D1103966

DISCARD

the

Weather
Almanac

ISSN 0731-5627

the
Weather
Almanac

A reference guide to weather, climate,
and related issues in the United States
and its key cities.

ELEVENTH EDITION

Richard A. Wood, Ph.D.
Editor

GALE®

THOMSON
GALE

Detroit • New York • San Diego • San Francisco • Cleveland • New Haven, Conn. • Waterville, Maine • London • Munich

Weather Almanac, Eleventh Edition
Richard A. Wood, Ph.D., Editor

Project Editor
Ryan L. Thomason

Editorial
Brigham Narins, Mark Springer

Permissions
Margaret Chamberlain

Imaging and Multimedia
Randy Bassett, Jeff Matlock, Christine O'Bryan, Barbara J. Yarrow

Product Design
Cynthia Baldwin

Manufacturing
Nekita McKee, Evi Seoud

© 2004 by Gale. Gale is an imprint of The Gale Group, Inc., a division of Thomson Learning, Inc.

Gale and Design™ and Thomson Learning™ are trademarks used herein under license.

For more information contact
The Gale Group, Inc.
27500 Drake Rd.
Farmington Hills, MI 48331-3535
Or you can visit our Internet site at
http://www.gale.com

ALL RIGHTS RESERVED
No part of this work covered by the copyright hereon may be reproduced or used in any form or by any means—graphic, electronic, or mechanical, including photocopying, recording, taping, Web distribution, or information storage retrieval systems—without the written permission of the publisher.

For permission to use material from this product, submit your request via Web at http://www.gale-edit.com/permissions, or you may download our Permissions Request form and submit your request by fax or mail to:

Permissions Department
The Gale Group, Inc.
27500 Drake Road
Farmington Hills, MI, 48331-3535
Permissions hotline:
248-699-8074 or 800-877-4253, ext. 8006
Fax: 248-699-8074 or 800-762-4058.

While every effort has been made to ensure the reliability of the information presented in this publication, The Gale Group, Inc. does not guarantee the accuracy of the data contained herein. The Gale Group, Inc. accepts no payment for listing; and inclusion in the publication of any organization, agency, institution, publication, service, or individual does not imply endorsement of the editors and publisher. Errors brought to the attention of the publisher and verified to the satisfaction of the publisher will be corrected in future editions.

ISBN 0-7876-7515-6
ISSN 0731-5627

Printed in the United States of America
10 9 8 7 6 5 4 3 2 1

Contents

Foreword

During the 53 years I have enjoyed working in the field of Meteorology, serving in the United States Navy during the early and mid 1950s (during the Korean War), 32 years in the National Weather Service (NWS), and the past 16 years as a Consulting Meteorologist, Educator, and Author/Editor of some 8 books and hundreds of articles in various publications, I have never found the weather to be the same on two different days. Serving as an observer, radar operator, pilot briefer, forecaster and manager of various NWS stations, plus serving as the Program Leader of the NWS National Disaster Preparedness Program at NWS Headquarters in Silver Spring, Maryland, during the 1980s, I had the opportunity to observe many strange and unusual weather events.

I decided to begin compiling and tabulating hundreds of these strange and unusual anecdotes and have placed several of these events in the following chapters; I hope you find them as interesting as I do. I am including in this book some of those weather-related incidents that produced the greatest reactions from my students. Many of the anecdotes point out that taking proper precautions can and does save lives, and many of the anecdotes have positive as well as negative results.

Whether a person is planning to move to a new location, looking for a vacation home, planning retirement in a new part of the country, thinking about a vacation trip, or if a business needs to know what kind of weather normally exists in a particular area—there are innumerable reasons and ways in which human activities are affected by the weather every day. These are some of the issues we hope to address in this 11th edition of *The Weather Almanac.*

What's happening in the Earth's atmosphere affects numerous aspects of our lives. More and more, individuals are working to increase their awareness, knowledge, and understanding of the weather and its related elements and phenomena.

SEVERE WEATHER OF THE 1990s AND EARLY 2000s

Category 4 Hurricane Andrew, upgraded to Category 5 in 2002, struck just south of Miami in August 1992 causing damage that may reach $35–40 billion; 126,000 homes were destroyed or damaged and 9,000 mobile homes were de-stroyed. Andrew left at least 160,000 people homeless in Dade County, Florida, alone. Despite the severe physical damages and crippling monetary losses, human casualties were surprisingly few. In Florida, 15 deaths were directly attributed to the storm, with another 29 fatalities indirectly related. Compare this to the 6,000 lives lost in the 1900 Galveston, Texas, hurricane.

The size and impact of the Great Midwest Flood of 1993 was unprecedented. Record river stages, areal extent of flooding, persons displaced, crop and property damage, and flood duration surpassed all floods in the United States in modern times. During the event, 95 forecast points in the Upper Midwest exceeded the previous floods of record, many by 6 feet or more. Approximately 500 forecast points on major rivers and tributary systems exceeded flood stage at some time during this flood.

Measured in terms of economic and human impacts, the Great Flood of 1993 will be recorded as the most devastating flood in modern U.S. history. Nine states, more than 15 percent of the contiguous United States, were catastrophically affected. Assessments of the economic damages of the flood indicate that losses ranged between $15–20 billion. More than 50,000 homes were damaged or destroyed and approximately 54,000 persons were evacuated from flooded areas. There were a total of 48 fatalities caused by the flood.

Included in this 11th edition of *The Weather Almanac* are details of the following weather related catastrophes that occurred in the last decade of the past millennium: Hurricanes Mitch, Georges, Charley, and Andrew, among others; the Great Midwest Flood of 1993; the winter superstorm of 1993; the Northridge earthquake of 1994; the major Pacific northwestern floods of December 1996/January 1997 in the states of Washington, Oregon, Idaho, Nevada, and California; the southeastern US 1998 tornadoes; and the January 1999 midwest blizzard.

THE WEATHER ALMANAC: A READY-REFERENCE TOOL

The Weather Almanac is a handy and comprehensive complement to usual sources of day-to-day weather information in newspapers, radio, and television information. This newly revised edition makes available to you a wide range of maps, charts, and safety rules based upon past records, research, and experience to inform you of what may be expected from our ever-changing atmosphere.

Separate sections in *The Weather Almanac* are devoted to tornadoes, hurricanes, thunderstorms and lightning, flash floods, and winter storms, among others, and they have been edited from official reports by the National Weather Service and Geological Survey, and other agencies, to serve as a popular basic reference for severe weather and extreme conditions. Official records and statistics are included wherever they are useful and available.

Being "weather-aware" can help save your life—being prepared to act is far better than having to react to a natural hazard. The flat cloud on top of a thunderstorm, called the anvil, can indicate by the way it is pointing which direction the storm is moving. If you can hear thunder you are at risk of being struck; normally thunder can be heard about 10 miles away, but in a calm period, particularly in a dry area, the distance can be up to 15 miles.

Information in this book will help you to better adapt yourself and your activities to the weather and let you prepare for a weather emergency at the best time: before it is upon you.

Reference to severe weather is only one feature of this book. *The Weather Almanac* offers you a large collection of data from hundreds of sources. *The Weather Almanac* begins with an overview of the various types of weather, then it examines some underlying weather principals, and finally there is a glossary of weather terms, a brief cloud atlas, and other guides to personal observations.

The Weather Almanac will take you around the world. Besides over 100 cities in the United States, about 160 key cities in every portion of the world are climatically covered in this book, including seasonal temperature and precipitation normals.

CHANGES IN THIS 11TH EDITION OF *THE WEATHER ALMANAC*

We have added a new chapter devoted just to the most recent year's weather, 2002, and early 2003. Earlier year highlights will continue in the individual chapters, and any other noteworthy items for early 2003. Recent drought conditions will be discussed in this new chapter.

All sections of the book have been revised as new data on particular subjects and phenomena have become available, but the greatest amount of change occurs in the El Niño and La Niña chapter; plus, there is a new section on Global Warming in the Heat and Humidity chapter; and a Significant Floods of the 20th Century section in the Flood chapter has been added, along with updated state-by-state flash flood/flood statistics for the period 1960–2001; updated in-

formation from the EPA in the Air Pollution chapter; the latest satellite information in the Weather Information and Communications Chapter; new items added to the Glossary of Weather Terms chapter, and the various other atmospheric natural hazards chapters have been completely updated. The Local Climatological Data statistics have been updated through the latest data available—2001.

Finely detailed weather statistics for each city typically include 40 years or more of weather history. Besides temperature and precipitation statistics, heating- and cooling-degree days are included in the data for the 108 cities.

We have included twelve "top 5" categories in the Record-Setting Weather section. Look for those locations that are the hottest, coldest, driest, wettest, windiest, snowiest, sunniest, cloudiest, most and least humid, and rainiest and least rainy cities in the United States. These charts will be popular with teachers and students looking for subjects for weather research and projects.

While earthquakes, tidal waves, and volcanoes are not manifestations of the weather, it is important that the general public be informed about them and about the precautions to be taken when, where, and before they occur. Thus these have special sections in the book.

In addition to these major information features, *The Weather Almanac* provides sections regarding Wind Chill; Summer Comfort Index; Livestock Safety; NOAA Weather Radio Warning Network; Marine Weather Advisories section; how to forecast the weather for yourself, Historic Presidential Disaster Declarations, Billion-Dollar Disasters, 1996–2002 Weather Events, and a potpourri of other weather-related subjects.

GENERAL NOTES ON SOURCES

Much of the information in this book is quoted directly from reports and records prepared by various United States Government departments, agencies, and services that share parts of the nation's great weather and environmental science efforts. For example, safety rules for various atmospheric natural hazards are taken directly from National Oceanic and Atmospheric Administration (NOAA), National Weather Service, publications. Similarly, the U.S. Geological Survey is the source for comparable safety advice for earthquakes and volcanic activity. These suggested safety rules have dramatically helped to minimize the impact of these hazards.

The city-by-city weather records are worthy of the same confidence, for they come directly from the cumulative records developed by the United States observers around the country and have been coordinated by the people of the National Climatic Data Center.

The core section of the information on Air Pollution has been compiled from a wide variety of sources drawn from the Environmental Protection Agency and the President's Council on Environmental Quality.

A special thanks to Joseph Lee Hudson, Research Meteorologist, Arlington, Virginia, for his expert review of the various chapters; Robert Waiter, Computer Programmer, Prescott, Arizona; and Rick Wood, Computer Specialist, Mesa, Arizona, for assisting in obtaining new photos, etc.

The Weather Almanac editorial staff hopes this new, revised, 11th edition, will be increasingly useful as a source of basic information about our atmosphere and Earth.

Comments or suggestions from readers are not only welcome but are solicited in our constant quest to improve the book.

Richard A. Wood, Ph.D.,
Editor

Introduction

The Weather Almanac Is Your Guide to Weather and Related Phenomena

Now in its eleventh edition, *The Weather Almanac* (WA) contains information on all types of weather and weather-related phenomena. It furnishes in-depth weather statistics and safety rules, as well as a climatic overview of the country.

Important Features of this Edition

The Weather Almanac has been modified to meet the needs of today's user:

- Contemporary Design. The almanac's design provides users with important information in an accessible, clean, and easy-to-use format.

- New chapter. A new "Recent Weather" chapter discusses major weather events of the last two years in detail.

- Statistics. Statistics throughout the book have been updated through 2002 (when available).

To further complement its thorough coverage, *The Weather Almanac* uses these important features to help users find information easily:

- Glossary—enables users to review, at a glance, fundamental weather terms.

- Further Resources—provides users with a number of books and other resources for additional weather information.

- General Index—a new, two-level index guides users to important terms and concepts throughout the book.

Acknowledgments

The editors wish to thank the following for making relevant data and its compilation available for use in this publication.

- Thirty-three U.S. Atlas maps. Illustrated by Mark Berger, Standley Publishing. Data courtesy of National Climatic Data Center/NOAA.

- Thirteen country climate charts. Compiled by Richard A. Wood and Robert Waiter from National Climatic Data Center/NOAA data.

- Local Climatological Data Reports. Courtesy of National Climatic Data Center/NOAA.

Suggestions Welcome

Comments and suggestions from users on any aspect of *The Weather Almanac* are welcome. Address correspondence to:

The Editor
The Weather Almanac
Gale Group
27500 Drake Rd.
Farmington Hills, MI 48331-3535

United States
Weather in Atlas
Format

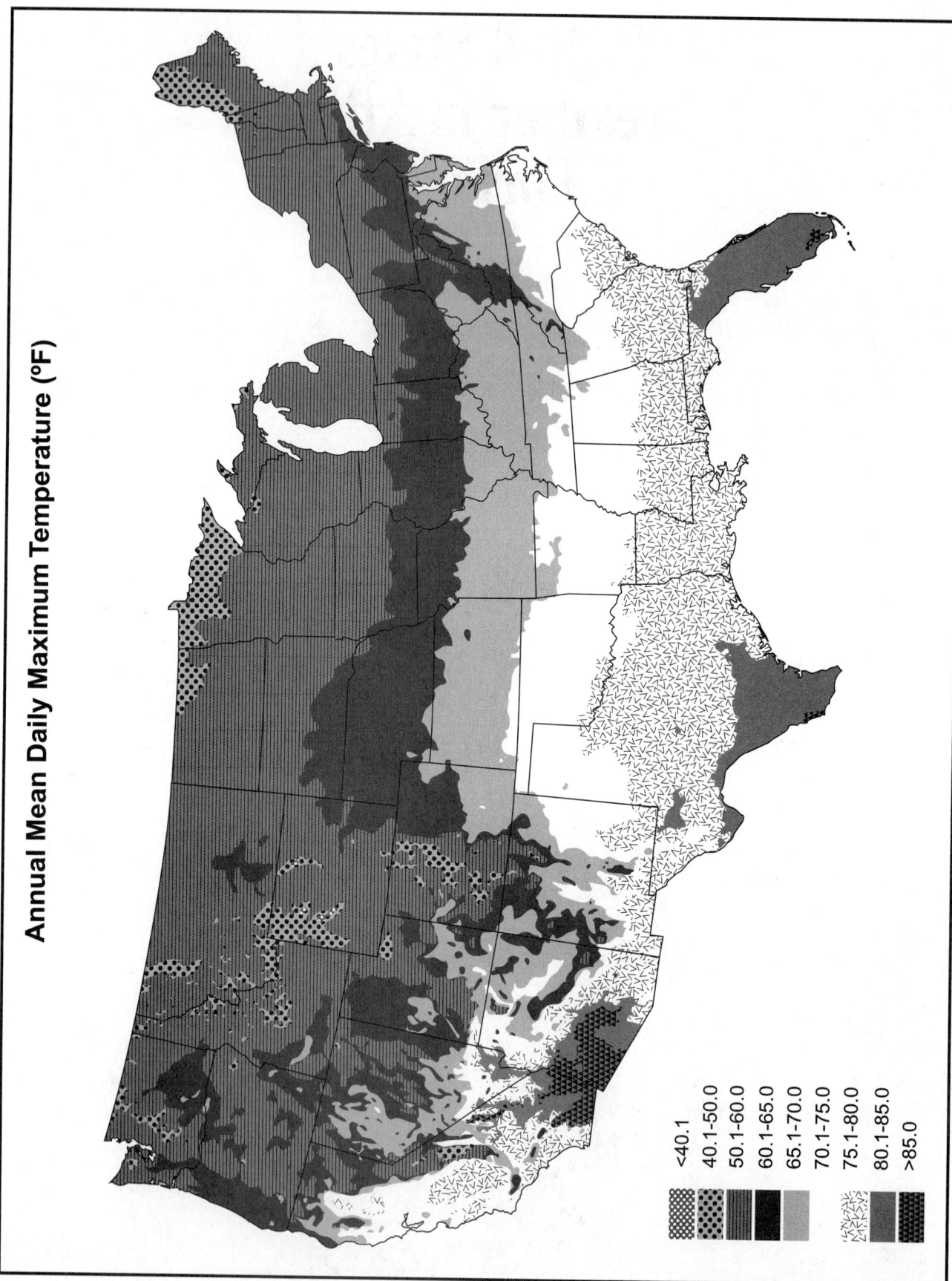

Annual Mean Daily Maximum Temperature (°F)

<40.1
40.1-50.0
50.1-60.0
60.1-65.0
65.1-70.0
70.1-75.0
75.1-80.0
80.1-85.0
>85.0

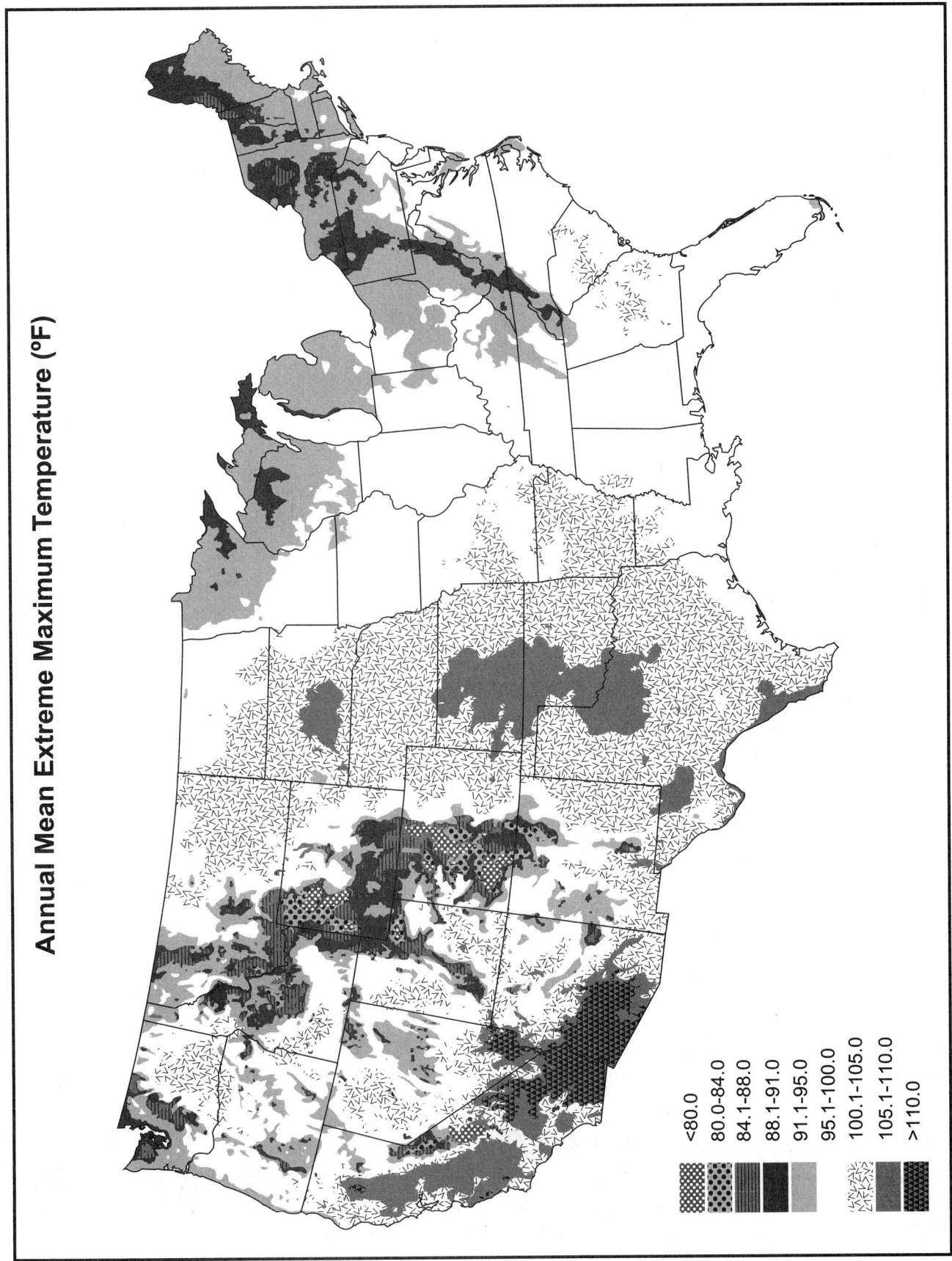

Annual Mean Extreme Maximum Temperature (°F)

<80.0
80.0-84.0
84.1-88.0
88.1-91.0
91.1-95.0
95.1-100.0
100.1-105.0
105.1-110.0
>110.0

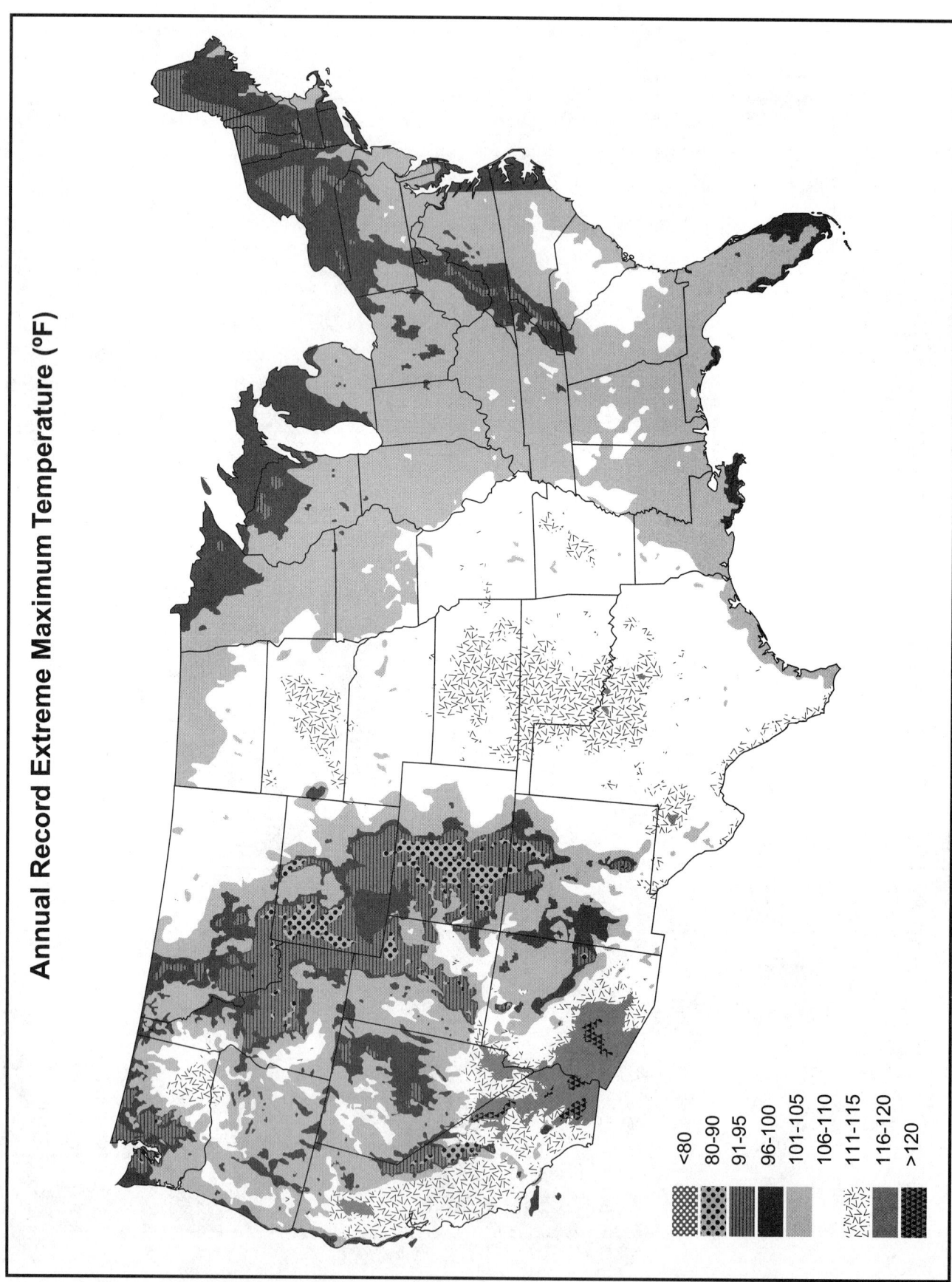

Annual Record Extreme Maximum Temperature (°F)

<80
80-90
91-95
96-100
101-105
106-110
111-115
116-120
>120

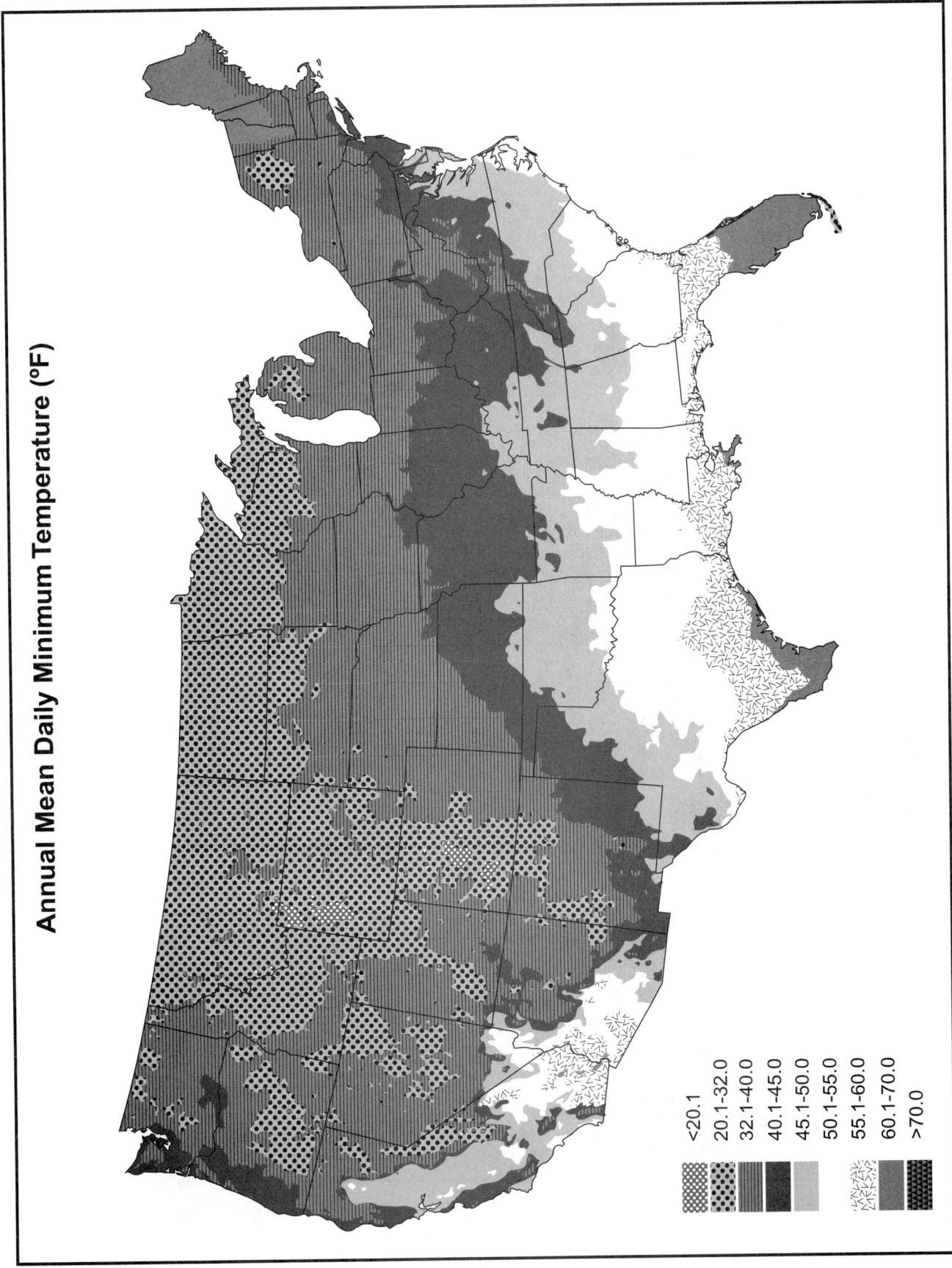

Annual Mean Daily Minimum Temperature (°F)

<20.1
20.1-32.0
32.1-40.0
40.1-45.0
45.1-50.0
50.1-55.0
55.1-60.0
60.1-70.0
>70.0

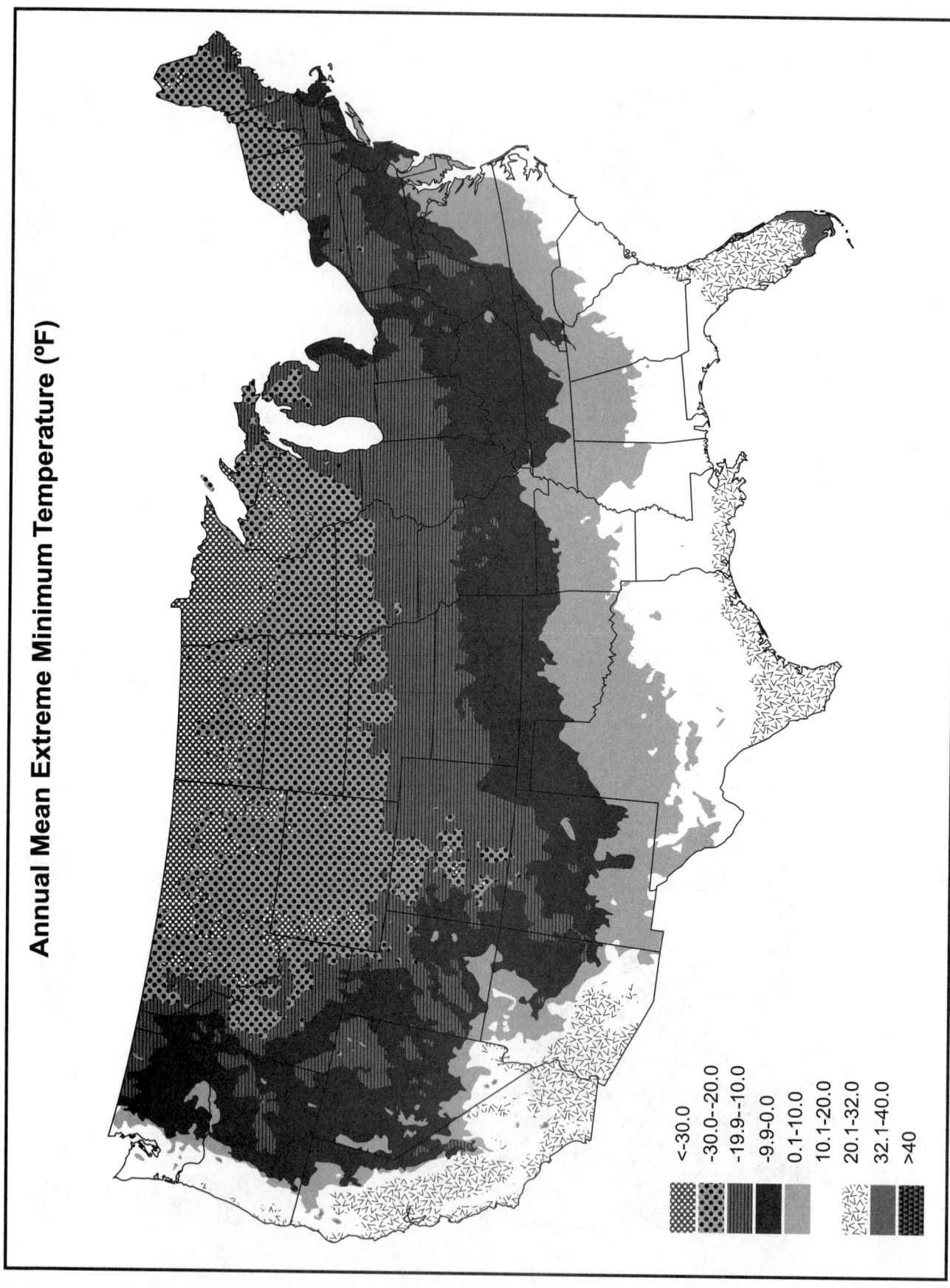

Annual Mean Extreme Minimum Temperature (°F)

<-30.0
-30.0–20.0
-19.9–10.0
-9.9-0.0
0.1-10.0
10.1-20.0
20.1-32.0
32.1-40.0
>40

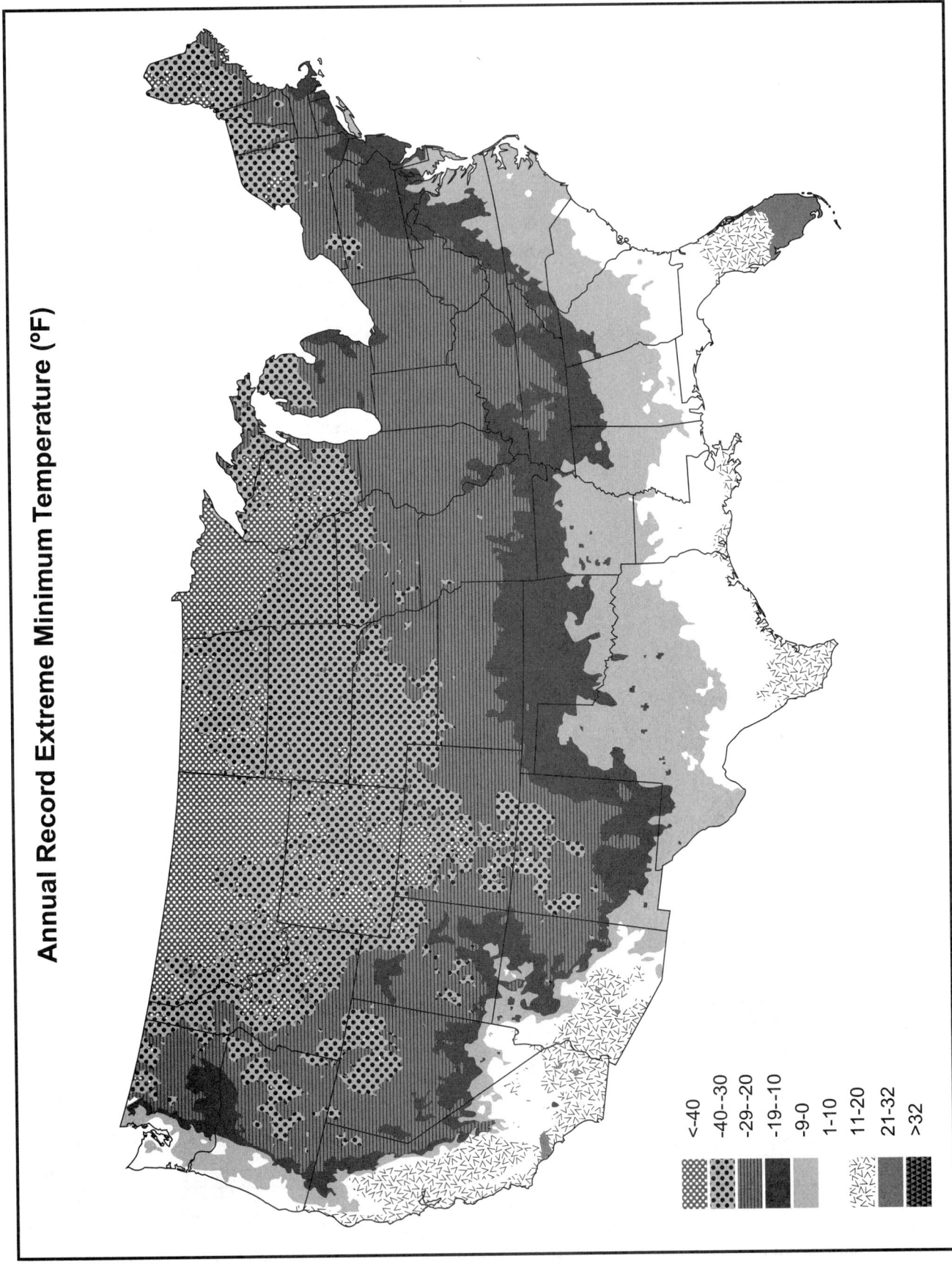

Annual Record Extreme Minimum Temperature (°F)

<40
-40--30
-29--20
-19--10
-9-0
1-10
11-20
21-32
>32

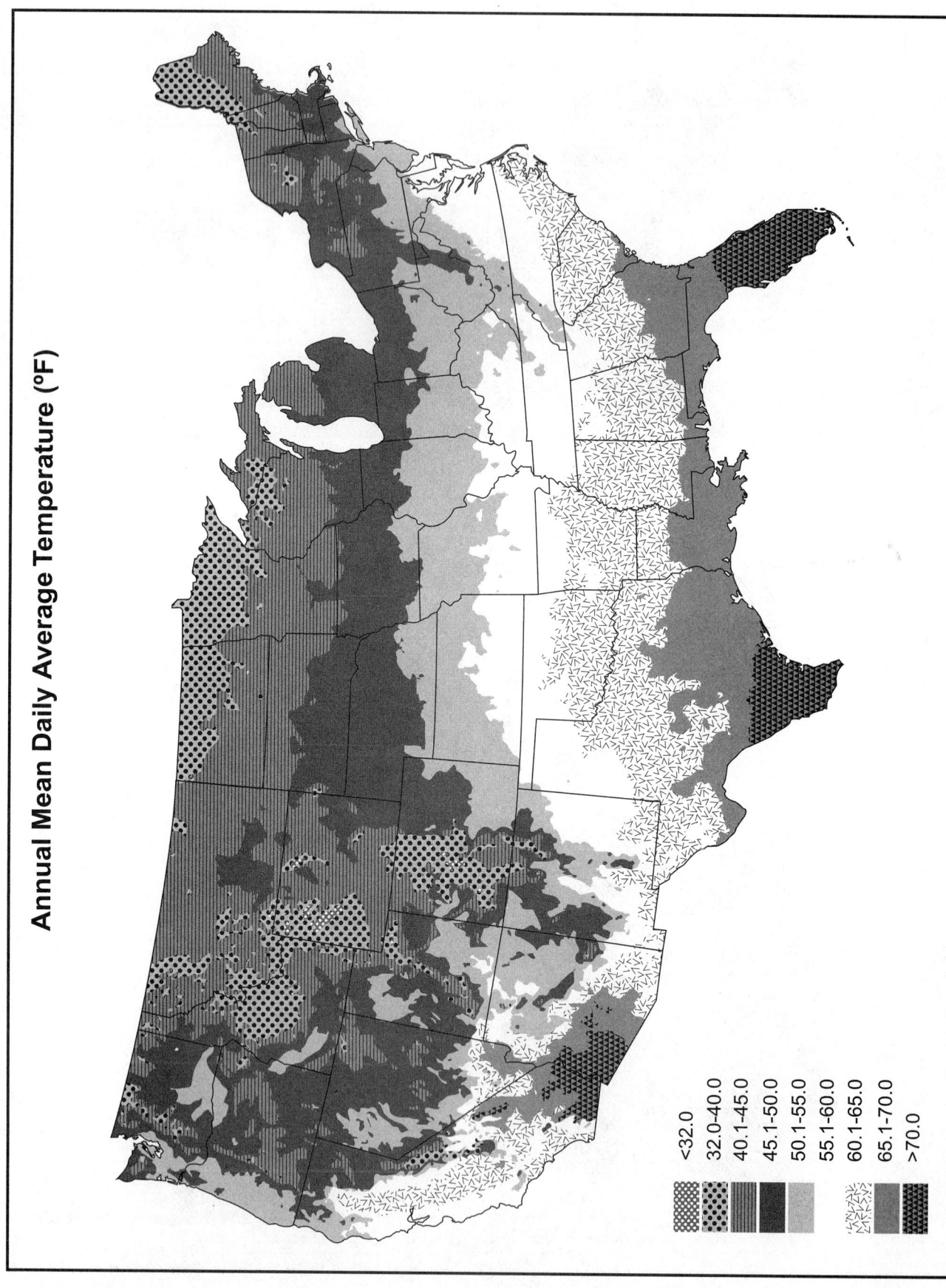

Annual Mean Daily Average Temperature (°F)

<32.0
32.0-40.0
40.1-45.0
45.1-50.0
50.1-55.0
55.1-60.0
60.1-65.0
65.1-70.0
>70.0

Annual Mean Daily Temperature Range (°F)

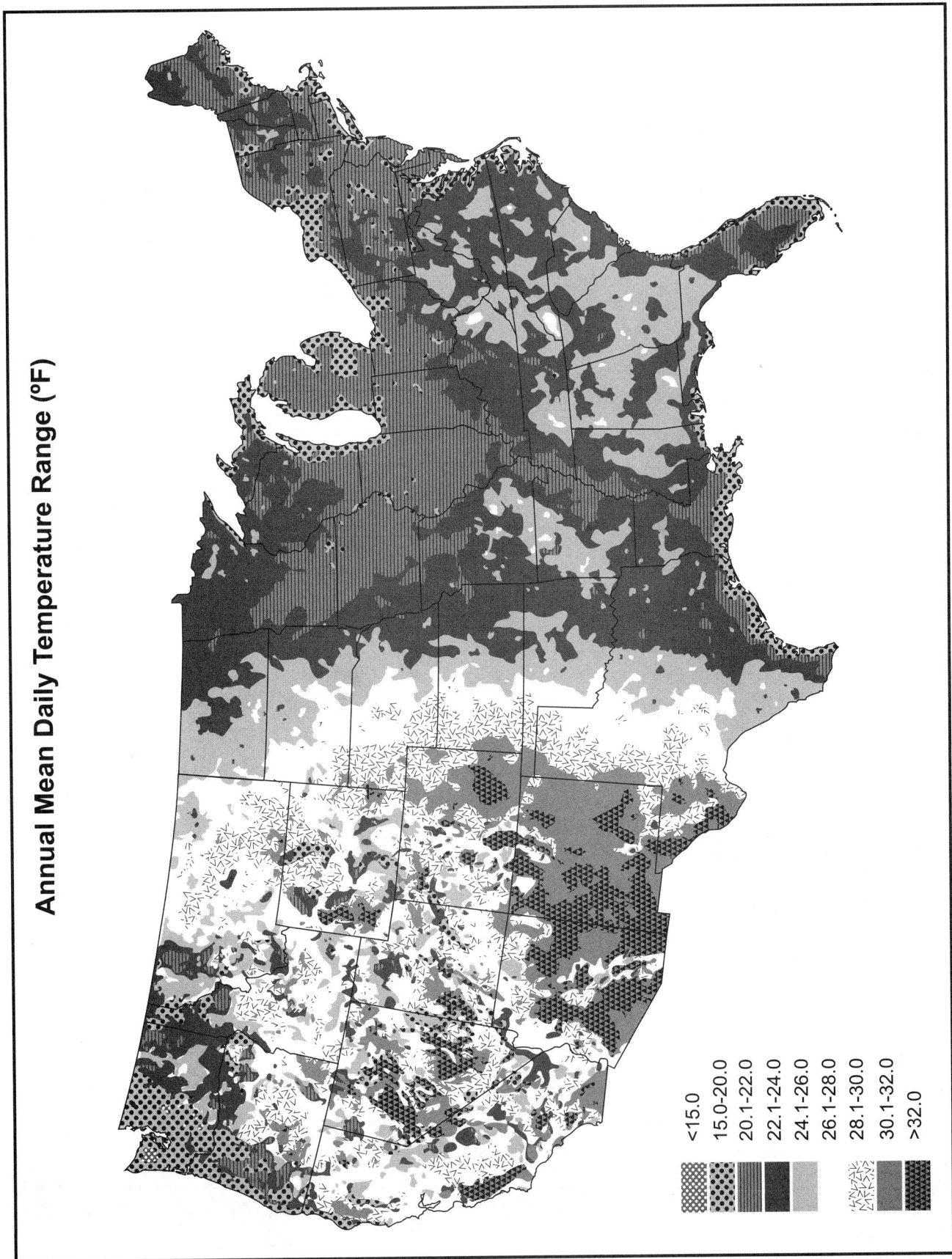

<15.0
15.0-20.0
20.1-22.0
22.1-24.0
24.1-26.0
26.1-28.0
28.1-30.0
30.1-32.0
>32.0

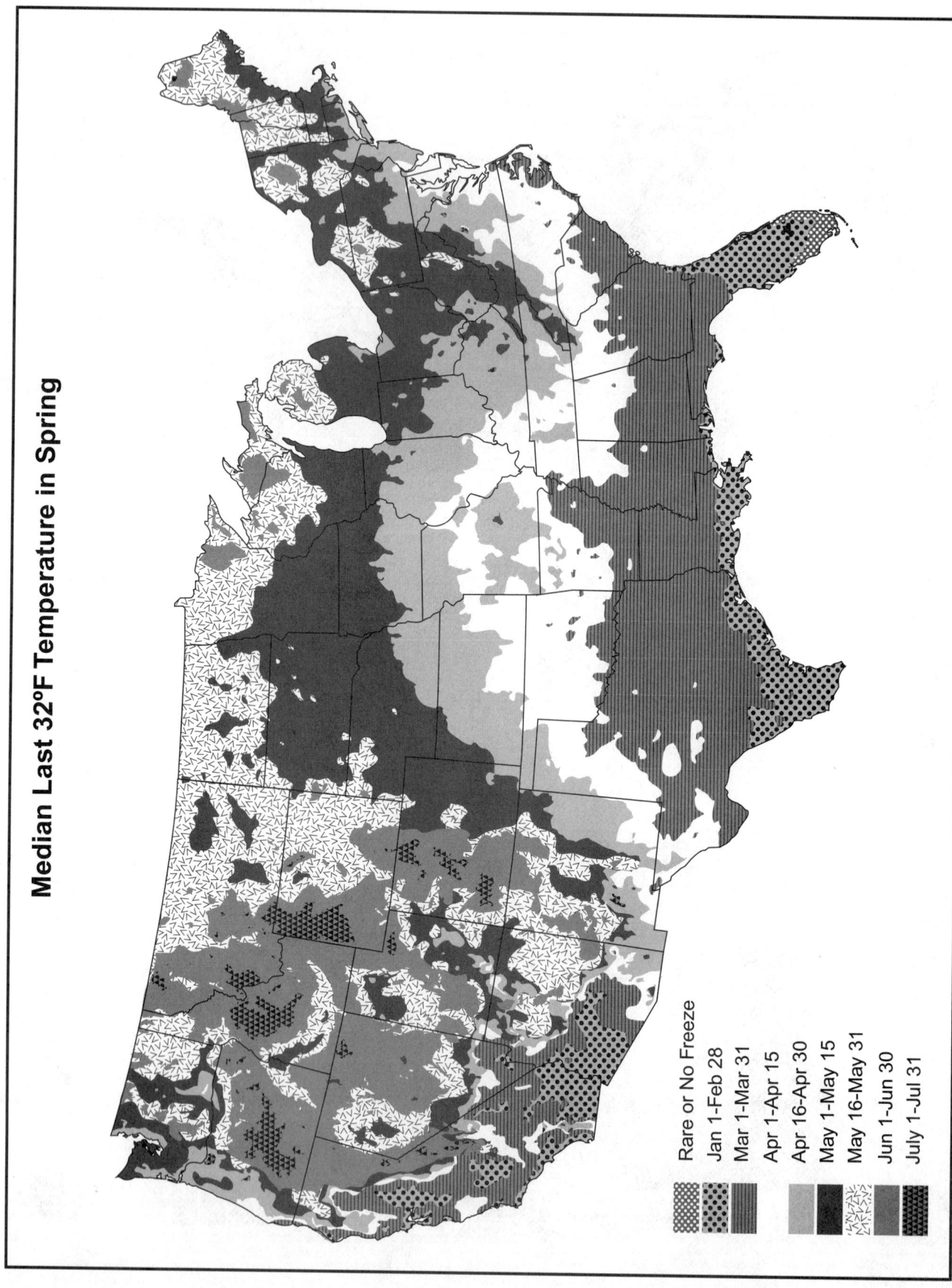

Median Last 32°F Temperature in Spring

Rare or No Freeze
Jan 1-Feb 28
Mar 1-Mar 31
Apr 1-Apr 15
Apr 16-Apr 30
May 1-May 15
May 16-May 31
Jun 1-Jun 30
July 1-Jul 31

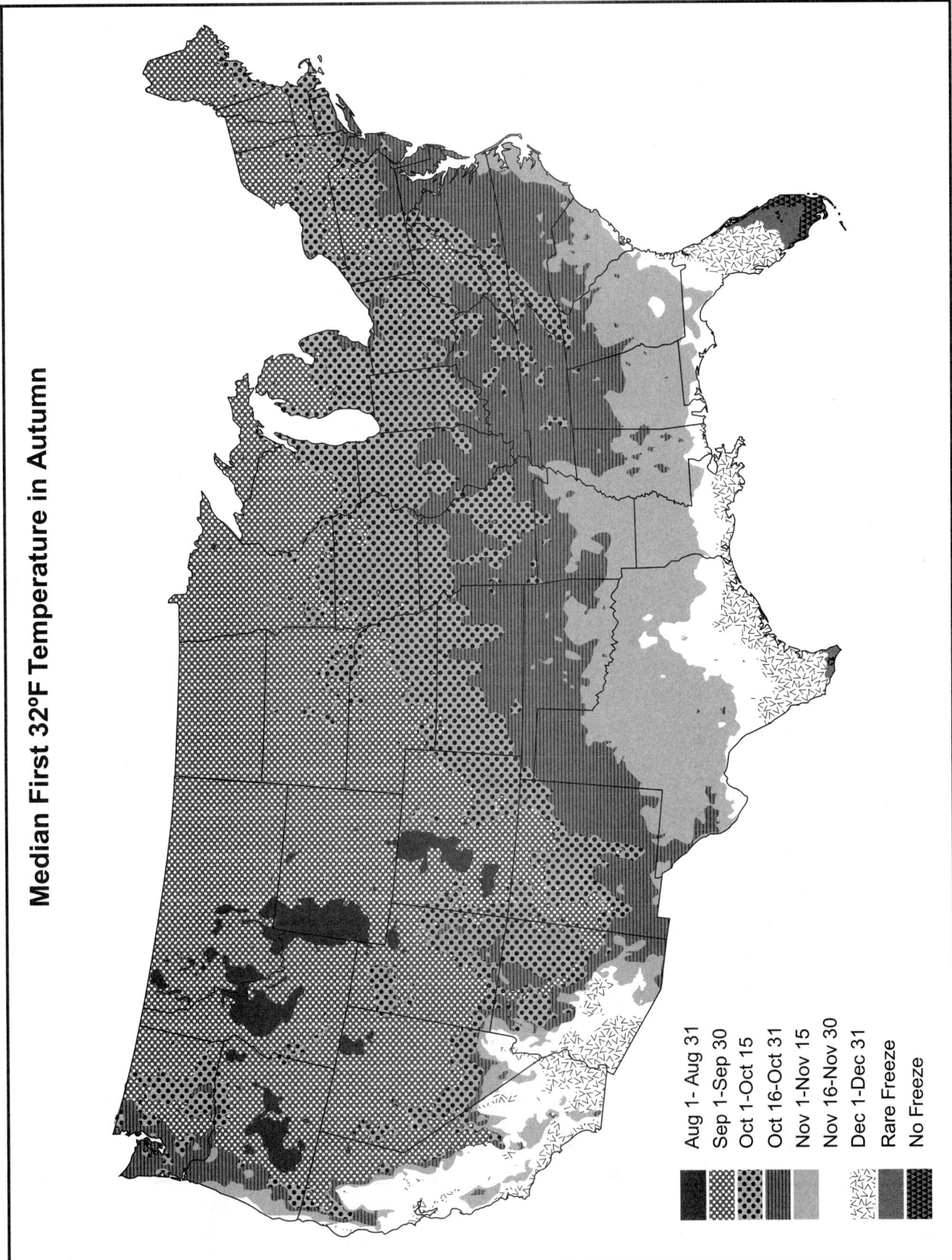

Median First 32°F Temperature in Autumn

Aug 1- Aug 31
Sep 1-Sep 30
Oct 1-Oct 15
Oct 16-Oct 31
Nov 1-Nov 15
Nov 16-Nov 30
Dec 1-Dec 31
Rare Freeze
No Freeze

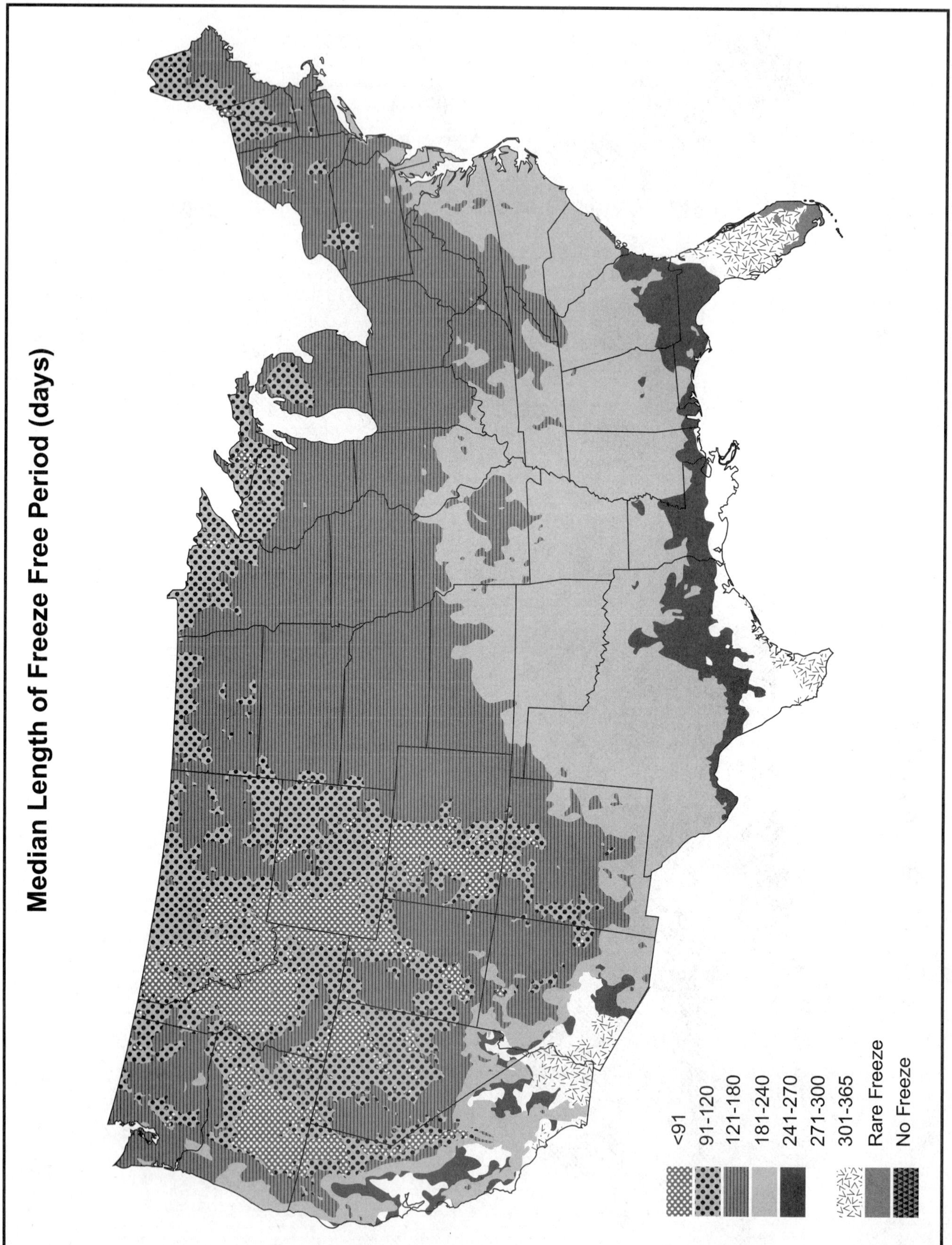

Median Length of Freeze Free Period (days)

<91
91-120
121-180
181-240
241-270
271-300
301-365
Rare Freeze
No Freeze

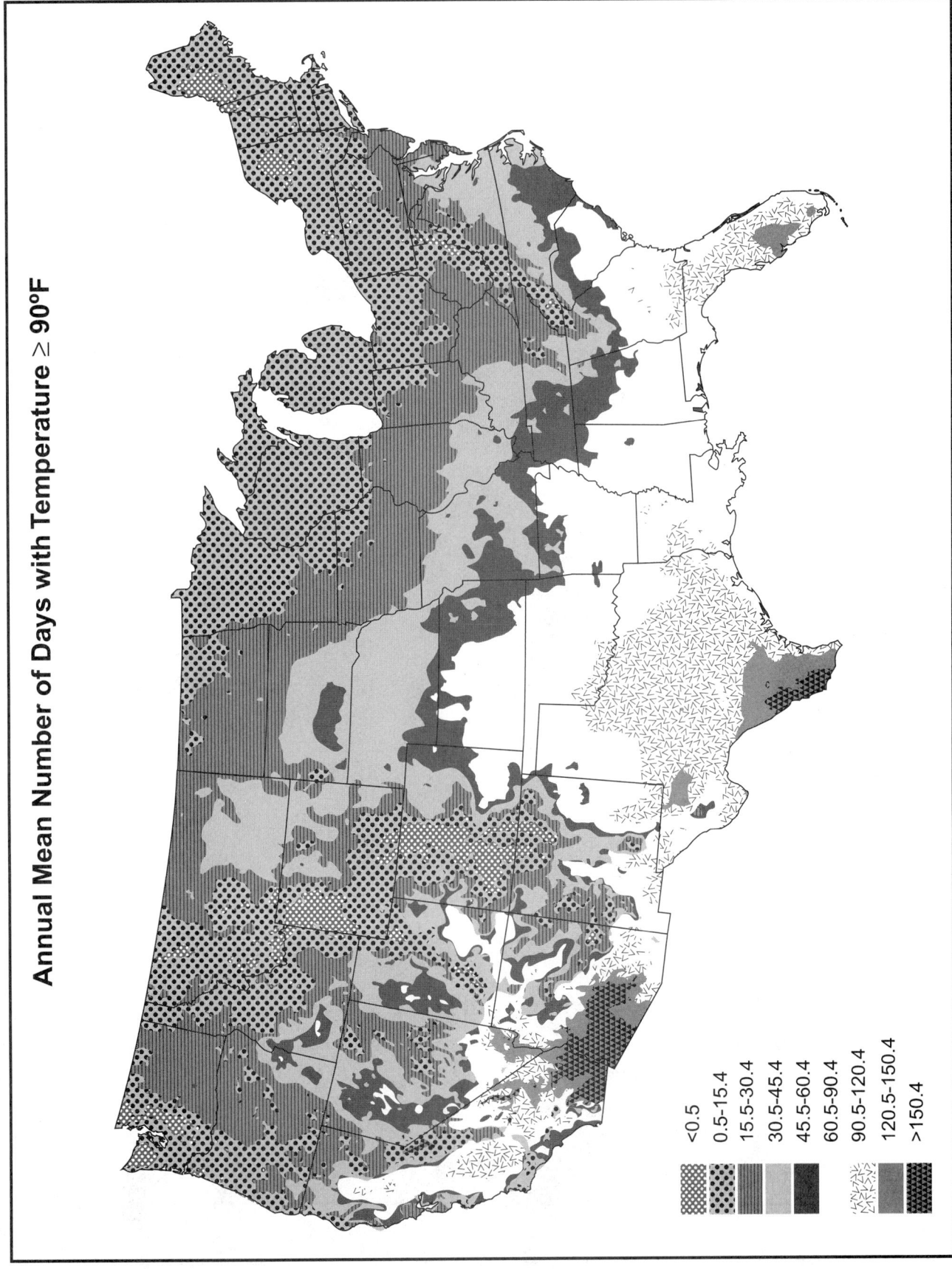

Annual Mean Number of Days with Temperature ≥ 90°F

<0.5
0.5-15.4
15.5-30.4
30.5-45.4
45.5-60.4
60.5-90.4
90.5-120.4
120.5-150.4
>150.4

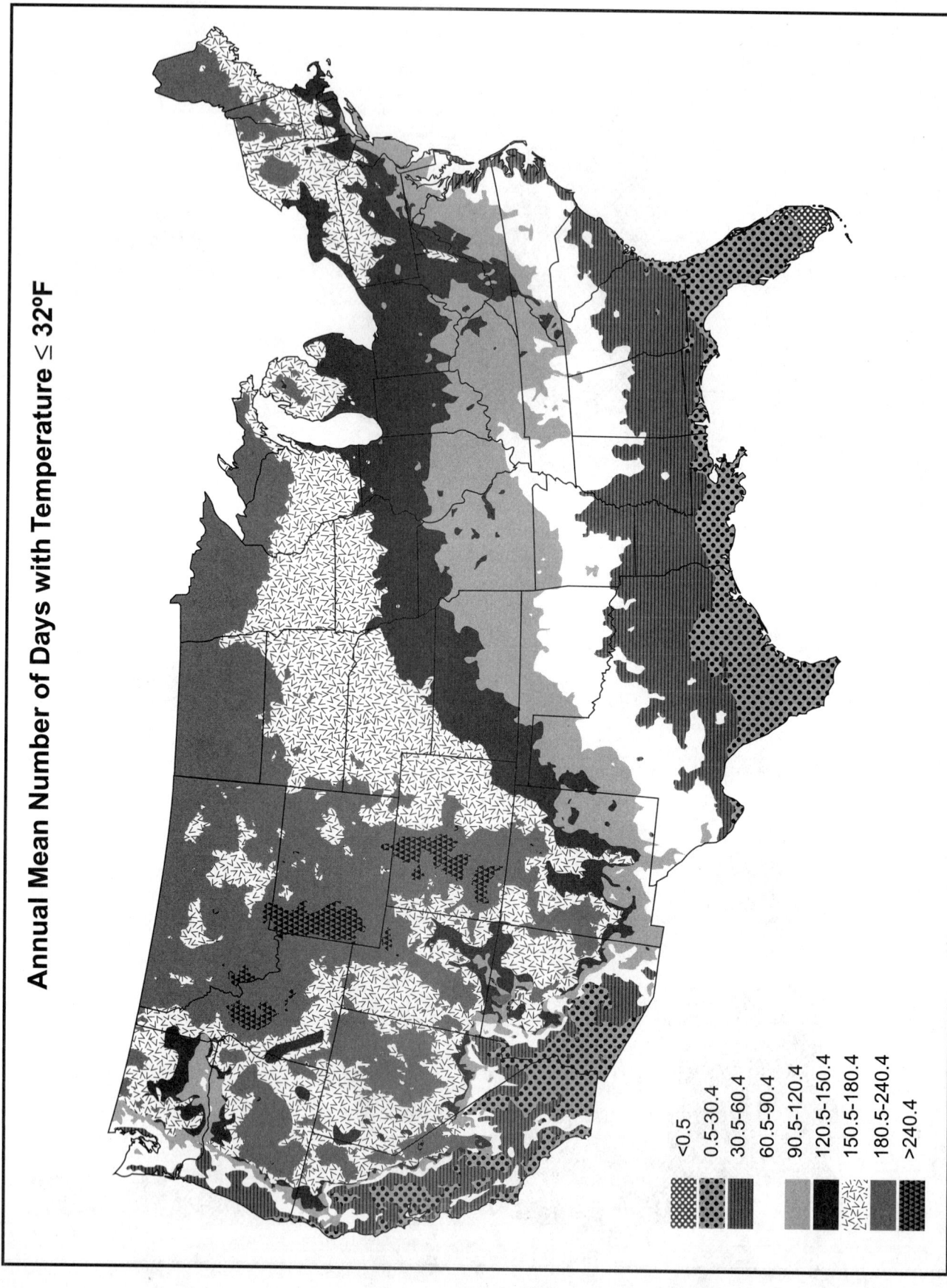

Annual Mean Number of Days with Temperature ≤ 32°F

<0.5
0.5-30.4
30.5-60.4
60.5-90.4
90.5-120.4
120.5-150.4
150.5-180.4
180.5-240.4
>240.4

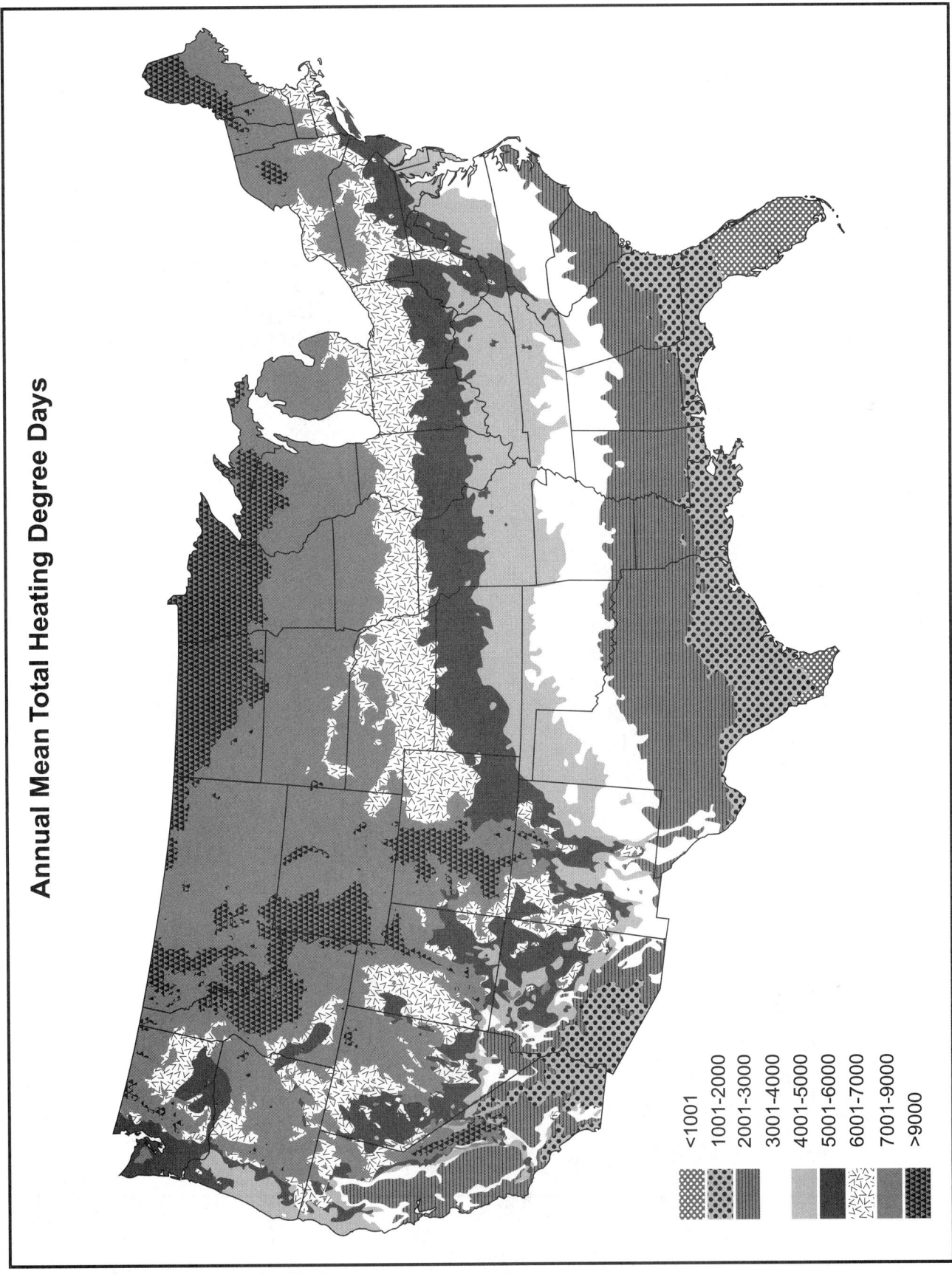

Annual Mean Total Heating Degree Days

<1001
1001-2000
2001-3000
3001-4000
4001-5000
5001-6000
6001-7000
7001-9000
>9000

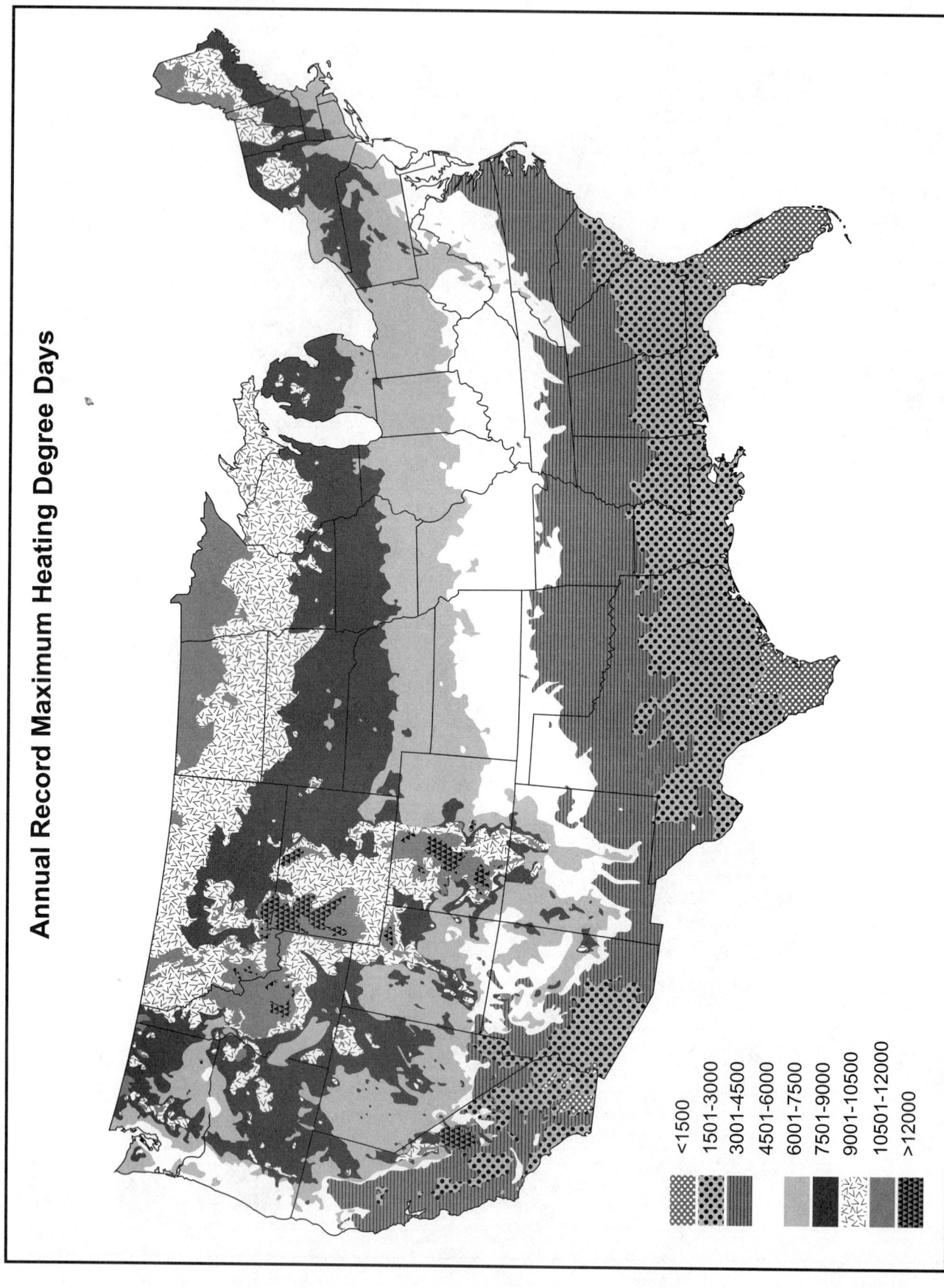

Annual Record Maximum Heating Degree Days

<1500
1501-3000
3001-4500
4501-6000
6001-7500
7501-9000
9001-10500
10501-12000
>12000

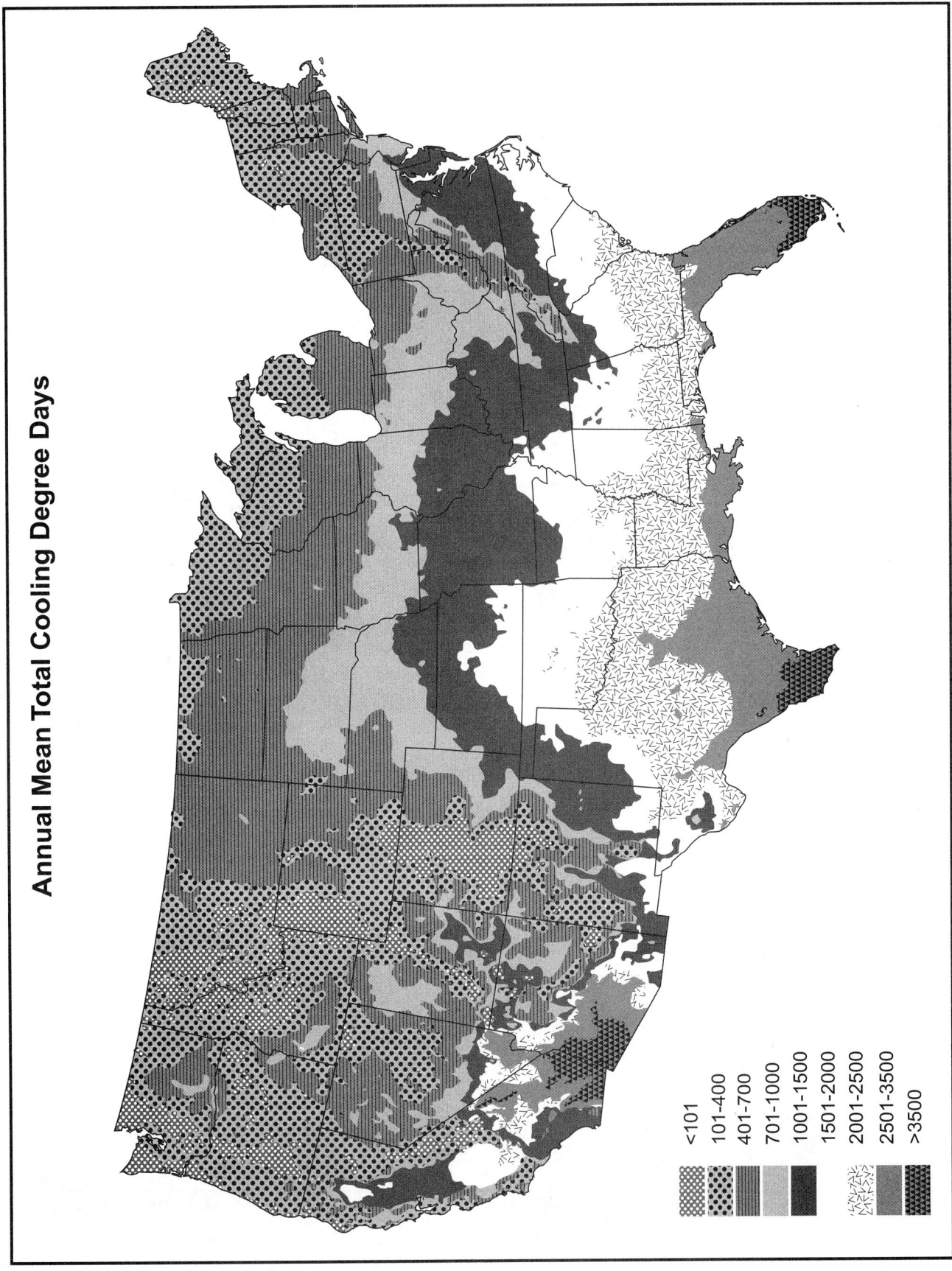

Annual Mean Total Cooling Degree Days

<101
101-400
401-700
701-1000
1001-1500
1501-2000
2001-2500
2501-3500
>3500

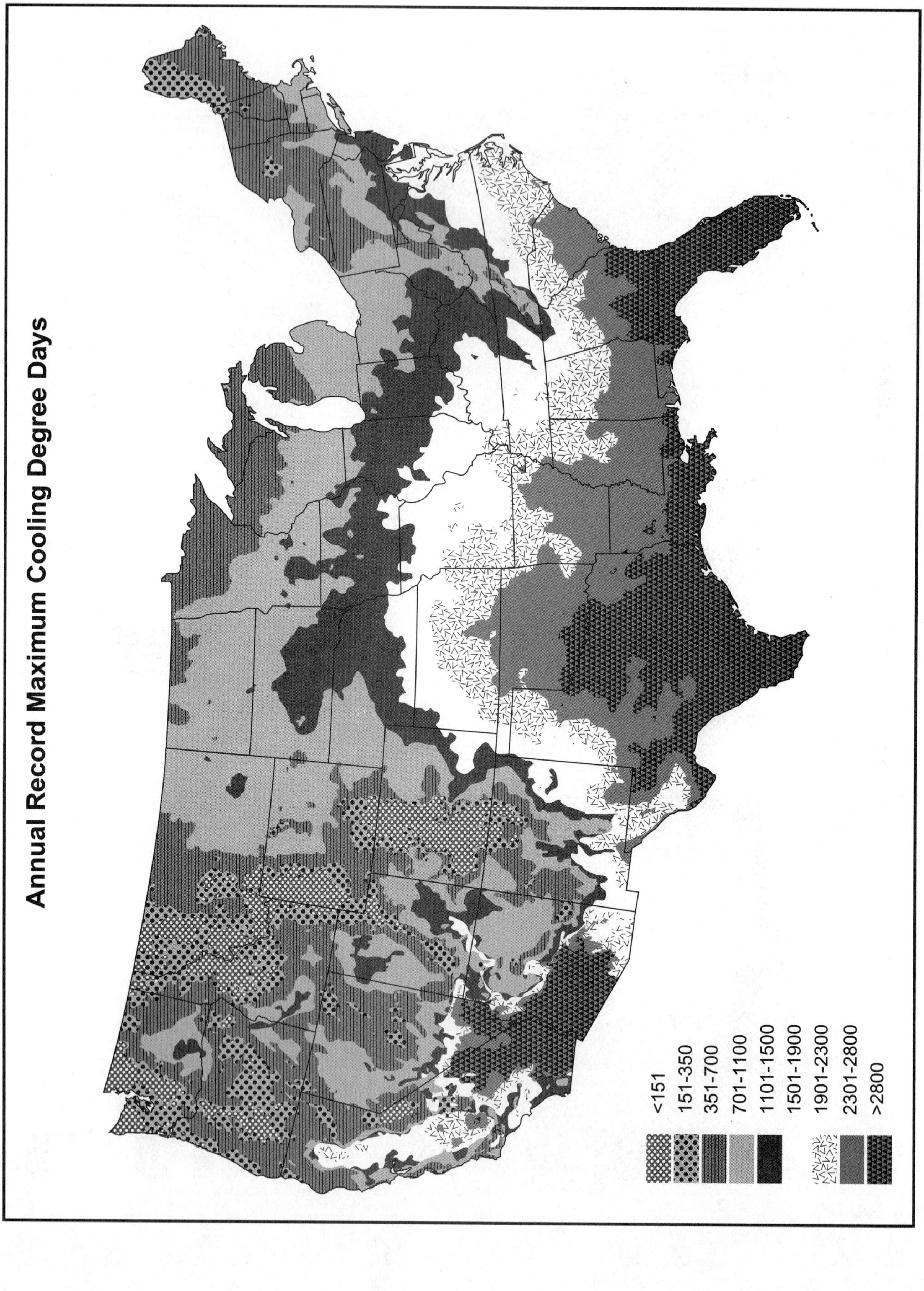

Annual Record Maximum Cooling Degree Days

<151
151-350
351-700
701-1100
1101-1500
1501-1900
1901-2300
2301-2800
>2800

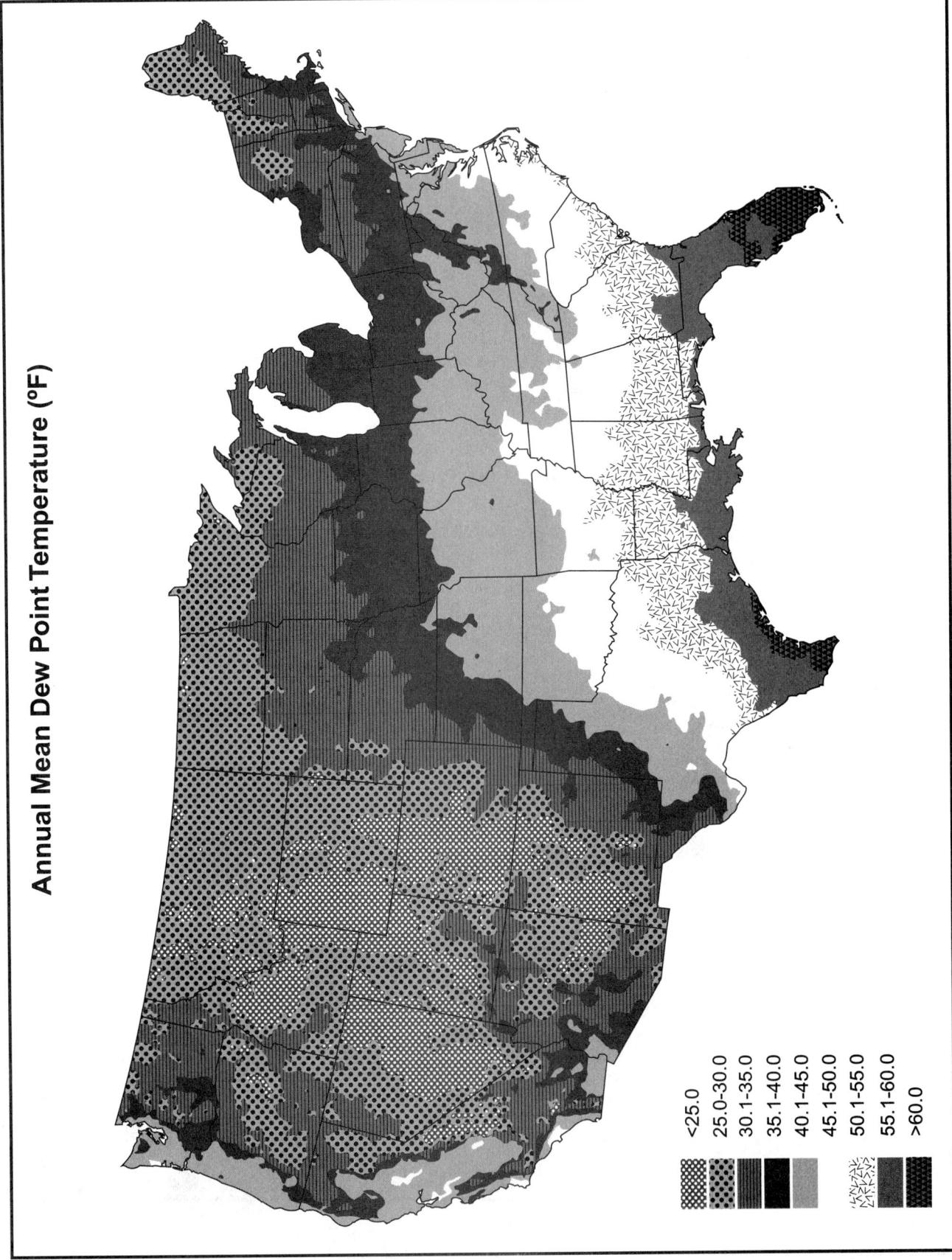

Annual Mean Dew Point Temperature (°F)

<25.0
25.0-30.0
30.1-35.0
35.1-40.0
40.1-45.0
45.1-50.0
50.1-55.0
55.1-60.0
>60.0

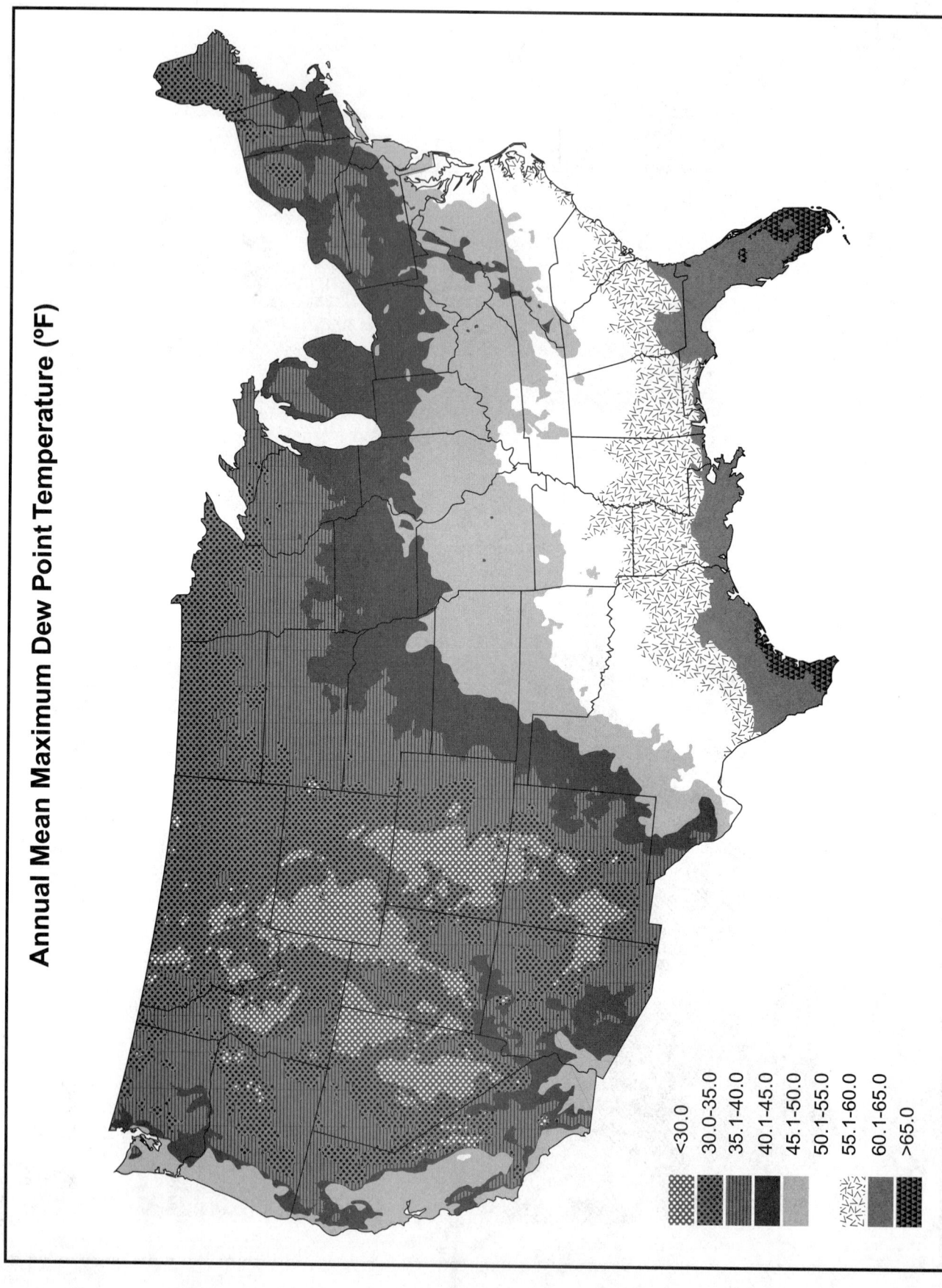

Annual Mean Maximum Dew Point Temperature (°F)

<30.0
30.0-35.0
35.1-40.0
40.1-45.0
45.1-50.0
50.1-55.0
55.1-60.0
60.1-65.0
>65.0

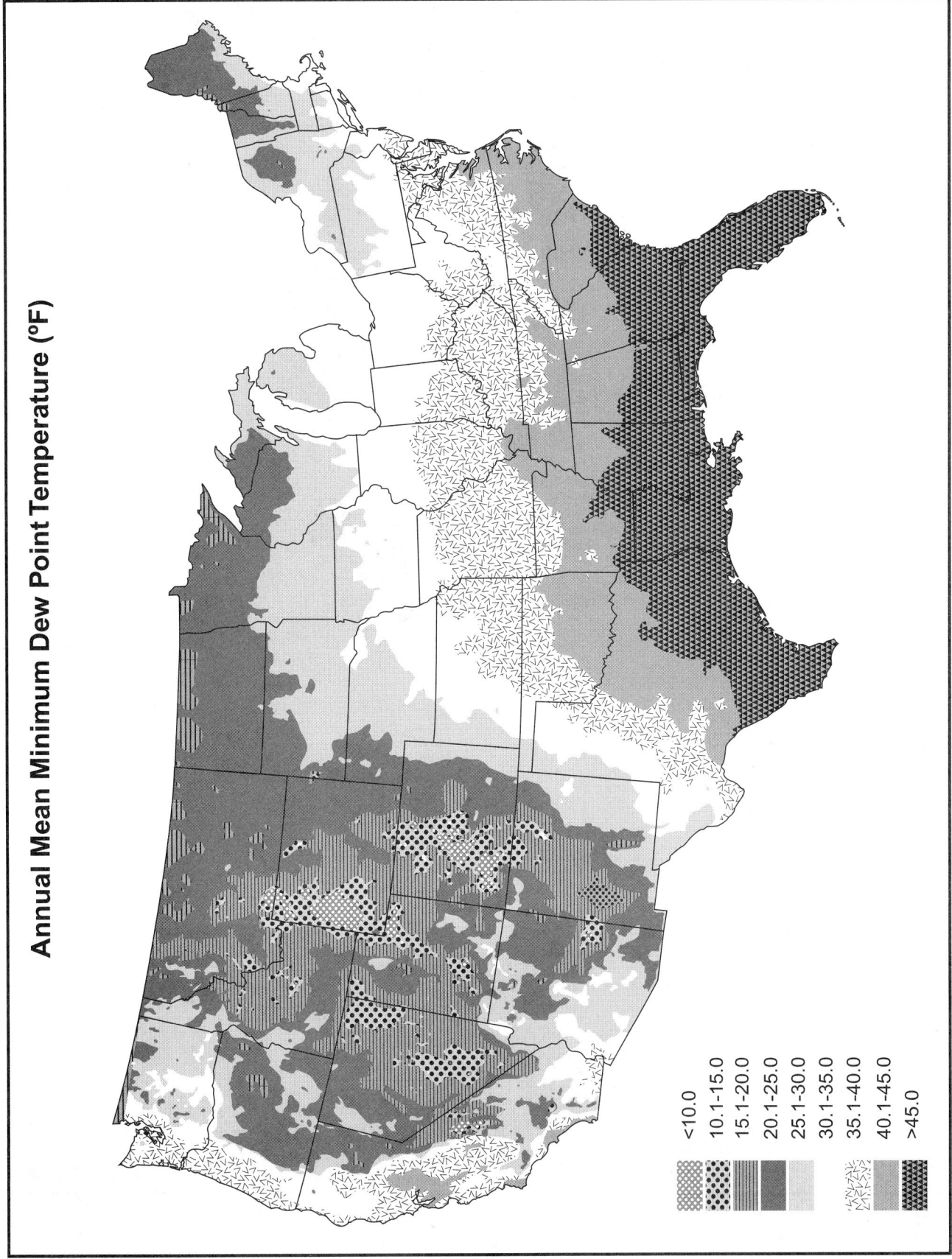

Annual Mean Minimum Dew Point Temperature (°F)

<10.0
10.1-15.0
15.1-20.0
20.1-25.0
25.1-30.0
30.1-35.0
35.1-40.0
40.1-45.0
>45.0

Annual Mean Total Precipitation (in)

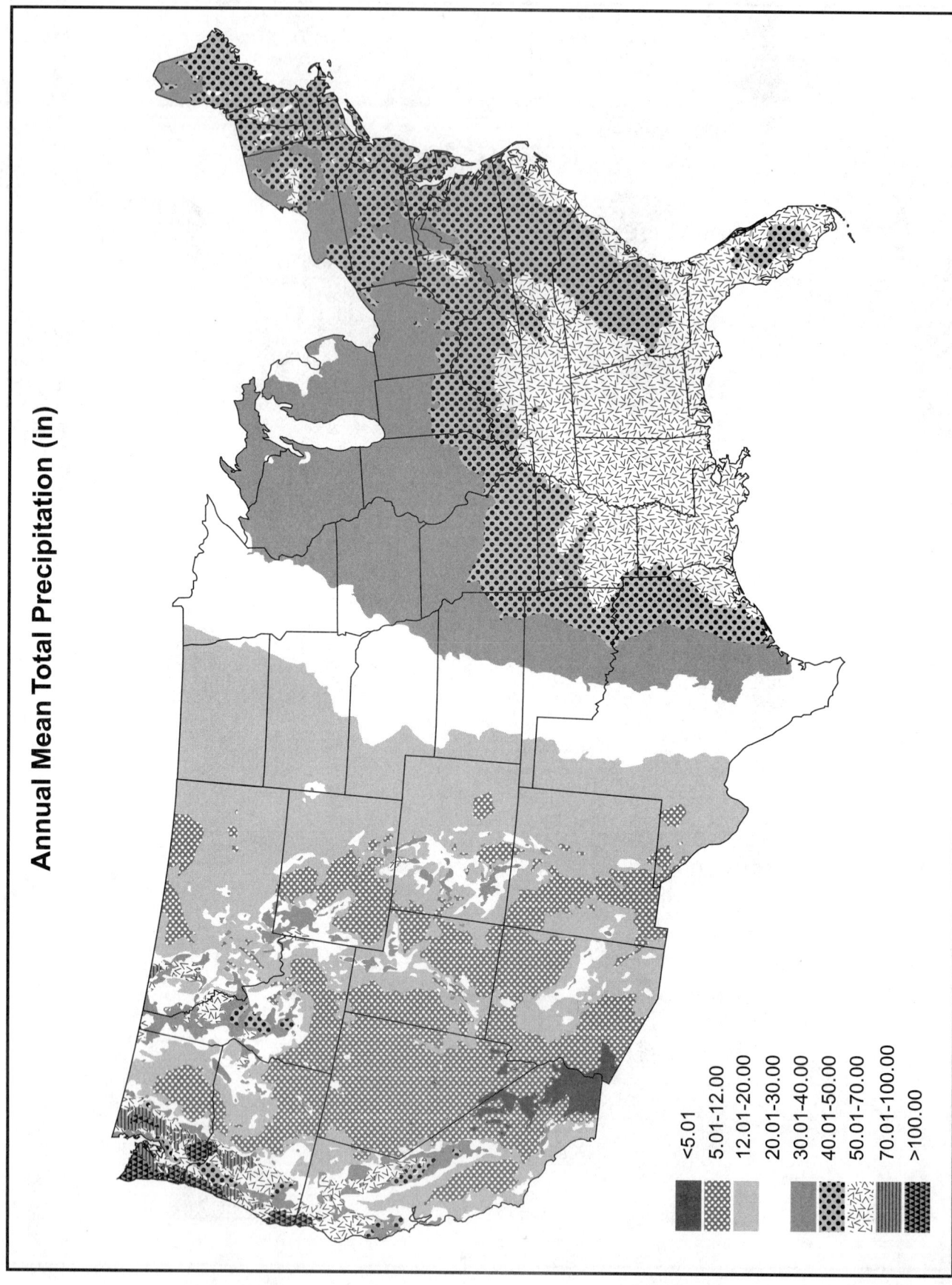

<5.01
5.01-12.00
12.01-20.00
20.01-30.00
30.01-40.00
40.01-50.00
50.01-70.00
70.01-100.00
>100.00

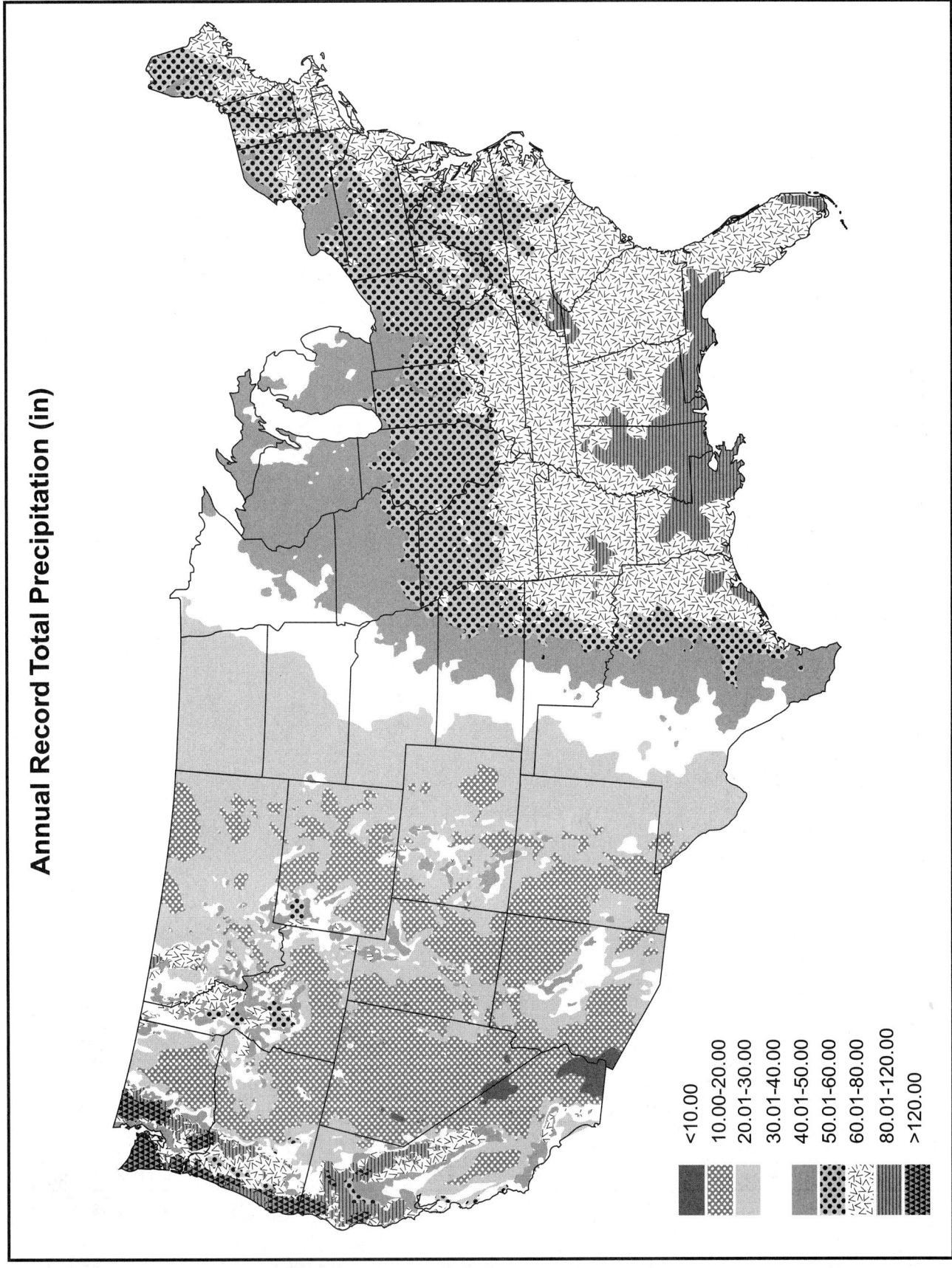

Annual Record Total Precipitation (in)

<10.00
10.00-20.00
20.01-30.00
30.01-40.00
40.01-50.00
50.01-60.00
60.01-80.00
80.01-120.00
>120.00

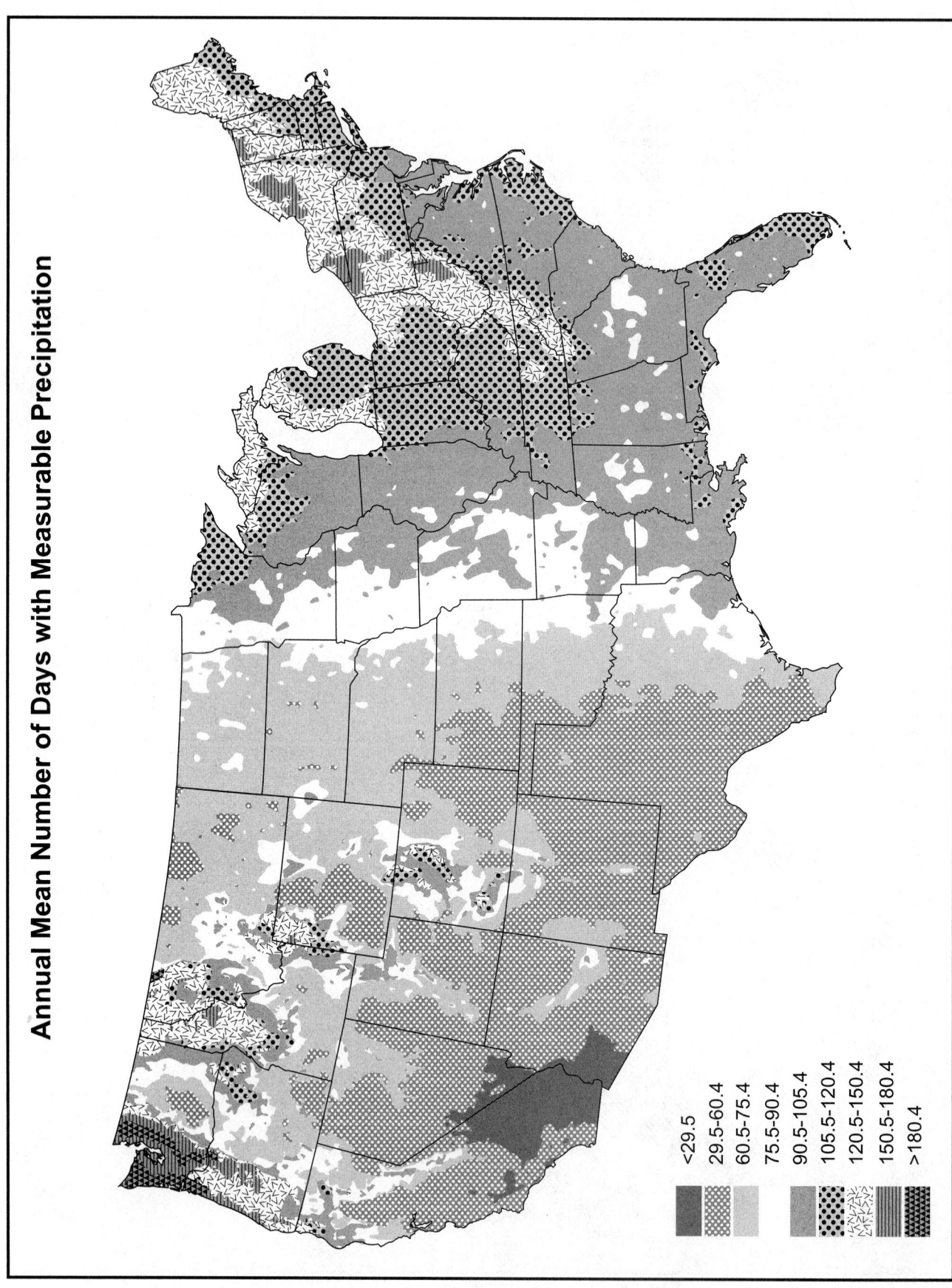

Annual Mean Number of Days with Measurable Precipitation

<29.5
29.5-60.4
60.5-75.4
75.5-90.4
90.5-105.4
105.5-120.4
120.5-150.4
150.5-180.4
>180.4

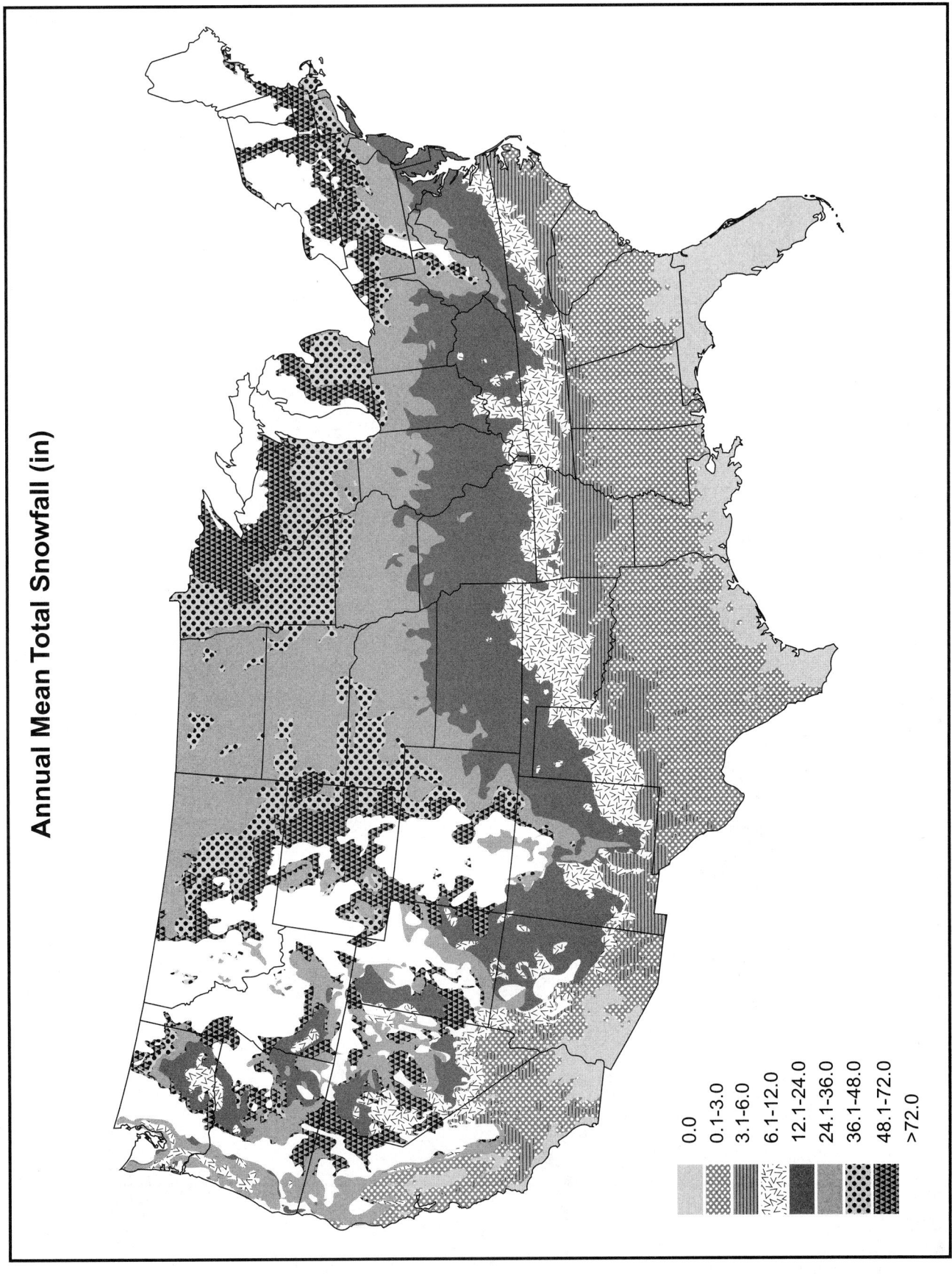

Annual Mean Total Snowfall (in)

0.0
0.1-3.0
3.1-6.0
6.1-12.0
12.1-24.0
24.1-36.0
36.1-48.0
48.1-72.0
>72.0

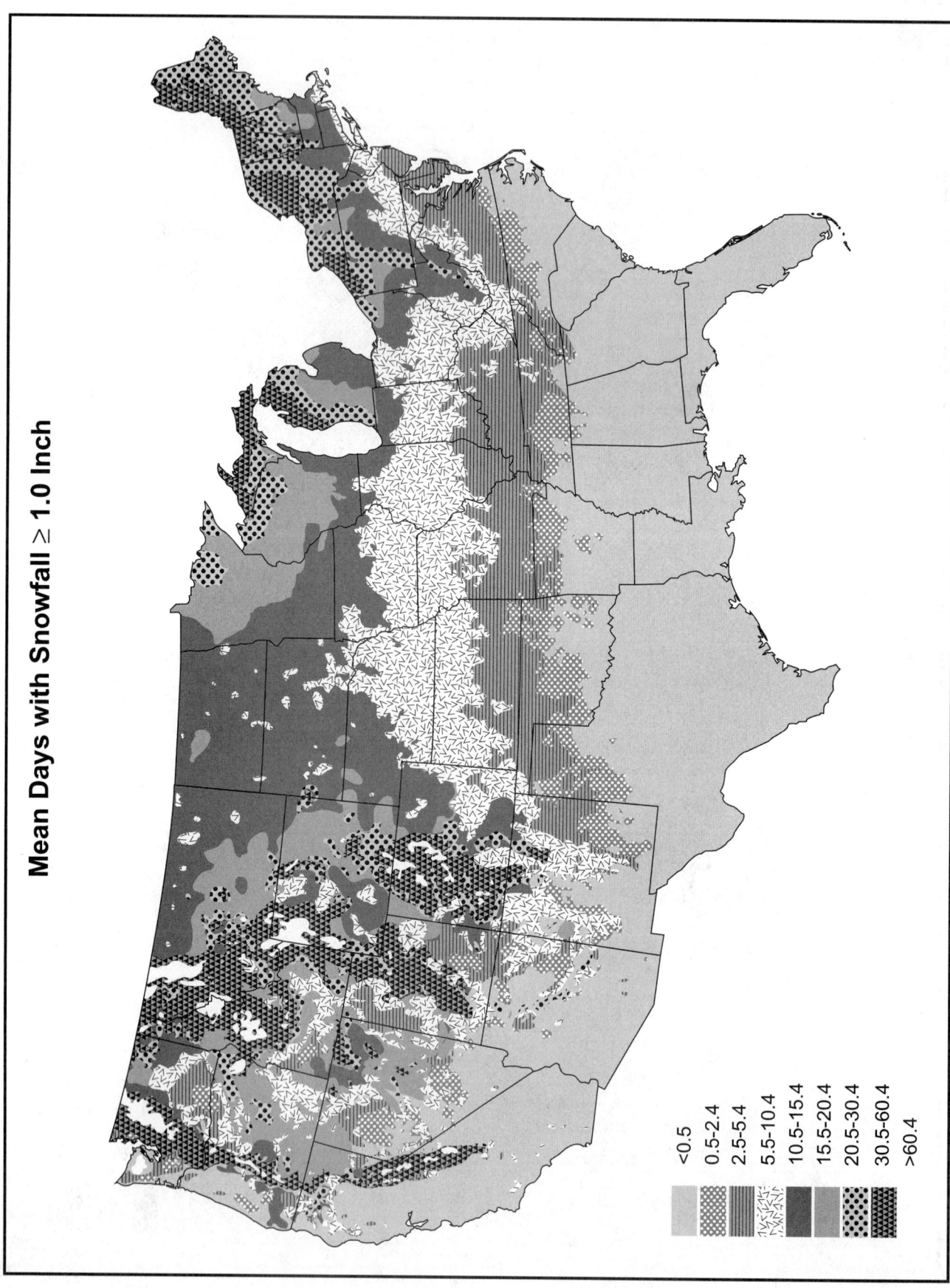

Mean Days with Snowfall ≥ 1.0 Inch

<0.5
0.5-2.4
2.5-5.4
5.5-10.4
10.5-15.4
15.5-20.4
20.5-30.4
30.5-60.4
>60.4

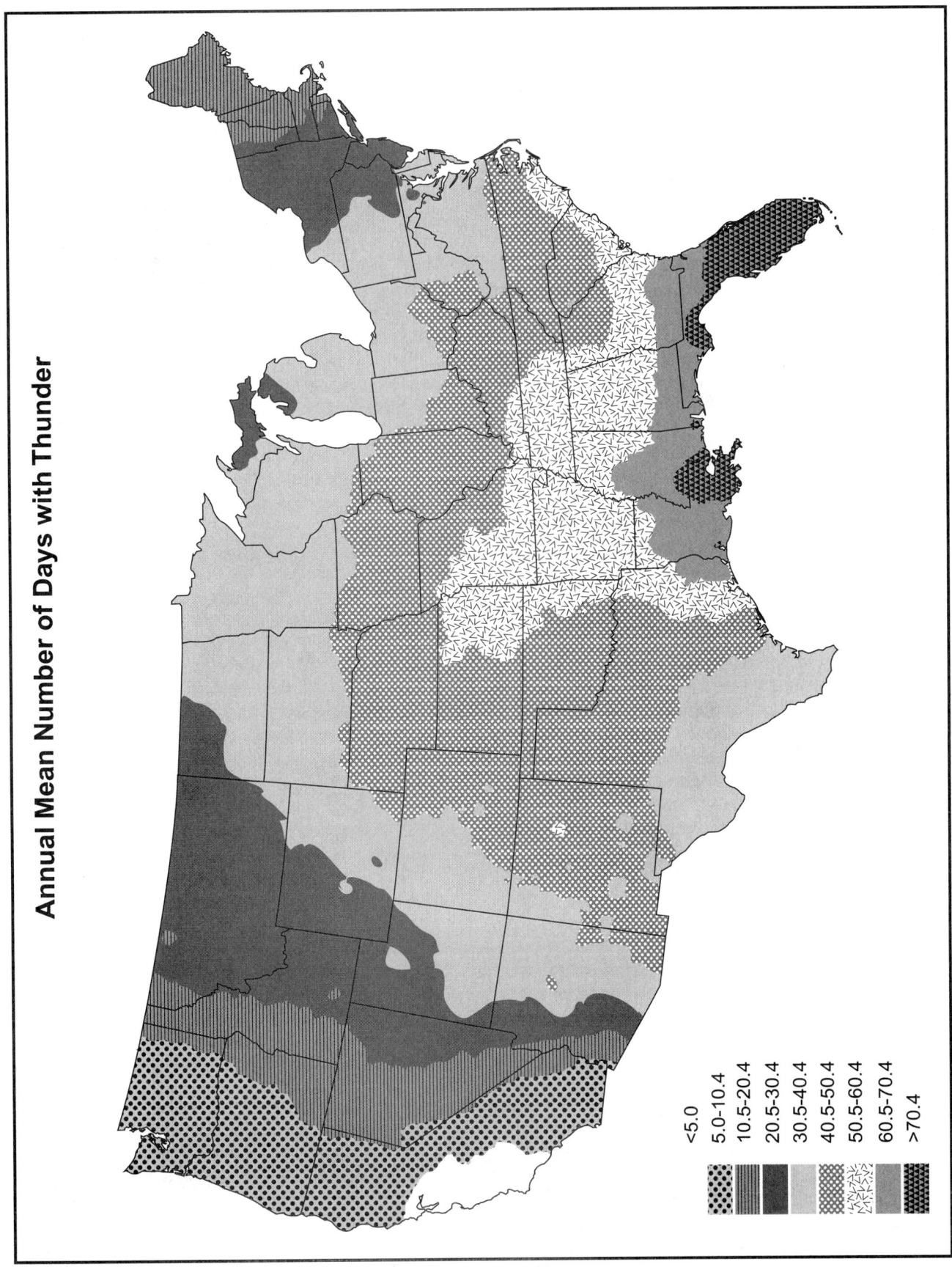

Annual Mean Number of Days with Thunder

<5.0
5.0-10.4
10.5-20.4
20.5-30.4
30.5-40.4
40.5-50.4
50.5-60.4
60.5-70.4
>70.4

Annual Mean Wind Speed (mph) and Prevailing Direction (arrows fly with the wind)

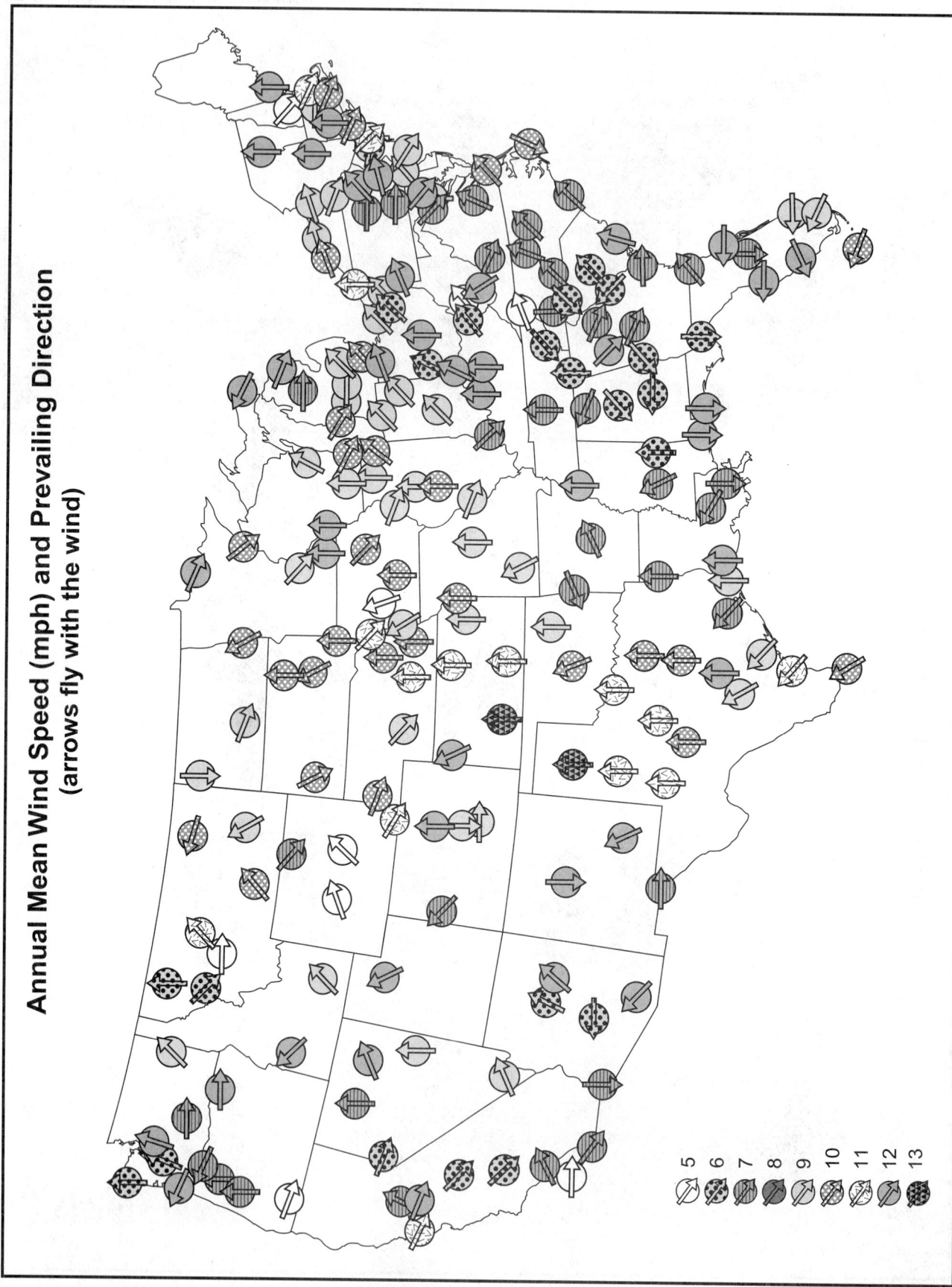

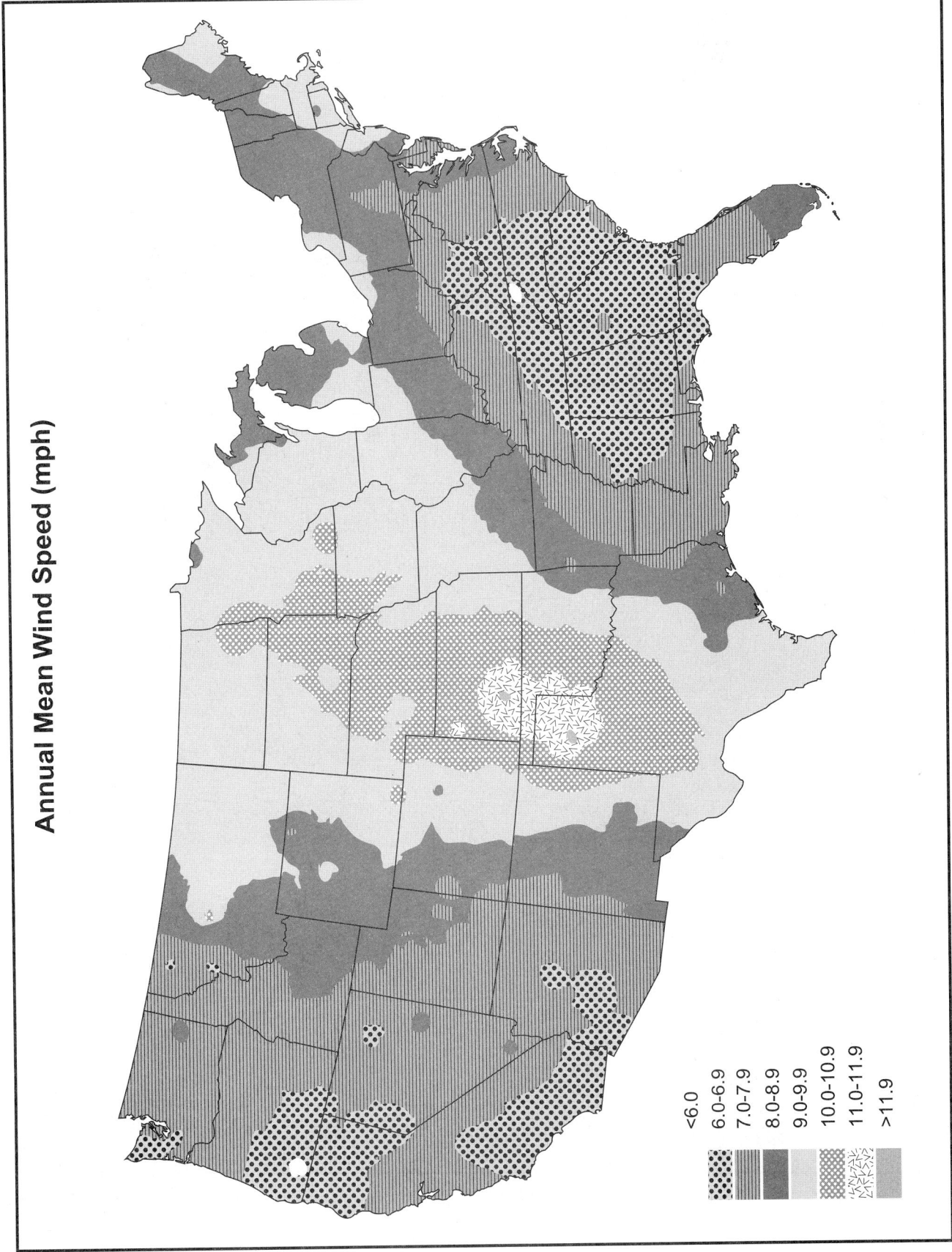

Annual Mean Wind Speed (mph)

<6.0
6.0-6.9
7.0-7.9
8.0-8.9
9.0-9.9
10.0-10.9
11.0-11.9
>11.9

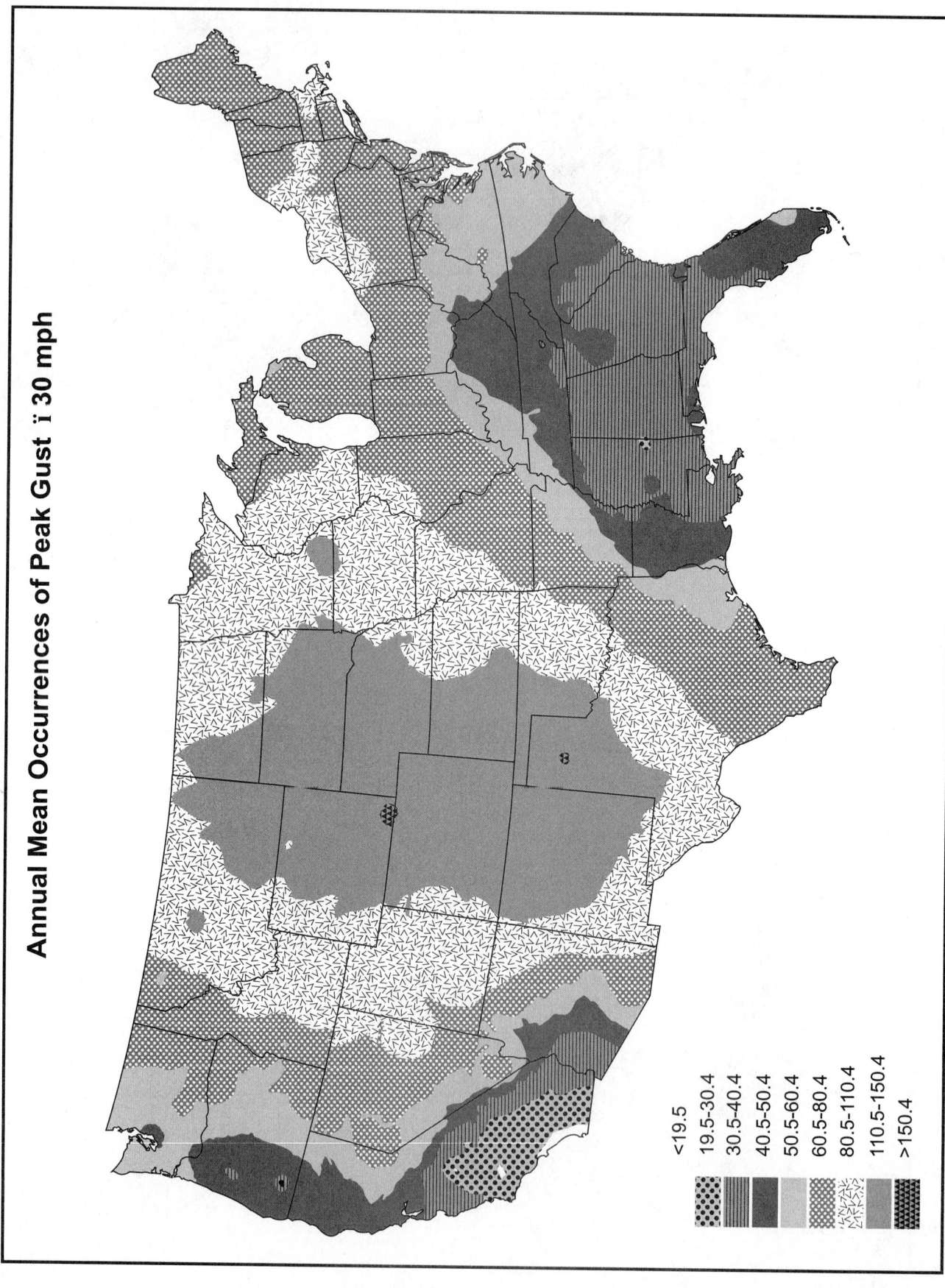

Annual Mean Occurrences of Peak Gust ï 30 mph

<19.5
19.5-30.4
30.5-40.4
40.5-50.4
50.5-60.4
60.5-80.4
80.5-110.4
110.5-150.4
>150.4

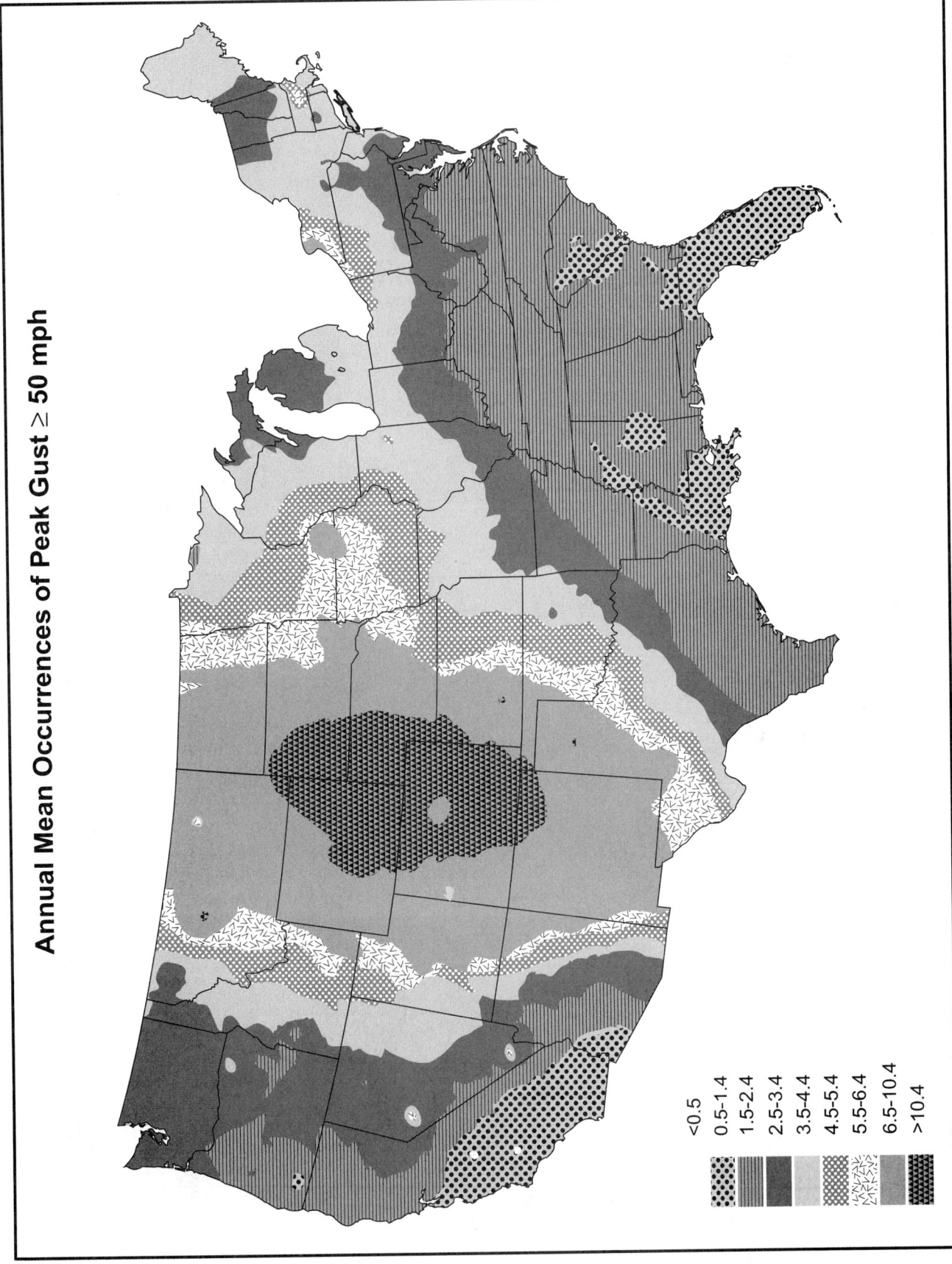

Annual Mean Occurrences of Peak Gust ≥ 50 mph

<0.5
0.5-1.4
1.5-2.4
2.5-3.4
3.5-4.4
4.5-5.4
5.5-6.4
6.5-10.4
>10.4

Annual Mean Sea Level Pressure (millibars)

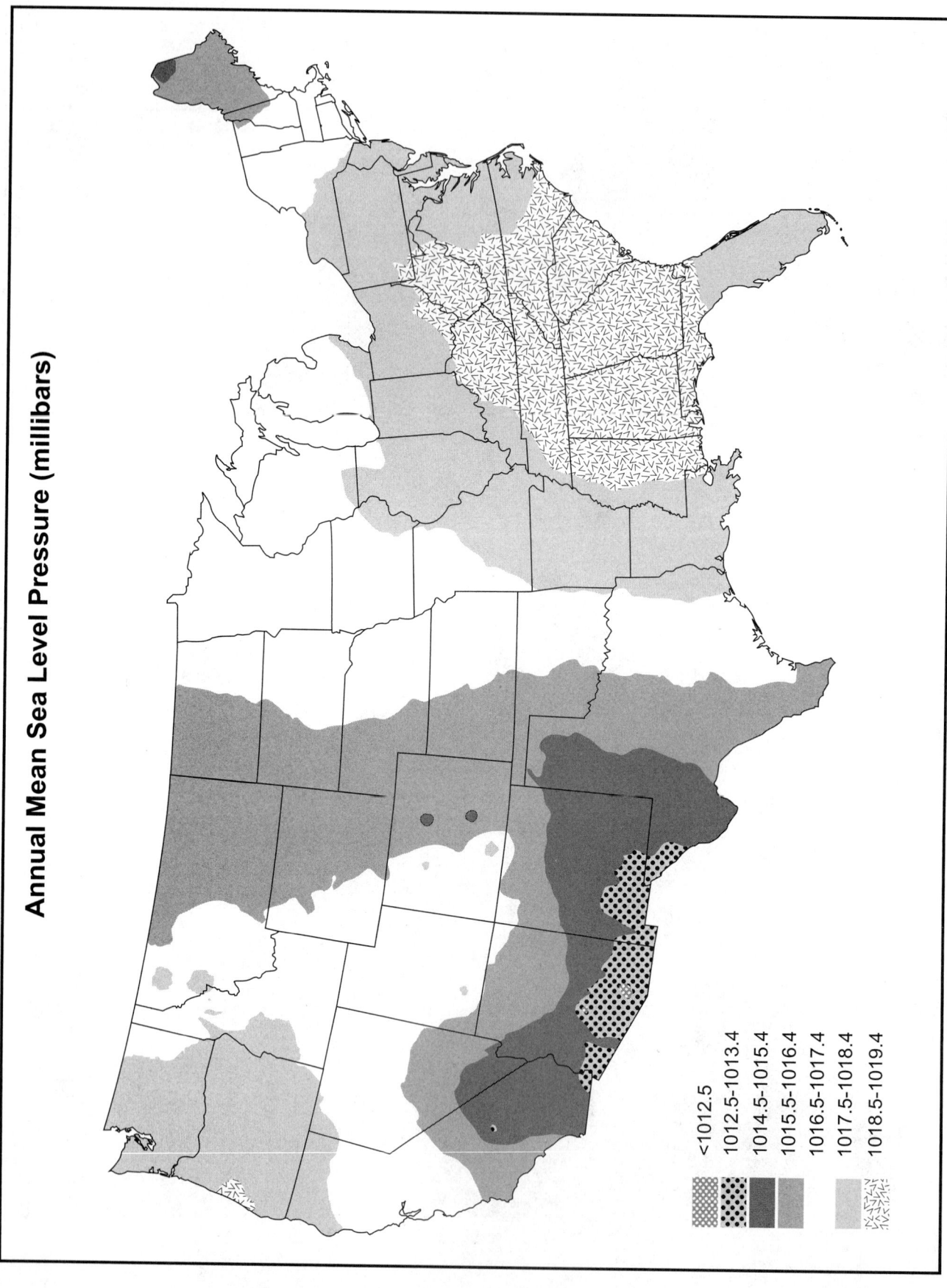

<1012.5
1012.5-1013.4
1014.5-1015.4
1015.5-1016.4
1016.5-1017.4
1017.5-1018.4
1018.5-1019.4

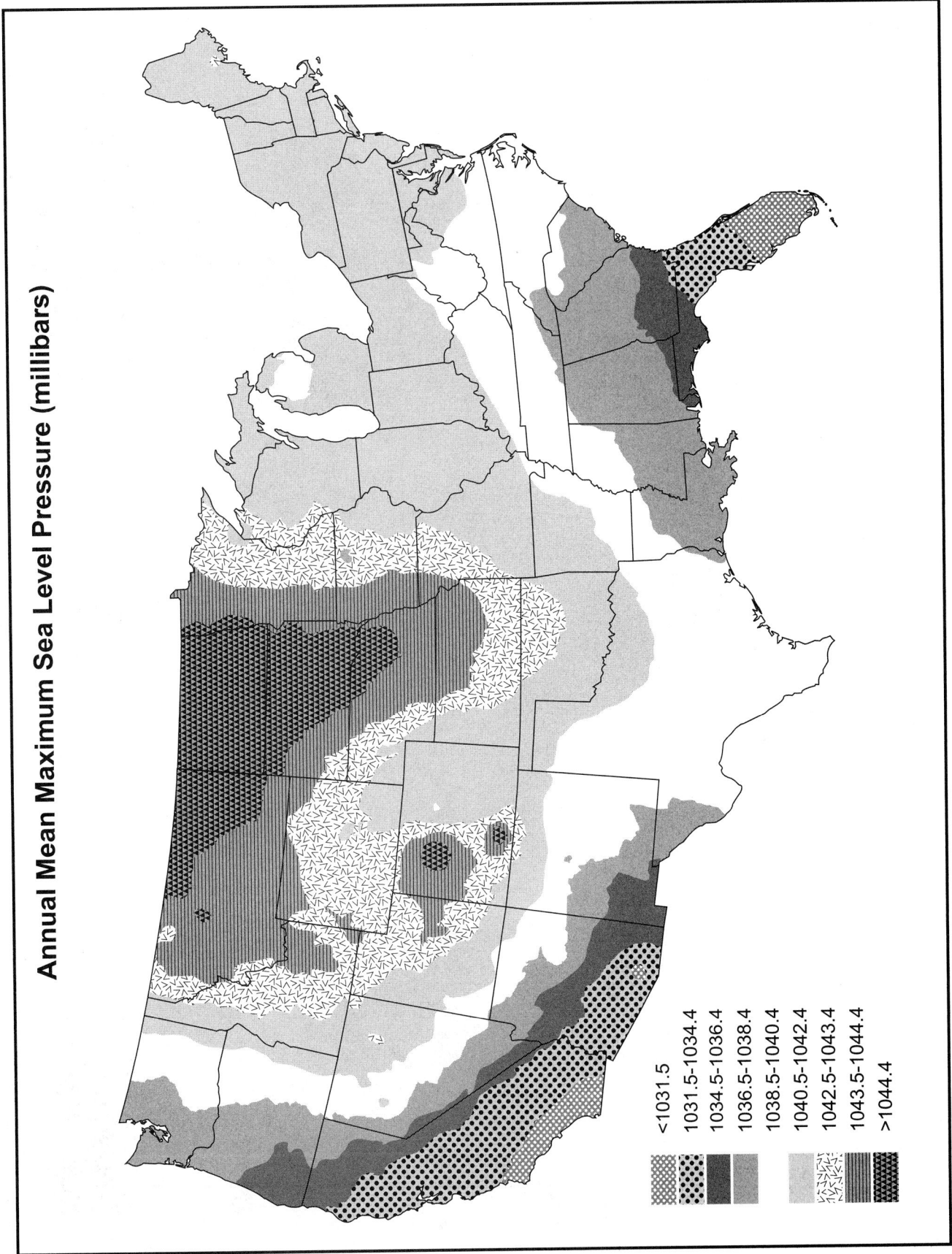

Annual Mean Maximum Sea Level Pressure (millibars)

<1031.5
1031.5-1034.4
1034.5-1036.4
1036.5-1038.4
1038.5-1040.4
1040.5-1042.4
1042.5-1043.4
1043.5-1044.4
>1044.4

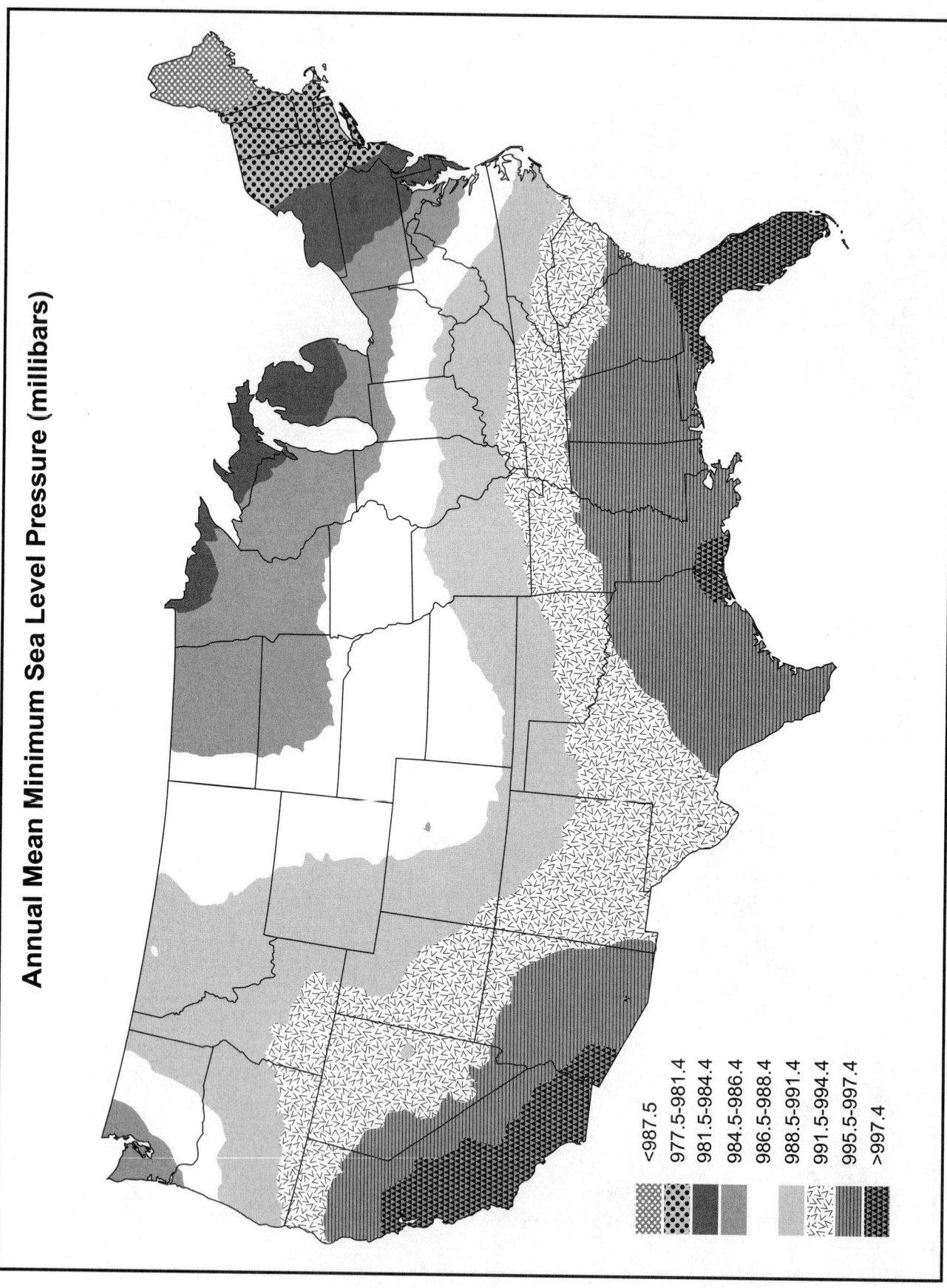

Annual Mean Minimum Sea Level Pressure (millibars)

<987.5
977.5-981.4
981.5-984.4
984.5-986.4
986.5-988.4
988.5-991.4
991.5-994.4
995.5-997.4
>997.4

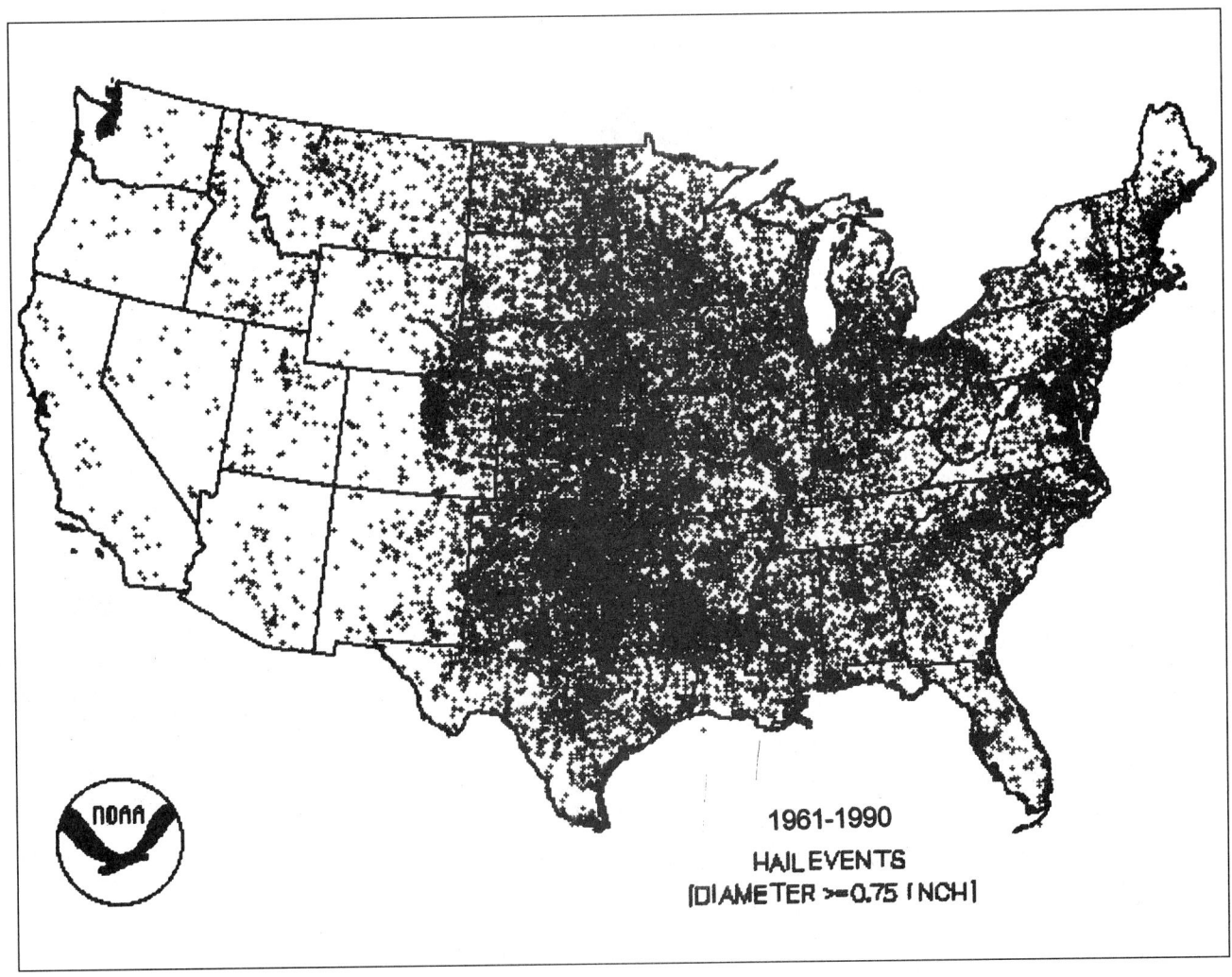

(Courtesy of National Oceanic and Atmospheric Administration (NOAA).)

Billion Dollar U.S. Weather Disasters 1980 - 2000

National Climatic Data Center
Asheville, NC   12/21/2000

The U.S. has sustained 48 weather-related disasters over the past 21 years in which overall damages/costs reached or exceeded $1 billion. 41 of these disasters occurred during the 1988-2000 period with total damages/costs exceeding $180 billion. Seven occurred during 1998 alone–the most for any year on record, though other years have recorded higher damage totals. Events are listed below beginning with the most recent.

Two damage figures are given for events prior to 1996–the first figure represents actual dollar costs at the time of the event and is not adjusted for inflation. Therefore, event costs over time should not be compared using this value. The second value in parenthesis (if given) is the dollar costs normalized to 1998 dollars using a GNP inflation/wealth index. The total normalized losses for the 48 events are nearly $280 billion.

1. **Drought/Heat Wave** *Spring-Summer 2000. Severe drought and persistent heat over south-central and southeastern states causing significant losses to agriculture and related industries; preliminary estimate of over $4.0 billion in damage/costs; estimated 140 deaths nationwide.*

2. **Western Fire Season** *Spring-Summer 2000. Severe fire season in western states due to drought and frequent winds, with nearly 7 million acres burned; estimate of over $2.0 billion in damage/costs (includes fire suppression); no deaths reported.*

Western Wildfires Summer 2000

3. **Hurricane Floyd** *September 1999. Large, category 2 hurricane makes landfall in eastern NC, causing 10-20 inch rains in 2 days, with severe flooding in NC and some flooding in SC, VA, MD, PA, NY, NJ, DE, RI, CT, MA, NH, and VT; estimate of at least $6.0 billion damage/costs; 77 deaths.*

4. **Eastern Drought/Heat Wave** *Summer 1999. Very dry summer and high temperatures, mainly in eastern U.S., with extensive agricultural losses; over $1.0 billion damage/costs; estimated 502 deaths.*

5. **Oklahoma-Kansas Tornadoes** *May 1999. Outbreak of F4-F5 tornadoes hit the states of Oklahoma and Kansas, along with Texas and Tennessee, Oklahoma City area hardest hit; over $1.1 billion damage/costs; 55 deaths.*

6. **Arkansas-Tennessee Tornadoes** *January 1999. Two outbreaks of tornadoes in 6-day period strike Arkansas and Tennessee; approximately $1.3 billion damage/costs; 17 deaths.*

7. **Texas Flooding** *October-November 1998. Severe flooding in southeast Texas from 2 heavy rain events, with 10-20 inch rainfall totals; approximately $1.0 billion damage/costs; 31 deaths.*

8. **Hurricane Georges** *September 1998. Category 2 hurricane strikes Puerto Rico, Florida Keys, and Gulf coasts of Louisiana, Mississippi, Alabama, and Florida panhandle, 15-30 inch 2-day rain totals in parts of AL/FL; estimated $5.9 billion damage/costs; 16 deaths.*

9. **Hurricane Bonnie** *August 1998. Category 3 hurricane strikes eastern North Carolina and Virginia, extensive agricultural damage due to winds and flooding, with 10-inch rains in 2 days in some locations; approximately $1.0 billion damage/costs; 3 deaths.*

10. **Southern Drought/Heat Wave** *Summer 1998. Severe drought and heat wave from Texas/Oklahoma eastward to the Carolinas; $6.0-$9.0 billion damage/costs to agriculture and ranching; at least 200 deaths.*

11. **Minnesota Severe Storms/Hail** *May 1998. Very damaging severe thunderstorms with large hail over wide areas of Minnesota; over $1.5 billion damage/costs; 1 death.*

12. **Southeast Severe Weather** *Winter-Spring 1998. Tornadoes and flooding related to El Nino in southeastern states; over $1.0 billion damage/costs; at least 132 deaths.*

13. **Northeast Ice Storm** *January 1998. Intense ice storm hits Maine, New Hampshire, Vermont, and New York, with extensive forestry losses; over $1.4 billion damage/costs; 16 deaths.*

14. **Northern Plains Flooding** *April-May 1997. Severe flooding in Dakotas and Minnesota due to heavy spring snowmelt; approximately $3.7 billion damage/costs; 11 deaths.*

15. **MS and OH Valleys Flooding & Tornadoes** *March 1997. Tornadoes and severe flooding hit the states of AR, MO, MS, TN, IL, IN, KY, OH, and WV, with over 10 inches of rain in 24 hours in Louisville; estimated $1.0 billion damage/costs; 67 deaths.*

16. **West Coast Flooding** *December 1996-January 1997. Torrential rains (10-40 inches in 2 weeks) and snowmelt produce severe flooding over portions of CA, WA, OR, ID, NV, and MT; approximately $3.0 billion damage/costs; 36 deaths.*

17. **Hurricane Fran** *September 1996. Category 3 hurricane strikes North Carolina and Virginia, over 10-inch 24-hour rains in some locations and extensive agricultural and other losses; over $5.0 billion damage/costs; 37 deaths.*

(Courtesy of National Climatic Data Center/NOAA.)

18. **Southern Plains Severe Drought** *Fall 1995 through Summer 1996. Severe drought in agricultural regions of southern plains--Texas and Oklahoma most severely affected; approximately $5.0 billion damage/costs; no deaths.*

19. **Pacific Northwest Severe Flooding** *February 1996. Very heavy, persistent rains (10-30 inches) and melting snow over OR, WA, ID, and western MT; approximately $1.0 billion damage/costs; 9 deaths.*

20. **Blizzard of '96 Followed by Flooding** *January 1996. Very heavy snowstorm (1-4 feet) over Appalachians, Mid-Atlantic, and Northeast; followed by severe flooding in parts of same area due to rain & snowmelt; approximately $3.0 billion damage/costs; 187 deaths.*

21. **Hurricane Opal** *October 1995. Category 3 hurricane strikes Florida panhandle, Alabama, western Georgia, eastern Tennessee, and the western Carolinas, causing storm surge, wind, and flooding damage; over $3.0 (3.3) billion damage/costs; 27 deaths.*

22. **Hurricane Marilyn** *September 1995. Category 2 hurricane devastates U.S. Virgin Islands; estimated $2.1 (2.3) billion damage/costs; 13 deaths*

23. **Texas/Oklahoma/Louisiana/Mississippi Severe Weather and Flooding** *May 1995. Torrential rains, hail, and tornadoes across Texas - Oklahoma and southeast Louisiana - southern Mississippi, with Dallas and New Orleans areas (10-25 inch rains in 5 days) hardest hit; $5.0-$6.0 (5.5-6.6) billion damage/costs; 32 deaths.*

24. **California Flooding** *January-March 1995. Frequent winter storms cause 20-70 inch rainfall and periodic flooding across much of California; over $3.0 (3.3) billion damage/costs; 27 deaths.*

25. **Western Fire Season** *Summer-Fall 1994. Severe fire season in western states due to dry weather; approximately $1.0 (1.1) billion damage/costs; death toll undetermined.*

26. **Texas Flooding** *October 1994. Torrential rain (10-25 inches in 5 days) and thunderstorms cause flooding across much of southeast Texas; approximately $1.0 (1.1) billion damage/costs; 19 deaths.*

27. **Tropical Storm Alberto** *July 1994. Remnants of slow-moving Alberto bring torrential 10-25 inch rains in 3 days, widespread flooding and agricultural damage in parts of Georgia, Alabama, and panhandle of Florida; approximately $1.0 (1.1) billion damage/costs; 32 deaths.*

28. **Southeast Ice Storm** *February 1994. Intense ice storm with extensive damage in portions of TX, OK, AR, LA, MS, AL, TN, GA, SC, NC, and VA; approximately $3.0 (3.3) billion damage/costs; 9 deaths.*

29. **California Wildfires** *Fall 1993. Dry weather, high winds and wildfires in Southern California; approximately $1.0 (1.1) billion damage/costs; 4 deaths.*

30. **Midwest Flooding** *Summer 1993. Severe, widespread flooding in central U.S. due to persistent heavy rains and thunderstorms; approximately $21.0 (23.1) billion damage/costs; 48 deaths.*

31. **Drought/Heat Wave** *Summer 1993. Southeastern U.S.; about $1.0 (1.1) billion damage/costs to agriculture; at least 16 deaths.*

32. **Storm/Blizzard** *March 1993. "Storm of the Century" hits entire eastern seaboard with tornadoes (FL), high winds, and heavy snows (2-4 feet); $3.0-$6.0 (3.3-6.6) billion damage/costs; approximately 270 deaths.*

33. **Nor'easter of 1992** *December 1992. Slow-moving storm batters northeast U.S. coast, New England hardest hit; $1.0-$2.0 (1.2-2.4) billion damage/costs; 19 deaths.*

34. **Hurricane Iniki** *September 1992. Category 4 hurricane hits Hawaiian island of Kauai; about $1.8 (2.2) billion damage/costs; 7 deaths.*

35. **Hurricane Andrew** *August 1992. Category 4 hurricane hits Florida and Louisiana, high winds damage or destroy over 125,000 homes; approximately $27.0 (32.4) billion damage/costs; 61 deaths.*

36. **Oakland Firestorm** *October 1991. Oakland, California firestorm due to low humidities and high winds; approximately $2.5 (3.3) billion damage/costs; 25 deaths.*

37. **Hurricane Bob** *August 1991. Category 2 hurricane--Mainly coastal North Carolina, Long Island, and New England; $1.5 (2.0) billion damage/costs; 18 deaths.*

38. **Texas/Oklahoma/Louisiana/Arkansas Flooding** *May 1990. Torrential rains cause flooding along the Trinity, Red, and Arkansas Rivers in TX, OK, LA, and AR; over $1.0 (1.3) billion damage/costs; 13 deaths.*

39. **Hurricane Hugo** *September 1989. Category 4 hurricane devastates South and North Carolina with ~ 20 foot storm surge and severe wind damage after hitting Puerto Rico and the U.S. Virgin Islands; over $9.0 (12.6) billion damage/costs (about $7.1 (9.9) billion in Carolinas); 86 deaths (57--U.S. mainland, 29--U.S. Islands).*

40. **Northern Plains Drought** *Summer 1989. Severe summer drought over much of the northern plains with significant losses to agriculture; at least $1.0 (1.4) billion in damage/costs; no deaths reported.*

41. **Drought/Heat Wave** *Summer 1988. 1988 drought in central and eastern U.S. with very severe losses to agriculture and related industries; estimated $40.0 (56.0) billion damage/costs; estimated 5,000 to 10,000 deaths (includes heat stress-related).*

42. **Southeast Drought/Heat Wave** *Summer 1986. Severe summer drought in parts of the southeastern U.S. with severe losses to agriculture; $1.0-$1.5 (1.6-2.4) billion in damage/costs; estimated 100 deaths.*

43. **Hurricane Juan** *October-November 1985. Category 1 hurricane-- Louisiana and Southeast U.S.--severe flooding; $1.5 (2.6) billion damage/costs; 63 deaths.*

44. **Hurricane Elena** *August-September 1985. Category 3 hurricane-- Florida to Louisiana; $1.3 (2.2) billion damage/costs; 4 deaths.*

45. **Florida Freeze** *January 1985. Severe freeze central/northern Florida; about $1.2 (2.0) billion damage to citrus industry; no deaths.*

46. **Florida Freeze** *December 1983. Severe freeze central/northern Florida; about $2.0 (3.6) billion damage to citrus industry; no deaths.*

47. **Hurricane Alicia** *August 1983. Category 3 hurricane--Texas; $3.0 (5.4) billion damage/costs; 21 deaths.*

48. **Drought/Heat Wave** *June-September 1980. Central and eastern U.S.; estimated $20.0 (44.0) billion damage/costs to agriculture and related industries; estimated 10,000 deaths (includes heat stress-related).*

These statistics were taken from a wide variety of sources and represent, to the best of our ability, the estimated total costs of these events---that is, the costs in terms of dollars and lives that would not have been incurred had the event not taken place. Insured and uninsured losses are included in damage estimates, and direct plus indirect deaths (i.e., related to the event, would not have occurred otherwise) are included in fatality totals. Economic costs are included for wide-scale, long-lasting events such as drought.

(Courtesy of National Climatic Data Center/NOAA.)

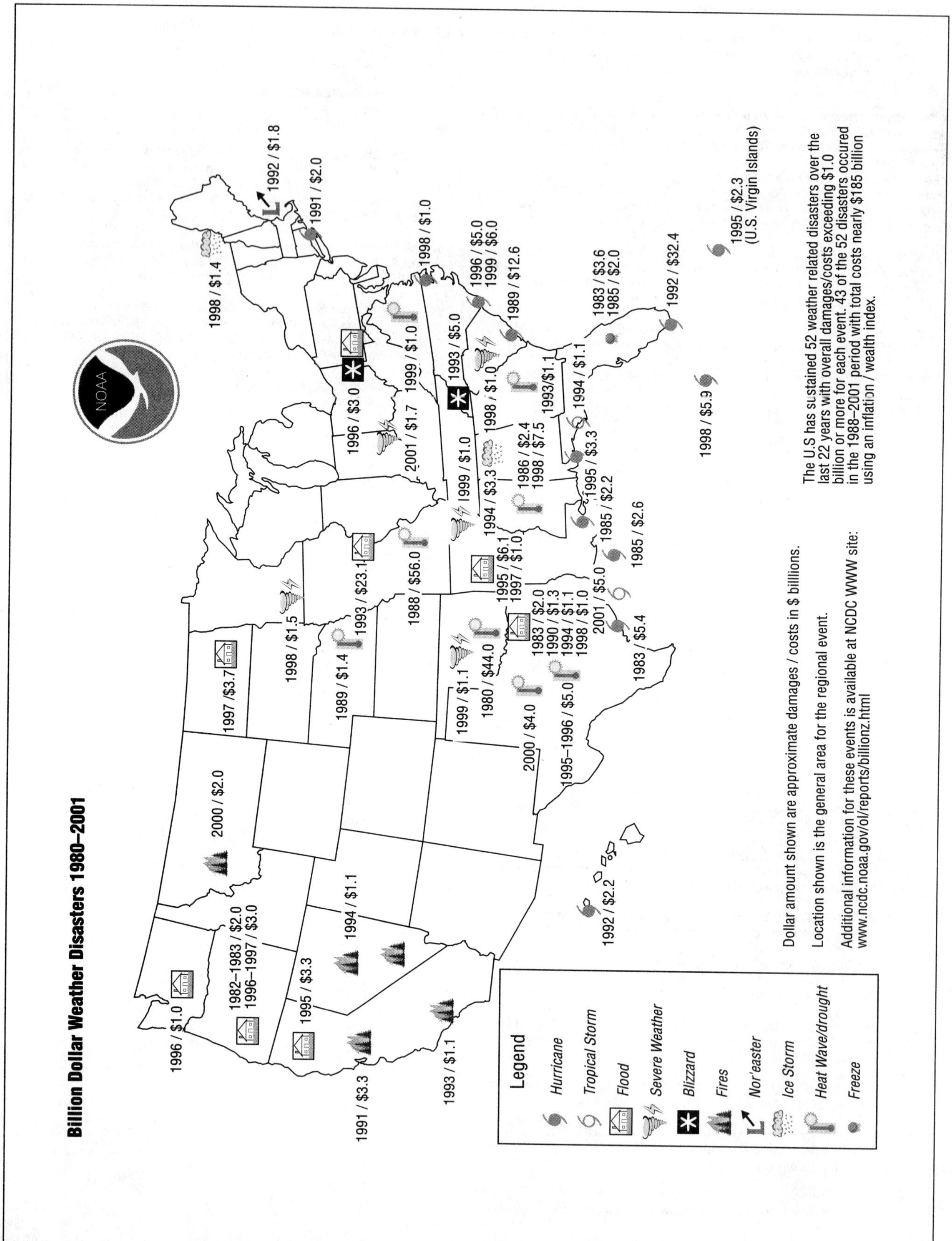

Billion Dollar Weather Disasters 1980–2001

1998 / $1.4

1992 / $1.8

1991 / $2.0

1998 / $1.0

1996 / $5.0
1999 / $6.0

1989 / $12.6

1995 / $2.3
(U.S. Virgin Islands)

1983 / $3.6
1985 / $2.0

1992 / $32.4

1996 / $3 0

2001 / $1.7   1999 / $1.0

1993 / $5.0

1998 / $1.0

1993/$1.1

1994 / $1.1

1999 / $1.0   1994 / $3.3   1986 / $2.4
1998 / $7.5

1995 / $3.3

1998 / $5.9

1997 / $3.7

1998 / $1.5

1989 / $1.4

1993 / $23.1

1988 / $56.0

1994 / $3.0

1995 / $6.1
1997 / $1.0

1995 / $2.2

1985 / $2.6

1983 / $2.0
1990 / $1.3
1994 / $1.1
1998 / $1.0

2001 / $5.0

1983 / $5.4

2000 / $2.0

1999 / $1.1

1980 / $44.0

2000 / $4.0

1995–1996 / $5.0

1982–1983 / $2.0
1996–1997 / $3.0

1994 / $1.1

1992 / $2.2

1996 / $1.0

1995 / $3.3

1993 / $1.1

1991 / $3.3

The U.S has sustained 52 weather related disasters over the last 22 years with overall damages/costs exceeding $1.0 billion or more for each event. 43 of the 52 disasters occured in the 1988–2001 period with total costs nearly $185 billion using an inflation / wealth index.

Dollar amount shown are approximate damages / costs in $ billlions.

Location shown is the general area for the regional event.

Additional information for these events is available at NCDC WWW site: www.ncdc.noaa.gov/ol/reports/billionz.html

Legend

Hurricane

Tropical Storm

Flood

Severe Weather

Blizzard

Fires

Nor'easter

Ice Storm

Heat Wave/drought

Freeze

(Courtesy of National Oceanic and Atmospheric Administration (NOAA).)

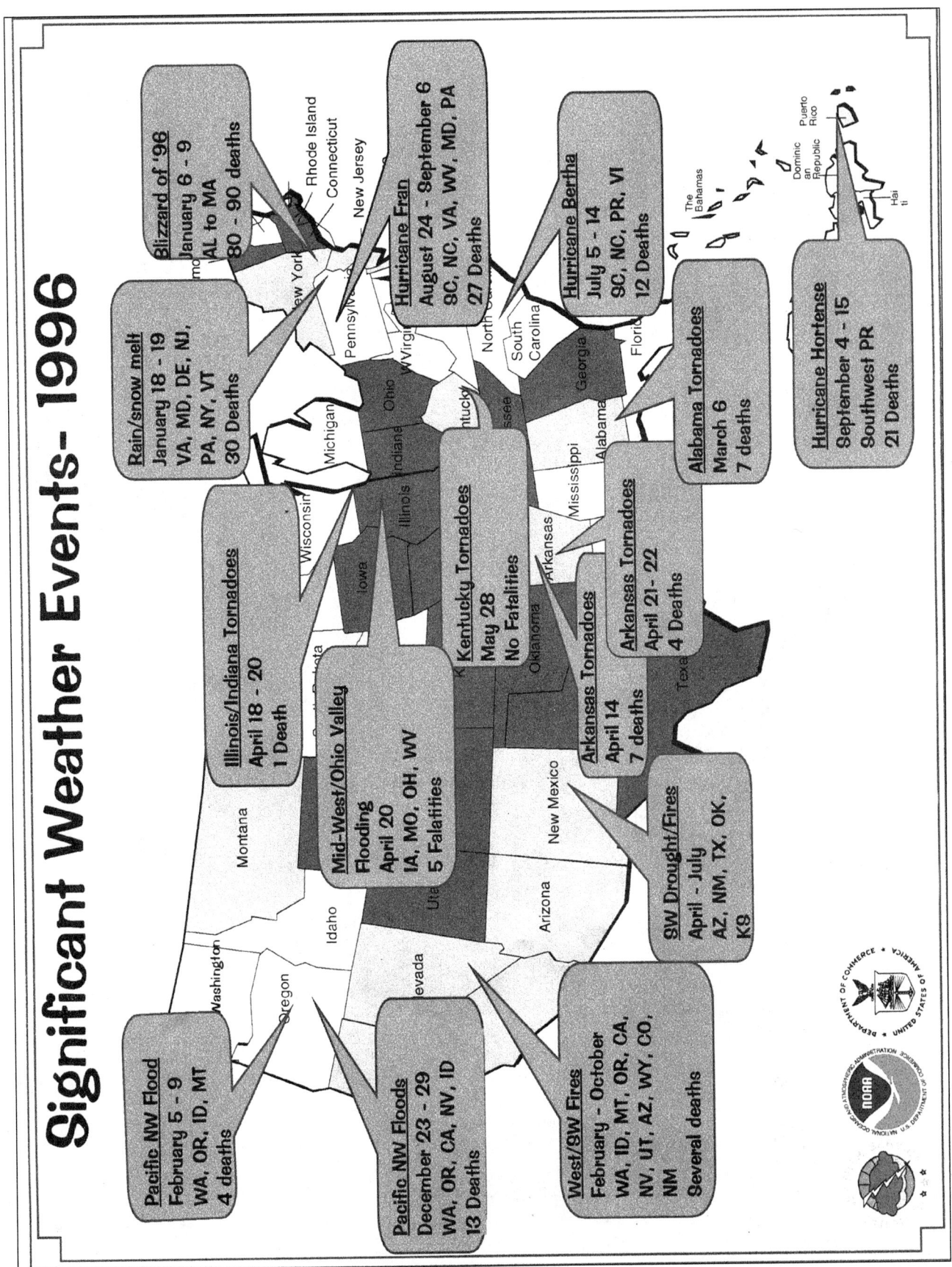

(Courtesy of National Oceanic and Atmospheric Administration (NOAA).)

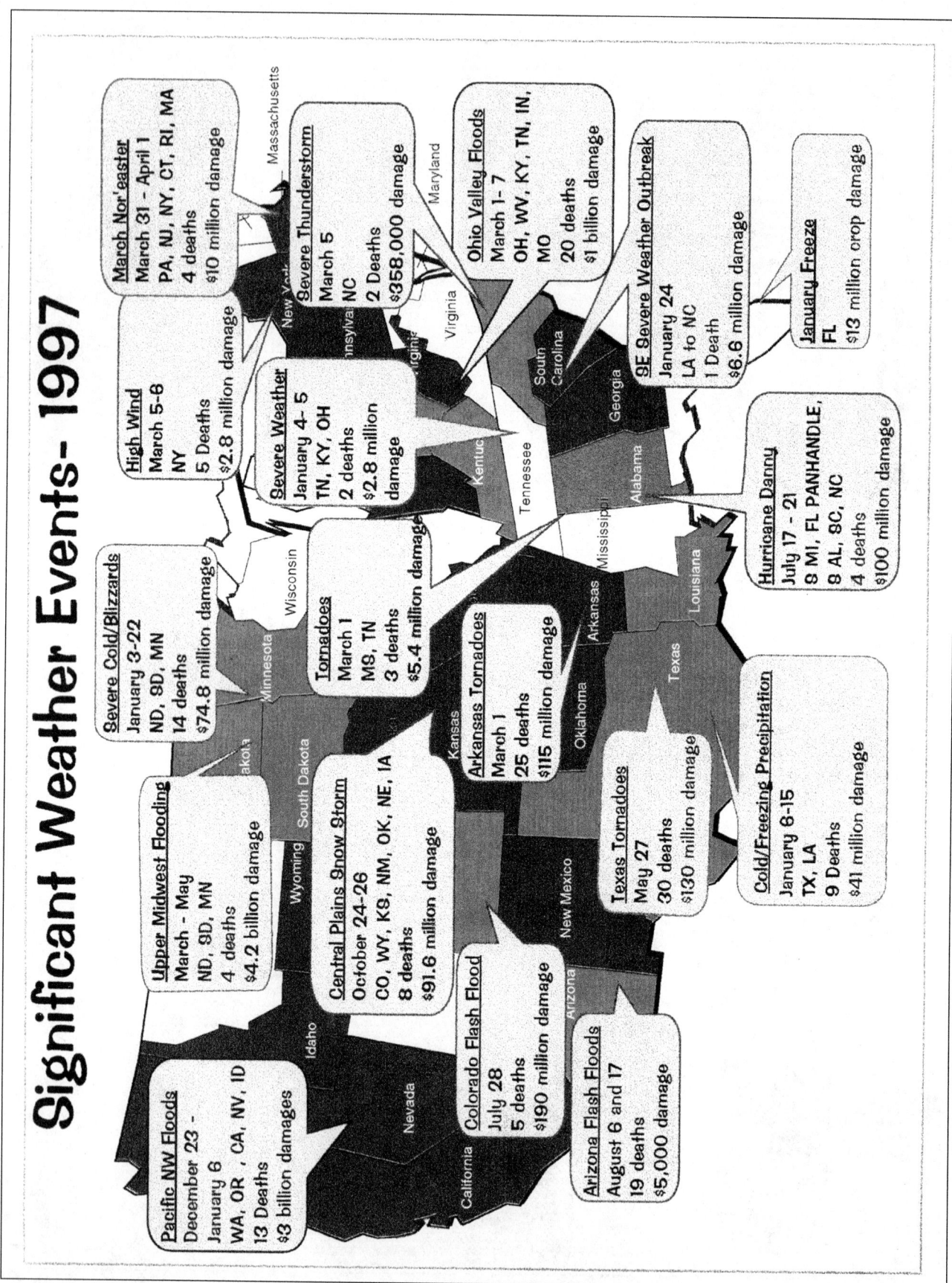

(Courtesy of National Oceanic and Atmospheric Administration (NOAA).)

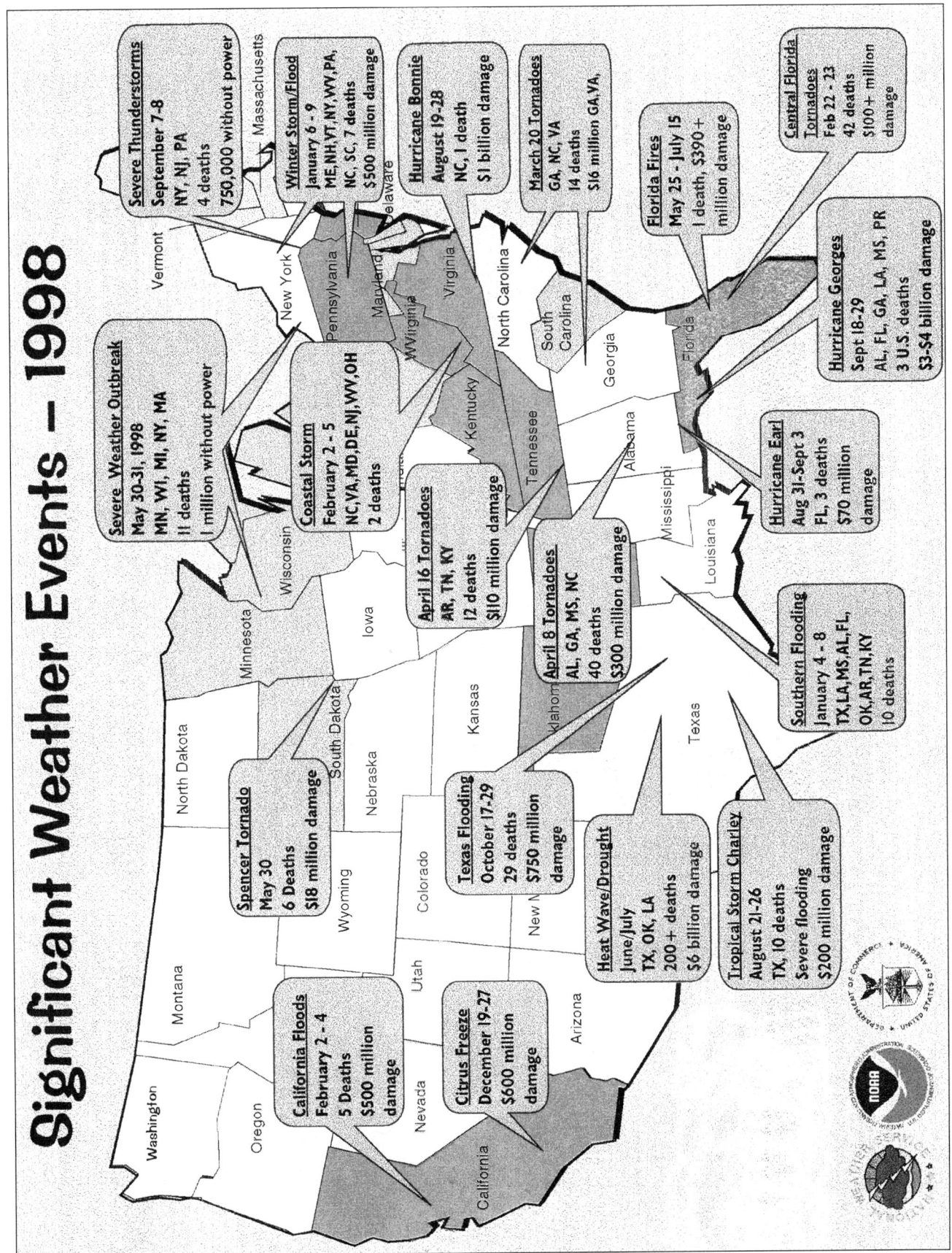

(Courtesy of National Oceanic and Atmospheric Administration (NOAA).)

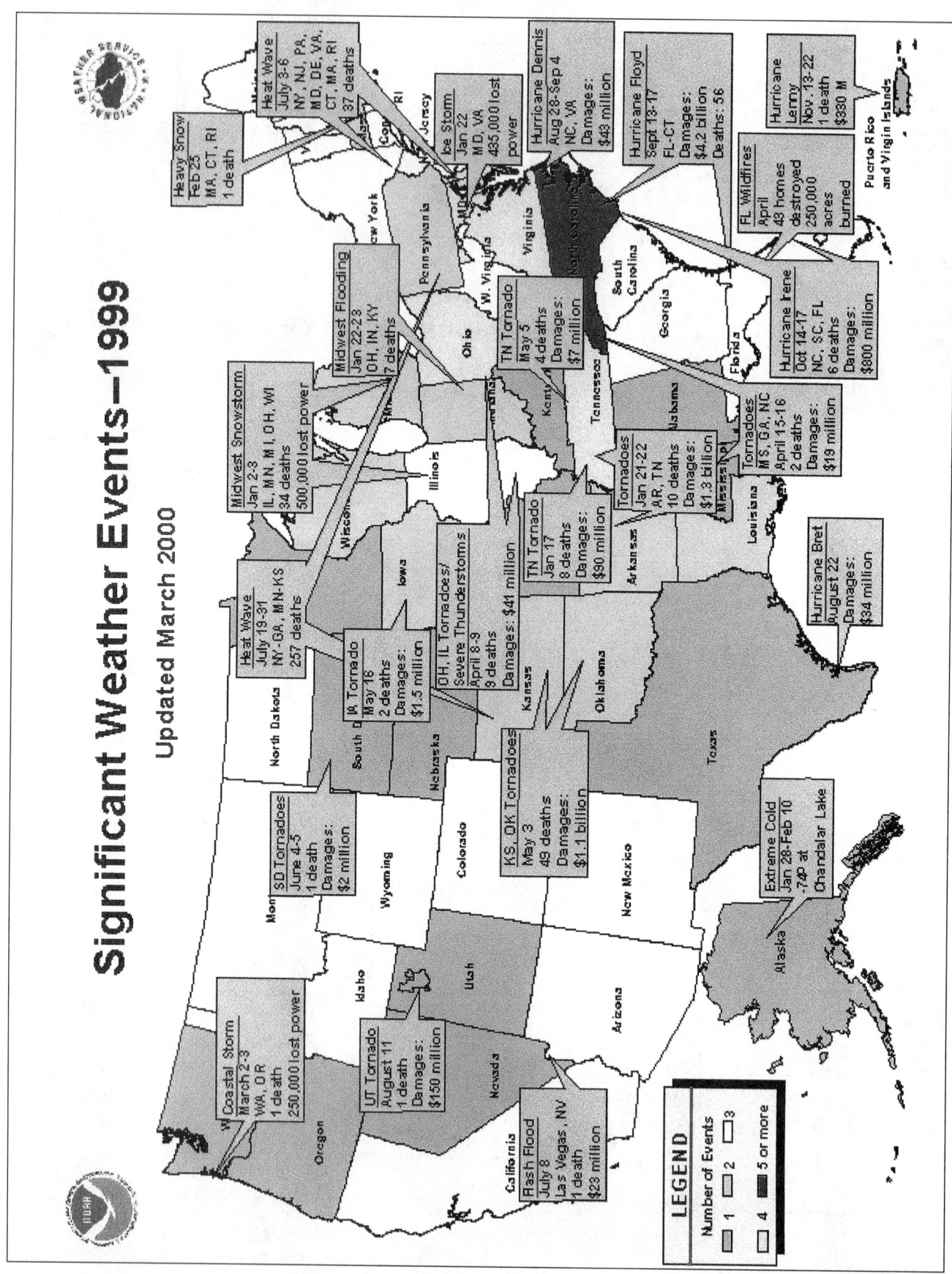

Significant Weather Events—1999

Updated March 2000

Heavy Snow
Feb 25
MA, CT, RI
1 death

Heat Wave
July 3-6
NY, NJ, PA,
MD, DE, VA,
CT, MA, RI
37 deaths

Ice Storm
Jan 22
MD, VA
435,000 lost power

Hurricane Dennis
Aug 28-Sep 4
NC, VA
Damages:
$43 million

Hurricane Floyd
Sept 13-17
FL-CT
Damages:
$4.2 billion
Deaths: 56

Hurricane Lenny
Nov. 13-22
1 death
$330 M

FL Wildfires
April
43 homes
destroyed
250,000
acres
burned

Hurricane Irene
Oct 14-17
NC, SC, FL
6 deaths
Damages:
$800 million

Midwest Flooding
Jan 22-23
OH, IN, KY
7 deaths

Midwest Snowstorm
Jan 2-3
IL, MN, MI, OH, WI
34 deaths
500,000 lost power

TN Tornado
May 5
4 deaths
Damages:
$7 million

Tornadoes
April 15-16
MS, GA, NC
2 deaths
Damages:
$19 million

Heat Wave
July 19-31
NY-GA, MN-KS
257 deaths

IA Tornado
May 16
2 deaths
Damages:
$1.5 million

**OH, IL Tornadoes/
Severe Thunderstorms**
April 8-9
9 deaths
Damages: $41 million

TN Tornado
Jan 17
8 deaths
Damages:
$90 million

Tornadoes
Jan 21-22
AR, TN
10 deaths
Damages:
$1.3 billion

Hurricane Bret
August 22
Damages:
$34 million

SD Tornadoes
June 4-5
1 death
Damages:
$2 million

KS, OK Tornadoes
May 3
49 deaths
Damages:
$1.1 billion

Extreme Cold
Jan 28-Feb 10
-74° at
Chandalar Lake

Coastal Storm
March 2-3
WA, OR
1 death
250,000 lost power

UT Tornado
August 11
1 death
Damages:
$150 million

Flash Flood
July 8
Las Vegas, NV
1 death
Damages:
$23 million

Puerto Rico
and Virgin Islands

LEGEND

Number of Events

1 2 3
4 5 or more

(Courtesy of National Weather Service/NOAA.)

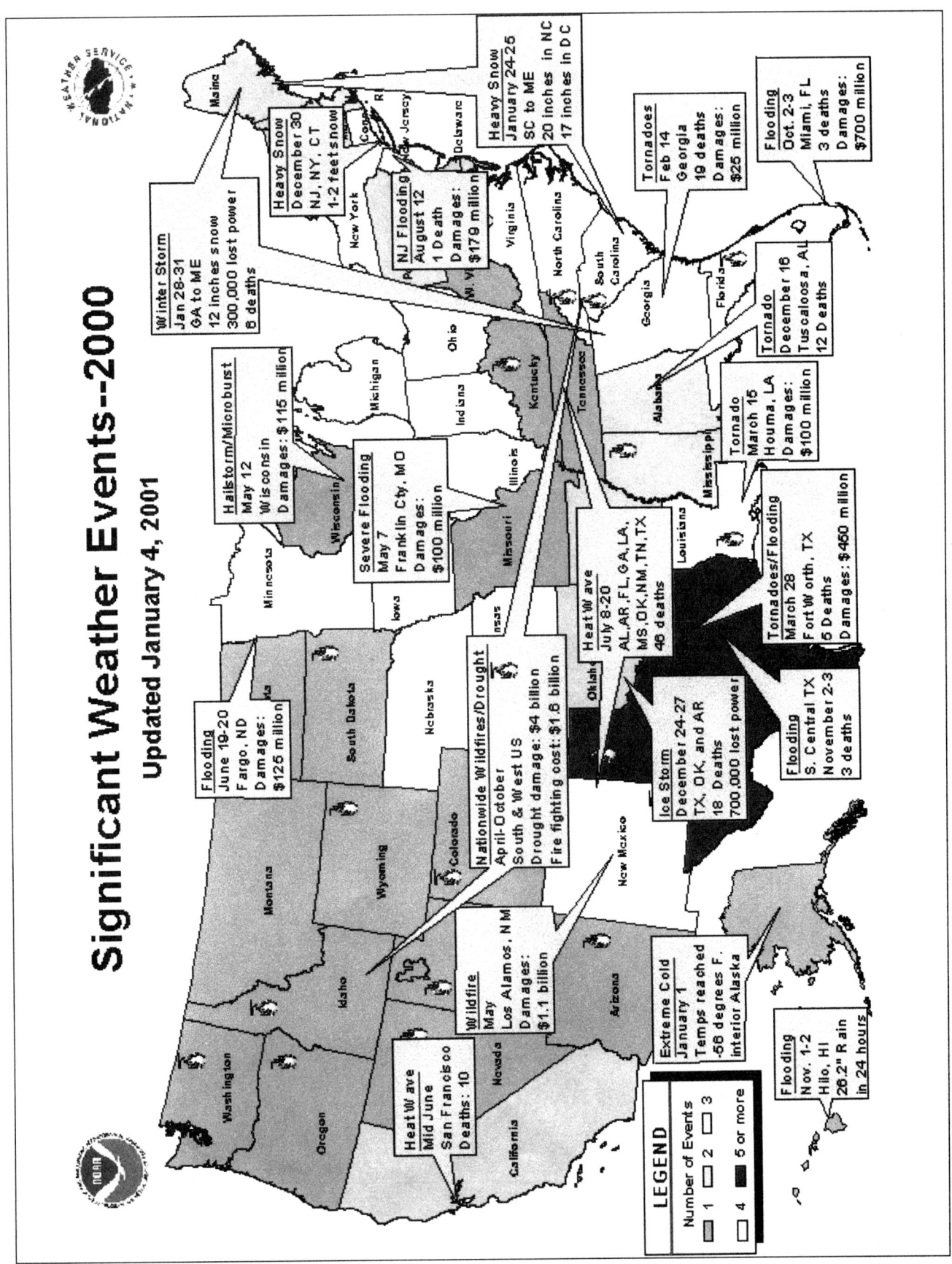

FAHRENHEIT TO CENTIGRADE

°F	°C	°F	°C	°F	°C	°F	°C	°F	°C
+130	+54.44	+80	+26.67	+30	-1.11	-20	-28.89	-70	-56.67
129	53.89	79	26.11	29	1.67	21	29.44	71	57.22
128	53.33	78	25.56	28	2.22	22	30.00	72	57.78
127	52.78	77	25.00	27	2.78	23	30.56	73	58.33
126	52.22	76	24.44	26	3.33	24	31.11	74	58.89
125	51.67	75	23.89	25	3.89	25	31.67	75	59.44
124	51.11	74	23.33	24	4.44	26	32.22	76	60.00
123	50.56	73	22.78	23	5.00	27	32.78	77	60.56
122	50.00	72	22.22	22	5.56	28	33.33	78	61.11
121	49.44	71	21.67	21	6.11	29	33.89	79	61.67
120	48.89	70	21.11	20	6.67	30	34.44	80	62.22
119	48.33	69	20.56	19	7.22	31	35.00	81	62.78
118	47.78	68	20.00	18	7.78	32	35.56	82	63.33
117	47.22	67	19.44	17	8.33	33	36.11	83	63.89
116	46.67	66	18.89	16	8.89	34	36.67	84	64.44
115	46.11	65	18.33	15	9.44	35	37.22	85	65.00
114	45.56	64	17.78	14	10.00	36	37.78	86	65.56
113	45.00	63	17.22	13	10.56	37	38.33	87	66.11
112	44.44	62	16.67	12	11.11	38	38.89	88	66.67
111	43.89	61	16.11	11	11.67	39	39.44	89	67.22
110	43.33	60	15.56	10	12.22	40	40.00	90	67.78
109	42.78	59	15.00	9	12.78	41	40.56	91	68.33
108	42.22	58	14.44	8	13.33	42	41.11	92	68.89
107	41.67	57	13.89	7	13.89	43	41.67	93	69.44
106	41.11	56	13.33	6	14.44	44	42.22	94	70.00
105	40.56	55	12.78	5	15.00	45	42.78	95	70.56
104	40.00	54	12.22	4	15.56	46	43.33	96	71.11
103	39.44	53	11.67	3	16.11	47	43.89	97	71.67
102	38.89	52	11.11	2	16.67	48	44.44	98	72.22
101	38.33	51	10.56	+1	17.22	49	45.00	99	72.78
100	37.78	50	10.00	0	17.78	50	45.56	100	73.33
99	37.22	49	9.44	-1	18.33	51	46.11	101	73.89
98	36.67	48	8.89	2	18.89	52	46.67	102	74.44
97	36.11	47	8.33	3	19.44	53	47.22	103	75.00
96	35.56	46	7.78	4	20.00	54	47.78	104	75.56
95	35.00	45	7.22	5	20.56	55	48.33	105	76.11
94	34.44	44	6.67	6	21.11	56	48.89	106	76.67
93	33.89	43	6.11	7	21.67	57	49.44	107	77.22
92	33.33	42	5.56	8	22.22	58	50.00	108	77.78
91	32.78	41	5.00	9	22.78	59	50.56	109	78.33
90	32.22	40	4.44	10	23.33	60	51.11	110	78.89
89	31.67	39	3.89	11	23.89	61	51.67	111	79.44
88	31.11	38	3.33	12	24.44	62	52.22	112	80.00
87	30.56	37	2.78	13	25.00	63	52.78	113	80.56
86	30.00	36	2.22	14	25.56	64	53.33	114	81.11
85	29.44	35	1.67	15	26.11	65	53.89	115	81.67
84	28.89	34	1.11	16	26.67	66	54.44	116	82.22
83	28.33	33	+0.56	17	27.22	67	55.00	117	82.78
82	27.78	32	0.00	18	27.78	68	55.56	118	83.33
81	27.22	31	-0.56	19	28.33	69	56.11	119	83.89
								120	84.44

(Chart by Datapage. Gale Group.)

INCHES TO MILLIMETERS

1 inch = 25.4 millimeters

in	mm	in	mm	in	mm	in	mm	in	mm	in	mm	in	mm
0.00	0.00	5.00	127.00	10.00	254.00	15.00	381.00	20.00	508.00	25.00	635.00	30.00	762.00
0.10	2.54	5.10	129.54	10.10	256.54	15.10	383.54	20.10	510.54	25.10	637.54	30.10	764.54
0.20	5.08	5.20	132.08	10.20	259.08	15.20	386.08	20.20	513.08	25.20	640.08	30.20	767.08
0.30	7.62	5.30	134.62	10.30	261.62	15.30	388.62	20.30	515.62	25.30	642.62	30.30	769.62
0.40	10.16	5.40	137.16	10.40	264.16	15.40	391.16	20.40	518.16	25.40	645.16	30.40	772.16
0.50	12.70	5.50	139.70	10.50	266.70	15.50	393.70	20.50	520.70	25.50	647.70	30.50	774.70
0.60	15.24	5.60	142.24	10.60	269.24	15.60	396.24	20.60	523.24	25.60	650.24	30.60	777.24
0.70	17.78	5.70	144.78	10.70	271.78	15.70	398.78	20.70	525.78	25.70	652.78	30.70	779.78
0.80	20.32	5.80	147.32	10.80	274.32	15.80	401.32	20.80	528.32	25.80	655.32	30.80	782.32
0.90	22.86	5.90	149.86	10.90	276.86	15.90	403.86	20.90	530.86	25.90	657.86	30.90	784.86
1.00	25.40	6.00	152.40	11.00	279.40	16.00	406.40	21.00	533.40	26.00	660.40	31.00	787.40
1.10	27.94	6.10	154.94	11.10	281.94	16.10	408.94	21.10	535.94	26.10	662.94	31.10	789.94
1.20	30.48	6.20	157.48	11.20	284.48	16.20	411.48	21.20	538.48	26.20	665.48	31.20	792.48
1.30	33.02	6.30	160.02	11.30	287.02	16.30	414.02	21.30	541.02	26.30	668.02	31.30	795.02
1.40	35.56	6.40	162.56	11.40	289.56	16.40	416.56	21.40	543.56	26.40	670.56	31.40	797.56
1.50	38.10	6.50	165.10	11.50	292.10	16.50	419.10	21.50	546.10	26.50	673.10	31.50	800.10
1.60	40.64	6.60	167.64	11.60	294.64	16.60	421.64	21.60	548.64	26.60	675.64	31.60	802.64
1.70	43.18	6.70	170.18	11.70	297.18	16.70	424.18	21.70	551.18	26.70	678.18	31.70	805.18
1.80	45.72	6.80	172.72	11.80	299.72	16.80	426.72	21.80	553.72	26.80	680.72	31.80	807.72
1.90	48.26	6.90	175.26	11.90	302.26	16.90	429.26	21.90	556.26	26.90	683.26	31.90	810.26
2.00	50.80	7.00	177.80	12.00	304.80	17.00	431.80	22.00	558.80	27.00	685.80		
2.10	53.34	7.10	180.34	12.10	307.34	17.10	434.34	22.10	561.34	27.10	688.34		
2.20	55.88	7.20	182.88	12.20	309.88	17.20	436.88	22.20	563.88	27.20	690.88		
2.30	58.42	7.30	185.42	12.30	312.42	17.30	439.42	22.30	566.42	27.30	693.42		
2.40	60.96	7.40	187.96	12.40	314.96	17.40	441.96	22.40	568.96	27.40	695.96		
2.50	63.50	7.50	190.50	12.50	317.50	17.50	444.50	22.50	571.50	27.50	698.50		
2.60	66.04	7.60	193.04	12.60	320.04	17.60	447.04	22.60	574.04	27.60	701.04		
2.70	68.58	7.70	195.58	12.70	322.58	17.70	449.58	22.70	576.58	27.70	703.58		
2.80	71.12	7.80	198.12	12.80	325.12	17.80	452.12	22.80	579.12	27.80	706.12		
2.90	73.66	7.90	200.66	12.90	327.66	17.90	454.66	22.90	581.66	27.90	708.66		
3.00	76.20	8.00	203.20	13.00	330.20	18.00	457.20	23.00	584.20	28.00	711.20		
3.10	78.74	8.10	205.74	13.10	332.74	18.10	459.74	23.10	586.74	28.10	713.74		
3.20	81.28	8.20	208.28	13.20	335.28	18.20	462.28	23.20	589.28	28.20	716.28		
3.30	83.82	8.30	210.82	13.30	337.82	18.30	464.82	23.30	591.82	28.30	718.82		
3.40	86.36	8.40	213.36	13.40	340.36	18.40	467.36	23.40	594.36	28.40	721.36		
3.50	88.90	8.50	215.90	13.50	342.90	18.50	469.90	23.50	596.90	28.50	723.90		
3.60	91.44	8.60	218.44	13.60	345.44	18.60	472.44	23.60	599.44	28.60	726.44		
3.70	93.98	8.70	220.98	13.70	347.98	18.70	474.98	23.70	601.98	28.70	728.98		
3.80	96.52	8.80	223.52	13.80	350.52	18.80	477.52	23.80	604.52	28.80	731.52		
3.90	99.06	8.90	226.06	13.90	353.06	18.90	480.06	23.90	607.06	28.90	734.06		
4.00	101.60	9.00	228.60	14.00	355.60	19.00	482.60	24.00	609.60	29.00	736.60		
4.10	104.14	9.10	231.14	14.10	358.14	19.10	485.14	24.10	612.14	29.10	739.14		
4.20	106.68	9.20	233.68	14.20	360.68	19.20	487.68	24.20	614.68	29.20	741.68		
4.30	109.22	9.30	236.22	14.30	363.22	19.30	490.22	24.30	617.22	29.30	744.22		
4.40	111.76	9.40	238.76	14.40	365.76	19.40	492.76	24.40	619.76	29.40	746.76		
4.50	114.30	9.50	241.30	14.50	368.30	19.50	495.30	24.50	622.30	29.50	749.30		
4.60	116.84	9.60	243.84	14.60	370.84	19.60	497.84	24.60	624.84	29.60	751.84		
4.70	119.38	9.70	246.38	14.70	373.38	19.70	500.38	24.70	627.38	29.70	754.38		
4.80	121.92	9.80	248.92	14.80	375.92	19.80	502.92	24.80	629.92	29.80	756.92		
4.90	124.46	9.90	251.46	14.90	378.46	19.90	505.46	24.90	632.46	29.90	759.46		

(Chart by Datapage. Gale Group.)

METRIC CONVERSION CARD

Approximate Conversions to Metric Measures

Symbol	When You Know	Multiply by	To Find	Symbol

LENGTH

Symbol	When You Know	Multiply by	To Find	Symbol
in	inches	2.5	centimeters	cm
ft	feet	30	centimeters	cm
yd	yards	0.9	meters	m
mi	miles	1.6	kilometers	km

AREA

Symbol	When You Know	Multiply by	To Find	Symbol
in^2	square inches	6.5	square centimeters	cm^2
ft^2	square feet	0.09	square meters	m^2
yd^2	square yards	0.8	square meters	m^2
mi^2	square miles	2.6	square kilometers	km^2
	acres	0.4	hectares	ha

MASS (weight)

Symbol	When You Know	Multiply by	To Find	Symbol
oz	ounces	28	grams	g
lb	pounds	0.45	kilograms	kg
	short tons (2000 lb)	0.9	metric ton	t

TEMPERATURE (exact)

Symbol	When You Know	Multiply by	To Find	Symbol
°F	degrees Fahrenheit	subtract 32, multiply by 5/9	degrees Celsius	°C

(Courtesy of National Institute of Standards and Technology, Office of Information Services.)

Knots to Miles Per Hour Conversion Chart

1 KNOT = 1.15155 MILES PER HOUR

KTS	0 MPH	1 MPH	2 MPH	3 MPH	4 MPH	5 MPH	6 MPH	7 MPH	8 MPH	9 MPH
0	0	1	2	3	5	6	7	8	9	10
10	12	13	14	15	16	17	18	20	21	22
20	23	24	25	26	28	29	30	31	32	33
30	35	36	37	38	39	40	41	43	44	45
40	46	47	48	49	51	52	53	54	55	56
50	58	59	60	61	62	63	64	66	67	68
60	69	70	71	72	74	75	76	77	78	79
70	81	82	83	84	85	86	87	89	90	91
80	92	93	94	96	97	98	99	100	101	102
90	104	105	106	107	108	109	110	112	113	114

(Courtesy of National Climatic Data Center/NOAA.)

Summary of 2002 Weather

- ## REVIEW OF U.S. EVENTS FOR 2002

January 2002. Moderate to heavy snows hit parts of Georgia, the Carolinas, and Virginia during the first week of January. Also, on the southeastern periphery of the snow, sleet and freezing rain fell, resulting in over 60,000 power outages. The hardest hit areas were in eastern North Carolina where 14 inches of snow were reported in parts of Nash, Halifax, Montgomery, Vance, Granville, and Person counties.

Interior Maine received heavy snows on the 15th and 16th of the month as a storm system moved across the area. Maine received a wide range of snowfall totals with one inch in the Princeton area to as high as 17 inches in Bucksport. This followed an earlier storm that rapidly deepened off the coast around the 13th of the month bringing heavy rains to coastal areas and over a foot of snow to down east Maine.

On the 29th, record rainfall fell at several Hawaii recording stations and severe storms caused millions of dollars in flood damage. The Hilo airport shattered their January 24-hour rainfall record of 9.51 inches. The new record is 12.47 inches.

A powerful late-January storm brought ice and heavy snows across the Plains, Great Lakes, and into New England. Heavy snow and freezing rain stranded airline passengers, shut down schools, and left thousands without electricity across the Midwest. The storm had dumped about a foot of snow in parts of Iowa and northern Illinois including the Chicago area. In Oklahoma, entire cities were without power, as heavy ice toppled trees and downed power lines. More than a foot of snow fell across New Mexico's high country. At least 15 people died in traffic accidents that were blamed on the weather.

February 2002. Farther south, a winter storm brought snowfall to portions of Texas, east Arkansas, north Mississippi, and west Tennessee on the 5th and 6th. Searcy and Polk Counties in Arkansas reported up to seven inches of snow. During a storm in the northern Plains on the 9th, blowing snow cut visibility nearly to zero and shut down hundreds of miles of major highways. Up to two feet of snow fell in the Black Hills of eastern Wyoming.

For the month of February, Marquette, Michigan, set a new all-time monthly snowfall record of 91.9 inches. The previous record was 91.7 inches in January 1997. The new record far surpassed the old February record of 63.6 inches set in 1995.

March 2002. A snowstorm that swept from Texas to Michigan was responsible for 21 deaths, according to media reports. On the 2nd, sleet, snow, and freezing rain contributed to more than 500 traffic accidents and about 100 cancelled flights at Dallas–Fort Worth International Airport.

The National Guard was called in to help evacuate residents affected by a storm that damaged or destroyed at least 250 homes in the worst flooding to hit eastern Kentucky in 25 years. At least seven deaths in Tennessee were also blamed on the storm, which dumped as much as six to eight inches of rain.

Anchorage, Alaska, received 28.6 inches of snow on the 16th and 17th. This far surpasses the previous snowfall record—15.6 inches—set on December 29, 1955. The snow fell at a rate of up to two inches per hour for much of the 16th, and schools were closed and flights canceled as a result of the storm.

Snow continued to fall through the end of March in Marquette, Michigan, leading to a seasonal total of over 300 inches. This surpassed the old seasonal snowfall record by 28 inches. Warmer than normal water in the Great Lakes enabled lake-effect snows to persist through late winter.

April 2002. On the 15th, temperatures soared into the mid to upper 90s in Nebraska. Temperatures were above 90°F in Omaha and reached 97°F at McCook, Nebraska. Temperatures were also high in parts of Kansas, northwest Missouri, western Iowa, and southwest Minnesota. The temperature reached 91°F in the St. Paul/Minneapolis area, breaking the old daily record of 82°F set in 1915 (which was equaled in 1976). This was the earliest in April that 90°F or higher was recorded. At the beginning of April 2002, there were four to seven inches of snow on the ground.

A spring storm on the 28th in the Tennessee and Ohio Valleys caused tornadoes, high wind, and hail. The storm then moved eastward causing additional damage and spawning killer tornadoes. The tornadoes were part of powerful storms carrying heavy rain and snow. On the system's

Scenes such as this were common across much of the United States during the winter of 2002 and 2003 as major storms blanketed many communities with record amounts of snow. *(Courtesy of National Oceanic and Atmospheric Administration (NOAA).)*

northern edge, up to 20 inches of snow fell overnight in Wisconsin. More than 40,000 people were without power in northern Wisconsin on the 28th and wet snow contributed to four traffic deaths in Minnesota. In Maryland, a powerful tornado killed three people and leveled parts of La Plata, a small town 25 miles south of Washington, D.C. In Kentucky and Illinois, tornadoes caused one death in each state. In Tennessee, a tornado injured 18 people 30 miles southeast of Nashville, and a tornado touched down in Ohio and caused widespread damage.

May 2002. Tug Fork River, which separates West Virginia and Kentucky, crested at nearly 20ft above flood stage on Friday, May 3, and led to the deaths of at least six people. Heavy rains on Tuesday, May 7, in western Indiana caused riverbanks to burst and prompted evacuations. Flash flooding also led to evacuations in south-central Ohio. Storms battered central West Virginia, killing at least one person and stranding approximately two-dozen people at a campground. In Webster County, West Virginia, 2,000 structures were damaged and 197 homes destroyed. At least two people died in the floods in Virginia. Four counties in West Virginia (McDowell, Mercer, Mingo, and Wyoming) and two in Virginia (Buchanan and Tazewell) were declared disaster areas by President George W. Bush. Hundreds of homes and other structures were also damaged and destroyed in Kentucky and more than 1,000 residents were evacuated from Kentucky and western Virginia.

A tornado swept through Happy, Texas, on the night of May 5, 2002. The tornado killed at least two people and injured others. The tornado was one of at least six reported in the state on the 5th. Happy, Texas, is located about 30 miles (48 km) south of Amarillo.

Drought intensified in much of the eastern U.S. during the month of May 2002.

June 2002. Heavy rain and flooding occurred in the upper Midwest on the 9th and 10th. Rainfall totals during the 48 hours beginning in the early morning of June 9 exceeded six inches in a broad area of northern Minnesota. However, local totals far exceeded 12 inches (305 mm) in places such as Roseau, Lake of the Woods, and Koochiching counties. Additional rainfall later in the month compounded flood problems in northern Minnesota.

By the end of June, nearly 2.8 million acres had been burned by wildfires in the U.S. This is over two-and-half times the 10-year average. Fires in Arizona burned over 550,000 acres and destroyed at least 400 homes. Separate fires near Durango and Denver, Colorado, consumed 70,000 and 140,000 acres respectively by the end of June, destroying dozens of homes. The latter (the Hayman Fire) was the largest fire in Colorado's history and prompted President George W. Bush to declare the area south of Denver a disaster area.

July 2002. During the first week of July, major flooding occurred in parts of Texas due to region-wide accumulations of 5 to 15 inches of rainfall in the San Antonio/Austin area. San Antonio airport received over 9.5 inches of rain on the 1st and over 10 inches on the 2nd. Around two dozen counties were declared disaster areas and costs from the flooding were in the hundreds of millions of dollars. At least nine deaths were attributed to the disaster. Additional rain fell in the middle of the month across parts of Texas adding to flooding problems and preventing rivers from receding.

During July 6–14, a heat wave affected the western United States, where numerous all-time high temperature records were broken. On July 13, Death Valley, California, recorded a high temperature of 127°F with a low of 100°F (37.8°C), or a mean temperature of 113.5°F. This is the second-warmest mean temperature at Death Valley since records began at the current station in 1961. Drought conditions intensified across parts of the Southeast, the High Plains, and much of the West during July 2002.

August 2002. Monmouth County, New Jersey, declared a state of emergency after storms ripped through the area on the 2nd. High winds and lightning damaged homes and property and left emergency crews clearing debris from roads and restoring power to around 140,000 homes.

In North Carolina on the evening of the 25th, up to eight inches of rain fell in the area north and east of Raleigh. This led to flash flooding in several counties along with some road closures. Heavy rain continued across eastern North Carolina through the 26th and 27th.

On the 2nd, a tornado ripped through the town of Ladysmith, in northwestern Wisconsin. A few dozen injuries were reported and significant structural damage occurred in the town of 4,000 residents.

The remnants of Tropical Storm Fay dumped heavy rains and spawned tornadoes on the 7th, as the storm moved inland over Texas. Up to a foot of rain fell in Freeport and

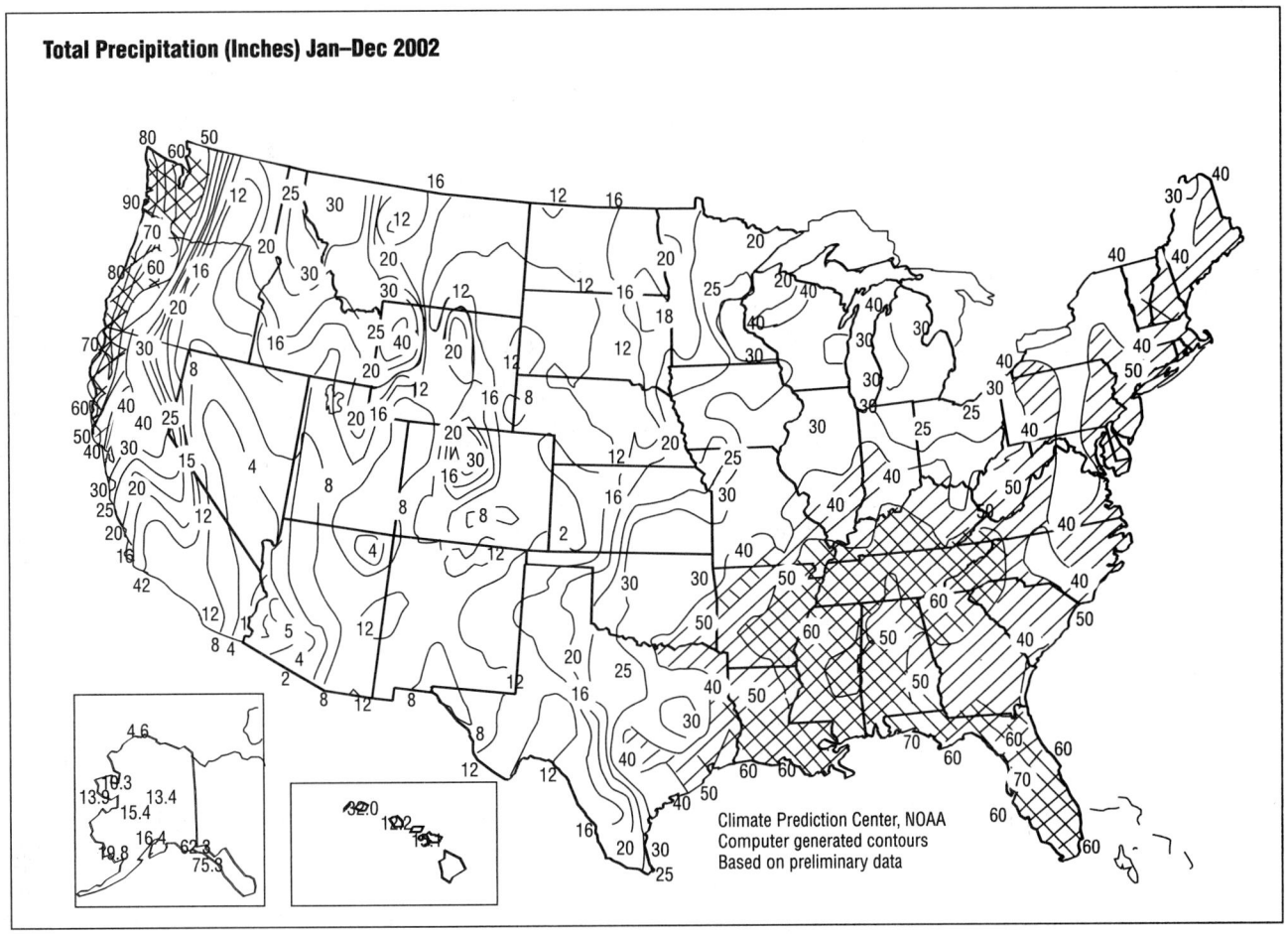

(Courtesy of Climate Prediction Center/NOAA.)

West Columbia, in coastal Brazoria County and five to eight inches were reported in Matagorda and Wharton counties.

The remnants of tropical storm Hanna brought heavy rain to much of the eastern third of the United States on the 13th and 15th, stretching from the Gulf Coast to the mid-Atlantic. Rainfall spread from the Florida Panhandle northward across the Appalachians into the Ohio Valley and the Northeast, with flash flooding reported in scattered parts of Georgia, South Carolina, and Florida. This storm brought much needed rains to parts of the Southeast that had been suffering from extended drought conditions. Two- to five-inch rainfall amounts were common across portions of western Florida, Georgia, and the Carolinas.

Thunderstorms preceding a strong cold front brought severe weather and tornadoes on the 20th to areas of Ohio and Indiana. Of at least 50 homes that were destroyed in Indiana, 20 of those were in the town of Martinsville, where a tornado struck. According to the state emergency management agency, this was the worst outbreak of severe weather in Indiana since June 1990.

October 2002. On October 3, Hurricane Lili came ashore in western Louisiana bringing additional rain to areas already affected by Tropical Storm Isidore in September. More rain-

fall later in October made this the second-wettest October on record for Louisiana and Texas. Further details on flooding and rainfall in the Southeast and parts of the Gulf Coast are given on the Atlantic Hurricane Season summary page.

Severe weather occurred in south Texas on October 24, as thunderstorms produced tornadoes that caused one fatality and 14 injuries in the Corpus Christi area. Storm damage in the city was estimated as high as $100 million, and Texas Governor Rick Perry declared Nueces County a disaster area.

An intrusion of Arctic air produced record-low temperatures across parts of the upper Midwest during the last week of October. Williston, North Dakota, dropped to –9°F on October 30, which is the coldest low temperature ever recorded during the month of October at Williston.

It was the warmest October in Alaska since 1938, and only the 8th time on record that Anchorage had no snow during the month.

November 2002. A major outbreak of severe weather and tornadoes occurred across the Tennessee and Ohio valley region on the 10th and 11th, producing damage in 13 states. A total of 75 tornadoes touched down on the 10th, resulting in at least 36 deaths. A tornado rated as F-4 on the Fujita Scale struck Van Wert county in Ohio. In Tennessee, the community of Mossy Grove was nearly destroyed by a mile-wide tornado that claimed 12 lives.

Severe to exceptional drought continued throughout much of the western United States. Two months of above-average precipitation in the eastern U.S. brought significant relief to long-term drought conditions.

December 2002. A significant snow and ice storm affected much of the eastern half of the United States during the 3rd through the 5th. In the Carolinas, electric utilities provider Duke Power characterized the ice storm as the worst in the company's history, with 1.2 million customers or nearly half its entire customer base without power on the morning of the 5th. This surpassed electrical outages inflicted by Hurricane Hugo as it swept through the central Carolinas in September 1989.

A strong Pacific storm struck the west coast of the United States at the beginning of the third week of the month with very heavy rains and strong hurricane-force winds. As many as two million homes were without power in California. At least 13 people lost their lives during this storm, which may have been a portion of the Super Typhoon Pongsona that struck Guam on December 8, 2002; it arrived along the west coast of the United States a week later and produced tornadoes in the midwest several days later. It might be that the ex-typhoon got caught and absorbed by a large longitudinal trough that migrated to the west coast then to the south and southeast states. During a 120-hour period ending on December 18, rainfall topped 20 inches at a few locations in California's Shasta Mountains, including Clear Creek (21.04 inches) and Brandy Creek (20.76 inches). Sierra Nevada snowfall during the same 5-day period reached or exceeded 100 inches at Kirkwood and several other sites. Flagstaff, Arizona, noted a daily record amount of 7.2 inches of snow on December 18 en route to a storm-snowfall of 11.8 inches. Yuma, Arizona, received 0.03 inch of rain on December 20, 2002, their first measurable precipitation since December 4, 2001.

Atlantic Hurricanes. There were 12 named tropical storms in 2002, four of which became hurricanes with two reaching major hurricane strength (categories three to five on the Saffir-Simpson Scale). On average, 10 named storms form with six growing to hurricane strength and two developing into major hurricanes. While there were more tropical storms than average, there were fewer strong storms in 2002. The development of an El Niño in the Eastern Equatorial Pacific led to the suppression of strong storms in 2002, and overall activity is considered somewhat lower than average for the season. However, seven storms made landfall in the U.S., which is the most since 1998.

Snow Season. Overall, the North American snow season for 2001–2002 was below average. However, due to above average warmth over the Great Lakes during the winter, the lakes remained largely unfrozen and provided ample moisture to fuel lake effect snowfall. Marquette, on Michigan's Upper Peninsula, received record seasonal snowfall amounts, and in Buffalo, New York, the monthly snowfall record was broken when more than 80 inches of snow fell between the 24th and 28th of December 2001. Elsewhere, the snow pack was mostly light and the lack of spring melt water from the snow added to the water shortages in much of the western U.S.

Drought—2002

The incidence of drought in the United States has varied greatly over the past century. From the dust bowl years of the 1930s to the major droughts of 1988 and 2000, much of the U.S. has suffered from the effects of drought during the past century. While annual and seasonal precipitation totals have generally increased in the United States since 1900, severe drought episodes continue to occur.

The nation's most devastating drought occurred in the 1930s during what many refer to as the "Dust Bowl" years. The drought affected almost the entire Plains and covered more than 60% of the U.S. during its peak in July 1934. It had a devastating economic impact and caused the migration of millions of people from the Plains to other parts of the country, many to the western U.S. Although the nation has not since experienced a drought as severe as the drought of the 1930s, subsequent droughts (e.g., those of the 1950s, 1988, and 2000) have also had serious economic and societal consequences.

Although a variety of weather-related phenomena have the potential to cause great economic and personal losses, in the U.S. it is drought that has historically had the greatest impact on the largest number of people. Since 1980, 48 weather-related disasters have each caused at least 1 billion dollars in economic losses. Of these 48 disasters, the greatest losses have been attributed to drought. Economic losses exceeded 40 billion dollars in the droughts of 1980 and 1988, and the combination of drought and heat-related deaths totaled more than 5,000 in each event. The drought of 2000 resulted in losses of 4 billion dollars and 140 deaths.

Although not as widespread as the droughts of the 1930s and 1950s, persistent above-normal temperatures and below-normal precipitation across much of the western and southern U.S. in 1999 and 2000 brought drought to these regions by the summer of 2000. More than one third of the country suffered from severe to extreme drought by August, leading to heavy agricultural losses, water rationing for many, and one of the worst wildfire seasons in the last 50 years. While some parts of the nation have received drought-ending precipitation since that time, parts of the nation continue to suffer from severe precipitation deficits through the spring season of 2001.

Defining Drought

The wide variety of disciplines affected by drought, its diverse geographical and temporal distribution, and the many scales that it operates on make it difficult to develop both a definition to describe drought and an index to measure it. Common to all types of drought is the fact that they originate from a deficiency of precipitation resulting from an unusual weather pattern. If the weather pattern lasts a short time (e.g., a few weeks or a couple of months), the drought is considered short-term. But if the weather or atmospheric circulation pattern becomes entrenched and the precipitation deficits last for several months to several years, the drought is described as long-term.

Many quantitative measures of drought have been developed in the United States, depending on the discipline affected, the region being considered, and the particular application. The most frequently used indicators of drought are those developed by Wayne Palmer in the 1960s. These include the Palmer Drought Severity Index (PDSI), the Palmer Hydrological Drought Index (PHDI), the Palmer Z Index, and the Crop Moisture Index (CMI). These indices have been used in countless research studies as well as in operational drought monitoring during the past 35 years. The Palmer drought index has proven to provide one of the best indications of drought for much of the United States. It is superior to other drought indices in many respects because it accounts not only for precipitation totals but also for temperature, evapotranspiration, soil runoff, and soil recharge.

The Z Index measures short-term drought on a monthly scale. The CMI measures short-term agricultural drought on a weekly scale. The PDSI measures drought duration and the intensity of long-term drought-inducing circulation patterns; it responds fairly quickly to meteorological patterns that often change rapidly from one regime to another. A measure of the hydrological impacts of drought, the PHDI measures the long-term effects of drought on systems affected by long-term precipitation deficits. These effects, such as reservoir levels, groundwater levels, etc., take longer to develop, and more time is needed to recover from them. It is from this index, the PHDI, that we calculate the precipitation amounts and probabilities of ending or ameliorating drought.

The Amount of Precipitation Needed to End a Drought (Based on the PHDI)

Because of the far-reaching societal and economic impacts of drought, there is considerable interest in determining 1) how much precipitation is required to end a drought, and 2) the probability that a region will receive the necessary amount of precipitation. Ending a hydrological drought requires that the moisture needs associated with recharge, demand, and runoff have been brought back to normal or above normal.

Many factors affect the quantity of precipitation required to end or ameliorate (reduce the severity of) a drought. Knowledge of the severity of the drought, as defined by the Palmer Hydrological Drought Index (PHDI), is the essential starting point for determining the needed precipitation. The typical conditions that a region experiences during each

month and season of the year (i.e., that region's climatology) is also essential. Given droughts of equal magnitude in dry and wet climates, the wetter region requires more precipitation to end the drought.

The season in which the precipitation falls can also greatly influence the quantity of precipitation required to end a drought. During a typically moist month (such as those experienced in the winter and spring along the West Coast) more precipitation may be required to end a drought than during the typically dry months of the summer. Because soil moisture conditions are generally lower in the dry months, the precipitation needed to bring soil conditions back to normal may be less than that required to return soil moisture conditions to normal during a generally wetter season. Nevertheless, regardless of a region's climate, a long period of near-normal precipitation is often sufficient for ending a drought with moisture conditions gradually returning to normal.

However, the quantity of precipitation needed to end a drought says nothing about the probability that a region will actually receive that amount of precipitation. A region, such as the West Coast, that does not typically experience excessively heavy precipitation during the summer season, may be less likely to receive a quantity sufficient for ending a drought than a region that has a record of experiencing extreme precipitation events during the same season. The months that have the greatest probability of receiving substantially more precipitation than normal would be those with precipitation distributions with the largest positive skew (that is, those subject to more extreme precipitation events), not necessarily those months that normally receive the greatest amount of precipitation.

The technical details associated with the calculation of precipitation totals needed to end or ameliorate drought and the probability of receiving the required precipitation can be found in *Drought Termination and Amelioration: Its Climatological Probability,* by Tom Karl, et al. 1987.

Maps of Precipitation Totals and Probability

More than 2,000 maps of the contiguous U.S. are provided that show the precipitation totals needed to end or ameliorate drought from periods of one month to six months based on PHDI values from –2 to –6. These data were calculated for each month of the year and include precipitation values for each of the 344 contiguous U.S. climate divisions. The end of a drought is defined by a PHDI value of –0.5, while drought amelioration is achieved when a PHDI value of –2.0 is reached. Maps showing the probability of receiving the necessary amount of precipitation are also provided.

Maps of precipitation needed to end or ameliorate a drought for those divisions currently experiencing drought are also available. Values are provided in all divisions with a monthly PHDI less than –2.0. Precipitation needed over

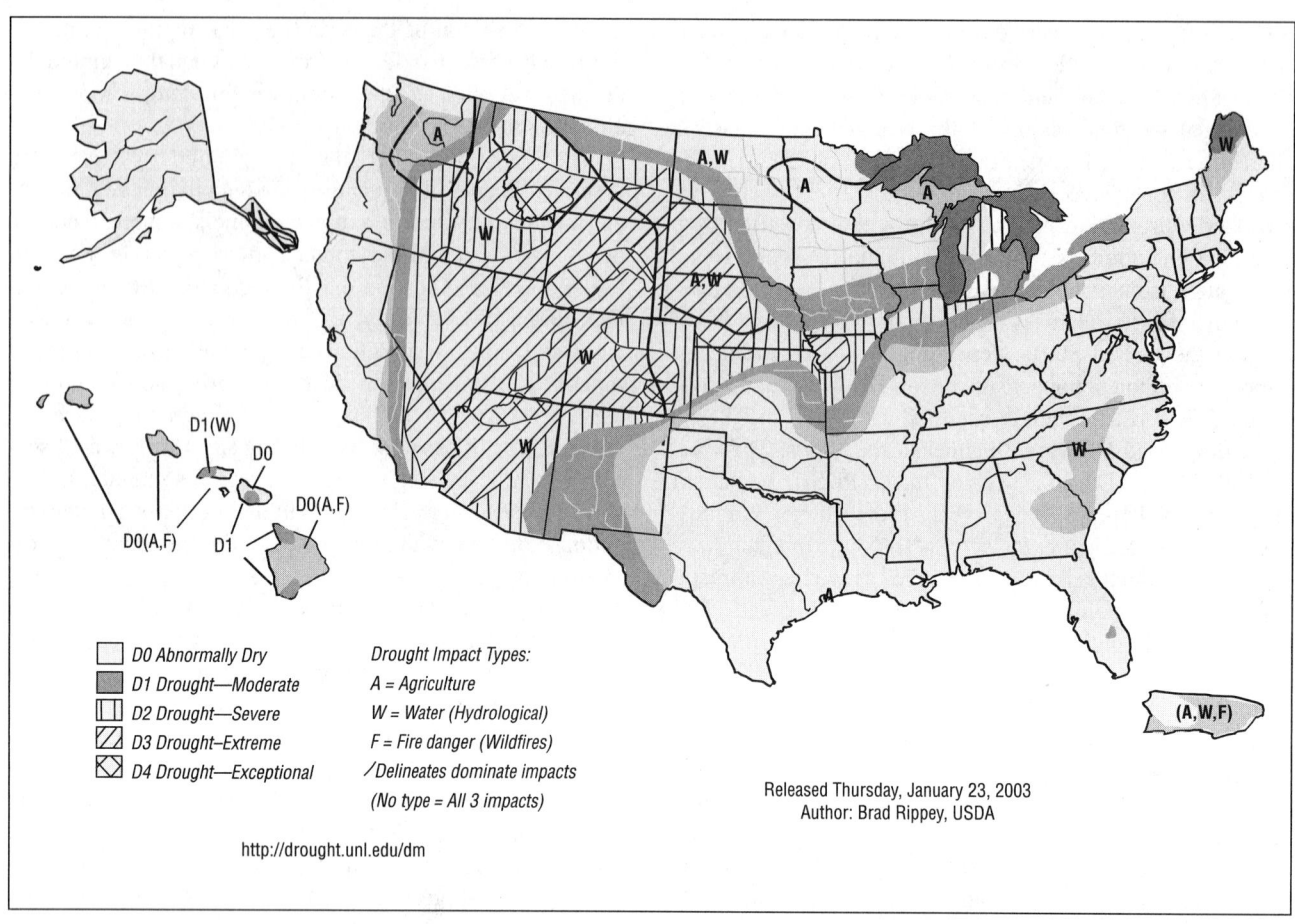

Drought Impact Types:
- A = Agriculture
- W = Water (Hydrological)
- F = Fire danger (Wildfires)
- /Delineates dominate impacts
- (No type = All 3 impacts)

- D0 Abnormally Dry
- D1 Drought—Moderate
- D2 Drought—Severe
- D3 Drought—Extreme
- D4 Drought—Exceptional

Released Thursday, January 23, 2003
Author: Brad Rippey, USDA

http://drought.unl.edu/dm

(Courtesy of National Climatic Data Center/NOAA.)

periods from one month to six months is included as well as the associated probability of receiving that quantity of precipitation. These maps are replaced on a monthly basis with the values reflecting conditions in the previous month. Maps are available from the National Climatic Data Center in Asheville, North Carolina.

A severe dry spell has its grasp on the Northwest

It has been five years since Portland has seen a late-year dry spell this severe. October and November, 2002, which mark the beginning of a 12-month cycle known as the water year, have been drier only three times in 64 years of record keeping.

The inland Northwest, much of it devoted to agriculture and ranching, also is parched. Precipitation in the two months across the Columbia River Basin east of the Dalles, a region making up much of Oregon, Washington, Idaho, and Montana, was about 50 percent below normal. That could hurt irrigators. 2001 posted a near-record drought and saw, as a result, huge rate increases for hydroelectric power sold by the Bonneville Power Administration, which sharply cut water availability to eastern Oregon farmers.

Of the 75 river-gauging stations in Oregon operated by U.S. Geological Survey, 67 percent are reporting below-normal flows. The low flows are delaying the migration of salmon. Portland's drinking-water reservoirs on Bull Run, typically full this time of year, were 62 percent below capacity recently. Portland has back-up wells, so the city's water managers said they expect no water restrictions next year.

• WINTER STORMS

Southeastern U.S. Ice Storm of December 2002

Some million-and-a-half homes were without power several days after the December 4 and 5, 2002 storm that moved from the southwestern U.S. to the eastern seaboard in about 48 hours. An armada of cherry-pickers helped to restore power to nearly 1.5 million people in the ice-coasted Carolinas. Six to eight inches or more of snow coated the mid-Atlantic region from Washington, D.C., northeastward into New England.

TEMPERATURES—HEAT AND HUMIDITY

Global Temperatures for 2002

Global temperatures in 2002 were 1.01°F above the long-term (1880–2001) average, which will place 2002 as the second warmest year on record. The only warmer year was 1998 in which a strong El Niño contributed to higher global temperatures. Land temperatures were 1.57°F above average and ocean temperatures 0.76°F above the 1880–2001 mean. Both land and ocean temperature ranks as second warmest on record.

Neutral ENSO conditions at the beginning of 2002 gave way to a strengthening El Niño episode during late boreal summer and continuing into early winter. Moderate positive anomalies of equatorial Pacific sea surface temperatures (El Niño conditions) were expected to persist through the early part of 2003.

The Northern Hemisphere temperature averaged near record levels in 2002 at 1.13°F above the long-term average. The Southern Hemisphere also reflected the globally warmer conditions, with a positive anomaly near 0.83°F.

In 2002 warmer temperatures and shifts in atmospheric circulation patterns contributed to the greatest surface melt on the Greenland Ice Sheet in the 24-year satellite record. There was also a record low level of Arctic sea ice extant in September, the lowest since satellite monitoring began in 1978, according to the National Snow and Ice Data Center.

Annual anomalies in excess of 1.8°F were widespread across much of North America and Asia.

The 1880–2001 average combined land and ocean annual temperature is 56.9°F, the annually averaged land temperature for the same period is 47.3°F, and the long-term annually averaged sea surface temperature is 60.9°F.

TEMPERATURE TRENDS

During the past century, global surface temperatures have increased at a rate near 1.1°F per century, but this trend has dramatically increased to a rate approaching 3.6°F per century during the past 25 years. There have been two sustained periods of warming, one beginning around 1910 and ending around 1945, and the most recent beginning about 1976. Temperatures during the latter period of warming have increased at a rate comparable to the rates of warming projected to occur during the next century, with continued increases of anthropogenic greenhouse gases.

Data collected by NOAA's polar orbiting satellites and analyzed for NOAA by the University of Alabama in Huntsville (UAH) and Remote Sensing Systems (RSS, Santa Rosa, California) indicate that temperatures centered in the middle troposphere at altitudes from two to six miles make 2002 the second warmest year for the globe.

The average lower troposphere temperature (surface to about five miles) for 2002 was second warmest on record.

Analysis of the satellite record that began in 1979 shows that the global average temperature in the middle troposphere has increased, but the differing analysis techniques of the two teams result in different trends. The UAH team found an increase of 0.06°F per decade, while a trend of 0.21°F per decade was found by the RSS team. This compares to surface temperature increases approaching 0.3°F per decade during the same period.

While lower tropospheric temperatures as measured by the MSU indicate increasing temperatures over the last two decades, stratospheric (nine to 14 miles) temperatures have been decreasing. This is consistent with the depletion of ozone in the lower stratosphere. The large increase in 1982 was caused by the volcanic eruption of El Chichon, and the increase in 1991 was caused by the eruption of Mt. Pinatubo in the Philippines.

United States Temperatures for 2002

2002 ranked as the 14th warmest year on record for the U.S., with an estimated preliminary temperature of 53.9°F, which is 1.1°F above the long-term average. After beginning the year with much above average warmth, especially in the northeast, 2002 ended with cooler-than-normal to near-average temperatures across much of the nation.

The 2001–2002 winter season (December–February) was ninth warmest on record for the U.S., with much of the warmth occurring in the northeast, which had its warmest winter on record. Spring (March–May) was near normal nationally with a warmer than average April compensating for a cooler March and May. However, the summer season (June–August) was one of the warmest in 108 years of national records. Summer 2002 was tied (with 1988) for 3rd warmest behind only 1936 and 1934. The 2002 fall season was near average, though September was the 7th warmest such month on record, followed by a cool October and near-average November.

The last three 5-year periods (1998–2002, 1997–2001, 1996–2000), have been the warmest 5-year periods in the last 108 years of national records, and the last 6 (1997–2002), 7 (1996–2002), 8 (1995–2002), 9 (1994–2002) and 10-year (1993–2002) periods have been the warmest on record for the U.S., illustrating the near persistent warmth of the last decade.

A tenth or more of the country averaged very warm for 8 months of the year, with 2 months, April and July, each exceeding 30 percent above average. More than 10 percent of the country was very cold in March and May, and more than a third of the country was very cold in October. Very warm and very cold conditions are defined as the warmest and coldest ten percent of recorded temperatures, respectively.

2002 was fourth warmest on record for Delaware, Maryland, and New Jersey. Nine other states were much warmer than normal during 2002. No state in the contiguous U.S. has averaged below normal for the year.

Annual temperatures averaged across the state of Alaska reached record levels for 2002, with every season in Alaska averaging above normal. Fall 2002 was the warmest September–November on record for the state. Nine states in the northeast and midwest had their warmest winters on record, and the southwest region had its warmest summer on record. Three states (Colorado, Maryland, and Delaware) also broke summer warmth records.

Data collected by NOAA's TIROS-N polar-orbiting satellites (adjusted for time-dependent biases by NASA and the Global Hydrology and Climate Center at the University of Alabama in Huntsville) indicate that temperatures in the lower half of the atmosphere (the lowest 8 km) were above the 20-year (1979–1998) average for 2002 for the fifth consecutive year and ranked as the eighth warmest such period since 1979.

• THUNDERSTORMS

Thunderstorms around the World and in the United States

Meteorologists estimate that, at any given moment, some 1,800 thunderstorms are in progress over Earth's surface, and about 18 million a year around the world. It is estimated that approximately 100,000 to 125,000 thunderstorms occur in the United States each year. Of that total anywhere from 10 to 20 percent may be severe. The National Weather Service considers a thunderstorm severe if it produces hail at least three-quarters of an inch in diameter, winds of 58 mph or stronger, or a tornado. From 1996 to 2001, a total of 134,005 severe thunderstorms were recorded (not associated with tornadoes), an average of 19,144 annually. The frequency with which these giant generators of local weather occur, along with the quantity of energy they release and the variety of forms this energy can take, make thunderstorms great destroyers of life and property.

• TORNADOES

Tornadoes of 2002

A line of thunderstorms that extended a thousand miles produced an unofficial 70 tornadoes from southern Mississippi to northern Ohio; at least 35 people were killed and more than 200 were injured. These 35 fatalities were the most since the May 3, 1999, tornado outbreak in Oklahoma and Kansas. There were at least 16 deaths in Tennessee; 12 deaths in Alabama; 5 deaths in Ohio; and 1 death each in Mississippi and Pennsylvania. This outbreak occurred in November 2002.

There were 21 very strong to violent tornadoes (wind speeds in excess of 158 mph, category F3–F5) during the 2002 tornado season (March–August). This is well below the long-term (1950–2001) mean of 38 and is the third consecutive season of much-below-normal activity. No trends have been observed since 1950 in very strong to violent tornadoes. Although the season was below average, there were several notable storm outbreaks in 2002, including the F4 that touched down in Maryland on April 28, the Corpus Christi (Texas) tornado in October, and the November outbreak in the Tennessee and Ohio valley region.

• HURRICANES

2002 Hurricane Season

The Atlantic hurricane season begins on June 1 and ends on November 30. The 2002 Atlantic hurricane season was again an active one, making it the fifth consecutive active season. There were 12 named storms, compared to a 1944–1996 annual average of 9.8. Four of those named storms became hurricanes, of which 2 were classified as major. This compares to an average of 5.8 hurricanes a year, 2.5 of which are major (based on a 53-year average). There were also 2 additional tropical depressions. So, there were more named tropical cyclones than the long-term mean, but fewer strong storms. The season was also slow to begin, extremely active in September, and less active than average at its end. Eight named systems developed in September as well as a tropical depression, making it the most active month of any month on record for tropical cyclone development in the Atlantic basin.

Seven tropical storms made landfall on the continental U.S. in 2002, the most since 1998 when seven was also the number of landfalling tropical systems for the U.S. Hurricane Lili was the first hurricane to hit the coast of the U.S. since Irene in October 1999.

Notable tropical systems of 2002 included Hurricanes Isidore and Lili in September, which were quite intense, and Hurricane Kyle, also in September, which had a fairly long life. Kyle was the third longest-lived tropical cyclone in the Atlantic basin after Ginger of 1971 and Inga of 1969. Kyle formed on September 20 and spent over three weeks in the North Atlantic before finally coming ashore in South Carolina on October 11. Isidore became a category 3 hurricane as it moved into the southeastern Gulf of Mexico and hit the Yucatan Peninsula on September 22 at category 3 strength. After weakening over the Yucatan Peninsula it never quite regained hurricane strength and came ashore in Louisiana as a tropical storm. Lili was the only hurricane to make landfall in the United States in 2002, and did so at category 2. Lili had been a category 4 hurricane shortly before landfall and quickly weakened before reaching the west Louisiana coastline on October 3.

Monthly Tornado Statistics

	Number of tornadoes							Number of tornado deaths					Killer tornadoes				
	2003 Prelim	Final	2002 Prelim	Final	2001 Final	2000 Final	3 year avg.	2003 Prelim	2002 Final	2001 Final	2000 Final	3 year avg.	2003 Prelim	2002 Final	2001 Final	2000 Final	3 year avg.
JAN	?	?	8	3	5	16	11	?	0	0	0	0	?	0	0	0	0
FEB	?	?	2	2	30	56	43	?	0	8	19	9	?	0	3	4	2
MAR	?	?	30	48	33	103	68	?	0	3	2	2	?	0	2	1	1
APR	?	?	114	115	135	136	136	?	7	5	2	4	?	6	4	2	4
MAY	?	?	177	202	241	241	241	?	4	1	2	2	?	2	1	2	2
JUN	?	?	87	97	248	135	192	?	0	5	0	2	?	0	3	0	1
JUL	?	?	58	?	120	148	134	?	0	0	1	0	?	0	0	1	0
AUG	?	?	72	?	69	52	61	?	0	0	0	0	?	0	0	0	0
SEP	?	?	66	?	84	47	66	?	0	2	2	1	?	0	1	2	1
OCT	?	?	56	?	116	63	90	?	4	2	0	2	?	3	2	0	2
NOV	?	?	115	?	110	48	79	?	37	14	0	17	?	14	7	0	7
DEC	?	?	69	?	22	26	24	?	3	0	12	5	?	3	0	2	2
Total	0	0	854	467	1213	1071	1142	0	55	40	40	44	0	28	23	14	22

Note: ? means final number not yet available.
Important! Prelim. numbers represent tornado segments. Columns marked Final represent total tornadoes.
2003 numbers updated through 6 AM CST 1/29/03

(Courtesy of National Weather Service/NOAA.)

EL NIÑO SUPPRESSED HURRICANES IN 2002 SEASON

The 2002 Atlantic hurricane season produced only four hurricanes due to a strengthening El Niño. However, twice the normal number of storm systems (eight) affected the nation, bringing storm surge and severe weather and rain to the nation, including Hurricane Lili, the first land-falling hurricane to strike the United States since the 1999 hurricane season.

Overall in 2002, there were 12 named storms, four of which became hurricanes. Hurricanes Lili and Isidore were classified as major (category 3 or higher on the Saffir-Simpson hurricane scale). Eight storms (Tropical Storms Bertha, Edouard, Fay, and Hanna; and Hurricanes Gustav, Isidore, Kyle, and Lili) affected the coastal United States. Hurricane Lili was the only storm to make landfall while still a hurricane. The other 2002 storms were Tropical Storms Arthur, Cristobal, Dolly, and Josephine.

Hurricane forecasters at NOAA Climate Prediction Center (CPC), Hurricane Research Division and National Hurricane Center (NHC) correctly forecast that climate conditions, including the El Niño, would reduce the overall hurricane activity this season. The forecast called for seven to 10 tropical storms, of which four to six could develop into hurricanes, with one to three classified as major.

Louisiana, the hardest hit area, was battered by four storms including the powerful Hurricane Lili and Tropical Storm Isidore. The 2002 season's storms caused 9 deaths in the United States and about $900 million in damages.

The public relied heavily on Internet access for lifesaving information from NOAA this season. The explosive use of the Internet to convey vital information to the public in near real time has been astonishing. Between August and September the NHC web site recorded almost 500 million hits. The peak day for the season was October 3 (Hurricane Lili), when the site recorded 35.9 million hits—doubling the previous record set in 1999 during Hurricane Floyd.

Typhoon Pongsona hits Guam, December 8, 2002

Guam, west of the international date line, about 3,700 miles southwest of Hawaii, was raked by wind gusts of more than 180 mph on this day, and winds the following day of 70 mph continued in this U.S. Territory. No deaths were reported but the entire island was without electricity, and water and sewer systems were not expected to be fully operational for weeks. Sustained winds of 150 mph around the eye of Pongsona gave the storm a "super typhoon" status. Gusts up to 184 mph were reported in some areas of Guam. A wind gust of 117 mph was clocked before the National Weather Service's sensor failed, along with the radar. A reported 2,000 homes were destroyed and many more were damaged. Schools became emergency shelters.

Cyclone Zoe Strikes the Solomon Islands

Southern Pacific Ocean Category 5 Cyclone Zoe struck the southernmost Solomon Islands December 29, 2002, and was reported to be the most powerful ever recorded in the area, with winds reaching 225 mph and producing 33-foot waves that destroyed houses, uprooted trees, and flooded entire villages. Those living on the island used the common sense gained from centuries of dealing with these storms and had taken shelter in the hills on the island. Amazingly no one was killed on Tikopia, some 1,400 miles northeast of Sydney, Australia.

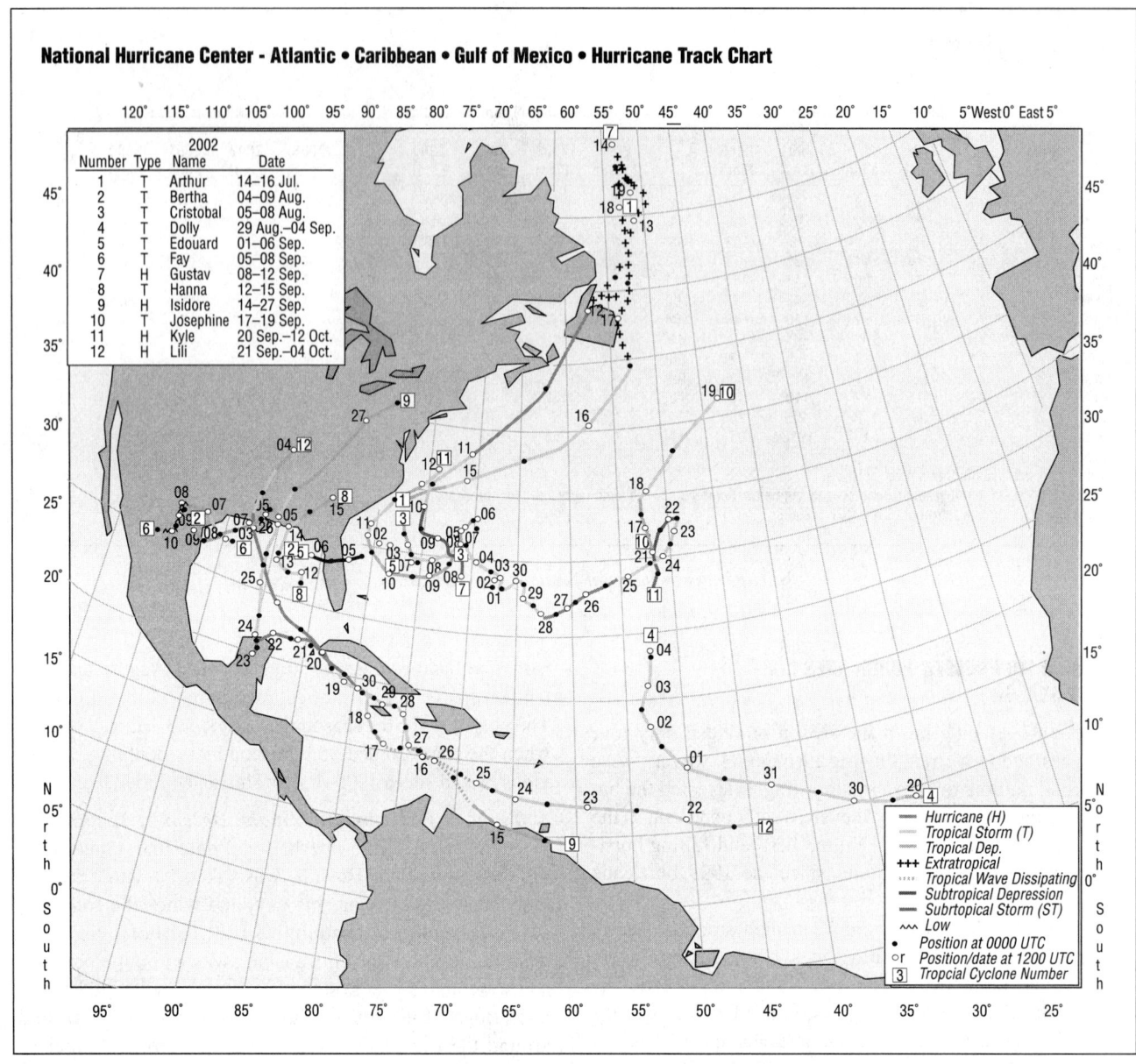

National Hurricane Center - Atlantic • Caribbean • Gulf of Mexico • Hurricane Track Chart

		2002	
Number	Type	Name	Date
1	T	Arthur	14–16 Jul.
2	T	Bertha	04–09 Aug.
3	T	Cristobal	05–08 Aug.
4	T	Dolly	29 Aug.–04 Sep.
5	T	Edouard	01–06 Sep.
6	T	Fay	05–08 Sep.
7	H	Gustav	08–12 Sep.
8	T	Hanna	12–15 Sep.
9	H	Isidore	14–27 Sep.
10	T	Josephine	17–19 Sep.
11	H	Kyle	20 Sep.–12 Oct.
12	H	Lili	21 Sep.–04 Oct.

(Courtesy of National Climatic Data Center/NOAA.)

• FLOODS AND FLASH FLOODS

Global Precipitation for 2002

Global precipitation was below the 1961–1990 average in 2002. Much of Australia experienced severe drought, with the eastern part of the country the worst affected. India monsoon rainfall was 19 percent below normal, with the resulting drought characterized as the worst since 1987. Other drought-affected areas included the western United States and portions of the north coast of China.

After a dry beginning to 2002, several typhoons brought excessive rains to parts of southeast Asia and Japan, the southeast coast of China, Taiwan, and the Philippines. In contrast to drought conditions during the first half of 2002, the onset of monsoon rains in southeast Asia promoted ex-

tensive flooding along the Mekong Delta. Seasonal flooding in much of south Asia (Nepal, Bangladesh, and northeastern India) during June–August claimed more than 1,000 lives. In the eastern United States, long-term drought was ameliorated by a turn to wetter weather, due in part to moisture from tropical systems.

National Precipitation for 2002

During 2002, more than a tenth of the country was very dry in eight of the 11 months, and in February and May, more than 20 percent of the country was very dry. Only in September and October did the percentage area of very wet conditions exceed ten percent, though in October more than 20 percent of the country was very wet. This was due in part to several land-falling tropical systems during these two months. Drought improvement generally occurred in the eastern U.S. during 2002, while in much of the western half of the country and the central and southern Great Lakes drought worsened.

Precipitation in the United States in 2002 was characterized by extreme dryness in the west, generally above-average wetness in the Mississippi Valley Region and dryness giving way to near average conditions for the east. Colorado had its driest year on record during 2002, and Wyoming, Nevada, and Nebraska had their third driest years. Six states were much drier than normal and the Southwest region as a whole was the fourth driest on record.

Louisiana had its wettest fall on record, in part due to two land-falling tropical systems that hit the state in September and October. It was the third driest July and August combined for the Northeast. Considerable improvement in drought conditions for the east coast occurred in the fall so that annual totals for much of the east coast were near normal.

• EARTHQUAKES

November 3, 2002 (UTC), Denali Fault, Alaska

This M7.9 shock is the largest earthquake on the Denali fault since at least 1912, when an M7.2 earthquake occurred in the general vicinity of the fault, 50 miles east of this latest epicenter. This M7.9 shock, which was also one of the largest ever recorded on U.S. soil, occurred on the Denali-Totschunda fault system, one of the longest strike-slip fault systems in the world and a rival in size to California's famed San Andreas strike-slip fault system, which spawned the destructive M7.8 San Francisco earthquake in 1906.

• VOLCANIC ACTIVITY

The Pu`u `O`o-Kupaianaha eruption of Kilauea, now in its nineteenth year and fifty-fifth eruptive episode, ranks as the most voluminous outpouring of lava on the volcano's east rift zone in the past five centuries. By September 2002, 2.3 km^3 of lava had covered 110 km^2 and added 220 hectares to Kilauea's southern shore. In the process, lava flows destroyed 189 structures and resurfaced 13 km of highway with as much as 25 m of lava.

Beginning in 1983, a series of short-lived lava fountains built the massive cinder-and-spatter cone of Pu`u` O`o. In 1986, the eruption migrated 3 km down the east rift zone to build a broad shield, Kupaianaha, which fed lava to the coast for the next five-and-a-half years. When the eruption

shifted back to Pu`u `O`o in 1992, a series of flank-vent eruptions formed a shield banked against the uprift side of the cone. Continuous eruption from these vents undermined the west and south flanks of the cone, resulting in large collapses of the west flank.

In May 2002 a new vent opened on the west side of the shield and fed flows down the western margin of the flow field, sparking the largest forest fire in the park in 15 years. These flows reached the ocean near the end of Chain of Craters Road in July, and as many as 4,000 visitors per day flocked to view flowing lava up close for the rest of the summer.

• 171 PEOPLE WERE RESCUED IN THE U.S. IN 2002 WITH HELP FROM NOAA SATELLITES

Thanks to environmental satellites with search and rescue tracking capability, the Commerce Department's National Oceanic and Atmospheric Administration (NOAA) helped save 171 lives in the United States in 2002.

The NOAA satellites, along with Russia's Cospas satellites, are part of an international Search-and-Rescue Satellite-Aided Tracking System known as Cospas-Sarsat. The system uses a constellation of satellites in geostationary and polar orbits to detect and locate emergency beacons from vessels and aircraft in distress and from hand-held Personal Locator Beacons (PLBs). India and the European Space Agency also provide geostationary satellites for the Cospas-Sarsat System.

Of the 171 rescues last year, 133 people were saved on the nation's seas, 27 in the Alaska wilderness, and 11 from downed aircraft in states around the country. Of the 69 separate Sarsat rescue events, a variety took place out at sea.

Engine fires, flooding, and rough seas all caused emergencies resulting in distress calls and rescues. In Alaska, stranded snowmobilers and lost persons were among those rescued. Downed aircraft incidents included those making emergency landings. In one such incident, a Piper Supercub had flipped after landing near Glenallen, Alaska. Both the pilot and passenger were uninjured.

More than 15,000 lives have been saved worldwide (and nearly 4,500 in the United States alone) since the system became operational in 1982. September 2002 marked the 20th anniversary of the first Sarsat rescue.

In one dramatic rescue, a father, a son, and their family dog were plucked from a life raft in the Gulf of Alaska about 90 miles south of Cordova, Alaska. They were in a dangerous predicament: their fishing vessel had struck an object and sustained uncontrollable flooding, causing them to abandon the craft. Yet, because there was an Emergency Position Indicating Radio Beacon, or EPIRB, on board that was manually activated, a U.S. Coast Guard search and rescue helicopter was able to respond to the distress call quickly once the alert information was received from the USMCC. On arrival the helicopter saw the situation unfolding and deployed a rescue swimmer to retrieve the three occupants and bring them to safety. This particular incident illustrates the importance of emergency beacon registration.

NOAA expects the number of worldwide rescues for 2002 will total about 1,500. The average number of distress alerts continues to rise internationally as more countries sign on to use the advantages and benefits of the Cospas-Sarsat system.

NOAA's Geostationary Operational Environmental Satellites (GOES) can instantly detect emergency distress signals. The polar-orbiting satellites in the system detect emergency signals as they circle the Earth from pole to pole. The signals are sent to the Mission Control Centers, then automatically sent to rescue forces around the world. There are 35 countries participating in the system as of 2002.

NOAA Satellites and Information is the nation's primary source of operational space-based meteorological and climate data. In addition to search and rescue, NOAA's environmental satellites are used for weather forecasting, climate monitoring, and other environmental applications such as volcanic eruptions, ozone monitoring, sea surface temperature measurements, and wild fire detection.

NOAA Satellites and Information also operates three data centers, which house global data bases in climatology, oceanography, solid earth geophysics, marine geology and geophysics, solar-terrestrial physics, and paleoclimatology.

NOAA is dedicated to enhancing economic security and national safety through the prediction and research of weather- and climate-related events and providing environmental stewardship of our nation's coastal and marine resources.

• 2002 UNITED STATES WEATHER REVIEW

According to the National Climatic Data Center (NCDC), most of the country was abnormally warm once again in 2002, resulting in this being one of the warmest 15 years since records began in 1895. Below-average precipitation led to persistent or worsening drought for much of the nation, although a series of storms ended drought across the East Coast by year's end, and wet conditions prevailed from the lower Mississippi Valley into the Tennessee and Ohio River Valleys for much of the year. Major flooding hit south central Texas this summer. Drought affected farm areas in the High Plains this spring and summer, but several timely frontal passages in July and August prevented drought from becoming widespread over the Corn Belt. Six states—Wyoming, Nebraska, Colorado,

Utah, Nevada, and Arizona—recorded one of the driest seven years on record, with Colorado recording its driest year ever.

• WINTER (DECEMBER 2001–FEBRUARY 2002)

Winter 2001–2002 was generally mild and tranquil, with less-than-normal snowfall, although there were some notable exceptions. A series of early winter storms crossed the Northwest from Washington and Oregon into Idaho and northern California, ending drought across most of the region. In contrast, precipitation was scarce across the plains of Montana and southward through Wyoming and Colorado. Extraordinarily dry weather covered the Southwest from southern California through Arizona and New Mexico. Cumulative precipitation from the Southwest through the Rockies into the High Plains totaled less than 50% of normal. Nationally, this was the ninth mildest winter (December–February) on record, as nearly the entire country east of the Continental Divide experienced above-normal warmth. Temperatures for the three-month period averaged 5–10°F above normal over the Midwest and Northeast. Ten states in the northeastern quadrant of the country measured their mildest winters on record.

For the central and eastern parts of the country, the pattern featured mild weather with little snow, the main exception being parts of the Great Lakes region, which saw heavy lake-effect snows. A marked dearth of winter storms led to near-record dryness from Maine to Georgia, resulting in unseasonably low groundwater, lake, and stream levels. Precipitation from the mid-Atlantic region to the Gulf Coast totaled less than 75% of normal, with several areas recording under 50% of normal. New Jersey and Maryland measured their driest winter of record.

Despite a major snowstorm that swept across the Southeast during the first few days of January, most locations east of the Continental Divide registered meager snowfall amounts this season. The New York City–Washington urban corridor saw only three to five inches of snow for the entire snow season. New York City's 3.5 inches was it second lowest snowfall total on record.

The most damaging and expensive storm of the winter season spread rain, ice, and snow from New Mexico to Maine from January 30 to February 1. A thick layer of ice toppled trees and power lines and left hundreds of thousands of customers without power in Missouri, Kansas, and Oklahoma. The storm did, however, bring much needed moisture to the Plains wheat crop. The cold air associated with the storm left the northern High Plains winter wheat crop exposed to temperatures as low as −20 F.

• SPRING (MARCH–MAY)

Wintry weather finally took hold over most of the nation during March, and an active storm pattern brought normal precipitation to many areas along the East Coast for the first time since August or September. This was the second-coldest March in the past 20 years nationally. Monthly temperatures averaged 10–20°F below normal across the northern Plains. Following March, extremes of temperature alternated during the rest of the spring, but Montana still ended up with its fourth-coldest spring since 1895. Record heat enveloped the country during the middle of April, sending mercury readings into the 90s across the Midwest and Northeast. Some 300 daily record high temperature records were set during April 14-20.

Only a few weeks later, a dramatic change in circulation brought polar air southward from Canada, resulting in frigid air covering a large expanse of the nation and nearly 500 low-temperature records during May 17-25. A number of locations from the Tennessee Valley into the mid-Atlantic region registered their latest freezes on record during this cold snap. The seesaw continued as, days later, a ridge of high pressure building up over the Southwest brought extreme heat to the West and Plains. Temperatures soaring into the 90s and 100s during May 30 to June 1 broke some 250 daily records and three dozen May monthly records.

Tornado alley was relatively quiet this spring, due to abnormally dry weather in the central parts of the country. But there were a number of outbreaks of severe weather from late spring into summer. About 50 tornadic thunderstorms hit central and eastern parts of the country during the last 10 days of April. One of the strongest East Coast tornadoes on record, an F4, struck La Plata, Maryland, on the 28th. The tornado was part of a storm system that brought heavy precipitation to many areas, including up to 20 inches of snow in Wisconsin.

Abnormally high pressure aloft kept southern Alaska unusually dry this spring, with the greatest deficits in the Panhandle. Although Anchorage recorded its all-time record 24-hour snowfall (26.7 inches) on March 16–17, most of the rest of the state saw dry conditions this month, especially across the south. Juneau saw its second driest April, with 0.47 inches (15% of normal) of precipitation. Spring precipitation totaled less than 50% of normal across the Panhandle.

Drought intensified over the Southwest, the Rockies, and the High Plains, as three-month precipitation totaled under 50% of normal across much of this region. Less than 25% of normal precipitation fell over the plains of Colorado, New Mexico, Arizona, southern California, and southern and eastern Utah. Colorado recorded its driest spring on record, and Arizona measured its second driest. The 12-month period ending in May was the driest ever for both states.

In contrast, the Midwest saw above-normal wetness, with spring precipitation 150% of normal across much of the Ohio Valley. Indiana saw its third wettest spring on record. Repeated rounds of heavy rain led to persistent lowland flooding from the southeastern Plains to the Ohio Valley during May, when more than a foot of rain fell on parts of Indiana, Illinois, and Missouri. Heavy rains led to significant fieldwork delays in the Corn Belt.

• SUMMER (JUNE–AUGUST)

Heat and dryness contributed to huge wildfires in Colorado and Arizona from late spring into early summer and an active fire season throughout the West this year. Five western states—Nevada, Utah, Arizona, California, and Colorado—measured their driest summers since 1895. This was the driest first half of a year (January–June) on record in Arizona, Utah, and Colorado.

By the end of June wildfires had burned 2.8 million acres across the country, with most of the acreage in the parched West. Record large fires burned in Arizona, Colorado, and Oregon this spring and summer. Nationally, fires burned 7.1 million acres by year's end, nearly double the 10-year average. This was the second worst fire year in the past 14 years.

Summer rainfall totaled less than one-half of normal from western South Dakota to eastern Kansas and over large parts of Colorado and the other western states. But heavy rain and snow relieved drought in northern Montana in June, resulting in summer rainfall more than twice normal.

In late June, drought indices showed some 50% of the contiguous United States in drought, with severe drought covering nearly 40% of the country. Over one-quarter of the nation endured extreme drought, primarily the Southwest and the southeastern Piedmont areas. In addition, abnormal dryness covered about one-half of Alaska and lingered over parts of Hawaii. The last time severe drought covered a larger area occurred during the mid-1950s mini-dust bowl era.

One area with a quickly disappearing drought this year was south-central Texas, as an upper-level low-pressure system delivered torrential rains from the end of June into the first week of July. Over a foot of rain brought devastating floods to the San Antonio region, with thousands of people displaced from their homes. San Antonio measured 16.16 inches of rain from June 30 to July 6, and the city's monthly total of 16.92 inches (833% of normal) was by far its wettest July total ever.

Strong thunderstorms also brought widespread flooding to North Dakota and western Minnesota in June, resulting in considerable crop and property damage.

A series of heat waves affected the country at various times this summer. Much of the nation sweltered from June 29 to July 4 as the Bermuda High pumped tropical air northward. Bismark, North Dakota, set an all-time high mark with 111°F on June 29. On Independence Day temperatures neared triple digit levels in the mid-Atlantic region. Both Baltimore and Richmond registered maximum readings of 100°F.

Cold fronts brought cooler air to central and eastern parts of the country temporarily after July 4, but the heat continued in the West, with July 10 entering the record books as one of the hottest days in recent history across the interior Pacific states and the western Great Basin. Readings reached 115°F in the Sacramento Valley and exceeded 100°F as far north as Washington. Reno, Nevada's, maximum of 108°F on July 10 was its all-time highest, and this record was tied just one day later. During July 7–14, triple-digit heat broke more than 500 daily high-temperature records and numerous all-time highs.

In the Plains, extreme heat further aggravated drought conditions in July. From the 15th to the 21st, thermometers from South Dakota to Kansas hit the century mark each day. Omaha, Nebraska, reached 104°F on the 22nd before a cold front brought temporary relief. Another heat wave covered central and northeastern parts of the country in late July and early August, followed by a return of the heat to the Northeast in mid-August. Washington, DC, recorded eight consecutive days of 95-degree or higher temperatures from August 12 to 19, tying a record for the longest stretch of 95-degree readings.

Nationally, this was the third hottest summer (June–August) in over 100 years of record, exceeded only by the summers of 1934 and 1936. The hot weather aggravated drought in many areas, especially the East and the western Plains states, significantly cutting crop yields. Although several bouts of showers eased dryness over most of the Corn Belt, summer rainfall totaled less than 75% of normal over southern parts of Illinois, Indiana, and Michigan, and across much of Ohio. A dry pocket in northwestern Ohio saw less than 50% of normal rainfall. Hot, dry conditions in July severely stressed reproductive to filling crops in the westernmost Corn Belt.

• AUTUMN (SEPTEMBER–NOVEMBER)

Twelve named tropical storms formed in the Atlantic basin during 2002, four becoming hurricanes. The two most notable storms took quite similar tracks from the Gulf of Mexico into the central Gulf states. In late September, Isidore slammed into southern Mississippi at tropical storm strength, its remains bringing tropical deluges of two to eight inches as far north as the Ohio Valley. On October 3, Hurricane Lili, the first storm to make landfall at hurricane strength since 1999, hit the central coast of Louisiana, bringing wind gusts to 92 mph and inundating low-lying areas. As with Isidore, tropical rains extended far northward into the Ohio Valley.

October brought a change in the weather pattern to most of the country, as the Bermuda High responsible for much of the summer heat retreated and the westerlies dropped southward, allowing cold Canadian air to penetrate the U.S. October 2002 was nearly opposite to October 2001, with below-normal temperatures and bouts of rain and snow affecting many states. One nor'easter early in the month ended many aspects of the long-term drought across the Eastern Seaboard, lifting water tables and adding substantially to reservoir levels. Tropical Storm Kyle, which had been meandering around the Atlantic for nearly three weeks, grazed the South Carolina coast on October 10–11, further eating away at the long-term drought.

October brought an early winter to many parts of the nation, as Canadian high pressure plunged southward. By the last day of the month, minimum temperatures hit sub-zero levels from Oregon to North Dakota, and dipped to –11°F in Montana.

The stormy weather pattern lasted through year's end, virtually ending the long-term drought over the Eastern Seaboard, but bringing violent weather to some areas.

Cold air behind an intense cold front sweeping across the central parts of the country clashed with unseasonably mild and humid air in the East on November 10, setting the stage for the year's deadliest outbreak of severe weather. Tornadoes in seven states from Mississippi, Alabama, and Georgia northward to Ohio and Indiana left 36 people dead. From late Sunday on the 10th through Monday the 11th there were more than 70 tornadoes, 250 damaging wind events, and 160 large hail occurrences from Louisiana across the Tennessee and Ohio Valleys to Pennsylvania and Georgia.

A few days earlier, a huge Pacific storm hit the Pacific states with a barrage of wind, waves, rain, and mountain snows. The first major storm of the season slammed the western states from November 7 to 9, bringing wind gusts of 55 mph to the San Francisco area and two to three feet of snow to the Sierra Nevada. The two- to four-inch rainfall amounts that covered large parts of California, Oregon, and Nevada constituted a big portion of the normal annual rainfall in some of the more arid locations. But the moisture was not entirely unwelcomed, as it put a big dent in the ongoing drought. Downtown Los Angeles recorded 2.31 inches of rain during the storm, more than the city received during the entire year-to-date through November 6 (1.61 inches). Nevertheless, even with the heavy rain, the city's year-to-date total through November 10 of 3.92 inches was just 31 percent of normal. The 12 months ending in October were coastal southern California's driest such period since at least 1895.

Nor'easters brought heavy rain, snow, or ice to the Eastern Seaboard on November 5–6 and 16–17, further eating away at any lingering drought. The freezing rain that fell on New England on Saturday the 16th turned into a major ice storm for Connecticut.

Conditions were abnormally dry north and west of the storm track this autumn. Reduced soil moisture and bouts of cool weather hindered winter wheat establishment across the northern and central Plains and the Northwest. Precipitation during the three-month period totaled under 50% of normal from Michigan into northeast Oklahoma. In Illinois, Peoria recorded only 34% of its normal September-November precipitation, setting a record for the city's driest autumn. Most of Washington and Oregon recorded less than 50% of normal precipitation.

• DECEMBER

One of the most damaging ice storms of the year took place on December 4. Freezing rain fell from southern Virginia to northern Georgia, with the Carolinas bearing the brunt of the ice storm. A layer of ice one-half to one-inch thick toppled trees and power lines, leaving 1.5 million customers without power in North Carolina and many others in the dark in South Carolina. The same storm spread five to eight inches of snow from Washington, D.C., to New York City on the 5th, resulting in more snow in one day than the I-95 corridor saw during the entire 2001–2002 winter season. A number of Pacific storms pelted the West Coast states with strong winds, heavy rain, and mountain snows during December, the largest bearing down on Washington, Oregon, and California from the 13th to the 16th. This storm dropped seven inches of rain near San Francisco and brought river flooding to northern California. Winds gusted to 90 mph along the Oregon coast. Winds measured at 82 mph caused major property damage to Reno, Nevada, on the 14th. Another Pacific storm a few days later brought street flooding to San Francisco.

The storm that brought flooding to California on December 19–20 developed into a massive winter storm that brought a lot of wintry weather from the Plains to the Northeast on December 23–25. The storm left six to 12 inches of snow from western Oklahoma and northern Texas to southern Missouri on the 23rd to 24th and triggered severe thunderstorms in eastern Texas and southern Georgia. Coastal development led to an intense nor'easter on Christmas day, resulting in one to two feet of snow from Pennsylvania through upstate New York into New England.

Still more storms struck the West Coast the last few days of the year. At Squaw Valley in California, snow depth rose to 122 inches on the last day of the year, and winds gusted on top to 111 mph. For the month, many Tahoe locations accumulated 10 to 15 feet of snow. Along the Pacific coast, rainfall exceeded two feet this month from northern California into southwestern Oregon.

Dry weather persisted across the upper Midwest, as precipitation totaled under one-half of normal from Nebraska to the Great Lakes. In Nebraska, Omaha recorded no precipi-

WEATHER ANECDOTES

March 26, 2002, Slidell, LA. Two teenage girls using curling irons were hospitalized after lightning traveled through their houses electrical wiring.

May 13, 2002, Iron Co., MO. A 43-year old man was trying to cross the flooding Stouts Creek on foot to get to his home to rescue his dogs. He was knocked down, but managed to grab hold of a tree. He was swept away and drowned by the rising water before workers could reach him.

August 23, 2002, Willard, MO. Three mourners at a rural graveside funeral were killed when they were struck by lightning that cracked down a tree under which they took shelter.

tation for the entire month. In Missouri, Kansas City measured only 0.03 inches, tying the record for the driest December. For the first time ever, Des Moines, Iowa, failed to receive measurable monthly precipitation.

• 2002 PRECIPITATION RECORDS FOR SELECTED LOCATIONS

Driest Year (Inches) on Record

Yuma, AZ 0.03
Palm Springs, CA 0.76
Phoenix, AZ 2.82
Pueblo, CO 3.94
Winslow, AZ 4.31
Laramie, WY 5.78
Denver, CO 7.48

Wettest Year on Record

Lake Charles, LA 85.17

• EARTHQUAKE IN WEST CENTRAL MEXICO, JANUARY 21, 2003.

A major earthquake occurred in Colima, Mexico, about 310 miles west of Mexico City near the coastal city of Manzanillo, at 7:06 PM MST, January 21, 2003 (8:06 PM CST in Mexico). A preliminary magnitude of 7.8 was computed for this earthquake. The magnitude and location may be revised when additional data and further analysis results are available. There were at least 28 deaths, 300 injuries, and considerable damage in the states of Colima, Michoacan, and Jalisco. The earthquake was felt strongly in Mexico City.

This shallow earthquake occurred in a seismically active zone near the coast of central Mexico. The earthquake occurred near the juncture of three tectonic plates: the North

American Plate to the northeast, the Rivera Plate to the northwest, and the Cocos Plate to the south. Both the Rivera Plate and the Cocos Plate are being consumed beneath the North American Plate. The slower moving Rivera Plate is moving northwest at about 2 cm per year relative to the North American Plate and the faster moving Cocos plate is moving in a similar direction at a rate of about 4.5 cm per year.

There have been several significant earthquakes near the recent event. In 1932, a magnitude 8.4 earthquake struck about 100 km to the northnorthwest. More recently, on October 9, 1995 a magnitude 8.0 earthquake struck about 50 km to the northwest killing at least 49 people and leaving 1,000 homeless. The most deadly earthquake in the region occurred about 170 km to the southeast on September 19, 1985. This magnitude 8.0 earthquake killed at least 9,500 people, injured about 30,000, and left 100,000 people homeless.

El Niño & La Niña

• BACKGROUND

The original definition of El Niño goes back to the eighteenth or nineteenth century when Peruvian sailors coined the term to describe a warm southward current that appeared annually near Christmas off the Peruvian coast. Hence the name El Niño, Spanish for "the Child," referring to the Christ Child. Throughout the year, a northward cool current prevails because of southeast trade winds, causing upwelling of cool, nutrient-rich water. However, during late December the upwelling relaxes, causing warmer and nutrient-poor water to appear, which signals the end of the local fishing season.

Over the years, the warm, southward current occasionally seemed more intense than usual and was associated with periods of extreme wetness along the normally very dry Peruvian coast. These events were called "years of abundance." In the early twentieth century, researchers found a strong inverse correlation, called the Southern Oscillation, between surface pressure over the Pacific and Indian Oceans, hence the saying, "When pressure is high in the Pacific and Indian Oceans." Researchers tried, but failed, to correlate the Southern Oscillation with Indian monsoon failures. In 1958–59, a strong "year of abundance" occurred, in which a large area of warm water in the Pacific Ocean extended from the South American coast westward to the International Date Line. Coinciding with the extensive warm water were wetness along the Peruvian coast, low surface pressure in the eastern Pacific, and high pressure in the western tropical Pacific. Consequently, scientists in the early 1960s concluded that these events were associated and occurred interannually. Since then, the term "El Niño" (or warm episode) has described not a local warm current, but warming of the tropical Pacific surface waters occurring every two to seven years and associated with changes in the atmospheric circulation in the tropical Pacific and worldwide.

• MECHANISMS

Figure 1 depicts the typical atmospheric and oceanic circulations that exist in the tropical Pacific. The prevailing

easterlies (NE and SE trades) converge over Indonesia in conjunction with the Asian monsoon, producing widespread convection. Additionally, warm water "piles up" in the western Pacific, due to the easterly winds. Further east, the SE trades and equatorial easterlies in the eastern and central Pacific produce upwelling of cool water along the equator and coast of South America.

As the El Niño event begins, the easterlies relax, reducing the amount of upwelling and allowing the western warm water to move eastward. As time goes on, the warm pool in the western Pacific grows and expands eastward toward the central Pacific (figure 2). Detailed monitoring of recorded El Niño episodes has revealed that once the warmest water reaches the International Date Line, anomalous convection usually appears in that region, accompanied by a weakening of the equatorial easterlies. This pattern typically occurs during the boreal winter (June–August) and may be preceded or followed by a warming that causes the Inter-Tropical Convergence Zone (ITCZ) to move farther south than normal, which contributes to enhanced rainfall across Ecuador and northern Peru, producing the "years of abundance."

In determining the atmospheric status of the tropical Pacific, climatologists devised the Southern Oscillation Index (SOI, Figure 3). It is the standardized sea level pressure difference between Darwin, Australia, and Tahiti, French Polynesia, in the central Pacific (Tahiti minus Darwin). Thus, when the surface pressure is high at Darwin and low at Tahiti, the SOI is negative (El Niño); conversely, when surface pressure is low at Darwin and high at Tahiti the SOI is positive. When the SOI is strongly positive, cooler than normal equatorial water appears throughout the central and eastern equatorial Pacific. This is called a cold episode or sometimes La Niña, "little girl." Climatologists prefer to use the acronym ENSO (El Niño/Southern Oscillation) to describe the warm (El Niño) and cold (La Niña) episodes that occur periodically across the tropical Pacific.

• IMPACTS

When El Niño or La Niña develops, several consistent weather anomalies typically occur around the world. Fig-

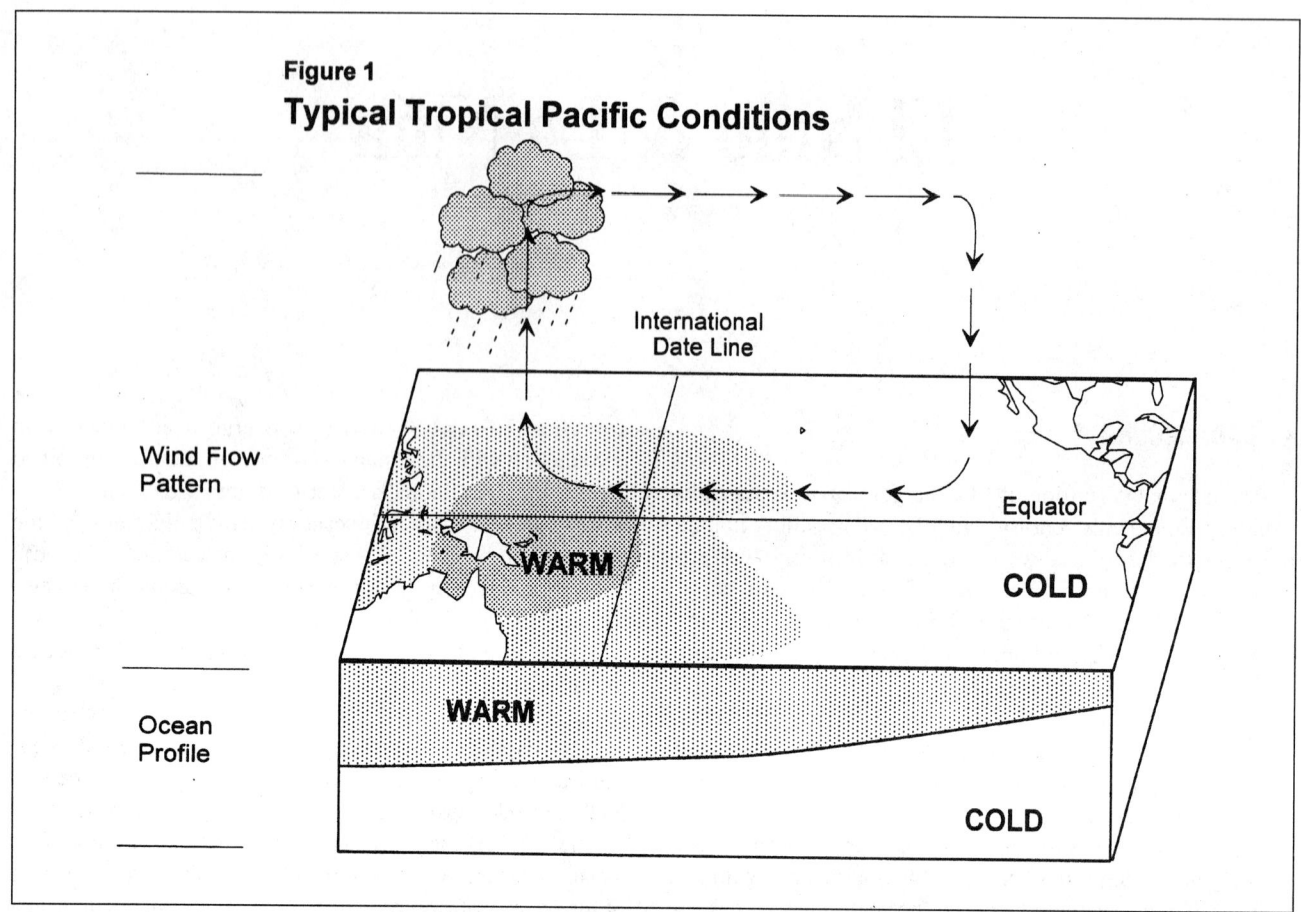

Figure 1
Typical Tropical Pacific Conditions

(Courtesy of National Oceanic and Atmospheric Administration (NOAA).)

ures 4 and 5 depict potential rainfall and temperature impacts from El Niño while figures 6 and 7 show potential rainfall and temperature impacts from La Niña. Most climate anomalies associated with El Niño are reversed during La Niña. In general, a majority of the impacts occur in climates that have significant oceanic influences and border the tropical Pacific. Thus, the regions of the world that show the highest correlation to warm or cold events are Indonesia, Australia, and the tropical Pacific islands. Weather anomalies (drought and excessive moisture) associated with El Niño and La Niña can have significant impact on agricultural production (i.e., poor crops due to failure of the Indian monsoon). However, several factors make the impacts on crop production less dramatic and sometimes nonexistent. These factors include the timing, duration, and intensity of ENSO events at various stages of crop development.

What is happening in the atmosphere during El Niño?

Several aspects of the atmosphere's behavior are remarkable and entirely unique to the ENSO phenomenon. Some normally arid tropical habitats are transformed into virtual gardens during El Niño. Abundant and reliable rains in other tropical areas become sparse and intermittent during El

Niño. Extreme climates have also been experienced in the higher latitudes during ENSO, though these are by no means unique to ENSO. One marvels that the atmosphere, especially thousands of miles away from the equatorial Pacific, "knows" about the modest warming of those waters during El Niño! Yet, all regions of the globe are not equally affected, nor is ENSO's impact uniform throughout the year. How do we understand these atmosphere manifestations of ENSO?

A meteorological view of the ENSO phenomenon offers some answers. However it does not explain ENSO itself; for that, one needs to account for the origin of the oceanic conditions, and the coupled interaction of the ocean and the atmosphere is central to that problem.

So how does the atmosphere "know" about El Niño? It is useful to imagine a chain of atmospheric processes, with each link in this chain carrying information from the local vicinity of El Niño sea surface temperature (SST) anomalies throughout the global climate system. The first link is the tropical response of rain-producing cumulonimbus; critical because deep convection is the principal agent for exchanging heat from Earth's surface and thereby communicating El Niño's presence to the free atmosphere. Wet tropical climates tend to coincide with warm pool SST area in the west-

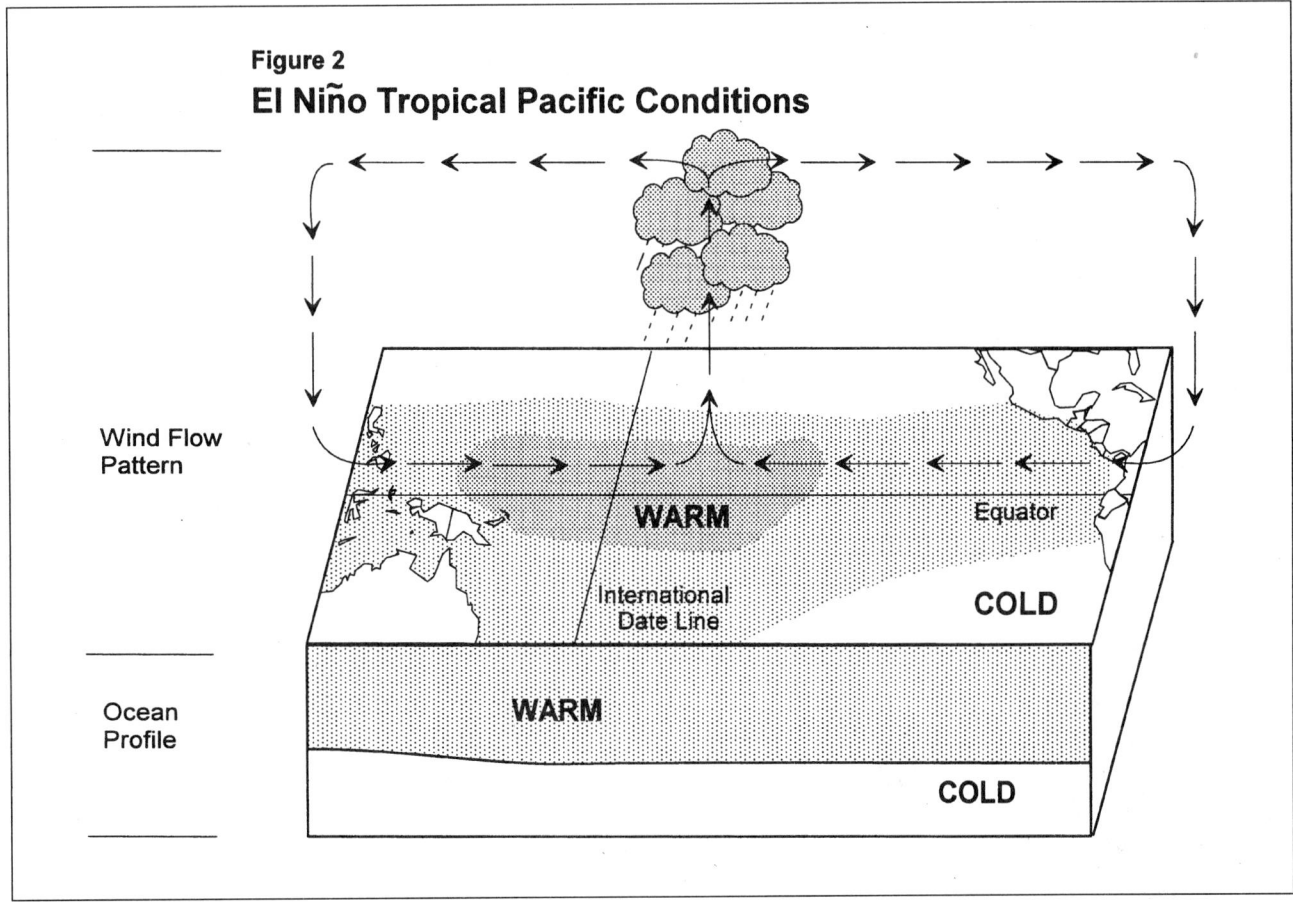

Figure 2
El Niño Tropical Pacific Conditions

Wind Flow Pattern

WARM

Equator

International Date Line

COLD

Ocean Profile

WARM

COLD

(Courtesy of National Oceanic and Atmospheric Administration (NOAA).)

ern Pacific, and the continental monsoons. During El Niño, rainfall increases over a distance of several thousand miles along the equator from the central to the eastern Pacific in response to the warming of the underlying SSTs. Reduced rainfall occurs on the periphery of this wet zone, and even the continental monsoons are not spared ENSO's influence. The opposite effect tends to be experienced during La Niña, although the west-east scale of rainfall anomalies over the equatorial Pacific is somewhat reduced compared to warm events.

The second link in the chain is the horizontal communication of El Niño's presence, and this involves the sensitivity of the atmosphere's circulation to shifts in organized cumulonimbus convection. Excited atmospheric wave motions adjust the climatological flow to the new tropical energy sources; not unlike the waves generated by a pebble dropped into a pond, although for ENSO the spatial scales of forcing are much larger and the atmosphere is readily forced. The major convection anomalies themselves are confined to within a few degrees of the equator during winter. However, associated with them is a circulation of mass and energy in the atmosphere that extends several thousand miles poleward into the subtropics. A deflecting force, due to the Earth's rotation, acts upon this outflow along its poleward

course, thereby initiating wave-like patterns in the perturbed flow. In addition, the climatological circulation in higher latitudes acts to channel the course of this poleward flowing energy. This flow is directed from west to east, has concentrated westerlies along the jet streams east of Asia and North America, and is characterized by stationary waves of alternating low and high pressure.

Through interactions with this background flow, the resulting atmospheric response to El Niño also consists of a wave train pattern having alternating low and high pressure. In the Northern Hemisphere, the wind circulates parallel to these contours, with low pressure to the left of the motion's direction. The wave paths follow "great circle" routes that arc toward the pole, into the higher reaches of the Pacific-North American region, and then curve toward the equator into the subtropical western Atlantic. A similar wave pattern in the atmosphere exists in the Southern Hemisphere with an attendant influence on the climate of South America.

The anomalous wave patterns are referred to as the atmospheric "teleconnections," linking equatorial and high latitudes during ENSO. These are based on a historical composite of how the atmosphere has behaved during the ten strongest El Niño and La Niña events of the 1950–96 period. Several features to note are the stronger Pacific jet during El

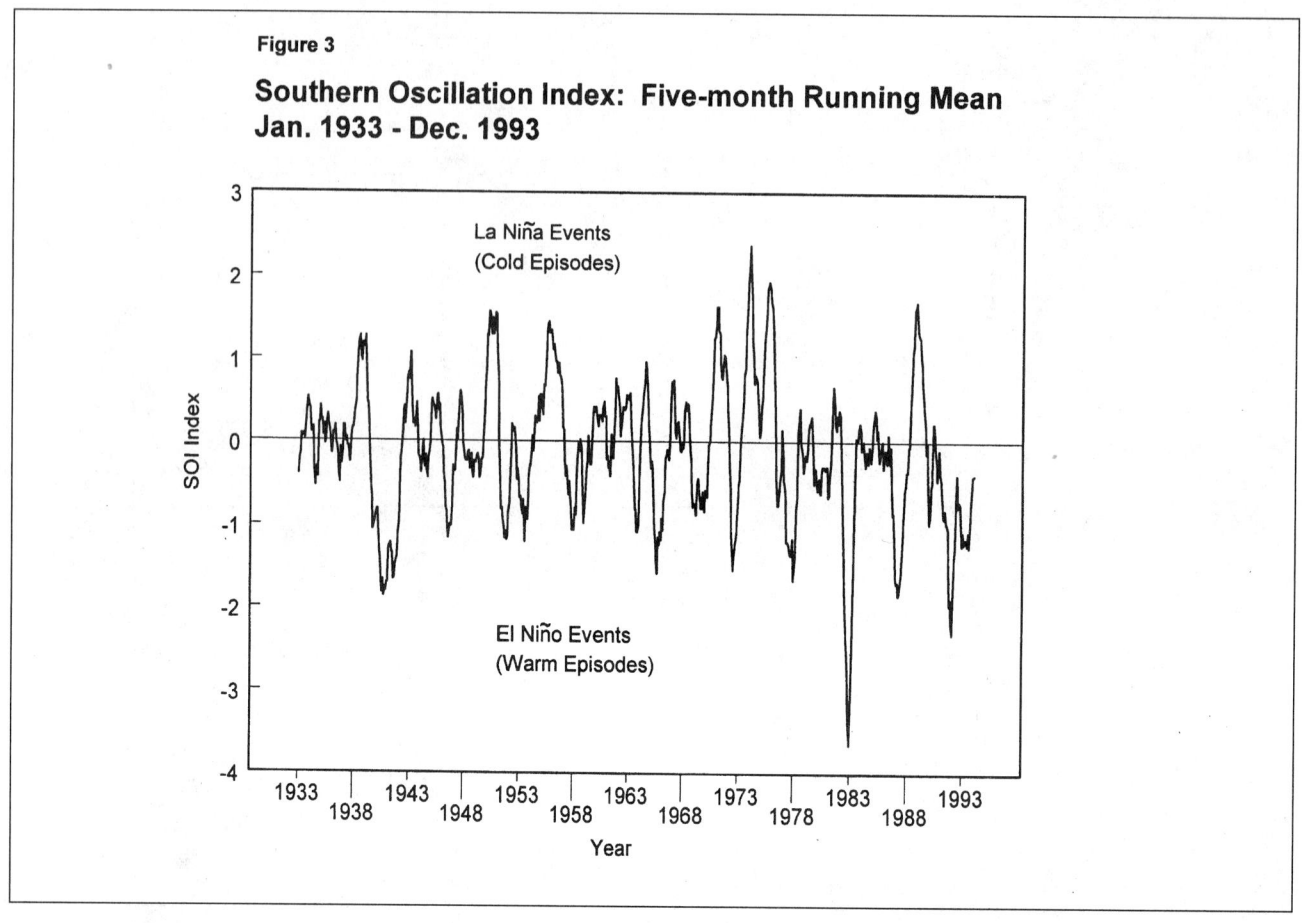

Figure 3

Southern Oscillation Index: Five-month Running Mean Jan. 1933 - Dec. 1993

(Courtesy of National Oceanic and Atmospheric Administration (NOAA).)

Niño and an eastward shift of the stationary wave pattern over the Pacific-North American region during El Niño. These upper tropospheric changes alter the course of storms (cyclone and anticyclones) that control the daily weather fluctuations in the higher latitudes. Changes in statistical properties of storms (for example, their frequency, strength, or origin and track) account for the bulk of ENSO's signal in precipitation and surface temperature in the higher latitudes. Such "storm track" feedback constitutes a third essential link along the chain that is initiated by the equatorial Pacific SST anomalies.

Each El Niño event has a unique signature in its SST lifecycle, and the question of the atmosphere's sensitivity to such inter-El Niño variations is a matter of intense research focusing on the implications for seasonal climate predictions. The atmospheric events are being actively monitored by the scientific community, and the expected climate response continues to be assessed.

Global consequences of El Niño

The twists and turns in the ongoing dialogue between ocean and atmosphere in the Pacific can have a ripple effect on climatic conditions in far flung regions of the globe. This worldwide message is conveyed by shifts in tropical rainfall, which affect wind patterns over much of the globe. Imagine a rushing stream flowing over and around a series of large boulders. The boulders create a train of waves that extend downstream, with crests and troughs that show up in fixed positions. If one of the boulders were to shift, the shape of the wave would also change and the crests and troughs might occur in different places.

Dense tropical rain clouds distort the air flow aloft (5–10 mi above sea level) much as rocks distort the flow of a stream, or islands distort the winds that blow over them, but on horizontal scale of thousands of miles. The waves in the air flow, in turn, determine the positions of the monsoons, and the storm tracks and belts of strong winds (commonly referred to as jet streams) that separate warm and cold regions at the Earth's surface. In El Niño years, when the rain area that is usually centered over Indonesia and the far western Pacific moves eastward into the central Pacific, the waves in the flow aloft are affected, causing unseasonable weather over many regions of the globe.

The impacts of El Niño upon climate in temperate latitudes show up most clearly during wintertime. For example,

Figure 4

Potential Rainfall Impacts from El Niño Events (Warm Episodes)

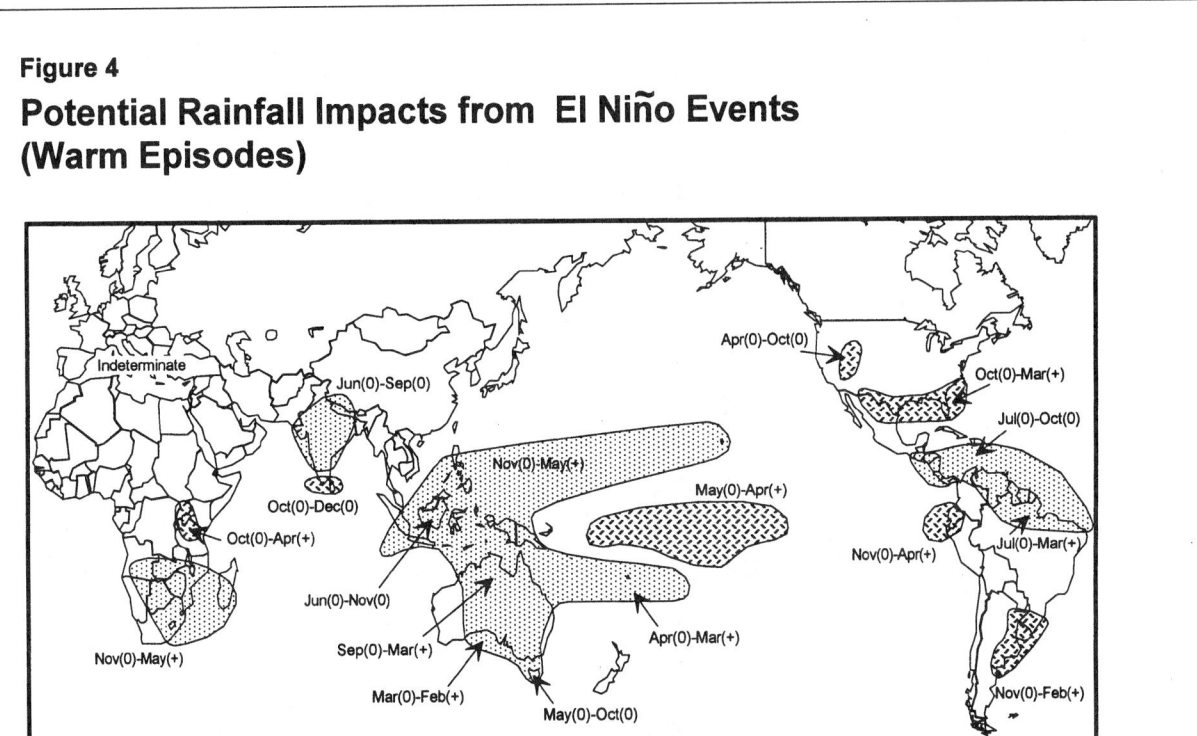

(Courtesy of National Oceanic and Atmospheric Administration (NOAA).)

most El Niño winters are mild over western Canada and parts of the northern United States, and wet over the southern United States from Texas to Florida. El Niño affects temperate climates in other seasons as well. But even during wintertime, El Niño is only one of a number of factors that influence temperate climates. El Niño years, therefore, are not always marked by "typical" El Niño conditions the way they are in parts of the tropics.

• BENEFITS OF EL NIÑO PREDICTION

Scientists are now taking our understanding of El Niño a step further by incorporating the descriptions of these events into numerical prediction models—computer programs designed to represent, in terms of equations, processes that occur in nature. Such models are fed information mostly in the form of sets of numbers, describing the present state of the atmosphere-ocean system (for example, observations of wind speeds, ocean currents, sea level, and the depth of the thermocline along the equator). Updated sets of numbers that the models produce indicate how the atmosphere-ocean system might evolve over the next few seasons or years. The

results thus far, though by no means perfect, give a better indication of the climatic conditions that will prevail during the next one or two seasons than simply assuming that rainfall and temperature will be "normal."

Peru provides a prime example of how even short term El Niño forecasts can be valuable. There, as in most developing countries in the tropics, the economy (and food production in particular) is highly sensitive to climate fluctuations. Warm (El Niño) years tend to be unfavorable for fishing and some of them have been marked by damaging floods along the coastal plain and in the western Andean foothills in the northern part of the country. Cold years are welcomed by fishermen, but not by farmers, because these years have frequently been marked by drought and crop failures. Such cold years often come on the heels of strong El Niño events.

Since 1983, forecasts of the upcoming rainy season have been issued each November based on observations of winds and water temperatures in the tropical Pacific region and the output of numerical prediction models. The forecasts are presented in terms of four possibilities: (1) near normal conditions, (2) a weak El Niño with a slightly wetter than normal growing season, (3) a full blown El Niño with flooding, and (4) cooler than normal waters offshore, with higher than normal chance of drought.

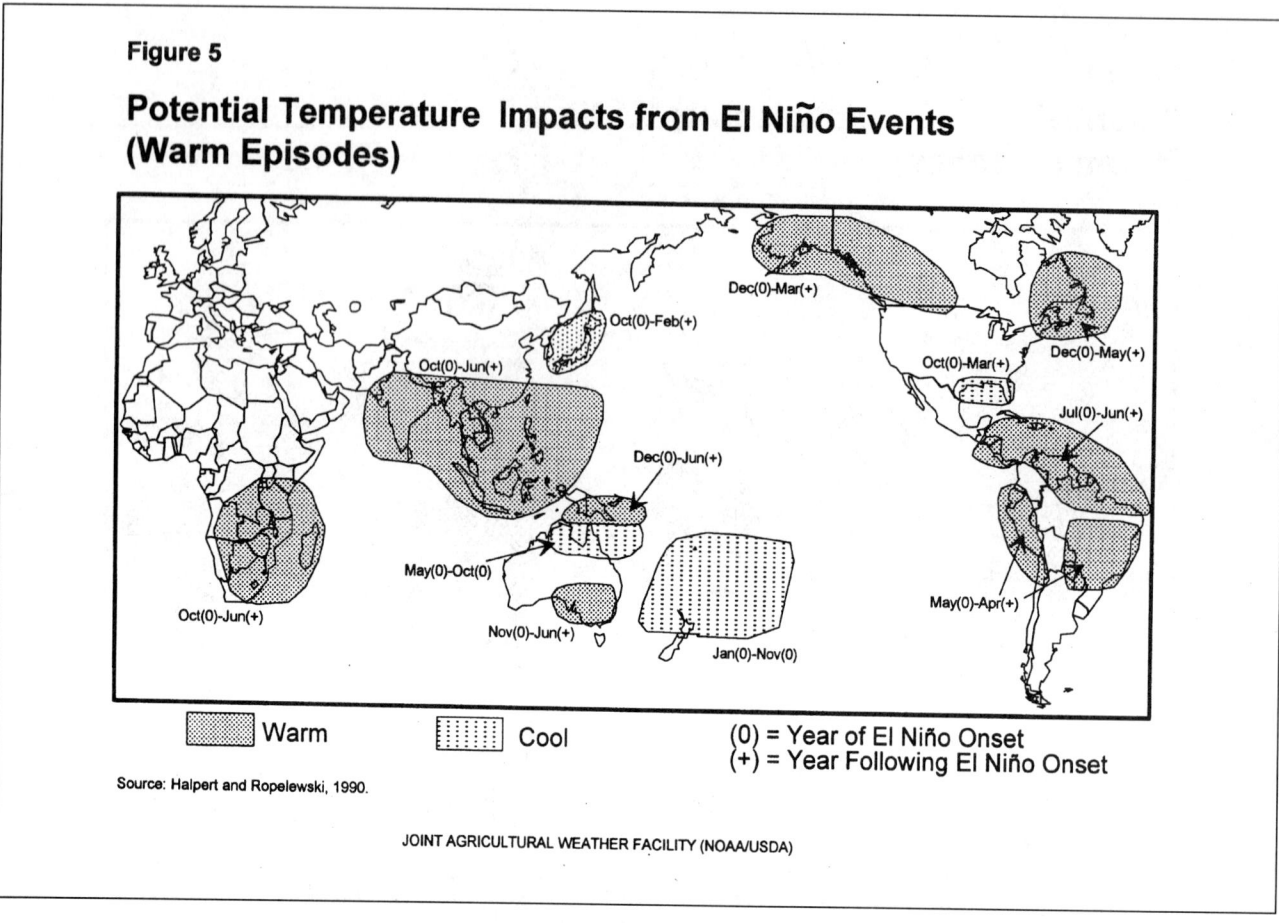

Figure 5

Potential Temperature Impacts from El Niño Events (Warm Episodes)

Source: Halpert and Ropelewski, 1990.

JOINT AGRICULTURAL WEATHER FACILITY (NOAA/USDA)

(Courtesy of National Oceanic and Atmospheric Administration (NOAA).)

Once the forecast is issued, farmer representatives and government officials meet to decide on the appropriate combination of crops to sow in order to maximize the overall yield. Rice and cotton, two of the primary crops grown in northern Peru, are highly sensitive to the quantities and timing of rainfall. Rice thrives on wet conditions during the growing season followed by drier conditions during the ripening phase. Cotton, with its deeper root system, can tolerate drier weather. Hence, a forecast of El Niño weather might induce farmers to sow more rice and less cotton than in a year without El Niño.

Countries that have taken similar initiatives include Australia, Brazil, Ethiopia, and India. Although tropical countries have the most to gain from successful prediction of El Niño, for many countries outside the tropics, such as Japan and the United States, more accurate prediction of El Niño will also benefit strategic planning in areas such as agriculture, and the management of water resources and reserves of grain and fuel oil.

Encouraged by the progress throughout the 1990s, scientists and governments in many countries are working together to design and build a global system for (1) observing the tropical oceans, (2) predicting El Niño and other irregular climate rhythms, and (3) making routine climate predic-

tions readily available to those who need them for planning purposes, much as weather forecasts are made available to the public today. The ability to anticipate how climate will change from one year to the next will lead to better management of agriculture, water supplies, fisheries, and other resources. By incorporating climate predictions into management decision, humankind is becoming better adapted to the irregular rhythms of climate.

• THE PACIFIC DECADAL OSCILLATION

The Pacific Decadal Oscillation (PDO) is a long-lived El Niño-like pattern of Pacific climate variability. While the two climate oscillations have similar spatial climate fingerprints, they have very different behavior in time. Fisheries scientist Steven Hare coined the term "Pacific Decadal Oscillation" in 1996 while researching connections between Alaska salmon production cycles and Pacific climate (his dissertation topic with advisor Robert Francis).

Two main characteristics distinguish PDO from El Niño/Southern Oscillation (ENSO): first, twentieth century PDO "events" persisted for 20–30 years, while typical

Figure 6

Potential Rainfall Impacts from La Niña Events (Cold Episodes)

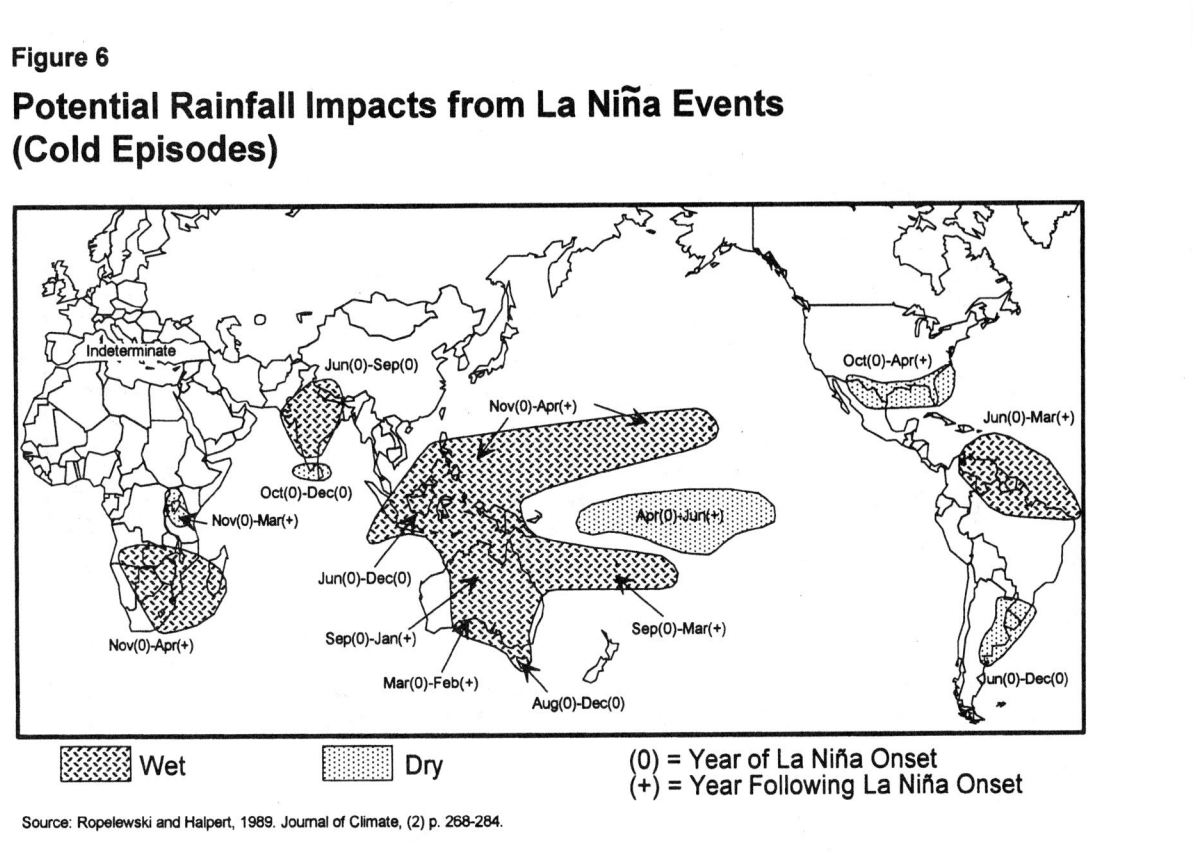

(Courtesy of National Oceanic and Atmospheric Administration (NOAA).)

ENSO events persisted for 6–18 months; second, the climatic fingerprints of the PDO are most visible in the North Pacific/North American sector, while secondary signatures exist in the tropics—the opposite is true for ENSO. Several independent studies find evidence for just two full PDO cycles in the past century: "cool" PDO regimes prevailed in 1890–1924 and again in 1947–1976, while "warm" PDO regimes dominated in 1925–1946 and from 1977 through (at least) the mid-1990s. Shoshiro Minobe has shown that twentieth century PDO fluctuations were most energetic in two general periodicities, one of 15–25 years, and the other of 50–70 years.

Major changes in northeast Pacific marine ecosystems have been correlated with phase changes in the PDO; warm eras have seen enhanced coastal ocean biological productivity in Alaska and inhibited productivity off the west coast of the contiguous United States, while cold PDO eras have seen the opposite north-south pattern of marine ecosystem productivity.

Causes for the PDO are not currently known. Likewise, the potential predictability for this climate oscillation is not known. Some climate simulation models produce PDO-like oscillations, although often for different reasons. The mechanisms giving rise to PDO will determine whether skillful decades-long PDO climate predictions are possible. For example, if PDO arises from air-sea interactions that require 10-year ocean adjustment times, then aspects of the phenomenon will (in theory) be predictable at lead times of up to 10 years. Even in the absence of a theoretical understanding, PDO climate information improves season-to-season and year-to-year climate forecasts for North America because of its strong tendency for multi-season and multi-year persistence. From a societal impacts perspective, recognition of PDO is important because it shows that "normal" climate conditions can vary over time periods comparable to the length of a human's lifetime.

• MADDEN-JULIAN OSCILLATION

Weather is not as predictable in the tropics as in mid–latitudes. This is because in mid–latitudes the weather variables (clouds, precipitation, wind, temperature, and pressure) are largely governed by the upper–tropospheric Rossby waves, which interact with surface weather in a process called baroclinic instability. In the tropics there is no such dominant instability or wave motion, and therefore the

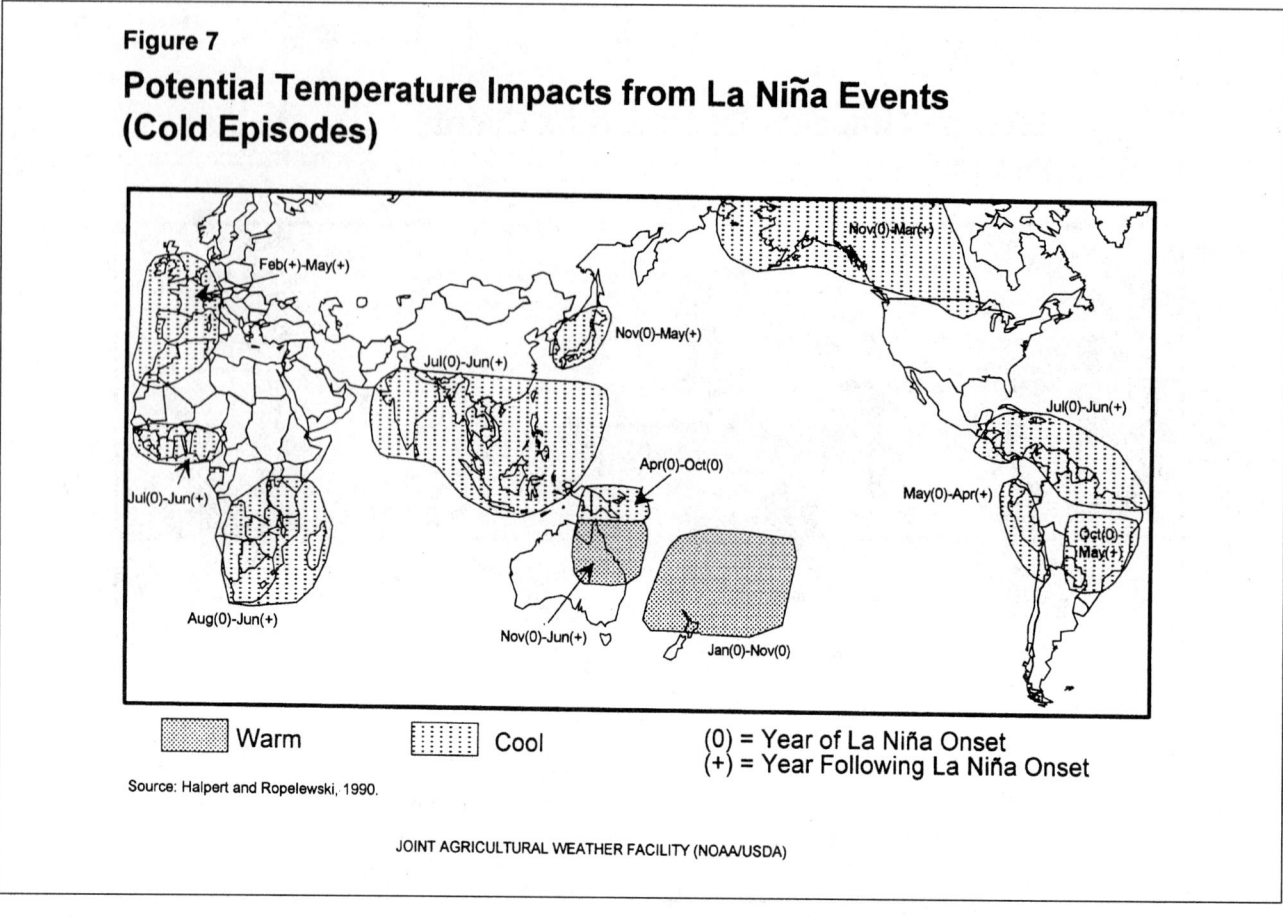

Figure 7

Potential Temperature Impacts from La Niña Events (Cold Episodes)

Warm Cool (0) = Year of La Niña Onset (+) = Year Following La Niña Onset

Source: Halpert and Ropelewski, 1990.

JOINT AGRICULTURAL WEATHER FACILITY (NOAA/USDA)

(Courtesy of National Oceanic and Atmospheric Administration (NOAA).)

weather is less predictable for the 1–10 day period. Until recently it was believed that tropical weather variations on time scales less than a year were essentially random.

In 1971 Roland Madden and Paul Julian stumbled upon a 40–50 day oscillation when analyzing zonal wind anomalies in the tropical Pacific. They used 10 years of pressure records at Canton (at 2.8°S in the Pacific) and upper level winds at Singapore.

The oscillation of surface and upper–level winds was remarkably clear in Singapore. Until the early 1980s little attention was paid to this oscillation, which became known as the Madden and Julian Oscillation (MJO), and some scientists questioned its global significance. Since the 1982-83 El Niño event, low-frequency variations in the tropics, both on intra-annual (less than a year) and inter-annual (more than a year) timescales, have received much more attention, and the number of MJO–related publications grew rapidly.

The MJO, also referred to as the 30–60 day or 40–50 day oscillation, turns out to be the main intra–annual fluctuation that explains weather variations in the tropics. The MJO affects the entire tropical troposphere, but is most evident in the Indian and western Pacific Oceans. The MJO involves variations in wind, sea surface temperature (SST), cloudiness, and rainfall. Because most tropical rainfall is convec-

tive, and convective cloud tops are very cold (emitting little longwave radiation), the MJO is most obvious in the variation of outgoing longwave radiation (OLR), as measured by an infrared sensor on a satellite.

The OLR signal in the Western Hemisphere is weaker, and the recurrence interval for the eastward propagating OLR anomalies in the Eastern Hemisphere is about 30–60 days. How exactly the anomaly propagates from the dateline to Africa (i.e. through the Western Hemisphere) is not well understood. It appears that near the dateline a weak Kelvin wave propagates eastward and poleward at a speed exceeding 10 m/s.

• NORTH ATLANTIC OSCILLATION

There are several prominent, recurring modes of low-frequency variability over the extra tropical North Atlantic and Europe. Perhaps the most well known of these patterns is the North Atlantic Oscillation. The NAO exhibits little variation in its climatological mean structure from month to month and consists of a north-south dipole of anomalies of opposite sign, with one center located over Greenland and the other spanning the central latitudes of the North Atlantic at 35–40°N. The positive phase of the NAO reflects below-normal heights and pressure across the high latitudes of the North Atlantic and above-normal heights and pressure over the central North Atlantic, the eastern United States, and western Europe. The negative phase reflects an opposite pattern of height and pressure anomalies. Both phases of the NAO are associated with basin-wide changes in the intensity and location of the North Atlantic jet stream and storm track, and in large-scale modulations of the normal patterns of zonal and meridianal heat and moisture transport, which in turn result in changes in temperature and precipitation patterns often extending from eastern North America to western and central Europe.

Strong positive phases of the NAO are often associated with above-normal temperatures in the eastern United States and across northern Europe and below-normal temperatures in Greenland and across southern Europe and the Middle East. The positive NAO phase is also associated with above-normal precipitation over northern Europe and Scandinavia and below-normal precipitation over southern and central Europe. Opposite patterns of temperature and precipitation anomalies are typically observed during strong negative phases of the NAO. During prolonged periods dominated by one phase of the NAO, abnormal height and temperature patterns are also often seen extending well into central Russia and north-central Siberia.

• NORTH ATLANTIC–ASIAN MONSOONS INFLUENCE EACH OTHER'S PATTERNS

Like ballroom dancers on a crowded floor, climatic phenomena like El Niño, the Asian monsoons, and the North Atlantic influence each other's patterns. Sediments from the floor of the Arabian Sea near Oman were studied by researchers looking for evidence of the strength of monsoons in the region over the past 10,000 years. There is a suggestion that the link between the North Atlantic climate and the Asian monsoon is a persistent aspect of global climate. The link was demonstrated previously by various researchers, but the new research examines a much longer time period (the past 10,000 years). The new study reveals substantial natural variation in climate and the monsoon in a time prior to any significant human influence. The new information may lead to improved predictions of the monsoon in the coming decades.

The significance of these results lies in demonstrating a pattern of persistent variability in monsoons throughout the Holocene (from 10,000 years ago to the present) that may be linked with episodic warming and cooling of the North Atlantic. The results highlight the need to improve our understanding of abrupt and difficult-to-predict weakening in monsoon strength, which could accompany major climate shifts in the North Atlantic in the future.

NOAA and university researchers used fossils of the plankton *Globigerina bulloides* to estimate wind intensity. During a monsoon, the seasonal reversal of winds brings moisture from the ocean onto land. The winds also blow surface waters off shore, causing an upwelling of colder, nutrient-rich water where the microscopic marine animals can thrive. By counting the amount of *G. bulloides* present in different layers of the sediment and using radiocarbon dating, the scientists were able to approximate monsoon strength from 10,500 years ago up to the present. The resulting record showed a natural variation in the monsoon from one century to the next. This provides new evidence that the strength of Asian monsoon varies substantially on century to millennial time scales, and the need to understand this if we're going to ensure human and ecological sustainability in Tibet, China, India, and the rest of Southeast Asia.

While researchers aren't sure of the exact causes of the link between the North Atlantic and the Asian monsoon, earlier research showed the amount of snow on the Tibetan plateau may play a critical role. As the land warms in the spring, the air rises above the land causing a pressure gradient that drives the monsoon. More snow on the plateau in spring or early summer uses up all the sun's heating because it has to be melted and evaporated before the land can warm. So the more snow you have in winter, the weaker the monsoon the following summers. There is speculation that when the North Atlantic is cold, areas downwind like the Tibetan plateau stay cold longer, allowing more snow to persist and setting up a weakened monsoon. The monsoon–snow cover link may lead to a stronger or more variable monsoon in the coming century as the Northern Hemisphere continues to warm faster than the tropics.

Other studies show that changes in the amount of sunlight correlate to variations in both the North Atlantic climate and the Asian monsoon. The researchers aren't certain if the sun affects each system directly or if solar radiation influences the North Atlantic circulation, which in turn affects the monsoon. In an earlier study, evidence from sediments in the same region showed an increase in monsoon strength in the past 400 years.

• YEAR 2000

A strong La Niña at the beginning of 2000 weakened during July and August, but was still evident at year's end. As a result, cooler than normal temperatures throughout the eastern equatorial Pacific held down temperatures in the tropics. However, temperatures in the non-tropical Northern Hemisphere continued to average near record levels. Temperatures north of 20°N were the second warmest on record during the December 1999–November 2000 period. In addition, annual anomalies in excess of 2°F were widespread across Canada, Scandinavia, much of Eastern Europe, and the Balkans.

• FREQUENTLY ASKED QUESTIONS ABOUT EL NIÑO AND LA NIÑA

What is the difference between La Niña and El Niño?

El Niño and La Niña are extreme phases of a naturally occurring climate cycle referred to as El Niño/Southern Oscillation. Both terms refer to large-scale changes in sea-surface temperature across the eastern tropical Pacific. Usually, sea-surface readings off South America's west coast range from the 60s to 70s°F, while they exceed 80°F in the "warm pool" located in the central and western Pacific. This warm pool expands to cover the tropics during El Niño, but during La Niña, the easterly trade winds strengthen and cold upwelling along the equator and the west coast of South America intensifies. Sea-surface temperatures along the equator can fall as low as 7°F below normal during La Niña. Both La Niña and El Niño impact global weather patterns.

How often does La Niña occur and how long does it last?

El Niño and La Niña occur on average every three to five years. However, in the historical record the interval between events has varied from two to seven years. According to the National Centers for Environmental Prediction, the twentieth century's previous La Niñas began in 1903, 1906, 1909, 1916, 1924, 1928, 1938, 1950, 1954, 1964, 1970, 1973, 1975, 1988, and 1995. These events typically continued into the following spring. Since 1975, La Niñas have been only half as frequent as El Niño.

La Niña conditions typically last approximately nine to 12 months, though some episodes may persist for as long as two years.

Does a La Niña typically follow an El Niño?

No, a La Niña episode may, but does not always, follow an El Niño.

Why do El Niño and La Niña only occur in the Pacific?

This question does not have a simple or straightforward answer, since this is not a settled issue. Fundamentally, no one is exactly sure why the Pacific should have an El Niño/La Niña cycle and the Atlantic not.

A principal difference between the Atlantic and Pacific is the width of the equatorial region. The Pacific is more than twice as wide at the equator. This is important to its capacity to sustain El Niño/La Niña because of the peculiar dynamics of equatorial waves. Equatorial waves are not the familiar surf or swell seen on the surface, but very large-scale motions that carry changes in currents and temperature over thousands of miles. The period of these waves is measured in months, and they take typically three months to more than a year to cross the Pacific. Surprisingly, these waves do not spread out equally in all directions like waves made by dropping a rock in a lake, but preferentially propagate eastward or westward. When winds blow over a large area of the ocean consistently for a month or more, equatorial waves are usually generated, and these then modify conditions over a very large region, including places far removed from where they were generated. For example, winds over the far western Pacific make waves that carry the signal to the coast of South America, even though the winds in the South American region may not change at all. The subsurface changes due to the arriving waves can then cause sea surface temperature changes, entirely due to winds occurring many thousands of miles to the west.

With the huge distances across the Pacific, one side of the ocean can be reacting to conditions due to one set of waves, while the other can be doing something completely different. As the waves propagate back and forth, a cycle can be set up that oscillates (El Niño/La Niña). The much smaller Atlantic, on the other hand, is not large enough to sustain much of an oscillation, since the waves cross it so quickly, often in only a month or so. This does not allow a cross-ocean contrast to be created, nor an oscillation to be set up. Some indications suggest that some kind of weak oscillation may in fact occur in the Atlantic, but it never reaches the amplitude of that in the Pacific.

A second reason that the Pacific is more important in this regard is that the fundamental driver of the whole ocean-atmosphere circulation is heat. The large width across the Pacific allows the existence of a huge pool of warm water in the west. The smaller distances across the Atlantic mean that the Atlantic warm pool is much smaller. The Pacific warm pool is a gigantic source of heat that is one of the main controls of the atmosphere. When the warm pool shifts east (during El Niño) or shrinks west (during La Niña), the effects reverberate around the world, causing the weather disruptions associated with this cycle. In the Atlantic, there is simply not enough of a warm pool to make that much difference to worldwide weather. So even if there is an analogue to El Niño in the Atlantic, it does not have the power to cause weather disturbances that affect more than local conditions.

Why do El Niño and La Niña occur?

El Niño and La Niña result from interaction between the surface of the ocean and the atmosphere in the tropical Pacific. Changes in the ocean impact the atmosphere and climate patterns around the globe. In turn, changes in the atmosphere impact the ocean temperatures and currents. The system oscillates between warm (El Niño) to neutral (or cold La Niña) conditions on an average of every three to five years.

Do volcanoes or sea-floor venting cause El Niño?

The idea that volcanoes cause El Niño events originally gained prominence because of the eruption of El Chichón in Mexico in February 1982 (preceding the El Niño of 1982–83), and the eruption of Mt. Pinatubo in the Philippines in June 1991 (preceding the El Niño of 1991–92). However, when the time series of El Niño is compared to the time series of volcanic eruptions, it becomes clear that the relationship is coincidental. There are numerous large volcanic eruptions around the world and almost as many El Niños. In that situation there is almost always an eruption at some time preceding any El Niño. Scientists are now convinced that this relationship is coincidental.

Certain experiments bear this out. For example, several computer models predicted the onset of the 1991–92 event as early as January 1991, based on the state of the ocean-atmosphere system at that time well before Pinatubo erupted. This indicates that the ocean-atmosphere system was already generating the El Niño, and Pinatubo erupted coincidentally. Computer models integrating the equations of fluid motion and the flow of heat routinely produce El Niño-like variability completely on their own. Of course, computer models are not reality, but these experiments suggest that El Niño is a natural mode of variability of the ocean-atmosphere system, as much as, for example, a thunderstorm. While we do not have a complete picture of how the El Niño cycle operates, these models (and a developing theoretical understanding) suggest that the fluid envelope of the Earth is prone to developing various kinds of instabilities, ranging from storm systems lasting a few hours or days, to El Niño, to longer-term fluctuations that we are just beginning to explore. There is no reason to think that external processes such as volcanoes are a necessary element.

None of this is to say that volcanoes do not affect the climate. They most certainly do, and since El Niños occur against the background existing climate, there is little doubt volcanic eruptions that eject large amounts of dust into the stratosphere must modify the frequency, character, and strength of El Niño events, possibly in important ways. The distinction is between "slowly modifying the background" and "causing" El Niño.

As far as deep-ocean vents modifying the ocean temperatures, researchers now think that this source of heat does contribute to the long-term evolution of the ocean state. The chemical signatures of undersea vents are of great interest as tracers of the slow deep circulation of the ocean, and therefore these signatures are studied carefully. (The deep circulation is so slow that its currents cannot be measured directly, so we look at tongues of chemical tracers to estimate the speed, direction, and transport of the flows.) Numerous scientific papers discuss these questions, studying a variety of chemical constituents. What is consistently found is that the traces spread extremely slowly through the water column and are vastly diluted. There is little doubt that over very long periods the effects of undersea venting on the ocean are large, both for their heat and for their contribution to the chemical makeup of the ocean. However, these effects occur on timescales of thousands of years, and certainly do not produce the kind of rapid signals that characterize El Niño. To trigger an El Niño event, one would look for a signal that produced surface variability on a month to month or year to year timescale, and undersea venting has never been observed to do that.

It is indeed tempting to look for nice clean causes for complex oscillations like the El Niño cycle. Unfortunately, it seems that the ocean-atmosphere system is capable of generating these oscillations on its own, and the task now is to understand how this happens. Volcanoes and sea-floor venting are part of the slowly changing background state against which phenomena like El Niño occur, and add to the complexity of the task.

Why isn't there much publicity about the causes of El Niño and La Niña?

The reason that there is not much publicity about the causes of El Niño and La Niña is that we do not understand the origins of the events. We do, however, have a pretty good understanding of how they evolve once they begin, and that allows us to make forecasts six to nine months ahead for some regions. This is the information that is publicized because it is reasonably secure knowledge. Of course, there are a variety of theories, and many scientists are working on various aspects of the genesis, which would presumably extend the predictive skill out another few months or even years.

The fact is, at several points over the past two decades scientists thought working theories of what causes El Niño had been firmly established. Unfortunately, nature has shown that those theories were at best incomplete. For example, during the mid-1980s, a group at Columbia University developed a fairly simple theory and wrote a computer model to produce predictions based on it. This was successful in predicting the 1986–87 and 1991–92 events almost a year in advance. Then along came the event of 1993, then another in 1994–95, the most prominent El Niño of the late 1990s, neither of which developed according to the ideas in their theory.

The main reason this is so difficult is that the processes that cause El Niño and La Niña involve the full complexity of ocean-atmosphere interaction on a global scale. Now that the sea surface temperature (SST) driving the atmospheric circulation is known, a reasonably accurate understanding of how the atmosphere works (at least in theory) has been developed. With a basic understanding, atmospheric models can make short-term weather forecasts, because the ocean changes rather slowly. However, when one considers longer-term phenomena like El Niño and La Niña, it is not enough to specify the SST; one must consider how the ocean will evolve under the winds, and then how the altered ocean will modify the winds, and so on, in many tricky and sensitive feedback loops. We are just beginning to be able to see how these fundamentally coupled disturbances work, and generally only in very idealized cases. Remember that for a long time meteorologists only talked to meteorologists, and oceanographers only to oceanographers. Now we are really at the initial stages of being able to think about these coupled problems.

Does El Niño have a purpose?

El Niño is part of the natural rhythm of the ocean-atmosphere system, as much as winter cold or summer thunderstorms or any other weather phenomenon.

In a complicated system like this, each feature fills a role in the grand scheme of things. The exact role cannot be pinpointed, but scientists do observe that these events drain the west Pacific of heat that is built up over several years by the trade winds. In any case, El Niño does not exist in isolation, and any changes in it would reverberate around the whole system in unpredictable ways. Further, as part of the natural environment of the Pacific basin that the animals, fish, birds, and plants have adapted to over the millennia (it is known that El Niños have occurred throughout history), it is not clear that stopping El Niño would even be desirable. Even if it was possible to make El Niño disappear, it is unclear what the outcome would be.

What is the relationship between El Niño/La Niña and global warming?

The jury is still out on this. Are we likely to see more El Niños because of global warming? Will they be more intense? These are the main research questions facing the scientific community today. Research will help us separate the natural climate variability from any trends due to human activities. We cannot figure out the "fingerprint" of global warming if we cannot sort out what the natural variability does. We also need to look at the link between decadal changes in natural variability and global warming. At this time we cannot preclude the possibility of links, but it is too early to say there is definitely a link.

Is El Niño or La Niña responsible for a specific hurricane/tropical storm/drought/fire/flood/winter storm?

It is inaccurate to label individual storms or events as La Niña or El Niño events. Rather, these climate extremes affect the position and intensity of the jet streams, which in turn affect the intensity and track of storms. During La Niña, the normal climate patterns are enhanced. For example, in areas that would normally experience a wet winter, conditions would likely be wetter than normal.

It is impossible to prove that El Niño or La Niña cause a particular event, just as it is impossible to say that winter caused a particular snowstorm—it is the likely suspect.

We cannot run experiments to see what a parallel Earth without El Niño or La Niña would do. But a group at NOAA's Climate Diagnostic Center in Boulder, Colorado, is trying something similar using numerical forecast models. First, they run the model with the actual conditions and produce weather forecasts, just like the regular ones. Then they make another run, in which everything is the same as the first one except that they change the Pacific SST to be like a "normal" year. The difference between these forecasts gives an indication of the effect of El Niño and La Niña conditions on the specific weather events being forecast.

What impacts do El Niño and La Niña have on tornado activity across the country?

Since a strong jet stream is an important ingredient for severe weather, the position of the jet stream determines the regions more likely to experience tornadoes. Contrasting El Niño and La Niña winters, the jet stream over the United States is considerably different. During El Niño the jet stream is oriented from west to east over the northern Gulf of Mexico and northern Florida. Thus this region is most susceptible to severe weather. During La Niña the jet stream extends from the central Rockies east-northeastward to the eastern Great Lakes. Thus severe weather is likely to be further north and west during La Niña than El Niño.

What are the impacts of La Niña?

Both El Niño and La Niña impact global and U.S. climate patterns. In many locations, especially in the tropics, La Niña (or cold episodes) produces the opposite climate variations from El Niño. For instance, parts of Australia and Indonesia are prone to drought during El Niño, but are typically wetter than normal during La Niña.

In the United States, La Niña often features drier than normal conditions in the Southwest in late summer through the subsequent winter. Drier than normal conditions also typically occur in the Central Plains in the fall and in the Southeast in the winter. In contrast, the Pacific Northwest is more likely to be wetter than normal in the late fall and early winter with the presence of a well-established La Niña. Additionally, on average, La Niña winters are warmer than normal in the Southeast and colder than normal in the Northwest.

How is La Niña influencing the Atlantic and Pacific hurricane seasons?

Dr. William Gray at the Colorado State University has pioneered research efforts leading to the discovery of La Niña impacts on Atlantic hurricane activity, and to the first—and presently only—operational long-range forecasts of Atlantic basin hurricane activity. According to this research, the chances for the continental United States and the Caribbean Islands to experience hurricane activity increase substantially during La Niña.

Does El Niño create dangerous conditions for marine life, and will it have a lasting effect on marine animals?

First, there is no question that El Niño has serious effects on life in many regions.

Second, it should be remembered that El Niño is part of the normal rhythm of Earth, and of the environment that marine life has evolved to face. A plant or creature will not last long in a place in which it can only handle ideal conditions. With natural variability, some winters are colder than others, some years drier, etc. El Niño is part of this normal climate, along with other influences that are less known. Living things may have varied success in these natural fluctuations. So El Niño may cause a temporary die-back of some forms of marine life in some regions, or reduce the survival rate of young, but it probably does not have a lasting effect. For example, El Niño devastates the population of seabirds off Peru by reducing the fish stock on which they live. But those birds will bounce back soon after El Niño is gone. If, however, El Niño became more frequent, then one might find the overall composition of marine life changing in that region. But the individual events themselves probably do not cause a permanent change.

Since normal variations can have large swings, such as those that occur during El Niño, the boundaries of where particular forms of life can survive are somewhat smaller than they would be if the climate were more constant. For example, palm trees can survive in Seattle, Washington, during most years, since it has generally mild winters. But every ten years or so a killing frost occurs, so natural palms do not occur there. But many people grow them in gardens and find that they thrive with only occasional protection.

Is there a scale for the intensity of El Niño?

The most widely used scale is known as the Southern Oscillation Index (SOI), which is based on the surface (atmospheric) pressure difference between Darwin, Australia and Tahiti, French Polynesia. It was noted as far back as the 1920s that these two stations were anti-correlated, so that when Tahiti pressure is high, Darwin pressure is low. This reflects the very large scale of the phenomenon, since one would not usually expect such a close relation between such

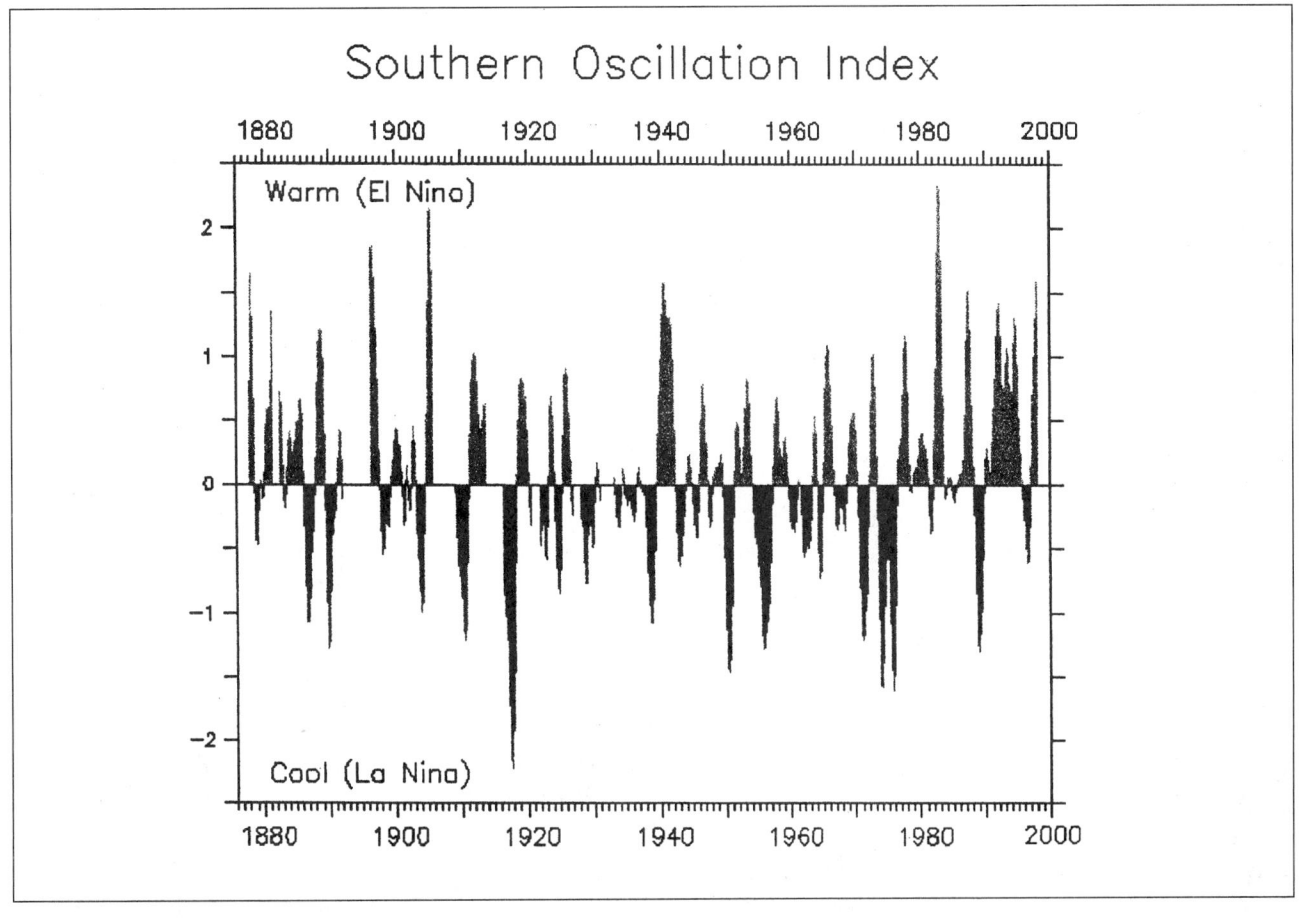

(Courtesy of National Oceanic and Atmospheric Administration (NOAA).)

Cold and Warm Episodes by Season

The following list of cold (La Niña) and warm (El Niño) episodes has been compiled to provide a season-by-season breakdown of conditions in the tropical Pacific. We have attempted to classify the intensity of each event by focusing on a key region of the tropical Pacific (along the equator from 150°W to the date line). The process of classification was primarily subjective using reanalyzed sea surface temperature analyses produced at the National Centers for Environmental Prediction/Climate Prediction Center and at the United Kingdom Meteorological Office. An objective procedure for classifying intensity is being explored at NCEP/CPC. In the following table, weak periods are designated as C- or W-, moderate strength periods as C or W strong periods as W+ or C+, and neutral periods as N.

	JFM	AMJ	JAS	OND		JFM	AMJ	JAS	OND
1950	C	C	C	C	1978	W-	N	N	N
1951	C	N	N	W-	1979	N	N	N	N
1952	N	N	N	N	1980	W-	N	N	N
1953	N	W-	W-	N	1981	N	N	N	N
1954	N	N	C-	C-	1982	N	W-	W	W+
1955	C	C-	C-	C+	1983	W+	W	N	C-
1956	C	C	C	C-	1984	C-	C-	N	C-
1957	N	W-	W	W	1985	C-	C-	N	N
1958	W+	W	W-	W-	1986	N	N	W-	W
1959	W-	N	N	N	1987	W	W	W+	W
1960	N	N	N	N	1988	W-	N	C-	C+
1961	N	N	N	N	1989	C+	C-	N	N
1962	N	N	N	N	1990	N	N	W-	W-
1963	N	N	W-	W	1991	W-	W-	W	W
1964	N	N	C-	C	1992	W+	W+	W-	W-
1965	C-	N	W	W+	1993	W-	W	W	W-
1966	W	W-	W-	N	1994	N	N	W	W
1967	N	N	N	N	1995	W	N	N	C-
1968	N	N	N	W-	1996	C-	N	N	N
1969	W	W-	W-	W-	1997	N	W	W+	W+
1970	W-	N	N	C	1998	W+	W	C-	C
1971	C	C-	C-	C-	1999	C+	C	C-	C
1972	N	W-	W	W+	2000	C	C-	N	C-
1973	W	N	C-	C+	2001	C-	N	N	N
1974	C+	C	C-	C-	2002	N	W-		
1975	C-	C-	C	C+	2003				
1976	C	N	N	W-	2004				
1977	N	N	N	W-	2005				

(Courtesy of National Centers for Environmental Prediction/Climate Prediction Center.)

faraway places. When Tahiti pressure is high, that indicates winds blowing towards the west (normal trade winds), and when it is low, winds blow to the east (El Niño).

Scientists use the Southern Oscillation Index for three main reasons, even though it is an indirect measure of El Niño and these locations are not ideally sited for this purpose. First, the time series at Darwin and Tahiti are more than 100 years long, and there is no other record that would allow us to categorize the El Niño cycle that far back. Second, the measurement of atmospheric pressure is simple (it is just the height of a column of mercury in a barometer) and not subject to calibration problems, as, for example, thermometers are. A column of mercury is just a measure of length, which is accurate and easily convertible between inches or centimeters, whereas thermometers are inherently less accurate since they rely on a carefully made glass tube that may be subject to expansion or irregularities; also the placement of the thermometer (in the sun or shade or breeze, near buildings, etc.) can have a large effect. For example, temperature measured in cities shows a long-term rise associated with the heat generated by urban activities and the increased absorption of solar heat by pavement compared with forest. This is one thing that makes it difficult to detect the signature of greenhouse warming. Therefore, pressure mea-

surements are highly desirable for interpreting long records. Third, pressure tends to be similar over wide regions, whereas more directly important quantities (SST and winds) can have many local effects that make it hard to interpret single point measurements as representative of the large-scale situation.

The SOI is given in normalized units of standard deviation. It can be used as an intensity scale. For example, SOI values for the 1982–83 El Niño were about 3.5 standard deviations, so by this measure that event was roughly twice as strong as the 1991–92 El Niño which measured only about 1.75 in SOI units. By this standard, the El Niño of the late 1990s is about as strong as 1991–92. However, the sea surface temperature anomaly, November 1997, was about as large as in 1982–83, and some might say that is a more important measure. This shows that there is no single number that summarizes the intensity of events.

Why is predicting these types of events so important?

Better predictions of the potential for extreme climate episodes like floods and droughts could save the United States billions of dollars in damage costs. Predicting the onset of a warm or cold phase is critical in helping farmers and water, energy, and transportation managers plan for, avoid, or mitigate potential losses. Advances in improved climate predictions will also result in significantly enhanced economic opportunities, particularly for the national agriculture, fishing, forestry, and energy sectors, as well as social benefits.

How do scientists detect La Niña and El Niño and predict their evolution?

Scientists from NOAA and other agencies use a variety of tools and techniques to monitor and forecast changes in the Pacific Ocean and the impact of those changes on global weather patterns. In the tropical Pacific Ocean, El Niño is detected by many methods, including satellites, moored buoys, drifting buoys, sea level analysis, and expendable buoys. Many of these ocean observing systems were part of the Tropical Ocean Global Atmosphere (TOGA) program, and are now evolving into an operational El Niño/Southern Oscillation (ENSO) observing system. NOAA also operates a research ship, the *KA'IMIMOANA*, which is dedicated to servicing the Tropical Atmosphere Ocean (TAO) buoy network component of the observing system. Large computer models of the global ocean and atmosphere, such as those at the National Centers for Environmental Prediction, use data from the ENSO observing system as input to predict El

Niño. Other models are used for El Niño research, such as those at NOAA's Geophysical Fluid Dynamics Laboratory, at the Center for Ocean-Land-Atmosphere Studies, and other research institutions.

How are sea surface temperatures monitored?

Sea surface temperatures in the tropical Pacific Ocean are monitored with data buoys and satellites. NOAA operates a network of 70 data buoys along the equatorial Pacific that provide important data about conditions at the ocean's surface. The data is complemented and calibrated with satellite data collected by NOAA's Polar Orbiting Environmental Satellites, NASA's TOPEX/POSEIDEN satellite, and others.

How are the data buoys used to monitor ocean temperatures?

Observations of conditions in the tropical Pacific are essential for the prediction of short term (a few months to one year) climate variations. To provide necessary data, NOAA operates a network of buoys that measure temperature, currents, and winds in the equatorial band. These buoys transmit data that are available to researchers and forecasters around the world in real time.

Winter Storms

• WHERE & WHY WINTER STORMS OCCUR

Winter storms are generated, as are many of the thunderstorms of summer, from disturbances along the boundary between cold polar and warm tropical air masses—the fronts where air masses of different temperatures and densities wage their perpetual war of instability and equilibrium. The disturbances may become intense low-pressure systems, churning over tens of thousands of square miles in a great counter-clockwise sweep. In order for disturbances to become winter storms there are three key factors: cold air (below-freezing temperatures facilitate the production of snow and ice); moisture (which forms clouds and precipitation); and lift (which raises the moist air to form clouds, precipitation, and fronts).

In the Pacific, these disturbances form along polar fronts off the east coast of Asia and travel northeastward toward Alaska. But some, particularly those forming along the mid-Pacific polar front, take a more southerly track, striking the United States as far south as southern California. Few Pacific disturbances cross the Rockies, but some do, redeveloping to the east. One region of such redevelopment lies east of the Colorado Rockies; the storms that come out of that region are called *Colorado cyclones*. Another region of storm redevelopment is east of the Canadian Rockies, from which come the so-called *Alberta cyclones*. Both types take an eastward path, and most frequently converge over the Great Lakes. The lakes themselves generate severe local winter storms, and forge other storms from northward-drifting disturbances that originate over the Gulf of Mexico and the southern plains.

On the East Coast, winter storms often form along the Atlantic polar front near the coast of Virginia and the Carolinas and in the general area east of the southern Appalachians. These are the notorious Cape Hatteras storms—nor'easters—that develop to great intensity as they move up the coast, then drift seaward toward Iceland, where they finally dissipate.

Because they form over water, these storms are difficult to forecast, and occasionally surprise the Atlantic megalopolis with paralyzing snows. In 1969, the U.S. Departments of Commerce, Transportation, and Defense tightened winter storm surveillance with reconnaissance aircraft, an ocean buoy, and a new weather ship. With better hour-to-hour information on the storms, weather forecasters ashore have begun to ease the burden of unexpected heavy snows in eastern cities.

For some parts of the United States—the Northern Rockies, for example—storms with snow followed by cold are a threat from mid-September to mid-May; during one of the colder months from November to March, it is not unusual for several separate storms to affect some area across the continent. Intense winter storms are frequently accompanied by cold waves, ice or glaze, heavy snow, blizzards, or a combination of these; often, in a single winter storm, precipitation type changes several times as the storm passes. The common feature of these storms is the ability to completely immobilize large areas and to isolate and kill persons and livestock in their path. In the north, the severity of these storms makes their threat a seasonal one. Farther south, the occasional penetration of severe winter storms into more moderate climates causes severe hardship and great loss of warm-weather crops.

• FREEZING RAIN & ICE STORMS

Freezing rain or freezing drizzle is rain or drizzle that occurs when surface temperatures are below freezing (32°F). The moisture falls in liquid form but freezes upon impact, resulting in a coating of ice glaze on all exposed objects. The occurrence of freezing rain or drizzle is often called an ice storm when a substantial glaze layer accumulates. Ice forming on exposed objects generally ranges from a thin glaze to coatings about 1 in thick, but much thicker deposits have been observed. For example, ice deposits to 8 in in diameter were reported on wires in northern Idaho in January 1961, and loadings of 11 lb per foot of telephone wire were found in Michigan in February 1922. It has been estimated that an evergreen tree 50 ft high with an average width of 20 ft may be coated with as much as five tons of ice during a severe ice storm. A heavy accumulation of ice, especially when accompanied by high winds, devastates trees and transmission lines. Sidewalks, streets, and highways become extremely hazardous to pedestrians and motorists; over 85% of ice-storm deaths are traffic related. Freezing rain and drizzle frequently occur for a short time as a transitory condition between the occurrence of rain or drizzle and snow, and therefore usually occur at temperatures slightly below freezing.

Some of the most destructive ice storms have occurred in the southern states, where warm-weather buildings and crops are not adapted to withstand severe winter conditions. The most damaging ice storm in the United States was probably that which struck the South from January 28–February 4, 1951, causing some $50 million damage in Mississippi, $15 million in Louisiana, and nearly $2 million in Arkansas; this storm also caused 22 deaths. The region of greatest incidence, however, is a broad belt from Nebraska, Kansas, and Oklahoma eastward through the middle Atlantic and New England states.

Sleet

Ice storms are sometimes incorrectly referred to as sleet storms. Sleet can be easily identified as frozen raindrops (ice pellets) that bounce when hitting the ground or other objects. Sleet does not stick to trees and wires; but sleet in sufficient depth does cause hazardous driving conditions.

• WINTER STORM IMPACT

Early winter storm records

Nearly everyone east of the Pacific coastal ranges remembers significant winter storms—days of heavy snow, interminable blizzards, inconvenience, economic loss, and sometimes, personal tragedy. For Wyoming, Kansas, and Texas, the blizzard of 1888 was one of the worst on record. January 11–13 of that year also brought the most disastrous blizzard ever known in Montana, the Dakotas, and Minnesota, combining gale winds, blowing snow, and extreme cold into a lethal, destructive push from the Rockies eastward. That same year, the eastern seaboard from Chesapeake Bay to Maine got its big storm of the century. On March 11–14, 1888, a blizzard dumped an average of 40 in of snow over southeastern New York and southern New England. The storm killed 200 in New York City alone; total deaths were over 400.

Effect of 1950s–60s storms

Large numbers of snow-related deaths—345 and 354—occurred in 1958 and 1960 respectively. About half of these

Heavy ice and snow can disrupt transportation, communication, and utilities. *(Photo courtesy of National Oceanic and Atmospheric Administration (NOAA) Central Library.)*

deaths occurred in New England, New York, and Pennsylvania. The 1966 season saw the eastern seaboard paralyzed by snow from Virginia through New England, with more than 50 deaths, and thousands marooned. A March storm buried the Dakotas, Minnesota, and Nebraska, with 30-ft drifts pushed up by winds gusting to more than 100 mph. The 1967 winter storm season was not much better, and included a May Day blizzard in the Dakotas and a nor'easter that brought snow and hurricane-force winds to northern New England late in May. Snowfall across middle America was as much as four times the normal amount in early 1968, and 1969 was called "the year of the big snows" in the Midwest.

The Blizzard of 1978

The winter of 1977–78 was unusually harsh, particularly in the Midwest and East. Its most devastating punch was the "Northeast Blizzard of 78," one of the worst of the twentieth century. On February 5–7, 1978, the blizzard created havoc along the eastern seaboard.

In New York City, the 17.7-in snowfall was the sixth largest since records began in 1869. Boston, Massachusetts, had over 2 ft of snow, as did Providence, Rhode Island. Winds of more than 55 mph caused massive snowdrifts, drove seas through seawalls, undermined homes, destroyed beaches (including Rocky Beaches on Long Island), breached protective dunes, and left many areas from Cape May, New Jersey, northward open to further damage from spring coastal storms. The American Red Cross reported 99 deaths and 4,587 injuries or illnesses attributable to the storm. Damage in Massachusetts exceeded $1.5 billion, while New York and New Jersey losses aggregated to about $94 million.

Superstorm, 1993

In March 1993, one of the nation's greatest non-tropical weather events occurred, adversely impacting 100 million individuals and severely crippling the commercial activity of the eastern one-third of the United States, as well as travel activities nationwide.

On March 12–14, 1993, 22 states in the eastern United States were subjected to the blizzard conditions (in the north) and high winds, coastal flooding, and convective weather (in the south) as the storm swept through the region. Record cold temperatures were noted in all affected areas.

Still recovering from Hurricane Andrew and a rash of tornadoes in Tampa Bay, Florida reported additional property damage costs of $1.6 billion; about one-third of the storm-related deaths occurred there. Thirty storm-related deaths and some $400 million in property damage were reported from the remainder of the southeastern United States. The mid-Atlantic and northeast, while sustaining significantly fewer storm-related deaths, suffered major economic misfortune due to the extensive slowdown of business activity.

Snowiest U.S. Cities

Listed below are the snowiest U.S. cities. Average annual snowfall is shown parenthetically.

1. Blue Canyon, CA (240.8 in)
2. Marquette, MI (128.6 in)
3. Sault Ste. Marie, MI (116.7 in)
4. Syracuse, NY (111.6 in)
5. Caribou, ME (110.4 in)

The Blizzard of 1996

In early December 1995, lake-effect snowstorms caused record 24-hour snowfalls in Buffalo, New York, of 38 in, and in Sault Ste. Marie, Michigan, which was hit with 28 in December 9–10.

In mid-December, an unusually intense storm struck the Pacific Northwest. Heavy rains of 5–20 in accompanied the system, which featured very low pressures and high winds. Record low sea-level pressure readings were recorded at Astoria, Oregon (28.5 in), Seattle, Washington (28.6 in), and Medford, Oregon (28.9 in). Winds gusted to 119 mph at Sea Lion Caves, Oregon, and 103 mph at Angel Island, California. Six deaths and over two million power outages were attributed to the storm.

In the latter part of the month, Miami, Florida, failed to exceed 65°F for eight consecutive days—December 21–28—an all-time record for the city.

January proved to be quite cold over much of the nation (below normal in 37 states), with unusually heavy snowfall over most of the East. On January 6–8, much of the eastern seaboard received 1–3 ft of snow during the "Blizzard of '96." A large area from the southern Appalachians to southern New Hampshire and Maine received a foot or more, with 20 in or more very common over the major metropolitan areas of the East. However, upstate New York received very little snow as the storm featured a very pronounced western edge where snowfall ended.

The sudden warm-up that followed proved to be almost as deadly and damaging as the blizzard itself. Moderate-to-heavy rains (3.03 in over 24 hours at Williamsport, Pennsylvania) and rapid snowmelt triggered serious flooding along the Delaware, Susquehanna, upper Ohio, Potomac, and James River basins, with crests as high as 20 ft above flood stage.

Nearly 200 deaths were attributed to the blizzard and ensuing flooding; damages and costs totaled approximately $3 billion.

New England/New York, 1998

A catastrophic ice storm and flood event struck northern New England and northern New York during the first two weeks of January 1998. Heavy rain associated with a warm, moist air mass overspread a shallow but dense layer of cold air producing ice accumulations in excess of 3 in. The heavy rainfall, exceeding 4 in in some areas, combined with significant runoff from the melting snowpack to produce record flooding. The ice coated all outdoor surfaces, destroying the electric power infrastructure, toppling trees, collapsing outdoor structures, and threatening the lives of a large, distributed population. The flooding exacerbated the icing problems by forcing the evacuation of more than 1,000 homes and the closure of numerous roads. A record crest was observed on the Black River at Watertown, New York, where the river crested 2 ft above the previous flood of record.

Conservative damage estimates approach $0.5 billion. More than three million people in four states and two Canadian provinces were without electricity. In Maine, 80% of the state's population lost electrical service, some for more than two weeks. Residents were forced to find alternative means of heating their homes, pumping water, traveling, and communicating. Tens of thousands of trees were downed or severely damaged. Agricultural losses exceeded one million dollars as farmers were unable to milk their cows without electricity. National Guard units were activated and many counties in Maine, New York, Vermont, and New Hampshire were declared federal disaster areas. Despite its severity, duration, and scope, only seven fatalities were directly attributed to the event.

The New Year's 1999 Blizzard in the Midwest

The second-worst blizzard of the twentieth century (ranking behind the blizzard in January 1967) struck portions of the Midwest on January 1–3, 1999. The storm, which developed over the Texas panhandle, produced 9–22 in of snow in Chicago, Illinois, and strong northeast winds gusting to over 30 mph—and over 60 mph along the Lake Michigan shoreline. Soon after the snow ended, record low temperatures occurred with values of −20°F or lower in parts of Illinois and surrounding states on January 3 and 4. The areas with the heaviest snows—15 inches or more—included central and northern Illinois, southern Wisconsin,

Coldest U.S. Cities

Listed below are the coldest cities in the United States. Average mean temperatures are listed parenthetically.
1. International Falls, MN (36.4°F)
2. Duluth, MN (38.2°F)
3. Caribou, ME (38.9°F)
4. Marquette, MI (39.2°F)
5. Sault Ste. Marie, MI (39.7°F)

central and northern Indiana, southern Michigan, and northern Ohio.

Estimates of losses and recovery costs were $3–4 billion with 73 dead as a result of the storm. The governor of Illinois declared the entire state a disaster area on January 4, and on January 20, President Clinton declared 45 Illinois counties disaster areas (half the state) and subject to receiving federal relief. Areas of Indiana were also declared disaster areas.

Many of the worst impacts were associated with the storm's effects on transportation. Every form of Midwestern transportation was either halted or delayed by two to four days, and transportation problems were the source of many accidents and deaths. Auto- and train-related deaths totaled 39 with five more dead due to snowmobile accidents.

Railroad trains in the storm's heart were stalled or delayed by 12–24 hours, and since Chicago is the nation's rail hub, many priority shipments for the East and West Coasts were delayed, at great expense, by one to four days. The suburban train service of Chicago, one of the nation's finest, was overwhelmed by travelers who were unable to use normal vehicular transport, and three separate train accidents killed three. These and many other operational problems greatly slowed train service for three days.

Major auto accidents involving numerous vehicles occurred on major highways and interstates, causing more than 2,500 auto and truck accidents. Snow removal on city streets buried thousands of cars and driveway entrances, trapping many motorists without transportation, and it took up to a week after the storm before most vehicles had been extracted from these man-made snowbanks. Many traveling motorists became trapped on highways and thousands were housed in emergency shelters (churches and city buildings) for one to three nights.

The storm's impact on commercial aviation was staggering at Chicago and Detroit. Northwest Airlines reported more than 1,100 canceled flights during January 2–4, and United Airlines canceled 60% of its flights at Chicago's O'Hare Airport during the two-day storm. O'Hare had 300,000 travelers stranded for periods of hours up to four days. Costs were in the millions of dollars, and the stress on travelers was immense, particularly since the storm occurred on the weekend a day after New Year's Eve.

Navigation on the major Midwestern rivers was reduced by 50%. The cold temperatures during and after the storm created large ice floes on the Illinois and Mississippi Rivers, and these limited safe barge movement and the operation of locks and dams.

Human health was another area of major impact. There were 39 known deaths due to vehicle and train accidents. Five died in snowmobile accidents, two froze to death, and another 32 died of heart attacks resulting from overexertion, due mainly to snow shoveling. The storm also induced numerous illnesses and created a blood shortage in the nation since the Midwest is the prime source of fresh blood supplies and many donors could not reach hospitals to give blood.

Cities in the storm's main track experienced enormous problems and costs in achieving snow removal. Fortunately, the storm was accurately predicted several days in advance and most cities, such as Chicago, made major preparations that lessened the storm's long-term impacts. Most Chicago streets had been cleared by January 3, whereas other cities with less advance preparation like Detroit, Michigan, were still digging out a week after the storm. Chicago put 850 snow removal trucks on the streets (240 is the normal number for heavy snow). The cost of the snow removal and salt in the Chicago metropolitan area was $44 million ($14 million for snow removal at O'Hare Airport). Communities with populations of 100,000 typically reported storm costs of $250,000.

The huge effect on transportation produced major impacts on retail business and school openings. Most retailers were closed for one or two days, and lost business for several days after the storm, but those selling snow removal equipment typically sold out their supplies. The blocked streets and country roads led to multi-day school closings throughout the five-state area where the storm struck. Even by January 9, a week after the storm, only 47% of the students in Chicago schools were able to attend classes. School closings in the storm area ranged from three to seven days.

Warm weather returned to the Midwest by mid-January and the melting of the deep snow cover began. However, heavy rains with thunderstorms in the heavy snow areas occurred on January 17 and 20–22. This precipitation coupled with above-freezing temperatures brought rapid melting, dense fogs, and major flooding.

This weather pattern took on a spring-like flavor later in the month as milder temperatures returned to the lower half of the country. This trend in combination with fast-moving storms coming in from the Pacific spawned a record number of January tornadoes. Three significant tornado outbreaks occurred, resulting in 19 deaths. The first outbreak occurred on January 2 in southeast Texas and along the western Gulf coast, the second on January 17 in Tennessee, and the third on January 21–22, mostly in and near Arkansas and Tennessee. There were 169 tornadoes reported in January 1999—the highest number of tornadoes ever reported for January.

• RECORD COLD GRIPS MUCH OF THE NATION IN NOVEMBER AND DECEMBER 2000: TWO-MONTH PERIOD IS THE COLDEST ON RECORD IN THE UNITED STATES

NOAA scientists announced that the United States national temperature during the November–December, 2000, two-month period was the coldest such period on record.

The scientists worked with data from the world's largest statistical weather database at NOAA's National Climatic Data Center in Asheville, North Carolina.

Following the second coldest November on record in the United States, below normal temperatures continued to grip much of the nation in December. With an average temperature of 28.9°F, December 2000 was the seventh coldest December since national records began in 1895. Jay Lawrimore, chief of the Climate Monitoring Branch at the National Climatic Data Center, said, "Two months in a row of much below average temperatures resulted in the coldest November–December U.S. temperature on record, 33.8°F." This broke the old record of 34.2°F set in 1898. Near record cold temperatures for the same period occurred most recently in 1985 and 1983, when the nation's average temperature was 34.6°F and 34.8°F respectively, the third and fifth coldest such two-month periods on record.

Forty-three states within the contiguous United States recorded below average temperatures during the November–December period. The only states with near-normal temperatures were Nevada, New Mexico, New Hampshire, Vermont, and Maine. Severe winter conditions hit the central and southern Plains particularly hard. The coldest November–December on record occurred in Oklahoma, Arkansas, and Missouri, while six states experienced the second coldest such two-month period (Illinois, Iowa, Kansas, Texas, Louisiana, and Mississippi). For Dallas-Fort Worth, December 2000 was 7.5°F below normal at 39.4°F.

Heavy snow also accompanied the cold in many areas, particularly throughout the Plains and Upper Midwest. In Buffalo, New York, snowfall records were set during the three month period of October–December, where a total of 95.9 in broke the previous record of 92.2 in. At Midway Airport in Chicago, Illinois, snowfall records were set for a 24-hour period, where a total of 14.5 in broke the previous record set in December 1960.

December snowfall records were set in Marquette, Michigan, where a total of 89.5 in broke the previous record of 82.6 inches set in December 1981. Cities such as Milwaukee, Wisconsin; Waterloo, Iowa; and Amarillo, Texas, also set records for the most snowfall in the month of December. While precipitation amounts were normal to above normal throughout the central and eastern United States—except for the mid-Atlantic region—the West and Northwest regions (composed of Washington, Oregon, Indiana, Nevada, and California) recorded their fourth driest November–December since records began in 1895.

This prolonged cold outbreak came at the end of a year that began with the warmest winter on record in the United States. Above normal temperatures continued through the month of October and made the January–October 2000 period the warmest such ten-month period since national temperature records began in 1895. Preliminary data indicate that 2000 was the thirteenth warmest year on record in the United States, 1.2°F above the long-term average of 52.8°F.

Even though average long-term United States and global temperatures are warmer than they were a century ago, dramatic short-term swings in temperature are to be expected due to variability in circulation patterns. This variability can lead to periods of record cold temperatures while long-term trends remain positive. Although the United States has experienced periods of below average temperatures throughout the past century, temperatures have risen approximately 1°F since 1900.

During the same period global temperatures have increased at a rate near 1.1°F/century. Global temperatures in 2000 are expected to be similar to those recorded in 1999, the fifth warmest year since records began in 1880. The only years warmer were 1998, 1997, 1995, and 1990. The ten warmest years on record have all occurred since 1983.

• SOUTHEASTERN U.S. ICE STORM OF DECEMBER 2002

Roughly 1.5 million homes were without power several days after the December 4–5, 2002, storm that moved from the southwestern U.S. to the eastern seaboard in about 48 hours. An armada of cherry–pickers helped to restore power to nearly 1.5 million people in the ice–coasted Carolinas. Six to eight inches or more of snow coated the mid-Atlantic region from Washington, D.C., northeastward into New England.

• DANGERS OF WINTER STORMS

Winter storms can kill without breaking climatological records. Their danger is persistent, year-to-year. Since 1936, snowstorms have caused, directly and indirectly, about 200 deaths a year—and a year of 300 deaths is not unusual. Of such deaths, usually 70% are attributed to snowstorm-related automobile and other accidents; about 25% are caused by overexertion, exhaustion, and consequent fatal heart attack resulting from shoveling snow, pushing cars, and other snow-related physical labor. The remaining number, about 5%, are deaths due to home fires, carbon monoxide poisoning in stalled cars, electrocution from downed wires, and building collapse. Of deaths directly related to exposure to the cold, 50% are individuals over the age of 60; 75% are males; and about 20% of these deaths overall occur in homes.

• WINTER WARNINGS

The terms *watch* and *warning* are used for winter storms, as for other natural hazards. The *watch* alerts the public that a storm has formed and is approaching the area. People in the alerted area should keep listening for the latest advisories over radio and television, and begin to take precautionary measures. The *warning* means that a storm is imminent and immediate action should be taken to protect life and property.

The word *snow* in a forecast, without a qualifying word such as *occasional* or *intermittent*, means that the fall of snow is of a steady nature and will probably continue for several hours without letup.

Heavy snow warnings are issued to the public when a fall of 4 in or more is expected in a 12-hour period, or a fall of 6 in or more is expected in a 24-hour period. Some variations on these rules may be used in different parts of the country. Where 4-in snowfalls are common, for example, the emphasis on heavy snow is generally associated with six or more inches of snow. In other parts of the country where heavy snow is infrequent or in metropolitan areas with heavy traffic, a snowfall of 2–3 in will justify a heavy snow warning.

Snow flurries are defined as snow falling for short durations at intermittent periods; however, snowfall during the flurries may reduce visibility to an eighth of a mile or less. Accumulations from snow flurries are generally small. *Snow squalls* are brief, intense falls of snow and are comparable to summer rain showers. They are accompanied by gusty surface winds.

Blowing and drifting snow generally occur together and result from strong winds and falling snow or loose snow on the ground. Blowing snow is defined as snow lifted from the surface by the wind and blown about to a degree that horizontal visibility is greatly restricted.

Drifting snow is used in forecasts to indicate that strong winds will blow falling snow or loose snow on the ground into significant drifts. In the northern Plains, the combination of blowing and drifting snow, after a substantial snowfall has ended, is often referred to as a ground blizzard.

Blizzards are the most dramatic and perilous of all winter storms, characterized by low temperatures and by strong winds bearing large amounts of snow. Most of the snow accompanying a blizzard is in the form of fine, powdery particles of snow that are whipped in such great quantities that at times visibility is only a few yards. *Blizzard warnings* are issued when winds with speeds of at least 35 mph are accompanied by considerable falling or blowing snow and temperatures of 20°F or lower are expected to prevail for an extended period of time.

Severe blizzard warnings are issued when blizzards of extreme proportions are expected and indicate wind with speeds of at least 45 mph plus a great density of falling or blowing snow and a temperature of 10°F or lower.

Hazardous driving (travelers') warnings are issued to indicate that falling, blowing, or drifting snow, freezing rain or drizzle, sleet, or strong winds will make driving difficult.

Livestock (stockmen's) warnings alert ranchers and farmers that livestock will require protection from a large accumulation of snow or ice, a rapid drop in temperature, or strong wind.

A *cold-wave warning* indicates an expected rapid fall in temperature within a 24-hour period that will require substantially increased protection to agricultural, industrial, commercial, and social activities. The temperature falls and minimum temperatures required to justify cold-wave warnings vary with the changing of the season and with geographic location. Regardless of the month or the section of the country, a cold-wave warning is a red-flag alert to the public that during a forthcoming forecast period a change to very cold weather will require greater-than-normal protective measures.

The terms *storm, freezing rain,* and *freezing drizzle* warn the public that a coating of ice is expected on the ground and on other exposed surfaces. The qualifying term *heavy* is used to indicate ice coating that, because of the extra weight of the ice, will cause significant damage to trees, overhead wires, and the like. Damage will be greater if the freezing rain or drizzle is accompanied by high winds.

NWS tests new winter weather warning index

On November 25, 1998, the National Weather Service's Cheyenne, Wyoming, office began testing a new storm-warning procedure that could result in a public-rating scale designed to better describe the potential impact of winter storms. The new rating system, created by forecasters in Cheyenne should let forecasters provide a winter watch and warning service more useful to those affected by taking into account variables such as wind and/or temperature—not just amount of snow—to rate a storm's impact.

Cheyenne forecasters devised the new index after an October 1997 storm dumped heavy snow on parts of Wyoming and Nebraska for 15 hours in gale-force winds. As the storm grew in strength, forecasts progressed from a winter storm watch to a snow advisory, a snow and blowing snow advisory and, finally, to a blizzard warning. More flexible locally adapted warning criteria could have saved time and made the frequent changes unnecessary.

The new index rates winter storms in five categories: one, a minor inconvenience; two, inconvenience; three, significant inconvenience; four, potentially life threatening; and five, life threatening. Winter storms are difficult to rate objectively on a number scale, as they do not have the same impact in all situations or locations; the NWS hopes that this more subjective index will be more useful to the public in preparing for a storm's impact.

• WIND CHILL INDEX

A reasonably satisfactory solution to that elusive characteristic of weather known as "coldness" was first proposed

---

Blizzard Safety for Livestock

Blizzards take a terrible toll on livestock. For both humane and economic reasons, stockmen should take necessary precautions in advance of severe winter storms.

MOVE LIVESTOCK, ESPECIALLY YOUNG LIVESTOCK, INTO SHELTERED AREAS. Shelter belts, properly oriented and laid out, provide better protection for range cattle than shed-type shelters, which may cause cattle to overcrowd, leading to overheating and respiratory disorders.

HAUL EXTRA FEED TO FEEDING AREAS before the storm arrives. Storm duration is the largest determinant of livestock losses; if the storm lasts more than 48 hours, emergency feed methods are required. Range cattle are hardy and can survive extreme winter weather providing they have some non-confining type of shelter from the wind and are able to feed at frequent intervals.

Autopsies of cattle killed by winter storms have shown the cause of death to be dehydration, not cold or suffocation. Because cattle cannot lick enough snow to satisfy their thirst, stockmen are advised to use heaters in water tanks to provide livestock with water and feed after prolonged exposure to winter storm conditions.

---

by Dr. Paul Siple in 1939. The term "wind chill" was used to describe the relative discomfort resulting from combinations of wind and temperature. The method used was not applicable to temperatures above 32°F and high wind speeds caused exaggerated windchill values. During the Antarctic winter of 1941, Siple and Passel developed a new formula to determine wind chill from experiments made at Little America. Measurements were made of the time required for the freezing of 8.8 oz of water in a plastic cylinder under a variety of conditions of wind and temperature. They assumed that the rate of heat loss was proportional to the difference in temperature between the cylinder and the temperature of the surrounding air. The results, expressed in kilocalories per square meter, per hour, per degree Celsius, were plotted against wind speed in meters per second.

Heat loss occurs by means of radiation, conduction, and convection. Combining all effects, the general equation for heat loss H is: $H = (A + B\sqrt{v} + Cv)\,\Delta t$. Within the equation, constants A, B, and C are equal to 10.45, 10.00, and -1.0 respectively. H is equal to heat loss (wind chill) in kilograms (kg) cals/m^2/hr; v is wind speed in meters per second; and Δt is the difference in degrees Celsius between neutral skin temperature of 33°F (0.6°C) and air temperature.

The constant A includes the cooling caused by radiation and conduction. The value of the constants A, B, and C varies widely in formulae presented by different investigators. This is to be expected since H also depends on certain properties of the body being cooled. The above formula measures the cooling power of the wind and temperature in

Winter Storm Safety Rules

Keep ahead of the winter storm by listening to the latest weather warnings and bulletins on radio and television.

CHECK BATTERY-POWERED EQUIPMENT BEFORE THE STORM ARRIVES. A portable radio or television set may be your only contact with the outside world during a winter storm. Also check emergency cooking facilities and flashlights.

CHECK YOUR SUPPLY OF HEATING FUEL. Fuel carriers may not be able to move if a winter storm buries your area in snow. Closing off unneeded rooms, stuffing towels or rags in cracks under doors, and covering windows at night will help retain heat when fuel supplies are exhausted.

CHECK YOUR FOOD and stock an extra supply. Your supplies should include food that requires no cooking or refrigeration in case of power failure. You will need to replenish your body with fluids to prevent dehydration; food provides the body with the energy necessary to generate its own heat.

PREVENT FIRE HAZARDS due to overheated coal or oil-burning stoves, fireplaces, heaters, or furnaces.

STAY INDOORS DURING STORMS unless you are in peak physical condition. If you must go out, avoid overexertion.

IF CAUGHT OUTSIDE find shelter, try to stay dry, and cover all exposed parts of the body. If no shelter is available, construct a lean-to, windbreak, or snow cave for protection. If possible, build a fire; it will draw attention to your location as well as provide warmth (especially if stones are placed around it to absorb and reflect the heat). Finally, avoid eating snow, as it lowers body temperature; melt it first.

DO NOT KILL YOURSELF SHOVELING SNOW. It is extremely hard work for anyone in less than prime physical condition, and can bring on a heart attack, a major cause of death during and after winter storms.

RURAL RESIDENTS: MAKE NECESSARY TRIPS FOR SUPPLIES BEFORE THE STORM DEVELOPS. Arrange for emergency heat supply in case of power failure; be sure camp stoves and lanterns are filled.

DRESS TO FIT THE SEASON. If you spend much time outdoors, wear loose-fitting, lightweight, warm clothing in several layers; layers can be removed to prevent perspiring and subsequent chill. Outer garments should be tightly woven, water repellent, and hooded. The hood should protect much of your face and cover your mouth to ensure warm breathing and protect your lungs from the extremely cold air. A warm, snug hat is very important since much of the body's heat loss is through the head. Remember that entrapped, insulating air, warmed by body heat, is the best protection against cold. Layers of protective clothing are more effective and efficient than single layers of thick clothing, and mittens, snug at the wrists, are better protection than fingered gloves.

AUTOMOBILE PREPARATIONS. Your automobile can be your best friend—or worst enemy—during winter storms, depending on your preparations. Get your car winterized before the storm season begins. Check the following before winter storms strike your area: ignition system; heater; battery; brakes; lights; tire treads; wiper blades; cooling system/defroster; fuel system; snow tires; lubrication; chains; exhaust system tight; and winter-grade oil. Keep water out of your fuel by maintaining a FULL tank of gasoline.

BE EQUIPPED FOR THE WORST. Carry a winter storm car kit, especially if cross-country travel is anticipated or if you live in the northern states. Suggested Winter Storm Car Kit: blankets or sleeping bags, matches and candles, empty 3-lb coffee can with plastic cover, facial tissue, paper towels, extra clothing, high-calorie, nonperishable food, bottled water, compass, road maps, knife, first-aid kit, shovel, sack of sand, flashlight or signal light, gas can, windshield scraper, booster cables, two tow chains, fire extinguisher, catalytic heater, and axe.

Safe Winter Auto Travel

Winter travel by automobile is serious business. Take your travel seriously. If possible, inform family members or friends of your travel plans (including routes, arrival/departure times, etc.).

1. If the storm exceeds or even tests your limitations, seek available refuge immediately.
2. Plan your travel and select primary and alternate routes.
3. Check latest weather information on your radio.
4. Try not to travel alone; two or three persons are preferable.
5. Travel in convoy with another vehicle, if possible.
6. Always fill gasoline tank before entering open country, even for a short distance.
7. Drive carefully, defensively, watching for "black ice" on roadways.

If a Blizzard Catches You in Your Car

AVOID OVEREXERTION AND EXPOSURE. Exertion from attempting to push your car, shovel heavy drifts, and perform other difficult chores during the strong winds, blinding snow, and bitter cold of a blizzard may cause a heart attack—even for persons in apparently good physical condition.

STAY IN YOUR VEHICLE. Do not attempt to walk out into a blizzard. Disorientation comes quickly in blowing and drifting snow. Being lost in open country during a blizzard is almost certain death. You are more likely to be found, and more likely to be sheltered, in your car.

DO NOT PANIC.

KEEP FRESH AIR IN YOUR VEHICLE. Freezing wet snow and wind-driven snow can completely seal the passenger compartment.

BEWARE THE SILENT KILLERS: CARBON MONOXIDE AND OXYGEN STARVATION. Run the motor and heater sparingly, and only with the downwind window open for ventilation.

EXERCISE by clapping hands and moving arms, fingers, legs, and toes vigorously from time to time. To keep blood circulating, do not stay in one position for long.

TURN ON DOME LIGHT AT NIGHT to make the vehicle visible to work crews.

TIE a piece of material (preferably red) to your car's antenna or door. Once the snow has stopped, raise the hood, indicating engine problems.

KEEP WATCH. Do not permit all occupants of the car to sleep at once.

WIND CHILL EQUIVALENT TEMPERATURE TABLE
DRY BULB TEMPERATURE (°F)

WIND VELOCITY (MPH)	45	40	35	30	25	20	15	10	5	0	−5	−10	−15	−20	−25	−30	−35	−40	−45	WIND VELOCITY (MPH)
4	45	40	35	30	25	20	15	10	5	0	−5	−10	−15	−20	−25	−30	−35	−40	−45	4
5	43	37	32	27	22	16	11	6	0	−5	−10	−15	−21	−26	−31	−36	−42	−47	−52	5
10	34	28	22	16	10	3	−3	−9	−15	−22	−27	−34	−40	−46	−52	−58	−64	−71	−77	10
15	29	23	16	9	2	−5	−11	−18	−25	−31	−38	−45	−51	−58	−65	−72	−78	−85	−92	15
20	26	19	12	4	−3	−10	−17	−24	−31	−39	−46	−53	−60	−67	−74	−81	−88	−95	−103	20
25	23	16	8	1	−7	−15	−22	−29	−36	−44	−51	−59	−66	−74	−81	−88	−96	−103	−110	25
30	21	13	6	−2	−10	−18	−25	−33	−41	−49	−56	−64	−71	−79	−86	−93	−107	−109	−116	30
35	20	12	4	−4	−12	−20	−27	−35	−43	−52	−58	−67	−74	−82	−89	−97	−105	−113	−120	35
40	19	11	3	−5	−13	−21	−29	−37	−45	−53	−60	−68	−76	−84	−92	−100	−107	−115	−123	40
45	18	10	2	−6	−14	−22	−30	−38	−46	−54	−62	−70	−78	−85	−93	−102	−109	−117	−125	45

VERY COLD
BITTER COLD
EXTREME COLD

(Chart by Hans & Cassidy. Gale Group.)

complete shade and does not consider the gain of heat from incoming radiation, either direct or diffuse. Under conditions of bright sunshine, the wind chill index should be reduced by about 200 kg/cals/m²/hr.

The wind chill index, or equivalent temperature, is based upon a neutral skin temperature of 91°F. With physical exertion, the body's heat production rises, perspiration begins, and heat is removed from the body by vaporization. The body also loses heat through conduction to cold surfaces with which it is in contact, and in breathing cold air, which results in the loss of heat from the lungs. The index, therefore, does not take into account all possible losses of heat from the body. It does, however, give a good measure of convective cooling, which is the major source of body heat loss.

• HEATING-DEGREE DAYS

Early in the twentieth century heating engineers developed the concept of heating-degree days as a useful index of heating fuel requirements. They found that when the daily mean temperature is lower than 65°F, most buildings require heat to maintain an inside temperature of 70°F. The daily mean temperature is obtained by adding together the maximum and minimum temperatures reported for the day and dividing the total by two. Each degree of mean temperature below 65 is counted as one heating-degree day. Thus, if the maximum temperature is 70°F and the minimum 52°F, four heating-degree days would be produced. (70 + 52 = 122; 122 divided by 2 = 61; 65 − 61 = 4). If the daily mean temperature is 65°F or higher, the heating-degree day total is zero.

For every additional heating-degree day, more fuel is needed to maintain a comfortable 70°F indoors. A day with a mean temperature of 35°F—30 heating-degree days—would require twice as much fuel as a day with a mean temperature of 50°F—15 heating-degree days—assuming, of course, similar meteorological conditions such as wind speed and cloudiness. Each degree below 70°F that is called for as indoor temperature will conserve one heating-degree day where the measure is used in calculating fuel unit requirements.

The heating-degree concept has become so valuable that daily, monthly, and seasonal totals are routinely computed for all temperature-observing stations in the National Weather Service's network. Daily figures are used by fuel

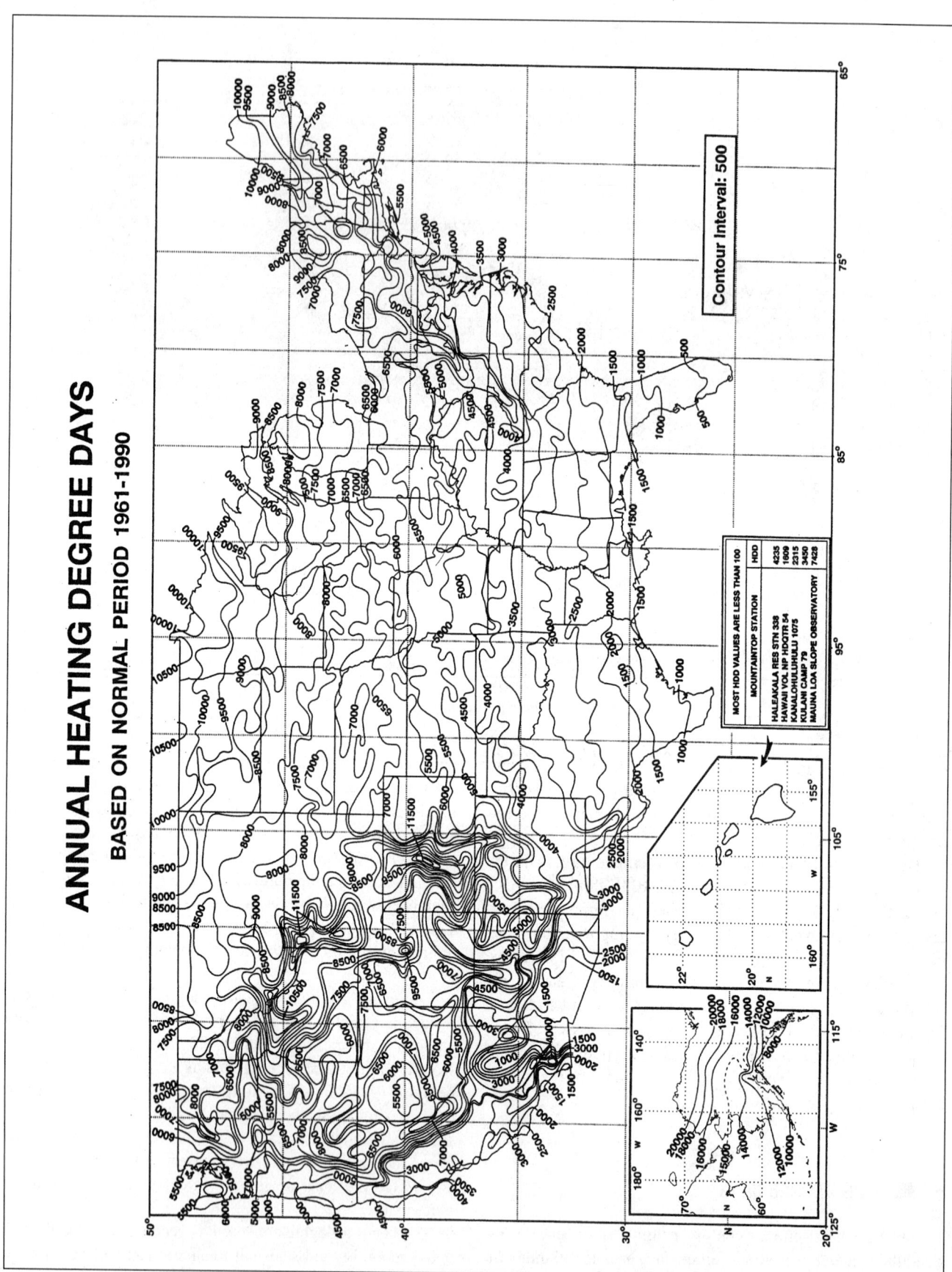

(Courtesy of U.S. government publication.)

companies for evaluation of fuel use rates and for efficient scheduling of deliveries. For example, if a heating system is known to use one gallon of fuel for every five heating-degree days, oil deliveries will be scheduled to meet this burning rate. Gas and electric company dispatchers use the data to anticipate demand and to implement priority procedures when demand exceeds capacity.

The amount of heat required to maintain a certain temperature level is proportional to the heating-degree days. A fuel bill usually will be twice as high for a month with 1,000 heating-degree days as for a month with 500. For example, it can be estimated that about four times as much fuel will be required to heat a building in Chicago, Illinois, where the annual average is 6,100 heating-degree days as it would to heat a building in New Orleans, Louisiana, where the average is about 1,500. All this is true only if building construction and living habits in these areas are similar. Since such factors are not constant, these ratios must be modified by actual experience. The use of heating-degree days has the advantage that consumption rates are fairly constant, i.e., fuel consumed for 100 degree days is about the same whether the 100 heating-degree days were accumulated on only three or four days or were spread over seven or eight days.

Accumulation of temperature data for a particular location has resulted in the establishment of "normal" values based on 30 years of record. NOAA's Environmental Data Service (EDS) publishes maps and tables of heating-degree day normals. The maps are useful only for broad general comparisons, because temperatures, even in a small area, vary considerably depending on differences in altitude, exposure, wind, and other circumstances. Tables of normal monthly and annual heating-degree days for U.S. cities provide a more accurate basis for comparison. The tables show, for instance, that Washington, D.C. (National Airport), has a normal annual total of 4,047 heating-degree days, while the normal for Boston, Massachusetts (Logan International Airport), is 5,641.

Heating-degree day comparisons within a single area are the most accurate. For example, March heating-degree day totals in the Midwest average about 70% of those for January. In Chicago, the coldest six months in order of decreasing coldness are January, December, February, March, November, and April. Annual heating-degree day data are published by heating season, which runs from July of one year through June of the next year. This enables direct comparison of seasonal heating-degree day data and seasonal heating fuel requirements.

• HUMIDITY & INDOOR WINTER COMFORT

Compared with summer, when the moisture content of the air (relative humidity) is an important factor of body discomfort, the amount of moisture in the air in the winter has a lesser effect on the human body during outdoor winter activ-

Weather Anecdotes

March 23, 1987 ... Colby Co., KS. An elderly couple became stranded and remained in their car through a second blizzard and were finally rescued April 4, 13 days later. They both suffered from exposure and were hospitalized. The man's feet were later amputated due to frostbite.

March 30, 1987 ... Oldham Co., KY. A 8-year-old boy with his dog became disoriented and lost in the storm and wandered away from their home while playing in the snow. Rescuers found them 14 hours later and the boy said he used the dog as a blanket to try and keep warm. Wind chill was about 8°F. His only injury was numbness to his extremities.

April 25, 1997 ... De Baca, Guadalupe, and Torrance Counties, NM. These three counties sustained losses of over 1 million dollars in lost sheep and cattle due to blizzard conditions. One ranch in eastern Torrance County lost 1,300 sheep from a herd of 1,700. Sheep loses totaled over 5,000 animals. Economic loses of sheep do not include the loss of over 4,000 unborn lambs.

ities. But moisture is a big factor for winter INDOOR comfort because of effects on health and energy consumption.

The colder the outdoor temperature, the more heat must be added indoors to be comfortable. That heat, however, dries the indoor air and lowers the indoor relative humidity. While a room temperature of 71–77°F may be comfortable for short periods under very dry conditions, prolonged exposure to dry air has varying effects on the human body and usually causes discomfort.

Dry air has been shown to have four main effects on the human body:
1. Breathing dry air can cause such respiratory ailments as asthma, bronchitis, sinusitis, and nosebleeds; it may also cause general dehydration since body fluids are depleted during respiration.
2. Skin moisture evaporation can cause skin irritations and eye itching.
3. Irritative effects, such as static electricity, which causes mild shocks when metal is touched, are common when moisture is low.
4. The "apparent temperature" of the air is lower than what the thermometer indicates, and the body "feels" colder.

These problems can be reduced by simply increasing the indoor relative humidity. By increasing the relative humidity to above 50% when the indoor temperature is between 71–77°F, most average-dressed persons will feel comfortable. This can be done by using humidifiers, vaporizers, steam generators, or large pans of water. Even wet towels or water in a bathtub will help. The lower the room temperature the easier the relative humidity can be brought to its desired level. A relative humidity indicator (hygrometer) may be of assistance in determining the humidity in the house.

How cold will you feel?

When the humidity is low, the body feels cooler than what the thermometer indicates. While the indoor temperature may read 75°F, the apparent temperature (what it feels like) may be warmer or cooler depending upon the moisture content (relative humidity) of the air. Apparent temperature can vary as much as 8°F when the relative humidity is in a range of 10–80%. Why? The human body cools when exposed to dry air because skin moisture evaporates, and the sense of coolness increases as humidity decreases. The drier the room, the cooler the skin feels. With a room temperature of 70°F, for example, a person will feel cooler if the humidity is low than if the humidity is high; this is especially noticeable when entering a dry room after bathing.

Heat & Humidity

• TEMPERATURE TRENDS

During the past century, global surface temperatures have increased at a rate near 1.1°F per century, but this trend has dramatically increased to a rate approaching 3.6°F per century during the past 25 years. There have been two sustained periods of warming, one beginning around 1910 and ending around 1945, and the most recent beginning about 1976. Temperatures during the latter period of warming have increased at a rate comparable to the rates of warming projected to occur during the next century with continued increases of anthropogenic greenhouse gases.

Data collected by NOAA's polar orbiting satellites and analyzed for NOAA by the University of Alabama in Huntsville (UAH) and Remote Sensing Systems (RSS, Santa Rosa, California) indicate that temperatures centered in the middle troposphere at altitudes from 2 to 6 miles are also on pace to make 2002 the second warmest year for the globe.

The average lower troposphere temperature (surface to about 5 miles) for 2002 will also very likely be the second warmest on record. Analysis of the satellite record that began in 1979 shows that the global average temperature in the middle troposphere has increased, but the differing analysis techniques of the two teams result in different trends. The UAH team found an increase of 0.06°F per decade while a trend of 0.21°F per decade was found by the RSS team. This compares to surface temperature increases approaching 0.3°F per decade during the same period.

While lower tropospheric temperatures as measured by the MSU indicate increasing temperatures over the last two decades, stratospheric (9–14 miles) temperatures have been decreasing. This is consistent with the depletion of ozone in the lower stratosphere. The large increase in 1982 was caused by the volcanic eruption of El Chichon, and the increase in 1991 was caused by the eruption of Mt. Pinatubo in the Philippines.

Preliminary global temperatures for 2002

Global temperatures in 2002 are likely to be 1.03°F above the long-term (1880–2001) average, which will place 2002 as the second warmest year on record. The only warmer year was 1998, in which a strong El Niño contributed to higher global temperatures. Land temperatures are likely to be 1.66°F above average and ocean temperatures 0.76°F above the 1880–2001 mean. Both land and ocean temperature will very likely rank as second warmest on record.

Neutral ENSO conditions at the beginning of 2002 gave way to a strengthening El Niño episode during late boreal summer and continuing into early winter. Moderate positive anomalies of equatorial Pacific sea surface temperatures (El Niño conditions) are expected to persist through the early part of 2003.

The Northern Hemisphere temperature will very likely average near record levels in 2002 at 1.19°F above the long-term average. The Southern Hemisphere also reflected the globally warmer conditions, with a positive anomaly expected near 0.83°F.

In 2002 warmer temperatures and shifts in atmospheric circulation patterns contributed to the greatest surface melt on the Greenland Ice Sheet in the 24-year satellite record. There was also a record low level of Arctic sea ice extant in September, the lowest since satellite monitoring began in 1978, according to the National Snow and Ice Data Center.

Annual anomalies in excess of 1.8°F were widespread across much of North America and Asia.

The 1880–2001 average combined land and ocean annual temperature is 56.9°F, the annually averaged land temperature for the same period is 47.3°F, and the long-term annually averaged sea surface temperature is 60.9°F.

Preliminary United States temperatures for 2002

2002 will likely rank as one of the top 20 warmest years on record for the U.S., with an estimated temperature of 53.6°F. After beginning the year with much-above-average warmth, especially in the Northeast, 2002 ended with cooler-than-normal to near-average temperatures across much of the nation.

The 2001–2002 winter season (December to February) was the 9th warmest on record for the U.S., with much of the warmth occurring in the Northeast, which had its warmest winter on record. Spring (March to May) was near normal nationally with a warmer-than-average April compensating for a cooler March and May. However, the summer season (June to August) was one of the warmest in 108 years of national records. Summer 2002 was tied (with 1988) for the 3rd warmest, behind only 1936 and 1934. The 2002 fall season was near average, though September was the 7th warmest such month on record, followed by a cool October and near-average November.

The last three 5-year periods (1998–2002, 1997–2001, 1996–2000), have been the warmest 5-year periods in the last 108 years of national records, and the last 6- (1997–2002), 7- (1996–2002), 8- (1995–2002), 9- (1994–2002), and 10-year (1993–2002) periods have been the warmest on record for the U.S., illustrating the near-persistent warmth of the last decade.

A tenth or more of the country experienced very warm averages for seven months of the year, with two months, April and July, each exceeding 30%. More than 10% of the country was very cold in March and May, and more than a third of the country was very cold in October. Very warm and very cold conditions are defined as the warmest and coldest 10% of recorded temperatures, respectively.

2002 is on pace to be the warmest on record for Delaware and the second warmest for Maryland and New Jersey. Fourteen other states will likely be much warmer than normal during 2002. No state in the contiguous U.S. has averaged below normal for the year.

Annual temperatures averaged across the state of Alaska will approach or exceed record levels for 2002, with every season in Alaska averaging above normal. Fall 2002 was the warmest September–November on record for the state. Nine states in the Northeast and Midwest had their warmest winter on record, and the Southwest region had its warmest summer on record. Three states (Colorado, Maryland, and Delaware) also broke summer warmth records.

Data collected by NOAA's TIROS–N polar-orbiting satellites and adjusted for time-dependent biases by NASA and the Global Hydrology and Climate Center at the University of Alabama in Huntsville, indicate that temperatures in the lower half of the atmosphere (lowest 8 km of the atmosphere) were above the 20-year (1979–1998) average for 2002 for the 5th consecutive year and ranked as the 8th warmest such period since 1979.

Global temperatures—2001

Global temperatures in 2001 were 0.92°F* above the long-term (1880–2000) average**, which makes 2001 the third warmest year on record. The only warmer years were 2002 and 1998, in which a strong El Niño contributed to higher global temperatures. Land temperatures were 1.35°F* above average and ocean temperatures 0.72°F* above the 1880–2000 mean.

*Updated anomalies using most recent station data and NCEP OI Version 2 anomalies.
**The 1880–2000 average combined land and ocean annual temperature is 56.9°F, the annually averaged land temperature for the same period is 47.3°F, and the long-term annually averaged sea surface temperature is 60.9°F.

A weak La Niña persisted into early 2001 in the tropical Pacific, but neutral ENSO conditions developed and were maintained throughout the latter half of the year. Near-normal or weak positive anomalies (El Niño) conditions are expected to persist through the early part of 2002.

The Northern Hemisphere temperature continued to average near record levels in 2001 at 1.08°F above the long-term average. The Southern Hemisphere also reflected the globally warmer conditions, with a positive anomaly of 0.77°F.

Annual anomalies in excess of 1.8°F were widespread across North America and much of Europe and the Middle East.

• 2000: A YEAR OF RECORD WARMTH

Temperature Trends—2000

The year 2000 began with record warmth and ended with colder than normal temperatures across much of the country. However, annual United States and global temperatures remained well above normal.

The U.S. national temperature was above average during 2000 according to statistics calculated by scientists working from the world's largest statistical weather database at NOAA's National Climatic Data Center in Asheville, North Carolina.

After beginning with record winter warmth, the year 2000 ended with colder than normal temperatures across much of the nation. This does not, however, alter the fact that the year 2000 was also one of the top 20 warmest years on record in 106 years. The average temperature in the United States was 54.0°F.

Heat waves and drought plagued much of the southern and western United States in 2000, while the Midwest and northeastern United States experienced prolonged periods of cooler and wetter than normal conditions. July 2000 was the coolest July on record in Pennsylvania and West Virginia, and the second coolest in New York. Precipitation was above average in 15 states throughout the Northeast and Midwest during the summer months (June–August).

By August 2000, 36% of the nation was in severe to extreme drought, although precipitation in the following months significantly reduced the severity of drought in many areas. The widespread drought contributed to one of the worst U.S. wildfire seasons in 50 years. More than 7 million acres of forests and grasslands were consumed by fire in 2000 with the greatest losses in western states, particularly Idaho and Montana, and estimated losses nationwide of more than $1 billion.

Global temperatures in 2000 were also warmer. Temperatures were 0.7°F above the long-term (1880–2000) average (56.9°F), the 8th warmest year on record. The only warmer years were 1998, 1997, 1995, 1990, 1999, 2001, and 2002.

Land temperatures were 1.1°F above the average, 47.3°F, and ocean temperatures 0.5°F above the 1880–2000 mean of 60.9°F.

A strong La Niña at the beginning of 2000 weakened during July and August, but was still evident at year's end.

• HOW ORDINARY SUMMER HEAT DEVELOPS

Given terrain and geographic situation, North American summers are bound to be hot. As the advancing Sun drives back the polar air, the land is opened up to light and solar heat, and occupied by masses of moist, warm air spun landward off the tropical ocean. With these rain-filled visitors come the tongues of dry desert air that flick northward out of Mexico, and, occasionally, the hot winds called chinooks that howl down the Rockies' eastern slopes.

Inequalities of atmospheric heating and cooling, of moistness and aridity, are regulated at middle latitudes by horizontal and vertical mixing. The mixing apparatus is the parade of cyclones (low-pressure centers, or lows) and anticyclones (high-pressure centers, or highs) that lie at the heart of most weather, good and bad.

The cyclones and anticyclones drift in the mid-latitude westerlies, the prevailing eastward-blowing winds that follow a scalloped path around the Northern Hemisphere. The large-scale undulations of these winds may extend for thousands of miles, and are called planetary waves. Their high-speed core is the jet stream, which snakes across the continent some 6–8 mi up, keeping mainly to the cool side of highs and lows as they form and spin and die below it.

The strength of the jet stream itself varies, and therefore, it rarely moves across a hemisphere as one continuous river of air. Generally, its segments span 1,000–3,000 mi, with a width of 100–400 mi, and a thickness of 3,000–7,000 ft. A jet stream's winds can be as low as 50 knots or as high as 300; frequently the jet stream wind speed settles at 100–150 knots.

The kind of weather predominating in an area over a period of time depends largely on the prevailing position and orientation of the jet stream. As the continent warms, the jet stream shifts northward, along with the tracks of surface weather disturbances. Cyclones like the ones that brought April rains to the Gulf states bring June thundershowers to the Plains; the humid spring of Georgia becomes the muggy summer of Illinois. These semi-regular alternations of instability and equilibrium, hot and cool, moist and dry, combine year-in and year-out to generate the average June–September climate for North America.

How heat waves occur

When these alternating processes are somehow interrupted, the climatic "norm" of summer is impacted by a heat wave. The anomaly is usually associated with a change in the planetary waves, so that the prevailing winds from the southwestern deserts sweep farther north than usual and blanket a large region with hot, often humid air at ground level. An upper-level high may settle over the mid-continent, destroying cloud cover with its descending, compression-heated currents, until the blessing of fair weather turns to the curse of drought. In addition, heat from the hot, dry ground feeds back into the atmosphere, tending to perpetuate the heat-wave circulation.

Whatever the cause, the effect is uncomfortable and dangerous. Continental heat waves live in human memory the way fierce winters do.

• THE HUMAN BODY'S REACTION TO HEAT

To keep on the cool side of their upper thermal limits, our bodies dissipate heat by varying the rate and depth of blood circulation, by losing water through the skin and sweat glands, and, as the last extremity is reached, by panting. Under normal conditions, these reflex activities are kept in balance and controlled by the brain's hypothalamus, a comparatively simple sensor of rising and falling environmental temperatures, and a sophisticated manager of temperatures inside the body.

Like the hot light in a car, the hypothalamus responds to the temperature of coolant, in this case, blood. A surge of blood heated above 98.6°F sends the hypothalamus into action. As its orders go out, the heart begins to pump more blood, blood vessels dilate to accommodate the increased

---

Most Humid U.S. Cities

1. Quillayute, WA
2. Olympia, WA
3. Port Arthur, TX; Lake Charles, LA; Gainesville, FL; Apalachicola, FL; Corpus Christi, TX (tie)
4. New Orleans, LA; Eugene, OR; Houston, TX (tie)

Least Humid U.S. Cities

1. Las Vegas, NV
2. Phoenix, AZ
3. Yuma, AZ
4. Tucson, AZ
5. El Paso, TX

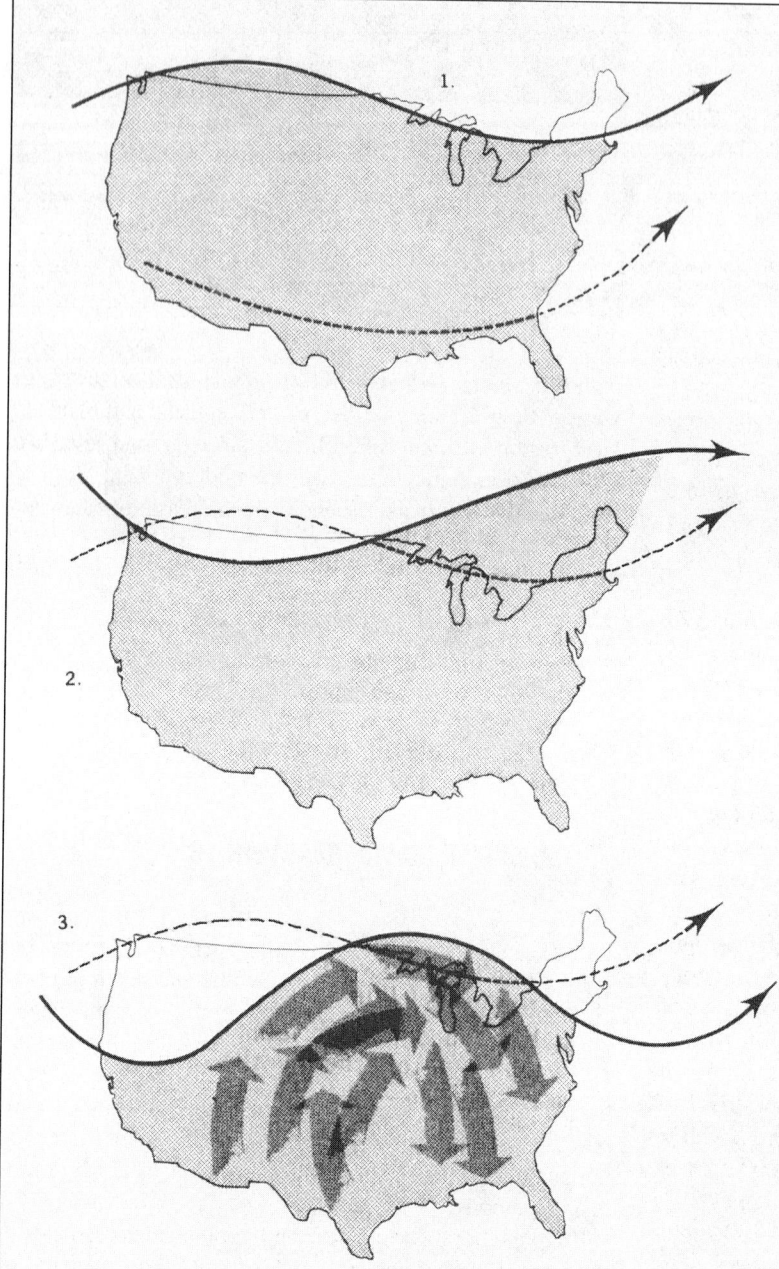

In summer the jet stream can be an ill wind indeed. As the sun drives the polar front back into Canada, the jet stream keeps to the cool side of the boundary, and shifts northward (1). The summer is a "normal" one—hot but not too hot; humid but not too humid.

But the polar front and jet stream may be oriented so that their eastern segment is displaced farther to the north (2), setting the stage for a midwestern and eastern heat wave. A persistent High can block the jet stream northward (3), its clockwise, sinking circulation drawing in hot dry air from the southwestern deserts, and dry air from the northwest. This classical Dust Bowl pattern brings hot, dry weather to the mid-continent, but often means cooler-than-normal conditions in New England and the far northwest. (Jet stream position at 500 mb level is shown here.)

(Courtesy of U.S. government publication.)

flow, and the bundles of tiny capillaries threading through the upper layers of the skin are put into operation. The body's blood is circulated closer to the skin's surface, and excess heat drains off into the cooler atmosphere. At the same time, water diffuses through the skin as insensible perspiration, so called because it evaporates before it becomes visible, and the skin seems dry to the touch.

Heat loss from increased circulation and insensible perspiration is a comparatively minor correction. If the hypothalamus continues to sense overheating, it calls upon the millions of sweat glands that perforate the outer layer of our skin. These tiny glands can shed great quantities of water

(and heat) in what is called sensible perspiration, or sweating. Between sweating and insensible perspiration, the skin handles about 90% of the body's heat-dissipating function.

As environmental temperature approaches normal body temperature, physical discomfort is replaced by physical danger. The body loses its ability to get rid of heat through the circulatory system, because there is no heat-drawing drop in temperature between the skin and the surrounding air. At this point, the skin's elimination of heat by sweating becomes virtually the only means of maintaining constant temperature.

Most water enters the atmosphere via the process of evaporation, the jump from liquid to vapor phase; to do this, a water molecule must absorb enough energy to break the tenacious clutch of its fellow molecules. Evaporation, consequently, has the effect of absorbing large quantities of energy in the form of latent heat, which cools the parent body. This is familiar to anyone who has stepped from a bath into a dry room. The breakdown of the evaporation process when one steps from a bath into a hot, moist room is just as familiar.

Sweating, by itself, does nothing to cool the body, unless the water is removed by evaporation—and high relative humidity retards evaporation. Under conditions of high temperature (above 90°F) and high relative humidity (above 75%), the body is doing everything it can to maintain 98.6°F inside. The heart is pumping a torrent of blood through dilated circulatory vessels, the sweat glands are pouring liquids—and essential dissolved chemicals, like sodium and chloride—onto the surface of the skin. And the body's metabolic heat production goes on in the vital organs. When thermal limits are exceeded by very much or for very long, the warm-blooded organism does not doze, reptile fashion—it dies.

A study of three September heat waves in Los Angeles and Orange County, California, shows what excessive temperature alone can do. Without the complicating factors of high humidity or air pollution, the heat waves were accompanied by an increased mortality, especially among the elderly. The California study agreed with other research that showed that increased mortality in a heat wave tends to follow maximum temperatures by about one day—the day it takes to overwork a tired circulatory system. The causes of "extra" deaths in September 1963 would seem to bear this out. Most were assigned to coronary and cerebrovascular disease. Heat syndrome was almost absent.

Heat syndrome refers to several clinically recognizable disturbances of the human thermoregulatory system. The disorders generally have to do with a reduction or collapse of the body's ability to shed heat by circulatory changes and sweating, or a chemical (salt) imbalance caused by too much sweating. Ranging in severity from the vague malaise of heat asthenia to the extremely lethal heat stroke, heat syndrome disorders share one common feature: the individual has overexposed or over-exercised for his or her age and physical condition for the thermal environment. Studies of heat syndrome and its victims indicate that it occurs at all ages, but, other things being equal, the severity of the disorder tends to increase with age—heat cramps in a 17-year-old may be heat exhaustion in someone age 40, and heat stroke in a person over 60.

Sunburn, while not categorized as heat syndrome, is pertinent here, for ultraviolet radiation burns can significantly retard the skin's ability to shed excess heat.

Acclimatization has to do with adjusting sweat-salt concentrations, among other things. In winter and summer, this concentration changes, just as it does when one moves from Boston to Panama. The idea is to lose enough water to regulate body temperature, with the least possible chemical disturbance. Because females appear to be better at this than males—females excrete less sweat and so less salt—heat syndrome usually strikes fewer females.

Climatic stress is worse for people with heart disease than for others. In a hot, humid environment, impaired evaporation and water loss hamper thermal regulation, while physical exertion and heart failure increase the body's rate of heat production. The ensuing cycle is vicious in the extreme.

Heat Index

It isn't the heat; it's the humidity. The job of keeping the body cool falls increasingly upon the evaporation of sweat as the temperature rises. Meanwhile, the other forms of heat dissipation such as radiation and convection, which depend upon temperature differences between the skin and surroundings, are reduced in effectiveness. In turn, the rate of evaporation of sweat is influenced by the humidity in the surrounding air. (Wind speed and thermal radiation are also factors.)

Discomfort is usually a complaint as soon as sweating begins, although, to be sure, the discomfort and heat stress on the body would be much greater if one could not sweat. Clothing reduces the effectiveness of sweating, but it is needed for protection from the Sun. In order to reflect heat

Rules to Heat Wave Safety

1. SLOW DOWN. Your body cannot do its best in high temperature and humidity, and might do its worst.

2. HEED YOUR BODY'S EARLY WARNINGS THAT HEAT SYNDROME IS ON THE WAY. Reduce your level of activities immediately and get to a cooler environment.

3. DRESS FOR SUMMER. Lightweight, light-colored clothing reflects heat and sunlight, and helps your thermoregulatory system maintain normal body temperature.

4. EAT SMALL MEALS AND EAT MORE OFTEN. Avoid foods that are high in protein, which increase metabolic heat production and increase water loss.

5. DO NOT DRY OUT. Heat wave weather can wring you out before you know it. Drink plenty of water or other non-alcoholic fluids while the hot spell lasts, even if you do not feel thirsty.

6. AVOID THERMAL SHOCK. Acclimatize yourself gradually to warmer weather. Be extra careful for those first critical two or three hot days.

7. VARY YOUR THERMAL ENVIRONMENT. Physical stress increases with exposure time in heat-wave weather. Try to get out of the heat for at least a few hours each day. If you cannot do this at home, drop into a cool store, restaurant, or theater—anything—to keep your exposure time down.

8. DO NOT GET TOO MUCH SUN. Sunburn makes the job of heat dissipation that much more difficult.

9. KNOW THE HEAT SYNDROME SYMPTOMS AND FIRST AID.

Heat Index Chart (Temperature & Relative Humidity)

RH (%)	Temperature (°F)															
	90	91	92	93	94	95	96	97	98	99	100	101	102	103	104	105
90	119	123	128	132	137	141	146	152	157	163	168	174	180	186	193	199
85	115	119	123	127	132	136	141	145	150	155	161	166	172	178	184	190
80	112	115	119	123	127	131	135	140	144	149	154	159	164	169	175	180
75	109	112	115	119	122	126	130	134	138	143	147	152	156	161	166	171
70	106	109	112	115	118	122	125	129	133	137	141	145	149	154	158	163
65	103	106	108	111	114	117	121	124	127	131	135	139	143	147	151	155
60	100	103	105	108	111	114	116	120	123	126	129	133	136	140	144	148
55	98	100	103	105	107	110	113	115	118	121	124	127	131	134	137	141
50	96	98	100	102	104	107	109	112	114	117	119	122	125	128	131	135
45	94	96	98	100	102	104	106	108	110	113	115	118	120	123	126	129
40	92	94	96	97	99	101	103	105	107	109	111	113	116	118	121	123
35	91	92	94	95	97	98	100	102	104	106	107	109	112	114	116	118
30	89	90	92	93	95	96	98	99	101	102	104	106	108	110	112	114

Note: Exposure to full sunshine can increase HI values by up to 15° F

(Courtesy of National Climatic Data Center/NOAA.)

and enhance circulation of air, hot-weather clothing should be light colored, lightweight, porous, and loose fitting. For most individuals, cotton or high-cotton blends are still the best hot weather fabrics.

Livestock weather safety index

Livestock, like human beings, are subject to heat stress that is variable not only with temperature but with different combinations of temperature and relative humidity. This is particularly true for animals that are confined or being loaded or transported.

Additional hazard: the effect of calm, cloudless days

Lack of cloud cover and little or no movement of air are additional hazards that can increase stress and should be considered. An emergency situation is most likely to develop when the temperature is 90–95°F early in the day, and higher temperatures are forecast for the period that the livestock will be in the marketing process. Additional stress created by handling livestock should be kept at an absolute minimum.

Reduction of hazard: the effect of wind

The cooling effect of wind can alleviate conditions by lowering temperature a few degrees in open areas. However, when the air temperature approaches the skin temperature of the animal, the cooling effect of wind becomes minimal.

Acclimatization

Hot, humid weather is more detrimental to livestock in the early summer than in mid- or late summer and during any season following an extended cool period. This heat tolerance has not been qualified but should be considered during periods of marginal danger or emergency categories.

Hogs are a special problem

Most livestock do not adjust readily to high temperatures (heat stress). Hogs are especially vulnerable when closely confined in a vehicle, building, or pen. A careful study by Livestock Conservation, Inc. (Chicago, Illinois) of the relationship of hog deaths during the marketing process shows that high temperatures, especially with high relative humidity, cause abnormally high losses.

Heat builds up internally in the hog's body if it cannot be thrown off by the lungs or skin. If the internal temperature reaches 105–106°F, heat exhaustion occurs and will be followed by death unless the situation is relieved.

Heat Index Chart (Temperature & Dewpoint)

Dewpoint (°F)	Temperature (°F)															
	90	91	92	93	94	95	96	97	98	99	100	101	102	103	104	105
65	94	95	96	97	98	100	101	102	103	104	106	107	108	109	110	112
66	94	95	97	98	99	100	101	103	104	105	106	108	109	110	111	112
67	95	96	97	98	100	101	102	103	105	106	107	108	110	111	112	113
68	95	97	98	99	100	102	103	104	105	107	108	109	110	112	113	114
69	96	97	99	100	101	103	104	105	106	108	109	110	113	114	115	
70	97	98	99	101	102	103	105	106	107	109	110	111	112	114	115	116
71	98	99	100	102	103	104	106	107	108	109	111	112	113	115	116	117
72	98	100	101	103	104	105	107	108	109	111	112	113	114	116	117	118
73	99	101	102	103	105	106	108	109	110	112	113	114	116	117	118	119
74	100	102	103	104	106	107	109	110	111	113	114	115	117	118	119	121
75	101	103	104	106	107	108	110	111	113	114	115	117	118	119	121	122
76	102	104	105	107	108	110	111	112	114	115	117	118	119	121	122	123
77	103	105	106	108	109	111	112	114	115	117	118	119	121	122	124	125
78	105	106	108	109	111	112	114	115	117	118	119	121	122	124	125	126
79	106	107	109	111	112	114	115	117	118	120	121	122	124	125	127	128
80	107	109	110	112	114	115	117	118	120	121	123	124	126	127	128	130
81	109	110	112	114	115	117	118	120	121	123	124	126	127	129	130	132
82	110	112	114	115	117	118	120	122	123	125	126	128	129	131	132	133

Note: Exposure to full sunshine can increase HI values by up to 15° F

(Courtesy of National Climatic Data Center/NOAA.)

Hogs lose about 80% of their body heat through the lungs when the environmental temperature is above 80°F; only 20% is lost from the skin by radiation and air movement. Hogs must breathe approximately 20 times as much air at 100°F as at 80°F to maintain a safe internal body temperature (around 102°F) when the environmental temperature is 100°F.

• HEAT WAVE DANGERS

In a "normal" year, about 151 Americans die from summer heat and too much Sun—"excessive heat and insolation" is the vital statistics category. Among our family of natural hazards, only the excessive cold of winter—not lightning, hurricanes, tornadoes, floods, earthquakes, or tsunamis—takes a greater average toll.

Tragic consequences of heat waves

In 1986–99, at least 2,115 persons were killed in the United States by the effects of heat and solar radiation. At least 1,021 people succumbed to heat in the Midwest in 1995. These are direct casualties. It is not known how many deaths are brought on by excessive heat or solar radiation—

for example, how many diseased or aging hearts surrender that would not have under better conditions. Heat waves bring great stresses to the human body; among the aged or infirm are many whose systems cannot withstand the extreme summer conditions.

Most summers see heat waves in one section or another of the United States. East of the Rockies, they tend to be periods of high temperatures and humidity—although the worst have been catastrophically dry. There is evidence that heat waves are worse in the airless brick and asphalt canyons of the inner cities than in the better-lawned, more open suburbs. Among the big ones are the hot summer of 1830, which scorched the north-central interior, and that of 1860, which dried up the Great Plains. July 1901 saw high temperatures in the Middle West that resulted in 9,508 heat deaths.

Heat waves, 1930s–1960s

There is nothing in American climatological annals to touch the heat waves that came with the Dust Bowl droughts of the 1930s. The years 1930, 1934, and 1936 brought progressively more severe summer weather. Record highs of 121°F in North Dakota and Kansas, and 120°F in South

Dakota, Oklahoma, Arkansas, and Texas were observed in the summer of 1936; July and August of that year saw record highs of 109°F or better tied or broken in Indiana, Louisiana, Maryland, Michigan, Minnesota, Nebraska, New Jersey, Pennsylvania, West Virginia, and Wisconsin.

These were cruel years in terms of heat deaths. From 1930 through 1936—ranging from a low of 678 deaths in 1932 to 4,768 in 1936—heat killed nearly 15,000 persons. The toll is consistently high, but tends sharply upward with increases in average July temperatures. This relationship between excessive July heat and significant jumps in heat deaths persists to the present day, despite the softening effects of modern consumer technology.

The second half of the century also reported a significant number of heat deaths. Many states had their hottest summer of record in 1952; that year's death toll—1,401—is the highest for the 1950–67 period. The summer of 1954, a year when heat killed 978, was almost as bad. The heat wave of July 1966 covered much of the eastern and middle continent with high temperatures and very high humidity. Over 400 people died as a result of the heat wave (as compared to nearly 100 in 1965).

Heat waves of 1980, 1993, 1995, and 1998–2000

One of the most tragic heat waves since the 1930s occurred in 1980, according to the National Center for Health Statistics. That year unusually high temperatures held the central and southwestern United States in their grip for the best part of 15 weeks, and directly or indirectly caused 1,700 deaths. This total, which also includes occupationally related heat deaths, is more than nine times that of an ordinary year.

The 1980 heat wave began about June 10, and—except for remissions from about July 21 to about August 10, and August 17 to about September 1—lasted continuously through September's third week. As an example of its intensity, the first week of September saw temperatures of 9°F above normal over much of the central and eastern states.

On the weekend of July 14, 1993, states throughout the Midwest reported record high temperatures. Chicago, Illinois, experiencing its worst heat emergency until 1995, attributed nearly 500 deaths (primarily senior citizens) to the heat or to heat-related activities. Later in the summer, the municipal government developed and implemented heat-crisis plans that included setting up "cooling centers" and contacting senior citizens (to offer fans or transportation to cooling centers, etc.).

On July 13–15, 1995, approximately 70 daily maximum temperature records were set at locations from the central and northern Great Plains to the Atlantic coast. The July 1995 heat wave in Chicago, Illinois, and Milwaukee, Wisconsin, was a highly rare, and in some respects, unprecedented event in terms of both unusually high maximum and minimum temperatures and the accompanying high relative humidity. Chicago experienced its worst weather-related disaster, with 465 heat-related deaths recorded during July

11–27. Milwaukee was also severely affected, with 85 heat-related deaths recorded during the same period. The number of deaths in Chicago and Milwaukee exceeds the average number of lives lost each year in the United States to all floods, hurricanes, and tornadoes.

Heat waves of the late 1990s were not without tragedy. In 1998, a heat wave in Texas killed more than 120 people. In 1999, there were 497 heat related deaths in the United States, more than five times the 92 deaths caused by tornadoes. Even in 2000, heat caused the death of 19 people in a San Francisco Bay area heat wave in June.

• COOLING-DEGREE DAYS

The cooling-degree day statistic—summer sister of the familiar heating-degree day—serves as an index of air-conditioning requirements during the year's warmest months.

According to experts, the need for air-conditioning begins to be felt when the daily maximum temperature climbs to 80°F and higher. The cooling-degree day is therefore a kind of mirror image of the heating-degree day. After obtaining the daily mean temperature—by adding together the day's high and low temperatures and dividing the total by two—the base 65 is subtracted from the resulting figure to determine the cooling-degree day total. For example, a day with a maximum temperature of 82°F and a minimum of 60°F would produce six cooling-degree days. (82 + 60 = 142; 142 divided by 2 = 71; 71 − 65 = 6). If the daily mean temperature is 65°F or lower, the cooling-degree day total is zero.

The greater the number of cooling-degree days, the more energy is required to maintain indoor temperatures at a comfortable level. However, the relationship between cooling-degree days and energy use is less precise than that between heating-degree days and fuel consumption. There is considerable controversy among meteorologists, as well as air-conditioning engineers, as to what meteorological variables are most closely related to energy consumption by air-conditioning systems. Many experts argue that because high-humidity levels make people feel more uncomfortable as temperatures rise, some measure of moisture should be included

Hottest U.S. Cities

Listed below are the hottest U.S. cities. Mean average temperature is shown parenthetically.
1. Key West, FL (77.7°F)
2. Miami, FL (75.6°F)
3. West Palm Beach, FL (74.6°F)
4. Fort Myers, FL and Yuma, AZ (73.9°F)
5. Brownsville, TX (73.6°F)

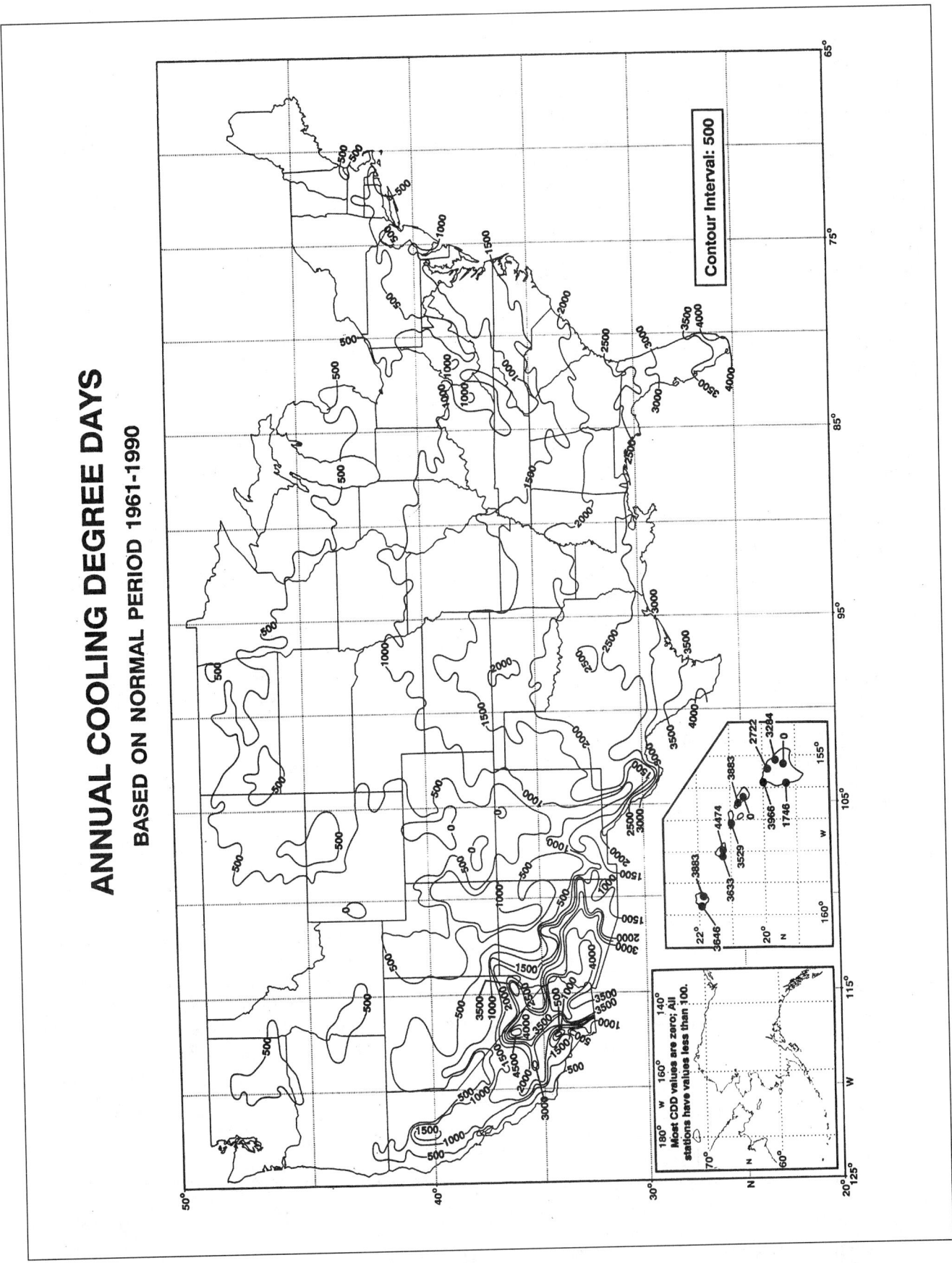

ANNUAL COOLING DEGREE DAYS
BASED ON NORMAL PERIOD 1961-1990

Contour Interval: 500

(Courtesy of U.S. government publication.)

in calculating energy needs for air-conditioning. In addition to humidity some experts feel there are other factors, such as cloudiness and wind speed, that should be included in computation of energy needs for air-conditioning. All agree, however, that there is a need for a more effective measure of the influence of weather on air-conditioning loads.

Until a definitive study of the problem is conducted, NOAA's EDS is continuing to use and publish statistics based on simple cooling-degree-day calculations, employing air temperatures measured at National Weather Service (NWS) offices and cooperating stations throughout the country. As with heating-degree days, normals of cooling-degree days have been established, based on 30 years of record. It should be noted that heating- and cooling-degree days do not cancel each other out. Totals for each are accumulated independently.

• NATIONAL WEATHER SERVICE WARNINGS

Many forecast offices of the NWS (NOAA, U.S. Department of Commerce) issue Danger and/or Emergency Warnings. If potential users are unaware of this program they should contact their local weather service office or agricultural extension agent for information on service in their region.

Global warming

This section is based on a brief synopsis of the 2001 report by the Intergovernmental Panel on Climate Change, the U.S. National Assessment, published by the U.S. Global Change Research Program in the year 2000, and the National Research Council's 2001 report "Climate Change Science: An Analysis of Some Key Questions," as well as the National Climatic Data Center's (NCDC) own data resources. It was prepared by David Easterling and Tom Karl, National Climatic Data Center, Asheville, N.C., 28801.

One of the most hotly debated topics on Earth is the issue of climate change, and the National Environmental Satellite, Data, and Information Service (NESDIS) data centers are central to answering some of the most pressing global change questions that remain unresolved. The National Climatic Data Center contains the instrumental records that can precisely define the nature of climatic fluctuations at time scales of a up to a century. Diverse kinds of data platforms contribute to NCDC's armamentarium; they include ships, buoys, weather stations, balloons, satellites, and aircraft. The National Oceanographic Data Center contains the subsurface data that reveal the ways that heat is distributed and redistributed over the planet. Knowing how these systems are changing and how they have changed in the past is crucial to understanding how they will change in the future. And, for climate information that extends from hundreds to

thousands of years, the paleoclimatology program, also at the National Climatic Data Center, helps to provide longer term perspectives.

Internationally, the Intergovernmental Panel on Climate Change (IPCC), under the auspices of the United Nations (UN), World Meteorological Organization (WMO), and the United Nations Environment Programme (UNEP), is the most senior and authoritative body providing scientific advice to global policy makers. The IPCC met in full session in 1990, 1995, and in 2001. They address issues such as the buildup of greenhouse gases; evidence, attribution, and prediction of climate change; impacts of climate change; and policy options.

Listed below are a number of questions commonly addressed to climate scientists, and brief replies (based on IPCC reports and other research) in common, understandable language.

What is the greenhouse effect, and is it affecting our climate?

The greenhouse effect is unquestionably real and helps to regulate the temperature of our planet. It is essential for life on Earth and is one of Earth's natural processes. It is the result of heat absorption by certain gases in the atmosphere (called greenhouse gases because they effectively "trap" heat in the lower atmosphere) and re-radiation downward of some of that heat. Water vapor is the most abundant greenhouse gas, followed by carbon dioxide and other trace gases. Without a natural greenhouse effect, the temperature of the Earth would be about 0°F instead of its present 57°F. So, the concern is not with the fact that we have a greenhouse effect, but whether human activities are leading to an enhancement of the greenhouse effect.

Are greenhouse gases increasing?

Human activity has been increasing the concentration of greenhouse gases in the atmosphere (mostly carbon dioxide from combustion of coal, oil, and gas; plus a few other trace gases). There is no scientific debate on this point. Pre-industrial levels of carbon dioxide (prior to the start of the Industrial Revolution) were about 280 parts per million by volume (ppmv), and current levels are about 370 ppmv. The concentration of CO_2 in our atmosphere today has not been exceeded in the last 420,000 years, and likely not in the last 20 million years. According to the IPCC Special Report on Emission Scenarios (SRES), by the end of the 21st century we could expect to see carbon dioxide concentrations of anywhere from 490 to 1260 ppm (75–350% above the pre–industrial concentration).

Is the climate warming?

Yes. Global surface temperatures have increased about 0.6°C (plus or minus 0.2°C) since the late-19th century, and about 0.4°F (0.2 to 0.3°C) over the past 25 years (the period with the most credible data). The warming has not been globally uniform. Some areas (including parts of the south-eastern U.S.) have, in fact, cooled over the last century. The recent warmth has been greatest over North America and Eurasia between 40 and 70°N. Warming, assisted by the record El Niño of 1997–1998, has continued right up to the present, with 2001 being the second warmest year on record after 1998.

Linear trends can vary greatly depending on the period over which they are computed. Temperature trends in the lower troposphere (between about 2,500 and 26,000 ft.) from 1979 to the present, the period for which Satellite Microwave Sounding Unit data exist, are small and may be unrepresentative of longer term trends and trends closer to the surface. Furthermore, there are small unresolved differences between radiosonde and satellite observations of tropospheric temperatures, though both data sources show slight warming trends. If one calculates trends beginning with the commencement of radiosonde data in the 1950s, the record shows a slight increase in warming due to temperature rises in the 1970s. There are statistical and physical reasons (e.g., short record lengths, the transient differential effects of volcanic activity and El Niño, and boundary layer effects) for expecting differences between recent trends in surface and lower tropospheric temperatures, but the exact causes for the differences are still under investigation (see National Research Council report Reconciling Observations of Global Temperature Change).

An enhanced greenhouse effect is expected to cause cooling in higher parts of the atmosphere because the increased "blanketing" effect in the lower atmosphere holds in more heat, allowing less to reach the upper atmosphere. Cooling of the lower stratosphere (about 49,000–79,500 ft.) since 1979 is shown by both satellite Microwave Sounding Unit and radiosonde data, but is larger in the radiosonde data.

Relatively cool surface and tropospheric temperatures, and a relatively warmer lower stratosphere, were observed in 1992 and 1993, following the 1991 eruption of Mt. Pinatubo. The warming reappeared in 1994. A dramatic global warming, at least partly associated with the record El Niño, took place in 1998. This warming episode is reflected from the surface to the top of the troposphere.

There has been a general, but not global, tendency toward reduced diurnal temperature range (DTR) (the difference between high and low daily temperatures) over about 50% of the global land mass since the middle of the 20th century. Cloud cover has increased in many of the areas with reduced diurnal temperature range. The overall positive trend for maximum daily temperature over the period of study (1950–1993) is 0.1°C per decade, whereas the trend for daily minimum temperatures is 0.2°C per decade. This results in a negative trend in the DTR of –0.1°C per decade.

Indirect indicators of warming such as borehole temperatures, snow cover, and glacier recession data, are in substantial agreement with the more direct indicators of recent warmth. Evidence such as changes in glacier length is useful because it not only provides qualitative support for existing meteorological data, but glaciers often exist in places too remote to support meteorological stations, the records of glacial advances and retreats often extend back further than weather station records, and glaciers are usually at much higher altitudes than weather stations, thus allowing more insight into temperature changes higher in the atmosphere.

Large-scale measurements of sea-ice have only been possible since the advent of satellites, but by looking at a number of different satellite estimates it has been determined that Arctic sea ice has decreased between 1973 and 1996 at a rate of –2.8% (+/– 0.3%) per decade. Although this seems to correspond to a general increase in temperature over the same period, there are lots of quasi-cyclic atmospheric dynamics (for example the Arctic Oscillation) that may also influence the extent and thickness of sea ice in the Arctic. Sea ice in the Antarctic has shown perhaps a slight increase since 1979, though extending the Antarctic sea-ice record back in time is more difficult due to the lack of direct observations in this part of the world.

Are El Niños related to global warming?

El Niños are not caused by global warming. Clear evidence exists from a variety of sources (including archaeological studies) that El Niños have been present for hundreds, and some indicators suggest maybe millions, of years. However, it has been hypothesized that warmer global sea-surface temperatures can enhance the El Niño phenomenon, and it is also true that El Niños have been more frequent and intense in recent decades. Recent climate model results that simulate the 21st century with increased greenhouse gases suggest that El Niño–like sea-surface temperature patterns in the tropical Pacific are likely to be more persistent.

Is the hydrological cycle (evaporation and precipitation) changing?

Overall, land precipitation for the globe has increased by roughly 2% since 1900; however, precipitation changes have been spatially variable over the last century. Instrumental records show that there has been a general increase in precipitation of about 0.5–1.0% per decade over land in northern mid–high latitudes, except in parts of eastern Russia. However, a decrease of about 0.3% per decade in precipitation has occurred during the 20th century over land in sub-tropical latitudes, though this trend has weakened in recent decades. Due to the difficulty in measuring precipitation, it has been important to constrain these observations by analyzing other related variables. The measured changes in precipitation are consistent with observed changes in streamflow, lake levels, and soil moisture (where data are available and have been analyzed).

The extent of snow cover in the Northern Hemisphere has consistently remained below average since 1987, and has decreased by about 10% since 1966. This is mostly due to a decrease in spring and summer snowfall over both the

Eurasian and North American continents since the mid-1980s. However, the extent of snow cover in the winter and autumn has shown no significant trend for the Northern Hemisphere over the same period.

Improved satellite data show that a general trend of increasing clouds over both land and ocean since the early 1980s seems to have reversed in the early 1990s, and the total amount of clouds over land and ocean now appears to be decreasing. However, there are several studies that suggest regional cloudiness, perhaps especially in the thick precipitating clouds, has increased over the 20th century.

Is the atmospheric/oceanic circulation changing?

A rather abrupt change in the El Niño–Southern Oscillation behavior occurred around 1976–1977 and the new regime has persisted. There have been relatively more frequent and persistent El Niño episodes rather than the cool La Niñas. This behavior is highly unusual in the last 120 years (the period of instrumental record). Changes in precipitation over the tropical Pacific are related to this change in the El Niño–Southern Oscillation, which has also affected the pattern and magnitude of surface temperatures. However, it is unclear whether this apparent change in the ENSO cycle is caused by global warming.

Is the climate becoming more variable or extreme?

On a global scale there is little evidence of sustained trends in climate variability or extremes. This perhaps reflects inadequate data and a dearth of analyses. However, on regional scales, there is clear evidence of changes in variability or extremes.

In areas where a drought or excessive wetness usually accompanies an El Niño, these dry or wet spells have been more intense in recent years. Other than these areas, little evidence is available of changes in drought frequency or intensity.

In some areas where overall precipitation has increased (i.e., the mid-high northern latitudes), there is evidence of increases in the heavy and extreme precipitation events. Even in areas such as eastern Asia, it has been found that extreme precipitation events have increased despite total precipitation remaining constant or even decreasing somewhat. This is related to a decrease in the frequency of precipitation in this region.

Many individual studies of various regions show that extra-tropical cyclone activity seems to have generally increased over the last half of the 20th century in the Northern Hemisphere; but it has decreased in the Southern Hemisphere. It is not clear whether these trends are multi-decadal fluctuations or part of a longer-term trend.

Where reliable data are available, tropical storm frequency and intensity show no significant long-term trend in any basin. There are apparent decadal-interdecadal fluctua-

tions, but nothing that is conclusive in suggesting a longer-term component.

Global temperature extremes have been found to exhibit no significant trend in interannual variability, but several studies suggest a significant decrease in intra-annual variability. There has been a clear trend to fewer extremely low minimum temperatures in several widely separated areas in recent decades. Widespread significant changes in extreme high temperature events have not been observed.

There is some indication of a decrease in day-to-day temperature variability in recent decades.

How important are these changes in a longer-term context?

Paleoclimatic data are critical for enabling us to extend our knowledge of climatic variability beyond what is measured by modern instruments. Many natural phenomena are climate dependent (the growth rate of a tree, for example), and, as such, they provide natural "archives" of climate information. Some useful paleoclimate data can be found in sources as diverse as tree rings, ice cores, corals, lake sediments (including fossil insects and pollen data), speleothems (stalactites, etc.), and ocean sediments. Some of these, including ice cores and tree rings, also provide chronological data because of the way in which they are formed; so high-resolution climate reconstruction is possible in these cases. However, there is not as comprehensive a network of paleoclimate data as there is with instrumental coverage, so global climate reconstructions are often difficult to obtain. Nevertheless, combining different types of paleoclimate records enables us to gain a near-global picture of climate changes in the past.

For the Northern Hemisphere summer temperature, recent decades appear to be the warmest since at least about 1000 A.D., and the warming trend since the late 19th century is unprecedented over the last 1,000 years. Older data are insufficient to provide reliable hemispheric temperature estimates. Ice-core data suggest that the 20th century has been warm in many parts of the globe; but they also suggest that the significance of this warming varies geographically, particularly when viewed in the context of climate variations of the last millennium.

Large and rapid climatic changes affecting the atmospheric and oceanic circulation and temperature, and the hydrological cycle, occurred during the last ice age and during the transition towards the present Holocene period (which began about 10,000 years ago). Based on the incomplete evidence available, the projected change of 1.5–4°C over the next century would be unprecedented in comparison with the best available records from the last several thousand years.

Is the sea level rising?

Global mean sea level has been rising at an average rate of 1–2 mm per year over the past 100 years, which is significantly larger than the rate averaged over the last several thousand years. The projected increase from 1990 to 2100 is anywhere from 0.09–0.88 meters, depending on which greenhouse gas scenario is used. There are many physical uncertainties that contribute to sea-level rise, including a variety of frozen and unfrozen water sources.

Can the observed changes be explained by natural variability, including changes in solar output?

Since our entire climate system is fundamentally driven by energy from the Sun, it stands to reason that if the Sun's energy output were to change, then so would the climate. Since the advent of space-borne measurements in the late 1970s, solar output has indeed been shown to vary. There appears to be confirmation of earlier suggestions of an 11- and 22-year cycle of irradiance. With only 20 years of reliable measurements however, it is difficult to deduce a trend. But, from the short record we have so far, the trend in solar irradiance is estimated at roughly 0.09 W/m^2 compared to 0.4 W/m^2 from well-mixed greenhouse gases. There are many indications that the Sun also has a longer-term variation that has potentially contributed to the century-scale forcing to a greater degree. There is, though, a great deal of uncertainty in estimates of solar irradiance beyond what can be measured by satellites, and still the contribution of direct solar irradiance forcing is small compared to the greenhouse gas component. However, our understanding of the indirect effects of changes in solar output and feedbacks in the climate system is minimal. There is much need to refine our understanding of key natural forcing mechanisms of the climate, including solar irradiance changes, in order to reduce uncertainty in our projections of future climate change.

In addition to changes in energy from the Sun itself, the Earth's position and orientation relative to the Sun (our orbit) also varies slightly, thereby bringing us closer and farther away from the Sun in predictable cycles (called Milankovitch cycles). Variations in these cycles are believed to be the cause of Earth's ice ages (glacials). Particularly important for the development of glacials is the radiation receipt at high northern latitudes. Diminishing radiation at these latitudes during the summer months would have enabled winter snow and ice cover to persist throughout the year, eventually leading to a permanent snow- or icepack. While Milankovitch cycles have tremendous value as a theory to explain ice ages and long-term changes in the climate, they are unlikely to have very much impact on the decade-century timescale. Over several centuries it may be possible to observe the effect of these orbital parameters; however, for the prediction of climate change in the 21st century, these changes will be far less important than radiative forcing from greenhouse gases.

What about the future?

Due to the enormous complexity of the atmosphere, the most useful tools for gauging future changes are "climate

Weather Anecdotes

May 21, 1996 ... Philadelphia, PA. A 74-year-old man died in a hospital parking lot when the defroster, instead of the air conditioner, was running in the car.

June 23, 1996 ... Spartanburg, SC. A 39-year-old man died from a heat stroke while relaxing beside his pool.

April 30, 1999 ... India. A record-breaking heat wave killed eight people, bringing the total deaths during the two-week heat to 89. The eastern state of Orissa is the worst hit area hit by the heat with 34 fatalities. Temperatures reached 117°F in the state's Balangir tribal district. More than 2,000 people died in a similar heat wave in Orissa the previous year, when a 50-year-old record high for April of nearly 110°F was reported.

models." These are computer-based mathematical models that simulate, in three dimensions, the climate's behavior, its components, and their interactions. Climate models are constantly improving, though by definition a model is a simplification and simulation of reality, an approximation of the climate system. The first step in any modeled projection of climate change is to simulate the present climate and compare it to observations. If the model is considered to do a good job at representing modern climate, then certain parameters can be changed, such as the concentration of greenhouse gases, which helps us understand how the climate would change in response. Projections of future climate change therefore depend on how well the computer climate model simulates the climate and on our understanding of how forcing functions will change in the future.

The IPCC Special Report on Emission Scenarios determines the range of future possible greenhouse gas concentrations (and other forcings) based on considerations such as population growth, economic growth, energy efficiency, and a host of other factors. This leads to a wide range of possible forcing scenarios, and consequently to a wide range of possible future climates.

According to the range of possible forcing scenarios, and taking into account uncertainty in climate model performance, the IPCC projects a global temperature increase of anywhere from 1.4 to 5.8°C from 1990 to 2100. However, this global average will integrate widely varying regional responses, such as the likelihood that land areas will warm much faster than ocean temperatures, particularly those land areas in northern high latitudes (and mostly in the cold season).

Precipitation is also expected to increase over the 21st century, particularly at northern mid-high latitudes; the trends may be more variable in the tropics.

Snow extent and sea-ice are also projected to decrease further in the Northern Hemisphere, and glaciers and ice-caps are expected to continue to retreat.

Thunderstorms

• THUNDERSTORMS AROUND THE WORLD AND THE UNITED STATES

Meteorologists estimate that, at any given moment, some 1,800 thunderstorms are in progress over Earth's surface, and about 18 million a year around the world. It is estimated that approximately 100,000 to 125,000 thunderstorms occur in the United States each year. Of that total anywhere from 10–20% may be severe. The National Weather Service considers a thunderstorm severe if it produces hail at least 0.75 inch in diameter, winds of 58 mph or stronger, or a tornado. From 1996-2001, a total of 134,005 severe thunderstorms were recorded (not associated with tornadoes), an average of 19,144 annually. The frequency with which these giant generators of local weather occur, the quantity of energy they release, and the variety of forms this energy may take, make thunderstorms great destroyers of life and property.

• HOW THUNDERSTORMS DEVELOP

Thunderstorms are generated by thermal instability in the atmosphere, and represent a violent example of convection—the vertical circulation produced in a fluid made thermally unstable by the local addition or subtraction of heat and the conversion of potential to kinetic energy. The convective overturning of atmospheric layers that sets up a thunderstorm is dynamically similar to convective circulations observed under laboratory conditions, where distinct patterns are generated in liquids by unequal heating.

The orderly circulations produced in a laboratory are rarely encountered in the atmosphere, where areas corresponding to the rising core of laboratory convective cells are marked by cumulus and cumulonimbus clouds. Clouds are parcels of air that have been lifted high enough to condense the water vapor they contain into very small, visible particles. These particles are too small and light to fall out as rain. As the lifting process continues, these particles grow in size by collision and coalescence until they are large enough to fall against the updrafts associated with any developing convective clouds. Cumulus (for accumulation) clouds begin their towering movement in response

---

Rainiest U.S. Cities

Listed below are the rainiest cities in the U.S. Average number of days with rain is shown parenthetically.
1. Quillayute, WA (210)
2. Astoria, OR (191)
3. Elkins, WV and Syracuse, NY (171)
4. Buffalo, NY (169)
5. Marquette, MI (168)

---

to atmospheric instability and convective overturning. Warmer and lighter than the surrounding air, they rise rapidly around a strong, central updraft. These elements grow vertically, appearing as rising mounds, domes, or towers.

The atmospheric instability in which thunderstorms begin may develop in several ways. Radiational cooling of cloud tops, heating of the cloud base from the ground, and frontal effects may produce an unstable condition. This is compensated in air, as in most fluids, by the convective overturning of layers to put denser layers below less-dense layers.

Mechanical processes are also at work. Warm, buoyant air may be forced upward by the wedge-like undercutting of a cold air mass, or lifted by a mountain slope. Convergence

---

Least Rainy U.S. Cities

Listed below are the U.S. cities with the least amount of rain. The average number of days with rain is shown parenthetically.
1. Yuma, AZ (17)
2. Las Vegas, NV (26)
3. Bishop, CA (29)
4. Santa Barbara, CA (30)
5. Long Beach, CA (32)

It is estimated that lightning flashes occur somewhere on Earth about 100 times each second. *(Photo by C. Clark. Courtesy of NOAA Photo Library, NOAA Central Library; OAR/ERL/National Severe Storms Laboratory (NSSL).)*

of horizontal winds into the center of a low-pressure area forces warm air near that center upward. Where these processes are sustained, and where lifting and cooling of the moist air continues, minor turbulence may generate a cumulus cloud, and then a towering cumulonimbus system.

The pattern of the vertical air movement in the center of the cumulus or cumulonimbus cloud system mimics the behavior of each convective cell. Most thunderstorms have, at maturity, a series of several cells, each following a life cycle characterized by changes in wind direction, development of precipitation and electrical charge, and other factors.

In the first stage of thunderstorm development, an updraft drives warm air up beyond condensation levels, where clouds form, and where continued upward movement produces cumulus formations. The updraft develops in a region of gently converging surface winds in which the atmospheric pressure is slightly lower than in surrounding areas. As the updraft continues, air flows in through the cloud's sides in a process called entrainment, mixing with and feeding the updraft. The updraft may be further augmented by a chimney effect produced by winds at high altitude.

Energy from water

A developing thunderstorm also feeds on another source of energy. Once the cloud has formed, the phase changes of water result in a release of heat energy, which increases the momentum of the storm's vertical development. The rate at which this energy is released is directly related to the amount of gaseous water vapor converted to liquid water.

As water vapor in the burgeoning cloud is raised to saturation levels, the air is cooled sufficiently to liberate solid and liquid particles of water, and rain and snow begin to fall within the cloud. The cloud tower rises above 1.8–3.1 mi where fibrous streamers of frozen precipitation elements appear; this apparent ice phase is thought to be a condition of thunderstorm precipitation. The formation and precipitation of particles large enough and in sufficient quantity to fall against the updraft marks the beginning of the second, mature stage of a thunderstorm cell.

A thunderstorm's mature stage is marked by a transition in wind direction within the storm cells. The prevailing updraft, which initiated the cloud's growth, is joined by a downdraft generated by precipitation. The downdraft is fed and strengthened, as the updraft was, by the addition of en-

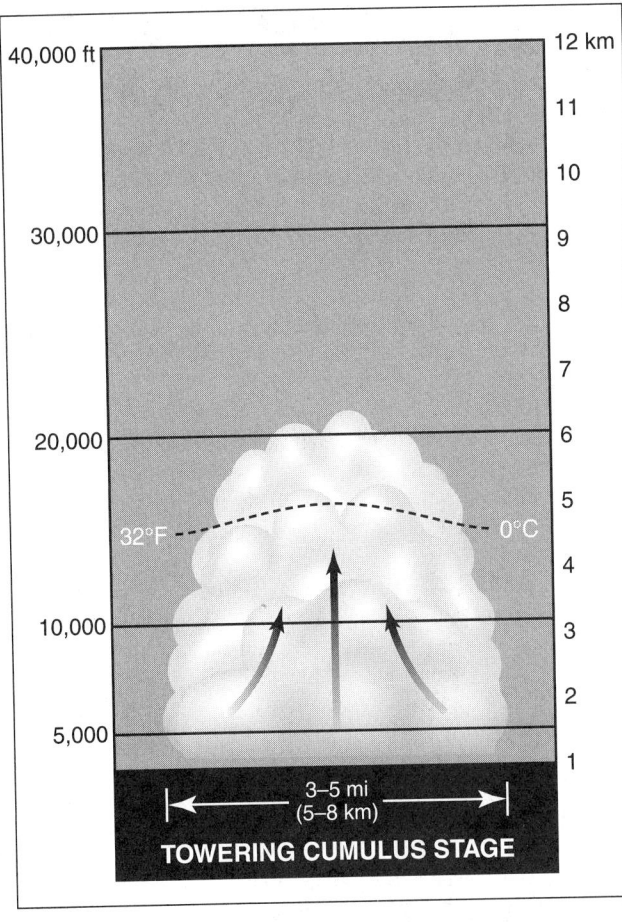

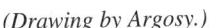

(Drawing by Argosy.)

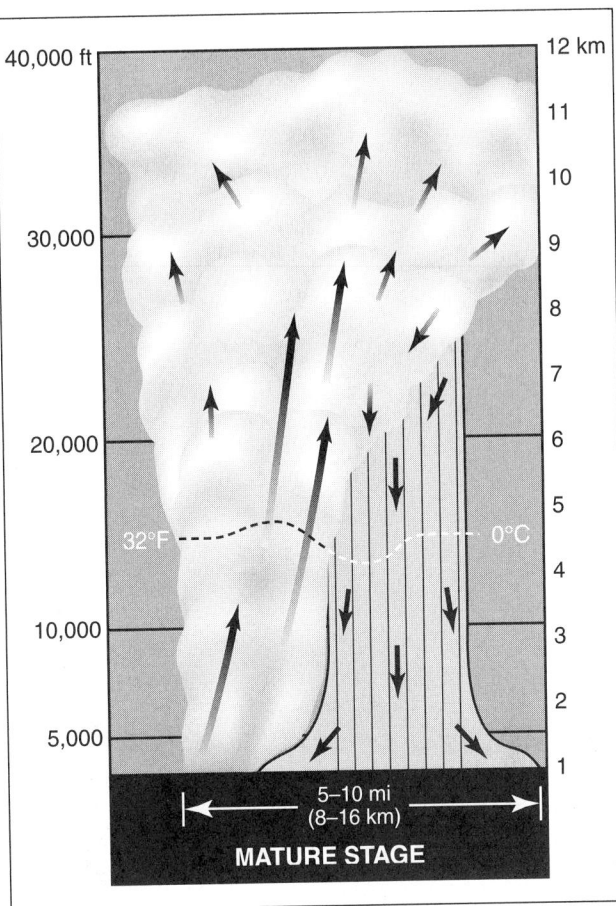

(Drawing by Argosy.)

trained air, and by evaporational cooling caused by interactions of entrained air and falling precipitation. The mature storm dominates the electrical field and atmospheric circulation for several miles around. Lightning—the discharge of electricity between large charges of opposite signs—occurs soon after precipitation begins, a clue to the relationship of thunderstorm electrification and formation of ice crystals and raindrops.

At maturity, a thunderstorm cloud is several miles across its base and may tower to altitudes of 40,000 ft or more. The swift winds of the upper troposphere shred the cloud top into the familiar anvil form, visible in dry regions as lonely giants, or as part of a squall line.

On the ground directly beneath a storm system, the mature stage is initially felt as rain, which is soon joined by the strong downdraft. The downdraft spreads out from the cloud in gusting, divergent winds, and brings a marked drop in temperature. Even where the rain has not reached the ground, this cold air stream flowing over the surface can identify a thunderstorm's mature stage. This is nature's warning that a thunderstorm is in its most violent phase. It is in this phase that a thunderstorm unleashes its lightning, hail, heavy rain, high wind, and—most destructive of all—its tornado. But even as it enters maturity, the storm has be-

gun to die. The violent downdraft initially shares the circulation with the sustaining updraft, but then strangles it. As the updraft is cut off from its converging low-level winds, the storm loses its source of moisture and heat energy. Precipitation weakens and stops, and the cold downdraft ceases. And the thunderstorm, violent creature of an instant, spreads and dies.

Types of thunderstorms

Storms, based on their physical characteristics, can be classified into four basic categories: single cell, multicell cluster, multicell line, and supercell.

Though single-cell storms are rare and relatively weak, they can produce brief bouts of severe weather lasting 20–30 minutes; these storms are not well organized and are seemingly random in occurrence. In the unstable single-cell environment, oftentimes pulse severe storms form. Pulse severe storms produce brief heavy rainfall, severe hail and/or microbursts, and occasionally, weak tornadoes. Single-cell storms are difficult to forecast.

Like a single-cell storm, each cell of a multicell cluster storm lasts usually only about 20 minutes; however, the cluster itself can last for several hours. The multicell cluster, the most common type of thunderstorm, comprises a num-

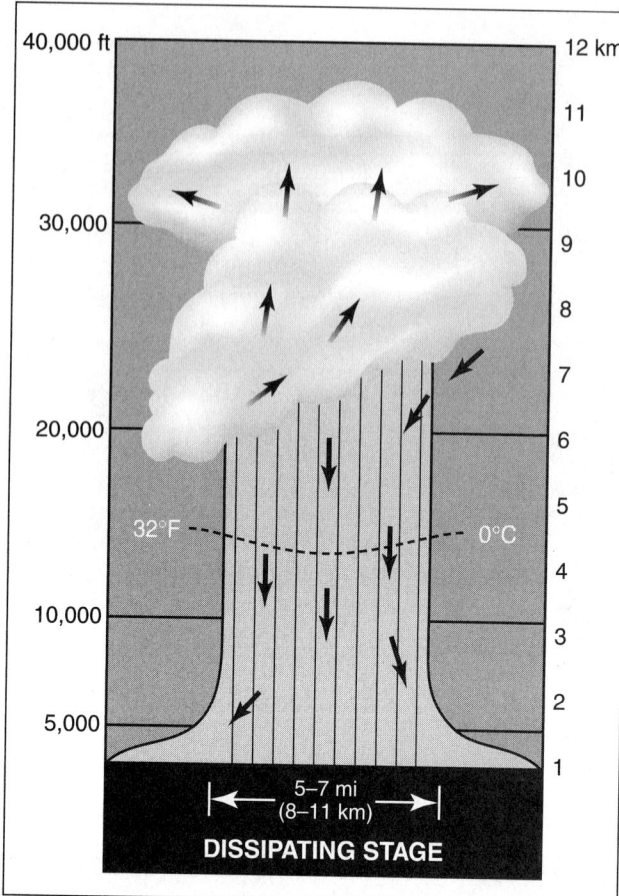

40,000 ft — 12 km

32°F — — — 0°C

5–7 mi
(8–11 km)

DISSIPATING STAGE

(Drawing by Argosy.)

Thunderstorm Safety Rules

1. KEEP AN EYE ON THE WEATHER DURING WARM PERIODS AND DURING THE PASSAGE OF COLD FRONTS. When cumulus clouds begin building up and darkening, you are probably in for a thunderstorm. Check the latest weather forecast.

2. KEEP CALM. Thunderstorms are usually of short duration; even squall lines pass in a matter of an hour or so. Be cautious, but do not be afraid. Stay indoors (away from windows and doors) and keep informed.

3. KNOW WHAT THE STORM IS DOING. Remember that the mature stage may be marked on the ground by a sudden reversal of wind direction, a noticeable rise in wind speed, and a sharp drop in temperature. Heavy rain, hail, tornadoes, and lightning generally occur only in the mature stage of the thunderstorm.

4. CONDITIONS MAY FAVOR TORNADO FORMATION. Tune in your radio or television receiver to determine whether there is a tornado watch or tornado warning out for your area. A tornado watch means tornado formation is likely in the area covered by the watch. A tornado warning means a tornado has been sighted or radar-indicated in your area. If you receive a tornado warning, seek inside shelter in a storm cellar, below ground level, or in reinforced concrete structures; stay away from windows and doors.

5. LIGHTNING IS THE THUNDERSTORM'S WORST KILLER. Stay indoors and away from electrical appliances while the storm is overhead. If lightning catches you outside, remember that it seeks the easiest—not necessarily the shortest—distance between positive and negative centers. Keep yourself lower than the nearest highly conductive object, and maintain a safe distance from it. If the object is a tree, twice its height is considered a safe distance.

6. THUNDERSTORM RAIN MAY PRODUCE FLASH FLOODS. Stay out of dry creek beds during thunderstorms. If you live along a river, listen for flash flood warnings from the National Weather Service.

ber of cells moving as one entity; the cells continuously roll through different storm cycles at different times. The most mature cells are found at the center of the cluster, new cells form at the upwind (usually the west or southwest) edge, while the dissipating cells are found at the downwind (usually east or northeast) edge. Multicell clusters are stronger than single-cell storms, and produce heavy rainfall, downbursts (wind speeds reaching 80 mph), medium-sized hail, and periodic tornadoes.

A long line of storms with a leading edge of strong wind gusts is called a multicell line storm, or squall line. Moving forward, the wind gusts of cold air force unstable warm air into the updraft at the stormfront's edge; heavy rain and large hail immediately follow. A large area behind this produces lighter rain. Squall lines produce golf-ball-size hail, heavy rains, tornadoes, and most notably, weak to strong downbursts.

The most severe (and rare) type of thunderstorm is the supercell. It is a highly organized storm consisting of one main updraft that can reach 150–175 mph. This rotating updraft is called a mesocyclone and works to produce extremely large hail (2 in), major downbursts (80 mph), and fierce tornadoes.

Microbursts

Microbursts are small-scale, hard-hitting downdrafts that result in both vertical and horizontal wind shears that can be extremely hazardous to low-altitude aircraft.

Microbursts most commonly occur during convective activity. They can appear at the point of heaviest rain during a thunderstorm or they can occur within weaker convective cells with far less precipitation. The downdraft's cold air (usually 1 mi in diameter) accelerates as it descends from the cloud base (about 1,000–3,000 ft above ground), reaching its highest speeds about five minutes after initially hitting the ground. The resulting "curl" (air moving away from the impact point) accelerates further, and can extend to approximately 2.5 mi in diameter, creating a serious threat to nearby aircraft. A downdraft can reach top speeds of 6,000

A supercell thunderstorm. *(Photo courtesy of NOAA Photo Library, NOAA Central Library; OAR/ERL/National Severe Storms Laboratory (NSSL).)*

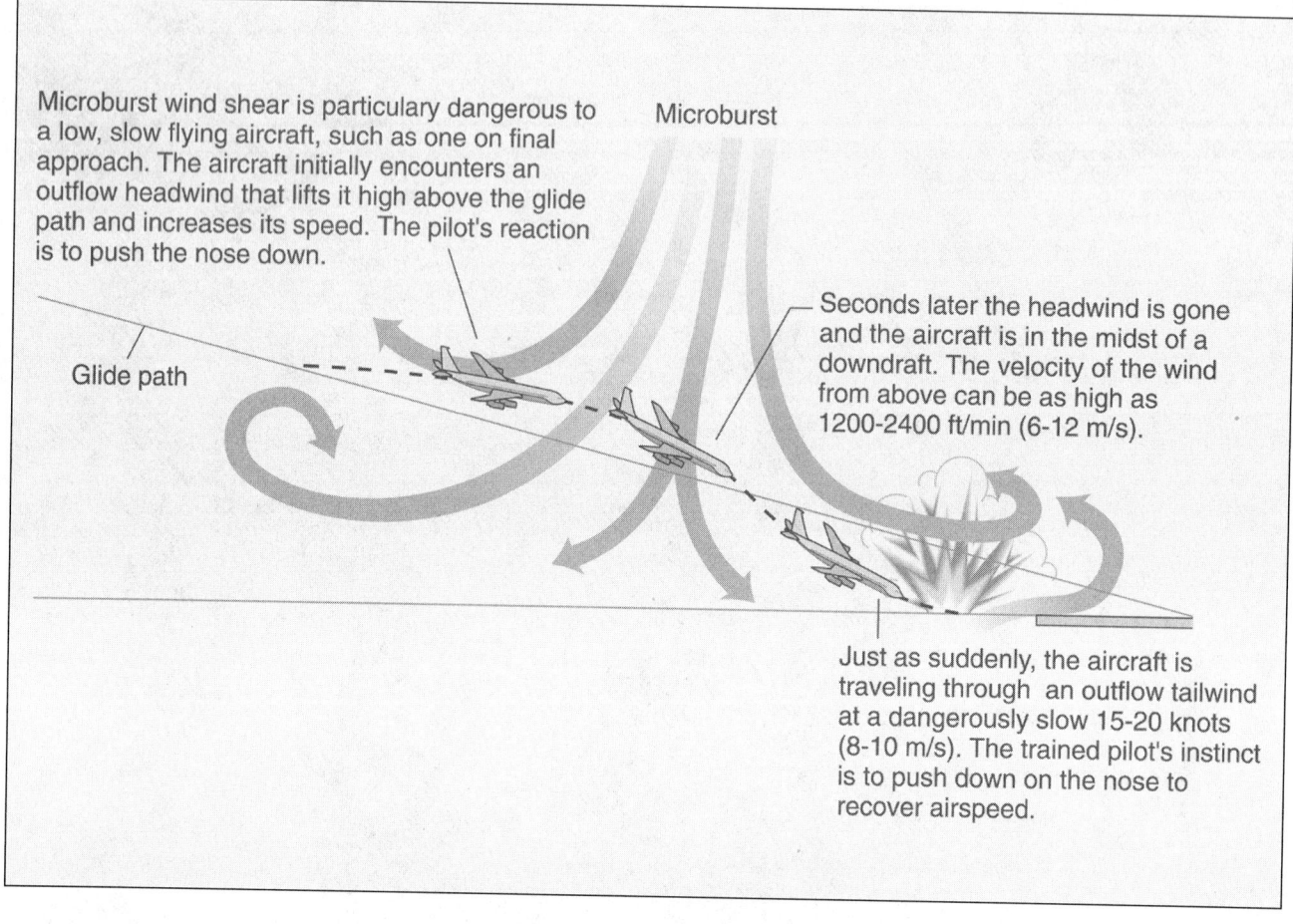

Microburst wind shear is particulary dangerous to a low, slow flying aircraft, such as one on final approach. The aircraft initially encounters an outflow headwind that lifts it high above the glide path and increases its speed. The pilot's reaction is to push the nose down.

Microburst

Glide path

Seconds later the headwind is gone and the aircraft is in the midst of a downdraft. The velocity of the wind from above can be as high as 1200-2400 ft/min (6-12 m/s).

Just as suddenly, the aircraft is traveling through an outflow tailwind at a dangerously slow 15-20 knots (8-10 m/s). The trained pilot's instinct is to push down on the nose to recover airspeed.

(Drawing by Hans & Cassidy. Gale Group.)

ft/min; the curl can be as strong as 45 knots, producing a 90-knot windshear.

Microbursts are not easily detectable by conventional radar due to their size, duration (no longer than 15 minutes), and because they can appear in areas without sur-face precipitation. Visual clues, however, provide proof of their existence. These clues include rings of blowing dust that often mark the impact point of a microburst; a rain foot—the "unfinished," outward distortion of the edge of an area of precipitation, suggesting the presence of a wet microburst; and a dust foot—the resulting plume of dust after the microburst hits the ground and moves away from its impact point. Multiple occurrences of downdrafts of this nature can continue for up to an hour; it is not un-common for more than one microburst to occur in one area.

Dust storms

Dust storms associated with summer thunderstorms are common in the southwestern United States and are found generally in the desert areas of western New Mexico, south-ern Arizona, and in the southeastern deserts of California. Dust storms develop due to the cool downdrafts of a thun-dershower that reach the ground and spread out in all direc-tions, picking up dust along the way. Dust storms associated with late winter and early spring storm systems are common during March and April.

Dust Storm Precautions for Motorists

Dust storms, so common to the southwestern United States, can wreak havoc for motorists traveling in the area. The sudden dark-brown cloud with strong winds and debris is often to blame for tragic, chain-reaction accidents.

If a dust storm does develop, motorists should observe the following:

● Pull off to the furthest edge of the shoulder, turn off lights and set emergency brake.

● If conditions prevent pulling off the roadway, a motorist should proceed at an appropriately reduced speed, turn lights on, and use the center line as a guide. A motorist should never stop on the roadway.

Average Number of Days with Thunderstorms Each Year.

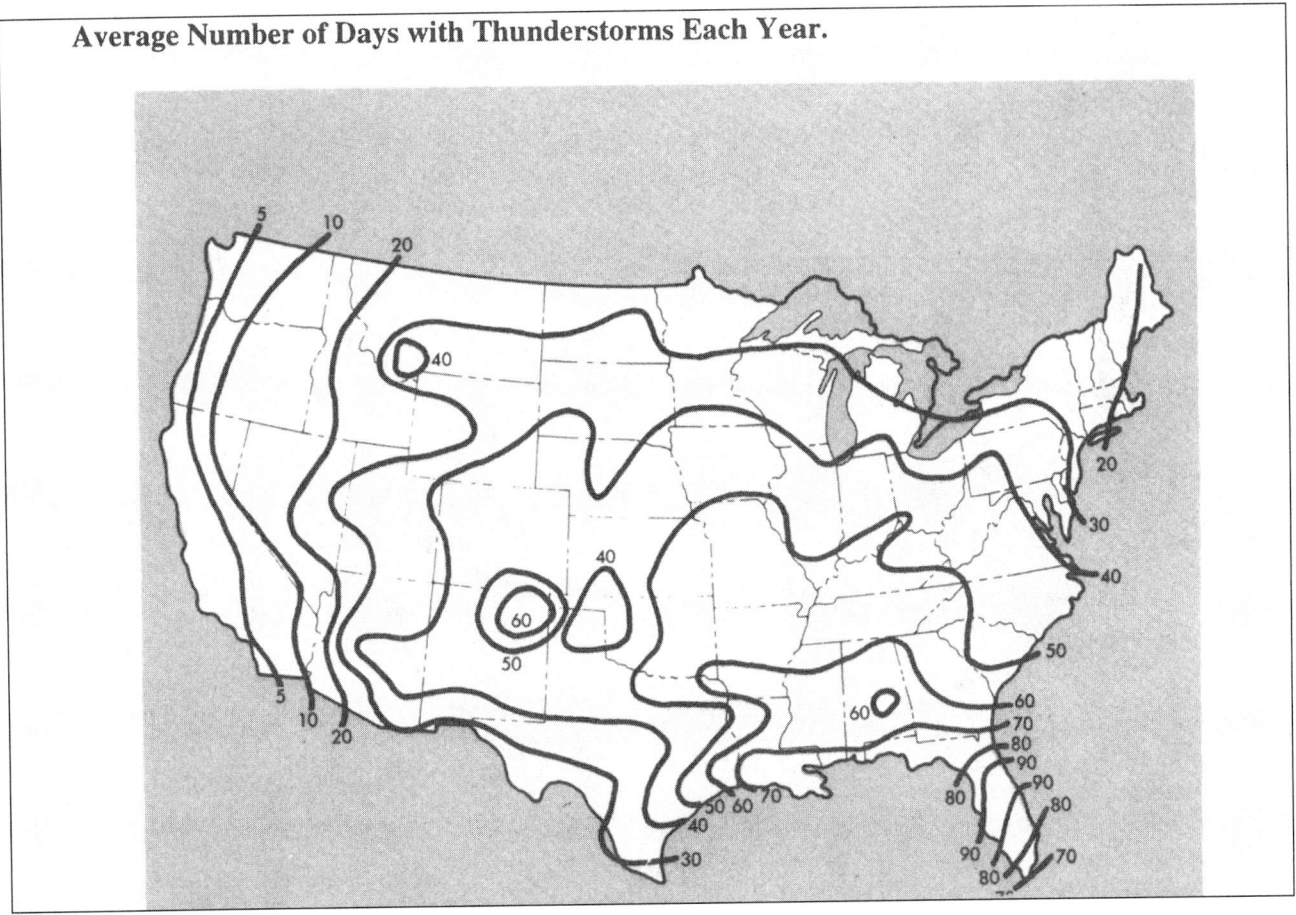

(Courtesy of United States Air Force.)

• LIGHTNING

Lightning strikes Earth an estimated 100 times per second. The average annual death toll for lightning is greater than for tornadoes or hurricanes.

According to data assembled by the National Oceanic and Atmospheric Administration (NOAA), lightning kills about 90 Americans per year and injures about 260. Property loss—fire and other damage to structures, aircraft damage, livestock deaths and injuries, forest fires, disruption of electromagnetic transmissions, and other effects—is estimated at more than $100 million annually.

What causes lightning?

Lightning is a secondary effect of electrification within a thunderstorm cloud system. Updrafts of warm, moist air rising into cold air can cause small cumulus clouds to grow into the large cumulonimbus cloud systems we associate with thunderstorms. These turbulent cloud systems tower about their companions, and dominate the atmospheric circulation and electrical field over a wide area. The transition from a small cloud to a turbulent, electrified giant can occur in as little as 30 minutes.

As a thunderstorm cumulonimbus develops, interactions of charged particles, external and internal electrical fields, and complex energy exchanges produce a large electrical field within the cloud. No completely acceptable theory explaining the complex processes of thunderstorm electrification has yet been advanced. But it is believed that electrical charge is important to formation of raindrops and ice crystals, and that thunderstorm electrification closely follows precipitation.

The distribution of electricity in a thunderstorm cloud is usually a concentration of positive charge in the frozen upper layers, and a large negative charge around a positive area in the lower portions of the cloud.

Earth is normally negatively charged with respect to the atmosphere. As the thunderstorm passes over the ground, the negative charge in the base of the cloud induces a positive charge on the ground below and several miles around the storm. The ground charge follows the storm like an electrical shadow, growing stronger as the negative cloud charge increases. The attraction between positive and negative charges makes the positive ground current flow up buildings, trees, and other elevated objects in an effort to establish a flow of current. But air, which is a poor conductor of

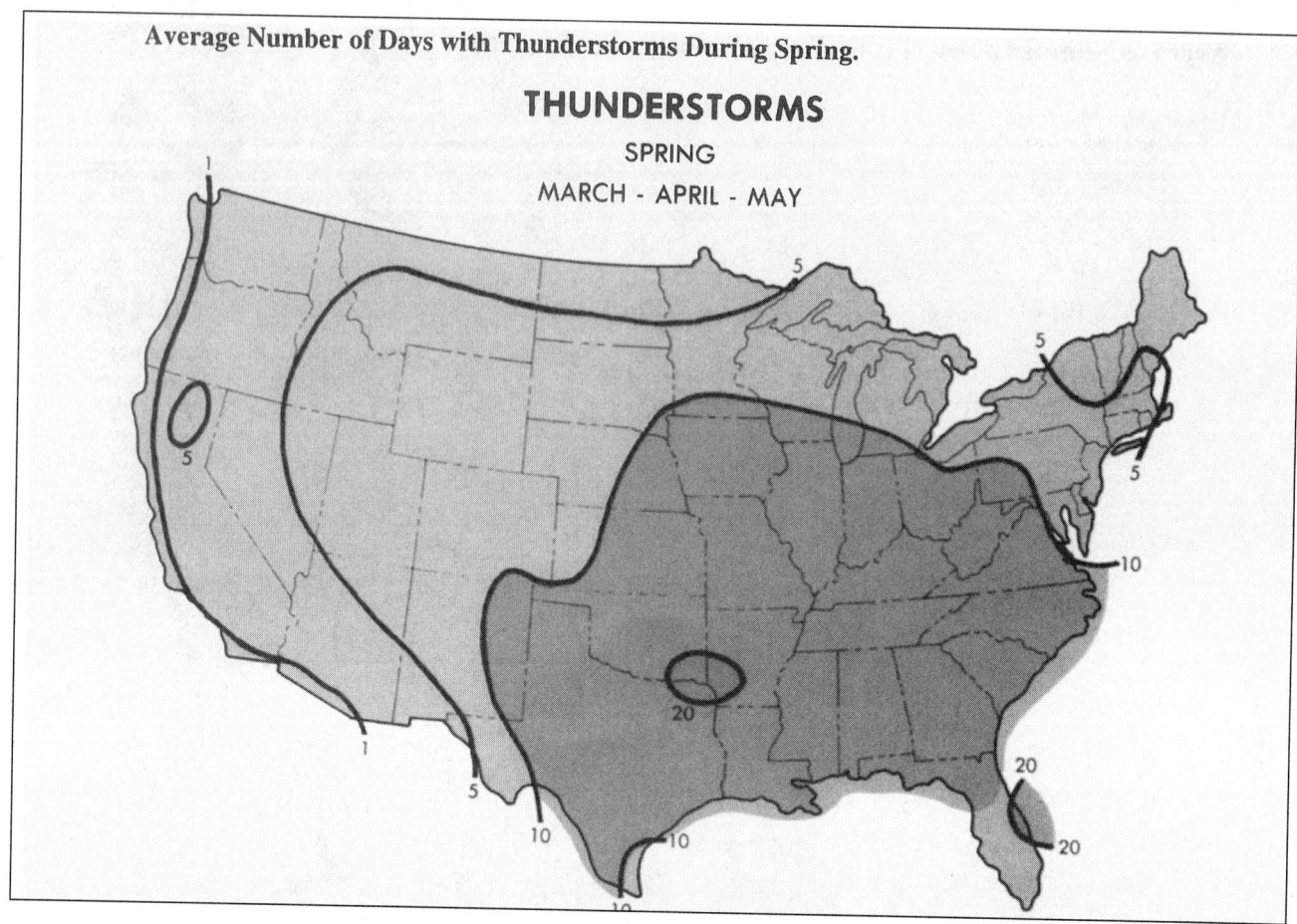

Average Number of Days with Thunderstorms During Spring.

THUNDERSTORMS

SPRING
MARCH - APRIL - MAY

(Courtesy of United States Air Force.)

electricity, insulates the cloud and ground charges, preventing a flow of current until large electrical charges are built up.

Lightning occurs when the difference between the positive and negative charges—the electrical potential—becomes great enough to overcome the resistance of the insulating air, and to force a conductive path for current to flow between the two charges. Potential in these cases can be as much as 100 million volts. Lightning strokes typically represent a flow of current from negative to positive, intra-cloud, and may proceed from cloud to cloud, cloud to air, cloud to ground, or, where high structures are involved, from ground to cloud.

The typical cloud-to-ground stroke we see most frequently begins as a pilot leader too faint to be visible, advances downward from the cloud, and sets up the initial portion of the stroke path. A surge of current called a step leader follows the pilot, moving 100 ft or more at a time toward the ground, pausing, then repeating the sequence until the conductive path of electrified (ionized) particles is near the ground. There, discharge streamers extending from the ground intercept the leader path and complete the conductive channel between ground and cloud charges. When this path is complete, a return stroke leaps upward at speeds approaching that of light, illuminating the branches of the de-

scending leader track. Because these tracks point downward, the stroke appears to come from the cloud. The bright light of the return stroke is the result of glowing atoms and molecules of air energized by the stroke.

Once the channel has been established and the return stroke has ended, dart leaders from the cloud initiate secondary returns, until the opposing charges are dissipated or the channel is gradually broken up by air movement. Even when luminous lightning is not visible, current may continue to flow along the ionized channel set up by the initial step leader.

Ground-to-cloud discharges are less frequently observed than the familiar cloud-to-ground stroke. In these cases, step leaders generally proceed from a tall conductive or semi-conductive structure to the clouds; the initial leader stroke is not followed by a return stroke from the cloud, possibly because charges are less mobile in the cloud than in the highly conducting Earth. Once the conductive path is established, however, current flow may set up cloud-to-ground sequences of dart leaders and returns.

Types of lightning

Lightning comes in many forms. *Streak lightning*, a single or multiple line from cloud to ground, is the form seen

Average Number of Days with Thunderstorms During Summer.

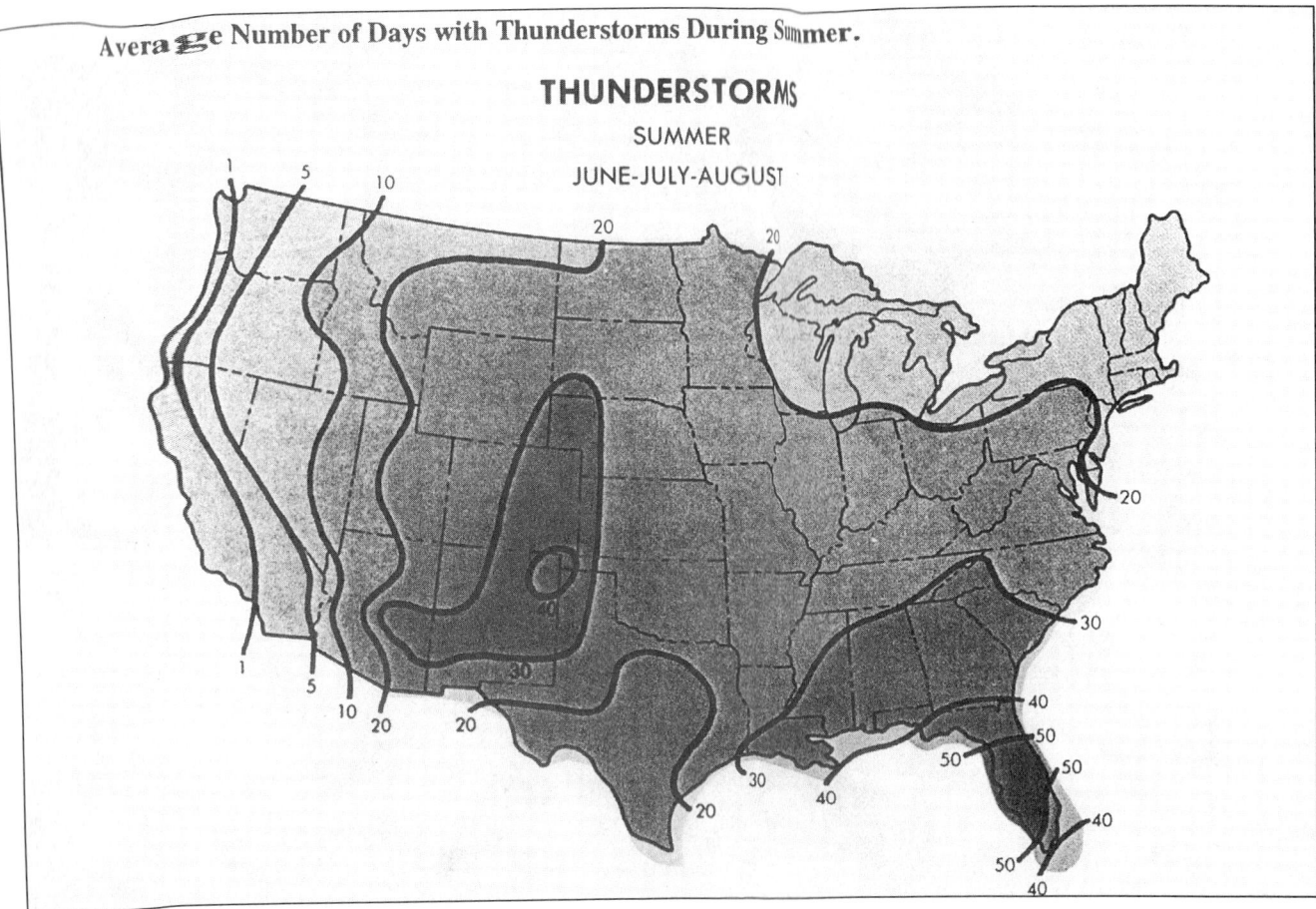

(Courtesy of United States Air Force.)

most frequently. *Forked lightning* shows the conductive channel. *Sheet lightning* is a shapeless flash covering a broad area, often seen in cloud-to-cloud discharges. *Heat lightning* is seen along the horizon during hot weather, and is believed to be the reflection of lightning occurring beyond the horizon. *Ribbon lightning* is streak lightning whose conductive channel is moved by high winds, making successive strokes seem to parallel one another. *Beaded lightning* appears as an interrupted stroke.

Ball lightning is in some ways the most interesting—and most controversial—form. Ball lightning has been reported in various shapes—from a luminous globe to a doughnut-shaped toroid to an ellipsoid. It hisses as it hurtles from cloud to Earth, maneuvers at high speeds, rolls along structures, or hangs suspended in the air.

The dual character of lightning—it carries high currents and produces destructive thermal effects—makes it doubly dangerous. The current peaks, which may reach magnitudes of 200,000 amperes or more, produce forces that have a crushing effect upon conductors, and which can build to explosive levels in non-conducting or semi-conducting materials like wood or brick. The continuous current produces heat, and is responsible for the numerous fires attributed to lightning. The peak temperature of lightning is greater than

50,000°F, about five times hotter than the visible surface of the Sun.

Lightning research

At the NOAA, lightning is the subject of considerable scientific interest. The severe storm warnings of NOAA's National Weather Service (NWS) carry implicit alerts that lightning can be expected—and avoided. U.S. Department of Commerce scientists at NOAA's Environmental Research Laboratories are experimenting with lightning suppression techniques, measuring atmospheric electricity over the open ocean, and studying the apparent but elusive connections between lightning and other events in the atmosphere, ionosphere, Earth, and geomagnetic field.

• THUNDER

Thunder is the crash and rumble associated with lightning and is caused by an explosive expansion of air heated by the stroke. When lightning is close by, its thunder makes a sharp explosive sound. More distant strokes produce the familiar growl and rumble of thunder, a result of sound be-

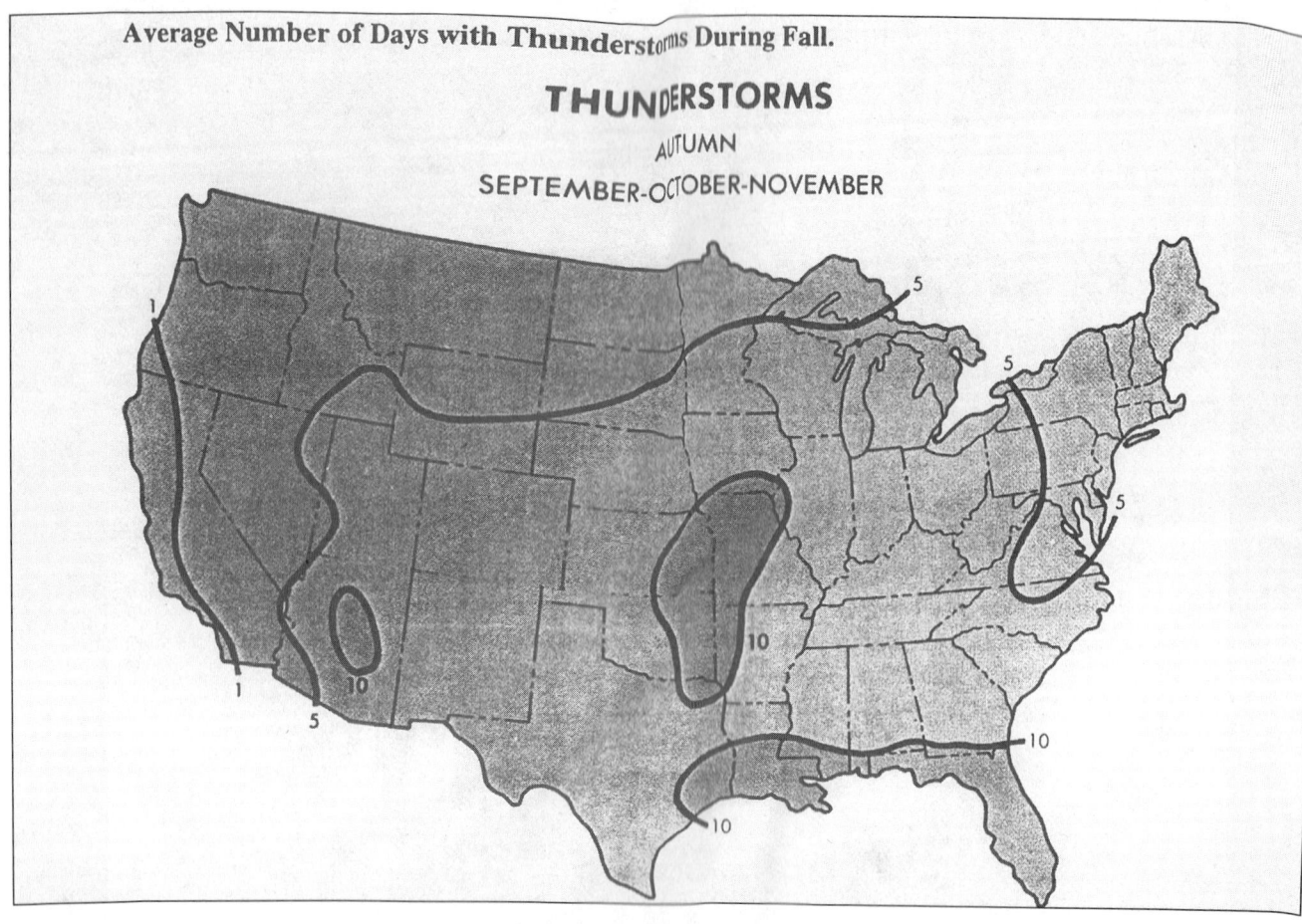

Average Number of Days with Thunderstorms During Fall.

THUNDERSTORMS
AUTUMN
SEPTEMBER-OCTOBER-NOVEMBER

(Courtesy of United States Air Force.)

Lightning Safety Rules

1. Stay indoors, and do not venture outside unless absolutely necessary.

2. Stay away from open doors and windows, fireplaces, radiators, stoves, metal pipes, sinks, and plugged-in electrical appliances.

3. Do not use plug-in electrical equipment such as hair dryers, electric toothbrushes, or electric razors during the storm.

4. Do not use the telephone during the storm—lightning may strike telephone lines outside.

5. Do not take laundry off the clothesline during the storm.

6. Do not work on fences, telephone or power lines, pipelines, or structural steel fabrication.

7. Do not use metal objects like fishing rods and golf clubs. Golfers wearing cleated shoes are particularly good lightning rods.

8. Do not handle flammable materials in open containers.

9. Stop tractor work, especially when the tractor is pulling metal equipment, and dismount. Tractors and other implements in metallic contact with the ground are often struck by lightning.

10. Get off the water and out of small boats.

11. Stay in your vehicle if you are traveling. Vehicles offer excellent lightning protection; however, avoid parking near large trees or power lines.

12. Seek shelter in buildings. If no buildings are available, your best protection is a cave, ditch, or canyon. If in a wooded area, take shelter under the shorter trees.

13. When there is no shelter, avoid the highest object in the area. If only isolated trees are nearby, your best protection is to crouch in the open, keeping twice as far away from isolated trees as the trees can act as lightning rods.

14. Avoid hilltops, open spaces, wire fences, metal clotheslines, exposed sheds, and any electrically conductive elevated objects.

15. When you feel the electrical charge—if your hair stands on end or your skin tingles—lightning may be about to strike you. Drop to the ground immediately.

Persons struck by lightning receive a severe electrical shock and may be burned, but they carry no electrical charge and can be handled safely. A person struck by lightning who shows no vital signs can often be revived by prompt mouth-to-mouth resuscitation, cardiac massage, and prolonged artificial respiration. In a group struck by lightning, the apparently dead should be treated first; those who show vital signs will probably recover spontaneously, although burns and other injuries may require treatment. Recovery from lightning strikes is usually complete except for possible impairment or loss of sight or hearing.

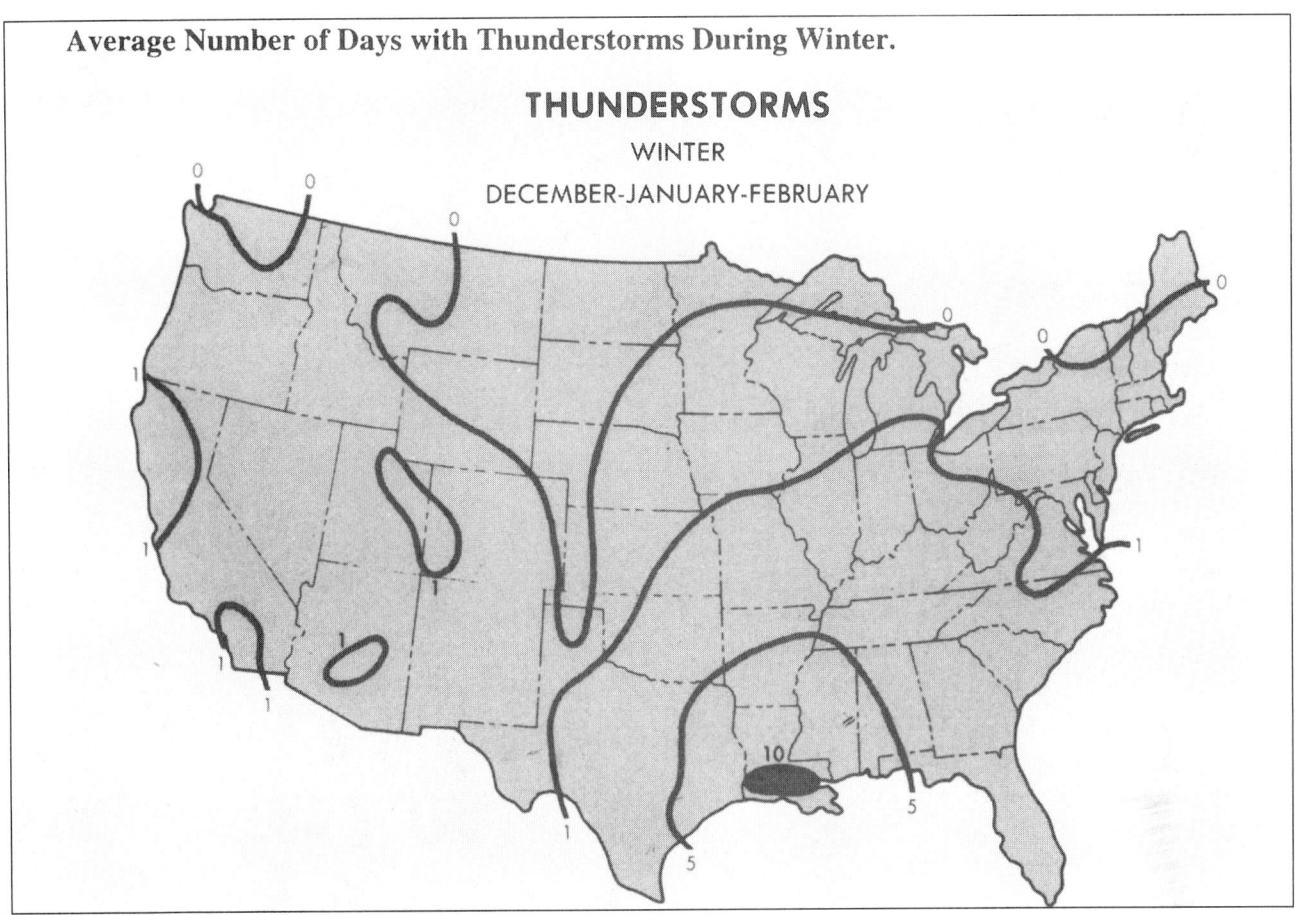

Average Number of Days with Thunderstorms During Winter.

THUNDERSTORMS
WINTER
DECEMBER-JANUARY-FEBRUARY

(Courtesy of United States Air Force.)

ing refracted and modified by the turbulent environment of a thunderstorm. Because the speed of light is about a million times that of sound, the distance (in miles) to a lightning's stroke can be estimated by counting the number of seconds between lightning and thunder, and dividing by five.

The electromagnetic impulses of a lightning stroke produce whistlers—gliding tones that travel along lines of force in Earth's magnetic field from their lightning source in one hemisphere to a similar point in the opposite hemisphere, often echoing back and forth several times. Their sound is something like the whistle of World War II bombs, occasionally modified in a way that produces musical variations.

Hail

Hailstones are precipitation in the form of lumps of ice that form during some thunderstorms. Hail can range in size from that of a pea to a softball. Hailstones are usually round, but may also be conical or irregular in shape, some with pointed projections. While it takes about one million cloud droplets to form a single raindrop, it takes about 10 billion cloud droplets to form a golf-ball-size hailstone.

Hail is formed as ice pellets (which were initially snowflakes or frozen raindrops) strike supercooled water droplets within a storm cloud. The supercooled water flows over the ice particles and part of it freezes instantly. Some of the unfrozen water remains attached to the growing hailstone until it freezes, and part of it slips away. This continues until the weight of the hailstone can no longer be supported by the updrafts, and it falls to the ground. The multiple trips through up- and downdrafts result in alternating bands of clear and cloudy ice within a hailstone; as many as 25 layers have been counted in one hailstone.

Of the thousands of thunderstorms that strike the United States each year, only about 10–15% produce potentially dangerous hailstones. Hail-producing thunderstorms are most frequently found in eastern Colorado, Nebraska, and Wyoming (the city of Cheyenne, Wyoming, observes the most hailstorm days per year, about 8–10); such storms also develop in the western plains, the Midwest, and the Ohio Valley.

Damage
Damage estimates from hailstorms alone reach up to nearly a billion dollars annually in the United States. The most costly single U.S. hailstorm struck on July 11, 1990, in Colorado Springs, Colorado, and resulted in damages of $625 million. Golf-ball- and baseball-size hailstones

pelted thousands of roofs, vehicles, windows, and other property.

Hail also causes injuries, but rarely death. In fact, during the twentieth century, only three deaths were reportedly due to hail—one was a farmer in Lubbock, Texas, in 1930; an infant in Fort Collins, Colorado, July 30, 1979, and a 19-year old man struck by softball-size hail in Lake Worth, TX, on March 28, 2000. Injuries are also sparse, but more common. In the last full year of statistical hail data (2001) there were 32 hail injuries reported in the United States.

Lightning myths and truths

MYTH: If it is not raining, then there is no danger from lightning.

TRUTH: Lightning often strikes outside of heavy rain and may occur as far as 10 miles away from any rainfall. This is especially true in the western United States where thunderstorms sometimes produce very little rain.

MYTH: The rubber soles of shoes or rubber tires on a car will protect you from being struck by lightning.

TRUTH: Rubber-soled shoes and rubber tires provide NO protection from lightning. The steel frame of a hard-topped vehicle provides increased protection if you are not touching metal. Although you may be injured if lightning strikes your car, you are much safer inside a vehicle than outside.

MYTH: People struck by lightning carry an electrical charge and should not be touched.

TRUTH: Lightning-strike victims carry no electrical charge and should be attended to immediately. Contact your local American Red Cross chapter for information on CPR and first aid classes.

MYTH: "Heat lightning" occurs after very hot summer days and poses no threat.

TRUTH: "Heat lightning" is a term used to describe lightning from a thunderstorm too far away for thunder to be heard.

Lightning struck the spot where this woman was standing minutes after she left. *(Photo courtesy of National Oceanic and Atmospheric Administration (NOAA).)*

Weather Anecdotes

Can lightning hit twice, from the same storm, at the same location? Yes. On July 19, 1993, at about 4:30 P.M., a mile and a half south of Winnsboro, SC, four people sought shelter during a thunderstorm. Lightning struck a nearby tree showering the car with debris. The storm weakened and the driver got out to clean the top of the car, saying "lightning never strikes twice in the same spot." As he touched the car handle to enter the car, lightning struck again. He was appeared to be dead, but people rushed over, administered CPR, and brought him back to life. The lightning blew the soles off his wife's shoes as she stood by the side of the car, but she was not seriously injured.

How rare was it that lightning would strike the same low ground location twice from the same storm and about the same time? A 50-foot structure will be struck about once every four to six years, and a quarter acre of flat land (a large residential lot) will be struck about once every 100 years or more. Any structure, no matter what its size, may be struck by lightning.

The moral of this anecdote is to stay indoors until you are sure the storm has passed. You can hear thunder when the storm is within 10 miles or so, and are at risk if you can hear thunder.

June 17, 1987 ... Harrison Co., MS. A couple was walking on a beach. Two Air Force sergeants were eating lunch and saw the lightning strike the two people. They ran to the man and woman and found the man conscious, but his wife was not breathing. One of the sergeants began CPR while the other called the police. She survived thanks to the two men.

May 31, 1998 ... WI. During the early morning hours, south central and southeast WI experienced an unprecedented and widespread downburst wind event known as a "derecho." Incredibly powerful, hurricane-force, straight-line winds, with peak gusts of 100–128 mph tore through 12 counties, and another eight counties had peak gusts to 60–80 mph. Newspaper headlines included: "Nature's Spring Cleaning," and "Traveling Back to the Dark Ages." A 48-year old Washington Co. woman was killed when a tree fell onto her house as she was sleeping. Another 32 people were injured during the storms.

July 28, 1999 ... Lake Powell, AZ. At least two fishing boats were swamped and other Lake Powell visitors had to take shelter when violent thunderstorms pounded the area, creating swells between six and eight feet high. No injuries were reported, but flash floods on the lake's shore roared through a campground and closed roads. In another emergency, three adults and four children on their way from Rainbow Bridge to Warm Creek were forced to beach their boat when waves made the going too treacherous. A nearby boat took them in. In another incident, a boat took in 10–15 people for the night when their boats proved too risky.

April 9, 1995 ... Cochise Co., AZ. A dust storm along Interstate 10 near Bowie caused the death of 10 people and injured another 20 people as vehicles piled into each other in near zero visibilities.

March 28, 2000 ... Tarrant Co., TX. A 19-year-old male was killed when struck by softball-size hail at Lake Worth while trying to move a new car. He died the following day from associated head injuries. He is only the third person directly killed by a hail strike since 1900.

May 23, 2001...Adams Co., CO. A strong, dry microburst swept a woman up into a swirl of dirt and carried her approximately 150 feet. The woman was in an open field, corralling a yearling horse, when the incident occurred. Fortunately, she received only minor injuries.

Standing under a tree is one of the most dangerous places to be during a thunderstorm. *(Photo by Johnny Autery. Reproduced by permission.)*

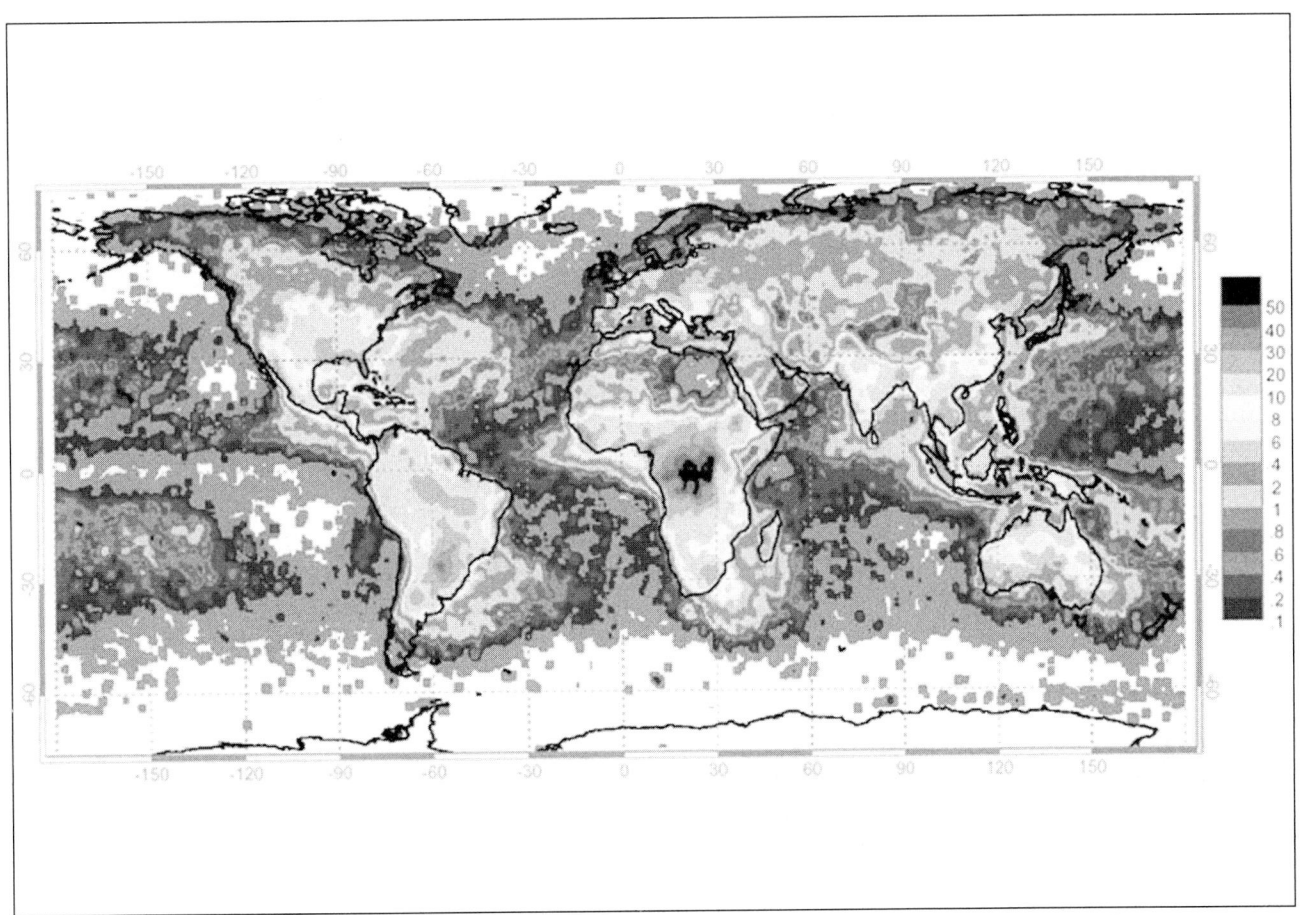

This map shows the incidence of lightning strikes annually around the world. *(Courtesy of National Oceanographic and Atmospheric Administration (NOAA).)*

National Total Lightning Injuries by Year for Period 1959–2001

Year	Jan	Feb	Mar	Apr	May	Jun	Jul	Aug	Sep	Oct	Nov	Dec	Ann
1959	0	0	0	5	27	52	110	103	23	3	1	1	325
1960	0	0	2	11	12	70	28	50	16	9	4	0	202
1961	0	0	7	14	15	49	83	50	31	5	1	1	256
1962	0	0	3	5	39	38	90	49	12	6	0	0	242
1963	7	0	0	6	14	64	5	44	18	1	0	0	209
1964	0	0	10	15	14	38	99	53	8	1	1	0	239
1965	3	2	2	4	26	42	59	59	19	1	0	0	217
1966	0	2	1	2	37	39	42	44	15	1	0	0	183
1967	0	0	0	4	7	35	59	33	4	2	0	1	145
1968	0	0	4	2	16	52	117	155	14	9	1	0	370
1969	0	0	0	4	19	75	39	23	12	0	0	1	173
1970	0	0	1	5	40	40	82	43	43	4	1	0	259
1971	0	1	0	1	24	71	79	54	22	1	1	0	254
1972	0	0	8	6	12	24	72	54	24	2	1	0	203
1973	0	0	10	2	20	23	74	59	29	9	2	0	228
1974	1	9	1	3	12	27	56	51	12	1	0	0	173
1975	0	3	0	1	30	60	107	154	42	1	0	1	399
1976	0	1	0	7	16	39	73	68	13	1	0	1	219
1977	0	0	0	3	35	58	58	67	62	4	4	0	291
1978	0	0	5	3	19	100	73	54	42	5	0	0	301
1979	0	2	4	26	32	73	55	49	9	2	2	0	254
1980	0	1	2	11	11	49	50	134	16	1	0	0	275
1981	1	0	2	9	34	60	108	52	9	3	13	0	291
1982	1	0	2	6	38	20	54	32	11	4	4	2	174
1983	0	0	24	3	25	24	87	113	30	31	0	0	337
1984	0	0	7	5	13	43	80	53	44	7	1	0	253
1985	0	0	29	4	42	48	61	33	27	4	0	0	248
1986	0	2	4	2	15	68	112	43	22	3	0	0	271
1987	0	0	2	8	66	49	121	70	43	3	1	1	364
1988	0	0	1	14	22	53	133	63	19	5	1	0	311
1989	1	0	8	16	23	70	135	51	12	4	2	0	322
1990	12	0	4	6	10	43	88	62	25	1	0	1	252
1991	0	0	2	30	48	111	94	107	37	3	0	0	432
1992	0	0	4	10	41	38	80	46	31	42	0	0	292
1993	1	0	4	4	15	75	103	65	20	5	0	3	295
1994	1	7	6	32	39	151	156	106	55	19	4	1	577
1995	4	4	10	18	33	81	229	102	26	0	3	0	510
1996	2	3	4	12	36	102	66	49	21	6	6	2	309
1997	3	12	7	8	8	69	109	48	27	11	4	0	306
1998	1	3	9	24	28	81	69	45	16	4	2	3	285
1999	3	0	3	8	34	42	67	67	15	3	1	0	243
2000	0	104	12	12	56	180	132	140	68	24	0	0	728
2001	9	0	6	10	45	86	105	77	24	7	1	0	370
Total	50	156	210	381	1148	2612	3749	2874	1068	258	62	19	12587
Mean	1	4	5	9	27	61	87	67	25	6	1	0	293

(Courtesy of National Climatic Data Center/NOAA.)

Total Lightning Injuries by State and Nation for Year 2001

State	Jan	Feb	Mar	Apr	May	Jun	Jul	Aug	Sep	Oct	Nov	Dec	Ann
Alabama	0	0	0	1	3	1	3	4	7	0	1	0	20
Alaska	0	0	0	0	0	0	0	0	0	0	0	0	0
Arizona	0	0	0	0	0	0	1	0	0	0	0	0	1
Arkansas	0	0	0	0	0	0	3	0	1	0	0	0	4
California	1	0	0	0	0	0	1	0	2	0	0	0	4
Colorado	0	0	0	1	5	5	3	5	0	0	0	0	19
Connecticut	0	0	0	0	0	0	2	2	0	0	0	0	4
Delaware	0	0	0	0	0	0	1	2	0	0	0	0	3
District of Columbia	0	0	0	0	0	0	0	0	0	0	0	0	0
Florida	0	0	4	0	3	34	20	9	4	1	0	0	75
Georgia	1	0	1	0	11	1	3	0	1	0	0	0	18
Hawaii	0	0	0	0	0	0	0	0	0	0	0	0	0
Idaho	0	0	0	0	0	0	0	0	0	0	0	0	0
Illinois	0	0	0	0	0	0	1	0	0	0	0	0	1
Indiana	0	0	0	0	0	1	1	2	2	0	0	0	6
Iowa	0	0	0	3	0	0	1	0	0	0	0	0	4
Kansas	0	0	0	0	0	0	0	0	0	0	0	0	0
Kentucky	0	0	0	0	0	0	1	0	0	0	0	0	1
Louisiana	0	0	0	0	0	0	0	0	1	0	0	0	1
Maine	0	0	0	0	0	1	8	0	0	0	0	0	9
Maryland	0	0	0	0	0	2	1	2	0	0	0	0	5
Massachusetts	0	0	0	0	2	0	0	0	0	0	0	0	2
Michigan	0	0	0	0	0	0	0	0	1	1	0	0	2
Minnesota	0	0	0	0	0	0	25	0	0	2	0	0	27
Mississippi	0	0	0	0	1	3	6	0	1	0	0	0	11
Missouri	0	0	0	0	0	0	0	0	0	0	0	0	0
Montana	0	0	0	0	3	0	0	0	0	0	0	0	3
Nebraska	0	0	0	0	0	0	1	0	0	0	0	0	1
Nevada	0	0	0	0	0	0	0	0	0	0	0	0	0
New Hampshire	0	0	0	0	1	0	0	0	0	0	0	0	1
New Jersey	0	0	0	0	2	1	2	14	0	0	0	0	19
New Mexico	0	0	0	0	0	0	1	0	0	0	0	0	1
New York	0	0	0	0	5	15	5	2	3	1	0	0	31
North Carolina	2	0	0	0	1	1	0	4	0	0	0	0	8
North Dakota	0	0	0	0	0	0	0	0	0	0	0	0	0
Ohio	0	0	1	0	1	2	1	0	0	0	0	0	5
Oklahoma	5	0	0	0	3	0	0	0	1	1	0	0	10
Oregon	0	0	0	0	0	0	0	0	0	0	0	0	0
Pennsylvania	0	0	0	0	0	9	0	1	0	0	0	0	10
Puerto Rico	0	0	0	0	0	0	0	0	0	0	0	0	0
Rhode Island	0	0	0	0	0	0	4	0	0	0	0	0	4
South Carolina	0	0	0	1	0	1	0	4	0	0	0	0	6
South Dakota	0	0	0	1	0	3	0	0	0	0	0	0	4
Tennessee	0	0	0	0	0	0	1	13	0	1	0	0	15
Texas	0	0	0	0	4	0	0	2	0	0	0	0	6
Utah	0	0	0	0	0	1	0	0	0	0	0	0	1
Vermont	0	0	0	0	0	0	0	0	0	0	0	0	0
Virginia	0	0	0	0	0	1	3	2	0	0	0	0	6
Washington	0	0	0	0	0	0	0	0	0	0	0	0	0
West Virginia	0	0	0	0	0	0	0	9	0	0	0	0	9
Wisconsin	0	0	0	3	0	0	4	0	0	0	0	0	7
Wyoming	0	0	0	0	0	4	2	0	0	0	0	0	6
Total: United States	9	0	6	10	45	86	105	77	24	7	1	0	370

(Courtesy of National Climatic Data Center/NOAA.)

Lightning Injuries by State, Rank, and Location of Occurrence

Column key (each location has a No. and % column):
OF = Open Fields, Ball Parks, and Open Spaces · UT = Under Trees · BF = Boating, Fishing and Water Related · NT = Near Tractors, Heavy Road Equipment · GC = Golf Courses · TP = At Telephones · VO = Various Other and Unknown Locations

State	Rank	OF No.	OF %	UT No.	UT %	BF No.	BF %	NT No.	NT %	GC No.	GC %	TP No.	TP %	VO No.	VO %	OF No.	OF %	UT No.	UT %	BF No.	BF %	NT No.	NT %	GC No.	GC %	TP No.	TP %	VO No.	VO %
		1959–2001														**2001**													
Alabama	15	81	29	66	23	5	2	8	3	1	0	15	5	107	38	8	40	5	25	0	0	1	5	0	0	1	5	5	25
Alaska	52	1	50	0	0	0	0	0	0	0	0	0	0	1	50	0	0	0	0	0	0	0	0	0	0	0	0	0	0
Arizona	26	97	51	17	9	4	2	11	6	1	1	4	2	55	29	0	0	0	0	0	0	0	0	0	0	0	0	1	100
Arkansas	16	78	28	40	14	15	5	11	4	4	1	16	6	119	42	3	75	0	0	0	0	0	0	0	0	1	25	0	0
California	36	35	35	16	16	8	8	3	3	0	0	2	2	35	35	3	75	0	0	0	0	0	0	0	0	0	0	1	25
Colorado	9	173	39	55	12	25	6	18	4	30	7	12	3	132	30	2	11	6	32	0	0	1	5	5	26	1	5	3	16
Connecticut	31	11	9	25	21	4	3	0	0	3	3	5	4	72	60	3	75	0	0	0	0	0	0	0	0	1	25	0	0
Delaware	46	9	28	10	31	0	0	0	0	0	0	0	0	10	31	0	0	0	0	0	0	0	0	0	0	0	0	3	100
Dist. of Columbia	44	22	67	6	18	0	0	1	3	1	3	0	0	3	9	0	0	0	0	0	0	0	0	0	0	0	0	0	0
Florida	1	474	29	164	10	210	13	73	5	53	3	38	2	598	37	15	20	18	24	0	0	0	0	0	0	0	0	26	35
Georgia	6	237	45	62	12	37	7	18	3	23	4	9	2	143	27	5	28	10	56	0	0	1	6	0	0	0	0	2	11
Hawaii	50	3	33	0	0	0	0	0	0	0	0	2	22	4	44	0	0	0	0	0	0	0	0	0	0	0	0	0	0
Idaho	38	16	18	7	8	4	5	3	3	2	2	4	5	52	59	0	0	0	0	0	0	0	0	0	0	0	0	0	0
Illinois	12	105	34	71	23	44	14	10	3	17	6	10	3	89	29	1	100	0	0	0	0	0	0	0	0	0	0	0	0
Indiana	27	30	16	33	18	16	9	11	6	9	5	7	4	82	44	0	0	0	0	0	0	0	0	0	0	0	0	6	100
Iowa	28	31	18	21	12	5	3	1	1	3	2	3	2	113	64	1	25	0	0	0	0	0	0	0	0	0	0	3	75
Kansas	24	29	15	14	7	6	3	9	5	10	5	8	4	117	61	0	0	0	0	0	0	0	0	0	0	0	0	0	0
Kentucky	21	57	27	24	11	9	4	4	2	14	7	9	4	97	45	0	0	0	0	0	0	0	0	0	0	0	0	1	100
Louisiana	14	108	37	41	14	26	9	9	3	1	1	2	1	102	35	0	0	0	0	0	0	0	0	0	0	0	0	1	100
Maine	30	17	10	60	37	4	2	1	1	2	1	1	1	77	47	0	0	5	56	0	0	0	0	0	0	0	0	3	33
Maryland	25	54	28	35	18	26	14	6	3	3	2	1	1	65	34	2	40	1	20	0	0	0	0	0	0	0	0	2	40
Massachusetts	11	81	21	20	5	13	3	5	1	12	3	9	2	249	64	0	0	0	0	0	0	0	0	0	0	0	0	1	100
Michigan	2	239	34	105	15	27	4	37	5	35	5	19	3	250	35	1	50	0	0	0	0	0	0	0	0	0	0	1	50
Minnesota	29	22	13	22	13	7	4	11	7	14	9	11	7	77	47	2	7	0	0	0	0	0	0	0	0	0	0	25	93
Mississippi	20	85	37	38	17	32	14	4	2	4	2	15	7	52	23	8	73	0	0	0	0	2	18	0	0	0	0	1	9
Missouri	32	40	35	25	22	3	3	2	2	4	4	3	3	37	32	0	0	0	0	0	0	0	0	0	0	0	0	0	0
Montana	42	15	29	6	12	8	16	5	10	4	8	0	0	13	25	0	0	0	0	0	0	0	0	0	0	0	0	0	0
Nebraska	39	26	33	2	3	1	1	5	6	6	8	5	6	34	43	0	0	0	0	0	0	3	100	0	0	0	0	0	0
Nevada	49	6	29	2	10	0	0	0	0	1	5	0	0	12	57	0	0	0	0	0	0	0	0	0	0	0	0	0	0
New Hampshire	37	23	23	3	3	0	0	1	1	4	4	3	3	65	66	0	0	0	0	0	0	0	0	0	0	0	0	1	100
New Jersey	23	71	36	29	15	9	5	1	1	7	4	6	3	76	38	5	26	2	11	1	5	0	0	0	0	0	0	11	58
New Mexico	22	103	52	26	13	3	2	5	3	11	6	1	1	50	25	0	0	0	0	0	0	0	0	0	0	0	0	1	100
New York	5	91	15	129	22	29	5	28	5	12	2	15	3	286	48	8	25	1	13	0	0	0	0	0	0	0	0	13	42
North Carolina	4	175	29	46	8	28	5	47	8	28	5	11	2	270	45	2	25	1	13	0	0	0	0	0	0	0	0	5	63
North Dakota	45	7	21	3	9	2	6	7	21	0	0	1	3	13	39	0	0	0	0	0	0	0	0	0	0	0	0	0	0
Ohio	7	110	21	93	18	18	3	5	1	43	8	13	3	232	45	2	40	0	0	0	0	0	0	0	0	1	20	2	40
Oklahoma	13	104	34	15	5	11	4	11	4	6	2	17	6	141	46	5	50	0	0	1	10	0	0	0	0	0	0	4	40
Oregon	47	10	40	0	0	0	0	1	4	0	0	2	4	13	52	0	0	0	0	0	0	0	0	0	0	0	0	0	0
Pennsylvania	3	207	33	51	8	6	1	10	2	14	2	9	1	324	52	1	10	9	90	0	0	0	0	0	0	0	0	0	0
Puerto Rico	51	1	14	1	14	0	0	0	0	0	0	0	0	5	71	0	0	0	0	0	0	0	0	0	0	0	0	0	0

(Courtesy of National Climatic Data Center/NOAA.)

Lightning Injuries by State, Rank, and Location of Occurrence [CONTINUED]

State	Rank	1959–2001														2001													
		Open Fields, Ball Parks, and Open Spaces		Under Trees		Boating, Fishing and Water Related		Near Tractors, Heavy Road Equipment		Golf Courses		At Telephones		Various Other and Unknown Locations		Open Fields, Ball Parks, and Open Spaces		Under Trees		Boating, Fishing and Water Related		Near Tractors, Heavy Road Equipment		Golf Courses		At Telephones		Various Other and Unknown Locations	
		No.	%	No.	%	No.	%	No.	%	No.	%	No.	%	No.	%	No.	%	No.	%	No.	%	No.	%	No.	%	No.	%	No.	%
Rhode Island	43	10	20	15	31	0	0	0	0	2	4	0	0	22	45	0	0	0	0	0	0	0	0	0	0	0	0	4	100
South Carolina	17	70	25	23	8	11	4	8	3	5	2	7	3	155	56	2	33	1	17	0	0	0	0	0	0	0	0	3	50
South Dakota	40	14	21	5	7	4	6	9	13	1	1	2	3	33	49	1	25	0	0	2	50	0	0	0	0	0	0	1	25
Tennessee	10	145	37	86	22	4	1	17	4	8	2	17	4	115	29	11	73	1	7	0	0	2	13	0	0	1	7	0	0
Texas	8	211	45	52	11	42	9	11	2	5	1	7	1	139	30	0	0	4	67	0	0	0	0	1	17	0	0	1	17
Utah	35	32	30	28	27	4	4	7	7	6	6	5	5	23	22	0	0	0	0	0	0	0	0	1	100	0	0	0	0
Vermont	48	7	30	1	4	0	0	0	0	0	0	0	0	15	65	0	0	0	0	0	0	0	0	0	0	0	0	0	0
Virginia	18	44	16	43	16	10	4	2	1	9	3	11	4	149	56	2	33	0	0	0	0	0	0	0	0	1	17	3	50
Washington	41	24	45	8	15	0	0	1	2	0	0	3	6	17	32	0	0	0	0	0	0	0	0	0	0	0	0	0	0
West Virginia	33	36	32	12	11	3	3	2	2	2	2	1	1	55	50	9	100	0	0	0	0	0	0	0	0	0	0	0	0
Wisconsin	19	96	37	8	3	8	3	7	3	10	4	6	2	126	48	7	100	0	0	0	0	0	0	0	0	0	0	0	0
Wyoming	34	46	43	3	3	23	21	13	12	4	4	0	0	19	18	2	33	0	0	0	0	4	67	0	0	0	0	0	0
United States	0	3819	30	1667	13	714	6	459	4	434	3	352	3	5140	41	119	32	65	18	22	6	14	4	8	2	7	2	135	36

(Courtesy of National Climatic Data Center/NOAA.)

National Total Lightning Deaths by Year for Period 1959–2001

Year	Jan	Feb	Mar	Apr	May	Jun	Jul	Aug	Sep	Oct	Nov	Dec	Ann
1959	1	0	1	4	18	25	50	39	13	7	0	0	158
1960	0	0	1	5	7	33	25	17	9	0	0	0	97
1961	0	0	1	2	9	23	47	20	10	1	0	0	113
1962	0	0	3	6	27	20	26	28	9	1	0	0	120
1963	0	0	4	3	11	37	42	20	10	2	0	81*	210
1964	0	0	9	6	15	21	29	19	7	1	1	0	108
1965	0	0	2	4	12	34	39	28	4	2	0	0	125
1966	0	0	1	1	8	15	21	16	11	3	0	0	76
1967	1	0	1	2	3	26	21	14	1	2	1	1	73
1968	0	0	0	1	5	24	30	29	9	3	1	1	103
1969	0	0	1	5	13	17	27	13	14	3	0	0	93
1970	0	0	0	1	17	25	27	19	21	1	0	0	111
1971	0	0	2	1	12	27	33	19	19	0	0	0	113
1972	0	0	1	1	5	21	31	28	3	1	0	0	91
1973	0	1	2	3	10	24	31	18	13	2	1	0	105
1974	0	2	0	7	12	21	28	24	6	0	2	0	102
1975	0	1	3	3	11	19	28	18	6	2	0	0	91
1976	0	0	0	1	9	19	19	19	3	2	0	0	72
1977	0	0	0	4	9	19	16	35	14	1	0	0	98
1978	0	0	1	1	9	26	24	22	3	1	0	1	88
1979	0	0	0	3	11	4	20	16	4	3	2	0	63
1980	0	0	0	0	7	16	27	20	5	1	0	0	76
1981	0	0	0	4	5	13	19	19	5	0	2	0	67
1982	1	0	0	3	5	14	29	18	4	3	0	0	77
1983	0	0	1	2	4	8	28	23	8	1	2	0	77
1984	0	0	1	3	10	14	20	10	7	1	1	0	67
1985	0	0	0	5	12	12	26	8	8	1	1	0	73
1986	0	0	0	2	9	13	21	17	5	1	0	0	68
1987	0	0	0	2	14	18	28	15	7	2	0	0	86
1988	0	0	0	3	9	17	21	14	2	1	2	0	69
1989	0	0	1	1	9	14	19	18	4	1	0	0	67
1990	1	0	3	1	3	18	22	15	10	0	0	1	74
1991	0	0	0	2	8	15	23	19	6	0	0	0	73
1992	0	0	0	2	6	6	9	10	8	0	0	0	41
1993	1	0	0	0	6	9	11	12	4	0	0	0	43
1994	0	2	2	3	7	24	17	8	10	1	0	0	74
1995	0	0	0	6	7	11	30	19	12	0	0	0	85
1996	1	0	0	4	3	18	8	13	4	0	0	1	52
1997	0	1	0	1	6	10	12	8	4	0	0	0	42
1998	0	1	2	2	3	12	9	12	2	0	0	1	44
1999	0	0	1	3	7	8	15	10	1	0	1	0	46
2000	0	0	8	0	2	26	30	24	7	4	0	0	101
2001	0	1	2	4	3	7	6	11	12	0	0	0	46
Total	6	9	54	117	378	783	1044	784	324	55	17	87	3658
Mean	0	0	1	3	9	18	24	18	8	1	0	2	85

*On December 8, 1963 the crash of a jetliner killing 81 people near Elkin, Maryland, was attributed to lightning by the Civil Aeronautics Board investigators.

(Courtesy of National Climatic Data Center/NOAA.)

Lightning Deaths by State, Rank, and Location of Occurrence

Column abbreviations — OF = Open Fields, Ball Parks, and Open Spaces; UT = Under Trees; BF = Boating, Fishing and Water Related; NT = Near Tractors, Heavy Road Equipment; GC = Golf Courses; Tel = At Telephones; VO = Various Other and Unknown Locations.

State	Rank	1959–2001 OF No.	%	UT No.	%	BF No.	%	NT No.	%	GC No.	%	Tel No.	%	VO No.	%	2001 OF No.	%	UT No.	%	BF No.	%	NT No.	%	GC No.	%	Tel No.	%	VO No.	%
Alabama	15	24	25	25	26	12	13	4	4	1	1	2	2	27	28	1	50	1	50	0	0	0	0	0	0	0	0	0	0
Alaska	52	0	0	0	0	0	0	0	0	0	0	0	0	0	0	0	0	0	0	0	0	0	0	0	0	0	0	0	0
Arizona	24	32	47	8	12	5	7	1	1	4	6	3	4	15	22	0	0	0	0	0	0	0	0	0	0	0	0	0	0
Arkansas	11	37	32	23	20	11	9	10	9	3	3	0	0	32	28	0	0	0	0	0	0	0	0	0	0	0	0	0	0
California	37	8	33	3	13	2	8	3	13	0	0	0	0	8	33	2	100	0	0	0	0	0	0	0	0	0	0	0	0
Colorado	10	58	49	22	18	5	4	5	4	7	6	0	0	22	18	1	33	1	33	0	0	0	0	0	0	0	0	1	33
Connecticut	41	3	19	2	13	0	0	0	0	3	19	0	0	8	50	0	0	0	0	0	0	0	0	0	0	0	0	0	0
Delaware	42	5	33	0	0	4	27	1	7	0	0	0	0	5	33	0	0	0	0	1	50	0	0	0	0	0	0	0	0
Dist. of Columbia	48	2	40	2	40	0	0	0	0	1	20	0	0	0	0	0	0	0	0	0	0	1	100	0	0	0	0	0	0
Florida	1	111	27	54	13	107	26	27	6	14	3	0	0	103	25	2	20	0	0	0	0	0	0	0	0	0	0	3	30
Georgia	16	25	26	23	24	14	15	7	7	5	5	2	2	19	20	0	0	0	0	0	0	0	0	0	0	0	0	0	0
Hawaii	53	0	0	0	0	0	0	0	0	0	0	0	0	0	0	0	0	0	0	0	0	0	0	0	0	0	0	0	0
Idaho	35	12	48	3	12	2	8	5	20	1	4	0	0	2	8	0	0	0	0	1	20	1	20	1	20	0	0	2	40
Illinois	13	23	24	19	20	4	4	9	9	8	8	1	1	32	33	0	0	0	0	0	0	0	0	1	100	0	0	0	0
Indiana	21	14	16	25	29	8	9	7	8	2	2	2	2	27	32	0	0	0	0	0	0	0	0	0	0	0	0	0	0
Iowa	23	9	13	9	13	3	4	7	10	2	3	0	0	40	57	0	0	0	0	0	0	0	0	0	0	0	0	0	0
Kansas	26	19	31	1	2	5	8	10	16	2	3	0	0	24	39	0	0	0	0	0	0	0	0	0	0	0	0	0	0
Kentucky	18	27	30	12	13	4	4	4	4	2	2	1	1	41	45	0	0	0	0	0	0	0	0	0	0	0	0	0	0
Louisiana	5	17	13	42	32	41	31	8	6	0	0	0	0	25	19	0	0	0	0	1	100	0	0	0	0	0	0	0	0
Maine	38	0	0	3	13	7	29	0	0	0	0	0	0	14	58	0	0	0	0	0	0	0	0	0	0	0	0	0	0
Maryland	8	13	10	9	7	16	13	1	1	1	1	2	2	84	67	0	0	0	0	0	0	0	0	0	0	0	0	0	0
Massachusetts	33	7	26	3	11	1	4	0	0	0	0	1	4	14	52	1	100	0	0	0	0	0	0	0	0	0	0	0	0
Michigan	12	29	29	26	26	12	12	3	3	10	10	2	2	17	17	1	100	0	0	0	0	0	0	0	0	0	0	0	0
Minnesota	28	19	32	15	25	5	8	6	10	2	3	2	3	10	17	0	0	0	0	0	0	0	0	0	0	0	0	0	0
Mississippi	14	30	31	22	23	14	15	6	6	0	0	0	0	24	25	0	0	0	0	0	0	0	0	0	0	0	0	0	0
Missouri	22	18	22	18	22	12	15	5	6	5	6	2	2	21	26	0	0	0	0	0	0	0	0	0	0	0	0	0	0
Montana	34	8	31	3	12	3	12	6	23	0	0	0	0	6	23	0	0	0	0	0	0	0	0	0	0	0	0	0	0
Nebraska	30	20	45	2	5	4	9	11	25	0	0	0	0	7	16	1	100	0	0	0	0	0	0	0	0	0	0	0	0
Nevada	47	0	0	0	0	1	17	0	0	1	17	0	0	4	67	0	0	0	0	0	0	0	0	0	0	0	0	0	0
New Hampshire	45	2	25	0	0	3	38	0	0	1	13	0	0	2	25	0	0	0	0	1	50	0	0	0	0	0	0	1	50
New Jersey	27	19	31	8	13	14	23	2	3	4	7	2	3	11	18	0	0	0	0	0	0	0	0	0	0	0	0	0	0
New Mexico	20	40	46	15	17	8	9	5	6	1	1	1	1	21	24	0	0	0	0	0	0	0	0	0	0	0	0	0	0
New York	6	22	17	31	23	17	13	5	4	7	5	1	1	50	38	0	0	0	0	0	0	0	0	0	0	0	0	0	0
North Carolina	3	43	24	25	14	23	13	6	3	8	5	1	1	71	40	0	0	0	0	0	0	0	0	0	0	0	0	0	0
North Dakota	44	2	17	0	0	0	0	4	33	0	0	0	0	6	50	0	0	1	50	0	0	0	0	0	0	0	0	1	50
Ohio	4	37	27	26	19	16	12	7	5	9	7	1	1	40	29	0	0	0	0	0	0	0	0	0	0	0	0	0	0
Oklahoma	17	33	35	11	12	17	18	7	7	3	3	2	2	22	23	0	0	0	0	0	0	0	0	0	0	0	0	0	0
Oregon	46	4	50	1	13	0	0	0	0	0	0	0	0	3	38	1	50	0	0	0	0	0	0	0	0	0	0	1	50
Pennsylvania	9	39	32	17	14	4	3	5	4	13	11	2	2	42	34	1	50	1	50	0	0	0	0	0	0	0	0	0	0
Puerto Rico	32	12	40	8	27	1	3	0	0	0	0	0	0	9	30	0	0	0	0	0	0	0	0	0	0	0	0	0	0

(Courtesy of National Climatic Data Center/NOAA.)

Lightning Deaths by State, Rank, and Location of Occurrence [CONTINUED]

State	Rank	1959–2001 Open Fields, Ball Parks, and Open Spaces No.	%	Under Trees No.	%	Boating, Fishing and Water Related No.	%	Near Tractors, Heavy Road Equipment No.	%	Golf Courses No.	%	At Telephones No.	%	Various Other and Unknown Locations No.	%	2001 Open Fields, Ball Parks, and Open Spaces No.	%	Under Trees No.	%	Boating, Fishing and Water Related No.	%	Near Tractors, Heavy Road Equipment No.	%	Golf Courses No.	%	At Telephones No.	%	Various Other and Unknown Locations No.	%
Rhode Island	50	0	0	0	0	1	25	0	0	0	0	0	0	3	75	0	0	0	0	0	0	0	0	0	0	0	0	0	0
South Carolina	19	17	19	22	24	10	11	10	11	2	2	4	4	25	28	0	0	0	0	0	0	0	0	0	0	0	0	1	100
South Dakota	40	7	32	1	5	3	14	8	36	1	5	1	5	2	9	1	50	0	0	1	50	0	0	0	0	0	0	0	0
Tennessee	7	35	27	33	25	9	7	12	9	7	5	2	2	32	25	0	0	1	100	0	0	0	0	0	0	0	0	0	0
Texas	2	78	40	29	15	30	15	12	6	5	3	0	0	40	21	0	0	0	0	0	0	0	0	0	0	0	0	1	100
Utah	31	21	48	11	25	2	5	2	5	1	2	1	2	6	14	0	0	0	0	0	0	0	0	0	0	0	0	0	0
Vermont	43	2	14	1	7	4	29	0	0	0	0	0	0	7	50	0	0	0	0	0	0	0	0	0	0	0	0	0	0
Virginia	25	15	24	18	29	7	11	5	8	2	3	0	0	16	25	0	0	0	0	1	33	2	67	0	0	0	0	0	0
Washington	49	3	60	1	20	0	0	1	20	0	0	0	0	0	0	0	0	0	0	0	0	0	0	0	0	0	0	0	0
West Virginia	36	7	28	6	24	2	8	1	4	1	4	0	0	8	32	0	0	0	0	0	0	0	0	0	0	0	0	0	0
Wisconsin	29	15	27	4	7	9	16	5	9	6	11	0	0	16	29	1	100	0	0	0	0	0	0	0	0	0	0	0	0
Wyoming	39	14	58	2	8	3	13	1	4	0	0	0	0	4	17	0	0	0	0	0	0	0	0	0	0	0	0	0	0
United States	0	1038	28	644	18	485	13	241	7	145	4	38	1	1067	29	13	28	5	11	10	22	5	11	2	4	0	0	11	24

*On December 8, 1963 the crash of a jetliner killing 81 people near Elton, Maryland, was attributed to lightning by the Civil Aeronautics investigators.

(Courtesy of National Climatic Data Center/NOAA.)

Tornadoes

• WHAT IS A TORNADO?

A tornado is a local storm of short duration (usually 5–10 minutes) formed of winds rotating at very high speeds, usually in a counter-clockwise direction (in the Northern Hemisphere). This storm is visible as a vortex—a whirlpool structure of winds rotating about a hollow cavity in which centrifugal forces produce a partial vacuum. As condensation occurs around the vortex, a pale cloud appears—the familiar and frightening tornado funnel. Funnels usually appear as an extension of the dark, heavy cumulonimbus clouds of thunderstorms, and stretch downward toward the ground. Some never reach the surface; others touch and rise again. Air surrounding the funnel is also part of the tornado vortex. As the storm moves along the ground, this outer ring of rotating winds becomes dark with dust and debris, which may eventually darken the entire funnel.

These small, severe storms form several thousand feet above Earth's surface, usually during warm, humid, unsettled weather, and usually in conjunction with a severe thunderstorm. Sometimes a series of two or more tornadoes is associated with a parent thunderstorm. As the thunderstorm moves, tornadoes may form at intervals along its path, travel for a few miles, then dissipate. The *forward* speed of tornadoes has been observed to range from almost no motion to 70 mph.

The winds of some tornadoes have been estimated to exceed 300 mph. (*Photo courtesy of NOAA Photo Library, NOAA Central Library; OAR/ERL/National Severe Storms Laboratory (NSSL).*)

• HOW A TORNADO IS FORMED

Tornado formation requires the presence of layers of air with contrasting characteristics of temperature, moisture, density, and wind flow. Complicated energy transformations produce the tornado vortex.

Many theories have been advanced as to the type of energy transformation necessary to generate a tornado, and none has won general acceptance. The two most frequently encountered theories visualize tornado generation as either the effect of thermally induced rotary circulations, or as the effect of converging rotary winds. Currently, scientists seem to agree that neither process generates tornadoes independently. It is more probable that tornadoes are produced by the combined effects of thermal and mechanical forces, with one or the other force being the stronger generating agent.

Numerous observations of lightning strokes and a variety of luminous features in and around tornado funnels have led scientists to speculate about the relationship between tornado formation and thunderstorm electrification. This hypothesis explores the alternative possibilities that atmospheric electricity accelerates rotary winds to tornado velocities, or that those high-speed rotary winds generate large electrical charges. Here, as in most attempts to understand complex atmospheric relationships, the reach of theory exceeds the grasp of proof.

• SIZE, SPEED, & DURATION

Tornadoes vary greatly in size, intensity, and appearance. Most (69%) of the tornadoes that occur each year fall into the "weak" category. Wind speeds are in the range of 110 mph or less. Weak tornadoes account for less than 5% of all tornado deaths.

About one out of every three tornadoes (29%) is classified as "strong." Strong tornadoes have wind speeds reaching about 205 mph, with an average path length of 9 mi, and a path width of 200 yd. Almost 30% of all tornado deaths occur each year from this type of storm. Nearly 70% of all tornado fatalities, however, result from "violent" tornadoes. Although very rare (only about 2% are violent), these extreme tornadoes can last for hours. Average path lengths and widths are 26 mi and 425 yd, respectively. The largest of these may exceed a mile or more in width, with wind speeds approaching 300 mph.

• WHERE DO TORNADOES OCCUR?

Tornadoes occur in many parts of the world and in all 50 states. But no area is more favorable to their formation than the continental plains of North America. The term "Tornado

---

Tornado Characteristics

Time of day during which tornadoes are most likely to occur is mid-afternoon, generally 3–7 P.M., but they have occurred at all times of day.

Direction of movement is usually from southwest to northeast. (Note: Tornadoes associated with hurricanes may move from an easterly direction.)

Length of path averages 4 mi, but may reach 300 mi. A tornado traveled 293 mi across Illinois and Indiana on May 26, 1917, and lasted seven hours and 20 minutes.

Width of path averages about 300–400 yd but tornadoes have cut paths a mile or more in width.

Speed of travel averages 25–40 mph, but speeds ranging from stationary to 68 mph have been reported.

The cloud directly associated with a tornado is a dark, heavy cumulonimbus (the familiar thunderstorm cloud) from which a whirling funnel-shaped pendant extends to the ground.

Precipitation associated with the tornado usually occurs first as rain just preceding the storm, frequently with hail, and as a heavy downpour immediately to the side of the tornado's path.

Sound occurring during a tornado has been described as a roaring, rushing noise, closely approximating that made by a train speeding through a tunnel or over a trestle, or the roar of many airplanes.

Tornado Intensity Rating System

The intensity of tornadoes is defined according to the Fujita Scale (or F scale), which ranges from F0 to F6 as outlined below.

F0: 40–72 mph winds. Damage is light and might include damage to tree branches, chimneys, and billboards. Shallow-rooted trees may be pushed over.

F1: 73–112 mph winds. Damage is moderate; mobile homes may be pushed off foundations and moving autos pushed off the road.

F2: 113–157 mph winds. Damage is considerable. Roofs can be torn off houses, mobile homes demolished, and large trees uprooted.

F3: 158–206 mph winds. Damage is severe. Even well-constructed houses may be torn apart, trees uprooted, and cars lifted off the ground.

F4: 207–260 mph winds. Damage is devastating. Houses can be leveled and cars thrown; objects become deadly missiles.

F5: 261–318 mph winds. Damage is incredible. Structures are lifted off foundations and carried away; cars become missiles. Fewer than 2% of all tornadoes reach an intensity of this magnitude.

F6: The maximum tornado wind speeds are not expected to exceed 318 mph.

Alley" is sometimes used to identify this section of the United States—a section in which more tornadoes strike than in any other place in the world. This broad band begins in Texas and covers many states northward including Oklahoma, Missouri, Kansas, and Ohio.

No season is free of these destructive storms. Normally, the number of tornadoes is at its lowest in the United States during December and January and at its (average) peak in May. The months of greatest total frequency are April, May, and June.

In February, when tornado frequency begins to increase, the center of maximum frequency lies over the central Gulf states. Then, during March, this center moves eastward to the southeast Atlantic states, where tornado frequency reaches a peak in April. During May, the center of maximum frequency moves to the southern plains states, and in June, northward to the northern plains and Great Lakes area, as far east as western New York state. The reason for this drift is the increasing penetration of warm, moist air while contrasting cool, dry air still surges in from the north and northwest; tornadoes are generated with greatest frequency where these air masses wage their wars. Thus, when the Gulf states are substantially "occupied" by warm air systems after May, there is no cold air intrusion to speak of, and

tornado frequency drops. This is the case across the nation after June, although in some states, a secondary tornado maximum occurs in the fall. Winter cooling permits fewer and fewer encounters between warm and overriding cold systems, and tornado frequency returns to its lowest level by December.

The mathematical chance that a specific location will be struck by a tornado in any one year is quite small. For example, the probability of a tornado striking a given point in the area most frequently subject to tornadoes is 0.0363, or about once in 250 years. In the far western states, the probability is close to zero.

But tornadoes have provided many un-mathematical exceptions. Oklahoma City, Oklahoma, has been struck by tornadoes 27 times since 1892. Baldwyn, Mississippi, was struck twice by tornadoes during a 25-minute period on March 16, 1942. One-third of Irving, Kansas, was left in ruins by two tornadoes that occurred 45 minutes apart on May 30, 1879. Austin, Texas, had two tornadoes in rapid succession on May 4, 1922; and Codell, Kansas, was struck three times in 1916, 1917, and 1918—all on May 20.

From 1953–99, an average of 847 tornadoes per year occurred in the United States, about half of them during three months of April, May, and June. For the 1953–97 period,

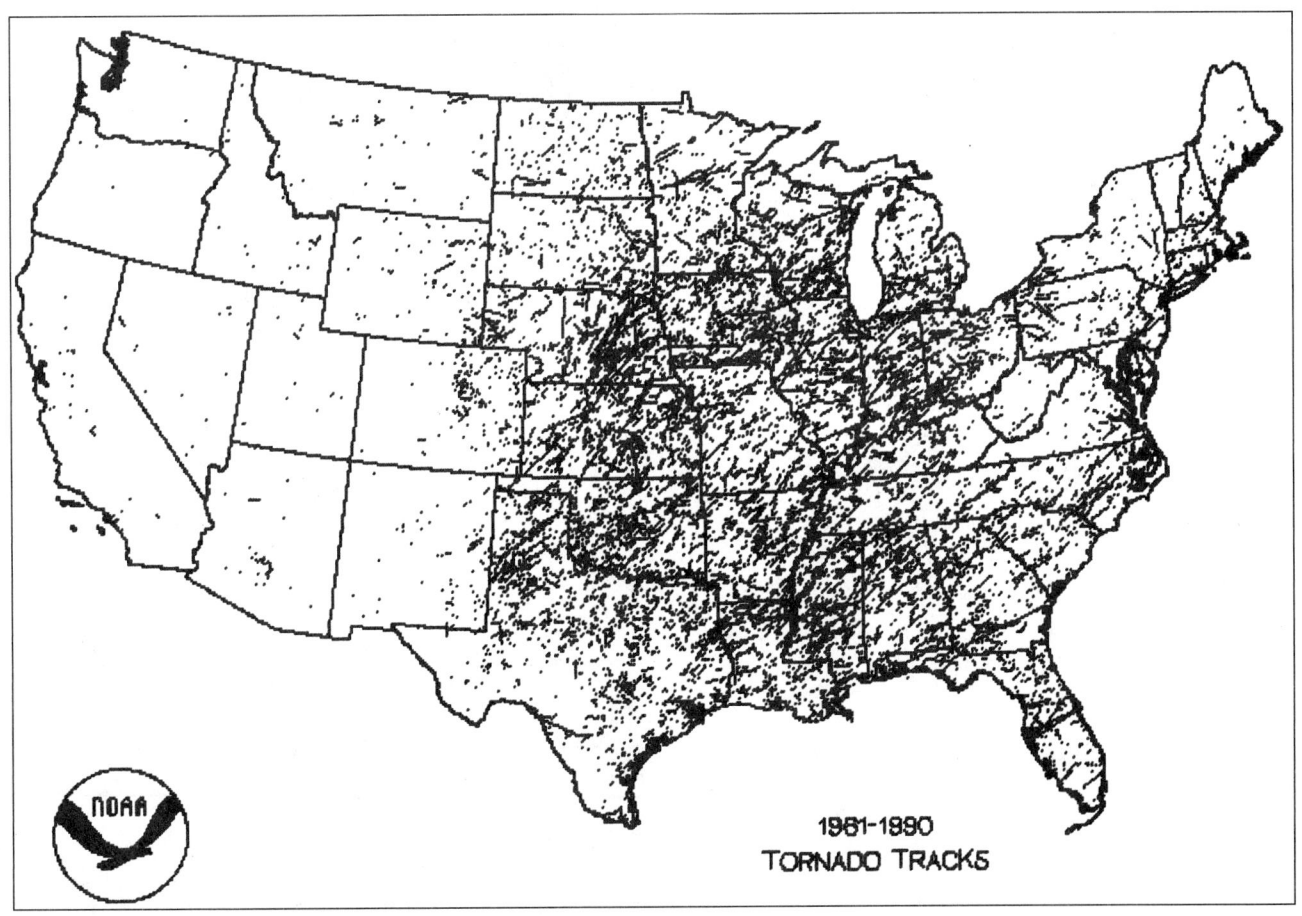

1961-1990
TORNADO TRACKS

(Courtesy of National Oceanic and Atmospheric Administration (NOAA).)

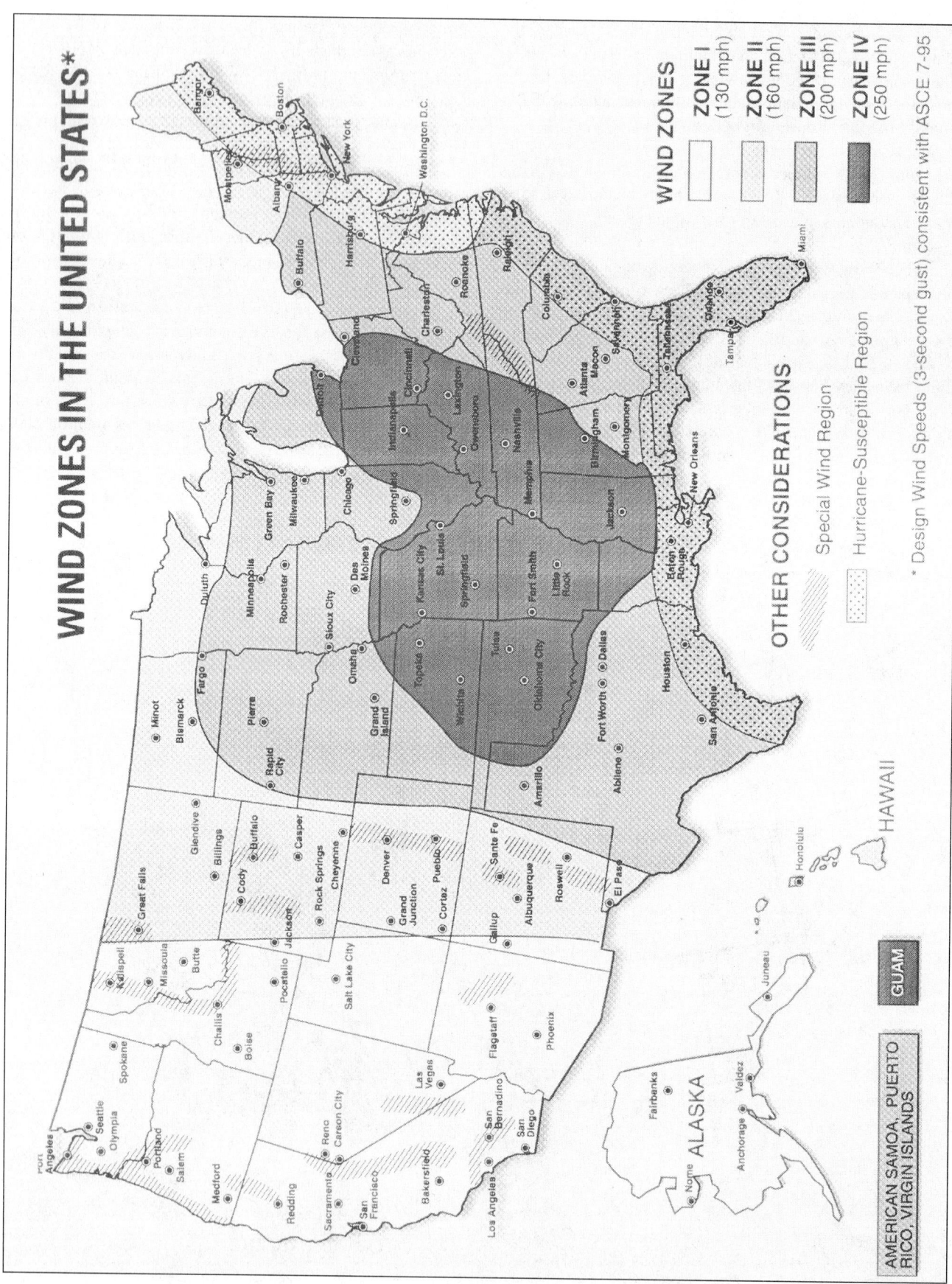

WIND ZONES IN THE UNITED STATES*

WIND ZONES

ZONE I (130 mph)
ZONE II (160 mph)
ZONE III (200 mph)
ZONE IV (250 mph)

OTHER CONSIDERATIONS

Special Wind Region

Hurricane-Susceptible Region

* Design Wind Speeds (3-second gust) consistent with ASCE 7-95

HAWAII

ALASKA

GUAM

AMERICAN SAMOA, PUERTO RICO, VIRGIN ISLANDS

(Courtesy of Federal Emergency Management Agency (FEMA).)

the annual average number of tornado days—days on which one or more tornadoes were reported—was 171. Average annual frequency by states for this period ranges from 129 tornadoes in Texas to fewer than three in most of the northeastern and far-western states.

Tornadoes may occur at any hour of the day or night, but, because of the meteorological combinations which create them, they form most readily during the warmest hours of the day. The greatest number of tornadoes—82% of the total—occurs between noon and midnight, and the greatest single concentration—23% of the total tornado activity—falls between 4 and 6 P.M.

• TORNADO TRENDS

From 1916–52, fewer than 300 tornadoes were reported in any one year. In 1953, when the U.S. Department of Commerce initiated its tornado forecasting effort, more than 421 tornadoes were observed and reported, beginning the first period of reliable statistical history. Since 1953, partly through improved equipment and techniques, partly through increasing public participation, essentially complete tornado records have been available.

Tornado statistics 1953–2001

The average number of tornadoes from 1953 through 2001 is 859. For the past 30 years there has been an average of 996 per year. The explosion of reports of tornadoes in recent years is due to changes in the system, not climate change.

The average number of tornado days from 1953 through 2001 was 174. This number has also grown with time, but because it has an upper bound, the growth has been slower.

The state that typically has the most tornadoes in a year is Texas: it's big and at the base of tornado alley. The annual Texas tornado count from 1953 through 2001 was 134. It ranged from 32 in 1953 to 232 in both 1967 and 1995.

More tornadoes occurred in 1998, a total of 1,424, in the United States, than in any year from the beginning of firm records from 1916 to 2001. In 1992, 1,297 tornadoes struck in 44 states, killing 39 and causing property damage of $794 million. A harsher year for tornado deaths, however, was 1953 when 515 people died from 422 recorded tornadoes.

• MAJOR TORNADOES

Early tornado records

The most death-dealing series of tornadoes on record occurred during the late afternoon on March 18, 1925, in portions of Missouri, Indiana, Illinois, Kentucky, and Tennessee. Eight separate and distinct tornadoes were observed.

Annual official total of tornadoes by year: 1953 through 2001:	
Year: Number of Tornadoes	
1953: 422	1978: 789
1954: 550	1979: 855
1955: 593	1980: 866
1956: 504	1981: 782
1957: 858	1982: 1047
1958: 564	1983: 931
1959: 604	1984: 907
1960: 616	1985: 684
1961: 697	1986: 765
1962: 657	1987: 656
1963: 463	1988: 702
1964: 704	1989: 856
1965: 897	1990: 1133
1966: 585	1991: 1132
1967: 926	1992: 1297
1968: 660	1993: 1173
1969: 608	1994: 1082
1970: 653	1995: 1234
1971: 889	1996: 1173
1972: 741	1997: 1148
1973: 1102	1998: 1424
1974: 945	1999: 1342
1975: 919	2000: 1071
1976: 834	2001: 1214
1977: 852	

One of these killed 689 persons, injured 1,890 and caused more than $16 million in property damage. The other seven tornadoes of the series killed 740 and also caused significant property damage. Another major series of tornadoes killed 268 people and injured 1,874 in Alabama on March 21, 1932. Property damage amounted to approximately $5 million.

Tornadoes of the 1990s

In March 1990, a series of four separate tornadoes occurred in central Kansas. Two of these tornadoes joined forces at one point and resulted in one of the three most intense to occur in the past decade. This series cut a path of over 100 mi in a two-and-one-half-hour period. The winds of another tornado in the Wichita, Kansas, area on April 26, 1991, reached 260 mph and traveled some 70 mi on the ground over a two-hour period.

On March 1, 1997, a severe weather system, with tornadoes and extremely heavy rainfall, erupted along a nearly stationary front from Texas to West Virginia. By mid-afternoon, an outbreak of strong tornadoes in Arkansas, northern Mississippi, and western Tennessee resulted in 27 deaths, including 25 in Arkansas. Several of the tornadoes have been estimated at F4 intensity, with winds ranging from 207–260 mph.

This was the deadliest U.S. outbreak since March 27, 1994, when 42 were killed in Alabama, Georgia, and South Carolina. Fortunately in this event, the National Weather Service (NWS) issued tornado warnings 10–32 minutes before the tornadoes struck, using NEXRAD radar to provide much more lead time than previously possible.

Tornadoes of 1998

During the late evening of February 22 and early morning of February 23, 1998, a series of tornadoes ripped across central Florida. At least one of the tornadoes reached an estimated F4 intensity. There were 42 fatalities; more than 3,500 residences were damaged, with more than 800 destroyed and another 700 left uninhabitable; and 135,000 utility customers lost power at the height of the storms. Damages from the tornado outbreak exceeded $60 million, and Florida's overall storm damage total was approximately $500 million. Hardest-hit locations in the tornado outbreak were Winter Garden, Altamonte Springs, Sanford, and Campbell. Overall, 54 of Florida's 67 counties were declared federal disaster areas due to storms over a period of months.

In the late afternoon of Wednesday, April 8, 1998, severe thunderstorms quickly developed over Mississippi, Arkansas, and southwest Tennessee. During the evening, some of the more powerful storms generated tornadoes that caused massive property damage and loss of life along a path from northeast Mississippi through central Alabama into northern Georgia. Hardest hit were Jefferson and St. Clair Counties in the Birmingham, Alabama metropolitan area. In the wake of these storms, 36 people were killed, 273 were injured, and property damage was estimated at over $300 million.

The path of destruction began about 6:30 P.M. CDT with one storm fatality in Pontotoc County, Mississippi, and ended near midnight with a death in De Kalb County, Georgia. The storms killed 34 people in Alabama—32 in Jefferson County and two in St. Clair County. Some of the tornadoes produced tracks more than 30 mi long. The most intense tornado, an F5—one of the worst in Alabama history—moved through Jefferson County, destroying more than 1,000 permanent homes and damaging almost 1,000 more. There was widespread catastrophic damage in Jefferson County, caused by winds in excess of 260 mph and a damage path up to 0.75 mi wide. The affected counties were declared disaster areas and drew emergency responses from the Federal Emergency Management Agency (FEMA), U.S. Department of Labor, Small Business Administration, National Guard, Red Cross, numerous volunteers, and dozens of state and local agencies.

Tornadoes of 1999

Throughout May 3–4, 1999, the state of Oklahoma and nearby areas of Kansas experienced about 76 violent tornadoes, with intensities ranging from F2 to F5 (the highest) on the Fujita Scale for Tornado Intensity—up to 318 mph. The largest of the series of twisters raged for four hours in Oklahoma and is estimated to have been a mile wide. This was the sixth deadliest tornado outbreak in Oklahoma history. (The deadliest was April 9, 1947, in Woodward where 116 people perished.)

Tornadoes occur every year, but what made this outbreak unique and deadly is that the tornadoes struck heavily populated areas. In Oklahoma and Kansas, 49 people were killed and another 898 injured as a result of the powerful storms. In Oklahoma, nearly 3,000 homes and 47 businesses were destroyed; in Kansas, estimates show that a total of 1,500 business and homes were destroyed. Insurance companies in Oklahoma City anticipated statewide damage claims of up to $750 million ($600 million of that in Oklahoma City alone).

During the same week, heavy storms left paths of destruction in Tennessee, Texas, and Arkansas.

There were 1,205 tornadoes observed in 1999, killing a total of 95 people.

Tornadoes of 2000

Figures show that there were fewer tornadoes in 2000 than average. Sixteen very strong to violent tornadoes (winds in excess of 158 mph) occurred between March and August 2000 in the United States. This is many fewer than the 1950–1999 average of 38. Throughout the past 50 years, little trend in very strong to violent tornado activity has been observed. A band of tornadoes pushed across Alabama in mid-December, killing at least 12 people. Hardest hit was the Tuscaloosa area.

There were 1,071 observed tornadoes during 2000, killing 41 people.

Tornadoes of 2001

There were eight very-strong to violent tornadoes (wind speeds in excess of 158 mph, category F3 to F5) during the 2001 tornado season (March-August). This is well below the long-term (1950–2000) mean of 38 and is the lowest such tornado count in the last 51 years. The previous record was nine violent tornadoes in 1987. Little trend in very-strong to violent tornadoes has been observed since 1950.

There were 1,214 observed tornadoes during 2001, killing 40 people.

Tornadoes of 2002

A line of thunderstorms, which that extended 1,000 miles, produced an unofficial 70 tornadoes from southern Mississippi to northern Ohio, killing at least 35 people and injuring more than 200 others. This total of 35 fatalities was the deadliest total since the May 3, 1999, tornado outbreak in Oklahoma and Kansas. There were at least 16 deaths in

Storm Prediction Center Monthly Tornado Statistics

	Number of Tornadoes							Number of Tornado Deaths					Killer Tornadoes				
	2003 Prelim	2003 Final	2002 Prelim	2002 Final	2001 Final	2000 Final	3 Year Avg.	2003 Prelim	2002 Final	2001 Final	2000 Final	3 Year Avg.	2003 Prelim	2002 Final	2001 Final	2000 Final	3 Year Avg.
JAN	0	?	8	3	5	16	8	0	0	0	0	0	0	0	0	0	0
FEB	11	?	2	2	30	56	29	2	0	8	19	9	1	0	3	4	2
MAR	49	?	30	47	33	103	61	8	0	3	2	2	4	0	2	1	1
APR	118	?	114	118	135	136	130	0	7	5	2	4	0	6	4	2	4
MAY	562	?	177	204	241	241	229	41	4	1	2	2	10	2	1	2	2
JUN	126	?	87	97	248	135	136	?	0	5	0	2	?	0	3	0	1
JUL	?	?	58	68	120	148	112	?	0	0	1	0	?	0	0	0	0
AUG	?	?	72	86	69	52	69	?	0	2	2	1	?	0	1	2	1
SEP	?	?	66	62	84	47	64	?	4	2	0	2	?	3	2	0	2
OCT	?	?	56	57	116	63	79	?	4	2	0	2	?	3	2	0	2
NOV	?	?	115	98	110	48	85	?	37	14	0	17	?	14	7	0	7
DEC	?	?	69	99	2	26	49	?	3	0	12	5	?	3	0	2	2
Total	866	?	854	941	1213	1071	1051	51	55	40	40	44	15	28	23	14	22

Note: ? means final number not available.
Important! Prelim. numbers represent tornado segments. Columns marked Final represent total tornadoes.
2003 numbers updated through 7AM CDT 06/16/2003

(Courtesy of Storm Prediction Center (NOAA/NWS/NCEP).)

Tennessee; 12 deaths in Alabama; 5 deaths in Ohio; and 1 death each in Mississippi and Pennsylvania. This outbreak occurred in November 2002.

• TORNADO DAMAGE

If there is some question as to the causes of tornadoes, there is none about the destructive effects of these violent storms. The dark funnel of a tornado can destroy solid buildings, make a deadly missile of a piece of straw, uproot large trees, and hurl people and animals for hundreds of yards. In 1931, a tornado in Minnesota carried an 83-ton railroad coach and its 117 passengers 80 ft through the air, and dropped them in a ditch.

As a tornado passes over a building, the winds twist, lift, and rip at the outside and roof, which causes walls to collapse and topple, and windows to explode; the debris of this destruction is driven through the air in a dangerous barrage. Heavy objects like machinery and railroad cars are lifted and carried by the wind for considerable distances.

Tornadoes do their destructive work through the combined action of their strong rotary winds and the impact of windborne debris. In the most simple case, the force of the tornado's winds pushes the windward wall of a building inward. The roof is lifted up and the other walls fall outward. Until recently, this damage pattern led to the incorrect belief that a structure explodes as a result of the atmospheric pressure drop associated with the tornado. Mobile homes are particularly vulnerable to strong winds and windborne debris; they should be evacuated for more substantial shelter when tornado warnings are in effect.

The Federal Emergency Management Agency has produced an excellent publication, FEMA 320, August 1999, "Taking Shelter From the Storm; Building a Safe Room Inside Your House," which includes construction plans and cost estimates.

• TORNADO WARNINGS

Although it is not possible to predict exactly where and when severe thunderstorms and tornadoes will occur, it is possible to predict general areas where the probability of severe thunderstorm and tornado development is greatest by detecting the larger-scale events that are usually associated with such storms.

This important function is performed by the Storm Prediction Center in Norman, Oklahoma. The Oklahoma facility is one of several environmental hazards centers of action operated by the National Weather Service (NWS), a major

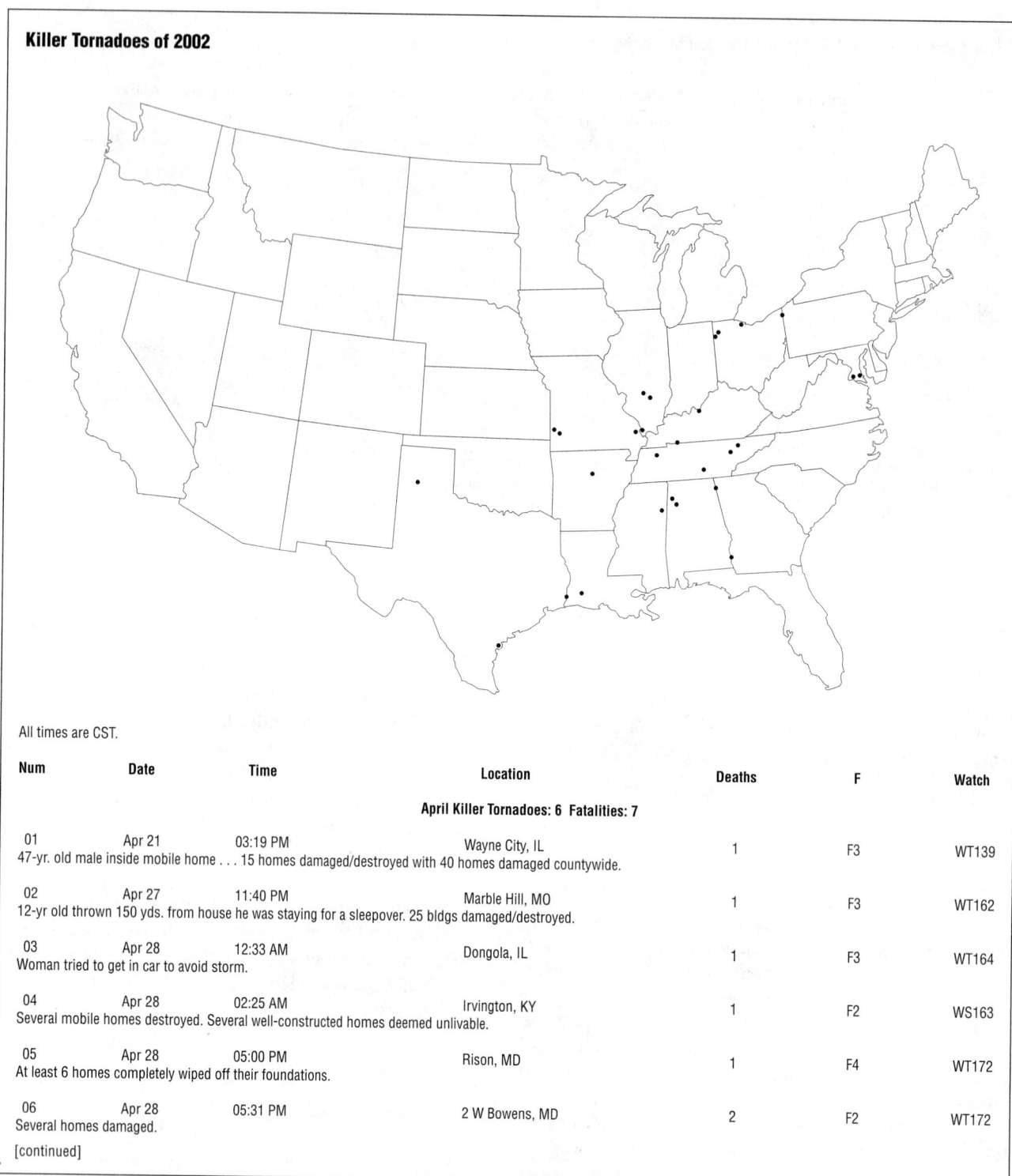

Killer Tornadoes of 2002

All times are CST.

Num	Date	Time	Location	Deaths	F	Watch
			April Killer Tornadoes: 6 Fatalities: 7			
01	Apr 21	03:19 PM	Wayne City, IL	1	F3	WT139
47-yr. old male inside mobile home . . . 15 homes damaged/destroyed with 40 homes damaged countywide.						
02	Apr 27	11:40 PM	Marble Hill, MO	1	F3	WT162
12-yr old thrown 150 yds. from house he was staying for a sleepover. 25 bldgs damaged/destroyed.						
03	Apr 28	12:33 AM	Dongola, IL	1	F3	WT164
Woman tried to get in car to avoid storm.						
04	Apr 28	02:25 AM	Irvington, KY	1	F2	WS163
Several mobile homes destroyed. Several well-constructed homes deemed unlivable.						
05	Apr 28	05:00 PM	Rison, MD	1	F4	WT172
At least 6 homes completely wiped off their foundations.						
06	Apr 28	05:31 PM	2 W Bowens, MD	2	F2	WT172
Several homes damaged.						

[continued]

(Courtesy of Storm Prediction Center (NOAA/NWS/NCEP).)

Killer Tornadoes of 2002

May Killer Tornadoes: 2 Fatalities: 4

07	May 5	05:45 PM	Happy TX	2	F2	WT227

Extensive damage reported . . . several homes destroyed.

08	May 9	01:10 AM	Centralia IL	2	F1	—

Twelve mobile homes destroyed and 1 home damaged.

October Killer Tornadoes: 3 Fatalities: 4

09	Oct 24	01:19 PM	Corpus Christi TX	1	F2	WS727

Extensive structural damage at Del Mar West Community College.

10	Oct 29	12:55 AM	DeQuincy LA	1	F?	WT731

Local sheriff reported 7 mobile homes damaged/destroyed.

11	Oct 29	02:55 AM	Chataignier LA	2	F2	WT732

Fire department reported 8 mobile homes destroyed, large house and barn badly damaged.

November Killer Tornadoes: 14 Fatalities: 37

12	Nov 5	05:45 PM	Abbeville AL	1	F2	WT736

Heavy damage to businesses, high school and adjacent residential area. At least 15 hospitalized.

13	Nov 10	12:45 am	Leach TN	2	F2	WT741

Five homes destroyed . . . 1 mobile home with major damage.

14	Nov 10	01:00 AM	Port Royal TN	2	F1	WT741

Several homes destroyed . . . a few automobiles totaled.

15	Nov 10	02:30 PM	Van Wert OH	2	F4	WS744

Numerous structures leveled.

16	Nov 10	03:00 PM	Continental OH	2	F3	WS744

Trailer destroyed.

17	Nov 10	04:25 PM	Republic OH	1	F3	WT747

Homes damaged.

18	Nov 10	06:25 PM	New Union TN	2	F2	WT750

Widespread damage.

19	Nov 10	06:52 PM	Carbon Hill AL	4	F3	WT748

Numerous buildings and homes destroyed.

20	Nov 10	06:57 PM	Clark PA	1	F2	WT750

Tornado travelled 7 miles on the ground.

21	Nov 10	07:08 PM	Crawford MS	1	F1	WT748

Fatality associated with tree falling onto mobile home.

22	Nov 10	07:40 PM	Mossy Grove/Petros TN	7	F3	WT746

Numerous buildings destroyed.

23	Nov 10	08:15 PM	Saragossa AL	7	F3	WT748

Numerous mobile homes destroyed.

24	Nov 10	09:45 PM	Crossville TN	4	F3	WT749

Mobile homes damaged.

25	Nov 10	11:20 PM	Peak's Crossing AL	1	F2	WT754

Damage to mobile homes.

December Killer Tornadoes: 3 Fatalities: 3

26	Dec 17	11:18 PM	Chesapeake MO	1	F2	WT778

Tornado hit Luck Lady trailer park and 3 homes east of trailer park.

27	Dec 18	01:59 AM	Kenoma MO	1	F1	WT779

At least three frame homes and one mobile home destroyed.

28	Dec 18	03:35 PM	Hamlet AR	1	F3	WT785

Number of residences destroyed. 84 yr. old woman thrown from mobile home.

State	Killer Tornado	Fatalities		F Scale	Killer Tornado	Fatalities		Circumstance	Fatalities
TN	5	17		F0	0	0		Mobile Home	37
AL	4	13		F1	4	6		Permanent Home	10
OH	3	5		F2	11	16		Vehicle	4
IL	3	4		F3	10	29		Outside Open	2
TX	2	3		F4	2	3		Total	53
LA	2	3		F5	0	0			
MO	3	3		F?	1	1			
MD	2	3		Total	28	55			
KY	1	1							
PA	1	1							
MS	1	1							
AR	1	1							
Total	28	55							

(Courtesy of Storm Prediction Center (NOAA/NWS/NCEP).)

Constructing a Shelter

In parts of the country where tornadoes are comparatively frequent, a form of shelter is vital for protection from tornadoes. The shelter may never be needed, but during a tornado emergency, it can be worth many times the effort and cost of preparing it. One of the safest tornado shelters is an underground excavation, known as a storm cellar.

1. **Location.** When possible, the storm cellar should be located outside and near the residence, but not so close that failing walls or debris could block the exit. If there is a rise in the ground, the cellar may be dug into it to make use of the rise for protection. The cellar should not be connected in any way with house drains, cesspools, or sewer and gas pipes.

2. **Size.** The size of the shelter depends on the number of persons to be accommodated and the storage needs. A structure 8 ft long by 6 ft wide and 7 ft high will protect eight people for a short time and provide limited storage space.

3. **Material.** Reinforced concrete is the best material for a tornado shelter. Other suitable building materials include: split logs, 2-in planks (treated with creosote and covered with tar paper), cinder block, hollow tile, and brick. The roof should be covered with a 3-ft mound of well-pounded dirt, sloped to divert surface water. The entrance door should be of heavy construction, hinged to open inward.

4. **Drainage.** The floor should slope to a drainage outlet if the terrain permits. If not, a dry well can be dug. An outside drain is better, because it will aid ventilation.

5. **Ventilation.** A vertical ventilating shaft about 1 ft^2 can extend from near the floor level through the ceiling. This can be converted into an emergency escape hatch if the opening through the ceiling is made 2 ft^2 and the 1-ft shaft below is made easily removable. Slat gratings of heavy wood on the floor also will improve air circulation.

6. **Emergency equipment.** A lantern and tools—crowbar, pick, shovel, hammer, pliers, screwdriver—should be stored in the cellar to ensure escape if cellar exits are blocked by debris. Stored metal tools should be greased to prevent rusting.

element of the National Oceanic and Atmospheric Administration (NOAA) of the U.S. Department of Commerce.

Meteorologists at the Storm Prediction Center monitor conditions in the North American atmosphere, using surface data from hundreds of points and radar summaries, satellite photographs, meteorological upper-air profile (obtained by sounding balloons), and reports from pilots. From these thousands of pieces of information, meteorologists determine the area that is most likely to experience severe thunderstorms or tornadoes. Information on this area is then issued to NWS offices and the public in the form of a *watch bulletin*. In addition, trained civilian volunteers, "spotters," in the program SKYWARN, work with their local communities to alert the areas to dangerous weather. The spotters use their personal equipment and vehicles to provide the NWS with timely and accurate reports of severe weather. Once the severe conditions are confirmed by NWS radar, local authorities are contacted.

A severe thunderstorm watch or tornado watch bulletin issued by the Center usually identifies an area about 140 mi wide by 240 mi long. Although the watch bulletin states approximately where and for how long the severe local storm threat will exist, it does not mean that severe local storms will not occur outside the watch area or time frame—the watch is only an indication of where and when the probabilities are highest.

Watch bulletins are transmitted to all NWS offices. Designated offices prepare and issue a redefining statement, which specifies the affected area in terms of counties, towns, and locally well-known geographic landmarks. These messages are disseminated to the public by all possible means, and are used to guide the activities of local government, law enforcement, and emergency agencies in preparing for severe weather.

Watches are not warnings. Until a severe thunderstorm or tornado warning is issued, persons in watch areas should maintain their normal routines, but watch for threatening weather and listen to the radio or television for further severe weather information.

A severe thunderstorm warning or tornado warning bulletin is issued by a local office of the NWS when a severe thunderstorm or tornado has actually been sighted in the area or indicated by radar. Warnings describe the location of the severe thunderstorm or tornado at the time of detection, the area (usually the counties) that could be affected, and the time period (usually one hour) covered by the warning. The length of this area is equal to the distance the storm is expected to travel in one hour.

When a warning is received, persons close to the storm should take cover immediately, especially in the case of a tornado warning. Persons farther away from the storm should be prepared to take cover if threatening conditions are sighted.

Severe weather statements are prepared by local offices of the NWS to keep the public fully informed of all current information, particularly when watch or warning bulletins are in effect. Statements are issued at least once each hour, and more frequently when the severe weather situation is changing rapidly. In this way, a close watch is kept on weather developments, and information is quickly disseminated to the counties for which the NWS office has responsibility.

All-clear bulletins are issued whenever the threat of severe thunderstorms or tornadoes has ended in the area previously warned in a tornado or severe thunderstorm warning bulletin. When a warning is canceled, but a watch continues in effect for the same area or a warning is in effect for an adjacent area, a *Severe Weather Bulletin* is issued. This qualified message is also issued when a portion, but not all, of a watch area is canceled. This permits a continuous alert in the path of the storm, with the alert being canceled as the severe weather moves through the watch area.

• TORNADO FORECASTING STUDIES

VORTEX

The Verification of the Origins of Rotation in Tornadoes Experiment (VORTEX) was an experiment that began in 1994 and concluded in June 1995. The National Severe Storms Laboratory (NSSL) and the Center for the Analysis and Prediction of Storms (CAPS) sponsored it. Using a dozen instrumented vehicles, a mobile Doppler radar, and two Doppler-equipped aircraft, data were collected through a field program in parts of Texas, Oklahoma, and Kansas. It is hoped that the data collected from more than 30 storms will provide meteorologists with a better understanding of different elements of the tornado: the connection between mid-level and surface rotations of the storm, wind speed, and airflow patterns. With this knowledge meteorologists will be able to more accurately predict tornado formation,

understand airflow patterns and wind speeds relevant to tornadic storms, advance the knowledge of the environment within a tornado—how it varies over time, pressure, humidity, and other tendencies—which can be used to comprehend how tornadoes cause damage. In the mid-1980s, the NSSL conducted tests to learn more about the internal workings of tornadoes by placing a 55-gal drum filled with sensors (called TOTO—TOtable Tornado Observatory) in the path of oncoming tornadoes. These efforts were met with minimal success. The 1996 movie *Twister* was based upon this work.

Tracking supercells

In 1998, a team of government and university scientists based at NOAA's National Severe Storms Laboratory and the University of Oklahoma in Norman, Oklahoma, began using mobile meteorological stations and truck-mounted radars in an attempt to intercept severe storms in the Great Plains to get close-up observations of a developing tornado. The radar data and other information collected attempt to offer researchers a three-dimensional view of the complete life cycle of a tornado and ultimately help improve NOAA fore-

---

Tornado Safety Rules

A **TORNADO WATCH** means tornadoes are expected to develop. Keep a battery-operated radio or television set nearby, and listen for weather advisories—even if the sky is blue. A **tornado warning** means a tornado has actually been sighted or indicated by weather radar. Seek shelter inside a storm cellar or reinforced building and stay away from windows. Curl up so that your head and eyes are protected. Keep a battery-operated radio or television nearby, and listen for further advisories.

ON THE STREET OR IN A CAR, leave your vehicle and take shelter in civil defense or other inside shelter areas with basements or storm cellars. Be sure to stay away from large glassed-in areas. If no building is available, or if caught out in the open countryside, take shelter in a ditch or ravine or lie flat on the ground upwind of your parked vehicle.

IN HIGH-RISE OFFICE BUILDINGS AND LARGE APARTMENT BUILDINGS, if possible post a trained spotter or lookout on the roof with a two-way radio. Go to the lower floors or the basement. Take shelter in small interior rooms such as rest rooms, closets, and utility rooms as well as interior corridors. Be sure to cover and protect the head from flying and falling debris.

IN HOMES, take shelter in the basement under sturdy items. Concrete laundry tubs, heavy-duty workbenches, pool tables, and staircases offer the greatest safety. If there is no basement, move to an interior bathroom or interior closet or, if necessary, an interior room or hallway on the lowest floor and take cover under heavily stuffed furniture in the center of the home. Take shelter away from all windows. Caution: Avoid bathrooms with an outside wall or window. Also do not lock yourself in a closet that has no inside latch or door handle.

IN SHOPPING CENTERS OR SHOPPING MALLS, if possible, post a trained security guard or lookout on the west or south side of the complex with a two-way radio. Take shelter in the basement or in shops. Be sure to protect your head from flying or falling debris. Caution: Avoid large open malls or walkways with glass or plastic skylights as well as large glass signs and display cases.

IN SCHOOLS, go to a storm cellar or underground shelter if available. If there is no underground shelter area move the pupils into interior hallways or small interior rooms on the lowest floor. Caution: Avoid auditoriums, gymnasiums, and other large rooms with long free-span roofs as well as southwest-to-northeast-oriented corridors with exposed entrances on the south and west side of the building. Also avoid glass display cases, glassed-in stairwells and all doorways.

IN FACTORIES, post a trained spotter or lookout on the roof with a two-way radio. Workers should move to sections of the plant that are below ground level. If this is not possible, have the workers take shelter in interior corridors or in small interior rooms such as rest rooms, closets, and storage rooms. Caution: Avoid large rooms or work areas with long free-span roofs. Stay away from all windows.

IN MOBILE HOMES, leave your trailer and take shelter in an administration building with a basement or an approved community shelter area. If no shelter is available, go to a ditch or ravine and lie down flat against the ground. Make sure you protect your head from flying debris.

casts and warnings of severe weather. The research is a follow-up to the VORTEX project of 1994 and 1995, in which scientists from NOAA, the University of Oklahoma, and other universities intercepted and studied 10 tornadoes from as close as 3 mi away; and 1997's SubVORTEX, in which the team intercepted a tornado for the first time using truck-mounted radar dishes called "Doppler on Wheels" or DOWs.

Using three DOWs and six sedans equipped with meteorological sensors, cameras, and communications gear, the scientists attempt to target a supercell—a large thunderstorm with a rotating updraft. When a likely storm system is forecast, the team sets out to intercept the supercell and position two 8-ft-diameter DOWs about a mile outside the storm to scan the entire area of rotating air every 30–90 seconds. Other team members use a third, shorter wavelength DOW to make finer-scale measurements. The team then attempts to send one or more of the three instrumented "probe" vehicles inside the *hook*—the rotating center of the supercell that looks like an inverted question mark in radar images and often signals a tornado is about to form.

One of the team's experiments, conducted in spring 1998, called SubVORTEX-RFD, attempted to photograph and measure the rear flank downdraft of winds of a supercell, since the scientists believe this is the region of a storm that often triggers a tornado. The scientists document the origin and evolution of the storm's rear flank downdraft by measuring variations in these winds just before and after a tornado forms. To do this, the team carefully positions some of the probe vehicles beneath the most strongly rotating part of the storm, something that has never been done before. Another experiment, called Radar Observations of Tornadoes and Thunderstorms (ROTATE), uses the dual-DOW network in an attempt to study the whole tornado genesis process, tornado maintenance, tornado death, and tornado structure.

• TORNADOES OF THE TWENTIETH CENTURY

Severe weather experts from NOAA's Storm Prediction Center in Norman, Oklahoma, have prepared a list of some of the more notable tornado outbreaks that occurred in the United States during the twentieth century. The summary lists the tornadoes by decade and notes the technological and policy improvements that resulted.

"For meteorologists who study tornadic storms either through forecasting or research or storm chasing, there are a number of memorable tornadoes or tornado outbreaks during the 1900s," says Dan McCarthy, warning coordination meteorologist for the Storm Prediction Center. "Many meteorologists are in the profession because of a certain outbreak or tornado that spurred their curiosity, driving them to the science."

Technological advancements in the second half of the century have contributed to better, more accurate severe weather watches and warnings from the National Weather Service, ultimately saving countless lives. The biggest advancement for severe weather forecasting was the development of Doppler radar. NOAA scientists and other researchers took the airborne radar developed by the U.S. military during World War II and applied it to weather forecasting and severe storm identification. The ultimate result was the Next Generation Radar (NEXRAD) Doppler weather radar system currently in use.

Developments in computer technology also have created continued advancements in numerical weather prediction, allowing meteorologists to apply physics in replicating motions of the atmosphere. This, combined with diligent analysis to recognize weather patterns, helped advance severe weather prediction to its current level of an average lead time of over 11 minutes for tornado warnings issued by National Weather Service forecasters.

The most impressive and devastating tornado outbreak in the twentieth century was the Super Outbreak of April 3–4, 1974. The outbreak lasted 16 hours and produced a total of 148 tornadoes across 13 states from Illinois, Indiana, and Michigan southward through the Ohio and Tennessee Valleys into Mississippi, Alabama, and Georgia. This outbreak produced more long-track tornadoes than any other, killing 315 people and injuring more than 5,000. The most notable individual tornado was the one that moved into Xenia, Ohio, just before 4:30 P.M. It destroyed much of the town, including the town square and high school, killing 34 people.

1900–1909: The outbreak of April 24–26, 1908, included violent tornadoes that moved through parts of Louisiana, Mississippi, and Alabama, killing 324 people and injuring 1,652 others. The worst damage took place in Amite, Louisiana, where 29 people died.

1910–1919: A long-track tornado on May 26, 1917, traveled across Illinois and Indiana for 293 miles, lasting seven hours and 20 minutes. The tornado killed 101 people and injured 638 others. Another tornado moved through the town of Mattoon, Illinois, destroying everything in a two-and-a-half block-wide path for 2.5 mi.

1920–1929: The Tri-State tornado of March 18, 1925, developed near Ellington, Missouri, and then for the next 3.5 hours killed more people and destroyed more schools, homes, and farmsteads than any other tornado to this point in history. The tornado cut across southern Illinois into southern Indiana, killing 695 people, 234 of them in the town of Murphysboro, Illinois, and injuring 2,027. Other tornadoes occurred in Kentucky, Tennessee, and Alabama. A total of nine tornadoes were reported, leaving 747 dead and nearly 2,300 people injured.

1930–1939: On March 21–22, 1932, a total of 330 people died as a result of tornadoes that touched down across northern Alabama. One tornado hit the northeast part of the state, killing 38 and injuring 500.

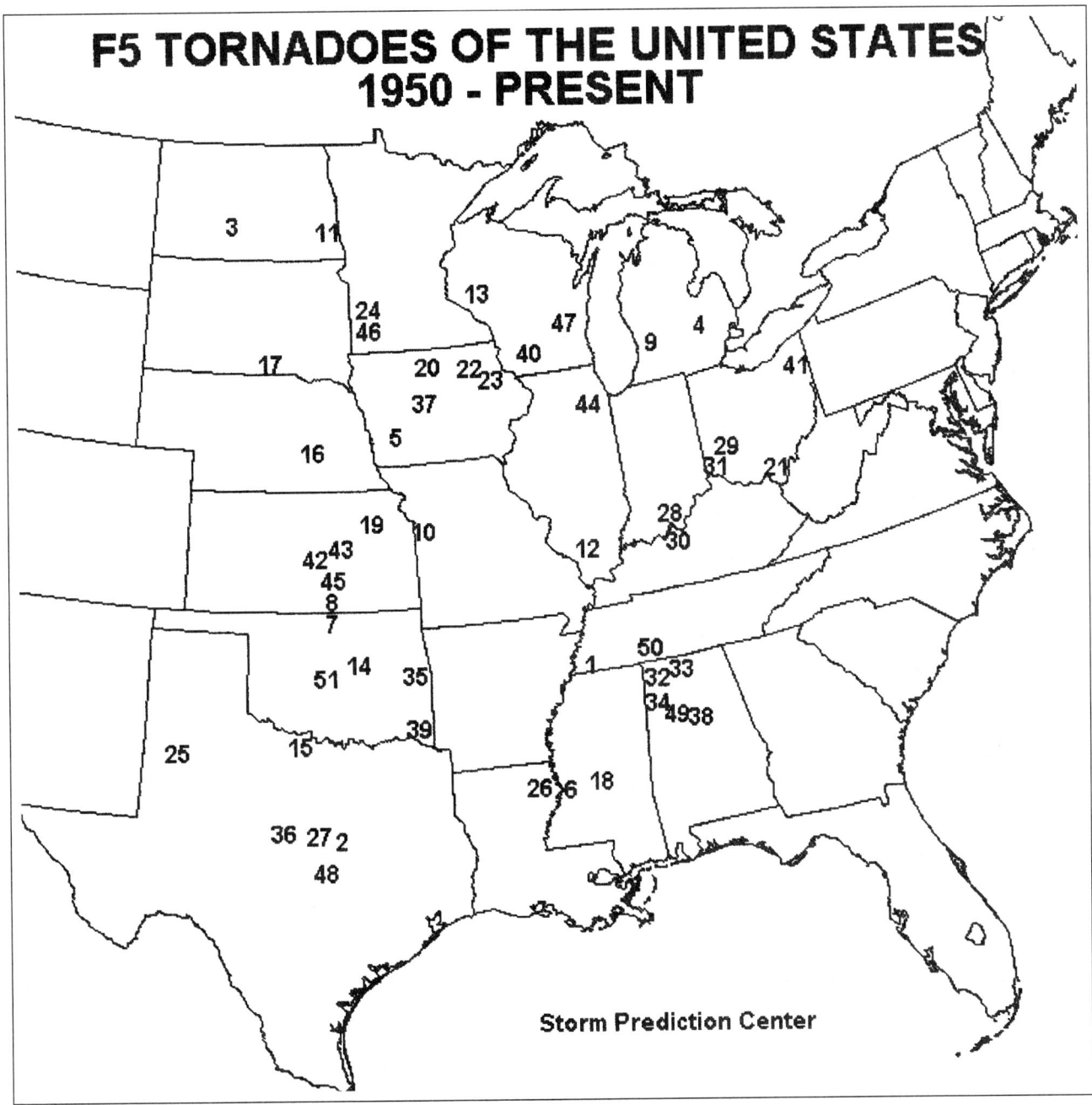

(Courtesy of Storm Prediction Center (NOAA/NWS/NCEP).)

During the Tupelo/Gainesville outbreak on April 5–6, 1936, 17 tornadoes were scattered across parts of northern Mississippi and northern Georgia. A massive pair of tornadoes hit Gainesville, Georgia, in the morning, killing 203 people and causing 1,600 injuries.

1940–1949: Three major outbreaks occurred during this decade. The first, on March 16, 1942, left 152 dead and 1,284 injured from tornadoes that raked across parts of Illinois, Mississippi, Tennessee, and Kentucky. As many as 63 people perished in a tornado northwest of Greenwood, Mississippi, that hit as buses carried school children home. Five hundred people were injured.

A total of 154 people died and nearly 1,000 were injured on June 23, 1944, as tornadoes struck parts of Ohio, Pennsylvania, West Virginia, Maryland, and Delaware. The worst areas affected were parts of northeast West Virginia and western Maryland, where a tornado family killed 30 and injured 300.

On April 9, 1947, a tornado outbreak that included eight tornadoes raked across parts of Texas, Oklahoma, and Kansas. One tornado killed 107 people in Woodward, Oklahoma. Devastation covered 100 city blocks and 1,000 homes were damaged or destroyed. Cost of the damage at that time was estimated at $6 million. Clean-up afterward was hampered by cold and snow.

```
==============================================================
NUMBER    DATE                     LOCATION
======    ====================     ==========================
51        May 3, 1999              Bridge Creek/Moore OK
50        April 16, 1998           Waynesboro TN
49        April 8, 1998            Pleasant Grove AL
48        May 27, 1997             Jarrell TX
47        July 18, 1996            Oakfield WI
46        June 16, 1992            Chandler MN
45        April 26, 1991           Andover KS
44        August 28, 1990          Plainfield IL
43        March 13, 1990           Goessel KS
42        March 13, 1990           Hesston KS
41        May 31, 1985             Niles OH
40        June 7, 1984             Barneveld WI
39        April 2, 1982            Broken Bow OK
38        April 4, 1977            Birmingham AL
37        June 13, 1976            Jordan IA
36        April 19, 1976           Brownwood TX
35        March 26, 1976           Spiro OK
34        April 3, 1974            Guin AL (#101)
33        April 3, 1974            Tanner AL (#98)
32        April 3, 1974            Mt. Hope AL (#96)
31        April 3, 1974            Saylor Park OH (#43)
30        April 3, 1974            Brandenburg KY (# 47)
29        April 3, 1974            Xenia OH   (# 37)
28        April 3, 1974            Daisy Hill IN   (# 40)
27        May 6, 1973              Valley Mills TX
26        February 21, 1971        Delhi LA
25        May 11, 1970             Lubbock TX
24        June 13, 1968            Tracy MN
23        May 15, 1968             Maynard IA
22        May 15, 1968             Charles City IA
21        April 23, 1968           Gallipolis OH
20        October 14, 1966         Belmond IA
19        June 8, 1966             Topeka KS
18        March 3, 1966            Jackson MS
17        May 8, 1965              Gregory SD
16        May 5, 1964              Bradshaw NE
15        April 3, 1964            Wichita Falls TX
14        May 5, 1960              Prague OK
13        June 4, 1958             Menomonie WI
12        December 18, 1957        Murphysboro IL
11        June 20, 1957            Fargo ND
10        May 20, 1957             Ruskin Heights MO
9         April 3, 1956            Grand Rapids MI
8         May 25, 1955             Udall KS
7         May 25, 1955             Blackwell OK
6         December 5, 1953         Vicksburg MS
5         June 27, 1953            Adair IA
4         June 8, 1953             Flint MI
3         May 29, 1953             Ft. Rice ND
2         May 11, 1953             Waco TX
1         March 21, 1952           Moscow TN
==============================================================
```

(Courtesy of Storm Prediction Center (NOAA/NWS/NCEP).)

1950–1959: On May 11, 1953, a violent tornado hit downtown Waco, Texas, killing 114 people and destroying about 200 business buildings. Heaps of bricks up to five feet high filled the streets. Survivors were buried for up to 14 hours.

A tornado outbreak in early June 1953 produced two major tornadoes. On June 8, a tornado hit in Flint, Michigan, leaving 116 people dead. The next day, June 9, a tornado described as "a huge cone of black smoke" carrying debris eastward over the Boston area and out over the Atlantic Ocean caused 94 deaths and nearly 1,300 injuries in Worcester, Massachusetts. In the United States, the death toll was 116 from tornadoes in Michigan, Ohio, Massachusetts, and New Hampshire. Other tornadoes occurred in Canada.

The hardest-hit area from a tornado outbreak in Oklahoma and Kansas on May 25, 1955, was Udall, Kansas. Eighty people were known dead and 270 were injured—more than half of the people in Udall—and the town was destroyed. For the entire outbreak, tornadoes killed 102 people and injured 563.

A tornado moved across southeast parts of Kansas City hitting the area of Ruskin Heights on May 20, 1957. Forty-four people were killed and 531 were injured. More than 825 homes and businesses were damaged or destroyed, including the local high school. The outbreak itself spread from northeast Kansas and northeast Oklahoma through Missouri into Iowa and Illinois. In all, 17 tornadoes killed 59 people and injured 665 others.

1960–1969: The second most damaging outbreak of the century, known as the Palm Sunday outbreak, occurred April 11–12, 1965. Nearly 50 tornadoes struck parts of the Great Lakes region from Wisconsin and Illinois eastward through lower Michigan and northern Ohio. The outbreak resulted in 256 deaths and 3,402 injuries. Twin tornadoes moved into Goshen, Indiana, destroying nearly 100 trailer homes. A large tornado hit Russiaville, Indiana, damaging or destroying 90% of the buildings. As many as 44 people died and 612 were injured as one tornado followed another tornado across Steuben County in northern Indiana and Monroe County and others in lower Michigan. Tornadoes devastated areas in northern Toledo, Ohio, killing 18 people. Other tornadoes moved through areas about 15 miles southwest of Cleveland just northeast of Strongsville. Six homes literally vanished; 18 people were killed and 200 others were injured.

On June 8, 1966, a tornado brought massive damage to Topeka, Kansas, causing $100 million in damage. This became the most expensive tornado to date.

1970–1979: The most prolific tornado outbreak of the twentieth century was the Super Outbreak of April 3–4, 1974. During a 16-hour period, 148 tornadoes occurred from Illinois and Indiana into Michigan and Ohio southward through the Tennessee Valley into Mississippi and Alabama. This outbreak produced the largest number of tornadoes, with 30 causing F4 damage or worse. On one occasion, as many as five large tornadoes were on the ground at one time. The outbreak killed 315 people and resulted in 6,142 injuries. One tornado hit Xenia, Ohio, at 4:30 P.M., moved through the center of town and demolished the high school. Thirty-four people died and 1,150 were injured in Xenia as 300 homes were destroyed and 2,100 homes were damaged.

Five years later, a tornado hit Wichita Falls, Texas, on April 10, 1979, killing 42 people and injuring 1,740.

1980–1989: Thirty tornadoes spread out across parts of northeast Ohio into western Pennsylvania on May 31, 1985. The outbreak killed 76 people and injured 876 others. Twelve people died from one tornado that moved from Ashtabula County, Ohio, into Erie County, Pennsylvania. Sixteen people were killed by a tornado that started over Trumbull County, Ohio, then moved east/northeast across parts of Pennsylvania.

Another outbreak moved across Iowa and Minnesota into Wisconsin on June 7–8, 1984. The town of Barneveld, Wisconsin, was hit by a tornado just before midnight. Everything but the water tower was demolished and nine people were killed. As many as 45 tornadoes in the entire outbreak killed 13 people.

1990–1999: The role of videotape and the advances in media technology provided many breathtaking views of tornadoes in the 1990s. People came from miles around to film the Hesston, Kansas tornado on March 13, 1990. One tornado started near Goshen, only to merge with a second near Hesston and track northeast to just southwest of Topeka, Kansas.

Another notable Palm Sunday tornado occurred on March 27, 1994, when 22 people died in Goshen, Alabama, after a tornado hit a church.

Most recently, a large tornado mowed through areas of southwest Oklahoma City and Moore, Oklahoma, on May 3, 1999, demolishing or damaging more than 8,000 homes and ringing up more than $1 billion in damage. This tornado was part of an outbreak of 74 tornadoes that affected parts of Oklahoma and southern Kansas, killing 48 people.

Advancements in communications through radio and television helped issue advanced watches and warnings to the public. Additionally, meteorological advancements from research in storm structure using Doppler radar helped forecasters identify tornadic storms, improving warnings from a few minutes to as many as 20 minutes and increasing public response.

• FREQUENTLY ASKED QUESTIONS ABOUT TORNADOES

Does hail always come before a tornado? Rain? Lightning? Utter silence?

None of these necessarily come before a tornado. Rain, wind, lightning, and hail characteristics vary from storm to storm, from one hour to the next, and even with the direction the storm is moving with respect to the observer. While large hail can indicate the presence of an unusually dangerous thunderstorm, and can happen before a tornado, do not depend on it. Hail, or any particular pattern of rain, lightning or calmness, is not a reliable predictor of tornado threat.

What does a tornado sound like?

That depends on what it is hitting, its size, intensity, closeness, and other factors. The most common tornado sound is a continuous rumble, like a close-by train. Sometimes a tornado produces a loud whooshing sound, like that of a waterfall or of open car windows while driving very fast. Tornadoes that are tearing through densely populated areas may produce all kinds of loud noises at once, which collectively may make a tremendous roar. Just because you may have heard a loud roar during a damaging storm does not necessarily mean it was a tornado. Any intense thunderstorm wind can produce damage and cause a roar.

How do tornadoes dissipate?

The details are still debated by tornado scientists. We do know tornadoes need a source of instability (heat, moisture, etc.) and a larger-scale property of rotation (vorticity) to keep going. There are a lot of processes around a thunderstorm that can possibly rob the area around a tornado of either instability or vorticity. One is relatively cold outflow—the flow of wind out of the precipitation area of a shower or thunderstorm. Many tornadoes have been observed to go away soon after being hit by outflow. For decades, storm observers have documented the death of numerous tornadoes when their parent circulations (mesocyclones) weaken after they become wrapped in outflow air—either from the same thunderstorm or a different one. The irony is that some kinds of thunderstorm outflow may help to cause tornadoes, while other forms of outflow may kill tornadoes.

Do tornadoes really skip?

No. There is no such thing as a "skipping" tornado, despite what you may have read in many older references, news stories, or even damage survey reports. By definition (above), a tornado must be in contact with the ground. When the vortex is not, it is literally no longer a tornado; and even if the same vortex or funnel cloud makes ground contact later, it is a separate tornado. Stories of skipping tornadoes usually mean either (1) there was continuous contact between vortex and ground in the path, but it was too weak to do damage; (2) multiple tornadoes occurred, but there was no survey done to precisely separate their paths (very common before the 1970s); or (3) there were multiple tornadoes with only short separation, but the survey erroneously classified them as one tornado.

Is a low-hanging cloud in a thunderstorm a tornado?

Many low-hanging clouds are not tornadoes, but sometimes are wrongly reported as tornadoes anyway. The most important things to look for when you see a suspicious cloud feature are: (1) rapid cloud-base rotation, if you are close enough to make out cloud movement, and (2) a concentrated, whirling debris or dust cloud at ground level under the thunderstorm base. (Imagine this spinning rapidly.) It is common to have one without the other. Many thunderstorms produce dust plumes in their outflow; these tend to move in one direction and not rotate. In gustnadoes there is spinning motion at ground level but not at cloud base (therefore, not a tornado). If the ground is wet enough, or the circulation weak enough, there may not be any debris under a rotating cloud base. But persistent rotation in the cloud base is potentially very dangerous and should be reported. At night, also look for persistent cloud lowering to ground, especially if accompanied by a power flash.

What is the difference between a funnel cloud and a tornado? What is a funnel cloud?

In a tornado, damaging circulation is on the ground—whether or not the cloud is. A true funnel cloud rotates, but has no ground contact or debris, and does no damage. If it is a low-hanging cloud with no rotation, it is not a funnel cloud. Caution: tornadoes can occur without a funnel, and what looks like "only" a funnel cloud may be doing damage that cannot be seen from a distance. Because it may quickly become a tornado, any funnel cloud (remember rotation!) should be reported by spotters.

Why are some tornadoes white, and others black, gray, or even red?

Tornadoes tend to look darkest when one looks southwest through northwest in the afternoon. In those cases, they are often silhouetted in front of a light source, such as brighter skies west of the thunderstorm. If there is heavy precipitation behind the tornado, it may be dark gray, blue, or even white—depending on where most of the daylight is coming from. This happens often when the spotter is looking north or east at a tornado, and part of the forward-flank and/or rear-flank cores. Tornadoes wrapped in rain may exhibit varieties of gray shades on gray, if they are visible at all. Lower parts of tornadoes also can assume the color of the dust and debris they are generating; for example, a tornado passing across dry fields in western or central Oklahoma may take on the hue of the red soil so prevalent there.

What is a "wedge" tornado? A "rope" tornado?

These are slang terms often used by storm observers to describe tornado shape and appearance. Remember, the size or shape of a tornado does not say anything certain about its strength! "Wedge" tornadoes simply appear to be at least as wide as they are tall (from ground to ambient cloud base). "Rope" tornadoes are very narrow, often sinuous or snake-like in form. Tornadoes often (but not always!) assume the "rope" shape in their last stage of life; and the cloud rope may even break up into segments. Again, tornado shape and size do not signal strength! Some rope tornadoes can still do violent damage of F4 or F5.

What is a multivortex tornado?

Multivortex (a.k.a. multiple-vortex) tornadoes contain two or more small, intense subvortices orbiting the center of the larger tornado circulation. When tornadoes do not contain too much dust and debris, they can sometimes be spectacularly visible. These vortices may form and die within a few seconds, sometimes appearing to train through the same part of the tornado one after another. They can happen in all sorts of tornado sizes, from huge "wedge" tornadoes to narrow "rope" tornadoes. Subvortices are the cause of most of the narrow, short, extreme swaths of damage that sometimes arc through tornado tracks. From the air, they can preferentially mow down crops and stack the stubble, leaving cycloidal marks in fields. Multivortex tornadoes are the source of most of the old stories from newspapers and other media before the late twentieth century that told of several tornadoes seen together at once.

What is a "satellite" tornado? Is it a kind of multivortex tornado?

No. There are important distinctions between satellite and multiple-vortex tornadoes. A satellite tornado develops independently from the primary tornado—not inside it as does a suction vortex. The tornadoes remain separate and distinct as the satellite tornado orbits its much larger companion within the same mesocyclone. Their cause is unknown; but they seem to form most often in the vicinity of exceptionally large and intense tornadoes.

What is a gustnado?

A gustnado is a small and usually weak whirlwind that forms as an eddy in thunderstorm outflows. They do not connect with any cloud-base rotation and are not tornadoes. But because gustnadoes often have a spinning dust cloud at ground level, they are sometimes wrongly reported as tornadoes. Gustnadoes can do minor damage (e.g., break windows and tree limbs, overturn trash cans, and toss lawn furniture), and should be avoided.

What is a landspout?

This is storm-chaser slang for a non-supercell tornado. So-called "landspouts" resemble waterspouts in that way, and also in their typically small size and weakness compared to the most intense tornadoes. But "landspouts" are tornadoes by definition; they are capable of doing significant damage and killing people.

What is a waterspout?

A waterspout is a tornado over water—usually meaning non-supercell tornadoes over water. Waterspouts are common along the southeast U.S. coast—especially off southern Florida and the Keys—and can happen over seas, bays, and lakes worldwide. Although waterspouts are always tornadoes by definition; they do not officially count in tornado records unless they hit land. They are smaller and weaker than the most intense Great Plains tornadoes, but still can be quite dangerous. Waterspouts can overturn small boats, damage ships, do significant damage when hitting land, and kill people. The National Weather Service will often issue special marine warnings when waterspouts are likely or have been sighted over coastal waters, or tornado warnings when waterspouts can move onshore.

How are tornadoes in the Northern Hemisphere different from tornadoes in the Southern Hemisphere?

The sense of rotation is usually the opposite. Most tornadoes—but not all—rotate cyclonically, which is coun-

Most waterspouts in the United States form in the Florida Keys area, although they may occur over inland bodies of water.
(Photo courtesy of NOAA Photo Library, NOAA Central Library; OAR/ERL/National Severe Storms Laboratory (NSSL).)

terclockwise in the Northern Hemisphere and clockwise south of the equator. Anticyclonic tornadoes (clockwise-spinning in the Northern Hemisphere) have been observed, however—usually in the form of waterspouts, non-supercell land tornadoes, or anticyclonic whirls around the rim of a supercell's mesocyclone. There have been several documented cases of cyclonic and anticyclonic tornadoes under the same thunderstorm at the same time. Anticyclonically rotating supercells with tornadoes are extremely rare; but one struck near Sunnyvale, California, in 1998. Remember, "cyclonic" tornadoes spin counter-clockwise in the Northern Hemisphere, and clockwise in the Southern Hemisphere.

How many tornadoes hit the United States yearly?

About 1,000. The actual average is unknown, because tornado spotting and reporting methods have changed so much in the last several decades that the officially recorded tornado climatologies are believed to be incomplete. Also, in the course of recording thousands of tornadoes, errors are bound to occur. Events can be missed or misclassified; and some nondamaging tornadoes in remote areas could still be unreported.

What is Tornado Alley?

Tornado Alley is a nickname in the popular media for a broad swath of relatively high tornado occurrence in the central United States. Various Tornado Alley maps that you may see can look different because tornado occurrence can be measured many ways—by all tornadoes, tornado county-segments, strong and violent tornadoes only, and databases with different time periods. Remember, this is only a map of greatest incidence. Violent or killer tornadoes do happen outside this Tornado Alley every year. Tornadoes can occur almost anywhere in the United States, and even overseas.

What city has been hit by the most tornadoes?

Oklahoma City. The exact count varies because city limits and tornado reporting practices have changed over the years; but the known total is now over 100.

Some towns have legends that they are protected from tornadoes by a hill, river, spirit, etc. Is there any truth to this?

No. Many towns that have not suffered a tornado strike contain well-meaning people who perpetuate these myths; but there is no basis for them besides the happenstance lack of a tornado. Many other towns used to have such myths before they were hit, including extreme examples like Topeka, Kansas (F5 damage, 16 killed, 1968) and Waco, Texas (F5 damage, 114 killed, 1953). Violent tornadoes have crossed rivers of all shapes and sizes. The deadliest tornado in U.S. history (Tri-state Tornado of March 18, 1925; F5 damage; 695 killed) roared undeterred across the Mississippi River, as have numerous other violent tornadoes. Almost every major river east of the Rockies has been crossed by a significant tornado, as have high elevations in the Appalachians and Rockies.

What is the highest-elevation tornado? Do they happen in the mountain West?

The highest elevation a tornado has ever occurred is unknown; but it is at least 10,000 ft above sea level. On July 21, 1987, there was a violent (F4 damage) tornado in the Tetons of Wyoming near that elevation; and there have probably been lesser ones at higher altitudes. On August 31, 2000, a supercell spawned a photogenic tornado in Nevada. Tornadoes are generally a lot less frequent west of the Rockies per unit area with a couple of exceptions. One exception is the Los Angeles Basin, where weak-tornado frequency over tens of square miles is on par with that in the Great Plains. Elsewhere, there are probably more high-elevation western tornadoes occurring than we have known about, just because many areas are so sparsely populated, and lack the density of spotters and storm chasers we have in the Plains.

Why does it seem that tornadoes avoid downtowns of major cities?

Simply, downtowns cover such tiny land areas relative to the entire nation. The chance of any particular tornado hitting a major downtown is quite low—not for any meteorological reason, but simply because downtowns are small targets. Even when tornadoes hit metro areas; their odds of hitting downtown are small. For example, downtown Dallas (inside the freeway loop) covers roughly three square miles—Dallas County, about 900 square miles. For a brief tornado in Dallas County, its odds of hitting downtown are only about one in 300. Still, downtown tornadoes have happened, including at least four hits on St. Louis alone. The idea of large buildings destroying or preventing a tornado is pure myth. Even the largest skyscrapers pale in size and volume when compared to the total circulation of a big tornado from ground through thunderhead.

What is the risk of another super-outbreak like April 3–4, 1974?

It is rare; but we do not know how rare, because an outbreak like that has only happened once since tornado records have been kept. There is no way to know if the odds are one in every 50 years, 100 years, or 1,000 years, since we just do not have the long climatology of reasonably accurate tornado numbers to use. So the bigger the outbreaks, the less reliably we can judge their potential to recur.

What are the chances of a tornado near my house?

The frequency that a tornado can hit any particular square mile of land is about every 1,000 years on average—but varies around the country. The reason this is not an exact number is because we do not have a long and accurate enough record of tornadoes to make more certain (statistically sound) calculations. The probability of any tornado hitting within sight of a spot (let's say 25 nautical miles) also varies during the year and across the country.

What is the F-scale?

Dr. T. Theodore Fujita developed a damage scale for winds, including tornadoes, that is supposed to relate the degree of damage to the intensity of the wind. This scale was the result. The F-scale should be used with great caution. Tornado wind speeds are still largely unknown; and the wind speeds on the F-scale have never been scientifically tested and proven. Different winds may be needed to cause the same damage depending on how well built a structure is, wind direction, wind duration, battering by flying debris, and many other factors. Also, the process of rating the damage itself is largely a judgment call—quite inconsistent and arbitrary. Even meteorologists and engineers highly experienced in damage survey techniques may come up with different F-scale ratings for the same damage. Even with all its flaws, the F-scale is the only widely used tornado rating method, and probably will remain so until ground-level winds can be measured in most tornadoes.

So if the F-scale winds are just guesses, why are they so specific?

Those winds were arbitrarily attached to the damage scale based on 12-step mathematical interpolation between the hurricane criteria of the Beaufort wind scale and the threshold for Mach 1 (738 mph). Though the F-scale actually peaks at F12 (Mach 1), only F1 through F5 are used in practice, with F0 attached for tornadoes of winds weaker than hurricane force. Again, F-scale wind-to-damage relationships are untested, unknown, and purely hypothetical. They have never been proven and may not represent real tornadoes. F-scale winds should not be taken literally.

It was noted that the Oklahoma City tornado was "almost F6." Is that a real level on the scale?

Only in untested theory. Fujita plotted hypothetical winds higher than F5; but as mentioned in the previous answer above, they were only guesses. Even if winds measured by portable Doppler radar (slightly above ground level) had been over 318 mph, the tornado would still be rated "only" F5 since F5 is the most intense possible damage level.

What is a "significant" tornado?

A tornado is classified as "significant" if it does F2 or greater damage on the F scale. It is important to know that those definitions are arbitrary, and exist for scientific research. No tornado is necessarily insignificant. Any tornado can kill or cause damage; and some tornadoes rated less than F2 probably could do F2 or greater damage if they hit a well-built house during peak intensity.

Big fat tornadoes are the strongest ones, right?

Not necessarily. The size or shape of a tornado does not say anything about its strength! Some small "rope" tornadoes can still do violent damage of F4 or F5; and some very large tornadoes over a quarter-mile wide have produced only weak damage of F0 to F1.

What was the strongest tornado? What is the highest wind speed in a tornado?

Nobody knows. Tornado wind speeds have only been directly recorded in the weaker ones, because strong and violent tornadoes destroy weather instruments. Mobile Doppler radars such as the University of Oklahoma's Doppler on Wheels have remotely sensed tornado wind speeds above ground level as high as 318 mph (May 3, 1999 near Bridge Creek, Oklahoma)—the highest winds ever found near Earth's surface by any means. (That tornado caused F5 damage.) But ground-level wind speeds in the most violent tornadoes have never been directly measured.

Does El Niño cause tornadoes?

No. Neither does La Niña. Both are major changes in sea surface temperature in the tropical Pacific, which occur over the span of months. U.S. tornadoes happen thousands of miles away on the order of seconds and minutes. El Niño does adjust large-scale weather patterns. But there are too many variables to say conclusively what role El Niño (or La Niña) has in changing tornado risk; and it certainly does not directly cause tornadoes. A few studies have shown some loose associations between El Niño years and regional trends in tornado numbers from year to year; but that still does not prove cause and effect. Weak associations by year may be as close as the El Niño-to-tornado connection can get—because there are so many things on the scales of states, counties, and individual thunderstorms that affect tornado formation.

How does cloud seeding affect tornadoes?

Nobody knows for certain. There is no proof that seeding can or cannot change tornado potential in a thunderstorm. This is because there is no way to know that the things a thunderstorm does after seeding would not have happened anyway. This includes any presence or lack of rain, hail, wind gusts, or tornadoes. Because the effects of seeding are impossible to prove or disprove, there is a great deal of controversy in meteorology about whether it works, and if so, under what conditions, and to what extent.

Do hurricanes and tropical storms produce tornadoes?

Often, but not always. There are great differences from storm to storm, not necessarily related to tropical cyclone size or intensity. Some landfalling hurricanes in the United States fail to produce any known tornadoes, while others cause major outbreaks. Though fewer tornadoes tend to occur with tropical depressions and tropical storms than with hurricanes, there are notable exceptions like TS Beryl in the Carolinas in 1994. Relatively weak hurricanes like Danny (1985) have spawned significant supercell tornadoes well inland, as have larger, more intense storms like Allen (1980) and Beulah (1967). Hurricane Beulah, in fact, caused the second biggest tornado outbreak on record in numbers, with 115. Hurricane-spawned tornadoes tend to occur in small, low-topped supercells within the outer bands, NNW through ESE of the center—mainly the northeast quadrant. There, the orientation and speed of the winds create vertical shear profiles somewhat resembling those around classic Great Plains supercells—but weaker and shallower. Because tornado-producing circulations in hurricane supercells tend to be smaller and shorter-lived than their Midwest counterparts, they are harder to detect on Doppler radar, and more difficult to warn for. But hurricane-spawned tornadoes can still be quite deadly and destructive, as shown by the F3 tornado from Hurricane Andrew at La Place, Louisiana (1992, 2 killed) and an F4 tornado at Galveston, Texas from Hurricane Carla (1961, 8 killed). We do not know how many tornadoes hurricanes produce over the water. But the similarity in Doppler radar velocity signatures over water to tornado-producing cells in landfalling hurricanes suggests that it does happen—and that they can be yet another good reason for ships to steer well clear of tropical cyclones.

Who forecasts tornadoes?

Only the National Weather Service (NWS) issues tornado forecasts nationwide. Warnings come from a local NWS office. The Storm Prediction Center (SPC) issues watches, general severe weather outlooks, and mesoscale discussions.

How are tornadoes forecast?

This is a very simple question with no simple answer! Here is a very generalized view from the perspective of a severe weather forecaster: when predicting severe weather (including tornadoes) a day or two in advance, we look for the development of temperature and wind flow patterns in the atmosphere that can cause enough moisture, instability, lift, and wind shear for tornadic thunderstorms. Those are the four needed ingredients. But it is not as easy as it sounds.

"How much is enough" of those is not a hard and fast number, but varies a lot from situation to situation—and sometimes is unknown! A large variety of weather patterns can lead to tornadoes; and often, similar patterns may produce no severe weather at all. To further complicate it, the various computer models we use days in advance can have major biases and flaws when the forecaster tries to interpret them on the scale of thunderstorms. As the event gets closer, the forecast usually (but not always) loses some uncertainty and narrows down to a more precise threat area. (At SPC, this is the transition from outlook to mesoscale discussion to watch.) Real-time weather observations—from satellites, weather stations, balloon packages, airplanes, wind profilers, and radar-derived winds—become more and more critical the sooner the thunderstorms are expected; and the models become less important. To figure out where the thunderstorms will form, we must do some hard, short-fuse detective work: find out the location, strength and movement of the fronts, drylines, outflows, and other boundaries between air masses that tend to provide lift. Figure out the moisture and temperatures—both near ground and aloft—that will help storms form and stay alive in this situation. Find the wind structures in the atmosphere that can make a thunderstorm rotate as a supercell, then produce tornadoes. (Many supercells never spawn a tornado!) Make an educated guess where the most favorable combination of ingredients will be and when; then draw the areas and type the forecast.

How is tornado damage rated?

The most widely used method worldwide is the F-scale. In Britain, there is a similar scale with more divisions. In both cases, the wind speeds are based on calculations of the Beaufort wind scale and have never been scientifically verified in real tornadoes. Because (1) nobody knows the "true" wind speeds at ground level in most tornadoes, and (2) the amount of wind needed to do similar-looking damage can vary greatly, even from block to block or building to building. Damage rating is (at best) an exercise in educated guessing. Even experienced damage-survey meteorologists and wind engineers can and often do disagree among themselves on a tornado's strength.

Who surveys tornado damage? What are the criteria for the National Weather Service to do a survey?

This varies from place to place; and there is no rigid criteria. The responsibility for damage survey decisions at each NWS office usually falls on the Warning-Coordination Meteorologist (WCM) and/or the Meteorologist in Charge (MIC). Budget constraints keep every tornado path from having a direct ground survey by NWS personnel; so spotter, chaser, and news accounts may be used to rate relatively weak, remote, or brief tornadoes. Killer tornadoes, those striking densely populated areas, or those generating reports of exceptional damage are given highest priority for ground surveys. Most ground surveys involve the WCM and/or forecasters not having shift responsibility the day of the survey. For outbreaks and unusually destructive events—usually only a few times a year—the NWS may support involvement by highly experienced damage survey experts and wind engineers from elsewhere in the country. Aerial surveys are expensive and usually reserved for tornado events with multiple casualties and/or massive amounts of damage. Sometimes, local NWS offices may have a cooperative agreement with local media or police to use their helicopters during surveys.

Do mobile homes attract tornadoes?

Of course not. It may seem that way, considering most tornado deaths occur in them, and that some of the most graphic reports of tornado damage come from mobile home communities. The reason for this is that mobile homes are, in general, much easier for a tornado to damage and destroy than well-built houses and office buildings. A brief, relatively weak tornado which may have gone undetected in the wilderness—or misclassified as severe straight-line thunderstorm winds while doing minor damage to sturdy houses—can blow a mobile home apart. Historically, mobile home parks have been reliable indicators, not attractors, of tornadoes.

How can a tornado destroy one house and leave the next one almost unscratched?

Most of the time, this happens either with multiple-vortex tornadoes or very small, intense single-vortex tornadoes. The winds in most of a multivortex tornado may only be strong enough to do minor damage to a particular house. But one of the smaller embedded subvortices, perhaps only a few dozen feet across, may strike the house next door with winds over 200 mph, causing complete destruction. Also, there can be great differences in construction from one building to the next, so that even in the same wind speed, one may be flattened while the other is barely nicked.

How do tornadoes do strange things, like drive straw into trees, strip road pavement, and drive splinters into bricks?

The list of bizarre things attributed to tornadoes is almost endless. Much of it is folklore; but there are some weird scenes in tornado damage. Asphalt pavement may strip when tornado winds sandblast the edges with gravel and other small detritus, eroding the edges and causing chunks to peel loose from the road base. Storm chasers and damage surveyors have observed this phenomenon often after the passage of a violent tornado. With a specially designed cannon, wind engineers at Texas Tech University have fired boards and other objects at over 100 mph into various types

of construction materials, duplicating some of the types of "bizarre" effects, such as wood splinters embedded in bricks. Intense winds can bend a tree or other objects, creating cracks in which debris (e.g., hay, straw) becomes lodged before the tree straightens and the crack tightens shut again. All bizarre damage effects have a physical cause inside the roiling maelstrom of tornado winds. We do not fully understand what some of those causes are yet, however; because much of it is almost impossible to simulate in a lab.

Do tornadoes really pick up objects and carry them for miles? Who does research on it?

Yes, numerous tornadoes have lofted (mainly light) debris many miles into the sky, which was then carried by middle- and upper-atmospheric winds for long distances. The vertical winds in tornadoes can be strong enough to temporarily levitate even heavy objects if they have a large face to the wind or flat sides (like roofs, walls, trees, and cars), and are strong enough to carry lightweight objects tens of thousands of feet high. Though the heaviest objects, such as railroad cars, can only be airborne for short distances, stories of checks and other papers found over 100 miles away are often true. The Worcester, Massachusetts, tornado on June 9, 1953 carried mattress pieces high into the thunderstorm, where they were coated in ice, before they fell into Boston Harbor. Pilots reported seeing debris fluttering through the air at high altitude near the thunderstorm which spawned the Ruskin Heights, Missouri tornado on May 20, 1957.

Is it helpful to open house windows to equalize pressure? Or is that a bad thing to do?

Opening the windows is absolutely useless, a waste of precious time, and can be very dangerous. Do not do it. You may be injured by flying glass while trying to do it. And if the tornado hits your home, it will blast the windows open anyway.

Is it safe to run under a bridge to ride out a tornado?

Absolutely not! Stopping under a bridge to take shelter from a tornado is a very dangerous idea, for several reasons:
1. Deadly flying debris can still be blasted into the spaces between bridge and grade—and impale in any people hiding there.
2. Even when strongly gripping the girders (if they exist), people may be blown loose, out from under the bridge and into the open—possibly well up into the tornado itself. Chances for survival are not good if that happens.
3. The bridge itself may fail, peeling apart and creating large flying objects, or even collapsing down onto people underneath. The structural integrity of many bridges in tornado winds is unknown—even for those that may look sturdy.

4. Whether or not the tornado hits, parking on traffic lanes is illegal and dangerous to yourself and others. It creates a potentially deadly hazard for others, who may plow into your vehicle at full highway speeds in the rain, hail, and/or dust. Also, it can trap people in the storm's path against their will, or block emergency vehicles from saving lives.

The people in that infamous video were extremely fortunate not to have been hurt or killed. They were actually not inside the tornado vortex itself, but instead in a surface inflow jet—a small belt of intense wind flowing into the base of the tornado a few dozen yards to their south. Even then, flying debris could have caused serious injury or death. More recently, on May 3, 1999, two people were killed and several others injured outdoors in Newcastle and Moore, Oklahoma, when a violent tornado blew them out from under bridges on I-44 and I-35. Another person was killed that night in his truck, which was parked under a bridge.

So if a person is in a car, which is supposed to be very unsafe, and should not get under a bridge, what can he or she do?

Vehicles are notorious as death traps in tornadoes, because they are easily tossed and destroyed. Either leave the vehicle for sturdy shelter or drive out of the tornado's path. When the traffic is jammed or the tornado is bearing down on you at close range, your only option may be to park safely off the traffic lanes, get out, and find a sturdy building for shelter, if possible. If not, lie flat in a low spot, as far from the road as possible (to avoid flying vehicles). However, in open country, the best option is to escape if the tornado is far away. If the traffic allows, and the tornado is distant, you probably have time to drive out of its path. Watch the tornado closely for a few seconds compared to a fixed object in the foreground (such as a tree, pole, or other landmark). If it appears to be moving to your right or left, it is not moving toward you. Still, you should escape at right angles to its track: to your right if it is moving to your left, and vice versa—just to put more distance between you and its path. If the tornado appears to stay in the same place, growing larger or getting closer—but not moving either right or left—it is headed right at you. You must take shelter away from the car or get out of its way fast!

Is it best to go to the southwest corner of a basement in a tornado?

Not necessarily. The southwest corner is no safer than any other part of the basement, because walls, floors, and furniture can collapse (or be blown) into any corner. The "safe southwest corner" is an old myth based on the belief that, since tornadoes usually come from the southwest, debris will preferentially fall into the northeast side of the basement. There are several problems with this concept, including: (1) tornadoes are not straight-line winds, even on the scale of a house, so the strongest wind may be blowing from any direction; and (2) tornadoes themselves may arrive from any direction. In a basement, the safest place is under a sturdy workbench, mattress, or other such protection—and out from under heavy furniture or appliances resting on top of the floor above.

What is a safe room?

So-called "safe rooms" are reinforced small rooms built in the interior of a home, which are fortified by concrete and/or steel to offer extra protection against tornadoes, hurricanes, and other severe windstorms. They can be built in a basement, or if no basement is available, on the ground floor. In existing homes, interior bathrooms or closets can be fortified into "safe rooms" also.

What about tornado safety in sports stadiums or outdoor festivals?

Excellent question—and a very, very disturbing one to many meteorologists. Tornadoes have passed close to such gatherings on a few occasions, including a horse race in Omaha on May 6, 1975, and a crowded dog track in West Memphis, Arkansas on December 14, 1987. A supercell without a tornado hit a riverside festival in Fort Worth in 1995, catching over 10,000 people outdoors and bashing many of them with hail larger than baseballs. Just in the last few years, tornadoes have hit the football stadium for the NFL Tennessee Titans, and the basketball arena for the NBA Utah Jazz. Fortunately, they were both nearly empty of people at the time. There is the potential for massive death tolls if a stadium or fairground is hit by a tornado during a concert, festival, or sporting event—even with a warning in effect. Fans may never know about the warning; and even if they do, mass-panic could ensue and result in casualties even if the tornado does not hit. Stadium and festival managers should work with local emergency management officials to develop a plan for tornado emergencies—both for crowd safety during the watch and warning stages, and (similar to a terrorism plan) for dealing with mass casualties after the tornado.

What would happen if a large, violent tornado hit a major city today?

This has happened on several occasions, most recently in parts of Oklahoma City on May 3, 1999. Because of excellent, timely watches and warnings and intense media coverage of the Oklahoma tornado long before it hit, only 36 people were killed. The damage toll exceeded $1 billion, making it the costliest tornado in U.S. history. Still, it did not strike downtown, and passed over many miles of undeveloped land. Moving the same path north or south in the same area may have led to much greater death and damage tolls. The threat exists for a far worse disaster. Placing the same tornado outbreak in the Dallas-Ft. Worth Metroplex,

Weather Anecdotes

June 19, 2001 ... Siren, WI. A tornado destroyed this town of 900 people some 65 miles northeast of St. Paul, MN. Three people were killed and 14 injured. A man and wife were playing cards with their son when they heard on the police scanner that a bad storm was coming. The wife reached the basement steps when a wall collapsed and pushed her into a counter. The husband curled into a fetal position and watched as linoleum tiles were ripped off the kitchen floor. Another couple huddled in their bath-room as the tornado tore down their house. They escaped with bruises. Siren has a village siren but it was damaged by lightning in late April or May. Repairs had been scheduled for the next week.

April 15, 1996 ... A woman rode her bathtub to safety when a tornado ripped through Pilot, NC. She ran into the bathroom and got in the tub, "hearing it was a good thing to do," and rode the tub out into the woods. She landed in the trees and crawled out through the briars.

February 23, 1998 ... Kissimmee, FL. In the Flamingo Lakes area, many said it was a miracle no one was killed. One woman grabbed her eight-year-old daughter out of her bed and returned to find the bed covered with shattered glass. Another woman, 69, recovering from hip surgery and plagued by old polio infirmities, was sitting in a big, overstuffed chair in her bedroom on the second floor watching TV, her daughter was asleep downstairs. The tornado hit with full force. In an instant, the roof and all four walls of the second floor were gone, and the older female was pinned in the chair by a wayward 18-inch satellite dish. Neighbors found her, after it was over, sitting in the chair, exposed to the night sky, without a scratch. She said "it was raining 250 miles per hour" and recalled "I said, If this is the time, don't let me go in water, I'm afraid of Mother Nature." In another home, a family found their cat trapped inside a washing machine. How the cat got inside the washing machine, which was covered with debris, remains a mystery.

The 10 Deadliest U.S. Tornado Years since 1950

	YEAR	DEATHS
1	1953	519
2	1974	366
3	1965	301
4	1952	230
5	1957	193
6	1971	159
7	1968	131
8	1998	130
9	1955	129
10	1984	122

(Courtesy of Storm Prediction Center (NOAA/NWS/NCEP).)

especially during rush hour gridlock (with up to 62,000 vehicles stuck in the path), the damage could triple what was done in Oklahoma. There could be staggering death tolls in the hundreds or thousands, and overwhelmed emergency services. Ponder the prospect of such a tornado's path in downtown Dallas, for example. The North Texas Council of Governments and NWS Ft. Worth has compiled a very detailed study of several such violent tornado disaster scenarios in the Metroplex, which could be adapted to other major metro areas as well.

Although they are common in the Great Plains, not all places have "tornado warning" sirens. Is this because tornadoes do not occur everywhere? Isn't every place required to have sirens?

Siren policy seems to vary a lot from place to place; and it is something over which the National Weather Ser-

vice has no control. There is no nationwide requirement for tornado sirens. The NWS issues watches and warnings; but it is up to local governments to have a community readiness system in place for their citizens. In conversations with emergency managers and spotter coordinators, it has been found that the two most common reasons for a lack of sirens are low budgets and the perception that tornadoes cannot happen in an area. The latter is false; and the former is a matter of fiscal priorities. Your city and/or county emergency manager is the first person to ask about the tornado preparedness program in your community.

How does a person become a storm spotter?

Local National Weather Service offices offer spotter-training sessions each year. Contact the Warning Coordination Meteorologist at the office that serves you for information on when and where they conduct these sessions, and how to become a spotter for them. There is also a national spotters' organization, SKYWARN, that can help you learn about storm spotting and get you in contact with spotting experts.

What is the difference between a spotter and a chaser?

The differences are in method and motivation. Chasers are more mobile than spotters, and unlike most spotters, travel hundreds of miles and across state lines to observe storms. Spotters' primary function is to report critical weather information, on a live basis, to the National Weather Service through some kind of local spotter coordinator. Chasers, on the other hand, may be doing it for any number of reasons, including scientific field programs, storm photography, self-education, commercial video opportunity, or news media coverage. Some storm spotters also do occasional chasing outside their home area; and some chasers are certified and equipped to do real-time spotting.

Who does scientific tornado research?

The National Severe Storms Laboratory has been the major force in tornado-related research for several decades. NSSL has been a leader in Doppler radar development, research, and testing, and has run numerous field programs to study tornadoes and other severe weather since the early 1970s. Others heavily involved with tornado research include UCAR/NCAR, the University of Oklahoma, the Tornado Project, and overseas, TORRO (UK). Members of the SELS/SPC staff have done research related to forecasting tornadoes for many years. Almost every university with an atmospheric science program, as well as many local National Weather Service offices, have also published some tornado-related studies.

The 10 Costliest U.S. Tornadoes

	DATE	LOCATION	PROPERTY DAMAGE ($)
1	6 May 1975	Omaha NE	1.132 billion
2	10 Apr 1979	Wichita Falls TX	840 million
3	11 May 1970	Lubbock TX	530 million
4	8 Jun 1966	Topeka KS	470 million
5	3 Oct 1979	Windsor Locks CT	420 million
6	31 Mar 1973	north-central GA	388 million
7	3 Apr 1974	Xenia OH	310 million
8	3 Jun 1980	Grand Island NE	260 million
9	27 May 1896	St. Louis MO	201 million
10	28 Aug 1990	Plainfield IL	192 million

(Courtesy of Storm Prediction Center (NOAA/NWS/NCEP).)

Deadliest Tornado Years for a State since 1950

Rank	Sum	Year	State	Major Event
1.	150	1953	Texas	11 May - Waco
2.	137	1965	Indiana	11 Apr - Palm Sunday Outbreak
3.	127	1953	Michigan	08 Jun - Flint
4.	118	1971	Mississippi	21 Feb - Delta Outbreak
5.	112	1952	Arkansas	21 Mar - Outbreak
6.	90	1953	Massachusetts	09 Jun - Worcester
7.	81	1955	Kansas	25 May - Udall
8.	79	1974	Alabama	03 Apr - Super-Outbreak
9.	75	1952	Tennessee	21 Apr - Outbreak
10.	72	1974	Kentucky	03 Apr - Super-Outbreak
11t.	65	1985	Pennsylvania	31 May - Outbreak
11t.	65	1968	Arkansas	15 May - Outbreak
13.	62	1965	Ohio	11 Apr - Palm Sunday Outbreak
14.	59	1967	Illinois	21 Apr - Outbreak
15.	58	1966	Mississippi	03 Mar - Vicksburg
16.	56	1957	Missouri	20 May - Ruskin Heights
17.	54	1979	Texas	10 Apr - Wichita Falls
18.	53	1965	Michigan	11 Apr - Palm Sunday Outbreak
19.	51	1970	Texas	11 May - Lubbock
20t.	47	1974	Tennessee	03 Apr - Super-Outbreak
20t.	47	1974	Indiana	03 Apr - Super-Outbreak
22t.	42	1984	North Carolina	28 Mar - Outbreak
22t.	42	1974	Ohio	03 Apr - Xenia (Super-Outbreak)
22t.	42	1998	Florida	22 Feb - Kissimmee
22t.	42	1999	Oklahoma	03 May - Bridge Creek/Moore

(Courtesy of Storm Prediction Center (NOAA/NWS/NCEP).)

Has there ever been anything done like "Dorothy" in the movie *Twister*? What was TOTO?

In *Twister*, "Dorothy" was a large, reinforced metal bin containing small instrument pods that, with help from refabricated Pepsi cans, were supposed to be drawn into a tornado when the tornado cracked "Dorothy" open. The idea for "Dorothy" was taken from a real device that University of Oklahoma and NSSL weather scientists used in the early to mid-1980s called TOTO—the TOtable TOrnado Observatory.

What are "turtles"?

Turtles are small, squat, heavy, aerodynamic instrument packages that were designed to withstand tornado wind speeds while measuring temperature, pressure, and humidity at ground level. During the VORTEX program, they were sometimes placed on the ground at 100–250 yard intervals in the path of tornadic mesocyclones. Scientists are still analyzing data from those deployments. (Turtles do not measure winds.)

What was Project VORTEX?

That was the acronym for Verification of the Origin of Rotation in Tornadoes EXperiment, conducted in the springs of 1994 and 1995 in the southern and central U.S. plains. The basic idea was to gather the densest possible set of observations in tornadic supercells, from sensors in cars, planes, balloons, "turtles" (small instrument packages that could be placed on the ground), and portable radars. The main goal is to better understand the cause of tornado formation in thunderstorms.

The 25 Deadliest U.S. Tornadoes

NOTE: Having happened before the era of comprehensive damage surveys, some of these events may have been composed of multiple tornadoes along a damage path. Death counts for events in the 1800s and early 1900s should be treated as estimates since recordkeeping of tornado deaths was erratic back then.

	DATE	LOCATION(S)	DEATHS
1	18 Mar 1925	Tri-State (MO/IL/IN)	695
2	06 May 1840	Natchez MS	317
3	27 May 1896	St. Louis MO	255
4	05 Apr 1936	Tupelo MS	216
5	06 Apr 1936	Gainesville GA	203
6	09 Apr 1947	Woodward OK	181
7	24 Apr 1908	Amite LA, Purvis MS	143
8	12 Jun 1899	New Richmond WI	117
9	8 Jun 1953	Flint MI	115
10	11 May 1953	Waco TX	114
10	18 May 1902	Goliad TX	114
12	23 Mar 1913	Omaha NE	103
13	26 May 1917	Mattoon IL	101
14	23 Jun 1944	Shinnston WV	100
15	18 Apr 1880	Marshfield MO	99
16	01 Jun 1903	Gainesville, Holland GA	98
16	09 May 1927	Poplar Bluff MO	98
18	10 May 1905	Snyder OK	97
19	24 Apr 1908	Natchez MS	91
20	09 Jun 1953	Worcester MA	90
21	20 Apr 1920	Starkville MS, Waco AL	88
22	28 Jun 1924	Lorain, Sandusky OH	85
23	25 May 1955	Udall KS	80
24	29 Sep 1927	St. Louis MO	79
25	27 Mar 1890	Louisville KY	76

(Courtesy of Storm Prediction Center (NOAA/NWS/NCEP).)

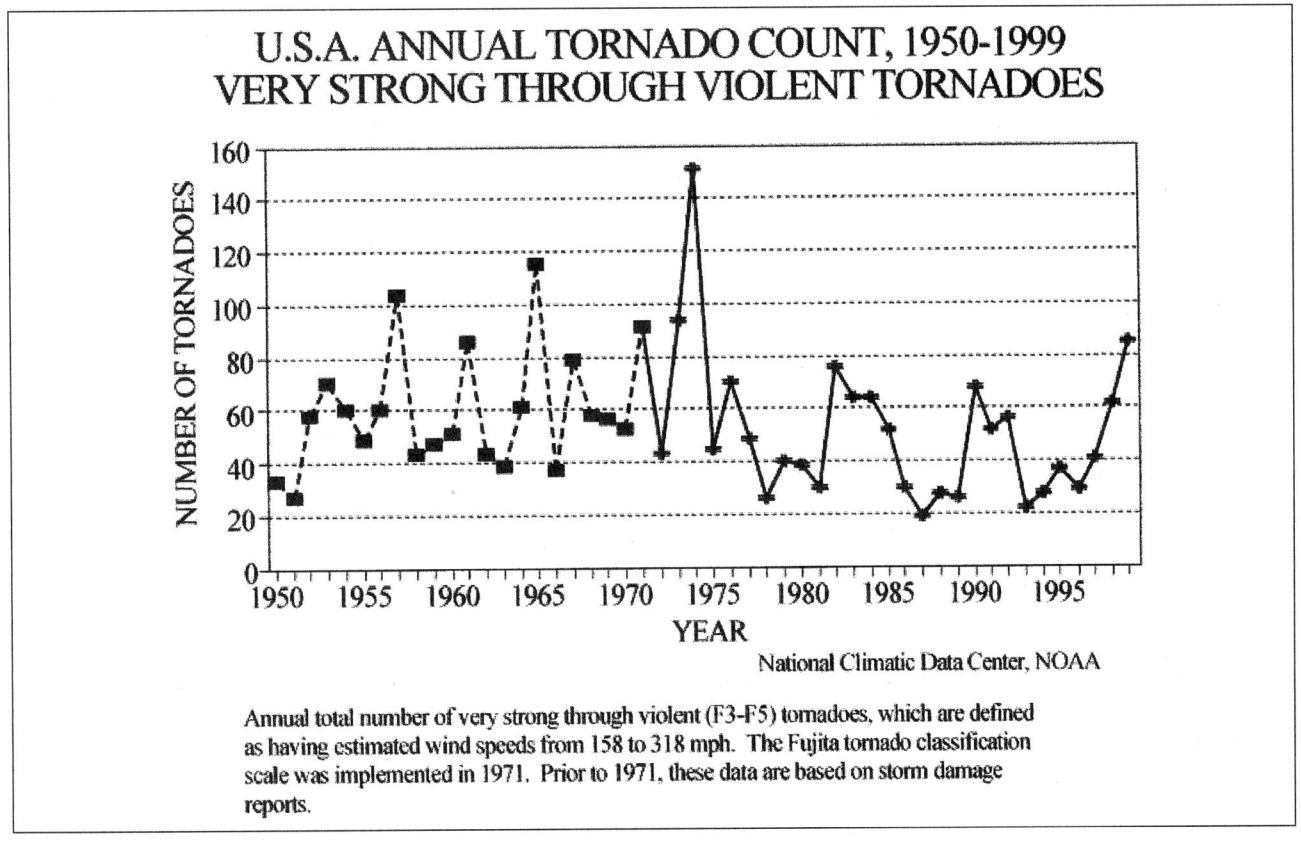

(Courtesy of National Climatic Data Center/NOAA.)

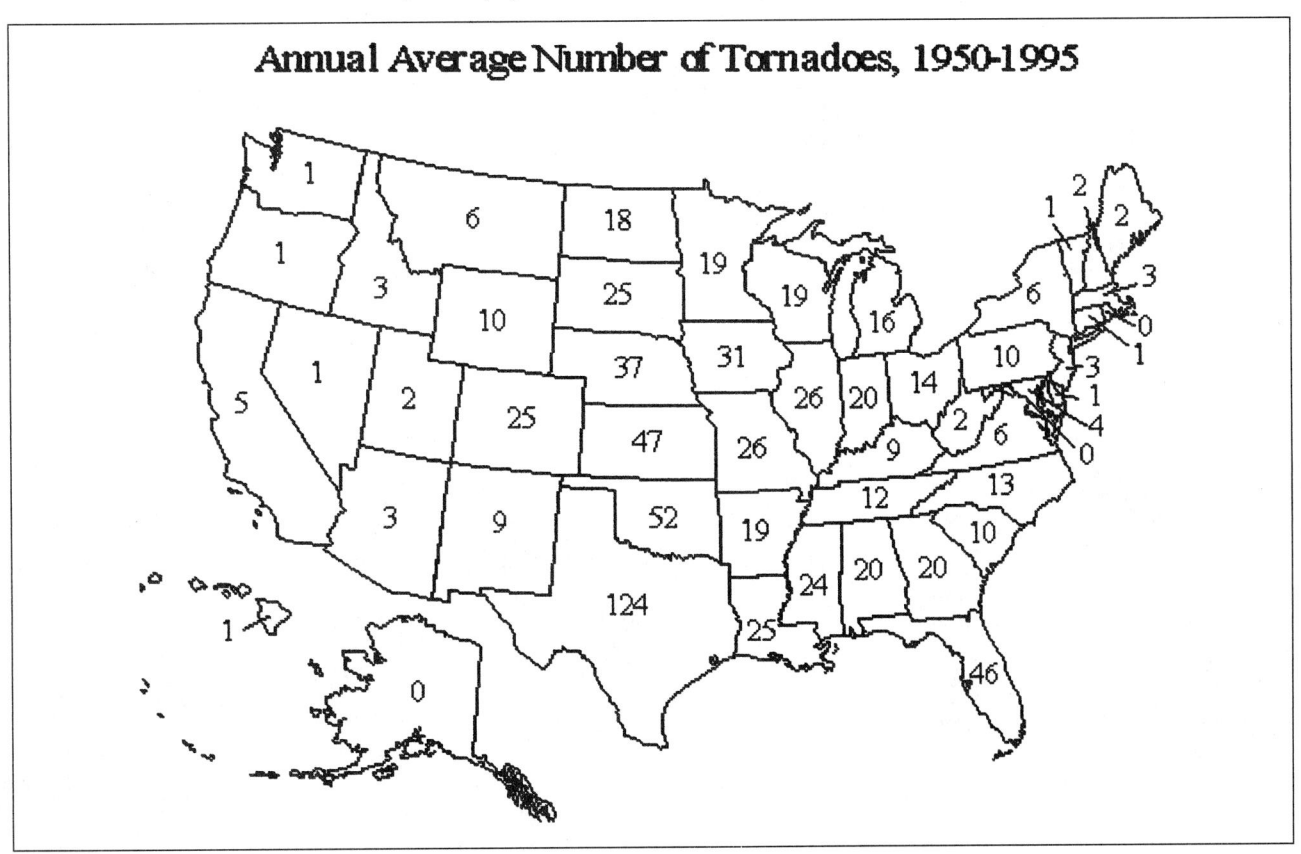

(Courtesy of Storm Prediction Center (NOAA/NWS/NCEP).)

Tornado Fatalities by Location

Year	Mobile Home	Permanent Home	Vehicle	Business	School/Church	Outdoors	Unknown	Total
1999	39	35	6	8	0	6	1	94
1998	65	40	15	7	0	3	0	130
1997	30	23	3	3	0	7	1	67
1996	14	8	2	0	0	0	1	25
1995	8	15	4	0	0	3	0	30
1994	26	14	3	0	20	6	0	69
1993	13	6	7	3	1	3	0	33
1992	20	18	0	0	0	1	0	39
1991	20	3	4	0	0	12	0	39
1990	7	11	14	15	5	1	0	53
1989	12	8	16	4	9	0	1	50
1988	21	6	3	2	0	0	0	32
1987	24	7	3	0	22	3	0	59
1986	7	3	3	0	0	0	2	15
1985	28	40	4	0	0	0	22	94

(Courtesy of Storm Prediction Center (NOAA/NWS/NCEP).)

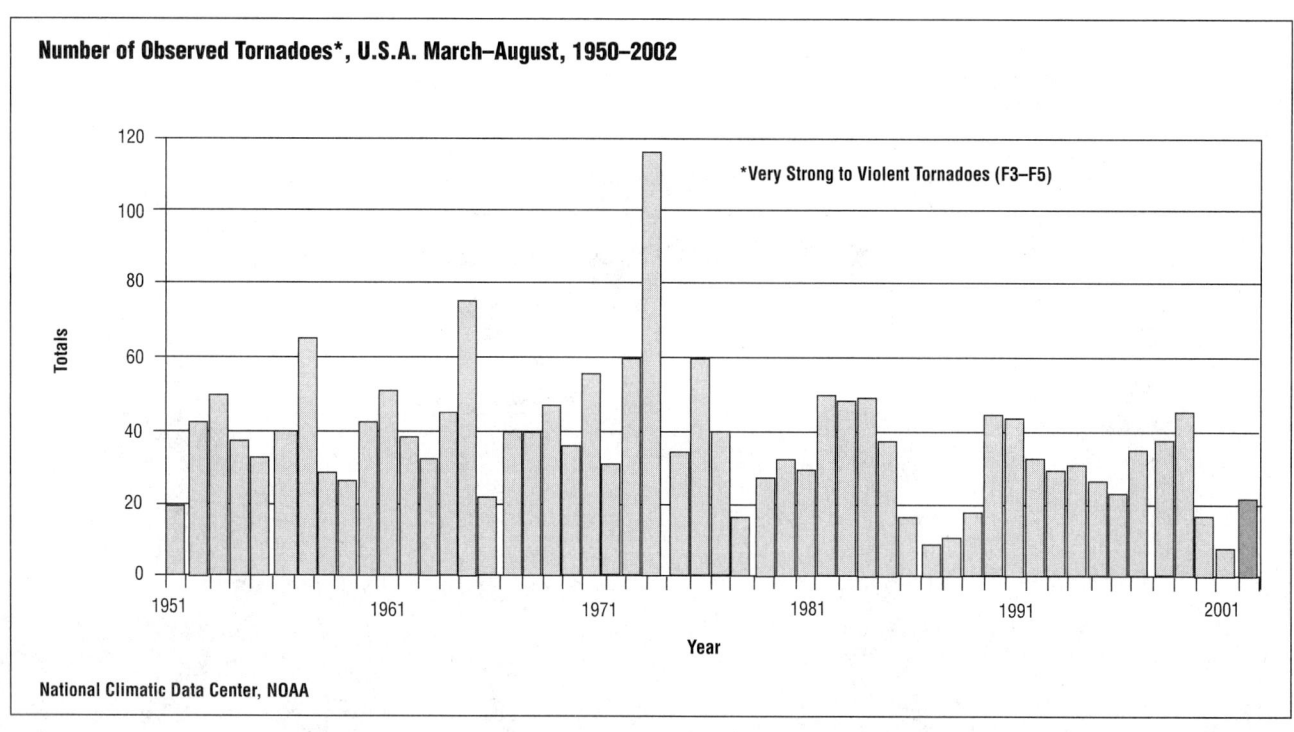

Number of Observed Tornadoes*, U.S.A. March–August, 1950–2002

*Very Strong to Violent Tornadoes (F3–F5)

National Climatic Data Center, NOAA

(Courtesy of National Climatic Data Center/NOAA.)

Nation Summary of Tornadoes, Tornado Days and Deaths by Month and Annual, 1953–2000

Year	January Number	Days	Deaths	February Number	Days	Deaths	March Number	Days	Deaths	April Number	Days	Deaths	May Number	Days	Deaths	June Number	Days	Deaths
1953	14	6	0	16	3	3	40	10	24	47	16	34	94	21	161	111	24	244
1954	2	1	0	17	9	2	63	13	10	112	22	3	101	22	9	107	26	5
1955	3	2	0	4	3	0	43	15	5	99	18	7	147	26	103	154	28	2
1956	2	2	0	47	12	8	31	7	1	85	15	67	79	24	4	65	21	0
1957	17	3	13	5	3	0	38	7	1	216	21	29	227	26	87	147	25	14
1958	12	7	0	20	5	13	15	10	0	76	19	4	68	21	0	127	27	42
1959	16	2	3	20	5	21	43	11	9	30	12	1	226	28	8	73	25	2
1960	9	4	0	28	10	0	28	10	0	70	20	7	201	26	34	124	27	3
1961	1	1	0	31	8	0	124	17	7	74	19	3	137	25	23	107	23	2
1962	12	3	1	25	7	0	37	9	17	41	8	1	200	22	3	171	29	0
1963	15	5	1	6	3	0	48	12	8	84	14	16	71	21	1	91	23	0
1964	14	3	10	2	2	0	36	11	6	157	23	15	135	20	16	136	24	0
1965	21	11	0	32	4	0	34	9	2	129	20	267	275	25	17	147	28	6
1966	1	1	0	28	5	0	12	6	58	80	20	12	98	17	0	126	28	19
1967	39	4	7	8	5	0	42	14	3	149	18	73	116	25	3	210	28	6
1968	5	3	0	7	3	0	28	8	0	102	15	40	145	26	72	136	27	11
1969	3	1	32	5	5	0	8	2	1	68	15	2	145	25	4	137	28	7
1970	9	5	0	16	3	0	25	12	2	117	16	29	88	19	26	134	24	6
1971	18	7	1	83	12	131	40	13	2	75	14	11	166	24	7	199	28	1
1972	33	10	5	7	4	0	69	17	0	96	20	16	140	27	0	114	25	2
1973	33	7	1	10	4	0	80	16	17	150	22	10	250	26	35	224	26	2
1974	24	8	2	23	9	0	36	12	1	269	22	313	144	28	10	194	26	31
1975	52	7	12	45	12	7	84	16	12	108	20	13	188	30	5	196	28	6
1976	12	5	0	37	6	5	180	18	21	113	23	1	155	24	8	169	26	3
1977	5	4	0	17	3	2	64	15	0	88	15	26	228	29	4	132	27	0
1978	23	7	2	6	3	0	17	8	0	107	17	4	213	27	7	148	28	17
1979	16	9	0	4	3	0	53	13	1	120	17	58	112	23	2	150	24	8
1980	5	4	0	11	9	0	41	15	2	137	16	4	203	25	8	217	30	7
1981	3	3	0	25	5	2	33	13	1	84	18	13	187	24	0	223	29	8
1982	18	8	1	3	2	0	60	15	6	150	20	30	327	28	14	198	30	4
1983	13	2	2	41	7	1	71	21	0	65	15	6	249	26	14	178	27	2
1984	1	1	0	27	4	0	73	15	64	176	22	33	169	27	6	242	25	14
1985	2	2	0	7	4	0	38	12	2	134	19	5	182	28	78	82	24	3
1986	0	0	0	30	11	2	75	9	6	84	17	2	173	25	1	134	25	0
1987	6	3	0	19	4	6	38	11	1	20	8	1	126	25	31	132	29	2
1988	17	3	5	4	3	0	28	10	1	58	16	4	132	24	3	63	21	0
1989	15	6	0	18	3	0	44	14	1	82	13	0	234	28	9	253	27	5
1990	11	7	0	57	10	1	88	8	3	108	17	0	243	27	5	353	28	11
1991	29	6	1	11	5	0	159	16	13	206	20	21	335	30	0	216	30	1
1992	15	6	0	29	9	0	55	17	5	53	11	0	137	21	0	400	29	1
1993	17	8	0	34	7	3	49	11	5	85	15	10	179	26	2	316	27	1
1994	13	5	0	9	7	0	61	12	40	206	23	12	161	27	0	236	29	3
1995	36	9	3	7	4	6	49	11	0	130	17	1	394	28	16	219	25	0
1996	35	10	1	14	5	1	72	10	6	180	16	12	236	29	1	127	27	0
1997	50	10	2	23	6	1	102	13	28	114	16	1	225	27	29	194	29	0
1998	49	12	0	78	18	42	80	12	16	208	19	55	327	29	10	401	29	3
1999	256	12	18	24	7	0	67	12	1	214	19	15	363	29	53	298	27	4
2000	16	4	0	58	12	19	103	20	2	136	19	2	241	29	3	136	27	0
Por	1018	249	123	1078	293	276	2704	588	411	5492	837	1289	8972	1219	932	8429	1277	508
Mean	21	5	3	22	6	6	56	12	9	114	17	27	187	25	19	176	27	11

(Courtesy of National Climatic Data Center/NOAA.)

Nation Summary of Tornadoes, Tornado Days and Deaths by Month and Annual, 1953–2000

Year	July Number	July Days	July Deaths	August Number	August Days	August Deaths	September Number	September Days	September Deaths	October Number	October Days	October Deaths	November Number	November Days	November Deaths	December Number	December Days	December Deaths	Annual Number	Annual Days	Annual Deaths
1953	31	19	0	24	15	0	5	4	0	6	4	0	12	6	0	21	8	49	421	136	515
1954	45	23	0	49	21	1	21	10	3	14	8	2	2	2	0	17	3	1	550	160	36
1955	49	21	5	33	18	0	15	8	2	23	7	1	20	4	1	3	2	0	593	152	126
1956	91	26	1	43	20	2	16	10	0	29	8	0	7	6	0	9	4	0	504	155	83
1957	55	19	0	20	14	0	17	10	2	18	11	2	58	11	25	38	4	19	856	154	192
1958	121	30	1	46	20	1	24	14	1	9	6	4	45	6	0	1	1	0	564	166	66
1959	63	24	0	38	18	0	58	15	14	24	10	0	11	4	0	2	2	0	604	156	58
1960	43	22	0	47	23	1	22	13	0	18	10	1	25	6	0	1	1	0	616	172	46
1961	77	27	0	27	16	0	53	16	15	14	5	0	36	7	1	16	5	0	697	169	51
1962	78	26	0	51	21	6	24	11	0	11	10	0	5	4	0	2	2	0	657	152	28
1963	62	26	0	26	13	2	33	13	3	13	5	0	15	6	0	0	0	0	464	141	31
1964	63	23	0	79	23	2	25	10	0	22	4	22	17	8	0	18	5	2	704	156	72
1965	86	26	0	61	23	1	64	21	0	16	4	1	34	6	5	7	4	0	906	181	299
1966	100	27	3	58	21	0	22	13	0	29	6	6	20	3	0	11	3	0	585	150	98
1967	90	25	1	28	16	2	139	16	5	36	7	4	8	5	0	61	10	10	926	173	110
1968	56	22	2	66	23	2	25	14	0	14	9	0	44	12	2	21	9	1	660	171	131
1969	99	27	0	69	21	19	20	11	0	26	10	0	5	3	0	23	7	1	608	155	66
1970	81	26	3	55	21	0	54	20	0	50	13	6	10	4	0	14	8	0	653	171	72
1971	100	30	1	50	21	0	47	15	0	38	12	0	16	7	0	56	9	2	888	192	156
1972	115	29	0	59	23	2	49	19	0	34	10	0	17	4	2	8	6	0	741	194	27
1973	80	26	0	51	23	4	69	22	3	25	11	0	81	11	12	49	12	3	1102	206	87
1974	59	19	0	107	26	0	25	11	0	45	10	4	13	8	0	8	5	0	947	184	361
1975	79	26	2	60	25	2	34	17	0	12	7	0	40	8	0	22	8	1	920	204	60
1976	84	28	2	38	18	1	35	15	3	11	5	0	0	0	0	1	1	0	835	169	44
1977	99	27	1	82	26	6	65	21	1	25	5	1	24	10	0	23	7	2	852	189	43
1978	143	30	11	65	24	1	20	10	6	7	5	0	9	5	0	30	9	5	788	173	53
1979	132	30	1	127	27	5	68	19	2	47	12	7	21	8	0	2	1	0	852	186	84
1980	95	26	5	73	27	0	37	14	1	43	7	1	3	2	0	1	1	0	866	176	28
1981	98	27	0	64	22	0	26	16	0	32	12	0	7	5	0	1	1	0	783	175	24
1982	95	29	0	34	15	0	38	12	2	9	4	0	19	6	0	95	13	7	1046	182	64
1983	99	27	4	76	21	0	20	15	0	12	5	0	49	11	0	58	13	5	931	190	34
1984	72	21	0	47	20	0	17	12	0	49	12	4	30	5	1	4	2	0	907	166	122
1985	51	19	0	108	26	3	40	16	0	18	8	0	19	8	3	3	2	0	684	168	94
1986	88	24	3	67	23	1	65	17	0	26	7	0	17	8	0	5	2	0	764	168	15
1987	163	28	0	63	24	1	19	10	0	1	1	0	55	5	11	14	3	6	656	151	59
1988	103	23	0	61	13	3	76	16	1	19	8	0	121	13	14	20	6	1	702	156	32
1989	59	19	0	36	20	0	31	12	0	30	7	4	58	10	31	3	1	0	856	160	50
1990	106	26	0	60	22	29	45	15	0	35	9	2	18	5	0	35	7	2	1133	181	53
1991	65	22	1	46	17	0	26	16	0	22	9	0	20	5	2	3	3	0	1132	179	39
1992	214	30	0	115	22	3	82	17	0	34	10	4	149	14	26	20	9	0	1298	195	39
1993	242	31	0	112	26	6	65	17	2	55	9	4	19	6	0	6	3	0	1176	186	33
1994	155	27	3	120	24	4	30	17	0	51	15	0	42	10	7	4	3	0	1082	199	69
1995	163	28	0	53	22	0	19	11	0	74	11	1	79	5	2	19	7	1	1235	178	30
1996	204	28	1	72	25	0	101	20	0	65	10	0	52	8	2	15	8	1	1170	196	25
1997	189	30	4	84	26	1	32	12	1	101	12	0	25	8	0	12	7	0	1148	196	67
1998	82	22	0	64	20	0	109	19	2	95	13	2	28	6	0	6	3	0	1525	202	130
1999	110	28	0	91	22	1	54	16	0	20	7	0	8	3	0	15	4	2	1520	186	94
2000	148	31	1	53	23	0	47	17	2	64	15	0	50	11	0	26	4	12	1076	212	41
Por	4782	1230	56	2958	1020	112	2028	695	71	1471	405	83	1463	318	148	840	238	133	41183	8369	4142
Mean	100	26	1	62	21	2	42	14	1	31	8	2	30	7	3	18	5	3	858	174	86

(Courtesy of National Climatic Data Center/NOAA.)

Tornadoes, Tornado Days and Deaths by State and Nation 1953–2000

State	Tornadoes							Days		Deaths		
	Total	Average	Greatest	Year	Least	Year	Per # 10,000 Sq. Mi.	Total	Average	Total	Average	Per[1] 10,000 Sq. Mi.
Alabama	1092	22	57	1998	5	1956	4.26	546	11	328	6	64
Alaska	1	0	1	1959	0	1989+	.00	1	0	0	0	0
Arizona	179	3	17	1972	0	1965+	.27	142	2	3	0	0
Arkansas	1131	23	140	1999	2	1987+	4.33	469	9	214	4	40
California	294	6	25	1998	0	1968+	.38	211	4	0	0	0
Colorado	1432	29	98	1996	1	1959	2.78	722	15	2	0	0
Connecticut	65	1	8	1973	0	1988+	2.00	56	1	4	0	8
Delaware	55	1	6	1992	0	1987+	4.86	44	0	2	0	10
District of Columbia	1	0	1	1995	0	2000+	.00	1	0	0	0	0
Florida	2513	52	115	1997+	10	1956	8.88	1541	32	128	2	22
Georgia	1018	21	52	1994	2	1987	3.57	531	11	147	3	25
Hawaii	27	0	4	1971	0	1987+	.00	23	0	0	0	0
Idaho	154	3	13	2000	0	1977+	.36	122	2	0	0	0
Illinois	1524	31	107	1974	4	1953	5.50	614	12	181	3	32
Indiana	1002	20	49	1990	4	1984	5.51	458	9	221	4	61
Iowa	1643	34	78	1998	7	1956	6.04	676	14	63	1	11
Kansas	2386	49	116	1991	14	1976	5.96	965	20	196	4	24
Kentucky	506	10	39	1997	0	1953	2.48	264	5	110	2	27
Louisiana	1277	26	79	1992	3	1955	5.36	689	14	111	2	23
Maine	90	1	11	1971	0	1987+	.30	80	1	1	0	0
Maryland	212	4	24	1995	0	1988+	3.78	126	2	2	0	2
Massachusetts	140	2	12	1958	0	1988+	2.42	99	2	102	2	124
Michigan	786	16	39	1974	2	1959	2.75	443	9	238	4	41
Minnesota	1069	22	65	1998	5	1988+	2.62	521	10	86	1	10
Mississippi	1202	25	62	1988	1	1979	5.24	566	11	371	7	78
Missouri	1319	27	79	1973	6	1987+	2.88	570	11	137	2	20
Montana	309	6	30	1991	0	1974+	.41	209	4	1	0	0
Nebraska	1999	41	106	1999	10	1966	5.31	846	17	51	1	7
Nevada	63	1	8	1987+	0	1985+	.09	57	1	0	0	0
New Hampshire	80	1	9	1963	0	1987+	1.07	68	1	0	0	0
New Jersey	123	2	17	1989	0	1984+	2.55	93	1	0	0	0
New Mexico	442	9	31	1991	0	1953	.74	317	6	3	0	0
New York	297	6	25	1992	0	1953	1.21	192	4	21	0	4
North Carolina	789	16	66	1998	2	1970	3.04	413	8	85	1	16
North Dakota	975	20	65	1999	2	1961	2.83	488	10	22	0	3
Ohio	744	15	61	1992	0	1988	3.64	374	7	177	3	43
Oklahoma	2706	56	171	1999	17	1988	8.01	992	20	255	5	36
Oregon	75	1	14	1997	0	1988+	.10	65	1	0	0	0
Pennsylvania	557	11	62	1998	0	1959	2.43	311	6	81	1	18
Puerto Rico	12	0	2	1979+	0	1989+	.00	11	0	0	0	0
Rhode Island	9	0	3	1986	0	1988+	.00	8	0	0	0	0
South Carolina	596	12	54	1995	1	1986+	3.86	340	7	50	1	16
South Dakota	1310	27	85	1993	1	1958	3.50	686	12	15	0	2
Tennessee	656	13	44	1974	1	1987+	3.08	318	6	117	2	28
Texas	6452	134	232	1967+	32	1953	5.01	2355	49	499	10	19
Utah	94	1	7	1998	0	1989+	.12	81	1	1	0	0
Vermont	33	0	5	1962	0	1985+	.00	30	0	0	0	0
Virginia	355	7	28	1993	1	1982+	1.72	217	4	25	0	6
Virgin Islands	2	0	1	1979+	0	1989+	.00	2	0	0	0	0
Washington	75	1	14	1997	0	1988+	.15	63	1	6	0	1
West Virginia	107	2	17	1998	0	1988+	.83	77	1	2	0	1
Wisconsin	927	19	43	1980	3	1953	3.44	469	9	82	1	15
Wyoming	513	10	42	1977	0	1970	1.02	339	7	2	0	0
Pacific Islands	2	0	1	1981+	0	1989+	.00	2	0	0	0	0
United States	41183*	857	1525	1998	421	1953	2.37	8369&	174	3142	86	11

+Also in earlier year(s).
*Corrected for Boundary-crossing tornadoes.
&Tornado days for country as a whole.
#Mean annual tornadoes per 10,000 square miles.
[1]Number of deaths per 10,000 square miles.

(Courtesy of National Climatic Data Center/NOAA.)

National Tornadoes, Tornado Days, Deaths and Resulting Losses by Years, 1916–2000

Year	Number Tornadoes	Tornado Days	Total Deaths	Most Deaths In Single Tornado	Total Property Losses $	Property Loss Frequency* Category 5	Category 6	Category 7 and over
1916	90	36	150	30	6	7	1	0
1917	121	38	551	101	7	21	9	0
1918	81	45	136	36	7	20	5	0
1919	64	35	206	59	7	10	2	0
1920	87	50	499	87	7	14	10	0
1921	105	55	202	61	7	22	3	0
1922	108	64	135	16	7	27	5	0
1923	102	59	110	23	6	21	1	0
1924	130	57	376	85	7	26	11	1
1925	119	65	794	689	7	34	2	1
1926	111	57	144	23	6	28	0	0
1927	163	62	540	92	7	42	9	1
1928	203	79	95	14	7	40	7	0
1929	197	74	274	40	7	48	4	0
1930	192	72	179	41	7	38	6	0
1931	94	57	36	6	6	14	1	0
1932	151	67	394	37	7	23	1	1
1933	258	96	362	34	7	46	9	0
1934	147	77	47	6	6	10	3	0
1935	180	77	71	11	6	29	0	0
1936	151	71	552	216	7	17	5	1
1937	147	75	29	5	6	24	0	0
1938	213	76	183	32	7	29	6	0
1939	152	75	91	27	7	21	3	0
1940	124	62	65	18	7	13	2	0
1941	118	57	53	25	6	24	1	0
1942	167	66	384	65	7	42	10	0
1943	152	61	58	5	7	28	8	0
1944	169	68	275	100	7	50	9	0
1945	121	66	210	69	7	21	10	1
1946	106	65	78	15	7	29	7	0
1947	165	78	313	169	7	46	7	1
1948	183	68	139	33	7	62	11	2
1949	249	80	211	58	7	54	13	0
1950	200	88	70	18	7	47	9	0
1951	262	113	34	6	7	35	11	2
1952	240	98	229	57	7	53	19	0
1953	421	136	515	116	8	63	18	7
1954	550	160	36	6	7	63	8	1
1955	593	152	126	80	7	74	13	1
1956	504	155	83	25	7	83	24	1
1957	856	154	192	44	8	129	26	3
1958	564	166	66	19	7	70	8	1
1959	604	156	58	21	7	70	4	1
1960	616	172	46	16	7	65	11	1
1961	697	169	51	16	7	103	21	1
1962	657	152	28	17	7	51	10	0
1963	464	141	31	5	7	77	15	1
1964	704	156	73	22	7	113	17	5
1965	906	181	299	44	8	126	30	11
1966	585	150	98	58	8	79	13	4
1967	926	173	114	33	8	125	33	8
1968	660	171	131	34	8	82	26	6
1969	608	155	66	32	8	98	16	3
1970	653	171	72	26	8	97	24	6
1971	888	192	156	58	8	71	30	5
1972	741	194	27	6	8	100	28	1
1973	1102	206	87	7	9	219	67	9
1974	947	184	361	34	9	166	82	25
1975	920	204	60	9	9	189	31	11
1976	835	169	44	5	8	145	41	5
1977	852	189	43	22	8	173	40	6
1978	788	173	53	16	9	153	53	6
1979	852	186	84	42	9	169	62	11
1980	866	176	28	5	9	201	79	13

(Courtesy of National Climatic Data Center/NOAA.)

National Tornadoes, Tornado Days, Deaths and Resulting Losses by Years, 1916–2000 [CONTINUED]

Year	Number Tornadoes	Tornado Days	Total Deaths	Most Deaths In Single Tornado	Total Property Losses $	Property Loss Frequency* Category 5	Category 6	Category 7 and over
1981	783	175	24	5	9	144	43	12
1982	1046	182	64	10	9	254	79	13
1983	931	190	34	3	9	211	85	10
1984	907	166	122	16	9	193	90	35
1985	684	168	94	18	9	114	55	14
1986	764	168	15	3	9	157	66	9
1987	656	151	59	30	8	112	32	6
1988	702	156	32	5	9	148	48	17
1989	856	160	50	21	9	133	60	18
1990	1133	181	53	29	8	215	91	18
1991	1132	179	39	17	8	194	49	15
1992	1298	195	39	12	8	212	83	25
1993	1176	186	33	7	8	186	59	13
1994	1082	199	69	22	8	194	68	15
1995	1235	178	30	6	8	203	52	11
1996	1170	196	25	5	8	224	86	22
1997	1148	196	67	27	8	202	63	10
1998	1525	202	130	32	9	327	134	46
1999	1520	186	94	12	9	260	115	24
2000	1076	212	41	11	8	182	53	15
Mean	858	174	86	—	—	146	47	10

Note: The above estimated losses are based on values at time of occurrence. Mean was derived from data for period 1953–1992
$ Storm damages in categories:
   5. $50,000 to $500,000
   6. $500,000 to $5 Million
   7. $5 Million to $50 Million
   8. $50 Million to $500 Million
   9. $500 Million and over
*Number of times property losses reported in Storm Data in categories 5, 6, 7 and over.

(Courtesy of National Climatic Data Center/NOAA.)

Tornado Summary by State and Nation, 1999

	Jan	Feb	Mar	Apr	May	Jun	Jul	Aug	Sep	Oct	Nov	Dec	Ann
Alabama													
Number	5	5	2	3	5	0	1	0	0	2	0	0	23
Days	1	2	2	2	3	0	1	0	0	1	0	0	12
Deaths	0	0	0	0	0	0	0	0	0	0	0	0	0
Injuries	0	1	0	0	0	0	0	0	0	0	0	0	1
Arizona													
Number	0	0	0	0	0	0	1	0	4	0	0	0	5
Days	0	0	0	0	0	0	1	0	3	0	0	0	4
Deaths	0	0	0	0	0	0	0	0	0	0	0	0	0
Injuries	0	0	0	0	0	0	0	0	0	0	0	0	0
Arkansas													
Number	91	2	17	10	12	2	1	0	0	0	0	5	140
Days	3	1	1	4	3	2	1	0	0	0	0	2	17
Deaths	8	0	0	0	0	0	0	0	0	0	0	0	8
Injuries	143	0	0	7	0	0	26	0	0	0	0	0	176
California													
Number	0	0	0	2	0	0	1	0	0	0	0	0	3
Days	0	0	0	2	0	0	1	0	0	0	0	0	3
Deaths	0	0	0	0	0	0	0	0	0	0	0	0	0
Injuries	0	0	0	0	0	0	0	0	0	0	0	0	0
Colorado													
Number	0	0	0	1	9	12	4	11	4	0	0	0	41
Days	0	0	0	1	4	6	4	4	2	0	0	0	21
Deaths	0	0	0	0	0	0	0	0	0	0	0	0	0
Injuries	0	0	0	0	0	1	0	0	0	0	0	0	1
Florida													
Number	13	3	6	0	11	4	9	4	5	5	1	0	61
Days	3	2	4	0	8	4	7	4	3	2	1	0	38
Deaths	0	0	0	0	0	0	0	0	0	0	0	0	0
Injuries	8	0	0	0	0	0	0	0	0	4	0	0	12
Georgia													
Number	1	0	0	5	4	2	0	0	0	0	1	0	13
Days	1	0	0	1	3	2	0	0	0	0	1	0	8
Deaths	0	0	0	0	0	0	0	0	0	0	0	0	0
Injuries	0	0	0	30	3	0	0	0	0	0	0	0	33
Idaho													
Number	0	0	0	0	0	0	1	0	0	0	0	0	1
Days	0	0	0	0	0	0	1	0	0	0	0	0	1
Deaths	0	0	0	0	0	0	0	0	0	0	0	0	0
Injuries	0	0	0	0	0	0	0	0	0	0	0	0	0
Illinois													
Number	4	5	0	21	5	39	0	2	0	0	0	0	76
Days	2	2	0	3	3	4	0	1	0	0	0	0	15
Deaths	0	0	0	2	0	1	0	0	0	0	0	0	3
Injuries	0	0	0	11	4	10	0	0	0	0	0	0	25
Indiana													
Number	0	0	0	4	7	0	0	0	1	0	0	0	12
Days	0	0	0	1	2	0	0	0	1	0	0	0	4
Deaths	0	0	0	0	0	0	0	0	0	0	0	0	0
Injuries	0	0	0	2	0	0	0	0	1	0	0	0	3
Iowa													
Number	0	0	0	29	15	19	6	3	1	0	0	0	73
Days	0	0	0	2	2	6	3	1	1	0	0	0	15
Deaths	0	0	0	0	2	0	0	0	0	0	0	0	2
Injuries	0	0	0	12	16	0	0	0	0	0	0	0	28
Kansas													
Number	0	0	0	11	36	13	6	0	2	0	0	0	68
Days	0	0	0	3	9	5	3	0	2	0	0	0	22
Deaths	0	0	0	0	6	0	0	0	0	0	0	0	6
Injuries	0	0	0	1	154	0	0	0	0	0	0	0	155
Kentucky													
Number	2	1	0	0	1	0	1	1	0	0	0	0	6
Days	1	1	0	0	1	0	1	1	0	0	0	0	5
Deaths	0	0	0	0	0	0	0	0	0	0	0	0	0
Injuries	8	2	0	0	0	0	0	0	0	0	0	0	10

(Courtesy of National Climatic Data Center/NOAA.)

Tornado Summary by State and Nation, 1999 [CONTINUED]

	Jan	Feb	Mar	Apr	May	Jun	Jul	Aug	Sep	Oct	Nov	Dec	Ann
Louisiana													
Number	51	1	9	5	3	0	1	0	1	0	0	1	72
Days	7	1	3	1	3	0	1	0	1	0	0	1	18
Deaths	0	0	0	7	0	0	0	0	0	0	0	0	7
Injuries	5	0	5	103	0	0	0	0	0	0	0	0	113
Maine													
Number	0	0	0	0	0	0	0	1	0	0	0	0	1
Days	0	0	0	0	0	0	0	1	0	0	0	0	1
Deaths	0	0	0	0	0	0	0	0	0	0	0	0	0
Injuries	0	0	0	0	0	0	0	0	0	0	0	0	0
Maryland													
Number	0	0	0	0	0	0	1	2	0	0	0	0	3
Days	0	0	0	0	0	0	1	2	0	0	0	0	3
Deaths	0	0	0	0	0	0	0	0	0	0	0	0	0
Injuries	0	0	0	0	0	0	0	0	0	0	0	0	0
Michigan													
Number	0	0	0	0	5	0	7	1	0	0	0	0	13
Days	0	0	0	0	3	0	4	1	0	0	0	0	8
Deaths	0	0	0	0	0	0	0	0	0	0	0	0	0
Injuries	0	0	0	0	0	0	2	0	0	0	0	0	2
Minnesota													
Number	0	0	0	0	0	9	15	16	0	0	0	0	40
Days	0	0	0	0	0	4	7	2	0	0	0	0	13
Deaths	0	0	0	0	0	0	0	0	0	0	0	0	0
Injuries0	0	0	0	0	0	0	2	0	0	0	0	0	2
Mississippi													
Number	33	2	3	6	5	0	0	0	0	0	0	1	50
Days	4	1	2	1	2	0	0	0	0	0	0	1	11
Deaths	0	0	0	1	0	0	0	0	0	0	0	0	1
Injuries	2	0	0	33	0	0	0	0	0	0	0	1	36
Missouri													
Number	13	2	0	19	2	11	0	0	0	0	0	0	47
Days	1	1	0	3	1	4	0	0	0	0	0	0	10
Deaths	0	0	0	0	0	0	0	0	0	0	0	0	0
Injuries	0	0	0	11	1	0	0	0	0	0	0	0	12
Montana													
Number	0	0	0	0	1	9	5	7	0	0	0	0	22
Days	0	0	0	0	1	6	3	2	0	0	0	0	12
Deaths	0	0	0	0	0	0	0	0	0	0	0	0	0
Injuries	0	0	0	0	0	0	0	3	0	0	0	0	3
Nebraska													
Number	0	0	0	17	27	49	4	4	5	0	0	0	106
Days	0	0	0	3	6	10	2	1	4	0	0	0	26
Deaths	0	0	0	0	0	0	0	0	0	0	0	0	0
Injuries	0	0	0	0	0	0	1	0	0	0	0	0	1
Nevada													
Number	0	0	0	0	0	0	1	0	0	0	0	0	1
Days	0	0	0	0	0	0	1	0	0	0	0	0	1
Deaths	0	0	0	0	0	0	0	0	0	0	0	0	0
Injuries	0	0	0	0	0	0	0	0	0	0	0	0	0
New Hampshire													
Number	0	0	0	0	0	0	3	2	0	0	0	0	5
Days	0	0	0	0	0	0	1	1	0	0	0	0	2
Deaths	0	0	0	0	0	0	0	0	0	0	0	0	0
Injuries	0	0	0	0	0	0	0	0	0	0	0	0	0
New Jersey													
Number	0	1	0	0	0	0	0	1	0	0	0	0	2
Days	0	1	0	0	0	0	0	1	0	0	0	0	2
Deaths	0	0	0	0	0	0	0	0	0	0	0	0	0
Injuries	0	0	0	0	0	0	0	1	0	0	0	0	1
New Mexico													
Number	0	0	0	1	4	3	0	0	0	0	0	0	8
Days	0	0	0	1	2	2	0	0	0	0	0	0	5
Deaths	0	0	0	0	0	0	0	0	0	0	0	0	0
Injuries	0	0	0	0	0	0	0	0	0	0	0	0	0

(Courtesy of National Climatic Data Center/NOAA.)

Tornado Summary by State and Nation, 1999 [CONTINUED]

	Jan	Feb	Mar	Apr	May	Jun	Jul	Aug	Sep	Oct	Nov	Dec	Ann
New York													
Number	0	0	0	0	0	0	0	1	0	0	0	0	1
Days	0	0	0	0	0	0	0	1	0	0	0	0	1
Deaths	0	0	0	0	0	0	0	0	0	0	0	0	0
Injuries	0	0	0	0	0	0	0	1	0	0	0	0	1
North Carolina													
Number	0	0	0	13	1	0	0	1	21	2	0	0	38
Days	0	0	0	2	1	0	0	1	2	1	0	0	7
Deaths	0	0	0	1	0	0	0	0	0	0	0	0	1
Injuries	0	0	0	37	0	0	0	0	0	1	0	0	38
North Dakota													
Number	0	0	0	0	0	41	9	15	0	0	0	0	65
Days	0	0	0	0	0	7	3	3	0	0	0	0	13
Deaths	0	0	0	0	0	0	0	0	0	0	0	0	0
Injuries	0	0	0	0	0	0	0	0	0	0	0	0	0
Ohio													
Number	0	0	0	6	0	0	11	2	0	3	0	0	22
Days	0	0	0	1	0	0	3	1	0	1	0	0	6
Deaths	0	0	0	4	0	0	0	0	0	0	0	0	4
Injuries	0	0	0	65	0	0	4	0	0	6	0	0	75
Oklahoma													
Number	0	1	6	22	113	14	0	1	2	4	5	3	171
Days	0	1	1	4	6	4	0	1	1	1	1	1	21
Deaths	0	0	0	0	40	2	0	0	0	0	0	0	42
Injuries	0	0	4	1	675	5	0	0	0	0	0	1	686
Oregon													
Number	0	0	0	0	0	0	0	1	0	2	0	1	4
Days	0	0	0	0	0	0	0	1	0	2	0	1	4
Deaths	0	0	0	0	0	0	0	0	0	0	0	0	0
Injuries	0	0	0	0	0	0	0	0	0	0	0	1	1
Pennsylvania													
Number	2	0	1	0	0	0	4	1	0	0	1	0	9
Days	1	0	1	0	0	0	2	1	0	0	1	0	6
Deaths	0	0	0	0	0	0	0	0	0	0	0	0	0
Injuries	18	0	0	0	0	0	0	0	0	0	12	0	30
South Carolina													
Number	0	1	0	11	3	1	0	0	1	1	0	0	18
Days	0	1	0	5	1	1	0	0	1	1	0	0	10
Deaths	0	0	0	0	0	0	0	0	0	0	0	0	0
Injuries	0	0	0	0	0	0	0	0	6	0	0	0	6
South Dakota													
Number	0	0	0	0	13	27	5	0	0	0	0	0	45
Days	0	0	0	0	3	8	3	0	0	0	0	0	14
Deaths	0	0	0	0	0	1	0	0	0	0	0	0	1
Injuries	0	0	0	0	0	54	0	0	0	0	0	0	54
Tennessee													
Number	18	0	0	0	6	0	0	1	0	0	0	0	25
Days	3	0	0	0	1	0	0	1	0	0	0	0	5
Deaths	9	0	0	0	3	0	0	0	0	0	0	0	12
Injuries	139	0	0	0	17	0	0	0	0	0	0	0	156
Texas													
Number	22	0	22	23	70	27	3	8	0	0	0	4	179
Days	4	0	6	7	14	10	3	5	0	0	0	2	51
Deaths	1	0	1	0	2	0	0	0	0	0	0	2	6
Injuries	23	0	6	1	41	0	0	0	0	0	0	3	74
Utah													
Number	0	0	0	0	1	0	0	1	1	0	0	0	3
Days	0	0	0	0	1	0	0	1	1	0	0	0	3
Deaths	0	0	0	0	0	0	0	1	0	0	0	0	1
Injuries	0	0	0	0	0	0	0	80	1	0	0	0	81
Virginia													
Number	1	0	1	2	0	0	6	0	3	0	0	0	13
Days	1	0	1	2	0	0	2	0	2	0	0	0	8
Deaths	0	0	0	0	0	0	0	0	0	0	0	0	0
Injuries	0	17	0	0	0	0	0	0	6	0	0	0	23
Washington													
Number	0	0	0	1	1	0	0	0	0	0	0	0	2
Days	0	0	0	1	1	0	0	0	0	0	0	0	2
Deaths	0	0	0	0	0	0	0	0	0	0	0	0	0

(Courtesy of National Climatic Data Center/NOAA.)

Tornado Summary by State and Nation, 1999 [CONTINUED]

	Jan	Feb	Mar	Apr	May	Jun	Jul	Aug	Sep	Oct	Nov	Dec	Ann
West Virginia													
Number	0	0	0	0	0	0	0	0	0	1	0	0	1
Days	0	0	0	0	0	0	0	0	0	1	0	0	1
Deaths	0	0	0	0	0	0	0	0	0	0	0	0	0
Injuries	0	0	0	0	0	0	0	0	0	0	0	0	0
Wisconsin													
Number	0	0	0	0	1	6	3	1	0	0	0	0	11
Days	0	0	0	0	1	3	2	1	0	0	0	0	7
Death	0	0	0	0	0	0	0	0	0	0	0	0	0
Injuries	0	0	0	0	0	0	3	0	0	0	0	0	3
Wyoming													
Number	0	0	0	2	2	10	1	3	3	0	0	0	21
Days	0	0	0	1	2	5	1	2	2	0	0	0	13
Deaths	0	0	0	0	0	0	0	0	0	0	0	0	0
Injuries	0	0	0	0	0	0	0	0	0	0	0	0	0
United States													
Number	256	24	67	214	363	298	110	91	54	20	8	15	1520
Days	12	7	12	19	29	27	28	22	16	7	3	4	186&
Deaths	18	0	1	15	53	4	0	1	0	0	0	2	94
Injuries	346	3	32	314	911	70	38	85	14	11	12	6	1842

*Corrected for boundary-crossing tornadoes.
&Tornado days for country as a whole.

(Courtesy of National Climatic Data Center/NOAA.)

Tornado Summary by State and Nation, 2000

	Jan	Feb	Mar	Apr	May	Jun	Jul	Aug	Sep	Oct	Nov	Dec	Ann
Alabama													
Number	3	3	7	11	0	0	0	0	0	0	8	12	44
Days	1	1	2	3	0	0	0	0	0	0	4	1	12
Deaths	0	0	0	1	0	0	0	0	0	0	0	12	13
Injuries	1	0	0	7	0	0	0	0	0	0	2	169	179
Arkansas													
Number	4	8	1	5	18	0	1	0	0	0	0	0	37
Days	1	3	1	1	6	0	1	0	0	0	0	0	13
Deaths	0	0	0	0	0	0	0	0	0	0	0	0	0
Injuries	0	11	0	0	1	0	0	0	0	0	0	0	12
California													
Number	0	5	0	0	0	0	1	2	0	0	1	0	9
Days	0	2	0	0	0	0	1	2	0	0	1	0	6
Deaths	0	0	0	0	0	0	0	0	0	0	0	0	0
Injuries	0	0	0	0	0	0	0	0	0	0	0	0	0
Colorado													
Number	0	0	0	2	28	8	19	0	3	0	0	0	60
Days	0	0	0	2	2	4	8	0	3	0	0	0	19
Deaths	0	0	0	0	0	0	0	0	0	0	0	0	0
Injuries	0	0	0	0	0	0	2	0	0	0	0	0	2
Connecticut													
Number	0	0	0	0	0	0	0	1	0	0	0	0	1
Days	0	0	0	0	0	0	0	1	0	0	0	0	1
Deaths	0	0	0	0	0	0	0	0	0	0	0	0	0
Injuries	0	0	0	0	0	0	0	0	0	0	0	0	0
Florida													
Number	1	3	8	6	2	10	13	7	19	4	1	3	77
Days	1	2	5	5	2	10	10	6	5	1	1	1	49
Deaths	0	0	0	0	0	0	0	0	0	0	0	0	0
Injuries	0	0	0	0	0	1	0	0	0	0	0	0	1
Georgia													
Number	0	6	4	7	0	1	4	0	3	0	0	3	28
Days	0	2	2	2	0	1	2	0	1	0	0	3	12
Deaths	0	19	0	0	0	0	0	0	0	0	0	2	19
Injuries	0	202	0	7	0	0	1	0	0	0	0	8	218
Idaho													
Number	0	5	0	0	0	0	2	4	1	0	0	1	13
Days	0	1	0	0	0	0	2	1	1	0	0	1	6
Deaths	0	0	0	0	0	0	0	0	0	0	0	0	0
Injuries	0	1	0	0	0	0	0	0	0	0	0	0	1
Illinois													
Number	0	0	0	6	35	7	4	1	1	1	0	0	55
Days	0	0	0	1	5	3	2	1	1	1	0	0	14
Deaths	0	0	0	0	0	0	0	0	0	0	0	0	0
Injuries	0	0	0	0	5	3	0	0	0	0	0	0	8
Indiana													
Number	1	0	0	0	3	7	2	0	0	0	0	0	13
Days	1	0	0	0	2	4	1	0	0	0	0	0	8
Deaths	0	0	0	0	0	0	0	0	0	0	0	0	0
Injuries	0	0	0	0	0	0	0	0	1	0	0	0	3
Iowa													
Number	0	0	0	0	24	11	7	0	1	2	0	0	45
Days	0	0	0	0	5	5	4	0	1	2	0	0	17
Deaths	0	0	0	0	1	0	0	0	0	0	0	0	1
Injuries	0	0	0	0	26	0	0	0	0	0	0	0	26
Kansas													
Number	0	0	8	6	8	15	3	0	0	19	0	0	59
Days	0	0	2	2	4	5	3	0	0	6	0	0	22
Deaths	0	0	0	0	0	0	0	0	0	0	0	0	0
Injuries	0	0	3	34	0	0	0	0	0	0	0	0	37
Kentucky													
Number	2	0	0	3	9	0	1	1	0	0	7	0	23
Days	1	0	0	1	3	0	1	1	0	0	1	0	8
Deaths	0	0	0	0	0	0	0	0	0	0	0	0	0
Injuries	21	0	0	4	20	0	1	0	0	0	1	0	47

(Courtesy of National Climatic Data Center/NOAA.)

Tornado Summary by State and Nation, 2000 [CONTINUED]

	Jan	Feb	Mar	Apr	May	Jun	Jul	Aug	Sep	Oct	Nov	Dec	Ann
Louisiana													
Number	0	0	3	24	1	2	3	6	0	0	4	0	43
Days	0	0	2	3	1	2	2	5	0	0	3	0	18
Deaths	0	0	0	0	0	0	0	0	0	0	0	0	0
Injuries	0	0	37	19	0	0	0	1	0	0	3	0	60
Maine													
Number	0	0	0	0	0	0	1	1	0	0	0	0	2
Days	0	0	0	0	0	0	1	1	0	0	0	0	2
Deaths	0	0	0	0	0	0	0	0	0	0	0	0	0
Injuries	0	0	0	0	0	0	0	0	0	0	0	0	0
Maryland													
Number	0	0	0	1	2	2	3	0	0	0	0	0	8
Days	0	0	0	1	1	1	3	0	0	0	0	0	6
Deaths	0	0	0	0	0	0	0	0	0	0	0	0	0
Injuries	0	0	0	0	1	0	0	0	0	0	0	0	1
Massachusetts													
Number	0	0	0	0	0	1	0	0	0	0	0	0	1
Days	0	0	0	0	0	1	0	0	0	0	0	0	1
Deaths	0	0	0	0	0	0	0	0	0	0	0	0	0
Injuries	0	0	0	0	0	0	0	0	0	0	0	0	0
Michigan													
Number	0	0	0	0	3	1	0	0	0	0	0	0	4
Days	0	0	0	0	1	1	0	0	0	0	0	0	2
Deaths	0	0	0	0	0	0	0	0	0	0	0	0	0
Injuries	0	0	0	0	0	0	0	0	0	0	0	0	0
Minnesota													
Number	0	0	0	2	5	0	18	6	0	0	1	0	32
Days	0	0	0	1	1	0	3	2	0	0	1	0	8
Deaths	0	0	0	0	0	0	1	0	0	0	0	0	1
Injuries0	0	0	0	0	0	0	19	0	0	0	0	0	19
Mississippi													
Number	4	5	1	4	0	0	7	1	2	0	2	1	27
Days	1	1	1	3	0	0	2	1	1	0	2	1	13
Deaths	0	0	0	0	0	0	0	0	0	0	0	0	0
Injuries	7	0	0	0	0	0	0	0	0	0	0	17	24
Missouri													
Number	0	8	1	2	10	4	3	0	0	0	0	0	28
Days	0	2	1	2	5	2	1	0	0	0	0	0	13
Deaths	0	0	0	0	0	0	0	0	0	0	0	0	0
Injuries	0	0	0	0	0	0	0	0	0	0	0	0	0
Montana													
Number	0	0	0	0	0	2	7	0	1	0	0	0	10
Days	0	0	0	0	0	1	5	0	1	0	0	0	7
Deaths	0	0	0	0	0	0	0	0	0	0	0	0	0
Injuries	0	0	0	0	0	0	0	0	0	0	0	0	0
Nebraska													
Number	0	0	0	5	9	21	6	0	0	18	1	0	60
Days	0	0	0	1	4	7	5	0	0	2	1	0	20
Deaths	0	0	0	0	0	0	0	0	0	0	0	0	0
Injuries	0	0	0	0	2	0	0	0	0	0	0	0	2
Nevada													
Number	0	1	0	0	0	0	0	1	0	0	0	0	2
Days	0	1	0	0	0	0	0	1	0	0	0	0	2
Deaths	0	0	0	0	0	0	0	0	0	0	0	0	0
Injuries	0	0	0	0	0	0	0	0	0	0	0	0	0
New Mexico													
Number	0	0	0	0	0	2	1	0	0	2	0	0	5
Days	0	0	0	0	0	1	1	0	0	1	0	0	3
Deaths	0	0	0	0	0	0	0	0	0	0	0	0	0
Injuries	0	0	0	0	0	0	0	0	0	0	0	0	0
New York													
Number	0	0	0	0	2	2	0	0	1	0	0	0	5
Days	0	0	0	0	1	1	0	0	1	0	0	0	3
Deaths	0	0	0	0	0	0	0	0	0	0	0	0	0
Injuries	0	0	0	0	0	2	0	0	0	0	0	0	2

(Courtesy of National Climatic Data Center/NOAA.)

Tornado Summary by State and Nation, 2000 [CONTINUED]

	Jan	Feb	Mar	Apr	May	Jun	Jul	Aug	Sep	Oct	Nov	Dec	Ann
North Carolina													
Number	0	2	3	5	6	2	1	0	1	0	0	3	23
Days	0	1	2	3	3	2	1	0	1	0	0	1	14
Deaths	0	0	0	0	0	0	0	0	0	0	0	0	0
Injuries	0	0	0	1	0	0	0	0	0	0	0	0	1
North Dakota													
Number	0	0	0	0	0	11	9	3	0	0	5	0	28
Days	0	0	0	0	0	3	3	2	0	0	1	0	9
Deaths	0	0	0	0	0	0	0	0	0	0	0	0	0
Injuries	0	0	0	0	0	1	0	0	0	0	2	0	3
Ohio													
Number	0	0	0	0	9	4	1	4	7	0	0	0	25
Days	0	0	0	0	2	4	1	2	2	0	0	0	11
Deaths	0	0	0	0	0	0	0	0	0	0	0	0	0
Injuries	0	0	0	0	0	0	0	4	103	0	0	0	107
Oklahoma													
Number	0	2	15	6	11	3	0	0	0	7	0	0	44
Days	0	2	3	1	3	2	0	0	0	2	0	0	13
Deaths	0	0	0	0	0	0	0	0	0	0	0	0	0
Injuries	0	0	00	7	0	0	0	0	0	0	0	0	7
Oregon													
Number	0	0	0	1	0	0	0	0	0	2	0	0	3
Days	0	0	0	1	0	0	0	0	0	1	0	0	2
Deaths	0	0	0	0	0	0	0	0	0	0	0	0	0
Injuries	0	0	0	1	0	0	0	0	0	0	0	0	1
Pennsylvania													
Number	0	0	0	0	1	3	1	0	0	0	0	0	5
Days	0	0	0	0	1	3	1	0	0	0	0	0	5
Deaths	0	0	0	0	0	0	0	0	0	0	0	0	0
Injuries	0	0	0	0	0	0	0	0	0	0	0	0	0
Puerto Rico													
Number	0	1	0	0	0	0	0	1	0	0	0	0	2
Days	0	1	0	0	0	0	0	1	0	0	0	0	2
Deaths	0	0	0	0	0	0	0	0	0	0	0	0	0
Injuries	0	0	0	0	0	0	0	0	0	0	0	0	0
Rhode Island													
Number	0	0	0	0	0	0	0	1	0	0	0	0	1
Days	0	0	0	0	0	0	0	1	0	0	0	0	1
Deaths	0	0	0	0	0	0	0	0	0	0	0	0	0
Injuries	0	0	0	0	0	0	0	0	0	0	0	0	0
South Carolina													
Number	0	1	4	1	1	3	3	1	5	0	0	1	20
Days	0	1	2	1	1	2	2	1	4	0	0	1	15
Deaths	0	0	0	0	0	0	0	0	1	0	0	0	1
Injuries	0	0	0	0	0	0	0	0	6	0	0	0	6
South Dakota													
Number	0	0	0	0	2	1	11	4	0	0	0	0	18
Days	0	0	0	0	2	1	4	3	0	0	0	0	10
Deaths	0	0	0	0	0	0	0	0	0	0	0	0	0
Injuries	0	0	0	0	0	0	0	0	0	0	0	0	0
Tennessee													
Number	0	1	1	4	19	0	1	0	0	0	0	1	27
Days	0	1	1	1	6	0	1	0	0	0	0	1	11
Deaths	0	0	0	1	0	0	0	0	0	0	0	0	1
Injuries	0	1	0	0	4	0	0	0	0	0	0	0	5
Texas													
Number	0	7	46	31	22	8	3	0	0	9	20	1	147
Days	0	2	8	4	7	5	3	0	0	5	5	1	40
Deaths	0	0	2	0	2	0	0	0	0	0	0	0	4
Injuries	0	0	87	3	0	0	0	0	0	0	3	0	93
Utah													
Number	0	0	0	1	0	0	0	0	2	0	0	0	3
Days	0	0	0	1	0	0	0	0	1	0	0	0	2
Deaths	0	0	0	0	0	0	0	0	0	0	0	0	0
Injuries	0	0	0	0	0	0	0	0	0	0	0	0	0
Virginia													
Number	0	0	0	2	3	0	4	2	0	0	0	0	11
Days	0	0	0	1	2	0	3	1	0	0	0	0	7
Deaths	0	0	0	0	0	0	0	0	0	0	0	0	0
Injuries	0	0	0	0	1	0	0	0	0	0	0	0	1

(Courtesy of National Climatic Data Center/NOAA.)

Tornado Summary by State and Nation, 2000 [CONTINUED]

	Jan	Feb	Mar	Apr	May	Jun	Jul	Aug	Sep	Oct	Nov	Dec	Ann
Washington													
Number	1	0	0	0	2	0	0	0	0	0	0	0	3
Days	1	0	0	0	2	0	0	0	0	0	0	0	3
Deaths	0	0	0	0	0	0	0	0	0	0	0	0	0
Injuries	0	0	0	0	0	0	0	0	0	0	0	0	0
West Virginia													
Number	0	0	0	0	1	0	0	3	0	0	0	0	4
Days	0	0	0	0	1	0	0	1	0	0	0	0	2
Deaths	0	0	0	0	0	0	0	0	0	0	0	0	0
Injuries	0	0	0	0	0	0	0	0	0	0	0	0	0
Wisconsin													
Number	0	0	1	0	1	5	8	3	0	0	0	0	18
Days	0	0	1	0	1	1	6	2	0	0	0	0	11
Death	0	0	0	0	0	0	0	0	0	0	0	0	0
Injuries	0	0	16	0	0	0	0	0	0	0	0	0	16
Wyoming													
Number	0	0	0	1	4	0	0	0	0	0	0	0	5
Days	0	0	0	1	2	0	0	0	0	0	0	0	3
Deaths	0	0	0	0	0	0	0	0	0	0	0	0	0
Injuries	0	0	0	0	0	0	0	0	0	0	0	0	0
United States													
Number	16	58*	103	136	241	136	148	53*	47	64	50	26	1076*
Days	4	12	20	19	29	27	31	23	17	15	11	4	212&
Deaths	0	19	2	2	3	0	1	0	2	0	0	12	41
Injuries	29	215	143	83	60	10	23	5	109	0	11	194	882

*Corrected for boundary-crossing tornadoes.
&Tornado days for country as a whole.

(Courtesy of National Climatic Data Center/NOAA.)

Hurricanes

There is nothing like the hurricane in the atmosphere. Even seen by sensors on satellites thousands of miles above Earth, the uniqueness of these powerful, tightly coiled storms is clear. Hurricanes are not the largest storm systems, nor the most violent—but they combine those qualities as no other phenomenon does, as if they were designed to be engines of death and destruction.

In the Northern Hemisphere, these storms are called hurricanes, a term that echoes colonial Spanish and Caribbean Indian words for evil spirits and big winds. The storms are products of the tropical ocean and atmosphere: powered by heat from the sea, steered by the easterly trades and temperate westerlies, and driven by their own fierce energy. Around their tranquil core, winds blow with lethal velocity and the ocean develops an inundating surge. In addition, as they move ashore, tornadoes may descend from the advancing bands of thunderclouds.

Hurricanes, as poorly understood as they are today, seem to have a single benefit—they are a major source of rain for those continental corners over which their unpredictable

Computer-enhanced photo of Hurricane Diana, September 11, 1984. *(Courtesy of National Aeronautics and Space Administration (NASA).)*

tracks carry them. Mostly they are seen as engines of tragedy, which still leave death and destruction in their paths, even though the effectiveness of warning systems have doubled and redoubled in recent decades.

• HURRICANE AND TROPICAL STORM SEASON

It is the coming of summer to the Northern Hemisphere that ushers in conditions that spawn tropical storms and hurricanes. The movement of the sun—which is not really movement, of course, but a positional shift relative to Earth caused by the planet's year-long orbit—brings the peak power of solar radiation northward. The sun's track moves first to the equator, and from the sun's position over the equator in late March its apparent movement is northward to the Tropic of Cancer in late June, (23.5° north latitude), when it begins to retreat southward again in July, and August. Behind this solar track the sea and air grow warmer, and the polar airflows make a steady retreat.

This northward shift of the sun brings the season of tropical cyclones to the Northern Hemisphere. This means it is time to look seaward, along our coasts. This is as true for Asia as it is for the United States and the Caribbean.

Over the western Pacific, the tropical cyclone season is never quite over, but varies greatly in intensity. Every year, conditions east of the Philippines send a score of violent storms howling toward Asia, but it is worst from June through October.

Southwest of Mexico, eastern Pacific hurricanes develop during the spring, summer, and fall. Most of these will die at sea as they move over colder ocean waters. But there are destructive exceptions when storms occasionally curve back toward Mexico and the southwestern United States, bringing flooding rains.

Along the U.S. Atlantic and Gulf coasts, the nominal hurricane season lasts from June through November. Early in this season, the western Caribbean and Gulf of Mexico are the principal areas of origin. In July and August, this spawning center begins to shift eastward, and by early September a few storms are being born as far east as the Cape Verde Islands off Africa's west coast. Again after mid-September, most storms begin in the western Caribbean and Gulf of Mexico.

In an average year, more than 100 disturbances with hurricane potential are observed in the Atlantic, Gulf, and Caribbean; on average, only 10 of these reach the tropical storm stage, and only about six mature into hurricanes. On average, two of these hurricanes strike the United States, where they are apt to kill 50–100 people, from Texas to Maine, and cause hundreds of millions of dollars in property damage. In a worse-than-average year, the same storms cause several hundred deaths, and property damage totaling billions of dollars. For the National Oceanic and Atmospheric Administration (NOAA), the hurricane season means another hazard from the atmosphere, at a time when tornadoes, floods, and severe storms are also playing seasonal havoc elsewhere on the continent.

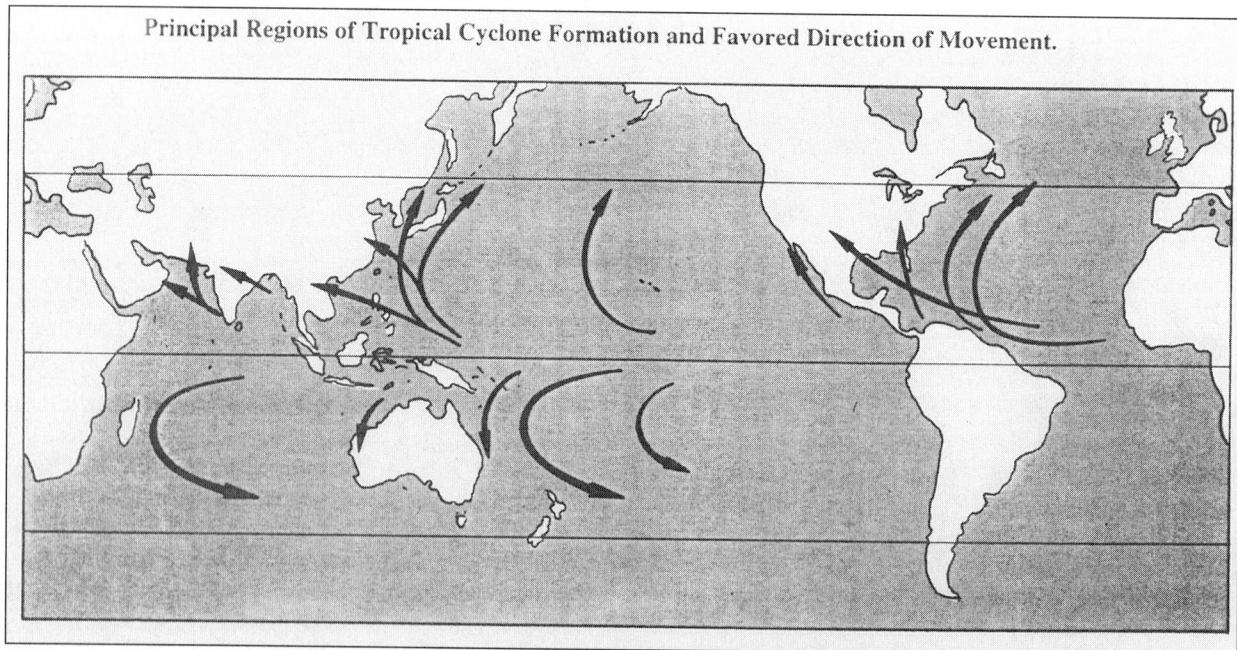

Principal Regions of Tropical Cyclone Formation and Favored Direction of Movement.

(Courtesy of National Oceanic and Atmospheric Administration (NOAA).)

• PORTRAIT OF A HURRICANE

Given that the hurricane, as an engine, is inefficient and hard to start and sustain, and most tropical storms will never reach hurricane proportions, some tropical storms still do. A certain number every season will manage to accumulate the complex combination of natural forces required. When one does, it is an awesome natural event indeed. The young storm stands upon the sea as a whirlwind of awful violence. Its hurricane-force winds (winds greater than 63 knots) cover thousands of square miles, and tropical storm-force winds (winds of 34–63 knots) cover an area ten times larger. Along the twisting contours of its spiral are "rain bands" of dense clouds from which torrential rains fall. These spiral rain bands ascend in "decks" of cumulus and cumulonimbus clouds to the high upper-atmosphere. There, condensing water vapor is swept off as ice-crystal wisps of cirrus clouds by high-altitude winds. Lightning glows in almost perpetual pulsations in the rain bands, and this cloudy terrain is whipped by turbulence.

In the lower few thousand feet, air flows in toward the center of the cyclone, and is whirled upward through ascending columns of air near the center. Above 40,000 ft, this cyclone pattern is replaced by an anticyclonic circulation—the high-level pump that functions as the "exhaust system" of the hurricane engine. (Anticyclonic circulation means, in the Northern Hemisphere, a system of winds rotating in a clockwise direction about a center of relatively low barometric pressure. Contrast this to "cyclonic circulation," which has a counterclockwise pattern. Both of these definitions are reversed in the Southern Hemisphere.)

At the lower levels, where the hurricane is most intense, winds on the rim of the storm follow a wide pattern, like the slower currents on the rim of a whirlpool; like those currents, these winds accelerate as they approach the central vortex. This inner band is the eyewall, where the storm's worst winds are found, and where moist air entering at the surface is "chimneyed" upward, releasing heat to drive the storm. In most hurricanes, these winds exceed 90 knots—in extreme cases they may double that velocity. Maximum winds run still higher in typhoons, the Pacific version of the same type of storm.

Hurricane winds are produced, as all winds are, by difference in atmospheric pressure, or density. The pressure gradient—the rate of pressure change with distance—produced in hurricanes is the sharpest pressure gradient in the atmosphere, excepting only the pressure change believed to exist across the narrow funnel of a tornado.

Hurricanes and barometric pressure

Atmospheric pressure is popularly expressed as the height of a column of mercury that can be supported by the weight of the overlying air at a given time. Weather maps show atmospheric pressure in millibars (mb), units equal to a thousandth of a bar. The bar is a unit of measure equal to 29.53 in of mercury in the English system, and to one million dynes per square centimeter in the metric system.

In North America, barometric measurements at sea level seldom go below 29 in of mercury, and in the tropics the barometer reading is generally close to 30 in under normal conditions. Hurricanes drop the bottom out of those normal categories. The Labor Day hurricane that struck the Florida Keys in 1935 had a central pressure of only 26.35 in. Hurricane Gilbert, which wrought havoc in the Caribbean and the Yucatan Peninsula during 1988, was measured at 26.22 in in its eye. And the pressure change is swift: pressure may drop a full inch per mile. Such pressure contrasts guarantee tremendous wind velocity.

At the center of the storm is a unique atmospheric entity, and a persistent metaphor for order in the midst of chaos—the eye of the hurricane. It is encountered suddenly. From the heated tower of maximum winds and thunderclouds, one bursts into the eye, where winds diminish to something less than 15 knots. Penetrating the opposite wall, one is abruptly in the worst of winds again.

A mature hurricane orchestrates more than a million cubic miles of atmosphere. Over the deep ocean, waves generated by hurricane winds can reach heights of 50 ft or more. Under the storm center, the ocean surface is drawn upward like water in a straw, forming a mound 1–3 ft or so higher than the surrounding ocean surface. This mound may translate into coastal surges of 20 ft or more. Besides this surge, massive swells pulse out through the upper levels of the sea.

Hurricane Eloise, which struck the Florida panhandle in September 1975, taught scientists something new about the influence of passing hurricanes on the marine environment. Expendable bathythermographs dropped from NOAA research aircraft ahead of, into, and in the wake of the storm showed that the ocean was disturbed to depths of hundreds of feet by a passing hurricane. Moreover, the ocean "remembered" hurricane passage with internal waves that persisted for weeks after the storm had gone. The same storm also demonstrated that a passing hurricane can be felt deep in the sea-floor sediments.

While a hurricane lives, the transaction of energy within its circulation is immense. The condensation heat energy released by a hurricane in one day can be the equivalent of energy released by fusion of 400 20-megaton hydrogen bombs. One day's released energy, converted to electricity, could supply the United States' electrical needs for about six months.

The fatal thrust toward land

From birth, the hurricane lives in an environment that constantly tries to kill it—and ultimately succeeds. The hurricane tends to survive while it is over warm water but forces drive the storm ashore or over colder water beyond the tropics. In these non-nourishing environments it will fall and die. This thrust away from the tropics is the clockwise curve that propels Atlantic hurricanes into the eastern

United States, and takes eastern Pacific typhoons across the coastlines of Japan and into the Asian mainland.

Even before a hurricane forms, the embryonic storm has forward motion, generally driven by the easterly flow of an air movement system of the tropic latitudes, featuring east to west flow of the atmosphere in which it is embedded. As long as this westerly drift is slow—less than about 20 knots—the young hurricane may intensify. More rapid forward motion generally inhibits intensification in the storm's early stages. Entering the temperate latitudes (north of the tropic of Cancer) some storms may move along at better than 50 knots, but such fast-moving storms soon weaken.

At middle latitudes, the hurricane's end usually comes swiftly. Colder air penetrates the cyclonic vortex; the warm core cools, and acts as a thermal brake on further intensification. Water below 80°F does not contribute much energy to a hurricane. Even though some large hurricanes may travel for days over cold North Atlantic water, all storms are doomed once they leave the warm tropical waters that sustain them. The farther they venture into higher latitudes, the less fuel they receive from the sea; this lack of fuel finally kills the storms. Over land, hurricanes break up rapidly. Cut off from their oceanic source of energy, and with the added effects of frictional drag, their circulation rapidly weakens and becomes more disorganized. Torrential rains, however, may continue even after the winds are much diminished. In the southeastern United States, about one-fourth of the annual rainfall comes from dissipating hurricanes, and the Asian mainland and Japan suffer typhoons to get water from the sky.

Hurricanes are often resurrected into extratropical cyclones at higher latitudes, or their dynamic forces combine with existing temperate-zone disturbances. Many storms moving up our Atlantic coast are in the throes of this transformation when they strike New England, and large continental lows are often invigorated by the remnants of storms born over the tropical sea.

Hurricane destruction

Hurricanes are the unstable, unreliable creatures of a moment in our planet's natural history. But their brief life ashore can leave scars that never quite heal. In the mid-1970s, the hand of 1969's Camille could still be seen along the Mississippi Gulf Coast and recovery from Florida's Hurricane Andrew in 1992 continued into the late 1990s. Most of a hurricane's destructive work is done by the general rise in the height of the seas that accompany the storm. This quick, tidal-like rise in sea level is called storm surge. Hurricane winds are a force to be reckoned with by coastal communities deciding how strong their structures should be. As winds increase, pressure against objects is added at a disproportionate rate. Pressure force against a wall mounts with the square of wind speed so that a threefold increase in wind speed results in a nine-fold increase in pressure. Thus, a 25

mph wind causes about 1.6 lb of pressure per square foot—a force of 50 lb. In 75 mph winds, that force becomes 450 lb and in 125 mph, it becomes 1,250 lb. For some structures this force is enough to cause failure. Tall structures like radio towers can be worried to destruction by gusty hurricane-force winds. Winds also carry a barrage of debris that can be extremely dangerous.

All the wind damage does not necessarily come from the hurricane. As the storm moves shoreward, interactions with other weather systems can produce tornadoes, which work around the fringes of the hurricane. Although hurricane-spawned tornadoes are not the most violent form of these whirlwinds, they add to the destruction.

Floods from hurricane rainfall are quite destructive. A typical hurricane brings an awesome 6–12 in of short-duration rainfall to the area it crosses, and some have brought much more. The resulting floods—often sudden flash floods—have caused great damage and loss of life, especially in mountainous areas, where heavy rains can mean flash floods. Rains from the dying Hurricane Agnes brought disastrous floods to the entire Atlantic tier of states, causing 129 deaths and some $2.1 billion in property damage.

Storm surge

The hurricane's worst comes from the sea, in the form of storm surge. This subtly approaching smash of tidal wave immensity actually claims nine of each 10 victims that fall to a hurricane.

As the storm crosses the continental shelf and moves close to the coast, mean water level may increase 15 ft or more. The advancing storm surge combines with the normal astronomical tide to create the hurricane storm tide. In addition, wind waves 5–10 ft high are superimposed on the storm tide. This buildup of water level can cause severe flooding in coastal areas—particularly when the storm surge coincides with normal high tides. Because much of the United States' densely populated coastline along the Atlantic and Gulf coasts lies less than 10 ft above mean sea level, the danger from storm surge is multiplied. Nearly every coastal location that is exposed to a hurricane is also a candidate for the smashing blow of storm surge.

Wave and current action associated with the surge also causes extensive damage. Water weighs some 1,700 lb per cubic yard; extended pounding by frequent waves can demolish any structures not specifically designed to withstand such forces.

Currents set up along the coast by the gradient in storm surge heights and wind combine with waves to severely erode beaches and coastal highways. Many buildings withstand hurricane winds until their foundations, undermined by erosion, are weakened and fail.

Storm tides, waves, and currents in confined harbors severely damage ships, marinas, and pleasure boats. In estuarine and bayou areas, intrusions of saltwater endanger the public health.

• HURRICANE CASUALTIES

In Asia, the price in life paid due to hurricanes has been enormous. As late as 1970, cyclone storm tides along the coast of what is now Bangladesh killed hundreds of thousands of persons. Eleven thousand people perished in a storm that struck that region in 1984, and even more in a storm seven years later.

The Western Hemisphere has not had such spectacular losses, but the toll has still been high. In August 1893, a storm surge drowned 1,000–2,000 people in Charleston, South Carolina. In October of that same year, nearly 2,000 more perished on the Gulf Coast of Louisiana. More than 1,800 perished along the south shore of Florida's Lake Okeechobee in 1928 when hurricane-driven waters broached an earthen levee. Cuba lost more than 2,000 to a storm in 1932. Four hundred died in Florida in an intense hurricane in September 1935—the Labor Day hurricane that, until Hugo hit in 1989, shared with 1969's Camille the distinction of being the most severe to strike the United States mainland during the years of record keeping.

Floods from 1974's Hurricane Fifi caused one of the Western Hemisphere's worst natural disasters, with an esti-

mated 5,000 persons dead in Honduras, El Salvador, Guatemala, and Belize.

• THE NATION'S WORST WEATHER DISASTER

Over one hundred years ago, in 1900, the great Galveston hurricane roared through the prosperous island city with winds in excess of 130 mph and a 15-ft storm surge. When it was finally over, at least 3,500 homes and buildings were destroyed and more than 8,000 people were killed. The number of people who lost their lives on that single day, September 8, 1900, represents more than the combined fatalities resulting from the 325 tropical storms and hurricanes that have struck the United States since then. In fact, that single event accounts for one third of all tropical storm or hurricane-related fatalities that have occurred in the United States since it was founded.

Since wireless ship-to-shore communication was not yet available in 1900, information was extremely sketchy and there was little if any knowledge that the hurricane was strengthening and heading toward Texas; even though hurricane warning flags were raised atop the Weather Bureau building on September 7, the day before the hurricane struck. Dr. Isaac Cline, Galveston's Weather Bureau Manager, was aware of the hurricane as it passed over Cuba on

Saffir-Simpson Hurricane Scale

All hurricanes are dangerous, but some are more so than others. The way storm surge, wind, and other factors combine determine the hurricane's destructive power. To make comparisons easier—and to make the predicted hazards of approaching hurricanes clearer to emergency forces—NOAA's hurricane forecasters use a disaster-potential scale, which assigns storms to five categories. Category 1 is a minimum hurricane; category 5 is the worst case. The criteria for each category are shown below.

This can be used to give an estimate of the potential property damage and flooding expected along the coast with a hurricane.

Category Definitions and Effects

ONE Winds 74–95 mph. No real damage to building structures. Damage primarily to unanchored mobile homes, shrubbery and trees. Also, some coastal road flooding and minor pier damage.

TWO Winds 96–110 mph. Some roofing material, door and window damage to buildings. Considerable damage to vegetation, mobile homes and piers. Coastal and low-lying escape routes flood two to four hours before arrival of center. Small craft in unprotected anchorages break moorings.

THREE Winds 111–130 mph. Some structural damage to small residences and utility buildings with a minor amount of curtain wall failures. Mobile homes are destroyed. Flooding near the coast destroys smaller structures with larger structures damaged by floating debris. Terrain continuously lower than 5 ft above sea level (ASL) may be flooded inland as far as 6 mi.

FOUR Winds 131–155 mph. More extensive curtain wall failures with some complete roof structure failure on small residences. Major erosion of beach areas. Major damage to lower floors of structures near the shore. Terrain continuously lower than 10 ft ASL may be flooded, requiring massive evacuation of residential areas inland as far as 6 mi.

FIVE Winds greater than 155 mph. Complete roof failure on many residences and industrial buildings. Some complete building failures with small utility buildings blown over or away. Major damage to lower floors of all structures located less than 15 ft ASL and within 500 yd of the shoreline. Massive evacuation of residential areas on low ground within 10 mi of the shoreline may be required.

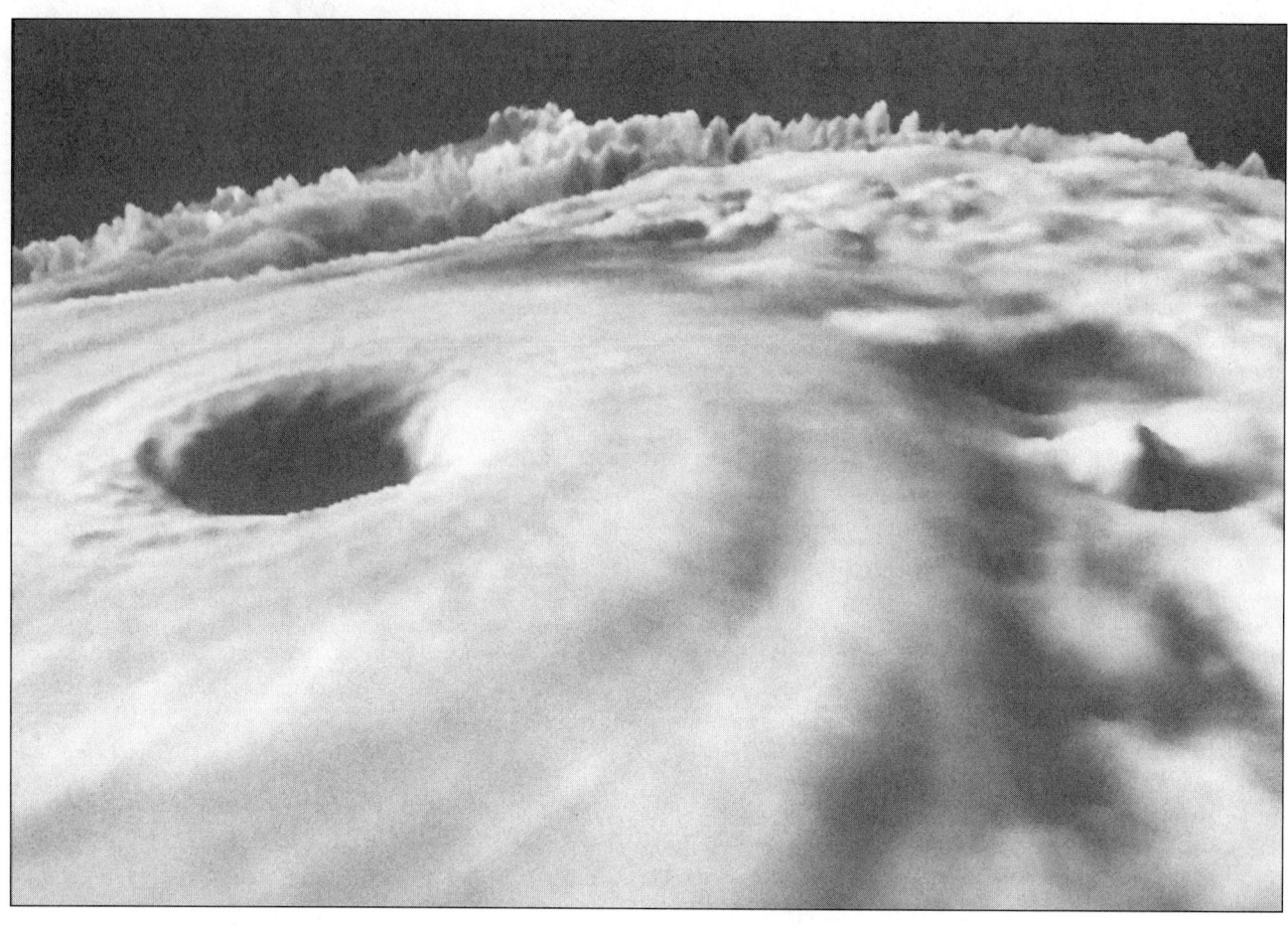

At their strongest, the winds of Diana reached 130 mph. *(Courtesy of National Aeronautics and Space Administration (NASA).)*

a northern track. Cline patrolled the beach and warned people to move to higher ground. With a population of more than 35,000 people, it is likely many more Galveston residents would have died without the warning.

In 1900, the highest point in Galveston was only 8.7 ft above sea level and the hurricane easily inundated the city with a storm surge of 15 ft. With the terrible memories of the 1900 hurricane in mind, the people of Galveston began an unprecedented effort to protect their city from the next "big one." In 1902, they began constructing a 16-ft thick, 17-ft high sea wall covering 3 mi of oceanfront. They also began the monumental task of raising the entire island by as much as 8 ft with sand dredged from Galveston Bay. Today's sea wall has been extended to a length of 10 mi of oceanfront to protect the heart of the city.

In the United States, the hurricane death toll has been greatly diminished by timely warnings of approaching storms. But damage to fixed property continues to mount. Camille, in 1969, caused some $1.42 billion in property damage. Floods from Agnes in 1972 cost an estimated $2.1 billion and damage from Frederic in 1979 hit $2.3 billion. Hugo, in 1989, wrought damage of more than $7 billion while coming ashore as a full-scale hurricane in South Car-

olina and moving hundreds of miles inland as a furious near-hurricane-strength storm.

• TROPICAL STORM/HURRICANE NAMES

The National Hurricane Center (NHC, Tropical Prediction Center) near Miami, Florida, keeps a constant watch on oceanic storm-breeding areas for tropical disturbances which may herald the formation of a hurricane. If a disturbance intensifies into a tropical storm with rotary circulation and sustained wind speeds above 38 mph, the Center will give the storm a name. The tropical disturbance may never reach hurricane intensity, nevertheless it is given a name in anticipation that it may. Experience shows that the use of short, distinctive given names in written, as well as in spoken communications, is quicker, and less subject to error than the older, more cumbersome latitude-longitude identification methods. This is especially important in exchanging detailed storm information between hundreds of widely scattered stations, airports, coastal bases, and ships at sea.

Six separate alphabetical lists of names (with alternating male and female names every other year) have been estab-

World-wide Tropical Cyclone Names

Atlantic names

2002	2003	2004	2005	2006	2007
Arthur	Ana	Alex	Arlene	Alberto	Andrea
Bertha	Bill	Bonnie	Bret	Beryl	Barry
Cristobal	Claudette	Charley	Cindy	Chris	Chantal
Dolly	Danny	Danielle	Dennis	Debby	Dean
Edouard	Erika	Earl	Emily	Ernesto	Erin
Fay	Fabian	Frances	Franklin	Florence	Felix
Gustav	Grace	Gaston	Gert	Gordon	Gabrielle
Hanna	Henri	Hermine	Harvey	Helene	Humberto
Isidore	Isabel	Ivan	Irene	Isaac	Ingrid
Josephine	Juan	Jeanne	Jose	Joyce	Jerry
Kyle	Kate	Karl	Katrina	Kirk	Karen
Lili	Larry	Lisa	Lee	Leslie	Lorenzo
Marco	Mindy	Matthew	Maria	Michael	Melissa
Nana	Nicholas	Nicole	Nate	Nadine	Noel
Omar	Odette	Otto	Ophelia	Oscar	Olga
Paloma	Peter	Paula	Philippe	Patty	Pablo
Rene	Rose	Richard	Rita	Rafael	Rebekah
Sally	Sam	Shary	Stan	Sandy	Sebastien
Teddy	Teresa	Tomas	Tammy	Tony	Tanya
Vicky	Victor	Virginie	Vince	Valerie	Van
Wilfred	Wanda	Walter	Wilma	William	Wendy

(Chart by GGS. Gale Group.)

lished for six-year periods to designate hurricanes and tropical storms. The first tropical storm or hurricane of the year is given the first alphabetical name from the set for that year. After the sets have all been used, they are used again. The 2002 set, for example, is the same set used to name storms in 1996. The letters Q, U, X, Y, and Z are not included because of the scarcity of names beginning with those letters. Furthermore, in cases when a land-falling storm results in economic or human disaster, the storm name is retired. If over 24 tropical cyclones occur in a year, the Greek alphabet is used.

The name lists have an international flavor because hurricanes affect other nations and are tracked by countries other than the United States. Names for these lists are selected from library sources and agreed upon by nations involved during international meetings of the World Meteorological Organization (WMO).

Problems for U.S. hurricane forecasters

The permanent populations of the hurricane-prone coastal counties of the United States continue to grow at a rapid rate. When weekend, seasonal, and holiday populations are considered, the number of people on barrier islands such as at Ocean City, Maryland; Gulf Shores, Alabama; and Padre Island, Texas, increases by 10- to 100-fold or more. Also, these areas are subject to inundation from the rapidly rising waters—the storm surge—associated with hurricanes that generally result in catastrophic damage and potentially large losses of life. Over the past several years, the warning system has provided adequate time for the great

majority of the people on barrier islands and along the immediate coast to move inland when hurricanes have threatened. However, it is becoming more difficult each year to evacuate people from these areas due to roadway systems that have not kept pace with the rapid population growth. This condition results in the requirement for longer and longer lead times for safe evacuation. Unfortunately, these extended forecasts suffer from increasing uncertainty. Furthermore, rates of improvements in forecast skills have been far out-paced by rates of population growth in areas vulnerable to hurricanes.

The combination of the growing populations on barrier islands and other vulnerable locations, and the uncertainties in the forecasts poses major dilemmas for forecasters and local and state emergency management officials alike, for example, how to prevent complacency caused by "false alarms" and yet provide adequate warning times.

Preparations for hurricanes are expensive. When a hurricane is forecast to move inland on a path nearly normal to the coasts, the area placed under warning is about 300 mi in length. The average cost of preparation, whether the hurricane strikes or not, is more than $50 million for the Gulf Coast. This estimate covers the cost of boarding up homes, closing down businesses and manufacturing plants, evacuating oilrigs, etc. It does not include economic losses due to disruption of commerce activities such as sales, tourists canceling reservations, etc.

In some locations, the loss for the Labor Day weekend alone can be a substantial portion of the yearly income of coastal businesses, for example, the losses experienced along the Florida panhandle during Hurricane Elena in 1985.

Eastern North Pacific Names

2002	2003	2004	2005	2006	2007
Alma	Andres	Agatha	Adrian	Aletta	Alvin
Boris	Blanca	Blas	Beatriz	Bud	Barbara
Cristina	Carlos	Celia	Calvin	Carlotta	Cosme
Douglas	Dolores	Darby	Dora	Daniel	Dalila
Elida	Enrique	Estelle	Eugene	Emilia	Erick
Fausto	Felicia	Frank	Fernanda	Fabio	Flossie
Genevieve	Guillermo	Georgette	Greg	Gilma	Gil
Hernan	Hilda	Howard	Hilary	Hector	Henriette
Iselle	Ignacio	Isis	Irwin	Ileana	Ivo
Julio	Jimena	Javier	Jova	John	Juliette
Kenna	Kevin	Kay	Kenneth	Kristy	Kiko
Lowell	Linda	Lester	Lidia	Lane	Lorena
Marie	Marty	Madeline	Max	Miriam	Manuel
Norbert	Nora	Newton	Norma	Norman	Narda
Odile	Olaf	Orlene	Otis	Olivia	Octave
Polo	Patricia	Paine	Pilar	Paul	Priscilla
Rachel	Rick	Roslyn	Ramon	Rosa	Raymond
Simon	Sandra	Seymour	Selma	Sergio	Sonia
Trudy	Terry	Tina	Todd	Tara	Tico
Vance	Vivian	Virgil	Veronica	Vicente	Velma
Winnie	Waldo	Winifred	Wiley	Willa	Wallis
Xavier	Xina	Xavier	Xina	Xavier	Xina
Yolanda	York	Yolanda	York	Yolanda	York
Zeke	Zelda	Zeke	Zelda	Zeke	Zelda

These lists are also re-cycled every six years (the 2002 list will be used again in 2008).

Central North Pacific Names

List 1	List 2	List 3	List 4
Akoni	Aka	Alika	Ana
Ema	Ekeka	Ele	Ela
Hana	Hali	Huko	Halola
Io	Iolana	Ioke	Iune
Keli	Keoni	Kika	Kimo
Lala	Li	Lana	Loke
Moke	Mele	Maka	Malia
Nele	Nona	Neki	Niala
Oka	Oliwa	Oleka	Oko
Peke	Paka	Peni	Pali
Uleki	Upana	Ulia	Ulika
Wila	Wene	Wali	Walaka

The names are used one after the other. When the bottom of one list is reached, the next name is the top of the next list.

Western North Pacific Names

Contributor	I	II	III	IV	V
Cambodia	Damrey	Kong-rey	Nakri	Krovanh	Sarika
China	Longwang	Yutu	Fengshen	Dujuan	Haima
DPR Korea	Kirogi	Toraji	Kalmaegi	Maemi	Meari
HK, China	Kai-Tak	Man-yi	Fung-wong	Choi-wan	Ma-on
Japan	Tenbin	Usagi	Kanmuri	Koppu	Tokage

(Chart by GGS. Gale Group.)

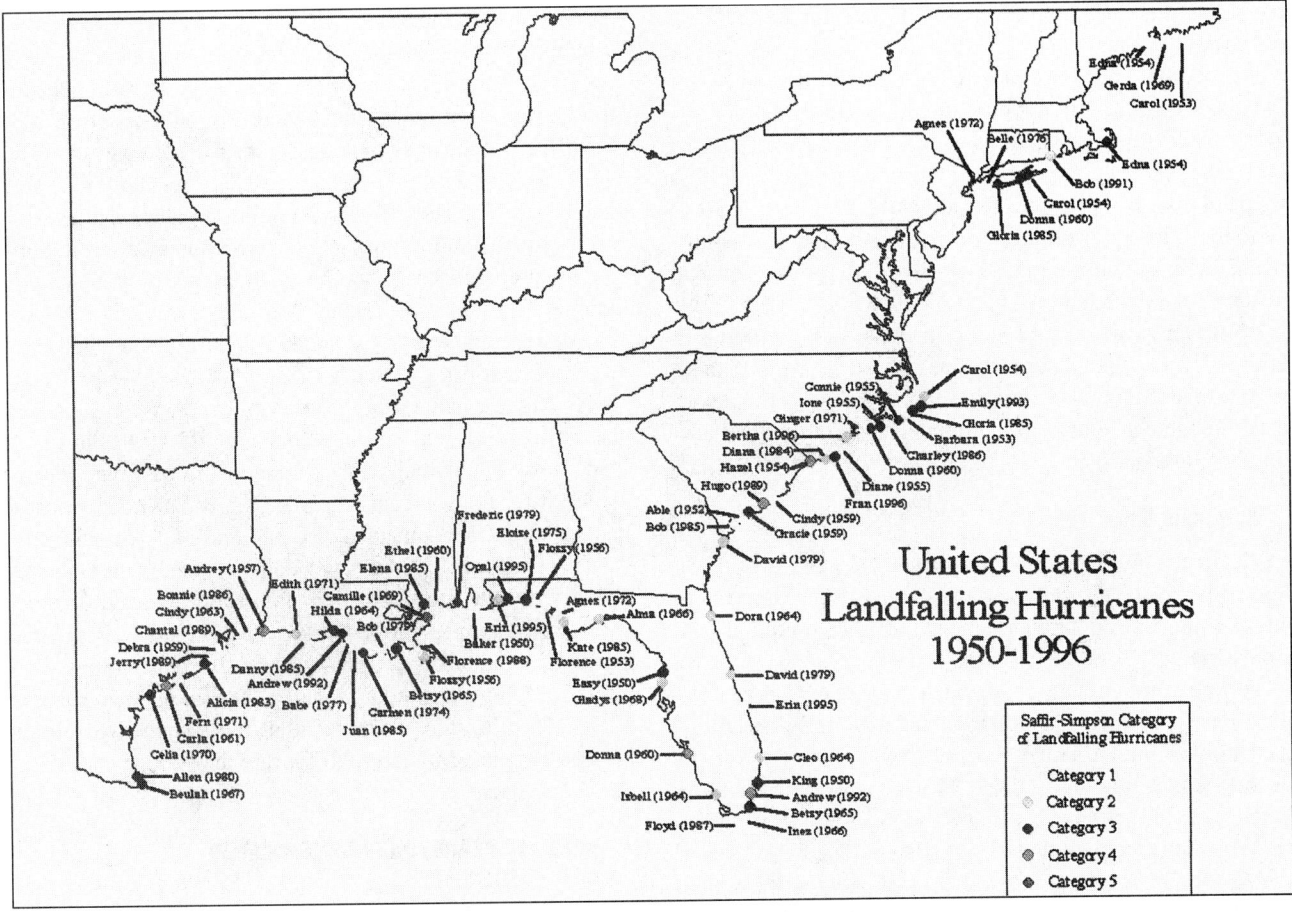

United States
Landfalling Hurricanes
1950-1996

Saffir-Simpson Category
of Landfalling Hurricanes

Category 1
Category 2
Category 3
Category 4
Category 5

(Courtesy of National Oceanic and Atmospheric Administration (NOAA).)

If the width of the warned area has to be increased by 20% because of greater uncertainties in the forecast, the additional cost for each would be $10 million. If uncertainties in the hurricane strength require warning for the next higher category of hurricane, then major increases in the number of people evacuated and preparation costs would be required.

Of course, if these uncertainties meant that major metropolitan areas such as Galveston/Houston, New Orleans, Tampa, Miami, or a number of other major coastal cities would or would not be included in the warning area, then the differences in preparation costs would be substantially more than the $10 million. Also, the number of people evacuated would be substantially more than tens of thousands of people. For instance, in the case of the Galveston/Houston area, an increase in storm strength from a category 2 hurricane to a category 3 hurricane on the Saffir-Simpson Scale would require the evacuation of an additional 200,000 people. Likewise, if major industrial areas such as Beaumont/Port Arthur, Texas, or tourist areas such as Atlantic City, New Jersey, were affected by these uncertainties, the financial impact would be quite significant.

Economic factors receive serious consideration from the National Hurricane Center (NHC), and local and state officials consider not only direct, but also indirect effects, on

people's response. People will not continually take expensive actions that, afterwards, prove to have been unnecessary. If we consistently overwarn by wide margins, people will not respond and such actions could result in large loss of life. To maintain credibility with the general public, NHC and local and state officials cannot treat all hurricanes as if they were Camilles, Hugos, or Andrews. Such an exaggerated approach may indeed provide maximum protection of life for a given event, but it endangers many more lives the next time when the threat may be even greater.

Finally, the hurricane problem is compounded by the fact that 80–90% of the people who now live in the hurricane-prone areas have never experienced the core of a major hurricane (a category 3 or stronger on the Saffir-Simpson Scale). Many of these people have been through weaker hurricanes or been brushed by the fringe of a major hurricane. The result is a false impression of the damage potential of these storms. This frequently breeds complacency and delayed action that could result in the loss of many lives. For example, people living on barrier islands might be reluctant to evacuate under "blue sky" conditions until they actually see the threat (water rising and winds increasing). The result could be people trapped in those areas as water cuts off escape routes. This situation nearly happened for about 200 people on western Galveston Island during Hurricane Alicia of 1983.

This type of response primarily results from three major factors. First, major hurricanes are infrequent events for any given location. Second, for the past three decades, major hurricanes striking the United States coast have been less frequent than previous decades, although that rate appears to be rising. Finally, it has been during this period of low hurricane activity that the great majority of the present coastal residents moved to the coast.

However, with the tremendous increase of populations in high-risk areas along our coastlines, the concern is that we may now not fare as well in the future when hurricane activity inevitably returns to the frequencies experienced during the 1940s–60s.

Hurricane Hugo, 1989

Hurricane Hugo, crossing the coast of South Carolina on September 21, 1989, at that point, was the strongest storm to strike the United States since Camille pounded the Louisiana and Mississippi coasts in 1969.

At one point east of Guadeloupe, a NOAA research aircraft measured winds of 160 mph and a central pressure of 27.1 in, which rated Hugo as a category 5 storm—the highest on the Saffir-Simpson Scale. It was somewhat less fierce when it reached the United States mainland.

When Hugo struck the Virgin Islands, Puerto Rico, and the Carolinas, it was classified as a category 4 hurricane. Storm tides of approximately 20 ft were experienced along part of the South Carolina coast, constituting record storm-tide heights for the U.S. East Coast. Although the highest surges struck sparsely populated areas north of Charleston, South Carolina, damage was extensive and lives were lost.

Forty-nine fatalities directly related to the storm were recorded; 26 in the United States and its Caribbean Islands, and 23 on other Leeward Islands. It is estimated that Hugo caused more than $9 billion in damage. The mainland of the United States alone accounted for $7 billion.

Hurricane Andrew, 1992

Hurricane Andrew slammed into heavily populated south Florida as the most destructive storm in U.S. history.

Andrew formed as a tropical wave off the African coast on August 14, 1992; by August 22, it was classified as a tropical storm. As it neared the Bahamas and Florida on August 23, Andrew had reached hurricane intensity.

Andrew annihilated homes and businesses along a 30-mi path through the Dade County, Florida, towns of Homestead, Leisure City, Goulds, Princeton, Naranja, and Florida City. When it was over, more than 60,000 homes were destroyed and 200,000 people were left homeless.

Andrew had a central pressure of 922 millibars (mb) at landfall making it among the three most intense hurricanes of the twentieth century. Only the infamous Labor Day hurricane that struck the Florida Keys in 1935 and Hurricane Camille in 1969 along the Mississippi/Louisiana coasts were as strong. Damage estimates have been as high as $25 billion.

Fifteen people died in Florida as a direct result of Andrew's fury. Another 29 lives were lost as a result of indirect effects of the hurricane within the next three weeks. The relatively low loss of life, compared to the hundreds that died in the 1935 storm and in Camille, stands as a testimony to the success and importance of hurricane awareness campaigns, preparedness planning, and actions by the joint efforts of federal, state, county, and city emergency forces. The news media played a major role in the life-saving actions before, during, and after Andrew hit.

As Andrew came ashore first in the northwest Bahamas, the storm surge reached an astonishing 23 ft. In Florida, a 17-ft storm tide, which headed inland from Biscayne Bay, is a record for the southeast Florida peninsula. Storm tides of more than 7 ft in Louisiana also caused severe flooding.

Evacuation from threatened coastal areas is the only defense from the storm surge's potential for death and destruction. After the National Hurricane Center (NHC) issued hurricane watches and warnings, massive evacuations were ordered in Florida and Louisiana by emergency management officials. It is estimated that more than two million people evacuated to safety in Florida and Louisiana as Andrew approached.

AFTER 10 YEARS, HURRICANE ANDREW GAINS STRENGTH

In the record books, it's still one of America's costliest hurricanes, and today National Oceanic and Atmospheric Administration (NOAA) scientists announced Hurricane Andrew was even stronger than originally believed when it made landfall in south Florida 10 years ago. Based on new research, scientists upgraded the storm from a Category 4, to a Category 5, the highest on the Saffir-Simpson Hurricane Scale.

In their re-analysis of Hurricane Andrew's maximum sustained surface-wind speeds, NOAA's National Hurricane Center Best Track Committee, a team of hurricane experts, concluded winds were 165 mph—20 mph faster than earlier estimated—as the storm made landfall. Herbert Saffir, a structural engineer who co-designed the Saffir-Simpson Hurricane Scale, joined the committee as an observer and reviewed the team's results.

The upgrade makes Andrew only the third Category 5 (wind speeds greater than 155 mph) hurricane on record to strike the continental United States. The other two Category 5 storms were the "Florida Keys 1935 Hurricane," and Hurricane Camille in 1969.

There is always some uncertainty in determining the maximum winds in a hurricane, and Andrew is no exception. The NHC's previous estimate was 145 mph, based on the science available in 1992. With advanced research techniques and technology, NHC now estimates the winds were stronger.

Hurricane Hugo prepares to strike Charleston, South Carolina, in 1989. *(Courtesy of National Oceanic and Atmospheric Administration (NOAA)/National Environmental Satellite, Data, and Information Service (NESDIS).)*

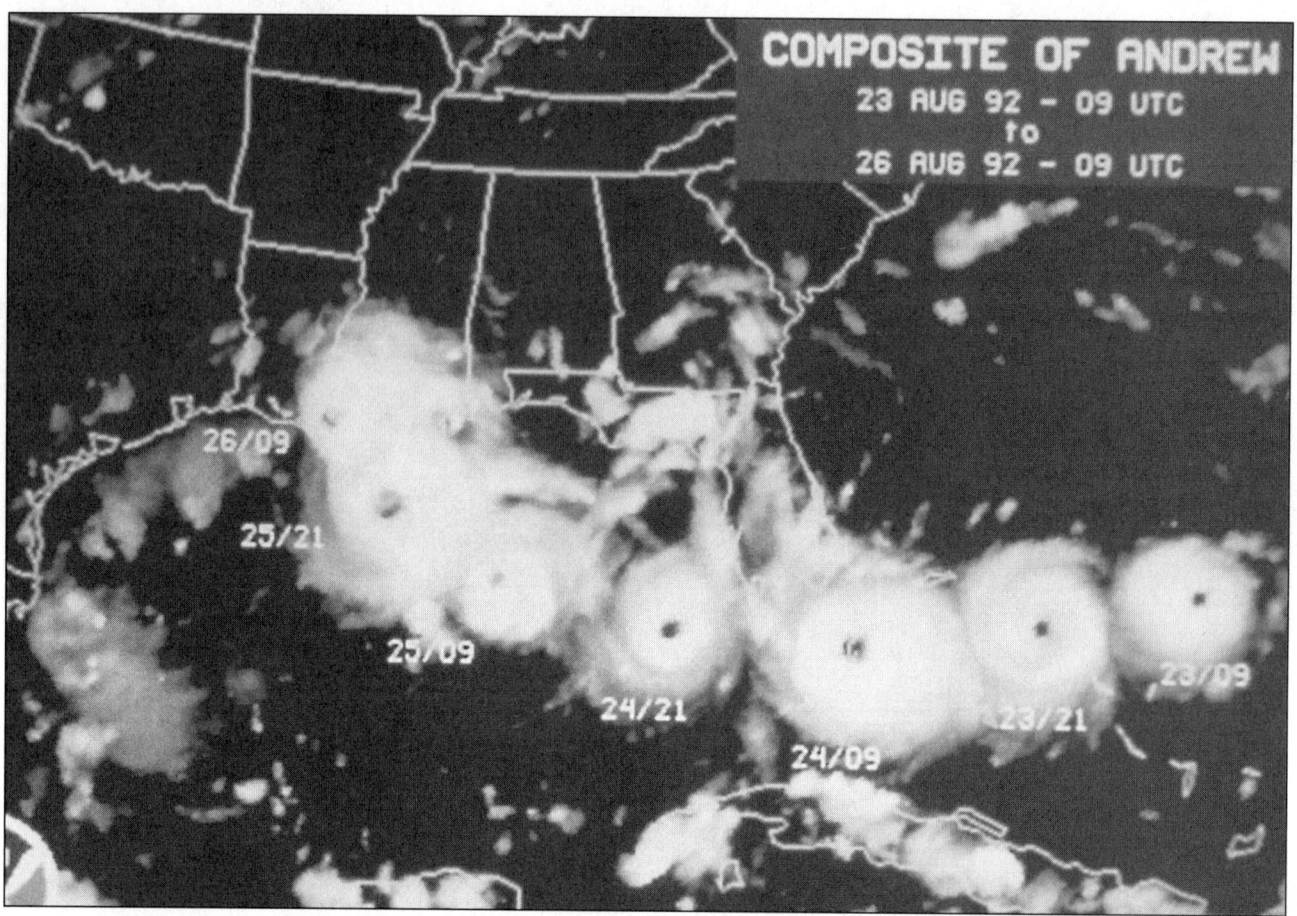

Hurricane Andrew's path of destruction, August 1992. *(Courtesy of National Oceanic and Atmospheric Administration (NOAA)/National Environmental Satellite, Data, and Information Service (NESDIS).)*

Andrew was directly responsible for 23 fatalities in Florida and Louisiana, and about $25 billion in damages (1992 dollars), according to NOAA. The National Hurricane Center has had an ongoing program to review the historical record of all storms. Scientists and other researchers note that society needs an accurate account of the frequency and intensity of past catastrophic events to best plan for the future.

The Best Track Committee at the NHC recently completed a review of a re-analysis of storms from 1851 to 1910. Since 1997, forecasters have used Global Positioning System dropwindsondes, a measuring device dropped from hurricane reconnaissance aircraft into the eyewall—the windiest part of the hurricane. The sonde system measures temperature, barometric pressure, water vapor, and wind data every 15 feet on its way down.

This new method gave meteorologists an important glimpse into the true strength of these devastating storms. The analyses of the dropwindsonde data indicated that, on average, the maximum sustained surface-wind speed was about 90 percent of the wind speed measured at the 10,000-foot aircraft level flown as Andrew approached south Florida. In 1992 Andrew's wind speed was estimated at 75

to 80 percent of the aircraft observations. The research findings resulted in an increase in the estimated wind speeds of Hurricane Andrew from 145 mph to 165 mph.

Hurricane Andrew was a Category 5 over open water on approach to South Florida. Hurricane Andrew was a Category 5 on the Saffir-Simpson Hurricane Scale at time of landfall, with Category 5 winds occurring in a small area on the immediate coast having open exposure to Biscayne Bay. Winds at specific locations over land in Miami-Dade County are unknown due to remaining scientific uncertainties.

There should be continuing research aimed at better determining hurricane winds immediately preceding, and during landfall. The "Hurricane Landfall" component of the U.S. Weather Research Program is structured to address such a question.

When Hurricane Andrew hit southeast Miami-Dade County, Florida, August 24, 1992, flying debris in the storm's winds knocked out most ground-based wind measuring instruments, and widespread power outages caused electric-based measuring equipment to fail. The winds were so strong many wind-measuring tools were incapable of registering the maximum winds. Surviving wind observations and measurements from aircraft reconnaissance, surface pressure, satellite analysis, radar, distribution of debris, and structural failures were used to estimate the surface winds.

Hurricane Fran, 1996

Fran was a Cape Verde hurricane that started as a tropical wave off the African west coast on August 22, 1996. The wave escalated to a tropical storm on August 27 (about 900 mi east of the Lesser Antilles), and finally made landfall on the North Carolina coast as a category 3 hurricane on September 6.

At its peak, Fran's minimum central pressure dropped to 946 mb and sustained surface winds reached 105 knots (on September 5); at landfall the minimum central pressure hit 954 mb and winds were approximately 100 knots. Fran eventually weakened to below-hurricane levels as it moved through Virginia and up through the eastern Great Lakes where it became extratropical.

Hitting at the peak of hurricane season, Fran's strength resulted in major storm surge flooding in the North Carolina coastal area, wind damage in North Carolina and Virginia, and significant flooding up through Pennsylvania. Storm-surge flood levels in North Carolina (primarily southwest of Cape Lookout) are estimated to have reached 8–12 ft; some outside water and debris marks are higher, due to breaking waves. And while Fran typically caused more than 6 in of rainfall in its path, up to 12 in were reported in two counties of North Carolina. Inland flooding was also extensive. In Alexandria, Virginia, for example, the Old Town historic district had to be partially evacuated as 3 ft of water covered the streets due to the rising Potomac River.

The casualty and damage figures for Fran are significantly higher than for those of Bertha (July 1996). Thirty-four people died in storm-related incidents, and 21 of those reported were in North Carolina.

In an effort to avoid the hazards of downed power lines, flash floods, and damaging winds, nearly a half-million people were evacuated from the North and South Carolina coasts. Fran's wrath wrought nearly $3.2 billion in total U.S. damage, with the bulk of that damage (about $1.275 billion) in North Carolina.

• THE 1998 ATLANTIC HURRICANE SEASON

The 1998 season will be remembered as being one of the deadliest in history—and for having the strongest October hurricane on record. It was a very active season with 14 named tropical storms of which 10 became hurricanes. Three of these were major hurricanes—category three, four, and five on the Saffir-Simpson Hurricane Scale. The four-year period of 1995–1998 had a total of 33 hurricanes—an all-time record.

Tropical cyclones claimed an estimated 11,629 lives in 1998. Of that total, 11,000 were due to hurricane Mitch in Central America. Not since 1780 has an Atlantic hurricane caused so many deaths.

Seven of the tropical storms and hurricanes hit the United Sates, which is more than twice the average. Total damages in the United States stands at $6.5 billion.

The season started a little late, but more than made up for lost time. There were no tropical cyclones in June and the first storm of the season, tropical storm Alex, developed on July 27. Then, in a hyperactive 35-day span from August 19 to September 23, 10 named tropical cyclones formed in October—including Lisa and powerful hurricane Mitch. The season concluded with hurricane Nicole in late November.

On September 25, there were four Atlantic hurricanes in progress at once. This is the first time such an event was observed since 1893.

Bonnie developed from a tropical wave over the Atlantic about 900 mi east of the Leeward Islands on August 19, and became a tropical storm a day later. It moved on a west-northwestward track, skirting the Leeward Islands. Late on August 21, the storm strengthened into a hurricane located about 200 mi north-northeast of eastern Hispaniola. Bonnie strengthened to its maximum wind speed of 115 mph late on August 23 while located about 175 mi east of San Salvador in the Bahamas. The hurricane turned toward the northwest and stayed east of the Bahamas.

Bonnie then headed toward the southeast U.S. coast in the general direction of the Carolinas, gradually turning toward the north-northwest and then north. As the center neared the coast, its forward speed slowed. The eye of Bonnie passed just east of Cape Fear, North Carolina, late on August 26. Bonnie made landfall near Wilmington as a category two hurricane early on August 27. It weakened to a tropical storm while moving slowly over eastern North Carolina. As the storm moved off the coast in the vicinity of the outer banks near Kitty Hawk, it re-strengthened into a hurricane. Bonnie soon weakened back to a tropical storm as it moved northeastward to eastward over the Atlantic into cooler waters, becoming extratropical about 240 mi southeast of Cape Race, Newfoundland. Three deaths were caused by Bonnie and the damage total is estimated to be $720 million.

The tropical depression that was to become Danielle formed early on August 24 about 700 mi west of the Cape Verde Islands. Tropical storm status was reached later that day. Moving west-northwestward, Danielle rapidly strengthened into a hurricane and reached the first of several peak intensities near 105 mph while centered about 1,040 mi east of the Leeward Islands. For the next several days, the hurricane continued west-northwestward, gradually slowing in forward speed. Danielle turned northwestward and northward on August 30–31, passing about 230 mi northwest of Bermuda early on September 2, and winds at Bermuda briefly reached tropical storm force. Danielle lost tropical characteristics about 260 mi east-southeast of Cape Race late on September 3.

Earl developed over the southwest Gulf of Mexico on August 31. The tropical depression became a tropical storm later that day but had a poorly defined center that was difficult to track. The general motion was north and then northeastward, becoming a hurricane located about 150 mi south-southeast of New Orleans, Louisiana. After briefly reaching 100 mph winds, Earl made landfall over the Florida panhandle near Panama City as a category one hurricane early on September 3. It weakened to below hurricane strength soon after making landfall and became extratropical on September 3 while moving northeastward through Georgia. Earl was directly responsible for three deaths and the total damage estimate is $79 million.

Georges formed in the far eastern Atlantic from a tropical wave early on September 15, and became a tropical storm on the morning of September 16. By late afternoon on September 17, satellite imagery indicated that Georges developed an eye and had become a hurricane. Georges moved on a general west to west-northwest course at 15–20 mph for the next several days. During this period, Georges is estimated to have reached a peak intensity of 155 mph category 4, and a minimum central pressure of 937 mb early on September 20 while located about 420 mi east of Guadeloupe in the lesser Antilles. Georges's first of many landfalls occurred at Antigua in the Leeward Islands late on September 20. After moving near or over other islands of the northeast Caribbean, including the U.S. Virgin Islands, it then hit Puerto Rico on the evening of September 21 with estimated maximum winds of 115 mph. Georges weakened very little while over Puerto Rico and was even stronger when it made landfall in the Dominican Republic on the afternoon of September 22, with estimated maximum winds of 120 mph. Georges weakened after crossing the mountainous terrain of Hispaniola and made landfall in eastern Cuba on the afternoon of September 23 with estimated maximum winds of 75 mph.

The hurricane continued along the northern coast of Cuba for most of September 24. Thereafter, Georges moved into the Florida Straits early on September 25 and re-intensified making landfall near Key West, Florida, on mid-morning of September 25 with estimated maximum winds of 105 mph. Georges continued on a general west-northwest to northwest track on September 26–27, turning to a north-northwest heading and gradually slowing down as it approached the coast of the central Gulf of Mexico. Georges made its final landfall near Biloxi, Mississippi, early on September 28 with 105 mph winds. Georges meandered over land and weakened to a tropical storm later that day. Georges was downgraded to a tropical depression by mid-morning on September 29 while located about 35 mi north-northeast of Mobile, Alabama. The remnant weak circulation center moved off the Georgia/South Carolina coast on October 1, becoming in-

volved with a frontal zone and dissipating. Georges caused one death and the total U.S. damage estimate is $5.1 billion.

Jeanne became a tropical storm unusually far to the east, close to the west coast of Africa early on September 21 and intensified into a hurricane the following day. It brushed the Cape Verde Islands while moving west-northwestward for a couple of days, reaching a peak intensity of 105 mph on September 24 about 650 mi west of those islands. The hurricane turned toward the north over the east-central Atlantic late on September 26, then toward the northeast and east-northeast. Jeanne weakened to a tropical storm on September 29, then dropped below storm strength in the vicinity of the Azores and lost its tropical characteristics late on September 30.

Mitch: The Deadliest Atlantic Hurricane Since 1780

In an awesome display of power and destruction, hurricane Mitch will be remembered as the most deadly hurricane to strike the Western Hemisphere in the last two centuries! Not since the Great Hurricane of 1780, which killed approximately 22,000 people in the eastern Caribbean, was there a more deadly hurricane. Mitch struck Central America with such viciousness that it was nearly a week before the magnitude of the disaster began to reach the outside world. The death toll has been reported as 11,000 with thousands of others missing. Though the final death toll will never be known, it is quite likely that Mitch directly killed more people than any Atlantic hurricane in over 200 years. More than three million people were either homeless or severely affected. In this extremely poor third world region of the globe, estimates of the total damage from the storm are at $5 billion and rising. The President of Honduras, Carlos Flores Facusse, claimed the storm destroyed 50 years of progress.

A Category 5 Monster

Within four days of its birth as a tropical depression on October 22, 1998, Mitch had grown into a category 5 storm on the Saffir-Simpson Scale. By 2100 UTC on October 26, the monster storm had deepened to a pressure of 905 mb with sustained winds of 180 mph (155 knots) and gusts well over 200 mph! Mitch thus became tied for the fourth strongest Atlantic hurricane on record based upon barometric pressure values. Though the pressure began rising six hours later, Mitch remained at category 5 status for a continuous period of 33 hours—the longest continuous period for a category 5 storm since the 36 consecutive hours by hurricane David in 1979. In addition, Mitch maintained sustained winds 155 knots for 15 hours—the third longest period of such winds on record after the continuous 18 hours of 155 knot winds or higher by Hurricane Camille in 1969 and Hurricane Dog in 1950. Though exact comparisons are suspect due to differing frequencies of observation times (3-hourly versus 6-hourly observations) and a bias in earlier years toward higher estimated wind speeds, it is quite ap-

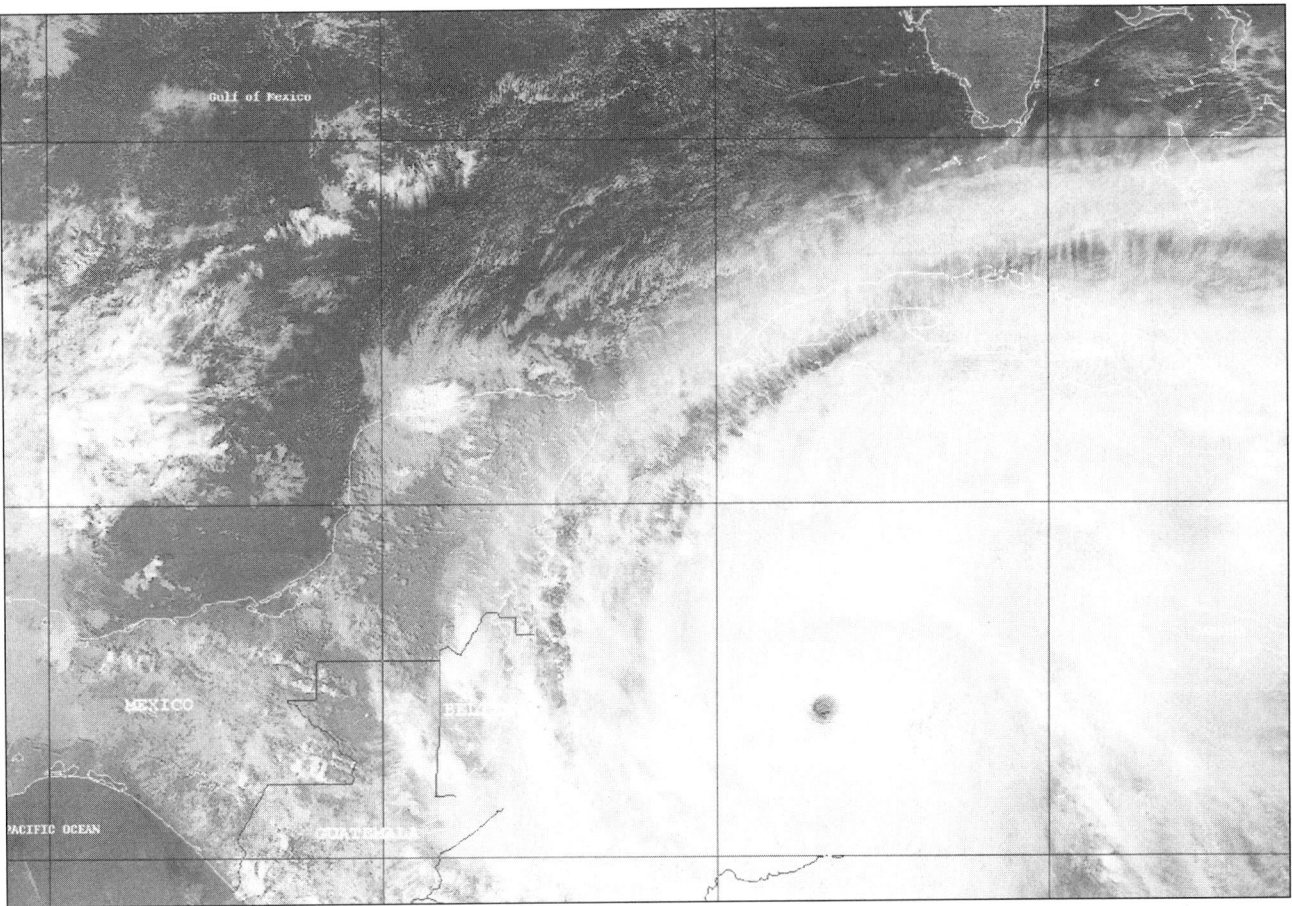

Hurricane Mitch (October 1998), a category 5 storm, was the deadliest hurricane since 1780. *(Courtesy of National Oceanic and Atmospheric Administration (NOAA).)*

parent that Mitch was one of the strongest storms ever recorded in the Atlantic.

Assault on Central America

After threatening Jamaica and the Cayman Islands, Mitch moved westward and by 2100 UTC on October 27, the category 5 storm was about 60 mi north of Trujillo on the north coast of Honduras. Preliminary wave height estimates north of Honduras during this time at the height of the hurricane are as high as 44 ft, according to one wave model. Although its ferocious winds began to abate slowly, it took Mitch two days to drift southward to make a landfall. Coastal regions and the offshore Honduras island of Guanaja were devastated. Mitch then began a slow westward drift through the mountainous interior of Honduras, finally reaching the border with Guatemala two days later on October 31.

Although the ferocity of the winds decreased during the westward drift, the storm produced enormous amounts of precipitation caused in part by the mountains of Central America. As Mitch's feeder bands swirled into its center from both the Caribbean and the Pacific Ocean to its south, the stage was set for a disaster of epic proportions. Taking into account the orographic effects by the volcanic peaks of Central America and Mitch's slow movement, rain fell at

the rate of 1–2 ft per day in many of the mountainous regions. Total rainfall has been reported as high as 75 in for the entire storm. The resulting floods and mudslides virtually destroyed the entire infrastructure of Honduras and devastated parts of Nicaragua, Guatemala, Belize, and El Salvador. Whole villages and their inhabitants were swept away in the torrents of floodwaters and deep mud that came rushing down the mountainsides. Hundreds of thousands of homes were destroyed.

Re-birth and Florida landfall

The remnants of Mitch drifted northwestward as a weak depression and entered the Bay of Campeche on November 2. Over the warm waters and favorable conditions aloft, Mitch once more regained tropical storm status and began moving rapidly northeastward. It struck the western side of Mexico's Yucatan Peninsula, which weakened it to tropical depression status once again. As Mitch moved back over the Gulf of Mexico, it regained tropical storm status for the third time. It raced northeastward and pounded Key West with tropical storm-force winds and heavy rains on November 4–5. Some of the roofs and buildings damaged by Hurricane Georges in September fell victim to Mitch. Rains of 6–8 in were common in southern Florida and several tornadoes

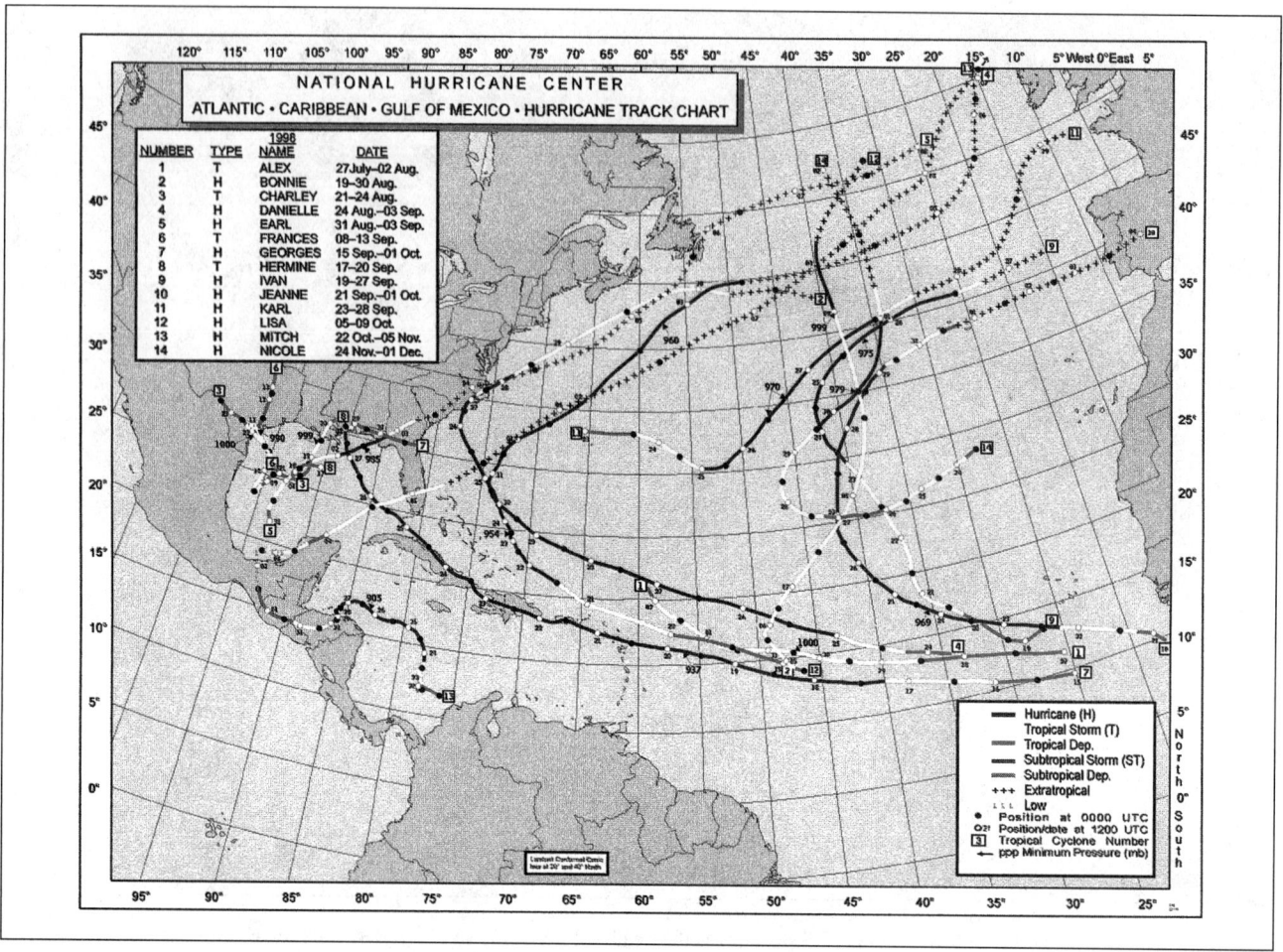

(Courtesy of National Hurricane Center/Tropical Prediction Center/NOAA.)

struck the region. At least seven were injured when a tornado swept from Marathon to Key Largo. A second tornado touched down at Miramar, north of Miami. At Fowey Rocks Lighthouse, just southeast of Miami, a wind gust of 73 mph was reported. Across south Florida, some 100,000 customers lost electrical power. One person was killed in the United States near Dry Tortugas when a fisherman died from a capsized boat. A second person was missing. Another person died as a result of an auto accident on a slick highway. Mitch passed through the Bahamas and finally became extratropical on November 5.

• NOTABLE TROPICAL CYCLONES OF 1999

The 1999 Atlantic hurricane season produced the most deadly hurricane to hit the United States in 28 years, Floyd, which killed 56 people. Three hurricanes hit the U.S. mainland, Bret, Floyd, and Irene, as category 3, 2, and 1 hurricanes, respectively. Hurricane Dennis produced near-hurricane conditions as it moved along the North Carolina coast but made landfall as a tropical storm. Tropical Storm

Harvey also made landfall in the United States. The U.S. Virgin Islands experienced Hurricane Lenny. In all, there were four tropical storms, and eight hurricanes during 1999, and five reached intense hurricane status. Four tropical storms were consistent with long term average (1950–98), the eight hurricanes were above the long-term average of six per season.

Only one system during the 1999 season developed prior to August 18, and four systems developed after October 12. Three of these late-season storms became hurricanes, with the last (Hurricane Lenny) reaching category 4 status. Lenny developed in mid-November and moved eastward across the central Caribbean Sea. This unusual track enabled it to become the first hurricane to strike the Lesser Antilles Islands from the west.

Hurricane Bret made landfall August 22, on Padre Island in sparsely populated Kenedy County (population under 500 people, about 60,000 cattle) in Texas, about halfway between Brownsville and Corpus Christi. Bret was only the sixteenth category 4 storm to ever hit the United States and the fourth category 4 storm to ever hit the Texas coast. Bret drifted westward dumping copious rainfall over south Texas, with over 20 in estimated by NEXRAD radar over a

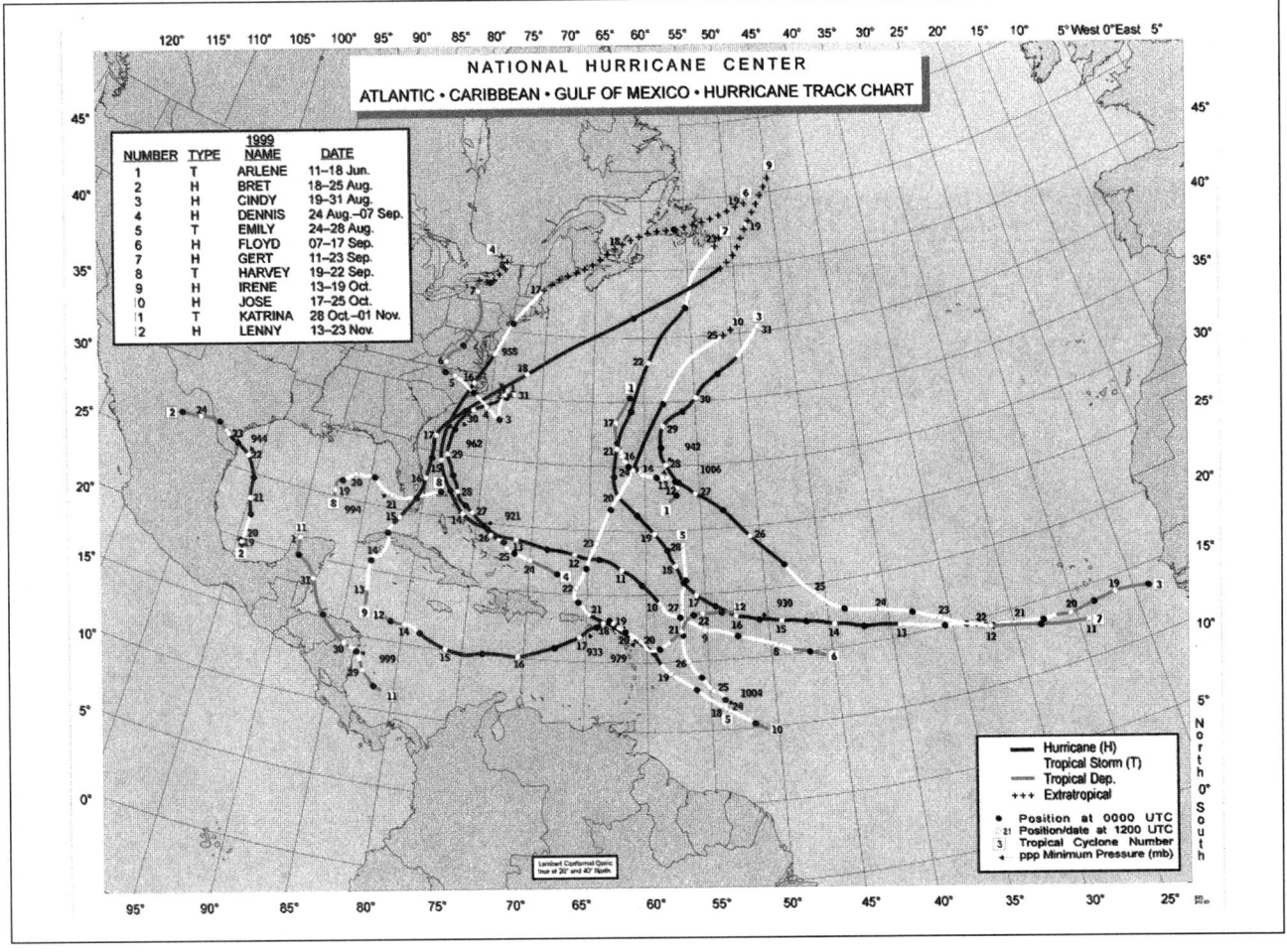

(Courtesy of National Hurricane Center/Tropical Prediction Center/NOAA.)

portion of Kenedy County. This region had been spared a land-falling hurricane in recent years, as the last hurricane to hit the Texas coast was Jerry in October, 1989.

Hurricane Dennis threatened the North Carolina coast, the fourth tropical storm scare in as many years, during August 29–30. The center of Dennis was within 60 mi of the North Carolina coast early on August 30 as a strong category 2 hurricane with highest sustained winds of 105 mph. Due to the fact that the hurricane never made landfall, damage was only moderate. Rainfall amounts approached 10 in in coastal southeastern North Carolina, and beach erosion was substantial. Four people were killed in Florida due to high surf, and a tornado in Hampton, Virginia, produced several serious injuries.

Hurricane Floyd brought flooding rains, high winds, and rough seas along a good portion of the Atlantic seaboard on September 14–18. The greatest damages were along the eastern Carolinas northeast into New Jersey, and adjacent areas northeastward along the East Coast into Maine. Several states had numerous counties declared disaster areas. Flooding caused major problems across the region, and at least 77 deaths were reported—57 directly related to the

hurricane. Damage estimates range from three to over six billion dollars. Although Hurricane Floyd reached category 4 intensity in the Bahamas, it weakened to category 2 intensity at landfall in North Carolina. Floyd's large size was a greater problem than its winds, as the heavy rainfall covered a larger area and lasted longer than with a typical category 2 hurricane. Approximately 2.6 million people evacuated their homes in Florida, Georgia, and the Carolinas—the largest peacetime evacuation in U.S. history. Ten states were declared major disaster areas as a result of Floyd, including Connecticut, Delaware, Florida, Maryland, New Jersey, New York, North Carolina, Pennsylvania, South Carolina, and Virginia. There were several reports from the Bahamas area northward of wave heights exceeding 50 feet. The maximum storm surge was estimated to be 10.3 ft on Masonborough Island in New Hanover County, North Carolina.

Hurricane Irene brought heavy rains to the Florida Keys northward to central Virginia during the middle of October. Some places in eastern North Carolina and eastern Virginia received over 12 in of additional rains, adding to the flooding problems. Eight people were killed from electrocution and drowning.

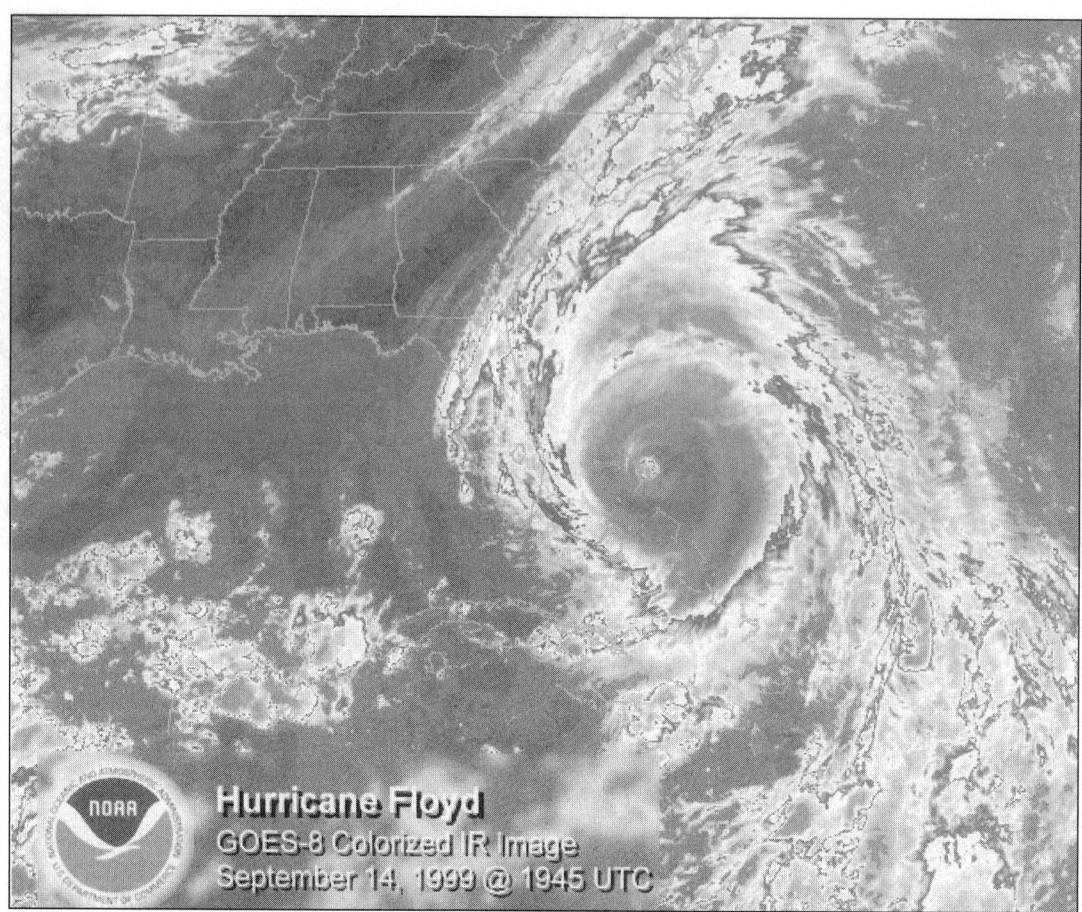

Hurricane Floyd, September 1999. *(Courtesy of National Oceanic and Atmospheric Administration (NOAA).)*

Hurricane Lenny, a very unusual west-to-east moving low latitude hurricane, battered portions of the Caribbean around mid-November. Lenny was a strong category 4 hurricane on November 17 with winds of 150 mph sustained and an estimated central pressure of 929 mb. Lenny was the second strongest storm of the twentieth century to hit the Virgin Islands, second only to Hurricane David in 1979. Lenny as of November 17 was slightly stronger than Hugo at landfall in South Carolina in 1989. Hurricane Lenny was responsible for 15 deaths directly associated with the storm, including three deaths in Dutch St. Maarten, two in Colombia, five in Guadeloupe, one in Martinique, and four offshore.

• NOTABLE TROPICAL CYCLONES OF 2000

The above-average 2000 Atlantic hurricane season continued the recent upturn in activity, but the United States was spared extensive hurricane damage. This was the third consecutive year of above-average activity. NOAA says that the increased activity was likely an indication that global climate variations on decadal time scales are again favoring more active Atlantic hurricane seasons. NOAA researchers are studying the decadal cycles and how storm paths this year may have been affected by a Bermuda high pressure system made weak without the influence of El Niño or La Niña.

There were 14 named storms in 2000, including eight hurricanes and three intense hurricanes. The current long-term averages are 9.3 named storms, 5.8 hurricanes and 2.2 intense storms. Only two named storms hit the mainland, both in Florida. Hurricane Gordon weakened into a tropical storm when it hit land September 17, and Helene came ashore as a weak tropical storm September 23. This is the first time since 1994 that the U.S. mainland was spared a landfall storm of hurricane strength.

Tropical Storm Beryl made landfall 150 mi south of Brownsville, Texas on August 14. Moisture from Beryl moved northward along the south Texas coast and brought 0.79 in of rain to Corpus Christi on August 15.

Hurricane/Tropical Storm Debby weakened as it moved across the Caribbean and was downgraded to a trough or open wave and then dissipated on August 24.

Tropical Storm/Hurricane Gordon brought heavy rains and isolated tornadoes to Florida on Sunday, September 17. It became extratropical on September 18, but brought heavy rains across parts of Georgia and the eastern Carolinas into southeast Virginia. The storm had dumped over 8 in of rain in Seabring, Florida, in the central part of the state. Gordon killed 23 people in Guatemala.

Hurricane Keith pounded much of Central America on October 2 with heavy rain and high winds. On October 2, Keith's center was located about 45 mi east of Belize City, Belize, and about 70 mi south-southeast of Chetumal, Mexico. With its slow motion, Keith was expected to dump as much as 20 in of rainfall on some parts of Central America. Keith killed 19 people in Belize, and caused $200 million in damages in the region.

Tropical weather systems do not have to reach the level of a hurricane to wreck havoc. Even before it became Tropical Storm Leslie, this system dropped 18 in of rain in south Florida, and caused massive urban flooding and $700 million in total damage. The United States recorded only modest tropical storm-damage and flooding in 2000 because none of the hurricanes made landfall.

• NOTABLE TROPICAL CYCLONES OF 2001

For the North Atlantic, Caribbean Sea, and the Gulf of Mexico

The 2001 hurricane season was an active one. There were 15 named storms of which 9 became hurricanes. Four of these became major hurricanes—Category 3 or higher on the Saffir-Simpson hurricane scale. The long-term averages are 10 tropical storms, 6 hurricanes, and 2 major hurricanes. The bulk of the activity occurred during the last 3 months of the season, during which 11 of the named storms, and all of the major hurricanes, formed. There were 3 hurricanes during November, which was the first such occurrence on record. Two major hurricanes—Iris and Michelle—struck land areas around the Caribbean, causing 48 deaths. Michelle also caused significant damage in the Bahamas. For the second consecutive year there were no U.S. hurricane landfalls, though 2 of the 3 tropical storms that hit the U.S. were almost hurricanes, and the third, Allison, caused enormous flooding, resulting in 41 deaths and billions of dollars in damage. There were 2 tropical depressions that did not become tropical storms.

Allison developed from a disturbance that moved from the eastern Pacific into the southwest Gulf of Mexico. On June 4, thunderstorms increased over the western Gulf, and on the morning of the 5th the system quickly developed into a tropical storm. Allison strengthened to a peak intensity of 60 mph on the afternoon of the 5th before moving inland over southeast Texas with 50 mph winds a few hours later.

After moving inland, Allison rapidly weakened before stalling over eastern Texas on June 7. The remnant circulation drifted southward and emerged over the northwestern Gulf of Mexico on the 9th, where the system reorganized as a subtropical cyclone before moving inland again over Louisiana early on the 11th. The subtropical low tracked east-northeastward before stalling over eastern North Car-

olina on the 14th. The low was nearly stationary for almost three days before finally moving northeastward off the Mid-Atlantic coast on the 17th. The subtropical cyclone merged with a cold front on the 17th and dissipated on the 19th southeast of Nova Scotia.

Allison produced extremely heavy rainfall from eastern Texas across the Gulf states and along the Mid-Atlantic coast, resulting in the most extensive flooding ever associated with a tropical storm. Damage estimates are $5.0 billion or more, and there were 41 direct deaths. Much of the damage and fatalities occurred in the Houston Metropolitan area, where more than 30 inches of rain were reported at several locations. The preliminary death toll by states is as follows: Texas 23, Florida 8, Pennsylvania 7, Louisiana 1, Mississippi 1, and Virginia 1.

Michelle started as a broad low-pressure area in the southwestern Caribbean Sea. It developed into a tropical depression on October 29 along the east coast of Nicaragua. The depression remained nearly stationary over northeastern Nicaragua for two days, producing extremely heavy rains with flooding over portions of Nicaragua and Honduras. Late on the 31st the depression moved into the northwestern Caribbean Sea just north of the Honduras-Nicaragua border and strengthened into Tropical Storm Michelle. Michelle moved slowly north-northwestward for the next two days as it strengthened into a hurricane. The hurricane turned slowly northward on the 3rd, then turned northeastward on the 4th as it reached a peak intensity of 140 mph. Later that day Michelle crossed the coastal islands of Cuba as a Category-4 hurricane, then it crossed the coast of the main island of Cuba as a Category-3 hurricane. A weakening Michelle continued northeastward through the Bahamas on the 5th, and the storm became extratropical over the southwestern Atlantic on the 6th.

Michelle left a trail of damage and death from central America to the Bahamas. So far 17 deaths are associated with the hurricane, including 6 in Honduras, 5 in Cuba, 4 in Nicaragua, and 2 in Jamaica. Michelle was the strongest hurricane to hit Cuba since 1952 and caused widespread damage over central and western Cuba. Additional damage was reported in the Bahamas. Widespread heavy rains over Central America and the northwest Caribbean caused extensive flooding and mud slides in Nicaragua, Honduras, and Jamaica. Minor damage was reported in the Cayman Islands and South Florida.

Five or more major hurricanes occurred three times in the 90's; 1995, 1996, and 1999. (A major hurricane is defined as category 3 or higher according to the Saffir-Simpson scale. A category 3 hurricane has winds of 111 to 130 mph.) Prior to 1995, five or more major Atlantic hurricanes had not occurred in one season since 1964. A new record number of hurricanes for November was set in 2001 as Michelle, Noel, and Olga all were active in the Atlantic Basin during the month. The contiguous U.S. has not been hit directly by a hurricane now for the past two years, although tropical storms have caused significant damage, as evidenced by Tropical Storm Allison. This storm, the costliest tropical storm on record ($5 billion in damage), caused severe flooding in Texas and Louisiana before moving across the Southeast and up the East Coast.

• NOTABLE TROPICAL CYCLONES OF 2002

The 2002 Atlantic hurricane season was again an active one, making it the 5th consecutive active season. There were 12 named storms, compared to a 1944–1996 annual average of 9.8. Four of those named storms became hurricanes, of which 2 were classified as major. This compares to an average of 5.8 hurricanes a year, 2.5 of which are major based on a 53-year average. There were 2 additional tropical depressions. So there were more named tropical cyclones than the long-term mean, but fewer strong storms. The season was also slow to begin, extremely active in September, and less active than average at its end. Eight named systems developed in September as well as a tropical depression, making it the most active month of any month on record for tropical cyclone development in the Atlantic basin.

Seven tropical storms made landfall on the continental U.S. in 2002—the most since 1998, when seven was also the number of landfalling tropical systems for the U.S. Hurricane Lili was the first hurricane to hit the coast of the U.S. since Irene in October 1999.

Notable tropical systems in 2002 include the intense Hurricanes Isidore and Lili in September and the long-lasting Hurricane Kyle, also in September. Kyle was the third-longest-lived tropical cyclone in the Atlantic basin after Ginger of 1971 and Inga of 1969. Kyle formed on September 20 and spent over 3 weeks in the North Atlantic before finally coming ashore in South Carolina on October 11. Isidore became a category-3 hurricane as it moved into the southeastern Gulf of Mexico and hit the Yucatan Peninsula on September 22 at category 3 strength. After weakening over the Yucatan Peninsula, it never quite regained hurricane strength and came ashore in Louisiana as a tropical storm. Lili was the only hurricane to make landfall in the United States in 2002, and did so at category 2. Lili had been a category 4 hurricane shortly before landfall and quickly weakened before reaching the west Louisiana coastline on October 3.

El Niño Suppressed Hurricanes in 2002 Season

The 2002 Atlantic hurricane season produced only four hurricanes due to a strengthening El Niño. However, twice the normal number of storm systems (eight) affected the nation, bringing storm surge and severe weather and rain to the nation, including Hurricane Lili, the first land-falling hurricane to strike the United States since the 1999 Hurricane Season.

Overall in 2002, there were 12 named storms, of which four became hurricanes. Hurricanes Lili and Isidore were classified as major (category 3 or higher on the Saffir-Simpson hurricane scale). Eight storms—Tropical Storms Bertha, Edouard, Fay, and Hanna; and Hurricanes Gustav, Isidore, Kyle, and Lili—affected the coastal United States. Hurricane Lili was the only storm to make landfall while still a hurricane. The other 2002 storms were: Tropical Storms Arthur, Cristobal, Dolly, and Josephine.

Hurricane forecasters at NOAA Climate Prediction Center (CPC), Hurricane Research Division, and National Hurricane Center (NHC) correctly forecast that climate conditions, including the El Niño, would reduce the overall hurricane activity this season. The forecast called for seven to 10 tropical storms, of which four to six could develop into hurricanes, with one to three classified as major.

Louisiana, the hardest hit area, was battered by four storms including the powerful Hurricane Lili and Tropical Storm Isidore. The 2002 season's storms caused nine deaths in the United States and about $900 million in damages.

The public relied heavily on Internet access for lifesaving information from NOAA this season. The explosive use of the Internet to convey vital information to the public in near real time has been astonishing. Between August and September the NHC web site recorded almost 500 million hits. The peak day for the season was October 3 (Hurricane Lili) when the site recorded 35.9 million hits—doubling the previous record set in 1999 during Hurricane Floyd.

• TYPHOON PONGSONA HITS GUAM, DECEMBER 8, 2002

Guam, west of the international date line, about 3,700 miles southwest of Hawaii, was raked by wind gusts of more than 180 mph, and winds the following day of 70 mph continued in this U.S. Territory. No deaths were reported but the entire island was without electricity, and water and sewer systems were not expected to be fully operational for weeks. Sustained winds of 150 mph around the eye of Pongsona gave the storm a "super typhoon" status. Gusts up to 184 mph were reported in some areas of Guam. A wind gust of 117 mph was clocked before the National Weather Service's sensor failed, along with the radar. A reported 2,000 homes were destroyed and many more were damaged. Schools became emergency shelters.

• CYCLONE ZOE STRIKES THE SOLOMON ISLANDS

Southern Pacific Ocean Category-5 Cyclone Zoe struck the southernmost Solomon Islands on December 29, 2002,

Hurricane Lili was the first hurricane to hit the United States in three years (Irene, October 1999). *(Courtesy of National Oceanographic and Atmospheric Administration (NOAA)/National Environmental Satellite, Data, and Information Service (NESDIS).)*

and was reported to be the most powerful ever recorded in the area, with winds reaching 225 mph and produced 33 foot waves that destroyed houses, uprooted trees, and flooded entire villages. Those living on the island used common sense that had come with centuries of dealing with these storms and had taken shelter in the hills on the island. It was considered a miracle that no one was killed on Tikopia, some 1,400 miles northeast of Sydney, Australia.

• HURRICANE WARNING SERVICE

A history of hurricane watching

The hurricane forecast and warning service stands as the finest of its kind in the world, distinguished by its character, credibility, and the confidence that our nation has in it. But that was not always the case.

The Weather Bureau was created as a civilian agency in 1890 mainly because of a general dissatisfaction with weather forecasting under the military. The hurricane of 1875 that destroyed Indianola, Texas, without much warning was a contributing factor.

It was not until the Spanish-American War of 1898 that an effort was made to establish a comprehensive hurricane warning service. President McKinley stated that he was more afraid of a hurricane than he was of the Spanish Navy. He extended the warning service to include warnings for shipping interests as well as the military. Before that, hurricane warnings were only issued for the United States coastal areas. Hurricane warning stations were established throughout the West Indies. A forecast center was established in Kingston, Jamaica, and later moved to Havana, Cuba, in 1899. The warning service was extended to Mexico and Central America. This recognition of the international responsibility for the United States hurricane warning service continues today under the auspices of the World Meteorological Organization (WMO) of the United Nations.

In 1900, the infamous Galveston, Texas, hurricane killed at least 8,000 people—the greatest natural disaster in United States history. There was no formal hurricane warning and this calamity prompted the transfer of the warning service to Washington, D.C., where it remained until 1935.

In the 1920s, there were several hurricanes that hit with little or no warning, leading to dissatisfaction with the hurricane service operating out of Washington. The coastal communities felt that Washington was insensitive to the hurricane problem. In 1926, a very strong hurricane (category 4 by today's standard) brought great devastation to southeast Florida, including Miami and Ft. Lauderdale, causing more than 200 deaths. The warnings for that storm were issued at night when most residents were asleep and unaware of the rapidly approaching hurricane. In 1928, another severe hurricane hit south Florida and killed an estimated 1,800 people, who drowned when Lake Okeechobee

overflowed. In 1933, the largest number of tropical storms—21—developed. Nine of them were hurricanes and two that affected the east coast of the United States, including Washington, were badly forecast and the public was inadequately warned. In 1934, a forecast and warning for an approaching hurricane in the very sensitive Galveston area was again badly flubbed by Washington.

These incidents led Congress and the President to revamp and decentralize the Hurricane Warning Service. Improvements included 24-hour operations with teletypewriter hookup along the Gulf and Atlantic coasts; weather observations at six hourly intervals; hurricane advisories at least four times a day; and a more adequate upper air observing network. New hurricane forecast centers were established at Jacksonville, Florida; New Orleans, Louisiana; San Juan, Puerto Rico; and Boston, Massachusetts (established in 1940).

In 1943, the primary hurricane forecast office at Jacksonville was moved to Miami, Florida, where the Weather Bureau established a joint hurricane warning service with the Army Air Corps and the Navy. It was also in 1943 that Col. Joseph Duckworth made the first intentional plane reconnaissance flight into the eye of a hurricane. The following year, regular aircraft reconnaissance was begun by the military, giving hurricane forecasters the location and intensity of the storms for the first time.

The Miami office was officially designated as the National Hurricane Center (NHC) in 1955. In the 1950s, a number of hurricanes, including Hazel, struck the East Coast, causing much damage and flooding. Congress responded with increased appropriations to strengthen the warning service and intensify research into hurricanes. The Weather Bureau organized the National Hurricane Research Project. The Air Force and Navy provided the first aircraft to be used by the Project to investigate the structure, characteristics, and movement of tropical storms.

In 1960, radar capable of "seeing" out to a distance of 200–250 mi from their coastal sites were established at strategic locations along the Atlantic and Gulf coasts from Maine to Brownsville, Texas. On April 1, 1960, the first weather satellite was placed in orbit, giving hurricane forecasters the ability to detect storms before they hit land.

• IMPROVEMENTS IN FORECASTING

Hurricane tracking & preparedness

The day is past when a hurricane could develop to maturity far out to sea and be unreported until it thrust toward land. The 1970–90s have seen greater emphasis on the need for hurricane preparedness among the hurricane-prone communities in the United States, as well as in the Caribbean. The rapid development of America's coastal areas has placed millions of people with little or no hurricane experience in the path of these lethal storms. For this vulnerable

Terms to Know

By international agreement, **TROPICAL CYCLONE** is the general term for all cyclone circulations originating over tropical waters, classified by form and intensity as follows:

TROPICAL WAVE: A trough of low pressure in the trade-wind easterlies.

TROPICAL DISTURBANCE: A moving area of thunderstorms in the tropics that maintains its identity for 24 hours or more. A common phenomenon in the tropics.

TROPICAL DEPRESSION: Rotary circulation at surface, highest constant wind speed 38 mph or less.

TROPICAL STORM: Distinct rotary circulation, constant wind speed ranges 39–73 mph.

HURRICANE: Pronounced rotary circulation, constant wind speed of 74 mph or greater.

SMALL CRAFT CAUTIONARY STATEMENTS: When a tropical cyclone threatens a coastal area, small craft operators are advised to remain in port or not to venture into the open sea.

TROPICAL STORM WATCH: Is issued for a coastal area when there is the threat of tropical storm conditions within 24–36 hours.

TROPICAL STORM WARNINGS: May be issued when winds of 39–73 mph are expected. If a hurricane is expected to strike a coastal area, tropical storm warnings will not usually precede hurricane warnings.

HURRICANE WATCH: Is issued for a coastal area when there is a threat of hurricane conditions within 24–36 hours.

HURRICANE WARNING: Is issued when hurricane conditions are expected in a specified coastal area in 24 hours or less.

STORM SURGE: An abnormal rise of the sea along a shore as the result, primarily, of the winds of a storm.

FLASH-FLOOD WATCH: Means a flash flood is possible in the area; stay alert.

FLASH-FLOOD WARNING: Means a flash flood is imminent; take immediate action.

---

coastal population, the answer must be community preparedness and public education in the hope that education and planning before the fact will save lives and lessen the impact of the hurricane and its effects. There has been increased national awareness of the hurricane threat through the cooperation of local and state emergency officials and the enlistment of the news media and other federal agencies in the campaign to substitute education and awareness for the lack of first-hand experience among the ever-increasing coastal populations. New technology and advances in the science under the weather service's modernization program now underway will lead to more improvement and effectiveness in the forecasting and warning of hurricanes.

Meteorologists with NOAA's National Weather Service (NWS) monitor the massive flow of data that might contain the early indications of a developing storm somewhere over the warm sea. Cloud images from Earth-orbiting satellites operated by NOAA keep Earth's atmosphere under virtually continuous surveillance, night and day. Meteorological data from hundreds of surface stations, radar and balloon probes of the atmosphere, and information from hurricane-hunting aircraft are other tools of the hurricane forecaster. Long before a storm has evolved even to the point of ruffling the easterly wave, scientists at NOAA's National Hurricane Center (NHC) in Miami, Florida, have begun to watch the disturbance.

In the satellite data coming in from both polar-orbiting and geostationary spacecraft, and in reports from ships and aircraft, the scientists look for subtle clues that mark the development of hurricanes—cumulus clouds covered by the cirrostratus deck of a highly organized convective system; showers that become steady rains; dropping atmospheric pressure; intensification of the trade winds, or a westerly wind component there. Then, if this hint of a disturbance blooms into a tropical storm, it receives a name. Naming the storm is a signal that brings the warning systems to readiness.

As an Atlantic hurricane drifts closer to land, it comes under surveillance by weather reconnaissance aircraft of the U.S. Air Force Reserve, the famous "Hurricane Hunters," who bump through the turbulent interiors of the storms to obtain precise fixes on the position of the eye, and measure winds and pressure fields. Despite the advent of satellites, the aircraft probes are the most detailed information hurricane forecasters receive. The hurricanes are also probed by the "flying laboratories" from NOAA's Aircraft Operations Center in Miami. Finally, the approaching storm comes within range of a radar network stretching from Texas to Maine, and from Miami to the Lesser Antilles.

Through the lifetime of the hurricane, advisories from the NHC warning give the storm's position and what the forecasters in Miami expect the storm to do. As the hurricane drifts to within a day or two of its predicted landfall, these advisories begin to carry watch and warning messages, telling people when and where the hurricane is expected to strike, and what its effects are likely to be. Not until the storm has decayed over land and its cloud elements and great cargo of moisture have blended with other brands of weather does the hurricane emergency end.

This system works well; the death toll in the United States from hurricanes has dropped steadily as NOAA's hurricane tracking and warning apparatus has matured.

Hurricane research & technology

Although the accuracy of hurricane forecasts has improved over the years, any significant improvements must come from quantum leaps in scientific understanding.

In NOAA's Environmental Research Laboratories, scientists follow eagerly as nature furnishes additional specimens of the great storms—specimens they can probe and analyze to gather ever-greater understanding of the mechanics of the storms. Such analyses assist the forecasters with their warnings.

The GOES-10 (Geostationary Operational Environmental Satellite) series of satellites has provided more accurate and higher-resolution sounding data than provided from geosynchronous satellites, and similar improvements can be expected from the polar orbit satellite systems.

Major improvements in longer-range hurricane forecasts (36–72 hours) will come through improved dynamical models. Global, hemispheric, and regional models show considerable promise.

Present operational reconnaissance aircraft provide invaluable data in the core of the hurricane. Doppler radar is now an integral part of NOAA's research aircraft operations, providing entire data fields within several miles of the aircraft's path.

Next Generation Radar (NEXRAD), also known as WSR-88D, has added new dimensions to hurricane warning capabilities. The NEXRAD stations use Doppler radar technology to provide much-needed information on tropical cyclone wind fields and the wind fields' changes as they move inland. Local offices are able to provide accurate short-term warnings as rain bands, high winds, and possible tornadoes move toward specific inland locations. Heavy rains and flooding frequently occur over widespread inland areas.

Improved observing systems and anticipated improvements in analysis, forecasting, and warning programs require efficient accessing, processing, and analysis of large quantities of data from numerous sources. These data also provide the opportunity for improved numerical forecasts. The Class VII computer at the National Meteorological Center permits operational implementation of next-generation hurricane prediction models.

The NWS's Advanced Weather Interactive Processing System (AWIPS) will be another tool for forecasting hurricanes. AWIPS is a highly automated and integrated weather information processing, communications, and display system that will be deployed to Weather Forecast Offices (WFO), River Forecast Centers (RFC), and the National Centers for Environmental Prediction (NCEP).

Critical high-resolution hurricane information needed by local, state, and other federal agencies as well as the private sector, will be displayed graphically and transmitted to the user faster and more completely than ever before, making more effective warning and evacuation response.

Aerial weather reconnaissance

Aerial weather reconnaissance is vitally important to the forecasters of the National Hurricane Center. Aircraft reports help the meteorologist determine what is going on inside a storm as it actually happens. This, along with the broader view provided by data from satellites, floating buoys, and land and ship reports, makes up the total "package" of information available to hurricane forecasters who must make forecasts of the speed, intensity, and direction of the storm. Reconnaissance aircraft penetrate to the core of the storm and provide detailed measurements of its strengths as well as accurate location of its center—information that is not available from any other source. The NHC is supported by specially modified aircraft of the U.S. Air Force Reserve (USAFR) and NOAA's Aircraft Operations Center (NOAA/AOC) The USAFR crews fly the Lockheed WC-130 Hercules, a giant four-engine turboprop aircraft that carries a crew of six people and can stay aloft for up to 14 hours. NOAA's AOC flies Lockheed WP-3 Orion, a four-engine turboprop aircraft that carries a crew of 7–17 persons and can stay aloft for up to 12 hours at a time. The NOAA/AOC aircraft and crews are based at Miami International Airport. Both units can be deployed as necessary in the Atlantic, Caribbean, Gulf of Mexico, and Central Pacific Ocean.

Meteorological information obtained from aerial reconnaissance includes measurements of the winds, atmospheric pressure, temperature, and the location of the center of the storm. In addition, these aircraft also drop instruments called *dropsondes* as they fly through the storm's center. These devices continuously radio back measurement of pressure, humidity, temperature, wind direction, and speed as they fall toward the sea. This information provides a detailed look at the structure of the storm and an indication of its intensity.

Aerial weather reconnaissance of nature's most powerful destructive force is not without risk. Since aircraft and crews first started flying into hurricanes and typhoons nearly 40 years ago, three have been lost, vanishing without a trace along with their crews. The first of these, a U.S. Navy P2V Neptune fell into the Caribbean Sea while flying into Hurricane Janet on September 26, 1955. Next came a U.S. Air Force WB-50 Super Fortress, which crashed into the Pacific Ocean on January 15, 1958 while penetrating Typhoon Ophelia. Also lost was a WC-130 Hercules that disappeared in the vicinity of Typhoon Bess in the Philippine Sea south of Taiwan on October 12, 1974.

Flying into a hurricane is like no other experience. Crew members who have flown combat missions say that their feelings before these flights and those involving hurricanes are very similar. There is a blend of excitement and apprehension that is difficult to describe. Adding to the tension is the fact that no two hurricanes are alike. Some are gentle while others seem like raging bulls. Preparations for flying into a hurricane are very thorough. All crew members are fully trained by highly trained specialists. The crew takes special precautions as they enter the hurricane. All loose objects are tied down or put away and crew members slip into safety harnesses and belts. When radar picks up the storm, the crew then determines how to get inside. The idea is to make the aircraft mesh with the storm rather than fight it. If it is a well-defined storm, getting inside can be a real experience. The winds at flight altitude

Weather Anecdotes

April 12, 2001 ...Gulfport, MS. As attractions go, "Big Boy" ranks decidedly lower than the Grand Canyon and Niagara Falls. But the cat—so-named for his impressive girth—has nonetheless attracted a cult following. Big Boy blew into town three years ago, riding the tail of Hurricane Georges. He was hurled off a bait shop roof and dropped onto a 60-foot oak tree in Jones Park. From his perch, the 4-year old tabby has since nurtured his celebrity status. Children and fishermen bring him treats. Big Boy has apparently never left his tree house since the 1998 storm, and he weighs at least 20 pounds and eats and drinks from dishes nailed to the tree—and he is not shy about demanding refills. When not sleeping or sharpening his claws, Big Boy moves from limb to limb for exercise.

November 9, 1998 ... Trujillo, Honduras. Swept out of her village by Hurricane Mitch, a 36-year old woman drifted alone for six days far into the Caribbean Sea. Her husband and three children had been killed in the storm. All she had was a makeshift raft, the sea below, the sun during the day, and the moon at night. No land was in sight. On the sixth day she spotted a duck nearby. Her desperation ended hours later. She was spotted by an airplane looking for a yacht that had disappeared during the storm. A British helicopter rescued her. She and her family had lived in the village of Barra de Aguan, near the mouth of the Aguan River. Normally her house was about 2 miles from the sea and more than a mile from the river on the other side. But when Mitch stalled over the Honduran coast on October 28, the sea and the river merged. Her house was quickly swept away and her family took refuge at a neighbor's home. Fourteen people climbed onto the roof, but soon the river tore through the house. One of her sons was ripped away from her, and she clung to some floating palm branches for four hours. Using debris in the water, she made a 4-by-4-foot raft out of tree roots, branches, and motorboard. While at sea she found coconuts, which gave her milk. She was found by the plane on November 2. She was picked up about 25 miles north of Guanaja Island, and about 75 miles from her home.

30 Costliest Mainland U.S. Hurricanes 1900–2000

Rank	Hurricane	Year	Category	Damage (U.S.)
1	Andrew (SE FL/SE LA)	1992	4	$26,500,000,000
2	Hugo (SC)	1989	4	7,000,000,000
3	Floyd (Mid Atlantic & NE U.S.)	1999	2	4,500,000,000
4	Fran (NC)	1996	3	3,200,000,000
5	Opal (NW FL/AL)	1995	3	3,000,000,000
6	Georges (FL Keys, MS, AL)	1998	2	2,310,000,000
7	Frederic (AL, MS)	1979	3	2,300,000,000
8	Agnes (FL/NE U.S.)	1972	1	2,100,000,000
9	Alicia (N TX)	1983	3	2,000,000,000
10	Bob (NC, NE U.S.)	1991	2	1,500,000,000
10	Juan (LA)	1985	1	1,500,000,000
12	Camille (MS/SE LA/VA)	1969	5	1,420,700,000
13	Betsy (SE FL/SE LA)	1965	3	1,420,500,000
14	Elena (MS/AL/NW FL)	1985	3	1,250,000,000
15	Gloria (Eastern U.S.)	1985	3*	900,000,000
16	Diane (NE U.S.)	1955	1	831,700,000
17	Bonnie (NC, VA)	1998	2	720,000,000
18	Erin (NW FL)	1995	2	700,000,000
19	Allison (N TX)	1989	TS@	500,000,000
19	Alberto (NW FL, GA, AL)	1994	TS@	500,000,000
19	Frances (TX)	1998	TS@	500,000,000
22	Eloise (NW FL)	1975	3	490,000,000
23	Carol (NE U.S.)	1954	3*	461,000,000
24	Celia (S TX)	1970	3	453,000,000
25	Carla (N & Central TX)	1961	4	408,000,000
26	Claudette (N TX)	1979	TS@	400,000,000
26	Gordon (S & Cent FL, NC)	1994	TS@	400,000,000
28	Donna (FL/Eastern U.S.)	1960	4	387,000,000
29	David (FL/Eastern U.S.)	1979	2	320,000,000
30	New England	1938	3*	306,000,000

ADDENDUM (Rank is independent of other events in group)

Rank	Hurricane	Year	Category	Damage (U.S.)
4	Georges (USVI, PR)	1998	3	3,600,000,000
10	Iniki (Kauai, HI)	1992	Unk.	1,800,000,000
10	Marilyn (USVI, PR)	1995	2	1,500,000,000
15	Hugo (USVI, PR)	1989	4	1,000,000,000
19	Hortense (PR)	1996	1	500,000,000
29	Lenny (USVI, PR)	1999	4	330,000,000
29	Olivia (CA)	1982	T.D.&	325,000,000
30	IWA (Kauai, HI)	1982	Unk.	312,000,000

Notes:
*Moving more than 30 miles per hour.
@Only of Tropical Storm intensity.
&Only a tropical Depression

(Courtesy of National Climatic Data Center/NOAA.)

often exceed 100 mph, and the wall cloud surrounding the center, or eye, can be several miles thick. Rain comes down in torrents, and the updrafts and downdrafts are usually strong and frequent. Inside the eye, however, the conditions are much different. The ocean is generally visible, and there is blue sky and sunshine. The flight level winds are nearly calm. After gathering all the information they need, the crew then exits the storm in the same manner they entered.

Making sure the NHC gets the aerial weather reconnaissance it needs is the job of a small group of Air Force people assigned to a liaison office in the Center. This office, under a former Chief, Aerial Reconnaissance Coordination, All Hurricanes (CARCAH), is responsible for coordinating requirements and arranging for the supporting flights. This office also records and monitors weather observations radioed back or received through direct satellite communication from the storm by the on-board meteorologists. These data are checked for accuracy and then transmitted to the worldwide meteorological community, through both military and civilian communications circuits.

Aircraft meteorological reconnaissance is a team effort. A host of different organizations, tied together by CARCAH, is dedicated to providing the NHC the vital information it needs to make accurate forecasts that help to ensure that communities in the path of a hurricane are adequately warned.

Science may never provide a full solution to the problems of hurricane safety. But warnings and forecasts help save countless lives and allow residents to take the necessary precautions to prevent enormous property damage each year.

30 Costliest Mainland U.S. Hurricanes 1900–2000 (adjusted)

Ranked Using 2000 Deflator**					Ranked Using 2000 Inflation, Population and Wealth Normalization				
Rank	Hurricane	Year	Category	Damage (U.S.)**	Rank	Hurricane	Year	Category	Damage (U.S.ˡ)
1	Andrew (SE FL/SE LA)	1992	4	34,954,825,000	1	SE Florida/Alabama	1926	4	87,167,000,000
2	Hugo (SC)	1989	4	9,739,820,675	2	Andrew (SE FL/SE LA)	1992	4	39,896,000,000
3	Agnes FL/NE U.S.)	1972	1	8,602,500,000	3	N Texas (Galveston)	1900	4	32,090,000,000
4	Betsy (SE FL/SE LA)	1965	3	8,516,866,023	4	N Texas (Galveston)	1915	4	27,190,000,000[1]
5	Camille (MS/SE LA/VA)	1969	5	6,992,441,549	5	SW Florida	1944	3	20,331,000,000
6	Diane (NE U.S.)	1955	1	5,540,676,187	6	New England	1938	3*	20,046,000,000
7	Frederic (AL/MS)	1979	3	4,965,327,332	7	SE Florida/Lake Okeechobee	1928	4	16,631,000,000
8	Floyd (Mid Atlantic & NE U.S.)	1999	2	4,666,817,360	8	Betsy (SE FL/SE LA)	1965	3	14,990,000,000
9	New England	1938	3*	4,748,580,000	9	Donna (FL/Eastern U.S.)	1960	4	14,526,000,000
10	Fran (NC)	1996	3	3,670,400,000	10	Camille (MS/SE LA/VA)	1969	5	13,219,000,000
11	Opal (NW FL/AL)	1995	3	3,520,596,085	11	Agnes (NW FL, NE U.S.)	1972	1	12,904,000,000
12	Alicia (N TX)	1983	3	3,421,660,182	12	Diane (NE U.S.)	1955	1	12,335,000,000
13	Carol (NE U.S>)	1954	3*	3,134,443,557	13	Hugo (SC)	1989	4	11,307,000,000
14	Carla (N & Central TX)	1961	4	2,550,580,095	14	Carol (NE U.S.)	1954	3*	10,929,000,000
15	Georges (FL Keys, MS, AL)	1998	2	2,494,800,000	15	SE Florida/Louisiana/Alabama	1947	4	10,015,000,000
16	Juan (LA)	1985	1	2,418,795,844	16	Carla (N & Central TX)	1961	4	8,522,000,000
17	Donna (FL/Eastern U.S.)	1960	4	2,407,888,443	17	Hazel (SC/NC)	1954	4*	8,486,000,000
18	Celia (S TX)	1970	3	2,015,663,203	18	NE U.S.	1944	3	7,790,000,000
19	Elena (MS/AL/NW FL)	1985	3	2,015,663,203	19	SE Florida	1945	3	7,611,000,000
20	Bob (NC, NE U.S.)	1991	2	2,004,635,258	20	Frederic (AL/MS)	1979	3	7,587,000,000
21	Hazel (SC/NC)	1954	4*	1,910,582,732	21	SE Florida	1949	3	7,038,000,000
22	FL (Miami, Pensacola)/MS/AL	1926	4	1,738,042,353	22	S Texas	1919	4	6,448,000,000
23	N TX (Galveston)	1915	4	1,544,253,659[1]	23	Alicia (N TX)	1983	3	4,890,000,000
24	Dora (NE FL)	1964	2	1,540,946,262	24	Floyd (NC)	1999	2	4,680,000,000
25	Eloise (NW FL)	1975	3	1,489,250,000	25	Celia (S TX)	1970	3	4,024,000,000
26	Gloria (Eastern U.S.)	1985	3*	1,451,277,506	26	Dora (NE FL)	1964	2	3,747,000,000
27	NE U.S.	1944	3*	1,221,342,593	27	Fran (NC)	1996	3	3,735,000,000
28	Beulah (S TX)	1967	3	1,113,122,363	28	Opal (NW FL/AL)	1995	3	3,617,000,000
29	SE FL/SE LA/MS	1947	4	930,099,359	29	Cleo (SE FL)	1964	2	2,936,000,000
30	N TX (Galveston)	1900	4	928,160,793[2]	30	Juan (LA)	1985	1	2,892,000,000
ADDENDUM									
10	Georges (USVI, PR)	1998	3	3,888,000,000	27	Hugo (USVI, PR)	1989	4	1,283,755,274
18	Iniki (Kauai, HI)	1992	Unk	2,190,600,000	28	San Felipe (PR)	1928	4	1,217,000,000
23	Marilyn (USVI, E, PR)	1995	2	1,624,110,320					

Notes:
**2000 $ based on U.S. DOC Implicit Price Deflator for Construction.
ˡ2000 Landsea normalization for population, wealth and inflation.
*Moving more than 30 miles per hour
[1]Damage estimate in 1915 reference is considered too high
[2]Using 1915 cost adjustment base - none available prior to 1915.

(Courtesy of National Climatic Data Center/NOAA.)

Most Intense Mainland U.S. Hurricanes 1900–2000

Rank	Hurricane	Year	Category (at landfall)	Minimum Pressure Millibars	Minimum Pressure Inches
1	FL (Keys)	1935	5	892	26.35
2	Camille (MS/SE LA/VA)	1969	5	909	26.84
3	Andrew (SE FL/SE LA)	1992	4	922	27.23
4	FL (Keys)/S TX	1919	4	927	27.37
5	FL (Lake Okeechobee)	1928	4	929	27.43
6	Donna (FL/Eastern U.S.)	1960	4	930	27.46
7	TX (Galveston)	1900	4	931	27.49
7	LA (Grand Isle)	1909	4	931	27.49
7	LA (New Orleans)	1915	4	931	27.49
7	Carla (N & Central TX)	1961	4	931	27.49
11	Hugo (SC)	1989	4	934	27.58
12	FL (Miami)/MS/AL/Pensacola	1926	4	935	27.61
13	Hazel (SC/NC)	1954	4*	938	27.70
14	SE FL/SE LA/MS	1947	4	940	27.76
15	N TX	1932	4	941	27.79
16	Gloria (Eastern U.S.)	1985	3*&	942	27.82
16	Opal (NW FL/AL)	1995	3&	942	27.82
18	Audrey (SW LA/N TX)	1957	4#	945	27.91
18	TX (Galveston)	1915	4#	945	27.91
18	Celia (S TX)	1970	3	945	27.91
18	Allen (S TX)	1980	3	945	27.91
22	New England	1938	3*	946	27.94
22	Frederic (AL/MS)	1979	3	946	27.94
24	NE U.S.	1944	3*	947	27.97
24	SC/NC	1906	3	947	27.97
26	Betsy (SE FL/SE LA)	1965	3	948	27.99
26	SE FL/NW FL	1929	3	948	27.99
26	SE FL	1933	3	948	27.99
26	S TX	1916	3	948	27.99
26	MS/AL	1916	3	948	27.99
31	Diana (NC)	1984	3+	949	28.02
31	S TX	1933	3	949	28.02
33	Beulah (S TX)	1967	3	950	28.05
33	Hilda (Central LA)	1964	3	950	28.05
33	TX (Central)	1942	3	950	28.05
33	TX (Central)	1942	3	950	28.05
37	SE FL	1945	3	951	28.08
37	Bret (S TX)	1999	3	951	28.08
39	FL (Tampa Bay)	1921	3	952	28.11
39	Carmen (Central LA)	1974	3*	952	28.11
41	Edna (New England)	1954	3	954	28.17
41	SE FL	1949	3	954	28.17
41	Fran (NC)	1996	3	954	28.17
44	Eloise (NW FL)	1975	3	955	28.20
44	King (SE FL)	1950	3	955	28.20
44	Central LA	1926	3	955	28.20
44	SW LA	1918	3	955	28.20
44	SW FL	1910	3	955	28.20
49	NC	1933	3	957	28.26
49	FL (Keys)	1909	3	957	28.26
51	Easy (NW FL)	1950	3	958	28.29
51	N TX	1941	3	958	28.29
51	NW FL	1917	3	958	28.29
51	N TX	1909	3	958	28.29
51	MS/AL	1906	3	958	28.29
56	Elena (MS/AL/NW FL)	1985	3	959	28.32
57	Carol (NE U.S.)	1954	3*	960	28.35
57	Ione (NC)	1955	3	960	28.35
57	Emily (NC)	1993	3	960	28.35
60	Alicia (N TX)	1983	3	962	28.41
60	Connie (NC/VA)	1955	3	962	28.41
60	SW FL/NE FL	1944	3	962	28.41
60	Central LA	1934	3	962	28.41
64	SW FL/SE FL	1948	3	963	28.44
65	NW FL	1936	3	964	28.47

ADDENDUM

Rank	Hurricane	Year	Category (at landfall)	Minimum Pressure Millibars	Minimum Pressure Inches
4	David (S of PR)	1979	4	924	27.29
7	San Felipe (PR)	1928	4	931	27.49
14	Hugo (USVI & PR)	1989	4	940	27.76
33	Iniki (Kauai, HI)	1992	UNK	950	27.91
43	Dot (Kauai, HI)	1959	UNK	955	28.11
50	Donna (St. Thomas, PR)	1960	4	948	28.29
64	IWA (Kauai, HI)	1982	UNK	964	28.47
65	Georges (USVI & PR)	1998	3	968	28.59

Notes:
*Moving more than 30 miles per hour.
&Highest category justified by winds.
#Classified 4 because of estimated winds.
+Cape Fear, NC area only; was a category 2 at final landfall.

(Courtesy of National Climatic Data Center/NOAA.)

Hurricane Direct Hits by State 1900–2000

Area	Category Number 1	2	3	4	5	All	Major Hurricanes
U.S. (Texas to Maine)	61	39	48	15	2	165	65
Texas	12	9	10	6	0	37	16
(North)	7	3	3	4	0	17	7
(Central)	2	2	1	1	0	6	2
(South)	3	4	6	1	0	14	7
Louisiana	9	5	8	3	1	26	12
Mississippi	1	2	5	0	1	9	6
Alabama	5	2	5	0	0	12	5
Florida	19	17	17	6	1	60	24
(Northwest)	10	8	7	0	0	25	7
(Northeast)	2	7	0	0	0	9	0
(Southwest)	7	3	6	2	1	19	9
(Southeast)	6	11	7	4	0	28	11
Georgia	1	4	0	0	0	5	0
South Carolina	6	4	2	2	0	14	4
North Carolina	10	6	10	1*	0	27	11
Virginia	2	1	1*	0	0	4	1*
Maryland	0	1*	0	0	0	1*	0
Delaware	0	0	0	0	0	0	0
New Jersey	1*	0	0	0	0	1*	0
New York	3	1*	5*	0	0	9	5*
Connecticut	2	3*	3*	0	0	8	3*
Rhode Island	0	2*	3*	0	0	5*	3*
Massachusetts	2	2*	2*	0	0	6	2*
New Hampshire	1*	1*	0	0	0	2*	0
Maine	5*	0	0	0	0	5	0

*Indicates all hurricanes in this group were moving faster than 30 mph. State totals will not equal U.S. totals, and Texas or Florida totals will not necessarily equal sum of sectional totals.

(Courtesy of National Climatic Data Center/NOAA.)

Maximun and Minimum Tropical Cyclone Activity Years 1871–2000

Maximum Activity

Tropical cyclones[1] Number	Years	Hurricanes[2] Number	Years
21	1933	12	1969
19	1995	11	1916, 1950, 1995
18	1969	10	1887, 1893, 1933, 1998
17	1887		
16	1936	9	1955, 1980, 1996

Maximum Activity

Tropical cyclones[1] Number	Years	Hurricanes[2] Number	Years
1	1890, 1914	0	1907, 1914
2	1925, 1930	1	1890, 1905, 1919, 1925
		2	1895, 1897, 1904, 1917, 1922, 1930, 1930, 1982

Notes:
[1]Includes subtropical storms after 1967; excludes depressions.
[2]Distinction of hurricanes recorded only after 1885.

(Courtesy of National Climatic Data Center/NOAA.)

Number of Hurricanes by Category to Hit Mainland U.S. by Decade

Decade	Category 1	2	3	4	5	All 1,2,3,4,5	Major 3,4,5
1900–1909	5	4	4	2	0	15	6
1910–1919	9	3	5	3	0	20	8
1920–1929	6	4	3	2	0	15	5
1930–1939	4	5	6	1	1	17	8
1940–1949	7	8	7	1	0	23	8
1950–1959	8	1	7	2	0	18	9
1960–1969	4	5	3	2	1	15	6
1970–1979	6	2	4	0	0	12	4
1980–1989	9	1	5	1	0	16	6
1990–1999	3	6	4	1	0	14	5
2000–2009	0	0	0	0	0	0	0
1900–1999	61	39	48	15	2	165	65

Note: Only the highest category to affect the U.S. has been used.

(Courtesy of National Climatic Data Center/NOAA.)

Named Atlantic Cyclones by Month of Origin 1900–2000

Month	1944–2000 Tropical Storms and Hurricanes		1944–2000 Hurricanes		1900–2000 U.S. Hurricanes	
	Total	Average	Total	Average	Total	Average
January–April	3	0.1	0	0.0	0	0.00
May	8	0.1	2	*	0	0.00
June	31	0.5	11	0.2	11	0.11
July	50	0.9	22	0.4	18	0.18
August	151	2.6	95	1.6	42	0.42
September	198	3.5	129	2.3	65	0.64
October	100	1.8	60	1.1	25	0.25
November	26	0.5	16	0.3	4	0.04
December	4	0.1	2	*	0	0.00
Year	571	10.0	337	5.9	165	1.63

[1]Includes subtropical storms after 1967. See Neumann et al. (1999) for details.
*Less than 0.05.

(Courtesy of National Climatic Data Center/NOAA.)

Incidence of Direct Hits on U.S.

Area	June	July	Aug.	Sept.	Oct.	All
U.S. (Texas to Maine)	2	3	16	36	8	65
Texas	1	1	8	6		16
(North)	1	1	3	2		7
(Central)			1	1		2
(South)			4	3		7
Louisiana	2		4	5	1	12
Mississippi		1	1	4		6
Alabama		1		4		5
Florida		1	2	15	6	24
(Northwest)		1		5	1	7
(Northeast)						0
(Southwest)			1	5	3	9
(Southeast)			2	7	2	11
Georgia						0
South Carolina				3	1	4
North Carolina			2	8	1	11
Virginia				1		1
Maryland						0
Delaware						0
New Jersey						0
New York				1	4	5
Connecticut				1	2	3
Rhode Island				1	2	3
Massachusetts				2		2
New Hampshire						0
Maine						0

Note: State totals do not equal U.S. totals and texas or Florida totals do not necessarily equal the sum of sectional entries.

(Courtesy of National Climatic Data Center/NOAA.)

Last Direct Hit by Hurricane

State	City	Direct Hits Last major	Direct Hits Last Any	Indirect Hits Last Any
Texas	Brownsville	1980(3) Allen	1980(3) Allen	
	Corpus Christi	1970(3) Celia	1971(1) Fern	1980(3) Allen
	Port Aransas	1970(3) Celia	1971(1) Fern	1980(3) Allen
	Matagorda	1961(4) Carla	1971(1) Fern	
	Freeport	1983(3) Alicia	1983(3) Alicia	1983(3) Alicia
	Galveston	1983(3) Alicia	1989(1) Jerry	
	Houston	1941(3)	1989(1) Jerry	
	Beaumont	<1900	1986(1) Bonnie	
Louisiana	Cameron	1957(4) Audrey	1985(1) Danny	1985(1) Juan
	Morgan City	1992(3) Andrew	1992(3) Andrew	
	Houma	1974(3) Carmen	1985(1) Juan	1992(3) Andrew
	New Orleans	1965(3) Betsy	1965(3) Betsy	1969(5) Camille
Mississippi	Bay St. Louis	1985(3) Elena	1985(3) Elena	
	Biloxi	1985(3) Elena	1998(2) Georges	
	Pascagoula	1985(3) Elena	1998(2) Georges	
Alabama	Mobile	1985(3) Elena	1998(2) Georges	
Florida	Pensacola	1926(3)	1995(1) Erin	1995(3) Opal
	Panama City	1995(3) Opal	1995(3) Opal	
	Apalachicola	1985(3) Elena	1985(2) Kate	1995(3) Opal
	Homosassa	1950(3) Easy	1968(2) Gladys	
	St. Petersburg	1921(3)	1946(1)	1968(2) Gladys
	Tampa	1921(3)	1946(1)	1968(2) Gladys
	Sarasota	1944(3)	1946(1)	1966(2) Alma
	Fort Myers	1960(3) Donna	1960(3) Donna	1966(2) Alma
	Naples	1960(4) Donna	1964(2) Isbell	1992(3) Andrew
	Key West	1948(3)	1999(1) Irene	
	Miami	1992(4) Andrew	1999(1) Irene	1992(4) Andrew
	Fort Lauderdale	1950(3) King	1999(1) Irene	
	West Palm Beach	1949(3)	1999(1) David	
	Stuart	1949(3) David	1979(2) David	
	Fort Pierce	1933(3) David	1979(2) David	
	Vero Beach	<1900	1995(1) Erin	
Florida	Cocoa	<1900	1995(1) Erin	
	Daytona Beach	<1900	1960(2) Donna	1979(2) David
	St. Augustine	<1900	1964(2) Dora	
	Jacksonville	<1900	1964(2) Dora	1964(2) Dora
	Fernandina Bch	<1900	1928(2) Dora	
Georgia	Brunswick	<1900	1928(1)	
	Savannah	<1900	1979(2) David	
S. Carolina	Hilton Head	1959(3) Gracie	1979(2) David	1985(1) Bob
	Charleston	1989(4) Hugo	1989(4) Hugo	
	Myrtle Beach	1954(4*) Hazel	1954(4*) Hazel	1989(4) Hugo
N. Carolina	Wilmington	1996(3) Fran	1999(1) Floyd	1999(2) Dennis
	Morehead City	1996(3) Fran	1999(2) Floyd	
	Cape Hatteras	1993(3) Emily	1993(3) Emily	1999(1) Dennis
Virginia	Virginia Beach	1944(3*)	1999(2) Floyd	
	Norfolk	<1900	1955(1) Connie	1999(1) Floyd
Maryland	Ocean City	<1900	<1900	1985(3*) Gloria
	Baltimore	<1900	<1900	1954(2*) Hazel
Delaware	Rehoboth Bch	<1900	<1900	1985(3*) Gloria
	Wilmington	<1900	<1900	1954(2*) Hazel
New Jersey	Cape May	<1900	1903(1)	1985(3*) Gloria
	Atlantic City	1903(1)	1903(1)	1985(3*) Gloria
New York	New York City	1903(1)	1903(1)	1976(1) Belle
	Westhampton	1903(1)	1985(3*) Gloria	
Connecticut	New London	1985(3*) Gloria	1991(2*) Bob	
	New Haven	1938(3*)	1985(2*) Gloria	
	Bridgeport	1938(3*)	1985(2*) Gloria	
Rhode Island	Providence	1954(3*) Carol	1991(2*) Bob	
Mass.	Cape Cod	1954(3*) Carol	1991(2*) Bob	
	Boston	1954(3*) Edna	1960(1*) Donna	1991(1*) Bob
N. Hampshire	Portsmouth	<1900	1985(2*) Gloria	
Maine	Portland	<1900	1985(1*) Gloria	
	Eastport	1969(1)	1969(1) Gerda	1985(1*) Gloria

Note: <1900 means before 1900
*Moving over 30 mph

(Courtesy of National Climatic Data Center/NOAA.)

Estimated Annual Deaths/Damage 1900–2000

Year	Deaths	Damage ($Millions)			Year	Deaths	Damage ($Millions)		
		Unadjusted	Adjusted[1]	Normalized[L]			Unadjusted	Adjusted[1]	Normalized[L]
1900	8,000*	30	965[2]	32,090	1951	0	2	13	219
1901	10	1	32[2]	773	1952	3	3	20	76
1902	0	Minor	Minor	0	1953	2	6	41	34
1903	15	1	32[2]	8,317	1954	193	756	5,140	21,121
1904	5	2	642	1,006	1955	218	985	6,562	15,906
1905	0	Minor	Minor	0	1956	19	27	170	422
1906	298	3*	972	4,906	1957	400	152	933	2,946
1907	0	Minor	Minor	0	1958	2	11	67	268
1908	0	Minor	Minor	0	1959	24	23	143	538
1909	406	8	257[2]	3,523	1960	65	396	2,464	14,717
1910	30	1	32[2]	1,360	1961	46	414	2,588	8,635
1911	17	1*	32[2]	260	1962	3	2	12	51
1912	1	Minor	Minor	0	1963	10	12	73	179
1913	5	3	97[2]	786	1964	49	515	3,174	8,499
1914	0	Minor	Minor	0	1965	75	1,445	8,664	15,308
1915	550	63	2,027[3]	28,503	1966	54	15	86	199
1916	107	33	888	4,340	1967	18	200	1,113	2,471
1917	5	Minor	Minor	0	1968	9	10	53	386
1918	34	5	87	441	1969	256	1,421	6,994	13,219
1919	287*	22	341	6,448	1970	11	454	2,109	4,024
1920	2	3	37	439	1971	8	213	927	1,461
1921	6	3	46	3,918	1972	122	2,100	8,603	12,923
1922	0	Minor	Minor	0	1973	5	18	68	114
1923	0	Minor	Minor	0	1974	1	150	498	863
1924	2	Minor	Minor	0	1975	21	490	1,489	2,117
1925	6	Minor	Minor	0	1976	9	100	290	370
1926	269	112	1,738	89,676	1977	0	10	27	39
1927	0	Minor	Minor	0	1978	36	20	48	92
1928	1,836	25	388	16,632	1979	22	3,045	6,574	10,414
1929	3	1	14	162	1980	2	300	584	1,043
1930	0	Minor	Minor	0	1981	0	25	45	94
1931	0	Minor	Minor	0	1982	0	Minor	Minor	0
1932	40	8	132	2,187	1983	22	2,000	3,422	4,890
1933	63	47	861	4,182	1984	4	66	109	157
1934	17	5	83	442	1985	30	4,000	6,450	7,921
1935	414	12	200	3,820	1986	9	17	26	35
1936	9	2	34	125	1987	0	8	12	16
1937	0	Minor	Minor	0	1988	6	59	86	106
1938	600	306	4,749	20,057	1989	56	7,670	10,672	12,422
1939	3	Minor	Minor	0	1990	13	57	77	89
1940	51	5	80	617	1991	16	1,500	2,005	2,065
1941	10	8	120	1,205	1992	24	26,500	34,955	39,896
1942	8	27	353	1,408	1993	4	57	72	77
1943	16	17	208	1,822	1994	38	973	1,187	1,238
1944	64[S]	165	2,015	28,322	1995	29	3,723	4,369	4,493
1945	7	80	951	8,512	1996	36	3,600	4,129	4,201
1946	0	5	50	2,703	1997	4	100	111	112
1947	53	136	1,150	12,990	1998	23	3,699	3,995	4,001
1948	3	18	139	2,037	1999	62	5,532	5,737	5,753
1949	4	59	455	7,443	2000	6	27	27	27
1950	19	36	273	3,383					

*1900 could have been as high as 10,000 to 12,000, other years means "more than."
[1]Adjusted to 2000 dollars based on U.S. Department of Commerce Implicit Price Deflator for Construction.
[2]Using 1915 cost adjustment - none available prior to 1915.
[3]Considered too high in 1915 reference.
[S]Figures do not agree with Table 2 because deaths at sea are not included here.
[L]Normalization reflects inflation, changes in personal wealth and coastal county population to 2000 (Pielke and Landsea 1998).

(Courtesy of National Climatic Data Center/NOAA.)

30 Deadliest/Costliest Hurricane Years 1900–2000

	Ranked on Deaths			Ranked on Unadjusted Damage			Ranked on Adjusted[1] Damage			Ranked by Normalized[L] Damage	
	Year	Deaths		Year	($ Millions)		Year	($ Millions)		Year	($ Millions)
1	1900	8,000*	1	1992	26,500	1	1992	34,955	1	1926	89,676
2	1928	1,836	2	1989	7,670	2	1989	10,672	2	1992	39,896
3	1938	600	3	1999	5,532	3	1965	8,664	3	1900	32,090
4	1915	550	4	1985	4,000	4	1972	8,603	4	1915	28,503
5	1935	414	5	1995	3,723	5	1969	6,994	5	1944	28,322
6	1909	406	6	1998	3,699	6	1979	6,574	6	1954	21,121
7	1957	400	7	1996	3,600	7	1955	6,562	7	1938	20,057
8	1906	298	8	1979	3,045	8	1985	6,450	8	1928	16,632
9	1919	287[s]	9	1972	2,100	9	1999	5,637	9	1955	15,906
10	1926	269	10	1983	2,000	10	1954	5,140	10	1965	15,308
11	1969	256	11	1991	1,500	11	1938	4,749	11	1960	14,717
12	1955	218	12	1965	1,445	12	1995	4,369	12	1969	13,219
13	1954	193	13	1969	1,421	13	1996	4,129	13	1947	12,990
14	1972	122	14	1955	985	14	1998	3,995	14	1972	12,923
15	1916	107	15	1994	973	15	1983	3,422	15	1989	12,422
16	1965	75	16	1954	756	16	1964	3,174	16	1979	10,414
17	1960	65	17	1964	515	17	1961	2,588	17	1961	8,635
18	1944	64[s]	18	1975	490	18	1960	2,464	18	1945	8,512
19	1933	63	19	1970	454	19	1970	2,109	19	1964	8,499
20	1999	62	20	1961	414	20	1915	2,027[3]	20	1903	8,317
21	1989	56	21	1960	396	21	1944	2,015	21	1985	7,921
22	1966	54	22	1938	306	22	1991	2,005	22	1949	7,443
23	1947	53	23	1980	300	23	1926	1,738	23	1919	6,448
24	1940	51	24	1971	213	24	1975	1,489	24	1999	5,753
25	1964	49	25	1967	200	25	1994	1,187	25	1906	4,906
26	1961	46	26	1944	165	26	1947	1,150	26	1983	4,890
27	1932	40	27	1957	152	27	1967	1,113	27	1995	4,493
28	1994	38	28	1974	150	28	1900	965[2]	28	1916	3,340
29	1978	36	29	1947	136	29	1945	951	29	1996	4,201
30	1996	36	30	1926	112	30	1957	933	30	1933	4,182

*Could have been as high as 10,000 to 12,000.
[1]Adjusted to 2000 dollars based on U.S. Department of Commerce Implicit Price Deflator for Construction.
[2]Using 1915 cost adjustment - none available prior to 1915.
[3]Considered too high in 1915 reference.
[s]Figures do not agree with Table 2 because deaths at sea are not included here.
[L]Landsea normalization reflects inflation, changes in personal wealth and coastal county population to 2000 (Pielke and Landsea 1998).

(Courtesy of National Climatic Data Center/NOAA.)

Major U.S. Hurricanes (≥ Category 3) 1901–1910

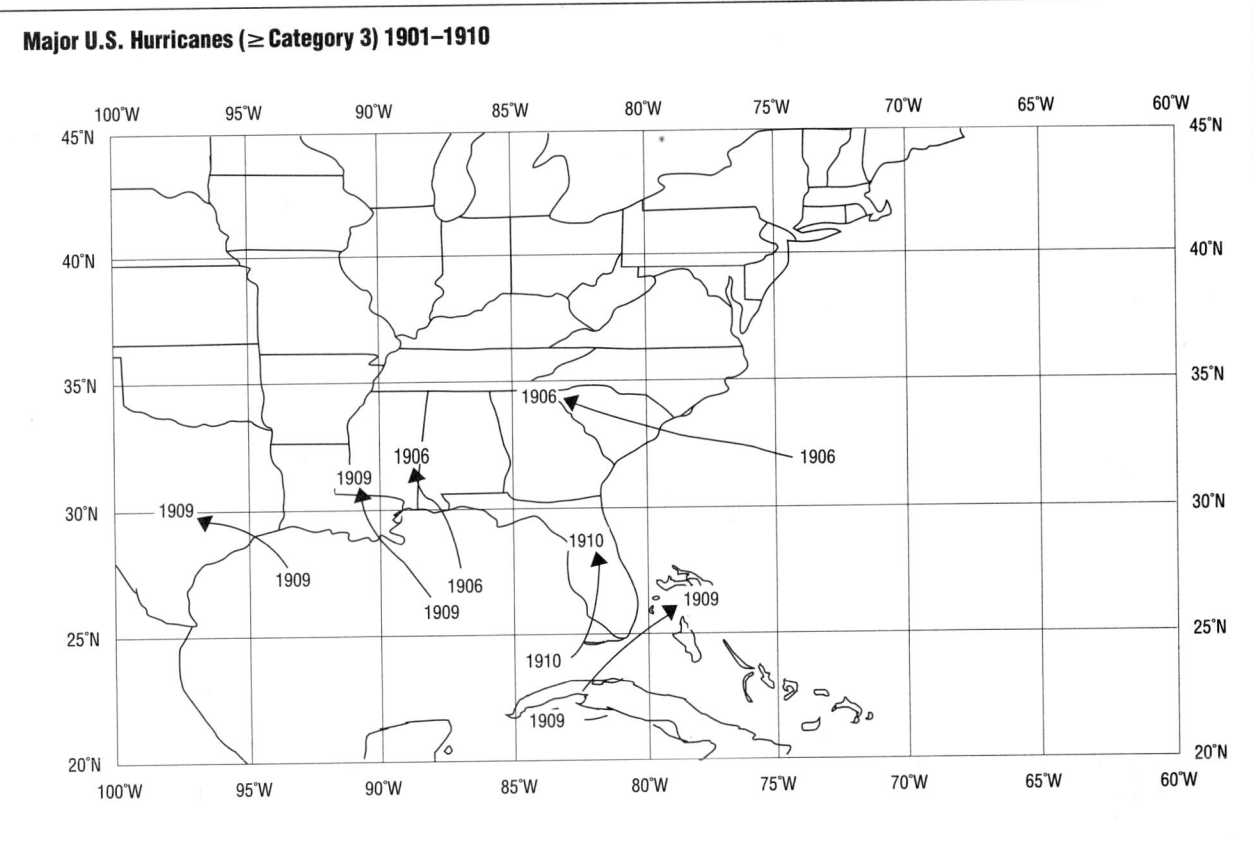

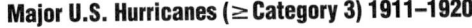

Major U.S. Hurricanes (≥ Category 3) 1911–1920

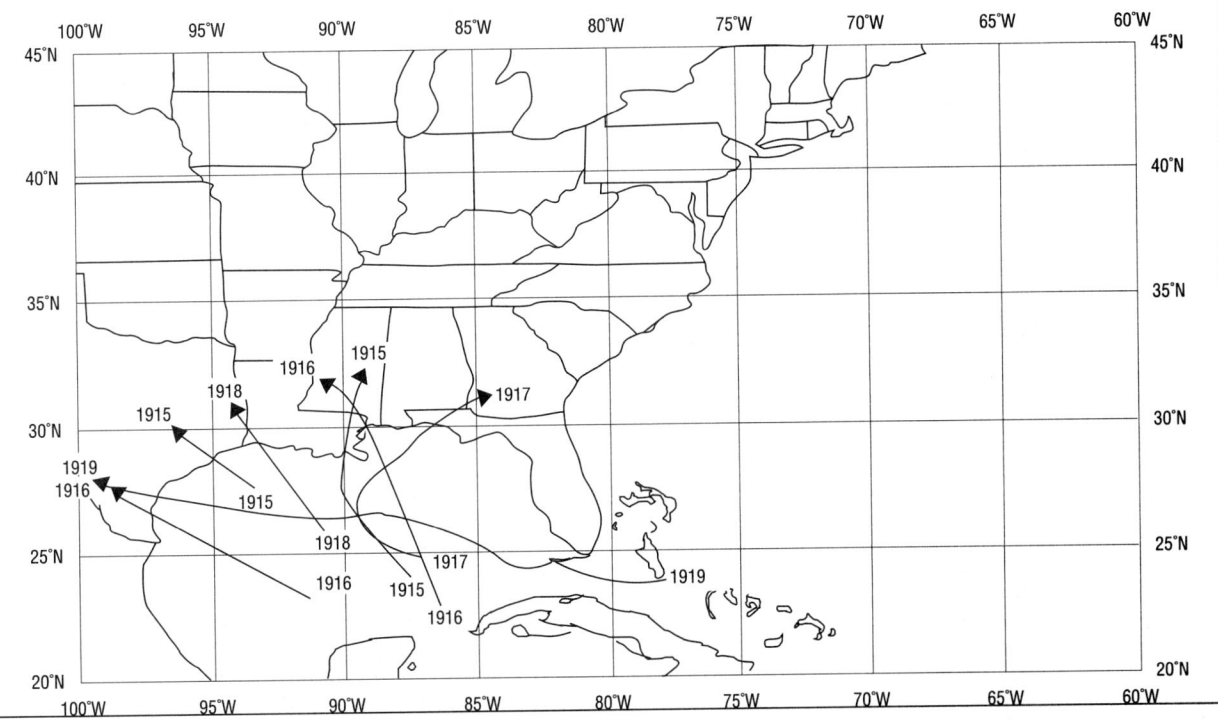

(Courtesy of National Climatic Data Center/NOAA.)

Major U.S. Hurricanes (≥ Category 3) 1921–1930

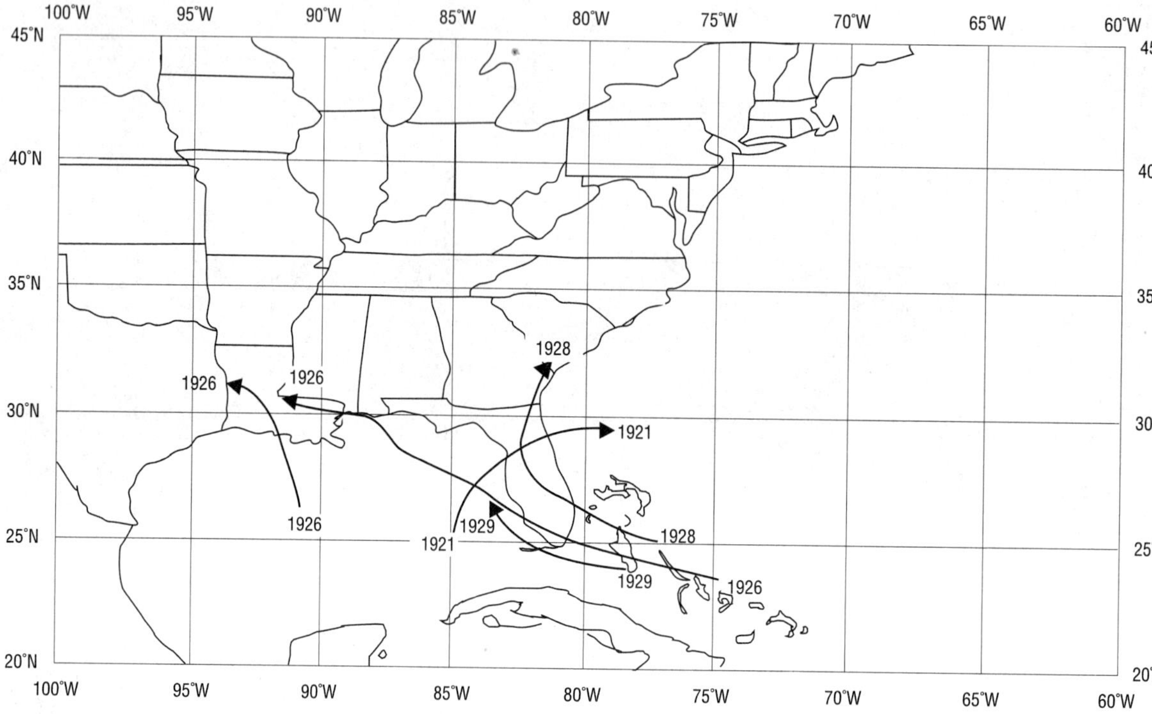

Major U.S. Hurricanes (≥ Category 3) 1931–1940

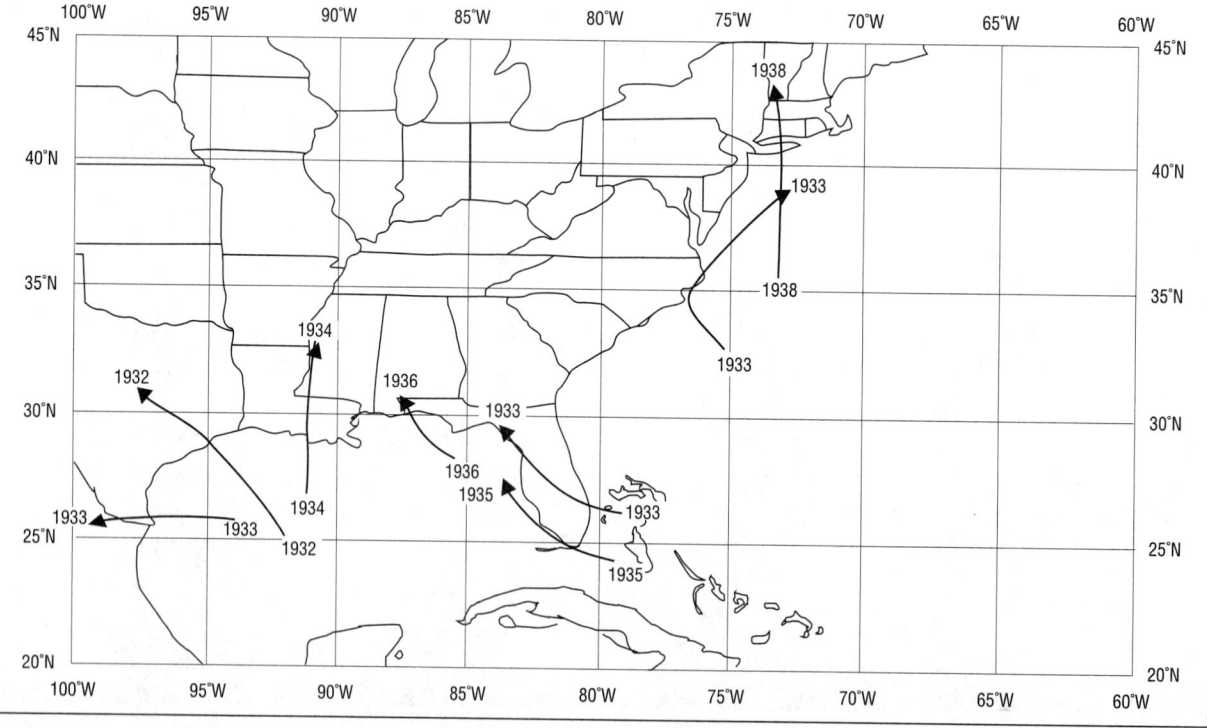

(Courtesy of National Climatic Data Center/NOAA.)

Major U.S. Hurricanes (≥ Category 3) 1941–1950

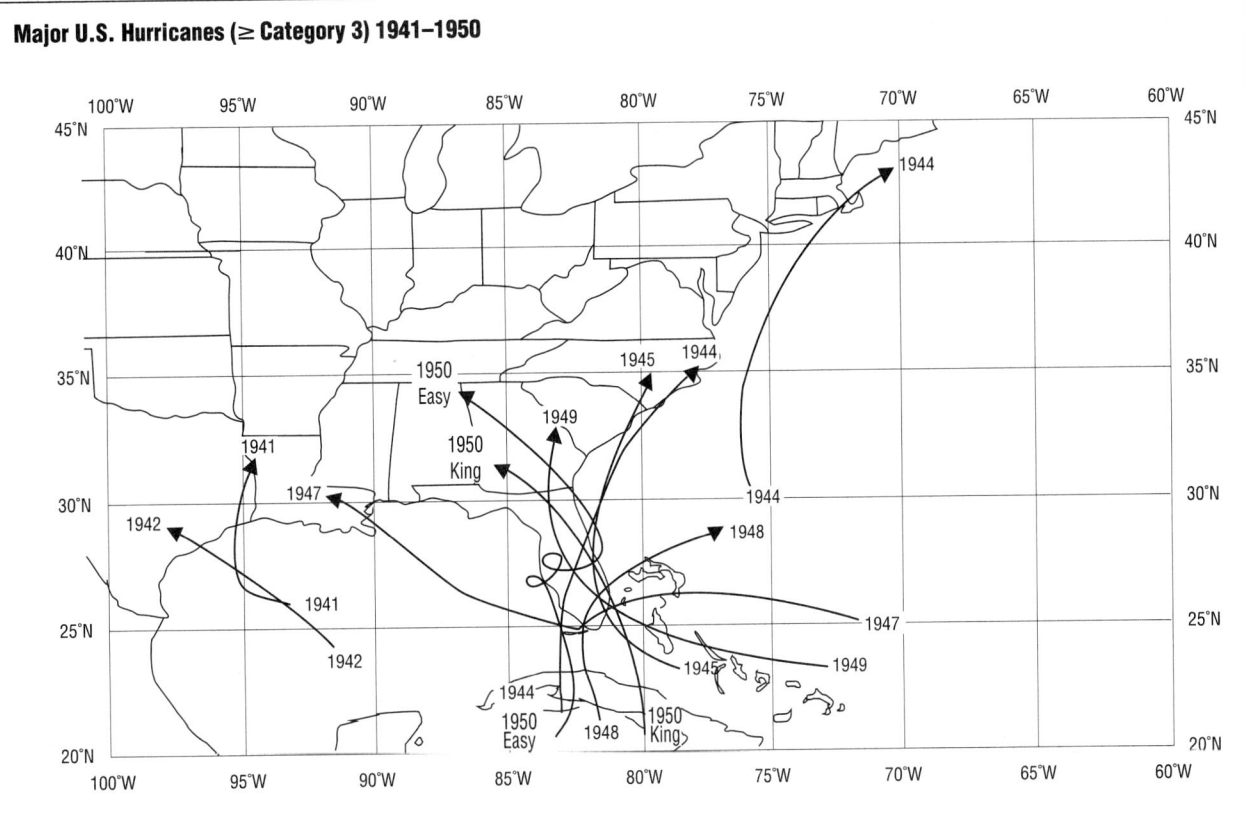

Major U.S. Hurricanes (≥ Category 3) 1951–1960

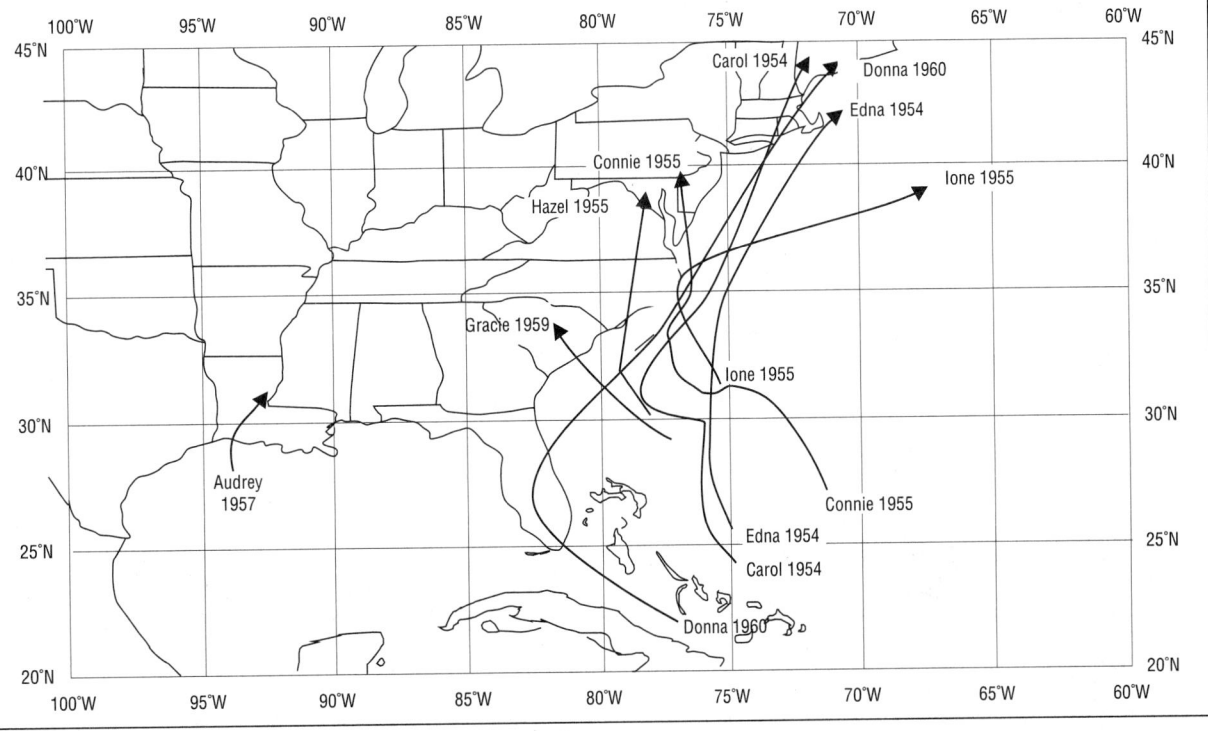

(Courtesy of National Climatic Data Center/NOAA.)

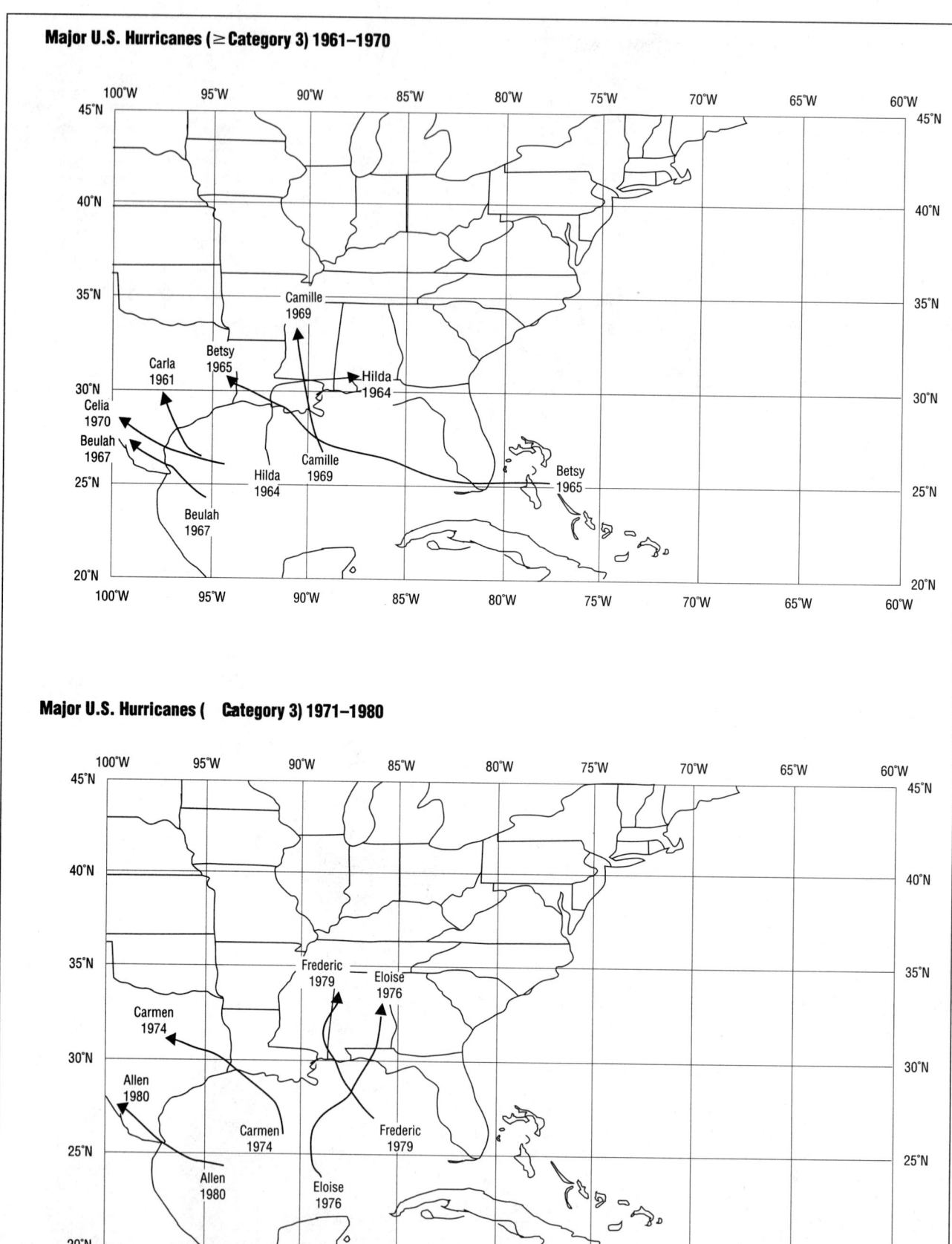

Major U.S. Hurricanes (≥ Category 3) 1961–1970

Major U.S. Hurricanes (Category 3) 1971–1980

(Courtesy of National Climatic Data Center/NOAA.)

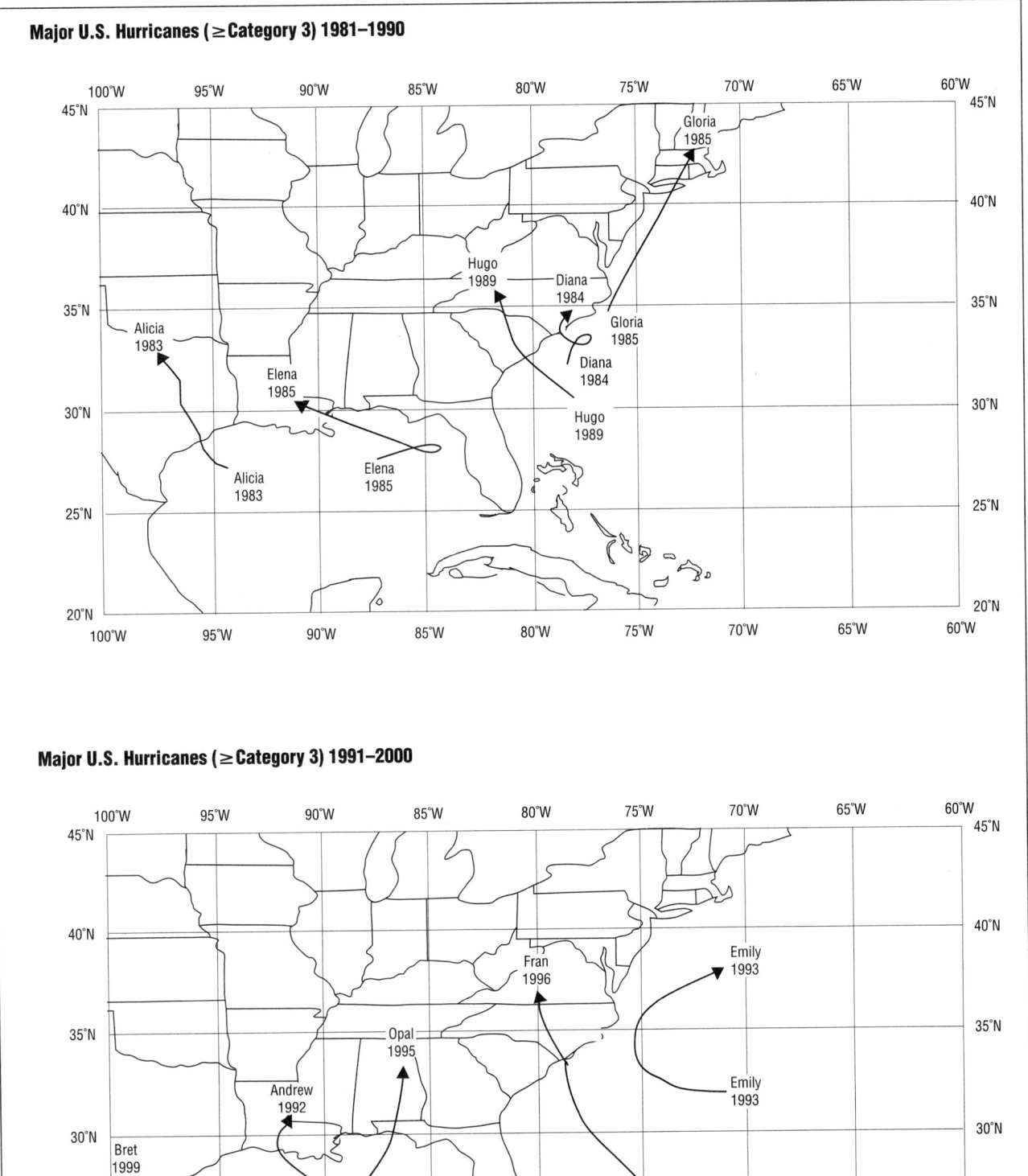

Major U.S. Hurricanes (≥ Category 3) 1981–1990

Major U.S. Hurricanes (≥ Category 3) 1991–2000

(Courtesy of National Climatic Data Center/NOAA.)

Floods & Flash Floods

The transformation of a tranquil river or normally dry wash into a destructive flood occurs hundreds of times each year, in every part of the United States. Every year, floods drive some 75,000 Americans from their homes; on the average, 127 persons are killed each year. These destructive overflows have caused property damage in some years estimated at more than $2 billion. During the years 1985–2001, the total annual number of floods or flash floods ranged from 361 (1988) to 3,376 in 1998. A total of 32,047 flash flood/flood events were recorded in 1985–2001.

• HOW FLOODS HAPPEN

Floods begin when soil and vegetation cannot absorb falling rain or melting snow, or when water runs off the land in such quantities that it cannot be carried in normal stream channels or retained in natural ponds and human-made reservoirs. Flash floods are the result of too much rain falling in too small an area, in too short a time. Flash floods frequently occur in seconds and minutes, while floods occur over hours and days. River Forecast Centers issue flood

Tropical Storm Agnes caused extensive flooding throughout the northeastern United States in 1972. (*Photo courtesy of United States Coast Guard.*)

forecasts and warnings when the rain that has fallen is enough to cause waterways to overflow their banks, and when melting snow combines with rainfall to produce similar effects.

• FLASH FLOODS

Typically, flash floods occur primarily at night and when there is an abundance of atmospheric moisture; in addition, there is usually little, if any, vertical wind shear present. Flash flooding can be produced by large, slow-moving storms or as a result of "train effect" storms (i.e., sequential mature storms that release precipitation over the same area). Train effect storms can be part of multicell cluster or squall line storm systems.

Flash-flood waves, moving at incredible speeds, can roll boulders, tear out trees, destroy buildings and bridges, and scour out new channels. Killing walls of water can reach 10–20 ft.

On small streams, especially near the headwaters of river basins, water levels may rise quickly in heavy rainstorms,

and flash floods can begin before the rain stops falling. There is little time between detection and flood crest. Swift action is essential for the protection of life and property.

• FLOOD EVENTS

The Great Flood of 1993

In terms of precipitation amounts, record river stages, extent of flooding, persons displaced, crop and property damage, and flood duration, the Great Flood of 1993 was perhaps the worst hydrometeorological event to occur since the United States started to provide weather services in the late 1800s.

Record and near-record precipitation during the spring of 1993, on soil saturated from previous seasonal precipitation, resulted in flooding along many of the major river systems and their tributaries in the Upper Midwest during the summer of 1993.

Spring flooding began in March as a result of a previous wet fall, normal to above-normal snow accumulation, and rapid spring snowmelt accompanied by heavy spring rain-

Flash floods are one of weather's most dangerous effects. *(Photo by Jack Shaffer, Arizona Daily Star. Reproduced by permission.)*

(Photo by Jack Shaffer, Arizona Daily Star. Reproduced by permission.)

Within a minute, a nearly dry creek bed can become a roaring torrent, totally submerging the road. *(Photo by Jack Shaffer, Arizona Daily Star. Reproduced by permission.)*

fall. By May 8, however, record flooding occurred in South Dakota on Split Rock Creek at Corson and in Minnesota on the Rock River in Luverne. On May 22–24, heavy thunderstorms produced 3–7 in of rain in three hours over Sioux Falls, South Dakota, resulting in major urban and residential flooding across the city. The Big Sioux and Vermillion Rivers in South Dakota went above flood stage in late May and remained in flood through mid-June. Major flooding continued throughout the summer along the Missouri and Mississippi Rivers—by September 1, 1993, for example, the towns of Hannibal, Louisiana, and Clarksville, Missouri, had experienced 153 consecutive days of flooding. Flooding at levels above the flood stages continued through the middle of September in many regions along the Mississippi River.

Records broken in 1993

In 1993, flood records were broken at 44 forecast points on the upper Mississippi River system, at 49 forecast points on the Missouri River system, and at two forecast points on the Red River of the North system. Within the Mississippi River system, 1993 floods of record include those set at 15 forecast points on the main stem, at four forecast points on

the Iowa River, at five forecast points on the Des Moines River, and at two forecast points on the Raccoon River.

Within the Missouri River system, 1993 records were set at 14 forecast points on the main stem and at four forecast points on each of the Saline, Smoky Hill, and Grand Rivers. Record 1993 flood stages surpassed old record stages by more than 6 ft in some cases. For example, in 1993, flood records set more than 42 years ago on the main stem of the Missouri were broken by more than 4 ft at multiple forecast points. In at least one case, a new record was established early on but was broken by higher water later. The historic record on the Mississippi at St. Louis was established on April 28, 1973, at 43.2 ft; reestablished on July 21, 1993, with a flood stage of 46.9 ft; and again reestablished 11 days later on August 1, 1993, with a record flood stage of 49.6 ft.

Flood damages of 1993

Damage caused by these record flood stages was massive. Over 17,000 mi^2 were inundated by the flood (the 1937 flood in the same region affected only 12,700 mi^2). It is estimated that the region suffered over $8 billion in damages; President Clinton declared more than 200 counties federal disaster areas, including all 99 counties in Iowa. Amazingly, only 26 individuals died as a direct result of the flood. Compare that figure to 23 killed in the 1973 flood and 250 in the 1937 flood of the same region and then to the 2,100 killed in the one-day Jamestown, Pennsylvania, flood in 1889. The decreased death tolls can be attributed to abundant warnings, well-executed evacuation plans, lack of flash floods, and foremost, the protection provided by the system of flood-controlling levees in the area.

The duration and magnitude of the Great Flood of 1993, as well as its leading conditions, strongly support the premise that this event was a significant climate variation rather than simply a sequence of meteorological events. It is quite possible that one or more climate-driving forces significantly contributed to this variation. (For example, a shift in the jet stream acted as a barrier to a cold front, bringing 150–200% more precipitation to the area than usual.) A more thorough analysis of this situation is expected to result in improved understanding of the roles contributing factors may have played.

Flooding in California, 1995

In January and March of 1995, large sections of northern and central California were hit with heavy precipitation and subsequent flooding, resulting in record losses of $2–3 billion. Over 10,000 homes were lost or destroyed and 27 lives were lost. In the end, all state counties but one were declared federal disaster areas.

Throughout the early winter and spring of 1995, a strong and displaced Pacific jet stream funneled storm systems and heavy moisture into California. These systems produced severe thunderstorms and sometimes tornadoes. The heavier-than-normal precipitation flooded small streams; this flooding spread to tributaries and larger rivers. Precipitation

The force of a flash flood carried these cars downstream. *(Photo courtesy of National Oceanic and Atmospheric Administration (NOAA) Central Library.)*

(which accumulated as snow) at altitudes of 5,000 ft or more also contributed to the flooding as melting and runoff. By March, the snow depth in some areas of the Sierra Nevada range were reported to be as high as 40 ft.

Floods of 1997

During the winter of 1996-97, substantial snow fell across north-central and western sections of the United States. As this snow began to melt, the potential for significant flooding existed across these areas. Heavy rainfall during February 1997 caused flooding across portions of the Southeast, middle and lower Mississippi Valley, and the Great Lakes regions, and added to the possibility of continued flooding in these areas with heavy spring rains. In early March, torrential rains of up to 12 in in 36 hours fell across the Ohio River Basin. This caused disastrous flooding throughout the basin, and the flood crest in March reached the confluence with the Mississippi.

Areas of elevated risk for spring flooding included the northern Plains, most of the Mississippi Valley, the western Great Lakes region, and much of the lower Missouri and Ohio Valleys, as well as a substantial area in the Southeast, extending from eastern Texas through South Carolina, ex-

cluding southern Alabama, Georgia, and all of Florida. Of particular concern were the Red River along the North Dakota-Minnesota state line, and the James River in South Dakota. There, the water content of the snow over the entire drainage south of Fargo, North Dakota, was exceptionally high and could cause serious flooding.

Flooding of March 1997

During February 1997, the middle third of the nation saw the passage of several very wet storm systems, and many of these areas accumulated more than 150% of normal precipitation for the month. In early March, this general pattern continued, and included the deluge that caused catastrophic flooding in portions of the Ohio Valley.

During late February or early March, a total of 18 states across the central and eastern United States were affected by river flooding, the most serious of which was observed along the south of the middle and lower Ohio River. Flow volume was reported four to 10 times the normal (for the date) along smaller rivers and creeks in Kentucky and Ohio alone. As of March 10, flooding had resulted in: 98 counties being declared federal disaster areas; damage to at least 100,000 structures; destruction of roads and bridges, and

Flood Safety Rules

Before the flood:

1. Become familiar with local flood areas and dams; know if floodwaters might affect your home and property. Know your flood risk and elevation above flood stage and flood plain. Do local waterways, or rivers, and washes flood easily? If so, be prepared to move to a place of safety.

2. Learn flood warning signals and community evacuation routes and shelters.

3. Keep a stock of food that requires no cooking or refrigeration; electric and gas services may be interrupted.

4. Keep a portable radio, emergency cooking equipment, lights, and flashlights in working order.

5. Keep first-aid supplies and any medicines your family may need on hand.

6. Store materials like sandbags, plywood, plastic sheeting, and lumber to protect your house from flood waters and to make quick repairs.

7. Keep your car fueled. In an emergency, filling stations may not be operating.

8. Contact your insurance agent or local government to discuss flood insurance coverage.

9. Install check valves in building sewer traps to prevent flood water from backing up in sewer drains.

10. Arrange for auxiliary electrical supplies for hospitals and other operations that are critically affected by power failure.

When you receive a flood warning:

11. Store drinking water in clean bathtubs and in various closed containers as water service may be disrupted.

12. In coastal areas, board up windows or protect them with storm shutters, or tape to prevent flying, broken glass.

13. Put sandbags or other protection in place, but away from outer walls. In the case of deep flooding you may opt to flood a basement with clean water.

14. If forced to leave your home and time permits: move essential items to safe ground or to upper levels of the house; turn off utilities at main switches, but do not touch electrical equipment if you are wet or standing in water; fill tanks to keep them from moving away; grease immovable machinery; leave a note on your house to advise authorities that you have evacuated.

15. Move to a safe area before access is cut off by flood water. Watch for mud slides, downed electrical lines, and areas with high or rising water levels.

During the flood:

16. Avoid areas subject to sudden flooding.

17. Do not drive into flooded areas. Even 2 ft of water will carry away most vehicles. If flood waters do rise around your car, abandon it and move to higher ground.

18. Do not swim or dive into the water.

After the flood:

19. Do not visit disaster areas; your presence might hamper rescue and other emergency operations.

20. If you have flood insurance, contact your agent that you have a loss.

21. Tune in to local radio and television for advice on where to obtain medical care and other assistance.

22. Do not enter structures if floodwaters have covered the first floor. Seek expert advice to determine if the building is safe to enter.

23. Use battery-powered lanterns or flashlights (not oil or gas lanterns); if the building may have a gas leak, do not use any kind of light.

24. Flood waters may have swollen doors tightly shut; use windows or other openings.

25. Check with local authorities before using any water; wells should be pumped out and water tested before drinking.

26. Do not use fresh food that has come into contact with flood waters.

27. Do not handle live electrical equipment in wet areas; have an expert check all equipment before returning to service.

28. Pump water out of basements gradually (one-third of the water per day) to lessen damage to walls and foundation.

29. Report broken utility lines to appropriate authorities; have the gas company check for leaks and to turn the gas back on.

30. Watch out for poisonous snakes in previously flooded areas.

Important flood terms:

FLOOD FORECASTS mean rainfall is heavy enough to cause rivers to overflow their banks, or melting snow is mixing with rainfall to produce similar effects.

FLOOD WARNINGS or forecasts of impending floods describe the affected river, lake, or tidewater, the severity of flooding (minor, moderate, or major), and when and where the flooding will begin.

---

contamination of public water supplies; and the deaths of approximately 30 people.

Pacific Northwest floods, 1997

In 1997, there was also serious concern for spring flooding in the northern and central Rocky Mountains—particularly in Idaho and Montana—and in parts of Wyoming, Utah, and Colorado, as snow pack totals were well above average. Most of the higher elevations of Washington and northern Oregon also had much above-average snow packs and were in some jeopardy of flooding. Finally, deep snow packs in the central and southern Sierra Nevada increased the possibility of snowmelt flooding in western Nevada and in the Sacramento and San Joaquin River drainages in California.

Precipitation and snow accumulation since the end of January was lighter than normal throughout much of the West. In particular, there was some optimism that areas along the West Coast, disastrously inundated at the start of the year, might escape the spring without additional serious flooding. This dry period allowed water managers to draw down reservoirs to provide storage capacity for at least some of the spring snowmelt. Many reservoirs had filled to very high levels earlier in the year.

• PRECIPITATION WORLDWIDE IN 2000

Global precipitation was also above-average in 2000. It is estimated that 2000 ended as one of the 10 wettest years on record. Precipitation in the tropics was heavily influenced by La Niña throughout much of the year, with above average precipitation in Indonesia and the western tropical Pacific, while drier than normal conditions were common in the central tropical Pacific. La Niña also contributed to above normal precipitation in northeast South America and southern Africa and enhanced monsoonal precipitation in southern Asia. Below normal precipitation across equatorial areas of East Africa and the Gulf Coast of the United States is also attributable to La Niña conditions.

• PRECIPITATION WORLDWIDE IN 2001

Global precipitation was below the 1961–1990 average in 2001. A drought that has persisted for nearly three years continued in 2001 across Afghanistan, Pakistan, and neighboring countries. In Iran, estimated agricultural losses approached $2.6 billion (USD). Much of Central America experienced drought during the middle part of the year, which is traditionally the rainy season. Late season tropical activity eased drought in these areas, although drought in Honduras and neighboring regions continued through the latter part of 2001. In Africa, drought in Kenya persisted during 2001, despite one of the wettest Januaries in 40 years. In Australia, drought affected much of Western Australia and parts of Queensland.

Several typhoons brought excessive rains to parts of Southeast Asia, including the southeast coast of China, Taiwan, the Philippines, and Vietnam. Monsoon rains in Southeast Asia promoted extensive flooding along the Mekong Delta with at least several hundred deaths between August and October. Across Siberia, spring rains and a rapid thaw brought flooding across a vast area from the Ural Mountains to the Russian Far East, with some of the worst flooding in the Sakha region along the Lena River. A third consecutive year of spring flooding along the Tisza River in Hungary displaced tens of thousands of people from their homes, with the Tisza rising to its highest level in 100 years by early March.

• PRECIPITATION IN THE UNITED STATES FOR 2001

Precipitation was below average for the conterminous United States in 2001. The estimated January–December anomaly was 0.7 inches below the long-term mean. This value could change based on conditions in December. This year marked the third consecutive year of below normal precipitation, following nine years of precipitation surpluses.

• PRELIMINARY GLOBAL PRECIPITATION FOR 2002

Global precipitation was below the 1961–1990 average in 2002. Much of Australia experienced severe drought, with the eastern part of the country the worst affected. India monsoon rainfall was 19% below normal, with the resulting drought characterized as the worst since 1987. Other drought-affected areas included the western United States and portions of the north coast of China.

After a dry beginning to 2002, several typhoons brought excessive rains to parts of Southeast Asia and Japan, the southeast coast of China, Taiwan, and the Philippines. In contrast to drought conditions during the first half of 2002, the onset of monsoon rains in Southeast Asia promoted extensive flooding along the Mekong Delta. Seasonal flooding in much of south Asia (Nepal, Bangladesh, and northeastern India) during June–August claimed more than 1,000 lives. In the eastern United States, long-term drought was ameliorated by a turn to wetter weather, due in part to moisture from tropical systems.

• PRELIMINARY NATIONAL PRECIPITATION FOR 2002

During 2002, more than a tenth of the country was very dry in 8 of the 11 months, and in February and May more than 20% of the country was very dry. Only in September and October did the percentage area of very wet conditions exceed 10%, though in October more than 20% of the country was very wet. This was due in part to several landfalling tropical systems during these two months. Drought improvement generally occurred in the eastern U.S. during 2002, while in much of the western half of the country and the central and southern Great Lakes drought worsened.

Precipitation in the United States in 2002 was characterized by extreme dryness in the West, above-average wetness in the Mississippi Valley region, and dryness giving way to near average conditions for the East. Colorado is on pace to have its driest year on record during 2002, and Wyoming and Arizona to have their second driest years. Eight states will likely be much drier than normal, and the West region as a whole will be the driest on record.

Louisiana had its wettest fall on record, in part due to two landfalling tropical systems that hit the state in September and October. It was the third driest July and August combined for the Northeast. Considerable improvement in drought conditions for the East Coast occurred in the fall.

• FLOOD WARNINGS

Early flood warnings allow time for residents to leave low-lying areas, and to move personal property, mobile equipment, and livestock to higher ground. Sometimes valuable crops can be harvested in advance of a destructive flood. Emergency and relief organizations can prepare to handle refugees and combat the inevitable health hazards caused by floods.

Flood warnings are forecasts of impending floods, and are distributed to the public by radio and television, and through local emergency forces. The warning message tells the expected severity of flooding (minor, moderate, or major), the affected river, and when and where flooding will begin. Careful preparation and prompt response will reduce property loss and ensure personal safety.

Flash-flood warnings can be issued in minutes and flood warnings can be issued hours to days in advance of the flood peak on major tributaries. Main river flood forecasts can be issued several days or even weeks in advance. In general, the time lapse between rainfall or snowmelt and the rise in river height increases with the size of the river.

• WARNING SYSTEMS

Thirteen regional river forecast centers are the first echelon offices that prepare river and flood forecasts and warnings for approximately 3,000 communities.

National Oceanic and Atmospheric Administration's National Weather Service (NWS) has helped set up additional flash-flood warning systems in about 100 communities. In these, a volunteer network of rainfall and river-observing stations is established in the area, and a local flood warning representative is appointed to collect reports from the network. The representative is authorized to issue official flash-flood warnings based on a series of graphs prepared by the NWS. These graphs show the local flooding that will occur under different conditions of soil moisture and rainfall. On the basis of reported rainfall, the representative can prepare a flood forecast from these graphs, and spread a warning within minutes. (Flash-flood warnings are the most urgent type of flood warning issued, and are transmitted to the public over radio, television, and by sirens and other signals.) Communities within range of a NWS radar have the additional protection of advance warning when flood-producing storms approach.

Flood Casualties

During 1960–2001, a total of 5,180 people died in the United States, Puerto Rico, and the Virgin Islands combined due to flash floods and floods. Most flash-flood fatalities (56%) are vehicle related and more than two-thirds (65%) are males.

The state with the most flash-flood/flood deaths during the past 42 years—with more than double the next states—is Texas, followed by totals of closely grouped states—California, West Virginia, Virginia, and South Dakota (Puerto Rico ranks just behind Texas).

All but 11 of the 249 fatalities in South Dakota occurred during the flash floods in the Rapid City area June 9–10, 1972, and more than half of the deaths in Virginia occurred during the August 1969 floods in the western portion of the state. Most of the Colorado deaths (83%) occurred during the Big Thompson Canyon flash flood, August 1, 1976.

States with Highest Flash-Flood/Flood Fatality Statistics, 1960–2001

1. Texas (736)
2. California (313)
3. West Virginia (270)
4. Virginia (256)
5. South Dakota (249)
6. Pennsylvania (228)

Totals by decades: 1960–69: 1,296; 1970–79: 1,837; 1980–89: 890; 1990–99: 1,042 2000–01: 115.

The flash-flood/flood totals include deaths during tropical cyclones. Years with the greatest totals can be directly tied to those years with major landfalling hurricanes (e.g., Betsy in 1965, Camille in 1969, and Agnes in 1972).

Sources: Richard A. Wood; National Weather Service, Office of Hydrology, and National Climatic Data Center ("Storm Data" publications).

Successful operation of a flash-flood warning system requires active community participation and planning, but very little financial outlay. Still, the communities with cooperative flash-flood warning systems are only a small fraction of the thousands of communities that need them.

• USGS FLOOD MEASUREMENTS

The USGS currently (2000) maintains more than 7,000 stream-gaging stations throughout the United States, Puerto Rico, and the Virgin Islands that monitor streamflow and provide data to various federal, state, and local cooperating agencies as well as the general public. Some of these stream-gauging stations have been in operation since before

Flash-Flood Safety Rules

BEFORE THE FLOOD know the elevation of your property in relation to nearby streams and other waterways. Investigate the flood history of your area and how human-made changes may affect future flooding. Make advance plans of what you will do and where you will go in a flash-flood emergency.

WHEN A FLASH-FLOOD WATCH IS ISSUED listen to area radio and television stations for possible Flash-Flood Warnings and reports of flooding in progress from the National Weather Service and public safety agencies. Be prepared to move out of danger at a moment's notice. If you are on the road, watch for flooding at highway dips, bridges, and low areas due to heavy rain not observable to you, but which may be indicated by thunder and lightning.

WHEN A FLASH-FLOOD WARNING IS ISSUED for your area act quickly to save yourself. You may have only seconds:

1. Get out of areas subject to flooding. Avoid already flooded areas. This includes dips, low spots, canyons, washes, etc.

2. Do not attempt to cross a flowing stream on foot where water is above your ankles.

3. If driving, know the depth of water in a dip before crossing. The road may not be intact under the water. If the vehicle stalls, abandon it immediately and seek higher ground—rapidly rising water may engulf the vehicle and its occupants and sweep them away. The depth of the water is not always obvious.

4. Be especially cautious at night when it is more difficult to recognize flood dangers.

5. When out of immediate danger, tune in area radio or television stations for additional information as conditions change and new reports are received.

6. Children should never play around high water, storm drains, viaducts, or arroyos.

7. Do not camp or park vehicles along streams or washes, particularly during threatening conditions.

AFTER THE FLASH-FLOOD WATCH OR WARNING IS CANCELED stay tuned to radio or television for follow-up information. Flash flooding may have ended, but general flooding may come later in headwater streams and major rivers.

Flash-flood terms used in forecasts and warnings:

FLASH FLOOD means the occurrence of a dangerous rise in water level of a stream or over a land area in a few hours or less caused by heavy rain, ice-jam breakup, earthquake, or dam failure.

FLASH-FLOOD WATCH means that heavy rains occurring or expected to occur may soon cause flash flooding in certain areas and citizens should be alert to the possibility of a flood emergency that will require immediate action.

FLASH-FLOOD WARNING means that flash flooding is occurring or imminent on certain streams or designated areas and immediate precautions should be taken by those threatened.

1900, providing more than a century of water information for the nation. In addition to providing critical information on flood heights and discharges, these stations provide data used in the effective management of water-supply and water-quality needs, protection of aquatic habitat, recreation, and water-resources research.

The basic building block for a stream-flow data network is the stage-discharge relation that is developed at each gauging-station location. Measurements of the flow (discharge) are related graphically to the respective water depths (stage), which then enables discharge to be determined from stage data.

Discharge measurements can either be direct, using a current meter, or indirect, using mathematical flow equations. Both methods require that an elevation of the floodwater surface be determined by a water-depth gauge or by a detailed survey of high-water marks. If time allows and conditions are safe, a direct measurement by USGS hydrographers is preferred. However, during major floods, direct measurements are often impossible or extremely dangerous, and indirect methods must be used.

Accurate identification and measurement of high-water marks from floods are very important in the accurate mapping of inundated areas as well as in the analysis of water-surface profiles for indirect discharge measurements. These

elevations, in combination with flood-frequency analysis using many years of annual flood maximums, are used by the Federal Emergency Management Agency (FEMA) to determine flood-insurance rates.

• SIGNIFICANT FLOODS OF THE TWENTIETH CENTURY

During the twentieth century, floods were the number-one natural disaster in the United States in terms of number of lives lost and property damaged. They can occur at any time of the year, in any part of the country, and at any time of the day or night. Most lives are lost when people are swept away by flood currents, whereas most property damage results from inundation by sediment-laden water. Flood currents also possess tremendous destructive power, as lateral forces can demolish buildings and erosion can undermine bridge foundations and footings leading to the collapse of structures. The accompanying map and table locate and describe 32 of the most significant floods of the twentieth century.

Floods are the result of a multitude of naturally occurring and human-induced factors, but they all can be defined as the accumulation of too much water in too little time in a

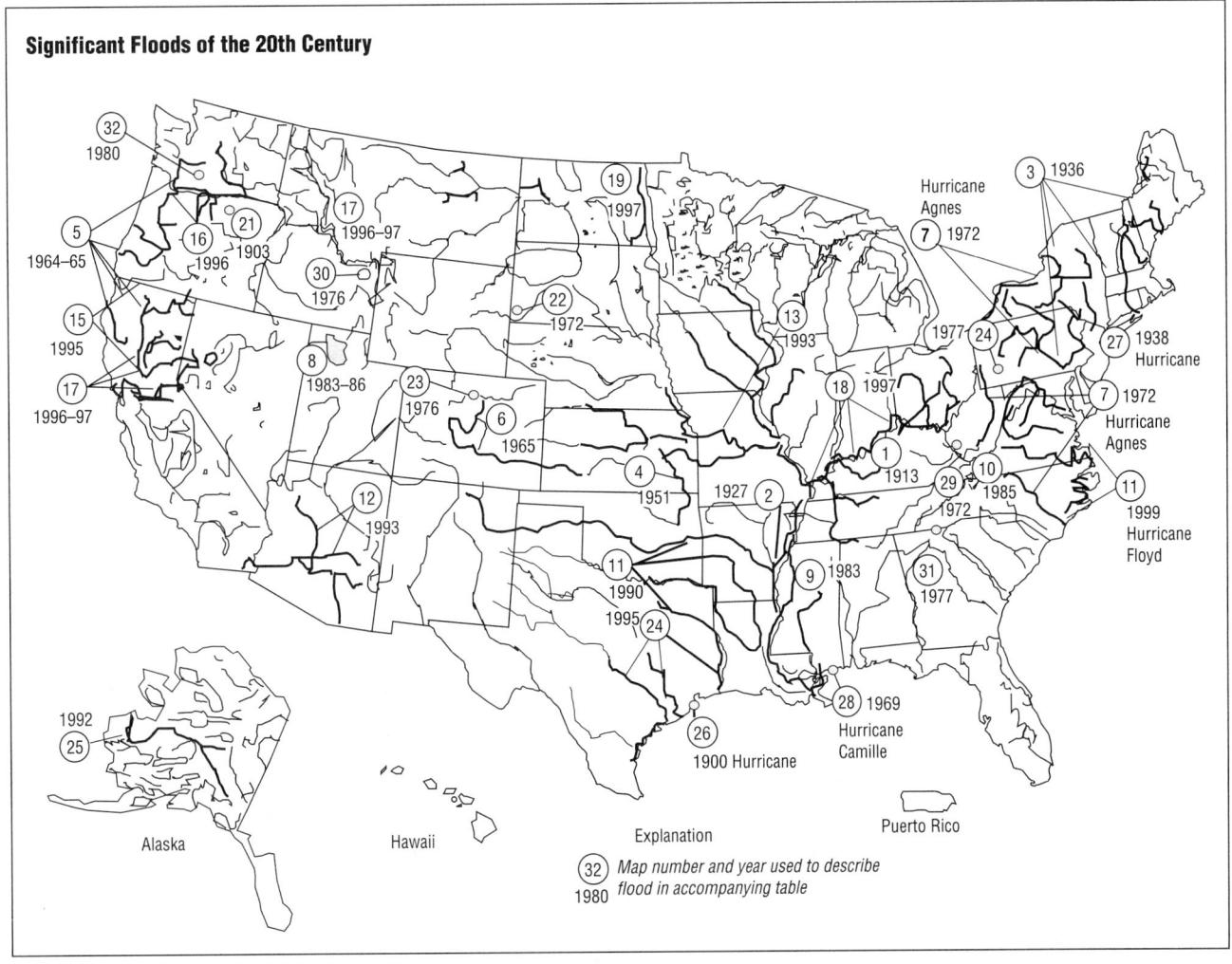

Significant Floods of the 20th Century

(Courtesy of National Climatic Data Center/NOAA.)

specific area. Types of floods include regional floods; flash floods; ice-jam floods; storm-surge floods; dam- and levee-failure floods; and debris, landslide, and mudflow floods.

Regional floods

Some regional floods occur seasonally when winter or spring rains, coupled with melting snow, fill river basins with too much water too quickly. The ground may be frozen, reducing infiltration into the soil and thereby increasing runoff. Such was the case for the New England flood of March 1936, in which more than 150 lives were lost and property damage totaled $300 million.

Extended wet periods during any part of the year can create saturated soil conditions, after which any additional rain runs off into streams and rivers, until river capacities are exceeded. Regional floods are many times associated with slow-moving, low-pressure, or frontal storm systems, including decaying hurricanes or tropical storms. Persistent wet meteorological patterns are usually responsible for very large regional floods, such as the Mississippi River Basin flood of 1993 wherein damages were $20 billion.

Flash floods

Flash floods can occur within several seconds to several hours, with little warning. Flash floods can be deadly because they produce rapid rises in water levels and have devastating flow velocities. Several factors can contribute to flash flooding. Among these are rainfall intensity, rainfall duration, surface conditions, and topography and slope of the receiving basin. Urban areas are susceptible to flash floods because a high percentage of the surface area is composed of impervious streets, roofs, and parking lots where runoff occurs very rapidly. Mountainous areas also are susceptible to flash floods, as steep topography may funnel runoff into a narrow canyon. Floodwaters accelerated by steep stream slopes can cause the floodwave to move downstream too fast to allow escape, resulting in many deaths. A

Significant Floods of the 20th Century

[M, million; B, billion]

Flood type	Map no.	Date	Area or stream with flooding	Reported deaths	Approximate cost (uninflated)	Comments
Regional flood	1	Mar.-Apr. 1913	Ohio, statewide	467	$143M	Excessive Regional rain.
	2	Apr.-May 1927	Mississippi River from Missouri to Louisiana	unknown	$230M	Record discharge downstream from Cairo, Illinois.
	3	Mar. 1936	New England	150+	$300M	Excessive rainfall on snow.
	4	July 1951	Kansas and Neosho River Basins in Kansas	15	$800M	Excessive regional rain.
	5	Dec. 1964-Jan. 1965	Pacific Northwest	47	$430M	Excessive rainfall on snow.
	6	June 1965	South Platte and Arkansas Rivers in Colorado	24	$570M	14 inches of rain in a few hours in eastern Colorado.
	7	June 1972	Northeastern United States	117	$3.2B	Extra tropical remnants of Hurricane Agnes.
	8	Apr.-June 1983	Shoreline of Great Salt Lake, Utah	unknown	$621M	In June 1986, the Great Salt Lake reached its highest elevation and caused $268M more in property damage.
	9	May 1983	Central and northeast Mississippi	1	$500M	Excessive regional rain.
	10	Nov. 1985	Shenandoah, James, and Roanoke Rivers in Virginia and West Virginia	69	$1.25B	Excessive regional rain.
	11	Apr. 1990	Trinity Arkansas, and Red Rivers in Texas, Arkansas, and Oklahoma	17	$1B	Recurring intense thunderstorms.
	12	Jan. 1993	Gila, Salt, and Santa Cruz Rivers in Arizona	unknown	$400M	Persistent winter precipitation.
	13	May-Sept. 1993	Mississippi River Basin in central United States	48	$20B	Long period of excessive rainfall.
	14	May 1995	South-central United States	32	$5-6B	Rain from recurring thunderstorms.
	15	Jan.-Mar. 1995	California	27	$3B	Frequent winter storms.
	16	Feb. 1996	Pacific Northwest and western Montana	9	$1B	Torrential rains and snowmelt
	17	Dec. 1996-Jan. 1997	Pacific Northwest and Montana	36	$2-3B	Torrential rains and snowmelt
	18	Mar. 1997	Ohio River and tributaries	50+	$500M	Slow-moving frontal system.
	19	Apr.-May 1997	Red River of the North in North Nakota and Minnesota	8	$2B	Very rapid snowmelt.
	20	Sept. 1999	Eastern North Carolina	42	$6B	Slow-moving Hurricane Floyd.
Flash flood	21	June 14, 1903	Willow Creek in Oregon	225	unknown	City of Heppner, Oregon, destroyed.
	22	June 9-10, 1972	Rapid City, South Dakota	237	$160M	15 inches of rain in 5 hours.
	23	July 31, 1976	Big Thompson and Cachela Poudre Rivers in Colorado	144	$39M	Flash flood in canyon after excessive rainfall.
	24	July 19-20, 1977	Conemaugh River in Pennsylvania	78	$300M	12 inches of rain in 6-8 hours.
Ice-jam flood	25	May 1992	Yukon River in Alaska	0	unknown	100-year flood on Yukon River.
Storm-surge flood	26	Sept. 1900	Galveston, Texas	6,000+	unknown	Hurricane
	27	Sept. 1938	Northeast United States	494	$306M	Hurricane
	28	Aug. 1969	Gulf Coast, Mississippi and Louisiana	259	$1.4B	Hurricane Camille
Dam-failure flood	29	Feb. 2, 1972	Buffalo Creek in West Virginia	125	$60M	Dam failure aafter excessive rainfall.
	30	June 5, 1976	Teton River in Idaho	11	$400M	Earthen dam breached.
	31	Nov. 8, 1977	Toccoa Creek in Georgia	39	$2.8M	Dam failure after excessive rainfall.
Mudflow flood	32	May 18, 1980	Toutle and lower Cowlitz Rivers in Washington	60	unknown	Result of eruption of Mt. St. Helens.

(Courtesy of National Climatic Data Center/NOAA.)

flash flood caused by 15 inches of rain in 5 hours from slow-moving thunderstorms killed 237 people in Rapid City, South Dakota, in 1972.

Floodwaves more than 30-feet high have occurred many miles from the rainfall area, catching people unaware. Even desert arroyos are not immune to flash floods, as distant thunderstorms can produce rapid rises in water levels in otherwise dry channels. Early-warning gauges upstream save lives by providing advanced notice of potential deadly floodwaves.

Ice-Jam Floods

Ice-jam floods occur on rivers that are totally or partially frozen. A rise in stream stage will break up a totally frozen river and create ice flows that can pile up on channel obstructions such as shallow riffles, log jams, or bridge piers. The jammed ice creates a dam across the channel over which the water and ice mixture continues to flow, allowing for more jamming to occur. Backwater upstream from the ice dam can rise rapidly and overflow the channel banks. Flooding moves downstream when the ice dam fails, and the water stored behind the dam is released. At this time the flood takes on the characteristics of a flash flood, with the added danger of ice flows that, when driven by the energy of the floodwave, can inflict serious damage on structures. An added danger of being caught in an ice-jam flood is hypothermia, which can kill quickly. Ice jams on the Yukon

River in Alaska contributed to severe flooding during the spring breakup of 1992.

Storm-Surge Floods

Storm-surge flooding is water that is pushed up onto otherwise dry land by onshore winds. Friction between the water and the moving air creates drag that, depending upon the distance of water (fetch) and the velocity of the wind, can pile water up to depths greater than 20 feet. Intense low-pressure systems and hurricanes can create storm-surge flooding. The storm surge is unquestionably the most dangerous kind of flood. Nine out of 10 hurricane fatalities are caused by the storm surge. Worst-case scenarios occur when the storm surge occurs concurrently with high tide. Stream flooding is much worse inland during the storm surge because of backwater effects. In September 1900, the hurricane and storm surge at Galveston, Texas, killed more than 6,000 people, making it the worst natural disaster in the nation's history.

Dam- and levee-failure floods

Dams and levees are built for flood protection. They usually are engineered to withstand a flood with a computed risk of occurrence. For example, a dam or levee may be designed to contain a flood at a location on a stream that has a certain probability of occurring in any one year. If a larger flood occurs, then that structure will be overtopped. If during the overtopping the dam or levee fails or is washed out, the water behind it is released to become a flash flood. Failed dams or levees can create floods that are catastrophic to life and property because of the tremendous energy of the released water. Warnings of the Teton Dam failure in Idaho in June 1976 reduced the loss of life to 11 people.

Debris, landslide, and mudflow floods

Debris or landslide floods are created by the accumulation of debris, mud, rocks, or logs in a channel, which form

Weather Anecdotes

October 16–18, 1994 ... The nation's number one atmospheric natural hazard killer is flash floods and floods. A total of 17 people lost their lives in a three-day period in southeastern Texas due to flash flooding. In an average year, about 56% of all deaths due to flash flooding are vehicle related.

November 28, 1993 ... Hamilton Township, PA. Two women were stranded in their vehicle when they drove onto a flooded roadway. According to them, "we hit some water—we thought it was a puddle." The puddle turned out to be the rain-swollen Back Creek. Both were swept down stream, but managed to climb on top of their car. One of them swam to safety, the other was rescued by fire fighters. When in doubt, stay out of waterways. Drive to a safe bridge or wait out the high water. If stuck, leave immediately and seek higher ground, probably the way you came in.

November 27, 1993 ... Two cars were stranded on a Blair County, PA, road near Geeseytown. A woman was forced to abandon her car after it became submerged. The swift current carried her downstream about 200 yards where she grabbed a tree limb and held on for about two hours. In an attempt to rescue the woman, two fire fighters were also stranded when their boat capsized. All three individuals were treated for exposure at local hospitals.

a temporary dam. Flooding occurs upstream as water becomes stored behind the temporary dam and then becomes a flash flood as the dam is breached and rapidly washes away. Landslides can create large waves on lakes or embayments and can be deadly. Mudflow floods can occur when volcanic activity rapidly melts mountain snow and glaciers, and the water mixed with mud and debris moves rapidly downslope. These mudflow events are also called lahars and, after the eruption of Mt. St. Helens in 1980, caused significant damage downstream along the Toutle and Cowlitz Rivers in southwest Washington.

U.S. Flash Flood/Flood Fatalities 1960–2001

Year	AL	AZ	AR	CA	CO	CT	DE	FL	GA	ID	IL	IN	IA	KS	KY	LA	ME	MD	MA	MI	MN	MS	MO	MT	NE	NV	NH
1960				7							1			3	2			2	3								
1961		4							1		1		5	5													
1962				2	1		7						1		1			4									
1963	1		1	3			1				1		1		2			2	1				1				
1964				19				4	5						10	28							1	36		1	
1965	2			8	22			5	2				4			58		1			12	1	5	2	1		
1966				6					1				1			2										1	
1967	7			4			3		1				1		3			1			1		1				
1968	4			1					1		1		5			2		2	2	4			1				
1969	1		2	58	1				2			4	1		3	3		4			6	132	2			1	
1970	1	23	1	21									1		1						3	1					
1971		1													1			16								1	
1972	2	8		1		1					1		2		11			19	4		3	3					
1973	2		4	5					1		1	11		4	3	3			3			19	12			1	8
1974		3	3	5			1	5			3		1	3	1	5		1			1	9	4	2		9	
1975	2		3	1	1	2	2				1	2	1	2	7	2		1		1		3	1			3	
1976	2	1		4	139				1	11				3													
1977			1						39		1			6	8	1							26				
1978	1	17	10	26							2	2			2	6		1			13				1		
1979	15		2	1		1		5	3		4	4			6	4		1				4	3				
1980	3	5		6	1			4							2						1	2				1	
1981		8		8							1	3	2	3	2	2		1		1							1
1982	1	3	10	25	4	5		3	1		5		2	6	2	6		1				1	10			1	
1983	1	15	2	7							2	2		6	1	6	2	1				5	1				
1984		1	1	2											7	1							4	2	2		8
1985		1	2								6				4			1					4				
1986			13								1		1		2			1	1	2	1		4	2		1	
1987		4							4		1	2		1	2		1				1	2					
1988	1	1		3				4				2			5						1	2					
1989		4	1				3	1			1	2	2	3	3		7						1				
1990	16	3	1	3					14		5	3	3		2					4	1	2	5		1	2	
1991	1	1	1	1								3	3		3	3	3	1				2			1	1	
1992		1		2				2	2			1		1	5	1							5				1
1993		3	2	17				2	2		1	1	3	2	2						2	1	30		4		
1994			2						28		2	1	1									1	3				
1995	1	2		7							1	2			2	6						1	2				
1996	1			5	3					1		2		2	2		1	3		2			7	4			1
1997	1	23	1	13	9	2		2		1	1	1	1		16	1		1							6		3
1998	8	4	2	19				2	1		1	2		3	1	1	1				2		13			1	1
1999	2			1	1	2					1	5	1		1							2	4			1	1
2000		1		6	1						4	1	2		1							1	2				
2001			6					1	1		1	2		1	2						3	3	1	3			
2002																											
2003																											
2004																											
2005																											
2006																											
2007																											
2008																											
2009																											
2010																											
Totals	76	125	67	313	181	12	19	37	114	13	52	59	41	54	116	151	9	68	14	18	52	190	162	49	15	32	11

(Chart by GGS. Data compiled by Richard A. Wood, Ph.D.)

U.S. Flash Flood/Flood Fatalities 1960–2001 [CONTINUED]

Year	NJ	NM	NY	NC	ND	OH	OK	OR	PA	RI	SC	SD	TN	TX	UT	VT	VA	WA	WV	WI	WY	AK	HI	PR	VI	Totals
1960	9		7	3				1	1					15		1	1		22	1				111		168
1961			1	1		2								36	6		1							8		93
1962	15		7	3			2							4			4		12		1					53
1963												2	4	4			2		12				3			41
1964				3		8	18				2			5				1					1			142
1965		3		1			1				9		6	28	7		1	2				1		6		188
1966		1			4									36									2	2		56
1967	1	7	3	1					3				3	10					3					5		53
1968	8					3							2	16												57
1969	1	2				30		1	2		2	3	2	12			153		4	2				10	1	445
1970								1	3					17		2								55	1	131
1971	3					2				17				23									2	1		68
1972	2	5	21	2				7	51			237	1	24			20	4	125					4		555
1973	9		2	16		1	10		7		3		11	27		5	3		3					9		178
1974	3	2		1	1		1	9	1				2	22									5	9		111
1975	6		6	4	5	2			4		1		5	13			2		3					41		127
1976		2	3	4	1		3		1		2		1	9		2	1	1						2		193
1977			6	11		1	5	78					11	5				4	1	2				4		210
1978		7	1			3	2		1					40			1		2	1			1	3		143
1979		3	5	6		1	3						2	28			8	2	1					9		121
1980			6			3	2	2	9				3	6	1				2	1			2			62
1981						2	4		4				3	36										1		84
1982		1				3	10	1	1		1	1	9	8				7			2			2		131
1983			4		1	1	4	2			3		5	19		1		5						1	1	97
1984		6	4	5			15		5		2		3	7		5			2					1		83
1985						8			3					16			19		38	12				52		166
1986		2	4			1	2		9		4		1	18	1	3	3	1	1	1				2		80
1987		11					3	1	1		1		3	17		1	3						3	9		71
1988		3												7									5			31
1989		2	4	6		4	1		1				12	17			6		4	1				2		85
1990		2	1			32	1		1		9		5	14		1		2	7					3		142
1991		4	1					1	1			1	2	16			3		3				7	3		61
1992	2		2	3		2	3							4			2									39
1993			8	2	1		5					5		6			1			1	1				1	103
1994	1		2			1	5		1				3	28					5	2			3			89
1995				9		5	2		1		3			26	1		5	1		1				1		79
1996	2	5	12	14	2	3	5	12	28		2			7	1	2	10		12					19		172
1997		2		3	2	5	2		4				5	23	1				6					1		135
1998			3			9	2	2			1		10	45	2		1		2	3				8		150
1999	6		2	22		2	2		6					2			2		1					4	1	72
2000			1		2	4	1				1			9			1		4	1				3		43
2001						3			1		1		1	31	1		1	1	5					3		72
2002																										
2003																										
2004																										
2005																										
2006																										
2007																										
2008																										
2009																										
2010																										
Totals	68	59	125	122	19	131	98	66	228	17	47	249	115	736	20	24	256	29	270	15	15	3	29	384	5	5180

(Chart by GGS. Data compiled by Richard A. Wood, Ph.D.)

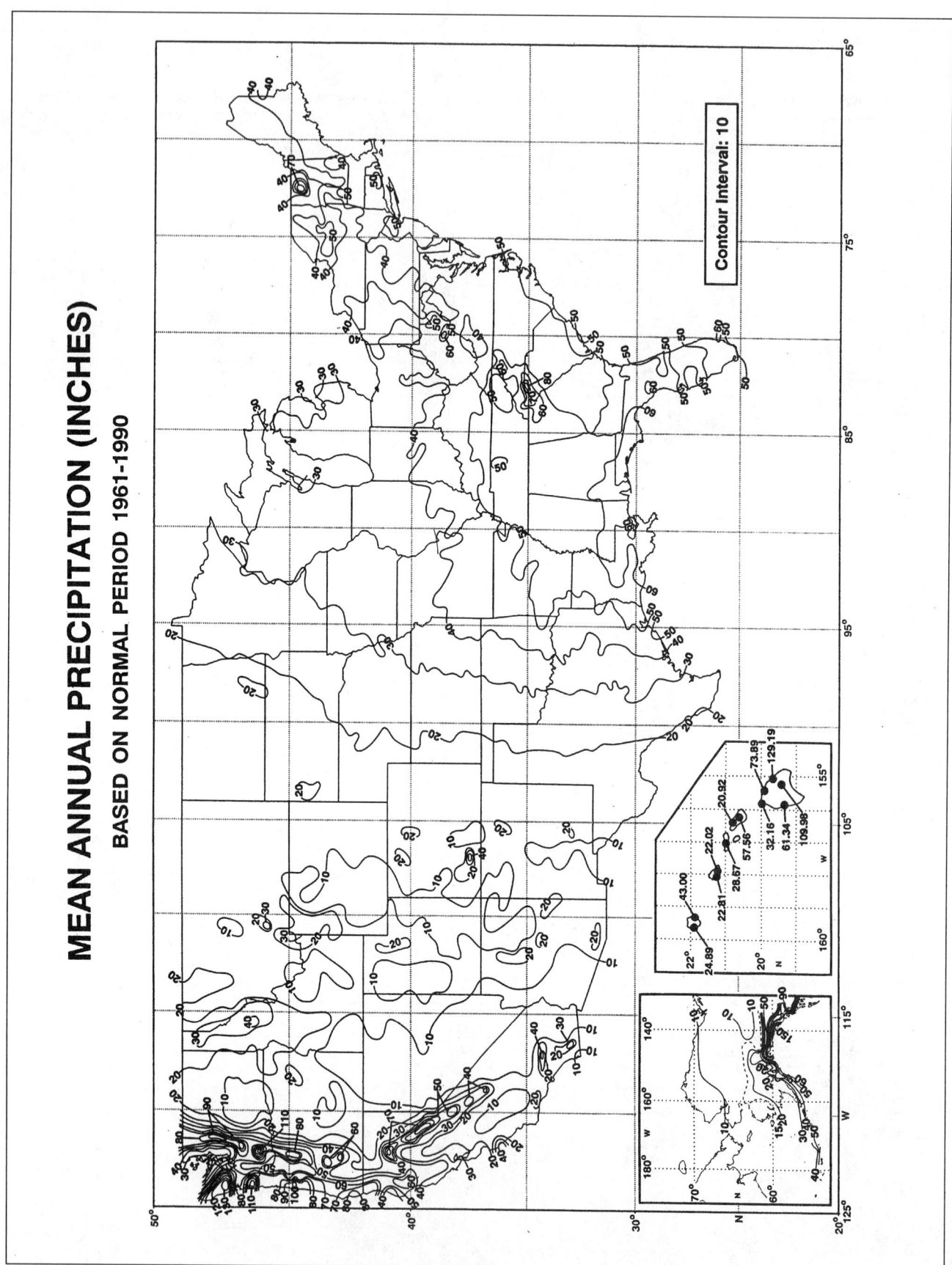

MEAN ANNUAL PRECIPITATION (INCHES)
BASED ON NORMAL PERIOD 1961-1990

Contour Interval: 10

(Courtesy of U.S. government publication.)

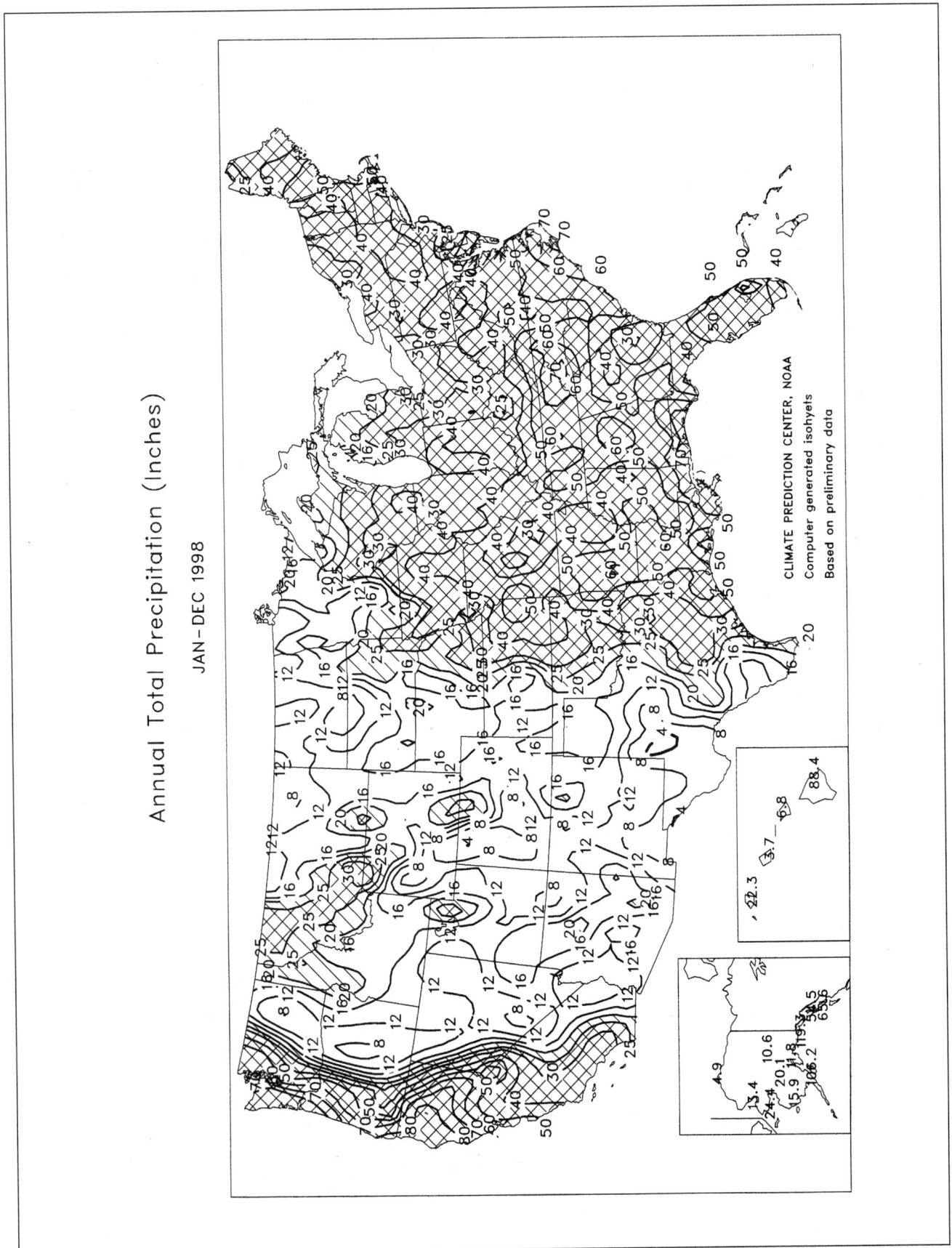

Annual Total Precipitation (Inches)

JAN-DEC 1998

CLIMATE PREDICTION CENTER, NOAA
Computer generated isohyets
Based on preliminary data

(Courtesy of U.S. government publication.)

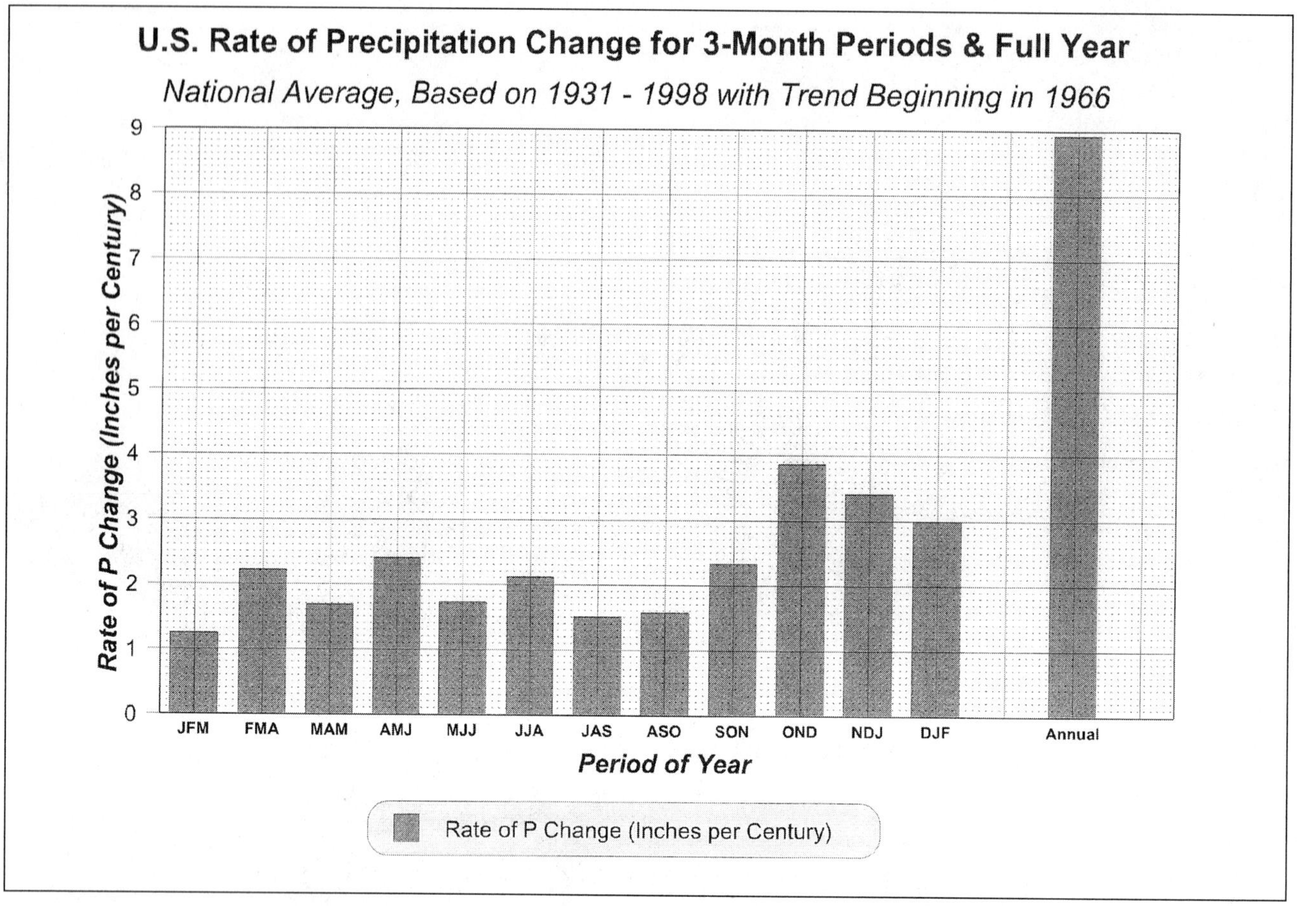

U.S. Rate of Precipitation Change for 3-Month Periods & Full Year
National Average, Based on 1931 - 1998 with Trend Beginning in 1966

(Courtesy of U.S. government publication.)

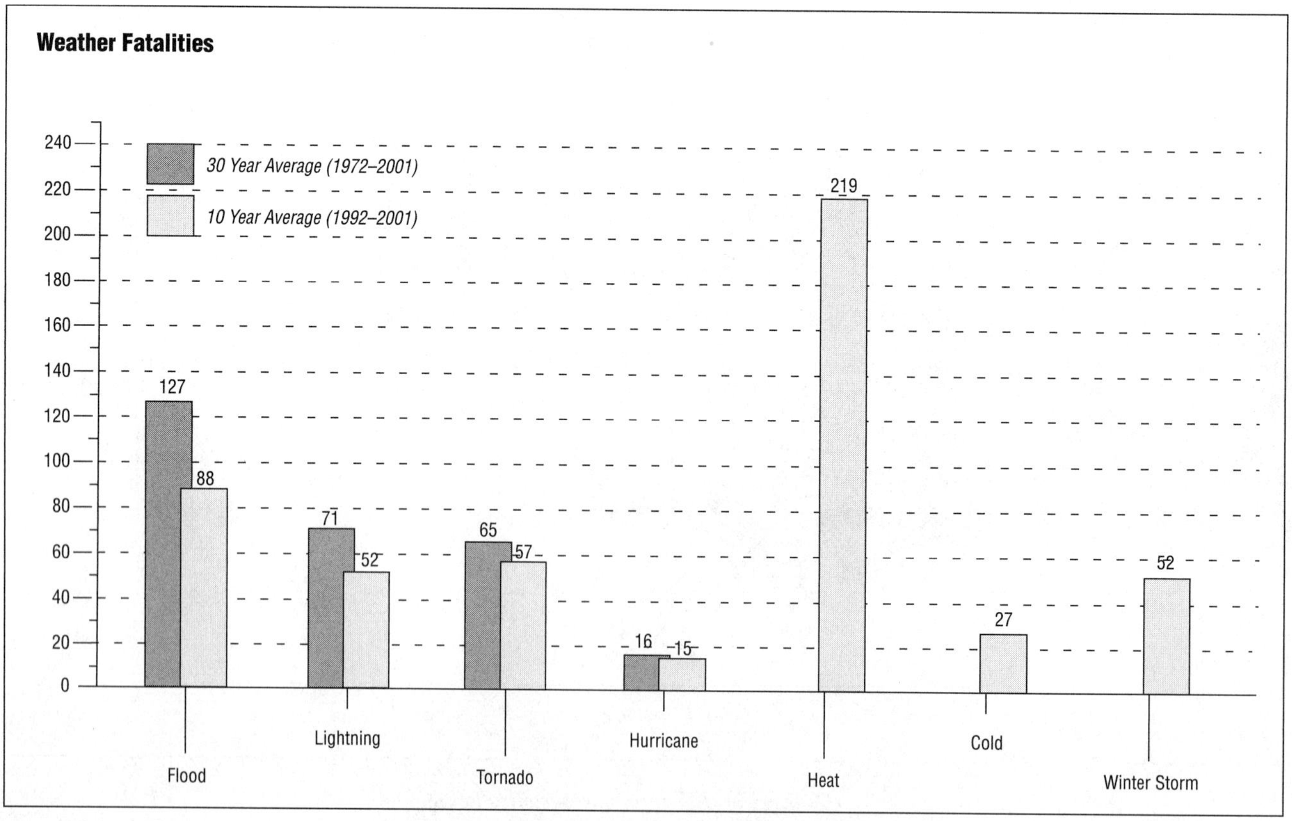

Weather Fatalities

(Courtesy of National Climatic Data Center/NOAA.)

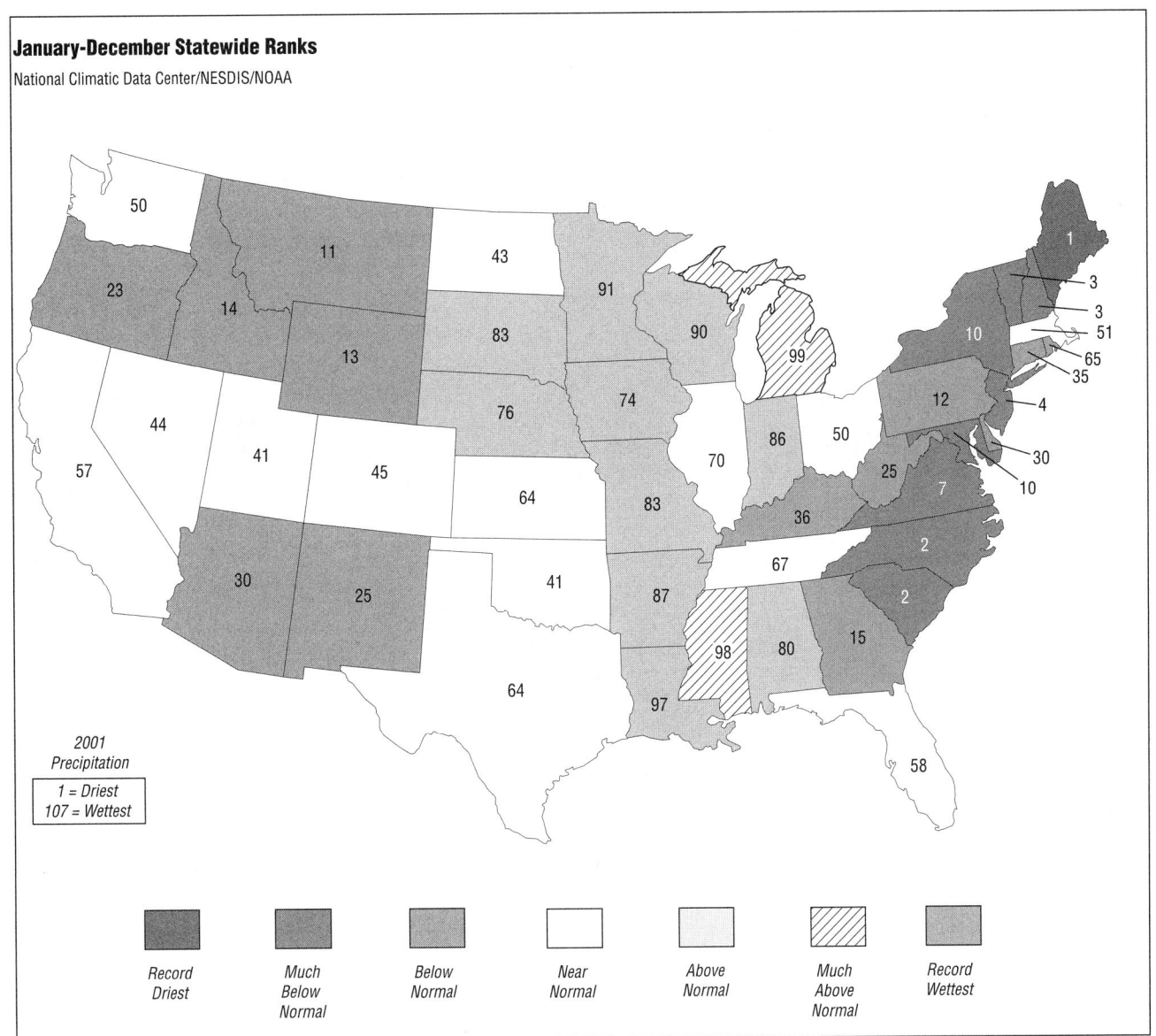

January-December Statewide Ranks

National Climatic Data Center/NESDIS/NOAA

50

11

43

91

23

14

90

99

1

13

83

74

10

3

3

51

65

35

44

76

86

50

12

4

41

45

64

70

83

25

30

57

87

36

7

30

30

25

41

67

2

10

98

80

2

15

64

97

58

2001
Precipitation

1 = Driest
107 = Wettest

| Record Driest | Much Below Normal | Below Normal | Near Normal | Above Normal | Much Above Normal | Record Wettest |

(Courtesy of National Climatic Data Center/NOAA.)

Earthquakes

• WHY EARTHQUAKES OCCUR

Earthquake conditions

The planet Earth is believed to consist of a thin crust 2–3 mi thick under the oceans and as much as 25 mi thick beneath the continents that covers the large, solid sphere of the rock mantle, which descends to about 1,800 mi. Below the mantle is the fluid outer core, and, at about 3,200 mi depth, the apparently solid inner core. The province of earthquakes recorded thus far is from the crust to a maximum depth of about 450 mi.

Conditions thought to prevail in this hot, dark, high-pressure land cannot be simulated in existing laboratories—at the base of the mantle, pressure is about 11,000 tons per square inch and temperature is 10,000°F. These diamond-smashing pressures produce a rigidity in mantle rock about four times that of ordinary steel, with an average density about that of titanium.

This very solid mantle rock seems to behave, over periods of millions of years, like a very sluggish fluid. Something, perhaps the temperature difference between the white-hot region near the core and the cooler region near the crust, drives slow-moving cycles of rising and descending currents in the mantle rock itself.

Evidently, these currents rise beneath the thin-crusted ocean floor, thrust up the mid-ocean ridges, and generate the stresses that produce their spine-like transverse cracks and shallow earthquakes. This is believed to be the force that causes material to well up through the crust, replacing and spreading the old sea floor, and pushing drifting continents apart.

Where the currents begin their descent at the edges of continents, they produce compressive pressures, and massive folding in the form of trenches and mountain ranges. These regions are the sites of the deeper earthquakes, and of most volcanism.

Earth stresses and strains and then releases

Stresses generated in the crust and upper mantle by convective currents are stored in the form of strain—physical deformation of the rock structure. Under normal circumstances, the "solid" rocks deform plastically, releasing pent-up energy before it builds to catastrophic levels. But, when stresses accumulate too rapidly to be removed by plastic flow, some structural compensation is necessary. Large blocks of material are slowly forced into highly strained positions along faults, and held in place by a supporting structure of stronger materials. These energy-absorbing zones of weakness continue to shift, like longbows being pulled to the breaking point. Finally, more stress causes the supporting rocks to rupture, triggering the "cocked" fracture back toward equilibrium. The sides of the rebounding fault move horizontally with respect to one another (strike-slip), vertically (dip-slip), or in combinations of such motion, as in the large-scale tilting that accompanied the Alaska earthquake in March 1964.

Foreshocks and aftershocks

Sometimes all the energy to be released goes out in one large wrench, followed by trains of smaller tremors, or aftershocks, produced by continuing collapse and slippage along the fracture. Sometimes the fault shift is preceded by the small structural failures we detect as foreshocks. The magnitude 5.9 earthquake that shook Fairbanks, Alaska, on June 21, 1967, was preceded by a magnitude 5.6 foreshock, followed by a magnitude 5.5 aftershock, and then, over the next 24 hours, by more than 2,000 smaller aftershocks. Small tremors were detected for days after the initial event. However, all small tremors or earthquake "swarms" do not necessarily indicate that a big one is on the way. The Matsushiro, Japan, swarm maintained an intermittent tremble for more than a year, probably doing more psychic than physical damage. Of more than 600,000 tremors recorded between August 3, 1965, and the end of 1966, 60,000 were strong enough to be felt, and 400 were damaging. During the most active period, in April and May 1966, Matsushiro felt hundreds of tremors daily, all under magnitude 5.

Whatever the time period involved, the energy of strain flows out through the shifted fault in the form of heat, sound, and earthquake waves.

How earthquake waves travel

There are four basic seismic waves: two preliminary "body" waves that travel through the earth, and two that travel only at the surface. Combinations, reflections, and diffractions produce a virtual infinity of other types. The behavior of these are well-enough understood that wave speed and amplitude have been the major means of describing Earth's interior. In addition, a large earthquake generates elastic waves that echo through the planet like vibrations in a ringing bell, which actually cause the planet to expand and contract infinitesimally.

The primary (P) wave is longitudinal, like a sound wave, propagates through both liquids and solids, and is usually the first signal that an earthquake has occurred. Where the disturbance is near enough or large enough to be felt, the P wave arrives at the surface like a hammer blow from the inside. This is the swiftest seismic wave, its speed varying with the material through which it passes. In the heterogeneous crustal structure, P-wave velocity is usually less than 4 mi per second—nearly 15,000 mph. Just below the crust, at a layer called the Mohorovicic discontinuity (the Moho), these speeds jump to 5 mi per second and subsequently increase to about 8.5 mi per second (more than 30,000 mph) through the core.

As the compressional phase of the P wave passes through the earth, particles are pushed together and displaced away from the disturbance. The rare factional phase dilates the particles and displaces them toward the earthquake source. For an object embedded in the ground, the result is a series of sharp pushes and pulls parallel to the wave path—

motions similar to those that passengers feel when a long train gets under way.

The secondary (S) wave is transverse, like a light or radio wave, and travels about half as fast as the primary wave. Because S waves require a rigid medium to travel in coherent rays, their apparent absence below the mantle gives credence to the theory of a fluid core. About twice the period and amplitude of the associated P waves, these shear waves displace particles at right angles to the direction of wave travel. The vertical component of this movement is somewhat dampened by the opposing force of gravity; but side-to-side shaking in the horizontal can be quite destructive. Where the motion is perceptible, the arrival of the S waves marks the beginning of a new series of shocks, often worse than the P-wave tremor.

Surface waves, named for their discoverers, Love and Rayleigh, are of much greater length and period, e.g., 30 seconds or more, versus less than one second for P waves. Love waves are shear in the horizontal dimension, and the Rayleigh wave induces a retrograde, elliptical motion, something like that in wind-driven ocean waves. The speed of the Love wave is about 2.5 mi per second; the Rayleigh

Many California freeways were subject to partial collapse during the Northridge quake in 1994, such as this section of the Golden Gate Freeway. *(Photo by Robert A. Eplett. Courtesy of Governor's Office of Emergency Services.)*

wave is about 10% slower. Despite the large proportion of earthquake energy represented by these waves, their long period smoothes out the motion they impart, reducing their destructiveness.

Wave motion is not considered in describing the travel of seismic waves through the earth. Instead, the P and S body waves, and their large family of reflected, combined, or resonated offspring, are treated as rays. If the planet were homogeneous, like a ball of wax, these rays would be straight lines. But in the heterogeneous Earth, the rays describe concavely spherical paths away from the earthquake source, and from points of reflection at the surface.

Since they travel at different speeds, seismic waves arrive at a given point on the Earth's surface at different times. Near the source, the ground will shake over a slightly longer interval of time than it took the fault to slip. At great distances, the same energy released by a single event may be detected instrumentally for days.

• MEASURING AN EARTHQUAKE

Intensity is an indication of an earthquake's apparent severity at a specified location, as determined by experienced observers. Through interviews with persons in the stricken area, damage surveys, and studies of Earth movement, an earthquake's regional effects can be systematically described. For seismologists and emergency workers, intensity becomes an efficient shorthand for describing what an earthquake has done to a given area.

The Modified Mercalli Intensity Scale generally used in the United States grades observed effects into 12 classes ranging from I, felt only under especially favorable circumstances, to XII, damage total. The older RossiForel Intensity Scale (RF) has 10 categories of observed effects, and is still used in Europe. Still other intensity scales are in use in Japan and the former Soviet Union.

Rating earthquakes by intensity has the disadvantage of being always relative. In recent years, an "objective" scale of earthquake magnitude has supplemented intensity ratings. Magnitude expresses the amount of energy released by an earthquake as determined by measuring the amplitudes produced on standardized recording instruments. The persistent misconception that the "Richter scale" rates the size of earthquakes on a "scale of 10" is extremely misleading, and has tended to mask the clear distinction between magnitude and intensity.

Earthquake magnitudes are similar to stellar magnitudes in that they describe the subject in absolute, not relative, terms, and that they refer to a logarithmic, not an arithmetic, scale. An earthquake of magnitude 8, for example, represents seismograph amplitudes 10 times larger than those of a magnitude 7 earthquake, 100 times larger than those of a magnitude 6 earthquake, and so on. There is no highest or lowest value, and it is possible here, as with temperature, to record

negative values. The largest earthquakes of record were rated at magnitude 8.9; the smallest, about minus 3. Preliminary magnitude determinations may vary with the observatory, equipment, and methods of estimating—the Alaska earthquake of March 1964, for example, was described variously as magnitude 8.4, 8.5, and 8.6 by different stations.

Magnitude also provides an indication of earthquake energy release, which intensity does not. In terms of ergs, (in the centimeter-gram-second system, an erg is the unit of work equal to a force of 1 dyne acting through a distance of 1 cm (0.39 in); a dyne is the force required to accelerate a freestanding gram mass 1 cm/second) a magnitude 1 earthquake releases about one billionth the energy of a magnitude 7 earthquake; a magnitude 5, about one thousandth that of a magnitude 7, etc.

• UNDERSTANDING EARTHQUAKES & WHY THEY ARE ALWAYS A SURPRISE

Most natural hazards can be detected before they strike. However *seisms* (from the Greek *seismos*, earthquake) have no known precursors, and so they come without warning. For this reason, they continue to kill, in some areas, at a level usually reserved for wars and epidemics.

Natural hazards worldwide, such as those caused by storms, earthquakes, volcanoes, and tsunamis cause $1 billion in damages daily. The years 1995–1998 have seen more destructive tsunamis than any other time period since the beginning of the twentieth century. Recent tsunamis alone have caused approximately $2 billion in damage, over 1,820 deaths, 1,500 serious injuries (not including the New Guinea Tsunami), and left more than 135,000 homeless. In addition, there have been at least 540,000 fatalities in 209 tsunamis from the year 684 to 1998.

It is estimated that there are 500,000 detectable earthquakes in the world each year—100,000 of those can be felt and 100 of them cause damage. Worldwide, 1,741,127 people have been killed in earthquakes during the twentieth century.

• WHERE EARTHQUAKES OCCUR

Our planet's most active earthquake-producing feature is the circum-Pacific seismic belt, which trends along the major geologic faults and the deep oceanic trenches of island arcs decorated here and there with the volcanic "Ring of Fire." The mid-Atlantic Ridge, with its fish-skeleton figure of transverse cracks, is also quite active. Other major seismic belts branch from the circum-Pacific system and arc across southeastern and southern Asia into southern Europe, through the Indian Ocean up through the eastern Mediterranean, and up through southern Asia into China.

In an average year, these belts will generate several million tremors, ranging in severity from barely detectable wiggles to great earthquakes of the size that ravaged San Francisco in 1906 and tilted a third of Alaska in 1964. There is always an earthquake in progress somewhere.

• ACTIVE FAULTS OF CALIFORNIA

The most earthquake-prone areas in the contiguous United States are those that are adjacent to the San Andreas fault system of coastal California and the fault system that separates the Sierra Nevada from the Great Basin. Many of the individual faults of these major systems are known to have been active during the last 150–200 years, and others are believed to have been active since the wane of the last great ice advance about 10,000 years ago. Parts of these earthquake-prone areas are among the most densely populated and rapidly urbanizing sections of the western states. A knowledge of the location of these active faults and an understanding of the nature of the earthquake activity that is related to them is necessary for people to accommodate themselves and their work to these hazards.

Earthquakes in California are relatively shallow and clearly related to movement along active faults. During historical times, at least 25 California earthquakes have been associated with movements that ruptured Earth's surface along these faults. On the San Andreas fault, eight moderate-to-severe earthquakes have been accompanied by movements on the fault at the earth's surface since 1838, and other faults in the California region have also experienced repeated earthquakes. The magnitude of shallow earthquakes can generally be correlated with the amount and length of the associated fault movement. Thus, the largest episode of fault movement (or fault slip) recorded in California accompanied the three great earthquakes of 1857, 1872, and 1906—all of which had estimated magnitudes that were over 8 on the Richter scale.

Many of the California faults have had one or more episodes of sudden slip or of slow movement, called creep, during historical time or a documented history of shallow earthquakes. For other faults, however, recent activity can only be inferred from geologic and topographic relations, which indicate that they have been active during the past several thousand years. Such activity suggests that some of these faults will, and that any of them might, slip or creep again.

In parts of California where relatively little geologic work has been done, evidence of other recently active faults will undoubtedly be found as research progresses. This is particularly true of large areas in northern California where topographic features by which recent fault movements can be recognized are commonly obscured by dense vegetation and rapid erosion. Further study may also reveal that some of the unknown faults have been recently active, and that some parts of faults thought to be active are actually dead.

Most of the active California faults are vertical or nearly vertical breaks, and movement along these breaks has been predominantly horizontal. If the block on the opposite side of the fault from the observer has moved to the right, the movement is termed right-lateral; movement of the opposite block to the left is termed left-lateral. Most of the faults trend northwesterly, and movement on these faults has been right-lateral. Notable exceptions to the predominantly northwesterly trend of faults are the west-trending Garlock and Big Pine faults; movement on these faults has been left-lateral.

A few reverse faults have also been active in California. The planes of such faults are inclined to the earth's surface, and the rocks above the fault have been thrust upward over the rocks below the fault plane. The magnitude 7.7 Arvin-Tehachapi earthquake of 1952 was associated with such movement along the White Horse reverse fault, and the magnitude 6.6 San Fernando earthquake of 1971 was caused by a sudden rupture along a reverse fault at the foot of the San Gabriel Mountains.

Studies of historical fault movement have shown that they occur in two ways. The first, and better known, is the sudden displacement, or slip, of the ground along a fault. Such displacement is accompanied by earthquakes and occasionally produces spectacular offsets of topographic and even of human-made features. During the 1906 earthquake, the ground was displaced as much as 21 ft along the San Andreas fault in northern California. During the 1857 earthquake, displacement of the ground along this fault was possibly as much as 30 ft in southern California. The second type of fault movement, termed creep, is now taking place on portions of several faults in California. This type of movement was well documented for the first time in 1956, and has since been found to be commonplace. It is characterized by continuous or intermittent slight slip without noticeable earthquakes. Recent fault creep on portions of the Hayward, Calaveras, and San Andreas faults has produced cumulative offsets ranging from a fraction of an inch to almost a foot in curbs, streets, and railroad tracks, and has caused some damage to buildings.

Most of the faults are, in reality, zones made up of a number of subsidiary faults or fault strands. These fault zones range in width from several feet to a mile or more. Slip along them during historical time and the recent geologic past has been found to recur repeatedly on only one or a few of the multiple strands that constitute these zones. Most of the strands commonly show no evidence of recent activity, although slip does at times recur on older strands or on entirely new ones. The strong tendency for fault slip and earthquakes to recur along the most recently active strands makes knowledge of the precise location of these strands essential to land-use planning.

The source of the stresses that cause the Earth's crust to break and slip in the California region is unknown, but the stresses appear to be related to crustal distortion on a global scale. Geologists have found abundant evidence that these stresses have been acting for millions of years. Whatever their source, the result is a continuing history of surface displacements and earthquakes along numerous faults in the California region.

The San Andreas fault

The most important of California's faults is the San Andreas, which is the "master fault" of the intricate network of faults that cuts through rocks of the coastal region of California. It is a fracture in the Earth's crust along which two parts of the crust have slipped with respect to each other.

The presence of the San Andreas fault was dramatically brought to the attention of the world on April 18, 1906, when displacement along the fault resulted in the great San Francisco earthquake and fire. This, however, was but one of many, many earthquakes that have resulted from displacement along the fault throughout its life of possibly 100 million years.

The fault is a huge fracture some 600 mi or more long, extending almost vertically into the earth to a depth of at least 20 mi. In detail, it is a complex zone of crushed and broken rock from a few hundred feet to a mile wide. Many smaller faults branch from and join the San Andreas fault zone, and if almost any road cut in the zone is examined, one will find a myriad of small fractures, fault gouge (pulverized rock), and a few solid pieces of rock.

Where is the San Andreas fault?

The San Andreas fault forms a continuous break from northern California southward to Cajon Pass. From Cajon Pass southeastward the identity of the fault becomes confused, because several branching faults such as the San Jacinto, Mission Creek, and Banning faults have similar characteristics. Nevertheless, the San Andreas type of faulting continues unabated southward to, and under, the Gulf of California.

Over much of its length, a linear trough reveals the presence of the fault; and from an airplane the linear arrangement of the lakes, bays, and valleys appears striking. Undoubtedly, however, many people driving near Crystal Springs Reservoir, along Tomales Bay, through Cajon or Tejon Passes, do not realize they are on the San Andreas fault zone. On the ground, the fault zone can be recognized by long straight escarpments, narrow ridges, and small-undrained ponds formed by the settling of small blocks within the fault zone. Characteristically, stream channels jog sharply along the fault trace.

Fault movement

Essentially, blocks on opposite sides of the San Andreas fault move horizontally, and if one were to stand on one side of the fault and look across it, the block on the opposite side would appear to be moved to the right. Geologists refer to this as a right-lateral strike-slip fault, or wrench fault.

During the 1906 earthquake, roads, fences, and rows of trees and bushes that crossed the fault were offset several

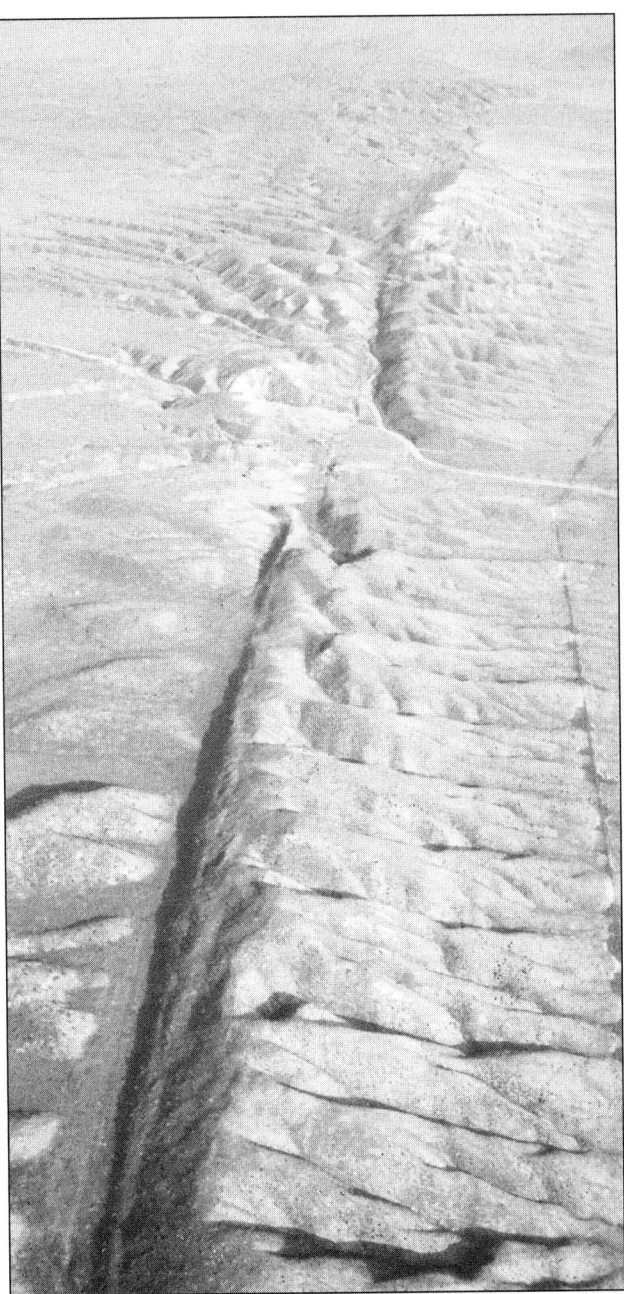

The San Andreas fault. *(Photo by Robert E. Wallace. Courtesy of U.S. Geological Survey (USGS).)*

feet, and the road across the head of Tomales Bay was offset 21 ft, the maximum offset recorded. In each case the ground west of the fault moved relatively northward.

Geologists who have studied in detail the fault between Los Angeles and San Francisco have suggested that the total accumulated displacement along the fault may be as much as 350 mi. Similarly, geologic study of a segment of the fault between Tejon Pass and the Salton Sea revealed geologically similar terrains on opposite sides of the fault now separated by 150 mi, indicating that the separation is a result of movement along the San Andreas and branching San Gabriel faults.

It is difficult to imagine this great amount of shifting of Earth's crust; yet the rate represented by these ancient offsets seems consistent with the rate measured in historical time. Precise surveying shows a slow drift at the rate of about 2 in per year. At that rate, if the fault has been uniformly active during its possible 100 million years of existence, over 300 mi of offset is indeed a possibility.

Since 1934, earthquake activity along the San Andreas fault system has been concentrated in the areas of three cities: Eureka, San Francisco, and Los Angeles/San Bernadino. These are areas where historical earthquakes and fault displacements of the Earth's surface have been most common and where fault creep is taking place today. The sections of the state intervening the three areas mentioned above, on the other hand, have had almost no earthquakes or known slip events since the great earthquakes of 1857 in the southernmost segment and 1906 in the segment between Eureka and San Francisco. This implies to some earth scientists that these two segments of the San Andreas fault are temporarily locked, whereas in the other areas stress is being continually relieved by slip, which produces small-to-moderate earthquakes, and by creep. The lack of such activity in the locked segments could mean that these segments are subject to less frequent but larger fault movements and correspondingly more severe earthquakes.

The recorded history of earthquakes along the San Andreas fault is an extremely small sample from which, however, a clear pattern of behavior can be determined. Judging from this short history, great earthquakes seem to occur only a few times a century, but smaller earthquakes recorded only on sensitive seismographs occur much more frequently.

It is a popular misconception that once there has been a small earthquake along a segment of the fault, strain is released and further earthquakes are not to be expected for many years. Seismologists have pointed out, however, that the really great earthquakes have been preceded by numerous strong shocks and that large earthquakes seem to cluster in 10–20 year periods. Furthermore, the energy released during small earthquakes is insignificant compared to that in earthquakes having the same magnitude as the one in 1906.

Different segments of the fault also behave differently. For example, in the vicinity of Hollister, frequent small shocks are recorded, and slow movement at the rate of 0.47 in per year has been recorded. In contrast, the segment near San Francisco, except for an earthquake of magnitude 5.3 in 1957, has been relatively quiet since 1906. Perhaps, as some believe, it is gradually bending or accumulating strain that will be adjusted all at once in one large "snap."

What can be done about the fault?

Much is yet to be learned about the nature and behavior of the San Andreas fault and the earthquakes it generates. Some questions geologists would like to answer are: How old is the fault? Has movement been uniform? What movement has there been on branching faults? What is the fundamental cause of the stresses that produced the San Andreas fault? Until these questions and others have been satisfactorily answered, the question "what can be done about the fault?" is best responded to, according to the U. S. Geological Survey (USGS), in this way: "Though man cannot stop earthquakes from happening, he can learn to live with the problems they cause. Of prime importance are adequate building codes, for experience shows that well-constructed buildings greatly lessen the hazards. In construction projects, greater consideration should be given to foundation conditions. Degree of damage will range widely, between construction on bedrock, water-saturated mud, filled ground, or landslide terrain. For example, in 1906, most buildings on filled or 'made' land near the foot of Market Street in San Francisco suffered particularly intense damage, whereas buildings on solid rock suffered little or no damage. Geologists are horrified to see land developers build rows of houses straddling the trace of the 1906 break."

Maps showing the most recently active strands or breaks along the San Andreas and related active faults are being prepared by the USGS. This governmental agency can be contacted at 804 National Center, Reston, Virginia 20192. The USGS also maintains Public Inquiries Offices in San Francisco and Los Angeles.

• RECENT QUAKE EVENTS

Bay Area's 1989 Earthquake Teaches Basic Lessons

There is no substitute for experience in understanding an earthquake. Most of us will never have a significant earthquake experience, but millions had the next thing to it when television was uniquely deployed in San Francisco (for other reasons) and shifted its focus to tell us first-hand how the violent shaking there felt and looked.

At 5:04 P.M., on a quiet, autumnal afternoon, October 17, 1989, the San Andreas fault upset life beyond description in the San Francisco-Oakland Bay area. It heaved its giant breast in the Santa Cruz Mountains and wrought havoc in widening circles that reached throughout the Bay Area and shook buildings as far as Reno, Nevada, 250 mi to the east, and rattled skyscrapers 400 mi south in Los Angeles. The internationally televised third game of baseball's World Series, about to begin, gave the world the word and picture as it was happening, beginning with views of apprehensive players and fans inside Candlestick Park stadium, and combining them with telephoto visuals of the fires and devastation some 8–10 mi to the north. However, virtually no one in the Bay Area needed the ABC television crew to tell them that the "big one" was happening. Fanning out in every direction from the epicenter, shock waves that reached 6.9 on the Richter scale sundered the quiet afternoon of uncounted

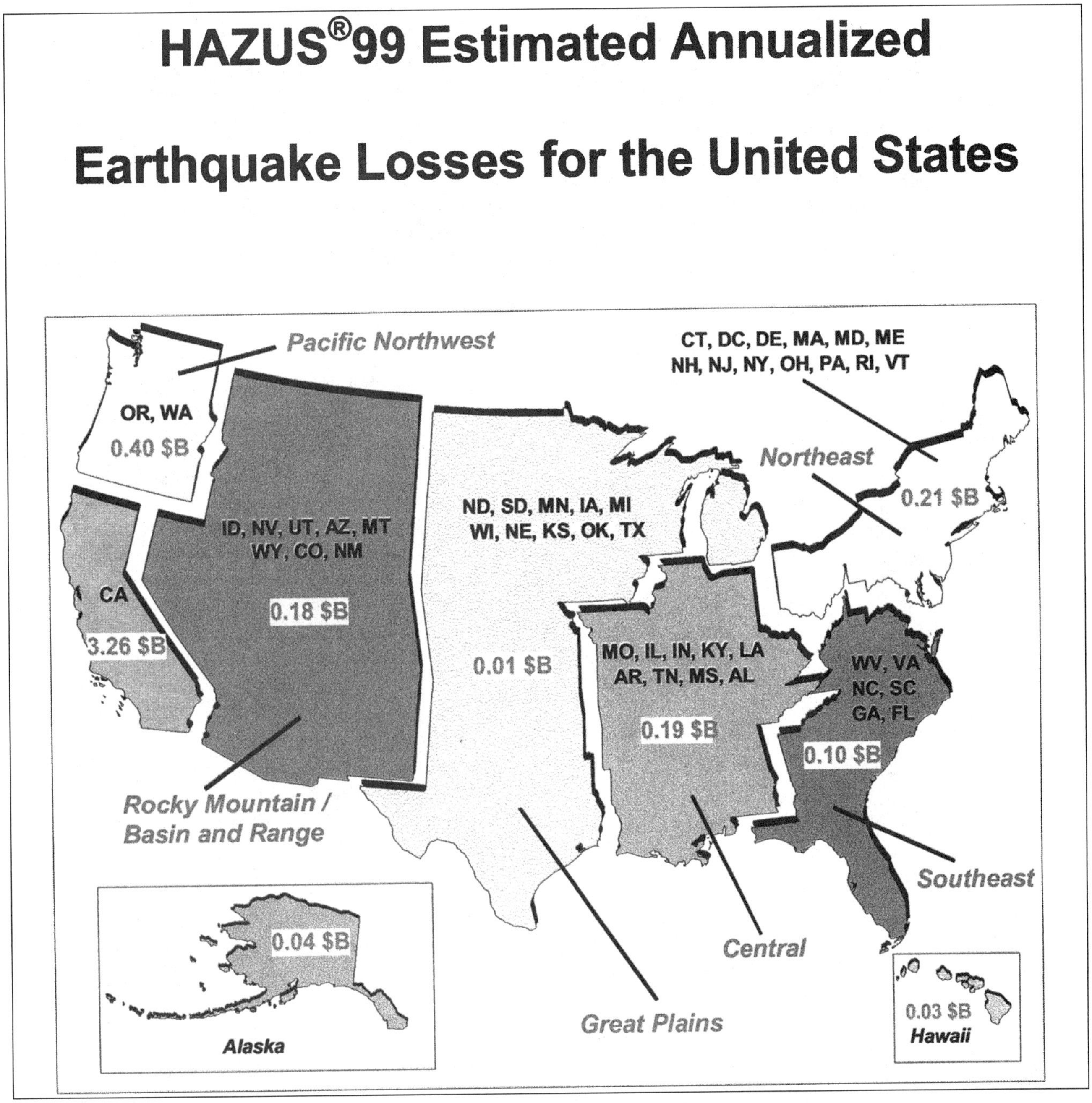

HAZUS®99 Estimated Annualized
Earthquake Losses for the United States

Pacific Northwest

CT, DC, DE, MA, MD, ME
NH, NJ, NY, OH, PA, RI, VT

OR, WA
0.40 $B

Northeast

0.21 $B

ID, NV, UT, AZ, MT
WY, CO, NM

ND, SD, MN, IA, MI
WI, NE, KS, OK, TX

0.18 $B

CA

3.26 $B

0.01 $B

MO, IL, IN, KY, LA
AR, TN, MS, AL

WV, VA
NC, SC
GA, FL

0.19 $B

0.10 $B

Rocky Mountain /
Basin and Range

Southeast

0.04 $B

Central

0.03 $B
Hawaii

Alaska

Great Plains

(Courtesy of Federal Emergency Management Agency (FEMA).)

thousands of people in the area, bringing fear, destruction, and, in the next minutes, death as only an earthquake can. There were more than 100 fatalities and 3,000 injuries.

As earthquakes go, the intensity, maintained for a short 15 seconds, was great. But it was nowhere near the 9.2 reading of the 1964 Alaskan quake that destroyed with tidal waves as much as with earth shaking. (Each whole number on the Richter scale equals ten times greater intensity than the previous whole number.) The toll in life and property of this 1989 cataclysm (the third most lethal of all time) resulted because the epicenter was so close to very large con-

centrations of population and technology-laced living styles. In the cities of San Francisco, Oakland, and other edge communities, the lives of millions, densely packed into a few square miles, involves structures of all types and descriptions. It also involves the steel and concrete double-deck freeway connecting Oakland and the mainland to San Francisco's peninsula.

Structures that survived, and some that did not

The safest structures proved to be high-rise office buildings in San Francisco, constructed since a 1971 tremor had spurred new standards; most vulnerable were the restored

Earthquake Safety Rules

An earthquake strikes your area and for a minute or two the "solid" earth moves like the deck of a ship. What you do during and immediately after the tremor may make life-and-death differences for you, your family, and your neighbors. These rules will help you survive.

Before an Earthquake

At home, bolt down water heaters and gas appliances. Place large, heavy objects and fragile items on securely fastened, lower shelves; brace or anchor heavy objects.

Keep a flashlight and battery-powered transistor radio in the home, ready for use.

During an Earthquake

1. Remain calm. Think through the consequences of any action you take. Try to calm and reassure others; prepare them for the certainty of aftershocks.

2. If indoors, watch for falling plaster, bricks, light fixtures, and other objects. Watch for high bookcases, china cabinets, shelves, and other furniture that might slide or topple. Stay away from windows, mirrors, and chimneys. If in danger, get under a table, desk, or bed; in a corner away from windows; or in a strong doorway. Encourage others to follow your example. Usually it is best not to run outside.

3. If in a high-rise building, get under a desk. Do not dash for exits, since stairways may be broken and jammed with people. Power for elevators may fail.

4. If in a crowded store, do not rush for a doorway since hundreds may have the same idea. If you must leave the building, choose your exit as carefully as possible.

5. If outside, avoid high buildings, walls, power poles, and other objects which could fall. Do not run through streets. If possible, move to an open area away from all hazards. If in an automobile, stop in the safest place available, preferably an open area.

After an Earthquake

1. Check for injuries in your family and neighborhood. Do not attempt to move seriously injured persons unless they are in immediate danger of further injury.

2. Check for fires or fire hazards.

3. Wear shoes in all areas near debris or broken glass.

4. Check utility lines and appliances for damage. If gas leaks exist, shut off the main gas valve. Shut off electrical power if there is damage to your house wiring. Report damage to the appropriate utility companies and follow their instructions. Do not use matches, lighters, or open-flame appliances until you are sure no gas leaks exist. Do not operate electrical switches or appliances if gas leaks are suspected. This creates sparks that can ignite gas from broken lines.

5. Do not touch downed power lines or objects touched by the downed wires.

6. Immediately clean up spilled medicines, drugs, and other potentially harmful materials.

7. If water is off, emergency water may be obtained from water heaters, toilet tanks, melted ice cubes, and canned vegetables.

8. Check to see that sewage lines are intact before permitting continued flushing of toilets.

9. Do not eat or drink anything from open containers near shattered glass. Liquids may be strained through a clean handkerchief or cloth if danger of glass contamination exists.

10. If power is off, check your freezer and plan meals to use foods that will spoil quickly.

11. Use outdoor charcoal broilers for emergency cooking.

12. Do not use your telephone except for genuine emergency calls. Turn on your radio for damage reports and information.

13. Check your chimney over its entire length for cracks and damage, particularly in the attic and at the roofline. Unnoticed damage could lead to a fire. The initial check should be made from a distance. Approach chimneys with caution.

14. Check closets and storage shelf areas. Open closets and cupboard doors carefully and watch out for objects falling from shelves.

15. Do not spread rumors. They often do great harm after disasters.

16. Do not go sightseeing immediately, particularly in beach and waterfront areas where seismic sea waves could strike. Keep the streets clear for passage of emergency vehicles.

17. Be prepared for additional earthquake shocks called "aftershocks." Although most of these are smaller than the main shock, some may be large enough to cause additional damage.

18. Respond to requests for help from police, fire fighting, civil defense, and relief organizations, but do not go into damaged areas unless your help has been requested. Cooperate fully with public-safety officials. In some areas, you may be arrested for getting in the way of disaster operations.

There are no rules that can eliminate all earthquake danger. However, damage and injury can be greatly reduced by following these simple rules.

---

single homes built 60–90 years ago on landfills in an area known as the Marina district of San Francisco. The former coped with the shocks with well-planned engineering provisions, resulting in minimal damage. Meanwhile, more than 50 of the latter (wood and brick structures) collapsed into their foundations, and many lives were lost as scores were trapped inside. The most shocking element of the catastrophe was reserved, however, for the freeway. Due to reasoning that seems strange in retrospect, the lifeline traffic artery known as the Nimitz Freeway (Interstate Route 80) had been constructed years before as a double-decker, with its supports assumed to be—but never tested to be—earthquake resistant. The supports failed this test. Cars and drivers alike were crushed in mid-cruise as the upper deck first undulated with the shock wave then dropped its millions of tons of steel and concrete on the deck below. Drivers were pinned and vehicles crushed as if made of cardboard. The San Francisco-Oakland Bay bridge fared only slightly better, with one end of an upper section falling to meet the lower level, closing it for days.

As terrifying as this quake was, carnage was in one sense light because the early start of the ball game had drained the streets of much rush-hour traffic. That is small consolation to the families of those who died, but a blessing to thousands of others whose route would have placed them directly under the collapsing concrete that repudiated its supports.

Northridge, California, 1994

Early on the morning of January 17, 1994, Martin Luther King Day, Los Angeles area residents were jolted awake by what was to be the most significant urban earthquake to occur in California since 1906. The initial 10 seconds of trembling ground resulted in massive property damage and loss of life.

The powerful quake struck at its epicenter of Northridge at 4:31 A.M. on January 17, reaching a magnitude of 6.8. Residents within an area of approximately 2,192 mi^2 experienced over 1,000 aftershocks of a magnitude of 1.5 for weeks after the quake.

Nearly 100 deaths and 9,000 injuries resulted from the incident. Over 50,000 people were displaced from their homes; thousands of individuals were forced to move to temporary shelters in schools and churches, or camp in city parks and endure the chilly, less-than-favorable temperatures and rain.

Nearly 114,039 residential and commercial structures were damaged in the area despite the fact that seismic building code provisions and other mitigations had been intensified, especially since the Loma Prieta earthquake in 1989. Total damage was estimated at $17 billion. This most recent seismic event brought to light the difficulties in protecting Californians from their seismically unsafe environment.

The 1999 Colombia Earthquake

On January 25, 1999, at 1:19 P.M., an earthquake shook the Armenia-Calarca-Pereira area of Colombia, causing extensive damage and killing over 2,000. It was followed by a magnitude 4.5 aftershock at 5:40 P.M. that day.

This region has had many earthquakes above magnitude 5.5 since 1973, most recently a magnitude 6.8 event on June 6, 1994, which killed at least 295 and caused extensive damage in Cauca, Tolima, and Valle Departments.

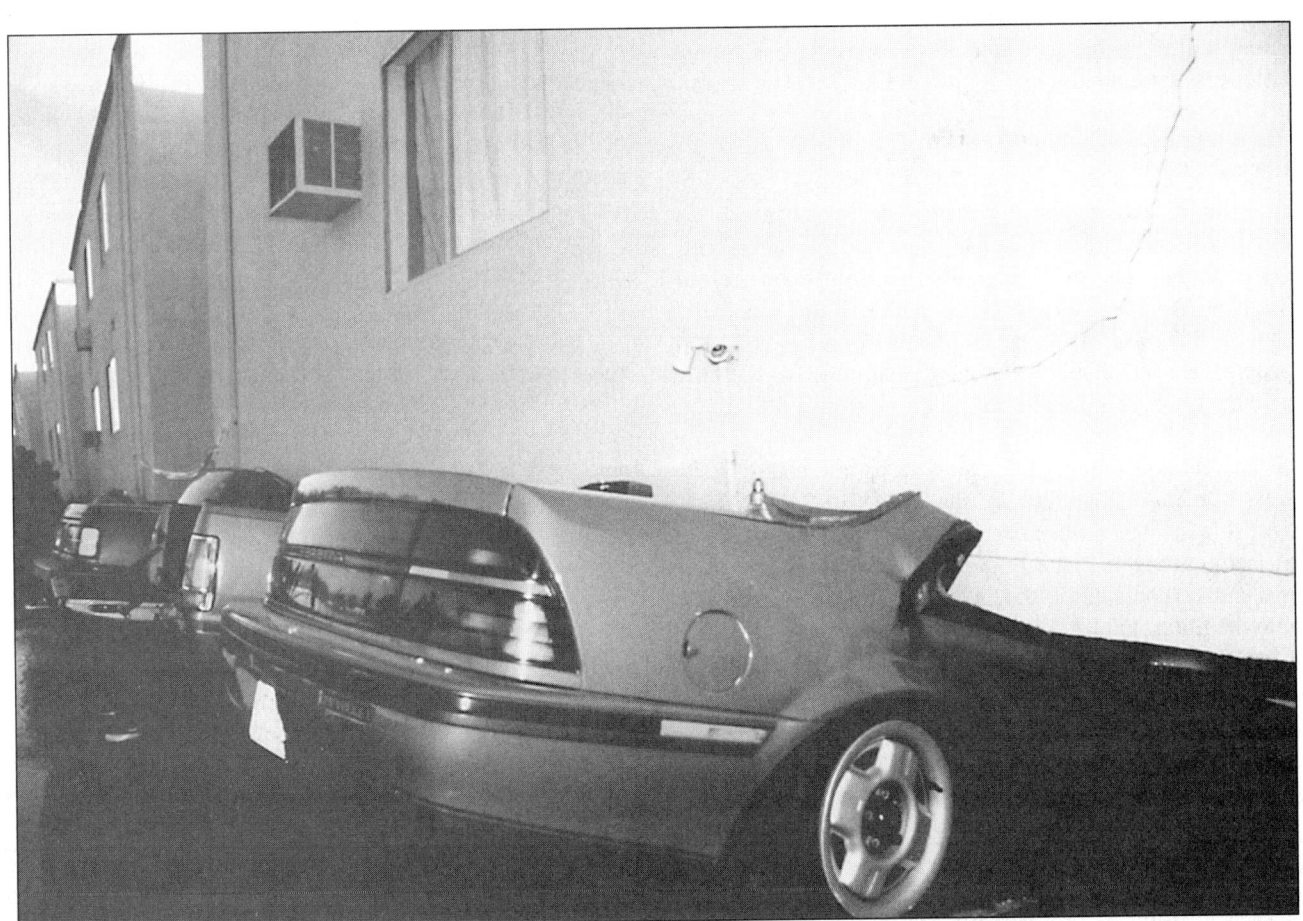

The intensity of the Northridge quake in 1994 caused the first floor of this building to collapse. *(Photo by Robert A. Eplett. Courtesy of Governor's Office of Emergency Services.)*

The January 1999 earthquake caused extraordinary damage for a very moderate earthquake, perhaps due to a phenomenon known as soil amplification, in which thick alluvial layers resonate and amplify seismic energy over what it would be in hard rock.

The quake occurred in the center mountain range of Cordillera Central, of three north-trending mountain ranges in western Colombia. This mountain range has active volcanoes and the earthquake occurred near the Ruiz-Tolima volcanic complex. The volcanoes exist here because the Nazca tectonic plate is subducting beneath South America; most earthquakes in this zone are in the subducting plate and thus at least 62 mi beneath the surface. This earthquake, however, had a source depth of about 11 mi, and was caused by some near-surface tectonic adjustment. This earthquake was a strike slip earthquake; that is, there was essentially no vertical motion in its faulting. Its faulting is similar to what typically occurs on the San Andreas fault.

January 13, 2001, El Salvador Earthquake

A major earthquake, 7.6 magnitude, struck about 105 miles south-southwest of San Miguel, El Salvador, in the Pacific Ocean. The death toll in Central America, as of early March 2001, is approximately 1,300 people, with 8,000 injured. An aftershock occurred in the same area a month later, killing hundreds.

January 26, 2001, southern India Earthquake

A major earthquake, 7.9 magnitude, struck about 65 miles north-northeast of Jamnagar, India. Buildings collapsed in the state of Gujarat. The earthquake was felt at Mumbai (Bombay) and Delhi, as well as Karachi and Peshawar, Pakistan, and in parts of Nepal. As many as 20,000 people were killed by this earthquake. On June 16, 1819, an earthquake in this same general area killed 1,500–2,000 people.

The earthquake occurred along an approximately east-west trending thrust fault at shallow (<15.5 mi) depth. Thrust faults occur when one portion of the Earth's crust is pushed up over an adjacent portion. The strain that caused this earthquake is due to the Indian plate pushing northward into the Eurasian plate.

February 28, 2001, western Washington State Earthquake

A major earthquake of 6.8 magnitude occurred in the Seattle-Tacoma area of Washington state on February 28, 2001. At least several hundred people were injured, but no deaths were directly connected to the earthquake. Damage is estimated at about $2 billion.

The location of this western Washington state earthquake is very near the locations of a 1949 magnitude 7.1 earthquake and a 1965 magnitude 6.5 earthquake. These events occurred on a normal fault within the Juan de Fuca plate where it subducts (goes under) the North America plate. The 2001 earthquake is called the "Nisqually earthquake," because of the proximity of the earthquake to the Nisqually River delta in Puget Sound. The name Nisqually is taken from a group of Native Americans who live in the area.

November 3, 2002 (UTC), Denali Fault, Alaska

This 7.9 magnitude shock is the largest earthquake on the Denali fault since at least 1912, when an 7.2 magnitude earthquake occurred in the general vicinity of the fault, 50 miles east of this latest epicenter. This 7.9 magnitude shock, one of the largest ever recorded on U.S. soil, occurred on the Denali-Totschunda fault system, which is one of the longest strike-slip fault systems in the world and rivals in size California's famed San Andreas strike-slip fault system that spawned the destructive San Francisco earthquake in 1906.

Earthquake in west central Mexico, January 21, 2003.

A major earthquake occurred in Colima, Mexico, about 30 miles southeast of Manzanillo, Colima, or about 310 miles west of Mexico City at 7:06 P.M. MST, January 21, 2003 (8:06 P.M. CST in Mexico). A preliminary magnitude of 7.8 was computed for this earthquake. The magnitude and location may be revised when additional data and further analysis results are available. There were at least 28 deaths, 300 injured, and considerable damage in the states of Colima, Michoacan, and Jalisco. The earthquake was felt strongly in Mexico City.

This shallow earthquake occurred in a seismically active zone near the coast of central Mexico. The earthquake occurred near the juncture of three tectonic plates: the North American Plate to the northeast, the Rivera Plate to the northwest, and the Cocos Plate to the south. Both the Rivera Plate and the Cocos Plate are being consumed beneath the North American Plate. The slower moving Rivera Plate is moving northwest at about 2 cm per year relative to the North American Plate, and the faster moving Cocos plate is moving in a similar direction at a rate of about 4.5 cm per year.

There have been several significant earthquakes near the recent event. In 1932, a magnitude 8.4 thrust earthquake struck about 100 km to the north-northwest. More recently, on October 9, 1995, a magnitude 8.0 earthquake struck about 50 km to the northwest killing at least 49 people and leaving 1,000 homeless. The most deadly earthquake in the region occurred about 170 km to the southeast on September 19, 1985. This magnitude 8.0 earthquake killed at least 9,500 people, injured about 30,000, and left 100,000 people homeless.

Largest Quakes in the United States (with Magnitudes and Dates)

Prince William Sound, AK (9.2; March 28, 1964)
Andreanof Islands, AK (8.8; March 9, 1957)
Rat Islands, AK (8.7; February 4, 1965)
east of Shumagin Islands, AK (8.3; November 10, 1938)
Lituya Bay, AK (8.3; July 10, 1958)
Yakutat Bay, AK (8.2; September 10, 1899)
Cape Yakataga, AK (8.2; September 4, 1899)
Andreanof Islands, AK (8.0; May 7, 1986)
New Madrid, MO (7.9; February 7, 1812)
Fort Tejon, CA (7.9; January 9, 1857)
Ka'u District, HI (7.9; April 3, 1868)
Kodiak Island, AK (7.9; October 9, 1900)
Gulf of Alaska (7.9; November 30, 1987)
Denali Fault, Alaska (7.9; November 3, 2002)
Owens Valley, CA (7.8; March 26, 1872)

Largest Quakes in the Contiguous United States

New Madrid, MO (7.9; February 7, 1812)
Fort Tejon, CA (7.9; January 9, 1857)
Owens Valley, CA (7.8; March 26, 1872)
Imperial Valley, CA (7.8; February 24, 1892)
New Madrid, MO area (7.7; December 16, 1811)
San Francisco, CA (7.7; April 18, 1906)
Pleasant Valley, NV (7.7; October 3, 1915)
New Madrid, MO (7.6; January 23, 1812)
Landers, CA (7.6; June 28, 1992)
Kern County, CA (7.5; July 21, 1952)
west of Lompoc, CA (7.3; November 4, 1927)
Dixie Valley, NV (7.3; December 16, 1954)
Hebgen Lake, MT (7.3; August 18, 1959)
Borah Peak, ID (7.3; October 28, 1983)

Widely differing magnitudes have been computed for some of these earthquakes; the values differ according to the methods and data used. For example, some sources list the magnitude of the 8.7 Rat Islands earthquake as low as 7.7. On the other hand, some sources list the magnitude of the February 7, 1812, New Madrid quake as high as 8.8. Similar variations exist for most events on this list, although generally not so large as for the examples given.

In general, the magnitudes given in the list above have been determined from the seismic moment, when available. For very large quakes, the moment magnitude is considered to be a more accurate determination than the traditional amplitude magnitude computation procedures. Note that all of these values can be called "magnitudes on the Richter scale," regardless of the method used to compute them.

Source: Stover, C.W., and J.L. Coffman. *Seismicity of the United States, 1968–1989.* Revised. U.S. Geological Survey Prof. Paper 1527, 1993.

• TSUNAMIS

What is a tsunami?

The phenomenon we call "tsunami" is a series of traveling ocean waves of great length and long period, generated by disturbances associated with earthquakes in oceanic and coastal regions. As the tsunami crosses the deep ocean, its length from crest to crest may be 100 mi or more, its height from trough to crest only a few feet. It cannot be felt aboard ships in deep water, and cannot be seen from the air. But in deep water, tsunami waves may reach forward speeds exceeding 600 mph.

As the tsunami enters the shoaling water of coastlines in its path, the velocity of its waves diminishes and wave height increases. It is in these shallow waters that tsunamis become a threat to life and property, for they can crest to heights of more than 100 ft, and strike with devastating force.

The tsunami of the century: Papua New Guinea, 1998

On the evening of Friday, July 17, at 7:30 P.M., a massive tsunami swept across the sandbar that forms the outer margin of Sissano Lagoon, West Sepik, Papua New Guinea, striking four villages west of the town of Aitape. The wave was reported to be 22.8–33 ft high; up to 3,000 persons were reported killed or missing. This was an unusually damaging tsunami, given the size of the earthquake (a magnitude 7) associated with it.

As of late 1998, scientists were continuing to examine this event, in an attempt to explain the unusually high run-ups, with the ultimate hope of mitigating such disasters in the future.

Scientists from the USGS participated in the second International Tsunami Survey Team to study the sedimentary deposits left by this tsunami. Animations of the tsunami have also been developed to graphically display how the tsunami evolved from an earthquake source.

Since 1992, the international community has responded to nine major tsunami disasters (Nicaragua, 1992; Flores, 1992; Okushiri, 1993; East Java, 1994; Mindoro, 1994; Kuril Islands, Russia, 1994; Manzanillo, 1995; Irian Jaya, Indonesia, 1996; and Peru, 1996) by dispatching this team of scientists, which has come to be known as the International Tsunami Survey Team (ITST), with more than 30 scientists and 20 students from Indonesia, Korea, Japan, Mexico, Peru, Russia, the United Kingdom, and the United States. The Papua New Guinea survey team was joined by scientists from Australia and New Zealand.

The tsunami warning system

Development of the National Oceanic and Atmospheric Administration (NOAA) Coast and Geodetic Survey's Pacific Tsunami Warning System was impelled by the disastrous waves of April 1946, which surprised Hawaii and took a heavy toll in life and property. The locally disastrous tsunami caused by the March 1964 Alaska earthquake impelled the development of another type of warning apparatus—the Regional Tsunami Warning System in Alaska.

The Regional Tsunami Warning System is headquartered at the Coast and Geodetic Survey's Seismological Observatory at Palmer, Alaska. This is the nerve center for an elaborate telemetry network linking Palmer with remote seismic and tidal stations along the Alaska coast and in the Aleutian Islands. Seismograph stations in the network are at Palmer Observatory and its two remote stations 25 mi south and west, and at Biorka, Sitka, Gilmore Creek, Kodiak, and Adak. Tide stations are at Seward, Sitka, Kodiak, Cold Bay, Unalaska, Adak, Yakutat, and Shemya. Data from these stations are recorded continuously at Palmer, where a 24-hour watch is kept.

When an earthquake occurs in the Alaska-Aleutian area, seismologists at Palmer Observatory rapidly determine its epicenter (the point on the Earth's surface above the underground source of the earthquake) and magnitude. If the epicenter falls in the Aleutian Island arc or near the Alaskan coastal area, and if the earthquake magnitude is great enough to generate a tsunami, Palmer Observatory issues a *tsunami warning* through the Alaska Disaster Office, Alaska Command, and Federal Aviation Administration (FAA) covering the area near the epicenter. A *tsunami watch* is issued for the rest of the Alaskan coastline, alerting the public to the possibility of a tsunami threat. If tide stations detect a tsunami, Palmer Observatory extends the tsunami warning to cover the entire coastline of Alaska. If no tsunami is observed, both the watch and warning bulletins are canceled.

Subsidiary warning centers have been established at Sitka and Adak Observatories. These facilities operate small seismic arrays and have a limited warning responsibility for local areas.

The Pacific Tsunami Warning System has its headquarters at the Coast and Geodetic Survey's Honolulu Observatory. There, seismologists monitor data received from seismic and tidal instruments in Hawaii and around the Pacific Ocean, and provide ocean-wide tsunami watches and warnings. The Pacific system works very closely with its regional counterpart in Alaska. Potentially tsunami-generating earthquakes in the Alaska-Aleutian area are detected and evaluated at Palmer Observatory, and the data relayed directly to the Honolulu Observatory. Where there is tidal evidence of a tsunami, the warning is extended by Honolulu to cover the Pacific Ocean basin. For tsunamis generated elsewhere in the Pacific area, tsunami watch and warning bulletins are prepared at the Honolulu Observatory and disseminated in Alaska by the Alaska Disaster Office, the military, and Federal Aviation Administration (FAA).

Tsunami Safety Rules

Tsunamis are generated by some earthquakes. When you hear a tsunami warning, you must assume a dangerous wave is on its way. History shows that when the great waves finally strike, they claim those who have ignored the warning.

REMEMBER:

1. Not all earthquakes cause tsunamis, but many do. When you hear that an earthquake has occurred, stand by for a tsunami emergency.

2. A strong earthquake felt in a low-lying coastal area is a natural warning of possible, immediate danger. Keep calm and move to higher ground, away from the coast.

3. A tsunami is not a single wave, but a series of waves. Stay out of danger areas until an "all-clear" is issued by competent authority.

4. Approaching tsunamis are sometimes heralded by a noticeable rise or fall of coastal water. This is nature's tsunami warning and should be heeded.

5. A small tsunami at one beach can be a giant a few miles away. Do not let the modest size of one make you lose respect for all.

6. All tsunamis—like hurricanes—are potentially dangerous, even though they may not damage every coastline they strike.

7. Never go down to the beach to watch for a tsunami. When you can see the wave you are too close to escape it.

8. During a tsunami emergency, your local Civil Defense, police, and other emergency organizations will try to save your life. Give them your fullest cooperation.

Stay tuned to your radio or television stations during a tsunami emergency—bulletins issued through Civil Defense and NOAA offices can help save your life.

• FREQUENTLY ASKED QUESTIONS ABOUT EARTHQUAKES

What is the biggest earthquake ever?

Since 1900, the earthquake in Chile on May 22, 1960, is the biggest in the world with magnitude 9.5 Mw.

What is the biggest earthquake in the United States?

Since 1900, the earthquake in Alaska on March 28, 1964, is the biggest earthquake in the United States, with magnitude 9.2 Mw. This earthquake is also the second biggest earthquake in the world.

Which states in the United States have the most earthquakes?

Alaska and California.

Which state has the most damaging earthquakes?

California.

Which states have the smallest number of earthquakes?

Florida and North Dakota.

What region has the fewest earthquakes?

Antarctica has the fewest earthquakes of any continent, but small earthquakes can occur anywhere in the world.

When will California slide into the ocean?

There is no scientific reason that indicates that California will ever fall into the ocean.

What is the difference between magnitude and intensity?

Magnitude measures the energy released at the source of the earthquake. The magnitude of an earthquake is determined from the logarithm of the amplitude of waves recorded on a seismogram at a certain period. *Intensity* measures the strength of shaking produced by the earthquake at a certain location. Intensity is determined from effects on people, human structures, and the natural environment. Intensity does not have a mathematical basis, but is based on observed effects.

Where can I buy a Richter scale?

The Richter scale is not a physical device, but a mathematical formula. The magnitude of an earthquake is determined from the logarithm of the amplitude of waves recorded on a seismogram at a certain period.

What is an aftershock?

Smaller earthquakes following the largest earthquake of a series, concentrated in a restricted crustal column.

How long can an earthquake shake?

Two to three minutes.

What is a fault?

A fracture or zone of fractures in rock along which the two sides have been displaced relative to each other parallel to the fracture.

What is liquefaction of soil?

The process of soil and sand behaving like dense fluid rather than a wet solid during an earthquake.

What is the Moho?

The Moho is the abbreviated form of Mohorovicic (pronounced Mo-ho-ro-vish-ich) discontinuity. This is a boundary surface or the sharp seismic velocity discontinuity that separates the Earth's crust from the underlying mantle. Its depth varies from about 3–6 mi beneath the ocean floor to about 20 mi below the continents. The discontinuity probably represents a change in chemical composition. It is named after its Croatian discoverer Andrija Mohorovicic.

When did the first instrument actually record an earthquake?

Probably the earliest seismoscope was invented by the Chinese philosopher Chang Heng in A.D. 132. This was a large urn on the outside of which were eight dragon heads facing the eight principal directions of the compass. Below each dragon head was a toad with its mouth opened toward the dragon. When an earthquake occurred, one or more of the eight dragon-mouths would release a ball into the open mouth of the toad sitting below. The direction of the shaking determined which of the dragons released its ball. The instrument is reported to have detected an earthquake 400 mi away that was not felt at the location of the seismoscope. The inside of the seismoscope is unknown: most speculations assume that the motion of some kind of pendulum would activate the dragons.

Where do earthquakes occur?

Earthquakes can strike any location at any time. But history shows they occur in the same general patterns year after year, principally in three large zones of the earth.

The world's largest earthquake belt, the circum-Pacific seismic belt, is found along the rim of the Pacific Ocean, where about 81% of the world's largest earthquakes occur. The belt extends from Chile, northward along the South American coast through Central America, Mexico, the West Coast of the United States, and the southern part of Alaska, through the Aleutian Islands to Japan, the Philippine Islands, New Guinea, the islands groups of the southwest Pacific, and to New Zealand. This earthquake belt was responsible for 70,000 deaths in Peru in May 1970, and 65 deaths and one billion dollars of damage in California in February 1971.

Why do so many earthquakes originate in this belt?

This is a region of young, growing mountains and deep ocean trenches that invariably parallel mountain chains. Earthquakes necessarily accompany elevation changes in mountains, the higher part of the Earth's crust, and changes in the ocean trenches, the lower part.

The second important belt, the Alpide, extends from Java to Sumatra through the Himalayas, the Mediterranean, and out into the Atlantic. This belt accounts for about 17% of the world's largest earthquakes, including some of the most destructive, such as the Iran shock that took 11,000 lives in August 1968, and the Turkey tremors in March 1970 and May 1971 that each killed over 1,000. All were near magnitude 7 on the Richter scale.

The third prominent belt follows the submerged mid-Atlantic Ridge. The remaining shocks are scattered in various areas of the world.

Earthquakes in these prominent seismic zones are taken for granted, but damaging shocks occur occasionally outside these areas. Examples in the United States are New Madrid, Missouri, and Charleston, South Carolina. Many years, however, usually elapse between such destructive shocks.

Can earthquakes be predicted?

It is not possible for scientist to predict earthquakes now and it may never be possible. Some people believe that animals and psychics can predict earthquakes, but that has not been proven.

• PROTECTING THE PUBLIC FROM EARTHQUAKE HAZARDS—ADVANCED NATIONAL SEISMIC SYSTEM COMES TO MEMPHIS

October 2002 marked a new milestone in the installation of modern seismic stations in seismically active urban areas across the country. These cities include Memphis, San Francisco, Seattle, Salt Lake City, Anchorage, and Reno. These new instruments are part of a nationwide network of sophisticated ground shaking measurement systems, both on the ground and in buildings, called the Advanced National Seismic System (ANSS). ANSS will become the first line of defense in the war on earthquake hazards—with the ultimate victory being public safety, lives saved, and major losses to the economy avoided.

ANSS stations will assist emergency responders within minutes of an event showing not only the magnitude and epicenter, but where damage is most likely to have occurred.

Ten new ANSS instruments have recently been installed in the Memphis area, 20 have been installed across the mid-America region, and more than 175 have been installed in other vulnerable urban areas to provide real-time information on how the ground responds when a strong earthquake happens.

The ultimate goal of ANSS is to save lives and ensure public safety, said Dr. John Filson, U. S. Geological Survey (USGS) Earthquake Program Coordinator. "This information, already available in Southern California, is generated by data from seismic instruments installed in urban areas and has revolutionized the response time of emergency managers to an earthquake, but its success depends on further deployment of instruments in other vulnerable cities."

In 1997, during the reauthorization of the National Earthquake Hazards Reduction Program, Congress asked for an assessment of the status and needs of earthquake monitoring. The result was the authorization of ANSS to be implemented by the USGS. The system, when implemented, would integrate all regional and national networks with 7,000 new seismic instruments, including 6,000 strong-motion sensors in 26 at-risk urban areas. To date, approximately 350 instruments have been installed.

Earthquakes pose one of the greatest risks for casualties and costly damage in the United States. California's Northridge earthquake in 1994, a magnitude 6.7 quake, took 57 lives when it struck a modern urban environment generally designed for seismic resistance. With losses estimated at $20 billion, this was the most expensive earthquake in U.S. history. During the 1989 World Series, as more than 62,000 fans filled Candlestick Park, a magnitude 7.1 earthquake struck about 60 miles south of San Francisco. The effects of the 20-second quake caused as much as $10 billion in damage. Sixty-two people died.

In March 1964, a magnitude 9.2 earthquake near Anchorage took 125 lives and caused about $311 million in property losses. Thirty blocks of dwellings and commercial buildings were damaged or destroyed in the downtown area of Anchorage. Landslides caused heavy damage, and an area of 130 acres broke the ground into blocks that were collapsed and tilted at all angles.

In 1811 and 1812, the central Mississippi Valley was struck by three of the most powerful earthquakes in U.S. history. Consider what the impact would be if these events happened today in this region that has more earthquakes than any area east of the Rocky Mountains.

The goal of USGS earthquake monitoring is to mitigate risk—using better instruments to understand the damage that shaking causes and to help engineers create stronger and sounder structures that ensure vital infrastructures, and keep utility, water, and communication networks operating safely and efficiently.

The ANSS "strong motion" instruments are critical in giving emergency response personnel real-time maps of severe ground shaking and providing engineers with information about building and site response.

ANSS provides the USGS with the capability to create tools to process earthquake information faster; for example, ShakeMap, a rapidly generated computer map that shows the location, severity, and extent of strong ground shaking within minutes after an earthquake. As it modernizes seismic networks, the USGS hopes to be able to provide the ANSS-generated ShakeMap capability for every seismically active urban area. A possibility USGS scientists have been keenly aware of throughout the development of ANSS is that an early warning of even a few seconds would give children enough time to get under their desks; could stop trains and subways; shut off pipelines; shut down nuclear facilities; and suspend medical procedures. Another new tool is the "Did you feel it?" website (http://pasadena .wr.usgs.gov/shake/). This allows citizens with internet access to record their observations of shaking. The result is a community intensity map (coded by zip code) across the region.

Volcanic Activity

Volcanic activity has played a dominant role in shaping the face of Earth. Much of the natural beauty of the land, its mineral wealth, and the fertility of the soil is owed to volcanism, especially in the western states. At one time or another during the last 70 million years, volcanic rocks covered nearly all of the western states of Washington, Oregon, California, Nevada, Arizona, Utah, Idaho, and large parts of Montana, Wyoming, Colorado, New Mexico, and Texas. Still older volcanic rocks, now largely deformed and metamorphosed, are found in nearly every state.

Since the Mount St. Helens volcano, located in the state of Washington, erupted violently on May 18, 1980, volcanism has become a much-talked-about part of the weather and climate pictures of both the United States and the world. Much of the continuing importance of this spectacular eruption is owed to the fact that it lifted a great volume of ash and debris into the upper atmosphere. All the ash particles, even the finest, fell out within a matter of days, at most a few weeks. What remained dispersed in the stratosphere—for up to a few years—were aerosols of sulfuric acid droplets formed by the volcanic SO_2 gas in the eruption plume reacting with moisture in the air.

Incidentally, Mount St. Helens only minimally affected climate (due to low levels of sulfur in the magma), compared with the 1982 eruption of El Chichón volcano (Mexico) and the 1991 eruption of Pinatubo (Philippines).

Mount St. Helens erupting in 1980. *(Photo from U.S. Geological Survey (USGS) Photographic Library. Reproduced by permission.)*

• WHAT ARE VOLCANOES?

Volcanoes are built by the accumulation of their own erup-tive products: lava, bombs, ash, and dust. Usually the volcano is a conical hill or mountain around a vent (the term "vol-cano" also is applied to the opening or vent) that connects with reservoirs of molten rock (known as magma) below Earth's surface. Forced upward by the pressure of contained gas, the molten rock, which is less dense than the surrounding solid rock, rises buoyantly and breaks through zones of weak-ness in Earth's crust. It spews from the vent as lava flows, or shoots into the air as dense clouds of lava fragments.

Larger fragments (bombs and cinders) fall back around the vent. Some of the finer materials (ash and dust) drift down and are blown by the wind to eventually fall to the ground many miles away.

The gas in lava can be compared to the behavior of the gas in a soda bottle that is shaken and then the top released. The violent separation of gas from lava may produce rock froth known as pumice, which is so light that it floats on wa-ter if the gas bubbles are not connected. If the gas cavities are interconnected, the pumice does not float.

Types of volcanoes

There are four main types of volcanoes—cinder cones, composite cones, shield volcanoes, and lava domes.

1. **Cinder cones** are built of lava fragments. They are very numerous in the western United States including Sun-set Crater, Arizona, and Craters of the Moon National Mon-ument, Idaho.

2. **Composite volcanoes** are built of alternating layers of lava flows, volcanic ash, and ash; they are sometimes called strato-volcanoes. Many of the world's large mountains are composite cones; for example, Mount Fuji in Japan; Mount Shasta in California; and Mount St. Helens and Mount Rainier in Washington State.

Crater Lake in Oregon is an interesting variation of a composite cone. Originally, like Mount Rainier, the volcano lost its summit in a series of tremendous explosions; the re-maining parts of the volcano eventually collapsed to form the depression or caldera, later filled by water, that is now Crater Lake.

3. **Shield volcanoes** are built almost entirely of very fluid lava flows. These flow out in all directions from the central vent in a group of vents. The Hawaiian Islands are clusters

Many small explosions since 1980 have built this lava dome (photographed in 1983) in Mount St. Helens crater. *(Photo courtesy of U.S. Geological Survey (USGS)/Cascades Volcano Observatory.)*

of shield volcanoes. Mauna Loa is the world's largest active volcano and rises 13,653 ft above sea level. In fact, Mauna Loa is the largest mountain (of any kind) in the world if its structure and height above the deep-sea floor, rather than sea level, is taken into account. Shield volcanoes are also very common in the Pacific Northwest (e.g., the "flood basalts" that constructed the Columbia River Plateau).

4. **Lava domes** are built of viscous or pasty lava extruded like toothpaste from a tube. Lassen Peak and Mono Dome in California are examples of lava domes. An excellent and photogenic example of a lava dome is that which formed within the new crater of Mount St. Helens following the May 18, 1980 explosive eruption.

• ANCIENT VOLCANOES

There are several areas in the United States where, for many thousands of years, the only activity has been that of hot springs and solfataras (steam vents). Yellowstone National Park is the most famous of these areas. It is visited each year by thousands of people who come to see the geysers, mud pots, boiling hot springs, steam vents, and beautiful carbonate and silica deposits formed by precipitation from the hot waters. Such phenomena are vestiges of a former period of very active volcanism. It cannot be determined with certainty, however, that someday more violent volcanic activity will not begin again.

The youngest and most dominant volcanic eruptions in Yellowstone Park produced lavas called rhyolite. Chemically, rhyolite is high in silica and is the volcanic equivalent of granite. Among active volcanoes rhyolite is exceedingly rare. In the geologic past, however, rhyolite eruptions were more common and among the most spectacular of natural phenomena. Over 600 mi^3 of rhyolite erupted from the Yellowstone volcanoes alone during their last active periods. These rhyolitic ash flows also form extensive plateaus (e.g., the Yellowstone Plateau). Two-thirds of this amount erupted as ash flows like the "river of sand" that produced the "Valley of Ten Thousand Smokes" in Alaska in 1912. The remaining 200 mi^3 of material issued as great sticky lava flows.

So great was the volume of flows of ash and pumice that valleys were completely filled and the intervening ridges were covered. The ash flows merged to form flat-topped plateaus thousands of square miles in area. When deposited, these great sheets of ash were so hot that the particles fused together to form rocks known as welded tuffs. Recent geologic studies have revealed that welded tuffs are perhaps the most abundant rhyolitic rocks in the western United States. Tens of thousands of cubic miles are known in Nevada alone, and great volumes are also recognized in the San Juan Mountains of Colorado, southwestern New Mexico, the Big Bend region of Texas, and southwestern Arizona.

In the Jemez Mountains of New Mexico, about one million years ago, eruptions similar to those at Yellowstone produced nearly 50 mi^3 of welded tuffs. Removal of this large volume of material from the abyssal storage chamber caused a great circular block of Earth's crust (more than 10 mi in diameter) to subside several thousand feet, producing a giant caldera at the surface. Formation of this caldera, the Valles Caldera, was similar to that of Crater Lake, but of special interest to geologists is the subsequent, more complex history of the Caldera. Renewed pressure from below uplifted and arched the subsided circular block (called the resurgent dome), so that its center rose even higher than its original elevation. During and after this uplift, new eruptions of rhyolitic lava broke out along the fracture system around the uplifted circular block and built a ring of 15 new volcanoes.

Detailed studies of the Valles Caldera have allowed geologists to relate this rhyolitic volcanism to more deeply seated processes of granite formation. Volcanism of this type is now known to have occurred in many other areas of the United States. The San Juan Mountains of Colorado are an outstanding example, as are the Mono Craters in California. They, like Little Glass and Big Glass Mountains in northeastern California, are famous for black volcanic glass (obsidian) and for the large blocks of pumice that have become popular as ornamental stones. The Mono Craters include some of the world's finest examples of rhyolitic pumice cones and lava domes. Their forms range from simple, almost perfectly symmetrical cones of pumice and ash, to cones whose craters are partly or completely filled or overflowing with lava. Some cones are completely covered with lava flows that have piled up to form steep-sided lava domes. Those cones containing lava that did not overflow the crater rim are similar to the plug dome of Lassen Peak, but are much smaller and formed by more fluid lava.

Recurrently, throughout geologic time, very fluid basaltic lava has erupted from swarms of fissures to form vast lava plateaus. The Columbia River Plateau of Washington and Oregon and the Snake River Plains of Idaho are among the finest examples of this type of volcanism. The Columbia River Plateau has an area of 100,000 mi^2, and the total volume of basaltic lava approximates 35,000 mi^3. Individual lava flows can be traced for distances of more than 100 mi. Such lava must have been almost as fluid as water to have covered such large areas so uniformly.

On the northern edge of the Snake River Plains is the Craters of the Moon National Monument. Here, basaltic lava erupted perhaps fewer than 2,000 years ago. The vents from which the lava issued are localized on great fissures and show a wide variety of forms, ranging from cinder cones built entirely on very frothy red and black lava cinders, to spatter cones formed by the piling up of liquid lava blobs and droplets around the vent. Both aa (pronounced ah-ah) and pahoehoe (pronounced pa-hoy-hoy) lavas are found at Craters of the Moon. Pahoehoe and aa are Hawaiian terms adopted the world over for two principal types of basaltic

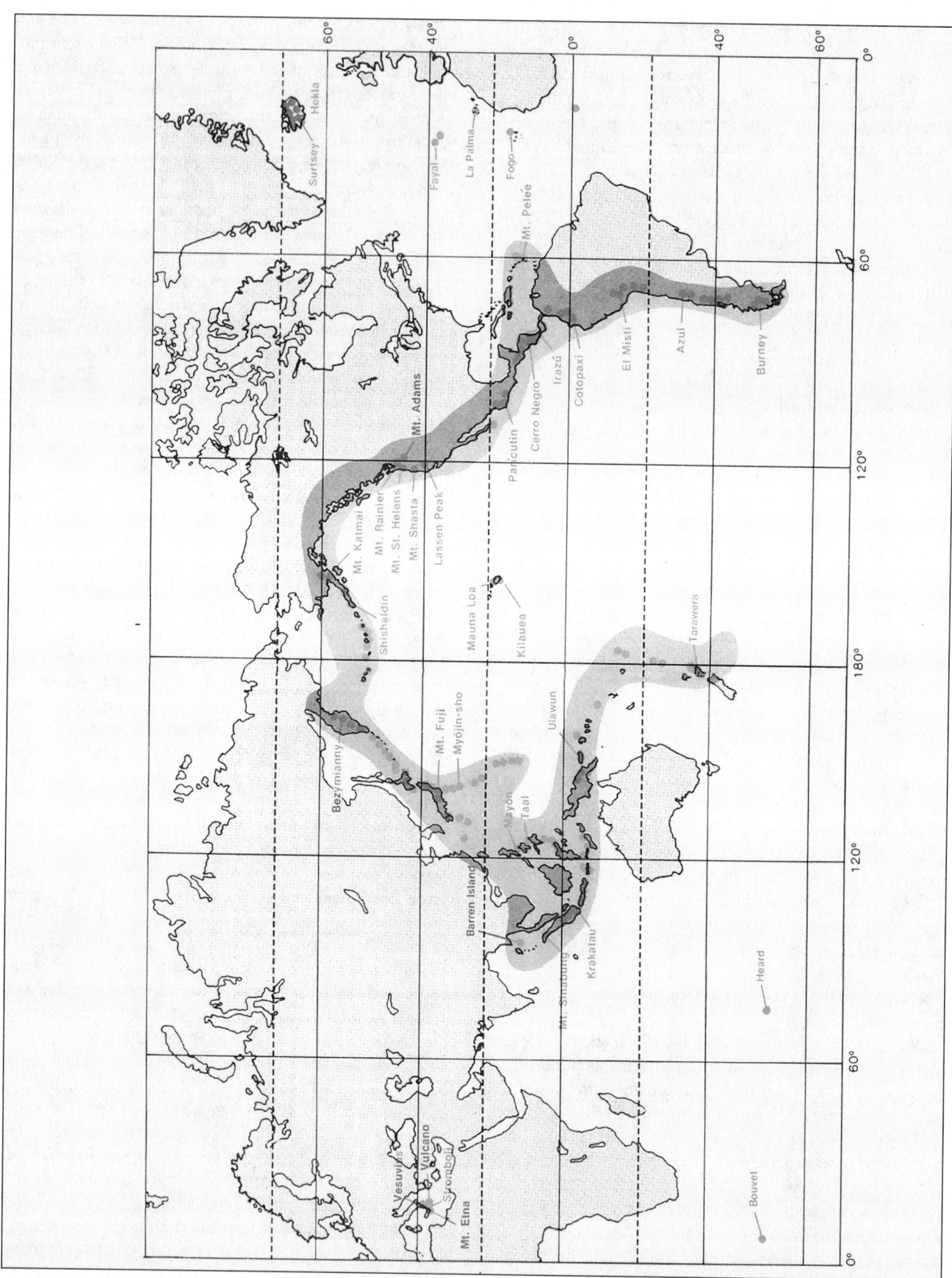

Known as the "Ring of Fire," this belt of high volcanic and seismic activity surrounds the Pacific Ocean. *(Courtesy of U.S. government publication.)*

lavas. Pahoehoe is very fluid lava with a smooth-to-ropy surface and flows that may travel long distances. Aa lava is a more viscous lava that forms steep-sided flows seemingly composed only of craggy blocks, but usually containing a continuous fluid interior.

Occasionally the surface and sides of these basaltic lava flows solidify to form a thick outer crust. Yet the hot lava inside continues its forward movement and eventually drains out of its own crust to form lava tunnels or lava tubes. Water entering such tunnels may freeze in the winter and, because of the excellent insulation provided by the basalt crust, may not thaw in the summer even under desert conditions. Excellent examples of these ice caves and lava tunnels may be seen in the Modoc Lava Beds National Monument, California.

Basaltic lava fields in cinder cones are numerous in many parts of the western United States. Some of these are isolated volcanic vents, but basaltic vents are commonly clustered near or around large volcanoes composed of andesitic and similar lavas. Fine examples of the clustered types are the volcanic fields of the San Francisco Mountains, Arizona, and Mount Taylor, New Mexico.

Hundreds of other localities for volcanoes and volcanic rocks are known, and this very brief discussion simply serves to emphasize that the United States, particularly the West, has been one of the most volcanically active areas in the world over geologic time. Geologists do know the "causes" of volcanism, but what they do not know with any certainty is the frequency of eruptions for volcanic systems. Does a given volcano, on average, erupt every few years, decades, centuries, or millennia? Or are the eruptions separated by long periods of inactivity (even millions of years)? For example, we know that the Yellowstone system during the past 2 million years had three huge eruptions about 600,000 years apart. So, from that average eruption frequency, we are about due for another huge one. It could erupt again in our lifetime, but the mathematical probability is very, very low. We can be sure, however, that in our lifetime volcanoes in Hawaii and Alaska will erupt many times again.

• ACTIVE VOLCANOES

"Active volcanism" is vague terminology and requires some kind of time context. Unfortunately, volcanologists have not agreed on the criteria to distinguish between "active," "dormant," and "extinct" volcanoes. A commonly used but inadequate definition of an "active" volcano is one that has erupted one or more times in recorded history. The problem is that "recorded history" of an Old World volcano-containing country (e.g., Japan, Italy, Iceland) differs from that of a New World country (e.g., United States, Mexico, Ecuador). For example, we know that eruptions have occurred within the last 200 years, but have not been chroni-

cled, in the Inyo-Mono chain in California. From the geologist's perspective, California has not had "active" volcanism.

Active volcanism in the United States is presently confined to the Hawaiian Islands, the Aleutian Islands, the Alaska Peninsula, and the Cascade Mountains.

The Hawaiian Islands consist entirely of volcanic rocks that form giant shield volcanoes; two (Mauna Kea and Mauna Loa) rise nearly 30,000 ft from the ocean floor. By far the dominant rock type is dark gray-to-black basalt in the form of lava flows, cinders, pumice, ash, and bombs.

On the island of Hawaii are Mauna Loa, the largest volcano in the world, and Kilauea, one of the most active. During an eruption at Kilauea in 1959–60, great fountains of lava, some as high as 1,900 ft, were observed near the volcano summit. Lava from the fountains ran into an old pit crater and filled it to a depth of 365 ft, forming a lava lake. Late in 1963, nearly four years after the eruption, the crust on top of the lake was nearly 50 ft thick and the temperature of the lava below the crust was still about 2,000°F.

The Aleutian Island arc, including the Alaska Peninsula, is more than 2,000 mi long and contains about 36 historically active and many extinct volcanoes. Among them are some of the world's most beautiful, but little-studied, volcanoes. The composite volcanoes Pavlof, Shishaldin, and Pavlof Sister are examples. Bogoslof Island is a disappearing volcano, having emerged and submerged in the sea more than once in historic time. Most of the Aleutian lavas consist of black-to-gray rocks called basalts and andesites.

Of the several active volcanoes on the Alaskan Peninsula, Mount Katmai (Katmai National Monument) is the most notable. In 1912, one of the most remarkable eruptions of historical time occurred near its base. The "river of sand," as the early explorers called it, flowed for more than 15 mi down a great glacial valley, filling it to a depth of more than 400 ft. This valley is known as the "Valley of Ten Thousand Smokes" because of the thousands of fumaroles (gas vents) that formed on the surface of the volcanic deposits and gave off steam and other vapors for many years.

In the Cascade Mountains of Washington, Oregon, and California are the well-known composite cones of Mounts Baker, Rainier, St. Helens, Adams, Hood, Mazama (Crater Lake), and Shasta. These high Cascade volcanoes are built primarily of rocks called andesites, which are intermediate in chemical composition between basalts and rhyolites. Andesitic volcanoes are usually great cones of rubble consisting of interlayered lava flows, ash deposits, and fragment deposits called mud flows. These form when eruptions take place through crater lakes, when fragmented lavas and landslides enter streams, or when water from melting snow or rain saturates and mobilizes previously erupted deposits. Volcanic mud flows may be hot or cold and may deposit boulders weighing many tons. Very young mud flow deposits are common on the flanks of Mount Rainier and other Cascade cones.

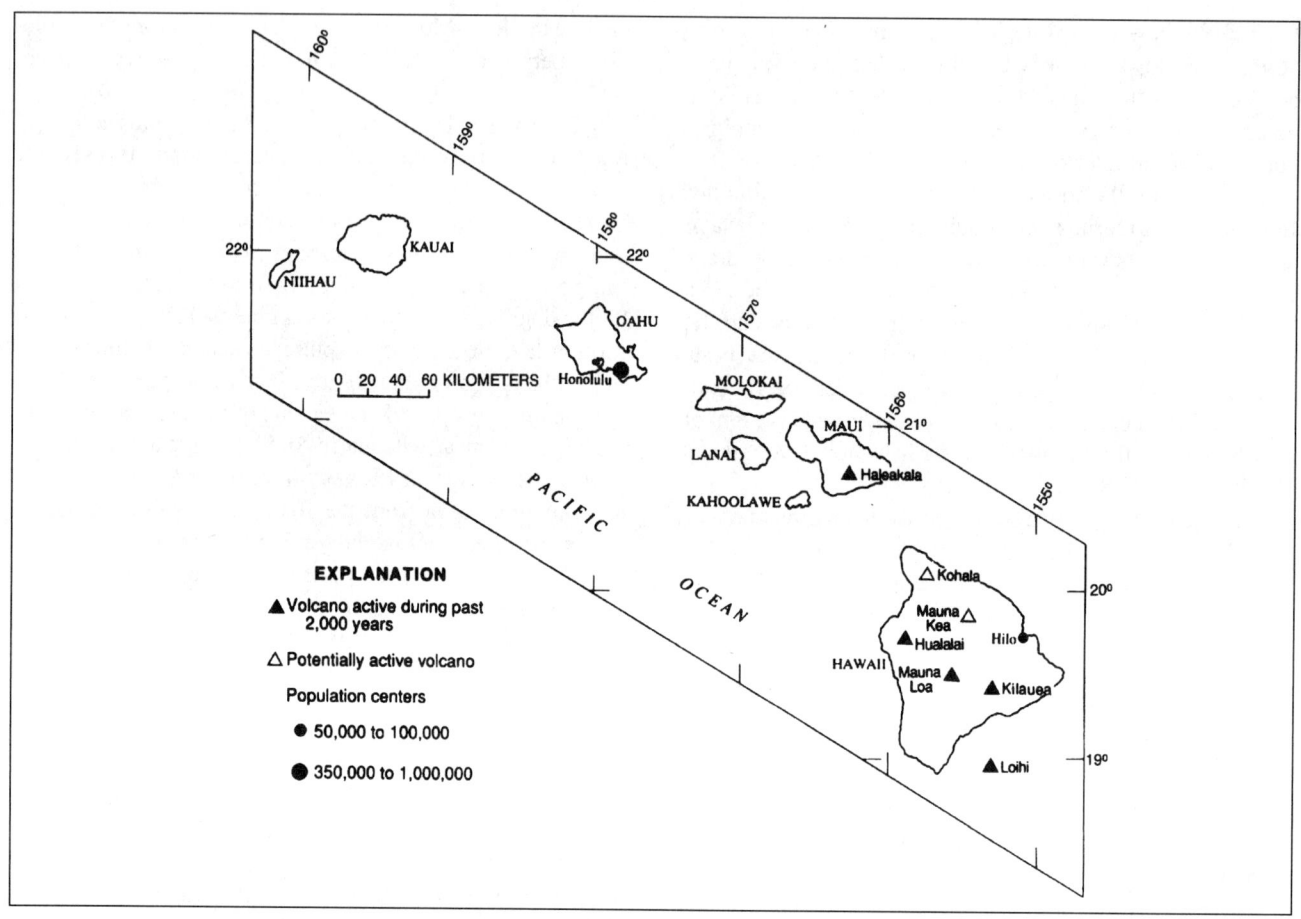

Volcanoes of the Hawaiian Islands. *(Courtesy of U.S. Geological Survey (USGS).)*

Mud flows—a major cause of destruction

In some areas of the world, mud flows have been the major cause of destruction and loss of life during volcanic catastrophes. Such an eruption in 1919 from Kelut volcano in Java covered 50 mi² of land with mud and lava blocks. More than 5,000 human lives were lost and 100 villages completely or partly damaged. This great devastation was caused by the ejection of over 1 billion cubic feet of water from Kelut's lake.

Mount Mazama (Crater Lake) is the volcano that lost most of its top about 7,000 years ago during a tremendous explosive eruption. Nearly 12 mi³ of ash and pumice erupted, producing under the volcano a void so large that the top caved in to form a great hole 6 mi across. Depressions formed in the top of volcanoes in this manner are known as calderas. They differ from craters that are produced by explosion rather than by collapse. The caldera of Mount Mazama is now filled with water to a depth of over 1,700 ft and is known as Crater Lake.

Lassen Peak is famous among volcanoes because it is one of the largest known plug domes. A plug dome is a part of a volcano formed by the vertical rise of a great sticky mass of lava that remains standing above the crater rim. Commonly,

great vertical grooves are formed in the margin of the dome as the material oozes from the orifice. The mechanism may be compared to the squeezing of toothpaste from a tube. Subsequent eruptions from plug domes are among the most dangerous known. Gasses dissolved in the lava may burst forth violently, shatter the dome, and cause the formation of extremely mobile avalanches of hot blocks, rock dust, and gas. Such avalanches may travel at speeds up to 100 mph and devastate everything in their paths. It was an eruption of this type in 1902 from Mount Pelee in the West Indies that destroyed the nearby town of St. Pierre and killed nearly 30,000 people.

Eruption at Mount St. Helens, 1980

Mount St. Helens is a symmetrical volcanic cone in southwestern Washington about 45 mi northeast of Portland, Oregon. Most of the cone that can be seen now was formed within the last thousand years—but this overlies an older volcanic center that probably has existed for at least 40,000 years. Mount St. Helens has had a long history of spasmodic explosive activity. It is an especially dangerous volcano because of its past behavior and the high frequency of its eruptions during the past 4,500 years.

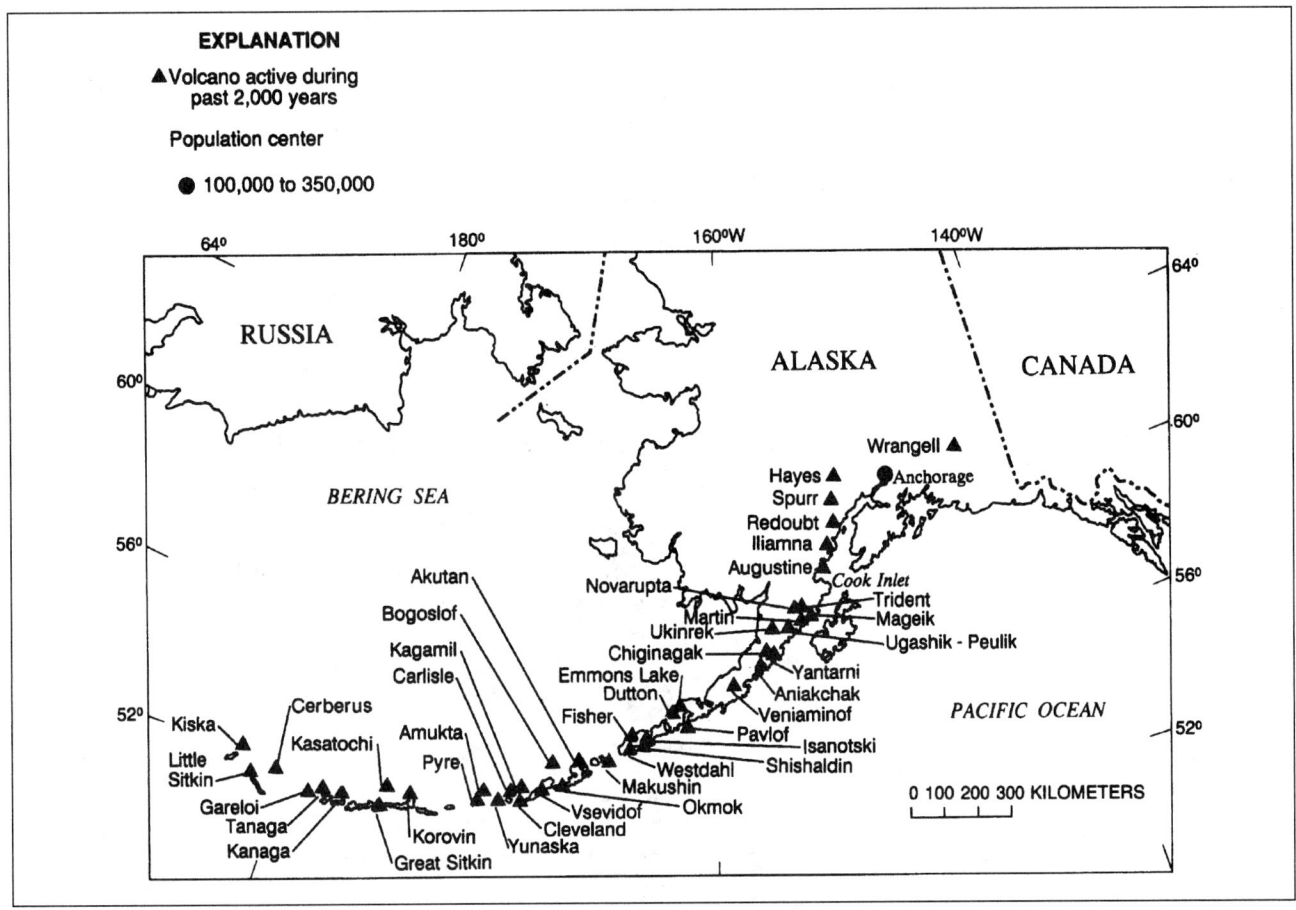

Volcanoes of the Aleutian Island arc. *(Courtesy of U.S. Geological Survey (USGS).)*

On May 18, 1980, a powerful explosion occurred from Mount St. Helens at 8:32 A.M. that was heard 200 mi away. The explosion climaxed a series of activities that began with an earthquake shock of magnitude 4.1 on March 20, 1980. Remarkable photographs, taken as the explosion began, show the north flank breaking away from the volcano as a large vertical cloud began to rise from the summit. The eruption column of ash and gas rose very rapidly to more than 10 mi above sea level; passing through the tropopause at 7 mi. Winds blew the cloud to the east. Ashfall at Yakima, 90 mi away, totaled as much as 4–5 in and caused respiratory problems for some residents. By mid-afternoon, the ash had reached Spokane, reducing visibility to only 10 ft, although only half an inch was deposited there. Almost 2 in of ash were reported from areas of Montana west of the Continental Divide, but only a dusting fell on the eastern slopes. Slight ashfall occurred in Denver on May 19. The ash blew generally eastward for the next several days, causing some problems for aircraft over the Midwest.

The U.S. Geological Survey (USGS) identified three components of the initial eruptive event in addition to the vertical cloud.

The first component was a directed blast that leveled the forest on the north and northwest flanks for a distance of up to 15 mi from the former summit. The blast swept over ridges and flowed down valleys, depositing significant quantities of ash. Although the blast was hot, it did not char fallen or buried trees. Many persons are known to have been killed by the blast, and others in the devastated zone went missing.

The second component was a combined pyroclastic flow and landslide that carried the remnants of the north flank uplift across the lower slopes and about 17 mi down the Toutle River valley, burying it to depths as great as 180 ft. Large quantities of mud, logs, and other debris clogged several valleys around Mount St. Helens and rendered some shipping lanes impassible in the Columbia River.

The third component was a pumiceous pyroclastic flow, funneled northward through the breach formed by the destruction of the north flank bulge. This flow dammed the outlet of Spirit Lake, trapping a large quantity of water.

The volcano maintained an eruption column 10 mi high until a relatively sudden diminution of activity occurred in the early morning of May 19. The altitude of the top of the column decreased to about 2.5 mi. Activity continued to weaken through May 22.

A new elliptical crater about a quarter mile deep was formed by the explosion.

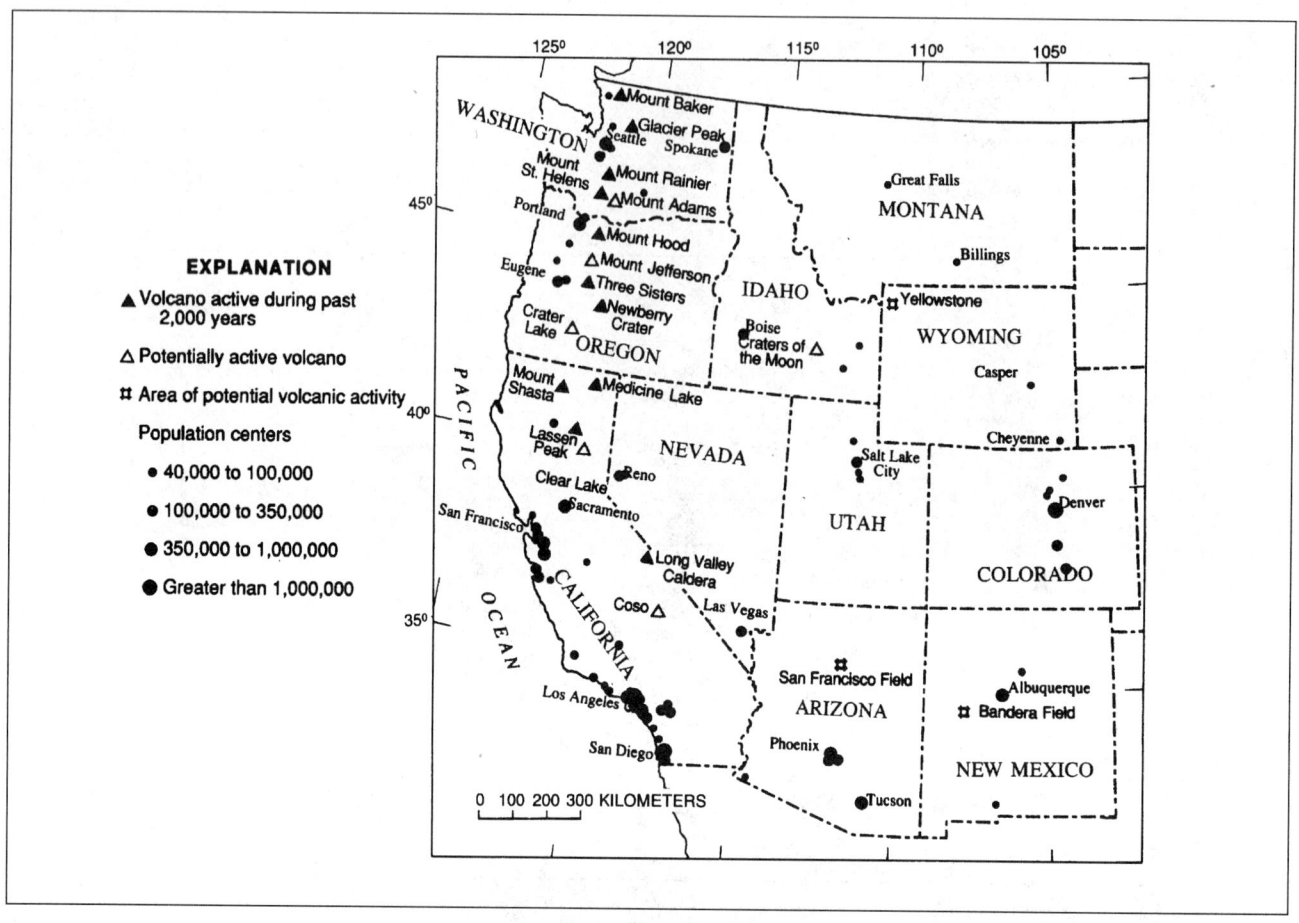

Volcanoes of the Cascades. *(Courtesy of U.S. Geological Survey (USGS).)*

Preliminary analysis of seismic and deformation data indicates that there was no immediate warning of the imminence of a large explosion. A magnitude 5.1 earthquake occurred essentially simultaneously with the explosion at 8:32 A.M. The tilt data were fragmentary and not diagnostic. Actually, the "inflation" began as early as March 20, 1980 with the onset of the first precursory earthquakes and the intrusion of magma "cryptodome" to begin forming the "bulge."

Although volume estimates for this eruption are very rough, comparison with previous eruptions in the Smithsonian Institution's Volcano Reference File (a computer data file of the world's volcanoes and their eruptions) indicates that explosions of this size occur only about once a decade. The May 18 eruption ejected about 0.3 mi³ of uncompacted ash.

Mount St. Helens has erupted 22 times since May 18, 1980. On 19 occasions the eruption has been predicted, as scientists use more sensitive equipment to pick up ground vibrations and super-accurate lasers to detect the most minute "bulges" that occur in the slope of the mountain.

• MOUNT ST. HELENS: WHAT WE HAVE LEARNED 20 YEARS LATER

Twenty years ago in late March, southwestern Washington's Mount St. Helens, a volcano in the Cascade Range, awoke from a 123-year slumber. Following two months of precursory activity including sustained energetic seismicity, phreatic (steam-blast) explosions, and rapid bulging of its north flank from magmatic intrusion, Mount St. Helens erupted cataclysmically on the morning of May 18, 1980.

This eruption caused the worst volcanic disaster in the recorded history of the United States, resulting in 57 deaths, scores of injuries, and economic losses exceeding $1 billion. Because it was thoroughly documented and received substantial media attention, the eruption and its aftermath ushered in two decades of heightened public awareness and expanded scientific studies, launching a veritable renaissance in volcanology that continues into the twenty-first century.

The reawakening of Mount St. Helens happened while Robert I. Tilling was head of the Volcano Hazards Program of the USGS and he was thus responsible for directing the survey's scientific response to the eruption. The 1980 and subsequent eruptions of Mount St. Helens furnish many lessons, not only scientific findings relevant to reducing volcano risk, but also lessons regarding the critical need for effective communication among scientists, emergency management officials, members of the media, and affected populations.

The eruption

Seconds after a 5.1 magnitude earthquake, the north flank of Mount St. Helens began to collapse, unleashing a powerful, laterally directed blast, comparable in some ways to the sudden removal of the cap from a vigorously shaken bottle of soda. This collapse produced a rockslide-debris avalanche of 0.5 mi^3, the world's largest in historical time. Although it lasted less than five minutes, the lateral blast traveled at speeds of up to 620 mph, extending out as far as 373 mi and devastating 230 square miles of land north of the volcano. This blast, much more powerful than any other in the volcano's history, was the principal cause of fatalities.

Ash fallout from the eruption affected more than 22,000 square miles in eastern Washington and neighboring states. The ash cloud drifted across the country in three days and ultimately circled the globe in about two weeks.

Within 10 minutes of the eruption's onset, the interaction of hot volcanic ejecta with snow and ice triggered lahars (volcanic mudflows) that caused widespread flooding and damaged roads, bridges, and other structures.

Several compelling lessons from the Mount St. Helens eruption must not be forgotten:

- Volcano hazards can be multiple (e.g., flank failure, debris avalanche, lahars, lateral blast, pyroclastic flow, tephra [solid material, usually ash] fall, flooding, and sedimentation), and they can happen in a very short time, from tens of seconds to a few hours.
- Communities far distant from a volcano can also be vulnerable to damaging hazards.
- Prolonged, increased sediment transport following explosive eruptions can have socioeconomic and environmental consequences that are more severe than the direct consequences of the actual eruption.

Anticipating the eruption

When activity began at Mount St. Helens in late March 1980, the USGS began intensive monitoring in cooperation with the Geophysics Program of the University of Washington. Data clearly indicated magmatic intrusion high into the volcanic edifice and the growing instability of the bulge on the north flank, but it was impossible to predict the onset of the events of May 18. Nevertheless, the monitoring data indicated to scientists that a flank failure might trigger a large magmatic eruption. Indeed, the on-site scientific team explained this scenario to emergency management officials before May 1.

Volcanologists now regard any significant rapid deformation of a volcano, as was well documented at Mount St.

Helens, as a warning of a potential sector collapse and lateral blast.

A successful but imperfect response

Besides monitoring the volcano around the clock, USGS scientists also worked daily with the U.S. Forest Service, the principal land manager for Mount St. Helens, as well as two counties and other government agencies, to provide updates of potential hazards and advice on mitigation and preparedness measures. By April 1, the USGS had developed a large-scale hazards zonation map and related hazards assessment information that were essential for preparing the Forest Service's Mount St. Helens Contingency Plan (which was completed April 9) and for locating roadblocks and restrictions on public access. Had these measures not been taken, the eruption would have caused considerably more casualties.

However, while the USGS response to Mount St. Helens was successful overall, it was hardly perfect. Lack of equipment, along with logistical difficulties, meant geodetic measurements of the north flank (bulging at an average rate of about 5 ft per day) did not begin until mid-April, several weeks after the bulge was first recognized. We will never know how much difference, if any, earlier measurements might have made.

Why didn't the USGS and Washington State take more action before 1980? By repeating the baseline monitoring measurements that had been initiated in the early 1970s and by developing protocols for working with other agencies, private organizations, and the public, they could have prepared both earlier and better for a reawakening of Mount St. Helens. After all, the USGS knew of the potential hazards as early as the late 1960s. Moreover, in 1975 three USGS geologists (Crandell, Mullineaux, and Rubin) published in *Science* a long-term forecast stating that Mount St. Helens would be the most likely volcano in the Cascade Range to reawaken, possibly even "before the end of the twentieth century." In 1978, Crandell and Mullineaux published a detailed volcano hazards assessment that received little notice, but after the onset of activity in 1980 this report was widely read by scientists and emergency management officials.

The Survey's lack of preparation in part stemmed from insufficient funds for additional work in the Cascades; and repeated efforts to obtain increased appropriations for volcano hazards studies had all failed. After the 1980 eruption, however, funding for the USGS Volcano Hazards Program increased significantly and was more than sufficient to establish the Cascades Volcano Observatory in Vancouver, Washington, to monitor Mount St. Helens and other Cascades volcanoes. The funding also expanded or initiated studies of other volcanoes in the United States. The lesson is obvious and disturbing: justification for increased funding of volcano hazards studies is greatly strengthened and perhaps only receives serious attention following a volcanic crisis or disaster.

Since 1980, scientists have made numerous advances in volcano monitoring. The scientific and public responses to two volcano crises outside the United States, Nevado del Ruiz (Colombia) in 1985 and Mount Pinatubo (Philippines) in 1991, underscore the lesson of Mount St. Helens regarding the need for effective communications among scientists, emergency responders, and the public.

On November 13, 1985, a very small amount of magma erupted from Colombia's Nevado del Ruiz volcano, triggering destructive lahars that killed more than 23,000 people. This tragedy could have been averted; a hazards zonation map had been prepared a month earlier and scientists had provided adequate warning that went unheeded.

The Ruiz disaster was the impetus for the International Association of Volcanology and Chemistry of the Earth's Interior to produce *Understanding Volcanic Hazards*, a video that depicts the deadly outcomes of volcano disasters, and *Reducing Volcanic Risk*, a video that shows what communities can do to mitigate volcanic hazards.

The Ruiz experience also launched the Volcanic Disaster Assistance Program (VDAP) in 1986, which is jointly funded by the USGS and the Office of Foreign Disaster Assistance of the U.S. Agency for International Development. Once officially invited, a VDAP team can quickly deploy a mobile volcano observatory to help host countries respond to volcanic crises.

In contrast to the 1985 Ruiz catastrophe, the response of scientists and emergency management officials to the 1991 eruption of Mount Pinatubo in the Philippines saved thousands of lives and reduced economic loss by hundreds of millions of dollars. Fatalities directly attributed to Pinatubo numbered fewer than 300, thanks to the timely evacuation of 250,000 people. Scientists from the Philippine Institute of Volcanology and Seismology and the VDAP team educated local authorities and populations about the eruption and its potential hazards using a draft version of the *Understanding Volcanic Hazards* video that convinced local officials to order evacuations and the people at risk to comply.

Averting volcanic disasters

Some trends of the past two decades continue into the twenty-first century, including the continued development and improvement of real-time volcano monitoring networks and methods, particularly the use of satellite technology. Worldwide eruption frequency (on average, about 60 volcanoes are active each year) is not likely to decrease in the foreseeable future. Thus, with continued growth in world population, economic development, and urbanization, the global risks due to volcanoes will become more acute. However, even as we look forward to continuing advances in volcanology and hazards studies, the 1985 Ruiz disaster provides a tragic reminder that good science alone is not enough. The greatest payoff in risk reduction will come from increased focus on the societal and human issues that emerge during volcanic crises and from developing or improving communication among scientists, emergency managers, representatives of the news media, educators, and the general public. The major challenge, indeed the goal, for volcanologists and other scientists is to prevent volcanic crises from turning into volcanic disasters.

• VOLCANIC RISKS IN THE UNITED STATES

Volcanic hazards (lava flows, mudflows, hot rock avalanches, ashfalls, and floods) are limited to areas in the western United States, principally in the Cascade Mountain Range in California, Oregon, and Washington; in Idaho's Snake River Plain; and in parts of Arizona, New Mexico, and Utah. No volcanic hazard areas are known east of New Mexico.

The short term risk from volcanic hazards is generally low because destructive eruptions are relatively infrequent. Severely destructive effects of eruptions, other than extremely rare ones of catastrophic scale, probably would be limited to areas within a few tens of miles down-valley or downwind from a volcano. Thus, the area seriously endangered by any one eruption would be only a very small part of the western United States.

Except for Mount St. Helens' renewed activity in 1980, the only explosive volcanic eruption in the contiguous states since the area was settled by Europeans was 10,457-ft-high Mount Lassen in northern California during a series of eruptions in 1914–15. The Mount Lassen eruption was moderate compared to major eruptions at other volcanoes in the world during recorded history. No one was killed in the Mount Lassen eruption, and damage was minor.

Eruptions of moderate volume may occur somewhere in the Cascade Range as often as once every 1,000–2,000 years, but very large eruptions may occur no more than once every 10,000 years. A few large cataclysmic eruptions have occurred during the last two million years in and near Yellowstone National Park, at Long Valley, California, and in the Jemez Mountains of New Mexico. These eruptions affected very large regions and deposited ash over much of the western United States. Such cataclysmic eruptions were not considered in outlining potential hazard zones. These eruptions are so infrequent that it is not possible to judge whether one might occur during the time for which planning is feasible.

Risk from volcanic hazards decreases as distance from an erupting volcano increases. Lava flows are nearly uniformly destructive to their outer limits. Some other volcanic hazards, especially ashfalls, become less destructive and less frequent with increasing distance. The boundary of such a hazard is indefinite and often dependent upon land use. For example, an ashfall a half-inch or so thick might cause little damage to structures, yet destroy crops.

• BRIEF SYNOPSIS OF HISTORICAL VOLCANIC ERUPTIONS IN THE UNITED STATES

Cascade Range of the Pacific Northwest

Mount Baker

1840-1870: Historical literature refers to several episodes of small tephra-producing events in the mid-1800s, and increased fumarolic activity began in Sherman Crater near the summit in 1975 and remains elevated through the late 1990s.

Glacier Peak

1600-1700s: Two hundred to three hundred years ago, small eruptions deposited pumice and ash east of the volcano, and may have been observed by Native Americans.

Mount Rainer

1894 and early 1800s: Several eyewitness accounts describe minor releases of steam and ash-laden steam during November and December 1894. The most recent eruption that formed a thin and discontinuous tephra layer, however, occurred during the first half of the nineteenth century.

Mount St. Helens

1980–1986: Large explosive eruption on May 18, 1980, followed by 21 smaller eruptive episodes. The last 17 episodes built a lava dome in the volcano's crater.

1800–1857: Large explosive eruption in 1800 was followed by eruptions of lava that formed a lava flow on the volcano's northwest flank (Floating Island lava flow) and a lava dome on the north flank (Goat Rocks lava dome).

Late 1700s: Layers of volcanic rocks record a variety of activity related to the growth of a lava dome at the volcano's summit, including pyroclastic flows, lahars, and tephra fall.

Oregon

Mount Hood

1856–1865 and late 1700s: According to eyewitnesses, small explosive eruptions occurred from the summit area between 1856 and 1865. In the latter half of the eighteenth century, however, a lava dome was erupted, which was accompanied by pyroclastic flows, lahars, and tephra fall.

California

Mount Shasta

1786: An eruption cloud was observed above the volcano from a ship passing by the north coast California, and the activity included pyroclastic flows.

Lassen Peak

1914–1917: A series of small explosions that began on May 30, 1914, was followed 12 months later by extrusion of lava from the summit and a destructive pyroclastic flow and lahars on May 21, 1915. Minor activity continued through the middle of 1917.

Hawaii

Kilauea Volcano, Hawaii

1790 to present: First written record of eruption at Kilauea was in 1823; strong explosive activity occurred in 1790.

Mauna Loa Volcano, Hawaii

1832 to present: First written record of eruption at Mauna Loa was in 1832 (eyewitness was on Maui, 118 mi away).

Hualalai Volcano, Hawaii

Late 1700s–1801: Six different vents erupted lava between the late 1700s and 1801, two of which generated lava flows that poured into the sea on the west coast of Hawaii.

East Maui Volcano (Haleakala), Maui

1790: Two lava flows erupted on the southwest flank and reached the sea.

• RECENT CHANGES IN U.S. VOLCANIC ACTIVITY AND UNREST

Mount Hood, Oregon

A swarm of a few dozen small earthquakes beneath Mount Hood that began January 11, 1999, prompted scientists to release an information statement on the afternoon of January 14. In the statement, scientists describe the earthquakes as most likely due to the regional tectonic stress rather than to the movement of magma. Since 1990, Mount Hood has generated about 15 earthquakes swarms similar to the January 1999 quake.

Kilauea Volcano, Hawaii

Lava continues to erupt from Pu'u O'o cone and travel through lava tubes into the sea. New land created by lava entering the sea collapses into the ocean repeatedly. An eruption update is available from the USGS Hawaiian Volcano Observatory.

Summary of the Pu`u `O`o-Kupaianaha Eruption, 1983–present

The Pu`u `O`o-Kupaianaha eruption of Kilauea, now in its 19th year and 55th eruptive episode, ranks as the most voluminous outpouring of lava on the volcano's east rift zone in the past five centuries. By September 2002, 2.3 km^3 of lava had covered 110 km^2 and added 220 hectares to Kilauea's southern shore. In the process, lava flows destroyed 189 structures and resurfaced 13 km of highway with as much as 25 m of lava.

Beginning in 1983, a series of short-lived lava fountains built the massive cinder-and-spatter cone of Pu`u `O`o. In 1986, the eruption migrated 3 km down the east rift zone to build a broad shield, Kupaianaha, which fed lava to the coast for the next 5.5 years. When the eruption shifted back to Pu`u `O`o in 1992, a series of flank-vent eruptions formed a shield banked against the uprift side of the cone. Continuous eruption from these vents undermined the west and south flanks of the cone, resulting in large collapses of the west flank.

In May 2002, a new vent opened on the west side of the shield and fed flows down the western margin of the flow field, sparking the largest forest fire in the park in 15 years. These flows reached the ocean near the end of Chain of Craters Road in July, and as many as 4,000 visitors per day flocked to view flowing lava up close for the rest of the summer.

Other recent U.S. activity

Mount St. Helens, Washington

Earthquake activity beneath the volcano has decreased since the small earthquake swarm in mid-1998 (May–July). Scientists interpret the swarm as representing a pulse of magma that moved in the volcano's magma reservoir.

Long Valley Caldera, California

The current condition of the caldera is GREEN (minor unrest posing no immediate risk). Earthquake activity is relatively low; the most recent significant earthquakes occurred on June 8, 1998, and July 14, 1998 (magnitude 5.1); a few aftershocks continue to occur. Deformation measurements within the caldera indicate that, following a period of little change from mid-April through mid-August, the resurgent dome has resumed slow inflation at a rate averaging 0.39–0.78 in per year.

Korovin Volcano, Alaska

Scientists of the Alaska Volcano Observatory (AVO) received reports on the morning of June 30, 1998, that a small eruption column above Korovin volcano had reached nearly 16,000 ft above sea level. Korovin volcano is located on the north end of Atka Island in the central Aleutians, 334 mi west of Dutch Harbor and 1,100 mi southwest of Anchorage. It is 13 mi north of the village of Atka, population about 100. Local winds at the time of the reports were light and to the south-southwest, and a dusting of ash was reported in Atka.

AVO does not maintain seismic monitoring equipment on Atka Island, and relies on satellite and field observations for monitoring the status of Korovin. No activity has been observed since June 1998.

Lake Becharof Seismic Swarm, Alaska

The intense swarm of earthquakes beneath Lake Becharof that shook the Alaska Peninsula beginning May 8, 1998, declined significantly. The swarm began with five earthquakes of magnitude 5.2–4.7. The earthquakes have been clustered on the southwest shore of Lake Becharof, several miles northwest of the 1977 Ukinrek Maars volcanoes. No volcanic activity (for example, fumarolic activity

---

Safety Rules—What to do When a Volcano Erupts

Most important, do not panic—keep calm. If volcanic ash begins to fall:

1. Stay indoors.
2. If you are outside, seek shelter such as a car or a building.
3. If you cannot find shelter, breathe through a cloth, such as a handkerchief, preferably a damp cloth to filter out the ash.
4. When air is full of ash, keep your eyes closed as much as possible.

● Heavy falls of ash seldom last more than a few hours—only rarely do they last a day or more.

● Heavy falls of ash may cause darkness during daylight hours and may temporarily interfere with telephone, radio, and television communications.

● Do not try to drive a car during heavy fall of ash—the chance of accident will be increased by poor visibility.

● The thick accumulation of ash could increase the load on roofs, and saturation of ash by rain could be an additional load. Ash should be removed from flat or low-pitched roofs to prevent a thick accumulation.

● Valleys that head on the volcano may be the routes of mudflows which carry boulders and resemble wet flowing concrete. Mudflows can move faster than you can walk or run, but you can drive a car down a valley faster than a mudflow will travel. When driving along a valley that heads on a volcano, watch up the river channel and parts of the valley floor for the occurrence of mudflows.

5. Before crossing a highway bridge, look upstream.
6. Do not cross a bridge while a mudflow is moving beneath it.
7. The danger from a mudflow increases as you approach a river channel and decreases as you move to higher ground.
8. Risk of mudflows also decreases with increasing distance from a volcano.
9. If you become isolated, do not stay near a river channel, move upslope.

DURING AN ERUPTION MOVE AWAY FROM A VOLCANO, NOT TOWARD IT. (In the immediate vicinity of an eruption the hazards to life are much greater than those listed above.)

Source: United States Geological Survey.

---

or steam-driven explosions) or ground cracks have been observed since the earthquake swarm began.

Chiginagak Volcano, Alaska

Reports of increased steaming, snowmelt, and sulfur smell at Chiginagok Volcano were received by AVO beginning October 22, 1997. Robust steam plumes have issued from an active fumarole at an elevation of about 5,500 ft on the north flank of the volcano since at least 1943 and sulfur deposition in the vicinity of the fumarole discolors the adjacent snow and ice. Reports of historic activity at Chiginagak are poorly documented. The volcano is not monitored by seismic instrumentation.

Fish and Wildlife Service personnel in the field and citizens of Pilot Point reported that on Thursday, August 13, 1998, the usual white steaming from Chiginagak fumaroles changed for a time to puffs of black ash accompanied by a greenish yellow gas and steam plume rising about 500–1,000 ft above the volcano.

Air and Water Pollution

Air pollution is a general term that covers a broad range of contaminants in the atmosphere. Pollution can occur from natural causes or from human activities. Discussions about the effects of air pollution have focused mainly on human health but attention is being directed to environmental quality and amenity as well. Air pollutants are found as gases or particles, and on a restricted scale they can be trapped inside buildings as indoor air pollutants. Urban air pollution has long been an important concern for civic administrators, but increasingly, air pollution has become an international problem.

The most characteristic sources of air pollution have always been combustion processes. Here the most obvious pollutant is smoke. However, the widespread use of fossil fuels has made sulfur and nitrogen oxides pollutants of great concern. With increasing use of petroleum-based fuels, a range of organic compounds have become widespread in the atmosphere.

In urban areas, air pollution has been a matter of concern since historical times. Indeed, there were complaints about smoke in ancient Rome. The use of coal throughout the centuries has caused cities to be very smoky places. Along with smoke, large concentrations of sulfur dioxide were produced. It was this mixture of smoke and sulfur dioxide that typified the foggy streets of Victorian London, paced by such figures as Sherlock Holmes and Jack the Ripper, whose images remain linked with smoke and fog. Such situations are far less common in the cities of North America and Europe today. However, until recently, they have been evident in other cities, such as Ankara, Turkey, and Shanghai, China, that rely heavily on coal.

Coal is still burned in large quantities to produce electricity or to refine metals, but these processes are frequently undertaken outside cities. Within urban areas, fuel use has shifted toward liquid and gaseous hydrocarbons (petroleum and natural gas). These fuels typically have a lower concentration of sulfur, so the presence of sulfur dioxide has declined in many urban areas. However, the widespread use of liquid fuels in automobiles has meant increased production of carbon monoxide, nitrogen oxides, and volatile organic compounds.

Primary pollutants such as sulfur dioxide or smoke are the direct emission products of the combustion process. Today, many of the key pollutants in the urban atmospheres are secondary pollutants, produced by processes initiated through photochemical reactions. The Los Angeles, California-type, photochemical smog is now characteristic of urban atmospheres dominated by secondary pollutants.

Although the automobile is the main source of air pollution in contemporary cities, there are other equally significant sources. Stationary sources are still important and the oil-burning furnaces that have replaced the older coal-burning ones are still responsible for a range of gaseous emissions and fly ash. Incineration is also an important source of complex combustion products, especially where this incineration burns a wide range of refuse. These emissions can include chlorinated hydrocarbons such as dioxin. When plastics, which often contain chlorine, are incinerated, hydrochloric acid is found in the waste gas stream. Metals, especially since they are volatile at high temperatures, can migrate to smaller, respirable particles. The accumulation of toxic metals, such as cadmium, on fly ash gives rise to concern over harmful effects from incinerator emissions. In specialized incinerators designed to destroy toxic compounds such as polychlorinated biphenyls (PCBs), many questions have been raised about the completeness of this destruction process. Even under optimum conditions when the furnace operation has been properly maintained, great care needs to be taken to control leaks and losses during transfer operations (fugitive emissions).

The enormous range of compounds used in modern manufacturing processes has also meant that there is an ever-widening range of emissions from both the industrial processes and the combustion of their wastes. Although the amounts of these toxic compounds are often rather small, they add to the complex range of compounds found in the urban atmosphere. Again, it is not only the deliberate loss of effluents through discharge from pipes and chimneys that needs attention. Fugitive emissions of volatile substances that leak from valves and seals often warrant careful control.

Air pollution control procedures are increasingly an important part of civic administration, although their goals are far from easy to achieve. It is also noticeable that although many urban concentrations of primary pollutants, for example, smoke and sulfur dioxide, are on the decline in developed countries, this is not always true in developing countries. Here the desire for rapid industrial growth has often lowered urban air quality. Secondary air pollutants are generally proving a more difficult problem to eliminate than primary pollutants like smoke.

Smog covers the Los Angeles basin. *(Photo by Walter A. Lyons, FMA Productions. Reproduced by permission.)*

• AIR POLLUTION AND HEALTH PROBLEMS

Urban air pollutants have a wide range of effects, with health problems being the most enduring concern. In the classical polluted atmospheres filled with smoke and sulfur dioxide, a range of bronchial diseases was enhanced. While respiratory diseases are still the principal problem, the issues are somewhat more subtle in atmospheres where the air pollutants are not so obvious. In photochemical smog, eye irritation from a secondary pollutant, peroxyacetyl nitrate (PAN), is one of the most characteristic direct effects of the smog. High concentrations of carbon monoxide in cities where automobiles operate at high density mean that the human heart has to work harder to make up for the oxygen displaced from the blood's hemoglobin by carbon monoxide. This extra stress appears to reveal itself through increased incidence of complaints among people with heart problems. There is a widespread belief that contemporary air pollutants are involved in the increases in asthma, but the links between asthma and air pollution are probably rather complex and related to a whole range of factors. Lead, from automotive exhausts, is thought by many to be a factor in lowering the IQs of urban children.

Air pollution also affects materials in the urban environment. Soiling has long been regarded as a problem, originally the result of the smoke from wood or coal fires, but now increasingly the result of fine black soot from diesel exhausts. The acid gases, particularly sulfur dioxide, increase the rate of destruction of building materials. This is most noticeable with calcareous stones, which are the predominant building material of many important historic structures. Metals also suffer from atmospheric acidity. In today's photochemical smog, natural rubbers crack and deteriorate rapidly.

Health problems relating to indoor air pollution are extremely ancient. Anthracosis, or black lung disease, has been found in mummified lung tissue. Recent decades have witnessed a shift from the predominance of concern about outdoor air pollution into a widening interest in indoor air quality.

The production of energy from combustion and the release of solvents is so large in the contemporary world that it causes air pollution problems of regional and global nature. Acid rain is now widely observed throughout the world. The sheer quantity of carbon dioxide emitted in combustion processes is increasing the concentration of carbon dioxide in the atmosphere and enhancing the greenhouse ef-

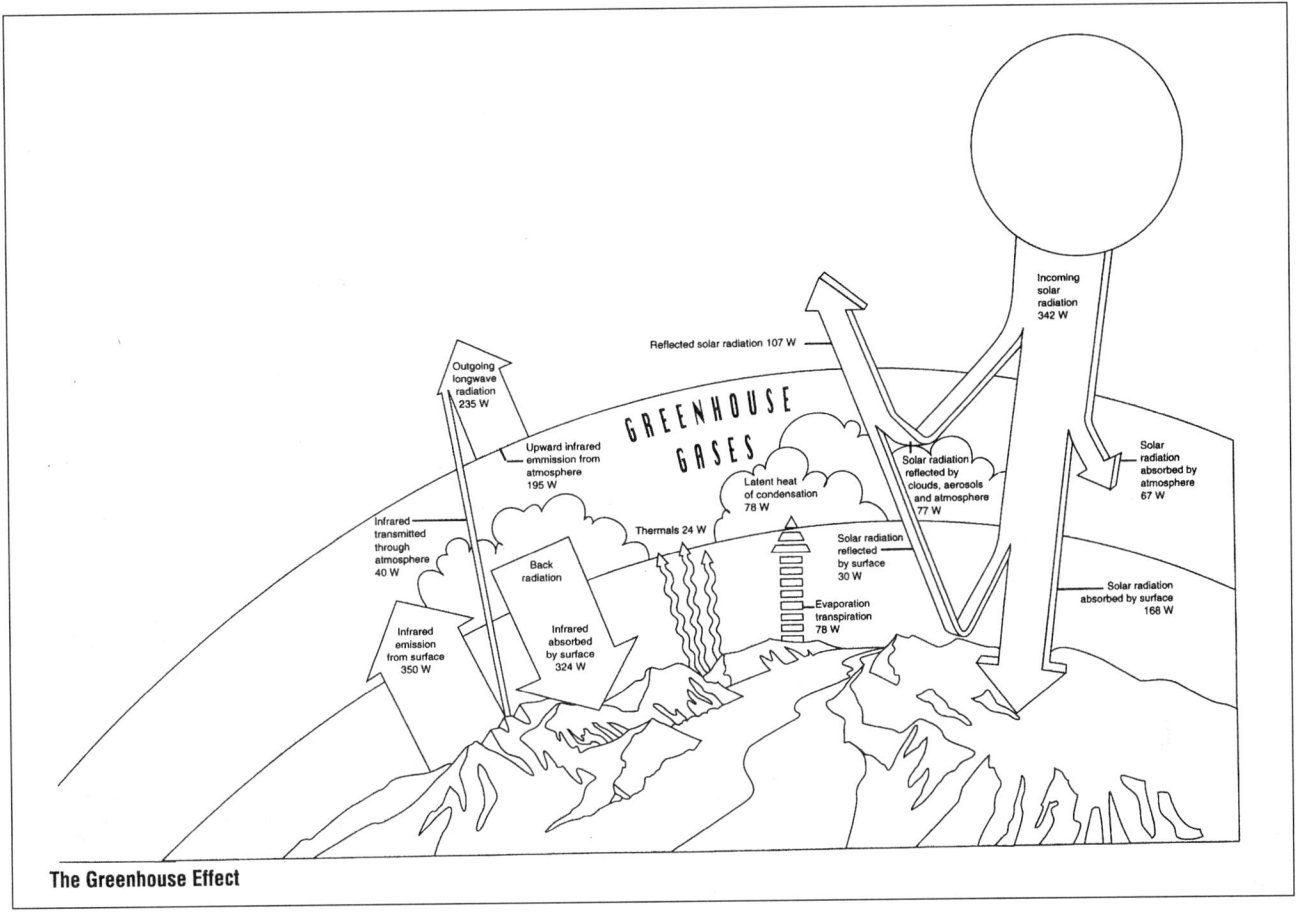

The Greenhouse Effect

(Courtesy of U.S. government publication.)

fect. Solvents, such as carbon tetrachloride and the aerosol propellants chlorofluorocarbons (CFCs) are now detectable all over the globe and responsible for problems such as ozone layer depletion.

At the other end of the scale, we need to remember that gases leak indoors from the polluted outdoor environment, but more often the serious pollutants arise from processes that take place indoors. Here there has been particular concern with regards to the generation of nitrogen oxides by sources such as gas stoves. Similarly, formaldehyde from insulating foams causes illnesses and adds to concerns about our exposure to a substance that may induce cancer in the long run. In the last decade it has become clear that radon leaks from the ground can expose some members of the public to high levels of this radioactive gas within their own homes. Cancers may also result from the emanation of solvents from consumer products—glues, paints, and mineral fibers (asbestos). More generally these compounds and a range of biological materials—animal hair, skin, pollen spores, and dusts—can cause allergic reactions in some people. At one end of the spectrum these simply cause annoyance, but in extreme cases, such as found with the bacterium *Legionella*, a large number of deaths can occur.

There are also important issues surrounding the effects of indoor air pollutants on materials. Many industries, especially the electronics industry, must take great care over the purity of indoor air where a speck of dust can destroy a microchip or low concentrations of air pollutants change the composition of surface films in component design. Museums must care for objects over long periods of time, so precautions must be taken to protect delicate dyes from the effects of photochemical smog, paper and books from sulfur dioxide, and metals from sulfide gases.

• AIR QUALITY

Air quality is determined with respect to the total air pollution in a given area as it interacts with meteorological conditions such as humidity, temperature, and wind to produce an overall atmospheric condition. Poor air quality can manifest itself aesthetically (as a displeasing odor, for example), and can also result in harm to plants, animals, and people, and even damage to objects.

As early as 1881, cities such as Chicago, Illinois, and Cincinnati, Ohio, passed laws to control some types of pollution, but it was not until several air pollution catastrophes occurred in the twentieth century that governments began to give more attention to air-quality problems. For instance, in 1930, smog trapped in the Meuse River Valley in Belgium caused 60 deaths. Similarly, in 1948, smog was blamed for 20 deaths in Donora, Pennsylvania. Most dramatically, in 1952, a sulfur-laden fog enshrouded London for five days and caused as many as 4,000 deaths over two weeks.

Disasters such as these prompted governments in a number of industrial countries to initiate programs to protect air quality. The year of the London tragedy, the United States passed the Air Pollution Control Act granting funds to assist the states in controlling airborne pollutants. In 1963, the Clean Air Act, which began to place authority for air quality into the hands of the federal government, was established. Today the Clean Air Act, with its 1970 and 1990 amendments, remains the principal air quality law in the United States.

The act established a National Ambient Air Quality Standard under which federal, state, and local monitoring stations at thousands of locations, together with temporary stations set up by the Environmental Protection Agency (EPA) and other federal agencies, directly measure pollutant concentrations in the air and compare those concentrations with national standards for six major pollutants: ozone, carbon monoxide, nitrogen oxides, lead, particulates, and sulfur dioxide. When the air we breathe contains amounts of these pollutants in excess of EPA standards, it is deemed unhealthy, and regulatory action is taken to reduce the pollution levels.

A December 1998 EPA report indicates that while air quality continues to improve, approximately 107 million Americans in 1997 lived in areas that did not meet the ambient air quality standards for at least one of the six major pollutants noted above. In general, though, improvements in air quality have been significant: carbon monoxide concentrations have decreased 38%; lead concentrations have decreased by 67%; nitrogen dioxide concentrations are down by 14%; ozone (smog) concentrations have been reduced by 19%; particulate matter concentrations decreased 26%; and sulfur dioxide concentrations decreased 39%. At the same time that air pollution has been decreasing significantly (1970–97), gross domestic product increased 114%, U.S. population increased 31%, and vehicle miles traveled increased 127%.

In addition, urban and industrial areas maintain an air pollution index. This scale, a composite of several pollutant levels recorded from a particular monitoring site or sites, yields an overall air quality value. Public warnings are given if the index exceeds certain values; in severe instances residents might be asked to stay indoors and factories might even be closed down.

While such air quality emergencies seem increasingly rare in the United States, developing countries, as well as Eastern European nations, continue to suffer poor air qual-

ity, especially in urban areas such as Bangkok, Thailand and Mexico City, Mexico. In Mexico City, for example, seven out of ten newborns have higher lead levels in their blood than the World Health Organization (WHO) considers acceptable. At present, many Third World countries place national economic development ahead of pollution control—and in many countries with rapid industrialization, high population growth, or increasing per capita income, the best efforts of governments to maintain air quality are outstripped by rapid proliferation of automobiles, escalating factory emissions, and runaway urbanization.

For all the progress the United States has made in reducing ambient air pollution, *indoor* air pollution may pose even greater risks than all of the pollutants we breathe outdoors. The Radon Gas and Indoor Air Quality Act of 1986 directed the EPA to research and implement a public information and technical assistance program on indoor air quality. From this program has come monitoring equipment to measure an individual's "total exposure" to pollutants both in indoor and outdoor air. Studies done using this equipment have shown indoor exposures to toxic air pollutants far exceed outdoor exposures for the simple reason that most people spend 90% of their time in office buildings, homes, and other enclosed spaces. Moreover, nationwide energy conservation efforts following the oil crisis of the 1970s led to building designs that trap pollutants indoors, thereby exacerbating the problem.

• AIR POLLUTION CONTROL

The need to control air pollution was recognized in the earliest cities. In the Mediterranean at the time of Christ, laws were developed to place objectionable sources of odor and smoke downwind or outside city walls. The adoption of fossil fuels in thirteenth-century England focused particular concern on the effect of coal smoke on health, with a number of attempts at regulation with regard to fuel type, chimney heights, and time of use. Given the complexity of the air pollution problem it is not surprising that these early attempts at control met with only limited success.

The nineteenth century was typified by a growing interest in urban public health. This developed against a background of continuing industrialization, which saw smoke abatement clauses incorporated into the growing body of sanitary legislation in both Europe and North America. However, a lack of both technology and political will doomed these early efforts to failure, except in the most blatantly destructive situations (for example, industrial settings such as those around alkali works in England).

The rise of environmental awareness in the current century has reminded us that air pollution ought not to be seen as a necessary product of industrialization. This has redirected responsibility for air pollution towards those who create it. The notion of "making the polluter pay" is seen as a central feature of air pollution control. The century has also seen the development of a range of broad air pollution control strategies, among them:

1. Air quality management strategies that set ambient air quality standards so that emissions from various sources can be monitored and controlled.

2. Emission standards strategy that sets limits for the amount of a pollutant that can be emitted from a given source. These may be set to meet air quality standards, but the strategy is optimally seen as one of adopting best available technology not entailing excessive costs (BATNEEC).

3. Economic strategies that involve charging the party responsible for the pollution. If the level of charge is set correctly, some polluters will find it more economical to install air pollution control equipment than continue to pollute. Other methods utilize a system of tradable pollution rights.

4. Cost-benefit analysis, which attempts to balance economic benefits with environmental costs. This is an appealing strategy but difficult to implement because of its controversial and imprecise nature.

In general, air pollution strategies have either been air quality or emission based. In the United Kingdom, emission strategy is frequently used; for example, the Alkali and Works Act of 1863 specifies permissible emissions of hydrochloric acid. By contrast, the United States has aimed to achieve air quality standards, as evidenced by the Clean Air Act. One criticism of using air quality strategy has been that while it improves air in poor areas it leads to degradation in areas with high air quality. Although the emission standards approach is relatively simple, it is criticized for failing to make explicit judgments about air quality and assumes that good practice will lead to an acceptable atmosphere.

Until the mid-twentieth century, legislation was primarily directed towards industrial sources, but the passage of the United Kingdom Clean Air Act (1956), which followed the disastrous smog of December 1952, directed attention towards domestic sources of smoke. While this particular act may have reinforced the improvements already under way, rather than initiating improvements, it has served as a catalyst for much subsequent legislative thinking. Its mode of operation was to initiate a change in fuel, perhaps one of the oldest methods of control. The other well-tried aspects were the creation of smokeless zones and an emphasis on tall chimneys to disperse the pollutants.

As simplistic as such passive control measures seem, they remain at the heart of much contemporary thinking. Changes from coal and oil to the less-polluting gas or electricity have contributed to the reduction in smoke and sulfur dioxide concentrations in cities all around the world. Industrial zoning has often kept power and large manufacturing plants away from centers of human population, and "superstacks," chimneys of enormous height, are now quite common. Successive changes in automotive fuels—lead-free gasoline, low-volatility gas, methanol,

or even the interest in the electric automobile—are further indications of continued use of these methods of control.

There are more active forms of air pollution control that seek to clean up the exhaust gases. The earliest of these were smoke and grit arresters that came into increasing use in large electrical stations during the twentieth century. Notable here were the cyclone collectors that removed large particles by driving the exhaust through a tight spiral that threw the grit outward where it could be collected. Finer particles could be removed by electrostatic precipitation. These methods were an important part of the development of the modern pulverized fuel power station. However, they failed to address the problem of gaseous emissions. Here it has been necessary to look at burning fuel in ways that reduce the production of nitrogen oxides. Control of sulfur dioxide emissions from large industrial plants can be achieved by desulfurization of the flue gases. This can be quite successful by passing the gas through towers of solid absorbers or spraying solutions through the exhaust gas stream. However, these are not necessarily cheap options.

The catalytic converter is also an important element of active attempts to control air pollutants. Although these can considerably reduce emissions, they have to be offset against the increasing use of the automobile. There is much talk of the development of zero pollution vehicles that do not emit any pollutants.

Legislation and control methods are often associated with monitoring networks that assess the effectiveness of the strategies and inform the general public about air quality where they live. A balanced approach to the control of air pollution in the future may have to look far more broadly than simply at technological controls. It will become necessary to examine the way we structure our lives in order to find more effective solutions to air pollution.

• AIR POLLUTION INDEX

The air pollution index is a value derived from an air quality scale that uses the measured or predicted concentrations of several criteria pollutants and other air-quality indicators, such as coefficient of haze (COH) or visibility. The best-known index of air pollution is the pollutant standard index (PSI).

The PSI has a scale that spans from 0 to 500. The index represents the highest value of several subindices; there is a subindex for each pollutant, or in some cases, for a product of pollutant concentrations and a product of pollutant concentrations and COH. If a pollutant is not monitored, its subindex is not used in deriving the PSI.

The subindex of each pollutant or pollutant product is derived from a PSI nomogram that matches concentrations with subindex values. The highest subindex value becomes the PSI. The PSI has five health-related categories: good (0–50); moderate (50–100); unhealthy (100–200); very unhealthy (200–300) hazardous (300–500).

• CLEAN AIR ACT (1963, 1970, 1990)

The 1970 Clean Air Act and major amendments to the act in 1977 and 1990 serve as the backbone of efforts to control air pollution in the United States. This law established one of the most complex regulatory programs in the country. Efforts to control air pollution in the United States date back to 1881, when Chicago and Cincinnati passed laws to control smoke and soot from factories in the cities. Other municipalities followed suit and the momentum continued to build. In 1952, Oregon became the first state to adopt a significant program to control air pollution, and three years later, the federal government became involved for the first time, when the Air Pollution Control Act was passed. This law granted funds to assist the states in their air pollution control activities.

In 1963, the first Clean Air Act was passed. This act provided permanent federal aid for research, support for the development of state pollution control agencies, and federal involvement in cross-boundary air pollution cases. An amendment to the act in 1965 directed the Department of Health, Education, and Welfare (HEW) to establish federal emission standards for motor vehicles. (At that time, HEW administered air pollution laws. The EPA was not created until 1970.) This represented a significant move by the federal government from a supportive to an active role in setting air-pollution policy. The 1967 Air Quality Act provided additional funding to the states, required the states to establish Air Quality Control Regions, and directed HEW to obtain and make available information on the health effects of air pollutants and to identify pollution control techniques. All of these components of the law were designed to assist the states, but they further demonstrated increasing federal involvement in the issue.

The Clean Air Act of 1970 marked a dramatic change in air pollution policy in the United States. Following the passage of this law, the federal government, not the states, would be the focal point for air pollution policy. This act established the framework that continues to be the foundation for air pollution control policy. The impetus for this change was the belief that the current state-based approach was not working. Public sentiment was growing so significantly that environmental issues demanded the attention of high-ranking officials. In fact, the leading policy entrepreneurs on the issue were President Richard Nixon and Senator Edmund Muskie of Maine.

These men and other leaders devised a plan with four key components. First, National Ambient Air Quality Standards (NAAQS) were established for six major pollutants: carbon monoxide, lead (in 1977), nitrogen dioxide, ground-level ozone (a key component of smog), particulate matter, and sulfur dioxide. For each of these pollutants, sometimes referred to as criteria pollutants, primary and secondary standards were set. The primary standards were designed to protect human health; the secondary standards were based on protecting crops, forests, and buildings if the primary standards were not capable of doing so. The act stipulated that these standards must apply to the entire country and be established by the EPA, based on the best available scientific information. Relatedly, the EPA was to establish standards for less common toxic air pollutants.

Second, New Source Performance Standards (NSPS) would be established by the EPA. These standards would determine how much air pollution would be allowed by new plants in the various industrial sectors. The standards are to be based on the best affordable technology available for the control of pollutants at sources such as power plants, steel factories, and chemical plants.

Third, mobile source emission standards were established to control automobile emissions. These standards were specified in the statute (rather than left to the EPA), and schedules for meeting these standards were also written into the law. It was thought that such an approach was crucial in having success with the powerful auto industry. The pollutants regulated were carbon monoxide, hydrocarbons, and nitrogen oxides, with goals of reducing the first two pollutants by 90% by 1975, and nitrogen oxides by 82% by 1975.

The final component of the air quality protection framework involved the implementation of the above procedures. Each state would be encouraged to devise a state implementation plan (SIP), which would indicate how the state would achieve the national standards. This gave each state some flexibility while still maintaining national standards. These plans had to be approved by the EPA; if a state did not have an approved SIP, the EPA would administer the Clean Air Act in that state. However, since the federal government is in charge of establishing pollution standards for new mobile and stationary sources, even the states with an SIP have limited flexibility. The main focal point for the states was the control of existing stationary sources, and if necessary, mobile sources. The states had to set limits in their SIPs that allowed them to achieve the NAAQS by a statutorily determined deadline (originally 1975, but subsequently delayed). One problem with this approach was the construction of tall smokestacks, which helped move pollution out of a particular airshed but did not reduce overall pollution levels. The states are also charged with monitoring and enforcing the Clean Air Act.

The 1977 amendments to the Clean Air Act dealt with three main issues: nonattainment, auto emissions, and the prevention of air quality deterioration in areas where the air was already relatively clean. The first two issues were re-

National Ambient Air Quality Standards

National Ambient Air Quality Standards (NAAQS) have been established by the U.S. Environmental Protection Agency for the following six criteria air pollutants:

NATIONAL AMBIENT AIR QUALITY STANDARDS			
Pollutant	**Averaging Time**	**Primary Standard**	**Secondary Standard**
CO	8 Hours	9 ppm	None
	1 Hour	35 ppm	None
Lead (Pb)	Calendar Quarter	1.5 $\mu g/m^3$	Same as Primary
NO_2	Annual	0.053 ppm	Same as Primary
O_3	1 Hour	0.12 ppm	Same as Primary
PM_{10}	Annual	50 $\mu g/m^3$	Same as Primary
	24 Hours	150 $\mu g/m^3$	Same as Primary
SO_2	Annual	0.03 ppm	None
	24 Hours	0.14 ppm	None
	3 Hours	None	0.5 ppm
The TSP NAAQS is no longer applicable. It was superseded by the PM_{10} NAAQS on 07/01/87. The old TSP NAAQS is provided for information only.			
TSP	Annual	75 $\mu g/m^3$	60 $\mu g/m^3$
	24 Hours	260 $\mu g/m^3$	150 $\mu g/m^3$

The NAAQS are the allowable ambient (outdoor) concentrations that must be maintained in order to protect public health and welfare. Limits have been set for carbon monoxide (CO), lead (Pb), nitrogen dioxide (NO_2), ozone (O_3), sulfur dioxide (SO_2), and particulate matter (PM_{10}). EPA is currently reviewing the adequacy of the ozone and PM_{10} standards.

(Courtesy of U.S. government publication.)

solved primarily by delaying deadlines and increasing penalties. Largely in response to a court decision in favor of environmentalists (*Sierra Club v. Ruckelshaus*, 1972), the 1977 amendments included a program for the prevention of significant deterioration (PSD) of air that was already clean. This program would prevent polluting the air up to the national levels in areas where the air was cleaner than the standards. In Class I areas, areas with near pristine air quality, no new significant air pollution would be allowed. Class I areas are airsheds over larger national parks and wilderness areas. In Class II areas a moderate degree of air quality deterioration would be allowed. And finally, in Class III areas, air deterioration up to the national secondary standards would be allowed. Most of the country that had area cleaner than the NAAQS was classified as Class II. Related to the prevention of significant deterioration is a provision to protect and enhance visibility in national parks and wilderness areas even if the air pollution is not a threat to human health. The impetus of this section of the bill was the growing visibility problem in parks, especially in the Southwest.

Throughout the 1980s efforts to further amend the Clean Air Act were stymied. President Ronald Reagan was opposed to any strengthening of the Act, which he argued would hurt the economy. In Congress, the controversy over acid rain between members from the Midwest and the Northeast further contributed to the stalemate. Gridlock on the issue broke with the election of George Bush, who supported amendments to the Act, and the rise of Senator George Mitchell of Maine to Senate Majority Leader. Over the next two years, the issues were hammered out between environmentalists and industry and between different regions of the country. Major players in Congress were Representatives John Dingell of Michigan and Henry Waxman of California and Senators Robert Byrd of West Virginia and Mitchell.

Major amendments to the Clean Air Act were finally passed in the fall of 1990. These amendments addressed four major topics: (1) acid rain, (2) toxic air pollutants, (3) nonattainment areas, and (4) ozone layer depletion. To address acid rain, a 10-million-ton reduction in annual sulfur dioxide emissions (a 40% reduction based on 1980 levels) and a 2-million-ton annual reduction in nitrogen oxides by the year 2000 was required. Most of this reduction came from old utility power plants. The law also creates marketable pollution allowances, so that a utility that reduces emissions more than required can sell those pollution rights to another source. Economists argue that such an approach should become more widespread for all pollution control, to

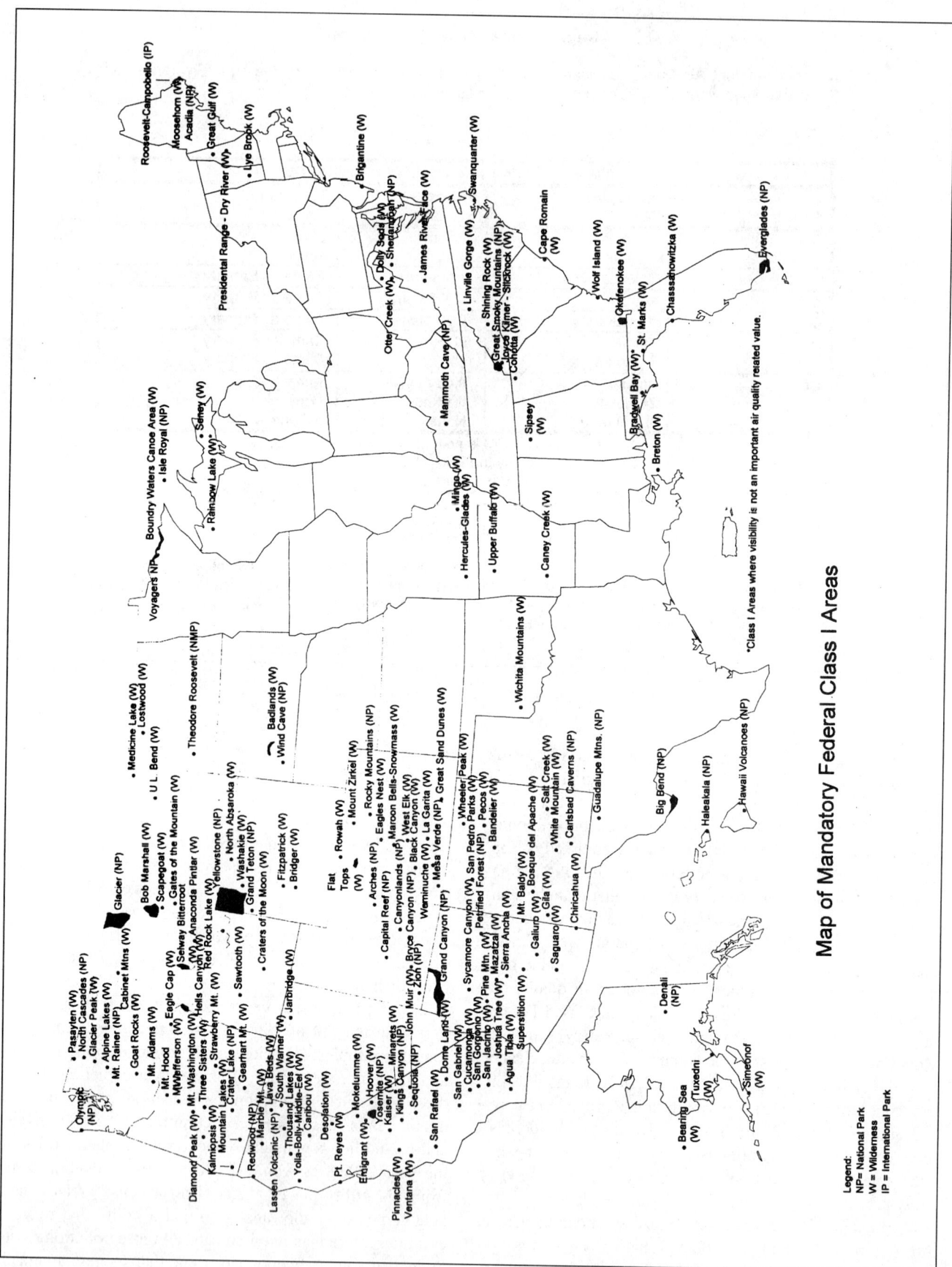

Map of Mandatory Federal Class I Areas

Legend:
NP= National Park
W = Wilderness
IP = International Park

*Class I Areas where visibility is not an important air quality related value.

(Courtesy of U.S. government publication.)

increase efficiency. Due to the failure of the toxic air pollutant provisions of the 1970 Clean Air Act, new, more stringent provisions were adopted requiring regulations for all major sources of 189 varieties of toxic air pollution within 10 years. Areas of the country still in nonattainment for criteria pollutants will be given from three to 20 years to meet these standards. These areas are also required to impose tighter controls to meet these standards. To help these areas and other parts of the country, the act requires stiffer motor vehicle emissions standards and cleaner gasoline. Finally, three chemical families that contribute to the destruction of the stratospheric ozone layer (chlorofluorocarbons [CFCs], hydrochlorofluorocarbons [HCFCs], and methyl chloroform) are to be phased out of production and use.

The Clean Air Act has met with mixed success. The national average pollutant levels for the criteria pollutants have decreased. Nevertheless, many localities have not achieved these standards and are in perpetual nonattainment. Not surprisingly, major urban areas are those most frequently in nonattainment. The pollutant for which standards are most often exceeded is ozone, or smog. The greatest successes have come with lead, which has been reduced by 96% (largely due to the phasing-out of leaded gasoline), and particulates, which were reduced by over 60%. Additionally, despite numerous delays, the carbon monoxide, hydrocarbon, and nitrogen oxides pollution from new cars has decreased by 96%, 96%, and 76% over the period from 1967 to 1990. A final point of caution concerning evaluating the Clean Air Act: due to the tremendous complexity of air quality, we cannot conclude that all changes in pollutant levels are due to the law. These changes may be due to shifts in the economy at large, changes in weather patterns, or other such variables.

• AIR QUALITY CONTROL REGION (AQCR)

The Clean Air Act defines an Air Quality Control Region as a contiguous area where air quality, and thus air pollution, is relatively uniform. In those cases where topography is a factor in air movement, AQCRs often correspond with airsheds. AQCRs may consist of two or more cities, counties, or other governmental entities, and each region is required to adopt consistent pollution control measures across the political jurisdictions involved. AQCRs may even cross state lines and, in these instances, the states must cooperate in developing pollution control strategies. Each AQCR is treated as a unit for the purposes of pollution reduction and achieving National Ambient Air Quality Standards. As of 1993, most AQCRs had achieved national air quality standards; however the remaining AQCRs where standards had not been achieved were a significant group, where a large percentage of the United States population dwelled. AQCRs involving major metro areas like Los Angeles, New York,

Houston, Denver, and Philadelphia were not achieving air quality standards because of smog, motor vehicle emissions, and other pollutants.

• OZONE

Ozone (O_3) is a toxic, colorless gas (but can be blue when in high concentration) with a characteristic acrid odor. A variant of normal oxygen, it has three oxygen atoms per molecule rather than the usual two. Ozone strongly absorbs ultraviolet radiation at wavelengths of 220–290 nanometers (nm) with peak absorption at 260.4 nm. Ozone will also absorb infrared radiation at wavelengths in the range 9–10 μm. Ozone occurs naturally in the ozonosphere (ozone layer), which surrounds Earth, protecting living organisms at Earth's surface from ultraviolet radiation. The ozonosphere is located in the stratosphere at 6–30 mi above Earth's surface, with the highest concentration at 7.5–12 mi. The concentration of ozone in the ozonosphere is 1 molecule per 100,000 molecules, or if the gas were at standard temperature and pressure, the ozone layer would be 0.12 in thick. However, the ozone layer absorbs over 90% of incident ultraviolet radiation.

Ozone in the stratosphere results from a chemical equilibrium between oxygen, ozone, and ultraviolet radiation. Ultraviolet radiation is absorbed by oxygen and produces ozone. Simultaneously, ozone absorbs ultraviolet radiation and decomposes to oxygen and other products. Ozone layer depletion occurs as a result of complex reactions in the atmosphere between organic compounds that react with ozone faster than the ozone is replenished. Compounds of most concern include the byproducts of ultraviolet degradation of chlorofluorocarbons (CFCs), chlorine, and fluorine.

Ozone is also a secondary air pollutant at Earth's surface as a result of complex chemical reactions between sunshine and primary pollutants, such as hydrocarbons and oxides of nitrogen. Ozone can also be generated in the presence of oxygen from equipment that gives off intense light, electrical sparks, or creates intense static electricity, such as photocopiers and laser printers. Human olfactory senses are very sensitive to ozone, being able to detect ozone odor at concentrations of 0.02–0.05 parts per million. Toxic symptoms for humans from exposure to ozone include headaches and drying of the throat and respiratory tracts. Ozone is highly toxic to many plant species and destroys or degrades many building materials, such as paint, rubber, and some plastics. The total losses in the United States each year due to ozone damage to crops, livestock, buildings, natural systems, and human health is estimated to be in the tens of billions of dollars. The threshold limit value (TLV) for air quality standards is 0.1 ppm, or 0.2 mg O_3 per m^3 of air.

Industrial uses of ozone include chemical manufacturing and air, water, and waste treatment. Industrial quantities of ozone are typically generated from air or pure oxygen by means of silent corona discharge. Ozone is used in water treatment as a disinfectant to kill pathogenic microorganisms or for oxidation of organic and inorganic compounds. Combinations of ozone and hydrogen peroxide or ultraviolet radiation in water can generate powerful oxidants useful in breaking down complex synthetic organic compounds. In wastewater treatment, ozone can be used to disinfect effluents, or decrease their color and odor. In some industrial applications, ozone can be used to enhance biodegradation of complex organic molecules. Industrial cooling tower treatment with ozone prevents transmission of airborne pathogenic organisms and can reduce odor.

• OZONE LAYER DEPLETION

The ozone layer in Earth's upper atmosphere helps make life on the planet possible by shielding it from 95–99% of the Sun's potentially deadly ultraviolet radiation. This radiation is harmful and sometimes lethal to wildlife, crops, and vegetation, and can cause fatal skin cancer, cataracts, and immune system damage in humans.

Destroying the ozone shield

Ozone, a form of oxygen consisting of three atoms of oxygen instead of two, is considered an air pollutant when found at ground levels and is a major component of smog. It is formed by the reaction of various air pollutants in the presence of sunlight. Ozone is also used commercially as a bleaching agent and to purify municipal water supplies. Since ozone is toxic, the gas is harmful to health when generated near Earth's surface. Because of its high rate of breakdown, such ozone never reaches the upper atmosphere.

But the ozone that shields Earth from the Sun's radiation is found in the stratosphere, a layer of the upper atmosphere found 9–30 mi above ground. This ozone layer is maintained as follows: the action of ultraviolet light breaks O_2 molecules into atoms of elemental oxygen (O). The elemental oxygen then attaches to other O_2 molecules to form O_3. When it absorbs ultraviolet radiation that would otherwise reach Earth, ozone is, in turn, broken down into $O_2 + O$. The elemental oxygen generated then finds another O_2 molecule to become O_3 once again.

In 1974, chemists F. Sherwood Rowland and Mario J. Molina realized that chlorine from chlorofluorocarbon (CFC) molecules was capable of breaking down ozone in the stratosphere. In time, evidence began to accumulate that the ozone layer was indeed being broken apart by these industrial chemicals, and to a lesser extent by nitrogen oxide emissions from jet airplanes as well as hydrogen chloride emissions from large volcanic eruptions.

When released into the environment, CFCs slowly rise into the upper atmosphere, where they are broken apart by solar radiation. This releases chlorine atoms that act as catalysts, breaking up molecules of ozone by stripping away one of their oxygen atoms. The chlorine atoms, unaltered by the reaction, are each capable of destroying ozone molecules repeatedly. Without a sufficient quantity of ozone to block its way, ultraviolet radiation from the Sun passes through the upper atmosphere and reaches Earth's surface.

When damage to the ozone layer first became apparent in 1974, propellants in aerosol spray cans were a major source of CFC emissions, and CFC aerosols were banned in the United States in 1978. Production of CFC-12 (also known by R-12 or the trade name Freon), used in cooling and refrigeration, ended in 1995, although use is allowed until supplies are depleted. However, CFCs have since remained in widespread use in thermal insulation, as cleaning solvents, and as foaming agents in plastics, resulting in continued and accelerating depletion of stratospheric ozone.

The Antarctic ozone hole

The most dramatic evidence of the destruction of the ozone layer has occurred over Antarctica, where a massive "hole" in the ozone layer appears each winter and spring, apparently exacerbated by the area's unique and violent climatological conditions. The destruction of ozone molecules begins during the long, completely dark, and extremely cold Antarctic winter, when swirling winds and ice clouds begin to form in the lower stratosphere. This ice reacts with chlorine compounds in the stratosphere (such as hydrogen chloride and chlorine nitrate) that come from the breakdown of CFCs, creating molecules of chlorine.

When spring returns in August and September, a seasonal vortex—a rotating air mass—causes the ozone to mix with certain chemicals in the presence of sunlight. This helps break down the chlorine molecules into chlorine atoms, which, in turn, react with and break up the molecules of ozone. A single chlorine, bromine, or nitrogen molecule can break up literally thousands of ozone molecules.

During December, the ozone-depleted air can move out of the Antarctic area, as happened in 1987, when levels of ozone over southern Australia and New Zealand sank by 10% over a three-week period, causing as much as a 20% increase in ultraviolet radiation reaching Earth. This may have been responsible for a reported increase in skin cancers and damage to some food crops.

The seasonal hole in the ozone layer over Antarctica has been monitored by scientists at the National Aeronautics and Space Administration's (NASA) Goddard Space Flight Center outside Washington, D.C. NASA's NIMBUS-7 satellite first discovered drastically reduced ozone levels over the Southern Hemisphere in 1985, and measurements are also being conducted with instruments on aircraft and balloons. Some of the data that has been gathered is alarming.

In October 1987, ozone levels within the Antarctic ozone hole were found to be 45% below normal, and similar reductions occurred in October 1989. A 1988 study revealed that since 1969, ozone levels had declined by 2% world-wide, and by as much as 3% or more over highly populated areas of North America, Europe, South America, Australia, and New Zealand.

In September 1992, the NIMBUS-7 satellite found that the depleted ozone area over the southern polar region had grown 15% from the previous year, to a size three times larger than the area of the United States, and was 80% thinner than usual. The ozone hole over Antarctica was measured at approximately 8.9 million mi^2, as compared to its usual size of 6.5 million mi^2. The contiguous 48 states are, by comparison, about 3 million mi^2, and all of North America covers 9.4 million mi^2. Researchers attributed the increased thinning not only to industrial chemicals but also to the 1991 volcanic eruptions of Mount Pinatubo in the Philippines and Mount Hudson in Chile, which emitted large amounts of sulfur dioxide into the atmosphere.

Dangers of ultraviolet radiation

The major consequence of the thinning of the ozone layer is the penetration of more solar radiation, especially ultraviolet-B (UV-B) rays, the most dangerous type, which can be extremely damaging to plants, wildlife, and human health. Because UV-B can penetrate the ocean's surface, it is potentially harmful to marine life forms and indeed to the entire chain of life in the seas as well.

UV-B can kill and affect the reproduction of fish, larvae, and other plants and animals, especially those found in shallow waters, including phytoplankton, which forms the basis of the oceanic food chain/web. The National Science Foundation reported in February 1992 that its research ship, on a six week Antarctic cruise, found that the production of phytoplankton decreases at least 6–12% during the period of greatest ozone layer depletion, and that the destructive effects of UV radiation could extend to depths of 90 ft.

A decrease in phytoplankton would affect all other creatures higher on the food chain and dependent on them, including zooplankton, microscopic ocean creatures that feed on phytoplankton and are also an essential part of the ocean food chain. And marine phytoplankton are the main food source for krill, tiny Antarctic shrimp that are the major food source for fish, squid, penguins, seals, whales, and other creatures in the Southern Hemisphere.

Moreover, phytoplankton are responsible for absorbing, through photosynthesis, great amounts of carbon dioxide (CO_2) and releasing oxygen. It is not known how a depletion of phytoplankton would affect the planet's supply of life-giving oxygen, but more CO_2 in the atmosphere would exacerbate the critical problem of global warming, the so-called greenhouse effect.

There are numerous reports, largely unconfirmed, of animals in the southern polar region being harmed by ultraviolet radiation. Rumors abound in Chile, for example, of pets, livestock, sheep, rabbits, and other wildlife getting cataracts, suffering reproductive irregularities, or even being blinded by solar radiation. Many residents of Chile believe these stories, and wear sunglasses, protective clothing, and sun-blocking lotion in the summer, or even stay indoors much of the day when the sun is out. If the ozone layer's thinning continues to spread, the lifestyles of people across the globe could be similarly disrupted for generations to come.

Particularly frightening have been incidents reported to have taken place in Punta Arenas, Chile's southernmost city, at the tip of Patagonia. After several days of record low levels of ozone were recorded in October 1992, people reported severe burns from short exposure to sunlight. Sheep and cattle became blind, and some starved because they could not find food. Trees wilted and died, and melanoma-type skin cancers seem to have increased dramatically. Similar stories have been reported from other areas of the Southern Hemisphere. And malignant melanoma, once a rare disorder, is now the fastest rising cancer in the world.

Ozone thinning spreads

Indeed, ozone layer depletion is spreading at an alarming rate. In the 1980s, scientists discovered that an ozone hole was also appearing over the Arctic region in the late winter months, and concern was expressed that similar thinning might begin to occur over, and threaten, heavily populated areas of the globe. These fears were confirmed in April 1991, when the Environmental Protection Agency (EPA) announced that satellite measurements had recorded an ominous decrease in atmospheric ozone, amounting to an average of 5% over the mid-latitudes (including the United States), almost double the loss previously thought to be occurring.

The data showed that ozone levels measured in the late fall, winter, and early spring over large areas of the United States, Europe, and the mid-latitudes of the Northern and Southern Hemispheres had dropped by 4–6% over the last decade—twice the amount estimated in earlier years. The greatest area of ozone thinning in the United States was found north of a line stretching from Philadelphia to Denver to Reno, Nevada. One of the most alarming aspects of the new findings was that the ozone depletion was continuing into April and May, a time when people spend more time outside, and crops are beginning to sprout, making both more vulnerable to ultraviolet radiation.

The new findings led the EPA to project that over the next 50 years, thinning of the ozone layer could cause Americans to suffer some 12 million cases of skin cancer, 200,000 of which would be fatal. Several years earlier, the agency had calculated that over the next century, there could be an additional 155 million cases of skin cancers and 3.2 million deaths if the ozone layer continued to thin at the then current rate. Another EPA projection made in the 1980s was that the increase in radiation could cause Americans to suffer 40 million cases of skin cancer and 800,000 deaths in the following 88 years, plus some 12 million eye cataracts.

No one can say how accurate such varying projections will turn out to be, but evidence of ozone layer thinning is well documented. In October 1991, additional data of spreading ozone layer destruction were made public. Dr. Robert Watson, a NASA scientist who co-chairs an 80-member panel of scientists from 80 countries, called the situation "extremely serious," saying that "we now see a significant decrease of ozone both in the Northern and Southern Hemispheres, not only in winter but in spring and summer, the time when people sunbathe, putting them at risk for skin cancer, and the time when we grow crops."

In February 1992, a team of NASA scientists announced that they had found record high levels of ozone-depleting chlorine over the Northern Hemisphere. This could, in turn, lead to an ozone "hole" similar to the one that appears over Antarctica developing over populated areas of the United States, Canada, and England. The areas over which increased levels of chlorine monoxide were found extended as far south as New England, France, Britain, and Scandinavia.

Action to protect the ozone layer

As evidence of the critical threats posed by ozone layer depletion has increased, the world community has begun to take steps to address the problem. In 1987, the United States and 22 other nations signed the Montreal Protocol, agreeing, by the year 2000, to cut CFC production in half, and to phase out two ozone-destroying gases, Halon 1301 and Halon 1211. Halons are human-made bromine compounds used mainly in fire extinguishers, and can destroy ozone at a rate 10 to 40 times more rapidly than CFCs. Fortunately, these restrictions appear to already be having an impact. In 1992, it was found that the rate at which these two Halon gases were accumulating in the atmosphere had fallen significantly since 1987. The rate of increase of levels of Halon 1301 was about 8% per year during 1989–1992, about half of the average annual rate of growth over previous years. Similarly, Halon 1211 was increasing at only 3% annually, much less than the previous growth of 15% a year.

Since the Montreal Protocol, other international treaties have been signed limiting the production and use of ozone-destroying chemicals. When alarming new evidence on the destruction of stratospheric ozone became available in 1988, the world's industrialized nations convened a series of conferences to plan remedial action. In March 1989, the 12-member European Economic Community (EEC) announced plans to end the use of CFCs by the turn of the century, and the United States agreed to join in the ban. A week later, 123 nations met in London to discuss ways to speed the CFC phase-out. The industrial nations agreed to cut their own domestic CFC production in half, while continuing to allow exports of CFCs, in order to accommodate Third World nations.

The large industrial nations, which have created the CFC problem, are now much more willing to take effective action to ban the compounds than are many developing nations, such as India and China. The latter nations resist restrictions on CFCs on the grounds that the chemicals are necessary for their own economic development.

After the meeting in London, leaders and representatives from 24 countries met in an environmental summit at The Hague, Netherlands, and agreed that the United Nations' authority to protect the world's ozone layer should be strengthened.

In May 1989, members of the EEC and 81 other nations that had signed the 1987 Montreal Protocol decided at a meeting in Helsinki, Finland, to try to achieve a total phase-out of CFCs by the year 2000, as well as phase-outs as soon as possible of other ozone-damaging chemicals like carbon tetrachloride, halons, and methyl chloroform. In London in June 1990, most of the Montreal Protocol's signatory nations formally adopted a deadline of the year 2000 for industrial nations to phase out the major ozone-destroying chemicals, with 2010 being the goal for developing countries.

Finally, in November 1992, 87 nations meeting in Copenhagen, Denmark, decided to strengthen the action agreed to under the Montreal Protocol and move up the phase-out deadline from 2000 to January 1, 1996 for CFCs, and to January 1, 1994 for halons. A timetable was also agreed to for eliminating hydrochlorofluorocarbons (HCFCs) by the year 2030. HCFCs are being used as substitutes for CFCs even though they also deplete ozone, albeit on a far lesser scale than CFCs. The conference failed to ban the production of the pesticide methyl bromide, which may account for 15% of ozone depletion by the year 2000, but did freeze production at 1991 levels.

Environmentalists were disappointed that stronger action was not taken to protect the ozone layer. But Environmental Protection Agency (EPA) Administrator William K. Reilly, who headed the United States delegation, estimated that the reductions agreed to could, by the year 2075, prevent a million cases of cancer and 20,000 deaths.

Although the restrictions apply to developed nations, which produce most of the ozone-damaging chemicals, it was also agreed to consider moving up a phase-out of such compounds by developing nations from 2010 to 1995. A month after the Copenhagen conference, the nations of the European Community agreed to push bans on the use of CFCs and carbon tetrachloride to 1995 and to cut CFC emissions by 85% by the end of 1993.

The private sector has also taken action to reduce CFC production. The world's largest manufacturer of the chemicals, DuPont Chemical Company, announced in 1988 that it was working on a variety of substitutes for CFCs, would phase out production of them by 1996, and would partially replace them with HCFCs. Environmentalists charge that DuPont has been moving too slowly to eliminate production of these chemicals.

There are many ways that individuals can help reduce the release of CFCs into the atmosphere, mainly by avoiding products that contain or are made from CFCs, and by recycling CFCs whenever possible. Although CFCs have not generally been used in spray cans in the United States since 1978, they are still used in many consumer and industrial products, such as styrofoam. Other products manufactured using CFCs include solvents and cleaning liquids used on electrical equipment, polystyrene foam products, and fire extinguishers that use halons.

Refrigerants in cars and home air conditioning units that still use CFCs must be poured into closed containers to be cleaned or recycled, or they will evaporate into the atmosphere. Using foam insulation to seal homes also releases CFCs. Many alternatives to foam insulation exist, such as cellulose fiber, gypsum, fiberboard, and fiberglass.

Unfortunately, whatever steps are taken in the next few years, the problem of ozone layer depletion will continue even after the release of ozone-destroying chemicals is limited or halted. It takes six to eight years for some of these compounds to reach the upper atmosphere, and once there, they will destroy ozone for another 20–25 years. Thus, even if all emissions of destructive chemicals were stopped, compounds already released would continue to damage the ozone layer for another quarter century.

Understanding ozone depletion

As detailed collection of data about interactions in the stratosphere progresses, the observational support for the ozone depletion theory continues to grow more compelling. Yet atmospheric scientists are beginning to realize that their understanding of the upper atmosphere is still quite crude. While certain key reactions that maintain and destroy ozone are theoretically and observationally supported, scientists will have to comprehend the interaction of dozens, if not hundreds, of reactions between natural and artificial species of hydrogen, nitrogen, bromine, chlorine, and oxygen before a complete picture of ozone-layer dynamics emerges. The eruption of Mt. Pinatubo, for example, made scientists aware that heterogenous processes—those reactions which require cloud surfaces to take place—may play a far greater role in causing ozone depletion than originally believed. Such reactions had previously been observed taking place only at Earth's poles, where stratospheric clouds form during the long winter darkness, but it is now thought that sulfur aerosols ejected by Pinatubo may be serving as catalysts to speed ozone depletion at nonpolar latitudes.

Ozone-depleting reactions are best understood around the thinly inhabited polar regions, where stable and isolated conditions over the winter allow scientists to understand stratospheric changes most easily. In contrast, at the temperate latitudes where constantly moving air masses undergo no seasonal isolation, it is difficult to determine whether a fluctuation in a given chemical's density is a result of local reactions or atmospheric turbulence. It is hoped that increasingly detailed measurements using a new generation of equipment (such as NASA's Perseus remote-control

aircraft) will begin to shed more light on the processes occurring away from the poles. Joe Waters of NASA's Jet Propulsion Laboratory summarizes the urgent task: "We must be able to lay out the catalytic cycles that are destroying ozone at all altitudes all over the globe—from its production region in the tropics to the higher latitudes and the polar regions."

• AIR POLLUTION AND WEATHER

Natural phenomena affecting air quality

The concentration of atmospheric pollutants observed at different locations depends on more than just the quantity of pollutants emitted at the various sources. The atmosphere is the agent that transports and disperses pollutants between sources and receptors. Consequently, the state of the atmosphere helps to determine the concentrations of pollutants observed at receptors. Unlike emissions sources, which can be controlled, the state of the atmosphere is not at present susceptible to human control.

Some skill has been attained, however, in predicting the future state of the atmosphere. Since meteorological conditions that favor high concentrations of pollutants are known, severe air pollution episodes can therefore be forecast.

In general, three parameters are used to describe atmospheric transport and dispersion processes. These are wind speed, wind direction, and atmospheric stability. For emissions at a given source, a higher wind speed provides the pollutants with a greater air volume within which to disperse. This causes ground level pollutant concentrations, other things being equal, to be inversely proportional to wind speed.

Horizontally, the wind direction is the strongest factor affecting pollutant concentrations. For a given wind direction, nearly all the pollutant transport and dispersion will be downwind. Wind direction determines which sector of the area surrounding a source will receive pollutants from that source.

Atmospheric stability directly affects the vertical dispersion of atmospheric pollutants. Unlike wind direction and wind speed, atmospheric stability cannot be measured directly. Atmospheric stability is a measure of air turbulence and may be defined in terms of the vertical atmospheric temperature profile. When the temperature decreases rapidly with height, vertical motions in the atmosphere are enhanced, and the atmosphere is called unstable. An unstable atmosphere, with its enhanced vertical motions, is more effective for dispersing pollutants, and because of the large volume of air available for the spread of pollutants, ground-level concentrations can be relatively low. When the temperature does not decrease rapidly with height, vertical motions are neither enhanced nor repressed and the stability is described as neutral. Under these conditions, pollutants are also allowed to disperse vertically in the atmosphere, although not as rapidly as when it is unstable.

When the temperature decreases very little, remains the same, or increases with increasing height, the atmosphere is called stable. Under these conditions, the atmosphere inhibits the upward spread of pollutants. Upward-moving smoke, which rapidly assumes the temperature of the surrounding air, reaches a point where it is colder, and hence denser, than the air above it, so it can rise no further. This suppression of upward motion effectively forms a lid beneath which pollutants can disperse freely. The weaker the temperature decrease with height, the higher the lid is. The extreme case is an inversion, when the temperature increases with height. Often, clouds are topped by a stable or inversion layer, which stops their vertical growth.

The well-mixed layer beneath a stable layer is called the mixing layer. When it extends to the ground its vertical extent is known as the mixing height or the mixing depth. Generally, turbulence is enhanced in the early morning hours as the sun heats the ground and temperature decreases with height, causing unstable conditions. At night, as the earth cools, temperature increases with height causing less turbulence and stable atmospheric conditions.

Wind speed, wind direction, and atmospheric stability will vary greatly with time. For a certain location, some combinations occur more frequently than others.

Where detailed meteorological records have been kept for a year or more, a stability wind rose can be calculated. This wind rose is a set of tables, one for each stability class (ranging from very stable to very unstable), listing the frequency of occurrence of all possible combinations of wind speed and wind direction. Such roses are available for many locations in the United States from the National Climatic Data Center in Asheville, North Carolina. It should be noted that topographical features such as mountains, hills, valleys, bodies of water, buildings, and other terrain features can change airflow patterns, resulting in unexpected pollution effects.

Near a large body of water, local sea breezes influence the spread of pollutants. Early in the morning, when the air is still or the wind is off the land, pollutants can accumulate over their sources or downwind of them. Later in the day, when a local sea breeze develops, a fresh breeze blows in the direction from the water toward land. This breeze brings with it not only the pollutants emitted from the sources at this time of day, but also those accumulated earlier in the day, because they are carried back from water to land. Unexpectedly high pollutant concentrations can occur near the shore when the high pollutant loading blows past. In addition to this effect, which generally occurs close to land, the sea breeze itself can penetrate as far inland as 40 mi or more.

Mountains and valleys have characteristic airflow patterns, too. In the evening, as the earth cools, the coldest air will sink into the lowest part of the valley. This creates a sta-

ble inversion layer because lighter, warmer air stays above the valley. In this way, pollutants are trapped in the valleys all night. During the daytime when heating occurs, the air in the valley is warmed and rises, permitting the pollutants to escape. Unfortunately, this heating and upward motion does not always occur. During periods when high pressure settles over a region and the air is stagnant, the atmosphere is stable all day long, and pollutants continue to accumulate in the valley. Some of the worst episodes of air pollution have occurred in mountain chains like the Appalachians, where industries are located in the valleys between adjacent hills.

In cities, buildings form the topography. Where rows of tall buildings front on narrow streets the air flows through the streets as though they were canyons. Since ventilation is determined by building configuration, many distortions in wind, and hence pollution flows, take place in a city. Air flows over a building and into a street downwind of it. The building, because the air cannot flow through it, creates an obstruction in the pattern of the smooth airflow. Downwind of the building, an eddy, or circular movement of air at variance with the main airflow, is formed in its wake. The eddy can trap pollutants emitted by cars in the street, and can cause concentrations of pollutants, for example, carbon monoxide, to be as much as three times higher on the side of the street further downwind than at the site of pollutant origin.

High air pollution potential advisories

High Air Pollution Potential Advisories (HAPPA) are prepared at the National Meteorological Center (NMC) in Suitland, Maryland, by meteorologists of the National Oceanographic and Atmospheric Administration (NOAA), U.S. Department of Commerce.

Advisories are based both on reports received hourly via teletype from National Weather Service stations in the United States and on numerous analyses and forecasts prepared by the NMC. With its electronic computer facilities, the NMC prepares mixing-depth and wind-speed data from all upper-air-observing stations in the contiguous United States (about 70 stations). These data are analyzed, interpreted, and integrated with other meteorological information.

National air pollution potential advisories based on these data are transmitted daily at 12:20 P.M., EST, to Weather Service stations. When meteorological conditions do not warrant issuance of a HAPPA, the teletype message is "none today." When the forecast indicates that an advisory of high air pollution potential should be issued, the message designates the affected areas. The daily message indicates significant changes in the boundaries of advisory areas, including termination of an episode.

Because conditions of atmospheric transport and dispersion typically vary with location and time, the forecasting staff cannot prepare advisories for each city in the United States. For this reason, the NOAA meteorologists limit their forecasts to areas at least as large as 75,000 mi^2; roughly the size of Oklahoma, in which stagnation conditions are expected to persist for at least 36 hours. Individual Weather Service stations may modify these generalized forecasts on the basis of local meteorological conditions.

Users of the service should realize that boundaries of the forecast areas of high air pollution potential cannot be delineated exactly. For practical purposes, the lines defining the advisory area should be interpreted as bands roughly 100 mi wide.

To be notified of these advisories, air pollution control or research officials must initiate arrangements with the nearest Weather Service station.

• EPA AND AGRICULTURE WORKING TOGETHER TO IMPROVE AMERICA'S WATERS

The U.S. Environmental Protection Agency (EPA) has announced that the agency is working with the agricultural community to control water pollution from the nation's largest livestock operations while keeping American agriculture viable. The EPA has joined the Agriculture Department in announcing a final rule that will require all large Concentrated Animal Feeding Operations (CAFOs) to obtain permits that will ensure they protect America's waters from wastewater and manure. The rule will control runoff from agricultural feeding operations, preventing billions of pounds of pollutants from entering America's waters.

The EPA looks forward to continuing to work with USDA and with the agricultural community to ensure that the goal we all share—cleaner, purer water—is being advanced by their efforts. The new rule is unique in that it comes after unprecedented cooperation between EPA and USDA to find a way to help producers meet their own and society's goals for environmental quality and profitability. USDA stands ready to provide assistance in an incentive-based approach combining information and education, research and technology transfer, direct technical assistance, and financial assistance through the Environmental Quality Incentives Program (EQIP) and other farm bill programs.

The December 2002 announcement finalizes a rule that will replace 25-year old technology requirements and permitting regulations that did not address today's environmental needs and did not keep pace with growth in the industry. Effective manure management practices required by this rule will maximize the use of manure as a resource for agriculture while reducing adverse impacts on the environment.

The new rule applies to about 15,500 livestock operations across the country. Under the new rule all large CAFOs will be required to apply for a permit, submit an annual report, and develop and follow a plan for handling manure and wastewater. In addition, the rule moves efforts to protect the environment forward by: placing controls on land application of manure and wastewater, covering all major animal agriculture sectors, and increasing public access to information through CAFO annual reports. The rule also eliminates current permitting exemptions and expands coverage over types of animals in three important ways: the rule eliminates the exemption that excuses CAFOs from applying for permits if they only discharge during large storms; second, the rule eliminates the exemption for operations that raise chickens with dry manure handling systems; and third, the rule extends coverage to immature swine and immature dairy cows.

Currently about 4,500 operations are covered by permits. Because of the new rule, EPA expects that up to 11,000 additional facilities will be required to apply for permits by 2006. This rule will enhance protection of the nation's waters from nutrient over-enrichment and eutrophication, which causes algal blooms, fish kills, and the expansion of the Gulf of Mexico dead zone. The rule will also reduce pathogens in drinking water and improve coastal water quality. The amount of phosphorus released into the environment will be reduced by 56 million pounds, while nitrogen releases will be slashed by more than 100 million pounds. In addition, over two billion pounds of sediments and nearly one million pounds of metals will not be released.

The new rule will affect large livestock operations including those with hundreds of thousands of hogs, cattle, and poultry. Large CAFOs are defined in the rule as operations raising more than 1,000 cattle, 700 dairy cows, 2,500 swine, 10,000 sheep, 125,000 chickens, 82,000 laying hens, and 55,000 turkeys in confinement. Approximately 500 million tons of manure are generated annually by an estimated 238,000 livestock operations. From 1982 to 1997 these large livestock operations have grown by 51%, with some of the largest facilities having capacities exceeding one million animals. Since 1978 the number of animals per confined-animal operation has increased significantly. The largest per operation increases have been: layers (176%), broilers (148%), swine (134%), turkeys (129%), dairy (93%), and beef cattle (56%).

To help these livestock operations meet the rule's requirements, Congress increased funding for land and water conservation programs in the 2002 Farm Bill by $20.9 billion, bringing total funding for these programs to $51 billion over the next decade. The Environmental Quality Incentives Program (EQIP) was authorized at $200 million in 2002 and will ultimately go up to $1.3 billion in 2007; 60% of those funds must go to livestock operations. New technology is also being perfected to aid farmers in meeting this new rule. States are being given significant flexibility to find geographically appropriate means of implementing the CAFO rule. For example, states retain the authority to determine the type of permit—general or individual—to be issued to a given operation. This enables states to develop permits that take into account the size, location, and environmental risks that may be posed by an operation. States will also have substantial flexibility to tailor nutrient management plans for CAFOs, and may authorize alternative performance standards for existing and new CAFOs that will help promote the use of innovative technologies.

• PROTECTING AND RESTORING AMERICA'S WETLANDS

The U.S. Army Corps of Engineers and the U.S. Environmental Protection Agency, in conjunction with the Departments of Agriculture, Commerce, Interior, and Transportation, have strengthened their commitment to achieve the goal of no net loss of our nation's wetlands with the release of a comprehensive action plan and improved guidance to ensure effective, scientifically based restoration of wetlands impacted by development activities. The Corps regulatory guidance and the multi-agency action plan will help advance technical capabilities for wetlands restoration and protection, as well as clarify policies to ensure ecologically sound, predictable, and enforceable wetlands restoration completed as part of Clean Water Act and related programs. Both actions are the result of extensive multi-agency collaboration.

The National Wetlands Mitigation Action Plan lists 17 action items that the agencies will undertake to improve the effectiveness of restoring wetlands that are impacted or lost to activities governed by clean water laws. Completing the actions in the plan will enable the agencies and the public to make better decisions regarding where and how to restore, enhance, and protect wetlands; improve their ability to measure and evaluate the success of mitigation efforts; and expand the public's access to information on these wetland restoration activities.

A revised Regulatory Guidance Letter leads the list of action items in the National Wetlands Mitigation Plan. Crafted with input from the federal agencies that play a role in wetlands protection, the Corps Regulatory Guidance Letter will improve wetlands restoration implemented under the Clean Water Act in support of the Administration's "no net loss of wetlands" goal.

In order to advance the goal of no net loss of wetlands, the guidance letter emphasizes the following:

- A watershed-wide approach to prospective mitigation efforts for proposed projects impacting wetlands and other waters;
- The increased use of functional assessment tools; and
- Improved performance standards.

In addition, the guidance letter emphasizes monitoring, long-term management, and financial assurances to help ensure that restored wetlands actually result in planned environmental gains. The guidance letter also provides greater consistency across the Corps' 38 district offices on issues such as the timing of mitigation activities and the party responsible for mitigation success.

Recent independent evaluations published in 2001 by the National Academy of Sciences (NAS) and the General Accounting Office (GAO) reviewed the effectiveness of wetlands compensatory mitigation for authorized losses of wetlands and other waters under Section 404 of the CWA. In its study the NAS concluded that, despite progress in the last 20 years, the goal of no net loss of wetlands is currently not being met for wetland functions by the compensatory mitigation programs of federal agencies. The action plan and guidance were developed in response to, and are consistent with, the recommendations made in those reports.

"Wetlands" is a collective term for marshes, swamps, bogs, and similar areas that filter and cleanse drinking water supplies, retain flood waters, harbor extensive fish and shellfish populations, and support a diverse array of wildlife. In performing these functions, wetlands provide invaluable ecosystem services. Consequently, their destruction increases flooding and runoff, harms neighboring property, causes stream and river pollution, and results in the loss of valuable habitat.

The agencies are committed to achieving the goal of no net loss of wetlands under the regulatory program and are hopeful of attaining in the near future an increase in the overall function and value of the nation's wetlands. This is especially important in light of the fact that, since the late 1700s, over half the nation's wetlands have been lost to development and other activities. These losses are widespread—almost half of all states have lost more than 50% of their historic wetland resources.

The CWA prohibits the discharge of dredged or fill material into regulated wetlands and other waters of the United States unless a permit is issued under Section 404 of the CWA authorizing such a discharge. The Corps makes decisions regarding Section 404 permit requests after it completes a careful environmental review of the impacts of proposed discharges, including the potential adverse effects on wetlands. This permit program is designed to avoid impacts to wetlands where possible and minimize these impacts when they are unavoidable. However, if a permit is issued for a project that will result in a loss of wetlands, compensatory mitigation is necessary to replace those lost wetlands. EPA leads the development of the environmental criteria used to evaluate proposed discharges under the CWA.

In addition to the Corps of Engineers and EPA, the Department of Commerce's National Oceanic and Atmospheric Administration, the Department of Interior, and the Department of Transportation implement programs involving the restoration of wetlands and other aquatic resources. In combination with the Department of Agriculture's Wetlands Reserve and Conservation Reserve Programs, these restoration efforts are expected to take the country from annual net wetlands loss to net wetlands gain.

• PROTECT WATER FOR LIFE

The U.S. has the safest drinking water in the world: 91% of people served by public water systems now drink water meeting all federal health standards—up from 79% in 1993. Even so, there is a lot we need to do to make sure water is safe for everybody.

No matter where we live, our drinking water originates in a watershed, a land area that drains to a single body of water that may be surface water or groundwater. These watersheds are constantly under siege from multiple threats. As rain washes over roofs, pavement, farms, and grassy areas, and as snow melts and soaks into the ground, it picks up pollution and deposits it into surface water and groundwater.

As our population expands, our need for food, shelter, clothing, electricity, and recreation places more demands on our water supply. As the number of households and businesses increase, so does the amount of natural resources we consume and the amount of waste we produce. These are just a few of the activities that create pollution that can enter our drinking water sources:

- Over-application and abuse of pesticides and fertilizers—67 million pounds of pesticides annually;
- Overburdened land fills—230 million tons annually; 5 pounds per person per day;
- Huge volumes of animal waste—half a million animal factory farms produce 130 times the amount of waste of the human population; and
- Careless or ignorant activities at home, work, and play—12 million recreational and house boats and 10,000 boat marinas release solvents, gasoline, detergents, and raw sewage directly into waterways.

This pollution is caused by humans, but choices we make in our communities and as individuals can help eliminate it and greatly reduce threats to our drinking water. Four basic protective barriers help keep water safe to drink:

Prevention: Keep contaminants out of the drinking water source to protect the environment and reduce the need for costly treatment.

Risk management: Support your local utilities. Your public water system makes sure pollution that has entered source water is removed before it is distributed to the community. Water utilities treat nearly 34 billion gallons of water daily. The total miles of water pipeline and aqueducts equal approximately one million miles—enough to circle the globe 40 times.

Risk and compliance monitoring: Learn about your drinking water quality. Our communities constantly monitor water quality—at the source, at the treatment plant, in the distribution system that delivers water to our homes, and, in some cases, at the tap. Your local water system can provide you with this information. If you receive water from a private well, make sure it is tested annually.

Individual action: The actions we take as individuals really do add up when it comes to protecting our water.

- Be informed! Read the annual Consumer Confidence Report provided by your water system.
- Be involved! Speak up at public hearings on land use and permitting.
- Be observant! Report any suspicious activities in or around your water supply to local authorities or call 911 immediately. Look for announcements in the local media for activities that could pollute your source water.
- Don't contaminate! Reduce or eliminate pesticide application. Reduce the amount of trash you create. Recycle used oil. Reduce paved areas. Keep pollutants away from boat marinas and waterways.

Weather & Climate of the Millennium

• WEATHER, CLIMATE, AND PALEOCLIMATOLOGY

What is weather?

Weather is the state of atmospheric conditions (i.e., hot/cold, wet/dry, calm/windy, sunny/cloudy) that exists over relatively short periods of time (hours to a couple of days). Weather includes the passing of a thunderstorm, hurricane, or blizzard, and the persistence of a heat wave, or a cold snap. Weather variability and extreme events may be an unpredictable response to climate change.

What is climate?

Climate is the weather we expect over the period of a month, a season, a decade, or a century. More technically, climate is defined as the weather conditions resulting from the mean state of the atmosphere-ocean-land system, often described in terms of "climate normals" or average weather conditions. Climate change is a departure from the expected average weather or climate normals.

What is paleoclimatology?

Paleoclimatology is the study of past climate. The word is derived from the Greek root *paleo-*, which means "ancient," and the term "climate." Paleoclimate is that which existed before humans began collecting instrumental measurements of weather (e.g., temperature from a thermometer, precipitation from a rain gauge, sea level pressure from a barometer, wind speed and direction from an anemometer). Instead of instrumental measurements of weather and climate, paleoclimatologists use natural environmental (or "proxy") records to infer past climate conditions.

How do scientists study paleoclimatology?

Paleoclimatology not only includes the collection of evidence of past climate conditions, but the investigation of the climate processes underlying these conditions.

Paleoclimatologists, people who study past climate, can learn about past climate by looking at human records such as sailing logs and diaries. Paleoclimatologists also study the natural record, such as tree rings and the evidence of glacial advancement and retreat, to determine what past climate was like. Evidence of ancient mammals living before the last ice age also indicate climate change.

The environment, including the atmosphere and water conditions, is reflected in the chemical and physical make up of the remains of plants and animals. In this way, past climate is locked in nature. By comparing today's climate conditions with past climatic conditions scientists can get an idea of the rate at which conditions have changed. Earth's average global temperature has risen and fallen many times due to astronomical and other influences. Earth has periods of cold climate—ice ages—and periods of warm climate, called interglacial periods. There have been times when the average global temperature on Earth has been warmer than our present average global temperature, which is around 53.6–55.4°F. Knowing the pattern of past climate change can help us predict how climate will change in the future.

There are several ways that scientists study how Earth's temperature is changing: satellites, instrumental records, and proxy data.

Some scientists look to satellites to reveal something about Earth's changing climate. However, the satellite record is very short (ca. 20 years) and hard to interpret due to changes in instruments and orbits.

The record of instrumental temperature measurements, extending back to the nineteenth century, provides data from thermometers, rain gauges, etc., since 1860.

Paleoclimatologists also find clues in natural records. Annual records of climate are preserved in tree-rings, locked in the skeletons of tropical coral reefs, frozen in glaciers and ice caps, and buried in the sediments of lakes and oceans. These natural recorders of climate are called proxy climate data—that is, they substitute for thermometers, rain gauges, and other modern instruments used to record climate. By analyzing records taken from trees, reefs, glaciers, sediments, and other proxy sources, scientists can extend our understanding far beyond the 140-year instrumental record provided by thermometers and rain gauges.

Recent changes in the natural record from environmental proxy data can be calibrated using the 140-year instrumental record of climate changes.

What do we know about the history of climate?

Good weather records extend back only about 125 years. In that time, the Earth's global average temperature has increased by approximately 0.9°F. Scientists are trying to determine if this warming is a natural fluctuation, or a result of greenhouse warming.

From paleo records, we know that the climate of the past million years has been dominated by the glacial cycle, a pattern of ice ages and glacial retreats lasting thousands of years. In the even more ancient past, changes in climate have been linked to the movement of continents and to the storage of vast amounts of carbon in oil and coal beds.

How is paleoclimatology important?

Climate variability, including changes in the frequency of climate extremes (like droughts, floods, and storms), has always had a large impact on humans. A particularly severe El Niño, or relatively short drought, can cost U.S. citizens billions of dollars. For this reason, scientists study past climate variability and change to gain clues that will help them anticipate future climate change. This scientific information then helps society plan for future climate change.

Unfortunately, records of past climate change from satellites and human measurements (thermometers, rain gauges) are too short, generally less than 150 years, to examine the full range of climate variability. For this reason, it is critical to examine climate change going back hundreds and thousands of years using paleoclimate records from trees, coral, sediments, glaciers, and other natural "proxy" sources.

The study of paleoclimates has been particularly helpful in the discovery that Earth's climate system can shift between dramatically different climate states in a matter of years and/or decades. Understanding "climate surprise" of the past is critical if we are to avoid being surprised in the future by abrupt climatic change.

The study of past climate change also helps us understand whether humans are affecting Earth's climate system. The study of climate change over the last thousand years clearly shows that global warming of this century is real; and that the recent record warm years are likely unprecedented in the last 1,200 years. The paleoclimate record also allows us to examine the causes of past climate change, and to help unravel the natural causes of such climate change (for example: volcanic eruptions and solar variability) that may explain twentieth century global warming.

Lastly, most state of the art climate prediction done in the world is accomplished using large sophisticated computer models of the climate system. There has been a great deal of research focused on ensuring that these models can simulate most aspects of the modern, present-day, climate. It is also important to know how these same models simulate climate change. This can only be accomplished by comparing modeled past climate change with changes that are observed using paleoclimate records. Thus, paleoclimatology helps us gain confidence that our computer model simulations of future climate are worth believing.

What can we predict about climate in the future?

Powerful computer models are used to try to predict the climate of the future. The 1995 best estimates of The Intergovernmental Panel on Climate Change (IPCC) are for global warming by 2.7–7.2°F by the end of the next century. It is also likely that there will be substantial climate variability over the next 100 years, quite possibly including shifts in the frequency of severe droughts, floods, and El Niño.

The paleoclimate perspective can help us answer many questions, including...

● Is the last century of climate change unprecedented relative to the last 500, 2,000, and 20,000 years?
● Do recent global temperatures represent new highs, or just part of a longer cycle of natural variability?
● Is the recent rate of climate change unique or commonplace in the past?
● Can we find evidence in the paleoclimate record for mechanisms or climate forcings that could be causing recent climate change?

• GLOBAL WARMING

What is global warming?

The term global warming refers to the perception that the atmosphere near Earth's surface is warming, without any implications for the causes or magnitude. This warming is one of many kinds of climate change that Earth has gone through in the past and will continue to go through in the future.

Temperature increases will have significant impacts on human activities; where we can live, what food we can grow and how or where we can grow food, and where organisms we consider pests can thrive. To be prepared for the effects of these potential impacts we need to know how much the Earth is warming, for how long the Earth is warming, and the cause of the warming. Answers to these questions will provide us with a better basis for making decisions related to issues such as water resource management and agricultural planning.

What is the greenhouse effect? How is it related to greenhouse warming and global warming?

The greenhouse effect is a term that describes how carbon dioxide, water vapor, and other gases in the atmosphere help maintain the temperature at Earth's surface. The atmosphere approximates the function of a greenhouse by first letting sunlight (solar or short wave radiation) pass through to warm Earth, while absorbing much of the heat (thermal or long wave radiation) radiated up from the surface of Earth and reradiating back to the surface.

Life on Earth would be very different without the greenhouse effect. The greenhouse effect serves to keep the long term annual average of Earth approximately 57°F higher than Earth's temperature would be without the greenhouse effect. It is reasonable to expect that Earth should warm as concentrations of greenhouse gases in the atmosphere increase above natural levels, much like what happens when the windows of a greenhouse are closed on a warm, sunny day. The additional warming is commonly referred to as greenhouse warming.

Greenhouse warming is global warming due to increases in atmospheric greenhouse gases (e.g., carbon dioxide, methane, chlorofluorocarbons, etc.) whereas global warming refers only to the observation that Earth is warming, without any indication of what might be causing the warming.

Global warming is accepted as fact by most of the scientific community. However, greenhouse warming is more controversial because it implies that we know what is causing the Earth to warm. It is certain that atmospheric concentrations of greenhouse gases are rising dramatically due to human activity, but exactly how these increases in greenhouse gases factor in the observed changes of Earth's climate and global temperatures is unknown.

How is the ozone issue different?

The ozone issue and greenhouse warming are related yet distinct scientific issues. In the lower atmosphere (the troposphere), ozone does act as a greenhouse gas, trapping outgoing radiation that would otherwise escape into space. Compared to carbon dioxide, ozone is a minor greenhouse gas. The significance of tropospheric ozone may be increasing however, due to the burning of fossil fuel, which generates ozone (commonly recognized as a component of smog) into the lower atmosphere.

Ozone also plays a very important, natural role in the upper atmosphere (the stratosphere). In the upper atmosphere, ozone acts as a shield against harmful ultraviolet (UV) radiation from the Sun. Reductions in stratospheric ozone result in the increase of harmful UV radiation reaching Earth's surface. Ninety percent of the atmospheric ozone is concentrated 6.2–24.9 miles above Earth's surface. Ironically, the biggest destroyer of the ozone in the stratosphere is the set of human produced chemical compounds—chlorofluorocarbons, or CFCs—that act as greenhouse gases in the lower atmosphere. Extremely cold stratospheric temperatures over the North and South Poles, combined with solar radiation and atmospheric circulation, amplify the impact of ozone destroying chemical reactions, resulting in "ozone holes" over Antarctica and the Arctic.

How do we study global warming?

There are several ways that scientists study how Earth's temperature is changing. Although each method has some uncertainties, they all suggest a similar story—that Earth has warmed dramatically over the last 140 years and that Earth is now warmer than it has been in the last 600 years.

As mentioned before, some scientists look at satellites to reveal something about Earth's changing climate. Although the satellite record is very short (ca. 20 years) and hard to interpret due to changes in instruments and orbits, the latest satellite studies confirm the same story—the globe is warming.

The record of instrumental temperature measurements, extending back to the nineteenth century, provides one clear indication: that the mean annual surface air temperature of the earth has risen approximately 0.9°F since 1860.

Paleoclimatic data provides an independent confirmation of this recent warming, and also places the nineteenth to twentieth century (1860 to the present) warming in the context of the last several centuries to millennia. The paleoclimatic record not only allows us to look at global temperature fluctuations over the last several centuries, it also permits scientists to examine past climate even further back in time. This perspective is an important capability in our quest to understand the possible causes of the twentieth century global warming. We can look at hypothesized warm periods in the distant past (e.g.,1,000; 6,000; 125,000; and even 165,000,000 years ago) to see if they provide clues for natural processes that could be causing the global warming we are now experiencing. So far, paleoclimatologists have been unable to find any natural climatic explanations for our present-day warming.

• THE INSTRUMENTAL RECORD OF PAST GLOBAL TEMPERATURES

Satellite-derived temperature records

Satellite measurements have been used to reconstruct global atmospheric temperatures for the past 21 years. Current calculations show the lower atmosphere to be warming at a rate of 0.12°F per decade—an amount that is less rapid than indicated by ground-based thermometer and paleoclimatic data. It is uncertain why this disagreement exists.

Thermometer-based temperature trends (global & hemispheric)

The earliest records of temperature measured by thermometers are from western Europe beginning in the late seventeenth and early eighteenth centuries. The network of temperature collection stations increased over time and by the early twentieth century, records were being collected in almost all regions, except for polar regions, where collection began in the 1940s and 1950s. A set of temperature records of over 7,000 stations around the world have been compiled by the NOAA National Climatic Data Center (NCDC). About 1,000 of these records extend back into the nineteenth century.

Three widely recognized research programs have used the available instrumental data to reconstruct global surface air temperature trends from the late 1800s through today. All use the same land-based thermometer measurement records, but the reconstructions contain some differences. These differences are due to different approaches to spatial averaging, the use of treatment of sea surface temperature data (from ship observations), and the handling of the influence of changes in land-cover (i.e., increases in urbanization). However, all three show the same basic trends over the last 100 years.

Paleoclimatic data of the last 1,000 years

Beginning in the 1970s, paleoclimatologists began constructing a blueprint of how Earth's temperature changed over the centuries before 1850 and the widespread use of thermometers. Out of this emerged a sketchy view of the past 1,000 years of climate, based on limited data from tree rings, historical documents, sediments, and other proxy data sources. Today, many more paleoclimate records are available from around the world, providing a much improved view of past changes in Earth's temperature.

In the last few years, there has been a major breakthrough in our understanding of global temperature change over the last 400–1,000 years. Although temperature reconstructions are different in various studies, they all show some similar patterns of temperature change over the last several centuries. Most striking is the fact that each record reveals that the twentieth century is the warmest of the entire record, and that warming was most dramatic after 1920.

The similar characteristics among the different paleoclimatic reconstructions provide greater confidence in concluding that there has been dramatic global warming since the nineteenth century and that the 1990s had the warmest temperatures on Earth in the last 1,000 years.

Paleoclimatic data before 1,000 years ago

To put the twentieth century in the perspective of the last 1,000 years, it is critical to look further back into Earth's history to see if previous periods of global warmth can provide clues about twentieth-century warming.

Several periods of warmth (listed below) are hypothesized to have occurred in the past. However, upon close examination of these periods, it becomes apparent that these periods of warmth are not similar to twentieth century warming for two specific reasons: (1) the periods of hypothesized past warming do not appear to be global in extent, or (2) the periods of warmth can be explained by known natural climatic forcing conditions that are uniquely different than those of the last 100 years.

Several commonly cited periods of warmth include:

- the so-called "Medieval Warm Period" (ca. ninth to fourteenth centuries)
- the so-called mid-Holocene "Warm Period" (ca. 6,000 years ago)
- the so-called penultimate interglacial period (ca. 125,000 years ago)
- the mid-Cretaceous Period (ca. 120–90 million years ago)

The latest paleoclimatic studies appear to confirm that the global warmth of the twentieth century may not necessarily be the warmest time in Earth's history. What is unique is that the warmth is global and cannot be explained by natural mechanisms.

What can paleoclimatology tell us about climate change relevant to society in the future?

To understand and predict changes in the climate system, we need a more complete understanding of seasonal to century scale climate variability than can be obtained from the instrumental climate record. The instrumental temperature record indicates that Earth has warmed by 0.9°F from 1860 to the present. However, this record is not long enough to determine if this warming should be expected under a naturally varying climate, or if it is unusual and perhaps due to human activities. Paleoclimatic proxy data can be used to extend climate records and provide a longer time frame (hundreds to tens of thousands of years) for evaluating the warming of the last 140 years.

The cause of global warming over the last century remains a heated debate with significant economic and societal implications. Many scientists attribute the current global warming to the greenhouse effect enhanced by human activities. Other scientists have suggested that other factors are responsible, such as natural changes in the number and size of volcanic eruptions or an increase in the Sun's output (such phenomena are referred to as climate forcings). A paleoclimate perspective provides information about long term changes in different climate forcings that may be the underlying cause of the observed climate change.

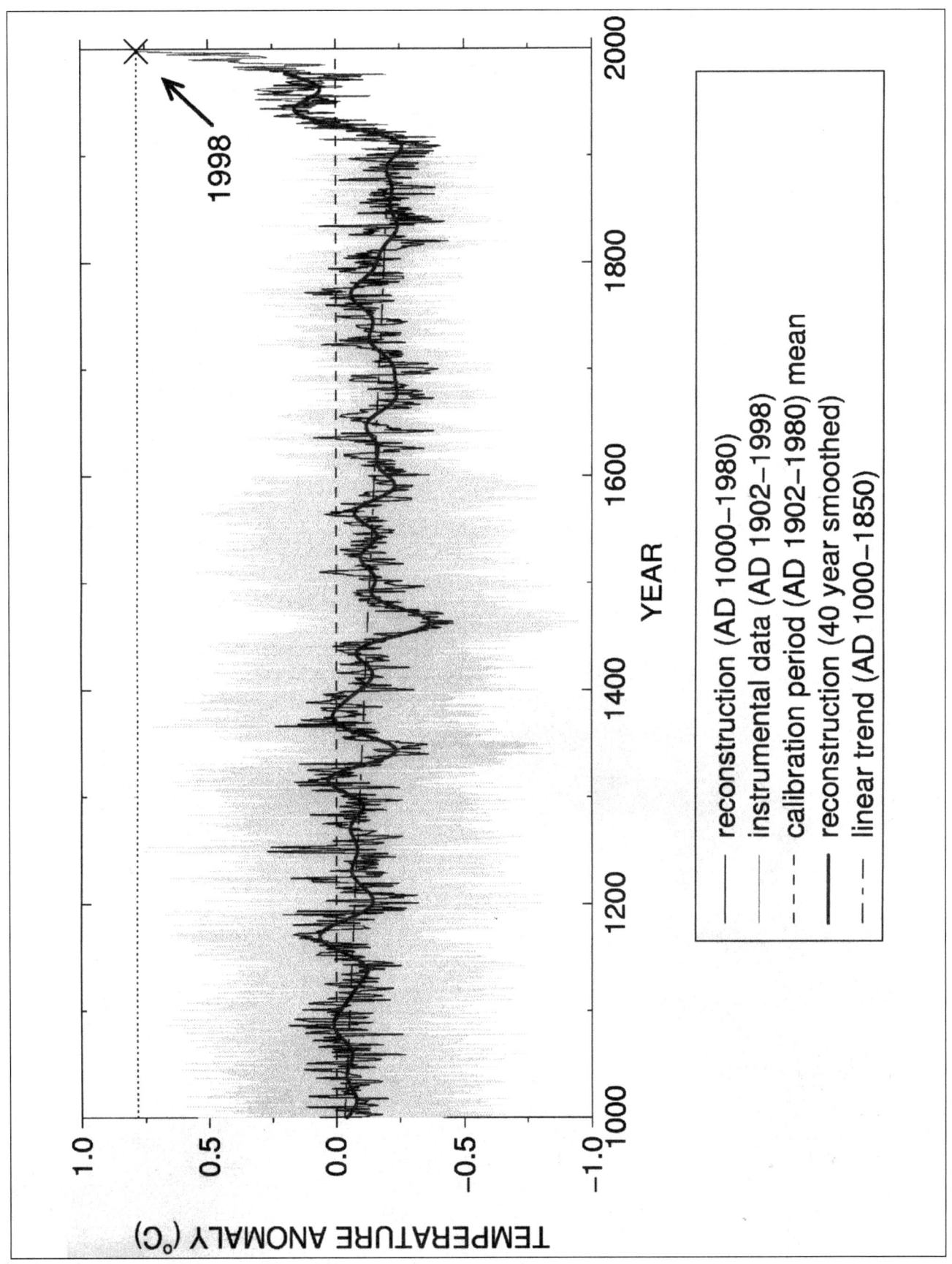

(Courtesy of Dr. Michael E. Mann, Univ. of Virginia. Reproduced by permission.)

An analogy of how paleoclimatic data improves our understanding of climate can be explained in terms of the stock market. Stock market analyses use longer term trends (one, two, three, or six months) in the stock market indices (Dow Jones, NASDAQ, etc.) rather than depending on changes from one day to the next or over a week to predict what the market will do next (i.e., bull or bear market). In much the same way, the paleoclimate perspective allows us to evaluate climate change many decades and centuries into the past, in order to develop a more reliable estimate of how climate may change the future.

Paleoclimatology and global warming

Researchers have tried to highlight what paleoclimatic data are, where they come from, and what the data contribute to the global warming debate. When one reviews all the data, both from thermometers and paleotemperature proxies, it becomes clear that Earth has warmed significantly over the last 140 years; global warming is a reality. Multiple paleoclimatic studies indicate that the recent year, decade, and century are all the warmest, on a global basis, of the last 600, and most likely 1,200 years. It appears that the global warming of the last century is unprecedented in the last 1,200 years.

There are, however, questions remaining concerning global warming. For instance, what is causing all the warming and what are the implications for the future? The answers to these questions are not simple.

There is considerable debate centered on the cause of twentieth century climate change. Few people contest the idea that some of the recent climate changes are likely due to natural processes, such as volcanic eruptions, changes in solar luminosity, and variations generated by natural interactions between parts of the climate system (for example, oceans and the atmosphere). There were significant climate changes before humans were around and there will be non-human causes of climate change in the future.

Just the same, with each year, more and more climate scientists are coming to the conclusion that human activity is also causing the climate of Earth to change. First on the list of likely human influences is greenhouse warming due to human-caused increases in atmospheric trace-gases. Other human activities are thought to drive climate as well. There is no doubt that humans are causing the level of atmospheric trace-gases to increase dramatically—the measurements match the predictions. There is also no doubt that these gases will contribute to global warming (since they warmed Earth before humans). However, there is uncertainty about some issues. For example, these questions remain to be answered with complete confidence:

- How much warming has occurred due to anthropogenic increases in atmospheric trace-gas levels?
- How much warming will occur in the future?
- How fast will this warming take place?

- What other kinds of climatic change will be associated with future warming?

Paleoclimatolgy offers to help answer each of these questions. The best estimate is that about 50% of the observed global warming is now due to greenhouse gas increases. Although this number will continue to be refined, it indicates that the climate modeling community is on target with their estimates that the Earth may warm an additional 3–7°F in the next century.

What future global warming means to society is beyond the scope of this discussion. However, studies have indicated that unprecedented twentieth century warming has affected the Arctic environment, and the warming already seems to be causing unprecedented changes in glaciers, permafrost, lakes, ecosystems, and the oceans, and it is likely that future changes will be even more dramatic as the warming continues.

• NORTH AMERICAN DROUGHT: A PALEO PERSPECTIVE

Droughts occur throughout North America, and in a given year, at least one region is experiencing drought conditions. The major drought of the twentieth century, in terms of duration and spatial extent, is considered to be the 1930s Dust Bowl drought, which lasted up to seven years in some areas of the Great Plains. The 1930s Dust Bowl drought, vividly portrayed in John Steinbeck's novel, *The Grapes of Wrath*, was so severe, widespread, and lengthy that it resulted in a mass migration of millions of people from the Great Plains to the western United States in search of jobs and better living conditions.

Just how unusual was the Dust Bowl drought? Was this a rare event or should we expect drought of similar magnitude to occur in the future? Rainfall records used to evaluate drought extend back 100 years, too short a time to answer these questions. However, these questions can be answered by analyzing records from tree rings, lake and dune sediments, archaeological remains, historical documents and other environmental indicators, which can extend our understanding of past climate far beyond the 100-year instrumental record.

What is drought?

The difficulty of recognizing the onset or end of a drought is compounded by the lack of any clear definition of drought. Drought can be defined by rainfall amounts, vegetation conditions, agricultural productivity, soil moisture, levels in reservoirs, stream flow, or economic impacts. In the most basic terms, a drought is simply a significant deficit in moisture availability due to lower than normal rainfall. However, even this simple definition is complicated when attempts are made to compare droughts in different regions.

For example, a drought in New Jersey would make for wet conditions in the deserts of Arizona!

Drought, as measured by scientists, is defined by evaluating precipitation, temperature, and soil moisture data, for the present and past months. A number of different indices of drought have been developed to quantify drought, each with its own strengths and weaknesses. Two of the most commonly used are the Palmer Drought Severity Index (PDSI) and the Standard Precipitation Index (SPI). Drought conditions are monitored constantly using these and other indices to provide current information on drought-impacted regions.

Because of the elusive nature of drought, we do not think of droughts in the same way as other weather-related catastrophes, such as floods, tornadoes, and hurricanes. However, although droughts may be less spectacular, they are often more costly than other types of natural disasters, and no region in North America is immune to periodic droughts.

Why are we concerned about drought?

Although the major droughts of the twentieth century, the 1930s Dust Bowl and the 1950s droughts, had the most severe impact on the central United States, droughts occur all across North America. Florida suffered from the 1998 drought along with the states of Oklahoma and Texas. Extensive drought-induced fires burned over 475,000 acres in Florida and cost $500 million in damages. In the same year, Canada suffered its fifth-highest fire occurrence season in 25 years. Starting in 1998, three years of record low rainfall plagued northern Mexico. 1998 was declared the worst drought in 70 years. It became worse as 1999 spring rains were 93% below normal. The government of Mexico declared five northern states disaster zones in 1999, and nine in 2000. The U.S. West Coast experienced a six-year drought in the late 1980s and early 1990s, causing Californians to take aggressive water conservation measures. Even the typically humid northeastern U.S. experienced a five-year drought in the 1960s that drained reservoirs in New York City down to 25% of capacity. In fact, almost every year, some region of North America experiences drought.

Drought is a natural hazard that cumulatively has affected more people in North America than any other natural hazard. The cost of losses due to drought in the United States averages $6–8 billion every year, but range as high as $39 billion for the three years drought of 1987–1989, which was the most costly natural disaster documented in U.S. history. Continuing uncertainty in drought prediction contributes to crop insurance payoffs of over $175 million per year in western Canada.

Beyond the monetary costs, the impacts of drought on society, the economy, and the natural environment are tremendous. Although measures such as development of irrigation systems, financial aid programs, and interbasin water transfers have been undertaken to mitigate the impacts of drought in recent decades, some regions of the United States are becoming more vulnerable to the impacts of drought.

Although irrigation has made it possible to grow crops on land that was once considered barren, this practice has led to a reliance on groundwater and surface storage in reservoirs. Increasing demands on water supplies have resulted in the depletion of groundwater reserves in many areas, which can make the removal of additional water uneconomical if not impossible, especially during a drought. In many urban areas of the semi-arid and arid western United States, population growth, expansion into marginal areas, and the subsequent development is overtaxing water supplies and heightening vulnerability to drought. Along with this increased vulnerability, concern exists because some research suggests that drought in the future may be amplified in certain areas due to changes in climate variability and extremes resulting from global warming.

Scientists have much to learn about the characteristics of drought and the conditions that lead to the persistence of drought. Although some progress has been made, (for instance, droughts that are related to El Niño and the Southern Oscillation (ENSO) are now more predictable on a seasonal scale), scientists still cannot predict longer, multi-year droughts.

The two major droughts of the twentieth century, the 1930s Dust Bowl drought and the 1950s drought, lasted five to seven years and covered large areas of the continental United States. Complete scientific understanding of how and why these two drought episodes occurred remains elusive. From a societal perspective, the important question is, how unusual are these events? Most instrumental records (from thermometers and rain gauges) are only about 100 years long, so they are too short to answer this question. However, paleoclimatic proxy data are a valuable tool to investigate this question by providing a longer context within which to evaluate the reoccurrence of these major droughts over hundreds to thousands of years.

Twentieth century drought

The Dust Bowl

The Dust Bowl drought was a natural disaster that severely affected much of the United States during the 1930s. The drought came in three waves, 1934, 1936, and 1939–40, but some regions of the High Plains experienced drought conditions for as long as eight years. The "dust bowl" effect was caused by sustained drought conditions compounded by years of land management practices that left topsoil susceptible to the forces of the wind. The soil, depleted of moisture, was lifted by the wind into great clouds of dust and sand that were so thick they concealed the sun for several days at a time. They were referred to as "black blizzards."

The agricultural and economic damage devastated residents of the Great Plains. The Dust Bowl drought worsened the already severe economic crisis that many Great Plains farmers faced. In the early 1930s, many farmers were trying to recover from economic losses suffered during the Great Depression. To compensate for these losses, they began to increase their crop yields. High production drove prices down, forcing farmers to keep increasing their production to pay for both their equipment and their land. When the drought hit, farmers could no longer produce enough crops to pay off loans or even pay for essential needs. Even with the federal emergency aid, many Great Plains farmers could not withstand the economic crisis of the drought. Many farmers were forced off their land, with one in ten farms changing possession at the peak of the farm transfer.

In the aftermath of the Dust Bowl, it was clear that many factors contributed to the severe impact of this drought. A better understanding of the interactions between the natural elements (climate, plants, and soil) and human-related elements (agricultural practices, economics, and social conditions) of the Great Plains was needed. Lessons were learned and because of this drought, farmers adopted new cultivation methods to help control soil erosion in dry land ecosystems. Subsequent droughts in this region have had less impact due to these cultivation practices.

The 1950s drought

Fueled by post-war economic stability and technological advancement, the 1950s represented a time of growth and prosperity for many Americans. While much of the country celebrated resurgence of well being, many residents of the Great Plains and the southwestern United States withstood a five-year drought, and in three of these years, drought conditions stretched coast to coast. The drought was first felt in the southwestern United States in 1950 and spread to Oklahoma, Kansas, and Nebraska by 1953. By 1954, the drought encompassed a 10-state area reaching from the Midwest to the Great Plains, and southwest into New Mexico. The area from the Texas panhandle to central and eastern Colorado, western Kansas, and central Nebraska experienced severe drought conditions. The drought maintained a stronghold in the Great Plains, reaching a peak in 1956. The drought subsided in most areas with the spring rains of 1957.

The 1950s drought was characterized by both low rainfall amounts and excessively high temperatures. Texas rainfall dropped by 40% in 1949–1951 and by 1953, 75% of Texas recorded below normal rainfall amounts. Excessive temperatures heated up cities like Dallas where temperatures exceeded 100°F on 52 days in the summer of 1953. Kansas experienced severe drought conditions during much of the five-year period, and recorded a negative Palmer Drought Severity Index from 1952 until March 1957, reaching a record low in September of 1956.

A drought of this magnitude creates severe social and economic repercussions and this was definitely the case in the southern Great Plains region. The drought devastated the region's agriculture. Crop yields in some areas dropped as much as 50%. Excessive temperatures and low rainfall scorched grasslands typically used for grazing. With grass scarce, hay prices became too costly, forcing some ranchers to feed their cattle a mixture of prickly pear cactus and molasses. By the time the drought subsided in 1957, many

counties across the region were declared federal drought disaster areas, including 244 of the 254 counties in Texas.

The 1987–1989 drought

The three-year drought of the late 1980s (1987–1989) covered 36% of the United States at its peak. Compared to the Dust Bowl drought, which covered 70% during its worst year, this does not seem significant. However, the 1980s drought was not only the costliest in U.S. history, but also the most expensive natural disaster of any kind to affect the United States. Combining the losses in energy, water, ecosystems, and agriculture, the total cost of the three-year drought was estimated at $39 billion. Drought-related losses in western Canada exceeded $1.8 billion dollars in 1988 alone.

The drought, beginning along the West Coast and extending into the northwestern United States, had its greatest impact in the northern Great Plains. By 1988, the drought intensified over the northern Great Plains and spread across much of the eastern half of the United States. The drought affected much of the nation's primary corn and soybean growing areas, where total precipitation for April through June of 1988 was even lower than during the Dust Bowl. The drought also encompassed the upper Mississippi River basin where low river levels caused major problems for barge navigation. The summer of 1988 is well known for the extensive forest fires that burned across western North America, including the catastrophic Yellowstone fire. In addition to dry conditions, heat waves during the summer of 1988 broke long-standing temperature records in many Midwestern and northeastern metropolitan areas.

The 1987–89 drought was the first widespread persistent drought since the 1950s and undoubtedly took people by surprise. Many had not experienced the 1950s drought and others had forgotten about the harsh realities of drought. The financial costs of this drought were an indication that many parts of the country are now more vulnerable to drought than ever before. This increased vulnerability was due in part to farming on marginally arable lands and pumping of ground water to the point of depletion. Although surplus grain and federal assistance programs offset the impacts of the 1987–89 drought, these types of assistance programs would be less feasible during a lengthier drought.

Another Dust Bowl?

What is the likelihood of another Dust Bowl-scale drought in the future? No one has yet been able to scientifically predict multi-year or decadal droughts, but the paleoclimatic record can tell us how frequently droughts such as the 1930s Dust Bowl occurred in the past or if droughts of this magnitude are indeed a rare event. If such droughts occurred with some regularity in the past, then we should expect them to occur in the future.

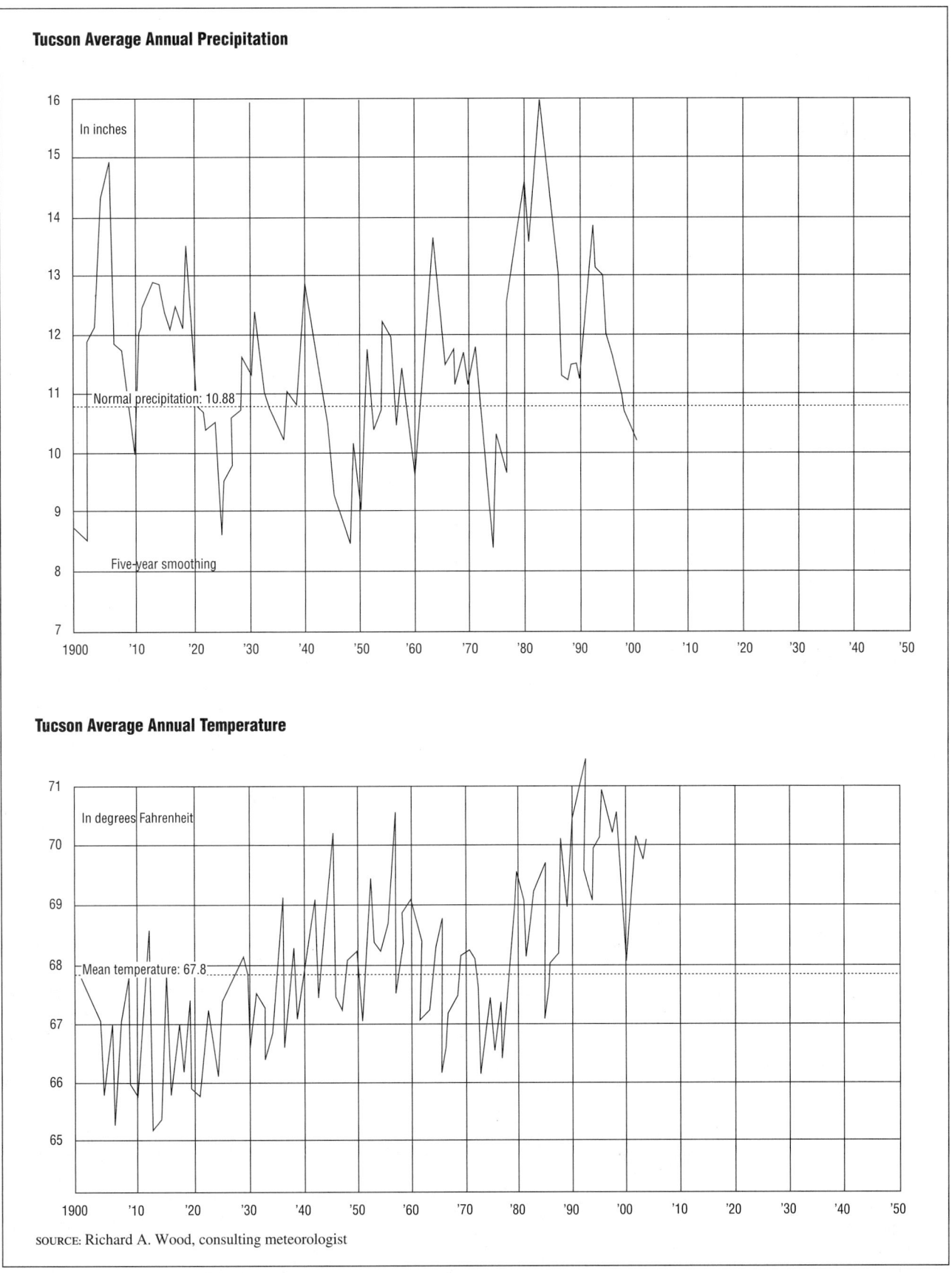

Tucson Average Annual Precipitation

Tucson Average Annual Temperature

SOURCE: Richard A. Wood, consulting meteorologist

Note the unusually dry 3-5 year cycles every 22-25 years during the past century in Tucson, Arizona. Using "five year smoothing," drought conditions occurred around 1900, 1925, 1950, and 2000. *(Chart by GGS. Data compiled by Richard A. Wood, Ph.D. Reproduced by permission.)*

• PALEOCLIMATOLOGY AND DROUGHT

How do we reconstruct drought from paleoclimatic data?

Records of rainfall (or other variables that reflect drought, such as changes in lake salinity, vegetation, or evidence of blowing sand) are preserved in tree-rings, buried in the sediments of sand dunes and lakes, contained within historical documents, and preserved in archaeological remains. By analyzing records taken from these proxy sources of paleodrought data, scientists can extend our records of drought far beyond the 100-year record provided by instruments.

To reconstruct drought or drought-related variables from environmental proxy data, the proxy data are calibrated with the instrumental record to determine how well the natural record estimates the climate record. The mathematical relationship between the proxy data and the climate record is defined, then used to produce a model. The model is then used to reconstruct the instrumental record from the proxy record for the length of the proxy.

Does paleoclimatic data help us understand drought?

Proxy records from tree rings, lake and dune sediments, historical records, and archaeological remains have all provided information about past droughts in the United States. Each record provides a piece of the puzzle, and together, they provide a more complete history than any one proxy would.

Historical records, such as diaries and newspaper accounts, can provide detailed information about droughts for the last 200 (Midwest and western U.S.) or 300 (eastern U.S.) years. Tree-ring records can extend back 300 years in most areas, and thousands of years in some regions. In trees that are sensitive to drought conditions, tree rings provide a record of drought for each year of the tree's growth. For records longer than those provided by trees and historical accounts (and for regions where we may not find trees and/or historical accounts), scientists turn to dune and lake sediments.

Lake sediments, if the core of the sediments is sampled at very frequent intervals, can provide information about variations occurring at frequencies less than a decade in length. Lake level fluctuations can be recorded as geologic bath tub rings as beach material sediments are deposited either high (further from the center under wetter conditions) or lower (closer to the center under drier conditions) within a basin as the water depth and thus lake level changes in response to drought. Droughts can increase the salinity of lakes, changing the species of small, lake-dwelling organisms that occur within a lake.

Pollen grains get washed or blown into lakes and accumulate in sediments. Different types of pollen in lake sediments reflect the vegetation around the lake and the climate conditions that are favorable for that vegetation. So, a change in the type of pollen found in sediments from, for example, an abundance of grass pollen to an abundance of sage pollen, can indicate a change from wet to dry conditions.

Records of more extreme environmental changes can be found by investigating the layers within sand dunes. The sand layers are interspersed among layers of soil material produced under wetter conditions, between the times when the sand dune was active. For a soil layer to develop, the climate needs to be wet for an extended period of time, so these layers reflect slower, longer-lasting changes.

Taken together, these different proxies record variations in drought conditions on the order of single seasons to decadal and century-scale changes, providing scientists with the information about both rapid and slow changes, and short and long periods of drought. These records are needed to put individual droughts in perspective, as well as to characterize droughts of the twentieth century.

The instrumental record

Instrumental records of drought (observed data measured by weather monitoring instruments) are a valuable resource that help detect the onset of drought. Although scientists have not yet refined drought prediction skills, these instrumental data provide us with information on current and developing droughts. The instrumental records also give us a picture of the short-term behavior (less than 100 years) and spatial patterns of drought, helping scientists learn more about the character of droughts. Actual observed data are important in paleoclimatic studies because they enable the calibration of the instrumental data with proxy data, a process needed to generate reconstructions of climate from proxy data, and also to allow researchers to determine how accurately different proxy records reconstruct climate.

Variations in the El Niño/Southern Oscillation (ENSO) in the equatorial Pacific are accomplished by changes in atmospheric flow and pressure systems in midlatitudes. These changes, in turn, affect climate across North America, especially in winter. Thus, certain phases of ENSO can increase the likelihood of more unusual and/or persistent weather conditions, such as drought, in some areas. For example, during El Niño, winters are wetter from California to the southeastern United States, while unusually warm conditions tend to persist from Alaska south through southwestern Canada and eastward to the Great Lakes. During La Niña, drought conditions are likely across the southwestern and southeastern U.S., while the northwestern United States can experience unusually wet winters, and cool conditions persist in a broad band from Alaska to western Canada and across the northern tier of the United States.

The last 500 years

A girded network of tree-ring reconstructions of Palmer Drought Severity Index (PDSI) for the last 300 years has been used to create a set of maps of the spatial pattern of PDSI for each year, back to A.D. 1700. This set of maps enables an assessment of the droughts of the twentieth century compared to droughts for the past 300 years. An inspection of the maps shows that droughts similar to the 1950s, in terms of duration and spatial extent, occurred once or twice a century for the three centuries (for example, during the 1860s, 1820, 1730s). However, there has not been another drought as extensive and prolonged as the 1930s drought in the past 300 years.

Longer records show strong evidence for a drought that appears to have been more severe in some areas of central North America than anything we have experienced in the twentieth century, including the 1930s drought. Tree-ring records from around North America document episodes of severe drought during the last half of the sixteenth century. Drought is reconstructed as far east as Jamestown, Virginia, where tree rings reflect several extended periods of drought that coincided with the disappearance of the Roanoke Island, North Carolina, colonists, and difficult times for the Jamestown colony. These droughts were extremely severe and lasted for three to six years, a long time for such severe drought conditions to persist in this region of North America.

Coincident droughts, or droughts during the same time, are apparent in tree-ring records from Mexico to British Columbia, and from California to the East Coast. Winter and spring drought conditions appear to have been particularly severe in the southwestern U.S., and northwestern Mexico, where this drought appears to have lasted several decades. In other areas, drought conditions were milder, suggesting drought impacts may have been tempered by seasonal variations.

The last 2,000 years

When records of drought for the last two millennia are examined, the major twentieth century droughts appear to be relatively mild in comparison with other droughts that occurred within this time frame. Even the sixteenth century drought appears to be fairly modest, when compared to some early periods of drought. Although there are still a few high resolution (offering data on annual to seasonal scales), precisely dated (to the calendar year), tree-ring records available that extend back 2,000 years, most of the paleo-drought data that extends back this far are less precisely dated and more coarsely resolved. These records reflect periods of more frequent drought, or drier overall conditions rather than single drought events, so it is difficult to compare droughts in these records with twentieth century drought events. However, the twentieth century can still be evaluated in this context, and we can assess whether parts of the twentieth century as a whole were wetter or drier than in the past with these records.

Even longer records

Data from a variety of paleoclimate sources document drought conditions across North America over the past 10,000 years. These records, with decade to century resolution, document extended periods of extremely dry conditions in different regions of North America. These periods of drought were severe enough and of long enough duration to impact vegetation composition, fire frequency, and to mobilize sand dunes in the Great Plains.

Evidence for these dry periods comes from paleoclimate proxy data such as pollen, charcoal, minerals and other materials within lake sediments, and sand dune sediments. Fossil pollen data provide information about changes in vegetation composition, used to reconstruct past changes in precipitation and temperature. Relative changes in charcoal abundance, indicative of fire, are interpreted in terms of climatic conditions favorable to fire. Changes in the composition and the chemistry of sediments of lakes provide information on regional aridity and drought. Lake level records are a direct measurement of moisture balance providing information on long-term hydrologic variability. The interpretation and dating of ancient soils and wind-blown sand/silt deposits constrain the timing and magnitude of sand dune mobilization associated with large-scale droughts.

How is the paleoclimatic record of drought relevant for understanding or predicting drought today, or in the future?

The North American record of past drought allows researchers to determine what has been the range of natural variability of drought over hundreds if not thousands of years. This long-term perspective is important because although severe droughts have occurred in the twentieth century, a more long-term look at past droughts, when climate conditions appear to have been similar to today, indicates that twentieth century droughts do not represent the possible range of drought variability.

The paleoclimatic record of past droughts is a better guide than the instrumental record alone for estimating the magnitude and duration of future droughts. For example, paleoclimatic data suggest that droughts as severe as the 1950s drought have occurred in central North America several times a century over the past 300–400 years, and thus we should expect (and plan for) similar droughts in the future. The paleoclimatic record also indicates that drought of a much greater duration than any in the twentieth century has occurred in parts of North America as recently as 500 years ago. These data indicate that we should be aware of the possibility of such droughts occurring in the future as well. The occurrence of such sustained drought conditions today would be a natural disaster of a magnitude unprecedented in the twentieth century.

In addition to establishing a baseline of drought variability over the long term, the paleoclimatic record of drought provides information about drought under a range of naturally varying climate conditions, some of which are the same as the climate of today and some which are quite different. This paleoclimatic perspective can be used to learn about the underlying process and characteristics of drought under very different future climate conditions.

The impact of droughts over the last few decades have shown that some regions and sectors of the population are becoming increasingly vulnerable to drought. Compounding these vulnerabilities is the uncertainly of the effects of human activities and global warming on climate in general and on drought in particular. A number of climate model simulations for doubled CO_2 conditions suggest an increased frequency of drought in midcontinental regions whereas other model simulations and recent decadal trends in the instrumental record suggest wetter conditions, at least in the short term, due to an intensification of the hydrologic cycle associated with warmer sea surface temperatures. Better constrained answers to the question of the severity of future droughts require improved understanding and modeling of the processes underlying the drought behavior exhibited in both the instrumental and the paleoclimatic records.

What can we do to better understand past droughts and predict future droughts?

Our understanding of what causes drought conditions to persist for years and decades is far from complete. Much is needed to comprehensively understand drought and the causes of drought, and to improve drought prediction capabilities. Putting together the pieces of past droughts through the use of paleoclimatic data is a vital part of building this understanding and developing an improved capacity to anticipate droughts in the future.

Focused efforts are needed to bring together paleoclimatic records of past droughts with scientists working to better understand the workings of the climate system. Currently, scientists are working on this sort of focused effort for western Canada. In the Prairie Drought Paleolimnology Project, paleoclimatic reconstructions will be incorporated into novel models specifically developed for use with long-term climatic data. The models will be used to predict drought frequency, duration, and intensity over the next five to 50 years. More such efforts are needed to understand the drought across all of North America.

• SUMMARY

In conclusion, paleoclimatology is an important tool in understanding the past and predicting future trends in Earth's climate. Paleoclimatology allows scientists to establish long-term norms and give current conditions a context for comparison. In this way scientists can determine whether or not humans are affecting Earth's climate, and if so, what the result of these effects might be. Some scientists note that a change of only a few degrees, of 6–8°F, in average global temperature can swing a moderate climate into an ice age or warm it up substantially. Swings of this magnitude have been occurring throughout Earth's history.

In addition, paleoclimatology lets scientists look at the past and determine how often a particular event, such as a drought, happens and plan accordingly. Paleoclimatology looks into the past to help explain the future.

Natural Disasters of the Millennium

- **1100s**

1138 Aleppo, Syria—Earthquake claimed 230,000 lives.

- **1200s**

1228 Netherlands—Sea flood killed 100,000 people.

1290 Chihli, China—Earthquake killed about 100,000 people.

- **1300s**

1300 North America—For about two decades in Arizona and New Mexico, rivers disappeared, crops failed, and towns declined, bringing death and destruction to that area.

1316 England—Long periods of heavy rains ruined crops, causing death for one-tenth of the population due to malnutrition or disease.

1346 Constantinople—The eastern arch of St. Sophia's crumbled during a strong earthquake that struck the Byzantine capital.

- **1500s**

1556 Shaanxi, China—In the deadliest earthquake in history, 830,000 people were killed.

1570 Northern Europe—Over 1,000 people were killed when a tidal wave in the North Sea destroyed sea walls from the Netherlands to Denmark.

- **1600s**

1667 Shemakha, Caucasia—Earthquake killed about 80,000 people.

- **1700s**

1727 Tabriz, Iran—Earthquake killed about 77,000 people.

1755 Lisbon, Portugal—More than 10,000 people were killed in an earthquake that devastated Lisbon. The earthquake occurred on All Saints' Day, when churches in this city of a quarter of million people were full, and the quake lasted about nine minutes. Floods and fires followed the event.

1776 Eastern Seaboard from North Carolina to Nova Scotia—At least 4,100 were killed during the storm called the Hurricane of Independence.

1780 Barbados, West Indies—Hurricane killed up to 22,000 people.

- **1800s**

1811 Mississippi Valley/New Madrid, Missouri, USA—Earthquake reversed the course of the Mississippi River. Due to the sparse population of the area at the time, the number of fatalities is unknown.

1815 Sumbawa, Indonesia—Tambora volcano erupted, throwing so much ash into the atmosphere that the year that followed was called the "Year without a Summer." In June and July of 1815, New England and northern Europe suffered frost and even snow.

1840 Natchez, Mississippi, USA—Tornado killed 317 people.

1842 China—Flooding killed 300,000 people.

1864 India—Cyclone killed 70,000; Calcutta was the most affected.

1883 Indonesia—Eruption of Krakatau, with sea waves sent as far away as Cape Horn, and possibly England. Possibly 36,000 killed.

1886 Charleston, South Carolina, USA—Earthquake killed 60 people.

1887 Huang He (Yellow River), China—Flood waters killed 900,000 people.

1888 East Coast of the United States—400 people died in the blizzard of 1888.

1889 Johnstown, Pennsylvania, USA—Flood killed more than 2,200 people.

1896 Sanriku, Japan—Earthquake and tidal wave killed 27,000 people.

• 1900s

1900 Galveston, Texas, USA—Hurricane killed more than 6,000–8,000 people.

1902 Martinique, West Indies—Volcano eruption killed 40,000 people.

1906 San Francisco, California, USA—Earthquake killed more than 1,000 people.

1908 Messina, Italy—Earthquake killed about 85,000 people.

1920 Gansu, China—Earthquake killed 200,000 people.

1923 Tokyo, Japan—Earthquake killed more than 132,000 people. Regional rivers burst their banks, bringing the total deaths to over 300,000.

1925 Missouri, Illinois, and Indiana, USA—Tri-state tornadoes killed 689 people.

1927 Xining, China—Earthquake killed about 200,000 people.

1930s New York through the Midwest to California, USA—Drought caused the "Great Dust Bowl" of the south central plains during the mid-1930s.

1931 Huang He (Yellow River), China—In what may be the greatest death toll due to a natural disaster, 3,700,000 people may have lost their lives in flooding.

1933 Long Beach, California, USA—Earthquake killed 117 people.

1935 Pakistan—Quetta earthquake left 30,000–60,000 dead.

1936 Mississippi and Georgia, USA—Tornadoes killed 455 people.

1939 Chile—Earthquake killed about 30,000 people.

1939 Northern Turkey—Earthquake caused about 100,000 deaths, mostly near Erzingan.

1954 Alaska, USA—Strongest earthquake in North America occurred east of Anchorage. Seismic wave 50 feet high traveled more than 8,000 miles at 450 mph.

1970 East Pakistan—Cyclone and tidal wave killed at least 300,000 people.

1970 Peru—Earthquake killed more than 50,000 people.

1972 Rapid City, South Dakota, USA—Flash flood caused 237 deaths.

1972 Mid Atlantic region, USA—Tropical Storm Agnes killed 129 people.

1974 Eastern and Central USA—Worst tornado outbreak in history killed at least 315 people.

1976 Big Thompson Canyon, Colorado, USA—Flash flood killed 139 people.

1976 Tangshan, China—Earthquake left 242,000–655,000 people dead.

1985 Mexico—Earthquake killed an estimated 25,000 people near and around Mexico City.

1985 Colombia—Earthquake killed about 25,000 people.

1989 San Francisco, California, USA—Earthquake killed 67 people.

1990 Northwest Iran—Earthquake killed at least 50,000 people.

1991 Bangladesh—Cyclone killed over 131,000 people.

1993 Midwestern USA—Major flooding killed almost 50 people.

1994 San Fernando Valley, California, USA—Earthquake killed 61 people.

1995 Osaka, Japan—Earthquake killed 5,100 people.

1995 Chicago, Illinois, USA—A July heat wave killed at least 465 people.

1998 China—At least 3,000 people were killed in flooding.

1998 Papua New Guinea—Tsunamis killed at least 2,000 people.

1999 Oklahoma and Kansas, USA—Tornadoes killed almost 50 people.

1999 Turkey—Earthquake killed 14,000 people.

Weather Fundamentals

• THE EARTH'S ATMOSPHERE

Composition

Air is a mixture of several gases. When completely dry, it is about 78% nitrogen and 21% oxygen. The remaining 1% is other gases such as argon, carbon dioxide, neon, helium, and others. However, in nature, air is never completely dry. It always contains some water vapor in amounts varying from almost none to 5% by volume. As water vapor content increases, the other gases decrease proportionately.

Vertical structure

The atmosphere is classified into layers, or spheres, by characteristics exhibited in these layers.

The *troposphere* is the layer from the surface to an average altitude of about 7 mi. It is characterized by an overall decrease of temperature with increasing altitude. The height of the troposphere varies with latitude and seasons. It slopes from about 20,000 ft over the poles to about 65,000 ft over the equator; and it is higher in summer than in winter.

At the top of the troposphere is the *tropopause*, a very thin layer marking the boundary between the troposphere and the layer above. A relationship between the height of the tropopause and certain weather phenomena has been documented.

Above the tropopause is the *stratosphere*. This layer is typified by relatively small changes in temperature with height except for a warming trend near the top.

Density

Air is matter and has weight. Since it is gaseous, it is compressible. Pressure the atmosphere exerts on the surface is the result of the weight of the air above. Thus, air near the surface is much more dense than air at high altitudes.

• TEMPERATURE

Temperature scales

Two commonly used temperature scales are Celsius (°C), or centigrade, and Fahrenheit (°F). The Celsius scale is used exclusively for upper air temperatures and is rapidly becoming the world standard for surface temperatures also.

Traditionally, two common temperature references are the melting point of pure ice and the boiling point of pure water at sea level. The melting point of ice is 32°F (0°C); the boiling point of water is 212°F (100°C). Thus, the difference between melting and boiling is 100°C, or 180°F; the ratio between degrees Celsius and Fahrenheit is 100/180 or 5/9. Since 0°F is 32°F colder than 0°C, you must apply this difference when comparing temperatures on the two scales. You can convert from one scale to the other using one of the following formulae: $C = 5/9 (F - 32)$ or $F = 9/5 C + 32$ where C is degrees Celsius and F is degrees Fahrenheit.

Heat and temperature

Heat is a form of energy. When a substance contains heat, it exhibits the property that is measured as temperature—the degree of "hotness" or "coldness." A specific amount of heat absorbed by or removed from a substance raises or lowers its temperature a definite amount. However, the amount of temperature change depends on characteristics of the substance. Each substance has its unique temperature change for the specific change in heat. For example, if a land surface and a water surface have the same temperature and an equal amount of heat is added, the land surface becomes hotter than the water surface. Conversely, with equal heat loss, the land becomes colder than the water.

The Earth receives energy from the Sun in the form of solar radiation. Earth and its atmosphere reflect about 55% of the radiation and absorb the remaining 45%, converting it to heat. The Earth, in turn, radiates energy, and this outgoing radiation is terrestrial radiation. It is evident that the average heat gained from incoming solar radiation must equal heat lost through terrestrial radiation in order to keep the Earth from getting progressively hotter or colder. However, this balance is worldwide; regional and local imbalances that create temperature variations should also be considered.

Temperature variations

The amount of solar energy received by any region varies with time of day, with seasons, and with latitude. These differences in solar energy create temperature variations. Tem-

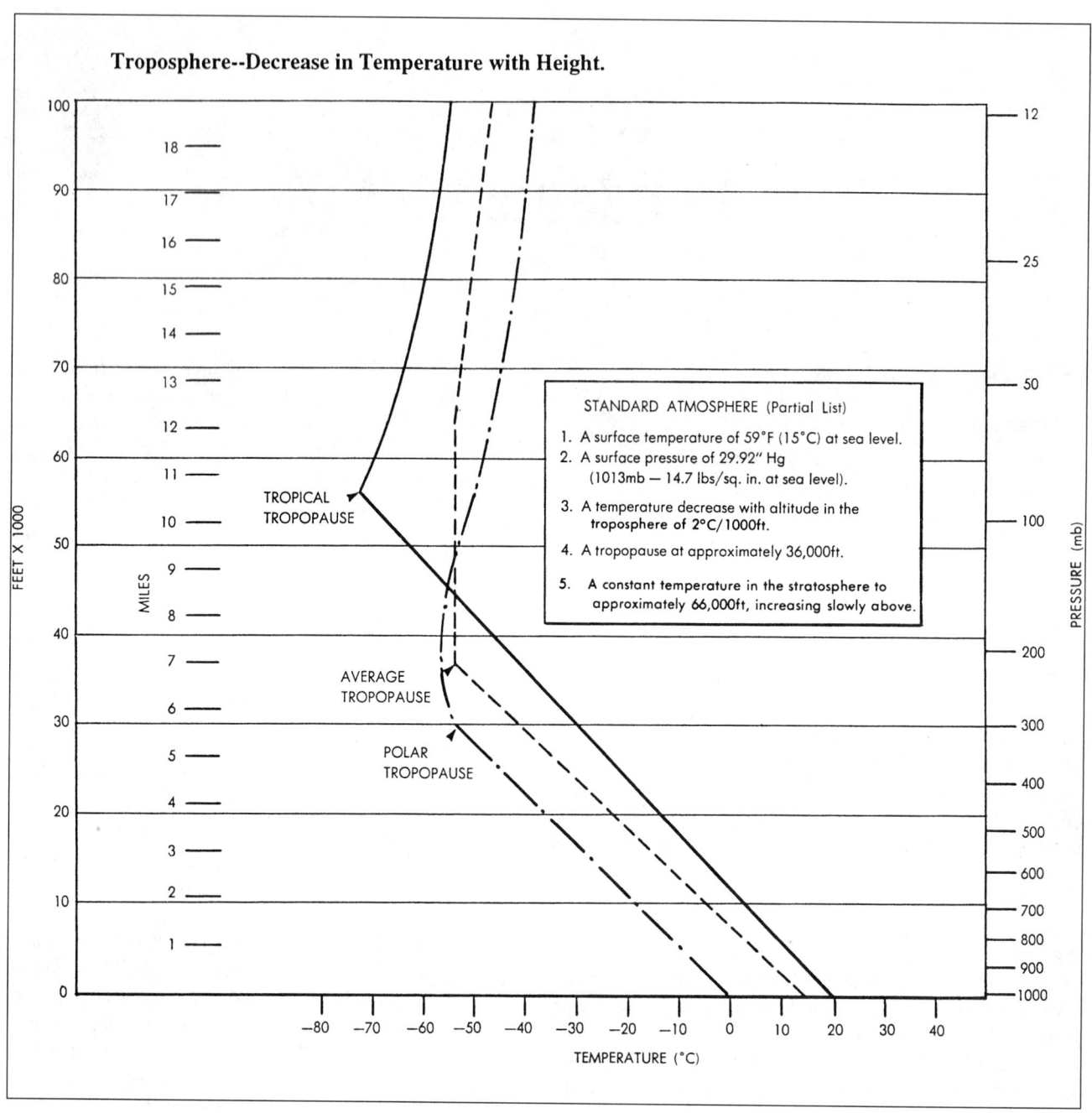

Troposphere--Decrease in Temperature with Height.

STANDARD ATMOSPHERE (Partial List)

1. A surface temperature of 59°F (15°C) at sea level.
2. A surface pressure of 29.92" Hg
   (1013mb — 14.7 lbs/sq. in. at sea level).
3. A temperature decrease with altitude in the troposphere of 2°C/1000ft.
4. A tropopause at approximately 36,000ft.
5. A constant temperature in the stratosphere to approximately 66,000ft, increasing slowly above.

(Courtesy of U.S. Air Force.)

peratures also vary with differences in topographical surface and with altitude. These temperature variations create forces that drive the atmosphere in its endless motions.

Day-to-night (diurnal) variation of temperature

Diurnal variation is the change in temperature from day to night brought about by the daily rotation of the Earth. Earth receives heat during the day from solar radiation but continually loses heat by terrestrial radiation. Warming and cooling depend on an imbalance of solar and terrestrial radiation. During the day, solar radiation exceeds terrestrial radiation and the surface becomes warmer. At night, solar radiation ceases, but terrestrial radiation continues and cools the surface. Cooling continues after sunrise until solar radiation again exceeds terrestrial radiation. Minimum temperature usually occurs after sunrise, sometimes as much as one hour after. The continued cooling after sunrise is one reason that fog sometimes forms shortly after the Sun is above the horizon.

Seasonal variation of temperature

In addition to its daily rotation, the Earth revolves in a complete orbit around the Sun once each year. Since the axis of the Earth tilts to the plane of orbit, the angle of incident solar radiation varies seasonally between hemispheres. The Northern Hemisphere is warmer in June, July, and August because it receives more solar energy than the Southern Hemisphere. During December, January, and February, the opposite is true—the Southern Hemisphere receives more solar energy and is warmer.

Temperature variation with latitude

The shape of the Earth causes a geographical variation in the angle of incident solar radiation. Since the Earth is essentially spherical, the Sun is closer to being overhead in equatorial regions than at higher latitudes. Equatorial regions, therefore, receive the most radiant energy and are warmest. Slanting rays of the Sun at higher latitudes deliver less energy over a given area with the least being received at the poles. Thus, temperature varies with latitude from the warm equator to the cold poles.

Temperature variations with topography

Not related to movement or shape of the Earth are temperature variations induced by water and terrain. Water absorbs and radiates energy with less temperature change than does land. Large, deep water bodies tend to minimize temperature changes, while continents favor large changes. Wet soil, such as in swamps and marshes, is almost as effective as water in suppressing temperature changes. Thick vegetation tends to control temperature changes since it contains some water and also insulates against heat transfer between the ground and the atmosphere. Arid, barren surfaces permit the greatest temperature changes.

These topographical influences are both diurnal and seasonal. For example, the difference between a daily maximum and minimum may be 10°F or less over water, near a shoreline, or over a swamp or marsh, while a difference of 50°F or more is common over rocky or sandy deserts. In the Northern Hemisphere in July, temperatures are warmer over continents than over oceans; in January they are colder over continents than over oceans. The opposite is true in the Southern Hemisphere, but not as pronounced because of more water surface in the Southern Hemisphere.

To compare land and water effect on seasonal temperature variation, consider northern Asia and southern California near San Diego. In the deep continental interior of northern Asia, July average temperature is about 50°F, and January average, about −30°F. Seasonal range is about 80°F. Near San Diego, due to the proximity of the Pacific Ocean, July average is about 70°F, and January average, about 50°F. Seasonal variation is only about 20°F.

Prevailing wind is also a factor in temperature controls. In an area where prevailing winds are from large water bodies, temperature changes are rather small. Most islands enjoy fairly constant temperatures. On the other hand, temperature changes are more pronounced where prevailing wind is from dry, barren regions.

The air slowly transfers heat from the surface upward. Thus, temperature changes aloft are more gradual than at the surface.

Temperature variation with altitude

Temperature normally decreases with increasing altitude throughout the troposphere. This decrease of temperature with altitude is defined as lapse rate. The average decrease of temperature—average lapse rate—in the troposphere is 3.6°F per 1,000 ft. But since this is an average, the exact value seldom exists. In fact, temperature sometimes increases with height through a layer. An increase in temperature with altitude is defined as an inversion, i.e., lapse rate is inverted.

An inversion often develops near the ground on clear, cool nights when wind is light. The ground radiates and cools much faster than the overlying air. Air in contact with the ground becomes cold while the temperature a few hundred feet above changes very little. Thus, temperature increases with height. Inversions may also occur at any altitude when conditions are favorable. For example, a current of warm air aloft overrunning cold air near the surface produces an inversion aloft. Inversions are common in the stratosphere.

• ATMOSPHERIC PRESSURE & THE BAROMETER

Atmospheric pressure

Atmospheric pressure is the force per unit area exerted by the weight of the atmosphere. Since air is not solid, it cannot be weighed with conventional scales. Yet, Toricelli proved three centuries ago that he could weigh the atmosphere by balancing it against a column of mercury. He actually measured pressure by converting it directly to weight.

Measuring pressure

The instrument Toricelli designed for measuring pressure is the barometer. Weather services and the aviation community use two types of barometers in measuring pressure—the mercurial and aneroid.

The mercurial barometer consists of an open dish of mercury that is placed into the open end of an evacuated glass tube. Atmospheric pressure forces mercury to rise in the tube. At stations near sea level, the column of mercury rises on the average to a height of 29.92 in. In other words, a column of mercury of that height weighs the same as a column of air having the same cross section as the column of mercury and extending from sea level to the top of the atmosphere.

Why is mercury used in the barometer? Mercury is the heaviest substance available that remains liquid at ordinary temperatures. It permits the instrument to be of manageable size. Water could be used but at sea level the water column would be about 34 ft high.

The aneroid barometer comprises the essential features of a flexible metal cell and the registering mechanism. The cell is partially evacuated and contracts or expands as pres-

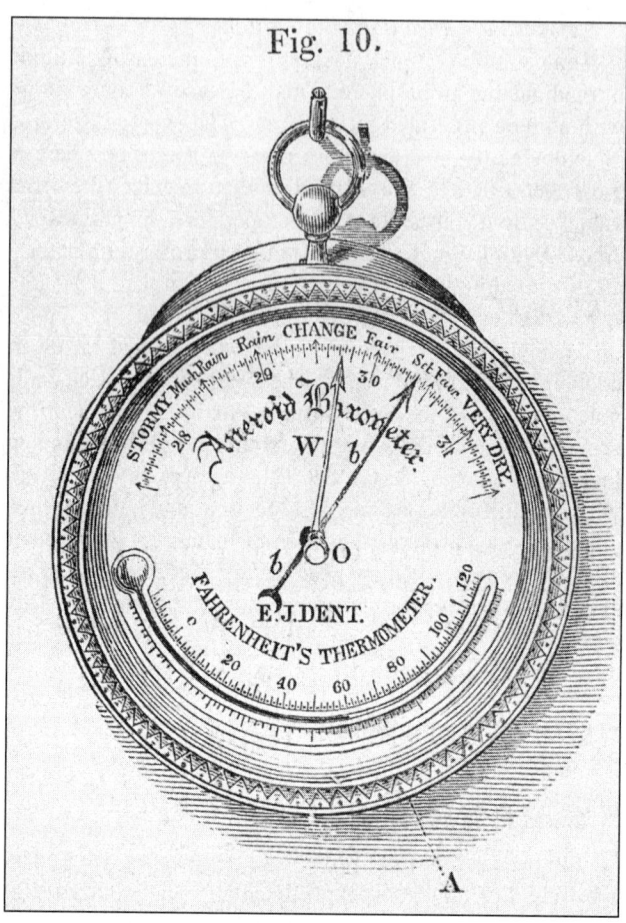

The aneroid consists of a partially evacuated metal cell that contracts and expands with changing pressure, and a coupling mechanism that drives the indicator along a scale graduated in pressure units. *(Photo by Steve Nicklas. Courtesy of National Oceanic and Atmospheric Administration (NOAA) Central Library.)*

sure changes. One end of the cell is fixed, while the other end moves the registering mechanism. The coupling mechanism magnifies movement of the cell, driving an indicator hand along a scale graduated in pressure units.

Pressure units

Pressure is expressed in many ways throughout the world. The term used depends somewhat on its application and the system of measurement. Two popular units are inches of mercury or millimeters of mercury. Since pressure is force per unit area, a more explicit expression of pressure is pounds per square inch (lb/in^2) or grams per square centimeter (g/cm^2). The term millibar (mb) precisely expresses pressure as a force per unit area, 1 mb being a force of 1,000 dynes per square centimeter. The millibar is rapidly becoming a universal pressure unit.

Station pressure

Pressure can be measured only at the point of measurement. The pressure measured at a station or airport is station pressure or the actual pressure at field elevation.

Pressure variation

Pressure varies with altitude and temperature of the air as well as with other minor influences.

Altitude

Moving upward through the atmosphere, weight of the air above becomes less and less. Within the lower few thousand feet of the troposphere, pressure decreases roughly 1 in of mercury for each 1,000-ft increase in altitude. At sea level, the average pressure is about 14.7 lb/in^2. It has been found that the pressure will decrease by half for each 18,000-ft increase in altitude. Thus, at 18,000 ft, we could expect an average pressure of about 7.4 lb/in^2 and at 36,000 ft, a pressure of only 3.7 lb/in^2, and so on.

Sea-level pressure

Since pressure varies with altitude, it is not easy to compare station pressures between stations at different altitudes. To make them comparable, pressure readings must be adjusted to some common level. Mean sea level seems the most feasible common reference. Pressure measured at a 5,000-ft station is 25 in; pressure increases about 1 in for each 1,000 ft or a total of 5 in. Sea-level pressure is approximately 25 + 5, or 30 in. The weather observer takes temperature and other effects into account, but this simplified example explains the basic principle of sea-level pressure reduction.

Sea-level pressure is usually expressed in millibars. Standard sea-level pressure is 1,013.2 mb, 29.9 in of mercury, or about 14.7 lb/in^2.

Pressure analyses (using isobars)

Sea-level pressure is commonly plotted on a map and lines are drawn connecting points of equal pressure. These lines of equal pressure are isobars. Hence, the surface map is an isobaric analysis showing identifiable, organized pressure patterns. Five pressure systems are defined as follows:

LOW—a center of pressure surrounded on all sides by higher pressure; also called a cyclone. Cyclonic curvature is the curvature of isobars to the left when you stand with lower pressure to your left.

HIGH—a center of pressure surrounded on all sides by lower pressure, also called an anticyclone. Anticyclonic curvature is the curvature of isobars to the right when you stand with lower pressure to your left.

TROUGH—an elongated area of low pressure with the lowest pressure along a line marking maximum cyclonic curvature.

RIDGE—an elongated area of high pressure with the highest pressure along a line marking maximum anticyclonic curvature.

COL—the neutral area between two highs and two lows. It is also the intersection of a trough and a ridge. The col on a pressure surface is analogous to a mountain pass on a topographic surface. We simply contour the heights of the pressure surface. For example, a 700-mb constant pressure analysis is a contour map of the heights of the 700-mb pressure surface. While the contour map is based on variations in height, these variations are small when compared to flight levels, and for all practical purposes, you may regard the 700-mb chart as a weather map at approximately 10,000 ft.

• WIND

What causes wind?

Differences in temperature create differences in pressure. For example, local winds along lake and ocean shores are the result of the temperature differences between land and water, which cause a pressure difference and wind. These pressure differences drive a complex system of winds in a never-ending attempt to reach equilibrium. Wind also transports water vapor and spreads fog, clouds, and precipitation.

Convection currents

When two surfaces are heated unequally, they heat the overlying air unevenly. The warmer air expands and becomes lighter or less dense than the cool air. The denser, cool air is drawn to the ground by its greater gravitational force lifting or forcing the warm air upward much as oil is forced to the top of water when the two are mixed. The rising air spreads and cools, eventually descending to complete the convective circulation. As long as the uneven heating persists, convection maintains a continuous convective current.

The horizontal air flow in a convective current is wind. Convection of both large and small scales accounts for systems ranging from hemispheric circulations down to local eddies. This horizontal flow, wind, is sometimes called advection. However, the term advection more commonly applies to the transport of atmospheric properties by the wind, i.e., warm advection; cold advection; advection of water vapor, etc.

Pressure gradient force of wind

Pressure differences must create a force in order to drive the wind. This force is the pressure gradient force. The force is from higher pressure to lower pressure and is perpendicular to isobars or contours. Whenever a pressure difference develops over an area, the pressure gradient force begins moving the air directly across the isobars. The closer the spacing of isobars, the stronger the pressure gradient force is. The stronger the pressure gradient force, the stronger the wind is. Thus, closely spaced isobars mean strong winds; widely spaced isobars mean lighter wind. From a pressure analysis, the reader can get a general idea of wind speed from contour or isobar spacing.

Because of uneven heating of the Earth, surface pressure is low in warm equatorial regions and high in cold polar regions. A pressure gradient develops from the poles to the equator. If the Earth did not rotate, this pressure gradient force would be the only force acting on the wind. Circulation would be two giant hemispheric convective currents. Cold air would sink at the poles; wind would blow straight from the poles to the equator; warm air at the equator would be forced upward; and high-level winds would blow directly toward the poles. However, the Earth does rotate; and because of its rotation, this simple circulation is greatly distorted.

Coriolis force: it modifies wind direction

A moving mass travels in a straight line until acted on by some outside force. However, if one views the moving mass from a rotating platform, the path of the moving mass relative to his platform appears to be deflected or curved. To illustrate, start rotating a potter's wheel. Then, using a piece of chalk and a ruler, draw a straight line from the center to the outer edge of the wheel. To you, the chalk traveled in a straight line. Now stop the turntable; on it, the line spirals outward from the center. To a viewer on the turntable, some apparent force deflected the chalk to the right.

A similar apparent force deflects moving particles on the earth. Because the earth is spherical, the deflective force is much more complex than the simple turntable example. This principle was first explained by a Frenchman, Coriolis, and carries his name—the Coriolis force.

The Coriolis force affects the paths of aircraft, missiles, flying birds, and ocean currents, and is most important to the study of weather and air currents. The force deflects air to the right in the Northern Hemisphere and to the left in the Southern Hemisphere. This text concentrates mostly on deflection to the right in the Northern Hemisphere.

Coriolis force is at a right angle to wind direction and directly proportional to wind speed. That is, as wind speed increases, Coriolis force increases. At a given latitude, double the wind speed and you double the Coriolis force.

Coriolis force varies with latitude from zero at the equator to a maximum at the poles. It influences wind direction everywhere except immediately at the equator, but the effects are more pronounced in middle and high latitudes.

Remember that the pressure gradient force drives the wind and is perpendicular to isobars. When a pressure gradient force is first established, wind begins to blow from higher to lower pressure directly across the isobars. However, the instant air begins moving, Coriolis force deflects it to the right. Soon the wind is deflected a full 90° and is parallel to the isobars or contours. At this time, Coriolis force exactly balances pressure gradient force. With the forces in balance, wind will remain parallel to isobars or contours. Surface friction disrupts this balance; Coriolis force distorts the fictitious global circulation.

• THE GENERAL CIRCULATION OF THE EARTH'S AIR

As air is forced aloft at the equator and begins its high-level trek northward, the Coriolis force turns it to the right or to the east. Wind becomes westerly at about 30° latitude, temporarily blocking further northward movement. Similarly, as air over the poles begins its low-level journey southward toward the equator, it likewise is deflected to the right and becomes an east wind, halting for a while its southerly progress. As a result, air literally "piles up" at about 30° and 60° latitude in both hemispheres. The added weight of the air increases the pressure into semipermanent high-pressure belts.

The building of these high-pressure belts creates a temporary impasse, disrupting the simple convective transfer between the equator and the poles. The restless atmosphere cannot live with this impasse in its effort to reach equilibrium. Something has to give. Huge masses of air begin overturning in middle latitudes to complete the exchange.

Large masses of cold air break through the northern barrier, plunging southward toward the tropics. Large midlatitude storms develop between cold outbreaks and carry warm air northward. The result is a midlatitude band of migratory storms with ever-changing weather.

Since pressure differences cause wind, seasonal pressure variations determine to a great extent the areas of these cold air outbreaks and midlatitude storms. But, seasonal pressure variations are largely due to seasonal temperature changes. It should be remembered that at the surface, warm temperatures to a great extent determine low pressure, and cold temperatures determine high pressure. It should also be recalled that seasonal temperature changes over continents are much greater than over oceans.

During summer, warm continents tend to be areas of low pressure and the relatively cool oceans tend to be areas of high pressure. In winter, the reverse is true: there is high pressure over the cold continents and low pressure over the relatively warm oceans. The same pressure variations occur in the warm and cold seasons of the Southern Hemisphere, although the effect is not as pronounced because of the much larger water areas of the Southern Hemisphere.

Cold outbreaks are strongest in the cold season and are predominantly from cold continental areas. Summer outbreaks are weaker and more likely to originate from cool water surfaces. Since these outbreaks are masses of cool, dense air, they characteristically are high-pressure areas.

As the air tries to blow outward from the high pressure, it is deflected to the right by the Coriolis force. Thus, the wind around a high blows clockwise. The high pressure with its associated wind system is an anticyclone.

The storms that develop between high-pressure systems are characterized by low pressure. As winds try to blow inward toward the center of low pressure, they also are deflected to the right. Thus, the wind around a low is counterclockwise. The low pressure and its wind system is a cyclone.

The high-pressure belt at about 30° north latitude forces air outward at the surface to the north and to the south. The

northbound air becomes entrained into the midlatitude storms. The southward moving air is again deflected by the Coriolis force, becoming the well-known subtropical northeast trade winds. In midlatitudes, high-level winds are predominantly from the west and are known as the prevailing westerlies. Polar easterlies dominate low-level circulation north of about 60° latitude.

There are three major wind belts. Northeasterly trade winds carry tropical storms from east to west. The prevailing westerlies drive midlatitude storms generally from west to east. Few major storm systems develop in the comparatively small Arctic region; the chief influence of the polar easterlies is their contribution to the development of midlatitude storms.

Friction effect on wind

Wind flow patterns aloft follow isobars or contours where friction has little effect. However, friction is a significant factor near the surface.

Friction between the wind and the terrain surface slows the wind. The rougher the terrain, the greater the frictional effect. Also, the stronger the wind speed, the greater the friction. One may not think of friction as a force, but it is a very real and effective force always acting opposite to wind direction.

As frictional force slows the wind speed, Coriolis force decreases. However, friction does not affect pressure gradient force. Pressure gradient and Coriolis forces are no longer in balance. The stronger pressure gradient force turns the wind at an angle across the isobars toward lower pressure until the three forces balance. Frictional and Coriolis forces combine to just balance pressure gradient force. Surface wind spirals outward from high pressure into low pressure, crossing isobars at an angle.

The angle of surface wind to isobars is about 10° over water, increasing with roughness of terrain. In mountainous regions, one often has difficulty relating surface wind to pressure gradient because of immense friction and also because of local terrain effects on pressure.

The jet stream

Winds, on the average, increase with height throughout the troposphere, culminating in a maximum near the level of the tropopause. These maximum winds tend to be further concentrated in narrow bands. A jet stream, then, is a narrow band of strong winds meandering through the atmosphere at a level near the tropopause. Further discussion of the jet stream is taken up later in this text.

• LOCAL & SMALL-SCALE WINDS

Local terrain features such as mountains and shore lines also influence local winds and weather.

Mountain and valley winds

In the daytime, air next to a mountain slope is heated by contact with the ground as it receives radiation from the sun. This air usually becomes warmer than air at the same altitude, but farther from the slope.

Surrounding colder, denser air settles downward and forces the warmer air near the ground up the mountain slope. This wind is a valley wind, so called because the air is flowing up out of the valley.

At night, the air in contact with the mountain slope is cooled by terrestrial radiation and becomes heavier than the surrounding air. It sinks along the slope, producing the mountain wind, which flows like water down the mountain slope. Mountain winds are usually stronger than valley winds, especially in winter. The mountain wind often continues down the more gentle slopes of canyons and valleys, and in such cases becomes drainage wind. It can become quite strong over some terrain conditions and in extreme cases can become hazardous when flowing through canyon restrictions.

Katabatic wind

A katabatic wind is any wind blowing down an incline when the incline is influential in causing the wind. Thus, the mountain wind is a katabatic wind. Any katabatic wind originates because cold, heavy air spills down sloping terrain, displacing warmer, less dense air ahead of it. Air is heated and dried as it flows downslope. Sometimes the descending air becomes warmer than the air it replaces.

Many katabatic winds recurring in local areas have been given colorful names to highlight their dramatic, local ef-fect. Some of these are the Bora, a cold northerly wind blowing from the Alps to the Mediterranean coast; the chinook, a warm wind down the east slope of the Rocky Mountains often reaching hundreds of miles into the high plains; the Taku, a cold wind in Alaska blowing off the Taku glacier; and the Santa Ana, a warm wind descending from the Sierras into the Santa Ana Valley of California.

Land and sea breezes

Land surfaces warm and cool more rapidly than do water surfaces; therefore, land is warmer than the sea during the day; wind blows from the cool water to warm land—the sea breeze, so called because it blows from the sea. At night, the wind reverses, blows from cool land to warmer water, and creates a land breeze.

Land and sea breezes develop only when the overall pressure gradient is weak. Wind with a stronger pressure gradient mixes the air so rapidly that local temperature and pressure gradients do not develop along the shoreline.

Wind shear

Rubbing two objects against each other creates friction. If the objects are solid, no exchange of mass occurs between the two. However, if the objects are fluid currents, friction creates eddies along a common shallow mixing zone, and a mass transfer takes place in the shallow mixing layer. This zone of induced eddies and mixing is called a shear zone.

Wind, pressure systems, and weather

Wind speed is proportional to the spacing of isobars or contours on a weather map. However, with the same spac-

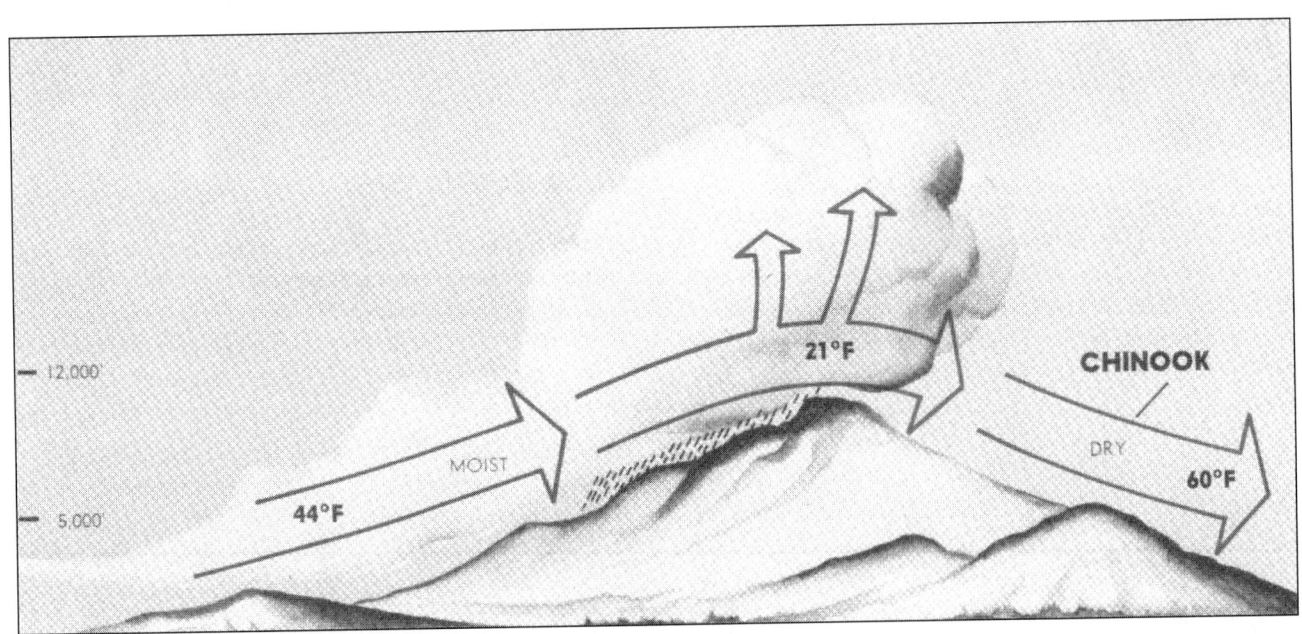

The "chinook" is a katabatic (downslope) wind. Air cools as it moves upslope and warms as it blows downslope. The chinook occasionally produces dramatic warming over the plains just east of the Rocky Mountains. *(Courtesy of U.S. government publication.)*

ing, wind speed at the surface will be less than aloft because of surface friction.

Wind direction can be determined from a weather map. If you face along an isobar or contour with lower pressure on your left, wind will be blowing in the direction you are facing. On a surface map, wind will cross the isobar at an angle toward lower pressure; on an upper-air chart, it will be parallel to the contour.

Wind blows counterclockwise (Northern Hemisphere) around a low, and clockwise around a high. At the surface where winds cross the isobars at an angle, the transport of air from high to low pressure can be seen. Although winds are virtually parallel to contours on an upper-air chart, there is still a slow transport of air from high to low pressure.

At the surface when air converges into a low, it cannot go outward against the pressure gradient, nor can it go downward into the ground; it must go upward. Therefore, a low or trough is an area of rising air.

Rising air is conducive to cloudiness and precipitation; thus we have the general association of low pressure—bad weather.

By similar reasoning, air moving out of a high or ridge depletes the quantity of air. Highs and ridges, therefore, are areas of descending air. Descending air favors dissipation of cloudiness; hence the association, high pressure—good weather.

Many times weather is more closely associated with an upper-air pattern than with features shown by the surface map. Although features on the two charts are related, they seldom are identical. A weak surface system often loses its identity in the upper-air pattern, while another system may be more evident on the upper-air chart than on the surface map.

Widespread cloudiness and precipitation often develop in advance of an upper trough or low. A line of showers and thunderstorms is not uncommon with a trough aloft even though the surface pressure pattern shows little or no cause for the development.

On the other hand, downward motion in a high or ridge places a "cap" on convection, preventing any upward motion. Air may become stagnant in a high, trap moisture and contamination in low levels, and restrict ceiling and visibility. Low stratus, fog, haze, and smoke are not uncommon in high-pressure areas. However, a high or ridge aloft with moderate surface winds most often produces good flying weather.

• MOISTURE, CLOUD FORMATION & PRECIPITATION

Water vapor

Water evaporates into the air and becomes an ever-present but variable constituent of the atmosphere. Water vapor is invisible just as oxygen and other gases are invisible. However, water vapor can be readily measured and expressed in different ways. Two commonly used terms are relative humidity and dew point.

Relative humidity

Relative humidity routinely is expressed as a percentage. It relates the actual water vapor present to that which could be present.

Temperature largely determines the maximum amount of water vapor air can hold. Warm air can hold more water vapor than cool air. Relative humidity expresses the degree of saturation. Air with 100% relative humidity is saturated; less than 100% is unsaturated.

Dew point

Dew point is the temperature to which air must be cooled to become saturated by the water vapor already present in the air. Aviation weather reports normally include the air temperature and dew-point temperature. Dew point when related to air temperature reveals qualitatively how close the air is to saturation.

Temperature dew-point spread

The difference between air temperature and dew-point temperature is popularly called the spread. As spread becomes less, relative humidity increases, and it is 100% when temperature and dew point are the same. Surface-temperature dew-point spread is important for anticipating fog, but has little bearing on precipitation. To support precipitation, air must be saturated through thick layers aloft.

Sometimes the spread at ground level may be quite large, yet at higher altitudes the air is saturated and clouds form. Some rain may reach the ground or it may evaporate as it falls into the drier air. Our never-ending weather cycle involves a continual reversible change of water from one state to another.

• CHANGE OF STATE

Evaporation, condensation, sublimation, freezing, and melting are changes of state. Evaporation is the changing of liquid water to invisible water vapor. Condensation is the reverse process. Sublimation is the changing of ice directly to water vapor, or water vapor to ice, bypassing the liquid state in each process. Snow or ice crystals result from the sublimation of water vapor directly to the solid state.

Latent heat

Any change of state involves a heat transaction with no change in temperature. Evaporation requires heat energy that comes from the nearest available heat source. This heat energy is known as the latent heat of vaporization, and its removal cools the source it comes from. An example is the cooling of your body by evaporation of perspiration.

What becomes of this heat energy used by evaporation? Energy cannot be created or destroyed, so it is hidden or stored in the invisible water vapor. When the water vapor condenses to liquid water or sublimates directly to ice, energy originally used in the evaporation reappears as heat and is released to the atmosphere. This energy is latent heat. Melting and freezing involve the exchange of "latent heat of fusion" in a similar manner. The latent heat of fusion is much less than that of condensation and evaporation; however, each in its own way plays an important role in weather.

Condensation nuclei

The atmosphere is never completely clean; an abundance of microscopic solid particles suspended in the air are condensation surfaces. These particles, such as salt, dust, and combustion by-products, are condensation nuclei. Some condensation nuclei have an affinity for water and can induce condensation or sublimation even when air is almost, but not completely, saturated.

As water vapor condenses or sublimates on condensation nuclei, liquid or ice particles begin to grow. Whether the particles are liquid or ice does not depend entirely on temperature. Liquid water may be present at temperatures well below freezing.

Supercooled water

Freezing is complex and liquid water droplets often condense or persist at temperatures colder than 32°F. Water droplets colder than 32°F are supercooled. When they strike an exposed object, the impact induces freezing. For example, impact freezing of supercooled water can result in aircraft icing.

Supercooled water drops are often in abundance in clouds at temperatures between 5°F and 32°F and, with decreasing amounts at colder temperatures. Usually, at temperatures colder than 5°F, sublimation is prevalent, and clouds and fog may be mostly ice crystals with a lesser amount of supercooled water. However, strong vertical currents may carry supercooled water to great heights where temperatures are much colder than 5°F. Supercooled water has been observed at temperatures colder than −40°F.

Dew and frost

During clear nights with little or no wind, vegetation often cools by radiation to a temperature at or below the dew point of the adjacent air. Moisture then collects on the leaves just as it does on a pitcher of ice water in a warm room. Heavy dew often collects on grass and plants while none collects on pavements or large solid objects. These more massive objects absorb abundant heat during the day, lose it slowly during the night, and cool below the dew point only in rather extreme cases.

Frost forms in much the same way as dew. The difference is that the dew point of surrounding air must be colder than freezing. Water vapor then sublimates directly as ice crystals or frost rather than condensing as dew. Sometimes dew forms and later freezes; however, frozen dew is easily distinguished from frost. Frozen dew is hard and transparent while frost is white and opaque.

Cloud formation

Normally, air must become saturated for condensation or sublimation to occur. Saturation may result from cooling temperature, increasing dew point, or both. Cooling is far more predominant.

Cooling processes

Three basic processes may cool air to saturation: (1) air moving over a colder surface, (2) stagnant air overlying a cooling surface, and (3) expansional cooling in upward moving air. Expansional cooling is the major cause of cloud formation.

Clouds and fog

A cloud is a visible aggregate of minute water or ice particles suspended in air. If the cloud is on the ground, it is fog. When entire layers of air cool to saturation, fog or sheet-like clouds result. Saturation of a localized updraft produces a towering cloud. A cloud may be composed entirely of liquid water, of ice crystals, or a mixture of the two.

Precipitation

Precipitation is an all-inclusive term denoting drizzle, rain, snow, ice pellets, hail, and ice crystals. Precipitation occurs when these particles grow in size and weight until the atmosphere no longer can suspend them and they fall. These particles grow primarily in two ways.

Particle growth

Once a water droplet or ice crystal forms, it continues to grow by added condensation or sublimation directly onto the particle. This is the slower of the two methods and usually results in drizzle or very light rain or snow.

Cloud particles collide and merge into a larger drop in the more rapid growth process. This process produces larger precipitation particles and does so more rapidly than the simple condensation growth process. Upward currents enhance the growth rate and also support larger drops. Precipitation formed by merging drops with mild upward currents can produce light to moderate rain and snow. Strong upward currents support the largest drops and build clouds to great heights. They can produce heavy rain, heavy snow, and hail.

Liquid, freezing, and frozen precipitation

Precipitation forming and remaining liquid falls as rain or drizzle. Sublimation forms snowflakes, and they reach the ground as snow if temperatures aloft remain below freezing.

Precipitation can change its state as the temperature of its environment changes. Falling snow may melt in warmer layers of air at lower altitudes to form rain. Rain falling through colder air may become supercooled, freezing on impact as freezing rain; or it may freeze during its descent, failing as ice pellets. Ice pellets always indicate freezing rain at higher altitude.

Sometimes strong upward currents sustain large supercooled water drops until some freeze; subsequently, other drops freeze to them, forming hailstones.

Precipitation versus cloud thickness

To produce significant precipitation, clouds usually are 4,000 ft thick or more. The heavier the precipitation, the thicker the clouds are likely to be.

Land and water effects on clouds

Land and water surfaces underlying the atmosphere greatly affect cloud and precipitation development. Large bodies of water such as oceans and large lakes add water vapor to the air.

The greatest frequency of low ceilings, fog, and precipitation can be expected in areas where prevailing winds have an over-water trajectory. The aviator should be especially alert for these hazards when moist winds are blowing upslope.

In winter, cold air frequently moves over relatively warm lakes. The warm water adds heat and water vapor to the air, causing showers. In other seasons, the air may be warmer than the lakes. When this occurs, the air may become saturated by evaporation from the water while also becoming cooler in the low levels by contact with the cool water. Fog often becomes extensive and dense to the lee of a lake. Strong cold winds across the Great Lakes often carry precipitation to the Appalachians.

A lake only a few miles across can influence convection and cause a diurnal fluctuation in cloudiness. During the day, cool air over the lake blows toward the land, and convective clouds form over the land. At night, the pattern reverses; clouds tend to form over the lake as cool air from the land flows over the lake, creating convective clouds over the water.

Water exists in three states—gaseous, liquid, and solid. Water vapor is an invisible gas. Condensation or sublimation of water vapor creates many common weather extremes. The following may be anticipated:

1. Fog when temperature dew-point spread is 5°F or less and decreasing.

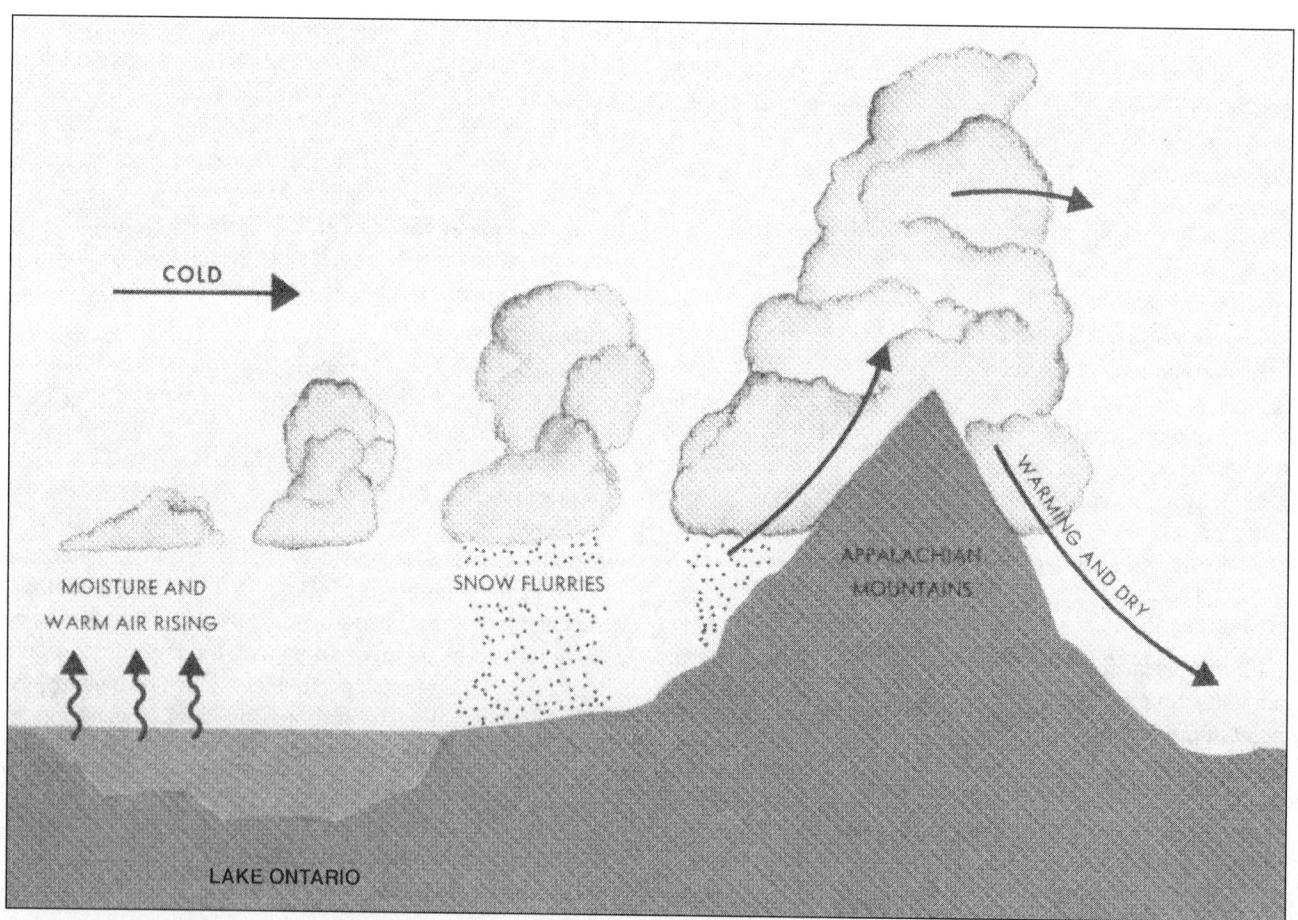

Strong cold winds across the Great Lakes absorb water vapor and may carry showers as far eastward as the Appalachians. *(Courtesy of U.S. government publication.)*

2. Lifting or clearing of low clouds and fog when temperature dew-point spread is increasing.

3. Frost on a clear night when temperature dew-point spread is 5°F or less, is decreasing, and dew point is colder than 32°F.

4. More cloudiness, fog, and precipitation when wind blows from water than when it blows from land.

5. Cloudiness, fog, and precipitation over higher terrain when moist winds are blowing uphill.

6. Showers to the lee of a lake when air is cold and the lake is warm. Expect fog to the lee of the lake when the air is warm and the lake is cold.

7. Clouds to be at least 4,000 ft thick when significant precipitation is reported. The heavier the precipitation, the thicker the clouds are likely to be.

• STABLE & UNSTABLE AIR

Changes within upward and downward moving air

Any time air moves upward, it expands because of decreasing atmospheric pressure. Conversely, downward-moving air is compressed by increasing pressure. But as pressure and volume change, temperature also changes.

When air expands, it cools; and when compressed, it warms. These changes are adiabatic, meaning that no heat is removed from or added to the air. We frequently use the terms expansional or adiabatic cooling and compressional or adiabatic heating. The adiabatic rate of change of temperature is virtually fixed in unsaturated air but varies in saturated air.

Unsaturated air

Unsaturated air moving upward and downward cools and warms at about 5.4°F per 1,000 ft. This rate is the dry adiabatic rate of temperature change and is independent of the temperature of the mass of air through which the vertical movements occur.

Saturated air

Condensation occurs when saturated air moves upward. Latent heat released through condensation partially offsets the expansional cooling. Therefore, the saturated adiabatic rate of cooling is slower than the dry adiabatic rate. The saturated rate depends on saturation temperature or dew point of the air. Condensation of copious moisture in saturated warm air releases more latent heat to offset expansional cooling than does the scant moisture in saturated cold air. Therefore, the saturated adiabatic rate of cooling is less in warm air than in cold air.

When saturated air moves downward, it heats at the same rate as it cools on ascent, provided liquid water evaporates rapidly enough to maintain saturation. Minute water droplets evaporate at virtually this rate. Larger drops evaporate more slowly and complicate the moist adiabatic process in downward-moving air.

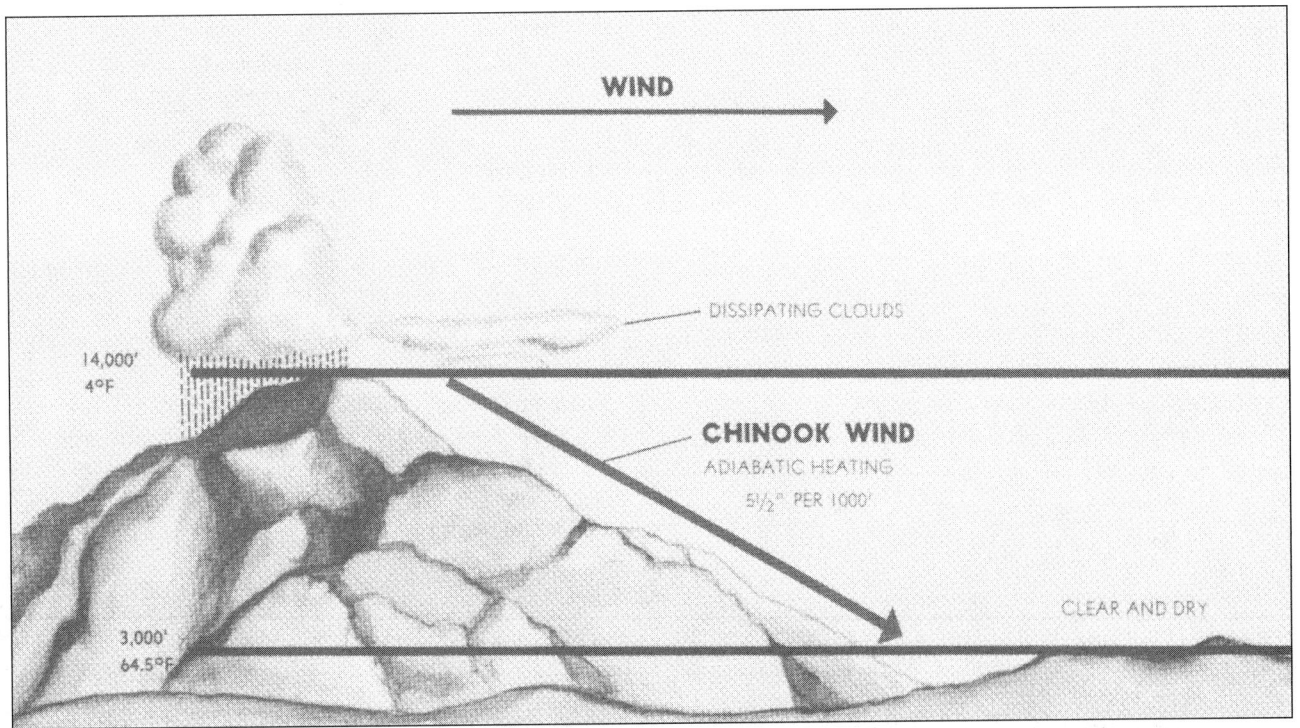

Adiabatic warming of downward-moving air produces the warm chinook wind. *(Courtesy of U.S. government publication.)*

Adiabatic cooling and vertical air movement

If a sample of air is forced upward into the atmosphere, two possibilities must be considered: (1) the air may become colder than the surrounding air, or (2) even though it cools, the air may remain warmer than the surrounding air.

If the upward-moving air becomes colder than surrounding air, it sinks; but if it remains warmer, it is accelerated upward as a convective current. Whether it sinks or rises depends on the ambient or existing temperature lapse rate.

Existing lapse rate should not be confused with adiabatic rates of cooling in vertically moving air. Sometimes the dry and moist adiabatic rates of cooling will be called the dry adiabatic lapse rate and the moist adiabatic lapse rate. Lapse rate refers exclusively to the existing, or actual, decrease of temperature with height in a real atmosphere. The dry or moist adiabatic lapse rate signifies a prescribed rate of expansional cooling or compressional heating. An adiabatic lapse rate becomes real only when it becomes a condition brought about by vertically moving air. The difference between the existing lapse rate of a given mass of air and the adiabatic rates of cooling in upward-moving air determines if the air is stable or unstable.

• CLOUDS

Clouds—stable or unstable?

When air is cooling and first becomes saturated, condensation, or sublimation, begins to form clouds. Whether the air is stable or unstable within a layer largely determines cloud structure.

Stratiform clouds

Since stable air resists convection, clouds in stable air form in horizontal, sheet-like layers or strata. Thus, within a stable layer, clouds are stratiform. Adiabatic cooling may be by upslope flow; by lifting over cold, denser air; or by converging winds. Cooling by an underlying cold surface is a stabilizing process and may produce fog. If clouds are to remain stratiform, the layer must remain stable after condensation occurs.

Cumuliform clouds

Unstable air favors convection. A cumulus cloud, meaning "heap," forms in a convective updraft and builds upward. Thus, within an unstable layer, clouds are cumuliform; and the vertical extent of the cloud depends on the depth of the unstable layer.

Initial lifting to trigger a cumuliform cloud may be the same as that for lifting stable air. In addition, convection may be set off by surface heating. Air may be unstable or slightly stable before condensation occurs; but for convective cumuliform clouds to develop, it must be unstable after saturation. Cooling in the updraft is now at the slower moist adiabatic rate because of the release of latent heat of condensation. Temperature in the saturated updraft is warmer than ambient temperature, and convection is spontaneous. Updrafts accelerate until temperature within the cloud cools below the ambient temperature. This condition occurs when a stable layer, which is often marked by a temperature inversion, caps the unstable layer. Vertical heights range from the shallow fair weather cumulus to the giant thunderstorm cumulonimbus—the ultimate in atmospheric instability capped by the tropopause.

When unstable air lies above stable air, convective currents aloft sometimes form middle- and high-level cumuliform clouds. In relatively shallow layers they occur as altocumulus and ice crystal cirrocumulus clouds. Altocumulus castellans clouds develop in deeper midlevel unstable layers.

Identification

The basic cloud types are divided into four families: high clouds, middle clouds, low clouds, and clouds with extensive vertical development. The first three families are further classified according to the way they are formed. Clouds formed by vertical currents in unstable air are cumulus, meaning accumulation or heap; they are characterized by their lumpy, billowy appearance. Clouds formed by the cooling of a stable layer are stratus, meaning stratified or layered; they are characterized by their uniform, sheet-like appearance. In addition to the above, the prefix *nimbo-*, and the suffix *-nimbus*, mean rain cloud. Thus, stratified clouds from which rain is falling are nimbostratus. A heavy, swelling cumulus-type cloud that produces precipitation is a cumulonimbus. Clouds broken into fragments are often identified by adding the suffix *-fractus*; for example, fragmentary cumulus is cumulus fractus.

High clouds

The high-cloud family is cirriform and includes cirrus, cirrocumulus, and cirrostratus. They are composed almost entirely of ice crystals. The height of the bases of these clouds is in the range of 16,500–45,000 ft in middle latitudes.

Middle clouds

In the middle-cloud family are the altostratus, altocumulus, and nimbostratus clouds. These clouds are primarily water, much of which may be supercooled. The height of the bases of these clouds is in the range of 6,500–23,000 ft in middle latitudes.

Low clouds

In the low-cloud family are the stratus, stratocumulus, and fair-weather cumulus clouds. Low clouds are almost entirely water, but at times the water may be supercooled. Low clouds at subfreezing temperatures can also contain snow and ice particles. The bases of these clouds range from near the surface to about 6,500 ft in middle latitudes.

Altocumulus clouds. *(Photo by Ralph F. Kresge. Courtesy of National Oceanic and Atmospheric Administration (NOAA) Central Library.)*

Clouds with extensive vertical development

The vertically developed family of clouds includes towering cumulus and cumulonimbus. These clouds usually contain supercooled water above the freezing level. But when a cumulus grows to great heights, water in the upper part of the cloud freezes into ice crystals, forming a cumulonimbus. The heights of cumuliform cloud bases range from 1,000 ft or lower to above 10,000 ft.

• FOG

Fog is a surface-based cloud composed of either water droplets or ice crystals.

Small temperature dew-point spread is essential for fog to form. Therefore, fog is prevalent in coastal areas where moisture is abundant. However, fog can occur anywhere. Abundant condensation nuclei enhance the formation of fog. Thus, fog is prevalent in industrial areas where by-products of combustion provide a high concentration of these nuclei. Fog occurs most frequently in the colder months, but the season and frequency of occurrence vary from one area to another.

Fog may form either by cooling air to its dew point, or by adding moisture to air near the ground. Fog is classified by the way it forms. Formation may involve more than one process.

Radiation fog

Radiation fog is relatively shallow fog. It may be dense enough to hide the entire sky or may conceal only part of the sky. "Ground fog" is a form of radiation fog. Conditions favorable for radiation fog are clear sky, little or no wind, and small temperature dew-point spread (high relative humidity). The fog forms almost exclusively at night or near daybreak. Terrestrial radiation cools the ground; in turn, the cool ground cools the air in contact with it. When the air is cooled to its dew point, fog forms. When rain soaks the ground, followed by clearing skies, radiation fog is not uncommon the following morning.

Radiation fog is restricted to land because water surfaces cool little from nighttime radiation. It is shallow when wind is calm. Winds up to about 5 knots mix the air slightly and tend to deepen the fog by spreading the cooling through a deeper layer. Stronger winds disperse the fog or mix the air through a still deeper layer with stratus clouds forming at the top of the mixing layer.

Ground fog usually "burns off" rather rapidly after sunrise. Other radiation fog generally clears before noon unless clouds move in over the fog.

Advection fog

Advection fog forms when moist air moves over colder ground or water. It is most common along coastal areas but often develops deep in continental areas. At sea it is called sea fog. Advection fog deepens as wind speed increases up to about 15 knots. Wind much stronger than 15 knots lifts the fog into a layer of low stratus or stratocumulus.

The west coast of the United States is quite vulnerable to advection fog. This fog frequently forms offshore as a result of cold water and then is carried inland by the wind. During the winter, advection fog over the central and eastern United States results when moist air from the Gulf of Mexico spreads northward over cold ground. The fog may extend as far north as the Great Lakes. Water areas in northern latitudes have frequent dense sea fog in summer as a result of warm, moist, tropical air flowing northward over colder arctic waters.

Advection fog is usually more extensive and much more persistent than radiation fog. Advection fog can move in rapidly regardless of the time of day or night.

Upslope fog

Upslope fog forms as a result of moist, stable air being cooled adiabatically as it moves up sloping terrain. Once the upslope wind ceases, the fog dissipates. Unlike radiation fog, it can form under cloudy skies. Upslope fog is common along the eastern slopes of the Rockies and somewhat less frequent east of the Appalachians. Upslope fog often is quite dense and extends to high altitudes.

Precipitation-induced fog

When relatively warm rain or drizzle falls through cool air, evaporation from the precipitation saturates the cool air and forms fog. Precipitation-induced fog can become quite dense and continue for an extended period of time. This fog may extend over large areas, completely suspending air operations. It is most commonly associated with warm fronts, but can occur with slow-moving cold fronts and with stationary fronts.

Ice fog

Ice fog occurs in cold weather when the temperature is much below freezing and water vapor sublimates directly as ice crystals. Conditions favorable for its formation are the same as for radiation fog except that it is associated with cold temperatures, usually $-25°F$ or colder. It occurs mostly in the arctic regions, but is not unknown in middle latitudes during the cold season.

Low stratus clouds

Stratus clouds, like fog, are composed of extremely small water droplets or ice crystals suspended in air. An observer on a mountain in a stratus layer would call it fog. Stratus and fog frequently exist together. In many cases there is no real line of distinction between the fog and stratus; rather, one gradually merges into the other. Stratus tends to be lowest during night and early morning, lifting or dissipating due to solar heating during the late morning or afternoon. Low stratus clouds often occur when moist air mixes with a colder air mass or in any situation where temperature dew-point spread is small.

Haze and smoke

Haze is a concentration of salt particles or other dry particles not readily classified as dust or other phenomena. It occurs in stable air, is usually only a few thousand feet thick, but sometimes may extend as high as 15,000 ft. Haze layers often have definite tops above which horizontal visibility is good. However, downward visibility from above a haze layer is poor, especially on a slant. Visibility in haze varies greatly, depending upon whether the observer is facing the sun.

Smoke concentrations form primarily in industrial areas when air is stable. It is most prevalent at night or early morning under a temperature inversion but it can persist throughout the day.

• AIR MASSES

Air masses

When a body of air comes to rest or moves slowly over an extensive area having fairly uniform properties of temperature and moisture, the air takes on those properties. Thus, the air over the area becomes somewhat of an entity and has fairly uniform horizontal distribution of its properties. The area over which the air mass acquires its identifying distribution of moisture and temperature is its source region.

Source regions are many and varied, but the best source regions for air masses are large snow- or ice-covered polar regions, cold northern oceans, tropical oceans, and large desert areas. Midlatitudes are poor source regions because transitional disturbances dominate these latitudes, giving little opportunity for air masses to stagnate and take on the properties of the underlying region.

Air-mass modification

Just as an air mass takes on the properties of its source region, it tends to also take on properties of the underlying surface when it moves away from its source region, thus becoming modified.

The degree of modification depends on the speed with which the air mass moves, the nature of the region over which it moves, and the temperature difference between the new surface and the air mass. Some ways air masses are modified are warming from below, cooling from below, addition of water vapor, and subtraction of water vapor:

1. Cool air moving over a warm surface is heated from below, generating instability and increasing the possibility of showers.
2. Warm air moving over a cool surface is cooled from below, increasing stability. If air is cooled to its dew point, stratus and/or fog forms.
3. Evaporation from water surfaces and failing precipitation adds water vapor to the air. When the water is warmer than the air, evaporation can raise the dew point sufficiently to saturate the air and form stratus or fog.
4. Water vapor is removed by condensation and precipitation.

Stability

Stability of an air mass determines its typical weather characteristics. When one type of air mass overlies another, conditions change with height. Characteristics typical of an unstable air mass are: cumuliform clouds, showery precipitation, rough air (turbulence), and good visibility. Characteristics of stable air include: stratiform clouds and fog, continuous precipitation, smooth air, and fair-to-poor visibility in haze and smoke.

• FRONTS

As air masses move out of their source regions, they come in contact with other air masses of different properties. The zone between two different air masses is a frontal zone or front. Across this zone, temperature, humidity, and wind often change rapidly over short distances.

Discontinuities

When one passes through a front, the change from the properties of one air mass to those of the other is sometimes quite abrupt. Abrupt changes indicate a narrow frontal zone. At other times, the change of properties is very gradual, indicating a broad and diffuse frontal zone.

Temperature—Temperature is one of the most easily recognized discontinuities across a front. At the surface, the passage of a front usually causes noticeable temperature change.

Dew point—Dew-point temperature is a measure of the amount of water vapor in the air. Temperature dew-point spread is a measure of the degree of saturation. Dew point and temperature dew-point spread usually differ across a front. The difference helps identify the front and may give a clue to differences of cloudiness and/or fog.

Wind—Wind always changes across a front. Wind discontinuity may be in direction, in speed, or in both.

Pressure—A front lies in a pressure trough, and pressure generally is higher in the cold air. Thus, when a front is crossed directly into colder air, pressure usually rises abruptly. When a front is approached toward warm air, pressure generally falls until the front is crossed, and then remains steady or falls slightly in the warm air. However, pressure patterns vary widely across fronts.

Types of fronts

The three principal types of fronts are the cold front, the warm front, and the stationary front.

Cold front

The leading edge of an advancing cold air mass is a cold front. At the surface, cold air is overtaking and replacing warmer air. Cold fronts move at about the speed of the wind component perpendicular to the front just above the frictional layer. A shallow cold air mass or a slow-moving cold front may have a frontal slope more like a warm front.

Warm front

The edge of an advancing warm air mass is a warm front—warmer air is overtaking and replacing colder air. Since the cold air is denser than the warm air, the cold air hugs the ground. The warm air slides up and over the cold air and lacks direct push on the cold air. Thus, the cold air is slow to retreat in advance of the warm air. This slowness of the cold air to retreat produces a frontal slope that is more gradual than the cold frontal slope. Consequently, warm fronts on the surface are seldom as well marked as cold fronts, and they usually move about half as fast when the general wind flow is the same in each case.

Stationary front

When neither air mass is replacing the other, the front is stationary. The opposing forces exerted by adjacent air masses of different densities are such that the frontal surface between them shows little or no movement. In such cases, the surface winds tend to blow parallel to the frontal zone. Slope of a stationary front is normally shallow, although it may be steep, depending on wind distribution and density difference.

Frontal waves and occlusion

Frontal waves and cyclones (areas of low pressure) usually form on slow-moving cold fronts or on stationary fronts. The life cycle and movement of a cyclone is dictated to a great extent by the upper wind flow.

In the initial condition of frontal wave development, the winds on both sides of the front are blowing parallel to the front. Small disturbances then may start a wavelike bend in the front.

If this tendency persists and the wave increases in size, a cyclonic (counterclockwise) circulation develops. One section of the front begins to move as a warm front, while the section next to it begins to move as a cold front. This deformation is a frontal wave.

The pressure at the peak of the frontal wave falls, and a low-pressure center forms. The cyclonic circulation becomes

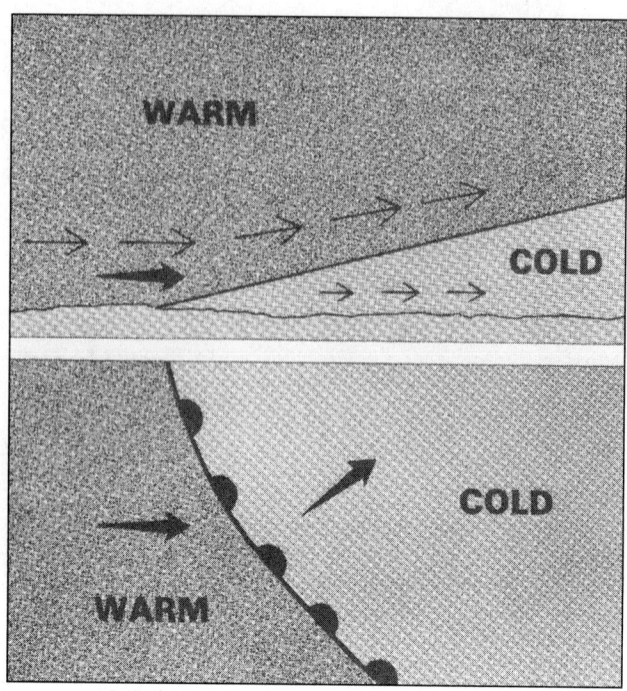

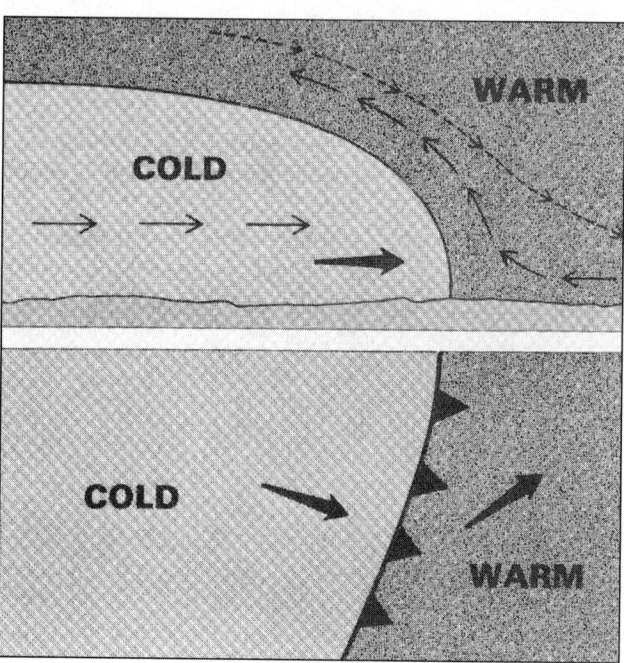

Cross section of a warm front (top) with the weather map symbol (bottom). The symbol is a line with rounded barbs pointing in the direction of movement. On a color map, a red line represents the warm front. The slope of a warm front is generally more shallow than that of a cold front. Movement of a warm front, shown by the heavy black arrow, is slower than the wind in the warm air, represented by the thin solid arrows. The warm air gradually erodes the cold air. *(Courtesy of U.S. government publication.)*

Cross section of a cold front (top) with the weather map symbol (bottom). The symbol is a line with pointed barbs pointing in the direction of movement. On a color map, a blue line represents the cold front. The vertical scale is expanded in the top illustration to show the frontal slope, which is steep near the leading edge as cold air replaces warm air. Warm air may descend over the front as indicated by the dashed arrows; but more often, the cold air forces warm air upward over the frontal surface as shown by the solid arrows. *(Courtesy of U.S. government publication.)*

stronger, and the surface winds are now strong enough to move the fronts; the cold front moves faster than the warm front. When the cold front catches up with the warm front, the two of them occlude (close together). The result is an occluded front or, for brevity, an occlusion. This is the time of maximum intensity for the wave cyclone. Note that the symbol depicting the occlusion is a combination of the symbols for the warm and cold fronts.

As the occlusion continues to grow in length, the cyclonic circulation diminishes in intensity and the frontal movement slows down. Sometimes a new frontal wave begins to form on the long westward-trailing portion of the cold front, or a secondary low-pressure system forms at the apex where the cold front and warm front come together to form the occlusion. In the final stage, the two fronts may have become a single stationary front again. The low center with its remnant of the occlusion is disappearing.

Nonfrontal lows

Since fronts are boundaries between air masses of different properties, fronts are not associated with lows lying solely in a homogeneous air mass. Nonfrontal lows are infrequent east of the Rocky Mountains in midlatitudes, but do occur occasionally during the warmer months. Small non-

frontal lows over the western mountains are common as is the semistationary thermal low in the extreme southwestern United States. Tropical lows are also nonfrontal.

Frontolysis

As adjacent air masses modify and as temperature and pressure differences equalize across a front, the front dissipates. This process is frontolysis, the generation of a front. It occurs when a relatively sharp zone of transition develops over an area between two air masses that have densities gradually becoming more and more in contrast with each other. The necessary wind flow pattern develops at the same time.

Frontal weather

Weather occurring with a front depends on the amount of moisture available, the degree of stability of the air that is forced upward, the slope of the front, the speed of frontal movement, and the upper wind flow.

Sufficient moisture must be available for clouds to form, or there will be no clouds. As an inactive front comes into an area of moisture, clouds and precipitation may develop rapidly. A good example of this is a cold front moving eastward from the dry slopes of the Rocky Mountains into a

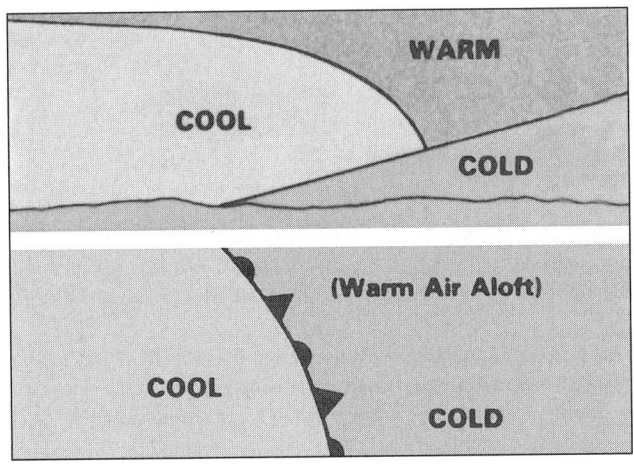

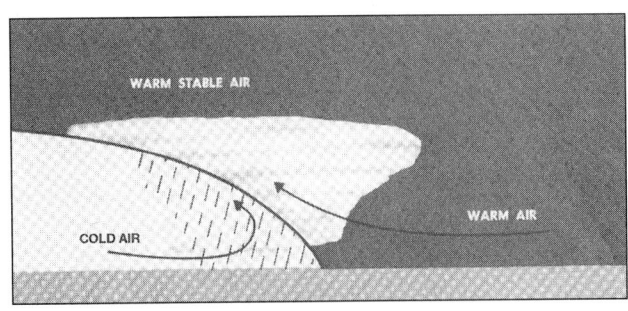

A cold front underrunning warm, moist, stable air. Clouds are stratified and precipitation is continuous. Precipitation induces stratus in the cold air. *(Courtesy of U.S. government publication.)*

Cross section of a warm-front occlusion (top) and its weather symbol (bottom). The symbol is a line with alternating pointed and rounded barbs on the same side of the line pointing in the direction of movement. On a color map, the line is purple. In the warm-front occlusion, air under the cold front is not as cold as air ahead of the warm front; and when the cold front overtakes the warm front, the cool air rides over the colder air. In a warm-front occlusion, cool air replaces cold air at the surface. *(Courtesy of U.S. government publication.)*

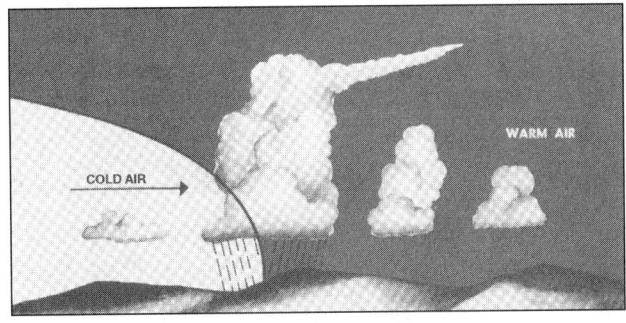

A cold front underrunning warm, moist, unstable air. Clouds are cumuliform with possible showers or thunderstorms near the surface position of the front. Convective clouds often develop in the warm air ahead of the front. The warm, wet ground behind the front generates low-level convection and fair-weather cumulus in the cold air. *(Courtesy of U.S. government publication.)*

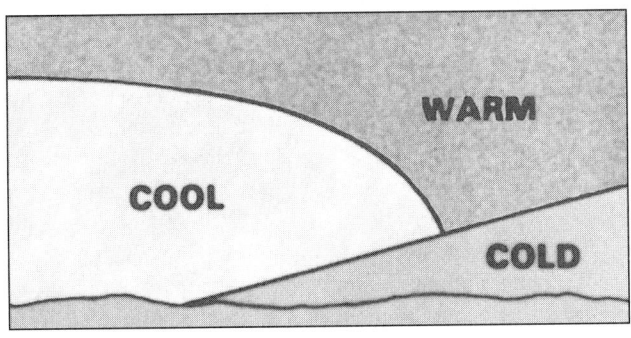

Cross section of a cold-front occlusion. Its weather map symbol is the same as that for a warm-front occlusion, and the coldest air is under the cold front. When it overtakes the warm front, it lifts the warm front aloft, and cold air replaces cool air at the surface. *(Courtesy of U.S. government publication.)*

tongue of moist air from the Gulf of Mexico over the Plains states. Thunderstorms may build rapidly.

The degree of stability of the lifted air determines whether cloudiness will be predominately stratiform or cumuliform. If the warm air overriding the front is stable, stratiform clouds develop. If the warm air is unstable, cumuliform clouds develop. Precipitation from stratiform clouds is usually steady and there is little or no turbulence. Precipitation from cumuliform clouds is of a shower type and the clouds are turbulent.

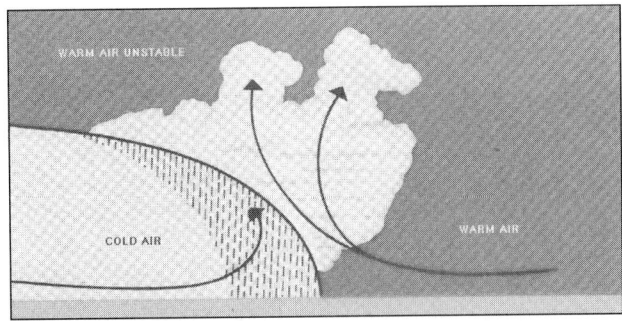

A slow-moving cold front underrunning warm, moist, unstable air. Clouds are stratified with embedded cumulonimbus and thunderstorms. This type of frontal weather is especially hazardous for aircraft, since the individual thunderstorms are hidden and cannot be avoided unless the aircraft is equipped with airborne radar. *(Courtesy of U.S. government publication.)*

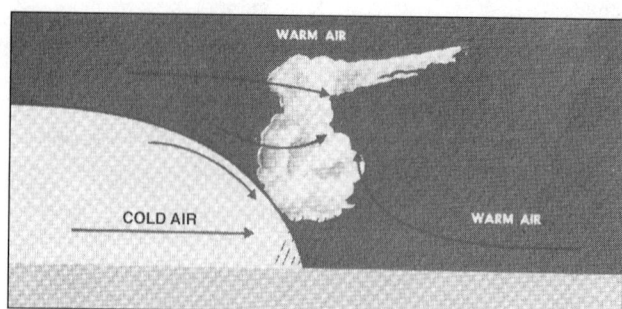

A fast-moving cold front underrunning warm, moist, unstable air. Showers and thunderstorms develop along the surface position of the front. *(Courtesy of U.S. government publication.)*

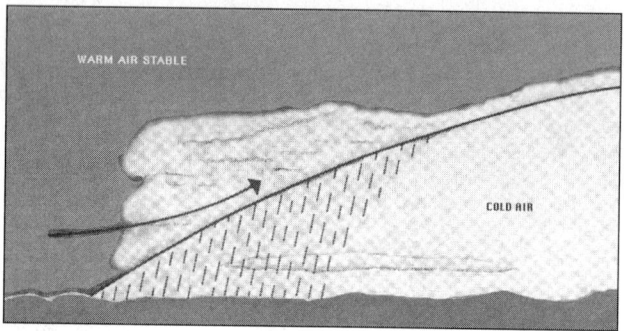

A warm front with overrunning moist, stable air. Clouds are stratiform and widespread over the shallow front. Precipitation is continuous and induces widespread stratus in the cold air. *(Courtesy of U.S. government publication.)*

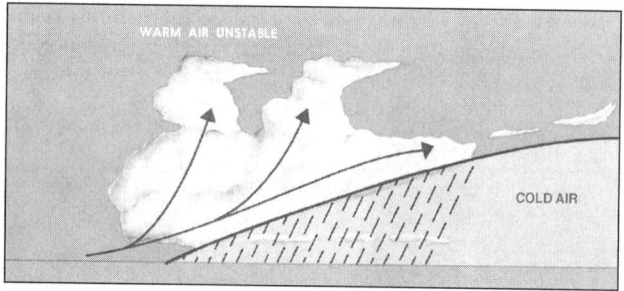

A warm front with overrunning warm, moist, unstable air. *(Courtesy of U.S. government publication.)*

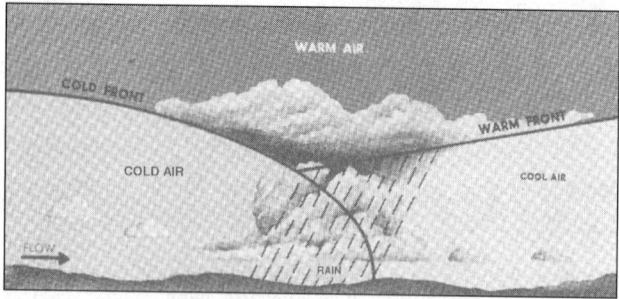

A cold-front occlusion lifting warm, moist, stable air. Associated weather encompasses that associated with both warm and cold fronts when air is moist and stable. *(Courtesy of U.S. government publication.)*

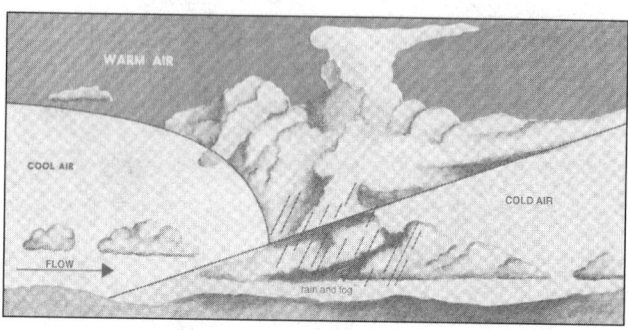

A warm-front occlusion lifting warm, moist, unstable air. The associated weather is complex and encompasses all types of weather related to both the warm and cold fronts when air is moist and unstable. *(Courtesy of U.S. government publication.)*

Shallow frontal surfaces tend to have extensive cloudiness with large precipitation areas. Widespread precipitation associated with a gradual sloping front often causes low stratus and fog. In this case, the rain raises the humidity of the cold air to saturation. This and related effects may produce low ceiling and poor visibility over thousands of square miles. If temperature of the cold air near the surface is below freezing but the warmer air aloft is above freezing, precipitation falls as freezing rain or ice pellets; however, if temperature of the warmer air aloft is well below freezing, precipitation forms as snow.

When the warm air overriding a shallow front is moist and unstable, the usual widespread cloud mass forms; but embedded in the cloud mass are altocumulus, cumulus, and even thunderstorms. These embedded storms are more common with warm and stationary fronts but may occur with a slow-moving, shallow cold front.

A fast-moving, steep cold front forces upward motion of the warm air along its leading edge. If the warm air is moist, precipitation occurs immediately along the surface position of the front.

Since an occluded front develops when a cold front overtakes a warm front, weather with an occluded front is a combination of both warm and cold frontal weather.

A front may have little or no cloudiness associated with it. Dry fronts occur when the warm air aloft is flowing down the frontal slope or the air is so dry that any cloudiness that occurs is at high levels.

The upper wind flow dictates to a great extent the amount of cloudiness and rain accompanying a frontal system as well as movement of the front itself. Systems tend to move with the upper winds. When winds aloft blow across a front, it tends to move with the wind. When winds aloft parallel a front, the front moves slowly, if at all. A deep, slow-moving trough aloft forms extensive cloudiness and precipitation, while a rapid-moving minor trough more often restricts weather to a rather narrow band. However, the latter often breeds severe, fast-moving, turbulent spring weather.

Instability line

An instability line is a narrow, nonfrontal line or band of convective activity. If the activity is fully developed in a thunderstorm, the line is a squall line. Instability lines form in moist, unstable air. An instability line may develop far from any front. More often, it develops ahead of a cold front, and sometimes a series of these lines move out ahead of the front. A favored location for instability lines which frequently erupt into severe thunderstorms is a dew-point front or dry line.

Dew-point front or dry line

During a considerable part of the year, dew-point fronts are common in western Texas and New Mexico northward over the Plains states. Moist air flowing north from the Gulf of Mexico abuts the dryer, and therefore slightly denser, air flowing from the southwest. Except for moisture differences, there is seldom any significant air mass contrast across this front, and therefore, it is commonly called a dry line. Nighttime and early morning fog and low-level clouds often prevail on the moist side of the line while generally clear skies mark the dry side. In spring and early summer over Texas, Oklahoma, and Kansas, and for some distance eastward, the dry line is a favored spawning area for squall lines and tornadoes.

• TURBULENCE

Convective currents

Convective currents are localized vertical air movements, both ascending and descending. For every rising current, there is a compensating downward current. The downward currents frequently occur over broader areas than do the upward currents, and therefore, they have a slower vertical speed than do the rising currents.

Convective currents are most active on warm summer afternoons when winds are light. Heated air at the surface creates a shallow, unstable layer, and the warm air is forced upward. Convection increases in strength and to greater heights as surface heating increases. Barren surfaces such as sandy or rocky wastelands and plowed fields become hotter than open water or ground covered by vegetation. Thus, air at and near the surface heats unevenly. Because of uneven heating, the strength of convective currents can vary considerably within short distances.

When cold air moves over a warm surface, it becomes unstable in lower levels. Convective currents extend several thousand feet above the surface, resulting in rough, choppy turbulence. This condition often occurs in any season after the passage of a cold front.

• HIGH-ALTITUDE WEATHER

The tropopause

The tropopause is a thin layer forming the boundary between the troposphere and stratosphere. Height of the tropopause varies from about 65,000 ft over the equator to 20,000 ft or lower over the poles. The tropopause is not continuous but generally descends step-wise from the equator to the poles. These steps occur as breaks.

An abrupt change in temperature lapse rate characterizes the tropopause.

Maximum winds generally occur at levels near the tropopause. These strong winds create narrow zones of wind shear that often generate hazardous turbulence for aircraft.

The jet stream

The jet stream is a narrow, shallow, meandering river of maximum winds extending around the globe in a wavelike pattern. A second jet stream is not uncommon, and three at one time are not unknown. A jet may be as far south as the northern tropics. A jet in midlatitudes generally is stronger than one in or near the tropics. The jet stream typically occurs in a break in the tropopause. Therefore, a jet stream occurs in an area of intensified temperature gradients characteristic of the break.

The concentrated winds, by arbitrary definition, must be 50 knots or greater to classify as a jet stream. The jet maximum is not constant; rather, it is broken into segments, shaped something like a boomerang.

Jet stream segments move with pressure ridges and troughs in the upper atmosphere. In general, they travel faster than pressure systems, and maximum wind speed varies as the segments progress through the systems. In midlatitude, wind speed in the jet stream averages considerably stronger in winter than in summer. Also, the jet shifts farther south in winter than in summer.

• CONDENSATION TRAILS

A condensation trail, or contrail, is generally defined as a cloudlike streamer that frequently is generated in the wake of aircraft flying in clear, cold, humid air. Two distinct types are observed—exhaust trails and aerodynamic trails.

Exhaust contrails

The exhaust contrail is formed by the addition to the atmosphere of sufficient water vapor from aircraft exhaust gases to cause saturation or super-saturation of the air. Since heat is also added to the atmosphere in the wake of an aircraft, the addition of water vapor must be of such magnitude that it saturates or supersaturates the atmosphere in spite of the added heat. There is evidence to support the idea that the nuclei, which are necessary for condensation or sublimation, may also be donated to the atmosphere in the exhaust gases of aircraft engines, further aiding contrail formation. These nuclei are relatively large. However, recent experiments have found that by adding very minute nuclei material (dust, for example) to the exhaust visible exhaust contrails could be prevented. Condensation and sublimation on these smaller nuclei result in contrail particles too small to be visible.

Aerodynamic contrails

In air that is almost saturated, aerodynamic pressure reduction around airfoils, engine nacelles, and propellers cools the air to saturation, leaving condensation trails from these components. This type of trail usually is neither as dense nor as persistent as exhaust trails. However, under critical atmospheric conditions, an aerodynamic contrail may trigger the formation and spreading of a deck of cirrus clouds.

Air travels in a corkscrew path around the jet core with upward motion on the equatorial side. Therefore, when high-level moisture is available, cirriform clouds form on the equatorial side of the jet. Jet stream cloudiness can form independently of well-defined pressure systems. Such cloudiness ranges primarily from scattered to broken coverage in shallow layers or streaks. Their sometimes fishhook and streamlined, wind-swept appearance always indicates very strong upper wind usually quite far from developing or intense weather systems.

The most dense cirriform clouds occur with well-defined systems. They appear in broad bands. Cloudiness is rather dense in an upper trough, thickens downstream, and becomes most dense at the crest of the downwind ridge. The clouds taper off after passing the ridge crest, in the area of descending air. The poleward boundary of the cirrus band often is quite abrupt and frequently casts a shadow on lower clouds, especially in an occluded frontal system.

The upper limit of dense, banded cirrus is near the tropopause; a band may be either a single layer of multiple layers 10,000–12,000 ft thick. Dense, jet stream cirriform cloudiness is most prevalent along midlatitude and polar jets. However, a cirrus band usually forms along the subtropical jet in winter, when a deep upper trough plunges southward into the tropics.

An important aspect of the jet stream cirrus shield is its association with turbulence. Extensive cirrus cloudiness often occurs with deepening surface and upper lows and these deepening systems produce the greatest turbulence.

Weather
Information &
Communications

- ## NATIONAL WEATHER SERVICE

The National Weather Service (NWS) continues full-scale redevelopment of its systems and its organization.

Applied research conducted in the National Oceanic and Atmospheric Administration's (NOAA) Environmental Research Laboratories and other federal laboratories has demonstrated that state-of-the-art laboratory techniques for analyzing and predicting severe weather and flood phenomena can be practicably applied to Weather Service operations. Because the scientific understanding of the atmosphere and the ability to forecast large- and small-scale weather phenomena has increased dramatically over the last two decades, the Department of Commerce has set an ambitious goal for NOAA's agency, the National Weather Service. The Service is to be modernized, to take full advantage of hundreds of new technological tools for upgrading weather forecasting.

In 1988, Public Law 100-685 was enacted and, in part, specifies conditions on the planning, reporting, and accomplishment of the modernization and associated restructuring of the NWS. This strategic plan was the first response to Congress required by Public Law 100-685.

Principles for the modernization and associated restructuring

Throughout the process of change, the NWS will continue to fulfill its mission, which is to provide weather and flood warnings, public forecasts, and advisories for all of the United States, its territories, adjacent waters, and ocean areas, primarily for the protection of life and property. NWS data and products will continue to be provided to private meteorologists for the provision of all specialized services. Certain principles are essential to meet the operational mission and will be continued during the modernization and associated restructuring transition period.

The need to implement new science and technology

The most deadly of our nation's weather events—tornadoes, severe thunderstorms, lightning, and flash floods—are also the most difficult to detect and forecast. The new systems being installed will enable earlier detection and permit the short-range prediction of destructive, violent local storms and floods. The new observational technologies implemented this decade will provide unprecedented amounts of complex data, thereby requiring that the operational forecasters have higher levels of analytical and interpretive skills. This will require training personnel and the deployment of proven, new observational information processing, and communications technologies.

New equipment, new methods

Until recently, the vintage technologies that composed part of earlier weather service infrastructure were in desperate need of replacement. As the equipment aged, it became costly to maintain. By replacing the equipment with more reliable technologies that support the new scientific capabilities, the nation can move into the future with strengthened confidence in its atmospheric prediction capabilities.

New technological systems are essential in providing the opportunity to improve warning and forecast services and for replacing obsolete and increasingly unreliable existing systems. Each of the new technologies to be installed plays a unique, but complementary role in the modernization process. New observational technologies will yield high-resolution, time-variant, three-dimensional representations of details on the state of the atmosphere.

At Weather Forecast Offices, new data processing systems aid the forecaster in the assimilation of changing data and numerical weather prediction outputs. The meteorologist and hydrologist are able to rapidly manipulate, display, and analyze information, thus enabling them to combine scientific principles and operational experience to produce more accurate and timely warning and forecast services for the nation. The new high-resolution data sets and derived information are an important input to business and economic decision making outside the NWS.

The NWS is joined in its acquisition of much of the major new technologies by the Department of Transportation's

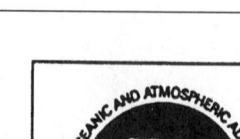

National Weather Service Milestones

1891 Weather Bureau becomes responsible for issuing flood warnings.

1894 William Eddy, using kites, makes first temperatures observations aloft.

1895 First Washington daily weather map published. Expanded version, covering the nation, is still published weekly.

1898 President McKinley orders establishment of hurricane warning network in the West Indies.

1900 Cable exchange of weather warnings/information begins with Europe. Devastating Galveston Hurricane kills 6000 including wife of Official-In-Charge and one employee. Hurricane was predicted 4 days in advance, but not the storm surge.

1901 Official three-day forecasts begin for the North Atlantic.

1904 Government begins using aircraft to conduct atmospheric research.

1905 The SS New York transmits first wireless weather report from a ship at sea.

1909 USWB begins free-rising balloon observations.

1910 USWB begins weekly agricultural planning forecasts and seasonal assessment of water availability for irrigation.

1912 First fire weather forecast issued.

1914 Aerological section established for aviation.

1917 Norwegian meteorologists begin air mass analysis techniques which will revolutionize meteorology.

1918 USWB begins forecasts for domestic military flights and new air mail routes.

1922 Histories of 500 river stations completed.

1927 West Coast prototype for an Airways Meteorological Service established. Lindbergh flies the Atlantic, ignoring a recommended Weather Bureau delay of 12 hours. The forecast was right on.

1928 Teletype replaces the telegraph. Telephone service becomes primary method of communicating weather information.

1931 Aircraft observations to 16,000 feet begin at Chicago, Cleveland, Dallas, and Omaha. The beginning of the end for "kite stations".

1933 A science advisory group apprises President Roosevelt that the work of the volunteer cooperative observer network is one of the most extraordinary services ever developed, netting the public more per dollar expended than any other government service in the world. By 1990 the 25 mile radius network encompasses nearly 10,000 stations.

1935 A hurricane warning service is established. The Smithsonian Institution begins making long-range weather forecasts based on solar cycles. Floating automatic weather instruments mounted on buoys begin collecting marine weather data.

1937 First official Weather Bureau radio meteorograph, or radiosonde sounding made at East Boston, Mass. This program spells the end for aircraft soundings. Twelve pilots die flying weather missions.

1939 USWB initiates automatic telephone weather service in New York City. Radiosondes replace all military and Weather Bureau aircraft observations.

1940 USWB transferred to Department of Commerce. Army and Navy establish weather centers. President Roosevelt orders Coast Guard to man ocean weather stations.

1942 A Central Analysis Center, forerunner of the National Meteorological Center (NMC), is created. Navy gives the USWB 25 surplus aircraft radars to be modified for ground meteorological use, marking the start of the first weather radar system in the U. S.

1948 Air Force meteorologists issue first tornado warning from a military installation. Princeton's Institute for Advanced Studies begins research into use of a computer for weather forecasting. Chicago Weather Bureau office demonstrates use of facsimile for map transmission.

1950 30-day outlooks and "tornado alerts" begin.

1951 Forerunner of the National Severe Storms Forecast Center begins at Tinker AFB, OK. National Weather Records Center opens at Asheville.

1952 USWB organizes Severe Local Storms Forecasting Unit in Washington and begins issuing tornado forecasts.

1954 The USWB, Navy, Air Force, MIT, and University of Chicago form a Joint Numerical Prediction Unit at Suitland, using an IBM 701. First radar designed for meteorological use, AN/CPS-9 is unveiled by USAF.

1955 Scheduled operational computer forecasts begin by the Joint Numerical Forecast Unit. USWB becomes a pioneer in civilian user of computers along with Census Bureau. Development of Barotropic Model begins.

1957 USWB accepts Dr. James Brantly's proposal of Cornell Aeronautical Laboratories to modify surplus Navy Doppler radars for severe storm observation. First endeavor to use Doppler radar in meteorology.

1958 NMC is established.

1959 Feasibility of a weather satellite is demonstrated. First WSR-57 weather radar is installed at Miami Hurricane Forecast Center.

1960 World's first weather satellite, TIROS I, launched. First advisories on air pollution issued.

1961 National Severe Storms Forecast Center established in Kansas City. First official forecast of clear air turbulence issued.

1963 TIROS III launched with automatic picture transmission (APT) capability, eventually provides images to over 100 nations.

1964 National Severe Storms Laboratory established.

1965 ESSA created.

1966 NMC introduces computer model capable of making sea-level predictions as accurate as those made manually.

1967 Fire weather forecasts extended to cover contiguous U. S.

1970 ESSA becomes NOAA.

1973 National Weather Service purchases its second generation radar (WSR-74).

1975 First Geostationary Operational Environmental Satellite (GOES) launched, for hurricane detection.

1976 Real-time operational forecasts and warnings using Doppler radar are evaluated by the Joint Doppler Operational Radar Project (JDOP), spawning third generation weather radar (WSR-88D).

1977 Success of weather satellites causes elimination of last U.S. weather observation ships. National centers have real-time access to satellite data.

1979 Nested Grid Model (NGM) becomes operational. Global Data Assimilation System (GDAS) developed. Automation of Field Operations and Services (AFOS) computer system deployed. System is the most ambitious computer network system yet created, setting records of volume of data and entry points while supporting word processing and other capabilities.

1980 Weather satellites spot eruption of Mt. St. Helens, beginning a trend of increasing usage for volcanic eruption detection. Mr. Edward Stoll, cooperative observer since 1905 honored at the Whitehouse.

1984 First official Air Transportable Mobile Unit ((ATMU) a remote observing and forecasting unit)) dispatched to the Shasta-Trinity National Forest wildfire. Removal of teletypewriters begins.

1989 Eight year national plan for modernization of the NWS is announced.

1990 Cray Y-MP8 supercomputer installed at NMC. Contract option exercised with UNISYS for full-scale production of 165 WSR-88D radars and 300 display subsystems. Installation begins immediately. Automatic Surface Observing System (ASOS) development and planning nears completion.

1991 ASOS contract awarded to AAI Corporation.

1992 First 10 Limited-Production WSR-88D/s are fully installed. Advanced Weather Information Processing System (AWIPS) Development Phase contract awarded to Planning Research Corporation (PRC). AWIPS is a UNIX-based communication and workstation replacement for AFOS.

1993 Two to three WSR-88D/s are installed per month as production ramps up. Practically each installation with a new or refurbished NWS office.

(Courtesy of National Oceanic and Atmospheric Administration (NOAA).)

Federal Aviation Administration (FAA) and the Department of Defense, which results in economies of scale and a reduction in purchase costs. Recently purchased geostationary meteorological satellites, newer radars, and automated surface observing systems provide data that are shared by all participating agencies.

Automated Surface Observing Systems

Automated Surface Observing Systems (ASOSs) relieve staff personnel from the time-consuming duty of collecting surface observations manually. Nearly 850 ASOSs nationwide serve as the primary surface weather-observing network, significantly expanding the information available to forecasters and the aviation community. The system works non-stop, updating observations every minute, every day.

Getting up-to-the-minute data to forecasters is crucial. This type of information will increase the accuracy and timeliness of forecasts and warnings. This same data will also address and alleviate the safety concerns of the aviation community.

Next Generation Weather Radars

Next Generation Weather Radar (NEXRAD) is a large step forward in early warnings of tornado and severe thunderstorms. Utilizing Doppler radar technology, the NEXRAD system (also known as the Weather Surveillance Radar [WSR-88D]) observes the presence and calculates the speed and direction of motion of severe weather elements. The nearly 161 NEXRAD radars also provide quantitative area precipitation measurements so important in hydrologic forecasting of potential flooding. For example, at present, currently limited (obsolescent) radar systems and tornado warnings are usually issued only when sightings have been reported. The advent of NEXRAD will not only allow for an earlier detection of the precursors to tornadic activity, but will also provide data on the direction and speed of tornado cells once they form.

The national network of NEXRAD systems, when fully deployed, will sharply upgrade uniform coverage way beyond the capability of present-day radar network. The NWS will operate the majority of NEXRAD systems; the remainder will be at FAA and Department of Defense locations.

• NOAA'S GEOSTATIONARY AND POLAR-ORBITING WEATHER SATELLITES

Operating the country's system of environmental (weather) satellites is one of the major responsibilities of the National Oceanic and Atmospheric Administration's (NOAA's) National Environmental Satellite, Data, and Information Service (NESDIS). NESDIS operates the satellites and manages the processing and distribution of the millions of bits of data and images these satellites produce daily. The primary customer is NOAA's National Weather Service, which uses satellite data to create forecasts for the

National Severe Storms Laboratory's first Doppler weather radar in Norman, Oklahoma. *(Photo courtesy of NOAA Photo Library, NOAA Central Library; OAR/ERL/National Severe Storms Laboratory (NSSL).)*

public, television, radio, and weather advisory services. Satellite information is also shared with various Federal agencies, such as the Departments of Agriculture, Interior, Defense, and Transportation; with other countries, such as Japan, India, and Russia, and members of the European Space Agency (ESA) and the United Kingdom Meteorological Office; and with the private sector.

NOAA's operational weather satellite system is composed of two types of satellites: geostationary operational environmental satellites (GOES) for short-range warning and "now-casting," and polar-orbiting satellites for longer-term forecasting. Both types of satellite are necessary for providing a complete global weather monitoring system.

A new series of GOES and polar-orbiting satellites has been developed for NOAA by the National Aeronautics and Space Administration (NASA). The new GOES-I through M series provide higher spatial and temporal resolution images and full-time operational soundings (vertical temperature and moisture profiles of the atmosphere). The newest polar-orbiting meteorological satellites (that began with NOAA-K in 1998) provide improved atmospheric temperature and moisture data in all weather situations. This new technology will help provide the National Weather Service with the most advanced weather forecast system in the world.

Geostationary Operational Environmental Satellites (GOES)

GOES satellites provide the kind of continuous monitoring necessary for intensive data analysis. They circle the Earth in a geosynchronous orbit, which means they orbit the equatorial plane of the Earth at a speed matching the Earth's rotation. This allows them to hover continuously over one position on the surface. The geosynchronous plane is about 22,300 miles above the Earth, high enough to allow the satellites a full-disc view of the Earth. Because they stay above a fixed spot on the surface, they provide a constant vigil for the atmospheric "triggers" for severe weather conditions such as tornadoes, flash floods, hail storms, and hurricanes. When these conditions develop the GOES satellites are able to monitor storm development and track their movements. GOES satellite imagery is also used to estimate rainfall during the thunderstorms and hurricanes (for flash flood warnings), as well as to estimate snowfall accumulations and the overall extent of snow cover. Such data help meteorologists issue winter storm warnings and spring snow melt advisories. Satellite sensors also detect ice fields and map the movements of sea and lake ice.

NASA launched the first GOES for NOAA in 1975 and followed it with another in 1977. Currently, the United States is operating GOES-8 and GOES-10. (GOES-9, which malfunctioned in 1998, is being stored in orbit as an emergency backup should either GOES-8 or GOES-10 fail.) GOES-11 was launched on May 3, 2000 and GOES-12 on July 23, 2001. Both are being stored in orbit as fully functioning replacements for GOES-8 or GOES-10 (should they fail).

GOES-8 and GOES-10

The United States normally operates two meteorological satellites in geostationary orbit over the equator. Each satellite views almost a third of the Earth's surface: one monitors North and South America and most of the Atlantic Ocean, the other North America and the Pacific Ocean basin. GOES-8 (or GOES-East) is positioned at 75° W longitude and the equator, while GOES-10 (or GOES-West) is positioned at 135° W longitude and the equator. The two operate together to produce a full-face picture of the Earth, day and night. Coverage extends approximately from 20° W longitude to 165° E longitude.

The main mission is carried out by the primary instruments, the Imager and the Sounder. The imager is a multichannel instrument that senses radiant energy and reflected solar energy from the Earth's surface and atmosphere. The Sounder provides data to determine the vertical temperature and moisture profile of the atmosphere, surface and cloud top temperatures, and ozone distribution.

Other instruments on board the spacecraft are a Search-and-Rescue transponder, a data collection and relay system for ground-based data platforms, and a space environment monitor. The latter consists of a magnetometer, an X-ray sensor, a high energy proton and alpha detector, and an energetic particles sensor. All are used for monitoring the near-Earth space environment or solar "weather".

The United States reaps many benefits from the new series of GOES satellites as they aid forecasters in providing better advanced warnings of thunderstorms, flash floods, hurricanes, and other severe weather. The GOES-I series provides meteorologists and hydrologists with detailed weather measurements, more frequent imagery, and new types of atmospheric soundings. The data gathered by the GOES satellites, combined with that from new Doppler radars and sophisticated communications systems, make for improved forecasts and weather warnings that save lives, protect property, and benefit agricultural and a variety of commercial interests.

For users who establish their own direct readout receiving station, the GOES satellites transmit low-resolution imagery in the WEFAX service. WEFAX can be received with an inexpensive receiver. Highest resolution Imager and Sounder data is found in the GVAR primary data user service, which requires more complex receiving equipment.

Polar-Orbiting Satellites

Complementing the geostationary satellites are two polar-orbiting satellites known as Advanced Television Infrared Observation Satellite (TIROS-N or ATN), constantly circling the Earth in an almost north-south orbit, passing close to both poles. The orbits are circular, with an altitude between 830 (morning orbit) and 870 (afternoon orbit) km, and are sun synchronous. One satellite crosses the equator at 7:30 a.m. local time, the other at 1:40 p.m. local time. The circular orbit permits uniform data acquisition by the satellite and efficient control of the satellite by the NOAA Command and Data Acquisition (CDA) stations located near Fairbanks, Alaska, and Wallops Island, Virginia. Operating as a pair, these satellites ensure that data for any region of the Earth are no more than six hours old.

A suite of instruments is able to measure many parameters of the Earth's atmosphere, its surface, cloud cover, incoming solar protons, positive ions, electron-flux density, and the energy spectrum at the satellite altitude. As a part of the mission, the satellites can receive, process, and retransmit data from Search-and-Rescue beacon transmitters and automatic data collection platforms on land, ocean buoys, or aboard free-floating balloons. The primary instrument aboard the satellite is the Advanced Very High Resolution Radiometer or AVHRR.

Data from all the satellite sensors is transmitted to the ground via a broadcast called the High Resolution Picture Transmission (HRPT). A second data transmission consists of only image data from two of the AVHRR channels, called

Automatic Picture Transmission (APT). For users who want to establish their own direct readout receiving station, low-resolution imagery data in the APT service can be received with inexpensive equipment, while the highest resolution data transmitted in the HRPT service utilizes a more complex receiver.

The polar orbiters are able to monitor the entire Earth, tracking atmospheric variables and providing atmospheric data and cloud images. They track weather conditions that eventually affect the weather and climate of the United States. The satellites provide visible and infrared radiometer data that are used for imaging purposes, radiation measurements, and temperature profiles. The polar orbiters ultraviolet sensors also provide ozone levels in the atmosphere and are able to detect the "ozone hole" over Antarctica during mid-September to mid-November. These satellites send more than 16,000 global measurements daily via NOAA's CDA station to NOAA computers, adding valuable information for forecasting models, especially for remote ocean areas where conventional data are lacking.

Currently, NOAA is operating five polar orbiters: NOAA-14 (classified as a stand-by satellite), launched in December 1994, and a new series of polar orbiters with improved sensors, which began with the launch of NOAA-15 in May 1998 and NOAA-16 on September 21, 2000. The newest, NOAA-17, was launched June 24, 2002. NOAA-12 continues transmitting HRPT data as a stand-by satellite. NOAA-15 and NOAA-16 are classified as the "operational" satellites.

How satellites are named

NOAA assigns a letter to the satellite before it is launched, and a number once it has achieved orbit. For example, GOES-H, once in orbit, was designated GOES-7, GOES-G, which was lost at launch, was never assigned a number. The same system is used for polar orbiters; for example, NOAA-11, still in orbit, was designated NOAA-H before launch. NOAA-J became NOAA-14.

Upgrading satellites

For severe weather and flood warnings and short-range forecasts, cloud imagery and atmospheric sounding data from the geostationary meteorological satellites will continue to be a major data source. The new Geostationary Operational Environmental Satellite (GOES) I–M system includes separate instrumentation that allows simultaneous image and sounding data to be observed and transmitted to ground stations. The GOES I–M system also provides visible and infrared imagery data updates as frequently as every six minutes during severe weather warning situations over selected areas of the United States. To date, three of the program's five scheduled satellites orbit the earth, covering the United States' east and west coasts.

For longer-range forecasting, soundings from the polar orbiting satellites are a primary data input into the National Meteorological Center (NMC) numerical forecast models.

• 171 PEOPLE WERE RESCUED IN THE U.S. IN 2002 WITH HELP FROM NOAA SATELLITES

Thanks to environmental satellites with search-and-rescue tracking capability, the Commerce Department's National Oceanic and Atmospheric Administration (NOAA) helped save 171 lives in the United States in 2002.

The NOAA satellites, along with Russia's Cospas satellites, are part of an international Search-and-Rescue Satellite–Aided Tracking System known as Cospas-Sarsat. Together the systems use a constellation of satellites in geostationary and polar orbits to detect and locate emergency beacons on vessels and aircraft in distress and from hand-held Personal Locator Beacons (PLBs). India and the European Space Agency also provide geostationary satellites for the Cospas-Sarsat System.

Of the 171 rescues last year, 133 people were saved on the nation's coastal waters, 27 in the Alaskan wilderness, and 11 from downed aircraft in states around the country. Of the 69 separate Sarsat-rescue events, a variety took place out at sea.

Engine fires, flooding, and rough seas all caused emergencies resulting in distress calls and rescues. In Alaska, stranded snowmobilers and lost persons were among those rescued. Downed aircraft incidents included those making emergency landings. In one such incident, a Piper Supercub had flipped after landing near Glenallen, Alaska. Both the pilot and passenger were uninjured.

More than 15,000 lives have been saved worldwide since the system became operational in 1982, and nearly 4,500 lives have been saved in the United States alone. September 2002 marked the 20th anniversary of the first Sarsat rescue.

In one dramatic rescue, a father, his son, and their family dog were plucked from a life raft in the Gulf of Alaska about 90 miles south of Cordova, Alaska. They were in a dangerous predicament, their fishing vessel had struck an object and sustained uncontrollable flooding causing them to abandon their vessel. Yet, because there was an Emergency Position Indicating Radio Beacon, or EPIRB, on board (which was manually activated), a U.S. Coast Guard search-and-rescue helicopter was able to respond quickly once the alert information was received from the USMCC. On arrival the helicopter saw the situation unfolding and deployed a rescue swimmer to retrieve the three occupants and bring them to safety. This particular incident illustrates the importance of emergency beacon registration.

NOAA expects the number of worldwide rescues for 2002 will total about 1,500. The average number of distress alerts continues to rise internationally as more countries sign on to use the advantages and benefits of the Cospas-Sarsat system.

NOAA's Geostationary Operational Environmental Satellites (GOES) can instantly detect emergency distress signals. The polar-orbiting satellites in the system detect emergency signals as they circle the Earth from pole to pole. The signals are sent to the Mission Control Centers, then automatically sent to rescue forces around the world. There are 35 countries participating in the system as of 2002.

NOAA Satellites and Information is the nation's primary source of operational space-based meteorological and climate data. In addition to search and rescue, NOAA's environmental satellites are used for weather forecasting, climate monitoring, and other environmental applications such as volcanic eruptions, ozone monitoring, sea surface temperature measurements, and wild fire detection.

NOAA Satellites and Information also operates three data centers, which house global data bases in climatology, oceanography, solid-earth geophysics, marine geology and geophysics, solar-terrestrial physics, and paleoclimatology.

NOAA is dedicated to enhancing economic security and national safety through the prediction and research of weather and climate related events and providing environmental stewardship of our nation's coastal and marine resources.

National Center advanced computer systems

Warnings and forecasts prepared by NWS offices rely heavily on the basic analyses and advisories provided by the National Meteorological Center, especially for periods of 36 hours and beyond. These analyses and guidance products result from numerical models of the atmosphere run on high-speed computers. These increased demands require the acquisition of dedicated next-generation Class VII computer capabilities with a processing capability that is a full order of magnitude greater than the present Class VI computer.

Advanced Weather Interactive Processing System

The revised system employed by the Advanced Weather Interactive Processing System (AWIPS) functions as the nerve center of the 118 National Meteorological Centers operations. AWIPS is the data integrator receiving the high-resolution data from the observation systems, the centrally collected data, and the centrally prepared analysis and guidance information developed by the National Meteorological Center. The integration of all of this data from multiple sources represents the information base from which all warning and forecast products will be prepared. The AWIPS system provides fast-response interactive analysis and display of the data to help support the meteorologists as they make rapid decisions, prepare warnings and forecasts, and disseminate information to users.

AWIPS includes the communications network that interconnects each Weather Forecast Office for exchange of locally generated data. NOAAPORT provides communications support for the operational distribution of the centrally collected data and centrally produced analysis and guidance products, as well as the satellite imagery and sounding data

processed by the National Environmental Satellite, Data, and Information Service (NESDIS).

The continued restructuring involves changing the number and location of field offices, a gradual transformation of the workforce to emphasize more professionalism in its makeup, and a reallocation of operational responsibilities between field offices and the National Centers.

The need to restructure is twofold: first, the combination of new operational concepts, new data sets, and an evolving scientific understanding of the dynamic processes associated with the most dangerous weather phenomena requires an increase in the number of meteorologists. The percentage increase of meteorologists in the NWS workforce will improve warnings and forecasts by taking advantage of the capabilities of the new technologies. Second, productivity and efficiency gains will occur as a result of increased integration of the new technological observation, information processing, and communication systems with the staff. Key trade-offs in the restructuring process exist between human capabilities, costs, and programmatic, scientific, and technological opportunities.

The Weather Forecast Office

How does the new Weather Forecast Office (WFO) fit into the scheme? Consider this analogy: think of the surface of a map of the United States. Now, consider a uniform arrangement of 118 mutually adjacent cylinders, each with a radius of approximately 125 mi. The cylinders would each extend upward from the earth's surface. Each cylinder represents the "area" of operational responsibility associated with the WFO. A WFO is located in the center of the base of the cylinder. Each section of the country and the coastal ocean area is contained in one of these cylinders, thus the whole of the country is theoretically uniformly covered.

The Geostationary Operational Environmental Satellite (GOES) satellite positioned over the United States provides to each of these cylinders an Earth image. The regularly updated image represents uniform coverage with visible and infrared imagery of each of the "cylinders." It also provides remote soundings that penetrate each cylinder from above. Associated with each WFO is one or more NEXRAD radars that scan the atmosphere from near Earth's surface to a height sufficient to detect the majority of meteorological events.

Also serving the WFOs are 1,000 ASOS units. They are spread across the surface of the country and each measures surface weather parameters as fast as once every minute. All of these data within the cylinder are sent directly to the AWIPS system in each WFO. The WFO's AWIPS system also receives the centrally produced guidance products from the National Centers, generated from globally exchanged data. Subsets of these data are available for all other WFOs through the AWIPS communication network.

Integrated operations within the WFO

These operations allow forecasters to comprehensively address the air-sea environment in their assigned area. The

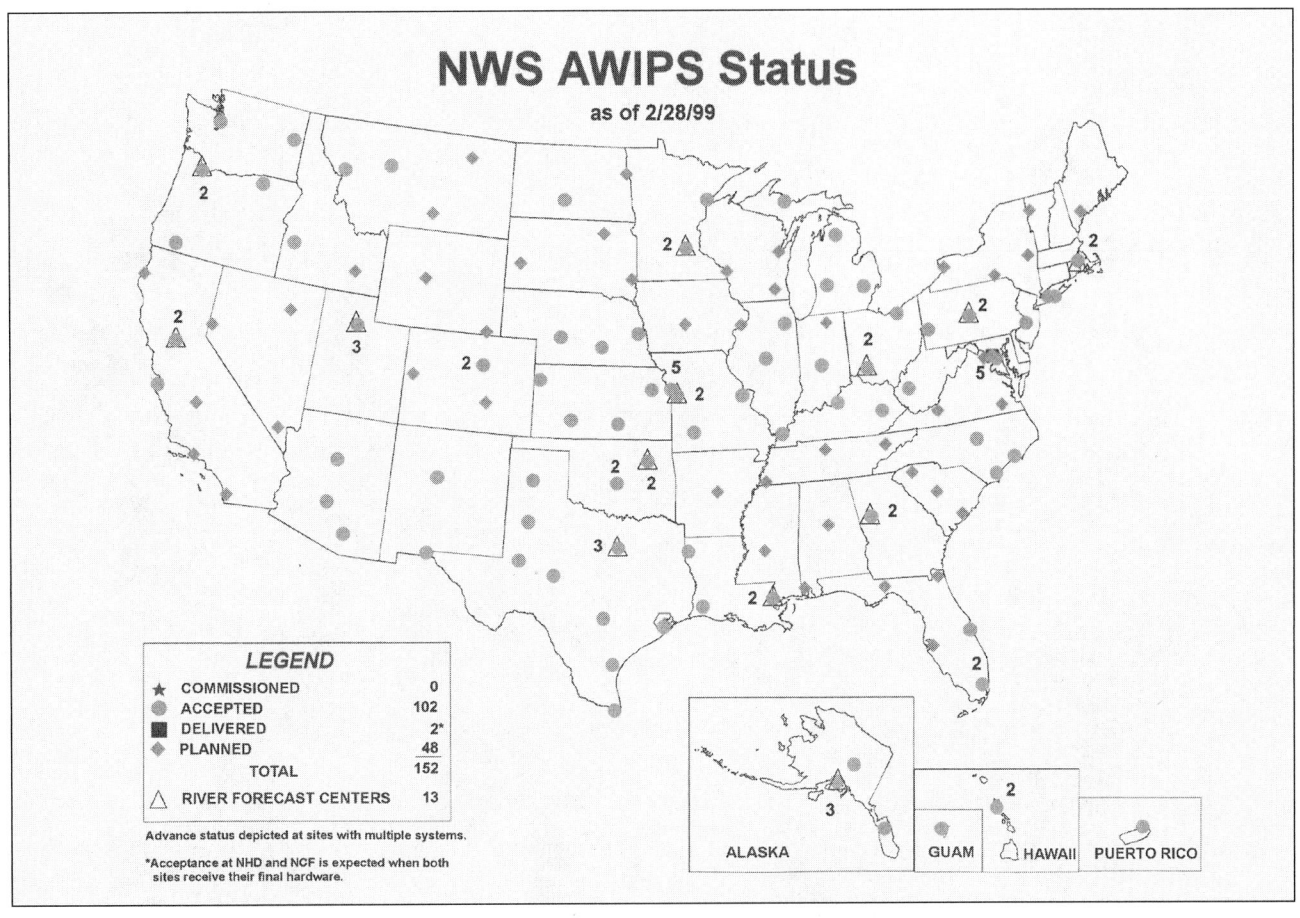

(Courtesy of National Oceanic and Atmospheric Administration (NOAA).)

observation and analysis of current and expected weather conditions can be quickly and reliably completed, and critical decisions can be made and translated into immediate warnings and forecasts. This is contrasted to past operations where a number of meteorologists and technicians were required to individually evaluate a limited data base and separately derive the various warnings and forecasts.

The concept of the local data base is central to future operations. The high volume of data from the local NEXRAD and geostationary meteorological satellites combined with the high-frequency observations from ASOS will flow directly to the Weather Forecast Office. The most complete data sets will only be available to the local WFO. However, summarized data from all NEXRADs and ASOSs in the nation are made available to all field offices.

The new observing systems are designed to provide data sets which can be immediately integrated into three-dimensional depictions of the rapidly changing state of the environment. Each system will contribute a critical part, combining with and complementing data from all other systems to form a complete set of information about that particular cylinder of space from the earth's surface to the upper atmosphere comprising the particular WFO's area of responsibility.

AWIPS work stations allow the forecaster to quickly update, quality control, and analyze current processes and events detailed within the area of concern. New dedicated supercomputer capabilities and high-resolution models running at the National Centers provide a stream of detailed, frequently updated guidance to forecasters, assisting in the prediction of future conditions. This represents a new, highly integrated mode of operation which greatly increases the productivity of personnel, and also holds the promise of increased accuracy and greater timeliness of forecast services for the nation.

The new structure

The WFO is the weather office that will provide all warning and forecast services for its assigned area of responsibility. The forecast and warnings operations at the WFO are supported by guidance products issued from the National Centers and River Forecast Center (RFC).

Weather Forecast Offices (WFOs)

A total of 118 WFOs provide weather and hydrologic services in four major areas: (1) Watches and warnings for the general public for severe local storms; floods, flash floods and winter storms; local and zone public forecasts;

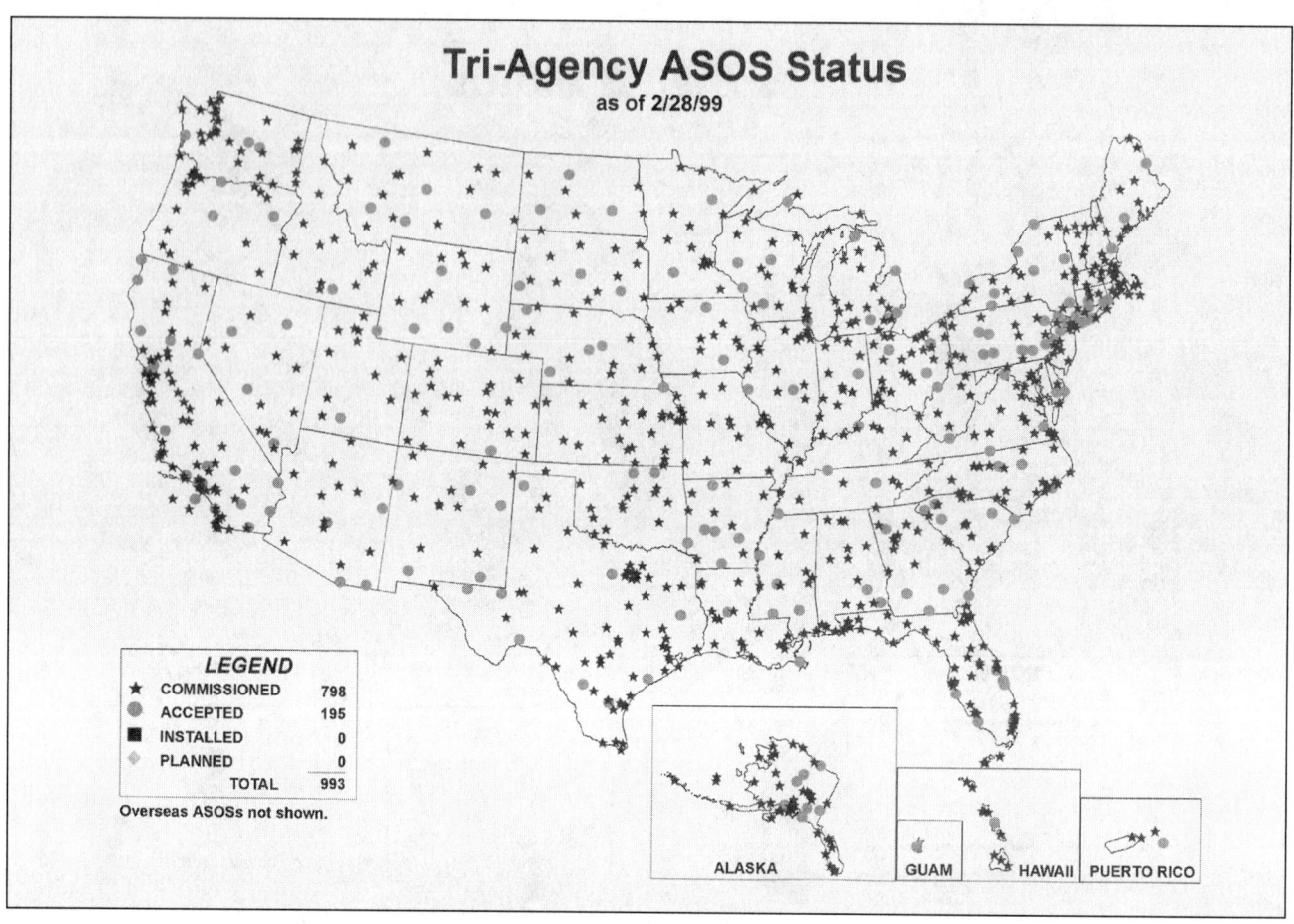

Tri-Agency ASOS Status
as of 2/28/99

LEGEND

★	COMMISSIONED	798
●	ACCEPTED	195
■	INSTALLED	0
◆	PLANNED	0
	TOTAL	993

Overseas ASOSs not shown.

ALASKA GUAM HAWAII PUERTO RICO

(Courtesy of National Oceanic and Atmospheric Administration (NOAA).)

and fire weather forecasts; (2) local aviation watches and warnings, terminal forecasts, and domestic aviation en-route forecasts; (3) marine warnings and forecasts for coastal areas of the nation and the Great Lakes; and (4) hydrologic services that identify flash-flood-prone areas and the development of community supported surveillance systems.

The foundation for the more accurate and timely warnings and forecasts will be the guidance products from the National Centers and RFCs and the data from the new observing systems: ASOS, NEXRAD, and geostationary meteorological satellites.

The basic tool for more accurate and timely warnings and forecasts from the WFO is AWIPS. It assembles, processes, and displays the observational data and guidance from National Centers. AWIPS helps meteorologists with the warning and forecast decision process through an interactive workstation. It pre-formats warnings and forecast products and disseminates these products to the users in a timely manner.

River Forecast Center

River Forecast Centers (RFCs) provide hydrologic forecasts and guidance information in three major cate-

gories: (1) mainstem river and flood forecasts regarding about 3,000 locations; (2) flash-flood and headwater guidance to WFOs for warning services; and (3) long-term, seasonal forecasts providing estimates of snowmelt and water supply outlooks (from excess to drought) at approximately 1,000 locations for periods up to several months in advance.

The operations of RFCs are expected to change in several ways. Each of the 13 RFCs will be colocated with a WFO. This will result in a more effective utilization of hydrological and meteorological information facilitated by a Hydrologic Analysis and Support Group in each colocated facility.

National Meteorological Center

The National Meteorological Center (NMC) has the responsibility for national and international data collection. This database is first employed for global atmospheric and oceanic analysis. The resultant analyses are distributed to international and domestic users that include the NWS, other government agencies, and private sector meteorologists. The database is then used as initial input to global atmospheric numerical models. These models produce international aviation forecast products, high-seas forecast products, long-range national forecasts, and forecast guidance

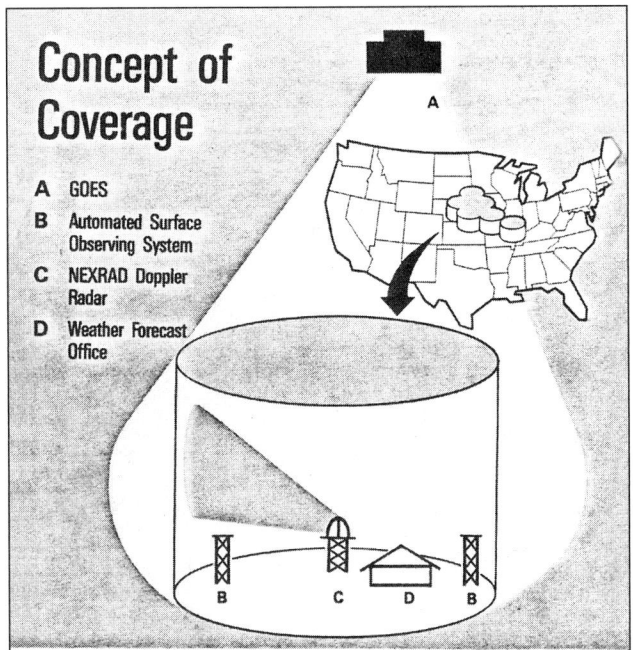

Concept of Coverage

A GOES

B Automated Surface
Observing System

C NEXRAD Doppler
Radar

D Weather Forecast
Office

The above schematic suggests how the individual weather forecast offices (WFO) will bear observation and forecast responsibility for an area which has a radius of 125 miles, and reaches to the top of the atmosphere, verically. Each office's "cylinder" of responsibility will abut cylinders around it so that 115 of these cylinders can cover the nation. Many formerly manual operations have been handed to automated gear.

(Courtesy of U.S. government publication.)

for local WFOs and RFCs. New dedicated Class VII computer capabilities will enable increases in the resolution of the models, resulting in improved forecast products and guidance. Traditionally, the long-range national forecasts have begun at three days and beyond. The new computers will reduce this threshold to beyond 36 hours. This will allow local forecasters to devote their attention to short-term weather events that are not aided by the use of centralized computer models.

Climate Analysis Center

This center is a specialized facility that is part of the National Meteorological Center and is colocated with it to take advantage of the facilities available there. The center's responsibilities are national and international in scope—collecting, organizing, and disseminating climate information for diagnosis of short-term climate change; researching the physical cause of short-term climate change; and issuing forecasts of departures of average weather conditions from climatological means.

National Hurricane Center (Tropical Prediction Center)

This special facility will continue to be responsible for providing the nation with its strongest measure of security

from tropical storms. It analyzes, predicts, and tracks tropical weather systems that often become hurricanes. The center provides leadership and coordination of storm-related preparedness. It uses geostationary meteorological satellites to track and monitor tropical storms 24 hours a day throughout the cycle of a storm. It will utilize coastal NEXRADs radar systems, which are becoming available to provide hurricane understanding well beyond that available at present. NMC's new Class VII computers will run new hurricane models which will greatly assist forecasters at the Hurricane Center. AWIPS at the National Hurricane Center will serve the center's mission as well as NMC's.

National Storm Prediction Center

The National Storm Prediction Center provides national severe weather guidance to WFOs and RFCs. It issues more timely and specific advisories necessary to support the severe weather and flood-warning activities of the WFOs.

National Data Buoy Center

The National Data Buoy Center will continue the operation of deep-sea, coastal buoys, and headland systems. Data from the buoys and these coastal systems are essential to marine warnings and forecasts, and numerical weather predictions.

Staffing

The new observing, data processing, and display systems provide forecasters with the opportunity to sample, observe, and analyze the environment to an extent never before possible. This means a better product, more efficiently produced. Field offices have a core staff of professional scientists at each WFO and RFC to take advantage of these new capabilities. These individuals will provide all warning and forecast services across their area of responsibility. They will have far better data with which to meet these tasks.

Taking a quantum leap such as this—improving product while using fewer people—places greater emphasis on a higher level of professional skills and on retraining many technicians. NWS has strategic, as well as tactical programs, to effect the implied transmutation.

For instance, each WFO operates 24 hours a day, and a certified meteorologist will be in charge at all times. Other such staffing upgrades will occur. The staffing level is determined by peak service demands and maximum weather activity, with reduced staff requirements at selected offices during hours of lower threat and service demands.

• NOAA WEATHER RADIO

The NOAA provides the service known as *Weather Radio*. This service provides continuous broadcasts of the latest weather information directly from National Weather Service

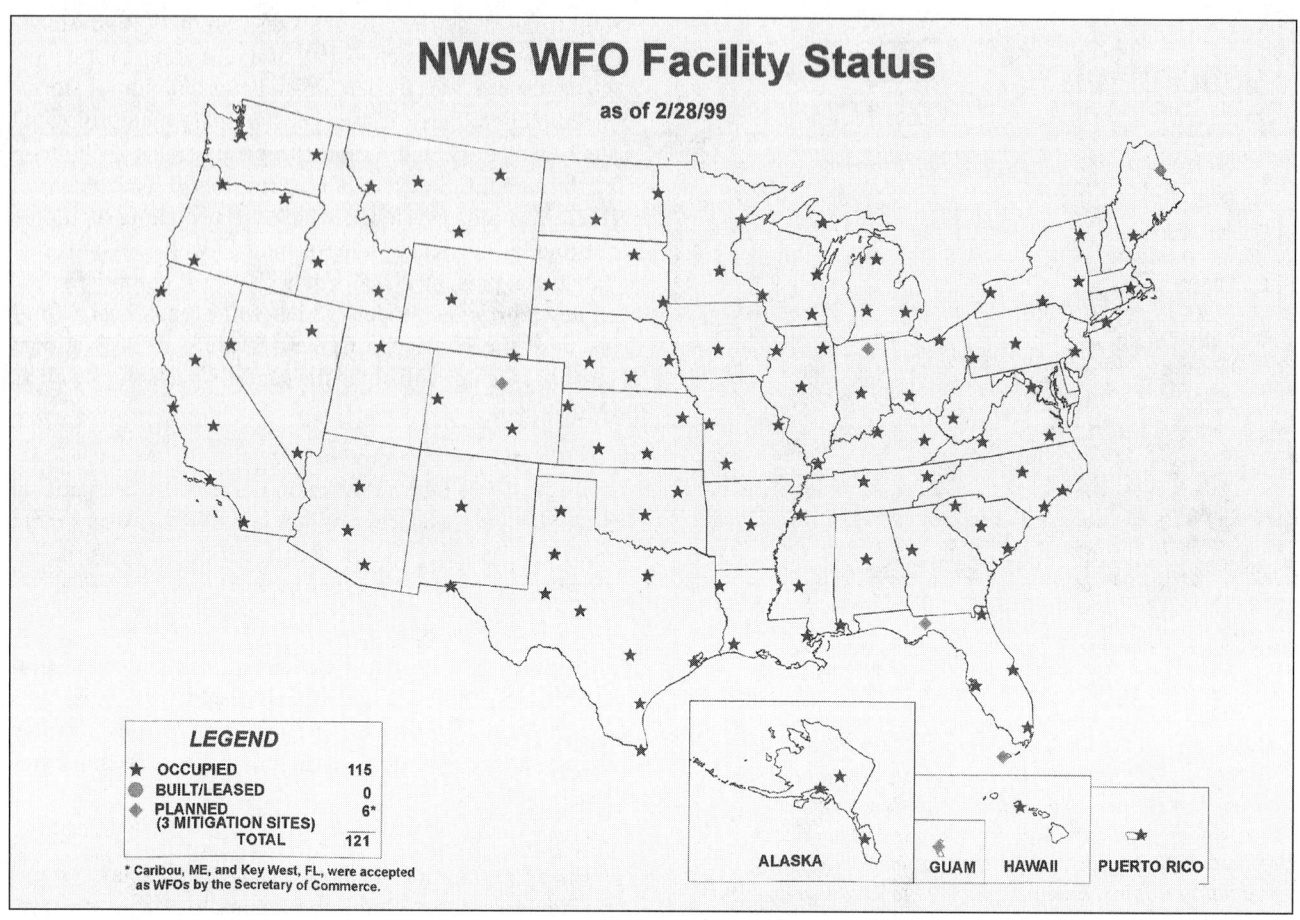

(Courtesy of National Oceanic and Atmospheric Administration (NOAA).)

offices. Taped weather messages are repeated every four to six minutes and are routinely revised every one to three hours, or more frequently if needed. Most of the stations operate 24 hours daily.

The broadcasts are tailored to weather information needs of people within the receiving area. For example, stations along the seacoasts and Great Lakes provide specialized weather information for boaters, fishers, and others engaged in marine activities, as well as general weather information.

During severe weather, National Weather Service forecasters can interrupt the routine weather broadcasts and substitute special warning messages. The forecasters can also activate specially designed warning receivers. Such receivers either sound an alarm indicating that an emergency exists, alerting the listener to turn the receiver up to an audible volume; or, when operated in a muted mode, are automatically turned on so that the warning message is heard. "Warning alarm" receivers are especially valuable for schools, hospitals, public-safety agencies, and news media offices.

Under a January, 1975, White House policy statement, NOAA Weather Radio was designated the sole government-operated radio system to provide direct warnings into private homes for both natural disasters and nuclear attack.

This capability is to supplement warnings by sirens and by commercial radio and television.

NOAA Weather Radio broadcasts are made on high-band FM frequencies. The 162.475 MHz frequency is used only in special cases where required to avoid channel interference. These frequencies are not found on the average home radio now in use. However, a number of radios manufacturers offer special weather radios to operate on these frequencies, with or without the emergency warning alarm. There are also many radios on the market that offer standard AM/FM frequencies plus the so-called "weather band" as an added feature.

NOAA Weather Radio broadcasts can usually be heard as far as 40 mi from the antenna site, sometimes more. The effective range depends on many factors, particularly the height of the broadcasting antenna, terrain, quality of the receiver, and type of receiving antenna. As a general rule, listeners close to or perhaps beyond the 40-mi range should have a good quality receiver system if they expect reliable reception.

Marine weather

Few people are affected more by weather than the mariner. An unexpected change in winds, seas, or visibility can reduce the efficiency of marine operations and threaten

NOAA Weather Radio . . . the Voice of the National Weather Service

BROADCAST FREQUENCIES: 162.400 MHZ, 162.425 MHZ, 162.450 MHZ, 162.475 MHZ, 162.500 MHZ, 162.525 MHZ, 162.550 MHZ

Broadcast range from the weather radio transmitter is approximately 40 miles. The effective range depends on such things as terrain and quality of the receiver and indoor/outdoor antenna. Before you buy a receiver, make sure your area is covered by one of these transmitters.

Alabama		Redding	162.550	Savannah	162.400	Harlan	162.450
Anniston	162.475	Sacramento	162.400	Valdosta	162.500	Hazard	162.475
Auburn	162.525	San Diego	162.400	Waycross	162.475	Lexington	162.400
Birmingham	162.550	San Francisco	162.400	Waynesboro	162.425	London	162.475
Cullman	162.450	San Luis Obispo	162.550			Louisville	162.475
Demopolis/Linden	162.475	Santa Ana	162.450	**Hawaii**		Manchester	162.400
Dozier	162.550	Santa Barbara	162.400	Hawaii (South Point)	162.550	Mayfield	162.475
Florence	162.475	Santa Barbara		Hawaii (Kulani Cone)	162.550	McKee	162.450
Fort Payne	162.500	Marine	162.475	Kaneohe		Monticello	162.425
Huntsville	162.400			(Puu Hawaiiloa)	162.400	Morehead	162.425
Jackson	162.500	**Colorado**		Kauai (Kokee)	162.400	Mt. Vernon	162.425
Louisville	162.475	Alamosa	162.475	Maui (Mt Haleakala)	162.400	Jackson	162.425
Mobile	162.550	Bethune	162.525	Oahu (Mt. Kaala)	162.550	Paintsville	162.525
Montgomery	162.400	Colorado Springs	162.475	Oahu (Hawaii Kai)	162.400	Phelps	162.500
Tuscaloosa	162.400	Denver	162.550			Pikeville	162.400
Winfield	162.525	Fort Collins	162.450	**Idaho**		Pineville	162.525
		Fowler	162.425	Boise	162.550	Richmond	162.525
Alaska		Glenwood Springs	162.500	Bonners Ferry	162.500	Somerset	162.550
Anchorage	162.550	Grand Junction	162.550	Lewiston	162.550	Stanton	162.550
Cordova	162.400	Greeley	162.400	McCall	162.475	West Liberty	162.450
Craig	162.475	La Junta	162.500	Pocatello	162.550	Williamsburg	162.500
Fairbanks	162.550	Mead/Longmont	162.475	Twin Falls	162.400		
Haines	162.400	Pueblo	162.400			**Louisiana**	
Homer	162.400	Sterling	162.400	**Illinois**		Alexandria	162.475
Juneau	162.550			Champaign	162.550	Baton Rouge	162.400
Ketchikan	162.550	**Connecticut**		Chicago	162.550	Buras	162.475
Kodi	162.550	Hartford	162.475	Marion	162.425	Lafayette	162.550
Nome	152.550	Meriden	162.400	Peoria	162.475	Lake Charles	162.400
Seward	162.550	New London	162.550	Rock Island/Moline	162.550	Monroe	162.550
Sitka	162.550			Rockford	162.475	Morgan City	162.475
Soldotna	162.475	**Delaware**		Springfield	162.400	New Orleans	162.550
Valdez	162.550	Lewes	162.550			Shreveport	162.400
Wrangell	162.400	Salisbury, Md	162.475	**Indiana**			
Yakutat	162.400			Bloomington	162.400	**Maine**	
		Florida		Edwardssport	162.425	Caribou	162.525
Arizona		Belle Glade	162.400	Evansville	162.550	Dresden	162.475
Flagstaff	162.400	Bethlehem	162.450	Fort Wayne	162.550	Ellsworth	162.400
Gila County North	162.425	Daytona Beach	162.400	Georgia	162.550	Falmouth	162.550
Gila County South	162.500	East Point	162.500	Indianapolis	162.550		
Grand Canyon		Fort Myers	162.475	Lafayette (Yeoman)	162.475	**Mariana Islands**	
(Hopi Point)	162.475	Fort Pierce	162.425	Marion	162.450	Guam (Nimitz Hill)	162.400
Phoenix	162.550	Gainesville	162.475	North Webster	162.450	Saipan	
Prescott	162.525	Inverness	162.400	Putnamville	162.400	(Mt. Tapoctobau)	162.550
Show Low (Porter Mt.)	162.400	Jacksonville	162.550	Seymour	162.525		
Tucson	162.400	Key West	162.400	South Bend	162.400	**Maryland**	
Window Rock	162.550	Live Oka	162.450			Baltimore	162.400
Yuma	162.550	Melbourne	162.550	**Iowa**		Hagerstown	162.475
		Miami	162.550	Cedar Rapids	162.475		
Arkansas		Naples	162.525	Des Moines	162.550	**Massachusetts**	
Fayetteville	162.475	Ocala	162.525	Dubuque	162.400	Boston	162.475
Fort Smith	162.550	Orlando	162.475	Sioux City	162.475	Hyannis	
Gurdon	162.475	Panama City	162.550	Waterloo	162.550	(Camp Edwards)	162.550
Jonesboro	162.550	Pensacola	162.400			Mt. Greylock	162.525
Little Rock	162.550	Salem	162.425	**Kansas**		Worcester	162.550
Mountain View	162.450	Sebring	162.500	Chanute	162.400		
Russellville	162.525	Tallahassee	162.400	Colby/Goodland	162.475	**Michigan**	
Star City	162.400	Tampa	162.550	Concordia	162.550	Alpena	162.550
Texarkana	162.550	Venice	162.400	Dodge City	162.475	Detroit	162.550
Yellville	162.500	West Palm Beach	162.475	Ellsworth	162.400	Flint	162.475
				Lenora	162.425	Gaylord	162.500
California		**Georgia**		Topeka	162.475	Grand Rapids	162.550
Bakersfield	162.550	Athens	162.400	Tribune	162.550	Hesperia	162.475
Coachella	162.400	Atlanta	162.550	Wichita	162.550	Houghton	162.400
Eureka	162.400	Augusta	162.550			Marquette	162.550
Fresno	162.400	Baxley	162.525	**Kentucky**		Onondaga	162.400
Grass Valley		Buchanan	162.425	Ashland	162.550	Oshtemo	162.475
(Wolf Mt.)	162.400	Chatsworth	162.400	Beattyville	162.500	Sault Ste Marie	162.550
Los Angeles	162.550	Columbus	162.400	Bowling Green	162.400	Traverse City	162.400
Monterey	162.550	Macon	162.475	Covington	162.550		
Pt. Arena/Ukiah	162.550	Pelham	162.550	Elizabethtown	162.550	**Minnesota**	
				Frenchburg	162.475	Bemidji	162.425
						Detroit Lakes	162.400

(Courtesy of National Weather Service/NOAA.)

NOAA Weather Radio . . . the Voice of the National Weather Service [CONTINUED]

BROADCAST FREQUENCIES: 162.400 MHZ, 162.425 MHZ, 162.450 MHZ, 162.475 MHZ, 162.500 MHZ, 162.525 MHZ, 162.550 MHZ

Broadcast range from the weather radio transmitter is approximately 40 miles. The effective range depends on such things as terrain and quality of the receiver and indoor/outdoor antenna. Before you buy a receiver, make sure your area is covered by one of these transmitters.

Minnesota cont'd

Duluth	152.550
International Falls	162.550
Mankato	162.400
Minneapolis/ St. Paul	162.550
Park Rapids	162.475
Rochester	162.475
Roosevelt	162.450
St. Cloud	162.400
Thief River Falls	162.550
Willmar	162.475

Mississippi

Ackerman	162.475
Booneville	162.550
Bude	162.550
Columbia	162.400
Gulfport	162.400
Hattiesburg	162.475
Inverness	162.550
Jackson	162.400
Kosciusko	162.425
Meridian	162.550
Oxford	162.400
Parchman	162.500

Missouri

Bourbon	162.525
Camdenton	162.550
Columbia	162.400
Doniphan	162.450
Fredericktown	162.500
Hannibal	162.475
Hermitage	162.450
Joplin	162.425
Kansas City	162.550
Sikeston	162.400
Springfield	162.400
St. Joseph	162.400
St. Louis	162.550
Summersville	162.475
Wardell	162.525

Montana

Billings	162.550
Butte	162.550
Glasgow	162.400
Glendive	162.475
Great Falls	162.550
Havre (Squaw Butte)	162.400
Helena	162.400
Kalispell	162.550
Malta	162.475
Miles City	162.400
Missoula	162.400
Plentywood	162.475
Pondera County	162.475
Scoby	162.450

Nebraska

Bassett	162.475
Grand Island	162.400
Holdrege	162.475
Lincoln	162.475
Merriman	162.400
Norfolk	162.550
North Platte	162.550
Omaha	162.400
Scottsbluff	162.550

Nevada

Elko	162.550
Ely (Cave Mtn)	162.400
Eureka	162.550
Hawthorne	162.475
Las Vegas (Boulder City)	162.550
Northwest Nevada	162.450
Reno	162.550
Winnemucca	162.400

New Hampshire

Concord	162.400

New Jersey

Atlantic City	162.400

New Mexico

Albuquerque	162.400
Carlsbad	162.475
Clovis	162.475
Des Moines	162.550
Farmington	162.475
Hobbs	162.400
Las Cruces	162.400
Roswell	162.450
Ruidoso	162.550
Sante Fe	162.550

New York

Albany	162.550
Binghamton	162.475
Buffalo	162.550
Elmira	162.400
Kingston	162.475
Little Valley	162.425
New York City	162.550
Riverhead	162.475
Rochester	162.400
Stamford	162.400
Syracuse	162.550
Watertown	162.475

North Carolina

Asheville	162.400
Badin	162.425
Cape Hatteras	162.475
Charlotte	162.475
Fayetteville	162.475
Lumber Bridge	162.525
Margaretsville	162.550
New Bern	162.400
Raleigh/Durham	162.550
Rocky Mount	162.475
Wilmington	162.550
Winston-Salem	162.400

North Dakota

Bismark	162.475
Devils Lake	162.425
Dickinson	162.400
Fargo	162.475
Grand Forks	162.475
Jamestown	162.550
Petersburg	162.400
Williston	162.550

Ohio

Akron	162.400
Bridgeport	162.525
Caldwell	162.475
Cleveland	162.550
Columbus	162.550
Dayton	162.475
Lima	162.400
Sandusky	162.400
Toledo	162.550

Oklahoma

Altus	162.425
Clinton	162.475
Enid	162.475
Lawton	162.550
McAlester	162.475
Oklahoma City	162.400
Ponca City	162.450
Tulsa	162.550
Woodward	162.500

Oregon

Astoria	162.400
Bend/Redmond	162.500
Coos Bay	162.400
Eugene	162.400
Glenwood	162.425
Klamath Falls	162.550
Medford	162.400
Mt. Ashland	162.475
Newport	162.550
Pendleton	162.400
Portland	162.550
Rosenburg	162.550
Salem	162.475
Tillamook	162.475
Umatilla	162.500

Pennsylvania

Allentown	162.400
Clearfield	162.550
Erie	162.400
Harrisburg	162.550
Johnstown	162.400
Parker	162.425
Philadelphia	162.475
Pittsburgh	162.550
State College	162.475
Three Springs	162.525
Towanda	162.550
Warren	162.450
Wellsboro	162.475
Wilkes-Barre	162.550
Williamsport	162.400

Puerto Rico

Maricao	162.550
San Juan	162.400

Rhode Island

Providence	162.400

South Carolina

Beaufort	162.475
Charleston	62.550
Columbia	162.400
Conway/Myrtle Beach	162.400
Cross	162.475
Florence	162.550
Greenville	162.550
Sumter	162.475

South Dakota

Aberdeen	162.475
Huron	162.550
Pierre	162.400
Rapid City	162.550
Sioux Falls	162.400

Tennessee

Bristol	162.550
Chattanooga	162.550
Cookeville	162.400
Jackson	162.550
Knoxville	162.475
Lawrenceburg	162.425
Memphis	162.475
Nashville	162.550
Shelbyville	162.475
Waverly	162.400

Texas

Abilene	162.400
Amarillo	162.550
Austin	162.400
Bay City	162.425
Beaumont	162.475
Big Spring	162.475
Brownsville	162.550
Bryan/College Station	162.550
Cedar Hill (Dallas)	162.400
Corpus Christi	162.550
Del Rio	162.400
El Paso	162.475
Ft. Worth	162.550
Galveston	162.550
Houston	162.400
Junction	162.475
Kerrville	162.450
La Grange	162.500
Laredo	162.475
Llano	162.420
Lubbock	162.400
Lufkin	162.550
Odessa/Midland	162.400
Paris	162.550
Pharr	162.400
Richland Springs	162.525
San Angelo	162.550
San Antonio	162.550
Sherman	162.475
Tyler	162.475
Victoria	162.400
Waco	162.475
Wichita Falls	162.475

Utah

Lake Powell	162.550
Logan	162.400
Milford/Cedar City	162.400
Salt Lake City	162.550
St. George (Utah Hill)	162.425
Tooele (South Mt.)	162.450
Tooele (Vernon Hills)	162.525
Vernal	162.400

Vermont

Burlington	162.400
Marlboro	162.425
Windsor	162.475

Virginia

Heathsville	162.400
Lynchburg	162.550
Norfolk	162.550
Richmond	162.475

(Courtesy of National Weather Service/NOAA.)

NOAA Weather Radio . . . the Voice of the National Weather Service [CONTINUED]

BROADCAST FREQUENCIES: 162.400 MHZ, 162.425 MHZ, 162.450 MHZ, 162.475 MHZ, 162.500 MHZ, 162.525 MHZ, 162.550 MHZ

Broadcast range from the weather radio transmitter is approximately 40 miles. The effective range depends on such things as terrain and quality of the receiver and indoor/outdoor antenna. Before you buy a receiver, make sure your area is covered by one of these transmitters.

Virginia cont'd		Seattle	162.550	Spencer	162.500	Park Falls	162.500
Roanoke	162.475	Spokane	162.400	Sutton	162.450	Prairie Du Chien	162.500
Washington, DC		Wenatchee	162.475	**Wisconsin**		Richland Center	162.450
(Manassas)	162.550	Yakima	162.550	Adams	162.400	Sheboygan	162.525
Virgin Islands		**West Virginia**		Crandon	162.450	Sister Bay	162.425
St. Thomas	162.475	Beckley	162.550	Fond Du Lac	162.500	Wausau	162.475
		Charleston	162.400	Green Bay	162.550	**Wyoming**	
Washington		Clarksburg	162.550	Janesville	162.425	Casper Mountain	162.550
Neah Bay	162.550	Gilbert	162.475	LaCrosse	162.550	Cheyenne	162.475
Okanagan (Tunk Mt.)	162.525	Hinton	162.425	Madison	162.550	Lander	162.475
Olympia	162.475	Moorefield	162.400	Menomonie	162.400	Sheridan	162.500
Port Angeles Marine	162.425	Ogden	162.425	Milwaukee	162.400		
Richland	162.450						

Note: NOAA Weather Radio coverage is expanding through our partnership programs with local commuties. For the latest list of frequencies, check the NOAA Weather Radio Web Site—http://www.nws.noaa.gov/nwr

(Courtesy of National Weather Service/NOAA.)

the very safety of a vessel and its crew. The National Weather Service provides marine warnings and forecasts to serve all who sail for livelihood or recreation.

Aviation weather

Fewer than 15 years after the Wright brothers' historic flights of December 1903, the National Weather Service, then the Signal Corps' Weather Bureau, issued the first official aviation forecast—to help deliver the mail. Now, 80 years later, the agency issues thousands of aviation forecasts, advisories, and warnings to make flying safe and efficient.

On December 1, 1918, Weather Bureau forecasters combined ground observations with data collected by instrumented kites and tethered balloons to provide a forecast for the "Aerial Mail Service" route from New York to Chicago. The reporting network for that historic forecast consisted of 18 kite stations; six were operated by the Weather Bureau and the rest by the military.

Since 1918, the NWS has made quantum leaps in aviation weather forecasting. Geostationary and polar-orbiting satellites enable the organization to cover the globe in its entirety. State-of-the-art computers facilitate fast and accurate analysis of the incredible amounts of collected data to provide forecasts of developing weather phenomena.

Growing from that fledgling, single forecast for a handful of aircraft in 1918, forecasters at the Aviation Weather Center, NWS Weather Forecast Offices, and Center Weather Service Units working in Federal Aviation Administration Air Route Traffic Control Centers issue thousands of forecasts every day. With the improved services, the commercial and general aviation industries are able to save on fuel costs and select routes that avoid hazardous weather.

The Aviation Weather Center, one of nine units in the National Centers for Environmental Prediction, was formed

Marine Advisories and Warnings

These advisories and warnings are "headlined" in marine forecasts. Small Craft Advisories can be issued up to 12 hours and warnings up to 24 hours prior to onset of adverse conditions.

Small Craft Advisory: Forecast winds of 18–33 knots. Small Craft Advisories may also be issued for hazardous sea conditions or lower wind speeds that may affect small craft operations.

Gale Warning: Forecast winds of 34–47 knots.

Storm Warning: Forecast winds of 48 knots or greater.

Tropical Storm Warning: Forecast winds of 34–63 knots associated with a tropical storm.

Hurricane Warning: Forecast winds of 64 knots or higher associated with a hurricane.

in 1995. More that 40 meteorologists at the center work closely with aviation weather research centers, the FAA, and the U.S. Air Force. The AWC issues en route warnings and forecasts over the continental United States. At Weather Forecast Offices, terminal and route forecasts are issued and updated continuously throughout the day. Aviation meteorologists at the Center Weather Services Units provide forecasts and consultation to FAA traffic managers at the FAA Air Route Traffic Control Centers. NWS offices in Anchorage, Alaska; Honolulu, Hawaii; and Guam provide special aviation products for their respective areas.

Data for pilots

The NWS, in cooperation with NOAA, also provides pilots with numerous weather briefings to prepare them for, or assist them with, their flights.

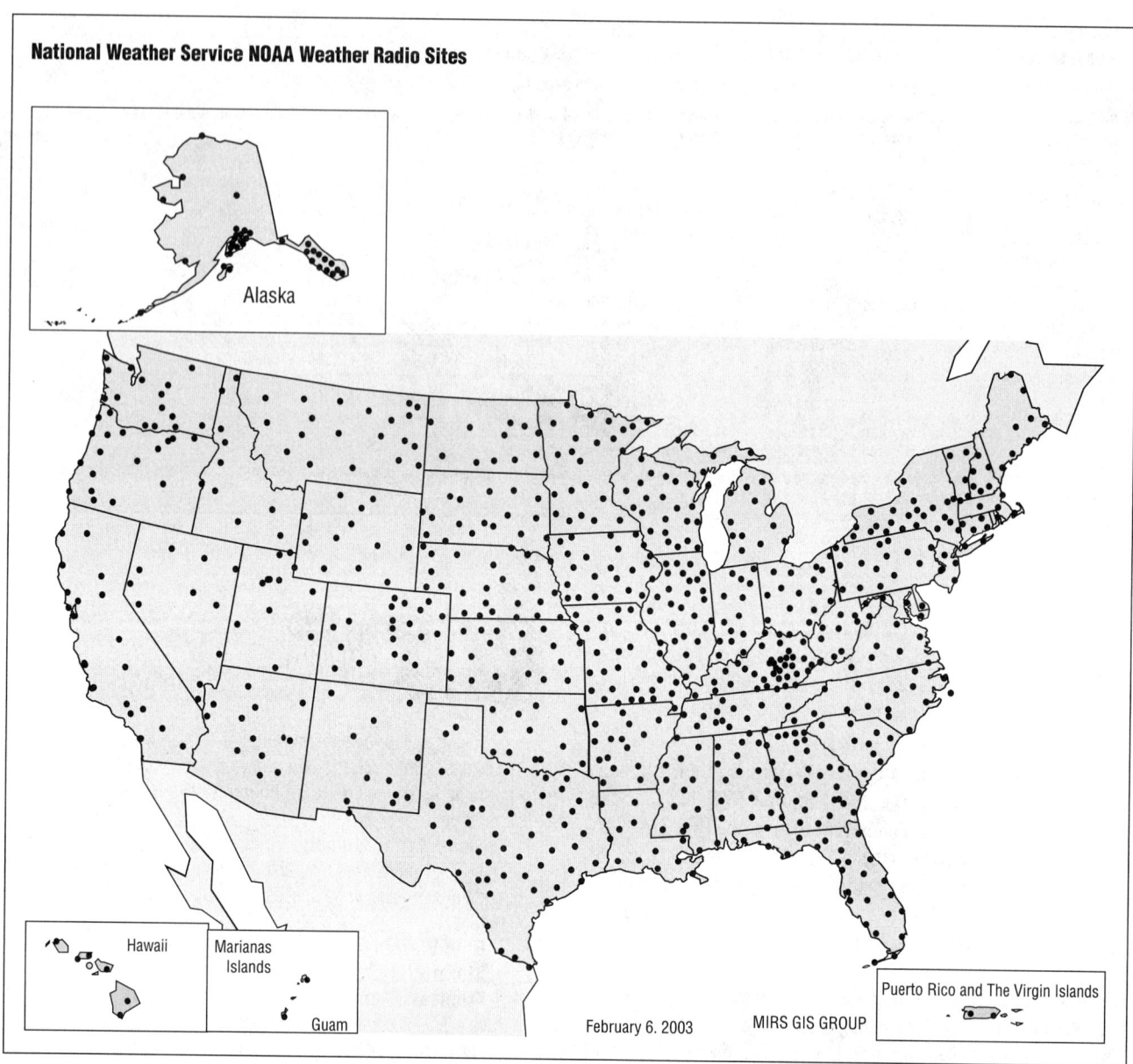

National Weather Service NOAA Weather Radio Sites

Alaska

Hawaii   Marianas Islands   Guam

February 6. 2003   MIRS GIS GROUP

Puerto Rico and The Virgin Islands

(Courtesy of National Weather Service/NOAA.)

Meteorological and aeronautical information is provided by continuous recorded Transcribed Weather Broadcasts (TWEB), the Pilot's Automatic Telephone Weather Answering Service (PATWAS), and the Telephone Information Briefing Service (TIBS). Complete weather information is available by phoning or visiting the nearest FAA Flight Service Station (FSS) or designated NOAA Weather Service Office. Information is also available from private commercial vendors. During periods of marginal weather, briefers are busy and telephone delays may occur. Pilots may get necessary information from TWEB, PATWAS, or TIBS, but it is recommended pilots wait to hear a briefer's information personally. The latest hourly aviation weather observation from distant stations are normally available by five minutes past each hour.

After providing briefers with necessary background information, pilots will receive either a standard, abbreviated, or outlook briefing.

Standard briefings automatically provide pilots with data on adverse conditions, synopsis of prevailing weather systems, current conditions, winds aloft, etc., and any other information the pilot has requested.

Abbreviated briefings are requested when the pilot has used prerecorded or mass-media weather information to make a go/no-go decision and only selected additional information is required.

Outlook briefing is requested for a long-range flight plan.

Marine Beaufort Scale

The Beaufort Scale was originally developed in 1805 by Sir Francis Beaufort as a system for estimating wind strengths without the use of instruments. It is currently still in use for this same purpose as well as to tie together various components of weather (wind strength, sea state, observable effects) into a unified picture.

Force	Speed		Marine Conditions
	knots	mph	
0	<1	<1	Calm, sea like a mirror.
1	1-3	1-3	Light air, ripples only.
2	4-6	4-7	Light breeze, small wavelets (0.2m). Crests have a glassy appearance.
3	7-10	8-12	Gentle breeze, large wavelets (0.6m), crests begin to break.
4	11-16	13-18	Moderate breeze, small waves (1m), some white horses.
5	17-21	19-24	Fresh breeze, moderate waves (1.8m), many white horses.
6	22-27	25-31	Strong breeze, large waves (3m), probably some spray.
7	28-33	32-38	Near gale, mounting sea (4m) with foam blown in streaks downwind.
8	34-40	39-46	Gale, moderately high waves (5.5m), crests break into spindrift.
9	41-47	47-54	Strong gale, high waves (7m), dense foam, visibility affected.
10	48-55	55-63	Storm, very high waves (9m), heavy sea roll, visibility impaired. Surface generally white.
11	56-63	64-73	Violent storm, exceptionally high waves (11m), visibility poor.
12	64+	74+	Hurricane, 14m waves, air filled with foam and spray, visibility bad.

(Courtesy of National Climatic Data Center/NOAA.)

Land Beaufort Scale

The Beaufort Scale was originally developed in 1805 by Sir Francis Beaufort as a system for estimating wind strengths without the use of instruments. It is currently still in use for this same purpose as well as to tie together various components of weather (wind strength, sea state, observable effects) into a unified picture.

Force	Speed		Land Conditions
	knots	mph	
0	<1	<1	Calm, smoke rises vertically
1	1-3	1-3	Light air, direction of wind shown by smoke drift only
2	4-6	4-7	Light breeze, wind felt on face, leaves rustle, vanes moved by wind
3	7-10	8-12	Gentle breeze, leaves and small twigs in constant motion, wind extends light flag
4	11-16	13-18	Moderate breeze, raises dust, loose paper, small branches move
5	17-21	19-24	Fresh breeze, small trees in leaf begin to sway
6	22-27	25-31	Strong breeze, large branches in motion, umbrellas used with difficulty
7	28-33	32-38	Near gale, whole trees in motion, inconvenience felt walking against the wind
8	34-40	39-46	Gale, breaks twigs off trees, impedes progress
9	41-47	47-54	Strong gale, slight structural damage occurs
10	48-55	55-63	Storm, trees uprooted, considerable damage occurs
11	56-63	64-73	Violent storm, widespread damage
12	64+	74+	Hurricane, extreme destruction

(Courtesy of National Climatic Data Center/NOAA.)

Record-Setting Weather

- ## TEMPERATURE EXTREMES, HIGHEST

Temperature extremes depend upon a number of factors, important among which are altitude, latitude, surface conditions, and the density and length of record of observing stations.

The world's highest temperatures, as well as the greatest range of extremes and the greatest and most rapid temperature fluctuations, occur over continental areas in the *temperate zones*.

A reading of 136°F, observed at Azizia (elevation about 380 ft, Tripolitania, Libya, North Africa) on September 13, 1922, is generally accepted as the world's highest temperature recorded under standard conditions.

The highest temperature ever observed in Canada was 115°F at Gleichen, Alberta on July 28, 1903. A high of 120°F or higher has been recorded on all the continents except Antarctica, where the high is only 58.3°F.

Greenland Ranch, California, with 134°F on July 10, 1913, holds the record for the highest temperature ever officially recorded in the United States. This station is located in barren Death Valley which is about 140 mi long and 4–16 mi wide and runs north and south in southeastern California and southwestern Nevada. The valley is below sea level and is flanked by towering mountain ranges with Mt. Whitney, the highest landmark in the 48 states, rising to 14,495 ft, less than 100 mi to the west. Death Valley has the hottest summers in the Western Hemisphere, and is the only known place in the United States where nighttime temperatures sometimes remain above 100°F.

The highest average annual temperature in the world, possibly a world record, is the 94°F, at Dalol (or Dallol), Ethiopia. The station is in a salt desert and is based on only 6 years of data, October 1960–November 1966. Lugh (or Luuq), Somalia, East Africa has an annual mean recorded temperature of 88°F. In the United States the station normally having the highest annual average is Key West, Florida, 77.8°F; the highest summer average, Death Valley, California, 98.2°F; and the highest winter average, Key West, Florida, 70.2°F.

Amazing temperature rises of 40–50°F in a few minutes occasionally may be brought about by chinook winds. Some outstanding extreme temperature rises in short periods are:

12 hours: 83°F, Granville, ND, Feb. 21, 1918, from −33°F to 50°F from early morning to late afternoon.
15 minutes: 42°F, Fort Assiniboine, MT, Jan. 19, 1892, from −5°F to 37°F.
7 minutes: 34°F, Kipp, MT, Dec. 1, 1896; observer also reported that a total rise of 80°F occurred in a few hours and that 30 in of snow disappeared in one-half day.
2 minutes: 49°F, Spearfish, SD, Jan. 22, 1943, from −4°F at 7:30 A.M., to 45°F at 7:32 A.M.

The range of temperature extremes over large bodies of water is much less than over land. Temperature extremes over the sea likely range from 100°F recorded by the SS *Titan* on August 8, 1920, in the Red Sea to −40°F observed by the SS *Baychino*, January 27, 1932, when beset by ice at latitude 70° 50′ N, longitude 159° 11′ W. Sea-surface temperatures in the Persian Gulf average as high as 88°F for July and August, and a high of 96°F was measured by the SS *Frankenfels* on August 5, 1924. These are among the highest—if not the highest—sea-surface temperatures ever observed. (Any official record of these over-the-sea temperatures has been lost over the years, so these measurements are not confirmed.)

- ## TEMPERATURE EXTREMES, LOWEST

Antarctica, a vast, elevated, snow-covered continent at the South Pole is one of the most favorable regions in the world for extremely low temperatures. Several stations there now have records dating back through 1957. A new

Sunniest U.S. Cities

1. Yuma, AZ
2. Redding, CA
3. Phoenix, AZ; Tucson, AZ; Las Vegas, NV
4. El Paso, TX
5. Fresno, CA; Reno, NV

Cloudiest U.S. Cities

1. Quillayute, WA
2. Astoria, OR
3. Olympia, WA
4. Seattle, WA
5. Portland, OR

world-record low temperature was observed at −128.6°F and was recorded at Vostok (Russian station) on July 21, 1983. At the Amundsen-Scott station (elevation 9,186 ft), located on a snow plain within a few hundred yards of the geographical South Pole, the average annual temperature from 1957 to 1964 was −59°F. For July, the average maximum temperature was −69°F, the minimum −80°F; and for January, these values were −17°F and −22°F, respectively. The average temperature at Vostok for the two-year period 1958–59 was −67°F. Even colder locations may exist on the continent.

Other regions favorable for unusually low winter extremes include Greenland, a high snow-covered area located mostly in the north polar regions; and north central Siberia, part of a great land mass at high latitudes. Minima of −90°F (Verhoyansk −89.7°F, February 5 and 7, 1892 and Oimekon −89.9°F, February 6, 1933) in the latter region stood as the world's lowest temperatures prior to observations in Antarctica. The lowest temperature on the Greenland Icecap, −86.8°F was observed at Northice January 9, 1954. Canada's lowest temperature, −81°F, was observed at Snag, Yukon Territory, near the border of Alaska at an altitude of 2,120 ft on February 3, 1947.

In the United States, the lowest temperature on record, −79.8°F, was recorded on January 23, 1971, at Prospect Creek Camp, which is located in the Endicott Mountains of Northern Alaska at latitude 66° 48′ N, longitude 150° 40′ W. The lowest temperature in the contiguous 48 states, −69.7°F, occurred on January 20, 1954, at Rogers Pass, in Lewis and Clark County, Montana. This location is in mountainous and heavily forested terrain, about 0.5 mi east of and 140 ft below the summit of the Continental Divide.

The lowest average annual temperature recorded in the United States is 9.4°F at Barrow, Alaska, which lies on the Arctic coast. Barrow also has the coolest summers (June, July, August) with an average of 41.9°F. The lowest average winter (December, January, February) temperature is −20.1°F at Barter Island on the Arctic coast of northeast Alaska. In Hawaii, average annual temperatures range from 44°F at Mauna Loa Slope Observatory (elevation 11,146 ft) on the island of Hawaii to 77.2°F at Honolulu on the island of Oahu.

In the contiguous 48 states, Mt. Washington, New Hampshire (elevation 6,262 ft) has the lowest mean annual temperature, 26.5°F, and the lowest mean summer (June, July, August) temperature, 51.6°F. A few stations in the Northeast and upper Rockies have mean annuals in the high 30s, and at the same stations in the latter area, summers may average in the high 40s. Winter (December, January, February) mean temperature are lowest in northeastern North Dakota where the average is 5.9°F at the Langdon Experiment Farm and northwestern Minnesota where the average is 6.1°F at Hallock.

In continental areas of the temperate zone, 40–50°F temperature falls in a few hours caused by advection of cold air masses are not uncommon. Sometimes, following these large drops due to advection, radiation may cause a further temperature fall resulting in remarkable changes. Some outstanding extreme temperature falls are:

24 hours: 100°F, Browning, MT, Jan. 23–24, 1916, from 44°F to −56°F.

12 hours: 84°F, Fairfield, MT, Dec. 24, 1924, from 63°F at noon to −21°F at midnight.

2 hours: 62°F, Rapid City, SD, Jan. 12, 1911, from 49°F at 6 A.M. to −13°F at 8 A.M.

27 minutes: 58°F, Spearfish, SD, Jan. 22, 1943, from 54°F at 9 A.M., to −4°F at 9:27 A.M.

15 minutes: 47°F, Rapid City, SD, Jan. 10, 1911, from 55°F at 7 A.M., to 8°F at 7:15 A.M.

Highest Temperature Extremes

Continent	Highest Temp. (deg F)	Place	Elevation (Feet)	Date
Africa	136	El Azizia, Libya	367	13 Sep 1922
North America	134	Death Valley, CA (Greenland Ranch)	−178	10 Jul 1913
Asia	129	Tirat Tsvi, Israel	−722	21 Jun 1942
Australia	128	Cloncurry, Queensland	622	16 Jan 1889
Europe	122	Seville, Spain	26	4 Aug 1881
South America	120	Rivadavia, Argentina	676	11 Dec 1905
Oceania	108	Tuguegarao, Philippines	72	29 Apr 1912
Antarctica	59	Vanda Station, Scott Coast	49	5 Jan 1974

Lowest Temperature Extremes

Continent	Lowest Temp. (deg F)	Place	Elevation (Feet)	Date
Antarctica	−129	Vostok	11220	21 Jul 1983
Asia	−90	Oimekon, Russia	2625	6 Feb 1933
Asia	−90	Verkhoyansk, Russia	350	7 Feb 1892
Greenland	−87	Northice	7687	9 Jan 1954
North America	−81.4	Snag, Yukon, Canada	2120	3 Feb 1947
Europe	−67	Ust'Shchugor, Russia	279	January*
South America	−27	Sarmiento, Argentina	879	1 Jun 1907
Africa	−11	Ifrane, Morocco	5364	11 Feb 1935
Australia	−9.4	Charlotte Pass, NSW	5758	29 Jun 1994
Oceania	14	Haleakala Summit, Maui, HI	9750	2 Jan 1961

*Exact date unknown, lowest in 15-year period

(Courtesy of National Climatic Data Center/NOAA.)

Highest Average Annual Precipitation Extremes

Continent	Highest Avg. (Inches)	Place	Elevation (Feet)	Years of Record
South America	523.6!^	Lloro, Colombia	520*	29
Asia	467.4!	Mawsynram, India	4597	38
Oceania	460.0!	Mt. Waialeale, Kauai, HI	5148	30
Africa	405.0	Debundscha, Cameroon	30	32
South America	354.0^	Quibdo, Colombia	120	16
Australia	340.0	Bellenden Ker, Queensland	5102	9
North America	256.0	Henderson Lake, British Colombia	12	14
Europe	183.0	Crkvica, Bosnia-Hercegovina	3337	22

!The value given is continent's highest and possibly the world's depending on measurement practices, procedures and period of record variations.
^The official greatest average annual precipitation for South America is 354 inches at Quibdo, Colombia. The 523.6 inches average at Lloro, Colombia [14 miles SE and at a higher elevation than Quibdo] is an estimated amount.
*Approximate elevation.

Lowest Average Annual Precipitation Extremes

Continent	Lowest Avg. (Inches)	Place	Elevation (Feet)	Years of Record
South America	0.03	Arica, Chile	95	59
Africa	<0.1	Wadi Halfa, Sudan	410	39
Antarctica	0.8~	Amundsen-Scott South Pole Station	9186	10
North America	1.2	Batagues, Mexico	16	14
Asia	1.8	Aden, Yemen	22	50
Australia	4.05	Mulka (Troudaninna), South Australia	160*	42
Europe	6.4	Astrakhan, Russia	45	25
Oceania	8.93	Puako, Hawaii, HI	5	13

~The value given is the average amount of solid snow accumulating in one year as indicated by snow markers. The liquid content of the snow is undetermined.
*Approximate elevation.

(Courtesy of National Climatic Data Center/NOAA.)

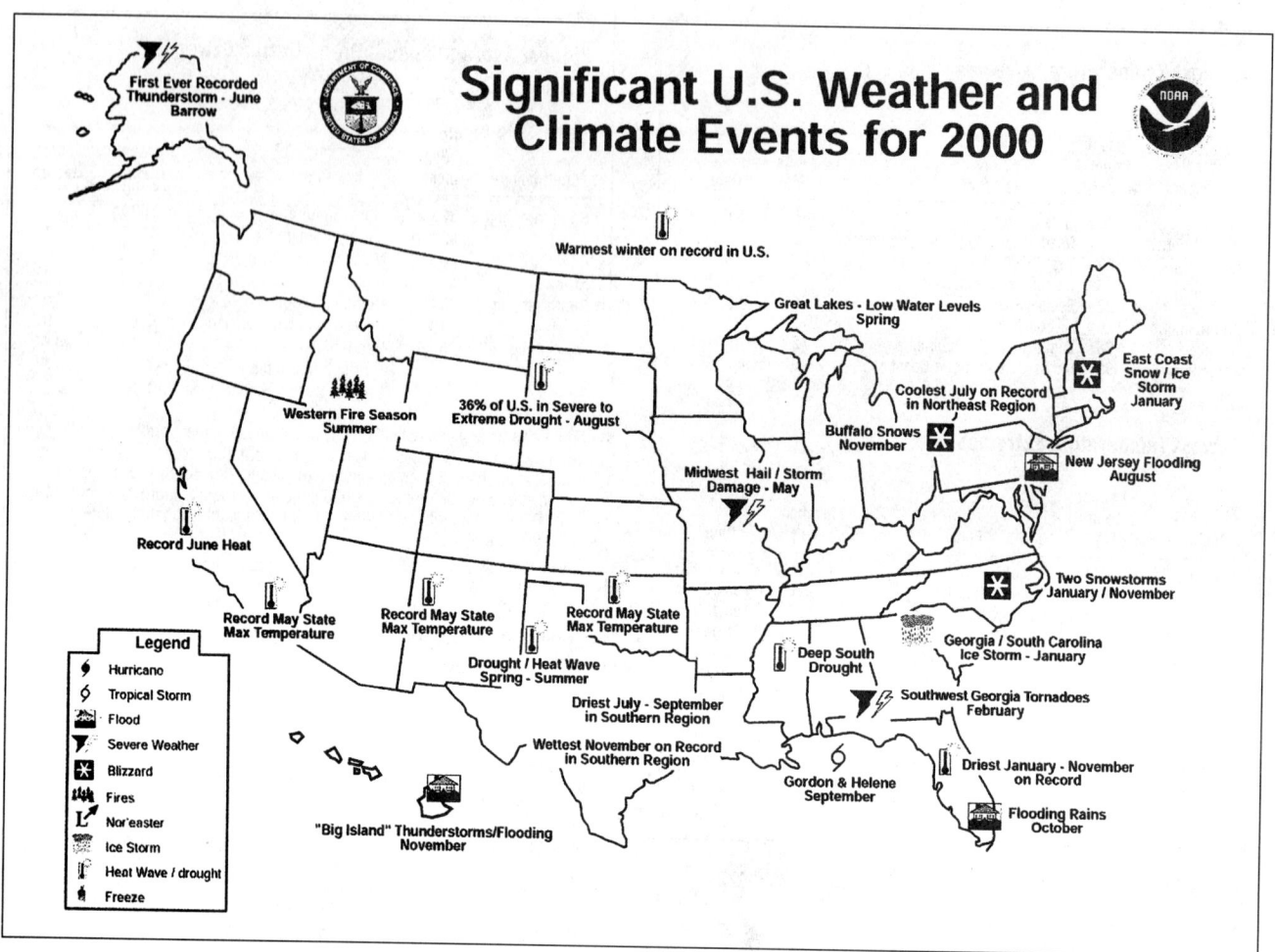

First Ever Recorded Thunderstorm - June Barrow

Significant U.S. Weather and Climate Events for 2000

Warmest winter on record in U.S.

Great Lakes - Low Water Levels Spring

East Coast Snow / Ice Storm January

Coolest July on Record in Northeast Region

Buffalo Snows November

Western Fire Season Summer

36% of U.S. in Severe to Extreme Drought - August

Midwest Hail / Storm Damage - May

New Jersey Flooding August

Record June Heat

Two Snowstorms January / November

Record May State Max Temperature

Record May State Max Temperature

Record May State Max Temperature

Georgia / South Carolina Ice Storm - January

Deep South Drought

Drought / Heat Wave Spring - Summer

Southwest Georgia Tornadoes February

Driest July - September in Southern Region

Driest January - November on Record

Wettest November on Record in Southern Region

Gordon & Helene September

Flooding Rains October

Legend

Hurricane	
Tropical Storm	
Flood	
Severe Weather	
Blizzard	
Fires	
Nor'easter	
Ice Storm	
Heat Wave / drought	
Freeze	

"Big Island" Thunderstorms/Flooding November

(Courtesy of National Oceanic and Atmospheric Administration (NOAA).)

Record Highest Temperatures by State

(Thru December 2000)

State	Temp. °F.	Date	Station	Elevation Feet
Alabama	112	Sep. 5, 1925	Centerville	345
Alaska	100	Jun. 27, 1915	Fort Yukon	est. 420
Arizona	128	Jun. 29, 1994	Lake Havasu City	505
Arkansas	120	Aug. 10, 1936	Ozark	396
California	134	Jul. 10, 1913	Greenland Ranch	−178
Colorado	118	Jul. 11, 1888	Bennett	5,484
Connecticut	106	Jul. 15, 1995	Danbury	450
Delaware	110	Jul. 21, 1930	Millsboro	20
Florida	109	Jun. 29, 1931	Monticello	207
Georgia	112	Aug. 20, 1983	Greenville	860
Hawaii	100	Apr. 27, 1931	Pahala	850
Idaho	118	Jul. 28, 1934	Orofino	1,027
Illinois	117	Jul. 14, 1954	East St. Louis	410
Indiana	116	Jul. 14, 1936	Collegeville	672
Iowa	118	Jul. 20, 1934	Keokuk	614
Kansas	121	Jul. 24, 1936*	Alton (near)	1,651
Kentucky	114	Jul. 28, 1930	Greensburg	581
Louisiana	114	Aug. 10, 1936	Plain Dealing	268
Maine	105	Jul. 10, 1911*	North Bridgton	450
Maryland	109	Jul. 10, 1936*	Cumberland & Frederick	623, 325
Massachusetts	107	Aug. 2, 1975	New Bedford & Chester	120; 640
Michigan	112	Jul. 13, 1936	Mio	963
Minnesota	114	Jul. 6, 1936	Moorhead	904
Mississippi	115	Jul. 29, 1930	Holly Springs	600
Missouri	118	Jul. 14, 1954*	Warsaw & Union	705; 560
Montana	117	Jul. 5, 1937	Medicine Lake	1,950
Nebraska	118	Jul. 24, 1936*	Minden	2,169
Nevada	125	Jun. 29, 1994*	Laughlin	605
New Hampshire	106	Jul. 4, 1911	Nashua	125
New Jersey	110	Jul. 10, 1936	Runyon	18
New Mexico	122	Jun. 27, 1994	Waste Isolat. Pilot Plt	3,418
New York	108	Jul. 22, 1926	Troy	35
North Carolina	110	Aug. 21, 1983	Fayetteville	213
North Dakota	121	Jul. 6, 1936	Steele	1,857
Ohio	113	Jul. 21, 1934*	Gallipolis (near)	673
Oklahoma	120	Jun. 27, 1994*	Tipton	1,350
Oregon	119	Aug. 10, 1898*	Pendleton	1,074
Pennsylvania	111	Jul. 10, 1936*	Phoenixville	100
Rhode Island	104	Aug. 2, 1975	Providence	51
South Carolina	111	Jun. 28, 1954*	Camden	170
South Dakota	120	Jul. 5, 1936	Gannvalley	1,750
Tennessee	113	Aug. 9, 1930*	Perryville	377
Texas	120	Jun. 28, 1994*	Monahans	2,660
Utah	117	Jul. 5, 1985	Saint George	2,880
Vermont	105	Jul. 4, 1911	Vernon	310
Virginia	110	Jul. 15, 1954	Balcony Falls	725
Washington	118	Aug. 5, 1961*	Ice Harbor Dam	475
West Virginia	112	Jul. 10, 1936*	Martinsburg	435
Wisconsin	114	Jul. 13, 1936	Wisconsin Dells	900
Wyoming	115	Aug. 8, 1983	Basin	3,500

*Also on earlier dates at the same time or other places.

(Courtesy of National Climatic Data Center/NOAA.)

Record Lowest Temperatures by State

(Thru December 2000)

State	Temp. °F.	Date	Station	Elevation Feet
Alabama	−27	Jan. 30, 1966	New Market	760
Alaska	−80	Jan. 23, 1971	Prospect Creek Camp	1,100
Arizona	−40	Jan. 7, 1971	Hawley Lake	8,180
Arkansas	−29	Feb. 13, 1905	Pond	1,250
California	−45	Jan. 20, 1937	Boca	5,532
Colorado	−61	Feb. 1, 1985	Maybell	5,920
Connecticut	−32	Jan. 22, 1961*	Coventry	480
Delaware	−17	Jan. 17, 1893	Millsboro	20
Florida	−2	Feb. 13, 1899	Tallahassee	193
Georgia	−17	Jan. 27, 1940	CCC Camp F-16	est. 1,000
Hawaii	12	May 17, 1979	Mauna Kea Obs 111.2	13,770
Idaho	−60	Jan. 18, 1943	Island Park Dam	6,285
Illinois	−36	Jan. 5, 1999	Congerville	635
Indiana	−36	Jan. 19, 1994	New Whiteland	785
Iowa	−47	Feb. 3, 1996*	Elkader	770
Kansas	−40	Feb. 13, 1905	Lebanon	1,812
Kentucky	−37	Jan. 19, 1994	Shelbyville	730
Louisiana	−16	Feb. 13, 1899	Minden	194
Maine	−48	Jan. 19, 1925	Van Buren	510
Maryland	−40	Jan. 13, 1912	Oakland	2,461
Massachusetts	−35	Jan. 12, 1981	Chester	640
Michigan	−51	Feb. 9, 1934	Vanderbilt	785
Minnesota	−60	Feb. 2, 1996	Tower	1,460
Mississippi	−19	Jan. 30, 1966	Corinth	420
Missouri	−40	Feb. 13, 1905	Warsaw	700
Montana	−70	Jan. 20, 1954	Rogers Pass	5,470
Nebraska	−47	Dec. 22, 1989*	Oshkosh	3,379
Nevada	−50	Jan. 8, 1937	San Jacinto	5,200
New Hampshire	−47	Jan. 29, 1934	Mt. Washington	6,262
New Jersey	−34	Jan. 5, 1904	River Vale	70
New Mexico	−50	Feb. 1, 1951	Gavilan	7,350
New York	−52	Feb. 18, 1979*	Old Forge	1,720
North Carolina	−34	Jan. 21, 1985	Mt. Mitchell	6,525
North Dakota	−60	Feb. 15, 1936	Parshall	1,929
Ohio	−39	Feb. 10, 1899	Milligan	800
Oklahoma	−27	Jan. 18, 1930*	Watts	958
Oregon	−54	Feb. 10, 1933*	Seneca	4,700
Pennsylvania	−42	Jan. 5, 1904	Smethport	est. 1,500
Rhode Island	−25	Feb. 5, 1996	Greene	425
South Carolina	−19	Jan. 21, 1985	Caesars Head	3,115
South Dakota	−58	Feb. 17, 1936	McIntosh	2,277
Tennessee	−32	Dec. 30, 1917	Mountain City	2,471
Texas	−23	Feb. 8, 1933*	Seminole	3,275
Utah	−69	Feb. 1, 1985	Peter's Sink	8,092
Vermont	−50	Dec. 30, 1933	Bloomfield	915
Virginia	−30	Jan. 22, 1985	Mtn. Lake Bio. Stn.	3,870
Washington	−48	Dec. 30, 1968	Mazama & Winthrop	2,120; 1,755
West Virginia	−37	Dec. 30, 1917	Lewisburg	2,200
Wisconsin	−55	Feb. 4, 1996	Couderay	1,300
Wyoming	−66	Feb. 9, 1933	Riverside R.S.	6,500

*Also on earlier dates at the same time or other places.

(Courtesy of National Climatic Data Center/NOAA.)

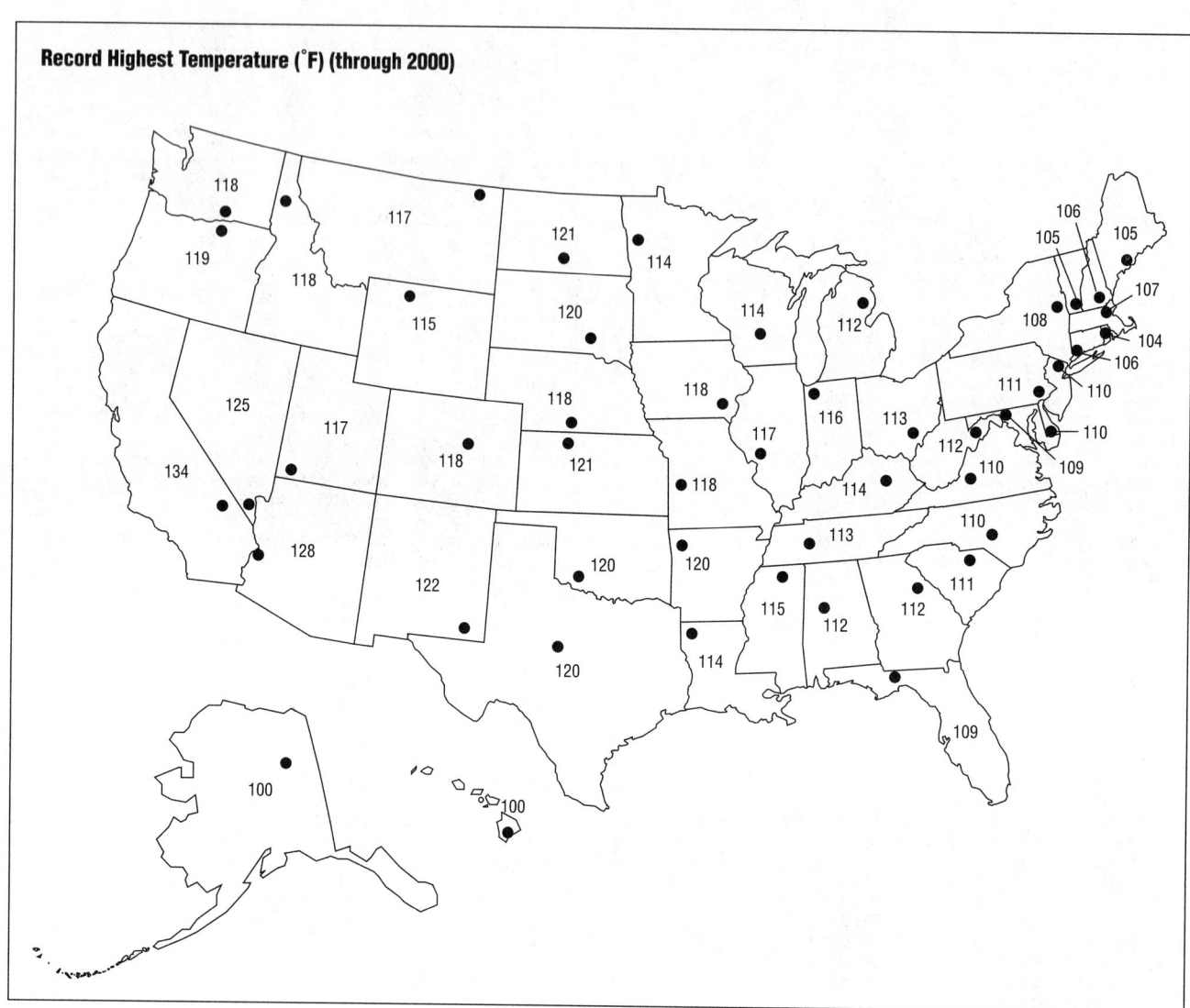

(Courtesy of National Climatic Data Center/NOAA.)

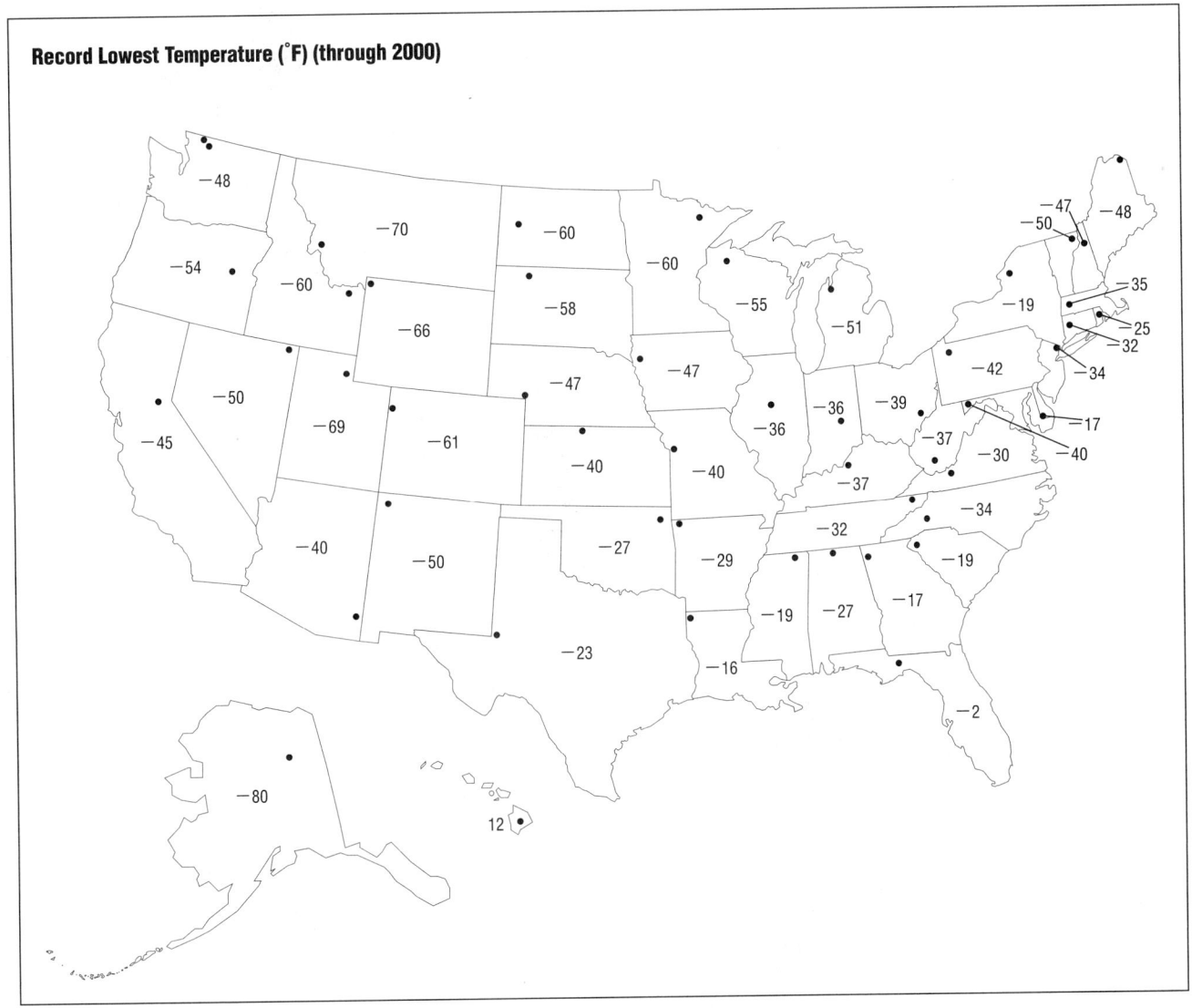

Record Lowest Temperature (˚F) (through 2000)

(Courtesy of National Climatic Data Center/NOAA.)

Round-the-World Weather

• CLIMATES OF THE WORLD

Temperature distribution

The distribution of temperature over the world and its variations through the year depend primarily on the amount of distribution of the radiant energy received from the sun in different regions. This in turn depends mainly on latitude but is greatly modified by the distribution of continents and oceans, prevailing winds, oceanic circulation, topography, and other factors.

In the winter of the Northern Hemisphere (see *Average January Temperature* map), the poleward temperature gradient (that is, the rate of fall in temperature) north of latitude 15° is very steep over the interior of North America. This is shown by the fact that the lines indicating changes in temperature come very close together. The temperature gradient is also steep toward the cold pole over Asia in the area marked −50°F. In western Europe, to the east of the Atlantic Ocean and the North Atlantic Drift, and in the region of prevailing westerly winds, the temperature gradient is much more gradual, as indicated by the fact that the isotherms, or lines of equal temperature, are far apart. In the winter of the Southern Hemisphere (see *Average July Temperature* map), the temperature gradient toward the South Pole is very gradual, and the isothermal deflections from the east-west direction (that is, the dipping of the isothermal lines) are of minor importance because continental effects are largely absent.

In the summers of the two hemispheres—July in the north and January in the south—the temperature gradients poleward are very much diminished as compared with those during the winter. This is especially marked over the middle and higher northern latitudes because of the greater warming of the extensive interiors of North America and Eurasia than of the smaller land areas in middle and higher southern latitudes.

Distribution of precipitation

Whether precipitation occurs as rain or snow or in the rarer forms of hail or sleet depends largely on the temperature climate, which may be influenced more by elevation than by latitude, as in the case of the perpetually snow-capped mountain peaks and glaciers on the equator in both South America and Africa.

The quantity of precipitation is governed by the amount of water vapor in the air and the nature of the process that leads to its condensation into liquid or solid form through cooling. Air may ascend to great elevations through local convection, as in thunderstorms and in tropical regions generally; it may be forced up over topographical elevations across the prevailing wind direction, as on the southern or

---

Coldest World Temperatures

1. Vostok, Antarctica (−129°F)
2. Oimekon, Russia (−90°F)
3. Northice, Greenland (−87°F)
4. Snag, Yukon, Canada (−81.4°F)
5. Ust'Shchugor, Russia (−67°F)

Highest World Temperatures

1. El Azizia, Libya (136°F)
2. Death Valley, CA (134°F)
3. Tirat Tsvi, Israel (129°F)
4. Cloncurry, Australia (128°F)
5. Seville, Spain (122°F)

Highest Annual Precipitation

1. Lloro, Columbia (523.6 in)
2. Mawsynram, India (467.4 in)
3. Mt. Waialeale, Kauai, Hawaii (460 in)
4. Debundscha, Cameroon (405 in)
5. Quibdo, Colombia (354 in)

Lowest Annual Precipitation

1. Arica, Chile (0.03 in)
2. Wadi Halfa, Sudan (0.1 in)
3. Amundsen-Scott, South Pole Station (0.8 in)
4. Batagues, Mexico (1.2 in)
5. Aden, Yemen (1.8 in)

windward slopes of the Himalayas in the path of the southwest monsoon in India; or it may ascend more or less gradually in migratory low-pressure formations such as those that govern the main features of weather in the United States.

The areas of heaviest precipitation on the map (*World Precipitation*) are generally located, as would be expected, in tropical regions, where because of high temperature the greatest amount of water vapor may be present in the atmosphere and the greatest evaporation takes place—although only where conditions favor condensation can rainfall occur. Outstanding exceptions are certain regions in high latitudes,

such as southern Alaska, western Norway, and southern Chile, where relatively warm, moist winds from the sea undergo forced ascent over considerable elevations.

In marked contrast to the rainy regions just named, are the dry polar regions, where the water vapor content of the air is always very low because of the low temperature and very limited evaporation. The dry areas in the subtropical belts of high atmospheric pressure in the vicinity of latitude 30° on all continents (especially from the extreme western Sahara over a broad, somewhat broken belt to the Desert of Gobi), and the arid strips on the lee sides of mountains on whose windward slopes precipitation is heavy to excessive, are caused by conditions which, even though the temperature may be high, are unfavorable to the condensation of whatever water vapor may be present in the atmosphere.

North America

North America is nearly all within middle and northern latitudes. Consequently, it has a large central area in which the continental type of climate with marked seasonal temperature is to be found. Along the coasts of northern Alaska, western Canada, and the northwestern part of the United States, moderate midsummer temperatures are in marked contrast to those prevailing in the interior east of the moun-

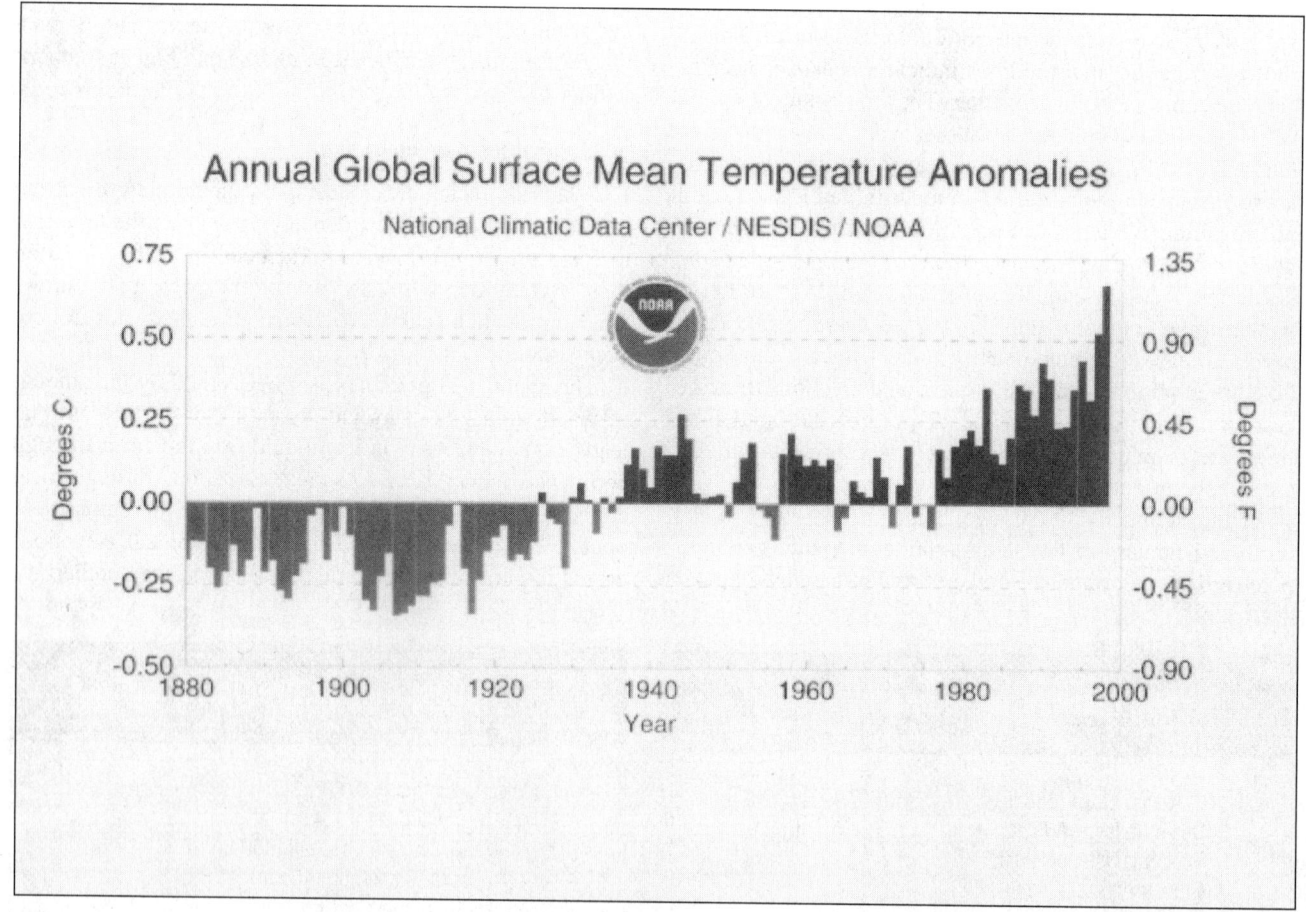

(Courtesy of National Oceanic and Atmospheric Administration (NOAA).)

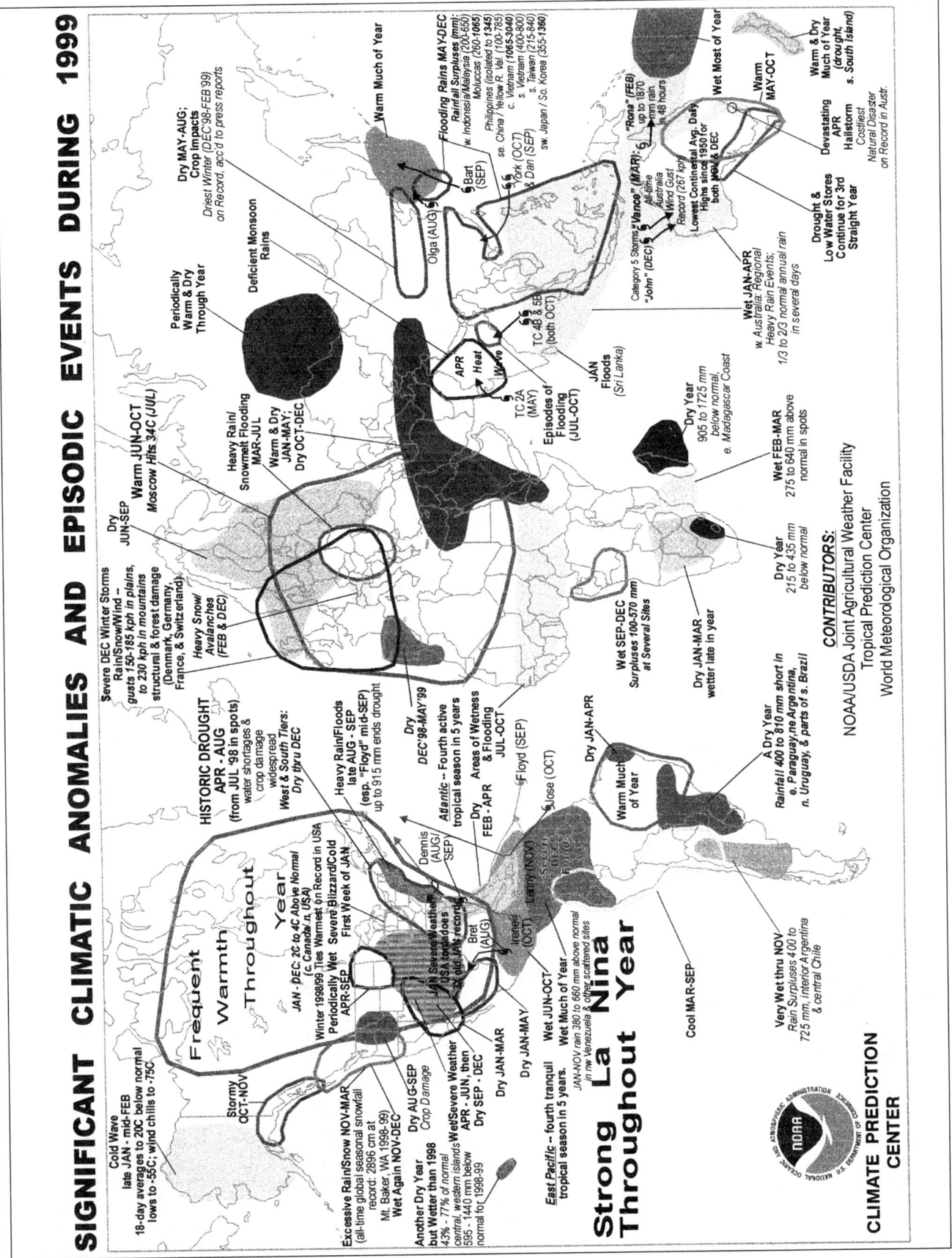

(Courtesy of National Oceanic and Atmospheric Administration (NOAA).)

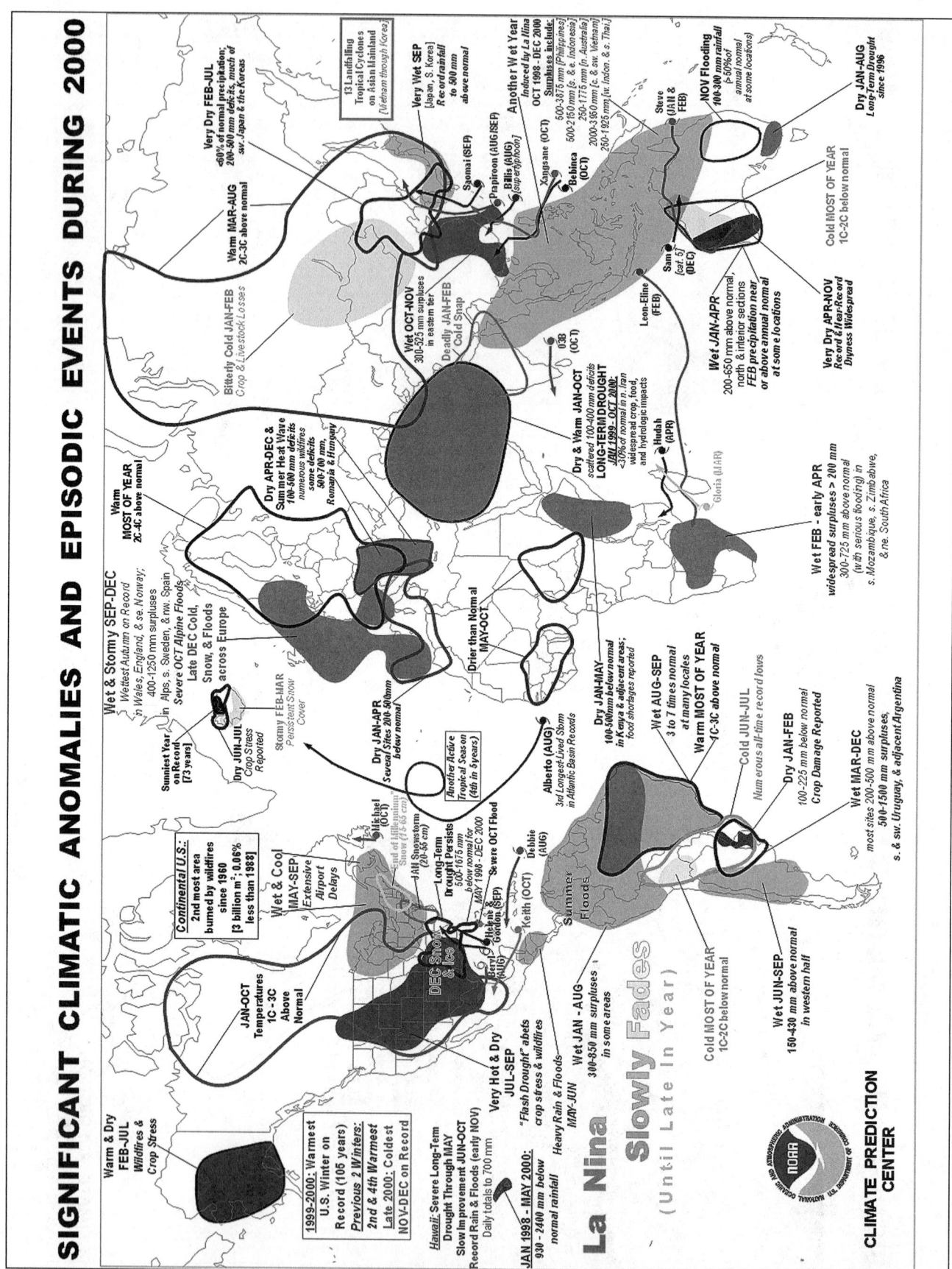

SIGNIFICANT CLIMATIC ANOMALIES AND EPISODIC EVENTS DURING 2000

(Courtesy of National Oceanic and Atmospheric Administration (NOAA).)

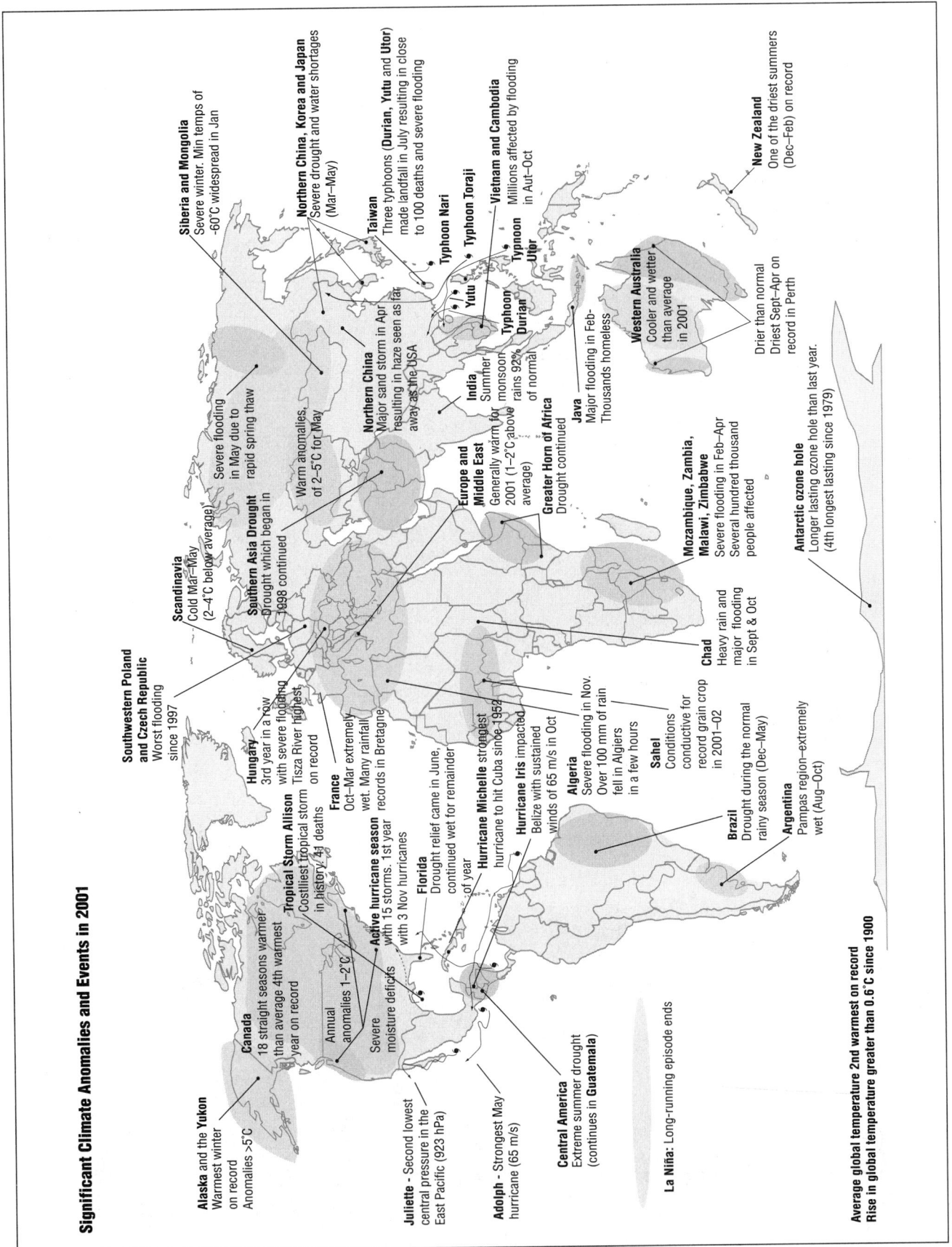

Significant Climate Anomalies and Events in 2001

Alaska and the Yukon Warmest winter on record Anomalies >5°C

Canada 18 straight seasons warmer than average 4th warmest year on record

Annual anomalies 1–2°C

Severe moisture deficits

Tropical Storm Allison Costliest Tropical storm in history, 41 deaths

Active hurricane season with 15 storms. 1st year with 3 Nov hurricanes

Juliette - Second lowest central pressure in the East Pacific (923 hPa)

Adolph - Strongest May hurricane (65 m/s)

Central America Extreme summer drought (continues in **Guatemala**)

Florida Drought relief came in June, continued wet for remainder of year

Hurricane Michelle strongest hurricane to hit Cuba since 1952

Hurricane Iris impacted Belize with sustained winds of 65 m/s in Oct

Hungary 3rd year in a row with severe flooding Tisza River highest on record

France Oct–Mar extremely wet. Many rainfall records in Bretagne

Scandinavia Cold Mar–May (2–4°C below average)

Southern Asia Drought Drought which began in 1998 continued

Southwestern Poland and Czech Republic Worst flooding since 1997

Severe flooding in May due to rapid spring thaw

Warm anomalies, of 2–5°C for May

Siberia and Mongolia Severe winter. Min temps of −60°C widespread in Jan

Northern China, Korea and Japan Severe drought and water shortages (Mar–May)

Northern China Major sand storm in Apr resulting in haze seen as far away as the USA

Taiwan Three typhoons (**Durian, Yutu** and **Utor**) made landfall in July resulting in close to 100 deaths and severe flooding

Typhoon Nari

Typhoon Toraji

Yutu

Typhoon Utor

Typhoon Durian

Vietnam and Cambodia Millions affected by flooding in Aut–Oct

India Summer monsoon rains 92% of normal

Europe and Middle East Generally warm for 2001 (1–2°C above average)

Greater Horn of Africa Drought continued

Java Major flooding in Feb. Thousands homeless

Mozambique, Zambia, Malawi, Zimbabwe Severe flooding in Feb–Apr Several hundred thousand people affected

Chad Heavy rain and major flooding in Sept & Oct

Algeria Severe flooding in Nov. Over 100 mm of rain fell in Algiers in a few hours

Sahel Conditions conductive for record grain crop in 2001–02

Brazil Drought during the normal rainy season (Dec–May)

Argentina Pampas region–extremely wet (Aug–Oct)

La Niña: Long-running episode ends

Western Australia Cooler and wetter than average in 2001

Drier than normal Driest Sept–Apr on record in Perth

New Zealand One of the driest summers (Dec–Feb) on record

Antarctic ozone hole Longer lasting ozone hole than last year. (4th longest lasting since 1979)

Average global temperature 2nd warmest on record
Rise in global temperature greater than 0.6°C since 1900

(Courtesy of National Oceanic and Atmospheric Administration (NOAA).)

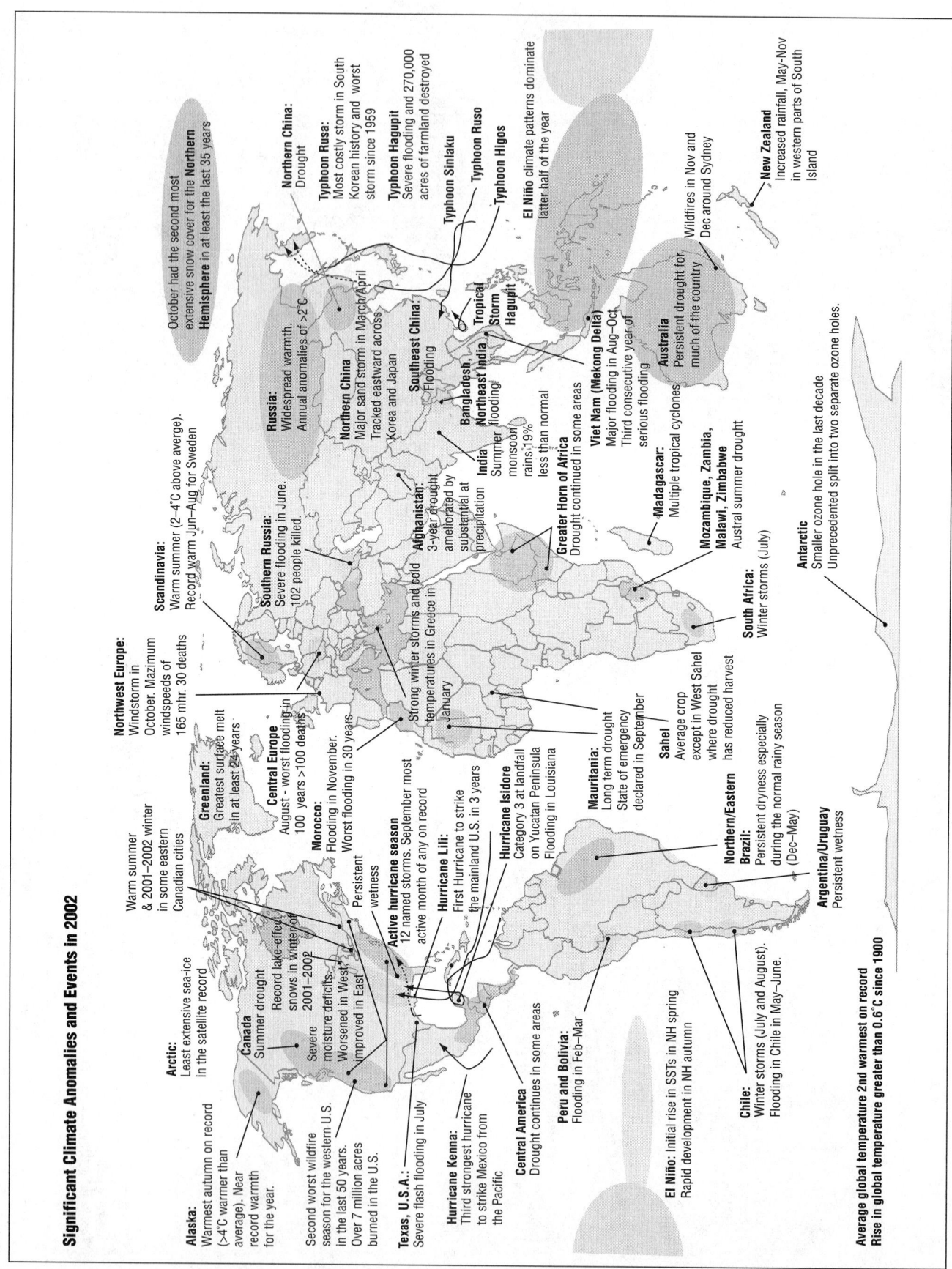

Significant Climate Anomalies and Events in 2002

October had the second most extensive snow cover for the **Northern Hemisphere** in at least the last 35 years

Northern China: Drought

Typhoon Rusa: Most costly storm in South Korean history and worst storm since 1959

Typhoon Hagupit Severe flooding and 270,000 acres of farmland destroyed

Typhoon Sinlaku

Typhoon Ruso

Typhoon Higos

El Niño climate patterns dominate latter half of the year

New Zealand Increased rainfall, May–Nov in western parts of South Island

Russia: Widespread warmth. Annual anomalies of >2°C

Northern China Major sand storm in March/April Tracked eastward across Korea and Japan

Southeast China: Flooding

Tropical Storm Hagupit

Bangladesh, Northeast India flooding

India Summer monsoon rains 19% less than normal

Viet Nam (Mekong Delta) Major flooding in Aug–Oct. Third consecutive year of serious flooding

Australia Persistent drought for much of the country

Wildfires in Nov and Dec around Sydney

Scandinavia: Warm summer (2–4°C above averge). Record warm Jun–Aug for Sweden

Southern Russia: Severe flooding in June. 102 people killed.

Afghanistan: 3-year drought ameliorated by substantial at precipitation

Greater Horn of Africa Drought continued in some areas

Madagascar: Multiple tropical cyclones

Mozambique, Zambia, Malawi, Zimbabwe Austral summer drought

South Africa: Winter storms (July)

Antarctic Smaller ozone hole in the last decade Unprecedented split into two separate ozone holes.

Northwest Europe: Windstorm in October. Maximum windspeeds of 165 mhr. 30 deaths

Strong winter storms and cold temperatures in Greece in January

Greenland: Greatest surface melt in at least 24 years

Central Europe August - worst flooding in 100 years >100 deaths

Morocco: Flooding in November. Worst flooding in 30 years

Persistent wetness

Mauritania: Long term drought State of emergency declared in September

Sahel Average crop except in West Sahel where drought has reduced harvest

Northern/Eastern Brazil: Persistent dryness especially during the normal rainy season (Dec–May)

Argentina/Uruguay Persistent wetness

Alaska: Warmest autumn on record (>4°C warmer than average). Near record warmth for the year.

Warm summer & 2001–2002 winter in some eastern Canadian cities

Canada Summer drought

Second worst wildfire season for the western U.S. in the last 50 years. Over 7 million acres burned in the U.S.

Severe moisture deficits. Worsened in West. Improved in East

Record lake-effect snows in winter 2001–2002

Persistent wetness

Active hurricane season 12 named storms. September most active month of any on record

Hurricane Lili: First Hurricane to strike the mainland U.S. in 3 years

Hurricane Isidore Category 3 at landfall on Yucatan Peninsula Flooding in Louisiana

Texas, U.S.A.: Severe flash flooding in July

Hurricane Kenna: Third strongest hurricane to strike Mexico from the Pacific

Central America Drought continues in some areas

Peru and Bolivia: Flooding in Feb–Mar

El Niño: Initial rise in SSTs in NH spring Rapid development in NH autumn

Chile: Winter storms (July and August). Flooding in Chile in May–June.

Arctic: Least extensive sea-ice in the satellite record

Average global temperature 2nd warmest on record Rise in global temperature greater than 0.6°C since 1900

(Courtesy of National Oceanic and Atmospheric Administration (NOAA).)

tains. Again, the mild midwinter temperatures in the coastal areas stand out against the severe conditions to be found from the Great Lakes region northward and northwestward.

In the West Indian region, temperature conditions are subtropical; and in Mexico and Central America, climatic zones depend on elevation, ranging from subtropical to temperate in the higher levels.

The prevailing westerly wind movement carries the continental type of climate eastward over the United States, so that the region of maritime climate along the Atlantic Ocean is very narrow.

From the Aleutian Peninsula to northern California west of the crests of the mountains, there is a narrow strip where annual precipitation is over 40 in; it exceeds 100 in locally on the coast of British Columbia. East of this belt there is an abrupt falloff in precipitation to less than 20 in annually over the western half of the continent from lower California northward, and to even less than 5 in in parts of what used to be called the "Great American Desert" in the southwestern part of the United States.

In the eastern part of the continent—that is, from the southeastern part of the United States northeastward to Newfoundland—the average annual precipitation is more than 40 in. Rainfall in the West Indies, southern Mexico, and Central America is generally abundant. It is very spotty, however, varying widely even within short distances, especially from the windward to the leeward sides of the mountains.

The northern areas are, of course, very cold; but the midwinter low temperatures fall far short of the records set in the cold pole area of northeastern Siberia, where the vast extent of land becomes much colder than the partly ice-covered area of northern Canada.

South America

A large part of South America lies within the tropics and has a characteristically tropical climate. The remaining rather narrow southern portion is not subject to the extremes of heat and cold that are found where wide land areas give full sway to the continental type of climate with its hot summers and cold winters, as in North America and Asia. Temperature anomalies unusual for a given latitude are to be found mainly at the elevated levels of the Andean region stretching from the Isthmus of Panama to Cape Horn.

The Antarctic Current and its cool Humboldt branch skirting the western shores northward to the equator, together with the prevailing onshore winds, exert a strong cooling influence over the coastal regions of all the western countries of South America except Colombia. On the east the southerly moving Brazilian current from tropical waters has the opposite, or warming, effect except along southern Argentina.

In the northern countries of South America the sharply contrasted dry and wet seasons are related to the regime of the trade winds. In the dry season (corresponding to winter in the Northern Hemisphere) these winds sweep the entire region, while the wet season (corresponding to summer in the Northern Hemisphere) calms and variable winds prevail. In the basin of the Amazon River the rainfall is related to the equatorial belt of low pressure and to the trade winds, which give the maximum amounts of rainfall in the extreme west, where they ascend the Andean slopes.

The desert areas on the west coast of South America, extending from the equator to the latitude of Santiago, Chile, are due primarily to the cold Humboldt or Peruvian Current and upwelling coastal water. The moist, cool ocean air is warmed while passing over the land, with a consequent decrease in relative humidity, so that the dew point is not reached and condensation of vapor does not occur until the incoming air has reached high elevations in the Andes, where temperatures are much lower than along the coast.

In southern Chile the summer has moderate rainfall, and winters are excessively wet. The conditions that prevail farther north are not present here, and condensation of moisture from the ocean progresses from the shores up to the crests of the Andes. By the time the air passes these elevations, however, the moisture has been so depleted that the winds in the leeward slopes are dry, becoming more and more so as they are warmed on reaching lower levels. The mountains can be looked upon as casting a great "rain shadow"—an area of little rain—over southern Argentina.

Europe

In Europe there is no extensive north-south mountain system such as is found in both the Americas, and the general east-west direction of the ranges in the south allows the conditions in the maritime west to change rather gradually toward Asia. Generally rainfall is heaviest on the western coast, where locally it exceeds 60 in annually, and diminishes toward the east—except in the elevated Alpine and Caucasus regions—to less than 20 in in eastern Russia. There is a well-defined rain shadow in Scandinavia, with over 60 in of rain in western Norway and less than 20 in in eastern Sweden.

Over much of Europe, rainfall is both abundant and rather evenly distributed throughout the year. The chief feature of seasonal distribution of precipitation is the marked winter maximum and the extremely dry, even droughty, summers in most of the Mediterranean lands.

Isothermal lines have the general direction of the parallels of latitude except in winter, when the waters of the western ocean, warmed by the Gulf Stream, give them a north-south trend. Generally there are no marked dips in isotherms due to elevation and continental type of climate such as are found in North America. In Scandinavia, however, the winter map shows an abrupt fall in temperature from the western coast of Norway to the eastern coast of Sweden and then a continued fall eastward, under a type of exposure more and more continental in contrast to the oceanic exposure on the west.

Asia

The vast extent of Asia gives full opportunity for continental conditions to develop a cold area of high barometric pressure in winter and a low-pressure, hot area in summer, the former northeast of the Himalayas and the latter stretching widely from west to east in the latitude of northern India. These distributions of pressure give India the well-known monsoon seasons, during which the wind comes from one direction for several months, and also affect the yearly distribution of rainfall over eastern Asia.

In winter, the air circulates outward over the land from the cold pole, and precipitation is very light over the entire continent. In summer, on the contrary, there is an inflow of air from the oceans; even the southeast trade winds flow across the equator and merge into the southwest monsoon that crosses India. This usually produces abundant rain over most of that country, with excessively heavy amounts when the air is forced to rise, even to moderate elevations, in its passage over the land. At Cherrapunji (4,590 ft), on the southern side of the Khasi Hills in Assam, the average rainfall in a winter month is about 1 in, while in both June and July it is approximately 100 in. However, this heavy summer rainfall meets an impassable barrier in the Himalayas, while the much lighter summer monsoon rainfall over Japan and eastern Asia does not extend far into China because of lesser elevations. Consequently, while the southeast quadrant of Asia, including the East Indies, also with monsoon winds, has heavy to excessive annual rainfall, the remainder of the continent is dry, with vast areas receiving less than 10 in annually.

North of the Himalayas the low plains are excessively cold in winter and temperatures rise rather high in summer. At Verkhoyansk in the cold pole area, and north of the Arctic Circle, the mean temperature in January is about −59°F, and in July approximately 64°F; the extreme records are a maximum of 98°F, from readings at 1 P.M., and a minimum of −90°F.

In southwestern Asia the winter temperature control is still the interior high-pressure area, and temperatures are generally low, especially at high elevations; in summer at low elevations excessively high maxima are recorded, as, for example, in the Tigris-Euphrates Valley.

Africa

Africa, like South America, lies very largely within the tropics. There too, temperature distribution is determined mainly by altitude. Moreover, along the southern portion of the western coast the cool Benguela Current moves northward, and on the eastern coast are the warm, tropical currents of the Indian Ocean, which create conditions closely paralleling those found around the South American continent. In the strictly tropical areas of Africa, conditions are characterized by prevailing low barometric pressure, with conventional rainfall and alternate northward and southward movement of the heat equator; while in both the north and the south the ruling influences are the belts of high barometric pressure.

Except in the Atlas Mountains in the northwest where the considerable elevations set up a barrier in the path of trade winds and produce moderate rainfall, the desert conditions typified by the Sahara extend from the Atlantic to the Red Sea and from the Mediterranean southward well beyond the tropic of Cancer to about the latitudes of southern Arabia.

South of the Sahara, rainfall increases rapidly, becoming abundant to heavy from the west coast to the central lakes, with annual maxima of over 80 in in the regions bordering the eastern and western extremes of the Guinea coast. This marked increase in precipitation does not extend to the eastern portion of the middle region of the continent, where the annual amounts received are below 40 in and decrease to less than 10 in on the coasts of Somalia. Also to the south of the central rainy area there is a rapid fall in precipitation toward the regions of southwest Africa, where conditions are similar to those in Somalia.

The heavy rainfall over sections of Ethiopia from June to October, when more than 40 in fall and bring the overflowing of the Nile Valley, is one of Earth's outstanding features of seasonal distribution of rainfall.

Moist equatorial climate is typified by conditions in the Democratic Republic of the Congo; arid torrid climate by those of the Arabian Peninsula and the Sahara; and moderate plateau climate by those found in parts of Ethiopia, Kenya, and Tanzania.

Australia

In the southern winter the high-pressure belt crosses the interior of Australia, and all except the southernmost parts of the continent are dry. In summer, on the other hand, this pressure belt has moved south of the continent, still giving dry conditions over the southern and western areas. Thus, the total annual precipitation is less than 20 in except in the extreme southwest and in a strip circling from southeast to northwest. The average annual precipitation is even less than 10 in in a large south-central area.

In the south, the winter precipitation is the cyclonic type; the heavy summer rains of the north are of monsoon origin; and those of the eastern borders are in large part orographic, owing to the presence of the highlands in the immediate vicinity of the coasts. In the outer border of the rainfall strip along the coastal region, the mean annual rainfall is over 40 in and in many localities over 60 in. This is true for the monsoon rains in the north.

Because of the location of Australia, on both sides of the tropic of Capricorn, temperatures far below freezing are found only in a small part of the continent, in the south at high elevations. In the arid interior, extreme maximum temperatures are very high, ranking with those of the hottest regions on Earth.

The following tables provide data on mean maximum and minimum temperatures for January, April, July, and October, with extremes recorded in the period of record, and monthly and annual precipitation for about 170 selected stations well-distributed worldwide.

COUNTRY AND STATION	TEMPERATURE (F) AVERAGE DAILY								AVERAGE PRECIPITATION INCHES OR *WET DAYS # LESS THAN .5 DAYS													
	JANUARY		APRIL		JULY		OCTOBER		JANUARY	FEBRUARY	MARCH	APRIL	MAY	JUNE	JULY	AUGUST	SEPTEMBER	OCTOBER	NOVEMBER	DECEMBER	YEAR
	MAXIMUM	MINIMUM	MAXIMUM	MINIMUM	MAXIMUM	MINIMUM	MAXIMUM	MINIMUM													
Mexico: Mazatlan	80	57	84	61	90	77	89	73	*3	*1	*1	#	*1	*3	*13	*13	*11	*3	*2	*3	*54
Mexico City	70	45	78	53	74	56	73	52	0.3	0.2	0.5	0.8	1.9	4.2	5.1	4.8	4.3	1.7	0.6	0.3	25.0
Tampico	71	59	82	71	88	77	84	73	*7	*7	*4	*4	*4	*6	*8	*7	*10	*6	*6	*6	*75
British Honduras: Belize	80	70	85	76	86	79	84	75	4.5	2.6	1.9	1.7	4.0	8.6	8.1	7.0	9.5	10.0	7.0	6.7	71.2
Canal Zone: Panama City	89	76	89	78	87	77	85	76	*24	*13	*10	*18	*26	*22	*22	*23	*24	*26	*26	*28	*262
Costa Rica: Limon	83	72	85	74	84	75	85	74	12.5	8.5	8.0	10.8	11.3	11.3	16.2	11.7	5.6	7.2	14.7	16.1	133.8
El Salvador: San Salvador	83	65	87	70	83	69	82	68	0.4	0	0.4	1.2	5.3	12.4	11.1	13.2	12.9	8.5	1.4	0.2	67.0
Guatemala: Guatemala City	72	55	78	60	74	61	73	61	*4	*3	*4	*5	*14	*23	*20	*20	*23	*15	*7	*5	*143
Honduras: Tela	80	70	86	74	87	75	84	74	9.9	7.6	4.6	3.5	3.6	5.7	8.4	9.0	9.1	16.0	15.9	16.1	110.0
Argentina: Buenos Aires	85	64	72	53	58	41	71	51	4.2	4.0	4.0	3.3	3.1	2.1	2.3	2.3	2.4	4.0	3.6	3.3	38.5
Bolivia: La Paz	55	39	57	36	55	27	58	36	5.1	4.1	2.8	1.4	0.5	0.2	0.3	0.6	1.2	1.6	2.0	3.3	23.1
Sucre	67	55	67	53	66	46	69	54	*7	*7	*6	*2	#	#	#	*1	*2	*4	*5	*6	*40

COUNTRY AND STATION	TEMPERATURE (F) AVERAGE DAILY								AVERAGE PRECIPITATION INCHES OR *WET DAYS # LESS THAN .5 DAYS												
	JANUARY		APRIL		JULY		OCTOBER		JANUARY	FEBRUARY	MARCH	APRIL	MAY	JUNE	JULY	AUGUST	SEPTEMBER	OCTOBER	NOVEMBER	DECEMBER	YEAR
	MAXIMUM	MINIMUM	MAXIMUM	MINIMUM	MAXIMUM	MINIMUM	MAXIMUM	MINIMUM													
Brazil:																					
Belem	87	74	87	75	89	74	89	74	13.8	16.2	17.4	14.6	11.1	6.5	6.1	4.8	5.1	4.1	4.0	7.9	110.0
Brasilia	81	64	82	62	79	52	83	63	*21	*18	*18	*12	*6	*2	*2	*3	*8	*16	*21	*24	*151
Rio de Janeiro	91	74	87	71	81	64	84	68	5.3	4.9	5.3	4.3	3.1	2.0	1.8	1.8	2.4	3.2	3.9	5.4	43.3
Chile:																					
Santiago	85	54	72	45	57	37	71	45	0	0.1	0.2	0.5	2.3	3.1	3.0	2.1	1.1	0.5	0.2	0.2	13.4
Colombia:																					
Bogota	66	43	66	48	64	47	65	47	1.9	2.0	3.2	4.7	4.0	2.4	1.8	1.9	2.3	5.6	4.5	2.7	37.8
Ecuador:																					
Quito	66	50	66	51	67	49	67	49	4.5	5.1	6.0	6.9	4.9	1.9	0.8	1.0	3.1	5.0	4.3	4.1	47.5
French Guiana:																					
Cayenne	82	74	84	75	85	73	87	73	15.2	12.6	15.2	15.2	20.2	15.7	8.0	4.3	1.7	2.0	5.0	11.8	126.2
Paraguay:																					
Asuncion	91	73	82	66	73	56	83	65	5.9	5.2	5.6	5.7	4.7	2.9	2.0	1.9	3.3	5.4	5.7	5.6	54.3
Peru:																					
Arequipa	69	51	70	49	69	46	71	48	1.1	1.6	0.7	0.1	0	0	0	0	0	0	0	0.3	4.1
Lima	79	68	76	66	67	60	69	61	0	0	0	0	0	0.1	0.2	0.1	0.1	0.1	0	0	0.3
Uruguay:																					
Salto	89	70	75	58	63	47	76	57	4.8	4.4	5.4	5.2	3.7	3.4	2.9	2.6	4.1	4.8	4.7	4.1	50.3
Venezuela:																					
Caracas	77	63	82	68	79	68	80	68	*5	*4	*2	*6	*9	*14	*12	*14	*11	*11	*9	*8	*105

COUNTRY AND STATION	TEMPERATURE (F) AVERAGE DAILY								AVERAGE PRECIPITATION INCHES OR *WET DAYS # LESS THAN .5 DAYS												
	JANUARY		APRIL		JULY		OCTOBER		JAN	FEB	MAR	APR	MAY	JUN	JUL	AUG	SEP	OCT	NOV	DEC	YEAR
	MAX	MIN	MAX	MIN	MAX	MIN	MAX	MIN													
Albania: Durres	52	44	62	55	80	73	69	61	*10	*9	*9	*8	*4	*5	*1	*2	*3	*8	*10	*11	*80
Austria: Vienna	36	27	57	41	77	59	57	43	*21	*17	*19	*19	*18	*20	*18	*16	*15	*15	*20	*22	*220
Bulgaria: Sofia	36	25	58	41	78	58	61	44	1.3	1.4	1.5	2.1	2.7	3.1	2.2	1.7	1.6	1.4	2.0	1.7	22.8
Cyprus: Larnaca	60	45	71	53	89	71	81	61	*15	*13	*12	*9	*6	*2	*1	#	*1	*6	*10	*15	*90
Czech Republic: Prague	34	24	54	36	72	54	54	39	0.8	0.7	1.0	1.4	2.3	2.7	2.6	2.5	1.6	1.2	1.1	0.9	18.8
Denmark: Copenhagen	37	30	49	36	69	55	53	44	1.7	1.0	1.4	1.6	1.7	2.1	2.6	2.9	2.0	2.1	2.1	2.0	25.4
Finland: Helsinki	26	16	45	31	70	53	46	36	1.8	1.4	1.4	1.5	1.7	1.8	2.4	2.9	2.6	2.7	2.6	2.2	24.9
France: Marseille	51	37	63	47	84	66	68	52	1.9	1.6	1.8	1.8	1.8	1.0	0.6	1.0	2.5	3.7	3.0	2.3	23.0
Paris	43	34	57	42	75	58	59	46	*20	*16	*18	*17	*16	*14	*13	*12	*14	*17	*17	*19	*193
Germany: Berlin	35	26	54	37	73	56	56	42	*23	*18	*20	*16	*15	*19	*18	*17	*17	*17	*22	*23	*225
Munich	36	24	53	36	72	54	55	40	1.9	1.7	2.1	2.8	4.0	4.9	5.0	4.4	3.3	2.4	2.1	2.0	36.5
Gibraltar: Gibraltar	61	52	66	56	81	68	71	62	4.8	4.2	4.2	2.6	1.5	0.4	0	0.1	1.0	3.0	5.9	5.2	32.8

Country and Station	Temperature (F) Average Daily								Average Precipitation Inches or *Wet Days / # Less Than .5 Days												
	January Maximum	January Minimum	April Maximum	April Minimum	July Maximum	July Minimum	October Maximum	October Minimum	January	February	March	April	May	June	July	August	September	October	November	December	Year
Greece: Athens	55	44	66	52	89	73	73	60	1.9	1.6	1.6	0.9	0.7	0.3	0.2	0.3	0.4	2.1	2.2	2.4	14.8
Hungary: Budapest	36	25	60	41	79	59	59	43	*19	*17	*20	*22	*25	*24	*23	*20	*15	*17	*19	*21	*242
Iceland: Akureyri	33	24	40	30	57	47	42	33	2.0	1.6	1.7	1.2	0.7	1.0	1.3	1.4	1.7	2.3	2.0	2.1	18.8
Ireland: Dublin	46	37	52	41	66	54	55	46	2.5	2.0	2.0	1.9	2.2	2.2	2.6	3.0	2.5	2.9	2.7	2.7	29.2
Italy: Rome	55	39	63	47	83	66	71	56	3.2	2.8	2.7	2.6	2.0	1.3	0.6	1.0	2.7	4.5	4.4	3.8	31.6
Venice	43	31	61	46	81	65	64	50	2.2	2.1	2.4	2.9	2.7	3.1	2.7	3.1	2.6	3.0	3.5	2.4	32.4
Luxembourg: Luxembourg	36	29	53	38	71	55	54	43	2.6	2.1	2.2	2.1	2.6	2.6	2.7	2.7	2.5	2.7	2.8	2.9	30.4
Netherlands: Amsterdam	41	34	53	40	69	55	57	46	3.1	1.7	3.5	1.5	2.0	2.4	2.9	2.4	3.2	4.1	3.0	2.8	32.1
Norway: Oslo	26	14	45	29	69	50	46	34	2.3	1.9	1.9	1.9	2.3	2.9	3.3	3.8	3.7	3.8	3.5	2.6	34.1
Poland: Warsaw	33	24	54	37	73	55	54	40	1.1	1.0	1.2	1.5	2.0	2.6	3.0	2.8	1.8	1.6	1.5	1.4	21.6
Portugal: Lisbon	57	45	65	51	82	63	70	57	*16	*16	*15	*16	*13	*9	*5	*4	*8	*13	*15	*17	*147
Romania: Bucharest	36	23	62	42	82	60	63	43	1.7	1.5	1.4	1.8	2.6	3.4	2.2	2.2	1.4	1.1	1.8	1.6	22.9

COUNTRY AND STATION	TEMPERATURE (F) AVERAGE DAILY								AVERAGE PRECIPITATION INCHES OR *WET DAYS # LESS THAN .5 DAYS												
	JANUARY		APRIL		JULY		OCTOBER		JANUARY	FEBRUARY	MARCH	APRIL	MAY	JUNE	JULY	AUGUST	SEPTEMBER	OCTOBER	NOVEMBER	DECEMBER	YEAR
	MAXIMUM	MINIMUM	MAXIMUM	MINIMUM	MAXIMUM	MINIMUM	MAXIMUM	MINIMUM													
Spain: Madrid	51	32	63	42	90	61	68	47	1.8	1.7	1.5	1.8	1.6	1.0	0.4	0.4	1.2	1.8	2.5	1.9	17.8
Sweden: Stockholm	31	22	47	31	70	54	48	38	*25	*20	*21	*20	*17	*18	*20	*19	*20	*20	*23	*25	*248
Switzerland: Geneva	39	29	56	39	77	56	57	43	2.2	2.1	2.4	2.5	3.0	3.2	2.8	3.5	3.6	3.6	3.2	2.6	34.5
Turkey: Istanbul	46	37	60	47	82	66	67	55	3.7	2.8	2.3	1.7	1.2	0.9	0.7	0.6	1.1	2.1	3.5	4.0	25.2
United Kingdom: London	45	36	55	41	72	56	58	46	2.4	1.4	2.0	1.7	1.8	1.8	1.8	1.7	1.7	2.9	1.8	2.3	23.3
Edinburgh	43	33	51	38	66	51	54	42	2.2	1.6	1.9	1.5	2.0	2.0	2.5	2.7	2.5	2.4	2.5	2.4	26.1
U.S.S.R.: Moscow	21	11	49	34	71	55	45	33	1.4	1.1	1.3	1.5	2.0	2.6	3.2	2.8	2.3	2.0	1.7	1.7	23.6
Volgogran	23	13	57	40	82	64	52	37	1.3	1.1	1.1	0.9	1.3	1.3	1.4	1.3	1.0	1.1	1.4	1.6	14.6
Yugoslavia: Belgrade	38	27	61	43	81	60	63	45	*16	*14	*14	*16	*16	*16	*11	*10	*10	*12	*14	*16	*165
Algeria: Annaba	60	45	67	50	85	67	76	59	4.0	3.0	2.8	1.6	1.2	0.6	0.1	0.4	1.3	2.8	2.6	4.3	24.9
Botswana: Gaborone	87	72	79	61	69	45	84	65	3.8	3.3	2.8	1.6	0.5	0.2	0.1	0.2	0.6	1.7	2.6	3.5	20.7
C. African Rep.: Bangui	84	68	88	74	82	72	83	72	*2	*3	*6	*10	*8	*11	*12	*14	*12	*11	*6	*2	*97

COUNTRY AND STATION	TEMPERATURE (F) AVERAGE DAILY								AVERAGE PRECIPITATION INCHES OR *WET DAYS # LESS THAN .5 DAYS												
	JANUARY		APRIL		JULY		OCTOBER		JANUARY	FEBRUARY	MARCH	APRIL	MAY	JUNE	JULY	AUGUST	SEPTEMBER	OCTOBER	NOVEMBER	DECEMBER	YEAR
	MAXIMUM	MINIMUM	MAXIMUM	MINIMUM	MAXIMUM	MINIMUM	MAXIMUM	MINIMUM													
Chad: N'Djamena	86	63	103	81	90	76	96	74	0	0	0	0.3	1.2	2.5	5.9	8.5	3.6	0.9	0	0	23.0
Dem. Rep. Congo: Kinshasa	84	75	86	75	78	68	84	74	*10	*8	*11	*12	*10	*2	*1	*3	*5	*10	*12	*11	*95
Rep. of Congo: Brazzaville	85	73	87	74	79	67	85	73	5.4	5.0	7.4	8.2	4.6	0.2	0	0.1	1.5	5.7	9.2	6.6	54.4
Egypt: Cairo	65	49	82	59	93	72	85	65	0.2	0.2	0.1	0.1	0	0	0	0	0	0	0.1	0.2	1.0
Alexandria	64	50	75	58	84	73	81	66	*10	*9	*5	*2	*1	*1	#	#	*1	*2	*5	*8	*44
Ethiopia: Addis Ababa	69	53	71	57	65	55	69	54	0.7	1.5	2.7	3.4	3.4	5.2	10.5	11.1	7.3	1.1	0.4	0.4	47.7
Ghana: Accra	87	77	87	79	81	75	84	76	0.6	1.1	2.2	3.5	5.3	7.8	2.0	0.7	1.7	2.5	1.3	0.8	29.8
Guinea: Conakry	84	76	86	79	80	76	83	77	*1	*1	*1	*1	*6	*13	*22	*22	*16	*9	*3	*1	*96
Ivory Coast: Abidjan	86	76	87	78	80	74	83	76	*2	*4	*6	*9	*15	*19	*11	*11	*12	*13	*10	*6	*118
Kenya: Mombasa	87	76	87	77	80	71	84	73	1.1	0.6	2.4	7.2	10.7	4.0	3.2	2.6	2.6	3.6	3.8	2.5	44.4
Nairobi	77	58	76	61	71	54	78	58	1.8	1.7	2.9	6.3	4.7	1.2	0.5	0.5	1.0	1.7	4.7	3.0	29.8
Libya: Tripoli	63	44	78	54	95	69	84	63	2.7	1.6	1.0	0.5	0.2	0.1	0	0	0.4	1.5	2.4	3.2	13.7

COUNTRY AND STATION	TEMPERATURE (F) AVERAGE DAILY								AVERAGE PRECIPITATION — INCHES OR *WET DAYS # LESS THAN .5 DAYS													
	JANUARY		APRIL		JULY		OCTOBER		JANUARY	FEBRUARY	MARCH	APRIL	MAY	JUNE	JULY	AUGUST	SEPTEMBER	OCTOBER	NOVEMBER	DECEMBER	YEAR
	MAXIMUM	MINIMUM	MAXIMUM	MINIMUM	MAXIMUM	MINIMUM	MAXIMUM	MINIMUM													
Morocco: Casablanca	62	47	66	53	77	68	72	60	2.2	2.1	2.0	1.5	0.8	0.2	0	0	0.2	1.3	2.6	2.9	16.1
Rabat	62	46	67	52	80	65	74	58	3.2	2.7	2.6	2.2	1.0	0.3	0	0	0.3	1.8	3.3	4.0	21.5
Niger: Niamey	88	64	104	81	92	76	99	76	#	#	*1	*1	*4	*6	*9	*11	*8	*2	*1	#	*43
Nigeria: Lagos	82	79	84	81	79	76	80	77	*1	*1	*3	*3	*5	*8	*6	*7	*9	*6	*2	*2	*53
Senegal: Dakar	75	65	75	66	84	77	86	77	*2	*1	*1	#	*1	*2	*7	*13	*9	*3	*1	*1	*41
Sierra Leone: Freetown/Lungi	84	76	86	79	81	76	82	76	0.3	0.3	0.9	3.1	9.3	16.6	23.5	22.0	22.5	11.9	5.4	1.3	118.0
Somalia: Mogadishu	85	79	87	82	82	78	84	79	0	0	0.3	2.3	2.3	3.1	2.6	1.7	0.8	1.2	1.6	0.3	16.2
South Africa: Capetown	77	63	72	56	62	47	69	54	*6	*5	*6	*9	*12	*13	*14	*13	*11	*9	*7	*7	*112
Tanzania: Dar es Salaam	87	77	85	74	82	67	86	70	2.8	2.5	5.0	10.6	7.2	1.3	1.1	1.0	1.1	1.9	3.3	3.7	41.6
Tunisia: Tunis	60	46	68	51	90	69	78	61	2.4	2.1	1.8	1.5	0.9	0.4	0.1	0.3	1.3	2.2	2.1	2.5	17.5
Zambia: Lusaka	78	68	77	63	72	51	85	68	*13	*12	*9	*4	*1	#	#	#	#	*2	*6	*13	*60
Zimbabwe: Harare	78	63	77	59	69	49	80	59	*19	*16	*15	*9	*4	*2	*1	*1	*2	*7	*12	*19	*107

COUNTRY AND STATION	TEMPERATURE (F) — AVERAGE DAILY								AVERAGE PRECIPITATION — INCHES OR *WET DAYS # LESS THAN .5 DAYS												
	JANUARY MAXIMUM	JANUARY MINIMUM	APRIL MAXIMUM	APRIL MINIMUM	JULY MAXIMUM	JULY MINIMUM	OCTOBER MAXIMUM	OCTOBER MINIMUM	JANUARY	FEBRUARY	MARCH	APRIL	MAY	JUNE	JULY	AUGUST	SEPTEMBER	OCTOBER	NOVEMBER	DECEMBER	YEAR
China: Beijing	34	17	67	47	86	72	66	47	0.2	0.2	0.3	0.7	1.3	3.1	8.8	6.7	2.3	0.7	0.4	0.1	25.1
Guangzhou	64	51	77	67	90	79	82	70	1.6	2.7	3.8	6.7	10.4	10.4	9.7	9.1	6.2	2.5	1.6	1.2	66.6
Hong Kong	67	58	77	69	89	81	83	75	1.1	1.7	2.9	5.5	11.2	15.7	14.3	14.8	11.7	4.7	1.5	1.0	86.2
Shanghai	45	34	65	52	88	77	72	59	1.8	2.4	3.3	3.7	4.1	6.8	5.7	5.4	5.4	2.7	2.1	1.5	45.2
Japan: Sapporo	28	17	50	37	76	63	60	44	*30	*27	*28	*23	*25	*22	*20	*22	*24	*26	*28	*29	*304
Tokyo	48	35	64	50	82	71	69	58	2.0	2.8	4.2	5.1	5.7	6.9	5.3	5.8	8.5	7.6	3.8	2.1	60.2
Korea: Seoul	33	21	62	46	82	71	66	51	*10	*9	*9	*10	*10	*13	*18	*15	*10	*9	*11	*11	*135
Taiwan: Taipei	66	55	77	65	92	78	80	70	*18	*17	*20	*18	*19	*17	*13	*15	*15	*16	*19	*17	*204
U.S.S.R: Vladivostok	15	4	46	36	68	60	52	42	0.4	0.5	0.9	1.7	2.6	3.6	4.2	5.8	4.8	2.2	1.2	0.6	28.3
Brunei: Brunei	85	75	90	75	89	74	88	74	*19	*14	*14	*2	*21	*17	*20	*19	*22	*23	*25	*23	*235
Burma: Rangoon	89	65	99	76	85	76	88	77	*1	*1	*2	*11	*17	*26	*28	*28	*22	*13	*6	*1	*147
Indonesia: Jakarta	83	75	88	76	88	74	89	76	*19	*14	*14	*11	*8	*5	*5	*5	*5	*8	*11	*14	*119

COUNTRY AND STATION	TEMPERATURE (F) AVERAGE DAILY								AVERAGE PRECIPITATION — INCHES OR *WET DAYS # LESS THAN .5 DAYS												
	JANUARY		APRIL		JULY		OCTOBER		JANUARY	FEBRUARY	MARCH	APRIL	MAY	JUNE	JULY	AUGUST	SEPTEMBER	OCTOBER	NOVEMBER	DECEMBER	YEAR
	MAXIMUM	MINIMUM	MAXIMUM	MINIMUM	MAXIMUM	MINIMUM	MAXIMUM	MINIMUM													
Laos: Vientiane	83	64	93	77	88	78	87	75	*2	*1	*3	*5	*13	*17	*17	*19	*14	*7	*2	*1	*101
Philippine Is.: Manila	86	71	93	76	88	76	87	75	0.8	0.4	0.6	1.2	4.9	10.3	15.9	14.4	13.5	7.8	5.3	2.6	77.8
Singapore: Singapore	85	73	88	76	86	76	86	75	9.4	6.5	6.8	6.6	6.7	6.4	5.9	6.7	6.4	7.5	9.8	10.6	89.7
Thailand: Bangkok	89	71	94	80	90	78	89	77	0.4	1.1	1.2	2.8	7.5	6.0	6.2	7.4	12.6	9.1	2.3	0.4	57.1
Viet Nam: Hanoi	66	58	80	71	90	80	82	73	*15	*16	*20	*17	*16	*16	*14	*15	*12	*12	*10	*6	*169
Afghanistan: Kabul	36	23	65	48	88	67	69	45	1.3	2.1	2.8	2.6	0.8	0	0.2	0.1	0.1	0.2	0.4	0.8	10.7
Bangladesh: Dhaka	76	58	89	77	87	81	87	77	0.3	0.8	2.3	4.6	10.5	14.1	15.7	12.5	10.1	6.4	1.2	0.2	77.9
India: Bombay	85	65	90	77	85	79	90	76	0	0	0	0	0.5	22.3	25.6	19.2	14.0	3.5	0.2	0	83.3
Calcutta	77	57	95	72	89	80	88	76	0.3	0.8	1.2	1.6	4.6	11.9	13.1	10.4	11.7	4.4	1.2	0.4	60.2
New Delhi	68	48	95	72	93	81	90	68	0.9	0.8	0.6	0.4	0.6	2.7	7.9	7.9	4.8	0.7	0.1	0.4	27.7
Iran: Tehran	41	29	71	54	97	78	74	58	1.7	1.5	1.5	1.3	0.6	0.1	0.1	0.1	0.1	0.4	1.0	1.2	9.3
Iraq: Baghdad	58	38	84	59	110	78	91	60	1.1	1.1	1.1	0.7	0.3	0	0	0	0	0.1	0.8	1.0	6.1

COUNTRY AND STATION	TEMPERATURE (F) AVERAGE DAILY								AVERAGE PRECIPITATION INCHES OR *WET DAYS # LESS THAN .5 DAYS												
	JANUARY		APRIL		JULY		OCTOBER		JANUARY	FEBRUARY	MARCH	APRIL	MAY	JUNE	JULY	AUGUST	SEPTEMBER	OCTOBER	NOVEMBER	DECEMBER	YEAR
	MAXIMUM	MINIMUM	MAXIMUM	MINIMUM	MAXIMUM	MINIMUM	MAXIMUM	MINIMUM													
Israel: Jerusalem	51	40	68	50	81	64	74	58	5.5	4.4	4.6	0.7	0.2	0	0	0	0	0.4	2.7	5.1	23.0
Tel Aviv	62	46	77	54	87	69	83	63	*15	*13	*11	*6	*3	*1	#	#	*1	*5	*9	*14	*78
Jordan: Amman	52	39	71	51	88	67	79	58	2.5	2.5	1.7	0.7	0.1	0	0	0	0	0.2	1.1	1.9	10.8
Kuwait: Kuwait	63	46	89	67	112	87	93	70	*7	*6	*9	*6	*3	#	#	#	#	*3	*4	*8	*46
Lebanon: Beirut	61	50	71	58	84	73	80	68	7.4	6.0	3.8	2.0	0.7	0.1	0	0	0.2	1.9	4.7	6.9	34.1
Oman/Muscat: Muscat	76	65	92	79	100	88	92	79	*3	*5	*4	*2	*1	*1	*1	*1	*1	*1	*1	*3	*24
Pakistan: Multan	66	45	93	72	98	86	90	69	0.3	0.3	0.7	0.4	0.5	0.4	2.2	1.3	0.5	0	0.1	0.3	7.0
Saudi Arabia: Riyadh	67	48	90	68	109	84	94	69	0.5	0.4	1.2	1.2	0.5	0	0	0	0	0	0.2	0.4	4.3
Sri Lanka: Colombo	86	74	88	79	85	80	85	78	3.3	2.5	4.5	10.0	13.2	7.5	5.1	3.8	6.2	13.9	12.1	6.0	87.8
Syria: Damascus	53	33	74	46	96	62	81	49	1.5	1.3	0.9	0.5	0.2	0	0	0	0	0.4	1.0	1.7	7.6
Turkey: Ankara	35	20	60	38	82	55	65	39	*16	*15	*15	*17	*17	*13	*7	*5	*5	*10	*12	*16	*148

COUNTRY AND STATION	TEMPERATURE (F) AVERAGE DAILY								AVERAGE PRECIPITATION INCHES OR *WET DAYS # LESS THAN .5 DAYS												
	JANUARY		APRIL		JULY		OCTOBER		JANUARY	FEBRUARY	MARCH	APRIL	MAY	JUNE	JULY	AUGUST	SEPTEMBER	OCTOBER	NOVEMBER	DECEMBER	YEAR
	MAXIMUM	MINIMUM	MAXIMUM	MINIMUM	MAXIMUM	MINIMUM	MAXIMUM	MINIMUM													
Australia: Adelaide	82	60	72	53	59	44	70	51	0.7	0.7	0.8	1.5	2.3	2.1	2.5	2.0	1.7	1.4	1.0	0.9	17.7
Alice Springs	97	70	82	55	67	39	87	59	1.4	1.6	1.5	0.5	0.7	0.6	0.6	0.4	0.3	0.8	1.0	1.4	10.9
Canberra	82	55	67	44	52	32	67	43	2.3	2.2	2.2	2.1	1.9	1.5	1.6	1.9	2.0	2.6	2.4	2.1	24.8
Melbourne	79	56	68	51	55	41	66	47	1.8	1.6	1.4	1.9	1.9	1.6	1.5	2.0	1.8	2.3	2.4	1.9	22.3
Perth	89	62	77	55	64	46	72	50	0.3	0.5	0.6	1.8	4.2	6.9	6.4	4.6	2.7	1.9	1.0	0.5	31.5
Sydney	79	65	73	57	62	44	72	55	4.0	4.5	5.2	4.2	3.9	5.2	2.5	3.2	2.2	3.1	3.4	3.1	44.5
Tasmania: Hobart	72	53	64	47	54	39	63	45	1.5	1.5	1.5	1.9	1.8	1.4	1.8	1.8	1.6	2.2	1.9	2.1	21.1
New Zealand: Auckland	73	62	66	56	56	47	62	52	2.8	3.4	3.1	3.8	4.5	5.0	5.2	4.4	3.7	3.7	3.2	3.1	45.6
Wellington	67	58	61	53	51	45	57	49	3.1	3.1	3.4	3.9	4.8	4.9	5.5	4.8	3.9	4.1	3.5	3.5	48.4
Pacific Islands: Honolulu, Hawaii	80	66	82	69	87	73	86	72	3.4	2.6	2.8	1.3	1.0	0.4	0.6	0.6	0.7	2.0	2.6	3.5	21.6
Pago Pago, Samoa	87	76	86	76	83	75	85	76	13.6	12.8	11.4	12.9	10.0	6.4	6.4	7.1	7.0	10.3	11.4	14.4	123.8
Tahiti, Society Is.:	86	75	86	75	82	71	83	73	11.9	9.2	6.9	5.2	4.1	2.7	2.3	1.9	2.6	3.5	6.1	10.6	67.2

COUNTRY AND STATION	TEMPERATURE (F) AVERAGE DAILY								AVERAGE PRECIPITATION INCHES OR *WET DAYS # LESS THAN .5 DAYS												
	JANUARY		APRIL		JULY		OCTOBER		JANUARY	FEBRUARY	MARCH	APRIL	MAY	JUNE	JULY	AUGUST	SEPTEMBER	OCTOBER	NOVEMBER	DECEMBER	YEAR
	MAXIMUM	MINIMUM	MAXIMUM	MINIMUM	MAXIMUM	MINIMUM	MAXIMUM	MINIMUM													
Bahamas:																					
Freeport	74	64	80	69	88	79	84	73	2.3	2.3	2.6	2.1	5.6	6.5	5.6	8.1	9.9	6.7	3.4	3.0	58.0
Nassau	77	64	81	69	89	77	85	74	1.9	1.7	1.6	2.6	5.2	7.0	6.0	6.7	7.1	6.7	2.8	1.7	50.6
Dominican Rep.:																					
Puerto Plata	81	70	82	72	87	77	86	76	7.6	6.1	4.8	6.2	5.0	2.4	2.8	3.2	3.6	5.4	11.5	11.0	70.6
Santo Domingo	82	70	84	73	86	76	86	75	2.2	1.7	1.9	3.0	7.0	6.1	6.1	6.4	6.8	6.5	4.4	2.5	54.5
Cayman Is.:																					
Grand Cayman	81	73	84	76	88	80	86	78	*4	*3	*3	*2	*5	*5	*6	*7	*7	*7	*6	*4	*59
Haiti:																					
Port-au-Prince	87	75	88	78	92	81	89	79	1.3	2.0	3.1	6.1	8.6	3.8	2.9	5.5	6.5	6.5	3.3	1.4	52.1
Jamaica:																					
Kingston	86	74	87	76	90	79	89	78	1.2	0.9	0.9	1.5	4.1	3.8	1.8	4.2	5.0	6.8	3.8	1.6	35.4
Montego Bay	82	72	85	74	88	77	87	76	2.7	1.8	2.3	2.5	5.9	5.5	2.9	5.2	6.1	7.4	5.5	3.6	50.7
Puerto Rico:																					
Ponce	83	71	85	74	87	78	86	76	*2	*2	*3	*2	*4	*3	*3	*4	*5	*6	*5	*2	*41
San Juan	83	70	86	73	88	76	88	75	3.0	2.2	2.3	3.7	6.1	4.4	4.5	5.3	5.3	5.5	5.8	4.7	52.9
Virgin Is.:																					
St. Thomas	84	73	86	76	89	80	88	78	*4	*4	*3	*3	*4	*4	*4	*5	*5	*5	*6	*5	*51
Christiansted	83	73	84	76	88	80	87	78	*4	*4	*4	*4	*5	*4	*5	*5	*6	*6	*7	*6	*60

Local Climatological Data Reports

• INTRODUCTION

Local Climatological Data (LCD) are data observed at principal meteorological stations by trained observers or automated equipment that has been tested and accepted by the controlling agency. The stations are located worldwide and are operated by agencies of the United States government. The controlling agencies are the National Weather Service (NWS), the U.S. Air Force/Air Weather Service (AWS), the U.S. Navy/National Oceanographic Command (NAVOC-FANCOM), and the Federal Aviation Administration (FAA).

The data are collected on a wide range of time scales from one minute for some current automated equipment to three observations per day for intermittent periods at some remote, part-time locations. The majority of stations collect hourly observations and special (between hour) observations when weather conditions warrant it. The data are used initially for aviation guidance and safety, weather forecasting, and severe weather warnings. The hourly data, special observations and summaries are generally transmitted over global telecommunication circuits.

In 1992, NWS began automating surface weather observations through the Automated Surface Observing System (ASOS). The measurement of all weather elements included in conventional observations have not been automated. Some of those elements that have not been automated include clouds above 12,000 ft, snowfall and snow depth, sunshine, and the identification of certain types of weather such as tornadoes, thunderstorms, hail, drizzle, smoke, and blowing phenomena. Until new automated sensors are designed to measure these elements, the data will be obtained from other sources. ASOS is considered to be only one source of the total surface observation. The total surface observation concept includes the ASOS system, augmentation of ASOS data by station personnel, supplementary data from networks distinct from ASOS (e.g., severe weather spotter networks, hydrological networks, and cooperative networks), and complementary data from other observing technologies such as satellite, radar, and lightning detection systems.

• HOW TO READ THESE REPORTS

Weather statistics on 108 U.S. cities

The information about the history of key cities is easy to find. The reports are planned to be informative, yet easy to read. The terminology used and the standard formats are explained below.

Narrative report

Typically, each report begins with a narrative description of the local area climate prepared by a local climatologist.

The narrative describes the area in terms of terrain, water bodies, and other topographical features as these features exercise key influences on the local weather. They are usually the cause if an area's weather differs sharply from weather in areas only a few miles away. For example, if a lake is near a city, it will always influence the city's weather, and, in fact, may even create climatic differences from one part of the city to another. Mountains, swamps, even plowed fields exercise their influences on air masses as these masses move toward a city.

The report generally discusses temperatures in the city, rainfall tendencies, snowfall history, and other points. It closes with notes about the area's agricultural adaptability. The history of first fall frost and last spring freeze is usually described, along with suggestions about the types of crops for which the area is climatologically suited.

Statistics

The statistics are of two different kinds. The first type distills many years of history to give a profile of the city's weather (e.g., *Normals, Means, and Extremes*). The second group offers data for individual years, allowing the user to see what variances and/or patterns have occurred.

There are several meteorological terms used in the table. The following notes are designed to clarify the information included in the tables.

Location

Precise geographical location is provided at the top of each *Normals, Means, and Extremes* table.

Normal

Term applied as to temperature, degree days, and precipitation refers to the value of that particular element averaged over the period of 1961–90. When the station does not have continuous records from an instrument site with the same "exposure," a difference factor between the old site and the new site is used to adjust the observed values to a common series. The difference factor is determined from a period of simultaneous measurements. The base period is revised every ten years by adding the averages for the most recent decade and dropping them for the first decade of the former normals for 1951–80. *Normal* does not refer to "normalcy" or "expectation," but only to the actual averages for a particular thirty-year period.

Means and extremes (note "a")

Means and extremes are based on the period of years in which observations have been made under comparable conditions of instrument exposure. Data are included for dates through 1995 unless otherwise noted. The Date of an Extreme is the most recent one in cases of repeated occurrences.

Length of observational record

Length of observational record for *Means and Extremes* is based on the length of January data for the present instrument site exposure (15 equals 15 years). The table does not give the all time high or low value if it was recorded at a different site within the area. The *Mean* (or average) values for relative humidity, wind sunshine, sky condition, and the mean number of days with the various other weather conditions listed are also based on length of record noted in each instance. Check the first column for each of these rows to read this length of record for each item.

Average of the highest temperature (°F)

Average of the highest temperature on each day of the month and year for the period 1961–90. This value is obtained by taking the sum of the highest temperature for each day of the period adjusted for site exposure if necessary) and dividing by the number of days included.

Average of the lowest temperature (°F)

Average of the lowest temperature on each day of the month and year for the period 1961–90.

Average of all daily temperatures (°F)

Average of all daily temperatures for the month and year for the period 1961–90; computed as being the sum of the minimum and maximum values divided by two. The monthly mean maximum or minimum temperatures are the sum of the daily maximums or minimums divided by the number of days in the month. The monthly average temperature is the sum of the monthly mean maximum and minimum temperatures divided by two.

Extremes—highest temperature (°F)

Highest temperature ever recorded during any month at present site exposure.

Extremes—lowest temperature (°F)

Lowest temperature ever recorded during any month at present site exposure.

Average number of heating degree days

The number of heating degree days (average) for each month and year for the period 1961–90. The statistic is based on the amount that the daily mean temperature falls below 65°F. Each degree of mean temperature below 65 is counted as one heating degree day. If the daily mean temperature is 65°F or higher, the heating degree day value for that day is zero. Monthly and annual sums are calculated for each period and averaged over the appropriate thirty years of record to establish these "normal" values. Compare this with cooling degree days.

Average number of cooling degree days

The number of cooling degree days (average) for each month and year for the period 1961–90. The concept of this statistic is the mirror image of the concept of heating degree days and is based on the amount that the daily mean temperature exceeds 65°F. Each degree of mean temperature above 65°F is counted as one cooling degree day. If the daily mean temperature is 65°F or below, the cooling degree day is zero. It should be noted that heating and cooling degree days are calculated independently and do not cancel each other out.

Sunshine

The average percent of daytime hours subject to direct radiation from the sun at the present site. The percentage is given without regard for the intensity of sunshine. That is, thin clouds, light haze, or other minor obstructions to direct solar rays may be present but would not mitigate the full counting of an hour.

Mean sky cover

Average amount of daytime sky obscured by any type of cover expressed in tenths (e.g., 4.8 or, 48%).

Activity limiting weather

Average number of days in month with specified weather conditions (precipitation, storms, etc.) based on present exposure. The symbol * indicates less than half a day.

Cloudiness

Average number of days in month at the present site with various amounts of cloud cover. Clear indicates average daytime cloudiness of 0.3 or less; partly cloudy indicates average daytime cloudiness between 0.4 and 0.7; cloudy indicates average daytime cloudiness of 0.8 or more.

Maximum ≥ 90° days

Average number of days in month and year when the temperatures at the present site is 90°F or above. (70°F or above at Alaskan stations.)

Maximum ≤ 32° days

Average number of days at the present site when the temperatures remained below 32°F at all times.

Minimum ≤ 32° days

Average number of days at the present site when the temperature dropped to a minimum of 32°F or below.

Minimum ≤ 0° days

Average number of days at present site when the minimum temperature was 0°F or below.

Average station pressure

Given in inches (in).

Average relative humidity (percent)

Relative humidity levels at various hours of the day. The time is expressed in terms of the 24-hour clock (00 is midnight, 06 is 6 A.M., 12 is noon, and 18 is 6 P.M.). Values are for present site only.

Average precipitation

Precipitation in inches of water equivalent for each month and year during the period 1961–90. As in the other precipitation data, the values are expressed in inches of depth of the liquid water content of all forms of precipitation even if initially frozen. As in all "normal" values, when the station does not have continuous records from the same instrument site, a "ratio factor" between the new and old exposure is used to adjust the observed values to a common series.

Maximum precipitation (month)

Greatest amount (in inches) of water equivalent ever recorded during *any month* at present site.

Minimum precipitation (month)

Least amount (in inches) of water equivalent ever recorded during any month at present site.

Maximum precipitation (in 24 hours)

Greatest amount (in inches) of water equivalent ever recorded during *any day* at present site.

Summaries for snow, ice pellets, hail

Figures include sleet and are similar to those for total precipitation. The values are expressed in inches of actual snow or ice fall. The water equivalent can be estimated roughly by using the rule-of-thumb that 10 in of snow equal 1 in of water.

Wind

Average speed of wind is expressed in miles-per-hour (mph) without regard to direction.

Prevailing wind direction

The most common single wind direction without regard for wind speed or any minimum amount of persistence. The aggregate total of wind from the other directions may be very much greater than that from the "prevailing" direction. Direction is coded in two different ways, some reports use letters, some use numbers. When letters are used, they have the usual meaning, such as WSW, indicating west-south-west. When numbers are used, they are given in tens of degrees clockwise from true north, so that 09 is 90° clockwise from north (east), 18 is 180° (south), 27 is 270° (west), and 36 is 360° (north).

High wind

The greatest speed in miles-per-hour of any "mile" of wind passing the station. The accompanying direction and year of occurrence are also given. A mile of wind passing the station one minute has an average speed of 60 mph; two minutes—30 mph; five minutes—12 mph, etc. The wind cups of the particular instrument involved operate much like the wheel of a car in actuating the car's odometer. The instrument does not record the strength of individual wind gusts, which usually last less than 20 seconds and may be very much greater than the value given here. The fastest mile however does give some idea of the extremes of wind that can be encountered.

Special symbols

Symbols that appear on many of the individual summaries: (*) less than one half; (**T**) trace, an amount too small to measure; (−) below zero temperatures are preceded by a minus sign.

Weather averages year-by-year

PRECIPITATION refers to the inches of water equivalent in the total of all forms of liquid or frozen precipitation that fell during each month. Snowfall refers to the actual amount of snow in inches that fell during the month. T (trace) is a precipitation amount of less than 0.005 in (note: in estimating the water equivalent of snow a ratio of 10 in of snow equal 1 in of water is customarily employed).

AVERAGE TEMPERATURE equals the average of the maximum and minimum temperatures for each day of the month for the given year; afternoon temperatures were typically higher than these values and late night/early morning temperatures were typically below them.

HEATING DEGREE DAYS (HDD) provide a well-established index of relative fuel consumption for space heating in a given place—a month of 2,000 HDD requires about twice the amount of space heating energy as one of 1,000 HDD, while 100 HDD will require about the same fuel whether accumulated in two or four days. Regional differences in the Heating Degree Day Index are only partially useful in estimating comparative fuel requirements because the building construction and cultural expectations tend to be different in different parts of the country (e.g., subjective ideas of comfort in relation to temperature will vary). For example, the average standards of efficiency in heating equipment and insulation are generally lower in warmer climates so that fuel requirements tend not to decrease as rapidly as the heating degree days decrease.

COOLING DEGREE DAYS provide only a rough guide to relative energy consumption in air conditioning. A proper air conditioning index will almost certainly require a factor for humidity variation and possibly factors for cloudiness or other weather variables. However, the Cooling Degree Days Index will have some usefulness in indicating relative outdoor comfort and relative indoor air conditioning requirements.

2001
BIRMINGHAM (MUNICIPAL AIRPORT), ALABAMA (BHM)

Birmingham is located in a hilly area of north-central Alabama in the foothills of the Appalachians about 300 miles inland from the Gulf of Mexico. There is a series of southwest to northeast valleys and ridges in the area.

The city is far enough inland to be protected from destructive tropical hurricanes, yet close enough that the Gulf has a pronounced modifying effect on the climate.

Although summers are long and hot, they are not generally excessively hot. On a typical mid-summer day, the temperature will be nearly 70 degrees at daybreak, approach 90 degrees at mid-day, and level off in the low 90s during the afternoon. It is not unusual for the temperature to remain below 100 degrees for several years in a row. However, every few years an extended heat wave will bring temperatures over 100 degrees. July is normally the hottest month but there is little difference from mid-June to mid-August. Rather persistent high humidity adds to the summer discomfort.

January is normally the coldest month but there is not much difference from mid-December to mid-February. Overall, winters are relatively mild. Even in cold spells, it is unusual for the temperature to remain below freezing all day. Sub-zero cold is extremely rare, occurring only a very few times this century. Extremely low temperatures almost always occur under clear skies after a snowfall.

Snowfall is erratic. Sometimes there is a two- or three-year span with no measurable snow. On rare occasions, there may be a 2 to 4 inch snowstorm. The snow usually melts quickly. Even 1 or 2 inches of snow can effectively shut down this sunbelt city because of the hilly terrain, the wetness of the snow and the unfamiliarity of motorists driving on snow and ice.

Birmingham is blessed with abundant rainfall. It is fairly well distributed throughout the year. However, some of the wetter winter months, plus March and July, have twice the rainfall of October, the driest month. Summer rainfall is almost entirely from scattered afternoon and early evening thunderstorms. Serious droughts are rare and most dry spells are not severe.

The stormiest time of the year with the greatest risk of severe thunderstorms and tornadoes is in spring, especially in March and April.

In a normal year, the last 32 degree minimum temperature in the spring is in mid to late March and the first in autumn is in early November.

NORMALS, MEANS, AND EXTREMES
BIRMINGHAM, AL (BHM)

LATITUDE:	LONGITUDE:	ELEVATION (FT):	TIME ZONE:	WBAN: 13876
33 33' 50" N	86 45' 16" W	GRND: 636 BARO: 639	CENTRAL (UTC + 6)	

	ELEMENT	POR	JAN	FEB	MAR	APR	MAY	JUN	JUL	AUG	SEP	OCT	NOV	DEC	YEAR
TEMPERATURE °F	NORMAL DAILY MAXIMUM	30	51.7	56.9	66.1	74.6	81.0	87.4	89.9	89.1	83.9	74.7	64.6	55.7	73.0
	MEAN DAILY MAXIMUM	54	53.3	58.3	65.6	74.6	81.8	87.8	90.6	90.0	84.6	75.1	64.2	55.9	73.5
	HIGHEST DAILY MAXIMUM	58	81	83	89	92	99	102	106	103	100	94	85	80	106
	YEAR OF OCCURRENCE		1949	1996	1982	1987	1962	1954	1980	1999	1990	1954	2000	1951	JUL 1980
	MEAN OF EXTREME MAXS.	54	70.7	75.2	81.5	86.2	90.5	95.3	97.5	96.1	94.1	86.3	78.6	72.1	85.3
	NORMAL DAILY MINIMUM	30	31.3	34.5	42.3	49.2	57.7	65.2	69.5	68.8	62.9	50.2	41.6	34.8	50.7
	MEAN DAILY MINIMUM	54	33.2	36.3	42.2	49.9	58.4	66.2	70.1	69.2	63.2	50.9	41.1	35.5	51.4
	LOWEST DAILY MINIMUM	58	-6	3	2	26	35	42	51	51	37	27	5	1	-6
	YEAR OF OCCURRENCE		1985	1958	1993	1973	1944	1966	1967	1946	1967	1956	1950	1989	JAN 1985
	MEAN OF EXTREME MINS.	54	14.2	18.0	24.7	32.8	43.4	54.1	61.7	60.4	48.2	34.1	24.3	17.4	36.1
	NORMAL DRY BULB	30	41.5	45.7	54.2	62.0	69.4	76.3	79.8	79.0	73.4	62.5	53.1	45.2	61.8
	MEAN DRY BULB	54	43.2	47.3	54.0	62.2	70.1	77.0	80.2	79.6	73.9	63.1	52.8	45.7	62.4
	MEAN WET BULB	18	39.3	43.2	48.1	54.8	63.3	69.8	68.9	67.7	66.5	56.5	45.8	39.1	55.2
	MEAN DEW POINT	18	33.3	36.9	40.9	48.9	59.0	66.5	70.0	68.6	62.5	51.9	40.7	33.6	51.1
	NORMAL NO. DAYS WITH:														
	MAXIMUM 90	30	0.0	0.0	0.0	0.1	2.5	11.1	16.6	15.5	6.7	0.1	0.0	0.0	52.6
	MAXIMUM 32	30	1.8	0.4	*	0.0	0.0	0.0	0.0	0.0	0.0	0.0	*	0.6	2.8
	MINIMUM 32	30	17.7	13.1	5.6	1.0	0.0	0.0	0.0	0.0	0.0	0.5	6.3	14.7	58.9
	MINIMUM 0	30	0.2	0.0	0.0	0.0	0.0	0.0	0.0	0.0	0.0	0.0	0.0	0.0	0.2
H/C	NORMAL HEATING DEG. DAYS	30	729	540	350	135	31	0	0	0	7	148	364	614	2918
	NORMAL COOLING DEG. DAYS	30	0	0	15	45	168	339	459	434	259	71	7	0	1797
RH	NORMAL (PERCENT)	30	70	66	64	65	70	72	74	75	74	72	71	71	70
	HOUR 00 LST	30	76	73	72	77	83	84	86	86	84	83	80	77	80
	HOUR 06 LST	30	80	79	79	83	86	86	88	89	88	86	84	81	84
	HOUR 12 LST	30	61	56	53	50	54	56	60	59	58	53	57	60	56
	HOUR 18 LST	30	65	58	53	51	57	60	65	65	68	69	68	68	62
S	PERCENT POSSIBLE SUNSHINE	34	42	50	55	63	66	65	59	63	61	66	55	46	58
W/O	MEAN NO. DAYS WITH:														
	HEAVY FOG(VISBY 1/4 MI)	59	1.2	0.6	0.6	0.3	0.3	0.6	0.4	0.5	0.5	0.7	1.0	1.2	7.9
	THUNDERSTORMS	59	2.0	2.4	4.5	5.1	7.1	8.8	11.2	8.6	4.0	1.3	1.9	1.2	58.1
CLOUDINESS	MEAN:														
	SUNRISE-SUNSET (OKTAS)														
	MIDNIGHT-MIDNIGHT (OKTAS)														
	MEAN NO. DAYS WITH:														
	CLEAR														
	PARTLY CLOUDY														
	CLOUDY														
PR	MEAN STATION PRESSURE(IN)	27	29.50	29.40	29.40	29.40	29.30	29.30	29.40	29.40	29.40	29.41	29.50	29.50	29.41
	MEAN SEA-LEVEL PRES. (IN)	18	30.15	30.12	30.06	30.02	30.00	30.00	30.04	30.03	30.04	30.10	30.13	30.17	30.07
WINDS	MEAN SPEED (MPH)	40	8.1	8.6	9.0	8.5	7.0	6.1	5.9	5.5	6.5	6.2	7.0	7.6	7.2
	PREVAIL.DIR(TENS OF DEGS)	24	36	36	36	18	15	04	04	04	09	36	36	36	36
	MAXIMUM 2-MINUTE:														
	SPEED (MPH)	3	32	51	32	38	30	36	48	38	25	29	32	32	51
	DIR. (TENS OF DEGS)		27	24	29	30	31	01	36	03	02	01	30	27	24
	YEAR OF OCCURRENCE		2000	2001	2000	2000	1999	1999	2000	2000	1999	2000	2001	2000	FEB 2001
	MAXIMUM 5-SECOND:														
	SPEED (MPH)	3	39	68	40	53	38	46	58	45	31	37	49	40	68
	DIR. (TENS OF DEGS)		16	24	13	29	34	01	36	04	34	32	28	21	24
	YEAR OF OCCURRENCE		1999	2001	1999	2000	2000	1999	2000	2000	1999	2001	2001	2000	FEB 2001
PRECIPITATION	NORMAL (IN)	30	5.10	4.72	6.19	4.96	4.85	3.73	5.25	3.59	3.93	2.81	4.33	5.12	54.58
	MAXIMUM MONTHLY (IN)	58	11.00	17.67	15.80	13.75	11.10	9.04	13.70	10.85	10.43	11.90	15.25	13.98	17.67
	YEAR OF OCCURRENCE		1949	1961	1980	1979	1969	1999	1950	1967	1977	1995	1948	1961	FEB 1961
	MINIMUM MONTHLY (IN)	58	1.09	1.20	1.71	0.42	0.88	0.67	0.30	0.38	T	0.07	0.42	0.81	T
	YEAR OF OCCURRENCE		1981	1968	1985	1986	2000	1968	1983	1989	1955	1991	1949	1980	SEP 1955
	MAXIMUM IN 24 HOURS (IN)	52	5.81	6.57	7.05	5.08	4.63	3.85	5.47	5.13	5.03	6.94	4.87	5.29	7.05
	YEAR OF OCCURRENCE		1949	1961	1970	1966	1969	1957	1985	1952	1977	1995	1948	1961	MAR 1970
	NORMAL NO. DAYS WITH:														
	PRECIPITATION 0.01	30	10.9	9.6	10.8	9.2	9.9	9.4	12.7	9.4	8.2	6.2	9.1	10.4	115.8
	PRECIPITATION 1.00	30	1.6	1.4	2.1	1.7	1.5	1.1	1.3	1.0	1.4	0.8	1.4	1.5	16.8
SNOWFALL	NORMAL (IN)	30	0.7	0.1	0.1	0.2	0.0	0.0	0.0	0.0	0.0	0.0	T	0.3	1.4
	MAXIMUM MONTHLY (IN)	58	6.6	2.3	13.0	5.0	T	T	0.0		T	T	1.4	8.0	13.0
	YEAR OF OCCURRENCE		1982	1960	1993	1987	1996	1992	1994		1992	1993	1950	1963	MAR 1993
	MAXIMUM IN 24 HOURS (IN)	54	4.5	2.3	13.0	5.0	T	T	0.0		T	T	1.4	8.4	13.0
	YEAR OF OCCURRENCE		1948	1960	1993	1987	1996	1992	1994		1992	1993	1950	1963	MAR 1993
	MAXIMUM SNOW DEPTH (IN)	54	8	2	13	5	0	0	0		0	0	1	11	13
	YEAR OF OCCURRENCE		1964	1960	1993	1987							1950	1958	MAR 1993
	NORMAL NO. DAYS WITH:														
	SNOWFALL 1.0	30	0.3	0.0	0.1	0.*	0.0	0.0	0.0	0.0	0.0	0.0	0.0	0.*	0.4

PRECIPITATION (inches) 2001 BIRMINGHAM (MUNICIPAL AIRPORT), AL (BHM)

YEAR	JAN	FEB	MAR	APR	MAY	JUN	JUL	AUG	SEP	OCT	NOV	DEC	ANNUAL
1972	9.30	2.15	4.79	2.56	3.82	2.70	3.55	2.01	8.09	3.35	4.47	5.76	52.55
1973	6.85	2.33	9.71	5.33	8.29	3.74	8.36	5.41	2.64	0.96	4.91	7.58	66.11
1974	6.85	4.94	2.43	5.43	5.43	1.42	4.69	8.28	4.94	1.49	4.13	5.97	56.00
1975	7.23	4.96	7.57	3.19	4.15	2.44	7.33	3.33	3.69	3.74	4.15	5.97	55.27
1976	4.12	1.80	14.15	1.99	9.00	2.75	4.92	3.34	4.91	1.59	2.23	4.35	55.15
1977	5.08	3.89	8.70	6.73	3.51	0.96	6.24	0.87	10.43	7.52	4.10	2.01	60.04
1978	4.54	1.31	3.07	2.64	8.51	5.04	5.09	2.09	1.14	0.22	2.71	5.43	41.79
1979	5.94	4.70	5.69	13.75	6.64	1.19	9.98	2.30	10.40	2.05	5.51	1.55	69.70
1980	6.63	2.36	15.80	9.10	7.30	3.01	2.11	2.84	5.26	3.21	3.04	0.81	61.47
1981	1.09	4.87	7.23	2.45	2.81	2.49	3.88	5.30	0.93	3.34	1.67	5.30	41.88
1982	5.19	6.29	2.71	7.86	3.19	5.39	3.53	2.68	0.66	3.73	7.11	9.51	57.85
1983	3.26	6.42	5.06	8.28	9.57	3.81	0.30	0.99	2.83	3.67	9.14	12.63	65.96
1984	3.96	2.79	4.07	8.61	6.07	1.41	5.06	3.86	0.16	3.91	5.42	2.30	47.62
1985	5.22	5.79	1.71	2.86	4.36	5.34	10.07	4.07	1.97	4.12	2.62	2.54	50.67
1986	1.21	1.79	2.45	0.42	3.66	3.87	1.61	5.56	2.52	5.24	9.66	3.08	41.07
1987	5.89	5.82	4.77	1.03	6.03	4.59	2.30	3.96	3.52	1.16	3.17	3.08	45.32
1988	5.55	2.52	3.18	3.18	1.22	0.79	2.95	3.43	8.57	3.41	6.33	2.84	43.97
1989	4.76	4.31	5.70	3.40	3.82	8.00	6.42	0.38	7.38	1.52	4.63	3.39	53.71
1990	7.38	7.43	5.81	2.38	4.12	2.08	3.16	0.59	2.04	2.98	4.02	5.47	47.46
1991	3.19	4.27	5.42	4.87	8.90	7.52	4.00	4.59	3.02	0.07	4.00	3.64	53.49
1992	3.22	3.96	3.36	2.61	1.18	5.34	7.41	7.43	5.70	2.18	7.94	5.27	55.60
1993	6.11	2.35	4.40	2.99	3.93	1.47	1.16	2.72	4.62	3.22	2.22	4.01	39.20
1994	5.10	4.78	7.56	3.77	3.74	5.41	7.75	4.05	4.67	5.39	3.93	4.10	60.25
1995	3.85	4.37	3.63	4.42	3.15	3.07	1.81	1.51	5.53	11.90	6.97	4.91	55.12
1996	9.59	3.09	10.59	2.70	5.16	2.17	8.50	4.45	5.83	3.55	4.00	3.28	62.91
1997	6.31	4.82	3.23	5.08	4.41	5.55	5.99	2.74	4.00	5.49	3.74	4.13	55.49
1998	8.06	8.52	6.36	7.99	4.03	3.25	7.75	8.98	0.52	1.17	4.37	6.27	67.27
1999	8.63	2.34	6.75	2.03	5.37	9.04	3.13	0.81	0.65	4.16	2.73	3.13	48.77
2000	5.72	2.17	10.67	8.19	0.88	2.89	4.62	2.20	1.66	1.26	8.14	1.84	50.24
2001	5.23	4.35	8.43	7.30	5.25	7.53	3.61	7.38	6.26	2.43	4.20	4.76	66.73
POR= 106 YRS	5.06	4.78	6.06	4.73	4.23	4.03	5.22	4.13	3.47	2.82	3.85	4.93	53.31

AVERAGE TEMPERATURE (F) 2001   BIRMINGHAM (MUNICIPAL AIRPORT), AL (BHM)

YEAR	JAN	FEB	MAR	APR	MAY	JUN	JUL	AUG	SEP	OCT	NOV	DEC	ANNUAL
1972	47.2	46.9	53.7	63.1	67.2	74.2	76.9	78.8	75.1	61.7	48.9	47.7	61.8
1973	40.7	42.3	58.6	58.1	67.5	75.9	79.4	77.2	77.2	66.6	56.2	45.0	62.1
1974	52.8	47.3	61.2	61.5	71.0	72.7	79.2	78.4	70.1	60.6	52.7	47.3	62.9
1975	49.1	50.1	53.9	60.5	72.4	76.4	78.1	79.1	70.2	63.5	54.1	44.6	62.7
1976	39.1	54.3	58.7	61.5	64.2	74.4	78.3	77.0	71.9	57.4	45.5	41.2	60.3
1977	31.6	44.8	57.3	65.2	72.4	79.8	83.4	82.1	75.7	59.9	55.0	44.1	62.6
1978	33.5	37.4	50.2	62.9	68.2	77.1	81.5	79.8	76.9	60.6	57.9	46.1	61.0
1979	37.7	44.0	55.2	63.4	69.8	75.1	79.9	78.4	71.4	63.9	52.8	45.9	61.5
1980	45.2	42.3	51.4	60.9	70.0	77.6	84.4	83.0	78.4	60.3	52.4	44.0	62.5
1981	39.2	48.2	51.3	67.1	67.8	80.6	82.2	79.2	71.9	61.8	54.6	42.7	62.2
1982	41.9	47.3	58.9	59.1	72.6	75.7	80.7	79.6	73.1	64.2	54.1	51.2	63.2
1983	40.6	44.8	50.8	56.6	67.3	74.1	80.3	81.7	71.7	63.0	51.9	39.8	63.2
1984	38.5	46.8	52.4	59.0	67.5	77.1	78.2	77.5	72.1	71.1	51.4	54.3	61.8
1985	35.5	43.2	52.7	63.7	69.5	76.2	78.1	78.2	72.0	67.3	61.0	40.2	61.8
1986	41.9	49.9	55.5	61.8	71.2	78.6	82.5	77.8	76.9	63.9	57.4	44.0	63.5
1987	42.1	46.9	54.6	59.5	73.8	76.7	80.5	82.0	73.1	56.8	54.5	49.1	62.5
1988	39.7	43.5	54.1	61.2	67.6	77.5	79.8	81.5	74.4	57.7	55.2	45.8	61.5
1989	48.9	45.7	56.4	60.3	67.7	75.5	79.2	79.5	72.5	61.9	52.9	38.0	61.5
1990	49.0	54.5	57.0	60.7	68.9	77.8	79.8	77.5	63.9	59.7	55.9	47.7	64.7
1991	44.8	49.7	55.8	65.7	73.1	76.9	81.4	80.8	75.4	65.1	50.0	49.3	64.0
1992	43.4	50.6	53.8	61.0	67.4	74.7	80.4	75.6	73.4	61.5	51.7	45.8	61.6
1993	47.0	44.9	50.4	58.8	68.7	78.5	83.8	81.2	73.2	61.9	51.7	44.4	62.0
1994	39.6	49.3	54.6	65.3	67.5	78.5	78.0	78.0	72.1	64.1	57.0	49.2	62.8
1995	44.4	46.2	57.4	63.7	72.0	75.6	83.2	84.8	74.5	62.5	48.0	43.6	63.0
1996	42.6	46.0	50.9	59.1	73.3	77.2	79.9	78.6	71.8	62.6	52.0	48.7	61.9
1997	44.7	50.5	59.5	58.2	66.0	74.4	80.3	77.9	74.9	62.7	49.7	43.6	61.8
1998	46.7	48.0	52.7	60.9	73.4	80.3	82.2	80.4	78.2	67.6	57.3	50.7	64.9
1999	49.3	51.2	51.6	67.2	69.9	77.1	81.7	83.6	74.8	64.4	55.5	46.7	64.4
2000	44.5	52.2	58.0	59.4	74.3	77.6	82.6	81.6	74.2	65.8	52.0	38.1	63.4
2001	40.1	51.0	49.7	64.5	70.5	75.3	80.1	78.5	71.6	60.6	58.6	49.3	62.5
POR= 106 YRS	44.7	47.1	54.9	62.6	70.4	77.6	80.1	79.6	74.9	64.1	53.5	46.0	63.0

REFERENCE NOTES:

PAGE 1:
THE TEMPERATURE GRAPH SHOWS NORMAL MAXIMUM AND NORMAL
MINIMUM DAILY TEMPERATURES (SOLID CURVES) AND THE
ACTUAL DAILY HIGH AND LOW TEMPERATURES (VERTICAL BARS).

PAGE 2 AND 3:
H/C INDICATES HEATING AND COOLING DEGREE DAYS.
RH INDICATES RELATIVE HUMIDITY
W/O INDICATES WEATHER AND OBSTRUCTIONS
S INDICATES SUNSHINE.
PR INDICATES PRESSURE.
CLOUDINESS ON PAGE 3 IS THE SUM OF THE CEILOMETER AND
SATELLITE DATA NOT TO EXCEED EIGHT EIGHTHS(OKTAS).

GENERAL:
T INDICATES TRACE PRECIPITATION, AN AMOUNT GREATER
THAN ZERO BUT LESS THAN THE LOWEST REPORTABLE VALUE.
+ INDICATES THE VALUE ALSO OCCURS ON EARLIER DATES.
BLANK ENTRIES DENOTE MISSING OR UNREPORTED DATA.
NORMALS ARE 30-YEAR AVERAGES (1961 - 1990).
ASOS INDICATES AUTOMATED SURFACE OBSERVING SYSTEM.
PM INDICATES THE LAST DAY OF THE PREVIOUS MONTH.
POR (PERIOD OF RECORD) BEGINS WITH THE JANUARY DATA
MONTH AND IS THE NUMBER OF YEARS USED TO COMPUTE
THE MEAN. INDIVIDUAL MONTHS WITHIN THE POR MAY
BE MISSING.
WHEN THE POR FOR A NORMAL IS LESS THAN 30 YEARS,
THE NORMAL IS PROVISIONAL AND IS BASED ON THE NUMBER
OF YEARS INDICATED.
0.* OR * INDICATES THE VALUE OR MEAN-DAYS-WITH
IS BETWEEN 0.00 AND 0.05.
CLOUDINESS FOR ASOS STATIONS DIFFERS FROM THE NON-ASOS
OBSERVATION TAKEN BY A HUMAN OBSERVER. ASOS STATION
CLOUDINESS IS BASED ON TIME-AVERAGED CEILOMETER DATA
FOR CLOUDS AT OR BELOW 12,000 FEET AND ON SATELLITE
DATA FOR CLOUDS ABOVE 12,000 FEET.
THE NUMBER OF DAYS WITH CLEAR, PARTLY CLOUDY, AND
CLOUDY CONDITIONS FOR ASOS STATIONS IS THE SUM
OF THE CEILOMETER AND SATELLITE DATA FOR THE
SUNRISE TO SUNSET PERIOD.

GENERAL CONTINUED:
CLEAR INDICATES 0 - 2 OKTAS, PARTLY CLOUDY INDICATES
3 - 6 OKTAS, AND CLOUDY INDICATES 7 OR 8 OKTAS.
WHEN AT LEAST ONE OF THE ELEMENTS (CEILOMETER OR
SATELLITE) IS MISSING, THE DAILY CLOUDINESS IS
NOT COMPUTED.
WIND DIRECTION IS RECORDED IN TENS OF DEGREES (2 DIGITS)
CLOCKWISE FROM TRUE NORTH. "00" INDICATES CALM. "36"
INDICATES TRUE NORTH.
RESULTANT WIND IS THE VECTOR AVERAGE OF THE SPEED AND
DIRECTION.
AVERAGE TEMPERATURE IS THE SUM OF THE MEAN DAILY MAXIMUM
AND MINIMUM TEMPERATURE DIVIDED BY 2.
SNOWFALL DATA COMPRISE ALL FORMS OF FROZEN
PRECIPITATION, INCLUDING HAIL.
A HEATING (COOLING) DEGREE DAY IS THE DIFFERENCE BETWEEN
THE AVERAGE DAILY TEMPERATURE AND 65 F.
DRY BULB IS THE TEMPERATURE OF THE AMBIENT AIR.
DEW POINT IS THE TEMPERATURE TO WHICH THE AIR MUST BE
COOLED TO ACHIEVE 100 PERCENT RELATIVE HUMIDITY.
WET BULB IS THE TEMPERATURE THE AIR WOULD HAVE IF THE
MOISTURE CONTENT WAS INCREASED TO 100 PERCENT RELATIVE
HUMIDITY.

ON JULY 1, 1996, THE NATIONAL WEATHER SERVICE BEGAN USING
THE "METAR" OBSERVATION CODE THAT WAS ALREADY EMPLOYED
BY MOST OTHER NATIONS OF THE WORLD. THE MOST NOTICEABLE
DIFFERENCE IN THIS ANNUAL PUBLICATION WILL BE THE CHANGE
IN UNITS FROM TENTHS TO EIGHTS(OKTAS) FOR REPORTING THE
AMOUNT OF SKY COVER.

HEATING DEGREE DAYS (base 65 F) 2001 BIRMINGHAM (MUNICIPAL AIRPORT), AL (BHM)

YEAR	JUL	AUG	SEP	OCT	NOV	DEC	JAN	FEB	MAR	APR	MAY	JUN	TOTAL
1972-73	0	0	9	135	486	534	749	629	217	228	48	0	3035
1973-74	0	0	3	73	285	615	374	491	161	147	9	0	2158
1974-75	0	0	19	159	378	542	493	412	361	190	2	0	2556
1975-76	0	0	49	105	349	626	626	314	213	129	65	0	2645
1976-77	0	0	7	241	580	732	1026	566	252	73	5	0	3482
1977-78	0	0	0	176	292	640	967	768	452	120	42	0	3457
1978-79	0	0	0	160	220	578	839	584	303	83	33	0	2800
1979-80	0	0	0	101	365	588	604	655	417	144	14	0	2888
1980-81	0	0	7	181	372	642	795	464	428	46	51	0	2986
1981-82	0	0	19	138	314	682	711	490	250	199	4	0	2807
1982-83	0	0	16	134	331	449	751	558	437	262	39	0	2977
1983-84	0	0	26	108	388	774	817	519	392	202	68	2	3296
1984-85	0	0	10	27	410	330	907	604	278	123	16	3	2708
1985-86	0	0	14	62	165	761	711	415	305	132	21	0	2586
1986-87	0	1	0	112	239	644	705	500	320	201	0	0	2722
1987-88	0	0	5	248	315	488	777	616	347	134	18	0	2948
1988-89	0	0	0	234	289	589	492	545	289	200	65	0	2703
1989-90	0	0	27	140	363	834	491	299	265	172	32	0	2623
1990-91	0	0	16	138	282	474	620	421	304	46	7	0	2308
1991-92	0	0	12	97	455	492	663	409	344	173	46	0	2691
1992-93	0	0	6	116	398	587	550	559	452	196	20	0	2884
1993-94	0	0	16	153	411	633	780	433	328	90	31	0	2875
1994-95	0	0	3	82	241	483	632	520	250	101	24	0	2336
1995-96	0	0	12	137	509	662	686	560	440	203	12	0	3221
1996-97	0	0	10	121	390	498	628	399	188	209	46	0	2489
1997-98	0	0	1	162	450	676	562	469	400	150	13	5	2888
1998-99	0	0	0	47	231	462	483	383	408	73	8	0	2095
1999-00	0	0	12	107	279	559	634	370	219	175	1	0	2356
2000-01	0	0	10	84	419	828	766	392	466	101	3	0	3069
2001-	0	0	32	183	200	478							

COOLING DEGREE DAYS (base 65 F) 2001 BIRMINGHAM (MUNICIPAL AIRPORT), AL (B

YEAR	JAN	FEB	MAR	APR	MAY	JUN	JUL	AUG	SEP	OCT	NOV	DEC	ANNUAL
1972	4	6	7	88	93	291	379	434	318	40	8	2	1670
1973	0	0	24	28	132	333	452	385	376	129	28	0	1887
1974	4	3	52	48	204	238	450	418	180	28	15	0	1640
1975	8	0	26	61	241	351	416	376	219	66	29	2	1858
1976	0	8	25	27	48	290	290	418	376	219	15	1	1427
1977	0	5	21	85	241	450	578	537	329	25	0	1	2272
1978	0	0	0	64	151	370	515	463	365	34	12	1	1975
1979	0	1	8	43	188	308	466	424	198	74	7	2	1719
1980	0	4	6	25	176	386	607	563	413	43	2	0	2225
1981	0	1	9	113	145	475	539	448	231	46	9	0	2016
1982	1	0	65	28	247	327	495	459	265	118	12	27	2044
1983	0	0	7	17	115	281	528	233	53	2	0	1717	
1984	0	0	11	30	151	373	421	395	230	221	6	6	1844
1985	0	1	27	90	163	349	413	417	233	139	51	1	1884
1986	0	0	16	43	222	413	547	406	364	84	18	0	2113
1987	0	0	5	46	280	357	487	532	254	1	7	3	1972
1988	0	0	13	26	105	382	467	516	292	14	1	0	1816
1989	0	14	27	65	155	323	447	459	257	51	3	0	1801
1990	1	14	21	49	160	393	466	543	398	110	9	4	2168
1991	0	0	28	74	266	364	515	496	333	106	9	10	2201
1992	0	0	2	59	129	298	484	333	265	14	5	0	1589
1993	0	0	5	19	140	410	591	511	271	63	18	0	2028
1994	0	0	10	109	115	408	410	411	221	58	8	3	1750
1995	2	0	23	67	247	323	571	620	301	67	3	3	2227
1996	0	16	10	34	280	374	470	428	219	54	8	1	1894
1997	6	1	23	14	82	287	481	404	303	99	0	0	1700
1998	1	0	27	34	277	470	538	484	403	134	6	26	2400
1999	7	5	0	146	168	371	524	580	312	83	0	0	2196
2000	3	8	9	12	298	385	553	521	294	117	37	0	2237
2001	0	6	0	95	182	314	473	423	234	53	17	0	1797

SNOWFALL (inches) 2001 BIRMINGHAM (MUNICIPAL AIRPORT), AL (BHM)

YEAR	JUL	AUG	SEP	OCT	NOV	DEC	JAN	FEB	MAR	APR	MAY	JUN	TOTAL
1972-73	0.0	0.0	0.0	0.0	0.0	0.0	T	0.0	0.0	0.0	0.0	0.0	T
1973-74	0.0	0.0	0.0	0.0	0.0	T	0.0	T	0.0	0.0	0.0	0.0	T
1974-75	0.0	0.0	0.0	0.0	T	0.4	T	0.0	0.0	0.0	0.0	0.0	0.4
1975-76	0.0	0.0	0.0	0.0	T	T	1.4	T	0.0	0.0	0.0	0.0	1.4
1976-77	0.0	0.0	0.0	0.0	T	T	1.4	T	0.0	0.0	0.0	0.0	1.4
1977-78	0.0	0.0	0.0	0.0	0.0	T	1.9	T	T	0.0	0.0	0.0	1.9
1978-79	0.0	0.0	0.0	0.0	0.0	0.0	T	T	0.0	0.0	0.0	0.0	T
1979-80	0.0	0.0	0.0	0.0	0.0	0.0	T	T	0.3	0.0	0.0	0.0	0.3
1980-81	0.0	0.0	0.0	0.0	T	T	0.0	0.0	T	0.0	0.0	0.0	T
1981-82	0.0	0.0	0.0	0.0	0.0	T	2.1	0.0	T	0.0	0.0	0.0	2.1
1982-83	0.0	0.0	0.0	0.0	0.0	T	1.0	T	1.5	0.0	0.0	0.0	2.5
1983-84	0.0	0.0	0.0	0.0	0.0	T	T	T	2.0	0.0	0.0	0.0	2.0
1984-85	0.0	0.0	0.0	0.0	0.0	T	T	0.3	0.0	0.0	0.0	0.0	0.3
1985-86	0.0	0.0	0.0	0.0	0.0	T	T	0.0	T	0.0	0.0	0.0	T
1986-87	0.0	0.0	0.0	0.0	0.0	T	2.6	0.0	T	5.0	0.0	0.0	7.6
1987-88	0.0	0.0	0.0	0.0	0.0	0.0	1.0	T	0.0	T	0.0	0.0	1.0
1988-89	0.0	0.0	0.0	0.0	0.0	T	0.0	T	0.0	0.0	T	0.0	T
1989-90	0.0	0.0	0.0	0.0	T	0.4	0.0	T	T	0.0	T	0.0	0.4
1990-91	T	0.0	0.0	0.0	0.0	T	T	T	T	0.0	T	T	4.4
1991-92	0.0	0.0	0.0	0.0	T	0.0	4.4	0.0	T	0.0	0.0	T	4.4
1992-93	0.0	0.0	T	T	0.0	0.0	0.0	T	13.0	0.0	T	0.0	13.0
1993-94	0.0	0.0	0.0	T	0.0	T	T	0.0	T	0.0	0.0	0.0	1.0
1994-95	T	0.0	0.0	0.0	0.0	0.0	T	1.0	T	0.0	T	0.0	1.5
1995-96	0.0	0.0	0.0	0.0	T	0.3	1.2	T	0.0	T	0.0	1.5	
1996-97	0.0	0.0	0.0	0.0	0.0	T	T	T	0.0	0.0	T	0.0	T
1997-98	0.0	0.0	0.0	0.0	T	1.7	T	T	T	T	0.0	T	1.7
1998-99	T	0.0	0.0	0.0	0.0	0.0	T	T	0.0	T	0.0	T	T
1999-00	0.0	0.0	0.0	0.0	0.0	0.0	3.0	T	T	T	0.0	0.0	3.0
2000-01	T	0.0	0.0	0.0	T	0.0	T	T	T	T	0.0	T	T
2001-	0.0	0.0	0.0	0.0	0.0	0.0							
POR= 57 YRS	T	0.0	T	T	0.0	0.3	0.6	0.2	0.3	0.1	T	T	1.5

2001
MOBILE,
ALABAMA (MOB)

Mobile is located at the head of Mobile Bay and approximately 30 miles from the Gulf of Mexico. Its weather is influenced to a considerable extent by the Gulf.

The summers are consistently warm, but temperatures are seldom as high as they are at inland stations. Normally, in summer, the day begins in the low 70s and the temperature rises rapidly before noon to the high 80s or low 90s, when it is checked by the onset of the sea breeze. On the rare occasions when northerly winds prevail throughout the day, temperatures may reach the high 90s or rise slightly above 100 degrees.

Winter weather is usually mild except for occasional invasions of cold air that last about three days. January is the coldest month in the year. Unusual winters may produce readings that require extensive protective measures as some citrus fruit is grown in the area and outdoor nurseries are numerous.

Based on the 1951–1980 period, the average first occurrence of 32 degrees Fahrenheit in the fall is November 26 and the average last occurrence in the spring is February 27.

The yearly rainfall is among the highest in the United States. It is fairly evenly distributed throughout the year with a slight maximum at the height of the summer thunderstorm season and a slight minimum during the late fall. Rainfall is usually of the shower type and long periods of continuous rain are rare.

Frontal thunderstorms may occur in any month of the year. There may be a thunderstorm every other day in July and August. The summer storms are usually not too violent and seldom produce hail.

The area is subject to hurricanes from the West Indies, the western Caribbean, and the Gulf of Mexico.

NORMALS, MEANS, AND EXTREMES
MOBILE, AL (MOB)

LATITUDE: 30 41' 18" N LONGITUDE: 88 14' 44" W ELEVATION (FT): GRND: 209 BARO: 212 TIME ZONE: CENTRAL (UTC + 6) WBAN: 13894

	ELEMENT	POR	JAN	FEB	MAR	APR	MAY	JUN	JUL	AUG	SEP	OCT	NOV	DEC	YEAR
TEMPERATURE F	NORMAL DAILY MAXIMUM	30	59.7	63.6	70.9	78.5	84.6	90.0	91.3	90.5	86.9	79.5	70.3	62.9	77.4
	MEAN DAILY MAXIMUM	54	60.9	64.6	70.4	77.8	84.7	89.8	91.1	90.8	86.9	79.3	69.9	63.1	77.4
	HIGHEST DAILY MAXIMUM	60	84	82	90	94	100	102	104	105	99	93	87	81	105
	YEAR OF OCCURRENCE		1949	1989	1946	1987	1953	1952	1952	2000	1990	1963	1971	1998	AUG 2000
	MEAN OF EXTREME MAXS.	54	74.7	77.4	82.1	86.6	92.4	95.9	97.0	96.3	93.8	88.3	81.8	76.5	86.9
	NORMAL DAILY MINIMUM	30	40.0	42.7	50.1	57.1	64.4	70.7	73.2	72.9	68.7	57.3	49.1	43.1	57.4
	MEAN DAILY MINIMUM	54	40.7	43.9	49.4	56.6	64.4	70.6	72.9	72.6	68.5	57.4	48.2	42.8	57.3
	LOWEST DAILY MINIMUM	60	3	11	21	32	43	49	60	59	42	30	22	8	3
	YEAR OF OCCURRENCE		1985	1996	1993	1987	1960	1984	1947	1956	1967	1993	1950	1983	JAN 1985
	MEAN OF EXTREME MINS.	54	22.4	25.8	32.3	41.5	52.1	62.3	68.5	67.4	56.6	40.8	31.4	24.8	43.8
	NORMAL DRY BULB	30	49.9	53.2	60.5	67.8	74.5	80.4	82.3	81.8	77.9	68.4	59.8	53.0	67.5
	MEAN DRY BULB	54	50.8	54.3	60.0	67.2	74.5	80.1	81.9	81.7	77.6	68.3	59.0	52.9	67.4
	MEAN WET BULB	18	46.4	50.1	54.3	60.1	67.7	72.7	75.1	74.6	70.8	61.5	54.8	48.9	61.4
	MEAN DEW POINT	18	41.4	44.9	49.3	55.3	64.0	70.0	72.6	72.2	67.7	57.2	50.4	44.4	57.5
	NORMAL NO. DAYS WITH:														
	MAXIMUM 90	30	0.0	0.0	0.0	0.3	4.6	17.0	22.4	20.7	10.7	1.0	0.0	0.0	76.7
	MAXIMUM 32	30	0.3	0.0	0.0	0.0	0.0	0.0	0.0	0.0	0.0	0.0	0.0	0.2	0.5
	MINIMUM 32	30	8.7	5.4	1.2	*	0.0	0.0	0.0	0.0	0.0	0.0	1.1	6.0	22.4
	MINIMUM 0	30	0.0	0.0	0.0	0.0	0.0	0.0	0.0	0.0	0.0	0.0	0.0	0.0	0.0
H/C	NORMAL HEATING DEG. DAYS	30	492	344	177	48	0	0	0	0	0	52	196	393	1702
	NORMAL COOLING DEG. DAYS	30	24	14	38	132	295	462	536	521	387	157	40	21	2627
RH	NORMAL (PERCENT)	30	72	70	70	70	71	73	76	78	76	71	73	74	73
	HOUR 00 LST	30	78	78	80	82	84	85	87	88	86	82	82	80	83
	HOUR 06 LST	30	81	82	84	87	87	88	89	91	89	85	85	83	86
	HOUR 12 LST	30	60	57	55	52	53	55	60	61	59	52	57	61	57
	HOUR 18 LST	30	68	64	63	62	63	66	71	73	71	67	71	71	68
S	PERCENT POSSIBLE SUNSHINE														
W/O	MEAN NO. DAYS WITH:														
	HEAVY FOG(VISBY 1/4 MI)	60	6.0	4.7	5.3	4.6	2.7	1.1	1.0	1.3	1.9	2.8	4.3	4.9	40.6
	THUNDERSTORMS	60	2.2	2.3	5.0	4.8	7.2	11.9	17.6	14.2	7.2	2.2	2.3	2.2	79.1
CLOUDINESS	MEAN:														
	SUNRISE-SUNSET (OKTAS)	48	5.3	5.0	4.9	4.6	4.6	4.6	5.1	4.6	4.4	3.5	4.2	4.9	4.6
	MIDNIGHT-MIDNIGHT (OKTAS)	32	5.0	4.6	4.7	4.2	4.2	4.0	4.5	4.2	3.9	3.3	4.0	4.7	4.3
	MEAN NO. DAYS WITH:														
	CLEAR	48	7.9	7.6	8.7	9.2	8.6	7.0	4.1	6.3	9.0	14.2	11.0	8.8	102.4
	PARTLY CLOUDY	48	6.2	6.6	7.8	8.7	11.3	13.8	15.0	14.8	10.1	7.6	7.2	6.4	115.5
	CLOUDY	48	16.9	14.0	14.5	12.0	11.1	9.2	11.9	9.9	11.0	9.1	11.8	15.9	147.3
PR	MEAN STATION PRESSURE(IN)	29	29.90	29.89	29.80	29.80	29.80	29.80	29.80	29.80	29.79	29.80	29.90	29.90	29.83
	MEAN SEA-LEVEL PRES. (IN)	18	30.13	30.11	30.05	30.02	29.99	30.01	30.04	30.01	30.00	30.07	30.11	30.15	30.06
WINDS	MEAN SPEED (MPH)	44	10.1	10.5	10.5	10.1	8.7	7.7	7.0	6.8	7.9	8.2	9.1	9.6	8.9
	PREVAIL.DIR(TENS OF DEGS)	27	36	36	14	14	18	18	22	04	04	04	36	36	36
	MAXIMUM 2-MINUTE:														
	SPEED (MPH)	5	43	34	39	31	30	45	37	40	51	40	38	33	51
	DIR. (TENS OF DEGS)		23	21	18	20	35	16	11	12	12	22	17	21	12
	YEAR OF OCCURRENCE		2001	1997	2001	2000	1998	2001	1999	1998	1998	2001	2000	2000	SEP 1998
	MAXIMUM 5-SECOND:														
	SPEED (MPH)	5	52	43	53	39	34	57	49	46	63	54	49	46	63
	DIR. (TENS OF DEGS)		22	22	24	20	20	34	06	12	12	21	18	21	12
	YEAR OF OCCURRENCE		2001	1998	1999	2000	1999	1998	1998	1998	1998	2001	2000	2000	SEP 1998
PRECIPITATION	NORMAL (IN)	30	4.76	5.46	6.41	4.48	5.74	5.04	6.85	6.96	5.91	2.94	4.10	5.31	63.96
	MAXIMUM MONTHLY (IN)	60	16.92	11.89	15.58	17.69	15.08	13.07	19.29	15.19	24.13	13.20	13.65	11.38	24.13
	YEAR OF OCCURRENCE		1998	1983	1946	1955	1980	1961	1949	1984	1998	1985	1948	1953	SEP 1998
	MINIMUM MONTHLY (IN)	60	0.98	1.09	0.59	0.08	0.36	1.19	1.72	1.04	0.58	T	0.25	1.29	T
	YEAR OF OCCURRENCE		1968	1999	1967	1999	1996	1966	1983	1997	1963	1978	1960	1980	OCT 1978
	MAXIMUM IN 24 HOURS (IN)	60	8.34	5.37	10.57	13.36	8.86	7.38	10.07	6.62	10.06	5.65	7.02	7.46	13.36
	YEAR OF OCCURRENCE		1965	1981	1990	1955	1995	1961	1997	1969	1998	1985	1975	1995	APR 1955
	NORMAL NO. DAYS WITH:														
	PRECIPITATION 0.01	30	10.8	9.2	10.0	6.8	8.4	11.1	15.1	14.5	10.2	5.5	8.0	10.2	119.8
	PRECIPITATION 1.00	30	1.4	1.7	2.2	1.5	1.8	1.4	2.1	2.4	1.7	1.1	1.0	1.9	20.2
SNOWFALL	NORMAL (IN)	30	0.1	0.2	T	T	0.0	0.0	0.0	0.0	0.0	0.0	T	0.1	0.4
	MAXIMUM MONTHLY (IN)	58	3.5	3.6	2.7	T	T	0.0	0.0	T	0.0	0.0	T	3.0	3.6
	YEAR OF OCCURRENCE		1955	1973	1993	1988	1991			1995			1966	1963	FEB 1973
	MAXIMUM IN 24 HOURS (IN)	58	3.5	3.6	2.7	T	T	0.0	0.0	T	0.0	0.0	T	3.0	3.6
	YEAR OF OCCURRENCE		1955	1973	1993	1988	1991			1995			1966	1963	FEB 1973
	MAXIMUM SNOW DEPTH (IN)	51	22	3	2	0	0	0	0	0	0	0	0	0	22
	YEAR OF OCCURRENCE		1964	1973	1993										JAN 1964
	NORMAL NO. DAYS WITH:														
	SNOWFALL 1.0	30	0.*	0.1	0.0	0.0	0.0	0.0	0.0	0.0	0.0	0.0	0.0	0.*	0.1

PRECIPITATION (inches) 2001 MOBILE, AL (MOB)

YEAR	JAN	FEB	MAR	APR	MAY	JUN	JUL	AUG	SEP	OCT	NOV	DEC	ANNUAL
1972	5.94	4.46	5.87	1.81	7.72	3.65	2.16	2.35	3.28	1.26	5.67	5.59	49.76
1973	1.96	3.40	11.63	9.91	2.97	5.33	6.24	5.46	12.68	2.15	4.52	4.57	70.82
1974	3.89	6.47	6.16	5.91	2.31	3.48	6.60	7.03	10.60	0.88	3.91	4.31	61.55
1975	3.43	3.75	7.45	9.05	7.12	3.76	11.82	8.50	7.37	6.72	12.63	4.98	86.58
1976	1.80	2.36	9.60	1.69	11.11	4.71	4.43	4.77	3.51	5.67	4.64	4.21	58.50
1977	5.54	1.86	6.06	2.73	5.30	1.26	7.77	5.42	9.61	4.57	9.51	4.94	64.57
1978	10.40	3.88	4.08	7.03	9.74	6.13	7.74	7.34	3.04	T	5.05	4.36	68.79
1979	5.14	9.14	8.37	4.05	7.02	3.70	10.80	6.10	11.73	2.80	6.46	3.94	79.25
1980	4.95	1.58	13.46	15.43	15.08	2.57	6.54	6.42	5.94	2.90	1.96	1.29	78.12
1981	1.23	8.75	3.00	0.96	12.51	6.50	6.53	5.06	3.97	0.94	0.85	6.82	57.12
1982	3.56	7.42	6.81	4.48	2.84	11.00	13.14	7.00	4.68	1.61	3.66	8.24	74.44
1983	5.82	11.89	6.89	12.53	1.53	8.09	1.72	15.19	5.97	3.81	5.30	8.34	83.46
1984	6.13	4.79	4.75	3.53	4.28	2.07	2.36	4.32	0.74	6.19	1.67	2.12	53.82
1985	5.06	6.39	5.49	1.22	5.77	3.94	8.31	5.60	10.32	13.20	1.61	4.34	69.97
1986	2.67	4.17	4.53	2.16	4.18	7.53	7.13	4.41	4.83	8.45	3.68	59.34
1987	5.81	8.64	6.18	0.83	10.69	7.68	10.33	10.33	0.02	5.54	3.60	67.12
1988	4.64	6.26	7.80	4.19	0.58	2.34	6.04	10.43	14.04	1.83	2.30	1.80	62.25
1989	2.13	1.47	5.57	3.55	6.47	9.82	7.16	3.82	4.55	0.90	11.33	7.23	64.00
1990	7.35	8.13	12.24	4.51	6.47	2.68	2.28	1.46	2.65	1.23	1.77	5.20	55.97
1991	16.07	3.35	4.83	10.43	15.03	7.04	8.15	5.17	1.87	0.86	5.78	3.09	81.67
1992	9.89	9.94	3.61	1.90	1.58	5.24	7.77	8.27	3.26	1.03	12.70	5.27	70.46
1993	8.03	4.48	10.45	4.80	5.24	3.01	5.64	3.52	3.35	5.32	2.87	3.69	60.40
1994	6.57	1.32	4.34	5.48	4.59	5.09	10.72	2.05	1.76	5.91	4.05	3.04	54.92
1995	7.09	3.01	8.53	7.46	12.08	3.32	4.63	7.60	1.16	8.70	8.05	8.86	80.49
1996	4.65	6.23	8.97	11.93	0.36	5.18	4.95	8.29	5.01	1.72	2.52	6.92	66.73
1997	5.67	5.98	5.42	5.66	8.50	8.49	18.52	1.04	1.98	5.83	8.68	4.37	80.14
1998	16.92	4.29	10.92	5.19	1.78	2.87	6.90	5.02	24.13	1.35	4.95	2.20	86.52
1999	5.19	1.09	9.44	0.08	4.71	7.58	4.30	2.87	4.41	4.75	3.05	3.43	50.90
2000	2.67	1.30	6.94	2.43	2.38	3.21	3.90	3.67	3.43	0.47	11.54	3.80	45.74
2001	4.08	2.70	11.04	0.88	1.52	6.20	8.55	9.49	2.59	3.53	1.24	2.83	54.65
POR= 131 YRS	4.99	5.05	6.70	5.02	4.84	5.53	7.44	6.54	5.48	3.20	3.87	5.06	63.72

AVERAGE TEMPERATURE (F) 2001 MOBILE, AL (MOB)

YEAR	JAN	FEB	MAR	APR	MAY	JUN	JUL	AUG	SEP	OCT	NOV	DEC	ANNUAL
1972	58.4	55.0	61.8	69.4	74.7	80.7	81.4	83.4	81.6	71.0	56.8	55.5	69.1
1973	50.5	51.7	64.6	65.6	74.6	81.6	84.1	81.4	80.2	73.6	65.3	52.9	68.8
1974	64.0	55.2	64.3	66.1	75.5	77.8	82.2	81.4	75.5	65.5	59.2	55.1	68.5
1975	56.6	59.8	60.7	66.2	76.5	80.7	81.8	81.7	75.3	70.3	60.5	52.1	68.5
1976	49.5	59.4	64.0	69.9	72.7	79.8	83.4	81.0	75.5	61.4	50.6	48.0	66.3
1977	40.9	52.2	62.6	68.2	75.6	82.9	83.5	83.4	80.9	65.2	60.9	50.8	67.3
1978	41.2	45.1	56.7	68.5	75.4	81.2	82.8	83.2	81.6	69.8	65.1	54.4	67.1
1979	45.1	50.7	60.8	69.3	73.3	79.7	81.3	81.1	76.8	68.1	57.2	51.3	66.2
1980	56.2	49.7	61.6	65.9	74.6	81.2	84.6	83.2	81.8	66.3	57.8	51.6	67.9
1981	46.0	53.4	59.5	71.0	71.6	82.6	84.0	82.7	76.8	69.4	63.0	49.9	67.5
1982	49.7	53.3	63.5	67.1	74.4	80.7	81.0	81.0	76.0	68.5	60.4	56.9	67.7
1983	47.1	51.1	55.0	61.8	72.1	76.4	81.8	81.4	73.6	67.4	57.7	47.7	64.4
1984	46.1	52.4	58.6	65.7	72.5	78.1	79.9	78.8	75.6	72.7	56.3	60.4	66.4
1985	43.5	50.7	64.5	67.7	73.7	79.9	79.7	81.0	76.0	71.4	66.1	48.8	66.9
1986	49.7	56.4	59.7	66.1	74.4	81.0	83.3	80.5	79.5	68.5	64.3	51.1	67.9
1987	48.8	53.6	59.2	64.6	75.7	78.9	81.8	82.2	77.2	62.5	59.2	56.5	66.7
1988	46.8	50.6	58.6	67.3	72.8	79.5	81.8	81.3	78.3	65.3	62.7	53.8	66.6
1989	57.1	54.0	61.7	65.9	74.1	79.0	81.0	81.6	76.7	66.1	59.1	44.4	66.7
1990	54.6	59.0	61.7	65.0	73.4	81.1	81.7	83.1	78.6	68.0	60.9	56.7	68.7
1991	52.4	55.9	61.8	69.3	76.0	79.7	81.9	81.3	77.1	69.4	54.1	54.5	67.8
1992	49.4	57.0	60.3	67.4	72.2	79.3	82.4	79.4	77.2	67.4	56.4	55.6	66.8
1993	55.1	52.7	57.0	62.1	71.9	79.9	82.6	82.8	78.3	68.1	56.8	49.8	66.4
1994	47.2	56.1	60.0	68.6	73.4	79.6	79.9	80.4	76.6	68.3	65.2	57.5	67.5
1995	51.0	55.1	61.8	67.1	75.6	78.6	82.5	83.2	78.6	68.5	56.7	51.6	67.5
1996	49.0	52.4	55.4	63.8	75.6	78.8	81.7	80.0	76.4	67.3	54.3	46.2	66.2
1997	51.0	54.9	66.3	63.5	72.2	77.4	81.5	81.4	79.2	67.2	55.1	49.4	66.6
1998	52.2	53.7	57.1	65.2	76.4	82.5	83.5	83.2	78.5	70.3	63.3	57.2	68.6
1999	55.3	57.3	58.6	70.8	73.7	79.1	81.4	84.4	76.0	68.6	60.0	51.6	68.1
2000	53.6	57.6	63.6	64.7	76.9	79.6	84.3	83.9	77.3	68.2	57.0	46.1	67.7
2001	47.2	58.3	56.9	69.6	74.5	79.1	81.6	80.7	76.0	65.1	63.9	55.1	67.3
POR= 129 YRS	51.4	54.2	60.2	67.0	74.2	80.2	81.8	81.5	78.0	68.7	59.0	52.9	67.4

REFERENCE NOTES:

PAGE 1:
THE TEMPERATURE GRAPH SHOWS NORMAL MAXIMUM AND NORMAL MINIMUM DAILY TEMPERATURES (SOLID CURVES) AND THE ACTUAL DAILY HIGH AND LOW TEMPERATURES (VERTICAL BARS).

PAGE 2 AND 3:
H/C INDICATES HEATING AND COOLING DEGREE DAYS.
RH INDICATES RELATIVE HUMIDITY
W/O INDICATES WEATHER AND OBSTRUCTIONS
S INDICATES SUNSHINE.
PR INDICATES PRESSURE.
CLOUDINESS ON PAGE 3 IS THE SUM OF THE CEILOMETER AND SATELLITE DATA NOT TO EXCEED EIGHT EIGHTHS(OKTAS).

GENERAL:
T INDICATES TRACE PRECIPITATION, AN AMOUNT GREATER THAN ZERO BUT LESS THAN THE LOWEST REPORTABLE VALUE.
+ INDICATES THE VALUE ALSO OCCURS ON EARLIER DATES.
BLANK ENTRIES DENOTE MISSING OR UNREPORTED DATA.
NORMALS ARE 30-YEAR AVERAGES (1961 - 1990).
ASOS INDICATES AUTOMATED SURFACE OBSERVING SYSTEM.
PM INDICATES THE LAST DAY OF THE PREVIOUS MONTH.
POR (PERIOD OF RECORD) BEGINS WITH THE JANUARY DATA MONTH AND IS THE NUMBER OF YEARS USED TO COMPUTE THE MEAN. INDIVIDUAL MONTHS WITHIN THE POR MAY BE MISSING.
WHEN THE POR FOR A NORMAL IS LESS THAN 30 YEARS, THE NORMAL IS PROVISIONAL AND IS BASED ON THE NUMBER OF YEARS INDICATED.
0.* OR * INDICATES THE VALUE OR MEAN-DAYS-WITH IS BETWEEN 0.00 AND 0.05.
CLOUDINESS FOR ASOS STATIONS DIFFERS FROM THE NON-ASOS OBSERVATION TAKEN BY A HUMAN OBSERVER. ASOS STATION CLOUDINESS IS BASED ON TIME-AVERAGED CEILOMETER DATA FOR CLOUDS AT OR BELOW 12,000 FEET AND ON SATELLITE DATA FOR CLOUDS ABOVE 12,000 FEET.
THE NUMBER OF DAYS WITH CLEAR, PARTLY CLOUDY, AND CLOUDY CONDITIONS FOR ASOS STATIONS IS THE SUM OF THE CEILOMETER AND SATELLITE DATA FOR THE SUNRISE TO SUNSET PERIOD.

GENERAL CONTINUED:
CLEAR INDICATES 0 - 2 OKTAS, PARTLY CLOUDY INDICATES 3 - 6 OKTAS, AND CLOUDY INDICATES 7 OR 8 OKTAS. WHEN AT LEAST ONE OF THE ELEMENTS (CEILOMETER OR SATELLITE) IS MISSING, THE DAILY CLOUDINESS IS NOT COMPUTED.
WIND DIRECTION IS RECORDED IN TENS OF DEGREES (2 DIGITS) CLOCKWISE FROM TRUE NORTH. "00" INDICATES CALM. "36" INDICATES TRUE NORTH.
RESULTANT WIND IS THE VECTOR AVERAGE OF THE SPEED AND DIRECTION.
AVERAGE TEMPERATURE IS THE SUM OF THE MEAN DAILY MAXIMUM AND MINIMUM TEMPERATURE DIVIDED BY 2.
SNOWFALL DATA COMPRISE ALL FORMS OF FROZEN PRECIPITATION, INCLUDING HAIL.
A HEATING (COOLING) DEGREE DAY IS THE DIFFERENCE BETWEEN THE AVERAGE DAILY TEMPERATURE AND 65 F.
DRY BULB IS THE TEMPERATURE OF THE AMBIENT AIR.
DEW POINT IS THE TEMPERATURE TO WHICH THE AIR MUST BE COOLED TO ACHIEVE 100 PERCENT RELATIVE HUMIDITY.
WET BULB IS THE TEMPERATURE THE AIR WOULD HAVE IF THE MOISTURE CONTENT WAS INCREASED TO 100 PERCENT RELATIVE HUMIDITY.

ON JULY 1, 1996, THE NATIONAL WEATHER SERVICE BEGAN USING THE "METAR" OBSERVATION CODE THAT WAS ALREADY EMPLOYED BY MOST OTHER NATIONS OF THE WORLD. THE MOST NOTICEABLE DIFFERENCE IN THIS ANNUAL PUBLICATION WILL BE THE CHANGE IN UNITS FROM TENTHS TO EIGHTS(OKTAS) FOR REPORTING THE AMOUNT OF SKY COVER.

HEATING DEGREE DAYS (base 65 F) 2001 MOBILE, AL (MOB)

YEAR	JUL	AUG	SEP	OCT	NOV	DEC	JAN	FEB	MAR	APR	MAY	JUN	TOTAL
1972-73	0	0	0	33	284	308	442	368	86	75	0	0	1596
1973-74	0	0	0	18	91	377	108	288	106	49	0	0	1037
1974-75	0	0	0	58	215	336	276	185	183	70	0	0	1323
1975-76	0	0	9	11	211	402	475	174	100	8	0	0	1390
1976-77	0	0	0	149	430	520	738	355	141	28	0	0	2361
1977-78	0	0	0	73	149	439	731	551	268	16	0	0	2227
1978-79	0	0	0	20	53	355	613	401	152	8	4	0	1606
1979-80	0	0	0	43	246	424	267	454	159	52	0	0	1645
1980-81	0	0	0	63	235	420	581	323	180	9	4	0	1815
1981-82	0	0	5	42	113	463	485	323	161	48	0	0	1640
1982-83	0	0	5	63	185	296	545	383	306	119	4	0	1906
1983-84	0	0	7	60	243	529	582	363	218	69	10	0	2081
1984-85	0	0	2	17	270	172	665	397	84	50	1	0	1658
1985-86	0	0	1	22	69	503	469	251	188	40	0	0	1543
1986-87	0	0	0	43	105	433	500	311	193	106	0	0	1691
1987-88	0	0	0	108	194	283	565	421	214	33	0	0	1818
1988-89	0	0	0	53	122	363	254	335	170	81	2	0	1380
1989-90	0	0	6	71	204	630	326	186	128	71	0	0	1622
1990-91	0	0	2	58	145	287	382	258	156	13	0	0	1301
1991-92	0	0	3	32	344	342	477	236	165	77	16	0	1692
1992-93	0	0	0	18	276	297	307	340	260	110	3	0	1611
1993-94	0	0	0	64	278	464	543	260	182	46	0	0	1837
1994-95	0	0	0	35	93	304	426	281	148	41	3	0	1331
1995-96	0	0	0	29	263	445	489	381	323	97	3	0	2030
1996-97	0	0	0	48	212	346	440	293	63	80	5	0	1487
1997-98	0	0	0	89	296	475	390	313	266	60	0	0	1889
1998-99	0	0	0	24	94	287	319	234	204	30	3	0	1195
1999-00	0	0	2	52	162	414	363	242	85	68	0	0	1388
2000-01	0	0	6	45	302	579	542	208	262	36	0	0	1980
2001-	0	0	4	95	89	312							

COOLING DEGREE DAYS (base 65 F) 2001 MOBILE, AL (MOB)

YEAR	JAN	FEB	MAR	APR	MAY	JUN	JUL	AUG	SEP	OCT	NOV	DEC	ANNUAL
1972	47	15	40	173	309	478	513	575	505	224	46	20	2945
1973	0	1	80	98	303	508	598	462	293	110	10		2981
1974	85	19	91	88	332	392	539	517	323	81	45	36	2548
1975	21	44	57	115	363	475	529	524	325	183	85	11	2732
1976	0	17	74	160	245	454	578	504	321	48	4	0	2405
1977	0	4	74	132	333	543	579	576	482	85	31	7	2846
1978	0	0	17	127	329	496	560	573	507	177	64	34	2884
1979	0	6	28	144	267	448	512	509	361	146	16	5	2442
1980	2	17	61	89	305	495	618	571	512	110	25	9	2814
1981	0	4	15	195	217	538	595	557	366	185	60	4	2736
1982	17	2	119	118	296	475	502	503	342	177	54	53	2658
1983	0	0	4	27	229	351	532	514	271	140	29	2	2099
1984	0	3	24	96	249	401	469	435	327	261	16	34	2315
1985	4	2	77	133	275	452	461	504	339	227	108	7	2589
1986	0	15	31	79	295	486	573	489	444	158	89	8	2667
1987	4	0	21	99	341	422	527	544	373	41	27	28	2427
1988	7	10	21	108	251	442	529	513	404	67	61	20	2433
1989	16	33	74	116	289	431	498	523	366	114	31	0	2491
1990	7	23	34	80	267	493	525	569	417	155	29	36	2635
1991	1	6	66	149	348	448	528	513	371	176	23	22	2651
1992	0	9	31	97	249	437	548	454	373	102	20	13	2333
1993	5	2	21	33	224	455	469	485	357	165	37	0	2463
1994	0	18	33	161	268	443	469	485	357	155	66	6	2461
1995	0	12	55	106	338	416	554	571	417	145	20	35	2669
1996	1	23	30	69	335	419	526	473	349	125	44	17	2411
1997	15	18	109	42	236	377	521	519	435	163	7	0	2442
1998	0	0	31	73	280	359	531	581	569	412	196	52	2853
1999	24	25	11	212	280	432	516	611	340	173	19	3	2646
2000	16	32	48	69	377	446	607	592	383	151	67	0	2788
2001	1	27	16	177	303	427	524	492	338	104	62	12	2483

SNOWFALL (inches) 2001 MOBILE, AL (MOB)

YEAR	JUL	AUG	SEP	OCT	NOV	DEC	JAN	FEB	MAR	APR	MAY	JUN	TOTAL
1972-73	0.0	0.0	0.0	0.0	0.0	0.0	T	3.6	0.0	0.0	0.0	0.0	3.6
1973-74	0.0	0.0	0.0	0.0	0.0	T	0.0	0.0	0.0	0.0	0.0	0.0	T
1974-75	0.0	0.0	0.0	0.0	0.0	0.0	0.0	0.0	0.0	0.0	0.0	0.0	0.0
1975-76	0.0	0.0	0.0	0.0	0.0	0.0	0.0	0.0	0.0	0.0	0.0	0.0	0.0
1976-77	0.0	0.0	0.0	0.0	0.0	T	1.9	0.0	0.0	0.0	0.0	0.0	1.9
1977-78	0.0	0.0	0.0	0.0	0.0	0.0	0.4	T	0.0	0.0	0.0	0.0	0.4
1978-79	0.0	0.0	0.0	0.0	0.0	0.0	0.0	0.0	0.0	0.0	0.0	0.0	T
1979-80	0.0	0.0	0.0	0.0	0.0	0.0	0.0	T	T	0.0	0.0	0.0	T
1980-81	0.0	0.0	0.0	0.0	0.0	0.0	T	0.0	0.0	0.0	0.0	0.0	T
1981-82	0.0	0.0	0.0	0.0	0.0	0.0	T	0.0	0.0	0.0	0.0	0.0	T
1982-83	0.0	0.0	0.0	0.0	0.0	0.0	0.0	0.0	0.0	0.0	0.0	0.0	0.0
1983-84	0.0	0.0	0.0	0.0	0.0	T	0.0	0.0	T	0.0	0.0	0.0	T
1984-85	0.0	0.0	0.0	0.0	0.0	0.0	T	T	0.0	0.0	0.0	0.0	T
1985-86	0.0	0.0	0.0	0.0	0.0	0.0	0.0	0.0	0.0	0.0	0.0	0.0	0.0
1986-87	0.0	0.0	0.0	0.0	0.0	0.0	T	0.0	0.0	T	0.0	0.0	T
1987-88	0.0	0.0	0.0	0.0	0.0	0.0	0.0	1.7	0.0	T	0.0	0.0	1.7
1988-89	0.0	0.0	0.0	0.0	0.0	0.0	0.0	0.0	T	0.0	0.0	0.0	T
1989-90	T	0.0	0.0	0.0	0.0	T	0.0	0.0	0.0	0.0	0.0	0.0	T
1990-91	0.0	0.0	0.0	0.0	0.0	0.0	0.0	0.0	0.0	0.0	T	0.0	0.0
1991-92	0.0	0.0	0.0	0.0	0.0	0.0	0.0	0.0	0.0	0.0	0.0	0.0	0.0
1992-93	T	0.0	0.0	0.0	0.0	0.0	0.0	0.0	2.7	0.0	0.0	0.0	2.7
1993-94	0.0	0.0	0.0	0.0	0.0	T	0.0	T	0.0	0.0	0.0	0.0	T
1994-95	0.0	0.0	0.0	0.0	0.0	0.0	T	T	0.0	0.0	0.0	0.0	T
1995-96	T	0.0	0.0	0.0	0.0	T	0.0						
1996-97						1.0							
1997-98				0.0	0.0		0.0	0.0	0.0	0.0	0.0		
1998-99	0.0	0.0	0.0	0.0	0.0	0.0	0.0	0.0	0.0	T	0.0	0.0	T
1999-00	0.0	0.0	0.0	0.0	0.0	0.0	T	0.0	0.0	0.0	0.0	0.0	T
2000-01	0.0	0.0	0.0	0.0	0.0	0.0							
2001-	0.0	0.0	0.0	0.0	0.0								
POR= 57 YRS	T	0.0	0.0	0.0	T	0.1	0.1	0.1	0.1	T	T	0.0	0.4

2001
ANCHORAGE,
ALASKA (ANC)

Anchorage is in a broad valley with adjacent narrow bodies of water. Cook Inlet, including Knik Arm and Turnagain Arm, lies approximately 2 miles to the west, north, and south. The terrain rises gradually to the east for about 10 miles, with marshes interspersed with glacial moraines, shallow depressions, small streams, and knolls. Beyond this area, the Chugach Mountains rise abruptly into a range oriented north-northeast to south-southwest, with average elevation 4,000 to 5,000 feet and some peaks to 8,000 or 10,000 feet. The Chugach Range acts as a barrier to the influx of warm, moist air from the Gulf of Alaska, so the average annual precipitation is only 10 to 15 percent of that at stations located on the Gulf of Alaska side of the Chugach Range. The Alaska Mountain Range lies in a long arc from southwest, through northwest, to northeast, approximately 100 miles distant from Anchorage. During the winter, this range is an effective barrier to the influx of very cold air from the north side of the range.

The four seasons are well marked in Anchorage. In the summer, high temperatures average about 60 degrees and low temperatures nearly 50 degrees. Temperatures in the 70s are considered very warm. On summer days, temperatures on the east side of Anchorage may be about 10 degrees warmer than the official airport readings. Rain increases after mid-June. About two-thirds of the days in July and August are cloudy and one-third have rain.

Autumn is brief, beginning in early September and ending in mid-October. Temperatures begin to fall in September with snow becoming more frequent in October.

Winter can be considered as mid-October to early April when streams and lakes are frozen. Temperatures steadily decrease into January when the highs are near 20 degrees and lows near 5 degrees. The coldest weather is normally in January, when very cold days have high temperatures below zero. Cold days generally have clear skies and calm wind. Mild days do occur with temperatures in the 30s. On cold winter nights, temperatures on the east side of Anchorage may be 10-20 degrees lower than airport readings on the west side. Most winter precipitation is snow, but rain may occur on a few days.

Annual snowfall varies from about 70 inches on the west side to about 90 inches on the east side of Anchorage at low elevations. Along the Chugach Mountains, snow totals increase steadily with increasing elevations and winter arrives a month earlier and stays a month longer at the 1,000 to 2,000 foot level. Most snow is light or dry, i.e., low in water content. Freezing rain is extremely rare. Fog, made of water droplets, occurs on about fifteen days. In general, ice-fog does not occur in Anchorage.

Spring begins in late April and May when days are warm and sunny, nights are cool, and precipitation is exceedingly small. Foliage turns green by late May.

The wind in Anchorage is generally light. However, on several days each winter, strong northerly winds, up to 90 mph, affect the entire Anchorage area. Also during the winter there are about eight occurrences of very strong southeast winds which affect only the east side of Anchorage and the slopes of the Chugach Mountains. These winds occur more often above the 800 feet elevation in the Chugach where winds are funneled thru creek canyons. On the east side of Anchorage, damaging winds of over 100 mph have been recorded.

The average occurrence of the first snow is mid-October, but has occurred as early as mid-September. The average date of the last snow is mid-April, but has occurred as late as early May. The growing season is about 125 days. Average occurrence of the last temperature of 32 degrees in spring is mid-May and the first in fall is mid-September. Daylight varies from about 19 hours in late June to 6 hours in late December with 12 hours of daylight occurring in late September and late March.

NORMALS, MEANS, AND EXTREMES
ANCHORAGE, AK (ANC)

LATITUDE:	LONGITUDE:	ELEVATION (FT):		TIME ZONE:	WBAN: 26451
61 10' 30" N	149 59' 36" W	GRND: 130	BARO: 133	ALASKA (UTC + 9)	

	ELEMENT	POR	JAN	FEB	MAR	APR	MAY	JUN	JUL	AUG	SEP	OCT	NOV	DEC	YEAR
TEMPERATURE °F	NORMAL DAILY MAXIMUM	30	21.4	25.8	33.1	42.8	54.4	61.6	65.2	63.0	55.2	40.5	27.2	22.5	42.7
	MEAN DAILY MAXIMUM	48	21.5	25.6	33.0	43.5	54.9	62.4	65.1	63.1	55.2	40.2	27.6	22.3	42.9
	HIGHEST DAILY MAXIMUM	48	50	48	51	65	77	85	82	82	73	61	53	48	85
	YEAR OF OCCURRENCE		1961	1991	1984	1976	1969	1969	1989	1978	1957	1993	1979	1999	JUN 1969
	MEAN OF EXTREME MAXS.	48	40.1	41.8	44.4	54.2	66.7	73.5	75.4	72.8	64.5	53.2	42.3	40.7	55.8
	NORMAL DAILY MINIMUM	30	8.4	11.5	18.1	28.6	38.8	47.2	51.7	49.5	41.6	28.7	15.1	10.0	29.1
	MEAN DAILY MINIMUM	48	8.3	11.4	17.5	28.6	38.9	47.3	51.5	49.4	41.2	28.1	15.8	9.6	29.0
	LOWEST DAILY MINIMUM	48	-34	-28	-24	-4	17	33	38	31	19	-5	-21	-30	-34
	YEAR OF OCCURRENCE		1975	1999	1971	1985	1964	1961	1964	1984	1992	1956	1956	1964	JAN 1975
	MEAN OF EXTREME MINS.	48	-13.2	-8.6	-.8	16.9	29.7	39.4	44.3	39.6	29.3	11.2	-3.0	-12.3	14.4
	NORMAL DRY BULB	30	14.9	18.7	25.7	35.8	46.6	54.4	58.4	56.3	48.4	34.6	21.2	16.3	35.9
	MEAN DRY BULB	48	14.9	18.6	25.4	36.1	46.9	54.9	58.4	56.2	48.2	34.1	21.6	15.9	35.9
	MEAN WET BULB	18	16.7	18.0	24.1	33.0	41.8	49.5	54.0	52.4	45.2	31.1	20.5	18.2	33.7
	MEAN DEW POINT	18	12.4	13.3	17.5	25.5	34.0	43.4	49.8	48.3	41.0	26.4	16.4	14.6	28.6
	NORMAL NO. DAYS WITH:														
	MAXIMUM 70	30	0.0	0.0	0.0	0.0	0.6	3.4	6.5	3.4	0.1	0.0	0.0	0.0	14.0
	MAXIMUM 32	30	24.8	19.9	11.5	2.0	0.0	0.0	0.0	0.0	0.0	4.6	20.8	24.5	108.1
	MINIMUM 32	30	30.5	27.3	28.3	20.3	2.7	0.0	0.0	0.1	3.2	19.8	28.2	30.2	190.6
	MINIMUM 0	30	9.8	7.2	2.3	0.0	0.0	0.0	0.0	0.0	0.0	0.1	3.2	7.2	29.8
H/C	NORMAL HEATING DEG. DAYS	30	1553	1296	1218	876	570	318	205	270	498	942	1314	1510	10570
	NORMAL COOLING DEG. DAYS	30	0	0	0	0	0	0	0	0	0	0	0	0	0
RH	NORMAL (PERCENT)														
	HOUR 03 LST	30	74	73	71	72	72	75	80	83	82	78	79	77	76
	HOUR 09 LST	30	74	74	70	66	64	68	73	78	80	78	78	77	73
	HOUR 15 LST	30	72	67	57	54	50	56	62	65	64	67	74	76	64
	HOUR 21 LST	30	73	71	68	64	59	62	69	76	78	76	78	77	71
S	PERCENT POSSIBLE SUNSHINE	40	34	42	50	50	50	46	42	38	38	35	31	26	40
W/O	MEAN NO. DAYS WITH:														
	HEAVY FOG(VISBY 1/4 MI)	48	5.7	4.3	1.6	0.7	0.3	0.1	0.2	0.9	1.3	2.1	3.5	4.7	25.4
	THUNDERSTORMS	48	0.0	0.0	0.0	0.0	0.1	0.1	0.5	0.2	0.1	0.0	0.0	0.0	1.0
CLOUDINESS	MEAN:														
	SUNRISE-SUNSET (OKTAS)	44	5.6	5.7	5.4	5.8	6.1	6.3	6.3	6.3	6.3	6.1	5.8	6.0	6.0
	MIDNIGHT-MIDNIGHT (OKTAS)	33	5.5	5.4	5.3	5.6	6.1	6.4	6.4	6.2	6.0	5.8	5.6	5.9	5.9
	MEAN NO. DAYS WITH:														
	CLEAR	45	7.1	6.3	7.7	5.7	3.9	2.8	3.2	3.2	3.6	5.0	5.5	5.6	59.6
	PARTLY CLOUDY	45	4.8	3.7	5.6	6.1	6.6	6.9	5.8	6.1	5.3	4.6	4.6	3.9	64.0
	CLOUDY	45	19.3	18.1	17.6	18.3	20.4	20.2	21.2	21.1	20.5	20.6	19.3	20.9	237.5
PR	MEAN STATION PRESSURE(IN)	29	29.48	29.58	29.58	29.60	29.71	29.80	29.80	29.79	29.60	29.50	29.49	29.48	29.62
	MEAN SEA-LEVEL PRES. (IN)	18	29.64	29.75	29.73	29.77	29.86	29.91	29.97	29.90	29.77	29.66	29.63	29.62	29.77
WINDS	MEAN SPEED (MPH)	48	6.4	6.9	7.1	7.3	8.5	8.4	7.3	6.9	6.7	6.8	6.4	6.2	7.1
	PREVAIL.DIR(TENS OF DEGS)	38	36	36	36	16	16	16	16	16	16	36	01	01	36
	MAXIMUM 2-MINUTE:														
	SPEED (MPH)	3	34	41	29	33	26	33	30	29	33	29	31	38	41
	DIR. (TENS OF DEGS)		16	15	02	15	16	18	17	04	15	18	17	16	15
	YEAR OF OCCURRENCE		1999	2000	1999	2001	1999	1999	2001	2000	2000	2000	2000	1999	FEB 2000
	MAXIMUM 5-SECOND:														
	SPEED (MPH)	3	47	58	41	45	36	44	40	38	47	40	39	49	58
	DIR. (TENS OF DEGS)		16	12	11	18	18	17	15	15	15	17	16	16	12
	YEAR OF OCCURRENCE		1999	2000	2000	2001	2001	1999	1999	2001	2000	2000	2000	1999	FEB 2000
PRECIPITATION	NORMAL (IN)	30	0.79	0.78	0.69	0.67	0.73	1.14	1.71	2.44	2.70	2.03	1.11	1.12	15.91
	MAXIMUM MONTHLY (IN)	48	2.13	3.07	2.76	1.91	1.93	3.40	4.49	9.77	6.64	4.11	2.84	2.67	9.77
	YEAR OF OCCURRENCE		1949	1955	1979	1977	1989	1962	2001	1989	1990	1986	1976	1955	AUG 1989
	MINIMUM MONTHLY (IN)	48	0.02	0.07	T	T	0.02	0.17	0.42	0.33	0.72	0.35	0.08	0.09	T
	YEAR OF OCCURRENCE		1982	1958	1983	1969	1957	1993	1972	1969	1998	1960	1985	1995	MAR 1983
	MAXIMUM IN 24 HOURS (IN)	48	1.19	1.16	1.25	0.78	1.18	1.84	2.37	4.12	1.92	1.60	1.66	1.62	4.12
	YEAR OF OCCURRENCE		1961	1956	1986	1989	1980	1962	2001	1989	1961	1986	1964	1955	AUG 1989
	NORMAL NO. DAYS WITH:														
	PRECIPITATION 0.01	30	7.7	8.0	7.3	5.8	7.1	7.9	11.5	13.4	14.5	12.2	9.6	11.0	116.0
	PRECIPITATION 1.00	30	*	0.0	*	0.0	0.0	0.1	0.1	0.1	0.1	0.2	*	0.0	0.5
SNOWFALL	NORMAL (IN)	30	8.8	11.0	9.1	5.8	0.2	0.0	0.0	0.0	0.0	0.3	8.0	10.5	13.9	67.6
	MAXIMUM MONTHLY (IN)	48	28.6	52.1	31.0	27.6	6.1			0.0	T	4.6	28.1	38.8	41.6	52.1
	YEAR OF OCCURRENCE		2000	1996	1979	1963	2001			1997	1965	1996	1994	1955	FEB 1996
	MAXIMUM IN 24 HOURS (IN)	48	10.5	13.9	14.5	9.1	5.0	0.0	0.0	T	3.5	14.6	16.4	17.7	17.7
	YEAR OF OCCURRENCE		1955	1996	1959	1955	2001			1997	1965	1996	1964	1955	DEC 1955
	MAXIMUM SNOW DEPTH (IN)	47	833	840	906	356	17	0	0	0	1	105	416	715	906
	YEAR OF OCCURRENCE		1956	1956	1959	1955	1955				1992	1991	1994	1994	MAR 1959
	NORMAL NO. DAYS WITH:														
	SNOWFALL 1.0	30	2.8	3.3	2.7	1.5	0.0	0.0	0.0	0.0	0.2	2.3	3.5	4.6	20.9

PRECIPITATION (inches) 2001 ANCHORAGE, AK (ANC)

YEAR	JAN	FEB	MAR	APR	MAY	JUN	JUL	AUG	SEP	OCT	NOV	DEC	ANNUAL
1972	0.56	0.63	0.68	0.73	0.81	0.61	0.42	1.40	4.42	2.89	0.76	0.72	14.63
1973	0.72	0.11	0.65	0.33	0.14	1.07	0.60	3.40	0.76	1.74	0.78	0.38	10.68
1974	0.02	1.15	0.60	0.61	0.34	0.69	1.22	1.62	1.53	2.63	1.01	2.00	13.42
1975	0.43	0.77	0.54	1.71	0.40	0.47	1.33	1.19	4.52	0.69	0.10	0.89	13.04
1976	0.98	0.33	1.77	0.74	0.16	0.33	0.60	0.97	3.50	1.29	2.84	1.03	14.54
1977	1.35	0.52	0.84	1.91	0.46	0.49	1.35	1.35	4.08	1.92	0.53	0.69	15.51
1978	0.39	1.19	0.45	0.02	0.03	3.09	1.78	0.54	2.16	1.65	0.85	2.60	14.75
1979	0.23	0.69	2.76	0.94	0.15	1.79	3.84	1.56	2.73	2.54	2.77	1.15	21.15
1980	1.28	1.18	0.30	0.19	1.68	2.73	2.27	3.06	2.53	3.05	0.49	0.41	19.17
1981	0.93	0.97	0.41	0.19	0.81	0.83	4.39	4.96	2.15	3.49	1.85	0.36	21.34
1982	0.02	0.69	0.42	0.27	0.54	1.56	2.41	2.33	4.66	2.95	1.72	0.11	17.68
1983	0.21	0.23	T	1.36	0.59	0.66	0.55	2.89	2.29	2.67	0.23	0.48	12.16
1984	1.30	1.08	0.08	0.93	0.96	1.10	1.11	3.21	2.59	1.38	0.15	1.08	14.97
1985	0.70	0.67	0.86	0.50	1.45	1.01	0.99	3.54	3.17	1.07	0.08	1.47	15.51
1986	0.20	0.55	1.70	0.42	0.50	0.33	2.02	3.62	2.85	4.11	1.23	1.42	18.95
1987	1.72	0.20	0.17	0.24	0.67	1.09	1.89	0.43	1.91	2.60	1.90	1.12	13.94
1988	0.38	0.32	0.65	0.37	0.56	0.79	0.64	3.77	1.26	2.96	1.11	1.51	14.32
1989	0.26	0.17	0.22	0.98	1.93	1.14	2.89	9.77	3.92	3.63	1.01	1.63	27.55
1990	1.42	1.46	0.46	0.27	0.71	1.52	0.81	1.90	6.64	0.73	1.31	1.78	19.01
1991	0.62	0.42	0.65	0.23	0.12	0.18	2.82	3.54	3.41	1.93	1.57	1.82	17.31
1992	1.17	1.04	0.31	0.08	0.58	1.21	0.79	2.49	2.83	2.08	1.17	0.69	14.44
1993	0.94	1.17	0.29	0.09	1.17	0.17	0.57	4.02	4.27	1.90	2.00	0.30	16.89
1994	0.59	0.28	1.51	0.45	0.51	1.34	0.57	1.02	1.66	1.21	2.47	1.51	13.12
1995	0.52	1.00	0.88	0.08	1.11	0.91	3.01	2.19	2.93	0.95	0.09	0.09	13.76
1996	0.11	2.40	0.42	0.08	0.20	0.50	2.04	2.53	1.93	2.63	1.38	0.24	14.46
1997	0.12	0.52	0.01	0.25	1.12	0.60	1.36	8.37	2.53	1.93	0.87	1.80	19.48
1998	0.45	0.24	0.07	0.39	0.63	2.70	1.01	3.25	0.72	0.54	0.18	1.47	11.65
1999	0.37	0.28	0.61	0.29	1.30	1.10	2.15	4.62	3.17	2.63	0.35	1.43	18.30
2000	1.04	0.54	0.48	0.39	0.69	1.43	2.58	1.68	3.24	0.59	1.13	0.58	14.37
2001	1.10	0.85	0.88	0.34	0.48	0.24	4.49	0.97	1.14	1.57	0.26	0.20	12.52
POR= 58 YRS	0.79	0.79	0.62	0.51	0.66	1.09	1.86	2.63	2.57	1.71	0.98	1.01	15.22

AVERAGE TEMPERATURE (F) 2001 ANCHORAGE, AK (ANC)

YEAR	JAN	FEB	MAR	APR	MAY	JUN	JUL	AUG	SEP	OCT	NOV	DEC	ANNUAL
1972	6.4	13.5	15.7	26.8	43.3	51.9	59.0	56.6	44.5	31.7	21.3	12.4	31.9
1973	2.9	13.1	24.2	35.8	43.6	51.4	57.8	53.8	45.7	32.2	13.6	18.3	32.7
1974	6.8	14.1	23.3	37.9	47.9	55.5	57.3	56.3	49.8	34.5	22.6	18.8	35.4
1975	11.9	12.9	22.5	32.9	46.3	53.0	58.6	56.6	49.3	34.6	14.2	11.6	33.7
1976	17.1	12.8	24.1	34.8	44.9	53.7	58.9	56.3	47.4	33.4	30.6	23.1	36.4
1977	32.0	32.7	24.7	35.7	46.9	57.8	62.6	60.3	50.7	38.3	15.3	11.3	39.0
1978	21.2	26.3	29.3	39.1	49.0	54.5	58.8	59.8	51.5	39.3	26.3	21.4	39.7
1979	22.3	10.6	31.6	38.8	50.2	55.9	60.4	58.8	52.0	41.1	33.5	10.0	38.8
1980	14.3	27.4	27.2	39.3	45.8	53.2	57.0	54.4	46.7	37.2	27.6	0.8	35.9
1981	31.5	24.8	34.4	36.0	50.7	53.8	57.4	54.8	47.9	36.0	21.8	15.9	38.8
1982	6.4	15.5	26.3	33.1	44.5	52.9	56.2	54.7	47.5	26.6	21.0	21.5	33.9
1983	16.2	21.4	28.7	37.4	48.7	55.9	58.5	56.1	45.3	34.4	24.9	16.7	37.0
1984	18.8	19.4	36.4	38.8	49.6	58.8	60.8	56.6	49.3	35.5	19.8	18.9	38.6
1985	30.3	13.5	26.7	28.4	45.1	51.9	58.5	55.2	47.6	30.3	14.0	27.5	35.8
1986	25.6	21.8	24.1	31.0	46.6	54.6	58.0	54.3	48.6	39.0	25.0	28.3	38.1
1987	22.8	25.3	26.8	37.9	47.2	51.9	57.1	57.3	48.0	38.9	26.9	18.2	38.2
1988	18.0	22.7	31.3	37.1	48.5	55.2	58.8	56.0	48.0	33.3	20.5	22.0	37.6
1989	3.5	17.6	23.6	39.3	46.3	55.3	59.4	59.0	50.6	34.0	17.2	24.2	35.8
1990	15.5	3.8	28.6	39.9	49.9	57.1	58.6	57.8	49.6	32.3	9.9	14.8	34.8
1991	15.9	19.6	23.7	37.7	46.6	55.7	57.5	55.5	51.0	33.0	25.0	20.5	36.8
1992	20.3	15.0	24.9	35.2	46.1	55.9	59.5	55.8	40.3	31.2	27.1	15.0	35.5
1993	14.5	21.0	28.8	40.6	50.7	56.3	61.1	58.8	48.8	38.7	25.2	24.0	39.0
1994	21.7	17.1	25.7	38.6	47.1	56.7	58.8	58.8	48.6	33.5	15.3	15.7	36.5
1995	15.6	20.7	18.6	40.4	48.8	56.0	59.7	57.9	53.7	38.1	21.0	19.0	37.4
1996	6.1	15.8	29.2	38.6	50.1	56.9	59.9	56.7	46.5	25.4	19.0	13.0	34.8
1997	15.9	30.7	24.6	38.2	48.0	56.7	60.8	58.1	50.4	29.6	28.0	16.8	38.2
1998	15.5	25.8	30.1	40.2	47.3	54.7	57.3	53.8	49.0	35.9	23.5	14.3	37.3
1999	11.9	7.2	24.1	34.5	45.7	55.3	58.4	56.9	48.9	34.0	19.7	14.9	34.3
2000	14.8	25.9	29.6	37.4	46.1	55.1	56.8	54.8	47.1	34.8	29.7	25.1	38.1
2001	27.5	23.1	28.8	37.7	44.8	58.1	57.7	58.4	49.2	30.1	19.6	10.8	37.2
POR= 58 YRS	14.2	18.1	24.8	35.6	46.5	54.5	58.2	56.0	48.0	34.3	21.7	15.5	35.6

REFERENCE NOTES:

PAGE 1:
THE TEMPERATURE GRAPH SHOWS NORMAL MAXIMUM AND NORMAL
MINIMUM DAILY TEMPERATURES (SOLID CURVES) AND THE
ACTUAL DAILY HIGH AND LOW TEMPERATURES (VERTICAL BARS).

PAGE 2 AND 3:
H/C INDICATES HEATING AND COOLING DEGREE DAYS.
RH INDICATES RELATIVE HUMIDITY.
W/O INDICATES WEATHER AND OBSTRUCTIONS
S INDICATES SUNSHINE.
PR INDICATES PRESSURE.
CLOUDINESS ON PAGE 3 IS THE SUM OF THE CEILOMETER AND
SATELLITE DATA NOT TO EXCEED EIGHT EIGHTHS (OKTAS).

GENERAL:
T INDICATES TRACE PRECIPITATION, AN AMOUNT GREATER
THAN ZERO BUT LESS THAN THE LOWEST REPORTABLE VALUE.
+ INDICATES THE VALUE ALSO OCCURS ON EARLIER DATES.
BLANK ENTRIES DENOTE MISSING OR UNREPORTED DATA.
NORMALS ARE 30-YEAR AVERAGES (1961 - 1990).
ASOS INDICATES AUTOMATED SURFACE OBSERVING SYSTEM.
PM INDICATES THE LAST DAY OF THE PREVIOUS MONTH.
POR (PERIOD OF RECORD) BEGINS WITH THE JANUARY DATA
MONTH AND IS THE NUMBER OF YEARS USED TO COMPUTE
THE MEAN. INDIVIDUAL MONTHS WITHIN THE POR MAY
BE MISSING.
WHEN THE POR FOR A NORMAL IS LESS THAN 30 YEARS,
THE NORMAL IS PROVISIONAL AND IS BASED ON THE NUMBER
OF YEARS INDICATED.
0.* OR * INDICATES THE VALUE OR MEAN-DAYS-WITH
IS BETWEEN 0.00 AND 0.05.
CLOUDINESS FOR ASOS STATIONS DIFFERS FROM THE NON-ASOS
OBSERVATION TAKEN BY A HUMAN OBSERVER. ASOS STATION
CLOUDINESS IS BASED ON TIME-AVERAGED CEILOMETER DATA
FOR CLOUDS AT OR BELOW 12,000 FEET AND ON SATELLITE
DATA FOR CLOUDS ABOVE 12,000 FEET.
THE NUMBER OF DAYS WITH CLEAR, PARTLY CLOUDY, AND
CLOUDY CONDITIONS FOR ASOS STATIONS IS THE SUM
OF THE CEILOMETER AND SATELLITE DATA FOR THE
SUNRISE TO SUNSET PERIOD.

GENERAL CONTINUED:
CLEAR INDICATES 0 - 2 OKTAS, PARTLY CLOUDY INDICATES
3 - 6 OKTAS, AND CLOUDY INDICATES 7 OR 8 OKTAS.
WHEN AT LEAST ONE OF THE ELEMENTS (CEILOMETER OR
SATELLITE) IS MISSING, THE DAILY CLOUDINESS IS
NOT COMPUTED.
WIND DIRECTION IS RECORDED IN TENS OF DEGREES (2 DIGITS)
CLOCKWISE FROM TRUE NORTH. "00" INDICATES CALM. "36"
INDICATES TRUE NORTH.
RESULTANT WIND IS THE VECTOR AVERAGE OF THE SPEED AND
DIRECTION.
AVERAGE TEMPERATURE IS THE SUM OF THE MEAN DAILY MAXIMUM
AND MINIMUM TEMPERATURE DIVIDED BY 2.
SNOWFALL DATA COMPRISE ALL FORMS OF FROZEN
PRECIPITATION, INCLUDING HAIL.
A HEATING (COOLING) DEGREE DAY IS THE DIFFERENCE BETWEEN
THE AVERAGE DAILY TEMPERATURE AND 65 F.
DRY BULB IS THE TEMPERATURE OF THE AMBIENT AIR.
DEW POINT IS THE TEMPERATURE TO WHICH THE AIR MUST BE
COOLED TO ACHIEVE 100 PERCENT RELATIVE HUMIDITY.
WET BULB IS THE TEMPERATURE THE AIR WOULD HAVE IF THE
MOISTURE CONTENT WAS INCREASED TO 100 PERCENT RELATIVE
HUMIDITY.

ON JULY 1, 1996, THE NATIONAL WEATHER SERVICE BEGAN USING
THE "METAR" OBSERVATION CODE THAT WAS ALREADY EMPLOYED
BY MOST OTHER NATIONS OF THE WORLD. THE MOST NOTICEABLE
DIFFERENCE IN THIS ANNUAL PUBLICATION WILL BE THE CHANGE
IN UNITS FROM TENTHS TO EIGHTS (OKTAS) FOR REPORTING THE
AMOUNT OF SKY COVER.

HEATING DEGREE DAYS (base 65 F) 2001 ANCHORAGE, AK (ANC)

YEAR	JUL	AUG	SEP	OCT	NOV	DEC	JAN	FEB	MAR	APR	MAY	JUN	TOTAL
1972-73	185	252	608	1025	1308	1627	1925	1448	1258	866	654	399	11555
1973-74	216	342	573	1012	1532	1440	1797	1416	1285	805	526	279	11223
1974-75	235	263	452	937	1263	1425	1643	1454	1313	954	575	354	10868
1975-76	192	252	463	937	1517	1654	1511	1260	1485	897	615	332	11115
1976-77	184	262	521	972	1028	1294	1017	897	1241	872	554	208	9050
1977-78	75	144	421	820	1486	1659	1349	1077	1100	771	491	308	9701
1978-79	186	160	400	792	1153	1344	1321	1029	1515	781	454	268	9403
1979-80	138	184	384	735	937	1704	1568	1083	1164	764	592	347	9600
1980-81	243	320	542	855	1115	1990	1032	1122	943	863	438	329	9792
1981-82	230	307	507	893	1290	1516	1813	1382	1191	949	625	356	11059
1982-83	261	313	520	1184	1315	1342	1507	1216	1117	821	500	267	10363
1983-84	194	269	585	945	1194	1491	1425	1319	880	778	471	179	9730
1984-85	129	254	464	906	1350	1423	1070	1437	1182	1091	610	388	10304
1985-86	193	298	516	1065	1523	1155	1215	1206	1260	1013	564	307	10315
1986-87	215	325	486	800	1194	1133	1303	1104	1176	805	543	386	9470
1987-88	243	232	506	801	1136	1444	1450	1221	1037	830	504	285	9689
1988-89	184	270	503	975	1331	1326	1908	1322	1277	765	573	286	10720
1989-90	173	181	423	956	1428	1255	1533	1715	1121	746	465	237	10233
1990-91	191	222	457	1006	1648	1552	1518	1265	1273	813	563	273	10781
1991-92	226	287	414	988	1193	1373	1380	1444	1240	891	579	268	10283
1992-93	161	280	735	1039	1131	1543	1563	1226	1117	725	436	252	10208
1993-94	125	187	477	808	1191	1267	1334	1335	1212	785	548	243	9512
1994-95	183	190	485	968	1488	1523	1526	1239	1433	734	496	265	10530
1995-96	172	214	335	826	1314	1425	1827	1423	1102	783	456	239	10116
1996-97	151	251	549	1220	1375	1608	1516	956	1246	796	520	249	10437
1997-98	123	207	432	1090	1103	1486	1530	1093	1073	739	540	302	9718
1998-99	232	340	475	895	1240	1566	1638	1611	1262	908	592	286	11045
1999-00	204	248	478	953	1352	1544	1548	1126	1090	821	579	291	10234
2000-01	245	312	534	931	1052	1230	1154	1168	1113	812	618	199	9368
2001-	220	200	466	1072	1356	1673							

COOLING DEGREE DAYS (base 65 F) 2001 ANCHORAGE, AK (ANC)

YEAR	JAN	FEB	MAR	APR	MAY	JUN	JUL	AUG	SEP	OCT	NOV	DEC	ANNUAL
1972	0	0	0	0	0	0	5	0	0	0	0	0	5
1973	0	0	0	0	0	0	0	0	0	0	0	0	0
1974	0	0	0	0	0	0	1	0	0	0	0	0	1
1975	0	0	0	0	0	0	2	0	0	0	0	0	2
1976	0	0	0	0	0	0	3	0	0	0	0	0	3
1977	0	0	0	0	0	0	8	3	0	0	0	0	11
1978	0	0	0	0	0	0	1	7	0	0	0	0	8
1979	0	0	0	0	0	0	4	0	0	0	0	0	4
1980	0	0	0	0	0	0	0	0	0	0	0	0	0
1981	0	0	0	0	0	0	0	0	0	0	0	0	0
1982	0	0	0	0	0	0	0	0	0	0	0	0	0
1983	0	0	0	0	0	0	0	0	0	0	0	0	0
1984	0	0	0	0	0	0	5	1	0	0	0	0	6
1985	0	0	0	0	0	0	0	0	0	0	0	0	0
1986	0	0	0	0	0	0	4	0	0	0	0	0	4
1987	0	0	0	0	0	0	2	0	0	0	0	0	2
1988	0	0	0	0	0	0	0	0	0	0	0	0	0
1989	0	0	0	0	0	0	5	2	0	0	0	0	7
1990	0	0	0	0	0	3	1	2	0	0	0	0	6
1991	0	0	0	0	0	0	0	0	0	0	0	0	0
1992	0	0	0	0	0	0	0	0	0	0	0	0	0
1993	0	0	0	0	0	0	11	0	0	0	0	0	11
1994	0	0	0	0	0	0	0	2	0	0	0	0	2
1995	0	0	0	0	0	1	0	0	0	0	0	0	1
1996	0	0	0	0	0	0	0	0	0	0	0	0	0
1997	0	0	0	0	0	5	0	1	0	0	0	0	6
1998	0	0	0	0	0	0	0	0	0	0	0	0	0
1999	0	0	0	0	0	0	7	0	0	0	0	0	7
2000	0	0	0	0	0	0	0	0	0	0	0	0	0
2001	0	0	0	0	0	0	0	0	0	0	0	0	0

SNOWFALL (inches) 2001 ANCHORAGE, AK (ANC)

YEAR	JUL	AUG	SEP	OCT	NOV	DEC	JAN	FEB	MAR	APR	MAY	JUN	TOTAL
1972-73	0.0	0.0	1.5	3.3	10.7	6.5	8.1	1.0	16.1	1.3	0.0	0.0	48.5
1973-74	0.0	0.0	0.0	6.6	10.6	6.7	0.5	23.3	8.2	1.9	0.0	0.0	57.8
1974-75	0.0	0.0	0.0	4.4	8.4	29.2	5.7	15.4	8.3	16.1	0.4	0.0	87.9
1975-76	0.0	0.0	0.0	T	2.0	11.5	9.7	1.8	30.7	5.6	T	0.0	61.3
1976-77	0.0	0.0	0.0	11.4	11.1	13.8	6.1	2.1	9.5	14.0	0.0	0.0	68.0
1977-78	0.0	0.0	1.0	13.2	12.6	10.6	7.3	20.8	T	2.8	0.0	0.0	75.0
1978-79	0.0	0.0	0.0	3.9	8.5	35.2	3.6	6.2	31.0	0.8	0.0	0.0	91.2
1979-80	0.0	0.0	0.0	4.3	13.7	16.0	12.0	18.7	3.4	4.4	1.1	0.0	68.9
1980-81	0.0	0.0	0.0	10.2	4.2	1.4	5.0	6.6	4.4	1.1	T	0.0	32.9
1981-82	0.0	0.0	1.5	6.3	20.0	7.6	0.5	0.6	5.6	3.5	0.7	0.0	46.3
1982-83	0.0	0.0	0.0	27.1	23.4	1.9	3.7	4.3	T	11.0	0.0	0.0	71.4
1983-84	0.0	0.0	T	23.7	2.1	10.5	15.0	18.9	0.2	9.8	0.0	0.0	80.2
1984-85	0.0	0.0	0.0	3.3	1.8	18.0	9.7	7.9	12.8	7.3	1.3	0.0	62.1
1985-86	0.0	0.0	0.0	0.8	1.5	6.1	5.1	6.1	21.0	5.4	0.1	0.0	46.1
1986-87	0.0	0.0	0.0	T	3.8	10.1	18.5	2.2	2.5	1.6	0.0	0.0	38.7
1987-88	0.0	0.0	0.0	T	29.2	26.3	4.7	9.2	8.5	2.0	0.0	0.0	79.9
1988-89	0.0	0.0	0.0	12.0	15.3	18.6	10.1	2.3	5.1	T	0.2	0.0	63.6
1989-90	0.0	0.0	0.0	16.3	10.1	20.0	27.5	23.0	4.7	0.8	T	0.0	102.4
1990-91	0.0	0.0	0.0	1.6	16.9	21.4	7.7	5.4	12.7	T	0.0	0.0	65.7
1991-92	0.0	0.0	0.0	11.6	19.3	26.2	21.4	18.3	2.7	T	0.2	0.0	99.7
1992-93	0.0	0.0	3.0	13.0	9.1	12.1	13.7	18.3	5.7	0.0	0.0	0.0	74.9
1993-94	0.0	0.0	T	4.4	11.9	5.1	7.5	1.7	29.9	6.0	0.0	0.0	66.5
1994-95	0.0	0.0	0.0	9.1	38.8	29.0	12.6	15.3	16.7	0.0	0.0	0.0	121.5
1995-96	0.0	0.0	0.0	4.0	0.9	2.5	2.5	52.1	6.1	0.9	0.0	0.0	69.0
1996-97	0.0	0.0	0.1	28.1	25.7	4.7	3.1	5.1	0.8	0.2	0.0	0.0	67.8
1997-98	0.0	T	0.0	11.6	6.4	26.6	6.8	3.1	1.2	2.9	0.0	0.0	79.3
1998-99	0.0	T	0.0	0.4	9.1	34.5	8.3	6.6	17.4	3.0	T	0.0	76.2
1999-00	0.0	0.0	0.0	5.7	8.5	18.6	28.6	4.6	7.4	2.8	0.0	0.0	76.2
2000-01	0.0	0.0	T	2.1	3.2	4.2	11.0	19.7	15.9	1.3	6.1	0.0	63.5
2001-	0.0	0.0	0.0	20.6	6.6	7.9							
POR= 58 YRS	0.0	0.0	0.3	7.3	10.7	15.3	10.4	11.5	9.2	4.9	1.0	0.0	70.6

2001
FAIRBANKS,
ALASKA (FAI)

Fairbanks is located in the Tanana Valley, in the interior of Alaska. It has a distinctly continental climate, with large variation of temperature from winter to summer.

The climate in Fairbanks is conditioned mainly by the response of the land mass to large changes in solar heat received by the area during the year. The sun is above the horizon from 18 to 21 hours during June and July. During this period, daily average maximum temperatures reach the lower 70s. Temperatures of 80 degrees or higher occur on about 10 days each summer. In contrast, from November to early March, when the period of daylight ranges from 10 to less than 4 hours per day, the lowest temperature readings normally fall below zero quite regularly. Low temperatures of -40 degrees or colder occur each winter. The range of temperatures in summer is comparatively low, from the lower 30s to the mid 90s. In winter, this range is larger, from about 65 below to 45 degrees above. This large winter range of temperature reflects the great difference between frigid weather associated with dry northerly airflow from the Arctic to mild temperatures associated with southerly airflow from the Gulf of Alaska, accompanied by chinook winds off the Alaska Range, 80 miles to the south of Fairbanks.

Snow cover is persistent in Fairbanks, without interruption, from October through April. Snowfalls of 4 inches or more in a day occur only three times during winter. Blizzard conditions are almost never seen, as winds in Fairbanks are above 20 miles an hour less than 1 percent of the time. Precipitation normally reaches a minimum in spring, and a maximum in August, when rainfall is common. During summer, thunderstorms occur in Fairbanks on an average of about eight days. Thunderstorms are about three times more frequent over the hills to the north and east of Fairbanks. Damaging hail or wind rarely accompany thunderstorms around Fairbanks.

There are rolling hills reaching elevations up to 2,000 feet above Fairbanks to the north and east of the city. During winter, the uplands are often warmer than Fairbanks, as cold air settles into the valley. In some months, temperatures in the uplands will average more than 10 degrees warmer than Fairbanks. During summer, the uplands are a few degrees cooler than the city. Precipitation in the uplands around Fairbanks is heavier than it is in the city by roughly 20 to 50 percent. Fairbanks exhibits an urban heat island, especially during winter. Low lying areas nearby, such as the community of North Pole, are often colder than the city, sometimes by as much as 15 degrees.

During winter, with temperatures of -20 degrees or colder, ice fog frequently forms in the city. Cold snaps accompanied by ice fog generally last about a week, but can last three weeks in unusual situations. The fog is almost always less than 300 feet deep, so that the surrounding uplands are usually in the clear, with warmer temperatures. Visibility in the ice fog is sometimes quite low, and this can hinder aircraft operations for as much as a day in severe cases. Aside from the low visibility in winter ice fog, flying weather in Fairbanks is quite favorable, especially from February through May, when crystal clear weather is common and the length of daylight is rapidly increasing.

Hardy vegetables and grains grow luxuriantly. Freezing of local rivers normally begins in the first week of October. The date when ice will normally support a persons weight is October 27. Rivers remain frozen and safe for travel until early April. Breakup of the river ice usually occurs in the first week of May.

NORMALS, MEANS, AND EXTREMES
FAIRBANKS, AK (FAI)

LATITUDE:	LONGITUDE:	ELEVATION (FT):		TIME ZONE:	WBAN: 26411
64 49' 00" N	147 51' 18" W	GRND: 461	BARO: 464	ALASKA (UTC + 9)	

	ELEMENT	POR	JAN	FEB	MAR	APR	MAY	JUN	JUL	AUG	SEP	OCT	NOV	DEC	YEAR
TEMPERATURE °F	NORMAL DAILY MAXIMUM	30	-1.6	7.2	23.8	41.0	59.3	70.1	72.3	66.3	54.8	32.0	10.9	1.8	36.5
	MEAN DAILY MAXIMUM	53	-1.1	7.5	24.1	42.1	59.6	70.5	72.2	66.1	54.4	31.9	11.2	1.0	36.6
	HIGHEST DAILY MAXIMUM	50	50	47	56	74	89	96	94	93	84	65	49	45	96
	YEAR OF OCCURRENCE		1981	1987	1994	1960	1960	1969	1975	1994	1957	1969	1997	1999	JUN 1969
	MEAN OF EXTREME MAXS.	53	29.0	33.0	44.6	59.7	74.6	83.9	85.3	80.5	68.8	51.5	34.3	29.9	56.3
	NORMAL DAILY MINIMUM	30	-18.5	-14.4	-1.7	20.4	38.0	49.5	52.6	47.2	36.2	18.1	-5.6	-14.8	17.2
	MEAN DAILY MINIMUM	53	-18.8	-14.6	-2.6	20.1	37.4	48.8	51.7	46.5	35.4	16.8	-5.1	-15.9	16.6
	LOWEST DAILY MINIMUM	50	-61	-58	-49	-24	-1	31	35	27	3	-27	-46	-62	-62
	YEAR OF OCCURRENCE		1969	1993	1956	1986	1964	1963	1959	1987	1992	1992	1990	1961	DEC 1961
	MEAN OF EXTREME MINS.	53	-43.5	-38.3	-26.6	-2.7	26.1	38.4	42.7	34.5	22.4	-5.9	-27.8	-39.7	-1.7
	NORMAL DRY BULB	30	-10.1	-3.6	11.0	30.7	48.6	59.8	62.5	56.8	45.5	25.1	2.7	-6.5	26.9
	MEAN DRY BULB	53	-10.0	-3.6	10.9	31.2	48.4	59.7	62.1	56.1	44.8	24.5	3.1	-7.6	26.6
	MEAN WET BULB	13	-2.9	-1.5	10.8	28.4	41.4	52.5	55.9	51.5	40.6	21.0	.2	-1.1	24.7
	MEAN DEW POINT	13	-6.0	-6.7	3.1	18.2	30.3	44.4	50.0	46.9	35.8	17.0	-3.9	-4.6	18.7
	NORMAL NO. DAYS WITH:														
	MAXIMUM 70	30	0.0	0.0	0.0	0.0	0.1	4.2	17.0	21.5	10.9	1.5	0.0	0.0	55.2
	MAXIMUM 32	30	29.8	25.8	21.5	6.6	0.1	0.0	0.0	0.0	0.1	15.9	28.1	29.5	157.4
	MINIMUM 32	30	31.0	28.0	30.9	26.8	6.2	*	0.0	0.6	8.7	28.4	30.0	31.0	221.6
	MINIMUM 0	30	25.6	22.2	16.7	2.5	*	0.0	0.0	0.0	0.0	3.6	19.2	24.2	114.0
H/C	NORMAL HEATING DEG. DAYS	30	2328	1921	1674	1029	508	182	123	267	585	1237	1869	2217	13940
	NORMAL COOLING DEG. DAYS	30	0	0	0	0	0	26	45	13	0	0	0	0	84
RH	NORMAL (PERCENT)	30	69	66	60	56	50	56	64	71	69	74	73	71	65
	HOUR 03 LST	30	69	67	66	66	65	73	79	84	79	77	73	71	72
	HOUR 09 LST	30	69	67	65	60	53	61	69	78	76	78	73	71	68
	HOUR 15 LST	30	69	63	52	45	37	42	50	54	54	66	72	71	56
	HOUR 21 LST	30	69	66	61	53	45	49	58	68	69	75	73	71	63
S	PERCENT POSSIBLE SUNSHINE														
W/O	MEAN NO. DAYS WITH:														
	HEAVY FOG(VISBY 1/4 MI)	50	4.3	2.4	0.5	0.3	0.2	0.3	0.8	1.7	1.4	1.7	1.2	2.8	17.6
	THUNDERSTORMS	50	0.0	0.0	0.0	0.0	0.4	3.0	2.4	0.9	0.1	0.0	0.0	0.0	6.8
CLOUDINESS	MEAN:														
	SUNRISE-SUNSET (OKTAS)	45	5.0	5.0	4.7	5.3	5.5	5.8	5.9	6.2	6.1	6.3	5.5	5.6	5.6
	MIDNIGHT-MIDNIGHT (OKTAS)	32	4.8	4.5	4.4	5.0	5.4	5.8	5.9	6.0	5.8	6.0	5.2	5.2	5.3
	MEAN NO. DAYS WITH:														
	CLEAR	46	9.1	8.1	9.9	6.7	4.6	3.0	3.3	2.8	4.1	3.8	6.8	6.7	68.9
	PARTLY CLOUDY	46	5.9	6.0	7.0	7.9	11.0	10.4	8.9	6.8	6.1	4.9	5.0	5.7	85.6
	CLOUDY	46	16.0	14.2	14.1	15.4	15.4	16.6	18.2	20.9	19.3	21.5	17.6	18.0	207.2
PR	MEAN STATION PRESSURE(IN)	27	29.39	29.37	29.31	29.30	29.32	29.32	29.40	29.40	29.30	29.22	29.31	29.31	29.33
	MEAN SEA-LEVEL PRES. (IN)	16	29.84	29.94	29.88	29.83	29.84	29.86	29.92	29.89	29.81	29.80	29.85	29.78	29.85
WINDS	MEAN SPEED (MPH)	45	3.1	3.8	5.1	6.4	7.5	6.9	6.5	6.0	6.0	5.1	3.6	3.0	5.2
	PREVAIL.DIR(TENS OF DEGS)	29	02	01	36	36	36	24	23	22	36	36	36	04	36
	MAXIMUM 2-MINUTE:														
	SPEED (MPH)	4	23	30	29	28	30	25	30	25	23	20	22	21	30
	DIR. (TENS OF DEGS)		13	07	26	26	12	08	26	01	26	28	05	25	07
	YEAR OF OCCURRENCE		2001	2001	1998	1999	1998	1999	1999	2000	1998	1998	1999	1999	FEB 2001
	MAXIMUM 5-SECOND:														
	SPEED (MPH)	4	36	38	34	36	37	32	40	32	28	24	25	24	40
	DIR. (TENS OF DEGS)		12	06	27	21	11	07	27	01	25	29	05	08	27
	YEAR OF OCCURRENCE		2001	2001	1998	1998	1998	2000	1999	2000	1998	1998	1999	2000	JUL 1999
PRECIPITATION	NORMAL (IN)	30	0.47	0.40	0.37	0.32	0.61	1.37	1.87	1.96	0.95	0.90	0.80	0.85	10.87
	MAXIMUM MONTHLY (IN)	50	2.40	1.75	2.24	0.93	1.67	3.52	4.87	6.20	3.05	2.19	3.32	3.23	6.20
	YEAR OF OCCURRENCE		1993	1966	1991	1982	1955	1955	1990	1967	1960	1983	1970	1984	AUG 1967
	MINIMUM MONTHLY (IN)	50	0.01	0.01	T	T	0.07	0.19	0.35	0.40	0.15	0.08	T	T	APR 1991
	YEAR OF OCCURRENCE		1966	2000	1987	1991	1957	1957	1966	1993	1957	1968	1954	1953	1969
	MAXIMUM IN 24 HOURS (IN)	50	0.75	0.97	1.17	0.47	0.88	1.52	1.73	3.42	1.21	2.22	0.84	1.25	3.42
	YEAR OF OCCURRENCE		1993	1966	1991	1979	1955	1955	1990	1967	1954	1976	1970	1968	AUG 1967
	NORMAL NO. DAYS WITH:														
	PRECIPITATION 0.01	30	7.2	6.1	5.8	5.4	7.5	11.1	12.9	12.4	9.4	11.6	10.8	9.5	109.7
	PRECIPITATION 1.00	30	0.0	0.0	0.0	0.0	0.0	0.1	0.2	0.1	0.0	0.0	0.0	0.0	0.4
SNOWFALL	NORMAL (IN)	30	8.9	8.4	6.5	3.8	0.5	0.0	0.0	T	1.0	11.7	15.1	14.9	70.8
	MAXIMUM MONTHLY (IN)	50	40.2	43.1	30.4	11.5	14.1	T	T	0.1	24.4	25.9	54.0	50.7	54.0
	YEAR OF OCCURRENCE		1993	1966	1991	1992	1992	1993	1990	1995	1992	1982	1970	1984	NOV 1970
	MAXIMUM IN 24 HOURS (IN)	50	10.1	20.1	12.6	6.3	9.4	T	T	0.1	9.0	10.4	14.6	14.7	20.1
	YEAR OF OCCURRENCE		1993	1966	1963	1992	1992	1993	1990	1995	1992	1974	1970	1968	FEB 1966
	MAXIMUM SNOW DEPTH (IN)	52	46	52	54	49	14	0	0	0	12	16	42	46	54
	YEAR OF OCCURRENCE		1993	1966	1991	1991	1991				1992	1982	1970	1990	MAR 1991
	NORMAL NO. DAYS WITH:														
	SNOWFALL 1.0	30	2.8	2.3	2.3	1.2	0.2	0.0	0.0	0.0	0.3	4.2	5.1	4.5	22.9

PRECIPITATION (inches) 2001 FAIRBANKS, AK (FAI)

YEAR	JAN	FEB	MAR	APR	MAY	JUN	JUL	AUG	SEP	OCT	NOV	DEC	ANNUAL
1972	0.73	0.16	0.27	0.20	0.35	0.55	0.63	1.09	2.08	0.86	0.46	1.13	8.51
1973	0.44	0.11	0.40	0.05	0.99	0.97	1.92	2.19	0.19	0.91	0.80	0.15	9.12
1974	0.14	0.33	0.27	0.21	0.11	1.22	1.17	1.14	0.47	1.08	1.03	0.55	7.72
1975	0.60	0.04	0.22	0.47	0.49	0.99	1.81	2.10	0.20	0.79	0.44	0.31	8.46
1976	0.22	0.01	0.55	0.08	0.94	1.08	1.60	0.69	1.05	0.89	0.13	0.08	7.32
1977	0.31	0.81	0.26	0.36	1.63	3.01	1.58	0.41	2.51	1.11	0.19	0.80	12.98
1978	0.39	0.19	0.09	0.16	0.44	1.71	1.19	1.24	0.98	0.59	1.02	1.40	9.40
1979	0.58	0.02	0.47	0.83	0.88	1.54	2.54	1.22	0.19	0.94	0.63	0.49	10.33
1980	0.52	0.22	0.13	0.10	0.31	1.38	1.37	1.68	0.76	0.39	0.70	0.32	7.88
1981	0.31	0.78	0.07	0.32	0.73	1.91	2.41	1.35	0.80	0.91	0.91	0.58	11.08
1982	0.34	0.38	0.39	0.93	0.96	1.96	2.33	1.67	0.77	1.48	1.49	0.23	12.93
1983	0.24	0.18	0.09	0.27	0.14	0.57	1.71	3.33	0.92	2.19	0.08	0.65	10.37
1984	0.89	0.64	0.03	0.47	1.17	0.48	2.95	1.15	0.22	0.70	0.42	3.23	12.35
1985	0.52	0.48	0.57	0.36	0.41	1.80	1.88	2.59	1.00	0.90	0.08	11.72
1986	0.13	0.19	0.32	0.07	0.54	0.87	2.12	2.36	0.65	1.79	0.48	0.34	9.86
1987	0.68	0.10	T	0.05	0.21	1.02	1.70	0.56	0.57	0.39	0.64	0.51	6.43
1988	0.32	0.13	0.13	0.21	1.51	2.26	1.02	1.95	0.73	1.07	0.68	0.46	10.47
1989	0.52	0.98	0.13	0.05	0.99	2.53	0.91	0.78	0.72	1.28	0.97	0.57	10.43
1990	0.52	0.72	0.11	0.07	0.40	1.73	4.87	3.60	1.74	0.31	1.51	2.94	18.52
1991	1.17	0.17	2.24	T	0.10	0.36	0.81	1.18	1.16	0.71	0.48	1.02	9.40
1992	0.85	0.66	0.07	0.47	1.23	2.15	2.32	0.59	1.34	0.91	0.93	1.21	12.73
1993	2.40	0.31	0.26	0.03	0.63	1.24	0.35	1.58	2.63	0.61	0.86	0.43	11.33
1994	0.47	0.32	0.17	0.07	0.22	2.41	1.11	1.38	0.61	0.82	1.67	0.48	9.73
1995	0.25	0.29	0.18	0.17	0.73	1.91	1.32	2.10	1.33	0.31	0.20	0.06	8.85
1996	0.32	1.40	0.53	0.05	0.14	1.56	1.07	2.83	1.06	1.13	0.81	0.46	11.36
1997	0.25	0.34	0.06	0.01	0.07	1.03	1.08	1.70	0.48	0.94	0.26	0.52	6.74
1998	0.08	0.11	T	0.05	0.41	1.33	3.35	3.18	1.19	0.27	0.21	0.56	10.74
1999	0.22	0.21	0.19	0.04	0.31	1.30	2.11	1.85	1.83	0.90	0.58	0.73	10.27
2000	1.97	T	0.10	T	0.74	0.72	1.29	3.04	1.48	0.91	0.37	0.16	10.78
2001	0.40	0.66	0.43	0.16	0.62	0.65	2.45	2.01	0.25	0.55	0.06	0.24	8.48
POR= 72 YRS	0.68	0.46	0.40	0.27	0.64	1.38	1.80	2.13	1.00	0.79	0.68	0.59	10.82

AVERAGE TEMPERATURE (F) 2001 FAIRBANKS, AK (FAI)

YEAR	JAN	FEB	MAR	APR	MAY	JUN	JUL	AUG	SEP	OCT	NOV	DEC	ANNUAL
1972	-16.3	-10.1	-2.8	20.8	47.4	59.4	64.5	58.9	40.1	26.8	7.0	-2.5	24.4
1973	-18.2	-1.5	11.9	35.3	50.6	60.3	62.0	55.0	47.3	25.1	-.7	-3.4	27.0
1974	-16.7	-17.8	7.6	34.9	51.4	58.7	63.5	59.1	51.4	21.4	0.6	-11.3	25.2
1975	-15.5	-3.4	12.6	30.5	53.5	63.4	68.4	56.1	45.8	23.9	-8.0	-16.1	25.9
1976	-11.5	-13.7	12.1	36.0	47.8	59.6	61.8	59.2	45.4	23.9	15.9	-3.9	27.7
1977	9.8	8.6	4.6	27.8	48.7	59.6	62.8	62.6	45.6	25.6	-7.6	-14.9	27.8
1978	0.1	3.9	14.0	34.8	50.2	54.6	63.5	59.5	46.8	23.3	8.6	3.3	30.2
1979	-7.7	-25.3	12.0	30.9	49.9	57.4	61.3	60.4	46.5	32.4	20.2	-10.2	27.3
1980	-9.5	16.0	17.3	35.9	50.7	56.5	60.9	53.6	43.0	33.0	11.5	-24.0	28.7
1981	18.1	5.1	27.1	31.4	51.3	58.9	56.5	53.6	44.1	29.5	12.3	-4.2	32.0
1982	-18.0	-3.9	13.1	27.7	46.8	58.5	63.1	56.6	49.3	18.5	4.4	2.2	26.5
1983	-11.0	3.4	13.8	37.4	50.4	62.3	64.2	53.5	41.2	23.6	8.6	-3.7	28.6
1984	-5.9	-13.3	21.6	30.3	47.2	61.6	60.9	53.8	46.8	25.9	0.1	-3.1	27.2
1985	11.1	-9.4	14.6	20.8	46.7	57.8	63.1	56.3	42.9	18.8	-4.7	7.7	27.1
1986	-2.1	4.7	6.0	24.0	47.8	62.6	63.6	54.7	46.3	27.1	0.2	7.3	28.5
1987	0.7	1.5	13.5	35.0	50.9	61.9	64.2	57.7	44.1	33.0	6.0	-3.2	30.4
1988	-5.3	3.9	17.8	33.6	52.8	62.9	65.8	58.4	44.4	17.4	-3.1	4.2	29.4
1989	-21.3	3.4	6.7	36.2	47.8	60.1	64.6	60.8	48.6	26.2	-7.1	4.5	27.5
1990	-12.9	-21.7	18.5	38.1	55.1	61.6	65.3	60.0	44.8	24.1	-4.9	-6.3	26.8
1991	-4.7	-1.2	11.9	35.4	51.2	63.8	60.6	54.4	48.0	24.9	0.6	-2.9	28.5
1992	-4.7	-8.7	14.0	26.1	41.8	60.0	64.1	56.5	31.7	17.5	10.4	-7.6	25.1
1993	-3.9	2.4	17.2	41.1	53.7	62.0	65.6	56.1	44.1	29.3	8.4	0.6	31.4
1994	-1.5	-6.4	10.5	34.7	51.6	58.4	64.7	58.3	43.7	21.0	0.3	-8.5	27.3
1995	-9.4	-.2	3.8	40.1	53.5	60.9	63.1	57.3	52.8	28.0	-2.7	-9.0	28.2
1996	-16.8	-2.6	15.5	33.3	49.1	59.5	63.4	53.4	42.0	13.2	-1.9	-13.4	24.6
1997	-16.2	13.0	4.7	35.4	49.0	63.1	64.9	58.7	49.7	17.5	10.6	-7.0	28.6
1998	-13.6	2.1	18.4	38.6	50.0	58.8	62.7	53.0	46.1	26.1	5.0	-5.8	28.5
1999	-17.0	-16.7	7.2	32.7	46.9	61.4	61.5	58.0	45.2	19.5	-4.7	-12.8	23.4
2000	-9.9	6.7	17.6	31.9	44.4	61.5	59.6	51.7	41.7	22.3	8.7	0.3	28.0
2001	7.8	7.0	9.9	33.3	44.5	61.1	60.0	57.3	47.5	22.0	0.6	-10.5	28.4
POR= 72 YRS	-10.1	-3.1	10.2	30.7	48.0	59.3	61.4	55.6	44.6	25.2	3.0	-7.8	26.4

REFERENCE NOTES:

PAGE 1:
THE TEMPERATURE GRAPH SHOWS NORMAL MAXIMUM AND NORMAL
MINIMUM DAILY TEMPERATURES (SOLID CURVES) AND THE
ACTUAL DAILY HIGH AND LOW TEMPERATURES (VERTICAL BARS).

PAGE 2 AND 3:
H/C INDICATES HEATING AND COOLING DEGREE DAYS.
RH INDICATES RELATIVE HUMIDITY
W/O INDICATES WEATHER AND OBSTRUCTIONS
S INDICATES SUNSHINE.
PR INDICATES PRESSURE.
CLOUDINESS ON PAGE 3 IS THE SUM OF THE CEILOMETER AND
SATELLITE DATA NOT TO EXCEED EIGHT EIGHTHS(OKTAS).

GENERAL:
T INDICATES TRACE PRECIPITATION, AN AMOUNT GREATER
THAN ZERO BUT LESS THAN THE LOWEST REPORTABLE VALUE.
+ INDICATES THE VALUE ALSO OCCURS ON EARLIER DATES.
BLANK ENTRIES DENOTE MISSING OR UNREPORTED DATA.
NORMALS ARE 30-YEAR AVERAGES (1961 - 1990).
ASOS INDICATES AUTOMATED SURFACE OBSERVING SYSTEM.
PM INDICATES THE LAST DAY OF THE PREVIOUS MONTH.
POR (PERIOD OF RECORD) BEGINS WITH THE JANUARY DATA
MONTH AND IS THE NUMBER OF YEARS USED TO COMPUTE
THE MEAN.   INDIVIDUAL MONTHS WITHIN THE POR MAY
BE MISSING.
WHEN THE POR FOR A NORMAL IS LESS THAN 30 YEARS,
THE NORMAL IS PROVISIONAL AND IS BASED ON THE NUMBER
OF YEARS INDICATED.
0.* OR * INDICATES THE VALUE OR MEAN-DAYS-WITH
IS BETWEEN 0.00 AND 0.05.
CLOUDINESS FOR ASOS STATIONS DIFFERS FROM THE NON-ASOS
OBSERVATION TAKEN BY A HUMAN OBSERVER. ASOS STATION
CLOUDINESS IS BASED ON TIME-AVERAGED CEILOMETER DATA
FOR CLOUDS AT OR BELOW 12,000 FEET AND ON SATELLITE
DATA FOR CLOUDS ABOVE 12,000 FEET.
THE NUMBER OF DAYS WITH CLEAR, PARTLY CLOUDY, AND
CLOUDY CONDITIONS FOR ASOS STATIONS IS THE SUM
OF THE CEILOMETER AND SATELLITE DATA FOR THE
SUNRISE TO SUNSET PERIOD.

GENERAL CONTINUED:
CLEAR INDICATES 0 - 2 OKTAS, PARTLY CLOUDY INDICATES
3 - 6 OKTAS, AND CLOUDY INDICATES 7 OR 8 OKTAS.
WHEN AT LEAST ONE OF THE ELEMENTS (CEILOMETER OR
SATELLITE) IS MISSING, THE DAILY CLOUDINESS IS
NOT COMPUTED.
WIND DIRECTION IS RECORDED IN TENS OF DEGREES (2 DIGITS)
CLOCKWISE FROM TRUE NORTH. "00" INDICATES CALM. "36"
INDICATES TRUE NORTH.
RESULTANT WIND IS THE VECTOR AVERAGE OF THE SPEED AND
DIRECTION.
AVERAGE TEMPERATURE IS THE SUM OF THE MEAN DAILY MAXIMUM
AND MINIMUM TEMPERATURE DIVIDED BY 2.
SNOWFALL DATA COMPRISE ALL FORMS OF FROZEN
PRECIPITATION, INCLUDING HAIL.
A HEATING (COOLING) DEGREE DAY IS THE DIFFERENCE BETWEEN
THE AVERAGE DAILY TEMPERATURE AND 65 F.
DRY BULB IS THE TEMPERATURE OF THE AMBIENT AIR.
DEW POINT IS THE TEMPERATURE TO WHICH THE AIR MUST BE
COOLED TO ACHIEVE 100 PERCENT RELATIVE HUMIDITY.
WET BULB IS THE TEMPERATURE THE AIR WOULD HAVE IF THE
MOISTURE CONTENT WAS INCREASED TO 100 PERCENT RELATIVE
HUMIDITY.

ON JULY 1, 1996, THE NATIONAL WEATHER SERVICE BEGAN USING
THE "METAR" OBSERVATION CODE THAT WAS ALREADY EMPLOYED
BY MOST OTHER NATIONS OF THE WORLD.   THE MOST NOTICEABLE
DIFFERENCE IN THIS ANNUAL PUBLICATION WILL BE THE CHANGE
IN UNITS FROM TENTHS TO EIGHTS(OKTAS) FOR REPORTING THE
AMOUNT OF SKY COVER.

HEATING DEGREE DAYS (base 65 F) 2001 FAIRBANKS, AK (FAI)

YEAR	JUL	AUG	SEP	OCT	NOV	DEC	JAN	FEB	MAR	APR	MAY	JUN	TOTAL
1972-73	63	184	738	1177	1739	2091	2582	1864	1637	883	439	150	13547
1973-74	111	302	523	1231	1968	2124	2535	2322	1778	895	414	188	14391
1974-75	85	195	402	1342	1935	2370	2497	1918	1620	1028	347	69	13808
1975-76	33	270	570	1270	2195	2513	2372	2285	1631	861	527	156	14683
1976-77	116	188	583	1269	1466	2136	1709	1574	1871	1107	500	155	12674
1977-78	101	124	573	1216	2184	2480	2013	1712	1576	898	454	304	13635
1978-79	65	176	542	1286	1689	1912	2260	2533	1638	1018	463	220	13802
1979-80	124	143	548	1004	1336	2335	2312	1415	1475	868	436	248	12244
1980-81	127	351	654	985	1599	2766	1447	1676	1168	999	418	188	12378
1981-82	255	347	622	1094	1573	2150	2581	1929	1602	1113	555	216	14037
1982-83	86	252	465	1434	1816	1946	2356	1725	1581	823	451	133	13068
1983-84	62	351	705	1280	1688	2126	2201	2277	1338	1035	549	120	13732
1984-85	140	344	538	1205	1950	2111	1666	2086	1558	1321	558	215	13692
1985-86	72	267	654	1430	2095	1776	2079	1686	1825	1224	527	113	13748
1986-87	110	312	559	1169	1943	1787	1994	1776	1594	893	428	128	12693
1987-88	61	218	620	987	1768	2111	2185	1768	1455	934	371	96	12574
1988-89	39	202	611	1469	2045	1883	2676	1722	1804	859	529	149	13988
1989-90	73	134	484	1195	2164	1875	2420	2433	1431	798	310	127	13444
1990-91	74	178	600	1261	2097	2212	2161	1849	1640	877	421	130	13500
1991-92	143	321	504	1234	1935	2105	2163	2140	1577	1160	711	157	14150
1992-93	45	259	995	1463	1636	2253	2134	1751	1478	711	343	103	13171
1993-94	43	273	620	1099	1694	1995	2061	2002	1689	900	408	210	12994
1994-95	56	217	634	1357	1941	2280	2310	1827	1896	739	354	143	13754
1995-96	92	235	365	1141	2033	2299	2541	1962	1528	946	486	171	13799
1996-97	71	353	681	1600	2010	2424	2509	1450	1862	880	485	106	14431
1997-98	40	204	453	1467	1624	2224	2429	1755	1437	786	460	199	13078
1998-99	103	365	563	1200	1793	2188	2535	2282	1784	963	553	131	14460
1999-00	147	221	584	1403	2084	2403	2315	1680	1467	987	631	112	14034
2000-01	165	407	693	1318	1682	2001	1764	1616	1701	940	629	126	13042
2001-	164	231	521	1326	1926	2333							

COOLING DEGREE DAYS (base 65 F) 2001 FAIRBANKS, AK (FAI)

YEAR	JAN	FEB	MAR	APR	MAY	JUN	JUL	AUG	SEP	OCT	NOV	DEC	ANNUAL
1972	0	0	0	0	0	40	55	0	0	0	0	0	95
1973	0	0	0	0	0	13	24	2	0	0	0	0	39
1974	0	0	0	0	1	9	44	20	0	0	0	0	74
1975	0	0	0	0	0	28	146	1	0	0	0	0	175
1976	0	0	0	0	0	2	23	14	0	0	0	0	39
1977	0	0	0	0	0	0	44	54	0	0	0	0	98
1978	0	0	0	0	0	0	27	10	0	0	0	0	37
1979	0	0	0	0	0	0	16	7	0	0	0	0	23
1980	0	0	0	0	0	0	8	6	0	0	0	0	14
1981	0	0	0	0	0	1	11	0	2	0	0	0	14
1982	0	0	0	0	0	27	36	0	0	0	0	0	63
1983	0	0	0	0	5	61	40	0	0	0	0	0	106
1984	0	0	0	0	0	22	21	2	0	0	0	0	45
1985	0	0	0	0	0	8	20	4	0	0	0	0	32
1986	0	0	0	0	0	46	74	0	0	0	0	0	120
1987	0	0	0	0	0	42	42	0	0	0	0	0	84
1988	0	0	0	0	0	41	72	2	0	0	0	0	115
1989	0	0	0	0	0	10	67	11	0	0	0	0	88
1990	0	0	0	0	11	32	91	35	0	0	0	0	169
1991	0	0	0	0	0	100	13	0	0	0	0	0	113
1992	0	0	0	0	0	17	21	4	0	0	0	0	42
1993	0	0	0	0	0	18	70	3	0	0	0	0	91
1994	0	0	0	0	0	19	55	55	0	0	0	0	129
1995	0	0	0	0	5	24	41	0	4	0	0	0	74
1996	0	0	0	0	0	11	29	0	0	0	0	0	40
1997	0	0	0	0	0	56	46	15	0	0	0	0	117
1998	0	0	0	0	0	18	37	1	0	0	0	0	56
1999	0	0	0	0	0	27	45	11	0	0	0	0	83
2000	0	0	0	0	0	16	5	0	0	0	0	0	21
2001	0	0	0	0	0	15	14	0	0	0	0	0	29

SNOWFALL (inches) 2001 FAIRBANKS, AK (FAI)

YEAR	JUL	AUG	SEP	OCT	NOV	DEC	JAN	FEB	MAR	APR	MAY	JUN	TOTAL
1972-73	0.0	0.0	7.8	6.9	12.2	26.9	14.2	1.8	8.7	1.0	T	0.0	79.5
1973-74	0.0	0.0	1.2	8.2	17.5	2.5	3.0	8.7	7.0	1.3	0.0	T	49.4
1974-75	0.0	0.0	0.3	24.4	22.5	17.9	14.1	1.6	5.0	4.4	T	0.0	90.2
1975-76	0.0	0.0	T	14.4	11.4	6.7	6.4	0.5	10.4	1.2	T	0.0	51.0
1976-77	0.0	0.0	T	14.7	4.3	2.6	6.7	19.0	5.5	6.8	T	0.0	59.6
1977-78	0.0	0.0	1.0	17.3	5.1	14.4	4.9	3.0	1.6	0.7	0.1	0.0	48.1
1978-79	0.0	0.0	0.3	7.6	15.1	21.7	9.7	0.5	7.7	3.1	0.0	0.0	65.7
1979-80	0.0	0.0	T	6.5	4.4	11.0	10.8	4.4	3.2	0.8	0.0	0.0	41.1
1980-81	0.0	0.0	3.4	5.1	10.6	4.8	6.2	10.2	0.9	1.9	0.0	0.0	43.1
1981-82	0.0	0.0	0.3	7.8	16.2	10.6	7.6	7.0	7.2	11.4	0.3	0.0	68.4
1982-83	0.0	0.0	0.6	25.9	27.8	3.8	4.8	3.6	2.1	1.9	0.4	0.0	70.9
1983-84	0.0	0.0	0.4	17.3	2.4	14.4	13.8	11.1	0.8	6.7	T	0.0	66.9
1984-85	0.0	0.0	0.0	11.3	8.5	50.7	8.1	8.0	7.4	5.9	1.0	0.0	100.9
1985-86	0.0	0.0	2.1	5.4	14.7	1.8	2.3	2.6	3.7	1.0	0.0	0.0	33.6
1986-87	0.0	0.0	T	11.8	7.4	6.0	12.0	1.5	T	1.0	T	0.0	39.7
1987-88	0.0	0.0	T	3.5	14.1	10.4	6.3	2.5	2.6	0.2	0.3	0.0	39.9
1988-89	0.0	0.0	0.2	12.9	15.2	10.5	10.8	13.4	3.9	0.4	0.8	T	68.1
1989-90	0.0	0.0	0.5	19.7	18.1	11.1	10.4	16.0	2.6	T	0.0	0.0	78.4
1990-91	T	0.0	1.6	6.9	37.3	47.5	20.6	3.0	30.4	T	0.0	0.0	147.3
1991-92	0.0	0.0	T	12.2	9.7	18.9	15.5	15.0	1.8	11.5	14.1	0.0	98.7
1992-93	0.0	0.0	24.4	16.7	18.7	28.5	40.2	5.4	5.2	T	0.0	T	139.1
1993-94	0.0	0.0	7.5	7.3	16.3	8.5	11.2	8.1	4.5	0.8	0.0	0.0	64.2
1994-95	0.0	0.0	0.0	15.3	35.3	13.1	5.9	7.4	3.1	1.3	0.0	0.0	81.4
1995-96	0.0	0.1	0.0	7.1	3.2	0.8	5.4	29.3	9.6	0.7	0.0	0.0	56.2
1996-97	0.0	T	7.1	16.2	18.4	9.4	4.9	8.8	2.3	0.4	0.0	T	67.5
1997-98	0.0	0.0	0.0	19.3	7.6	13.1	1.3	3.2	0.9	0.5	0.1	0.0	46.0
1998-99	0.0	0.0	T	6.2	4.6	8.5	4.2	2.7	3.5	0.9	0.4	0.0	31.0
1999-00	0.0	0.0	3.0	11.1	11.2	13.8	27.9	0.2	3.0	T	T	0.0	70.2
2000-01	0.0	0.0	1.4	9.8	7.5	2.7	6.7	10.7	8.6	2.2	7.0	0.0	56.6
2001-	0.0	0.0	0.3	6.4	2.8	4.6							
POR= 49 YRS	T	0.0	1.5	10.9	13.2	13.0	10.3	8.6	6.2	3.1	1.3	T	68.1

2001
JUNEAU,
ALASKA (JNU)

Juneau lies well within the area of maritime influences which prevail over the coastal areas of southeastern Alaska, and is in the path of most storms that cross the Gulf of Alaska. Consequently, the area has little sunshine, generally moderate temperatures, and abundant precipitation. In contrast with the characteristic lack of sunshine there are greatly appreciated intervals, sometimes lasting for several days at a stretch, during which clear skies prevail. The rugged terrain exerts a fundamental influence upon local temperatures and the distribution of precipitation, creating considerable variations in both weather elements within relatively short distances.

Temperature variations, both daily and seasonal, are usually confined to relatively narrow limits by the dominant maritime influences. There are, however, periods of comparatively severe cold, which usually start with strong northerly winds, and are most often caused by the flow of cold air from northwestern Canada through nearby mountain passes and over the Juneau ice field. These are generally of brief duration. During such periods strong, gusty winds, known locally as Taku Winds, often occur especially in downtown Juneau, Douglas, and other local areas, but generally they are not felt in the Mendenhall Valley. At times these are strong enough to cause considerable damage. During periods of calm or light winds, temperature differences within short distances are frequently very pronounced. Variations in local sunlight and air drainage patterns produce wide differences in temperatures particularly between upland or sloping areas and areas of low, flat terrain. Juneau International Airport, located on low, flat terrain formed by the Mendenhall River delta, and in the path of drainage air from the Mendenhall Glacier, averages about 10 days a year with minimum readings below zero. Downtown Juneau, located on a sloping portion of a rugged mountain area, experiences on the average only about one day each year with minimum readings below zero. At the airport the growing season averages 146 days, from May 4 to September 28, while the downtown average is 181 days, from April 22 to October 21.

The months of February to June mark the period of lightest precipitation, with monthly averages of about 3 inches. After June the monthly amounts increase gradually, reaching an average of 7.71 inches in October. Due to the rugged topography, precipitation throughout the year tends to vary greatly within short distances. At the Juneau Airport, yearly precipitation is 53 inches while downtown, only 8 miles away, it is 93 inches. The maximum yearly amount received in the city is almost double the maximum received at the airport.

Although a trace of snow has fallen as early as September 9, first falls usually occur in the latter part of October, and sometimes not until the first part of December. On the average there is very little accumulation on the ground at low levels until the last of November, although at higher elevations, and particularly on mountain tops, a cover is usually established in early October. Snow accumulation usually reaches its greatest depth during the middle of February. Individual storms may produce heavy falls as late as the first half of May. However, snow cover is usually gone before the middle of April. Ice accumulations due to alternating thawing and freezing of snow or due to freezing precipitation are frequent problems in the Juneau area during the winter months.

NORMALS, MEANS, AND EXTREMES
JUNEAU, AK (JNU)

LATITUDE: 58 21' 18" N LONGITUDE: 134 34' 30" W ELEVATION (FT): GRND: 37 BARO: 40 TIME ZONE: ALASKA (UTC + 9) WBAN: 25309

	ELEMENT	POR	JAN	FEB	MAR	APR	MAY	JUN	JUL	AUG	SEP	OCT	NOV	DEC	YEAR
TEMPERATURE F	NORMAL DAILY MAXIMUM	30	29.4	34.1	38.7	47.2	55.1	60.9	63.9	62.7	55.9	47.1	36.7	31.6	46.9
	MEAN DAILY MAXIMUM	52	29.1	34.1	38.7	47.3	55.2	61.7	63.8	62.8	56.1	47.0	37.7	32.5	47.2
	HIGHEST DAILY MAXIMUM	57	57	57	61	72	82	86	90	83	73	61	56	54	90
	YEAR OF OCCURRENCE		1958	1992	1998	1995	1947	1969	1975	1999	1996	1987	1949	1999	JUL 1975
	MEAN OF EXTREME MAXS.	52	42.7	43.9	48.2	59.4	69.7	75.9	78.4	75.6	65.6	55.6	48.1	43.9	58.9
	NORMAL DAILY MINIMUM	30	19.0	22.7	26.7	32.1	38.9	45.0	48.1	47.3	42.9	37.2	27.2	22.6	34.1
	MEAN DAILY MINIMUM	52	18.6	22.7	26.4	32.4	39.3	45.3	48.3	47.4	43.1	37.0	28.6	23.4	34.4
	LOWEST DAILY MINIMUM	57	-22	-22	-15	6	25	31	36	27	23	11	-5	-21	-22
	YEAR OF OCCURRENCE		1972	1968	1972	1963	1972	1971	1950	1948	1972	1984	1966	1949	JAN 1972
	MEAN OF EXTREME MINS.	52	-.7	4.9	10.3	23.4	30.0	37.2	41.5	39.2	31.8	24.6	12.2	3.1	21.5
	NORMAL DRY BULB	30	24.2	28.4	32.7	39.7	47.0	53.0	56.0	55.0	49.4	42.2	32.0	27.1	40.6
	MEAN DRY BULB	52	24.0	28.4	32.7	39.9	47.2	53.5	56.1	55.1	49.6	42.0	33.2	28.0	40.8
	MEAN WET BULB	14	27.1	28.2	32.5	35.8	44.7	50.4	50.2	49.7	45.4	38.7	30.9	28.1	38.7
	MEAN DEW POINT	14	24.6	24.3	28.6	34.5	40.5	46.7	47.6	47.3	43.9	36.9	28.6	26.0	35.8
	NORMAL NO. DAYS WITH:														
	MAXIMUM 70	30	0.0	0.0	0.0	0.1	1.4	5.1	7.2	5.7	0.2	0.0	0.0	0.0	19.7
	MAXIMUM 32	30	16.3	8.7	4.3	0.1	0.0	0.0	0.0	0.0	0.0	0.3	6.8	12.9	49.4
	MINIMUM 32	30	25.4	21.8	22.0	14.4	3.5	0.1	0.0	*	1.4	7.2	19.1	23.9	138.8
	MINIMUM 0	30	4.7	1.6	0.5	0.0	0.0	0.0	0.0	0.0	0.0	0.0	0.1	1.8	8.7
H/C	NORMAL HEATING DEG. DAYS	30	1265	1025	1001	759	558	360	279	310	468	707	990	1175	8897
	NORMAL COOLING DEG. DAYS	30	0	0	0	0	0	0	0	0	0	0	0	0	0
RH	NORMAL (PERCENT)	30	80	81	79	77	76	78	81	84	88	88	85	83	82
	HOUR 03 LST	30	81	83	86	87	89	90	90	92	93	91	87	84	88
	HOUR 09 LST	30	80	82	80	76	76	78	82	85	88	89	86	83	82
	HOUR 15 LST	30	77	75	70	63	63	67	72	75	79	82	81	82	74
	HOUR 21 LST	30	81	82	82	80	78	78	81	85	91	90	87	83	83
S	PERCENT POSSIBLE SUNSHINE	33	32	32	37	39	39	34	31	32	26	19	23	20	30
W/O	MEAN NO. DAYS WITH:														
	HEAVY FOG(VISBY 1/4 MI)	57	2.1	2.4	1.8	1.0	0.8	0.3	0.2	1.1	2.6	2.8	3.2	2.4	20.7
	THUNDERSTORMS	57	0.0	0.0	0.0	0.0	0.0	0.1	0.1	0.0	0.1	0.0	0.0	0.0	0.3
CLOUDINESS	MEAN:														
	SUNRISE-SUNSET (OKTAS)	47	6.2	6.4	6.4	6.4	6.3	6.5	6.6	6.3	6.8	7.0	6.7	6.7	6.5
	MIDNIGHT-MIDNIGHT (OKTAS)	15	5.2	5.7	6.1	5.9	6.3	6.1	6.2	5.9	6.4	6.6	6.0	6.1	6.0
	MEAN NO. DAYS WITH:														
	CLEAR	47	5.7	4.3	4.4	3.7	3.8	3.4	2.8	3.9	2.7	2.4	3.3	3.2	43.6
	PARTLY CLOUDY	47	2.6	3.0	3.2	4.2	4.4	4.2	4.5	4.7	3.3	2.1	2.3	2.0	40.5
	CLOUDY	47	22.5	21.0	23.4	22.1	22.6	22.4	23.0	21.9	23.4	26.0	24.0	25.2	277.5
PR	MEAN STATION PRESSURE(IN)	16	29.87	29.88	29.79	29.90	29.90	29.99	30.01	29.98	29.89	29.79	29.78	29.80	29.88
	MEAN SEA-LEVEL PRES. (IN)	4	29.87	29.91	29.82	29.91	29.93	30.00	30.04	29.98	29.98	29.91	29.79	29.78	29.83	29.90
WINDS	MEAN SPEED (MPH)	35	7.8	8.1	8.1	8.5	8.2	7.9	7.5	7.4	7.9	9.3	8.7	8.9	8.2
	PREVAIL.DIR(TENS OF DEGS)	20	09	11	11	11	11	08	08	09	11	11	11	11	11
	MAXIMUM 2-MINUTE:														
	SPEED (MPH)	3	45	44	39	37	36	24	26	38	41	41	47	44	47
	DIR. (TENS OF DEGS)		12	12	11	11	12	12	12	12	12	12	12	11	12
	YEAR OF OCCURRENCE		2001	2001	2000	1999	2001	1999	2001	1999	2001	1999	2001	1999	NOV 2001
	MAXIMUM 5-SECOND:														
	SPEED (MPH)	3	56	51	49	48	47	32	32	48	53	51	61	54	61
	DIR. (TENS OF DEGS)		12	12	12	12	15	10	12	13	20	12	12	12	12
	YEAR OF OCCURRENCE		2001	2001	2000	1999	1999	2001	2001	1999	2000	1999	2001	1999	NOV 2001
PRECIPITATION	NORMAL (IN)	30	4.54	3.75	3.28	2.77	3.42	3.15	4.16	5.32	6.73	7.84	4.91	4.44	54.31
	MAXIMUM MONTHLY (IN)	57	9.11	8.48	6.50	7.48	9.20	6.22	10.36	12.31	15.14	15.25	11.22	13.61	15.25
	YEAR OF OCCURRENCE		1993	1964	1994	1999	1992	1996	1997	1961	1991	1974	1956	1997	OCT 1974
	MINIMUM MONTHLY (IN)	57	0.94	0.07	0.59	0.27	1.25	1.08	1.15	0.56	2.34	2.71	1.15	0.49	0.07
	YEAR OF OCCURRENCE		1969	1989	1983	1948	1946	1950	1972	1979	1965	1950	1983	1983	FEB 1989
	MAXIMUM IN 24 HOURS (IN)	57	2.74	2.71	1.81	2.05	2.30	2.26	2.46	2.62	3.35	4.66	3.40	3.56	4.66
	YEAR OF OCCURRENCE		1948	1993	1992	1997	1992	1996	2001	1974	1996	1946	1988	1956	OCT 1946
	NORMAL NO. DAYS WITH:														
	PRECIPITATION 0.01	30	18.1	16.5	17.9	16.7	16.8	15.9	16.1	17.3	20.0	23.7	19.2	20.2	218.4
	PRECIPITATION 1.00	30	0.6	0.4	0.1	0.1	0.2	0.2	0.6	0.8	1.3	1.2	0.8	0.4	6.7
SNOWFALL	NORMAL (IN)	30	30.6	18.6	15.4	3.5	0.*	0.0	0.0	0.0	T	0.9	13.3	22.2	104.5
	MAXIMUM MONTHLY (IN)	57	69.2	86.3	52.6	46.3	1.2	T	0.0	0.0	T	15.6	69.8	54.7	86.3
	YEAR OF OCCURRENCE		1982	1965	1948	1963	1964	1970			1974	1974	1956	1964	FEB 1965
	MAXIMUM IN 24 HOURS (IN)	57	20.1	23.7	31.0	24.2	0.7	T	0.0	0.0	T	8.8	19.4	25.6	31.0
	YEAR OF OCCURRENCE		1975	1949	1948	1963	1945	1970			1974	1956	1994	1962	MAR 1948
	MAXIMUM SNOW DEPTH (IN)	51	38	41	40	33	0	0	0	0	0	7	25	36	41
	YEAR OF OCCURRENCE		1966	1949	1972	1963						1956	1994	1962	FEB 1949
	NORMAL NO. DAYS WITH:														
	SNOWFALL 1.0	30	7.8	5.4	4.1	0.8	0.0	0.0	0.0	0.0	0.0	0.3	3.6	6.1	28.1

PRECIPITATION (inches) 2001 JUNEAU, AK (JNU)

YEAR	JAN	FEB	MAR	APR	MAY	JUN	JUL	AUG	SEP	OCT	NOV	DEC	ANNUAL
1972	3.73	2.71	4.19	3.62	4.03	3.98	1.15	8.62	6.24	8.49	3.35	3.56	53.67
1973	4.37	3.94	3.01	2.41	4.09	2.80	3.65	6.64	4.95	6.07	1.63	2.30	45.86
1974	2.37	6.23	1.15	2.59	1.66	4.92	3.12	5.78	5.96	15.25	7.79	7.03	63.85
1975	4.10	3.76	2.17	3.04	3.59	2.48	4.96	2.78	7.25	3.55	2.83	5.81	46.32
1976	8.19	4.82	3.61	2.14	3.42	3.37	2.48	3.16	8.32	6.19	5.15	5.56	56.41
1977	4.59	4.56	3.31	4.02	1.56	3.47	3.19	3.03	5.57	7.14	4.58	2.16	47.18
1978	1.71	1.50	1.84	2.19	2.86	3.18	3.98	4.39	3.07	13.00	3.90	4.46	46.08
1979	2.19	0.91	3.98	0.98	2.45	2.74	5.44	0.56	4.89	9.06	8.36	7.73	49.29
1980	3.44	2.83	2.75	5.32	2.53	4.37	6.49	5.61	7.91	11.26	7.10	2.27	61.88
1981	4.66	2.57	1.88	2.11	3.27	2.44	4.25	6.19	11.61	6.18	6.93	2.24	54.33
1982	3.74	1.42	2.52	2.44	5.10	1.86	1.73	5.97	5.10	7.97	2.10	1.17	41.12
1983	4.00	1.69	0.59	2.53	5.37	2.69	3.16	9.52	6.13	4.24	1.15	0.49	41.56
1984	6.06	5.40	3.75	2.11	1.84	4.17	6.92	6.26	3.39	6.69			
1985												8.33	
1986	7.00	3.25	6.08	2.98	2.54	2.76	2.38	6.89	2.40	12.33	5.96	6.42	60.99
1987	3.99	3.13	2.12	2.08	2.60	6.02	2.54	4.54	8.92	10.36	7.17	5.32	58.79
1988	2.58	6.55	4.15	2.25	3.91	2.05	5.21	5.53	5.46	9.71	8.62	4.75	60.77
1989	6.77	0.07	1.33	0.87	3.44	1.10	3.81	2.82	7.29	6.37	6.23	6.78	46.88
1990	3.72	4.54	4.86	1.06	1.72	3.32	4.65	5.35	10.63	6.59	4.89	6.03	57.36
1991	4.16	6.55	4.41	4.73	4.72	3.41	4.85	9.60	15.14	8.63	9.63	9.32	85.15
1992	8.69	7.24	6.37	3.63	9.20	2.98	5.18	5.02	11.45	5.90	7.91	5.73	79.30
1993	9.11	8.09	3.50	1.94	2.19	4.92	2.25	3.20	8.44	9.00	11.06	7.89	71.59
1994	7.05	2.52	6.50	3.68	4.20	1.83	4.32	2.68	11.17	9.15	9.57	6.22	68.89
1995	1.78	2.83	3.01	2.08	2.85	3.45	4.36	5.01	7.43	6.04	2.93	4.58	46.35
1996	2.26	8.43	4.12	2.19	1.80	6.22	3.16	7.91	10.68	6.20	2.75	4.73	60.45
1997	2.73	8.17	3.91	4.41	3.25	3.51	10.36	3.93	8.26	7.85	4.63	13.61	74.62
1998	2.54	1.90	3.71	3.12	2.21	2.50	4.95	6.80	6.17	12.13	1.72	5.45	53.20
1999	8.14	2.66	2.58	7.48	5.69	2.69	4.10	6.77	10.62	12.19	5.77	10.30	78.99
2000	4.82	1.56	5.75	4.40	3.25	5.72	6.65	6.12	10.05	10.11	6.37	4.17	68.97
2001	7.43	4.40	3.33	2.19	5.19	1.65	7.26	3.66	8.37	7.80	3.62	4.49	59.39
POR= 57 YRS	4.29	3.72	3.55	2.92	3.49	3.12	4.42	5.19	7.22	7.97	5.45	4.87	56.21

AVERAGE TEMPERATURE (F) 2001   JUNEAU, AK (JNU)

YEAR	JAN	FEB	MAR	APR	MAY	JUN	JUL	AUG	SEP	OCT	NOV	DEC	ANNUAL
1972	15.8	19.5	26.5	34.6	44.8	50.4	58.0	55.4	47.0	38.7	34.5	23.6	37.4
1973	18.9	24.6	32.9	39.6	45.9	51.3	53.7	51.8	48.0	41.2	23.0	28.0	38.2
1974	14.8	28.8	24.7	39.3	46.8	50.2	53.5	54.7	50.1	42.5	36.4	33.8	39.6
1975	23.1	24.5	30.6	38.3	47.4	51.4	55.9	53.9	51.4	41.8	28.5	24.7	39.3
1976	28.3	25.8	32.4	41.1	45.5	52.0	55.7	55.9	50.6	42.8	40.9	34.4	42.1
1977	35.0	40.1	36.0	42.3	47.7	54.1	57.0	58.5	50.5	42.4	29.4	18.9	42.7
1978	25.1	31.8	33.9	42.0	47.8	54.2	55.2	56.3	50.6	45.1	30.3	28.3	41.7
1979	20.6	11.0	35.6	41.1	47.7	52.2	56.7	58.2	51.0	45.2	37.2	26.5	40.3
1980	19.5	33.8	34.1	42.2	49.4	55.6	55.6	54.7	49.0	44.6	38.7	21.7	41.6
1981	37.6	32.7	39.4	39.1	52.1	54.3	56.1	55.9	49.2	42.8	36.8	26.9	43.6
1982	13.8	21.3	31.8	37.1	45.4	56.3	57.7	54.8	50.3	42.2	30.6	31.6	39.4
1983	30.2	31.8	34.8	42.6	49.7	55.6	57.1	54.6	48.0	42.1	31.8	18.9	41.4
1984	32.0		39.6	43.1	49.1	53.5	55.5	55.3	50.0	40.5			
1985												32.5	
1986	34.3	28.7	35.3	37.3	46.6	54.3	56.8	54.4	50.7	45.8	30.5	36.0	42.6
1987	33.1	34.7	31.8	41.6	47.8	51.6	58.6	57.5	50.1	44.4	40.0	34.3	43.8
1988	27.0	32.6	37.6	41.7	48.5	54.0	53.8	53.9	48.2	44.1	36.2	31.3	42.4
1989	25.5	23.9	29.4	42.7	49.0	55.4	60.1	58.2	52.4	41.7	32.9	36.0	42.3
1990	26.4	25.1	36.4	42.7	49.7	55.0	59.3	58.1	51.2	40.7	26.4	23.7	41.2
1991	24.9	35.3	33.1	41.7	48.2	55.0	55.2	55.0	50.3	40.1	36.3	33.6	42.4
1992	35.7	32.7	36.9	41.1	46.4	55.2	57.0	56.0	46.8	40.5	38.4	25.6	42.7
1993	24.5	30.2	35.8	44.4	52.1	55.8	59.6	57.1	51.2	45.8	36.8	35.9	44.1
1994	27.7	18.4	36.7	43.3	48.1	54.9	54.9	59.3	50.2	43.3	29.4	27.7	41.4
1995	27.2	29.0	30.3	43.4	51.4	55.4	57.0	54.8	53.8		32.4	25.2	
1996	16.6	29.0	32.8	40.7	48.7	54.1	57.7	54.7	49.0	40.2	31.3	26.5	40.1
1997	26.1	35.3	31.5	41.5	51.5	56.6	58.2	58.8	53.3	41.4	37.3	36.7	44.0
1998	24.5	37.1	35.0	42.4	50.1	55.9	57.5	54.2	49.4	42.8	33.1	27.9	42.5
1999	26.6	28.6	33.1	39.5	44.8	53.8	57.4	56.4	49.7	43.5	36.1	35.9	42.5
2000	26.5	31.7	36.3	39.7	46.9	53.0	55.3	54.4	49.0	41.9	37.7	35.9	42.0
2001	36.7	28.2	33.4	40.3	45.2	54.2	55.4	57.2	50.3	41.9	32.9	29.0	42.1
POR= 57 YRS	24.4	28.2	32.7	39.6	47.4	53.5	56.0	54.9	49.5	42.0	33.1	28.1	40.8

REFERENCE NOTES:

PAGE 1:
THE TEMPERATURE GRAPH SHOWS NORMAL MAXIMUM AND NORMAL
MINIMUM DAILY TEMPERATURES (SOLID CURVES) AND THE
ACTUAL DAILY HIGH AND LOW TEMPERATURES (VERTICAL BARS).

PAGE 2 AND 3:
H/C INDICATES HEATING AND COOLING DEGREE DAYS.
RH INDICATES RELATIVE HUMIDITY
W/O INDICATES WEATHER AND OBSTRUCTIONS
S INDICATES SUNSHINE.
PR INDICATES PRESSURE.
CLOUDINESS ON PAGE 3 IS THE SUM OF THE CEILOMETER AND
SATELLITE DATA NOT TO EXCEED EIGHT EIGHTHS (OKTAS).

GENERAL:
T INDICATES TRACE PRECIPITATION, AN AMOUNT GREATER
THAN ZERO BUT LESS THAN THE LOWEST REPORTABLE VALUE.
+ INDICATES THE VALUE ALSO OCCURS ON EARLIER DATES.
BLANK ENTRIES DENOTE MISSING OR UNREPORTED DATA.
NORMALS ARE 30-YEAR AVERAGES (1961 - 1990).
ASOS INDICATES AUTOMATED SURFACE OBSERVING SYSTEM.
PM INDICATES THE LAST DAY OF THE PREVIOUS MONTH.
POR (PERIOD OF RECORD) BEGINS WITH THE JANUARY DATA
MONTH AND IS THE NUMBER OF YEARS USED TO COMPUTE
THE MEAN. INDIVIDUAL MONTHS WITHIN THE POR MAY
BE MISSING.
WHEN THE POR FOR A NORMAL IS LESS THAN 30 YEARS,
THE NORMAL IS PROVISIONAL AND IS BASED ON THE NUMBER
OF YEARS INDICATED.
0.* OR * INDICATES THE VALUE OR MEAN-DAYS-WITH
IS BETWEEN 0.00 AND 0.05.
CLOUDINESS FOR ASOS STATIONS DIFFERS FROM THE NON-ASOS
OBSERVATION TAKEN BY A HUMAN OBSERVER. ASOS STATION
CLOUDINESS IS BASED ON TIME-AVERAGED CEILOMETER DATA
FOR CLOUDS AT OR BELOW 12,000 FEET AND ON SATELLITE
DATA FOR CLOUDS ABOVE 12,000 FEET.
THE NUMBER OF DAYS WITH CLEAR, PARTLY CLOUDY, AND
CLOUDY CONDITIONS FOR ASOS STATIONS IS THE SUM
OF THE CEILOMETER AND SATELLITE DATA FOR THE
SUNRISE TO SUNSET PERIOD.

GENERAL CONTINUED:
CLEAR INDICATES 0 - 2 OKTAS, PARTLY CLOUDY INDICATES
3 - 6 OKTAS, AND CLOUDY INDICATES 7 OR 8 OKTAS.
WHEN AT LEAST ONE OF THE ELEMENTS (CEILOMETER OR
SATELLITE) IS MISSING, THE DAILY CLOUDINESS IS
NOT COMPUTED.
WIND DIRECTION IS RECORDED IN TENS OF DEGREES (2 DIGITS)
CLOCKWISE FROM TRUE NORTH. "00" INDICATES CALM. "36"
INDICATES TRUE NORTH.
RESULTANT WIND IS THE VECTOR AVERAGE OF THE SPEED AND
DIRECTION.
AVERAGE TEMPERATURE IS THE SUM OF THE MEAN DAILY MAXIMUM
AND MINIMUM TEMPERATURE DIVIDED BY 2.
SNOWFALL DATA COMPRISE ALL FORMS OF FROZEN
PRECIPITATION, INCLUDING HAIL.
A HEATING (COOLING) DEGREE DAY IS THE DIFFERENCE BETWEEN
THE AVERAGE DAILY TEMPERATURE AND 65 F.
DRY BULB IS THE TEMPERATURE OF THE AMBIENT AIR.
DEW POINT IS THE TEMPERATURE TO WHICH THE AIR MUST BE
COOLED TO ACHIEVE 100 PERCENT RELATIVE HUMIDITY.
WET BULB IS THE TEMPERATURE THE AIR WOULD HAVE IF THE
MOISTURE CONTENT WAS INCREASED TO 100 PERCENT RELATIVE
HUMIDITY.

ON JULY 1, 1996, THE NATIONAL WEATHER SERVICE BEGAN USING
THE "METAR" OBSERVATION CODE THAT WAS ALREADY EMPLOYED
BY MOST OTHER NATIONS OF THE WORLD. THE MOST NOTICEABLE
DIFFERENCE IN THIS ANNUAL PUBLICATION WILL BE THE CHANGE
IN UNITS FROM TENTHS TO EIGHTS (OKTAS) FOR REPORTING THE
AMOUNT OF SKY COVER.

HEATING DEGREE DAYS (base 65 F) 2001 JUNEAU, AK (JNU)

YEAR	JUL	AUG	SEP	OCT	NOV	DEC	JAN	FEB	MAR	APR	MAY	JUN	TOTAL
1972-73	207	293	535	810	907	1275	1423	1126	986	752	584	404	9302
1973-74	343	404	505	732	1253	1143	1550	1006	1242	765	556	437	9936
1974-75	349	315	437	690	851	957	1296	1129	1063	791	541	402	8821
1975-76	281	337	402	712	1088	1244	1132	1131	1006	706	597	384	9020
1976-77	280	275	427	679	717	938	918	690	893	673	531	320	7341
1977-78	243	196	428	695	1062	1423	1233	922	954	683	525	317	8681
1978-79	298	262	427	612	1037	1134	1370	1505	904	712	528	378	9167
1979-80	251	205	415	609	830	1187	1404	895	949	678	477	278	8178
1980-81	283	308	472	628	783	1333	843	899	786	772	516	316	7815
1981-82	269	275	469	682	841	1175	1579	1214	1021	830	601	257	9213
1982-83	220	310	435	699	1027	1029	1073	924	931	663	470	275	8056
1983-84	237	317	502	701	991	1423	1014		780	649	486	338	
1984-85	286	291	444	754									
1985-86							1001	943	1011	913	822	564	316	
1986-87	249	319	423	587	1028	891	982	841	1020	697	526	394	7957
1987-88	199	222	440	635	744	944	1169	935	844	691	508	324	7655
1988-89	338	338	497	641	855	1040	1217	1144	1097	663	491	283	8604
1989-90	159	210	370	713	959	890	1191	1109	879	661	467	295	7903
1990-91	180	210	407	748	1152	1274	1238	823	981	694	516	298	8521
1991-92	294	303	435	764	855	966	902	930	865	712	571	292	7889
1992-93	240	274	540	750	791	1217	1250	966	899	612	394	271	8204
1993-94	166	240	406	588	838	891	1147	1298	869	641	514	299	7897
1994-95	229	171	437	668	1061	1150	1165	1003	1066	641	416	285	8292
1995-96	241	309	329		971	1229	1493	1038	993	719	498	322	
1996-97	218	312	476	764	1004	1187	1196	826	1031	698	414	252	8378
1997-98	203	183	345	725	823	869	1247	776	926	673	457	266	7493
1998-99	224	331	463	681	951	1141	1181	1012	978	758	619	329	8668
1999-00	230	264	452	660	858	895	1184	959	882	751	552	354	8041
2000-01	295	322	471	710	813	1040	873	1026	972	734	605	320	8181
2001-	292	235	434	709	956	1111							

COOLING DEGREE DAYS (base 65 F) 2001 JUNEAU, AK (JNU)

YEAR	JAN	FEB	MAR	APR	MAY	JUN	JUL	AUG	SEP	OCT	NOV	DEC	ANNUAL
1972	0	0	0	0	0	0	0	0	0	0	0	0	0
1973	0	0	0	0	0	0	0	0	0	0	0	0	0
1974	0	0	0	0	0	0	0	0	0	0	0	0	0
1975	0	0	0	0	0	0	7	0	0	0	0	0	7
1976	0	0	0	0	0	0	3	3	0	0	0	0	6
1977	0	0	0	0	0	0	0	1	0	0	0	0	1
1978	0	0	0	0	0	0	0	0	0	0	0	0	0
1979	0	0	0	0	0	0	0	1	0	0	0	0	1
1980	0	0	0	0	0	1	0	0	0	0	0	0	1
1981	0	0	0	0	0	0	0	0	0	0	0	0	0
1982	0	0	0	0	0	2	0	0	0	0	0	0	2
1983	0	0	0	0	0	0	0	0	0	0	0	0	0
1984	0		0	0	0	0	0	0	0	0	0		
1985											0		
1986	0		0	0	0	2	0	0	0	0	0	0	2
1987	0	0	0	0	0	0	5	0	0	0	0	0	5
1988	0	0	0	0	0	0	0	0	0	0	0	0	0
1989	0	0	0	0	0	0	14	0	0	0	0	0	14
1990	0	0	0	0	0	1	8	3	0	0	0	0	12
1991	0	0	0	0	0	6	0	0	0	0	0	0	6
1992	0	0	0	0	0	6	0	0	0	0	0	0	6
1993	0	0	0	0	0	0	5	0	0	0	0	0	5
1994	0	0	0	0	0	0	0	2	0	0	0	0	2
1995	0	0	0	0	0	1	0	0	0	0	0		
1996	0	0	0	0	0	0	0	2	0	0	0	0	2
1997	0	0	0	0	0	6	0	0	0	0	0	0	6
1998	0	0	0	0	0	0	0	0	0	0	0	0	0
1999	0	0	0	0	0	0	0	2	3	0	0	0	5
2000	0	0	0	0	0	0	0	0	0	0	0	0	0
2001	0	0	0	0	0	0	0	0	0	0	0	0	0

SNOWFALL (inches) 2001 JUNEAU, AK (JNU)

YEAR	JUL	AUG	SEP	OCT	NOV	DEC	JAN	FEB	MAR	APR	MAY	JUN	TOTAL
1972-73	0.0	0.0	0.0	2.2	2.8	31.6	63.8	20.9	9.6	0.5	0.0	0.0	131.4
1973-74	0.0	0.0	0.0	T	18.6	15.7	36.0	32.8	15.3	0.5	0.0	0.0	118.9
1974-75	0.0	0.0	T	0.9	3.5	17.3	41.5	16.5	18.9	4.5	0.0	0.0	103.1
1975-76	0.0	0.0	0.0	5.3	32.5	51.0	32.9	34.1	25.5	2.3	0.0	0.0	183.6
1976-77	0.0	0.0	0.0	1.2	1.1	26.3	5.0	T	1.7	0.4	0.0	0.0	46.3
1977-78	0.0	0.0	0.0	T	27.4	16.6	5.2	1.1	1.7	0.4	0.0	0.0	52.4
1978-79	0.0	0.0	0.0	0.0	14.9	21.6	24.2	21.4	5.4	T	0.0	0.0	87.5
1979-80	0.0	0.0	0.0	0.0	1.6	48.4	41.6	4.1	2.4	T	0.0	0.0	98.1
1980-81	0.0	0.0	0.0	0.0	0.5	40.5	2.4	16.4	0.5	2.2	0.0	0.0	62.5
1981-82	0.0	0.0	0.0	0.0	4.0	6.0	69.2	29.6	8.4	1.1	T	0.0	118.3
1982-83	0.0	0.0	0.0	2.0	0.4	10.8	40.1	15.7	0.2	T	0.0	0.0	69.2
1983-84	0.0	0.0	0.0	0.0	8.1	13.3	43.1	0.7	1.0	T	T	0.0	66.2
1984-85	0.0	0.0	0.0	0.0			2.0	10.3	7.4	30.4	4.4	T	
1985-86													
1986-87	0.0	0.0	0.0	T	22.1	1.4	3.3	1.4	7.3	T	0.0	0.0	35.5
1987-88	0.0	0.0	0.0	T	4.6	6.8	3.5	8.0	1.0	0.5	0.0	0.0	24.4
1988-89	0.0	0.0	0.0	0.0	4.8	11.3	44.7	0.2	10.0	T	0.0	0.0	71.0
1989-90	0.0	0.0	0.0	0.6	32.5	6.4	36.5	39.4	0.6	0.0	0.0	0.0	116.0
1990-91	0.0	0.0	0.0	0.0	48.8	33.2	31.8	15.5	9.4	0.8	T	0.0	139.5
1991-92	0.0	0.0	0.0	5.4	7.7	49.3	14.3	12.4	4.1	T	0.0	0.0	93.2
1992-93	0.0	0.0	0.0	1.1	4.4	25.0	32.5	36.8	2.6	T	0.0	0.0	102.4
1993-94	0.0	0.0	0.0	0.0	4.3	10.4	61.2	32.4	22.9	T	0.0	0.0	131.2
1994-95	0.0	0.0	0.0	0.0	69.8	25.8	8.6	14.0	28.4	T	0.0	0.0	146.6
1995-96	0.0	0.0	0.0	0.0	2.9	23.0	20.8	33.1	5.5	T	0.0	0.0	85.3
1996-97	0.0	0.0	0.0	0.5	2.5	4.6	12.3	17.3	19.2	0.4	0.0	0.0	56.8
1997-98	0.0	0.0	0.0	1.9	1.2	14.9	12.9	0.5		T	0.0		
1998-99						22.8	53.1	34.2	7.3	1.1	T	0.0	
1999-00	0.0	0.0	0.0	T	5.1	19.9	13.6	4.6	T	T	0.0	0.0	43.2
2000-01	0.0	0.0	T	2.3	1.0	2.3	7.6	14.9	0.6	T	T	0.0	28.7
2001-	0.0	0.0	0.0	3.4	3.1	25.5							
POR= 55 YRS	0.0	0.0	T	1.0	12.2	22.3	25.2	18.6	14.2	2.9	0.0	T	96.4

2001
FLAGSTAFF,
ARIZONA (FLG)

Flagstaff, elevation 7,000 feet, is situated on a volcanic plateau at the base of the highest mountains in Arizona. The climate may be classified as vigorous with cold winters, mild, pleasantly cool summers, moderate humidity, and considerable diurnal temperature change. Only limited farming exists due to the short growing season. The stormy months are January, February, March, July, and August.

Based on the 1951-1980 period, the average first occurrence of 32 degrees Fahrenheit in the fall is September 21 and the average last occurrence in the spring is June 13.

Temperatures in Flagstaff are characteristic of high altitude climates. The average daily range of temperature is relatively high, especially in the winter months, October to March, as a result of extensive snow cover and clear skies. Winter minimum temperatures frequently reach zero or below and temperatures of -25 degrees or less have occurred. Summer maximum temperatures are often above 80 degrees and occasionally, temperatures have exceeded 95 degrees.

The Flagstaff area is semi-arid. Several months have recorded little or no precipitation. Over 90 consecutive days without measurable precipitation have occurred. Annual precipitation ranges from less than 10 inches to more than 35 inches. Winter snowfalls can be heavy, exceeding 100 inches during one month and over 200 inches during the winter season. However, accumulations are quite variable from year to year. Some winter months may experience little or no snow and the winter season has produced total snow accumulations of less than 12 inches.

NORMALS, MEANS, AND EXTREMES
FLAGSTAFF, AZ (FLG)

LATITUDE: 35 08' 25" N　　LONGITUDE: 111 40' 20" W　　ELEVATION (FT): GRND: 7000　BARO: 7003　　TIME ZONE: MOUNTAIN (UTC + 7)　　WBAN: 03103

	ELEMENT	POR	JAN	FEB	MAR	APR	MAY	JUN	JUL	AUG	SEP	OCT	NOV	DEC	YEAR
TEMPERATURE °F	NORMAL DAILY MAXIMUM	30	42.2	45.3	49.2	57.8	67.4	78.2	81.9	79.3	73.2	63.4	51.1	43.3	61.0
	MEAN DAILY MAXIMUM	51	42.4	45.0	49.7	57.9	67.4	77.9	81.7	79.0	74.0	63.5	50.9	43.5	61.1
	HIGHEST DAILY MAXIMUM	52	66	71	73	80	87	96	97	92	90	85	74	68	97
	YEAR OF OCCURRENCE		1971	1986	1988	1992	1974	1970	1973	1978	1950	1980	1977	1950	JUL 1973
	MEAN OF EXTREME MAXS.	52	57.3	59.5	63.9	71.7	78.4	88.6	90.0	87.5	83.3	76.0	65.2	58.4	73.3
	NORMAL DAILY MINIMUM	30	15.2	17.7	21.3	26.7	33.3	41.4	50.5	48.9	41.2	31.0	22.4	15.8	30.4
	MEAN DAILY MINIMUM	51	15.8	17.9	21.7	27.0	33.2	41.1	50.2	49.1	41.3	30.8	21.4	16.3	30.5
	LOWEST DAILY MINIMUM	52	-22	-23	-16	-2	0	22	32	24	23	-2	-13	-23	-23
	YEAR OF OCCURRENCE		1971	1985	1966	1975	1996	1955	1955	1968	1971	1971	1958	1990	DEC 1990
	MEAN OF EXTREME MINS.	52	-5.0	-.9	4.7	13.9	21.2	29.1	40.8	40.4	29.4	18.4	5.3	-2.9	16.2
	NORMAL DRY BULB	30	28.7	31.5	35.3	42.3	50.4	59.8	66.3	64.1	57.3	47.2	36.8	29.6	45.8
	MEAN DRY BULB	52	29.2	31.6	35.8	42.6	50.5	59.6	66.0	64.0	57.5	47.3	36.5	30.0	45.9
	MEAN WET BULB	48	24.2	26.7	29.4	33.8	39.5	45.1	53.6	53.4	47.5	38.1	30.0	24.6	37.2
	MEAN DEW POINT	48	15.7	17.7	19.1	20.9	24.6	28.5	44.0	46.1	38.7	27.4	20.3	15.4	26.5
	NORMAL NO. DAYS WITH:														
	MAXIMUM 90	30	0.0	0.0	0.0	0.0	0.0	1.7	1.8	0.4	0.0	0.0	0.0	0.0	3.9
	MAXIMUM 32	30	4.5	2.6	1.6	0.2	0.0	0.0	0.0	0.0	0.0	0.2	1.1	5.2	15.4
	MINIMUM 32	30	30.4	27.5	29.9	25.2	13.9	2.9	0.0	0.1	2.8	18.7	28.0	30.3	209.7
	MINIMUM 0	30	3.2	1.2	0.7	0.1	0.0	0.0	0.0	0.0	0.0	*	0.4	2.4	8.0
H/C	NORMAL HEATING DEG. DAYS	30	1125	938	921	681	453	187	29	65	237	552	846	1097	7131
	NORMAL COOLING DEG. DAYS	30	0	0	0	0	0	31	70	38	6	0	0	0	145
RH	NORMAL (PERCENT)	30	62	60	55	46	39	34	51	58	55	53	57	61	53
	HOUR 05 LST	30	72	73	71	65	62	53	68	77	74	70	70	71	69
	HOUR 11 LST	30	52	49	45	35	29	23	35	40	39	38	44	50	40
	HOUR 17 LST	30	50	45	41	32	26	21	39	43	37	36	44	50	39
	HOUR 23 LST	30	69	68	63	55	47	40	60	68	66	63	65	68	61
S	PERCENT POSSIBLE SUNSHINE	15	77	73	76	82	88	86	75	76	81	79	75	73	78
W/O	MEAN NO. DAYS WITH:														
	HEAVY FOG (VISBY 1/4 MI)	37	2.5	2.1	2.0	1.6	0.3	0.0	0.3	0.3	0.5	1.2	1.3	2.0	14.1
	THUNDERSTORMS	37	0.0	0.4	0.7	1.4	2.3	3.3	15.0	15.2	6.6	2.0	0.6	0.2	47.7
CLOUDINESS	MEAN:														
	SUNRISE-SUNSET (OKTAS)	2		5.6	4.4	4.8	1.6	2.4	3.2	4.8				4.4	
	MIDNIGHT-MIDNIGHT (OKTAS)	1		5.6	5.6		0.8	2.4	3.2	4.8		0.0	3.2	4.0	
	MEAN NO. DAYS WITH:														
	CLEAR	1		10.0	6.0	9.0	1.0		3.0	2.0	6.0	14.0		6.0	
	PARTLY CLOUDY	1		5.0	8.0	5.0				2.0	3.0	7.0	3.0	5.0	
	CLOUDY	1	4.0	8.0	8.0	5.0					6.0	3.0		7.0	
PR	MEAN STATION PRESSURE (IN)	11	23.23	23.22	23.16	23.18	23.23	23.29	23.38	23.38	23.33	23.31	23.25	23.26	23.27
	MEAN SEA-LEVEL PRES. (IN)	48	30.13	30.08	29.98	29.94	29.92	29.91	30.00	30.01	30.00	30.06	30.10	30.12	30.02
WINDS	MEAN SPEED (MPH)	22	5.9	6.1	6.4	7.0	7.1	6.6	5.2	4.3	4.8	5.2	5.8	5.8	5.9
	PREVAIL.DIR (TENS OF DEGS)	6	21	21	22	21	21	21	21	21	21	21	21	21	21
	MAXIMUM 2-MINUTE:														
	SPEED (MPH)	7	36	33	38	38	36	34	32	30	31	38	37	34	38
	DIR. (TENS OF DEGS)		22	07	19	21	20	20	07	09	21	04	06	19	04
	YEAR OF OCCURRENCE		1996	1999	1995	2001	2001	1995	2000	1998	2000	1999	2000	1996	OCT 1999
	MAXIMUM 5-SECOND:														
	SPEED (MPH)	7	45	45	49	48	44	44	40	37	44	53	52	46	53
	DIR. (TENS OF DEGS)		24	18	19	21	23	21	07	33	20	04	26	19	04
	YEAR OF OCCURRENCE		1996	1998	1995	2001	2001	1995	2000	1998	2000	1999	2001	2001	OCT 1999
PRECIPITATION	NORMAL (IN)	30	2.04	2.09	2.55	1.48	0.72	0.40	2.78	2.75	2.03	1.61	1.95	2.40	22.80
	MAXIMUM MONTHLY (IN)	52	9.55	10.05	6.75	5.62	4.14	2.92	6.62	8.06	6.75	9.86	6.64	7.30	10.05
	YEAR OF OCCURRENCE		1993	1993	1970	1965	1992	1955	1986	1986	1983	1972	1985	1967	FEB 1993
	MINIMUM MONTHLY (IN)	52	0.00	T	T	T	T	0.00	T	0.26	0.00	T	0.00	T	NOV 1999
	YEAR OF OCCURRENCE		1972	1967	1972	1991	1974	1971	1993	1962	1995	1952	1999	1958	
	MAXIMUM IN 24 HOURS (IN)	52	2.71	4.48	2.96	1.79	1.11	2.79	2.55	3.04	3.43	2.73	3.69	3.11	4.48
	YEAR OF OCCURRENCE		1993	1993	1970	1985	1965	1956	1964	1986	1965	1972	1978	1951	FEB 1993
	NORMAL NO. DAYS WITH:														
	PRECIPITATION 0.01	30	7.7	7.1	9.2	6.0	4.8	2.8	11.9	11.6	7.4	5.0	5.5	7.4	86.4
	PRECIPITATION 1.00	30	0.3	0.3	0.3	0.2	*	0.0	0.5	0.5	0.5	0.4	0.4	0.6	4.0
SNOWFALL	NORMAL (IN)	30	20.5	19.0	24.4	11.9	1.9	0.0	0.0	0.0	0.1	2.5	10.7	17.8	108.8
	MAXIMUM MONTHLY (IN)	46	63.4	45.5	79.4	58.3	8.2	T	T	0.3	2.0	24.7	40.7	86.0	86.0
	YEAR OF OCCURRENCE		1980	1990	1991	1965	1975	1993	1992	1992	1965	1971	1985	1967	DEC 1967
	MAXIMUM IN 24 HOURS (IN)	46	23.1	23.1	26.3	17.2	6.6	T	T	0.3	2.0	13.5	19.6	27.3	27.3
	YEAR OF OCCURRENCE		1980	1987	1970	1977	1965	1993	1992	1992	1965	1974	1991	1967	DEC 1967
	MAXIMUM SNOW DEPTH (IN)	45	34	40	37	31	8	0	0	0	T	11	23	66	66
	YEAR OF OCCURRENCE		1979	1979	1969	1973	1954				1965	1974	1975	1967	DEC 1967
	NORMAL NO. DAYS WITH:														
	SNOWFALL 1.0	30	4.1	4.2	5.4	2.6	0.6	0.0	0.0	0.0	0.*	0.6	2.2	4.1	23.8

PRECIPITATION (inches) 2001 FLAGSTAFF, AZ (FLG)

YEAR	JAN	FEB	MAR	APR	MAY	JUN	JUL	AUG	SEP	OCT	NOV	DEC	ANNUAL
1972	0.00	0.02	T	0.72	0.14	1.93	1.90	2.82	0.81	9.86	2.34	4.13	24.67
1973	1.89	3.69	6.18	1.21	1.17	0.40	1.87	1.25	T	0.03	1.90	0.12	19.71
1974	3.63	0.26	1.01	0.57	T	T	3.00	0.93	3.64	1.03	1.18	1.18	17.41
1975	1.76	1.90	2.92	2.20	1.16	0.05	2.24	0.74	1.89	0.33	2.96	1.95	20.10
1976	0.17	5.96	2.06	3.09	1.65	T	3.82	0.58	1.16	0.73	0.10	0.80	20.12
1977	1.85	0.84	0.92	1.47	0.96	0.91	3.60	3.72	1.52	1.04	0.76	1.18	18.77
1978	4.09	4.67	5.58	1.60	0.27	0.09	1.17	0.68	0.46	0.56	6.16	5.39	30.72
1979	5.54	1.73	2.52	0.31	2.16	0.18	0.79	2.38	0.13	1.30	1.14	1.50	19.68
1980	6.52	7.81	4.16	1.21	1.79	0.25	2.49	2.19	0.65	1.08	T	1.15	29.30
1981	1.31	1.16	4.04	1.50	0.72	1.09	2.87	3.73	2.54	1.81	2.43	0.17	23.37
1982	4.62	2.55	5.69	0.25	0.86	T	1.89	2.32	3.17	0.71	5.36	3.67	31.09
1983	1.61	3.04	4.36	2.18	0.06	0.28	2.86	3.53	6.75	0.75	1.53	2.52	29.47
1984	0.36	0.13	0.89	0.73	0.21	0.13	4.20	3.86	1.75	1.43	1.40	5.00	20.09
1985	2.38	1.67	2.54	3.39	0.26	0.09	2.36	1.07	3.68	2.44	6.64	0.15	26.67
1986	0.31	1.76	2.60	1.23	0.72	1.16	6.62	8.06	4.80	2.02	1.68	1.43	32.39
1987	2.51	2.54	1.69	0.21	0.70	0.25	1.93	3.74	1.49	4.64	2.91	1.37	23.98
1988	1.64	2.30	0.14	3.83	0.14	1.86	3.48	4.77	0.15	1.23	1.14	1.00	21.68
1989	1.84	1.35	2.08	0.01	0.62	0.23	2.28	3.40	0.30	1.28	T	1.05	14.44
1990	1.54	3.20	2.17	2.32	0.73	0.24	4.32	1.71	6.18	0.49	1.09	1.68	25.67
1991	1.76	2.08	6.00	T	0.14	0.59	1.04	1.64	0.26	1.12	4.47	2.73	21.83
1992	2.03	3.69	4.40	0.78	3.14	0.32	2.67	5.80	T	3.64	0.46	6.78	34.71
1993	9.55	10.05	1.54	0.26	0.44	0.55	T	4.19	1.95	3.29	3.02	0.76	35.60
1994	0.38	2.47	3.03	2.48	1.01		1.70	3.61	2.75	1.12	1.91	1.43	
1995	2.39	3.77	3.99	1.49	0.44	0.06	0.61	2.30	0.00	0.01	0.31	0.35	15.72
1996	0.19	1.36	0.54	0.07	T	T	1.79	0.92	3.73	0.93		0.64	
1997	3.21	0.99	0.03	0.60	0.08	0.18	0.21	2.83	3.68	1.15	0.80	1.85	15.61
1998	1.30	2.15	3.74	1.52	1.31	0.00	4.72	2.82	4.45	3.11	1.76	0.42	27.30
1999	0.28	0.48	0.53	2.80	0.42	0.95	3.27	2.45	4.54	T	0.00	T	15.72
2000	0.62	1.61	3.12	0.19	0.12	1.11	0.29	2.83	0.36	3.85	1.07	0.21	15.38
2001	2.60	1.68	1.28	1.40	0.82	0.03	2.80	3.46	0.68	1.21	0.43	1.16	17.55
POR= 102 YRS	2.00	2.05	2.12	1.32	0.71	0.50	2.68	2.88	1.86	1.56	1.47	1.87	21.02

AVERAGE TEMPERATURE (F) 2001 FLAGSTAFF, AZ (FLG)

YEAR	JAN	FEB	MAR	APR	MAY	JUN	JUL	AUG	SEP	OCT	NOV	DEC	ANNUAL
1972	28.4	31.8	40.0	39.8	46.4	57.6	66.5	62.9	56.7	45.9	29.6	21.9	44.0
1973	22.9	28.7	26.8	38.0	52.8	60.5	65.7	64.6	57.7	50.0	37.1	33.0	44.8
1974	27.8	30.5	40.4	43.2	54.8	66.5	66.3	64.7	58.7	47.9	36.8	26.8	47.0
1975	27.7	27.6	33.2	36.2	47.5	57.1	65.9	63.5	57.9	46.5	37.3	29.2	44.1
1976	30.6	35.1	35.4	41.4	52.8	58.9	66.5	62.8	57.1	47.0	39.5	31.6	46.6
1977	26.8	34.2	33.5	44.7	47.4	62.2	67.2	66.2	60.2	50.8	41.0	37.6	47.7
1978	31.5	30.7	40.3	42.2	49.4	61.4	66.6	63.7	57.3	49.8	34.6	24.3	46.0
1979	22.6	25.9	32.3	40.7	47.9	57.8	64.1	60.5	58.8	47.8	31.2	31.2	43.4
1980	30.7	32.8	32.3	41.9	45.6	60.9	69.0	66.3	59.7	49.0	41.4	39.9	47.5
1981	36.2	36.3	36.6	48.5	52.1	66.1	68.2	65.4	58.7	46.7	41.8	36.4	49.4
1982	28.3	30.5	35.3	43.5	49.9	57.1	63.8	65.6	57.6	44.0	35.2	28.1	44.9
1983	31.0	32.3	36.2	37.0	49.9	57.3	65.5	63.8	60.7	48.6	36.9	34.0	46.1
1984	31.7	32.8	38.3	40.7	56.8	58.5	65.7	63.8	59.3	42.4	35.3	29.0	46.2
1985	27.5	28.9	35.5	46.2	51.5	62.6	67.2	65.3	53.1	47.7	33.8	32.4	46.0
1986	37.0	34.2	39.7	44.1	52.0	61.2	64.0	65.7	53.0	43.5	37.6	30.2	46.9
1987	27.6	31.1	32.8	45.9	50.7	61.3	62.8	65.2	56.5	50.4	36.2	27.1	45.4
1988	29.1	34.2	37.4	44.0	50.8	61.5	67.4	64.3	55.9	52.5	36.9	27.9	46.8
1989	26.2	32.3	41.8	50.4	54.2	60.8	68.1	63.6	58.7	47.5	38.4	31.6	47.8
1990	28.6	29.3	38.3	46.1	50.3	64.4	66.9	62.6	59.8	48.9	37.6	25.1	46.5
1991	30.5	37.2	32.2	43.1	49.4	57.3	67.1	66.5	59.6	50.8	36.9	30.2	46.7
1992	27.3	35.2	38.1	49.1	53.3	59.0	65.1	64.2	59.3	50.4	33.8	26.3	46.8
1993	32.2	32.5	40.2	46.4	53.9	60.4	65.5	64.8	57.9	47.4	35.0	30.1	47.2
1994	32.8	29.8	40.6	44.3	51.7		65.4	66.4	58.1	44.2	32.2	32.1	
1995	28.4	37.2	38.2	40.4		55.1	64.6	66.9		47.7	42.3	31.9	
1996	31.4	36.9	37.8	44.9		61.9							
1997	28.0	29.6	39.1	40.5	54.1	57.1	63.8	63.8	59.4	45.0	37.7	26.6	45.4
1998	31.4	27.2	33.7	37.2	46.8	55.0	66.2	64.9	57.3	44.6	37.1	31.1	44.4
1999	33.9	35.1	39.0	37.5	49.4	58.0	64.2	62.1	56.1	47.7	40.5	29.3	46.1
2000	32.9	35.6	35.4	46.6	55.4	62.2	66.5	65.0	60.1	46.2	30.8	34.7	47.6
2001	27.2	29.7	39.0	43.5	55.3	62.2	65.6	64.3	59.9	50.2	38.8	27.0	46.9
POR= 102 YRS	28.4	31.2	35.8	42.8	50.6	59.4	65.8	63.9	57.4	47.1	36.3	29.8	45.7

REFERENCE NOTES:

PAGE 1:
THE TEMPERATURE GRAPH SHOWS NORMAL MAXIMUM AND NORMAL MINIMUM DAILY TEMPERATURES (SOLID CURVES) AND THE ACTUAL DAILY HIGH AND LOW TEMPERATURES (VERTICAL BARS).

PAGE 2 AND 3:
H/C INDICATES HEATING AND COOLING DEGREE DAYS.
RH INDICATES RELATIVE HUMIDITY
W/O INDICATES WEATHER AND OBSTRUCTIONS
S INDICATES SUNSHINE.
PR INDICATES PRESSURE.
CLOUDINESS ON PAGE 3 IS THE SUM OF THE CEILOMETER AND SATELLITE DATA NOT TO EXCEED EIGHT EIGHTHS(OKTAS).

GENERAL:
T INDICATES TRACE PRECIPITATION, AN AMOUNT GREATER THAN ZERO BUT LESS THAN THE LOWEST REPORTABLE VALUE.
+ INDICATES THE VALUE ALSO OCCURS ON EARLIER DATES.
BLANK ENTRIES DENOTE MISSING OR UNREPORTED DATA.
NORMALS ARE 30-YEAR AVERAGES (1961 - 1990).
ASOS INDICATES AUTOMATED SURFACE OBSERVING SYSTEM.
PM INDICATES THE LAST DAY OF THE PREVIOUS MONTH.
POR (PERIOD OF RECORD) BEGINS WITH THE JANUARY DATA MONTH AND IS THE NUMBER OF YEARS USED TO COMPUTE THE MEAN. INDIVIDUAL MONTHS WITHIN THE POR MAY BE MISSING.
WHEN THE POR FOR A NORMAL IS LESS THAN 30 YEARS, THE NORMAL IS PROVISIONAL AND IS BASED ON THE NUMBER OF YEARS INDICATED.
0.* OR * INDICATES THE VALUE OR MEAN-DAYS-WITH IS BETWEEN 0.00 AND 0.05.
CLOUDINESS FOR ASOS STATIONS DIFFERS FROM THE NON-ASOS OBSERVATION TAKEN BY A HUMAN OBSERVER. ASOS STATION CLOUDINESS IS BASED ON TIME-AVERAGED CEILOMETER DATA FOR CLOUDS AT OR BELOW 12,000 FEET AND ON SATELLITE DATA FOR CLOUDS ABOVE 12,000 FEET.
THE NUMBER OF DAYS WITH CLEAR, PARTLY CLOUDY, AND CLOUDY CONDITIONS FOR ASOS STATIONS IS THE SUM OF THE CEILOMETER AND SATELLITE DATA FOR THE SUNRISE TO SUNSET PERIOD.

GENERAL CONTINUED:
CLEAR INDICATES 0 - 2 OKTAS, PARTLY CLOUDY INDICATES 3 - 6 OKTAS, AND CLOUDY INDICATES 7 OR 8 OKTAS. WHEN AT LEAST ONE OF THE ELEMENTS (CEILOMETER OR SATELLITE) IS MISSING, THE DAILY CLOUDINESS IS NOT COMPUTED.
WIND DIRECTION IS RECORDED IN TENS OF DEGREES (2 DIGITS) CLOCKWISE FROM TRUE NORTH. "00" INDICATES CALM. "36" INDICATES TRUE NORTH.
RESULTANT WIND IS THE VECTOR AVERAGE OF THE SPEED AND DIRECTION.
AVERAGE TEMPERATURE IS THE SUM OF THE MEAN DAILY MAXIMUM AND MINIMUM TEMPERATURE DIVIDED BY 2.
SNOWFALL DATA COMPRISE ALL FORMS OF FROZEN PRECIPITATION, INCLUDING HAIL.
A HEATING (COOLING) DEGREE DAY IS THE DIFFERENCE BETWEEN THE AVERAGE DAILY TEMPERATURE AND 65 F.
DRY BULB IS THE TEMPERATURE OF THE AMBIENT AIR.
DEW POINT IS THE TEMPERATURE TO WHICH THE AIR MUST BE COOLED TO ACHIEVE 100 PERCENT RELATIVE HUMIDITY.
WET BULB IS THE TEMPERATURE THE AIR WOULD HAVE IF THE MOISTURE CONTENT WAS INCREASED TO 100 PERCENT RELATIVE HUMIDITY.

ON JULY 1, 1996, THE NATIONAL WEATHER SERVICE BEGAN USING THE "METAR" OBSERVATION CODE THAT WAS ALREADY EMPLOYED BY MOST OTHER NATIONS OF THE WORLD. THE MOST NOTICEABLE DIFFERENCE IN THIS ANNUAL PUBLICATION WILL BE THE CHANGE IN UNITS FROM TENTHS TO EIGHTS(OKTAS) FOR REPORTING THE AMOUNT OF SKY COVER.

HEATING DEGREE DAYS (base 65 F) 2001 FLAGSTAFF, AZ (FLG)

YEAR	JUL	AUG	SEP	OCT	NOV	DEC	JAN	FEB	MAR	APR	MAY	JUN	TOTAL
1972-73	22	78	241	584	1055	1330	1297	1006	1176	801	370	164	8124
1973-74	39	45	212	458	829	987	1147	960	754	650	310	67	6458
1974-75	19	34	195	522	841	1178	1150	1041	978	856	536	231	7581
1975-76	23	66	207	566	821	1101	1059	862	912	702	370	190	6879
1976-77	17	71	230	553	757	1027	1177	858	970	603	540	102	6905
1977-78	9	19	157	432	715	843	1032	954	756	677	478	128	6200
1978-79	33	73	227	465	907	1254	1307	1089	1005	722	524	219	7825
1979-80	68	157	186	526	1008	1042	1056	930	1009	684	596	158	7420
1980-81	6	43	153	491	703	774	886	794	873	486	398	50	5657
1981-82	1	39	182	558	689	880	1130	963	911	640	458	230	6681
1982-83	65	22	218	643	888	1136	1046	911	888	835	461	222	7335
1983-84	26	64	134	502	837	952	1025	929	820	722	247	204	6462
1984-85	21	51	165	695	884	1109	1155	1005	911	557	411	102	7066
1985-86	26	31	351	532	931	1005	862	855	777	619	392	119	6500
1986-87	58	28	353	661	816	1069	1150	942	990	566	435	114	7182
1987-88	82	86	246	447	859	1167	1103	885	848	621	434	124	6902
1988-89	9	41	266	381	836	1141	1196	910	710	432	330	139	6391
1989-90	7	67	184	540	792	1030	1124	996	822	558	449	84	6653
1990-91	10	93	167	494	816	1231	1060	772	1008	652	476	225	7004
1991-92	7	12	159	438	835	1071	1159	854	830	470	356	180	6371
1992-93	38	80	163	444	932	1191	1014	903	761	551	339	158	6574
1993-94	29	45	207	538	893	1073	990	979	747	614	406		
1994-95	35	11	203	637	974	1014	1128	772	821	731	290		
1995-96	60	10		531	671	1015	1035	807	834	597	108		
1996-97	2	36	291	620		966	1141	984	797	729	330	231	
1997-98	61	45	161	613	813	1182	1037	1050	964	828	559	300	7613
1998-99	15	27	226	624	832	1044	957	829	798	817	480	220	6869
1999-00	45	86	260	529	726	1099	987	846	911	547	293	98	6427
2000-01	26	49	148	577	1020	933	1165	983	801	641	295	109	6747
2001-	25	40	149	452	781	1172							

COOLING DEGREE DAYS (base 65 F) 2001 FLAGSTAFF, AZ (FLG)

YEAR	JAN	FEB	MAR	APR	MAY	JUN	JUL	AUG	SEP	OCT	NOV	DEC	ANNUAL
1972	0	0	0	0	0	4	76	20	0	0	0	0	100
1973	0	0	0	0	0	34	69	40	0	0	0	0	143
1974	0	0	0	0	0	120	66	33	13	0	0	0	232
1975	0	0	0	0	0	0	60	28	0	0	0	0	88
1976	0	0	0	0	0	15	70	13	0	0	0	0	98
1977	0	0	0	0	0	28	82	61	20	0	0	0	191
1978	0	0	0	0	0	25	87	38	2	0	0	0	152
1979	0	0	0	0	0	10	47	24	4	0	0	0	85
1980	0	0	0	0	0	43	133	91	2	0	0	0	269
1981	0	0	0	0	0	91	108	58	0	0	0	0	257
1982	0	0	0	0	0	1	36	46	6	0	0	0	89
1983	0	0	0	0	0	0	45	31	12	0	0	0	88
1984	0	0	0	0	0	2	14	49	23	2	0	0	90
1985	0	0	0	0	0	36	102	47	0	0	0	0	185
1986	0	0	0	0	0	11	33	55	0	0	0	0	99
1987	0	0	0	0	0	13	21	26	0	0	0	0	60
1988	0	0	0	0	0	28	94	28	1	0	0	0	151
1989	0	0	0	0	0	22	111	29	0	0	0	0	162
1990	0	0	0	0	0	76	76	28	18	0	0	0	198
1991	0	0	0	0	0	0	79	65	4	0	0	0	148
1992	0	0	0	0	0	6	48	62	0	0	0	0	116
1993	0	0	0	0	0	26	50	46	1	0	0	0	123
1994	0	0	0	0	0		53	57	0	0	0	0	
1995	0	0	0	0	0		53	76	0	0			
1996	0	0	0	0		18							
1997	0	0	0	0	1	29	14	2	0	0	0		46
1998	0	0	0	0	0	5	60	28	0	0	0	0	93
1999	0	0	0	0	0	15	29	5	0	0	0	0	49
2000	0	0	0	0	3	18	81	55	8	0	0	0	165
2001	0	0	0	0	0	30	50	26	0	0	0	0	106

SNOWFALL (inches) 2001 FLAGSTAFF, AZ (FLG)

YEAR	JUL	AUG	SEP	OCT	NOV	DEC	JAN	FEB	MAR	APR	MAY	JUN	TOTAL
1972-73	0.0	0.0	0.0	11.8	23.2	28.9	21.0	33.8	77.4	10.9	3.0	0.0	210.0
1973-74	0.0	0.0	0.0	T	20.7	1.2	35.3	2.4	8.8	1.6	0.0	0.0	70.0
1974-75	0.0	0.0	0.0	16.6	8.2	15.6	20.1	18.2	29.1	25.1	8.2	0.0	141.1
1975-76	0.0	0.0	0.0	T	25.2	18.9	1.5	31.2	20.4	3.2	0.0	0.0	131.6
1976-77	0.0	0.0	0.0	T	0.7	9.0	21.1	11.2	8.8	17.8	1.6	0.0	70.2
1977-78	0.0	0.0	0.0	0.0	7.0	1.6	40.4	32.1	24.5	9.3	1.3	0.0	116.2
1978-79	0.0	0.0	0.0	0.0	16.5	19.8	59.4	18.1	22.8	4.1	4.8	0.0	145.5
1979-80	0.0	0.0	0.0	0.5	5.5	20.7	63.4	32.9	42.5	11.6	T	0.0	177.1
1980-81	0.0	0.0	0.0	6.9	T	6.8	11.7	11.9	45.6	9.5	0.0	0.0	92.4
1981-82	0.0	0.0	0.0	0.0			47.5	20.4	26.7	1.9	0.4	0.0	
1982-83	0.0	0.0	T	T	22.6	27.1	15.3	25.0	38.5	13.6	0.5	0.0	142.6
1983-84	0.0	0.0	0.0	0.0	14.3	5.8	4.3	1.8	0.3	5.5	0.0	0.0	32.0
1984-85	0.0	0.0	0.0	0.6	9.4	28.7	26.2	31.3	21.3	18.5	0.4	0.0	136.0
1985-86	0.0	0.0	0.0	0.0	40.7	2.6	0.4	26.9	32.8	1.6	0.4	0.0	105.4
1986-87	0.0	0.0	0.9	0.6	4.8	9.5	38.6	40.7	25.0	1.5	0.0	0.0	121.6
1987-88	0.0	0.0	0.0	0.0	2.9	16.7	28.9	21.0	1.5	33.1	0.4	0.0	104.5
1988-89	0.0	0.0	0.0	0.0	11.9	15.0	21.7	12.0	16.6	0.5	0.0		77.7
1989-90	0.0	T	T	T	T	13.1	24.2	45.5	25.0	4.2	1.4	0.0	113.4
1990-91	T	T	T	T	9.6	22.3	3.1	13.5	79.4	T	T		127.9
1991-92	0.0	0.0	T	5.9	39.5	24.0	24.7	24.9	35.9	4.0	T	T	158.9
1992-93	T	0.3	0.0	0.5	2.1	41.7	55.7	35.2	11.9	2.3	0.3	T	150.0
1993-94	0.0	T	T	0.0	23.0	13.4	6.9	26.1	12.3	27.8	T		
1994-95	0.0	T	0.0	0.0							0.0		
1995-96	0.0	0.0	0.0	0.0	0.0								
1996-97													
1997-98													
1998-99													
1999-00													
2000-01													
2001-													
POR= 46 YRS	T	T	0.1	2.0	9.7	15.6	20.0	18.0	21.5	9.7	1.7	T	98.3

2001
PHOENIX,
ARIZONA (PHX)

Phoenix is located in the Salt River Valley at an elevation of about 1,100 feet. The valley is oval shaped and flat except for scattered precipitous mountains rising a few hundred to as much as 1,500 feet above the valley floor. Sky Harbor Airport, where the weather observations are taken, is in the southern part of the city. Six miles to the south of the airport are the South Mountains rising to 2,500 feet. Eighteen miles southwest, the Estrella Mountains rise to 4,500 feet, and 30 miles to the west are the White Tank Mountains rising to 4,100 feet. The Superstition Mountains, over 30 miles to the east, rise to as much as 5,000 feet. The valley, though located in the Sonora Desert, supports large acreages of cotton, citrus, and other agriculture along with one of the largest urban populations in the United States. The water supply for this complex desert community is partly from reservoirs on the impounded Salt and Verde Rivers, and partly from a large underground water table.

Temperatures range from very hot in summer to mild in winter. Many winter days reach over 70 degrees and typical high temperatures in the middle of the winter are in the 60s. The climate becomes less attractive in the summer. The normal high temperature is over 90 degrees from early May through early October, and over 100 degrees from early June through early September. Many days each summer will exceed 110 degrees in the afternoon and remain above 85 degrees all night. When temperatures are extremely high, the low humidity does not provide much comfort.

Indeed, the climate is very dry. Annual precipitation is only about 7 inches, and afternoon humidities range from about 30 percent in winter to only about 10 percent in June. Rain comes mostly in two seasons. From about Thanksgiving to early April there are periodic rains from Pacific storms. Moisture from the south and southeast results in a summer thunderstorm peak in July and August. Usually the break from extreme dryness in June to the onset of thunderstorms in early July is very abrupt. Afternoon humidities suddenly double to about 20 percent, which with the great heat, gives a feeling of mugginess. Fog is rare, occurring about once per winter, and is unknown in the other seasons.

The valley is characterized by light winds. High winds associated with thunderstorms occur periodically in the summer. These occasionally create duststorms which move large distances across the deserts. Strong thunderstorm winds occur any month of the year, but are rare outside the summer months. Persistent strong winds of 30 mph or more are rare except for two or three events in an average spring due to Pacific storms. Winter storms rarely bring high winds due to the relatively stable air in the valley during that season.

Based on the 1951-1980 period, the average first occurrence of 32 degrees Fahrenheit in the fall is December 13 and the average last occurrence in the spring is February 7.

NORMALS, MEANS, AND EXTREMES
PHOENIX, AZ (PHX)

LATITUDE: 33 26' 35" N LONGITUDE: 111 59' 25" W ELEVATION (FT): GRND: 1103 BARO: 1106 TIME ZONE: MOUNTAIN (UTC + 7) WBAN: 23183

ELEMENT	POR	JAN	FEB	MAR	APR	MAY	JUN	JUL	AUG	SEP	OCT	NOV	DEC	YEAR
TEMPERATURE F														
NORMAL DAILY MAXIMUM	30	65.9	70.7	75.5	84.5	93.6	103.5	105.9	103.7	98.3	88.1	74.9	66.2	85.9
MEAN DAILY MAXIMUM	51	66.1	70.3	75.4	84.2	93.4	102.9	105.2	103.2	98.7	88.2	74.7	66.3	85.7
HIGHEST DAILY MAXIMUM	64	88	92	100	105	113	122	121	116	118	107	95	88	122
YEAR OF OCCURRENCE		1971	1986	1988	1992	1984	1990	1995	1975	1950	1980	2001	1950	JUN 1990
MEAN OF EXTREME MAXS.	54	77.6	82.2	87.9	97.2	104.7	112.3	113.4	110.6	107.4	99.4	86.8	77.6	96.4
NORMAL DAILY MINIMUM	30	41.2	44.7	48.8	55.3	63.9	72.9	81.0	79.2	72.8	60.8	48.9	41.8	59.3
MEAN DAILY MINIMUM	51	41.9	45.1	49.4	56.0	64.6	73.2	80.9	79.7	73.1	61.3	49.0	42.3	59.7
LOWEST DAILY MINIMUM	64	17	22	25	32	40	50	61	60	47	34	25	22	17
YEAR OF OCCURRENCE		1950	1948	1966	1945	1967	1944	1944	1942	1965	1971	1938	1948	JAN 1950
MEAN OF EXTREME MINS.	54	30.8	34.5	38.4	44.6	52.9	63.0	71.6	71.0	62.7	48.8	37.3	31.9	49.0
NORMAL DRY BULB	30	53.6	57.7	62.2	69.9	78.8	88.2	93.5	91.5	85.6	74.5	61.9	54.1	72.6
MEAN DRY BULB	54	53.7	57.5	62.3	69.9	78.9	88.0	92.8	91.3	85.9	74.4	61.8	54.0	72.5
MEAN WET BULB	49	43.9	46.1	48.6	52.3	56.7	62.0	70.0	70.5	65.7	57.4	49.0	44.0	55.5
MEAN DEW POINT	49	33.1	33.3	33.6	33.7	36.0	41.2	57.1	59.4	53.1	43.6	36.1	32.9	41.1
NORMAL NO. DAYS WITH:														
MAXIMUM 90	30	0.0	0.1	2.2	9.7	22.8	29.3	30.9	30.7	27.4	14.8	0.5	0.0	168.4
MAXIMUM 32	30	0.0	0.0	0.0	0.0	0.0	0.0	0.0	0.0	0.0	0.0	0.0	0.0	0.0
MINIMUM 32	30	3.7	1.4	0.4	0.0	0.0	0.0	0.0	0.0	0.0	0.0	0.2	2.0	7.7
MINIMUM 0	30	0.0	0.0	0.0	0.0	0.0	0.0	0.0	0.0	0.0	0.0	0.0	0.0	0.0
H/C														
NORMAL HEATING DEG. DAYS	30	362	227	182	75	8	0	0	0	0	17	134	345	1350
NORMAL COOLING DEG. DAYS	30	8	22	95	222	436	696	884	822	618	311	41	7	4162
RH														
NORMAL (PERCENT)	30	51	44	39	28	22	19	32	36	36	37	44	52	37
HOUR 05 LST	30	66	60	56	43	35	31	45	51	50	51	58	66	51
HOUR 11 LST	30	45	39	34	23	18	16	28	33	31	31	37	46	32
HOUR 17 LST	30	32	26	24	16	13	11	20	23	23	22	27	33	22
HOUR 23 LST	30	56	48	42	29	22	20	33	38	38	41	49	57	39
S PERCENT POSSIBLE SUNSHINE	11	79	82	85	89	94	95	86	86	89	88	83	77	86
W/O MEAN NO. DAYS WITH:														
HEAVY FOG (VISBY 1/4 MI)	64	0.5	0.1	0.1	0.0	0.0	0.0	0.0	0.0	0.0	0.0	0.2	0.5	1.4
THUNDERSTORMS	62	0.4	0.6	0.9	0.7	1.0	1.0	5.9	7.0	3.4	1.4	0.5	0.6	23.4
CLOUDINESS MEAN:														
SUNRISE-SUNSET (OKTAS)	1		2.4	2.0	2.4	1.2	1.2	1.6	0.8				1.6	
MIDNIGHT-MIDNIGHT (OKTAS)	1		2.4		2.4	1.2	1.2	1.6	0.8		0.0			
MEAN NO. DAYS WITH:														
CLEAR	1	5.0	17.5	18.0	14.0	26.0	17.0	6.0	10.0	15.0	15.0		13.0	
PARTLY CLOUDY	1	1.0	5.0	5.0	4.0	1.5	4.0		1.0		2.0		5.0	
CLOUDY	1		6.0	4.0	1.0	1.0			1.0					
PR MEAN STATION PRESSURE (IN)	28	28.89	28.87	28.78	28.72	28.66	28.64	28.68	28.69	28.69	28.76	28.84	28.89	28.76
MEAN SEA-LEVEL PRES. (IN)	48	30.06	30.01	29.93	29.86	29.80	29.75	29.79	29.80	29.80	29.89	30.00	30.05	29.89
WINDS MEAN SPEED (MPH)	46	5.1	5.7	6.5	6.8	6.9	6.6	6.9	6.5	6.1	5.6	5.2	4.9	6.1
PREVAIL.DIR (TENS OF DEGS)	30	10	10	10	10	27	27	27	09	10	10	10	10	10
MAXIMUM 2-MINUTE:														
SPEED (MPH)	7	36	30	43	51	32	33	41	37	37	36	30	29	51
DIR. (TENS OF DEGS)		25	21	24	30	26	04	17	14	15	24	25	31	30
YEAR OF OCCURRENCE		1995	2001	1998	2001	2001	2000	1995	1999	1999	1998	1997	2000	APR 2001
MAXIMUM 5-SECOND:														
SPEED (MPH)	7	39	39	53	59	40	40	52	53	51	48	36	36	59
DIR. (TENS OF DEGS)		24	26	24	29	27	04	18	25	15	24	25	31	29
YEAR OF OCCURRENCE		1995	2000	1998	2001	2001	2000	1995	2000	1999	1998	1997	2000	APR 2001
PRECIPITATION NORMAL (IN)	30	0.67	0.68	0.88	0.22	0.12	0.13	0.83	0.96	0.86	0.65	0.66	1.00	7.66
MAXIMUM MONTHLY (IN)	64	5.22	2.93	4.16	2.10	1.06	1.70	5.15	5.56	4.23	4.40	3.04	3.98	5.56
YEAR OF OCCURRENCE		1993	1998	1941	1941	1976	1972	1984	1951	1939	1972	1952	1967	AUG 1951
MINIMUM MONTHLY (IN)	64	0.00	0.00	0.00	0.00	0.00	0.00	T	T	0.00	0.00	0.00	0.00	0.00
YEAR OF OCCURRENCE		1972	1967	1959	1962	1983	1983	1995	1975	1973	1973	1980	1981	JUN 1983
MAXIMUM IN 24 HOURS (IN)	64	1.84	1.49	2.04	1.38	0.96	1.64	2.75	3.07	2.43	2.32	2.16	1.89	3.07
YEAR OF OCCURRENCE		1993	1987	1983	1941	1976	1972	1984	1943	1970	1988	1993	1967	AUG 1943
NORMAL NO. DAYS WITH:														
PRECIPITATION 0.01	30	4.0	3.8	4.1	1.6	0.8	0.7	4.3	4.7	3.2	2.7	2.9	4.0	36.8
PRECIPITATION 1.00	30	0.0	*	0.1	0.0	0.0	*	0.1	0.2	0.2	0.1	0.0	0.1	0.8
SNOWFALL NORMAL (IN)	30	T	T	0.0	0.0	0.0	0.0	0.0	0.0	0.0	0.0	0.0	0.*	0.0
MAXIMUM MONTHLY (IN)	62	T	0.6	T	T	T	0.0	0.0	0.0	0.0	T	0.0	0.4	0.6
YEAR OF OCCURRENCE		1993	1939	1991	1949	1992					1992		1990	FEB 1939
MAXIMUM IN 24 HOURS (IN)	62	T	0.6	T	T	T	0.0	0.0	0.0	0.0	T	0.0	0.4	0.6
YEAR OF OCCURRENCE		1993	1939	1991	1949	1992					1992		1990	FEB 1939
MAXIMUM SNOW DEPTH (IN)	48	0	0	0	0	0	0	0	0	0	0	0	0	0
YEAR OF OCCURRENCE														
NORMAL NO. DAYS WITH:														
SNOWFALL 1.0	30	0.0	0.0	0.0	0.0	0.0	0.0	0.0	0.0	0.0	0.0	0.0	0.0	0.0

PRECIPITATION (inches) 2001 PHOENIX, AZ (PHX)

YEAR	JAN	FEB	MAR	APR	MAY	JUN	JUL	AUG	SEP	OCT	NOV	DEC	ANNUAL
1972	0.00	T	T	T	T	1.70	0.72	1.20	0.28	4.40	1.01	1.56	10.87
1973	0.13	1.36	1.69	0.07	0.10	T	1.30	T	0.00	0.00	1.36	0.00	6.01
1974	0.57	0.02	1.37	0.01	0.00	0.00	0.84	1.15	1.07	2.12	0.44	0.59	8.18
1975	0.02	0.33	0.63	0.43	T	T	0.38	T	0.82	0.23	0.55	1.12	4.51
1976	T	0.47	0.40	0.67	1.06	0.09	1.48	0.12	1.69	0.70	0.43	0.85	7.96
1977	0.35	0.06	0.27	0.06	0.16	0.10	0.30	0.18	0.53	0.61	T	0.54	3.16
1978	2.33	2.21	2.14	0.20	T	0.01	1.44	1.79	T	0.35	2.30	2.46	15.23
1979	2.16	0.09	1.78	0.02	0.76	0.04	0.34	1.18	0.09	0.09	0.12	0.13	6.80
1980	1.58	2.09	0.86	0.44	0.21	0.03	0.56	0.06	0.13	0.02	0.00	0.08	6.06
1981	0.71	1.08	0.98	0.20	0.03	T	1.14	0.11	0.18	1.34	0.95	0.00	6.72
1982	0.81	0.67	1.30	T	0.50	T	0.43	1.97	0.12	T	2.50	1.64	9.94
1983	0.70	1.17	3.17	0.18	0.00	0.00	0.38	2.48	2.43	0.71	0.43	1.16	12.81
1984	0.31	0.00	0.00	0.91	0.18	0.18	5.15	0.87	3.36	0.31	0.71	2.93	14.91
1985	0.95	0.18	0.46	0.17	T	0.00	0.00	0.21	1.60	0.92	1.59	0.86	7.92
1986	0.07	1.19	1.58	0.01	T	0.01	1.19	1.27	0.47	0.41	0.03	1.38	7.61
1987	0.67	2.06	0.28	0.09	0.06	0.01	1.08	0.45	0.57	0.47	1.04	1.62	8.40
1988	0.90	0.23	0.17	1.09	0.00	0.02	0.87	0.63	0.00	2.38	0.78	0.14	7.21
1989	1.19	T	1.25	0.00	T	0.00	0.13	1.11	0.47	0.46	0.14	0.19	4.94
1990	0.80	0.70	0.35	0.17	0.16	0.04	1.05	2.70	1.11	0.04	0.15	0.46	7.73
1991	0.63	0.56	2.05	0.00	0.00	T	0.14	0.12	0.81	1.16	1.25	1.63	8.35
1992	1.62	0.90	2.49	0.49	1.05	0.04	2.95	1.30	0.03	0.26	3.08		14.24
1993	5.22	1.72	1.62	0.00	0.08	0.01	T	0.55	0.06	1.27	2.79	0.02	13.34
1994	0.13	0.54	1.36	0.09	0.39	T	0.25	0.02	1.74	0.55	0.68	3.03	8.78
1995	1.41	0.34	1.04	0.29	0.09	T	T	3.50	1.08	0.00	1.75	0.01	9.51
1996	0.25	1.03	0.55	T	T	T	1.04	0.34	0.54	0.37	0.23	0.02	4.37
1997	0.90	0.55	0.01	0.24	T	T	0.17	1.39	0.45	0.07	0.06	0.83	4.67
1998	0.35	2.93	1.31	0.43	0.04	0.00	1.94	1.05	0.58	1.03	0.19	0.68	10.53
1999	0.01	0.17	0.11	1.13	0.00	T	2.96	0.92	1.31	0.00	0.00	0.00	6.61
2000	0.01	T	2.98	0.00	0.00	0.34	0.28	0.58	0.01	3.17	0.50	0.00	7.87
2001	1.77	0.86	0.77	1.06	0.02	0.01	0.67	0.46	0.00	0.02	0.20	0.88	6.72
POR= 106 YRS	0.80	0.76	0.80	0.35	0.14	0.10	0.94	1.02	0.80	0.55	0.64	0.90	7.80

AVERAGE TEMPERATURE (F) 2001 PHOENIX, AZ (PHX)

YEAR	JAN	FEB	MAR	APR	MAY	JUN	JUL	AUG	SEP	OCT	NOV	DEC	ANNUAL
1972	51.4	59.1	70.6	71.4	78.3	87.8	94.4	89.9	84.8	71.9	58.1	52.1	72.5
1973	51.2	57.5	56.6	67.2	80.9	88.1	93.5	93.4	84.7	74.4	60.8	55.4	72.0
1974	54.0	56.7	64.5	70.6	80.2	92.2	92.4	91.2	87.2	75.9	61.5	50.6	73.1
1975	52.3	54.0	59.0	62.6	76.7	86.6	94.3	91.9	86.2	72.9	60.9	54.8	71.0
1976	55.4	60.7	61.5	68.7	80.7	87.9	91.6	90.7	83.0	74.0	64.1	55.6	72.8
1977	53.8	61.7	60.8	73.5	75.7	95.0	94.1	94.1	87.6	78.7	65.8	51.7	74.8
1978	56.6	58.7	65.6	69.2	78.5	90.9	94.6	91.4	86.3	78.6	61.5	51.7	73.6
1979	50.1	55.7	60.4	70.1	78.1	89.5	93.8	89.4	90.2	77.2	58.2	55.9	72.4
1980	56.6	60.6	60.7	69.8	76.0	88.9	95.6	92.2	87.3	75.6	64.1	61.3	74.1
1981	59.2	61.4	63.8	76.0	80.5	93.4	95.2	95.8	89.2	73.6	66.1	54.8	74.1
1982	53.9	60.1	62.4	72.5	80.4	88.1	93.7	93.7	86.7	73.5	61.9	54.1	73.4
1983	56.0	58.4	62.2	66.6	80.6	88.6	95.5	92.6	91.0	77.2	62.4	57.0	74.0
1984	57.4	60.1	67.6	70.7	87.0	88.9	91.7	91.2	87.5	71.4	61.9	53.7	74.1
1985	54.3	57.4	62.8	75.1	84.2	92.4	94.9	94.5	82.3	75.1	65.0	56.4	74.2
1986	61.4	61.0	69.3	74.2	82.3	92.8	92.3	94.5	84.1	74.7	65.0	56.4	75.7
1987	54.7	59.7	63.4	77.9	82.6	93.0	93.1	92.2	86.9	80.9	63.1	52.7	75.0
1988	55.1	62.5	66.3	73.0	81.4	93.1	96.2	93.9	87.4	82.4	64.4	54.7	75.6
1989	54.4	61.9	70.1	80.1	83.1	92.1	97.4	93.7	89.9	77.3	66.4	57.0	77.0
1990	55.6	56.6	67.2	76.2	81.1	93.8	93.6	90.8	87.6	78.7	63.5	53.6	75.1
1991	55.9	66.0	60.3	72.2	79.7	87.8	95.1	94.5	88.5	80.2	63.5	57.3	75.1
1992	56.4	62.1	64.7	77.0	83.1	90.1	92.8	92.3	90.5	79.8	61.5	53.8	75.3
1993	58.2	58.2	65.7	73.8	83.7	89.6	92.9	91.6	87.9	76.7	61.4	56.2	74.7
1994	56.8	58.1	65.0	71.8	78.3	92.2	93.9	95.3	87.1	78.1	58.0	53.7	73.7
1995	54.4	63.1	64.3	68.4	76.2	86.2	94.5	94.6	89.2	76.2	66.6	56.9	74.2
1996	55.8	62.3	65.1	74.1	83.0	91.9	95.2	94.1	84.3	74.9	66.5	56.5	75.1
1997	55.7	57.8	69.4	70.9	86.3	87.7	93.8	92.9	90.0	74.7	64.1	52.9	74.7
1998	56.6	54.1	61.9	66.7	76.2	85.4	94.5	94.7	87.9	74.1	62.7	54.4	72.4
1999	56.8	60.3	64.9	65.9	79.8	88.8	91.3	93.0	87.4	79.3	68.0	55.3	74.2
2000	58.1	60.9	63.1	75.2	84.8	91.0	95.0	92.5	90.3	72.5	56.9	57.3	74.8
2001	54.1	56.9	65.6	71.7	86.9	92.0	94.4	94.7	92.2	79.4	68.4	53.8	75.8
POR= 106 YRS	52.5	56.5	61.2	68.9	77.3	86.4	91.5	89.9	84.6	72.9	60.8	53.1	71.3

REFERENCE NOTES:

PAGE 1:
THE TEMPERATURE GRAPH SHOWS NORMAL MAXIMUM AND NORMAL MINIMUM DAILY TEMPERATURES (SOLID CURVES) AND THE ACTUAL DAILY HIGH AND LOW TEMPERATURES (VERTICAL BARS).

PAGE 2 AND 3:
H/C INDICATES HEATING AND COOLING DEGREE DAYS.
RH INDICATES RELATIVE HUMIDITY
W/O INDICATES WEATHER AND OBSTRUCTIONS
S INDICATES SUNSHINE.
PR INDICATES PRESSURE.
CLOUDINESS ON PAGE 3 IS THE SUM OF THE CEILOMETER AND SATELLITE DATA NOT TO EXCEED EIGHT EIGHTHS(OKTAS).

GENERAL:
T INDICATES TRACE PRECIPITATION, AN AMOUNT GREATER THAN ZERO BUT LESS THAN THE LOWEST REPORTABLE VALUE.
+ INDICATES THE VALUE ALSO OCCURS ON EARLIER DATES.
BLANK ENTRIES DENOTE MISSING OR UNREPORTED DATA.
NORMALS ARE 30-YEAR AVERAGES (1961 - 1990).
ASOS INDICATES AUTOMATED SURFACE OBSERVING SYSTEM.
PM INDICATES THE LAST DAY OF THE PREVIOUS MONTH.
POR (PERIOD OF RECORD) BEGINS WITH THE JANUARY DATA MONTH AND IS THE NUMBER OF YEARS USED TO COMPUTE THE MEAN. INDIVIDUAL MONTHS WITHIN THE POR MAY BE MISSING.
WHEN THE POR FOR A NORMAL IS LESS THAN 30 YEARS, THE NORMAL IS PROVISIONAL AND IS BASED ON THE NUMBER OF YEARS INDICATED.
0.* OR * INDICATES THE VALUE OR MEAN-DAYS-WITH IS BETWEEN 0.00 AND 0.05.
CLOUDINESS FOR ASOS STATIONS DIFFERS FROM THE NON-ASOS OBSERVATION TAKEN BY A HUMAN OBSERVER. ASOS STATION CLOUDINESS IS BASED ON TIME-AVERAGED CEILOMETER DATA FOR CLOUDS AT OR BELOW 12,000 FEET AND ON SATELLITE DATA FOR CLOUDS ABOVE 12,000 FEET.
THE NUMBER OF DAYS WITH CLEAR, PARTLY CLOUDY, AND CLOUDY CONDITIONS FOR ASOS STATIONS IS THE SUM OF THE CEILOMETER AND SATELLITE DATA FOR THE SUNRISE TO SUNSET PERIOD.

GENERAL CONTINUED:
CLEAR INDICATES 0 - 2 OKTAS, PARTLY CLOUDY INDICATES 3 - 6 OKTAS, AND CLOUDY INDICATES 7 OR 8 OKTAS.
WHEN AT LEAST ONE OF THE ELEMENTS (CEILOMETER OR SATELLITE) IS MISSING, THE DAILY CLOUDINESS IS NOT COMPUTED.
WIND DIRECTION IS RECORDED IN TENS OF DEGREES (2 DIGITS) CLOCKWISE FROM TRUE NORTH. "00" INDICATES CALM. "36" INDICATES TRUE NORTH.
RESULTANT WIND IS THE VECTOR AVERAGE OF THE SPEED AND DIRECTION.
AVERAGE TEMPERATURE IS THE SUM OF THE MEAN DAILY MAXIMUM AND MINIMUM TEMPERATURE DIVIDED BY 2.
SNOWFALL DATA COMPRISE ALL FORMS OF FROZEN PRECIPITATION, INCLUDING HAIL.
A HEATING (COOLING) DEGREE DAY IS THE DIFFERENCE BETWEEN THE AVERAGE DAILY TEMPERATURE AND 65 F.
DRY BULB IS THE TEMPERATURE OF THE AMBIENT AIR.
DEW POINT IS THE TEMPERATURE TO WHICH THE AIR MUST BE COOLED TO ACHIEVE 100 PERCENT RELATIVE HUMIDITY.
WET BULB IS THE TEMPERATURE THE AIR WOULD HAVE IF THE MOISTURE CONTENT WAS INCREASED TO 100 PERCENT RELATIVE HUMIDITY.

ON JULY 1, 1996, THE NATIONAL WEATHER SERVICE BEGAN USING THE "METAR" OBSERVATION CODE THAT WAS ALREADY EMPLOYED BY MOST OTHER NATIONS OF THE WORLD. THE MOST NOTICEABLE DIFFERENCE IN THIS ANNUAL PUBLICATION WILL BE THE CHANGE IN UNITS FROM TENTHS TO EIGHTS(OKTAS) FOR REPORTING THE AMOUNT OF SKY COVER.

HEATING DEGREE DAYS (base 65 F) 2001 PHOENIX, AZ (PHX)

YEAR	JUL	AUG	SEP	OCT	NOV	DEC	JAN	FEB	MAR	APR	MAY	JUN	TOTAL
1972-73	0	0	0	38	205	395	422	200	254	39	0	0	1553
1973-74	0	0	0	2	156	291	333	229	77	5	0	0	1093
1974-75	0	0	0	21	112	439	388	301	191	107	4	0	1563
1975-76	0	0	0	15	159	310	296	123	134	52	0	0	1089
1976-77	0	0	0	2	112	285	339	122	149	33	0	0	1042
1977-78	0	0	0	0	42	155	254	172	67	25	0	0	715
1978-79	0	0	0	1	148	405	455	254	143	30	0	0	1436
1979-80	0	0	0	11	204	277	254	130	129	35	0	0	1040
1980-81	0	0	0	12	108	122	181	131	74	8	0	0	636
1981-82	0	0	0	1	56	196	335	151	99	4	0	0	842
1982-83	0	0	0	1	103	331	272	181	120	53	0	0	1061
1983-84	0	0	0	0	154	236	228	139	16	23	0	0	796
1984-85	0	0	0	7	126	345	328	222	102	5	0	0	1135
1985-86	0	0	0	0	149	274	110	158	66	2	1	0	760
1986-87	0	0	0	0	43	260	318	172	95	4	0	0	892
1987-88	0	0	0	0	98	375	311	100	60	20	2	0	966
1988-89	0	0	0	0	135	284	321	133	46	0	0	0	919
1989-90	0	0	0	1	36	243	291	253	76	0	0	0	900
1990-91	0	0	0	0	65	348	275	27	161	5	0	0	881
1991-92	0	0	0	34	107	233	260	89	56	7	0	0	786
1992-93	0	0	0	0	127	340	205	184	61	0	0	0	917
1993-94	0	0	0	0	125	264	246	192	54	13	0	0	894
1994-95	0	0	0	15	211	293	320	61	74	48	6	0	1028
1995-96	0	0	0	0	22	243	278	111	63	0	0	0	717
1996-97	0	0	0	46	67	260	284	202	39	36	0	0	934
1997-98	0	0	0	9	83	369	250	299	138	77	2	0	1227
1998-99	0	0	0	4	73	322	245	127	65	104	2	0	942
1999-00	0	0	0	0	53	297	214	121	110	8	0	0	803
2000-01	0	0	0	19	238	234	330	222	87	45	0	0	1175
2001-	0	0	0	0	68	341							

COOLING DEGREE DAYS (base 65 F) 2001 PHOENIX, AZ (PHX)

YEAR	JAN	FEB	MAR	APR	MAY	JUN	JUL	AUG	SEP	OCT	NOV	DEC	ANNUAL
1972	0	11	200	212	419	691	919	780	599	259	4	0	4094
1973	0	0	0	109	499	701	894	885	598	302	36	0	4024
1974	0	2	69	182	477	825	858	821	673	365	13	0	4285
1975	0	0	12	42	374	654	913	839	640	265	45	1	3785
1976	6	4	34	169	495	692	833	804	548	289	91	0	3965
1977	0	36	25	295	334	797	936	907	683	434	73	1	4521
1978	3	1	92	158	422	787	928	828	644	431	49	0	4343
1979	0	0	11	191	411	741	901	763	764	397	7	0	4186
1980	0	5	2	187	344	724	956	852	675	346	88	13	4192
1981	5	36	40	345	489	857	943	961	731	277	95	5	4784
1982	0	21	24	234	481	697	899	897	658	272	12	0	4195
1983	2	1	38	112	489	715	951	861	787	388	85	0	4429
1984	0	2	107	203	688	724	836	821	681	208	41	0	4311
1985	0	17	40	316	603	826	934	920	525	319	47	0	4547
1986	3	52	209	282	543	844	853	921	582	307	51	1	4648
1987	3	30	51	396	553	846	879	850	665	499	48	0	4820
1988	10	31	108	265	520	851	972	904	678	543	124	3	5009
1989	1	49	210	459	566	820	1013	897	751	392	87	0	5245
1990	5	25	150	339	506	873	895	806	683	431	101	0	4814
1991	0	61	22	227	465	691	937	920	712	514	70	0	4619
1992	0	13	54	373	567	762	868	851	771	464	30	0	4753
1993	1	1	89	271	585	746	871	832	695	369	24	0	4484
1994	0	3	59	223	421	821	904	943	664	273	10	0	4321
1995	0	19	58	158	365	642	923	927	733	354	79	0	4258
1996	0	41	70	278	564	817	943	908	583	358	60	4	4626
1997	2	8	185	217	665	689	899	873	756	315	62	0	4671
1998	0	0	50	136	361	619	924	928	691	295	13	2	4019
1999	0	7	68	138	468	724	820	875	679	453	151	1	4384
2000	6	7	60	321	624	789	939	860	764	259	0	0	4629
2001	0	0	112	250	685	816	920	928	823	452	177	0	5163

SNOWFALL (inches) 2001 PHOENIX, AZ (PHX)

YEAR	JUL	AUG	SEP	OCT	NOV	DEC	JAN	FEB	MAR	APR	MAY	JUN	TOTAL
1972-73	0.0	0.0	0.0	0.0	0.0	0.0	0.0	0.0	0.0	0.0	0.0	0.0	0.0
1973-74	0.0	0.0	0.0	0.0	0.0	0.0	0.0	0.0	0.0	0.0	0.0	0.0	0.0
1974-75	0.0	0.0	0.0	0.0	0.0	T	0.0	0.0	0.0	0.0	0.0	0.0	T
1975-76	0.0	0.0	0.0	0.0	0.0	0.0	0.0	0.0	T	0.0	0.0	0.0	T
1976-77	0.0	0.0	0.0	0.0	0.0	0.0	0.0	0.0	0.0	0.0	0.0	0.0	0.0
1977-78	0.0	0.0	0.0	0.0	0.0	0.0	0.0	0.0	0.0	0.0	0.0	0.0	0.0
1978-79	0.0	0.0	0.0	0.0	0.0	0.0	0.0	0.0	0.0	0.0	0.0	0.0	0.0
1979-80	0.0	0.0	0.0	0.0	0.0	0.0	0.0	0.0	0.0	0.0	0.0	0.0	0.0
1980-81	0.0	0.0	0.0	0.0	0.0	0.0	0.0	0.0	0.0	0.0	0.0	0.0	0.0
1981-82	0.0	0.0	0.0	0.0	0.0	0.0	0.0	0.0	0.0	0.0	0.0	0.0	0.0
1982-83	0.0	0.0	0.0	0.0	0.0	0.0	0.0	0.0	0.0	0.0	0.0	0.0	0.0
1983-84	0.0	0.0	0.0	0.0	0.0	0.0	0.0	0.0	0.0	0.0	0.0	0.0	0.0
1984-85	0.0	0.0	0.0	0.0	0.0	0.0	0.0	0.0	T	0.0	0.0	0.0	T
1985-86	0.0	0.0	0.0	0.0	0.0	0.1	0.0	0.0	0.0	0.0	0.0	0.0	0.1
1986-87	0.0	0.0	0.0	0.0	0.0	0.0	T	0.0	0.0	0.0	0.0	0.0	T
1987-88	0.0	0.0	0.0	0.0	0.0	0.0	0.0	0.0	0.0	0.0	0.0	0.0	0.0
1988-89	0.0	0.0	0.0	0.0	0.0	0.0	0.0	0.0	0.0	0.0	0.0	0.0	0.0
1989-90	0.0	0.0	0.0	0.0	0.0	0.0	0.0	0.0	0.0	0.0	0.0	0.0	0.0
1990-91	0.0	0.0	0.0	0.0	0.0	0.4	0.0	0.0	T	0.0	T	0.0	0.4
1991-92	0.0	0.0	0.0	0.0	0.0	0.0	0.0	0.0	0.0	0.0	0.0	0.0	T
1992-93	0.0	0.0	0.0	T	0.0	0.0	T	0.0	0.0	0.0	0.0	0.0	T
1993-94	0.0	0.0	0.0	0.0	0.0	0.0	0.0	T	0.0	0.0	0.0	0.0	T
1994-95	0.0	0.0	0.0	0.0	0.0	0.0	0.0	0.0	0.0	0.0	0.0	0.0	0.0
1995-96	0.0	0.0	0.0	0.0	0.0	0.0	0.0	0.0	0.0	0.0	0.0	0.0	0.0
1996-97	0.0	0.0	0.0	0.0	0.0	0.0	0.0	0.0	0.0	0.0	0.0	0.0	0.0
1997-98	0.0	0.0	0.0	0.0	0.0	0.0	0.0	0.0	0.0	0.0	0.0	0.0	0.0
1998-99	0.0	0.0	0.0	0.0	0.0	T	0.0	0.0	0.0	0.0	0.0	0.0	T
1999-00	0.0	0.0	0.0	0.0	0.0	0.0							
2000-01													
2001-													
POR= 62 YRS	0.0	0.0	0.0	T	0.0	T	T	T	T	T	T	0.0	T

2001
TUCSON,
ARIZONA (TUS)

Tucson lies at the foot of the Catalina Mountains, north of the airport. The area within about 15 miles of the airport station is flat or gently rolling, with many dry washes. The soil is sandy, and vegetation is mostly brush, cacti, and small trees. Rugged mountains encircle the valley. The mountains to the north, east, and south rise to over 5,000 feet above the airport. The western hills and mountains range from 500 to 4,000 feet.

The climate of Tucson is characterized by a long hot season, from April to October. Temperatures above 90 degrees prevail from May through September. Temperatures of 100 degrees or higher average 41 days annually, including 14 days each for June and July, but these extreme temperatures are moderated by low relative humidities. The temperature range is large, averaging 30 degrees or more a day.

More than 50 percent of the annual precipitation falls between July 1 and September 15, and over 20 percent falls from December through March. During the summer, scattered convective or orographic showers and thunderstorms often fill dry washes to overflowing. On occasion, brief, torrential downpours cause destructive flash floods in the Tucson area. Hail rarely occurs in thunderstorms. The December through March precipitation occurs as prolonged rainstorms that replenish the ground water. During these storms, snow often falls on the higher mountains, but snow in Tucson is infrequent, particularly in accumulations exceeding an inch in depth.

From the first of the year, the humidity decreases steadily until the summer thunderstorm season, when it shows a marked increase. From mid-September, the end of the thunderstorm season, the humidity decreases again until late November. Occasionally during the summer, humidities are high enough to produce discomfort, but only for short periods. During the hot season, humidity values sometimes fall below 5 percent.

Tucson lies in the zone receiving more sunshine than any other section of the United States. Cloudless days are commonplace, and average cloudiness is low.

Surface winds are generally light, with no major seasonal changes in velocity or direction. Occasional duststorms occur in areas where the ground has been disturbed. During the spring, winds may briefly be strong enough to cause some damage to trees and buildings. Wind velocities and directions are influenced by the surrounding mountains, and the general slope of the terrain. Usually local winds tend to be in the southeast quadrant during the night and early morning hours, veering to northwest during the day. Highest velocities usually occur with winds from the southwest and east to south.

While dust and haze are frequently visible, their effect on the general clarity of the atmosphere is not great. Visibility is normally high.

Based on the 1951-1980 period, the average first occurrence of 32 degrees Fahrenheit in the fall is November 29 and the average last occurrence in the spring is February 28.

NORMALS, MEANS, AND EXTREMES
TUCSON, AZ (TUS)

LATITUDE: 32 07' 53" N   LONGITUDE: 110 57' 19" W   ELEVATION (FT): GRND: 2578   BARO: 2581   TIME ZONE: MOUNTAIN (UTC + 7)   WBAN: 23160

	ELEMENT	POR	JAN	FEB	MAR	APR	MAY	JUN	JUL	AUG	SEP	OCT	NOV	DEC	YEAR
TEMPERATURE F	NORMAL DAILY MAXIMUM	30	63.9	67.8	72.8	81.2	89.9	99.6	99.4	96.8	93.3	84.3	72.7	64.3	82.2
	MEAN DAILY MAXIMUM	53	64.6	68.3	73.1	81.4	90.1	99.6	99.0	96.8	94.1	84.6	72.9	65.3	82.5
	HIGHEST DAILY MAXIMUM	61	87	92	99	104	108	117	114	112	107	102	93	84	117
	YEAR OF OCCURRENCE		1999	1957	1988	1989	2000	1990	1995	1993	2000	1993	1999	1954	JUN 1990
	MEAN OF EXTREME MAXS.	53	78.0	81.3	86.3	93.5	100.8	107.9	107.7	104.4	102.1	95.5	84.8	77.6	93.3
	NORMAL DAILY MINIMUM	30	38.6	41.0	44.6	50.4	58.0	67.9	73.6	72.1	67.5	56.6	45.6	39.8	54.6
	MEAN DAILY MINIMUM	53	38.6	40.9	44.5	50.5	58.4	67.8	73.7	72.2	67.7	56.7	45.1	39.1	54.6
	LOWEST DAILY MINIMUM	61	16	20	20	27	38	47	59	61	44	26	24	16	16
	YEAR OF OCCURRENCE		1949	1955	1965	1945	1950	1955	1992	1956	1965	1971	1979	1974	DEC 1974
	MEAN OF EXTREME MINS.	53	26.6	29.2	32.4	38.9	46.4	56.8	67.0	66.5	58.7	43.5	32.1	27.1	43.8
	NORMAL DRY BULB	30	51.3	54.4	58.7	65.8	74.0	83.8	86.6	84.5	80.4	70.4	59.2	52.0	68.4
	MEAN DRY BULB	53	51.5	54.6	58.9	66.0	74.3	83.7	86.3	84.5	80.9	70.7	58.9	52.0	68.5
	MEAN WET BULB	18	41.8	43.9	46.5	49.5	54.6	59.8	67.2	68.9	63.4	54.7	46.1	41.8	53.2
	MEAN DEW POINT	18	29.0	29.9	29.5	29.0	32.6	37.9	55.0	60.0	51.4	40.2	31.9	29.9	38.0
	NORMAL NO. DAYS WITH:														
	MAXIMUM 90	30	0.0	0.0	0.7	4.0	16.8	28.1	29.3	28.6	22.3	7.5	*	0.0	137.3
	MAXIMUM 32	30	0.0	0.0	0.0	0.0	0.0	0.0	0.0	0.0	0.0	0.0	0.0	0.0	0.0
	MINIMUM 32	30	5.9	3.3	1.1	0.0	0.0	0.0	0.0	0.0	0.0	*	1.1	4.7	16.1
	MINIMUM 0	30	0.0	0.0	0.0	0.0	0.0	0.0	0.0	0.0	0.0	0.0	0.0	0.0	0.0
H/C	NORMAL HEATING DEG. DAYS	30	425	302	229	97	7	0	0	0	0	27	188	403	1678
	NORMAL COOLING DEG. DAYS	30	0	5	33	121	286	564	670	605	462	194	14	0	2954
RH	NORMAL (PERCENT)	30	48	43	37	27	22	21	42	47	42	38	43	50	38
	HOUR 05 LST	30	62	58	53	41	35	33	58	64	57	53	56	63	53
	HOUR 11 LST	30	40	35	29	21	17	17	34	38	34	31	34	41	31
	HOUR 17 LST	30	31	26	22	16	13	13	28	32	28	24	28	34	25
	HOUR 23 LST	30	56	49	41	30	24	23	47	52	46	44	49	58	43
S	PERCENT POSSIBLE SUNSHINE	52	80	81	86	92	93	92	78	80	87	88	85	79	85
W/O	MEAN NO. DAYS WITH:														
	HEAVY FOG(VISBY 1/4 MI)	61	0.3	0.2	0.0	0.0	0.0	0.0	0.0	0.0	0.0	0.0	0.2	0.4	1.1
	THUNDERSTORMS	61	0.4	0.3	0.5	0.7	1.6	2.7	13.7	13.7	5.5	2.0	0.5	0.3	41.9
CLOUDINESS	MEAN:														
	SUNRISE-SUNSET (OKTAS)	1			2.4		0.0	1.6							
	MIDNIGHT-MIDNIGHT (OKTAS)	1					0.0	1.6							
	MEAN NO. DAYS WITH:														
	CLEAR	1	1.0	5.0	12.0		27.0	15.0							
	PARTLY CLOUDY	1	1.0	3.0	3.0		1.0	2.0							
	CLOUDY	1	4.0	1.0			1.0								
PR	MEAN STATION PRESSURE(IN)	29	27.40	27.38	27.32	27.28	27.25	27.25	27.31	27.31	27.30	27.33	27.37	27.40	27.32
	MEAN SEA-LEVEL PRES. (IN)	18	30.04	29.98	29.92	29.85	29.78	29.76	29.82	29.84	29.83	29.89	29.98	30.03	29.89
WINDS	MEAN SPEED (MPH)	36	7.8	7.9	8.4	8.9	8.7	8.6	8.4	7.8	8.0	8.0	8.0	7.6	8.2
	PREVAIL.DIR(TENS OF DEGS)	22	14	14	14	14	14	14	14	14	14	14	14	14	14
	MAXIMUM 2-MINUTE:														
	SPEED (MPH)	6	33	37	33	35	33	38	55	48	38	38	32	32	55
	DIR. (TENS OF DEGS)		28	20	20	12	19	09	07	12	11	11	12	12	07
	YEAR OF OCCURRENCE		1996	1998	1998	2000	1997	1999	2001	2000	1999	2000	1996	1999	JUL 2001
	MAXIMUM 5-SECOND:														
	SPEED (MPH)	6	40	48	43	46	43	51	60	60	43	53	37	40	60
	DIR. (TENS OF DEGS)		27	22	21	25	24	07	07	12	11	26	12	10	07
	YEAR OF OCCURRENCE		1996	1998	1998	2001	1997	2001	2001	2000	1998	1996	1996	1998	JUL 2001
PRECIPITATION	NORMAL (IN)	30	0.87	0.70	0.72	0.30	0.18	0.20	2.37	2.19	1.67	1.06	0.67	1.07	12.00
	MAXIMUM MONTHLY (IN)	61	4.81	3.20	2.26	1.66	1.11	1.56	6.17	7.93	5.11	4.98	1.90	5.02	7.93
	YEAR OF OCCURRENCE		1993	1998	1952	1951	1992	2000	1981	1955	1964	1983	1952	1965	AUG 1955
	MINIMUM MONTHLY (IN)	61	T	0.00	0.00	0.00	0.00	0.00	0.04	0.23	0.00	0.00	0.00	0.00	0.00
	YEAR OF OCCURRENCE		1970	1972	1956	1972	1974	1983	1995	1976	1953	1982	1980	1981	JUN 1983
	MAXIMUM IN 24 HOURS (IN)	61	1.46	1.49	1.19	1.28	0.89	1.27	3.93	2.48	3.05	3.58	1.86	2.12	3.93
	YEAR OF OCCURRENCE		1993	1942	1952	1999	1943	1954	1958	1961	1964	1983	1968	1994	JUL 1958
	NORMAL NO. DAYS WITH:														
	PRECIPITATION 0.01	30	4.8	3.6	4.3	2.0	1.7	2.0	10.6	9.5	5.6	3.2	3.3	5.0	55.6
	PRECIPITATION 1.00	30	0.0	0.1	0.0	0.0	0.0	0.0	0.5	0.3	0.4	0.3	*	0.1	1.7
SNOWFALL	NORMAL (IN)	30	0.3	0.3	0.3	0.1	0.0	0.0	0.0	0.0	0.0	0.0	0.*	0.4	1.4
	MAXIMUM MONTHLY (IN)	61	4.7	3.9	5.7	2.0	T	0.0	T	T	T	T	6.4	6.8	6.8
	YEAR OF OCCURRENCE		1987	1965	1964	1976	1992		1995	1995	1996	1991	1958	1971	DEC 1971
	MAXIMUM IN 24 HOURS (IN)	60	4.3	3.9	5.7	2.0	T	0.0	T	T	T	T	6.4	6.8	6.8
	YEAR OF OCCURRENCE		1987	1965	1964	1976	1992		1995	1995	1990	1991	1958	1971	DEC 1971
	MAXIMUM SNOW DEPTH (IN)	52	1	4	5	0	0	0	0	0	0	0	1	5	5
	YEAR OF OCCURRENCE		1987	1965	1964								1958	1971	DEC 1971
	NORMAL NO. DAYS WITH:														
	SNOWFALL 1.0	30	0.1	0.2	0.1	0.*	0.0	0.0	0.0	0.0	0.0	0.0	0.0	0.1	0.5

PRECIPITATION (inches) 2001 TUCSON, AZ (TUS)

YEAR	JAN	FEB	MAR	APR	MAY	JUN	JUL	AUG	SEP	OCT	NOV	DEC	ANNUAL
1972	0.00	0.00	0.01	0.00	0.24	0.68	3.49	2.93	1.09	4.51	1.30	0.61	14.86
1973	0.06	1.60	2.20	0.02	0.09	0.50	1.74	0.54	T	0.00	0.47	0.00	7.22
1974	0.93	T	0.55	T	0.00	0.01	4.44	1.04	1.69	2.12	0.81	0.33	11.92
1975	0.36	0.13	0.95	0.27	0.11	0.00	2.38	0.32	1.26	T	0.34	0.52	6.64
1976	0.06	0.53	0.38	0.57	0.23	0.10	1.18	0.23	1.68	0.37	0.48	0.47	6.28
1977	1.83	0.04	0.74	0.43	0.08	0.06	0.76	0.80	1.41	2.36	0.33	1.33	10.17
1978	2.05	1.75	0.89	0.01	0.61	0.22	0.78	1.59	1.66	1.86	1.58	2.73	15.73
1979	2.94	0.42	0.64	0.04	0.67	0.53	2.04	2.60	0.02	0.33	0.15	0.15	10.39
1980	0.73	2.90	1.22	0.08	T	0.23	1.78	1.95	2.93	0.22	0.00	0.19	12.23
1981	1.29	0.71	1.98	0.56	0.26	0.16	6.17	0.80	1.10	0.06	0.61	0.00	13.70
1982	1.56	0.06	1.26	0.05	0.51	0.13	2.13	2.51	2.69	0.00	1.30	1.59	13.79
1983	1.70	0.94	1.28	0.14	T	0.00	1.98	4.28	4.28	4.98	1.71	0.61	21.86
1984	0.62	0.00	0.00	0.36	0.06	1.05	2.92	4.19	1.81	0.77	0.45	3.30	15.53
1985	1.71	1.08	0.20	0.45	T	0.07	3.14	1.97	1.13	2.03	0.95	0.15	12.88
1986	0.98	1.13	1.30	T	0.44	0.06	1.82	3.56	0.31	0.50	0.42	1.28	11.80
1987	0.59	1.64	0.83	0.80	0.74	0.16	0.37	2.79	2.30	0.34	0.44	1.50	12.50
1988	0.41	0.53	0.35	1.15	0.02	0.15	1.69	3.64	0.80	2.09	0.75	0.05	11.63
1989	0.96	0.23	0.62	0.00	0.13	0.06	1.42	0.90	0.02	1.84	0.12	0.18	6.48
1990	0.96	0.71	0.38	0.10	0.03	0.64	5.45	2.70	1.63	0.58	0.23	1.54	14.95
1991	1.15	0.91	1.40	0.00	0.00	0.20	0.44	2.17	1.54	0.73	0.80	1.44	10.78
1992	1.21	1.80	2.12	0.19	1.11	0.07	0.93	4.55	0.94	0.03	T	3.47	16.42
1993	4.81	1.50	0.49	0.00	0.59	0.02	0.26	4.93	0.46	0.81	0.98	0.14	14.99
1994	0.02	1.03	1.14	0.04	0.52	0.26	0.41	0.45	1.46	0.76	1.83	3.71	11.63
1995	1.41	1.32	0.54	0.28	0.15	T	0.04	3.71	2.29	0.36	0.86	0.22	11.18
1996	0.01	0.82	0.32	T	0.00	T	1.88	1.87	3.68	1.74	0.19	T	10.51
1997	0.93	0.67	0.02	0.47	0.44	0.02	0.51	2.32	1.43	0.38	0.49	2.88	10.56
1998	0.17	3.20	1.64	0.39	T	0.00	4.06	1.70	1.10	0.24	0.67	0.45	13.62
1999	0.01	T	T	1.34	0.00	0.16	4.15	3.05	0.97	T	0.00	T	9.68
2000	0.10	0.19	0.93	T	0.00	1.56	1.59	1.70	0.02	4.98	1.36	T	12.43
2001	1.24	0.46	0.88	0.84	0.24	0.54	1.09	0.85	0.33	0.69	0.05	0.60	7.81
POR= 102 YRS	0.87	0.85	0.75	0.35	0.22	0.28	2.11	2.19	1.31	0.75	0.69	0.51	10.88

AVERAGE TEMPERATURE (F) 2001   TUCSON, AZ (TUS)

YEAR	JAN	FEB	MAR	APR	MAY	JUN	JUL	AUG	SEP	OCT	NOV	DEC	ANNUAL
1972	50.4	55.8	65.0	65.8	72.3	81.6	86.6	82.9	78.6	66.5	53.0	49.0	67.3
1973	47.6	53.4	51.6	59.7	73.0	81.4	84.3	84.7	79.6	70.7	58.4	52.3	66.4
1974	50.2	51.9	60.1	66.1	74.3	86.9	83.5	83.0	77.8	69.1	57.5	47.0	67.3
1975	49.8	50.7	55.3	57.9	69.8	80.5	84.2	85.8	80.0	69.5	59.3	53.0	66.3
1976	52.6	58.4	58.2	64.8	74.5	83.4	83.9	85.3	77.7	67.8	60.0	52.2	68.2
1977	50.7	56.9	55.7	67.0	70.8	84.7	87.0	86.4	82.0	73.3	61.7	56.9	69.4
1978	53.1	53.6	61.8	65.2	73.1	85.8	88.1	84.7	80.9	73.8	58.5	49.7	69.0
1979	48.4	53.8	56.4	65.6	72.2	83.1	87.5	83.4	84.2	73.0	56.6	55.0	68.3
1980	54.3	57.9	57.5	65.6	71.5	84.9	88.6	84.6	80.5	69.6	59.5	58.1	69.4
1981	54.8	57.1	57.1	69.1	73.4	86.1	85.2	86.4	80.7	68.1	62.2	55.0	69.5
1982	50.7	54.7	57.7	66.1	72.3	80.5	84.8	83.9	79.2	67.0	57.7	50.1	67.1
1983	52.9	53.8	57.3	60.4	73.8	81.6	86.9	84.0	82.2	69.5	57.4	53.5	67.8
1984	51.8	53.7	60.5	64.0	79.9	83.1	84.2	82.9	81.5	66.3	57.8	51.5	68.1
1985	50.3	53.1	58.7	68.7	75.9	85.8	87.5	86.1	77.4	70.0	58.2	52.9	68.7
1986	58.7	56.9	63.8	69.0	76.8	86.6	85.5	86.0	79.0	69.6	59.8	52.3	70.3
1987	50.9	54.2	57.9	70.1	74.3	86.3	87.4	85.1	79.9	75.1	58.9	50.3	69.2
1988	53.0	59.4	61.4	68.0	76.4	86.8	87.9	85.9	80.4	75.3	59.2	51.9	70.5
1989	49.9	58.2	65.0	73.8	77.4	85.4	90.0	86.6	84.5	71.1	61.7	53.0	71.4
1990	51.8	52.8	61.8	69.7	75.2	88.7	85.0	82.6	82.2	71.1	61.6	51.1	69.6
1991	52.3	59.8	55.4	65.2	73.5	81.5	87.5	86.6	80.7	74.0	58.9	54.3	69.1
1992	51.6	57.3	59.4	70.8	76.7	84.5	86.8	85.1	83.6	74.2	56.1	51.4	69.8
1993	55.2	54.0	61.3	68.6	78.1	85.0	88.0	85.5	81.4	72.6	58.8	53.4	70.2
1994	53.7	55.2	62.9	68.6	75.6	89.2	90.4	90.3	84.2	70.5	56.7	53.9	70.9
1995	52.6	60.7	61.2	64.8	72.6	83.3	88.4	87.3	82.9	72.4	63.1	54.0	70.3
1996	53.6	58.8	61.1	68.9	79.0	87.4	88.6	86.4	77.7	70.4	61.0	53.7	70.6
1997	52.4	53.6	64.8	65.8	79.7	83.1	88.0	85.8	84.2	70.0	60.7	48.6	69.4
1998	53.2	50.8	57.9	61.4	72.9	81.8	86.5	86.5	82.6	70.7	60.5	52.0	68.1
1999	53.6	56.9	61.4	62.3	74.7	83.8	84.0	85.2	81.8	74.7	65.4	51.3	69.6
2000	55.0	57.4	58.8	70.3	80.2	84.5	88.2	84.9	84.8	68.2	52.8	54.3	70.0
2001	49.7	52.8	60.2	66.6	79.3	85.6	86.2	85.9	84.4	73.2	62.9	49.4	69.7
POR= 102 YRS	50.7	53.7	58.2	65.1	73.3	82.5	86.1	84.2	80.2	69.8	58.4	51.5	67.8

REFERENCE NOTES:

PAGE 1:
THE TEMPERATURE GRAPH SHOWS NORMAL MAXIMUM AND NORMAL
MINIMUM DAILY TEMPERATURES (SOLID CURVES) AND THE
ACTUAL DAILY HIGH AND LOW TEMPERATURES (VERTICAL BARS).

PAGE 2 AND 3:
H/C INDICATES HEATING AND COOLING DEGREE DAYS.
RH INDICATES RELATIVE HUMIDITY
W/O INDICATES WEATHER AND OBSTRUCTIONS
S INDICATES SUNSHINE.
PR INDICATES PRESSURE.
CLOUDINESS ON PAGE 3 IS THE SUM OF THE CEILOMETER AND
SATELLITE DATA NOT TO EXCEED EIGHT EIGHTHS (OKTAS).

GENERAL:
T INDICATES TRACE PRECIPITATION, AN AMOUNT GREATER
THAN ZERO BUT LESS THAN THE LOWEST REPORTABLE VALUE.
+ INDICATES THE VALUE ALSO OCCURS ON EARLIER DATES.
BLANK ENTRIES DENOTE MISSING OR UNREPORTED DATA.
NORMALS ARE 30-YEAR AVERAGES (1961 - 1990).
ASOS INDICATES AUTOMATED SURFACE OBSERVING SYSTEM.
PM INDICATES THE LAST DAY OF THE PREVIOUS MONTH.
POR (PERIOD OF RECORD) BEGINS WITH THE JANUARY DATA
MONTH AND IS THE NUMBER OF YEARS USED TO COMPUTE
THE MEAN. INDIVIDUAL MONTHS WITHIN THE POR MAY
BE MISSING.
WHEN THE POR FOR A NORMAL IS LESS THAN 30 YEARS,
THE NORMAL IS PROVISIONAL AND IS BASED ON THE NUMBER
OF YEARS INDICATED.
0.* OR * INDICATES THE VALUE OR MEAN-DAYS-WITH
IS BETWEEN 0.00 AND 0.05.
CLOUDINESS FOR ASOS STATIONS DIFFERS FROM THE NON-ASOS
OBSERVATION TAKEN BY A HUMAN OBSERVER. ASOS STATION
CLOUDINESS IS BASED ON TIME-AVERAGED CEILOMETER DATA
FOR CLOUDS AT OR BELOW 12,000 FEET AND ON SATELLITE
DATA FOR CLOUDS ABOVE 12,000 FEET.
THE NUMBER OF DAYS WITH CLEAR, PARTLY CLOUDY, AND
CLOUDY CONDITIONS FOR ASOS STATIONS IS THE SUM
OF THE CEILOMETER AND SATELLITE DATA FOR THE
SUNRISE TO SUNSET PERIOD.

GENERAL CONTINUED:
CLEAR INDICATES 0 - 2 OKTAS, PARTLY CLOUDY INDICATES
3 - 6 OKTAS, AND CLOUDY INDICATES 7 OR 8 OKTAS.
WHEN AT LEAST ONE OF THE ELEMENTS (CEILOMETER OR
SATELLITE) IS MISSING, THE DAILY CLOUDINESS IS
NOT COMPUTED.
WIND DIRECTION IS RECORDED IN TENS OF DEGREES (2 DIGITS)
CLOCKWISE FROM TRUE NORTH. "00" INDICATES CALM. "36"
INDICATES TRUE NORTH.
RESULTANT WIND IS THE VECTOR AVERAGE OF THE SPEED AND
DIRECTION.
AVERAGE TEMPERATURE IS THE SUM OF THE MEAN DAILY MAXIMUM
AND MINIMUM TEMPERATURE DIVIDED BY 2.
SNOWFALL DATA COMPRISE ALL FORMS OF FROZEN
PRECIPITATION, INCLUDING HAIL.
A HEATING (COOLING) DEGREE DAY IS THE DIFFERENCE BETWEEN
THE AVERAGE DAILY TEMPERATURE AND 65 F.
DRY BULB IS THE TEMPERATURE OF THE AMBIENT AIR.
DEW POINT IS THE TEMPERATURE TO WHICH THE AIR MUST BE
COOLED TO ACHIEVE 100 PERCENT RELATIVE HUMIDITY.
WET BULB IS THE TEMPERATURE THE AIR WOULD HAVE IF THE
MOISTURE CONTENT WAS INCREASED TO 100 PERCENT RELATIVE
HUMIDITY.

ON JULY 1, 1996, THE NATIONAL WEATHER SERVICE BEGAN USING
THE "METAR" OBSERVATION CODE THAT WAS ALREADY EMPLOYED
BY MOST OTHER NATIONS OF THE WORLD. THE MOST NOTICEABLE
DIFFERENCE IN THIS ANNUAL PUBLICATION WILL BE THE CHANGE
IN UNITS FROM TENTHS TO EIGHTS (OKTAS) FOR REPORTING THE
AMOUNT OF SKY COVER.

HEATING DEGREE DAYS (base 65 F) 2001 TUCSON, AZ (TUS)

YEAR	JUL	AUG	SEP	OCT	NOV	DEC	JAN	FEB	MAR	APR	MAY	JUN	TOTAL
1972-73	0	0	0	96	358	489	533	320	410	174	19	0	2399
1973-74	0	0	0	23	216	390	451	362	161	49	5	0	1657
1974-75	0	0	0	53	218	552	465	393	299	217	29	0	2226
1975-76	0	0	0	38	191	365	378	180	221	88	5	0	1466
1976-77	0	0	0	45	178	390	435	221	287	65	9	0	1630
1977-78	0	0	0	1	117	242	365	313	144	64	24	0	1270
1978-79	0	0	0	15	213	470	511	311	260	76	20	0	1876
1979-80	0	0	0	26	252	302	323	202	227	84	3	0	1419
1980-81	0	0	0	66	197	310	220	220	244	31	0	0	1278
1981-82	0	0	0	34	106	304	437	291	223	46	10	0	1451
1982-83	0	0	0	41	211	456	371	309	239	168	6	0	1801
1983-84	0	0	0	0	232	348	402	323	140	110	0	0	1555
1984-85	0	0	0	49	221	413	448	328	200	41	0	0	1700
1985-86	0	0	0	9	217	369	193	244	117	22	6	0	1177
1986-87	0	0	0	11	154	387	429	299	225	24	0	0	1529
1987-88	0	0	0	0	188	452	366	171	161	46	12	0	1396
1988-89	0	0	0	0	220	402	461	199	82	9	4	0	1377
1989-90	0	0	0	25	107	361	402	340	156	16	3	0	1410
1990-91	0	0	0	5	152	427	384	140	296	47	3	0	1454
1991-92	0	0	0	56	195	325	408	215	169	24	0	0	1392
1992-93	0	0	0	0	261	418	298	299	129	28	0	0	1433
1993-94	0	0	0	5	186	355	345	272	94	42	0	0	1299
1994-95	0	0	0	24	255	335	377	123	143	84	17	0	1358
1995-96	0	0	0	3	64	332	344	173	135	23	0	0	1074
1996-97	0	0	0	91	147	346	386	315	66	77	0	0	1428
1997-98	0	0	0	52	146	502	360	392	232	153	3	0	1840
1998-99	0	0	0	20	136	398	346	220	126	155	7	0	1408
1999-00	0	0	0	0	83	416	308	217	186	26	0	0	1236
2000-01	0	0	0	60	357	323	470	335	178	95	1	0	1819
2001-	0	0	0	0	134	479							

COOLING DEGREE DAYS (base 65 F) 2001 TUCSON, AZ (TUS)

YEAR	JAN	FEB	MAR	APR	MAY	JUN	JUL	AUG	SEP	OCT	NOV	DEC	ANNUAL
1972	0	1	82	82	236	506	678	563	414	150	1	0	2713
1973	0	0	0	21	272	495	603	615	445	206	26	2	2685
1974	0	0	18	87	301	664	581	564	387	185	1	0	2788
1975	0	0	4	11	184	471	604	651	458	182	27	0	2592
1976	2	0	14	89	306	557	597	636	386	139	34	0	2760
1977	0	0	5	133	198	597	691	669	517	266	23	0	3099
1978	0	0	54	76	283	630	721	616	483	293	28	0	3184
1979	0	0	1	101	249	551	706	576	580	282	6	0	3052
1980	0	4	1	109	211	606	742	615	474	216	37	3	3018
1981	0	8	4	159	267	639	633	670	476	137	27	2	3022
1982	0	4	4	82	244	471	622	594	437	112	0	0	2570
1983	0	0	8	36	288	503	688	600	523	145	10	0	2801
1984	0	0	6	87	469	549	601	562	503	96	12	0	2885
1985	0	1	7	159	345	633	704	660	379	173	14	0	3075
1986	2	23	88	150	378	653	643	657	431	158	3	0	3186
1987	0	2	12	184	297	644	702	630	452	325	12	0	3260
1988	2	13	58	142	374	658	716	657	471	327	51	1	3470
1989	0	16	89	281	397	619	780	676	592	221	16	0	3687
1990	0	6	63	164	327	719	625	553	522	262	56	0	3297
1991	0	1	6	58	274	501	703	675	479	345	21	0	3063
1992	0	0	4	204	372	590	683	627	563	291	1	0	3335
1993	1	0	22	142	413	604	721	641	500	250	11	3	3308
1994	0	4	34	156	332	733	794	788	584	203	11	0	3639
1995	0	8	31	85	261	558	733	698	544	241	12	0	3171
1996	0	1	24	148	440	679	740	670	387	266	33	2	3390
1997	0	2	65	105	462	549	718	651	581	212	26	0	3371
1998	1	0	19	54	255	512	672	671	536	203	9	1	2933
1999	0	1	18	83	318	573	598	635	512	306	103	1	3148
2000	3	1	1	194	479	595	725	624	599	164	0	0	3385
2001	0	0	37	150	450	623	667	655	586	265	79	0	3512

SNOWFALL (inches) 2001 TUCSON, AZ (TUS)

YEAR	JUL	AUG	SEP	OCT	NOV	DEC	JAN	FEB	MAR	APR	MAY	JUN	TOTAL
1972-73	0.0	0.0	0.0	0.0	0.0	0.0	T	0.0	0.0	0.0	0.0	0.0	T
1973-74	0.0	0.0	0.0	0.0	0.0	0.0	0.4	0.0	T	0.0	0.0	0.0	0.4
1974-75	0.0	0.0	0.0	0.0	0.0	T	0.0	T	0.5	0.0	0.0	0.0	0.5
1975-76	0.0	0.0	0.0	0.0	T	T	0.0	0.0	3.8	2.0	0.0	0.0	5.8
1976-77	0.0	0.0	0.0	0.0	0.0	0.0	0.0	0.0	0.0	0.0	0.0	0.0	0.0
1977-78	0.0	0.0	0.0	0.0	0.0	0.0	0.0	0.0	0.0	0.0	0.0	0.0	0.0
1978-79	0.0	0.0	0.0	0.0	0.0	T	1.2	0.0	0.0	0.0	0.0	0.0	1.2
1979-80	0.0	0.0	0.0	0.0	0.0	0.0	0.0	0.0	T	0.0	0.0	0.0	T
1980-81	0.0	0.0	0.0	0.0	0.0	0.0	0.0	0.0	T	0.0	0.0	0.0	T
1981-82	0.0	0.0	0.0	0.0	0.0	0.0	T	0.0	T	0.0	0.0	0.0	T
1982-83	0.0	0.0	0.0	0.0	0.0	T	0.0	0.0	0.0	0.0	0.0	0.0	T
1983-84	0.0	0.0	0.0	0.0	0.0	0.0	0.0	0.0	0.0	0.0	0.0	0.0	0.0
1984-85	0.0	0.0	0.0	0.0	0.0	T	0.0	2.2	0.0	0.0	0.0	0.0	2.2
1985-86	0.0	0.0	0.0	0.0	0.0	T	0.0	T	0.0	0.0	0.0	0.0	T
1986-87	0.0	0.0	0.0	0.0	0.0	0.0	4.7	0.0	T	0.0	0.0	0.0	4.7
1987-88	0.0	0.0	0.0	0.0	0.0	3.6	0.0	0.0	0.0	0.0	0.0	0.0	3.6
1988-89	0.0	0.0	0.0	0.0	0.0	T	0.0	0.0	T	0.0	0.0	0.0	T
1989-90	0.0	0.0	0.0	0.0	0.0	0.6	0.0	2.7	2.3	0.0	0.0	0.0	5.0
1990-91	0.0	T	T	0.0	0.0	0.0	0.0	T	0.3	T	0.0	0.0	0.9
1991-92	0.0	0.0	0.0	T	0.0	T	0.0	0.0	0.0	T	0.0	0.0	T
1992-93	0.0	T	0.0	0.0	0.0	0.0	0.0	0.0	0.0	0.0	0.0	0.0	T
1993-94	0.0	T	0.0	0.0	0.0	0.0	0.0	0.0	0.0	0.0	0.0	0.0	T
1994-95	0.0	0.0	0.0	0.0	T	0.0	0.0	T	0.0	0.0	0.0	0.0	T
1995-96	T	T	0.0	0.0	0.0	0.0	0.0	0.0	T	0.0	0.0	0.0	T
1996-97	0.0	0.0	T	0.0	0.0	0.0	T	0.0	0.0	T	0.0	0.0	T
1997-98	0.0	0.0	0.0	T	0.0	0.0	0.0	0.0	T	T	0.0	0.0	T
1998-99	0.0	0.0	0.0	0.0	0.0	0.0	T	0.0	0.0	0.0	0.0	0.0	T
1999-00	0.0	0.0	0.0	0.0	0.0	0.0	0.0	0.0	0.0	0.0	0.0	0.0	0.0
2000-01	0.0	0.0	0.0	0.0	0.0	T	0.0	T	0.0	0.0	0.0	0.0	T
2001-	0.0	0.0	0.0	0.0	0.0	T							
POR= 60 YRS	T	T	T	T	0.1	0.3	0.3	0.2	0.2	0.1	T	0.0	1.2

2001
LITTLE ROCK,
ARKANSAS (LIT)

Little Rock is located on the Arkansas River near the geographical center of the state. It is situated on the dividing line between the Ouachita Mountains to the west and the flat lowlands comprising the Mississippi River Valley to the east. Elevations range from 222 feet at the river level to 257 feet over much of the flat land, including the airport in the southeast, to near 600 feet in the hilly residential area of the western portions of the city. Two minor temperature variations are observed due to the terrain; somewhat lower minimum temperatures are observed in the airport vicinity and a slight downslope adiabatic heating effect accompanies airflow from the ridges and hills in the west and northwest.

The modified continental climate of Little Rock includes exposure to all of the North American air mass types. However, with its proximity to the Gulf of Mexico, the summer season is marked by prolonged periods of warm and humid weather. The growing season averages 233 days in which 62 percent of the normal precipitation occurs. Winters are mild, but polar and Arctic outbreaks are not uncommon.

Precipitation is fairly well distributed throughout the year. Summer rainfall is almost completely of the convective type. The driest period usually occurs in the late summer and early fall. Snow is almost negligible. Glaze and ice storms, although infrequent, are at times severe. Warm front weather in the winter and early spring, characterized by shallow surface cold air flow from the north under warm moist Gulf air, results in excellent conditions for the production of freezing precipitation.

NORMALS, MEANS, AND EXTREMES
LITTLE ROCK, AR (LIT)

LATITUDE: 34 44' 48" N   LONGITUDE: 92 13' 59" W   ELEVATION (FT): GRND: 289 BARO: 292   TIME ZONE: CENTRAL (UTC + 6)   WBAN: 13963

	ELEMENT	POR	JAN	FEB	MAR	APR	MAY	JUN	JUL	AUG	SEP	OCT	NOV	DEC	YEAR
TEMPERATURE °F	NORMAL DAILY MAXIMUM	30	49.0	53.9	64.0	73.4	81.3	89.3	92.4	91.4	84.6	75.1	62.7	52.5	72.5
	MEAN DAILY MAXIMUM	54	49.8	55.0	63.1	73.3	81.5	89.2	92.4	91.9	85.0	75.2	62.2	52.8	72.6
	HIGHEST DAILY MAXIMUM	60	83	85	91	95	98	105	112	109	106	97	86	80	112
	YEAR OF OCCURRENCE		1950	1986	1974	1987	1998	1988	1986	2000	1947	1963	1955	1956	JUL 1986
	MEAN OF EXTREME MAXS.	54	71.9	75.9	81.5	86.2	91.5	96.9	100.5	100.1	95.8	88.6	79.5	72.7	86.8
	NORMAL DAILY MINIMUM	30	29.1	33.2	42.2	50.7	59.0	67.4	71.5	69.8	63.5	50.9	41.5	33.1	51.0
	MEAN DAILY MINIMUM	54	31.0	34.9	42.3	51.3	59.9	68.1	71.8	70.4	63.4	51.4	41.2	33.7	51.6
	LOWEST DAILY MINIMUM	60	−4	−5	11	28	40	46	54	52	37	29	17	−1	−5
	YEAR OF OCCURRENCE		1962	1951	1951	1971	1971	1969	1972	1986	1942	1989	1976	1989	FEB 1951
	MEAN OF EXTREME MINS.	54	13.7	18.8	26.1	35.7	46.4	56.5	63.7	61.5	48.8	36.4	25.0	17.7	37.5
	NORMAL DRY BULB	30	39.1	43.6	53.1	62.1	70.2	78.4	81.9	80.6	74.1	63.0	52.1	42.8	61.7
	MEAN DRY BULB	54	40.4	45.0	52.6	62.4	70.7	78.8	82.1	81.1	74.1	63.2	51.7	43.3	62.1
	MEAN WET BULB	16	36.8	41.5	47.3	56.0	64.5	71.2	69.7	68.6	62.2	53.1	45.0	37.3	54.5
	MEAN DEW POINT	16	31.0	35.0	40.6	50.3	60.4	67.6	67.6	65.2	58.5	48.4	40.2	32.8	49.7
	NORMAL NO. DAYS WITH:														
	MAXIMUM 90	30	0.0	0.0	*	0.3	3.8	16.3	22.1	19.8	9.0	1.2	0.0	0.0	72.5
	MAXIMUM 32	30	3.6	1.1	0.1	0.0	0.0	0.0	0.0	0.0	0.0	0.0	*	1.6	6.4
	MINIMUM 32	30	20.5	13.7	4.5	0.5	0.0	0.0	0.0	0.0	0.0	0.2	5.3	15.6	60.3
	MINIMUM 0	30	0.1	0.0	0.0	0.0	0.0	0.0	0.0	0.0	0.0	0.0	0.0	0.1	0.2
H/C	NORMAL HEATING DEG. DAYS	30	803	599	384	133	25	0	0	0	8	128	387	688	3155
	NORMAL COOLING DEG. DAYS	30	0	0	15	46	187	402	524	484	281	66	0	0	2005
RH	NORMAL (PERCENT)	30	70	68	65	67	71	70	72	72	74	70	71	71	70
	HOUR 00 LST	30	76	74	72	74	82	82	83	84	85	82	78	76	79
	HOUR 06 LST	30	80	80	79	82	87	86	88	88	89	86	83	80	84
	HOUR 12 LST	30	61	59	56	56	58	55	56	56	58	53	58	62	57
	HOUR 18 LST	30	64	59	55	55	59	57	60	60	64	63	65	65	60
S	PERCENT POSSIBLE SUNSHINE	32	46	54	57	62	68	73	71	73	68	69	56	48	62
W/O	MEAN NO. DAYS WITH:														
	HEAVY FOG (VISBY 1/4 MI)	59	2.7	1.9	1.2	0.7	0.7	0.3	0.5	0.7	1.0	1.7	2.0	2.7	16.1
	THUNDERSTORMS	59	1.9	2.3	4.9	6.5	7.4	7.8	8.6	6.3	3.6	2.6	2.9	1.8	56.6
CLOUDINESS	MEAN:														
	SUNRISE-SUNSET (OKTAS)	35	5.2	4.8	5.0	4.9	4.8	4.3	4.3	3.9	4.0	3.5	4.3	4.8	4.5
	MIDNIGHT-MIDNIGHT (OKTAS)	12	4.5	4.0	4.4	4.2	3.9	3.4	3.6	3.4	3.7	3.2	4.0	4.4	3.9
	MEAN NO. DAYS WITH:														
	CLEAR	35	8.6	9.1	8.6	8.7	8.0	9.5	8.8	11.6	11.2	14.4	11.0	9.2	118.7
	PARTLY CLOUDY	35	6.1	5.7	7.0	7.5	10.8	11.6	12.9	10.9	8.6	7.1	5.9	5.9	100.0
	CLOUDY	35	16.3	13.5	15.4	13.8	12.3	8.9	8.9	8.5	10.1	9.5	13.1	16.0	146.7
PR.	MEAN STATION PRESSURE (IN)	27	29.89	29.84	29.74	29.71	29.68	29.70	29.73	29.74	29.77	29.81	29.83	29.87	29.78
	MEAN SEA-LEVEL PRES. (IN)	18	30.16	30.12	30.05	29.98	29.96	29.96	30.01	30.02	30.04	30.10	30.13	30.17	30.06
WINDS	MEAN SPEED (MPH)	41	8.4	8.7	9.4	8.8	7.5	7.1	6.7	6.3	6.7	6.8	7.9	8.0	7.7
	PREVAIL.DIR (TENS OF DEGS)	27	24	05	18	18	18	18	18	23	24	06	24	24	24
	MAXIMUM 2-MINUTE:														
	SPEED (MPH)	3	31	41	32	32	33	46	35	29	35	29	33	36	46
	DIR. (TENS OF DEGS)		30	25	33	19	31	30	34	28	32	30	26	29	30
	YEAR OF OCCURRENCE		1999	2001	2000	2001	1999	2001	2000	2001	2000	2001	2001	2000	JUN 2001
	MAXIMUM 5-SECOND:														
	SPEED (MPH)	3	39	51	39	43	41	87	46	36	47	38	44	43	87
	DIR. (TENS OF DEGS)		21	25	32	18	31	32	34	27	30	31	26	27	32
	YEAR OF OCCURRENCE		1999	2001	1999	2001	1999	1999	2000	2001	2000	2001	2001	2000	JUN 1999
PRECIPITATION	NORMAL (IN)	30	3.42	3.61	4.91	5.49	5.17	3.57	3.60	3.26	4.05	3.75	5.20	4.83	50.86
	MAXIMUM MONTHLY (IN)	60	12.53	11.02	10.40	14.20	12.74	7.82	7.95	14.46	10.17	15.35	13.14	16.48	16.48
	YEAR OF OCCURRENCE		1950	1956	1990	1973	1968	1974	1988	1966	1978	1984	1988	1987	DEC 1987
	MINIMUM MONTHLY (IN)	60	0.50	0.51	0.73	0.50	0.69	T	0.14	T	0.28	0.01	0.26	1.26	T
	YEAR OF OCCURRENCE		1986	1947	1966	1987	1970	1952	1986	1995	1956	1944	1999	1958	AUG 1995
	MAXIMUM IN 24 HOURS (IN)	54	5.18	5.15	4.56	7.96	7.71	4.61	3.58	7.32	4.05	5.67	7.81	7.01	7.96
	YEAR OF OCCURRENCE		1969	1950	1990	1974	1955	1960	1988	1966	1967	1990	1988	1987	APR 1974
	NORMAL NO. DAYS WITH:														
	PRECIPITATION 0.01	30	9.3	8.8	10.1	9.9	10.0	8.5	8.5	7.1	7.7	7.1	8.4	9.4	104.8
	PRECIPITATION 1.00	30	0.9	1.2	1.7	1.8	1.7	0.8	1.0	1.0	1.1	1.2	1.7	1.5	15.6
SNOWFALL	NORMAL (IN)	30	2.5	1.7	0.7	T	0.0	0.0	0.0	0.0	0.0	0.0	0.3	0.5	5.7
	MAXIMUM MONTHLY (IN)	57	13.6	9.8	7.0	T	T	T	0.0	0.0	0.0	T	4.8	9.8	13.6
	YEAR OF OCCURRENCE		1988	1979	1971	1994	1997	1998				1993	1971	1963	JAN 1988
	MAXIMUM IN 24 HOURS (IN)	51	12.1	9.6	6.7	T	T	T	0.0	0.0	0.0	T	4.8	9.8	12.1
	YEAR OF OCCURRENCE		1988	1966	1971	1994	1988	1998				1993	1971	1963	JAN 1988
	MAXIMUM SNOW DEPTH (IN)	50	13	5	4	0	0	0	0	0	0	0	3	8	13
	YEAR OF OCCURRENCE		1988	1985	1984								1971	1963	JAN 1988
	NORMAL NO. DAYS WITH:														
	SNOWFALL 1.0	30	1.1	0.6	0.2	0.0	0.0	0.0	0.0	0.0	0.0	0.0	0.1	0.1	2.1

PRECIPITATION (inches) 2001 LITTLE ROCK, AR (LIT)

YEAR	JAN	FEB	MAR	APR	MAY	JUN	JUL	AUG	SEP	OCT	NOV	DEC	ANNUAL
1972	1.71	1.55	3.32	1.81	2.07	2.62	1.77	3.58	6.43	7.63	7.38	5.14	45.01
1973	5.64	2.95	7.89	14.20	3.96	2.66	6.59	1.26	9.09	5.93	9.03	5.19	74.39
1974	5.77	2.60	2.07	9.76	6.26	7.82	4.09	3.20	4.31	3.36	5.73	2.99	57.96
1975	4.64	4.38	7.67	4.14	5.87	1.56	3.98	2.73	1.86	1.62	3.68	2.92	45.05
1976	3.00	5.12	5.43	1.06	4.88	5.69	1.97	0.70	1.82	6.04	1.79	2.30	39.80
1977	2.70	1.96	6.75	4.47	2.89	4.70	5.07	1.37	6.38	0.63	9.34	1.40	47.66
1978	5.44	1.52	3.56	4.22	6.27	5.39	2.70	6.38	10.17	1.01	6.64	11.56	64.86
1979	4.05	5.67	3.10	9.64	11.54	4.45	4.27	6.51	4.35	3.36	4.02	3.53	64.49
1980	2.73	0.89	6.60	5.85	4.57	0.53	0.99	0.19	5.09	2.64	6.28	1.86	38.22
1981	1.11	3.89	4.00	2.75	9.73	7.80	3.15	2.91	1.37	6.11	1.64	1.34	45.80
1982	8.74	3.37	2.87	9.32	5.63	4.10	1.01	4.52	1.47	2.26	9.72	8.28	61.29
1983	2.25	1.49	4.19	6.72	7.58	3.34	1.07	0.79	0.41	3.73	4.47	9.07	45.11
1984	1.31	3.52	5.58	3.77	8.22	1.06	4.15	5.69	3.28	15.35	8.49	3.54	63.96
1985	3.11	2.78	5.27	8.63	2.99	2.40	3.30	3.52	4.36	3.91	5.78	2.97	49.02
1986	0.50	3.45	3.68	7.33	4.07	6.42	0.14	4.56	1.94	6.05	5.67	3.86	47.67
1987	2.07	7.07	3.52	0.50	4.56	4.63	1.60	2.12	7.56	1.37	10.96	16.48	62.44
1988	3.71	3.41	3.50	3.82	2.05	1.04	7.95	2.54	2.19	1.95	13.14	2.91	48.21
1989	3.01	9.55	7.64	2.57	4.04	3.95	7.87	1.21	3.57	1.70	1.95	2.19	49.25
1990	6.50	4.82	10.40	7.73	7.71	0.80	4.63	1.57	4.08	8.75	3.29	6.79	67.07
1991	6.88	3.03	3.56	12.44	2.87	2.28	2.03	6.78	3.01	7.00	5.18	4.59	59.65
1992	1.75	2.05	6.48	1.86	3.67	5.07	6.76	2.14	2.90	0.67	4.71	3.85	41.91
1993	5.06	2.44	3.05	5.40	5.49	2.04	1.24	2.77	1.44	4.10	6.33	4.41	43.77
1994	4.87	3.21	5.60	5.20	3.97	5.57	4.28	4.00	2.11	3.88	6.13	4.58	53.40
1995	3.94	2.40	3.74	4.95	4.56	1.85	2.99	T	1.88	5.54	2.33	2.83	37.01
1996	2.60	2.14	3.57	4.24	3.98	2.82	3.56	1.23	6.39	6.35	7.41	2.81	47.10
1997	1.88	4.65	6.48	7.73	3.92	5.41	1.88	2.18	3.75	4.35	3.87	3.66	49.76
1998	4.70	4.11	4.80	3.31	2.86	2.16	2.96	3.23	3.45	3.39	2.30	4.41	41.68
1999	6.11	1.10	4.85	5.33	3.40	6.10	2.41	0.91	1.56	4.97	0.26	5.25	42.25
2000	1.03	3.93	3.86	2.94	5.80	5.66	0.94	0.04	2.39	0.79	10.99	3.42	41.79
2001	2.98	8.46	3.94	1.40	4.04	2.05	1.56	1.76	2.54	5.24	5.30	8.31	47.58
POR= 122 YRS	4.45	3.90	4.71	5.13	4.96	3.67	3.35	3.12	3.32	3.17	4.45	4.24	48.47

AVERAGE TEMPERATURE (F) 2001 LITTLE ROCK, AR (LIT)

YEAR	JAN	FEB	MAR	APR	MAY	JUN	JUL	AUG	SEP	OCT	NOV	DEC	ANNUAL
1972	43.6	46.7	53.3	62.5	69.8	79.4	80.4	81.1	75.8	62.6	47.3	41.0	62.0
1973	39.7	42.1	58.2	59.9	68.2	78.6	81.1	80.4	75.7	67.6	56.6	42.8	62.6
1974	42.4	45.7	58.1	60.7	71.3	74.3	83.2	79.0	69.0	62.3	51.9	44.4	61.9
1975	44.6	44.6	48.7	60.7	72.5	78.6	80.2	79.5	69.2	63.0	51.4	42.9	61.3
1976	39.7	52.5	56.5	61.1	64.6	74.4	80.2	78.7	72.1	57.8	45.9	41.9	60.5
1977	31.3	46.9	56.4	64.7	73.7	80.0	82.1	80.4	77.3	62.6	52.9	42.0	62.5
1978	31.7	34.0	51.0	65.9	71.4	78.9	84.1	83.0	76.7	62.0	54.6	43.3	61.4
1979	29.9	38.7	55.5	62.7	70.2	77.9	81.0	79.0	72.7	65.3	50.3	45.8	60.8
1980	44.0	40.9	50.3	61.5	70.6	79.4	88.1	87.0	78.6	60.4	50.4	43.1	62.9
1981	39.7	44.6	52.5	67.5	67.4	80.1	83.5	79.9	75.3	61.4	55.5	43.2	62.6
1982	37.5	41.3	57.1	58.0	72.7	76.6	83.1	82.1	74.2	64.7	53.2	48.3	62.4
1983	39.2	43.8	51.1	54.4	67.7	77.4	82.5	86.1	76.0	64.0	50.8	30.9	60.3
1984	36.7	46.6	50.1	59.7	68.0	79.8	79.7	78.1	71.0	65.0	49.0	52.1	61.3
1985	33.7	39.3	57.6	63.0	70.0	78.2	81.2	80.8	72.5	66.2	58.1	38.1	61.4
1986	42.5	48.2	55.4	63.6	71.4	79.7	86.3	78.1	77.6	63.1	49.7	42.4	63.2
1987	40.2	47.1	53.4	62.4	76.3	79.9	82.2	84.4	74.9	59.1	53.0	45.2	63.2
1988	35.6	42.9	52.2	61.5	70.4	78.8	81.7	82.4	76.0	60.4	53.3	44.4	61.6
1989	46.3	38.4	52.5	62.6	69.6	76.2	79.3	80.2	71.2	63.3	55.3	35.7	60.9
1990	48.1	51.3	55.4	62.0	68.1	80.6	83.2	82.2	77.5	61.5	56.3	43.1	64.1
1991	39.0	49.2	56.2	64.4	74.1	79.5	82.8	80.0	73.7	64.3	49.2	46.8	63.3
1992	42.7	50.6	54.4	62.6	69.0	76.5	81.0	76.5	72.7	64.4	50.1	43.9	62.0
1993	40.4	43.2	51.3	58.3	68.8	78.7	86.0	83.7	73.5	61.4	48.3	44.9	61.5
1994	38.3	45.2	53.9	64.4	68.3	81.8	80.0	78.9	72.7	64.3	56.1	46.3	62.5
1995	43.0	46.5	55.5	62.4	70.7	77.6	83.1	86.7	72.5	64.1	50.1	42.1	62.9
1996	40.0	46.2	48.2	60.4	74.3	79.4	81.9	80.8	73.5	63.6	48.4	46.4	61.9
1997	41.0	47.2	56.7	58.2	68.3	77.1	84.0	80.4	76.4	63.9	49.9	42.6	62.1
1998	46.7	48.9	51.7	62.0	75.9	83.1	87.2	83.9	80.6	65.8	54.8	44.9	65.5
1999	43.8	51.0	50.0	65.1	70.1	78.3	83.6	83.3	74.0	64.0	56.8	44.1	62.5
2000	43.0	50.7	55.9	61.1	72.3	76.5	82.7	86.5	75.4	65.7	48.0	32.0	62.5
2001	38.6	46.4	49.5	67.3	71.3	76.9	83.2	82.0	72.7	60.6	55.7	45.9	62.5
POR= 122 YRS	41.3	44.9	53.0	62.4	70.3	78.2	81.6	80.7	74.3	63.6	51.9	43.6	62.1

REFERENCE NOTES:

PAGE 1:
THE TEMPERATURE GRAPH SHOWS NORMAL MAXIMUM AND NORMAL
MINIMUM DAILY TEMPERATURES (SOLID CURVES) AND THE
ACTUAL DAILY HIGH AND LOW TEMPERATURES (VERTICAL BARS).

PAGE 2 AND 3:
H/C INDICATES HEATING AND COOLING DEGREE DAYS.
RH INDICATES RELATIVE HUMIDITY
W/O INDICATES WEATHER AND OBSTRUCTIONS
S INDICATES SUNSHINE.
PR INDICATES PRESSURE.
CLOUDINESS ON PAGE 3 IS THE SUM OF THE CEILOMETER AND
SATELLITE DATA NOT TO EXCEED EIGHT EIGHTHS(OKTAS).

GENERAL:
T INDICATES TRACE PRECIPITATION, AN AMOUNT GREATER
THAN ZERO BUT LESS THAN THE LOWEST REPORTABLE VALUE.
+ INDICATES THE VALUE ALSO OCCURS ON EARLIER DATES.
BLANK ENTRIES DENOTE MISSING OR UNREPORTED DATA.
NORMALS ARE 30-YEAR AVERAGES (1961 - 1990).
ASOS INDICATES AUTOMATED SURFACE OBSERVING SYSTEM.
PM INDICATES THE LAST DAY OF THE PREVIOUS MONTH.
POR (PERIOD OF RECORD) BEGINS WITH THE JANUARY DATA
MONTH AND IS THE NUMBER OF YEARS USED TO COMPUTE
THE MEAN. INDIVIDUAL MONTHS WITHIN THE POR MAY
BE MISSING.
WHEN THE POR FOR A NORMAL IS LESS THAN 30 YEARS,
THE NORMAL IS PROVISIONAL AND IS BASED ON THE NUMBER
OF YEARS INDICATED.
0.* OR * INDICATES THE VALUE OR MEAN-DAYS-WITH
IS BETWEEN 0.00 AND 0.05.
CLOUDINESS FOR ASOS STATIONS DIFFERS FROM THE NON-ASOS
OBSERVATION TAKEN BY A HUMAN OBSERVER. ASOS STATION
CLOUDINESS IS BASED ON TIME-AVERAGED CEILOMETER DATA
FOR CLOUDS AT OR BELOW 12,000 FEET AND ON SATELLITE
DATA FOR CLOUDS ABOVE 12,000 FEET.
THE NUMBER OF DAYS WITH CLEAR, PARTLY CLOUDY, AND
CLOUDY CONDITIONS FOR ASOS STATIONS IS THE SUM
OF THE CEILOMETER AND SATELLITE DATA FOR THE
SUNRISE TO SUNSET PERIOD.

GENERAL CONTINUED:
CLEAR INDICATES 0 - 2 OKTAS, PARTLY CLOUDY INDICATES
3 - 6 OKTAS, AND CLOUDY INDICATES 7 OR 8 OKTAS.
WHEN AT LEAST ONE OF THE ELEMENTS (CEILOMETER OR
SATELLITE) IS MISSING, THE DAILY CLOUDINESS IS
NOT COMPUTED.
WIND DIRECTION IS RECORDED IN TENS OF DEGREES (2 DIGITS)
CLOCKWISE FROM TRUE NORTH. "00" INDICATES CALM. "36"
INDICATES TRUE NORTH.
RESULTANT WIND IS THE VECTOR AVERAGE OF THE SPEED AND
DIRECTION.
AVERAGE TEMPERATURE IS THE SUM OF THE MEAN DAILY MAXIMUM
AND MINIMUM TEMPERATURE DIVIDED BY 2.
SNOWFALL DATA COMPRISE ALL FORMS OF FROZEN
PRECIPITATION, INCLUDING HAIL.
A HEATING (COOLING) DEGREE DAY IS THE DIFFERENCE BETWEEN
THE AVERAGE DAILY TEMPERATURE AND 65 F.
DRY BULB IS THE TEMPERATURE OF THE AMBIENT AIR.
DEW POINT IS THE TEMPERATURE TO WHICH THE AIR MUST BE
COOLED TO ACHIEVE 100 PERCENT RELATIVE HUMIDITY.
WET BULB IS THE TEMPERATURE THE AIR WOULD HAVE IF THE
MOISTURE CONTENT WAS INCREASED TO 100 PERCENT RELATIVE
HUMIDITY.

ON JULY 1, 1996, THE NATIONAL WEATHER SERVICE BEGAN USING
THE "METAR" OBSERVATION CODE THAT WAS ALREADY EMPLOYED
BY MOST OTHER NATIONS OF THE WORLD. THE MOST NOTICEABLE
DIFFERENCE IN THIS ANNUAL PUBLICATION WILL BE THE CHANGE
IN UNITS FROM TENTHS TO EIGHTS(OKTAS) FOR REPORTING THE
AMOUNT OF SKY COVER.

HEATING DEGREE DAYS (base 65 F) 2001 LITTLE ROCK, AR (LIT)

YEAR	JUL	AUG	SEP	OCT	NOV	DEC	JAN	FEB	MAR	APR	MAY	JUN	TOTAL
1972-73	0	0	8	142	530	736	777	637	216	186	28	0	3260
1973-74	0	0	2	61	261	680	690	533	255	163	4	0	2649
1974-75	0	0	23	111	401	634	630	566	499	196	5	0	3065
1975-76	0	0	48	130	414	681	777	359	284	146	70	0	2909
1976-77	0	0	1	257	567	710	1041	502	265	70	10	0	3423
1977-78	0	0	0	105	370	709	1025	862	436	68	48	0	3623
1978-79	0	0	0	118	321	667	1083	732	314	110	13	0	3358
1979-80	0	0	0	80	436	588	645	693	450	142	15	0	3049
1980-81	0	0	16	184	437	673	774	565	388	37	45	0	3119
1981-82	0	0	4	186	278	668	847	656	298	223	6	0	3166
1982-83	0	0	12	119	369	536	795	587	425	332	33	0	3208
1983-84	0	0	19	89	422	1050	872	530	460	190	24	0	3656
1984-85	0	0	44	81	476	408	962	713	251	101	8	0	3044
1985-86	0	0	31	82	283	825	691	467	298	91	7	0	2775
1986-87	0	1	0	112	454	694	762	496	353	145	0	0	3017
1987-88	0	0	0	182	358	609	904	637	388	123	4	0	3205
1988-89	0	0	1	163	358	633	573	738	395	156	39	0	3056
1989-90	0	0	23	112	313	898	516	380	316	152	31	0	2741
1990-91	0	0	8	173	260	675	798	438	294	69	10	0	2725
1991-92	0	0	23	97	483	560	682	413	324	134	40	0	2756
1992-93	0	0	14	69	441	647	755	606	423	215	15	1	3186
1993-94	0	0	13	174	506	613	824	548	339	105	49	0	3171
1994-95	0	0	22	111	276	573	678	512	318	119	27	0	2636
1995-96	0	0	24	93	445	703	767	540	520	180	7	0	3279
1996-97	0	0	11	103	492	576	741	488	261	217	28	0	2917
1997-98	0	0	1	149	451	685	562	445	446	123	0	0	2862
1998-99	0	0	0	67	301	627	650	387	458	73	1	0	2564
1999-00	0	0	4	100	250	581	672	418	288	131	8	0	2452
2000-01	0	0	16	97	513	1014	811	515	477	64	7	0	3514
2001-	0	0	17	165	274	585							

COOLING DEGREE DAYS (base 65 F) 2001 LITTLE ROCK, AR (LIT)

YEAR	JAN	FEB	MAR	APR	MAY	JUN	JUL	AUG	SEP	OCT	NOV	DEC	ANNUAL
1972	2	0	3	81	174	440	484	507	341	76	3	0	2111
1973	0	0	14	41	135	415	485	530	330	147	17	0	2090
1974	0	0	45	37	206	288	572	441	148	36	14	2	1787
1975	6	0	4	73	245	416	475	455	178	73	14	0	1941
1976	0	5	27	37	65	292	480	434	220	42	0	0	1602
1977	0	0	7	67	287	455	537	482	377	40	14	0	2266
1978	0	0	9	104	253	424	599	565	357	31	16	0	2358
1979	0	0	24	48	178	396	502	442	240	95	1	0	1926
1980	0	1	1	42	196	439	725	688	432	50	5	0	2579
1981	0	0	9	117	131	458	580	470	318	78	2	0	2163
1982	0	0	57	21	253	355	570	540	294	114	20	24	2248
1983	0	0	0	21	121	381	550	660	355	63	6	0	2157
1984	0	1	4	38	126	451	462	416	234	88	3	17	1840
1985	0	0	31	48	167	404	508	501	265	125	21	0	2070
1986	0	5	7	56	211	446	668	415	385	63	0	0	2256
1987	0	0	3	74	359	456	540	610	304	8	3	0	2357
1988	0	0	2	24	177	423	523	546	332	25	15	0	2067
1989	0	0	13	91	189	345	450	479	213	70	29	0	1879
1990	0	4	26	67	135	475	571	540	392	73	8	0	2291
1991	0	1	30	56	299	445	558	472	290	82	14	2	2249
1992	0	0	4	70	170	352	505	364	249	60	0	0	1774
1993	0	0	4	22	138	419	656	591	276	69	14	0	2189
1994	0	3	3	96	161	510	472	435	256	100	15	0	2051
1995	5	0	31	47	207	383	568	683	254	73	4	0	2255
1996	0	5	6	49	303	437	529	499	273	62	1	6	2170
1997	5	0	13	21	138	369	597	485	351	122	4	0	2105
1998	0	0	40	40	344	548	697	592	473	98	3	11	2846
1999	0	2	0	85	168	406	582	571	280	74	11	1	2180
2000	0	7	10	21	237	353	556	676	334	128	9	0	2331
2001	0	0	0	138	209	366	574	533	254	35	6	0	2115

SNOWFALL (inches) 2001 LITTLE ROCK, AR (LIT)

YEAR	JUL	AUG	SEP	OCT	NOV	DEC	JAN	FEB	MAR	APR	MAY	JUN	TOTAL
1972-73	0.0	0.0	0.0	0.0	T	0.7	2.6	T	0.0	T	0.0	0.0	3.3
1973-74	0.0	0.0	0.0	0.0	0.0	T	0.3	T	0.0	0.0	0.0	0.0	0.3
1974-75	0.0	0.0	0.0	0.0	T	T	1.4	0.4	2.4	0.0	0.0	0.0	4.2
1975-76	0.0	0.0	0.0	0.0	0.2	1.0	T	0.0	0.0	0.0	0.0	0.0	1.2
1976-77	0.0	0.0	0.0	0.0	1.0	0.0	3.8	0.0	0.0	0.0	0.0	0.0	4.8
1977-78	0.0	0.0	0.0	0.0	0.0	0.0	10.0	3.4	T	0.0	0.0	0.0	13.4
1978-79	0.0	0.0	0.0	0.0	0.0	T	1.4	9.8	0.0	0.0	0.0	0.0	11.2
1979-80	0.0	0.0	0.0	0.0	0.0	0.0	0.9	0.5	T	T	0.0	0.0	1.4
1980-81	0.0	0.0	0.0	0.0	1.8	T	T	0.0	0.0	0.0	0.0	0.0	1.8
1981-82	0.0	0.0	0.0	0.0	0.0	0.0	5.0	6.3	T	0.0	0.0	0.0	11.3
1982-83	0.0	0.0	0.0	0.0	0.0	T	T	T	T	T	0.0	0.0	T
1983-84	0.0	0.0	0.0	0.0	0.0	0.8	1.5	0.2	4.5	0.0	0.0	0.0	7.0
1984-85	0.0	0.0	0.0	0.0	0.0	T	6.3	1.5	0.0	0.0	0.0	0.0	7.8
1985-86	0.0	0.0	0.0	0.0	0.0	T	0.0	1.5	T	0.0	0.0	0.0	1.5
1986-87	0.0	0.0	0.0	0.0	T	0.0	1.0	0.8		0.0	0.0	0.0	1.8
1987-88	0.0	0.0	0.0	0.0	0.0	T	13.6	2.5	T	0.0	0.0	0.0	16.1
1988-89	0.0	0.0	0.0	0.0	0.0	T	2.0	T	1.0	0.0	0.0	0.0	3.0
1989-90	0.0	0.0	0.0	0.0	0.0	T	T	0.0	0.0	0.0	0.0	0.0	T
1990-91	0.0	0.0	0.0	0.0	0.0	T	T	T	T	0.0	0.0	0.0	T
1991-92	0.0	0.0	0.0	0.0	T	T	0.0	0.0	0.0	0.0	0.0	0.0	T
1992-93	0.0	0.0	0.0	0.0	T	T	0.0	T	T	0.0	0.0	0.0	T
1993-94	0.0	0.0	0.0	T	0.0	0.0	0.1	0.7	T	T	0.0	0.0	0.8
1994-95	0.0	0.0	0.0	0.0	0.0	T	7.0	1.0	T	0.0	0.0	0.0	8.0
1995-96	0.0	0.0	0.0	0.0	0.1	T	T	0.6	0.7	T	T	0.0	1.4
1996-97	0.0	0.0	0.0	0.0	0.0	T	T	4.7	0.0	T	T	0.0	4.7
1997-98	0.0	0.0	0.0	0.0	T	T	0.0	0.0	T	T	0.0	T	T
1998-99	0.0	0.0	0.0										
1999-00													
2000-01													
2001-													
POR= 55 YRS	0.0	0.0	0.0	T	0.2	0.6	2.3	1.4	0.5	T	0.0	0.0	5.0

2001
EUREKA,
CALIFORNIA (EKA)

Humboldt Bay is one-quarter mile north and one mile west of the station. There are no hills in Eureka of any consequence. The land slopes upward gently from the Bay toward the Coast Range, which begins about 3 miles east of the station and reaches the top of its first ridge approximately 10 miles to the east. The elevation of the ridge is 2,000 feet and extends in a semicircle from a point 20 miles north of Eureka to a point 25 miles south.

The climate of Eureka is completely maritime with high humidity prevailing the entire year. There are definite rainy and dry seasons. The rainy season begins in October and continues through April, accounting for about 90 percent of the annual precipitation. The dry season from May through September is marked by considerable fog or low cloudiness that usually clears in the late morning and sunny weather is generally the case during the early afternoon hours.

Temperatures are moderate the entire year. Although record highs have reached the mid 80s and record lows near 20 degrees, the usual yearly range is from lows in the mid 30s to highs in the mid 70s.

The principal industries are lumbering, fishing, tourism, and dairy farming. There is very little truck farming due to the low temperatures and lack of sunshine, however, the climate is nearly ideal for berries and flowers.

Based on the 1951-1980 period, the average first occurrence of 32 degrees Fahrenheit in the fall is December 10 and the average last occurrence in the spring is February 6.

NORMALS, MEANS, AND EXTREMES
EUREKA, CA (EKA)

LATITUDE: LONGITUDE: ELEVATION (FT):   TIME ZONE: WBAN: 24213
40 48' 0 " N   124 09' 0 " W   GRND: 19   BARO: 20   PACIFIC (UTC + 8)

	ELEMENT	POR	JAN	FEB	MAR	APR	MAY	JUN	JUL	AUG	SEP	OCT	NOV	DEC	YEAR
TEMPERATURE °F	NORMAL DAILY MAXIMUM	30	54.4	55.6	55.4	55.9	57.9	60.3	61.8	62.6	63.0	60.8	58.1	54.8	58.4
	MEAN DAILY MAXIMUM	60	54.3	55.3	55.2	56.2	58.3	60.4	61.6	62.5	62.8	60.8	58.0	54.9	58.4
	HIGHEST DAILY MAXIMUM	91	78	85	78	80	84	85	76	82	86	87	78	77	87
	YEAR OF OCCURRENCE		1986	1930	1914	1989	1939	1945	1992	1991	1983	1993	1987	1963	OCT 1993
	MEAN OF EXTREME MAXS.	60	65.4	66.4	65.4	66.2	68.5	68.3	67.7	69.8	73.3	73.7	68.1	64.8	68.1
	NORMAL DAILY MINIMUM	30	41.5	42.9	43.4	44.3	47.5	50.6	52.3	53.1	51.6	48.6	45.2	42.0	46.9
	MEAN DAILY MINIMUM	60	41.3	42.6	43.0	44.6	47.8	50.6	52.3	53.0	51.5	48.5	45.0	41.9	46.8
	LOWEST DAILY MINIMUM	91	25	27	29	32	36	40	45	44	41	32	29	21	21
	YEAR OF OCCURRENCE		1937	1990	1917	2001	1954	1999	1924	1935	1946	1971	1994	1972	DEC 1972
	MEAN OF EXTREME MINS.	60	31.7	33.3	34.7	37.2	41.0	45.5	48.2	49.0	45.6	40.6	35.4	32.2	39.5
	NORMAL DRY BULB	30	48.0	49.3	49.4	50.2	52.8	55.5	57.0	57.9	57.3	54.7	51.7	48.4	52.7
	MEAN DRY BULB	60	47.9	49.0	49.1	50.3	53.0	55.5	56.9	57.8	57.1	54.7	51.5	48.4	52.6
	MEAN WET BULB														
	MEAN DEW POINT														
	NORMAL NO. DAYS WITH:														
	MAXIMUM 90	30	0.0	0.0	0.0	0.0	0.0	0.0	0.0	0.0	0.0	0.0	0.0	0.0	0.0
	MAXIMUM 32	30	0.0	0.0	0.0	0.0	0.0	0.0	0.0	0.0	0.0	0.0	0.0	0.0	0.0
	MINIMUM 32	30	1.6	0.9	0.2	0.0	0.0	0.0	0.0	0.0	0.0	*	0.4	1.9	5.0
	MINIMUM 0	30	0.0	0.0	0.0	0.0	0.0	0.0	0.0	0.0	0.0	0.0	0.0	0.0	0.0
H/C	NORMAL HEATING DEG. DAYS	30	527	440	484	444	378	285	248	223	234	319	399	515	4496
	NORMAL COOLING DEG. DAYS	30	0	0	0	0	0	0	0	0	0	0	0	0	0
RH	NORMAL (PERCENT)														
	HOUR 04 LST														
	HOUR 10 LST														
	HOUR 16 LST														
	HOUR 22 LST														
S	PERCENT POSSIBLE SUNSHINE	84	43	46	52	57	58	59	55	51	55	50	44	41	51
W/O	MEAN NO. DAYS WITH:														
	HEAVY FOG(VISBY 1/4 MI)	78	4.0	2.7	1.9	1.6	1.2	2.1	3.5	5.4	7.5	9.5	6.0	4.3	49.7
	THUNDERSTORMS	78	0.7	0.6	0.4	0.2	0.2	0.2	0.2	0.1	0.3	0.4	0.6	0.6	4.5
CLOUDINESS	MEAN:														
	SUNRISE-SUNSET (OKTAS)	57	5.8	5.9	5.8	5.6	5.4	5.2	5.1	5.3	4.8	5.1	5.7	5.6	5.4
	MIDNIGHT-MIDNIGHT (OKTAS)	2	2.6	3.2	0.0	0.0	2.6	0.0	0.0	0.0	2.8	2.4	0.0	2.4	1.3
	MEAN NO. DAYS WITH:														
	CLEAR	91	5.9	5.1	5.6	6.1	6.5	7.2	6.4	5.4	8.6	8.0	6.1	6.3	77.2
	PARTLY CLOUDY	91	6.3	5.8	7.9	8.4	9.9	9.7	10.8	10.8	8.6	8.4	6.6	6.5	99.7
	CLOUDY	91	18.9	17.4	17.6	15.5	14.7	13.1	13.5	14.3	12.4	14.2	16.9	17.9	186.4
PR	MEAN STATION PRESSURE(IN)														
	MEAN SEA-LEVEL PRES. (IN)														
WINDS	MEAN SPEED (MPH)	54	6.9	7.2	7.6	8.0	7.9	7.4	6.8	5.8	5.5	5.6	6.0	6.4	6.8
	PREVAIL.DIR(TENS OF DEGS)														
	FASTEST MILE:														
	SPEED (MPH)	84	54	48	48	49	40	39	35	34	44	56	55	56	56
	DIR.		S	SW	SW	N	NW	NW	N	N	N	SW	S	S	SW
	YEAR OF OCCURRENCE		1955	1960	1953	1915	1955	1949	1986	1920	1941	1962	1981	1931	OCT 1962
	PEAK GUST :														
	SPEED (MPH)														
	DIR. (TENS OF DEGS)														
	YEAR OF OCCURRENCE														
PRECIPITATION	NORMAL (IN)	30	6.00	4.73	5.32	2.88	1.44	0.51	0.13	0.48	0.89	2.67	6.44	6.04	37.53
	MAXIMUM MONTHLY (IN)	91	13.92	13.95	13.97	10.68	6.05	2.57	1.34	3.42	3.56	13.04	16.58	21.26	21.26
	YEAR OF OCCURRENCE		1969	1998	1938	1963	1960	1954	1916	1983	1925	1950	1973	1996	DEC 1996
	MINIMUM MONTHLY (IN)	91	0.66	0.50	0.07	0.31	0.03	0.00	0.00	0.00	0.00	0.00	T	0.52	0.00
	YEAR OF OCCURRENCE		1985	1923	1926	1956	1955	1917	1967	1940	1929	1917	1976	1976	JUL 1967
	MAXIMUM IN 24 HOURS (IN)	91	4.42	4.88	4.02	2.56	2.23	1.73	1.18	2.21	1.54	5.83	5.21	4.86	5.83
	YEAR OF OCCURRENCE		1912	1959	1975	1983	1943	1943	1916	1983	1977	1950	1998	1996	OCT 1950
	NORMAL NO. DAYS WITH:														
	PRECIPITATION 0.01	30	14.3	13.8	16.2	11.4	7.6	5.0	2.1	3.0	4.2	8.7	15.1	15.4	116.8
	PRECIPITATION 1.00	30	1.6	0.9	0.8	0.4	0.0	0.1	0.0	0.1	0.1	0.5	1.7	1.3	7.5
SNOWFALL	NORMAL (IN)	30	0.1	0.2	0.*	T	0.0	0.0	0.0	0.0	0.0	0.0	0.*	0.1	0.4
	MAXIMUM MONTHLY (IN)	91	3.0	3.5	1.0	T	0.0	0.0	0.0	0.0	0.0	0.0	0.1	1.9	3.5
	YEAR OF OCCURRENCE		1935	1989	1966	1995							1977	1972	FEB 1989
	MAXIMUM IN 24 HOURS (IN)	91	3.0	2.0	1.0	T	0.0	0.0	0.0	0.0	0.0	0.0	0.1	1.9	3.0
	YEAR OF OCCURRENCE		1935	1989	1966	1995							1977	1972	JAN 1935
	MAXIMUM SNOW DEPTH (IN)	59	0	1	1	0	0	0	0	0	0	0	0	0	1
	YEAR OF OCCURRENCE			1989	1999										MAR 1999
	NORMAL NO. DAYS WITH:														
	SNOWFALL 1.0	30	0.0	0.1	0.0	0.0	0.0	0.0	0.0	0.0	0.0	0.0	0.0	0.1	0.2

PRECIPITATION (inches) 2001 EUREKA, CA (EKA)

YEAR	JAN	FEB	MAR	APR	MAY	JUN	JUL	AUG	SEP	OCT	NOV	DEC	ANNUAL
1972	7.96	5.93	5.08	2.27	1.11	0.88	0.01	0.07	1.06	1.97	5.41	7.42	39.17
1973	6.47	3.85	7.10	0.35	0.85	0.23	T	0.08	2.35	4.14	16.58	7.02	49.02
1974	6.02	5.98	6.98	3.15	0.42	0.33	0.11	0.32	T	1.76	2.75	6.40	34.22
1975	5.20	7.68	10.73	3.29	1.05	0.58	0.10	0.58	0.01	6.77	4.72	5.38	46.09
1976	1.88	7.51	3.12	2.80	0.54	0.14	0.20	1.70	0.04	0.28	2.98	0.52	21.71
1977	1.90	2.24	4.33	1.20	2.10	0.07	T	0.20	3.35	2.79	4.51	6.60	29.29
1978	4.52	6.06	2.88	4.10	0.82	0.34	0.03	0.59	2.72	0.04	2.39	1.16	25.65
1979	3.82	6.26	1.70	3.94	2.25	0.05	0.31	0.13	1.15	6.14	6.19	3.75	35.69
1980	3.19	4.67	6.14	4.18	1.70	0.42	T	0.07	0.14	1.38	2.49	6.10	30.48
1981	7.67	3.72	4.64	0.71	2.02	0.57	T	0.01	0.97	3.71	9.39	9.88	43.29
1982	4.75	5.76	7.06	5.97	0.07	0.78	0.08	0.03	0.62	4.89	7.83	10.30	48.14
1983	8.48	9.18	10.73	5.47	1.12	0.65	0.89	3.42	0.87	1.87	10.40	14.13	67.21
1984	0.76	5.18	4.70	2.76	2.51	1.07	0.03	0.05	0.55	3.67	15.15	4.27	40.70
1985	0.66	3.69	4.68	0.45	1.14	0.89	0.15	0.52	1.06	4.07	2.98	2.78	23.07
1986	7.19	10.08	6.12	1.46	2.34	0.21	0.02	T	2.70	1.75	1.85	3.83	37.55
1987	6.48	3.38	6.10	1.15	0.41	0.26	0.20	0.06	0.02	1.05	4.23	10.92	34.26
1988	7.13	0.54	1.18	2.06	2.70	2.22	0.05	T	0.12	0.41	8.93	6.26	31.60
1989	4.71	2.88	7.63	2.01	1.67	0.21	0.08	0.13	0.85	2.90	1.60	0.80	25.47
1990	7.20	4.50	3.30	1.41	3.74	0.32	0.22	0.71	0.19	1.73	3.07	2.95	29.34
1991	1.65	2.75	6.94	2.52	2.16	0.26	1.13	0.37	T	1.06	1.95	2.36	23.15
1992	3.99	3.80	3.51	2.42	0.06	1.27	0.25	0.01	0.33	2.08	2.21	9.33	29.26
1993	7.15	5.93	4.72	5.94	4.44	1.23	0.37	0.54	0.03	0.56	1.35	7.12	39.38
1994	5.09	7.12	2.06	3.30	1.10	0.71	0.08	T	0.06	0.54	8.21	7.00	35.27
1995	12.74	1.40	11.18	7.47	1.21	1.85	0.08	0.22	0.69	0.53	2.26	11.56	51.19
1996	10.74	8.11	3.51	4.64	2.40	0.05	0.03	T	1.21	3.50	5.16	21.26	60.61
1997	8.81	2.55	2.73	3.06	0.90	1.25	T	0.84	2.05	2.73	7.39	4.73	37.04
1998	13.42	13.95	7.83	2.23	3.12	0.33	0.16	0.01	0.08	3.06	14.09	5.40	63.68
1999	4.37	10.32	8.94	1.79	1.62	0.15	0.04	0.30	0.05	1.60	7.36	3.02	39.56
2000	9.71	7.00	2.81	2.15	1.86	0.54	0.04	T	0.55	2.99	3.51	1.97	33.13
2001	3.79	3.60	2.45	2.54	0.71	0.69	0.20	0.21	0.28	1.00	7.71	11.56	34.74
POR= 115 YRS	6.76	5.73	5.25	3.06	1.86	0.72	0.11	0.21	0.80	2.61	5.42	6.43	38.96

AVERAGE TEMPERATURE (F) 2001 EUREKA, CA (EKA)

YEAR	JAN	FEB	MAR	APR	MAY	JUN	JUL	AUG	SEP	OCT	NOV	DEC	ANNUAL
1972	44.6	49.0	51.2	49.1	51.2	54.6	57.8	58.2	55.8	54.2	51.8	45.7	51.9
1973	47.3	50.7	47.4	50.1	52.0	55.1	55.6	54.8	57.1	52.1	51.4	51.2	52.1
1974	46.7	46.1	50.1	49.7	51.1	54.1	57.7	57.7	55.6	53.2	51.2	48.7	51.8
1975	45.8	48.0	47.6	46.5	51.4	53.2	56.9	55.3	55.2	54.2	48.5	47.4	50.8
1976	46.4	46.9	45.7	48.4	51.1	52.8	57.5	57.9	56.2	54.4	52.2	47.4	51.4
1977	47.4	50.5	46.2	49.2	51.5	54.2	55.1	58.3	57.1	53.8	51.2	51.0	52.1
1978	51.8	50.3	53.5	51.0	53.3	56.0	55.9	57.0	57.6	54.9	48.0	43.3	52.7
1979	46.6	48.0	50.0	51.0	52.7	54.0	58.1	59.4	62.3	57.3	52.4	51.3	53.6
1980	48.3	53.5	48.6	51.9	52.3	55.5	57.3	55.0	56.6	54.3	51.3	51.2	53.0
1981	52.5	51.3	50.2	50.6	53.3	56.3	58.1	58.1	56.9	53.6	52.1	51.3	53.2
1982	44.9	49.5	48.3	50.8	52.8	56.3	58.3	59.6	58.7	57.3	52.1	49.6	53.2
1983	51.3	53.4	53.4	51.8	54.4	58.0	60.5	61.9	60.7	57.9	53.6	50.6	55.6
1984	49.4	50.0	52.8	51.3	55.0	55.2	57.6	60.2	58.3	55.3	52.1	46.8	53.7
1985	48.4	47.9	47.4	51.6	53.9	56.6	58.6	58.5	56.7	54.5	46.6	47.7	52.4
1986	54.3	52.7	53.1	50.7	53.9	59.0	57.4	57.2	57.2	55.6	53.5	51.3	54.7
1987	49.1	51.4	52.7	54.2	56.6	57.7	59.5	58.5	57.3	57.8	54.8	49.7	54.9
1988	50.4	50.2	50.3	52.8	56.2	57.7	58.8	57.7	55.8	53.3	54.7	47.7	53.9
1989	46.0	45.6	51.8	54.6	55.9	57.6	59.4	59.3	56.6	54.9	52.6	49.2	53.6
1990	48.3	45.5	50.2	52.8	54.0	58.2	59.4	60.4	61.6	54.3	49.7	42.8	53.1
1991	47.8	52.8	48.1	50.9	52.2	53.8	57.3	59.9	57.2	56.1	51.4	48.3	53.0
1992	50.3	53.9	54.1	56.4	56.7	58.3	60.6	59.1	57.5	58.0	51.7	46.9	55.3
1993	46.4	49.4	53.4	53.6	57.9	57.9	61.4	59.8	55.8	56.9	49.9	49.5	54.0
1994	50.5	48.5	51.1	52.8	55.5	57.0	57.1	61.4	59.5	54.4	46.7	53.4	53.4
1995	52.4	51.7	50.4	51.1	53.7	56.0	60.2	58.4	60.3	54.9	53.9	52.0	54.6
1996	49.5	52.3	50.7	52.9	53.7	57.5	57.4	58.1	55.8	54.5	51.3	50.9	53.6
1997	48.5	47.9	49.5	51.4	57.8	57.8	59.5	61.4	62.2	55.1	53.4	47.7	54.4
1998	52.0	50.1	50.2	50.4	53.9	56.7	58.5	58.8	57.4	55.0	51.7	44.6	53.3
1999	47.4	47.3	46.8	48.6	51.3	55.0	56.5	60.1	57.4	54.9	54.7	46.0	51.8
2000	48.3	51.6	48.4	52.3	55.3	56.7	55.0	56.5	60.1	54.9	54.7	46.0	51.8
2001	46.9	47.0	50.2	49.2	54.0	56.0	57.4	59.5	56.2	54.4	51.8	49.2	52.7
POR= 115 YRS	47.5	48.3	48.8	50.3	52.7	55.1	56.3	56.9	56.5	54.2	51.3	48.3	52.2

REFERENCE NOTES:

PAGE 1:
THE TEMPERATURE GRAPH SHOWS NORMAL MAXIMUM AND NORMAL MINIMUM DAILY TEMPERATURES (SOLID CURVES) AND THE ACTUAL DAILY HIGH AND LOW TEMPERATURES (VERTICAL BARS).

PAGE 2 AND 3:
H/C INDICATES HEATING AND COOLING DEGREE DAYS.
RH INDICATES RELATIVE HUMIDITY.
W/O INDICATES WEATHER AND OBSTRUCTIONS
S INDICATES SUNSHINE.
PR INDICATES PRESSURE.
CLOUDINESS ON PAGE 3 IS THE SUM OF THE CEILOMETER AND SATELLITE DATA NOT TO EXCEED EIGHT EIGHTHS(OKTAS).

GENERAL:
T INDICATES TRACE PRECIPITATION, AN AMOUNT GREATER THAN ZERO BUT LESS THAN THE LOWEST REPORTABLE VALUE.
+ INDICATES THE VALUE ALSO OCCURS ON EARLIER DATES.
BLANK ENTRIES DENOTE MISSING OR UNREPORTED DATA.
NORMALS ARE 30-YEAR AVERAGES (1961 - 1990).
ASOS INDICATES AUTOMATED SURFACE OBSERVING SYSTEM.
PM INDICATES THE LAST DAY OF THE PREVIOUS MONTH.
POR (PERIOD OF RECORD) BEGINS WITH THE JANUARY DATA MONTH AND IS THE NUMBER OF YEARS USED TO COMPUTE THE MEAN. INDIVIDUAL MONTHS WITHIN THE POR MAY BE MISSING.
WHEN THE POR FOR A NORMAL IS LESS THAN 30 YEARS, THE NORMAL IS PROVISIONAL AND IS BASED ON THE NUMBER OF YEARS INDICATED.
0.* OR * INDICATES THE VALUE OR MEAN-DAYS-WITH IS BETWEEN 0.00 AND 0.05.
CLOUDINESS FOR ASOS STATIONS DIFFERS FROM THE NON-ASOS OBSERVATION TAKEN BY A HUMAN OBSERVER. ASOS STATION CLOUDINESS IS BASED ON TIME-AVERAGED CEILOMETER DATA FOR CLOUDS AT OR BELOW 12,000 FEET AND ON SATELLITE DATA FOR CLOUDS ABOVE 12,000 FEET.
THE NUMBER OF DAYS WITH CLEAR, PARTLY CLOUDY, AND CLOUDY CONDITIONS FOR ASOS STATIONS IS THE SUM OF THE CEILOMETER AND SATELLITE DATA FOR THE SUNRISE TO SUNSET PERIOD.

GENERAL CONTINUED:
CLEAR INDICATES 0 - 2 OKTAS, PARTLY CLOUDY INDICATES 3 - 6 OKTAS, AND CLOUDY INDICATES 7 OR 8 OKTAS. WHEN AT LEAST ONE OF THE ELEMENTS (CEILOMETER OR SATELLITE) IS MISSING, THE DAILY CLOUDINESS IS NOT COMPUTED.
WIND DIRECTION IS RECORDED IN TENS OF DEGREES (2 DIGITS) CLOCKWISE FROM TRUE NORTH. "00" INDICATES CALM. "36" INDICATES TRUE NORTH.
RESULTANT WIND IS THE VECTOR AVERAGE OF THE SPEED AND DIRECTION.
AVERAGE TEMPERATURE IS THE SUM OF THE MEAN DAILY MAXIMUM AND MINIMUM TEMPERATURE DIVIDED BY 2.
SNOWFALL DATA COMPRISE ALL FORMS OF FROZEN PRECIPITATION, INCLUDING HAIL.
A HEATING (COOLING) DEGREE DAY IS THE DIFFERENCE BETWEEN THE AVERAGE DAILY TEMPERATURE AND 65 F.
DRY BULB IS THE TEMPERATURE OF THE AMBIENT AIR.
DEW POINT IS THE TEMPERATURE TO WHICH THE AIR MUST BE COOLED TO ACHIEVE 100 PERCENT RELATIVE HUMIDITY.
WET BULB IS THE TEMPERATURE THE AIR WOULD HAVE IF THE MOISTURE CONTENT WAS INCREASED TO 100 PERCENT RELATIVE HUMIDITY.

ON JULY 1, 1996, THE NATIONAL WEATHER SERVICE BEGAN USING THE "METAR" OBSERVATION CODE THAT WAS ALREADY EMPLOYED BY MOST OTHER NATIONS OF THE WORLD. THE MOST NOTICEABLE DIFFERENCE IN THIS ANNUAL PUBLICATION WILL BE THE CHANGE IN UNITS FROM TENTHS TO EIGHTS(OKTAS) FOR REPORTING THE AMOUNT OF SKY COVER.

HEATING DEGREE DAYS (base 65 F) 2001 EUREKA, CA (EKA)

YEAR	JUL	AUG	SEP	OCT	NOV	DEC	JAN	FEB	MAR	APR	MAY	JUN	TOTAL
1972-73	217	204	269	328	389	590	542	395	537	439	393	293	4596
1973-74	287	308	230	393	399	419	559	523	455	454	423	321	4771
1974-75	222	220	274	360	407	501	587	469	532	547	417	347	4883
1975-76	244	290	286	328	486	542	569	516	590	490	424	360	5125
1976-77	226	213	258	324	375	535	537	400	577	468	415	317	4645
1977-78	302	200	231	342	408	427	403	404	347	415	357	264	4100
1978-79	274	241	215	307	503	667	562	467	459	412	374	323	4804
1979-80	208	165	92	231	369	404	511	330	500	386	388	280	3864
1980-81	230	303	246	328	402	422	384	377	451	423	357	249	4172
1981-82	299	205	203	324	339	396	616	430	512	419	373	258	4374
1982-83	204	158	181	232	381	468	415	317	355	391	320	203	3625
1983-84	133	90	129	215	336	443	475	429	369	403	301	285	3608
1984-85	222	142	195	295	378	556	507	472	532	396	338	243	4276
1985-86	195	194	244	316	546	533	329	337	365	422	338	169	3988
1986-87	227	236	227	290	341	417	487	372	377	316	260	213	3763
1987-88	163	196	226	221	302	470	446	423	453	358	265	215	3738
1988-89	187	218	274	297	345	529	582	535	403	309	278	214	4171
1989-90	164	171	243	306	365	482	513	541	453	332	198	124	4124
1990-91	154	141	95	325	451	680	528	336	516	418	387	330	4361
1991-92	229	164	225	274	401	511	450	316	329	250	248	193	3590
1992-93	135	176	220	211	393	556	570	433	354	337	214	203	3802
1993-94	229	152	268	253	446	472	442	455	426	357	286	233	4019
1994-95	241	102	155	321	576	559	383	368	445	412	341	263	4166
1995-96	142	196	133	308	325	394	472	360	435	352	344	278	3739
1996-97	231	205	268	323	407	431	501	476	475	401	212	209	4139
1997-98	163	106	82	300	341	529	397	409	450	432	337	242	3788
1998-99	192	186	223	305	393	624	541	490	556	485	417	296	4708
1999-00	254	146	295	382	303	580	511	382	506	378	298	242	4277
2000-01	191	179	181	326	484	506	555	500	453	467	337	264	4443
2001-	228	168	257	322	389	481							

COOLING DEGREE DAYS (base 65 F) 2001 EUREKA, CA (EKA)

YEAR	JAN	FEB	MAR	APR	MAY	JUN	JUL	AUG	SEP	OCT	NOV	DEC	ANNUAL
1972	0	0	0	0	0	0	0	1	0	0	0	0	1
1973	0	0	0	0	0	1	0	0	0	0	0	0	1
1974	0	0	0	0	0	0	0	0	0	0	0	0	0
1975	0	0	0	0	0	0	0	0	0	0	0	0	0
1976	0	0	0	0	0	0	0	0	0	0	0	0	0
1977	0	0	0	0	0	0	0	0	0	1	0	0	1
1978	0	0	0	0	0	0	0	0	1	0	0	0	1
1979	0	0	0	0	0	0	0	0	15	0	0	2	17
1980	0	2	0	0	0	0	0	0	0	3	0	0	5
1981	4	0	0	0	0	0	0	0	0	0	0	0	4
1982	0	0	0	0	0	3	0	0	0	2	0	0	5
1983	0	0	0	0	0	0	0	2	7	0	0	0	9
1984	0	0	0	0	0	0	0	0	4	1	0	0	5
1985	0	0	0	0	0	0	1	0	0	2	0	0	3
1986	0	0	0	0	0	0	0	0	0	5	0	0	5
1987	0	0	0	0	3	0	0	0	0	5	0	0	8
1988	0	0	0	0	0	0	0	0	4	0	0	0	4
1989	0	0	0	1	0	0	0	0	0	0	0	0	1
1990	0	0	0	0	0	0	0	0	4	0	0	0	4
1991	0	0	0	0	0	0	0	10	0	4	0	0	14
1992	0	0	0	0	0	0	2	0	3	2	0	0	7
1993	0	3	0	0	3	0	0	0	0	8	0	0	14
1994	0	0	0	0	0	0	0	0	0	0	0	0	0
1995	0	0	0	0	0	0	0	0	0	0	1	0	1
1996	0	0	0	0	0	0	0	0	0	0	0	0	0
1997	0	0	1	0	4	0	0	3	5	0	0	0	13
1998	0	0	0	0	0	0	0	0	0	0	0	0	0
1999	0	0	0	0	0	0	0	0	0	0	1	0	1
2000	0	0	0	0	0	0	0	0	0	0	0	0	0
2001	0	0	0	0	0	0	0	3	0	0	0	0	3

SNOWFALL (inches) 2001 EUREKA, CA (EKA)

YEAR	JUL	AUG	SEP	OCT	NOV	DEC	JAN	FEB	MAR	APR	MAY	JUN	TOTAL
1972-73	0.0	0.0	0.0	0.0	0.0	1.9	0.0	0.0	T	0.0	0.0	0.0	1.9
1973-74	0.0	0.0	0.0	0.0	0.0	0.0	T	0.0	0.0	0.0	0.0	0.0	T
1974-75	0.0	0.0	0.0	0.0	0.0	T	0.0	0.0	0.0	T	0.0	0.0	T
1975-76	0.0	0.0	0.0	0.0	0.0	0.0	0.0	0.0	0.1	0.0	0.0	0.0	0.1
1976-77	0.0	0.0	0.0	0.0	0.0	0.0	0.0	0.0	T	0.0	0.0	0.0	T
1977-78	0.0	0.0	0.0	0.0	0.1	0.0	0.0	0.0	0.0	0.0	0.0	0.0	0.1
1978-79	0.0	0.0	0.0	0.0	0.0	0.0	0.0	0.0	0.0	0.0	0.0	0.0	0.0
1979-80	0.0	0.0	0.0	0.0	0.0	0.0	0.0	0.0	0.0	0.0	0.0	0.0	0.0
1980-81	0.0	0.0	0.0	0.0	0.0	0.0	T	0.0	T	T	0.0	0.0	T
1981-82	0.0	0.0	0.0	0.0	0.0	0.0	0.0	0.0	0.0	0.0	0.0	0.0	T
1982-83	0.0	0.0	0.0	0.0	0.0	0.0	T	0.0	T	0.0	0.0	0.0	T
1983-84	0.0	0.0	0.0	0.0	T	1.0	0.0	0.0	0.0	0.0	0.0	0.0	1.0
1984-85	0.0	0.0	0.0	0.0	0.0	0.0	0.0	0.0	0.0	0.0	0.0	0.0	0.0
1985-86	0.0	0.0	0.0	0.0	0.0	0.0	0.0	0.0	0.0	0.0	0.0	0.0	0.0
1986-87	0.0	0.0	0.0	0.0	0.0	0.0	0.0	0.0	0.0	0.0	0.0	0.0	0.0
1987-88	0.0	0.0	0.0	0.0	0.0	T	0.0	0.0	0.0	0.0	0.0	0.0	T
1988-89	0.0	0.0	0.0	0.0	0.0	T	0.0	3.5	0.0	0.0	0.0	0.0	3.5
1989-90	0.0	0.0	0.0	0.0	0.0	0.0	0.0	1.0	0.0	0.0	0.0	0.0	1.0
1990-91	0.0	0.0	0.0	0.0	0.0	T	0.0	0.0	T	0.0	0.0	0.0	T
1991-92	0.0	0.0	0.0	0.0	0.0	T	0.0	0.0	0.0	0.0	0.0	0.0	T
1992-93	0.0	0.0	0.0	0.0	0.0	T	0.0	T	T	0.0	0.0	0.0	T
1993-94	0.0	0.0	0.0	0.0	0.0	0.0	0.0	T	T	T	0.0	0.0	T
1994-95	0.0	0.0	0.0	0.0	T	0.0	0.0	0.0	0.0	T	0.0	0.0	T
1995-96	0.0	0.0	0.0	0.0	0.0	0.0	T	T	0.0	T	0.0	0.0	T
1996-97	0.0	0.0	0.0	0.0	0.0	T	T	0.0	0.0	0.0	0.0	0.0	T
1997-98	0.0	0.0	0.0	0.0	0.0	0.0	T	T	T	0.0	0.0	0.0	T
1998-99	0.0	0.0	0.0	0.0	0.0	T	T	0.6	T	0.0	0.0	0.0	0.6
1999-00	0.0	0.0	0.0	0.0	0.0	T	T	0.0	0.0	0.0	0.0	0.0	T
2000-01	0.0	0.0	0.0	0.0	0.0	0.0	0.0	0.0	T	0.0	0.0	0.0	T
2001-	0.0	0.0	0.0	0.0	0.0	T							
POR= 90 YRS	0.0	0.0	0.0	0.0	0.0	0.0	0.1	0.1	T	T	0.0	0.0	0.2

2001
FRESNO,
CALIFORNIA (FAT)

Fresno is located about midway and toward the eastern edge of the San Joaquin Valley, which is oriented northwest to southeast and has a length of about 225 miles and an average width of 50 miles. The San Joaquin Valley is generally flat. About 15 miles east of Fresno the terrain slopes upward with the foothills of the Sierra Nevada. The Sierra Nevada attain an elevation of more than 14,000 feet 50 miles east of Fresno. West of the city 45 miles lie the foothills of the Coastal Range.

The climate of Fresno is dry and mild in winter and hot in summer. Nearly nine-tenths of the annual precipitation falls in the six months from November to April.

Due to clear skies during the summer and the protection of the San Joaquin Valley from marine effects, the normal daily maximum temperature reaches the high 90s during the latter part of July. The daily maximum temperature during the warmest month has ranged from 76 to 115 degrees. Low relative humidities and some wind movement substantially lower the sensible temperature during periods of high readings. Humidity readings of 15 percent are common on summer afternoons, and readings as low as 8 percent have been recorded. In contrast to this, humidity readings average 90 percent during the morning hours of December and January.

Winds flow with the major axis of the San Joaquin Valley, generally from the northwest. This feature is especially beneficial since, during the warmest months, the northwest winds increase during the evenings. These refreshing breezes and the normally large temperature variation of about 35 degrees between the highest and lowest readings of the day, generally result in comfortable evening and night temperatures.

Winter temperatures are usually mild with infrequent cold spells dropping the readings below freezing. Heavy frost occurs almost every year, and the first frost usually occurs during the last week of November. The last frost in spring is usually in early March, however, one year in five will have the last frost after the first of April. The growing season is 291 days.

Although the heaviest rains recorded at Fresno for short periods have occurred in June, usually any rainfall during the summer is very light. Snow is a rare occurrence in Fresno.

Fresno enjoys a very high percentage of sunshine, receiving more than 80 percent of the possible amounts during all but the four months of November, December, January, and February. Reduction of sunshine during these months is caused by fog and short periods of stormy weather.

During foggy periods, at times lasting nearly two weeks, sunshine is reduced to a minimum. This fog frequently lifts to a few hundred feet above the surface of the valley and presents the appearance of a heavy, solid cloud layer.

Spring and autumn are very enjoyable seasons in Fresno, with clear skies, light rainfall and winds and mild temperatures.

NORMALS, MEANS, AND EXTREMES
FRESNO, CA (FAT)

LATITUDE:	LONGITUDE:	ELEVATION (FT):		TIME ZONE:	WBAN: 93193
36 46' 48" N	119 43' 10" W	GRND: 372	BARO: 375	PACIFIC (UTC + 8)	

	ELEMENT	POR	JAN	FEB	MAR	APR	MAY	JUN	JUL	AUG	SEP	OCT	NOV	DEC	YEAR
TEMPERATURE F	NORMAL DAILY MAXIMUM	30	54.1	61.7	66.6	75.1	84.2	92.7	98.6	96.7	90.1	79.7	64.7	53.7	76.5
	MEAN DAILY MAXIMUM	52	54.4	61.5	66.8	74.6	83.4	91.5	98.0	96.2	90.4	79.9	65.3	54.6	76.4
	HIGHEST DAILY MAXIMUM	52	78	80	90	100	107	110	112	112	111	102	89	76	112
	YEAR OF OCCURRENCE		1986	1991	1972	1981	1984	1964	1991	1996	1955	1980	1949	1958	AUG 1996
	MEAN OF EXTREME MAXS.	52	67.5	73.2	79.9	89.7	98.4	104.5	106.7	105.5	101.9	93.9	79.5	67.2	89.0
	NORMAL DAILY MINIMUM	30	37.4	40.5	43.4	47.3	53.7	60.4	65.1	63.8	58.8	50.7	42.5	37.1	50.1
	MEAN DAILY MINIMUM	52	37.6	40.5	43.5	47.7	54.0	60.1	65.1	63.7	59.2	50.9	42.2	37.0	50.1
	LOWEST DAILY MINIMUM	52	19	24	26	32	36	44	50	49	37	27	26	18	18
	YEAR OF OCCURRENCE		1963	1990	1966	1982	1975	1955	1955	1966	1950	1972	1975	1990	DEC 1990
	MEAN OF EXTREME MINS.	52	27.4	31.0	33.9	38.5	44.0	50.9	56.4	56.0	50.5	40.8	32.2	27.7	40.8
	NORMAL DRY BULB	30	45.7	51.2	55.1	61.2	69.0	76.6	81.9	80.3	74.5	65.2	53.6	45.4	63.3
	MEAN DRY BULB	52	46.1	51.2	55.1	61.2	68.7	75.7	81.5	79.9	74.8	65.5	53.8	45.6	63.3
	MEAN WET BULB	17	43.9	47.3	50.9	53.4	57.1	61.8	65.5	60.8	61.9	56.1	48.6	42.5	54.1
	MEAN DEW POINT	17	41.0	43.0	45.5	44.8	43.8	47.3	54.5	54.5	50.5	52.3	48.2	43.8	46.2
	NORMAL NO. DAYS WITH:														
	MAXIMUM 90	30	0.0	0.0	*	2.1	9.4	19.5	28.5	26.3	16.9	4.2	0.0	0.0	106.9
	MAXIMUM 32	30	0.0	0.0	0.0	0.0	0.0	0.0	0.0	0.0	0.0	0.0	0.0	0.1	0.1
	MINIMUM 32	30	8.5	3.1	0.7	0.1	0.0	0.0	0.0	0.0	0.0	0.1	2.0	8.9	23.4
	MINIMUM 0	30	0.0	0.0	0.0	0.0	0.0	0.0	0.0	0.0	0.0	0.0	0.0	0.0	0.0
H/C	NORMAL HEATING DEG. DAYS	30	598	386	314	182	34	0	0	0	8	84	342	608	2556
	NORMAL COOLING DEG. DAYS	30	0	0	7	68	158	352	524	474	293	91	0	0	1967
RH	NORMAL (PERCENT)	30	83	77	69	57	47	42	39	45	50	58	74	84	60
	HOUR 04 LST	30	91	90	86	80	72	64	60	67	72	78	87	92	78
	HOUR 10 LST	30	84	77	65	51	43	39	37	42	45	52	71	85	58
	HOUR 16 LST	30	67	56	46	34	26	23	21	25	28	34	54	69	40
	HOUR 22 LST	30	88	83	74	62	50	43	40	46	52	64	81	89	64
S	PERCENT POSSIBLE SUNSHINE	46	47	65	77	85	90	95	97	96	94	88	66	46	79
W/O	MEAN NO. DAYS WITH:														
	HEAVY FOG(VISBY 1/4 MI)	52	11.7	5.8	1.7	0.3	0.1	0.0	0.0	0.0	0.1	0.9	5.4	11.5	37.5
	THUNDERSTORMS	52	0.3	0.4	0.9	0.5	0.6	0.4	0.3	0.2	0.7	0.5	0.2	0.3	5.3
CLOUDINESS	MEAN:														
	SUNRISE-SUNSET (OKTAS)														
	MIDNIGHT-MIDNIGHT (OKTAS)														
	MEAN NO. DAYS WITH:														
	CLEAR	1		3.0	5.0		6.0	4.0							
	PARTLY CLOUDY	1		1.0			1.0								
	CLOUDY	1	2.0	3.0	1.0		1.0								
PR	MEAN STATION PRESSURE(IN)	28	29.80	29.71	29.70	29.70	29.60	29.50	29.50	29.50	29.60	29.60	29.79	29.80	29.65
	MEAN SEA-LEVEL PRES. (IN)	17	30.15	30.09	30.04	30.00	29.93	29.99	29.88	29.88	29.89	29.98	30.10	30.16	30.01
WINDS	MEAN SPEED (MPH)	40	5.2	5.7	6.7	7.4	8.2	8.3	7.4	6.8	6.1	5.2	4.7	4.8	6.4
	PREVAIL.DIR(TENS OF DEGS)	26	13	13	32	31	31	31	30	30	30	31	30	13	31
	MAXIMUM 2-MINUTE:														
	SPEED (MPH)	6	29	36	29	36	32	28	23	25	26	26	30	26	36
	DIR. (TENS OF DEGS)		09	13	30	29	32	30	08	30	30	31	28	30	29
	YEAR OF OCCURRENCE		2001	1998	1997	1999	1998	1999	1999	1999	1996	2000	2001	2001	APR 1999
	MAXIMUM 5-SECOND:														
	SPEED (MPH)	6	41	43	37	40	37	32	30	30	31	35	37	31	43
	DIR. (TENS OF DEGS)		31	29	29	29	32	30	30	31	30	31	29	31	29
	YEAR OF OCCURRENCE		1998	1999	1997	1999	1998	1998	2001	1999	1996	1998	2001	2001	FEB 1999
PRECIPITATION	NORMAL (IN)	30	1.96	1.80	1.89	0.97	0.30	0.08	0.01	0.03	0.24	0.53	1.37	1.42	10.60
	MAXIMUM MONTHLY (IN)	52	8.56	6.12	7.24	4.41	1.65	1.93	0.25	0.25	1.19	2.45	3.50	6.73	8.56
	YEAR OF OCCURRENCE		1969	2000	1991	1967	1990	1998	1992	1964	1976	2000	1972	1955	JAN 1969
	MINIMUM MONTHLY (IN)	52	0.04	T	0.00	0.02	0.00	0.00	0.00	0.00	0.00	0.00	0.00	0.00	0.00
	YEAR OF OCCURRENCE		1976	1964	1972	1997	1982	1983	1983	1981	1981	1978	1959	1989	DEC 1989
	MAXIMUM IN 24 HOURS (IN)	52	2.59	1.99	2.43	1.39	1.42	1.80	0.22	0.25	0.97	1.76	1.35	1.76	2.59
	YEAR OF OCCURRENCE		1969	1969	1995	1983	1990	1998	1992	1964	1978	1992	1953	1955	JAN 1969
	NORMAL NO. DAYS WITH:														
	PRECIPITATION 0.01	30	6.8	7.0	7.3	4.2	1.6	0.7	0.2	0.4	1.1	2.3	6.2	6.7	44.5
	PRECIPITATION 1.00	30	0.2	0.3	0.1	0.1	*	0.0	0.0	0.0	0.0	0.0	*	0.0	0.1	0.8
SNOWFALL	NORMAL (IN)	30	0.1	T	0.0	0.0	0.0	0.0	0.0	0.0	0.0	0.0	0.0	0.*	0.1
	MAXIMUM MONTHLY (IN)	47	2.2	T	T	0.0	0.0	T	0.0	0.0	0.0	T	0.0	1.2	2.2
	YEAR OF OCCURRENCE		1962	1994	1991			1995				1974		1968	JAN 1962
	MAXIMUM IN 24 HOURS (IN)	47	1.5	T	T	0.0	0.0	T	0.0	0.0	0.0	T	0.0	1.2	1.5
	YEAR OF OCCURRENCE		1962	1994	1991			1995				1974		1968	JAN 1962
	MAXIMUM SNOW DEPTH (IN)	46	0	0	0	0	0	0	0	0	0	0	0	1	1
	YEAR OF OCCURRENCE													1968	DEC 1968
	NORMAL NO. DAYS WITH:														
	SNOWFALL 1.0	30	0.*	0.0	0.0	0.0	0.0	0.0	0.0	0.0	0.0	0.0	0.0	0.*	0.0

PRECIPITATION (inches) 2001 FRESNO, CA (FAT)

YEAR	JAN	FEB	MAR	APR	MAY	JUN	JUL	AUG	SEP	OCT	NOV	DEC	ANNUAL
1972	0.37	0.67	0.00	0.27	0.15	0.60	T	0.00	0.29	0.22	3.50	1.40	7.47
1973	1.91	3.69	2.84	0.09	T	T	0.00	T	0.00	1.02	1.39	1.74	12.68
1974	2.82	0.25	2.56	0.64	0.00	0.00	T	0.00	1.44	0.34	1.26		9.31
1975	0.69	0.97	2.44	0.55	T	0.00	T	0.05	0.22	1.07	0.20	0.14	6.33
1976	0.04	4.72	0.44	0.93	T	0.00	0.37	0.21	1.19	1.55	0.87	0.71	11.04
1977	0.68	0.09	1.04	0.04	1.16	0.06	T	T		0.01	0.46	3.02	6.56
1978	3.16	4.41	4.25	2.85	0.00	0.00	T	T	1.05	0.00	1.34	0.62	17.68
1979	2.71	2.53	2.27	0.07	0.06	T	0.08	0.00	T	0.48	1.01	0.74	9.95
1980	3.83	3.30	2.05	0.25	0.18	T	0.01	0.00	T	0.03	0.14	0.49	10.28
1981	2.67	1.29	2.59	1.01	T	0.00	0.00	0.00	0.00	0.58	1.22	0.65	10.01
1982	2.11	0.58	4.76	0.89	0.00	0.31	0.00	0.00	1.10	1.58	3.16	1.59	16.08
1983	5.14	3.70	4.53	2.76	0.01	0.00	0.00	0.09	1.03	0.09	2.51	1.75	21.61
1984	0.15	1.05	0.48	0.25	0.02	0.20	T	0.00	0.00	0.70	1.94	1.98	6.77
1985	0.43	0.71	1.73	0.12	0.00	0.33	0.04	0.02	0.43	0.85	3.02	0.72	8.40
1986	2.12	3.66	3.42	0.36	0.16	0.00	T	0.00	0.38	0.00	0.01	2.30	12.41
1987	1.93	1.36	2.39	0.07	0.87	0.01	0.00	0.00	T	0.85	0.52	1.19	9.19
1988	1.52	0.83	0.27	2.41	0.45	0.03	0.00	0.00	0.00	0.00	1.42	2.46	9.39
1989	0.48	1.18	2.25	0.05	0.89	0.00	0.00	0.03	1.11	0.42	0.50	0.00	6.91
1990	2.82	1.33	0.67	0.92	1.65	0.00	T	0.00	0.15	0.05	0.46	0.68	8.73
1991	0.13	1.01	7.24	0.18	0.03	T	0.00	T	T	0.80	0.04	1.22	10.49
1992	1.94	4.73	2.14	0.18	T	T	0.22	T	T	2.19	T	2.68	14.08
1993	5.18	2.44	1.76	0.20	0.25	1.61	0.00	0.00	0.00	0.12	1.16	1.03	13.75
1994	1.15	1.92	0.52	1.36	1.30	0.00	T	0.00	0.20	0.77	1.57	1.33	10.12
1995	5.42	0.93	5.88	1.08	1.19	0.66	0.01	0.00	0.00	0.00	T	2.12	17.29
1996	2.07	3.57	1.52	1.17	0.38	0.08	T	0.00	0.00	1.97	1.94	4.27	16.97
1997	3.53	0.17	0.10	T	0.01	T	0.00	0.00	0.15	0.07	2.66	0.99	7.68
1998	3.40	4.89	3.44	1.26	1.37	1.93	0.00	0.00	0.15	0.16	0.43	0.62	17.65
1999	2.82	1.18	0.49	0.93	0.03	0.20	0.00	0.01	0.00	T	0.48	0.03	6.17
2000	3.15	6.12	1.35	1.16	0.05	0.56	0.00	T	0.32	T	2.45	0.01	15.24
2001	2.66	2.22	0.96	1.87	0.00	0.00	0.08	0.00	T	0.29	1.99	1.95	12.02
POR= 124 YRS	2.05	1.94	1.88	0.97	0.34	0.17	0.02	0.02	0.16	0.57	1.09	1.64	10.85

AVERAGE TEMPERATURE (°F) 2001 FRESNO, CA (FAT)

YEAR	JAN	FEB	MAR	APR	MAY	JUN	JUL	AUG	SEP	OCT	NOV	DEC	ANNUAL
1972	40.6	52.5	60.7	61.1	69.9	77.5	81.5	79.7	71.8	62.6	50.2	40.9	62.4
1973	45.1	51.9	50.4	61.2	72.9	78.6	80.4	78.5	72.0	63.3	52.9	47.2	62.9
1974	47.9	49.1	56.3	60.0	69.5	77.7	81.3	79.3	77.5	66.0	53.1	44.5	63.5
1975	43.4	49.9	51.5	53.9	68.4	74.7	78.1	75.9	75.8	61.4	49.5	43.9	60.5
1976	44.3	49.6	52.4	57.2	69.7	72.9	79.4	72.7	72.2	65.1	53.4	46.5	61.3
1977	44.3	53.5	52.4	65.5	63.6	79.8	81.5	80.6	74.0	66.8	54.6	51.3	64.0
1978	51.4	52.6	60.3	58.9	69.9	76.3	82.4	81.4	73.0	70.0	52.1	42.8	64.3
1979	47.0	51.4	57.4	62.7	71.1	77.9	82.2	79.9	79.5	67.8	54.0	46.9	64.8
1980	49.4	53.8	53.7	61.8	67.2	73.7	84.0	80.7	75.6	68.4	54.2	46.8	64.1
1981	47.9	52.0	54.5	63.2	70.9	72.8	84.9	82.9	76.5	61.4	55.5	47.5	64.2
1982	41.7	50.5	51.4	58.0	69.3	72.9	81.0	80.4	72.3	65.0	51.1	45.4	61.6
1983	45.2	53.1	55.9	57.9	69.7	76.3	79.0	82.1	78.8	68.5	54.6	51.1	64.3
1984	47.8	50.7	58.4	60.8	74.8	77.5	87.0	83.5	81.0	62.4	53.6	46.5	65.3
1985	43.3	51.3	53.1	67.2	69.4	81.8	86.0	80.5	72.3	65.0	52.5	43.8	63.9
1986	53.6	55.7	60.3	62.7	71.2	79.4	81.9	84.2	71.3	66.9	56.7	47.5	66.0
1987	45.3	52.8	55.6	66.7	71.8	78.4	77.0	80.2	75.5	70.1	52.3	44.2	64.2
1988	46.0	52.2	56.8	61.6	67.0	75.6	85.5	81.2	76.4	64.3	54.3	44.5	64.2
1989	42.9	48.8	57.9	67.3	69.6	77.0	82.5	79.3	74.3	65.3	54.3	43.8	63.6
1990	45.5	48.0	57.3	65.7	68.1	76.8	84.0	80.6	75.8	67.7	52.9	41.5	63.7
1991	47.0	55.8	51.5	59.5	66.1	74.7	83.8	78.6	79.9	70.5	55.8	47.0	64.2
1992	42.7	55.5	58.8	66.8	76.0	77.0	81.3	83.2	77.0	68.6	54.3	45.3	65.5
1993	47.1	51.9	60.3	61.7	69.9	75.7	80.2	79.7	75.7	67.8	53.9	45.6	64.1
1994	46.9	49.9	59.3	63.2	68.5	77.7	82.3	83.3	74.8	66.8	48.1	45.3	63.7
1995	51.9	54.1	56.2	60.7	66.2	73.3	80.7	82.6	76.3	66.8	58.7	50.5	64.8
1996	48.3	54.2	57.2	63.6	69.9	77.8	85.4	83.4	74.8	64.1	53.9	49.1	65.1
1997	48.7	50.3	60.0	63.5	75.3	75.8	81.3	80.6	77.3	63.8	56.9	44.7	64.9
1998	49.0	50.0	55.5	59.0	62.0	71.5	82.1	84.1	75.8	63.1	53.1	42.8	62.3
1999	44.7	49.9	53.5	58.5	68.0	75.9	80.6	78.4	75.8	63.1	53.1	42.8	62.3
2000	50.2	53.8	56.5	62.4	71.0	79.8	78.4	81.2	77.3	68.7	56.9	47.0	63.3
2001	46.2	48.7	58.8	58.6	77.3	79.7	81.6	81.9	77.0	68.5	56.4	47.4	65.2
POR= 114 YRS	45.9	51.2	55.2	61.1	68.2	75.5	81.8	80.0	74.2	65.0	54.2	46.2	63.2

REFERENCE NOTES:

PAGE 1:
THE TEMPERATURE GRAPH SHOWS NORMAL MAXIMUM AND NORMAL MINIMUM DAILY TEMPERATURES (SOLID CURVES) AND THE ACTUAL DAILY HIGH AND LOW TEMPERATURES (VERTICAL BARS).

PAGE 2 AND 3:
H/C INDICATES HEATING AND COOLING DEGREE DAYS.
RH INDICATES RELATIVE HUMIDITY.
W/O INDICATES WEATHER AND OBSTRUCTIONS
S INDICATES SUNSHINE.
PR INDICATES PRESSURE.
CLOUDINESS ON PAGE 3 IS THE SUM OF THE CEILOMETER AND SATELLITE DATA NOT TO EXCEED EIGHT EIGHTHS (OKTAS).

GENERAL:
T INDICATES TRACE PRECIPITATION, AN AMOUNT GREATER THAN ZERO BUT LESS THAN THE LOWEST REPORTABLE VALUE.
+ INDICATES THE VALUE ALSO OCCURS ON EARLIER DATES.
BLANK ENTRIES DENOTE MISSING OR UNREPORTED DATA.
NORMALS ARE 30-YEAR AVERAGES (1961 - 1990).
ASOS INDICATES AUTOMATED SURFACE OBSERVING SYSTEM.
PM INDICATES THE LAST DAY OF THE PREVIOUS MONTH.
POR (PERIOD OF RECORD) BEGINS WITH THE JANUARY DATA MONTH AND IS THE NUMBER OF YEARS USED TO COMPUTE THE MEAN. INDIVIDUAL MONTHS WITHIN THE POR MAY BE MISSING.
WHEN THE POR FOR A NORMAL IS LESS THAN 30 YEARS, THE NORMAL IS PROVISIONAL AND IS BASED ON THE NUMBER OF YEARS INDICATED.
0.* OR * INDICATES THE VALUE OR MEAN-DAYS-WITH IS BETWEEN 0.00 AND 0.05.
CLOUDINESS FOR ASOS STATIONS DIFFERS FROM THE NON-ASOS OBSERVATION TAKEN BY A HUMAN OBSERVER. ASOS STATION CLOUDINESS IS BASED ON TIME-AVERAGED CEILOMETER DATA FOR CLOUDS AT OR BELOW 12,000 FEET AND ON SATELLITE DATA FOR CLOUDS ABOVE 12,000 FEET.
THE NUMBER OF DAYS WITH CLEAR, PARTLY CLOUDY, AND CLOUDY CONDITIONS FOR ASOS STATIONS IS THE SUM OF THE CEILOMETER AND SATELLITE DATA FOR THE SUNRISE TO SUNSET PERIOD.

GENERAL CONTINUED:
CLEAR INDICATES 0 - 2 OKTAS, PARTLY CLOUDY INDICATES 3 - 6 OKTAS, AND CLOUDY INDICATES 7 OR 8 OKTAS. WHEN AT LEAST ONE OF THE ELEMENTS (CEILOMETER OR SATELLITE) IS MISSING, THE DAILY CLOUDINESS IS NOT COMPUTED.
WIND DIRECTION IS RECORDED IN TENS OF DEGREES (2 DIGITS) CLOCKWISE FROM TRUE NORTH. "00" INDICATES CALM. "36" INDICATES TRUE NORTH.
RESULTANT WIND IS THE VECTOR AVERAGE OF THE SPEED AND DIRECTION.
AVERAGE TEMPERATURE IS THE SUM OF THE MEAN DAILY MAXIMUM AND MINIMUM TEMPERATURE DIVIDED BY 2.
SNOWFALL DATA COMPRISE ALL FORMS OF FROZEN PRECIPITATION, INCLUDING HAIL.
A HEATING (COOLING) DEGREE DAY IS THE DIFFERENCE BETWEEN THE AVERAGE DAILY TEMPERATURE AND 65 F.
DRY BULB IS THE TEMPERATURE OF THE AMBIENT AIR.
DEW POINT IS THE TEMPERATURE TO WHICH THE AIR MUST BE COOLED TO ACHIEVE 100 PERCENT RELATIVE HUMIDITY.
WET BULB IS THE TEMPERATURE THE AIR WOULD HAVE IF THE MOISTURE CONTENT WAS INCREASED TO 100 PERCENT RELATIVE HUMIDITY.

ON JULY 1, 1996, THE NATIONAL WEATHER SERVICE BEGAN USING THE "METAR" OBSERVATION CODE THAT WAS ALREADY EMPLOYED BY MOST OTHER NATIONS OF THE WORLD. THE MOST NOTICEABLE DIFFERENCE IN THIS ANNUAL PUBLICATION WILL BE THE CHANGE IN UNITS FROM TENTHS TO EIGHTS (OKTAS) FOR REPORTING THE AMOUNT OF SKY COVER.

HEATING DEGREE DAYS (base 65 F) 2001 FRESNO, CA (FAT)

YEAR	JUL	AUG	SEP	OCT	NOV	DEC	JAN	FEB	MAR	APR	MAY	JUN	TOTAL
1972-73	0	0	2	108	437	740	610	358	444	140	12	2	2853
1973-74	0	0	0	94	360	544	522	438	260	160	33	0	2411
1974-75	0	0	0	59	350	628	661	419	409	325	53	3	2907
1975-76	0	0	0	154	455	648	636	440	385	242	10	9	2979
1976-77	0	1	5	63	342	566	636	313	386	42	98	0	2452
1977-78	0	0	0	46	302	417	415	343	143	182	19	0	1867
1978-79	0	0	6	30	382	682	549	372	234	96	34	0	2385
1979-80	0	0	0	56	323	555	473	318	343	129	46	0	2243
1980-81	0	0	0	69	318	553	521	359	316	114	9	0	2259
1981-82	0	0	0	118	278	530	711	398	412	217	21	4	2689
1982-83	0	0	13	62	411	602	607	327	276	206	55	0	2559
1983-84	0	0	1	3	304	421	530	408	198	149	6	0	2020
1984-85	0	0	0	128	335	566	664	378	361	39	8	3	2482
1985-86	0	0	0	63	369	651	345	258	156	98	30	0	1970
1986-87	0	0	13	22	242	537	602	337	282	56	26	0	2117
1987-88	0	0	0	7	374	636	583	366	251	124	69	12	2422
1988-89	0	0	0	20	316	629	679	450	213	52	14	0	2373
1989-90	0	0	7	73	310	649	598	470	236	35	19	1	2398
1990-91	0	0	0	17	356	722	549	253	412	163	65	0	2537
1991-92	0	0	0	81	276	551	683	267	183	25	0	1	2067
1992-93	0	0	0	18	316	602	549	359	145	113	9	12	2123
1993-94	0	0	0	12	326	595	553	414	168	97	37	0	2202
1994-95	0	0	0	58	500	602	398	298	269	146	60	16	2347
1995-96	0	0	0	30	184	444	513	304	238	99	8	0	1820
1996-97	0	0	0	148	329	486	500	405	169	97	2	0	2136
1997-98	0	0	0	92	246	621	490	412	293	226	104	7	2491
1998-99	0	0	7	79	351	682	619	418	348	227	35	12	2778
1999-00	0	0	0	14	235	550	452	317	259	72	27	3	1929
2000-01	0	0	0	103	466	526	577	451	208	222	0	0	2553
2001-	0	0	0	23	251	538							

COOLING DEGREE DAYS (base 65 F) 2001 FRESNO, CA (FAT)

YEAR	JAN	FEB	MAR	APR	MAY	JUN	JUL	AUG	SEP	OCT	NOV	DEC	ANNUAL
1972	0	0	17	18	195	383	518	464	213	42	0	0	1850
1973	0	0	0	32	264	419	484	423	218	47	4	0	1891
1974	0	0	0	20	179	384	512	448	381	96	0	0	2020
1975	0	0	0	0	164	303	413	344	329	49	0	0	1602
1976	0	0	2	16	162	254	456	246	228	73	0	0	1437
1977	0	0	0	62	60	451	518	494	275	108	0	0	1968
1978	0	0	3	6	179	342	546	516	250	187	0	0	2029
1979	0	0	2	37	229	396	541	471	442	149	0	0	2267
1980	0	0	0	39	120	265	541	493	326	181	0	0	2018
1981	0	0	0	67	200	545	622	562	352	14	0	0	2362
1982	0	0	0	12	162	251	501	483	240	70	0	0	1719
1983	0	0	0	0	207	343	440	537	422	119	0	0	2068
1984	0	0	1	30	318	382	688	581	487	55	0	0	2542
1985	0	0	0	111	153	516	657	487	227	69	2	0	2222
1986	0	1	18	34	231	440	530	603	206	87	0	0	2150
1987	0	0	0	114	243	409	379	480	323	172	0	0	2120
1988	0	0	3	28	139	338	642	511	349	143	3	0	2156
1989	0	0	4	129	166	366	546	449	291	90	0	0	2041
1990	0	0	2	61	122	360	595	490	333	108	0	0	2071
1991	0	0	0	6	107	298	588	428	454	259	5	0	2145
1992	0	0	0	88	350	366	511	572	365	135	0	0	2387
1993	0	0	3	20	168	342	476	462	331	105	0	0	1907
1994	0	0	1	52	151	389	576	547	318	59	0	0	2093
1995	0	0	0	25	104	273	494	551	347	91	0	0	1885
1996	0	0	4	66	162	389	640	579	300	125	0	0	2265
1997	0	0	18	61	330	334	514	492	373	61	11	0	2194
1998	0	0	6	50	18	210	536	600	338	25	0	0	1783
1999	0	0	0	39	135	348	487	423	373	135	0	0	1940
2000	0	0	0	54	217	454	434	509	291	81	0	0	2040
2001	0	0	20	37	389	447	521	533	365	137	0	0	2449

SNOWFALL (inches) 2001 FRESNO, CA (FAT)

YEAR	JUL	AUG	SEP	OCT	NOV	DEC	JAN	FEB	MAR	APR	MAY	JUN	TOTAL
1972-73	0.0	0.0	0.0	0.0	0.0	T	0.0	0.0	T	0.0	0.0	0.0	T
1973-74	0.0	0.0	0.0	0.0	0.0	0.0	0.0	0.0	0.0	0.0	0.0	0.0	0.0
1974-75	0.0	0.0	0.0	T	0.0	0.0	0.0	0.0	0.0	0.0	0.0	0.0	T
1975-76	0.0	0.0	0.0	0.0	0.0	0.0	0.0	0.0	0.0	0.0	0.0	0.0	T
1976-77	0.0	0.0	0.0	0.0	0.0	0.0	0.0	0.0	0.0	0.0	0.0	0.0	0.0
1977-78	0.0	0.0	0.0	0.0	0.0	0.0	0.0	0.0	0.0	0.0	0.0	0.0	0.0
1978-79	0.0	0.0	0.0	0.0	0.0	0.0	0.0	T	T	0.0	0.0	0.0	T
1979-80	0.0	0.0	0.0	0.0	0.0	0.0	0.0	0.0	0.0	0.0	0.0	0.0	0.0
1980-81	0.0	0.0	0.0	0.0	0.0	0.0	0.0	0.0	0.0	0.0	0.0	0.0	0.0
1981-82	0.0	0.0	0.0	0.0	0.0	0.0	0.0	0.0	0.0	0.0	0.0	0.0	0.0
1982-83	0.0	0.0	0.0	0.0	0.0	0.0	0.0	0.0	0.0	0.0	0.0	0.0	0.0
1983-84	0.0	0.0	0.0	0.0	0.0	0.0	0.0	0.0	0.0	0.0	0.0	0.0	0.0
1984-85	0.0	0.0	0.0	0.0	0.0	0.0	0.0	0.0	0.0	0.0	0.0	0.0	0.0
1985-86	0.0	0.0	0.0	0.0	0.0	0.0	0.0	0.0	0.0	0.0	0.0	0.0	0.0
1986-87	0.0	0.0	0.0	0.0	0.0	0.0	0.0	0.0	0.0	0.0	0.0	0.0	0.0
1987-88	0.0	0.0	0.0	0.0	0.0	0.0	0.0	0.0	0.0	0.0	0.0	0.0	0.0
1988-89	0.0	0.0	0.0	0.0	0.0	0.0	0.0	T	0.0	0.0	0.0	0.0	T
1989-90	0.0	0.0	0.0	0.0	0.0	0.0	0.0	T	0.0	0.0	0.0	0.0	T
1990-91	0.0	0.0	0.0	0.0	0.0	T	0.0	0.0	T	0.0	0.0	0.0	T
1991-92	0.0	0.0	0.0	0.0	0.0	0.0	0.0	T	0.0	0.0	0.0	0.0	T
1992-93	0.0	0.0	0.0	0.0	0.0	0.0	0.0	0.0	0.0	0.0	0.0	0.0	0.0
1993-94	0.0	0.0	0.0	0.0	0.0	0.0	0.0	T	0.0	0.0	0.0	0.0	T
1994-95	0.0	0.0	0.0	0.0	0.0	0.0	T	0.0	0.0	0.0	0.0	T	T
1995-96	0.0	0.0	0.0	0.0	0.0	0.0	0.0						
1996-97													
1997-98													
1998-99						0.5							
1999-00													
2000-01													
2001-													
POR= 46 YRS	0.0	0.0	0.0	T	0.0	T	0.0	T	T	0.0	0.0	T	T

2001
LOS ANGELES, CALIFORNIA
Downtown L.A./USC Campus (CQT)

The climate of Los Angeles is normally pleasant and mild through the year. The Pacific Ocean is the primary moderating influence. The coastal mountain ranges lying along the north and east sides of the Los Angeles coastal basin act as a buffer against extremes of summer heat and winter cold occurring in desert and plateau regions in the interior. A variable balance between mild sea breezes, and either hot or cold winds from the interior, results in some variety in weather conditions, but temperature and humidity are usually well within the limits of human comfort. An important, and somewhat unusual, aspect of the climate of the Los Angeles metropolitan area is the pronounced difference in temperature, humidity, cloudiness, fog, rain, and sunshine over fairly short distances.

These differences are closely related to the distance from, and elevation above, the Pacific Ocean. Both high and low temperatures become more extreme and the average relative humidity becomes lower as one goes inland and up foothill slopes. Relative humidity is frequently high near the coast, but may be quite low along the foothills. During periods of high temperatures, the relative humidity is usually below normal so that discomfort is rare, except for infrequent periods when high temperatures and high humidities occur together.

Like other Pacific Coast areas, most rainfall comes during the winter with nearly 85 percent of the annual total occurring from November through March, while summers are practically rainless. As in many semi-arid regions, there is a marked variability in monthly and seasonal totals. Precipitation generally increases with distance from the ocean, from a yearly total of around 12 inches in coastal sections to the south of the city to over 20 inches in foothill areas. Destructive flash floods occasionally develop in and below some mountain canyons. Snow is often visible on nearby mountains in the winter, but is extremely rare in the coastal basin. Thunderstorms are infrequent.

Prevailing winds are from the west during the spring, summer, and early autumn, with northeasterly wind predominating the remainder of the year. At times, the lack of air movement, combined with a frequent and persistent temperature inversion, is associated with concentrations of air pollution in the Los Angeles coastal basin and some adjacent areas. In fall, winter, and early spring months, occasional foehn-like descending Santa Ana winds come from the northeast over ridges and through passes in the coastal mountains. These Santa Ana winds may pick up considerable amounts of dust and reach speeds of 35 to 50 mph in north and east sections of the city, with higher speeds in outlying areas to the north and east, but rarely reach coastal portions of the city.

Sunshine, fog, and clouds depend a great deal on topography and distance from the ocean. Low clouds are common at night and in the morning along the coast during spring and summer, but form later and clear earlier near the foothills so that annual cloudiness and fog frequencies are greatest near the ocean, and sunshine totals are highest on the inland side of the city. The sun shines about 75 percent of daytime hours at the Civic Center. Light fog may accompany the usual night and morning low clouds, but dense fog is more likely to occur during the night and early morning hours of the winter months.

NORMALS, MEANS, AND EXTREMES
LOS ANGELES, CA (CQT)

LATITUDE:	LONGITUDE:	ELEVATION (FT):		TIME ZONE:	WBAN: 93134
34 01' 40" N	118 17' 45" W	GRND: 182	BARO: 185	PACIFIC (UTC + 8)	

	ELEMENT	POR	JAN	FEB	MAR	APR	MAY	JUN	JUL	AUG	SEP	OCT	NOV	DEC	YEAR
TEMPERATURE F	NORMAL DAILY MAXIMUM	30	67.7	69.4	69.5	72.3	73.9	78.3	84.0	84.5	82.7	79.0	72.4	67.8	75.1
	MEAN DAILY MAXIMUM	81	66.5	67.6	68.7	71.2	73.4	77.2	82.7	83.5	82.1	77.8	73.2	67.7	74.3
	HIGHEST DAILY MAXIMUM	61	95	95	98	106	102	112	107	105	110	108	100	91	112
	YEAR OF OCCURRENCE		1971	1995	1988	1989	1967	1990	1985	1983	1988	1987	1966	1979	JUN 1990
	MEAN OF EXTREME MAXS.	81	80.9	82.4	84.0	87.9	89.0	89.9	93.0	94.2	96.7	94.4	87.5	81.7	88.5
	NORMAL DAILY MINIMUM	30	48.9	50.6	51.8	54.2	57.7	61.1	64.5	65.7	64.6	60.3	53.5	48.8	56.8
	MEAN DAILY MINIMUM	81	48.4	49.9	51.1	53.7	56.8	59.8	63.3	64.3	63.0	59.0	53.5	49.5	56.0
	LOWEST DAILY MINIMUM	61	28	34	35	39	46	49	54	53	51	41	38	30	28
	YEAR OF OCCURRENCE		1949	1989	1976	1975	1964	1999	1952	1943	1948	1971	1978	1978	JAN 1949
	MEAN OF EXTREME MINS.	81	39.9	42.3	43.8	46.8	51.2	54.7	58.8	59.8	57.2	52.0	45.1	40.9	49.4
	NORMAL DRY BULB	30	58.3	60.1	60.7	63.3	65.8	69.7	74.3	75.1	73.7	69.7	63.0	58.3	66.0
	MEAN DRY BULB	81	57.4	58.7	60.0	62.5	65.1	68.4	73.0	73.9	72.6	68.4	63.3	58.6	65.2
	MEAN WET BULB	2	49.4	50.2	53.9	55.2	60.2	62.9	63.9	65.4		61.9	55.0	48.1	
	MEAN DEW POINT	2	42.5	44.9	49.4	50.2	55.7	59.1	60.2	62.0		59.1	49.8	38.3	
	NORMAL NO. DAYS WITH:														
	MAXIMUM 90	30	0.1	0.2	0.3	1.1	1.3	2.1	4.1	5.3	5.9	3.4	0.7	*	24.5
	MAXIMUM 32	30	0.0	0.0	0.0	0.0	0.0	0.0	0.0	0.0	0.0	0.0	0.0	0.0	0.0
	MINIMUM 32	30	*	0.0	0.0	0.0	0.0	0.0	0.0	0.0	0.0	0.0	0.0	0.1	0.1
	MINIMUM 0	30	0.0	0.0	0.0	0.0	0.0	0.0	0.0	0.0	0.0	0.0	0.0	0.0	0.0
H/C	NORMAL HEATING DEG. DAYS	30	222	170	169	128	72	35	0	0	10	17	105	226	1154
	NORMAL COOLING DEG. DAYS	30	14	32	36	77	97	176	293	316	271	162	45	18	1537
RH	NORMAL (PERCENT)														
	HOUR 04 LST														
	HOUR 10 LST														
	HOUR 16 LST														
	HOUR 22 LST														
S	PERCENT POSSIBLE SUNSHINE	32	69	72	73	70	66	65	82	83	79	73	74	71	73
W/O	MEAN NO. DAYS WITH:														
	HEAVY FOG(VISBY 1/4 MI)	26	1.6	1.7	1.1	1.2	0.6	0.6	0.5	0.8	1.3	2.4	2.4	2.1	16.3
	THUNDERSTORMS	26	0.5	1.1	0.9	0.8	0.2	0.1	0.2	0.4	0.4	0.3	0.6	0.7	6.2
CLOUDINESS	MEAN:														
	SUNRISE-SUNSET (OKTAS)	34	3.5	3.8	3.8	3.8	3.8	3.4	2.2	2.1	2.4	3.0	3.0	3.4	3.2
	MIDNIGHT-MIDNIGHT (OKTAS)														
	MEAN NO. DAYS WITH:														
	CLEAR	34	14.3	12.4	12.9	12.0	11.4	13.6	20.9	22.4	18.4	16.1	16.5	15.0	185.9
	PARTLY CLOUDY	34	8.1	6.9	9.3	9.8	11.8	10.5	8.9	7.4	8.4	9.3	7.4	8.0	105.8
	CLOUDY	34	8.5	9.0	8.7	8.2	7.8	5.9	1.1	1.2	3.3	5.6	6.1	8.0	73.4
PR	MEAN STATION PRESSURE(IN)	2	29.92	29.87	29.80	29.81	29.73	29.73	29.74	29.72	29.69	29.77	29.86	29.90	29.79
	MEAN SEA-LEVEL PRES. (IN)	2	30.11	30.07	29.99	30.01	29.92	29.92	29.94	29.92		29.97	30.06	30.10	
WINDS	MEAN SPEED (MPH)	26	6.4	6.6	6.6	6.4	6.1	5.5	5.1	5.0	5.2	5.3	5.8	6.0	5.8
	PREVAIL.DIR(TENS OF DEGS)														
	MAXIMUM 2-MINUTE:														
	SPEED (MPH)	2	21	15	20	16	16	13	12	12	13	15	15	16	21
	DIR. (TENS OF DEGS)		33	20	03	28	25	26	26	26	23	25	25	33	33
	YEAR OF OCCURRENCE		2000	2000	2000	2001	2000	2001	2001	2001	2000	2000	2000	2001	JAN 2000
	MAXIMUM 5-SECOND:														
	SPEED (MPH)	2	29	28	30	25	25	20	18	18	18	24	24	23	30
	DIR. (TENS OF DEGS)		33	25	03	25	27	25	26	26	25	25	27	24	03
	YEAR OF OCCURRENCE		2000	2000	2000	2001	2000	2000	2001	2000	2000	2000	2000	2001	MAR 2000
PRECIPITATION	NORMAL (IN)	30	2.92	3.07	2.61	1.03	0.19	0.03	0.01	0.14	0.45	0.31	1.98	2.03	14.77
	MAXIMUM MONTHLY (IN)	61	14.94	13.68	8.37	6.02	3.10	0.98	0.18	2.26	2.82	2.37	9.68	6.57	14.94
	YEAR OF OCCURRENCE		1969	1998	1983	1965	1998	1999	1986	1977	1976	1987	1965	1971	JAN 1969
	MINIMUM MONTHLY (IN)	61	0.00	T	0.00	0.00	0.00	0.00	0.00	0.00	0.00	0.00	0.00	0.00	DEC 1990
	YEAR OF OCCURRENCE		1976	1951	1959	1979	1981	1982	1983	1982	1980	1980	1980	1990	
	MAXIMUM IN 24 HOURS (IN)	61	6.11	4.02	3.79	2.05	2.41	0.76	0.18	2.22	1.95	1.77	4.07	3.92	6.11
	YEAR OF OCCURRENCE		1956	1944	1978	1956	1977	1993	1986	1986	1977	1986	1983	1970	1965	JAN 1956
	NORMAL NO. DAYS WITH:														
	PRECIPITATION 0.01	30	5.3	5.3	6.2	3.4	1.0	0.5	0.2	0.7	1.8	1.8	3.6	4.4	34.2
	PRECIPITATION 1.00	30	1.0	1.2	0.7	0.2	0.1	0.0	0.0	*	0.1	*	0.6	0.6	4.5
SNOWFALL	NORMAL (IN)	30	0.0	0.0	0.0	0.0	0.0	0.0	0.0	0.0	0.0	0.0	0.0	0.0	0.0
	MAXIMUM MONTHLY (IN)	56	0.3	T	0.0	0.0	0.0	0.0	0.0	0.0	0.0	0.0	0.0	T	0.3
	YEAR OF OCCURRENCE		1949	1951										1947	JAN 1949
	MAXIMUM IN 24 HOURS (IN)	56	0.3	T	0.0	0.0	0.0	0.0	0.0	0.0	0.0	0.0	0.0	T	0.3
	YEAR OF OCCURRENCE		1949	1951										1947	JAN 1949
	MAXIMUM SNOW DEPTH (IN)	56	0	0	0	0	0	0	0	0	0	0	0	0	0
	YEAR OF OCCURRENCE														
	NORMAL NO. DAYS WITH:														
	SNOWFALL 1.0	30	0.0	0.0	0.0	0.0	0.0	0.0	0.0	0.0	0.0	0.0	0.0	0.0	0.0

PRECIPITATION (inches) 2001 LOS ANGELES, CALIFORNIA CA (CQT)

YEAR	JAN	FEB	MAR	APR	MAY	JUN	JUL	AUG	SEP	OCT	NOV	DEC	ANNUAL
1972	0.00	0.13	T	0.03	0.03	0.07	0.00	0.35	0.02	0.29	3.26	2.36	6.54
1973	4.39	7.89	2.70	0.00	T	0.00	0.00	0.00	0.00	0.12	1.68	0.67	17.45
1974	8.35	0.14	3.78	0.10	0.08	0.00	0.00	0.00	0.00	0.58	0.07	3.59	16.69
1975	0.12	3.54	4.83	1.53	0.09	0.00	0.00	0.00	0.00	0.27	0.00	0.32	10.70
1976	0.00	3.71	1.81	0.84	0.05	0.22	0.00	0.08	2.82	0.24	0.49	0.75	11.01
1977	2.84	0.17	1.89	0.00	3.03	0.00	0.00	2.26	0.00	0.00	0.08	4.70	14.97
1978	7.70	8.91	8.02	1.77	0.00	0.00	0.00	0.39	0.05	2.28	1.45	30.57
1979	6.59	3.06	5.85	0.00	0.00	0.00	0.00	0.01	T	0.77	0.21	0.51	17.00
1980	7.50	12.75	4.79	0.31	0.13	0.00	0.00	0.00	0.00	0.00	0.00	0.85	26.33
1981	2.02	1.48	4.10	0.53	0.00	0.00	0.00	0.00	0.02	0.49	1.80	0.48	10.92
1982	2.17	0.70	3.54	1.39	0.12	0.00	0.00	0.00	0.84	0.19	4.41	1.05	14.41
1983	6.49	4.37	8.37	5.16	0.00	0.01	0.00	0.79	1.99	0.75	2.52	3.23	34.04
1984	0.17	0.00	0.28	0.69	0.00	0.01	0.00	0.40	0.23	0.15	1.44	5.53	8.90
1985	0.71	2.84	1.29	0.00	0.23	0.00	0.00	0.00	0.19	0.42	2.91	0.33	8.92
1986	2.19	6.10	5.27	0.45	0.00	0.00	0.18	0.00	1.97	0.53	0.94	0.37	18.00
1987	1.39	1.22	0.95	0.06	0.00	0.05	0.01	0.00	0.09	2.37	1.13	1.84	9.11
1988	1.65	1.72	0.26	3.41	0.00	0.00	0.00	0.05	0.04	0.00	0.70	3.80	11.63
1989	0.73	1.90	0.81	0.00	0.05	0.00	0.00	0.00	0.35	0.43	0.29	0.00	4.56
1990	1.24	3.12	0.17	0.58	1.17	0.00	0.00	0.02	0.00	0.00	0.19	0.00	6.49
1991	1.69	4.13	5.92	0.03	0.00	0.01	0.13	0.00	0.09	0.37	0.00	3.22	15.59
1992	1.74	7.96	7.12	0.33	0.04	0.00	0.08	0.00	0.00	0.70	0.00	4.68	22.65
1993	11.77	6.61	2.74	0.00	0.02	0.76	0.00	0.00	0.00	0.16	0.66	0.78	23.50
1994	0.33	3.21	1.86	0.83	0.28	0.00	0.00	0.00	0.00	0.00	0.61	1.35	8.66
1995	12.56	1.30	6.98	0.58	0.18	0.60	0.02	0.00	0.00	0.00	0.09	1.34	23.65
1996	3.16	4.94	2.16	0.71	0.04	0.00	0.00	0.00	0.00	1.06	1.59	4.09	17.75
1997	5.58	0.08	0.00	0.00	0.00	0.00	0.00	0.00	0.45	0.00	2.06	2.52	10.69
1998	4.12	13.68	4.06	0.97	3.10	0.05	0.00	0.00	0.01	0.00	1.32	0.54	27.85
1999	1.85	0.56	1.24	2.57	0.02	0.98	T	0.00	T	0.00	0.44	0.40	8.06
2000	0.88	5.54	2.82	1.49	T	0.00	0.00	0.07	0.15	0.98	T	T	11.93
2001	5.59	8.87	1.17	1.11	0.00	T	T	0.00	0.00	0.06	1.42	1.38	19.60
POR= 124 YRS	3.49	3.23	2.45	1.09	0.25	0.05	0.01	0.10	0.31	0.30	1.67	1.89	14.84

AVERAGE TEMPERATURE (F) 2001 LOS ANGELES, CALIFORNIA CA (CQT)

YEAR	JAN	FEB	MAR	APR	MAY	JUN	JUL	AUG	SEP	OCT	NOV	DEC	ANNUAL
1972	55.5	60.3	63.7	63.9	67.6	72.2	78.0	77.4	72.3	67.2	62.2	58.1	66.5
1973	56.4	60.0	57.9	63.1	65.8	72.0	72.4	73.6	70.0	68.8	60.0	59.9	65.0
1974	55.2	59.2	59.6	64.7	65.7	72.2	74.1	72.3	73.2	67.6	64.0	56.2	65.3
1975	57.6	55.8	55.7	56.0	62.7	65.7	72.5	71.9	74.0	66.4	61.2	57.0	63.0
1976	59.4	56.4	58.4	57.8	64.3	71.1	72.6	71.6	72.6	70.7	66.9	60.4	65.2
1977	58.1	63.1	56.9	63.7	61.9	69.2	74.2	75.6	71.8	69.0	66.3	60.8	65.9
1978	58.1	58.9	63.2	60.8	68.6	71.8	73.4	73.7	76.0	70.3	58.4	53.2	65.5
1979	53.3	55.0	57.9	62.7	65.4	71.5	72.1	72.9	77.4	68.7	64.6	63.2	65.4
1980	60.9	64.6	60.9	64.8	63.2	71.8	77.1	76.3	72.6	71.5	65.3	63.7	67.7
1981	61.8	64.3	62.0	66.0	68.9	77.4	77.2	78.3	75.0	68.5	65.0	62.1	68.9
1982	57.1	64.0	59.3	62.2	64.4	65.3	74.0	75.1	73.9	70.9	61.7	58.1	65.5
1983	61.9	63.0	63.9	63.5	70.7	70.7	75.9	80.8	79.1	74.2	63.5	59.8	68.9
1984	61.2	61.9	65.6	65.3	72.4	72.2	78.7	76.4	81.3	68.5	61.0	57.2	68.5
1985	57.5	60.4	59.3	66.8	66.3	73.5	79.2	75.7	71.3	64.4	61.7	67.0
1986	65.9	62.4	64.5	66.4	68.1	71.2	73.2	76.0	68.8	69.4	66.4	60.1	67.7
1987	57.2	60.3	61.2	67.8	68.1	69.7	70.8	73.0	75.2	71.9	62.9	54.4	66.0
1988	58.3	62.9	64.9	64.1	67.2	67.9	74.3	72.9	72.2	69.7	61.9	57.1	66.1
1989	56.3	56.4	62.4	67.9	66.2	69.8	75.1	72.8	74.5	69.2	66.7	62.7	66.7
1990	59.4	58.0	61.7	65.7	66.9	74.3	77.3	74.0	76.0	73.2	65.6	57.5	67.5
1991	59.2	63.5	56.8	64.2	63.9	67.1	71.0	73.1	73.6	72.1	66.2	59.6	65.9
1992	60.3	62.3	60.8	69.6	69.0	70.4	75.9	78.9	76.6	70.4	65.1	56.5	68.0
1993	57.3	58.3	64.5	67.0	68.9	72.4	73.0	74.4	74.3	71.2	64.6	60.8	67.2
1994	62.2	59.3	64.7	64.1	64.5	74.4	73.6	80.5	76.5	70.4	59.9	59.9	67.2
1995	58.4	65.3	62.6	64.8	64.0	69.0	75.8	77.5	77.0	71.5	67.1	60.9	67.8
1996	60.9	61.4	63.3	66.8	68.8	71.9	75.1	77.2	73.5	67.0	64.3	59.7	67.6
1997	58.7	61.0	65.1	65.7	72.7	71.0	73.2	77.6	79.8	71.1	65.2	58.9	68.3
1998	58.8	57.1	61.9	62.2	64.2	68.8	76.2	79.9	73.6	69.0	62.4	59.1	66.1
1999	60.8	59.9	56.9	59.4	63.3	66.8	71.8	71.4	69.2	71.5	61.8	58.1	64.2
2000	58.5	57.6	59.5	64.1	67.7	71.2	72.3	74.7	72.7	65.1	59.0	58.9	65.1
2001	54.4	54.7	60.4	59.9	67.3	70.8	71.0	72.0	71.4	68.2	62.0	56.4	64.0
POR= 124 YRS	56.4	57.6	59.1	61.4	64.0	67.7	72.0	72.8	71.4	67.2	62.4	57.8	64.2

REFERENCE NOTES:

PAGE 1:
THE TEMPERATURE GRAPH SHOWS NORMAL MAXIMUM AND NORMAL
MINIMUM DAILY TEMPERATURES (SOLID CURVES) AND THE
ACTUAL DAILY HIGH AND LOW TEMPERATURES (VERTICAL BARS).

PAGE 2 AND 3:
H/C INDICATES HEATING AND COOLING DEGREE DAYS.
RH INDICATES RELATIVE HUMIDITY
W/O INDICATES WEATHER AND OBSTRUCTIONS
S INDICATES SUNSHINE.
PR INDICATES PRESSURE.
CLOUDINESS ON PAGE 3 IS THE SUM OF THE CEILOMETER AND
SATELLITE DATA NOT TO EXCEED EIGHT EIGHTHS(OKTAS).

GENERAL:
T INDICATES TRACE PRECIPITATION, AN AMOUNT GREATER
THAN ZERO BUT LESS THAN THE LOWEST REPORTABLE VALUE.
+ INDICATES THE VALUE ALSO OCCURS ON EARLIER DATES.
BLANK ENTRIES DENOTE MISSING OR UNREPORTED DATA.
NORMALS ARE 30-YEAR AVERAGES (1961 - 1990).
ASOS INDICATES AUTOMATED SURFACE OBSERVING SYSTEM.
PM INDICATES THE LAST DAY OF THE PREVIOUS MONTH.
POR (PERIOD OF RECORD) BEGINS WITH THE JANUARY DATA
MONTH AND IS THE NUMBER OF YEARS USED TO COMPUTE
THE MEAN. INDIVIDUAL MONTHS WITHIN THE POR MAY
BE MISSING.
WHEN THE POR FOR A NORMAL IS LESS THAN 30 YEARS,
THE NORMAL IS PROVISIONAL AND IS BASED ON THE NUMBER
OF YEARS INDICATED.
0. OR * INDICATES THE VALUE OR MEAN-DAYS-WITH
IS BETWEEN 0.00 AND 0.05.
CLOUDINESS FOR ASOS STATIONS DIFFERS FROM THE NON-ASOS
OBSERVATION TAKEN BY A HUMAN OBSERVER. ASOS STATION
CLOUDINESS IS BASED ON TIME-AVERAGED CEILOMETER DATA
FOR CLOUDS AT OR BELOW 12,000 FEET AND ON SATELLITE
DATA FOR CLOUDS ABOVE 12,000 FEET.
THE NUMBER OF DAYS WITH CLEAR, PARTLY CLOUDY, AND
CLOUDY CONDITIONS FOR ASOS STATIONS IS THE SUM
OF THE CEILOMETER AND SATELLITE DATA FOR THE
SUNRISE TO SUNSET PERIOD.

GENERAL CONTINUED:
CLEAR INDICATES 0 - 2 OKTAS, PARTLY CLOUDY INDICATES
3 - 6 OKTAS, AND CLOUDY INDICATES 7 OR 8 OKTAS.
WHEN AT LEAST ONE OF THE ELEMENTS (CEILOMETER OR
SATELLITE) IS MISSING, THE DAILY CLOUDINESS IS
NOT COMPUTED.
WIND DIRECTION IS RECORDED IN TENS OF DEGREE (2 DIGITS)
CLOCKWISE FROM TRUE NORTH. "00" INDICATES CALM. "36"
INDICATES TRUE NORTH.
RESULTANT WIND IS THE VECTOR AVERAGE OF THE SPEED AND
DIRECTION.
AVERAGE TEMPERATURE IS THE SUM OF THE MEAN DAILY MAXIMUM
AND MINIMUM TEMPERATURE DIVIDED BY 2.
SNOWFALL DATA COMPRISE ALL FORMS OF FROZEN
PRECIPITATION, INCLUDING HAIL.
A HEATING (COOLING) DEGREE DAY IS THE DIFFERENCE BETWEEN
THE AVERAGE DAILY TEMPERATURE AND 65 F.
DRY BULB IS THE TEMPERATURE OF THE AMBIENT AIR.
DEW POINT IS THE TEMPERATURE TO WHICH THE AIR MUST BE
COOLED TO ACHIEVE 100 PERCENT RELATIVE HUMIDITY.
WET BULB IS THE TEMPERATURE THE AIR WOULD HAVE IF THE
MOISTURE CONTENT WAS INCREASED TO 100 PERCENT RELATIVE
HUMIDITY.

ON JULY 1, 1996, THE NATIONAL WEATHER SERVICE BEGAN USING
THE "METAR" OBSERVATION CODE THAT WAS ALREADY EMPLOYED
BY MOST OTHER NATIONS OF THE WORLD. THE MOST NOTICEABLE
DIFFERENCE IN THIS ANNUAL PUBLICATION WILL BE THE CHANGE
IN UNITS FROM TENTHS TO EIGHTS(OKTAS) FOR REPORTING THE
AMOUNT OF SKY COVER.

HEATING DEGREE DAYS (base 65 F) 2001 LOS ANGELES, CALIFORNIA CA (CQT)

YEAR	JUL	AUG	SEP	OCT	NOV	DEC	JAN	FEB	MAR	APR	MAY	JUN	TOTAL
1972-73	0	0	0	14	97	230	266	136	214	77	32	2	1068
1973-74	0	0	1	8	156	174	300	160	171	54	32	1	1057
1974-75	0	0	0	25	73	268	243	254	283	262	75	21	1504
1975-76	0	0	0	31	132	247	190	246	215	215	50	12	1338
1976-77	0	0	0	0	62	138	215	86	247	57	108	1	914
1977-78	0	0	0	14	51	132	209	174	102	122	24	0	828
1978-79	0	0	0	7	209	361	354	274	226	80	46	6	1563
1979-80	0	0	0	1	59	114	128	60	123	79	75	4	643
1980-81	0	0	0	2	41	85	103	91	97	43	1	0	463
1981-82	0	0	0	11	58	102	238	58	184	113	41	16	821
1982-83	0	0	0	3	117	205	134	73	68	68	2	0	670
1983-84	0	0	0	0	99	158	140	99	29	59	5	0	589
1984-85	0	0	0	4	129	239	225	162	179	40	21	0	999
1985-86	0	0	0	0	163	131	42	125	92	32	7	0	592
1986-87	0	0	8	2	14	151	241	140	131	31	19	0	737
1987-88	0	0	0	3	91	323	216	82	81	88	30	11	925
1988-89	0	0	1	2	98	258	270	271	104	36	27	5	1072
1989-90	0	0	0	2	27	102	173	206	130	26	16	2	684
1990-91	0	0	0	0	42	244	183	63	248	74	72	2	928
1991-92	0	0	0	23	46	168	159	114	125	1	0	0	636
1992-93	0	0	0	1	49	256	235	181	68	9	0	6	805
1993-94	0	0	0	0	45	136	106	153	61	60	35	0	596
1994-95	0	0	0	0	158	160	211	90	62	91	59	52	14	807
1995-96	0	0	0	1	6	132	146	125	83	16	3	1	513
1996-97	0	0	0	31	76	158	196	120	61	49	0	0	691
1997-98	0	0	0	0	71	193	185	216	118	130	36	3	952
1998-99	0	0	0	1	79	200	137	144	245	211	66	32	1115
1999-00	0	0	2	1	95	207	197	211	166	61	9	0	949
2000-01	0	0	0	36	179	186	321	293	147	164	5	0	1331
2001-	0	0	0	3	97	260							

COOLING DEGREE DAYS (base 65 F) 2001 LOS ANGELES, CALIFORNIA CA (CQT)

YEAR	JAN	FEB	MAR	APR	MAY	JUN	JUL	AUG	SEP	OCT	NOV	DEC	ANNUAL
1972	0	4	27	34	122	223	409	391	225	89	21	25	1570
1973	9	2	0	25	64	220	236	272	157	133	9	21	1148
1974	3	5	8	51	58	223	288	235	254	115	53	2	1295
1975	21	0	0	0	11	48	241	221	277	82	26	4	931
1976	22	3	20	10	32	203	245	212	233	185	123	2	1290
1977	7	39	3	23	18	135	293	334	210	148	96	8	1314
1978	0	8	52	2	145	212	269	277	338	177	17	0	1497
1979	0	0	14	17	67	209	229	252	379	124	53	62	1406
1980	10	54	3	82	26	215	380	357	233	210	56	53	1679
1981	12	75	13	81	132	380	387	422	306	124	67	17	2016
1982	0	36	15	36	33	32	286	322	275	194	25	0	1254
1983	44	23	41	21	185	174	342	495	432	292	60	4	2113
1984	29	14	56	73	240	222	433	360	496	123	13	6	2065
1985	0	41	10	100	68	264	447	339	210	203	31	35	1748
1986	77	56	83	80	110	194	261	349	132	145	65	6	1558
1987	6	18	21	120	121	147	186	257	312	221	36	1	1446
1988	13	30	84	68	107	107	297	252	223	154	14	20	1369
1989	8	37	31	131	73	154	318	251	290	139	85	41	1558
1990	10	16	36	54	81	291	388	287	336	262	68	20	1849
1991	8	28	0	54	43	72	191	260	265	250	88	6	1265
1992	17	44	1	146	132	168	347	435	353	175	59	0	1877
1993	6	0	59	77	129	232	254	299	287	199	38	12	1592
1994	25	0	59	47	47	288	272	488	354	178	9	10	1777
1995	13	76	24	61	31	142	341	394	369	212	77	12	1752
1996	25	26	28	134	134	215	320	386	266	101	60	0	1695
1997	7	14	72	79	245	188	260	402	450	196	85	11	2009
1998	0	0	29	53	18	125	356	468	266	132	11	24	1482
1999	14	8	0	49	23	90	217	202	136	208	7	0	954
2000	5	2	4	39	98	192	233	308	239	45	2	3	1170
2001	1	13	9	18	85	180	191	226	200	110	14	3	1050

SNOWFALL (inches) 2001 LOS ANGELES, CALIFORNIA CA (CQT)

YEAR	JUL	AUG	SEP	OCT	NOV	DEC	JAN	FEB	MAR	APR	MAY	JUN	TOTAL
1970-71	0.0	0.0	0.0	0.0	0.0	0.0	0.0	0.0	0.0	0.0	0.0	0.0	0.0
1971-72	0.0	0.0	0.0	0.0	0.0	0.0	0.0	0.0	0.0	0.0	0.0	0.0	0.0
1972-73	0.0	0.0	0.0	0.0	0.0	0.0	0.0	0.0	0.0	0.0	0.0	0.0	0.0
1973-74	0.0	0.0	0.0	0.0	0.0	0.0	0.0	0.0	0.0	0.0	0.0	0.0	0.0
1974-75	0.0	0.0	0.0	0.0	0.0	0.0	0.0	0.0	0.0	0.0	0.0	0.0	0.0
1975-76	0.0	0.0	0.0	0.0	0.0	0.0	0.0	0.0	0.0	0.0	0.0	0.0	0.0
1976-77	0.0	0.0	0.0	0.0	0.0	0.0	0.0	0.0	0.0	0.0	0.0	0.0	0.0
1977-78	0.0	0.0	0.0	0.0	0.0	0.0	0.0	0.0	0.0	0.0	0.0	0.0	0.0
1978-79	0.0	0.0	0.0	0.0	0.0	0.0	0.0	0.0	0.0	0.0	0.0	0.0	0.0
1979-80	0.0	0.0	0.0	0.0	0.0	0.0	0.0	0.0	0.0	0.0	0.0	0.0	0.0
1980-81	0.0	0.0	0.0	0.0	0.0	0.0	0.0	0.0	0.0	0.0	0.0	0.0	0.0
1981-82	0.0	0.0	0.0	0.0	0.0	0.0	0.0	0.0	0.0	0.0	0.0	0.0	0.0
1982-83	0.0	0.0	0.0	0.0	0.0	0.0	0.0	0.0	0.0	0.0	0.0	0.0	0.0
1983-84	0.0	0.0	0.0	0.0	0.0	0.0	0.0	0.0	0.0	0.0	0.0	0.0	0.0
1984-85													
1996-97													
1997-98													
1998-99													
1999-00													
2000-01													
2001-													
POR= 43 YRS	0.0	0.0	0.0	0.0	0.0	T	0.0	T	0.0	0.0	0.0	0.0	T

2001
SACRAMENTO,
CALIFORNIA (SAC)

Sacramento, and the lower Sacramento Valley, has a mild climate with abundant sunshine most of the year. A nearly cloud-free sky prevails throughout the summer months, and in much of the spring and fall. The summers are usually dry with warm to hot afternoons and mostly mild nights. The rainy season generally is November through March. About 75 percent of the annual precipitation occurs then, but measurable rain falls only on an average of nine days per month during that period. The shielding effect of mountains to the north, east, and west usually modifies winter storms. The Sierra Nevada snow fields, only 70 miles east of Sacramento, usually provide an adequate water supply during the dry season, and an important recreational area in winter. Heavy snowfall and torrential rains frequently fall on the western Sierra slopes, and may produce flood conditions along the Sacramento River and its tributaries. In the valley, however, excessive rainfall as well as damaging winds are rare.

The prevailing wind at Sacramento is southerly every month but November, when it is northerly. Topographic effects, the north-south alignment of the valley, the coast range, and the Sierra Nevada strongly influence the wind flow in the valley. A sea level gap in the coast range permits cool, oceanic air to flow, occasionally, into the valley during the summer season with a marked lowering of temperature through the Sacramento-San Joaquin River Delta to the capital. In the spring and fall, a large north-to-south pressure gradient develops over the northern part of the state. Air flowing over the Siskiyou mountains to the north warms and dries as it descends to the valley floor. This gusty, blustery north wind is a local variation of the chinook. It apparently carries a form of pollen which may cause allergic responses by susceptible individuals.

As is well known, relative humidity has a marked influence on the reactions of plants and animals to temperature. The extremely low relative humidity that ordinarily accompanies high temperatures in this valley should be considered when comparing temperatures here with those of cities in more humid regions. The extreme hot spells, with temperatures exceeding 100 degrees, are usually caused by air flow from a sub-tropical high pressure area that brings light to nearly calm winds and humidities below 20 percent.

Thunderstorms are few in number, usually mild in character, and occur mainly in the spring. An occasional thunderstorm may drift over the valley from the Sierra Nevada in the summer. Snow falls so rarely, and in such small amounts, that its occurrence may be disregarded as a climatic feature. Heavy fog occurs mostly in midwinter, never in summer, and seldom in spring or autumn. An occasional winter fog, under stagnant atmospheric conditions, may continue for several days. Light and moderate fogs are more frequent, and may come anytime during the wet, cold season. The fog is the radiational cooling type, and is usually confined to the early morning hours.

Sacramento is the geographical center of the great interior valley of California that reaches from Red Bluff in the north to Bakersville in the south. This predominantly agricultural region produces an extremely wide and abundant variety of fruits, grains, and vegetables ranging from the semi-tropical to the hardier varieties.

Based on the 1951-1980 period, the average first occurrence of 32 degrees Fahrenheit in the fall is December 1 and the average last occurrence in the spring is February 14.

NORMALS, MEANS, AND EXTREMES
SACRAMENTO, CA (SAC)

LATITUDE: 38 30' 45" N LONGITUDE: 121 29' 33" W ELEVATION (FT): GRND: 38 BARO: 41 TIME ZONE: PACIFIC (UTC + 8) WBAN: 23232

ELEMENT	POR	JAN	FEB	MAR	APR	MAY	JUN	JUL	AUG	SEP	OCT	NOV	DEC	YEAR
TEMPERATURE (F)														
NORMAL DAILY MAXIMUM	30	52.7	60.0	64.0	71.1	80.3	87.8	93.2	92.1	87.3	77.9	63.1	52.7	73.5
MEAN DAILY MAXIMUM	54	53.1	59.7	64.3	71.4	80.0	87.2	90.9	91.5	87.6	77.8	63.6	53.4	73.4
HIGHEST DAILY MAXIMUM	51	70	76	88	95	105	115	114	110	108	104	87	72	115
YEAR OF OCCURRENCE		1991	1992	1988	1996	1984	1961	1972	1996	1988	2001	1960	1999	JUN 1961
MEAN OF EXTREME MAXS.	54	63.3	69.7	76.2	85.5	95.5	102.8	102.8	103.4	100.2	92.0	76.3	64.0	86.0
NORMAL DAILY MINIMUM	30	37.7	41.4	43.2	45.5	50.3	55.3	58.1	58.0	55.7	50.4	43.4	37.8	48.1
MEAN DAILY MINIMUM	54	38.2	41.4	43.0	45.8	50.5	55.1	56.7	57.7	55.9	50.2	42.8	37.9	47.9
LOWEST DAILY MINIMUM	51	23	23	26	31	36	41	48	49	43	36	26	18	18
YEAR OF OCCURRENCE		1979	1989	1971	1999	1974	1990	1983	1978	1978	1989	1993	1990	DEC 1990
MEAN OF EXTREME MINS.	54	27.7	31.5	34.1	37.7	42.1	48.0	51.3	52.0	48.7	41.5	32.3	28.2	39.6
NORMAL DRY BULB	30	45.2	50.7	53.6	58.3	65.3	71.6	75.7	75.1	71.5	64.2	53.3	45.3	60.8
MEAN DRY BULB	54	45.8	50.5	53.8	58.7	65.1	71.1	74.0	74.6	71.7	64.0	53.2	45.6	60.7
MEAN WET BULB	16	43.8	47.1	50.0	52.3	55.7	59.7	58.5	56.1	55.5	51.5	45.0	39.8	51.2
MEAN DEW POINT	16	40.9	43.1	45.2	45.9	48.3	51.7	51.4	49.3	48.5	44.7	40.8	36.8	45.5
NORMAL NO. DAYS WITH:														
MAXIMUM 90	30	0.0	0.0	0.0	0.4	5.8	12.5	22.5	20.0	12.4	2.8	0.0	0.0	76.4
MAXIMUM 32	30	*	0.0	0.0	0.0	0.0	0.0	0.0	0.0	0.0	0.0	0.0	*	0.0
MINIMUM 32	30	7.0	1.8	0.5	0.0	0.0	0.0	0.0	0.0	0.0	0.0	1.2	6.9	17.4
MINIMUM 0	30	0.0	0.0	0.0	0.0	0.0	0.0	0.0	0.0	0.0	0.0	0.0	0.0	0.0
H/C														
NORMAL HEATING DEG. DAYS	30	614	400	357	230	80	12	0	0	16	78	351	611	2749
NORMAL COOLING DEG. DAYS	30	0	0	0	29	89	210	332	313	211	53	0	0	1237
RH														
NORMAL (PERCENT)	30	83	77	72	64	59	55	53	56	57	63	76	83	66
HOUR 04 LST	30	90	87	84	82	81	78	76	78	77	80	86	90	82
HOUR 10 LST	30	85	77	69	58	50	47	47	50	50	57	74	84	62
HOUR 16 LST	30	70	59	53	43	35	31	28	29	31	38	58	70	45
HOUR 22 LST	30	86	81	77	73	69	64	61	64	65	70	81	86	73
S — PERCENT POSSIBLE SUNSHINE	47	48	65	74	82	90	94	97	96	93	86	66	49	78
W/O MEAN NO. DAYS WITH:														
HEAVY FOG(VISBY 1/4 MI)	53	10.0	5.2	1.7	0.3	0.2	0.0	0.0	0.0	0.2	1.4	5.2	9.4	33.6
THUNDERSTORMS	53	0.4	0.6	0.8	0.7	0.4	0.3	0.2	0.1	0.5	0.3	0.2	0.2	4.7
CLOUDINESS MEAN:														
SUNRISE-SUNSET (OKTAS)	48	5.7	5.0	4.5	3.8	2.9	1.8	0.9	1.1	1.5	2.6	4.5	5.5	3.3
MIDNIGHT-MIDNIGHT (OKTAS)	31	5.3	4.6	3.9	3.3	2.3	1.7	0.8	1.0	1.2	2.1	3.9	4.8	2.9
MEAN NO. DAYS WITH:														
CLEAR	49	6.5	7.6	9.9	11.9	16.9	21.7	26.9	25.1	23.2	18.9	9.8	7.6	186.0
PARTLY CLOUDY	49	5.9	5.9	8.7	9.6	8.6	5.9	3.2	4.1	4.2	6.0	7.4	5.8	76.4
CLOUDY	49	18.7	13.6	12.4	8.4	5.5	2.4	1.0	1.3	2.1	5.4	12.4	17.0	100.2
PR														
MEAN STATION PRESSURE(IN)	24	30.10	30.05	29.99	29.98	29.90	29.86	29.85	29.86	29.86	29.96	30.06	30.12	29.97
MEAN SEA-LEVEL PRES. (IN)	16	30.13	30.07	30.03	30.00	29.93	29.87	29.87	29.87	29.88	29.97	30.09	30.15	29.99
WINDS														
MEAN SPEED (MPH)	36	7.1	7.8	8.5	8.6	9.0	9.5	9.1	8.7	7.3	6.6	6.1	6.4	7.9
PREVAIL.DIR(TENS OF DEGS)	21	14	14	22	22	22	21	20	20	21	20	14	14	21
MAXIMUM 2-MINUTE:														
SPEED (MPH)	3	35	33	36	33	38	29	26	22	25	36	38	31	38
DIR. (TENS OF DEGS)		14	24	32	35	33	33	23	22	21	35	15	15	15
YEAR OF OCCURRENCE		2001	2000	2000	1999	2001	2000	1999	2001	2000	2000	2001	2001	NOV 2001
MAXIMUM 5-SECOND:														
SPEED (MPH)	3	48	40	44	43	46	35	31	29	32	44	49	44	49
DIR. (TENS OF DEGS)		13	23	14	35	33	33	22	21	22	34	16	14	16
YEAR OF OCCURRENCE		2001	2000	2001	1999	2001	2000	2001	2000	2000	2000	2001	2001	NOV 2001
PRECIPITATION														
NORMAL (IN)	30	3.73	2.87	2.57	1.16	0.27	0.12	0.05	0.07	0.37	1.08	2.72	2.51	17.52
MAXIMUM MONTHLY (IN)	62	9.69	9.95	8.13	4.76	3.13	1.26	0.79	0.65	2.78	7.51	7.41	12.64	12.64
YEAR OF OCCURRENCE		1995	1998	1995	1941	1948	1993	1974	1976	1989	1962	1970	1955	DEC 1955
MINIMUM MONTHLY (IN)	62	0.16	0.15	0.05	0.00	T	0.00	0.00	0.00	0.00	0.00	0.00	0.00	0.00
YEAR OF OCCURRENCE		1984	1964	1994	1949	1992	1981	1983	1982	1980	1966	1995	1989	DEC 1989
MAXIMUM IN 24 HOURS (IN)	53	3.41	3.01	2.30	2.22	1.38	1.21	T	0.32	0.50	5.59	2.95	3.64	5.59
YEAR OF OCCURRENCE		1967	1986	1982	1958	1994	1993	2001	1997	2001	1962	1970	1955	OCT 1962
NORMAL NO. DAYS WITH:														
PRECIPITATION 0.01	30	9.6	8.1	8.6	5.1	2.1	1.1	0.3	0.5	1.7	3.6	7.9	8.7	57.3
PRECIPITATION 1.00	30	0.8	0.5	0.3	0.1	*	0.0	0.0	0.0	*	0.2	0.6	0.3	2.8
SNOWFALL														
NORMAL (IN)	30	T	0.1	0.0	0.0	0.0	0.0	0.0	0.0	0.0	0.0	0.0	T	0.1
MAXIMUM MONTHLY (IN)	50	T	2.0	T	0.0	T	0.0	0.0	0.0	0.0	0.0	0.0	T	2.0
YEAR OF OCCURRENCE		1974	1976	1982		1994						1995		FEB 1976
MAXIMUM IN 24 HOURS (IN)	50	T	2.0	T	0.0	T	0.0	0.0	0.0	0.0	0.0	0.0	T	2.0
YEAR OF OCCURRENCE		1974	1976	1982		1994						1995		FEB 1976
MAXIMUM SNOW DEPTH (IN)	48	0	0	0	0	0	0	0	0	0	0	0	0	0
YEAR OF OCCURRENCE														
NORMAL NO. DAYS WITH:														
SNOWFALL 1.0	30	0.0	0.*	0.0	0.0	0.0	0.0	0.0	0.0	0.0	0.0	0.0	0.0	0.0

PRECIPITATION (inches) 2001 SACRAMENTO, CA (SAC)

YEAR	JAN	FEB	MAR	APR	MAY	JUN	JUL	AUG	SEP	OCT	NOV	DEC	ANNUAL
1972	0.81	1.28	0.29	1.39	0.28	0.19	0.00	0.00	0.90	1.75	5.14	1.88	13.91
1973	6.87	5.64	2.76	0.05	0.13	0.00	0.00	0.00	0.33	1.64	6.27	2.79	26.48
1974	3.58	1.37	3.27	0.96	0.01	0.50	0.79	T	0.00	1.16	0.66	2.86	15.16
1975	0.73	4.59	4.28	0.81	T	T	0.04	0.23	T	2.03	0.29	0.18	13.18
1976	0.36	1.49	0.44	1.53	0.00	0.04	0.00	0.65	0.52	0.02	0.55	0.65	6.25
1977	1.17	1.17	1.27	0.30	0.73	0.00	T	0.00	0.76	0.12	1.92	4.27	11.71
1978	9.14	4.46	3.38	2.31	T	T	0.00	T	0.30	T	3.20	0.95	23.74
1979	5.66	4.55	2.47	0.76	0.14	0.00	0.25	0.00	T	1.62	1.48	3.41	20.34
1980	5.64	7.12	2.62	1.06	0.49	0.04	0.40	0.00	0.00	0.06	0.12	1.79	19.34
1981	4.56	0.87	3.55	0.66	0.50	0.00	0.00	0.00	0.25	2.57	6.09	3.28	22.33
1982	5.50	2.35	7.12	3.07	T	0.15	0.00	0.00	1.81	2.61	5.74	3.25	31.60
1983	4.92	5.56	6.75	4.21	0.25	0.40	0.00	0.11	0.66	0.40	4.91	5.26	33.43
1984	0.16	1.22	1.35	0.34	0.01	0.10	0.00	0.01	0.07	1.39	3.61	1.23	9.49
1985	0.66	1.52	2.01	T	0.01	0.15	T	0.06	0.56	0.53	3.72	2.34	11.56
1986	3.67	8.60	3.20	0.91	0.07	0.00	0.00	0.00	0.60	0.19	0.14	0.76	18.14
1987	2.29	3.23	3.05	0.20	T	T	0.00	0.00	0.00	1.28	2.53	3.25	15.83
1988	2.96	0.99	0.17	1.58	0.89	0.19	0.00	0.00	0.00	0.19	1.68	2.73	11.38
1989	0.71	1.25	6.29	0.31	0.06	0.43	0.00	0.20	2.78	1.76	1.32	0.00	15.11
1990	4.97	2.91	0.93	0.73	2.10	0.00	T	0.00	0.00	0.09	0.43	1.60	13.76
1991	0.36	3.10	6.14	0.29	0.25	0.53	T	0.14	0.04	1.25	0.19	1.60	13.89
1992	1.39	5.47	2.05	0.92	T	0.15	0.00	T	1.31	0.28	4.94		16.51
1993	8.63	4.94	2.39	0.63	1.14	1.26	0.00	0.00	0.00	0.47	2.28	1.75	23.49
1994	2.12	3.15	0.05	0.67	1.68	0.00	0.00	0.00			2.68		
1995	9.69	0.20	8.13	1.46	1.06	0.47	0.00	0.00	0.00	T	T	5.49	26.50
1996	4.16	5.49	1.73	1.25	0.79	0.00	0.00	T	T	0.67	1.97	6.39	22.45
1997	9.05	0.28	0.34	0.18	0.35	0.59	0.00	0.32	0.16	0.82	4.56	2.91	19.56
1998	6.40	9.95	2.47	1.05	2.98	0.58	0.00	0.00	0.23	0.76	2.84	0.58	27.84
1999	2.63	4.45	1.50	0.89	0.07	0.03	0.00	0.00	0.00	0.18	1.63	0.06	11.44
2000	6.49	8.49	2.03	1.39	1.17	0.04	0.00	T	0.00	0.09	0.68	0.59	22.59
2001	3.75	4.57	2.04	1.50	T	0.08	T	0.00	0.50	0.36	2.43	6.27	21.50
POR= 62 YRS	3.83	3.13	2.52	1.78	0.50	0.15	0.03	0.07	0.36	0.90	2.28	2.99	18.54

AVERAGE TEMPERATURE (F) 2001   SACRAMENTO, CA (SAC)

YEAR	JAN	FEB	MAR	APR	MAY	JUN	JUL	AUG	SEP	OCT	NOV	DEC	ANNUAL
1972	41.0	51.4	58.6	58.6	66.9	72.5	76.0	75.9	69.5	62.0	49.7	40.6	60.2
1973	44.3	53.1	51.1	60.8	69.4	74.6	76.8	74.2	70.8	63.4	51.9	46.9	61.4
1974	46.3	48.4	54.1	56.6	63.8	70.5	74.1	74.0	72.2	66.3	53.2	46.4	60.5
1975	43.4	49.2	50.3	51.8	68.2	73.2	77.3	76.9	77.4	65.1	54.6	47.4	61.2
1976	47.2	51.9	54.6	57.9	70.1	73.9	76.5	73.4	71.8	66.1	56.7	46.5	61.2
1977	43.8	52.4	51.0	62.2	59.4	72.2	74.2	73.9	68.8	63.9	54.5	49.6	60.5
1978	50.3	51.9	57.2	55.8	66.4	70.0	75.1	75.0	69.6	65.9	49.8	41.7	60.7
1979	45.3	48.8	54.6	56.9	66.7	71.9	75.6	73.1	74.6	64.0	51.7	46.7	60.8
1980	46.9	51.9	51.6	59.6	62.7	66.8	75.0	71.4	69.4	63.7	53.5	45.4	59.8
1981	46.8	50.4	51.2	57.9	64.7	74.8	75.1	74.5	69.7	63.5	60.3	48.6	61.5
1982	42.0	50.5	50.8	55.5	64.6	66.2	72.1	71.7	68.2	61.0	46.9	43.0	57.7
1983	43.1	52.2	53.4	54.7	64.2	70.8	72.2	76.6	74.9	67.5	53.7	51.0	61.2
1984	48.2	50.2	58.1	58.7	70.0	71.7	78.3	75.3	75.5	62.8	53.6	45.1	62.3
1985	42.4	51.4	50.8	61.5	63.2	75.1	77.0	72.9	68.5	63.3	49.8	42.6	59.9
1986	51.4	54.7	58.8	58.4	65.5	71.6	75.0	75.2	66.2	64.8	55.5	45.7	61.9
1987	44.9	51.3	53.8	62.7	69.1	72.4	74.9	74.9	71.8	67.6	53.4	47.2	61.8
1988	48.0	54.2	58.0	60.9	64.7	72.9	80.4	75.9	72.5	66.5	53.8	46.2	62.8
1989	44.1	47.1	55.6	63.2	65.8	71.7	76.2	73.8	69.6	62.4	54.3	44.3	60.7
1990	47.5	48.6	55.4	63.4	65.5	72.4	77.7	76.6	74.0	66.6	53.0	41.0	61.8
1991	47.3	55.0	51.0	58.0	63.3	70.2	77.1	73.2	74.3	68.8	55.9	46.3	61.7
1992	43.6	54.1	56.2	62.1	70.6	70.9	75.3	77.0	72.4	66.6	53.4	44.1	62.2
1993	45.2	49.5	57.9	58.4	64.6	71.7	74.3	74.1	71.5	65.0	51.5	44.3	60.7
1994	47.0	48.8	56.7	60.1	65.3	71.6	74.0	75.2	71.7		47.6	43.7	
1995	51.3	52.1	53.0	57.7	63.1	69.0	74.2	75.1	72.3		59.3	51.1	
1996	48.2	54.3	56.7	61.1	67.0	73.3	78.7	78.3	71.0	63.9	55.0	51.1	63.2
1997	48.3	52.7	57.9	62.4	71.9	72.9	76.5	75.9	75.1	64.1	56.9	46.2	63.4
1998	49.7	50.4	55.1	57.5	58.7	67.0	74.8	76.8	72.5	62.5	52.5	42.5	59.9
1999	44.7	48.1	50.8	57.4	62.8	69.9	72.0	73.0	72.3	65.1	54.6	46.8	59.8
2000	48.8	51.4	55.5	60.7	65.7	73.2	72.1	74.0	71.2	61.9	48.6	47.0	60.8
2001	45.8	48.8	57.2	55.9	71.7	73.4	73.4	74.5	71.1	65.8	55.9	48.8	61.9
POR= 61 YRS	45.7	50.4	53.8	58.6	65.0	70.9	74.1	74.4	71.6	63.9	53.1	45.8	60.6

REFERENCE NOTES:

PAGE 1:
    THE TEMPERATURE GRAPH SHOWS NORMAL MAXIMUM AND NORMAL
    MINIMUM DAILY TEMPERATURES (SOLID CURVES) AND THE
    ACTUAL DAILY HIGH AND LOW TEMPERATURES (VERTICAL BARS).

PAGE 2 AND 3:
    H/C INDICATES HEATING AND COOLING DEGREE DAYS.
    RH INDICATES RELATIVE HUMIDITY.
    W/O INDICATES WEATHER AND OBSTRUCTIONS
    S INDICATES SUNSHINE.
    PR INDICATES PRESSURE.
    CLOUDINESS ON PAGE 3 IS THE SUM OF THE CEILOMETER AND
    SATELLITE DATA NOT TO EXCEED EIGHT EIGHTHS(OKTAS).

GENERAL:
    T INDICATES TRACE PRECIPITATION, AN AMOUNT GREATER
    THAN ZERO BUT LESS THAN THE LOWEST REPORTABLE VALUE.
    + INDICATES THE VALUE ALSO OCCURS ON EARLIER DATES.
    BLANK ENTRIES DENOTE MISSING OR UNREPORTED DATA.
    NORMALS ARE 30-YEAR AVERAGES (1961 - 1990).
    ASOS INDICATES AUTOMATED SURFACE OBSERVING SYSTEM.
    PM INDICATES THE LAST DAY OF THE PREVIOUS MONTH.
    POR (PERIOD OF RECORD) BEGINS WITH THE JANUARY DATA
    MONTH AND IS THE NUMBER OF YEARS USED TO COMPUTE
    THE MEAN. INDIVIDUAL MONTHS WITHIN THE POR MAY
    BE MISSING.
    WHEN THE POR FOR A NORMAL IS LESS THAN 30 YEARS,
    THE NORMAL IS PROVISIONAL AND IS BASED ON THE NUMBER
    OF YEARS INDICATED.
    0.* OR * INDICATES THE VALUE OR MEAN-DAYS-WITH
    IS BETWEEN 0.00 AND 0.05.
    CLOUDINESS FOR ASOS STATIONS DIFFERS FROM THE NON-ASOS
    OBSERVATION TAKEN BY A HUMAN OBSERVER. ASOS STATION
    CLOUDINESS IS BASED ON TIME-AVERAGED CEILOMETER DATA
    FOR CLOUDS AT OR BELOW 12,000 FEET AND ON SATELLITE
    DATA FOR CLOUDS ABOVE 12,000 FEET.
    THE NUMBER OF DAYS WITH CLEAR, PARTLY CLOUDY, AND
    CLOUDY CONDITIONS FOR ASOS STATIONS IS THE SUM
    OF THE CEILOMETER AND SATELLITE DATA FOR THE
    SUNRISE TO SUNSET PERIOD.

GENERAL CONTINUED:
    CLEAR INDICATES 0 - 2 OKTAS, PARTLY CLOUDY INDICATES
    3 - 6 OKTAS, AND CLOUDY INDICATES 7 OR 8 OKTAS.
    WHEN AT LEAST ONE OF THE ELEMENTS (CEILOMETER OR
    SATELLITE) IS MISSING, THE DAILY CLOUDINESS IS
    NOT COMPUTED.
    WIND DIRECTION IS RECORDED IN TENS OF DEGREES (2 DIGITS)
    CLOCKWISE FROM TRUE NORTH. "00" INDICATES CALM. "36"
    INDICATES TRUE NORTH.
    RESULTANT WIND IS THE VECTOR AVERAGE OF THE SPEED AND
    DIRECTION.
    AVERAGE TEMPERATURE IS THE SUM OF THE MEAN DAILY MAXIMUM
    AND MINIMUM TEMPERATURE DIVIDED BY 2.
    SNOWFALL DATA COMPRISE ALL FORMS OF FROZEN
    PRECIPITATION, INCLUDING HAIL.
    A HEATING (COOLING) DEGREE DAY IS THE DIFFERENCE BETWEEN
    THE AVERAGE DAILY TEMPERATURE AND 65 F.
    DRY BULB IS THE TEMPERATURE OF THE AMBIENT AIR.
    DEW POINT IS THE TEMPERATURE TO WHICH THE AIR MUST BE
    COOLED TO ACHIEVE 100 PERCENT RELATIVE HUMIDITY.
    WET BULB IS THE TEMPERATURE THE AIR WOULD HAVE IF THE
    MOISTURE CONTENT WAS INCREASED TO 100 PERCENT RELATIVE
    HUMIDITY.

    ON JULY 1, 1996, THE NATIONAL WEATHER SERVICE BEGAN USING
    THE "METAR" OBSERVATION CODE THAT WAS ALREADY EMPLOYED
    BY MOST OTHER NATIONS OF THE WORLD. THE MOST NOTICEABLE
    DIFFERENCE IN THIS ANNUAL PUBLICATION WILL BE THE CHANGE
    IN UNITS FROM TENTHS TO EIGHTS(OKTAS) FOR REPORTING THE
    AMOUNT OF SKY COVER.

HEATING DEGREE DAYS (base 65 F) 2001 SACRAMENTO, CA (SAC)

YEAR	JUL	AUG	SEP	OCT	NOV	DEC	JAN	FEB	MAR	APR	MAY	JUN	TOTAL
1972-73	0	0	6	115	451	749	636	325	424	141	15	1	2863
1973-74	0	0	0	77	384	553	571	456	332	251	93	9	2726
1974-75	7	0	0	44	347	569	661	435	449	389	69	1	2971
1975-76	0	0	0	72	306	539	547	374	315	211	1	3	2368
1976-77	0	0	1	44	252	567	650	345	424	92	187	9	2571
1977-78	0	0	17	68	309	472	451	362	235	269	46	0	2229
1978-79	0	0	11	51	449	715	606	446	313	236	57	2	2886
1979-80	0	0	0	100	391	558	551	373	408	164	107	29	2681
1980-81	2	0	4	134	339	596	557	405	420	229	81	2	2769
1981-82	0	0	9	66	145	498	708	398	434	282	70	40	2650
1982-83	3	0	31	125	532	675	670	353	354	303	99	4	3149
1983-84	3	0	0	7	333	425	514	421	206	191	22	11	2133
1984-85	0	0	0	115	335	611	693	377	433	122	89	11	2786
1985-86	0	2	15	95	450	689	411	284	192	200	73	0	2411
1986-87	0	0	53	47	277	593	614	377	340	95	37	0	2433
1987-88	1	0	0	11	339	544	522	307	212	138	94	27	2195
1988-89	0	0	3	38	329	576	640	496	285	106	50	3	2526
1989-90	0	0	11	107	316	634	536	453	289	71	53	6	2476
1990-91	0	0	0	24	356	739	543	274	427	205	104	6	2678
1991-92	0	0	0	82	267	572	657	310	265	104	0	9	2266
1992-93	0	0	0	24	340	643	605	426	214	202	55	21	2530
1993-94	0	0	5	33	399	634	550	449	248	147	48	1	2514
1994-95	0	0	0		515	654	415	354	364	210	105	26	
1995-96	0	0	0		166	421	513	302	250	154	21	1	
1996-97	0	0	0	121	294	423	511	336	214	104	7	1	2011
1997-98	0	0	0	56	248	577	465	404	299	233	190	13	2485
1998-99	0	0	8	113	367	689	621	465	431	239	92	29	3054
1999-00	1	0	0	57	303	556	496	389	291	138	83	2	2316
2000-01	0	4	5	123	484	551	588	445	240	276	4	0	2720
2001-	0	0	0	44	264	496							

COOLING DEGREE DAYS (base 65 F) 2001 SACRAMENTO, CA (SAC)

YEAR	JAN	FEB	MAR	APR	MAY	JUN	JUL	AUG	SEP	OCT	NOV	DEC	ANNUAL
1972	0	0	6	5	129	245	351	349	147	30	0	0	1262
1973	0	0	0	19	156	295	373	293	181	34	0	0	1351
1974	0	0	0	5	61	180	296	285	222	89	0	0	1138
1975	0	0	0	0	177	258	388	375	375	81	0	0	1654
1976	0	0	1	8	167	278	363	270	213	83	8	0	1391
1977	0	0	0	12	19	230	290	284	139	40	0	0	1014
1978	0	0	0	0	98	157	318	315	157	87	0	0	1132
1979	0	0	0	0	117	214	336	260	295	72	0	0	1294
1980	0	0	0	8	42	91	317	207	145	99	0	0	909
1981	0	0	0	26	78	303	318	301	155	28	7	0	1216
1982	0	0	0	2	67	83	230	213	133	9	0	0	737
1983	0	0	0	0	81	183	235	368	304	92	0	0	1263
1984	0	0	0	6	183	216	419	327	320	57	0	0	1528
1985	0	0	0	22	41	319	380	254	128	48	0	0	1192
1986	0	0	10	9	95	207	315	321	95	47	0	0	1099
1987	0	0	0	34	171	234	220	314	212	100	0	0	1285
1988	0	0	5	22	88	269	484	346	233	92	0	0	1539
1989	0	0	1	60	83	211	354	280	158	32	0	0	1179
1990	0	0	0	33	75	236	399	367	276	82	0	0	1468
1991	0	0	0	3	54	171	379	261	300	208	0	0	1376
1992	0	0	0	23	180	193	330	381	231	81	0	0	1419
1993	0	0	1	9	49	227	294	291	207	38	0	0	1116
1994	0	0	0	9	67	205	285	320	209		0	0	
1995	0	0	0	0	54	152	294	322	228		0	0	
1996	0	0	0	42	91	258	430	422	187	96	0	0	1526
1997	0	0	1	31	227	244	362	348	312	33	11	0	1569
1998	0	0	0	12	2	78	311	371	241	8	0	0	1023
1999	0	0	0	17	30	184	225	253	225	69	0	0	1003
2000	0	0	2	15	111	255	227	290	198	32	0	0	1130
2001	0	0	4	9	219	259	266	300	190	75	0	0	1322

SNOWFALL (inches) 2001 SACRAMENTO, CA (SAC)

YEAR	JUL	AUG	SEP	OCT	NOV	DEC	JAN	FEB	MAR	APR	MAY	JUN	TOTAL
1972-73	0.0	0.0	0.0	0.0	0.0	T	0.0	0.0	0.0	0.0	0.0	0.0	T
1973-74	0.0	0.0	0.0	0.0	0.0	0.0	0.0	T	0.0	0.0	0.0	0.0	T
1974-75	0.0	0.0	0.0	0.0	0.0	0.0	0.0	0.0	0.0	0.0	0.0	0.0	0.0
1975-76	0.0	0.0	0.0	0.0	0.0	0.0	0.0	0.0	2.0	0.0	0.0	0.0	2.0
1976-77	0.0	0.0	0.0	0.0	0.0	0.0	0.0	0.0	0.0	0.0	0.0	0.0	0.0
1977-78	0.0	0.0	0.0	0.0	0.0	0.0	0.0	0.0	0.0	0.0	0.0	0.0	0.0
1978-79	0.0	0.0	0.0	0.0	0.0	0.0	0.0	0.0	0.0	0.0	0.0	0.0	0.0
1979-80	0.0	0.0	0.0	0.0	0.0	0.0	0.0	0.0	0.0	0.0	0.0	0.0	0.0
1980-81	0.0	0.0	0.0	0.0	0.0	0.0	0.0	0.0	0.0	0.0	0.0	0.0	0.0
1981-82	0.0	0.0	0.0	0.0	0.0	0.0	0.0	0.0	T	0.0	0.0	0.0	T
1982-83	0.0	0.0	0.0	0.0	0.0	0.0	0.0	0.0	0.0	0.0	0.0	0.0	0.0
1983-84	0.0	0.0	0.0	0.0	0.0	0.0	0.0	0.0	0.0	0.0	0.0	0.0	0.0
1984-85	0.0	0.0	0.0	0.0	0.0	0.0	0.0	0.0	0.0	0.0	0.0	0.0	0.0
1985-86	0.0	0.0	0.0	0.0	0.0	0.0	0.0	0.0	0.0	0.0	0.0	0.0	0.0
1986-87	0.0	0.0	0.0	0.0	0.0	0.0	0.0	0.0	0.0	0.0	0.0	0.0	0.0
1987-88	0.0	0.0	0.0	0.0	0.0	0.0	0.0	0.0	0.0	0.0	0.0	0.0	T
1988-89	0.0	0.0	0.0	0.0	0.0	T	0.0	0.0	0.0	0.0	0.0	0.0	T
1989-90	0.0	0.0	0.0	0.0	0.0	0.0	0.0	0.0	0.0	0.0	0.0	0.0	0.0
1990-91	0.0	0.0	0.0	0.0	0.0	0.0	0.0	0.0	0.0	0.0	0.0	0.0	0.0
1991-92	0.0	0.0	0.0	0.0	0.0	0.0	0.0	0.0	0.0	0.0	0.0	0.0	0.0
1992-93	0.0	0.0	0.0	0.0	0.0	T	0.0	0.0	0.0	0.0	0.0	0.0	T
1993-94	0.0	0.0	0.0	0.0	0.0	T	0.0	T	0.0	T	0.0	0.0	T
1994-95	0.0	0.0		0.0		0.0	0.0	0.0	0.0	0.0	0.0	0.0	0.0
1995-96	0.0	0.0		0.0	0.0	T	0.0	T	0.0	0.0	0.0	0.0	T
1996-97	0.0	0.0		0.0	0.0	0.0	0.0	0.0	0.0	0.0	0.0	0.0	0.0
1997-98	0.0	0.0	0.0	0.0	0.0	0.0	0.0	0.0	0.0				
1998-99													
1999-00													
2000-01													
2001-													
POR= 48 YRS	0.0	0.0	0.0	0.0	0.0	T	T	0.0	T	0.0	T	0.0	T

2001
SAN DIEGO,
CALIFORNIA (SAN)

The city of San Diego is located on San Diego Bay in the southwest corner of southern California. The prevailing winds and weather are tempered by the Pacific Ocean, with the result that summers are cool and winters warm in comparison with other places along the same general latitude. Temperatures of freezing or below have rarely occurred at the station since the record began in 1871, but hot weather, 90 degrees or above, is more frequent.

Dry easterly winds sometimes blow in the vicinity for several days at a time, bringing temperatures in the 90s and at times even in the 100s in the eastern sections of the city and outlying suburbs. At the National Weather Service station itself, however, there have been relatively few days on which 100 degrees or higher was reached.

As these hot winds are predominant in the fall, highest temperatures occur in the months of September and October. Records show that over 60 percent of the days with 90 degrees or higher have occurred in these two months. High temperatures are almost invariably accompanied by very low relative humidities, which often drop below 20 percent and occasionally below 10 percent.

A marked feature of the climate is the wide variation in temperature within short distances. In nearby valleys daytimes are much warmer in summer and nights noticeably cooler in winter, and freezing occurs much more frequently than in the city. Although records show unusually small daily temperature ranges, only about 15 degrees between the highest and lowest readings, a few miles inland these ranges increase to 30 degrees or more.

Strong winds and gales associated with Pacific, or tropical storms, are infrequent due to the latitude.

The seasonal rainfall is about 10 inches in the city, but increases with elevation and distance from the coast. In the mountains to the north and east the average is between 20 and 40 inches, depending on slope and elevation. Most of the precipitation falls in winter, except in the mountains where there is an occasional thunderstorm. Eighty-five percent of the rainfall occurs from November through March, but wide variations take place in monthly and seasonal totals. Infrequent measurable amounts of hail occur in San Diego, but snow is practically unknown at the Weather Service Office location. In each occurrence of snowfall only a trace was recorded officially, but in some locations amounts up to or slightly exceeding a half-inch fell, and remained on the ground for an hour or more.

As on the rest of the Pacific Coast, a dominant characteristic of spring and summer is the nighttime and early morning cloudiness. Low clouds form regularly and frequently extend inland over the coastal valleys and foothills, but they usually dissipate during the morning and the afternoons are generally clear.

Considerable fog occurs along the coast, but the amount decreases with distance inland. The fall and winter months are usually the foggiest. Thunderstorms are rare, averaging about three a year in the city. Visibilities are good as a rule. The sunshine is plentiful for a marine location, with a marked increase toward the interior.

NORMALS, MEANS, AND EXTREMES
SAN DIEGO, CA   (SAN)

LATITUDE:	LONGITUDE:	ELEVATION (FT):		TIME ZONE:	WBAN: 23188
32 44' 05" N	117 10' 07" W	GRND: 78 BARO: 81		PACIFIC (UTC + 8)	

	ELEMENT	POR	JAN	FEB	MAR	APR	MAY	JUN	JUL	AUG	SEP	OCT	NOV	DEC	YEAR
TEMPERATURE °F	NORMAL DAILY MAXIMUM	30	65.9	66.5	66.3	68.4	69.1	71.6	76.2	77.8	77.1	74.6	69.9	66.1	70.8
	MEAN DAILY MAXIMUM	54	65.3	66.0	66.3	68.2	69.2	71.7	75.6	77.3	76.8	74.0	70.1	66.2	70.6
	HIGHEST DAILY MAXIMUM	61	88	90	93	98	96	101	95	98	111	107	97	88	111
	YEAR OF OCCURRENCE		1953	1995	1988	1989	1953	1979	1985	1955	1963	1961	1976	1963	SEP 1963
	MEAN OF EXTREME MAXS.	54	78.3	78.5	79.0	81.2	78.9	81.4	83.3	85.8	90.3	88.4	83.1	78.2	82.2
	NORMAL DAILY MINIMUM	30	48.9	50.7	52.8	55.6	59.1	61.9	65.7	67.3	65.6	60.9	53.9	48.8	57.6
	MEAN DAILY MINIMUM	54	48.5	50.2	52.2	55.3	58.6	61.4	65.1	66.4	64.9	60.2	53.3	48.6	57.1
	LOWEST DAILY MINIMUM	61	29	36	39	41	48	51	55	57	51	43	38	34	29
	YEAR OF OCCURRENCE		1949	1949	1971	1945	1967	1967	1948	1944	1948	1971	1964	1987	JAN 1949
	MEAN OF EXTREME MINS.	54	40.4	43.4	45.5	49.2	53.3	57.4	61.3	62.5	59.2	52.8	45.3	40.9	50.9
	NORMAL DRY BULB	30	57.4	58.6	59.6	62.0	64.1	66.8	71.0	72.6	71.4	67.7	62.0	57.4	64.2
	MEAN DRY BULB	54	56.8	58.0	59.2	61.7	63.9	66.5	70.5	71.9	70.9	67.1	61.8	57.5	63.8
	MEAN WET BULB	18	51.3	52.5	54.2	56.3	58.8	61.2	64.5	66.1	65.0	61.1	54.8	50.4	58.0
	MEAN DEW POINT	18	44.9	46.8	49.1	51.5	55.1	58.0	61.7	63.2	61.8	56.9	48.8	43.4	53.4
	NORMAL NO. DAYS WITH:														
	MAXIMUM 90	30	0.0	0.0	0.1	0.2	0.1	0.5	0.3	0.2	1.4	0.9	0.2	0.0	3.9
	MAXIMUM 32	30	0.0	0.0	0.0	0.0	0.0	0.0	0.0	0.0	0.0	0.0	0.0	0.0	0.0
	MINIMUM 32	30	*	0.0	0.0	0.0	0.0	0.0	0.0	0.0	0.0	0.0	0.0	0.0	0.0
	MINIMUM 0	30	0.0	0.0	0.0	0.0	0.0	0.0	0.0	0.0	0.0	0.0	0.0	0.0	0.0
H/C	NORMAL HEATING DEG. DAYS	30	245	189	177	113	73	51	13	0	19	24	109	243	1256
	NORMAL COOLING DEG. DAYS	30	9	10	9	23	45	105	199	240	211	107	19	7	984
RH	NORMAL (PERCENT)	30	63	66	67	67	71	74	75	74	73	69	66	64	69
	HOUR 04 LST	30	70	73	75	75	77	81	82	81	80	76	73	70	76
	HOUR 10 LST	30	55	58	60	60	65	69	69	68	66	61	56	54	62
	HOUR 16 LST	30	56	58	59	59	64	66	66	66	65	63	61	58	62
	HOUR 22 LST	30	70	72	72	72	75	78	80	79	78	75	73	71	75
S	PERCENT POSSIBLE SUNSHINE	56	72	71	70	68	59	58	68	70	69	68	75	73	68
W/O	MEAN NO. DAYS WITH:														
	HEAVY FOG(VISBY 1/4 MI)	61	2.9	2.5	1.5	1.1	0.6	0.6	0.6	0.6	2.2	3.1	3.4	3.9	23.0
	THUNDERSTORMS	61	0.2	0.3	0.4	0.2	0.1	0.1	0.2	0.3	0.3	0.3	0.3	0.4	3.1
CLOUDINESS	MEAN:														
	SUNRISE-SUNSET (OKTAS)	56	4.1	4.2	4.2	4.2	4.6	4.4	3.6	3.2	3.2	3.5	3.3	3.7	3.9
	MIDNIGHT-MIDNIGHT (OKTAS)	32	3.9	4.4	4.5	4.1	5.1	5.0	4.3	4.0	4.0	3.8	3.4	3.5	4.2
	MEAN NO. DAYS WITH:														
	CLEAR	56	12.3	10.1	10.8	10.3	8.5	9.3	12.7	15.1	15.0	13.7	14.7	13.6	146.1
	PARTLY CLOUDY	56	7.6	7.6	9.7	10.0	11.3	11.8	12.8	11.5	9.5	9.7	8.0	7.7	117.0
	CLOUDY	56	11.1	10.5	10.7	9.7	11.2	9.0	5.0	4.4	5.5	7.6	7.3	9.6	101.6
PR	MEAN STATION PRESSURE(IN)	29	30.04	30.03	29.98	29.96	29.92	29.89	29.90	29.89	29.87	29.93	30.00	30.03	29.95
	MEAN SEA-LEVEL PRES. (IN)	18	30.06	30.06	30.02	29.97	29.95	29.92	29.93	29.92	29.90	29.95	30.03	30.07	29.98
WINDS	MEAN SPEED (MPH)	38	6.0	6.6	7.4	7.8	7.8	7.8	7.6	7.5	7.2	6.6	6.0	5.6	7.0
	PREVAIL.DIR(TENS OF DEGS)	23	30	30	30	30	30	30	30	30	30	30	30	30	30
	MAXIMUM 2-MINUTE:														
	SPEED (MPH)	5	31	38	38	32	26	23	20	20	21	23	26	34	38
	DIR. (TENS OF DEGS)		14	17	25	26	17	25	28	29	14	29	13	04	25
	YEAR OF OCCURRENCE		2001	1998	2000	1999	1998	2000	1999	1997	2001	1997	1998	1997	MAR 2000
	MAXIMUM 5-SECOND:														
	SPEED (MPH)	5	37	46	45	37	32	25	23	23	24	28	34	43	46
	DIR. (TENS OF DEGS)		14	14	25	26	18	24	27	28	33	29	19	05	14
	YEAR OF OCCURRENCE		2001	2001	2000	1999	1998	2000	1997	2000	1997	1997	1998	1997	FEB 2001
PRECIPITATION	NORMAL (IN)	30	1.80	1.53	1.77	0.79	0.19	0.07	0.02	0.10	0.24	0.37	1.45	1.57	9.90
	MAXIMUM MONTHLY (IN)	61	9.09	7.65	6.96	3.71	1.79	0.87	0.24	2.13	1.90	2.90	5.82	7.60	9.09
	YEAR OF OCCURRENCE		1993	1998	1991	1988	1977	1990	1991	1977	1963	1941	1965	1943	JAN 1993
	MINIMUM MONTHLY (IN)	61	T	0.00	0.00	0.00	0.00	0.00	0.00	0.00	0.00	0.00	0.00	0.01	0.00
	YEAR OF OCCURRENCE		1976	1967	1997	1993	1952	1981	1982	1981	1979	1967	1980	2000	MAR 1997
	MAXIMUM IN 24 HOURS (IN)	61	2.65	2.61	2.40	1.98	1.50	0.82	0.23	2.13	1.00	1.39	2.44	3.07	3.07
	YEAR OF OCCURRENCE		1978	1979	1952	1988	1977	1990	1991	2013	1986	1986	1944	1945	DEC 1945
	NORMAL NO. DAYS WITH:														
	PRECIPITATION 0.01	30	6.3	5.3	6.6	4.3	2.2	1.1	0.5	0.6	1.5	2.3	5.0	5.5	41.2
	PRECIPITATION 1.00	30	0.3	0.2	0.3	0.1	*	0.0	0.0	*	0.0	0.0	0.3	0.3	1.5
SNOWFALL	NORMAL (IN)	30	0.0	0.0	0.0	0.0	0.0	0.0	0.0	0.0	0.0	0.0	0.0	T	0.0
	MAXIMUM MONTHLY (IN)	59	T	0.0	T	T	0.0	0.0	0.0	0.0	0.0	0.0	T	T	T
	YEAR OF OCCURRENCE		1949		1985	1999							1985	1967	APR 1999
	MAXIMUM IN 24 HOURS (IN)	59	T	0.0	T	T	0.0	0.0	0.0	0.0	0.0	0.0	T	T	T
	YEAR OF OCCURRENCE		1949		1985	1999							1985	1967	APR 1999
	MAXIMUM SNOW DEPTH (IN)	51	0	0	0	0	0	0	0	0	0	0	0	0	0
	YEAR OF OCCURRENCE														
	NORMAL NO. DAYS WITH:														
	SNOWFALL 1.0	30	0.0	0.0	0.0	0.0	0.0	0.0	0.0	0.0	0.0	0.0	0.0	0.0	0.0

PRECIPITATION (inches) 2001 SAN DIEGO, CA (SAN)

YEAR	JAN	FEB	MAR	APR	MAY	JUN	JUL	AUG	SEP	OCT	NOV	DEC	ANNUAL
1972	0.07	0.10	T	0.02	0.10	0.38	T	0.02	0.44	0.58	3.16	1.61	6.48
1973	1.68	1.63	2.26	0.05	T	T	T	T	0.02	0.01	1.63	0.19	7.47
1974	2.96	0.04	1.70	0.02	0.01	0.02	0.01	T	T	1.03	0.14	2.20	8.13
1975	0.49	0.96	3.79	2.00	0.01	0.02	T	T	T	0.09	0.64	0.37	8.37
1976	T	5.40	0.99	1.33	0.27	0.02	0.02	0.01	1.00	0.38	0.75	1.06	11.23
1977	2.36	0.06	0.61	0.01	1.79	0.03	T	2.13	T	0.50	0.05	1.67	9.21
1978	5.95	2.64	5.00	0.73	0.04	T	0.00	T	0.72	0.05	2.09	2.19	19.41
1979	5.82	0.85	3.71	0.02	0.09	0.01	0.09	0.01	0.00	0.73	0.27	0.02	11.62
1980	5.58	4.47	2.71	1.18	0.65	0.01	T	0.00	T	0.05	0.00	0.31	14.96
1981	1.48	2.26	3.74	0.22	0.04	0.00	T	0.00	0.03	0.14	1.79	0.54	10.24
1982	2.71	0.88	4.74	0.62	0.01	0.04	0.00	T	0.38	0.05	2.10	1.43	12.96
1983	2.10	3.88	6.57	1.74	0.01	T	0.01	0.39	0.21	0.40	1.94	1.53	18.78
1984	0.46	0.09	0.04	0.62	0.00	0.04	0.19	0.06	T	0.29	2.37	4.55	8.71
1985	0.52	0.77	0.58	0.32	T	T	0.00	T	0.20	0.29	4.92	1.06	8.66
1986	0.75	2.59	3.12	1.17	0.00	T	0.01	0.00	1.04	1.39	1.16	0.95	12.18
1987	1.68	1.53	1.04	0.78	0.03	T	0.03	0.01	0.70	1.74	1.33	2.73	11.60
1988	0.89	1.37	0.59	3.71	0.08	0.00	T	T	T	T	1.39	2.23	10.26
1989	0.42	0.70	0.69	0.12	0.04	0.06	0.00	T	0.23	0.47	0.09	1.01	3.83
1990	2.52	1.13	0.25	0.76	0.51	0.87	T	0.01	T	0.65	0.59	0.59	7.29
1991	1.06	2.46	6.96	0.05	0.01	T	0.24	0.01	0.28	0.69	0.05	1.70	13.51
1992	1.81	3.34	4.42	0.28	0.07	0.04	0.03	0.05	0.00	0.18	0.03	2.56	12.81
1993	9.09	4.73	1.22	0.00	0.01	0.41	0.03	T	T	0.22	0.77	0.78	17.26
1994	0.70	2.75	3.67	0.93	0.07	T	0.03	0.01	T	0.01	0.46	0.80	9.43
1995	8.06	1.93	3.81	0.96	0.59	0.46	0.05	0.00	T	T	0.30	0.88	17.04
1996	1.52	0.88	1.10	0.36	0.02	0.00	0.09	T	0.03	0.94	1.70	0.63	7.27
1997	3.02	0.31	0.00	0.28	T	T	T	0.00	0.85	0.02	1.17	1.35	7.00
1998	2.68	7.65	2.21	1.11	0.64	0.10	0.20	T	0.03	0.08	0.69	0.66	16.05
1999	1.54	0.70	1.09	1.62	0.06	0.04	T	0.00	0.02	0.00	0.04	0.32	5.43
2000	0.17	3.67	1.00	0.54	T	T	0.00	0.01	T	1.24	0.00	0.01	6.90
2001	3.28	2.38	0.63	0.76	0.01	0.00	T	0.00	0.00	0.00	0.95	0.46	8.47
POR= 152 YRS	2.19	1.72	1.90	0.79	0.19	0.08	0.03	0.08	0.19	0.35	1.14	1.34	10.00

AVERAGE TEMPERATURE (F) 2001 SAN DIEGO, CA (SAN)

YEAR	JAN	FEB	MAR	APR	MAY	JUN	JUL	AUG	SEP	OCT	NOV	DEC	ANNUAL
1972	54.9	57.8	60.2	62.3	64.7	67.0	72.7	72.2	68.7	65.6	59.8	57.5	63.6
1973	55.6	59.9	58.1	61.5	63.4	68.0	69.1	70.5	68.8	66.8	60.6	58.2	63.4
1974	56.9	58.2	59.1	62.0	63.3	66.9	71.4	70.2	70.3	66.8	62.2	56.3	63.6
1975	56.1	56.4	57.5	58.7	62.2	65.0	69.4	68.9	71.5	65.9	60.4	56.9	62.4
1976	58.9	59.6	60.3	61.0	65.2	69.7	71.1	72.4	73.8	71.2	66.8	60.7	65.9
1977	60.3	61.7	57.5	61.4	61.9	65.8	71.6	73.1	72.2	68.9	64.9	63.3	65.2
1978	61.0	60.9	64.3	63.4	68.2	71.3	71.6	72.9	74.0	70.1	61.7	55.2	66.2
1979	56.9	56.9	60.1	63.4	65.6	70.2	71.8	73.9	76.3	68.7	62.4	60.6	65.6
1980	61.1	63.5	61.5	63.9	63.8	68.5	72.9	74.2	70.4	67.3	62.7	60.8	65.9
1981	61.3	62.2	61.1	64.4	67.3	72.9	75.6	75.8	73.7	67.1	63.5	60.3	67.1
1982	56.6	60.7	60.5	63.8	65.8	66.7	71.9	73.5	73.1	70.1	62.1	57.4	65.2
1983	60.7	60.9	62.0	62.4	66.2	68.1	72.6	77.4	76.8	72.2	64.4	60.6	67.0
1984	61.2	60.2	63.7	64.3	68.1	69.9	77.2	76.6	78.9	68.5	61.4	56.7	67.2
1985	57.0	57.2	58.9	63.6	64.8	69.0	75.3	72.4	69.8	67.9	60.1	58.0	64.5
1986	61.0	58.9	60.5	62.8	64.6	67.4	69.6	71.8	66.9	65.5	62.8	57.6	64.1
1987	55.4	58.0	59.1	63.4	64.7	65.8	67.1	69.9	69.9	69.5	61.8	53.9	63.2
1988	56.7	59.9	61.6	62.4	63.9	64.9	70.4	71.0	70.0	66.7	60.1	56.0	63.6
1989	54.7	56.7	59.8	65.6	63.7	66.0	70.1	71.0	70.4	66.3	63.1	58.7	63.8
1990	56.6	55.2	58.7	63.2	64.3	69.0	72.3	71.6	71.7	68.6	62.7	55.6	64.1
1991	57.4	59.4	56.5	61.7	62.1	64.1	67.4	68.9	69.4	68.0	62.3	57.3	62.9
1992	57.4	61.1	60.4	67.0	68.0	68.1	71.8	74.9	72.4	68.2	62.6	55.3	65.6
1993	56.9	58.0	61.3	63.8	66.0	68.6	69.8	70.2	69.0	67.3	61.6	57.0	64.1
1994	57.9	56.5	60.4	61.0	62.1	68.1	69.5	74.0	72.5	66.8	56.4	55.8	63.4
1995	56.9	61.4	60.4	61.5	62.0	64.8	69.0	71.9	71.5	67.1	63.2	58.3	64.0
1996	57.6	58.8	60.1	64.6	64.8	67.8	70.0	72.8	70.6	64.3	61.6	57.8	64.4
1997	58.0	58.4	61.6	62.5	68.7	67.5	69.3	72.9	75.2	68.7	64.0	57.4	65.4
1998	58.2	57.3	59.3	59.6	62.8	65.7	68.8	72.8	70.3	65.7	59.7	55.4	63.0
1999	57.5	57.8	58.3	58.9	60.5	62.8	68.5	68.0	67.0	68.8	60.5	57.8	62.2
2000	58.2	59.0	58.1	62.4	64.6	68.2	69.2	72.0	70.5	65.1	58.4	58.1	63.7
2001	54.7	54.9	58.9	58.5	63.7	67.6	69.0	69.6	68.2	66.3	61.4	55.5	62.4
POR= 127 YRS	55.6	56.6	58.0	60.3	62.5	65.2	68.8	70.2	69.0	65.2	60.8	57.0	62.4

REFERENCE NOTES:

PAGE 1:
THE TEMPERATURE GRAPH SHOWS NORMAL MAXIMUM AND NORMAL MINIMUM DAILY TEMPERATURES (SOLID CURVES) AND THE ACTUAL DAILY HIGH AND LOW TEMPERATURES (VERTICAL BARS).

PAGE 2 AND 3:
H/C INDICATES HEATING AND COOLING DEGREE DAYS.
RH INDICATES RELATIVE HUMIDITY
W/O INDICATES WEATHER AND OBSTRUCTIONS
S INDICATES SUNSHINE.
PR INDICATES PRESSURE.
CLOUDINESS ON PAGE 3 IS THE SUM OF THE CEILOMETER AND SATELLITE DATA NOT TO EXCEED EIGHT EIGHTHS(OKTAS).

GENERAL:
T INDICATES TRACE PRECIPITATION, AN AMOUNT GREATER THAN ZERO BUT LESS THAN THE LOWEST REPORTABLE VALUE.
+ INDICATES THE VALUE ALSO OCCURS ON EARLIER DATES.
BLANK ENTRIES DENOTE MISSING OR UNREPORTED DATA.
NORMALS ARE 30-YEAR AVERAGES (1961 - 1990).
ASOS INDICATES AUTOMATED SURFACE OBSERVING SYSTEM.
PM INDICATES THE LAST DAY OF THE PREVIOUS MONTH.
POR (PERIOD OF RECORD) BEGINS WITH THE JANUARY DATA MONTH AND IS THE NUMBER OF YEARS USED TO COMPUTE THE MEAN. INDIVIDUAL MONTHS WITHIN THE POR MAY BE MISSING.
WHEN THE POR FOR A NORMAL IS LESS THAN 30 YEARS, THE NORMAL IS PROVISIONAL AND IS BASED ON THE NUMBER OF YEARS INDICATED.
0.* OR * INDICATES THE VALUE OR MEAN-DAYS-WITH IS BETWEEN 0.00 AND 0.05.
CLOUDINESS FOR ASOS STATIONS DIFFERS FROM THE NON-ASOS OBSERVATION TAKEN BY A HUMAN OBSERVER. ASOS STATION CLOUDINESS IS BASED ON TIME-AVERAGED CEILOMETER DATA FOR CLOUDS AT OR BELOW 12,000 FEET AND ON SATELLITE DATA FOR CLOUDS ABOVE 12,000 FEET.
THE NUMBER OF DAYS WITH CLEAR, PARTLY CLOUDY, AND CLOUDY CONDITIONS FOR ASOS STATIONS IS THE SUM OF THE CEILOMETER AND SATELLITE DATA FOR THE SUNRISE TO SUNSET PERIOD.

GENERAL CONTINUED:
CLEAR INDICATES 0 - 2 OKTAS, PARTLY CLOUDY INDICATES 3 - 6 OKTAS, AND CLOUDY INDICATES 7 OR 8 OKTAS. WHEN AT LEAST ONE OF THE ELEMENTS (CEILOMETER OR SATELLITE) IS MISSING, THE DAILY CLOUDINESS IS NOT COMPUTED.
WIND DIRECTION IS RECORDED IN TENS OF DEGREES (2 DIGITS) CLOCKWISE FROM TRUE NORTH. "00" INDICATES CALM. "36" INDICATES TRUE NORTH.
RESULTANT WIND IS THE VECTOR AVERAGE OF THE SPEED AND DIRECTION.
AVERAGE TEMPERATURE IS THE SUM OF THE MEAN DAILY MAXIMUM AND MINIMUM TEMPERATURE DIVIDED BY 2.
SNOWFALL DATA COMPRISE ALL FORMS OF FROZEN PRECIPITATION, INCLUDING HAIL.
A HEATING (COOLING) DEGREE DAY IS THE DIFFERENCE BETWEEN THE AVERAGE DAILY TEMPERATURE AND 65 F.
DRY BULB IS THE TEMPERATURE OF THE AMBIENT AIR.
DEW POINT IS THE TEMPERATURE TO WHICH THE AIR MUST BE COOLED TO ACHIEVE 100 PERCENT RELATIVE HUMIDITY.
WET BULB IS THE TEMPERATURE THE AIR WOULD HAVE IF THE MOISTURE CONTENT WAS INCREASED TO 100 PERCENT RELATIVE HUMIDITY.

ON JULY 1, 1996, THE NATIONAL WEATHER SERVICE BEGAN USING THE "METAR" OBSERVATION CODE THAT WAS ALREADY EMPLOYED BY MOST OTHER NATIONS OF THE WORLD. THE MOST NOTICEABLE DIFFERENCE IN THIS ANNUAL PUBLICATION WILL BE THE CHANGE IN UNITS FROM TENTHS TO EIGHTS(OKTAS) FOR REPORTING THE AMOUNT OF SKY COVER.

HEATING DEGREE DAYS (base 65 F) 2001 SAN DIEGO, CA (SAN)

YEAR	JUL	AUG	SEP	OCT	NOV	DEC	JAN	FEB	MAR	APR	MAY	JUN	TOTAL
1972-73	0	0	0	29	149	224	286	131	208	107	61	1	1196
1973-74	0	0	0	6	132	205	243	184	176	85	55	4	1090
1974-75	0	0	0	14	97	265	273	237	225	182	83	10	1386
1975-76	0	0	0	19	141	246	196	150	148	115	16	0	1031
1976-77	0	0	0	0	39	129	143	94	224	103	88	3	823
1977-78	0	0	0	0	37	55	117	117	52	43	8	0	429
1978-79	0	0	0	0	102	297	244	219	153	45	20	6	1086
1979-80	0	0	0	4	75	136	117	50	104	61	43	1	591
1980-81	0	0	0	6	75	133	113	101	116	40	1	0	585
1981-82	0	0	0	9	57	136	258	119	139	64	9	2	793
1982-83	0	0	0	1	93	228	137	110	88	83	9	0	749
1983-84	0	0	0	0	66	130	123	134	51	43	4	0	551
1984-85	0	0	0	4	104	250	238	219	183	60	18	2	1078
1985-86	0	0	0	3	145	211	118	173	132	85	29	0	896
1986-87	0	0	7	10	66	223	291	197	178	72	21	6	1071
1987-88	0	0	0	0	98	338	250	147	125	85	53	22	1118
1988-89	0	0	0	4	141	275	313	237	158	37	40	14	1219
1989-90	0	0	1	13	67	188	252	268	185	52	39	1	1066
1990-91	0	0	0	3	88	284	227	152	254	104	96	28	1236
1991-92	0	0	0	24	96	231	228	115	136	8	1	0	839
1992-93	0	0	0	0	84	294	240	191	111	44	7	6	977
1993-94	0	0	0	4	103	242	214	235	147	116	85	1	1147
1994-95	0	0	0	12	249	276	242	106	136	105	87	24	1237
1995-96	0	0	0	3	49	200	223	171	146	54	6	0	852
1996-97	0	0	0	45	115	216	217	185	124	81	0	1	984
1997-98	0	0	0	2	65	227	203	208	172	155	69	9	1110
1998-99	1	0	0	13	152	293	224	198	200	189	132	64	1466
1999-00	4	0	3	3	130	217	212	171	206	82	23	0	1051
2000-01	0	0	0	27	191	208	314	281	183	191	39	5	1439
2001-	0	0	0	4	107	289							

COOLING DEGREE DAYS (base 65 F) 2001 SAN DIEGO, CA (SAN)

YEAR	JAN	FEB	MAR	APR	MAY	JUN	JUL	AUG	SEP	OCT	NOV	DEC	ANNUAL
1972	0	0	0	4	33	68	247	230	117	53	0	1	753
1973	0	0	0	10	17	97	133	176	121	70	8	1	633
1974	0	0	0	2	9	69	204	169	164	75	19	0	711
1975	0	0	0	0	1	18	142	124	201	54	8	0	548
1976	14	0	10	3	31	147	196	240	269	200	102	0	1212
1977	5	9	0	2	1	34	212	258	224	128	40	8	921
1978	1	7	38	4	115	194	213	251	276	166	11	0	1276
1979	0	0	10	6	46	169	216	283	348	124	5	8	1215
1980	2	13	3	35	15	110	253	289	170	86	15	7	998
1981	7	29	0	26	81	244	335	343	265	80	21	0	1431
1982	0	7	6	32	42	58	219	271	250	164	12	0	1061
1983	11	0	1	9	51	99	242	392	364	231	55	0	1455
1984	13	0	15	31	107	156	387	366	422	119	4	0	1620
1985	0	7	0	22	19	128	325	235	153	104	6	0	999
1986	2	11	4	29	23	78	152	218	73	31	9	0	630
1987	0	6	5	29	17	35	71	158	154	147	10	0	632
1988	0	5	28	16	25	26	176	193	161	64	0	5	699
1989	0	13	2	63	5	48	165	193	168	58	17	0	732
1990	0	0	2	6	21	127	233	211	209	123	25	0	957
1991	0	0	0	9	11	10	80	130	138	122	21	0	521
1992	0	12	1	72	99	100	216	313	228	106	20	0	1167
1993	0	0	4	12	47	119	155	170	126	81	8	0	722
1994	0	0	10	3	1	100	146	289	231	76	0	0	856
1995	0	14	1	8	1	27	133	223	199	75	2	0	683
1996	0	0	4	53	70	92	164	251	175	30	19	0	858
1997	6	4	23	12	119	83	138	252	314	123	40	2	1116
1998	0	0	1	0	6	34	127	250	167	40	0	0	625
1999	0	0	0	14	0	5	118	100	69	126	3	0	435
2000	8	5	0	8	19	101	137	226	170	39	0	1	714
2001	0	3	2	5	4	90	132	149	102	53	4	0	544

SNOWFALL (inches) 2001 SAN DIEGO, CA (SAN)

YEAR	JUL	AUG	SEP	OCT	NOV	DEC	JAN	FEB	MAR	APR	MAY	JUN	TOTAL
1972-73	0.0	0.0	0.0	0.0	0.0	0.0	0.0	0.0	0.0	0.0	0.0	0.0	0.0
1973-74	0.0	0.0	0.0	0.0	0.0	0.0	0.0	0.0	0.0	0.0	0.0	0.0	0.0
1974-75	0.0	0.0	0.0	0.0	0.0	0.0	0.0	0.0	0.0	0.0	0.0	0.0	0.0
1975-76	0.0	0.0	0.0	0.0	0.0	0.0	0.0	0.0	0.0	0.0	0.0	0.0	0.0
1976-77	0.0	0.0	0.0	0.0	0.0	0.0	0.0	0.0	0.0	0.0	0.0	0.0	0.0
1977-78	0.0	0.0	0.0	0.0	0.0	0.0	0.0	0.0	0.0	0.0	0.0	0.0	0.0
1978-79	0.0	0.0	0.0	0.0	0.0	0.0	0.0	0.0	0.0	0.0	0.0	0.0	0.0
1979-80	0.0	0.0	0.0	0.0	0.0	0.0	0.0	0.0	0.0	0.0	0.0	0.0	0.0
1980-81	0.0	0.0	0.0	0.0	0.0	0.0	0.0	0.0	0.0	0.0	0.0	0.0	0.0
1981-82	0.0	0.0	0.0	0.0	0.0	0.0	0.0	0.0	0.0	0.0	0.0	0.0	0.0
1982-83	0.0	0.0	0.0	0.0	0.0	0.0	0.0	0.0	0.0	0.0	0.0	0.0	0.0
1983-84	0.0	0.0	0.0	0.0	0.0	0.0	0.0	0.0	0.0	0.0	0.0	0.0	0.0
1984-85	0.0	0.0	0.0	0.0	0.0	0.0	0.0	0.0	T	0.0	0.0	0.0	T
1985-86	0.0	0.0	0.0	0.0	0.0	T	0.0	0.0	0.0	0.0	0.0	0.0	T
1986-87	0.0	0.0	0.0	0.0	0.0	0.0	0.0	0.0	0.0	0.0	0.0	0.0	0.0
1987-88	0.0	0.0	0.0	0.0	0.0	0.0	0.0	0.0	0.0	0.0	0.0	0.0	0.0
1988-89	0.0	0.0	0.0	0.0	0.0	0.0	0.0	0.0	0.0	0.0	0.0	0.0	0.0
1989-90	0.0	0.0	0.0	0.0	0.0	0.0	0.0	0.0	0.0	0.0	0.0	0.0	0.0
1990-91	0.0	0.0	0.0	0.0	0.0	0.0	0.0	0.0	0.0	0.0	0.0	0.0	0.0
1991-92	0.0	0.0	0.0	0.0	0.0	0.0	0.0	0.0	0.0	0.0	0.0	0.0	0.0
1992-93	0.0	0.0	0.0	0.0	0.0	0.0	0.0	0.0	0.0	0.0	0.0	0.0	0.0
1993-94	0.0	0.0	0.0	0.0	0.0	0.0	0.0	0.0	0.0	0.0	0.0	0.0	0.0
1994-95	0.0	0.0	0.0	0.0	0.0	0.0	0.0	0.0	0.0	0.0	0.0	0.0	0.0
1995-96	0.0	0.0	0.0	0.0	0.0	0.0	0.0	0.0	0.0	0.0	0.0	0.0	0.0
1996-97	0.0												
1997-98				0.0	0.0	0.0	0.0	0.0	T	0.0	0.0		T
1998-99	0.0	0.0	0.0	0.0	0.0	0.0	0.0	0.0	0.0	T	0.0	0.0	T
1999-00	0.0	0.0	0.0	0.0	0.0	0.0	0.0	0.0	0.0	0.0			
2000-01													
2001-													
POR= 57 YRS	0.0	0.0	0.0	0.0	T	T	T	0.0	T	0.0	0.0	0.0	T

2001
SAN FRANCISCO, CALIFORNIA
INTERNATIONAL AIRPORT (SFO)

The station is located in the central Terminal Building of the San Francisco International Airport, which is on flat filled tideland on the west shore of San Francisco Bay. The bay borders the airport from the north to the south-southeast. San Bruno Mountain, 5 miles to the north-northwest, rises to 1,300 feet. A north-south trending ridge of coastal mountains, 4 miles to the west, varies in elevation from 700 to 1,900 feet, being highest southward along the peninsula. The Pacific Ocean west of the ridge is 6 miles from the airport. A broad gap to the northwest of the station, between San Bruno Mountain and the coastal mountains, allows a strong flow of marine air over the station and dominate the local climate.

San Francisco Airport enjoys a marine-type climate characterized by mild and moderately wet winters and by dry, cool summers. Winter rains, occurring from November through March, account for over 80 percent of the annual rainfall, and measurable precipitation occurs on an average of 10 days per month during this period. However, there are frequent dry periods lasting well over a week. Severe winter storms with gale winds and heavy rains occur only occasionally. Thunderstorms average two a year and may occur in any month.

The daily and annual range in temperature is small. A few frosty mornings occur during the winter but the temperature seldom drops below freezing. Winter temperatures generally rise to the high 50s in the early afternoon.

The summer weather is dominated by a cool sea breeze resulting in an average summer wind speed of nearly 15 mph. Winds are light in the early morning but normally reach 20 to 25 mph in the afternoon.

A sea fog, arriving over the station during the late evening or night as a low cloud, is another persistent feature of the summer weather. This high fog, occasionally producing drizzle or mist, usually disappears during the late forenoon. Despite the morning overcast, summer days are sunny. On the average a total of only 14 days during the four months from June through September are classified as cloudy.

Daytime temperatures are held down both by the morning low overcast and the afternoon strengthening sea breeze, resulting in daily maximum readings averaging about 70 degrees from May through August. However, during these months occasional hot spells, lasting a few days, are experienced without the usual high fog and sea breeze. September, when the sea breeze becomes less pronounced, is the warmest month with highs in the 70s. Low temperatures during the summer are in the mid-50s.

A strong temperature inversion with its base usually about 1,500 feet persists throughout the summer. Inversions close to the ground are infrequent in summer but rather common in fall and winter. As a consequence of these factors and the continued population and economic growth of the area, atmospheric pollution has become a problem of increasing importance.

NORMALS, MEANS, AND EXTREMES
SAN FRANCISCO, CA (SFO)

LATITUDE:	LONGITUDE:	ELEVATION (FT):	TIME ZONE:	WBAN: 23234
37 37' 11" N	122 23' 53" W	GRND: 86 BARO: 89	PACIFIC (UTC + 8)	

	ELEMENT	POR	JAN	FEB	MAR	APR	MAY	JUN	JUL	AUG	SEP	OCT	NOV	DEC	YEAR
TEMPERATURE °F	NORMAL DAILY MAXIMUM	30	55.6	59.4	60.8	63.9	66.5	70.3	71.6	72.3	73.6	70.1	62.4	56.1	65.2
	MEAN DAILY MAXIMUM	56	55.6	59.0	60.9	63.8	66.4	69.9	71.2	71.9	73.3	70.1	62.7	56.3	65.1
	HIGHEST DAILY MAXIMUM	74	72	78	85	92	97	106	105	100	103	99	85	75	106
	YEAR OF OCCURRENCE		1948	1930	1952	1989	1984	1961	1988	1993	1971	1987	1967	1958	JUN 1961
	MEAN OF EXTREME MAXS.	56	64.5	68.3	72.9	78.9	84.1	88.4	86.6	86.0	90.1	85.9	74.0	64.6	78.7
	NORMAL DAILY MINIMUM	30	41.8	45.0	45.8	47.2	49.7	52.6	53.9	55.0	55.2	51.8	47.1	42.7	49.0
	MEAN DAILY MINIMUM	56	42.0	44.6	45.8	47.4	49.9	52.4	53.7	54.6	54.4	51.6	47.1	42.8	48.9
	LOWEST DAILY MINIMUM	74	24	25	30	31	36	41	43	42	38	34	25	20	20
	YEAR OF OCCURRENCE		1928	1929	1929	1929	1929	1932	1928	1935	1929	1931	1929	1932	DEC 1932
	MEAN OF EXTREME MINS.	56	33.5	36.8	38.7	41.3	44.7	48.1	50.0	50.6	49.4	45.0	39.0	34.7	42.7
	NORMAL DRY BULB	30	48.7	52.2	53.3	55.6	58.1	61.5	62.7	63.7	64.5	61.0	54.8	49.4	57.1
	MEAN DRY BULB	56	48.9	51.8	53.4	55.6	58.3	61.2	62.5	63.3	63.9	60.9	54.9	49.5	57.0
	MEAN WET BULB	17	47.3	48.7	50.2	51.2	53.2	55.4	57.0	57.8	57.3	55.2	50.8	46.7	52.6
	MEAN DEW POINT	17	44.0	45.1	46.2	46.8	48.7	50.9	53.0	54.2	53.7	50.9	46.4	42.9	48.6
	NORMAL NO. DAYS WITH:														
	MAXIMUM 90	30	0.0	0.0	0.0	0.1	0.4	0.9	0.7	0.3	1.4	0.4	0.0	0.0	4.2
	MAXIMUM 32	30	0.0	0.0	0.0	0.0	0.0	0.0	0.0	0.0	0.0	0.0	0.0	0.0	0.0
	MINIMUM 32	30	1.2	0.1	*	0.0	0.0	0.0	0.0	0.0	0.0	0.0	0.0	1.0	2.3
	MINIMUM 0	30	0.0	0.0	0.0	0.0	0.0	0.0	0.0	0.0	0.0	0.0	0.0	0.0	0.0
H/C	NORMAL HEATING DEG. DAYS	30	505	358	363	287	218	121	92	68	79	135	306	484	3016
	NORMAL COOLING DEG. DAYS	30	0	0	0	5	0	16	21	28	64	11	0	0	145
RH	NORMAL (PERCENT)	30	78	76	73	71	71	71	73	74	72	72	75	78	74
	HOUR 04 LST	30	86	84	81	81	83	84	86	86	83	82	84	85	84
	HOUR 10 LST	30	79	75	70	65	63	63	65	67	65	68	72	77	69
	HOUR 16 LST	30	66	64	62	59	59	58	59	61	58	59	63	67	61
	HOUR 22 LST	30	80	78	77	76	78	79	82	82	78	77	77	80	79
S	PERCENT POSSIBLE SUNSHINE														
W/O	MEAN NO. DAYS WITH:														
	HEAVY FOG (VISBY 1/4 MI)	64	3.3	2.6	0.4	0.1	0.1	0.0	0.0	0.2	0.6	1.3	2.1	3.0	13.7
	THUNDERSTORMS	74	0.4	0.4	0.3	0.2	0.1	0.1	0.1	0.1	0.2	0.3	0.1	0.2	2.5
CLOUDINESS	MEAN:														
	SUNRISE-SUNSET (OKTAS)	55	5.0	5.0	4.6	4.2	3.6	3.0	2.4	2.6	2.6	3.1	4.2	4.7	3.8
	MIDNIGHT-MIDNIGHT (OKTAS)	32	4.7	4.9	4.6	3.9	3.4	3.2	2.6	2.8	2.6	3.0	4.0	4.3	3.7
	MEAN NO. DAYS WITH:														
	CLEAR	69	8.5	7.8	9.5	10.8	13.5	16.2	20.3	18.8	17.9	15.5	11.1	9.4	159.3
	PARTLY CLOUDY	69	7.5	7.3	8.7	9.3	9.7	8.6	7.5	8.6	8.2	8.9	8.3	7.5	100.1
	CLOUDY	69	15.0	13.1	12.8	9.9	7.8	5.2	2.8	3.3	3.6	6.5	10.6	14.1	104.7
PR	MEAN STATION PRESSURE (IN)	28	30.09	30.05	30.03	30.03	29.97	29.95	29.95	29.94	29.93	30.00	30.08	30.11	30.01
	MEAN SEA-LEVEL PRES. (IN)	17	30.12	30.08	30.06	30.05	30.00	29.96	29.96	29.96	29.94	30.01	30.11	30.13	30.03
WINDS	MEAN SPEED (MPH)	43	7.7	9.0	11.0	12.5	13.9	14.1	13.7	12.8	11.4	9.7	8.1	7.7	11.0
	PREVAIL.DIR (TENS OF DEGS)	27	14	30	29	29	28	28	30	30	30	30	30	14	30
	MAXIMUM 2-MINUTE:														
	SPEED (MPH)	5	43	44	46	45	46	43	38	37	32	40	46	45	46
	DIR. (TENS OF DEGS)		17	27	26	26	26	26	26	26	26	26	24	16	24
	YEAR OF OCCURRENCE		1997	2001	2000	2001	1999	1997	1997	1999	1997	1997	2001	2001	NOV 2001
	MAXIMUM 5-SECOND:														
	SPEED (MPH)	5	56	59	51	54	54	53	44	46	39	49	55	55	59
	DIR. (TENS OF DEGS)		16	18	26	28	26	26	26	26	25	29	24	16	18
	YEAR OF OCCURRENCE		1997	1999	2000	1999	1999	1997	1997	2001	1997	1997	2001	2001	FEB 1999
PRECIPITATION	NORMAL (IN)	30	4.35	3.17	3.06	1.37	0.19	0.11	0.03	0.05	0.20	1.22	2.86	3.09	19.70
	MAXIMUM MONTHLY (IN)	74	11.26	13.64	9.01	6.36	3.81	0.86	0.35	0.66	2.30	7.30	7.94	12.30	13.64
	YEAR OF OCCURRENCE		1993	1998	1958	1958	1957	1967	1977	1976	1959	1962	1973	1955	FEB 1998
	MINIMUM MONTHLY (IN)	74	0.24	T	T	T	0.00	0.00	0.00	0.00	T	0.00	0.00	0.01	0.00
	YEAR OF OCCURRENCE		1991	1953	1934	1977	2001	1928	1930	1996	1995	1978	1929	1989	MAY 2001
	MAXIMUM IN 24 HOURS (IN)	74	5.71	3.41	2.46	2.66	1.54	0.83	0.35	0.60	2.30	3.74	2.43	3.33	5.71
	YEAR OF OCCURRENCE		1982	1998	1982	1958	1957	1967	1977	1997	1959	1962	1994	1955	JAN 1982
	NORMAL NO. DAYS WITH:														
	PRECIPITATION 0.01	30	9.8	8.9	10.4	6.1	2.3	0.7	0.4	0.6	1.4	3.6	8.1	9.4	61.7
	PRECIPITATION 1.00	30	1.4	0.8	0.5	0.2	0.0	0.0	0.0	0.0	0.0	0.4	0.6	0.7	4.6
SNOWFALL	NORMAL (IN)	30	0.1	T	0.0	0.0	0.0	0.0	0.0	0.0	0.0	0.0	0.0	T	0.1
	MAXIMUM MONTHLY (IN)	69	1.5	T	T	0.0	0.0	0.0	0.0	0.0	0.0	0.0	0.0	1.0	1.5
	YEAR OF OCCURRENCE		1962	1996	1995									1932	JAN 1962
	MAXIMUM IN 24 HOURS (IN)	69	1.5	T	T	T	0.0	0.0	0.0	0.0	0.0	0.0	0.0	1.0	1.5
	YEAR OF OCCURRENCE		1962	1996	1995	1987								1932	JAN 1962
	MAXIMUM SNOW DEPTH (IN)	48	0	0	0	0	0	0	0	0	0	0	0	0	0
	YEAR OF OCCURRENCE														
	NORMAL NO. DAYS WITH:														
	SNOWFALL 1.0	30	0.*	0.0	0.0	0.0	0.0	0.0	0.0	0.0	0.0	0.0	0.0	0.0	0.0

PRECIPITATION (inches) 2001 SAN FRANCISCO, CALIFORNIA CA (SFO)

YEAR	JAN	FEB	MAR	APR	MAY	JUN	JUL	AUG	SEP	OCT	NOV	DEC	ANNUAL
1972	1.09	1.35	0.18	1.20	T	0.06	T	T	0.30	5.24	5.15	2.40	16.97
1973	8.32	6.82	2.93	0.11	0.07	T	T	T	0.04	1.60	7.94	3.55	31.38
1974	3.21	1.70	4.21	2.32	T	0.14	0.23	T	0.93	0.50	2.36	15.60
1975	2.60	3.94	5.91	1.66	0.02	0.04	0.13	0.21	T	2.27	0.26	0.21	17.25
1976	0.37	2.13	1.22	0.92	T	0.01	T	0.66	0.30	0.34	1.37	2.70	10.02
1977	2.22	1.04	2.01	T	0.41	T	0.35	T	0.47	0.15	2.20	3.69	12.54
1978	8.90	4.92	4.90	4.50	0.02	T	T	T	0.26	T	1.67	0.64	25.81
1979	6.61	5.87	2.74	0.69	0.13	T	0.09	T	T	2.20	1.94	4.30	24.57
1980	4.85	7.62	2.65	0.90	0.24	0.03	0.10	T	T	0.10	0.12	1.73	18.34
1981	5.92	2.21	3.60	0.24	0.07	T	T	T	0.28	2.35	4.89	3.91	23.47
1982	8.81	2.82	7.63	3.25	T	0.06	T	T	0.96	1.95	5.34	3.99	34.81
1983	6.83	6.64	8.50	3.11	0.32	T	0.01	T	0.57	0.10	6.03	6.23	38.34
1984	0.46	1.47	1.36	0.68	T	0.03	T	0.11	0.05	1.96	6.12	1.89	14.13
1985	0.74	2.35	3.30	0.12	0.05	0.29	0.03	0.02	0.18	0.69	3.19	1.61	12.57
1986	4.04	8.09	5.84	0.39	0.15	T	0.01	T	0.47	0.02	0.06	1.66	20.73
1987	2.80	3.52	1.98	0.16	0.06	T	T	T	0.93	1.64	4.51	15.60
1988	3.92	0.38	0.05	2.02	0.29	0.60	T	0.03	0.42	2.31	3.65	13.67
1989	1.25	1.28	4.00	0.78	0.04	0.01	T	1.24	1.40	1.34	0.01	11.35
1990	3.06	2.28	0.79	0.20	1.55	T	0.01	T	0.20	0.19	0.28	1.79	10.35
1991	0.24	3.76	6.07	0.61	0.21	0.11	T	0.27	0.04	1.73	0.23	2.70	15.97
1992	2.04	6.44	4.12	0.25	T	0.39	0.00	0.14	T	1.12	0.15	6.04	20.69
1993	11.26	4.68	2.34	0.41	0.55	0.16	T	T	T	0.45	1.47	2.19	23.51
1994	2.50	5.26	0.24	1.12	1.52	0.03	T	T	0.10	0.33	5.73	2.49	19.32
1995	8.89	0.38	8.75	1.41	0.93	0.60	T	T	T	0.03	0.02	6.41	27.42
1996	6.92	6.03	2.89	1.40	1.24	T	T	0.00	T	0.76	2.56	6.97	28.77
1997	7.52	0.31	0.25	0.30	0.21	0.24	T	0.60	T	0.68	6.41	3.87	20.39
1998	8.20	13.64	2.05	2.24	2.37	0.03	T	0.00	0.09	0.62	2.43	0.96	32.63
1999	2.96	4.59	2.80	2.18	0.10	0.18	0.00	0.06	0.27	0.46	1.47	0.43	15.50
2000	5.83	8.46	1.74	1.30	0.53	0.14	0.00	0.01	0.07	2.14	0.91	0.44	21.57
2001	3.87	6.12	1.02	1.56	0.00	0.10	0.00	0.00	0.11	0.31	4.51	8.54	26.14
POR= 74 YRS	4.47	3.50	2.88	1.33	0.39	0.13	0.02	0.05	0.20	0.97	2.40	3.42	19.76

AVERAGE TEMPERATURE (F) 2001 SAN FRANCISCO, CALIFORNIA CA (SFO)

YEAR	JAN	FEB	MAR	APR	MAY	JUN	JUL	AUG	SEP	OCT	NOV	DEC	ANNUAL
1972	45.6	52.0	55.3	55.4	57.6	60.9	64.3	64.0	62.6	61.0	53.2	44.9	56.4
1973	48.0	52.9	51.4	56.7	58.7	63.5	62.1	60.7	63.5	60.7	53.9	50.0	56.8
1974	48.6	49.7	53.4	54.8	56.2	60.4	63.0	63.7	62.5	62.0	53.8	48.6	56.4
1975	47.4	50.9	51.4	50.6	58.4	59.8	61.8	63.0	61.4	58.8	52.2	49.3	55.4
1976	48.5	50.5	51.0	53.3	58.3	63.2	62.5	64.3	63.3	61.3	57.0	48.8	56.8
1977	47.0	53.2	50.9	55.5	55.8	60.4	62.4	64.1	63.5	60.5	55.3	52.3	56.7
1978	52.5	52.8	57.0	54.9	60.4	60.4	61.4	63.2	65.8	61.1	52.5	46.0	57.3
1979	47.5	50.3	54.5	55.5	60.4	61.0	63.8	63.7	67.3	62.6	54.1	50.9	57.6
1980	50.5	54.4	53.0	55.9	56.3	59.9	63.0	61.5	63.2	61.2	55.6	50.7	57.1
1981	51.1	54.0	53.2	56.2	59.0	65.0	61.7	63.0	62.7	58.6	56.3	52.2	57.8
1982	45.1	51.7	51.3	54.6	57.4	59.7	61.7	63.3	64.0	61.1	52.3	48.9	55.9
1983	48.0	53.4	54.1	54.7	57.8	61.5	65.1	66.9	68.3	64.4	55.5	53.4	58.6
1984	51.3	52.9	57.0	56.0	61.8	61.3	65.6	64.4	69.7	60.4	53.8	47.7	58.5
1985	46.4	51.6	51.4	59.0	58.6	65.2	64.8	64.0	63.2	60.7	52.0	47.1	57.0
1986	53.7	56.3	57.1	56.2	58.6	62.5	62.4	61.2	62.9	61.4	57.1	50.3	58.3
1987	49.3	53.3	54.9	59.2	61.6	62.4	65.1	65.1	64.0	63.9	57.0	50.5	58.7
1988	50.6	54.5	56.5	58.1	59.5	62.5	65.3	65.0	63.1	61.4	56.5	50.4	58.6
1989	48.3	48.4	54.9	60.8	59.8	62.7	62.8	64.0	61.4	60.8	56.4	50.1	57.5
1990	49.9	49.2	53.3	58.5	59.0	62.4	64.5	66.3	66.4	63.1	56.0	46.4	57.9
1991	50.1	55.3	52.2	55.7	56.9	59.4	63.7	64.4	62.9	62.8	57.5	50.8	57.6
1992	48.9	56.1	57.4	60.6	63.4	63.6	65.8	63.8	65.3	65.4	56.8	49.8	59.7
1993	49.3	52.5	57.5	58.1	62.2	64.8	64.8	67.3	63.1	63.4	56.3	50.1	59.1
1994	51.5	50.6	56.0	56.8	58.5	61.5	62.1	64.5	64.1	60.1	49.9	48.0	57.0
1995	52.2	54.1	53.9	55.0	57.0	60.8	64.8	63.0	63.2	62.5	58.4	53.7	58.2
1996	51.8	54.9	56.0	59.1	60.4	61.5	63.2	62.7	63.0	60.9	56.3	53.9	58.6
1997	50.9	53.4	56.1	58.2	64.4	62.3	64.0	66.9	68.3	61.8	57.7	51.1	59.6
1998	52.8	52.3	55.0	55.7	57.7	61.6	62.8	64.3	64.3	60.3	54.2	47.0	57.3
1999	49.3	50.4	51.0	54.6	55.5	59.6	61.9	63.8	63.6	62.5	57.1	51.1	56.7
2000	52.4	53.5	54.4	58.2	60.2	62.9	61.4	63.6	66.2	60.0	52.6	51.6	58.1
2001	48.8	51.0	55.7	53.6	62.6	62.6	62.7	63.2	62.9	62.4	57.3	51.7	57.9
POR= 74 YRS	48.7	51.5	53.3	55.2	58.0	60.8	62.1	62.7	63.4	60.5	54.6	49.6	56.7

REFERENCE NOTES:

PAGE 1:
THE TEMPERATURE GRAPH SHOWS NORMAL MAXIMUM AND NORMAL
   MINIMUM DAILY TEMPERATURES (SOLID CURVES) AND THE
   ACTUAL DAILY HIGH AND LOW TEMPERATURES (VERTICAL BARS).

PAGE 2 AND 3:
H/C INDICATES HEATING AND COOLING DEGREE DAYS.
RH INDICATES RELATIVE HUMIDITY
W/O INDICATES WEATHER AND OBSTRUCTIONS
S INDICATES SUNSHINE.
PR INDICATES PRESSURE.
CLOUDINESS ON PAGE 3 IS THE SUM OF THE CEILOMETER AND
   SATELLITE DATA NOT TO EXCEED EIGHT EIGHTHS(OKTAS).

GENERAL:
T INDICATES TRACE PRECIPITATION, AN AMOUNT GREATER
   THAN ZERO BUT LESS THAN THE LOWEST REPORTABLE VALUE.
+ INDICATES THE VALUE ALSO OCCURS ON EARLIER DATES.
BLANK ENTRIES DENOTE MISSING OR UNREPORTED DATA.
NORMALS ARE 30-YEAR AVERAGES (1961 - 1990).
ASOS INDICATES AUTOMATED SURFACE OBSERVING SYSTEM.
PM INDICATES THE LAST DAY OF THE PREVIOUS MONTH.
POR (PERIOD OF RECORD) BEGINS WITH THE JANUARY DATA
   MONTH AND IS THE NUMBER OF YEARS USED TO COMPUTE
   THE MEAN. INDIVIDUAL MONTHS WITHIN THE POR MAY
   BE MISSING.
WHEN THE POR FOR A NORMAL IS LESS THAN 30 YEARS,
   THE NORMAL IS PROVISIONAL AND IS BASED ON THE NUMBER
   OF YEARS INDICATED.
0.* OR * INDICATES THE VALUE OR MEAN-DAYS-WITH
   IS BETWEEN 0.00 AND 0.05.
CLOUDINESS FOR ASOS STATIONS DIFFERS FROM THE NON-ASOS
   OBSERVATION TAKEN BY A HUMAN OBSERVER. ASOS STATION
   CLOUDINESS IS BASED ON TIME-AVERAGED CEILOMETER DATA
   FOR CLOUDS AT OR BELOW 12,000 FEET AND ON SATELLITE
   DATA FOR CLOUDS ABOVE 12,000 FEET.
THE NUMBER OF DAYS WITH CLEAR, PARTLY CLOUDY, AND
   CLOUDY CONDITIONS FOR ASOS STATIONS IS THE SUM
   OF THE CEILOMETER AND SATELLITE DATA FOR THE
   SUNRISE TO SUNSET PERIOD.

GENERAL CONTINUED:
CLEAR INDICATES 0 - 2 OKTAS, PARTLY CLOUDY INDICATES
   3 - 6 OKTAS, AND CLOUDY INDICATES 7 OR 8 OKTAS.
   WHEN AT LEAST ONE OF THE ELEMENTS (CEILOMETER OR
   SATELLITE) IS MISSING, THE DAILY CLOUDINESS IS
   NOT COMPUTED.
WIND DIRECTION IS RECORDED IN TENS OF DEGREES (2 DIGITS)
   CLOCKWISE FROM TRUE NORTH. "00" INDICATES CALM. "36"
   INDICATES TRUE NORTH.
RESULTANT WIND IS THE VECTOR AVERAGE OF THE SPEED AND
   DIRECTION.
AVERAGE TEMPERATURE IS THE SUM OF THE MEAN DAILY MAXIMUM
   AND MINIMUM TEMPERATURE DIVIDED BY 2.
SNOWFALL DATA COMPRISE ALL FORMS OF FROZEN
   PRECIPITATION, INCLUDING HAIL.
A HEATING (COOLING) DEGREE DAY IS THE DIFFERENCE BETWEEN
   THE AVERAGE DAILY TEMPERATURE AND 65 F.
DRY BULB IS THE TEMPERATURE OF THE AMBIENT AIR.
DEW POINT IS THE TEMPERATURE TO WHICH THE AIR MUST BE
   COOLED TO ACHIEVE 100 PERCENT RELATIVE HUMIDITY.
WET BULB IS THE TEMPERATURE THE AIR WOULD HAVE IF THE
   MOISTURE CONTENT WAS INCREASED TO 100 PERCENT RELATIVE
   HUMIDITY.

ON JULY 1, 1996, THE NATIONAL WEATHER SERVICE BEGAN USING
   THE "METAR" OBSERVATION CODE THAT WAS ALREADY EMPLOYED
   BY MOST OTHER NATIONS OF THE WORLD. THE MOST NOTICEABLE
   DIFFERENCE IN THIS ANNUAL PUBLICATION WILL BE THE CHANGE
   IN UNITS FROM TENTHS TO EIGHTS(OKTAS) FOR REPORTING THE
   AMOUNT OF SKY COVER.

HEATING DEGREE DAYS (base 65 F) 2001 SAN FRANCISCO, CALIFORNIA CA (SFO)

YEAR	JUL	AUG	SEP	OCT	NOV	DEC	JAN	FEB	MAR	APR	MAY	JUN	TOTAL
1972-73	64	42	75	131	350	613	521	334	416	241	197	103	3087
1973-74	100	129	76	128	327	459	501	424	354	298	270	145	3211
1974-75	83	54	101	117	329	499	540	387	415	424	220	151	3320
1975-76	109	75	120	188	377	480	504	415	427	344	222	136	3397
1976-77	72	38	79	127	231	494	549	326	432	278	278	141	3045
1977-78	103	48	55	139	284	385	381	335	238	295	161	135	2559
1978-79	111	65	32	143	371	581	536	406	319	277	148	132	3121
1979-80	55	56	13	85	320	431	441	298	366	269	261	155	2750
1980-81	76	109	74	145	275	436	424	301	358	279	180	53	2710
1981-82	112	65	71	197	252	389	611	364	416	307	241	154	3179
1982-83	100	63	47	130	376	491	521	322	330	301	225	113	3019
1983-84	33	1	18	43	281	354	415	342	242	269	124	115	2237
1984-85	43	34	5	147	328	527	570	370	416	180	192	27	2839
1985-86	49	51	60	158	382	546	343	236	239	260	191	78	2593
1986-87	77	113	62	122	228	447	477	320	309	168	128	85	2536
1987-88	60	16	40	70	233	445	440	296	259	212	184	78	2333
1988-89	40	29	71	128	246	447	511	455	308	162	160	94	2651
1989-90	70	38	103	138	249	454	459	437	356	189	185	94	2772
1990-91	33	13	8	77	262	570	454	265	387	273	244	166	2752
1991-92	57	44	65	96	223	434	494	252	226	133	52	58	2134
1992-93	20	49	34	40	235	466	480	346	201	201	100	54	2250
1993-94	33	11	71	67	256	455	408	396	273	238	200	118	2526
1994-95	91	43	46	154	446	522	388	301	338	293	242	143	3007
1995-96	48	79	73	98	192	344	398	289	270	187	149	128	2255
1996-97	75	73	71	155	255	333	431	317	270	194	76	74	2324
1997-98	32	8	2	109	222	422	370	348	301	269	218	109	2410
1998-99	81	59	57	147	320	550	483	403	428	315	289	170	3302
1999-00	104	47	75	92	229	423	381	327	328	207	169	94	2476
2000-01	106	56	32	151	367	409	497	384	282	334	107	95	2820
2001-	76	60	74	100	225	408							

COOLING DEGREE DAYS (base 65 F) 2001 SAN FRANCISCO, CALIFORNIA CA (SFO)

YEAR	JAN	FEB	MAR	APR	MAY	JUN	JUL	AUG	SEP	OCT	NOV	DEC	ANNUAL
1972	0	0	1	0	7	14	49	16	8	14	0	0	109
1973	0	0	0	0	7	64	18	2	37	1	0	0	129
1974	0	0	0	0	6	11	27	22	31	30	0	0	127
1975	0	0	0	0	21	4	15	19	19	2	0	0	80
1976	0	0	0	0	21	88	4	23	33	23	0	0	192
1977	0	0	0	0	0	10	30	26	17	5	0	0	88
1978	0	0	0	0	24	0	7	18	62	33	0	0	144
1979	0	0	0	0	11	19	25	21	88	18	0	0	182
1980	0	0	0	0	0	10	22	7	30	33	1	0	103
1981	0	0	0	17	1	61	17	7	7	3	0	0	113
1982	0	0	0	1	12	5	7	15	23	12	0	0	75
1983	0	0	0	0	7	16	42	66	119	32	0	0	282
1984	0	0	0	4	33	10	70	24	152	9	0	0	302
1985	0	0	0	8	1	38	50	28	11	33	0	0	169
1986	0	0	1	2	0	11	5	0	7	16	0	0	42
1987	0	0	0	4	29	15	9	26	17	43	0	0	143
1988	0	0	0	11	19	8	55	34	23	24	0	0	174
1989	0	0	0	0	40	6	35	8	15	2	16	0	122
1990	0	0	0	1	3	23	58	58	13	24	0	0	190
1991	0	0	0	0	0	6	21	29	13	33	2	0	104
1992	0	0	0	9	10	21	53	18	49	58	0	0	218
1993	0	0	0	0	18	58	33	91	19	26	1	0	246
1994	0	0	0	0	4	18	4	30	27	8	0	0	91
1995	0	0	0	0	0	26	48	26	25	29	0	0	154
1996	0	0	0	18	16	30	22	9	21	36	0	0	152
1997	0	0	1	0	65	1	5	78	107	15	7	0	279
1998	0	0	0	0	0	13	23	44	45	6	0	0	131
1999	0	0	0	9	0	15	17	19	41	24	0	0	125
2000	0	0	2	10	27	36	15	2	20	73	1	0	171
2001	0	0	0	0	38	30	13	10	16	28	0	0	135

SNOWFALL (inches) 2001 SAN FRANCISCO, CALIFORNIA CA (SFO)

YEAR	JUL	AUG	SEP	OCT	NOV	DEC	JAN	FEB	MAR	APR	MAY	JUN	TOTAL
1972-73	0.0	0.0	0.0	0.0	0.0	T	0.0	0.0	T	0.0	0.0	0.0	T
1973-74	0.0	0.0	0.0	0.0	0.0	0.0	0.0	0.0	0.0	0.0	0.0	0.0	0.0
1974-75	0.0	0.0	0.0	0.0	0.0	0.0	0.0	T	0.0	0.0	0.0	0.0	T
1975-76	0.0	0.0	0.0	0.0	0.0	0.0	0.0	T	T	0.0	0.0	0.0	T
1976-77	0.0	0.0	0.0	0.0	0.0	0.0	0.0	0.0	0.0	0.0	0.0	0.0	0.0
1977-78	0.0	0.0	0.0	0.0	0.0	0.0	0.0	0.0	0.0	0.0	0.0	0.0	0.0
1978-79	0.0	0.0	0.0	0.0	0.0	0.0	T	0.0	0.0	0.0	0.0	0.0	T
1979-80	0.0	0.0	0.0	0.0	0.0	0.0	0.0	0.0	T	0.0	0.0	0.0	T
1980-81	0.0	0.0	0.0	0.0	0.0	0.0	0.0	0.0	T	0.0	0.0	0.0	T
1981-82	0.0	0.0	0.0	0.0	0.0	0.0	0.0	T	0.0	T	0.0	0.0	T
1982-83	0.0	0.0	0.0	0.0	0.0	0.0	0.0	0.0	T	T	0.0	0.0	T
1983-84	0.0	0.0	0.0	0.0	0.0	0.0	0.0	0.0	0.0	0.0	0.0	0.0	0.0
1984-85	0.0	0.0	0.0	0.0	0.0	0.0	0.0	0.0	0.0	0.0	0.0	0.0	0.0
1985-86	0.0	0.0	0.0	0.0	0.0	0.0	0.0	0.0	0.0	0.0	0.0	0.0	0.0
1986-87	0.0	0.0	0.0	0.0	0.0	0.0	0.0	0.0	T	0.0	0.0	0.0	T
1987-88	0.0	0.0	0.0	0.0	0.0	0.0	0.0	0.0	0.0	0.0	0.0	0.0	0.0
1988-89	0.0	0.0	0.0	0.0	0.0	T	T	0.0	T	0.0	0.0	0.0	T
1989-90	0.0	0.0	0.0	0.0	0.0	0.0	0.0	0.0	0.0	0.0	0.0	0.0	0.0
1990-91	0.0	0.0	0.0	0.0	0.0	0.0	0.0	0.0	0.0	0.0	0.0	0.0	0.0
1991-92	0.0	0.0	0.0	0.0	0.0	0.0	0.0	0.0	T	0.0	0.0	0.0	T
1992-93	0.0	0.0	0.0	0.0	0.0	T	0.0	0.0	0.0	0.0	0.0	0.0	T
1993-94	0.0	0.0	0.0	0.0	0.0	0.0	0.0	T	0.0	0.0	0.0	0.0	T
1994-95	0.0	0.0	0.0	0.0	0.0	0.0	0.0	0.0	T	0.0	0.0	0.0	T
1995-96	0.0	0.0	0.0	0.0	0.0	0.0	0.0	T	0.0	0.0	0.0	0.0	T
1996-97	0.0	0.0	0.0										
1997-98													
1998-99													
1999-00													
2000-01													
2001-													
POR= 68 YRS	0.0	0.0	0.0	0.0	0.0	0.0	0.0	T	T	0.0	0.0	0.0	T

2001
COLORADO SPRINGS,
COLORADO (COS)

At an elevation near 6,200 feet above sea level, Colorado Springs is located in relatively flat semi-arid country on the eastern slope of the Rocky Mountains. Immediately to the west the mountains rise abruptly to heights ranging from 10,000 to 14,000 feet but generally averaging near 11,000 feet. To the east lie gently undulating prairie lands. The land slopes upward to the north, reaching an average height of about 8,000 feet in 20 miles at the top of Palmer Lake Divide.

Colorado Springs is in the Arkansas River drainage basin. The principal tributary feeding the Arkansas from this area is Fountain Creek which rises in the high mountains west of the city and is fed by Monument Creek originating to the north in the Palmer Lake Divide area.

Other topographical features of the area, and particularly its wide range of elevations, help to give Colorado Springs the various and altogether delightful plains and mountain mixture of climate that has established the locality as a highly desirable place to live. The higher elevations immediately to the west and north of the city produce significant differences in temperature and precipitation. Precipitation amounts at these higher elevations are approximately twice those at nearby lower elevations and the number of rainy days is almost triple.

In Colorado Springs itself, precipitation is relatively sparse. Over 80 percent of it falls between April 1 and September 30, mostly as heavy downpours accompanying summer thunderstorms. Temperatures, in view of the station latitude and elevation, are mild. Uncomfortable extremes, in either summer or winter, are comparatively rare and of short duration. Relative humidity is normally low and wind movement moderately high. This is notably true of the west-to-east movement of the chinook winds, that cause rapid rises in winter temperatures and remind us that the Indian meaning of CHINOOK is SNOW EATER.

Colorado Springs is best known as a resort city, but is also important to the high-tech industry and military community. Several military installations, including the United States Air Force Academy and the Space Command are located within or near the city. The surrounding prairie is also important for cattle raising and a considerable amount of grazing land is used for sheep in the summer months. The growing season varies considerably in length but averages from the first week in May to the first week of October.

NORMALS, MEANS, AND EXTREMES
COLORADO SPRINGS, CO (COS)

	LATITUDE:	LONGITUDE:	ELEVATION (FT):	TIME ZONE:	WBAN: 93037
	38 48' 43" N	104 42' 40" W	GRND: 6180 BARO: 6183	MOUNTAIN (UTC + 7)	

	ELEMENT	POR	JAN	FEB	MAR	APR	MAY	JUN	JUL	AUG	SEP	OCT	NOV	DEC	YEAR
TEMPERATURE °F	NORMAL DAILY MAXIMUM	30	41.4	44.6	50.0	59.8	68.7	79.0	84.4	81.3	73.6	63.5	50.7	42.2	61.6
	MEAN DAILY MAXIMUM	52	42.2	45.3	50.1	59.3	68.7	79.1	84.7	82.1	74.6	64.1	50.7	43.6	62.0
	HIGHEST DAILY MAXIMUM	52	73	76	81	87	94	100	100	99	94	86	78	77	100
	YEAR OF OCCURRENCE		1997	1963	1971	1992	2000	1954	1954	1954	1995	1979	1981	1955	JUL 1954
	MEAN OF EXTREME MAXS.	24	63.3	65.2	70.1	78.8	84.3	92.5	95.3	92.1	88.2	80.1	70.7	64.1	78.7
	NORMAL DAILY MINIMUM	30	16.1	19.3	24.6	33.0	42.1	51.1	57.1	55.2	47.1	36.3	24.9	17.4	35.4
	MEAN DAILY MINIMUM	52	16.3	19.6	24.3	32.7	42.3	51.2	56.8	55.3	47.3	36.4	24.7	17.8	35.4
	LOWEST DAILY MINIMUM	52	-26	-27	-11	-3	21	32	42	39	22	5	-8	-24	-27
	YEAR OF OCCURRENCE		1951	1951	1956	1959	1954	1951	1952	1992	1985	1969	1976	1990	FEB 1951
	MEAN OF EXTREME MINS.	24	-1.7	-.2	7.9	18.4	30.4	40.9	50.3	48.7	33.0	20.4	8.0	-1.4	21.2
	NORMAL DRY BULB	30	28.8	32.0	37.3	46.4	55.4	65.0	70.8	68.3	60.4	49.9	37.8	29.8	48.5
	MEAN DRY BULB	53	29.5	32.5	37.3	46.0	55.7	65.3	70.8	68.6	60.9	50.3	37.7	30.8	48.8
	MEAN WET BULB	56	22.9	25.6	29.5	36.2	44.5	52.1	56.9	56.0	48.9	39.1	29.5	23.9	38.8
	MEAN DEW POINT	56	10.9	13.5	17.5	23.7	34.1	41.7	48.4	48.4	38.9	27.0	18.2	12.3	27.9
	NORMAL NO. DAYS WITH:														
	MAXIMUM 90	30	0.0	0.0	0.0	0.0	0.2	4.3	10.8	3.3	0.8	0.0	0.0	0.0	19.4
	MAXIMUM 32	30	8.3	6.6	2.7	0.7	0.0	0.0	0.0	0.0	0.1	0.3	3.3	7.8	29.8
	MINIMUM 32	30	30.3	26.2	25.1	12.8	2.1	0.0	0.0	0.0	1.1	9.1	24.4	29.3	160.4
	MINIMUM 0	30	2.6	2.3	0.3	0.0	0.0	0.0	0.0	0.0	0.0	0.0	0.0	2.3	7.5
H/C	NORMAL HEATING DEG. DAYS	30	1122	924	859	558	302	87	6	18	164	468	816	1091	6415
	NORMAL COOLING DEG. DAYS	30	0	0	0	0	0	87	186	120	26	0	0	0	419
RH	NORMAL (PERCENT)	30	51	50	50	46	49	48	50	54	51	46	50	51	50
	HOUR 05 LST	30	56	59	61	61	66	66	68	70	66	58	59	56	62
	HOUR 11 LST	30	41	40	40	35	37	36	36	40	39	35	39	42	38
	HOUR 17 LST	30	46	41	38	34	36	34	39	41	37	36	44	48	40
	HOUR 23 LST	30	57	58	58	55	59	57	60	64	60	56	58	57	58
S	PERCENT POSSIBLE SUNSHINE														
W/O	MEAN NO. DAYS WITH:														
	HEAVY FOG(VISBY 1/4 MI)	52	2.4	2.9	2.9	2.3	1.9	0.8	0.6	1.0	2.0	1.9	2.7	2.2	23.6
	THUNDERSTORMS	52	0.0	0.0	0.5	2.4	8.5	11.1	15.9	13.7	4.8	0.8	0.0	0.0	57.7
CLOUDINESS	MEAN:														
	SUNRISE-SUNSET (OKTAS)	1	3.2	3.7	5.6	6.0	4.8	3.6	3.6	5.6	3.2	4.0	3.2	2.4	4.1
	MIDNIGHT-MIDNIGHT (OKTAS)	1	4.0	3.6	6.4	6.0	4.8	3.2	3.6	5.6	3.2	4.4	2.8	2.4	4.2
	MEAN NO. DAYS WITH:														
	CLEAR	2	5.3	8.3	8.0	5.0	1.0	3.0	7.0	6.0	11.0	10.5	6.0	11.0	82.1
	PARTLY CLOUDY	2	2.0	5.0	5.7	5.5	5.0	4.0	4.0	8.5	4.0	1.0	2.0	4.5	51.2
	CLOUDY	2	1.7	6.7	7.0	8.0	2.0	2.0	1.5	7.5	6.0	5.0	1.5	1.0	49.9
PR	MEAN STATION PRESSURE(IN)	28	23.89	23.88	23.82	23.88	23.90	23.98	24.06	24.07	24.04	24.00	23.92	23.91	23.95
	MEAN SEA-LEVEL PRES. (IN)	55	30.09	30.06	29.96	29.92	29.89	29.88	29.95	29.97	29.99	30.04	30.08	30.09	29.99
WINDS	MEAN SPEED (MPH)	35	9.5	10.0	11.0	11.5	11.2	10.5	9.3	8.8	9.4	9.6	9.6	9.5	10.0
	PREVAIL.DIR(TENS OF DEGS)	20	36	36	36	36	36	16	36	36	36	36	36	36	36
	MAXIMUM 2-MINUTE:														
	SPEED (MPH)	8	49	61	45	61	51	47	49	45	44	59	47	53	61
	DIR. (TENS OF DEGS)		27	28	30	28	36	27	22	34	30	27	30	29	28
	YEAR OF OCCURRENCE		1996	1999	2000	1996	2001	1998	2000	2000	2001	2001	1998	1993	FEB 1999
	MAXIMUM 5-SECOND:														
	SPEED (MPH)	8	55	78	53	68	62	62	60	55	55	70	57	63	78
	DIR. (TENS OF DEGS)		26	28	31	30	27	28	23	25	33	30	26	28	28
	YEAR OF OCCURRENCE		1996	1999	2000	1996	2001	2001	2000	1995	1996	1996	1995	2000	FEB 1999
PRECIPITATION	NORMAL (IN)	30	0.29	0.40	0.94	1.19	2.15	2.25	2.90	3.02	1.33	0.84	0.47	0.46	16.24
	MAXIMUM MONTHLY (IN)	52	1.17	2.45	2.42	7.50	5.67	8.00	5.27	7.04	4.28	5.01	2.21	1.05	8.00
	YEAR OF OCCURRENCE		1987	1987	1998	1999	1957	1965	1968	1999	1976	1984	1957	1988	JUN 1965
	MINIMUM MONTHLY (IN)	52	T	T	0.01	0.01	0.33	0.13	0.67	0.15	T	0.01	T	T	DEC 1995
	YEAR OF OCCURRENCE		1995	1991	1966	1964	1974	1990	1987	1962	1953	1980	1995	1995	
	MAXIMUM IN 24 HOURS (IN)	52	0.79	1.49	1.63	3.30	2.57	3.09	3.66	4.11	1.73	1.60	1.45	0.69	4.11
	YEAR OF OCCURRENCE		1987	1987	1998	1999	1955	1954	1997	1999	1959	1960	1979	1981	AUG 1999
	NORMAL NO. DAYS WITH:														
	PRECIPITATION 0.01	30	6.2	5.8	8.3	7.7	12.5	9.7	11.0	14.0	6.8	5.2	4.3	6.4	97.9
	PRECIPITATION 1.00	30	0.0	0.1	0.1	0.2	0.4	0.2	0.4	0.5	0.2	0.1	0.0	0.0	2.2
SNOWFALL	NORMAL (IN)	30	6.7	7.1	12.1	4.5	2.5	0.0	0.0	0.0	0.2	2.8	5.3	10.2	51.4
	MAXIMUM MONTHLY (IN)	53	28.7	23.2	23.2	42.7	19.4	1.1	T	T	27.9	25.9	26.3	18.2	42.7
	YEAR OF OCCURRENCE		1987	1987	1984	1957	1978	1975	1992	1992	1959	1984	1991	1983	APR 1957
	MAXIMUM IN 24 HOURS (IN)	53	22.0	14.8	15.0	18.0	17.4	1.1	T	T	17.1	19.9	14.5	9.6	22.0
	YEAR OF OCCURRENCE		1987	1987	1998	1957	1978	1975	1992	1992	1959	1997	1972	1979	JAN 1987
	MAXIMUM SNOW DEPTH (IN)	25	16	12	15	12	11	0	0	0	2	20	11	10	20
	YEAR OF OCCURRENCE		1987	1987	1998	1997	1978				1985	1997	1979	1979	OCT 1997
	NORMAL NO. DAYS WITH:														
	SNOWFALL 1.0	30	1.9	1.9	3.1	1.3	0.4	0.0	0.0	0.0	0.1	0.6	1.5	3.0	13.8

PRECIPITATION (inches) 2001 COLORADO SPRINGS, CO (COS)

YEAR	JAN	FEB	MAR	APR	MAY	JUN	JUL	AUG	SEP	OCT	NOV	DEC	ANNUAL
1972	0.27	0.25	0.55	0.42	1.46	2.07	4.08	3.55	4.13	1.34	1.08	0.83	20.03
1973	0.06	0.06	1.16	1.72	4.27	0.47	3.31	0.89	1.03	0.35	0.15	0.64	14.11
1974	0.26	0.18	0.52	1.92	0.33	1.29	1.42	1.14	0.43	1.36	0.23	0.42	9.50
1975	0.13	0.29	0.24	0.68	1.00	2.97	2.65	2.06	0.16	0.52	1.00	0.07	11.77
1976	0.32	0.23	0.63	1.63	2.09	2.46	1.75	5.94	4.28	0.49	0.40	0.12	20.34
1977	0.29	0.20	1.18	2.57	1.12	3.87	3.02	5.11	0.45	0.19	0.60	0.18	18.78
1978	0.25	0.38	0.40	1.15	3.58	0.54	2.14	2.51	0.05	0.90	0.37	1.01	13.28
1979	0.53	0.04	2.38	1.83	3.13	1.58	2.73	2.50	0.92	0.55	1.82	1.02	19.03
1980	0.25	0.54	1.30	3.64	4.99	1.60	1.69	4.59	0.65	0.01	0.35	0.05	19.66
1981	0.07	0.12	0.93	0.13	3.14	1.98	3.64	5.24	0.52	0.37	0.03	0.82	16.99
1982	0.25	0.27	0.73	0.76	3.07	3.81	3.64	5.37	3.02	0.22	0.10	0.70	21.94
1983	0.43	0.09	1.79	0.97	3.08	2.41	0.99	2.59	0.37	0.28	1.09	0.70	14.79
1984	0.32	0.09	1.93	1.66	0.74	1.54	3.97	4.03	0.93	5.01	0.14	0.64	21.00
1985	0.42	0.24	1.68	2.07	3.36	0.78	4.92	1.56	1.49	0.52	0.42	0.55	18.01
1986	0.01	0.30	0.31	0.65	1.89	2.47	1.63	6.06	0.61	1.41	0.64	0.28	16.26
1987	1.17	2.45	1.79	0.50	3.82	2.89	0.67	2.77	0.55	0.54	0.44	0.64	18.23
1988	0.43	0.68	0.90	0.27	1.01	1.69	2.07	2.88	1.19	0.08	0.36	1.05	12.61
1989	0.23	1.23	0.49	1.06	1.11	3.42	2.26	2.63	2.30	0.28	0.02	0.41	15.44
1990	0.53	0.59	1.77	2.04	3.90	0.13	5.13	1.45	1.50	1.46	0.30	0.27	19.07
1991	0.09	T	0.42	1.76	0.80	3.07	2.87	4.57	0.56	0.88	2.05	0.45	17.52
1992	0.06	0.02	2.36	0.92	2.07	3.91	0.76	3.37	0.13	0.30	0.75	0.11	14.76
1993	0.52	0.21	0.79	1.02	1.60	1.27	2.38	2.17	1.44	0.91	0.97	0.11	13.39
1994	0.18	0.28	0.54	1.49	4.10	4.32	1.29	3.92	1.52	2.67	0.32	0.13	20.76
1995	T	0.21	0.71	3.05	4.81	7.78	1.91	1.77	1.87	0.02	T	T	22.13
1996	0.16	0.34	0.82	0.39	2.22	1.58	1.46	3.46	2.04	0.89	0.17	0.04	16.57
1997	0.11	0.18	0.34	3.30	1.16	5.44	4.63	4.70	1.78	0.98	0.22	0.10	22.94
1998	0.03	0.34	2.42	1.38	0.72	1.27	5.26	2.75	0.51	0.93	0.44	0.15	16.20
1999	0.12	0.05	0.41	7.50	3.57	1.36	4.70	7.04	0.52	1.10	1.01	0.20	27.58
2000	0.68	0.23	1.97	0.62	1.27	1.73	2.72	5.82	0.55	0.86	0.19	0.25	16.89
2001	0.82	0.26	1.38	0.98	3.21	2.14	3.25	1.47	1.01	0.02	0.37	0.09	15.00
POR= 53 YRS	0.30	0.33	0.92	1.39	2.30	2.26	2.95	2.97	1.25	0.82	0.48	0.34	16.31

AVERAGE TEMPERATURE (F) 2001 COLORADO SPRINGS, CO (COS)

YEAR	JAN	FEB	MAR	APR	MAY	JUN	JUL	AUG	SEP	OCT	NOV	DEC	ANNUAL
1972	30.0	36.1	43.3	49.1	55.9	67.8	68.9	67.7	60.8	49.4	29.8	23.5	48.5
1973	25.4	32.3	36.5	40.9	53.3	65.6	68.4	70.6	58.8	52.6	39.6	31.2	47.9
1974	27.0	33.9	42.1	46.0	59.7	66.2	72.6	67.9	58.0	38.5	28.0	29.4	49.4
1975	29.1	29.8	35.6	44.4	53.4	63.8	71.0	70.2	59.4	52.3	36.4	35.3	48.4
1976	30.1	37.8	36.1	47.7	54.7	64.3	72.1	68.1	59.3	45.7	36.2	32.9	48.8
1977	26.8	35.0	37.1	48.3	58.6	68.0	71.4	68.6	64.0	51.5	38.6	34.6	50.2
1978	25.1	27.7	40.9	48.8	52.5	66.2	72.8	67.5	62.8	51.9	36.5	21.9	47.9
1979	16.9	32.5	38.1	48.3	54.0	64.3	70.6	67.5	64.4	51.7	31.2	33.5	47.8
1980	26.7	34.3	35.7	44.3	53.4	69.2	75.3	70.4	62.3	49.9	39.5	39.8	50.1
1981	34.9	34.4	39.3	53.8	54.5	69.4	71.9	67.3	63.4	50.8	43.3	32.7	51.3
1982	29.4	29.0	38.1	45.7	52.7	60.1	70.2	68.5	59.0	47.6	35.4	29.8	47.1
1983	32.5	34.4	35.7	40.1	50.5	61.1	72.3	71.9	63.8	50.3	37.7	18.4	47.5
1984	26.1	33.3	35.3	41.4	58.2	65.1	71.5	68.4	59.5	42.8	38.4	33.1	47.8
1985	25.1	26.3	37.9	48.7	57.2	65.0	70.3	69.8	58.0	49.0	32.1	27.9	47.3
1986	38.2	34.7	44.3	48.5	54.6	65.8	70.4	67.6	59.1	48.0	37.7	29.8	49.9
1987	29.4	33.0	35.3	48.5	56.0	65.1	70.7	66.1	60.0	50.4	39.2	29.0	48.6
1988	24.3	31.8	36.2	48.2	56.4	68.7	70.6	70.4	60.8	53.0	39.2	29.2	49.1
1989	32.8	21.8	43.7	49.0	57.6	62.1	71.8	68.4	61.0	49.7	41.5	27.3	48.9
1990	33.6	31.7	38.9	47.1	53.8	69.5	68.0	68.0	64.5	49.5	42.7	24.3	49.3
1991	28.0	38.0	39.9	45.8	58.0	67.0	69.4	68.1	60.5	50.3	33.0	31.0	49.1
1992	32.6	37.6	41.6	52.1	57.8	62.5	68.4	66.6	62.9	52.4	31.9	29.3	49.6
1993	26.7	29.4	39.8	46.1	55.6	65.0	70.7	67.3	58.1	48.0	32.3	32.3	47.6
1994	31.5	31.6	40.6	45.5	57.7	69.1	69.8	70.2	62.7	49.1	37.4	33.6	49.8
1995	31.4	35.8	38.7	41.3	49.4	60.6	67.4	71.9	59.0	48.9	41.4	32.1	48.2
1996	27.2	34.0	35.5	47.2	59.6	65.7	69.8	67.6	57.4	49.4	38.1	33.0	48.7
1997	27.1	30.2	40.7	39.5	53.8	64.3	70.5	67.1	62.2	48.5	33.5	30.5	47.3
1998	32.1	31.6	35.4	43.0	58.9	63.8	71.0	67.6	65.8	50.0	41.7	28.3	49.1
1999	33.7	38.1	42.1	42.2	53.4	62.9	71.1	68.7	57.7	50.8	44.9	33.6	49.9
2000	32.1	38.0	39.0	48.8	59.4	64.2	71.4	71.4	62.0	49.7	30.3	28.0	49.6
2001	27.1	30.3	37.4	49.4	55.1	66.5	73.5	69.8	63.1	50.9	41.0	31.6	49.6
POR= 53 YRS	29.4	32.5	37.4	46.1	55.7	65.3	70.8	68.9	60.8	50.3	37.8	30.9	48.8

REFERENCE NOTES:

PAGE 1:
THE TEMPERATURE GRAPH SHOWS NORMAL MAXIMUM AND NORMAL MINIMUM DAILY TEMPERATURES (SOLID CURVES) AND THE ACTUAL DAILY HIGH AND LOW TEMPERATURES (VERTICAL BARS).

PAGE 2 AND 3:
H/C INDICATES HEATING AND COOLING DEGREE DAYS.
RH INDICATES RELATIVE HUMIDITY
W/O INDICATES WEATHER AND OBSTRUCTIONS
S INDICATES SUNSHINE.
PR INDICATES PRESSURE.
CLOUDINESS ON PAGE 3 IS THE SUM OF THE CEILOMETER AND SATELLITE DATA NOT TO EXCEED EIGHT EIGHTHS(OKTAS).

GENERAL:
T INDICATES TRACE PRECIPITATION, AN AMOUNT GREATER THAN ZERO BUT LESS THAN THE LOWEST REPORTABLE VALUE.
+ INDICATES THE VALUE ALSO OCCURS ON EARLIER DATES.
BLANK ENTRIES DENOTE MISSING OR UNREPORTED DATA.
NORMALS ARE 30-YEAR AVERAGES (1961 - 1990).
ASOS INDICATES AUTOMATED SURFACE OBSERVING SYSTEM.
PR INDICATES THE LAST DAY OF THE PREVIOUS MONTH.
POR (PERIOD OF RECORD) BEGINS WITH THE JANUARY DATA MONTH AND IS THE NUMBER OF YEARS USED TO COMPUTE THE MEAN. INDIVIDUAL MONTHS WITHIN THE POR MAY BE MISSING.
WHEN THE POR FOR A NORMAL IS LESS THAN 30 YEARS, THE NORMAL IS PROVISIONAL AND IS BASED ON THE NUMBER OF YEARS INDICATED.
0.* OR * INDICATES THE VALUE OR MEAN-DAYS-WITH IS BETWEEN 0.00 AND 0.05.
CLOUDINESS FOR ASOS STATIONS DIFFERS FROM THE NON-ASOS OBSERVATION TAKEN BY A HUMAN OBSERVER. ASOS STATION CLOUDINESS IS BASED ON TIME-AVERAGED CEILOMETER DATA FOR CLOUDS AT OR BELOW 12,000 FEET AND ON SATELLITE DATA FOR CLOUDS ABOVE 12,000 FEET.
THE NUMBER OF DAYS WITH CLEAR, PARTLY CLOUDY, AND CLOUDY CONDITIONS FOR ASOS STATIONS IS THE SUM OF THE CEILOMETER AND SATELLITE DATA FOR THE SUNRISE TO SUNSET PERIOD.

GENERAL CONTINUED:
CLEAR INDICATES 0 - 2 OKTAS, PARTLY CLOUDY INDICATES 3 - 6 OKTAS, AND CLOUDY INDICATES 7 OR 8 OKTAS. WHEN AT LEAST ONE OF THE ELEMENTS (CEILOMETER OR SATELLITE) IS MISSING, THE DAILY CLOUDINESS IS NOT COMPUTED.
WIND DIRECTION IS RECORDED IN TENS OF DEGREES (2 DIGITS) CLOCKWISE FROM TRUE NORTH. "00" INDICATES CALM. "36" INDICATES TRUE NORTH.
RESULTANT WIND IS THE VECTOR AVERAGE OF THE SPEED AND DIRECTION.
AVERAGE TEMPERATURE IS THE SUM OF THE MEAN DAILY MAXIMUM AND MINIMUM TEMPERATURE DIVIDED BY 2.
SNOWFALL DATA COMPRISE ALL FORMS OF FROZEN PRECIPITATION, INCLUDING HAIL.
A HEATING (COOLING) DEGREE DAY IS THE DIFFERENCE BETWEEN THE AVERAGE DAILY TEMPERATURE AND 65 F.
DRY BULB IS THE TEMPERATURE OF THE AMBIENT AIR.
DEW POINT IS THE TEMPERATURE TO WHICH THE AIR MUST BE COOLED TO ACHIEVE 100 PERCENT RELATIVE HUMIDITY.
WET BULB IS THE TEMPERATURE THE AIR WOULD HAVE IF THE MOISTURE CONTENT WAS INCREASED TO 100 PERCENT RELATIVE HUMIDITY.

ON JULY 1, 1996, THE NATIONAL WEATHER SERVICE BEGAN USING THE "METAR" OBSERVATION CODE THAT WAS ALREADY EMPLOYED BY MOST OTHER NATIONS OF THE WORLD. THE MOST NOTICEABLE DIFFERENCE IN THIS ANNUAL PUBLICATION WILL BE THE CHANGE IN UNITS FROM TENTHS TO EIGHTS(OKTAS) FOR REPORTING THE AMOUNT OF SKY COVER.

HEATING DEGREE DAYS (base 65 F) 2001 COLORADO SPRINGS, CO (COS)

YEAR	JUL	AUG	SEP	OCT	NOV	DEC	JAN	FEB	MAR	APR	MAY	JUN	TOTAL
1972-73	41	34	136	476	1049	1281	1221	912	877	715	359	76	7177
1973-74	32	1	194	378	754	1041	1172	866	700	566	176	88	5968
1974-75	1	17	229	376	789	1143	1102	980	904	608	350	88	6587
1975-76	0	10	200	391	852	916	1075	782	891	512	314	80	6023
1976-77	0	11	191	593	859	988	1181	837	858	494	192	5	6209
1977-78	2	22	73	413	784	938	1231	1036	741	479	386	98	6203
1978-79	3	44	119	400	848	1329	1484	906	825	494	336	97	6885
1979-80	6	41	88	407	1005	969	1180	883	901	615	351	32	6478
1980-81	0	7	113	463	759	776	928	850	789	335	321	38	5379
1981-82	5	30	70	433	643	993	1095	1001	827	571	374	163	6205
1982-83	8	11	198	532	880	1084	1001	851	904	742	444	159	6814
1983-84	2	0	101	417	811	1438	1198	911	912	700	220	58	6768
1984-85	0	6	200	684	790	982	1233	1077	830	481	242	77	6602
1985-86	5	8	253	487	978	1142	822	840	635	487	315	49	6021
1986-87	4	14	174	519	813	1081	1096	888	912	491	272	50	6314
1987-88	17	74	150	445	767	1108	1256	958	886	499	273	25	6458
1988-89	7	8	154	366	767	1099	989	1207	655	475	247	134	6108
1989-90	0	4	172	473	699	1164	966	928	805	526	345	24	6106
1990-91	28	21	83	473	663	1258	1142	750	773	568	219	33	6011
1991-92	16	16	145	453	954	1048	998	788	717	383	219	96	5833
1992-93	21	53	91	383	990	1101	1179	991	776	558	286	84	6513
1993-94	0	40	212	519	972	1008	1032	926	749	576	223	14	6271
1994-95	10	14	98	486	821	969	1035	811	808	702	477	152	6383
1995-96	38	3	231	490	700	1011	1162	890	908	527	192	48	6200
1996-97	2	17	237	490	800	986	1167	967	747	758	341	70	6582
1997-98	6	28	111	506	937	1060	1012	928	911	653	208	116	6476
1998-99	3	6	43	458	691	1129	965	748	702	674	357	101	5877
1999-00	0	5	239	434	600	968	1010	773	798	478	206	83	5594
2000-01	0	2	150	473	1033	1137	1169	964	850	462	305	54	6599
2001-	0	6	97	431	714	1029							

COOLING DEGREE DAYS (base 65 F) 2001 COLORADO SPRINGS, CO (COS)

YEAR	JAN	FEB	MAR	APR	MAY	JUN	JUL	AUG	SEP	OCT	NOV	DEC	ANNUAL
1972	0	0	0	2	0	96	168	124	17	0	0	0	407
1973	0	0	0	0	0	104	145	180	13	0	0	0	442
1974	0	0	0	0	18	130	241	109	26	0	0	0	524
1975	0	0	0	0	0	59	195	180	41	5	0	0	480
1976	0	0	0	0	0	66	227	114	28	0	0	0	435
1977	0	0	0	0	0	103	204	142	49	0	0	0	498
1978	0	0	0	0	4	143	255	127	59	1	0	0	589
1979	0	0	0	0	1	84	185	124	77	2	0	0	473
1980	0	0	0	0	0	169	327	180	41	0	0	0	717
1981	0	0	0	4	2	176	226	105	27	0	0	0	540
1982	0	0	0	0	0	23	176	127	26	0	0	0	352
1983	0	0	0	0	1	48	236	219	71	0	0	0	575
1984	0	0	0	0	17	68	207	119	42	0	0	0	453
1985	0	0	0	0	5	83	179	163	51	0	0	0	481
1986	0	0	0	0	1	82	180	102	3	0	0	0	368
1987	0	0	0	0	0	62	199	113	6	0	0	0	380
1988	0	0	0	0	12	143	190	181	33	0	0	0	559
1989	0	0	0	3	25	54	220	117	57	3	0	0	479
1990	0	0	0	0	6	168	128	121	73	0	0	0	496
1991	0	0	0	0	8	101	161	120	15	4	0	0	409
1992	0	0	0	4	3	28	131	106	32	0	0	0	304
1993	0	0	0	0	2	89	183	117	11	1	0	0	403
1994	0	0	0	0	5	143	165	182	33	0	0	0	528
1995	0	0	0	0	0	30	120	226	60	0	0	0	436
1996	0	0	0	4	31	77	159	106	16	1	0	0	394
1997	0	0	0	0	0	55	182	101	33	0	0	0	371
1998	0	0	0	0	23	88	194	95	61	0	0	0	461
1999	0	0	0	0	0	42	196	130	26	0	0	0	394
2000	0	0	0	0	39	65	220	208	67	5	0	0	604
2001	0	0	0	0	5	106	272	164	45	0	0	0	592

SNOWFALL (inches) 2001 COLORADO SPRINGS, CO (COS)

YEAR	JUL	AUG	SEP	OCT	NOV	DEC	JAN	FEB	MAR	APR	MAY	JUN	TOTAL
1972-73	0.0	0.0	0.0	14.4	16.4	11.4	3.9	2.3	15.6	11.2	0.8	0.0	76.0
1973-74	0.0	0.0	0.0	3.9	1.8	9.1	2.4	2.4	6.8	10.0	0.0	T	36.4
1974-75	0.0	0.0	T	T	1.5	6.3	2.8	4.8	3.5	6.9	0.2	1.1	27.1
1975-76	0.0	0.0	0.0	4.3	9.8	0.9	6.9	5.2	8.6	10.1	0.0	0.0	45.8
1976-77	0.0	0.0	T	2.5	4.9	2.6	4.8	2.5	13.8	4.3	0.0	0.0	35.4
1977-78	0.0	0.0	0.0	0.9	1.9	3.0	4.2	8.6	3.3	1.1	19.4	0.0	42.4
1978-79	0.0	0.0	T	0.5	4.0	15.2	9.9	1.2	20.0	14.6	4.1	0.0	69.5
1979-80	0.0	0.0	0.0	1.3	19.1	17.6	4.7	5.9	12.7	11.3	0.0	0.0	72.6
1980-81	0.0	0.0	0.0	0.2	4.4	1.4	1.0	1.7	9.0	0.3	0.2	0.0	18.2
1981-82	0.0	0.0	0.0	0.4	0.5	9.1	3.6	6.2	8.4	2.3	3.9	0.0	34.4
1982-83	0.0	0.0	0.0	0.2	0.9	8.2	4.0	1.1	16.3	4.8	0.8	0.0	36.3
1983-84	0.0	0.0	0.0	0.0	10.3	18.2	7.8	1.4	23.2	9.0	0.8	0.0	70.7
1984-85	0.0	0.0	0.9	25.9	2.0	10.9	8.0	4.7	22.3	0.8	T	0.0	75.5
1985-86	0.0	0.0	1.9	1.7	8.3	6.3	0.2	4.6	2.9	4.0	T	0.0	29.9
1986-87	0.0	0.0	0.0	1.4	7.3	4.4	28.7	23.2	14.9	3.3	T	0.0	83.2
1987-88	0.0	0.0	0.0	0.0	4.8	9.5	4.9	11.5	12.6	1.0	0.3	0.0	44.6
1988-89	0.0	0.0	0.0	0.0	1.6	13.6	3.0	18.9	T	1.0	0.0	T	38.1
1989-90	T	T	T	2.1	0.2	7.5	8.7	9.3	9.7	11.3	4.2	0.0	53.0
1990-91	0.0	0.0	0.0	8.2	2.7	4.1	0.9	T	5.2	5.9	1.5	0.0	28.5
1991-92	0.0	0.0	0.0	7.5	26.3	4.5	1.2	0.2	3.0	T	0.0	T	42.7
1992-93	T	T	0.0	T	11.4	2.6	7.5	3.3	2.3	1.2	T	T	28.3
1993-94	T	T	0.1	0.5	12.3	0.5	2.2	1.5	6.8	8.7	T	T	32.6
1994-95	0.0	0.0	T	0.2	6.1	3.9	4.5	6.5	7.7	0.0	0.0	T	28.9
1995-96	0.0	0.0	0.5	0.2	T	T	2.9	4.7	6.9	3.2			
1996-97				8.7			9.3	3.4	2.4	12.3	T	0.0	
1997-98	0.0	T	0.0	19.9	3.5	1.8	0.3	4.3	17.5	9.5	0.3	T	56.8
1998-99	T	T	0.0	0.0	1.1	1.7	7.3	2.9	0.3	2.7	17.6	0.3	33.9
1999-00	0.0	0.0	0.2	3.0	10.6	2.3	7.4	2.0	12.6	2.0	T	0.0	40.1
2000-01	T	0.0	1.5	1.5	2.9	3.7	14.6	5.9	12.0	8.1	6.5	T	56.7
2001-	0.0	T	2.4	0.0	3.7	1.2							
POR= 54 YRS	T	T	1.0	3.3	5.4	5.0	5.2	4.8	9.4	6.7	1.5	T	42.3

2001
DENVER,
COLORADO (DEN)

Denver enjoys the invigorating climate that prevails over much of the central Rocky Mountain region, without the extremely cold mornings of the high elevations during winter, or the hot afternoons of summer at lower altitudes. Extremely warm or cold weather in Denver is usually of short duration.

Situated a long distance from any moisture source, and separated from the Pacific Ocean by several high mountain barriers, Denver enjoys low relative humidity, light precipitation, and abundant sunshine.

Air masses from four different sources influence Denver weather. These include arctic air from Canada and Alaska, warm, moist air from the Gulf of Mexico, warm, dry air from Mexico and the southwestern deserts, and Pacific air modified by its passage over mountains to the west.

In winter, the high altitude and mountains to the west combine to moderate temperatures in Denver. Invasions of cold air from the north, intensified by the high altitude, can be abrupt and severe. However, many of the cold air masses that spread southward out of Canada never reach the altitude of Denver, but move off over the lower plains to the east. Surges of air from the west are moderated in their descent down the east face of the Rockies, and reach Denver in the form of chinook winds that often raise temperatures into the 60s, even in midwinter.

In spring, polar air often collides with warm, moist air from the Gulf of Mexico and these collisions result in frequent, rapid and drastic weather changes. Spring is the cloudiest, windiest, and wettest season in the city. Much of the precipitation falls as snow, especially in March and early April. Stormy periods are interspersed with stretches of mild, sunny weather that quickly melt previous snow cover.

Summer precipitation falls mainly from scattered thunderstorms during the afternoon and evening. Mornings are usually clear and sunny, with clouds forming during early afternoon to cut off the sunshine at what would otherwise be the hottest part of the day. Severe thunderstorms, with large hail and heavy rain occasionally occur in the city, but these conditions are more common on the plains to the east.

Autumn is the most pleasant season. Few thunderstorms occur and invasions of cold air are infrequent. As a result, there is more sunshine and less severe weather than at any other time of the year.

Based on the 1951-1980 period, the average first occurrence of 32 degrees Fahrenheit in the fall is October 8 and the average last occurrence in the spring is May 3.

NORMALS, MEANS, AND EXTREMES
DENVER, CO (DEN)

LATITUDE:	LONGITUDE:	ELEVATION (FT):	TIME ZONE:	WBAN: 03017
39 49' 58" N	104 39' 27" W	GRND: 5379　BARO: 5382	MOUNTAIN　(UTC + 7)	

	ELEMENT	POR	JAN	FEB	MAR	APR	MAY	JUN	JUL	AUG	SEP	OCT	NOV	DEC	YEAR
TEMPERATURE °F	NORMAL DAILY MAXIMUM	30	42.5	46.3	52.2	61.4	70.7	81.2	88.1	85.6	76.6	66.3	52.0	43.9	63.9
	MEAN DAILY MAXIMUM	6	43.4	47.1	53.9	58.9	70.0	80.7	88.6	86.6	77.7	64.7	52.3	43.2	63.9
	HIGHEST DAILY MAXIMUM	6	72	69	79	84	93	98	101	99	97	87	78	72	101
	YEAR OF OCCURRENCE		1997	1999	1997	1996	1996	2001	2001	1995	1995	1997	1999	1998	JUL 2001
	MEAN OF EXTREME MAXS.	6	65.7	65.5	74.5	78.0	86.9	93.7	99.0	95.6	92.4	83.4	71.7	62.7	80.8
	NORMAL DAILY MINIMUM	30	13.8	18.3	23.7	32.0	41.2	50.1	56.0	54.2	45.1	34.2	23.4	15.3	33.9
	MEAN DAILY MINIMUM	6	17.9	20.9	25.6	32.1	42.7	51.2	59.6	58.4	49.3	36.3	25.7	18.5	36.5
	LOWEST DAILY MINIMUM	6	-14	-16	-2	6	23	34	44	42	25	3	-3	-19	-19
	YEAR OF OCCURRENCE		1997	1996	1996	1997	2000	1998	1997	1995	1996	1997	1997	1998	DEC 1998
	MEAN OF EXTREME MINS.	6	-2.4	2.7	7.7	16.8	30.0	39.9	50.7	49.6	32.3	20.1	4.9	0.3	21.1
	NORMAL DRY BULB	30	28.2	32.3	38.0	46.8	55.9	65.6	72.1	69.8	61.0	50.2	37.6	29.6	48.9
	MEAN DRY BULB	6	30.7	34.0	39.7	45.4	56.5	66.0	74.1	72.5	63.5	50.5	39.0	30.9	50.2
	MEAN WET BULB	6	25.1	27.6	27.6	31.9	40.2	46.0	59.3	58.9	51.0	40.4	31.5	25.4	38.7
	MEAN DEW POINT	6	15.1	16.3	21.1	27.2	38.8	44.3	49.5	50.2	41.1	28.9	21.1	15.5	30.8
	NORMAL NO. DAYS WITH:														
	MAXIMUM　90														
	MAXIMUM　32														
	MINIMUM　32														
	MINIMUM　0														
H/C	NORMAL HEATING DEG. DAYS	30	1141	916	837	546	288	84	0	11	162	459	822	1097	6363
	NORMAL COOLING DEG. DAYS	30	0	0	0	0	6	102	224	160	42	0	0	0	534
RH	NORMAL (PERCENT)														
	HOUR 05 LST														
	HOUR 11 LST														
	HOUR 17 LST														
	HOUR 23 LST														
S	PERCENT POSSIBLE SUNSHINE														
W/O	MEAN NO. DAYS WITH:														
	HEAVY FOG(VISBY 1/4 MI)	6	2.3	3.0	3.5	2.8	2.2	1.2	1.7	1.5	1.5	2.2	1.9	1.2	25.0
	THUNDERSTORMS	6	0.0	0.2	0.8	2.4	7.3	10.8	14.5	11.4	5.6	0.7	0.0	0.0	53.7
CLOUDINESS	MEAN:														
	SUNRISE-SUNSET (OKTAS)	1		5.0	5.3	7.2	5.6	2.5		2.5				2.5	
	MIDNIGHT-MIDNIGHT (OKTAS)	1		5.3	7.2	6.4	3.0		2.0						
	MEAN NO. DAYS WITH:														
	CLEAR	1	3.0	10.0	9.0	6.0	10.0	12.0	2.0	7.0	6.0	9.0		13.0	
	PARTLY CLOUDY	1	4.0	2.0	6.0	4.0	5.5	9.0	2.0	9.0	6.0			1.0	
	CLOUDY	1	3.0	6.0	10.0	13.0	5.5	5.0	1.0	3.0	3.0	2.0		2.0	
PR	MEAN STATION PRESSURE(IN)	6	24.53	24.56	24.56	24.54	24.56	24.61	24.69	24.70	24.66	24.61	24.60	24.57	24.60
	MEAN SEA-LEVEL PRES. (IN)	6	29.99	29.99	29.93	29.88	29.84	29.83	29.88	29.92	29.94	29.95	30.01	30.04	29.93
WINDS	MEAN SPEED (MPH)	6	9.4	9.7	10.3	11.3	10.5	10.4	9.7	9.4	9.3	9.8	9.3	9.9	9.9
	PREVAIL.DIR(TENS OF DEGS)	5	20	21	21	01	21	16	21	21	22	21	22	21	21
	MAXIMUM 2-MINUTE:														
	SPEED (MPH)	6	41	46	53	53	49	49	54	49	39	46	45	47	54
	DIR. (TENS OF DEGS)		28	26	28	33	36	30	13	28	02	29	27	30	13
	YEAR OF OCCURRENCE		1996	2000	1995	2001	2001	1999	1999	2001	2001	2001	2000	1997	JUL 1999
	MAXIMUM 5-SECOND:														
	SPEED (MPH)	6	48	54	56	60	58	63	64	61	49	54	54	53	64
	DIR. (TENS OF DEGS)		29	26	32	33	01	29	13	29	02	29	28	29	13
	YEAR OF OCCURRENCE		1996	2000	1997	2001	2001	1999	1999	2001	2001	2001	2000	2000	JUL 1999
PRECIPITATION	NORMAL (IN)	30	0.50	0.54	1.26	1.68	2.62	2.05	1.99	1.65	1.34	0.99	0.89	0.60	16.11
	MAXIMUM MONTHLY (IN)	6	0.78	0.64	1.96	5.86	4.67	3.07	5.92	3.52	2.34	1.87	0.72	0.50	5.92
	YEAR OF OCCURRENCE		2001	2001	2000	1999	1995	1995	1998	1997	1996	1997	2001	1997	JUL 1998
	MINIMUM MONTHLY (IN)	6	0.05	0.09	0.19	0.33	1.57	0.73	1.01	0.56	0.73	0.08	0.31	0.06	0.05
	YEAR OF OCCURRENCE		1998	1996	1999	1996	1997	1998	1996	1996	1996	1998	2001	1995	JAN 1998
	MAXIMUM IN 24 HOURS (IN)	6	0.51	0.30	0.63	2.06	2.00	1.15	3.06	1.46	1.22	1.12	0.47	0.19	3.06
	YEAR OF OCCURRENCE		2001	2001	1996	1999	2000	1999	1997	2000	1996	1997	1999	1997	JUL 1997
	NORMAL NO. DAYS WITH:														
	PRECIPITATION　0.01														
	PRECIPITATION　1.00														
SNOWFALL	NORMAL (IN)														
	MAXIMUM MONTHLY (IN)														
	YEAR OF OCCURRENCE														
	MAXIMUM IN 24 HOURS (IN)														
	YEAR OF OCCURRENCE														
	MAXIMUM SNOW DEPTH (IN)														
	YEAR OF OCCURRENCE														
	NORMAL NO. DAYS WITH:														
	SNOWFALL　1.0														

PRECIPITATION (inches) 2001 DENVER, CO (DEN)

YEAR	JAN	FEB	MAR	APR	MAY	JUN	JUL	AUG	SEP	OCT	NOV	DEC	ANNUAL
1995			0.28	2.44	4.67	3.07	2.31	1.04	2.28	0.72	0.31	0.06	
1996	0.29	0.09	0.77	0.33	2.40	1.77	1.01	0.56	2.34	0.39	0.38	0.06	10.39
1997	0.26	0.54	0.26	1.30	1.57	2.57	5.60	3.52	0.97	1.87	0.61	0.50	19.57
1998	0.05	0.23	0.86	2.47	1.73	0.73	5.92	1.19	0.73	1.20	0.40	0.42	15.93
1999	0.38	0.15	0.19	5.86	2.37	2.52	3.84	3.37	1.20	0.31	0.47	0.29	20.95
2000	0.24	0.23	1.96	0.71	3.09	0.79	1.42	3.06	1.52	0.52	0.61	0.27	14.42
2001	0.78	0.64	1.10	1.20	3.80	1.53	4.76	0.71	1.00	0.08	0.72	0.14	16.46
POR= 6 YRS	0.34	0.32	0.82	2.16	2.69	1.76	3.87	2.12	1.30	0.85	0.52	0.29	17.04

AVERAGE TEMPERATURE (F) 2001 DENVER, CO (DEN)

YEAR	JAN	FEB	MAR	APR	MAY	JUN	JUL	AUG	SEP	OCT	NOV	DEC	ANNUAL
1995			39.3	42.9	50.0	62.2	70.9	75.3	61.7	48.5	41.8	33.0	
1996	27.0	33.9	36.0	48.0	58.1	68.2	73.4	71.6	60.8	50.9	37.2	33.0	49.8
1997	27.9	30.0	42.1	40.5	56.6	67.8	73.1	69.7	64.3	49.7	34.8	27.9	48.7
1998	32.7	33.9	36.9	44.8	59.1	63.0	74.3	71.7	68.0	50.2	42.1	28.9	50.5
1999	33.7	38.6	43.7	42.6	54.8	64.2	73.9	71.2	59.2	52.5	47.3	33.8	51.3
2000	33.0	39.2	40.4	49.8	59.2	67.0	76.7	74.5	63.6	50.5	28.9	28.3	50.9
2001	30.0	28.3	39.8	49.6	57.1	69.4	76.7	73.5	66.8	51.5	40.9	31.7	51.3
POR= 6 YRS	30.7	34.0	40.1	45.5	56.9	66.1	74.5	72.3	63.9	50.8	38.9	30.7	50.4

REFERENCE NOTES:

PAGE 1:
THE TEMPERATURE GRAPH SHOWS NORMAL MAXIMUM AND NORMAL
   MINIMUM DAILY TEMPERATURES (SOLID CURVES) AND THE
   ACTUAL DAILY HIGH AND LOW TEMPERATURES (VERTICAL BARS).

PAGE 2 AND 3:
H/C INDICATES HEATING AND COOLING DEGREE DAYS.
RH INDICATES RELATIVE HUMIDITY
W/O INDICATES WEATHER AND OBSTRUCTIONS
S INDICATES SUNSHINE.
PR INDICATES PRESSURE.
CLOUDINESS ON PAGE 3 IS THE SUM OF THE CEILOMETER AND
   SATELLITE DATA NOT TO EXCEED EIGHT EIGHTHS(OKTAS).

GENERAL:
T INDICATES TRACE PRECIPITATION, AN AMOUNT GREATER
   THAN ZERO BUT LESS THAN THE LOWEST REPORTABLE VALUE.
+ INDICATES THE VALUE ALSO OCCURS ON EARLIER DATES.
BLANK ENTRIES DENOTE MISSING OR UNREPORTED DATA.
NORMALS ARE 30-YEAR AVERAGES (1961 - 1990).
ASOS INDICATES AUTOMATED SURFACE OBSERVING SYSTEM.
PM INDICATES THE LAST DAY OF THE PREVIOUS MONTH.
POR (PERIOD OF RECORD) BEGINS WITH THE JANUARY DATA
   MONTH AND IS THE NUMBER OF YEARS USED TO COMPUTE
   THE MEAN.   INDIVIDUAL MONTHS WITHIN THE POR MAY
   BE MISSING.
WHEN THE POR FOR A NORMAL IS LESS THAN 30 YEARS,
   THE NORMAL IS PROVISIONAL AND IS BASED ON THE NUMBER
   OF YEARS INDICATED.
0.* OR * INDICATES THE VALUE OR MEAN-DAYS-WITH
   IS BETWEEN 0.00 AND 0.05.
CLOUDINESS FOR ASOS STATIONS DIFFERS FROM THE NON-ASOS
   OBSERVATION TAKEN BY A HUMAN OBSERVER.   ASOS STATION
   CLOUDINESS IS BASED ON TIME-AVERAGED CEILOMETER DATA
   FOR CLOUDS AT OR BELOW 12,000 FEET AND ON SATELLITE
   DATA FOR CLOUDS ABOVE 12,000 FEET.
THE NUMBER OF DAYS WITH CLEAR, PARTLY CLOUDY, AND
   CLOUDY CONDITIONS FOR ASOS STATIONS IS THE SUM
   OF THE CEILOMETER AND SATELLITE DATA FOR THE
   SUNRISE TO SUNSET PERIOD.

GENERAL CONTINUED:
CLEAR INDICATES 0 - 2 OKTAS, PARTLY CLOUDY INDICATES
   3 - 6 OKTAS, AND CLOUDY INDICATES 7 OR 8 OKTAS.
   WHEN AT LEAST ONE OF THE ELEMENTS (CEILOMETER OR
   SATELLITE) IS MISSING, THE DAILY CLOUDINESS IS
   NOT COMPUTED.
WIND DIRECTION IS RECORDED IN TENS OF DEGREES (2 DIGITS)
   CLOCKWISE FROM TRUE NORTH. "00" INDICATES CALM. "36"
   INDICATES TRUE NORTH.
RESULTANT WIND IS THE VECTOR AVERAGE OF THE SPEED AND
   DIRECTION.
AVERAGE TEMPERATURE IS THE SUM OF THE MEAN DAILY MAXIMUM
   AND MINIMUM TEMPERATURE DIVIDED BY 2.
SNOWFALL DATA COMPRISE ALL FORMS OF FROZEN
   PRECIPITATION, INCLUDING HAIL.
A HEATING (COOLING) DEGREE DAY IS THE DIFFERENCE BETWEEN
   THE AVERAGE DAILY TEMPERATURE AND 65 F.
DRY BULB IS THE TEMPERATURE OF THE AMBIENT AIR.
DEW POINT IS THE TEMPERATURE TO WHICH THE AIR MUST BE
   COOLED TO ACHIEVE 100 PERCENT RELATIVE HUMIDITY.
WET BULB IS THE TEMPERATURE THE AIR WOULD HAVE IF THE
   MOISTURE CONTENT WAS INCREASED TO 100 PERCENT RELATIVE
   HUMIDITY.

ON JULY 1, 1996, THE NATIONAL WEATHER SERVICE BEGAN USING
   THE "METAR" OBSERVATION CODE THAT WAS ALREADY EMPLOYED
   BY MOST OTHER NATIONS OF THE WORLD.   THE MOST NOTICEABLE
   DIFFERENCE IN THIS ANNUAL PUBLICATION WILL BE THE CHANGE
   IN UNITS FROM TENTHS TO EIGHTS(OKTAS) FOR REPORTING THE
   AMOUNT OF SKY COVER.

HEATING DEGREE DAYS (base 65 F) 2001 DENVER, CO (DEN)

YEAR	JUL	AUG	SEP	OCT	NOV	DEC	JAN	FEB	MAR	APR	MAY	JUN	TOTAL
1994-95									788	655	457	132	5600
1995-96	26	2	188	505	686	981	1166	894	893	230	29	0	6297
1996-97	0	4	192	444	824	985	1142	975	704	728	264	35	
1997-98	2	11	92	475	895	1142	996	865	865	597	186	137	6263
1998-99	1	1	46	453	680	1113	962	731	654	666	311	85	5703
1999-00	1	3	194	383	528	962	984	744	754	446	215	61	5275
2000-01	0	5	149	447	1074	1131	1079	1021	775	455	256	46	6438
2001-	0	4	65	416	717	1026							

COOLING DEGREE DAYS (base 65 F) 2001 DENVER, CO (DEN)

YEAR	JAN	FEB	MAR	APR	MAY	JUN	JUL	AUG	SEP	OCT	NOV	DEC	ANNUAL
1995			0	0	0	55	212	327	98	0	0	0	730
1996	0	0	0	3	26	133	269	215	71	13	0	0	
1997	0	0	0	0	11	126	260	160	77	8	0	0	642
1998	0	0	0	0	13	88	296	215	143	0	0	0	755
1999	0	0	0	0	2	69	283	203	30	2	0	0	589
2000	0	0	0	0	43	127	368	305	115	5	0	0	963
2001	0	0	0	0	18	184	373	274	125	5	0	0	979

SNOWFALL (inches) 2001 DENVER, CO (DEN)

YEAR	JUL	AUG	SEP	OCT	NOV	DEC	JAN	FEB	MAR	APR	MAY	JUN	TOTAL
POR=													

2001
GRAND JUNCTION,
COLORADO (GJT)

Grand Junction is located at the junction of the Colorado and Gunnison Rivers. It is on the west slope of the Rockies, in a large mountain valley. The area has a climate marked by the wide seasonal range usual to interior localities at this latitude. Thanks, however, to the protective topography of the vicinity, sudden and severe weather changes are very infrequent. The valley floor slopes from 4,800 feet near Palisade to 4,400 feet at the west end near Fruita. Mountains are on all sides at distances of from 10 to 60 miles and reach heights of 9,000 to over 12,000 feet.

This mountain valley location, with attendant valley breezes, provides protection from spring and fall frosts. This results in a growing season averaging 191 days in the city. This varies considerably in the outlying districts. It is about the same in the upper valley around Palisade, and 3 to 4 weeks shorter near the river west of Grand Junction. The growing season is sufficiently long to permit commercial growth of almost all fruits except citrus varieties. Summer grazing of cattle and sheep on nearby mountain ranges is extensive.

The interior, continental location, ringed by mountains on all sides, results in quite low precipitation in all seasons. Consequently, agriculture is dependent on irrigation. Adequate supplies of water are available from mountain snows and rains. Summer rains occur chiefly as scattered light showers and thunderstorms which develop over nearby mountains. Winter snows are fairly frequent, but are mostly light and quick to melt. Even the infrequent snows of from 4 to 8 inches seldom remain on the ground for prolonged periods. Blizzard conditions in the valley are extremely rare.

Temperatures above 100 degrees are infrequent, and about one-third of the winters have no readings below zero. Summer days with maximum temperatures in the middle 90s and minimums in the low 60s are common. Relative humidity is very low during the summer, with values similar to other dry locations such as the southern parts of New Mexico and Arizona. Spells of cold winter weather are sometimes prolonged due to cold air becoming trapped in the valley. Winds are usually very light during the coldest weather. Changes in winter are normally gradual, and abrupt changes are much less frequent than in eastern Colorado. Cold waves are rare. Sunny days predominate in all seasons.

The prevailing wind is from the east-southeast due to the valley breeze effect. The strongest winds are associated with thunderstorms or with pre-frontal weather. They usually are from the south or southwest.

NORMALS, MEANS, AND EXTREMES
GRAND JUNCTION, CO (GJT)

LATITUDE: 39 08' 03" N　LONGITUDE: 108 32' 15" W　ELEVATION (FT): GRND: 4823　BARO: 4826　TIME ZONE: MOUNTAIN (UTC + 7)　WBAN: 23066

	ELEMENT	POR	JAN	FEB	MAR	APR	MAY	JUN	JUL	AUG	SEP	OCT	NOV	DEC	YEAR
TEMPERATURE F	NORMAL DAILY MAXIMUM	30	35.5	45.4	55.6	65.8	76.0	87.7	93.6	90.5	81.1	67.7	51.4	38.7	65.8
	MEAN DAILY MAXIMUM	54	36.4	44.8	54.8	65.0	75.8	87.3	93.1	89.8	81.1	67.8	50.6	39.1	65.5
	HIGHEST DAILY MAXIMUM	55	60	68	81	89	101	105	105	103	100	88	75	64	105
	YEAR OF OCCURRENCE		1971	1986	1971	1992	2000	1990	1976	2000	1995	1963	1977	1980	JUN 1990
	MEAN OF EXTREME MAXS.	54	49.3	58.5	70.7	80.5	89.4	98.4	100.8	98.1	92.6	81.6	65.1	52.5	78.1
	NORMAL DAILY MINIMUM	30	14.5	23.5	31.3	38.5	47.9	57.1	63.9	62.2	52.8	41.6	29.4	18.7	40.1
	MEAN DAILY MINIMUM	54	16.4	23.3	30.9	38.7	48.0	57.0	63.8	61.8	52.9	41.1	28.5	19.0	40.1
	LOWEST DAILY MINIMUM	55	-23	-18	5	11	26	34	44	43	29	18	-2	-17	-23
	YEAR OF OCCURRENCE		1963	1989	1948	1975	1970	1976	1993	1968	1978	1976	1990		JAN 1963
	MEAN OF EXTREME MINS.	54	1.5	8.0	17.7	25.5	34.7	44.4	55.5	53.2	40.2	28.2	15.2	5.2	27.4
	NORMAL DRY BULB	30	25.0	34.5	43.4	52.2	62.0	72.4	78.8	76.4	67.0	54.7	40.4	28.7	53.0
	MEAN DRY BULB	54	26.4	34.0	42.8	51.8	61.8	72.2	78.4	75.8	67.0	54.4	39.7	28.9	52.8
	MEAN WET BULB	18	24.0	30.2	36.2	41.3	48.1	53.4	58.6	58.8	51.4	42.7	32.8	25.2	41.9
	MEAN DEW POINT	18	18.2	22.7	24.9	27.9	33.1	35.6	43.7	46.1	37.7	30.3	24.2	18.8	30.3
	NORMAL NO. DAYS WITH:														
	MAXIMUM 90	30	0.0	0.0	0.0	0.0	1.2	14.3	24.8	19.4	4.2	0.0	0.0	0.0	63.9
	MAXIMUM 32	30	11.7	2.3	0.2	0.0	0.0	0.0	0.0	0.0	0.0	0.0	0.6	7.2	22.0
	MINIMUM 32	30	30.3	24.7	17.1	6.6	0.4	0.0	0.0	0.0	0.1	2.9	20.2	29.5	131.8
	MINIMUM 0	30	4.0	0.6	0.0	0.0	0.0	0.0	0.0	0.0	0.0	0.0	*	1.4	6.0
H/C	NORMAL HEATING DEG. DAYS	30	1240	854	670	389	132	13	0	0	55	332	738	1125	5548
	NORMAL COOLING DEG. DAYS	30	0	0	0	0	39	235	428	353	115	13	0	0	1183
RH	NORMAL (PERCENT)	30	70	60	50	40	36	29	34	37	39	46	58	68	47
	HOUR 05 LST	30	77	71	63	55	51	44	48	51	52	58	70	76	60
	HOUR 11 LST	30	64	53	42	33	30	24	28	32	34	38	50	61	41
	HOUR 17 LST	30	61	47	36	27	24	19	22	24	26	33	46	59	35
	HOUR 23 LST	30	75	66	56	45	40	32	36	39	42	51	64	73	52
S	PERCENT POSSIBLE SUNSHINE	55	61	65	65	70	73	81	79	77	79	74	63	61	71
W/O	MEAN NO. DAYS WITH:														
	HEAVY FOG(VISBY 1/4 MI)	56	2.7	1.8	0.6	0.1	0.0	0.0	0.0	0.0	0.0	0.2	0.8	2.0	8.2
	THUNDERSTORMS	56	0.1	0.3	0.8	2.1	4.8	4.8	7.6	8.2	5.3	1.5	0.4	0.1	36.0
CLOUDINESS	MEAN:														
	SUNRISE-SUNSET (OKTAS)	50	4.9	5.0	5.0	4.8	4.4	3.2	3.3	3.4	2.9	3.4	4.2	4.5	4.1
	MIDNIGHT-MIDNIGHT (OKTAS)	32	4.6	4.4	4.6	4.3	4.0	3.0	3.2	3.3	2.7	3.2	4.0	4.3	3.8
	MEAN NO. DAYS WITH:														
	CLEAR	51	9.1	7.6	8.0	7.9	9.6	14.9	13.7	13.1	16.0	14.7	10.4	7.4	134.6
	PARTLY CLOUDY	51	7.0	7.4	8.6	9.4	10.7	9.4	11.2	11.2	8.1	7.6	7.4	7.6	105.6
	CLOUDY	51	14.8	13.3	14.4	12.7	10.7	5.7	5.5	6.1	5.3	8.0	11.6	13.1	121.2
PR	MEAN STATION PRESSURE(IN)	29	25.20	25.19	25.10	25.10	25.10	25.10	25.20	25.20	25.20	25.20	25.21	25.30	25.17
	MEAN SEA-LEVEL PRES. (IN)	18	30.23	30.10	29.97	29.89	29.83	29.81	29.87	29.91	29.94	30.03	30.12	30.22	29.99
WINDS	MEAN SPEED (MPH)	55	5.8	6.7	8.4	9.4	9.6	9.8	9.6	9.3	9.2	8.1	6.9	6.1	8.2
	PREVAIL.DIR(TENS OF DEGS)														
	MAXIMUM 2-MINUTE:														
	SPEED (MPH)	5	36	41	53	49	44	48	41	45	38	44	37	35	53
	DIR. (TENS OF DEGS)		25	20	34	27	25	30	15	24	30	21	13	31	34
	YEAR OF OCCURRENCE		1999	1998	2001	2000	2000	1999	2000	1999	2000	1997	1997	2000	MAR 2001
	MAXIMUM 5-SECOND:														
	SPEED (MPH)	5	45	55	60	60	54	55	52	64	48	54	45	43	64
	DIR. (TENS OF DEGS)		23	19	34	19	25	31	15	26	29	22	26	31	26
	YEAR OF OCCURRENCE		1999	1998	2001	2000	1999	1999	2000	1999	2001	1997	1999	2000	AUG 1999
PRECIPITATION	NORMAL (IN)	30	0.56	0.48	0.90	0.75	0.87	0.50	0.65	0.81	0.82	0.98	0.71	0.61	8.64
	MAXIMUM MONTHLY (IN)	55	2.46	1.56	2.02	2.15	2.04	2.07	1.92	3.48	2.84	3.45	2.00	1.89	3.48
	YEAR OF OCCURRENCE		1957	1948	1979	1997	1995	1969	1983	1957	1997	1972	1983	1951	AUG 1957
	MINIMUM MONTHLY (IN)	55	T	T	0.02	0.06	T	T	0.01	0.04		0.00		0.01	0.00
	YEAR OF OCCURRENCE		1961	1972	1972	1958	1970	1980	1994	1956		1952		1976	OCT 1952
	MAXIMUM IN 24 HOURS (IN)	55	0.72	0.81	1.15	1.33	1.13	1.57	1.42	1.68	1.35	1.24	0.83	1.16	1.68
	YEAR OF OCCURRENCE		2000	1996	1993	1965	1983	1969	1974	1997	1965	1957	1983	1951	AUG 1997
	NORMAL NO. DAYS WITH:														
	PRECIPITATION 0.01	30	6.4	5.2	7.7	6.6	6.3	4.2	5.8	6.1	6.3	5.6	5.6	6.4	72.2
	PRECIPITATION 1.00	30	0.0	0.0	0.0	0.0	0.0	*	*	0.0	*	0.0	0.0	0.0	0.0
SNOWFALL	NORMAL (IN)	30	6.6	3.3	3.7	1.2	0.2	0.0	0.0	0.0	0.1	0.7	2.5	5.7	24.0
	MAXIMUM MONTHLY (IN)	55	33.7	18.4	14.9	14.3	5.0	T	T	T	3.1	6.1	12.1	19.0	33.7
	YEAR OF OCCURRENCE		1957	1948	1948	1975	1979	1997	2001	1993	1965	1975	1964	1983	JAN 1957
	MAXIMUM IN 24 HOURS (IN)	55	9.1	9.0	6.1	8.9	5.0	T	T	T	3.1	6.1	8.4	6.3	9.1
	YEAR OF OCCURRENCE		1957	1989	1948	1975	1979	1997	2001	1993	1965	1975	1954	1998	JAN 1957
	MAXIMUM SNOW DEPTH (IN)	53	16	12	8	7	1	0	0	0		2	5	8	16
	YEAR OF OCCURRENCE		1957	1957	1960	1975	1979				1965	1975	1954	1983	JAN 1957
	NORMAL NO. DAYS WITH:														
	SNOWFALL 1.0	30	2.5	1.0	1.2	0.3	0.1	0.0	0.0	0.0	0.*	0.2	0.8	2.1	8.2

PRECIPITATION (inches) 2001 GRAND JUNCTION, CO (GJT)

YEAR	JAN	FEB	MAR	APR	MAY	JUN	JUL	AUG	SEP	OCT	NOV	DEC	ANNUAL
1972	0.20	T	0.02	0.11	0.44	0.64	0.03	0.29	0.72	3.45	0.69	0.74	7.33
1973	0.79	0.12	0.65	0.86	1.45	0.87	0.52	0.62	0.33	0.20	0.91	0.62	7.94
1974	1.20	0.40	0.81	1.03	0.01	0.14	1.53	0.48	0.38	0.72	1.18	0.32	8.20
1975	0.53	0.49	1.74	1.38	1.23	0.43	1.39	0.09	0.16	0.85	0.39	0.50	9.18
1976	0.13	0.81	0.75	0.40	1.49	0.14	0.20	0.31	0.67	0.32	0.04	0.01	5.27
1977	0.37	0.06	0.50	0.54	0.59	0.04	0.89	0.59	0.52	0.50	0.70	0.38	5.68
1978	1.08	0.64	1.19	1.19	0.55	0.01	0.25	0.54	0.49	0.03	0.62	1.30	7.89
1979	1.36	0.63	2.02	0.42	1.45	0.78	0.08	0.61	0.01	0.25	1.02	0.27	8.90
1980	0.57	1.10	1.77	0.53	1.17	T	0.96	1.39	0.58	1.31	0.52	0.24	10.14
1981	0.44	0.16	1.35	0.56	1.49	0.17	0.41	0.82	0.25	2.06	0.47	0.60	8.78
1982	0.29	0.41	0.79	0.09	0.75	0.21	0.35	0.94	2.81	0.83	0.48	0.27	8.22
1983	0.50	0.64	1.59	0.90	1.68	1.54	1.92	0.73	1.11	0.36	2.00	1.85	14.82
1984	0.28	0.11	1.57	1.21	0.55	1.68	0.62	1.77	0.34	2.65	0.38	0.43	11.59
1985	0.51	0.26	0.92	1.78	1.09	0.39	1.21	0.24	1.67	2.32	1.10	0.73	12.22
1986	0.13	0.33	0.25	0.71	1.15	0.15	0.94	0.97	1.52	1.22	1.02	0.47	8.86
1987	0.30	1.21	1.95	0.46	1.51	0.23	1.51	0.83	0.13	0.65	1.92	0.83	11.53
1988	1.07	0.21	0.72	0.99	1.10	0.21	0.18	1.37	0.76	0.02	1.02	0.20	7.85
1989	0.98	1.33	0.51	0.23	0.39	0.24	0.27	1.01	0.33	0.14	T	0.08	5.51
1990	0.59	0.55	1.07	0.71	0.05	0.26	0.96	0.49	1.23	0.95	0.57	0.98	8.41
1991	0.92	0.13	0.70	0.87	0.20	0.30	0.40	0.57	2.30	1.10	1.10	0.54	9.23
1992	0.24	0.35	1.71	0.15	1.81	0.17	1.03	0.84	0.33	1.45	0.76	0.35	9.19
1993	1.36	1.09	1.72	1.30	1.99	0.03	0.04	1.42	0.41	1.34	0.41	0.57	11.68
1994	0.23	0.56	0.25	1.81	0.19	0.04	0.01	0.48	1.50	0.58	0.69	0.64	6.98
1995	0.62	0.52	1.74	0.96	2.04	1.32	0.87	0.47	0.66	0.24	0.20	0.55	10.19
1996	0.65	1.07	0.53	0.90	0.99	0.58	0.77	0.15	1.53		1.35	0.53	10.06
1997	0.63	0.34	0.53	2.15	1.53	0.29	0.28	2.67	2.84	1.20	0.62	0.14	13.22
1998	0.47	0.48	1.36	0.75	0.21	0.55	1.21	0.61	1.44	1.36	0.42	0.26	9.12
1999	0.09	0.28	0.03	2.06	0.68	0.60	0.51	2.22	1.01	0.17	0.18	0.26	8.09
2000	1.35	0.70	1.26	0.32	0.43	0.34	0.20	0.60	0.60	1.18	0.35	0.18	7.51
2001	0.43	0.67	0.98	0.58	0.57	T	1.20	1.46	0.15	0.99	1.06	0.31	8.40
POR= 109 YRS	0.61	0.59	0.84	0.78	0.79	0.44	0.62	1.03	0.84	0.90	0.61	0.50	8.55

AVERAGE TEMPERATURE (F) 2001 GRAND JUNCTION, CO (GJT)

YEAR	JAN	FEB	MAR	APR	MAY	JUN	JUL	AUG	SEP	OCT	NOV	DEC	ANNUAL
1972	30.0	36.6	46.6	53.3	63.1	74.3	80.2	77.1	68.1	54.0	37.1	22.7	53.6
1973	11.5	29.1	42.1	48.1	61.5	70.5	78.1	77.4	65.7	56.4	41.2	30.1	51.0
1974	16.9	19.9	48.2	51.0	65.0	74.9	75.6	66.4	56.2	39.6	27.1	51.6	
1975	20.0	33.0	41.0	46.4	57.1	67.5	78.3	75.4	67.1	53.5	36.2	27.4	50.2
1976	21.7	38.2	38.7	51.9	61.8	70.4	79.6	75.3	66.9	49.3	39.1	27.4	51.9
1977	23.9	37.1	40.8	56.9	63.7	79.1	80.2	78.3	70.2	58.2	40.3	33.3	55.2
1978	29.4	34.4	46.7	52.3	58.8	73.0	78.4	74.5	65.7	54.7	40.2	16.0	52.0
1979	16.6	23.5	41.1	52.5	60.3	71.0	78.7	74.6	72.0	58.9	33.2	26.9	50.8
1980	32.6	39.2	40.9	51.4	59.1	74.0	78.1	75.3	67.9	53.8	42.3	40.1	54.6
1981	36.8	37.9	44.0	56.8	60.9	76.4	79.4	76.8	69.2	50.7	41.6	31.3	55.2
1982	26.0	34.7	46.0	51.3	61.4	72.2	79.0	78.2	67.5	51.9	41.2	33.0	53.5
1983	34.3	40.9	45.9	48.7	58.7	69.8	78.3	80.4	71.4	58.2	42.1	30.5	54.9
1984	20.7	31.7	44.4	49.0	66.2	70.1	78.0	76.6	68.3	50.2	40.8	32.6	52.4
1985	31.1	32.0	43.9	54.5	63.6	73.0	77.6	77.1	62.4	52.8	38.8	31.9	53.2
1986	34.2	40.4	49.0	52.5	60.6	74.3	76.0	75.6	63.2	51.5	40.8	32.4	54.2
1987	27.4	36.8	40.1	54.5	61.2	73.5	75.2	72.9	66.2	56.8	39.6	27.9	52.7
1988	17.4	29.2	41.0	53.1	60.9	76.5	80.6	76.2	64.5	58.9	40.7	30.0	52.4
1989	20.2	27.7	47.5	57.0	63.8	71.0	80.5	74.4	68.3	54.9	40.6	29.2	52.9
1990	28.5	35.4	46.8	55.7	61.3	75.2	78.0	76.6	69.7	53.2	39.5	20.6	53.4
1991	17.6	32.0	42.0	48.8	62.0	72.5	77.6	76.2	69.0	55.0	37.7	26.2	51.2
1992	19.9	37.6	47.2	59.0	64.5	71.2	75.1	75.1	67.7	57.8	35.9	24.5	53.0
1993	31.9	36.2	45.6	50.0	61.9	70.1	76.7	73.2	65.9	51.7	35.6	29.2	52.3
1994	31.7	34.2	47.4	53.2	65.3	77.8	81.3	79.4	68.3	52.9	37.1	33.1	55.1
1995	33.7	43.4	44.2	50.6	56.9	67.8	76.1	79.0	68.5	53.3	43.6	35.3	54.5
1996	30.1	40.2	44.9	51.0	64.2	73.3	79.2	77.6	63.0	51.3	39.9	30.7	53.8
1997	30.6	34.2	45.0	46.5	62.2	72.7	76.2	74.0	66.4	51.7	38.2	28.2	52.2
1998	33.5	35.8	41.8	48.8	62.1	67.8	79.1	76.3	70.0	53.3	40.8	25.6	53.0
1999	32.9	37.4	48.5	47.8	58.7	70.6	77.6	73.6	63.5	54.2	43.0	30.3	53.2
2000	32.8	39.5	43.4	56.4	65.8	73.1	80.1	78.4	71.9	53.3	32.8	30.3	54.6
2001	28.0	36.2	45.3	53.8	64.3	74.1	78.9	75.7	69.7	55.0	43.7	28.3	54.4
POR= 109 YRS	26.2	33.7	43.1	52.2	61.8	72.0	78.2	75.6	66.9	54.2	39.7	28.6	52.7

REFERENCE NOTES:

PAGE 1:
THE TEMPERATURE GRAPH SHOWS NORMAL MAXIMUM AND NORMAL MINIMUM DAILY TEMPERATURES (SOLID CURVES) AND THE ACTUAL DAILY HIGH AND LOW TEMPERATURES (VERTICAL BARS).

PAGE 2 AND 3:
H/C INDICATES HEATING AND COOLING DEGREE DAYS.
RH INDICATES RELATIVE HUMIDITY
W/O INDICATES WEATHER AND OBSTRUCTIONS
S INDICATES SUNSHINE.
PR INDICATES PRESSURE.
CLOUDINESS ON PAGE 3 IS THE SUM OF THE CEILOMETER AND SATELLITE DATA NOT TO EXCEED EIGHT EIGHTHS(OKTAS).

GENERAL:
T INDICATES TRACE PRECIPITATION, AN AMOUNT GREATER THAN ZERO BUT LESS THAN THE LOWEST REPORTABLE VALUE.
+ INDICATES THE VALUE ALSO OCCURS ON EARLIER DATES.
BLANK ENTRIES DENOTE MISSING OR UNREPORTED DATA.
NORMALS ARE 30-YEAR AVERAGES (1961 - 1990).
ASOS INDICATES AUTOMATED SURFACE OBSERVING SYSTEM.
PM INDICATES THE LAST DAY OF THE PREVIOUS MONTH.
POR (PERIOD OF RECORD) BEGINS WITH THE JANUARY DATA MONTH AND IS THE NUMBER OF YEARS USED TO COMPUTE THE MEAN. INDIVIDUAL MONTHS WITHIN THE POR MAY BE MISSING.
WHEN THE POR FOR A NORMAL IS LESS THAN 30 YEARS, THE NORMAL IS PROVISIONAL AND IS BASED ON THE NUMBER OF YEARS INDICATED.
0.* OR * INDICATES THE VALUE OR MEAN-DAYS-WITH IS BETWEEN 0.00 AND 0.05.
CLOUDINESS FOR ASOS STATIONS DIFFERS FROM THE NON-ASOS OBSERVATION TAKEN BY A HUMAN OBSERVER. ASOS STATION CLOUDINESS IS BASED ON TIME-AVERAGED CEILOMETER DATA FOR CLOUDS AT OR BELOW 12,000 FEET AND ON SATELLITE DATA FOR CLOUDS ABOVE 12,000 FEET.
THE NUMBER OF DAYS WITH CLEAR, PARTLY CLOUDY, AND CLOUDY CONDITIONS FOR ASOS STATIONS IS THE SUM OF THE CEILOMETER AND SATELLITE DATA FOR THE SUNRISE TO SUNSET PERIOD.

GENERAL CONTINUED:
CLEAR INDICATES 0 - 2 OKTAS, PARTLY CLOUDY INDICATES 3 - 6 OKTAS, AND CLOUDY INDICATES 7 OR 8 OKTAS.
WHEN AT LEAST ONE OF THE ELEMENTS (CEILOMETER OR SATELLITE) IS MISSING, THE DAILY CLOUDINESS IS NOT COMPUTED.
WIND DIRECTION IS RECORDED IN TENS OF DEGREES (2 DIGITS) CLOCKWISE FROM TRUE NORTH. "00" INDICATES CALM. "36" INDICATES TRUE NORTH.
RESULTANT WIND IS THE VECTOR AVERAGE OF THE SPEED AND DIRECTION.
AVERAGE TEMPERATURE IS THE SUM OF THE MEAN DAILY MAXIMUM AND MINIMUM TEMPERATURE DIVIDED BY 2.
SNOWFALL DATA COMPRISE ALL FORMS OF FROZEN PRECIPITATION, INCLUDING HAIL.
A HEATING (COOLING) DEGREE DAY IS THE DIFFERENCE BETWEEN THE AVERAGE DAILY TEMPERATURE AND 65 F.
DRY BULB IS THE TEMPERATURE OF THE AMBIENT AIR.
DEW POINT IS THE TEMPERATURE TO WHICH THE AIR MUST BE COOLED TO ACHIEVE 100 PERCENT RELATIVE HUMIDITY.
WET BULB IS THE TEMPERATURE THE AIR WOULD HAVE IF THE MOISTURE CONTENT WAS INCREASED TO 100 PERCENT RELATIVE HUMIDITY.

ON JULY 1, 1996, THE NATIONAL WEATHER SERVICE BEGAN USING THE "METAR" OBSERVATION CODE THAT WAS ALREADY EMPLOYED BY MOST OTHER NATIONS OF THE WORLD. THE MOST NOTICEABLE DIFFERENCE IN THIS ANNUAL PUBLICATION WILL BE THE CHANGE IN UNITS FROM TENTHS TO EIGHTS(OKTAS) FOR REPORTING THE AMOUNT OF SKY COVER.

HEATING DEGREE DAYS (base 65 F) 2001 GRAND JUNCTION, CO (GJT)

YEAR	JUL	AUG	SEP	OCT	NOV	DEC	JAN	FEB	MAR	APR	MAY	JUN	TOTAL
1972-73	0	0	31	335	832	1303	1651	999	705	499	139	49	6543
1973-74	0	0	72	266	708	1075	1487	1260	513	415	66	32	5894
1974-75	0	0	60	266	756	1167	1387	888	736	551	249	51	6111
1975-76	0	0	35	358	858	1161	1335	775	807	386	122	25	5862
1976-77	0	0	41	421	769	1153	1267	775	743	250	94	0	5513
1977-78	0	1	17	214	736	975	1098	852	561	373	210	9	5046
1978-79	0	6	95	313	737	1510	1493	1154	732	377	192	37	6646
1979-80	0	3	0	209	945	1175	999	741	740	405	195	4	5416
1980-81	0	2	21	359	674	765	864	754	645	247	153	15	4499
1981-82	0	0	12	439	696	1039	1203	841	581	405	136	6	5358
1982-83	2	0	61	397	704	983	946	668	586	482	238	22	5089
1983-84	0	0	27	208	678	1064	1366	959	631	474	89	44	5540
1984-85	0	0	54	452	719	996	1044	919	646	310	81	12	5233
1985-86	0	0	139	371	779	1018	949	685	489	366	168	3	4967
1986-87	0	0	130	414	718	1001	1159	785	765	314	143	0	5429
1987-88	0	6	34	248	754	1147	1469	1031	741	350	172	8	5960
1988-89	0	0	106	183	724	1078	1379	1038	534	258	113	8	5421
1989-90	0	0	40	316	729	1103	1124	820	557	271	139	20	5119
1990-91	0	0	28	360	759	1371	1464	919	706	478	136	18	6239
1991-92	0	2	37	304	815	1193	1390	788	540	195	53	8	5325
1992-93	0	6	25	222	868	1245	1018	799	597	446	144	33	5403
1993-94	4	0	59	410	875	1102	1025	853	540	360	64	0	5292
1994-95	0	0	24	368	832	984	962	596	578	425	256	47	5072
1995-96	8	0	73	357	634	914	1073	712	614	415	88	2	4890
1996-97	0	0	135	421	748	1055	1056	857	613	547	122	4	5558
1997-98	0	0	42	412	799	1138	970	813	709	478	137	55	5553
1998-99	0	0	9	355	715	1217	987	767	501	508	219	33	5311
1999-00	0	0	88	331	651	1067	992	732	663	256	100	1	4881
2000-01	0	0	58	339	959	1053	1140	802	605	333	105	31	5425
2001-	0	0	23	316	633	1130							

COOLING DEGREE DAYS (base 65 F) 2001 GRAND JUNCTION, CO (GJT)

YEAR	JAN	FEB	MAR	APR	MAY	JUN	JUL	AUG	SEP	OCT	NOV	DEC	ANNUAL
1972	0	0	0	0	86	288	479	381	130	3	0	0	1367
1973	0	0	0	0	35	222	410	393	101	6	0	0	1167
1974	0	0	0	1	73	335	420	335	109	0	0	0	1273
1975	0	0	0	0	9	133	419	328	106	9	0	0	1004
1976	0	0	0	0	32	195	460	324	103	0	0	0	1114
1977	0	0	0	16	60	429	477	420	180	10	0	0	1592
1978	0	0	0	0	25	258	420	308	123	1	0	0	1135
1979	0	0	0	6	52	225	428	310	215	27	0	0	1263
1980	0	0	0	1	16	280	427	325	115	19	0	0	1183
1981	0	0	0	9	31	367	456	375	143	0	0	0	1381
1982	0	0	0	0	33	229	443	415	144	0	0	0	1264
1983	0	0	0	0	49	171	421	483	226	3	0	0	1353
1984	0	0	0	0	134	200	408	368	159	0	0	0	1269
1985	0	0	0	4	45	261	396	382	67	0	0	0	1155
1986	0	0	0	0	39	289	348	334	82	0	0	0	1092
1987	0	0	0	5	30	262	324	256	76	2	0	0	955
1988	0	0	0	0	51	360	489	357	98	4	0	0	1359
1989	0	0	0	26	85	195	489	300	145	11	0	0	1251
1990	0	0	0	1	34	331	412	368	174	3	0	0	1323
1991	0	0	0	0	50	247	398	356	88	0	0	0	1139
1992	0	0	0	21	43	203	319	328	114	7	0	0	1035
1993	0	0	0	0	56	193	371	260	92	6	0	0	978
1994	0	0	0	13	82	388	514	454	126	0	0	0	1577
1995	0	0	0	0	9	138	362	444	185	3	0	0	1141
1996	0	0	0	1	67	256	448	396	82	4	0	0	1254
1997	0	0	0	0	44	240	353	285	93	8	0	0	1023
1998	0	0	0	0	51	148	445	375	163	0	0	0	1182
1999	0	0	0	0	27	204	401	274	45	4	0	0	955
2000	0	0	0	5	131	249	476	421	129	15	0	0	1426
2001	0	0	0	3	89	311	437	341	170	13	0	0	1364

SNOWFALL (inches) 2001 GRAND JUNCTION, CO (GJT)

YEAR	JUL	AUG	SEP	OCT	NOV	DEC	JAN	FEB	MAR	APR	MAY	JUN	TOTAL
1972-73	0.0	0.0	0.0	5.7	1.3	9.7	12.8	1.2	1.3	2.0	T	0.0	34.0
1973-74	0.0	0.0	0.0	0.0	7.7	5.7	17.0	5.5	T	1.2	0.0	0.0	37.1
1974-75	0.0	0.0	0.0	T	0.1	4.6	7.9	4.4	8.8	14.3	1.3	0.0	41.4
1975-76	0.0	0.0	0.0	6.1	3.9	7.2	1.7	4.0	6.8	0.2	0.0	0.0	29.9
1976-77	0.0	0.0	0.0	0.0	T	0.1	4.2	T	2.3	1.7	0.0	0.0	8.3
1977-78	0.0	0.0	0.0	T	3.3	2.5	12.0	2.5	0.6	T	T	0.0	20.9
1978-79	0.0	0.0	0.0	0.0	2.9	11.8	18.7	9.6	3.4	1.1	5.0	0.0	52.5
1979-80	0.0	0.0	0.0	0.0	8.2	3.5	2.2	0.5	7.3	0.2	0.0	0.0	21.9
1980-81	0.0	0.0	0.0	0.0	0.0	T	0.0	3.9	0.8	1.2	T	0.0	5.9
1981-82	0.0	0.0	0.0	0.5	3.3	3.4	3.4	4.0	0.8	T	0.0	0.0	15.4
1982-83	0.0	0.0	0.0	T		1.9	6.1	3.1	1.5	2.2	T	0.0	14.8
1983-84	0.0	0.0	0.0	0.0	4.2	19.0	3.7	0.6	6.1	2.9	0.0	0.0	36.5
1984-85	0.0	0.0	0.0	0.7	2.0	2.7	5.0	2.7	5.6	0.1	0.0	0.0	18.8
1985-86	0.0	0.0	0.0	0.0	4.6	4.4	1.8	0.7	T	0.2	T	0.0	11.7
1986-87	0.0	0.0	0.0	2.2	1.2	1.0	3.0	5.5	9.4	0.6	T	0.0	22.9
1987-88	0.0	0.0	0.0	0.0	1.1	7.1	12.2	2.2	4.3	0.0	0.0	0.0	26.9
1988-89	0.0	0.0	0.0	0.0	0.9	3.1	10.2	16.0	1.1	T	T	0.0	31.3
1989-90	0.0	0.0	0.0	0.0	0.0	1.1	6.2	8.6	1.8	0.8	0.0	0.0	18.5
1990-91	0.0	0.0	0.0	0.0	1.5	5.1	12.7	0.3	3.7	4.7	0.0	0.0	28.0
1991-92	0.0	0.0	T	2.5	1.9	7.6	2.7	1.9	T	0.0	T	0.0	16.6
1992-93	0.0	0.0	0.0	0.0	2.0	4.4	6.0	8.4	0.0	0.4	T	0.0	21.2
1993-94	0.0	T	0.0	0.0	0.6	5.9	1.1	4.6	1.1	0.3	T	0.0	12.5
1994-95	0.0	0.0	0.0	T	5.4	2.2	4.0	1.1	3.7	T	T	0.0	16.4
1995-96	0.0	0.0	T	1.2	T	3.7	2.1	0.5	3.0	T	0.0	0.0	10.5
1996-97	0.0	0.0	0.0	1.1	4.2	3.5	4.9	3.2	1.2	6.5	T	T	24.6
1997-98	0.0	T	0.0	0.5	T	1.7	3.8	0.5	3.8	0.1	0.0	T	10.4
1998-99	0.0	0.0	T	0.0	2.5	7.0	0.5	1.4	0.0	3.8	T	0.0	15.2
1999-00	0.0	T	0.0	0.0	T	3.3	3.1	0.6	6.2	0.0	0.0	0.0	13.2
2000-01	0.0	0.0	0.0	0.0	3.6	1.5	4.3	3.3	3.7	0.0	0.0	0.0	16.4
2001-	0.0	T	0.0	0.0	2.8	3.3							
POR= 54 YRS	0.0	T	0.1	0.5	2.5	5.0	6.6	3.8	3.6	1.2	0.2	0.0	23.5

2001
HARTFORD,
CONNECTICUT (BDL)

Bradley International Airport is located about 3 miles west of the Connecticut River on a slight rise of ground in a broad portion of the Connecticut River Valley between north-south mountain ranges whose heights do not exceed 1,200 feet.

The station is in the northern temperate climate zone. The prevailing west to east movement of air brings the majority of weather systems into Connecticut from the west. The average wintertime position of the Polar Front boundary between cold, dry polar air and warm, moist tropical air is just south of New England, which helps to explain the extensive winter storm activity and day to day variability of local weather. In summer, the Polar Front has an average position along the New England-Canada border with this station in a warm and pleasant atmosphere.

The location of Hartford, relative to continent and ocean, is also significant. Rapid weather changes result when storms move northward along the mid-Atlantic coast, frequently producing strong and persistent northeast winds associated with storms known locally as coastals or northeasters. Seasonally, weather characteristics vary from the cold and dry continental-polar air of winter to the warm and humid maritime air of summer.

Summer thunderstorms develop in the Berkshire Mountains to the west and northwest, move over the Connecticut Valley, and when accompanied by wind and hail, sometimes cause considerable damage to crops, particularly tobacco. During the winter, rain often falls through cold air trapped in the valley, creating extremely hazardous ice conditions. On clear nights in the late summer or early autumn, cool air drainage into the valley, and moisture from the Connecticut River, produce steam and/or ground fog which becomes quite dense throughout the valley, hampering ground and air transportation.

The mean date of the last springtime temperature of 32 degrees or lower is April 22, and the mean date of the first autumn temperature of 32 degrees is October 15.

NORMALS, MEANS, AND EXTREMES
HARTFORD, CT (BDL)

LATITUDE: 41 56' 17" N LONGITUDE: 72 40' 57" W ELEVATION (FT): GRND: 162 BARO: 165 TIME ZONE: EASTERN (UTC + 5) WBAN: 14740

ELEMENT	POR	JAN	FEB	MAR	APR	MAY	JUN	JUL	AUG	SEP	OCT	NOV	DEC	YEAR
TEMPERATURE F														
NORMAL DAILY MAXIMUM	30	33.2	36.4	46.8	59.9	71.6	80.0	85.0	82.7	74.8	63.7	51.0	37.5	60.2
MEAN DAILY MAXIMUM	53	34.2	37.3	46.3	59.7	71.4	80.0	83.0	82.4	74.3	63.9	50.9	38.5	60.2
HIGHEST DAILY MAXIMUM	47	65	73	89	96	99	100	102	102	99	91	81	76	102
YEAR OF OCCURRENCE		1967	1985	1998	1976	1996	1964	1966	2001	1983	1963	1974	1998	AUG 2001
MEAN OF EXTREME MAXS.	53	53.7	55.1	68.3	80.8	89.0	93.6	93.9	93.8	89.2	80.9	70.0	58.7	77.3
NORMAL DAILY MINIMUM	30	15.8	18.6	28.1	37.5	47.6	56.9	62.2	60.4	51.8	40.7	32.8	21.3	39.5
MEAN DAILY MINIMUM	53	17.3	19.7	27.8	37.6	47.5	56.8	60.9	60.3	51.7	41.0	32.6	21.9	39.6
LOWEST DAILY MINIMUM	47	-26	-21	-6	9	28	37	44	36	30	17	1	-14	-26
YEAR OF OCCURRENCE		1961	1961	1967	1970	2001	1986	1962	1965	2000	1978	1989	1980	JAN 1961
MEAN OF EXTREME MINS.	53	-2.8	-.3	10.8	24.8	34.1	43.6	49.4	47.1	35.6	26.0	17.4	3.1	24.1
NORMAL DRY BULB	30	24.6	27.5	37.5	48.7	59.6	68.5	73.7	71.6	63.3	52.2	41.9	29.5	49.9
MEAN DRY BULB	53	25.6	28.5	36.9	48.7	59.4	68.5	72.1	71.4	63.2	52.4	41.7	30.1	49.9
MEAN WET BULB	16	24.2	26.2	32.6	42.3	52.6	61.2	61.7	61.0	57.6	47.2	35.3	26.9	44.1
MEAN DEW POINT	16	16.9	17.9	23.7	33.4	45.8	56.1	57.5	57.2	53.4	41.7	29.2	20.1	37.7
NORMAL NO. DAYS WITH:														
MAXIMUM 90	30	0.0	0.0	0.0	0.3	1.1	3.6	8.0	4.8	1.3	*	0.0	0.0	19.1
MAXIMUM 32	30	14.1	9.4	1.6	0.1	0.0	0.0	0.0	0.0	0.0	0.0	0.5	9.5	35.2
MINIMUM 32	30	28.8	25.4	21.4	8.5	0.8	0.0	0.0	0.0	0.3	6.4	16.6	26.9	135.1
MINIMUM 0	30	3.2	1.7	*	0.0	0.0	0.0	0.0	0.0	0.0	0.0	0.0	1.1	6.0
H/C														
NORMAL HEATING DEG. DAYS	30	1252	1050	853	489	194	20	0	6	96	397	693	1101	6151
NORMAL COOLING DEG. DAYS	30	0	0	0	0	27	125	270	210	45	0	0	0	677
RH														
NORMAL (PERCENT)	30	64	63	60	58	63	67	68	71	73	69	68	68	66
HOUR 01 LST	30	69	68	68	68	76	80	82	84	85	80	75	73	76
HOUR 07 LST	30	71	72	71	69	73	77	78	82	86	83	78	75	76
HOUR 13 LST	30	56	54	50	45	47	51	51	53	54	51	56	59	52
HOUR 19 LST	30	62	60	56	52	56	60	61	65	70	67	66	66	62
S PERCENT POSSIBLE SUNSHINE	42	53	56	57	55	57	60	62	62	59	57	45	47	56
W/O MEAN NO. DAYS WITH:														
HEAVY FOG(VISBY 1/4 MI)	47	2.5	2.4	2.3	1.4	1.7	2.2	2.0	2.3	2.8	3.2	2.2	2.7	27.7
THUNDERSTORMS	47	0.1	0.2	0.7	1.2	2.5	3.9	4.3	3.7	2.1	1.0	0.4	0.1	20.2
CLOUDINESS MEAN:														
SUNRISE-SUNSET (OKTAS)	42	5.2	5.2	5.4	5.4	5.4	5.3	5.1	4.9	4.8	4.6	5.4	5.2	5.2
MIDNIGHT-MIDNIGHT (OKTAS)	32	5.0	4.9	5.1	5.0	5.1	5.0	4.8	4.7	4.6	4.3	5.0	5.0	4.9
MEAN NO. DAYS WITH:														
CLEAR	42	7.7	6.6	6.7	6.3	5.6	5.7	5.7	6.9	8.4	9.4	5.7	6.9	81.6
PARTLY CLOUDY	42	7.8	7.7	8.4	8.5	9.7	10.3	12.1	10.8	8.9	8.5	8.2	7.5	108.4
CLOUDY	42	15.5	13.9	16.0	15.2	15.7	14.0	13.2	13.3	12.7	13.0	16.1	16.6	175.2
PR MEAN STATION PRESSURE(IN)	13	29.82	29.83	29.78	29.79	29.79	29.79	29.81	29.81	29.85	29.90	29.89	29.87	29.83
MEAN SEA-LEVEL PRES. (IN)	17	30.06	30.05	30.01	29.96	29.96	29.93	29.98	30.01	30.05	30.09	30.07	30.07	30.02
WINDS MEAN SPEED (MPH)	41	8.9	9.5	10.1	9.9	8.8	8.1	7.3	7.0	7.4	7.8	8.5	8.6	8.5
PREVAIL.DIR(TENS OF DEGS)	27	31	31	32	18	18	18	18	18	18	18	36	36	18
MAXIMUM 2-MINUTE:														
SPEED (MPH)	5	46	38	43	41	39	45	39	34	38	38	36	41	46
DIR. (TENS OF DEGS)		24	30	20	20	28	36	27	28	02	29	15	26	24
YEAR OF OCCURRENCE		1999	2001	1997	2000	1998	2000	1999	1997	1999	1998	1999	2000	JAN 1999
MAXIMUM 5-SECOND:														
SPEED (MPH)	5	56	54	54	53	61	52	60	41	46	46	54	55	61
DIR. (TENS OF DEGS)		24	29	29	20	31	34	27	28	31	28	24	25	31
YEAR OF OCCURRENCE		1999	2001	1997	2000	1998	2000	1999	1997	1999	1998	1998	2000	MAY 1998
PRECIPITATION NORMAL (IN)	30	3.41	3.23	3.63	3.85	4.12	3.75	3.19	3.65	3.79	3.57	4.04	3.91	44.14
MAXIMUM MONTHLY (IN)	47	9.61	7.27	6.86	9.90	12.00	13.60	8.43	21.87	11.22	11.61	8.53	8.36	21.87
YEAR OF OCCURRENCE		1978	1981	1983	1983	1989	1982	1988	1955	1999	1955	1972	1969	AUG 1955
MINIMUM MONTHLY (IN)	47	0.38	0.45	0.27	1.10	0.73	0.67	0.97	0.54	0.84	0.35	0.51	0.78	0.27
YEAR OF OCCURRENCE		1981	1987	1981	1999	1959	1988	2001	1981	1986	1963	1976	1955	MAR 1981
MAXIMUM IN 24 HOURS (IN)	47	2.56	2.16	2.62	3.01	4.90	6.14	3.48	12.12	5.72	4.45	2.90	3.12	12.12
YEAR OF OCCURRENCE		1979	1965	1987	1979	1989	1982	1960	1955	1999	1959	1988	1973	AUG 1955
NORMAL NO. DAYS WITH:														
PRECIPITATION 0.01	30	10.7	10.1	11.2	10.7	11.8	11.5	9.4	9.6	9.3	8.2	11.2	12.3	126.0
PRECIPITATION 1.00	30	0.7	0.9	0.9	1.0	1.1	0.9	0.7	1.1	1.2	1.0	1.0	1.2	11.7
SNOWFALL NORMAL (IN)	30	12.8	12.2	7.7	1.3	0.*	0.0	0.0	0.0	0.0	0.1	2.3	11.0	47.4
MAXIMUM MONTHLY (IN)	43	42.8	32.2	43.3	14.3	1.3	T	0.0	0.0	0.0	1.7	8.7	35.4	43.3
YEAR OF OCCURRENCE		1996	1969	1956	1982	1977	1993				1979	1986	1969	MAR 1956
MAXIMUM IN 24 HOURS (IN)	43	14.9	21.0	14.8	14.1	1.3	T	0.0	0.0	0.0	1.7	8.6	13.9	21.0
YEAR OF OCCURRENCE		1996	1983	1993	1982	1977	1993				1979	1980	1969	FEB 1983
MAXIMUM SNOW DEPTH (IN)	48	25	29	20	14	0	0	0	0	0	1	8	16	29
YEAR OF OCCURRENCE		1961	1961	1956	1982						1979	1971	1969	FEB 1961
NORMAL NO. DAYS WITH:														
SNOWFALL 1.0	30	3.0	2.6	2.0	0.4	0.*	0.0	0.0	0.0	0.0	0.*	0.6	3.4	12.0

PRECIPITATION (inches) 2001 HARTFORD, CT (BDL)

YEAR	JAN	FEB	MAR	APR	MAY	JUN	JUL	AUG	SEP	OCT	NOV	DEC	ANNUAL
1972	2.02	5.12	6.71	4.61	7.49	9.66	3.84	3.45	1.84	4.20	8.53	7.08	64.55
1973	3.28	3.05	3.22	6.59	5.95	5.07	1.77	4.50	3.73	3.47	2.14	8.31	51.08
1974	4.10	1.95	4.49	3.64	3.03	2.38	2.39	3.36	8.57	2.34	2.62	4.52	43.39
1975	4.30	3.22	3.82	2.99	3.29	3.83	6.11	4.60	9.02	5.28	4.57	4.31	55.34
1976	5.57	3.11	2.86	3.93	4.45	2.86	3.51	5.76	2.55	4.10	0.51	2.97	42.18
1977	2.41	2.81	6.57	4.89	3.70	3.99	3.37	2.44	8.17	5.45	4.38	5.68	53.86
1978	9.61	1.42	3.63	1.51	4.61	2.94	2.51	3.61	2.67	1.75	2.12	4.23	40.61
1979	9.12	2.83	4.25	5.88	3.48	0.91	1.97	4.44	2.95	4.76	3.46	2.57	46.62
1980	0.72	0.98	5.87	5.39	1.65	3.81	2.65	1.60	1.40	2.58	4.22	0.82	31.69
1981	0.38	7.27	0.27	2.92	2.17	1.37	4.21	0.54	4.49	5.19	2.34	4.37	35.15
1982	4.76	2.83	2.23	4.12	3.30	13.60	2.60	4.41	2.41	3.31	3.12	1.32	48.01
1983	4.68	3.83	6.86	9.90	4.82	2.61	1.07	2.55	2.10	5.52	6.09	5.97	56.00
1984	1.80	4.72	3.93	4.24	11.55	2.16	4.22	1.32	1.20	2.76	2.49	2.46	42.85
1985	0.73	1.72	2.16	1.54	2.77	3.55	4.55	6.44	3.83	2.27	6.04	1.28	36.88
1986	5.34	3.02	2.72	1.55	2.28	6.79	4.44	3.44	0.84	2.18	5.57	6.15	44.32
1987	6.20	0.45	4.44	5.23	2.18	2.27	3.66	4.25	7.19	3.67	3.66	1.57	44.77
1988	3.36	3.99	2.06	2.35	3.46	0.67	8.43	2.12	1.88	2.29	7.84	1.35	39.80
1989	0.88	1.85	3.02	3.33	12.00	6.65	3.40	6.81	4.67	7.62	2.89	1.49	54.61
1990	4.03	3.37	2.46	4.55	6.38	3.59	2.09	8.32	2.13	7.63	3.76	4.86	53.17
1991	2.45	1.78	4.52	3.54	5.18	2.37	2.90	8.69	5.67	3.17	4.03	2.96	47.26
1992	2.73	2.23	3.79	3.13	3.21	5.77	4.62	3.60	2.43	1.95	4.19	4.33	41.98
1993	2.63	2.90	6.67	4.71	1.92	2.63	4.90	1.80	5.35	4.15	3.27	4.16	45.09
1994	5.83	3.38	5.70	2.51	4.12	3.84	5.32	5.33	5.47	1.53	4.57	5.38	52.98
1995	3.84	3.24	1.89	2.60	2.63	1.02	2.58	3.81	3.15	9.46	4.38	2.32	40.92
1996	6.99	2.86	2.45	6.29	2.98	2.39	6.97	1.67	7.53	5.25	4.14	5.69	55.21
1997	3.15	1.38	3.60	2.43	3.37	1.90	3.92	7.33	0.97	1.65	5.87	2.18	37.75
1998	3.37	3.12	4.87	3.35	7.84	7.18	2.23	1.98	2.33	5.67	2.34	0.83	45.11
1999	5.26	3.50	4.28	1.10	3.23	0.72	2.59	2.66	11.22	3.54	3.54	2.47	44.11
2000	2.83	2.24	3.69	4.21	4.45	6.74	5.48	3.33	3.88	1.07	0.95	2.47	41.34
2001	1.35	2.90	6.13	1.22	4.71	5.12	0.97	3.71	3.10	0.76	0.86	2.20	33.03
POR= 50 YRS	3.53	3.13	3.79	3.68	3.75	3.65	3.43	3.88	3.67	3.15	3.69	3.63	42.98

AVERAGE TEMPERATURE (F) 2001 HARTFORD, CT (BDL)

YEAR	JAN	FEB	MAR	APR	MAY	JUN	JUL	AUG	SEP	OCT	NOV	DEC	ANNUAL
1972	27.9	26.0	34.6	44.3	60.1	65.9	73.8	71.1	64.3	48.9	38.7	30.8	48.9
1973	29.4	27.2	43.3	50.4	58.0	71.8	75.0	76.4	64.7	54.7	43.6	32.9	52.3
1974	28.2	27.0	36.8	50.7	56.6	67.3	73.7	72.7	63.3	47.3	40.7	30.7	49.6
1975	31.2	29.7	35.9	45.8	64.6	68.3	76.1	71.8	61.7	55.7	48.2	28.5	51.5
1976	19.5	34.8	40.3	53.3	58.4	72.7	72.4	71.0	62.4	49.9	38.3	24.8	49.8
1977	18.7	27.6	42.9	51.2	63.6	68.0	74.5	72.8	64.0	52.0	44.4	28.0	50.6
1978	23.6	22.1	35.1	48.1	59.9	69.2	71.9	70.0	58.6	49.0	38.6	29.3	48.0
1979	26.6	18.0	41.2	49.0	64.1	69.0	74.6	70.8	61.6	50.7	43.5	36.0	50.4
1980	27.9	34.2	38.2	49.3	61.0	66.4	74.2	73.2	64.3	50.3	37.9	24.6	49.1
1981	17.8	35.3	38.1	52.0	61.6	69.6	74.8	70.6	62.5	49.3	43.7	31.0	50.5
1982	18.8	29.2	36.7	45.8	61.4	65.0	74.4	69.5	63.0	51.5	45.8	36.0	49.8
1983	27.1	29.1	39.2	48.9	56.8	69.9	74.9	72.7	66.5	52.5	42.7	28.1	50.7
1984	21.8	34.3	31.4	48.0	56.0	69.8	71.8	73.2	59.8	55.2	41.5	35.7	49.9
1985	21.5	29.9	39.7	50.7	60.6	63.7	72.4	70.2	63.4	51.9	43.2	27.5	49.6
1986	27.4	26.2	38.7	51.0	61.7	66.0	72.3	69.5	61.8	51.4	38.3	33.1	49.8
1987	25.0	26.7	39.8	49.7	60.8	68.8	74.2	69.0	62.9	49.2	41.4	33.2	50.1
1988	23.1	28.3	38.5	47.4	59.7	66.7	75.2	74.5	62.2	47.6	42.3	29.3	49.6
1989	30.8	28.6	37.4	46.5	60.4	68.3	72.6	71.4	63.9	53.4	40.9	18.1	49.4
1990	34.7	33.0	40.2	49.2	56.7	69.0	74.4	73.3	64.0	57.4	44.5	36.7	52.8
1991	27.0	33.9	40.5	53.3	65.8	70.5	73.7	73.1	62.1	55.1	42.7	32.8	52.5
1992	28.6	30.3	34.6	46.4	58.5	66.4	69.9	69.1	62.6	49.2	40.6	31.2	49.0
1993	28.8	54.3	34.5	49.6	61.2	68.6	74.3	73.4	63.0	49.8	40.8	30.9	52.4
1994	18.8	23.2	36.1	50.9	58.3	71.1	77.1	70.0	63.7	52.6	46.1	35.0	50.2
1995	32.4	25.8	41.1	46.8	58.0	69.3	76.5	72.1	62.2	55.8	38.1	27.1	50.4
1996	25.1	27.7	34.3	49.6	58.3	69.3	71.2	71.9	62.9	51.1	38.1	35.8	49.6
1997	27.1	34.7	36.2	47.2	56.1	68.7	72.3	70.1	63.2	50.6	38.8	31.4	49.7
1998	32.6	36.2	40.3	49.9	62.9	67.0	72.9	73.6	65.7	52.6	41.6	36.8	52.7
1999	25.9	31.8	38.5	49.2	59.9	71.0	76.5	71.3	66.0	51.1	46.3	34.6	51.8
2000	24.0	31.0	43.7	47.8	59.6	67.7	69.6	69.6	62.1	51.8	40.9	25.4	49.4
2001	25.4	28.4	34.3	49.7	59.5	69.6	69.7	75.2	63.4	53.5	45.9	36.9	51.0
POR= 97 YRS	26.7	28.0	37.2	48.2	59.2	67.9	72.4	71.0	63.5	52.9	42.1	30.7	50.0

REFERENCE NOTES:

PAGE 1:
THE TEMPERATURE GRAPH SHOWS NORMAL MAXIMUM AND NORMAL MINIMUM DAILY TEMPERATURES (SOLID CURVES) AND THE ACTUAL DAILY HIGH AND LOW TEMPERATURES (VERTICAL BARS).

PAGE 2 AND 3:
H/C INDICATES HEATING AND COOLING DEGREE DAYS.
RH INDICATES RELATIVE HUMIDITY
W/O INDICATES WEATHER AND OBSTRUCTIONS
S INDICATES SUNSHINE.
PR INDICATES PRESSURE.
CLOUDINESS ON PAGE 3 IS THE SUM OF THE CEILOMETER AND SATELLITE DATA NOT TO EXCEED EIGHT EIGHTHS(OKTAS).

GENERAL:
T INDICATES TRACE PRECIPITATION, AN AMOUNT GREATER THAN ZERO BUT LESS THAN THE LOWEST REPORTABLE VALUE.
+ INDICATES THE VALUE ALSO OCCURS ON EARLIER DATES.
BLANK ENTRIES DENOTE MISSING OR UNREPORTED DATA.
NORMALS ARE 30-YEAR AVERAGES (1961 - 1990).
ASOS INDICATES AUTOMATED SURFACE OBSERVING SYSTEM.
PM INDICATES THE LAST DAY OF THE PREVIOUS MONTH.
POR (PERIOD OF RECORD) BEGINS WITH THE JANUARY DATA MONTH AND IS THE NUMBER OF YEARS USED TO COMPUTE THE MEAN. INDIVIDUAL MONTHS WITHIN THE POR MAY BE MISSING.
WHEN THE POR FOR A NORMAL IS LESS THAN 30 YEARS, THE NORMAL IS PROVISIONAL AND IS BASED ON THE NUMBER OF YEARS INDICATED.
0.* OR * INDICATES THE VALUE OR MEAN-DAYS-WITH IS BETWEEN 0.00 AND 0.05.
CLOUDINESS FOR ASOS STATIONS DIFFERS FROM THE NON-ASOS OBSERVATION TAKEN BY A HUMAN OBSERVER. ASOS STATION CLOUDINESS IS BASED ON TIME-AVERAGED CEILOMETER DATA FOR CLOUDS AT OR BELOW 12,000 FEET AND ON SATELLITE DATA FOR CLOUDS ABOVE 12,000 FEET.
THE NUMBER OF DAYS WITH CLEAR, PARTLY CLOUDY, AND CLOUDY CONDITIONS FOR ASOS STATIONS IS THE SUM OF THE CEILOMETER AND SATELLITE DATA FOR THE SUNRISE TO SUNSET PERIOD.

GENERAL CONTINUED:
CLEAR INDICATES 0 - 2 OKTAS, PARTLY CLOUDY INDICATES 3 - 6 OKTAS, AND CLOUDY INDICATES 7 OR 8 OKTAS.
WHEN AT LEAST ONE OF THE ELEMENTS (CEILOMETER OR SATELLITE) IS MISSING, THE DAILY CLOUDINESS IS NOT COMPUTED.
WIND DIRECTION IS RECORDED IN TENS OF DEGREES (2 DIGITS) CLOCKWISE FROM TRUE NORTH. "00" INDICATES CALM. "36" INDICATES TRUE NORTH.
RESULTANT WIND IS THE VECTOR AVERAGE OF THE SPEED AND DIRECTION.
AVERAGE TEMPERATURE IS THE SUM OF THE MEAN DAILY MAXIMUM AND MINIMUM TEMPERATURE DIVIDED BY 2.
SNOWFALL DATA COMPRISE ALL FORMS OF FROZEN PRECIPITATION, INCLUDING HAIL.
A HEATING (COOLING) DEGREE DAY IS THE DIFFERENCE BETWEEN THE AVERAGE DAILY TEMPERATURE AND 65 F.
DRY BULB IS THE TEMPERATURE OF THE AMBIENT AIR.
DEW POINT IS THE TEMPERATURE TO WHICH THE AIR MUST BE COOLED TO ACHIEVE 100 PERCENT RELATIVE HUMIDITY.
WET BULB IS THE TEMPERATURE THE AIR WOULD HAVE IF THE MOISTURE CONTENT WAS INCREASED TO 100 PERCENT RELATIVE HUMIDITY.

ON JULY 1, 1996, THE NATIONAL WEATHER SERVICE BEGAN USING THE "METAR" OBSERVATION CODE THAT WAS ALREADY EMPLOYED BY MOST OTHER NATIONS OF THE WORLD. THE MOST NOTICEABLE DIFFERENCE IN THIS ANNUAL PUBLICATION WILL BE THE CHANGE IN UNITS FROM TENTHS TO EIGHTS (OKTAS) FOR REPORTING THE AMOUNT OF SKY COVER.

HEATING DEGREE DAYS (base 65 F) 2001 HARTFORD, CT (BDL)

YEAR	JUL	AUG	SEP	OCT	NOV	DEC	JAN	FEB	MAR	APR	MAY	JUN	TOTAL
1972-73	6	9	82	494	782	1054	1097	1051	665	444	229	13	5926
1973-74	0	0	102	322	635	988	1134	1056	868	435	287	37	5864
1974-75	2	1	121	542	725	1057	1040	986	894	567	111	43	6089
1975-76	0	11	121	292	503	1125	1403	869	759	391	213	20	5707
1976-77	0	16	118	467	794	1242	1429	1038	684	419	130	45	6382
1977-78	1	8	112	399	610	1141	1276	1192	920	500	220	25	6404
1978-79	9	15	209	489	790	1102	1184	1310	730	473	81	26	6418
1979-80	16	30	152	442	578	965	1151	1174	916	466	146	68	6104
1980-81	0	0	99	449	808	1246	1456	824	828	380	149	10	6249
1981-82	0	9	115	481	635	1048	1427	996	871	569	128	64	6343
1982-83	1	30	96	416	575	894	1170	1002	793	483	261	24	5745
1983-84	0	7	106	404	662	1135	1332	884	1035	503	286	32	6386
1984-85	3	3	186	298	698	896	1341	975	776	428	167	76	5847
1985-86	0	14	119	401	648	1157	1159	1081	809	413	174	63	6038
1986-87	14	32	135	422	793	981	1230	1065	773	452	191	29	6117
1987-88	1	31	100	481	700	981	1292	1057	817	523	186	75	6244
1988-89	9	23	112	539	672	1101	1054	1012	847	553	175	31	6128
1989-90	0	22	103	354	715	1444	935	890	763	478	251	21	5976
1990-91	5	0	112	276	608	873	1170	863	755	373	107	16	5158
1991-92	1	0	156	311	663	990	1122	1002	936	553	218	37	5989
1992-93	9	16	138	486	722	1042	1114	1148	935	454	139	43	6246
1993-94	3	4	142	464	722	1049	1424	1163	888	417	226	15	6517
1994-95	0	8	77	379	561	923	1005	1088	737	539	224	13	5554
1995-96	0	4	130	283	802	1169	1227	1078	944	465	241	19	6362
1996-97	1	2	120	424	802	900	1167	841	883	526	274	49	5989
1997-98	3	6	104	445	777	1036	998	801	769	447	101	59	5546
1998-99	1	3	61	378	694	873	1207	923	815	466	168	9	5598
1999-00	0	11	67	422	555	934	1264	978	657	512	210	65	5675
2000-01	3	17	154	398	716	1221	1217	1023	946	466	202	27	6390
2001-	12	0	101	361	568	863							

COOLING DEGREE DAYS (base 65 F) 2001 HARTFORD, CT (BDL)

YEAR	JAN	FEB	MAR	APR	MAY	JUN	JUL	AUG	SEP	OCT	NOV	DEC	ANNUAL
1972	0	0	0	3	30	83	286	203	70	1	0	0	676
1973	0	0	0	11	19	221	318	362	99	8	0	0	1038
1974	0	0	0	11	34	110	282	247	77	0	3	0	764
1975	0	0	0	0	106	147	348	229	27	7	6	0	870
1976	0	0	0	47	18	257	236	208	47	6	0	0	819
1977	0	0	4	13	93	144	303	259	88	1	0	0	905
1978	0	0	0	0	71	159	228	173	26	0	0	0	657
1979	0	0	0	0	60	151	320	218	56	6	0	0	811
1980	0	0	0	0	31	117	296	263	107	1	0	0	815
1981	0	0	0	0	53	152	311	190	48	0	0	0	754
1982	0	0	0	0	22	70	298	176	45	2	3	0	616
1983	0	0	0	5	16	177	313	253	158	23	0	0	945
1984	0	0	0	0	11	182	218	265	38	4	0	0	718
1985	0	0	0	3	37	44	234	182	78	3	0	0	581
1986	0	0	0	0	79	103	249	179	48	7	0	0	665
1987	0	0	0	3	70	150	292	161	42	0	0	0	718
1988	0	0	0	0	0	27	134	331	326	37	6	0	861
1989	0	0	0	0	0	37	136	240	224	77	0	0	714
1990	0	0	0	0	13	1	146	305	263	89	48	1	866
1991	0	0	0	0	29	139	191	278	257	76	12	0	982
1992	0	0	0	0	0	21	85	170	151	74	2	0	503
1993	0	0	0	0	0	28	155	297	271	89	0	0	840
1994	0	0	0	1	25	205	381	173	43	0	0	0	828
1995	0	0	0	0	0	11	151	363	230	57	1	0	813
1996	0	0	0	9	42	155	200	221	63	0	0	0	690
1997	0	0	0	0	5	168	237	172	58	6	0	0	646
1998	0	0	13	1	42	127	254	273	89	0	0	0	799
1999	0	0	0	0	17	194	368	210	101	0	0	0	890
2000	0	0	0	0	46	154	152	166	77	1	0	0	596
2001	0	0	0	11	40	170	166	322	60	12	0	0	781

SNOWFALL (inches) 2001 HARTFORD, CT (BDL)

YEAR	JUL	AUG	SEP	OCT	NOV	DEC	JAN	FEB	MAR	APR	MAY	JUN	TOTAL
1972-73	0.0	0.0	0.0	0.4	2.1	12.0	14.1	5.9	0.4	0.3	0.0	0.0	35.2
1973-74	0.0	0.0	0.0	0.0	T	3.1	14.3	5.8	4.8	2.1	0.0	0.0	30.1
1974-75	0.0	0.0	0.0	T	0.8	8.5	10.2	16.0	2.5	0.3	0.0	0.0	38.3
1975-76	0.0	0.0	0.0	0.0	0.3	13.4	15.6	5.0	12.3	0.0	0.0	0.0	46.6
1976-77	0.0	0.0	0.0	0.0	0.4	7.3	20.0	9.1	11.0	0.3	1.3	0.0	49.4
1977-78	0.0	0.0	0.0	0.0	1.3	12.6	37.0	18.1	13.3	T	0.0	0.0	82.3
1978-79	0.0	0.0	0.0	0.0	4.3	10.3	8.6	9.2	T	3.6	0.0	0.0	36.0
1979-80	0.0	0.0	0.0	1.7	0.0	0.9	0.2	7.7	5.9	T	0.0	0.0	16.4
1980-81	0.0	0.0	0.0	0.0	8.6	3.9	4.1	0.9	0.2	0.0	0.0	0.0	17.7
1981-82	0.0	0.0	0.0	T	T	13.1	16.7	5.8	6.5	14.3	0.0	0.0	56.4
1982-83	0.0	0.0	0.0	0.0	T	5.7	10.2	29.4	0.2	0.9	0.0	0.0	46.4
1983-84	0.0	0.0	0.0	0.0	T	7.9	14.7	1.3	19.3	T	0.0	0.0	43.2
1984-85	0.0	0.0	0.0	0.0	0.1	3.8	6.9	9.4	2.1	1.4	0.0	0.0	23.7
1985-86	0.0	0.0	0.0	0.0	2.0	5.4	5.1	11.8	0.2	0.8	0.0	0.0	25.3
1986-87	0.0	0.0	0.0	0.0	8.7	4.9	34.0	1.6	1.7	0.4	0.0	0.0	51.3
1987-88	0.0	0.0	0.0	T	8.6	5.8	22.6	17.6	4.9	T	T	0.0	59.5
1988-89	0.0	0.0	0.0	0.0	0.0	6.3	0.6	4.6	3.4	T	0.0	0.0	14.9
1989-90	0.0	0.0	0.0	0.0	5.3	12.4	10.5	0.0	4.3	1.5	0.0	0.0	43.0
1990-91	0.0	0.0	0.0	0.0	T	8.1	10.2	5.8	5.7	0.0	0.0	0.0	29.8
1991-92	0.0	0.0	0.0	0.0	0.7	6.0	1.7	5.3	7.3	2.6	0.0	0.0	23.6
1992-93	0.0	0.0	0.0	T	T	6.7	10.5	13.8	31.1	T	0.0	T	62.1
1993-94	0.0	0.0	0.0	0.0	T	6.7	31.3	29.4	17.5	0.0	T	0.0	84.9
1994-95	0.0	0.0	0.0	0.0	3.9	1.1	5.7	10.1	0.0	1.5	0.0	0.0	22.3
1995-96	0.0	0.0	0.0	0.0	5.6	20.3	42.8	20.6	17.8				
1996-97													
1997-98													
1998-99													
1999-00													
2000-01						8.0	10.4	21.4	13.5	0.0			
2001-						3.0							
POR= 41 YRS	0.0	0.0	0.0	0.1	2.1	10.3	12.7	11.7	9.8	1.5	0.0	T	48.2

2001
WILMINGTON,
DELAWARE (ILG)

Delaware is part of the Atlantic Coastal Plain consisting mainly of flat low land with many marshes. Small streams and tidal estuaries comprise the drainage of the State. Wilmington, at the northern end of the State, marks the beginning of low rolling hills extending northward and northwestward into Pennsylvania. The Delaware River, the Delaware Bay, and the Atlantic Ocean are along the eastern boundary of the State. The broad Chesapeake Bay lies 35 miles, or less, to the west of the western boundary of nearly the entire State. These large water areas considerably influence the climate of the Wilmington, Delaware region.

Summers are warm and humid, winters are usually mild. During the summer maximum temperatures are usually in the 80s. The temperature reaches 100 degrees on the average once in six years. During January, the coldest month of the year, the daily average temperature is 32 degrees. Temperatures of zero may be expected once in four years. Most of the winter precipitation falls as rain. Seasonal snowfall has been as little as 1 inch, and as much as 50 inches. Snow is frequently mixed with rain and sleet, and seldom remains on the ground more than a few days.

The proximity of large water areas and the inflow of southerly winds cause the relative humidity to be quite high all year. During the summer months the relative humidity is approximately 75 percent. Fog is relatively frequent and may occur in any month. Light southeast winds blowing up the Delaware Bay favor the formation of fog. Light north-northeast winds bring in smoke from Philadelphia and from the heavy industry area located along the Delaware River north of Wilmington.

Rainfall distribution throughout the year is fairly uniform, however, the greatest amounts normally come during the summer months. Mostly, the summer rainfall comes in the form of thunderstorms. Moisture deficiencies for crops occur occasionally, but severe droughts are rare. During the fall, winter, and spring seasons, much of the rainfall comes from storms forming over the southern states or the South Atlantic and moving northward along the coast. During the late summer and early fall, hurricanes occasionally cause heavy rainfall, but winds seldom reach hurricane force in Wilmington. Heavy rains occasionally cause minor flooding, but the streams and rivers of northern Delaware are not subject to major flooding. Strong easterly and southeasterly winds sometimes cause high tides in the Delaware Bay and the Delaware River, resulting in the flooding of lowlands and damage to bay front and river front properties.

Based on the 1951–1980 period, the average first occurrence of 32 degrees Fahrenheit in the fall is October 29 and the average last occurrence in the spring is April 13.

NORMALS, MEANS, AND EXTREMES
WILMINGTON, DE (ILG)

LATITUDE:	LONGITUDE:	ELEVATION (FT):	TIME ZONE:	WBAN: 13781
39 40' 22" N	75 36' 03" W	GRND: 92 BARO: 95	EASTERN (UTC + 5)	

	ELEMENT	POR	JAN	FEB	MAR	APR	MAY	JUN	JUL	AUG	SEP	OCT	NOV	DEC	YEAR
TEMPERATURE °F	NORMAL DAILY MAXIMUM	30	38.7	41.9	52.1	62.6	72.9	81.4	85.6	84.1	77.7	66.6	55.5	43.9	63.6
	MEAN DAILY MAXIMUM	45	39.3	42.4	51.4	63.0	72.8	81.5	85.5	84.1	77.6	66.3	55.4	44.1	63.6
	HIGHEST DAILY MAXIMUM	54	75	78	78	86	94	96	100	102	101	100	91	85	75	102
	YEAR OF OCCURRENCE		1950	1985	1998	1985	1996	1994	1966	1955	1983	1951	1950	1998	JUL 1966
	MEAN OF EXTREME MAXS.	54	59.8	62.4	73.0	81.9	88.4	93.3	95.4	93.5	90.0	81.8	72.6	63.9	79.7
	NORMAL DAILY MINIMUM	30	22.4	24.8	33.1	41.8	52.2	61.6	67.1	65.9	58.2	45.7	37.0	27.6	44.8
	MEAN DAILY MINIMUM	45	23.2	25.4	32.9	42.0	52.2	61.6	66.9	65.6	57.9	45.6	36.9	27.8	44.8
	LOWEST DAILY MINIMUM	54	-14	-6	2	18	30	41	48	43	36	24	14	-7	-14
	YEAR OF OCCURRENCE		1985	1979	1984	1982	1978	1972	1988	1982	1974	1976	1955	1983	JAN 1985
	MEAN OF EXTREME MINS.	54	7.0	9.4	18.0	28.8	38.5	48.9	55.7	53.4	42.5	31.8	22.3	12.5	30.7
	NORMAL DRY BULB	30	30.6	33.4	42.7	52.2	62.5	71.5	76.4	75.0	68.0	56.2	46.3	35.8	54.2
	MEAN DRY BULB	54	32.0	34.1	42.2	52.6	62.4	71.5	76.3	74.7	67.6	56.4	46.0	36.0	54.3
	MEAN WET BULB	49	29.0	30.5	37.2	46.1	55.7	64.2	68.7	67.8	61.6	51.3	41.5	32.6	48.8
	MEAN DEW POINT	49	22.2	23.3	29.2	38.7	50.2	60.0	64.6	64.2	57.7	47.3	35.8	26.4	43.3
	NORMAL NO. DAYS WITH:														
	MAXIMUM 90	30	0.0	0.0	0.0	0.2	0.7	3.3	7.6	4.9	1.9	0.0	0.0	0.0	18.6
	MAXIMUM 32	30	8.8	5.0	0.6	0.0	0.0	0.0	0.0	0.0	0.0	0.0	0.1	3.9	18.4
	MINIMUM 32	30	26.0	22.0	14.5	3.3	0.1	0.0	0.0	0.0	0.0	1.9	9.6	22.2	99.6
	MINIMUM 0	30	0.6	0.3	0.0	0.0	0.0	0.0	0.0	0.0	0.0	0.0	0.0	0.1	1.0
H/C	NORMAL HEATING DEG. DAYS	30	1066	885	691	384	122	0	0	0	35	288	561	905	4937
	NORMAL COOLING DEG. DAYS	30	0	0	0	0	44	199	353	310	125	15	0	0	1046
RH	NORMAL (PERCENT)	30	68	65	62	62	67	68	70	72	72	71	69	69	68
	HOUR 01 LST	30	72	71	70	71	77	80	81	83	83	81	76	74	77
	HOUR 07 LST	30	74	74	73	72	75	77	79	83	84	84	80	76	78
	HOUR 13 LST	30	59	56	51	49	53	54	54	56	56	54	56	59	55
	HOUR 19 LST	30	66	63	60	57	62	63	64	68	70	69	68	68	65
S	PERCENT POSSIBLE SUNSHINE														
W/O	MEAN NO. DAYS WITH:														
	HEAVY FOG(VISBY 1/4 MI)	54	4.2	3.5	2.7	2.1	2.2	1.7	1.6	2.3	2.4	3.8	3.3	3.4	33.2
	THUNDERSTORMS	54	0.2	0.3	1.1	2.3	3.8	5.5	6.1	5.4	2.4	0.9	0.6	0.2	28.8
CLOUDINESS	MEAN:														
	SUNRISE-SUNSET (OKTAS)	1			6.0			5.0							
	MIDNIGHT-MIDNIGHT (OKTAS)														
	MEAN NO. DAYS WITH:														
	CLEAR	1	2.0	2.0	6.0		8.0	9.0	3.0	7.0	5.0	9.0		5.0	
	PARTLY CLOUDY	1	1.0	1.0	6.0		4.0	5.0	1.0	4.0	3.0	2.0		2.0	
	CLOUDY	1	4.0	5.0	11.0		6.0	8.0		2.0	7.0	3.0		8.0	
PR	MEAN STATION PRESSURE(IN)	29	30.00	29.99	29.94	29.90	29.90	29.90	29.92	29.96	29.99	30.02	30.02	30.01	29.96
	MEAN SEA-LEVEL PRES. (IN)	49	30.08	30.07	30.02	29.99	30.00	29.98	30.00	30.02	30.08	30.10	30.07	30.10	30.04
WINDS	MEAN SPEED (MPH)	42	9.8	10.1	11.0	10.4	8.8	8.2	7.7	7.3	7.7	8.0	9.0	9.1	8.9
	PREVAIL.DIR(TENS OF DEGS)	26	30	30	30	30	30	18	30	18	32	31	30	30	30
	MAXIMUM 2-MINUTE:														
	SPEED (MPH)	7	51	43	45	46	48	41	45	40	39	36	44	43	51
	DIR. (TENS OF DEGS)		15	29	28	33	24	27	32	14	35	29	14	27	15
	YEAR OF OCCURRENCE		1999	1996	1996	1995	1999	1998	1995	1997	2001	1995	1999	2000	JAN 1999
	MAXIMUM 5-SECOND:														
	SPEED (MPH)	7	61	54	56	53	61	52	55	53	52	46	57	53	61
	DIR. (TENS OF DEGS)		23	29	27	33	23	27	32	15	36	26	14	28	23
	YEAR OF OCCURRENCE		1999	1996	1996	1995	1999	1998	1995	1997	2001	1999	1999	2000	MAY 1999
PRECIPITATION	NORMAL (IN)	30	3.03	2.91	3.43	3.39	3.84	3.55	4.23	3.40	3.43	2.88	3.27	3.48	40.84
	MAXIMUM MONTHLY (IN)	54	8.41	7.02	9.17	6.80	7.38	7.49	12.63	12.09	12.68	8.01	7.84	7.96	12.68
	YEAR OF OCCURRENCE		1978	1979	2000	1983	1983	1972	1989	1955	1999	1995	1972	1996	SEP 1999
	MINIMUM MONTHLY (IN)	54	0.52	0.83	0.81	0.35	0.22	0.21	0.16	0.25	0.82	0.08	0.49	0.19	0.08
	YEAR OF OCCURRENCE		1981	1980	1966	1985	1964	1988	1955	1972	1970	2000	1976	1955	OCT 2000
	MAXIMUM IN 24 HOURS (IN)	54	2.53	2.29	4.87	2.56	2.72	4.35	6.83	4.11	8.43	3.88	3.83	2.33	8.43
	YEAR OF OCCURRENCE		1998	1966	2000	1961	1990	1972	1989	1971	1999	1966	1956	1996	SEP 1999
	NORMAL NO. DAYS WITH:														
	PRECIPITATION 0.01	30	10.3	9.5	9.9	10.6	11.2	9.8	9.1	8.8	7.8	7.3	9.4	9.8	113.5
	PRECIPITATION 1.00	30	0.7	0.6	1.0	0.7	0.9	0.8	1.2	0.9	1.1	0.7	0.8	1.0	10.4
SNOWFALL	NORMAL (IN)	30	7.6	6.9	2.4	0.3	T	0.0	0.0	0.0	0.0	0.*	0.7	3.4	21.3
	MAXIMUM MONTHLY (IN)	50	26.2	27.5	20.3	2.6	T	T	T	0.0	0.0	2.5	11.9	21.5	27.5
	YEAR OF OCCURRENCE		1996	1979	1958	1982	1991	1992	1990			1979	1953	1966	FEB 1979
	MAXIMUM IN 24 HOURS (IN)	50	22.0	16.5	15.6	2.4	T	T	T	0.0	0.0	2.5	11.9	12.4	22.0
	YEAR OF OCCURRENCE		1996	1979	1958	1987	1991	1992	1990			1979	1953	1966	JAN 1996
	MAXIMUM SNOW DEPTH (IN)	46	13	20	8	2	0	0	0	0	0	T	9	20	20
	YEAR OF OCCURRENCE		1987	1979	1956	1987						1962	1953	1966	FEB 1979
	NORMAL NO. DAYS WITH:														
	SNOWFALL 1.0	30	2.3	1.7	0.8	0.1	0.0	0.0	0.0	0.0	0.0	0.0	0.2	1.0	6.1

PRECIPITATION (inches) 2001 WILMINGTON, DE (ILG)

YEAR	JAN	FEB	MAR	APR	MAY	JUN	JUL	AUG	SEP	OCT	NOV	DEC	ANNUAL
1972	2.50	5.43	2.40	4.47	3.85	7.49	2.07	0.25	1.64	4.20	7.84	5.99	48.13
1973	3.81	3.42	4.02	6.57	5.56	5.19	2.82	2.44	3.02	2.22	0.67	7.31	47.05
1974	2.92	1.73	4.56	3.08	3.96	3.97	1.49	5.65	6.19	1.77	1.19	4.18	39.61
1975	4.23	2.95	4.63	3.03	5.65	6.16	5.53	2.55	6.19	3.06	2.63	3.00	49.61
1976	4.21	1.70	2.25	1.40	5.05	2.14	4.33	2.00	2.11	6.12	0.49	1.79	33.59
1977	2.18	1.09	4.55	3.91	0.96	4.41	1.38	4.82	1.29	3.59	6.14	5.81	40.13
1978	8.41	1.77	5.59	2.16	6.94	3.00	5.53	5.97	2.18	1.48	2.69	5.56	51.28
1979	7.61	7.02	2.61	4.03	3.10	4.01	4.76	6.11	5.94	3.45	3.23	1.44	53.31
1980	2.44	0.83	6.22	4.55	2.40	4.23	3.49	1.09	1.44	3.99	2.41	0.83	33.92
1981	0.52	3.23	1.26	3.54	5.05	4.50	2.52	3.38	3.82	2.84	0.67	3.95	35.28
1982	3.75	2.71	2.87	5.41	3.72	4.70	2.70	4.68	2.30	1.97	3.87	2.39	41.07
1983	2.98	3.55	6.84	6.80	7.38	3.94	2.33	1.29	3.44	3.87	5.48	6.80	54.70
1984	1.25	4.27	5.40	4.24	5.03	4.54	6.53	1.56	2.02	3.31	1.63	1.94	41.72
1985	1.56	2.05	2.03	0.35	5.52	1.37	6.91	2.28	4.56	1.84	4.46	0.80	33.73
1986	4.21	2.77	1.19	2.77	1.69	4.05	3.99	2.88	2.75	4.04	6.42	6.11	42.87
1987	4.35	1.52	1.16	2.63	3.15	2.31	4.09	4.21	4.85	2.31	3.50	1.90	35.98
1988	2.46	4.14	1.82	2.59	4.95	0.21	8.29	3.03	2.18	1.94	5.29	0.90	37.80
1989	2.48	2.75	3.69	2.76	6.57	5.43	12.63	1.97	4.31	3.92	1.99	1.27	49.77
1990	3.56	1.35	2.15	3.42	7.03	3.94	4.27	6.15	2.64	2.85	1.61	5.16	44.13
1991	4.30	0.97	4.64	3.28	1.98	3.41	3.71	5.38	5.36	1.27	1.26	4.26	39.82
1992	1.05	1.81	4.36	1.76	4.48	3.14	4.34	2.21	4.30	1.11	4.27	4.21	37.04
1993	2.64	3.11	7.50	5.87	3.95	1.60	4.04	2.65	6.26	2.77	2.85	3.51	46.75
1994	5.00	3.55	7.36	2.85	3.69	2.11	7.01	5.68	2.10	0.85	2.96	2.24	45.40
1995	3.08	2.28	2.47	2.10	3.50	1.26	2.89	2.03	5.17	8.01	4.31	2.17	39.27
1996	4.58	1.19	3.63	4.98	3.27	5.00	6.25	3.04	4.05	4.70	3.25	7.96	51.90
1997	1.83	1.83	3.49	1.49	0.82	1.75	3.08	3.66	1.93	2.33	3.24	2.57	28.02
1998	4.80	2.95	4.86	2.91	4.13	4.66	2.18	3.14	1.76	2.80	1.26	1.01	36.46
1999	5.41	3.51	3.96	3.36	3.56	1.62	0.89	4.24	12.68	3.42	2.09	2.94	47.68
2000	3.83	2.00	9.17	3.43	2.94	4.83	4.64	2.47	7.30	0.08	2.54	2.80	46.03
2001	3.13	2.81	5.62	1.43	5.33	4.28	2.35	2.65	2.57	0.74	0.99	1.95	33.85
POR= 108 YRS	3.34	3.01	3.81	3.46	3.67	3.70	4.50	4.44	3.72	2.99	3.19	3.41	43.24

AVERAGE TEMPERATURE (F) 2001 WILMINGTON, DE (ILG)

YEAR	JAN	FEB	MAR	APR	MAY	JUN	JUL	AUG	SEP	OCT	NOV	DEC	ANNUAL
1972	36.1	32.6	41.3	50.0	62.5	68.6	76.9	75.5	69.6	53.5	45.2	41.6	54.5
1973	35.6	35.5	49.4	54.7	61.4	75.8	78.1	78.1	70.4	59.9	49.1	37.7	57.1
1974	36.2	33.1	44.3	55.5	62.4	70.4	76.7	76.1	66.5	53.2	46.6	38.9	55.0
1975	37.5	35.9	40.6	47.5	65.2	71.5	75.7	76.3	65.4	60.0	51.0	36.2	55.2
1976	27.7	41.2	46.5	54.7	59.9	72.4	74.2	73.7	66.9	52.5	40.4	31.0	53.4
1977	20.8	33.3	47.2	54.4	63.9	68.9	76.4	75.4	69.3	53.4	46.6	33.4	53.6
1978	27.2	22.8	37.6	50.3	60.3	70.6	73.4	77.1	66.7	54.1	46.8	37.2	52.0
1979	31.4	22.1	45.4	50.6	63.8	67.9	75.3	75.0	68.2	55.0	49.6	38.1	53.5
1980	32.4	29.9	40.0	54.6	64.9	68.9	77.7	78.2	71.3	54.8	43.3	33.1	54.1
1981	23.4	37.9	40.2	55.0	62.5	72.3	77.1	75.5	66.9	53.0	45.3	34.2	53.6
1982	24.2	34.2	41.8	50.6	65.0	69.9	77.3	72.0	67.4	56.0	47.5	41.3	53.9
1983	35.2	35.3	45.9	53.1	61.0	71.8	77.6	77.0	69.3	56.9	46.7	32.1	55.2
1984	24.8	38.6	35.6	50.7	61.2	73.8	75.2	75.2	63.7	61.2	43.3	42.1	53.8
1985	27.5	37.4	47.1	58.0	65.9	70.6	76.6	74.4	69.1	58.3	51.0	32.9	55.7
1986	32.2	31.6	43.6	52.4	65.7	72.1	77.1	72.5	67.5	57.2	44.1	37.3	54.4
1987	31.4	31.9	44.6	52.3	63.1	73.5	79.1	74.3	68.3	51.7	47.4	38.6	54.7
1988	27.4	34.8	44.2	50.8	62.9	71.6	79.4	77.3	65.8	51.0	46.7	35.1	53.9
1989	36.0	34.3	42.1	51.6	62.1	74.3	75.9	74.4	65.2	57.6	44.6	25.0	53.9
1990	40.5	41.1	46.0	53.7	61.5	72.1	77.4	74.6	66.7	60.0	48.4	41.0	56.9
1991	34.3	39.7	44.9	54.7	69.1	73.6	77.3	76.7	67.2	57.6	46.8	39.5	56.8
1992	35.2	37.2	41.5	52.0	60.7	69.6	76.3	72.2	67.3	53.2	47.2	38.3	54.2
1993	37.6	31.0	39.0	52.8	65.3	72.8	79.4	77.8	68.7	55.7	47.4	36.4	55.3
1994	26.3	31.6	41.8	58.4	60.0	75.8	79.8	73.2	67.2	53.9	49.7	40.0	54.8
1995	36.6	30.3	45.3	51.5	62.1	72.0	78.7	77.1	68.2	59.3	41.3	31.3	54.5
1996	29.9	33.3	38.4	52.6	60.1	73.0	74.1	74.0	67.7	55.1	40.1	39.2	53.1
1997	31.8	38.9	43.2	50.3	58.6	69.6	75.9	72.8	65.5	55.6	43.6	37.6	53.6
1998	39.8	41.0	44.7	54.2	65.3	70.5	75.1	75.5	70.9	57.0	46.9	41.0	56.8
1999	34.8	36.9	41.7	52.7	62.9	71.6	80.0	75.8	68.1	53.8	49.4	38.8	55.5
2000	31.5	36.6	47.3	51.9	63.8	72.0	72.9	72.7	65.0	56.4	44.2	30.1	53.7
2001	31.6	36.2	39.8	53.3	62.7	73.0	72.7	77.3	65.6	56.7	50.8	42.5	55.2
POR= 107 YRS	32.3	33.5	42.2	52.3	62.7	71.4	76.1	74.3	67.8	56.5	45.8	35.5	54.2

REFERENCE NOTES:

PAGE 1:
THE TEMPERATURE GRAPH SHOWS NORMAL MAXIMUM AND NORMAL MINIMUM DAILY TEMPERATURES (SOLID CURVES) AND THE ACTUAL DAILY HIGH AND LOW TEMPERATURES (VERTICAL BARS).

PAGE 2 AND 3:
H/C INDICATES HEATING AND COOLING DEGREE DAYS.
RH INDICATES RELATIVE HUMIDITY
W/O INDICATES WEATHER AND OBSTRUCTIONS
S INDICATES SUNSHINE.
PR INDICATES PRESSURE.
CLOUDINESS ON PAGE 3 IS THE SUM OF THE CEILOMETER AND SATELLITE DATA NOT TO EXCEED EIGHT EIGHTHS(OKTAS).

GENERAL:
T INDICATES TRACE PRECIPITATION, AN AMOUNT GREATER THAN ZERO BUT LESS THAN THE LOWEST REPORTABLE VALUE.
+ INDICATES THE VALUE ALSO OCCURS ON EARLIER DATES.
BLANK ENTRIES DENOTE MISSING OR UNREPORTED DATA.
NORMALS ARE 30-YEAR AVERAGES (1961 - 1990).
ASOS INDICATES AUTOMATED SURFACE OBSERVING SYSTEM.
PM INDICATES THE LAST DAY OF THE PREVIOUS MONTH.
POR (PERIOD OF RECORD) BEGINS WITH THE JANUARY DATA MONTH AND IS THE NUMBER OF YEARS USED TO COMPUTE THE MEAN. INDIVIDUAL MONTHS WITHIN THE POR MAY BE MISSING.
WHEN THE POR FOR A NORMAL IS LESS THAN 30 YEARS, THE NORMAL IS PROVISIONAL AND IS BASED ON THE NUMBER OF YEARS INDICATED.
0.* OR * INDICATES THE VALUE OR MEAN-DAYS-WITH IS BETWEEN 0.00 AND 0.05.
CLOUDINESS FOR ASOS STATIONS DIFFERS FROM THE NON-ASOS OBSERVATION TAKEN BY A HUMAN OBSERVER. ASOS STATION CLOUDINESS IS BASED ON TIME-AVERAGED CEILOMETER DATA FOR CLOUDS AT OR BELOW 12,000 FEET AND ON SATELLITE DATA FOR CLOUDS ABOVE 12,000 FEET.
THE NUMBER OF DAYS WITH CLEAR, PARTLY CLOUDY, AND CLOUDY CONDITIONS FOR ASOS STATIONS IS THE SUM OF THE CEILOMETER AND SATELLITE DATA FOR THE SUNRISE TO SUNSET PERIOD.

GENERAL CONTINUED:
CLEAR INDICATES 0 - 2 OKTAS, PARTLY CLOUDY INDICATES 3 - 6 OKTAS, AND CLOUDY INDICATES 7 OR 8 OKTAS. WHEN AT LEAST ONE OF THE ELEMENTS (CEILOMETER OR SATELLITE) IS MISSING, THE DAILY CLOUDINESS IS NOT COMPUTED.
WIND DIRECTION IS RECORDED IN TENS OF DEGREES (2 DIGITS) CLOCKWISE FROM TRUE NORTH. "00" INDICATES CALM. "36" INDICATES TRUE NORTH.
RESULTANT WIND IS THE VECTOR AVERAGE OF THE SPEED AND DIRECTION.
AVERAGE TEMPERATURE IS THE SUM OF THE MEAN DAILY MAXIMUM AND MINIMUM TEMPERATURE DIVIDED BY 2.
SNOWFALL DATA COMPRISE ALL FORMS OF FROZEN PRECIPITATION, INCLUDING HAIL.
A HEATING (COOLING) DEGREE DAY IS THE DIFFERENCE BETWEEN THE AVERAGE DAILY TEMPERATURE AND 65 F.
DRY BULB IS THE TEMPERATURE OF THE AMBIENT AIR.
DEW POINT IS THE TEMPERATURE TO WHICH THE AIR MUST BE COOLED TO ACHIEVE 100 PERCENT RELATIVE HUMIDITY.
WET BULB IS THE TEMPERATURE THE AIR WOULD HAVE IF THE MOISTURE CONTENT WAS INCREASED TO 100 PERCENT RELATIVE HUMIDITY.

ON JULY 1, 1996, THE NATIONAL WEATHER SERVICE BEGAN USING THE "METAR" OBSERVATION CODE THAT WAS ALREADY EMPLOYED BY MOST OTHER NATIONS OF THE WORLD. THE MOST NOTICEABLE DIFFERENCE IN THIS ANNUAL PUBLICATION WILL BE THE CHANGE IN UNITS FROM TENTHS TO EIGHTS(OKTAS) FOR REPORTING THE AMOUNT OF SKY COVER.

HEATING DEGREE DAYS (base 65 F) 2001 WILMINGTON, DE (ILG)

YEAR	JUL	AUG	SEP	OCT	NOV	DEC	JAN	FEB	MAR	APR	MAY	JUN	TOTAL
1972-73	0	0	20	356	586	716	902	820	477	322	143	0	4342
1973-74	0	0	15	179	469	839	886	887	635	300	137	7	4354
1974-75	0	0	65	362	553	805	847	808	753	520	84	5	4802
1975-76	0	0	56	177	418	888	1149	681	567	341	178	31	4486
1976-77	0	4	44	387	734	1047	1361	884	546	333	107	32	5479
1977-78	0	1	31	353	550	975	1165	1179	842	433	191	17	5737
1978-79	6	0	60	337	542	854	1037	1197	605	424	89	28	5179
1979-80	4	7	31	318	458	827	1004	1009	768	307	83	35	4851
1980-81	0	0	20	322	645	985	1222	752	763	299	135	4	5147
1981-82	0	0	57	370	585	947	1259	855	715	426	69	12	5295
1982-83	0	14	29	305	519	724	919	822	587	368	163	7	4457
1983-84	0	0	74	275	542	1013	1240	758	904	422	162	5	5395
1984-85	0	2	113	149	641	701	1154	766	550	248	73	7	4404
1985-86	0	0	45	213	411	986	1011	930	653	373	99	11	4732
1986-87	0	27	36	276	619	848	1032	923	628	374	143	2	4908
1987-88	0	2	22	406	521	811	1159	869	637	419	121	38	5005
1988-89	3	0	52	434	541	923	893	854	710	395	142	0	4947
1989-90	0	2	54	236	605	1231	749	661	593	368	127	6	4632
1990-91	2	1	69	214	494	734	943	700	617	320	61	5	4160
1991-92	0	0	64	244	541	785	914	799	723	386	169	12	4637
1992-93	0	1	57	363	527	817	843	945	799	360	64	12	4788
1993-94	0	0	55	286	526	879	1193	929	715	223	189	2	4997
1994-95	0	0	29	343	454	770	875	967	598	401	122	0	4559
1995-96	0	0	41	208	703	1033	1083	912	816	378	206	10	5390
1996-97	0	0	43	304	741	794	1025	729	672	436	201	56	5001
1997-98	2	0	71	324	635	844	775	667	645	317	89	27	4396
1998-99	0	0	22	241	536	738	926	782	715	362	95	7	4424
1999-00	0	1	36	340	462	807	1032	816	544	390	117	15	4560
2000-01	0	2	99	274	619	1074	1029	801	776	361	110	10	5155
2001-	1	0	73	271	418	691							

COOLING DEGREE DAYS (base 65 F) 2001 WILMINGTON, DE (ILG)

YEAR	JAN	FEB	MAR	APR	MAY	JUN	JUL	AUG	SEP	OCT	NOV	DEC	ANNUAL
1972	0	0	2	0	35	143	376	334	165	8	0	0	1063
1973	0	0	0	19	42	332	416	413	183	29	0	0	1434
1974	0	0	0	24	64	175	370	355	113	2	6	0	1109
1975	0	0	0	0	97	207	337	355	73	27	5	0	1101
1976	0	0	0	37	25	260	291	278	106	6	0	0	1003
1977	0	0	5	20	80	156	360	328	166	0	5	0	1120
1978	0	0	0	0	48	188	273	383	117	7	0	0	1016
1979	0	0	4	1	57	123	327	324	138	16	0	0	990
1980	0	0	0	0	83	159	400	417	214	10	0	0	1283
1981	0	0	0	9	62	228	381	270	120	3	0	0	1073
1982	0	0	0	2	75	163	391	238	107	32	1	0	1009
1983	0	0	0	17	47	218	398	378	209	29	0	0	1296
1984	0	0	0	0	50	276	321	327	80	34	0	0	1088
1985	0	0	4	47	106	181	366	300	174	13	0	0	1191
1986	0	0	0	0	129	227	379	267	116	40	0	0	1158
1987	0	0	0	3	91	264	446	295	129	0	0	0	1228
1988	0	0	0	0	0	62	242	455	389	80	5	0	1233
1989	0	0	6	0	61	287	345	299	162	17	0	0	1177
1990	0	0	0	10	36	23	227	395	304	127	66	0	1188
1991	0	0	1	18	197	271	390	371	135	23	2	0	1408
1992	0	0	0	3	39	156	358	230	134	3	0	0	923
1993	0	0	0	0	1	79	252	452	405	176	3	4	1372
1994	0	0	0	32	41	334	466	260	99	4	0	0	1236
1995	0	0	0	0	4	42	218	434	381	148	39	1	1267
1996	0	0	0	13	64	254	291	285	129	2	0	0	1038
1997	0	0	0	0	9	204	348	248	92	40	0	0	941
1998	0	0	22	0	106	195	320	337	204	2	0	0	1186
1999	0	0	0	0	40	211	476	342	139	1	0	0	1209
2000	0	0	0	3	89	230	253	247	106	11	0	0	939
2001	0	0	0	16	46	257	247	388	100	21	0	0	1075

SNOWFALL (inches) 2001 WILMINGTON, DE (ILG)

YEAR	JUL	AUG	SEP	OCT	NOV	DEC	JAN	FEB	MAR	APR	MAY	JUN	TOTAL
1972-73	0.0	0.0	0.0	T	T	T	T	1.2	T	T	0.0	0.0	1.2
1973-74	0.0	0.0	0.0	0.0	0.0	6.1	2.4	11.5	T	T	0.0	0.0	20.0
1974-75	0.0	0.0	0.0	0.0	T	T	5.1	4.8	1.1	T	0.0	0.0	11.0
1975-76	0.0	0.0	0.0	0.0	0.0	0.3	4.5	1.7	6.7	T	0.0	0.0	13.2
1976-77	0.0	0.0	0.0	0.0	0.1	3.2	14.5	T	0.0	T	0.0	0.0	17.8
1977-78	0.0	0.0	0.0	0.0	0.4	0.9	16.0	18.4	9.9	T	0.0	0.0	45.6
1978-79	0.0	0.0	0.0	0.0	4.5	T	12.0	27.5	0.2	T	0.0	0.0	44.2
1979-80	0.0	0.0	0.0	2.5	0.0	1.4	6.1	0.8	5.1	0.0	0.0	0.0	15.9
1980-81	0.0	0.0	0.0	0.0	0.5	1.4	6.5	T	3.7	0.0	0.0	0.0	12.1
1981-82	0.0	0.0	0.0	0.0	T	2.8	14.6	4.5	0.4	2.6	0.0	0.0	24.9
1982-83	0.0	0.0	0.0	0.0	T	5.8	T	18.5	0.3	0.5	0.0	0.0	25.1
1983-84	0.0	0.0	0.0	0.0	T	T	9.7	T	5.2	T	0.0	0.0	14.9
1984-85	0.0	0.0	0.0	0.0	T	0.3	14.2	0.7	T	0.4	0.0	0.0	15.6
1985-86	0.0	0.0	0.0	0.0	0.0	1.4	3.1	9.7	T	T	0.0	0.0	14.2
1986-87	0.0	0.0	0.0	0.0	T	0.3	21.4	15.7	0.2	2.4	0.0	0.0	40.0
1987-88	0.0	0.0	0.0	0.0	0.7	2.1	10.8	1.1	T	T	0.0	0.0	14.7
1988-89	0.0	0.0	0.0	0.0	T	0.2	6.7	2.9	1.2	0.0	0.0	0.0	11.0
1989-90	0.0	0.0	0.0	0.0	5.6	8.9	1.5	1.0	1.3	1.6	0.0	0.0	19.9
1990-91	T	0.0	0.0	0.0	0.0	6.4	5.2	0.6	0.9	T	T	0.0	13.1
1991-92	0.0	0.0	0.0	0.0	T	0.2	1.5	1.3	0.5	T	0.0	T	3.5
1992-93	0.0	0.0	0.0	0.0	T	0.1	1.4	10.0	13.9	0.0	0.0	0.0	25.4
1993-94	0.0	0.0	0.0	0.0	T	T	2.4	2.7	9.2	3.4	0.0	0.0	17.7
1994-95	0.0	0.0	0.0	0.0	T	T	0.0	T	8.3	T	0.0	0.0	8.3
1995-96	0.0	0.0	0.0	0.0	3.2	7.1	26.2	7.5	6.0	5.9	T	0.0	55.9
1996-97	0.0	0.0	0.0	0.0	0.0	T	0.8	5.6	6.5	2.8	0.0	0.0	15.7
1997-98	0.0	0.0	0.0	0.0	0.0	T	T	0.0	T	0.0	0.0	0.0	T
1998-99	0.0	0.0	0.0	0.0	0.0	2.0	4.5	0.0	2.0	0.0	0.0	0.0	8.5
1999-00	0.0	0.0	0.0	0.0	0.0	0.0	14.2	4.0	0.0	2.1	0.0	0.0	20.3
2000-01	0.0	0.0	0.0	0.0	0.0	2.1	4.8	5.0	0.7	T	0.0	0.0	12.6
2001-	0.0	0.0	0.0	0.0	0.0	0.0							
POR= 51 YRS	T	0.0	0.0	0.1	0.9	3.1	6.9	6.0	3.1	0.2	T	T	20.3

2001
WASHINGTON, D.C.
RONALD REAGAN NATIONAL AIRPORT (DCA)

Washington lies at the western edge of the mid Atlantic Coastal Plain, about 50 miles east of the Blue Ridge Mountains and 35 miles west of Chesapeake Bay, adjacent to the Potomac and Anacostia Rivers. Elevations range from a few feet above sea level to about 400 feet in parts of the northwest section of the city.

Observations have been kept continuously since November 1870. Since June 1941 the official observations have been taken at Washington National Airport.

National Airport is located at the center of the urban heat island. As a result, low temperatures are the highest for the area. Differences between the airport and suburban locations are often 10 to 15 degrees. There is less variation in the high temperatures.

Summers are warm and humid and winters are cold, but not severe. Periods of pleasant weather often occur in the spring and fall. The summertime temperature is in the upper 80s and the winter is in the upper 20s. Precipitation is rather uniformly distributed throughout the year.

Thunderstorms can occur at any time but are most frequent during the late spring and summer. The storms are most often accompanied by downpours and gusty winds, but are not usually severe.

Tornadoes, which infrequently occur, have resulted in significant damage. Severe hailstorms have occurred in the spring.

Tropical storms can bring heavy rain, high winds and flooding, but extensive damage from wind and tidal flooding is rare. Wind gusts of nearly 100 mph and rainfall over 7 inches have occurred during the passage of tropical storms and hurricanes.

Major flooding of the Potomac River can result from heavy rains over the basin, occasionally augmented by snowmelt, and above normal tides associated with hurricanes or severe storms along the coast. Flooding may also occur after a cold winter when the Potomac may be blocked with ice.

Although a snowfall of 10 inches or more in 24 hours is unusual, several notable falls of more than 25 inches have occurred. Normal snowfall during the winter season is 18 inches.

The average date of the last freezing temperature in the spring is April 1 and the average date for the first freezing temperature in the fall is November 10.

NORMALS, MEANS, AND EXTREMES
WASHINGTON DC, DC　(DCA)

LATITUDE:	LONGITUDE:	ELEVATION (FT):	TIME ZONE:	WBAN: 13743
38 51' 54" N	77 02' 03" W　GRND:　10　BARO:　3		EASTERN　(UTC + 5)	

ELEMENT	POR	JAN	FEB	MAR	APR	MAY	JUN	JUL	AUG	SEP	OCT	NOV	DEC	YEAR
TEMPERATURE °F														
NORMAL DAILY MAXIMUM	30	42.3	45.9	56.5	66.7	76.2	84.7	88.5	86.9	80.1	69.1	58.3	47.0	66.8
MEAN DAILY MAXIMUM	56	43.4	46.8	55.3	66.6	75.7	83.9	87.9	86.1	79.6	68.9	57.6	46.8	66.5
HIGHEST DAILY MAXIMUM	60	79	82	89	95	99	101	104	105	101	94	86	79	105
YEAR OF OCCURRENCE		1950	1948	1990	1976	1991	1994	1988	1997	1980	1954	1974	1998	AUG 1997
MEAN OF EXTREME MAXS.	56	64.5	67.4	77.5	85.7	90.3	95.3	96.8	95.5	92.3	83.8	75.3	66.3	82.6
NORMAL DAILY MINIMUM	30	26.8	29.1	37.7	46.4	56.6	66.5	71.4	70.0	62.5	50.3	41.1	31.7	49.2
MEAN DAILY MINIMUM	56	28.2	30.0	37.0	46.3	56.2	65.4	70.2	68.8	62.0	50.2	40.4	31.8	48.9
LOWEST DAILY MINIMUM	60	-5	4	11	24	34	47	54	49	39	29	16	1	-5
YEAR OF OCCURRENCE		1982	1961	1943	1982	1947	1972	1988	1986	1963	1969	1955	1942	JAN 1982
MEAN OF EXTREME MINS.	56	12.3	15.3	22.9	33.2	43.7	53.5	60.8	58.6	48.3	36.6	26.8	17.4	35.8
NORMAL DRY BULB	30	34.6	37.5	47.2	56.5	66.4	75.6	80.0	78.5	71.3	59.7	49.8	39.4	58.0
MEAN DRY BULB	56	35.7	38.4	46.2	56.5	65.9	74.7	79.1	77.4	70.7	59.6	49.0	39.3	57.7
MEAN WET BULB	18	32.2	34.8	40.2	49.5	58.9	67.4	71.2	70.2	63.9	53.7	43.7	35.5	51.8
MEAN DEW POINT	18	24.7	27.1	31.6	41.9	53.3	62.7	66.9	66.1	59.7	48.4	37.3	28.5	45.7
NORMAL NO. DAYS WITH:														
MAXIMUM　90	30	0.0	0.0	0.0	0.4	1.5	7.4	13.8	10.1	3.9	0.1	0.0	0.0	37.2
MAXIMUM　32	30	5.2	2.3	0.2	0.0	0.0	0.0	0.0	0.0	0.0	0.0	*	2.1	9.8
MINIMUM　32	30	22.3	18.5	8.4	0.9	0.0	0.0	0.0	0.0	0.0	0.4	4.2	15.7	70.4
MINIMUM　0	30	0.1	0.0	0.0	0.0	0.0	0.0	0.0	0.0	0.0	0.0	0.0	0.0	0.1
H/C														
NORMAL HEATING DEG. DAYS	30	942	770	552	264	60	0	0	0	14	195	456	794	4047
NORMAL COOLING DEG. DAYS	30	0	0	0	9	104	318	465	419	203	31	0	0	1549
RH														
NORMAL (PERCENT)	30	62	60	59	58	64	66	67	69	70	67	65	64	64
HOUR 01 LST	30	66	66	64	66	74	76	77	79	79	76	71	68	72
HOUR 07 LST	30	69	69	70	69	74	75	76	80	81	79	75	71	74
HOUR 13 LST	30	55	52	49	48	52	52	53	55	55	53	54	56	53
HOUR 19 LST	30	59	56	53	51	59	60	61	64	66	64	61	61	60
S PERCENT POSSIBLE SUNSHINE	50	46	50	55	57	58	64	62	62	61	59	51	46	56
W/O MEAN NO. DAYS WITH:														
HEAVY FOG(VISBY 1/4 MI)	53	1.7	1.5	0.9	0.8	0.4	0.2	0.1	0.1	0.3	1.3	1.3	1.7	10.3
THUNDERSTORMS	53	0.2	0.2	1.2	2.6	4.6	5.9	6.6	5.2	2.4	1.1	0.6	0.0	30.6
CLOUDINESS MEAN:														
SUNRISE-SUNSET (OKTAS)	49	5.4	5.2	5.2	5.1	5.1	4.8	4.8	4.6	4.6	4.3	5.0	5.2	4.9
MIDNIGHT-MIDNIGHT (OKTAS)	33	5.1	5.0	5.0	4.8	5.0	4.7	4.8	4.5	4.5	4.2	4.7	5.1	4.8
MEAN NO. DAYS WITH:														
CLEAR	50	7.2	7.0	7.5	7.1	6.9	7.3	7.4	8.6	9.4	10.7	7.7	7.9	94.7
PARTLY CLOUDY	50	7.4	6.7	8.3	9.1	10.1	11.4	11.8	10.2	8.4	7.9	7.8	6.6	105.7
CLOUDY	50	16.4	14.5	15.2	13.9	14.0	11.4	11.4	11.7	11.6	11.9	13.8	15.9	161.7
PR MEAN STATION PRESSURE(IN)	29	30.03	30.02	29.97	29.92	29.92	29.92	29.94	29.98	30.01	30.04	30.03	30.05	29.99
MEAN SEA-LEVEL PRES. (IN)	18	30.11	30.11	30.05	29.99	29.99	29.98	30.00	30.03	30.07	30.11	30.13	30.13	30.06
WINDS														
MEAN SPEED (MPH)	54	10.2	10.7	11.1	10.7	9.6	9.0	8.5	8.3	8.5	8.8	9.5	9.6	9.5
PREVAIL.DIR(TENS OF DEGS)	32	32	32	32	32	18	18	18	18	18	18	18	33	18
MAXIMUM 2-MINUTE:														
SPEED (MPH)	3	41	36	37	39	37	49	47	33	39	32	34	34	49
DIR. (TENS OF DEGS)		29	28	28	31	26	31	05	31	32	35	16	32	31
YEAR OF OCCURRENCE		2000	2001	2001	2000	1999	1999	2000	2000	2000	1999	1999	1999	JUN 1999
MAXIMUM 5-SECOND:														
SPEED (MPH)	3	55	51	48	51	55	60	53	45	55	41	44	45	60
DIR. (TENS OF DEGS)		31	27	29	31	25	31	05	35	32	36	16	28	31
YEAR OF OCCURRENCE		2000	2001	2001	2000	2000	1999	2000	2000	1999	1999	1999	2000	JUN 1999
PRECIPITATION														
NORMAL (IN)	30	2.72	2.71	3.17	2.71	3.66	3.38	3.80	3.91	3.31	3.02	3.12	3.12	38.63
MAXIMUM MONTHLY (IN)	60	7.11	5.71	8.45	6.88	10.69	11.53	11.06	14.31	12.36	8.65	6.70	6.54	14.31
YEAR OF OCCURRENCE		1978	1961	1994	1983	1953	1972	1945	1955	1975	1995	1963	1969	AUG 1955
MINIMUM MONTHLY (IN)	60	0.31	0.42	0.64	0.03	0.75	0.95	0.93	0.55	0.20	T	0.29	0.22	T
YEAR OF OCCURRENCE		1955	1978	1945	1985	1986	1988	1966	1962	1967	1963	1981	1955	OCT 1963
MAXIMUM IN 24 HOURS (IN)	58	2.13	2.13	3.43	3.08	4.32	7.19	4.69	6.39	5.31	4.98	4.03	2.86	7.19
YEAR OF OCCURRENCE		1976	1998	1958	1970	1953	1972	1970	1955	1975	1955	1993	1977	JUN 1972
NORMAL NO. DAYS WITH:														
PRECIPITATION　0.01	30	9.8	9.5	10.1	9.6	10.8	9.7	9.6	8.9	7.2	7.2	8.5	8.9	109.8
PRECIPITATION　1.00	30	0.4	0.5	0.6	0.4	0.9	0.8	1.0	1.3	1.0	1.0	0.8	0.7	9.4
SNOWFALL														
NORMAL (IN)	30	6.2	6.3	1.6	0.*	0.0	0.0	0.0	0.0	0.0	0.*	1.0	3.1	18.2
MAXIMUM MONTHLY (IN)	58	23.8	30.6	17.1	0.6	T	T	T	T	0.0	0.3	11.5	16.2	30.6
YEAR OF OCCURRENCE		1996	1979	1960	1972	1993	1998	1990	1996		1979	1987	1962	FEB 1979
MAXIMUM IN 24 HOURS (IN)	58	13.8	18.7	8.4	0.6	T	T	T	T	0.0	0.3	11.5	11.4	18.7
YEAR OF OCCURRENCE		1966	1979	1999	1972	1993	1998	1990	1992		1979	1987	1957	FEB 1979
MAXIMUM SNOW DEPTH (IN)	55	18	22	8	1	0	0	0	0	0	0	12	8	22
YEAR OF OCCURRENCE		1987	1979	1960	1972							1987	1973	FEB 1979
NORMAL NO. DAYS WITH:														
SNOWFALL　1.0	30	1.6	1.5	0.5	0.0	0.0	0.0	0.0	0.0	0.0	0.0	0.2	0.7	4.5

PRECIPITATION (inches) 2001 WASHINGTON, D.C. DC (DCA)

YEAR	JAN	FEB	MAR	APR	MAY	JUN	JUL	AUG	SEP	OCT	NOV	DEC	ANNUAL
1972	2.45	5.27	2.27	3.99	4.78	11.53	3.43	2.82	1.27	3.56	6.05	4.55	51.97
1973	2.26	2.68	2.97	4.19	3.39	2.11	2.68	4.41	1.58	1.71	0.97	6.03	34.98
1974	2.66	0.95	4.21	2.26	4.37	3.40	1.15	5.77	4.39	1.13	1.24	4.43	35.96
1975	3.09	1.56	5.33	2.13	4.71	2.15	7.16	3.54	12.36	2.38	2.05	4.04	50.50
1976	3.56	1.55	2.51	1.17	3.57	1.21	4.54	2.13	7.23	7.76	0.85	1.99	38.07
1977	1.50	0.66	2.17	2.66	1.73	3.28	4.06	4.74	0.32	5.35	4.81	4.86	36.14
1978	7.11	0.42	4.48	1.38	5.13	2.43	4.28	5.85	1.01	1.16	2.31	4.00	39.56
1979	6.64	5.62	2.45	1.88	3.55	2.99	3.43	5.41	6.64	5.54	2.33	0.85	47.33
1980	2.85	1.16	5.04	3.28	2.64	1.68	3.86	1.11	1.90	2.59	2.56	0.65	29.32
1981	0.38	2.82	1.49	2.63	3.42	2.55	5.69	3.02	1.94	3.64	0.29	2.80	30.67
1982	2.27	3.33	2.64	3.19	5.11	5.41	2.98	2.68	1.71	1.75	2.96	1.74	35.77
1983	1.69	3.09	4.84	6.88	4.62	7.09	1.78	3.11	2.90	4.87	5.09	5.91	51.87
1984	1.71	3.43	6.14	3.71	3.80	2.01	4.09	2.30	2.51	3.18	3.66	1.19	37.73
1985	2.11	3.07	1.88	0.03	5.79	2.05	2.91	2.35	6.67	3.85	4.47	0.68	35.86
1986	2.38	3.49	0.74	1.98	0.75	1.29	3.79	5.33	0.60	2.01	5.23	4.98	32.57
1987	4.90	2.11	1.54	2.28	2.54	3.90	2.59	2.07	5.11	2.53	4.49	2.57	36.63
1988	3.14	2.52	2.27	2.00	4.50	0.95	3.74	2.39	1.85	1.75	5.33	1.30	31.74
1989	2.49	2.80	4.30	3.50	7.77	6.02	5.66	1.15	6.68	5.48	2.37	2.10	50.32
1990	2.95	1.30	2.57	4.09	5.20	3.14	3.78	6.74	0.87	3.30	2.17	4.73	40.84
1991	2.90	0.83	4.42	1.39	1.57	1.27	3.76	2.03	3.50	2.03	0.85	5.07	29.62
1992	2.78	2.23	3.48	2.55	3.41	2.35	5.34	2.48	3.49	2.03	3.38	2.86	36.38
1993	2.90	2.27	6.82	3.62	3.40	1.73	1.36	3.87	3.68	2.62	4.38	4.76	41.41
1994	4.28	4.20	8.45	1.58	1.56	1.59	3.61	4.35	2.84	1.19	1.57	2.35	37.57
1995	3.22	1.71	2.14	1.89	4.19	2.42	4.03	0.88	3.73	8.65	4.77	2.17	39.80
1996	5.01	1.99	3.60	3.17	4.96	3.14	5.60	2.63	7.79	4.04	3.58	5.51	51.02
1997	2.55	2.43	4.15	2.41	3.04	2.94	1.14	3.51	1.59	3.72	4.60	1.74	33.82
1998	5.43	5.23	5.40	3.96	4.05	4.42	1.79	0.59	1.83	0.59	0.91	1.74	35.94
1999	5.42	2.54	3.87	2.09	1.28	2.26	1.01	5.02	10.27	2.16	1.82	2.49	40.23
2000	3.66	2.06	3.98	5.13	3.08	4.93	5.51	3.77	4.91	0.02	1.60	2.01	40.66
2001	2.22	1.83	3.88	1.68	3.71	4.69	4.78	2.98	1.41	0.69	0.55	1.53	29.95
POR= 60 YRS	2.90	2.56	3.49	2.81	3.80	3.30	3.84	4.01	3.43	2.90	2.92	2.96	38.92

AVERAGE TEMPERATURE (F) 2001 WASHINGTON, D.C. DC (DCA)

YEAR	JAN	FEB	MAR	APR	MAY	JUN	JUL	AUG	SEP	OCT	NOV	DEC	ANNUAL
1972	38.5	36.5	45.6	54.1	64.6	70.2	77.5	75.9	71.0	56.0	46.8	43.6	56.7
1973	37.6	37.0	51.1	56.0	62.8	77.1	79.2	79.9	74.3	63.3	51.6	41.9	59.3
1974	42.9	39.2	49.2	58.3	65.1	71.5	79.0	78.4	70.2	57.3	50.9	43.1	58.8
1975	40.9	40.6	45.2	53.6	69.7	76.4	79.3	80.1	68.5	63.2	54.4	40.5	59.4
1976	33.9	46.9	51.3	59.9	65.0	77.6	78.4	76.7	70.4	55.4	43.0	35.5	57.8
1977	25.4	38.8	52.7	60.1	69.4	74.3	80.9	78.8	73.9	59.0	51.8	38.1	58.6
1978	32.5	31.4	44.4	57.7	65.8	76.7	78.8	81.3	73.6	59.4	52.2	43.1	58.1
1979	35.1	28.4	51.5	56.0	67.7	72.4	78.6	78.5	71.6	58.6	54.4	43.7	58.0
1980	37.2	36.1	46.2	60.1	69.5	74.8	82.3	82.8	77.1	59.9	48.6	39.8	59.5
1981	33.0	43.7	47.6	62.1	66.2	78.7	80.2	77.0	71.0	58.3	51.4	38.5	59.0
1982	28.1	38.3	45.7	54.0	69.0	72.8	80.3	75.4	70.6	60.2	51.8	45.5	57.6
1983	38.1	38.7	48.8	53.3	64.9	75.0	81.2	81.0	72.6	60.5	50.3	36.0	58.4
1984	32.2	43.8	41.8	54.9	64.9	76.9	76.5	77.8	68.3	65.2	46.0	45.6	57.8
1985	30.8	37.8	47.7	61.6	68.1	72.3	79.0	76.7	71.9	61.2	54.3	36.4	58.2
1986	35.4	35.3	47.4	56.2	68.1	76.6	81.1	74.6	70.9	61.1	46.5	39.8	57.8
1987	34.7	37.0	47.7	54.8	67.2	76.4	82.6	78.7	72.1	54.4	49.9	41.5	58.1
1988	31.0	37.3	47.2	54.4	65.8	74.4	81.9	80.7	68.9	54.4	49.9	38.7	57.1
1989	39.9	37.8	46.1	55.4	64.1	76.8	78.3	77.1	71.4	60.5	48.0	27.9	56.9
1990	43.6	45.2	50.2	56.8	64.3	75.0	79.4	76.5	69.6	62.8	52.0	44.5	60.0
1991	38.6	43.0	48.8	58.2	73.0	76.8	81.4	80.0	71.0	60.4	48.8	42.3	60.2
1992	38.2	41.2	45.0	55.4	62.3	71.7	79.5	74.0	69.2	56.0	48.8	39.6	56.7
1993	39.7	34.3	42.2	54.8	67.4	75.3	83.1	79.6	71.0	58.1	48.8	38.1	57.7
1994	28.8	36.3	45.4	62.0	63.1	79.4	81.8	75.5	70.2	59.1	53.4	44.2	58.3
1995	39.6	34.4	49.2	56.3	65.8	74.6	81.5	81.3	71.0	62.3	43.1	35.6	57.9
1996	32.9	37.3	42.9	56.8	63.9	77.2	77.6	76.5	70.6	59.4	44.1	43.0	56.9
1997	37.0	44.7	48.7	54.0	62.9	73.1	80.4	77.6	70.5	59.6	46.2	41.0	58.0
1998	43.0	43.4	46.9	57.4	67.5	73.0	78.9	79.4	75.4	60.1	50.3	44.4	60.0
1999	38.2	41.0	44.5	56.5	67.3	74.7	83.0	79.7	70.0	57.2	53.1	42.0	58.9
2000	35.9	42.5	51.7	55.6	67.8	74.7	74.7	75.1	67.6	60.2	46.7	31.8	57.0
2001	35.5	40.9	43.8	57.6	65.9	75.2	75.3	78.9	68.9	59.7	54.8	45.5	58.5
POR= 60 YRS	35.7	38.3	46.3	56.4	66.0	74.7	79.0	77.3	70.7	59.5	49.0	39.1	57.7

REFERENCE NOTES:

PAGE 1:
THE TEMPERATURE GRAPH SHOWS NORMAL MAXIMUM AND NORMAL MINIMUM DAILY TEMPERATURES (SOLID CURVES) AND THE ACTUAL DAILY HIGH AND LOW TEMPERATURES (VERTICAL BARS).

PAGE 2 AND 3:
H/C INDICATES HEATING AND COOLING DEGREE DAYS.
RH INDICATES RELATIVE HUMIDITY
W/O INDICATES WEATHER AND OBSTRUCTIONS
S INDICATES SUNSHINE.
PR INDICATES PRESSURE.
CLOUDINESS ON PAGE 3 IS THE SUM OF THE CEILOMETER AND SATELLITE DATA NOT TO EXCEED EIGHT EIGHTHS (OKTAS).

GENERAL:
T INDICATES TRACE PRECIPITATION, AN AMOUNT GREATER THAN ZERO BUT LESS THAN THE LOWEST REPORTABLE VALUE.
+ INDICATES THE VALUE ALSO OCCURS ON EARLIER DATES.
BLANK ENTRIES DENOTE MISSING OR UNREPORTED DATA.
NORMALS ARE 30-YEAR AVERAGES (1961 - 1990).
ASOS INDICATES AUTOMATED SURFACE OBSERVING SYSTEM.
PM INDICATES THE LAST DAY OF THE PREVIOUS MONTH.
POR (PERIOD OF RECORD) BEGINS WITH THE JANUARY DATA MONTH AND IS THE NUMBER OF YEARS USED TO COMPUTE THE MEAN. INDIVIDUAL MONTHS WITHIN THE POR MAY BE MISSING.
WHEN THE POR FOR A NORMAL IS LESS THAN 30 YEARS, THE NORMAL IS PROVISIONAL AND IS BASED ON THE NUMBER OF YEARS INDICATED.
0.* OR * INDICATES THE VALUE OR MEAN-DAYS-WITH IS BETWEEN 0.00 AND 0.05.
CLOUDINESS FOR ASOS STATIONS DIFFERS FROM THE NON-ASOS OBSERVATION TAKEN BY A HUMAN OBSERVER. ASOS STATION CLOUDINESS IS BASED ON TIME-AVERAGED CEILOMETER DATA FOR CLOUDS AT OR BELOW 12,000 FEET AND ON SATELLITE DATA FOR CLOUDS ABOVE 12,000 FEET.
THE NUMBER OF DAYS WITH CLEAR, PARTLY CLOUDY, AND CLOUDY CONDITIONS FOR ASOS STATIONS IS THE SUM OF THE CEILOMETER AND SATELLITE DATA FOR THE SUNRISE TO SUNSET PERIOD.

GENERAL CONTINUED:
CLEAR INDICATES 0 - 2 OKTAS, PARTLY CLOUDY INDICATES 3 - 6 OKTAS, AND CLOUDY INDICATES 7 OR 8 OKTAS. WHEN AT LEAST ONE OF THE ELEMENTS (CEILOMETER OR SATELLITE) IS MISSING, THE DAILY CLOUDINESS IS NOT COMPUTED.
WIND DIRECTION IS RECORDED IN TENS OF DEGREES (2 DIGITS) CLOCKWISE FROM TRUE NORTH. "00" INDICATES CALM. "36" INDICATES TRUE NORTH.
RESULTANT WIND IS THE VECTOR AVERAGE OF THE SPEED AND DIRECTION.
AVERAGE TEMPERATURE IS THE SUM OF THE MEAN DAILY MAXIMUM AND MINIMUM TEMPERATURE DIVIDED BY 2.
SNOWFALL DATA COMPRISE ALL FORMS OF FROZEN PRECIPITATION, INCLUDING HAIL.
A HEATING (COOLING) DEGREE DAY IS THE DIFFERENCE BETWEEN THE AVERAGE DAILY TEMPERATURE AND 65 F.
DRY BULB IS THE TEMPERATURE OF THE AMBIENT AIR.
DEW POINT IS THE TEMPERATURE TO WHICH THE AIR MUST BE COOLED TO ACHIEVE 100 PERCENT RELATIVE HUMIDITY.
WET BULB IS THE TEMPERATURE THE AIR WOULD HAVE IF THE MOISTURE CONTENT WAS INCREASED TO 100 PERCENT RELATIVE HUMIDITY.

ON JULY 1, 1996, THE NATIONAL WEATHER SERVICE BEGAN USING THE "METAR" OBSERVATION CODE THAT WAS ALREADY EMPLOYED BY MOST OTHER NATIONS OF THE WORLD. THE MOST NOTICEABLE DIFFERENCE IN THIS ANNUAL PUBLICATION WILL BE THE CHANGE IN UNITS FROM TENTHS TO EIGHTS (OKTAS) FOR REPORTING THE AMOUNT OF SKY COVER.

HEATING DEGREE DAYS (base 65 F) 2001 WASHINGTON, D.C. DC (DCA)

YEAR	JUL	AUG	SEP	OCT	NOV	DEC	JAN	FEB	MAR	APR	MAY	JUN	TOTAL
1972-73	0	0	8	278	543	654	843	777	423	286	109	0	3921
1973-74	0	0	4	103	399	708	677	716	490	228	85	4	3414
1974-75	0	0	26	250	446	674	740	677	608	345	24	0	3790
1975-76	0	0	20	102	328	752	956	524	415	236	80	0	3413
1976-77	0	0	11	306	652	907	1221	729	389	188	32	3	4438
1977-78	0	0	1	196	406	829	1001	933	633	219	86	0	4304
1978-79	0	0	9	192	378	671	918	1019	425	273	30	0	3915
1979-80	0	0	5	231	313	654	857	830	573	149	28	0	3640
1980-81	0	0	4	189	487	774	984	592	536	133	75	0	3774
1981-82	0	0	19	219	399	818	1135	743	592	328	19	3	4275
1982-83	0	2	9	193	402	597	827	730	497	365	77	0	3699
1983-84	0	0	32	177	433	890	1009	610	710	302	95	4	4262
1984-85	0	0	54	59	561	594	1053	757	533	166	30	6	3813
1985-86	0	0	14	147	320	879	913	824	542	267	61	3	3970
1986-87	0	13	18	180	548	775	931	777	527	304	68	0	4141
1987-88	0	0	4	325	448	719	1047	796	544	317	69	25	4294
1988-89	0	0	18	330	442	807	771	755	596	297	112	0	4128
1989-90	0	0	35	167	507	1144	656	550	481	285	64	4	3893
1990-91	0	0	38	153	381	630	810	608	502	237	22	0	3381
1991-92	0	0	27	175	486	696	824	686	614	295	127	5	3935
1992-93	0	0	43	282	477	781	779	855	700	307	32	3	4259
1993-94	0	0	33	217	487	825	1115	796	599	135	132	0	4339
1994-95	0	0	5	190	348	639	782	853	485	274	59	0	3635
1995-96	0	0	30	144	651	901	993	798	675	268	131	0	4591
1996-97	0	0	9	182	617	677	861	564	499	322	110	30	3871
1997-98	0	0	14	232	557	737	675	596	581	229	50	11	3682
1998-99	0	0	4	153	433	638	824	667	630	251	34	3	3637
1999-00	0	0	22	240	348	707	896	649	408	280	46	2	3598
2000-01	0	0	60	171	544	1022	908	670	654	246	54	4	
2001-	0	0	44	207	302	598							

COOLING DEGREE DAYS (base 65 F) 2001 WASHINGTON, D.C. DC (DCA)

YEAR	JAN	FEB	MAR	APR	MAY	JUN	JUL	AUG	SEP	OCT	NOV	DEC	ANNUAL
1972	0	0	3	5	50	184	393	346	195	8	2	0	1186
1973	0	0	2	21	47	371	448	469	288	57	3	0	1706
1974	0	0	4	33	96	205	441	422	192	17	27	0	1437
1975	1	0	0	12	177	344	448	475	132	50	15	0	1654
1976	0	4	1	92	86	383	424	370	179	15	0	0	1554
1977	0	0	10	49	177	289	496	434	274	18	15	0	1762
1978	0	0	0	10	117	358	434	514	274	25	0	0	1732
1979	0	0	14	9	120	231	431	425	208	39	2	0	1479
1980	0	0	0	9	174	301	546	563	374	38	1	0	2006
1981	0	0	6	49	118	417	478	380	204	18	0	0	1670
1982	0	0	0	6	155	244	479	330	185	51	13	1	1464
1983	0	0	0	21	81	310	510	504	269	42	0	0	1737
1984	0	0	0	4	99	368	365	404	157	73	0	0	1470
1985	0	0	6	70	135	232	444	373	228	37	6	0	1531
1986	0	0	5	10	162	358	503	318	202	70	1	0	1629
1987	0	0	0	8	146	347	554	431	222	0	0	0	1708
1988	0	0	1	4	101	313	534	490	144	11	0	0	1598
1989	0	0	16	14	91	362	417	381	233	33	1	0	1548
1990	0	0	30	46	50	309	451	364	183	88	0	0	1521
1991	0	0	5	38	278	362	517	472	214	41	5	0	1932
1992	0	0	0	16	53	214	457	285	175	8	0	0	1208
1993	0	0	0	5	114	319	569	460	218	9	8	0	1702
1994	0	0	1	53	82	439	530	333	166	11	6	0	1621
1995	0	0	0	20	91	297	515	511	216	67	1	0	1718
1996	0	0	0	29	105	373	397	364	181	14	0	0	1463
1997	0	0	1	0	52	282	484	399	184	72	0	0	1474
1998	0	0	27	11	132	257	438	456	322	9	0	5	1657
1999	0	0	0	4	111	301	565	462	180	6	0	0	1629
2000	0	0	4	5	143	300	307	319	146	29	0	0	1253
2001	0	0	0	31	88	318	323	441	171	47	5	0	1424

SNOWFALL (inches) 2001 WASHINGTON, D.C. DC (DCA)

YEAR	JUL	AUG	SEP	OCT	NOV	DEC	JAN	FEB	MAR	APR	MAY	JUN	TOTAL
1972-73	0.0	0.0	0.0	T	T	0.0	T	0.1	T	T	0.0	0.0	0.1
1973-74	0.0	0.0	0.0	0.0	0.0	11.0	1.5	4.2	T	T	0.0	0.0	16.7
1974-75	0.0	0.0	0.0	T	T	0.1	6.6	5.8	0.3	T	0.0	0.0	12.8
1975-76	0.0	0.0	0.0	0.0	0.0	0.4	0.1	0.9	0.8	0.0	0.0	0.0	2.2
1976-77	0.0	0.0	0.0	0.0	0.8	0.6	9.7	0.0	0.0	T	0.0	0.0	11.1
1977-78	0.0	0.0	0.0	0.0	0.1	0.2	10.3	3.8	8.3	0.0	0.0	0.0	22.7
1978-79	0.0	0.0	0.0	0.0	3.1	T	4.0	30.6	T	0.0	0.0	0.0	37.7
1979-80	0.0	0.0	0.0	0.3	0.0	T	8.6	5.1	6.1	0.0	0.0	0.0	20.1
1980-81	0.0	0.0	0.0	0.0	T	0.3	4.2	T	T	0.0	0.0	0.0	4.5
1981-82	0.0	0.0	0.0	0.0	T	1.7	15.3	5.3	0.2	T	0.0	0.0	22.5
1982-83	0.0	0.0	0.0	0.0	0.0	6.6	T	21.0	0.0	T	0.0	0.0	27.6
1983-84	0.0	0.0	0.0	0.0	0.3	T	6.5	T	1.8	T	0.0	0.0	8.6
1984-85	0.0	0.0	0.0	0.0	T	0.3	10.0	T	T	T	0.0	0.0	10.3
1985-86	0.0	0.0	0.0	0.0	0.0	0.7	1.8	12.9	T	T	0.0	0.0	15.4
1986-87	0.0	0.0	0.0	0.0	T	20.8	10.3	T	T	T	0.0	0.0	31.1
1987-88	0.0	0.0	0.0	0.0	11.5	T	13.1	T	0.4	T	0.0	0.0	25.0
1988-89	0.0	0.0	0.0	0.0	0.0	1.2	2.9	1.2	0.4	0.0	0.0	0.0	5.7
1989-90	0.0	0.0	0.0	0.0	3.5	9.0	0.2	T	2.4	0.2	0.0	0.0	15.3
1990-91	T	0.0	0.0	0.0	0.0	3.0	4.8	0.3	T	0.0	T	0.0	8.1
1991-92	0.0	0.0	0.0	0.0	T	0.0	4.0	2.6	T	0.0	0.0	0.0	6.6
1992-93	0.0	T	0.0	0.0	T	1.0	T	4.1	6.6	0.0	T	0.0	11.7
1993-94	0.0	0.0	0.0	0.0	T	2.6	3.5	3.1	4.0	0.0	0.0	0.0	13.2
1994-95	0.0	0.0	0.0	0.0	T	0.0	3.9	5.8	0.4	0.0	0.0	0.0	10.1
1995-96	0.0	0.0	0.0	0.0	0.5	1.3	23.8	15.2	5.2	T	0.0	0.0	46.0
1996-97	0.0	T	0.0	0.0	0.2	0.2	2.3	4.0	T	0.0	0.0	0.0	6.7
1997-98	0.0	0.0	0.0	0.0	T	0.1	T	T	T	0.0	T	0.0	0.1
1998-99	0.0	0.0	0.0	0.0	0.0	0.0	0.5	2.2	0.2	8.7	0.0	0.0	11.6
1999-00	0.0	0.0	0.0	0.0	0.0	T	14.5	0.9	0.0	T	0.0	0.0	15.4
2000-01	0.0	T	0.0	0.0	0.0	2.0	2.4	2.8	0.2	T	0.0	0.0	7.4
2001-	0.0	0.0	0.0	0.0	0.0								
POR= 57 YRS	T	T	0.0	0.0	0.8	2.8	5.3	5.1	2.2	0.0	T	0.0	16.2

2001
DAYTONA BEACH,
FLORIDA (DAB)

Daytona Beach is located on the Atlantic Ocean. The Halifax River, part of the Florida Inland Waterway, runs through the city. The terrain in the area is flat and the soil is mostly sandy. Elevations in the area range from 3 to 15 feet above mean sea level near the ocean to about 31 feet at the airport and on a ridge running along the western city limits.

Nearness to the ocean results in a climate tempered by the effect of land and sea breezes. In the summer, while maximum temperatures reach 90 degrees or above during the late morning or early afternoon, the number of hours of 90 degrees or above is relatively small due to the beginning of the sea breeze near midday and the occurrence of local afternoon convective thunderstorms which lower the temperature to the comfortable 80s. Winters, although subject to invasions of cold air, are relatively mild due to the nearness of the ocean and latitudinal location.

The rainy season from June through mid-October produces 60 percent of the annual rainfall. The major portion of the summer rainfall occurs in the form of local convective thunderstorms which are occasionally heavy and produce as much as 2 or 3 inches of rain. The more severe thunderstorms may be attended by strong gusty winds. Almost all rainfall during the winter months is associated with frontal passages.

Long periods of cloudiness and rain are infrequent, usually not lasting over 2 or 3 days. These periods are usually associated with a stationary front, a so-called northeaster, or a tropical disturbance.

Tropical disturbances or hurricanes are not considered a great threat to this area of the state. Generally hurricanes in this latitude tend to pass well offshore or lose much of their intensity while crossing the state before reaching this area. Only in gusts have hurricane-force winds been recorded at this station.

Heavy fog occurs mostly during the winter and early spring. These fogs usually form by radiational cooling at night and dissipate soon after sunrise. On rare occasions sea fog moves in from the ocean and persists for two or three days. There is no significant source in the area for air pollution.

NORMALS, MEANS, AND EXTREMES
DAYTONA BEACH, FL (DAB)

LATITUDE: 29 10' 38" N LONGITUDE: 81 03' 36" W ELEVATION (FT): GRND: 31 BARO: 34 TIME ZONE: EASTERN (UTC + 5) WBAN: 12834

	ELEMENT	POR	JAN	FEB	MAR	APR	MAY	JUN	JUL	AUG	SEP	OCT	NOV	DEC	YEAR
TEMPERATURE °F	NORMAL DAILY MAXIMUM	30	68.0	69.5	74.8	80.0	84.5	88.0	89.8	89.1	86.8	81.5	75.5	70.4	79.8
	MEAN DAILY MAXIMUM	54	68.8	70.6	74.8	79.7	84.9	88.4	90.1	89.3	86.8	81.6	75.5	70.3	80.1
	HIGHEST DAILY MAXIMUM	58	87	89	92	96	100	102	102	100	99	95	89	88	102
	YEAR OF OCCURRENCE		1991	1985	1994	1968	1953	1944	1981	1999	1944	1959	1948	1990	JUL 1981
	MEAN OF EXTREME MAXS.	54	81.7	83.4	87.1	90.1	93.0	95.0	95.6	94.6	92.2	89.1	84.6	82.1	89.0
	NORMAL DAILY MINIMUM	30	46.9	48.4	53.9	58.6	64.9	70.8	72.5	72.9	71.9	65.2	56.3	49.7	61.0
	MEAN DAILY MINIMUM	54	47.5	49.4	53.6	58.6	65.0	70.5	72.3	72.7	71.9	65.4	56.1	49.7	61.1
	LOWEST DAILY MINIMUM	58	15	24	26	35	44	52	60	65	52	41	27	19	15
	YEAR OF OCCURRENCE		1985	1958	1980	1950	1971	1984	1981	1984	1956	1993	1950	1983	JAN 1985
	MEAN OF EXTREME MINS.	54	30.3	33.3	37.3	45.1	54.4	63.6	68.0	68.8	65.3	50.7	40.3	32.3	49.1
	NORMAL DRY BULB	30	57.5	59.0	64.3	69.3	74.7	79.4	81.2	81.0	79.4	73.4	65.9	60.1	70.4
	MEAN DRY BULB	54	58.1	59.9	64.2	69.2	74.9	79.5	81.2	81.0	79.4	73.5	65.9	60.0	70.6
	MEAN WET BULB	18	54.1	55.9	59.0	62.5	68.7	73.5	75.1	71.3	74.1	68.4	62.5	56.4	65.1
	MEAN DEW POINT	18	50.0	51.8	54.6	58.0	64.9	70.9	72.9	69.4	71.9	65.5	59.6	52.9	61.9
	NORMAL NO. DAYS WITH:														
	MAXIMUM 90	30	0.0	0.0	0.1	1.7	4.4	9.7	16.1	13.4	5.3	0.7	0.0	0.0	51.4
	MAXIMUM 32	30	0.0	0.0	0.0	0.0	0.0	0.0	0.0	0.0	0.0	0.0	0.0	0.0	0.0
	MINIMUM 32	30	2.8	1.1	0.3	0.0	0.0	0.0	0.0	0.0	0.0	0.0	0.1	1.6	5.9
	MINIMUM 0	30	0.0	0.0	0.0	0.0	0.0	0.0	0.0	0.0	0.0	0.0	0.0	0.0	0.0
H/C	NORMAL HEATING DEG. DAYS	30	282	205	112	21	0	0	0	0	0	0	82	207	909
	NORMAL COOLING DEG. DAYS	30	50	37	90	150	301	432	502	496	432	265	109	55	2919
RH	NORMAL (PERCENT)	30	75	72	71	69	72	77	78	80	79	75	76	76	75
	HOUR 01 LST	30	84	82	82	81	83	86	87	89	87	83	84	84	84
	HOUR 07 LST	30	85	84	85	84	84	86	88	90	88	85	86	85	86
	HOUR 13 LST	30	59	56	54	51	56	63	64	66	66	61	60	59	60
	HOUR 19 LST	30	75	71	69	65	69	75	76	79	78	75	78	78	74
S	PERCENT POSSIBLE SUNSHINE														
W/O	MEAN NO. DAYS WITH:														
	HEAVY FOG(VISBY 1/4 MI)	58	5.1	3.4	2.9	1.8	1.5	1.1	1.1	1.3	0.7	1.5	2.7	4.5	27.6
	THUNDERSTORMS	58	1.2	1.8	3.4	3.4	7.4	13.4	17.2	15.1	8.9	3.0	1.3	1.2	77.3
CLOUDINESS	MEAN:														
	SUNRISE-SUNSET (OKTAS)	0			5.6		3.2	4.0							
	MIDNIGHT-MIDNIGHT (OKTAS)	0			5.6			4.0							
	MEAN NO. DAYS WITH:														
	CLEAR	1	1.0	4.0	7.0		13.0	4.0							
	PARTLY CLOUDY	1	2.0	4.0	4.0		5.0	11.0							
	CLOUDY	1	2.0	2.0	8.0		3.0	6.0							
PR	MEAN STATION PRESSURE(IN)	29	30.10	30.09	30.00	30.00	30.00	30.00	30.00	30.00	30.00	30.00	30.00	30.09	30.02
	MEAN SEA-LEVEL PRES. (IN)	18	30.13	30.12	30.06	30.03	30.01	30.03	30.07	30.03	30.01	30.03	30.10	30.13	30.06
WINDS	MEAN SPEED (MPH)	37	8.7	9.3	9.7	9.4	8.8	7.7	7.1	6.9	8.0	8.9	8.4	8.1	8.4
	PREVAIL.DIR(TENS OF DEGS)	22	36	36	24	24	12	08	24	12	06	07	31	32	08
	MAXIMUM 2-MINUTE:														
	SPEED (MPH)	6	36	44	38	31	46	36	35	41	43	46	39	29	46
	DIR. (TENS OF DEGS)		23	23	07	31	28	22	24	28	05	03	05	26	03
	YEAR OF OCCURRENCE		1998	1998	2001	2000	1996	1997	2000	1996	2001	1999	2001	1997	OCT 1999
	MAXIMUM 5-SECOND:														
	SPEED (MPH)	6	44	56	53	40	55	47	54	51	64	56	49	38	64
	DIR. (TENS OF DEGS)		23	22	33	20	28	21	36	32	36	01	04	26	36
	YEAR OF OCCURRENCE		1998	1998	1996	1997	1996	1997	1998	1998	1999	1999	2001	1997	SEP 1999
PRECIPITATION	NORMAL (IN)	30	2.75	3.11	2.90	2.23	3.45	5.99	5.40	6.16	6.34	4.13	2.84	2.59	47.89
	MAXIMUM MONTHLY (IN)	58	7.16	9.13	12.15	7.12	12.33	15.19	14.58	19.89	16.11	13.00	12.91	11.98	19.89
	YEAR OF OCCURRENCE		1986	1960	1996	1949	1976	1966	1944	1953	2001	1950	1994	1983	AUG 1953
	MINIMUM MONTHLY (IN)	58	0.15	0.29	0.25	T	0.08	0.83	0.16	2.01	0.42	0.19	T	0.06	T
	YEAR OF OCCURRENCE		1950	1944	1956	1967	1967	1965	1998	1992	1963	1972	1967	1967	NOV 1967
	MAXIMUM IN 24 HOURS (IN)	58	5.73	4.39	7.45	4.03	4.22	6.28	4.21	4.76	6.34	9.29	10.15	5.22	10.15
	YEAR OF OCCURRENCE		1989	1971	2000	1982	1947	1966	1986	1974	1964	1953	1994	1983	NOV 1994
	NORMAL NO. DAYS WITH:														
	PRECIPITATION 0.01	30	7.7	8.0	7.8	4.8	8.1	12.2	12.6	13.3	12.2	9.9	7.3	7.4	111.3
	PRECIPITATION 1.00	30	0.7	0.9	1.0	0.7	1.0	1.6	1.6	1.7	2.2	1.3	0.7	0.6	14.0
SNOWFALL	NORMAL (IN)	30	T	0.0	0.0	0.0	0.0	0.0	0.0	0.0	0.0	0.0	0.0	T	0.0
	MAXIMUM MONTHLY (IN)	53	T	T	T	0.0	0.0	T	0.0	T	0.0	0.0	0.0	T	T
	YEAR OF OCCURRENCE		1977	1951	1993			1989		1994				1989	AUG 1994
	MAXIMUM IN 24 HOURS (IN)	53	T	T	T	0.0	0.0	T	0.0	T	0.0	0.0	0.0	T	T
	YEAR OF OCCURRENCE		1977	1951	1993			1989		1994				1989	AUG 1994
	MAXIMUM SNOW DEPTH (IN)	48	0	0	0	0	0	0	0	0	0	0	0	0	0
	YEAR OF OCCURRENCE														
	NORMAL NO. DAYS WITH:														
	SNOWFALL 1.0	30	0.0	0.0	0.0	0.0	0.0	0.0	0.0	0.0	0.0	0.0	0.0	0.0	0.0

PRECIPITATION (inches) 2001 DAYTONA BEACH, FL (DAB)

YEAR	JAN	FEB	MAR	APR	MAY	JUN	JUL	AUG	SEP	OCT	NOV	DEC	ANNUAL
1972	2.37	3.97	6.66	1.41	4.02	7.06	3.22	8.29	0.42	3.08	10.96	2.48	53.94
1973	4.66	2.02	2.63	3.09	2.41	4.32	4.69	7.58	5.14	4.40	0.75	2.54	44.23
1974	0.30	1.10	3.19	0.44	2.66	8.65	6.31	9.96	10.50	1.42	0.48	2.20	47.21
1975	1.66	2.27	1.52	2.96	2.99	9.00	6.89	3.16	6.61	5.84	1.46	0.83	45.19
1976	0.60	0.70	2.03	4.27	12.33	11.14	1.07	3.80	5.10	1.90	3.38	6.00	52.32
1977	4.69	2.45	1.43	0.41	4.61	1.15	2.23	7.91	6.55	1.46	3.04	4.74	40.67
1978	2.89	5.98	2.31	3.30	0.56	7.48	5.53	7.99	4.63	8.31	0.07	4.89	53.94
1979	7.10	1.94	4.08	3.96	6.13	3.03	11.69	5.24	15.20	2.13	7.96	0.56	69.02
1980	3.75	0.76	2.41	2.54	3.62	5.57	5.82	4.13	1.83	2.42	3.12	1.39	37.36
1981	0.32	5.54	3.00	0.29	1.74	1.03	4.69	7.19	7.59	1.08	2.57	4.64	39.68
1982	2.46	2.08	5.81	6.04	4.68	8.29	5.31	3.21	4.96	3.23	1.58	2.53	50.18
1983	2.51	5.96	7.71	6.17	3.86	6.37	1.92	6.82	8.57	10.11	2.01	11.98	73.99
1984	1.46	3.44	1.31	5.29	6.04	2.84	6.77	4.02	10.73	1.09	3.52	0.20	46.71
1985	0.79	0.58	1.49	3.14	3.42	6.81	2.16	9.83	10.62	4.08	0.41	2.05	45.38
1986	7.16	1.28	1.85	0.44	0.99	3.50	14.43	3.47	3.58	3.47	5.08	2.76	48.01
1987	2.21	6.64	7.94	0.28	2.65	3.81	2.78	4.89	5.63	2.77	5.87	0.25	45.72
1988	5.36	1.72	4.57	1.68	1.78	2.39	2.94	4.79	6.81	1.24	6.70	0.93	40.91
1989	6.82	0.64	2.01	2.92	2.02	1.84	2.44	4.47	5.04	11.64	0.88	3.93	44.65
1990	1.42	5.61	1.94	1.48	1.45	2.71	5.85	7.00	1.61	5.88	0.83	0.34	36.12
1991	2.25	1.65	8.11	5.57	6.79	12.67	11.97	7.60	5.52	2.94	0.61	1.51	67.19
1992	2.42	1.71	2.28	2.81	3.13	10.64	0.16	8.86	6.57	5.21	2.15	0.47	46.41
1993	4.29	3.02	5.56	0.33	0.65	2.19	5.05	2.66	2.74	5.53	1.83	1.86	35.71
1994	5.60	2.66	3.44	5.05	3.09	6.54	6.91	7.08	5.93	4.72	12.91	2.71	66.64
1995	1.53	1.39	2.01	1.34	1.26	6.60	6.59	10.71	14.13	3.99	1.44	3.44	54.43
1996	5.53	1.32	12.15	2.22	2.28	11.35	1.90	5.70	3.92	11.15	0.96	2.01	60.49
1997	2.03	0.46	2.30	3.30	3.77	6.38	7.69	7.91	4.78	5.29	3.02	7.76	54.69
1998	4.33	7.25	3.97	0.14	0.16	0.83	5.63	7.56	5.79	1.84	1.66	1.35	40.51
1999	4.88	1.81	1.01	1.48	1.47	8.54	4.03	3.58	7.05	7.84	3.12	1.56	46.37
2000	1.80	0.65	8.48	1.15	0.32	3.08	5.09	3.17	13.55	0.93	1.14	0.80	40.16
2001	0.88	0.38	9.98	0.28	1.77	5.26	9.55	3.57	16.11	3.22	6.92	0.35	58.27
POR= 67 YRS	2.50	2.87	3.56	2.44	2.92	6.01	6.07	6.22	6.96	4.83	2.63	2.32	49.33

AVERAGE TEMPERATURE (F) 2001 DAYTONA BEACH, FL (DAB)

YEAR	JAN	FEB	MAR	APR	MAY	JUN	JUL	AUG	SEP	OCT	NOV	DEC	ANNUAL
1972	65.5	59.0	64.7	70.8	74.4	79.6	80.7	80.2	78.7	75.0	67.3	63.5	71.6
1973	58.9	56.3	68.2	68.7	75.2	79.8	81.9	80.3	80.8	75.0	69.6	58.4	71.1
1974	69.5	59.5	68.5	69.3	76.0	78.6	79.2	80.3	80.5	72.0	65.5	59.5	71.5
1975	63.6	65.7	65.6	70.1	77.2	79.5	80.5	80.5	79.6	75.2	66.8	58.6	71.9
1976	54.4	61.1	68.1	67.7	73.3	77.4	80.9	80.0	78.3	70.4	60.5	59.4	69.3
1977	50.6	55.5	68.9	70.0	74.8	82.3	82.6	82.6	80.8	70.8	67.1	58.3	70.4
1978	53.9	52.1	62.3	71.3	77.3	81.5	82.7	82.3	80.7	74.2	71.1	65.6	71.3
1979	56.7	57.1	64.4	72.3	75.5	78.7	82.1	80.0	80.2	72.9	66.1	59.6	70.5
1980	57.7	55.3	66.3	68.8	74.9	79.2	82.8	82.1	80.3	72.7	65.3	57.0	70.2
1981	48.8	59.2	60.4	70.5	73.5	82.0	82.8	81.5	77.7	73.8	62.7	57.1	69.2
1982	56.6	64.4	66.8	69.4	72.6	79.5	80.0	79.9	77.9	71.5	68.8	64.0	71.0
1983	53.9	57.2	60.2	64.3	72.4	77.0	81.7	81.1	77.8	73.5	62.6	58.1	68.3
1984	55.1	58.0	61.8	66.8	72.4	76.3	79.0	81.4	79.5	75.7	66.5	65.2	69.8
1985	53.7	61.0	66.6	69.4	76.1	81.7	80.6	81.4	78.3	76.8	71.2	56.1	71.1
1986	56.7	62.4	63.1	66.3	73.8	79.9	81.4	81.4	79.6	75.3	72.5	64.7	71.4
1987	55.8	59.7	63.3	65.1	74.3	79.7	81.9	82.3	79.6	70.0	66.5	61.9	70.0
1988	55.1	56.8	62.8	69.1	72.6	79.0	81.2	81.5	80.6	70.7	67.5	59.8	69.7
1989	64.8	61.9	67.8	69.5	75.4	80.3	82.7	81.8	80.5	73.4	65.9	53.3	71.4
1990	62.7	67.5	66.4	69.6	77.3	80.7	81.9	81.9	80.5	76.0	67.4	65.1	73.1
1991	63.4	61.9	65.6	73.7	78.5	80.3	82.5	82.4	80.8	73.8	64.4	63.5	72.6
1992	56.5	61.7	63.3	67.2	72.5	80.5	83.4	80.7	79.8	72.3	69.1	61.4	70.7
1993	64.5	57.5	62.9	65.5	72.7	80.0	82.5	81.9	79.6	73.7	66.9	56.4	70.3
1994	58.1	63.8	65.8	72.3	75.0	80.2	80.5	79.9	78.2	74.9	70.1	62.9	71.8
1995	56.5	58.7	66.0	70.3	79.0	79.0	81.2	81.3	79.5	76.4	62.8	58.1	70.7
1996	56.8	59.2	60.7	66.7	75.6	78.2	81.2	79.5	78.9	72.9	66.2	60.8	69.7
1997	59.0	64.7	69.9	67.8	74.0	78.4	81.6	81.5	79.3	73.2	64.3	59.4	71.1
1998	60.6	60.2	61.2	69.9	76.6	84.5	83.5	82.4	80.7	76.5	70.8	66.0	72.7
1999	62.0	62.1	62.2	72.4	74.5	79.6	82.4	82.9	79.0	74.4	67.3	60.0	71.6
2000	59.0	59.7	67.4	68.2	76.3	79.5	81.4	81.0	80.5	71.6	63.2	56.6	70.4
2001	53.9	64.1	64.5	69.1	74.6	79.9	81.2	81.3	77.6	72.9	68.9	65.4	71.1
POR= 67 YRS	58.3	59.8	64.5	69.3	74.7	79.4	81.0	80.9	79.3	73.4	65.9	60.1	70.6

REFERENCE NOTES:

PAGE 1:
THE TEMPERATURE GRAPH SHOWS NORMAL MAXIMUM AND NORMAL
MINIMUM DAILY TEMPERATURES (SOLID CURVES) AND THE
ACTUAL DAILY HIGH AND LOW TEMPERATURES (VERTICAL BARS).

PAGE 2 AND 3:
H/C INDICATES HEATING AND COOLING DEGREE DAYS.
RH INDICATES RELATIVE HUMIDITY
W/O INDICATES WEATHER AND OBSTRUCTIONS
S INDICATES SUNSHINE.
PR INDICATES PRESSURE.
CLOUDINESS ON PAGE 3 IS THE SUM OF THE CEILOMETER AND
SATELLITE DATA NOT TO EXCEED EIGHT EIGHTHS(OKTAS).

GENERAL:
T INDICATES TRACE PRECIPITATION, AN AMOUNT GREATER
THAN ZERO BUT LESS THAN THE LOWEST REPORTABLE VALUE.
+ INDICATES THE VALUE ALSO OCCURS ON EARLIER DATES.
BLANK ENTRIES DENOTE MISSING OR UNREPORTED DATA.
NORMALS ARE 30-YEAR AVERAGES (1961 - 1990).
ASOS INDICATES AUTOMATED SURFACE OBSERVING SYSTEM.
PM INDICATES THE LAST DAY OF THE PREVIOUS MONTH.
POR (PERIOD OF RECORD) BEGINS WITH THE JANUARY DATA
MONTH AND IS THE NUMBER OF YEARS USED TO COMPUTE
THE MEAN.   INDIVIDUAL MONTHS WITHIN THE POR MAY
BE MISSING.
WHEN THE POR FOR A NORMAL IS LESS THAN 30 YEARS,
THE NORMAL IS PROVISIONAL AND IS BASED ON THE NUMBER
OF YEARS INDICATED.
0.* OR * INDICATES THE VALUE OR MEAN-DAYS-WITH
IS BETWEEN 0.00 AND 0.05.
CLOUDINESS FOR ASOS STATIONS DIFFERS FROM THE NON-ASOS
OBSERVATION TAKEN BY A HUMAN OBSERVER.   ASOS STATION
CLOUDINESS IS BASED ON TIME-AVERAGED CEILOMETER DATA
FOR CLOUDS AT OR BELOW 12,000 FEET AND ON SATELLITE
DATA FOR CLOUDS ABOVE 12,000 FEET.
THE NUMBER OF DAYS WITH CLEAR, PARTLY CLOUDY, AND
CLOUDY CONDITIONS FOR ASOS STATIONS IS THE SUM
OF THE CEILOMETER AND SATELLITE DATA FOR THE
SUNRISE TO SUNSET PERIOD.

GENERAL CONTINUED:
CLEAR INDICATES 0 - 2 OKTAS, PARTLY CLOUDY INDICATES
3 - 6 OKTAS, AND CLOUDY INDICATES 7 OR 8 OKTAS.
WHEN AT LEAST ONE OF THE ELEMENTS (CEILOMETER OR
SATELLITE) IS MISSING, THE DAILY CLOUDINESS IS
NOT COMPUTED.
WIND DIRECTION IS RECORDED IN TENS OF DEGREES (2 DIGITS)
CLOCKWISE FROM TRUE NORTH. "00" INDICATES CALM. "36"
INDICATES TRUE NORTH.
RESULTANT WIND IS THE VECTOR AVERAGE OF THE SPEED AND
DIRECTION.
AVERAGE TEMPERATURE IS THE SUM OF THE MEAN DAILY MAXIMUM
AND MINIMUM TEMPERATURE DIVIDED BY 2.
SNOWFALL DATA COMPRISE ALL FORMS OF FROZEN
PRECIPITATION, INCLUDING HAIL.
A HEATING (COOLING) DEGREE DAY IS THE DIFFERENCE BETWEEN
THE AVERAGE DAILY TEMPERATURE AND 65 F.
DRY BULB IS THE TEMPERATURE OF THE AMBIENT AIR.
DEW POINT IS THE TEMPERATURE TO WHICH THE AIR MUST BE
COOLED TO ACHIEVE 100 PERCENT RELATIVE HUMIDITY.
WET BULB IS THE TEMPERATURE THE AIR WOULD HAVE IF THE
MOISTURE CONTENT WAS INCREASED TO 100 PERCENT RELATIVE
HUMIDITY.

ON JULY 1, 1996, THE NATIONAL WEATHER SERVICE BEGAN USING
THE "METAR" OBSERVATION CODE THAT WAS ALREADY EMPLOYED
BY MOST OTHER NATIONS OF THE WORLD.   THE MOST NOTICEABLE
DIFFERENCE IN THIS ANNUAL PUBLICATION WILL BE THE CHANGE
IN UNITS FROM TENTHS TO EIGHTS(OKTAS) FOR REPORTING THE
AMOUNT OF SKY COVER.

HEATING DEGREE DAYS (base 65 F) 2001 DAYTONA BEACH, FL (DAB)

YEAR	JUL	AUG	SEP	OCT	NOV	DEC	JAN	FEB	MAR	APR	MAY	JUN	TOTAL
1972-73	0	0	0	0	62	139	226	251	38	29	0	0	745
1973-74	0	0	0	10	15	239	0	197	30	24	0	0	515
1974-75	0	0	0	0	69	201	108	69	91	30	0	0	568
1975-76	0	0	0	0	102	222	334	143	34	16	0	0	851
1976-77	0	0	0	11	168	209	444	273	53	20	0	0	1178
1977-78	0	0	0	23	63	241	352	356	132	5	0	0	1172
1978-79	0	0	0	0	4	71	279	244	79	5	0	0	682
1979-80	0	0	0	0	75	183	234	297	84	16	0	0	889
1980-81	0	0	0	11	93	247	497	184	171	0	1	0	1204
1981-82	0	0	0	0	127	284	273	72	63	26	0	0	845
1982-83	0	0	0	24	21	125	345	220	167	74	2	0	978
1983-84	0	0	0	2	126	255	323	215	148	37	3	0	1109
1984-85	0	0	0	0	63	77	372	173	44	21	0	0	750
1985-86	0	0	0	0	24	303	261	119	141	30	0	0	878
1986-87	0	0	0	0	11	84	301	160	99	81	0	0	736
1987-88	0	0	0	10	74	146	316	259	120	23	0	0	948
1988-89	0	0	0	1	39	187	70	154	68	20	1	0	540
1989-90	0	0	0	31	59	369	120	47	37	14	0	0	677
1990-91	0	0	0	9	35	96	114	126	83	9	0	0	472
1991-92	0	0	0	0	110	114	264	139	106	43	15	0	791
1992-93	0	0	0	0	70	136	104	210	107	49	0	0	676
1993-94	0	0	0	12	62	281	230	100	70	10	0	0	765
1994-95	0	0	0	0	10	117	269	200	41	16	0	0	653
1995-96	0	0	0	3	135	239	272	212	197	55	0	0	1113
1996-97	0	0	0	11	62	155	204	80	12	23	1	0	548
1997-98	0	0	0	8	69	215	171	159	174	22	0	0	818
1998-99	0	0	0	0	10	88	141	126	105	17	6	0	493
1999-00	0	0	0	6	30	174	201	164	16	32	0	0	623
2000-01	0	0	0	6	118	293	355	87	88	26	0	0	973
2001-	0	0	0	14	4	108							

COOLING DEGREE DAYS (base 65 F) 2001 DAYTONA BEACH, FL (DAB)

YEAR	JAN	FEB	MAR	APR	MAY	JUN	JUL	AUG	SEP	OCT	NOV	DEC	ANNUAL
1972	107	16	67	193	299	447	494	477	415	317	138	98	3068
1973	45	14	148	148	324	450	527	482	483	326	156	40	3143
1974	147	50	145	159	349	414	447	480	469	223	90	40	3013
1975	71	95	118	191	386	441	488	487	445	325	163	29	3239
1976	12	36	136	104	264	378	497	471	406	186	39	40	2569
1977	4	12	181	177	310	527	553	554	478	212	131	41	3180
1978	14	0	56	198	388	499	553	543	477	295	192	94	3309
1979	26	28	68	231	332	419	538	471	462	252	111	23	2961
1980	12	21	131	135	315	435	559	538	467	258	109	5	2985
1981	0	25	37	172	269	516	559	521	385	282	65	47	2878
1982	19	61	127	166	240	440	472	470	392	234	141	97	2859
1983	6	6	28	57	238	369	521	504	391	270	62	46	2498
1984	22	20	55	96	238	345	442	515	441	338	114	91	2717
1985	29	67	101	160	348	506	490	511	405	373	217	35	3242
1986	13	50	89	79	280	452	516	515	444	324	246	82	3090
1987	20	17	52	92	297	449	530	543	442	171	125	58	2796
1988	17	27	62	155	242	425	509	518	474	185	121	32	2767
1989	71	74	162	162	331	468	553	530	474	299	94	11	3229
1990	55	124	85	161	385	478	531	528	470	355	114	107	3393
1991	71	45	113	278	426	465	548	548	480	278	98	74	3424
1992	8	51	60	118	256	472	577	492	448	231	196	35	2944
1993	95	7	48	71	245	456	551	534	446	289	126	23	2891
1994	23	73	101	237	318	466	487	468	402	315	170	57	3117
1995	12	29	81	182	442	426	507	511	441	365	74	32	3102
1996	24	50	72	114	338	402	505	459	427	262	106	29	2788
1997	26	77	170	113	283	405	522	519	436	269	55	47	2922
1998	42	29	64	176	369	592	581	546	479	363	191	124	3556
1999	54	47	26	249	307	442	548	561	429	304	106	29	3102
2000	21	18	98	134	355	440	514	503	471	216	71	40	2881
2001	19	69	82	156	305	455	511	512	383	263	129	126	3010

SNOWFALL (inches) 2001 DAYTONA BEACH, FL (DAB)

YEAR	JUL	AUG	SEP	OCT	NOV	DEC	JAN	FEB	MAR	APR	MAY	JUN	TOTAL
1972-73	0.0	0.0	0.0	0.0	0.0	0.0	0.0	0.0	0.0	0.0	0.0	0.0	0.0
1973-74	0.0	0.0	0.0	0.0	0.0	0.0	0.0	0.0	0.0	0.0	0.0	0.0	0.0
1974-75	0.0	0.0	0.0	0.0	0.0	0.0	0.0	0.0	0.0	0.0	0.0	0.0	0.0
1975-76	0.0	0.0	0.0	0.0	0.0	0.0	0.0	0.0	0.0	0.0	0.0	0.0	0.0
1976-77	0.0	0.0	0.0	0.0	0.0	0.0	T	0.0	0.0	0.0	0.0	0.0	T
1977-78	0.0	0.0	0.0	0.0	0.0	0.0	0.0	0.0	0.0	0.0	0.0	0.0	0.0
1978-79	0.0	0.0	0.0	0.0	0.0	0.0	0.0	0.0	0.0	0.0	0.0	0.0	0.0
1979-80	0.0	0.0	0.0	0.0	0.0	0.0	0.0	0.0	0.0	0.0	0.0	0.0	0.0
1980-81	0.0	0.0	0.0	0.0	0.0	0.0	0.0	0.0	0.0	0.0	0.0	0.0	0.0
1981-82	0.0	0.0	0.0	0.0	0.0	0.0	0.0	0.0	0.0	0.0	0.0	0.0	0.0
1982-83	0.0	0.0	0.0	0.0	0.0	0.0	0.0	0.0	0.0	0.0	0.0	0.0	0.0
1983-84	0.0	0.0	0.0	0.0	0.0	0.0	0.0	0.0	0.0	0.0	0.0	0.0	0.0
1984-85	0.0	0.0	0.0	0.0	0.0	0.0	0.0	0.0	0.0	0.0	0.0	0.0	0.0
1985-86	0.0	0.0	0.0	0.0	0.0	0.0	0.0	0.0	0.0	0.0	0.0	0.0	0.0
1986-87	0.0	0.0	0.0	0.0	0.0	0.0	0.0	0.0	0.0	0.0	0.0	0.0	0.0
1987-88	0.0	0.0	0.0	0.0	0.0	0.0	0.0	0.0	0.0	0.0	0.0	0.0	0.0
1988-89	0.0	0.0	0.0	0.0	0.0	0.0	0.0	0.0	0.0	0.0	T		T
1989-90	0.0	0.0	0.0	0.0	0.0	T	0.0	0.0	0.0	0.0	0.0	0.0	T
1990-91	0.0	0.0	0.0	0.0	0.0	0.0	0.0	0.0	0.0	0.0	0.0	0.0	0.0
1991-92	0.0	0.0	0.0	0.0	0.0	0.0	0.0	0.0	0.0	0.0	0.0	0.0	0.0
1992-93	0.0	0.0	0.0	0.0	0.0	0.0	0.0	0.0	T	0.0	0.0	0.0	T
1993-94	0.0	0.0	0.0	0.0	0.0	0.0	0.0	0.0	0.0	0.0	0.0	0.0	0.0
1994-95	0.0	T	0.0	0.0	0.0	0.0	0.0	0.0	0.0	0.0	0.0	0.0	T
1995-96	0.0	0.0	0.0	0.0	0.0	0.0	0.0						
1996-97													
1997-98													
1998-99													
1999-00													
2000-01													
2001-													
POR= 52 YRS	0.0	T	0.0	0.0	0.0	T	T	T	T	0.0	0.0	T	T

2001
MIAMI,
FLORIDA (MIA)

Miami is located on the lower east coast of Florida. To the east of the city lies Biscayne Bay, an arm of the ocean, about 15 miles long and 3 miles wide. East of the bay is the island of Miami Beach, a mile or less wide and about 10 miles long, and beyond Miami Beach is the Atlantic Ocean. The surrounding countryside is level and sparsely wooded.

The climate of Miami is essentially subtropical marine, featured by a long and warm summer, with abundant rainfall, followed by a mild, dry winter. The marine influence is evidenced by the low daily range of temperature and the rapid warming of cold air masses which pass to the east of the state. The Miami area is subject to winds from the east or southeast about half the time, and in several specific respects has a climate whose features differ from those farther inland.

One of these features is the annual precipitation for the area. During the early morning hours more rainfall occurs at Miami Beach than at the airport, while during the afternoon the reverse is true. The airport office is about 9 miles inland.

An even more striking difference appears in the annual number of days with temperatures reaching 90 degrees or higher, with inland stations having about four times more than the beach. Minimum temperature contrasts also are particularly marked under proper conditions, with the difference between inland locations and the Miami Beach station frequently reaching to 15 degrees or more, especially in winter.

Freezing temperatures occur occasionally in the suburbs and farming districts southwest, west, and northwest of the city, but rarely near the ocean.

Hurricanes occasionally affect the area. The months of greatest frequency are September and October. Destructive tornadoes are very rare. Funnel clouds are occasionally sighted and a few touch the ground briefly but significant damage is seldom reported. Waterspouts are often visible from the beaches during the summer months, however, significant damage is seldom reported. June, July, and August have the highest frequency of dangerous lightning events.

NORMALS, MEANS, AND EXTREMES
MIAMI, FL (MIA)

LATITUDE:	LONGITUDE:	ELEVATION (FT)		TIME ZONE:	WBAN: 12839
25 49' 26" N	80 17' 59" W	GRND: 26	BARO: 29	EASTERN (UTC + 5)	

	ELEMENT	POR	JAN	FEB	MAR	APR	MAY	JUN	JUL	AUG	SEP	OCT	NOV	DEC	YEAR
TEMPERATURE °F	NORMAL DAILY MAXIMUM	30	75.2	76.5	79.1	82.4	85.3	87.6	89.0	89.0	87.8	84.5	80.4	76.7	82.8
	MEAN DAILY MAXIMUM	54	75.5	77.0	79.7	82.7	85.7	88.1	89.3	89.7	88.2	84.8	80.4	76.9	83.2
	HIGHEST DAILY MAXIMUM	59	88	89	92	96	96	98	98	98	97	95	89	87	98
	YEAR OF OCCURRENCE		1987	1994	1977	1971	1995	1985	1998	1990	1987	1980	1997	1989	JUL 1998
	MEAN OF EXTREME MAXS.	54	83.5	85.3	87.4	89.9	91.2	92.9	93.4	93.7	92.2	89.7	86.0	83.6	89.1
	NORMAL DAILY MINIMUM	30	59.2	60.4	64.2	67.8	72.1	75.1	76.2	76.7	75.9	72.1	66.7	61.5	69.0
	MEAN DAILY MINIMUM	54	59.6	60.7	64.3	67.8	71.9	74.9	76.2	76.5	75.7	72.1	66.4	61.5	69.0
	LOWEST DAILY MINIMUM	59	30	32	32	46	53	60	69	68	68	51	39	30	30
	YEAR OF OCCURRENCE		1985	1980	1971	1971	1945	1984	1985	1950	1943	1950	1989	1989	DEC 1989
	MEAN OF EXTREME MINS.	54	42.1	45.5	48.8	56.5	63.9	69.8	72.0	72.3	71.8	63.0	52.6	44.9	58.6
	NORMAL DRY BULB	30	67.2	68.5	71.7	75.2	78.7	81.4	82.6	82.8	81.9	78.3	73.6	69.1	75.9
	MEAN DRY BULB	54	67.5	68.9	72.0	75.2	78.8	81.5	82.8	83.1	82.0	78.4	73.5	69.3	76.1
	MEAN WET BULB	17	62.7	63.9	65.2	67.4	71.5	75.4	76.5	76.6	76.1	72.7	69.1	64.4	70.1
	MEAN DEW POINT	17	58.6	59.6	60.8	62.7	67.7	72.5	73.5	74.0	73.5	69.6	65.6	60.6	66.6
	NORMAL NO. DAYS WITH:														
	MAXIMUM 90	30	0.0	0.0	0.2	1.5	3.0	8.0	13.8	14.0	9.2	1.9	0.0	0.0	51.6
	MAXIMUM 32	30	0.0	0.0	0.0	0.0	0.0	0.0	0.0	0.0	0.0	0.0	0.0	0.0	0.0
	MINIMUM 32	30	0.1	0.0	*	0.0	0.0	0.0	0.0	0.0	0.0	0.0	0.0	0.1	0.2
	MINIMUM 0	30	0.0	0.0	0.0	0.0	0.0	0.0	0.0	0.0	0.0	0.0	0.0	0.0	0.0
H/C	NORMAL HEATING DEG. DAYS	30	88	51	14	0	0	0	0	0	0	0	6	41	200
	NORMAL COOLING DEG. DAYS	30	156	149	221	306	425	492	546	552	507	412	264	168	4198
RH	NORMAL (PERCENT)	30	73	71	70	67	72	76	75	76	78	75	74	72	73
	HOUR 01 LST	30	81	79	77	76	79	83	82	83	85	82	81	80	81
	HOUR 07 LST	30	84	83	82	80	81	84	84	86	88	86	85	83	84
	HOUR 13 LST	30	59	57	56	53	59	65	63	65	66	62	61	59	60
	HOUR 19 LST	30	69	67	66	64	69	74	72	74	76	72	72	70	70
S	PERCENT POSSIBLE SUNSHINE	20	66	68	74	76	72	68	72	71	70	70	67	63	70
W/O	MEAN NO. DAYS WITH:														
	HEAVY FOG (VISBY 1/4 MI)	53	1.3	0.8	0.6	0.6	0.3	0.0	0.1	0.1	0.1	0.2	0.8	0.8	5.7
	THUNDERSTORMS	52	0.9	1.3	1.8	2.7	6.4	12.4	14.7	15.5	11.5	4.5	1.2	0.7	73.6
CLOUDINESS	MEAN:														
	SUNRISE-SUNSET (OKTAS)	48	4.3	4.2	4.3	4.2	4.6	5.4	5.1	5.1	5.3	4.6	4.3	4.2	4.6
	MIDNIGHT-MIDNIGHT (OKTAS)	32	3.8	3.8	3.8	3.5	4.1	4.9	4.4	4.4	4.7	4.0	3.8	3.6	4.1
	MEAN NO. DAYS WITH:														
	CLEAR	47	9.2	8.6	8.5	8.4	6.3	3.1	2.6	2.5	2.1	6.6	7.5	8.9	74.3
	PARTLY CLOUDY	47	13.1	12.1	14.1	14.9	15.3	14.3	17.4	17.8	15.5	14.3	14.0	12.9	175.7
	CLOUDY	47	8.7	7.6	8.3	6.7	9.3	12.6	11.0	10.7	12.4	10.1	8.5	9.1	115.0
PR	MEAN STATION PRESSURE (IN)	29	30.10	30.07	30.05	30.02	29.99	30.01	30.06	30.02	29.98	29.98	30.04	30.09	30.03
	MEAN SEA-LEVEL PRES. (IN)	17	30.10	30.09	30.06	30.03	30.00	30.01	30.07	30.03	29.98	29.99	30.05	30.10	30.04
WINDS	MEAN SPEED (MPH)	49	9.5	10.0	10.6	10.6	9.4	8.2	8.1	7.9	8.3	9.3	9.7	9.2	9.2
	PREVAIL.DIR (TENS OF DEGS)	33	34	11	13	10	09	11	11	10	10	06	09	34	10
	MAXIMUM 2-MINUTE:														
	SPEED (MPH)	5	30	55	29	28	43	32	36	34	43	59	36	29	59
	DIR. (TENS OF DEGS)		09	19	09	11	10	03	31	10	10	09	18	22	09
	YEAR OF OCCURRENCE		1998	1998	1998	1997	1999	1998	1997	1998	1998	1999	1998	1997	OCT 1999
	MAXIMUM 5-SECOND:														
	SPEED (MPH)	5	37	104	36	37	63	41	47	44	51	70	44	40	104
	DIR. (TENS OF DEGS)		26	19	11	35	33	02	23	11	09	09	31	23	19
	YEAR OF OCCURRENCE		2000	1998	1997	1999	1998	1998	1997	1998	1998	1999	1998	1997	FEB 1998
PRECIPITATION	NORMAL (IN)	30	2.01	2.08	2.39	2.85	6.21	9.33	5.70	7.58	7.63	5.64	2.66	1.83	55.91
	MAXIMUM MONTHLY (IN)	59	6.66	8.07	10.57	17.29	18.54	22.36	13.51	16.88	24.40	21.64	13.84	6.39	24.40
	YEAR OF OCCURRENCE		1969	1983	1986	1979	1968	1968	1947	1943	1960	1991	1992	1958	SEP 1960
	MINIMUM MONTHLY (IN)	59	0.04	0.01	0.02	0.05	0.44	1.81	1.77	1.65	2.63	0.09	0.12	0.01	0.01
	YEAR OF OCCURRENCE		1951	1944	1956	1981	1965	1945	1963	1954	1951	1977	1970	1988	FEB 1944
	MAXIMUM IN 24 HOURS (IN)	59	2.68	5.73	7.07	16.21	11.59	8.20	4.55	6.92	7.58	12.66	8.01	5.26	16.21
	YEAR OF OCCURRENCE		1973	1966	1949	1979	1977	1977	1952	1964	1960	2000	1992	2000	APR 1979
	NORMAL NO. DAYS WITH:														
	PRECIPITATION 0.01	30	7.0	6.2	6.2	5.3	10.4	15.7	15.6	17.9	16.6	13.4	8.9	6.1	129.3
	PRECIPITATION 1.00	30	0.5	0.5	0.7	0.7	1.9	3.2	1.5	2.0	2.3	1.8	0.6	0.5	16.2
SNOWFALL	NORMAL (IN)	30	0.0	0.0	0.0	0.0	0.0	0.0	0.0	0.0	0.0	0.0	0.0	0.0	0.0
	MAXIMUM MONTHLY (IN)	5	0.0	0.0	0.0	0.0	T	0.0	0.0	0.0	0.0	0.0	0.0	0.0	T
	YEAR OF OCCURRENCE						1998								MAY 1998
	MAXIMUM IN 24 HOURS (IN)	59	0.0	0.0	0.0	0.0	T	0.0	0.0	0.0	0.0	0.0	0.0	0.0	T
	YEAR OF OCCURRENCE						1998								MAY 1998
	MAXIMUM SNOW DEPTH (IN)	53	0	0	0	0	0	0	0	0	0	0	0	0	0
	YEAR OF OCCURRENCE														
	NORMAL NO. DAYS WITH:														
	SNOWFALL 1.0	30	0.0	0.0	0.0	0.0	0.0	0.0	0.0	0.0	0.0	0.0	0.0	0.0	0.0

PRECIPITATION (inches) 2001 MIAMI, FL (MIA)

YEAR	JAN	FEB	MAR	APR	MAY	JUN	JUL	AUG	SEP	OCT	NOV	DEC	ANNUAL
1972	1.60	2.71	3.01	2.67	13.71	10.90	7.13	6.49	5.08	2.86	2.77	4.18	63.11
1973	3.41	2.21	1.76	2.24	1.08	8.93	6.14	14.60	6.59	3.36	0.46	2.46	53.24
1974	2.54	0.10	2.27	2.11	2.63	8.12	6.09	9.29	6.38	3.68	4.62	1.17	49.00
1975	1.39	0.90	0.61	0.53	4.94	6.37	4.99	5.19	4.69	6.25	2.80	0.44	39.10
1976	0.95	3.54	0.23	4.17	10.45	9.45	3.83	7.75	4.42	2.69	1.61	55.90
1977	1.44	2.10	0.91	1.97	15.82	12.42	5.23	8.28	7.04	1.25	5.94	2.55	64.95
1978	2.07	3.44	2.92	3.50	5.66	5.29	2.69	3.93	3.42	7.68	3.17	2.06	45.83
1979	1.28	0.57	0.30	17.29	5.29	4.06	5.06	4.81	13.36	3.63	1.62	2.84	60.11
1980	1.89	0.88	3.17	10.20	2.14	3.02	9.40	11.32	5.60	6.05	3.47	0.20	57.34
1981	0.61	4.66	1.32	0.05	4.94	5.49	2.78	12.25	14.79	1.62	2.14	0.14	50.79
1982	0.44	1.22	4.22	9.27	8.80	10.82	3.84	5.79	7.62	7.12	7.09	1.18	67.41
1983	5.36	8.07	2.82	1.79	1.44	8.66	6.20	5.88	7.48	3.52	2.01	4.19	57.42
1984	0.18	0.70	6.12	4.51	10.91	7.24	7.38	5.44	10.45	2.35	4.04	0.70	60.02
1985	0.35	0.06	1.35	3.27	3.19	6.33	11.23	11.88	8.59	5.17	1.37	3.47	56.26
1986	5.04	1.72	10.57	0.71	8.24	9.06	7.81	7.67	4.38	3.96	4.75	2.21	66.12
1987	0.87	2.62	3.82	0.38	4.99	5.48	5.17	3.24	10.17	4.33	4.92	4.28	50.27
1988	1.88	0.61	0.39	1.82	5.28	10.36	10.90	7.89	3.09	1.49	0.76	0.12	44.59
1989	0.67	0.71	0.89	2.14	0.99	10.83	3.53	12.78	5.83	2.65	0.99	0.62	42.63
1990	0.24	1.19	2.28	6.96	7.79	6.84	4.31	11.06	3.52	4.82	1.67	1.03	51.71
1991	1.59	2.04	2.32	5.16	2.50	7.51	7.29	8.84	11.17	21.64	1.18	0.18	71.42
1992	1.80	1.49	2.67	2.43	0.55	13.17	4.21	7.22	6.48	2.02	13.84	1.94	57.82
1993	5.04	2.14	5.98	3.08	4.13	3.64	7.28	5.13	12.59	7.23	6.06	0.49	62.79
1994	3.59	5.66	1.94	2.14	4.72	4.97	3.03	16.64	13.50	9.50	8.92	4.95	79.56
1995	3.13	1.41	4.60	3.73	2.94	20.33	6.36	13.13	10.37	9.91	2.53	0.86	79.30
1996	2.33	0.80	1.40	3.37	8.30	11.67	5.25	5.55	7.21	10.10	0.69	1.04	57.71
1997	1.71	1.57	2.06	5.16	9.80	13.18	7.62	6.28	12.47	2.60	2.89	5.27	70.61
1998	1.04	6.62	5.97	0.66	3.45	6.67	5.41	11.66	14.41	5.70	6.66	1.98	70.23
1999	2.98	0.27	0.25	1.46	4.89	11.08	3.60	13.87	7.01	14.55	1.45	2.68	64.09
2000	0.52	1.24	0.35	3.36	1.80	5.19	5.29	7.42	10.58	18.65	0.50	6.15	61.05
2001	0.60	0.05	4.76	1.79	6.10	8.94	6.92	7.27	17.99	13.16	1.42	3.03	72.03
POR= 62 YRS	2.02	1.96	2.39	3.47	5.88	8.81	6.43	7.71	8.82	7.29	2.93	1.87	59.58

AVERAGE TEMPERATURE (F) 2001   MIAMI, FL (MIA)

YEAR	JAN	FEB	MAR	APR	MAY	JUN	JUL	AUG	SEP	OCT	NOV	DEC	ANNUAL
1972	73.0	68.4	72.1	75.0	77.6	79.9	80.9	81.7	80.4	77.9	73.3	70.8	75.9
1973	70.3	65.3	74.5	75.6	79.6	81.3	81.8	81.3	81.8	77.6	76.2	67.0	76.0
1974	74.3	68.9	75.6	76.2	80.0	82.1	82.6	84.0	84.1	78.1	72.9	69.0	77.3
1975	72.7	73.1	73.4	77.5	79.4	81.5	81.1	82.6	82.0	79.2	72.3	69.0	77.0
1976	64.7	68.8	75.8	75.1	78.5	79.1	83.1	81.9	80.4	76.3	71.5	68.2	75.3
1977	61.1	66.1	74.9	74.8	77.0	81.7	83.7	83.2	83.0	76.5	74.0	69.1	75.4
1978	64.0	63.2	68.9	74.0	79.2	81.9	82.5	82.6	82.0	78.8	75.7	73.0	75.5
1979	65.0	64.9	69.2	77.8	80.6	81.9	83.2	82.1	80.7	77.9	75.4	70.2	75.7
1980	67.5	64.0	73.2	75.4	79.0	81.4	82.6	82.8	82.1	80.1	74.3	67.3	75.8
1981	59.7	69.5	70.1	77.8	79.6	83.7	85.0	83.2	81.2	79.7	71.4	67.8	75.7
1982	67.8	74.4	74.7	77.9	77.2	82.0	84.3	84.0	82.7	77.9	75.0	72.6	77.5
1983	67.2	67.5	67.6	71.9	78.2	81.8	85.0	83.3	81.6	78.3	72.5	69.8	75.4
1984	67.0	68.6	70.4	73.2	77.1	79.8	81.9	82.6	80.1	78.2	71.5	71.1	75.1
1985	62.1	68.4	72.5	74.2	79.1	82.4	81.0	82.4	80.6	80.5	75.6	66.0	75.4
1986	65.2	69.4	68.6	71.7	77.5	81.3	83.1	83.5	83.3	80.3	79.3	73.6	76.4
1987	66.1	70.8	71.9	70.6	78.7	84.2	84.2	85.4	83.6	77.6	75.3	69.8	76.5
1988	67.9	67.7	70.7	76.1	77.9	82.0	83.1	83.6	84.0	79.1	76.9	70.5	76.6
1989	72.7	70.8	73.6	77.1	81.0	82.7	83.3	84.3	84.0	79.0	76.2	65.0	77.5
1990	73.6	74.0	73.7	75.2	80.3	83.0	83.5	83.7	83.1	80.4	74.4	72.9	78.2
1991	72.9	69.7	73.9	78.4	81.5	82.9	83.5	84.6	82.4	78.9	73.1	72.2	77.8
1992	67.4	70.5	71.9	74.0	77.8	81.5	84.9	84.4	83.2	79.5	76.8	71.6	77.0
1993	73.2	68.9	71.5	74.0	79.2	83.3	84.6	84.8	83.0	80.8	75.9	68.9	77.3
1994	69.4	73.3	74.0	78.2	81.0	83.6	83.7	82.8	81.9	80.3	77.2	72.0	78.1
1995	67.3	67.9	73.5	77.5	82.1	81.8	84.5	84.2	83.6	81.5	73.8	68.2	77.2
1996	68.1	66.7	69.7	76.0	81.2	82.5	84.5	83.1	83.2	78.4	74.3	70.2	76.5
1997	68.3	74.3	76.3	75.7	80.6	82.2	84.1	84.3	81.5	78.6	74.1	68.9	77.4
1998	70.1	74.2	69.5	76.0	80.7	85.4	84.8	84.9	83.2	80.8	76.3	73.4	77.9
1999	70.0	69.6	70.5	77.8	78.6	80.9	84.0	83.6	81.9	79.2	74.3	70.0	76.7
2000	68.6	69.5	75.0	75.1	80.2	82.1	83.4	83.4	83.0	78.1	73.5	68.8	76.8
2001	63.2	74.2	73.5	76.0	77.7	82.6	82.6	84.4	81.6	79.1	74.2	73.3	76.9
POR= 62 YRS	67.3	68.4	71.7	75.1	78.3	81.4	82.6	82.9	81.8	78.2	73.2	69.2	75.8

REFERENCE NOTES:

PAGE 1:
THE TEMPERATURE GRAPH SHOWS NORMAL MAXIMUM AND NORMAL
MINIMUM DAILY TEMPERATURES (SOLID CURVES) AND THE
ACTUAL DAILY HIGH AND LOW TEMPERATURES (VERTICAL BARS).

PAGE 2 AND 3:
H/C INDICATES HEATING AND COOLING DEGREE DAYS.
RH INDICATES RELATIVE HUMIDITY
W/O INDICATES WEATHER AND OBSTRUCTIONS.
S INDICATES SUNSHINE.
PR INDICATES PRESSURE.
CLOUDINESS ON PAGE 3 IS THE SUM OF THE CEILOMETER AND
SATELLITE DATA NOT TO EXCEED EIGHT EIGHTHS(OKTAS).

GENERAL:
T INDICATES TRACE PRECIPITATION, AN AMOUNT GREATER
THAN ZERO BUT LESS THAN THE LOWEST REPORTABLE VALUE.
+ INDICATES THE VALUE ALSO OCCURS ON EARLIER DATES.
BLANK ENTRIES DENOTE MISSING OR UNREPORTED DATA.
NORMALS ARE 30-YEAR AVERAGES (1961 - 1990).
ASOS INDICATES AUTOMATED SURFACE OBSERVING SYSTEM.
PM INDICATES THE LAST DAY OF THE PREVIOUS MONTH.
POR (PERIOD OF RECORD) BEGINS WITH THE JANUARY DATA
MONTH AND IS THE NUMBER OF YEARS USED TO COMPUTE
THE MEAN. INDIVIDUAL MONTHS WITHIN THE POR MAY
BE MISSING.
WHEN THE POR FOR A NORMAL IS LESS THAN 30 YEARS,
THE NORMAL IS PROVISIONAL AND IS BASED ON THE NUMBER
OF YEARS INDICATED.
0.* OR * INDICATES THE VALUE OR MEAN-DAYS-WITH
IS BETWEEN 0.00 AND 0.05.
CLOUDINESS FOR ASOS STATIONS DIFFERS FROM THE NON-ASOS
OBSERVATION TAKEN BY A HUMAN OBSERVER. ASOS STATION
CLOUDINESS IS BASED ON TIME-AVERAGED CEILOMETER DATA
FOR CLOUDS AT OR BELOW 12,000 FEET AND ON SATELLITE
DATA FOR CLOUDS ABOVE 12,000 FEET.
THE NUMBER OF DAYS WITH CLEAR, PARTLY CLOUDY, AND
CLOUDY CONDITIONS FOR ASOS STATIONS IS THE SUM
OF THE CEILOMETER AND SATELLITE DATA FOR THE
SUNRISE TO SUNSET PERIOD.

GENERAL CONTINUED:
CLEAR INDICATES 0 - 2 OKTAS, PARTLY CLOUDY INDICATES
3 - 6 OKTAS, AND CLOUDY INDICATES 7 OR 8 OKTAS.
WHEN AT LEAST ONE OF THE ELEMENTS (CEILOMETER OR
SATELLITE) IS MISSING, THE DAILY CLOUDINESS IS
NOT COMPUTED.
WIND DIRECTION IS RECORDED IN TENS OF DEGREES (2 DIGITS)
CLOCKWISE FROM TRUE NORTH. "00" INDICATES CALM. "36"
INDICATES TRUE NORTH.
RESULTANT WIND IS THE VECTOR AVERAGE OF THE SPEED AND
DIRECTION.
AVERAGE TEMPERATURE IS THE SUM OF THE MEAN DAILY MAXIMUM
AND MINIMUM TEMPERATURE DIVIDED BY 2.
SNOWFALL DATA COMPRISE ALL FORMS OF FROZEN
PRECIPITATION, INCLUDING HAIL.
A HEATING (COOLING) DEGREE DAY IS THE DIFFERENCE BETWEEN
THE AVERAGE DAILY TEMPERATURE AND 65 F.
DRY BULB IS THE TEMPERATURE OF THE AMBIENT AIR.
DEW POINT IS THE TEMPERATURE TO WHICH THE AIR MUST BE
COOLED TO ACHIEVE 100 PERCENT RELATIVE HUMIDITY.
WET BULB IS THE TEMPERATURE THE AIR WOULD HAVE IF THE
MOISTURE CONTENT WAS INCREASED TO 100 PERCENT RELATIVE
HUMIDITY.

ON JULY 1, 1996, THE NATIONAL WEATHER SERVICE BEGAN USING
THE "METAR" OBSERVATION CODE THAT WAS ALREADY EMPLOYED
BY MOST OTHER NATIONS OF THE WORLD. THE MOST NOTICEABLE
DIFFERENCE IN THIS ANNUAL PUBLICATION WILL BE THE CHANGE
IN UNITS FROM TENTHS TO EIGHTS(OKTAS) FOR REPORTING THE
AMOUNT OF SKY COVER.

HEATING DEGREE DAYS (base 65 F) 2001 MIAMI, FL (MIA)

YEAR	JUL	AUG	SEP	OCT	NOV	DEC	JAN	FEB	MAR	APR	MAY	JUN	TOTAL
1972-73	0	0	0	0	3	30	41	64	0	0	0	0	138
1973-74	0	0	0	0	1	93	0	37	0	0	0	0	131
1974-75	0	0	0	0	2	32	14	1	10	0	0	0	59
1975-76	0	0	0	0	33	49	93	27	0	0	0	0	202
1976-77	0	0	0	0	9	32	165	62	3	0	0	0	271
1977-78	0	0	0	0	6	58	123	99	34	0	0	0	320
1978-79	0	0	0	0	0	1	84	82	13	0	0	0	180
1979-80	0	0	0	0	6	10	50	95	39	0	0	0	200
1980-81	0	0	0	0	7	59	168	25	12	0	0	0	271
1981-82	0	0	0	0	1	80	65	1	3	0	0	0	150
1982-83	0	0	0	0	0	22	50	25	38	2	0	0	137
1983-84	0	0	0	0	4	69	54	37	17	0	0	0	181
1984-85	0	0	0	0	9	18	135	61	4	1	0	0	228
1985-86	0	0	0	0	2	78	76	22	54	0	0	0	232
1986-87	0	0	0	0	0	0	83	15	6	27	0	0	131
1987-88	0	0	0	0	3	29	49	38	26	0	0	0	145
1988-89	0	0	0	0	0	36	1	49	18	0	0	0	104
1989-90	0	0	0	0	1	0	110	7	4	0	0	0	122
1990-91	0	0	0	0	0	4	2	31	5	0	0	0	42
1991-92	0	0	0	0	7	0	38	7	6	0	0	0	58
1992-93	0	0	0	0	2	10	5	7	21	0	0	0	45
1993-94	0	0	0	0	4	31	26	15	1	0	0	0	77
1994-95	0	0	0	0	0	14	39	51	1	0	0	0	105
1995-96	0	0	0	0	3	77	65	77	41	0	0	0	263
1996-97	0	0	0	0	0	26	58	2	0	0	0	0	86
1997-98	0	0	0	0	2	49	20	29	25	0	0	0	125
1998-99	0	0	0	0	0	8	35	19	5	0	0	0	67
1999-00	0	0	0	0	0	26	37	19	0	0	0	0	82
2000-01	0	0	0	0	4	51	121	0	6	0	0	0	182
2001-	0	0	0	0	0	11							

COOLING DEGREE DAYS (base 65 F) 2001 MIAMI, FL (MIA)

YEAR	JAN	FEB	MAR	APR	MAY	JUN	JUL	AUG	SEP	OCT	NOV	DEC	ANNUAL
1972	262	144	227	307	398	454	498	523	471	408	261	217	4170
1973	212	81	301	324	459	499	531	516	511	394	343	163	4334
1974	294	150	335	342	471	518	551	596	578	414	257	178	4570
1975	261	233	276	382	456	501	508	553	517	448	209	141	4014
1976	92	144	336	309	424	429	569	530	470	361	209	141	4014
1977	50	97	318	299	381	508	587	574	549	364	284	191	4202
1978	97	54	163	273	449	515	547	552	513	437	329	254	4183
1979	90	81	149	391	492	516	572	537	481	407	324	178	4218
1980	138	75	296	321	441	501	501	555	563	519	476	292	135	4312
1981	10	154	177	389	460	568	568	625	570	492	460	198	173	4276
1982	161	270	311	394	385	518	606	596	537	406	304	264	4752
1983	125	101	124	213	417	514	628	576	503	419	236	221	4077
1984	124	144	194	252	380	452	532	554	460	416	213	214	3935
1985	55	164	244	285	445	529	505	546	476	488	329	114	4180
1986	86	150	175	207	395	495	569	582	556	483	432	272	4402
1987	122	186	227	202	430	580	603	639	565	401	314	182	4451
1988	145	123	209	339	408	516	571	584	578	445	364	216	4498
1989	247	219	292	367	502	540	576	603	578	442	346	114	4826
1990	279	262	276	314	479	547	547	587	552	486	287	254	4901
1991	254	167	288	408	515	547	583	614	531	437	255	231	4830
1992	121	173	226	277	404	503	624	609	553	454	366	222	4532
1993	269	123	227	277	449	557	613	622	550	497	338	159	4681
1994	167	252	288	403	503	566	589	557	512	479	374	239	4929
1995	119	138	274	377	537	509	612	602	565	519	272	178	4702
1996	168	134	193	335	512	532	609	566	555	421	284	195	4504
1997	164	270	360	326	489	523	600	607	502	427	282	179	4729
1998	186	152	174	337	493	616	623	621	555	498	348	277	4880
1999	195	152	184	390	429	482	594	583	513	447	285	190	4444
2000	152	154	317	312	480	518	592	579	545	411	267	178	4505
2001	71	266	277	335	400	532	552	605	505	444	283	276	4546

SNOWFALL (inches) 2001 MIAMI, FL (MIA)

YEAR	JUL	AUG	SEP	OCT	NOV	DEC	JAN	FEB	MAR	APR	MAY	JUN	TOTAL
1972-73	0.0	0.0	0.0	0.0	0.0	0.0	0.0	0.0	0.0	0.0	0.0	0.0	0.0
1973-74	0.0	0.0	0.0	0.0	0.0	0.0	0.0	0.0	0.0	0.0	0.0	0.0	0.0
1974-75	0.0	0.0	0.0	0.0	0.0	0.0	0.0	0.0	0.0	0.0	0.0	0.0	0.0
1975-76	0.0	0.0	0.0	0.0	0.0	0.0	0.0	0.0	0.0	0.0	0.0	0.0	0.0
1976-77	0.0	0.0	0.0	0.0	0.0	0.0	0.0	0.0	0.0	0.0	0.0	0.0	0.0
1977-78	0.0	0.0	0.0	0.0	0.0	0.0	0.0	0.0	0.0	0.0	0.0	0.0	0.0
1978-79	0.0	0.0	0.0	0.0	0.0	0.0	0.0	0.0	0.0	0.0	0.0	0.0	0.0
1979-80	0.0	0.0	0.0	0.0	0.0	0.0	0.0	0.0	0.0	0.0	0.0	0.0	0.0
1980-81	0.0	0.0	0.0	0.0	0.0	0.0	0.0	0.0	0.0	0.0	0.0	0.0	0.0
1981-82	0.0	0.0	0.0	0.0	0.0	0.0	0.0	0.0	0.0	0.0	0.0	0.0	0.0
1982-83	0.0	0.0	0.0	0.0	0.0	0.0	0.0	0.0	0.0	0.0	0.0	0.0	0.0
1983-84	0.0	0.0	0.0	0.0	0.0	0.0	0.0	0.0	0.0	0.0	0.0	0.0	0.0
1984-85	0.0	0.0	0.0	0.0	0.0	0.0	0.0	0.0	0.0	0.0	0.0	0.0	0.0
1985-86	0.0	0.0	0.0	0.0	0.0	0.0	0.0	0.0	0.0	0.0	0.0	0.0	0.0
1986-87	0.0	0.0	0.0	0.0	0.0	0.0	0.0	0.0	0.0	0.0	0.0	0.0	0.0
1987-88	0.0	0.0	0.0	0.0	0.0	0.0	0.0	0.0	0.0	0.0	0.0	0.0	0.0
1988-89	0.0	0.0	0.0	0.0	0.0	0.0	0.0	0.0	0.0	0.0	0.0	0.0	0.0
1989-90	0.0	0.0	0.0	0.0	0.0	0.0	0.0	0.0	0.0	0.0	0.0	0.0	0.0
1990-91	0.0	0.0	0.0	0.0	0.0	0.0	0.0	0.0	0.0	0.0	0.0	0.0	0.0
1991-92	0.0	0.0	0.0	0.0	0.0	0.0	0.0	0.0	0.0	0.0	0.0	0.0	0.0
1992-93	0.0	0.0	0.0	0.0	0.0	0.0	0.0	0.0	0.0	0.0	0.0	0.0	0.0
1993-94	0.0	0.0	0.0	0.0	0.0	0.0	0.0	0.0	0.0	0.0	0.0	0.0	0.0
1994-95	0.0	0.0	0.0	0.0	0.0	0.0	0.0	0.0	0.0	0.0	0.0	0.0	0.0
1995-96	0.0	0.0	0.0	0.0	0.0	0.0	0.0	0.0	0.0	0.0	0.0	0.0	0.0
1996-97	0.0	0.0	0.0	0.0	0.0	0.0	0.0	0.0	0.0	0.0	0.0	0.0	0.0
1997-98	0.0	0.0	0.0	0.0	0.0	0.0	0.0	0.0	0.0	0.0	T	0.0	T
1998-99	0.0	0.0	0.0	0.0	0.0	0.0	0.0	0.0	0.0	0.0	T	0.0	T
1999-00	0.0	0.0	0.0	0.0	0.0	0.0	0.0	0.0	0.0	0.0	T	0.0	T
2000-01	0.0	0.0	0.0	0.0	0.0	0.0	0.0	0.0	0.0				
2001-													
POR= 57 YRS	0.0	0.0	0.0	0.0	0.0	0.0	0.0	0.0	0.0	0.0	0.0	0.0	0.0

2001
ORLANDO,
FLORIDA (MCO)

Orlando is located in the central section of the Florida peninsula, surrounded by many lakes. Relative humidities remain high the year-round, with values near 90 percent at night and 40 to 50 percent in the afternoon. On some winter days, the humidity may drop to 20 percent.

The rainy season extends from June through September, sometimes through October when tropical storms are near. During this period, scattered afternoon thunderstorms are an almost daily occurrence, and these bring a drop in temperature to make the climate bearable. Summer temperatures above 95 degrees are rather rare. There is usually a breeze which contributes to the general comfort.

During the winter months rainfall is light. While temperatures, on infrequent occasion, may drop at night to near freezing, they rise rapidly during the day and, in brilliant sunshine, afternoons are pleasant.

Frozen precipitation in the form of snowflakes, snow pellets, or sleet is rare. However, hail is occasionally reported during thunderstorms.

Hurricanes are usually not considered a great threat to Orlando, since, to reach this area, they must pass over a substantial stretch of land and, in so doing, lose much of their punch. Sustained hurricane winds of 75 mph or higher rarely occur. Orlando, being inland, is relatively safe from high water, although heavy rains sometimes briefly flood sections of the city.

NORMALS, MEANS, AND EXTREMES
ORLANDO, FL (MCO)

LATITUDE: 28 26' 02" N LONGITUDE: 81 19' 30" W ELEVATION (FT): GRND: 95 BARO: 98 TIME ZONE: EASTERN (UTC + 5) WBAN: 12815

	ELEMENT	POR	JAN	FEB	MAR	APR	MAY	JUN	JUL	AUG	SEP	OCT	NOV	DEC	YEAR
TEMPERATURE °F	NORMAL DAILY MAXIMUM	30	70.8	72.7	78.0	83.0	87.8	90.5	91.5	91.5	89.7	84.6	78.5	72.9	82.6
	MEAN DAILY MAXIMUM	45	70.3	72.9	77.1	82.3	87.5	90.0	91.1	90.7	88.8	83.5	77.6	72.1	82.0
	HIGHEST DAILY MAXIMUM	59	87	90	92	96	102	100	101	100	98	95	89	90	102
	YEAR OF OCCURRENCE		1991	1962	1994	1968	1945	1998	1998	1980	1988	1986	1992	1978	MAY 1945
	MEAN OF EXTREME MAXS.	45	82.7	84.3	87.4	90.5	93.7	95.3	95.5	95.1	93.5	90.5	85.8	83.0	89.8
	NORMAL DAILY MINIMUM	30	48.6	49.7	55.2	59.4	65.9	71.8	73.1	73.4	72.4	65.8	57.5	51.3	62.0
	MEAN DAILY MINIMUM	45	49.0	51.4	55.5	60.4	66.4	71.5	73.2	73.6	72.4	65.7	57.9	51.5	62.4
	LOWEST DAILY MINIMUM	59	19	26	25	38	48	53	64	64	56	43	29	20	19
	YEAR OF OCCURRENCE		1985	1996	1980	1987	1992	1984	1981	1957	1956	1957	1950	1983	JAN 1985
	MEAN OF EXTREME MINS.	45	32.0	35.7	40.3	48.4	57.7	66.3	69.5	67.0	52.7	42.8	34.7	51.4	
	NORMAL DRY BULB	30	59.7	61.2	66.7	71.2	76.9	81.1	82.3	82.5	81.0	75.2	68.0	62.1	72.3
	MEAN DRY BULB	45	59.6	62.0	66.4	71.5	76.9	80.8	82.1	82.2	80.7	74.7	67.8	61.9	72.2
	MEAN WET BULB	18	55.8	57.5	60.0	63.4	69.1	73.8	75.2	75.5	74.3	68.7	63.3	57.8	66.2
	MEAN DEW POINT	18	51.6	52.9	55.2	58.6	64.9	71.1	72.7	73.2	72.0	65.8	60.3	54.1	62.7
	NORMAL NO. DAYS WITH:														
	MAXIMUM 90	30	0.0	0.0	0.4	2.8	9.5	17.6	23.1	22.8	15.7	3.3	0.0	*	95.2
	MAXIMUM 32	30	0.0	0.0	0.0	0.0	0.0	0.0	0.0	0.0	0.0	0.0	0.0	0.0	0.0
	MINIMUM 32	30	1.7	0.5	0.1	0.0	0.0	0.0	0.0	0.0	0.0	0.0	*	0.7	3.0
	MINIMUM 0	30	0.0	0.0	0.0	0.0	0.0	0.0	0.0	0.0	0.0	0.0	0.0	0.0	0.0
H/C	NORMAL HEATING DEG. DAYS	30	234	164	65	5	0	0	0	0	0	0	54	164	686
	NORMAL COOLING DEG. DAYS	30	70	58	117	191	369	483	536	543	480	316	144	74	3381
RH	NORMAL (PERCENT)	30	73	71	70	67	70	76	78	79	79	75	75	74	74
	HOUR 01 LST	30	84	84	84	83	86	89	89	91	90	87	87	86	87
	HOUR 07 LST	30	87	87	88	87	88	90	91	92	91	88	89	87	89
	HOUR 13 LST	30	56	53	51	46	49	57	59	60	60	55	55	56	55
	HOUR 19 LST	30	68	63	62	58	63	73	75	78	78	74	73	72	70
S	PERCENT POSSIBLE SUNSHINE														
W/O	MEAN NO. DAYS WITH:														
	HEAVY FOG (VISBY 1/4 MI)	53	5.1	3.4	2.5	1.4	1.4	0.8	0.5	0.9	1.1	1.7	2.6	4.5	25.9
	THUNDERSTORMS	57	1.1	1.7	3.0	3.4	7.5	14.7	18.8	17.5	9.6	2.6	1.2	1.2	82.3
CLOUDINESS	MEAN:														
	SUNRISE-SUNSET (OKTAS)	48	4.6	4.6	4.5	4.2	4.3	5.2	5.1	5.1	5.1	4.4	4.2	4.5	4.7
	MIDNIGHT-MIDNIGHT (OKTAS)	22	4.4	4.3	4.2	3.6	4.1	4.9	4.6	4.6	4.5	3.9	4.1	4.3	4.3
	MEAN NO. DAYS WITH:														
	CLEAR	48	8.9	8.6	9.1	10.0	8.7	4.1	3.3	3.2	3.7	9.4	10.1	9.6	88.7
	PARTLY CLOUDY	48	10.2	8.5	10.5	11.4	13.6	14.1	16.7	16.7	14.6	11.1	10.2	9.2	146.8
	CLOUDY	48	11.9	11.1	11.5	8.6	8.7	11.8	11.1	11.1	11.7	10.6	9.7	12.2	130.0
PR	MEAN STATION PRESSURE (IN)	29	30.00	30.00	30.00	29.90	29.90	29.90	30.00	29.90	29.90	29.90	30.00	29.99	29.95
	MEAN SEA-LEVEL PRES. (IN)	17	30.13	30.12	30.06	30.04	30.02	30.02	30.08	30.04	30.00	30.03	30.09	30.14	30.06
WINDS	MEAN SPEED (MPH)	47	8.3	8.7	9.0	8.5	7.7	6.9	6.2	5.9	6.5	7.5	7.7	7.6	7.5
	PREVAIL.DIR (TENS OF DEGS)	36	36	36	18	10	09	18	18	18	06	36	36	36	36
	MAXIMUM 2-MINUTE:														
	SPEED (MPH)	5	31	38	36	37	51	40	34	36	36	41	32	28	51
	DIR. (TENS OF DEGS)		28	18	23	19	35	04	14	31	13	01	02	29	35
	YEAR OF OCCURRENCE		2000	1998	2000	2000	1997	1999	1997	1998	2001	1999	2001	2000	MAY 1997
	MAXIMUM 5-SECOND:														
	SPEED (MPH)	5	45	47	54	49	70	49	52	53	48	48	39	36	70
	DIR. (TENS OF DEGS)		28	20	23	20	34	12	29	19	01	01	01	29	34
	YEAR OF OCCURRENCE		2000	1998	1998	2000	1997	2000	2000	1998	1999	1999	2001	1997	MAY 1997
PRECIPITATION	NORMAL (IN)	30	2.30	3.02	3.21	1.80	3.55	7.32	7.25	6.78	6.01	2.42	2.30	2.15	48.11
	MAXIMUM MONTHLY (IN)	59	7.23	8.74	11.38	9.10	10.36	18.28	19.57	16.11	15.87	14.51	10.29	12.63	19.57
	YEAR OF OCCURRENCE		1986	1998	1987	1992	1976	1968	1960	1972	1945	1950	1987	1997	JUL 1960
	MINIMUM MONTHLY (IN)	59	0.15	0.10	0.16	0.14	0.43	1.58	2.60	2.83	0.43	0.35	0.03	T	T
	YEAR OF OCCURRENCE		1950	1944	1956	1977	1961	1998	1992	2001	1972	1967	1967	1944	DEC 1944
	MAXIMUM IN 24 HOURS (IN)	59	4.19	4.38	5.03	5.65	3.18	8.40	8.19	5.29	9.67	7.74	5.87	3.61	9.67
	YEAR OF OCCURRENCE		1986	1970	1960	1992	1980	1945	1960	1949	1945	1950	1988	1969	SEP 1945
	NORMAL NO. DAYS WITH:														
	PRECIPITATION 0.01	30	7.0	7.4	7.4	4.3	8.8	14.3	16.3	15.7	13.4	7.7	7.0	6.0	115.3
	PRECIPITATION 1.00	30	0.6	0.8	0.8	0.4	1.0	2.3	2.0	1.9	1.7	0.4	1.6	0.6	14.1
SNOWFALL	NORMAL (IN)	30	T	0.0	0.0	0.0	0.0	0.0	0.0	0.0	0.0	0.0	0.0	0.0	0.0
	MAXIMUM MONTHLY (IN)	29	T	0.0	T	T	T	0.0	T	T	0.0	0.0	0.0	0.0	T
	YEAR OF OCCURRENCE		1977		1992	1992	1997		1993	1989					MAY 1997
	MAXIMUM IN 24 HOURS (IN)	29	T	0.0	T	T	T	0.0	T	T	0.0	0.0	0.0	0.0	T
	YEAR OF OCCURRENCE		1977		1992	1992	1997		1993	1989					MAY 1997
	MAXIMUM SNOW DEPTH (IN)	44	0	0	0	0	0	0	0	0	0	0	0	0	0
	YEAR OF OCCURRENCE														
	NORMAL NO. DAYS WITH:														
	SNOWFALL 1.0	30	0.0	0.0	0.0	0.0	0.0	0.0	0.0	0.0	0.0	0.0	0.0	0.0	0.0

PRECIPITATION (inches) 2001 ORLANDO, FL (MCO)

YEAR	JAN	FEB	MAR	APR	MAY	JUN	JUL	AUG	SEP	OCT	NOV	DEC	ANNUAL
1972	0.99	4.96	5.06	1.39	3.76	6.33	3.98	16.11	0.43	2.34	4.11	1.89	51.35
1973	4.82	2.73	4.13	2.82	4.74	6.63	6.24	7.33	11.53	1.10	0.74	2.56	55.37
1974	0.18	0.63	3.67	1.17	2.69	15.28	6.01	6.56	5.78	0.48	0.31	1.62	44.38
1975	0.98	1.49	1.10	1.36	7.52	9.70	9.26	4.75	4.97	4.74	0.66	0.51	47.04
1976	0.37	0.83	1.72	2.16	10.36	9.93	7.05	3.25	5.87	0.74	2.03	2.77	47.08
1977	1.81	1.76	1.82	0.14	1.47	4.47	6.61	6.28	7.03	0.43	2.60	3.70	38.12
1978	2.49	5.45	2.14	0.61	3.16	10.00	11.92	5.13	4.31	1.51	0.18	3.69	50.59
1979	6.48	1.45	3.24	1.08	7.66	4.00	7.95	5.88	9.19	0.43	1.93	0.94	50.23
1980	2.45	1.64	1.51	4.07	6.96	5.25	5.14	2.92	3.70	0.55	6.55	0.47	41.21
1981	0.21	4.36	1.85	0.18	2.02	12.49	3.53	5.60	8.26	3.13	2.50	2.97	47.10
1982	1.72	1.34	4.85	6.27	5.29	6.06	11.81	5.03	6.96	0.74	0.53	1.01	51.61
1983	2.08	8.32	5.37	3.21	1.77	7.82	6.49	4.83	5.16	3.78	1.36	5.33	55.52
1984	2.01	2.73	1.85	6.21	3.20	5.32	6.19	7.89	6.19	0.56	2.10	0.19	44.44
1985	0.91	1.27	4.59	1.69	3.00	4.54	7.28	11.63	5.45	2.55	0.82	3.46	47.19
1986	7.23	1.84	2.63	0.49	0.88	9.50	5.85	5.99	4.50	5.63	1.69	3.60	49.83
1987	1.27	1.74	11.38	0.59	1.40	3.54	7.95	6.07	8.64	3.41	10.29	0.51	56.79
1988	3.12	1.38	6.07	2.02	2.82	4.17	9.44	7.94	5.67	1.42	7.44	1.00	52.49
1989	3.80	0.15	1.35	2.28	2.38	6.79	4.74	6.20	10.29	1.75	1.44	4.49	45.66
1990	0.23	4.13	1.92	1.73	0.55	6.22	6.68	3.78	2.46	2.10	1.05	0.83	31.68
1991	2.37	0.98	6.66	7.72	9.48	5.98	10.78	7.13	4.53	4.76	0.27	0.24	60.90
1992	1.35	2.42	3.67	9.10	1.19	8.68	2.60	8.03	7.13	5.17	2.74	0.88	52.96
1993	4.89	1.48	6.26	1.78	2.32	4.47	6.49	5.95	5.35	4.61	0.17	0.76	44.53
1994	4.00	3.58	1.21	3.03	2.87	10.28	13.27	6.23	7.84	5.18	7.32	3.04	67.85
1995	1.50	1.13	2.12	0.81	4.24	8.23	5.10	9.48	3.59	4.35	1.74	0.76	43.05
1996	5.39	1.52	9.87	0.68	5.12	6.51	4.06	11.33	6.04	3.28	0.72	2.14	56.66
1997	1.13	2.44	3.46	4.02	3.17	8.20	11.51	7.99	2.59	4.22	3.15	12.63	64.51
1998	1.99	8.74	5.26	0.52	3.17	1.58	8.61	5.59	5.36	0.64	1.67	0.62	43.75
1999	2.99	0.36	0.56	2.40	5.43	13.84	5.14	4.50	4.50	6.40	8.40	2.65	54.80
2000	1.23	0.36	0.45	2.22	1.00	6.19	4.07	4.48	6.37	1.33	1.10	1.58	30.38
2001	0.66	0.22	3.72	2.09	5.84	7.65	18.27	2.83	10.47	1.01	1.68	0.48	54.92
POR= 59 YRS	2.25	2.65	3.52	2.55	3.48	7.10	7.83	6.66	6.67	3.38	1.89	2.06	50.04

AVERAGE TEMPERATURE (F) 2001   ORLANDO, FL (MCO)

YEAR	JAN	FEB	MAR	APR	MAY	JUN	JUL	AUG	SEP	OCT	NOV	DEC	ANNUAL
1972	68.9	62.0	68.7	72.7	77.4	82.2	83.2	82.8	81.8	76.8	68.9	66.1	74.3
1973	62.4	59.7	71.1	71.1	78.3	83.1	84.2	81.8	81.4	75.6	70.9	64.4	73.3
1974	71.6	60.5	70.2	71.4	78.0	80.3	80.7	82.0	81.8	72.6	67.6	60.9	73.1
1975	65.8	67.6	67.4	72.4	79.1	80.8	80.5	82.3	80.7	76.6	67.4	60.2	73.4
1976	56.5	63.7	70.4	71.3	76.8	79.7	82.4	81.9	80.5	72.6	63.0	60.1	71.6
1977	55.7	57.4	69.7	70.6	75.2	82.6	82.0	81.5	82.6	69.6	66.1	61.0	71.3
1978	56.8	55.8	66.3	73.4	79.3	82.9	82.6	82.6	81.7	75.0	72.3	66.8	73.0
1979	58.2	58.4	64.6	73.4	75.4	80.7	83.3	82.4	81.7	74.4	68.3	62.6	71.9
1980	60.5	57.2	68.2	70.4	76.4	80.1	83.6	83.6	81.7	75.4	67.1	59.0	71.9
1981	51.3	61.7	64.0	73.1	76.7	83.2	84.1	82.0	80.6	76.4	65.3	60.5	71.7
1982	60.0	68.4	70.4	72.6	75.3	82.0	82.6	82.2	80.2	74.1	70.8	66.7	73.8
1983	58.0	59.9	63.5	68.6	76.4	80.5	83.2	83.5	80.6	76.5	65.8	61.2	71.5
1984	57.8	61.2	64.7	69.2	75.6	78.4	80.7	81.5	78.9	75.4	65.8	66.0	71.3
1985	54.7	62.2	68.4	70.7	77.2	82.4	82.1	82.3	79.8	79.4	73.0	58.8	72.6
1986	59.8	64.3	65.4	69.3	76.7	81.7	82.3	83.3	81.7	77.5	75.8	67.3	73.8
1987	58.8	62.7	65.9	66.8	76.8	83.1	83.5	85.0	82.7	72.2	69.0	64.2	72.6
1988	58.5	60.4	65.5	72.0	75.5	80.3	80.7	82.8	83.9	73.7	70.5	62.4	72.2
1989	66.9	64.5	69.7	71.9	77.9	81.9	83.2	83.3	82.2	75.3	69.0	55.5	73.4
1990	65.8	69.1	69.3	71.5	79.4	81.9	82.8	83.5	82.0	77.1	69.3	64.3	74.8
1991	66.3	64.2	67.7	75.3	79.5	81.1	82.6	83.0	81.7	75.3	65.8	65.5	74.0
1992	59.7	64.8	66.4	69.9	74.9	81.2	84.5	83.2	81.6	73.4	70.4	63.5	72.7
1993	66.3	60.6	64.8	68.3	75.5	81.9	83.9	83.2	81.2	75.1	68.3	58.9	72.3
1994	60.9	66.4	68.2	74.1	77.4	81.3	81.5	81.0	79.2	76.1	71.9	64.9	73.6
1995	58.8	61.1	68.2	72.6	80.2	79.9	82.4	83.0	82.1	77.3	65.6	61.0	72.7
1996	59.1	61.3	62.9	71.6	80.2	81.8	82.9	81.6	80.6	73.8	67.5	62.3	72.1
1997	60.9	67.7	72.3	69.8	76.1	79.9	82.4	82.6	80.5	73.9	66.1	61.4	72.8
1998	62.4	61.7	63.8	71.3	77.7	85.0	84.0	83.5	80.9	77.4	71.2	67.3	73.9
1999	63.9	64.0	64.6	74.2	75.5	80.1	82.8	83.4	80.5	77.4	68.4	61.5	72.8
2000	60.6	62.6	69.9	70.4	77.9	81.3	82.6	82.3	81.4	72.9	65.7	59.8	72.3
2001	56.2	67.2	66.8	71.7	76.2	80.6	81.4	82.3	78.7	73.8	69.4	66.4	72.6
POR= 59 YRS	60.5	62.4	66.9	71.8	77.3	81.3	81.0	82.6	81.0	74.9	67.7	62.3	72.5

REFERENCE NOTES:

PAGE 1:
THE TEMPERATURE GRAPH SHOWS NORMAL MAXIMUM AND NORMAL MINIMUM DAILY TEMPERATURES (SOLID CURVES) AND THE ACTUAL DAILY HIGH AND LOW TEMPERATURES (VERTICAL BARS).

PAGE 2 AND 3:
H/C INDICATES HEATING AND COOLING DEGREE DAYS.
RH INDICATES RELATIVE HUMIDITY
W/O INDICATES WEATHER AND OBSTRUCTIONS
S INDICATES SUNSHINE.
PR INDICATES PRESSURE.
CLOUDINESS ON PAGE 3 IS THE SUM OF THE CEILOMETER AND SATELLITE DATA NOT TO EXCEED EIGHT EIGHTHS (OKTAS).

GENERAL:
T INDICATES TRACE PRECIPITATION, AN AMOUNT GREATER THAN ZERO BUT LESS THAN THE LOWEST REPORTABLE VALUE.
+ INDICATES THE VALUE ALSO OCCURS ON EARLIER DATES.
BLANK ENTRIES DENOTE MISSING OR UNREPORTED DATA.
NORMALS ARE 30-YEAR AVERAGES (1961 - 1990).
ASOS INDICATES AUTOMATED SURFACE OBSERVING SYSTEM.
PM INDICATES THE LAST DAY OF THE PREVIOUS MONTH.
POR (PERIOD OF RECORD) BEGINS WITH THE JANUARY DATA MONTH AND IS THE NUMBER OF YEARS USED TO COMPUTE THE MEAN. INDIVIDUAL MONTHS WITHIN THE POR MAY BE MISSING.
WHEN THE POR FOR A NORMAL IS LESS THAN 30 YEARS, THE NORMAL IS PROVISIONAL AND IS BASED ON THE NUMBER OF YEARS INDICATED.
0.* OR * INDICATES THE VALUE OR MEAN-DAYS-WITH IS BETWEEN 0.00 AND 0.05.
CLOUDINESS FOR ASOS STATIONS DIFFERS FROM THE NON-ASOS OBSERVATION TAKEN BY A HUMAN OBSERVER. ASOS STATION CLOUDINESS IS BASED ON TIME-AVERAGED CEILOMETER DATA FOR CLOUDS AT OR BELOW 12,000 FEET AND ON SATELLITE DATA FOR CLOUDS ABOVE 12,000 FEET.
THE NUMBER OF DAYS WITH CLEAR, PARTLY CLOUDY, AND CLOUDY CONDITIONS FOR ASOS STATIONS IS THE SUM OF THE CEILOMETER AND SATELLITE DATA FOR THE SUNRISE TO SUNSET PERIOD.

GENERAL CONTINUED:
CLEAR INDICATES 0 - 2 OKTAS, PARTLY CLOUDY INDICATES 3 - 6 OKTAS, AND CLOUDY INDICATES 7 OR 8 OKTAS. WHEN AT LEAST ONE OF THE ELEMENTS (CEILOMETER OR SATELLITE) IS MISSING, THE DAILY CLOUDINESS IS NOT COMPUTED.
WIND DIRECTION IS RECORDED IN TENS OF DEGREES (2 DIGITS) CLOCKWISE FROM TRUE NORTH. "00" INDICATES CALM. "36" INDICATES TRUE NORTH.
RESULTANT WIND IS THE VECTOR AVERAGE OF THE SPEED AND DIRECTION.
AVERAGE TEMPERATURE IS THE SUM OF THE MEAN DAILY MAXIMUM AND MINIMUM TEMPERATURE DIVIDED BY 2.
SNOWFALL DATA COMPRISE ALL FORMS OF FROZEN PRECIPITATION, INCLUDING HAIL.
A HEATING (COOLING) DEGREE DAY IS THE DIFFERENCE BETWEEN THE AVERAGE DAILY TEMPERATURE AND 65 F.
DRY BULB IS THE TEMPERATURE OF THE AMBIENT AIR.
DEW POINT IS THE TEMPERATURE TO WHICH THE AIR MUST BE COOLED TO ACHIEVE 100 PERCENT RELATIVE HUMIDITY.
WET BULB IS THE TEMPERATURE THE AIR WOULD HAVE IF THE MOISTURE CONTENT WAS INCREASED TO 100 PERCENT RELATIVE HUMIDITY.

ON JULY 1, 1996, THE NATIONAL WEATHER SERVICE BEGAN USING THE "METAR" OBSERVATION CODE THAT WAS ALREADY EMPLOYED BY MOST OTHER NATIONS OF THE WORLD. THE MOST NOTICEABLE DIFFERENCE IN THIS ANNUAL PUBLICATION WILL BE THE CHANGE IN UNITS FROM TENTHS TO EIGHTS (OKTAS) FOR REPORTING THE AMOUNT OF SKY COVER.

HEATING DEGREE DAYS (base 65 F) 2001 ORLANDO, FL (MCO)

YEAR	JUL	AUG	SEP	OCT	NOV	DEC	JAN	FEB	MAR	APR	MAY	JUN	TOTAL
1972-73	0	0	0	0	54	105	160	169	12	9	0	0	509
1973-74	0	0	0	6	13	193	0	173	15	8	0	0	408
1974-75	0	0	0	0	40	163	73	44	57	10	0	0	387
1975-76	0	0	0	0	85	174	278	104	18	1	0	0	660
1976-77	0	0	0	4	118	197	440	218	41	8	0	0	1026
1977-78	0	0	0	6	38	179	275	255	71	0	0	0	824
1978-79	0	0	0	0	0	56	230	214	71	0	0	0	571
1979-80	0	0	0	0	47	119	161	245	61	4	0	0	637
1980-81	0	0	0	1	67	190	416	119	76	1	0	0	870
1981-82	0	0	0	0	75	205	204	21	33	7	0	0	545
1982-83	0	0	0	14	16	94	233	148	105	13	0	0	623
1983-84	0	0	0	0	63	188	252	137	86	18	0	0	744
1984-85	0	0	0	0	68	71	340	146	22	12	0	0	659
1985-86	0	0	0	0	14	228	180	82	105	4	0	0	613
1986-87	0	0	0	0	0	42	216	97	48	66	0	0	469
1987-88	0	0	0	0	39	97	221	169	71	7	0	0	604
1988-89	0	0	0	0	11	135	32	119	59	4	0	0	360
1989-90	0	0	0	21	27	308	71	34	11	5	0	0	477
1990-91	0	0	0	6	14	69	75	88	52	0	0	0	304
1991-92	0	0	0	0	85	76	187	79	51	19	8	0	505
1992-93	0	0	0	0	47	102	73	131	80	12	0	0	445
1993-94	0	0	0	10	45	201	158	61	42	3	0	0	520
1994-95	0	0	0	0	6	88	205	153	24	5	0	0	481
1995-96	0	0	0	0	76	191	208	173	149	12	0	0	809
1996-97	0	0	0	6	38	122	164	43	0	7	0	0	380
1997-98	0	0	0	6	45	170	130	124	127	8	0	0	610
1998-99	0	0	0	0	7	64	109	92	61	12	5	0	350
1999-00	0	0	0	3	20	140	164	105	2	18	0	0	452
2000-01	0	0	0	4	70	217	300	49	55	11	0	0	706
2001-	0	0	0	11	0	87							

COOLING DEGREE DAYS (base 65 F) 2001 ORLANDO, FL (MCO)

YEAR	JAN	FEB	MAR	APR	MAY	JUN	JUL	AUG	SEP	OCT	NOV	DEC	ANNUAL
1972	181	44	146	243	391	524	570	561	509	374	179	148	3870
1973	88	28	207	198	421	548	602	529	501	341	199	58	3720
1974	213	51	183	207	410	463	492	536	510	241	125	43	3474
1975	105	121	141	237	442	481	489	541	479	366	167	32	3601
1976	18	75	194	196	374	449	549	529	474	247	65	49	3219
1977	1	13	192	182	324	534	537	521	536	257	185	62	3344
1978	26	3	116	259	449	541	550	553	508	321	225	115	3666
1979	26	31	65	260	330	479	575	546	498	299	153	53	3315
1980	27	25	169	172	362	459	586	582	508	331	138	12	3371
1981	0	34	52	253	372	552	602	559	458	359	89	73	3403
1982	56	123	211	241	325	518	550	542	465	303	196	152	3682
1983	22	11	68	129	361	473	573	582	476	362	95	77	3229
1984	37	35	84	151	332	411	490	520	426	331	99	107	3023
1985	27	74	137	191	386	531	539	548	451	454	262	45	3645
1986	25	69	124	139	372	506	543	573	507	392	333	121	3704
1987	32	38	82	127	376	549	582	627	540	230	163	78	3424
1988	26	43	95	223	336	466	496	559	573	275	182	61	3335
1989	101	111	213	216	408	509	573	579	523	346	153	19	3751
1990	102	156	156	206	453	514	559	581	518	388	149	116	3898
1991	121	71	143	315	455	490	553	565	508	326	118	98	3763
1992	28	79	101	175	325	496	612	540	507	265	217	62	3407
1993	120	14	80	118	334	514	591	573	492	331	146	18	3331
1994	39	107	149	284	393	495	517	501	436	352	218	89	3580
1995	19	52	132	241	476	455	546	567	518	391	103	74	3574
1996	33	72	93	217	477	512	561	524	478	283	124	46	3420
1997	46	125	234	155	353	452	546	552	475	290	82	69	3379
1998	55	38	96	206	402	605	599	579	485	390	201	139	3795
1999	82	69	54	297	336	460	560	578	471	321	130	38	3396
2000	35	42	159	188	410	496	553	542	499	257	97	63	3341
2001	32	117	117	217	352	474	515	542	419	289	140	134	3348

SNOWFALL (inches) 2001 ORLANDO, FL (MCO)

YEAR	JUL	AUG	SEP	OCT	NOV	DEC	JAN	FEB	MAR	APR	MAY	JUN	TOTAL
1972-73	0.0	0.0	0.0	0.0	0.0	0.0	0.0	0.0	0.0	0.0	0.0	0.0	0.0
1973-74	0.0	0.0	0.0	0.0	0.0	0.0	0.0	0.0	0.0	0.0	0.0	0.0	0.0
1974-75	0.0	0.0	0.0	0.0	0.0	0.0	0.0	0.0	0.0	0.0	0.0	0.0	0.0
1975-76	0.0	0.0	0.0	0.0	0.0	0.0	0.0	0.0	0.0	0.0	0.0	0.0	0.0
1976-77	0.0	0.0	0.0	0.0	0.0	0.0	T	0.0	0.0	0.0	0.0	0.0	T
1977-78	0.0	0.0	0.0	0.0	0.0	0.0	0.0	0.0	0.0	0.0	0.0	0.0	0.0
1978-79	0.0	0.0	0.0	0.0	0.0	0.0	0.0	0.0	0.0	0.0	0.0	0.0	0.0
1979-80	0.0	0.0	0.0	0.0	0.0	0.0	0.0	0.0	0.0	0.0	0.0	0.0	0.0
1980-81	0.0	0.0	0.0	0.0	0.0	0.0	0.0	0.0	0.0	0.0	0.0	0.0	0.0
1981-82	0.0	0.0	0.0	0.0	0.0	0.0	0.0	0.0	0.0	0.0	0.0	0.0	0.0
1982-83	0.0	0.0	0.0	0.0	0.0	0.0	0.0	0.0	0.0	0.0	0.0	0.0	0.0
1983-84	0.0	0.0	0.0	0.0	0.0	0.0	0.0	0.0	0.0	0.0	0.0	0.0	0.0
1984-85	0.0	0.0	0.0	0.0	0.0	0.0	0.0	0.0	0.0	0.0	0.0	0.0	0.0
1985-86	0.0	0.0	0.0	0.0	0.0	0.0	0.0	0.0	0.0	0.0	0.0	0.0	0.0
1986-87	0.0	0.0	0.0	0.0	0.0	0.0	0.0	0.0	0.0	0.0	0.0	0.0	0.0
1987-88	0.0	0.0	0.0	0.0	0.0	0.0	0.0	0.0	0.0	0.0	0.0	0.0	0.0
1988-89	0.0	0.0	0.0	0.0	0.0	0.0	0.0	0.0	0.0	0.0	0.0	0.0	0.0
1989-90	0.0	T	0.0	0.0	0.0	0.0	0.0	0.0	0.0	0.0	0.0	0.0	T
1990-91	0.0	0.0	0.0	0.0	0.0	0.0	0.0	0.0	0.0	0.0	0.0	0.0	0.0
1991-92	T	0.0	0.0	0.0	0.0	0.0	0.0	0.0	T	T	0.0	0.0	T
1992-93	0.0	0.0	0.0	0.0	0.0	0.0	0.0	0.0	0.0	0.0	0.0	0.0	0.0
1993-94	T	0.0	0.0	0.0	0.0	0.0	0.0	0.0	0.0	0.0	0.0	0.0	T
1994-95	0.0	0.0	0.0	0.0	0.0	0.0	0.0	0.0	0.0	0.0	0.0	0.0	0.0
1995-96	0.0	0.0	0.0	0.0	0.0	0.0	0.0	0.0	0.0	0.0	0.0	0.0	0.0
1996-97	0.0	0.0	0.0	0.0	0.0	0.0	0.0	0.0	0.0	T	T	0.0	T
1997-98	0.0	0.0	0.0	0.0	0.0	0.0	0.0	0.0	0.0	0.0	0.0	0.0	0.0
1998-99	0.0	0.0	0.0	0.0	0.0	0.0	0.0	0.0	0.0	0.0	0.0	0.0	0.0
1999-00	0.0	0.0	0.0	0.0	0.0	0.0	0.0	0.0	0.0	0.0	0.0	0.0	0.0
2000-01	T	0.0	0.0	0.0	0.0	0.0	0.0	0.0	0.0	0.0	0.0	0.0	T
2001-	0.0	0.0	0.0	0.0	0.0	0.0							
POR= 58 YRS	T	T	0.0	0.0	0.0	0.0	T	0.0	T	T	0.0	0.0	T

2001
TAMPA,
FLORIDA (TPA)

Tampa is on west central coast of the Florida Peninsula. Very near the Gulf of Mexico at the upper end of Tampa Bay, land and sea breezes modify the subtropical climate. Major rivers flowing into the area are the Hillsborough, the Alafia, and the Little Manatee.

Winters are mild. Summers are long, rather warm, and humid. Low temperatures are about 50 degrees in the winter and 70 degrees during the summer. Afternoon highs range from the low 70s in the winter to around 90 degrees from June through September. Invasions of cold northern air produce an occasional cool winter morning. Freezing temperatures occur on one or two mornings per year during December, January, and February. In some years no freezing temperatures occur. Temperatures rarely fail to recover to the 60s on the cooler winter days. Temperatures above the low 90s are uncommon because of the afternoon sea breezes and thunderstorms. An outstanding feature of the Tampa climate is the summer thunderstorm season. Most of the thunderstorms occur in the late afternoon hours from June through September. The resulting sudden drop in temperature from about 90 degrees to around 70 degrees makes for a pleasant change. Between a dry spring and a dry fall, some 30 inches of rain, about 60 percent of the annual total, falls during the summer months. Snowfall is very rare. Measurable snows under 1/2 inch have occurred only a few times in the last one hundred years.

A large part of the generally flat sandy land near the coast has an elevation of under 15 feet above sea level. This does make the area vulnerable to tidal surges. Tropical storms threaten the area on a few occasions most years. The greatest risk of hurricanes has been during the months of June and October. Many hurricanes, by replenishing the soil moisture and raising the water table, do far more good than harm. The heaviest rains in a 24-hour period, around 12 inches, have been associated with hurricanes.

Fittingly named the Suncoast, the sun shines more than 65 percent of the possible, with the sunniest months being April and May. Afternoon humidities are usually 60 percent or higher in the summer months, but range from 50 to 60 percent the remainder of the year.

Night ground fogs occur frequently during the cooler winter months. Prevailing winds are easterly, but westerly afternoon and early evening sea breezes occur most months of the year. Winds in excess of 25 mph are not common and usually occur only with thunderstorms or tropical disturbances.

Based on the 1951–1980 period, the average first occurrence of 32 degrees Fahrenheit in the fall is December 26 and the average last occurrence in the spring is February 3.

NORMALS, MEANS, AND EXTREMES
TAMPA, FL (TPA)

LATITUDE:	LONGITUDE:	ELEVATION (FT):		TIME ZONE:	WBAN: 12842
27 57' 41" N	82 32' 25" W	GRND: 8	BARO: 40	EASTERN (UTC + 5)	

	ELEMENT	POR	JAN	FEB	MAR	APR	MAY	JUN	JUL	AUG	SEP	OCT	NOV	DEC	YEAR
TEMPERATURE °F	NORMAL DAILY MAXIMUM	30	69.8	71.4	76.6	81.7	87.2	89.5	90.2	90.2	89.0	84.3	77.7	72.1	81.6
	MEAN DAILY MAXIMUM	54	70.5	72.3	76.4	81.7	87.4	89.9	90.3	90.4	89.0	84.0	77.5	72.2	81.8
	HIGHEST DAILY MAXIMUM	55	86	88	91	93	98	99	97	98	96	94	90	86	99
	YEAR OF OCCURRENCE		1991	1971	1949	1975	1975	1985	1995	1975	1991	1990	1971	1994	JUN 1985
	MEAN OF EXTREME MAXS.	54	80.6	82.3	85.2	88.7	93.0	94.7	94.6	94.4	93.4	90.3	85.2	82.1	88.7
	NORMAL DAILY MINIMUM	30	50.0	51.6	56.5	60.8	67.5	72.9	74.5	74.5	72.8	65.2	57.2	52.3	63.0
	MEAN DAILY MINIMUM	54	50.3	52.2	56.7	61.4	67.7	72.7	74.3	74.4	72.9	65.9	57.6	52.2	63.2
	LOWEST DAILY MINIMUM	55	21	24	29	40	49	53	63	67	57	40	23	18	18
	YEAR OF OCCURRENCE		1985	1958	1980	1987	1992	1984	1970	1973	1981	1964	1970	1962	DEC 1962
	MEAN OF EXTREME MINS.	54	32.5	36.0	41.1	48.2	57.8	66.5	70.4	70.5	67.3	52.5	41.5	34.6	51.6
	NORMAL DRY BULB	30	59.9	61.5	66.4	71.2	77.2	81.0	82.1	82.1	81.0	74.9	67.6	62.2	72.3
	MEAN DRY BULB	54	60.4	62.2	66.7	71.5	77.5	81.3	82.3	82.3	81.0	75.0	67.7	62.1	72.5
	MEAN WET BULB	18	56.4	58.4	60.9	64.7	70.5	74.9	76.1	76.3	74.9	69.1	63.6	55.1	66.7
	MEAN DEW POINT	18	52.6	54.0	56.6	60.0	66.4	72.0	73.8	74.1	72.6	65.8	60.1	51.5	63.3
	NORMAL NO. DAYS WITH:														
	MAXIMUM 90	30	0.0	0.0	0.0	0.6	8.2	15.8	20.5	21.4	15.4	2.9	0.1	0.0	84.9
	MAXIMUM 32	30	0.0	0.0	0.0	0.0	0.0	0.0	0.0	0.0	0.0	0.0	0.0	0.0	0.0
	MINIMUM 32	30	1.8	0.6	0.1	0.0	0.0	0.0	0.0	0.0	0.0	0.0	0.1	1.0	3.6
	MINIMUM 0	30	0.0	0.0	0.0	0.0	0.0	0.0	0.0	0.0	0.0	0.0	0.0	0.0	0.0
H/C	NORMAL HEATING DEG. DAYS	30	234	160	81	7	0	0	0	0	0	0	72	171	725
	NORMAL COOLING DEG. DAYS	30	76	62	130	196	384	489	539	539	477	304	147	84	3427
RH	NORMAL (PERCENT)	30	75	73	72	69	70	74	77	78	78	74	75	75	74
	HOUR 01 LST	30	84	83	82	82	82	84	85	87	86	85	86	84	84
	HOUR 07 LST	30	86	86	87	86	86	87	88	90	91	89	88	87	88
	HOUR 13 LST	30	58	56	54	50	52	60	63	64	62	56	57	58	58
	HOUR 19 LST	30	73	69	67	62	62	69	73	75	75	71	74	74	70
S	PERCENT POSSIBLE SUNSHINE	49	63	65	71	75	75	67	62	61	61	65	64	61	66
W/O	MEAN NO. DAYS WITH:														
	HEAVY FOG(VISBY 1/4 MI)	55	5.1	2.8	2.5	1.1	0.4	0.3	0.1	0.2	0.3	1.0	2.5	3.6	19.9
	THUNDERSTORMS	55	1.1	1.7	2.6	2.6	5.1	13.6	20.3	19.7	11.3	2.8	1.2	1.2	83.2
CLOUDINESS	MEAN:														
	SUNRISE-SUNSET (OKTAS)														
	MIDNIGHT-MIDNIGHT (OKTAS)														
	MEAN NO. DAYS WITH:														
	CLEAR	1	1.0	2.0	9.0		9.0	9.0							
	PARTLY CLOUDY	1	1.0	2.0	3.0		10.0	9.0							
	CLOUDY	1	3.0	1.0	8.0		2.0	5.0							
PR	MEAN STATION PRESSURE(IN)	28	30.12	30.09	30.05	30.02	29.99	30.01	30.05	30.02	29.99	30.01	30.07	30.11	30.04
	MEAN SEA-LEVEL PRES. (IN)	18	30.12	30.11	30.06	30.03	30.01	30.01	30.06	30.02	29.99	30.02	30.08	30.13	30.05
WINDS	MEAN SPEED (MPH)	49	8.6	8.9	9.5	9.2	8.6	7.9	7.2	7.0	7.6	8.3	8.3	8.2	8.3
	PREVAIL.DIR(TENS OF DEGS)	33	06	06	09	09	27	27	27	09	09	06	06	06	09
	MAXIMUM 2-MINUTE:														
	SPEED (MPH)	6	44	36	31	44	36	29	32	29	39	40	35	37	44
	DIR. (TENS OF DEGS)		32	28	32	28	25	22	19	32	35	21	19	29	32
	YEAR OF OCCURRENCE		1999	1998	1996	1997	1999	1999	2001	2000	2001	1996	2000	1997	JAN 1999
	MAXIMUM 5-SECOND:														
	SPEED (MPH)	6	51	44	38	49	47	47	47	43	49	53	41	47	53
	DIR. (TENS OF DEGS)		32	09	21	28	34	13	20	12	34	21	17	30	21
	YEAR OF OCCURRENCE		1999	1998	2001	1997	1997	1996	2001	1996	2001	1996	2000	1997	OCT 1996
PRECIPITATION	NORMAL (IN)	30	1.99	3.08	3.01	1.15	3.10	5.48	6.58	7.61	5.98	2.02	1.77	2.15	43.92
	MAXIMUM MONTHLY (IN)	55	8.02	10.82	12.64	10.71	17.64	13.75	20.59	18.59	13.98	7.36	6.12	15.57	20.59
	YEAR OF OCCURRENCE		1948	1998	1959	1997	1979	1974	1960	1949	1979	1952	1963	1997	JUL 1960
	MINIMUM MONTHLY (IN)	55	T	0.21	0.06	T	0.02	1.46	1.65	2.35	1.28	0.06	T	0.07	T
	YEAR OF OCCURRENCE		1950	1950	1956	1981	2001	1997	1981	1952	1972	2000	1960	1984	APR 1981
	MAXIMUM IN 24 HOURS (IN)	55	3.81	4.41	5.20	5.44	11.84	5.53	12.11	5.37	8.45	2.93	4.48	4.76	12.11
	YEAR OF OCCURRENCE		1996	1998	1960	1997	1979	1974	1960	1949	1997	1985	1988	1997	JUL 1960
	NORMAL NO. DAYS WITH:														
	PRECIPITATION 0.01	30	6.7	6.7	6.2	3.6	6.2	11.7	14.8	16.7	11.9	6.2	5.7	6.3	102.7
	PRECIPITATION 1.00	30	0.4	0.9	1.0	0.2	0.7	1.8	1.9	2.6	2.1	0.6	0.3	0.6	13.1
SNOWFALL	NORMAL (IN)	30	0.*	0.0	T	0.0	0.0	0.0	0.0	0.0	0.0	0.0	0.0	T	0.0
	MAXIMUM MONTHLY (IN)	55	0.2	T	T	T	0.0	0.0	T	0.0	0.0	0.0	0.0	T	0.2
	YEAR OF OCCURRENCE		1977	1951	1980	1997			1998					1989	JAN 1977
	MAXIMUM IN 24 HOURS (IN)	55	0.2	T	T	T	0.0	0.0	T	0.0	0.0	0.0	0.0	T	0.2
	YEAR OF OCCURRENCE		1977	1951	1980	1997			1998					1989	JAN 1977
	MAXIMUM SNOW DEPTH (IN)	53	0	0	0	0	0	0	0	0	0	0	0	0	0
	YEAR OF OCCURRENCE														
	NORMAL NO. DAYS WITH:														
	SNOWFALL 1.0	30	0.0	0.0	0.0	0.0	0.0	0.0	0.0	0.0	0.0	0.0	0.0	0.0	0.0

PRECIPITATION (inches) 2001 TAMPA, FL (TPA)

YEAR	JAN	FEB	MAR	APR	MAY	JUN	JUL	AUG	SEP	OCT	NOV	DEC	ANNUAL
1972	0.54	4.44	3.01	0.38	1.88	5.24	6.65	9.78	1.28	3.29	3.53	2.16	42.18
1973	3.75	2.54	4.21	2.42	0.17	4.19	4.77	9.43	8.91	0.98	2.82	5.52	49.71
1974	0.17	0.89	2.35	0.38	1.11	13.75	3.43	4.67	4.00	0.23	0.12	2.80	33.90
1975	0.91	1.56	1.09	0.91	2.07	8.73	6.65	4.24	11.25	4.94	0.22	0.87	43.44
1976	0.40	0.49	1.64	1.83	8.13	7.22	4.58	7.02	6.04	1.30	1.59	2.05	42.29
1977	2.75	2.41	0.73	0.86	0.73	2.66	5.36	5.98	4.28	0.42	1.89	3.40	31.47
1978	2.82	5.17	2.44	0.94	5.00	2.03	5.85	5.97	3.08	3.42	0.01	3.12	39.85
1979	5.72	2.87	2.43	0.55	17.64	2.07	5.93	12.76	13.98	0.16	0.83	1.52	66.46
1980	1.72	2.01	3.09	4.38	3.94	3.81	5.66	7.62	4.05	1.27	2.68	0.37	40.60
1981	0.44	5.34	1.70	T	1.68	9.37	1.65	7.71	5.87	0.87	0.43	3.58	38.64
1982	1.86	2.09	2.99	1.87	5.90	8.34	10.49	7.20	10.76	2.17	0.85	1.29	55.81
1983	1.25	7.35	7.59	2.76	4.10	7.17	6.37	8.89	6.61	1.74	2.33	4.71	60.87
1984	1.62	3.32	1.31	1.51	3.19	3.24	7.15	5.68	4.21	0.29	0.72	0.07	32.31
1985	2.06	2.07	1.80	0.96	0.22	6.43	6.48	8.65	9.04	4.77	0.99	1.13	44.60
1986	2.37	1.49	4.27	0.95	2.46	5.00	6.24	5.46	3.87	6.21	1.33	1.95	41.60
1987	3.29	1.50	12.01	0.39	2.86	3.39	6.06	8.50	4.76	1.46	4.36	0.50	49.08
1988	2.76	1.44	4.09	1.83	1.27	5.19	3.40	11.09	13.56	0.09	5.97	1.64	52.33
1989	1.54	0.41	1.79	0.71	0.24	7.41	8.86	7.90	6.11	1.89	2.05	4.72	43.63
1990	0.53	4.58	1.71	1.47	1.76	5.16	10.01	3.27	2.42	2.63	0.66	0.19	34.39
1991	2.41	0.41	4.73	1.54	6.88	3.78	9.92	7.35	3.43	0.78	1.26	0.67	43.16
1992	1.47	3.67	0.95	2.17	0.10	7.03	2.80	8.22	2.95	2.20	2.43	0.99	34.98
1993	3.60	2.32	3.93	2.45	1.74	3.18	2.92	5.06	6.60	4.23	0.22	1.28	37.53
1994	3.68	0.43	0.66	3.43	0.07	5.98	11.31	8.37	8.20	3.29	0.24	1.57	47.23
1995	3.51	2.02	2.02	1.48	1.67	9.79	10.12	13.75	2.80	4.71	1.24	1.02	54.13
1996	5.42	3.04	4.65	4.20	1.45	8.96	2.72	7.39	5.44	3.12	0.91	2.11	49.41
1997	0.95	0.66	1.28	10.71	1.70	1.46	6.73	8.20	12.84	4.20	3.41	15.57	67.71
1998	4.64	10.82	5.16	0.41	1.96	2.65	12.95	6.55	8.42	0.47	0.40	0.92	55.35
1999	3.04	0.29	0.72	0.40	1.52	4.65	3.65	8.35	6.05	2.85	1.78	1.02	34.32
2000	1.95	0.30	0.41	0.43	0.02	4.53	8.14	5.44	5.14	0.06	2.04	1.39	29.85
2001	1.03	1.18	6.73	0.02	T	6.81	6.01	2.83	11.76	2.39	0.10	0.89	39.75
POR= 111 YRS	2.29	2.75	3.02	2.05	2.91	6.69	7.53	7.83	6.51	2.57	1.60	2.10	47.85

AVERAGE TEMPERATURE (F) 2001 TAMPA, FL (TPA)

YEAR	JAN	FEB	MAR	APR	MAY	JUN	JUL	AUG	SEP	OCT	NOV	DEC	ANNUAL
1972	67.0	60.7	66.8	71.4	76.6	81.0	81.9	82.1	81.3	76.1	68.2	65.1	73.2
1973	61.9	57.4	70.2	69.3	76.8	81.8	83.2	81.8	81.7	75.8	70.8	60.1	72.6
1974	71.1	61.1	70.9	70.9	78.2	80.1	81.1	82.8	82.8	72.9	67.9	61.9	73.5
1975	65.1	66.6	67.6	74.0	81.5	82.6	83.2	83.7	81.9	77.9	68.3	60.2	74.4
1976	56.6	63.1	70.5	70.6	76.1	79.0	81.7	81.6	79.3	71.0	62.8	59.6	71.0
1977	51.2	57.5	70.9	71.5	76.5	83.7	82.9	83.0	82.3	72.5	67.7	58.7	71.5
1978	55.0	53.2	64.2	72.3	78.7	82.4	83.0	82.8	81.4	75.1	71.7	66.2	72.2
1979	57.8	59.3	65.4	74.2	75.9	80.8	83.9	82.2	81.9	75.2	68.7	63.0	72.4
1980	62.0	56.6	68.1	70.1	77.2	81.6	84.0	83.0	81.3	74.0	66.4	57.5	71.8
1981	50.4	61.4	62.8	72.4	75.4	81.5	82.5	81.7	78.6	74.5	64.4	59.2	70.4
1982	59.8	67.9	68.1	71.4	74.4	81.5	82.1	82.1	80.2	74.3	70.8	67.6	73.4
1983	58.9	60.3	63.3	68.6	76.8	80.9	82.2	82.2	79.4	75.8	65.9	59.9	71.2
1984	58.0	62.6	66.0	71.0	78.0	81.5	82.4	82.4	79.9	75.7	64.9	67.3	72.3
1985	55.9	63.6	69.4	72.5	79.8	83.7	82.4	83.1	80.5	79.2	73.6	59.0	73.6
1986	59.3	65.0	65.4	69.1	77.4	81.8	83.0	82.6	82.3	76.8	76.3	65.5	73.8
1987	59.2	63.2	66.4	66.4	77.8	82.7	83.1	83.7	81.4	71.3	68.9	64.3	72.4
1988	58.6	59.1	65.6	70.6	75.3	81.0	82.7	82.9	82.0	73.5	70.8	63.0	72.1
1989	67.1	64.9	69.8	72.0	78.4	82.4	83.3	83.0	82.4	75.4	68.9	56.2	73.7
1990	66.1	69.2	69.7	72.1	80.5	82.7	82.5	83.0	81.9	77.6	70.2	66.9	75.4
1991	66.7	64.2	68.4	76.8	81.2	81.3	82.3	83.2	81.9	75.3	65.8	64.6	74.3
1992	59.8	63.6	64.8	69.4	74.3	82.1	81.8	83.8	82.0	72.7	70.0	64.3	72.4
1993	67.0	60.2	64.3	67.2	76.1	81.8	83.8	83.7	81.9	75.8	69.1	59.5	72.5
1994	60.6	66.9	68.0	75.4	78.2	82.4	81.7	81.7	80.2	76.5	72.4	65.1	74.1
1995	58.8	61.4	68.7	73.5	81.8	80.2	83.0	83.3	82.0	77.8	65.2	61.0	73.1
1996	59.2	60.0	62.4	70.4	79.3	80.9	83.7	83.2	82.0	75.8	68.1	63.5	72.4
1997	62.6	68.6	73.9	71.6	77.7	81.8	82.7	82.8	81.7	74.7	66.8	61.3	73.9
1998	63.6	62.5	64.6	72.3	79.2	85.6	83.5	83.6	81.7	77.5	72.5	68.0	74.6
1999	63.8	64.2	65.1	74.3	77.8	81.2	81.3	83.6	83.7	80.9	68.9	63.2	73.2
2000	61.3	63.7	71.0	71.6	80.3	82.4	82.1	82.8	82.3	74.3	66.7	60.3	73.2
2001	55.2	68.2	65.7	72.7	77.4	82.2	82.1	83.2	79.3	75.1	71.3	68.8	73.4
POR= 111 YRS	60.8	62.4	66.8	71.4	77.0	80.8	81.1	82.1	80.7	74.7	67.4	62.2	72.3

REFERENCE NOTES:

PAGE 1:
THE TEMPERATURE GRAPH SHOWS NORMAL MAXIMUM AND NORMAL MINIMUM DAILY TEMPERATURES (SOLID CURVES) AND THE ACTUAL DAILY HIGH AND LOW TEMPERATURES (VERTICAL BARS).

PAGE 2 AND 3:
H/C INDICATES HEATING AND COOLING DEGREE DAYS.
RH INDICATES RELATIVE HUMIDITY
W/O INDICATES WEATHER AND OBSTRUCTIONS
S INDICATES SUNSHINE.
PR INDICATES PRESSURE.
CLOUDINESS ON PAGE 3 IS THE SUM OF THE CEILOMETER AND SATELLITE DATA NOT TO EXCEED EIGHT EIGHTHS (OKTAS).

GENERAL:
T INDICATES TRACE PRECIPITATION, AN AMOUNT GREATER THAN ZERO BUT LESS THAN THE LOWEST REPORTABLE VALUE.
+ INDICATES THE VALUE ALSO OCCURS ON EARLIER DATES.
BLANK ENTRIES DENOTE MISSING OR UNREPORTED DATA.
NORMALS ARE 30-YEAR AVERAGES (1961 - 1990).
ASOS INDICATES AUTOMATED SURFACE OBSERVING SYSTEM.
PM INDICATES THE LAST DAY OF THE PREVIOUS MONTH.
POR (PERIOD OF RECORD) BEGINS WITH THE JANUARY DATA MONTH AND IS THE NUMBER OF YEARS USED TO COMPUTE THE MEAN. INDIVIDUAL MONTHS WITHIN THE POR MAY BE MISSING.
WHEN THE POR FOR A NORMAL IS LESS THAN 30 YEARS, THE NORMAL IS PROVISIONAL AND IS BASED ON THE NUMBER OF YEARS INDICATED.
0.* OR * INDICATES THE VALUE OR MEAN-DAYS-WITH IS BETWEEN 0.00 AND 0.05.
CLOUDINESS FOR ASOS STATIONS DIFFERS FROM THE NON-ASOS OBSERVATION TAKEN BY A HUMAN OBSERVER. ASOS STATION CLOUDINESS IS BASED ON TIME-AVERAGED CEILOMETER DATA FOR CLOUDS AT OR BELOW 12,000 FEET AND ON SATELLITE DATA FOR CLOUDS ABOVE 12,000 FEET.
THE NUMBER OF DAYS WITH CLEAR, PARTLY CLOUDY, AND CLOUDY CONDITIONS FOR ASOS STATIONS IS THE SUM OF THE CEILOMETER AND SATELLITE DATA FOR THE SUNRISE TO SUNSET PERIOD.

GENERAL CONTINUED:
CLEAR INDICATES 0 - 2 OKTAS, PARTLY CLOUDY INDICATES 3 - 6 OKTAS, AND CLOUDY INDICATES 7 OR 8 OKTAS. WHEN AT LEAST ONE OF THE ELEMENTS (CEILOMETER OR SATELLITE) IS MISSING, THE DAILY CLOUDINESS IS NOT COMPUTED.
WIND DIRECTION IS RECORDED IN TENS OF DEGREES (2 DIGITS) CLOCKWISE FROM TRUE NORTH. "00" INDICATES CALM. "36" INDICATES TRUE NORTH.
RESULTANT WIND IS THE VECTOR AVERAGE OF THE SPEED AND DIRECTION.
AVERAGE TEMPERATURE IS THE SUM OF THE MEAN DAILY MAXIMUM AND MINIMUM TEMPERATURE DIVIDED BY 2.
SNOWFALL DATA COMPRISE ALL FORMS OF FROZEN PRECIPITATION, INCLUDING HAIL.
A HEATING (COOLING) DEGREE DAY IS THE DIFFERENCE BETWEEN THE AVERAGE DAILY TEMPERATURE AND 65 F.
DRY BULB IS THE TEMPERATURE OF THE AMBIENT AIR.
DEW POINT IS THE TEMPERATURE TO WHICH THE AIR MUST BE COOLED TO ACHIEVE 100 PERCENT RELATIVE HUMIDITY.
WET BULB IS THE TEMPERATURE THE AIR WOULD HAVE IF THE MOISTURE CONTENT WAS INCREASED TO 100 PERCENT RELATIVE HUMIDITY.
ON JULY 1, 1996, THE NATIONAL WEATHER SERVICE BEGAN USING THE "METAR" OBSERVATION CODE THAT WAS ALREADY EMPLOYED BY MOST OTHER NATIONS OF THE WORLD. THE MOST NOTICEABLE DIFFERENCE IN THIS ANNUAL PUBLICATION WILL BE THE CHANGE IN UNITS FROM TENTHS TO EIGHTS (OKTAS) FOR REPORTING THE AMOUNT OF SKY COVER.

HEATING DEGREE DAYS (base 65 F) 2001 TAMPA, FL (TPA)

YEAR	JUL	AUG	SEP	OCT	NOV	DEC	JAN	FEB	MAR	APR	MAY	JUN	TOTAL
1972-73	0	0	0	0	65	130	166	223	18	21	0	0	623
1973-74	0	0	0	6	24	200	0	159	17	12	0	0	418
1974-75	0	0	0	0	39	138	84	64	61	5	0	0	391
1975-76	0	0	0	0	88	183	268	109	18	2	0	0	668
1976-77	0	0	0	11	122	208	422	214	28	6	0	0	1011
1977-78	0	0	0	18	53	222	320	323	99	4	0	0	1039
1978-79	0	0	0	0	2	75	245	190	53	0	0	0	565
1979-80	0	0	0	0	47	112	136	262	64	8	0	0	629
1980-81	0	0	0	1	65	233	447	127	103	2	0	0	978
1981-82	0	0	0	0	83	223	209	24	53	8	0	0	600
1982-83	0	0	0	12	18	95	218	148	103	20	0	0	614
1983-84	0	0	0	0	57	214	252	115	68	5	0	0	711
1984-85	0	0	0	0	87	61	306	119	17	5	0	0	595
1985-86	0	0	0	0	9	238	185	78	105	7	0	0	622
1986-87	0	0	0	0	0	53	202	88	42	64	0	0	449
1987-88	0	0	0	4	46	107	221	195	85	14	0	0	672
1988-89	0	0	0	0	9	127	41	116	45	7	0	0	345
1989-90	0	0	0	17	27	285	70	32	13	5	0	0	449
1990-91	0	0	0	7	11	70	72	84	46	0	0	0	290
1991-92	0	0	0	0	93	94	179	90	69	32	5	0	562
1992-93	0	0	0	0	57	83	58	137	84	24	0	0	443
1993-94	0	0	0	6	44	185	158	62	50	5	0	0	510
1994-95	0	0	0	0	7	81	200	151	23	1	0	0	463
1995-96	0	0	0	0	83	180	198	188	152	16	0	0	817
1996-97	0	0	0	3	36	101	132	39	0	7	0	0	318
1997-98	0	0	0	7	36	163	108	103	113	4	0	0	534
1998-99	0	0	0	0	4	56	118	97	44	6	5	0	330
1999-00	0	0	0	5	20	110	154	99	3	10	0	0	401
2000-01	0	0	0	3	61	212	318	48	60	8	0	0	710
2001-	0	0	0	9	0	65							

COOLING DEGREE DAYS (base 65 F) 2001 TAMPA, FL (TPA)

YEAR	JAN	FEB	MAR	APR	MAY	JUN	JUL	AUG	SEP	OCT	NOV	DEC	ANNUAL
1972	131	20	90	204	365	487	528	536	497	352	168	141	3519
1973	75	15	188	158	374	510	574	529	510	348	205	55	3541
1974	196	55	204	197	413	460	506	562	540	250	130	48	3561
1975	93	115	151	282	521	536	572	585	512	407	191	39	4004
1976	17	63	199	175	348	424	525	517	440	202	63	45	3018
1977	2	9	218	210	364	567	559	565	526	258	139	36	3453
1978	18	0	79	232	431	529	565	557	500	319	208	121	3559
1979	28	36	73	283	344	482	592	543	515	322	164	55	3437
1980	45	22	164	165	386	506	506	564	493	284	115	7	3349
1981	0	32	43	230	331	501	552	525	414	303	71	49	3051
1982	56	114	156	208	299	499	537	537	467	311	197	182	3563
1983	36	24	57	137	369	487	541	546	439	342	91	64	3127
1984	42	52	104	190	410	468	517	546	454	337	92	135	3347
1985	30	88	163	237	464	569	566	475	526	445	275	58	3917
1986	15	85	123	139	391	510	565	551	526	374	348	107	3734
1987	27	43	91	114	405	538	567	583	497	207	169	91	3332
1988	30	32	110	188	326	489	554	562	517	271	191	74	3344
1989	112	120	202	224	425	528	575	564	529	344	151	18	3792
1990	107	154	164	225	487	537	549	592	541	406	176	139	4077
1991	131	68	158	361	509	498	543	572	515	326	126	88	3895
1992	25	55	70	170	301	519	589	544	518	248	212	72	3323
1993	126	12	72	95	352	511	593	587	513	347	174	21	3403
1994	32	118	148	321	415	535	524	524	461	364	233	91	3766
1995	15	59	145	264	526	464	561	574	518	405	95	63	3689
1996	27	50	76	182	450	482	589	570	516	348	135	63	3488
1997	62	145	283	213	397	510	556	559	508	314	97	55	3699
1998	75	41	107	229	446	623	582	583	506	394	236	160	3982
1999	85	84	52	294	409	492	583	589	481	363	144	62	3638
2000	48	68	197	213	482	529	537	558	524	300	121	71	3648
2001	22	141	92	245	393	522	536	571	434	326	196	187	3665

SNOWFALL (inches) 2001 TAMPA, FL (TPA)

YEAR	JUL	AUG	SEP	OCT	NOV	DEC	JAN	FEB	MAR	APR	MAY	JUN	TOTAL
1972-73	0.0	0.0	0.0	0.0	0.0	0.0	0.0	0.0	0.0	0.0	0.0	0.0	0.0
1973-74	0.0	0.0	0.0	0.0	0.0	0.0	0.0	0.0	0.0	0.0	0.0	0.0	0.0
1974-75	0.0	0.0	0.0	0.0	0.0	0.0	0.0	0.0	0.0	0.0	0.0	0.0	0.0
1975-76	0.0	0.0	0.0	0.0	0.0	0.0	0.0	0.0	0.0	0.0	0.0	0.0	0.0
1976-77	0.0	0.0	0.0	0.0	0.0	0.0	0.2	0.0	0.0	0.0	0.0	0.0	0.2
1977-78	0.0	0.0	0.0	0.0	0.0	0.0	0.0	0.0	0.0	0.0	0.0	0.0	0.0
1978-79	0.0	0.0	0.0	0.0	0.0	0.0	0.0	0.0	T	0.0	0.0	0.0	T
1979-80	0.0	0.0	0.0	0.0	0.0	0.0	0.0	0.0	0.0	0.0	0.0	0.0	0.0
1980-81	0.0	0.0	0.0	0.0	0.0	0.0	0.0	0.0	0.0	0.0	0.0	0.0	0.0
1981-82	0.0	0.0	0.0	0.0	0.0	0.0	0.0	0.0	0.0	0.0	0.0	0.0	0.0
1982-83	0.0	0.0	0.0	0.0	0.0	0.0	0.0	0.0	0.0	0.0	0.0	0.0	0.0
1983-84	0.0	0.0	0.0	0.0	0.0	0.0	0.0	0.0	0.0	0.0	0.0	0.0	0.0
1984-85	0.0	0.0	0.0	0.0	0.0	0.0	0.0	0.0	0.0	0.0	0.0	0.0	0.0
1985-86	0.0	0.0	0.0	0.0	0.0	0.0	0.0	0.0	0.0	0.0	0.0	0.0	0.0
1986-87	0.0	0.0	0.0	0.0	0.0	0.0	0.0	0.0	0.0	0.0	0.0	0.0	0.0
1987-88	0.0	0.0	0.0	0.0	0.0	0.0	0.0	0.0	0.0	0.0	0.0	0.0	0.0
1988-89	0.0	0.0	0.0	0.0	0.0	0.0	0.0	0.0	0.0	0.0	0.0	0.0	0.0
1989-90	0.0	0.0	0.0	0.0	0.0	T	0.0	0.0	0.0	0.0	0.0	0.0	T
1990-91	0.0	0.0	0.0	0.0	0.0	0.0	0.0	0.0	0.0	0.0	0.0	0.0	0.0
1991-92	0.0	0.0	0.0	0.0	0.0	0.0	0.0	0.0	0.0	0.0	0.0	0.0	0.0
1992-93	0.0	0.0	0.0	0.0	0.0	0.0	0.0	0.0	0.0	0.0	0.0	0.0	0.0
1993-94	0.0	0.0	0.0	0.0	0.0	0.0	0.0	0.0	0.0	0.0	0.0	0.0	0.0
1994-95	0.0	0.0	0.0	0.0	0.0	0.0	0.0	0.0	0.0	0.0	0.0	0.0	0.0
1995-96	0.0	0.0	0.0	0.0	0.0	0.0	0.0	0.0	0.0	0.0	0.0	0.0	0.0
1996-97	0.0	0.0	0.0	0.0	0.0	0.0	T	0.0	0.0	T	0.0	0.0	T
1997-98	0.0	0.0	0.0	0.0	0.0	0.0	0.0	0.0	0.0	0.0	0.0	0.0	0.0
1998-99	T	0.0	0.0	0.0	0.0	0.0	0.0	0.0	0.0	0.0	0.0	0.0	T
1999-00	0.0	0.0	0.0	0.0	0.0	0.0	0.0	0.0	0.0	0.0	0.0	0.0	0.0
2000-01	0.0	0.0	0.0	0.0	0.0	0.0	0.0	0.0	0.0	0.0	0.0	0.0	0.0
2001-	0.0	0.0	0.0	0.0	0.0								
POR= 54 YRS	0.0	0.0	0.0	0.0	0.0	T	0.0	T	T	0.0	0.0	0.0	T

2001
ATLANTA,
GEORGIA (ATL)

Atlanta is located in the foothills of the southern Appalachians in north-central Georgia. The terrain is rolling to hilly and slopes downward toward the east, west, and south so that drainage of the major river systems is generally into the Gulf of Mexico from the western and southern sections of the city and to the Atlantic from the eastern portions of the city.

The Gulf of Mexico and the Atlantic Ocean are approximately 250 miles south and southeast of the city, respectively. Both the Appalachian chain of mountains and the two nearby maritime bodies exert an important influence on the Atlanta climate. Temperatures are moderated throughout the year while abundant precipitation fosters natural vegetation and growth of crops. Summer temperatures in Atlanta are moderated somewhat by elevation but are still rather warm. However, prolonged periods of hot weather are unusual and 100 degree heat is rarely experienced.

With the mountains to the north tending to retard the southward movement of Polar air masses, Atlanta winters are rather mild. Cold spells are not unusual but they are rather short-lived and seldom disrupt outdoor activities for an extended period of time. Late March is the average date of the last temperature of 32 degrees in the spring and mid-November is the average date of the first temperature of 32 degrees in the fall, which gives an average growing season of about 234 days.

Minimum dry precipitation periods occur mainly during the late summer and early autumn. Maximum thunderstorm activity occurs during July, but severe local thunderstorms occur most frequently in March, April, and May, some spawning highly damaging tornadoes.

The average annual snowfall varies widely from year to year. A fall of 4 inches or more occurs about once every five years. Most snows melt in a short period of time due to the rapid warming which often follows the storm. Ice storms, freezing rain or glaze, occur about two out of every three years, causing hazardous travel and disruption of utilities. Severe ice storms occur about once in ten years, causing major disruption of utilities and significant property damage.

The Bermuda High pressure area has a dominant effect on Atlanta weather, particularly in the summer months. East or northeast winds produce the most unpleasant weather although southerly winds are quite humid during the summer. The generally light wind conditions contribute to the formation of an occasional early morning fog.

NORMALS, MEANS, AND EXTREMES
ATLANTA, GA (ATL)

LATITUDE: 33 38' 25" N LONGITUDE: 84 25' 37" W ELEVATION (FT): GRND: 971 BARO: 974 TIME ZONE: EASTERN (UTC + 5) WBAN: 13874

	ELEMENT	POR	JAN	FEB	MAR	APR	MAY	JUN	JUL	AUG	SEP	OCT	NOV	DEC	YEAR
TEMPERATURE °F	NORMAL DAILY MAXIMUM	30	50.4	55.0	64.3	72.7	79.6	85.8	88.0	87.1	81.8	72.7	63.4	54.0	71.2
	MEAN DAILY MAXIMUM	56	52.0	56.5	63.8	72.7	80.3	86.2	88.7	87.8	82.0	72.9	62.8	54.0	71.6
	HIGHEST DAILY MAXIMUM	53	79	80	89	93	95	101	105	102	98	95	84	79	105
	YEAR OF OCCURRENCE		1949	1996	1995	1986	1996	1952	1980	1995	1954	1954	1961	1991	JUL 1980
	MEAN OF EXTREME MAXS.	56	69.4	72.9	80.0	85.0	89.6	94.3	95.7	94.6	91.7	84.2	77.2	70.8	83.8
	NORMAL DAILY MINIMUM	30	31.5	34.5	42.5	50.2	58.7	66.2	69.5	69.0	63.5	51.9	42.8	35.0	51.3
	MEAN DAILY MINIMUM	56	33.7	36.3	42.6	50.8	59.5	66.6	69.9	69.3	64.0	52.5	42.6	35.7	52.0
	LOWEST DAILY MINIMUM	53	-8	5	10	26	37	46	53	55	36	28	3	0	-8
	YEAR OF OCCURRENCE		1985	1958	1960	1973	1971	1956	1967	1992	1967	1950	1983		JAN 1985
	MEAN OF EXTREME MINS.	56	14.8	19.2	25.9	35.1	46.5	56.6	63.9	62.5	51.0	37.6	26.6	18.7	38.2
	NORMAL DRY BULB	30	41.0	44.8	53.5	61.5	69.2	76.0	78.8	78.1	72.7	62.3	53.1	44.5	61.3
	MEAN DRY BULB	56	42.9	46.4	53.1	61.8	70.0	76.5	79.2	78.6	73.0	62.8	52.8	44.9	61.8
	MEAN WET BULB	17	38.9	42.6	47.4	53.9	62.5	69.0	72.1	71.2	65.9	53.3	48.6	38.8	55.3
	MEAN DEW POINT	17	32.0	35.0	39.3	46.6	57.2	64.6	68.5	67.8	61.8	48.2	42.7	32.9	49.7
	NORMAL NO. DAYS WITH:														
	MAXIMUM 90	30	0.0	0.0	0.0	0.1	1.0	8.0	11.6	9.2	3.0	0.0	0.0	0.0	32.9
	MAXIMUM 32	30	1.7	0.3	*	0.0	0.0	0.0	0.0	0.0	0.0	0.0	*	0.5	2.5
	MINIMUM 32	30	16.7	12.8	5.1	0.5	0.0	0.0	0.0	0.0	0.0	0.2	4.9	13.2	53.4
	MINIMUM 0	30	0.2	0.0	0.0	0.0	0.0	0.0	0.0	0.0	0.0	0.0	0.0	*	0.2
H/C	NORMAL HEATING DEG. DAYS	30	744	566	365	138	27	0	0	0	10	138	367	636	2991
	NORMAL COOLING DEG. DAYS	30	0	0	8	33	157	330	428	406	241	54	10	0	1667
RH	NORMAL (PERCENT)	30	68	63	62	61	67	70	74	75	74	68	68	68	68
	HOUR 01 LST	30	73	69	69	69	77	80	85	85	83	78	75	74	76
	HOUR 07 LST	30	78	76	78	78	82	84	88	90	88	84	81	79	82
	HOUR 13 LST	30	59	54	51	49	53	56	60	60	60	53	55	58	56
	HOUR 19 LST	30	62	56	54	50	57	60	66	66	67	63	63	64	61
S	PERCENT POSSIBLE SUNSHINE	63	49	54	58	66	68	67	62	64	62	66	58	50	60
W/O	MEAN NO. DAYS WITH:														
	HEAVY FOG (VISBY 1/4 MI)	67	4.8	3.3	2.7	1.4	1.3	1.0	1.5	1.6	2.0	2.2	3.1	4.2	29.1
	THUNDERSTORMS	67	1.2	2.0	3.5	4.2	5.9	8.4	9.9	7.5	3.1	1.0	1.0	0.7	48.4
CLOUDINESS	MEAN:														
	SUNRISE-SUNSET (OKTAS)	1			7.2		2.4	3.2							
	MIDNIGHT-MIDNIGHT (OKTAS)	1			6.4			3.2							
	MEAN NO. DAYS WITH:														
	CLEAR	1	2.0	1.0	10.0		12.0	9.0							
	PARTLY CLOUDY	1			1.0		6.0	9.0							
	CLOUDY	1	3.0	2.0	11.0		2.0	3.0							
PR	MEAN STATION PRESSURE (IN)	28	29.02	28.99	28.93	28.93	28.92	28.93	28.97	28.98	28.98	29.02	29.02	29.02	28.98
	MEAN SEA-LEVEL PRES. (IN)	17	30.13	30.10	30.05	30.01	30.00	30.00	30.04	30.02	30.04	30.10	30.13	30.15	30.06
WINDS	MEAN SPEED (MPH)	56	10.4	10.6	10.7	10.1	8.8	8.1	7.7	7.2	8.1	8.5	9.2	9.8	9.1
	PREVAIL.DIR (TENS OF DEGS)	37	32	32	32	30	30	27	29	08	08	08	32	32	32
	MAXIMUM 2-MINUTE:														
	SPEED (MPH)	6	37	48	38	44	37	38	46	39	32	28	33	32	48
	DIR. (TENS OF DEGS)		26	26	06	03	01	32	31	07	35	32	30	29	26
	YEAR OF OCCURRENCE		1997	2001	2001	1996	1996	1998	2001	1996	1999	1999	1996	2000	FEB 2001
	MAXIMUM 5-SECOND:														
	SPEED (MPH)	6	48	70	46	52	45	51	61	59	40	36	43	43	70
	DIR. (TENS OF DEGS)		25	26	28	32	27	28	31	25	33	32	29	26	26
	YEAR OF OCCURRENCE		1996	2001	2000	1996	1997	1998	2001	1997	1999	1999	1996	2000	FEB 2001
PRECIPITATION	NORMAL (IN)	30	4.75	4.81	5.77	4.26	4.29	3.56	5.01	3.66	3.42	3.05	3.86	4.33	50.77
	MAXIMUM MONTHLY (IN)	67	10.82	12.77	11.66	11.86	8.37	9.99	17.71	8.69	11.64	11.04	15.72	9.92	17.71
	YEAR OF OCCURRENCE		1936	1961	1980	1979	1980	1991	1994	1967	1989	1995	1948	1961	JUL 1994
	MINIMUM MONTHLY (IN)	67	0.84	0.77	1.86	0.49	0.32	0.16	0.57	0.50	0.04	T	0.41	0.69	T
	YEAR OF OCCURRENCE		1981	1978	1985	1986	1936	1988	1995	1976	1984	1963	1939	1979	OCT 1963
	MAXIMUM IN 24 HOURS (IN)	67	3.91	5.67	5.74	5.58	5.13	4.22	6.47	5.05	5.87	7.27	4.11	3.85	7.27
	YEAR OF OCCURRENCE		1973	1961	1990	1979	1948	1991	1994	1940	1992	1995	1935	1961	OCT 1995
	NORMAL NO. DAYS WITH:														
	PRECIPITATION 0.01	30	11.7	9.8	10.7	8.4	9.4	9.9	12.0	9.3	7.8	6.3	9.1	10.0	114.4
	PRECIPITATION 1.00	30	1.3	1.5	1.9	1.2	1.0	1.0	1.5	0.9	1.1	1.1	1.2	1.3	15.0
SNOWFALL	NORMAL (IN)	30	1.0	0.6	0.4	T	0.0	0.0	0.0	0.0	0.0	0.0	0.1	0.2	2.3
	MAXIMUM MONTHLY (IN)	64	8.3	4.4	7.9	T	0.0	0.0	T	0.0	0.0	T	1.0	3.0	8.3
	YEAR OF OCCURRENCE		1940	1979	1983	1990			2001			1993	1968	2000	JAN 1940
	MAXIMUM IN 24 HOURS (IN)	64	8.3	4.2	7.9	T	0.0	0.0	T	0.0	0.0	T	1.0	2.8	8.3
	YEAR OF OCCURRENCE		1940	1979	1983	1990			2001			1993	1968	1993	JAN 1940
	MAXIMUM SNOW DEPTH (IN)	53	9	4	4	0	0	0	0	0	0	0	1	2	9
	YEAR OF OCCURRENCE		1948	1979	1993								1975	2000	JAN 1948
	NORMAL NO. DAYS WITH:														
	SNOWFALL 1.0	30	0.3	0.2	0.1	0.0	0.0	0.0	0.0	0.0	0.0	0.0	0.*	0.1	0.7

PRECIPITATION (inches) 2001 ATLANTA, GA (ATL)

YEAR	JAN	FEB	MAR	APR	MAY	JUN	JUL	AUG	SEP	OCT	NOV	DEC	ANNUAL
1972	9.26	3.16	4.49	2.31	4.28	4.04	3.81	2.78	3.04	3.96	7.62	50.61
1973	8.89	3.44	9.53	4.03	7.14	3.35	2.10	1.35	4.16	0.75	2.31	8.11	55.16
1974	5.36	6.37	2.44	3.72	3.83	3.20	4.64	6.26	1.06	1.22	3.89	5.31	47.30
1975	6.19	8.98	8.31	4.28	4.62	5.52	8.52	3.30	2.99	5.31	4.62	3.36	66.00
1976	5.15	1.84	10.95	1.49	6.99	2.36	4.29	0.50	0.72	3.55	4.11	4.01	45.96
1977	3.49	2.14	6.28	1.77	2.04	3.03	4.26	4.23	4.90	5.00	7.18	2.36	46.68
1978	7.03	0.77	2.63	3.49	7.28	2.86	2.56	5.66	0.94	1.42	2.96	3.75	41.35
1979	5.03	5.71	3.19	11.86	2.43	1.46	3.62	7.28	6.08	2.17	5.19	0.69	54.71
1980	5.69	2.69	11.66	1.88	8.37	4.49	0.76	1.59	4.77	1.61	2.14	1.29	46.94
1981	0.84	6.62	3.93	2.06	3.89	2.69	2.74	2.76	5.27	3.01	1.85	6.25	41.91
1982	4.75	6.99	3.79	6.02	2.60	6.09	6.31	1.45	3.00	5.83	4.15	5.23	56.21
1983	3.09	4.99	6.68	4.79	1.42	1.52	1.85	1.06	7.52	1.97	7.46	9.27	51.62
1984	4.66	5.97	5.83	6.62	6.57	0.74	11.21	6.46	0.04	1.54	2.10	3.65	55.39
1985	4.11	4.98	1.86	2.75	4.69	2.04	9.92	4.57	2.63	5.74	4.23	2.28	49.80
1986	0.88	2.46	4.13	0.49	2.95	2.18	3.27	6.08	3.68	5.15	6.20	3.03	40.50
1987	5.63	6.13	5.44	1.16	2.74	6.36	7.35	1.22	3.02	0.70	2.36	4.13	46.24
1988	4.64	3.32	2.57	6.06	1.71	0.16	5.04	4.92	6.35	5.00	4.87	1.21	45.85
1989	2.57	4.30	3.85	5.24	6.42	9.34	7.65	2.13	11.64	1.71	3.97	4.49	63.31
1990	8.47	9.75	8.36	2.76	5.26	1.39	3.49	4.64	3.01	6.12	1.27	3.04	57.56
1991	4.66	3.10	6.98	5.28	7.35	5.82	4.37	2.03	0.39	3.19	2.69	55.85
1992	3.58	3.94	3.81	1.03	1.73	4.14	9.03	5.04	8.55	2.84	6.38	60.11
1993	3.94	4.43	5.73	2.77	4.87	6.01	3.05	2.96	3.91	3.83	4.01	2.54	48.05
1994	5.11	3.76	5.77	3.68	2.16	2.44	17.71	4.16	5.86	4.51	3.27	1.59	60.02
1995	3.36	6.74	2.66	3.00	2.12	3.97	0.57	5.82	2.52	11.04	7.40	3.57	52.77
1996	8.26	3.82	6.42	2.91	2.12	1.70	2.14	4.66	4.32	0.89	3.22	4.14	44.60
1997	5.65	7.93	2.18	4.28	3.36	3.91	4.71	1.32	4.83	5.12	3.34	5.05	51.68
1998	5.83	7.10	6.25	5.12	1.23	3.58	2.93	5.54	4.45	0.26	1.97	1.90	46.16
1999	5.33	1.97	3.32	1.14	4.42	5.83	3.43	1.26	4.19	2.41	3.34	2.21	38.85
2000	4.89	1.26	3.63	2.63	1.86	1.11	2.70	4.03	4.93	0.88	5.02	2.62	35.56
2001	2.77	3.61	9.08	3.29	3.31	6.69	2.54	1.03	2.19	0.79	0.87	2.22	38.39
POR= 53 YRS	4.74	4.68	5.47	3.94	3.62	3.83	4.74	3.86	3.41	2.68	3.33	4.24	48.54

AVERAGE TEMPERATURE (F) 2001   ATLANTA, GA (ATL)

YEAR	JAN	FEB	MAR	APR	MAY	JUN	JUL	AUG	SEP	OCT	NOV	DEC	ANNUAL
1972	46.8	42.6	52.4	61.3	66.5	72.2	76.8	77.7	74.6	61.1	49.6	48.3	60.8
1973	41.4	42.8	57.4	57.6	65.0	75.6	78.9	77.4	75.6	64.7	55.6	44.2	61.4
1974	53.2	45.8	57.8	61.1	71.0	72.5	77.9	76.7	70.2	61.3	52.4	44.3	62.0
1975	47.2	47.1	50.5	59.8	71.0	75.3	76.4	77.9	70.3	63.3	54.0	43.4	61.4
1976	38.5	51.5	56.4	61.7	65.4	73.8	76.4	76.0	69.8	56.2	44.2	39.8	59.1
1977	29.3	42.0	55.3	63.0	69.9	77.1	79.5	77.7	73.5	59.6	54.3	42.1	60.3
1978	33.7	39.3	51.6	61.5	67.6	76.3	78.6	78.3	76.3	62.5	58.5	46.1	60.9
1979	37.3	41.7	56.2	62.7	70.1	75.7	78.8	80.1	72.7	62.4	54.3	46.7	61.6
1980	44.9	41.9	52.1	62.6	72.0	79.1	85.1	83.8	78.9	61.5	51.9	44.9	63.2
1981	39.5	46.8	51.8	67.7	67.6	81.3	82.2	77.7	72.4	60.2	54.5	39.1	61.7
1982	38.5	47.4	56.5	58.4	72.5	76.3	79.1	77.5	70.5	62.7	53.7	49.9	61.9
1983	40.4	44.4	51.3	56.4	67.8	74.0	81.4	81.4	71.3	62.1	51.5	39.7	60.1
1984	39.6	47.5	51.7	58.1	67.5	78.3	76.8	77.5	71.3	69.8	50.6	53.7	61.9
1985	36.3	44.2	56.8	64.0	69.9	77.5	78.4	77.2	66.9	62.0	41.4	62.3
1986	43.4	49.8	54.4	62.9	71.0	80.0	84.1	77.4	74.6	64.0	57.9	45.1	63.7
1987	41.9	45.7	53.2	60.3	73.2	77.8	81.0	82.0	74.1	59.7	55.7	48.8	62.8
1988	39.2	45.5	54.9	63.0	70.0	78.6	80.5	81.0	73.4	59.3	55.0	46.6	62.3
1989	49.7	47.5	56.8	68.8	76.9	79.8	79.4	72.9	64.2	54.3	39.1	62.7
1990	49.8	54.4	57.7	61.9	70.4	78.6	80.0	80.6	75.7	64.4	56.5	49.1	65.0
1991	44.3	49.2	56.2	65.9	72.8	76.7	81.0	79.2	74.8	64.6	51.0	63.7
1992	45.1	51.8	54.0	61.9	68.1	74.5	80.2	76.1	73.2	61.9	51.5	44.5	61.9
1993	47.0	45.2	51.7	59.4	71.1	79.3	85.4	82.1	76.7	63.3	53.9	45.3	63.4
1994	40.5	50.1	57.4	67.6	69.4	80.5	79.1	79.1	73.9	64.1	58.3	50.4	64.2
1995	46.3	46.4	58.9	65.8	74.4	77.0	84.3	80.7	71.0	62.0	47.7	42.8	63.1
1996	41.1	47.1	50.6	61.2	74.9	79.1	81.8	79.5	73.2	63.4	52.0	48.8	62.1
1997	46.7	51.1	60.6	57.7	64.8	71.5	78.9	76.6	73.3	61.8	47.8	42.9	61.1
1998	46.0	47.0	50.3	60.3	72.7	78.9	80.7	77.8	75.6	65.9	56.7	50.0	63.4
1999	48.0	49.8	50.5	65.1	68.9	74.9	79.2	81.9	73.1	62.5	56.7	47.1	63.4
2000	43.1	50.9	57.3	58.6	72.9	77.9	81.4	79.7	70.8	64.2	56.7	47.1	62.1
2001	41.7	50.8	50.4	63.5	70.0	74.4	78.6	78.8	71.3	60.7	59.8	50.1	62.5
POR= 53 YRS	43.3	46.2	53.1	61.4	69.7	76.5	78.9	78.1	73.3	62.9	52.5	44.9	61.7

REFERENCE NOTES:

PAGE 1:
THE TEMPERATURE GRAPH SHOWS NORMAL MAXIMUM AND NORMAL
MINIMUM DAILY TEMPERATURES (SOLID CURVES) AND THE
ACTUAL DAILY HIGH AND LOW TEMPERATURES (VERTICAL BARS).

PAGE 2 AND 3:
H/C INDICATES HEATING AND COOLING DEGREE DAYS.
RH INDICATES RELATIVE HUMIDITY
W/O INDICATES WEATHER AND OBSTRUCTIONS
S INDICATES SUNSHINE.
PR INDICATES PRESSURE.
CLOUDINESS ON PAGE 3 IS THE SUM OF THE CEILOMETER AND
SATELLITE DATA NOT TO EXCEED EIGHT EIGHTHS(OKTAS).

GENERAL:
T INDICATES TRACE PRECIPITATION, AN AMOUNT GREATER
THAN ZERO BUT LESS THAN THE LOWEST REPORTABLE VALUE.
+ INDICATES THE VALUE ALSO OCCURS ON EARLIER DATES.
BLANK ENTRIES DENOTE MISSING OR UNREPORTED DATA.
NORMALS ARE 30-YEAR AVERAGES (1961 - 1990).
ASOS INDICATES AUTOMATED SURFACE OBSERVING SYSTEM.
PM INDICATES THE LAST DAY OF THE PREVIOUS MONTH.
POR (PERIOD OF RECORD) BEGINS WITH THE JANUARY DATA
MONTH AND IS THE NUMBER OF YEARS USED TO COMPUTE
THE MEAN.   INDIVIDUAL MONTHS WITHIN THE POR MAY
BE MISSING.
WHEN THE POR FOR A NORMAL IS LESS THAN 30 YEARS,
THE NORMAL IS PROVISIONAL AND IS BASED ON THE NUMBER
OF YEARS INDICATED.
0.* OR * INDICATES THE VALUE OR MEAN-DAYS-WITH
IS BETWEEN 0.00 AND 0.05.
CLOUDINESS FOR ASOS STATIONS DIFFERS FROM THE NON-ASOS
OBSERVATION TAKEN BY A HUMAN OBSERVER. ASOS STATION
CLOUDINESS IS BASED ON TIME-AVERAGED CEILOMETER DATA
FOR CLOUDS AT OR BELOW 12,000 FEET AND ON SATELLITE
DATA FOR CLOUDS ABOVE 12,000 FEET.
THE NUMBER OF DAYS WITH CLEAR, PARTLY CLOUDY, AND
CLOUDY CONDITIONS FOR ASOS STATIONS IS THE SUM
OF THE CEILOMETER AND SATELLITE DATA FOR THE
SUNRISE TO SUNSET PERIOD.

GENERAL CONTINUED:
CLEAR INDICATES 0 - 2 OKTAS, PARTLY CLOUDY INDICATES
3 - 6 OKTAS, AND CLOUDY INDICATES 7 OR 8 OKTAS.
WHEN AT LEAST ONE OF THE ELEMENTS (CEILOMETER OR
SATELLITE) IS MISSING, THE DAILY CLOUDINESS IS
NOT COMPUTED.
WIND DIRECTION IS RECORDED IN TENS OF DEGREES (2 DIGITS)
CLOCKWISE FROM TRUE NORTH. "00" INDICATES CALM. "36"
INDICATES TRUE NORTH.
RESULTANT WIND IS THE VECTOR AVERAGE OF THE SPEED AND
DIRECTION.
AVERAGE TEMPERATURE IS THE SUM OF THE MEAN DAILY MAXIMUM
AND MINIMUM TEMPERATURE DIVIDED BY 2.
SNOWFALL DATA COMPRISE ALL FORMS OF FROZEN
PRECIPITATION, INCLUDING HAIL.
A HEATING (COOLING) DEGREE DAY IS THE DIFFERENCE BETWEEN
THE AVERAGE DAILY TEMPERATURE AND 65 F.
DRY BULB IS THE TEMPERATURE OF THE AMBIENT AIR.
DEW POINT IS THE TEMPERATURE TO WHICH THE AIR MUST BE
COOLED TO ACHIEVE 100 PERCENT RELATIVE HUMIDITY.
WET BULB IS THE TEMPERATURE THE AIR WOULD HAVE IF THE
MOISTURE CONTENT WAS INCREASED TO 100 PERCENT RELATIVE
HUMIDITY.

ON JULY 1, 1996, THE NATIONAL WEATHER SERVICE BEGAN USING
THE "METAR" OBSERVATION CODE THAT WAS ALREADY EMPLOYED
BY MOST OTHER NATIONS OF THE WORLD.   THE MOST NOTICEABLE
DIFFERENCE IN THIS ANNUAL PUBLICATION WILL BE THE CHANGE
IN UNITS FROM TENTHS TO EIGHTS(OKTAS) FOR REPORTING THE
AMOUNT OF SKY COVER.

HEATING DEGREE DAYS (base 65 F) 2001 ATLANTA, GA (ATL)

YEAR	JUL	AUG	SEP	OCT	NOV	DEC	JAN	FEB	MAR	APR	MAY	JUN	TOTAL
1972-73	0	0	3	136	465	511	725	617	240	230	72	0	2999
1973-74	0	0	1	86	295	639	357	531	241	155	5	0	2310
1974-75	0	0	26	148	381	633	547	493	451	192	2	0	2873
1975-76	0	0	28	113	342	665	814	384	265	124	48	4	2787
1976-77	0	0	10	277	618	775	1099	640	300	102	11	0	3832
1977-78	0	0	4	178	313	701	966	714	412	137	57	0	3482
1978-79	0	0	0	112	194	580	853	646	279	97	16	0	2777
1979-80	0	0	5	122	320	559	616	668	399	113	3	0	2805
1980-81	0	0	18	154	391	618	786	502	410	36	43	0	2958
1981-82	0	0	17	179	314	795	819	486	282	204	2	0	3098
1982-83	0	0	16	139	341	466	755	571	423	261	24	0	2996
1983-84	0	0	32	123	400	770	780	503	409	221	50	0	3288
1984-85	0	0	13	22	426	346	882	576	265	111	14	1	2656
1985-86	0	0	15	71	131	725	663	422	331	133	14	0	2505
1986-87	0	11	2	107	243	609	709	534	359	191	6	0	2771
1987-88	0	0	0	172	279	494	791	559	310	104	6	0	2715
1988-89	0	0	0	188	291	566	468	490	284	160	44	0	2491
1989-90	0	0	29	103	318	797	462	297	250	150	20	0	2426
1990-91	0	0	12	109	252	488	636	437	281	54	8	0	2277
1991-92	0	0	8	76	419	499	611	377	345	161	50	2	2548
1992-93	0	0	12	110	398	627	548	549	417	184	13	0	2858
1993-94	0	0	6	129	346	604	753	412	245	55	20	0	2570
1994-95	0	0	0	79	207	446	573	515	207	59	11	0	2097
1995-96	0	0	25	135	514	680	735	517	437	159	8	0	3210
1996-97	0	0	4	104	389	497	561	385	156	226	63	18	2403
1997-98	0	0	5	159	508	679	580	499	465	175	12	3	3085
1998-99	0	0	0	53	242	463	519	421	443	92	8	0	2241
1999-00	0	0	7	110	246	548	669	401	232	186	0	0	2399
2000-01	0	0	16	86	428	856	715	391	446	108	5	0	3051
2001-		0	26	172	163	452							

COOLING DEGREE DAYS (base 65 F) 2001 ATLANTA, GA (ATL)

YEAR	JAN	FEB	MAR	APR	MAY	JUN	JUL	AUG	SEP	OCT	NOV	DEC	ANNUAL
1972	0	0	3	56	72	227	370	396	297	25	7	0	1453
1973	0	0	11	15	78	322	438	388	323	79	20	0	1674
1974	0	1	24	42	198	229	405	368	187	41	11	0	1506
1975	0	0	6	44	195	313	359	406	193	63	21	0	1600
1976	0	1	8	30	67	273	359	346	159	11	0	0	1254
1977	0	0	3	51	171	367	456	403	266	17	1	0	1735
1978	0	0	2	40	144	346	428	420	345	40	7	1	1773
1979	0	0	13	33	181	327	436	475	243	49	5	0	1762
1980	0	4	4	49	227	428	632	589	440	51	0	0	2424
1981	0	0	9	124	131	494	540	398	246	36	4	0	1982
1982	2	0	25	13	243	346	446	394	192	73	8	6	1748
1983	0	0	3	10	118	278	515	512	212	40	0	2	1688
1984	0	0	2	21	132	405	372	397	210	178	1	2	1720
1985	0	0	18	88	172	381	423	401	248	119	49	0	1899
1986	0	0	11	74	208	455	599	401	300	83	34	0	2165
1987	0	0	2	60	266	391	502	531	281	12	6	2	2053
1988	0	0	5	49	169	364	490	467	273	85	6	0	1907
1989	0	0	7	36	101	170	364	452	273	98	2	0	1961
1990	0	5	26	66	194	415	490	488	341	98	2	0	2125
1991	0	0	20	89	258	358	502	446	305	70	4	13	2065
1992	0	1	13	73	155	292	478	349	265	20	0	0	1646
1993	0	0	11	22	208	435	639	536	364	80	19	0	2314
1994	0	2	19	141	167	470	445	447	274	60	13	0	2038
1995	0	0	23	91	311	368	608	494	211	47	2	0	2155
1996	0	8	2	54	322	429	527	456	255	64	7	2	2126
1997	3	3	28	13	62	221	438	366	264	68	0	0	1466
1998	0	0	16	4	257	426	494	407	323	88	3	5	2023
1999	0	0	0	103	136	305	446	531	256	39	1	0	1817
2000	0	0	3	4	253	396	515	464	196	64	16	0	1911
2001	0	0	0	69	166	288	429	435	224	45	13	0	1669

SNOWFALL (inches) 2001 ATLANTA, GA (ATL)

YEAR	JUL	AUG	SEP	OCT	NOV	DEC	JAN	FEB	MAR	APR	MAY	JUN	TOTAL
1972-73	0.0	0.0	0.0	0.0	0.0	0.0	1.0	T	0.0	0.0	0.0	0.0	1.0
1973-74	0.0	0.0	0.0	0.0	0.0	T	0.0	T	T	0.0	0.0	0.0	T
1974-75	0.0	0.0	0.0	0.0	0.0	T	T	T	T	0.0	0.0	0.0	T
1975-76	0.0	0.0	0.0	0.0	0.6	0.0	T	T	T	0.0	0.0	0.0	0.6
1976-77	0.0	0.0	0.0	0.0	0.0	0.0	1.0	0.0	0.0	0.0	0.0	0.0	1.0
1977-78	0.0	0.0	0.0	0.0	0.0	T	T	0.3	T	0.0	0.0	0.0	0.3
1978-79	0.0	0.0	0.0	0.0	0.0	0.0	0.2	4.4	0.0	0.0	0.0	0.0	4.6
1979-80	0.0	0.0	0.0	0.0	0.0	0.0	T	1.7	2.7	0.0	0.0	0.0	4.4
1980-81	0.0	0.0	0.0	0.0	0.0	0.0	0.0	T	T	0.0	0.0	0.0	T
1981-82	0.0	0.0	0.0	0.0	0.0	T	7.0	0.7	0.0	0.0	0.0	0.0	7.7
1982-83	0.0	0.0	0.0	0.0	0.0	0.0	1.9	0.5	7.9	0.0	0.0	0.0	10.3
1983-84	0.0	0.0	0.0	0.0	0.0	T	T	1.3	T	0.0	0.0	0.0	1.3
1984-85	0.0	0.0	0.0	0.0	0.0	T	0.4	1.5	0.0	0.0	0.0	0.0	1.9
1985-86	0.0	0.0	0.0	0.0	0.0	T	0.4	T	0.0	0.0	0.0	0.0	0.4
1986-87	0.0	0.0	0.0	0.0	0.0	0.0	3.6	T	1.2	T	0.0	0.0	4.8
1987-88	0.0	0.0	0.0	0.0	0.0	0.0	4.2	T	0.0	0.0	0.0	0.0	4.2
1988-89	0.0	0.0	0.0	0.0	0.0	T	0.0	0.7	0.0	T	0.0	0.0	0.7
1989-90	0.0	0.0	0.0	0.0	0.0	1.3	0.0	0.0	0.0	0.0	0.0	0.0	1.3
1990-91	0.0	0.0	0.0	0.0	0.0	0.0	2.1	T	T	0.0	0.0	0.0	2.1
1991-92	0.0	0.0	0.0	0.0	0.0	0.0	5.0	0.0	T	0.0	0.0	0.0	5.0
1992-93	0.0	0.0	0.0	0.0	0.0	T	0.0	T	4.2	0.0	0.0	0.0	4.2
1993-94	0.0	0.0	0.0	T	0.0	2.8	T	0.0	0.0	0.0	0.0	0.0	2.8
1994-95	0.0	0.0	0.0	0.0	0.0	0.0	T	0.4	0.0	0.0	0.0	0.0	0.4
1995-96	0.0	0.0	0.0	0.0	0.0	0.0	1.4						
1996-97													
1997-98													
1998-99													
1999-00						T	0.1						
2000-01	0.0	0.0	0.0	0.0	T	3.0		0.0	0.0	0.0	0.0	0.0	3.1
2001-	T	0.0	0.0	0.0	0.0	0.0							
POR= 51 YRS	0.0	0.0	0.0	T	T	0.2	0.9	0.5	0.4	T	0.0	0.0	2.0

2001
SAVANNAH,
GEORGIA (SAV)

Savannah is surrounded by flat terrain, low and marshy to the north and east, and rising to several feet above sea level to the west and south. About half the land to the west and south is cleared and the other half is wooded and swampy.

The area has a temperate climate, with a seasonal low temperature of 51 degrees in winter, 66 degrees in spring, 80 degrees in summer, and 66 degrees in autumn. The lowest temperatures are below 10 degrees and the highest temperatures are about 100 degrees.

The normal annual rainfall is about 49 inches. About half falls in the thunderstorm season of June 15 through September 15. The remainder, produced principally by squall-line and frontal showers, is spread over the other nine months with a minor peak in March. Considerable periods of fair, mild weather are experienced in October, November, April, and to a less extent, in May. Snow is a rarity and even a trace does not occur on an average of once a year. The heaviest snowfalls are under 5 inches. Severe tropical storms affect this area about once in ten years. Rainfall from these storms constitute the heaviest sustained precipitation. Accumulations exceeding 22 inches have occurred.

The present exposure of the thermometers gives readings more nearly commensurate with those of suburban street levels of Savannah than was the case of previous locations atop various buildings. During that time, especially on still, clear nights, temperatures near the ground and in lower inland areas were as much as 15 degrees lower than the official low temperature. Present differences on comparable nights range from 3 - 8 degrees.

Sunshine is adequate at all seasons and seldom are there two or more days in succession without it. Sea- and land-breeze effect is usually not felt in Savannah, though it is a daily feature on the nearby islands. Dry, continental air masses reach this area in summer mostly by sliding down the Atlantic coast and giving cooler northeast winds. Such masses reaching this area from the northwest or west in summer bring mostly clear skies and high temperatures.

Based on the 1951-1980 period, the average first occurrence of 32 degrees Fahrenheit in the fall is November 15 and the average last occurrence in the spring is March 10.

NORMALS, MEANS, AND EXTREMES
SAVANNAH, GA (SAV)

LATITUDE: 32 07' 08" N   LONGITUDE: 81 12' 08" W   ELEVATION (FT): GRND: 48   BARO: 51   TIME ZONE: EASTERN (UTC + 5)   WBAN: 03822

	ELEMENT	POR	JAN	FEB	MAR	APR	MAY	JUN	JUL	AUG	SEP	OCT	NOV	DEC	YEAR
TEMPERATURE °F	NORMAL DAILY MAXIMUM	30	59.7	62.4	70.1	77.5	84.0	88.8	91.1	89.7	85.2	77.5	70.0	62.3	76.5
	MEAN DAILY MAXIMUM	51	60.3	63.9	70.3	77.7	84.5	89.1	91.6	90.3	85.6	78.1	70.1	62.5	77.0
	HIGHEST DAILY MAXIMUM	51	84	86	91	95	100	104	105	104	98	97	89	83	105
	YEAR OF OCCURRENCE		1957	1989	1974	1986	1953	1985	1986	1954	1986	1961	1971		JUL 1986
	MEAN OF EXTREME MAXS.	51	76.0	79.6	84.4	89.3	93.7	97.4	98.3	97.0	93.7	88.5	82.4	77.8	88.2
	NORMAL DAILY MINIMUM	30	38.1	41.1	48.3	54.5	62.9	69.2	72.4	72.2	67.8	56.9	48.1	41.0	56.0
	MEAN DAILY MINIMUM	51	38.3	41.4	47.5	54.0	62.3	69.2	72.2	71.9	67.7	56.6	46.8	40.2	55.7
	LOWEST DAILY MINIMUM	51	3	14	20	32	39	51	61	57	43	28	15	9	3
	YEAR OF OCCURRENCE		1985	1958	1980	1987	1963	1984	1972	1986	1967	1952	1970	1983	JAN 1985
	MEAN OF EXTREME MINS.	51	21.0	24.6	30.4	38.6	49.0	60.3	66.7	65.7	56.2	39.3	29.5	23.1	42.0
	NORMAL DRY BULB	30	48.9	51.8	59.2	66.0	73.5	79.1	81.8	81.0	76.6	67.3	59.1	51.7	66.3
	MEAN DRY BULB	51	49.3	52.6	58.8	65.9	73.2	79.2	81.9	81.1	76.7	67.3	58.2	51.3	66.3
	MEAN WET BULB	17	45.5	48.7	53.2	58.2	66.0	72.2	75.0	74.6	70.7	62.1	54.7	44.7	60.5
	MEAN DEW POINT	17	39.6	42.4	46.9	52.2	61.4	69.1	72.2	72.2	68.1	58.1	50.4	39.6	56.0
	NORMAL NO. DAYS WITH:														
	MAXIMUM 90	30	0.0	0.0	0.1	1.5	5.5	14.2	21.1	18.1	7.4	0.7	0.0	0.0	68.6
	MAXIMUM 32	30	0.2	0.0	*	0.0	0.0	0.0	0.0	0.0	0.0	0.0	0.0	0.1	0.3
	MINIMUM 32	30	10.8	7.2	1.9	0.1	0.0	0.0	0.0	0.0	0.0	0.0	0.1	2.4	8.4	30.9
	MINIMUM 0	30	0.0	0.0	0.0	0.0	0.0	0.0	0.0	0.0	0.0	0.0	0.0	0.0	0.0
H/C	NORMAL HEATING DEG. DAYS	30	516	378	204	47	0	0	0	0	0	63	213	426	1847
	NORMAL COOLING DEG. DAYS	30	16	9	24	77	266	423	521	496	348	135	36	14	2365
RH	NORMAL (PERCENT)	30	70	67	67	65	70	74	76	79	78	73	72	71	72
	HOUR 01 LST	30	78	77	78	79	84	86	88	90	88	84	83	79	83
	HOUR 07 LST	30	82	81	83	83	85	87	89	91	91	87	86	82	86
	HOUR 13 LST	30	54	51	49	46	50	55	58	61	60	53	52	54	54
	HOUR 19 LST	30	66	61	60	58	63	68	72	76	76	72	72	69	68
S	PERCENT POSSIBLE SUNSHINE	46	54	57	62	71	68	65	63	62	57	63	61	55	62
W/O	MEAN NO. DAYS WITH:														
	HEAVY FOG(VISBY 1/4 MI)	51	4.5	3.2	3.1	2.6	3.0	2.4	1.1	2.0	3.5	3.3	4.6	4.7	38.0
	THUNDERSTORMS	51	1.1	1.4	2.9	3.6	6.8	10.4	14.8	12.0	5.5	1.6	0.6	0.6	61.3
CLOUDINESS	MEAN:														
	SUNRISE-SUNSET (OKTAS)	46	5.0	4.9	4.8	4.3	4.6	4.9	4.9	4.9	4.9	4.1	4.2	4.8	4.7
	MIDNIGHT-MIDNIGHT (OKTAS)	32	4.6	4.5	4.3	3.7	4.1	4.4	4.5	4.4	4.3	3.7	3.8	4.3	4.2
	MEAN NO. DAYS WITH:														
	CLEAR	46	9.3	8.5	9.0	10.6	9.2	6.8	5.6	5.8	7.0	11.8	11.0	9.2	103.8
	PARTLY CLOUDY	46	6.0	6.2	8.5	8.6	10.0	11.2	13.6	13.6	10.3	7.9	7.0	7.1	110.0
	CLOUDY	46	15.7	13.7	13.5	10.7	11.8	12.0	11.9	11.6	12.6	11.4	12.0	14.7	151.6
PR	MEAN STATION PRESSURE(IN)	28	30.09	30.05	30.00	29.98	29.96	29.96	29.99	29.99	29.99	30.03	30.07	30.09	30.02
	MEAN SEA-LEVEL PRES. (IN)	17	30.14	30.12	30.06	30.03	30.01	30.01	30.05	30.02	30.02	30.08	30.12	30.15	30.07
WINDS	MEAN SPEED (MPH)	38	8.3	9.0	9.0	8.5	7.7	7.2	6.9	6.5	7.2	7.5	7.4	7.7	7.7
	PREVAIL.DIR(TENS OF DEGS)	25	27	30	27	18	18	18	23	23	04	04	03	27	18
	MAXIMUM 2-MINUTE:														
	SPEED (MPH)	5	29	31	43	30	29	43	45	37	40	32	26	29	45
	DIR. (TENS OF DEGS)		32	09	16	31	30	05	04	34	01	36	02	01	04
	YEAR OF OCCURRENCE		2000	1998	2001	2001	1997	1998	1997	1999	1999	1999	2001	2000	JUL 1997
	MAXIMUM 5-SECOND:														
	SPEED (MPH)	5	40	37	58	41	43	56	51	54	54	44	31	38	58
	DIR. (TENS OF DEGS)		32	28	16	26	27	06	04	06	17	01	01	27	16
	YEAR OF OCCURRENCE		2000	1999	2001	2001	1999	1998	1997	1997	1998	1999	2001	2000	MAR 2001
PRECIPITATION	NORMAL (IN)	30	3.59	3.22	3.78	3.03	4.09	5.66	6.38	7.46	4.47	2.39	2.19	2.96	49.22
	MAXIMUM MONTHLY (IN)	51	8.98	7.92	9.57	10.57	10.08	14.39	20.10	17.03	13.47	19.84	5.26	5.80	20.10
	YEAR OF OCCURRENCE		1991	1964	1959	1991	1957	1963	1964	1995	1953	1994	1993	1977	JUL 1964
	MINIMUM MONTHLY (IN)	51	0.45	0.26	0.18	0.38	0.36	0.84	1.35	1.02	0.35	0.02	0.15	0.12	0.02
	YEAR OF OCCURRENCE		1989	1991	1955	1986	2001	1954	1972	1980	1991	2000	1966	1984	OCT 2000
	MAXIMUM IN 24 HOURS (IN)	51	3.74	3.46	4.65	5.62	5.67	6.77	6.36	8.71	6.80	8.86	5.02	3.47	8.86
	YEAR OF OCCURRENCE		1998	1964	1959	1976	1976	1999	1957	1995	1979	1994	1969	1964	OCT 1994
	NORMAL NO. DAYS WITH:														
	PRECIPITATION 0.01	30	9.6	8.8	9.1	6.9	8.7	11.1	13.7	13.3	9.7	6.0	6.4	8.5	111.8
	PRECIPITATION 1.00	30	0.9	0.6	1.2	0.9	1.2	1.7	1.9	2.2	1.4	0.5	0.6	0.7	13.8
SNOWFALL	NORMAL (IN)	30	0.1	0.3	0.*	0.0	0.0	0.0	0.0	0.0	0.0	0.0	0.0	0.1	0.5
	MAXIMUM MONTHLY (IN)	47	2.0	3.6	1.1	0.0	0.0	T	0.0	0.0	0.0	0.0	0.0	3.6	3.6
	YEAR OF OCCURRENCE		1977	1968	1986			1989						1989	DEC 1989
	MAXIMUM IN 24 HOURS (IN)	47	1.3	3.6	1.1	0.0	0.0	T	0.0	0.0	0.0	0.0	0.0	3.6	3.6
	YEAR OF OCCURRENCE		1977	1968	1986			1989						1989	FEB 1968
	MAXIMUM SNOW DEPTH (IN)	45	0	4	0	0	0	0	0	0	0	0	0	4	4
	YEAR OF OCCURRENCE			1968										1989	DEC 1989
	NORMAL NO. DAYS WITH:														
	SNOWFALL 1.0	30	0.*	0.1	0.*	0.0	0.0	0.0	0.0	0.0	0.0	0.0	0.0	0.*	0.1

PRECIPITATION (inches) 2001 SAVANNAH, GA (SAV)

YEAR	JAN	FEB	MAR	APR	MAY	JUN	JUL	AUG	SEP	OCT	NOV	DEC	ANNUAL
1972	3.99	4.61	3.84	1.20	5.84	6.54	1.35	12.62	0.36	0.54	4.91	2.77	48.57
1973	3.61	4.46	5.36	4.43	1.23	9.19	2.89	6.45	3.65	0.19	0.68	3.26	45.40
1974	1.37	2.79	1.87	2.75	7.25	6.00	6.48	7.90	2.61	0.10	0.96	1.85	41.93
1975	3.17	3.01	3.99	4.71	6.00	2.08	11.55	3.13	8.01	1.25	1.09	3.19	51.18
1976	2.19	1.24	2.51	5.62	6.33	7.49	7.56	7.28	10.07	4.75	4.83	3.87	63.74
1977	3.14	1.83	2.72	1.94	1.03	2.00	5.62	8.01	6.52	1.16	2.07	5.80	41.84
1978	4.02	3.14	1.93	3.68	4.50	2.19	3.61	4.43	2.61	0.60	1.85	2.85	35.41
1979	3.96	4.14	2.42	3.83	8.49	7.37	10.78	2.65	12.20	0.70	2.70	2.68	61.92
1980	2.95	1.29	7.75	3.68	4.50	3.47	2.38	1.02	5.81	1.62	2.04	1.33	37.84
1981	1.03	2.94	3.91	1.75	2.10	3.01	5.42	10.91	2.88	1.29	1.65	3.17	40.06
1982	3.47	2.94	1.64	6.25	4.18	9.15	6.70	9.18	2.98	1.74	0.40	3.63	52.26
1983	5.90	5.23	9.01	5.15	1.07	5.81	5.30	3.67	3.39	1.03	4.18	4.77	54.51
1984	8.87	3.21	5.13	3.41	5.29	1.48	7.88	3.46	7.43	1.23	3.15	0.12	50.66
1985	0.51	1.37	1.65	1.37	2.18	6.72	5.00	9.42	0.76	3.37	4.28	2.01	38.64
1986	2.03	5.28	2.85	0.38	2.06	2.98	5.49	12.31	0.49	1.99	4.40	5.07	45.33
1987	8.62	4.39	5.33	0.50	3.82	8.03	4.37	9.46	8.16	0.33	2.06	1.41	56.48
1988	3.44	4.09	2.11	5.05	3.52	2.63	1.80	10.68	9.62	2.81	1.43	0.99	48.17
1989	0.45	0.67	1.41	3.59	3.10	7.30	4.91	6.29	7.98	4.71	1.26	5.20	46.87
1990	3.91	3.08	3.79	1.75	2.07	0.97	1.92	7.25	1.26	12.50	2.48	2.10	43.08
1991	8.98	0.26	5.48	10.57	7.13	5.12	15.41	10.51	0.35	1.60	1.26	1.75	68.42
1992	6.60	2.24	3.97	2.08	2.04	13.01	2.03	7.69	8.42	3.60	5.15	1.53	58.36
1993	5.52	3.35	7.96	3.22	1.31	2.48	4.34	3.04	6.84	2.58	5.26	2.15	48.05
1994	4.70	0.78	3.75	2.13	6.17	7.80	6.90	2.64	6.60	19.84	3.67	4.46	69.44
1995	2.44	4.47	0.72	0.69	3.52	5.89	6.65	17.03	4.10	3.42	1.29	0.89	51.11
1996	2.01	1.35	4.29	2.43	1.44	2.44	5.23	7.12	2.84	3.45	0.88	2.72	36.20
1997	3.12	2.34	1.54	3.62	2.99	5.90	11.64	4.56	5.32	5.65	5.04	3.98	55.70
1998	7.51	6.88	3.98	5.68	1.97	1.93	8.27	3.79	5.56	1.28	0.30	2.32	49.47
1999	4.73	1.95	1.25	1.68	2.54	14.25	7.15	4.26	6.50	2.04	0.49	1.94	48.78
2000	2.71	1.56	4.30	2.84	0.96	5.47	3.38	4.18	7.45	T	1.77	2.82	37.44
2001	1.71	0.77	6.65	0.71	0.36	6.41	4.85	4.65	4.72	0.16	0.16	0.49	31.64
POR= 132 YRS	3.03	3.10	3.59	2.93	3.36	5.58	6.61	7.00	5.51	2.86	1.98	2.70	48.25

AVERAGE TEMPERATURE (F) 2001 SAVANNAH, GA (SAV)

YEAR	JAN	FEB	MAR	APR	MAY	JUN	JUL	AUG	SEP	OCT	NOV	DEC	ANNUAL
1972	57.5	51.3	59.3	66.7	71.6	75.7	81.0	81.0	77.3	68.1	58.1	57.1	67.1
1973	49.8	49.6	63.6	64.0	73.5	79.3	82.2	80.0	79.0	68.9	61.6	51.2	66.9
1974	62.9	53.0	63.8	65.6	74.5	76.8	79.2	79.6	76.3	64.6	57.7	53.1	67.3
1975	55.3	57.7	59.5	65.4	76.7	79.5	78.4	81.7	76.5	70.0	59.3	50.5	67.5
1976	45.9	56.4	62.6	64.3	69.7	75.8	81.7	78.2	74.9	62.5	51.5	49.0	64.4
1977	39.9	49.2	62.1	67.6	74.0	82.0	83.3	81.1	78.9	64.1	61.2	50.0	66.1
1978	43.9	43.6	56.3	68.0	73.7	80.0	82.2	82.3	78.2	67.0	64.7	53.4	66.1
1979	45.4	49.0	59.8	67.7	73.9	76.8	82.1	81.4	77.4	67.5	60.4	50.7	66.0
1980	50.7	48.5	57.1	66.2	73.3	79.7	84.4	83.4	80.4	65.4	56.7	48.1	66.2
1981	43.5	52.6	56.8	69.0	71.6	84.5	84.4	79.2	75.2	65.4	57.7	48.3	65.7
1982	48.7	55.9	62.1	64.8	74.3	80.2	81.3	81.0	75.5	67.6	61.8	57.4	67.6
1983	46.1	50.8	58.0	62.8	73.0	78.2	84.1	82.9	75.7	69.7	57.7	48.5	65.6
1984	47.7	53.7	59.3	65.7	72.9	79.2	80.6	81.5	75.0	73.2	55.7	59.5	67.0
1985	45.3	53.0	61.3	66.9	74.6	81.3	82.7	80.7	76.6	72.2	67.5	48.9	67.6
1986	47.6	56.9	59.1	66.8	75.0	82.5	85.7	81.6	79.7	70.0	67.5	53.9	68.7
1987	49.0	50.4	58.2	63.9	74.1	80.6	83.6	84.5	77.8	61.5	60.4	54.7	66.6
1988	45.0	50.2	58.2	66.1	72.4	77.7	82.7	82.5	77.4	64.2	61.3	51.1	65.7
1989	56.6	56.1	60.6	65.6	72.3	81.0	82.9	80.7	76.8	68.6	59.5	43.7	67.0
1990	55.7	60.0	62.6	65.0	74.3	81.5	84.4	82.4	78.9	70.7	60.8	55.6	69.4
1991	52.2	56.2	62.4	70.0	77.5	79.4	83.2	82.2	77.9	68.4	56.6	55.6	68.5
1992	49.9	55.7	59.2	64.5	71.7	78.6	84.3	81.0	77.1	66.6	60.6	52.0	66.8
1993	54.8	50.3	57.0	62.4	73.6	81.3	86.7	83.1	79.6	68.1	59.9	49.7	67.2
1994	47.7	56.0	62.9	69.6	72.4	80.7	82.2	81.2	76.6	67.8	63.7	54.9	68.0
1995	50.7	52.4	62.5	69.1	76.5	79.2	83.8	82.4	76.1	71.5	55.6	50.1	67.5
1996	50.1	54.6	56.0	64.0	75.2	78.8	82.4	79.3	76.1	66.5	55.9	52.9	66.0
1997	50.9	55.1	64.7	63.1	68.9	75.4	81.2	79.3	76.9	66.6	55.1	50.5	65.6
1998	52.9	54.1	56.4	65.3	75.6	83.0	83.4	81.2	77.5	68.9	62.6	55.5	68.0
1999	52.4	54.3	55.8	69.0	70.8	77.1	82.7	83.1	75.7	67.8	59.7	50.8	66.5
2000	48.1	53.2	61.7	62.8	75.3	79.2	82.0	81.0	75.5	64.8	55.6	43.7	65.2
2001	46.8	56.0	57.4	65.1	73.0	78.8	80.8	81.3	74.2	64.9	63.0	55.9	66.4
POR= 128 YRS	51.2	53.3	59.4	66.1	73.5	79.4	81.7	81.0	77.0	67.9	58.8	52.2	66.8

REFERENCE NOTES:

PAGE 1:
THE TEMPERATURE GRAPH SHOWS NORMAL MAXIMUM AND NORMAL
MINIMUM DAILY TEMPERATURES (SOLID CURVES) AND THE
ACTUAL DAILY HIGH AND LOW TEMPERATURES (VERTICAL BARS).

PAGE 2 AND 3:
H/C INDICATES HEATING AND COOLING DEGREE DAYS.
RH INDICATES RELATIVE HUMIDITY
W/O INDICATES WEATHER AND OBSTRUCTIONS
S INDICATES SUNSHINE.
PR INDICATES PRESSURE.
CLOUDINESS ON PAGE 3 IS THE SUM OF THE CEILOMETER AND
SATELLITE DATA NOT TO EXCEED EIGHT EIGHTHS(OKTAS).

GENERAL:
T INDICATES TRACE PRECIPITATION, AN AMOUNT GREATER
THAN ZERO BUT LESS THAN THE LOWEST REPORTABLE VALUE.
+ INDICATES THE VALUE ALSO OCCURS ON EARLIER DATES.
BLANK ENTRIES DENOTE MISSING OR UNREPORTED DATA.
NORMALS ARE 30-YEAR AVERAGES (1961 - 1990).
ASOS INDICATES AUTOMATED SURFACE OBSERVING SYSTEM.
PM INDICATES THE LAST DAY OF THE PREVIOUS MONTH.
POR (PERIOD OF RECORD) BEGINS WITH THE JANUARY DATA
MONTH AND IS THE NUMBER OF YEARS USED TO COMPUTE
THE MEAN. INDIVIDUAL MONTHS WITHIN THE POR MAY
BE MISSING.
WHEN THE POR FOR A NORMAL IS LESS THAN 30 YEARS,
THE NORMAL IS PROVISIONAL AND IS BASED ON THE NUMBER
OF YEARS INDICATED.
0.* OR * INDICATES THE VALUE OR MEAN-DAYS-WITH
IS BETWEEN 0.00 AND 0.05.
CLOUDINESS FOR ASOS STATIONS DIFFERS FROM THE NON-ASOS
OBSERVATION TAKEN BY A HUMAN OBSERVER. ASOS STATION
CLOUDINESS IS BASED ON TIME-AVERAGED CEILOMETER DATA
FOR CLOUDS AT OR BELOW 12,000 FEET AND ON SATELLITE
DATA FOR CLOUDS ABOVE 12,000 FEET.
THE NUMBER OF DAYS WITH CLEAR, PARTLY CLOUDY, AND
CLOUDY CONDITIONS FOR ASOS STATIONS IS THE SUM
OF THE CEILOMETER AND SATELLITE DATA FOR THE
SUNRISE TO SUNSET PERIOD.

GENERAL CONTINUED:
CLEAR INDICATES 0 - 2 OKTAS, PARTLY CLOUDY INDICATES
3 - 6 OKTAS, AND CLOUDY INDICATES 7 OR 8 OKTAS.
WHEN AT LEAST ONE OF THE ELEMENTS (CEILOMETER OR
SATELLITE) IS MISSING, THE DAILY CLOUDINESS IS
NOT COMPUTED.
WIND DIRECTION IS RECORDED IN TENS OF DEGREES (2 DIGITS)
CLOCKWISE FROM TRUE NORTH. "00" INDICATES CALM. "36"
INDICATES TRUE NORTH.
RESULTANT WIND IS THE VECTOR AVERAGE OF THE SPEED AND
DIRECTION.
AVERAGE TEMPERATURE IS THE SUM OF THE MEAN DAILY MAXIMUM
AND MINIMUM TEMPERATURE DIVIDED BY 2.
SNOWFALL DATA COMPRISE ALL FORMS OF FROZEN
PRECIPITATION, INCLUDING HAIL.
A HEATING (COOLING) DEGREE DAY IS THE DIFFERENCE BETWEEN
THE AVERAGE DAILY TEMPERATURE AND 65 F.
DRY BULB IS THE TEMPERATURE OF THE AMBIENT AIR.
DEW POINT IS THE TEMPERATURE TO WHICH THE AIR MUST BE
COOLED TO ACHIEVE 100 PERCENT RELATIVE HUMIDITY.
WET BULB IS THE TEMPERATURE THE AIR WOULD HAVE IF THE
MOISTURE CONTENT WAS INCREASED TO 100 PERCENT RELATIVE
HUMIDITY.

ON JULY 1, 1996, THE NATIONAL WEATHER SERVICE BEGAN USING
THE "METAR" OBSERVATION CODE THAT WAS ALREADY EMPLOYED
BY MOST OTHER NATIONS OF THE WORLD. THE MOST NOTICEABLE
DIFFERENCE IN THIS ANNUAL PUBLICATION WILL BE THE CHANGE
IN UNITS FROM TENTHS TO EIGHTS(OKTAS) FOR REPORTING THE
AMOUNT OF SKY COVER.

HEATING DEGREE DAYS (base 65 F) 2001 SAVANNAH, GA (SAV)

YEAR	JUL	AUG	SEP	OCT	NOV	DEC	JAN	FEB	MAR	APR	MAY	JUN	TOTAL
1972-73	0	0	0	25	239	261	462	423	98	85	4	0	1597
1973-74	0	0	0	47	148	435	107	340	124	73	1	0	1275
1974-75	0	0	2	79	243	368	315	232	212	86	0	0	1537
1975-76	0	0	0	25	228	446	586	256	134	69	11	0	1755
1976-77	0	0	0	142	405	490	771	437	152	42	2	0	2441
1977-78	0	0	0	96	165	457	645	594	283	35	2	0	2277
1978-79	0	0	0	45	53	378	602	448	181	17	1	0	1725
1979-80	0	0	0	41	183	438	436	489	257	43	7	0	1894
1980-81	0	0	0	72	252	518	659	342	263	25	8	0	2139
1981-82	0	0	3	59	231	513	501	258	149	76	0	0	1790
1982-83	0	0	0	73	139	266	579	392	228	115	0	0	1792
1983-84	0	0	2	19	232	513	531	320	200	68	7	0	1892
1984-85	0	0	1	8	299	185	615	360	157	60	2	0	1687
1985-86	0	0	1	16	51	504	531	240	215	59	4	0	1621
1986-87	0	5	0	48	101	349	491	401	231	110	5	0	1741
1987-88	0	0	0	122	185	332	612	426	218	52	1	0	1948
1988-89	0	0	0	84	141	423	268	289	193	110	13	0	1521
1989-90	0	0	1	59	191	653	286	175	135	81	0	0	1581
1990-91	0	0	0	55	143	279	395	260	139	14	0	0	1285
1991-92	0	0	0	34	270	315	461	274	204	101	22	0	1681
1992-93	0	0	1	45	194	401	320	406	254	113	0	0	1734
1993-94	0	0	1	45	193	470	530	260	123	20	7	0	1649
1994-95	0	0	0	32	95	317	433	358	114	30	1	0	1380
1995-96	0	0	6	25	313	457	457	313	288	109	7	0	1975
1996-97	0	0	0	51	287	374	435	290	88	107	24	5	1661
1997-98	0	0	0	80	300	442	373	303	283	63	0	0	1844
1998-99	0	0	0	44	107	305	390	294	283	60	17	0	1500
1999-00	0	0	0	56	179	436	518	340	122	97	1	0	1749
2000-01	0	0	1	72	304	655	560	262	248	88	0	0	2190
2001-	0	0	4	107	103	294							

COOLING DEGREE DAYS (base 65 F) 2001 SAVANNAH, GA (SAV)

YEAR	JAN	FEB	MAR	APR	MAY	JUN	JUL	AUG	SEP	OCT	NOV	DEC	ANNUAL
1972	28	8	14	130	214	328	505	503	379	125	38	23	2295
1973	0	0	63	63	275	434	540	473	429	177	52	13	2519
1974	52	9	96	99	302	360	448	462	347	187	32	8	2289
1975	20	33	48	104	374	439	423	528	351	187	65	2	2574
1976	0	12	68	53	164	331	526	416	306	70	7	1	1954
1977	0	1	71	130	289	517	574	508	421	74	58	0	2643
1978	0	0	22	132	281	456	538	543	403	115	51	29	2570
1979	0	5	28	105	282	360	537	516	378	126	51	2	2390
1980	0	16	19	87	270	449	607	579	468	90	8	2	2595
1981	0	1	14	150	219	589	609	444	319	80	17	4	2446
1982	6	10	67	79	292	463	514	503	323	158	48	42	2505
1983	0	0	16	55	253	400	598	562	332	171	18	6	2411
1984	0	0	31	94	261	431	486	517	309	270	24	19	2442
1985	11	31	58	123	307	496	557	496	355	248	133	7	2822
1986	0	18	40	121	321	532	651	525	449	212	107	13	2989
1987	0	0	26	85	292	474	583	611	392	22	50	21	2556
1988	0	3	11	92	238	386	555	547	378	66	35	1	2312
1989	13	45	64	135	248	488	563	493	362	177	33	0	2621
1990	4	41	69	90	296	503	608	549	422	237	25	24	2868
1991	7	18	64	169	397	441	572	540	393	143	25	29	2798
1992	0	11	29	94	234	414	603	504	370	101	69	7	2436
1993	12	0	10	43	271	497	678	567	447	149	46	1	2721
1994	0	14	62	165	243	476	542	509	355	125	63	14	2568
1995	1	9	46	160	366	433	591	547	348	235	36	4	2776
1996	3	20	19	84	331	419	549	451	339	105	20	2	2342
1997	4	20	84	57	155	322	510	452	363	137	11	0	2115
1998	10	5	24	77	336	546	575	508	383	173	44	18	2699
1999	7	1	4	187	205	368	556	568	328	153	26	1	2404
2000	3	5	27	37	325	435	538	503	325	73	28	0	2299
2001	2	17	20	99	253	422	499	515	287	106	49	19	2288

SNOWFALL (inches) 2001 SAVANNAH, GA (SAV)

YEAR	JUL	AUG	SEP	OCT	NOV	DEC	JAN	FEB	MAR	APR	MAY	JUN	TOTAL
1972-73	0.0	0.0	0.0	0.0	0.0	0.0	T	3.2	0.0	0.0	0.0	0.0	3.2
1973-74	0.0	0.0	0.0	0.0	0.0	0.0	0.0	0.0	0.0	0.0	0.0	0.0	0.0
1974-75	0.0	0.0	0.0	0.0	0.0	0.0	0.0	0.0	0.0	0.0	0.0	0.0	0.0
1975-76	0.0	0.0	0.0	0.0	0.0	0.0	T	0.0	0.0	0.0	0.0	0.0	T
1976-77	0.0	0.0	0.0	0.0	0.0	0.0	2.0	T	0.0	0.0	0.0	0.0	2.0
1977-78	0.0	0.0	0.0	0.0	0.0	0.0	0.0	0.0	0.0	0.0	0.0	0.0	0.0
1978-79	0.0	0.0	0.0	0.0	0.0	0.0	0.0	T	0.0	0.0	0.0	0.0	T
1979-80	0.0	0.0	0.0	0.0	0.0	0.0	0.0	0.0	T	0.0	0.0	0.0	T
1980-81	0.0	0.0	0.0	0.0	0.0	T	0.0	0.0	0.0	0.0	0.0	0.0	T
1981-82	0.0	0.0	0.0	0.0	0.0	0.0	0.0	0.0	0.0	0.0	0.0	0.0	0.0
1982-83	0.0	0.0	0.0	0.0	0.0	0.0	T	0.0	T	0.0	0.0	0.0	T
1983-84	0.0	0.0	0.0	0.0	0.0	0.0	T	0.0	0.0	0.0	0.0	0.0	T
1984-85	0.0	0.0	0.0	0.0	0.0	0.0	0.0	0.0	0.0	0.0	0.0	0.0	0.0
1985-86	0.0	0.0	0.0	0.0	0.0	0.0	0.3	0.0	1.1	0.0	0.0	0.0	1.4
1986-87	0.0	0.0	0.0	0.0	0.0	0.0	T	0.0	0.0	0.0	0.0	0.0	T
1987-88	0.0	0.0	0.0	0.0	0.0	0.0	T	T	0.0	0.0	0.0	0.0	T
1988-89	0.0	0.0	0.0	0.0	0.0	0.0	T	1.0	0.0	0.0	0.0	T	1.0
1989-90	0.0	0.0	0.0	0.0	0.0	3.6	0.0	0.0	0.0	0.0	0.0	0.0	3.6
1990-91	0.0	0.0	0.0	0.0	0.0	0.0	0.0	T	0.0	0.0	0.0	0.0	T
1991-92	0.0	0.0	0.0	0.0	0.0	0.0	0.0	0.0	0.0	0.0	0.0	0.0	0.0
1992-93	0.0	0.0	0.0	0.0	0.0	0.0	0.0	0.0	0.2	0.0	0.0	0.0	0.2
1993-94	0.0	0.0	0.0	0.0	0.0	0.0	T	0.0	T	0.0	0.0	0.0	T
1994-95	0.0	0.0	0.0	0.0	0.0	0.0	0.0	T	0.0	0.0	0.0	0.0	T
1995-96	0.0	0.0	0.0	0.0	0.0	0.0	T	0.2	0.0				
1996-97													
1997-98													
1998-99													
1999-00													
2000-01						T							
2001-													
POR= 45 YRS	0.0	0.0	0.0	0.0	0.0	0.1	0.1	0.2	0.0	0.0	0.0	T	0.4

2001
HILO,
HAWAII (ITO)

The city of Hilo is located near the midpoint of the eastern shore of the Island of Hawaii. This island is by far the largest of the Hawaiian group, with an area of 4,038 square miles, more than twice that of all the other islands combined. Its topography is dominated by the great volcanic masses of Mauna Loa (13,653 feet), Mauna Kea (13,796 feet), and of Haulalai, the Kohala Mountains, and Kilauea. In fact, the island consists entirely of the slopes of these mountains and of the broad saddles between them. Mauna Loa and Kilauea, which occupy the southern half of the island, are still active volcanoes.

Hawaii lies well within the belt of northeasterly trade winds generated by the semi-permanent Pacific high pressure cell to the north and east. The climate provides equable temperatures from day to day and season to season. In Hilo, July and August are the warmest months, with average daily highs and lows of 83 and 68 degrees. January and February, the coolest months, have highs of 80 degrees and lows of 63 degrees. Greater variations occur in localities with less rain and cloud, but temperatures in the mid-90s and low 50s are uncommon anywhere on the island near sea level.

Over the windward slopes of Hawaii, rainfall occurs principally as orographic showers within the ascending moist trade winds. Mean annual rainfall, except for the semi-sheltered Hamakua district, increases from 100 inches or more along the coasts to a maximum of over 300 inches at elevations of 2,000 to 3,000 feet, and then declines to about 15 inches at the summits of Mauna Kea and Mauna Loa. Leeward areas are topographically sheltered from the trades and are therefore drier, although sea breezes created by daytime heating of the land move onshore and upslope, causing afternoon and evening cloudiness and showers. The driest locality on the island, and in the State, with an annual rainfall of less than 10 inches, is the coastal strip just leeward of the southern portion of the Kohala Mountains and of the saddle between the Kohalas and Mauna Kea.

Within the city of Hilo, average rainfall varies from about 130 inches a year near the shore to as much as 200 upslope. The wettest part of the island, with a mean annual rainfall exceeding 300 inches, lies about 6 miles upslope from the city limits. Relative humidity at Hilo is in the moderate range, however, due to the natural ventilation provided by the prevailing winds, the weather is seldom oppressive.

The trade winds prevail throughout the year and profoundly influence the climate. The islands entire western coast is sheltered from the trades by high mountains, except that unusually strong trade winds may sweep through the saddle between the Kohala Mountains and Mauna Kea and reach the areas to the lee. But even places exposed to the trades may be affected by local mountain circulations. Except for heavy rain, really severe weather seldom occurs. During the winter, cold fronts or the cyclonic storms of subtropical origin may bring blizzards to the upper slopes of Mauna Loa and Mauna Kea, with snow extending at times to 9,000 feet or below and icing nearer the summit.

Storms crossing the Pacific a thousand miles to the north, low pressure or tropical storms, may generate seas that cause heavy swell and surf.

NORMALS, MEANS, AND EXTREMES
HILO, HI (ITO)

LATITUDE: 19 43' 24" N LONGITUDE: 155 03' 05" W ELEVATION (FT): GRND: 44 BARO: 47 TIME ZONE: HAWAII (UTC + 10) WBAN: 21504

	ELEMENT	POR	JAN	FEB	MAR	APR	MAY	JUN	JUL	AUG	SEP	OCT	NOV	DEC	YEAR
TEMPERATURE °F	NORMAL DAILY MAXIMUM	30	79.8	79.8	79.5	79.8	81.2	82.7	83.0	83.6	83.8	83.2	81.4	80.0	81.5
	MEAN DAILY MAXIMUM	52	79.4	79.2	79.2	79.7	80.8	82.4	82.8	83.4	83.6	83.0	80.9	79.5	81.2
	HIGHEST DAILY MAXIMUM	55	92	92	93	89	94	90	89	93	92	91	92	93	94
	YEAR OF OCCURRENCE		1997	1968	1972	1978	1966	1969	1995	1950	1951	1979	1996	1980	MAY 1966
	MEAN OF EXTREME MAXS.	52	85.5	85.1	84.4	83.7	84.6	85.5	85.9	86.8	86.8	87.2	85.5	84.5	85.5
	NORMAL DAILY MINIMUM	30	63.6	63.6	64.4	65.5	66.5	67.6	68.6	68.9	68.6	68.1	66.8	64.8	66.4
	MEAN DAILY MINIMUM	52	63.5	63.3	64.2	65.3	66.4	67.6	68.6	69.0	68.5	67.9	66.7	64.6	66.3
	LOWEST DAILY MINIMUM	55	54	53	54	56	58	60	62	63	61	62	58	55	53
	YEAR OF OCCURRENCE		1995	1962	1983	1949	1947	1946	1970	1955	1970	1999	1985	1977	FEB 1962
	MEAN OF EXTREME MINS.	52	58.6	58.0	59.6	61.6	62.6	64.3	65.0	65.3	64.9	64.1	62.4	59.5	62.2
	NORMAL DRY BULB	30	71.7	71.7	72.0	72.7	73.9	75.2	75.8	76.3	76.2	75.7	74.2	72.4	74.0
	MEAN DRY BULB	52	71.4	71.3	71.7	72.4	73.7	75.0	75.7	76.2	76.1	75.4	73.8	72.1	73.7
	MEAN WET BULB	17	66.0	65.7	66.6	67.5	68.7	70.0	71.2	71.8	71.4	70.8	69.8	67.2	68.9
	MEAN DEW POINT	17	63.1	62.6	63.5	64.8	66.1	67.5	68.9	69.5	69.1	68.5	67.7	64.7	66.3
	NORMAL NO. DAYS WITH:														
	MAXIMUM 90	30	0.1	0.1	*	0.0	0.1	*	0.0	0.1	0.2	0.3	*	0.1	1.0
	MAXIMUM 32	30	0.0	0.0	0.0	0.0	0.0	0.0	0.0	0.0	0.0	0.0	0.0	0.0	0.0
	MINIMUM 32	30	0.0	0.0	0.0	0.0	0.0	0.0	0.0	0.0	0.0	0.0	0.0	0.0	0.0
	MINIMUM 0	30	0.0	0.0	0.0	0.0	0.0	0.0	0.0	0.0	0.0	0.0	0.0	0.0	0.0
H/C	NORMAL HEATING DEG. DAYS	30	0	0	0	0	0	0	0	0	0	0	0	0	0
	NORMAL COOLING DEG. DAYS	30	208	188	217	231	276	306	335	350	336	332	276	229	3284
RH	NORMAL (PERCENT)	30	77	76	78	80	79	77	80	80	79	80	80	79	79
	HOUR 02 LST	30	83	83	85	88	87	86	88	87	87	87	86	85	86
	HOUR 08 LST	30	78	77	80	81	79	78	81	80	79	79	81	80	79
	HOUR 14 LST	30	66	65	67	70	67	65	68	68	69	70	70	68	68
	HOUR 20 LST	30	82	81	82	83	82	81	82	82	84	85	85	84	83
S	PERCENT POSSIBLE SUNSHINE	49	46	46	42	37	37	44	41	41	43	38	33	37	40
W/O	MEAN NO. DAYS WITH:														
	HEAVY FOG(VISBY 1/4 MI)	56	0.0	0.0	0.0	0.0	0.0	0.0	0.0	0.0	0.0	0.0	0.1	0.0	0.1
	THUNDERSTORMS	56	0.9	1.3	1.5	1.0	0.6	0.1	0.3	0.3	0.5	1.1	1.2	0.9	9.7
CLOUDINESS	MEAN:														
	SUNRISE-SUNSET (OKTAS)	51	5.0	5.3	6.0	6.4	6.2	5.9	6.1	5.9	5.6	5.7	5.9	5.5	5.8
	MIDNIGHT-MIDNIGHT (OKTAS)	33	5.0	5.2	5.9	6.4	6.2	6.1	6.3	5.9	5.6	5.8	6.0	5.5	5.8
	MEAN NO. DAYS WITH:														
	CLEAR	51	6.5	5.3	2.7	1.2	1.2	1.7	1.3	1.8	2.9	2.7	3.2	5.0	35.5
	PARTLY CLOUDY	51	11.4	10.3	10.2	9.2	10.6	11.3	11.5	12.2	12.0	11.8	10.0	10.8	131.3
	CLOUDY	51	13.1	12.7	18.0	19.7	19.1	17.1	17.7	16.5	14.5	16.1	16.2	14.6	195.3
PR	MEAN STATION PRESSURE(IN)	29	29.96	29.98	30.02	30.02	30.02	30.01	29.99	29.97	29.94	29.95	29.96	29.97	29.98
	MEAN SEA-LEVEL PRES. (IN)	18	30.02	30.03	30.05	30.07	30.05	30.05	30.02	30.00	29.97	29.98	30.00	30.02	30.02
WINDS	MEAN SPEED (MPH)	35	7.6	7.9	8.0	7.8	7.6	7.3	7.2	7.2	7.1	7.0	7.0	7.3	7.4
	PREVAIL.DIR(TENS OF DEGS)	22	22	22	22	22	22	22	23	23	23	22	22	22	22
	MAXIMUM 2-MINUTE:														
	SPEED (MPH)	4	29	31	26	23	23	24	21	24	22	24	24	25	31
	DIR. (TENS OF DEGS)		01	35	08	08	08	06	07	10	08	02	04	07	35
	YEAR OF OCCURRENCE		1998	1998	2001	2001	1999	2001	2000	1998	1998	2001	1999	2001	FEB 1998
	MAXIMUM 5-SECOND:														
	SPEED (MPH)	4	35	38	29	32	29	28	25	32	25	31	29	32	38
	DIR. (TENS OF DEGS)		36	34	08	08	11	06	10	08	10	03	04	08	34
	YEAR OF OCCURRENCE		1998	1998	2001	2001	1999	2001	1998	1998	1998	2001	1999	2001	FEB 1998
PRECIPITATION	NORMAL (IN)	30	9.88	10.29	13.92	15.26	9.91	6.20	9.71	9.34	8.53	9.60	14.51	12.04	129.19
	MAXIMUM MONTHLY (IN)	59	32.24	45.55	49.93	43.24	25.01	22.70	28.59	26.92	21.82	26.10	45.90	50.82	50.82
	YEAR OF OCCURRENCE		1979	1979	1980	1986	1964	1997	1982	1991	1994	1951	2000	1954	DEC 1954
	MINIMUM MONTHLY (IN)	59	0.13	0.52	0.88	2.93	1.18	1.80	3.54	2.66	1.59	2.40	1.01	0.28	0.13
	YEAR OF OCCURRENCE		1998	2000	1972	1962	1945	1985	1999	1971	1974	1962	1989	1980	JAN 1998
	MAXIMUM IN 24 HOURS (IN)	59	10.90	22.30	17.05	11.07	10.26	4.72	7.11	11.57	9.49	8.88	27.36	11.45	27.36
	YEAR OF OCCURRENCE		1990	1979	1980	1971	1965	1997	1982	1991	1994	1951	2000	1987	NOV 2000
	NORMAL NO. DAYS WITH:														
	PRECIPITATION 0.01	30	16.4	16.6	23.1	25.1	25.5	24.3	27.2	25.9	23.4	23.6	22.4	19.8	273.3
	PRECIPITATION 1.00	30	2.6	2.6	3.7	3.7	2.2	1.0	1.7	1.4	2.2	2.2	3.9	3.1	30.3
SNOWFALL	NORMAL (IN)	30	0.0	0.0	0.0	0.0	0.0	0.0	0.0	0.0	0.0	0.0	0.0	0.0	0.0
	MAXIMUM MONTHLY (IN)	1	0.0	0.0	0.0	0.0	0.0	0.0	0.0	0.0	0.0	0.0	0.0	0.0	0.0
	YEAR OF OCCURRENCE														
	MAXIMUM IN 24 HOURS (IN)	55	0.0	0.0	0.0	0.0	0.0	0.0	0.0	0.0	0.0	0.0	0.0	0.0	0.0
	YEAR OF OCCURRENCE														
	MAXIMUM SNOW DEPTH (IN)	47	0	0	0	0	0	0	0	0	0	0	0	0	0
	YEAR OF OCCURRENCE														
	NORMAL NO. DAYS WITH:														
	SNOWFALL 1.0	30	0.0	0.0	0.0	0.0	0.0	0.0	0.0	0.0	0.0	0.0	0.0	0.0	0.0

PRECIPITATION (inches) 2001 HILO, HI (ITO)

YEAR	JAN	FEB	MAR	APR	MAY	JUN	JUL	AUG	SEP	OCT	NOV	DEC	ANNUAL
1972	10.96	10.13	0.88	17.79	4.71	4.58	9.07	8.77	5.20	9.52	13.23	4.01	98.85
1973	3.45	5.51	18.84	7.34	8.34	3.69	4.40	3.54	8.07	9.72	26.88	8.19	107.97
1974	5.88	7.57	13.47	19.11	8.07	4.76	7.81	4.25	1.59	6.65	14.56	19.20	112.92
1975	19.62	9.28	10.40	10.23	3.01	4.20	3.83	8.13	2.73	8.88	11.15	8.47	99.93
1976	15.62	11.63	25.00	11.58	6.01	2.97	5.46	5.13	5.31	11.35	7.24	7.37	114.67
1977	1.22	9.56	15.49	10.90	10.86	2.46	6.36	7.60	4.19	10.30	8.78	2.66	90.38
1978	5.41	4.26	12.95	6.53	9.64	10.99	11.19	13.53	5.44	10.12	20.21	8.82	119.09
1979	32.24	45.55	5.32	9.90	4.10	10.45	6.54	7.04	3.64	5.03	21.56	7.40	158.77
1980	0.91	4.14	49.93	11.01	5.88	9.66	9.17	8.24	13.70	7.69	7.13	0.28	127.74
1981	1.51	4.95	5.66	4.63	4.16	2.43	4.32	8.97	12.79	10.23	11.73	18.53	89.91
1982	13.58	1.35	48.50	12.00	6.89	6.03	28.59	25.45	9.92	6.53	4.74	6.78	170.36
1983	0.90	0.83	1.98	10.31	9.60	3.94	7.21	7.48	12.08	8.06	2.33	3.37	68.09
1984	10.76	10.06	3.37	12.08	6.59	4.28	6.63	9.36	4.05	2.52	18.38	12.00	100.08
1985	2.25	16.14	21.28	10.61	17.04	1.80	9.86	6.71	11.78	8.19	4.71	2.59	112.96
1986	4.95	0.58	15.37	43.24	8.61	9.11	11.17	10.64	14.36	11.53	35.72	5.75	171.03
1987	9.02	5.06	4.79	9.24	15.65	12.91	18.26	11.56	14.21	15.83	22.19		142.41
1988	10.31	9.95	13.09	12.90	7.77	5.11	5.50	16.56	11.30	8.50	25.74	13.46	140.19
1989	27.46	6.54	7.33	37.19	19.80	7.03	22.93	8.82	9.73	13.16	1.01	5.71	166.71
1990	29.13	15.24	10.80	4.02	8.13	10.04	10.78	7.80	18.47	20.96	45.75	30.10	211.22
1991	3.81	9.32	37.88	11.02	8.08	9.85	9.82	26.92	9.41	5.15	6.74	15.04	153.04
1992	1.33	1.29	3.90	6.62	2.99	9.36	17.63	13.62	17.59	3.38	25.16	17.02	119.89
1993	2.17	2.67	11.96	9.04	7.54	6.63	18.43	11.38	4.99	12.83	10.74	16.11	114.49
1994	10.39	25.52	18.48	8.59	7.18	13.29	11.71	14.58	21.82	8.73	35.91	6.61	182.81
1995	4.52	1.56	4.17	8.14	8.68	5.35	15.13	13.93	4.20	7.62	8.52	4.10	85.92
1996	14.29	11.81	16.66	6.27	3.65	10.33	13.22	4.77	7.03	11.07	14.22	6.89	120.21
1997	2.33	7.84	19.25	6.03	10.75	22.70	19.38	4.75	8.98	12.64	8.86	8.10	131.61
1998	0.13	2.40	3.67	8.86	15.65	11.27	8.48	6.09	10.76	16.01	15.57	3.37	108.78
1999	16.68	19.34	12.13	16.04	2.84	4.66	3.54	10.14	5.65	3.61	7.74	14.41	116.78
2000	17.87	0.52	5.81	7.25	3.36	8.19	13.16	10.54	9.20	17.65	45.90	4.59	144.04
2001	2.28	12.47	8.35	12.56	2.94	3.64	6.54	7.90	9.01	13.16	19.89	13.77	112.51
POR= 59 YRS	9.48	11.18	13.46	12.75	8.84	6.89	10.13	10.15	8.15	10.07	15.38	13.20	129.68

AVERAGE TEMPERATURE (F) 2001   HILO, HI (ITO)

YEAR	JAN	FEB	MAR	APR	MAY	JUN	JUL	AUG	SEP	OCT	NOV	DEC	ANNUAL
1972	70.1	70.7	73.8	72.7	73.0	75.3	75.4	76.5	76.4	76.0	73.3	71.4	73.7
1973	72.2	71.1	72.5	72.2	72.9	74.6	75.7	76.3	76.3	75.8	75.4	73.8	74.1
1974	74.5	72.6	73.1	73.5	73.8	75.3	76.1	76.9	77.3	76.9	73.6	72.3	74.7
1975	71.0	71.9	71.2	72.4	73.1	74.4	74.8	75.7	75.5	74.7	73.5	72.2	73.4
1976	71.3	71.2	71.6	72.1	73.0	73.6	74.5	76.2	76.9	76.2	74.5	73.2	73.7
1977	73.9	74.0	73.3	74.2	74.7	76.2	77.1	78.1	77.5	76.9	75.2	73.7	75.4
1978	71.7	72.0	73.2	74.2	76.2	76.5	77.1	76.8	76.2	75.5	74.1	71.1	74.6
1979	69.8	70.4	71.5	73.8	73.8	74.2	74.6	75.6	76.5	76.1	73.0	72.8	73.5
1980	71.6	72.6	72.3	74.5	77.3	77.6	77.8	75.0	75.7	74.8	73.8	74.2	74.8
1981	73.5	72.7	71.6	72.8	74.2	76.0	76.1	76.1	76.2	74.6	73.9	72.0	74.1
1982	71.9	71.8	70.3	71.2	72.9	76.3	76.7	76.9	76.1	74.9	74.6	71.8	73.8
1983	71.4	71.9	72.5	71.9	72.6	74.3	74.8	75.2	74.9	74.1	73.8	72.9	73.4
1984	72.4	71.5	73.8	73.0	74.0	74.7	75.2	75.3	75.4	76.5	73.6	71.1	73.9
1985	69.8	70.5	69.4	69.8	71.4	74.4	75.4	75.7	75.7	74.3	73.0	71.6	72.6
1986	71.1	73.6	74.7	73.6	75.4	76.6	77.8	78.5	77.9	76.4	75.1	72.8	75.3
1987	71.8	70.7	71.6	72.2	72.5	75.4	76.7	77.9	77.8	76.6	74.7	73.1	74.3
1988	71.9	72.3	72.2	72.6	74.2	74.7	75.7	76.0	76.6	77.9	76.3	74.9	74.6
1989	72.2	71.4	72.4	71.1	72.7	74.7	75.2	75.0	74.6	75.6	73.6	71.3	73.3
1990	72.1	70.4	71.2	73.5	74.1	75.0	76.0	77.0	77.?	76.2	76.1	72.5	74.2
1991	72.0	72.8	70.8	72.6	74.2	74.8	76.0	76.9	76.9	76.2	75.8	72.9	74.3
1992	71.2	71.4	72.3	72.4	74.8	76.2	76.2	77.2	77.8	77.7	75.2	73.6	74.7
1993	71.1	70.1	71.6	73.5	73.3	75.4	75.8	77.0	77.1	76.0	73.4	71.7	73.8
1994	70.0	71.3	71.7	73.4	74.9	76.0	78.1	78.6	78.1	77.4	74.9	74.9	74.8
1995	72.6	72.9	74.8	74.1	75.5	76.9	77.9	77.7	78.2	76.2	75.6	74.6	75.6
1996	73.4	70.9	71.5	74.2	76.1	76.9	77.5	77.5	77.5	77.3	75.8	73.2	75.2
1997	73.1	72.7	73.3	74.0	75.0	76.7	77.4	78.0	77.7	76.8	74.3	72.4	75.1
1998	71.8	71.7	72.9	71.8	72.2	73.8	74.5	76.1	76.1	74.2	72.4	69.9	73.0
1999	69.7	69.0	70.5	71.1	72.9	73.4	74.3	74.3	74.0	74.0	71.3	71.3	72.2
2000	69.1	71.3	71.8	71.3	73.9	75.5	75.5	76.1	75.4	75.3	73.0	71.7	73.3
2001	71.6	71.0	71.2	71.5	72.3	73.9	75.4	76.1	76.2	75.1	74.0	72.7	73.4
POR= 55 YRS	71.3	71.2	71.5	72.3	73.5	74.9	75.5	76.1	75.9	75.4	73.8	72.0	73.6

REFERENCE NOTES:

PAGE 1:
THE TEMPERATURE GRAPH SHOWS NORMAL MAXIMUM AND NORMAL
MINIMUM DAILY TEMPERATURES (SOLID CURVES) AND THE
ACTUAL DAILY HIGH AND LOW TEMPERATURES (VERTICAL BARS).

PAGE 2 AND 3:
H/C INDICATES HEATING AND COOLING DEGREE DAYS.
RH INDICATES RELATIVE HUMIDITY
W/O INDICATES WEATHER AND OBSTRUCTIONS
S INDICATES SUNSHINE.
PR INDICATES PRESSURE.
CLOUDINESS ON PAGE 3 IS THE SUM OF THE CEILOMETER AND
SATELLITE DATA NOT TO EXCEED EIGHT EIGHTHS(OKTAS).

GENERAL:
T INDICATES TRACE PRECIPITATION, AN AMOUNT GREATER
THAN ZERO BUT LESS THAN THE LOWEST REPORTABLE VALUE.
+ INDICATES THE VALUE ALSO OCCURS ON EARLIER DATES.
BLANK ENTRIES DENOTE MISSING OR UNREPORTED DATA.
NORMALS ARE 30-YEAR AVERAGES (1961 - 1990).
ASOS INDICATES AUTOMATED SURFACE OBSERVING SYSTEM.
PM INDICATES THE LAST DAY OF THE PREVIOUS MONTH.
POR (PERIOD OF RECORD) BEGINS WITH THE JANUARY DATA
MONTH AND IS THE NUMBER OF YEARS USED TO COMPUTE
THE MEAN. INDIVIDUAL MONTHS WITHIN THE POR MAY
BE MISSING.
WHEN THE POR FOR A NORMAL IS LESS THAN 30 YEARS,
THE NORMAL IS PROVISIONAL AND IS BASED ON THE NUMBER
OF YEARS INDICATED.
0.0 OR * INDICATES THE VALUE OR MEAN-DAYS-WITH
IS BETWEEN 0.00 AND 0.05.
CLOUDINESS FOR ASOS STATIONS DIFFERS FROM THE NON-ASOS
OBSERVATION TAKEN BY A HUMAN OBSERVER. ASOS STATION
CLOUDINESS IS BASED ON TIME-AVERAGED CEILOMETER DATA
FOR CLOUDS AT OR BELOW 12,000 FEET AND ON SATELLITE
DATA FOR CLOUDS ABOVE 12,000 FEET.
THE NUMBER OF DAYS WITH CLEAR, PARTLY CLOUDY, AND
CLOUDY CONDITIONS FOR ASOS STATIONS IS THE SUM
OF THE CEILOMETER AND SATELLITE DATA FOR THE
SUNRISE TO SUNSET PERIOD.

GENERAL CONTINUED:
CLEAR INDICATES 0 - 2 OKTAS, PARTLY CLOUDY INDICATES
3 - 6 OKTAS, AND CLOUDY INDICATES 7 OR 8 OKTAS.
WHEN AT LEAST ONE OF THE ELEMENTS (CEILOMETER OR
SATELLITE) IS MISSING, THE DAILY CLOUDINESS IS
NOT COMPUTED.
WIND DIRECTION IS RECORDED IN TENS OF DEGREES (2 DIGITS)
CLOCKWISE FROM TRUE NORTH. "00" INDICATES CALM. "36"
INDICATES TRUE NORTH.
RESULTANT WIND IS THE VECTOR AVERAGE OF THE SPEED AND
DIRECTION.
AVERAGE TEMPERATURE IS THE SUM OF THE MEAN DAILY MAXIMUM
AND MINIMUM TEMPERATURE DIVIDED BY 2.
SNOWFALL DATA COMPRISE ALL FORMS OF FROZEN
PRECIPITATION, INCLUDING HAIL.
A HEATING (COOLING) DEGREE DAY IS THE DIFFERENCE BETWEEN
THE AVERAGE DAILY TEMPERATURE AND 65 F.
DRY BULB IS THE TEMPERATURE OF THE AMBIENT AIR.
DEW POINT IS THE TEMPERATURE TO WHICH THE AIR MUST BE
COOLED TO ACHIEVE 100 PERCENT RELATIVE HUMIDITY.
WET BULB IS THE TEMPERATURE THE AIR WOULD HAVE IF THE
MOISTURE CONTENT WAS INCREASED TO 100 PERCENT RELATIVE
HUMIDITY.

ON JULY 1, 1996, THE NATIONAL WEATHER SERVICE BEGAN USING
THE "METAR" OBSERVATION CODE THAT WAS ALREADY EMPLOYED
BY MOST OTHER NATIONS OF THE WORLD. THE MOST NOTICEABLE
DIFFERENCE IN THIS ANNUAL PUBLICATION WILL BE THE CHANGE
IN UNITS FROM TENTHS TO EIGHTS(OKTAS) FOR REPORTING THE
AMOUNT OF SKY COVER.

HEATING DEGREE DAYS (base 65 F) 2001 HILO, HI (ITO)

YEAR	JUL	AUG	SEP	OCT	NOV	DEC	JAN	FEB	MAR	APR	MAY	JUN	TOTAL
1983-84	0	0	0	0	0	0	0	0	0	0	0	0	0
1984-85	0	0	0	0	0	0	0	0	0	0	0	0	0
1985-86	0	0	0	0	0	0	0	0	0	0	0	0	0
1986-87	0	0	0	0	0	0	0	0	0	0	0	0	0
1987-88	0	0	0	0	0	0	0	0	0	0	0	0	0
1988-89	0	0	0	0	0	0	0	0	0	0	0	0	0
1989-90	0	0	0	0	0	0	0	0	0	0	0	0	0
1990-91	0	0	0	0	0	0	0	0	0	0	0	0	0
1991-92	0	0	0	0	0	0	0	0	0	0	0	0	0
1992-93	0	0	0	0	0	0	0	0	0	0	0	0	0
1993-94	0	0	0	0	0	0	0	0	0	0	0	0	0
1994-95	0	0	0	0	0	0	0	0	0	0	0	0	0
1995-96	0	0	0	0	0	0	0	0	0	0	0	0	0
1996-97	0	0	0	0	0	0	0	0	0	0	0	0	0
1997-98	0	0	0	0	0	0	0	0	0	0	0	0	0
1998-99	0	0	0	0	0	0	0	0	0	0	0	0	0
1999-00	0	0	0	0	0	0	0	0	0	0	0	0	0
2000-01	0	0	0	0	0	0	0						
2001-	0												

COOLING DEGREE DAYS (base 65 F) 2001 HILO, HI (ITO)

YEAR	JAN	FEB	MAR	APR	MAY	JUN	JUL	AUG	SEP	OCT	NOV	DEC	ANNUAL
1972	163	171	281	236	256	316	330	365	349	348	256	207	3278
1973	233	180	239	222	253	294	341	358	345	340	321	278	3404
1974	299	219	261	263	276	315	351	375	375	376	262	235	3607
1975	192	201	197	232	257	288	311	339	323	309	262	233	3144
1976	201	186	214	222	255	268	302	355	364	357	291	261	3276
1977	280	260	264	281	307	343	379	415	382	374	312	274	3871
1978	216	215	263	283	353	351	383	375	341	332	279	195	3586
1979	155	160	210	271	278	280	302	338	351	350	246	248	3189
1980	213	227	234	293	390	385	405	316	328	313	269	295	3668
1981	271	220	210	242	293	338	350	348	345	302	274	225	3418
1982	220	196	170	194	252	348	369	379	340	317	293	219	3297
1983	207	200	239	214	240	287	313	324	303	288	272	250	3137
1984	236	194	282	247	284	298	324	326	320	363	261	195	3330
1985	154	161	142	152	204	290	329	339	329	294	248	211	2853
1986	196	246	308	264	329	356	404	423	396	363	309	250	3844
1987	218	163	212	226	241	319	369	407	389	365	299	259	3467
1988	221	216	233	238	293	298	338	349	353	405	345	315	3604
1989	227	188	238	189	248	297	327	315	294	335	264	202	3124
1990	227	157	200	260	290	308	349	379	376	353	317	237	3453
1991	223	222	188	234	296	301	348	378	365	351	333	251	3490
1992	197	192	235	229	312	343	355	384	387	402	315	275	3626
1993	193	148	213	263	263	318	343	380	370	350	260	217	3318
1994	161	183	214	261	312	338	412	427	401	389	305	254	3657
1995	241	228	310	281	335	364	410	402	401	355	325	303	3955
1996	266	178	209	291	349	360	394	394	379	387	331	265	3805
1997	261	221	261	277	317	358	388	407	389	373	287	239	3778
1998	216	194	253	211	230	269	311	351	293	289	228	162	3007
1999	149	117	179	189	252	257	292	296	278	284	228	200	2721
2000	133	191	221	194	280	322	332	350	317	328	247	213	3128
2001	209	174	201	204	234	276	330	350	341	319	278	246	3162

SNOWFALL (inches) 2001 HILO, HI (ITO)

YEAR	JUL	AUG	SEP	OCT	NOV	DEC	JAN	FEB	MAR	APR	MAY	JUN	TOTAL
1972-73	0.0	0.0	0.0	0.0	0.0	0.0	0.0	0.0	0.0	0.0	0.0	0.0	0.0
1973-74	0.0	0.0	0.0	0.0	0.0	0.0	0.0	0.0	0.0	0.0	0.0	0.0	0.0
1974-75	0.0	0.0	0.0	0.0	0.0	0.0	0.0	0.0	0.0	0.0	0.0	0.0	0.0
1975-76	0.0	0.0	0.0	0.0	0.0	0.0	0.0	0.0	0.0	0.0	0.0	0.0	0.0
1976-77	0.0	0.0	0.0	0.0	0.0	0.0	0.0	0.0	0.0	0.0	0.0	0.0	0.0
1977-78	0.0	0.0	0.0	0.0	0.0	0.0	0.0	0.0	0.0	0.0	0.0	0.0	0.0
1978-79	0.0	0.0	0.0	0.0	0.0	0.0	0.0	0.0	0.0	0.0	0.0	0.0	0.0
1979-80	0.0	0.0	0.0	0.0	0.0	0.0	0.0	0.0	0.0	0.0	0.0	0.0	0.0
1980-81	0.0	0.0	0.0	0.0	0.0	0.0	0.0	0.0	0.0	0.0	0.0	0.0	0.0
1981-82	0.0	0.0	0.0	0.0	0.0	0.0	0.0	0.0	0.0	0.0	0.0	0.0	0.0
1982-83	0.0	0.0	0.0	0.0	0.0	0.0	0.0	0.0	0.0	0.0	0.0	0.0	0.0
1983-84	0.0	0.0	0.0	0.0	0.0	0.0	0.0	0.0	0.0	0.0	0.0	0.0	0.0
1984-85	0.0	0.0	0.0	0.0	0.0	0.0	0.0	0.0	0.0	0.0	0.0	0.0	0.0
1985-86	0.0	0.0	0.0	0.0	0.0	0.0	0.0	0.0	0.0	0.0	0.0	0.0	0.0
1986-87	0.0	0.0	0.0	0.0	0.0	0.0	0.0	0.0	0.0	0.0	0.0	0.0	0.0
1987-88	0.0	0.0	0.0	0.0	0.0	0.0	0.0	0.0	0.0	0.0	0.0	0.0	0.0
1988-89	0.0	0.0	0.0	0.0	0.0	0.0	0.0	0.0	0.0	0.0	0.0	0.0	0.0
1989-90	0.0	0.0	0.0	0.0	0.0	0.0	0.0	0.0	0.0	0.0	0.0	0.0	0.0
1990-91	0.0	0.0	0.0	0.0	0.0	0.0	0.0	0.0	0.0	0.0	0.0	0.0	0.0
1991-92	0.0	0.0	0.0	0.0	0.0	0.0	0.0	0.0	0.0	0.0	0.0	0.0	0.0
1992-93	0.0	0.0	0.0	0.0	0.0	0.0	0.0	0.0	0.0	0.0	0.0	0.0	0.0
1993-94	0.0	0.0	0.0	0.0	0.0	0.0	0.0	0.0	0.0	0.0	0.0	0.0	0.0
1994-95	0.0	0.0	0.0	0.0	0.0	0.0	0.0	0.0	0.0	0.0	0.0	0.0	0.0
1995-96	0.0	0.0	0.0	0.0	0.0	0.0	0.0	0.0	0.0	0.0	0.0	0.0	0.0
1996-97	0.0	0.0	0.0	0.0	0.0	0.0	0.0	0.0	0.0	0.0	0.0	0.0	0.0
1997-98	0.0	0.0	0.0	0.0	0.0	0.0							
1998-99													
1999-00													
2000-01													
2001-													
POR= 54 YRS	0.0	0.0	0.0	0.0	0.0	0.0	0.0	0.0	0.0	0.0	0.0	0.0	0.0

2001
HONOLULU,
HAWAII (HNL)

Oahu, on which Honolulu is located, is the third largest of the Hawaiian Islands. The Koolau Range, at an average elevation of 2,000 feet parallels the northeastern coast. The Waianae Mountains, somewhat higher in elevation, parallel the west coast. Honolulu Airport, the business and Waikiki districts, and a number of the residential areas of Honolulu lie along the southern coastal plain.

The climate of Hawaii is unusually pleasant for the tropics. Its outstanding features are the persistence of the trade winds, the remarkable variability in rainfall over short distances, the sunniness of the leeward lowlands in contrast to the persistent cloudiness over nearby mountain crests, the equable temperature, and the general infrequency of severe storms.

The prevailing wind throughout the year is the northeasterly trade wind, although its average frequency varies from more than 90 percent during the summer to only 50 percent in January.

Heavy mountain rainfall sustains extensive irrigation of cane fields and the water supply for Honolulu. Oahu is driest along the coast west of the Waianaes where rainfall drops to about 20 inches a year. Daytime showers, usually light, often occur while the sun continues to shine, a phenomenon referred to locally as liquid sunshine.

The moderate temperature range is associated with the small seasonal variation in the energy received from the sun and the tempering effect of the surrounding ocean. Honolulu Airport has recorded as high as the lower 90s and as low as the lower 50s.

Because of the trade winds, even the warmest months are usually comfortable. But when the trades diminish or give way to southerly winds, a situation known locally as kona weather, or kona storms when stormy, the humidity may become oppressively high.

Intense rains of the October to April winter season sometimes cause serious, flash flooding. Thunderstorms are infrequent and usually mild and hail seldom occurs. Infrequently, a small tornado or a waterspout may do some damage. Only a few tropical cyclones have struck Hawaii, although others have come near enough for their outlying winds, waves, clouds, and rain to affect the Islands.

NORMALS, MEANS, AND EXTREMES
HONOLULU, HI (HNL)

LATITUDE: 21 19' 39" N LONGITUDE: 157 56' 35" W ELEVATION (FT): GRND: 15 BARO: 18 TIME ZONE: HAWAII (UTC + 10) WBAN: 22521

	ELEMENT	POR	JAN	FEB	MAR	APR	MAY	JUN	JUL	AUG	SEP	OCT	NOV	DEC	YEAR
TEMPERATURE F	NORMAL DAILY MAXIMUM	30	80.1	80.5	81.6	82.8	84.7	86.5	87.5	88.7	88.5	86.9	84.1	81.2	84.4
	MEAN DAILY MAXIMUM	52	80.0	80.0	81.0	82.4	84.2	86.1	86.9	88.0	87.9	86.5	83.7	81.0	84.0
	HIGHEST DAILY MAXIMUM	32	88	88	88	91	93	92	94	93	95	94	93	89	95
	YEAR OF OCCURRENCE		1996	1984	1998	1996	1988	1996	1995	1997	1994	1984	1986	1995	SEP 1994
	MEAN OF EXTREME MAXS.	52	83.9	83.9	84.8	86.0	87.4	88.5	89.4	90.6	90.7	89.7	87.3	85.0	87.3
	NORMAL DAILY MINIMUM	30	65.6	65.4	67.2	68.7	70.3	72.2	73.5	74.2	73.5	72.3	70.3	67.0	70.0
	MEAN DAILY MINIMUM	52	65.7	65.5	67.0	68.7	70.2	72.2	73.4	74.2	73.5	72.3	70.4	67.6	70.1
	LOWEST DAILY MINIMUM	32	53	53	55	57	60	65	66	67	66	61	57	54	53
	YEAR OF OCCURRENCE		1998	1983	1976	1985	1989	1982	1990	1984	1985	1993	1990	1962	JAN 1998
	MEAN OF EXTREME MINS.	52	58.8	58.9	60.8	63.5	65.5	68.1	69.8	70.2	69.2	67.0	64.4	60.2	64.7
	NORMAL DRY BULB	30	72.9	73.0	74.4	75.8	77.5	79.4	80.5	81.4	81.0	79.6	77.2	74.1	77.2
	MEAN DRY BULB	52	72.8	72.8	74.0	75.5	77.1	79.1	80.2	81.0	80.7	79.3	77.0	74.2	77.0
	MEAN WET BULB	17	66.7	66.3	67.0	67.7	68.8	70.4	71.7	72.3	72.6	72.0	70.3	68.2	69.5
	MEAN DEW POINT	17	62.9	62.5	62.6	63.2	64.3	65.7	67.1	67.9	68.7	68.2	66.7	64.6	65.4
	NORMAL NO. DAYS WITH:														
	MAXIMUM 90	30	0.0	0.0	0.0	0.0	0.3	1.4	4.4	11.2	9.6	4.1	0.4	0.0	31.4
	MAXIMUM 32	30	0.0	0.0	0.0	0.0	0.0	0.0	0.0	0.0	0.0	0.0	0.0	0.0	0.0
	MINIMUM 32	30	0.0	0.0	0.0	0.0	0.0	0.0	0.0	0.0	0.0	0.0	0.0	0.0	0.0
	MINIMUM 0	30	0.0	0.0	0.0	0.0	0.0	0.0	0.0	0.0	0.0	0.0	0.0	0.0	0.0
H/C	NORMAL HEATING DEG. DAYS	30	0	0	0	0	0	0	0	0	0	0	0	0	0
	NORMAL COOLING DEG. DAYS	30	245	224	291	324	388	432	481	508	480	453	366	282	4474
RH	NORMAL (PERCENT)	30	73	71	69	67	66	64	65	64	66	68	70	72	68
	HOUR 02 LST	30	81	79	76	75	74	73	73	72	74	76	78	80	76
	HOUR 08 LST	30	81	78	73	70	67	66	67	67	68	70	74	78	72
	HOUR 14 LST	30	62	59	57	56	54	52	52	51	52	56	59	61	56
	HOUR 20 LST	30	74	71	70	70	69	67	68	68	68	70	72	74	70
S	PERCENT POSSIBLE SUNSHINE	46	65	68	72	70	72	74	76	77	77	71	64	63	71
W/O	MEAN NO. DAYS WITH:														
	HEAVY FOG (VISBY 1/4 MI)	52	0.0	0.0	0.0	0.0	0.0	0.0	0.0	0.0	0.0	0.0	0.0	0.0	0.0
	THUNDERSTORMS	52	0.7	1.1	0.8	0.5	0.3	0.1	0.2	0.1	0.5	0.8	0.9	0.7	6.7
CLOUDINESS	MEAN:														
	SUNRISE-SUNSET (OKTAS)	51	4.3	4.4	4.6	4.9	4.6	4.4	4.2	4.1	4.2	4.5	4.6	4.4	4.4
	MIDNIGHT-MIDNIGHT (OKTAS)	33	4.1	4.1	4.3	4.5	4.3	4.2	4.2	3.9	3.9	4.1	4.5	4.3	4.2
	MEAN NO. DAYS WITH:														
	CLEAR	48	9.5	8.1	7.4	5.9	6.7	6.5	7.4	8.0	7.9	7.5	7.2	7.9	90.0
	PARTLY CLOUDY	48	12.9	12.5	14.1	14.6	15.6	17.4	18.0	16.9	15.9	14.8	13.4	13.7	179.8
	CLOUDY	49	8.5	7.6	9.3	9.6	8.7	6.2	5.1	5.7	5.7	8.1	8.8	8.7	92.0
PR	MEAN STATION PRESSURE (IN)	28	29.98	29.99	30.04	30.03	30.03	30.02	30.00	29.98	29.99	29.96	29.97	29.98	30.00
	MEAN SEA-LEVEL PRES. (IN)	17	30.00	30.01	30.06	30.06	30.04	30.02	30.01	30.00	29.99	29.97	29.99	30.01	30.01
WINDS	MEAN SPEED (MPH)	52	9.4	10.1	11.3	11.7	11.6	12.6	13.4	13.0	11.4	10.7	10.9	10.6	11.4
	PREVAIL.DIR (TENS OF DEGS)	5	06	06	06	06	06	06	06	06	06	06	06	06	06
	MAXIMUM 2-MINUTE:														
	SPEED (MPH)	3	32	33	31	35	26	30	30	31	26	29	30	35	35
	DIR. (TENS OF DEGS)		06	06	06	05	06	07	06	07	06	06	07	07	07
	YEAR OF OCCURRENCE		2000	1999	1999	2000	2000	1999	1999	2001	1999	2001	2000	2001	DEC 2001
	MAXIMUM 5-SECOND:														
	SPEED (MPH)	3	38	43	38	44	35	37	36	37	33	37	37	40	44
	DIR. (TENS OF DEGS)		05	07	06	07	08	07	08	07	05	07	06	08	07
	YEAR OF OCCURRENCE		2000	1999	1999	2000	2000	1999	2001	2001	2001	2001	2000	2001	APR 2000
PRECIPITATION	NORMAL (IN)	30	3.55	2.21	2.20	1.54	1.13	0.50	0.59	0.44	0.78	2.28	3.00	3.80	22.02
	MAXIMUM MONTHLY (IN)	55	14.74	13.68	20.79	8.92	7.23	2.46	2.33	3.08	2.74	11.15	18.79	17.29	20.79
	YEAR OF OCCURRENCE		1949	1955	1951	1963	1965	1971	1989	1959	1947	1978	1996	1987	MAR 1951
	MINIMUM MONTHLY (IN)	55	0.18	0.06	0.01	0.01	0.03	T	0.03	T	0.05	0.07	0.03	0.06	T
	YEAR OF OCCURRENCE		1986	1983	1957	1960	2000	1959	1950	1974	1977	1996	1962	1976	AUG 1974
	MAXIMUM IN 24 HOURS (IN)	52	6.72	6.88	17.07	4.21	3.44	2.28	2.20	2.35	1.40	7.57	9.15	8.25	17.07
	YEAR OF OCCURRENCE		1963	1955	1958	1972	1965	1967	1989	1959	1963	1978	1954	1987	MAR 1958
	NORMAL NO. DAYS WITH:														
	PRECIPITATION 0.01	30	9.7	8.6	9.2	9.5	7.9	5.8	7.4	6.1	7.4	8.8	9.8	10.5	100.7
	PRECIPITATION 1.00	30	0.9	0.6	0.8	0.2	0.3	0.1	*	0.1	0.1	0.6	0.9	1.0	5.6
SNOWFALL	NORMAL (IN)	30	0.0	0.0	0.0	0.0	0.0	0.0	0.0	0.0	0.0	0.0	0.0	0.0	0.0
	MAXIMUM MONTHLY (IN)	2	0.0	0.0	0.0	0.0	0.0	0.0	0.0	0.0	0.0	0.0	0.0	0.0	0.0
	YEAR OF OCCURRENCE														
	MAXIMUM IN 24 HOURS (IN)	52	0.0	0.0	0.0	0.0	0.0	0.0	0.0	0.0	0.0	0.0	0.0	0.0	0.0
	YEAR OF OCCURRENCE														
	MAXIMUM SNOW DEPTH (IN)	48	0	0	0	0	0	0	0	0	0	0	0	0	0
	YEAR OF OCCURRENCE														
	NORMAL NO. DAYS WITH:														
	SNOWFALL 1.0	30	0.0	0.0	0.0	0.0	0.0	0.0	0.0	0.0	0.0	0.0	0.0	0.0	0.0

PRECIPITATION (inches) 2001 HONOLULU, HI (HNL)

YEAR	JAN	FEB	MAR	APR	MAY	JUN	JUL	AUG	SEP	OCT	NOV	DEC	ANNUAL
1972	5.28	5.00	2.45	5.15	0.12	0.79	0.20	0.46	0.92	2.39	0.59	3.59	26.94
1973	0.67	0.60	0.40	0.72	0.89	0.09	0.46	0.32	0.64	1.78	3.73	3.94	14.24
1974	4.21	1.28	3.49	4.13	0.82	1.52	0.44	T	2.08	2.77	2.69	0.59	24.02
1975	6.42	2.36	2.02	0.51	0.19	0.03	0.40	0.03	0.11	0.18	11.54	0.60	24.39
1976	1.29	6.08	2.67	0.71	0.26	0.18	0.24	0.17	0.33	0.45	0.46	0.06	12.90
1977	0.52	0.32	2.36	1.81	4.76	0.11	0.14	0.08	0.05	0.15	0.61	1.45	12.36
1978	0.34	0.75	1.37	2.07	3.39	1.06	0.20	0.83	0.28	11.15	1.55	2.06	25.05
1979	4.57	7.21	0.77	0.55	0.21	0.32	0.13	0.15	0.47	0.53	0.52	1.50	16.93
1980	8.91	2.26	3.04	1.13	0.78	1.76	0.37	0.36	0.41	0.30	0.21	7.37	26.90
1981	0.81	0.97	0.71	1.01	0.94	0.14	0.42	0.70	0.39	1.84	1.01	4.47	13.41
1982	12.82	2.16	3.73	1.28	0.13	0.35	0.20	1.98	0.52	7.24	1.32	3.19	34.92
1983	0.32	0.06	0.53	0.42	0.35	0.26	0.22	0.29	1.16	0.23	0.13	1.06	5.03
1984	0.21	0.60	1.08	2.41	0.16	0.08	0.23	0.04	1.36	1.89	3.58	5.44	17.08
1985	1.46	3.87	1.26	0.20	1.11	0.13	0.53	0.16	1.28	5.08	2.11	0.19	17.38
1986	0.18	1.38	0.17	0.35	0.81	0.36	1.54	0.90	2.00	1.23	4.23	0.78	13.93
1987	0.42	0.86	0.31	0.65	0.73	0.46	0.33	0.22	1.13	0.20	0.93	17.29	23.53
1988	3.05	1.31	0.67	0.50	1.25	0.04	0.12	0.34	0.86	0.23	1.39	6.71	16.47
1989	2.07	6.48	2.58	1.23	0.29	0.11	2.33	0.08	0.15	10.37	0.51	1.32	27.52
1990	4.32	4.15	0.86	0.30	0.30	0.08	0.49	0.01	0.98	0.47	2.96	4.92	19.84
1991	0.80	2.09	6.24	1.00	0.48	0.26	0.16	0.16	0.56	3.43	1.52	1.24	17.94
1992	0.43	1.35	0.72	0.11	1.13	0.10	2.01	0.97	2.08	2.57	1.04	6.49	19.00
1993	0.70	0.41	0.02	0.23	0.19	0.11	0.69	1.03	0.10	1.63	0.46	0.27	5.84
1994	0.78	7.04	3.77	0.46	0.09	0.23	0.48	0.14	0.39	1.56	0.20	0.45	15.59
1995	0.54	6.53	1.33	1.15	0.45	0.05	0.19	0.44	0.33	0.36	0.92	1.31	13.60
1996	3.52	1.00	2.68	0.34	0.13	1.01	0.74	2.17	0.78	0.07	18.79	1.89	33.12
1997	6.92	0.95	4.90	1.02	0.63	0.40	0.52	0.13	0.67	2.57	0.84	0.44	19.99
1998	0.77	0.21	0.03	0.54	0.16	0.29	0.19	0.15	0.05	0.13	0.85	1.15	4.52
1999	2.12	0.73	0.46	0.68	2.13	0.11	0.57	0.12	0.20	2.01	0.21	2.65	11.99
2000	1.26	0.07	0.38	0.46	0.03	0.03	0.41	1.17	0.78	0.25	2.09	0.17	7.10
2001	0.18	0.57	0.62	0.29	0.14	1.16	0.13	0.05	0.28	1.05	3.91	0.76	9.14
POR= 55 YRS	3.51	2.46	2.55	1.23	0.90	0.40	0.49	0.52	0.69	1.81	2.72	3.17	20.45

AVERAGE TEMPERATURE (F) 2001 HONOLULU, HI (HNL)

YEAR	JAN	FEB	MAR	APR	MAY	JUN	JUL	AUG	SEP	OCT	NOV	DEC	ANNUAL
1972	70.4	70.6	72.8	75.0	77.3	78.9	80.4	81.1	80.5	79.3	76.7	71.6	76.2
1973	72.9	72.6	76.1	75.5	77.1	79.2	80.5	81.2	81.0	79.4	77.0	73.8	77.2
1974	74.5	74.4	74.0	77.4	78.2	79.3	79.9	81.2	80.0	79.5	75.7	75.8	77.5
1975	72.4	72.8	73.0	74.5	75.7	78.1	79.4	80.1	79.4	79.1	77.1	73.0	76.2
1976	73.7	72.0	73.6	75.1	77.5	78.2	79.8	80.8	80.7	79.1	75.3	75.3	76.8
1977	73.7	75.6	76.2	76.3	77.6	79.5	80.9	82.2	81.6	81.1	78.6	75.1	78.2
1978	74.2	73.2	75.7	76.8	78.2	78.7	79.0	80.5	80.5	77.8	74.7	76.8	76.8
1979	69.9	72.1	72.8	74.8	78.0	80.0	80.9	80.4	81.1	81.0	77.4	75.3	77.0
1980	71.9	72.4	75.0	76.1	78.3	79.5	80.9	81.0	81.6	80.1	78.0	74.4	77.4
1981	73.2	73.6	74.7	75.9	77.3	80.6	79.7	80.1	80.7	78.3	76.7	74.0	77.1
1982	73.2	71.7	74.0	74.6	78.3	79.6	80.6	81.4	81.4	79.4	75.7	72.0	76.9
1983	71.9	71.3	73.5	74.6	75.7	78.9	79.7	82.4	82.3	81.1	80.1	75.1	77.2
1984	74.6	74.6	75.8	77.0	78.7	79.3	81.0	81.7	81.3	80.2	79.0	74.1	78.1
1985	71.4	73.9	74.5	74.5	76.5	79.2	81.6	81.9	81.1	79.8	75.1	73.3	76.9
1986	72.8	72.6	76.5	77.5	78.3	80.0	81.6	81.6	82.1	80.6	79.2	75.1	78.3
1987	73.4	71.2	74.0	76.0	75.7	80.4	82.1	82.7	82.9	81.4	78.8	75.8	78.5
1988	73.1	74.7	76.0	77.3	78.9	80.8	81.8	82.1	82.1	80.1	79.9	75.6	78.5
1989	74.5	73.6	75.3	74.5	78.4	80.9	81.6	81.4	81.9	78.6	76.7	72.9	77.5
1990	74.7	71.5	73.1	76.6	78.1	80.0	80.8	82.3	82.3	80.9	77.3	74.1	77.6
1991	72.4	73.4	72.9	75.9	77.8	79.4	81.2	82.4	81.5	80.0	79.5	76.2	77.7
1992	72.9	73.2	74.9	75.6	77.8	81.3	81.5	82.2	81.3	79.4	77.0	76.7	77.8
1993	70.9	71.1	74.0	77.4	77.2	80.2	80.6	81.3	81.1	79.7	76.3	75.0	77.1
1994	72.0	73.6	73.2	76.0	79.3	80.9	82.9	84.3	84.0	82.5	80.8	76.6	78.8
1995	74.2	73.4	75.6	76.4	78.6	81.3	83.3	83.4	83.2	82.7	80.2	77.3	79.3
1996	76.2	74.0	74.3	79.9	79.0	81.3	82.1	82.8	81.4	81.7	76.9	73.1	78.6
1997	72.3	74.7	75.3	76.3	76.2	81.1	81.5	82.6	82.7	80.6	76.4	74.0	77.8
1998	72.5	72.8	75.1	75.1	76.6	78.3	79.7	81.1	81.0	79.8	77.8	74.8	77.1
1999	73.3	73.7	74.6	75.4	77.1	78.8	79.3	80.8	80.2	78.3	76.9	74.1	76.9
2000	72.5	73.6	75.4	75.3	78.3	80.5	81.1	81.4	80.5	80.4	77.6	74.7	76.9
2001	75.4	74.1	75.0	76.6	78.2	79.7	81.5	82.2	82.1	79.9	77.4	76.4	78.2
POR= 49 YRS	72.7	72.7	73.8	75.3	76.9	79.0	80.0	80.7	80.6	79.2	76.7	74.2	76.8

REFERENCE NOTES:

PAGE 1:
THE TEMPERATURE GRAPH SHOWS NORMAL MAXIMUM AND NORMAL MINIMUM DAILY TEMPERATURES (SOLID CURVES) AND THE ACTUAL DAILY HIGH AND LOW TEMPERATURES (VERTICAL BARS).

PAGE 2 AND 3:
H/C INDICATES HEATING AND COOLING DEGREE DAYS.
RH INDICATES RELATIVE HUMIDITY
W/O INDICATES WEATHER AND OBSTRUCTIONS
S INDICATES SUNSHINE.
PR INDICATES PRESSURE.
CLOUDINESS ON PAGE 3 IS THE SUM OF THE CEILOMETER AND SATELLITE DATA NOT TO EXCEED EIGHT EIGHTHS(OKTAS).

GENERAL:
T INDICATES TRACE PRECIPITATION, AN AMOUNT GREATER THAN ZERO BUT LESS THAN THE LOWEST REPORTABLE VALUE.
+ INDICATES THE VALUE ALSO OCCURS ON EARLIER DATES.
BLANK ENTRIES DENOTE MISSING OR UNREPORTED DATA.
NORMALS ARE 30-YEAR AVERAGES (1961 - 1990).
ASOS INDICATES AUTOMATED SURFACE OBSERVING SYSTEM.
PM INDICATES THE LAST DAY OF THE PREVIOUS MONTH.
POR (PERIOD OF RECORD) BEGINS WITH THE JANUARY DATA MONTH AND IS THE NUMBER OF YEARS USED TO COMPUTE THE MEAN. INDIVIDUAL MONTHS WITHIN THE POR MAY BE MISSING.
WHEN THE POR FOR A NORMAL IS LESS THAN 30 YEARS, THE NORMAL IS PROVISIONAL AND IS BASED ON THE NUMBER OF YEARS INDICATED.
0.* OR * INDICATES THE VALUE OR MEAN-DAYS-WITH IS BETWEEN 0.00 AND 0.05.
CLOUDINESS FOR ASOS STATIONS DIFFERS FROM THE NON-ASOS OBSERVATION TAKEN BY A HUMAN OBSERVER. ASOS STATION CLOUDINESS IS BASED ON TIME-AVERAGED CEILOMETER DATA FOR CLOUDS AT OR BELOW 12,000 FEET AND ON SATELLITE DATA FOR CLOUDS ABOVE 12,000 FEET.
THE NUMBER OF DAYS WITH CLEAR, PARTLY CLOUDY, AND CLOUDY CONDITIONS FOR ASOS STATIONS IS THE SUM OF THE CEILOMETER AND SATELLITE DATA FOR THE SUNRISE TO SUNSET PERIOD.

GENERAL CONTINUED:
CLEAR INDICATES 0 - 2 OKTAS, PARTLY CLOUDY INDICATES 3 - 6 OKTAS, AND CLOUDY INDICATES 7 OR 8 OKTAS.
WHEN AT LEAST ONE OF THE ELEMENTS (CEILOMETER OR SATELLITE) IS MISSING, THE DAILY CLOUDINESS IS NOT COMPUTED.
WIND DIRECTION IS RECORDED IN TENS OF DEGREES (2 DIGITS) CLOCKWISE FROM TRUE NORTH. "00" INDICATES CALM. "36" INDICATES TRUE NORTH.
RESULTANT WIND IS THE VECTOR AVERAGE OF THE SPEED AND DIRECTION.
AVERAGE TEMPERATURE IS THE SUM OF THE MEAN DAILY MAXIMUM AND MINIMUM TEMPERATURE DIVIDED BY 2.
SNOWFALL DATA COMPRISE ALL FORMS OF FROZEN PRECIPITATION, INCLUDING HAIL.
A HEATING (COOLING) DEGREE DAY IS THE DIFFERENCE BETWEEN THE AVERAGE DAILY TEMPERATURE AND 65 F.
DRY BULB IS THE TEMPERATURE OF THE AMBIENT AIR.
DEW POINT IS THE TEMPERATURE TO WHICH THE AIR MUST BE COOLED TO ACHIEVE 100 PERCENT RELATIVE HUMIDITY.
WET BULB IS THE TEMPERATURE THE AIR WOULD HAVE IF THE MOISTURE CONTENT WAS INCREASED TO 100 PERCENT RELATIVE HUMIDITY.

ON JULY 1, 1996, THE NATIONAL WEATHER SERVICE BEGAN USING THE "METAR" OBSERVATION CODE THAT WAS ALREADY EMPLOYED BY MOST OTHER NATIONS OF THE WORLD. THE MOST NOTICEABLE DIFFERENCE IN THIS ANNUAL PUBLICATION WILL BE THE CHANGE IN UNITS FROM TENTHS TO EIGHTS(OKTAS) FOR REPORTING THE AMOUNT OF SKY COVER.

HEATING DEGREE DAYS (base 65 F) 2001 HONOLULU, HI (HNL)

YEAR	JUL	AUG	SEP	OCT	NOV	DEC	JAN	FEB	MAR	APR	MAY	JUN	TOTAL
1983-84	0	0	0	0	0	0	0	0	0	0	0	0	0
1984-85	0	0	0	0	0	0	0	0	0	0	0	0	0
1985-86	0	0	0	0	0	0	0	0	0	0	0	0	0
1986-87	0	0	0	0	0	0	0	0	0	0	0	0	0
1987-88	0	0	0	0	0	0	0	0	0	0	0	0	0
1988-89	0	0	0	0	0	0	0	0	0	0	0	0	0
1989-90	0	0	0	0	0	0	0	0	0	0	0	0	0
1990-91	0	0	0	0	0	0	0	0	0	0	0	0	0
1991-92	0	0	0	0	0	0	0	0	0	0	0	0	0
1992-93	0	0	0	0	0	0	0	0	0	0	0	0	0
1993-94	0	0	0	0	0	0	0	0	0	0	0	0	0
1994-95	0	0	0	0	0	0	0	0	0	0	0	0	0
1995-96	0	0	0	0	0	0	0	0	0	0	0	0	0
1996-97							0	0	0	0	0	0	
1996-97	0	0	0	0	0	0	0	0	0	0	0	0	0
1997-98	0	0	0	0	0	0	0	0	0	0	0	0	0
1998-99	0	0	0	0	0	0		0	0	0	0	0	0
1999-00	0	0	0	0	0	0		0	0	0	0	0	0
2000-01	0	0	0	0	0	0	0	0	0	0	0	0	0
2001-	0	0	0	0	0	0							

COOLING DEGREE DAYS (base 65 F) 2001 HONOLULU, HI (HNL)

YEAR	JAN	FEB	MAR	APR	MAY	JUN	JUL	AUG	SEP	OCT	NOV	DEC	ANNUAL
1972	178	170	249	307	386	425	484	506	470	450	357	209	4191
1973	252	219	353	322	382	432	485	512	455	457	367	279	4542
1974	300	270	285	378	415	434	468	509	458	457	328	341	4643
1975	235	224	256	292	337	400	438	475	438	442	372	257	4166
1976	278	209	275	311	393	402	464	498	479	446	315	325	4395
1977	276	305	355	344	396	441	498	541	505	507	417	320	4905
1978	292	238	336	361	417	418	439	489	473	401	298	239	4401
1979	159	209	250	299	412	458	500	485	489	504	378	326	4469
1980	222	220	317	340	418	442	501	504	503	476	395	295	4633
1981	263	249	311	335	385	474	463	477	477	419	355	284	4492
1982	261	195	288	318	421	442	493	514	499	452	326	225	4434
1983	223	182	270	295	335	425	461	544	508	461	318	318	4547
1984	304	285	340	366	432	438	501	527	494	475	425	291	4878
1985	205	256	300	293	364	437	521	532	491	464	310	264	4437
1986	251	217	366	384	421	457	519	561	521	491	433	318	4939
1987	267	178	285	337	337	465	537	556	544	516	418	342	4782
1988	260	289	346	373	437	482	527	537	520	478	455	336	5040
1989	301	244	325	291	425	482	521	517	512	431	358	252	4659
1990	306	189	258	354	412	456	498	543	525	525	377	289	4708
1991	236	243	250	333	404	442	508	547	501	474	443	353	4734
1992	251	244	316	327	404	495	518	540	495	452	369	368	4779
1993	190	180	285	381	387	463	564	606	575	547	345	314	4507
1994	227	248	262	337	451	486	564	606	575	547	481	367	5151
1995	294	239	334	351	428	498	572	578	555	554	464	442	5309
1996	354	266	297	453	441	494	538	560	502	525	365	257	5052
1997	231	279	326	346	355	489	521	553	537	492	350	288	4767
1998	238	227	322	309	368	407	463	506	489	464	390	310	4493
1999	267	248	305	317	383	417	452	500	460	421	361	289	4420
2000	241	256	331	318	419	470	508	517	476	482	382	307	4707
2001	332	262	319	356	415	451	515	541	522	467	375	357	4912

SNOWFALL (inches) 2001 HONOLULU, HI (HNL)

YEAR	JUL	AUG	SEP	OCT	NOV	DEC	JAN	FEB	MAR	APR	MAY	JUN	TOTAL
1972-73	0.0	0.0	0.0	0.0	0.0	0.0	0.0	0.0	0.0	0.0	0.0	0.0	0.0
1973-74	0.0	0.0	0.0	0.0	0.0	0.0	0.0	0.0	0.0	0.0	0.0	0.0	0.0
1974-75	0.0	0.0	0.0	0.0	0.0	0.0	0.0	0.0	0.0	0.0	0.0	0.0	0.0
1975-76	0.0	0.0	0.0	0.0	0.0	0.0	0.0	0.0	0.0	0.0	0.0	0.0	0.0
1976-77	0.0	0.0	0.0	0.0	0.0	0.0	0.0	0.0	0.0	0.0	0.0	0.0	0.0
1977-78	0.0	0.0	0.0	0.0	0.0	0.0	0.0	0.0	0.0	0.0	0.0	0.0	0.0
1978-79	0.0	0.0	0.0	0.0	0.0	0.0	0.0	0.0	0.0	0.0	0.0	0.0	0.0
1979-80	0.0	0.0	0.0	0.0	0.0	0.0	0.0	0.0	0.0	0.0	0.0	0.0	0.0
1980-81	0.0	0.0	0.0	0.0	0.0	0.0	0.0	0.0	0.0	0.0	0.0	0.0	0.0
1981-82	0.0	0.0	0.0	0.0	0.0	0.0	0.0	0.0	0.0	0.0	0.0	0.0	0.0
1982-83	0.0	0.0	0.0	0.0	0.0	0.0	0.0	0.0	0.0	0.0	0.0	0.0	0.0
1983-84	0.0	0.0	0.0	0.0	0.0	0.0	0.0	0.0	0.0	0.0	0.0	0.0	0.0
1984-85	0.0	0.0	0.0	0.0	0.0	0.0	0.0	0.0	0.0	0.0	0.0	0.0	0.0
1985-86	0.0	0.0	0.0	0.0	0.0	0.0	0.0	0.0	0.0	0.0	0.0	0.0	0.0
1986-87	0.0	0.0	0.0	0.0	0.0	0.0	0.0	0.0	0.0	0.0	0.0	0.0	0.0
1987-88	0.0	0.0	0.0	0.0	0.0	0.0	0.0	0.0	0.0	0.0	0.0	0.0	0.0
1988-89	0.0	0.0	0.0	0.0	0.0	0.0	0.0	0.0	0.0	0.0	0.0	0.0	0.0
1989-90	0.0	0.0	0.0	0.0	0.0	0.0	0.0	0.0	0.0	0.0	0.0	0.0	0.0
1990-91	0.0	0.0	0.0	0.0	0.0	0.0	0.0	0.0	0.0	0.0	0.0	0.0	0.0
1991-92	0.0	0.0	0.0	0.0	0.0	0.0	0.0	0.0	0.0	0.0	0.0	0.0	0.0
1992-93	0.0	0.0	0.0	0.0	0.0	0.0	0.0	0.0	0.0	0.0	0.0	0.0	0.0
1993-94	0.0	0.0	0.0	0.0	0.0	0.0	0.0	0.0	0.0	0.0	0.0	0.0	0.0
1994-95	0.0	0.0	0.0	0.0	0.0	0.0	0.0	0.0	0.0	0.0	0.0	0.0	0.0
1995-96	0.0	0.0	0.0	0.0	0.0	0.0	0.0	0.0	0.0	0.0	0.0	0.0	0.0
1996-97	0.0	0.0	0.0	0.0	0.0	0.0	0.0	0.0	0.0	0.0	0.0	0.0	0.0
1997-98	0.0	0.0	0.0	0.0	0.0	0.0	0.0						
1998-99													
1999-00													
2000-01													
2001-													
POR= 50 YRS	0.0	0.0	0.0	0.0	0.0	0.0	0.0	0.0	0.0	0.0	0.0	0.0	0.0

2001
BOISE,
IDAHO (BOI)

Boise is situated in the Boise River Valley about 8 miles below the mouth of a mountain canyon where the valley proper begins. Sheltered by large shade trees and averaging 2,710 feet in elevation, the denser part of the city covers a gentle alluvial slope about 2 miles wide, stretching southwest from the foothills of the Boise Mountains to the river. The Boise Mountains immediately north of the city rise 5,000 to 6,000 feet above sea level in about 8 miles, the slopes partly mantled with sagebrush and then chaparral giving way near the summit to ridges of fir, spruce, and pine. Across the river, the land rises in two irregular steps, or benches, for several miles, finally reaching the low divide between the Boise and Snake Rivers. Downstream the valley widens, merging with the valley of the Snake about 40 miles to the northwest. Once semi-arid, the entire area is now irrigated from the upstream reservoirs.

Although air masses from the Pacific are considerably modified by the time they reach Boise, their influence, particularly in winter, alternates with that of atmospheric developments from other directions. The result is almost a typical upland continental type of climate in summer, while winters are usually tempered by periods of cloudy or stormy and mild weather. Autumns have prolonged periods of near ideal weather, while springtime is noted by changeable weather and varied temperatures. The Boise climate in general may be described as dry and temperate, with sufficient variation to be stimulating.

Summer hot periods rarely last longer than a few days. Temperatures of 100 degrees or higher occur nearly every year.

Winter cold spells with temperatures of 10 degrees or lower generally last longer than the summer hot spells. During cold weather, however, there is ordinarily little wind to add to the discomfort.

The normal precipitation pattern in the Boise area shows a winter high and a very pronounced summer low. Total amounts and intensity are generally greatest near the foothills, dwindling to westward and southward.

Tornadoes are very rare as are destructive force winds. Northwesterly winds, drying and rather raw in character, although of moderate velocity, are common from March through May. Diurnal southeasterly winds, descending from nearby foothills at night, frequently have a moderating effect on winter temperatures. There is an occasional, but moderate, duststorm during the warmer months, usually occurring at times of cold frontal passage.

Relative humidity is low but widespread irrigation maintains humidity several percent above the general dryness of western arid conditions in summer. Thunderstorms occur primarily during spring and summer, with less frequency during fall and occasionally during winter. December and January are the months of heavy fog or low stratus cloud conditions. Only a moderate amount of sunshine is received in the average winter, but protracted periods of clear, sunny weather are the rule in summer. Ice storms are practically unknown.

Based on the 1951-1980 period, the average first occurrence of 32 degrees Fahrenheit in the fall is October 9 and the average last occurrence in the spring is May 8.

NORMALS, MEANS, AND EXTREMES
BOISE, ID (BOI)

LATITUDE: 43 33' 54" N LONGITUDE: 116 13' 12" W ELEVATION (FT): GRND: 2858 BARO: 2861 TIME ZONE: MOUNTAIN (UTC + 7) WBAN: 24131

ELEMENT	POR	JAN	FEB	MAR	APR	MAY	JUN	JUL	AUG	SEP	OCT	NOV	DEC	YEAR
TEMPERATURE °F														
NORMAL DAILY MAXIMUM	30	36.4	44.2	52.9	61.4	71.0	80.9	90.2	88.1	77.0	64.6	48.7	37.7	62.8
MEAN DAILY MAXIMUM	54	37.0	44.4	52.7	61.7	71.1	80.2	90.1	88.3	77.9	64.8	48.7	38.3	62.9
HIGHEST DAILY MAXIMUM	62	63	71	81	92	98	109	111	110	102	94	78	65	111
YEAR OF OCCURRENCE		1953	1992	1978	1987	1986	1940	1960	1961	1945	1997	1999	1964	JUL 1960
MEAN OF EXTREME MAXS.	54	51.3	58.5	68.4	78.8	89.4	97.5	102.4	100.7	93.8	82.4	65.1	53.2	78.5
NORMAL DAILY MINIMUM	30	21.6	27.5	31.9	36.7	43.9	52.1	57.7	56.8	48.2	39.0	31.1	22.5	39.1
MEAN DAILY MINIMUM	54	22.1	27.3	31.8	37.0	44.3	51.7	58.0	57.3	48.9	39.0	30.5	23.3	39.3
LOWEST DAILY MINIMUM	62	-17	-15	6	19	22	31	35	34	23	11	-3	-25	-25
YEAR OF OCCURRENCE		1950	1989	1971	1968	1982	1995	1986	1992	1970	1971	1985	1990	DEC 1990
MEAN OF EXTREME MINS.	54	3.3	11.5	19.6	24.4	30.3	38.8	45.9	44.9	34.7	24.7	16.5	6.8	25.1
NORMAL DRY BULB	30	29.0	35.9	42.4	49.1	57.5	66.5	74.0	72.5	62.6	51.8	39.9	30.1	50.9
MEAN DRY BULB	54	29.6	35.8	42.2	49.3	57.8	66.0	74.1	72.8	63.4	51.8	39.6	30.7	51.1
MEAN WET BULB	18	27.2	31.7	37.6	42.7	48.4	53.6	57.5	56.4	50.7	42.5	34.7	26.4	42.5
MEAN DEW POINT	18	23.4	26.3	29.5	32.9	37.9	41.7	43.8	42.3	38.1	31.8	28.9	22.3	33.2
NORMAL NO. DAYS WITH:														
MAXIMUM 90	30	0.0	0.0	0.0	0.1	1.3	6.4	18.7	16.0	3.2	0.1	0.0	0.0	45.8
MAXIMUM 32	30	10.2	3.1	0.2	0.0	0.0	0.0	0.0	0.0	0.0	0.0	1.0	8.0	22.5
MINIMUM 32	30	25.7	19.5	17.0	8.9	2.1	*	0.0	0.0	0.6	5.8	16.5	25.3	121.4
MINIMUM 0	30	1.7	0.5	0.0	0.0	0.0	0.0	0.0	0.0	0.0	0.0	0.1	1.8	4.1
H/C														
NORMAL HEATING DEG. DAYS	30	1116	815	701	477	242	75	6	20	160	414	753	1082	5861
NORMAL COOLING DEG. DAYS	30	0	0	0	0	9	120	285	252	88	0	0	0	754
RH														
NORMAL (PERCENT)	30	75	70	60	52	49	45	36	37	45	54	68	75	56
HOUR 05 LST	30	79	78	72	68	67	65	54	53	60	66	76	79	68
HOUR 11 LST	30	73	66	54	46	43	39	32	33	39	47	64	72	51
HOUR 17 LST	30	68	57	43	35	31	28	21	23	29	37	57	68	41
HOUR 23 LST	30	78	75	66	59	56	50	40	41	50	60	74	78	61
S PERCENT POSSIBLE SUNSHINE	59	40	50	62	68	72	76	87	85	82	69	43	38	64
W/O MEAN NO. DAYS WITH:														
HEAVY FOG(VISBY 1/4 MI)	63	5.8	3.1	0.8	0.3	0.2	0.1	0.0	0.0	0.1	0.5	2.9	5.5	19.3
THUNDERSTORMS	63	0.0	0.3	0.6	0.9	2.8	2.8	2.6	2.4	1.6	0.6	0.3	0.1	15.0
CLOUDINESS MEAN:														
SUNRISE-SUNSET (OKTAS)														
MIDNIGHT-MIDNIGHT (OKTAS)														
MEAN NO. DAYS WITH:														
CLEAR														
PARTLY CLOUDY														
CLOUDY														
PR MEAN STATION PRESSURE(IN)	29	27.19	27.09	27.00	27.00	27.00	27.00	27.00	27.00	27.00	27.10	27.10	27.19	27.06
MEAN SEA-LEVEL PRES. (IN)	18	30.22	30.12	30.02	29.97	29.91	29.99	29.90	29.91	29.97	30.07	30.14	30.24	30.04
WINDS MEAN SPEED (MPH)	47	7.7	8.7	9.8	9.7	9.2	8.9	8.3	8.0	7.9	8.0	8.1	7.8	8.5
PREVAIL.DIR(TENS OF DEGS)	31	13	13	13	31	31	31	31	13	13	13	13	13	13
MAXIMUM 2-MINUTE:														
SPEED (MPH)	6	33	36	32	46	32	40	36	39	33	35	39	45	46
DIR. (TENS OF DEGS)		31	30	32	31	27	31	34	30	19	28	11	31	31
YEAR OF OCCURRENCE		1996	1999	2000	1998	1997	1997	1997	1999	1997	2001	2001	2000	APR 1998
MAXIMUM 5-SECOND:														
SPEED (MPH)	6	38	43	40	52	41	51	47	48	41	44	45	63	63
DIR. (TENS OF DEGS)		31	26	32	32	27	30	17	14	18	28	11	30	30
YEAR OF OCCURRENCE		1996	2000	1997	1998	1997	1997	1998	2000	1997	2001	2001	2000	DEC 2000
PRECIPITATION NORMAL (IN)	30	1.45	1.07	1.29	1.24	1.08	0.81	0.35	0.43	0.80	0.75	1.48	1.36	12.11
MAXIMUM MONTHLY (IN)	62	3.87	3.70	3.46	3.40	4.40	3.41	1.62	2.37	2.93	2.59	3.36	4.23	4.40
YEAR OF OCCURRENCE		1970	1986	1989	1955	1998	1941	1982	1968	1986	2000	1988	1983	MAY 1998
MINIMUM MONTHLY (IN)	62	0.12	0.18	0.17	0.09	T	0.01	0.00	0.00	0.00	0.00	0.14	0.09	0.00
YEAR OF OCCURRENCE		1949	1997	1994	1949	1992	1966	1947	1998	1987	1988	1976	1976	AUG 1998
MAXIMUM IN 24 HOURS (IN)	62	1.48	1.00	1.65	1.27	2.05	2.24	0.94	1.61	1.74	1.06	0.88	1.16	2.24
YEAR OF OCCURRENCE		1953	1951	1981	1969	1990	1958	1960	1979	1976	2000	1971	1955	JUN 1958
NORMAL NO. DAYS WITH:														
PRECIPITATION 0.01	30	11.8	9.7	9.7	8.2	7.2	5.6	2.8	3.3	4.2	5.6	10.4	11.4	89.9
PRECIPITATION 1.00	30	0.0	0.0	*	0.1	0.1	0.1	0.0	*	*	0.0	0.0	0.0	0.3
SNOWFALL NORMAL (IN)	30	6.3	2.7	1.4	1.0	0.2	0.0	0.0	0.0	0.0	0.1	2.7	6.9	21.3
MAXIMUM MONTHLY (IN)	62	21.4	25.2	11.9	8.0	4.0	T	T	T	T	2.7	18.6	26.2	26.2
YEAR OF OCCURRENCE		1964	1949	1951	1967	1964	1995	1995	1989	1998	1971	1985	1983	DEC 1983
MAXIMUM IN 24 HOURS (IN)	62	8.5	13.0	6.4	7.2	4.0	T	T	T	T	1.7	6.5	9.8	13.0
YEAR OF OCCURRENCE		1950	1949	1952	1969	1964	1995	1995	1989	1998	1971	1964	1996	FEB 1949
MAXIMUM SNOW DEPTH (IN)	53	12	9	6	1	0	0	0	0	0	0	11	13	13
YEAR OF OCCURRENCE		1982	1949	1952	1975							1985	1985	DEC 1985
NORMAL NO. DAYS WITH:														
SNOWFALL 1.0	30	2.1	0.9	0.4	0.3	0.*	0.0	0.0	0.0	0.0	0.1	1.0	2.7	7.5

PRECIPITATION (inches)   2001   BOISE, ID (BOI)

YEAR	JAN	FEB	MAR	APR	MAY	JUN	JUL	AUG	SEP	OCT	NOV	DEC	ANNUAL
1972	2.15	0.91	1.50	0.62	0.32	0.90	0.21	0.05	1.11	0.64	1.11	1.79	11.31
1973	1.14	0.42	0.65	1.49	0.74	0.19	0.07	0.03	0.82	1.15	2.44	2.23	11.37
1974	1.35	0.66	1.50	0.67	0.10	0.60	0.53	0.22	T	1.45	0.67	1.71	9.46
1975	0.59	2.62	1.92	1.53	0.88	0.78	0.82	0.48	0.01	1.99	0.78	1.29	13.69
1976	1.49	1.31	0.72	1.60	0.46	1.66	1.15	0.95	2.11	0.52	0.14	0.09	12.20
1977	0.65	0.57	0.86	0.19	1.80	1.26	0.41	0.73	1.20	0.21	1.86	2.46	12.20
1978	2.37	1.50	1.43	2.34	0.36	0.56	0.48	0.24	0.89	T	1.06	0.60	11.83
1979	1.93	1.20	0.48	1.60	1.28	0.18	0.01	1.81	0.04	1.50	1.30	0.74	12.07
1980	1.56	1.29	2.14	1.20	3.77	0.58	0.03	T	1.59	0.30	1.26	1.49	15.21
1981	1.20	1.02	2.76	1.93	0.95	0.77	0.23	0.13	0.36	0.97	2.24	2.72	15.28
1982	1.42	1.54	1.39	0.79	0.39	0.35	1.62	0.19	1.38	1.74	1.10	1.92	13.83
1983	1.67	1.26	2.70	2.29	1.93	0.17	1.16	0.28	0.65	0.56	1.87	4.23	18.77
1984	0.80	0.86	1.43	1.62	1.06	1.47	0.23	1.24	0.69	0.85	2.36	0.63	13.24
1985	0.20	0.55	0.97	0.90	1.52	0.37	0.85	0.04	1.81	0.84	1.85	1.24	11.14
1986	0.98	3.70	2.01	1.55	1.10	0.35	0.17	0.07	2.93	0.33	1.00	0.12	14.31
1987	0.73	1.24	2.01	0.38	0.69	0.58	0.70	0.11	0.00	T	1.00	1.05	8.49
1988	1.30	0.43	1.45	1.80	1.33	0.47	0.02	0.09	0.24	0.00	3.36	0.81	11.30
1989	1.14	1.15	3.46	0.46	0.21	0.08	0.03	0.78	1.20	1.24	0.59	0.10	10.44
1990	0.84	0.79	0.77	2.14	4.07	0.11	0.42	0.39	0.50	0.45	0.61	0.98	12.07
1991	0.96	0.46	0.55	1.65	1.57	0.64	0.37	0.04	0.21	0.91	1.76	0.35	9.47
1992	0.36	0.92	0.17	0.66	T	2.07	0.03	T	0.30	0.90	1.37	0.89	7.67
1993	1.65	0.96	2.45	2.09	0.92	2.10	0.52	0.24	T	0.47	0.38	0.98	12.76
1994	1.28	0.90	0.17	1.25	1.01	0.24	0.10	T	0.10	0.78	1.78	1.79	9.40
1995	2.10	0.51	1.51	1.03	2.36	0.88	0.59	0.06	0.22	0.42	2.20	2.14	14.02
1996	1.33	1.07	1.97	1.47	1.67	0.21	0.12	T	0.46	0.68	1.71	3.43	14.12
1997	2.74	0.18	0.52	1.89	1.14	1.35	0.45	0.27	0.67	0.55	0.68	0.65	11.09
1998	2.73	1.39	0.99	0.81	4.40	1.21	0.49	0.00	1.96	0.11	0.97	1.65	16.71
1999	1.40	1.96	0.75	0.61	1.10	0.47	T	0.29	0.00	0.11	1.00	0.90	8.59
2000	1.51	2.06	1.69	1.01	0.83	0.14	0.03	0.10	0.60	2.59	0.68	0.80	12.04
2001	1.07	0.49	1.07	1.20	0.27	0.32	0.15	T	0.44	0.86	1.52	1.15	8.54
POR= 102 YRS	1.47	1.25	1.30	1.21	1.25	0.87	0.20	0.20	0.50	0.90	1.29	1.30	11.74

AVERAGE TEMPERATURE (F) 2001   BOISE, ID (BOI)

YEAR	JAN	FEB	MAR	APR	MAY	JUN	JUL	AUG	SEP	OCT	NOV	DEC	ANNUAL
1972	30.9	36.5	45.5	46.7	60.6	67.6	73.5	74.5	58.8	51.7	40.5	23.8	50.9
1973	30.7	39.4	43.0	49.1	60.5	67.6	75.6	73.0	63.1	52.4	41.7	38.4	52.9
1974	29.4	38.7	42.6	49.7	55.4	71.7	73.8	71.7	65.3	52.2	41.7	33.2	52.1
1975	28.2	36.9	41.5	44.5	56.2	64.4	78.3	70.0	65.5	52.3	39.5	31.4	50.7
1976	32.2	34.2	37.1	47.3	59.3	62.6	73.1	67.7	64.1	50.8	40.9	29.3	49.9
1977	19.0	33.8	39.8	54.1	53.7	70.2	72.7	73.6	62.0	53.2	39.6	37.3	50.8
1978	37.0	38.2	48.7	48.5	54.3	64.5	72.8	69.9	61.4	52.8	36.7	27.0	51.0
1979	16.2	34.3	43.2	48.7	57.8	66.7	74.1	71.2	67.3	54.4	34.7	35.8	50.4
1980	30.3	39.8	41.1	52.6	57.2	62.6	72.9	67.2	62.6	51.6	39.8	33.1	50.9
1981	33.9	36.8	44.5	50.3	54.7	63.2	71.0	74.2	63.7	48.7	44.0	35.3	51.7
1982	24.8	30.0	41.3	45.2	54.9	65.8	70.1	72.7	60.4	50.9	36.0	31.5	48.6
1983	35.9	41.5	44.6	47.1	56.5	63.8	69.4	74.9	60.8	53.8	42.0	23.2	51.1
1984	21.2	30.2	41.9	45.9	54.8	61.7	74.2	75.5	60.1	46.9	39.1	22.9	47.9
1985	19.1	25.8	36.0	51.6	58.5	67.2	77.7	69.2	56.2	48.4	27.7	12.6	45.8
1986	29.4	41.0	48.0	48.3	59.1	72.0	69.6	75.9	57.4	52.6	40.4	28.0	51.8
1987	27.8	37.6	44.2	56.0	62.2	70.2	71.2	70.0	65.6	54.9	40.7	32.9	52.8
1988	26.1	37.8	42.6	52.7	57.9	70.7	74.6	71.6	61.7	59.9	40.7	27.0	51.9
1989	24.7	23.2	43.7	53.1	56.3	68.5	77.0	70.0	63.3	51.2	39.4	30.4	50.1
1990	34.2	34.1	14.3	54.7	55.7	66.3	76.2	74.0	69.9	51.0	41.2	18.1	51.7
1991	24.4	41.6	43.0	48.3	54.9	62.6	75.7	76.6	65.8	51.8	37.4	31.7	51.2
1992	33.1	42.1	48.9	54.9	64.3	69.8	71.4	73.9	63.4	54.4	34.5	28.5	53.3
1993	24.8	28.9	41.7	48.1	63.1	61.9	65.0	68.6	63.0	53.2	32.8	33.3	48.7
1994	32.7	34.7	46.1	52.6	61.0	67.9	77.0	76.1	66.8	50.3	32.6	30.4	52.4
1995	36.2	41.5	43.6	48.2	56.9	63.9	73.4	71.2	65.7	49.2	44.4	33.1	52.3
1996	32.5	35.1	44.2	50.0	55.4	64.7	76.7	74.8	61.8	52.4	41.3	36.3	52.3
1997	32.1	36.8	45.4	48.9	62.9	66.0	72.8	75.6	67.7	52.0	43.0	31.2	52.9
1998	38.9	40.2	44.3	49.9	56.3	63.4	79.2	76.7	69.8	51.7	43.7	30.7	53.7
1999	35.3	37.0	43.8	47.5	55.8	66.3	73.7	75.1	64.0	54.3	46.9	31.9	52.6
2000	34.2	41.7	43.4	54.3	59.7	68.6	76.1	75.9	63.0	52.3	32.7	31.1	52.8
2001	27.2	33.9	45.9	48.1	61.4	67.3	74.3	78.7	68.4	53.5	44.0	30.7	52.8
POR= 102 YRS	29.7	35.6	42.4	49.8	57.8	65.5	74.3	72.4	63.0	52.3	40.1	31.5	51.2

REFERENCE NOTES:

PAGE 1:
THE TEMPERATURE GRAPH SHOWS NORMAL MAXIMUM AND NORMAL
   MINIMUM DAILY TEMPERATURES (SOLID CURVES) AND THE
   ACTUAL DAILY HIGH AND LOW TEMPERATURES (VERTICAL BARS).

PAGE 2 AND 3:
H/C INDICATES HEATING AND COOLING DEGREE DAYS.
RH INDICATES RELATIVE HUMIDITY
W/O INDICATES WEATHER AND OBSTRUCTIONS
S INDICATES SUNSHINE.
PR INDICATES PRESSURE.
CLOUDINESS ON PAGE 3 IS THE SUM OF THE CEILOMETER AND
   SATELLITE DATA NOT TO EXCEED EIGHT EIGHTHS(OKTAS).

GENERAL:
T INDICATES TRACE PRECIPITATION, AN AMOUNT GREATER
   THAN ZERO BUT LESS THAN THE LOWEST REPORTABLE VALUE.
+ INDICATES THE VALUE ALSO OCCURS ON EARLIER DATES.
BLANK ENTRIES DENOTE MISSING OR UNREPORTED DATA.
NORMALS ARE 30-YEAR AVERAGES (1961 - 1990).
ASOS INDICATES AUTOMATED SURFACE OBSERVING SYSTEM.
PM INDICATES THE LAST DAY OF THE PREVIOUS MONTH.
POR (PERIOD OF RECORD) BEGINS WITH THE JANUARY DATA
   MONTH AND IS THE NUMBER OF YEARS USED TO COMPUTE
   THE MEAN.   INDIVIDUAL MONTHS WITHIN THE POR MAY
   BE MISSING.
WHEN THE POR FOR A NORMAL IS LESS THAN 30 YEARS,
   THE NORMAL IS PROVISIONAL AND IS BASED ON THE NUMBER
   OF YEARS INDICATED.
0.* OR * INDICATES THE VALUE OR MEAN-DAYS-WITH
   IS BETWEEN 0.00 AND 0.05.
CLOUDINESS FOR ASOS STATIONS DIFFERS FROM THE NON-ASOS
   OBSERVATION TAKEN BY A HUMAN OBSERVER. ASOS STATION
   CLOUDINESS IS BASED ON TIME-AVERAGED CEILOMETER DATA
   FOR CLOUDS AT OR BELOW 12,000 FEET AND ON SATELLITE
   DATA FOR CLOUDS ABOVE 12,000 FEET.
THE NUMBER OF DAYS WITH CLEAR, PARTLY CLOUDY, AND
   CLOUDY CONDITIONS FOR ASOS STATIONS IS THE SUM
   OF THE CEILOMETER AND SATELLITE DATA FOR THE
   SUNRISE TO SUNSET PERIOD.

GENERAL CONTINUED:
CLEAR INDICATES 0 - 2 OKTAS, PARTLY CLOUDY INDICATES
   3 - 6 OKTAS, AND CLOUDY INDICATES 7 OR 8 OKTAS.
   WHEN AT LEAST ONE OF THE ELEMENTS (CEILOMETER OR
   SATELLITE) IS MISSING, THE DAILY CLOUDINESS IS
   NOT COMPUTED.
WIND DIRECTION IS RECORDED IN TENS OF DEGREES (2 DIGITS)
   CLOCKWISE FROM TRUE NORTH. "00" INDICATES CALM. "36"
   INDICATES TRUE NORTH.
RESULTANT WIND IS THE VECTOR AVERAGE OF THE SPEED AND
   DIRECTION.
AVERAGE TEMPERATURE IS THE SUM OF THE MEAN DAILY MAXIMUM
   AND MINIMUM TEMPERATURE DIVIDED BY 2.
SNOWFALL DATA COMPRISE ALL FORMS OF FROZEN
   PRECIPITATION, INCLUDING HAIL.
A HEATING (COOLING) DEGREE DAY IS THE DIFFERENCE BETWEEN
   THE AVERAGE DAILY TEMPERATURE AND 65 F.
DRY BULB IS THE TEMPERATURE OF THE AMBIENT AIR.
DEW POINT IS THE TEMPERATURE TO WHICH THE AIR MUST BE
   COOLED TO ACHIEVE 100 PERCENT RELATIVE HUMIDITY.
WET BULB IS THE TEMPERATURE THE AIR WOULD HAVE IF THE
   MOISTURE CONTENT WAS INCREASED TO 100 PERCENT RELATIVE
   HUMIDITY.

ON JULY 1, 1996, THE NATIONAL WEATHER SERVICE BEGAN USING
   THE "METAR" OBSERVATION CODE THAT WAS ALREADY EMPLOYED
   BY MOST OTHER NATIONS OF THE WORLD.   THE MOST NOTICEABLE
   DIFFERENCE IN THIS ANNUAL PUBLICATION WILL BE THE CHANGE
   IN UNITS FROM TENTHS TO EIGHTS(OKTAS) FOR REPORTING THE
   AMOUNT OF SKY COVER.

HEATING DEGREE DAYS (base 65 F) 2001 BOISE, ID (BOI)

YEAR	JUL	AUG	SEP	OCT	NOV	DEC	JAN	FEB	MAR	APR	MAY	JUN	TOTAL
1972-73	11	0	222	406	727	1270	1056	708	673	470	182	91	5816
1973-74	4	13	103	382	692	817	1099	728	687	452	304	42	5323
1974-75	10	11	53	391	689	983	1132	782	721	607	275	76	5730
1975-76	6	21	74	399	759	1035	1013	886	858	524	189	132	5896
1976-77	3	36	76	434	720	1097	1418	868	772	342	358	7	6131
1977-78	8	32	145	362	758	853	859	744	500	488	329	74	5152
1978-79	5	38	173	370	841	1171	1503	855	668	481	241	72	6418
1979-80	5	2	26	326	903	899	1070	725	736	367	257	133	5449
1980-81	0	41	104	409	750	983	957	783	631	432	315	97	5502
1981-82	15	5	137	497	624	915	1240	974	729	586	312	86	6120
1982-83	27	2	182	432	863	1030	897	653	622	530	309	82	5629
1983-84	38	0	145	338	682	1290	1353	1004	710	566	328	162	6616
1984-85	0	8	204	557	771	1299	1412	1093	895	398	226	53	6916
1985-86	0	26	259	509	1113	1619	1097	668	522	499	280	15	6607
1986-87	35	2	259	376	733	1141	1149	761	639	287	140	41	5563
1987-88	23	18	86	306	722	990	1198	780	686	359	261	59	5488
1988-89	4	5	157	178	724	1169	1242	1166	656	356	276	30	5963
1989-90	0	29	97	421	759	1064	951	858	633	303	286	82	5483
1990-91	6	10	26	430	710	1449	1252	651	676	493	306	100	6109
1991-92	0	0	55	409	822	1026	982	657	492	308	92	54	4897
1992-93	6	40	118	340	907	1124	1239	1004	715	501	140	155	6289
1993-94	53	45	124	367	960	975	993	839	579	378	148	64	5525
1994-95	9	1	34	449	967	1066	887	652	657	499	255	111	5587
1995-96	5	23	86	485	610	984	1000	859	636	447	292	55	5482
1996-97	1	6	162	405	703	883	1013	783	601	473	128	41	5199
1997-98	18	0	61	410	651	1040	803	684	635	452	270	85	5109
1998-99	0	0	51	408	635	1058	914	778	652	520	311	95	5422
1999-00	8	13	101	324	535	1017	949	673	663	315	179	41	4818
2000-01	4	0	128	389	960	1044	1165	867	585	504	191	69	5906
2001-	6	0	23	355	626	1056							

COOLING DEGREE DAYS (base 65 F) 2001 BOISE, ID (BOI)

YEAR	JAN	FEB	MAR	APR	MAY	JUN	JUL	AUG	SEP	OCT	NOV	DEC	ANNUAL
1972	0	0	0	0	62	129	283	303	40	0	0	0	817
1973	0	0	0	0	51	177	341	269	51	1	0	0	890
1974	0	0	0	0	13	252	289	226	71	0	0	0	851
1975	0	0	0	0	10	64	426	182	96	11	0	0	789
1976	0	0	0	0	19	66	263	130	55	2	0	0	535
1977	0	0	0	20	10	170	255	306	61	0	0	0	822
1978	0	0	1	0	6	64	254	200	72	0	0	0	597
1979	0	0	0	0	27	129	293	199	101	3	0	0	752
1980	0	0	0	3	25	68	251	117	38	2	0	0	504
1981	0	0	0	1	3	52	205	296	101	0	0	0	658
1982	0	0	0	0	2	117	194	248	50	0	0	0	611
1983	0	0	0	0	55	50	180	313	26	0	0	0	624
1984	0	0	0	0	19	70	291	340	64	2	0	0	786
1985	0	0	0	2	28	125	402	165	4	0	0	0	726
1986	0	0	0	1	103	235	184	348	37	0	0	0	908
1987	0	0	0	23	61	202	223	180	111	0	0	0	800
1988	0	0	0	0	46	237	308	215	66	24	0	0	896
1989	0	0	0	6	14	140	376	191	56	0	0	0	783
1990	0	0	0	3	5	145	357	293	180	6	0	0	989
1991	0	0	0	0	0	36	337	368	85	6	0	0	832
1992	0	0	0	11	76	202	208	321	77	20	0	0	915
1993	0	0	0	0	86	68	61	162	71	8	0	0	456
1994	0	0	0	15	30	156	386	354	96	0	0	0	1037
1995	0	0	0	0	12	83	272	225	114	0	0	0	706
1996	0	0	0	0	4	111	372	314	73	22	0	0	896
1997	0	0	0	0	71	80	266	337	151	13	0	0	918
1998	0	0	0	0	4	9	42	449	371	201	3	0	1079
1999	0	0	0	0	32	142	282	334	80	1	0	0	871
2000	0	0	0	3	22	155	354	347	74	2	0	0	957
2001	0	0	0	4	85	146	303	433	131	7	0	0	1109

SNOWFALL (inches) 2001 BOISE, ID (BOI)

YEAR	JUL	AUG	SEP	OCT	NOV	DEC	JAN	FEB	MAR	APR	MAY	JUN	TOTAL
1972-73	0.0	0.0	0.0	T	T	12.6	7.3	0.3	0.6	T	0.0	0.0	20.8
1973-74	0.0	0.0	0.0	0.0	8.8	4.5	6.0	2.4	5.7	T	0.0	0.0	27.4
1974-75	0.0	0.0	0.0	0.0	0.9	4.6	5.4	6.3	2.2	2.6	0.9	0.0	22.9
1975-76	0.0	0.0	0.0	T	3.9	4.2	6.3	6.7	3.9	0.2	0.0	0.0	25.2
1976-77	0.0	0.0	0.0	0.0	T	1.1	7.2	3.5	2.9	T	T	0.0	14.7
1977-78	0.0	0.0	0.0	T	4.7	4.4	1.6	7.3	T	T	T	0.0	18.0
1978-79	0.0	0.0	0.0	0.0	0.2	3.1	11.9	4.3	0.8	0.2	T	0.0	20.5
1979-80	0.0	0.0	0.0	0.0	6.6	1.4	3.8	0.8	2.7	T	0.0	0.0	15.3
1980-81	0.0	0.0	0.0	0.0	3.2	1.7	3.6	0.7	T	0.5	0.0	0.0	9.7
1981-82	0.0	0.0	0.0	0.0	2.8	11.1	12.2	1.4	3.6	1.4	0.0	0.0	32.5
1982-83	0.0	0.0	0.0	0.0	2.1	6.4	1.6	0.9	T	T	0.8	0.0	11.8
1983-84	0.0	0.0	0.0	0.0	2.2	26.2	4.3	4.4	T	0.3	T	0.0	37.4
1984-85	0.0	0.0	0.0	T	0.2	7.7	2.6	5.3	1.5	1.0	0.0	0.0	18.3
1985-86	0.0	0.0	0.0	T	18.6	12.6	3.9	4.4	0.0	T	0.0	0.0	39.5
1986-87	0.0	0.0	0.0	0.0	5.9	0.9	0.5	0.6	T	T	0.0	0.0	7.9
1987-88	0.0	0.0	0.0	0.0	0.5	3.0	3.9	0.3	2.9	1.2	0.0	0.0	11.8
1988-89	0.0	0.0	0.0	0.0	2.5	10.8	8.0	0.7	T	T	0.0	0.0	22.0
1989-90	0.0	T	0.0	T	0.4	T	5.2	6.5	0.4	T	T	0.0	12.5
1990-91	0.0	0.0	0.0	0.0	0.1	15.7	1.5	T	1.2	T	0.0	0.0	18.5
1991-92	0.0	0.0	0.0	1.1	2.8	T	0.3	T	0.0	T	0.0	0.0	4.2
1992-93	0.0	0.0	0.0	T	3.6	5.7	14.6	10.4	0.2	T	T	0.0	34.5
1993-94	0.0	0.0	0.0	0.0	1.9	0.2	3.8	T	T	T	0.0	0.0	8.7
1994-95	0.0	0.0	0.0	0.0	8.8	8.5	1.5	5.7	1.2	T	T	T	25.7
1995-96	T	0.0	0.0	0.2	5.9	6.7	6.9	1.8	T	T	0.0	0.0	21.5
1996-97	0.0	0.0	0.0	T	T	13.0	3.4	T	0.3	T	0.0	0.0	16.7
1997-98	0.0	0.0	0.0	0.0	0.0	0.6	4.2	1.4	1.1	T	0.0	0.0	7.3
1998-99	T	0.0	0.0	0.0	T	8.8	1.7	5.7	6.7	0.4	0.0	0.0	23.3
1999-00	0.0	0.0	0.0	0.0	0.9	2.7	6.2	2.1	2.3	0.0	T	0.0	14.2
2000-01	0.0	0.0	0.0	0.0	4.0	1.6	8.9	3.1	1.3	T	0.0	0.0	18.9
2001-	0.0	0.0	0.0	0.0	3.5	9.9							
POR= 61 YRS	T	T	0.0	0.1	2.3	5.7	6.3	3.5	1.8	0.6	0.1	T	20.4

2001
CHICAGO, O'HARE INTERNATIONAL AIRPORT, ILLINOIS (ORD)

Chicago is located along the southwest shore of Lake Michigan and occupies a plain which, for the most part, is only some tens of feet above the lake. Lake Michigan averages 579 feet above sea level. Natural water drainage over most of the city would be into Lake Michigan, and from areas west of the city is into the Mississippi River System. But actual drainage over most of the city is artificially channeled also into the Mississippi system. Topography does not significantly affect air flow in or near the city except that lesser frictional drag over Lake Michigan causes winds to be frequently stronger along the lakeshore, and often permits air masses moving from the north to reach shore areas an hour or more before affecting western parts of the city.

Chicago is in a region of frequently changeable weather. The climate is predominately continental, ranging from relatively warm in summer to relatively cold in winter. However, the continentality is partially modified by Lake Michigan, and to a lesser extent by other Great Lakes. In late autumn and winter, air masses that are initially very cold often reach the city only after being tempered by passage over one or more of the lakes. Similarly, in late spring and summer, air masses reaching the city from the north, northeast, or east are cooler because of movement over the Great Lakes. Very low winter temperatures most often occur in air that flows southward to the west of Lake Superior before reaching the Chicago area. In summer the higher temperatures are with south or southwest flow and are therefore not influenced by the lakes, the only modifying effect being a local lake breeze. Strong south or southwest flow may overcome the lake breeze and cause high temperatures to extend over the entire city.

During the warm season, when the lake is cold relative to land, there is frequently a lake breeze that reduces daytime temperature near the shore, sometimes by 10 degrees or more below temperatures farther inland. When the breeze off the lake is light this effect usually reaches inland only a mile or two, but with stronger on-shore winds the whole city is cooled. On the other hand, temperatures at night are warmer near the lake so that 24-hour averages on the whole are only slightly different in various parts of the city and suburbs.

At the O'Hare International Airport temperatures of 96 degrees or higher occur in about half the summers, while about half the winters have a minimum as low as -15 degrees. The average occurrence of the first temperature as low as 32 degrees in the fall is mid-October and the average occurrence of the last temperature as low as 32 degrees in the spring is late April.

Precipitation falls mostly from air that has passed over the Gulf of Mexico. But in winter there is sometimes snowfall, light inland but locally heavy near the lakeshore, with Lake Michigan as the principal moisture source. The heavy lakeshore snow occurs when initially colder air moves from the north with a long trajectory over Lake Michigan and impinges on the Chicago lakeshore. In this situation the air mass is warmed and its moisture content increased up to a height of several thousand feet. Snowfall is produced by upward currents that become stronger, because of frictional effects, when the air moves from the lake onto land. This type of snowfall therefore tends to be heavier and to extend farther inland in south-shore areas of the city and in Indiana suburbs, where the angle between wind-flow and shoreline is greatest. The effect of Lake Michigan, both on winter temperatures and lake-produced snowfall, is enhanced by non-freezing of much of the lake during the winter, even though areas and harbors are often ice-choked.

Summer thunderstorms are often locally heavy and variable, parts of the city may receive substantial rainfall and other parts none. Longer periods of continuous precipitation are mostly in autumn, winter, and spring. About one-half the precipitation in winter, and about 10 percent of the yearly total precipitation, falls as snow. Snowfall from month to month and year to year is greatly variable. There is a 50 percent likelihood that the first and last 1-inch snowfall of a season will occur by December 5 and March 20, respectively.

Channeling of winds between tall buildings often causes locally stronger gusts in the central business area. However, the nickname, windy city, is a misnomer as the average wind speed is not greater than in many other parts of the U.S.

NORMALS, MEANS, AND EXTREMES
CHICAGO, IL (ORD)

LATITUDE:	LONGITUDE:	ELEVATION (FT):		TIME ZONE:	WBAN: 94846
41 59' 10" N	87 54' 51" W	GRND: 655	BARO: 658	CENTRAL (UTC + 6)	

	ELEMENT	POR	JAN	FEB	MAR	APR	MAY	JUN	JUL	AUG	SEP	OCT	NOV	DEC	YEAR
TEMPERATURE °F	NORMAL DAILY MAXIMUM	30	29.0	33.5	45.8	58.6	70.1	79.6	83.7	81.8	74.8	63.3	48.4	34.0	58.5
	MEAN DAILY MAXIMUM	43	29.5	34.3	45.4	58.4	70.1	79.5	83.6	81.9	74.7	63.0	48.0	34.7	58.6
	HIGHEST DAILY MAXIMUM	43	65	72	88	91	93	104	104	101	99	91	78	71	104
	YEAR OF OCCURRENCE		1989	2000	1986	1980	1977	1988	1995	1991	1985	1963	1978	1982	JUL 1995
	MEAN OF EXTREME MAXS.	43	49.5	54.5	72.0	80.5	87.5	93.0	95.0	92.9	89.9	81.7	68.1	55.9	76.7
	NORMAL DAILY MINIMUM	30	12.9	17.2	28.5	38.6	47.7	57.5	62.6	61.6	53.9	42.2	31.6	19.1	39.5
	MEAN DAILY MINIMUM	43	14.0	18.7	28.0	38.6	48.2	57.4	62.9	62.1	53.9	42.4	31.8	20.1	39.8
	LOWEST DAILY MINIMUM	43	-27	-19	-8	7	24	36	40	41	28	17	1	-25	-27
	YEAR OF OCCURRENCE		1985	1996	1962	1982	1966	1972	1965	1965	1974	1981	1976	1983	JAN 1985
	MEAN OF EXTREME MINS.	43	-9.0	-2.9	10.0	23.1	34.1	43.5	50.2	50.8	38.6	27.2	15.3	-1.9	23.2
	NORMAL DRY BULB	30	21.0	25.4	37.2	48.6	58.9	68.6	73.2	71.7	64.4	52.8	40.0	26.6	49.0
	MEAN DRY BULB	43	21.7	26.5	36.8	48.4	59.2	68.5	73.2	71.9	64.3	52.8	39.7	27.4	49.2
	MEAN WET BULB	18	22.5	26.2	34.0	43.0	53.3	62.1	66.7	65.9	58.3	47.7	34.8	26.6	45.1
	MEAN DEW POINT	18	17.8	21.0	27.9	36.2	47.2	57.0	62.3	61.9	53.5	42.3	30.0	21.8	39.9
	NORMAL NO. DAYS WITH:														
	MAXIMUM 90	30	0.0	0.0	0.0	*	0.8	3.8	6.7	4.2	1.5	0.1	0.0	0.0	17.1
	MAXIMUM 32	30	17.6	12.7	3.9	0.1	0.0	0.0	0.0	0.0	0.0	0.0	1.7	11.9	47.9
	MINIMUM 32	30	28.7	25.0	21.0	7.8	0.9	0.0	0.0	0.0	0.2	5.3	16.5	26.7	132.1
	MINIMUM 0	30	7.0	3.2	0.2	0.0	0.0	0.0	0.0	0.0	0.0	0.0	0.0	2.9	13.3
H/C	NORMAL HEATING DEG. DAYS	30	1364	1109	862	492	235	35	5	19	84	391	750	1190	6536
	NORMAL COOLING DEG. DAYS	30	0	0	0	0	46	143	259	226	66	12	0	0	752
RH	NORMAL (PERCENT)	30	72	72	70	65	64	66	68	71	71	69	72	76	70
	HOUR 00 LST	30	75	76	76	72	73	75	79	81	81	76	77	78	77
	HOUR 06 LST	30	76	77	79	77	77	78	82	85	85	81	80	80	80
	HOUR 12 LST	30	67	65	61	55	53	55	57	57	57	55	64	70	60
	HOUR 18 LST	30	71	69	64	57	54	55	58	61	63	64	70	74	63
S	PERCENT POSSIBLE SUNSHINE	16	44	49	51	50	58	67	66	62	59	55	38	43	54
W/O	MEAN NO. DAYS WITH:														
	HEAVY FOG(VISBY 1/4 MI)	44	1.7	1.8	2.2	0.9	1.2	0.6	0.5	0.7	0.4	0.8	1.3	1.9	14.0
	THUNDERSTORMS	44	0.3	0.5	2.0	4.0	5.1	6.6	5.9	6.0	4.2	1.7	1.1	0.6	38.0
CLOUDINESS	MEAN:														
	SUNRISE-SUNSET (OKTAS)	38	5.5	5.4	5.8	5.5	5.0	4.7	4.4	4.4	4.5	4.7	5.7	5.6	5.1
	MIDNIGHT-MIDNIGHT (OKTAS)	32	5.3	5.0	5.1	5.0	4.5	4.3	4.0	4.0	4.2	4.3	5.4	5.5	4.7
	MEAN NO. DAYS WITH:														
	CLEAR	39	6.8	6.0	4.9	6.0	7.2	7.3	8.0	8.6	8.5	8.6	5.2	5.7	82.8
	PARTLY CLOUDY	39	6.2	6.5	8.5	7.6	9.9	11.5	12.1	11.2	9.5	8.5	6.2	5.9	103.6
	CLOUDY	39	18.0	15.7	17.6	16.4	13.9	11.2	10.1	10.4	11.2	13.3	17.7	18.5	174.0
PR	MEAN STATION PRESSURE(IN)	29	29.30	29.30	29.30	29.20	29.20	29.20	29.30	29.30	29.30	29.31	29.31	29.31	29.28
	MEAN SEA-LEVEL PRES. (IN)	18	30.10	30.10	30.06	29.96	29.97	30.01	30.00	30.04	30.06	30.08	30.06	30.11	30.05
WINDS	MEAN SPEED (MPH)	38	11.6	11.4	11.9	12.0	10.5	9.4	8.6	8.4	9.0	10.2	11.1	11.0	10.4
	PREVAIL.DIR(TENS OF DEGS)	33	27	29	29	04	04	21	23	23	18	19	21	27	21
	MAXIMUM 2-MINUTE:														
	SPEED (MPH)	5	37	38	41	43	37	39	37	41	37	43	41	32	43
	DIR. (TENS OF DEGS)		09	26	02	26	28	30	01	24	28	07	23	33	07
	YEAR OF OCCURRENCE		1999	2001	1998	1997	1998	1998	1997	2000	1997	1997	1998	2000	OCT 1997
	MAXIMUM 5-SECOND:														
	SPEED (MPH)	5	47	56	47	56	54	48	61	59	48	52	57	41	61
	DIR. (TENS OF DEGS)		09	24	02	25	27	30	20	25	21	07	24	27	20
	YEAR OF OCCURRENCE		1999	1999	1998	1997	1998	1998	1999	2000	2001	1997	1998	2001	JUL 1999
PRECIPITATION	NORMAL (IN)	30	1.53	1.36	2.69	3.64	3.32	3.78	3.66	4.22	3.82	2.41	2.92	2.47	35.82
	MAXIMUM MONTHLY (IN)	43	4.47	5.56	5.91	7.69	7.14	9.96	8.33	17.10	11.44	8.54	8.22	8.56	17.10
	YEAR OF OCCURRENCE		1999	1997	1976	1983	1970	1993	1982	1987	1961	2001	1985	1982	AUG 1987
	MINIMUM MONTHLY (IN)	43	0.10	0.12	0.63	0.97	0.30	0.95	1.18	0.51	0.02	0.16	0.44	0.23	0.02
	YEAR OF OCCURRENCE		1981	1969	1981	1971	1992	1991	1977	1969	1979	1964	1999	1962	SEP 1979
	MAXIMUM IN 24 HOURS (IN)	43	2.00	3.78	2.39	2.78	3.45	3.79	2.90	9.35	3.00	4.62	2.99	4.53	9.35
	YEAR OF OCCURRENCE		1960	1997	1985	1983	1981	1994	1993	1987	1978	1969	1990	1982	AUG 1987
	NORMAL NO. DAYS WITH:														
	PRECIPITATION 0.01	30	11.1	9.4	12.4	12.6	10.9	10.1	9.8	9.5	9.5	9.2	10.9	11.6	127.0
	PRECIPITATION 1.00	30	0.2	0.1	0.4	0.7	0.5	0.9	1.0	1.2	1.1	0.3	0.5	0.5	7.4
SNOWFALL	NORMAL (IN)	30	10.7	8.4	6.5	1.8	0.1	0.0	0.0	0.0	0.0	0.5	1.9	8.8	38.7
	MAXIMUM MONTHLY (IN)	42	34.3	26.2	24.7	11.1	1.6	T	T	T	T	6.6	10.4	35.3	35.3
	YEAR OF OCCURRENCE		1979	1994	1965	1975	1966	1992	1995	1989	1967	1967	1959	1978	DEC 1978
	MAXIMUM IN 24 HOURS (IN)	42	18.6	11.1	10.6	10.9	1.6	T	T	T	T	6.6	5.8	11.0	18.6
	YEAR OF OCCURRENCE		1999	2000	1970	1975	1966	1992	1995	1989	1967	1967	1975	1969	JAN 1999
	MAXIMUM SNOW DEPTH (IN)	41	28	27	20	11	1	0	0	0	0	3	6	17	28
	YEAR OF OCCURRENCE		1979	1967	1965	1975	1966					1989	1975	2000	JAN 1979
	NORMAL NO. DAYS WITH:														
	SNOWFALL 1.0	30	3.3	2.7	2.0	0.4	0.*	0.0	0.0	0.0	0.0	0.2	0.6	2.6	11.8

PRECIPITATION (inches) 2001 CHICAGO, O'HARE INTERNATIONAL AIRPORT, IL (ORD)

YEAR	JAN	FEB	MAR	APR	MAY	JUN	JUL	AUG	SEP	OCT	NOV	DEC	ANNUAL
1972	1.01	0.73	3.45	4.77	3.02	3.55	4.97	6.97	8.14	2.92	3.05	2.89	45.47
1973	1.24	1.38	3.91	4.99	3.69	2.87	5.27	0.67	6.01	2.86	1.50	3.71	38.10
1974	3.29	2.11	2.40	4.27	5.09	4.69	2.96	2.60	1.47	1.88	2.47	2.12	35.35
1975	3.69	2.48	2.02	5.50	3.02	5.07	2.19	7.37	0.80	1.90	2.53	3.05	39.62
1976	0.85	1.87	5.91	4.05	4.03	2.93	1.44	1.29	1.49	1.41	0.65	0.64	26.56
1977	0.55	0.71	3.67	2.62	1.88	5.12	1.18	5.39	6.07	1.36	2.05	1.96	32.56
1978	1.48	0.43	1.16	3.94	2.80	6.36	4.61	1.96	6.88	1.08	2.24	4.41	37.35
1979	2.81	1.02	4.49	4.92	2.58	4.63	2.19	7.57	0.02	1.49	2.80	2.58	37.10
1980	1.04	1.24	1.96	3.41	3.22	3.42	3.56	8.54	5.65	2.09	1.10	3.43	38.66
1981	0.10	2.35	0.63	6.14	5.85	4.46	4.50	6.60	3.25	1.80	2.46	1.05	39.19
1982	2.90	0.41	4.15	2.78	2.08	1.56	8.33	3.93	1.15	1.88	6.95	8.56	44.68
1983	0.66	2.06	3.56	7.69	6.26	4.11	4.25	2.08	5.41	4.41	5.87	2.99	49.35
1984	1.15	1.39	3.00	4.11	4.49	2.02	3.19	2.10	3.84	3.15	2.64	2.92	34.00
1985	1.48	3.46	4.73	1.48	2.79	1.97	3.75	3.90	1.82	4.98	8.22	1.49	40.07
1986	0.39	2.58	1.49	1.85	3.11	3.49	4.30	1.15	7.12	3.75	1.41	1.09	31.73
1987	1.67	0.99	1.59	2.34	2.21	2.19	4.19	17.10	0.94	1.59	2.77	3.77	41.35
1988	1.88	1.29	2.15	2.08	1.19	1.05	2.74	3.29	3.79	5.05	6.45	2.40	33.36
1989	0.82	0.77	1.67	1.37	1.59	2.01	5.89	7.31	3.91	1.49	2.16	0.46	29.45
1990	1.97	2.25	3.09	1.79	6.85	4.50	2.25	7.75	1.03	4.10	5.60	1.94	43.12
1991	1.41	0.62	3.54	4.00	5.20	0.95	1.32	2.81	2.51	7.36	3.59	1.71	35.02
1992	0.87	1.39	2.67	2.21	0.30	1.35	3.77	3.56	4.31	1.79	5.41	2.49	30.12
1993	3.83	0.82	4.52	4.57	1.83	9.96	4.45	5.74	4.47	2.19	1.52	1.00	44.90
1994	1.77	2.56	1.09	2.20	0.58	6.09	1.62	4.05	1.04	3.23	3.75	1.61	29.59
1995	3.21	0.41	1.43	5.79	4.47	1.40	3.17	3.49	1.04	4.20	3.68	0.59	32.88
1996	1.58	0.71	0.95	2.59	6.95	4.80	3.95	1.45	2.73	2.32	1.48	1.21	30.72
1997	1.38	5.56	1.57	1.76	2.69	3.81	3.04	4.50	1.69	2.75	1.46	1.50	31.71
1998	2.67	1.70	4.29	3.56	3.02	2.90	1.75	6.88	2.34	5.22	2.00	1.20	37.53
1999	4.47	1.64	1.73	7.51	4.46	4.95	3.73	2.30	3.27	1.07	0.44	2.68	38.25
2000	1.35	1.97	1.18	5.15	4.02	4.32	3.58	2.26	3.59	1.12	2.71	2.11	33.36
2001	1.12	2.57	1.30	2.82	3.34	2.61	2.96	12.25	6.05	8.54	1.22	0.99	45.77
POR= 43 YRS	1.75	1.53	2.55	3.66	3.32	3.79	3.55	4.23	3.48	2.73	2.71	2.12	35.42

AVERAGE TEMPERATURE (F) 2001   CHICAGO, O'HARE INTERNATIONAL AIRPORT, IL (

YEAR	JAN	FEB	MAR	APR	MAY	JUN	JUL	AUG	SEP	OCT	NOV	DEC	ANNUAL
1972	19.6	23.6	34.0	44.8	61.0	65.7	73.6	73.8	63.5	49.3	37.7	23.9	47.5
1973	28.2	28.7	44.0	48.1	54.8	71.1	74.7	74.6	66.0	57.9	41.9	28.1	51.5
1974	24.8	27.4	38.6	52.3	56.8	65.5	73.6	70.0	60.5	52.8	40.6	30.2	49.4
1975	27.3	26.2	34.1	43.3	62.3	70.5	75.5	76.3	61.4	55.8	47.2	31.5	51.0
1976	19.9	35.2	42.8	52.3	55.9	70.1	74.0	70.8	62.7	48.3	32.4	19.4	48.7
1977	10.7	26.9	44.9	55.0	67.2	69.3	77.5	71.9	66.0	51.5	40.0	24.2	50.4
1978	15.7	16.8	31.9	47.5	58.3	67.6	72.0	72.4	68.8	51.4	40.8	25.8	47.4
1979	12.5	16.2	36.4	45.5	59.3	69.2	72.0	71.0	66.1	53.3	40.6	33.7	48.0
1980	23.4	21.5	32.6	46.5	59.7	65.3	75.7	75.7	66.0	48.4	39.9	28.0	48.6
1981	22.6	28.0	37.6	51.8	55.3	69.8	72.5	71.2	61.7	49.1	40.2	24.9	48.8
1982	12.2	21.5	35.1	44.5	64.3	62.1	74.1	68.8	62.1	53.2	39.1	36.0	47.8
1983	26.3	30.5	37.4	43.4	53.2	69.7	76.7	77.3	64.6	52.8	41.1	14.3	48.9
1984	17.1	33.9	29.5	45.8	55.5	70.3	70.3	72.8	61.1	54.7	37.9	31.0	48.3
1985	14.4	20.4	39.4	52.6	60.2	63.6	71.4	69.2	65.4	52.5	37.8	17.0	47.0
1986	22.8	24.0	40.4	51.5	59.5	66.3	74.9	68.5	66.8	53.7	36.0	30.6	49.6
1987	25.9	33.9	40.8	50.6	63.4	72.4	76.7	71.9	65.1	47.3	43.9	32.2	52.0
1988	19.8	22.7	38.1	48.2	61.0	71.7	76.8	76.8	65.9	46.1	41.7	25.7	49.7
1989	32.4	19.6	36.6	46.8	57.8	67.5	73.9	71.4	62.0	54.0	37.7	17.4	49.1
1990	33.9	31.3	41.3	49.9	56.2	69.6	71.7	71.9	64.5	51.6	44.7	28.6	51.4
1991	20.8	31.0	40.4	52.0	65.6	71.9	75.5	73.6	63.7	53.2	35.2	30.3	51.1
1992	28.1	33.3	37.5	46.1	56.9	64.9	69.3	67.0	62.7	50.4	38.3	28.6	48.6
1993	26.2	24.4	34.2	45.0	59.7	66.4	74.3	73.3	59.2	49.5	29.8	27.7	48.4
1994	15.9	22.1	38.5	51.1	58.2	70.2	73.4	68.7	66.8	54.7	44.4	34.8	49.9
1995	24.0	26.5	40.2	46.0	58.8	72.3	77.6	79.0	62.5	53.7	32.8	26.3	50.0
1996	23.4	26.0	30.8	45.2	58.0	68.0	69.9	72.3	63.5	51.9	33.4	27.7	47.3
1997	19.3	29.0	37.9	45.2	53.8	68.3	73.2	69.5	64.2	53.2	36.4	31.5	48.5
1998	29.6	38.7	39.0	49.8	64.8	69.3	74.5	73.5	67.7	55.5	44.8	34.7	53.5
1999	22.6	34.0	35.6	49.6	61.7	70.4	78.4	70.3	63.4	52.9	45.1	29.9	51.2
2000	25.3	34.1	44.2	47.2	62.0	67.3	71.1	72.4	64.7	56.1	37.0	16.0	49.8
2001	24.6	26.1	34.2	52.5	60.0	67.4	74.6	71.2	61.9	52.1	48.2	33.4	50.7
POR= 43 YRS	21.7	26.5	36.8	48.5	59.4	68.5	73.2	71.8	64.3	52.8	39.7	27.3	49.2

REFERENCE NOTES:

PAGE 1:
THE TEMPERATURE GRAPH SHOWS NORMAL MAXIMUM AND NORMAL MINIMUM DAILY TEMPERATURES (SOLID CURVES) AND THE ACTUAL DAILY HIGH AND LOW TEMPERATURES (VERTICAL BARS).

PAGE 2 AND 3:
H/C INDICATES HEATING AND COOLING DEGREE DAYS.
RH INDICATES RELATIVE HUMIDITY
W/O INDICATES WEATHER AND OBSTRUCTIONS
S INDICATES SUNSHINE.
PR INDICATES PRESSURE.
CLOUDINESS ON PAGE 3 IS THE SUM OF THE CEILOMETER AND SATELLITE DATA NOT TO EXCEED EIGHT EIGHTHS(OKTAS).

GENERAL:
T INDICATES TRACE PRECIPITATION, AN AMOUNT GREATER THAN ZERO BUT LESS THAN THE LOWEST REPORTABLE VALUE.
+ INDICATES THE VALUE ALSO OCCURS ON EARLIER DATES.
BLANK ENTRIES DENOTE MISSING OR UNREPORTED DATA.
NORMALS ARE 30-YEAR AVERAGES (1961 - 1990).
ASOS INDICATES AUTOMATED SURFACE OBSERVING SYSTEM.
PM INDICATES THE LAST DAY OF THE PREVIOUS MONTH.
POR (PERIOD OF RECORD) BEGINS WITH THE JANUARY DATA MONTH AND IS THE NUMBER OF YEARS USED TO COMPUTE THE MEAN. INDIVIDUAL MONTHS WITHIN THE POR MAY BE MISSING.
WHEN THE POR FOR A NORMAL IS LESS THAN 30 YEARS, THE NORMAL IS PROVISIONAL AND IS BASED ON THE NUMBER OF YEARS INDICATED.
0.* OR * INDICATES THE VALUE OR MEAN-DAYS-WITH IS BETWEEN 0.00 AND 0.05.
CLOUDINESS FOR ASOS STATIONS DIFFERS FROM THE NON-ASOS OBSERVATION TAKEN BY A HUMAN OBSERVER. ASOS STATION CLOUDINESS IS BASED ON TIME-AVERAGED CEILOMETER DATA FOR CLOUDS AT OR BELOW 12,000 FEET AND ON SATELLITE DATA FOR CLOUDS ABOVE 12,000 FEET.
THE NUMBER OF DAYS WITH CLEAR, PARTLY CLOUDY, AND CLOUDY CONDITIONS FOR ASOS STATIONS IS THE SUM OF THE CEILOMETER AND SATELLITE DATA FOR THE SUNRISE TO SUNSET PERIOD.

GENERAL CONTINUED:
CLEAR INDICATES 0 - 2 OKTAS, PARTLY CLOUDY INDICATES 3 - 6 OKTAS, AND CLOUDY INDICATES 7 OR 8 OKTAS. WHEN AT LEAST ONE OF THE ELEMENTS (CEILOMETER OR SATELLITE) IS MISSING, THE DAILY CLOUDINESS IS NOT COMPUTED.
WIND DIRECTION IS RECORDED IN TENS OF DEGREES (2 DIGITS) CLOCKWISE FROM TRUE NORTH. "00" INDICATES CALM. "36" INDICATES TRUE NORTH.
RESULTANT WIND IS THE VECTOR AVERAGE OF THE SPEED AND DIRECTION.
AVERAGE TEMPERATURE IS THE SUM OF THE MEAN DAILY MAXIMUM AND MINIMUM TEMPERATURE DIVIDED BY 2.
SNOWFALL DATA COMPRISE ALL FORMS OF FROZEN PRECIPITATION, INCLUDING HAIL.
A HEATING (COOLING) DEGREE DAY IS THE DIFFERENCE BETWEEN THE AVERAGE DAILY TEMPERATURE AND 65 F.
DRY BULB IS THE TEMPERATURE OF THE AMBIENT AIR.
DEW POINT IS THE TEMPERATURE TO WHICH THE AIR MUST BE COOLED TO ACHIEVE 100 PERCENT RELATIVE HUMIDITY.
WET BULB IS THE TEMPERATURE THE AIR WOULD HAVE IF THE MOISTURE CONTENT WAS INCREASED TO 100 PERCENT RELATIVE HUMIDITY.

ON JULY 1, 1996, THE NATIONAL WEATHER SERVICE BEGAN USING THE "METAR" OBSERVATION CODE THAT WAS ALREADY EMPLOYED BY MOST OTHER NATIONS OF THE WORLD. THE MOST NOTICEABLE DIFFERENCE IN THIS ANNUAL PUBLICATION WILL BE THE CHANGE IN UNITS FROM TENTHS TO EIGHTS(OKTAS) FOR REPORTING THE AMOUNT OF SKY COVER.

HEATING DEGREE DAYS (base 65 F) 2001 CHICAGO, O'HARE INTERNATIONAL AIRPORT, IL (OR

YEAR	JUL	AUG	SEP	OCT	NOV	DEC	JAN	FEB	MAR	APR	MAY	JUN	TOTAL
1972-73	15	10	109	481	811	1269	1135	1012	645	503	311	0	6301
1973-74	0	0	72	244	687	1139	1240	1046	812	383	266	63	5952
1974-75	0	1	176	384	724	1072	1160	1078	951	643	152	30	6371
1975-76	1	0	147	303	531	1033	1392	859	681	411	285	17	5660
1976-77	0	9	119	522	973	1408	1679	1060	616	332	115	41	6874
1977-78	0	8	42	413	741	1254	1521	1346	1020	518	264	46	7173
1978-79	1	4	59	418	718	1206	1622	1360	879	580	233	30	7110
1979-80	16	19	62	382	722	967	1281	1254	995	558	198	83	6537
1980-81	0	3	71	511	746	1140	1308	1031	846	397	313	6	6372
1981-82	8	6	135	489	719	1236	1632	1213	922	608	93	118	7179
1982-83	7	37	152	372	772	891	1194	961	847	643	364	38	6278
1983-84	16	0	125	383	714	1568	1479	894	1095	575	300	18	7167
1984-85	19	1	189	320	807	1046	1563	1245	787	418	183	103	6681
1985-86	0	6	141	380	813	1480	1302	1142	765	417	202	74	6722
1986-87	3	29	64	343	863	1060	1205	866	742	432	162	14	5783
1987-88	4	19	74	541	629	1011	1396	1221	828	503	176	40	6442
1988-89	0	9	63	583	693	1149	1003	1265	882	540	261	43	6491
1989-90	0	5	131	344	813	1471	956	938	733	491	271	33	6186
1990-91	10	5	103	425	605	1120	1365	945	756	393	142	13	5882
1991-92	0	0	163	367	887	1066	1137	913	847	560	284	77	6301
1992-93	9	37	136	449	795	1122	1196	1133	948	595	184	69	6673
1993-94	0	3	185	479	784	1084	1516	1197	817	433	253	51	6802
1994-95	1	23	63	322	611	932	1262	1074	760	561	199	25	5833
1995-96	1	0	150	349	958	1193	1284	1124	1054	589	343	58	7103
1996-97	9	0	119	399	940	1148	1410	1003	832	587	344	53	6844
1997-98	9	4	77	406	852	1030	1091	732	813	449	87	69	5619
1998-99	0	0	35	289	598	933	1309	860	903	456	149	34	5566
1999-00	1	4	110	368	591	1081	1224	892	640	528	148	57	5644
2000-01	6	3	112	286	833	1512	1248	1085	948	374	205	88	6700
2001-	5	0	128	394	496	973							

COOLING DEGREE DAYS (base 65 F) 2001 CHICAGO, O'HARE INTERNATIONAL AIRPORT

YEAR	JAN	FEB	MAR	APR	MAY	JUN	JUL	AUG	SEP	OCT	NOV	DEC	ANNUAL
1972	0	0	0	0	64	106	289	289	72	0	0	0	820
1973	0	0	0	5	3	189	308	301	108	32	0	0	946
1974	0	0	0	10	21	83	274	162	48	12	0	0	610
1975	0	0	0	0	76	203	332	358	46	24	1	0	1040
1976	0	0	0	36	6	178	286	196	56	8	0	0	766
1977	0	0	0	39	191	178	395	229	76	0	0	0	1108
1978	0	0	0	0	60	132	227	243	181	2	0	0	845
1979	0	0	0	2	61	164	241	213	99	26	0	0	806
1980	0	0	0	10	43	101	338	342	107	2	0	0	943
1981	0	0	0	9	20	157	248	204	44	0	0	0	682
1982	0	0	0	0	79	38	295	161	69	14	0	0	656
1983	0	0	1	0	4	189	385	388	122	10	0	0	1099
1984	0	0	0	5	11	184	190	254	77	8	0	0	729
1985	0	0	0	53	42	71	204	142	158	0	0	0	670
1986	0	0	7	17	37	118	318	145	123	3	0	0	768
1987	0	0	0	6	116	241	377	238	83	0	1	0	1062
1988	0	0	0	5	59	247	373	383	96	1	0	0	1164
1989	0	0	2	0	44	121	282	207	48	11	0	0	715
1990	0	0	7	43	8	179	226	224	137	11	1	0	836
1991	0	0	0	11	167	226	334	273	132	10	0	0	1153
1992	0	0	0	1	40	79	152	106	75	4	0	0	457
1993	0	0	0	0	28	118	294	266	19	5	0	0	730
1994	0	0	0	23	47	212	268	143	126	10	0	0	829
1995	0	0	0	0	13	254	398	445	81	8	0	0	1199
1996	0	0	0	0	41	154	166	235	79	2	0	0	677
1997	0	0	0	4	158	265	154	59	44	0	0	0	684
1998	0	0	13	0	88	205	301	267	123	5	0	0	1002
1999	0	0	0	0	52	201	422	176	70	2	2	0	925
2000	0	0	3	0	63	131	199	240	112	18	0	0	766
2001	0	0	0	7	58	168	309	262	44	1	0	0	849

SNOWFALL (inches) 2001 CHICAGO, O'HARE INTERNATIONAL AIRPORT, IL (ORD)

YEAR	JUL	AUG	SEP	OCT	NOV	DEC	JAN	FEB	MAR	APR	MAY	JUN	TOTAL
1972-73	0.0	0.0	0.0	0.1	0.9	11.2	0.5	9.3	3.4	0.2	T	0.0	25.6
1973-74	0.0	0.0	0.0	0.0	T	18.8	7.4	9.6	1.4	T	0.0	0.0	37.2
1974-75	0.0	0.0	0.0	0.0	1.0	9.4	3.5	8.2	4.5	11.1	0.0	0.0	37.7
1975-76	0.0	0.0	0.0	0.0	6.4	6.8	10.0	1.6	1.9	0.8	T	0.0	27.5
1976-77	0.0	0.0	0.0	1.6	0.5	6.5	7.2	4.0	4.9	T	0.0	0.0	24.7
1977-78	0.0	0.0	0.0	0.0	5.2	12.7	21.9	7.9	4.5	0.2	0.0	0.0	52.4
1978-79	0.0	0.0	0.0	0.0	5.2	35.3	34.3	6.8	2.0	0.1	0.0	0.0	83.7
1979-80	0.0	0.0	0.0	0.0	4.0	0.9	6.2	14.7	11.6	4.2	0.0	0.0	41.6
1980-81	0.0	0.0	0.0	T	5.1	9.7	2.0	15.9	2.3	0.0	0.0	0.0	35.0
1981-82	0.0	0.0	0.0	T	3.6	4.9	21.1	4.8	14.3	10.6	0.0	0.0	59.3
1982-83	0.0	0.0	0.0	0.0	0.4	2.1	5.0	8.9	9.0	1.2	0.0	0.0	26.6
1983-84	0.0	0.0	0.0	0.0	1.0	16.5	17.2	1.9	9.7	2.7	0.0	0.0	49.0
1984-85	0.0	0.0	0.0	0.0	T	6.6	18.9	13.3	0.3	T	0.0	0.0	39.1
1985-86	0.0	0.0	0.0	0.0	1.1	5.2	6.9	10.9	4.1	0.8	0.0	0.0	29.0
1986-87	0.0	0.0	0.0	T	3.8	0.4	17.3	4.7	T	0.0	0.0		26.2
1987-88	0.0	0.0	0.0	0.1	1.0	18.7	5.4	15.5	1.9	T	0.0	0.0	42.6
1988-89	0.0	0.0	0.0	T	0.9	5.0	0.4	15.1	2.0	0.6	0.5	0.0	24.5
1989-90	0.0	0.0	T	6.3	3.9	5.4	3.2	13.6	1.3	0.1	T	0.0	33.8
1990-91	0.0	0.0	0.0	T	T	3.2	11.1	3.3	3.3	5.9	1.1	0.0	28.4
1991-92	0.0	0.0	0.0	T	1.2	7.6	5.6	1.3	11.6	1.1	0.0	T	28.4
1992-93	T	0.0	0.0	0.3	0.2	5.7	15.2	8.0	13.8	3.7	0.0	0.0	46.9
1993-94	0.0	0.0	0.0	T	0.2	1.2	14.2	26.2	T	T	0.0	0.0	41.8
1994-95	0.0	0.0	0.0	0.0	T	7.0	13.1	0.4	3.5	0.1	0.0	0.0	24.1
1995-96	T	0.0	0.0	0.0	3.9	9.9	5.9	0.3	3.9	T	T	0.0	23.9
1996-97	0.0	0.0											
1997-98							11.0	T	8.2	0.0	0.0	T	50.9
1998-99	0.0	0.0	0.0	0.0	0.2	1.0	29.6	1.9	18.2	0.0	0.0	T	50.9
1999-00	0.0	0.0	0.0	0.0	0.0	3.5	13.6	11.6	T	1.6	0.0	T	30.3
2000-01	0.0	0.0	0.0	T	0.1	30.9	1.5	2.2	4.2	0.3	T	T	39.2
2001-	T	0.0	0.0	T	0.0	1.6							
POR= 41 YRS	T	T	T	0.4	1.9	8.4	10.8	7.8	6.5	1.6	0.1	T	37.5

2001
PEORIA,
ILLINOIS (PIA)

The airport station is situated on a rather level tableland surrounded by well-drained and gently rolling terrain. It is set back a mile from the rim of the Illinois River Valley and is almost 200 feet above the river bed. Exposures of all instruments are good. The climate of this area is typically continental as shown by its changeable weather and the wide range of temperature extremes.

June and September are usually the most pleasant months of the year. Then during October or the first of November, Indian Summer is often experienced with an extended period of warm, dry weather.

Precipitation is normally heaviest during the growing season and lowest during midwinter.

The earliest snowfalls have occurred in September and the latest in the spring have occurred as late as May. Heavy snowfalls have rarely exceeded 20 inches.

Based on the 1951-1980 period, the average first occurrence of 32 degrees Fahrenheit in the fall is October 20 and the average last occurrence in the spring is April 24.

NORMALS, MEANS, AND EXTREMES
PEORIA, IL (PIA)

LATITUDE: 40 40' 03" N   LONGITUDE: 89 41' 02" W   ELEVATION (FT): GRND: 713 BARO: 716   TIME ZONE: CENTRAL (UTC + 6)   WBAN: 14842

	ELEMENT	POR	JAN	FEB	MAR	APR	MAY	JUN	JUL	AUG	SEP	OCT	NOV	DEC	YEAR
TEMPERATURE °F	NORMAL DAILY MAXIMUM	30	29.9	34.9	48.1	62.0	72.8	82.2	85.7	83.1	76.9	64.8	49.8	34.6	60.4
	MEAN DAILY MAXIMUM	96	32.1	36.3	48.3	61.6	72.5	82.2	86.4	84.4	77.2	65.2	49.3	36.1	61.0
	HIGHEST DAILY MAXIMUM	62	70	72	86	92	93	105	103	103	100	90	81	71	105
	YEAR OF OCCURRENCE		1989	1976	1986	1986	1987	1988	1940	1988	1953	1963	1950	1982	JUN 1988
	MEAN OF EXTREME MAXS.	96	52.9	57.9	72.9	81.8	88.2	94.2	96.5	94.9	91.5	83.0	70.4	57.3	78.5
	NORMAL DAILY MINIMUM	30	13.2	17.7	29.8	40.8	50.9	60.7	65.4	63.1	55.2	43.1	32.5	19.3	41.0
	MEAN DAILY MINIMUM	96	15.9	19.8	29.5	40.7	50.9	60.5	64.8	63.0	55.1	43.4	31.7	20.9	41.3
	LOWEST DAILY MINIMUM	62	-25	-19	-10	14	25	39	47	41	26	19	-2	-23	-25
	YEAR OF OCCURRENCE		1977	1996	1960	1982	1966	1993	1972	1986	1942	1972	1977	1989	JAN 1977
	MEAN OF EXTREME MINS.	96	-7.0	-1.7	10.8	25.4	35.7	46.6	53.3	50.4	38.4	27.1	14.1	-1.5	24.3
	NORMAL DRY BULB	30	21.6	26.3	39.0	51.4	61.9	71.5	75.5	73.1	66.1	54.0	41.2	27.0	50.7
	MEAN DRY BULB	96	24.0	28.0	38.8	51.2	61.8	71.3	75.6	73.7	66.1	54.3	40.5	28.5	51.2
	MEAN WET BULB	18	23.2	27.8	36.1	46.2	56.5	64.8	69.0	67.5	59.2	48.5	37.4	26.9	46.9
	MEAN DEW POINT	18	19.1	23.0	29.8	39.5	50.9	60.4	65.5	64.2	54.9	43.2	32.6	22.8	42.2
	NORMAL NO. DAYS WITH:														
	MAXIMUM 90	30	0.0	0.0	0.0	0.1	0.5	4.7	8.4	5.2	1.6	*	0.0	0.0	20.5
	MAXIMUM 32	30	17.4	11.7	3.2	0.1	0.0	0.0	0.0	0.0	0.0	0.0	1.6	12.6	46.6
	MINIMUM 32	30	29.4	25.3	19.5	5.9	0.4	0.0	0.0	0.0	0.1	4.7	16.7	26.7	128.7
	MINIMUM 0	30	6.7	3.4	0.1	0.0	0.0	0.0	0.0	0.0	0.0	0.0	0.1	2.9	13.2
H/C	NORMAL HEATING DEG. DAYS	30	1345	1084	806	408	183	11	0	9	54	356	714	1178	6148
	NORMAL COOLING DEG. DAYS	30	0	0	0	0	87	206	326	261	87	15	0	0	982
RH	NORMAL (PERCENT)	30	74	74	70	65	66	67	72	74	73	70	74	78	71
	HOUR 00 LST	30	77	78	76	71	74	76	82	84	82	78	79	81	78
	HOUR 06 LST	30	79	80	81	78	80	81	86	89	88	85	83	83	83
	HOUR 12 LST	30	68	66	62	55	56	56	59	60	59	58	66	72	61
	HOUR 18 LST	30	72	70	64	55	55	56	60	63	64	63	71	76	64
S	PERCENT POSSIBLE SUNSHINE	52	47	50	51	55	60	67	69	67	64	61	43	42	56
W/O	MEAN NO. DAYS WITH:														
	HEAVY FOG(VISBY 1/4 MI)	59	3.2	2.8	2.2	0.9	0.9	0.6	1.0	1.4	1.4	1.7	2.0	3.2	21.3
	THUNDERSTORMS	59	0.5	0.6	2.6	5.0	6.9	8.2	7.6	6.4	4.8	2.5	1.4	0.6	47.1
CLOUDINESS	MEAN:														
	SUNRISE-SUNSET (OKTAS)	1						5.6						7.2	
	MIDNIGHT-MIDNIGHT (OKTAS)														
	MEAN NO. DAYS WITH:														
	CLEAR	1		1.0	5.0		5.0	9.0							
	PARTLY CLOUDY	1	1.0		3.0		1.0	4.0							
	CLOUDY	1	5.0	6.0	10.0		11.0	9.0							
PR	MEAN STATION PRESSURE(IN)	29	29.40	29.39	29.30	29.30	29.29	29.29	29.30	29.30	29.30	29.40	29.31	29.40	29.33
	MEAN SEA-LEVEL PRES. (IN)	18	30.12	30.12	30.06	29.96	29.96	29.96	29.99	30.03	30.05	30.08	30.08	30.15	30.05
WINDS	MEAN SPEED (MPH)	45	10.9	10.8	11.7	11.4	9.6	8.5	7.5	7.1	8.0	9.2	10.6	10.3	9.6
	PREVAIL.DIR(TENS OF DEGS)	33	18	18	18	18	18	18	18	18	18	18	18	18	18
	MAXIMUM 2-MINUTE:														
	SPEED (MPH)	6	38	39	54	45	36	39	30	30	37	35	44	33	54
	DIR. (TENS OF DEGS)		23	25	23	24	28	24	26	24	30	26	27	23	23	23
	YEAR OF OCCURRENCE		1996	2001	1998	1997	2001	2001	1997	2001	2001	1996	1998	1998	MAR 1998
	MAXIMUM 5-SECOND:														
	SPEED (MPH)	6	47	52	64	56	44	59	41	49	46	60	58	38	64
	DIR. (TENS OF DEGS)		27	28	20	24	28	22	22	29	26	23	24	28	20
	YEAR OF OCCURRENCE		1996	2001	1998	1997	2001	2001	1997	2001	2001	1996	1998	2001	MAR 1998
PRECIPITATION	NORMAL (IN)	30	1.51	1.42	2.91	3.77	3.70	3.99	4.20	3.10	3.87	2.65	2.69	2.44	36.25
	MAXIMUM MONTHLY (IN)	62	8.11	5.37	6.95	8.66	10.19	11.69	10.15	8.61	13.09	10.80	7.62	6.34	13.09
	YEAR OF OCCURRENCE		1965	1997	1973	1947	1995	1974	1993	1965	1961	1941	1985	1949	SEP 1961
	MINIMUM MONTHLY (IN)	62	0.22	0.33	0.39	0.71	0.82	0.60	0.33	0.25	0.03	0.03	0.22	0.32	0.03
	YEAR OF OCCURRENCE		1986	1947	1958	1971	1992	1988	1988	1992	1979	1964	1999	1995	SEP 1979
	MAXIMUM IN 24 HOURS (IN)	58	4.45	3.34	3.39	5.06	3.62	4.44	4.22	4.32	4.15	3.70	4.32	3.38	5.06
	YEAR OF OCCURRENCE		1965	1997	1944	1950	1956	1974	1993	1955	1961	1969	1990	1949	APR 1950
	NORMAL NO. DAYS WITH:														
	PRECIPITATION 0.01	30	9.5	8.2	11.0	11.6	10.7	9.0	9.1	8.7	9.0	8.4	9.5	10.6	115.3
	PRECIPITATION 1.00	30	0.3	0.2	0.4	1.0	0.8	1.0	1.4	0.8	1.0	0.5	0.6	0.5	8.5
SNOWFALL	NORMAL (IN)	30	7.3	5.9	3.4	1.2	0.*	0.0	0.0	0.0	0.0	0.1	1.9	6.4	26.2
	MAXIMUM MONTHLY (IN)	57	24.7	15.2	16.9	13.4	0.1	T	T	0.0	T	1.8	9.1	21.7	24.7
	YEAR OF OCCURRENCE		1979	1989	1960	1982	1966	2000	1990		1992	1967	1974	1977	JAN 1979
	MAXIMUM IN 24 HOURS (IN)	57	12.2	7.6	9.0	6.1	0.1	T	T	0.0	T	1.8	7.2	10.2	12.2
	YEAR OF OCCURRENCE		1979	1944	1946	1982	1966	2000	1990		1992	1967	1951	1973	JAN 1979
	MAXIMUM SNOW DEPTH (IN)	94	20	17	10	10	0	0	0	0	1	3	7	13	20
	YEAR OF OCCURRENCE		1979	1979	1960	1982					1942	1929	1975	1973	JAN 1979
	NORMAL NO. DAYS WITH:														
	SNOWFALL 1.0	30	2.4	1.7	1.2	0.4	0.0	0.0	0.0	0.0	0.0	0.*	0.4	2.1	8.2

PRECIPITATION (inches) 2001 PEORIA, IL (PIA)

YEAR	JAN	FEB	MAR	APR	MAY	JUN	JUL	AUG	SEP	OCT	NOV	DEC	ANNUAL
1972	0.81	0.74	2.48	4.38	1.30	5.97	3.54	4.26	5.21	2.50	2.56	2.48	36.23
1973	1.76	0.99	6.95	4.26	4.51	6.46	6.04	0.90	7.58	5.18	1.48	2.48	50.22
1974	3.09	1.65	2.69	4.11	6.26	11.69	2.63	0.81	1.45	2.07	4.13	1.93	42.51
1975	2.59	2.85	1.73	3.92	5.19	3.90	4.26	5.62	2.74	3.63	2.75	2.04	41.22
1976	0.78	2.56	4.25	4.86	5.11	2.92	2.98	2.30	1.78	2.48	0.83	0.38	31.23
1977	1.22	0.95	4.41	1.24	3.54	2.06	3.43	7.28	6.26	4.00	1.77	2.25	38.41
1978	0.69	0.59	1.56	4.69	7.72	1.96	3.47	1.28	2.32	1.73	2.54	3.54	32.09
1979	2.48	1.37	4.42	4.48	1.96	1.77	4.81	0.87	0.03	1.70	2.76	2.33	28.98
1980	0.59	1.06	2.79	2.78	2.05	8.94	1.43	6.16	4.09	2.44	0.67	2.25	35.25
1981	0.48	2.41	0.92	5.71	5.77	6.22	7.08	5.61	1.31	1.37	1.64	1.24	39.76
1982	2.88	1.13	4.80	5.40	3.15	3.15	7.53	3.97	1.24	1.47	4.95	5.45	45.12
1983	0.53	1.01	2.84	7.06	6.66	4.48	1.99	1.09	5.08	3.01	5.58	2.65	41.98
1984	0.59	2.28	3.95	5.18	4.84	2.90	5.02	0.78	2.38	5.07	3.95	3.82	40.76
1985	0.99	2.62	5.77	1.14	3.14	5.11	3.43	3.70	3.43	4.61	7.62	2.24	43.80
1986	0.22	1.79	0.87	1.39	2.95	6.53	7.00	1.74	6.39	4.64	1.32	2.60	37.44
1987	1.49	0.84	1.98	1.84	1.69	3.27	2.90	4.02	1.62	0.73	2.88	4.15	27.41
1988	1.99	0.71	2.83	1.59	1.68	0.60	0.33	2.11	2.82	1.08	4.19	2.23	22.16
1989	1.00	1.17	1.14	4.39	2.23	1.28	2.22	2.86	2.87	1.57	0.93	0.87	22.53
1990	1.73	3.59	3.95	2.32	6.19	7.99	9.18	5.31	1.03	3.17	7.19	3.70	55.35
1991	1.19	0.57	3.67	2.97	5.94	1.50	0.35	3.41	3.59	7.31	3.57	2.06	36.13
1992	1.06	1.55	2.58	1.61	0.82	0.80	8.19	0.25	5.81	1.33	5.58	2.99	32.57
1993	3.55	1.68	4.08	4.89	3.25	5.70	10.15	7.38	7.56	2.42	2.22	1.19	54.07
1994	0.92	1.98	0.66	3.86	1.52	1.81	1.37	2.92	1.27	3.19	3.49	2.21	25.20
1995	2.83	0.54	1.56	4.77	10.19	1.57	2.83	1.95	1.47	3.23	2.48	0.32	33.74
1996	1.44	0.82	2.04	2.55	7.60	2.08	4.57	1.03	1.62	2.51	2.16	1.06	29.48
1997	1.05	5.37	1.68	2.76	2.65	1.38	0.89	6.10	3.13	2.29	2.85	1.69	31.84
1998	2.55	2.64	4.67	4.96	5.50	5.19	1.64	5.26	2.30	3.21	2.63	1.75	42.30
1999	3.07	1.16	0.94	4.31	4.92	3.21	4.22	2.78	1.54	1.35	0.22	2.55	30.27
2000	0.80	1.82	1.63	2.53	4.04	3.76	1.95	0.91	2.64	1.95	3.22	0.96	26.21
2001	3.29	2.82	1.09	4.21	5.33	2.79	1.46	4.24	4.15	5.08	1.90	1.35	37.71
POR= 146 YRS	1.79	1.81	2.77	3.55	3.99	3.82	3.75	3.03	3.58	2.51	2.39	1.99	34.98

AVERAGE TEMPERATURE (F) 2001 PEORIA, IL (PIA)

YEAR	JAN	FEB	MAR	APR	MAY	JUN	JUL	AUG	SEP	OCT	NOV	DEC	ANNUAL
1972	19.7	24.5	36.9	48.9	63.2	68.0	73.2	72.4	66.2	50.0	36.2	23.8	48.6
1973	27.4	28.9	46.5	50.9	57.9	71.8	75.0	74.7	67.0	57.7	42.4	24.8	52.1
1974	23.1	29.4	41.1	53.6	59.3	66.7	76.9	72.2	61.1	53.6	39.6	29.9	50.5
1975	27.5	26.1	33.8	47.0	64.6	72.1	73.8	74.7	60.9	53.6	44.9	30.2	50.9
1976	19.5	34.9	43.9	54.1	58.2	70.5	75.3	70.3	63.3	47.3	32.0	21.0	49.2
1977	8.6	27.0	44.6	57.1	68.6	70.2	78.2	71.4	66.4	51.3	40.3	32.8	50.5
1978	13.3	15.4	32.4	51.2	60.3	71.3	75.0	73.2	70.5	52.1	41.3	26.8	48.6
1979	9.4	14.7	37.9	47.7	60.5	71.2	73.0	72.5	65.6	52.6	38.0	31.8	47.9
1980	23.6	19.9	35.6	49.2	63.0	69.3	78.5	76.9	67.9	49.6	40.7	28.9	50.3
1981	23.7	27.8	40.7	55.4	58.7	73.2	75.4	72.4	65.9	53.4	45.0	27.3	51.6
1982	15.8	24.8	37.7	46.5	68.4	67.2	75.7	71.3	64.9	55.0	41.5	37.4	50.5
1983	28.2	33.5	40.3	46.8	58.6	72.6	80.2	80.8	67.9	55.3	44.9	15.2	52.0
1984	20.6	35.5	31.3	49.6	58.8	74.3	73.0	74.9	64.0	57.6	40.3	13.9	51.2
1985	16.8	23.2	43.8	57.0	64.3	68.7	73.7	70.2	66.6	55.5	39.0	18.6	49.8
1986	26.6	23.5	43.2	55.4	64.0	72.3	77.4	68.9	68.9	55.2	36.2	30.9	51.9
1987	25.2	35.7	44.1	54.1	68.4	74.0	79.0	73.7	65.7	48.0	44.6	32.4	53.7
1988	22.7	23.5	39.9	51.1	65.4	73.1	78.6	73.5	68.2	48.2	41.5	29.7	51.7
1989	33.9	18.7	39.2	50.7	59.1	69.8	75.3	72.4	62.2	54.9	39.8	16.2	49.4
1990	34.8	33.7	43.4	50.0	57.8	71.0	72.6	72.1	66.1	51.9	45.2	27.7	52.2
1991	21.5	32.4	42.8	55.8	68.0	74.7	75.4	74.3	65.4	53.9	36.0	32.0	52.7
1992	30.1	35.6	42.3	50.0	61.0	69.7	72.4	68.6	64.2	53.1	39.8	29.5	51.4
1993	27.4	25.1	36.4	49.8	63.6	70.4	76.1	75.0	61.4	51.7	39.4	31.4	50.6
1994	18.2	26.2	41.0	53.2	61.7	74.4	74.4	71.6	66.7	56.2	44.8	34.7	51.9
1995	24.1	28.3	42.6	49.3	58.3	72.1	76.2	79.4	62.7	54.2	34.4	27.5	50.8
1996	21.9	28.9	34.0	48.1	60.0	71.1	72.0	73.4	64.4	54.0	33.7	28.7	49.2
1997	19.3	31.9	41.4	47.1	57.1	70.7	71.5	71.5	66.1	54.9	38.1	30.9	50.4
1998	29.7	39.6	39.5	52.1	66.8	70.7	75.7	75.2	70.4	55.5	45.2	33.3	54.5
1999	22.9	36.1	38.2	53.7	63.9	72.0	78.1	71.2	64.1	54.4	47.2	31.5	52.8
2000	25.7	36.7	46.5	51.6	64.5	69.2	73.3	75.9	66.5	57.4	37.6	16.1	51.8
2001	24.2	28.6	36.4	57.8	63.8	69.7	76.8	74.6	64.0	53.0	50.0	34.5	52.8
POR= 97 YRS	24.0	28.1	38.8	51.1	61.8	71.3	75.6	73.6	66.2	54.3	40.5	28.5	51.1

REFERENCE NOTES:

PAGE 1:
THE TEMPERATURE GRAPH SHOWS NORMAL MAXIMUM AND NORMAL MINIMUM DAILY TEMPERATURES (SOLID CURVES) AND THE ACTUAL DAILY HIGH AND LOW TEMPERATURES (VERTICAL BARS).

PAGE 2 AND 3:
H/C INDICATES HEATING AND COOLING DEGREE DAYS.
RH INDICATES RELATIVE HUMIDITY
W/O INDICATES WEATHER AND OBSTRUCTIONS
S INDICATES SUNSHINE.
PR INDICATES PRESSURE.
CLOUDINESS ON PAGE 3 IS THE SUM OF THE CEILOMETER AND SATELLITE DATA NOT TO EXCEED EIGHT EIGHTHS (OKTAS).

GENERAL:
T INDICATES TRACE PRECIPITATION, AN AMOUNT GREATER THAN ZERO BUT LESS THAN THE LOWEST REPORTABLE VALUE.
+ INDICATES THE VALUE ALSO OCCURS ON EARLIER DATES.
BLANK ENTRIES DENOTE MISSING OR UNREPORTED DATA.
NORMALS ARE 30-YEAR AVERAGES (1961 - 1990).
ASOS INDICATES AUTOMATED SURFACE OBSERVING SYSTEM.
PM INDICATES THE LAST DAY OF THE PREVIOUS MONTH.
POR (PERIOD OF RECORD) BEGINS WITH THE JANUARY DATA MONTH AND IS THE NUMBER OF YEARS USED TO COMPUTE THE MEAN. INDIVIDUAL MONTHS WITHIN THE POR MAY BE MISSING.
WHEN THE POR FOR A NORMAL IS LESS THAN 30 YEARS, THE NORMAL IS PROVISIONAL AND IS BASED ON THE NUMBER OF YEARS INDICATED.
0.* OR * INDICATES THE VALUE OR MEAN-DAYS-WITH IS BETWEEN 0.00 AND 0.05.
CLOUDINESS FOR ASOS STATIONS DIFFERS FROM THE NON-ASOS OBSERVATION TAKEN BY A HUMAN OBSERVER. ASOS STATION CLOUDINESS IS BASED ON TIME-AVERAGED CEILOMETER DATA FOR CLOUDS AT OR BELOW 12,000 FEET AND ON SATELLITE DATA FOR CLOUDS ABOVE 12,000 FEET.
THE NUMBER OF DAYS WITH CLEAR, PARTLY CLOUDY, AND CLOUDY CONDITIONS FOR ASOS STATIONS IS THE SUM OF THE CEILOMETER AND SATELLITE DATA FOR THE SUNRISE TO SUNSET PERIOD.

GENERAL CONTINUED:
CLEAR INDICATES 0 - 2 OKTAS, PARTLY CLOUDY INDICATES 3 - 6 OKTAS, AND CLOUDY INDICATES 7 OR 8 OKTAS. WHEN AT LEAST ONE OF THE ELEMENTS (CEILOMETER OR SATELLITE) IS MISSING, THE DAILY CLOUDINESS IS NOT COMPUTED.
WIND DIRECTION IS RECORDED IN TENS OF DEGREES (2 DIGITS) CLOCKWISE FROM TRUE NORTH. "00" INDICATES CALM. "36" INDICATES TRUE NORTH.
RESULTANT WIND IS THE VECTOR AVERAGE OF THE SPEED AND DIRECTION.
AVERAGE TEMPERATURE IS THE SUM OF THE MEAN DAILY MAXIMUM AND MINIMUM TEMPERATURE DIVIDED BY 2.
SNOWFALL DATA COMPRISE ALL FORMS OF FROZEN PRECIPITATION, INCLUDING HAIL.
A HEATING (COOLING) DEGREE DAY IS THE DIFFERENCE BETWEEN THE AVERAGE DAILY TEMPERATURE AND 65 F.
DRY BULB IS THE TEMPERATURE OF THE AMBIENT AIR.
DEW POINT IS THE TEMPERATURE TO WHICH THE AIR MUST BE COOLED TO ACHIEVE 100 PERCENT RELATIVE HUMIDITY.
WET BULB IS THE TEMPERATURE THE AIR WOULD HAVE IF THE MOISTURE CONTENT WAS INCREASED TO 100 PERCENT RELATIVE HUMIDITY.

ON JULY 1, 1996, THE NATIONAL WEATHER SERVICE BEGAN USING THE "METAR" OBSERVATION CODE THAT WAS ALREADY EMPLOYED BY MOST OTHER NATIONS OF THE WORLD. THE MOST NOTICEABLE DIFFERENCE IN THIS ANNUAL PUBLICATION WILL BE THE CHANGE IN UNITS FROM TENTHS TO EIGHTS (OKTAS) FOR REPORTING THE AMOUNT OF SKY COVER.

HEATING DEGREE DAYS (base 65 F) 2001 PEORIA, IL (PIA)

YEAR	JUL	AUG	SEP	OCT	NOV	DEC	JAN	FEB	MAR	APR	MAY	JUN	TOTAL
1972-73	9	19	87	461	855	1271	1161	1005	566	423	216	0	6073
1973-74	0	0	54	253	671	1237	1292	991	736	348	214	42	5838
1974-75	0	1	157	354	756	1081	1156	1085	959	534	92	17	6192
1975-76	13	1	171	306	596	1069	1404	868	648	351	221	5	5653
1976-77	0	9	106	556	981	1357	1747	1061	623	273	60	17	6790
1977-78	0	6	39	418	734	1301	1595	1383	1006	405	222	14	7123
1978-79	0	4	49	390	704	1174	1722	1403	833	510	194	10	6993
1979-80	3	19	70	401	804	1022	1279	1300	907	474	123	26	6428
1980-81	0	0	65	470	722	1112	1273	1037	748	295	221	1	5944
1981-82	1	0	60	360	594	1163	1520	1119	839	548	29	28	6261
1982-83	0	13	94	325	697	849	1133	875	758	537	206	17	5504
1983-84	2	0	92	311	595	1541	1371	849	1038	467	206	1	6473
1984-85	1	1	153	246	734	956	1489	1164	656	284	71	38	5793
1985-86	0	6	111	287	774	1432	1184	1156	683	314	92	7	6046
1986-87	0	26	37	305	858	1048	1228	814	640	340	53	2	5351
1987-88	0	16	68	520	609	1001	1306	1198	772	409	64	12	5975
1988-89	0	4	38	517	698	1090	958	1290	796	442	231	25	6089
1989-90	0	3	134	317	749	1509	929	871	672	475	226	16	5901
1990-91	8	3	99	409	589	1148	1341	909	683	285	86	0	5560
1991-92	0	0	143	347	863	1017	1078	847	698	444	180	19	5636
1992-93	0	23	109	367	748	1094	1159	878	758	450	96	48	6082
1993-94	0	0	135	408	761	1035	1448	1078	734	371	157	12	6139
1994-95	0	8	67	282	602	932	1258	1022	687	466	207	7	5538
1995-96	2	0	133	332	911	1154	1328	1041	954	499	210	28	6592
1996-97	2	0	97	338	930	1120	1412	922	723	530	252	18	6344
1997-98	5	3	54	364	802	1051	1088	707	793	381	50	44	5342
1998-99	0	0	19	297	587	975	1300	806	823	333	91	18	5249
1999-00	0	0	102	330	529	1031	1214	814	566	394	103	22	5105
2000-01	0	0	92	252	818	1507	1256	1013	881	248	120	36	6223
2001-	1	0	102	371	444	938							

COOLING DEGREE DAYS (base 65 F) 2001 PEORIA, IL (PIA)

YEAR	JAN	FEB	MAR	APR	MAY	JUN	JUL	AUG	SEP	OCT	NOV	DEC	ANNUAL
1972	0	0	0	0	103	140	273	253	130	3	0	0	902
1973	0	0	0	7	3	211	315	306	121	35	0	0	998
1974	0	0	0	12	43	99	377	233	47	6	0	0	817
1975	0	0	0	0	83	237	292	307	55	14	0	0	988
1976	0	0	0	28	14	176	326	180	62	14	0	0	800
1977	0	0	0	42	176	179	416	209	88	0	0	0	1110
1978	0	0	0	0	81	208	316	267	221	0	0	0	1093
1979	0	0	0	0	62	206	259	258	95	23	0	0	903
1980	0	0	0	6	70	160	425	378	156	1	0	0	1196
1981	0	0	0	13	33	250	331	250	93	6	1	0	977
1982	0	0	0	0	141	101	338	215	96	26	0	0	917
1983	0	0	0	0	14	250	479	494	188	17	0	0	1442
1984	0	0	0	12	24	285	256	315	129	22	0	0	1043
1985	0	0	3	48	54	155	279	173	164	0	0	0	876
1986	0	0	15	30	68	234	392	157	161	6	0	0	1063
1987	0	0	0	19	166	278	440	293	92	0	3	0	1291
1988	0	0	0	2	84	266	431	428	140	5	0	0	1356
1989	0	0	2	20	56	177	324	240	57	11	0	0	887
1990	0	0	9	34	8	204	251	230	138	9	0	0	883
1991	0	0	4	16	189	296	331	293	159	11	0	0	1299
1992	0	0	0	1	62	171	237	140	90	6	0	0	707
1993	0	0	0	2	60	218	350	321	33	6	0	0	990
1994	0	0	0	25	61	300	307	218	123	16	0	0	1050
1995	0	0	0	0	6	225	355	453	73	5	0	0	1117
1996	0	0	0	0	63	217	224	267	88	5	0	0	864
1997	0	0	0	12	199	334	210	94	59	0	0	0	908
1998	0	0	9	0	114	222	337	318	187	10	0	0	1197
1999	0	0	2	0	68	233	414	197	82	8	0	0	1004
2000	0	0	2	0	92	155	262	345	146	25	4	0	1031
2001	0	0	0	38	89	184	372	307	78	4	0	0	1072

SNOWFALL (inches) 2001 PEORIA, IL (PIA)

YEAR	JUL	AUG	SEP	OCT	NOV	DEC	JAN	FEB	MAR	APR	MAY	JUN	TOTAL
1972-73	0.0	0.0	0.0	0.3	7.3	4.7	0.8	2.7	1.2	0.7	0.0	0.0	17.7
1973-74	0.0	0.0	0.0	0.0	0.1	18.9	6.7	1.9	0.9	1.2	0.0	0.0	29.7
1974-75	0.0	0.0	0.0	0.0	9.1	6.2	8.8	12.8	3.8	1.6	0.0	0.0	42.3
1975-76	0.0	0.0	0.0	0.0	8.1	1.4	8.6	2.5	2.0	0.0	0.0	0.0	22.6
1976-77	0.0	0.0	0.0	T	0.4	4.1	16.3	4.1	2.4	1.4	0.0	0.0	28.7
1977-78	0.0	0.0	0.0	0.0	6.4	21.7	7.0	7.0	5.0	0.0	0.0	0.0	47.1
1978-79	0.0	0.0	0.0	0.0	2.6	14.2	24.7	3.6	6.5	T	0.0	0.0	51.6
1979-80	0.0	0.0	0.0	0.0	0.8	0.1	3.7	11.3	5.3	6.3	0.0	0.0	27.5
1980-81	0.0	0.0	0.0	T	4.1	3.3	5.9	10.5	T	0.0	0.0	0.0	23.8
1981-82	0.0	0.0	0.0	0.0	0.1	9.8	11.0	6.1	6.5	13.4	0.0	0.0	46.9
1982-83	0.0	0.0	0.0	0.0	0.9	2.0	5.6	4.8	5.7	0.1	0.0	0.0	19.1
1983-84	0.0	0.0	0.0	0.0	3.3	15.9	6.9	4.1	6.0	0.0	0.0	0.0	36.2
1984-85	0.0	0.0	0.0	0.0	T	0.9	9.8	6.2	T	T	0.0	0.0	16.9
1985-86	0.0	0.0	0.0	0.0	1.0	6.2	1.3	13.9	0.4	0.1	0.0	0.0	22.9
1986-87	0.0	0.0	0.0	0.0	1.0	T	18.0	0.1	T	0.0	0.0	0.0	19.1
1987-88	0.0	0.0	0.0	T	0.3	9.8	1.9	9.7	1.9	0.0	0.0	0.0	23.6
1988-89	0.0	0.0	0.0	0.0	0.7	4.7	0.3	15.2	T	0.9	T	0.0	21.8
1989-90	0.0	0.0	0.0	0.6	T	10.5	4.8	6.2	T	T	0.0	0.0	22.1
1990-91	T	0.0	0.0	T	T	3.8	9.8	1.0	6.9	T	0.0	0.0	21.5
1991-92	0.0	0.0	0.0	0.0	3.6	5.1	6.4	0.4	4.1	T	0.0	0.0	19.6
1992-93	0.0	0.0	T	T	0.7	6.1	6.8	12.4	3.5	0.7	0.0	0.0	30.2
1993-94	0.0	0.0	0.0	T	3.6	2.4	9.4	12.6	T	0.1	0.0	0.0	28.1
1994-95	0.0	0.0	0.0	0.0	T	0.5	3.8	2.8	0.7	T	T	0.0	7.8
1995-96	0.0	0.0	0.0	0.0	0.2	3.6	5.9	1.9	1.1	0.5	0.0	0.0	13.2
1996-97	0.0	0.0	0.0								0.0		
1997-98		0.0		T	1.9	13.0	7.1	T	7.3	7.3	0.0	0.0	31.7
1998-99	0.0	0.0	0.0	0.0	0.5	5.2	18.1	1.0	6.9	T	T	T	14.1
1999-00	0.0	0.0	0.0	0.0	0.0	4.7	8.8	0.6	1.8	2.1	T	0.0	30.2
2000-01	0.0	0.0	0.0	0.0	T	21.2	5.1	1.8	2.1	T	0.0	0.0	30.2
2001-	0.0	0.0	0.0	0.0	T	0.4							
POR= 56 YRS	T	0.0	T	0.1	2.0	6.1	6.7	4.9	3.9	0.8	0.0	0.0	24.5

2001
SPRINGFIELD,
ILLINOIS (SPI)

The location of Springfield near the center of North America gives it a typical continental climate with warm summers and fairly cold winters. The surrounding country is nearly level. There are no large hills in the vicinity, but rolling terrain is found near the Sangamon River and Spring Creek.

Monthly temperatures range from the upper 20s for January to the upper 70s for July. Considerable variation may take place within the seasons. Temperatures of 70 degrees or higher may occur in winter and temperatures near 50 degrees are sometimes recorded during the summer months.

There are no wet and dry seasons. Monthly precipitation ranges from a little over 4 inches in May and June to about 2 inches in January. There is some variation in rainfall totals from year to year. Thunderstorms are common during hot weather, and these are sometimes locally severe with brief but heavy showers. The average year has about fifty thunderstorms of which two-thirds occur during the months of May through August. Damaging hail accompanies only a few of the thunderstorms and the areas affected are usually small.

Sunshine is particularly abundant during the summer months when days are long and not very cloudy. January is the cloudiest month, with only about a third as much sunshine as July or August. March is the windiest month, and August the month with the least wind. Velocities of more than 40 mph are not unusual for brief periods in most months of the year. The prevailing wind direction is southerly during most of the year with northwesterly winds during the late fall and early spring months.

An overall description of the climate of Springfield would be one indicating pleasant conditions with sharp seasonal changes, but no extended periods of severely cold weather. Summer weather is often uncomfortably warm and humid.

Based on the 1951-1980 period, the average first occurrence of 32 degrees Fahrenheit in the fall is October 19 and the average last occurrence in the spring is April 17.

NORMALS, MEANS, AND EXTREMES
SPRINGFIELD, IL (SPI)

LATITUDE:	LONGITUDE:	ELEVATION (FT):		TIME ZONE:	WBAN: 93822
39 50' 43" N	89 41' 02" W	GRND: 591	BARO: 594	CENTRAL (UTC + 6)	

	ELEMENT	POR	JAN	FEB	MAR	APR	MAY	JUN	JUL	AUG	SEP	OCT	NOV	DEC	YEAR
TEMPERATURE F	NORMAL DAILY MAXIMUM	30	32.5	37.2	50.0	63.9	74.7	83.9	86.9	84.2	78.7	66.8	51.7	37.3	62.3
	MEAN DAILY MAXIMUM	101	34.5	38.7	50.4	63.4	74.3	83.7	87.6	85.4	78.7	67.0	51.2	38.3	62.8
	HIGHEST DAILY MAXIMUM	54	71	74	87	90	95	103	112	103	101	93	83	74	112
	YEAR OF OCCURRENCE		1950	1996	1981	1986	1967	1954	1954	1964	1984	1954	1950	1984	JUL 1954
	MEAN OF EXTREME MAXS.	101	56.5	61.1	74.9	82.9	89.2	95.2	97.5	95.6	92.3	84.5	72.1	60.4	80.2
	NORMAL DAILY MINIMUM	30	15.9	20.2	31.6	42.5	52.3	61.9	66.1	63.3	55.9	44.4	34.0	21.9	42.5
	MEAN DAILY MINIMUM	101	19.0	22.5	32.0	42.6	53.1	62.6	66.7	64.7	56.7	45.7	33.9	23.4	43.6
	LOWEST DAILY MINIMUM	54	-21	-22	-12	19	28	40	48	43	32	17	-3	-21	-22
	YEAR OF OCCURRENCE		1999	1963	1960	1997	1966	1966	1975	1986	1995	1952	1964	1989	FEB 1963
	MEAN OF EXTREME MINS.	101	-3.9	0.8	13.4	27.6	38.0	48.8	55.1	52.5	40.5	29.3	16.4	1.9	26.7
	NORMAL DRY BULB	30	24.2	28.7	40.8	53.2	63.5	72.9	76.5	73.7	67.3	55.6	42.9	29.7	52.4
	MEAN DRY BULB	101	26.8	30.5	41.2	53.1	63.6	73.1	77.2	75.0	67.8	56.4	42.6	30.8	53.2
	MEAN WET BULB	16	25.5	29.1	37.7	48.0	57.6	65.8	65.3	68.2	60.2	46.4	37.3	26.9	47.3
	MEAN DEW POINT	16	21.0	24.2	31.6	41.4	48.0	61.4	62.1	64.9	55.5	40.9	32.4	22.8	42.2
	NORMAL NO. DAYS WITH:														
	MAXIMUM　90	30	0.0	0.0	0.0	*	2.1	7.6	11.1	6.7	3.1	0.1	0.0	0.0	30.7
	MAXIMUM　32	30	14.8	9.7	2.5	*	0.0	0.0	0.0	0.0	0.0	0.0	1.0	9.9	37.9
	MINIMUM　32	30	28.1	23.6	17.1	4.3	0.2	0.0	0.0	0.0	*	3.6	14.5	25.3	116.7
	MINIMUM　0	30	4.9	2.6	0.1	0.0	0.0	0.0	0.0	0.0	0.0	0.0	*	2.1	9.7
H/C	NORMAL HEATING DEG. DAYS	30	1265	1016	750	361	166	7	0	8	44	314	663	1094	5688
	NORMAL COOLING DEG. DAYS	30	0	0	0	7	120	244	357	278	113	22	0	0	1141
RH	NORMAL (PERCENT)	30	74	74	71	65	66	67	70	74	72	68	74	78	71
	HOUR 00 LST	30	77	78	77	72	75	77	80	84	82	76	78	80	78
	HOUR 06 LST	30	78	80	81	79	80	82	85	89	88	83	83	83	83
	HOUR 12 LST	30	68	67	63	56	54	54	57	60	56	54	65	72	60
	HOUR 18 LST	30	71	71	65	56	55	55	59	64	63	62	70	76	64
S	PERCENT POSSIBLE SUNSHINE	48	48	52	51	56	63	68	71	70	68	63	48	44	58
W/O	MEAN NO. DAYS WITH:														
	HEAVY FOG(VISBY 1/4 MI)	54	2.8	2.6	1.9	0.9	0.9	0.4	0.7	1.3	1.1	1.1	1.5	2.5	17.7
	THUNDERSTORMS	54	0.5	0.7	2.5	5.3	6.9	7.7	8.3	6.7	4.2	2.4	1.6	0.6	47.4
CLOUDINESS	MEAN:														
	SUNRISE-SUNSET (OKTAS)														
	MIDNIGHT-MIDNIGHT (OKTAS)														
	MEAN NO. DAYS WITH:														
	CLEAR														
	PARTLY CLOUDY														
	CLOUDY														
PR	MEAN STATION PRESSURE(IN)	28	29.46	29.44	29.35	29.31	29.30	29.30	29.34	29.40	29.40	29.42	29.41	29.46	29.38
	MEAN SEA-LEVEL PRES. (IN)	16	30.12	30.12	30.04	29.95	29.96	29.95	29.98	30.02	30.04	30.07	30.08	30.13	30.04
WINDS	MEAN SPEED (MPH)	42	12.3	12.1	13.1	12.8	10.9	9.4	8.1	7.7	8.7	10.2	12.4	12.1	10.8
	PREVAIL.DIR(TENS OF DEGS)	27	18	18	18	18	18	18	18	18	18	18	18	18	18
	MAXIMUM 2-MINUTE:														
	SPEED (MPH)	6	39	51	46	51	46	43	52	44	32	38	40	35	52
	DIR. (TENS OF DEGS)		25	29	25	25	28	16	04	33	28	26	24	28	04
	YEAR OF OCCURRENCE		1996	2001	1996	1997	2001	1996	2001	1999	1997	1996	1998	2001	JUL 2001
	MAXIMUM 5-SECOND:														
	SPEED (MPH)	6	48	62	53	60	56	53	62	51	44	53	53	43	62
	DIR. (TENS OF DEGS)		14	27	25	25	28	34	06	32	28	25	17	24	06
	YEAR OF OCCURRENCE		1996	2001	1996	1997	2001	1998	2001	1999	1997	1996	1998	1996	JUL 2001
PRECIPITATION	NORMAL (IN)	30	1.51	1.77	3.24	3.68	3.62	3.43	3.52	3.29	3.33	2.60	2.53	2.73	35.25
	MAXIMUM MONTHLY (IN)	54	5.67	4.89	7.89	9.91	10.72	9.22	10.76	8.37	8.57	6.41	6.94	8.94	10.76
	YEAR OF OCCURRENCE		1949	1990	1973	1964	1996	1990	1981	1981	1986	1991	1985	1982	JUL 1981
	MINIMUM MONTHLY (IN)	54	0.04	0.51	0.63	0.73	0.52	0.23	0.89	0.46	T	0.16	0.25	0.15	T
	YEAR OF OCCURRENCE		1986	1995	1956	1971	1992	1959	1997	1992	1979	1964	1999	1955	SEP 1979
	MAXIMUM IN 24 HOURS (IN)	54	2.78	2.54	2.84	4.45	3.95	4.73	4.43	4.79	5.12	3.51	2.46	6.12	6.12
	YEAR OF OCCURRENCE		1975	1990	1972	1979	1990	1958	1981	1956	1959	1973	1964	1982	DEC 1982
	NORMAL NO. DAYS WITH:														
	PRECIPITATION　0.01	30	8.8	8.4	11.8	11.1	9.9	9.4	8.5	8.3	8.4	8.1	9.5	10.5	112.7
	PRECIPITATION　1.00	30	0.2	0.3	0.8	1.0	0.9	0.7	0.9	0.9	1.0	0.7	0.5	0.6	8.5
SNOWFALL	NORMAL (IN)	30	7.0	7.2	3.8	0.7	T	0.0	0.0	0.0	0.0	0.*	1.8	5.7	26.2
	MAXIMUM MONTHLY (IN)	54	21.1	16.0	20.3	7.3	T	0.0	T	0.0	0.0	0.3	9.2	22.7	22.7
	YEAR OF OCCURRENCE		1977	1993	1960	1980	1996		1994			1989	1951	1973	DEC 1973
	MAXIMUM IN 24 HOURS (IN)	54	8.8	10.0	8.2	6.1	T	0.0	T	0.0	0.0	0.3	8.0	10.9	10.9
	YEAR OF OCCURRENCE		1964	1965	1978	1980	1996		1994			1989	1951	1973	DEC 1973
	MAXIMUM SNOW DEPTH (IN)	100	16	12	16	5	0	0	0	0	0	2	7	15	16
	YEAR OF OCCURRENCE		1918	1965	1978	1920						1929	1951	1973	MAR 1978
	NORMAL NO. DAYS WITH:														
	SNOWFALL　1.0	30	2.2	1.9	1.1	0.3	0.0	0.0	0.0	0.0	0.0	0.0	0.6	1.8	7.9

PRECIPITATION (inches) 2001 SPRINGFIELD, IL (SPI)

YEAR	JAN	FEB	MAR	APR	MAY	JUN	JUL	AUG	SEP	OCT	NOV	DEC	ANNUAL
1972	1.03	0.82	4.03	3.35	1.88	2.72	1.70	4.52	3.95	1.40	3.27	3.36	32.03
1973	1.31	0.84	7.89	5.29	2.62	7.29	3.36	1.66	3.28	5.46	1.43	3.86	44.29
1974	2.61	3.15	3.39	3.11	6.37	5.00	0.91	7.70	2.17	1.39	3.58	1.44	40.82
1975	4.28	3.63	1.91	2.89	5.90	4.38	2.71	3.34	2.84	1.37	2.50	1.91	37.66
1976	0.98	3.67	5.60	1.07	1.96	1.41	2.29	2.33	2.20	0.53	0.66	0.66	25.70
1977	1.51	1.21	5.09	2.78	5.78	4.26	1.16	5.95	5.94	5.16	1.63	2.24	42.71
1978	0.72	0.83	4.20	2.84	5.81	1.73	2.99	4.04	1.58	1.57	2.13	3.39	31.83
1979	1.90	1.09	3.75	7.17	1.32	0.94	4.63	2.85	T	1.34	1.98	2.36	29.33
1980	0.72	1.42	4.29	2.22	2.22	3.23	2.08	3.91	4.95	1.47	0.57	1.99	29.07
1981	0.43	2.12	2.27	4.57	1.67	5.80	10.76	8.37	1.13	1.94	2.21	2.35	48.12
1982	4.48	1.81	3.04	3.40	4.12	2.54	2.53	3.68	2.75	2.69	4.50	8.94	44.48
1983	0.46	0.96	3.44	5.02	4.53	1.60	0.84	1.36	3.63	4.71	3.50	3.22	32.67
1984	0.70	1.97	4.00	5.45	6.32	2.26	3.46	0.63	4.80	4.74	4.36	3.91	42.60
1985	0.65	2.96	4.19	1.46	1.75	5.82	2.95	6.03	3.08	6.94	2.43	38.90	
1986	0.04	1.80	1.45	1.57	2.56	6.23	5.39	1.13	8.57	3.63	1.95	1.40	35.72
1987	1.46	0.73	2.08	2.59	0.56	4.08	4.12	3.23	0.99	1.26	3.25	5.00	29.35
1988	2.17	1.39	2.69	1.27	1.76	0.62	1.74	1.56	2.84	1.68	4.37	3.22	25.31
1989	0.88	1.27	1.68	5.50	4.18	0.89	3.13	2.57	5.49	1.02	0.84	0.58	28.03
1990	1.49	4.89	3.41	1.28	8.84	9.22	5.48	2.68	1.91	5.03	3.47	4.97	52.67
1991	1.29	0.71	2.64	3.57	5.74	1.26	3.22	4.03	4.35	6.41	3.31	1.38	37.91
1992	1.03	1.59	2.39	2.82	0.52	2.31	6.86	0.46	1.54	1.47	6.14	2.36	29.49
1993	3.97	1.48	2.48	4.59	2.05	7.26	9.46	3.21	6.05	2.92	3.04	1.07	47.58
1994	1.15	1.01	1.27	8.16	3.99	6.05	1.93	3.09	0.93	2.73	3.09	1.99	35.49
1995	4.01	0.51	3.41	2.71	7.54	1.29	2.70	3.67	0.65	2.24	1.38	1.34	31.45
1996	1.54	1.04	1.93	3.93	10.72	1.95	3.32	1.61	1.59	2.96	0.72	32.63	
1997	1.58	2.74	2.50	1.48	3.10	1.54	0.89	4.64	3.53	1.79	4.50	1.75	30.04
1998	2.43	2.71	4.63	4.05	5.65	8.81	3.32	5.30	1.27	3.30	2.81	0.64	44.92
1999	1.94	2.15	0.97	4.61	2.90	2.95	2.08	4.64	2.42	1.78	0.25	2.20	28.89
2000	0.54	1.27	2.80	1.94	1.35	7.46	3.16	3.33	2.92	2.55	2.99	0.91	31.22
2001	2.06	3.01	1.11	1.99	3.50	4.42	3.41	3.34	2.50	4.96	2.61	2.09	35.00
POR= 122 YRS	1.92	2.01	3.03	3.54	4.08	3.99	3.18	3.05	3.25	2.60	2.52	2.09	35.26

AVERAGE TEMPERATURE (F) 2001 SPRINGFIELD, IL (SPI)

YEAR	JAN	FEB	MAR	APR	MAY	JUN	JUL	AUG	SEP	OCT	NOV	DEC	ANNUAL
1972	24.4	29.0	40.8	52.5	64.7	70.5	75.9	73.8	68.7	52.9	38.3	26.7	51.5
1973	29.7	30.7	48.3	51.9	59.6	73.2	76.0	69.8	60.7	45.5	28.0	54.1	
1974	26.1	32.6	44.2	54.8	61.6	68.3	78.8	72.9	61.9	55.8	42.2	32.7	52.7
1975	31.1	29.4	37.4	50.4	66.5	73.6	74.7	76.6	63.8	57.7	47.1	33.4	53.5
1976	23.5	39.6	45.4	55.0	59.7	72.4	77.4	71.6	65.2	49.4	34.6	24.1	51.5
1977	10.3	29.5	47.6	59.9	71.0	71.6	79.4	72.8	68.6	53.8	42.8	27.0	52.9
1978	15.6	16.7	33.6	53.9	62.3	74.2	76.3	73.1	70.9	53.7	45.2	31.4	50.6
1979	12.4	17.1	40.3	50.4	64.0	74.9	75.4	74.0	67.7	55.7	42.0	35.8	50.8
1980	27.8	21.8	37.1	51.1	64.6	72.0	81.4	79.3	68.7	52.8	42.3	31.7	52.6
1981	26.4	32.4	44.1	60.4	60.3	74.9	77.1	73.8	67.2	54.6	45.3	26.8	53.6
1982	17.0	24.2	40.3	48.1	70.0	68.3	77.1	72.5	66.1	55.4	43.0	38.8	51.7
1983	28.7	34.9	40.6	47.8	59.9	73.8	80.4	80.0	69.1	57.2	45.7	16.1	52.9
1984	22.3	35.9	31.8	50.4	60.3	75.2	74.5	76.9	65.7	59.9	41.7	35.9	52.5
1985	18.5	24.9	45.1	57.8	65.2	69.7	74.3	74.8	67.9	58.1	42.8	21.5	51.4
1986	29.1	26.9	45.3	57.4	65.9	74.7	78.8	70.0	70.4	56.1	37.9	32.6	53.8
1987	26.1	36.6	45.4	54.7	70.5	75.4	78.8	75.3	67.6	50.6	46.2	34.6	55.2
1988	25.6	25.6	41.5	52.8	65.8	73.0	78.6	75.7	68.7	49.7	43.0	31.6	53.0
1989	36.0	20.3	40.6	52.3	59.3	70.5	75.4	73.3	63.8	56.9	42.0	18.7	50.8
1990	37.1	36.1	45.5	50.8	59.9	73.1	75.2	74.0	68.4	53.5	47.8	30.0	54.3
1991	24.0	35.3	45.1	57.3	69.2	75.6	75.3	74.5	66.8	55.8	38.5	34.7	54.3
1992	32.1	38.1	43.7	52.2	62.3	69.6	74.3	70.9	67.0	56.1	42.4	32.8	53.5
1993	29.8	27.6	38.6	50.9	64.3	72.0	77.2	76.0	62.4	52.9	40.5	33.1	52.1
1994	21.4	29.4	42.9	53.9	62.3	75.2	75.4	72.8	66.7	57.8	47.8	37.3	53.6
1995	26.4	30.3	45.1	51.6	61.2	73.7	78.3	80.2	64.7	56.0	37.5	28.0	52.8
1996	23.2	31.4	36.0	49.3	63.3	72.5	72.1	73.6	64.3	56.3	35.3	31.5	50.7
1997	20.4	34.4	43.4	48.2	58.8	70.9	75.6	71.8	66.7	56.4	39.2	32.4	51.5
1998	32.7	40.4	40.4	53.0	68.3	71.7	75.4	74.9	71.1	56.7	46.0	34.0	55.4
1999	25.1	37.3	38.7	55.0	64.7	72.3	77.9	71.5	64.5	54.9	48.4	33.9	53.7
2000	27.5	39.0	46.2	52.2	65.9	69.8	73.5	75.4	67.1	58.1	38.1	17.4	52.5
2001	25.8	31.4	37.4	59.4	65.2	70.4	76.6	74.1	64.4	54.1	49.9	35.5	53.7
POR= 122 YRS	26.9	30.5	40.9	53.3	63.7	73.1	77.2	74.9	67.8	56.4	42.5	31.2	53.2

REFERENCE NOTES:

PAGE 1:
THE TEMPERATURE GRAPH SHOWS NORMAL MAXIMUM AND NORMAL
MINIMUM DAILY TEMPERATURES (SOLID CURVES) AND THE
ACTUAL DAILY HIGH AND LOW TEMPERATURES (VERTICAL BARS).

PAGE 2 AND 3:
H/C INDICATES HEATING AND COOLING DEGREE DAYS.
RH INDICATES RELATIVE HUMIDITY
W/O INDICATES WEATHER AND OBSTRUCTIONS
S INDICATES SUNSHINE.
PR INDICATES PRESSURE.
CLOUDINESS ON PAGE 3 IS THE SUM OF THE CEILOMETER AND
SATELLITE DATA NOT TO EXCEED EIGHT EIGHTHS(OKTAS).

GENERAL:
T INDICATES TRACE PRECIPITATION, AN AMOUNT GREATER
THAN ZERO BUT LESS THAN THE LOWEST REPORTABLE VALUE.
+ INDICATES THE VALUE ALSO OCCURS ON EARLIER DATES.
BLANK ENTRIES DENOTE MISSING OR UNREPORTED DATA.
NORMALS ARE 30-YEAR AVERAGES (1961 - 1990).
ASOS INDICATES AUTOMATED SURFACE OBSERVING SYSTEM.
PM INDICATES THE LAST DAY OF THE PREVIOUS MONTH.
POR (PERIOD OF RECORD) BEGINS WITH THE JANUARY DATA
MONTH AND IS THE NUMBER OF YEARS USED TO COMPUTE
THE MEAN. INDIVIDUAL MONTHS WITHIN THE POR MAY
BE MISSING.
WHEN THE POR FOR A NORMAL IS LESS THAN 30 YEARS,
THE NORMAL IS PROVISIONAL AND IS BASED ON THE NUMBER
OF YEARS INDICATED.
0.* OR * INDICATES THE VALUE OR MEAN-DAYS-WITH
IS BETWEEN 0.00 AND 0.05.
CLOUDINESS FOR ASOS STATIONS DIFFERS FROM THE NON-ASOS
OBSERVATION TAKEN BY A HUMAN OBSERVER. ASOS STATION
CLOUDINESS IS BASED ON TIME-AVERAGED CEILOMETER DATA
FOR CLOUDS AT OR BELOW 12,000 FEET AND ON SATELLITE
DATA FOR CLOUDS ABOVE 12,000 FEET.
THE NUMBER OF DAYS WITH CLEAR, PARTLY CLOUDY, AND
CLOUDY CONDITIONS FOR ASOS STATIONS IS THE SUM
OF THE CEILOMETER AND SATELLITE DATA FOR THE
SUNRISE TO SUNSET PERIOD.

GENERAL CONTINUED:
CLEAR INDICATES 0 - 2 OKTAS, PARTLY CLOUDY INDICATES
3 - 6 OKTAS, AND CLOUDY INDICATES 7 OR 8 OKTAS.
WHEN AT LEAST ONE OF THE ELEMENTS (CEILOMETER OR
SATELLITE) IS MISSING, THE DAILY CLOUDINESS IS
NOT COMPUTED.
WIND DIRECTION IS RECORDED IN TENS OF DEGREES (2 DIGITS)
CLOCKWISE FROM TRUE NORTH. "00" INDICATES CALM. "36"
INDICATES TRUE NORTH.
RESULTANT WIND IS THE VECTOR AVERAGE OF THE SPEED AND
DIRECTION.
AVERAGE TEMPERATURE IS THE SUM OF THE MEAN DAILY MAXIMUM
AND MINIMUM TEMPERATURE DIVIDED BY 2.
SNOWFALL DATA COMPRISE ALL FORMS OF FROZEN
PRECIPITATION, INCLUDING HAIL.
A HEATING (COOLING) DEGREE DAY IS THE DIFFERENCE BETWEEN
THE AVERAGE DAILY TEMPERATURE AND 65 F.
DRY BULB IS THE TEMPERATURE OF THE AMBIENT AIR.
DEW POINT IS THE TEMPERATURE TO WHICH THE AIR MUST BE
COOLED TO ACHIEVE 100 PERCENT RELATIVE HUMIDITY.
WET BULB IS THE TEMPERATURE THE AIR WOULD HAVE IF THE
MOISTURE CONTENT WAS INCREASED TO 100 PERCENT RELATIVE
HUMIDITY.

ON JULY 1, 1996, THE NATIONAL WEATHER SERVICE BEGAN USING
THE "METAR" OBSERVATION CODE THAT WAS ALREADY EMPLOYED
BY MOST OTHER NATIONS OF THE WORLD. THE MOST NOTICEABLE
DIFFERENCE IN THIS ANNUAL PUBLICATION WILL BE THE CHANGE
IN UNITS FROM TENTHS TO EIGHTS(OKTAS) FOR REPORTING THE
AMOUNT OF SKY COVER.

HEATING DEGREE DAYS (base 65 F) 2001   SPRINGFIELD, IL (SPI)

YEAR	JUL	AUG	SEP	OCT	NOV	DEC	JAN	FEB	MAR	APR	MAY	JUN	TOTAL
1972-73	3	11	58	375	795	1181	1089	957	512	398	174	0	5553
1973-74	0	0	30	191	578	1140	1200	901	642	316	167	29	5194
1974-75	0	2	135	299	680	996	1044	992	848	435	63	7	5501
1975-76	10	0	125	253	528	973	1280	732	601	321	188	2	5013
1976-77	0	3	84	497	905	1260	1693	989	532	222	44	12	6241
1977-78	0	2	21	340	661	1172	1521	1348	968	334	186	4	6557
1978-79	0	0	36	349	591	1035	1627	1336	758	435	119	0	6286
1979-80	0	9	50	323	684	898	1146	1249	857	421	95	8	5740
1980-81	0	0	49	395	675	1028	1193	906	648	181	184	0	5259
1981-82	0	0	51	332	581	1175	1483	1139	760	502	9	20	6052
1982-83	0	5	86	325	656	806	1117	836	749	510	169	14	5273
1983-84	0	0	75	269	574	1512	1320	835	1023	446	170	1	6225
1984-85	0	0	127	194	691	899	1437	1116	601	262	67	32	5426
1985-86	0	11	99	223	658	1340	1105	1060	618	261	74	1	5450
1986-87	0	19	25	284	807	999	1199	788	599	325	34	0	5079
1987-88	0	3	44	440	565	934	1216	1139	720	360	59	11	5491
1988-89	0	5	33	475	654	1027	890	1242	749	405	224	19	5723
1989-90	0	4	106	269	683	1431	856	803	613	459	168	10	5402
1990-91	7	2	72	360	512	1080	1262	827	612	243	69	20	5046
1991-92	0	0	120	295	787	932	1007	771	651	382	154	20	5119
1992-93	0	8	82	296	671	991	1083	1041	813	419	85	35	5524
1993-94	0	0	117	382	730	979	1347	992	679	351	142	12	5731
1994-95	0	4	64	242	512	851	1187	964	610	395	141	7	4977
1995-96	0	0	109	289	820	1140	1289	973	892	462	144	19	6137
1996-97	1	0	97	305	884	1031	1375	852	664	500	212	20	5941
1997-98	2	5	57	342	768	1003	993	681	768	354	39	37	5049
1998-99	0	0	17	266	567	953	1231	771	807	305	77	21	5015
1999-00	0	0	100	319	496	955	1155	748	581	376	89	24	4843
2000-01	0	1	78	244	804	1470	1208	933	852	220	109	34	5953
2001-	0	0	103	339	443	907							

COOLING DEGREE DAYS (base 65 F) 2001   SPRINGFIELD, IL (SPI)

YEAR	JAN	FEB	MAR	APR	MAY	JUN	JUL	AUG	SEP	OCT	NOV	DEC	ANNUAL
1972	0	0	0	6	114	199	347	291	178	10	0	0	1145
1973	0	0	0	12	12	251	350	348	184	64	2	0	1221
1974	0	0	7	14	70	136	437	252	49	17	0	0	984
1975	0	0	0	2	115	271	315	367	94	36	0	0	1200
1976	0	0	3	28	34	228	393	214	98	23	0	0	1021
1977	0	0	0	76	238	217	452	251	134	0	1	0	1369
1978	0	0	1	8	107	289	358	258	222	5	5	0	1253
1979	0	0	0	2	93	305	327	295	138	41	0	0	1201
1980	0	0	0	10	88	228	515	452	169	20	0	0	1482
1981	0	0	8	50	43	303	380	280	121	13	0	0	1198
1982	0	0	0	1	173	131	379	244	126	32	1	1	1088
1983	0	0	1	1	17	289	483	471	205	34	0	1	1501
1984	0	0	0	16	30	312	303	375	155	44	0	0	1236
1985	0	0	7	52	80	178	293	202	191	17	0	0	1020
1986	0	0	14	38	109	299	434	182	193	14	0	0	1283
1987	0	0	0	21	212	318	436	329	130	0	8	0	1454
1988	0	0	0	3	93	281	430	437	150	6	0	0	1400
1989	0	0	2	30	57	190	332	268	77	24	0	0	980
1990	0	0	14	39	18	261	328	289	182	12	1	0	1144
1991	0	0	5	20	207	327	326	302	184	18	0	0	1389
1992	0	0	0	4	75	166	294	199	150	28	0	0	916
1993	0	0	0	1	71	253	386	352	46	13	0	0	1122
1994	0	0	0	26	67	327	327	257	123	27	2	0	1158
1995	0	0	0	0	33	276	418	479	105	16	0	0	1327
1996	0	0	0	0	99	250	231	274	80	20	0	0	954
1997	0	0	0	0	28	204	341	223	115	81	0	1	992
1998	0	0	11	0	145	246	327	312	207	15	0	0	1264
1999	0	0	0	12	74	249	409	210	94	14	1	0	1063
2000	0	0	4	0	123	174	273	332	148	37	3	0	1094
2001	0	0	0	60	122	203	365	288	88	8	0	0	1134

SNOWFALL (inches) 2001   SPRINGFIELD, IL (SPI)

YEAR	JUL	AUG	SEP	OCT	NOV	DEC	JAN	FEB	MAR	APR	MAY	JUN	TOTAL
1972-73	0.0	0.0	0.0	0.0	5.4	3.1	0.6	3.1	0.3	0.9	0.0	0.0	13.4
1973-74	0.0	0.0	0.0	0.0	1.3	22.7	7.3	5.3	4.4	0.4	0.0	0.0	41.4
1974-75	0.0	0.0	0.0	0.0	6.8	2.5	4.5	14.2	4.5	0.4	0.0	0.0	32.9
1975-76	0.0	0.0	0.0	0.0	8.5	4.3	8.5	2.1	2.2	0.0	0.0	0.0	25.6
1976-77	0.0	0.0	0.0	T	T	6.9	21.1	8.9	0.6	1.1	0.0	0.0	38.6
1977-78	0.0	0.0	0.0	0.0	6.4	8.9	8.9	9.4	18.5	T	0.0	0.0	52.1
1978-79	0.0	0.0	0.0	0.0	T	3.2	15.9	4.6	8.1	T	0.0	0.0	31.8
1979-80	0.0	0.0	0.0	0.0	0.8	0.1	5.1	11.7	5.5	7.3	0.0	0.0	30.5
1980-81	0.0	0.0	0.0	T	3.5	2.1	2.7	8.6	0.6	0.0	0.0	0.0	17.5
1981-82	0.0	0.0	0.0	0.0	0.1	21.6	12.0	11.4	0.7	4.6	0.0	0.0	50.4
1982-83	0.0	0.0	0.0	0.0	0.2	0.9	2.4	1.3	5.4	0.2	0.0	0.0	10.4
1983-84	0.0	0.0	0.0	T	T	16.2	3.9	9.8	5.7	0.0	0.0	0.0	35.6
1984-85	0.0	0.0	0.0	0.0	1.8	0.7	9.3	2.5	T	0.0	0.0	0.0	14.3
1985-86	0.0	0.0	0.0	0.0	T	4.3	0.2	15.1	0.6	0.1	0.0	0.0	20.3
1986-87	0.0	0.0	0.0	0.0	0.5	T	20.3	T	T	0.0	0.0	0.0	20.8
1987-88	0.0	0.0	0.0	0.0	0.1	5.7	0.4	10.2	5.0	0.0	0.0	0.0	21.4
1988-89	0.0	0.0	0.0	0.0	T	5.5	0.3	13.7	5.2	T	T	0.0	24.7
1989-90	0.0	0.0	0.0	0.3	T	6.7	0.6	0.5	1.2	0.0	0.0	0.0	9.3
1990-91	0.0	0.0	0.0	T	T	9.4	6.8	0.8	3.1	0.0	0.0	0.0	20.1
1991-92	0.0	0.0	0.0	T	3.2	0.5	4.6	0.9	2.2	T	0.0	0.0	11.4
1992-93	0.0	0.0	0.0	0.0	1.2	2.5	7.0	16.0	2.9	0.4	0.0	0.0	30.0
1993-94	0.0	0.0	0.0	0.2	0.8	1.7	5.8	4.4	0.9	2.7	T	0.0	16.5
1994-95	T	0.0	0.0	0.0	T	0.2	3.7	4.4	0.2	0.0	0.0	0.0	8.5
1995-96	0.0	0.0	0.0	0.0	1.2	5.9	11.2	1.1	1.6	0.0	T	0.0	21.0
1996-97	0.0	0.0	0.0	0.0	0.2	2.4	19.3	T	T	T	0.0	0.0	21.9
1997-98	0.0	0.0	0.0	T	1.2	7.6	3.5	0.0	6.9	0.0	0.0	0.0	19.2
1998-99	0.0	0.0	0.0	0.0	T	2.5	16.1	3.2	6.4	T	0.0	0.0	28.2
1999-00	0.0	0.0	0.0	0.0	0.0	2.4	6.8	0.3	5.5	T	0.0	0.0	15.0
2000-01	0.0	0.0	0.0	0.0	T	10.0	2.1	0.9	0.4	T	0.0	0.0	13.4
2001-	0.0	0.0	0.0	0.0	T	3.3							
POR= 53 YRS	T	0.0	0.0	0.0	1.7	5.0	6.1	5.6	3.7	0.7	T	0.0	22.8

2001
INDIANAPOLIS,
INDIANA (IND)

Indianapolis is located in the central part of the state and is situated on level or slightly rolling terrain. The greater part of the city lies east of the White River which flows in a general north to south direction.

The National Weather Service Forecast Office is located approximately 7 miles southwest of the central part of the city at the Indianapolis International Airport. From a field elevation of 797 feet above sea level at the Indianapolis International Airport the terrain slopes gradually downward to a little below 645 feet at the White River, then upward to just over 910 feet in the northwest corner and eastern sections of the county. The street elevation at the former city office located in the Old Federal Building is 718 feet.

Indianapolis has a temperate climate, with very warm summers and without a dry season. Very cold temperatures may be produced by the invasion of continental polar air in the winter from northern latitudes. The polar air can be quite frigid with very low humidity. The arrival of maritime tropical air from the Gulf in the summer brings warm temperatures and moderate humidity. One of the longest and most severe heat waves brought temperatures of 100 degrees or more for nine consecutive days.

Precipitation is distributed fairly evenly throughout the year, and therefore there is no pronounced wet or dry season. Rainfall in the spring and summer is produced mostly by showers and thunderstorms. A rainfall of about 2 1/2 inches in a 24-hour period can be expected about once a year. Snowfalls of 3 inches or more occur on an average of two or three times in the winter.

Local levees and/or channel improvements now protect some formerly flood-prone areas.

Based on the 1951-1980 period, the average first occurrence of 32 degrees Fahrenheit in the fall is October 20 and the average last occurrence in the spring is April 22.

NORMALS, MEANS, AND EXTREMES
INDIANAPOLIS, IN (IND)

LATITUDE: LONGITUDE: ELEVATION (FT):   TIME ZONE: WBAN: 93819
39 42' 36" N   86 16' 20" W   GRND: 794   BARO: 797   EASTERN (UTC + 5)

	ELEMENT	POR	JAN	FEB	MAR	APR	MAY	JUN	JUL	AUG	SEP	OCT	NOV	DEC	YEAR
TEMPERATURE °F	NORMAL DAILY MAXIMUM	30	33.7	38.3	50.9	63.3	73.8	82.7	85.5	83.6	77.6	65.8	51.9	38.5	62.1
	MEAN DAILY MAXIMUM	54	35.0	39.9	50.2	62.9	73.3	82.3	85.4	83.7	77.7	66.1	51.4	39.3	62.3
	HIGHEST DAILY MAXIMUM	62	71	76	85	89	93	102	104	102	100	90	81	74	104
	YEAR OF OCCURRENCE		1950	2000	1981	1970	1988	1988	1954	1988	1954	1954	1950	1982	JUL 1954
	MEAN OF EXTREME MAXS.	54	57.5	62.5	73.9	81.4	86.8	92.4	93.9	92.4	90.2	82.2	71.1	61.2	78.8
	NORMAL DAILY MINIMUM	30	17.2	20.9	31.9	41.5	51.7	61.0	65.2	62.8	55.6	43.5	34.1	23.2	42.4
	MEAN DAILY MINIMUM	54	18.8	22.6	31.3	41.6	51.7	61.2	65.3	63.0	55.2	43.9	33.6	23.9	42.7
	LOWEST DAILY MINIMUM	62	-27	-21	-7	16	28	37	44	41	28	17	-2	-23	-27
	YEAR OF OCCURRENCE		1994	1982	1980	1940	1966	1992	1942	1965	1942	1958	1989	JAN 1994
	MEAN OF EXTREME MINS.	54	-4.2	1.4	12.6	25.7	36.6	47.5	53.6	51.7	39.3	28.2	17.0	2.2	26.0
	NORMAL DRY BULB	30	25.5	29.6	41.4	52.4	62.8	71.9	75.4	73.2	66.6	54.7	43.0	30.9	52.3
	MEAN DRY BULB	54	27.0	31.2	40.6	52.3	62.5	71.8	75.4	73.3	66.4	55.0	42.4	31.6	52.5
	MEAN WET BULB	17	26.5	30.4	37.2	46.9	56.8	65.1	68.8	67.5	60.2	49.7	39.7	28.1	48.1
	MEAN DEW POINT	17	22.1	25.0	30.7	40.5	51.4	60.6	65.2	64.1	55.8	44.6	34.9	24.1	43.2
	NORMAL NO. DAYS WITH:														
	MAXIMUM ≥ 90	30	0.0	0.0	0.0	0.0	0.6	3.9	7.2	4.3	1.6	0.0	0.0	0.0	17.6
	MAXIMUM ≤ 32	30	14.0	9.0	2.0	*	0.0	0.0	0.0	0.0	0.0	0.0	0.9	8.7	34.6
	MINIMUM ≤ 32	30	27.8	23.4	16.9	5.5	0.4	0.0	0.0	0.0	0.0	4.2	14.2	24.9	117.3
	MINIMUM ≤ 0	30	4.4	2.2	0.2	0.0	0.0	0.0	0.0	0.0	0.0	0.0	0.0	1.8	8.6
H/C	NORMAL HEATING DEG. DAYS	30	1225	991	732	378	165	5	0	6	58	338	660	1057	5615
	NORMAL COOLING DEG. DAYS	30	0	0	0	0	96	212	322	260	106	18	0	0	1014
RH	NORMAL (PERCENT)	30	75	74	70	66	67	68	73	75	74	72	76	78	72
	HOUR 01 LST	30	78	77	75	73	77	79	84	86	85	80	80	81	80
	HOUR 07 LST	30	80	81	80	78	81	82	86	90	90	86	84	83	83
	HOUR 13 LST	30	70	67	62	55	55	56	60	61	58	57	66	72	62
	HOUR 19 LST	30	72	70	65	58	57	58	62	66	67	65	72	76	66
S	PERCENT POSSIBLE SUNSHINE	52	40	49	50	54	60	65	66	68	65	61	41	38	55
W/O	MEAN NO. DAYS WITH:														
	HEAVY FOG(VISBY 1/4 MI)	60	3.3	2.5	1.7	0.6	0.8	0.7	1.1	1.7	1.4	1.2	1.6	2.8	19.4
	THUNDERSTORMS	60	0.7	0.7	2.7	4.4	6.2	7.3	7.6	5.9	3.5	1.8	1.1	0.4	42.3
CLOUDINESS	MEAN:														
	SUNRISE-SUNSET (OKTAS)														
	MIDNIGHT-MIDNIGHT (OKTAS)														
	MEAN NO. DAYS WITH:														
	CLEAR	1			1.0		1.0								
	PARTLY CLOUDY	1	1.0		2.0		1.0								
	CLOUDY	1		2.0	6.0		1.0	1.0							
PR	MEAN STATION PRESSURE(IN)	29	29.20	29.19	29.11	29.10	29.10	29.10	29.20	29.20	29.20	29.20	29.20	29.21	29.17
	MEAN SEA-LEVEL PRES. (IN)	17	30.12	30.11	30.06	29.97	29.98	30.01	30.01	30.04	30.06	30.10	30.09	30.14	30.06
WINDS	MEAN SPEED (MPH)	47	11.2	10.9	12.0	11.6	9.9	8.7	7.7	7.3	8.1	9.1	10.7	10.7	9.8
	PREVAIL.DIR(TENS OF DEGS)	31	24	23	30	23	23	23	23	23	23	24	24	24	23
	MAXIMUM 2-MINUTE:														
	SPEED (MPH)	5	39	46	38	47	40	40	49	45	46	44	44	41	49
	DIR. (TENS OF DEGS)		22	28	26	25	30	28	28	29	26	27	23	26	28
	YEAR OF OCCURRENCE		1999	2001	2001	1997	1997	1998	1998	2000	2000	2001	1998	2000	JUL 1998
	MAXIMUM 5-SECOND:														
	SPEED (MPH)	5	47	56	48	58	48	61	60	54	55	51	52	49	61
	DIR. (TENS OF DEGS)		22	23	28	25	21	28	30	29	26	27	22	26	28
	YEAR OF OCCURRENCE		1999	1999	1998	1997	1999	1998	2001	2000	2000	2001	1998	2000	JUN 1998
PRECIPITATION	NORMAL (IN)	30	2.32	2.46	3.79	3.70	4.00	3.49	4.47	3.64	2.87	2.63	3.23	3.34	39.94
	MAXIMUM MONTHLY (IN)	62	12.69	5.35	10.74	8.09	10.10	10.26	11.79	8.34	8.06	8.36	8.50	7.72	12.69
	YEAR OF OCCURRENCE		1950	1971	1963	1964	1943	1998	1992	1980	1989	1941	1985	1990	JAN 1950
	MINIMUM MONTHLY (IN)	62	0.21	0.36	0.64	0.98	1.06	0.36	0.55	0.68	0.24	0.17	0.69	0.45	0.17
	YEAR OF OCCURRENCE		1944	1978	2001	1976	1988	1997	1964	1964	1963	1963	1999	1976	OCT 1963
	MAXIMUM IN 24 HOURS (IN)	59	3.47	2.50	3.05	2.56	3.53	3.80	5.32	4.72	3.07	3.90	4.15	2.83	5.32
	YEAR OF OCCURRENCE		1950	1977	1963	1961	1961	1963	1987	1976	1961	1959	1993	1990	JUL 1987
	NORMAL NO. DAYS WITH:														
	PRECIPITATION ≥ 0.01	30	11.5	10.1	13.4	11.9	11.3	9.3	9.9	8.9	8.1	8.5	10.2	13.2	126.3
	PRECIPITATION ≥ 1.00	30	0.4	0.5	0.8	0.9	1.0	0.8	1.2	1.1	0.7	0.5	0.8	0.6	9.3
SNOWFALL	NORMAL (IN)	30	8.2	7.8	3.2	0.5	0.*	0.0	0.0	0.0	0.0	0.4	1.6	5.8	27.5
	MAXIMUM MONTHLY (IN)	70	30.6	18.0	12.5	4.0	0.2	T	0.0	T	0.0	9.3	8.3	27.5	30.6
	YEAR OF OCCURRENCE		1978	1979	1996	1940	1989	1995		1989		1989	1966	1973	JAN 1978
	MAXIMUM IN 24 HOURS (IN)	59	12.2	12.5	11.3	3.1	0.2	T	0.0	T	0.0	7.5	8.2	11.5	12.5
	YEAR OF OCCURRENCE		1978	1965	1996	1953	1989	1995		1989		1989	1966	1973	FEB 1965
	MAXIMUM SNOW DEPTH (IN)	53	20	15	9	3	0	0	0	0	0	2	8	13	20
	YEAR OF OCCURRENCE		1978	1978	1960	1953						1989	1966	1973	JAN 1978
	NORMAL NO. DAYS WITH:														
	SNOWFALL ≥ 1.0	30	2.4	2.4	1.0	0.2	0.0	0.0	0.0	0.0	0.0	0.1	0.5	1.9	8.5

PRECIPITATION (inches) 2001 INDIANAPOLIS, IN (IND)

YEAR	JAN	FEB	MAR	APR	MAY	JUN	JUL	AUG	SEP	OCT	NOV	DEC	ANNUAL
1972	1.57	1.15	2.48	5.81	1.89	6.04	2.01	2.94	5.65	2.25	5.65	2.83	40.27
1973	2.27	1.11	5.63	2.76	1.79	5.91	6.67	2.74	2.43	3.11	3.62	4.27	42.31
1974	3.39	2.58	3.60	3.45	6.27	5.15	1.20	5.63	3.25	0.99	2.99	2.81	41.31
1975	4.37	4.13	4.16	4.14	2.42	5.73	4.63	4.68	2.32	2.80	3.63	3.71	46.72
1976	2.29	2.90	3.46	0.98	3.10	3.97	3.09	7.95	2.02	2.79	0.82	0.45	33.82
1977	1.50	3.62	3.83	1.91	2.78	3.86	2.57	4.47	3.40	2.79	3.01	4.31	38.05
1978	3.80	0.36	3.54	3.59	4.21	4.43	5.04	6.89	0.85	3.82	2.38	4.03	42.94
1979	3.24	2.86	2.43	3.14	2.23	3.93	11.06	6.09	0.36	2.32	4.37	2.57	44.60
1980	1.67	1.84	4.26	2.10	2.26	4.15	2.87	8.34	3.31	1.87	1.41	0.78	34.86
1981	0.36	2.88	1.22	5.81	9.23	1.64	5.75	1.69	2.04	2.35	1.12	3.41	37.49
1982	5.64	1.62	4.73	2.40	5.94	5.16	3.44	1.00	1.20	0.91	4.16	5.78	41.98
1983	1.05	1.03	2.94	4.47	4.68	4.53	1.58	2.79	1.28	3.87	4.55	3.43	36.20
1984	0.97	3.16	3.14	3.90	4.35	1.51	4.83	3.27	4.69	2.60	5.38	4.33	42.13
1985	1.37	3.73	5.94	2.60	4.60	3.06	4.06	5.29	2.71	1.82	8.50	3.30	46.98
1986	0.73	2.84	3.93	4.34	7.37	3.58	4.88	1.18	5.68	7.84	2.32	1.71	46.40
1987	1.55	1.28	1.84	2.68	1.77	4.11	9.22	0.86	1.41	1.36	2.60	4.77	33.45
1988	2.35	3.04	3.22	4.02	1.06	0.36	4.71	1.46	1.14	3.07	4.39	2.50	31.32
1989	1.75	1.32	3.72	4.32	5.79	3.80	6.15	8.05	8.06	2.92	2.79	1.90	50.57
1990	1.79	5.17	3.93	2.44	7.59	3.11	3.68	4.46	2.68	4.64	3.23	7.72	50.44
1991	1.59	1.94	6.41	4.34	4.64	0.91	2.17	3.54	1.12	5.47	3.95	1.45	37.53
1992	1.40	1.15	2.61	4.17	1.56	4.07	11.79	1.42	3.40	2.84	7.88	1.89	44.18
1993	3.29	2.31	3.72	3.75	2.76	5.15	4.94	6.42	5.69	2.25	8.30	2.18	50.76
1994	2.68	1.39	0.92	5.62	2.03	7.00	1.27	2.59	2.19	0.86	3.40	1.66	31.61
1995	2.44	0.90	2.99	2.79	7.37	3.49	2.61	4.16	1.27	2.08	2.64	2.72	35.46
1996	4.08	1.16	3.51	7.02	8.89	5.16	6.26	2.27	7.61	1.88	6.22	2.75	56.81
1997	4.47	3.92	4.59	1.97	4.93	3.17	0.55	2.96	1.51	1.55	1.94	1.33	32.89
1998	2.51	1.23	5.49	4.84	6.13	10.26	3.97	3.66	0.48	4.89	2.52	1.00	46.98
1999	6.35	3.57	1.71	4.09	3.75	2.57	2.96	1.50	0.75	1.82	0.69	2.61	32.37
2000	2.07	2.86	1.64	3.80	5.00	4.55	2.95	4.26	4.82	3.08	2.67	2.76	40.46
2001	0.74	1.95	0.64	1.82	4.10	4.46	8.34	2.47	4.66	7.01	2.69	3.01	41.89
POR= 130 YRS	2.88	2.51	3.72	3.71	4.04	4.08	3.92	3.29	3.13	2.74	3.28	2.89	40.19

AVERAGE TEMPERATURE (F) 2001 INDIANAPOLIS, IN (IND)

YEAR	JAN	FEB	MAR	APR	MAY	JUN	JUL	AUG	SEP	OCT	NOV	DEC	ANNUAL
1972	26.5	28.4	39.8	52.1	65.0	68.8	74.6	73.1	68.3	51.5	39.9	31.9	51.7
1973	30.7	31.5	49.4	51.3	59.5	73.3	76.0	74.4	69.9	59.4	46.4	30.9	54.4
1974	31.6	32.4	45.3	55.0	62.1	68.6	76.0	71.8	60.6	52.5	42.5	32.9	52.6
1975	32.0	32.0	36.9	48.9	65.5	71.5	73.5	76.1	62.6	55.9	46.4	32.8	52.8
1976	23.9	38.8	46.6	53.4	58.8	71.3	73.9	71.3	63.6	48.7	34.7	24.6	50.8
1977	10.3	28.2	46.5	57.1	70.6	71.2	78.0	74.0	69.6	54.2	45.9	29.2	52.9
1978	18.2	17.8	36.7	55.2	63.5	74.0	77.2	75.0	70.4	52.7	45.7	34.4	51.7
1979	18.0	18.8	43.4	49.8	61.1	71.1	73.0	72.4	64.9	53.1	41.3	34.9	50.2
1980	28.5	22.5	36.0	49.1	64.0	69.1	78.5	76.6	67.9	50.9	40.3	31.8	51.3
1981	23.5	33.0	39.9	57.4	59.9	73.3	75.4	72.8	64.6	53.1	44.2	27.8	52.1
1982	20.1	26.4	42.5	48.6	68.6	67.8	76.2	71.7	64.4	55.7	44.2	40.2	52.2
1983	30.6	35.5	42.9	48.2	58.2	71.8	79.7	80.1	69.4	57.5	44.1	20.2	53.2
1984	22.8	37.3	33.0	50.0	58.8	75.3	72.4	74.3	64.2	61.3	43.0	38.9	52.6
1985	20.4	26.1	44.6	57.1	64.9	70.6	74.3	72.5	66.2	57.6	46.6	22.5	51.8
1986	28.5	31.4	43.9	53.8	63.5	72.8	77.5	70.1	69.8	55.6	39.8	32.3	53.3
1987	27.6	35.5	44.5	52.5	68.1	73.7	75.8	73.6	68.1	48.6	46.3	35.8	54.2
1988	25.9	27.2	41.4	51.9	64.4	73.4	78.3	77.5	67.3	48.2	44.2	31.7	52.6
1989	36.3	27.2	42.5	51.4	59.4	71.4	75.7	71.8	64.4	55.4	40.9	18.8	51.3
1990	37.3	37.6	46.2	51.4	60.1	71.3	73.9	72.5	66.9	53.9	47.5	34.6	54.4
1991	26.9	35.3	43.7	55.3	68.3	74.6	76.7	75.2	67.9	57.5	40.2	35.7	54.8
1992	31.5	38.4	43.2	52.1	60.5	68.1	73.5	69.5	65.1	53.4	43.2	33.3	52.7
1993	31.6	27.4	39.1	50.7	63.5	71.0	77.2	75.5	62.2	51.6	42.2	32.4	52.0
1994	21.8	29.7	41.0	54.1	60.4	75.0	75.5	71.8	66.6	54.3	48.9	38.9	53.4
1995	28.6	30.0	45.0	51.9	61.6	72.8	76.6	79.6	64.7	55.7	37.0	28.6	52.7
1996	25.1	31.2	35.6	49.3	61.9	72.1	73.1	74.9	65.5	55.5	37.3	34.4	51.3
1997	24.3	36.0	43.0	48.3	56.7	69.6	75.4	71.6	66.1	54.6	39.0	33.4	51.5
1998	36.6	40.8	42.0	53.3	67.1	71.5	74.9	75.3	71.4	57.0	45.5	36.5	56.0
1999	28.5	37.5	38.4	54.9	64.3	73.7	79.2	72.4	67.1	55.2	48.8	34.6	54.6
2000	27.9	38.7	47.0	51.4	64.4	71.0	73.0	73.6	65.2	57.8	40.7	19.8	52.6
2001	28.0	34.6	38.4	58.5	64.4	70.2	74.4	75.4	65.1	54.3	49.5	38.2	54.3
POR= 130 YRS	28.1	31.0	40.5	52.1	62.5	71.7	75.7	73.6	66.8	55.2	42.3	31.8	52.6

REFERENCE NOTES:

PAGE 1:
THE TEMPERATURE GRAPH SHOWS NORMAL MAXIMUM AND NORMAL MINIMUM DAILY TEMPERATURES (SOLID CURVES) AND THE ACTUAL DAILY HIGH AND LOW TEMPERATURES (VERTICAL BARS).

PAGE 2 AND 3:
H/C INDICATES HEATING AND COOLING DEGREE DAYS.
RH INDICATES RELATIVE HUMIDITY
W/O INDICATES WEATHER AND OBSTRUCTIONS
S INDICATES SUNSHINE.
PR INDICATES PRESSURE.
CLOUDINESS ON PAGE 3 IS THE SUM OF THE CEILOMETER AND SATELLITE DATA NOT TO EXCEED EIGHT EIGHTHS(OKTAS).

GENERAL:
T INDICATES TRACE PRECIPITATION, AN AMOUNT GREATER THAN ZERO BUT LESS THAN THE LOWEST REPORTABLE VALUE.
+ INDICATES THE VALUE ALSO OCCURS ON EARLIER DATES.
BLANK ENTRIES DENOTE MISSING OR UNREPORTED DATA.
NORMALS ARE 30-YEAR AVERAGES (1961 - 1990).
ASOS INDICATES AUTOMATED SURFACE OBSERVING SYSTEM.
PM INDICATES THE LAST DAY OF THE PREVIOUS MONTH.
POR (PERIOD OF RECORD) BEGINS WITH THE JANUARY DATA MONTH AND IS THE NUMBER OF YEARS USED TO COMPUTE THE MEAN. INDIVIDUAL MONTHS WITHIN THE POR MAY BE MISSING.
WHEN THE POR FOR A NORMAL IS LESS THAN 30 YEARS, THE NORMAL IS PROVISIONAL AND IS BASED ON THE NUMBER OF YEARS INDICATED.
0.* OR * INDICATES THE VALUE OR MEAN-DAYS-WITH IS BETWEEN 0.00 AND 0.05.
CLOUDINESS FOR ASOS STATIONS DIFFERS FROM THE NON-ASOS OBSERVATION TAKEN BY A HUMAN OBSERVER. ASOS STATION CLOUDINESS IS BASED ON TIME-AVERAGED CEILOMETER DATA FOR CLOUDS AT OR BELOW 12,000 FEET AND ON SATELLITE DATA FOR CLOUDS ABOVE 12,000 FEET.
THE NUMBER OF DAYS WITH CLEAR, PARTLY CLOUDY, AND CLOUDY CONDITIONS FOR ASOS STATIONS IS THE SUM OF THE CEILOMETER AND SATELLITE DATA FOR THE SUNRISE TO SUNSET PERIOD.

GENERAL CONTINUED:
CLEAR INDICATES 0 - 2 OKTAS, PARTLY CLOUDY INDICATES 3 - 6 OKTAS, AND CLOUDY INDICATES 7 OR 8 OKTAS. WHEN AT LEAST ONE OF THE ELEMENTS (CEILOMETER OR SATELLITE) IS MISSING, THE DAILY CLOUDINESS IS NOT COMPUTED.
WIND DIRECTION IS RECORDED IN TENS OF DEGREES (2 DIGITS) CLOCKWISE FROM TRUE NORTH. "00" INDICATES CALM. "36" INDICATES TRUE NORTH.
RESULTANT WIND IS THE VECTOR AVERAGE OF THE SPEED AND DIRECTION.
AVERAGE TEMPERATURE IS THE SUM OF THE MEAN DAILY MAXIMUM AND MINIMUM TEMPERATURE DIVIDED BY 2.
SNOWFALL DATA COMPRISE ALL FORMS OF FROZEN PRECIPITATION, INCLUDING HAIL.
A HEATING (COOLING) DEGREE DAY IS THE DIFFERENCE BETWEEN THE AVERAGE DAILY TEMPERATURE AND 65 F.
DRY BULB IS THE TEMPERATURE OF THE AMBIENT AIR.
DEW POINT IS THE TEMPERATURE TO WHICH THE AIR MUST BE COOLED TO ACHIEVE 100 PERCENT RELATIVE HUMIDITY.
WET BULB IS THE TEMPERATURE THE AIR WOULD HAVE IF THE MOISTURE CONTENT WAS INCREASED TO 100 PERCENT RELATIVE HUMIDITY.

ON JULY 1, 1996, THE NATIONAL WEATHER SERVICE BEGAN USING THE "METAR" OBSERVATION CODE THAT WAS ALREADY EMPLOYED BY MOST OTHER NATIONS OF THE WORLD. THE MOST NOTICEABLE DIFFERENCE IN THIS ANNUAL PUBLICATION WILL BE THE CHANGE IN UNITS FROM TENTHS TO EIGHTS(OKTAS) FOR REPORTING THE AMOUNT OF SKY COVER.

HEATING DEGREE DAYS (base 65 F) 2001 INDIANAPOLIS, IN (IND)

YEAR	JUL	AUG	SEP	OCT	NOV	DEC	JAN	FEB	MAR	APR	MAY	JUN	TOTAL
1972-73	8	4	36	413	746	1018	1059	937	477	416	184	0	5298
1973-74	0	0	20	211	552	1052	1028	905	617	314	158	18	4875
1974-75	0	5	163	380	671	988	1016	918	866	481	78	22	5588
1975-76	8	0	137	288	551	992	1265	754	363	363	203	1	5126
1976-77	0	2	79	503	904	1249	1693	1025	567	276	45	19	6362
1977-78	0	0	30	326	575	1104	1443	1313	873	292	150	4	6110
1978-79	0	0	36	377	571	944	1453	1288	665	455	170	5	5964
1979-80	3	13	87	384	705	929	1123	1224	893	474	93	36	5964
1980-81	0	0	45	438	734	1022	1279	889	769	244	180	1	5601
1981-82	1	0	94	368	621	1146	1388	1075	690	486	26	18	5913
1982-83	0	2	110	325	621	764	1062	819	681	498	211	21	5114
1983-84	1	0	64	246	619	1386	1304	796	987	447	211	1	6062
1984-85	1	0	119	138	653	803	1375	1082	631	269	85	16	5172
1985-86	0	1	97	245	544	1311	1126	935	652	344	114	7	5376
1986-87	0	24	30	312	750	1009	1150	820	627	378	51	0	5151
1987-88	2	5	37	504	553	900	1205	1090	726	390	84	15	5511
1988-89	0	2	37	517	618	1024	882	1052	693	419	226	11	5481
1989-90	0	8	106	313	718	1426	851	760	588	432	153	19	5374
1990-91	4	1	86	348	518	932	1175	824	651	292	71	0	4902
1991-92	0	0	103	254	739	902	1035	766	670	390	185	34	5078
1992-93	2	7	88	353	645	973	1028	1046	796	424	103	35	5500
1993-94	0	1	129	409	680	1005	1332	984	738	336	192	3	5809
1994-95	0	4	61	256	483	805	1118	973	615	392	131	6	4844
1995-96	1	0	94	288	835	1125	1230	971	908	469	162	7	6090
1996-97	0	0	85	296	823	942	1254	807	676	494	261	30	5668
1997-98	3	5	47	366	772	973	876	670	720	344	48	44	4868
1998-99	0	0	14	257	579	878	1125	767	814	297	81	8	4820
1999-00	0	0	70	298	481	936	1143	755	553	401	83	18	4738
2000-01	0	0	101	237	724	1394	1141	845	816	246	100	31	5635
2001-	5	0	91	331	457	826							

COOLING DEGREE DAYS (base 65 F) 2001 INDIANAPOLIS, IN (IND)

YEAR	JAN	FEB	MAR	APR	MAY	JUN	JUL	AUG	SEP	OCT	NOV	DEC	ANNUAL
1972	0	0	0	8	95	156	313	266	141	0	0	0	979
1973	0	0	0	11	20	256	349	302	172	44	1	0	1155
1974	0	0	11	19	73	131	346	225	40	3	2	0	850
1975	0	0	0	3	100	222	281	355	71	14	0	0	1046
1976	0	0	0	21	15	198	284	205	43	4	0	0	770
1977	0	0	3	45	226	212	410	286	175	0	6	0	1363
1978	0	0	0	4	110	282	382	318	203	1	0	0	1300
1979	0	0	0	7	57	197	255	250	92	24	0	0	882
1980	0	0	0	3	68	168	425	368	139	6	0	0	1177
1981	0	0	1	23	29	256	332	249	88	5	0	0	983
1982	0	0	0	0	146	109	356	214	98	40	3	1	967
1983	0	0	1	3	9	231	464	474	202	18	0	0	1402
1984	0	0	0	6	25	318	237	291	103	28	0	0	1008
1985	0	0	5	36	90	190	296	199	143	21	0	0	980
1986	0	0	6	12	74	249	395	189	181	24	0	0	1130
1987	0	0	0	6	156	265	343	279	137	0	0	0	1186
1988	0	0	1	3	72	274	422	395	114	2	0	0	1283
1989	0	0	1	20	57	215	338	227	94	21	0	0	973
1990	0	0	13	30	10	215	289	241	147	10	1	0	956
1991	0	0	0	7	178	294	367	323	196	30	0	0	1395
1992	0	0	0	10	53	137	274	152	99	1	0	0	726
1993	0	0	0	0	64	223	389	328	51	0	0	0	1055
1994	0	0	0	16	56	307	330	221	114	13	0	0	1057
1995	0	0	0	4	32	250	371	459	92	6	0	0	1214
1996	0	0	0	7	71	229	261	313	107	3	0	0	991
1997	0	0	0	0	12	178	327	217	87	49	0	0	870
1998	0	0	15	0	120	246	312	331	214	16	0	1	1255
1999	0	0	0	5	67	278	450	236	138	3	1	0	1178
2000	0	0	2	0	88	205	256	274	113	22	0	0	960
2001	0	0	0	60	90	194	302	328	101	6	0	0	1081

SNOWFALL (inches) 2001 INDIANAPOLIS, IN (IND)

YEAR	JUL	AUG	SEP	OCT	NOV	DEC	JAN	FEB	MAR	APR	MAY	JUN	TOTAL
1972-73	0.0	0.0	0.0	T	1.9	1.1	0.4	1.4	2.0	1.1	0.0	0.0	7.9
1973-74	0.0	0.0	0.0	0.0	0.4	27.5	3.8	8.0	3.0	2.1	0.0	0.0	44.8
1974-75	0.0	0.0	0.0	0.0	3.8	5.8	6.8	4.8	10.5	0.1	0.0	0.0	31.8
1975-76	0.0	0.0	0.0	0.0	4.5	8.1	5.6	0.6	2.3	0.0	0.0	0.0	21.1
1976-77	0.0	0.0	0.0	0.0	0.4	3.1	20.9	3.6	1.6	0.4	0.0	0.0	30.0
1977-78	0.0	0.0	0.0	0.0	2.8	15.2	30.6	3.9	5.4	0.0	0.0	0.0	57.9
1978-79	0.0	0.0	0.0	0.0	T	0.7	19.1	18.0	0.2	0.4	0.0	0.0	38.4
1979-80	0.0	0.0	0.0	0.0	0.8	0.2	5.0	14.5	3.6	0.7	0.0	0.0	24.8
1980-81	0.0	0.0	0.0	T	3.4	2.1	3.9	7.2	0.7	0.0	0.0	0.0	17.3
1981-82	0.0	0.0	0.0	T	0.4	15.6	21.8	13.6	3.5	3.3	0.0	0.0	58.2
1982-83	0.0	0.0	0.0	0.0	0.1	0.4	2.8	2.5	1.3	T	0.0	0.0	7.1
1983-84	0.0	0.0	0.0	0.0	0.1	8.3	7.2	17.1	9.2	T	0.0	0.0	41.9
1984-85	0.0	0.0	0.0	0.0	2.5	3.6	10.6	11.0	T	0.1	0.0	0.0	27.8
1985-86	0.0	0.0	0.0	0.0	T	8.1	1.7	9.5	1.1	T	0.0	0.0	20.4
1986-87	0.0	0.0	0.0	0.0	T	1.6	11.6	5.3	1.4	T	0.0	0.0	19.9
1987-88	0.0	0.0	0.0	0.0	T	1.1	2.9	4.8	2.5	T	0.0	0.0	11.3
1988-89	0.0	0.0	0.0	T	0.4	6.8	0.1	2.4	1.9	1.7	0.2	0.0	13.5
1989-90	0.0	0.0	T	9.3	0.6	8.2	4.1	2.5	1.3	T	0.0	0.0	26.0
1990-91	0.0	0.0	0.0	0.0	0.0	9.6	5.4	2.4	0.1	0.0	T	0.0	17.5
1991-92	0.0	0.0	0.0	0.0	3.0	0.2	7.1	0.2	2.9	1.3	0.0	T	14.7
1992-93	0.0	0.0	0.0	0.1	1.4	3.7	3.1	17.2	3.0	T	0.0	0.0	28.5
1993-94	0.0	0.0	0.0	2.4	1.1	5.6	14.0	3.9	3.7	0.8	0.0	T	31.5
1994-95	0.0	0.0	0.0	0.0	0.0	2.5	13.1	3.6	0.6	0.0	0.0	0.0	19.8
1995-96	0.0	0.0	0.0	0.0	1.4	11.0	25.2	1.6	12.5	T	0.0	0.0	51.7
1996-97	0.0	0.0	0.0	0.0	0.9	14.6	12.9	3.1	T	T	0.0	0.0	31.5
1997-98	0.0	0.0	0.0	0.0	5.8	2.5	0.8	0.6	0.7	0.0	0.0	0.0	10.4
1998-99	0.0	0.0	0.0	0.0	0.0	3.3	18.3	2.9	5.2	T	0.0	0.0	29.7
1999-00	0.0	0.0	0.0	0.0	0.0	3.1	11.3	1.6	8.1	T	0.0	0.0	24.1
2000-01	0.0	0.0	T	0.0	0.2	16.3	2.1	0.8	T	0.2	T	0.0	19.6
2001-	0.0	0.0	0.0	T	T	0.7							
POR= 69 YRS	0.0	T	0.0	0.2	1.9	5.1	6.7	5.3	3.4	0.5	0.0	T	23.1

2001
SOUTH BEND,
INDIANA (SBN)

South Bend is located on the Saint Joseph River in the northern portion of Saint Joseph County, situated on mostly level to gently rolling terrain and some former marshland. Drainage for the area is through the Saint Joseph River and Kankakee River.

South Bend is under the climatic influence of Lake Michigan with its nearest shore 20 miles to the northwest. The lake has a moderating effect on the temperature. Temperatures of 100 degrees or higher are rare and cold waves are less severe than at many locations at the same latitude. This results in favorable conditions for orchard and vegetable growth.

Based on the 1951-1980 period, the average first occurrence of 32 degrees Fahrenheit in the fall is October 18 and the average last occurrence in the spring is May 1.

Precipitation is fairly evenly distributed throughout the year with the greatest amounts during the growing season. The predominant snow season is from November through March, although there are also generally lighter amounts in October and April.

Winter is marked by considerable cloudiness and rather high humidity along with frequent periods of snow. Heavy snowfalls, resulting from a cold northwest wind passing over Lake Michigan are not uncommon.

NORMALS, MEANS, AND EXTREMES
SOUTH BEND, IN   (SBN)

LATITUDE:	LONGITUDE:	ELEVATION (FT):		TIME ZONE:	WBAN: 14848
41 42' 26" N	86 19' 59" W	GRND: 777	BARO: 780	EASTERN (UTC + 5)	

	ELEMENT	POR	JAN	FEB	MAR	APR	MAY	JUN	JUL	AUG	SEP	OCT	NOV	DEC	YEAR
TEMPERATURE F	NORMAL DAILY MAXIMUM	30	30.4	34.1	45.6	58.7	70.0	79.5	82.9	80.7	74.1	62.3	48.5	35.4	58.5
	MEAN DAILY MAXIMUM	54	31.1	35.0	45.2	58.6	70.2	79.6	82.9	81.2	74.4	62.9	48.0	35.8	58.7
	HIGHEST DAILY MAXIMUM	62	68	74	85	91	95	104	102	103	99	92	82	70	104
	YEAR OF OCCURRENCE		1950	2000	1981	1942	1942	1988	1999	1988	1953	1963	1950	2001	JUN 1988
	MEAN OF EXTREME MAXS.	54	51.0	54.3	70.5	79.3	86.5	92.5	93.8	92.0	89.0	80.5	68.4	56.0	76.2
	NORMAL DAILY MINIMUM	30	16.1	18.7	29.1	38.7	48.8	58.6	63.0	61.1	53.7	42.8	33.4	22.3	40.5
	MEAN DAILY MINIMUM	54	16.6	19.6	28.0	38.4	48.4	58.4	62.8	61.2	53.6	43.2	33.0	22.3	40.5
	LOWEST DAILY MINIMUM	62	-22	-17	-13	11	24	35	42	40	29	20	-7	-16	-22
	YEAR OF OCCURRENCE		1943	1951	1943	1972	1968	1972	2001	1965	1942	1988	1950	1960	JAN 1943
	MEAN OF EXTREME MINS.	54	-6.1	-1.5	10.1	22.9	33.5	43.8	50.6	49.0	38.3	29.0	17.2	1.5	24.0
	NORMAL DRY BULB	30	23.3	26.4	37.4	48.7	59.4	69.1	72.9	70.9	63.9	52.6	40.9	28.9	49.5
	MEAN DRY BULB	54	23.9	27.3	36.5	48.5	59.4	69.0	72.8	71.1	64.0	53.1	40.5	29.1	49.6
	MEAN WET BULB	18	23.4	26.6	33.4	43.5	53.5	62.1	66.5	65.2	57.9	47.2	37.3	27.5	45.3
	MEAN DEW POINT	18	19.4	21.6	27.2	36.4	47.3	57.0	62.3	61.9	53.8	42.1	32.7	23.7	40.5
	NORMAL NO. DAYS WITH:														
	MAXIMUM 90	30	0.0	0.0	0.0	0.0	0.6	3.0	4.8	2.9	0.8	*	0.0	0.0	12.1
	MAXIMUM 32	30	17.0	12.6	3.9	0.1	0.0	0.0	0.0	0.0	0.0	0.0	1.5	11.1	46.2
	MINIMUM 32	30	28.3	24.5	19.9	8.1	0.8	0.0	0.0	0.0	0.0	3.2	14.0	25.6	124.4
	MINIMUM 0	30	4.6	2.5	0.1	0.0	0.0	0.0	0.0	0.0	0.0	0.0	0.0	1.7	8.9
H/C	NORMAL HEATING DEG. DAYS	30	1293	1081	856	489	233	28	7	22	85	395	723	1119	6331
	NORMAL COOLING DEG. DAYS	30	0	0	0	0	59	151	251	205	52	10	0	0	728
RH	NORMAL (PERCENT)	30	77	75	71	66	66	67	70	73	74	72	76	79	72
	HOUR 01 LST	30	79	78	77	73	75	77	81	84	83	79	79	81	79
	HOUR 07 LST	30	80	81	80	78	79	80	84	88	88	84	82	83	82
	HOUR 13 LST	30	72	69	63	56	54	54	57	59	59	60	68	76	62
	HOUR 19 LST	30	75	72	66	59	56	56	59	63	67	68	74	78	66
S	PERCENT POSSIBLE SUNSHINE														
W/O	MEAN NO. DAYS WITH:														
	HEAVY FOG(VISBY 1/4 MI)	63	2.7	2.3	2.0	1.3	1.4	1.1	1.2	2.1	1.9	2.1	1.9	3.3	23.3
	THUNDERSTORMS	63	0.4	0.4	2.4	4.4	5.0	7.9	7.1	6.5	4.4	1.9	1.0	0.4	41.8
CLOUDINESS	MEAN:														
	SUNRISE-SUNSET (OKTAS)	51	6.4	6.2	6.0	5.5	5.0	4.9	4.5	4.5	4.6	4.9	6.1	6.4	5.4
	MIDNIGHT-MIDNIGHT (OKTAS)	32	6.3	5.9	5.7	5.3	4.8	4.6	4.2	4.2	4.4	4.7	6.0	6.4	5.2
	MEAN NO. DAYS WITH:														
	CLEAR	57	3.4	3.7	4.7	5.8	6.9	7.2	8.2	8.6	8.8	8.7	3.6	3.0	72.6
	PARTLY CLOUDY	57	5.9	5.6	7.3	7.6	9.6	10.4	12.7	12.1	9.0	8.0	6.0	5.6	99.8
	CLOUDY	57	21.7	18.9	19.1	16.6	14.5	12.4	10.1	10.3	12.2	14.4	20.4	22.5	193.1
PR	MEAN STATION PRESSURE(IN)	29	29.22	29.23	29.17	29.13	29.14	29.14	29.18	29.21	29.23	29.23	29.20	29.23	29.19
	MEAN SEA-LEVEL PRES. (IN)	18	30.08	30.09	30.05	29.96	29.98	29.96	29.99	30.04	30.05	30.07	30.06	30.10	30.04
WINDS	MEAN SPEED (MPH)	44	11.6	11.2	11.8	11.6	10.1	9.0	8.1	7.6	8.5	9.5	11.1	11.1	10.1
	PREVAIL.DIR(TENS OF DEGS)	27	23	23	22	31	21	21	22	23	23	21	23	23	23
	MAXIMUM 2-MINUTE:														
	SPEED (MPH)	5	33	45	40	43	44	36	44	39	36	56	41	37	56
	DIR. (TENS OF DEGS)		21	20	33	24	31	34	32	31	25	25	23	29	25
	YEAR OF OCCURRENCE		1999	1999	1998	1997	1997	2001	1997	1997	1997	2001	1998	1998	OCT 2001
	MAXIMUM 5-SECOND:														
	SPEED (MPH)	5	43	56	48	54	54	49	55	51	44	90	52	49	90
	DIR. (TENS OF DEGS)		09	20	33	23	31	34	32	32	28	26	24	30	26
	YEAR OF OCCURRENCE		1999	1999	1998	1997	1997	2001	1998	1997	1997	2001	1998	1998	OCT 2001
PRECIPITATION	NORMAL (IN)	30	2.23	1.90	3.10	3.82	3.22	4.11	3.82	3.67	3.62	3.08	3.27	3.30	39.14
	MAXIMUM MONTHLY (IN)	62	5.28	5.23	7.96	9.20	8.09	10.86	7.47	8.30	9.01	9.75	6.72	5.50	10.86
	YEAR OF OCCURRENCE		1959	1976	1976	1947	1996	1993	1982	1979	1977	1954	1985	1965	JUN 1993
	MINIMUM MONTHLY (IN)	62	0.44	0.54	0.54	0.50	0.80	0.48	0.02	0.32	0.01	0.42	1.24	0.60	0.01
	YEAR OF OCCURRENCE		1945	1969	1958	1971	1994	1988	1946	1950	1979	1950	1998	1943	SEP 1979
	MAXIMUM IN 24 HOURS (IN)	62	2.81	2.64	2.33	3.14	2.99	4.70	3.64	4.88	3.00	3.49	3.95	3.33	4.88
	YEAR OF OCCURRENCE		1960	1954	1972	1947	1976	1968	1989	1995	1977	1988	1990	1965	AUG 1995
	NORMAL NO. DAYS WITH:														
	PRECIPITATION 0.01	30	16.0	12.5	14.5	13.3	10.6	10.4	10.0	9.9	9.6	11.1	13.0	16.6	147.5
	PRECIPITATION 1.00	30	0.3	0.2	0.4	0.9	0.4	1.1	1.1	1.0	0.9	0.6	0.5	0.5	7.9
SNOWFALL	NORMAL (IN)	30	22.8	16.0	9.7	2.7	0.*	0.0	0.0	0.0	0.0	1.0	8.6	21.0	81.8
	MAXIMUM MONTHLY (IN)	62	86.1	35.1	33.9	14.0	0.6	T	T	T	1.2	8.8	30.3	44.6	86.1
	YEAR OF OCCURRENCE		1978	1958	1960	1982	1966	1992	1996	1996	1942	1989	1977	2000	JAN 1978
	MAXIMUM IN 24 HOURS (IN)	62	16.7	14.4	14.8	8.7	0.6	T	T	T	1.0	8.8	17.5	13.7	17.5
	YEAR OF OCCURRENCE		1978	1993	1960	1982	1966	1992	1994	1989	1994	1989	1977	1981	NOV 1977
	MAXIMUM SNOW DEPTH (IN)	53	41	35	21	6	0	0	0	0	0	6	20	23	41
	YEAR OF OCCURRENCE		1978	1978	1960	1982						1989	1977	1962	JAN 1978
	NORMAL NO. DAYS WITH:														
	SNOWFALL 1.0	30	6.6	5.4	3.3	0.9	0.0	0.0	0.0	0.0	0.0	0.3	2.4	6.3	25.2

PRECIPITATION (inches) 2001 SOUTH BEND, IN (SBN)

YEAR	JAN	FEB	MAR	APR	MAY	JUN	JUL	AUG	SEP	OCT	NOV	DEC	ANNUAL
1972	1.82	1.39	3.63	3.23	3.01	2.72	4.52	3.85	7.67	3.76	2.90	4.78	43.28
1973	1.64	1.02	3.85	3.88	3.61	4.85	3.33	1.29	2.11	3.49	1.45	4.30	34.82
1974	3.24	2.20	2.81	4.17	4.82	4.08	1.17	1.70	4.65	2.46	3.21	3.00	37.51
1975	4.58	3.26	2.96	6.02	2.08	5.46	2.58	7.55	1.15	1.31	4.73	3.72	45.40
1976	2.21	5.23	7.96	5.20	6.67	6.60	5.96	2.44	3.34	3.23	3.23	2.21	54.28
1977	1.63	1.27	7.05	2.75	1.92	5.71	2.68	6.03	9.01	3.08	4.27	3.65	49.05
1978	4.03	0.86	2.37	4.35	3.35	3.79	5.21	3.80	3.07	3.99	2.79	4.43	42.04
1979	3.22	1.51	4.03	5.80	3.03	4.66	1.75	8.30	0.01	4.79	4.88	3.66	45.64
1980	1.52	1.51	3.74	3.44	1.65	5.97	3.29	7.84	5.64	3.35	1.47	3.91	43.33
1981	0.68	1.92	0.88	5.28	6.79	3.71	2.30	3.81	1.23	2.23	1.81	37.61
1982	2.95	1.17	4.54	1.46	5.51	3.12	7.47	2.84	2.51	0.91	4.52	3.40	40.40
1983	0.77	0.79	2.46	5.36	4.83	2.04	2.45	1.28	2.81	1.66	2.60	3.23	30.28
1984	0.86	1.45	2.10	4.22	4.02	3.43	1.76	1.47	4.02	4.38	2.73	4.42	34.86
1985	2.58	4.32	3.86	1.93	1.50	2.88	3.80	3.82	1.88	3.36	6.72	2.51	39.16
1986	1.24	2.46	2.09	1.87	3.42	5.06	6.15	1.90	4.27	3.81	2.90	1.67	36.84
1987	2.31	1.32	1.18	2.67	3.50	3.57	3.34	3.34	3.64	3.20	2.11	4.12	34.57
1988	2.21	1.98	3.03	2.91	1.40	0.48	1.28	5.63	4.42	6.68	5.72	2.91	38.65
1989	1.58	1.05	2.27	2.83	2.72	3.49	5.90	5.65	3.78	1.45	3.55	1.83	36.10
1990	2.36	3.66	2.79	2.91	6.86	4.40	5.45	4.60	3.76	7.09	6.69	5.04	55.61
1991	1.64	1.79	2.79	4.58	4.01	0.62	1.32	3.68	2.71	8.75	2.75	1.67	36.31
1992	1.64	1.73	2.93	2.19	1.17	1.74	5.24	2.07	8.84	1.60	5.54	3.99	38.68
1993	3.35	1.20	2.62	3.64	2.34	10.86	1.51	4.38	7.76	4.09	2.39	1.50	45.64
1994	2.46	1.45	0.80	2.80	0.80	5.10	4.97	4.19	3.49	4.68	4.36	2.50	37.60
1995	2.46	2.00	1.84	4.56	3.67	2.36	6.50	8.29	0.89	3.22	4.40	1.89	42.08
1996	1.66	2.09	1.24	3.60	8.09	7.20	6.69	1.75	3.30	3.27	3.70	2.90	45.49
1997	2.63	3.86	2.02	1.36	3.77	3.16	1.98	5.05	2.86	2.16	2.45	2.14	33.44
1998	3.76	1.69	3.62	3.78	2.49	3.98	2.27	5.84	1.54	2.50	1.24	2.04	34.75
1999	3.06	1.58	1.18	7.48	1.64	2.60	2.39	4.12	1.25	1.37	1.32	2.66	30.65
2000	2.57	1.63	1.81	3.69	4.60	7.75	2.88	1.49	3.22	2.28	3.07	2.29	37.28
2001	0.83	3.47	1.11	3.65	4.31	4.25	2.97	3.75	3.65	7.06	2.66	2.25	39.96
POR= 108 YRS	2.28	1.86	2.80	3.46	3.56	3.74	3.40	3.53	3.36	3.01	2.89	2.59	36.48

AVERAGE TEMPERATURE (F) 2001 SOUTH BEND, IN (SBN)

YEAR	JAN	FEB	MAR	APR	MAY	JUN	JUL	AUG	SEP	OCT	NOV	DEC	ANNUAL
1972	23.9	26.5	33.4	45.4	60.0	63.6	71.4	69.4	62.9	49.7	38.1	28.8	47.8
1973	29.9	30.0	46.7	50.1	56.1	72.3	74.4	74.1	67.0	58.3	44.4	29.2	52.7
1974	27.9	28.1	39.6	51.7	57.3	66.8	75.0	71.9	62.0	53.0	42.4	32.9	50.7
1975	30.0	27.8	34.4	43.6	63.2	70.3	71.8	73.6	60.1	55.9	48.1	32.7	51.0
1976	21.6	35.3	43.6	52.5	57.0	70.9	72.8	69.0	61.3	47.9	33.2	22.8	49.0
1977	12.3	26.8	44.1	55.4	68.7	67.1	76.2	69.8	65.3	50.5	42.6	25.9	50.4
1978	18.5	14.8	31.2	48.7	59.5	69.2	71.7	71.9	68.9	51.9	42.9	29.5	48.2
1979	17.9	16.3	39.2	46.2	58.5	70.1	72.3	71.3	65.4	54.2	42.3	34.8	49.0
1980	27.1	23.6	35.5	49.0	62.1	67.8	76.6	75.0	66.3	50.4	41.6	31.1	50.5
1981	23.4	32.1	40.7	52.5	56.6	69.2	71.4	70.6	62.8	50.0	41.5	27.6	49.9
1982	15.6	22.7	35.1	44.7	66.0	64.5	73.3	68.9	63.3	54.0	42.5	39.0	49.1
1983	29.3	33.2	40.2	45.5	55.3	72.0	78.7	78.3	66.5	53.8	43.8	18.4	51.3
1984	18.5	35.9	30.0	48.2	56.1	72.3	71.7	74.4	63.8	57.1	41.0	34.4	50.3
1985	19.6	23.7	41.3	55.2	63.1	66.8	73.1	69.8	65.6	54.6	41.6	20.1	49.5
1986	25.8	24.8	40.4	51.3	59.9	67.7	74.8	67.6	65.9	53.0	36.1	31.2	49.9
1987	25.3	30.8	40.0	50.4	64.7	72.8	75.8	71.7	64.6	46.8	43.7	32.8	51.6
1988	21.3	22.7	37.3	48.4	62.2	72.0	76.4	75.9	64.5	45.9	42.5	28.8	49.8
1989	33.4	22.2	37.0	47.3	57.2	68.0	73.9	70.6	61.8	52.9	38.8	17.7	48.4
1990	34.0	31.3	40.9	48.7	56.7	68.8	71.2	69.9	65.0	52.3	45.9	31.8	51.4
1991	24.2	32.3	41.5	52.6	67.5	74.0	75.7	73.2	63.2	54.9	36.6	32.1	52.3
1992	29.1	33.2	37.4	46.1	58.4	65.1	70.0	67.4	62.3	51.4	40.0	30.9	49.3
1993	28.3	24.4	35.0	46.9	60.7	67.7	74.5	73.6	58.9	50.5	39.4	29.7	49.1
1994	16.3	22.5	36.6	51.0	57.4	71.4	72.9	67.9	61.8	54.6	44.6	35.1	49.6
1995	25.3	25.3	40.3	46.5	57.9	71.3	74.9	77.2	62.2	54.3	34.8	26.0	49.7
1996	24.1	27.8	32.3	46.0	57.3	68.9	68.5	71.1	62.6	52.4	33.8	29.4	47.9
1997	20.6	29.6	37.6	43.7	52.5	68.3	72.2	67.9	62.4	51.6	36.2	30.7	47.8
1998	31.6	37.0	38.6	49.5	65.0	68.3	72.5	73.0	67.4	53.8	43.6	34.6	52.9
1999	22.7	33.0	33.5	51.0	62.7	71.3	78.0	70.1	64.9	52.6	45.9	31.9	51.5
2000	25.5	35.0	44.1	46.7	61.4	67.4	70.0	70.9	61.8	52.7	48.6	35.1	51.1
2001	24.7	28.1	34.1	53.7	61.2	67.3	72.4	72.9			37.4	17.1	49.5
POR= 108 YRS	24.5	26.3	36.5	48.3	59.3	68.9	73.3	71.5	64.6	53.2	40.2	28.7	49.6

REFERENCE NOTES:

PAGE 1:
THE TEMPERATURE GRAPH SHOWS NORMAL MAXIMUM AND NORMAL
  MINIMUM DAILY TEMPERATURES (SOLID CURVES) AND THE
  ACTUAL DAILY HIGH AND LOW TEMPERATURES (VERTICAL BARS).

PAGE 2 AND 3:
  H/C INDICATES HEATING AND COOLING DEGREE DAYS.
  RH INDICATES RELATIVE HUMIDITY
  W/O INDICATES WEATHER AND OBSTRUCTIONS
  S INDICATES SUNSHINE.
  PR INDICATES PRESSURE.
  CLOUDINESS ON PAGE 3 IS THE SUM OF THE CEILOMETER AND
  SATELLITE DATA NOT TO EXCEED EIGHT EIGHTHS (OKTAS).

GENERAL:
  T INDICATES TRACE PRECIPITATION, AN AMOUNT GREATER
  THAN ZERO BUT LESS THAN THE LOWEST REPORTABLE VALUE.
  + INDICATES THE VALUE ALSO OCCURS ON EARLIER DATES.
  BLANK ENTRIES DENOTE MISSING OR UNREPORTED DATA.
  NORMALS ARE 30-YEAR AVERAGES (1961 - 1990).
  ASOS INDICATES AUTOMATED SURFACE OBSERVING SYSTEM.
  PM INDICATES THE LAST DAY OF THE PREVIOUS MONTH.
  POR (PERIOD OF RECORD) BEGINS WITH THE JANUARY DATA
  MONTH AND IS THE NUMBER OF YEARS USED TO COMPUTE
  THE MEAN. INDIVIDUAL MONTHS WITHIN THE POR MAY
  BE MISSING.
  WHEN THE POR FOR A NORMAL IS LESS THAN 30 YEARS,
  THE NORMAL IS PROVISIONAL AND IS BASED ON THE NUMBER
  OF YEARS INDICATED.
  0.* OR * INDICATES THE VALUE OR MEAN-DAYS-WITH
  IS BETWEEN 0.00 AND 0.05.
  CLOUDINESS FOR ASOS STATIONS DIFFERS FROM THE NON-ASOS
  OBSERVATION TAKEN BY A HUMAN OBSERVER. ASOS STATION
  CLOUDINESS IS BASED ON TIME-AVERAGED CEILOMETER DATA
  FOR CLOUDS AT OR BELOW 12,000 FEET AND ON SATELLITE
  DATA FOR CLOUDS ABOVE 12,000 FEET.
  THE NUMBER OF DAYS WITH CLEAR, PARTLY CLOUDY, AND
  CLOUDY CONDITIONS FOR ASOS STATIONS IS THE SUM
  OF THE CEILOMETER AND SATELLITE DATA FOR THE
  SUNRISE TO SUNSET PERIOD.

GENERAL CONTINUED:
  CLEAR INDICATES 0 - 2 OKTAS, PARTLY CLOUDY INDICATES
  3 - 6 OKTAS, AND CLOUDY INDICATES 7 OR 8 OKTAS.
  WHEN AT LEAST ONE OF THE ELEMENTS (CEILOMETER OR
  SATELLITE) IS MISSING, THE DAILY CLOUDINESS IS
  NOT COMPUTED.
  WIND DIRECTION IS RECORDED IN TENS OF DEGREES (2 DIGITS)
  CLOCKWISE FROM TRUE NORTH. "00" INDICATES CALM. "36"
  INDICATES TRUE NORTH.
  RESULTANT WIND IS THE VECTOR AVERAGE OF THE SPEED AND
  DIRECTION.
  AVERAGE TEMPERATURE IS THE SUM OF THE MEAN DAILY MAXIMUM
  AND MINIMUM TEMPERATURE DIVIDED BY 2.
  SNOWFALL DATA COMPRISE ALL FORMS OF FROZEN
  PRECIPITATION, INCLUDING HAIL.
  A HEATING (COOLING) DEGREE DAY IS THE DIFFERENCE BETWEEN
  THE AVERAGE DAILY TEMPERATURE AND 65 F.
  DRY BULB IS THE TEMPERATURE OF THE AMBIENT AIR.
  DEW POINT IS THE TEMPERATURE TO WHICH THE AIR MUST BE
  COOLED TO ACHIEVE 100 PERCENT RELATIVE HUMIDITY.
  WET BULB IS THE TEMPERATURE THE AIR WOULD HAVE IF THE
  MOISTURE CONTENT WAS INCREASED TO 100 PERCENT RELATIVE
  HUMIDITY.

  ON JULY 1, 1996, THE NATIONAL WEATHER SERVICE BEGAN USING
  THE "METAR" OBSERVATION CODE THAT WAS ALREADY EMPLOYED
  BY MOST OTHER NATIONS OF THE WORLD. THE MOST NOTICEABLE
  DIFFERENCE IN THIS ANNUAL PUBLICATION WILL BE THE CHANGE
  IN UNITS FROM TENTHS TO EIGHTS (OKTAS) FOR REPORTING THE
  AMOUNT OF SKY COVER.

HEATING DEGREE DAYS (base 65 F) 2001 SOUTH BEND, IN (SBN)

YEAR	JUL	AUG	SEP	OCT	NOV	DEC	JAN	FEB	MAR	APR	MAY	JUN	TOTAL
1972-73	24	35	112	468	801	1115	1080	977	561	446	268	0	5887
1973-74	0	4	60	229	611	1102	1142	1028	780	403	250	51	5660
1974-75	0	3	144	378	672	985	1077	1035	944	634	136	37	6045
1975-76	13	1	175	300	503	992	1338	856	658	407	253	9	5505
1976-77	0	19	139	525	949	1302	1628	1063	640	327	85	68	6745
1977-78	2	18	58	443	669	1206	1436	1401	1041	480	233	41	7028
1978-79	5	4	57	399	656	1095	1453	1356	795	560	241	22	6643
1979-80	3	17	73	353	672	928	1172	1195	908	482	147	58	6008
1980-81	0	1	62	449	694	1047	1282	915	749	374	271	6	5850
1981-82	4	8	132	460	700	1154	1523	1178	922	604	64	72	6821
1982-83	2	30	114	353	668	798	1100	886	760	581	298	30	5620
1983-84	6	0	94	352	628	1440	1431	838	1080	503	290	4	6666
1984-85	7	0	128	244	714	940	1401	1153	727	339	116	47	5816
1985-86	0	4	121	319	694	1381	1208	1117	760	425	192	46	6267
1986-87	3	48	81	369	858	1038	1224	950	766	436	139	16	5928
1987-88	5	25	78	558	638	993	1347	1220	851	498	158	37	6408
1988-89	1	11	72	581	670	1116	972	1190	865	528	271	40	6317
1989-90	0	12	147	381	779	1462	954	936	751	521	257	46	6246
1990-91	6	14	110	408	565	1021	1260	910	722	378	106	2	5502
1991-92	0	1	167	318	844	1012	1104	915	850	561	247	75	6094
1992-93	6	25	139	417	742	1049	1130	1130	925	536	157	55	6311
1993-94	0	7	198	450	765	1087	1505	1186	875	441	273	28	6815
1994-95	0	30	92	336	606	921	1224	1105	759	548	221	20	5862
1995-96	7	0	135	337	896	1202	1258	1072	1005	566	276	23	6777
1996-97	22	1	140	384	928	1098	1368	983	844	631	383	39	6821
1997-98	12	20	120	437	858	1057	1026	775	824	458	74	80	5741
1998-99	0	0	52	346	632	936	1304	893	968	412	134	41	5718
1999-00	0	3	92	377	566	1020	1218	863	646	540	167	59	5551
2000-01	7	6	153	291	820	1476	1242	1026	951	342	168	87	6569
2001-	22	0	139	381	489	923							

COOLING DEGREE DAYS (base 65 F) 2001 SOUTH BEND, IN (SBN)

YEAR	JAN	FEB	MAR	APR	MAY	JUN	JUL	AUG	SEP	OCT	NOV	DEC	ANNUAL
1972	0	0	0	0	47	74	226	180	55	0	0	0	582
1973	0	0	0	7	1	226	297	292	126	29	0	0	978
1974	0	0	0	10	17	114	321	222	60	11	0	0	755
1975	0	0	0	0	86	203	232	274	32	25	1	0	853
1976	0	0	1	37	13	192	249	150	36	4	0	0	682
1977	0	0	0	47	206	137	355	173	74	0	2	0	994
1978	0	0	0	0	70	173	218	227	179	0	1	0	868
1979	0	0	0	0	2	48	181	236	220	93	21	0	801
1980	0	0	0	9	65	145	367	319	107	6	0	0	1018
1981	0	0	2	6	15	137	211	191	72	0	0	0	634
1982	0	0	0	3	105	65	266	159	71	17	0	0	686
1983	0	0	0	1	3	247	440	417	146	13	0	0	1267
1984	0	0	0	5	19	228	226	298	98	7	0	0	881
1985	0	0	0	52	64	109	260	159	147	3	0	0	794
1986	0	0	4	22	41	135	312	136	113	2	0	0	765
1987	0	0	0	3	136	256	345	240	71	0	3	0	1054
1988	0	0	0	3	77	254	362	357	65	0	0	0	1118
1989	0	0	1	6	37	137	283	194	55	11	0	0	724
1990	0	0	9	37	8	167	203	171	117	21	0	0	733
1991	0	0	0	13	189	281	338	264	121	11	0	0	1217
1992	0	0	0	2	50	90	168	103	66	2	0	0	481
1993	0	0	0	0	29	144	302	284	18	9	0	0	786
1994	0	0	0	26	45	226	255	124	98	2	0	0	776
1995	0	0	0	0	9	216	320	382	57	10	0	0	994
1996	0	0	0	1	44	147	136	201	74	0	0	0	603
1997	0	0	0	0	0	146	241	117	48	29	0	0	581
1998	0	0	9	0	82	185	238	253	131	8	0	0	906
1999	0	0	0	0	67	236	410	166	96	3	0	0	978
2000	0	0	4	0	62	138	171	197	102	9	0	0	683
2001	0	0	0	11	56	164	259	253	53	6	0	0	802

SNOWFALL (inches) 2001 SOUTH BEND, IN (SBN)

YEAR	JUL	AUG	SEP	OCT	NOV	DEC	JAN	FEB	MAR	APR	MAY	JUN	TOTAL
1972-73	0.0	0.0	0.0	1.5	12.6	19.7	5.5	10.6	4.7	1.7	0.0	0.0	56.3
1973-74	0.0	0.0	0.0	0.0	1.0	22.6	14.4	11.9	9.4	1.2	0.0	0.0	60.5
1974-75	0.0	0.0	0.0	0.6	7.9	19.9	9.2	13.9	17.8	5.4	0.0	0.0	74.7
1975-76	0.0	0.0	0.0	0.0	10.7	14.0	31.2	13.9	3.8	0.6	0.1	0.0	74.3
1976-77	0.0	0.0	0.0	0.8	21.6	37.6	37.2	13.9	15.8	2.3	0.0	0.0	129.2
1977-78	0.0	0.0	0.0	0.3	30.3	33.6	86.1	16.6	5.1	T	0.0	0.0	172.0
1978-79	0.0	0.0	0.0	0.0	7.5	26.4	45.1	15.9	6.3	0.1	0.0	0.0	101.3
1979-80	0.0	0.0	0.0	T	7.5	13.6	11.5	22.3	9.8	1.7	0.0	0.0	66.4
1980-81	0.0	0.0	0.0	1.1	8.8	24.3	23.8	20.7	6.3	T	0.0	0.0	85.0
1981-82	0.0	0.0	0.0	0.1	9.1	41.3	41.3	19.2	10.2	14.0	0.0	0.0	135.2
1982-83	0.0	0.0	0.0	0.0	2.1	2.5	8.0	9.6	12.0	1.1	0.0	0.0	35.3
1983-84	0.0	0.0	0.0	0.0	1.4	35.6	16.7	15.9	11.1	0.4	0.0	0.0	81.1
1984-85	0.0	0.0	0.0	0.0	0.6	14.1	40.0	28.9	1.6	3.1	0.0	0.0	88.3
1985-86	0.0	0.0	0.0	0.0	2.2	40.4	26.3	11.3	3.6	0.2	0.0	0.0	84.0
1986-87	0.0	0.0	0.0	9.7	4.8	31.4	5.9	2.0	1.5	0.0	0.0	0.0	55.3
1987-88	0.0	0.0	T	T	1.6	13.1	11.4	22.9	12.1	T	0.0	0.0	61.1
1988-89	0.0	0.0	0.0	0.3	7.8	14.8	3.1	16.3	2.5	1.7	T	T	46.5
1989-90	T	T	0.0	8.8	15.2	29.4	1.2	13.9	3.4	1.1	T	0.0	59.0
1990-91	0.0	0.0	0.0	T	T	17.5	15.7	23.4	2.4	T	T	0.0	67.7
1991-92	0.0	0.0	0.0	T	7.6	11.7	19.9	10.9	16.7	0.9	0.0	T	67.7
1992-93	0.0	0.0	0.0	1.2	7.7	12.6	8.1	32.2	19.0	1.4	0.0	0.0	82.2
1993-94	0.0	0.0	0.1	T	2.2	11.3	21.5	20.0	0.1	0.8	T	0.0	56.0
1994-95	T	0.0	1.0	0.0	T	2.4	17.3	25.8	2.8	0.6	0.0	0.0	49.9
1995-96	0.0	0.0	0.0	T	6.2	22.1	12.2	7.9	8.3	0.9	T	0.0	57.6
1996-97	T	T	0.0	0.0	15.3	10.5	36.4	10.2	3.3	0.4	T	0.0	76.1
1997-98	0.0	0.0	0.0	0.2	5.6	18.1	7.8	T	20.8	T	0.0	T	52.5
1998-99	T	0.0	0.0	0.0	0.0	17.1	37.0	10.4	14.2	T	T	0.0	78.7
1999-00	0.0	0.0	0.0	0.0	1.9	7.4	37.1	11.4	T	1.1	0.0	0.0	58.9
2000-01	T	0.0	0.0	T	10.4	44.6	2.9	5.4	10.6	2.7	0.0	T	76.6
2001-	T	0.0	0.0	T	0.0	19.5							
POR= 61 YRS	T	T	0.0	0.7	7.5	17.8	19.2	14.4	8.7	1.9	0.3	T	70.5

2001
DES MOINES,
IOWA (DSM)

Located in the heart of North America, Des Moines has a climate which is continental in character. This results in a marked seasonal contrast in both temperature and precipitation. There is a gently rolling terrain in and around the Des Moines metropolitan area. Drainage of the area is generally to the southeast to the Des Moines River and its tributaries.

Since agriculture and services for it are the mainstay of the area, it is convenient to separate the year into arbitrary seasons corresponding to the growing seasons of the principal crops of the section. The winter season, when most plant life is dormant, is from mid-November to late March. The summer season, when corn and soybeans can be grown, lasts from early May to early October. The spring growing season, including part of the growing season of oats and forage crops, and the fall harvest season, each runs about 6 weeks. There is a large variation in annual precipitation from a minimum of about 17 inches to a maximum of about 56 inches. The average annual snowfall is 32 inches. Annual variation of snowfall is also large, ranging from a minimum of about 8 inches to as much as 72 inches.

The winter is a season of cold dry air, interrupted by occasional storms of short duration. At the beginning and the end of the season, the precipitation may occur as rain, but during the major portion of the season it falls as snow. Drifting snow may be extensive and impede transportation. The average precipitation for this season is approximately 20 percent of the annual amount. Although occasional cold waves follow the storms, bitterly cold days on which the temperatures fail to rise above zero occur on an average of only 3 days in 4 years.

The average growing season with temperatures above 32 degrees normally spans 160 to 165 days between late April and mid-October. The growing season is characterized by prevailing southerly winds and precipitation falling primarily as showers and thunderstorms, occasionally with damaging wind, erosive downpours or hail. Some 60 percent of the annual precipitation falls during the crop season with the maximum rate normally in late May and June. The autumn is characteristically sunny with diminishing precipitation, a condition favorable for drying and harvesting crops.

NORMALS, MEANS, AND EXTREMES

DES MOINES, IA (DSM)

LATITUDE: 41 32' 16" N   LONGITUDE: 93 39' 58" W   ELEVATION (FT): GRND: 968   BARO: 971   TIME ZONE: CENTRAL (UTC + 6)   WBAN: 14933

	ELEMENT	POR	JAN	FEB	MAR	APR	MAY	JUN	JUL	AUG	SEP	OCT	NOV	DEC	YEAR
TEMPERATURE °F	NORMAL DAILY MAXIMUM	30	28.1	33.7	46.9	61.8	73.0	82.2	86.7	84.2	75.6	64.3	48.0	32.6	59.8
	MEAN DAILY MAXIMUM	56	28.9	34.6	46.1	61.3	72.4	81.7	86.0	83.9	74.4	64.7	47.3	33.6	59.6
	HIGHEST DAILY MAXIMUM	62	65	73	91	93	98	103	105	108	101	95	81	69	108
	YEAR OF OCCURRENCE		1989	1972	1986	1980	1967	1988	1955	1983	1939	1963	1999	1984	AUG 1983
	MEAN OF EXTREME MAXS.	56	51.2	56.8	72.9	83.3	87.7	93.5	97.0	95.7	89.5	83.6	68.3	56.9	78.0
	NORMAL DAILY MINIMUM	30	10.7	15.6	27.6	40.0	51.5	61.2	66.5	63.6	54.5	42.7	29.9	16.1	40.0
	MEAN DAILY MINIMUM	56	11.4	16.9	27.2	39.8	51.2	61.1	65.9	63.7	53.3	43.1	29.3	17.4	40.0
	LOWEST DAILY MINIMUM	62	-24	-26	-22	9	30	38	47	40	0	14	-4	-22	-26
	YEAR OF OCCURRENCE		1970	1996	1962	1975	1967	1945	1971	1950	1996	1972	1991	1989	FEB 1996
	MEAN OF EXTREME MINS.	56	-11.6	-5.2	7.2	23.5	36.4	48.0	55.0	52.0	37.0	26.1	11.0	-4.4	22.9
	NORMAL DRY BULB	30	19.4	24.7	37.3	50.9	62.3	71.8	76.6	73.9	65.1	53.5	39.0	24.4	49.9
	MEAN DRY BULB	56	20.2	25.7	36.5	50.4	61.8	71.3	75.9	73.9	64.0	53.9	38.3	25.5	49.8
	MEAN WET BULB	16	20.8	25.2	34.1	44.5	54.4	64.2	68.5	68.5	63.2	54.7	34.3	23.8	44.6
	MEAN DEW POINT	16	15.9	19.7	27.4	37.0	49.3	55.9	64.8	59.8	50.3	41.2	28.8	19.4	39.1
	NORMAL NO. DAYS WITH:														
	MAXIMUM 90	30	0.0	0.0	*	0.2	0.5	4.5	10.0	7.1	1.8	0.1	0.0	0.0	24.2
	MAXIMUM 32	30	17.4	13.0	4.6	0.2	0.0	0.0	0.0	0.0	0.0	0.0	3.1	14.4	52.7
	MINIMUM 32	30	30.0	25.6	20.7	6.6	0.2	0.0	0.0	0.0	0.2	4.7	18.6	28.9	135.5
	MINIMUM 0	30	8.3	4.0	0.2	0.0	0.0	0.0	0.0	0.0	0.0	0.0	0.2	4.1	16.8
H/C	NORMAL HEATING DEG. DAYS	30	1414	1128	859	428	165	10	0	11	71	372	780	1259	6497
	NORMAL COOLING DEG. DAYS	30	0	0	0	5	81	214	360	287	74	15	0	0	1036
RH	NORMAL (PERCENT)	30	71	71	68	63	63	65	68	70	71	66	71	75	68
	HOUR 00 LST	30	74	76	73	69	70	72	76	79	79	73	75	78	74
	HOUR 06 LST	30	75	78	78	77	77	79	82	85	85	79	79	79	79
	HOUR 12 LST	30	66	65	61	55	54	55	57	58	59	56	63	69	60
	HOUR 18 LST	30	68	66	60	52	52	53	56	58	60	58	66	72	60
S	PERCENT POSSIBLE SUNSHINE	48	51	54	57	56	61	68	72	70	66	62	49	46	59
W/O	MEAN NO. DAYS WITH:														
	HEAVY FOG (VISBY 1/4 MI)	52	2.6	2.3	2.0	1.0	0.9	0.7	0.7	1.3	1.2	1.2	1.7	2.8	18.4
	THUNDERSTORMS	62	0.3	0.4	2.1	4.3	7.0	9.0	8.3	7.0	4.9	2.6	1.1	0.3	47.3
CLOUDINESS	MEAN:														
	SUNRISE-SUNSET (OKTAS)	0						4.8							
	MIDNIGHT-MIDNIGHT (OKTAS)														
	MEAN NO. DAYS WITH:														
	CLEAR	1	1.0	1.0	6.0		3.0	10.0							
	PARTLY CLOUDY	1	2.0		3.0		2.0	1.0							
	CLOUDY	1	3.0	3.0	7.0		12.0	10.0							
PR	MEAN STATION PRESSURE (IN)	27	29.10	29.09	29.00	28.90	28.90	28.90	29.00	29.00	29.00	29.01	29.00	29.10	29.00
	MEAN SEA-LEVEL PRES. (IN)	16	30.11	30.12	30.05	29.94	29.94	30.01	29.97	30.01	30.03	30.05	30.06	30.13	30.03
WINDS	MEAN SPEED (MPH)	49	11.4	11.3	12.5	12.8	11.0	10.2	8.9	8.7	9.5	10.4	11.7	11.2	10.8
	PREVAIL.DIR (TENS OF DEGS)	30	32	32	32	32	18	18	18	15	18	18	32	32	18
	MAXIMUM 2-MINUTE:														
	SPEED (MPH)	6	43	41	45	51	52	53	49	53	52	46	52	40	53
	DIR. (TENS OF DEGS)		30	30	31	29	32	22	02	28	29	29	25	28	22
	YEAR OF OCCURRENCE		2000	1999	1996	1996	2000	2000	2000	1998	2001	1996	1998	2001	JUN 2000
	MAXIMUM 5-SECOND:														
	SPEED (MPH)	6	52	51	53	64	59	68	61	70	60	57	62	47	70
	DIR. (TENS OF DEGS)		33	29	29	28	32	21	01	29	28	28	26	28	29
	YEAR OF OCCURRENCE		1996	1996	1996	1996	2000	2000	2000	1998	2001	1996	1998	2001	AUG 1998
PRECIPITATION	NORMAL (IN)	30	0.96	1.11	2.33	3.36	3.66	4.46	3.78	4.20	3.53	2.62	1.79	1.32	33.12
	MAXIMUM MONTHLY (IN)	62	4.38	2.99	5.82	7.76	12.13	14.19	10.51	13.68	10.19	7.29	6.52	3.43	14.19
	YEAR OF OCCURRENCE		1960	1951	1990	1976	1996	1947	1958	1977	1961	1941	1983	1982	JUN 1947
	MINIMUM MONTHLY (IN)	62	0.04	0.13	0.17	0.23	1.23	1.02	0.04	0.25	0.41	0.03	0.03	0.03	0.03
	YEAR OF OCCURRENCE		1997	1968	1994	1985	1949	1992	1975	1984	1950	1952	1969	1998	DEC 1998
	MAXIMUM IN 24 HOURS (IN)	62	2.97	1.77	2.42	3.80	3.23	5.50	5.14	6.18	4.47	2.81	3.35	1.69	6.18
	YEAR OF OCCURRENCE		1960	1961	1945	1974	1996	1947	1958	1975	1961	1947	1952	1982	AUG 1975
	NORMAL NO. DAYS WITH:														
	PRECIPITATION 0.01	30	7.7	7.2	10.1	10.7	11.5	10.2	9.6	9.1	9.2	8.0	7.6	8.8	109.7
	PRECIPITATION 1.00	30	0.1	0.1	0.3	0.7	0.8	1.2	1.0	1.3	0.8	0.7	0.3	0.1	7.4
SNOWFALL	NORMAL (IN)	30	7.6	7.9	5.6	2.4	T	0.0	0.0	0.0	T	0.4	3.2	7.7	34.8
	MAXIMUM MONTHLY (IN)	57	22.3	21.3	18.8	15.6	0.2	T	T	0.0	T	7.4	14.7	23.9	23.9
	YEAR OF OCCURRENCE		1996	1962	1948	1982	1944	1993	1992		1992	1980	1991	1961	DEC 1961
	MAXIMUM IN 24 HOURS (IN)	57	19.8	12.1	8.5	10.4	0.2	T	T	0.0	T	7.4	11.8	11.0	19.8
	YEAR OF OCCURRENCE		1942	1950	1957	1973	1944	1993	1992		1992	1980	1968	1961	JAN 1942
	MAXIMUM SNOW DEPTH (IN)	51	16	17	18	12	0	0	0	0	0	5	10	30	30
	YEAR OF OCCURRENCE		2001	1979	1960	1973						1980	1968	1962	DEC 1962
	NORMAL NO. DAYS WITH:														
	SNOWFALL 1.0	30	2.4	2.4	1.6	0.6	0.0	0.0	0.0	0.0	0.0	0.1	0.9	2.4	10.4

PRECIPITATION (inches) 2001 DES MOINES, IA (DSM)

YEAR	JAN	FEB	MAR	APR	MAY	JUN	JUL	AUG	SEP	OCT	NOV	DEC	ANNUAL
1972	0.44	0.63	1.05	3.56	3.05	2.58	5.86	6.65	5.45	2.36	2.43	1.96	36.02
1973	2.09	2.21	4.15	4.67	5.01	2.04	9.17	1.37	7.07	3.26	1.49	2.65	45.18
1974	1.51	0.84	1.99	6.31	7.19	4.62	1.33	2.81	2.08	3.96	1.20	1.83	35.67
1975	1.41	1.48	1.90	2.65	3.41	5.98	0.04	9.73	1.70	0.63	2.20	0.48	31.61
1976	0.23	2.43	3.04	7.76	2.84	7.25	1.87	2.22	1.00	1.11	0.10	0.12	30.01
1977	0.50	0.36	3.57	2.45	2.29	1.25	2.63	13.68	2.82	5.10	0.69	1.81	37.15
1978	0.28	1.27	0.90	4.57	3.49	2.74	2.95	3.10	6.39	1.14	3.16	1.37	31.36
1979	1.72	0.52	4.23	3.23	2.50	5.78	2.96	5.07	0.97	3.23	1.43	0.20	31.84
1980	1.80	0.64	1.15	0.86	1.94	5.56	1.52	7.24	1.03	1.90	0.45	1.00	25.09
1981	0.25	0.97	0.39	2.00	2.46	5.02	5.76	6.32	2.30	2.06	2.63	1.14	31.30
1982	2.63	0.78	3.30	5.03	5.79	2.59	7.00	5.25	2.94	3.44	2.62	3.43	44.80
1983	1.17	1.95	3.72	3.80	3.93	3.65	2.44	3.01	3.87	5.54	6.52	1.57	41.17
1984	0.99	0.82	1.65	5.85	5.58	7.81	6.22	0.25	2.76	6.28	1.16	2.41	41.78
1985	0.64	1.98	3.37	0.23	1.56	3.72	2.04	2.83	5.42	3.75	1.65	1.31	28.50
1986	0.12	1.76	2.92	5.66	4.35	7.08	3.90	4.52	6.41	3.89	0.99	0.98	42.58
1987	0.42	1.38	2.99	2.92	3.75	2.10	5.08	10.04	1.40	1.03	3.27	2.59	36.97
1988	0.37	0.59	0.66	0.75	1.46	2.75	4.78	3.05	2.89	0.59	3.38	0.84	22.11
1989	1.30	1.05	0.37	1.95	3.62	2.22	3.65	6.53	5.41	2.28	0.19	0.57	29.14
1990	1.43	0.89	5.82	3.43	4.36	9.52	8.75	1.83	1.40	1.80	2.52	2.18	43.93
1991	0.95	0.17	3.90	7.54	7.88	2.87	1.14	3.65	0.90	4.96	3.61	2.20	39.77
1992	0.97	2.12	2.13	3.99	1.45	1.02	7.76	1.39	4.99	0.51	5.20	1.98	33.51
1993	1.59	1.52	3.22	2.96	7.51	7.68	9.75	12.24	5.79	1.70	1.06	0.86	55.88
1994	1.22	1.71	0.17	2.71	1.76	5.29	4.30	2.81	3.11	1.02	1.68	2.42	28.20
1995	0.98	0.68	2.75	6.04	6.29	2.88	4.36	1.04	3.12	1.24	1.57	0.08	31.03
1996	2.99	0.57	1.38	2.12	12.13	2.95	4.02	2.54	4.05	2.87	2.40	0.52	38.54
1997	0.04	0.77	0.85	4.11	3.95	4.90	2.89	2.36	2.14	4.62	1.13	0.77	28.53
1998	0.60	1.23	3.03	1.79	4.97	9.95	4.28	5.65	0.69	3.39	2.09	0.03	37.70
1999	0.26	0.63	0.90	4.43	5.82	3.46	2.78	4.56	2.32	0.36	1.28	0.35	27.15
2000	0.27	1.32	0.35	2.36	3.37	7.60	4.12	1.71	1.84	1.07	1.57	0.56	26.14
2001	1.31	2.08	1.08	2.20	7.46	2.61	1.04	2.22	4.77	2.12	0.86	0.70	28.45
POR= 50 YRS	1.10	1.15	1.96	2.98	4.32	4.66	3.46	3.65	3.24	2.38	1.60	1.16	31.66

AVERAGE TEMPERATURE (F) 2001 DES MOINES, IA (DSM)

YEAR	JAN	FEB	MAR	APR	MAY	JUN	JUL	AUG	SEP	OCT	NOV	DEC	ANNUAL
1972	16.6	19.9	37.1	48.9	62.5	70.2	73.9	72.4	64.6	48.0	33.9	18.4	47.2
1973	22.0	27.7	45.6	49.2	59.8	73.6	76.4	76.9	65.7	59.0	40.7	22.1	51.6
1974	19.5	28.0	39.8	53.0	60.3	69.1	80.9	71.1	60.9	55.0	39.6	28.2	50.5
1975	22.7	22.4	29.7	46.9	65.9	72.3	77.9	77.1	61.8	57.2	43.5	29.9	50.6
1976	22.9	34.1	39.7	55.4	60.7	71.4	77.2	73.8	65.7	48.5	32.6	21.8	50.3
1977	10.1	29.8	44.7	58.5	69.5	74.7	81.0	72.1	66.5	52.3	39.1	23.2	51.8
1978	11.0	13.3	33.2	50.5	61.6	72.8	75.8	75.1	70.5	52.4	39.0	23.0	48.2
1979	7.5	13.8	35.2	47.0	61.1	70.9	74.5	74.3	66.8	54.3	38.0	31.6	47.9
1980	23.4	21.3	34.7	52.0	63.5	71.2	79.9	76.2	66.8	49.6	41.6	26.6	50.6
1981	25.7	29.8	42.7	57.6	60.4	72.9	76.1	72.4	66.0	51.7	42.8	25.5	52.0
1982	9.6	22.9	35.4	46.9	64.7	67.1	76.9	73.2	64.9	54.6	38.4	31.6	48.9
1983	27.3	32.3	39.4	45.4	58.4	73.2	80.9	83.3	67.7	52.6	40.7	9.8	50.9
1984	19.7	35.5	31.1	48.8	58.7	73.2	75.9	77.5	62.4	52.8	39.5	27.6	50.2
1985	15.8	22.2	42.0	55.2	65.1	68.4	76.5	71.9	64.8	52.6	30.0	13.4	48.2
1986	26.9	21.7	42.5	53.9	62.6	73.3	77.2	69.3	67.4	52.6	33.3	28.8	50.8
1987	26.6	35.7	42.7	54.6	66.8	74.7	76.5	72.1	65.2	48.2	43.1	30.0	53.2
1988	19.6	21.6	40.2	51.4	67.4	75.6	78.5	78.8	66.9	47.9	39.8	28.8	51.4
1989	32.5	15.4	37.1	52.3	61.0	68.9	77.1	73.3	62.2	54.1	36.0	16.9	48.9
1990	31.7	31.1	42.0	50.0	57.9	71.6	73.8	74.4	68.1	52.4	43.7	22.9	51.6
1991	16.8	33.5	42.5	53.6	66.7	74.8	76.1	73.7	65.0	52.5	30.9	30.6	51.4
1992	30.7	34.9	42.4	48.9	62.0	70.3	71.2	68.3	63.1	53.4	34.9	21.2	50.6
1993	21.3	21.5	34.1	47.1	61.0	69.0	73.8	73.8	59.2	50.4	36.4	28.4	48.0
1994	13.0	21.3	40.5	50.9	62.5	72.4	73.0	71.3	66.4	55.5	42.1	30.3	49.9
1995	20.8	28.8	39.8	47.4	58.9	71.0	76.4	79.1	63.2	52.9	33.3	26.0	49.8
1996	17.3	25.0	32.5	47.8	57.7	70.9	72.7	72.3	0.0	53.4	32.1	22.2	42.0
1997	17.5	27.6	40.0	45.9	55.9	72.1	75.6	72.1	66.4	53.7	34.5	28.8	49.2
1998	25.9	36.3	33.6	50.8	66.5	68.1	75.4	75.2	69.4	54.0	40.2	30.3	52.4
1999	20.2	35.2	38.7	51.4	61.3	70.7	79.7	72.4	62.9	53.7	47.4	29.9	52.0
2000	25.4	36.5	45.1	52.0	64.8	69.3	73.8	75.5	67.2	57.7	33.9	11.6	51.1
2001	23.6	20.3	32.2	56.2	61.7	70.6	78.0	75.2	63.5	52.2	49.8	32.4	51.3
POR= 123 YRS	20.7	25.2	36.8	50.6	61.6	71.2	76.1	73.8	64.9	53.8	38.4	26.0	49.9

REFERENCE NOTES:

PAGE 1:
THE TEMPERATURE GRAPH SHOWS NORMAL MAXIMUM AND NORMAL
MINIMUM DAILY TEMPERATURES (SOLID CURVES) AND THE
ACTUAL DAILY HIGH AND LOW TEMPERATURES (VERTICAL BARS).

PAGE 2 AND 3:
H/C INDICATES HEATING AND COOLING DEGREE DAYS.
RH INDICATES RELATIVE HUMIDITY
W/O INDICATES WEATHER AND OBSTRUCTIONS
S INDICATES SUNSHINE.
PR INDICATES PRESSURE.
CLOUDINESS ON PAGE 3 IS THE SUM OF THE CEILOMETER AND
SATELLITE DATA NOT TO EXCEED EIGHT EIGHTHS(OKTAS).

GENERAL:
T INDICATES TRACE PRECIPITATION, AN AMOUNT GREATER
THAN ZERO BUT LESS THAN THE LOWEST REPORTABLE VALUE.
+ INDICATES THE VALUE ALSO OCCURS ON EARLIER DATES.
BLANK ENTRIES DENOTE MISSING OR UNREPORTED DATA.
NORMALS ARE 30-YEAR AVERAGES (1961 - 1990).
ASOS INDICATES AUTOMATED SURFACE OBSERVING SYSTEM.
PM INDICATES THE LAST DAY OF THE PREVIOUS MONTH.
POR (PERIOD OF RECORD) BEGINS WITH THE JANUARY DATA
MONTH AND IS THE NUMBER OF YEARS USED TO COMPUTE
THE MEAN. INDIVIDUAL MONTHS WITHIN THE POR MAY
BE MISSING.
WHEN THE POR FOR A NORMAL IS LESS THAN 30 YEARS,
THE NORMAL IS PROVISIONAL AND IS BASED ON THE NUMBER
OF YEARS INDICATED.
0.* OR * INDICATES THE VALUE OR MEAN-DAYS-WITH
IS BETWEEN 0.00 AND 0.05.
CLOUDINESS FOR ASOS STATIONS DIFFERS FROM THE NON-ASOS
OBSERVATION TAKEN BY A HUMAN OBSERVER. ASOS STATION
CLOUDINESS IS BASED ON TIME-AVERAGED CEILOMETER DATA
FOR CLOUDS AT OR BELOW 12,000 FEET AND ON SATELLITE
DATA FOR CLOUDS ABOVE 12,000 FEET.
THE NUMBER OF DAYS WITH CLEAR, PARTLY CLOUDY, AND
CLOUDY CONDITIONS FOR ASOS STATIONS IS THE SUM
OF THE CEILOMETER AND SATELLITE DATA FOR THE
SUNRISE TO SUNSET PERIOD.

GENERAL CONTINUED:
CLEAR INDICATES 0 - 2 OKTAS, PARTLY CLOUDY INDICATES
3 - 6 OKTAS, AND CLOUDY INDICATES 7 OR 8 OKTAS.
WHEN AT LEAST ONE OF THE ELEMENTS (CEILOMETER OR
SATELLITE) IS MISSING, THE DAILY CLOUDINESS IS
NOT COMPUTED.
WIND DIRECTION IS RECORDED IN TENS OF DEGREES (2 DIGITS)
CLOCKWISE FROM TRUE NORTH. "00" INDICATES CALM. "36"
INDICATES TRUE NORTH.
RESULTANT WIND IS THE VECTOR AVERAGE OF THE SPEED AND
DIRECTION.
AVERAGE TEMPERATURE IS THE SUM OF THE MEAN DAILY MAXIMUM
AND MINIMUM TEMPERATURE DIVIDED BY 2.
SNOWFALL DATA COMPRISE ALL FORMS OF FROZEN
PRECIPITATION, INCLUDING HAIL.
A HEATING (COOLING) DEGREE DAY IS THE DIFFERENCE BETWEEN
THE AVERAGE DAILY TEMPERATURE AND 65 F.
DRY BULB IS THE TEMPERATURE OF THE AMBIENT AIR.
DEW POINT IS THE TEMPERATURE TO WHICH THE AIR MUST BE
COOLED TO ACHIEVE 100 PERCENT RELATIVE HUMIDITY.
WET BULB IS THE TEMPERATURE THE AIR WOULD HAVE IF THE
MOISTURE CONTENT WAS INCREASED TO 100 PERCENT RELATIVE
HUMIDITY.

ON JULY 1, 1996, THE NATIONAL WEATHER SERVICE BEGAN USING
THE "METAR" OBSERVATION CODE THAT WAS ALREADY EMPLOYED
BY MOST OTHER NATIONS OF THE WORLD. THE MOST NOTICEABLE
DIFFERENCE IN THIS ANNUAL PUBLICATION WILL BE THE CHANGE
IN UNITS FROM TENTHS TO EIGHTS(OKTAS) FOR REPORTING THE
AMOUNT OF SKY COVER.

HEATING DEGREE DAYS (base 65 F) 2001 DES MOINES, IA (DSM)

YEAR	JUL	AUG	SEP	OCT	NOV	DEC	JAN	FEB	MAR	APR	MAY	JUN	TOTAL
1972-73	7	10	112	523	925	1442	1326	1039	594	470	173	0	6621
1973-74	0	0	68	209	719	1325	1406	1033	775	363	189	20	6107
1974-75	0	12	168	307	755	1131	1308	1185	1090	539	81	11	6587
1975-76	0	0	148	267	637	1085	1297	890	780	302	160	4	5570
1976-77	0	1	76	527	964	1333	1700	981	624	234	25	2	6467
1977-78	0	3	35	388	769	1289	1667	1442	988	427	181	9	7198
1978-79	0	0	48	385	776	1293	1779	1433	912	532	163	13	7334
1979-80	1	10	57	339	801	1031	1281	1263	932	408	124	9	6256
1980-81	0	0	77	473	695	1182	1214	979	684	241	177	0	5722
1981-82	6	2	57	406	660	1218	1713	1175	911	536	73	30	6787
1982-83	0	6	113	326	791	1026	1162	908	787	587	219	17	5942
1983-84	0	0	96	394	720	1709	1401	851	1043	488	217	1	6920
1984-85	0	0	172	376	759	1154	1520	1192	707	335	59	27	6301
1985-86	0	0	172	378	1046	1596	1181	1208	702	344	116	2	6745
1986-87	0	25	51	378	941	1114	1184	813	687	336	58	6	5593
1987-88	0	24	54	513	648	1083	1399	1254	764	400	33	3	6175
1988-89	0	6	35	524	749	1114	1002	1384	866	417	170	31	6298
1989-90	0	6	140	345	865	1489	1025	943	703	469	221	21	6227
1990-91	3	0	85	394	633	1301	1491	875	699	355	112	0	5948
1991-92	0	1	144	395	1019	1058	1054	868	691	486	152	9	5877
1992-93	0	26	135	363	897	1169	1344	1210	953	528	151	42	6818
1993-94	0	2	185	460	851	1129	1606	1221	751	437	132	17	6791
1994-95	0	11	74	292	681	1071	1363	1007	774	521	191	20	6005
1995-96	3	0	136	381	945	1201	1471	1155	1002	514	264	31	7103
1996-97	0			364	982	1322	1468	1043	764	568	288	0	
1997-98	4	0	52	409	908	1118	1207	799	967	419	45	67	5995
1998-99	0	0	26	339	674	1067	1384	824	808	402	147	27	5698
1999-00	0	2	130	355	522	1080	1218	821	615	388	90	20	5241
2000-01	0	0	85	241	930	1647	1274	1244	1009	286	154	33	6903
2001-	0	0	105	399	449	1005							

COOLING DEGREE DAYS (base 65 F) 2001 DES MOINES, IA (DSM)

YEAR	JAN	FEB	MAR	APR	MAY	JUN	JUL	AUG	SEP	OCT	NOV	DEC	ANNUAL
1972	0	0	0	3	77	184	289	247	109	0	0	0	909
1973	0	0	0	2	19	267	358	378	98	30	0	0	1152
1974	0	0	0	9	52	149	499	209	52	4	0	0	974
1975	0	0	0	3	116	237	408	383	58	32	0	0	1237
1976	0	0	0	19	36	203	385	283	104	20	0	0	1050
1977	0	0	0	48	172	298	505	232	86	1	0	0	1342
1978	0	0	6	0	83	251	341	321	221	3	0	0	1226
1979	0	0	0	0	48	194	304	305	118	15	0	0	984
1980	0	0	0	22	83	200	469	353	138	2	0	0	1267
1981	0	0	1	27	42	243	358	239	95	2	0	0	1007
1982	0	0	0	0	71	101	374	269	120	11	0	0	946
1983	0	0	0	4	20	272	502	574	183	19	0	0	1574
1984	0	0	0	8	27	254	345	397	101	2	0	0	1134
1985	0	0	0	46	69	134	361	221	174	0	0	0	1005
1986	0	0	11	17	46	258	386	162	130	0	0	0	1010
1987	0	0	0	31	121	304	426	250	65	0	0	0	1197
1988	0	0	0	1	112	329	425	444	101	3	0	0	1415
1989	0	0	6	44	53	156	379	269	61	15	0	0	983
1990	0	0	1	26	5	226	283	297	184	11	2	0	1035
1991	0	0	0	9	19	172	302	349	279	150	14	0	1294
1992	0	0	0	0	10	64	174	197	134	86	10	0	675
1993	0	0	0	0	0	31	167	281	280	19	15	0	793
1994	0	0	0	22	65	245	253	213	125	4	0	0	927
1995	0	0	0	0	0	9	206	361	440	90	11	0	1117
1996	0	0	0	4	44	212	245	231		10	0	0	
1997	0	0	0	0	14	223	338	228	101	65	0	0	969
1998	0	0	0	2	1	99	165	330	323	190	3	0	1113
1999	0	0	0	0	40	204	463	236	72	11	1	0	1027
2000	0	0	3	4	90	156	278	334	157	25	0	0	1047
2001	0	0	0	28	61	207	409	322	67	6	0	0	1100

SNOWFALL (inches) 2001 DES MOINES, IA (DSM)

YEAR	JUL	AUG	SEP	OCT	NOV	DEC	JAN	FEB	MAR	APR	MAY	JUN	TOTAL
1972-73	0.0	0.0	0.0	T	10.2	9.5	15.3	4.6	T	15.1	0.0	0.0	54.7
1973-74	0.0	0.0	0.0	0.0	T	9.6	10.6	6.1	1.9	1.2	0.0	0.0	29.4
1974-75	0.0	0.0	0.0	0.0	9.3	9.1	13.3	17.9	5.6	4.5	0.0	0.0	59.7
1975-76	0.0	0.0	0.0	0.0	7.0	0.5	2.4	11.8	0.9	0.0	T	0.0	22.6
1976-77	0.0	0.0	0.0	T	1.1	2.8	7.6	0.9	4.8	1.9	0.0	0.0	19.1
1977-78	0.0	0.0	0.0	0.0	2.2	22.0	2.7	18.6	9.9	0.3	0.0	0.0	55.7
1978-79	0.0	0.0	0.0	0.0	5.8	11.0	16.6	8.2	4.1	8.0	0.0	0.0	53.7
1979-80	0.0	0.0	0.0	T	0.7	0.4	7.9	5.7	5.8	2.8	0.0	0.0	23.3
1980-81	0.0	0.0	0.0	7.4	T	2.7	3.7	6.6	T	0.0	0.0	0.0	20.4
1981-82	0.0	0.0	0.0	0.4	3.2	10.7	18.5	2.6	11.9	15.6	0.0	0.0	62.9
1982-83	0.0	0.0	0.0	0.8	0.3	3.6	4.2	16.8	13.2	12.6	0.0	0.0	51.5
1983-84	0.0	0.0	0.0	T	9.8	19.6	12.5	1.3	13.7	0.1	0.0	0.0	57.0
1984-85	0.0	0.0	0.0	T	2.7	7.6	7.7	6.9	6.7	T	0.0	0.0	31.6
1985-86	0.0	0.0	T	0.0	8.1	15.9	1.3	6.7	0.2	0.1	0.0	0.0	32.3
1986-87	0.0	0.0	0.0	T	3.1	2.9	5.1	2.8	5.5	0.0	0.0	0.0	19.4
1987-88	0.0	0.0	0.0	T	0.1	13.0	1.2	7.6	0.8	0.0	0.0	0.0	22.7
1988-89	0.0	0.0	0.0	0.0	1.1	1.0	0.1	16.3	1.2	0.6	T	0.0	20.3
1989-90	0.0	0.0	0.0	T	1.7	6.6	11.3	7.4	0.1	T	0.0	0.0	27.1
1990-91	0.0	0.0	0.0	T	1.3	12.1	9.9	0.4	0.2	T	T	0.0	23.9
1991-92	0.0	0.0	0.0	T	14.7	1.2	1.9	4.5	0.2	0.1	0.0	0.0	22.6
1992-93	T	0.0	T	T	10.0	3.7	11.1	10.1	5.1	1.5	T	T	41.5
1993-94	0.0	0.0	0.0	0.2	0.9	2.0	8.8	15.2	2.0	0.5	0.7	0.0	28.3
1994-95	0.0	0.0	0.0	0.0	0.2	12.4	8.7	2.0	5.3	0.4	T	0.0	29.0
1995-96	0.0	0.0	0.0	1.8	6.5		22.3		1.6				
1996-97													
1997-98													
1998-99													
1999-00													
2000-01													
2001-													
POR= 56 YRS	T	0.0	T	0.3	3.1	6.7	8.3	7.1	5.9	1.8	0.0	T	33.2

2001
TOPEKA,
KANSAS (TOP)

Topeka, is located near the geographical center of the United States, and the middle of the temperate zone. The city straddles the Kansas River about 60 miles above its junction with the Missouri River. The Kansas River flows in an easterly direction through northeastern Kansas. Near Topeka, the river valley ranges from 2 to 4 miles wide, and is bordered on both sides by rolling prairie uplands of some 200 to 300 feet. The city is built on both banks of the Kansas River and along two tributaries, Soldier Creek in north Topeka and Shunganunga Creek in the south and east part of town. Flooding is always a threat following periods of heavy rains but protective construction has reduced the problem.

Seventy percent of the annual precipitation normally falls during the six crop-growing months, April through September. The rains of this period are usually of short duration, predominantly of the thunderstorm type. They occur more frequently during the nighttime and early morning hours than at other times of the day. Excessive precipitation rates may occur with warm-season thunderstorms. Rainfall accumulations over 8 inches in 24 hours have occurred in Topeka. Tornadoes have occurred in the area on several occasions and caused severe damage and numerous injuries.

Individual summers show wide departures from average conditions. Hottest summers may produce temperatures of 100 degrees or higher on more than 50 days. On the other hand, 25 percent of the summers pass with two or fewer 100 degree days. Similarly, precipitation has shown a wide range for June, July, and August, varying from under 3 inches to more than 27 inches during the 3 months. Summers are hot with low relative humidity and persistent southerly winds. Oppressively warm periods with high relative humidity are usually of short duration.

Winter temperatures average about 45 degrees cooler than summer. Cold spells are seldom prolonged. Only on rare occasions do daytime temperatures fail to rise above freezing. Winter precipitation is often in the form of snow, sleet, or glaze, but storms of such severity to prevent normal movement of traffic or to interfere with scheduled activity are not common.

In the transitional spring and fall seasons, the numerous days of fair weather are interspersed with short intervals of stormy weather. Strong, blustery winds are quite common in late winter and spring. Autumn is characteristically a season of warm days, cool nights, and infrequent precipitation, with cold air invasions gradually increasing in intensity as the season progresses.

Nearly all crops of the temperate zone can be produced in the vicinity of Topeka. Wheat and other small grains, clover, soybeans, fruit, and berries do well, and the area supports an extensive dairy industry.

Based on the 1951-1980 period, the average first occurrence of 32 degrees Fahrenheit in the fall is October 14 and the average last occurrence in the spring is April 21.

NORMALS, MEANS, AND EXTREMES
TOPEKA, KS (TOP)

LATITUDE: 39 04' 21" N   LONGITUDE: 95 37' 33" W   ELEVATION (FT): GRND: 880 BARO: 883   TIME ZONE: CENTRAL (UTC + 6)   WBAN: 13996

	ELEMENT	POR	JAN	FEB	MAR	APR	MAY	JUN	JUL	AUG	SEP	OCT	NOV	DEC	YEAR
TEMPERATURE F	NORMAL DAILY MAXIMUM	30	37.0	42.6	55.0	66.9	75.8	84.2	89.3	87.5	79.7	69.0	54.0	40.5	65.1
	MEAN DAILY MAXIMUM	54	37.5	43.4	54.0	66.5	76.0	84.5	89.3	88.3	80.6	69.7	53.8	41.9	65.5
	HIGHEST DAILY MAXIMUM	54	73	84	89	95	97	107	110	110	109	96	85	73	110
	YEAR OF OCCURRENCE		1967	1972	1986	1987	1998	1953	1980	1984	2000	1963	1980	2001	AUG 1984
	MEAN OF EXTREME MAXS.	53	61.4	66.9	78.1	86.4	89.8	95.4	99.3	99.5	94.8	87.1	74.1	64.0	83.1
	NORMAL DAILY MINIMUM	30	16.3	21.7	32.1	42.8	53.3	62.9	67.6	64.9	55.7	43.6	32.0	21.0	42.8
	MEAN DAILY MINIMUM	54	17.3	22.6	31.2	42.7	53.6	63.1	67.6	65.5	55.9	44.7	32.0	22.1	43.2
	LOWEST DAILY MINIMUM	54	-20	-23	-7	10	26	43	43	41	29	19	2	-26	-26
	YEAR OF OCCURRENCE		1974	1979	1978	1975	1963	1993	1972	1988	1984	1993	1976	1989	DEC 1989
	MEAN OF EXTREME MINS.	53	-3.5	1.8	12.6	26.4	37.3	49.1	55.3	52.5	38.4	27.0	15.4	2.0	26.2
	NORMAL DRY BULB	30	26.7	32.2	43.6	54.9	64.6	73.6	78.5	76.2	67.8	56.3	43.0	30.8	54.0
	MEAN DRY BULB	54	27.4	33.1	42.6	54.6	64.8	73.9	78.5	77.0	68.3	57.2	43.0	32.0	54.4
	MEAN WET BULB	56	24.8	29.8	37.6	48.1	58.1	66.9	70.6	68.8	60.9	50.5	38.2	28.9	48.6
	MEAN DEW POINT	56	19.0	23.6	30.9	41.7	53.6	63.1	67.0	64.9	56.5	44.8	32.7	23.5	43.4
	NORMAL NO. DAYS WITH:														
	MAXIMUM 90	30	0.0	0.0	0.0	0.4	1.1	6.7	15.5	12.8	4.7	0.6	0.0	0.0	41.8
	MAXIMUM 32	30	10.9	6.8	1.4	*	0.0	0.0	0.0	0.0	0.0	0.0	1.1	7.7	27.9
	MINIMUM 32	30	29.0	23.1	16.7	4.3	0.3	0.0	0.0	0.0	0.2	4.0	16.1	27.1	120.8
	MINIMUM 0	30	4.1	2.0	0.1	0.0	0.0	0.0	0.0	0.0	0.0	0.0	0.0	1.9	8.1
H/C	NORMAL HEATING DEG. DAYS	30	1187	918	663	312	121	5	0	0	56	283	660	1060	5265
	NORMAL COOLING DEG. DAYS	30	0	0	0	9	109	263	419	351	140	13	0	0	1304
RH	NORMAL (PERCENT)	30	70	69	66	64	68	71	70	71	72	68	71	72	69
	HOUR 00 LST	30	75	75	73	72	78	81	78	80	82	77	78	77	77
	HOUR 06 LST	30	77	78	79	80	84	87	86	87	88	83	81	80	82
	HOUR 12 LST	30	63	62	57	54	57	60	59	58	58	54	60	65	59
	HOUR 18 LST	30	64	60	53	51	55	58	56	57	60	57	64	68	59
S	PERCENT POSSIBLE SUNSHINE	50	57	55	57	58	61	67	71	71	66	64	56	51	61
W/O	MEAN NO. DAYS WITH:														
	HEAVY FOG (VISBY 1/4 MI)	54	2.2	1.9	1.0	0.9	0.9	0.6	0.5	1.2	1.1	1.7	1.3	2.2	15.5
	THUNDERSTORMS	54	0.4	0.7	2.5	5.4	8.8	10.0	8.6	7.9	5.8	3.3	1.2	0.4	55.0
CLOUDINESS	MEAN:														
	SUNRISE-SUNSET (OKTAS)	1	7.2	5.6	4.0	6.4	4.8	4.0	5.2	3.2	2.8	4.4	4.4	3.6	4.6
	MIDNIGHT-MIDNIGHT (OKTAS)	1	7.2	5.6	4.0	6.4	4.8	4.0	5.2	3.2	2.8	4.8	4.4	2.4	4.6
	MEAN NO. DAYS WITH:														
	CLEAR	3	4.3	8.7	7.0	8.0	11.0	11.0	6.5	12.0	7.0	10.0	6.5	10.5	102.5
	PARTLY CLOUDY	3	4.0	3.3	5.0	4.0	4.3	6.0	9.0	7.0	1.0	2.0	2.5	3.0	51.1
	CLOUDY	3	9.7	7.3	6.0	9.5	6.3	5.3	5.5	4.5	4.5	7.5	8.5	8.0	82.6
PR	MEAN STATION PRESSURE (IN)	28	29.19	29.14	29.03	29.00	28.99	29.00	29.04	29.06	29.09	29.11	29.10	29.18	29.08
	MEAN SEA-LEVEL PRES. (IN)	56	30.16	30.12	30.00	29.95	29.93	29.91	29.96	29.98	30.02	30.05	30.07	30.13	30.02
WINDS	MEAN SPEED (MPH)	46	9.7	10.1	11.5	11.8	10.3	9.6	8.5	8.2	8.4	9.1	9.7	9.5	9.7
	PREVAIL. DIR (TENS OF DEGS)	29	36	36	36	18	18	18	18	18	18	18	18	30	18
	MAXIMUM 2-MINUTE:														
	SPEED (MPH)	8	39	36	37	43	47	48	44	38	43	37	45	37	48
	DIR. (TENS OF DEGS)		31	17	21	21	34	34	34	27	32	14	31	33	34
	YEAR OF OCCURRENCE		1996	1996	2000	2001	1999	1996	1996	1996	2000	2001	1997	2000	JUN 1996
	MAXIMUM 5-SECOND:														
	SPEED (MPH)	8	49	47	48	53	59	66	62	65	52	52	54	47	66
	DIR. (TENS OF DEGS)		31	16	23	21	34	35	33	26	32	23	30	31	35
	YEAR OF OCCURRENCE		1996	1994	2000	2001	1999	1994	1994	1994	2000	1996	1997	1997	JUN 1994
PRECIPITATION	NORMAL (IN)	30	0.95	1.04	2.46	3.08	4.45	5.54	3.59	3.89	3.81	3.06	1.93	1.43	35.23
	MAXIMUM MONTHLY (IN)	54	5.24	3.49	8.44	8.69	9.39	15.20	12.02	11.18	12.71	7.24	6.27	4.30	15.20
	YEAR OF OCCURRENCE		1949	1971	1973	1999	1982	1967	1950	1977	1973	1980	1964	1973	JUN 1967
	MINIMUM MONTHLY (IN)	54	T	0.02	0.10	0.62	0.41	0.56	0.59	0.26	0.66	0.04	T	0.04	15.20
	YEAR OF OCCURRENCE		1986	1991	1966	1989	1966	1980	1983	1971	1952	1952	1989	1996	NOV 1989
	MAXIMUM IN 24 HOURS (IN)	54	1.55	2.33	3.76	3.59	3.62	5.52	4.19	4.48	4.80	4.10	4.66	2.65	5.52
	YEAR OF OCCURRENCE		1988	1971	1987	1967	1978	1967	1951	1962	1989	1985	1964	1980	JUN 1967
	NORMAL NO. DAYS WITH:														
	PRECIPITATION 0.01	30	5.9	5.9	8.7	9.8	11.0	10.3	7.9	8.3	8.3	6.8	6.5	6.6	96.0
	PRECIPITATION 1.00	30	0.1	0.1	0.5	0.7	1.3	1.8	1.2	1.3	1.2	1.1	0.3	0.2	9.8
SNOWFALL	NORMAL (IN)	30	5.8	4.7	2.8	0.6	0.0	0.0	0.0	0.0	0.0	0.*	1.2	5.6	20.7
	MAXIMUM MONTHLY (IN)	54	23.0	22.4	22.1	6.8	T	T	T	T	0.0	8.0	9.4	18.8	23.0
	YEAR OF OCCURRENCE		1993	1971	1960	1970	1991	1992	1992	1994		1996	1972	1983	JAN 1993
	MAXIMUM IN 24 HOURS (IN)	54	15.2	15.2	8.4	7.6	T	T	T	T	0.0	8.0	7.4	9.0	15.2
	YEAR OF OCCURRENCE		1993	1971	1960	1970	1991	1992	1992	1994	1994	1996	1975	1973	JAN 1993
	MAXIMUM SNOW DEPTH (IN)	56	12	12	18	4	0	T	0	T	0	6	8	9	18
	YEAR OF OCCURRENCE		1979	1971	1960	1970		1993		1949		1996	1975	1983	MAR 1960
	NORMAL NO. DAYS WITH:														
	SNOWFALL 1.0	30	1.9	1.5	1.0	0.2	0.0	0.0	0.0	0.0	0.0	0.0	0.4	1.8	6.8

PRECIPITATION (inches) 2001 TOPEKA, KS (TOP)

YEAR	JAN	FEB	MAR	APR	MAY	JUN	JUL	AUG	SEP	OCT	NOV	DEC	ANNUAL
1972	0.47	0.56	1.37	3.93	2.90	1.14	4.81	3.26	4.89	2.11	3.99	1.78	31.21
1973	2.67	1.71	8.44	4.03	4.37	2.96	10.16	2.83	12.71	4.57	2.14	4.30	60.89
1974	0.99	1.20	1.22	2.78	3.59	3.72	2.90	4.89	1.40	5.16	2.19	1.18	31.22
1975	1.50	1.67	1.66	3.26	3.88	4.85	0.68	1.69	4.35	0.05	4.44	1.12	29.15
1976	0.41	0.51	1.38	4.85	4.63	1.69	2.04	1.12	0.56	0.04	0.21		20.75
1977	0.90	0.22	2.06	2.46	7.83	10.91	1.37	11.18	3.22	4.92	3.38	0.26	48.71
1978	0.19	0.84	1.63	2.35	5.75	4.57	2.26	2.89	6.65	0.36	3.22	0.55	31.26
1979	1.81	0.63	3.95	2.37	2.25	5.63	5.84	4.05	2.17	4.15	1.80	0.05	34.70
1980	1.34	0.91	4.15	1.03	4.85	0.56	0.87	5.86	1.19	7.24	0.25	3.86	32.11
1981	0.32	0.21	1.61	1.98	5.93	9.40	7.63	3.92	2.03	3.72	3.63	0.22	40.60
1982	1.67	0.59	1.14	1.58	9.39	5.99	5.08	4.53	1.17	1.25	2.26	3.61	38.26
1983	0.69	0.63	4.39	6.29	4.93	6.08	0.59	0.62	2.25	5.19	3.61	1.34	36.61
1984	0.11	1.35	4.57	4.26	3.45	10.17	1.66	1.04	4.24	4.10	0.72	2.36	38.03
1985	0.70	2.02	2.38	3.60	3.79	5.15	2.90	7.97	8.16	5.20	2.02	0.71	44.60
1986	T	1.55	1.35	3.15	7.53	2.51	4.21	5.50	6.21	3.30	0.87	1.20	37.38
1987	1.09	2.71	5.92	2.33	3.89	4.86	2.78	5.90	1.81	1.86	1.94	1.87	36.96
1988	2.04	0.48	0.73	2.93	3.08	3.13	1.74	1.34	1.94	0.26	0.86	0.86	19.39
1989	1.24	0.86	3.11	0.62	4.05	4.76	5.21	6.22	8.65	3.44	T	0.61	38.77
1990	1.22	2.31	3.75	1.01	4.45	5.57	3.01	5.69	0.83	2.71	2.91	0.97	34.43
1991	0.76	0.02	2.98	3.63	7.09	1.49	1.47	1.76	2.15	3.20	2.20	2.44	29.19
1992	0.89	1.18	5.29	3.25	1.75	3.35	6.37	1.24	3.92	1.41	5.27	2.01	35.93
1993	1.11	1.61	2.56	5.43	6.95	2.18	10.98	5.32	7.03	1.37	1.12	0.90	46.56
1994	0.42	0.82	0.19	4.31	0.95	4.63	3.16	7.87	1.46	1.30	2.87	1.52	29.50
1995	1.50	0.71	2.11	3.32	11.82	3.43	5.10	4.29	2.90	0.21	0.66	0.57	36.62
1996	0.76	0.19	1.48	1.57	7.72	7.97	2.65	6.09	3.60	2.79	2.66	0.04	37.52
1997	0.24	2.67	0.26	4.99	3.54	1.36	2.59	4.65	2.15	3.58	2.14	2.41	30.58
1998	0.79	0.77	2.88	2.16	2.08	7.22	9.32	0.88	4.19	5.01	5.64	1.22	42.16
1999	1.17	0.94	0.99	8.69	6.38	6.20	0.59	1.09	4.43	0.87	1.60	1.76	34.71
2000	0.19	2.00	2.62	1.07	2.08	7.25	2.77	0.61	2.97	3.52	1.91	0.35	27.34
2001	1.22	2.90	3.56	4.27	3.85	6.39	2.31	5.95	7.46	3.51	1.13	0.13	42.68
POR= 55 YRS	0.95	1.15	2.40	3.16	4.39	5.03	4.24	3.94	3.44	2.80	1.87	1.34	34.71

AVERAGE TEMPERATURE (F) 2001 TOPEKA, KS (TOP)

YEAR	JAN	FEB	MAR	APR	MAY	JUN	JUL	AUG	SEP	OCT	NOV	DEC	ANNUAL
1972	25.6	31.8	46.7	54.7	63.6	74.3	74.4	74.7	68.5	54.6	39.4	27.7	53.0
1973	27.5	33.9	47.5	52.6	61.3	74.9	77.4	77.1	66.6	60.4	44.8	29.9	54.5
1974	22.3	35.9	46.9	56.8	67.2	70.0	80.6	74.1	61.9	58.4	43.1	32.7	54.2
1975	30.7	28.8	37.8	54.4	67.3	74.2	77.3	79.3	64.0	59.3	45.5	34.5	54.4
1976	27.8	42.7	45.4	57.0	60.4	72.7	78.0	76.9	69.0	50.3	35.4	28.5	53.7
1977	15.2	37.4	49.6	60.2	70.1	75.2	79.4	76.4	71.6	56.7	42.7	30.1	55.4
1978	17.3	20.4	38.4	55.9	63.0	74.6	77.3	75.7	72.9	54.6	43.0	30.0	51.9
1979	11.8	19.2	42.6	51.6	63.1	72.4	77.8	76.9	68.0	57.0	40.0	35.5	51.3
1980	28.6	26.0	40.8	53.7	62.8	76.5	86.4	80.7	70.0	53.9	45.0	32.6	54.8
1981	31.4	35.5	46.1	60.6	60.9	75.5	79.5	73.1	68.0	56.1	47.2	30.1	55.3
1982	21.9	28.5	43.2	50.2	63.7	69.0	78.7	75.5	66.5	55.9	42.0	35.8	52.6
1983	32.5	36.1	44.9	49.4	62.5	73.5	81.1	83.0	72.2	58.7	45.8	14.4	54.5
1984	26.0	40.2	38.1	51.7	62.4	73.9	77.0	78.0	66.5	56.6	45.5	36.8	54.4
1985	19.9	25.6	48.6	58.7	66.5	72.0	79.7	72.8	66.8	56.6	36.7	25.1	52.4
1986	35.8	32.5	49.8	57.7	65.9	77.0	80.4	72.3	71.6	56.6	38.3	34.6	56.0
1987	29.7	40.3	46.7	57.1	70.4	76.2	78.1	75.5	68.2	52.6	47.4	35.9	56.5
1988	28.1	30.8	43.4	53.9	68.8	75.1	76.7	79.5	70.3	52.8	45.2	35.3	55.0
1989	38.0	22.9	44.4	57.9	64.2	71.4	77.6	74.8	62.3	57.1	42.3	21.0	52.8
1990	37.3	36.2	45.5	51.9	60.3	77.2	77.7	76.5	71.6	57.0	49.1	29.6	52.8
1991	25.2		48.2	57.7	69.4	77.1	80.2	77.3	69.3	58.6	37.9	37.4	55.8
1992	37.2	41.5	47.8	54.7	62.5	69.1	75.9	71.7	67.9	56.5	39.1	32.6	54.7
1993	26.4	29.9	40.8	50.2	63.1	72.9	78.2	77.8	63.4	54.0	39.3	34.8	52.6
1994	26.1	29.9	47.0	54.1	64.5	76.4	76.1	75.9	58.3	45.8	36.0		54.8
1995	29.3	37.0	45.1	52.0	59.3	72.5	80.2	80.9	65.8	57.2	40.1	30.6	54.2
1996	24.5	35.0	38.5	54.0	65.7	75.3	76.4	74.6	64.6	56.2	37.6	30.1	52.7
1997	26.2	35.0	46.2	49.9	60.5	73.5	78.5	75.5	69.7	58.1	40.3	32.8	53.9
1998	32.4	39.8	38.3	53.0	70.3	74.0	79.2	78.9	74.0	58.1	48.7	36.0	57.0
1999	28.6	42.9	43.3	55.1	64.8	73.4	82.4	77.6	65.6	56.9	51.3	36.2	56.5
2000	32.3	41.9	47.5	55.2	68.1	72.1	79.2	85.4	72.3	60.5	37.3	20.7	56.0
2001	30.3	30.5	40.8	60.6	67.6	73.4	82.9	79.0	66.5	56.7	51.1	37.8	56.4
POR= 55 YRS	27.5	33.2	42.7	54.5	64.8	74.0	78.5	77.1	68.4	57.2	43.0	32.1	54.4

REFERENCE NOTES:

PAGE 1:
THE TEMPERATURE GRAPH SHOWS NORMAL MAXIMUM AND NORMAL MINIMUM DAILY TEMPERATURES (SOLID CURVES) AND THE ACTUAL DAILY HIGH AND LOW TEMPERATURES (VERTICAL BARS).

PAGE 2 AND 3:
H/C INDICATES HEATING AND COOLING DEGREE DAYS.
RH INDICATES RELATIVE HUMIDITY
W/O INDICATES WEATHER AND OBSTRUCTIONS
S INDICATES SUNSHINE.
PR INDICATES PRESSURE.
CLOUDINESS ON PAGE 3 IS THE SUM OF THE CEILOMETER AND SATELLITE DATA NOT TO EXCEED EIGHT EIGHTHS(OKTAS).

GENERAL:
T INDICATES TRACE PRECIPITATION, AN AMOUNT GREATER THAN ZERO BUT LESS THAN THE LOWEST REPORTABLE VALUE.
+ INDICATES THE VALUE ALSO OCCURS ON EARLIER DATES.
BLANK ENTRIES DENOTE MISSING OR UNREPORTED DATA.
NORMALS ARE 30-YEAR AVERAGES (1961 - 1990).
ASOS INDICATES AUTOMATED SURFACE OBSERVING SYSTEM.
PM INDICATES THE LAST DAY OF THE PREVIOUS MONTH.
POR (PERIOD OF RECORD) BEGINS WITH THE JANUARY DATA MONTH AND IS THE NUMBER OF YEARS USED TO COMPUTE THE MEAN. INDIVIDUAL MONTHS WITHIN THE POR MAY BE MISSING.
WHEN THE POR FOR A NORMAL IS LESS THAN 30 YEARS, THE NORMAL IS PROVISIONAL AND IS BASED ON THE NUMBER OF YEARS INDICATED.
0. OR * INDICATES THE VALUE OR MEAN-DAYS-WITH IS BETWEEN 0.00 AND 0.05.
CLOUDINESS FOR ASOS STATIONS DIFFERS FROM THE NON-ASOS OBSERVATION TAKEN BY A HUMAN OBSERVER. ASOS STATION CLOUDINESS IS BASED ON TIME-AVERAGED CEILOMETER DATA FOR CLOUDS AT OR BELOW 12,000 FEET AND ON SATELLITE DATA FOR CLOUDS ABOVE 12,000 FEET.
THE NUMBER OF DAYS WITH CLEAR, PARTLY CLOUDY, AND CLOUDY CONDITIONS FOR ASOS STATIONS IS THE SUM OF THE CEILOMETER AND SATELLITE DATA FOR THE SUNRISE TO SUNSET PERIOD.

GENERAL CONTINUED:
CLEAR INDICATES 0 - 2 OKTAS, PARTLY CLOUDY INDICATES 3 - 6 OKTAS, AND CLOUDY INDICATES 7 OR 8 OKTAS. WHEN AT LEAST ONE OF THE ELEMENTS (CEILOMETER OR SATELLITE) IS MISSING, THE DAILY CLOUDINESS IS NOT COMPUTED.
WIND DIRECTION IS RECORDED IN TENS OF DEGREES (2 DIGITS) CLOCKWISE FROM TRUE NORTH. "00" INDICATES CALM. "36" INDICATES TRUE NORTH.
RESULTANT WIND IS THE VECTOR AVERAGE OF THE SPEED AND DIRECTION.
AVERAGE TEMPERATURE IS THE SUM OF THE MEAN DAILY MAXIMUM AND MINIMUM TEMPERATURE DIVIDED BY 2.
SNOWFALL DATA COMPRISE ALL FORMS OF FROZEN PRECIPITATION, INCLUDING HAIL.
A HEATING (COOLING) DEGREE DAY IS THE DIFFERENCE BETWEEN THE AVERAGE DAILY TEMPERATURE AND 65 F.
DRY BULB IS THE TEMPERATURE OF THE AMBIENT AIR.
DEW POINT IS THE TEMPERATURE TO WHICH THE AIR MUST BE COOLED TO ACHIEVE 100 PERCENT RELATIVE HUMIDITY.
WET BULB IS THE TEMPERATURE THE AIR WOULD HAVE IF THE MOISTURE CONTENT WAS INCREASED TO 100 PERCENT RELATIVE HUMIDITY.

ON JULY 1, 1996, THE NATIONAL WEATHER SERVICE BEGAN USING THE "METAR" OBSERVATION CODE THAT WAS ALREADY EMPLOYED BY MOST OTHER NATIONS OF THE WORLD. THE MOST NOTICEABLE DIFFERENCE IN THIS ANNUAL PUBLICATION WILL BE THE CHANGE IN UNITS FROM TENTHS TO EIGHTS(OKTAS) FOR REPORTING THE AMOUNT OF SKY COVER.

HEATING DEGREE DAYS (base 65 F) 2001 TOPEKA, KS (TOP)

YEAR	JUL	AUG	SEP	OCT	NOV	DEC	JAN	FEB	MAR	APR	MAY	JUN	TOTAL
1972-73	10	0	59	337	764	1152	1158	864	537	378	129	0	5388
1973-74	0	0	58	191	603	1082	1317	807	558	258	64	7	4945
1974-75	0	3	134	213	649	991	1056	1008	839	352	46	7	5298
1975-76	0	2	137	230	581	941	1148	639	599	269	178	5	4729
1976-77	0	0	45	471	881	1126	1537	767	469	180	11	0	5487
1977-78	0	0	6	263	662	1075	1473	1240	824	280	156	6	5985
1978-79	0	0	34	319	655	1078	1643	1277	693	401	129	9	6238
1979-80	0	4	45	267	741	908	1123	1123	744	344	129	3	5431
1980-81	0	0	65	344	591	1001	1035	822	579	175	176	0	4788
1981-82	0	2	46	283	529	1076	1329	1014	664	449	76	32	5500
1982-83	0	0	93	303	683	896	1002	804	615	466	120	13	4995
1983-84	0	0	56	223	570	1565	1204	713	830	405	137	0	5703
1984-85	0	0	145	276	578	871	1389	1098	501	228	35	8	5129
1985-86	0	0	127	259	844	1228	899	906	491	252	49	0	5055
1986-87	0	9	27	263	792	934	1084	688	560	292	16	0	4665
1987-88	0	3	24	376	531	893	1136	988	662	331	16	5	4965
1988-89	2	4	24	383	587	912	832	1174	641	296	125	5	4985
1989-90	0	2	155	276	672	1360	851	801	600	413	176	4	5310
1990-91	1	1	39	276	477	1093	1227		523	233	48	0	
1991-92	0	0	95	262	808	849	855	673	528	326	132	7	4535
1992-93	0	2	68	278	770	995	1189	979	744	440	101	22	5588
1993-94	0	1	108	356	763	930	1202	974	553	347	94	0	5328
1994-95	0	1	64	237	568	892	1097	774	613	382	193	1	4822
1995-96	0	0	107	246	740	1059	1250	867	814	347	93	6	5529
1996-97	0	0	98	294	813	1076	1193	833	577	450	173	0	5507
1997-98	1	0	21	286	737	994	1000	699	830	361	27	27	4983
1998-99	0	0	9	203	485	893	1122	613	667	295	58	12	4357
1999-00	0	0	92	261	408	886	1006	666	536	295	51	10	4211
2000-01	0	0	61	174	824	1368	1069	958	742	182	53	9	5440
2001-	0	0	58	262	415	837							

COOLING DEGREE DAYS (base 65 F) 2001 TOPEKA, KS (TOP)

YEAR	JAN	FEB	MAR	APR	MAY	JUN	JUL	AUG	SEP	OCT	NOV	DEC	ANNUAL
1972	0	0	5	22	74	297	308	309	169	20	0	0	1204
1973	0	0	13	21		304	394	384	115	52	0	0	1283
1974	0	0	6	21	140	165	490	292	47	12	0	0	1173
1975	0	0		38	129	289	390	448	116	61	3	0	1474
1976	0	0	1	34	40	242	410	376	171	20	0	0	1294
1977	0	0	0	40	176	311	453	360	209	14	0	0	1563
1978	0	0	6	15	101	298	390	339	277	5	3	0	1434
1979	0	0	4	7	76	237	401	379	144	27	0	0	1275
1980	0	0	0	9	69	356	670	496	220	9	0	0	1829
1981	0	0	0	53	58	321	457	260	143	17	0	0	1309
1982	0	0	0	11	43	157	432	334	147	28	0	0	1152
1983	0	0	0	7	50	274	509	564	278	33	2	0	1717
1984	0	0	0	14	67	274	379	407	196	20	0	3	1360
1985	0	0	0	46	88	225	461	249	188	6	0	0	1263
1986	0	0	26	42	85	363	488	243	233	9	0	0	1489
1987	0	0	0	61	192	344	410	335	126	0	9	0	1477
1988	0	0	0	4	140	314	458	375	191	11	0	0	1493
1989	0	0	11	90	107	206	399	311	81	41	0	0	1246
1990	0	0	1	26	39	377	403	366	241	37	7	0	1497
1991	0	0	11	22	192	371	478	387	229	69	0	0	1759
1992	0	0	0	25	61	134	344	217	162	20	0	0	963
1993	0	0	0	0	48	269	417	405	64	22	0	0	1225
1994	0	0		26	86	348	351	345	140	36	0	0	1332
1995	0	0	4	1	22	237	481	502	140	16	0	0	1403
1996	0	0	0	25	125	321	358	302	92	28	0	0	1251
1997	0	0	0	1	38	262	424	330	166	81	0	0	1302
1998	0	0	10	9	196	304	445	440	287	26	1	0	1718
1999	0	0	0	5	57	269	545	396	120	20	1	0	1413
2000	0	0	2	6	153	231	446	640	284	44	0	0	1806
2001	0	0	0	59	140	267	563	441	109	12	5	0	1596

SNOWFALL (inches) 2001 TOPEKA, KS (TOP)

YEAR	JUL	AUG	SEP	OCT	NOV	DEC	JAN	FEB	MAR	APR	MAY	JUN	TOTAL
1972-73	0.0	0.0	0.0	0.0	9.4	4.0	13.5	1.7	0.0	0.1	0.0	0.0	28.7
1973-74	0.0	0.0	0.0	0.0	T	15.2	7.8	2.1	1.5	1.3	0.0	0.0	27.9
1974-75	0.0	0.0	0.0	0.0	1.3	1.4	5.0	6.7	7.8	3.6	0.0	0.0	25.8
1975-76	0.0	0.0	0.0	0.0	8.3	2.7	6.4	0.7	2.9	0.0	0.0	0.0	21.0
1976-77	0.0	0.0	0.0	T	0.3	T	13.6	0.1	T	0.2	0.0	0.0	14.2
1977-78	0.0	0.0	0.0	0.0	0.1	0.2	3.9	12.4	6.2	0.0	0.0	0.0	22.8
1978-79	0.0	0.0	0.0	0.0	T	11.1	20.1	3.1	7.5	1.1	0.0	0.0	42.9
1979-80	0.0	0.0	0.0	0.0	T	T	3.5	11.4	3.4	0.0	0.0	0.0	18.3
1980-81	0.0	0.0	0.0	T	0.0	3.8	2.6	2.5	0.0	0.0	0.0	0.0	8.9
1981-82	0.0	0.0	0.0	0.0	T	1.4	3.2	8.0	0.3	0.5	0.0	0.0	13.4
1982-83	0.0	0.0	0.0	0.0	1.1	5.0	6.1	10.1	0.6	4.5	0.0	0.0	27.4
1983-84	0.0	0.0	0.0	0.0	4.1	18.8	2.6	T	4.2	0.0	0.0	0.0	29.7
1984-85	0.0	0.0	0.0	0.0	T	9.8	18.2	7.9	0.5	0.0	0.0	0.0	36.4
1985-86	0.0	0.0	0.0	0.0	3.3	5.8	T	1.5	T	0.0	0.0	0.0	10.6
1986-87	0.0	0.0	0.0	T	0.7	1.7	15.1	2.3	0.5	0.0	0.0	0.0	20.3
1987-88	0.0	0.0	0.0	0.0	0.9	9.6	0.6	6.0	4.7	0.0	0.0	0.0	21.8
1988-89	0.0	0.0	0.0	0.0	0.7	T	0.8	T	9.0	1.6	0.0	T	12.1
1989-90	0.0	0.0	0.0	0.0	T	9.5	1.0	0.1	7.6	0.0	0.0	0.0	18.2
1990-91	0.0	0.0	0.0	0.0	0.0	2.9	9.6	T	T	T	0.0		
1991-92	0.0	0.0	0.0	T	6.2	0.1	T	T	0.9	T	0.0	T	7.2
1992-93	T	0.0	0.0	T	4.5	0.9	23.0	14.2	0.6	T		T	
1993-94	0.0	0.0	0.0	0.0	T	3.3	2.0	6.0	1.9	1.4	0.0	0.0	14.6
1994-95	0.0	T	0.0	0.0	T	0.8	2.0	0.1	5.0				14.5
1995-96	0.0	0.0	0.0	0.0	0.7	5.5	8.3	T	T	T	0.0	T	14.5
1996-97	0.0	0.0	0.0	8.0	1.1	0.8	3.9	5.8	T	1.6	T	0.0	21.2
1997-98	0.0	0.0	0.0	T	T	8.2	2.0	T	4.4	T	0.0	T	14.6
1998-99	0.0	0.0	0.0	T	0.0	0.5	4.0	2.0	1.8	T	T	0.0	8.3
1999-00	0.0	0.0	0.0	0.0	0.0	7.6	3.7	0.2	T	0.0	0.0	0.0	11.5
2000-01	0.0	0.0	0.0	0.0	T	8.3	1.2	8.3	1.3	T	T	0.0	19.1
2001-	0.0	0.0	0.0	0.0	0.0	T							
POR= 55 YRS	T	T	0.0	0.2	1.2	4.7	5.7	0.3	0.2	T	T	T	12.3

2001
WICHITA,
KANSAS (ICT)

Wichita is in the Central Great Plains where masses of warm, moist air from the Gulf of Mexico collide with cold, dry air from the Arctic region to create a wide range of weather the year around. Summers are usually warm and humid, and can be very hot and dry. The winters are usually mild, with brief periods of very cold weather.

The elevation is just over 1,300 feet above sea level. The terrain is basically flat with natural tree areas mainly along the Arkansas River and its tributaries.

The temperature extremes for the period of weather records at Wichita range from more than 110 degrees to less than -20 degrees. Temperatures above 90 degrees occur an average of 63 days per year, while very cold temperatures below zero occur about 2 days per year.

Precipitation averages about 30 inches per year, with 70 percent of that falling from April through September during the growing season. The wettest years have recorded over 50 inches. The driest years less than 15 inches.

Thunderstorms occur mainly during the spring and early summer. They can be severe and cause damage from heavy rain, large hail, strong winds and tornadoes.

The city of Wichita is protected against floods from the Arkansas River and its local tributaries by the Wichita-Vally Center Flood Control Project, which is designed to protect against floods up to the 75 to 100 year frequency class.

Snowfall normally is 15 inches per year, falling from December through March. Monthly snowfalls in excess of 20 inches and 24-hour snowfalls of more than 13 inches have occurred.

The prevailing wind direction is south with the windiest months March and April. July has the least wind. Strong north winds often occur with the passage of cold fronts from late fall through early spring. Extremely low wind chill factors are experienced with very cold outbreaks during the mid winter. On rare occasions during the summer, strong, hot, dry southwest winds can do considerable damage to crops.

NORMALS, MEANS, AND EXTREMES
WICHITA, KS (ICT)

LATITUDE: 37 38' 50" N　　LONGITUDE: 97 25' 46" W　　ELEVATION (FT): GRND: 1338　BARO: 1341　　TIME ZONE: CENTRAL (UTC + 6)　　WBAN: 03928

ELEMENT	POR	JAN	FEB	MAR	APR	MAY	JUN	JUL	AUG	SEP	OCT	NOV	DEC	YEAR
TEMPERATURE F														
NORMAL DAILY MAXIMUM	30	39.8	45.9	57.2	68.3	76.9	86.8	92.8	90.7	81.4	70.6	55.3	43.0	67.4
MEAN DAILY MAXIMUM	48	39.6	45.7	55.5	66.7	75.4	85.1	91.1	89.5	80.3	69.2	54.2	44.8	66.4
HIGHEST DAILY MAXIMUM	49	75	87	89	96	100	110	113	110	108	95	85	83	113
YEAR OF OCCURRENCE		1967	1996	1989	1972	1996	1980	1954	1984	2000	1979	1980	1955	JUL 1954
MEAN OF EXTREME MAXS.	49	61.2	67.9	77.8	83.9	89.5	97.5	101.7	101.1	95.5	86.3	72.1	63.3	83.1
NORMAL DAILY MINIMUM	30	19.2	23.7	33.6	44.5	54.3	64.6	69.9	67.9	59.2	46.6	33.9	23.0	45.0
MEAN DAILY MINIMUM	48	19.7	24.3	32.6	43.6	53.6	63.2	68.4	66.7	58.2	46.2	33.3	24.2	44.5
LOWEST DAILY MINIMUM	49	-12	-21	-2	15	31	43	51	48	31	18	1	-16	-21
YEAR OF OCCURRENCE		1962	1982	1960	1975	1976	1969	1975	1967	1984	1993	1975	1989	FEB 1982
MEAN OF EXTREME MINS.	49	1.5	5.8	14.9	27.4	38.6	50.6	58.3	55.9	42.0	29.9	17.6	6.6	29.1
NORMAL DRY BULB	30	29.5	34.8	45.4	56.4	65.6	75.7	81.4	79.3	70.3	58.6	44.7	33.0	56.2
MEAN DRY BULB	48	29.6	35.1	44.0	55.1	64.6	74.2	79.8	78.3	69.4	57.6	43.6	34.0	55.4
MEAN WET BULB	46	29.2	34.4	42.7	53.7	64.0	72.5	76.0	74.6	67.1	56.2	43.1	33.1	53.9
MEAN DEW POINT	46	22.9	27.4	34.8	46.5	58.8	67.3	70.2	68.6	61.6	49.5	36.8	27.1	47.6
NORMAL NO. DAYS WITH:														
MAXIMUM 90	30	0.0	0.0	0.0	0.3	2.0	12.1	21.9	18.9	7.0	0.8	0.0	0.0	63.0
MAXIMUM 32	30	9.4	5.5	1.0	*	0.0	0.0	0.0	0.0	0.0	0.0	0.7	6.0	22.6
MINIMUM 32	30	28.4	22.5	14.6	2.8	0.1	0.0	0.0	0.0	*	1.3	14.3	26.8	110.8
MINIMUM 0	30	2.2	1.0	0.1	0.0	0.0	0.0	0.0	0.0	0.0	0.0	0.0	0.9	4.2
H/C														
NORMAL HEATING DEG. DAYS	30	1101	846	608	278	102	5	0	0	29	221	609	992	4791
NORMAL COOLING DEG. DAYS	30	0	0	0	20	121	326	508	443	188	22	0	0	1628
RH														
NORMAL (PERCENT)	30	70	68	64	63	67	64	59	61	67	65	70	72	66
HOUR 00 LST	30	75	74	71	71	76	74	67	69	75	73	76	76	73
HOUR 06 LST	30	78	78	77	78	82	82	78	79	83	80	80	80	80
HOUR 12 LST	30	62	60	55	52	56	53	48	50	56	53	59	63	56
HOUR 18 LST	30	64	59	52	50	54	49	44	46	53	54	64	67	55
S PERCENT POSSIBLE SUNSHINE	46	59	60	63	65	64	69	76	77	70	65	58	57	65
W/O MEAN NO. DAYS WITH:														
HEAVY FOG(VISBY 1/4 MI)	48	3.0	2.7	1.5	0.9	0.7	0.3	0.2	0.2	1.0	1.3	2.0	3.3	17.1
THUNDERSTORMS	48	0.3	0.9	2.9	5.2	8.7	9.8	8.2	7.2	5.6	3.1	1.2	0.4	53.5
CLOUDINESS MEAN:														
SUNRISE-SUNSET (OKTAS)	1	6.4	6.0	5.6	6.0	4.4	3.2	2.4	2.1	3.2	3.6	4.0	5.6	4.4
MIDNIGHT-MIDNIGHT (OKTAS)	1	6.4	6.0	6.4	5.6	4.4	2.8	2.4	2.4	3.2	4.8	4.0	4.0	4.4
MEAN NO. DAYS WITH:														
CLEAR	3	6.3	6.7	7.0	6.5	11.0	10.5	15.0	16.5	6.0	10.5	7.0	11.0	114.0
PARTLY CLOUDY	2	5.0	6.5	3.5	2.5	8.0	7.0	4.5	5.0	3.0	4.0	2.5	3.5	55.0
CLOUDY	3	8.3	8.0	8.0	9.0	8.0	5.5	3.0	3.0	2.0	2.5	5.0	8.0	69.8
PR MEAN STATION PRESSURE(IN)	29	28.69	28.65	28.53	28.53	28.51	28.52	28.57	28.59	28.61	28.63	28.62	28.68	28.59
MEAN SEA-LEVEL PRES. (IN)	47	32.08	32.02	31.90	31.84	31.81	31.82	31.84	31.87	31.91	32.00	32.03	32.05	31.93
WINDS MEAN SPEED (MPH)	40	11.9	12.5	13.6	13.9	12.2	11.9	11.3	11.0	11.5	11.9	12.3	11.8	12.2
PREVAIL.DIR(TENS OF DEGS)	29	36	36	18	18	18	18	18	18	18	18	18	36	18
MAXIMUM 2-MINUTE:														
SPEED (MPH)	9	48	44	49	56	61	48	70	52	40	49	48	44	70
DIR. (TENS OF DEGS)		35	02	24	23	18	17	34	04	18	31	33	32	34
YEAR OF OCCURRENCE		1996	1997	2000	2001	1998	1998	1996	1998	2001	1993	1997	1997	JUL 1996
MAXIMUM 5-SECOND:														
SPEED (MPH)	9	59	55	60	66	70	59	101	72	53	53	57	53	101
DIR. (TENS OF DEGS)		35	33	24	23	18	36	04	04	19	31	35	35	04
YEAR OF OCCURRENCE		1996	1999	2000	2001	1998	1996	1993	1998	2001	1993	1997	2000	JUL 1993
PRECIPITATION NORMAL (IN)	30	0.79	0.96	2.43	2.38	3.81	4.31	3.13	3.02	3.49	2.22	1.59	1.20	29.33
MAXIMUM MONTHLY (IN)	48	2.73	3.89	9.17	6.02	9.62	10.46	9.22	7.91	10.69	9.42	5.88	4.71	10.69
YEAR OF OCCURRENCE		1973	2001	1973	1999	1993	1957	1962	1960	1999	1998	1964	1984	SEP 1999
MINIMUM MONTHLY (IN)	48	T	T	0.01	0.22	0.52	0.40	0.05	0.14	0.03	T	T	0.03	T
YEAR OF OCCURRENCE		1986	1991	1971	1971	1963	1973	1998	1975	2000	1956	1958	1989	FEB 1991
MAXIMUM IN 24 HOURS (IN)	48	1.72	2.05	2.65	2.51	4.70	4.98	3.86	4.50	7.93	5.84	4.33	2.60	7.93
YEAR OF OCCURRENCE		1980	2001	1961	1988	1963	1965	1983	1991	1999	1998	1964	1984	SEP 1999
NORMAL NO. DAYS WITH:														
PRECIPITATION 0.01	30	5.1	5.3	7.9	8.0	10.6	9.3	7.1	7.7	7.9	6.2	5.2	5.8	86.1
PRECIPITATION 1.00	30	0.1	0.1	0.7	0.5	1.1	1.3	0.9	0.8	1.1	0.6	0.5	0.2	7.9
SNOWFALL NORMAL (IN)	30	4.7	4.5	2.1	0.3	0.0	0.0	0.0	0.0	0.0	T	1.4	3.8	16.8
MAXIMUM MONTHLY (IN)	48	19.7	16.7	16.5	4.6	T	T	T	T	T	1.5	7.1	13.8	19.7
YEAR OF OCCURRENCE		1987	1971	1970	1979	1991	1992	1993	1994	1992	1991	1972	1983	JAN 1987
MAXIMUM IN 24 HOURS (IN)	48	13.0	11.9	13.5	4.6	T	T	T	T	T	1.5	6.8	9.0	13.5
YEAR OF OCCURRENCE		1962	1971	1970	1979	1991	1992	1993	1994	1992	1991	1984	1983	MAR 1970
MAXIMUM SNOW DEPTH (IN)	49	17	13	13	3	0	0	0	0	0	T	5	8	17
YEAR OF OCCURRENCE		1962	1971	1970	1979						1993	1987	1987	JAN 1962
NORMAL NO. DAYS WITH:														
SNOWFALL 1.0	30	1.4	1.3	0.4	0.1	0.0	0.0	0.0	0.0	0.0	0.0	0.4	1.2	4.8

PRECIPITATION (inches) 2001 WICHITA, KS (ICT)

YEAR	JAN	FEB	MAR	APR	MAY	JUN	JUL	AUG	SEP	OCT	NOV	DEC	ANNUAL
1972	0.15	0.28	0.56	3.32	2.47	2.02	3.86	3.31	1.31	2.00	3.06	0.97	23.31
1973	2.73	1.20	9.17	3.78	0.52	1.21	6.07	0.68	9.46	3.43	0.91	2.80	41.96
1974	0.56	0.25	2.36	4.29	4.65	2.79	0.09	4.11	1.08	3.44	2.69	2.22	28.53
1975	1.28	2.12	1.72	1.57	8.60	6.88	0.05	2.77	1.19	0.08	2.89	0.48	29.63
1976	0.04	0.25	1.50	5.57	2.69	3.12	6.13	0.31	2.02	1.82	0.06	0.07	23.58
1977	0.54	0.08	1.42	3.32	8.85	3.15	3.98	6.31	4.35	1.19	2.38	0.19	35.76
1978	0.49	1.71	2.10	2.71	2.24	3.19	1.49	1.90	3.58	0.05	2.21	0.59	22.26
1979	1.57	0.23	4.47	1.46	3.05	6.54	2.18	0.67	2.96	2.05	1.99		28.71
1980	1.82	0.81	3.99	1.07	2.66	1.34	0.47	3.76	0.67	1.25	0.54	2.11	20.49
1981	0.25	0.22	2.15	0.38	6.33	4.25	1.27	2.65			2.93	0.29	27.66
1982	1.68	0.77	2.05	0.73	7.82	8.28	0.56	1.51	1.08	0.41	0.73	1.51	27.13
1983	1.66	1.23	4.26	3.80	4.08	7.38	3.86	1.39	2.53	2.97	2.39	1.13	36.68
1984	0.20	1.23	7.57	3.71	1.15	2.30	0.30	0.75	2.18	2.78	1.44	4.71	28.32
1985	0.26	2.07	1.64	2.28	2.01	4.79	3.97	2.86	5.97	5.58	1.60	0.61	33.64
1986	T	1.26	1.22	1.80	2.98	5.39	3.42	6.00	3.81	3.61	0.58	1.22	31.29
1987	1.40	3.33	4.13	0.61	8.01	4.50	2.14	7.69	2.10	0.90	1.50	2.25	38.56
1988	0.51	0.18	2.91	4.45	2.40	1.86	0.91	1.10	0.53	0.94	0.77	0.50	17.07
1989	0.79	0.39	2.38	0.23	4.96	7.96	4.07	5.72	7.38	0.37	T	0.44	34.69
1990	1.73	2.19	2.68	0.80	1.29	1.91	1.72	2.01	1.95	0.64	2.01	0.78	19.71
1991	0.58	T	0.71	2.27	4.09	1.34	2.65	7.57	2.08	1.00	2.38	2.07	26.74
1992	0.67	0.47	3.60	1.42	3.44	7.26	4.77	1.94	2.94	2.80	4.91	1.16	35.38
1993	1.12	2.25	1.75	2.11	9.62	4.44	6.21	1.31	1.77	1.68	0.55	0.30	33.11
1994	0.03	0.27	0.25	3.80	0.95	2.72	5.89	2.30	1.08	3.45	3.08	1.04	24.86
1995	0.44	0.41	2.45	3.40	5.99	8.86	4.20	5.51	2.00	0.28	0.08	0.68	34.30
1996	0.10	0.10	1.82	1.48	4.42	2.06	3.97	5.24	3.80	1.90	5.63	0.03	28.48
1997	0.36	2.18	0.46	3.81	4.78	4.35	5.89	4.36	3.50	3.39	1.30	2.65	37.03
1998	1.04	0.31	4.22	3.37	1.64	0.40	5.17	1.58	3.25	9.42	3.20	1.06	34.66
1999	1.35	0.39	1.89	6.02	7.17	7.55	3.48	1.25	1.06	0.16	1.45	4.06	45.46
2000	0.92	2.72	5.99	1.21	3.00	7.00	3.66	0.14	1.02	4.82	0.98	0.36	31.82
2001	1.47	3.89	2.25	1.42	3.12	4.31	1.24	1.90	3.10	1.06	0.65	0.08	24.49
POR= 48 YRS	0.75	1.01	2.30	2.37	4.03	4.30	3.50	2.96	3.35	2.54	1.57	1.14	29.82

AVERAGE TEMPERATURE (F) 2001 WICHITA, KS (ICT)

YEAR	JAN	FEB	MAR	APR	MAY	JUN	JUL	AUG	SEP	OCT	NOV	DEC	ANNUAL
1972	27.0	34.8	47.9	55.3	64.1	76.7	77.1	78.4	71.0	55.6	39.8	28.5	54.7
1973	27.4	35.6	48.4	52.2	61.9	76.8	80.3	79.8	67.8	61.1	46.7	31.9	55.8
1974	24.9	38.7	47.9	56.4	68.3	72.5	84.4	76.7	64.1	60.5	44.9	35.1	56.2
1975	33.4	28.5	40.4	54.7	64.0	73.6	79.2	81.4	65.8	60.3	44.8	36.0	55.2
1976	32.0	45.5	46.3	57.5	60.2	73.7	78.4	79.7	69.8	52.4	38.3	33.7	55.6
1977	24.5	41.7	50.2	59.8	69.2	78.3	73.2	77.9		59.8	46.1	35.0	58.3
1978	20.5	23.8	43.6	59.0	65.0	76.8	85.1	81.2	76.3	58.6	45.0	32.0	55.6
1979	16.7	24.0	46.9	54.1	64.1	74.5	80.4	79.8	72.7	61.9	41.8	37.7	54.6
1980	31.4	28.2	41.5	54.3	63.5	79.9	90.5	85.3	75.2	58.8	46.9	36.9	57.7
1981	34.1	40.1	47.6	63.7	62.6	77.9	83.6	78.1	72.0	55.9	47.0	32.9	58.0
1982	25.5	28.0	46.0	53.5	65.3	70.4	81.5	82.0	71.9	58.0	43.0	36.1	55.1
1983	31.7	35.5	43.3	48.0	60.5	71.5	81.6	85.0	72.7	58.5	45.5	16.4	54.2
1984	26.7	41.4	40.7	51.8	63.6	77.7	81.6	82.8	70.3	58.3	45.5	37.2	56.5
1985	25.2	31.2	49.1	59.9	67.7	74.0	81.7	77.5	69.8	57.4	39.4	28.8	55.1
1986	38.1	37.8	51.9	58.7	66.7	78.7	83.0	75.9	73.6	57.6	39.9	35.8	58.1
1987	29.3	42.3	47.0	57.4	69.7	76.4	80.1	78.7	70.4	55.7	47.7	34.9	57.5
1988	27.0	34.1	44.2	53.9	68.3	78.6	80.9	83.0	72.0	56.7	47.5	38.5	57.1
1989	38.5	27.5	47.0	59.6	66.0	72.0	79.0	77.0	65.7	60.7	45.5	25.2	53.3
1990	39.5	38.8	46.9	54.3	63.7	81.7	81.7	80.8	74.3	58.4	50.4	30.2	58.4
1991	28.9	44.6	49.7	58.9	70.4	79.3	83.8	80.5	69.6	59.4	39.5	39.1	58.6
1992	39.0	44.9	50.0	56.7	63.3	71.2	73.8	73.8	71.0		41.3	33.3	56.9
1993	28.6	32.2	43.1	52.5	63.5	74.5	81.7	81.0	67.4	55.3	40.8	37.8	54.9
1994	31.0	33.9	50.3	54.7	66.4	79.3	77.6	79.0	70.5	60.5	46.7	37.1	57.3
1995	32.9	40.6	45.4	52.3	60.2	72.3	80.1	80.9	68.5	59.1	43.8	33.5	55.8
1996	28.3	38.0	40.6	55.2	69.6	77.2	79.1	77.4	67.7	58.4	40.8	34.0	55.5
1997	30.9	37.5	47.4	51.0	62.9	73.9	79.3	76.7	72.6	59.2	42.1	34.6	55.7
1998	34.5	41.2	40.6	54.3	69.9	78.4	81.8	81.6	78.6	60.8	49.5	36.0	58.9
1999	32.7	45.3	44.6	56.0	65.3	73.0	82.3	81.2	67.1	59.2	53.3	37.7	58.1
2000	33.8	42.0	48.5	55.1	68.9	73.1	80.3	86.6	74.4	61.6	40.2	23.9	57.4
2001	32.3	33.5	42.8	60.2	67.5	75.3	86.3	82.7	70.0	59.1	50.9	39.0	58.3
POR= 48 YRS	30.3	36.0	45.1	56.4	65.8	75.7	81.5	79.9	70.7	58.9	44.6	34.1	56.6

REFERENCE NOTES:

PAGE 1:
THE TEMPERATURE GRAPH SHOWS NORMAL MAXIMUM AND NORMAL
MINIMUM DAILY TEMPERATURES (SOLID CURVES) AND THE
ACTUAL DAILY HIGH AND LOW TEMPERATURES (VERTICAL BARS).

PAGE 2 AND 3:
H/C INDICATES HEATING AND COOLING DEGREE DAYS.
RH INDICATES RELATIVE HUMIDITY
W/O INDICATES WEATHER AND OBSTRUCTIONS
S INDICATES SUNSHINE.
PR INDICATES PRESSURE.
CLOUDINESS ON PAGE 3 IS THE SUM OF THE CEILOMETER AND
SATELLITE DATA NOT TO EXCEED EIGHT EIGHTHS(OKTAS).

GENERAL:
T INDICATES TRACE PRECIPITATION, AN AMOUNT GREATER
THAN ZERO BUT LESS THAN THE LOWEST REPORTABLE VALUE.
+ INDICATES THE VALUE ALSO OCCURS ON EARLIER DATES.
BLANK ENTRIES DENOTE MISSING OR UNREPORTED DATA.
NORMALS ARE 30-YEAR AVERAGES (1961 - 1990).
ASOS INDICATES AUTOMATED SURFACE OBSERVING SYSTEM.
PM INDICATES THE LAST DAY OF THE PREVIOUS MONTH.
POR (PERIOD OF RECORD) BEGINS WITH THE JANUARY DATA
MONTH AND IS THE NUMBER OF YEARS USED TO COMPUTE
THE MEAN. INDIVIDUAL MONTHS WITHIN THE POR MAY
BE MISSING.
WHEN THE POR FOR A NORMAL IS LESS THAN 30 YEARS,
THE NORMAL IS PROVISIONAL AND IS BASED ON THE NUMBER
OF YEARS INDICATED.
0.* OR * INDICATES THE VALUE OR MEAN-DAYS-WITH
IS BETWEEN 0.00 AND 0.05.
CLOUDINESS FOR ASOS STATIONS DIFFERS FROM THE NON-ASOS
OBSERVATION TAKEN BY A HUMAN OBSERVER. ASOS STATION
CLOUDINESS IS BASED ON TIME-AVERAGED CEILOMETER DATA
FOR CLOUDS AT OR BELOW 12,000 FEET AND ON SATELLITE
DATA FOR CLOUDS ABOVE 12,000 FEET.
THE NUMBER OF DAYS WITH CLEAR, PARTLY CLOUDY, AND
CLOUDY CONDITIONS FOR ASOS STATIONS IS THE SUM
OF THE CEILOMETER AND SATELLITE DATA FOR THE
SUNRISE TO SUNSET PERIOD.

GENERAL CONTINUED:
CLEAR INDICATES 0 - 2 OKTAS, PARTLY CLOUDY INDICATES
3 - 6 OKTAS, AND CLOUDY INDICATES 7 OR 8 OKTAS.
WHEN AT LEAST ONE OF THE ELEMENTS (CEILOMETER OR
SATELLITE) IS MISSING, THE DAILY CLOUDINESS IS
NOT COMPUTED.
WIND DIRECTION IS RECORDED IN TENS OF DEGREES (2 DIGITS)
CLOCKWISE FROM TRUE NORTH. "00" INDICATES CALM. "36"
INDICATES TRUE NORTH.
RESULTANT WIND IS THE VECTOR AVERAGE OF THE SPEED AND
DIRECTION.
AVERAGE TEMPERATURE IS THE SUM OF THE MEAN DAILY MAXIMUM
AND MINIMUM TEMPERATURE DIVIDED BY 2.
SNOWFALL DATA COMPRISE ALL FORMS OF FROZEN
PRECIPITATION, INCLUDING HAIL.
A HEATING (COOLING) DEGREE DAY IS THE DIFFERENCE BETWEEN
THE AVERAGE DAILY TEMPERATURE AND 65 F.
DRY BULB IS THE TEMPERATURE OF THE AMBIENT AIR.
DEW POINT IS THE TEMPERATURE TO WHICH THE AIR MUST BE
COOLED TO ACHIEVE 100 PERCENT RELATIVE HUMIDITY.
WET BULB IS THE TEMPERATURE THE AIR WOULD HAVE IF THE
MOISTURE CONTENT WAS INCREASED TO 100 PERCENT RELATIVE
HUMIDITY.

ON JULY 1, 1996, THE NATIONAL WEATHER SERVICE BEGAN USING
THE "METAR" OBSERVATION CODE THAT WAS ALREADY EMPLOYED
BY MOST OTHER NATIONS OF THE WORLD. THE MOST NOTICEABLE
DIFFERENCE IN THIS ANNUAL PUBLICATION WILL BE THE CHANGE
IN UNITS FROM TENTHS TO EIGHTS(OKTAS) FOR REPORTING THE
AMOUNT OF SKY COVER.

HEATING DEGREE DAYS (base 65 F) 2001　WICHITA, KS (ICT)

YEAR	JUL	AUG	SEP	OCT	NOV	DEC	JAN	FEB	MAR	APR	MAY	JUN	TOTAL
1972-73	2	0	46	319	750	1123	1159	816	506	386	125	0	5232
1973-74	0	0	58	159	541	1021	1237	732	529	263	42	1	4583
1974-75	0	0	92	156	596	920	974	1016	757	333	72	9	4925
1975-76	0	0	98	201	596	892	1015	562	575	238	173	2	4352
1976-77	0	0	38	409	794	966	1253	646	456	170	12	0	4744
1977-78	0	0	1	176	558	926	1375	1149	663	194	112	6	5160
1978-79	0	0	18	210	598	1016	1491	1143	560	333	104	3	5476
1979-80	0	0	10	156	690	838	1038	1063	723	318	116	0	4952
1980-81	0	0	28	239	535	864	954	692	533	104	126	0	4075
1981-82	0	0	24	292	537	990	1214	1033	583	356	54	17	5100
1982-83	0	0	37	239	653	889	1022	818	664	507	168	20	5017
1983-84	0	0	47	221	582	1504	1180	680	747	394	95	0	5450
1984-85	0	0	103	237	576	856	1224	938	487	184	33	8	4646
1985-86	0	0	111	230	762	1116	826	755	416	220	41	0	4477
1986-87	0	3	11	233	747	899	1099	631	551	263	14	0	4451
1987-88	0	2	7	282	523	924	1170	891	637	330	33	0	4799
1988-89	0	0	16	265	519	813	817	1044	556	238	90	8	4366
1989-90	0	0	105	193	578	1228	783	728	555	332	112	0	4614
1990-91	0	0	18	238	445	1074	1114	567	475	201	53	0	4185
1991-92	0	0	71	242	759	793	802	574	461	259	113	10	4084
1992-93	0	2	25	209	703	975	1122	913	670	368	100	10	5097
1993-94	0	0	59	321	718	837	1046	862	458	321	66	0	4688
1994-95	0	0	33	186	542	856	989	675	603	375	164	2	4425
1995-96	0	0	90	205	627	969	1129	776	749	308	49	5	4907
1996-97	0	0	48	231	719	953	1050	763	541	415	106	0	4826
1997-98	0	0	9	263	680	934	941	660	754	322	30	21	4614
1998-99	0	0	4	157	459	892	994	545	625	271	64	10	4021
1999-00	0	0	69	198	345	838	960	659	506	289	46	5	3915
2000-01	0	0	55	154	740	1266	1009	878	679	167	46	0	4994
2001-	0	0	23	199	419	796							

COOLING DEGREE DAYS (base 65 F) 2001　WICHITA, KS (ICT)

YEAR	JAN	FEB	MAR	APR	MAY	JUN	JUL	AUG	SEP	OCT	NOV	DEC	ANNUAL
1972	0	0	2	26	88	358	385	421	234	34	0	0	1548
1973	0	0	0	7	39	360	482	468	149	47	0	0	1552
1974	0	0	4	12	149	233	608	368	69	23	0	0	1466
1975	0	0	0	32	49	275	450	515	128	63	0	0	1512
1976	0	0	2	19	32	270	420	464	187	23	0	0	1417
1977	0	0	3	19	152	404	581	404	254	24	0	0	1841
1978	0	0	6	20	122	366	631	510	364	23	5	0	2047
1979	0	0	5	14	81	294	488	465	249	67	0	0	1663
1980	0	0	0	3	75	456	796	635	340	52	1	0	2358
1981	0	0	0	72	56	393	582	412	240	17	0	0	1772
1982	0	0	1	16	70	186	516	534	253	29	0	0	1605
1983	0	0	0	2	34	220	521	628	286	28	5	0	1724
1984	0	0	0	6	61	388	520	558	272	35	0	0	1840
1985	0	0	0	36	122	285	523	394	262	1	0	0	1623
1986	0	0	17	40	102	419	563	349	275	10	0	0	1775
1987	0	0	0	42	166	350	473	434	177	3	10	0	1655
1988	0	0	0	3	140	415	497	566	235	15	0	0	1871
1989	0	0	5	81	129	225	442	379	132	70	0	0	1463
1990	0	0	0	17	79	508	525	497	306	41	11	0	1984
1991	0	0	9	25	225	433	592	484	214	77	0	0	2059
1992	0	0	0	17	66	203	441	281	213	37	0	0	1258
1993	0	0	0	0	59	304	526	502	134	27	0	0	1552
1994	0	0	7	22	117	434	399	441	202	55	0	0	1677
1995	0	0	6	4	24	231	478	502	204	32	0	0	1481
1996	0	0	0	19	197	377	444	393	136	35	0	0	1601
1997	0	0	0	0	47	274	448	369	244	89	0	0	1471
1998	0	0	5	7	189	427	528	522	416	37	0	0	2131
1999	0	0	0	7	78	259	543	511	136	28	1	0	1563
2000	0	0	0	2	174	256	483	675	341	57	0	0	1988
2001	0	0	0	31	129	317	670	557	178	25	2	0	1909

SNOWFALL (inches) 2001　WICHITA, KS (ICT)

YEAR	JUL	AUG	SEP	OCT	NOV	DEC	JAN	FEB	MAR	APR	MAY	JUN	TOTAL
1972-73	0.0	0.0	0.0	0.0	7.1	2.7	17.7	0.2	T	2.3	0.0	0.0	30.0
1973-74	0.0	0.0	0.0	0.0	T	8.7	4.1	0.7	2.0	0.3	0.0	0.0	15.8
1974-75	0.0	0.0	0.0	0.0	1.8	2.2	7.6	15.2	7.6	T	0.0	0.0	34.4
1975-76	0.0	0.0	0.0	0.0	5.5	T	0.6	0.9	T	0.0	0.0	0.0	7.0
1976-77	0.0	0.0	0.0	T	0.3	T	3.6	T	T	T	0.0	0.0	3.9
1977-78	0.0	0.0	0.0	0.0	T	T	7.4	7.8	0.3	0.0	0.0	0.0	15.5
1978-79	0.0	0.0	0.0	0.0	T	6.7	13.9	1.9	1.6	4.6	0.0	0.0	28.7
1979-80	0.0	0.0	0.0	0.0	T	T	T	12.3	0.4	0.0	0.0	0.0	12.7
1980-81	0.0	0.0	0.0	T	T	0.4	T	2.5	0.2	0.0	0.0	0.0	3.1
1981-82	0.0	0.0	0.0	0.0	T	1.2	0.0	12.7	T	0.0	0.0	0.0	13.9
1982-83	0.0	0.0	0.0	0.0	T	1.4	13.0	8.9	1.5	0.7	0.0	0.0	25.5
1983-84	0.0	0.0	0.0	0.0	4.1	13.8	4.3	T	6.9	0.0	0.0	0.0	29.1
1984-85	0.0	0.0	0.0	0.0	6.8	7.6	3.5	3.6	0.2	0.0	0.0	0.0	21.7
1985-86	0.0	0.0	0.0	0.0	1.0	3.0	0.0	7.5	0.0	0.0	0.0	0.0	11.5
1986-87	0.0	0.0	0.0	0.0	T	0.4	19.7	6.0	T	0.0	0.0	0.0	26.1
1987-88	0.0	0.0	0.0	0.0	6.2	12.6	8.5	1.1	11.0	0.0	0.0	0.0	39.4
1988-89	0.0	0.0	0.0	0.0	3.0	0.7	0.3	0.8	T	T	T	0.0	4.8
1989-90	0.0	0.0	0.0	0.0	T	4.4	0.3	7.0	0.7	0.1	0.0	0.0	12.5
1990-91	0.0	0.0	0.0	0.0	0.0	4.8	2.1	0.0	T	T	T	0.0	6.9
1991-92	0.0	0.0	0.0	1.5	0.3	0.0	T	T	T	T	0.0	0.0	1.8
1992-93	0.0	0.0	T	T	3.8	3.3	7.7	5.1	2.2	T	0.0	0.0	22.1
1993-94	T	0.0	0.0	T	T	0.1	T	1.2	T	T	0.0	0.0	1.3
1994-95	0.0	T	0.0	0.0	0.0	0.0	1.9	2.6	5.6	0.0	0.0	0.0	10.1
1995-96	0.0	0.0	0.0	0.0	0.3	5.7	1.9	0.6	T	T	0.0	0.0	8.5
1996-97	0.0	0.0	0.0	0.2	4.6	0.4	3.8	5.5	T	2.0	0.0	0.0	16.5
1997-98	T	0.0	0.0	0.1	0.8	5.4	0.6	T	13.6	T	0.0	0.0	20.5
1998-99	0.0	0.0	0.0	0.0	0.0	0.9	0.1	T	8.3	T	T	0.0	9.3
1999-00	0.0	0.0	T	0.0	0.0	7.5	10.2	T	4.2	0.0	0.0	0.0	21.9
2000-01	0.0	0.0	T	0.0	T	5.5	2.5	1.1	T	T	T	0.0	9.1
2001-	0.0	0.0	T	0.0	0.0	0.0							
POR= 48 YRS	T	T	T	T	1.3	3.1	4.3	3.6	2.6	0.3	T	T	15.2

2001
LEXINGTON,
KENTUCKY (LEX)

Lexington, County Seat of Fayette County, is located in the heart of the famed Kentucky Blue Grass Region. Fayette County is a gently rolling plateau with the elevation varying between 900 and 1,050 feet above sea level. It is noted for its beauty, the fertility of its soil, excellent grass, stock farms, and burley tobacco. The soil has a high phosphorus content and this is very valuable in growing pasture grasses for the grazing of cattle and horses. Lexington has a decided continental climate with a rather large diurnal temperature range. The climate is temperate and well suited to a varied plant and animal life. There are no bodies of water close enough to have any effect on the climate. The closest river is the Kentucky which makes an arc about 15 to 20 miles to the southeast, south, and southwest on its course to the Ohio River. There are numerous small creeks that rise in the county and flow into the river. The reservoirs of the Lexington Water Company are about 5 miles southeast of the city and are the largest bodies of water in the area.

Lexington is subject to rather sudden and large changes in temperature with the spells generally of rather short duration. Temperatures above 100 degrees and below zero degrees are relatively rare. The average temperature for the winter is 35 degrees, spring 62 degrees, fall 50 degrees, and summer 74 degrees.

Precipitation is evenly distributed throughout the winter, spring, and summer, with about 12 inches recorded on the average for each of these seasons. The fall season averages nearly 8 1/2 inches. Snowfall amounts are variable and the ground does not retain snow cover more than a few days at a time.

The months of September and October are the most pleasant of the year. They have the least amount of precipitation, the greatest number of clear days, and generally comfortable temperatures are the rule during these months.

Based on the 1951-1980 period, the average first occurrence of 32 degrees Fahrenheit in the fall is October 25 and the average last occurrence in the spring is April 17.

NORMALS, MEANS, AND EXTREMES
LEXINGTON, KY (LEX)

LATITUDE: 38 02' 27" N LONGITUDE: 84 36' 21" W ELEVATION (FT): GRND: 977 BARO: 980 TIME ZONE: EASTERN (UTC + 5) WBAN: 93820

	ELEMENT	POR	JAN	FEB	MAR	APR	MAY	JUN	JUL	AUG	SEP	OCT	NOV	DEC	YEAR
TEMPERATURE °F	NORMAL DAILY MAXIMUM	30	39.1	43.6	55.3	65.5	74.3	82.7	85.8	84.9	78.3	67.2	54.9	44.2	64.7
	MEAN DAILY MAXIMUM	54	40.4	45.1	54.2	65.5	74.6	82.8	86.0	85.0	78.8	67.7	54.5	44.4	64.9
	HIGHEST DAILY MAXIMUM	57	76	80	83	88	92	101	103	103	103	91	83	75	103
	YEAR OF OCCURRENCE		1950	1996	1945	1962	1987	1988	1999	1983	1954	1959	1987	1982	JUL 1999
	MEAN OF EXTREME MAXS.	54	63.0	66.5	75.4	82.1	86.7	91.9	93.8	93.2	90.4	82.5	73.8	64.8	80.3
	NORMAL DAILY MINIMUM	30	22.4	25.3	35.3	44.2	53.5	61.5	65.7	64.4	58.0	46.0	37.0	27.6	45.1
	MEAN DAILY MINIMUM	54	24.0	27.0	34.3	44.1	53.8	62.1	66.3	64.7	57.7	46.2	36.2	27.9	45.4
	LOWEST DAILY MINIMUM	57	−21	−15	−2	18	26	39	47	42	34	20	−3	−19	−21
	YEAR OF OCCURRENCE		1963	1951	1960	1982	1966	1966	1972	1965	1993	1976	1950	1989	JAN 1963
	MEAN OF EXTREME MINS.	54	0.9	4.1	16.7	27.6	37.9	48.7	55.4	53.3	41.9	29.8	18.7	7.4	28.6
	NORMAL DRY BULB	30	30.8	34.5	45.3	54.8	64.0	72.2	75.8	74.7	68.2	56.7	46.0	35.9	54.9
	MEAN DRY BULB	54	32.3	36.0	44.2	54.8	64.2	72.4	76.3	74.9	68.3	56.9	45.3	36.2	55.2
	MEAN WET BULB	18	30.2	34.0	39.8	48.5	58.3	66.2	69.5	68.1	61.5	51.2	41.8	34.2	50.3
	MEAN DEW POINT	18	25.4	28.1	33.0	41.9	53.6	62.3	66.0	64.7	57.4	46.1	36.5	29.3	45.4
	NORMAL NO. DAYS WITH:														
	MAXIMUM ≥ 90	30	0.0	0.0	0.0	0.0	0.4	3.3	6.8	5.9	1.8	0.0	0.0	0.0	18.2
	MAXIMUM ≤ 32	30	9.9	5.8	0.8	0.0	0.0	0.0	0.0	0.0	0.0	0.0	0.5	5.4	22.4
	MINIMUM ≤ 32	30	24.3	20.2	13.3	3.6	0.1	0.0	0.0	0.0	0.0	2.4	10.9	20.8	95.6
	MINIMUM ≤ 0	30	2.0	0.7	0.1	0.0	0.0	0.0	0.0	0.0	0.0	0.0	0.0	0.5	3.3
H/C	NORMAL HEATING DEG. DAYS	30	1060	854	611	312	135	5	0	0	47	287	570	902	4783
	NORMAL COOLING DEG. DAYS	30	0	0	0	6	104	221	335	301	143	30	0	0	1140
RH	NORMAL (PERCENT)	30	73	71	66	63	68	69	72	72	73	69	72	74	70
	HOUR 01 LST	30	76	75	72	69	76	80	83	83	83	77	76	77	77
	HOUR 07 LST	30	80	79	77	75	80	82	85	87	88	84	81	81	82
	HOUR 13 LST	30	68	64	58	54	56	56	58	58	58	56	62	68	60
	HOUR 19 LST	30	70	66	60	55	59	60	63	64	66	64	68	72	64
S	PERCENT POSSIBLE SUNSHINE														
W/O	MEAN NO. DAYS WITH:														
	HEAVY FOG(VISBY ≤ 1/4 MI)	57	2.4	2.0	1.4	0.8	1.1	1.1	1.5	2.1	2.1	1.8	1.3	2.0	19.6
	THUNDERSTORMS	57	0.8	0.9	2.7	3.8	6.4	7.8	8.5	6.3	2.9	1.4	1.0	0.4	42.9
CLOUDINESS	MEAN:														
	SUNRISE-SUNSET (OKTAS)	52	5.9	5.7	5.6	5.3	5.0	4.8	4.6	4.3	4.2	4.2	5.3	5.7	5.0
	MIDNIGHT-MIDNIGHT (OKTAS)	32	5.8	5.5	5.3	4.9	4.7	4.6	4.3	4.0	4.1	4.1	5.1	5.6	4.8
	MEAN NO. DAYS WITH:														
	CLEAR	52	5.5	5.7	5.8	6.3	7.0	6.9	8.0	9.5	10.5	11.6	6.7	5.8	89.3
	PARTLY CLOUDY	52	5.8	5.6	7.3	8.5	10.0	11.8	12.3	11.9	8.4	7.2	6.7	5.7	101.2
	CLOUDY	52	19.7	17.0	15.2	14.0	14.0	11.3	10.7	9.6	11.1	12.1	16.6	19.5	174.7
PR	MEAN STATION PRESSURE(IN)	29	29.05	29.02	28.96	28.94	28.94	28.96	28.99	29.01	29.03	29.05	29.04	29.04	29.00
	MEAN SEA-LEVEL PRES. (IN)	17	30.13	30.11	30.05	29.97	29.99	30.01	30.01	30.03	30.03	30.07	30.10	30.12	30.16	30.06
WINDS	MEAN SPEED (MPH)	46	10.7	10.5	10.8	10.4	8.5	7.8	7.1	6.8	7.5	8.0	9.9	10.2	9.0
	PREVAIL.DIR(TENS OF DEGS)	30	18	18	18	18	18	18	18	18	18	18	18	18	18
	MAXIMUM 2-MINUTE:														
	SPEED (MPH)	5	47	40	32	36	33	44	37	36	32	32	37	39	47
	DIR. (TENS OF DEGS)		18	24	26	24	31	30	31	27	21	26	25	22	18
	YEAR OF OCCURRENCE		1999	1999	2001	1999	1997	1998	1999	1997	2000	2001	2000	2001	JAN 1999
	MAXIMUM 5-SECOND:														
	SPEED (MPH)	5	56	53	46	48	47	56	51	45	38	44	49	49	56
	DIR. (TENS OF DEGS)		19	23	21	20	19	30	32	01	32	26	24	23	19
	YEAR OF OCCURRENCE		1999	1999	1998	1999	2001	1998	1999	2000	2000	2001	2000	2001	JAN 1999
PRECIPITATION	NORMAL (IN)	30	2.86	3.21	4.40	3.88	4.47	3.66	5.00	3.93	3.20	2.57	3.39	3.98	44.55
	MAXIMUM MONTHLY (IN)	57	16.65	10.12	13.82	9.30	10.84	11.69	10.64	11.18	9.69	6.13	6.87	10.17	16.65
	YEAR OF OCCURRENCE		1950	1989	1997	1970	1983	1960	1958	1974	1979	1983	1951	1990	JAN 1950
	MINIMUM MONTHLY (IN)	57	0.37	0.67	0.99	0.79	1.20	0.61	1.26	0.29	0.24	0.33	0.45	0.61	0.24
	YEAR OF OCCURRENCE		1981	1978	1966	1946	1965	1988	1995	1998	1959	1963	1976	1965	SEP 1959
	MAXIMUM IN 24 HOURS (IN)	57	2.98	3.79	5.56	4.39	3.24	5.88	4.73	3.56	4.35	3.21	2.71	3.77	5.88
	YEAR OF OCCURRENCE		1951	1989	1997	1948	1983	1960	1978	1968	1979	1962	1988	1978	JUN 1960
	NORMAL NO. DAYS WITH:														
	PRECIPITATION ≥ 0.01	30	11.3	11.2	12.5	11.9	12.0	10.2	11.2	9.2	8.7	8.4	10.7	11.8	129.1
	PRECIPITATION ≥ 1.00	30	0.6	0.7	1.1	0.9	1.1	0.7	1.5	1.0	0.7	0.5	0.9	0.9	10.6
SNOWFALL	NORMAL (IN)	30	6.4	5.2	2.2	0.4	T	0.0	0.0	0.0	0.0	0.*	0.8	2.5	17.5
	MAXIMUM MONTHLY (IN)	52	21.9	16.4	17.7	5.9	T	T	T	T	0.0	0.2	9.7	10.7	21.9
	YEAR OF OCCURRENCE		1978	1960	1960	1987	1995	1993	1989	1989		1972	1950	1967	JAN 1978
	MAXIMUM IN 24 HOURS (IN)	52	10.2	7.3	9.5	4.9	T	T	T	T	0.0	0.2	7.5	7.8	10.2
	YEAR OF OCCURRENCE		1994	1971	1947	1987	1995	1993	1989	1989		1972	1966	1967	JAN 1994
	MAXIMUM SNOW DEPTH (IN)	48	14	9	12	2	0	0	0	0	0	0	8	5	14
	YEAR OF OCCURRENCE		1978	1985	1960	1961							1950	1984	JAN 1978
	NORMAL NO. DAYS WITH:														
	SNOWFALL ≥ 1.0	30	2.0	1.7	0.7	0.1	0.0	0.0	0.0	0.0	0.0	0.0	0.3	0.7	5.5

PRECIPITATION (inches) 2001 LEXINGTON, KY (LEX)

YEAR	JAN	FEB	MAR	APR	MAY	JUN	JUL	AUG	SEP	OCT	NOV	DEC	ANNUAL
1972	4.10	5.60	4.04	8.75	3.84	3.61	5.58	3.95	4.30	2.71	4.21	6.92	57.61
1973	1.53	1.58	5.08	5.67	8.22	6.06	5.15	3.58	1.40	2.65	6.58	3.42	50.92
1974	6.39	2.24	5.89	3.33	5.52	7.21	4.82	11.18	4.18	1.53	4.08	3.72	60.09
1975	3.66	5.70	10.38	6.17	2.69	2.23	5.60	3.96	6.46	5.09	2.93	4.24	59.11
1976	3.59	4.67	3.72	1.24	3.16	3.34	6.74	1.26	4.23	0.45	1.22	37.48
1977	2.30	1.03	4.21	3.42	1.51	4.80	4.59	4.83	2.71	3.77	3.95	3.04	40.16
1978	6.38	0.67	2.87	3.15	5.74	1.94	7.60	10.00	3.10	3.20	3.11	9.97	57.73
1979	4.07	2.92	3.22	4.92	4.17	2.80	4.72	6.20	9.69	2.96	4.52	3.81	54.00
1980	1.63	1.17	6.04	2.82	2.27	1.88	5.55	5.10	2.47	2.07	2.02	1.67	34.69
1981	0.37	4.76	1.76	4.88	5.10	2.29	5.27	2.72	1.97	2.44	1.99	3.10	36.65
1982	5.48	2.16	3.89	2.19	2.51	3.95	3.82	4.01	1.21	1.56	3.45	4.53	38.76
1983	1.29	1.61	1.48	5.18	10.84	2.18	2.41	1.26	1.33	6.13	3.59	3.46	40.76
1984	1.64	3.31	4.09	5.02	5.34	2.20	4.80	0.56	1.36	3.87	5.19	4.89	42.27
1985	1.91	1.11	3.69	2.34	4.34	4.98	3.37	3.76	1.93	4.23	4.96	1.13	37.75
1986	0.53	2.48	2.43	1.65	3.24	1.29	5.64	2.67	3.08	2.06	6.49	3.30	34.86
1987	1.30	3.62	3.13	2.23	1.80	6.59	3.48	4.18	0.91	0.55	2.72	6.17	36.68
1988	2.94	3.06	2.34	2.93	3.02	0.61	3.51	4.18	5.96	1.34	5.39	3.62	38.90
1989	3.99	10.12	6.08	2.60	5.39	4.26	4.20	3.98	4.98	3.38	2.38	1.80	53.16
1990	4.17	3.43	1.89	2.37	5.41	4.59	6.45	4.36	2.12	4.49	2.69	10.17	52.14
1991	2.57	3.91	5.80	2.70	3.95	2.91	3.60	3.08	2.09	2.70	1.27	7.22	41.80
1992	3.63	1.84	4.70	2.11	4.68	7.74	10.27	4.73	3.44	0.65	3.50	1.80	49.09
1993	2.42	4.15	3.77	3.53	2.43	5.46	3.38	4.52	3.00	4.19	5.42	3.31	45.58
1994	4.50	4.42	6.83	5.18	4.86	3.84	2.29	3.72	1.19	2.11	2.89	3.87	45.70
1995	5.01	2.26	3.32	3.90	8.97	8.17	1.26	4.89	2.76	3.64	3.19	2.71	50.08
1996	4.51	1.86	4.62	4.84	8.98	5.10	5.30	2.30	4.15	2.10	4.79	5.26	53.81
1997	3.70	3.97	13.82	1.89	8.85	9.54	3.29	2.58	2.38	2.37	4.06	2.68	59.13
1998	3.99	2.58	3.40	6.20	6.14	10.81	7.98	0.29	0.61	2.41	1.96	3.23	49.60
1999	5.77	2.38	3.80	2.23	1.31	5.38	2.47	0.99	1.39	1.63	1.82	2.70	31.87
2000	3.40	4.81	3.89	4.52	2.99	3.82	3.36	3.50	5.32	0.74	2.00	3.75	42.10
2001	1.35	3.56	3.27	1.14	6.00	2.58	5.78	2.93	2.46	3.71	3.30	2.89	38.97
POR= 118 YRS	3.96	3.25	4.50	3.63	4.06	4.24	4.43	3.44	2.81	2.40	3.20	3.70	43.62

AVERAGE TEMPERATURE (F) 2001   LEXINGTON, KY (LEX)

YEAR	JAN	FEB	MAR	APR	MAY	JUN	JUL	AUG	SEP	OCT	NOV	DEC	ANNUAL
1972	35.4	33.2	42.2	53.7	63.2	67.6	73.8	72.9	69.8	53.0	44.5	40.9	54.2
1973	35.1	35.2	53.8	52.9	60.1	73.6	75.8	74.9	72.2	60.9	48.5	36.4	56.6
1974	40.7	37.4	48.5	55.8	63.6	67.2	74.4	73.3	62.6	54.3	45.7	37.8	55.1
1975	37.3	39.5	41.4	52.2	67.4	73.5	76.4	77.5	63.4	57.4	48.5	35.8	55.9
1976	28.4	43.9	50.0	54.5	67.6	71.1	73.0	71.2	64.1	49.7	37.1	30.9	52.9
1977	17.8	34.9	50.8	59.6	69.8	72.3	78.4	75.7	72.2	56.1	49.0	33.4	55.8
1978	21.6	21.3	40.4	57.0	61.1	73.1	76.0	74.1	70.9	53.7	48.4	37.9	53.0
1979	23.6	26.9	48.0	53.8	63.1	70.9	74.0	74.0	66.9	55.6	45.7	37.9	53.4
1980	32.4	28.3	41.5	52.5	64.8	71.2	78.8	78.2	70.6	53.7	43.6	36.0	54.3
1981	27.5	37.0	42.6	59.6	60.4	73.8	75.8	73.4	65.8	45.9	43.0	34.0	54.2
1982	28.2	34.9	47.1	50.6	69.8	68.5	77.1	72.9	65.6	58.3	48.4	44.2	55.5
1983	33.8	37.2	46.3	50.8	60.7	72.8	79.8	80.5	70.2	46.5	28.4	55.5
1984	27.6	41.2	39.7	53.1	60.6	76.0	72.9	74.9	66.5	63.5	41.8	45.4	55.3
1985	23.8	30.5	48.5	58.5	64.7	70.3	75.1	72.8	67.6	53.0	29.6	54.4
1986	33.2	38.6	46.8	57.3	65.5	74.2	78.6	72.9	71.0	58.0	44.9	35.7	56.4
1987	31.9	38.0	46.8	53.7	70.3	75.0	77.1	77.5	70.0	52.0	50.0	38.9	56.8
1988	29.8	33.7	44.8	54.2	64.5	74.3	79.1	77.9	67.8	49.6	46.1	36.5	54.9
1989	40.5	33.1	47.3	54.2	60.6	71.6	76.5	74.1	67.6	57.0	45.0	23.0	54.2
1990	41.6	43.1	48.9	53.2	61.6	72.2	75.3	73.8	68.5	56.4	49.9	40.4	57.1
1991	33.9	39.1	47.4	58.1	70.5	74.3	77.6	75.8	68.8	58.7	43.4	40.1	57.3
1992	35.2	41.3	45.4	56.1	62.0	69.4	75.6	70.0	67.0	55.8	37.0	55.1
1993	37.4	33.1	42.0	52.7	65.0	72.1	80.1	76.5	66.0	55.1	44.6	35.0	55.0
1994	25.2	36.8	43.6	58.1	60.2	75.1	76.8	73.7	66.2	57.8	50.6	41.2	55.4
1995	33.9	33.5	47.2	55.5	63.3	72.8	77.3	79.5	66.5	56.9	39.7	33.0	54.9
1996	31.0	35.7	39.1	50.9	66.4	72.1	73.1	73.6	66.3	59.8	39.5	53.6
1997	31.6	40.8	46.2	49.1	58.0	69.3	76.1	72.7	66.6	56.1	41.6	35.6	53.6
1998	40.7	41.1	45.8	53.7	67.5	72.7	74.5	76.2	74.1	57.9	47.4	39.7	57.6
1999	36.2	40.2	40.3	56.5	65.3	73.5	79.6	75.7	68.8	56.7	50.3	37.3	56.7
2000	31.9	42.7	48.3	53.3	66.8	73.5	73.8	73.7	66.3	59.4	43.5	25.1	54.9
2001	31.1	39.9	40.6	59.7	66.5	71.3	75.3	76.1	66.1	56.7	52.0	40.9	56.4
POR= 118 YRS	33.1	35.4	44.1	54.4	64.1	72.7	76.2	74.8	69.0	57.2	45.2	35.9	55.2

REFERENCE NOTES:

PAGE 1:
THE TEMPERATURE GRAPH SHOWS NORMAL MAXIMUM AND NORMAL MINIMUM DAILY TEMPERATURES (SOLID CURVES) AND THE ACTUAL DAILY HIGH AND LOW TEMPERATURES (VERTICAL BARS).

PAGE 2 AND 3:
H/C INDICATES HEATING AND COOLING DEGREE DAYS.
RH INDICATES RELATIVE HUMIDITY
W/O INDICATES WEATHER AND OBSTRUCTIONS
S INDICATES SUNSHINE.
PR INDICATES PRESSURE.
CLOUDINESS ON PAGE 3 IS THE SUM OF THE CEILOMETER AND SATELLITE DATA NOT TO EXCEED EIGHT EIGHTHS (OKTAS).

GENERAL:
T INDICATES TRACE PRECIPITATION, AN AMOUNT GREATER THAN ZERO BUT LESS THAN THE LOWEST REPORTABLE VALUE.
+ INDICATES THE VALUE ALSO OCCURS ON EARLIER DATES.
BLANK ENTRIES DENOTE MISSING OR UNREPORTED DATA.
NORMALS ARE 30-YEAR AVERAGES (1961 - 1990).
ASOS INDICATES AUTOMATED SURFACE OBSERVING SYSTEM.
PM INDICATES THE LAST DAY OF THE PREVIOUS MONTH.
POR (PERIOD OF RECORD) BEGINS WITH THE JANUARY DATA MONTH AND IS THE NUMBER OF YEARS USED TO COMPUTE THE MEAN. INDIVIDUAL MONTHS WITHIN THE POR MAY BE MISSING.
WHEN THE POR FOR A NORMAL IS LESS THAN 30 YEARS, THE NORMAL IS PROVISIONAL AND IS BASED ON THE NUMBER OF YEARS INDICATED.
0.* OR * INDICATES THE VALUE OR MEAN-DAYS-WITH IS BETWEEN 0.00 AND 0.05.
CLOUDINESS FOR ASOS STATIONS DIFFERS FROM THE NON-ASOS OBSERVATION TAKEN BY A HUMAN OBSERVER. ASOS STATION CLOUDINESS IS BASED ON TIME-AVERAGED CEILOMETER DATA FOR CLOUDS AT OR BELOW 12,000 FEET AND ON SATELLITE DATA FOR CLOUDS ABOVE 12,000 FEET.
THE NUMBER OF DAYS WITH CLEAR, PARTLY CLOUDY, AND CLOUDY CONDITIONS FOR ASOS STATIONS IS THE SUM OF THE CEILOMETER AND SATELLITE DATA FOR THE SUNRISE TO SUNSET PERIOD.

GENERAL CONTINUED:
CLEAR INDICATES 0 - 2 OKTAS, PARTLY CLOUDY INDICATES 3 - 6 OKTAS, AND CLOUDY INDICATES 7 OR 8 OKTAS. WHEN AT LEAST ONE OF THE ELEMENTS (CEILOMETER OR SATELLITE) IS MISSING, THE DAILY CLOUDINESS IS NOT COMPUTED.
WIND DIRECTION IS RECORDED IN TENS OF DEGREES (2 DIGITS) CLOCKWISE FROM TRUE NORTH. "00" INDICATES CALM. "36" INDICATES TRUE NORTH.
RESULTANT WIND IS THE VECTOR AVERAGE OF THE SPEED AND DIRECTION.
AVERAGE TEMPERATURE IS THE SUM OF THE MEAN DAILY MAXIMUM AND MINIMUM TEMPERATURE DIVIDED BY 2.
SNOWFALL DATA COMPRISE ALL FORMS OF FROZEN PRECIPITATION, INCLUDING HAIL.
A HEATING (COOLING) DEGREE DAY IS THE DIFFERENCE BETWEEN THE AVERAGE DAILY TEMPERATURE AND 65 F.
DRY BULB IS THE TEMPERATURE OF THE AMBIENT AIR.
DEW POINT IS THE TEMPERATURE TO WHICH THE AIR MUST BE COOLED TO ACHIEVE 100 PERCENT RELATIVE HUMIDITY.
WET BULB IS THE TEMPERATURE THE AIR WOULD HAVE IF THE MOISTURE CONTENT WAS INCREASED TO 100 PERCENT RELATIVE HUMIDITY.

ON JULY 1, 1996, THE NATIONAL WEATHER SERVICE BEGAN USING THE "METAR" OBSERVATION CODE THAT WAS ALREADY EMPLOYED BY MOST OTHER NATIONS OF THE WORLD. THE MOST NOTICEABLE DIFFERENCE IN THIS ANNUAL PUBLICATION WILL BE THE CHANGE IN UNITS FROM TENTHS TO EIGHTS (OKTAS) FOR REPORTING THE AMOUNT OF SKY COVER.

HEATING DEGREE DAYS (base 65 F) 2001 LEXINGTON, KY (LEX)

YEAR	JUL	AUG	SEP	OCT	NOV	DEC	JAN	FEB	MAR	APR	MAY	JUN	TOTAL
1972-73	10	1	20	366	612	739	920	827	353	371	167	0	4386
1973-74	0	1	21	172	490	880	744	767	514	289	128	37	4043
1974-75	0	0	125	338	578	836	852	705	726	387	51	4	4602
1975-76	0	0	128	249	488	895	1130	606	468	339	158	2	4463
1976-77	1	4	64	474	829	1050	1457	836	444	208	52	19	5438
1977-78	0	0	6	277	498	972	1338	1219	755	254	179	6	5504
1978-79	0	0	20	348	492	834	1277	1061	522	337	110	15	5016
1979-80	0	5	40	307	574	833	1005	1057	721	371	88	17	5018
1980-81	0	0	23	358	633	892	1156	777	687	182	180	0	4888
1981-82	0	0	77	286	568	985	1134	840	549	429	14	9	4891
1982-83	0	1	75	259	500	646	961	772	580	422	151	7	4374
1983-84	0	0	59	201	550	1128	1152	685	778	370	178	3	5104
1984-85	2	0	89	84	689	601	1275	959	510	228	66	23	4526
1985-86	0	0	72	179	360	1092	978	735	561	259	94	2	4332
1986-87	0	15	14	250	595	903	1016	749	559	342	39	0	4482
1987-88	0	0	17	399	447	804	1085	901	620	328	90	18	4709
1988-89	0	3	30	474	560	877	750	887	548	351	196	8	4684
1989-90	0	6	61	267	592	1297	720	608	505	378	128	17	4579
1990-91	0	3	57	288	453	757	955	719	544	215	34	0	4025
1991-92	0	0	77	230	642	765	915	682	600	293	159	17	4380
1992-93	0	5	64	288	566	863	847	884	705	363	64	27	4676
1993-94	0	0	67	313	608	922	1231	783	658	229	185	3	4999
1994-95	0	3	37	232	425	730	960	877	545	298	108	2	4217
1995-96	0	0	65	254	755	983	1048	844	799	426	96	8	5278
1996-97	0	0	64	268	750	784	1025	670	576	471	233	31	4872
1997-98	1	3	39	321	698	904	745	666	614	331	46	24	4392
1998-99	0	0	8	237	522	781	887	690	758	254	51	1	4189
1999-00	0	0	43	254	433	855	1019	640	513	346	53	6	4162
2000-01	0	0	84	215	642	1227	1043	692	752	226	56	20	4957
2001-	0	0	84	275	387	741							

COOLING DEGREE DAYS (base 65 F) 2001 LEXINGTON, KY (LEX)

YEAR	JAN	FEB	MAR	APR	MAY	JUN	JUL	AUG	SEP	OCT	NOV	DEC	ANNUAL
1972	0	0	0	11	47	130	287	250	171	0	4	0	900
1973	0	0	12	18	21	266	342	314	245	51	0	0	1269
1974	0	0	10	21	94	108	296	264	60	11	4	0	868
1975	0	0	0	11	130	267	357	394	86	18	0	0	1263
1976	0	0	9	30	26	193	257	205	46	4	0	0	770
1977	0	0	11	52	206	241	422	337	232	8	23	0	1532
1978	0	0	0	19	69	257	349	290	202	4	0	0	1190
1979	0	0	2	8	57	199	287	292	102	21	0	0	968
1980	0	0	0	4	87	210	438	415	199	17	0	0	1370
1981	0	0	1	29	43	270	341	267	109	2	0	0	1062
1982	0	0	0	4	171	121	383	252	101	62	9	7	1110
1983	0	0	4	3	27	248	465	487	219	21	0	0	1474
1984	0	0	0	17	50	340	254	312	141	44	1	0	1159
1985	0	0	5	40	67	189	317	245	155	49	4	0	1071
1986	0	0	4	34	115	285	427	269	197	42	0	0	1373
1987	0	0	0	10	212	304	383	395	173	2	5	0	1484
1988	0	0	1	8	81	306	442	407	120	5	0	0	1370
1989	0	0	8	34	66	214	362	296	146	27	0	0	1153
1990	0	0	13	29	32	239	326	285	168	26	5	0	1123
1991	0	0	3	15	210	285	398	341	198	41	0	0	1491
1992	0	0	0	35	76	155	340	192	132	8	0	0	938
1993	0	0	0	2	74	247	474	365	103	13	3	0	1281
1994	0	0	0	31	43	312	372	280	80	16	0	0	1134
1995	0	0	0	22	65	246	391	457	117	10	0	0	1308
1996	0	0	0	7	122	228	258	274	109	10	0	0	1008
1997	0	0	0	2	24	166	354	249	94	51	0	0	940
1998	0	0	25	0	132	264	303	356	291	24	1	1	1396
1999	0	0	0	5	67	261	459	339	161	6	1	0	1299
2000	0	1	0	2	113	268	279	278	126	45	3	0	1115
2001	0	0	0	75	111	212	327	351	124	24	2	0	1226

SNOWFALL (inches) 2001 LEXINGTON, KY (LEX)

YEAR	JUL	AUG	SEP	OCT	NOV	DEC	JAN	FEB	MAR	APR	MAY	JUN	TOTAL
1972-73	0.0	0.0	0.0	0.2	1.7	0.4	1.0	1.4	0.4	0.2	0.0	0.0	5.3
1973-74	0.0	0.0	0.0	0.0	0.0	4.3	T	3.6	0.7	T	0.0	0.0	8.6
1974-75	0.0	0.0	0.0	T	1.4	2.8	4.6	1.5	5.6	T	0.0	0.0	15.9
1975-76	0.0	0.0	0.0	0.0	T	0.9	6.5	0.4	2.2	0.0	0.0	0.0	10.0
1976-77	0.0	0.0	0.0	0.0	2.9	1.6	18.5	3.5	0.1	0.8	0.0	0.0	27.4
1977-78	0.0	0.0	0.0	0.0	1.7	3.0	21.9	7.1	8.4	0.0	0.0	0.0	42.1
1978-79	0.0	0.0	0.0	0.0	0.0	0.7	11.4	11.6	0.1	T	0.0	0.0	23.8
1979-80	0.0	0.0	0.0	0.0	0.1	T	11.9	4.0	4.2	0.3	0.0	0.0	20.5
1980-81	0.0	0.0	0.0	0.0	0.1	0.5	2.2	0.4	0.5	0.0	0.0	0.0	3.7
1981-82	0.0	0.0	0.0	0.0	0.4	1.7	5.6	3.9	0.3	0.7	0.0	0.0	12.6
1982-83	0.0	0.0	0.0	0.0	T	0.2	7.5	0.3	T	0.0	0.0	0.0	8.0
1983-84	0.0	0.0	0.0	0.0	T	1.7	8.4	4.6	0.3	0.0	0.0	0.0	15.0
1984-85	0.0	0.0	0.0	0.0	T	4.9	10.2	10.7	T	0.5	0.0	0.0	26.3
1985-86	0.0	0.0	0.0	0.0	0.0	3.5	1.2	8.9	0.7	T	0.0	0.0	14.3
1986-87	0.0	0.0	0.0	0.0	0.2	T	3.6	3.5	2.1	5.9	0.0	0.0	15.3
1987-88	0.0	0.0	0.0	0.0	1.0	1.8	3.3	3.4	0.7	0.0	0.0	0.0	10.2
1988-89	0.0	0.0	0.0	0.0	T	0.7	T	1.5	T	T	0.0	0.0	2.2
1989-90	T	T	0.0	T	1.1	9.3	0.2	T	3.7	T	0.0	0.0	14.3
1990-91	0.0	0.0	0.0	0.0	0.0	0.8	0.1	3.2	1.7	0.0	0.0	0.0	5.8
1991-92	0.0	0.0	0.0	0.0	T	0.2	0.7	0.6	1.3	0.4	0.0	0.0	3.2
1992-93	0.0	0.0	0.0	T	1.8	2.5	0.4	11.5	7.1	T	0.0	T	23.3
1993-94	0.0	0.0	0.0	T	0.1	7.4	16.4	2.8	5.0	0.0	0.0	0.0	31.7
1994-95	0.0	0.0	0.0	0.0	0.0	T	2.0	3.7	3.4	T	0.0	0.0	9.1
1995-96	0.0	0.0	0.0	T	1.1	2.2	16.0	3.8		1.5		0.0	
1996-97	0.0		0.0		T	T							
1997-98													
1998-99													
1999-00													
2000-01													
2001-													
POR= 51 YRS	T	T	0.0	0.0	0.6	1.9	5.7	4.6	2.6	0.3	T	T	15.7

2001
LOUISVILLE,
KENTUCKY (SDF)

Louisville is located on the south bank of the Ohio River, 604 miles below Pittsburgh, Pennsylvania, and 377 miles above the mouth of the river at Cairo, Illinois. The city is divided by Beargrass Creek and its south fork into two portions with entirely different types of topography. The eastern portion is rolling, containing several creeks, and consists of plateaus and rolling hillsides. The highest elevation in this area is 565 feet. The western portion is mostly flat with an average elevation about 100 feet lower than the eastern area. Much of the western section lies in the flood plain of the Ohio River. Nearly all of the industries in the city are located in the western portion, while the eastern portion is almost entirely residential. A range of low hills about five miles northwest of Louisville, on the Indiana side of the Ohio River, present a partial barrier to arctic blasts in the winter months. During colder months, snow is frequently observed on the summits of these hills when there is no snow in the city of Louisville or in riverside communities on the Indiana side of the Ohio River.

The climate of Louisville, while continental in type, is of a variable nature because of its position with respect to the paths of high and low pressure systems and the occasional influx of warm moist air from the Gulf of Mexico. In winter and summer there are occasional cold and hot spells of short duration. As a whole, winters are moderately cold and summers are quite warm. Temperatures of 100 degrees or more in summer and zero degrees or less in winter are rare.

Thunderstorms with high rainfall intensities are common during the spring and summer months. The precipitation in Louisville is nonseasonal and varies from year to year. The fall months are usually the driest. Generally, March has the most rainfall and October the least. Snowfall usually occurs from November through March. As with rainfall, amounts vary from year to year and month to month. Some snow has also been recorded in the months of October and April. Mean total amounts for the months of January, February, and March are about the same with January showing a slight edge in total amount. Relative humidity remains rather high throughout the summer months. Cloud cover is about equally distributed throughout the year with the winter months showing somewhat of an increase in amount. The percentage of possible sunshine at Louisville varies from month to month with the greatest amount during the summer months as a result of the decreasing sky cover during that season. Heavy fog is unusual and there is only an average of 10 days during the year with heavy fog and these occur generally in the months of September through March.

The average date for the last occurrence in the spring of temperatures as low as 32 degrees is mid April, and the first occurrence in the fall is generally in late October.

The prevailing direction of the wind has a southerly component and the velocity averages under 10 mph. The strongest winds are usually associated with thunderstorms.

NORMALS, MEANS, AND EXTREMES
LOUISVILLE, KY (SDF)

LATITUDE: 38 10' 38" N   LONGITUDE: 85 43' 47" W   ELEVATION (FT): GRND: 481 BARO: 484   TIME ZONE: EASTERN (UTC + 5)   WBAN: 93821

	ELEMENT	POR	JAN	FEB	MAR	APR	MAY	JUN	JUL	AUG	SEP	OCT	NOV	DEC	YEAR
TEMPERATURE F	NORMAL DAILY MAXIMUM	30	40.3	44.8	56.3	67.3	76.0	83.5	87.0	85.7	80.3	69.2	56.8	45.1	66.0
	MEAN DAILY MAXIMUM	51	41.4	46.3	55.7	67.6	76.4	84.5	88.1	87.0	80.5	69.3	56.1	45.4	66.5
	HIGHEST DAILY MAXIMUM	54	77	77	86	91	95	102	106	101	104	92	84	76	106
	YEAR OF OCCURRENCE		1950	2000	1981	1960	1959	1952	1999	1988	1954	1959	1958	1982	JUL 1999
	MEAN OF EXTREME MAXS.	54	64.2	67.9	77.4	84.4	88.6	93.8	94.0	93.7	90.6	84.2	74.9	66.1	81.7
	NORMAL DAILY MINIMUM	30	23.2	26.5	36.2	45.4	54.7	62.9	67.3	65.8	58.7	45.8	37.3	28.6	46.0
	MEAN DAILY MINIMUM	51	24.8	28.1	35.9	45.8	55.2	64.0	66.9	65.3	57.9	46.8	37.4	29.1	46.4
	LOWEST DAILY MINIMUM	54	-22	-19	-1	22	31	42	50	50	32	23	-1	-15	-22
	YEAR OF OCCURRENCE		1994	1951	1960	1982	1966	1966	1996	1996	1996	1952	1950	1989	JAN 1994
	MEAN OF EXTREME MINS.	54	3.6	8.6	19.3	30.2	40.1	51.2	56.4	54.7	42.8	31.4	20.8	9.7	30.7
	NORMAL DRY BULB	30	31.7	35.7	46.3	56.3	65.3	73.2	77.2	75.8	69.5	57.6	47.1	36.9	56.0
	MEAN DRY BULB	54	33.4	37.3	45.7	56.7	66.0	74.2	78.2	76.7	69.7	58.2	46.6	37.3	56.7
	MEAN WET BULB	49	30.0	33.4	40.3	49.5	58.8	66.8	70.3	69.0	62.8	52.0	41.8	33.8	50.7
	MEAN DEW POINT	49	23.9	27.1	33.2	42.5	53.7	62.7	66.5	65.2	58.5	47.0	36.1	28.2	45.4
	NORMAL NO. DAYS WITH:														
	MAXIMUM 90	30	0.0	0.0	0.0	*	0.4	5.9	11.0	9.2	3.0	0.0	0.0	0.0	29.5
	MAXIMUM 32	30	9.2	5.1	0.6	0.0	0.0	0.0	0.0	0.0	0.0	0.0	0.3	4.5	19.7
	MINIMUM 32	30	24.4	19.9	11.8	2.3	0.1	0.0	0.0	0.0	0.0	1.6	8.9	20.3	89.3
	MINIMUM 0	30	1.4	0.2	0.0	0.0	0.0	0.0	0.0	0.0	0.0	0.0	0.0	0.4	2.0
H/C	NORMAL HEATING DEG. DAYS	30	1032	820	580	273	105	6	0	0	36	254	537	871	4514
	NORMAL COOLING DEG. DAYS	30	0	0	0	12	115	252	378	335	171	25	0	0	1288
RH	NORMAL (PERCENT)	30	69	68	64	62	67	69	71	72	73	70	69	70	69
	HOUR 01 LST	30	72	72	69	68	77	80	81	82	83	79	74	73	76
	HOUR 07 LST	30	76	77	75	76	82	83	85	87	88	85	79	77	81
	HOUR 13 LST	30	63	62	56	52	55	56	58	58	58	55	60	64	58
	HOUR 19 LST	30	64	62	56	52	56	58	60	61	63	62	64	67	60
S	PERCENT POSSIBLE SUNSHINE	48	41	49	51	56	61	67	68	67	65	61	46	40	56
W/O	MEAN NO. DAYS WITH:														
	HEAVY FOG(VISBY 1/4 MI)	54	0.9	0.9	0.5	0.2	0.3	0.3	0.5	0.9	1.0	1.5	0.7	0.8	8.5
	THUNDERSTORMS	54	0.9	1.1	3.1	4.6	6.9	7.6	8.3	6.9	3.4	1.7	1.5	0.6	46.6
CLOUDINESS	MEAN:														
	SUNRISE-SUNSET (OKTAS)	1									3.2				
	MIDNIGHT-MIDNIGHT (OKTAS)	1									3.2				
	MEAN NO. DAYS WITH:														
	CLEAR	1					1.0		3.0	4.0	3.0	8.0		1.0	
	PARTLY CLOUDY	1			2.0						1.0				
	CLOUDY	1	1.0	4.0	6.0			3.0	1.0		3.0	4.0		2.0	
PR	MEAN STATION PRESSURE(IN)	28	29.61	29.57	29.49	29.47	29.46	29.46	29.49	29.52	29.54	29.57	29.57	29.59	29.53
	MEAN SEA-LEVEL PRES. (IN)	48	30.14	30.10	30.03	29.99	29.98	29.98	30.00	30.02	30.05	30.09	30.09	30.13	30.05
WINDS	MEAN SPEED (MPH)	49	9.6	9.6	10.1	9.8	8.0	7.4	6.9	6.4	6.8	7.2	9.0	9.1	8.3
	PREVAIL.DIR(TENS OF DEGS)	32	28	30	31	19	18	18	18	18	18	18	18	18	18
	MAXIMUM 2-MINUTE:														
	SPEED (MPH)	7	38	44	40	56	40	54	37	47	38	40	44	40	56
	DIR. (TENS OF DEGS)		21	23	20	22	23	04	21	21	27	29	23	27	22
	YEAR OF OCCURRENCE		1999	1999	1997	1999	1996	1998	1999	1999	2001	2001	1998	2000	APR 1999
	MAXIMUM 5-SECOND:														
	SPEED (MPH)	7	43	51	47	56	53	60	44	56	44	46	49	49	60
	DIR. (TENS OF DEGS)		26	23	20	24	22	04	34	22	27	31	22	27	04
	YEAR OF OCCURRENCE		2000	1999	1998	1999	1996	1998	2001	1999	2001	2001	1998	2000	JUN 1998
PRECIPITATION	NORMAL (IN)	30	2.86	3.30	4.66	4.23	4.62	3.46	4.51	3.54	3.16	2.71	3.70	3.64	44.39
	MAXIMUM MONTHLY (IN)	54	11.38	9.02	14.91	11.10	11.57	10.11	10.05	8.79	10.49	6.47	9.12	8.86	14.91
	YEAR OF OCCURRENCE		1950	1989	1964	1970	1990	1960	1979	1974	1979	1983	1957	1990	MAR 1964
	MINIMUM MONTHLY (IN)	54	0.45	0.76	1.02	0.76	1.37	0.49	0.34	0.23	0.01	0.39	0.72	0.65	0.01
	YEAR OF OCCURRENCE		1981	1978	1966	1976	1977	1984	1999	1953	1995	1987	1976	1976	SEP 1995
	MAXIMUM IN 24 HOURS (IN)	54	3.99	3.66	7.22	4.85	4.60	5.14	5.46	3.13	4.97	3.25	3.58	2.79	7.22
	YEAR OF OCCURRENCE		2000	1990	1997	1970	1961	1960	1979	1992	1979	1977	1948	1978	MAR 1997
	NORMAL NO. DAYS WITH:														
	PRECIPITATION 0.01	30	10.3	10.5	13.1	11.4	11.8	9.6	10.7	8.6	8.4	7.7	10.7	11.9	124.7
	PRECIPITATION 1.00	30	0.5	0.8	1.0	1.0	1.1	0.9	1.0	0.8	0.6	0.6	1.0	0.9	10.2
SNOWFALL	NORMAL (IN)	30	5.9	5.0	3.1	0.2	T	0.0	0.0	0.0	0.0	0.*	1.0	2.2	17.4
	MAXIMUM MONTHLY (IN)	54	28.4	19.3	22.9	1.6	T	T	T	T	0.0	2.4	13.2	9.7	28.4
	YEAR OF OCCURRENCE		1978	1998	1960	1973	1989	1993	1994	1999		1993	1966	2000	JAN 1978
	MAXIMUM IN 24 HOURS (IN)	54	15.9	11.0	12.1	1.6	T	T	T	T	0.0	2.4	13.0	5.0	15.9
	YEAR OF OCCURRENCE		1994	1966	1968	1973	1989	1993	1994	1999		1993	1966	1961	JAN 1994
	MAXIMUM SNOW DEPTH (IN)	50	19	11	11	2	0	0	0	0	0	T	8	5	19
	YEAR OF OCCURRENCE		1978	1966	1968	1987						1989	1966	1984	JAN 1978
	NORMAL NO. DAYS WITH:														
	SNOWFALL 1.0	30	1.7	1.3	0.7	0.1	0.0	0.0	0.0	0.0	0.0	0.*	0.3	0.7	4.8

PRECIPITATION (inches) 2001 LOUISVILLE, KY (SDF)

YEAR	JAN	FEB	MAR	APR	MAY	JUN	JUL	AUG	SEP	OCT	NOV	DEC	
1972	2.87	3.94	4.07	8.48	4.46	1.08	3.64	2.45	4.24	2.55	6.31	5	
1973	1.96	1.60	6.26	5.77	7.04	6.20	9.38	0.91	2.34	2.28	7.59	·	
1974	4.38	1.64	5.41	2.74	3.86	2.58	2.04	8.79	3.52	2.09	3.03		
1975	4.87	4.53	9.65	6.47	4.50	3.15	1.91	3.89	2.64	6.12	3.69	4.8	
1976	3.85	3.13	2.87	0.76	5.09	4.71	3.10	3.99	0.72	0.65			
1977	2.33	1.45	4.69	3.40	1.37	7.59	3.29	6.12	3.67	4.76	4.32		49.10
1978	5.90	0.76	3.76	3.33	4.76	2.67	3.77	5.50	0.96	2.26	5.14	7.64	46.45
1979	3.81	4.49	2.71	7.32	3.59	3.03	10.05	2.37	10.49	2.27	5.85	3.82	59.80
1980	1.71	1.09	4.80	2.63	4.58	3.70	5.41	3.76	3.17	3.37	2.42	1.25	37.89
1981	0.45	3.23	1.54	4.44	4.63	3.23	3.98	3.21	3.22	1.60	2.40	2.02	33.95
1982	5.28	1.55	5.89	3.05	2.96	3.86	3.72	3.74	3.46	1.26	5.50	5.11	45.38
1983	1.63	1.52	2.16	7.10	10.58	4.42	0.99	2.39	1.13	6.47	5.03	3.96	47.38
1984	0.92	1.68	4.41	5.53	6.78	0.49	6.94	5.08	3.70	2.12	5.87	5.86	49.38
1985	2.20	2.08	4.43	1.69	3.93	4.37	3.45	4.49	1.48	4.24	0.96		37.75
1986	0.91	3.90	2.69	1.04	4.28	2.32	7.04	2.19	2.75	3.08	4.62	2.69	37.51
1987	0.81	4.42	3.05	2.35	1.61	3.58	3.58	2.66	1.15	0.39	2.62	4.70	32.65
1988	4.00	3.58	2.97	3.52	2.68	0.87	4.68	3.00	1.48	1.54	5.76	3.45	37.53
1989	3.68	9.02	5.50	4.93	4.39	5.26	6.90	2.20	2.42	2.65	2.57	1.45	50.97
1990	3.90	6.72	2.78	3.46	11.57	6.13	1.96	3.21	2.57	3.97	2.34	8.86	57.47
1991	3.29	3.72	4.79	2.61	4.02	1.23	2.99	3.35	2.74	2.31	1.87	5.23	38.15
1992	1.97	1.74	5.88	2.66	3.51	3.04	6.51	4.71	3.50	0.96	4.71	1.60	40.79
1993	3.50	4.20	5.20	3.57	2.80	4.05	4.58	5.74	3.90	4.03	3.26	2.56	47.39
1994	4.08	2.96	3.90	5.32	2.12	1.85	2.50	1.58	2.90	1.96	3.57	3.24	35.98
1995	3.20	2.00	2.17	2.64	9.48	2.84	3.39	4.07	0.01	5.42	2.39	3.28	40.89
1996	4.44	2.03	4.99	5.65	9.18	3.84	1.31	5.66	2.59	3.35	4.56		
1997	3.35	3.39	12.58	2.01	6.01	8.11	1.74	3.70	1.28	1.41	3.63	2.50	49.71
1998	2.88	2.88	4.07	6.69	4.53	5.73	6.89	2.92	1.00	2.76	2.74	3.24	46.33
1999	7.23	2.20	3.47	3.04	3.12	6.36	0.34	0.97	1.74	2.46	1.61	4.81	37.35
2000	6.22	5.80	3.56	2.95	2.91	3.88	3.50	2.87	5.36	0.89	2.97	4.31	45.22
2001	1.46	3.42	2.27	1.04	5.19	2.61	2.61	4.47	3.42	4.08	6.39	5.16	43.99
POR= 129 YRS	3.78	3.38	4.52	3.89	4.08	3.86	3.79	3.28	2.81	2.63	3.51	3.61	43.14

AVERAGE TEMPERATURE (F) 2001 LOUISVILLE, KY (SDF)

YEAR	JAN	FEB	MAR	APR	MAY	JUN	JUL	AUG	SEP	OCT	NOV	DEC	ANNUAL
1972	35.2	34.9	44.8	56.2	65.5	70.6	77.1	76.1	72.3	55.3	44.0	39.1	55.9
1973	35.0	36.4	53.7	54.4	61.5	75.6	78.4	77.0	73.6	62.3	49.8	37.1	57.9
1974	39.8	39.3	49.8	57.2	65.1	68.7	75.9	75.0	63.2	54.9	47.0	39.1	56.3
1975	38.1	40.2	43.3	54.4	69.0	75.4	77.7	79.3	66.2	59.4	50.6	39.1	57.7
1976	31.3	45.4	52.4	57.5	62.9	72.9	76.8	74.2	66.8	52.5	39.5	33.1	55.4
1977	18.6	36.9	51.7	60.3	71.2	73.9	80.2	77.5	72.5	55.5	49.6	34.6	56.9
1978	22.9	23.8	41.7	58.0	63.8	75.7	78.5	77.1	73.7	55.5	50.0	40.0	55.1
1979	24.6	28.0	48.3	55.0	64.2	73.9	75.3	76.1	69.4	58.2	46.9	39.2	54.9
1980	33.5	29.6	41.8	53.6	66.8	73.4	81.5	81.0	73.5	55.8	46.3	38.3	56.3
1981	30.4	38.8	45.7	62.4	62.9	76.2	78.8	76.1	67.7	56.5	47.4	33.8	56.4
1982	28.6	34.9	47.1	51.3	70.3	69.3	78.0	73.5	66.8	59.0	48.7	44.9	56.0
1983	34.7	37.5	46.7	51.7	62.1	73.4	81.1	81.7	71.0	59.1	47.8	28.4	56.3
1984	28.9	41.5	40.4	55.0	62.6	77.7	75.5	76.0	67.2	63.9	44.0	45.9	56.6
1985	25.4	32.8	50.2	60.3	66.5	72.1	77.2	74.8	69.2	61.4	53.7	30.4	56.2
1986	34.5	39.9	48.3	58.5	67.0	75.7	80.3	74.3	73.1	59.5	45.9	36.7	57.8
1987	33.7	39.5	47.9	55.4	71.5	76.2	78.9	78.2	71.2	52.6	50.8	40.2	58.0
1988	31.0	34.7	46.1	57.0	67.1	75.6	80.3	80.0	70.1	52.3	47.8	38.0	56.7
1989	41.6	34.0	48.4	56.7	62.6	73.5	78.1	76.6	69.4	58.4	46.7	25.3	55.9
1990	43.1	44.3	51.2	55.5	64.2	75.1	78.5	77.5	71.8	58.7	52.0	40.8	59.4
1991	34.1	40.5	49.4	60.3	73.1	78.3	81.3	79.2	71.7	61.5	45.0	41.4	59.7
1992	37.1	43.7	47.9	58.2	63.9	72.1	78.5	73.2	69.1	58.1	47.9	38.6	57.4
1993	38.5	34.0	44.0	54.9	66.8	74.5	82.0	79.0	68.0	55.9	45.8	36.5	56.7
1994	26.8	38.0	45.4	59.9	63.0	77.5	79.0	76.1	68.7	59.4	52.2	42.4	57.4
1995	35.6	36.2	49.5	57.7	65.6	74.9	79.7	82.2	68.7	59.2	41.7	35.2	57.2
1996	32.4	37.7	41.0	53.6	68.5	74.6				58.9	41.8	40.9	
1997	32.3	42.2	49.5	52.6	61.4	72.2	78.7	75.4	70.3	58.8	43.9	37.2	56.2
1998	42.4	43.6	48.1	56.2	70.3	75.0	78.3	78.8	76.2	62.6	51.3	42.0	60.4
1999	36.8	42.6	43.0	59.3	67.6	76.3	83.2	78.5	72.1	58.9	53.5	39.1	59.2
2000	34.4	45.1	51.4	55.8	69.2	75.0	76.8	77.1	68.0	61.4	44.6	26.2	57.1
2001	33.0	41.1	43.0	62.3	68.6	73.8	78.8	79.1	68.8	58.9	53.5	42.6	58.6
POR= 129 YRS	34.2	37.0	45.8	56.5	66.0	74.6	77.8	76.2	69.8	58.7	46.8	37.1	56.7

REFERENCE NOTES:

PAGE 1:
THE TEMPERATURE GRAPH SHOWS NORMAL MAXIMUM AND NORMAL MINIMUM DAILY TEMPERATURES (SOLID CURVES) AND THE ACTUAL DAILY HIGH AND LOW TEMPERATURES (VERTICAL BARS).

PAGE 2 AND 3:
H/C INDICATES HEATING AND COOLING DEGREE DAYS.
RH INDICATES RELATIVE HUMIDITY
W/O INDICATES WEATHER AND OBSTRUCTIONS
S INDICATES SUNSHINE.
PR INDICATES PRESSURE.
CLOUDINESS ON PAGE 3 IS THE SUM OF THE CEILOMETER AND SATELLITE DATA NOT TO EXCEED EIGHT EIGHTHS(OKTAS).

GENERAL:
T INDICATES TRACE PRECIPITATION, AN AMOUNT GREATER THAN ZERO BUT LESS THAN THE LOWEST REPORTABLE VALUE.
+ INDICATES THE VALUE ALSO OCCURS ON EARLIER DATES.
BLANK ENTRIES DENOTE MISSING OR UNREPORTED DATA.
NORMALS ARE 30-YEAR AVERAGES (1961 - 1990).
ASOS INDICATES AUTOMATED SURFACE OBSERVING SYSTEM.
PM INDICATES THE LAST DAY OF THE PREVIOUS MONTH.
POR (PERIOD OF RECORD) BEGINS WITH THE JANUARY DATA MONTH AND IS THE NUMBER OF YEARS USED TO COMPUTE THE MEAN. INDIVIDUAL MONTHS WITHIN THE POR MAY BE MISSING.
WHEN THE POR FOR A NORMAL IS LESS THAN 30 YEARS, THE NORMAL IS PROVISIONAL AND IS BASED ON THE NUMBER OF YEARS INDICATED.
0.* OR * INDICATES THE VALUE OR MEAN-DAYS-WITH IS BETWEEN 0.00 AND 0.05.
CLOUDINESS FOR ASOS STATIONS DIFFERS FROM THE NON-ASOS OBSERVATION TAKEN BY A HUMAN OBSERVER. ASOS STATION CLOUDINESS IS BASED ON TIME-AVERAGED CEILOMETER DATA FOR CLOUDS AT OR BELOW 12,000 FEET AND ON SATELLITE DATA FOR CLOUDS ABOVE 12,000 FEET.
THE NUMBER OF DAYS WITH CLEAR, PARTLY CLOUDY, AND CLOUDY CONDITIONS FOR ASOS STATIONS IS THE SUM OF THE CEILOMETER AND SATELLITE DATA FOR THE SUNRISE TO SUNSET PERIOD.

GENERAL CONTINUED:
CLEAR INDICATES 0 - 2 OKTAS, PARTLY CLOUDY INDICATES 3 - 6 OKTAS, AND CLOUDY INDICATES 7 OR 8 OKTAS.
WHEN AT LEAST ONE OF THE ELEMENTS (CEILOMETER OR SATELLITE) IS MISSING, THE DAILY CLOUDINESS IS NOT COMPUTED.
WIND DIRECTION IS RECORDED IN TENS OF DEGREES (2 DIGITS) CLOCKWISE FROM TRUE NORTH. "00" INDICATES CALM. "36" INDICATES TRUE NORTH.
RESULTANT WIND IS THE VECTOR AVERAGE OF THE SPEED AND DIRECTION.
AVERAGE TEMPERATURE IS THE SUM OF THE MEAN DAILY MAXIMUM AND MINIMUM TEMPERATURE DIVIDED BY 2.
SNOWFALL DATA COMPRISE ALL FORMS OF FROZEN PRECIPITATION, INCLUDING HAIL.
A HEATING (COOLING) DEGREE DAY IS THE DIFFERENCE BETWEEN THE AVERAGE DAILY TEMPERATURE AND 65 F.
DRY BULB IS THE TEMPERATURE OF THE AMBIENT AIR.
DEW POINT IS THE TEMPERATURE TO WHICH THE AIR MUST BE COOLED TO ACHIEVE 100 PERCENT RELATIVE HUMIDITY.
WET BULB IS THE TEMPERATURE THE AIR WOULD HAVE IF THE MOISTURE CONTENT WAS INCREASED TO 100 PERCENT RELATIVE HUMIDITY.

ON JULY 1, 1996, THE NATIONAL WEATHER SERVICE BEGAN USING THE "METAR" OBSERVATION CODE THAT WAS ALREADY EMPLOYED BY MOST OTHER NATIONS OF THE WORLD. THE MOST NOTICEABLE DIFFERENCE IN THIS ANNUAL PUBLICATION WILL BE THE CHANGE IN UNITS FROM TENTHS TO EIGHTS(OKTAS) FOR REPORTING THE AMOUNT OF SKY COVER.

HEATING DEGREE DAYS (base 65 F) 2001 LOUISVILLE, KY (SDF)

YEAR	JUL	AUG	SEP	OCT	NOV	DEC	JAN	FEB	MAR	APR	MAY	JUN	TOTAL
1972-73	0	0	16	298	628	793	927	796	349	343	129	0	4279
1973-74	0	0	13	144	450	860	772	714	487	257	99	19	3815
1974-75	0	0	122	314	543	794	830	688	665	333	22	0	4311
1975-76	0	0	73	205	431	801	1040	562	405	266	111	1	3895
1976-77	0	0	29	393	757	982	1435	780	421	183	36	7	5023
1977-78	0	0	6	295	472	935	1294	1145	720	221	142	1	5231
1978-79	0	0	4	293	442	765	1246	1030	514	301	94	5	4694
1979-80	0	0	19	244	534	792	969	1021	713	342	68	8	4710
1980-81	0	0	12	309	555	821	1065	728	595	142	122	0	4349
1981-82	0	0	61	268	523	960	1124	837	549	408	13	3	4746
1982-83	0	1	56	246	495	624	933	763	571	399	121	5	4214
1983-84	0	0	54	196	509	1128	1115	673	757	315	141	0	4888
1984-85	0	0	73	84	623	584	1222	896	458	180	52	16	4188
1985-86	0	0	53	160	347	1067	941	696	516	224	69	0	4073
1986-87	0	12	5	210	570	869	962	706	526	294	21	0	4175
1987-88	0	0	9	377	423	762	1048	872	580	244	38	7	4360
1988-89	0	0	13	398	510	833	720	860	513	291	156	4	4298
1989-90	0	0	49	230	539	1222	672	574	445	320	82	13	4146
1990-91	0	0	34	229	387	745	949	677	482	167	27	0	3697
1991-92	0	0	52	168	590	725	855	610	523	244	124	8	3899
1992-93	0	0	40	219	505	813	819	859	644	299	44	18	4260
1993-94	0	0	48	289	572	875	1180	752	602	189	122	3	4632
1994-95	0	0	20	186	384	696	904	800	471	236	72	0	3769
1995-96	0	0	48	192	693	915	1002	782	738	353	66	2	4791
1996-97				202	689	741	1005	634	472	366	140	12	
1997-98	0	0	9	263	621	854	696	594	561	261	27	15	3901
1998-99	0	0	0	119	405	711	866	620	676	183	16	0	3596
1999-00	0	0	20	197	346	796	942	575	416	270	26	2	3590
2000-01	0	0	60	174	608	1198	980	664	674	175	31	8	4572
2001-	0	0	52	224	339	689							

COOLING DEGREE DAYS (base 65 F) 2001 LOUISVILLE, KY (SDF)

YEAR	JAN	FEB	MAR	APR	MAY	JUN	JUL	AUG	SEP	OCT	NOV	DEC	ANNUAL
1972	0	0	3	25	81	193	386	351	242	2	4	0	1287
1973	0	0	7	29	28	325	422	380	280	71	2	0	1544
1974	0	0	22	31	109	136	345	319	75	8	10	0	1055
1975	0	0	0	24	152	320	402	451	116	36	5	0	1506
1976	0	0	21	47	51	243	372	294	92	10	0	0	1130
1977	0	0	14	50	234	281	479	396	238	5	20	0	1717
1978	0	0	0	20	110	323	425	383	270	6	2	0	1539
1979	0	0	5	10	73	279	326	350	154	39	0	0	1236
1980	0	0	0	8	134	266	519	504	276	31	1	0	1739
1981	0	0	5	68	63	343	435	348	150	10	0	0	1422
1982	0	0	1	2	183	139	408	274	118	68	13	8	1214
1983	0	0	7	8	39	264	504	524	240	19	0	0	1605
1984	0	0	0	20	69	386	333	349	145	56	0	1	1359
1985	0	2	8	48	106	233	387	311	185	55	14	0	1349
1986	0	0	5	37	138	330	481	306	255	46	0	0	1598
1987	0	0	0	14	232	342	439	416	203	1	4	0	1651
1988	0	0	4	10	111	333	481	472	173	10	0	0	1594
1989	0	0	0	6	48	88	264	412	364	188	30	0	1400
1990	0	0	22	44	65	323	427	392	244	42	7	0	1566
1991	0	0	8	31	286	406	514	445	262	68	0	0	2020
1992	0	0	2	48	100	229	424	262	169	14	0	0	1248
1993	0	0	0	4	106	310	534	442	146	12	3	0	1557
1994	0	0	0	42	63	384	443	349	138	21	4	0	1444
1995	0	0	0	29	100	304	466	544	165	21	0	0	1629
1996	0	0	0	18	183	298				19	0		
1997	0	0	0	2	35	237	435	330	174	79	0	0	1292
1998	0	0	44	4	200	321	416	436	345	52	0	7	1825
1999	0	0	0	18	103	345	573	427	238	12	6	0	1722
2000	0	0	5	2	162	308	375	385	156	68	2	0	1463
2001	0	0	0	103	150	280	433	445	174	40	1	0	1626

SNOWFALL (inches) 2001 LOUISVILLE, KY (SDF)

YEAR	JUL	AUG	SEP	OCT	NOV	DEC	JAN	FEB	MAR	APR	MAY	JUN	TOTAL
1972-73	0.0	0.0	0.0	0.0	2.0	2.2	1.1	1.1	0.5	1.6	0.0	0.0	8.5
1973-74	0.0	0.0	0.0	0.0	0.0	4.5	1.0	0.9	2.8	T	0.0	0.0	9.2
1974-75	0.0	0.0	0.0	0.0	1.0	1.2	3.0	1.3	10.0	T	0.0	0.0	16.5
1975-76	0.0	0.0	0.0	0.0	0.1	0.7	2.5	0.1	0.7	0.0	0.0	0.0	4.1
1976-77	0.0	0.0	0.0	0.0	1.6	1.1	19.6	0.8	0.1	0.8	0.0	0.0	24.0
1977-78	0.0	0.0	0.0	0.0	4.8	2.2	28.4	5.3	9.4	T	0.0	0.0	50.1
1978-79	0.0	0.0	0.0	0.0	0.0	T	8.5	10.9	0.9	T	0.0	0.0	20.3
1979-80	0.0	0.0	0.0	0.0	0.1	T	10.7	3.6	3.9	T	0.0	0.0	18.3
1980-81	0.0	0.0	0.0	T	T	T	2.5	0.3	0.1	0.0	0.0	0.0	2.9
1981-82	0.0	0.0	0.0	0.0	0.1	3.6	2.7	2.9	0.3	1.4	0.0	0.0	11.0
1982-83	0.0	0.0	0.0	0.0	T	0.0	0.6	4.5	0.1	T	0.0	0.0	5.2
1983-84	0.0	0.0	0.0	0.0	0.0	0.6	3.1	8.8	1.0	0.0	0.0	0.0	13.5
1984-85	0.0	0.0	0.0	0.0	T	4.8	7.4	6.7	T	T	0.0	0.0	18.9
1985-86	0.0	0.0	0.0	0.0	0.0	1.6	1.1	8.8	0.1	0.0	0.0	0.0	11.6
1986-87	0.0	0.0	0.0	0.0	T	T	2.2	6.7	9.3	T	0.0	0.0	18.2
1987-88	0.0	0.0	0.0	0.0	T	T	3.0	5.0	0.5	0.0	0.0	0.0	8.5
1988-89	0.0	0.0	0.0	0.0	T	T	0.3	T	0.6	T	0.0	0.0	0.9
1989-90	0.0	0.0	0.0	1.4	T	6.5	1.9	0.8	4.1	T	0.0	0.0	14.7
1990-91	0.0	0.0	0.0	0.0	0.0	4.1	0.3	1.5	0.2	0.0	0.0	0.0	6.1
1991-92	0.0	0.0	0.0	0.0	0.5	0.1	0.9	0.1	0.9	0.7	0.0	0.0	3.2
1992-93	0.0	0.0	0.0	0.0	0.9	1.0	T	15.9	1.1	T	0.0	T	18.9
1993-94	0.0	0.0	0.0	2.4	T	3.6	17.7	1.5	4.7	T	0.0	0.0	29.9
1994-95	T	0.0	0.0	0.0	0.0	T	0.1	2.8	1.1	0.0	0.0	0.0	4.0
1995-96	0.0	0.0	0.0	0.0	T	1.1	13.8	1.3	8.0	T			
1996-97					0.0	T	0.4	3.0	1.9	T	T	0.0	
1997-98	0.0	0.0	0.0	0.0	0.4	0.0	1.9	19.3	1.2	T	T	0.0	22.8
1998-99	0.0	0.0	0.0	0.0	0.0	2.8	5.5	3.7	1.3	0.0	0.0	T	13.3
1999-00	0.0	0.0	T	0.0	0.0	4.9	6.7	0.1	0.3	T	0.0	0.0	12.0
2000-01	0.0	0.0	0.0	0.0	T	9.7	4.8	0.3	T	T	0.0	0.0	14.8
2001-	0.0	0.0	0.0	T	0.0	0.6							
POR= 54 YRS	T	0.0	0.0	0.1	T	2.3	5.3	4.4	3.0	T	T	T	15.1

2001
NEW ORLEANS,
LOUISIANA (MSY)

The New Orleans metropolitan area is virtually surrounded by water. Lake Pontchartrain, some 610 square miles in area, borders the city on the north and is connected to the Gulf of Mexico through Lake Borgne on the east. In other directions there are bayous, lakes, and marshy delta land. The proximity of the Gulf of Mexico also has a great influence on the climate. Elevations in the city vary from a few feet below to a few feet above mean sea level. A massive levee system surrounding the city and along the Mississippi River offers protection against flooding from the river and tidal surges. The New Orleans International Airport is located 12 miles west of downtown New Orleans, between the Mississippi River and Lake Pontchartrain.

The climate of the city can best be described as humid with the surrounding water modifying the temperature and decreasing the range between the extremes. Almost daily sporadic afternoon thunderstorms from mid-June through September keep the temperature from rising much above 90 degrees. From about mid-November to mid-March, the area is subjected alternately to the southerly flow of warm tropical air and to the northerly flow of cold continental air in periods of varying lengths. The usual track of winter storms is to the north of New Orleans, but occasionally one moves this far south, bringing large and rather sudden drops in temperature. However, the cold spells seldom last over three or four days. The lowest temperatures observed are below 10 degrees. In about two-thirds of the years, the lowest temperature is about 24 degrees or warmer. The lowest temperatures in some years are entirely above freezing.

During the winter and spring, the cold Mississippi River water enhances the formation of river fogs, particularly when light southerly winds bring warm, moist air into the area from the Gulf of Mexico. The nearby lakes and marshes also contribute to fog formation. Even so, the fog usually does not seriously affect automobile traffic except for brief periods. However, air travel will be suspended for several hours and river traffic, at times, will be unable to move between New Orleans and the Gulf for several days.

Rather frequent and sometimes very heavy rains are typical for this area. There are an average of 120 days of measurable rain per year and an annual average accumulation of over 60 inches. A fairly definite rainy period occurs from mid-December to mid-March. Precipitation during this period is most likely to be steady rain for two to three day periods. April, May, October, and November are generally dry, but there have been some extremely heavy showers in those months. The greatest 24-hour amounts have exceeded 14 inches. Snowfall is rather infrequent and light. However, on rare occasions, snowstorms have produced accumulations over 8 inches.

While thunder occurs with most of the showers in the area, thunderstorms with damaging winds are infrequent. Hail of a damaging nature seldom occurs, and tornadoes are extremely rare. However, waterspouts are observed quite often on nearby lakes. Hurricanes have effected the area.

The lower Mississippi River floods result from runoff upstream. If the water level in the river becomes dangerously high, the spillways upriver can be opened to divert the floodwaters. Rainfall in the New Orleans area is pumped into the surrounding lakes and bayous. Local street and minor urban flooding of short duration result from occasional downpours.

Air pollution is not a serious problem. The area is not highly industrialized, and long periods of air stagnation are rare.

Based on the 1951-1980 period, the average first occurrence of 32 degrees Fahrenheit in the fall is December 5 and the average last occurrence in the spring is February 20.

NORMALS, MEANS, AND EXTREMES
NEW ORLEANS, LA (MSY)

LATITUDE: 29 59' 34" N   LONGITUDE: 90 15' 03" W   ELEVATION (FT): GRND: 4   BARO: 7   TIME ZONE: CENTRAL (UTC + 6)   WBAN: 12916

	ELEMENT	POR	JAN	FEB	MAR	APR	MAY	JUN	JUL	AUG	SEP	OCT	NOV	DEC	YEAR
TEMPERATURE °F	NORMAL DAILY MAXIMUM	30	60.8	64.1	71.6	78.5	84.4	89.2	90.6	90.2	86.6	79.4	71.1	64.3	77.6
	MEAN DAILY MAXIMUM	54	62.0	65.3	71.2	78.3	84.9	89.4	90.9	90.6	86.9	79.6	70.7	64.4	77.8
	HIGHEST DAILY MAXIMUM	55	83	85	89	92	96	100	101	102	101	94	87	84	102
	YEAR OF OCCURRENCE		1982	1972	1982	1987	1953	1954	1981	1980	1980	1998	1997	1995	AUG 1980
	MEAN OF EXTREME MAXS.	54	77.4	79.4	82.5	86.5	91.4	94.6	95.8	95.6	93.2	88.6	83.2	79.9	87.3
	NORMAL DAILY MINIMUM	30	41.8	44.4	51.6	58.4	65.2	70.8	73.1	72.8	69.5	58.7	51.0	44.8	58.5
	MEAN DAILY MINIMUM	54	43.6	46.2	51.9	58.7	65.8	71.3	73.5	73.4	70.0	59.9	50.8	45.5	59.2
	LOWEST DAILY MINIMUM	55	14	16	25	32	41	50	60	60	42	35	24	11	11
	YEAR OF OCCURRENCE		1985	1996	1980	1971	1960	1984	1967	1968	1967	1993	1970	1989	DEC 1989
	MEAN OF EXTREME MINS.	54	26.4	29.9	35.8	43.6	54.1	63.6	68.2	68.4	59.7	44.3	34.8	28.4	46.5
	NORMAL DRY BULB	30	51.3	54.3	61.6	68.5	74.8	80.0	81.9	81.5	78.1	69.1	61.1	54.5	68.1
	MEAN DRY BULB	54	52.8	55.7	61.7	68.5	75.4	80.4	82.2	82.0	78.5	69.7	60.8	55.0	68.6
	MEAN WET BULB	17	48.7	52.5	56.8	62.5	69.9	74.5	76.0	76.0	72.6	64.3	57.4	48.6	63.3
	MEAN DEW POINT	17	44.2	47.9	52.3	58.2	66.4	72.0	73.7	73.7	69.6	60.4	53.3	44.7	59.7
	NORMAL NO. DAYS WITH:														
	MAXIMUM ≥ 90	30	0.0	0.0	0.0	0.2	3.1	14.4	20.4	19.8	8.9	0.9	0.0	0.2	67.7
	MAXIMUM ≤ 32	30	0.2	0.0	0.0	0.0	0.0	0.0	0.0	0.0	0.0	0.0	0.0	0.2	0.4
	MINIMUM ≤ 32	30	6.2	3.5	0.6	*	0.0	0.0	0.0	0.0	0.0	0.0	0.8	4.7	15.8
	MINIMUM ≤ 0	30	0.0	0.0	0.0	0.0	0.0	0.0	0.0	0.0	0.0	0.0	0.0	0.0	0.0
H/C	NORMAL HEATING DEG. DAYS	30	450	316	162	28	0	0	0	0	0	30	178	349	1513
	NORMAL COOLING DEG. DAYS	30	25	17	56	133	304	450	524	512	393	157	61	23	2655
RH	NORMAL (PERCENT)	30	76	73	73	73	74	76	79	79	78	75	77	77	76
	HOUR 00 LST	30	81	80	81	84	85	87	88	88	86	84	85	83	84
	HOUR 06 LST	30	84	83	85	88	89	90	92	92	89	88	87	86	88
	HOUR 12 LST	30	66	63	61	59	60	64	66	67	65	59	63	66	63
	HOUR 18 LST	30	71	66	64	64	65	68	72	73	74	72	76	74	70
S	PERCENT POSSIBLE SUNSHINE	22	46	50	56	62	62	63	59	61	61	64	54	48	57
W/O	MEAN NO. DAYS WITH:														
	HEAVY FOG(VISBY 1/4 MI)	53	5.7	4.2	3.5	1.6	0.9	0.2	0.1	0.2	0.3	1.6	3.5	4.5	26.3
	THUNDERSTORMS	53	2.2	2.9	4.0	4.2	5.8	9.8	14.7	12.4	6.6	1.9	2.0	2.1	68.6
CLOUDINESS	MEAN:														
	SUNRISE-SUNSET (OKTAS)	48	5.4	5.0	5.0	4.6	4.9	4.9	5.1	4.6	4.3	3.6	4.3	5.1	4.7
	MIDNIGHT-MIDNIGHT (OKTAS)	32	5.2	4.8	4.9	4.4	4.2	4.0	4.6	4.3	4.0	3.3	4.0	4.8	4.4
	MEAN NO. DAYS WITH:														
	CLEAR	48	6.9	7.5	7.8	7.9	8.9	8.3	4.6	7.2	9.6	14.3	10.2	7.7	100.9
	PARTLY CLOUDY	48	7.1	6.4	8.0	10.4	11.2	12.5	14.6	13.8	10.6	7.9	8.2	7.4	118.1
	CLOUDY	48	16.9	14.3	15.2	11.7	10.9	9.2	11.8	10.0	9.8	8.9	11.5	15.9	146.1
PR	MEAN STATION PRESSURE(IN)	28	30.11	30.08	30.00	29.98	29.94	29.96	30.00	29.98	29.96	30.03	30.08	30.11	30.02
	MEAN SEA-LEVEL PRES. (IN)	17	30.15	30.11	30.05	30.01	29.98	30.01	30.04	30.00	29.99	30.06	30.11	30.16	30.06
WINDS	MEAN SPEED (MPH)	39	9.4	9.9	9.9	9.5	8.1	6.9	6.2	6.1	7.5	7.8	8.8	9.2	8.3
	PREVAIL.DIR(TENS OF DEGS)	23	36	36	16	16	18	18	23	05	04	05	06	36	18
	MAXIMUM 2-MINUTE:														
	SPEED (MPH)	5	48	43	36	40	33	36	39	40	46	32	36	39	48
	DIR. (TENS OF DEGS)		27	21	25	01	28	28	24	02	02	03	26	24	27
	YEAR OF OCCURRENCE		1998	1998	2001	1997	2001	2001	1998	1997	1998	2000	2001	2000	JAN 1998
	MAXIMUM 5-SECOND:														
	SPEED (MPH)	5	63	51	44	53	44	48	47	48	55	38	44	51	63
	DIR. (TENS OF DEGS)		33	21	10	21	26	25	24	04	01	02	26	24	33
	YEAR OF OCCURRENCE		1998	1998	2001	1997	1999	2001	1998	1997	1998	2000	2001	2001	JAN 1998
PRECIPITATION	NORMAL (IN)	30	5.05	6.01	4.90	4.50	4.56	5.84	6.12	6.17	5.51	3.05	4.42	5.75	61.88
	MAXIMUM MONTHLY (IN)	55	19.28	12.59	19.09	16.12	21.18	17.62	13.15	16.12	18.98	13.20	19.81	10.77	21.18
	YEAR OF OCCURRENCE		1998	1983	1948	1980	1995	2001	1991	1977	1998	1985	1989	1967	MAY 1995
	MINIMUM MONTHLY (IN)	55	0.54	0.15	0.24	0.28	0.07	0.23	1.38	1.68	0.24	0.00	0.21	1.46	0.00
	YEAR OF OCCURRENCE		1968	1989	1955	1976	2000	1979	2000	1980	1953	1978	1949	1958	OCT 1978
	MAXIMUM IN 24 HOURS (IN)	55	6.08	5.60	7.87	8.08	12.40	7.40	4.43	4.96	6.50	4.51	12.66	6.81	12.66
	YEAR OF OCCURRENCE		1978	1961	1948	1988	1995	1988	1996	1992	1971	1985	1989	1990	NOV 1989
	NORMAL NO. DAYS WITH:														
	PRECIPITATION ≥ 0.01	30	10.3	8.9	8.8	6.7	7.6	10.9	14.3	13.5	10.0	5.5	7.9	9.9	114.3
	PRECIPITATION ≥ 1.00	30	1.5	2.0	1.7	1.2	1.5	2.0	1.7	1.6	1.9	1.0	1.5	1.8	19.4
SNOWFALL	NORMAL (IN)	30	0.*	0.*	T	0.0	0.0	0.0	0.0	0.0	0.0	0.0	0.0	0.1	0.1
	MAXIMUM MONTHLY (IN)	51	0.4	2.0	T	T	T	0.0	0.0	0.0	0.0	0.0	T	2.7	2.7
	YEAR OF OCCURRENCE		1985	1958	1993	1996	1989						1950	1963	DEC 1963
	MAXIMUM IN 24 HOURS (IN)	50	0.4	2.0	T	T	T	0.0	0.0	0.0	0.0	0.0	T	2.7	2.7
	YEAR OF OCCURRENCE		1985	1958	1993	1996	1989						1950	1963	DEC 1963
	MAXIMUM SNOW DEPTH (IN)	48	2	2	0	0	0	0	0	0	0	0	0	1	2
	YEAR OF OCCURRENCE		1964	1958										1989	FEB 1958
	NORMAL NO. DAYS WITH:														
	SNOWFALL ≥ 1.0	30	0.0	0.0	0.0	0.0	0.0	0.0	0.0	0.0	0.0	0.0	0.0	0.*	0.0

PRECIPITATION (inches) 2001 NEW ORLEANS, LA (MSY)

YEAR	JAN	FEB	MAR	APR	MAY	JUN	JUL	AUG	SEP	OCT	NOV	DEC	ANNUAL
1972	6.98	6.03	6.07	1.64	6.31	3.10	3.90	4.92	3.29	4.64	8.45	8.65	63.98
1973	2.68	5.40	12.17	10.47	4.68	6.08	5.94	3.37	11.07	5.07	4.04	8.31	79.28
1974	8.46	5.53	6.64	5.52	9.84	3.83	5.66	6.70	7.58	2.26	5.88	4.89	72.79
1975	2.95	3.64	5.32	6.69	8.03	12.28	8.35	10.11	3.97	4.00	11.35	3.81	80.50
1976	2.61	3.85	3.08	0.28	5.58	3.36	5.67	1.69	1.57	5.08	5.80	8.81	47.38
1977	5.62	2.75	3.96	6.38	2.59	1.74	2.91	16.12	13.48	4.33	8.77	4.15	72.80
1978	13.63	2.53	2.67	3.44	9.72	7.82	10.34	14.68	2.98	0.00	4.67	4.42	76.90
1979	5.55	12.49	3.31	4.90	4.38	0.23	11.43	4.57	4.55	1.49	4.27	3.07	60.24
1980	6.37	3.09	10.08	16.12	9.65	3.69	4.84	1.68	6.31	5.87	3.85	1.54	73.09
1981	0.94	8.34	2.70	2.28	5.35	8.47	1.92	11.10	4.78	2.03	1.10	5.50	54.51
1982	2.76	7.88	2.56	5.86	1.19	5.43	13.07	1.92	5.40	3.84	5.45	10.26	65.62
1983	3.31	12.59	4.88	14.86	3.71	10.64	2.95	6.29	5.72	4.88	6.32	9.15	85.30
1984	4.10	5.27	4.90	1.72	3.54	7.21	3.86	9.51	3.79	2.84	2.80	2.53	52.07
1985	4.83	9.28	7.07	2.11	1.16	4.56	6.92	6.37	5.74	13.20	0.96	4.78	66.98
1986	3.49	2.93	1.88	1.50	1.61	8.87	3.60	6.74	1.42	2.87	7.90	5.05	47.86
1987	8.88	7.38	4.39	2.27	3.46	15.01		5.05	1.29	0.72	2.92	2.88	60.63
1988	3.74	11.31	8.90	9.25	1.68	11.28	6.78	7.53	5.86	2.87	1.26	3.94	74.40
1989	2.47	0.15	7.14	3.20	3.50	8.22	8.34	3.31	4.53	0.51	3.50	6.28	67.46
1990	7.59	11.45	5.98	4.59	5.87	1.01	2.30	2.45	4.55	2.38	3.21	9.67	61.05
1991	19.25	5.42	6.27	15.29	14.28	10.71	13.15	7.86	3.44	1.88	2.19	2.63	102.37
1992	9.94	8.73	6.69	2.52	0.95	9.52	5.75	9.64	6.63	0.55	15.27	5.68	81.87
1993	6.21	2.34	5.65	6.82	7.23	4.96	5.77	2.26	2.47	3.67	2.43	2.90	52.71
1994	3.25	0.54	4.82	2.83	3.67	9.35	8.95	4.59	5.61	2.30	1.39	4.61	51.91
1995	3.66	4.94	7.89	3.81	21.18	2.84	6.44	3.26	0.69	1.31	4.24	5.07	65.33
1996	4.66	1.56	2.97	3.87	1.37	8.60	10.32	8.76	3.96	2.59	3.10	5.55	57.31
1997	6.32	6.88	2.57	4.91	5.03	6.97	3.94	2.25	0.81	1.36	8.09	2.55	51.68
1998	19.28	4.28	5.97	4.39	0.43	3.38	6.56	8.30	18.98	1.82	3.40	2.25	79.04
1999	3.20	0.92	4.60	0.30	3.37	12.20	4.05	5.21	2.87	5.46	0.28	3.85	46.31
2000	2.25	1.81	2.41	1.13	0.07	5.46	1.38	2.35	6.50	1.10	11.72	2.70	38.88
2001	3.05	1.59	8.07	1.08	6.85	17.62	6.97	7.41	6.30	5.13	2.54	2.90	69.51
POR= 72 YRS	5.12	4.92	5.26	4.64	4.84	5.66	6.61	5.89	5.33	3.01	4.45	4.86	60.59

AVERAGE TEMPERATURE (F) 2001 NEW ORLEANS, LA (MSY)

YEAR	JAN	FEB	MAR	APR	MAY	JUN	JUL	AUG	SEP	OCT	NOV	DEC	ANNUAL
1972	58.6	56.1	61.9	69.8	73.9	80.8	79.4	81.1	79.6	70.4	56.6	55.3	68.6
1973	50.3	52.6	65.5	64.2	72.5	81.7	84.4	81.7	79.7	73.4	66.6	54.1	68.9
1974	63.3	55.9	67.3	69.0	75.8	78.1	80.3	80.4	77.0	66.9	59.8	55.3	69.1
1975	57.2	58.9	61.4	66.8	75.1	79.4	80.2	80.5	75.1	69.9	61.2	51.9	68.1
1976	50.6	58.2	64.8	68.5	72.3	78.3	81.2	81.5	77.8	64.0	52.7	50.6	66.7
1977	43.4	53.8	65.0	69.0	75.9	82.4	83.9	81.9	80.2	68.2	62.9	54.3	68.4
1978	44.1	45.0	59.9	71.4	76.9	81.2	82.6	83.2	81.1	69.7	67.1	55.9	68.2
1979	45.9	52.9	62.6	71.1	74.6	81.1	83.6	82.9	79.3	71.0	58.0	52.5	68.0
1980	56.1	52.0	62.0	66.2	77.9	83.3	85.8	85.5	83.5	68.8	60.0	53.6	69.6
1981	48.5	55.3	61.9	71.4	74.8	85.0	83.1		77.9	71.1	64.9	54.5	69.4
1982	54.5	55.0	65.9	69.8	76.5	81.5	81.4	82.2	76.8	70.4	62.5	59.4	69.7
1983	50.2	53.8	58.3	64.3	74.0	77.6	81.5	82.4	75.3	69.1	60.0	49.5	66.3
1984	46.6	53.9	59.3	67.5	73.7	77.4	78.8	79.2	76.3	73.5	58.8	42.4	67.3
1985	45.2	52.3	65.3	69.0	74.7	79.3	80.2	81.6	77.0	72.6	67.3	51.0	68.0
1986	51.2	59.2	60.6	67.2	76.7	81.0	83.2	81.6	81.0	70.4	66.3	53.2	69.3
1987	50.0	56.3	60.3	66.2	76.8	79.9	82.9	83.5	78.2	64.4	61.6	59.0	68.3
1988	49.4	53.2	60.9	68.4	73.3	78.5	81.6	81.4	79.8	68.0	65.6	56.0	68.0
1989	60.2	55.9	62.8	67.0	76.4	79.4	81.4	81.7	76.9	67.5	62.4	46.9	68.2
1990	57.2	61.3	63.3	67.6	76.2	82.6	82.3	83.0	79.6	68.1	62.3	59.0	70.2
1991	52.8	58.2	64.2	71.2	77.5	81.3	83.5	81.7	78.2	71.6	55.9	57.6	69.5
1992	51.1	58.3	62.0	66.4	72.8	80.6	83.1	79.6	78.5	69.5	57.4	58.2	68.1
1993	57.2	54.7	58.8	64.0	71.9	80.4	83.3	83.6	79.7	69.9	58.0	52.2	67.8
1994	50.0	56.9	60.8	69.6	75.1	81.4	80.8	81.3	77.8	70.8	65.3	56.5	68.9
1995	53.0	56.3	63.0	68.8	77.6	79.3	83.9	84.7	79.5	71.3	59.8	55.2	69.4
1996	52.9	55.9	58.4	67.1	78.1	80.6	82.8	81.1	79.0	64.3	63.3	57.3	68.9
1997	53.8	57.1	66.4	65.4	74.5	79.6	83.4	83.6	81.1	71.3	58.6	52.8	69.0
1998	56.1	55.9	59.8	67.5	78.8	83.7	85.3	84.6	81.1	73.3	65.4	59.1	70.9
1999	57.7	61.1	61.6	73.1	76.7	81.4	82.2	85.5	78.2	70.7	62.0	55.2	70.5
2000	56.4	60.6	66.0	68.9	80.1	81.7	84.6	84.4	79.3	69.8	59.9	49.6	70.1
2001	50.2	60.3	59.4	72.2	76.1	80.3	82.9	82.5	78.5	68.3	65.0	57.6	69.4
POR= 72 YRS	53.3	56.3	61.9	68.8	75.5	80.7	82.3	82.2	78.7	70.3	61.1	55.3	68.9

REFERENCE NOTES:

PAGE 1:
THE TEMPERATURE GRAPH SHOWS NORMAL MAXIMUM AND NORMAL MINIMUM DAILY TEMPERATURES (SOLID CURVES) AND THE ACTUAL DAILY HIGH AND LOW TEMPERATURES (VERTICAL BARS).

PAGE 2 AND 3:
H/C INDICATES HEATING AND COOLING DEGREE DAYS.
RH INDICATES RELATIVE HUMIDITY
W/O INDICATES WEATHER AND OBSTRUCTIONS
S INDICATES SUNSHINE.
PR INDICATES PRESSURE.
CLOUDINESS ON PAGE 3 IS THE SUM OF THE CEILOMETER AND SATELLITE DATA NOT TO EXCEED EIGHT EIGHTHS(OKTAS).

GENERAL:
T INDICATES TRACE PRECIPITATION, AN AMOUNT GREATER THAN ZERO BUT LESS THAN THE LOWEST REPORTABLE VALUE.
+ INDICATES THE VALUE ALSO OCCURS ON EARLIER DATES.
BLANK ENTRIES DENOTE MISSING OR UNREPORTED DATA.
NORMALS ARE 30-YEAR AVERAGES (1961 - 1990).
ASOS INDICATES AUTOMATED SURFACE OBSERVING SYSTEM.
PM INDICATES THE LAST DAY OF THE PREVIOUS MONTH.
POR (PERIOD OF RECORD) BEGINS WITH THE JANUARY DATA MONTH AND IS THE NUMBER OF YEARS USED TO COMPUTE THE MEAN. INDIVIDUAL MONTHS WITHIN THE POR MAY BE MISSING.
WHEN THE POR FOR A NORMAL IS LESS THAN 30 YEARS, THE NORMAL IS PROVISIONAL AND IS BASED ON THE NUMBER OF YEARS INDICATED.
0.* OR * INDICATES THE VALUE OR MEAN-DAYS-WITH IS BETWEEN 0.00 AND 0.05.
CLOUDINESS FOR ASOS STATIONS DIFFERS FROM THE NON-ASOS OBSERVATION TAKEN BY A HUMAN OBSERVER. ASOS STATION CLOUDINESS IS BASED ON TIME-AVERAGED CEILOMETER DATA FOR CLOUDS AT OR BELOW 12,000 FEET AND ON SATELLITE DATA FOR CLOUDS ABOVE 12,000 FEET.
THE NUMBER OF DAYS WITH CLEAR, PARTLY CLOUDY, AND CLOUDY CONDITIONS FOR ASOS STATIONS IS THE SUM OF THE CEILOMETER AND SATELLITE DATA FOR THE SUNRISE TO SUNSET PERIOD.

GENERAL CONTINUED:
CLEAR INDICATES 0 - 2 OKTAS, PARTLY CLOUDY INDICATES 3 - 6 OKTAS, AND CLOUDY INDICATES 7 OR 8 OKTAS. WHEN AT LEAST ONE OF THE ELEMENTS (CEILOMETER OR SATELLITE) IS MISSING, THE DAILY CLOUDINESS IS NOT COMPUTED.
WIND DIRECTION IS RECORDED IN TENS OF DEGREES (2 DIGITS) CLOCKWISE FROM TRUE NORTH. "00" INDICATES CALM. "36" INDICATES TRUE NORTH.
RESULTANT WIND IS THE VECTOR AVERAGE OF THE SPEED AND DIRECTION.
AVERAGE TEMPERATURE IS THE SUM OF THE MEAN DAILY MAXIMUM AND MINIMUM TEMPERATURE DIVIDED BY 2.
SNOWFALL DATA COMPRISE ALL FORMS OF FROZEN PRECIPITATION, INCLUDING HAIL.
A HEATING (COOLING) DEGREE DAY IS THE DIFFERENCE BETWEEN THE AVERAGE DAILY TEMPERATURE AND 65 F.
DRY BULB IS THE TEMPERATURE OF THE AMBIENT AIR.
DEW POINT IS THE TEMPERATURE TO WHICH THE AIR MUST BE COOLED TO ACHIEVE 100 PERCENT RELATIVE HUMIDITY.
WET BULB IS THE TEMPERATURE THE AIR WOULD HAVE IF THE MOISTURE CONTENT WAS INCREASED TO 100 PERCENT RELATIVE HUMIDITY.

ON JULY 1, 1996, THE NATIONAL WEATHER SERVICE BEGAN USING THE "METAR" OBSERVATION CODE THAT WAS ALREADY EMPLOYED BY MOST OTHER NATIONS OF THE WORLD. THE MOST NOTICEABLE DIFFERENCE IN THIS ANNUAL PUBLICATION WILL BE THE CHANGE IN UNITS FROM TENTHS TO EIGHTS(OKTAS) FOR REPORTING THE AMOUNT OF SKY COVER.

HEATING DEGREE DAYS (base 65 F) 2001 NEW ORLEANS, LA (MSY)

YEAR	JUL	AUG	SEP	OCT	NOV	DEC	JAN	FEB	MAR	APR	MAY	JUN	TOTAL
1972-73	0	0	0	28	293	314	447	351	72	114	9	0	1628
1973-74	0	0	0	18	80	355	117	274	71	16	0	0	931
1974-75	0	0	0	24	194	341	270	210	183	73	0	0	1295
1975-76	0	0	6	16	222	417	445	205	98	21	0	0	1430
1976-77	0	0	0	93	375	438	664	318	117	18	0	0	2023
1977-78	0	0	0	43	113	342	646	556	191	2	0	0	1893
1978-79	0	0	0	16	39	324	586	347	128	8	2	0	1450
1979-80	0	0	0	13	230	396	278	385	154	38	0	0	1494
1980-81	0	0	0	35	195	363	504	275	123	12	0	0	1507
1981-82	0	0	0	36	100	333	365	278	127	29	0	0	1268
1982-83	0	0	0	31	146	234	453	309	217	81	1	0	1472
1983-84	0	0	1	37	183	483	564	321	197	48	2	0	1836
1984-85	0	0	2	14	214	146	605	359	62	28	0	0	1430
1985-86	0	0	0	12	49	443	421	195	160	28	0	0	1308
1986-87	0	0	0	28	85	370	464	242	168	75	0	0	1432
1987-88	0	0	0	58	149	222	490	351	166	23	0	0	1459
1988-89	0	0	0	12	92	301	186	292	155	60	0	0	1098
1989-90	0	0	0	53	142	559	253	136	101	41	0	0	1285
1990-91	0	0	0	62	122	244	371	196	105	8	0	0	1108
1991-92	0	0	0	22	312	262	426	203	128	54	5	0	1412
1992-93	0	0	0	2	240	218	248	285	209	82	0	0	1284
1993-94	0	0	0	42	259	399	464	263	177	49	0	0	1653
1994-95	0	0	0	16	72	268	375	257	123	33	0	0	1144
1995-96	0	0	0	16	186	358	380	307	248	54	1	0	1550
1996-97	0	0	0	17	116	253	373	248	58	44	0	0	1109
1997-98	0	0	0	38	202	383	273	251	210	25	0	0	1382
1998-99	0	0	0	2	48	243	263	150	118	21	0	0	845
1999-00	0	0	0	23	112	318	295	181	63	33	0	0	1025
2000-01	0	0	4	34	227	469	453	181	184	16	0	0	1568
2001-	0	0	0	44	60	249							

COOLING DEGREE DAYS (base 65 F) 2001 NEW ORLEANS, LA (MSY)

YEAR	JAN	FEB	MAR	APR	MAY	JUN	JUL	AUG	SEP	OCT	NOV	DEC	ANNUAL
1972	50	26	38	175	281	479	453	507	446	200	48	19	2722
1973	0	9	96	99	247	507	607	524	448	289	136	24	2986
1974	71	27	147	144	345	402	484	484	368	93	45	45	2655
1975	34	45	80	132	321	440	479	491	314	171	114	16	2637
1976	4	18	100	132	234	404	509	518	390	68	13	0	2390
1977	0	10	123	145	345	528	593	532	463	151	56	16	2962
1978	5	0	39	203	380	493	553	569	489	169	110	49	3059
1979	0	14	63	198	307	491	581	559	435	206	25	16	2895
1980	10	13	70	85	409	554	653	640	561	160	51	17	3223
1981	0	12	35	210	311	570	627	565	396	231	102	12	3071
1982	49	6	160	182	366	504	517	541	363	208	78	66	3040
1983	0	0	16	67	286	385	518	545	317	171	42	10	2357
1984	0	6	31	130	281	379	436	448	351	286	33	71	2452
1985	0	10	78	154	308	437	480	521	366	251	124	13	2742
1986	0	40	32	99	370	487	573	524	488	203	127	9	2952
1987	3	4	30	120	373	456	562	580	402	48	53	42	2673
1988	14	15	49	131	263	411	523	513	448	113	118	30	2628
1989	46	43	95	124	363	439	515	525	365	137	70	6	2728
1990	17	40	56	127	353	538	545	567	448	166	50	62	2969
1991	2	12	88	202	396	496	580	524	402	233	47	40	3022
1992	0	18	41	103	258	472	568	459	408	149	20	14	2510
1993	13	4	26	61	222	468	572	585	447	200	57	8	2663
1994	4	40	54	191	320	501	495	513	390	204	94	12	2818
1995	10	20	67	155	399	434	594	615	444	216	39	59	3052
1996	13	50	47	124	412	473	558	507	427	189	70	21	2891
1997	34	31	107	67	301	445	576	583	489	238	19	9	2899
1998	4	1	59	108	435	568	635	613	403	267	68	70	3317
1999	41	48	20	273	369	497	543	643	610	209	29	23	3098
2000	35	58	103	159	477	506	616	610	439	192	79	0	3274
2001	0	56	19	239	353	469	564	550	410	153	69	29	2911

SNOWFALL (inches) 2001 NEW ORLEANS, LA (MSY)

YEAR	JUL	AUG	SEP	OCT	NOV	DEC	JAN	FEB	MAR	APR	MAY	JUN	TOTAL
1972-73	0.0	0.0	0.0	0.0	0.0	0.0	0.1	0.6	0.0	0.0	0.0	0.0	0.7
1973-74	0.0	0.0	0.0	0.0	0.0	T	0.0	0.0	0.0	0.0	0.0	0.0	T
1974-75	0.0	0.0	0.0	0.0	0.0	0.0	0.0	0.0	0.0	0.0	0.0	0.0	0.0
1975-76	0.0	0.0	0.0	0.0	0.0	0.0	0.0	0.0	0.0	0.0	0.0	0.0	0.0
1976-77	0.0	0.0	0.0	0.0	0.0	0.0	T	0.0	0.0	0.0	0.0	0.0	T
1977-78	0.0	0.0	0.0	0.0	0.0	0.0	T	T	0.0	0.0	0.0	0.0	T
1978-79	0.0	0.0	0.0	0.0	0.0	0.0	T	0.0	0.0	0.0	0.0	0.0	T
1979-80	0.0	0.0	0.0	0.0	0.0	0.0	0.0	0.0	T	0.0	0.0	0.0	T
1980-81	0.0	0.0	0.0	0.0	0.0	0.0	0.0	0.0	0.0	0.0	0.0	0.0	0.0
1981-82	0.0	0.0	0.0	0.0	0.0	0.0	0.0	T	0.0	0.0	0.0	0.0	T
1982-83	0.0	0.0	0.0	0.0	0.0	0.0	0.0	0.0	0.0	0.0	0.0	0.0	0.0
1983-84	0.0	0.0	0.0	0.0	0.0	0.0	0.0	0.0	0.0	0.0	0.0	0.0	0.0
1984-85	0.0	0.0	0.0	0.0	0.0	0.0	0.4	0.0	0.0	0.0	0.0	0.0	0.4
1985-86	0.0	0.0	0.0	0.0	0.0	0.0	0.0	0.0	0.0	0.0	0.0	0.0	0.0
1986-87	0.0	0.0	0.0	0.0	0.0	0.0	0.0	0.0	0.0	0.0	0.0	0.0	0.0
1987-88	0.0	0.0	0.0	0.0	0.0	0.0	0.0	0.0	T	0.0	0.0	0.0	T
1988-89	0.0	0.0	0.0	0.0	0.0	0.0	0.0	0.0	T	0.0	T	0.0	T
1989-90	0.0	0.0	0.0	0.0	0.0	0.5	0.0	0.0	0.0	T	0.0	0.0	0.5
1990-91	0.0	0.0	0.0	0.0	0.0	0.0	0.0	0.0	0.0	T	0.0	0.0	T
1991-92	0.0	0.0	0.0	0.0	0.0	0.0	0.0	0.0	T	0.0	0.0	0.0	T
1992-93	0.0	0.0	0.0	0.0	0.0	0.0	0.0	0.0	T	0.0	0.0	0.0	T
1993-94	0.0	0.0	0.0	0.0	0.0	0.0	0.0	0.0	T	0.0	0.0	0.0	T
1994-95	0.0	0.0	0.0	0.0	0.0	0.0	0.0	0.0	T	0.0	0.0	0.0	T
1995-96	0.0	0.0	0.0	0.0	0.0	T	0.0	0.0	0.0	T			
1996-97													
1997-98													
1998-99													
1999-00													
2000-01						T							
2001-													
POR= 49 YRS	0.0	0.0	0.0	0.0	T	0.1	0.0	0.1	T	T	T	0.0	0.2

2001
SHREVEPORT,
LOUISIANA (SHV)

Shreveport is located on the west side of the Red River, opposite Bossier City, in the northwestern section of Louisiana, some 30 miles south of Arkansas and 15 miles east of Texas. A portion of the city is situated in the Red River bottom lands and the remainder in gently rolling hills that begin about 1 mile west of the river. The NOAA National Weather Service Office is at the Shreveport Regional Airport, about 8 miles southwest of the downtown area. Elevations in the Shreveport area range from about 170 to 280 feet above sea level.

The climate of Shreveport is transitional between the subtropical humid type prevalent to the south and the continental climates of the Great Plains and Middle West to the north. During winter, masses of moderate to severely cold air move periodically through the area. The spring and fall seasons are usually mild, while the summer months are consistently quite warm and humid with high pressure and a moist southerly flow being the dominant feature. Rainfall is abundant with the normal annual just over 46 inches, with monthly averages ranging less than 3 inches in August to more than 5 inches in May. The average growing season for northwest Louisiana ranges between 230 and 240 days in length.

The majority of rainfall is of convective and air mass types-showery and brief-except during winter when nearly continuous frontal rains may persist for a few days. Extremes of precipitation occur in all seasons. While torrential rainfall is the exception in the Shreveport area, some heavy rainfall events of notes are 12.44 inches in a 24-hour peroid on July 24-25, 1933, and 19.08 inches over a three-day peroid on July 23-25, 1933. The July 1933 total of 25.44 inches was the greatest monthly total. The greatest annual rainfall of record was in 1991 with 81.99 inches, and the driest year of record was 1899 with 23.10 inches. The months with the fewest days of rain are August and October, with August having the least average precipitation.

The winter months are normally mild with cold spells generally of short duration. The typical pattern is turning cold one day, reaching the lowest temperature on the second day, and a warming trend on the third day. The coldest reading on record is -5 degrees F on February 12, 1899. Temperatures of freezing or below occur each winter with an average of 39 days during the year. Temperatures drop below 15 degrees F only about one out of every two winters. The average date of the first 32 degrees F in the fall is November 15 and the average date of the last freeze in the spring is March 10. Freezing temperatures have been recorded as early as October 19 and as late as April 11. Temperatures recorded at the NWS Office on clear, calm nights are normally 2 to 5 degrees warmer than those in the low-lying river bottom lands of the area.

Measurable snowfall amounts occur on an average of only once every other year; many consecuitive years may pass with no measurable snowfall. The heaviest snowstorm of record in the Shreveport area is 11.0 inches in december of 1929. This fell on the 21st and 22nd, and one-half inch remained on the ground December 25th, making this the only Christmas Day of record with snow on the ground. In 1948, 12.4 inches of snow was measured for the month of January for the greatest monthly amount on record. Occasional ice and sleet storms do considerable damage to trees, power and telephone lines, as well as make travel very difficult.

The summer months are consistently quite warm, with maximum temperatures exceeding 100 degrees about 6 days per year, exceeding 95 degrees about 32 days per year, and exceeding 90 degrees about 87 days per year. The highest temperature on record is 110 degrees F on August 18, 1909. Showers and thunderstorms at any one location in the area give about eight days in a month of measurable rainfall. The resulting point rainfall totals are usually less than one-half inch except on two or three days per month when heavier amounts are recorded.

Thunderstorms occur each month, but are most frequent in spring and summer months. The showers and thunderstorms during the spring and autumn months are most often produced by squall lines and fronts, and are generally heavier than the air mass showers which occur in the summer months. Severe local storms, including hailstorms, tornadoes, and local windstorms have occurred over small areas in all seasons, but are most frequent during the spring months, with a secondary peak from November to early January. large hail of a damaging nature is infrequent, although hail as large as grapefruit fell in March 1961, and baseball size hail fell in May 1974 and April 1995.

The average relative humidity is rather high in all seasons. These high humidity values may be experienced at any hour but occur mainly during the early morning hours, with two-thirds of the hours shortly before sunrise having relative humidity of 90 percent or higher. In contrast, more than half of the mid-afternoon hours have had relative humidity values of less than 50 percent.

Tropical cyclones are in the dissipating stages by the time they reach this portion of the state and winds from them are usually not a destructive factor. rainfall accompanying these systems can be heavy and can contribute to local flooding.

NORMALS, MEANS, AND EXTREMES
SHREVEPORT, LA (SHV)

LATITUDE: 32 26' 49" N　　LONGITUDE: 93 49' 27" W　　ELEVATION (FT): GRND: 271　BARO: 274　　TIME ZONE: CENTRAL (UTC + 6)　　WBAN: 13957

ELEMENT	POR	JAN	FEB	MAR	APR	MAY	JUN	JUL	AUG	SEP	OCT	NOV	DEC	YEAR
TEMPERATURE (F)														
NORMAL DAILY MAXIMUM	30	55.4	60.6	69.2	77.1	83.2	89.7	93.0	93.1	87.3	78.7	68.0	58.5	76.2
MEAN DAILY MAXIMUM	54	56.2	61.4	68.6	76.9	83.5	90.1	93.3	93.4	87.6	78.5	67.1	58.8	76.3
HIGHEST DAILY MAXIMUM	49	84	89	92	94	102	102	107	109	109	97	88	84	109
YEAR OF OCCURRENCE		1972	1986	1974	1987	1998	1998	1998	2000	2000	1954	1984	1955	SEP 2000
MEAN OF EXTREME MAXS.	54	76.5	79.6	84.1	87.2	91.8	95.7	98.9	100.1	96.5	90.7	82.8	76.9	88.4
NORMAL DAILY MINIMUM	30	34.8	38.0	45.8	54.1	62.0	69.0	72.3	71.3	66.0	54.3	45.3	37.3	54.2
MEAN DAILY MINIMUM	54	36.4	39.8	46.0	54.2	62.8	69.6	72.7	71.6	66.0	54.6	44.8	38.3	54.7
LOWEST DAILY MINIMUM	49	3	12	20	31	42	52	58	53	42	28	16	5	3
YEAR OF OCCURRENCE		1962	1978	1980	1989	1960	1977	1972	1992	1984	1993	1976	1989	JAN 1962
MEAN OF EXTREME MINS.	54	18.9	23.3	28.8	38.2	49.5	59.5	66.6	64.2	51.9	38.4	28.2	21.4	40.8
NORMAL DRY BULB	30	45.1	49.3	57.5	65.6	72.6	79.4	82.7	82.2	76.7	66.5	56.7	48.0	65.2
MEAN DRY BULB	54	46.3	50.5	57.3	65.6	73.2	79.9	83.0	82.6	76.8	66.5	55.9	48.5	65.5
MEAN WET BULB	17	42.0	46.8	51.6	58.7	67.1	72.6	75.0	74.1	68.9	60.1	51.6	44.5	59.4
MEAN DEW POINT	17	36.4	40.8	45.6	53.5	63.5	69.4	71.8	70.3	65.0	55.9	47.0	39.5	54.9
NORMAL NO. DAYS WITH:														
MAXIMUM 90	30	0.0	0.0	*	0.3	3.7	17.2	25.0	25.0	13.3	2.5	0.0	0.0	87.0
MAXIMUM 32	30	1.3	0.4	*	0.0	0.0	0.0	0.0	0.0	0.0	0.0	0.0	0.6	2.3
MINIMUM 32	30	14.1	8.3	2.5	0.2	0.0	0.0	0.0	0.0	0.0	0.2	2.7	10.8	38.8
MINIMUM 0	30	0.0	0.0	0.0	0.0	0.0	0.0	0.0	0.0	0.0	0.0	0.0	0.0	0.0
H/C														
NORMAL HEATING DEG. DAYS	30	623	448	262	69	0	0	0	0	0	63	264	535	2264
NORMAL COOLING DEG. DAYS	30	6	8	30	87	239	432	549	533	351	110	15	8	2368
RH														
NORMAL (PERCENT)	30	73	70	68	70	73	73	72	72	74	72	74	74	72
HOUR 00 LST	30	78	76	76	79	84	84	83	83	84	82	82	81	81
HOUR 06 LST	30	84	84	84	87	90	91	90	91	91	89	87	86	88
HOUR 12 LST	30	63	60	56	56	59	59	58	56	58	54	59	63	58
HOUR 18 LST	30	65	58	54	56	60	60	58	58	62	62	68	68	61
S PERCENT POSSIBLE SUNSHINE	48	52	57	58	60	64	72	75	75	70	70	60	54	64
W/O MEAN NO. DAYS WITH:														
HEAVY FOG (VISBY 1/4 MI)	49	3.3	2.2	1.4	1.3	1.0	0.5	0.3	0.4	0.9	2.2	2.8	3.1	19.4
THUNDERSTORMS	49	2.2	2.9	5.2	5.5	7.2	7.5	8.1	6.5	4.3	3.0	3.1	2.2	57.7
CLOUDINESS MEAN:														
SUNRISE-SUNSET (OKTAS)	1			2.4			1.6							
MIDNIGHT-MIDNIGHT (OKTAS)	1			2.4										
MEAN NO. DAYS WITH:														
CLEAR	1	6.0	5.0	10.0			6.0	7.0						
PARTLY CLOUDY	1		2.0	3.0			7.0	8.0						
CLOUDY	1	3.0	2.0	7.0			3.0	4.0						
PR MEAN STATION PRESSURE (IN)	28	29.86	29.82	29.73	29.70	29.67	29.69	29.73	29.73	29.74	29.79	29.82	29.86	29.76
MEAN SEA-LEVEL PRES. (IN)	17	30.15	30.09	30.03	29.97	29.94	29.95	30.01	29.99	30.01	30.07	30.11	30.15	30.04
WINDS														
MEAN SPEED (MPH)	42	9.3	9.6	9.9	9.8	8.6	7.7	7.0	6.7	7.2	7.3	8.5	9.0	8.4
PREVAIL.DIR (TENS OF DEGS)	27	18	18	18	18	18	18	18	18	12	14	16	13	18
MAXIMUM 2-MINUTE:														
SPEED (MPH)	6	38	43	54	45	63	41	41	40	32	37	40	36	63
DIR. (TENS OF DEGS)		28	34	29	26	32	29	15	11	32	25	29	30	32
YEAR OF OCCURRENCE		1999	2001	1999	1999	2000	1998	1998	1998	1996	2001	1998	2000	MAY 2000
MAXIMUM 5-SECOND:														
SPEED (MPH)	6	46	52	68	55	81	55	67	59	41	46	51	45	81
DIR. (TENS OF DEGS)		26	28	31	26	32	31	18	34	33	25	29	30	32
YEAR OF OCCURRENCE		2000	1999	1999	1999	2000	1998	1998	1997	1996	2001	1996	2000	MAY 2000
PRECIPITATION														
NORMAL (IN)	30	3.88	3.92	3.59	3.75	5.18	4.29	3.67	2.43	3.12	3.73	4.45	4.10	46.11
MAXIMUM MONTHLY (IN)	49	12.96	8.57	8.72	21.84	11.78	17.11	9.46	9.23	9.59	12.05	10.81	10.00	21.84
YEAR OF OCCURRENCE		1999	1983	1997	1991	1967	1989	1972	1991	1968	1984	1987	1982	APR 1991
MINIMUM MONTHLY (IN)	49	0.27	0.42	0.56	0.43	0.15	0.13	0.15	0.35	0.08	0.00	0.52	0.59	0.00
YEAR OF OCCURRENCE		1971	1999	1966	1987	1998	1988	1964	2000	1994	1963	1999	1981	OCT 1963
MAXIMUM IN 24 HOURS (IN)	49	7.00	3.53	3.63	10.44	5.27	7.28	4.40	4.64	5.39	3.88	6.51	3.94	10.44
YEAR OF OCCURRENCE		1999	1965	1979	1991	1978	1993	1995	1955	1961	1957	1987	2001	APR 1991
NORMAL NO. DAYS WITH:														
PRECIPITATION 0.01	30	9.4	8.1	9.3	8.2	9.0	7.9	7.9	6.6	7.0	6.6	8.1	9.5	97.6
PRECIPITATION 1.00	30	1.2	1.2	1.1	1.2	1.8	1.3	1.2	1.0	1.0	1.2	1.4	1.1	14.7
SNOWFALL														
NORMAL (IN)	30	1.0	0.6	0.2	0.*	0.0	0.0	0.0	0.0	0.0	0.0	0.*	0.3	2.1
MAXIMUM MONTHLY (IN)	49	5.9	4.4	4.0	0.3	T	0.0	0.0	T	0.0	T	1.3	5.4	5.9
YEAR OF OCCURRENCE		1978	1985	1965	1987	1994			1997		1992	1980	1983	JAN 1978
MAXIMUM IN 24 HOURS (IN)	49	5.6	4.4	4.0	0.3	T	0.0	0.0	T	0.0	T	1.3	5.4	5.6
YEAR OF OCCURRENCE		1982	1985	1965	1987	1994			1997		1992	1980	1983	JAN 1982
MAXIMUM SNOW DEPTH (IN)	53	9	4	2	0	0	0	0	0	0	0	1	3	9
YEAR OF OCCURRENCE		1948	1951	1965								1980	1963	JAN 1948
NORMAL NO. DAYS WITH:														
SNOWFALL 1.0	30	0.3	0.2	0.1	0.0	0.0	0.0	0.0	0.0	0.0	0.0	0.*	0.1	0.7

PRECIPITATION (inches) 2001 SHREVEPORT, LA (SHV)

YEAR	JAN	FEB	MAR	APR	MAY	JUN	JUL	AUG	SEP	OCT	NOV	DEC	ANNUAL
1972	5.97	0.94	2.45	2.06	4.13	2.76	9.46	1.27	2.10	6.32	5.32	4.18	46.96
1973	5.65	1.52	5.01	6.44	2.00	5.84	7.63	0.77	6.39	5.38	5.16	6.37	58.16
1974	10.09	3.67	3.60	3.09	4.58	6.29	7.73	3.84	6.64	3.79	5.80	2.34	61.46
1975	4.55	4.51	5.84	3.91	5.31	3.48	3.45	1.65	0.98	3.87	4.44	1.88	43.87
1976	2.07	2.45	6.67	1.75	5.95	4.42	3.47	2.96	6.28	2.08	1.63	3.77	43.50
1977	3.00	3.68	4.94	2.05	2.40	2.41	3.89	4.28	0.53	0.31	2.11	2.58	32.18
1978	4.89	1.90	2.66	2.79	7.92	1.21	1.74	3.90	2.40	2.74	4.18	5.13	41.46
1979	9.22	4.98	5.74	7.42	7.99	3.04	7.50	1.86	4.33	3.96	4.76	3.12	63.92
1980	4.67	3.10	3.75	5.34	4.42	2.60	1.83	0.42	1.63	2.48	3.59	0.74	34.57
1981	1.43	3.83	3.33	1.97	9.96	6.45	2.36	0.94	3.32	5.63	1.49	0.59	41.30
1982	3.59	3.19	2.59	2.72	2.32	1.84	4.25	2.20	1.11	5.19	5.72	10.00	44.72
1983	2.45	8.57	3.68	1.47	8.22	6.60	1.18	1.67	3.12	0.79	4.90	7.18	49.83
1984	2.10	5.66	3.58	2.52	5.86	3.56	2.20	0.87	2.61	12.05	4.46	2.88	48.35
1985	2.38	4.42	4.28	3.05	1.96	4.57	8.40	0.35	4.40	9.87	4.25	3.37	51.30
1986	0.49	3.48	0.75	3.50	6.60	14.67	2.92	1.68	3.51	6.63	9.19	4.69	58.11
1987	2.26	7.80	1.48	0.43	6.67	5.43	1.21	3.50	0.94	5.49	10.81	8.12	54.14
1988	2.06	3.59	3.89	3.45	0.42	0.13	3.12	3.52	1.61	4.44	5.44	4.71	36.38
1989	7.20	4.06	3.41	2.41	10.07	17.11	4.46	3.94	1.08	1.50	2.32	3.34	60.90
1990	10.02	6.92	4.90	4.29	10.48	2.56	3.53	2.88	2.93	4.33	8.81	3.99	65.64
1991	7.70	5.13	2.89	21.84	10.71	2.53	3.47	9.23	3.45	3.59	3.94	7.51	81.99
1992	4.63	6.41	5.94	3.26	2.81	3.95	3.36	1.24	5.15	4.13	4.69	5.84	51.41
1993	4.63	4.80	5.94	4.19	3.30	15.73	0.27	4.09	3.51	4.43	4.85	1.44	57.18
1994	3.63	5.02	3.67	3.67	5.85	2.81	6.43	2.34	3.80	0.08	9.16	2.50	54.60
1995	5.44	3.75	4.05	7.80	3.26	1.09	5.68	0.83	3.36	1.65	1.94	5.11	43.96
1996	2.12	0.64	2.33	3.86	0.93	6.50	5.70	5.78	7.17	1.66	5.87	2.24	44.80
1997	4.47	8.09	8.72	11.93	3.19	6.14	1.73	5.48	2.41	7.50	3.44	6.10	69.20
1998	5.84	7.19	4.28	0.79	0.15	1.35	2.84	3.83	7.79	5.72	4.58	6.24	50.60
1999	12.96	0.42	5.10	7.88	3.96	7.98	2.80	1.47	4.90	3.21	0.52	3.82	55.02
2000	2.60	2.31	7.90	5.67	10.76	7.32	1.05	T	1.13	1.65	9.93	7.56	57.88
2001	5.76	6.52	6.47	0.86	4.31	7.33	1.75	4.10	6.84	5.17	4.16	6.10	59.37
POR= 131 YRS	4.24	3.77	4.23	4.69	4.62	3.70	3.47	2.65	2.92	3.25	3.96	4.54	46.04

AVERAGE TEMPERATURE (F) 2001 SHREVEPORT, LA (SHV)

YEAR	JAN	FEB	MAR	APR	MAY	JUN	JUL	AUG	SEP	OCT	NOV	DEC	ANNUAL
1972	49.4	51.9	59.8	66.9	72.3	80.8	81.0	82.6	80.2	67.2	50.9	45.5	65.7
1973	44.7	49.5	60.9	62.0	71.7	78.5	81.1	78.2	74.9	68.2	61.5	46.8	64.8
1974	47.8	51.2	63.6	64.3	74.0	76.6	82.1	80.1	71.0	66.2	55.6	47.3	65.0
1975	50.2	48.4	55.6	63.7	72.4	78.2	80.9	80.7	73.7	68.0	56.3	49.3	64.8
1976	47.1	59.2	60.2	67.3	67.9	75.7	78.6	78.7	74.3	59.9	49.3	46.3	63.7
1977	37.3	50.8	59.6	65.5	74.1	79.5	83.5	80.3	78.7	66.3	56.5	47.3	65.0
1978	34.9	38.1	52.8	65.5	73.6	80.5	85.4	83.3	77.8	66.0	60.0	48.0	63.8
1979	37.4	46.6	59.1	66.0	70.2	78.0	81.2	80.3	74.2	67.0	52.8	48.9	63.5
1980	48.3	47.9	54.9	63.0	74.3	83.4	86.9	85.5	82.1	63.5	54.2	49.1	66.1
1981	44.7	49.8	56.0	70.0	69.2	80.2	82.8	81.3	73.9	65.0	56.8	46.9	64.7
1982	46.1	45.6	61.5	63.3	74.4	78.6	83.3	83.2	75.6	64.6	55.4	51.2	65.2
1983	44.6	48.8	55.0	59.7	70.0	77.4	82.3	84.0	75.7	66.7	55.9	37.5	63.1
1984	40.6	50.4	58.0	64.9	72.1	79.3	81.1	82.1	75.0	70.6	56.4	60.0	65.9
1985	40.0	46.2	61.4	67.0	72.8	79.3	83.1	84.8	76.3	68.5	61.5	44.2	65.4
1986	49.0	54.4	59.7	66.5	72.3	79.9	83.7	80.8	79.4	65.2	55.6	46.1	66.1
1987	44.8	51.7	55.7	64.2	75.3	79.1	82.3	83.5	76.7	64.1	56.1	50.2	65.5
1988	42.2	49.3	56.3	65.2	71.8	79.8	83.3	83.7	77.8	64.1	58.6	49.2	65.1
1989	51.5	45.8	56.7	65.6	73.8	76.7	81.2	81.0	73.7	66.5	58.6	40.8	64.3
1990	52.5	56.4	59.4	65.6	72.6	82.7	82.7	82.2	83.3	79.7	65.0	58.9	67.2
1991	44.2	52.0	60.0	68.6	75.5	80.6	82.6	81.2	75.7	68.7	52.1	51.2	66.0
1992	47.2	54.7	59.5	65.1	71.4	78.5	82.9	78.6	76.0	67.0	52.0	49.5	65.2
1993	46.4	49.3	55.2	61.3	70.7	80.2	84.6	84.8	77.3	64.4	52.0	49.1	64.6
1994	46.4	51.2	58.0	66.7	70.9	81.4	82.1	81.1	76.3	67.0	60.0	51.1	66.0
1995	48.1	53.1	59.4	64.6	73.9	79.0	84.2	86.6	77.3	66.2	55.0	49.1	66.4
1996	46.2	52.4	53.4	68.3	77.0	78.3	82.3	80.0	74.1	66.0	55.7	51.5	65.1
1997	45.6	51.2	61.4	60.3	70.3	78.1	83.6	81.0	78.2	66.8	51.9	46.0	64.5
1998	51.7	51.6	56.1	63.5	77.3	84.9	88.5	84.6	81.7	68.5	58.0	49.3	68.0
1999	51.6	57.0	56.4	69.4	72.0	79.8	82.9	85.9	75.3	65.5	58.8	49.6	67.0
2000	50.4	57.2	61.7	63.6	75.8	79.1	81.8	83.9	86.9	78.1	67.8	52.6	39.8	66.4
2001	43.4	53.2	53.4	69.2	74.6	79.0	84.5	82.3	74.6	63.3	59.6	51.1	65.7
POR= 127 YRS	47.0	50.7	57.9	65.9	73.2	80.3	83.0	82.6	77.1	67.0	56.1	48.9	65.8

REFERENCE NOTES:

PAGE 1:
THE TEMPERATURE GRAPH SHOWS NORMAL MAXIMUM AND NORMAL MINIMUM DAILY TEMPERATURES (SOLID CURVES) AND THE ACTUAL DAILY HIGH AND LOW TEMPERATURES (VERTICAL BARS).

PAGE 2 AND 3:
H/C INDICATES HEATING AND COOLING DEGREE DAYS.
RH INDICATES RELATIVE HUMIDITY
W/O INDICATES WEATHER AND OBSTRUCTIONS
S INDICATES SUNSHINE.
PR INDICATES PRESSURE.
CLOUDINESS ON PAGE 3 IS THE SUM OF THE CEILOMETER AND SATELLITE DATA NOT TO EXCEED EIGHT EIGHTHS(OKTAS).

GENERAL:
T INDICATES TRACE PRECIPITATION, AN AMOUNT GREATER THAN ZERO BUT LESS THAN THE LOWEST REPORTABLE VALUE.
+ INDICATES THE VALUE ALSO OCCURS ON EARLIER DATES.
BLANK ENTRIES DENOTE MISSING OR UNREPORTED DATA.
NORMALS ARE 30-YEAR AVERAGES (1961 – 1990).
ASOS INDICATES AUTOMATED SURFACE OBSERVING SYSTEM.
PM INDICATES THE LAST DAY OF THE PREVIOUS MONTH.
POR (PERIOD OF RECORD) BEGINS WITH THE JANUARY DATA MONTH AND IS THE NUMBER OF YEARS USED TO COMPUTE THE MEAN. INDIVIDUAL MONTHS WITHIN THE POR MAY BE MISSING.
WHEN THE POR FOR A NORMAL IS LESS THAN 30 YEARS, THE NORMAL IS PROVISIONAL AND IS BASED ON THE NUMBER OF YEARS INDICATED.
0.* OR * INDICATES THE VALUE OR MEAN-DAYS-WITH IS BETWEEN 0.00 AND 0.05.
CLOUDINESS FOR ASOS STATIONS DIFFERS FROM THE NON-ASOS OBSERVATION TAKEN BY A HUMAN OBSERVER. ASOS STATION CLOUDINESS IS BASED ON TIME-AVERAGED CEILOMETER DATA FOR CLOUDS AT OR BELOW 12,000 FEET AND ON SATELLITE DATA FOR CLOUDS ABOVE 12,000 FEET.
THE NUMBER OF DAYS WITH CLEAR, PARTLY CLOUDY, AND CLOUDY CONDITIONS FOR ASOS STATIONS IS THE SUM OF THE CEILOMETER AND SATELLITE DATA FOR THE SUNRISE TO SUNSET PERIOD.

GENERAL CONTINUED:
CLEAR INDICATES 0 – 2 OKTAS, PARTLY CLOUDY INDICATES 3 – 6 OKTAS, AND CLOUDY INDICATES 7 OR 8 OKTAS.
WHEN AT LEAST ONE OF THE ELEMENTS (CEILOMETER OR SATELLITE) IS MISSING, THE DAILY CLOUDINESS IS NOT COMPUTED.
WIND DIRECTION IS RECORDED IN TENS OF DEGREES (2 DIGITS) CLOCKWISE FROM TRUE NORTH. "00" INDICATES CALM. "36" INDICATES TRUE NORTH.
RESULTANT WIND IS THE VECTOR AVERAGE OF THE SPEED AND DIRECTION.
AVERAGE TEMPERATURE IS THE SUM OF THE MEAN DAILY MAXIMUM AND MINIMUM TEMPERATURE DIVIDED BY 2.
SNOWFALL DATA COMPRISE ALL FORMS OF FROZEN PRECIPITATION, INCLUDING HAIL.
A HEATING (COOLING) DEGREE DAY IS THE DIFFERENCE BETWEEN THE AVERAGE DAILY TEMPERATURE AND 65 F.
DRY BULB IS THE TEMPERATURE OF THE AMBIENT AIR.
DEW POINT IS THE TEMPERATURE TO WHICH THE AIR MUST BE COOLED TO ACHIEVE 100 PERCENT RELATIVE HUMIDITY.
WET BULB IS THE TEMPERATURE THE AIR WOULD HAVE IF THE MOISTURE CONTENT WAS INCREASED TO 100 PERCENT RELATIVE HUMIDITY.

ON JULY 1, 1996, THE NATIONAL WEATHER SERVICE BEGAN USING THE "METAR" OBSERVATION CODE THAT WAS ALREADY EMPLOYED BY MOST OTHER NATIONS OF THE WORLD. THE MOST NOTICEABLE DIFFERENCE IN THIS ANNUAL PUBLICATION WILL BE THE CHANGE IN UNITS FROM TENTHS TO EIGHTS(OKTAS) FOR REPORTING THE AMOUNT OF SKY COVER.

HEATING DEGREE DAYS (base 65 F) 2001 SHREVEPORT, LA (SHV)

YEAR	JUL	AUG	SEP	OCT	NOV	DEC	JAN	FEB	MAR	APR	MAY	JUN	TOTAL
1972-73	0	0	6	92	419	597	621	429	135	145	9	0	2453
1973-74	0	0	0	40	164	557	533	386	152	78	2	0	1912
1974-75	0	0	14	32	312	541	473	457	305	124	0	0	2258
1975-76	0	0	4	39	286	492	551	186	202	38	17	0	1815
1976-77	0	0	0	199	471	574	851	399	188	46	0	0	2728
1977-78	0	0	0	72	260	549	933	746	374	61	30	0	3025
1978-79	0	0	0	57	181	528	849	517	216	50	11	0	2409
1979-80	0	0	0	52	366	498	508	494	312	96	4	0	2330
1980-81	0	0	3	128	340	488	620	425	279	14	20	0	2317
1981-82	0	0	8	129	246	554	588	537	202	125	4	0	2393
1982-83	0	0	9	120	309	457	624	449	308	186	12	0	2474
1983-84	0	0	15	69	305	848	747	421	247	81	11	0	2744
1984-85	0	0	19	36	286	208	770	528	151	42	1	0	2041
1985-86	0	0	11	49	174	638	490	331	176	44	1	0	1914
1986-87	0	0	0	86	299	579	618	366	286	117	0	0	2351
1987-88	0	0	0	79	279	456	701	453	278	54	1	0	2301
1988-89	0	0	0	76	218	482	418	535	295	93	2	0	2119
1989-90	0	0	17	85	244	743	382	243	216	92	3	0	2025
1990-91	0	0	6	126	208	509	634	357	197	26	6	0	2069
1991-92	0	0	9	36	401	422	544	297	185	84	23	0	2001
1992-93	0	0	2	33	381	480	569	435	307	154	3	0	2364
1993-94	0	0	5	132	404	490	577	392	239	81	20	0	2340
1994-95	0	0	5	74	179	440	528	334	224	80	10	0	1874
1995-96	0	0	10	56	311	507	576	387	372	115	2	0	2336
1996-97	0	0	10	59	277	427	604	383	133	157	5	0	2055
1997-98	0	0	0	95	392	580	406	368	324	88	0	0	2253
1998-99	0	0	0	37	215	501	422	250	261	41	1	0	1728
1999-00	0	0	2	79	189	477	462	247	149	103	0	0	1708
2000-01	0	0	10	70	388	774	662	332	352	35	0	0	2623
2001-	0	0	5	118	175	441							

COOLING DEGREE DAYS (base 65 F) 2001 SHREVEPORT, LA (SHV)

YEAR	JAN	FEB	MAR	APR	MAY	JUN	JUL	AUG	SEP	OCT	NOV	DEC	ANNUAL
1972	22	10	31	128	235	480	501	553	467	167	1	0	2595
1973	0	2	16	64	223	412	504	417	305	145	66	0	2154
1974	7	6	115	63	288	355	541	477	200	77	35	0	2164
1975	23	0	21	91	238	403	501	493	271	141	34	12	2228
1976	2	26	58	116	112	326	428	432	288	48	7	0	1843
1977	0	7	28	69	289	443	580	479	419	119	12	9	2454
1978	5	0	3	84	303	472	637	570	391	96	39	8	2608
1979	0	8	39	86	178	395	509	483	284	124	8	2	2116
1980	1	6	6	43	298	560	686	643	522	86	22	1	2874
1981	0	5	10	171	157	463	558	511	284	135	6	0	2300
1982	14	0	99	81	300	413	573	573	333	115	24	32	2557
1983	0	0	7	34	176	381	540	595	343	126	39	0	2241
1984	0	5	38	83	235	436	511	540	329	219	35	61	2492
1985	0	8	49	109	252	436	568	620	356	163	78	0	2639
1986	2	41	17	95	236	454	586	494	438	101	24	0	2488
1987	1	0	7	99	327	431	544	634	357	57	19	5	2481
1988	3	3	14	67	220	449	575	587	390	53	37	1	2399
1989	8	7	43	121	283	358	538	536	572	132	30	3	2684
1990	2	9	50	115	244	475	550	509	336	158	20	5	2576
1991	0	1	47	141	334	475	550	509	336	158	20	5	2576
1992	0	4	21	93	227	412	561	428	339	99	1	4	2189
1993	0	0	10	50	189	461	614	620	381	121	20	4	2470
1994	3	12	29	141	209	500	538	506	355	145	36	13	2487
1995	11	6	58	72	290	427	600	676	385	100	16	19	2660
1996	3	31	23	85	384	403	543	471	291	96	6	16	2352
1997	9	4	29	25	180	398	583	504	399	157	6	0	2294
1998	1	0	56	50	392	602	737	613	508	152	15	22	3148
1999	12	29	0	180	224	451	562	655	317	101	9	6	2546
2000	15	28	55	67	341	429	593	689	409	166	22	0	2814
2001	0	12	2	166	308	426	611	542	301	69	23	16	2476

SNOWFALL (inches) 2001 SHREVEPORT, LA (SHV)

YEAR	JUL	AUG	SEP	OCT	NOV	DEC	JAN	FEB	MAR	APR	MAY	JUN	TOTAL
1972-73	0.0	0.0	0.0	0.0	0.0	0.0	0.6	T	0.0	0.0	0.0	0.0	0.6
1973-74	0.0	0.0	0.0	0.0	0.0	T	0.0	0.0	0.0	0.0	0.0	0.0	T
1974-75	0.0	0.0	0.0	0.0	0.0	0.0	3.4	T	T	0.0	0.0	0.0	3.4
1975-76	0.0	0.0	0.0	0.0	T	0.0	0.0	0.0	0.0	0.0	0.0	0.0	T
1976-77	0.0	0.0	0.0	0.0	T	0.0	5.4	0.0	0.0	0.0	0.0	0.0	5.4
1977-78	0.0	0.0	0.0	0.0	0.0	0.0	5.9	2.0	0.3	0.0	0.0	0.0	8.2
1978-79	0.0	0.0	0.0	0.0	0.0	0.0	T	1.5	0.0	0.0	0.0	0.0	1.5
1979-80	0.0	0.0	0.0	0.0	T	0.0	0.0	1.4	T	0.0	0.0	0.0	1.4
1980-81	0.0	0.0	0.0	0.0	0.0	1.3	0.0	0.7	0.1	0.0	0.0	0.0	2.1
1981-82	0.0	0.0	0.0	0.0	0.0	0.0	5.6	T	T	0.0	0.0	0.0	5.6
1982-83	0.0	0.0	0.0	0.0	0.0	T	T	T	0.0	0.0	0.0	0.0	T
1983-84	0.0	0.0	0.0	0.0	0.0	5.4	T	T	0.0	0.0	0.0	0.0	5.4
1984-85	0.0	0.0	0.0	0.0	0.0	0.0	0.4	4.4	0.0	0.0	0.0	0.0	4.8
1985-86	0.0	0.0	0.0	0.0	0.0	0.0	T	T	T	0.3	0.0	0.0	0.3
1986-87	0.0	0.0	0.0	0.0	0.0	T	0.0	T	T	0.0	0.0	0.0	T
1987-88	0.0	0.0	0.0	0.0	0.0	0.0	1.2	0.8	0.0	0.0	0.0	0.0	2.0
1988-89	0.0	0.0	0.0	0.0	0.0	0.0	T	T	T	0.0	0.0	0.0	T
1989-90	0.0	0.0	0.0	0.0	0.0	T	T	0.0	T	0.0	0.0	0.0	T
1990-91	0.0	0.0	0.0	0.0	0.0	T	0.0	T	T	T	0.0	0.0	T
1991-92	0.0	0.0	0.0	0.0	0.0	0.0	0.0	T	T	0.0	0.0	0.0	T
1992-93	0.0	0.0	0.0	T	T	0.0	T	T	1.0	T	0.0	0.0	1.0
1993-94	0.0	0.0	0.0	0.0	0.0	T	0.0	0.4	T	T	0.0	0.0	0.4
1994-95	0.0	0.0	0.0	0.0	T	0.0	T	T	0.0	0.0	0.0	0.0	T
1995-96	0.0	0.0	0.0	0.0	0.0	0.0	T	0.4	T	0.0	0.0	0.0	0.4
1996-97	0.0	0.0	0.0	0.0	0.0	0.3	0.1	T	0.0	T	0.0	0.0	0.4
1997-98	0.0	T	0.0	0.0	T	0.0	T	T	T	0.0	0.0	0.0	T
1998-99	0.0	0.0	0.0	0.0	0.0	0.1	0.0	0.0	T	0.0	0.0	0.0	0.1
1999-00	0.0	0.0	0.0	0.0	0.0	T	1.1	T	T	T	0.0	0.0	1.1
2000-01	0.0	0.0	0.0	0.0	T	2.2	T	0.0	0.0	0.0	0.0	0.0	2.2
2001-	0.0	0.0	0.0	0.0	T	T							
POR= 48 YRS	0.0	0.0	0.0	T	0.0	0.2	0.7	0.4	0.2	0.0	T	0.0	1.5

2001
CARIBOU,
MAINE (CAR)

The Caribou Municipal Airport is located in Aroostook County, the largest and northernmost county in the state. The airport lies on top of high land which is about on the same level as most of the surrounding gently rolling hills. The Aroostook River, which runs about 1 mile to the east and southeast of the station, has little effect on the local weather. Even though Caribou is located only 150 miles from the Atlantic coast, its climate can be justly classed as a severe typical continental type. Winters are particularly long and windy, and seasonal snowfalls averaging over 100 inches are not unusual. While the extreme low temperatures may be less severe than one might expect, temperatures of zero or lower normally occur over 40 times per year. A study of heating degree day data will show the outstanding part that cold weather plays here.

Summers are cool and generally favored with abundant rainfall, which is one of the most important factors in the high yield of the potato and grain crops throughout the county. Our location high up in the St. Lawrence Valley allows Aroostook County to come under the influence of the Summer Polar Front, resulting in practically no dry periods of more than 3 or 4 days in the growing season. The growing season at Caribou averages more than 120 days, with the average last freeze in the spring in mid-May and the average first freeze in autumn in late September.

Autumn climate is nearly ideal, with mostly sunny warm days and crisp cool nights predominating. Aroostook County, even with its relatively short growing season, provides profitable farming. The principal crops are potatoes, peas, a variety of grains, and some hardy vegetables.

Probably unknown to many victims of hay fever and similar afflictions, the immediate Caribou area offers sparkling visibility and relatively pollen-free air in the late summer months. This latter condition is principally due to the extremely high degree of cultivation of all available land.

NORMALS, MEANS, AND EXTREMES
CARIBOU, ME (CAR)

LATITUDE: 46 52' 01" N   LONGITUDE: 68 01' 58" W   ELEVATION (FT): GRND: 627 BARO: 630   TIME ZONE: EASTERN (UTC + 5)   WBAN: 14607

	ELEMENT	POR	JAN	FEB	MAR	APR	MAY	JUN	JUL	AUG	SEP	OCT	NOV	DEC	YEAR
TEMPERATURE °F	NORMAL DAILY MAXIMUM	30	19.4	23.0	34.3	46.7	61.7	71.9	76.5	73.6	64.0	52.0	37.6	24.0	48.7
	MEAN DAILY MAXIMUM	54	19.7	23.1	33.6	46.5	61.7	71.2	75.8	73.7	64.3	52.0	38.1	25.0	48.7
	HIGHEST DAILY MAXIMUM	62	53	59	73	86	96	96	95	95	91	79	68	58	96
	YEAR OF OCCURRENCE		1995	1994	1962	1990	1977	1944	1991	1975	1945	1968	1956	1950	MAY 1977
	MEAN OF EXTREME MAXS.	54	41.3	41.0	51.1	66.1	81.3	87.1	88.5	86.5	80.9	70.2	57.3	44.7	66.3
	NORMAL DAILY MINIMUM	30	-1.6	.7	14.9	29.0	40.1	49.1	54.5	52.1	43.2	34.4	23.7	5.5	28.8
	MEAN DAILY MINIMUM	54	1.0	2.8	15.0	29.0	40.2	49.5	54.6	52.1	43.5	34.4	24.4	8.7	29.6
	LOWEST DAILY MINIMUM	62	-33	-41	-28	-2	18	30	36	34	23	14	-8	-31	-41
	YEAR OF OCCURRENCE		1995	1955	2001	1964	1974	1958	1969	1982	1980	1972	1995	1989	FEB 1955
	MEAN OF EXTREME MINS.	54	-22.5	-19.3	-8.2	14.0	27.9	36.4	43.6	39.7	29.8	21.2	5.9	-14.3	12.8
	NORMAL DRY BULB	30	8.9	11.9	24.6	37.9	50.9	60.5	65.5	62.8	53.7	43.2	30.7	14.8	38.8
	MEAN DRY BULB	54	10.4	13.0	24.2	37.7	50.9	60.3	65.3	62.9	53.9	43.3	31.3	16.9	39.2
	MEAN WET BULB	17	10.4	13.0	23.1	34.8	46.8	55.9	57.3	59.6	51.2	40.6	28.1	17.3	36.5
	MEAN DEW POINT	17	4.8	5.9	14.9	25.0	36.7	47.2	53.2	55.3	46.9	35.7	23.8	12.6	30.2
	NORMAL NO. DAYS WITH:														
	MAXIMUM 90	30	0.0	0.0	0.0	0.0	0.2	0.4	1.0	0.4	0.0	0.0	0.0	0.0	2.0
	MAXIMUM 32	30	26.0	22.3	12.4	1.7	*	0.0	0.0	0.0	0.0	0.3	9.5	23.5	95.7
	MINIMUM 32	30	30.6	27.5	28.3	20.6	5.2	*	0.0	0.0	2.7	13.6	24.8	30.4	183.7
	MINIMUM 0	30	16.6	12.7	4.5	*	0.0	0.0	0.0	0.0	0.0	0.0	0.3	10.3	44.4
H/C	NORMAL HEATING DEG. DAYS	30	1739	1487	1252	813	437	143	61	115	343	676	1029	1556	9651
	NORMAL COOLING DEG. DAYS	30	0	0	0	0	0	8	76	47	0	0	0	0	131
RH	NORMAL (PERCENT)	30	70	71	70	74	68	68	77	79	82	82	84	79	75
	HOUR 01 LST	30	74	74	76	82	80	82	90	91	92	89	88	82	83
	HOUR 07 LST	30	75	74	76	75	74	78	83	86	87	86	85	80	80
	HOUR 13 LST	30	65	61	59	56	53	55	59	60	61	63	71	70	61
	HOUR 19 LST	30	71	68	65	63	60	63	69	73	75	74	79	77	70
S	PERCENT POSSIBLE SUNSHINE														
W/O	MEAN NO. DAYS WITH:														
	HEAVY FOG (VISBY 1/4 MI)	31	2.3	1.9	2.0	2.2	1.0	1.6	2.6	2.5	3.0	2.1	3.4	2.7	27.3
	THUNDERSTORMS	31	0.0	0.0	0.2	0.5	1.8	4.3	6.7	4.0	1.2	1.0	0.0	0.0	19.7
CLOUDINESS	MEAN:														
	SUNRISE-SUNSET (OKTAS)	51	5.4	5.4	5.4	5.8	5.8	5.8	5.5	5.3	5.3	5.6	6.3	5.7	5.6
	MIDNIGHT-MIDNIGHT (OKTAS)														
	MEAN NO. DAYS WITH:														
	CLEAR	56	7.0	6.2	7.0	5.1	4.1	3.3	3.0	4.5	5.3	4.8	2.8	5.5	58.6
	PARTLY CLOUDY	56	6.9	6.3	6.8	6.8	9.0	9.8	12.7	11.2	8.9	7.7	6.3	6.5	98.9
	CLOUDY	56	17.1	15.8	17.3	18.0	17.9	16.9	14.7	14.6	15.1	18.1	20.4	18.5	204.4
PR	MEAN STATION PRESSURE (IN)	15	29.22	29.24	29.21	29.21	29.21	29.21	29.21	29.30	29.28	29.30	29.23	29.20	29.24
	MEAN SEA-LEVEL PRES. (IN)	18	29.97	29.97	29.96	29.93	29.94	29.89	29.91	29.97	30.01	30.03	30.00	30.00	29.96
WINDS	MEAN SPEED (MPH)	21	11.3	11.1	11.9	11.2	10.4	9.5	8.8	8.2	9.2	10.0	10.2	10.5	10.2
	PREVAIL.DIR (TENS OF DEGS)	6	32	31	32	32	32	18	18	24	18	32	32	32	32
	MAXIMUM 2-MINUTE:														
	SPEED (MPH)	7	36	41	37	33	37	30	30	28	26	40	41	39	41
	DIR. (TENS OF DEGS)		31	25	33	31	33	32	31	31	31	30	32	31	32
	YEAR OF OCCURRENCE		1998	2001	1999	1996	1994	1997	1999	1999	2000	1996	2001	1994	NOV 2001
	MAXIMUM 5-SECOND:														
	SPEED (MPH)	5	43	56	46	41	41	36	36	37	43	52	52	51	56
	DIR. (TENS OF DEGS)		30	26	30	12	22	35	30	32	13	30	31	22	26
	YEAR OF OCCURRENCE		2000	2001	1999	2000	2000	1997	2000	1998	1999	1996	2001	2000	FEB 2001
PRECIPITATION	NORMAL (IN)	30	2.42	1.92	2.43	2.45	3.07	2.91	4.01	4.07	3.45	3.10	3.55	3.22	36.60
	MAXIMUM MONTHLY (IN)	62	5.60	4.13	5.13	5.26	6.27	7.11	6.83	12.09	8.81	8.73	8.15	7.97	12.09
	YEAR OF OCCURRENCE		1995	1955	1953	1973	1947	1940	1957	1981	1999	1990	1983	1973	AUG 1981
	MINIMUM MONTHLY (IN)	62	0.12	0.26	0.66	0.54	0.47	0.88	0.96	0.93	0.86	0.63	0.45	0.74	0.12
	YEAR OF OCCURRENCE		1944	1978	1965	1967	1982	1983	1991	1957	1968	1955	1939	1963	JAN 1944
	MAXIMUM IN 24 HOURS (IN)	62	1.81	1.94	1.82	2.11	2.25	2.37	2.92	6.89	6.23	4.07	2.27	2.80	6.89
	YEAR OF OCCURRENCE		1995	2000	2001	1958	1948	1957	1957	1981	1954	1970	1983	1973	AUG 1981
	NORMAL NO. DAYS WITH:														
	PRECIPITATION 0.01	30	14.8	11.3	12.8	12.6	13.5	12.8	13.5	13.3	12.5	12.6	14.7	14.8	159.2
	PRECIPITATION 1.00	30	0.2	0.1	0.2	0.2	0.3	0.2	1.0	0.7	0.6	0.5	0.5	0.4	4.9
SNOWFALL	NORMAL (IN)	30	25.3	18.8	19.3	10.1	1.0	T	0.0	0.0	T	1.7	13.1	26.3	115.6
	MAXIMUM MONTHLY (IN)	61	44.5	41.0	47.1	36.4	10.9	T	T	T	2.5	12.1	34.9	59.9	59.9
	YEAR OF OCCURRENCE		1994	1960	1955	1982	1967	1991	1992	1995	1991	1963	1974	1972	DEC 1972
	MAXIMUM IN 24 HOURS (IN)	61	19.0	21.2	28.6	21.1	5.8	T	T	T	2.5	9.4	21.0	19.7	28.6
	YEAR OF OCCURRENCE		1994	1995	1984	1982	1967	1991	1992	1995	1991	1963	1986	1989	MAR 1984
	MAXIMUM SNOW DEPTH (IN)	53	57	62	51	42	4		0	0	0	1	9	28	39	62
	YEAR OF OCCURRENCE		1977	1977	1977	1982	1966					1991	1963	1974	1978	FEB 1977
	NORMAL NO. DAYS WITH:														
	SNOWFALL 1.0	30	5.9	5.0	4.7	2.8	0.3	0.0	0.0	0.0	0.0	0.5	3.2	6.7	29.1

PRECIPITATION (inches) 2001 CARIBOU, ME (CAR)

YEAR	JAN	FEB	MAR	APR	MAY	JUN	JUL	AUG	SEP	OCT	NOV	DEC	ANNUAL
1972	1.32	2.38	4.72	1.09	3.69	4.97	4.21	5.07	3.88	4.03	2.76	5.28	43.40
1973	2.60	2.89	2.48	5.26	5.03	2.13	4.62	2.95	2.62	1.47	2.63	7.97	42.65
1974	1.89	1.37	3.56	3.81	3.61	2.82	3.39	3.98	3.40	1.15	4.16	2.05	35.19
1975	2.71	1.66	1.94	1.95	3.04	2.40	4.28	1.39	2.98	1.51	3.40	4.05	31.31
1976	3.51	3.43	2.57	3.31	5.11	2.85	6.74	6.17	2.52	5.46	2.33	4.63	48.63
1977	3.42	2.99	2.36	1.83	0.74	6.44	1.75	7.86	3.54	5.30	1.63	4.45	41.45
1978	5.10	0.26	2.69	2.33	1.98	3.70	5.56	1.95	2.74	1.94	1.84	3.04	33.13
1979	4.49	2.22	3.70	3.08	4.27	3.39	3.36	4.98	4.68	1.68	2.90	3.05	41.80
1980	1.55	0.82	3.15	2.57	2.05	2.19	5.42	2.28	4.06	2.51	3.21	2.74	32.55
1981	1.68	2.39	3.43	2.17	3.18	4.15	2.62	12.09	2.38	6.28	2.51	3.81	46.69
1982	2.46	2.17	2.72	4.01	0.47	3.08	4.25	4.78	3.70	1.61	5.50	2.51	37.26
1983	2.95	1.77	3.84	4.20	5.28	0.88	5.92	3.86	2.70	1.81	8.15	5.01	46.37
1984	2.10	3.06	2.55	1.74	5.72	5.90	4.52	1.65	1.54	1.81	2.01	3.00	35.60
1985	0.99	2.77	1.87	1.80	2.64	2.89	5.05	1.74	2.30	1.42	3.50	2.24	29.21
1986	4.86	1.13	2.32	2.29	2.13	1.96	4.21	4.97	3.58	1.47	3.96	1.66	34.54
1987	2.29	0.33	1.24	1.75	2.46	3.59	3.16	1.82	4.37	2.18	2.33	2.56	28.08
1988	2.79	2.65	1.23	1.99	1.84	2.37	2.28	5.65	1.82	3.09	4.10	1.00	30.81
1989	1.88	1.43	1.40	2.24	4.13	2.29	2.63	5.41	3.52	1.62	3.88	2.35	32.78
1990	3.36	1.84	1.16	2.28	3.53	4.56	4.15	3.23	3.78	8.73	4.22	5.60	46.44
1991	2.06	1.13	4.71	2.62	3.51	1.98	0.96	6.72	3.45	4.46	1.61	2.10	35.31
1992	3.76	2.68	1.87	2.57	1.83	4.58	4.49	6.28	1.51	3.52	1.60	1.58	36.27
1993	2.18	2.10	1.72	3.48	3.58	4.54	2.48	3.57	5.11	4.31	2.94	3.99	40.00
1994	3.68	1.39	2.44	3.24	4.75	4.64	4.64	2.82	3.16	0.88	3.75	3.01	38.40
1995	5.60	2.70	2.23	2.12	2.46	1.18	1.48	2.94	1.90	5.13	4.88	1.79	34.41
1996	4.05	2.69	1.74	3.59	3.52	3.42	6.32	2.66	3.81	3.41	1.49	3.72	40.42
1997	3.60	2.52	2.47	1.68	5.10	4.37	2.64	4.12	2.67	1.31	2.08	2.81	35.37
1998	4.08	2.62	3.51	2.23	3.61	3.22	5.35	2.29	3.22	2.22	2.10	1.57	36.02
1999	3.59	1.38	2.30	1.43	2.40	3.20	2.94	3.69	8.81	3.48	2.70	2.76	38.68
2000	2.94	2.96	1.86	4.58	4.52	2.88	4.39	4.23	1.59	2.59	2.09	3.73	38.36
2001	0.81	2.66	3.15	0.96	2.01	2.77	5.82	1.51	3.65	2.69	2.08	0.99	29.10
POR= 61 YRS	2.49	2.09	2.42	2.53	3.08	3.40	3.97	3.83	3.37	3.15	3.19	2.87	36.39

AVERAGE TEMPERATURE (F) 2001   CARIBOU, ME (CAR)

YEAR	JAN	FEB	MAR	APR	MAY	JUN	JUL	AUG	SEP	OCT	NOV	DEC	ANNUAL
1972	8.2	6.4	18.3	34.7	52.4	61.5	64.7	60.5	54.2	38.3	27.1	7.6	36.2
1973	9.7	12.6	29.6	37.9	48.9	63.0	68.6	66.4	52.9	43.9	27.5	23.9	40.4
1974	6.7	10.8	21.0	36.8	45.2	62.7	64.7	64.3	52.7	39.3	31.4	18.5	37.8
1975	9.6	11.5	23.1	35.1	53.7	61.4	68.6	64.4	54.5	42.3	32.8	10.9	39.0
1976	5.3	13.5	22.7	38.4	51.2	64.0	63.7	62.6	51.5	39.2	25.6	7.9	37.1
1977	6.0	12.9	32.3	36.4	53.3	58.8	65.2	64.2	51.2	43.5	33.3	16.9	39.5
1978	9.9	12.5	20.8	34.9	55.3	60.8	66.2	64.9	50.4	41.6	27.7	16.3	38.4
1979	15.4	10.4	31.8	41.2	54.6	63.0	68.6	61.9	54.4	45.2	35.8	19.2	41.8
1980	13.3	11.9	24.0	42.6	51.2	59.4	64.8	66.0	50.8	40.9	30.1	9.0	38.7
1981	5.6	27.6	28.0	39.5	54.2	61.2	66.3	64.5	53.1	40.4	31.9	23.1	41.3
1982	4.0	11.3	24.5	35.1	53.6	60.3	66.4	58.7	54.9	44.4	32.7	22.1	39.0
1983	14.9	15.3	27.0	41.1	48.9	61.9	64.5	64.3	57.2	43.6	32.2	14.5	40.5
1984	6.7	22.4	19.3	39.8	49.2	59.2	65.8	65.5	51.7	44.5	33.4	18.4	39.7
1985	6.0	16.2	23.5	35.3	49.5	58.2	66.0	62.2	56.1	43.8	28.3	12.1	38.1
1986	11.8	12.2	24.4	42.8	52.2	56.9	62.6	60.7	50.5	41.2	25.9	16.2	38.1
1987	9.9	12.5	28.0	43.9	50.6	60.3	65.8	61.1	54.5	44.2	29.0	18.7	39.9
1988	12.1	13.6	23.1	39.4	54.6	58.5	67.4	64.4	52.4	40.5	33.5	13.8	39.4
1989	12.3	10.4	19.5	36.7	56.7	60.1	64.7	63.5	55.4	45.0	27.6	3.5	38.0
1990	17.3	11.7	24.7	39.6	49.0	62.9	66.2	66.6	53.5	44.7	31.0	20.5	40.6
1991	7.1	15.1	27.0	39.6	52.5	61.4	65.8	65.3	52.1	44.8	32.8	13.5	39.8
1992	10.4	10.9	20.2	36.4	53.2	60.2	60.3	63.7	56.2	41.1	29.3	19.6	38.5
1993	8.6	4.2	23.3	39.9	51.2	59.9	64.8	65.8	54.1	39.2	29.7	19.4	38.3
1994	-.7	7.7	25.3	37.0	48.7	63.6	68.1	61.6	53.5	46.1	33.9	20.1	38.7
1995	15.5	9.1	25.5	34.3	50.8	63.0	69.1	65.1	51.8	47.8	28.2	14.7	39.6
1996	11.2	13.5	23.7	38.8	48.7	62.1	64.6	63.7	55.0	42.2	29.1	26.2	39.9
1997	10.6	12.6	18.8	35.5	47.3	60.3	65.3	61.5	54.2	41.4	29.2	15.9	37.7
1998	13.8	19.7	27.4	40.1	56.2	61.2	66.3	64.1	55.5	43.4	30.0	21.4	41.6
1999	10.5	17.5	30.2	39.2	57.2	63.9	67.2	62.9	61.7	40.6	34.7	22.2	42.3
2000	11.1	14.5	29.9	37.2	49.4	58.7	63.4	62.4	53.6	42.7	35.0	15.5	39.5
2001	10.8	12.2	23.6	37.0	56.4	61.9	64.4	65.8	57.1	46.7	35.6	26.5	41.5
POR= 62 YRS	10.1	13.0	24.0	37.4	50.8	60.1	65.2	63.0	53.9	43.3	31.1	16.4	39.0

REFERENCE NOTES:

PAGE 1:
THE TEMPERATURE GRAPH SHOWS NORMAL MAXIMUM AND NORMAL
    MINIMUM DAILY TEMPERATURES (SOLID CURVES) AND THE
    ACTUAL DAILY HIGH AND LOW TEMPERATURES (VERTICAL BARS).

PAGE 2 AND 3:
    H/C INDICATES HEATING AND COOLING DEGREE DAYS.
    RH INDICATES RELATIVE HUMIDITY
    W/O INDICATES WEATHER AND OBSTRUCTIONS
    S INDICATES SUNSHINE.
    PR INDICATES PRESSURE.
    CLOUDINESS ON PAGE 3 IS THE SUM OF THE CEILOMETER AND
        SATELLITE DATA NOT TO EXCEED EIGHT EIGHTHS(OKTAS).

GENERAL:
    T INDICATES TRACE PRECIPITATION, AN AMOUNT GREATER
        THAN ZERO BUT LESS THAN THE LOWEST REPORTABLE VALUE.
    + INDICATES THE VALUE ALSO OCCURS ON EARLIER DATES.
    BLANK ENTRIES DENOTE MISSING OR UNREPORTED DATA.
    NORMALS ARE 30-YEAR AVERAGES (1961 - 1990).
    ASOS INDICATES AUTOMATED SURFACE OBSERVING SYSTEM.
    PM INDICATES THE LAST DAY OF THE PREVIOUS MONTH.
    POR (PERIOD OF RECORD) BEGINS WITH THE JANUARY DATA
        MONTH AND IS THE NUMBER OF YEARS USED TO COMPUTE
        THE MEAN. INDIVIDUAL MONTHS WITHIN THE POR MAY
        BE MISSING.
    WHEN THE POR FOR A NORMAL IS LESS THAN 30 YEARS,
        THE NORMAL IS PROVISIONAL AND IS BASED ON THE NUMBER
        OF YEARS INDICATED.
    0.* OR * INDICATES THE VALUE OR MEAN-DAYS-WITH
        IS BETWEEN 0.00 AND 0.05.
    CLOUDINESS FOR ASOS STATIONS DIFFERS FROM THE NON-ASOS
        OBSERVATION TAKEN BY A HUMAN OBSERVER. ASOS STATION
        CLOUDINESS IS BASED ON TIME-AVERAGED CEILOMETER DATA
        FOR CLOUDS AT OR BELOW 12,000 FEET AND ON SATELLITE
        DATA FOR CLOUDS ABOVE 12,000 FEET.
    THE NUMBER OF DAYS WITH CLEAR, PARTLY CLOUDY, AND
        CLOUDY CONDITIONS FOR ASOS STATIONS IS THE SUM
        OF THE CEILOMETER AND SATELLITE DATA FOR THE
        SUNRISE TO SUNSET PERIOD.

GENERAL CONTINUED:
    CLEAR INDICATES 0 - 2 OKTAS, PARTLY CLOUDY INDICATES
        3 - 6 OKTAS, AND CLOUDY INDICATES 7 OR 8 OKTAS.
        WHEN AT LEAST ONE OF THE ELEMENTS (CEILOMETER OR
        SATELLITE) IS MISSING, THE DAILY CLOUDINESS IS
        NOT COMPUTED.
    WIND DIRECTION IS RECORDED IN TENS OF DEGREES (2 DIGITS)
        CLOCKWISE FROM TRUE NORTH. "00" INDICATES CALM. "36"
        INDICATES TRUE NORTH.
    RESULTANT WIND IS THE VECTOR AVERAGE OF THE SPEED AND
        DIRECTION.
    AVERAGE TEMPERATURE IS THE SUM OF THE MEAN DAILY MAXIMUM
        AND MINIMUM TEMPERATURE DIVIDED BY 2.
    SNOWFALL DATA COMPRISE ALL FORMS OF FROZEN
        PRECIPITATION, INCLUDING HAIL.
    A HEATING (COOLING) DEGREE DAY IS THE DIFFERENCE BETWEEN
        THE AVERAGE DAILY TEMPERATURE AND 65 F.
    DRY BULB IS THE TEMPERATURE OF THE AMBIENT AIR.
    DEW POINT IS THE TEMPERATURE TO WHICH THE AIR MUST BE
        COOLED TO ACHIEVE 100 PERCENT RELATIVE HUMIDITY.
    WET BULB IS THE TEMPERATURE THE AIR WOULD HAVE IF THE
        MOISTURE CONTENT WAS INCREASED TO 100 PERCENT RELATIVE
        HUMIDITY.

    ON JULY 1, 1996, THE NATIONAL WEATHER SERVICE BEGAN USING
        THE "METAR" OBSERVATION CODE THAT WAS ALREADY EMPLOYED
        BY MOST OTHER NATIONS OF THE WORLD. THE MOST NOTICEABLE
        DIFFERENCE IN THIS ANNUAL PUBLICATION WILL BE THE CHANGE
        IN UNITS FROM TENTHS TO EIGHTS(OKTAS) FOR REPORTING THE
        AMOUNT OF SKY COVER.

HEATING DEGREE DAYS (base 65 F) 2001 CARIBOU, ME (CAR)

YEAR	JUL	AUG	SEP	OCT	NOV	DEC	JAN	FEB	MAR	APR	MAY	JUN	TOTAL
1972-73	55	145	326	818	1130	1775	1711	1464	1092	807	491	122	9936
1973-74	9	44	373	649	1117	1270	1810	1512	1359	837	609	81	9670
1974-75	40	66	369	789	998	1433	1717	1498	1295	890	343	158	9596
1975-76	19	93	310	700	960	1674	1849	1489	1302	793	426	123	9738
1976-77	84	132	405	793	1175	1765	1828	1453	1002	850	406	203	10096
1977-78	56	93	408	657	944	1489	1702	1467	1364	894	333	156	9563
1978-79	53	86	434	717	1114	1503	1534	1527	1018	708	326	104	9124
1979-80	34	146	327	612	870	1413	1593	1534	1265	664	420	200	9078
1980-81	71	41	425	740	1042	1733	1839	1040	1141	757	333	125	9287
1981-82	37	77	355	757	984	1292	1891	1499	1251	890	355	145	9533
1982-83	60	199	310	632	961	1325	1544	1387	1171	712	493	148	8942
1983-84	75	78	257	656	978	1562	1804	1229	1412	748	493	199	9491
1984-85	38	57	396	630	940	1441	1822	1363	1281	884	472	198	9522
1985-86	43	118	272	650	1094	1636	1646	1473	1255	660	394	246	9487
1986-87	105	147	427	733	1167	1505	1704	1465	1140	629	442	142	9606
1987-88	79	152	314	641	1071	1430	1636	1485	1289	760	321	232	9410
1988-89	47	114	373	752	939	1583	1627	1524	1402	841	257	181	9640
1989-90	66	101	303	613	1116	1905	1471	1490	1241	759	490	107	9662
1990-91	57	47	337	623	1014	1370	1791	1392	1172	755	383	158	9099
1991-92	55	66	383	619	959	1595	1691	1564	1382	852	381	163	9710
1992-93	145	73	273	732	1064	1402	1747	1699	1284	748	423	175	9765
1993-94	73	53	327	791	1051	1407	2040	1601	1222	832	500	107	10004
1994-95	24	124	337	578	926	1388	1529	1565	1218	912	433	131	9165
1995-96	12	74	388	522	1098	1551	1666	1490	1271	782	496	116	9466
1996-97	48	83	304	702	1070	1196	1678	1462	1427	879	544	161	9554
1997-98	62	123	318	723	1068	1516	1579	1263	1159	740	269	151	8971
1998-99	42	69	275	660	1045	1346	1681	1323	1073	767	248	103	8632
1999-00	40	105	163	750	903	1320	1662	1456	1079	828	478	216	9000
2000-01	75	93	337	681	893	1525	1672	1474	1277	833	265	147	9272
2001-	65	64	260	563	874	1186							

COOLING DEGREE DAYS (base 65 F) 2001 CARIBOU, ME (CAR)

YEAR	JAN	FEB	MAR	APR	MAY	JUN	JUL	AUG	SEP	OCT	NOV	DEC	ANNUAL
1972	0	0	0	0	21	35	51	11	9	0	0	0	127
1973	0	0	0	0	0	67	126	97	17	0	0	0	307
1974	0	0	0	0	0	2	21	38	50	5	0	0	116
1975	0	0	0	0	0	54	137	80	0	0	0	0	271
1976	0	0	0	0	6	104	52	65	4	0	0	0	231
1977	0	0	0	0	52	24	69	77	1	0	0	0	223
1978	0	0	0	0	38	35	97	93	1	0	0	0	264
1979	0	0	0	0	8	50	153	57	16	6	0	0	290
1980	0	0	0	0	0	37	71	78	8	0	0	0	194
1981	0	0	0	0	3	17	83	68	4	0	0	0	175
1982	0	0	0	0	11	10	110	9	12	0	0	0	152
1983	0	0	0	0	0	62	65	66	32	1	0	0	226
1984	0	0	0	0	9	33	72	79	0	0	0	0	193
1985	0	0	0	0	0	0	82	40	10	0	0	0	132
1986	0	0	0	0	5	10	39	20	1	0	0	0	75
1987	0	0	0	0	2	10	111	38	8	0	0	0	169
1988	0	0	0	0	6	45	131	103	1	0	0	0	286
1989	0	0	0	0	9	37	64	63	19	0	0	0	192
1990	0	0	0	2	0	51	102	106	0	0	0	0	261
1991	0	0	0	0	2	55	89	85	4	0	0	0	235
1992	0	0	0	0	23	26	5	40	12	0	0	0	106
1993	0	0	0	0	0	31	72	83	6	0	0	0	192
1994	0	0	0	0	4	71	128	26	0	0	0	0	229
1995	0	0	0	0	0	76	147	82	0	0	0	0	305
1996	0	0	0	0	0	36	40	51	11	0	0	0	138
1997	0	0	0	0	0	30	77	21	1	0	0	0	129
1998	0	0	0	0	3	43	91	44	0	0	0	0	181
1999	0	0	0	0	12	76	115	48	70	0	0	0	321
2000	0	0	0	0	0	33	33	20	1	0	0	0	87
2001	0	0	0	0	5	59	56	94	31	0	0	0	245

SNOWFALL (inches) 2001 CARIBOU, ME (CAR)

YEAR	JUL	AUG	SEP	OCT	NOV	DEC	JAN	FEB	MAR	APR	MAY	JUN	TOTAL
1972-73	0.0	0.0	0.0	1.7	13.3	59.9	20.3	27.8	7.8	22.2	0.0	0.0	153.0
1973-74	0.0	0.0	T	14.5	23.4	20.7	11.9	21.6	13.1	4.2	0.0	0.0	109.4
1974-75	0.0	0.0	T	0.4	34.9	12.6	31.2	9.8	18.2	15.1	0.0	0.0	122.2
1975-76	0.0	0.0	0.0	10.4	36.4	30.1	30.2	23.3	1.5	0.5	0.0	0.0	132.4
1976-77	0.0	0.0	0.0	3.7	10.2	31.9	39.1	34.4	16.0	10.6	T	0.0	145.9
1977-78	0.0	0.0	0.0	0.0	5.5	37.7	41.4	4.4	13.7	15.3	0.8	0.0	118.8
1978-79	0.0	0.0	0.0	0.0	T	9.2	40.3	32.1	18.1	8.6	14.9	T	123.2
1979-80	0.0	0.0	0.0	0.7	3.0	18.0	7.9	12.6	27.6	0.8	0.0	T	70.6
1980-81	0.0	0.0	T	0.2	12.0	28.9	35.2	8.6	36.4	1.4	0.2	0.0	122.9
1981-82	0.0	0.0	T	2.6	4.8	34.3	30.9	25.1	23.7	36.4	1.0	0.0	158.8
1982-83	0.0	0.0	0.0	T	14.6	4.0	22.8	22.8	13.9	4.8	T	0.0	82.9
1983-84	0.0	0.0	0.0	T	21.0	26.6	27.3	20.1	35.4	3.9	0.2	0.0	134.5
1984-85	0.0	0.0	0.0	0.8	3.8	30.7	10.5	26.8	11.9	5.0	1.3	0.0	90.8
1985-86	0.0	0.0	0.0	T	14.8	18.3	30.0	11.9	18.6	11.3	T	0.0	104.9
1986-87	0.0	0.0	T	0.4	28.1	8.5	26.4	4.1	12.3	5.2	T	0.0	85.0
1987-88	0.0	0.0	T	T	5.2	22.6	32.1	28.2	3.9	6.4	T	0.0	98.4
1988-89	0.0	0.0	0.0	1.2	13.8	11.5	18.4	16.6	11.9	9.4	0.0	0.0	82.8
1989-90	0.0	0.0	T	0.1	14.3	38.0	28.5	18.3	7.6	10.7	0.6	T	118.1
1990-91	0.0	0.0	0.0	T	12.0	22.6	21.5	12.2	24.0	2.5	0.0	T	94.8
1991-92	0.0	0.0	2.5	T	7.1	19.4	14.8	38.6	4.2	7.4	T	0.0	94.0
1992-93	T	0.0	T	0.4	5.3	15.9	5.6	22.8	23.4	7.1	T	0.0	80.5
1993-94	0.0	0.0	0.0	0.2	10.5	13.9	44.5	20.3	28.8	9.5	T	0.0	127.7
1994-95	0.0	0.0	T	T	9.8	25.9	25.9	40.9	15.7	7.9	1.1	0.0	
1995-96	0.0	T	0.0	T	8.9	25.1	18.4	22.7	16.5	13.1	5.7	0.0	110.4
1996-97	0.0	0.0	0.0	0.8	8.5	14.4	35.5	30.2	28.9	11.7	0.9	0.0	130.9
1997-98	0.0	0.0	0.0	8.9	9.6	37.3	39.9	8.8	11.0	8.5	T	0.0	124.0
1998-99	0.0	0.0	0.0	T	12.6	21.9	36.7	18.5	30.3	8.8	T	0.0	128.8
1999-00	0.0	0.0	0.0	0.2	4.1	6.7	43.7	32.5	15.4	9.8	T	0.0	112.4
2000-01	T	0.0	0.0	10.3	5.2	32.7	14.2	30.5	30.6	8.6	0.0	0.0	132.1
2001-	0.0	0.0	0.0	0.0	9.0	4.2							
POR= 61 YRS	T	T	0.0	1.8	11.8	22.9	24.0	21.7	19.3	8.5	0.9	T	110.9

2001
PORTLAND,
MAINE (PWM)

The Portland City Airport is located 2 3/4 miles west of the site of the former city office. The surrounding country is mostly open, rolling and sloping generally toward the Fore River, a body of brackish water about 1,000 feet wide at a distance of about 1/2 mile from the station and forming one boundary (north through east) of the field. The airport is about 5 1/2 miles west-northwest of the open ocean. A slight rise reaching an elevation of 100 feet, lying northwest of the field, cuts down the wind slightly from that direction. The older portion of the city is situated on a hill rising abruptly from sea level to 170 feet, 1 1/2 miles east of the airport and on the opposite side of the Fore River. A line of low hills southeast of the airport, near the ocean, which reach a maximum height of 160 feet, shuts off sight of the ocean from the airport. Sebago Lake with an area of 44 square miles is situated about 15 miles to the northwest and 45 miles farther are the White Mountains, averaging 3,000 to 5,000 feet in height.

As a rule, Portland has very pleasant summers and falls, cold winters with frequent thaws, and disagreeable springs. Very few summer nights are too warm and humid for comfortable sleeping. Autumn has the greatest number of sunny days and the least cloudiness. Winters are quite severe, but begin late and then extend deeply into the normal springtime.

Heavy seasonal snowfalls, over 100 inches, normally occur about each 10 years. True blizzards are very rare. The White Mountains, to the northwest, keep considerable snow from reaching the Portland area and also moderate the temperature. Normal monthly precipitation is remarkably uniform throughout the year.

Winds are generally quite light with the highest velocities being confined mostly to March and November. Even in these months the occasional northeasterly gales have usually lost much of their severity before reaching the coast of Maine.

Temperatures well below zero are recorded frequently each winter. Cold waves sometimes come in on strong winds, but extremely low temperatures are generally accompanied by light winds.

The average freeze-free season at the airport station is 139 days. Mid-May is the average occurrence of the last freeze in spring, and the average occurrence of the first freeze in fall is late September. The freeze-free period is longer in the city proper, but may be even shorter at susceptible places further inland.

Daily maximum temperatures at the present airport site agree closely with those near the former intown office, but minimum temperatures on clear, quiet mornings range as much as 15 degrees lower at the airport.

NORMALS, MEANS, AND EXTREMES

PORTLAND, ME　(PWM)

LATITUDE:	LONGITUDE:	ELEVATION (FT):		TIME ZONE:	WBAN: 14764
43　38' 32" N	70　18' 16" W	GRND:　69	BARO:　72	EASTERN　(UTC + 5)	

	ELEMENT	POR	JAN	FEB	MAR	APR	MAY	JUN	JUL	AUG	SEP	OCT	NOV	DEC	YEAR
TEMPERATURE °F	NORMAL DAILY MAXIMUM	30	30.3	33.1	41.4	52.3	63.2	72.7	78.8	77.4	69.3	58.7	47.0	35.1	54.9
	MEAN DAILY MAXIMUM	45	30.7	33.3	41.4	52.7	63.5	73.2	78.7	77.5	69.5	58.8	47.4	35.9	55.2
	HIGHEST DAILY MAXIMUM	61	64	64	88	85	94	98	99	103	95	88	74	71	103
	YEAR OF OCCURRENCE		1950	1957	1998	1957	1987	1991	1977	1975	1983	1963	1987	2001	AUG 1975
	MEAN OF EXTREME MAXS.	54	49.1	50.6	59.4	71.7	83.3	89.2	92.0	90.2	86.0	75.8	64.7	54.5	72.2
	NORMAL DAILY MINIMUM	30	11.4	13.5	24.5	34.1	43.4	52.1	58.3	57.1	48.9	38.3	30.4	17.8	35.8
	MEAN DAILY MINIMUM	45	12.6	14.0	24.5	33.6	43.1	52.3	58.2	56.9	48.8	38.1	30.2	18.2	35.9
	LOWEST DAILY MINIMUM	61	-26	-39	-21	8	23	33	40	33	23	15	3	-21	-39
	YEAR OF OCCURRENCE		1971	1943	1950	1954	1956	1944	1965	1965	1941	1976	1989	1963	FEB 1943
	MEAN OF EXTREME MINS.	54	-9.1	-7.0	3.9	21.7	30.6	40.6	47.6	44.5	33.3	23.6	14.9	-2.7	20.2
	NORMAL DRY BULB	30	20.8	23.3	33.0	43.3	53.3	62.4	68.6	67.3	59.1	48.5	38.7	26.5	45.4
	MEAN DRY BULB	54	21.8	23.9	32.5	43.1	53.3	62.7	68.6	67.0	58.8	48.6	38.8	27.3	45.5
	MEAN WET BULB	49	20.4	22.1	29.5	38.6	48.3	57.3	62.9	61.8	54.8	45.0	36.0	25.0	41.8
	MEAN DEW POINT	49	13.4	14.6	22.3	31.9	43.0	53.4	59.3	58.5	51.4	40.5	30.8	18.8	36.5
	NORMAL NO. DAYS WITH:														
	MAXIMUM　90	30	0.0	0.0	0.0	0.0	0.2	0.9	1.9	1.2	0.3	0.0	0.0	0.0	4.5
	MAXIMUM　32	30	17.2	12.9	4.1	0.1	0.0	0.0	0.0	0.0	0.0	0.0	1.0	12.4	47.7
	MINIMUM　32	30	29.9	26.5	25.2	13.5	2.0	0.0	0.0	0.0	0.8	8.6	19.3	28.9	154.7
	MINIMUM　0	30	6.1	3.9	0.5	0.0	0.0	0.0	0.0	0.0	0.0	0.0	0.0	2.8	13.3
H/C	NORMAL HEATING DEG. DAYS	30	1370	1168	992	651	363	100	11	39	189	512	789	1194	7378
	NORMAL COOLING DEG. DAYS	30	0	0	0	0	0	22	123	111	12	0	0	0	268
RH	NORMAL (PERCENT)	30	67	65	66	67	71	75	75	76	77	74	73	70	71
	HOUR 01 LST	30	72	70	72	77	82	87	88	88	88	83	79	75	80
	HOUR 07 LST	30	73	73	73	72	75	79	81	84	85	84	80	77	78
	HOUR 13 LST	30	58	56	56	56	54	58	60	60	60	60	58	61	58
	HOUR 19 LST	30	66	64	66	66	69	73	73	76	78	75	73	69	71
S	PERCENT POSSIBLE SUNSHINE	55	56	60	56	55	54	60	63	64	63	57	48	53	57
W/O	MEAN NO. DAYS WITH:														
	HEAVY FOG(VISBY 1/4 MI)	61	2.1	2.2	3.4	3.0	4.9	5.0	6.3	5.6	5.0	4.6	3.5	2.0	47.6
	THUNDERSTORMS	61	0.0	0.1	0.4	0.5	1.9	4.1	4.5	3.5	1.5	0.6	0.4	0.1	17.6
CLOUDINESS	MEAN:														
	SUNRISE-SUNSET (OKTAS)	1					5.6	6.4							
	MIDNIGHT-MIDNIGHT (OKTAS)														
	MEAN NO. DAYS WITH:														
	CLEAR	1	4.0	2.0	7.0		7.0	6.0	1.0	9.0	3.0	8.0		5.0	
	PARTLY CLOUDY	1		1.0	3.0		4.0	5.0	1.0	1.0	2.0			4.0	
	CLOUDY	1	4.0	5.0	7.0		10.0	9.0	2.0	3.0	6.0	3.0		8.0	
PR	MEAN STATION PRESSURE(IN)	29	29.93	29.93	29.90	29.88	29.89	29.87	29.88	29.94	29.97	29.99	29.96	29.94	29.92
	MEAN SEA-LEVEL PRES. (IN)	48	29.98	29.99	29.96	29.94	29.97	29.94	29.94	29.96	29.99	30.05	30.06	30.00	30.01	29.99
WINDS	MEAN SPEED (MPH)	41	9.2	9.4	10.1	10.1	9.1	8.4	7.7	7.7	8.1	8.5	8.9	8.8	8.8
	PREVAIL.DIR(TENS OF DEGS)	24	36	34	32	18	18	18	18	18	18	34	36	34	18
	MAXIMUM 2-MINUTE:														
	SPEED (MPH)	7	38	45	41	40	32	34	32	57	33	37	41	40	57
	DIR. (TENS OF DEGS)		16	08	11	29	36	26	02	28	29	36	10	30	28
	YEAR OF OCCURRENCE		1996	1995	1999	1995	1995	2000	1999	1998	1995	2000	1997	1997	AUG 1998
	MAXIMUM 5-SECOND:														
	SPEED (MPH)	7	53	56	54	47	44	44	44	61	41	47	72	51	72
	DIR. (TENS OF DEGS)		16	07	13	29	32	26	25	28	02	12	15	17	15
	YEAR OF OCCURRENCE		1996	1995	2000	1995	1995	2000	1997	1998	1996	1995	1995	2000	NOV 1995
PRECIPITATION	NORMAL (IN)	30	3.53	3.33	3.67	4.08	3.62	3.44	3.09	2.87	3.09	3.90	5.17	4.55	44.34
	MAXIMUM MONTHLY (IN)	61	11.92	7.10	9.97	9.90	9.64	9.01	7.48	15.22	9.81	16.83	13.50	9.69	16.83
	YEAR OF OCCURRENCE		1979	1981	1953	1973	1984	1998	1976	1991	1954	1996	1983	1969	OCT 1996
	MINIMUM MONTHLY (IN)	61	0.76	0.04	0.81	0.28	0.49	0.70	0.61	0.27	0.30	0.26	0.90	0.98	0.04
	YEAR OF OCCURRENCE		1970	1987	1965	1999	1965	1941	1965	1947	1948	1947	1976	1955	FEB 1987
	MAXIMUM IN 24 HOURS (IN)	61	3.56	3.41	3.47	5.26	4.66	5.58	2.68	7.83	7.49	13.32	4.70	3.82	13.32
	YEAR OF OCCURRENCE		1977	1981	1951	1973	1989	1967	1996	1991	1954	1996	1990	1969	OCT 1996
	NORMAL NO. DAYS WITH:														
	PRECIPITATION　0.01	30	10.5	9.5	11.4	11.7	12.0	11.4	10.1	9.7	8.9	9.5	11.8	12.2	128.7
	PRECIPITATION　1.00	30	0.7	0.8	1.0	0.9	0.7	0.8	0.7	0.6	0.8	1.2	1.7	1.4	11.3
SNOWFALL	NORMAL (IN)	30	19.3	16.5	11.6	3.4	0.1	0.0	0.0	0.0	T	0.4	3.2	16.4	70.9
	MAXIMUM MONTHLY (IN)	60	62.4	61.2	49.0	15.9	7.0	0.0	0.0	0.0	T	3.8	20.5	54.8	62.4
	YEAR OF OCCURRENCE		1979	1969	1993	1982	1945				1992	1969	1997	1970	JAN 1979
	MAXIMUM IN 24 HOURS (IN)	60	27.1	21.5	18.6	15.9	7.0	0.0	0.0	0.0	T	3.6	11.1	22.8	27.1
	YEAR OF OCCURRENCE		1979	1969	1993	1982	1945				1992	1969	1972	1970	JAN 1979
	MAXIMUM SNOW DEPTH (IN)	49	31	31	33	17	1	0	0	0	0	2	10	40	40
	YEAR OF OCCURRENCE		1979	1967	1967	1956	1966					1969	1972	1970	DEC 1970
	NORMAL NO. DAYS WITH:														
	SNOWFALL　1.0	30	4.4	3.4	3.1	0.9	0.*	0.0	0.0	0.0	0.0	0.1	1.0	4.3	17.2

PRECIPITATION (inches) 2001 PORTLAND, ME (PWM)

YEAR	JAN	FEB	MAR	APR	MAY	JUN	JUL	AUG	SEP	OCT	NOV	DEC	ANNUAL
1972	2.09	5.14	6.01	2.53	3.17	4.24	2.05	0.80	4.31	3.92	7.87	6.49	48.62
1973	2.58	2.57	3.33	9.90	6.28	4.87	1.70	3.48	2.23	3.38	2.40	9.57	52.29
1974	3.41	2.07	3.82	3.82	4.20	4.69	3.66	1.45	5.43	1.74	4.85	4.41	43.55
1975	4.40	2.51	3.20	3.71	1.09	4.87	2.06	3.89	4.34	4.50	6.01	8.14	48.72
1976	4.44	2.84	2.47	2.42	3.99	1.53	7.48	4.87	1.86	5.37	0.90	3.22	41.39
1977	6.46	3.75	6.92	3.46	2.04	3.38	2.83	2.79	4.63	8.30	6.46	6.61	57.63
1978	6.91	0.87	4.19	4.46	4.36	2.42	1.67	2.36	0.59	3.23	2.26	3.15	36.47
1979	11.92	3.50	4.17	6.48	5.15	1.97	5.90	5.53	3.28	6.71	3.95	2.59	61.15
1980	0.98	1.36	4.54	5.78	1.83	3.34	1.99	2.14	3.00	2.99	4.75	1.18	33.88
1981	0.93	7.10	1.44	3.46	2.27	4.59	5.44	2.31	6.14	4.71	2.80	4.51	45.70
1982	5.17	2.53	3.20	4.54	2.91	6.75	2.61	3.35	1.90	1.93	3.61	1.18	39.68
1983	4.59	3.94	9.75	6.82	5.98	1.35	4.31	2.58	1.35	3.38	13.50	8.78	66.33
1984	2.56	4.99	5.12	4.81	9.64	3.87	3.86	2.09	0.84	3.26	3.69	3.44	48.17
1985	1.03	1.54	3.15	1.25	2.03	2.74	3.30	3.18	2.97	4.07	6.36	2.39	34.01
1986	6.58	2.61	4.21	3.49	2.51	3.91	3.44	1.87	2.64	2.09	5.18	5.91	44.44
1987	5.21	0.04	4.29	6.33	2.62	5.01	1.79	2.48	4.64	2.54	3.82	2.01	40.78
1988	1.97	3.34	1.85	3.68	4.27	2.36	5.89	5.24	1.50	3.47	8.84	1.21	43.62
1989	1.15	2.37	2.14	2.94	8.74	4.49	2.50	1.73	4.48	4.81	3.97	2.23	41.55
1990	3.19	2.49	1.42	5.16	5.23	4.12	3.21	1.89	3.12	7.46	7.50	7.90	52.69
1991	2.91	1.85	6.19	6.71	3.77	1.47	2.35	15.22	5.44	2.83	4.34	4.06	57.14
1992	4.80	3.42	3.93	2.47	1.15	3.99	4.06	2.59	3.11	2.48	4.39	2.12	38.51
1993	2.80	3.84	6.26	5.69	1.14	2.89	2.73	1.21	4.10	3.74	4.05	5.40	43.85
1994	5.34	1.29	6.71	2.94	4.87	1.25	1.73	2.82	6.60	0.63	3.67	6.20	44.05
1995	4.92	3.55	2.14	2.36	3.31	2.60	3.14	0.47	2.40	4.78	7.27	4.35	41.29
1996	5.30	3.03	2.68	6.44	3.72	4.35	6.19	0.50	3.54	16.83	1.65	6.53	60.76
1997	4.59	2.48	4.05	5.21	2.41	0.76	2.01	4.15	2.68	1.37	5.21	2.57	37.49
1998	4.83	5.72	4.23	3.36	3.83	9.01	2.92	3.00	3.00	10.45	1.80	1.59	54.77
1999	6.02	3.87	4.52	0.28	4.98	0.95	1.62	1.53	2.46	3.89	2.25	2.00	40.70
2000	3.41	2.88	3.66	5.44	3.07	2.06	4.03	1.81	4.14	3.17	4.19	4.49	40.67
2001	1.33	2.58	8.01	1.26	1.14	5.37	2.05	1.28	4.14	1.72	2.20	2.03	33.11
POR= 130 YRS	3.91	3.65	3.99	3.65	3.43	3.25	3.12	3.03	3.28	3.51	4.10	3.95	42.87

AVERAGE TEMPERATURE (F) 2001   PORTLAND, ME (PWM)

YEAR	JAN	FEB	MAR	APR	MAY	JUN	JUL	AUG	SEP	OCT	NOV	DEC	ANNUAL
1972	22.2	21.1	28.9	40.6	52.6	59.6	67.5	65.3	58.7	45.2	35.0	24.0	43.4
1973	23.1	23.4	37.7	45.5	51.3	63.8	70.8	70.9	58.4	49.2	37.7	33.5	47.1
1974	23.2	24.5	33.9	45.1	50.6	61.3	68.2	67.9	58.8	44.7	39.3	30.2	45.6
1975	26.7	25.0	29.8	40.0	55.6	61.5	70.1	66.8	57.2	49.3	43.0	24.3	45.8
1976	15.7	28.2	31.5	45.0	52.3	65.9	65.9	65.4	57.3	43.6	36.1	19.4	43.9
1977	14.6	21.9	36.2	42.4	54.4	59.0	68.2	66.6	57.6	48.1	39.4	25.4	44.5
1978	21.1	19.2	30.2	40.6	52.7	61.0	68.0	68.6	57.8	48.2	36.4	26.1	44.2
1979	23.7	15.6	35.6	42.2	55.3	62.7	69.3	64.8	57.4	47.5	42.4	29.8	45.5
1980	22.7	20.5	32.0	44.3	53.7	61.0	69.6	71.2	61.5	45.9	36.2	21.3	45.0
1981	13.8	32.3	35.7	45.2	55.3	63.9	68.8	66.0	58.6	46.6	38.9	29.2	46.2
1982	15.0	23.3	32.1	42.0	54.6	58.3	69.2	64.4	59.2	48.1	41.3	32.0	45.0
1983	25.2	26.2	35.8	44.4	52.0	63.8	69.7	67.6	63.0	47.9	40.2	26.1	46.8
1984	19.8	32.0	28.2	43.3	53.0	63.9	69.6	69.1	57.8	49.6	39.2	31.2	46.4
1985	16.2	26.4	34.8	44.1	53.7	61.7	69.8	66.6	60.7	50.8	39.9	24.6	45.8
1986	25.0	23.4	34.6	46.5	53.6	60.9	66.3	66.2	57.4	47.9	36.3	29.6	45.6
1987	21.6	23.0	34.0	44.9	54.6	63.8	68.0	66.2	59.6	47.1	38.1	30.2	45.9
1988	21.6	25.7	34.2	43.5	54.7	63.8	71.1	71.1	59.4	46.5	41.0	26.0	46.6
1989	26.8	24.0	31.6	41.1	55.6	63.9	69.4	68.0	60.5	49.8	37.3	14.1	45.2
1990	30.2	25.7	34.7	44.6	51.8	62.4	70.3	69.8	59.8	52.4	41.8	33.7	48.1
1991	23.4	29.8	37.0	45.9	58.1	65.5	70.1	69.9	58.2	50.8	40.4	26.4	48.0
1992	23.9	25.9	30.9	41.4	51.9	63.4	65.8	66.8	58.9	47.3	37.6	28.9	45.2
1993	23.4	16.8	29.8	43.3	54.8	64.5	69.8	69.5	59.6	46.2	39.0	29.9	45.6
1994	14.0	19.6	33.5	44.5	52.3	66.0	72.4	66.4	58.4	49.5	41.5	31.5	45.8
1995	27.1	21.2	34.8	41.2	52.1	64.1	70.0	67.4	56.4	52.4	35.7	24.5	45.6
1996	21.4	23.9	29.8	42.9	52.5	63.0	66.4	67.4	59.8	46.8	35.1	34.5	45.3
1997	23.4	28.7	30.4	41.8	51.1	63.3	69.0	67.3	58.7	47.0	36.2	29.5	45.5
1998	27.1	31.3	36.7	45.8	56.8	60.7	69.5	69.5	61.2	49.2	39.4	32.5	48.3
1999	22.3	29.0	35.6	44.9	54.5	66.3	71.3	67.7	63.3	46.6	42.0	31.9	48.0
2000	21.5	26.4	37.8	43.3	53.2	63.7	67.1	67.3	59.9	48.9	40.4	24.6	46.2
2001	20.6	24.3	30.6	43.3	54.9	67.2	66.5	69.4	61.0	50.2	41.4	34.8	47.0
POR= 127 YRS	22.2	23.6	32.4	42.9	53.3	62.4	68.3	66.7	59.4	49.3	38.6	27.1	45.5

REFERENCE NOTES:

PAGE 1:
* THE TEMPERATURE GRAPH SHOWS NORMAL MAXIMUM AND NORMAL MINIMUM DAILY TEMPERATURES (SOLID CURVES) AND THE ACTUAL DAILY HIGH AND LOW TEMPERATURES (VERTICAL BARS).

PAGE 2 AND 3:
H/C INDICATES HEATING AND COOLING DEGREE DAYS.
RH INDICATES RELATIVE HUMIDITY
W/O INDICATES WEATHER AND OBSTRUCTIONS
S INDICATES SUNSHINE.
PR INDICATES PRESSURE.
CLOUDINESS ON PAGE 3 IS THE SUM OF THE CEILOMETER AND SATELLITE DATA NOT TO EXCEED EIGHTHS (OKTAS).

GENERAL:
T INDICATES TRACE PRECIPITATION, AN AMOUNT GREATER THAN ZERO BUT LESS THAN THE LOWEST REPORTABLE VALUE.
+ INDICATES THE VALUE ALSO OCCURS ON EARLIER DATES.
BLANK ENTRIES DENOTE MISSING OR UNREPORTED DATA.
NORMALS ARE 30-YEAR AVERAGES (1961 - 1990).
ASOS INDICATES AUTOMATED SURFACE OBSERVING SYSTEM.
PM INDICATES THE LAST DAY OF THE PREVIOUS MONTH.
POR (PERIOD OF RECORD) BEGINS WITH THE JANUARY DATA MONTH AND IS THE NUMBER OF YEARS USED TO COMPUTE THE MEAN. INDIVIDUAL MONTHS WITHIN THE POR MAY BE MISSING.
WHEN THE POR FOR A NORMAL IS LESS THAN 30 YEARS, THE NORMAL IS PROVISIONAL AND IS BASED ON THE NUMBER OF YEARS INDICATED.
0.* OR * INDICATES THE VALUE OR MEAN-DAYS-WITH IS BETWEEN 0.00 AND 0.05.
CLOUDINESS FOR ASOS STATIONS DIFFERS FROM THE NON-ASOS OBSERVATION TAKEN BY A HUMAN OBSERVER. ASOS STATION CLOUDINESS IS BASED ON TIME-AVERAGED CEILOMETER DATA FOR CLOUDS AT OR BELOW 12,000 FEET AND ON SATELLITE DATA FOR CLOUDS ABOVE 12,000 FEET.
THE NUMBER OF DAYS WITH CLEAR, PARTLY CLOUDY, AND CLOUDY CONDITIONS FOR ASOS STATIONS IS THE SUM OF THE CEILOMETER AND SATELLITE DATA FOR THE SUNRISE TO SUNSET PERIOD.

GENERAL CONTINUED:
CLEAR INDICATES 0 - 2 OKTAS, PARTLY CLOUDY INDICATES 3 - 6 OKTAS, AND CLOUDY INDICATES 7 OR 8 OKTAS. WHEN AT LEAST ONE OF THE ELEMENTS (CEILOMETER OR SATELLITE) IS MISSING, THE DAILY CLOUDINESS IS NOT COMPUTED.
WIND DIRECTION IS RECORDED IN TENS OF DEGREES (2 DIGITS) CLOCKWISE FROM TRUE NORTH. "00" INDICATES CALM. "36" INDICATES TRUE NORTH.
RESULTANT WIND IS THE VECTOR AVERAGE OF THE SPEED AND DIRECTION.
AVERAGE TEMPERATURE IS THE SUM OF THE MEAN DAILY MAXIMUM AND MINIMUM TEMPERATURE DIVIDED BY 2.
SNOWFALL DATA COMPRISE ALL FORMS OF FROZEN PRECIPITATION, INCLUDING HAIL.
A HEATING (COOLING) DEGREE DAY IS THE DIFFERENCE BETWEEN THE AVERAGE DAILY TEMPERATURE AND 65 F.
DRY BULB IS THE TEMPERATURE OF THE AMBIENT AIR.
DEW POINT IS THE TEMPERATURE TO WHICH THE AIR MUST BE COOLED TO ACHIEVE 100 PERCENT RELATIVE HUMIDITY.
WET BULB IS THE TEMPERATURE THE AIR WOULD HAVE IF THE MOISTURE CONTENT WAS INCREASED TO 100 PERCENT RELATIVE HUMIDITY.

ON JULY 1, 1996, THE NATIONAL WEATHER SERVICE BEGAN USING THE "METAR" OBSERVATION CODE THAT WAS ALREADY EMPLOYED BY MOST OTHER NATIONS OF THE WORLD. THE MOST NOTICEABLE DIFFERENCE IN THIS ANNUAL PUBLICATION WILL BE THE CHANGE IN UNITS FROM TENTHS TO EIGHTS (OKTAS) FOR REPORTING THE AMOUNT OF SKY COVER.

HEATING DEGREE DAYS (base 65 F) 2001 PORTLAND, ME (PWM)

YEAR	JUL	AUG	SEP	OCT	NOV	DEC	JAN	FEB	MAR	APR	MAY	JUN	TOTAL
1972-73	27	53	190	607	893	1264	1292	1157	842	575	419	99	7418
1973-74	0	9	231	480	813	970	1290	1126	958	595	444	131	7047
1974-75	15	17	206	624	762	1071	1177	1112	1086	746	286	146	7248
1975-76	13	59	230	480	653	1258	1522	1061	1030	594	389	94	7383
1976-77	45	73	229	660	858	1404	1559	1199	888	674	354	184	8127
1977-78	29	54	233	518	761	1219	1353	1276	1071	724	377	134	7749
1978-79	39	32	230	513	852	1201	1272	1380	905	677	311	97	7509
1979-80	21	82	240	539	672	1083	1305	1284	1018	613	346	163	7366
1980-81	16	6	163	584	855	1349	1578	910	901	588	312	54	7316
1981-82	16	45	189	566	778	1102	1543	1161	1014	684	320	198	7616
1982-83	20	78	185	519	704	1015	1225	1080	895	612	393	101	6827
1983-84	8	38	139	527	738	1198	1397	949	1132	642	368	110	7246
1984-85	11	13	223	469	767	1043	1506	1076	930	620	347	115	7120
1985-86	4	32	157	433	747	1245	1236	1161	935	548	354	138	6990
1986-87	47	52	242	523	855	1092	1336	1172	955	597	343	77	7291
1987-88	20	58	171	548	798	1070	1339	1130	950	641	323	112	7160
1988-89	13	32	180	569	713	1201	1174	1141	1028	708	286	91	7136
1989-90	6	25	167	464	824	1573	1071	1093	935	607	402	107	7274
1990-91	12	24	170	388	690	964	1283	979	861	568	236	76	6251
1991-92	11	16	228	433	730	1191	1267	1127	1051	700	414	84	7252
1992-93	47	30	208	543	814	1112	1280	1344	1084	645	311	81	7499
1993-94	15	5	192	575	774	1080	1577	1266	968	609	389	61	7511
1994-95	1	35	201	472	697	1028	1167	1216	929	706	391	85	6928
1995-96	9	35	265	380	873	1248	1343	1181	1083	659	384	90	7550
1996-97	23	17	179	556	889	940	1284	1011	1065	689	424	135	7212
1997-98	11	21	192	554	857	1093	1164	938	872	569	264	143	6678
1998-99	3	9	124	482	761	1002	1314	1002	903	599	322	61	6582
1999-00	6	23	105	563	686	1020	1341	1113	838	645	361	112	6813
2000-01	11	18	184	491	729	1247	1371	1134	1060	643	330	57	7275
2001-	32	15	140	456	702	928							

COOLING DEGREE DAYS (base 65 F) 2001 PORTLAND, ME (PWM)

YEAR	JAN	FEB	MAR	APR	MAY	JUN	JUL	AUG	SEP	OCT	NOV	DEC	ANNUAL
1972	0	0	0	0	0	3	114	71	8	0	0	0	196
1973	0	0	0	0	1	71	189	201	40	0	0	0	502
1974	0	0	0	2	4	26	121	115	28	0	0	0	296
1975	0	0	0	0	1	49	179	120	2	0	0	0	351
1976	0	0	0	0	2	128	80	93	5	0	0	0	308
1977	0	0	0	1	32	12	135	109	19	0	0	0	308
1978	0	0	0	0	6	22	138	150	20	0	0	0	336
1979	0	0	0	0	15	34	162	83	19	3	0	0	316
1980	0	0	0	0	1	50	163	205	67	0	0	0	486
1981	0	0	0	0	20	30	138	84	5	0	0	0	277
1982	0	0	0	0	6	4	158	66	17	0	0	0	251
1983	0	0	0	0	0	73	161	125	84	3	0	0	446
1984	0	0	0	0	0	84	162	147	12	0	0	0	405
1985	0	0	0	0	5	25	161	92	35	1	0	0	319
1986	0	0	0	0	8	23	93	96	19	0	0	0	239
1987	0	0	0	0	28	47	121	103	17	0	0	0	316
1988	0	0	0	0	11	85	209	227	17	2	0	0	551
1989	0	0	0	0	2	67	151	126	38	0	0	0	384
1990	0	0	0	0	0	34	181	179	19	7	0	0	420
1991	0	0	0	0	29	96	178	176	31	0	0	0	510
1992	0	0	0	0	13	43	76	93	30	0	0	0	255
1993	0	0	0	0	1	72	170	150	37	0	0	0	430
1994	0	0	0	0	1	101	237	89	10	0	0	0	438
1995	0	0	0	0	3	67	174	118	14	0	0	0	376
1996	0	0	0	0	5	38	75	99	29	0	0	0	246
1997	0	0	0	0	0	90	145	100	8	2	0	0	345
1998	0	0	2	0	15	21	152	154	17	0	0	0	361
1999	0	0	0	0	4	109	208	116	59	0	0	0	496
2000	0	0	0	0	1	76	81	97	37	1	0	0	293
2001	0	0	0	0	23	127	85	161	25	3	0	0	424

SNOWFALL (inches) 2001 PORTLAND, ME (PWM)

YEAR	JUL	AUG	SEP	OCT	NOV	DEC	JAN	FEB	MAR	APR	MAY	JUN	TOTAL
1972-73	0.0	0.0	0.0	T	15.6	35.1	9.6	6.6	0.5	2.3	0.0	0.0	69.7
1973-74	0.0	0.0	0.0	0.0	0.0	7.2	15.0	4.3	6.2	8.3	0.0	0.0	41.0
1974-75	0.0	0.0	0.0	T	3.2	8.2	15.4	11.6	6.3	1.7	0.0	0.0	46.4
1975-76	0.0	0.0	0.0	T	3.6	25.3	18.1	4.9	22.2	T	0.0	0.0	74.1
1976-77	0.0	0.0	0.0	0.0	1.5	23.3	35.2	7.9	19.3	1.4	T	0.0	88.6
1977-78	0.0	0.0	0.0	0.0	1.9	23.1	30.7	8.2	12.5	0.8	0.0	0.0	77.2
1978-79	0.0	0.0	0.0	0.0	3.6	18.9	62.4	4.5	T	2.9	0.0	0.0	92.3
1979-80	0.0	0.0	0.0	1.7	T	1.8	6.0	11.2	6.8	T	0.0	0.0	27.5
1980-81	0.0	0.0	0.0	0.0	8.9	13.0	9.2	4.6	3.1	T	0.0	0.0	38.8
1981-82	0.0	0.0	0.0	0.0	T	24.0	25.9	11.0	8.5	15.9	0.0	0.0	85.3
1982-83	0.0	0.0	0.0	0.0	0.6	5.7	12.4	24.5	2.1	T	0.0	0.0	45.3
1983-84	0.0	0.0	0.0	0.0	T	12.6	28.3	3.3	26.4	T	0.0	0.0	70.6
1984-85	0.0	0.0	0.0	0.0	T	17.0	12.1	7.2	13.1	2.4	0.0	0.0	51.8
1985-86	0.0	0.0	0.0	0.0	3.1	11.2	18.6	12.0	6.4	T	0.0	0.0	51.3
1986-87	0.0	0.0	0.0	0.0	5.2	4.0	50.7	0.8	14.3	3.4	0.0	0.0	78.4
1987-88	0.0	0.0	T	0.0	5.4	9.1	19.8	20.8	3.0	4.2	0.0	0.0	62.3
1988-89	0.0	0.0	0.0	T	T	3.5	4.0	13.8	8.9	0.7	0.0	0.0	30.9
1989-90	0.0	0.0	0.0	0.0	5.0	15.6	20.4	25.6	3.2	T	0.0	0.0	69.8
1990-91	0.0	0.0	T	0.0	0.2	6.8	13.4	6.3	5.7	T	0.0	0.0	32.4
1991-92	0.0	0.0	0.0	0.0	T	22.5	2.4	10.3	13.4	10.0	0.0	0.0	58.6
1992-93	0.0	0.0	T	T	2.8	2.1	17.1	33.5	49.0	11.1	0.0	0.0	115.6
1993-94	0.0	0.0	0.0	0.2	T	12.3	39.3	12.2	12.2	T	0.0	0.0	76.2
1994-95	0.0	0.0	0.0				15.5	17.3	1.4	0.0	0.0		
1995-96	0.0	0.0	0.0	0.0	2.0	37.3	37.1	13.1	25.0	8.5			
1996-97						1.4							
1997-98					20.5	6.3	17.1	0.9	9.7		0.0	0.0	
1998-99	0.0	0.0	0.0	0.0	T	11.7	19.2	5.1	17.5	T	0.0	0.0	53.5
1999-00	0.0	0.0	0.0	0.0	T	14.9	14.6	11.6	T	0.0	0.0		41.1
2000-01	0.0	0.0	0.0	T	T	18.8	15.5	24.2	40.5	0.3	0.0		99.3
2001-	0.0	0.0	0.0	T	T	5.0							
POR= 60 YRS	0.0	0.0	T	0.2	3.0	14.3	19.4	16.8	13.5	2.9	0.2	0.0	70.3

2001
BALTIMORE,
MARYLAND (BWI)

Baltimore-Washington International Airport lies in a region about midway between the rigorous climates of the North and the mild climates of the South, and adjacent to the modifying influences of the Chesapeake Bay and Atlantic Ocean to the east and the Appalachian Mountains to the west. Since this region is near the average path of the low pressure systems which move across the country, changes in wind direction are frequent and contribute to the changeable character of the weather. The net effect of the mountains to the west and the bay and ocean to the east is to produce a more equable climate compared with other continental locations farther inland at the same latitude.

Rainfall distribution throughout the year is rather uniform, however, the greatest intensities are confined to the summer and early fall months, the season for hurricanes and severe thunderstorms. Moisture deficiencies for crops occur occasionally during the growing season, but severe droughts are rare. Rainfall during the growing season occurs principally in the form of thunderstorms, and rainfall totals during these months vary appreciably.

The average date for the last occurrence in spring of temperatures as low as 32 degrees is mid-April. The average date for the first occurrence in fall of temperatures as low as 32 degrees is late October. The freeze-free period is approximately 194 days.

In summer, the area is under the influence of the large semi-permanent high pressure system commonly known as the Bermuda High and centered over the Atlantic Ocean near 30 degrees N Latitude. This pressure system brings warm humid air to the area. The proximity of large water areas and the inflow of southerly winds contribute to high relative humidities during much of the year.

January is the coldest month, and July, the warmest. Snowfall occurs on about eleven days per year on the average, however, an average of only about six days annually produces snowfalls of 1 inch or greater. Snow is frequently mixed with rain and sleet, and snow seldom remains on the ground more than a few days.

Glaze or freezing rain which is hazardous to highway traffic occurs on an average of two to three times per year, generally in January or February. Some years pass without the occurrence of freezing rain, while in others it occurs on as many as eight to ten days. Sleet is observed on about five days annually with the greatest frequency of occurrence in January.

The annual prevailing wind direction is from the west. Winter and spring months have the highest average wind speed. Destructive velocities are rare and occur mostly during summer thunderstorms. Only rarely have hurricanes in the vicinity caused widespread damage, then primarily through flooding.

NORMALS, MEANS, AND EXTREMES
BALTIMORE, MD (BWI)

LATITUDE: 39 10' 20" N   LONGITUDE: 76 41' 02" W   ELEVATION (FT): GRND: 193 BARO: 196   TIME ZONE: EASTERN (UTC + 5)   WBAN: 93721

ELEMENT	POR	JAN	FEB	MAR	APR	MAY	JUN	JUL	AUG	SEP	OCT	NOV	DEC	YEAR
TEMPERATURE °F														
NORMAL DAILY MAXIMUM	30	40.2	43.7	54.0	64.3	74.2	83.2	87.2	85.4	78.5	67.3	56.9	45.2	65.0
MEAN DAILY MAXIMUM	51	41.5	44.8	53.4	65.0	74.4	83.2	87.3	85.3	78.5	67.6	56.2	45.4	65.2
HIGHEST DAILY MAXIMUM	51	75	79	89	94	98	101	104	105	100	92	83	77	105
YEAR OF OCCURRENCE		1975	2000	1998	1960	1991	1994	1988	1983	1983	1954	1974	1998	AUG 1983
MEAN OF EXTREME MAXS.	51	62.3	66.1	76.3	85.0	90.0	95.1	97.0	95.4	92.2	83.2	74.8	65.7	81.9
NORMAL DAILY MINIMUM	30	23.4	25.9	34.1	42.5	52.6	61.8	66.8	65.7	58.4	45.9	37.1	28.2	45.2
MEAN DAILY MINIMUM	51	24.6	26.5	33.8	43.0	52.6	61.7	66.8	65.4	58.1	46.1	36.6	28.4	45.3
LOWEST DAILY MINIMUM	51	-7	-3	6	20	32	40	50	45	35	25	13	0	-7
YEAR OF OCCURRENCE		1984	1979	1960	1965	1966	1972	2001	1986	1963	1969	1955	1983	JAN 1984
MEAN OF EXTREME MINS.	51	6.9	10.7	19.0	28.9	38.6	48.9	55.9	53.6	42.9	31.6	21.5	12.6	30.9
NORMAL DRY BULB	30	31.8	34.8	44.1	53.4	63.4	72.5	77.0	75.6	68.5	56.6	46.8	36.7	55.1
MEAN DRY BULB	51	33.0	35.7	43.6	53.9	63.4	72.4	77.1	75.5	68.4	56.8	46.5	36.9	55.3
MEAN WET BULB	17	30.6	33.0	38.5	47.7	57.1	65.9	70.0	68.5	62.2	51.8	40.2	32.0	49.8
MEAN DEW POINT	17	23.3	25.2	30.0	40.1	51.3	61.3	65.9	64.7	58.3	46.9	34.3	25.4	43.9
NORMAL NO. DAYS WITH:														
MAXIMUM 90	30	0.0	0.0	0.0	0.4	1.4	5.8	11.3	8.0	3.4	0.0	0.0	0.0	30.3
MAXIMUM 32	30	7.2	4.2	0.4	0.0	0.0	0.0	0.0	0.0	0.0	0.0	0.1	3.6	15.5
MINIMUM 32	30	25.3	21.1	14.0	3.4	*	0.0	0.0	0.0	0.0	1.9	10.2	21.1	97.0
MINIMUM 0	30	0.5	0.1	0.0	0.0	0.0	0.0	0.0	0.0	0.0	0.0	0.0	*	0.6
H/C														
NORMAL HEATING DEG. DAYS	30	1029	846	648	348	108	0	0	0	29	276	546	877	4707
NORMAL COOLING DEG. DAYS	30	0	0	0	0	59	227	372	329	134	16	0	0	1137
RH														
NORMAL (PERCENT)	30	63	61	59	59	66	68	69	71	71	70	66	66	66
HOUR 01 LST	30	67	66	65	67	76	81	81	82	82	79	73	70	74
HOUR 07 LST	30	70	70	70	71	77	79	81	83	84	82	77	73	76
HOUR 13 LST	30	56	54	50	48	53	52	53	55	55	53	55	57	53
HOUR 19 LST	30	61	58	54	52	60	62	63	66	68	67	63	63	61
S PERCENT POSSIBLE SUNSHINE	40	51	55	56	56	56	62	64	62	60	58	51	49	57
W/O MEAN NO. DAYS WITH:														
HEAVY FOG (VISBY 1/4 MI)	52	3.1	3.2	2.5	1.8	1.6	0.9	0.8	1.0	1.3	2.5	2.6	3.1	24.4
THUNDERSTORMS	52	0.3	0.2	0.8	2.4	4.0	5.4	5.9	4.9	2.0	1.0	0.4	0.1	27.4
CLOUDINESS MEAN:														
SUNRISE-SUNSET (OKTAS)	46	5.1	5.0	5.0	5.0	5.0	4.6	4.4	4.4	4.2	4.1	4.8	5.0	4.7
MIDNIGHT-MIDNIGHT (OKTAS)	32	4.9	4.7	4.8	4.6	4.7	4.4	4.3	4.2	4.2	3.9	4.5	4.8	4.5
MEAN NO. DAYS WITH:														
CLEAR	47	8.0	7.7	7.9	7.7	7.7	8.4	8.9	9.2	10.3	11.6	8.1	8.1	103.6
PARTLY CLOUDY	47	7.5	6.8	8.7	9.0	10.3	11.4	11.5	10.5	8.3	7.9	8.0	6.9	106.8
CLOUDY	47	15.5	13.8	14.4	13.4	13.0	10.2	9.9	10.6	10.7	10.9	13.2	15.3	150.9
PR MEAN STATION PRESSURE (IN)	28	29.90	29.91	29.89	29.80	29.80	29.80	29.80	29.90	29.90	29.90	29.91	29.91	29.87
MEAN SEA-LEVEL PRES. (IN)	17	30.10	30.09	30.04	29.98	29.99	29.97	30.00	30.03	30.06	30.11	30.11	30.13	30.05
WINDS MEAN SPEED (MPH)	51	9.5	9.9	10.6	10.2	8.9	8.3	7.7	7.6	7.9	8.2	9.0	9.0	8.9
PREVAIL.DIR (TENS OF DEGS)	16	29	29	30	29	28	25	25	25	27	27	28	29	28
MAXIMUM 2-MINUTE:														
SPEED (MPH)	5	40	40	41	36	39	38	38	34	34	31	36	41	41
DIR. (TENS OF DEGS)		30	29	28	32	29	28	27	28	26	29	27	28	28
YEAR OF OCCURRENCE		2000	2001	1997	2001	1997	1997	1997	1997	1999	2001	2000	2000	DEC 2000
MAXIMUM 5-SECOND:														
SPEED (MPH)	5	53	48	53	52	49	46	44	53	44	40	44	54	54
DIR. (TENS OF DEGS)		28	29	28	23	32	31	26	27	27	27	27	28	28
YEAR OF OCCURRENCE		1999	2001	1997	1998	1997	2000	1997	1997	1999	2001	2000	2000	DEC 2000
PRECIPITATION NORMAL (IN)	30	3.05	3.12	3.38	3.09	3.72	3.67	3.69	3.92	3.41	2.98	3.32	3.41	40.76
MAXIMUM MONTHLY (IN)	51	7.84	7.16	8.64	8.15	8.71	9.95	8.18	18.35	11.50	8.09	7.68	7.44	18.35
YEAR OF OCCURRENCE		1979	1979	1994	1952	1989	1972	1960	1955	1999	1976	1952	1969	AUG 1955
MINIMUM MONTHLY (IN)	51	0.29	0.56	0.93	0.39	0.37	0.15	0.30	0.77	0.21	T	0.31	0.20	T
YEAR OF OCCURRENCE		1955	1978	1966	1985	1986	1954	1955	1951	1967	1963	1981	1955	OCT 1963
MAXIMUM IN 24 HOURS (IN)	51	3.11	3.26	3.18	2.80	3.64	5.23	5.86	8.35	6.04	3.49	3.43	3.39	8.35
YEAR OF OCCURRENCE		1976	1983	1958	1952	1960	1972	1952	1955	1985	1955	1952	1977	AUG 1955
NORMAL NO. DAYS WITH:														
PRECIPITATION 0.01	30	10.2	9.4	10.0	10.5	10.9	9.2	9.6	9.4	7.2	7.4	9.0	9.2	112.0
PRECIPITATION 1.00	30	0.5	0.7	0.7	0.7	0.7	1.1	1.1	1.1	1.0	0.9	0.9	0.9	10.3
SNOWFALL NORMAL (IN)	30	6.6	7.5	3.0	0.*	T	0.0	0.0	0.0	0.0	0.*	1.1	3.6	21.8
MAXIMUM MONTHLY (IN)	51	32.6	33.1	21.6	0.7	T	0.0	T	0.0	0.0	0.3	8.4	20.4	33.1
YEAR OF OCCURRENCE		1996	1979	1960	1985	1963		1992			1979	1967	1966	FEB 1979
MAXIMUM IN 24 HOURS (IN)	51	16.8	22.8	13.0	0.7	T	0.0	T	0.0	0.0	0.3	8.4	14.1	22.8
YEAR OF OCCURRENCE		1996	1983	1962	1985	1963		1992			1979	1967	1960	FEB 1983
MAXIMUM SNOW DEPTH (IN)	50	30	23	70	0	0	0	0	0	0	0	6	12	70
YEAR OF OCCURRENCE		1957	1983	1960								1987	1960	MAR 1960
NORMAL NO. DAYS WITH:														
SNOWFALL 1.0	30	1.8	1.8	0.8	0.0	0.0	0.0	0.0	0.0	0.0	0.0	0.3	0.9	5.6

PRECIPITATION (inches) 2001 BALTIMORE, MD (BWI)

YEAR	JAN	FEB	MAR	APR	MAY	JUN	JUL	AUG	SEP	OCT	NOV	DEC	ANNUAL
1972	2.82	6.01	2.38	5.30	4.11	9.95	2.81	2.22	1.15	3.51	7.05	5.02	52.33
1973	2.81	2.82	3.96	6.41	3.73	3.16	4.22	3.35	4.87	2.86	1.28	6.36	45.83
1974	2.92	0.94	4.12	2.59	3.58	2.84	0.85	5.85	5.45	1.53	1.39	5.70	37.76
1975	3.47	2.47	5.17	2.73	4.63	3.82	7.15	4.23	8.62	2.89	2.03	4.61	51.82
1976	4.10	2.16	2.23	1.27	5.03	2.49	5.56	2.98	6.93	8.09	0.56	2.04	43.44
1977	1.36	0.63	3.93	3.05	1.49	3.44	2.62	3.31	0.62	5.17	5.01	5.76	36.39
1978	7.34	0.56	4.74	1.26	5.49	2.81	6.83	3.39	1.03	0.71	2.70	4.63	41.49
1979	7.84	7.16	2.05	3.37	4.15	5.74	3.71	9.38	6.73	5.53	2.45	0.87	58.98
1980	2.58	1.06	5.46	4.24	3.58	3.04	3.25	4.00	1.00	3.08	2.72	0.70	34.71
1981	0.49	2.93	1.14	2.04	3.63	5.40	4.59	1.93	2.89	2.57	0.31	3.30	31.22
1982	3.37	4.04	3.03	3.61	1.85	5.70	2.16	0.95	3.63	2.31	3.13	2.39	36.17
1983	2.21	4.81	6.80	6.55	5.47	5.23	1.31	1.57	3.58	5.02	6.72	51.03	
1984	1.96	3.90	5.79	2.95	4.29	1.65	3.27	4.11	2.38	1.94	3.01	1.71	36.96
1985	2.03	3.03	2.37	0.39	6.01	2.44	2.53	3.72	6.22	2.48	4.71	0.84	36.77
1986	2.16	3.78	0.96	2.64	0.37	1.46	4.12	4.26	0.58	1.86	5.96	5.52	33.67
1987	5.85	2.22	0.99	1.86	4.16	2.63	5.05	1.61	7.34	2.25	5.05	2.07	41.08
1988	3.24	3.25	2.35	2.44	4.37	0.84	3.78	2.64	2.05	1.59	4.78	0.97	32.30
1989	3.07	3.36	4.24	3.16	8.71	5.98	7.35	3.38	3.64	4.90	1.97	2.12	51.88
1990	3.71	1.48	2.54	4.23	4.92	2.55	5.68	6.17	1.07	2.57	2.10	4.86	41.88
1991	3.54	0.73	5.65	1.68	1.16	1.08	1.76	2.54	3.05	3.20	1.69	4.08	30.16
1992	1.27	2.49	4.58	1.76	2.92	1.89	5.07	2.19	5.96	2.73	3.44	4.63	38.93
1993	2.73	2.84	8.12	3.68	3.66	2.56	1.71	2.55	4.09	3.02	3.09	4.45	42.50
1994	4.59	4.07	8.64	2.53	3.02	2.84	4.54	3.44	3.93	1.82	1.95	1.95	43.32
1995	2.87	1.88	2.12	1.92	3.40	1.80	3.65	2.98	3.29	6.24	4.12	2.66	36.93
1996	6.80	2.36	3.57	3.76	5.68	4.08	7.38	4.17	5.65	4.32	3.77	6.77	58.31
1997	2.83	2.23	5.67	2.40	3.03	3.74	1.49	4.21	1.47	3.43	5.79	2.05	38.34
1998	5.65	6.40	5.56	3.02	3.46	3.22	1.42	0.91	1.27	1.06	1.13	1.27	34.37
1999	4.70	2.65	3.46	2.27	1.73	2.04	2.06	6.14	11.50	2.48	1.95	2.96	43.94
2000	3.64	2.01	4.35	5.06	2.82	5.54	5.64	3.18	5.55	0.08	1.73	2.31	41.91
2001	2.68	2.35	4.76	1.32	5.34	3.58	3.85	5.74	1.43	0.78	1.01	1.73	34.57
POR= 51 YRS	3.13	2.98	3.90	3.14	3.60	3.52	3.79	4.00	3.54	2.89	3.04	3.25	40.78

AVERAGE TEMPERATURE (F) 2001 BALTIMORE, MD (BWI)

YEAR	JAN	FEB	MAR	APR	MAY	JUN	JUL	AUG	SEP	OCT	NOV	DEC	ANNUAL
1972	37.6	34.3	43.6	51.6	62.7	68.1	76.9	75.4	69.8	53.5	43.2	40.4	54.8
1973	34.6	34.3	48.3	53.1	59.6	73.5	75.9	76.9	69.8	58.2	47.3	37.3	55.7
1974	37.9	33.8	45.2	55.3	61.9	68.5	76.5	75.0	67.5	55.3	48.2	40.3	55.5
1975	38.5	39.1	42.1	50.4	66.3	73.0	76.1	77.9	66.0	60.7	51.9	37.2	56.6
1976	30.8	44.1	48.1	56.9	62.1	74.8	75.0	73.9	67.5	52.9	40.9	32.6	55.0
1977	22.9	36.5	50.0	57.9	66.7	71.4	79.0	77.7	72.1	56.0	49.2	35.6	56.3
1978	29.2	27.3	41.7	54.2	62.4	73.1	75.9	78.1	69.7	56.1	48.7	40.2	54.7
1979	33.1	25.6	48.5	53.1	64.7	70.7	75.9	75.7	68.8	55.7	50.6	40.3	55.2
1980	33.8	31.5	41.5	55.7	65.5	71.3	78.2	78.7	72.2	55.3	44.2	35.5	55.3
1981	27.9	38.8	41.9	57.0	62.2	74.3	77.3	74.4	67.7	53.2	46.2	34.5	54.6
1982	25.5	35.8	42.9	50.7	66.1	69.4	77.1	73.0	67.3	56.3	48.4	42.0	54.5
1983	34.6	34.7	45.4	51.8	61.5	72.1	78.7	78.0	69.5	57.3	47.1	33.2	55.3
1984	28.5	41.7	38.2	51.5	61.3	73.4	73.9	75.0	64.8	62.2	43.9	44.1	54.9
1985	29.3	38.7	46.0	57.9	65.1	70.4	76.4	74.5	69.4	58.8	52.4	33.8	56.1
1986	33.2	32.9	45.0	53.5	66.7	74.4	79.4	73.1	68.9	58.9	44.8	38.2	55.8
1987	32.5	34.3	46.2	53.1	65.0	74.5	80.0	76.1	69.3	51.5	47.8	39.8	55.8
1988	28.7	35.9	45.1	52.0	64.0	73.0	80.3	78.5	66.8	51.3	48.1	36.3	55.0
1989	37.9	36.5	43.8	52.5	62.0	73.9	76.0	74.4	69.0	58.3	44.8	25.4	54.5
1990	42.0	42.3	47.6	54.8	62.3	73.3	78.4	74.6	67.3	60.7	49.6	42.2	57.9
1991	35.5	40.7	46.7	55.9	70.6	74.6	79.5	77.8	69.0	57.8	45.8	38.7	57.7
1992	34.6	37.1	41.3	52.0	60.8	70.1	77.4	72.3	67.7	54.3	47.2	38.9	54.5
1993	37.9	31.4	39.4	52.5	65.0	72.2	80.2	76.7	68.8	55.5	46.5	36.2	55.2
1994	27.1	34.0	43.0	59.6	60.6	77.2	80.1	74.1	68.1	56.8	51.9	42.6	56.3
1995	39.0	33.2	47.8	55.2	64.5	74.5	81.5	80.1	70.4	61.1	42.6	33.9	57.0
1996	31.7	35.7	39.9	54.0	66.6	73.3	74.3	73.2	67.8	55.6	40.2	39.6	53.8
1997	32.8	41.0	45.5	51.6	59.5	70.1	77.3	74.0	67.3	56.5	43.7	38.4	54.8
1998	40.9	41.7	45.9	55.2	66.5	71.7	76.6	75.7	71.8	56.3	46.1	41.1	57.5
1999	35.1	37.6	41.8	53.2	64.2	71.5	80.0	75.7	68.2	53.9	49.9	39.1	55.9
2000	32.5	38.1	48.5	52.9	64.7	72.8	72.7	73.4	65.3	57.1	44.2	30.0	54.4
2001	33.1	38.5	41.8	55.4	63.4	74.1	72.8	77.0	65.2	56.0	50.7	42.1	55.8
POR= 51 YRS	32.9	35.7	43.6	53.8	63.4	72.4	77.0	75.4	68.5	56.7	46.4	36.9	55.2

REFERENCE NOTES:

PAGE 1:
THE TEMPERATURE GRAPH SHOWS NORMAL MAXIMUM AND NORMAL MINIMUM DAILY TEMPERATURES (SOLID CURVES) AND THE ACTUAL DAILY HIGH AND LOW TEMPERATURES (VERTICAL BARS).

PAGE 2 AND 3:
H/C INDICATES HEATING AND COOLING DEGREE DAYS.
RH INDICATES RELATIVE HUMIDITY
W/O INDICATES WEATHER AND OBSTRUCTIONS
S INDICATES SUNSHINE.
PR INDICATES PRESSURE.
CLOUDINESS ON PAGE 3 IS THE SUM OF THE CEILOMETER AND SATELLITE DATA NOT TO EXCEED EIGHT EIGHTHS(OKTAS).

GENERAL:
T INDICATES TRACE PRECIPITATION, AN AMOUNT GREATER THAN ZERO BUT LESS THAN THE LOWEST REPORTABLE VALUE.
+ INDICATES THE VALUE ALSO OCCURS ON EARLIER DATES.
BLANK ENTRIES DENOTE MISSING OR UNREPORTED DATA.
NORMALS ARE 30-YEAR AVERAGES (1961 - 1990).
ASOS INDICATES AUTOMATED SURFACE OBSERVING SYSTEM.
PM INDICATES THE LAST DAY OF THE PREVIOUS MONTH.
POR (PERIOD OF RECORD) BEGINS WITH THE JANUARY DATA MONTH AND IS THE NUMBER OF YEARS USED TO COMPUTE THE MEAN. INDIVIDUAL MONTHS WITHIN THE POR MAY BE MISSING.
WHEN THE POR FOR A NORMAL IS LESS THAN 30 YEARS, THE NORMAL IS PROVISIONAL AND IS BASED ON THE NUMBER OF YEARS INDICATED.
0.* OR * INDICATES THE VALUE OR MEAN-DAYS-WITH IS BETWEEN 0.00 AND 0.05.
CLOUDINESS FOR ASOS STATIONS DIFFERS FROM THE NON-ASOS OBSERVATION TAKEN BY A HUMAN OBSERVER. ASOS STATION CLOUDINESS IS BASED ON TIME-AVERAGED CEILOMETER DATA FOR CLOUDS AT OR BELOW 12,000 FEET AND ON SATELLITE DATA FOR CLOUDS ABOVE 12,000 FEET.
THE NUMBER OF DAYS WITH CLEAR, PARTLY CLOUDY, AND CLOUDY CONDITIONS FOR ASOS STATIONS IS THE SUM OF THE CEILOMETER AND SATELLITE DATA FOR THE SUNRISE TO SUNSET PERIOD.

GENERAL CONTINUED:
CLEAR INDICATES 0 - 2 OKTAS, PARTLY CLOUDY INDICATES 3 - 6 OKTAS, AND CLOUDY INDICATES 7 OR 8 OKTAS. WHEN AT LEAST ONE OF THE ELEMENTS (CEILOMETER OR SATELLITE) IS MISSING, THE DAILY CLOUDINESS IS NOT COMPUTED.
WIND DIRECTION IS RECORDED IN TENS OF DEGREES (2 DIGITS) CLOCKWISE FROM TRUE NORTH. "00" INDICATES CALM. "36" INDICATES TRUE NORTH.
RESULTANT WIND IS THE VECTOR AVERAGE OF THE SPEED AND DIRECTION.
AVERAGE TEMPERATURE IS THE SUM OF THE MEAN DAILY MAXIMUM AND MINIMUM TEMPERATURE DIVIDED BY 2.
SNOWFALL DATA COMPRISE ALL FORMS OF FROZEN PRECIPITATION, INCLUDING HAIL.
A HEATING (COOLING) DEGREE DAY IS THE DIFFERENCE BETWEEN THE AVERAGE DAILY TEMPERATURE AND 65 F.
DRY BULB IS THE TEMPERATURE OF THE AMBIENT AIR.
DEW POINT IS THE TEMPERATURE TO WHICH THE AIR MUST BE COOLED TO ACHIEVE 100 PERCENT RELATIVE HUMIDITY.
WET BULB IS THE TEMPERATURE THE AIR WOULD HAVE IF THE MOISTURE CONTENT WAS INCREASED TO 100 PERCENT RELATIVE HUMIDITY.

ON JULY 1, 1996, THE NATIONAL WEATHER SERVICE BEGAN USING THE "METAR" OBSERVATION CODE THAT WAS ALREADY EMPLOYED BY MOST OTHER NATIONS OF THE WORLD. THE MOST NOTICEABLE DIFFERENCE IN THIS ANNUAL PUBLICATION WILL BE THE CHANGE IN UNITS FROM TENTHS TO EIGHTS(OKTAS) FOR REPORTING THE AMOUNT OF SKY COVER.

HEATING DEGREE DAYS (base 65 F) 2001 BALTIMORE, MD (BWI)

YEAR	JUL	AUG	SEP	OCT	NOV	DEC	JAN	FEB	MAR	APR	MAY	JUN	TOTAL
1972-73	2	0	16	357	649	759	935	854	511	365	191	1	4640
1973-74	0	0	24	221	524	852	830	868	613	309	148	14	4403
1974-75	0	0	49	303	509	759	818	720	702	436	66	2	4364
1975-76	0	0	50	156	397	853	1050	603	518	293	133	11	4064
1976-77	0	0	34	377	716	1001	1296	790	469	245	62	18	5008
1977-78	0	0	9	278	476	904	1101	1048	715	318	141	9	4999
1978-79	0	0	33	280	483	763	984	1100	520	354	75	6	4598
1979-80	0	3	22	311	425	757	962	967	723	273	74	6	4525
1980-81	2	0	20	311	620	908	1145	727	706	252	148	1	4838
1981-82	0	0	51	363	557	940	1218	808	677	422	58	20	5114
1982-83	0	5	42	289	495	707	936	842	602	410	152	6	4486
1983-84	0	0	70	257	530	979	1123	671	825	397	169	9	5030
1984-85	0	1	96	123	625	643	1101	731	589	252	79	10	4250
1985-86	0	0	41	201	378	962	980	892	613	342	86	6	4501
1986-87	0	23	34	236	598	822	1002	853	576	357	106	1	4608
1987-88	0	1	15	412	511	774	1120	838	613	389	96	27	4796
1988-89	2	0	39	424	504	882	834	792	663	374	145	0	4659
1989-90	0	0	51	229	600	1221	707	631	552	341	102	5	4439
1990-91	1	0	63	195	454	701	907	674	562	289	55	4	3905
1991-92	0	0	49	246	570	809	936	802	730	387	161	8	4698
1992-93	0	1	51	328	529	801	834	934	787	369	61	11	4706
1993-94	0	0	52	292	553	886	1169	861	677	190	180	1	4861
1994-95	0	0	13	256	391	684	798	885	525	307	77	0	3936
1995-96	0	0	30	176	669	958	1024	840	772	345	199	12	5025
1996-97	0	0	42	283	736	778	778	667	597	394	182	53	4726
1997-98	0	0	49	307	633	815	737	647	625	295	59	22	4189
1998-99	0	1	25	263	560	734	919	762	714	349	71	9	4407
1999-00	0	0	37	336	445	794	999	774	508	362	102	8	4365
2000-01	0	1	97	254	616	1079	984	736	715	309	99	12	4902
2001-	2	0	76	289	424	706							

COOLING DEGREE DAYS (base 65 F) 2001 BALTIMORE, MD (BWI)

YEAR	JAN	FEB	MAR	APR	MAY	JUN	JUL	AUG	SEP	OCT	NOV	DEC	ANNUAL
1972	0	0	5	1	29	140	379	331	166	7	0	0	1058
1973	0	0	0	15	29	263	344	376	173	19	0	0	1219
1974	0	0	4	24	57	126	361	317	130	8	11	0	1038
1975	0	0	0	4	112	252	351	404	85	27	10	0	1245
1976	0	1	0	58	51	315	317	284	114	9	0	0	1149
1977	0	0	10	37	124	217	439	401	229	7	10	0	1474
1978	0	0	0	0	63	260	344	413	182	12	0	0	1274
1979	0	0	15	4	72	183	348	341	145	28	1	0	1137
1980	0	0	0	0	97	203	415	431	245	17	0	0	1408
1981	0	0	0	19	69	287	389	296	141	5	0	0	1206
1982	0	0	0	4	99	160	381	259	119	26	4	1	1053
1983	0	0	0	18	51	228	430	410	214	24	0	2	1375
1984	0	0	0	0	59	268	281	316	98	41	0	0	1065
1985	0	2	7	43	89	179	363	298	178	17	5	0	1181
1986	0	0	0	1	143	295	452	281	158	54	0	0	1384
1987	0	0	0	7	115	292	473	352	152	0	0	0	1391
1988	0	0	2	4	71	274	485	427	100	8	0	0	1371
1989	0	0	14	5	58	276	351	298	178	25	1	0	1206
1990	0	0	19	38	26	261	422	303	137	68	0	0	1274
	0	0	2	24	233	303	462	402	177	29	2	0	1634
	0	0	0	6	39	168	392	232	139	4	0	0	980
	0	0	0	0	70	235	476	371	175	3	5	0	1335
			0	38	49	374	476	292	112	6	3	0	1350
			0	20	72	289	520	475	199	60	3	0	1638
			0	19	70	265	295	259	135	1	0	0	1044
			0	20	211	385	287	124	51	0	0		1078
			9	115	228	367	341	235	0	0	0		1334
				54	210	471	340	138	0	0	0		1213
				102	248	245	269	115	17	0	0		1005
				54	290	249	381	90	12	2	0		1105

...T)							
N	FEB	MAR	APR	MAY	JUN	TOTAL
	1.2	T	T	0.0	0.0	1.2
7.6	T	T	T	0.0	0.0	17.1
	1.2	1.2	T	0.0	0.0	12.2
	7.8	T	0.0	0.0	0.0	11.5
	T	T	T	0.0	0.0	11.1
		T	T	0.0	0.0	34.3
		T	T	0.0	0.0	42.5
			0	0.0	0.0	14.6
				0.0	0.0	4.6
				0.0	0.0	25.5
		T	0.0	0.0	0.0	35.6
		T	T	0.0	0.0	14.5
	.7	T	T	0.0	0.0	10.3
		T	T	0.0	0.0	15.6
		T	T	0.0	0.0	35.2
		T	T	0.0	0.0	20.4
		0.0	0.0	0.0	0.0	8.3
	.7	0.1	0.0	0.0	0.0	17.3
	0.3	0.0	0.0	0.0	0.0	9.4
		T	T	0.0	0.0	4.1
		12.7	T	0.0	0.0	24.4
		4.2	T	0.0	0.0	17.3
	.5	0.2	0.0	0.0	0.0	8.2
.0		7.6				
7.1		2.7	2.7	T	0.0	0.0
		T	2.1	0.0	T	0.0	3.2
		0.6	7.6	0.0	0.0	0.0	15.2
		2.6	0.0	0.2	0.0	0.0	26.1
.7	3.7	3.7	T	T	0.0	0.0	8.7
6.3	6.3	3.6	0.1	T	0.0	20.4

2001
BOSTON,
MASSACHUSETTS (BOS)

Climate is the composite of numerous weather elements. Three important influences are responsible for the main features of the Boston climate. First, the latitude places the city in the zone of prevailing west to east atmospheric flow. Both polar and tropical air masses influence the region. Secondly, Boston is situated on or near several tracks frequently followed by low pressure storm systems. The weather fluctuates regularly from fair to cloudy to stormy conditions and assures an adequate amount of precipitation. The third factor is the east-coast location of Boston. The ocean has a moderating influence on temperature extremes of winter and summer.

Hot summer afternoons are frequently relieved by the locally celebrated sea breeze, as air flows inland from the cool water surface to displace the warm air over the land. This refreshing east wind is more commonly experienced along the shore than in the interior of the city or the western suburbs. In winter, under appropriate conditions, the severity of cold waves is reduced by the nearness of the relatively warm ocean. The average last occurrence of freezing temperature in spring is early April and the first occurrence of freezing temperature in autumn is early November. In suburban areas, especially away from the coast, these dates are later in spring and earlier in autumn by up to one month in the more susceptible localities.

Boston has no dry season. Most growing seasons have several shorter dry spells during which irrigation for high-value crops may be useful. Much of the rainfall from June to September comes from showers and thunderstorms. During the rest of the year, low pressure systems pass more or less regularly and produce precipitation on an average of roughly one day in three. Coastal storms, or northeasters, are prolific producers of rain and snow. The main snow season extends from December through March. Periods when the ground is bare or nearly bare of snow may occur at any time in the winter.

Relative humidity has been known to fall as low as 5 percent but such desert dryness is very rare. Heavy fog occurs on an average of about two days per month with its prevalence increasing eastward from the interior of Boston Bay to the open waters beyond.

Although winds of 30 mph or higher may be expected on at least one day in every month of the year, gales are both more common and more severe in winter.

NORMALS, MEANS, AND EXTREMES

BOSTON, MA (BOS)

LATITUDE:	LONGITUDE:	ELEVATION (FT):		TIME ZONE:	WBAN: 14739
42 21' 38" N	71 00' 38" W	GRND: 19　　BARO: 180		EASTERN (UTC + 5)	

	ELEMENT	POR	JAN	FEB	MAR	APR	MAY	JUN	JUL	AUG	SEP	OCT	NOV	DEC	YEAR
TEMPERATURE °F	NORMAL DAILY MAXIMUM	30	35.7	37.5	45.8	55.9	66.6	76.3	81.8	79.8	72.8	62.7	52.2	40.4	59.0
	MEAN DAILY MAXIMUM	56	36.6	38.3	45.5	56.1	66.8	76.8	81.8	79.9	72.5	62.7	52.0	40.9	59.2
	HIGHEST DAILY MAXIMUM	50	66	70	89	94	95	100	102	102	100	90	79	76	102
	YEAR OF OCCURRENCE		1995	1985	1998	1976	1979	1952	1977	1975	1953	1963	1994	1998	JUL 1977
	MEAN OF EXTREME MAXS.	56	56.0	56.2	66.9	77.8	87.3	92.7	95.2	92.9	88.9	80.1	69.9	60.8	77.1
	NORMAL DAILY MINIMUM	30	21.6	23.0	31.3	40.2	49.8	59.1	65.1	64.0	56.8	46.9	38.3	26.7	43.6
	MEAN DAILY MINIMUM	56	22.0	23.6	31.2	40.3	49.8	59.3	65.2	64.0	56.7	47.0	38.1	27.2	43.7
	LOWEST DAILY MINIMUM	50	-12	-4	6	16	34	45	50	47	38	28	15	-7	-12
	YEAR OF OCCURRENCE		1957	1961	1984	1982	1956	1986	1988	1986	2000	1976	1989	1980	JAN 1957
	MEAN OF EXTREME MINS.	56	3.4	6.1	15.4	29.6	40.3	49.4	57.1	54.7	44.2	34.6	24.3	9.5	30.7
	NORMAL DRY BULB	30	28.6	30.3	38.6	48.1	58.2	67.7	73.5	71.9	64.8	54.8	45.3	33.6	51.3
	MEAN DRY BULB	56	29.3	31.0	38.3	48.2	58.3	68.0	73.5	72.0	64.5	54.9	45.1	34.1	51.4
	MEAN WET BULB	18	26.7	28.2	33.9	42.1	51.6	61.0	65.8	65.4	58.8	46.4	40.1	29.1	45.8
	MEAN DEW POINT	18	18.8	19.7	24.6	34.7	43.3	55.8	61.3	61.3	54.5	40.9	33.4	21.5	39.1
	NORMAL NO. DAYS WITH:														
	MAXIMUM　　90	30	0.0	0.0	0.0	0.1	0.5	2.5	5.4	2.9	0.8	*	0.0	0.0	12.2
	MAXIMUM　　32	30	11.4	8.0	2.0	*	0.0	0.0	0.0	0.0	0.0	0.0	0.3	6.5	28.2
	MINIMUM　　32	30	26.1	23.5	16.8	2.7	0.0	0.0	0.0	0.0	0.0	0.6	7.0	22.3	99.0
	MINIMUM　　 0	30	0.5	0.5	0.0	0.0	0.0	0.0	0.0	0.0	0.0	0.0	0.0	0.2	1.2
H/C	NORMAL HEATING DEG. DAYS	30	1128	972	818	507	221	32	0	6	72	321	591	973	5641
	NORMAL COOLING DEG. DAYS	30	0	0	0	0	10	113	264	220	66	5	0	0	678
RH	NORMAL (PERCENT)	30	62	62	63	63	67	68	68	71	72	68	68	65	66
	HOUR 01 LST	30	65	65	67	69	74	77	77	79	80	76	72	68	72
	HOUR 07 LST	30	67	68	69	68	71	73	74	77	79	77	74	71	72
	HOUR 13 LST	30	57	56	56	55	58	58	57	59	59	57	59	59	58
	HOUR 19 LST	30	61	60	62	61	64	66	66	69	71	67	66	64	65
S	PERCENT POSSIBLE SUNSHINE	61	53	56	57	56	58	63	65	65	63	60	50	52	58
W/O	MEAN NO. DAYS WITH:														
	HEAVY FOG(VISBY 1/4 MI)	67	1.8	1.7	2.0	1.7	2.7	2.1	2.3	1.8	1.9	2.2	1.8	1.3	23.3
	THUNDERSTORMS	67	0.2	0.1	0.6	1.0	2.3	3.4	4.2	3.4	1.7	0.7	0.4	0.2	18.2
CLOUDINESS	MEAN:														
	SUNRISE-SUNSET (OKTAS)	61	5.0	5.0	5.1	5.3	5.3	5.0	4.9	4.5	4.4	4.4	5.0	5.0	4.9
	MIDNIGHT-MIDNIGHT (OKTAS)	32	4.8	4.7	4.9	4.9	5.0	4.7	4.7	4.3	4.3	4.1	4.8	4.7	4.7
	MEAN NO. DAYS WITH:														
	CLEAR	62	9.0	8.2	7.7	7.0	6.3	6.6	6.6	9.1	10.0	10.7	7.7	8.5	97.4
	PARTLY CLOUDY	62	6.8	6.6	8.0	8.1	9.9	10.4	12.0	10.5	7.9	7.7	7.2	7.3	102.4
	CLOUDY	62	15.2	13.4	15.3	14.9	14.8	12.9	11.9	10.9	11.6	12.1	14.7	14.9	162.6
PR	MEAN STATION PRESSURE(IN)	29	29.98	30.00	29.99	29.90	29.90	29.90	29.90	29.99	29.98	30.00	29.98	29.99	29.96
	MEAN SEA-LEVEL PRES. (IN)	18	30.03	30.03	30.01	29.97	29.97	29.95	29.97	30.02	30.06	30.08	30.06	30.04	30.02
WINDS	MEAN SPEED (MPH)	53	13.7	13.8	14.0	13.5	12.2	11.6	11.0	10.8	11.3	11.9	12.9	13.5	12.5
	PREVAIL.DIR(TENS OF DEGS)	34	30	29	29	29	30	20	20	23	24	24	29	30	30
	MAXIMUM 2-MINUTE:														
	SPEED (MPH)	5	39	40	46	41	40	45	45	30	47	41	48	46	48
	DIR. (TENS OF DEGS)		30	27	03	09	27	30	30	09	23	31	11	18	11
	YEAR OF OCCURRENCE		2000	2001	1997	2000	2001	2000	1999	1997	1998	1999	1997	2000	NOV 1997
	MAXIMUM 5-SECOND:														
	SPEED (MPH)	5	54	54	56	51	49	62	74	37	63	52	57	55	74
	DIR. (TENS OF DEGS)		30	21	06	01	32	29	25	10	23	32	11	18	25
	YEAR OF OCCURRENCE		2000	2000	2001	1997	2001	2000	1999	1997	1998	1999	1997	1997	JUL 1999
PRECIPITATION	NORMAL (IN)	30	3.59	3.62	3.69	3.60	3.25	3.09	2.84	3.24	3.06	3.30	4.22	4.01	41.51
	MAXIMUM MONTHLY (IN)	50	10.55	7.81	11.00	9.46	13.38	13.20	8.12	17.09	9.86	10.66	8.89	9.74	17.09
	YEAR OF OCCURRENCE		1979	1984	1953	1987	1954	1982	1959	1955	1999	1996	1983	1969	AUG 1955
	MINIMUM MONTHLY (IN)	50	0.61	0.72	0.62	0.83	0.53	0.48	0.52	0.82	0.35	0.41	0.64	0.81	0.35
	YEAR OF OCCURRENCE		1989	1987	1981	1999	1964	1999	1952	1995	1957	1994	1976	1989	SEP 1957
	MAXIMUM IN 24 HOURS (IN)	50	2.72	2.68	4.13	3.32	5.74	5.69	3.36	8.40	5.64	6.63	3.76	5.14	8.40
	YEAR OF OCCURRENCE		1979	1969	1968	1991	1954	1998	1996	1955	1954	1996	1992	1992	AUG 1955
	NORMAL NO. DAYS WITH:														
	PRECIPITATION　0.01	30	11.0	9.9	11.6	11.0	11.2	10.7	8.9	9.9	8.4	8.6	10.9	12.0	124.1
	PRECIPITATION　1.00	30	0.9	1.2	0.8	1.0	0.7	0.6	0.7	0.7	0.7	0.7	1.2	0.9	10.1
SNOWFALL	NORMAL (IN)	30	11.9	12.3	6.9	1.1	0.*	0.0	0.0	0.0	0.0	0.*	1.3	8.1	41.6
	MAXIMUM MONTHLY (IN)	64	39.8	41.3	38.9	13.3	0.5	0.0	0.0	T	0.0	0.2	10.0	27.9	41.3
	YEAR OF OCCURRENCE		1996	1969	1993	1982	1977			1994		1979	1938	1970	FEB 1969
	MAXIMUM IN 24 HOURS (IN)	64	21.0	23.6	17.7	13.2	0.5	0.0	0.0	T	0.0	0.2	8.0	13.0	23.6
	YEAR OF OCCURRENCE		1978	1978	1960	1982	1977			1994		1979	1987	1960	FEB 1978
	MAXIMUM SNOW DEPTH (IN)	51	26	29	20	12	0	0	0	0	0	0	6	14	29
	YEAR OF OCCURRENCE		1978	1978	1978	1982							1987	1975	FEB 1978
	NORMAL NO. DAYS WITH:														
	SNOWFALL　　1.0	30	2.9	2.7	2.1	0.3	0.0	0.0	0.0	0.0	0.0	0.0	0.5	2.5	11.0

PRECIPITATION (inches) 2001 BOSTON, MA (BOS)

YEAR	JAN	FEB	MAR	APR	MAY	JUN	JUL	AUG	SEP	OCT	NOV	DEC	ANNUAL
1972	2.05	5.29	5.37	3.34	5.26	6.76	2.19	0.83	5.94	2.98	7.02	6.08	53.11
1973	3.12	2.13	2.20	5.65	3.76	4.68	4.83	2.78	1.95	2.71	1.74	7.20	42.75
1974	3.22	3.24	4.01	3.86	2.87	2.29	1.54	2.31	1.79	1.92	1.73	3.92	40.24
1975	5.70	3.37	2.74	2.40	1.78	2.10	2.35	5.52	5.49	4.41	5.13	4.80	45.79
1976	5.29	2.45	2.42	2.00	1.98	0.58	4.30	7.99	1.56	0.64	3.35	36.72	
1977	4.41	2.40	4.76	4.07	3.52	2.49	2.21	2.91	4.03	4.63	2.54	6.20	44.17
1978	8.12	2.87	2.46	1.79	4.50	1.53	1.48	4.62	1.30	3.13	2.21	3.63	37.64
1979	10.55	3.46	3.03	3.19	4.24	0.86	2.36	5.02	3.61	3.14	3.29	1.42	44.17
1980	0.74	0.88	5.37	4.36	2.30	3.05	2.20	1.55	0.82	4.14	3.01	0.97	29.39
1981	0.95	6.65	0.62	3.14	1.17	1.65	3.47	1.04	2.54	3.43	4.78	6.27	35.71
1982	4.69	2.66	2.17	3.42	2.58	13.20	4.22	2.22	1.57	3.19	3.42	1.27	44.61
1983	5.03	5.00	9.72	6.86	2.94	1.07	1.07	3.28	1.06	3.74	8.89	4.94	53.60
1984	2.31	7.81	6.82	4.43	8.77	3.06	4.43	1.60	1.22	5.18	1.68	2.93	50.24
1985	1.12	1.83	2.29	1.62	3.36	3.94	1.51	6.67	3.00	1.65	6.39	1.21	36.59
1986	3.42	2.83	3.42	1.59	1.31	7.74	3.96	3.32	1.08	3.27	6.01	6.38	44.33
1987	7.28	0.72	4.27	9.46	1.75	2.62	0.82	2.93	7.29	2.73	3.49	2.12	45.48
1988	2.50	3.93	3.52	1.47	2.86	1.29	7.62	1.11	1.29	1.60	6.57	1.02	34.78
1989	0.61	2.51	3.07	3.58	3.54	2.84	5.09	5.92	4.61	5.71	4.13	0.81	42.42
1990	3.78	3.60	1.71	5.94	6.53	0.69	4.08	6.57	1.67	7.36	1.39	3.18	46.50
1991	3.24	1.58	4.33	4.84	0.92	2.89	1.95	5.27	6.32	4.27	4.06	2.58	42.25
1992	3.11	2.28	3.59	2.34	1.40	4.61	2.66	4.25	3.46	1.62	6.14	8.26	43.72
1993	2.17	4.94	7.67	4.86	1.04	1.75	1.75	1.32	4.64	3.61	2.86	6.60	43.21
1994	5.22	2.95	7.49	2.25	5.35	0.86	1.80	7.03	4.58	0.41	4.31	5.37	47.62
1995	4.33	2.57	2.20	1.40	1.82	1.55	2.06	0.82	3.60	6.42	5.13	3.20	35.10
1996	7.44	3.17	2.36	4.38	2.73	1.03	5.23	1.54	6.09	10.66	2.29	5.76	52.68
1997	2.34	1.28	4.68	3.46	2.63	1.41	0.63	3.01	1.02	1.78	5.86	2.29	30.39
1998	4.76	5.54	4.15	3.58	6.84	11.58	2.47	3.37	3.03	5.38	1.38	1.59	53.67
1999	5.69	3.51	2.52	0.83	2.70	T	3.51	1.33	9.86	4.30	2.14	1.52	37.91
2000	2.62	2.55	3.59	5.02	2.88	6.61	5.20	2.22	2.87	2.86	4.51	4.67	45.60
2001	1.58	1.37	7.57	0.88	1.23	4.99	2.13	4.14	2.29	0.98	0.73	2.83	30.72
POR= 131 YRS	3.67	3.35	3.89	3.56	3.24	3.20	3.12	3.56	3.25	3.27	3.86	3.58	41.55

AVERAGE TEMPERATURE (F) 2001 BOSTON, MA (BOS)

YEAR	JAN	FEB	MAR	APR	MAY	JUN	JUL	AUG	SEP	OCT	NOV	DEC	ANNUAL
1972	33.0	29.6	36.3	44.9	57.6	65.4	73.8	71.5	65.7	51.8	42.3	33.0	50.4
1973	31.4	30.1	43.3	49.9	57.0	70.0	74.3	74.8	64.4	55.6	45.8	39.6	53.0
1974	31.7	29.1	38.7	50.9	54.7	64.8	72.4	72.0	63.7	50.1	45.3	37.8	50.9
1975	34.9	32.1	36.9	45.1	61.5	67.5	75.9	72.9	63.9	57.3	51.8	34.4	52.9
1976	26.1	37.3	41.2	55.1	60.2	73.4	72.9	72.0	64.9	52.3	41.9	29.0	52.2
1977	23.3	30.7	44.7	51.3	62.6	67.4	74.9	73.4	64.4	55.3	48.1	34.2	52.5
1978	28.5	27.1	36.2	48.8	59.3	68.3	72.1	71.6	61.4	52.5	43.6	35.3	50.4
1979	32.5	23.1	42.5	48.7	61.1	68.2	74.5	71.7	64.9	52.7	48.6	36.7	52.1
1980	29.4	27.9	36.9	48.7	59.4	66.3	75.8	74.2	67.0	52.4	41.2	28.6	50.7
1981	21.4	36.4	39.1	51.7	60.4	70.7	74.6	72.1	63.7	51.2	43.9	33.2	51.5
1982	22.9	30.8	38.7	48.2	57.8	63.3	74.9	70.3	64.1	54.2	47.6	39.6	51.0
1983	31.2	32.8	40.6	49.1	58.2	70.7	78.0	73.6	70.6	55.2	46.1	32.1	53.2
1984	26.7	37.6	31.9	46.1	58.0	70.5	74.7	74.6	62.1	53.3	44.6	39.5	51.6
1985	24.4	32.8	40.4	49.3	59.3	64.8	73.5	70.4	65.4	55.4	45.4	31.3	51.0
1986	31.4	28.9	40.7	48.4	58.4	66.1	71.0	70.5	63.2	54.0	42.3	35.5	50.9
1987	28.9	29.1	38.5	45.1	57.2	65.1	71.7	70.3	65.4	54.3	43.9	36.1	50.5
1988	27.8	32.2	39.2	46.8	57.6	68.5	73.7	75.5	64.6	50.8	46.7	32.8	51.4
1989	34.5	30.5	37.3	45.9	59.4	67.8	72.8	71.6	64.7	55.3	42.8	21.7	50.4
1990	36.4	34.1	40.1	47.6	54.9	66.6	73.1	73.3	64.6	58.3	48.5	40.7	53.2
1991	29.4	36.1	41.6	51.3	63.3	70.0	74.6	73.8	63.7	56.4	45.2	36.0	53.5
1992	31.0	32.4	35.4	46.4	55.6	67.8	69.5	70.4	63.9	52.5	42.9	34.8	50.2
1993	32.4	27.1	36.4	48.3	60.3	69.5	74.7	73.6	64.8	52.3	45.6	34.2	51.6
1994	22.2	26.9	38.2	51.4	58.4	71.9	77.5	72.4	64.2	55.5	49.0	38.5	52.2
1995	34.6	28.5	38.8	46.1	57.2	68.6	75.9	72.8	63.1	58.4	41.9	31.7	51.5
1996	30.1	30.9	36.5	47.9	57.4	68.1	71.8	70.9	64.2	53.2	40.3	39.3	50.9
1997	29.2	36.0	36.7	46.3	56.1	68.2	73.7	71.2	64.2	52.8	41.7	35.2	50.9
1998	33.9	35.3	41.5	49.4	60.3	64.7	74.4	72.5	66.3	54.5	44.6	39.1	53.0
1999	29.5	33.6	39.4	49.2	58.2	71.0	75.7	71.3	67.1	53.0	48.0	37.3	52.8
2000	27.5	34.2	43.3	47.3	57.2	67.3	70.0	70.3	63.5	53.9	43.8	29.2	50.6
2001	30.0	31.8	35.3	48.7	59.6	71.1	69.9	73.9	65.3	56.2	48.3	40.5	52.6
POR= 129 YRS	28.8	29.6	37.3	47.2	57.9	67.1	72.7	70.9	64.1	54.1	43.7	33.1	50.5

REFERENCE NOTES:

PAGE 1:
THE TEMPERATURE GRAPH SHOWS NORMAL MAXIMUM AND NORMAL MINIMUM DAILY TEMPERATURES (SOLID CURVES) AND THE ACTUAL DAILY HIGH AND LOW TEMPERATURES (VERTICAL BARS).

PAGE 2 AND 3:
H/C INDICATES HEATING AND COOLING DEGREE DAYS.
RH INDICATES RELATIVE HUMIDITY
W/O INDICATES WEATHER AND OBSTRUCTIONS
S INDICATES SUNSHINE.
PR INDICATES PRESSURE.
CLOUDINESS ON PAGE 3 IS THE SUM OF THE CEILOMETER AND SATELLITE DATA NOT TO EXCEED EIGHT EIGHTHS(OKTAS).

GENERAL:
T INDICATES TRACE PRECIPITATION, AN AMOUNT GREATER THAN ZERO BUT LESS THAN THE LOWEST REPORTABLE VALUE.
+ INDICATES THE VALUE ALSO OCCURS ON EARLIER DATES.
BLANK ENTRIES DENOTE MISSING OR UNREPORTED DATA.
NORMALS ARE 30-YEAR AVERAGES (1961 - 1990).
ASOS INDICATES AUTOMATED SURFACE OBSERVING SYSTEM.
PM INDICATES THE LAST DAY OF THE PREVIOUS MONTH.
POR (PERIOD OF RECORD) BEGINS WITH THE JANUARY DATA MONTH AND IS THE NUMBER OF YEARS USED TO COMPUTE THE MEAN. INDIVIDUAL MONTHS WITHIN THE POR MAY BE MISSING.
WHEN THE POR FOR A NORMAL IS LESS THAN 30 YEARS, THE NORMAL IS PROVISIONAL AND IS BASED ON THE NUMBER OF YEARS INDICATED.
0.* OR * INDICATES THE VALUE OR MEAN-DAYS-WITH IS BETWEEN 0.00 AND 0.05.
CLOUDINESS FOR ASOS STATIONS DIFFERS FROM THE NON-ASOS OBSERVATION TAKEN BY A HUMAN OBSERVER. ASOS STATION CLOUDINESS IS BASED ON TIME-AVERAGED CEILOMETER DATA FOR CLOUDS AT OR BELOW 12,000 FEET AND ON SATELLITE DATA FOR CLOUDS ABOVE 12,000 FEET.
THE NUMBER OF DAYS WITH CLEAR, PARTLY CLOUDY, AND CLOUDY CONDITIONS FOR ASOS STATIONS IS THE SUM OF THE CEILOMETER AND SATELLITE DATA FOR THE SUNRISE TO SUNSET PERIOD.

GENERAL CONTINUED:
CLEAR INDICATES 0 - 2 OKTAS, PARTLY CLOUDY INDICATES 3 - 6 OKTAS, AND CLOUDY INDICATES 7 OR 8 OKTAS. WHEN AT LEAST ONE OF THE ELEMENTS (CEILOMETER OR SATELLITE) IS MISSING, THE DAILY CLOUDINESS IS NOT COMPUTED.
WIND DIRECTION IS RECORDED IN TENS OF DEGREES (2 DIGITS) CLOCKWISE FROM TRUE NORTH. "00" INDICATES CALM. "36" INDICATES TRUE NORTH.
RESULTANT WIND IS THE VECTOR AVERAGE OF THE SPEED AND DIRECTION.
AVERAGE TEMPERATURE IS THE SUM OF THE MEAN DAILY MAXIMUM AND MINIMUM TEMPERATURE DIVIDED BY 2.
SNOWFALL DATA COMPRISE ALL FORMS OF FROZEN PRECIPITATION, INCLUDING HAIL.
A HEATING (COOLING) DEGREE DAY IS THE DIFFERENCE BETWEEN THE AVERAGE DAILY TEMPERATURE AND 65 F.
DRY BULB IS THE TEMPERATURE OF THE AMBIENT AIR.
DEW POINT IS THE TEMPERATURE TO WHICH THE AIR MUST BE COOLED TO ACHIEVE 100 PERCENT RELATIVE HUMIDITY.
WET BULB IS THE TEMPERATURE THE AIR WOULD HAVE IF THE MOISTURE CONTENT WAS INCREASED TO 100 PERCENT RELATIVE HUMIDITY.

ON JULY 1, 1996, THE NATIONAL WEATHER SERVICE BEGAN USING THE "METAR" OBSERVATION CODE THAT WAS ALREADY EMPLOYED BY MOST OTHER NATIONS OF THE WORLD. THE MOST NOTICEABLE DIFFERENCE IN THIS ANNUAL PUBLICATION WILL BE THE CHANGE IN UNITS FROM TENTHS TO EIGHTS(OKTAS) FOR REPORTING THE AMOUNT OF SKY COVER.

HEATING DEGREE DAYS (base 65 F) 2001 BOSTON, MA (BOS)

YEAR	JUL	AUG	SEP	OCT	NOV	DEC	JAN	FEB	MAR	APR	MAY	JUN	TOTAL
1972-73	3	4	51	405	673	985	1033	971	666	450	258	24	5523
1973-74	0	2	94	289	570	782	1023	1000	809	429	335	77	5410
1974-75	0	2	102	458	587	836	925	918	866	590	162	59	5505
1975-76	0	8	70	239	395	941	1198	800	733	331	166	16	4897
1976-77	1	10	55	393	688	1108	1290	956	623	414	158	43	5739
1977-78	0	4	85	304	498	948	1127	1057	885	480	209	18	5615
1978-79	11	11	150	381	635	916	1002	1169	691	481	149	19	5615
1979-80	2	15	80	390	484	873	1096	1071	866	481	185	66	5609
1980-81	2	5	72	387	706	1120	1344	794	796	393	200	7	5826
1981-82	2	6	91	419	628	979	1300	948	811	496	231	113	6024
1982-83	2	19	71	338	515	783	1040	896	749	478	223	22	5136
1983-84	0	8	42	327	561	1012	1182	790	1020	563	239	36	5780
1984-85	3	0	142	359	605	781	1255	897	758	471	204	71	5546
1985-86	3	11	65	298	580	1035	1035	1008	746	490	258	66	5595
1986-87	21	16	98	344	674	904	1112	997	814	588	285	76	5929
1987-88	8	18	57	326	626	888	1145	945	792	541	253	61	5660
1988-89	9	10	64	443	541	992	938	959	853	565	196	51	5621
1989-90	2	4	88	294	660	1336	880	857	762	524	307	60	5774
1990-91	4	5	84	236	496	744	1096	803	721	407	126	35	4757
1991-92	1	8	111	273	586	894	1049	937	913	552	317	37	5678
1992-93	21	14	109	386	656	930	1002	880		493	167	31	5745
1993-94	3	1	89	387	579	946	1320	1059	827	404	226	7	5848
1994-95	1	3	61	288	479	813	932	1016	804	561	258	30	5246
1995-96	1	2	113	214	685	1023	1074	981	875	510	260	20	5758
1996-97	1	6	80	358	739	792	1104	806	868	551	269	87	5661
1997-98	5	3	83	383	693	917	955	826	736	462	192	66	5321
1998-99	0	2	36	321	606	798	1092	872	790	468	215	29	5229
1999-00	2	7	39	363	505	851	1152	887	665	523	250	74	5318
2000-01	6	2	108	341	628	1103	1079	925	914	488	217	18	5829
2001-	2	1	67	286	492	753							

COOLING DEGREE DAYS (base 65 F) 2001 BOSTON, MA (BOS)

YEAR	JAN	FEB	MAR	APR	MAY	JUN	JUL	AUG	SEP	OCT	NOV	DEC	ANNUAL
1972	0	0	0	0	26	74	279	213	79	0	0	0	671
1973	0	0	0	7	18	180	296	316	84	3	0	0	904
1974	0	0	0	0	10	22	81	235	226	68	1	3	646
1975	0	0	0	0	60	139	345	261	44	9	4	0	862
1976	0	0	0	43	25	276	251	231	61	8	0	0	895
1977	0	0	1	13	92	124	314	272	75	6	0	0	897
1978	0	0	0	0	40	122	237	221	48	0	0	0	668
1979	0	0	0	0	35	122	304	226	85	17	0	0	789
1980	0	0	0	0	18	114	347	299	137	1	0	0	916
1981	0	0	0	0	67	185	306	232	60	0	0	0	850
1982	0	0	0	0	15	67	314	192	49	10	2	0	649
1983	0	0	0	7	18	200	410	283	217	27	0	0	1162
1984	0	0	0	3	31	207	312	306	62	3	0	0	924
1985	0	0	0	5	30	72	271	183	83	8	0	0	652
1986	0	0	0	0	60	105	211	190	55	10	0	0	631
1987	0	0	0	0	48	87	221	189	76	0	2	0	623
1988	0	0	0	0	31	173	287	342	59	11	0	0	903
1989	0	0	1	0	29	142	248	214	89	0	0	0	723
1990	0	0	0	10	2	116	261	268	77	34	8	0	776
1991	0	0	0	3	79	189	304	287	79	15	0	0	956
1992	0	0	0	0	30	126	165	189	83	5	0	0	598
1993	0	0	0	0	28	173	310	273	89	0	2	0	875
1994	0	0	0	1	29	221	395	241	44	1	4	0	936
1995	0	0	0	0	24	147	344	252	64	15	0	0	846
1996	0	0	0	3	30	120	222	196	63	0	1	0	635
1997	0	0	0	0	0	189	280	202	66	12	0	0	749
1998	0	0	16	2	52	59	298	240	83	1	0	0	751
1999	0	0	0	0	11	215	344	213	110	0	0	0	893
2000	0	0	0	0	15	151	167	174	71	5	0	0	583
2001	0	0	0	9	55	210	160	283	83	20	0	0	820

SNOWFALL (inches) 2001 BOSTON, MA (BOS)

YEAR	JUL	AUG	SEP	OCT	NOV	DEC	JAN	FEB	MAR	APR	MAY	JUN	TOTAL
1972-73	0.0	0.0	0.0	T	0.6	3.3	3.6	2.5	0.3	T	0.0	0.0	10.3
1973-74	0.0	0.0	0.0	0.0	T	16.0	17.8	0.1	3.0	0.0	0.0	0.0	36.9
1974-75	0.0	0.0	0.0	0.0	2.0	3.6	2.2	17.0	1.8	1.0	0.0	0.0	27.6
1975-76	0.0	0.0	0.0	T	0.1	19.3	15.0	1.4	10.8	T	0.0	0.0	46.6
1976-77	0.0	0.0	0.0	0.0	1.0	17.2	23.2	5.9	10.7	T	0.5	0.0	58.5
1977-78	0.0	0.0	0.0	0.0	0.7	5.2	35.9	27.2	16.1	T	0.0	0.0	85.1
1978-79	0.0	0.0	0.0	0.0	4.2	5.8	10.5	6.6	T	0.4	0.0	0.0	27.5
1979-80	0.0	0.0	0.0	0.2	T	2.0	0.4	6.5	3.6	T	0.0	0.0	12.7
1980-81	0.0	0.0	0.0	0.0	2.4	5.6	11.9	1.9	0.5	0.0	0.0	0.0	22.3
1981-82	0.0	0.0	0.0	0.0	T	17.6	18.0	7.6	5.3	13.3	0.0	0.0	61.8
1982-83	0.0	0.0	0.0	0.0	T	5.5	4.7	22.3	0.2	T	0.0	0.0	32.7
1983-84	0.0	0.0	0.0	0.0	T	2.6	21.1	0.3	19.0	T	0.0	0.0	43.0
1984-85	0.0	0.0	0.0	0.0	0.0	3.7	7.0	10.2	3.7	2.0	0.0	0.0	26.6
1985-86	0.0	0.0	0.0	0.0	3.0	1.3	0.8	10.4	2.6	T	0.0	0.0	18.1
1986-87	0.0	0.0	0.0	0.0	3.5	3.4	24.3	3.7	3.5	4.1	0.0	0.0	42.5
1987-88	0.0	0.0	0.0	0.0	9.0	7.5	17.0	14.1	5.0	T	0.0	0.0	52.6
1988-89	0.0	0.0	0.0	T	0.0	3.7	1.5	6.7	3.2	0.4	0.0	0.0	15.5
1989-90	0.0	0.0	0.0	0.0	4.5	6.2	7.0	16.9	4.1	0.5	0.0	0.0	39.2
1990-91	0.0	0.0	0.0	0.0	T	1.2	11.7	2.8	3.4	0.0	0.0	0.0	19.1
1991-92	0.0	0.0	0.0	0.0	T	5.8	0.4	4.0	10.8	1.0	0.0	0.0	22.0
1992-93	0.0	0.0	0.0	0.0	0.6	9.7	12.9	19.6	38.9	2.2	0.0	0.0	83.9
1993-94	0.0	0.0	0.0	0.0	T	11.6	33.7	36.2	14.8	0.0	0.0	0.0	96.3
1994-95	0.0	T	0.0	0.0	0.1	1.5	4.4	8.5	0.4	T	0.0	0.0	14.9
1995-96	0.0	0.0	0.0	0.0	4.1	24.1	39.8	15.5	16.8				
1996-97					1.6								
1997-98													
1998-99							16.8	7.4	11.4	0.0	0.0	0.0	
1999-00	0.0	0.0	0.0	0.0	0.0	0.0	13.7	9.2	2.0	T	0.0	0.0	24.9
2000-01	0.0	0.0	0.0	T	T	4.5	12.4	9.8	19.2	T	0.0	0.0	45.9
2001-	0.0	0.0	0.0	T	0.0	5.0							
POR= 63 YRS	0.0	T	0.0	0.0	1.3	7.5	12.6	11.4	8.1	0.9	0.0	0.0	41.8

2001
WORCESTER,
MASSACHUSETTS (ORH)

Worcester Municipal Airport is located on the crest of a hill, 1,000 feet above sea level. It is about 500 feet above and 3 1/2 miles northwest of the city proper. The airport is surrounded by ridges and valleys with many of the valleys containing reservoirs. Only two of the ridges extend above the airport elevation. One is 400 feet higher and 2 1/2 miles to the northwest, and the other is 1,000 feet higher and 15 miles to the north.

The proximity to the Atlantic Ocean, Long Island Sound, and the Berkshire Hills plays an important part in determining the weather and, hence, the climate of Worcester. Rapid weather changes occur when storms move up the east coast after developing off the Carolina Coast. In the majority of these cases, they pass to the south and east, resulting in northeast and easterly winds with rain or snow and fog. Storms developing in the Texas-Oklahoma area normally travel up the St. Lawrence River Valley and, depending on the movement and intensity, usually deposit little precipitation over the area. However, they do bring an influx of warm air into the region. Wintertime cold snaps are quite frequent, but temperatures are usually modified by the passage of the air over land and mountains before reaching the county. Summertime thunderstorms develop over the hills to the west, with a majority moving toward the northeast. From the use of radar, we find many break up just before reaching Worcester, or pass either north or south of the city proper.

Airport site temperatures are moderate. The normal mean for the warmest month, July, is around 70 degrees. Though winters are reasonably cold, prolonged periods of severe cold weather are extremely rare. The three coldest months, December through February, have an average temperature of over 25 degrees. A review of Worcester Cooperative records since 1901 shows maximum temperatures above 100 degrees and minimum temperatures below -24 degrees.

Precipitation is usually plentiful and well distributed throughout the year. The annual snowfall for all Worcester sites since 1901, averages slightly less than 60 inches. The airport location averages slightly higher.

Based on the 1951-1980 period, the average first occurrence of 32 degrees Fahrenheit in the fall is October 17 and the average last occurrence in the spring is April 27.

NORMALS, MEANS, AND EXTREMES
WORCESTER, MA (ORH)

LATITUDE:	LONGITUDE:	ELEVATION (FT):	TIME ZONE:	WBAN: 94746
42 16' 02" N	71 52' 34" W	GRND: 963 BARO: 966	EASTERN (UTC + 5)	

ELEMENT	POR	JAN	FEB	MAR	APR	MAY	JUN	JUL	AUG	SEP	OCT	NOV	DEC	YEAR
TEMPERATURE °F														
NORMAL DAILY MAXIMUM	30	30.7	33.0	42.4	54.0	65.9	74.5	79.3	77.3	69.7	59.5	47.4	34.7	55.7
MEAN DAILY MAXIMUM	47	31.6	33.9	42.1	54.5	66.1	74.7	77.6	77.1	69.3	59.3	47.2	35.8	55.8
HIGHEST DAILY MAXIMUM	45	60	67	84	91	92	94	96	96	91	85	78	72	96
YEAR OF OCCURRENCE		1995	1985	1998	1976	1962	1988	1988	1975	1983	1963	1982	1998	JUL 1988
MEAN OF EXTREME MAXS.	48	51.3	52.2	65.1	76.0	84.1	88.1	88.2	87.7	84.0	76.0	66.5	56.9	73.0
NORMAL DAILY MINIMUM	30	15.0	16.5	24.9	34.7	45.1	53.9	60.1	58.6	50.5	40.5	31.3	20.1	37.6
MEAN DAILY MINIMUM	48	15.6	17.2	25.5	35.6	46.0	55.3	59.5	59.3	51.5	41.4	32.2	21.5	38.4
LOWEST DAILY MINIMUM	46	-19	-12	-4	11	28	36	43	38	30	20	6	-13	-19
YEAR OF OCCURRENCE		1957	1967	1986	1982	1970	1986	1988	1965	1992	1969	1989	1962	JAN 1957
MEAN OF EXTREME MINS.	48	-2.7	-1.6	7.7	22.7	34.4	43.2	50.1	48.1	37.6	27.3	17.2	2.6	23.9
NORMAL DRY BULB	30	22.8	24.8	33.7	44.4	55.5	64.2	69.7	68.0	60.2	50.0	39.4	27.4	46.7
MEAN DRY BULB	47	23.8	25.8	33.8	45.0	55.9	65.0	68.6	68.2	60.3	50.4	39.6	28.5	47.1
MEAN WET BULB	15	22.3	23.9	30.4	39.5	49.7	58.7	59.4	62.4	50.7	45.4	36.1	24.9	42.0
MEAN DEW POINT	15	15.1	15.9	22.0	31.1	42.4	53.6	55.4	58.3	46.4	39.3	29.9	18.4	35.6
NORMAL NO. DAYS WITH:														
MAXIMUM ≥ 90	30	0.0	0.0	0.0	0.1	0.1	0.7	1.5	0.8	*	0.0	0.0	0.0	3.2
MAXIMUM ≤ 32	30	17.6	13.9	5.1	0.2	0.0	0.0	0.0	0.0	0.0	0.0	1.7	13.6	52.1
MINIMUM ≤ 32	30	29.5	26.4	24.9	11.1	0.6	0.0	0.0	0.0	*	4.9	17.0	27.9	142.3
MINIMUM ≤ 0	30	3.1	1.9	0.2	0.0	0.0	0.0	0.0	0.0	0.0	0.0	0.0	0.9	6.1
H/C														
NORMAL HEATING DEG. DAYS	30	1308	1126	970	618	301	69	6	25	157	465	768	1166	6979
NORMAL COOLING DEG. DAYS	30	0	0	0	0	6	45	151	118	13	0	0	0	333
RH														
NORMAL (PERCENT)	30	65	64	63	60	64	68	70	73	75	69	71	70	68
HOUR 01 LST	30	68	68	68	68	73	78	79	82	83	77	75	74	74
HOUR 07 LST	30	70	71	70	68	70	74	76	79	81	78	77	75	74
HOUR 13 LST	30	58	57	54	50	52	57	57	60	61	56	61	62	57
HOUR 19 LST	30	64	62	60	57	60	66	68	72	75	68	70	69	66
S PERCENT POSSIBLE SUNSHINE														
W/O MEAN NO. DAYS WITH:														
HEAVY FOG(VISBY 1/4 MI)	29	6.0	5.8	7.8	6.6	7.1	7.7	6.0	6.1	7.6	6.8	7.4	6.8	81.7
THUNDERSTORMS	29	0.0	0.1	0.7	1.2	2.5	3.6	4.4	3.5	1.3	1.0	0.4	0.1	18.8
CLOUDINESS MEAN:														
SUNRISE-SUNSET (OKTAS)														
MIDNIGHT-MIDNIGHT (OKTAS)														
MEAN NO. DAYS WITH:														
CLEAR	1			1.0			5.9							
PARTLY CLOUDY	1	1.0		1.0		2.0	12.1							
CLOUDY	1	1.0	1.0	1.0			13.0							
PR MEAN STATION PRESSURE(IN)	28	28.87	28.87	28.86	28.85	28.88	28.88	28.91	28.95	28.97	28.96	28.92	28.89	28.90
MEAN SEA-LEVEL PRES. (IN)	16	30.03	30.00	29.98	29.94	29.98	29.95	29.99	30.05	30.08	30.08	30.06	30.02	30.01
WINDS MEAN SPEED (MPH)	29	12.2	12.4	12.6	11.6	10.6	9.9	9.2	8.9	9.5	10.2	11.2	11.8	10.8
PREVAIL.DIR(TENS OF DEGS)	16	28	28	29	29	28	28	26	26	25	27	27	29	28
MAXIMUM 2-MINUTE:														
SPEED (MPH)	5	41	49	47	43	51	36	33	41	41	38	41	45	51
DIR. (TENS OF DEGS)		30	28	27	21	31	31	30	31	32	31	29	27	31
YEAR OF OCCURRENCE		2000	1996	1997	2000	1998	2001	2001	1996	1999	1999	1997	2000	MAY 1998
MAXIMUM 5-SECOND:														
SPEED (MPH)	5	57	70	61	58	94	47	43	49	52	52	52	59	94
DIR. (TENS OF DEGS)		30	29	27	19	26	05	28	30	29	29	29	19	26
YEAR OF OCCURRENCE		2000	1996	1997	2000	1998	2000	1996	1996	1999	1998	1997	2000	MAY 1998
PRECIPITATION NORMAL (IN)	30	3.68	3.46	3.95	3.91	4.33	3.88	3.85	3.82	4.01	4.32	4.49	4.05	47.75
MAXIMUM MONTHLY (IN)	46	11.16	8.37	7.96	8.79	9.94	12.17	8.11	8.01	13.13	10.19	10.40	9.83	13.13
YEAR OF OCCURRENCE		1979	1981	1972	1987	1984	1982	1959	1991	1974	1990	1972	1973	SEP 1974
MINIMUM MONTHLY (IN)	46	0.89	0.25	0.74	0.75	0.86	0.32	0.74	1.03	0.69	0.70	0.67	0.74	0.25
YEAR OF OCCURRENCE		1970	1987	1981	2001	1959	1999	1987	1981	1986	2001	1976	1989	FEB 1987
MAXIMUM IN 24 HOURS (IN)	46	2.97	2.46	4.56	3.34	3.03	3.98	3.87	5.00	4.79	3.91	2.98	3.00	5.00
YEAR OF OCCURRENCE		1978	1973	1987	2000	1967	1986	3087	1991	1960	1990	1972	1986	AUG 1991
NORMAL NO. DAYS WITH:														
PRECIPITATION ≥ 0.01	30	11.5	10.6	12.0	11.4	12.2	11.6	9.5	10.1	9.2	8.7	11.9	12.7	131.4
PRECIPITATION ≥ 1.00	30	0.9	1.0	1.0	1.1	1.2	0.9	1.2	1.1	1.2	1.4	1.2	1.2	13.4
SNOWFALL NORMAL (IN)	30	16.6	16.4	10.9	3.6	0.4	0.0	0.0	0.0	0.0	0.5	4.2	12.9	65.5
MAXIMUM MONTHLY (IN)	40	46.8	45.2	44.1	21.0	12.7	0.1	0.0	T	T	7.5	20.7	37.0	46.8
YEAR OF OCCURRENCE		1987	1962	1993	1987	1977	1992		1994	1994	1979	1971	1992	JAN 1987
MAXIMUM IN 24 HOURS (IN)	40	18.7	24.0	19.7	17.0	12.7	0.1	0.0	T	T	7.5	14.8	28.1	28.1
YEAR OF OCCURRENCE		1961	1962	1993	1987	1977	1992		1994	1994	1979	1971	1992	DEC 1992
MAXIMUM SNOW DEPTH (IN)	42	35	42	33	17	10	0		0	0	8	15	29	42
YEAR OF OCCURRENCE		1966	1961	1969	1987	1977					1979	1971	1992	FEB 1961
NORMAL NO. DAYS WITH:														
SNOWFALL ≥ 1.0	30	4.2	3.5	2.7	0.9	0.1	0.0	0.0	0.0	0.0	0.1	1.0	3.6	16.1

PRECIPITATION (inches) 2001 WORCESTER, MA (ORH)

YEAR	JAN	FEB	MAR	APR	MAY	JUN	JUL	AUG	SEP	OCT	NOV	DEC	ANNUAL
1972	2.34	4.91	7.96	4.29	7.83	9.25	6.39	2.89	4.98	4.93	10.40	5.49	71.66
1973	4.04	3.30	3.50	6.33	4.73	6.98	3.82	4.29	3.80	4.63	2.00	9.83	57.25
1974	3.54	2.75	5.05	3.24	5.15	4.35	3.22	3.50	13.13	3.45	3.06	6.04	56.48
1975	5.35	3.52	3.37	2.66	2.07	3.32	3.92	4.59	7.21	6.06	5.28	4.61	51.96
1976	6.03	2.57	2.74	2.45	3.46	2.95	3.28	5.72	2.38	4.99	0.67	3.20	40.44
1977	2.76	2.46	5.75	3.69	2.20	4.00	3.84	2.73	6.54	6.33	3.69	4.99	48.98
1978	9.90	2.08	3.22	2.24	3.69	1.57	3.57	5.00	1.02	3.85	2.07	3.56	41.77
1979	11.16	2.64	3.71	4.49	4.14	0.79	5.74	7.39	3.80	4.36	3.58	1.89	53.69
1980	0.95	0.73	6.86	4.77	2.23	4.55	3.59	1.95	1.82	6.16	4.58	1.06	39.25
1981	0.93	8.37	0.74	3.85	4.48	2.45	7.90	1.03	4.66	5.49	3.13	5.94	48.97
1982	5.00	3.22	3.67	4.30	2.96	12.17	3.61	3.36	2.69	2.67	4.32	1.70	49.67
1983	4.85	4.67	7.84	8.59	5.97	2.56	1.32	6.26	1.38	5.77	8.75	6.37	64.33
1984	2.44	5.78	5.47	4.23	9.94	2.85	5.69	1.17	1.68	3.99	2.71	2.84	48.79
1985	1.16	2.72	2.89	1.26	5.46	5.24	6.35	3.74	3.12	6.41	1.93	44.05	
1986	5.56	3.14	2.93	1.59	3.14	7.21	4.83	3.20	0.69	2.72	5.63	7.25	47.89
1987	5.52	0.25	6.57	8.79	1.55	4.55	0.74	4.61	6.37	4.18	2.77	1.85	47.75
1988	2.71	2.78	3.46	3.45	4.47	1.25	6.27	2.19	2.70	3.66	7.91	1.42	42.27
1989	1.18	2.47	2.66	4.25	6.17	5.27	5.67	5.65	4.71	8.21	4.00	0.74	50.98
1990	3.75	3.88	1.52	4.78	7.65	1.74	2.44	6.84	1.73	10.19	2.41	5.46	52.39
1991	2.98	2.08	4.92	5.04	4.16	3.06	2.78	8.01	6.40	3.44	5.47	2.89	51.23
1992	3.01	2.51	4.15	2.59	2.54	4.68	5.25	4.83	3.58	2.36	4.94	4.61	45.05
1993	2.56	2.38	5.46	4.00	1.79	2.36	3.34	1.90	8.85	3.88	4.85	5.11	46.48
1994	4.78	1.86	5.38	2.73	5.87	2.48	3.09	7.64	4.84	1.24	4.54	4.81	49.26
1995	3.71	2.86	1.85	2.19	2.39	1.51	4.33	2.02	3.15	8.65	4.61	1.04	38.31
1996	6.30	2.66	2.20	6.70	3.34	3.01	6.51	3.99	6.07	5.81	2.93	6.91	56.42
1997	3.25	1.71	4.66	3.22	2.72	1.60	2.97	4.34	1.44	2.11	5.50	2.32	35.84
1998	4.59	3.17	5.82	3.30	5.89	9.68	1.76	2.38	1.69	4.93	2.28	1.46	46.95
1999	6.01	3.38	4.09	0.92	2.77	0.32	4.14	1.87	8.81	3.57	3.38	2.55	41.81
2000	3.11	2.59	3.82	6.85	3.51	5.84	4.04	2.09	3.01	2.05	3.61	3.62	44.14
2001	1.64	2.40	6.53	0.75	2.26	6.27	1.91	2.31	3.42	0.70	1.36	2.77	32.32
POR= 53 YRS	3.73	3.21	4.10	3.93	3.95	3.67	3.56	4.06	3.95	4.09	4.37	3.83	46.45

AVERAGE TEMPERATURE (F) 2001 WORCESTER, MA (ORH)

YEAR	JAN	FEB	MAR	APR	MAY	JUN	JUL	AUG	SEP	OCT	NOV	DEC	ANNUAL
1972	26.4	23.3	30.4	40.4	55.7	62.2	71.1	66.9	59.7	44.7	34.9	27.6	45.3
1973	26.1	24.7	40.9	46.3	54.2	67.4	71.1	71.7	60.1	51.3	39.8	31.9	48.8
1974	26.4	23.9	33.5	47.5	52.4	64.0	69.7	69.7	59.7	45.8	39.4	30.4	46.9
1975	27.5	25.2	31.3	40.8	60.1	63.6	71.8	67.5	57.3	51.5	45.2	27.3	47.4
1976	19.2	31.2	35.1	49.1	55.0	68.0	67.8	67.2	58.9	46.4	34.6	21.9	46.2
1977	16.2	24.9	39.1	45.9	58.1	62.4	69.3	68.6	59.6	49.3	41.1	26.0	46.7
1978	20.9	20.0	30.8	42.4	56.6	64.3	68.2	68.1	57.6	48.4	39.2	28.8	45.4
1979	25.3	16.5	38.2	44.7	57.7	63.6	71.2	67.7	60.0	49.0	44.3	31.9	47.5
1980	25.0	22.2	33.1	46.0	57.0	61.5	71.3	69.9	61.1	47.3	36.4	24.2	46.3
1981	15.9	32.8	34.7	47.4	57.3	66.0	70.7	67.4	58.2	46.7	39.1	27.8	47.1
1982	17.1	25.7	33.8	43.6	58.3	61.5	71.5	65.9	61.3	50.8	43.8	34.8	47.3
1983	26.1	29.1	36.6	46.6	54.7	67.9	72.3	70.2	64.9	50.3	42.0	27.3	49.0
1984	22.7	34.0	29.2	45.6	55.4	68.1	70.0	71.4	59.6	53.9	41.1	34.9	48.8
1985	20.1	28.6	37.2	47.4	57.8	61.4	70.4	67.9	62.0	51.7	40.9	25.3	47.6
1986	25.9	23.1	36.4	48.1	57.9	62.6	68.2	66.8	59.4	49.6	37.1	30.3	47.1
1987	24.0	24.5	36.8	45.7	57.6	65.8	70.7	66.6	60.6	48.8	39.4	30.8	47.6
1988	23.0	26.6	35.0	43.8	57.3	64.6	72.6	72.3	60.2	45.5	41.7	27.2	47.5
1989	28.5	24.4	33.0	42.4	57.0	64.0	68.7	67.3	60.6	50.7	36.5	15.1	45.7
1990	31.3	28.8	36.2	43.9	51.5	64.6	69.2	68.8	59.3	53.0	42.1	34.0	48.6
1991	23.8	30.3	36.5	48.2	61.2	64.7	70.0	69.1	58.7	51.9	40.2	29.7	48.7
1992	25.3	27.1	30.7	42.3	55.5	63.5	65.9	65.5	59.2	46.8	37.8	28.7	45.7
1993	26.0	19.6	31.3	45.7	57.8	64.5	70.4	70.0	59.5	47.2	39.7	28.9	46.7
1994	15.9	21.1	33.5	47.7	54.3	67.2	72.9	66.3	59.0	50.9	43.5	33.3	47.1
1995	29.8	23.0	36.1	42.6	54.4	65.5			59.4	53.5	35.5	24.1	
1996	23.4	24.9	31.6	45.3	54.9	65.9		68.9	61.1	49.7	35.9	34.2	
1997	23.6	0.0	31.9	43.5	52.2	65.8	69.1	67.6	60.3	48.8	36.2	30.0	44.1
1998		32.9	37.0	47.2	59.7	62.9	70.0	70.3	63.1	50.3	40.2	35.7	
1999	26.0	31.3	36.3	46.7	58.1	68.2	72.4	68.5	64.6	49.1	44.5	32.9	49.9
2000	23.0	30.0	40.4	44.7	56.2	64.6	66.5	66.9	59.5	50.1	39.2	23.7	47.1
2001	24.7	26.1	31.1	46.8	58.1	67.0	66.6	72.2	62.2	52.5	44.5	35.4	48.9
POR= 53 YRS	23.2	25.3	33.7	45.1	56.0	64.9	68.6	68.3	60.4	50.3	39.6	28.3	47.0

REFERENCE NOTES:

PAGE 1:
THE TEMPERATURE GRAPH SHOWS NORMAL MAXIMUM AND NORMAL
MINIMUM DAILY TEMPERATURES (SOLID CURVES) AND THE
ACTUAL DAILY HIGH AND LOW TEMPERATURES (VERTICAL BARS).

PAGE 2 AND 3:
H/C INDICATES HEATING AND COOLING DEGREE DAYS.
RH INDICATES RELATIVE HUMIDITY
W/O INDICATES WEATHER AND OBSTRUCTIONS
S INDICATES SUNSHINE.
PR INDICATES PRESSURE.
CLOUDINESS ON PAGE 3 IS THE SUM OF THE CEILOMETER AND
SATELLITE DATA NOT TO EXCEED EIGHT EIGHTHS(OKTAS).

GENERAL:
T INDICATES TRACE PRECIPITATION, AN AMOUNT GREATER
THAN ZERO BUT LESS THAN THE LOWEST REPORTABLE VALUE.
+ INDICATES THE VALUE ALSO OCCURS ON EARLIER DATES.
BLANK ENTRIES DENOTE MISSING OR UNREPORTED DATA.
NORMALS ARE 30-YEAR AVERAGES (1961 - 1990).
ASOS INDICATES AUTOMATED SURFACE OBSERVING SYSTEM.
PM INDICATES THE LAST DAY OF THE PREVIOUS MONTH.
POR (PERIOD OF RECORD) BEGINS WITH THE JANUARY DATA
MONTH AND IS THE NUMBER OF YEARS USED TO COMPUTE
THE MEAN. INDIVIDUAL MONTHS WITHIN THE POR MAY
BE MISSING.
WHEN THE POR FOR A NORMAL IS LESS THAN 30 YEARS,
THE NORMAL IS PROVISIONAL AND IS BASED ON THE NUMBER
OF YEARS INDICATED.
0.* OR * INDICATES THE VALUE OR MEAN-DAYS-WITH
IS BETWEEN 0.00 AND 0.05.
CLOUDINESS FOR ASOS STATIONS DIFFERS FROM THE NON-ASOS
OBSERVATION TAKEN BY A HUMAN OBSERVER. ASOS STATION
CLOUDINESS IS BASED ON TIME-AVERAGED CEILOMETER DATA
FOR CLOUDS AT OR BELOW 12,000 FEET AND ON SATELLITE
DATA FOR CLOUDS ABOVE 12,000 FEET.
THE NUMBER OF DAYS WITH CLEAR, PARTLY CLOUDY, AND
CLOUDY CONDITIONS FOR ASOS STATIONS IS THE SUM
OF THE CEILOMETER AND SATELLITE DATA FOR THE
SUNRISE TO SUNSET PERIOD.

GENERAL CONTINUED:
CLEAR INDICATES 0 - 2 OKTAS, PARTLY CLOUDY INDICATES
3 - 6 OKTAS, AND CLOUDY INDICATES 7 OR 8 OKTAS.
WHEN AT LEAST ONE OF THE ELEMENTS (CEILOMETER OR
SATELLITE) IS MISSING, THE DAILY CLOUDINESS IS
NOT COMPUTED.
WIND DIRECTION IS RECORDED IN TENS OF DEGREES (2 DIGITS)
CLOCKWISE FROM TRUE NORTH. "00" INDICATES CALM. "36"
INDICATES TRUE NORTH.
RESULTANT WIND IS THE VECTOR AVERAGE OF THE SPEED AND
DIRECTION.
AVERAGE TEMPERATURE IS THE SUM OF THE MEAN DAILY MAXIMUM
AND MINIMUM TEMPERATURE DIVIDED BY 2.
SNOWFALL DATA COMPRISE ALL FORMS OF FROZEN
PRECIPITATION, INCLUDING HAIL.
A HEATING (COOLING) DEGREE DAY IS THE DIFFERENCE BETWEEN
THE AVERAGE DAILY TEMPERATURE AND 65 F.
DRY BULB IS THE TEMPERATURE OF THE AMBIENT AIR.
DEW POINT IS THE TEMPERATURE TO WHICH THE AIR MUST BE
COOLED TO ACHIEVE 100 PERCENT RELATIVE HUMIDITY.
WET BULB IS THE TEMPERATURE THE AIR WOULD HAVE IF THE
MOISTURE CONTENT WAS INCREASED TO 100 PERCENT RELATIVE
HUMIDITY.

ON JULY 1, 1996, THE NATIONAL WEATHER SERVICE BEGAN USING
THE "METAR" OBSERVATION CODE THAT WAS ALREADY EMPLOYED
BY MOST OTHER NATIONS OF THE WORLD. THE MOST NOTICEABLE
DIFFERENCE IN THIS ANNUAL PUBLICATION WILL BE THE CHANGE
IN UNITS FROM TENTHS TO EIGHTS(OKTAS) FOR REPORTING THE
AMOUNT OF SKY COVER.

HEATING DEGREE DAYS (base 65 F) 2001 WORCESTER, MA (ORH)

YEAR	JUL	AUG	SEP	OCT	NOV	DEC	JAN	FEB	MAR	APR	MAY	JUN	TOTAL
1972-73	16	41	171	622	897	1152	1200	1120	738	555	338	37	6887
1973-74	6	8	202	420	750	1017	1191	1145	969	520	391	82	6701
1974-75	6	9	190	592	760	1063	1158	1110	1037	721	180	104	6930
1975-76	2	47	223	413	586	1161	1412	972	920	511	310	60	6617
1976-77	17	46	192	571	905	1331	1508	1115	798	571	244	115	7413
1977-78	16	32	189	481	711	1202	1359	1255	1054	674	288	69	7330
1978-79	33	38	229	511	767	1116	1225	1354	824	601	243	80	7021
1979-80	25	53	180	500	614	1019	1235	1233	983	562	246	148	6798
1980-81	2	19	165	540	853	1259	1516	894	934	521	241	42	6986
1981-82	1	29	204	562	772	1145	1478	1094	961	639	212	127	7224
1982-83	7	55	140	436	632	929	1199	996	871	548	318	44	6175
1983-84	5	20	115	459	682	1163	1303	892	1099	577	300	54	6669
1984-85	9	1	184	338	713	928	1382	1011	855	518	230	121	6290
1985-86	2	25	128	406	713	1223	1206	1166	879	501	262	119	6630
1986-87	48	49	182	471	830	1069	1266	1130	867	571	273	58	6814
1987-88	9	56	152	495	761	1052	1298	1107	922	629	253	115	6849
1988-89	19	32	155	597	693	1166	1123	1129	984	674	257	90	6919
1989-90	7	38	172	436	846	1540	1035	1006	884	632	413	58	7067
1990-91	22	30	192	379	683	957	1270	965	877	501	167	80	6123
1991-92	13	16	222	402	739	1090	1222	1092	1056	676	321	82	6931
1992-93	40	55	203	558	810	1117	1204	1259	1039	574	229	74	7162
1993-94	10	5	201	545	754	1115	1515	1226	968	515	337	49	7240
1994-95	5	48	182	431	636	975	1086	1170	887	667	329	58	6474
1995-96			191	352	877	1260	1279	1158	1031	587	331	36	
1996-97		14	159	467	863	980	1277		1020	638	394	93	
1997-98	15	18	158	498	858	1079		891	872	527	180	111	6023
1998-99	3	8	98	450	736	902	1201	938	882	540	226	39	6261
1999-00	7	27	91	486	609	989	1295	1008	758	602	289	100	
2000-01	26	38	190	455	769	1273	1243	1081	1045	544	258	44	6966
2001-	38	1	117	391	608	910							

COOLING DEGREE DAYS (base 65 F) 2001 WORCESTER, MA (ORH)

YEAR	JAN	FEB	MAR	APR	MAY	JUN	JUL	AUG	SEP	OCT	NOV	DEC	ANNUAL
1972	0	0	0	0	9	41	215	109	20	0	0	0	394
1973	0	0	0	1	11	118	201	224	62	0	0	0	617
1974	0	0	0	4	8	59	159	164	36	0	1	0	430
1975	0	0	0	0	33	69	218	132	0	0	0	0	453
1976	0	0	0	37	7	154	110	119	16	0	0	0	443
1977	0	0	2	4	37	44	156	150	32	0	0	0	425
1978	0	0	0	0	32	57	138	137	14	0	0	0	378
1979	0	0	0	0	24	44	225	142	38	10	0	0	483
1980	0	0	0	0	8	48	206	178	57	0	0	0	497
1981	0	0	0	0	29	78	184	110	10	0	0	0	411
1982	0	0	0	0	8	29	216	92	34	0	0	0	379
1983	0	0	0	4	4	138	238	188	118	11	0	0	701
1984	0	0	0	2	10	156	171	209	30	0	0	0	578
1985	0	0	0	0	15	22	177	120	47	0	0	0	381
1986	0	0	0	0	50	54	151	111	21	2	0	0	389
1987	0	0	0	0	49	90	193	110	25	0	0	0	467
1988	0	0	0	0	24	112	260	266	17	1	0	0	680
1989	0	0	0	0	16	70	131	116	47	0	0	0	380
1990	0	0	0	6	0	52	159	153	25	10	0	0	405
1991	0	0	0	4	57	79	173	149	40	2	0	0	504
1992	0	0	0	0	30	44	75	78	37	0	0	0	264
1993	0	0	0	0	12	67	181	163	44	0	0	0	467
1994	0	0	0	0	12	122	260	97	9	0	0	0	500
1995	0	0	0	0	6	77			29	3	0	0	
1996	0	0	0	2	23	70		143	47	0	0	0	
1997	0		0	0	0	123	148	106	25	2	0	0	
1998		0	10	0	21	58	181	47	0	0	0		
1999	0	0	0	0	19	142	245	140	86	0	0	0	632
2000	0	0	0	0	22	95	82	104	31	1	0	0	335
2001	0	0	0	4	51	109	95	232	41	7	0	0	539

SNOWFALL (inches) 2001 WORCESTER, MA (ORH)

YEAR	JUL	AUG	SEP	OCT	NOV	DEC	JAN	FEB	MAR	APR	MAY	JUN	TOTAL
1972-73	0.0	0.0	0.0	T	6.1	13.8	17.9	5.8	0.4	0.4	0.0	0.0	44.4
1973-74	0.0	0.0	0.0	0.0	T	0.9	12.5	15.0	1.7	3.7	0.0	0.0	33.8
1974-75	0.0	0.0	0.0	T	1.2	13.1	22.6	21.9	4.9	1.4	0.0	0.0	65.1
1975-76	0.0	0.0	0.0	T	1.5	18.1	21.6	4.7	16.4	T	T	0.0	62.3
1976-77	0.0	0.0	0.0	T	3.0	13.5	21.7	13.8	21.5	1.0	12.7	0.0	87.2
1977-78	0.0	0.0	0.0	0.0	2.2	13.7	34.2	20.8	15.0	T	T	0.0	85.9
1978-79	0.0	0.0	0.0	0.0	5.4	13.1	16.0	6.5	1.3	5.4	0.0	0.0	47.7
1979-80	0.0	0.0	0.0	7.5	0.0	2.1	0.8	6.5	9.7	T	0.0	0.0	26.6
1980-81	0.0	0.0	0.0	0.0	9.0	6.8	12.5	11.4	3.3	T	0.0	0.0	43.0
1981-82	0.0	0.0	0.0	T	T	24.6	16.7	6.5	11.0	15.1	0.0	0.0	73.9
1982-83	0.0	0.0	0.0	0.0	0.5	6.4	18.6	32.1	3.5	2.3	T	0.0	63.4
1983-84	0.0	0.0	0.0	T	1.1	17.2	24.1	3.3	30.9	T	T	0.0	76.6
1984-85	0.0	0.0	0.0	0.0	T	7.0	9.7	11.0	7.2	4.9	0.0	0.0	39.8
1985-86	0.0	0.0	0.0	0.0	6.9	9.1	5.8	14.5	2.3	0.1	0.3	0.0	39.0
1986-87	0.0	0.0	0.0	0.0	11.5	4.9	46.8	3.0	6.4	21.0	0.0	0.0	93.6
1987-88	0.0	0.0	0.0	0.0	10.2	12.9	25.2	15.8	6.4	0.6	0.0	0.0	71.1
1988-89	0.0	0.0	0.0	0.4	T	5.0	2.8	7.7	8.5	3.7	0.0	0.0	28.1
1989-90	0.0	0.0	0.0	0.0	7.9	10.2	11.3	15.2	6.4	2.1	0.0	T	53.1
1990-91	0.0	0.0	0.0	0.0	0.7	5.0	11.3	9.1	9.3	0.2	0.0	0.0	35.6
1991-92	0.0	0.0	0.0	0.0	5.8	14.5	2.7	8.4	11.8	2.7	0.0	0.1	46.0
1992-93	0.0	0.0	T	0.0	1.9	37.0	14.6	19.7	44.1	2.8	0.0	T	120.1
1993-94	0.0	0.0	0.0	T	0.2	12.9	34.1	25.9	27.1	0.0	T	0.0	100.2
1994-95	0.0	T	T	0.0	2.9	3.2	4.5	14.3	T	T	0.0	0.0	24.9
1995-96	0.0	0.0	0.0	0.0									
1996-97													
1997-98													
1998-99													
1999-00													
2000-01													
2001-													
POR= 40 YRS	0.0	T	T	0.5	3.6	13.2	15.9	15.7	13.7	3.5	0.3	0.0	66.4

2001
DETROIT, METROPOLITAN AIRPORT
MICHIGAN (DTW)

Detroit and the immediate suburbs, including nearby urban areas in Canada, occupy an area approximately 25 miles in radius. The waterway, consisting of the Detroit and St. Clair Rivers, Lake St. Clair, and the west end of Lake Erie, lies at an elevation of 568 to 580 feet above sea level. Nearly flat land slopes up gently from the waters edge northwestward for about 10 miles and then gives way to increasingly rolling terrain. The Irish Hills, parallel to and about 40 miles northwest of the waterway, have tops 1,000 to 1,250 feet above sea level. On the Canadian side of the waterway the land is relatively level.

Northwest winds in winter bring snow flurry accumulations to all of Michigan except in the Detroit Metropolitan area while summer showers moving from the northwest weaken and sometimes dissipate as they approach Detroit. On the other hand, much of the heaviest precipitation in winter comes from southeast winds, especially to the northwest suburbs of the city.

The climate of Detroit is influenced by its location with respect to major storm tracks and the influence of the Great Lakes. The normal wintertime storm track is south of the city, which brings on the average, about 3 inch snowfalls. Winter storms can bring combinations of rain, snow, freezing rain, and sleet with heavy snowfall accumulations possible at times. In summer, most storms pass to the north allowing for intervals of warm, humid, sunny skies with occasional thunderstorms followed by days of mild, dry, and fair weather. Temperatures of 90 degrees or higher are reached during each summer.

The most pronounced lake effect occurs in the winter when arctic air moving across the lakes is warmed and moistened. This produces an excess of cloudiness but a moderation of cold wave temperatures.

Local climatic variations are due largely to the immediate effect of Lake St. Clair and the urban heat island. On warm days in late spring or early summer, lake breezes often lower temperatures by 10 to 15 degrees in the eastern part of the city and the northeastern suburbs. The urban heat island effect shows up mainly at night where minimum temperatures at the Metropolitan Airport average 4 degrees lower than downtown Detroit. On humid summer nights or on very cold winter nights, this difference can exceed 10 degrees.

The growing season averages 180 days and has ranged from 145 days to 205 days. On average, the last freezing temperature occurs in late April while the average first freezing temperature occurs in late October. A freeze has occurred as late as mid-May and as early as late September.

Air pollution comes primarily from heavy industry spread along both shores of the waterway from Port Huron to Toledo. However, wind dispersion is usually sufficient to keep it from becoming a major hazard.

NORMALS, MEANS, AND EXTREMES
DETROIT, MI (DTW)

LATITUDE: 42 12' 55" N LONGITUDE: 83 20' 55" W ELEVATION (FT): GRND: 628 BARO: 631 TIME ZONE: EASTERN (UTC + 5) WBAN: 94847

ELEMENT	POR	JAN	FEB	MAR	APR	MAY	JUN	JUL	AUG	SEP	OCT	NOV	DEC	YEAR
TEMPERATURE °F														
NORMAL DAILY MAXIMUM	30	30.3	33.3	44.4	57.7	69.6	78.9	83.3	81.3	73.9	61.5	48.1	35.2	58.1
MEAN DAILY MAXIMUM	43	30.8	34.0	44.6	58.0	69.8	79.1	83.2	81.3	73.9	61.8	48.1	35.5	58.3
HIGHEST DAILY MAXIMUM	43	62	70	81	89	93	104	102	100	98	91	77	69	104
YEAR OF OCCURRENCE		1995	1999	1998	1977	1988	1988	1988	1988	1976	1963	1968	1998	JUN 1988
MEAN OF EXTREME MAXS.	43	49.6	52.6	69.2	79.0	86.1	92.0	93.5	91.9	88.8	79.7	67.2	55.4	75.4
NORMAL DAILY MINIMUM	30	15.6	17.6	27.0	36.8	47.1	56.3	61.3	59.6	52.5	40.9	32.2	21.4	39.0
MEAN DAILY MINIMUM	43	16.5	18.7	27.2	37.4	47.9	57.3	61.8	60.5	52.9	41.6	32.6	22.4	39.7
LOWEST DAILY MINIMUM	43	-21	-15	-4	10	25	36	41	38	29	17	9	-10	-21
YEAR OF OCCURRENCE		1984	1985	1978	1982	1966	1972	1965	1982	1974	1974	1969	1983	JAN 1984
MEAN OF EXTREME MINS.	43	-3.8	-.1	9.7	23.1	34.2	44.1	50.1	48.9	37.6	26.8	17.8	2.6	24.2
NORMAL DRY BULB	30	22.9	25.3	35.7	47.3	58.4	67.6	72.3	70.5	63.2	51.2	40.2	28.3	48.6
MEAN DRY BULB	43	23.6	26.4	35.8	47.7	58.9	68.1	72.4	71.0	63.5	51.7	40.2	29.1	49.0
MEAN WET BULB	18	23.8	26.1	32.6	42.6	53.2	62.0	66.0	65.2	57.8	47.3	37.4	28.2	45.2
MEAN DEW POINT	18	18.9	20.4	26.0	35.4	46.7	56.6	61.4	61.2	53.5	42.3	32.4	23.5	39.9
NORMAL NO. DAYS WITH:														
MAXIMUM 90	30	0.0	0.0	0.0	0.0	0.4	2.7	5.3	2.7	0.7	0.1	0.0	0.0	11.9
MAXIMUM 32	30	17.1	13.2	4.1	0.2	0.0	0.0	0.0	0.0	0.0	0.0	1.3	11.7	47.6
MINIMUM 32	30	29.2	25.3	22.6	10.0	0.9	0.0	0.0	0.0	0.1	5.0	16.4	26.5	136.0
MINIMUM 0	30	3.8	2.2	0.1	0.0	0.0	0.0	0.0	0.0	0.0	0.0	0.0	1.4	7.5
H/C														
NORMAL HEATING DEG. DAYS	30	1305	1109	908	531	243	38	0	16	102	435	744	1138	6569
NORMAL COOLING DEG. DAYS	30	0	0	0	0	38	116	231	186	48	7	0	0	626
RH														
NORMAL (PERCENT)	30	75	73	70	66	65	67	68	72	73	72	75	77	71
HOUR 01 LST	30	78	77	76	74	76	79	81	83	84	80	79	80	79
HOUR 07 LST	30	80	79	79	78	78	79	82	86	87	84	82	81	81
HOUR 13 LST	30	69	65	61	54	53	54	54	56	58	57	66	71	60
HOUR 19 LST	30	73	70	66	59	56	58	58	63	68	68	73	76	66
S PERCENT POSSIBLE SUNSHINE	31	40	46	52	53	60	65	68	67	61	51	35	31	52
W/O MEAN NO. DAYS WITH:														
HEAVY FOG(VISBY 1/4 MI)	44	2.5	2.4	2.3	0.9	0.8	0.7	0.6	1.3	1.6	2.0	1.7	3.1	19.9
THUNDERSTORMS	44	0.2	0.4	1.6	3.2	3.8	6.0	5.9	5.2	4.1	1.3	0.6	0.4	32.7
CLOUDINESS MEAN:														
SUNRISE-SUNSET (OKTAS)														
MIDNIGHT-MIDNIGHT (OKTAS)														
MEAN NO. DAYS WITH:														
CLEAR														
PARTLY CLOUDY														
CLOUDY														
PR MEAN STATION PRESSURE(IN)	29	29.30	29.31	29.30	29.30	29.30	29.29	29.30	29.30	29.30	29.30	29.32	29.31	29.30
MEAN SEA-LEVEL PRES. (IN)	17	30.07	30.08	30.05	29.96	29.98	30.01	29.99	30.04	30.04	30.08	30.06	30.09	30.04
WINDS MEAN SPEED (MPH)	39	11.8	11.4	11.7	11.4	10.1	9.2	8.6	8.3	8.8	9.8	11.3	11.2	10.3
PREVAIL.DIR (TENS OF DEGS)	34	24	23	29	23	23	23	22	23	23	23	24	23	23
MAXIMUM 2-MINUTE:														
SPEED (MPH)	6	44	51	46	47	43	32	53	35	35	47	45	49	53
DIR. (TENS OF DEGS)		22	22	21	22	22	27	28	29	27	24	24	29	28
YEAR OF OCCURRENCE		1996	1997	1996	2001	1999	1998	1998	2001	2001	1996	1998	1998	JUL 1998
MAXIMUM 5-SECOND:														
SPEED (MPH)	6	53	60	53	57	55	41	67	41	45	56	58	60	67
DIR. (TENS OF DEGS)		24	24	24	24	25	27	28	29	28	24	25	31	28
YEAR OF OCCURRENCE		1996	2001	1996	1997	2000	1998	1998	2001	1997	1996	1998	1998	JUL 1998
PRECIPITATION NORMAL (IN)	30	1.76	1.74	2.55	2.95	2.92	3.61	3.18	3.43	2.89	2.10	2.67	2.82	32.62
MAXIMUM MONTHLY (IN)	43	3.92	5.02	4.48	5.40	6.20	7.04	6.02	7.83	7.52	6.76	5.68	6.00	7.83
YEAR OF OCCURRENCE		1993	1990	1973	1961	1991	1987	1969	1975	1986	2001	1982	1965	AUG 1975
MINIMUM MONTHLY (IN)	43	0.27	0.15	0.82	0.92	0.87	0.97	0.59	0.43	0.43	0.35	0.79	0.46	0.15
YEAR OF OCCURRENCE		1961	1969	1981	1971	1988	1988	1974	1996	1960	1964	1960	1960	FEB 1969
MAXIMUM IN 24 HOURS (IN)	43	1.72	2.41	1.82	3.58	2.87	2.84	4.34	3.21	4.08	2.57	2.20	3.71	4.34
YEAR OF OCCURRENCE		1967	1998	1997	2000	1968	1983	1998	1964	2000	1985	1982	1965	JUL 1998
NORMAL NO. DAYS WITH:														
PRECIPITATION 0.01	30	13.0	11.0	13.2	12.3	11.0	10.5	9.2	9.3	9.8	9.6	11.6	14.1	134.6
PRECIPITATION 1.00	30	0.1	0.2	0.2	0.4	0.6	0.9	0.9	1.0	0.5	0.4	0.3	0.4	5.9
SNOWFALL NORMAL (IN)	30	10.0	9.2	6.6	1.8	T	0.0	0.0	0.0	0.0	0.2	3.0	10.9	41.7
MAXIMUM MONTHLY (IN)	42	29.6	20.8	16.1	9.0	T	0.0	0.0	0.0	T	2.9	11.8	34.9	34.9
YEAR OF OCCURRENCE		1978	1986	1965	1982	1994				1994	1980	1966	1974	DEC 1974
MAXIMUM IN 24 HOURS (IN)	42	11.1	10.3	9.2	7.4	T	0.0	0.0	0.0	T	2.9	5.6	19.2	19.2
YEAR OF OCCURRENCE		1992	1965	1973	1982	1994				1994	1980	1977	1974	DEC 1974
MAXIMUM SNOW DEPTH (IN)	41	24	18	9	6	0	0	0	0	0	1	6	19	24
YEAR OF OCCURRENCE		1999	1982	1982	1982						1980	1966	1974	JAN 1999
NORMAL NO. DAYS WITH:														
SNOWFALL 1.0	30	3.1	3.1	2.1	0.6	0.0	0.0	0.0	0.0	0.0	0.1	1.1	3.3	13.4

PRECIPITATION (inches) 2001 DETROIT, METROPOLITAN AIRPORT MI (DTW)

YEAR	JAN	FEB	MAR	APR	MAY	JUN	JUL	AUG	SEP	OCT	NOV	DEC	ANNUAL
1972	1.28	1.00	2.55	3.63	2.68	3.30	2.21	3.07	3.40	2.24	3.19	3.11	31.66
1973	1.65	1.08	4.48	1.42	3.72	4.86	4.66	1.67	1.82	2.01	3.21	3.51	34.09
1974	3.26	2.37	4.20	2.75	3.49	2.38	0.59	2.95	2.22	0.81	2.86	4.00	31.88
1975	2.90	2.65	1.66	2.50	2.82	2.39	1.98	7.83	3.18	1.29	2.39	3.00	34.59
1976	1.91	2.87	1.47	4.24	3.15	3.26	3.26	1.47	1.68	3.66	2.01	0.79	29.09
1977	0.98	1.64	3.57	4.17	2.40	3.16	3.28	2.23	4.23	1.37	2.88	2.97	32.88
1978	3.16	0.45	2.05	2.49	3.58	2.69	1.97	1.73	1.82	2.49	2.41	2.81	27.65
1979	1.52	0.57	2.44	4.97	2.82	4.04	4.96	2.99	0.94	1.24	4.19	2.36	33.04
1980	0.69	1.00	3.88	4.23	3.22	6.42	4.33	6.09	2.94	1.26	0.88	2.30	37.24
1981	0.57	3.13	0.82	3.44	2.60	3.33	4.29	2.32	5.47	3.92	1.26	2.38	33.53
1982	3.43	1.10	3.14	1.60	2.83	4.11	4.78	0.72	2.55	1.01	5.68	3.29	34.24
1983	0.84	0.89	1.87	4.20	5.47	4.88	4.53	1.57	2.49	2.85	4.28	3.78	37.65
1984	0.78	1.31	3.12	2.48	3.62	1.04	0.95	3.00	2.30	2.28	2.49	2.90	26.27
1985	2.63	3.83	4.42	2.11	3.11	1.62	3.96	4.88	2.59	3.91	5.51	1.51	40.08
1986	1.30	3.46	2.29	2.73	1.36	5.75	2.47	3.52	7.52	3.05	1.88	2.28	37.61
1987	2.35	0.53	2.19	2.14	2.50	7.04	2.20	6.87	2.69	2.00	3.17	4.60	38.28
1988	1.30	2.02	1.16	1.50	0.87	0.97	2.43	3.13	3.65	3.57	4.29	1.97	26.86
1989	1.28	0.77	2.16	2.22	4.16	3.79	4.21	2.14	3.03	1.73	2.53	1.24	29.26
1990	1.80	5.02	1.91	2.72	3.74	4.92	1.47	3.85	6.06	4.14	2.64	4.37	42.64
1991	1.44	0.94	1.41	2.66	6.20	1.89	1.23	4.31	0.90	4.14	2.61	1.91	29.64
1992	1.78	1.54	3.34	4.34	1.33	2.35	5.91	2.50	5.55	2.01	4.33	2.35	37.33
1993	3.92	1.27	2.12	3.32	1.24	6.05	2.17	1.60	4.26	2.21	1.69	0.78	30.63
1994	2.79	1.38	2.29	4.04	1.18	3.97	3.20	3.30	2.38	1.35	2.74	2.39	31.01
1995	2.47	0.89	1.73	3.44	3.55	1.55	3.40	3.71	0.62	3.53	3.08	0.85	28.82
1996	1.85	1.76	1.56	3.39	2.82	2.37	2.64	0.43	4.42	1.59	1.99	2.57	27.39
1997	1.57	3.90	3.22	1.56	5.23	3.17	2.68	3.22	3.41	1.91	0.94	1.61	32.42
1998	2.60	3.56	3.62	3.86	2.46	2.69	5.72	4.19	1.50	1.41	1.36	1.16	34.13
1999	3.00	1.98	1.12	5.13	2.20	5.46	3.62	1.31	3.11	1.56	1.49	2.22	32.20
2000	1.29	0.84	1.55	4.35	5.11	4.90	5.40	4.63	6.71	3.05	1.69	2.63	42.15
2001	0.69	2.88	0.93	3.20	3.70	3.40	1.16	2.87	4.28	6.76	2.35	2.23	34.45
POR= 43 YRS	1.92	1.81	2.40	3.13	3.00	3.61	3.18	3.23	2.97	2.28	2.48	2.53	32.54

AVERAGE TEMPERATURE (F) 2001 DETROIT, METROPOLITAN AIRPORT MI (DTW)

YEAR	JAN	FEB	MAR	APR	MAY	JUN	JUL	AUG	SEP	OCT	NOV	DEC	ANNUAL
1972	23.8	24.6	32.6	44.6	60.3	64.2	71.2	69.1	63.0	47.3	37.4	29.4	47.3
1973	28.8	25.3	43.3	48.8	55.5	69.9	72.6	72.9	64.9	56.2	41.4	28.7	50.7
1974	26.5	23.6	35.7	49.2	55.2	65.9	72.5	72.3	59.7	48.8	40.6	28.6	48.2
1975	28.3	27.5	32.5	40.9	62.8	69.0	72.2	72.3	59.1	52.9	46.8	29.1	49.5
1976	19.2	33.3	40.4	50.0	56.4	70.6	72.7	70.2	62.1	47.4	33.5	21.5	48.1
1977	12.8	25.2	41.5	52.4	64.4	65.5	75.8	70.6	65.5	47.9	40.5	25.5	48.9
1978	19.6	16.3	30.0	45.5	59.3	66.8	70.6	71.9	67.5	50.2	40.5	28.9	47.3
1979	18.6	16.5	37.7	44.6	56.5	66.6	70.4	67.9	67.6	45.7	39.5	31.9	46.9
1980	24.5	22.2	31.3	45.9	59.8	63.7	72.7	72.7	63.0	46.3	37.4	26.0	47.2
1981	19.0	28.8	36.5	49.8	55.9	68.0	72.4	70.0	60.9	47.4	41.1	27.8	48.2
1982	17.1	20.7	33.0	43.2	64.3	64.2	72.4	67.7	61.8	52.6	41.6	37.3	48.0
1983	28.7	31.6	38.4	44.2	54.4	68.2	74.5	73.6	64.0	51.6	41.2	20.8	49.3
1984	18.0	33.3	28.9	47.8	54.5	70.8	70.8	72.7	61.2	54.9	38.6	34.0	48.8
1985	20.4	23.5	38.4	51.0	60.1	62.8	71.3	69.2	64.3	53.0	42.4	22.2	48.2
1986	23.9	24.6	37.6	50.6	61.3	67.3	75.0	68.9	65.9	52.6	37.3	31.7	49.7
1987	26.1	29.6	39.8	50.8	63.3	71.3	76.1	71.6	64.6	46.6	43.5	33.6	51.4
1988	23.8	23.4	36.9	48.5	62.0	70.4	77.1	75.1	63.3	46.0	42.2	28.7	49.8
1989	32.8	24.1	35.2	45.1	57.5	67.5	73.0	69.9	61.9	52.1	38.2	18.0	47.9
1990	33.6	30.7	39.5	49.0	56.6	68.5	72.2	71.2	64.5	52.8	44.2	32.8	51.3
1991	25.0	31.2	40.3	52.0	66.5	72.4	74.9	73.4	63.1	54.8	38.5	32.1	52.0
1992	28.3	30.8	35.5	46.3	58.3	65.5	68.8	66.7	61.4	49.7	40.5	33.2	48.8
1993	29.4	24.2	34.7	47.8	60.2	67.5	75.5	74.5	61.0	51.9	41.2	30.8	49.9
1994	17.3	23.5	37.1	51.2	58.7	72.3	74.2	69.6	66.2	51.8	45.5	35.4	50.4
1995	28.4	24.9	39.2	45.7	59.7	71.6	74.8	77.2	62.8	55.1	35.5	25.6	50.0
1996	24.3	26.0	30.8	45.2	56.7	70.7	70.6	72.9	64.1	52.0	34.3	31.5	48.3
1997	23.0	30.6	37.2	45.8	52.0	69.5	72.2	68.1	62.7	51.8	37.1	32.2	48.5
1998	32.8	36.7	39.5	50.4	65.6	69.1	73.4	73.2	68.0	53.8	43.8	35.3	53.5
1999	23.1	32.8	34.8	50.7	62.4	70.8	76.8	70.2	65.5	51.6	45.2	32.0	51.3
2000	24.6	31.9	44.0	48.0	61.8	69.4	70.3	70.8	62.5	55.1	40.2	19.3	49.8
2001	26.2	29.7	35.1	51.2	61.2	69.6	73.6	74.1	62.3	52.5	47.6	35.9	51.6
POR= 43 YRS	23.6	26.5	35.8	47.8	59.0	67.9	72.5	71.0	63.5	51.6	40.2	29.2	49.1

REFERENCE NOTES:

PAGE 1:
THE TEMPERATURE GRAPH SHOWS NORMAL MAXIMUM AND NORMAL MINIMUM DAILY TEMPERATURES (SOLID CURVES) AND THE ACTUAL DAILY HIGH AND LOW TEMPERATURES (VERTICAL BARS).

PAGE 2 AND 3:
H/C INDICATES HEATING AND COOLING DEGREE DAYS.
RH INDICATES RELATIVE HUMIDITY
W/O INDICATES WEATHER AND OBSTRUCTIONS
S INDICATES SUNSHINE.
PR INDICATES PRESSURE.
CLOUDINESS ON PAGE 3 IS THE SUM OF THE CEILOMETER AND SATELLITE DATA NOT TO EXCEED EIGHT EIGHTHS (OKTAS).

GENERAL:
T INDICATES TRACE PRECIPITATION, AN AMOUNT GREATER THAN ZERO BUT LESS THAN THE LOWEST REPORTABLE VALUE.
+ INDICATES THE VALUE ALSO OCCURS ON EARLIER DATES.
BLANK ENTRIES DENOTE MISSING OR UNREPORTED DATA.
NORMALS ARE 30-YEAR AVERAGES (1961 - 1990).
ASOS INDICATES AUTOMATED SURFACE OBSERVING SYSTEM.
PM INDICATES THE LAST DAY OF THE PREVIOUS MONTH.
POR (PERIOD OF RECORD) BEGINS WITH THE JANUARY DATA MONTH AND IS THE NUMBER OF YEARS USED TO COMPUTE THE MEAN. INDIVIDUAL MONTHS WITHIN THE POR MAY BE MISSING.
WHEN THE POR FOR A NORMAL IS LESS THAN 30 YEARS, THE NORMAL IS PROVISIONAL AND IS BASED ON THE NUMBER OF YEARS INDICATED.
0.* OR * INDICATES THE VALUE OR MEAN-DAYS-WITH IS BETWEEN 0.00 AND 0.05.
CLOUDINESS FOR ASOS STATIONS DIFFERS FROM THE NON-ASOS OBSERVATION TAKEN BY A HUMAN OBSERVER. ASOS STATION CLOUDINESS IS BASED ON TIME-AVERAGED CEILOMETER DATA FOR CLOUDS AT OR BELOW 12,000 FEET AND ON SATELLITE DATA FOR CLOUDS ABOVE 12,000 FEET.
THE NUMBER OF DAYS WITH CLEAR, PARTLY CLOUDY, AND CLOUDY CONDITIONS FOR ASOS STATIONS IS THE SUM OF THE CEILOMETER AND SATELLITE DATA FOR THE SUNRISE TO SUNSET PERIOD.

GENERAL CONTINUED:
CLEAR INDICATES 0 - 2 OKTAS, PARTLY CLOUDY INDICATES 3 - 6 OKTAS, AND CLOUDY INDICATES 7 OR 8 OKTAS.
WHEN AT LEAST ONE OF THE ELEMENTS (CEILOMETER OR SATELLITE) IS MISSING, THE DAILY CLOUDINESS IS NOT COMPUTED.
WIND DIRECTION IS RECORDED IN TENS OF DEGREES (2 DIGITS) CLOCKWISE FROM TRUE NORTH. "00" INDICATES CALM. "36" INDICATES TRUE NORTH.
RESULTANT WIND IS THE VECTOR AVERAGE OF THE SPEED AND DIRECTION.
AVERAGE TEMPERATURE IS THE SUM OF THE MEAN DAILY MAXIMUM AND MINIMUM TEMPERATURE DIVIDED BY 2.
SNOWFALL DATA COMPRISE ALL FORMS OF FROZEN PRECIPITATION, INCLUDING HAIL.
A HEATING (COOLING) DEGREE DAY IS THE DIFFERENCE BETWEEN THE AVERAGE DAILY TEMPERATURE AND 65 F.
DRY BULB IS THE TEMPERATURE OF THE AMBIENT AIR.
DEW POINT IS THE TEMPERATURE TO WHICH THE AIR MUST BE COOLED TO ACHIEVE 100 PERCENT RELATIVE HUMIDITY.
WET BULB IS THE TEMPERATURE THE AIR WOULD HAVE IF THE MOISTURE CONTENT WAS INCREASED TO 100 PERCENT RELATIVE HUMIDITY.

ON JULY 1, 1996, THE NATIONAL WEATHER SERVICE BEGAN USING THE "METAR" OBSERVATION CODE THAT WAS ALREADY EMPLOYED BY MOST OTHER NATIONS OF THE WORLD. THE MOST NOTICEABLE DIFFERENCE IN THIS ANNUAL PUBLICATION WILL BE THE CHANGE IN UNITS FROM TENTHS TO EIGHTS (OKTAS) FOR REPORTING THE AMOUNT OF SKY COVER.

HEATING DEGREE DAYS (base 65 F) 2001 DETROIT, METROPOLITAN AIRPORT MI (DTW)

YEAR	JUL	AUG	SEP	OCT	NOV	DEC	JAN	FEB	MAR	APR	MAY	JUN	TOTAL
1972-73	24	28	113	539	822	1096	1115	1103	667	480	289	3	6279
1973-74	0	10	98	276	702	1119	1189	1152	901	476	308	54	6285
1974-75	0	2	189	495	726	1123	1129	1043	996	714	142	41	6600
1975-76	4	0	178	375	537	1107	1413	914	757	473	269	6	6033
1976-77	1	15	133	540	938	1341	1609	1106	721	395	122	85	7006
1977-78	1	17	85	524	729	1218	1400	1357	1077	580	235	65	7288
1978-79	17	0	73	452	728	1112	1432	1355	843	604	291	55	6962
1979-80	12	29	126	471	758	1019	1249	1233	1036	568	191	104	6796
1980-81	0	0	110	578	822	1201	1418	1008	878	452	293	19	6779
1981-82	3	9	167	534	710	1144	1477	1237	985	647	75	70	7058
1982-83	2	39	145	383	696	852	1119	928	816	618	323	59	5980
1983-84	6	0	125	418	708	1367	1450	912	1112	507	334	9	6948
1984-85	11	4	164	310	785	955	1377	1154	818	435	177	93	6283
1985-86	2	8	129	366	672	1317	1271	1125	842	435	166	48	6381
1986-87	1	33	76	380	824	1028	1198	984	776	423	158	11	5892
1987-88	4	30	69	566	639	969	1273	1201	864	486	138	46	6285
1988-89	2	3	90	590	679	1118	991	1138	916	591	254	33	6405
1989-90	0	11	151	400	797	1451	966	955	785	506	258	27	6307
1990-91	1	1	112	380	618	994	1234	939	761	394	125	5	5564
1991-92	0	0	151	319	788	1013	1129	985	906	555	224	59	6129
1992-93	15	30	153	469	725	976	1097	1138	931	506	161	51	6252
1993-94	0	2	155	404	710	1050	1472	1156	858	433	238	24	6502
1994-95	0	9	55	345	579	910	1129	1115	793	573	170	17	5695
1995-96	2	0	125	306	877	1215	1253	1122	1055	589	289	8	6841
1996-97	3	0	102	397	915	1033	1297	959	855	566	394	25	6546
1997-98	3	13	103	435	830	1008	991	787	791	431	69	80	5541
1998-99	0	2	44	350	629	914	1293	898	927	424	115	33	5629
1999-00	0	3	81	413	585	1016	1246	958	645	504	159	31	5641
2000-01	1	13	148	307	737	1412	1194	983	920	415	146	45	6321
2001-	9	0	131	382	514	898							

COOLING DEGREE DAYS (base 65 F) 2001 DETROIT, METROPOLITAN AIRPORT MI (DTW

YEAR	JAN	FEB	MAR	APR	MAY	JUN	JUL	AUG	SEP	OCT	NOV	DEC	ANNUAL
1972	0	0	0	0	36	73	222	160	59	0	0	0	550
1973	0	0	0	3	2	156	241	261	104	11	0	0	778
1974	0	0	0	8	10	91	237	237	36	1	0	0	620
1975	0	0	0	0	82	171	233	232	7	6	0	0	731
1976	0	0	0	30	10	182	246	182	53	3	0	0	706
1977	0	0	0	25	108	107	341	198	94	0	0	0	873
1978	0	0	0	0	63	122	200	221	154	0	0	0	760
1979	0	0	0	0	32	109	184	124	57	16	0	0	522
1980	0	0	0	3	38	69	246	248	79	3	0	0	686
1981	0	0	0	1	17	118	241	168	51	0	0	0	596
1982	0	0	0	0	58	55	237	129	57	5	0	0	541
1983	0	0	0	2	0	160	306	272	104	6	0	0	850
1984	0	0	0	0	15	189	197	252	55	2	0	0	710
1985	0	0	0	25	32	32	201	146	116	0	0	0	552
1986	0	0	0	10	55	120	319	160	110	3	0	0	777
1987	0	0	0	4	111	207	355	245	64	0	1	0	987
1988	0	0	0	0	52	214	385	322	46	8	0	0	1027
1989	0	0	0	0	29	114	256	171	64	5	0	0	639
1990	0	0	0	1	32	8	139	234	200	101	11	0	726
1991	0	0	0	0	10	179	233	315	268	104	9	0	1118
1992	0	0	0	0	27	80	143	91	51	0	0	0	392
1993	0	0	0	0	21	135	334	302	41	5	0	0	838
1994	0	0	0	24	49	248	290	160	97	2	0	0	870
1995	0	0	0	0	11	222	312	383	67	7	0	0	1002
1996	0	0	0	2	37	185	184	251	82	0	0	0	741
1997	0	0	0	0	0	165	231	114	40	29	0	0	579
1998	0	0	0	7	95	209	268	261	136	9	0	0	985
1999	0	0	0	0	43	213	372	171	104	2	0	0	905
2000	0	0	0	3	0	66	172	170	199	78	4	0	692
2001	0	0	0	6	36	190	279	290	58	4	0	0	863

SNOWFALL (inches) 2001 DETROIT, METROPOLITAN AIRPORT MI (DTW)

YEAR	JUL	AUG	SEP	OCT	NOV	DEC	JAN	FEB	MAR	APR	MAY	JUN	TOTAL
1972-73	0.0	0.0	0.0	T	7.1	12.5	2.4	12.8	10.1	0.1	T	0.0	45.0
1973-74	0.0	0.0	0.0	0.0	0.1	16.4	14.1	11.2	5.7	1.7	T	0.0	49.2
1974-75	0.0	0.0	0.0	T	7.7	34.9	4.9	7.5	4.5	3.6	0.0	0.0	63.1
1975-76	0.0	0.0	0.0	0.0	6.5	19.8	15.1	4.9	7.5	2.1	T	0.0	55.9
1976-77	0.0	0.0	0.0	T	1.4	9.8	14.7	5.0	12.3	0.7	0.0	0.0	43.9
1977-78	0.0	0.0	0.0	0.0	7.4	16.6	29.6	5.3	2.5	0.3	0.0	0.0	61.7
1978-79	0.0	0.0	0.0	0.0	6.1	6.6	13.3	3.9	2.7	3.0	0.0	0.0	35.6
1979-80	0.0	0.0	0.0	T	3.2	2.3	2.8	5.5	11.7	1.4	0.0	0.0	26.9
1980-81	0.0	0.0	0.0	2.9	3.4	10.5	7.6	13.4	0.6	0.0	0.0	0.0	38.4
1981-82	0.0	0.0	0.0	0.1	0.7	17.3	20.0	13.3	13.6	9.0	0.0	0.0	74.0
1982-83	0.0	0.0	0.0	T	1.8	1.4	1.5	4.3	7.6	3.4	0.0	0.0	20.8
1983-84	0.0	0.0	0.0	0.0	3.5	19.9	9.9	8.7	9.7	0.1	0.0	0.0	51.8
1984-85	0.0	0.0	0.0	0.0	4.1	6.2	20.9	16.9	6.1	0.9	0.0	0.0	55.1
1985-86	0.0	0.0	0.0	0.0	2.0	14.1	8.6	20.8	7.4	1.3	0.0	0.0	54.2
1986-87	0.0	0.0	0.0	T	3.3	6.0	24.0	2.0	13.3	1.1	0.0	0.0	49.7
1987-88	0.0	0.0	0.0	T	0.7	15.3	7.0	19.2	2.7	0.2	0.0	0.0	45.1
1988-89	0.0	0.0	0.0	T	1.0	6.3	5.3	9.6	2.4	0.5	T	0.0	25.1
1989-90	0.0	0.0	0.0	2.7	2.4	11.8	4.0	11.1	7.8	2.0	0.0	0.0	41.8
1990-91	0.0	0.0	T	0.0	T	13.2	8.8	9.2	0.2	T	T	0.0	31.4
1991-92	0.0	0.0	0.0	T	2.2	8.6	18.4	2.4	11.7	0.2	0.0	0.0	43.5
1992-93	0.0	0.0	0.0	0.4	0.9	5.0	11.0	15.2	15.7	4.0	0.0	0.0	52.2
1993-94	0.0	0.0	0.0	0.4	0.6	1.9	17.9	17.1	3.7	4.2	T	0.0	45.8
1994-95	0.0	0.0	T	0.0	T	9.6	13.1	5.7	3.5	1.6	0.0	0.0	33.5
1995-96	0.0	0.0	0.0	0.0	1.3	4.5	6.3	11.8					
1996-97													
1997-98					4.6	6.0	8.3	T	4.4	0.0	T	0.0	49.5
1998-99	0.0	0.0	T	0.0	0.0	1.2	27.3	7.8	13.2	0.0	0.0	0.0	23.4
1999-00	0.0	0.0	0.0	T	T	4.0	9.6	8.1	1.1	0.6	T	0.0	
2000-01	0.0	0.0	0.0	T	1.3	25.1	3.4	2.9	5.4	0.9	0.0	0.0	39.0
2001-	0.0	0.0	0.0	T	0.0	4.9							
POR= 41 YRS	0.0	0.0	T	0.2	2.6	9.9	10.3	8.7	6.7	1.7	T	0.0	40.1

2001
GRAND RAPIDS,
MICHIGAN (GRR)

Grand Rapids, Michigan, is located in the west-central part of Kent County, in the picturesque Grand River valley about 30 air miles east of Lake Michigan. The Grand River, the longest stream in Michigan, flows through the city and bisects it into east and west sections. High hills rise on either side of the valley. Elevations range from 602 feet on the valley floor to 1,020 feet in the extreme southern part of Kent County, southwest of the airport.

Grand Rapids is under the natural climatic influence of Lake Michigan. In spring the cooling effect of Lake Michigan helps retard the growth of vegetation until the danger of frost has passed. The warming effect in the fall retards frost until most of the crops have matured. Fall is a colorful time of year in western Michigan, compensating for the late spring. During the winter, excessive cloudiness and numerous snow flurries occur with strong westerly winds. The tempering effect of Lake Michigan on cold waves coming in from the west and northwest is quite evident.

The tempering effect of the lake promotes the growth of a great variety of fruit trees and berries, especially apples, peaches, cherries, and blueberries. The intense cold of winter is modified, thus reducing winter kill of fruit trees. Summer days are pleasantly warm and most summer nights are quite comfortable, although there are about three weeks of hot, humid weather during most summers. Prolonged severe cold waves with below-zero temperatures are infrequent. The temperature usually rises to above zero during the daytime hours regardless of early morning readings.

July is the sunniest month and December is the month with the least sunshine. November through January is usually a period of excessive cloudiness and minimal sunshine.

Precipitation is usually ample for the growth and development of all vegetation. About one-half of the annual precipitation falls during the growing season, May through September. Droughts occur occasionally, but are seldom of protracted length. The snowfall season extends from mid-November to mid-March. Some winters have had continuous snow cover throughout this period, although there is usually a mid-winter thaw. The Grand River flows through the city and reaches critical heights a couple of times each year, generally once in January-February and again in March-April. Overflow is generally limited to the lowlands of the flood plain.

November is one of the windiest months and although violent windstorms are infrequent, gusts have on occasion exceeded 65 mph. Summer thunderstorms occasionally produce gusty winds over 60 mph.

NORMALS, MEANS, AND EXTREMES
GRAND RAPIDS, MI　(GRR)

LATITUDE:　　　　LONGITUDE:　　　ELEVATION (FT):　　　　TIME ZONE:　　　　WBAN: 94860
42　52' 56" N　　85　31' 23" W　　GRND:　785　　BARO:　788　　EASTERN　(UTC + 5)

	ELEMENT	POR	JAN	FEB	MAR	APR	MAY	JUN	JUL	AUG	SEP	OCT	NOV	DEC	YEAR
TEMPERATURE °F	NORMAL DAILY MAXIMUM	30	29.0	31.6	42.8	56.6	69.3	78.7	82.8	80.5	72.0	59.8	45.8	33.5	56.9
	MEAN DAILY MAXIMUM	54	29.7	32.5	42.3	56.8	69.3	78.6	82.5	80.4	72.5	60.9	46.1	34.1	57.1
	HIGHEST DAILY MAXIMUM	38	62	69	78	88	92	98	100	100	93	87	77	69	100
	YEAR OF OCCURRENCE		1967	1999	2000	1970	1978	1995	1988	1964	1973	1975	1975	2001	JUL 1988
	MEAN OF EXTREME MAXS.	54	47.7	50.0	67.0	78.4	85.4	91.4	92.5	89.7	87.5	78.8	66.5	53.2	74.0
	NORMAL DAILY MINIMUM	30	14.7	15.8	25.4	35.4	45.6	55.3	60.4	58.4	49.9	39.1	30.2	20.7	37.6
	MEAN DAILY MINIMUM	54	15.8	17.1	25.1	35.9	46.2	55.9	60.4	59.0	50.9	40.8	31.3	21.3	38.3
	LOWEST DAILY MINIMUM	38	-22	-19	-8	3	22	33	41	39	27	18	5	-18	-22
	YEAR OF OCCURRENCE		1994	1973	1978	1982	1966	1972	1983	1976	1991	1988	1977	1983	JAN 1994
	MEAN OF EXTREME MINS.	54	-6.1	-3.8	5.6	20.7	31.1	41.1	47.6	45.2	34.8	26.2	15.9	0.9	21.6
	NORMAL DRY BULB	30	21.8	23.7	34.1	46.0	57.5	67.0	71.6	69.5	61.0	49.5	38.0	27.1	47.2
	MEAN DRY BULB	54	22.8	24.8	33.7	46.3	57.7	67.2	71.4	69.7	61.8	50.8	38.8	27.7	47.7
	MEAN WET BULB	17	22.7	24.5	31.3	41.6	52.1	60.9	64.9	64.1	56.6	43.5	35.9	26.9	43.8
	MEAN DEW POINT	17	18.9	20.0	25.4	34.8	45.9	55.8	60.9	60.7	52.9	39.5	31.7	23.3	39.2
	NORMAL NO. DAYS WITH:														
	MAXIMUM　　90	30	0.0	0.0	0.0	0.0	0.5	2.3	5.1	2.6	0.4	0.0	0.0	0.0	10.9
	MAXIMUM　　32	30	19.3	15.3	5.8	0.4	0.0	0.0	0.0	0.0	0.0	0.0	2.3	14.3	57.4
	MINIMUM　　32	30	29.4	26.1	23.9	11.9	2.2	0.0	0.0	0.0	0.3	6.1	17.3	27.9	145.1
	MINIMUM　　0	30	4.1	3.0	0.5	0.0	0.0	0.0	0.0	0.0	0.0	0.0	0.0	1.6	9.2
H/C	NORMAL HEATING DEG. DAYS	30	1339	1156	958	570	273	50	0	18	139	485	810	1175	6973
	NORMAL COOLING DEG. DAYS	30	0	0	0	0	40	110	208	157	19	0	0	0	534
RH	NORMAL (PERCENT)	30	77	74	71	67	65	68	70	73	76	75	77	80	73
	HOUR 01 LST	30	80	78	76	74	76	80	82	86	86	82	81	82	80
	HOUR 07 LST	30	81	80	80	79	79	81	84	88	89	85	83	83	83
	HOUR 13 LST	30	72	67	63	56	53	55	56	58	61	62	69	75	62
	HOUR 19 LST	30	76	72	66	59	55	57	57	62	71	72	76	79	67
S	PERCENT POSSIBLE SUNSHINE	36	28	39	46	51	56	61	64	61	54	44	27	23	46
W/O	MEAN NO. DAYS WITH:														
	HEAVY FOG(VISBY 1/4 MI)	39	2.3	2.2	2.5	1.7	1.5	1.5	1.4	1.9	2.0	1.9	2.1	3.3	24.3
	THUNDERSTORMS	39	0.2	0.2	1.6	3.4	3.7	5.5	5.8	5.0	4.1	1.7	1.2	0.4	32.8
CLOUDINESS	MEAN:														
	SUNRISE-SUNSET (OKTAS)														
	MIDNIGHT-MIDNIGHT (OKTAS)														
	MEAN NO. DAYS WITH:														
	CLEAR	1			1.0										
	PARTLY CLOUDY	1			1.0		2.0								
	CLOUDY	1	1.0		1.0										
PR	MEAN STATION PRESSURE(IN)	28	29.20	29.20	29.11	29.10	29.10	29.10	29.10	29.20	29.20	29.21	29.20	29.19	29.16
	MEAN SEA-LEVEL PRES. (IN)	15	30.06	30.06	30.05	29.95	29.97	30.01	29.98	30.03	30.04	30.07	30.04	30.07	30.03
WINDS	MEAN SPEED (MPH)	26	11.4	10.6	11.2	11.0	9.7	8.8	8.4	7.8	8.1	9.3	10.5	10.5	9.8
	PREVAIL.DIR(TENS OF DEGS)	27	25	24	25	25	25	24	24	24	18	18	24	20	24
	MAXIMUM 2-MINUTE:														
	SPEED (MPH)	6	45	55	43	52	47	38	51	32	39	47	49	38	55
	DIR. (TENS OF DEGS)		24	24	24	19	26	27	26	27	25	24	23	24	24
	YEAR OF OCCURRENCE		1996	1999	1999	1999	1998	2000	1999	2000	1997	1996	1998	2001	FEB 1999
	MAXIMUM 5-SECOND:														
	SPEED (MPH)	6	55	64	54	61	60	52	58	38	48	60	66	47	66
	DIR. (TENS OF DEGS)		24	25	23	25	27	22	29	27	25	25	23	24	23
	YEAR OF OCCURRENCE		1996	1999	1996	1997	1998	1998	1999	2000	1997	1996	1998	2001	NOV 1998
PRECIPITATION	NORMAL (IN)	30	1.83	1.42	2.63	3.37	3.13	3.68	3.19	3.57	4.24	2.81	3.32	2.85	36.04
	MAXIMUM MONTHLY (IN)	38	4.36	4.80	5.12	6.69	10.01	8.21	8.83	8.46	11.85	7.38	7.81	6.63	11.85
	YEAR OF OCCURRENCE		1975	1997	1974	1999	2001	1967	1992	1987	1986	2001	1966	1971	SEP 1986
	MINIMUM MONTHLY (IN)	38	0.47	0.33	0.54	1.70	0.94	0.25	0.81	0.14	T	0.60	0.95	T	T
	YEAR OF OCCURRENCE		1981	1969	2001	1997	1987	1988	1976	1969	1979	1964	1986	1952	SEP 1979
	MAXIMUM IN 24 HOURS (IN)	38	2.16	3.05	1.78	2.66	5.48	4.03	3.68	3.68	4.55	3.19	3.00	2.79	5.48
	YEAR OF OCCURRENCE		1993	1997	1985	1999	1981	1994	1994	1987	1986	1993	1990	1982	MAY 1981
	NORMAL NO. DAYS WITH:														
	PRECIPITATION　0.01	30	16.2	11.9	12.8	13.0	10.4	10.0	9.2	9.4	10.2	11.1	13.1	16.5	143.8
	PRECIPITATION　1.00	30	0.1	0.1	0.3	0.6	0.8	0.9	0.8	0.9	1.1	0.4	0.6	0.6	7.2
SNOWFALL	NORMAL (IN)	30	21.3	11.9	9.9	3.1	0.*	0.0	0.0	0.0	T	0.7	7.2	18.0	72.1
	MAXIMUM MONTHLY (IN)	38	46.8	29.6	36.0	12.4	0.2	0.0	T	T	T	8.4	25.3	59.2	59.2
	YEAR OF OCCURRENCE		1999	1994	1965	1982	1990		1990	1993	1967	1967	1991	2000	DEC 2000
	MAXIMUM IN 24 HOURS (IN)	38	16.1	9.1	13.2	9.8	0.2	0.0	T	T	T	8.4	10.4	15.1	16.1
	YEAR OF OCCURRENCE		1978	1985	1970	1975	1990		1990	1993	1967	1967	1991	1970	JAN 1978
	MAXIMUM SNOW DEPTH (IN)	53	27	21	15	10	0	0	0	0	0	5	11	22	27
	YEAR OF OCCURRENCE		1978	1985	1965	1975						1967	1951	1951	JAN 1978
	NORMAL NO. DAYS WITH:														
	SNOWFALL　1.0	30	6.9	4.1	2.9	0.9	0.0	0.0	0.0	0.0	0.0	0.1	2.5	6.0	23.4

PRECIPITATION (inches) 2001 GRAND RAPIDS, MI (GRR)

YEAR	JAN	FEB	MAR	APR	MAY	JUN	JUL	AUG	SEP	OCT	NOV	DEC	ANNUAL
1972	1.26	0.90	2.11	3.85	1.99	4.64	3.72	5.01	3.96	2.92	2.06	4.96	37.38
1973	1.66	1.15	3.34	3.47	4.31	3.58	2.06	1.45	2.47	4.12	3.54	3.28	34.43
1974	3.23	2.09	5.12	2.93	4.01	4.43	0.97	4.61	2.05	2.44	3.11	1.83	36.82
1975	4.36	1.92	2.28	4.07	2.08	5.97	2.31	7.38	2.00	1.04	3.82	4.02	41.25
1976	1.67	2.13	4.99	4.75	6.63	2.79	0.81	1.03	1.21	2.00	1.51	1.05	30.57
1977	1.59	1.35	3.81	4.04	1.33	3.50	5.16	4.77	4.26	2.28	2.34	3.35	37.78
1978	2.22	0.54	2.01	2.55	2.91	4.65	2.83	5.00	5.62	3.05	1.83	3.32	36.53
1979	2.09	0.61	3.72	3.56	1.37	4.16	2.27	4.33	T	2.10	5.46	2.97	32.64
1980	1.76	1.76	1.74	3.64	3.19	4.00	5.90	3.18	4.57	1.99	1.57	3.60	36.90
1981	0.47	2.03	1.29	6.11	8.29	4.22	3.74	2.95	9.52	2.54	2.58	1.10	44.84
1982	2.98	0.36	3.36	2.11	3.63	2.45	3.81	3.07	1.92	1.42	5.36	6.49	36.96
1983	1.33	1.12	3.30	5.06	4.64	2.09	4.76	1.49	4.87	2.66	3.00	2.79	37.11
1984	0.94	1.15	2.77	2.10	4.77	0.62	2.12	1.49	2.16	3.34	2.85	4.37	28.68
1985	1.94	3.26	4.20	2.54	1.36	1.68	3.09	6.48	4.26	4.64	5.45	2.00	40.90
1986	1.07	3.34	2.35	2.58	3.88	7.14	5.27	5.30	11.85	2.76	0.95	1.04	47.53
1987	0.67	0.37	1.15	2.40	0.94	3.56	2.93	8.46	4.47	2.33	2.49	3.29	33.06
1988	2.39	1.14	2.12	3.11	1.07	0.25	3.69	3.04	7.49	5.37	4.82	1.88	36.37
1989	0.95	1.01	2.47	1.79	4.33	5.02	1.29	4.78	4.90	1.53	4.86	0.97	33.90
1990	2.39	2.08	1.96	2.23	4.39	3.00	3.73	3.40	4.22	5.05	7.14	2.97	42.56
1991	1.32	0.64	3.58	5.58	4.44	1.76	6.24	3.79	2.93	5.61	6.41	2.63	44.93
1992	1.52	1.06	3.51	3.98	1.45	1.61	8.83	3.55	5.60	2.34	5.64	3.27	42.36
1993	4.21	1.13	2.31	4.93	2.17	6.05	1.83	7.73	8.20	4.32	2.12	1.47	46.47
1994	2.68	1.70	1.46	3.25	2.64	7.17	8.07	7.39	2.38	3.29	5.44	1.11	46.58
1995	2.73	0.94	1.49	3.85	2.85	3.97	6.11	3.10	1.54	2.78	4.37	1.52	35.25
1996	1.18	0.90	0.96	2.41	4.83	6.33	1.28	0.33	2.72	2.76	1.40	2.45	27.55
1997	2.63	4.80	1.40	1.70	3.05	2.76	1.95	1.99	4.85	2.02	1.49	0.98	29.62
1998	4.09	1.50	4.93	2.75	1.86	2.11	2.49	1.70	2.49	2.98	2.27	1.20	30.37
1999	3.54	1.49	0.96	6.69	2.46	3.81	2.88	3.22	3.21	1.00	0.95	2.31	32.52
2000	1.02	1.09	1.33	3.98	8.65	4.67	4.06	2.31	6.31	1.85	2.82	2.07	40.16
2001	0.71	2.58	0.54	2.05	10.01	3.35	2.00	3.48	3.76	7.38	2.26	2.37	40.49
POR= 132 YRS	2.19	1.85	2.47	3.04	3.39	3.58	2.99	2.87	3.53	2.82	2.78	2.38	33.89

AVERAGE TEMPERATURE (F) 2001 GRAND RAPIDS, MI (GRR)

YEAR	JAN	FEB	MAR	APR	MAY	JUN	JUL	AUG	SEP	OCT	NOV	DEC	ANNUAL
1972	20.8	21.9	29.7	42.3	59.3	63.5	70.3	68.7	62.0	46.1	35.9	26.6	45.6
1973	27.4	22.8	41.9	46.8	53.7	69.9	72.4	72.7	63.4	55.0	39.9	26.2	49.3
1974	25.7	21.1	33.6	47.2	53.2	63.0	71.1	67.3	56.4	46.9	37.6	28.4	46.0
1975	25.2	24.3	28.8	39.8	60.7	67.6	70.6	70.5	57.0	53.0	44.7	27.3	47.5
1976	18.9	31.0	37.9	48.2	53.9	69.5	72.5	68.9	60.2	45.7	31.5	19.1	46.4
1977	12.7	22.8	40.0	52.1	65.8	64.5	74.0	67.9	62.9	48.6	39.5	25.9	48.1
1978	19.3	14.4	27.9	45.3	59.9	66.8	70.4	68.1	63.7	47.2	37.6	26.6	45.6
1979	17.1	15.0	36.9	44.5	57.7	67.6	71.1	68.5		51.5	40.5	33.0	47.2
1980	25.1	22.7	32.5	47.0	59.9	64.4	72.7	72.6	62.7	45.8	37.5	25.1	47.3
1981	20.8	30.2	35.1	47.3	55.5	68.6	72.1	70.9	60.8	48.2	39.8	28.7	48.2
1982	17.2	22.1	32.5	41.8	65.0	62.8	73.1	68.6	61.9	53.6	40.8	36.2	48.0
1983	27.6	30.9	36.9	42.6	52.8	67.7	74.7	72.2	62.4	50.7	40.7	19.2	48.2
1984	17.1	34.0	28.9	47.3	53.5	70.1	70.1	73.0	60.5	54.2	40.3	32.2	48.4
1985	18.6	21.3	36.2	51.7	60.4	65.3	70.7	67.4	63.6	50.6	38.4	22.3	47.1
1986	22.7	22.5	36.7	49.7	58.8	64.4	72.5	66.0	62.8	50.2	35.0	29.8	47.6
1987	25.4	29.9	37.3	50.1	62.6	71.2	74.1	69.3	62.2	45.0	41.1	31.5	50.0
1988	20.8	20.7	34.1	47.1	60.9	68.5	74.7	73.4	61.3	44.2	40.7	27.3	47.8
1989	30.5	19.7	31.8	43.9	55.8	65.9	71.9	68.4	59.4	51.1	35.7	17.2	45.9
1990	32.1	28.1	37.1	47.7	55.0	67.1	70.3	69.0	62.8	49.9	43.3	30.0	49.4
1991	22.2	29.2	37.9	50.8	64.6	71.2	72.4	70.5	60.1	52.0	35.7	29.4	49.7
1992	27.5	30.3	33.7	44.0	57.4	64.3	67.2	65.0	59.7	48.2	38.2	30.5	47.2
1993	25.9	21.9	32.4	44.0	58.2	65.2	72.7	71.4	56.5	48.0	37.9	27.8	46.8
1994	14.5	19.2	34.1	47.9	56.1	68.6	71.1	66.8	61.8	51.7	42.3	33.2	47.4
1995	26.5	23.2	37.0	42.8	56.4	69.8	72.5	74.8	59.1	51.4	32.0	24.2	47.5
1996	21.7	24.4	28.4	43.0	55.6	67.7	68.2	72.2	62.3	50.2	33.2	28.7	46.3
1997	21.6	27.3	34.6	44.1	50.3	68.3	70.8	66.1	61.1	49.9	35.6	31.1	46.7
1998	29.6	34.1	36.1	48.6	64.3	67.4	72.2	72.2	66.0	52.7	42.1	33.3	51.6
1999	21.2	31.4	32.9	48.8	61.0	69.4	74.7	67.8	62.7	49.5	43.6	30.4	49.5
2000	23.4	31.4	42.2	45.9	59.6	66.1	68.6	69.1	60.4	53.6	37.6	19.0	48.1
2001	25.5	25.9	32.5	50.0	59.6	66.3	71.3	71.6	59.3	50.4	46.8	33.9	49.4
POR= 108 YRS	24.0	24.7	34.1	46.6	58.0	67.8	72.5	70.4	62.8	51.5	39.1	28.3	48.3

REFERENCE NOTES:

PAGE 1:
THE TEMPERATURE GRAPH SHOWS NORMAL MAXIMUM AND NORMAL MINIMUM DAILY TEMPERATURES (SOLID CURVES) AND THE ACTUAL DAILY HIGH AND LOW TEMPERATURES (VERTICAL BARS).

PAGE 2 AND 3:
H/C INDICATES HEATING AND COOLING DEGREE DAYS.
RH INDICATES RELATIVE HUMIDITY
W/O INDICATES WEATHER AND OBSTRUCTIONS
S INDICATES SUNSHINE.
PR INDICATES PRESSURE.
CLOUDINESS ON PAGE 3 IS THE SUM OF THE CEILOMETER AND SATELLITE DATA NOT TO EXCEED EIGHT EIGHTHS(OKTAS).

GENERAL:
T INDICATES TRACE PRECIPITATION, AN AMOUNT GREATER THAN ZERO BUT LESS THAN THE LOWEST REPORTABLE VALUE.
+ INDICATES THE VALUE ALSO OCCURS ON EARLIER DATES.
BLANK ENTRIES DENOTE MISSING OR UNREPORTED DATA.
NORMALS ARE 30-YEAR AVERAGES (1961 - 1990).
ASOS INDICATES AUTOMATED SURFACE OBSERVING SYSTEM.
PM INDICATES THE LAST DAY OF THE PREVIOUS MONTH.
POR (PERIOD OF RECORD) BEGINS WITH THE JANUARY DATA MONTH AND IS THE NUMBER OF YEARS USED TO COMPUTE THE MEAN. INDIVIDUAL MONTHS WITHIN THE POR MAY BE MISSING.
WHEN THE POR FOR A NORMAL IS LESS THAN 30 YEARS, THE NORMAL IS PROVISIONAL AND IS BASED ON THE NUMBER OF YEARS INDICATED.
0.* OR * INDICATES THE VALUE OR MEAN-DAYS-WITH IS BETWEEN 0.00 AND 0.05.
CLOUDINESS FOR ASOS STATIONS DIFFERS FROM THE NON-ASOS OBSERVATION TAKEN BY A HUMAN OBSERVER. ASOS STATION CLOUDINESS IS BASED ON TIME-AVERAGED CEILOMETER DATA FOR CLOUDS AT OR BELOW 12,000 FEET AND ON SATELLITE DATA FOR CLOUDS ABOVE 12,000 FEET.
THE NUMBER OF DAYS WITH CLEAR, PARTLY CLOUDY, AND CLOUDY CONDITIONS FOR ASOS STATIONS IS THE SUM OF THE CEILOMETER AND SATELLITE DATA FOR THE SUNRISE TO SUNSET PERIOD.

GENERAL CONTINUED:
CLEAR INDICATES 0 - 2 OKTAS, PARTLY CLOUDY INDICATES 3 - 6 OKTAS, AND CLOUDY INDICATES 7 OR 8 OKTAS. WHEN AT LEAST ONE OF THE ELEMENTS (CEILOMETER OR SATELLITE) IS MISSING, THE DAILY CLOUDINESS IS NOT COMPUTED.
WIND DIRECTION IS RECORDED IN TENS OF DEGREES (2 DIGITS) CLOCKWISE FROM TRUE NORTH. "00" INDICATES CALM. "36" INDICATES TRUE NORTH.
RESULTANT WIND IS THE VECTOR AVERAGE OF THE SPEED AND DIRECTION.
AVERAGE TEMPERATURE IS THE SUM OF THE MEAN DAILY MAXIMUM AND MINIMUM TEMPERATURE DIVIDED BY 2.
SNOWFALL DATA COMPRISE ALL FORMS OF FROZEN PRECIPITATION, INCLUDING HAIL.
A HEATING (COOLING) DEGREE DAY IS THE DIFFERENCE BETWEEN THE AVERAGE DAILY TEMPERATURE AND 65 F.
DRY BULB IS THE TEMPERATURE OF THE AMBIENT AIR.
DEW POINT IS THE TEMPERATURE TO WHICH THE AIR MUST BE COOLED TO ACHIEVE 100 PERCENT RELATIVE HUMIDITY.
WET BULB IS THE TEMPERATURE THE AIR WOULD HAVE IF THE MOISTURE CONTENT WAS INCREASED TO 100 PERCENT RELATIVE HUMIDITY.

ON JULY 1, 1996, THE NATIONAL WEATHER SERVICE BEGAN USING THE "METAR" OBSERVATION CODE THAT WAS ALREADY EMPLOYED BY MOST OTHER NATIONS OF THE WORLD. THE MOST NOTICEABLE DIFFERENCE IN THIS ANNUAL PUBLICATION WILL BE THE CHANGE IN UNITS FROM TENTHS TO EIGHTS(OKTAS) FOR REPORTING THE AMOUNT OF SKY COVER.

HEATING DEGREE DAYS (base 65 F) 2001 GRAND RAPIDS, MI (GRR)

YEAR	JUL	AUG	SEP	OCT	NOV	DEC	JAN	FEB	MAR	APR	MAY	JUN	TOTAL
1972-73	28	41	126	577	867	1185	1162	1178	708	552	341	4	6769
1973-74	1	7	136	314	748	1196	1212	1225	967	532	364	113	6815
1974-75	5	22	275	555	816	1127	1227	1133	1113	750	191	61	7275
1975-76	16	5	242	382	602	1161	1420	980	835	526	342	18	6529
1976-77	0	32	185	591	999	1415	1616	1177	765	401	106	91	7378
1977-78	3	48	99	501	759	1204	1407	1413	1140	584	213	62	7433
1978-79	15	22	124	545	816	1182	1477	1393	863	607	264	42	7350
1979-80	7	31	109	431	731	984	1230	1221	1000	541	189	106	6580
1980-81	0	4	115	588	821	1227	1363	969	919	525	298	8	6837
1981-82	7	5	173	513	749	1116	1475	1196	998	689	91	98	7110
1982-83	6	34	140	361	717	884	1151	949	866	663	373	61	6205
1983-84	16	2	149	440	721	1413	1480	892	1107	532	353	13	7118
1984-85	16	4	189	333	735	1014	1431	1216	886	428	173	81	6506
1985-86	9	17	157	436	790	1316	1302	1186	871	466	214	75	6839
1986-87	11	56	118	452	894	1084	1220	978	849	440	178	20	6300
1987-88	18	36	118	610	712	1032	1364	1277	950	531	169	60	6877
1988-89	3	21	135	639	722	1162	1062	1263	1023	625	297	51	7003
1989-90	2	23	203	424	874	1477	1014	1026	865	549	306	45	6808
1990-91	11	15	139	475	645	1077	1323	997	830	430	157	18	6117
1991-92	5	4	211	404	871	1098	1155	1001	964	626	262	75	6676
1992-93	25	66	197	512	794	1060	1204	1203	1004	624	227	84	7000
1993-94	1	16	259	524	810	1150	1561	1274	950	520	297	42	7404
1994-95	0	46	94	405	675	980	1186	1165	860	659	265	28	6363
1995-96	14	0	212	417	982	1260	1337	1166	1130	653	320	29	7520
1996-97	10	2	147	451	944	1119	1338	1048	935	616	448	24	7082
1997-98	8	43	136	473	876	1046	1091	858	899	486	76	98	6090
1998-99	0	2	58	380	682	975	1353	940	989	481	156	49	6065
1999-00	1	17	116	474	634	1066	1282	971	706	567	213	64	6111
2000-01	15	19	199	348	815	1420	1218	1089	1001	444	189	80	6837
2001-	20	1	194	449	538	955							

COOLING DEGREE DAYS (base 65 F) 2001 GRAND RAPIDS, MI (GRR)

YEAR	JAN	FEB	MAR	APR	MAY	JUN	JUL	AUG	SEP	OCT	NOV	DEC	ANNUAL
1972	0	0	0	0	37	61	200	163	42	0	0	0	503
1973	0	0	0	10	0	157	238	255	96	12	0	0	768
1974	0	0	0	4	6	60	201	103	23	3	0	0	400
1975	0	0	0	0	65	148	199	184	8	15	0	0	619
1976	0	0	0	27	4	158	239	161	49	0	0	0	638
1977	0	0	0	19	137	85	286	143	44	0	0	0	714
1978	0	0	0	0	62	123	188	125	90	0	0	0	588
1979	0	0	0	0	46	129	204	147	69	19	0	0	614
1980	0	0	0	6	34	96	247	245	53	0	0	0	681
1981	0	0	0	0	13	124	236	190	55	0	0	0	618
1982	0	0	0	0	99	40	263	153	55	12	0	0	622
1983	0	0	0	0	2	146	325	234	76	5	0	0	788
1984	0	0	0	7	5	174	179	259	60	6	0	0	690
1985	0	0	0	36	39	52	193	99	119	0	0	0	538
1986	0	0	0	14	26	64	252	94	61	0	0	0	511
1987	0	0	0	2	110	213	306	174	38	0	0	0	843
1988	0	0	0	0	50	170	310	289	29	2	0	0	850
1989	0	0	0	0	19	84	223	135	40	0	0	0	501
1990	0	0	0	6	37	6	185	144	83	13	0	0	589
1991	0	0	0	12	155	211	240	182	71	7	0	0	878
1992	0	0	0	2	36	60	99	74	44	1	0	0	316
1993	0	0	0	0	21	94	244	222	10	5	0	0	596
1994	0	0	0	13	29	159	196	102	65	1	0	0	565
1995	0	0	0	0	5	179	252	315	40	6	0	0	797
1996	0	0	0	0	33	118	116	230	74	0	0	0	571
1997	0	0	0	0	0	131	194	81	26	13	0	0	445
1998	0	0	0	9	62	177	230	233	95	4	0	0	810
1999	0	0	0	3	39	186	308	110	55	0	0	0	701
2000	0	0	0	2	0	48	102	132	154	66	2	0	506
2001	0	0	0	0	27	128	223	212	30	3	0	0	623

SNOWFALL (inches) 2001 GRAND RAPIDS, MI (GRR)

YEAR	JUL	AUG	SEP	OCT	NOV	DEC	JAN	FEB	MAR	APR	MAY	JUN	TOTAL
1972-73	0.0	0.0	0.0	0.9	11.0	19.8	7.0	13.2	8.5	5.0	0.1	0.0	65.5
1973-74	0.0	0.0	0.0	0.4	20.0	13.3	18.4	11.3	1.0	T	0.0	0.0	64.4
1974-75	0.0	0.0	0.0	0.4	8.9	16.5	10.7	10.6	11.8	10.0	0.0	0.0	68.9
1975-76	0.0	0.0	0.0	0.0	6.6	23.3	25.0	6.5	3.5	4.2	0.1	0.0	69.2
1976-77	0.0	0.0	0.0	2.0	8.5	17.7	26.1	5.0	9.5	2.0	0.0	0.0	70.8
1977-78	0.0	0.0	0.0	0.0	10.6	23.2	35.8	8.8	6.2	T	0.0	0.0	84.6
1978-79	0.0	0.0	0.0	T	6.2	30.0	45.5	5.3	7.2	1.8	0.0	0.0	96.0
1979-80	0.0	0.0	0.0	T	9.4	2.6	13.3	12.6	6.6	4.0	T	0.0	48.5
1980-81	0.0	0.0	0.0	0.4	5.5	17.3	8.1	18.8	1.4	T	0.0	0.0	51.5
1981-82	0.0	0.0	0.0	T	4.4	8.9	30.3	6.7	11.8	12.4	0.0	0.0	74.5
1982-83	0.0	0.0	0.0	0.0	5.2	8.2	5.7	2.9	13.2	0.7	0.0	0.0	35.9
1983-84	0.0	0.0	0.0	T	4.7	34.8	19.6	1.6	10.6	0.1	T	0.0	71.4
1984-85	0.0	0.0	0.0	T	15.7	22.6	21.3	6.7	3.3	0.0	0.0	69.6
1985-86	0.0	0.0	0.0	0.0	3.5	30.7	18.4	20.2	6.1	0.2	0.0	0.0	79.1
1986-87	0.0	0.0	0.0	0.0	5.3	12.7	19.2	0.9	5.7	3.8	0.0	0.0	47.6
1987-88	0.0	0.0	0.0	1.6	0.7	18.2	21.9	18.1	3.4	0.3	0.0	0.0	64.2
1988-89	0.0	0.0	0.0	0.0	0.2	5.5	14.4	8.7	25.1	6.3	2.2	T	62.4
1989-90	0.0	0.0	0.0	5.8	19.4	25.2	10.6	23.8	2.7	2.1	0.2	0.0	89.8
1990-91	T	0.0	0.0	T	2.0	18.6	27.7	9.5	2.8	T	0.0	0.0	60.6
1991-92	0.0	0.0	0.0	0.3	25.3	27.9	13.4	3.5	15.1	2.3	0.0	0.0	87.8
1992-93	0.0	0.0	0.0	2.3	4.2	14.2	11.1	18.6	11.6	3.3	0.0	0.0	65.3
1993-94	0.0	T	0.0	T	1.9	17.7	25.3	29.6	1.9	T	0.1	0.0	76.5
1994-95	0.0	0.0	0.0	0.0	0.6	8.9	21.6	18.6	4.4	0.8	0.0	0.0	54.9
1995-96	0.0	0.0	0.0	T	20.8	17.4	13.5	6.9	19.3	1.8	0.0	0.0	79.7
1996-97	0.0	0.0	0.0	0.0	5.6	22.8	45.5	14.0	5.2	3.5	T	0.0	96.6
1997-98	0.0	0.0	0.0	2.4	10.1	11.5	20.3	0.5	13.8	0.0	0.0	0.0	58.6
1998-99	0.0	0.0	0.0	0.0	0.2	7.5	46.8	8.0	14.2	T	T	0.0	76.7
1999-00	0.0	0.0	0.0	0.0	0.1	18.4	15.7	11.5	0.6	6.8	T	0.0	
2000-01	0.0	0.0	0.0	0.0	19.0	59.2	4.1	7.4	4.1	0.2	0.0		94.0
2001-	0.0	0.0	0.0	0.5	T	53.9							
POR= 37 YRS	T	T	T	0.8	7.7	19.5	20.5	11.7	9.3	2.5	0.2	0.0	72.2

2001
HOUGHTON LAKE,
MICHIGAN (HTL)

Houghton Lake is located in north-central lower Michigan. The present station is on the northeast shore of Houghton Lake, the largest inland lake in Michigan, with a circumference of about 32 miles. The Muskegon River source is Higgins Lake, 8 miles to the north. It flows through Houghton Lake, then southwestward to Lake Michigan. The station lies within an elongated bowl shaped 1,000-foot plateau, which extends roughly 50 miles north, 75 miles southwest, and about 20 miles southeast of Houghton Lake. In the immediate area, the land is level to rolling, but there are hills and ridges from 100 to 300 feet higher in elevation surrounding the station. Soils are generally sand, or sandy loam supporting little agricultural production, but the area is rich in natural resources of forests, lakes, and streams.

The interior location diminishes the influence of the larger Great Lakes, which lie 70 to 80 miles east and west of Houghton Lake. Hence, the daily temperature range is larger, especially in summer, and temperature extremes are greater than are found nearer the shores of either Lake Michigan or Lake Huron. Temperatures reach the 100 degree mark about one summer out of ten, and at the other extreme, fall below zero an average of twenty-two times during the winter season.

Precipitation is normally a little heavier during the summer season. About 60 percent of the annual total falls in the six-month period from April through September. The heaviest precipitation occurs with summertime thunderstorms.

Snowfall averages above 80 inches per year at Houghton Lake, with considerable variation from year to year. Much heavier snows, averaging over 100 inches a season, fall within a 30- to 60-mile radius to the north and west of Houghton Lake. Seasonal totals have ranged from 24 inches to over 124 inches. Measurable amounts of snow have occurred in nine of the twelve months, and the average number of months with measurable snowfall is six.

Cloudiness is greatest in the late fall and early winter, while sunshine percentage is highest in the spring and summer. Cloudiness is increased in the late fall due to the moisture and warmth picked up by the westerly and northwesterly winds while crossing Lake Michigan.

The growing season is normally quite short, averaging about 90 days between spring and fall freezes.

NORMALS, MEANS, AND EXTREMES
HOUGHTON LAKE, MI (HTL)

LATITUDE:　44 22' 04" N　　LONGITUDE:　84 41' 27" W　　ELEVATION (FT): GRND: 1148　BARO: 1151　　TIME ZONE: EASTERN (UTC + 5)　　WBAN: 94814

	ELEMENT	POR	JAN	FEB	MAR	APR	MAY	JUN	JUL	AUG	SEP	OCT	NOV	DEC	YEAR
TEMPERATURE °F	NORMAL DAILY MAXIMUM	30	25.3	28.1	38.2	52.7	66.1	74.8	79.6	76.4	67.8	55.8	42.0	29.5	53.0
	MEAN DAILY MAXIMUM	37	25.7	29.0	38.9	52.9	66.3	75.3	79.5	76.8	68.1	55.8	41.8	30.6	53.4
	HIGHEST DAILY MAXIMUM	37	54	59	76	86	90	103	98	96	92	85	70	64	103
	YEAR OF OCCURRENCE		1996	2000	1990	1980	1998	1995	1995	2001	1985	1971	1990	2001	JUN 1995
	MEAN OF EXTREME MAXS.	37	42.4	45.7	60.0	74.5	82.9	87.9	90.0	87.9	82.8	73.8	60.5	46.9	69.6
	NORMAL DAILY MINIMUM	30	8.4	8.2	18.2	31.6	42.0	50.6	55.4	53.8	46.9	37.5	28.4	15.7	33.1
	MEAN DAILY MINIMUM	37	9.4	9.8	18.7	31.5	42.2	51.0	55.4	54.0	46.7	37.2	28.4	17.2	33.5
	LOWEST DAILY MINIMUM	37	-26	-34	-23	3	21	29	33	29	21	16	-5	-21	-34
	YEAR OF OCCURRENCE		1981	1979	1967	1982	1966	1998	1965	1982	1989	1969	1995	1976	FEB 1979
	MEAN OF EXTREME MINS.	37	-14.6	-15.8	-5.7	15.8	28.3	36.2	40.8	39.2	29.6	22.8	12.2	-5.1	15.3
	NORMAL DRY BULB	30	16.9	18.2	28.2	42.2	54.1	62.7	67.5	65.1	57.3	46.7	35.2	22.6	43.1
	MEAN DRY BULB	37	17.6	19.3	29.0	42.2	54.5	63.2	67.5	65.2	57.4	46.6	35.1	23.7	43.4
	MEAN WET BULB	18	19.2	21.0	28.3	38.9	49.6	58.7	59.3	61.9	54.5	41.5	33.2	24.0	40.8
	MEAN DEW POINT	18	14.5	15.2	20.4	30.6	39.5	52.8	54.4	57.8	50.3	37.1	29.1	20.4	35.2
	NORMAL NO. DAYS WITH:														
	MAXIMUM 90	30	0.0	0.0	0.0	0.0	*	0.8	1.9	0.5	*	0.0	0.0	0.0	3.2
	MAXIMUM 32	30	23.6	18.5	9.3	1.0	0.0	0.0	0.0	0.0	0.0	0.0	5.3	19.3	77.0
	MINIMUM 32	30	30.8	27.6	27.5	17.2	5.1	0.3	0.0	*	1.7	9.8	21.9	29.6	171.5
	MINIMUM 0	30	8.8	8.1	3.2	0.0	0.0	0.0	0.0	0.0	0.0	0.0	0.1	3.3	23.5
H/C	NORMAL HEATING DEG. DAYS	30	1491	1310	1141	684	358	120	35	70	234	567	894	1314	8218
	NORMAL COOLING DEG. DAYS	30	0	0	0	0	20	51	112	73	0	0	0	0	256
RH	NORMAL (PERCENT)	30	80	76	74	68	65	70	71	76	78	77	82	83	75
	HOUR 01 LST	30	83	81	80	77	78	84	85	88	88	85	85	85	83
	HOUR 07 LST	30	82	82	83	80	78	81	85	90	92	88	87	85	84
	HOUR 13 LST	30	72	69	64	55	50	55	54	60	63	65	73	77	63
	HOUR 19 LST	30	77	72	67	58	53	57	58	66	73	74	79	81	68
S	PERCENT POSSIBLE SUNSHINE														
W/O	MEAN NO. DAYS WITH:														
	HEAVY FOG(VISBY 1/4 MI)	38	2.3	2.0	2.8	1.5	1.5	1.8	2.6	4.0	3.6	2.6	2.6	2.5	29.8
	THUNDERSTORMS	38	0.2	0.1	0.8	2.1	3.9	5.4	5.4	5.6	3.6	1.4	0.5	0.2	29.2
CLOUDINESS	MEAN:														
	SUNRISE-SUNSET (OKTAS)	31	6.4	5.9	5.6	5.5	5.1	4.8	4.5	4.6	5.0	5.6	6.5	6.4	5.5
	MIDNIGHT-MIDNIGHT (OKTAS)	16	5.9	5.2	5.1	4.7	4.4	4.4	3.9	3.9	4.7	5.0	6.1	6.2	5.0
	MEAN NO. DAYS WITH:														
	CLEAR	31	2.9	4.2	5.9	6.2	7.0	7.0	7.3	7.8	6.3	5.3	2.4	2.4	64.7
	PARTLY CLOUDY	31	6.3	6.7	7.2	7.2	9.3	11.2	13.2	11.1	9.3	7.7	5.0	5.9	100.1
	CLOUDY	31	21.8	17.4	17.8	16.6	14.1	11.8	10.1	12.1	14.4	18.0	22.6	22.8	199.5
PR	MEAN STATION PRESSURE(IN)	12	28.72	28.74	28.72	28.72	28.70	28.71	28.79	28.81	28.78	28.80	28.74	28.71	28.74
	MEAN SEA-LEVEL PRES. (IN)	18	30.02	30.06	30.04	29.95	29.97	30.01	29.99	30.04	30.03	30.05	30.02	30.04	30.02
WINDS	MEAN SPEED (MPH)	10	9.5	8.8	9.4	9.3	8.7	7.8	7.3	6.9	7.6	8.5	9.3	8.9	8.5
	PREVAIL.DIR(TENS OF DEGS)	12	27	27	32	32	27	27	23	23	23	27	27	27	27
	MAXIMUM 2-MINUTE:														
	SPEED (MPH)	5	31	37	32	37	40	51	37	28	30	35	39	33	51
	DIR. (TENS OF DEGS)		32	24	28	24	26	19	13	32	29	16	22	26	19
	YEAR OF OCCURRENCE		2000	2001	2000	1997	1998	1999	2000	2001	1998	2001	1998	2001	JUN 1999
	MAXIMUM 5-SECOND:														
	SPEED (MPH)	5	44	48	45	52	54	61	46	40	39	46	51	46	61
	DIR. (TENS OF DEGS)		33	23	30	22	23	19	13	29	29	16	21	26	19
	YEAR OF OCCURRENCE		2000	1999	1999	1997	1998	1999	2000	1997	1998	2001	1998	2001	JUN 1999
PRECIPITATION	NORMAL (IN)	30	1.50	1.16	2.02	2.22	2.57	3.02	2.58	3.37	3.41	2.18	2.27	1.95	28.25
	MAXIMUM MONTHLY (IN)	37	3.13	3.36	5.67	4.73	5.99	6.67	5.33	7.18	9.49	8.08	5.10	4.48	9.49
	YEAR OF OCCURRENCE		1974	1971	1974	1991	1983	1969	1994	1975	1986	1991	1988	1971	SEP 1986
	MINIMUM MONTHLY (IN)	37	0.59	0.29	0.19	0.86	0.40	0.85	0.55	0.85	0.01	0.47	0.45	0.34	0.01
	YEAR OF OCCURRENCE		2001	1982	2001	1998	1966	1988	1989	1969	1979	1971	1986	1997	SEP 1979
	MAXIMUM IN 24 HOURS (IN)	37	1.39	1.52	2.18	1.81	1.94	3.28	3.83	3.12	2.55	3.47	1.82	1.70	3.83
	YEAR OF OCCURRENCE		1974	1997	1976	1991	1973	1996	1984	1981	1985	1998	1988	1971	JUL 1984
	NORMAL NO. DAYS WITH:														
	PRECIPITATION 0.01	30	14.9	11.3	12.2	11.7	10.3	10.7	9.0	10.0	11.6	11.9	13.2	15.7	142.5
	PRECIPITATION 1.00	30	0.0	*	0.1	0.3	0.3	0.7	0.5	0.7	0.8	0.1	0.1	0.2	3.8
SNOWFALL	NORMAL (IN)	30	19.3	12.9	12.0	4.2	0.3	0.0	0.0	0.0	0.*	0.9	9.7	16.9	76.2
	MAXIMUM MONTHLY (IN)	33	38.0	23.6	28.7	11.6	2.3	0.0	T	T	0.1	4.4	41.9	30.4	41.9
	YEAR OF OCCURRENCE		1982	1971	1971	1979	1979		1996	1993	1967	1980	1995	1968	NOV 1995
	MAXIMUM IN 24 HOURS (IN)	33	15.4	8.5	11.7	7.6	3.2	0.0	T	T	0.1	3.5	14.4	13.2	15.4
	YEAR OF OCCURRENCE		1978	1974	1970	1979	1994		1970	1993	1967	1980	1981	1980	JAN 1978
	MAXIMUM SNOW DEPTH (IN)	32	24	21	22	7	3	0	0	0	0	1	17	14	24
	YEAR OF OCCURRENCE		1979	1979	1978	1973	1994					1992	1995	1972	JAN 1979
	NORMAL NO. DAYS WITH:														
	SNOWFALL 1.0	30	6.3	4.7	3.5	1.4	0.1	0.0	0.0	0.0	0.0	0.3	3.0	5.6	24.9

PRECIPITATION (inches) 2001 HOUGHTON LAKE, MI (HTL)

YEAR	JAN	FEB	MAR	APR	MAY	JUN	JUL	AUG	SEP	OCT	NOV	DEC	ANNUAL
1972	0.82	0.99	2.36	1.40	1.79	2.00	2.52	4.63	2.67	2.50	0.99	3.48	26.15
1973	1.22	1.33	1.95	1.66	4.88	2.84	2.34	2.29	1.95	3.11	1.81	26.72
1974	3.13	1.14	1.44	3.47	2.92	4.60	4.70	2.78	2.53	1.40	1.41	1.43	30.95
1975	1.97	1.14	1.50	2.63	2.79	3.79	4.96	7.18	1.52	0.85	2.20	1.33	31.86
1976	1.77	2.49	5.67	1.86	2.86	2.77	1.22	1.07	0.99	1.31	0.75	0.65	23.41
1977	0.60	1.46	2.40	2.29	1.39	0.94	1.72	5.70	1.89	2.14	2.20	26.92
1978	1.93	0.55	1.20	1.35	2.42	2.38	0.91	4.10	6.70	1.43	1.28	2.08	26.33
1979	1.51	0.63	3.05	3.18	1.85	4.46	0.87	3.90	0.01	2.43	2.47	1.71	26.07
1980	1.61	0.69	1.10	3.30	1.55	3.44	2.29	2.01	3.75	1.91	1.55	1.95	25.15
1981	0.79	2.10	0.88	3.88	1.73	4.02	1.89	7.06	1.89	2.61	2.07	1.08	30.00
1982	2.43	0.29	2.41	2.46	2.98	3.21	3.31	3.25	3.58	1.86	2.52	3.47	31.77
1983	1.20	0.79	3.11	1.86	5.99	0.95	1.40	3.89	4.63	3.66	1.60	1.69	30.77
1984	1.06	0.88	2.28	1.96	2.60	3.01	4.30	2.95	2.74	2.17	1.99	2.93	28.87
1985	1.64	1.99	3.55	2.42	1.87	1.71	2.28	4.76	6.14	1.63	3.56	2.14	33.69
1986	1.06	1.73	2.20	1.73	3.20	5.43	4.38	1.76	9.49	1.75	0.45	0.85	34.03
1987	1.06	0.61	0.78	0.97	1.56	1.04	1.62	6.69	4.35	2.21	2.63	2.45	25.97
1988	2.09	0.75	2.39	2.37	0.56	0.85	2.49	4.50	3.63	3.38	5.10	1.86	29.97
1989	0.97	0.70	2.99	0.98	3.19	2.90	0.55	2.62	1.03	1.30	2.08	0.92	20.23
1990	2.43	1.09	1.47	1.77	4.15	2.57	3.82	3.28	3.00	2.94	2.90	1.55	30.97
1991	1.22	0.62	3.02	4.73	5.03	1.48	4.43	2.36	3.05	8.08	1.88	1.76	37.66
1992	1.29	1.35	1.84	3.13	0.49	1.91	2.86	3.45	3.63	2.75	4.65	1.98	29.33
1993	1.73	0.89	0.62	2.96	2.50	4.84	1.28	5.67	2.10	1.78	1.58	0.74	26.69
1994	1.88	1.44	1.57	2.72	1.45	1.72	5.33	5.52	1.56	1.04	2.86	0.53	27.62
1995	2.10	0.65	1.61	2.96	1.51	3.83	2.52	5.64	1.67	2.20	4.82	1.38	30.89
1996	1.71	1.46	0.96	2.99	1.59	5.84	2.76	3.56	4.41	2.84	0.72	2.29	31.13
1997	2.46	2.60	1.51	1.30	3.33	1.26	2.63	3.58	3.45	1.85	1.40	0.34	25.71
1998	2.12	0.89	4.40	0.86	2.31	2.24	0.65	1.30	2.10	4.00	2.53	1.04	24.44
1999	1.86	1.29	0.40	2.04	2.06	6.31	4.48	2.21	3.19	1.77	0.58	1.63	27.82
2000	1.38	1.45	1.17	1.39	5.15	2.79	4.98	2.15	2.00	0.64	2.32	0.60	26.02
2001	0.59	1.60	0.19	3.22	5.31	2.53	0.81	3.15	3.61	4.61	1.91	0.59	28.12
POR= 84 YRS	1.48	1.27	1.89	2.41	2.79	3.07	2.70	2.99	3.11	2.51	2.35	1.66	28.23

AVERAGE TEMPERATURE (F) 2001 HOUGHTON LAKE, MI (HTL)

YEAR	JAN	FEB	MAR	APR	MAY	JUN	JUL	AUG	SEP	OCT	NOV	DEC	ANNUAL
1972	16.3	16.4	22.3	36.6	56.9	59.1	66.5	64.4	56.8	41.7	33.6	22.0	41.1
1973	22.1	18.1	38.1	43.1	50.6	64.9	67.7	68.9	57.4	51.3	35.8	24.7	45.1
1974	20.3	14.4	27.0	43.2	50.0	61.6	67.8	65.0	53.9	44.3	36.3	26.1	42.5
1975	22.0	20.7	24.3	37.0	59.7	64.1	68.1	65.0	53.7	49.6	40.8	23.8	44.1
1976	14.0	24.1	30.8	45.5	50.8	66.1	67.0	64.6	55.6	42.3	28.3	13.8	41.9
1977	8.7	17.8	34.8	46.6	60.3	61.0	70.1	62.1	58.3	45.1	35.1	21.8	43.5
1978	14.5	10.8	23.6	39.1	56.9	61.6	65.1	66.4	58.6	44.3	35.9	21.8	41.6
1979	11.5	10.4	30.9	40.0	52.6	62.3	66.9	62.7	59.2	45.7	35.0	27.6	42.1
1980	19.1	15.7	25.0	41.9	56.0	59.2	67.4	68.0	56.7	41.6	33.9	18.9	42.0
1981	14.8	24.0	32.5	44.3	53.0	63.6	65.6	65.5	55.9	42.9	35.6	25.3	43.7
1982	11.4	17.9	25.8	37.2	60.9	57.4	68.6	61.9	57.2	48.9	36.2	31.2	42.9
1983	21.8	25.9	32.4	40.3	48.5	65.3	71.5	68.6	59.7	46.4	36.6	16.8	44.3
1984	12.4	28.2	24.1	44.8	50.1	65.2	66.4	68.8	56.1	19.7	35.4	27.0	44.0
1985	14.9	17.9	30.3	46.5	57.0	59.5	66.1	63.9	59.8	47.5	34.5	18.0	43.0
1986	17.3	18.4	30.9	47.3	56.9	60.5	69.7	62.8	58.1	46.8	32.3	26.8	44.0
1987	21.6	23.4	33.6	47.3	58.1	67.1	71.4	65.6	59.5	43.9	37.7	28.7	46.5
1988	17.3	16.3	28.0	43.7	58.6	64.8	71.4	68.8	57.9	42.4	36.6	23.7	44.1
1989	24.9	15.5	25.0	41.1	54.5	62.2	69.5	64.8	56.2	47.9	30.7	12.9	42.1
1990	26.2	21.3	32.1	44.9	51.9	63.4	66.8	65.0	57.9	45.3	38.9	25.8	45.0
1991	16.1	24.5	33.1	46.9	61.1	67.7	68.6	68.0	55.8	47.9	33.4	25.4	45.7
1992	22.8	24.1	28.6	39.4	54.3	60.4	63.0	62.0	56.8	44.8	33.6	26.1	43.0
1993	21.0	16.2	28.7	40.2	54.6	61.9	70.2	68.8	52.9	44.6	34.7	25.9	44.3
1994	9.3	12.9	30.1	43.2	55.0	65.3	68.1	64.4	61.9	50.9	39.3	30.7	44.3
1995	23.1	18.2	34.3	40.1	54.5	68.9	70.7	72.3	56.9	50.1	29.1	19.5	44.8
1996	15.9	19.8	25.7	36.8	50.8	64.0	64.0	66.3	58.6	47.0	30.6	24.8	42.0
1997	17.0	20.4	27.2	40.3	46.6	64.8	65.8	61.6	57.0	46.0	32.3	28.2	42.3
1998	24.1	30.2	31.2	45.2	60.3	63.0	66.7	66.7	60.4	48.8	37.9	28.7	46.9
1999	17.1	26.1	29.1	44.7	56.1	64.9	69.8	63.1	56.8	44.9	39.7	26.2	44.9
2000	17.8	25.0	38.1	41.4	55.9	62.7	64.7	64.7	57.6	49.6	35.8	16.0	44.1
2001	22.8	20.1	27.7	46.1	56.8	63.4	67.0	67.0	55.5	46.5	42.2	31.4	45.5
POR= 83 YRS	18.8	19.7	28.6	42.4	54.5	63.6	67.6	65.6	58.0	47.6	35.3	24.0	43.8

REFERENCE NOTES:

PAGE 1:
THE TEMPERATURE GRAPH SHOWS NORMAL MAXIMUM AND NORMAL
   MINIMUM DAILY TEMPERATURES (SOLID CURVES) AND THE
   ACTUAL DAILY HIGH AND LOW TEMPERATURES (VERTICAL BARS).

PAGE 2 AND 3:
H/C INDICATES HEATING AND COOLING DEGREE DAYS.
RH INDICATES RELATIVE HUMIDITY
W/O INDICATES WEATHER AND OBSTRUCTIONS
S INDICATES SUNSHINE.
PR INDICATES PRESSURE.
CLOUDINESS ON PAGE 3 IS THE SUM OF THE CEILOMETER AND
   SATELLITE DATA NOT TO EXCEED EIGHT EIGHTHS (OKTAS).

GENERAL:
T INDICATES TRACE PRECIPITATION, AN AMOUNT GREATER
   THAN ZERO BUT LESS THAN THE LOWEST REPORTABLE VALUE.
+ INDICATES THE VALUE ALSO OCCURS ON EARLIER DATES.
BLANK ENTRIES DENOTE MISSING OR UNREPORTED DATA.
NORMALS ARE 30-YEAR AVERAGES (1961 - 1990).
ASOS INDICATES AUTOMATED SURFACE OBSERVING SYSTEM.
PM INDICATES THE LAST DAY OF THE PREVIOUS MONTH.
POR (PERIOD OF RECORD) BEGINS WITH THE JANUARY DATA
   MONTH AND IS THE NUMBER OF YEARS USED TO COMPUTE
   THE MEAN. INDIVIDUAL MONTHS WITHIN THE POR MAY
   BE MISSING.
WHEN THE POR FOR A NORMAL IS LESS THAN 30 YEARS,
   THE NORMAL IS PROVISIONAL AND IS BASED ON THE NUMBER
   OF YEARS INDICATED.
0. OR * INDICATES THE VALUE OR MEAN-DAYS-WITH
   IS BETWEEN 0.00 AND 0.05.
CLOUDINESS FOR ASOS STATIONS DIFFERS FROM THE NON-ASOS
   OBSERVATION TAKEN BY A HUMAN OBSERVER. ASOS STATION
   CLOUDINESS IS BASED ON TIME-AVERAGED CEILOMETER DATA
   FOR CLOUDS AT OR BELOW 12,000 FEET AND ON SATELLITE
   DATA FOR CLOUDS ABOVE 12,000 FEET.
THE NUMBER OF DAYS WITH CLEAR, PARTLY CLOUDY, AND
   CLOUDY CONDITIONS FOR ASOS STATIONS IS THE SUM
   OF THE CEILOMETER AND SATELLITE DATA FOR THE
   SUNRISE TO SUNSET PERIOD.

GENERAL CONTINUED:
CLEAR INDICATES 0 - 2 OKTAS, PARTLY CLOUDY INDICATES
   3 - 6 OKTAS, AND CLOUDY INDICATES 7 OR 8 OKTAS.
   WHEN AT LEAST ONE OF THE ELEMENTS (CEILOMETER OR
   SATELLITE) IS MISSING, THE DAILY CLOUDINESS IS
   NOT COMPUTED.
WIND DIRECTION IS RECORDED IN TENS OF DEGREES (2 DIGITS)
   CLOCKWISE FROM TRUE NORTH. "00" INDICATES CALM. "36"
   INDICATES TRUE NORTH.
RESULTANT WIND IS THE VECTOR AVERAGE OF THE SPEED AND
   DIRECTION.
AVERAGE TEMPERATURE IS THE SUM OF THE MEAN DAILY MAXIMUM
   AND MINIMUM TEMPERATURE DIVIDED BY 2.
SNOWFALL DATA COMPRISE ALL FORMS OF FROZEN
   PRECIPITATION, INCLUDING HAIL.
A HEATING (COOLING) DEGREE DAY IS THE DIFFERENCE BETWEEN
   THE AVERAGE DAILY TEMPERATURE AND 65 F.
DRY BULB IS THE TEMPERATURE OF THE AMBIENT AIR.
DEW POINT IS THE TEMPERATURE TO WHICH THE AIR MUST BE
   COOLED TO ACHIEVE 100 PERCENT RELATIVE HUMIDITY.
WET BULB IS THE TEMPERATURE THE AIR WOULD HAVE IF THE
   MOISTURE CONTENT WAS INCREASED TO 100 PERCENT RELATIVE
   HUMIDITY.

ON JULY 1, 1996, THE NATIONAL WEATHER SERVICE BEGAN USING
   THE "METAR" OBSERVATION CODE THAT WAS ALREADY EMPLOYED
   BY MOST OTHER NATIONS OF THE WORLD. THE MOST NOTICEABLE
   DIFFERENCE IN THIS ANNUAL PUBLICATION WILL BE THE CHANGE
   IN UNITS FROM TENTHS TO EIGHTS (OKTAS) FOR REPORTING THE
   AMOUNT OF SKY COVER.

HEATING DEGREE DAYS (base 65 F) 2001 HOUGHTON LAKE, MI (HTL)

YEAR	JUL	AUG	SEP	OCT	NOV	DEC	JAN	FEB	MAR	APR	MAY	JUN	TOTAL
1972-73	77	92	242	716	933	1326	1324	1307	829	653	442	54	7995
1973-74	26	25	280	420	871	1298	1380	1411	1171	649	463	135	8129
1974-75	27	56	337	633	856	1197	1327	1234	1254	836	215	101	8073
1975-76	44	68	333	472	720	1272	1575	1180	1054	590	434	53	7795
1976-77	24	92	296	698	1094	1582	1743	1315	932	544	199	159	8678
1977-78	29	134	200	610	869	1347	1556	1513	1276	769	289	144	8736
1978-79	79	46	220	636	869	1333	1655	1530	1053	742	396	135	8694
1979-80	48	109	196	597	893	1153	1416	1424	1233	690	286	209	8254
1980-81	18	27	258	716	928	1423	1553	1143	1001	613	368	80	8128
1981-82	50	44	280	676	872	1222	1658	1315	1209	824	148	224	8522
1982-83	12	139	255	494	855	1040	1331	1087	1003	733	505	116	7570
1983-84	25	29	209	569	846	1487	1628	1062	1262	599	458	53	8227
1984-85	42	28	270	481	881	1168	1547	1313	1071	571	248	172	7792
1985-86	46	88	209	535	908	1451	1474	1297	1048	532	274	148	8010
1986-87	23	109	212	557	975	1179	1335	1159	963	523	271	57	7363
1987-88	41	60	169	649	813	1117	1473	1406	1140	636	225	106	7835
1988-89	7	75	218	692	847	1276	1236	1381	1234	711	333	126	8136
1989-90	19	79	277	520	1025	1607	1196	1217	1009	627	398	97	8071
1990-91	32	51	248	604	773	1208	1509	1127	985	540	219	33	7329
1991-92	37	30	295	525	939	1219	1298	1178	1123	763	341	161	7909
1992-93	83	132	257	619	934	1196	1359	1358	1119	737	331	123	8248
1993-94	4	37	358	627	904	1205	1722	1457	1073	649	320	79	8435
1994-95	26	75	132	431	763	1059	1293	1307	947	741	320	35	7129
1995-96	22	2	260	461	1072	1403	1515	1307	1214	840	440	91	8627
1996-97	65	37	212	552	1021	1240	1481	1246	1162	737	561	64	8378
1997-98	79	134	239	582	972	1135	1260	968	1041	592	175	146	7323
1998-99	34	48	176	499	805	1118	1477	1083	1105	602	282	100	7329
1999-00	17	91	249	617	754	1193	1456	1154	826	701	299	114	7471
2000-01	83	69	256	473	867	1512	1303	1250	1149	558	262	127	7909
2001-	62	62	296	567	677	1032							

COOLING DEGREE DAYS (base 65 F) 2001 HOUGHTON LAKE, MI (HTL)

YEAR	JAN	FEB	MAR	APR	MAY	JUN	JUL	AUG	SEP	OCT	NOV	DEC	ANNUAL
1972	0	0	0	0	11	19	131	78	4	0	0	0	243
1973	0	0	0	1	0	58	119	155	60	2	0	0	395
1974	0	0	0	1	6	41	120	62	11	0	0	0	241
1975	0	0	0	0	58	83	146	78	2	2	0	0	369
1976	0	0	0	13	0	92	93	86	24	0	0	0	308
1977	0	0	0	0	61	46	192	50	5	0	0	0	354
1978	0	0	0	0	44	49	85	96	34	0	0	0	308
1979	0	0	0	0	18	58	112	45	31	7	0	0	271
1980	0	0	0	2	17	43	98	124	16	0	0	0	300
1981	0	0	0	0	3	47	114	70	13	0	0	0	247
1982	0	0	0	0	27	3	132	52	29	3	0	0	246
1983	0	0	0	0	0	73	235	148	56	0	0	0	512
1984	0	0	0	1	2	65	90	151	11	0	0	0	320
1985	0	0	0	22	8	11	86	61	60	0	0	0	248
1986	0	0	0	8	29	22	175	46	15	0	0	0	295
1987	0	0	0	1	64	125	250	86	12	0	0	0	538
1988	0	0	0	0	35	108	213	199	11	0	0	0	566
1989	0	0	0	0	10	49	164	83	19	0	0	0	325
1990	0	0	0	29	0	55	94	56	41	0	0	0	275
1991	0	0	0	2	104	120	157	130	26	0	0	0	539
1992	0	0	0	0	15	31	28	46	19	0	0	0	139
1993	0	0	0	0	16	34	175	161	4	0	0	0	390
1994	0	0	0	0	19	97	130	64	45	0	0	0	355
1995	0	0	0	0	4	160	202	233	24	5	0	0	628
1996	0	0	0	0	10	67	41	85	25	0	0	0	228
1997	0	0	0	0	0	63	112	39	7	0	0	0	221
1998	0	0	0	0	34	93	98	110	45	0	0	0	380
1999	0	0	0	0	15	107	174	41	10	0	0	0	347
2000	0	0	0	0	24	51	79	65	42	1	0	0	262
2001	0	0	0	0	12	87	132	130	19	0	0	0	380

SNOWFALL (inches) 2001 HOUGHTON LAKE, MI (HTL)

YEAR	JUL	AUG	SEP	OCT	NOV	DEC	JAN	FEB	MAR	APR	MAY	JUN	TOTAL
1972-73	0.0	0.0	0.0	0.6	4.1	29.2	8.3	15.2	5.3	7.9	0.8	0.0	71.4
1973-74	0.0	0.0	0.0	T	4.8	13.0	11.2	17.5	12.3	3.1	1.0	0.0	62.9
1974-75	0.0	0.0	0.0	T	6.0	18.0	11.8	12.4	14.0	4.5	0.0	0.0	66.7
1975-76	0.0	0.0	T	T	5.9	14.6	29.6	20.8	17.6	6.0	0.6	0.0	95.1
1976-77	0.0	0.0	0.0	0.7	13.6	18.6	16.3	11.4	3.9	2.7	0.0	0.0	67.2
1977-78	0.0	0.0	0.0	T	16.0	14.0	33.3	13.1	12.5	0.3	0.0	0.0	89.2
1978-79	0.0	0.0	0.0	T	10.5	29.0	21.8	8.3	5.7	11.6	2.3	0.0	89.2
1979-80	0.0	0.0	0.0	0.7	5.0	7.1	14.4	12.3	12.1	7.7	0.0	0.0	59.3
1980-81	0.0	0.0	0.0	4.4	3.2	25.6	19.3	16.3	5.4	0.2	0.0	0.0	74.4
1981-82	0.0	0.0	0.0	3.0	16.1	15.7	38.0	6.1	14.3	5.5	0.0	0.0	98.7
1982-83	0.0	0.0	0.0	0.2	7.7	4.8	15.4	6.4	13.3	2.7	1.0	0.0	51.5
1983-84	0.0	0.0	0.0	0.0	5.0	19.8	17.0	6.3	6.5	5.1	0.4	0.0	60.1
1984-85	0.0	0.0	0.0	0.0	0.6	11.7	25.3	18.7	14.1	10.0	0.0	0.0	80.4
1985-86	0.0	0.0	0.0	0.0	12.6	23.1	12.7	14.0	8.9	0.3	0.0	0.0	71.6
1986-87	0.0	0.0	0.0	T	3.4	10.3	14.1	6.4	2.3	2.0	0.0	0.0	38.5
1987-88	0.0	0.0	0.0	2.4	8.7	18.6	13.4	12.9	10.3	1.4	0.0	0.0	67.7
1988-89	0.0	0.0	0.0	0.0	7.8	12.3	9.3	13.5	18.7	3.1	T	0.0	
1989-90	0.0	0.0	T	0.6	12.2	15.0	21.7	11.0	2.4	2.1	0.9	0.0	65.9
1990-91	0.0	0.0	0.0	T	9.1	15.3	22.6	8.5	4.5	1.8	0.0	0.0	61.8
1991-92	0.0	0.0	T	T	4.9	17.5	15.0	15.8	9.5	3.6	0.0	0.0	66.3
1992-93	0.0	0.0	0.0	3.2	11.8	14.3	14.7	13.1	8.1	2.9	0.0	0.0	68.1
1993-94	0.0	T	0.0	0.1	2.3	9.0	26.2	12.7	9.0	3.9	0.1	0.0	63.3
1994-95	0.0	0.0	0.0	0.0	6.6	7.6	12.2	10.2	7.4	1.9	0.0	0.0	45.9
1995-96	0.0	0.0	T	T	41.9	16.4	26.0	8.7	13.4	10.6		0.0	
1996-97	T	0.0	T	T	8.9								
1997-98													
1998-99													
1999-00						1.6	3.7	16.1	10.4	T			
2000-01													
2001-													
POR= 31 YRS	T	T	0.0	0.7	10.2	16.0	19.3	12.3	11.3	4.2	0.5	0.0	74.5

2001
MARQUETTE COUNTY AIRPORT, MICHIGAN (MQT)

The Marquette County Airport lies about 7.5 miles southwest of the nearest shoreline of Lake Superior and about 8 miles west of the city of Marquette. Lake Superior is the largest body of fresh water in the world and the deepest and coldest of the Great Lakes. An irregular northwest-southeast ridge line lies just to the east of the airport. There are several water storage basins in the vicinity of the station. One basin, about 20 miles long, is 3 miles northwest and another, about 8 miles in diameter, is 3 miles west.

The climate is influenced considerably by the proximity of Lake Superior. As a consequence of the cool expanse of water in the summer, there is rarely a long period of sweltering hot weather. Periods of drought are extremely rare. In the winter, cold outbreaks are tempered considerably by the waters of Lake Superior if the lake is unfrozen. However, winds blowing across these relatively warmer waters pick up moisture and cause cloudy weather throughout the winter, as well as frequent periods of light snow. Lake-formed snow showers and snow squalls are intensified near the station by upslope winds, especially from the northwest through northeast. With a northeast through east wind, especially in autumn, the upslope condition will cause light snow at the airport, while along the lakeshore, only drizzle or no precipitation may occur.

The growing season averages 117 days. Precipitation is rather evenly distributed throughout the year, with an average precipitation of 4 inches or more in June and September and less than 2 inch averages only in January and February. One hundred inches or more of snow occur in nine of ten winter seasons.

NORMALS, MEANS, AND EXTREMES
MARQUETTE, MI (MQT)

LATITUDE:	LONGITUDE:	ELEVATION (FT):		TIME ZONE:	WBAN: 94850
46 32' 0 " N	87 34' 0 " W	GRND: 1415	BARO: 1415	EASTERN (UTC + 5)	

	ELEMENT	POR	JAN	FEB	MAR	APR	MAY	JUN	JUL	AUG	SEP	OCT	NOV	DEC	YEAR
TEMPERATURE F	NORMAL DAILY MAXIMUM	30	20.5	23.9	33.5	47.5	61.9	70.9	76.6	73.3	63.8	52.3	36.9	24.6	48.8
	MEAN DAILY MAXIMUM	39	20.9	24.6	34.1	47.2	62.2	71.3	76.2	73.6	64.0	52.3	36.9	25.4	49.1
	HIGHEST DAILY MAXIMUM	23	46	61	71	92	93	96	99	96	92	87	73	59	99
	YEAR OF OCCURRENCE		1981	1981	2000	1980	1986	1995	1988	2001	1998	1992	1999	1998	JUL 1988
	MEAN OF EXTREME MAXS.	39	37.7	43.6	55.0	72.0	84.3	87.8	89.7	88.1	82.0	73.6	56.5	41.3	67.6
	NORMAL DAILY MINIMUM	30	3.3	4.0	13.6	27.0	38.4	47.4	53.5	51.7	44.1	34.8	23.3	9.9	29.2
	MEAN DAILY MINIMUM	39	4.0	5.0	14.0	27.3	38.9	48.4	53.5	52.0	44.5	34.9	23.4	11.3	29.8
	LOWEST DAILY MINIMUM	23	-27	-34	-23	-5	17	28	36	34	24	14	-5	-28	-34
	YEAR OF OCCURRENCE		1996	1979	1982	1979	1983	1986	2000	1992	1993	1984	1989	1983	FEB 1979
	MEAN OF EXTREME MINS.	39	-18.1	-18.4	-10.6	8.6	24.6	33.5	39.9	37.9	29.1	21.6	4.6	-11.0	11.8
	NORMAL DRY BULB	30	11.9	14.0	23.6	37.3	50.2	59.2	65.1	62.5	54.0	43.6	30.1	17.3	39.1
	MEAN DRY BULB	39	12.6	14.7	23.9	37.4	50.5	59.8	64.9	62.9	54.3	43.6	30.1	18.2	39.4
	MEAN WET BULB														
	MEAN DEW POINT														
	NORMAL NO. DAYS WITH:														
	MAXIMUM 90	30	0.0	0.0	0.0	*	0.2	0.5	1.6	0.6	0.1	0.0	0.0	0.0	3.0
	MAXIMUM 32	30	26.9	22.2	14.7	2.9	0.1	0.0	0.0	0.0	0.0	0.5	10.8	24.9	103.0
	MINIMUM 32	30	30.9	27.9	29.3	22.4	9.8	0.9	0.0	0.1	2.9	14.8	26.2	30.7	195.9
	MINIMUM 0	30	12.3	11.7	6.1	0.3	0.0	0.0	0.0	0.0	0.0	0.0	0.7	7.5	38.6
H/C	NORMAL HEATING DEG. DAYS	30	1646	1428	1283	831	471	193	74	122	330	663	1047	1479	9567
	NORMAL COOLING DEG. DAYS	30	0	0	0	0	13	19	78	45	0	0	0	0	155
RH	NORMAL (PERCENT)														
	HOUR 01 LST														
	HOUR 07 LST														
	HOUR 13 LST														
	HOUR 19 LST														
S	PERCENT POSSIBLE SUNSHINE	20	34	40	48	50	60	63	65	61	53	44	34	32	49
W/O	MEAN NO. DAYS WITH:														
	HEAVY FOG(VISBY 1/4 MI)	21	0.9	1.0	1.8	2.3	2.3	1.2	2.1	3.0	2.8	2.4	1.0	1.4	22.2
	THUNDERSTORMS	21	0.0	0.1	0.4	1.0	2.4	4.6	5.3	4.5	3.2	1.3	0.1	0.0	22.9
CLOUDINESS	MEAN:														
	SUNRISE-SUNSET (OKTAS)														
	MIDNIGHT-MIDNIGHT (OKTAS)														
	MEAN NO. DAYS WITH:														
	CLEAR														
	PARTLY CLOUDY														
	CLOUDY														
PR	MEAN STATION PRESSURE(IN)														
	MEAN SEA-LEVEL PRES. (IN)							30.01							
WINDS	MEAN SPEED (MPH)														
	PREVAIL.DIR(TENS OF DEGS)														
	FASTEST MILE:														
	SPEED (MPH)	6	44	31	40	44	34	38	35	37	35	38	31	35	44
	DIR.		NW	NW	NW	NW	N	NW	NW	NW	NW	SE	NW	SW	NW
	YEAR OF OCCURRENCE		1980	1985	1982	1982	1981	1984	1982	1984	1983	1984	1979	1982	APR 1982
	PEAK GUST :														
	SPEED (MPH)														
	DIR. (TENS OF DEGS)														
	YEAR OF OCCURRENCE														
PRECIPITATION	NORMAL (IN)	30	2.17	1.73	2.77	2.64	3.03	3.48	2.88	3.41	4.08	3.61	2.89	2.61	35.30
	MAXIMUM MONTHLY (IN)	23	6.61	3.68	6.08	6.56	6.49	6.61	5.40	8.59	6.94	7.59	8.25	4.45	8.59
	YEAR OF OCCURRENCE		1997	1984	1979	1985	1983	1981	1991	1988	1980	1979	1988	1996	AUG 1988
	MINIMUM MONTHLY (IN)	23	0.92	0.48	0.56	0.90	0.06	0.61	0.57	0.81	1.21	1.65	1.00	0.37	0.06
	YEAR OF OCCURRENCE		1991	1994	1980	1998	1986	1992	1981	1991	1989	1994	1990	1994	MAY 1986
	MAXIMUM IN 24 HOURS (IN)	22	2.23	2.05	2.40	3.09	3.44	2.80	2.64	2.34	2.34	3.66	2.97	2.48	3.66
	YEAR OF OCCURRENCE		1988	1983	1986	1985	1983	1989	1985	1988	1993	1985	1988	1985	OCT 1985
	NORMAL NO. DAYS WITH:														
	PRECIPITATION 0.01	30	16.4	12.6	13.5	10.9	10.6	12.5	9.5	11.9	14.0	14.1	15.0	17.6	158.6
	PRECIPITATION 1.00	30	0.2	0.1	0.6	0.4	0.8	0.7	0.6	0.8	0.8	0.9	0.5	0.3	6.7
SNOWFALL	NORMAL (IN)	30	35.2	24.5	28.9	9.8	1.5	T	0.0	0.0	0.1	5.2	19.8	36.4	161.4
	MAXIMUM MONTHLY (IN)	23	91.7	63.6	59.1	43.4	22.6	T	T	T	1.7	18.6	48.9	89.5	91.7
	YEAR OF OCCURRENCE		1997	1995	1985	1996	1990	1994	2000	1989	1993	1979	1991	2000	JAN 1997
	MAXIMUM IN 24 HOURS (IN)	22	23.3	20.6	28.0	20.0	17.2	T	T	T	1.7	12.7	24.4	25.8	28.0
	YEAR OF OCCURRENCE		1988	1983	1997	1990	1990	1994	2000	1989	1993	1989	2001	1985	MAR 1997
	MAXIMUM SNOW DEPTH (IN)	38	54	56	63	41	12	0	0	0	1	10	27	47	63
	YEAR OF OCCURRENCE		1969	1971	1997	1997	1990				1974	1989	1975	1983	MAR 1997
	NORMAL NO. DAYS WITH:														
	SNOWFALL 1.0	30	9.0	6.2	5.8	2.6	0.4	0.0	0.0	0.0	0.*	1.6	5.8	8.3	39.7

PRECIPITATION (inches) 2001 MARQUETTE COUNTY AIRPORT, MI (MQT)

YEAR	JAN	FEB	MAR	APR	MAY	JUN	JUL	AUG	SEP	OCT	NOV	DEC	ANNUAL
1972	1.53	1.43	3.25	2.62	2.50	1.70	2.85	4.12	5.32	1.96	3.39	3.03	33.70
1973	1.47	0.99	2.37	3.01	3.42	2.16	2.93	2.03	2.53	1.16	2.43	31.66
1974	1.31	1.42	0.61	3.28	2.49	3.57	1.80	3.87	4.01	3.05	2.87	0.81	29.09
1975	2.85	2.10	2.11	2.41	3.53	4.59	1.06	2.97	2.80	1.34	3.16	1.87	30.79
1976	2.65	2.22	3.95	1.83	2.95	1.63	1.52	0.50	1.32	2.24	1.68	1.79	24.28
1977	1.01	1.57	4.09	3.19	2.04	3.10	3.89	3.55	5.17	2.65	2.52	3.24	36.02
1978	3.03	0.89	0.56	1.49	3.34	2.68	4.10	4.47	4.54	1.72	2.72	2.18	31.72
1979	2.43	1.99	6.08	1.84	2.70	6.13	4.63	1.41	2.09	7.59	2.47	1.56	40.92
1980	2.95	1.47	0.56	4.11	1.87	2.51	3.18	3.49	6.94	2.71	1.92	2.21	33.92
1981	1.96	2.18	2.30	3.10	2.43	6.61	0.57	2.28	2.11	4.63	2.00	4.33	34.50
1982	3.80	0.59	1.89	4.45	2.76	1.24	4.65	3.45	5.00	3.87	2.53	2.83	37.06
1983	2.67	3.14	4.85	3.17	6.49	1.62	1.45	3.21	4.98	5.75	5.68	3.35	46.36
1984	1.27	3.68	3.99	2.46	0.79	2.81	1.92	4.29	4.29	2.35	2.60	32.48
1985	2.96	2.91	4.63	6.56	3.59	1.49	3.76	4.78	6.32	4.88	5.74	3.97	51.59
1986	2.94	1.21	4.77	2.36	0.06	2.45	3.07	4.62	2.85	4.50	1.04	0.52	30.39
1987	1.42	1.53	2.07	2.30	3.93	2.01	4.88	3.58	2.66	4.31	4.05	3.56	36.30
1988	4.02	0.95	3.66	1.48	1.34	0.71	1.46	8.59	3.97	5.01	8.25	2.36	41.80
1989	1.82	1.60	2.89	1.48	2.15	5.36	0.91	2.61	1.21	3.49	3.43	2.58	29.53
1990	1.74	1.96	1.88	2.32	4.25	3.13	2.01	2.70	4.76	5.27	1.00	1.71	32.73
1991	0.92	1.78	3.75	3.50	2.64	3.06	5.40	0.81	3.10	4.73	5.87	1.51	37.07
1992	1.28	2.66	2.57	2.12	1.76	0.61	4.93	3.28	3.35	1.98	4.27	2.46	31.27
1993	2.39	0.68	0.99	4.11	4.61	2.42	1.95	3.98	4.41	3.32	2.84	1.31	33.01
1994	1.65	0.48	2.27	2.27	2.31	2.60	1.48	3.98	2.92	1.65	2.17	0.37	24.15
1995	1.45	2.89	3.31	2.31	3.85	1.49	3.93	4.31	3.60	6.37	2.97	2.48	38.96
1996	5.22	2.91	2.09	6.50	1.70	3.94	4.43	2.11	3.58	3.64	1.87	4.45	42.44
1997	6.61	0.97	3.65	1.03	2.96	3.44	1.60	3.01	1.36	2.38	3.48	2.40	32.89
1998	2.71	1.12	5.35	0.90	1.93	3.06	3.01	2.70	3.56	1.93	2.89	1.60	28.34
1999	4.93	3.15	1.04	2.21	5.74	3.05	4.78	2.35	3.09	3.47	2.27	1.54	37.62
2000	2.34	1.26	3.34	1.11	1.22	3.63	2.10	3.50	1.90	2.80	3.66	3.02	29.88
2001	2.80	2.00	1.72	3.18	2.37	4.21	3.03	1.26	3.93	3.19	4.89	1.66	34.24
POR= 129 YRS	2.14	1.72	2.22	2.50	2.89	3.33	2.99	2.79	3.48	2.91	2.92	2.23	32.12

AVERAGE TEMPERATURE (F) 2001 MARQUETTE COUNTY AIRPORT, MI (MQT)

YEAR	JAN	FEB	MAR	APR	MAY	JUN	JUL	AUG	SEP	OCT	NOV	DEC	ANNUAL
1972	13.6	14.9	22.8	34.9	52.3	57.0	63.9	63.3	54.6	43.6	33.3	18.9	39.4
1973	22.1	20.5	35.2	39.8	45.4	60.3	67.0	69.1	58.6	53.4	35.1	23.0	44.1
1974	17.1	16.4	26.7	39.9	47.7	59.7	69.1	64.5	52.9	46.7	36.6	29.4	42.2
1975	20.4	23.2	26.5	35.4	54.7	61.1	70.3	67.1	54.2	50.1	39.4	23.1	43.8
1976	15.6	26.1	28.4	43.1	48.5	64.1	66.5	67.6	56.8	43.4	29.1	14.6	42.0
1977	10.1	19.9	34.3	43.8	58.4	59.2	67.3	61.8	56.6	48.4	35.0	22.0	43.1
1978	17.9	18.6	27.4	37.3	53.3	60.5	65.1	66.4	58.7	47.0	34.2	20.1	42.2
1979	5.6	6.9	24.4	34.4	46.2	58.1	64.7	60.8	55.6	39.5	28.8	23.5	37.4
1980	13.1	12.8	21.3	39.1	54.5	56.9	65.7	64.5	52.9	38.6	29.8	15.4	38.7
1981	13.9	18.4	27.3	39.0	48.1	59.6	65.7	63.8	52.0	40.1	34.5	18.7	40.1
1982	4.8	12.1	23.1	32.8	54.9	54.0	66.0	59.8	53.8	45.4	29.1	23.3	38.3
1983	18.0	21.9	26.0	33.8	44.1	60.5	70.1	67.9	57.5	43.2	31.9	8.7	40.3
1984	9.0	25.3	17.5	40.3	48.2	62.0	64.4	65.4	52.4	46.7	30.5	18.7	40.0
1985	11.3	11.8	26.7	40.9	52.1	56.8	62.6	61.4	54.7	44.0	26.2	9.7	38.2
1986	14.4	14.8	26.4	42.0	54.8	57.1	66.3	60.0	53.2	42.7	25.2	21.3	39.9
1987	18.0	22.9	30.1	44.6	53.7	63.8	67.8	62.7	56.9	39.0	32.6	23.5	43.0
1988	11.6	10.8	23.1	38.5	54.7	62.3	68.6	65.0	54.9	38.0	31.8	16.3	39.6
1989	19.1	8.4	19.4	34.7	50.8	57.5	66.7	63.2	55.7	45.3	24.8	9.2	37.9
1990	22.0	18.1	28.5	42.3	46.9	59.9	63.7	63.2	57.3	41.3	34.0	18.0	41.0
1991	10.0	20.1	26.7	42.2	55.7	62.9	64.9	65.2	52.4	41.0	27.0	19.8	40.7
1992	17.0	19.8	22.6	34.6	52.6	56.3	58.4	59.2	53.5	41.4	28.4	20.2	38.7
1993	16.9	13.9	25.6	34.4	49.8	57.8	66.1	65.4	49.9	40.3	27.0	20.8	39.0
1994	2.8	7.7	27.3	36.8	50.2	61.9	64.0	60.7	58.2	47.3	33.7	27.5	39.8
1995	17.4	15.0	28.2	32.1	50.4	66.3	65.6	68.1	53.9	44.3	22.8	16.1	40.0
1996	10.5	11.9	19.5	31.5	46.6	62.3	61.6	64.6	56.3	43.8	26.8	19.5	37.9
1997	12.9	16.9	21.2	35.8	44.8	64.6	63.6	59.6	56.0	44.0	27.3	24.7	39.3
1998	17.8	28.4	24.7	41.7	55.0	59.5	65.5	65.8	58.3	47.4	34.1	22.4	43.4
1999	11.7	21.5	26.8	39.3	54.0	62.3	68.8	62.2	56.7	41.8	37.2	20.7	41.9
2000	12.9	20.1	32.9	35.8	53.9	58.0	62.8	62.4	54.4	47.3	31.4	12.1	40.3
2001	20.8	12.6	23.7	41.8	53.3	61.4	65.2	66.2	55.4	43.2	39.9	25.9	42.5
POR= 127 YRS	16.7	17.6	25.9	38.6	49.5	59.3	65.7	64.3	57.0	46.6	33.1	22.3	41.4

REFERENCE NOTES:

PAGE 1:
THE TEMPERATURE GRAPH SHOWS NORMAL MAXIMUM AND NORMAL
MINIMUM DAILY TEMPERATURES (SOLID CURVES) AND THE
ACTUAL DAILY HIGH AND LOW TEMPERATURES (VERTICAL BARS).

PAGE 2 AND 3:
H/C INDICATES HEATING AND COOLING DEGREE DAYS.
RH INDICATES RELATIVE HUMIDITY
W/O INDICATES WEATHER AND OBSTRUCTIONS
S INDICATES SUNSHINE.
PR INDICATES PRESSURE.
CLOUDINESS ON PAGE 3 IS THE SUM OF THE CEILOMETER AND
SATELLITE DATA NOT TO EXCEED EIGHT EIGHTHS(OKTAS).

GENERAL:
T INDICATES TRACE PRECIPITATION, AN AMOUNT GREATER
THAN ZERO BUT LESS THAN THE LOWEST REPORTABLE VALUE.
+ INDICATES THE VALUE ALSO OCCURS ON EARLIER DATES.
BLANK ENTRIES DENOTE MISSING OR UNREPORTED DATA.
NORMALS ARE 30-YEAR AVERAGES (1961 - 1990).
ASOS INDICATES AUTOMATED SURFACE OBSERVING SYSTEM.
PM INDICATES THE LAST DAY OF THE PREVIOUS MONTH.
POR (PERIOD OF RECORD) BEGINS WITH THE JANUARY DATA
MONTH AND IS THE NUMBER OF YEARS USED TO COMPUTE
THE MEAN. INDIVIDUAL MONTHS WITHIN THE POR MAY
BE MISSING.
WHEN THE POR FOR A NORMAL IS LESS THAN 30 YEARS,
THE NORMAL IS PROVISIONAL AND IS BASED ON THE NUMBER
OF YEARS INDICATED.
0.* OR * INDICATES THE VALUE OR MEAN-DAYS-WITH
IS BETWEEN 0.00 AND 0.05.
CLOUDINESS FOR ASOS STATIONS DIFFERS FROM THE NON-ASOS
OBSERVATION TAKEN BY A HUMAN OBSERVER. ASOS STATION
CLOUDINESS IS BASED ON TIME-AVERAGED CEILOMETER DATA
FOR CLOUDS AT OR BELOW 12,000 FEET AND ON SATELLITE
DATA FOR CLOUDS ABOVE 12,000 FEET.
THE NUMBER OF DAYS WITH CLEAR, PARTLY CLOUDY, AND
CLOUDY CONDITIONS FOR ASOS STATIONS IS THE SUM
OF THE CEILOMETER AND SATELLITE DATA FOR THE
SUNRISE TO SUNSET PERIOD.

GENERAL CONTINUED:
CLEAR INDICATES 0 - 2 OKTAS, PARTLY CLOUDY INDICATES
3 - 6 OKTAS, AND CLOUDY INDICATES 7 OR 8 OKTAS.
WHEN AT LEAST ONE OF THE ELEMENTS (CEILOMETER OR
SATELLITE) IS MISSING, THE DAILY CLOUDINESS IS
NOT COMPUTED.
WIND DIRECTION IS RECORDED IN TENS OF DEGREES (2 DIGITS)
CLOCKWISE FROM TRUE NORTH. "00" INDICATES CALM. "36"
INDICATES TRUE NORTH.
RESULTANT WIND IS THE VECTOR AVERAGE OF THE SPEED AND
DIRECTION.
AVERAGE TEMPERATURE IS THE SUM OF THE MEAN DAILY MAXIMUM
AND MINIMUM TEMPERATURE DIVIDED BY 2.
SNOWFALL DATA COMPRISE ALL FORMS OF FROZEN
PRECIPITATION, INCLUDING HAIL.
A HEATING (COOLING) DEGREE DAY IS THE DIFFERENCE BETWEEN
THE AVERAGE DAILY TEMPERATURE AND 65 F.
DRY BULB IS THE TEMPERATURE OF THE AMBIENT AIR.
DEW POINT IS THE TEMPERATURE TO WHICH THE AIR MUST BE
COOLED TO ACHIEVE 100 PERCENT RELATIVE HUMIDITY.
WET BULB IS THE TEMPERATURE THE AIR WOULD HAVE IF THE
MOISTURE CONTENT WAS INCREASED TO 100 PERCENT RELATIVE
HUMIDITY.

ON JULY 1, 1996, THE NATIONAL WEATHER SERVICE BEGAN USING
THE "METAR" OBSERVATION CODE THAT WAS ALREADY EMPLOYED
BY MOST OTHER NATIONS OF THE WORLD. THE MOST NOTICEABLE
DIFFERENCE IN THIS ANNUAL PUBLICATION WILL BE THE CHANGE
IN UNITS FROM TENTHS TO EIGHTS(OKTAS) FOR REPORTING THE
AMOUNT OF SKY COVER.

HEATING DEGREE DAYS (base 65 F) 2001 MARQUETTE COUNTY AIRPORT, MI (MQT)

YEAR	JUL	AUG	SEP	OCT	NOV	DEC	JAN	FEB	MAR	APR	MAY	JUN	TOTAL
1972-73	103	121	305	653	950	1423	1323	1239	916	752	598	159	8542
1973-74	59	37	244	360	889	1296	1482	1354	1180	747	535	192	8375
1974-75	27	77	360	563	843	1097	1376	1165	1188	880	320	176	8072
1975-76	49	45	317	455	762	1290	1525	1120	1128	653	506	94	7944
1976-77	36	79	293	668	1069	1556	1695	1253	946	630	239	193	8657
1977-78	50	126	244	504	894	1329	1455	1295	1156	826	381	183	8443
1978-79	64	57	220	554	918	1385	1840	1626	1254	914	577	214	9623
1979-80	93	153	295	785	1080	1282	1604	1508	1350	772	340	271	9533
1980-81	53	68	361	813	1050	1532	1581	1303	1160	776	519	168	9384
1981-82	73	78	384	766	907	1425	1864	1480	1292	960	313	326	9868
1982-83	37	190	350	598	1069	1289	1446	1201	1205	929	638	207	9159
1983-84	35	38	264	672	989	1740	1737	1145	1469	733	517	124	9463
1984-85	71	72	369	560	1028	1431	1658	1486	1178	724	400	249	9226
1985-86	111	146	323	645	1156	1709	1564	1400	1191	684	334	250	9513
1986-87	71	169	349	684	1185	1348	1450	1171	1075	606	387	106	8601
1987-88	58	130	247	800	964	1278	1650	1569	1295	787	347	166	9291
1988-89	37	101	305	833	988	1504	1416	1584	1412	902	432	245	9759
1989-90	65	117	283	610	1199	1727	1324	1308	1123	702	556	176	9190
1990-91	104	111	347	726	921	1454	1703	1250	1182	678	334	115	8925
1991-92	74	112	385	739	1133	1394	1483	1308	1306	905	398	276	9513
1992-93	208	202	340	732	1092	1383	1485	1424	1215	910	474	233	9698
1993-94	38	78	446	760	1130	1366	1927	1601	1164	844	470	135	9959
1994-95	84	153	220	538	932	1155	1471	1395	1137	978	452	106	8621
1995-96	68	21	336	637	1259	1509	1688	1538	1407	997	566	141	10167
1996-97	126	73	294	649	1140	1402	1607	1343	1350	867	620	89	9560
1997-98	111	193	276	649	1127	1243	1455	1019	1243	690	326	203	8535
1998-99	65	51	221	538	920	1314	1646	1213	1178	764	346	153	8409
1999-00	40	111	289	711	827	1364	1609	1295	990	867	349	208	8660
2000-01	117	108	317	543	1000	1632	1365	1460	1268	693	357	189	9049
2001-	94	63	302	672	745	1205							

COOLING DEGREE DAYS (base 65 F) 2001 MARQUETTE COUNTY AIRPORT, MI (MQT)

YEAR	JAN	FEB	MAR	APR	MAY	JUN	JUL	AUG	SEP	OCT	NOV	DEC	ANNUAL
1972	0	0	0	0	14	26	76	73	0	0	0	0	189
1973	0	0	0	0	0	25	127	172	59	6	0	0	389
1974	0	0	0	0	5	40	162	70	3	0	0	0	280
1975	0	0	0	0	9	67	218	118	1	0	0	0	413
1976	0	0	0	3	0	75	92	166	53	4	0	0	393
1977	0	0	0	0	44	24	126	35	0	0	0	0	229
1978	0	0	0	0	25	53	76	108	38	2	0	0	302
1979	0	0	0	0	0	15	92	29	19	0	0	0	155
1980	0	0	0	2	21	36	83	61	5	0	0	0	208
1981	0	0	0	0	1	13	103	50	2	0	0	0	169
1982	0	0	0	0	7	2	75	35	23	0	0	0	142
1983	0	0	0	0	0	78	200	135	47	1	0	0	461
1984	0	0	0	0	3	42	59	89	0	0	0	0	193
1985	0	0	0	7	7	10	44	41	21	0	0	0	130
1986	0	0	0	1	21	17	119	21	0	0	0	0	179
1987	0	0	0	3	41	77	150	65	13	0	0	0	349
1988	0	0	0	0	32	89	157	106	8	0	0	0	392
1989	0	0	0	0	1	27	123	70	13	1	0	0	235
1990	0	0	0	24	0	29	70	63	14	0	0	0	200
1991	0	0	0	0	51	60	77	126	13	0	0	0	327
1992	0	0	0	0	21	22	8	30	3	5	0	0	89
1993	0	0	0	0	10	22	79	97	0	0	0	0	208
1994	0	0	0	0	18	49	62	29	24	0	0	0	182
1995	0	0	0	0	8	152	93	123	8	5	0	0	389
1996	0	0	0	0	3	70	29	67	40	0	0	0	209
1997	0	0	0	0	0	80	74	33	9	3	0	0	199
1998	0	0	0	0	25	47	86	83	26	0	0	0	267
1999	0	0	0	0	13	77	162	31	47	0	0	0	330
2000	0	0	0	0	14	7	57	35	6	1	0	0	120
2001	0	0	0	5	1	91	110	109	21				337

SNOWFALL (inches) 2001 MARQUETTE COUNTY AIRPORT, MI (MQT)

YEAR	JUL	AUG	SEP	OCT	NOV	DEC	JAN	FEB	MAR	APR	MAY	JUN	TOTAL
1972-73	0.0	0.0	0.0	2.2	6.5	37.7	14.8	12.8	0.4	2.5	3.1	0.0	80.0
1973-74	0.0	0.0	0.0	T	7.4	27.9	15.5	18.2	5.2	13.2	2.3	0.0	89.7
1974-75	0.0	0.0	5.1	5.8	10.3	10.6	34.7	28.7	29.0	0.2	0.0	0.0	124.4
1975-76	0.0	0.0	T	0.1	16.9	15.3	45.6	25.9	44.3	1.4	4.3	0.0	153.8
1976-77	0.0	0.0	T	17.5	17.6	33.3	16.7	17.5	16.7	13.5	T	0.0	132.8
1977-78	0.0	0.0	0.0	T	12.2	38.3	43.3	16.3	6.1	1.0	0.0	0.0	177.2
1978-79	0.0	0.0	0.0	0.4	13.0	24.2	39.6	23.1	43.9	11.8	T	0.0	156.0
1979-80	0.0	0.0	0.0	18.6	24.8	18.9	33.9	30.2	7.0	11.3	1.4	0.0	146.1
1980-81	0.0	0.0	0.1	11.9	13.1	41.5	41.9	29.5	34.0	4.1	T	0.0	176.1
1981-82	0.0	0.0	0.0	14.2	15.3	82.6	68.8	9.6	24.1	29.2	0.0	0.0	243.8
1982-83	0.0	0.0	0.0	7.2	15.1	17.4	42.4	42.9	54.1	20.2	T	0.0	199.3
1983-84	0.0	0.0	T	2.5	31.6	54.3	30.2	38.1	46.3	1.1	T	0.0	204.1
1984-85	0.0	0.0	T	T	13.6	30.6	56.2	51.4	59.1	18.1	0.0	0.0	229.0
1985-86	0.0	0.0	0.0	T	28.1	56.4	49.8	17.4	49.1	6.8	0.1	0.0	207.7
1986-87	0.0	T	0.0	2.3	15.7	10.5	21.9	27.7	19.3	11.3	T	0.0	108.8
1987-88	0.0	0.0	T	8.3	21.4	36.1	62.8	28.0	49.0	5.8	0.0	0.0	211.4
1988-89	0.0	0.0	0.0	10.9	42.0	41.7	43.1	33.7	44.3	7.8	0.2	0.0	223.7
1989-90	0.0	T	0.5	16.6	41.7	58.4	26.4	35.8	16.3	17.1	22.6	T	235.4
1990-91	0.0	0.0	T	9.6	2.5	27.8	25.7	35.2	21.1	9.0	0.6	0.0	131.5
1991-92	0.0	0.0	T	2.9	48.9	30.7	33.3	49.6	23.3	16.7	T	0.0	205.4
1992-93	0.0	0.0	T	11.5	30.0	25.7	40.9	19.9	18.1	29.1	T	0.0	175.2
1993-94	0.0	0.0	1.7	8.7	29.0	22.9	46.8	12.4	28.3	2.7	0.1	T	152.6
1994-95	0.0	0.0	0.0	T	14.6	7.4	23.8	63.6	45.9	10.2	0.0	0.0	165.5
1995-96	0.0	0.0	T	3.3	31.7	45.1	66.6	37.8	22.9	43.4	0.6	0.0	251.4
1996-97	0.0	0.0	0.0	1.5	15.6	79.8	91.7	18.7	50.4	4.8	9.1	0.0	271.6
1997-98	0.0	0.0	0.0	8.3	37.2	38.8	41.7	15.3	35.4	5.7	0.0	T	182.4
1998-99	0.0	0.0	T	0.3	9.3	28.0	67.7	34.4	20.9	1.8	0.0	0.0	162.4
1999-00	0.0	0.0	0.0	3.5	7.5	30.0	44.4	30.8	27.4	8.8	0.0	T	152.4
2000-01	T	0.0	0.0	12.1	38.0	89.5	44.5	41.5	40.2	2.2	0.0	0.0	268.0
2001-	0.0	0.0	0.0	13.7	39.3	37.4							
POR= 63 YRS	0.0	T	0.2	4.2	18.2	28.1	29.8	23.7	22.4	9.0	1.6	T	137.2

2001
DULUTH,
MINNESOTA (DLH)

Duluth, Minnesota is located at the western tip of Lake Superior. The city, about 20 miles long, lies at the base of a range of hills that rise abruptly to 600 - 800 feet above the level of Lake Superior. The range runs in a northeast and southwest direction. Two or 3 miles from the lake the land becomes a slightly rolling plateau.

Duluth in the summer is known as the Air Conditioned City. Being situated below high terrain and along the lake, any easterly component winds automatically cool the city. However, with westerly flow in the summer, the wind generally abates at night, thus, allowing cool lake air to move back into the city area near the lake.

An important influence on the climate is the passage of a succession of high and low pressure systems west and east. The proximity of Lake Superior, which is the largest and coldest of the Great Lakes, modifies the local weather. Summer temperatures are cooler and winter temperatures are warmer. The lake effect at Duluth is most prevalent when low pressure systems pass to the south creating easterly winds. In the summer, warm, moist air flowing over the cold lake surface has a stabilizing effect that results in cool, cloudy weather over Duluth. However, during the winter cold air flowing over the warm open lake surface absorbs moisture that is later precipitated over Duluth as snow. The lake effect is further reflected from the low frequency of severe storms such as wind, hail, tornadoes, freezing rain (glaze), and blizzards when compared to other areas that are a further distance from the lake.

Easterly component winds at Duluth occur 40 to 50 percent of the time from March through August and 20 to 25 percent of the time from November through February. During the winter 60 to 70 percent of the winds are from a westerly component.

The climate of Duluth is predominantly continental with significant local Lake Superior effects. Duluth averages 143 days between the last occurrence of 32 degrees in mid-May and the first in early October. At the Duluth Airport about six miles away from the lake, the average first and last occurrences of 32 degrees are late May and late September, giving a freeze-free period of 123 days.

Fall colors throughout this area are outstanding. Reds, yellows, browns, and combinations of these are an experience to see. Recreation is superb from December through March for cross-country and down-hill skiing and snowmobiling. The snow is dry.

Ice in the harbor forms about mid-November and generally is gone by mid-April. The shipping season can vary from year to year depending on temperatures and the winds that move the ice around. In most years there is little or no shipping during February and March on Lake Superior.

NORMALS, MEANS, AND EXTREMES
DULUTH, MN (DLH)

LATITUDE:	LONGITUDE:	ELEVATION (FT):		TIME ZONE:	WBAN: 14913
46 50' 38" N	92 11' 39" W	GRND: 1426 BARO: 1429		CENTRAL (UTC + 6)	

	ELEMENT	POR	JAN	FEB	MAR	APR	MAY	JUN	JUL	AUG	SEP	OCT	NOV	DEC	YEAR
TEMPERATURE °F	NORMAL DAILY MAXIMUM	30	16.2	21.7	32.9	48.2	61.9	71.0	77.1	73.9	63.8	52.3	35.2	20.7	47.9
	MEAN DAILY MAXIMUM	54	16.8	22.5	32.6	47.9	61.8	70.6	76.1	74.0	64.1	52.6	35.0	21.7	48.0
	HIGHEST DAILY MAXIMUM	60	52	55	78	88	90	94	97	97	95	86	71	55	97
	YEAR OF OCCURRENCE		1942	2000	1946	1952	1986	1995	1988	1947	1976	1953	1999	1962	JUL 1988
	MEAN OF EXTREME MAXS.	54	36.2	41.2	51.3	70.7	81.8	85.9	88.6	87.5	81.4	73.4	54.8	39.5	66.0
	NORMAL DAILY MINIMUM	30	-2.2	2.8	15.7	28.9	39.6	48.5	55.1	53.3	44.5	35.1	21.5	4.9	29.0
	MEAN DAILY MINIMUM	54	-1.5	3.6	15.1	29.0	39.9	48.3	54.5	53.2	44.8	35.1	21.2	6.0	29.1
	LOWEST DAILY MINIMUM	60	-39	-39	-29	-5	17	27	35	32	22	8	-23	-34	-39
	YEAR OF OCCURRENCE		1972	1996	1989	1975	1967	1972	1988	1942	1976	1964	1983		FEB 1996
	MEAN OF EXTREME MINS.	54	-26.8	-21.1	-9.8	12.1	26.7	35.6	42.7	40.8	29.4	20.0	0.1	-17.8	11.0
	NORMAL DRY BULB	30	7.0	12.3	24.4	38.6	50.8	59.8	66.1	63.7	54.2	43.7	28.4	12.8	38.5
	MEAN DRY BULB	54	7.7	13.1	23.8	38.5	50.7	59.6	65.3	63.7	54.4	43.9	28.1	14.0	38.6
	MEAN WET BULB	16	10.1	14.4	23.6	34.2	45.8	54.6	59.6	59.1	50.8	38.8	24.5	13.2	35.7
	MEAN DEW POINT	16	5.4	8.6	17.1	26.1	38.5	49.9	55.8	55.6	46.8	33.3	20.2	9.1	30.5
	NORMAL NO. DAYS WITH:														
	MAXIMUM 90	30	0.0	0.0	0.0	0.0	0.1	0.2	1.1	0.8	0.2	0.0	0.0	0.0	2.4
	MAXIMUM 32	30	27.9	22.3	15.1	2.0	0.0	0.0	0.0	0.0	0.0	0.6	11.8	26.5	106.2
	MINIMUM 32	30	31.0	27.8	28.9	20.2	5.6	0.4	0.0	*	2.6	12.4	25.6	30.8	185.3
	MINIMUM 0	30	17.4	12.9	4.6	0.1	0.0	0.0	0.0	0.0	0.0	0.0	1.5	12.4	48.9
H/C	NORMAL HEATING DEG. DAYS	30	1798	1476	1259	792	445	170	60	113	329	660	1098	1618	9818
	NORMAL COOLING DEG. DAYS	30	0	0	0	0	0	14	94	72	0	0	0	0	180
RH	NORMAL (PERCENT)	30	72	70	69	64	63	70	71	74	76	71	75	76	71
	HOUR 00 LST	30	74	73	74	70	71	79	82	84	84	77	78	78	77
	HOUR 06 LST	30	76	76	78	76	76	81	84	88	87	82	81	79	80
	HOUR 12 LST	30	69	65	63	56	53	59	59	62	64	62	70	74	63
	HOUR 18 LST	30	69	64	62	54	52	58	59	64	68	66	72	74	64
S	PERCENT POSSIBLE SUNSHINE	47	48	53	55	56	57	58	65	61	52	46	35	39	52
W/O	MEAN NO. DAYS WITH:														
	HEAVY FOG(VISBY 1/4 MI)	53	2.3	2.5	3.6	3.6	5.2	6.6	5.4	6.4	5.5	4.5	3.3	3.2	52.1
	THUNDERSTORMS	59	0.1	0.0	0.6	1.7	3.5	6.7	7.9	6.9	3.9	1.3	0.4	0.1	33.1
CLOUDINESS	MEAN:														
	SUNRISE-SUNSET (OKTAS)	48	5.4	5.2	5.4	5.4	5.3	5.3	4.7	4.7	5.2	5.3	6.0	5.6	5.3
	MIDNIGHT-MIDNIGHT (OKTAS)	32	5.0	5.0	5.2	5.0	4.8	4.8	4.3	4.3	4.7	5.0	5.7	5.4	4.9
	MEAN NO. DAYS WITH:														
	CLEAR	49	7.4	7.3	7.0	6.2	6.2	5.1	6.7	7.2	6.3	6.4	4.4	6.0	76.2
	PARTLY CLOUDY	49	7.0	6.2	6.9	7.9	9.6	11.0	12.7	11.9	8.4	7.6	5.5	5.8	100.5
	CLOUDY	49	16.7	14.7	17.0	15.9	15.2	13.9	10.9	11.3	14.6	16.4	19.5	18.6	184.7
PR	MEAN STATION PRESSURE(IN)	28	28.50	28.50	28.50	28.40	28.40	28.40	28.50	28.50	28.50	28.50	28.41	28.51	28.47
	MEAN SEA-LEVEL PRES. (IN)	16	30.06	30.07	30.06	29.97	29.96	30.00	29.96	30.01	30.01	30.01	30.02	30.08	30.02
WINDS	MEAN SPEED (MPH)	43	11.6	11.6	12.0	12.6	11.9	10.5	9.7	9.5	10.6	11.5	12.2	11.6	11.3
	PREVAIL.DIR(TENS OF DEGS)	27	30	31	09	09	09	09	11	10	11	30	31	30	30
	MAXIMUM 2-MINUTE:														
	SPEED (MPH)	5	37	37	47	43	39	33	39	46	33	38	44	41	47
	DIR. (TENS OF DEGS)		33	09	08	20	28	09	28	29	26	35	08	28	08
	YEAR OF OCCURRENCE		2001	1999	1998	1998	1998	2000	1999	2001	1998	1999	1998	2001	MAR 1998
	MAXIMUM 5-SECOND:														
	SPEED (MPH)	5	44	46	57	57	49	46	47	62	40	49	55	56	62
	DIR. (TENS OF DEGS)		33	07	08	20	30	23	28	29	27	01	09	31	29
	YEAR OF OCCURRENCE		2001	1999	1998	1998	2000	1999	1999	2001	1998	1999	1998	1999	AUG 2001
PRECIPITATION	NORMAL (IN)	30	1.22	0.80	1.91	2.25	3.03	3.82	3.61	3.99	3.84	2.49	1.80	1.24	30.00
	MAXIMUM MONTHLY (IN)	60	4.70	2.72	5.12	8.18	7.67	8.04	8.74	10.31	9.38	7.53	5.08	3.70	10.31
	YEAR OF OCCURRENCE		1969	1998	1965	2001	1962	1986	1999	1972	1991	1949	2000	1968	AUG 1972
	MINIMUM MONTHLY (IN)	60	0.14	0.13	0.22	0.24	0.15	0.55	0.97	0.71	0.19	0.13	0.19	0.13	0.13
	YEAR OF OCCURRENCE		1961	1988	1959	1987	1976	1995	1947	1970	1952	1944	1976	1997	DEC 1997
	MAXIMUM IN 24 HOURS (IN)	52	1.74	1.38	2.38	2.60	3.25	4.05	3.68	5.79	3.77	2.90	2.64	2.12	5.79
	YEAR OF OCCURRENCE		1975	1965	1977	2001	1979	1958	1987	1978	1972	1973	1968	1950	AUG 1978
	NORMAL NO. DAYS WITH:														
	PRECIPITATION 0.01	30	12.1	9.4	10.9	10.4	11.8	12.3	11.2	11.8	12.1	9.8	10.5	11.7	134.0
	PRECIPITATION 1.00	30	0.1	0.0	0.2	0.2	0.4	0.8	1.1	1.2	0.9	0.6	0.3	0.1	5.9
SNOWFALL	NORMAL (IN)	30	17.2	10.3	14.1	6.3	0.6	0.0	0.0	0.0	0.*	1.3	11.4	16.1	77.3
	MAXIMUM MONTHLY (IN)	58	46.8	32.1	45.5	31.5	8.1	0.2	T	T	2.4	9.7	50.1	44.3	50.1
	YEAR OF OCCURRENCE		1969	2001	1965	1950	1954	1945		1992	1995	1995	1991	1950	NOV 1991
	MAXIMUM IN 24 HOURS (IN)	58	16.3	17.0	19.4	11.6	4.3	0.2	T	T	2.4	7.9	24.1	25.4	25.4
	YEAR OF OCCURRENCE		1994	1948	1965	1983	1954	1945		1992	1995	1991	1966	1991	1950	DEC 1950
	MAXIMUM SNOW DEPTH (IN)	53	42	38	48	41	9	0	0	0	0	6	30	32	48
	YEAR OF OCCURRENCE		1969	1969	1965	1965	1950					1966	1991	1983	MAR 1965
	NORMAL NO. DAYS WITH:														
	SNOWFALL 1.0	30	4.2	3.3	4.0	1.8	0.2	0.0	0.0	0.0	0.0	0.4	2.9	4.5	21.3

PRECIPITATION (inches)   2001   DULUTH, MN (DLH)

YEAR	JAN	FEB	MAR	APR	MAY	JUN	JUL	AUG	SEP	OCT	NOV	DEC	ANNUAL
1972	2.28	1.47	1.45	2.17	2.00	3.70	6.71	10.31	5.30	0.83	1.37	2.02	39.61
1973	0.67	0.29	1.54	1.35	3.81	2.43	2.36	8.46	4.28	4.56	1.61	0.69	32.05
1974	0.97	0.82	0.78	2.07	3.09	4.07	4.85	3.79	0.98	1.57	1.36	1.15	25.50
1975	3.69	0.76	2.59	2.21	1.44	5.59	2.26	2.52	2.32	1.20	4.19	0.64	29.41
1976	1.57	1.05	3.67	0.73	0.15	6.16	2.60	1.84	0.48	0.19	0.19	0.39	20.67
1977	0.36	0.47	4.43	1.27	3.50	3.97	3.91	3.26	5.97	3.20	2.37	1.31	34.02
1978	0.52	0.35	0.47	1.96	3.49	2.96	7.67	7.49	1.52	0.77	1.27	1.19	29.66
1979	0.76	1.89	3.58	1.15	6.01	4.33	5.45	2.10	2.01	3.01	0.47	0.16	30.92
1980	1.55	0.56	1.02	0.41	0.82	2.35	3.94	5.34	6.61	1.64	0.70	0.63	25.57
1981	0.32	1.50	1.05	4.48	1.15	5.83	3.26	2.84	2.42	3.59	0.96	0.97	28.37
1982	2.02	0.48	2.06	2.06	4.30	1.97	6.21	1.60	4.19	5.07	3.08	1.19	34.23
1983	1.34	0.49	2.05	2.28	2.12	2.00	3.51	3.37	5.57	2.32	5.01	1.97	32.03
1984	0.78	0.61	0.54	2.34	1.83	5.70	1.33	1.96	3.82	5.19	0.82	1.91	26.83
1985	0.39	0.66	1.85	2.35	4.44	3.18	4.16	3.91	6.02	1.76	2.33	0.78	31.83
1986	0.66	0.74	0.88	4.11	2.59	8.04	4.58	5.29	6.26	0.66	2.01	0.45	36.27
1987	0.69	0.31	0.60	0.24	4.02	0.83	5.46	1.87	2.93	0.96	1.26	0.67	19.84
1988	0.78	0.13	2.55	0.44	3.96	4.56	1.14	6.82	6.18	1.05	3.44	1.12	32.17
1989	1.87	0.34	1.49	2.11	3.50	3.81	1.09	5.02	4.40	1.02	1.01	0.63	26.29
1990	0.51	0.51	3.35	3.76	1.48	4.83	2.42	5.39	6.49	3.51	0.65	0.49	33.39
1991	0.52	0.55	1.17	3.90	6.11	5.64	5.33	2.49	9.38	2.85	4.89	0.61	43.44
1992	0.60	0.58	0.84	2.87	2.87	5.04	3.64	4.13	3.90	1.13	1.84	1.23	28.67
1993	1.79	0.38	0.44	2.42	4.74	6.95	5.75	3.63	1.81	0.56	2.60	1.28	32.35
1994	1.85	0.62	1.11	4.11	1.91	4.54	2.20	3.45	5.61	2.27	2.53	0.33	30.53
1995	1.38	1.04	1.97	2.19	4.13	0.55	5.73	6.67	3.02	4.24	1.84	1.54	34.30
1996	1.35	1.10	0.50	1.47	1.61	5.10	8.37	2.92	5.34	3.08	3.95	0.86	35.65
1997	0.94	0.23	1.48	1.07	1.90	5.17	4.33	1.26	1.75	2.29	0.43	0.13	20.98
1998	0.36	2.72	1.84	1.42	2.21	6.13	1.72	2.59	3.33	4.21	3.42	1.57	31.52
1999	0.87	0.64	0.76	2.96	2.93	4.72	8.74	7.41	5.80	2.35	0.70	0.23	38.11
2000	0.77	1.08	2.64	1.38	3.07	4.18	3.50	4.42	2.02	2.27	5.08	0.93	31.34
2001	1.25	1.76	0.69	8.18	3.49	3.06	1.89	3.31	1.41	2.42	2.21	0.55	30.22
POR= 98 YRS	1.12	0.93	1.64	2.27	3.10	4.05	3.78	3.51	3.39	2.25	1.77	1.09	28.90

AVERAGE TEMPERATURE (F) 2001   DULUTH, MN (DLH)

YEAR	JAN	FEB	MAR	APR	MAY	JUN	JUL	AUG	SEP	OCT	NOV	DEC	ANNUAL
1972	0.2	6.4	17.9	33.3	53.5	58.4	61.2	61.9	50.0	38.9	25.9	7.2	34.6
1973	11.6	13.7	31.8	38.3	47.8	59.1	64.8	65.5	53.6	48.2	27.9	11.7	39.5
1974	5.6	11.0	20.6	37.9	46.6	58.6	67.1	60.1	47.9	42.6	29.2	19.5	37.2
1975	9.6	12.0	18.2	31.0	53.3	57.9	68.7	62.0	50.9	46.7	31.3	11.8	37.8
1976	4.7	20.7	22.0	42.6	50.6	63.2	66.1	64.9	55.3	37.1	22.1	4.5	37.8
1977	-.2	17.2	31.4	44.4	57.4	60.0	65.9	58.6	53.2	44.3	28.0	11.3	39.3
1978	5.1	10.5	25.6	38.3	54.8	60.2	64.7	63.6	57.3	44.9	26.2	10.1	38.4
1979	0.6	5.0	23.0	34.5	47.2	59.8	66.1	62.0	56.9	43.0	29.4	22.2	37.5
1980	8.7	10.7	19.7	41.4	54.9	59.5	67.6	63.6	52.9	38.9	30.1	11.6	38.3
1981	11.9	16.3	28.3	39.2	50.2	58.7	65.5	64.3	52.9	40.5	34.7	14.8	39.8
1982	-3.2	10.6	20.8	35.7	52.8	54.8	64.5	60.8	54.3	45.1	24.4	19.6	36.7
1983	13.2	21.1	25.6	35.1	46.7	59.9	69.6	69.7	56.6	44.9	31.6	1.8	39.7
1984	7.1	23.0	18.7	42.4	50.5	61.4	66.6	67.7	51.9	45.7	28.4	13.2	39.7
1985	6.5	11.6	30.3	42.1	54.9	56.5	64.6	59.7	52.1	42.7	19.9	3.0	37.0
1986	11.7	11.5	28.5	42.1	51.8	58.9	64.7	60.8	52.7	43.4	24.0	18.8	39.1
1987	15.3	23.7	31.6	46.1	53.1	63.1	67.6	63.5	57.2	39.4	32.2	20.4	42.8
1988	4.8	6.2	24.4	40.1	56.0	62.8	70.0	64.5	55.4	38.6	28.9	13.5	38.8
1989	14.3	3.4	18.9	36.8	51.7	58.4	68.6	64.8	55.9	44.4	23.2	4.1	37.0
1990	18.7	14.9	27.1	40.2	48.3	61.5	64.6	64.1	56.2	42.1	31.4	12.6	40.1
1991	5.9	19.4	26.1	42.7	54.7	61.7	63.8	66.3	53.0	40.4	22.0	15.9	39.3
1992	16.7	21.2	25.9	36.2	53.8	56.3	59.4	60.5	53.8	41.7	27.3	15.9	39.1
1993	11.9	14.5	27.1	37.2	50.3	57.1	63.6	65.0	49.5	40.0	24.4	17.0	38.1
1994	-2.1	6.5	28.6	39.6	53.8	62.5	64.5	62.5	58.5	47.2	33.6	21.3	39.7
1995	13.7	11.7	28.2	34.7	52.4	65.1	66.8	67.4	55.1	43.7	20.6	12.1	39.3
1996	2.2	11.9	19.4	33.7	48.2	60.4	62.1	65.4	56.2	43.6	21.6	11.2	36.3
1997	7.0	14.5	23.5	35.7	45.8	62.2	63.8	61.0	57.5	43.9	24.9	22.7	38.5
1998	16.4	29.0	25.4	44.1	55.9	58.8	65.7	67.1	59.3	45.9	32.1	18.7	43.2
1999	9.3	21.3	27.8	41.4	53.6	60.6	67.6	63.5	54.5	42.3	36.2	19.7	41.5
2000	10.3	21.4	33.4	38.1	52.9	56.0	64.6	64.7	53.9	46.9	28.6	4.4	39.6
2001	17.4	7.8	24.5	39.8	52.8	60.9	65.7	67.0	55.4	43.4	39.5	22.2	41.4
POR= 98 YRS	8.6	12.9	24.3	38.4	49.6	58.7	65.2	63.7	55.1	44.1	28.6	14.7	38.7

REFERENCE NOTES:

PAGE 1:
THE TEMPERATURE GRAPH SHOWS NORMAL MAXIMUM AND NORMAL
   MINIMUM DAILY TEMPERATURES (SOLID CURVES) AND THE
   ACTUAL DAILY HIGH AND LOW TEMPERATURES (VERTICAL BARS).

PAGE 2 AND 3:
H/C INDICATES HEATING AND COOLING DEGREE DAYS.
RH INDICATES RELATIVE HUMIDITY
W/O INDICATES WEATHER AND OBSTRUCTIONS
S INDICATES SUNSHINE.
PR INDICATES PRESSURE.
CLOUDINESS ON PAGE 3 IS THE SUM OF THE CEILOMETER AND
   SATELLITE DATA NOT TO EXCEED EIGHT EIGHTHS(OKTAS).

GENERAL:
T INDICATES TRACE PRECIPITATION, AN AMOUNT GREATER
   THAN ZERO BUT LESS THAN THE LOWEST REPORTABLE VALUE.
+ INDICATES THE VALUE ALSO OCCURS ON EARLIER DATES.
BLANK ENTRIES DENOTE MISSING OR UNREPORTED DATA.
NORMALS ARE 30-YEAR AVERAGES (1961 - 1990).
ASOS INDICATES AUTOMATED SURFACE OBSERVING SYSTEM.
PM INDICATES THE LAST DAY OF THE PREVIOUS MONTH.
POR (PERIOD OF RECORD) BEGINS WITH THE JANUARY DATA
   MONTH AND IS THE NUMBER OF YEARS USED TO COMPUTE
   THE MEAN. INDIVIDUAL MONTHS WITHIN THE POR MAY
   BE MISSING.
WHEN THE POR FOR A NORMAL IS LESS THAN 30 YEARS,
   THE NORMAL IS PROVISIONAL AND IS BASED ON THE NUMBER
   OF YEARS INDICATED.
0.* OR * INDICATES THE VALUE OR MEAN-DAYS-WITH
   IS BETWEEN 0.00 AND 0.05.
CLOUDINESS FOR ASOS STATIONS DIFFERS FROM THE NON-ASOS
   OBSERVATION TAKEN BY A HUMAN OBSERVER. ASOS STATION
   CLOUDINESS IS BASED ON TIME-AVERAGED CEILOMETER DATA
   FOR CLOUDS AT OR BELOW 12,000 FEET AND ON SATELLITE
   DATA FOR CLOUDS ABOVE 12,000 FEET.
THE NUMBER OF DAYS WITH CLEAR, PARTLY CLOUDY, AND
   CLOUDY CONDITIONS FOR ASOS STATIONS IS THE SUM
   OF THE CEILOMETER AND SATELLITE DATA FOR THE
   SUNRISE TO SUNSET PERIOD.

GENERAL CONTINUED:
CLEAR INDICATES 0 - 2 OKTAS, PARTLY CLOUDY INDICATES
   3 - 6 OKTAS, AND CLOUDY INDICATES 7 OR 8 OKTAS.
   WHEN AT LEAST ONE OF THE ELEMENTS (CEILOMETER OR
   SATELLITE) IS MISSING, THE DAILY CLOUDINESS IS
   NOT COMPUTED.
WIND DIRECTION IS RECORDED IN TENS OF DEGREES (2 DIGITS)
   CLOCKWISE FROM TRUE NORTH. "00" INDICATES CALM. "36"
   INDICATES TRUE NORTH.
RESULTANT WIND IS THE VECTOR AVERAGE OF THE SPEED AND
   DIRECTION.
AVERAGE TEMPERATURE IS THE SUM OF THE MEAN DAILY MAXIMUM
   AND MINIMUM TEMPERATURE DIVIDED BY 2.
SNOWFALL DATA COMPRISE ALL FORMS OF FROZEN
   PRECIPITATION, INCLUDING HAIL.
A HEATING (COOLING) DEGREE DAY IS THE DIFFERENCE BETWEEN
   THE AVERAGE DAILY TEMPERATURE AND 65 F.
DRY BULB IS THE TEMPERATURE OF THE AMBIENT AIR.
DEW POINT IS THE TEMPERATURE TO WHICH THE AIR MUST BE
   COOLED TO ACHIEVE 100 PERCENT RELATIVE HUMIDITY.
WET BULB IS THE TEMPERATURE THE AIR WOULD HAVE IF THE
   MOISTURE CONTENT WAS INCREASED TO 100 PERCENT RELATIVE
   HUMIDITY.

ON JULY 1, 1996, THE NATIONAL WEATHER SERVICE BEGAN USING
   THE "METAR" OBSERVATION CODE THAT WAS ALREADY EMPLOYED
   BY MOST OTHER NATIONS OF THE WORLD.   THE MOST NOTICEABLE
   DIFFERENCE IN THIS ANNUAL PUBLICATION WILL BE THE CHANGE
   IN UNITS FROM TENTHS TO EIGHTS(OKTAS) FOR REPORTING THE
   AMOUNT OF SKY COVER.

HEATING DEGREE DAYS (base 65 F) 2001 ALBUQUERQUE, NM (ABQ)

YEAR	JUL	AUG	SEP	OCT	NOV	DEC	JAN	FEB	MAR	APR	MAY	JUN	TOTAL
1972-73	0	3	14	244	740	925	1020	811	607	440	113	3	4920
1973-74	0	0	43	257	606	955	963	754	373	255	29	4	4239
1974-75	0	2	68	212	593	1020	1051	748	614	449	143	6	4906
1975-76	0	0	47	256	664	905	979	622	634	304	99	1	4511
1976-77	0	0	35	367	726	985	1084	675	669	250	61	0	4852
1977-78	0	0	1	192	551	757	870	713	454	215	175	2	3930
1978-79	0	0	20	167	521	945	988	665	509	241	100	12	4168
1979-80	0	0	23	148	715	840	763	595	577	379	139	2	4181
1980-81	0	0	6	335	640	752	827	611	575	197	62	2	4007
1981-82	0	0	3	280	534	754	895	709	538	268	94	0	4075
1982-83	0	0	23	314	658	941	922	703	556	439	127	0	4683
1983-84	0	0	11	198	592	875	948	714	559	362	22	3	4284
1984-85	0	0	51	411	631	903	960	744	536	220	74	7	4537
1985-86	0	0	61	228	581	842	727	610	431	249	80	8	3817
1986-87	0	0	51	313	680	882	1004	717	653	300	81	2	4683
1987-88	0	0	2	133	589	914	937	605	551	290	103	2	4126
1988-89	0	5	39	118	579	959	909	640	373	133	31	0	3786
1989-90	0	0	10	260	551	918	934	735	501	233	103	0	4245
1990-91	0	0	14	202	595	1013	903	563	581	263	60	12	4206
1991-92	0	0	21	188	645	851	994	651	493	170	53	5	4071
1992-93	0	0	8	128	752	991	778	624	496	238	69	3	4087
1993-94	0	0	15	284	642	853	827	676	453	218	52	0	4020
1994-95	0	0	1	251	610	741	793	435	433	320	67	0	3651
1995-96	0	0	37	165	419	741	787	540	558	239	12	0	3498
1996-97	0	0	66	297	575	809	973	676	403	366	48	12	4225
1997-98	0	0	9	278	635	994	834	727	558	376	51	4	4466
1998-99	0	0	0	225	549	816	744	584	435	332	98	5	3788
1999-00	0	0	23	203	447	902	740	571	528	184	32	0	3630
2000-01	0	0	16	302	760	858	961	619	509	218	35	3	4281
2001-	0	0	1	134	508	881							

COOLING DEGREE DAYS (base 65 F) 2001 DULUTH, MN (DLH)

YEAR	JAN	FEB	MAR	APR	MAY	JUN	JUL	AUG	SEP	OCT	NOV	DEC	ANNUAL
1972	0	0	0	0	8	22	24	59	0	0	0	0	113
1973	0	0	0	0	0	1	72	84	18	0	0	0	175
1974	0	0	0	0	0	19	115	15	0	0	0	0	149
1975	0	0	0	0	2	18	168	41	0	0	0	0	229
1976	0	0	0	0	0	53	75	117	26	0	0	0	271
1977	0	0	0	0	12	24	80	6	0	0	0	0	122
1978	0	0	0	0	17	34	70	64	39	0	0	0	224
1979	0	0	0	0	4	25	95	30	15	0	0	0	169
1980	0	0	0	0	25	46	126	40	3	0	0	0	240
1981	0	0	0	0	2	2	97	48	6	0	0	0	155
1982	0	0	0	0	0	0	58	36	18	0	0	0	112
1983	0	0	0	0	0	42	179	165	42	0	0	0	428
1984	0	0	0	0	4	13	96	133	4	0	0	0	250
1985	0	0	0	0	9	4	54	20	15	0	0	0	102
1986	0	0	0	0	13	18	74	26	0	0	0	0	131
1987	0	0	0	0	13	62	121	60	7	0	0	0	263
1988	0	0	0	0	27	83	183	89	4	0	0	0	386
1989	0	0	0	0	0	17	147	80	7	0	0	0	251
1990	0	0	0	7	0	32	70	73	16	0	0	0	198
1991	0	0	0	0	20	35	69	117	8	0	0	0	249
1992	0	0	0	0	13	18	9	37	3	3	0	0	83
1993	0	0	0	0	0	8	36	76	0	0	0	0	120
1994	0	0	0	0	11	36	52	43	8	0	0	0	150
1995	0	0	0	0	7	107	105	99	14	4	0	0	336
1996	0	0	0	0	0	44	25	58	21	0	0	0	148
1997	0	0	0	0	0	27	67	37	8	0	0	0	139
1998	0	0	0	0	8	12	81	93	32	0	0	0	226
1999	0	0	0	0	12	33	127	35	20	0	0	0	227
2000	0	0	0	0	2	10	69	51	9	0	0	0	141
2001	0	0	0	0	0	49	96	117	17	0	0	0	279

SNOWFALL (inches) 2001 DULUTH, MN (DLH)

YEAR	JUL	AUG	SEP	OCT	NOV	DEC	JAN	FEB	MAR	APR	MAY	JUN	TOTAL
1972-73	0.0	0.0	T	0.5	6.7	20.0	9.2	3.0	2.1	2.8	1.5	0.0	45.8
1973-74	0.0	0.0	0.0	T	7.6	15.7	13.4	15.3	15.1	5.2	1.0	0.0	73.3
1974-75	0.0	0.0	0.3	0.8	4.5	19.1	32.7	12.3	30.3	0.4	T	0.0	100.4
1975-76	0.0	0.0	0.0	0.1	25.7	7.8	20.6	8.8	26.4	0.0	T	0.0	89.4
1976-77	0.0	0.0	0.0	0.8	2.7	7.8	8.8	5.1	15.3	0.1	0.0	0.0	40.6
1977-78	0.0	0.0	0.0	1.7	16.0	12.0	13.5	11.4	8.3	6.8	0.0	0.0	69.7
1978-79	0.0	0.0	0.0	T	10.1	20.8	11.9	23.3	17.0	4.2	1.4	0.0	88.7
1979-80	0.0	0.0	0.0	0.8	6.0	2.0	21.9	10.3	12.6	1.5	T	0.0	55.1
1980-81	0.0	0.0	0.0	1.3	1.6	7.1	4.7	13.3	4.2	4.3	T	0.0	36.5
1981-82	0.0	0.0	T	1.4	12.5	15.5	34.2	8.0	17.8	6.3	0.0	0.0	95.7
1982-83	0.0	0.0	0.0	0.8	16.0	21.4	15.7	9.0	9.9	23.7	T	0.0	96.5
1983-84	0.0	0.0	T	T	37.7	32.1	20.0	4.0	9.9	3.3	0.3	0.0	107.3
1984-85	0.0	0.0	T	3.5	2.4	8.2	15.3	12.0	26.7	0.1	0.0	0.0	68.2
1985-86	0.0	0.0	0.7	2.3	34.1	18.8	11.5	10.3	11.4	0.2	T	0.0	89.3
1986-87	0.0	0.0	0.0	1.8	8.2	7.3	11.2	5.1	6.7	0.3	T	0.0	40.6
1987-88	0.0	0.0	0.0	3.9	5.8	11.3	16.7	2.3	13.7	0.1	0.0	0.0	53.8
1988-89	0.0	0.0	0.0	0.3	24.3	20.7	31.2	5.3	17.6	19.0	0.7	0.0	119.1
1989-90	0.0	T	0.0	0.6	8.4	14.2	5.5	15.0	11.6	2.4	0.6	T	58.3
1990-91	0.0	0.0	0.0	3.2	5.1	13.5	11.4	9.7	9.5	8.5	2.9	0.0	63.8
1991-92	0.0	0.0	2.4	4.3	50.1	12.9	9.3	10.6	0.7	9.7	0.0	0.0	100.0
1992-93	T	0.0	T	5.1	29.4	15.5	24.9	10.0	5.8	3.5	T	0.0	94.2
1993-94	0.0	0.0	T	0.6	28.5	15.1	35.3	7.9	13.0	10.0	0.0	T	110.4
1994-95	0.0	0.0	0.0	0.2	15.5	6.2	12.8	17.6	20.9	16.2	T	0.0	89.4
1995-96	0.0	T	0.1	9.7	20.4	25.9	34.0	21.3	11.4	12.6	T		
1996-97					15.9	41.7	35.5	8.9	23.6	1.7	0.1	0.0	
1997-98	0.0	0.0	0.0	1.2	16.9	6.0	29.9	9.6	14.2	2.3	0.0	T	80.1
1998-99	0.0	T	0.0	T	15.7	12.8	18.9	13.5	17.8	11.5	0.0	0.0	90.2
1999-00	0.0	0.0	0.0	0.0	6.2	4.4	16.5	13.3	9.5	5.6	T	0.0	55.5
2000-01	0.0	0.0	0.0	T	17.5	19.2	14.9	32.1	7.9	7.7	0.0	0.0	99.3
2001-	0.0	0.0	0.0	2.0	11.4	7.3							
POR= 57 YRS	T	T	0.1	1.5	12.9	15.0	17.6	11.7	13.3	6.6	1.0	T	79.7

2001
MINNEAPOLIS - ST. PAUL,
MINNESOTA (MSP)

The Twin Cities of Minneapolis and St. Paul are located at the confluence of the Mississippi and Minnesota Rivers over the heart of an artesian water basin. Its flat or gently rolling terrain varies little in elevation from that of the official observation station at International Airport. Numerous lakes dot the surrounding area. Minneapolis alone boasts of 22 lakes within the city park system. The largest body of water, nearly 15,000 acres, is Lake Minnetonka, located about 15 miles west of the airport. Most bodies of water are relatively small and shallow and are ice covered during winter.

The climate of the Minneapolis-St. Paul area is predominantly continental. Seasonal temperature variations are quite large. Temperatures range from less than -30 degrees to over 100 degrees. The growing season is 166 days. Because of this favorable growing season, all crops generally mature before the autumn freeze occurs.

The Twin Cities lie near the northern edge of the influx of moisture from the Gulf of Mexico. Severe storms such as blizzards, freezing rain (glaze), tornadoes, wind and hail storms do occur. The total annual precipitation is important. Even more significant is its proper distribution during the growing season. During the five month growing season, May through September, the major crops produced are corn, soybeans, small grains, and hay. During this period, the normal rainfall is over 16 inches, approximately 65 percent of the annual precipitation. Winter snowfall is nearly 48 inches. Winter recreational weather is excellent because of the dry snow. These conditions exist from about Christmas into early March. Snow depths average 6 to 8 inches in the city and 8 to 10 inches in the suburbs during this period.

Floods occur along the Mississippi River due to spring snow melt, excessive rainfall, or both. Occasionally an ice jam forms and creates a local flood condition. The flood problem at St. Paul is complicated because the Minnesota River empties into the Mississippi River between the two cities. Consequently, high water or flooding on the Minnesota River creates a greater flood potential at St. Paul. Flood stage at St. Paul can be expected on the average once in every eight years.

NORMALS, MEANS, AND EXTREMES
MINNEAPOLIS-ST.PAUL, MN (MSP)

LATITUDE:	LONGITUDE:	ELEVATION (FT):	TIME ZONE:	WBAN: 14922
44 52' 59" N	93 13' 44" W	GRND: 871 BARO: 874	CENTRAL (UTC + 6)	

	ELEMENT	POR	JAN	FEB	MAR	APR	MAY	JUN	JUL	AUG	SEP	OCT	NOV	DEC	YEAR
TEMPERATURE F	NORMAL DAILY MAXIMUM	30	20.7	26.6	39.2	56.5	69.4	78.8	84.0	80.7	70.7	58.8	41.0	25.5	54.3
	MEAN DAILY MAXIMUM	57	21.6	27.2	39.0	56.1	69.0	78.5	83.3	80.8	70.9	59.5	40.6	26.5	54.4
	HIGHEST DAILY MAXIMUM	63	58	61	83	95	96	102	105	102	98	90	77	68	105
	YEAR OF OCCURRENCE		1944	2000	1986	1980	1978	1985	1988	1947	1976	1997	1999	1998	JUL 1988
	MEAN OF EXTREME MAXS.	57	40.7	45.3	61.8	78.7	87.6	93.3	94.9	93.6	88.7	80.1	61.2	46.0	72.7
	NORMAL DAILY MINIMUM	30	2.8	9.2	22.7	36.2	47.6	57.6	63.1	60.3	50.3	38.8	25.2	10.2	35.3
	MEAN DAILY MINIMUM	57	4.0	9.8	22.1	36.0	48.0	57.7	62.9	60.8	50.5	39.6	25.0	11.3	35.6
	LOWEST DAILY MINIMUM	63	-34	-32	-32	2	18	34	43	39	26	13	-17	-29	-34
	YEAR OF OCCURRENCE		1970	1996	1962	1962	1967	1945	1972	1967	1974	1997	1964	1983	JAN 1970
	MEAN OF EXTREME MINS.	57	-19.5	-13.3	0.0	20.6	32.6	43.5	51.5	48.8	34.8	24.6	5.1	-11.7	18.1
	NORMAL DRY BULB	30	11.8	17.9	31.0	46.4	58.5	68.2	73.6	70.5	60.5	48.8	33.2	17.9	44.9
	MEAN DRY BULB	57	12.8	18.6	30.6	46.0	58.3	68.0	73.1	70.9	60.7	49.6	33.8	19.0	45.0
	MEAN WET BULB	17	14.4	19.1	29.1	40.5	52.1	61.3	65.4	64.0	55.3	43.6	29.6	18.2	41.0
	MEAN DEW POINT	17	9.2	13.7	22.3	31.4	44.6	55.5	60.7	59.9	50.6	37.3	24.3	13.4	35.2
	NORMAL NO. DAYS WITH:														
	MAXIMUM 90	30	0.0	0.0	0.0	0.1	0.9	2.9	6.6	3.8	0.8	0.0	0.0	0.0	15.1
	MAXIMUM 32	30	23.6	17.8	8.6	0.4	0.0	0.0	0.0	0.0	0.0	0.0	6.7	21.6	78.7
	MINIMUM 32	30	30.9	27.0	25.1	11.1	1.1	0.0	0.0	0.0	0.5	7.4	23.4	30.1	156.6
	MINIMUM 0	30	14.1	8.5	1.8	0.0	0.0	0.0	0.0	0.0	0.0	0.0	0.6	8.1	33.1
H/C	NORMAL HEATING DEG. DAYS	30	1649	1319	1054	558	244	41	11	22	167	502	954	1460	7981
	NORMAL COOLING DEG. DAYS	30	0	0	0	0	43	137	278	192	32	0	0	0	682
RH	NORMAL (PERCENT)	30	70	70	67	60	60	64	65	68	71	68	73	74	68
	HOUR 00 LST	30	72	73	72	66	67	72	74	77	79	74	76	76	73
	HOUR 06 LST	30	74	75	76	74	76	78	80	83	85	81	80	78	78
	HOUR 12 LST	30	66	65	61	52	51	54	53	56	59	58	66	70	59
	HOUR 18 LST	30	68	65	61	50	49	52	52	56	61	60	69	72	60
S	PERCENT POSSIBLE SUNSHINE	58	53	59	57	58	61	66	72	69	62	55	39	42	58
W/O	MEAN NO. DAYS WITH:														
	HEAVY FOG(VISBY 1/4 MI)	63	1.2	1.4	1.3	0.4	0.5	0.5	0.3	0.6	0.8	0.9	1.1	1.3	10.3
	THUNDERSTORMS	63	0.0	0.2	1.0	2.6	5.1	7.7	7.6	6.5	4.2	1.8	0.6	0.1	37.4
CLOUDINESS	MEAN:														
	SUNRISE-SUNSET (OKTAS)	58	5.1	5.0	5.4	5.3	5.1	4.9	4.2	4.2	4.4	4.6	5.7	5.5	5.0
	MIDNIGHT-MIDNIGHT (OKTAS)	32	4.9	4.8	5.1	5.0	4.8	4.5	4.0	4.0	4.2	4.6	5.3	5.2	4.7
	MEAN NO. DAYS WITH:														
	CLEAR	58	8.3	7.7	7.1	6.9	7.0	7.2	9.8	10.0	9.8	9.8	5.5	6.4	95.5
	PARTLY CLOUDY	58	7.3	6.9	7.4	7.7	9.1	10.4	11.8	11.2	8.5	7.5	6.4	6.4	100.6
	CLOUDY	58	15.4	13.7	16.5	15.3	14.8	12.4	9.5	9.8	11.8	13.7	18.1	18.2	169.2
PR	MEAN STATION PRESSURE(IN)	28	29.17	29.18	29.10	29.06	29.04	29.03	29.08	29.11	29.12	29.12	29.11	29.17	29.11
	MEAN SEA-LEVEL PRES. (IN)	17	30.11	30.12	30.06	29.96	29.93	29.92	29.96	30.00	30.01	30.02	30.05	30.11	30.02
WINDS	MEAN SPEED (MPH)	52	10.5	10.3	11.3	12.3	11.3	10.5	9.5	9.3	10.0	10.6	10.9	10.3	10.6
	PREVAIL.DIR(TENS OF DEGS)	33	31	31	31	12	14	14	18	14	18	31	31	31	31
	MAXIMUM 2-MINUTE:														
	SPEED (MPH)	5	31	36	37	45	49	48	40	35	37	38	38	38	49
	DIR. (TENS OF DEGS)		33	32	32	28	22	33	22	26	34	31	29	32	22
	YEAR OF OCCURRENCE		1997	1999	2000	2000	1998	1998	1999	2000	1999	2001	1999	1999	MAY 1998
	MAXIMUM 5-SECOND:														
	SPEED (MPH)	5	38	44	49	59	64	56	52	46	48	47	47	51	64
	DIR. (TENS OF DEGS)		32	31	32	28	22	22	21	26	20	29	29	33	26
	YEAR OF OCCURRENCE		1997	1999	2000	2000	1998	2001	1999	2000	1999	2001	1998	1999	MAY 1998
PRECIPITATION	NORMAL (IN)	30	0.95	0.88	1.94	2.42	3.39	4.05	3.53	3.62	2.72	2.19	1.55	1.08	28.32
	MAXIMUM MONTHLY (IN)	63	3.63	2.14	4.75	7.00	8.03	9.82	17.90	9.31	7.53	5.68	5.29	4.27	17.90
	YEAR OF OCCURRENCE		1967	1981	1965	2001	1962	1990	1987	1977	1942	1971	1991	1982	JUL 1987
	MINIMUM MONTHLY (IN)	63	0.10	0.06	0.32	0.16	0.61	0.22	0.58	0.43	0.41	0.01	0.02	T	T
	YEAR OF OCCURRENCE		1990	1964	1994	1987	1967	1988	1975	1946	1940	1952	1939	1943	DEC 1943
	MAXIMUM IN 24 HOURS (IN)	63	1.21	1.10	1.66	2.23	3.03	3.00	10.00	7.36	3.55	2.95	2.91	2.47	10.00
	YEAR OF OCCURRENCE		1967	1966	1965	1975	1965	1986	1987	1977	1942	1966	1940	1982	JUL 1987
	NORMAL NO. DAYS WITH:														
	PRECIPITATION 0.01	30	9.2	7.4	9.9	11.2	11.4	11.1	9.2	9.6	10.0	8.2	8.4	9.8	115.4
	PRECIPITATION 1.00	30	0.1	0.1	0.2	0.3	0.7	1.0	0.9	1.0	0.4	0.5	0.2	0.1	5.5
SNOWFALL	NORMAL (IN)	30	12.5	9.2	11.6	3.6	0.1	0.0	0.0	0.0	0.*	0.4	7.3	11.3	56.0
	MAXIMUM MONTHLY (IN)	61	46.4	26.5	40.0	21.8	3.0	T	T	T	1.7	8.2	46.9	33.2	46.9
	YEAR OF OCCURRENCE		1982	1962	1951	1983	1946	1995	1994	1992	1942	1991	1991	1969	NOV 1991
	MAXIMUM IN 24 HOURS (IN)	61	18.5	9.3	14.7	13.6	3.0	T	T	T	1.7	8.2	21.0	16.5	21.0
	YEAR OF OCCURRENCE		1982	1939	1985	1983	1946	1995	1994	1992	1942	1991	1991	1982	NOV 1991
	MAXIMUM SNOW DEPTH (IN)	54	38	30	27	10	2	0	0	0	0	1	23	21	38
	YEAR OF OCCURRENCE		1982	1967	1965	1985	1984					1969	1991	1991	JAN 1982
	NORMAL NO. DAYS WITH:														
	SNOWFALL 1.0	30	3.9	3.1	3.2	0.9	0.*	0.0	0.0	0.0	0.0	0.1	2.2	3.7	17.1

PRECIPITATION (inches) 2001 MINNEAPOLIS - ST. PAUL, MN (MSP)

YEAR	JAN	FEB	MAR	APR	MAY	JUN	JUL	AUG	SEP	OCT	NOV	DEC	ANNUAL
1972	0.84	0.49	1.25	1.69	2.18	3.31	5.12	2.48	1.96	1.77	1.11	1.57	23.77
1973	0.92	0.84	1.12	2.32	2.48	1.06	2.90	3.05	2.08	1.29	1.97	1.10	21.13
1974	0.17	1.06	1.00	2.42	2.08	5.21	1.14	2.75	0.58	1.69	0.66	0.35	19.11
1975	2.82	0.79	1.67	5.40	3.81	7.99	0.58	4.92	1.31	0.27	4.80	0.79	35.15
1976	0.87	0.59	2.83	0.80	1.13	3.86	2.45	1.39	1.42	0.49	0.16	0.51	16.50
1977	0.65	0.93	2.66	1.84	2.86	3.57	3.72	9.31	4.43	2.34	1.42	1.15	34.88
1978	0.38	0.24	0.79	3.63	3.79	7.09	3.19	5.77	2.47	0.19	1.84	0.88	30.26
1979	1.09	1.39	2.55	0.66	4.55	4.78	2.34	7.04	2.20	3.16	0.98	0.33	31.07
1980	0.94	0.67	1.12	0.83	2.29	5.52	2.30	3.26	3.68	0.66	0.26	0.24	21.77
1981	0.30	2.14	0.71	2.17	2.18	4.42	4.09	4.73	1.46	2.69	2.16	0.92	27.97
1982	2.45	0.43	2.09	1.62	4.99	1.44	0.92	3.80	1.50	3.45	3.27	4.27	30.23
1983	0.67	1.19	3.22	3.97	6.20	5.22	3.07	3.12	3.34	2.61	4.93	1.53	39.07
1984	0.88	1.64	1.47	3.86	2.29	7.95	3.03	5.15	2.65	5.48	0.31	2.24	36.95
1985	0.87	0.50	4.48	1.81	3.65	2.18	2.20	5.02	4.37	3.66	1.72	1.20	31.66
1986	0.90	0.84	2.03	5.88	3.48	5.34	4.11	4.44	6.90	1.77	0.62	0.31	36.62
1987	0.63	0.13	0.64	0.16	1.88	1.95	17.90	3.67	1.28	0.60	2.07	1.25	32.16
1988	1.37	0.30	1.33	1.58	1.70	0.22	1.17	4.29	2.79	0.80	2.86	0.67	19.08
1989	0.52	1.04	2.19	2.66	3.38	3.50	2.92	1.28	0.53	1.38	0.42	23.32
1990	0.10	0.77	3.66	3.80	3.36	9.82	5.06	1.71	1.88	1.23	0.65	1.01	33.05
1991	0.49	1.03	2.29	3.58	6.35	2.57	2.95	3.14	5.43	2.52	5.29	1.05	36.69
1992	0.66	0.57	1.56	1.99	1.15	3.68	5.21	4.54	5.20	2.11	1.95	1.05	29.67
1993	1.25	0.39	1.25	1.99	4.02	6.28	5.58	6.50	2.04	0.79	1.57	0.55	32.21
1994	1.17	0.78	0.32	3.77	2.21	3.09	4.12	2.90	4.74	4.65	1.39	0.53	29.67
1995	0.36	0.25	2.11	1.90	2.43	3.38	2.72	4.59	2.21	3.68	0.88	1.15	25.66
1996	1.87	0.24	1.39	0.76	2.37	4.76	2.09	1.43	1.30	3.01	5.08	1.75	26.05
1997	1.71	0.30	1.18	1.01	1.70	3.70	12.60	6.01	3.19	2.03	0.69	0.31	34.43
1998	1.64	0.80	4.56	1.56	4.40	6.52	2.63	5.99	1.32	2.19	1.32	0.46	33.39
1999	2.67	0.40	1.86	3.43	6.56	3.68	4.55	2.64	2.73	0.92	0.77	0.33	30.54
2000	0.90	1.08	1.12	1.12	4.56	4.56	6.10	3.19	2.15	1.09	3.38	1.23	30.48
2001	1.21	1.33	1.09	7.00	4.53	6.35	2.12	2.31	3.50	1.28	2.77	0.74	34.23
POR= 111 YRS	0.90	0.85	1.62	2.21	3.41	4.20	3.60	3.41	2.88	2.00	1.45	0.90	27.43

AVERAGE TEMPERATURE (F) 2001   MINNEAPOLIS - ST. PAUL, MN (MSP)

YEAR	JAN	FEB	MAR	APR	MAY	JUN	JUL	AUG	SEP	OCT	NOV	DEC	ANNUAL
1972	5.5	10.5	26.5	41.9	61.3	66.0	68.5	69.8	57.9	43.7	32.2	11.3	41.3
1973	17.4	21.6	40.2	44.4	55.2	69.5	73.8	73.4	60.1	53.8	34.3	16.7	46.7
1974	11.9	16.9	29.5	47.1	54.4	65.5	76.6	67.3	55.3	49.8	33.7	24.4	44.4
1975	14.5	15.5	22.1	38.9	60.9	68.8	76.3	71.7	57.7	52.8	37.5	21.3	44.8
1976	11.6	27.8	31.4	51.8	58.9	71.7	76.1	73.1	60.3	44.6	28.3	13.6	45.9
1977	0.3	22.7	37.5	53.0	66.9	68.4	74.8	66.1	60.5	47.1	30.8	14.4	45.2
1978	5.5	11.6	30.0	45.2	61.8	67.8	71.1	72.2	67.3	49.8	32.5	15.2	44.2
1979	3.2	10.0	28.9	44.0	55.5	67.3	73.6	69.9	63.4	46.6	31.7	26.0	43.3
1980	15.3	15.3	27.3	49.2	61.5	67.6	75.2	70.7	59.5	45.1	36.6	19.8	45.3
1981	18.0	23.4	37.7	49.1	57.1	67.0	70.9	69.3	60.0	46.7	38.0	17.5	46.2
1982	2.3	15.8	29.0	43.8	62.5	63.7	75.6	71.8	60.9	50.3	31.5	25.7	44.4
1983	19.6	26.9	34.2	42.3	54.6	68.0	77.2	76.8	62.6	48.4	34.0	3.7	45.7
1984	12.0	27.5	24.8	47.1	56.0	69.7	72.2	73.5	57.2	50.7	33.3	17.9	45.2
1985	10.1	16.5	35.6	52.1	62.2	63.9	73.9	67.6	59.9	47.5	24.8	7.7	43.5
1986	17.5	15.7	33.9	49.6	59.4	68.6	73.9	67.1	59.8	49.2	28.2	24.7	45.6
1987	21.2	31.6	38.7	53.5	63.5	72.8	76.0	69.0	62.5	44.6	37.9	25.0	49.7
1988	10.4	13.9	33.8	47.4	65.4	74.4	78.1	73.9	62.4	44.0	32.7	20.5	46.4
1989	21.2	8.6	26.6	45.3	57.5	69.8	68.4	76.4	70.8	60.9	49.9	28.0	10.6	43.7
1990	26.3	23.7	35.7	46.8	56.3	69.5	71.3	70.6	64.4	48.1	37.4	16.9	47.3
1991	12.5	24.4	34.3	49.1	61.9	72.9	72.3	71.1	59.0	47.2	24.5	21.2	45.9
1992	21.9	28.0	33.1	43.6	60.5	65.6	65.8	65.9	59.6	47.4	31.4	21.2	45.3
1993	14.6	17.2	29.5	44.2	57.2	64.5	70.3	70.4	55.0	46.5	30.6	22.2	43.5
1994	4.4	13.2	34.7	45.9	60.7	69.9	70.1	67.4	64.3	52.2	38.0	24.5	45.4
1995	18.5	19.3	35.0	42.2	56.9	71.2	73.1	74.7	60.2	48.6	27.4	19.1	45.5
1996	10.2	18.0	25.3	41.4	55.6	67.4	70.0	70.5	62.2	48.8	25.4	13.7	42.4
1997	10.3	19.9	29.3	43.0	53.4	70.0	71.0	68.8	62.4	50.2	28.1	26.9	44.4
1998	19.1	31.9	31.9	50.7	63.4	64.9	72.6	71.6	66.6	51.2	37.2	24.6	48.8
1999	12.4	27.9	33.8	49.0	60.1	67.3	76.2	70.1	61.1	49.6	41.8	25.6	47.9
2000	15.9	27.9	41.1	46.7	60.9	66.1	72.4	72.2	61.6	53.3	31.2	7.6	46.4
2001	20.0	11.8	27.5	48.4	59.7	69.1	75.9	74.2	60.9	48.6	46.4	27.6	47.5
POR= 111 YRS	13.3	17.5	30.2	46.0	58.3	67.9	73.1	70.6	61.5	49.6	32.9	19.4	45.0

REFERENCE NOTES:

PAGE 1:
THE TEMPERATURE GRAPH SHOWS NORMAL MAXIMUM AND NORMAL
MINIMUM DAILY TEMPERATURES (SOLID CURVES) AND THE
ACTUAL DAILY HIGH AND LOW TEMPERATURES (VERTICAL BARS).

PAGE 2 AND 3:
H/C INDICATES HEATING AND COOLING DEGREE DAYS.
RH INDICATES RELATIVE HUMIDITY
W/O INDICATES WEATHER AND OBSTRUCTIONS
S INDICATES SUNSHINE.
PR INDICATES PRESSURE.
CLOUDINESS ON PAGE 3 IS THE SUM OF THE CEILOMETER AND
SATELLITE DATA NOT TO EXCEED EIGHT EIGHTHS(OKTAS).

GENERAL:
T INDICATES TRACE PRECIPITATION, AN AMOUNT GREATER
THAN ZERO BUT LESS THAN THE LOWEST REPORTABLE VALUE.
+ INDICATES THE VALUE ALSO OCCURS ON EARLIER DATES.
BLANK ENTRIES DENOTE MISSING OR UNREPORTED DATA.
NORMALS ARE 30-YEAR AVERAGES (1961 - 1990).
ASOS INDICATES AUTOMATED SURFACE OBSERVING SYSTEM.
PM INDICATES THE LAST DAY OF THE PREVIOUS MONTH.
POR (PERIOD OF RECORD) BEGINS WITH THE JANUARY DATA
MONTH AND IS THE NUMBER OF YEARS USED TO COMPUTE
THE MEAN. INDIVIDUAL MONTHS WITHIN THE POR MAY
BE MISSING.
WHEN THE POR FOR A NORMAL IS LESS THAN 30 YEARS,
THE NORMAL IS PROVISIONAL AND IS BASED ON THE NUMBER
OF YEARS INDICATED.
0.* OR * INDICATES THE VALUE OR MEAN-DAYS-WITH
IS BETWEEN 0.00 AND 0.05.
CLOUDINESS FOR ASOS STATIONS DIFFERS FROM THE NON-ASOS
OBSERVATION TAKEN BY A HUMAN OBSERVER. ASOS STATION
CLOUDINESS IS BASED ON TIME-AVERAGED CEILOMETER DATA
FOR CLOUDS AT OR BELOW 12,000 FEET AND ON SATELLITE
DATA FOR CLOUDS ABOVE 12,000 FEET.
THE NUMBER OF DAYS WITH CLEAR, PARTLY CLOUDY, AND
CLOUDY CONDITIONS FOR ASOS STATIONS IS THE SUM
OF THE CEILOMETER AND SATELLITE DATA FOR THE
SUNRISE TO SUNSET PERIOD.

GENERAL CONTINUED:
CLEAR INDICATES 0 - 2 OKTAS, PARTLY CLOUDY INDICATES
3 - 6 OKTAS, AND CLOUDY INDICATES 7 OR 8 OKTAS.
WHEN AT LEAST ONE OF THE ELEMENTS (CEILOMETER OR
SATELLITE) IS MISSING, THE DAILY CLOUDINESS IS
NOT COMPUTED.
WIND DIRECTION IS RECORDED IN TENS OF DEGREES (2 DIGITS)
CLOCKWISE FROM TRUE NORTH. "00" INDICATES CALM. "36"
INDICATES TRUE NORTH.
RESULTANT WIND IS THE VECTOR AVERAGE OF THE SPEED AND
DIRECTION.
AVERAGE TEMPERATURE IS THE SUM OF THE MEAN DAILY MAXIMUM
AND MINIMUM TEMPERATURE DIVIDED BY 2.
SNOWFALL DATA COMPRISE ALL FORMS OF FROZEN
PRECIPITATION, INCLUDING HAIL.
A HEATING (COOLING) DEGREE DAY IS THE DIFFERENCE BETWEEN
THE AVERAGE DAILY TEMPERATURE AND 65 F.
DRY BULB IS THE TEMPERATURE OF THE AMBIENT AIR.
DEW POINT IS THE TEMPERATURE TO WHICH THE AIR MUST BE
COOLED TO ACHIEVE 100 PERCENT RELATIVE HUMIDITY.
WET BULB IS THE TEMPERATURE THE AIR WOULD HAVE IF THE
MOISTURE CONTENT WAS INCREASED TO 100 PERCENT RELATIVE
HUMIDITY.

ON JULY 1, 1996, THE NATIONAL WEATHER SERVICE BEGAN USING
THE "METAR" OBSERVATION CODE THAT WAS ALREADY EMPLOYED
BY MOST OTHER NATIONS OF THE WORLD. THE MOST NOTICEABLE
DIFFERENCE IN THIS ANNUAL PUBLICATION WILL BE THE CHANGE
IN UNITS FROM TENTHS TO EIGHTS(OKTAS) FOR REPORTING THE
AMOUNT OF SKY COVER.

HEATING DEGREE DAYS (base 65 F) 2001 MINNEAPOLIS - ST. PAUL, MN (MSP)

YEAR	JUL	AUG	SEP	OCT	NOV	DEC	JAN	FEB	MAR	APR	MAY	JUN	TOTAL
1972-73	34	52	218	651	974	1664	1474	1208	761	611	299	13	7959
1973-74	1	3	185	350	915	1493	1642	1344	1092	535	338	72	7970
1974-75	0	48	289	467	933	1252	1561	1379	1324	775	188	39	8255
1975-76	15	7	231	387	818	1346	1650	1074	1031	405	195	11	7170
1976-77	0	4	162	632	1092	1590	2005	1180	844	365	75	17	7966
1977-78	0	35	145	548	1016	1565	1842	1488	1080	584	162	46	8511
1978-79	5	7	89	464	968	1538	1914	1537	1112	623	307	38	8602
1979-80	0	24	105	566	992	1203	1536	1436	1165	484	184	34	7729
1980-81	0	12	194	611	845	1396	1453	1160	838	472	249	28	7258
1981-82	11	11	172	564	803	1466	1945	1374	1111	629	117	71	8274
1982-83	0	14	168	448	997	1212	1400	1061	947	673	313	49	7282
1983-84	2	0	161	514	923	1901	1641	1082	1240	531	284	7	8286
1984-85	5	12	251	435	943	1453	1694	1355	904	403	123	104	7682
1985-86	0	28	240	537	1201	1774	1466	1377	957	454	212	30	8276
1986-87	0	43	177	480	1096	1243	1352	929	809	347	134	13	6623
1987-88	2	29	106	623	804	1236	1688	1479	962	523	76	4	7532
1988-89	1	16	116	646	963	1373	1353	1576	1184	583	251	44	8106
1989-90	0	6	159	470	1105	1683	1194	1151	899	569	274	37	7547
1990-91	2	5	136	516	820	1484	1624	1130	945	481	197	3	7343
1991-92	7	8	228	548	1206	1354	1333	1067	981	636	190	72	7630
1992-93	32	52	182	542	1003	1351	1557	1335	1096	617	243	70	8080
1993-94	3	18	302	566	1025	1322	1879	1445	932	569	180	27	8268
1994-95	2	45	99	390	802	1250	1434	1274	924	678	247	47	7192
1995-96	6	0	201	511	1123	1416	1697	1360	1222	699	304	62	8601
1996-97	3	2	167	500	1182	1583	1688	1255	1100	653	351	6	8490
1997-98	27	26	113	483	1101	1173	1414	917	1019	423	104	107	6907
1998-99	0	0	74	422	829	1249	1625	1034	958	473	171	76	6911
1999-00	0	2	174	471	690	1214	1515	1070	734	542	176	72	6660
2000-01	12	1	146	364	1008	1771	1386	1483	1155	497	197	54	8074
2001-	8	2	162	505	552	1152							

COOLING DEGREE DAYS (base 65 F) 2001 MINNEAPOLIS - ST. PAUL, MN (MSP)

YEAR	JAN	FEB	MAR	APR	MAY	JUN	JUL	AUG	SEP	OCT	NOV	DEC	ANNUAL
1972	0	0	0	0	94	109	148	208	13	0	0	0	572
1973	0	0	0	1	4	158	280	271	47	8	0	0	769
1974	0	0	0	5	18	93	369	127	6	1	0	0	619
1975	0	0	0	0	66	159	371	220	18	16	0	0	850
1976	0	0	0	14	14	223	351	269	72	7	0	0	950
1977	0	0	0	12	145	129	310	76	19	0	0	0	691
1978	0	0	0	0	72	138	201	236	164	0	0	0	811
1979	0	0	0	0	17	113	275	181	65	0	0	0	651
1980	0	0	0	16	82	121	322	194	38	1	0	0	774
1981	0	0	0	0	10	96	200	151	28	0	0	0	485
1982	0	0	0	0	46	40	338	232	53	0	0	0	709
1983	0	0	0	0	0	145	389	368	98	8	0	0	1008
1984	0	0	0	0	13	155	237	280	24	0	0	0	709
1985	0	0	0	22	43	77	284	118	93	0	0	0	637
1986	0	0	0	1	45	148	286	115	32	0	0	0	627
1987	0	0	0	11	95	253	348	159	37	0	0	0	903
1988	0	0	0	1	96	296	412	302	45	0	0	0	1152
1989	0	0	0	0	26	153	359	192	41	8	0	0	779
1990	0	0	0	28	11	178	206	191	125	1	0	0	740
1991	0	0	0	8	109	246	238	205	51	0	0	0	857
1992	0	0	0	3	56	96	64	88	28	2	0	0	337
1993	0	0	0	0	12	60	176	195	8	0	0	0	451
1994	0	0	0	3	52	183	167	126	86	0	0	0	617
1995	0	0	0	0	3	240	264	308	63	9	0	0	887
1996	0	0	0	0	20	142	168	181	87	4	0	0	602
1997	0	0	0	0	1	163	222	150	41	33	0	0	610
1998	0	0	0	0	62	111	243	212	130	0	0	0	758
1999	0	0	0	0	28	151	357	166	64	0	0	0	766
2000	0	0	0	0	55	111	249	228	53	8	0	0	704
2001	0	0	0	8	38	184	351	293	46	2	0	0	922

SNOWFALL (inches) 2001 MINNEAPOLIS - ST. PAUL, MN (MSP)

YEAR	JUL	AUG	SEP	OCT	NOV	DEC	JAN	FEB	MAR	APR	MAY	JUN	TOTAL
1972-73	0.0	0.0	T	T	1.1	15.3	11.6	11.3	0.4	2.0	0.0	0.0	41.7
1973-74	0.0	0.0	0.0	0.0	0.1	17.9	2.5	15.7	7.7	7.3	0.0	0.0	51.2
1974-75	0.0	0.0	0.0	0.0	1.2	6.1	27.4	9.0	18.3	2.2	0.0	0.0	64.2
1975-76	0.0	0.0	0.0	0.0	16.2	5.6	12.8	5.1	13.6	0.0	1.2	0.0	54.5
1976-77	0.0	0.0	0.0	2.3	1.4	8.3	13.4	1.8	14.6	1.8	0.0	0.0	43.6
1977-78	0.0	0.0	0.0	3.0	11.7	14.2	6.8	4.6	8.5	1.9	0.0	0.0	50.7
1978-79	0.0	0.0	0.0	0.0	16.5	15.1	14.2	13.5	8.4	0.7	0.0	0.0	68.4
1979-80	0.0	0.0	0.0	T	7.7	1.7	12.9	8.8	13.7	8.5	0.0	0.0	53.3
1980-81	0.0	0.0	0.0	0.0	0.9	2.8	4.6	11.0	0.1	1.7	0.0	0.0	21.1
1981-82	0.0	0.0	0.0	0.9	14.0	10.6	46.4	7.4	10.9	4.8	0.0	0.0	95.0
1982-83	0.0	0.0	0.0	1.4	3.6	19.3	3.2	10.8	14.3	21.8	0.0	0.0	74.4
1983-84	0.0	0.0	0.0	T	30.4	21.0	10.6	9.3	17.3	9.8	0.0	0.0	98.4
1984-85	0.0	0.0	0.0	0.3	2.0	16.3	13.1	4.2	36.8	T	0.0	0.0	72.7
1985-86	0.0	0.0	0.4	T	23.9	13.5	10.3	12.3	8.7	0.4	0.0	0.0	69.5
1986-87	0.0	0.0	0.0	T	4.4	4.2	5.5	1.2	2.1	T	0.0	0.0	17.4
1987-88	0.0	0.0	0.0	0.3	4.5	7.5	19.5	4.5	3.7	2.4	0.0	0.0	42.4
1988-89	0.0	0.0	0.0	0.2	15.8	7.2	6.0	17.3	22.7	0.8	0.1	T	70.1
1989-90	0.0	0.0	0.0	0.0	11.3	7.0	1.1	10.7	3.2	2.2	0.0	0.0	35.5
1990-91	0.0	0.0	0.0	T	5.0	11.7	6.5	14.2	4.4	1.5	0.3	0.0	43.6
1991-92	0.0	T	0.0	8.2	46.9	6.7	5.0	5.9	10.8	0.6	0.0	0.0	84.1
1992-93	0.0	T	T	1.3	12.2	9.2	12.0	5.3	6.9	0.5	0.0	0.0	47.4
1993-94	0.0	0.0	0.0	T	7.7	4.5	24.3	12.0	1.7	5.5	0.0	T	55.7
1994-95	T	0.0	0.0	T	6.2	6.5	4.2	2.1	10.4	0.2	0.0	T	29.6
1995-96	0.0	0.0	T	0.7	6.6	16.1	14.5	1.2	14.1	2.3	T	0.0	55.5
1996-97	0.0	0.0	0.0	T	15.3	23.5	14.2	4.0	14.3	0.6	T	T	71.9
1997-98	0.0	T	0.0	T	8.6	3.3	20.4	1.1	11.6	T	T	T	45.0
1998-99	0.0	0.0	T	0.0	0.1	3.1	33.1	4.2	16.0	T	0.0	0.0	56.5
1999-00	0.0	0.0	0.0	T	0.7	7.3	18.2	7.7	1.0	1.3	0.0	0.0	36.2
2000-01	0.0	0.0	0.0	0.0	0.0								
2001-													
POR=60 YRS	T	T	0.0	0.5	7.7	9.2	10.5	8.0	10.3	2.8	0.4	T	49.4

2001
JACKSON,
MISSISSIPPI (JAN)

Jackson is located on the west bank of the Pearl River, about 45 miles east of the Mississippi River and 150 miles north of the Gulf of Mexico. The nearby terrain is gently rolling with no topographic features that appreciably influence the weather. The National Weather Service Office is nearly 7 miles east-northeast of the Jackson Post Office and over 5 miles southwest of the Ross Barnett Reservoir. Alluvial plains up to 3 miles wide extend along the river near Jackson, where some levees have been built on both sides of the river.

The climate is significantly humid during most of the year, with relatively short mild winters and long warm summers. The Gulf of Mexico has a moderating effect on the climate. Cold spells are fairly frequent in winter, but are usually of short duration. Sub-zero temperatures rarely occur. Temperatures occasionally exceed 80 degrees in mid-winter. In summer, temperatures reach 90 degrees or higher on about two-thirds of the days, 100 degree readings are infrequent. Extended periods of very hot weather are rare. On unusual occasions, temperatures at night may drop into the 50s, even in July or August.

Snowfall averages less than two inches per season, with nearly two-thirds of the seasons having only a trace of snow or none at all. Ice storms occasionally cause major damage to trees and power lines during the winter or early spring season. Rainfall is abundant and fairly well-distributed throughout the year. The area does not have a true dry season. However, the six-month period, June through November, is relatively dry in comparison with the December through May period when 60 percent of the annual precipitation can be expected.

Excessive rainfall may occur in any season. In spite of the normally abundant rainfall, fairly serious droughts occasionally occur during the summer or fall season. Tropical disturbances, including hurricanes and their remnants, are infrequent. However, those that pass near or visit the Mississippi Coast in the summer or early fall may bring several days of heavy rain to the Jackson area.

Thunderstorms can be expected on an average of 65 days a year, usually occurring in each month. They are most frequent in summer when they occur on about one-third of the days. At other times of the year, thunderstorms are usually associated with passing weather systems and are likely to be attended by higher winds than in summer. Severe thunderstorms normally affect portions of the metropolitan area a few times each year.

NORMALS, MEANS, AND EXTREMES
JACKSON, MS (JAN)

LATITUDE: 32 19' 11" N   LONGITUDE: 90 04' 39" W   ELEVATION (FT): GRND: 293   BARO: 296   TIME ZONE: CENTRAL (UTC + 6)   WBAN: 03940

	ELEMENT	POR	JAN	FEB	MAR	APR	MAY	JUN	JUL	AUG	SEP	OCT	NOV	DEC	YEAR
TEMPERATURE F	NORMAL DAILY MAXIMUM	30	55.6	60.1	69.3	77.4	84.0	90.6	92.4	92.0	88.0	79.1	69.2	59.5	76.4
	MEAN DAILY MAXIMUM	37	56.0	60.7	68.5	76.9	83.7	90.3	92.2	91.7	87.0	77.7	68.0	59.5	76.0
	HIGHEST DAILY MAXIMUM	37	82	85	89	94	99	105	106	107	104	95	88	84	107
	YEAR OF OCCURRENCE		1972	1989	1982	1987	1964	1988	1980	2000	2000	1998	1971	1978	AUG 2000
	MEAN OF EXTREME MAXS.	38	74.2	78.2	83.6	87.2	91.9	96.1	97.9	97.4	94.7	89.0	82.5	76.8	87.5
	NORMAL DAILY MINIMUM	30	32.7	35.7	44.1	51.9	60.0	67.1	70.5	69.7	63.7	50.3	42.3	36.1	52.0
	MEAN DAILY MINIMUM	37	34.8	37.4	44.5	52.5	60.8	68.0	71.4	70.2	64.8	51.6	43.4	37.5	53.1
	LOWEST DAILY MINIMUM	37	2	10	15	27	38	47	51	55	35	26	17	4	2
	YEAR OF OCCURRENCE		1985	1996	1980	1987	1971	1984	1967	1992	1967	1993	1976	1989	JAN 1985
	MEAN OF EXTREME MINS.	38	16.5	19.8	26.5	35.1	46.2	55.7	64.0	62.0	49.4	34.8	26.1	19.7	38.0
	NORMAL DRY BULB	30	44.1	47.9	56.7	64.6	72.0	78.8	81.5	80.9	75.9	64.7	55.8	47.8	64.2
	MEAN DRY BULB	38	45.3	49.0	56.6	64.6	72.2	79.0	81.8	81.1	75.8	64.8	55.7	48.2	64.5
	MEAN WET BULB	35	41.5	44.2	50.9	58.6	65.5	71.5	74.3	73.4	69.0	58.5	51.2	44.2	58.6
	MEAN DEW POINT	35	36.9	39.1	45.7	53.8	61.9	68.3	71.6	70.7	65.9	54.4	47.1	40.0	54.6
	NORMAL NO. DAYS WITH:														
	MAXIMUM 90	30	0.0	0.0	0.0	0.3	4.8	18.7	23.9	22.8	12.1	1.3	0.0	0.0	83.9
	MAXIMUM 32	30	0.9	0.3	0.0	0.0	0.0	0.0	0.0	0.0	0.0	0.0	*	0.4	1.6
	MINIMUM 32	30	15.3	11.3	4.2	0.4	0.0	0.0	0.0	0.0	0.0	0.5	5.7	12.6	50.0
	MINIMUM 0	30	0.0	0.0	0.0	0.0	0.0	0.0	0.0	0.0	0.0	0.0	0.0	0.0	0.0
H/C	NORMAL HEATING DEG. DAYS	30	656	485	285	87	7	0	0	0	6	104	295	542	2467
	NORMAL COOLING DEG. DAYS	30	8	6	28	75	224	414	512	493	333	94	19	9	2215
RH	NORMAL (PERCENT)	30	76	73	71	72	74	74	77	77	77	75	76	76	75
	HOUR 00 LST	30	84	82	82	84	87	88	90	90	90	88	87	84	86
	HOUR 06 LST	30	87	87	88	90	92	92	94	94	94	93	90	87	91
	HOUR 12 LST	30	64	60	56	54	56	55	59	58	58	53	57	62	58
	HOUR 18 LST	30	70	63	58	57	60	60	66	66	70	71	73	74	66
S	PERCENT POSSIBLE SUNSHINE	31	49	55	61	66	63	71	64	65	63	67	57	47	61
W/O	MEAN NO. DAYS WITH:														
	HEAVY FOG(VISBY 1/4 MI)	37	3.3	2.0	1.7	1.5	1.1	0.7	1.3	1.5	1.5	2.2	2.4	3.3	22.5
	THUNDERSTORMS	37	2.3	2.7	5.8	5.6	7.0	8.9	12.6	10.2	5.1	2.2	2.5	2.3	67.2
CLOUDINESS	MEAN:														
	SUNRISE-SUNSET (OKTAS)	1	6.4	5.6	4.8	4.0	4.0	3.6	5.6	3.2	2.4	5.2	5.6	6.4	4.7
	MIDNIGHT-MIDNIGHT (OKTAS)	1	6.4	5.6	4.8	4.0	4.0	3.2	3.6	2.8	2.4	5.6	5.6	6.4	4.5
	MEAN NO. DAYS WITH:														
	CLEAR	2	8.5	12.5	10.0	12.0	16.0	14.5	9.0	9.0	5.0	6.0	5.0	3.0	110.5
	PARTLY CLOUDY	2	3.5	3.5	4.0	6.0	8.0	12.0	4.0	7.0			1.0	3.0	
	CLOUDY	2	15.0	13.0	10.0	5.0	11.5	9.0	14.0	5.0	1.0	12.0	7.0	5.0	107.5
PR	MEAN STATION PRESSURE(IN)	28	29.80	29.76	29.68	29.66	29.63	29.64	29.68	29.68	29.68	29.74	29.76	29.80	29.71
	MEAN SEA-LEVEL PRES. (IN)	34	30.16	30.12	30.04	30.01	29.98	29.99	30.02	30.02	30.02	30.08	30.11	30.15	30.06
WINDS	MEAN SPEED (MPH)	30	7.9	8.1	8.4	7.8	6.7	6.0	5.2	5.1	6.0	6.0	6.9	7.6	6.8
	PREVAIL.DIR(TENS OF DEGS)	31	15	34	16	16	15	15	15	14	14	14	15	15	15
	MAXIMUM 2-MINUTE:														
	SPEED (MPH)	8	32	43	37	43	28	39	44	33	55	26	29	41	55
	DIR. (TENS OF DEGS)		34	13	14	12	10	34	02	04	06	35	26	15	06
	YEAR OF OCCURRENCE		1996	1998	1999	1997	2001	1994	1998	2000	1998	2001	2000	1994	SEP 1998
	MAXIMUM 5-SECOND:														
	SPEED (MPH)	8	40	51	47	54	39	47	53	45	63	34	39	46	63
	DIR. (TENS OF DEGS)		34	14	33	29	26	33	02	01	06	26	27	15	06
	YEAR OF OCCURRENCE		1996	1998	2000	1996	1997	1994	1998	1999	1998	1996	2000	1994	SEP 1998
PRECIPITATION	NORMAL (IN)	30	5.24	4.70	5.82	5.57	5.05	3.18	4.51	3.77	3.55	3.26	4.81	5.91	55.37
	MAXIMUM MONTHLY (IN)	37	14.10	10.28	15.13	15.95	10.82	8.45	13.25	8.33	9.61	9.13	9.98	17.70	17.70
	YEAR OF OCCURRENCE		1979	1987	1976	1991	1967	1997	1979	1992	1965	1970	1977	1982	DEC 1982
	MINIMUM MONTHLY (IN)	37	0.75	1.29	2.05	1.21	0.29	0.10	1.04	0.26	0.56	0.00	0.51	0.91	0.00
	YEAR OF OCCURRENCE		1986	2000	1966	1987	1988	1988	1987	2000	1969	1963	1985	1980	OCT 1963
	MAXIMUM IN 24 HOURS (IN)	37	5.63	3.46	4.69	8.42	3.43	6.49	5.49	4.79	5.86	6.99	4.34	6.71	8.42
	YEAR OF OCCURRENCE		1979	1974	1991	1979	1989	1997	2001	1992	1965	1975	1983	1982	APR 1979
	NORMAL NO. DAYS WITH:														
	PRECIPITATION 0.01	30	10.8	9.3	10.1	8.3	9.4	8.2	10.3	9.5	8.3	5.9	8.1	10.1	108.3
	PRECIPITATION 1.00	30	1.7	1.4	2.0	1.9	1.9	0.9	1.0	1.0	0.9	0.9	1.6	2.1	17.3
SNOWFALL	NORMAL (IN)	30	0.6	0.2	0.2	0.*	0.0	0.0	0.0	0.0	0.0	0.0	0.*	0.1	1.1
	MAXIMUM MONTHLY (IN)	37	6.3	3.6	5.3	1.1	0.0	0.0	0.0	T	0.0	0.0	0.2	4.8	6.3
	YEAR OF OCCURRENCE		1982	1968	1968	1987				2001			1976	1997	JAN 1982
	MAXIMUM IN 24 HOURS (IN)	37	6.0	3.6	5.3	1.1	0.0	0.0	0.0	T	0.0	0.0	0.2	4.8	6.0
	YEAR OF OCCURRENCE		1982	1968	1968	1987				2001			1976	1997	JAN 1982
	MAXIMUM SNOW DEPTH (IN)	36	6	1	2	1	0	0	0	0	0	0	T	5	6
	YEAR OF OCCURRENCE		1982	1985	1968	1987							1976	1997	JAN 1982
	NORMAL NO. DAYS WITH:														
	SNOWFALL 1.0	30	0.2	0.1	0.*	0.0	0.0	0.0	0.0	0.0	0.0	0.0	0.0	0.1	0.4

PRECIPITATION (inches) 2001 JACKSON, MS (JAN)

YEAR	JAN	FEB	MAR	APR	MAY	JUN	JUL	AUG	SEP	OCT	NOV	DEC	ANNUAL
1972	5.94	3.09	5.57	2.44	4.52	2.01	3.31	2.84	5.04	2.08	3.52	9.67	50.03
1973	4.59	4.23	6.12	9.44	5.96	0.32	1.99	2.38	4.44	2.72	6.15	6.71	55.05
1974	11.00	6.72	3.50	6.74	3.01	3.39	1.54	6.17	5.06	1.74	4.12	7.22	60.21
1975	4.57	6.18	4.86	5.07	6.53	7.44	9.81	6.21	2.68	8.25	4.34	4.29	70.23
1976	3.64	1.43	15.13	2.08	8.01	2.80	4.96	5.26	3.78	3.52	3.34	3.44	57.39
1977	6.18	2.26	6.41	7.98	0.74	2.17	6.22	1.45	3.93	2.79	9.98	3.47	53.58
1978	5.32	2.36	3.37	3.54	10.48	1.03	3.65	1.80	2.90	1.21	3.30	8.37	47.33
1979	14.10	8.35	4.67	14.38	5.52	4.38	13.25	7.44	5.93	1.76	8.79	4.18	92.75
1980	7.53	3.19	13.57	14.33	6.60	1.74	2.91	1.45	3.25	3.47	4.11	0.91	63.06
1981	1.41	2.63	6.19	1.26	6.64	3.66	6.51	3.51	5.12	1.97	4.90	3.66	46.61
1982	4.48	5.22	5.13	6.59	0.77	6.27	9.29	4.97	1.05	6.73	7.43	17.70	75.63
1983	8.17	6.55	6.00	15.53	9.41	2.93	1.70	3.70	2.70	1.52	8.11	6.95	73.27
1984	2.64	4.64	4.84	3.96	5.61	3.18	3.07	4.56	0.93	7.68	6.48	2.17	49.76
1985	4.05	7.55	3.13	3.31	0.86	1.74	4.43	3.94	7.06	7.17	0.51	3.61	47.36
1986	0.75	1.53	3.34	1.75	10.00	3.72	4.78	2.03	2.63	5.10	9.40	4.98	50.01
1987	4.66	10.28	5.47	1.21	4.98	6.17	1.04	4.03	1.50	0.27	4.20	3.50	47.31
1988	2.25	3.89	7.46	5.37	0.29	0.10	2.73	3.02	2.28	6.14	5.66	4.80	43.99
1989	4.38	2.52	4.53	2.13	7.92	8.17	4.47	1.74	5.40	0.23	6.86	4.20	52.55
1990	12.17	8.30	3.55	3.66	6.34	1.46	2.84	0.61	4.83	1.24	3.33	5.71	54.04
1991	4.98	5.30	7.50	15.95	7.11	3.35	3.32	1.49	3.45	2.95	2.39	5.27	63.06
1992	4.27	3.83	2.11	1.48	1.22	4.75	4.48	8.33	3.80	1.55	8.27	4.32	48.41
1993	4.72	3.40	3.84	4.09	4.31	3.79	7.00	2.47	2.04	4.08	5.39	2.89	48.02
1994	8.31	6.14	5.63	4.32	2.25	7.37	5.09	4.29	1.81	5.12	2.57	3.31	56.21
1995	3.68	2.01	6.63	8.59	4.30	3.60	5.29	5.26	1.52	6.51	6.20	5.44	59.03
1996	8.85	2.01	8.58	6.01	0.72	3.85	6.99	6.70	1.49	1.59	4.84	3.33	54.96
1997	4.46	6.75	2.26	7.77	6.83	8.45	1.15	5.02	2.69	4.26	3.53	5.77	58.94
1998	9.52	5.51	6.17	3.68	1.47	3.26	6.44	2.02	2.41	0.51	4.32	6.24	51.55
1999	8.47	2.04	4.65	2.09	2.66	4.34	4.41	1.71	2.48	5.74	1.88	2.76	43.23
2000	1.88	1.29	4.41	7.82	2.67	5.82	1.86	0.26	3.50	1.46	7.75	3.87	42.59
2001	5.41	3.84	9.41	1.73	4.71	6.50	9.45	4.71	4.52	3.58	6.29	4.07	64.22
POR= 93 YRS	5.07	4.69	5.58	5.40	4.53	3.90	4.58	3.63	2.85	2.63	4.13	5.39	52.38

AVERAGE TEMPERATURE (F) 2001 JACKSON, MS (JAN)

YEAR	JAN	FEB	MAR	APR	MAY	JUN	JUL	AUG	SEP	OCT	NOV	DEC	ANNUAL
1972	51.5	51.2	58.8	66.4	71.9	79.6	80.4	82.8	81.5	68.0	52.0	50.1	66.2
1973	44.3	46.9	62.1	62.5	70.9	81.3	83.7	80.1	77.6	68.6	61.7	49.3	65.8
1974	55.1	50.1	62.4	63.1	73.8	74.8	80.4	79.3	72.0	64.1	55.6	50.4	65.1
1975	51.8	52.9	56.8	63.5	74.1	78.3	81.2	80.6	72.3	66.2	56.6	47.4	65.1
1976	44.3	56.7	60.0	65.4	67.9	76.7	80.8	80.2	74.8	59.9	47.2	45.0	63.2
1977	35.3	49.3	59.7	66.3	75.5	81.7	82.7	82.0	78.4	62.0	57.1	47.5	64.8
1978	36.5	39.6	52.5	65.4	71.9	80.0	83.4	82.0	78.4	63.6	60.3	48.8	63.5
1979	38.3	46.0	57.4	64.8	70.3	76.7	80.2	79.7	73.8	64.1	50.9	46.2	62.4
1980	47.4	45.5	54.3	61.6	72.8	80.7	85.8	84.7	82.4	61.4	53.2	46.2	64.7
1981	41.8	48.7	55.1	71.0	69.9	81.8	83.6	82.6	73.2	63.4	58.6	46.6	64.9
1982	47.2	48.8	62.1	63.8	74.7	79.5	82.4	81.8	74.1	65.1	56.4	52.9	65.7
1983	43.7	47.4	53.9	60.1	70.7	76.6	82.9	82.7	74.1	65.8	55.1	41.8	62.9
1984	39.9	49.7	56.5	63.9	71.0	78.9	80.6	79.9	74.6	71.2	53.6	58.3	64.8
1985	38.0	44.8	61.3	65.2	72.2	78.8	80.5	80.3	74.0	68.0	62.5	42.7	64.0
1986	45.3	52.6	57.0	64.1	72.9	80.4	83.2	80.5	79.8	66.1	60.1	46.6	65.7
1987	44.4	51.3	55.5	62.2	75.9	79.0	82.0	82.9	75.2	59.9	56.6	52.1	64.8
1988	41.7	47.5	55.8	64.6	70.8	79.8	82.5	82.6	77.9	61.6	59.3	49.3	64.5
1989	52.5	48.2	58.8	63.2	72.0	78.4	80.6	80.8	74.1	63.9	57.1	40.4	64.2
1990	51.0	56.2	60.0	63.7	71.8	81.3	81.5	82.3	78.4	63.7	58.0	51.8	66.6
1991	46.6	51.7	59.3	68.0	75.6	79.9	82.9	81.4	76.0	67.2	52.4	52.0	66.1
1992	45.6	53.7	57.6	64.1	71.1	77.7	82.0	77.8	75.6	65.0	52.7	50.1	64.4
1993	49.9	49.2	54.4	61.0	71.0	80.9	82.5	82.9	75.9	62.3	52.1	45.8	64.0
1994	41.9	49.4	55.8	66.2	70.6	79.4	78.9	79.0	74.0	66.2	58.5	49.7	64.1
1995	46.9	49.5	58.5	64.7	73.5	76.0	82.4	83.5	75.2	63.8	52.1	47.6	64.5
1996	45.2	49.8	51.9	61.5	75.7	78.6	81.3	78.9	73.6	65.2	55.3	50.9	64.0
1997	46.0	51.5	61.5	59.6	69.8	77.1	82.3	79.5	76.7	65.1	50.7	44.9	63.7
1998	49.2	50.0	55.7	62.6	74.4	82.2	83.5	82.7	80.6	68.4	58.7	51.4	66.6
1999	51.9	53.9	54.4	69.3	71.5	79.5	82.0	84.0	75.2	65.0	55.8	48.4	65.9
2000	48.8	54.7	60.4	61.8	75.3	78.7	83.5	85.0	76.5	65.7	52.1	38.5	65.1
2001	41.6	53.5	52.1	67.8	72.3	77.2	81.6	80.5	74.0	62.1	59.8	52.0	64.5
POR= 93 YRS	47.3	50.6	57.2	65.1	72.5	79.5	81.9	81.4	76.7	66.0	55.7	49.0	65.2

REFERENCE NOTES:

PAGE 1:
THE TEMPERATURE GRAPH SHOWS NORMAL MAXIMUM AND NORMAL MINIMUM DAILY TEMPERATURES (SOLID CURVES) AND THE ACTUAL DAILY HIGH AND LOW TEMPERATURES (VERTICAL BARS).

PAGE 2 AND 3:
H/C INDICATES HEATING AND COOLING DEGREE DAYS.
RH INDICATES RELATIVE HUMIDITY
W/O INDICATES WEATHER AND OBSTRUCTIONS
S INDICATES SUNSHINE.
PR INDICATES PRESSURE.
CLOUDINESS ON PAGE 3 IS THE SUM OF THE CEILOMETER AND SATELLITE DATA NOT TO EXCEED EIGHT EIGHTHS(OKTAS).

GENERAL:
T INDICATES TRACE PRECIPITATION, AN AMOUNT GREATER THAN ZERO BUT LESS THAN THE LOWEST REPORTABLE VALUE.
+ INDICATES THE VALUE ALSO OCCURS ON EARLIER DATES.
BLANK ENTRIES DENOTE MISSING OR UNREPORTED DATA.
NORMALS ARE 30-YEAR AVERAGES (1961 - 1990).
ASOS INDICATES AUTOMATED SURFACE OBSERVING SYSTEM.
PM INDICATES THE LAST DAY OF THE PREVIOUS MONTH.
POR (PERIOD OF RECORD) BEGINS WITH THE JANUARY DATA MONTH AND IS THE NUMBER OP YEARS USED TO COMPUTE THE MEAN. INDIVIDUAL MONTHS WITHIN THE POR MAY BE MISSING.
WHEN THE POR FOR A NORMAL IS LESS THAN 30 YEARS, THE NORMAL IS PROVISIONAL AND IS BASED ON THE NUMBER OF YEARS INDICATED.
0.* OR * INDICATES THE VALUE OR MEAN-DAYS-WITH IS BETWEEN 0.00 AND 0.05.
CLOUDINESS FOR ASOS STATIONS DIFFERS FROM THE NON-ASOS OBSERVATION TAKEN BY A HUMAN OBSERVER. ASOS STATION CLOUDINESS IS BASED ON TIME-AVERAGED CEILOMETER DATA FOR CLOUDS AT OR BELOW 12,000 FEET AND ON SATELLITE DATA FOR CLOUDS ABOVE 12,000 FEET.
THE NUMBER OF DAYS WITH CLEAR, PARTLY CLOUDY, AND CLOUDY CONDITIONS FOR ASOS STATIONS IS THE SUM OF THE CEILOMETER AND SATELLITE DATA FOR THE SUNRISE TO SUNSET PERIOD.

GENERAL CONTINUED:
CLEAR INDICATES 0 - 2 OKTAS, PARTLY CLOUDY INDICATES 3 - 6 OKTAS, AND CLOUDY INDICATES 7 OR 8 OKTAS.
WHEN AT LEAST ONE OF THE ELEMENTS (CEILOMETER OR SATELLITE) IS MISSING, THE DAILY CLOUDINESS IS NOT COMPUTED.
WIND DIRECTION IS RECORDED IN TENS OF DEGREES (2 DIGITS) CLOCKWISE FROM TRUE NORTH. "00" INDICATES CALM. "36" INDICATES TRUE NORTH.
RESULTANT WIND IS THE VECTOR AVERAGE OF THE SPEED AND DIRECTION.
AVERAGE TEMPERATURE IS THE SUM OF THE MEAN DAILY MAXIMUM AND MINIMUM TEMPERATURE DIVIDED BY 2.
SNOWFALL DATA COMPRISE ALL FORMS OF FROZEN PRECIPITATION, INCLUDING HAIL.
A HEATING (COOLING) DEGREE DAY IS THE DIFFERENCE BETWEEN THE AVERAGE DAILY TEMPERATURE AND 65 F.
DRY BULB IS THE TEMPERATURE OF THE AMBIENT AIR.
DEW POINT IS THE TEMPERATURE TO WHICH THE AIR MUST BE COOLED TO ACHIEVE 100 PERCENT RELATIVE HUMIDITY.
WET BULB IS THE TEMPERATURE THE AIR WOULD HAVE IF THE MOISTURE CONTENT WAS INCREASED TO 100 PERCENT RELATIVE HUMIDITY.

ON JULY 1, 1996, THE NATIONAL WEATHER SERVICE BEGAN USING THE "METAR" OBSERVATION CODE THAT WAS ALREADY EMPLOYED BY MOST OTHER NATIONS OF THE WORLD. THE MOST NOTICEABLE DIFFERENCE IN THIS ANNUAL PUBLICATION WILL BE THE CHANGE IN UNITS FROM TENTHS TO EIGHTS(OKTAS) FOR REPORTING THE AMOUNT OF SKY COVER.

HEATING DEGREE DAYS (base 65 F) 2001 JACKSON, MS (JAN)

YEAR	JUL	AUG	SEP	OCT	NOV	DEC	JAN	FEB	MAR	APR	MAY	JUN	TOTAL
1972-73	0	0	4	71	400	466	634	503	135	146	19	0	2378
1973-74	0	0	0	61	173	486	327	419	165	115	1	0	1747
1974-75	0	0	6	79	308	465	429	348	293	138	0	0	2066
1975-76	0	0	26	59	304	543	635	252	201	59	18	0	2097
1976-77	0	0	0	197	528	614	913	435	196	43	1	0	2927
1977-78	0	0	0	136	246	538	881	706	377	73	14	0	2971
1978-79	0	0	0	98	164	519	821	531	262	69	20	0	2484
1979-80	0	0	0	98	421	580	540	570	337	124	3	0	2673
1980-81	0	0	0	158	363	575	711	454	312	20	18	0	2611
1981-82	0	0	16	104	217	568	569	451	199	123	4	0	2251
1982-83	0	0	19	120	286	421	653	487	347	176	12	0	2521
1983-84	0	0	21	83	315	716	770	440	289	113	29	0	2776
1984-85	0	0	12	27	350	249	832	562	158	92	2	0	2284
1985-86	0	0	14	54	148	685	606	360	261	90	8	0	2226
1986-87	0	0	0	79	183	566	632	378	297	150	0	0	2285
1987-88	0	0	0	161	271	408	716	502	292	74	4	0	2428
1988-89	0	0	0	134	208	484	392	495	239	135	20	0	2107
1989-90	0	0	11	109	265	754	429	261	209	125	9	0	2172
1990-91	0	0	5	145	227	427	566	368	229	35	2	0	2004
1991-92	0	0	4	67	404	403	593	323	233	118	18	0	2163
1992-93	0	0	1	45	364	452	461	437	327	163	5	0	2255
1993-94	0	0	10	161	401	588	709	437	286	88	13	0	2693
1994-95	0	0	7	67	211	470	558	430	229	80	14	0	2066
1995-96	0	0	4	106	389	533	603	460	410	161	4	0	2670
1996-97	0	0	11	80	298	435	592	381	143	170	16	0	2126
1997-98	0	0	0	129	420	617	481	414	336	124	0	1	2522
1998-99	0	0	0	46	197	442	412	320	324	53	3	0	1797
1999-00	0	0	8	87	268	509	496	308	165	119	0	0	1960
2000-01	0	0	7	73	404	813	717	330	392	65	1	0	2802
2001-	0	0	16	149	168	408							

COOLING DEGREE DAYS (base 65 F) 2001 JACKSON, MS (JAN)

YEAR	JAN	FEB	MAR	APR	MAY	JUN	JUL	AUG	SEP	OCT	NOV	DEC	ANNUAL
1972	20	17	30	126	221	445	487	561	506	169	15	16	2613
1973	1	0	51	77	210	498	583	477	382	181	81	5	2546
1974	25	8	91	64	279	298	487	451	223	57	34	19	2036
1975	26	17	46	96	290	407	509	487	255	101	59	7	2300
1976	0	17	52	76	116	357	497	481	300	47	0	0	1943
1977	0	3	39	89	331	508	555	534	409	47	16	4	2535
1978	2	0	0	92	235	456	576	536	409	60	31	24	2421
1979	0	2	35	69	190	357	482	461	269	76	4	2	1947
1980	0	11	11	28	253	476	652	619	527	52	14	1	2644
1981	0	3	12	207	174	514	582	555	267	127	30	2	2473
1982	25	3	116	93	310	443	544	530	297	130	35	51	2577
1983	0	0	13	35	194	354	564	555	300	115	24	6	2160
1984	0	5	35	87	221	422	489	470	310	227	19	49	2334
1985	0	3	54	105	233	424	486	480	290	153	79	0	2307
1986	0	21	18	70	259	469	572	487	446	122	44	2	2510
1987	0	0	9	72	344	429	533	560	311	14	22	15	2309
1988	0	1	14	67	193	452	552	552	395	34	45	3	2308
1989	8	30	54	87	245	406	489	498	293	84	36	0	2230
1990	3	25	60	92	226	494	515	544	412	113	24	24	2532
1991	0	3	59	129	337	456	562	515	341	142	37	7	2588
1992	0	1	10	97	213	383	532	403	324	51	2	0	2016
1993	0	1	6	48	197	485	547	562	340	82	19	0	2287
1994	0	6	9	132	196	437	437	439	283	110	24	4	2077
1995	5	2	38	79	282	339	545	581	318	76	9	4	2278
1996	0	28	12	64	343	418	512	439	275	92	13	5	2201
1997	13	9	40	15	175	368	540	459	357	139	0	0	2115
1998	2	0	55	60	298	522	582	554	474	156	12	29	2744
1999	12	14	0	188	213	441	533	596	319	96	1	1	2414
2000	2	17	28	29	328	417	581	628	357	102	23	0	2512
2001	0	16	0	156	238	375	522	487	290	67	17	10	2178

SNOWFALL (inches) 2001 JACKSON, MS (JAN)

YEAR	JUL	AUG	SEP	OCT	NOV	DEC	JAN	FEB	MAR	APR	MAY	JUN	TOTAL
1972-73	0.0	0.0	0.0	0.0	0.0	0.0	T	T	0.0	0.0	0.0	0.0	T
1973-74	0.0	0.0	0.0	0.0	0.0	0.3	0.0	0.0	0.0	0.0	0.0	0.0	0.3
1974-75	0.0	0.0	0.0	0.0	0.0	0.0	T	0.0	T	0.0	0.0	0.0	T
1975-76	0.0	0.0	0.0	0.0	T	T	T	0.0	0.0	0.0	0.0	0.0	T
1976-77	0.0	0.0	0.0	0.0	0.2	0.0	5.8	0.0	0.0	0.0	0.0	0.0	6.0
1977-78	0.0	0.0	0.0	0.0	0.0	0.0	1.1	T	0.1	0.0	0.0	0.0	1.2
1978-79	0.0	0.0	0.0	0.0	0.0	T	T	T	0.0	0.0	0.0	0.0	T
1979-80	0.0	0.0	0.0	0.0	0.0	0.0	0.0	T	T	T	0.0	0.0	T
1980-81	0.0	0.0	0.0	0.0	0.0	0.0	T	T	0.0	0.0	0.0	0.0	T
1981-82	0.0	0.0	0.0	0.0	0.0	T	6.3	T	T	0.0	0.0	0.0	6.3
1982-83	0.0	0.0	0.0	0.0	0.0	0.0	T	T	0.0	T	0.0	0.0	T
1983-84	0.0	0.0	0.0	0.0	0.0	T	T	T	0.0	0.0	0.0	0.0	T
1984-85	0.0	0.0	0.0	0.0	0.0	T	0.3	1.4	0.0	0.0	0.0	0.0	1.7
1985-86	0.0	0.0	0.0	0.0	0.0	T	T	0.0	0.0	0.0	0.0	0.0	T
1986-87	0.0	0.0	0.0	0.0	0.0	0.0	T	0.0	T	1.1	0.0	0.0	1.1
1987-88	0.0	0.0	0.0	0.0	0.0	0.0	T	T	0.0	0.0	0.0	0.0	T
1988-89	0.0	0.0	0.0	0.0	0.0	T	0.0	T	T	T	0.0	0.0	T
1989-90	0.0	0.0	0.0	0.0	T	0.1	0.0	0.0	0.0	0.0	0.0	0.0	0.1
1990-91	0.0	0.0	0.0	0.0	0.0	T	0.0	0.0	0.0	T	0.0	0.0	T
1991-92	0.0	0.0	0.0	0.0	T	0.0	0.2	0.0	0.0	0.0	0.0	0.0	0.2
1992-93	0.0	0.0	0.0	0.0	T	0.0	T	0.0	1.6	0.0	0.0	0.0	1.6
1993-94	0.0	0.0	0.0	0.0	0.0	T	0.0	0.0	0.0	0.0	0.0	0.0	T
1994-95	0.0	0.0	0.0	0.0	0.0	0.0	0.0	0.0	0.0	0.0	0.0	0.0	0.0
1995-96	0.0	0.0	0.0	0.0	0.0	0.0	T	T	T	0.0	0.0	0.0	
1996-97	0.0	0.0	0.0	0.0	0.0	0.7	T	T	0.0	0.0	0.0	0.0	0.7
1997-98	0.0	0.0	0.0	0.0	0.0	4.8	0.0	0.0	0.0	0.0	0.0	0.0	4.8
1998-99	0.0	0.0	0.0	0.0	0.0	T	0.0	0.0	T	0.0	0.0	0.0	T
1999-00	0.0	0.0	0.0	0.0	0.0	0.0	0.4	0.0	0.0	T	0.0	0.0	0.4
2000-01	0.0	0.0	0.0	0.0	T	1.0	T	0.0	0.0	0.0	0.0	0.0	1.0
2001-	0.0	T	0.0	0.0	0.0	0.0							
POR= 38 YRS	0.0	0.0	0.0	0.0	T	0.1	0.5	0.2	0.2	T	0.0	0.0	1.0

2001
INTERNATIONAL AIRPORT,
KANSAS CITY, MISSOURI (MCI)

The National Weather Service Office at Kansas City is very near the geographical center of the United States. The surrounding terrain is gently rolling. It has a modified continental climate. There are no natural topographic obstructions to prevent the free sweep of air from all directions. The influx of moist air from the Gulf of Mexico, or dry air from the semi-arid regions of the southwest, determine whether wet or dry conditions will prevail. There is often conflict between the warm moist gulf air and the cold polar continental air from the north in this area.

Early spring brings a period of frequent and rapid fluctuations in weather, with the fluctuations generally less frequent as spring progresses. The summer season is characterized by warm days and mild nights, with moderate humidities. July is the warmest month. The fall season is normally mild and usually includes a period near the middle of the season characterized by mild, sunny days, and cool nights. Winters are not severely cold. January is the coldest month. Falls of snow to a depth of 10 inches or more are comparatively rare. The distribution of measurable snow normally extends from November to April.

Nearly 60 percent of the annual precipitation occurs during the six months from April through September. More than 75 percent of the annual moisture normally falls during the growing season. The frequency and distribution of precipitation over a normal day is also important. The maximum frequency of precipitation, from April through October, occurs during the six hours following midnight and the minimum frequency occurs during the six hours following noon.

NORMALS, MEANS, AND EXTREMES
KANSAS CITY, MO (MCI)

LATITUDE: 39 17' 57" N   LONGITUDE: 94 43' 04" W   ELEVATION (FT): GRND: 1005   BARO: 1008   TIME ZONE: CENTRAL (UTC + 6)   WBAN: 03947

	ELEMENT	POR	JAN	FEB	MAR	APR	MAY	JUN	JUL	AUG	SEP	OCT	NOV	DEC	YEAR
TEMPERATURE °F	NORMAL DAILY MAXIMUM	30	34.7	40.6	52.8	65.1	74.3	83.3	88.7	86.4	78.1	67.5	52.6	38.8	63.6
	MEAN DAILY MAXIMUM	29	35.7	42.3	53.8	64.8	74.4	83.3	88.7	86.9	78.5	66.9	51.8	39.8	63.9
	HIGHEST DAILY MAXIMUM	29	69	77	86	93	95	105	107	109	106	92	82	74	109
	YEAR OF OCCURRENCE		1989	1995	1986	1987	1998	1980	1974	1984	2000	1976	1990	2001	AUG 1984
	MEAN OF EXTREME MAXS.	29	57.0	65.9	77.3	83.9	87.7	93.8	98.7	98.4	92.7	85.0	71.6	61.7	81.1
	NORMAL DAILY MINIMUM	30	16.7	21.8	32.6	43.8	53.9	63.1	68.2	65.7	56.9	45.7	33.6	21.9	43.7
	MEAN DAILY MINIMUM	29	18.0	23.4	33.2	43.8	54.1	63.1	68.6	66.3	57.0	45.7	33.7	22.4	44.1
	LOWEST DAILY MINIMUM	29	−17	−19	−10	12	30	42	51	43	31	17	1	−23	−23
	YEAR OF OCCURRENCE		1982	1982	1978	1975	1976	1990	1997	1986	1995	1993	1991	1989	DEC 1989
	MEAN OF EXTREME MINS.	29	−2.6	2.6	13.0	26.8	39.8	49.2	57.6	54.9	40.1	28.1	16.5	0.7	27.2
	NORMAL DRY BULB	30	25.7	31.2	42.7	54.5	64.1	73.2	78.5	76.1	67.5	56.6	43.1	30.4	53.6
	MEAN DRY BULB	29	26.9	32.9	43.5	54.2	64.2	73.2	78.8	76.5	67.8	56.3	42.7	31.1	54.0
	MEAN WET BULB	18	26.2	30.6	38.4	47.9	58.7	66.6	70.8	69.1	61.0	50.6	38.1	27.6	48.8
	MEAN DEW POINT	18	20.7	24.6	31.3	41.0	53.7	62.6	67.1	65.6	56.8	44.8	32.7	22.6	43.6
	NORMAL NO. DAYS WITH:														
	MAXIMUM ≥ 90	30	0.0	0.0	0.0	0.4	0.3	5.8	16.3	11.7	3.9	0.2	0.0	0.0	38.6
	MAXIMUM ≤ 32	30	12.2	8.8	1.8	0.1	0.0	0.0	0.0	0.0	0.0	0.0	1.9	8.7	33.5
	MINIMUM ≤ 32	30	28.1	21.9	14.3	3.8	0.1	0.0	0.0	0.0	0.0	2.2	13.5	26.4	110.3
	MINIMUM ≤ 0	30	4.1	2.6	0.2	0.0	0.0	0.0	0.0	0.0	0.0	0.0	0.0	2.3	9.2
H/C	NORMAL HEATING DEG. DAYS	30	1218	946	691	325	135	7	0	6	56	279	657	1073	5393
	NORMAL COOLING DEG. DAYS	30	0	0	0	10	107	253	419	350	131	18	0	0	1288
RH	NORMAL (PERCENT)	30	69	70	67	63	68	69	67	70	70	67	70	71	68
	HOUR 00 LST	30	72	74	72	68	75	78	75	78	78	73	74	74	74
	HOUR 06 LST	30	76	77	78	77	82	84	84	86	86	80	79	78	81
	HOUR 12 LST	30	63	64	59	55	58	58	57	59	59	56	62	65	60
	HOUR 18 LST	30	64	63	57	52	56	57	54	58	59	59	65	67	59
S	PERCENT POSSIBLE SUNSHINE	23	58	55	58	62	61	66	72	67	66	60	49	49	60
W/O	MEAN NO. DAYS WITH:														
	HEAVY FOG (VISBY ≤ 1/4 MI)	30	3.0	2.9	2.0	1.1	1.2	0.8	0.6	1.3	1.3	1.5	1.9	3.1	20.7
	THUNDERSTORMS	30	0.4	0.9	2.6	4.9	8.0	9.2	8.1	6.7	5.6	3.0	1.3	0.5	51.2
CLOUDINESS	MEAN:														
	SUNRISE-SUNSET (OKTAS)	1			5.6			4.0							
	MIDNIGHT-MIDNIGHT (OKTAS)	1			6.4										
	MEAN NO. DAYS WITH:														
	CLEAR	1	4.0	5.0	10.0		6.0	7.0							
	PARTLY CLOUDY	1	1.0	1.0	5.0			9.0							
	CLOUDY	1	3.0		9.0		9.0	5.0							
PR	MEAN STATION PRESSURE (IN)	29	29.00	28.99	28.90	28.90	28.88	28.89	28.90	28.90	28.99	29.00	29.00	29.01	28.95
	MEAN SEA-LEVEL PRES. (IN)	17	30.14	30.11	30.03	29.95	29.92	30.01	29.97	30.00	30.02	30.06	30.07	30.16	30.04
WINDS	MEAN SPEED (MPH)	29	11.2	11.1	12.2	12.3	10.3	9.9	9.2	8.8	9.6	10.5	11.2	10.9	10.6
	PREVAIL.DIR (TENS OF DEGS)	29	20	19	19	18	18	18	19	19	18	18	18	19	18
	MAXIMUM 2-MINUTE:														
	SPEED (MPH)	6	39	40	46	48	43	51	58	37	41	40	37	37	58
	DIR. (TENS OF DEGS)		32	20	23	20	27	01	02	36	14	21	22	20	02
	YEAR OF OCCURRENCE		2000	2000	2000	2001	1996	2000	2000	2001	1996	1996	1998	2001	JUL 2000
	MAXIMUM 5-SECOND:														
	SPEED (MPH)	6	48	52	58	59	55	61	74	55	49	52	52	45	74
	DIR. (TENS OF DEGS)		32	19	23	18	20	36	01	14	14	23	33	33	01
	YEAR OF OCCURRENCE		2000	2000	2000	2001	1999	2000	2000	2001	1996	1996	1997	1997	JUL 2000
PRECIPITATION	NORMAL (IN)	30	1.09	1.10	2.51	3.12	5.04	4.72	4.38	4.01	4.86	3.29	1.92	1.58	37.62
	MAXIMUM MONTHLY (IN)	29	2.66	3.25	9.08	8.43	12.75	11.86	15.47	9.58	11.34	8.15	5.12	5.42	15.47
	YEAR OF OCCURRENCE		1982	2001	1973	1999	1995	2001	1992	1982	1977	1998	1992	1980	JUL 1992
	MINIMUM MONTHLY (IN)	29	0.02	0.20	0.33	0.66	1.05	1.80	0.25	0.50	1.13	0.21		0.03	0.00
	YEAR OF OCCURRENCE		1986	1991	1994	2000	1992	1988	1975	2000	1988	1988		1996	NOV
	MAXIMUM IN 24 HOURS (IN)	29	1.83	1.85	3.07	4.69	4.26	4.48	5.08	6.19	8.82	4.92	2.15	3.67	8.82
	YEAR OF OCCURRENCE		1982	1997	2001	1975	1974	2001	1986	1982	1977	1973	1998	1980	SEP 1977
	NORMAL NO. DAYS WITH:														
	PRECIPITATION ≥ 0.01	30	7.1	7.2	10.3	10.3	11.3	10.3	7.3	8.9	8.1	7.8	7.7	7.6	103.9
	PRECIPITATION ≥ 1.00	30	0.1	0.1	0.4	0.4	1.6	1.2	1.1	1.1	1.5	1.0	0.4	0.4	9.3
SNOWFALL	NORMAL (IN)	30	5.8	5.2	3.1	0.8	0.0	0.0	0.0	0.0	0.0	T	1.2	4.5	20.6
	MAXIMUM MONTHLY (IN)	29	14.2	15.7	11.4	7.2	T	T	T	T	T	6.5	7.1	13.2	15.7
	YEAR OF OCCURRENCE		1977	1993	1978	1983	1990	1994	1992	2001	1992	1996	1975	1983	FEB 1993
	MAXIMUM IN 24 HOURS (IN)	29	9.5	10.8	9.2	4.0	T	T	T	T	T	6.5	6.1	10.8	10.8
	YEAR OF OCCURRENCE		1993	1993	1990	1983	1990	1994	1992	2001	1992	1996	1975	1987	FEB 1993
	MAXIMUM SNOW DEPTH (IN)	28	12	11	9	2	0	0	0	0	0	0	7	11	12
	YEAR OF OCCURRENCE		1979	1979	1990	1994							1975	1987	JAN 1979
	NORMAL NO. DAYS WITH:														
	SNOWFALL ≥ 1.0	30	2.1	1.8	0.9	0.3	0.0	0.0	0.0	0.0	0.0	0.0	0.5	1.2	6.8

PRECIPITATION (inches) 2001 INTERNATIONAL AIRPORT, MO (MCI)

YEAR	JAN	FEB	MAR	APR	MAY	JUN	JUL	AUG	SEP	OCT	NOV	DEC	ANNUAL
1972	0.56	0.41	2.13	3.98	2.18	2.71	3.43	2.90	2.69	3.00	1.47	27.75
1973	2.05	1.35	9.08	2.91	5.65	2.84	8.71	1.60	10.32	5.80	2.36	2.59	55.26
1974	1.05	1.12	1.18	2.94	10.07	2.16	1.13	4.98	1.13	7.22	1.62	1.52	36.12
1975	2.14	1.59	1.49	6.61	3.45	2.46	0.25	4.85	6.10	0.35	2.75	2.03	34.07
1976	0.53	0.66	2.53	3.30	5.49	5.08	0.77	0.76	1.41	2.84	0.21	0.10	23.68
1977	1.15	0.57	2.59	2.35	5.43	6.18	2.74	7.99	11.34	7.67	1.36	0.37	49.74
1978	0.39	1.32	1.77	5.36	4.98	2.64	5.49	2.66	4.04	0.33	3.93	1.05	33.96
1979	2.35	0.78	2.96	2.35	3.00	5.37	4.68	3.66	1.16	3.56	1.83	0.05	31.75
1980	1.60	1.44	3.64	1.02	3.06	2.52	1.99	4.89	1.63	4.13	0.45	5.42	31.79
1981	0.49	0.31	1.43	1.94	9.46	7.44	8.43	2.43	2.71	4.14	2.84	0.45	42.07
1982	2.66	1.13	2.94	1.55	9.81	6.04	2.73	9.58	1.58	3.04	2.21	3.94	47.21
1983	0.58	0.57	2.93	5.52	6.03	5.03	0.26	0.86	1.89	3.85	3.94	1.42	32.88
1984	0.14	1.96	4.52	6.82	2.26	4.14	3.91	0.75	3.42	6.04	1.24	3.57	38.77
1985	0.94	2.69	2.05	1.75	7.00	3.56	5.82	6.98	9.23	7.51	3.95	1.24	52.72
1986	0.02	1.25	1.34	2.12	4.76	2.48	8.36	3.16	10.40	3.17	1.18	1.20	39.44
1987	0.77	2.26	2.85	2.24	4.74	4.58	3.00	4.64	3.66	1.32	1.88	2.05	33.99
1988	1.40	0.72	1.43	2.15	2.14	1.80	1.21	1.87	8.48	0.21	1.96	0.85	24.22
1989	0.98	0.59	2.13	1.50	4.56	3.44	4.76	7.38	8.87	2.88	T	0.55	37.64
1990	1.20	2.11	3.90	2.47	7.36	6.27	4.40	5.04	1.28	2.46	3.01	1.11	40.61
1991	1.37	0.20	2.36	4.99	3.69	3.06	1.72	1.35	2.12	3.71	2.05	2.08	28.70
1992	1.21	2.01	3.79	4.92	1.05	3.84	15.47	2.37	5.69	1.38	5.12	3.78	50.63
1993	1.96	1.28	2.21	5.59	7.30	5.67	10.90	3.98	7.63	1.75	2.07	1.12	51.46
1994	0.63	1.47	0.33	6.98	1.29	2.45	2.79	3.54	2.65	1.27	3.18	1.76	28.34
1995	1.42	1.35	1.12	2.12	12.75	3.36	4.64	4.00	1.85	0.50	1.18	0.40	34.69
1996	1.12	0.35	1.28	1.80	10.29	7.51	4.83	2.97	3.44	3.67	3.15	0.03	40.44
1997	0.69	2.94	1.16	4.13	4.63	2.90	3.53	2.49	3.34	2.98	1.95	2.33	33.07
1998	0.97	1.10	3.44	2.15	1.75	9.22	4.97	3.61	8.69	8.15	4.29	1.19	49.53
1999	2.30	1.71	1.49	8.43	5.62	8.67	0.51	1.56	5.32	0.67	1.63	2.18	40.09
2000	0.46	2.21	2.93	0.66	4.55	7.55	6.02	0.50	3.13	3.55	2.59	0.81	34.96
2001	2.08	3.25	3.88	4.03	4.81	11.86	6.26	5.48	7.98	2.56	0.56	0.75	53.50
POR= 111 YRS	1.31	1.45	2.57	3.41	4.85	4.90	4.04	3.85	4.36	2.95	1.93	1.41	37.03

AVERAGE TEMPERATURE (F) 2001 INTERNATIONAL AIRPORT, MO (MCI)

YEAR	JAN	FEB	MAR	APR	MAY	JUN	JUL	AUG	SEP	OCT	NOV	DEC	ANNUAL
1972	27.3	33.2	47.1	57.4	67.1	76.9	77.2	77.3	70.9	53.9	39.6	27.5	54.6
1973	27.3	33.8	47.7	51.8	61.5	74.5	76.6	77.1	66.7	60.7	45.5	30.1	54.4
1974	23.8	35.3	46.9	56.2	65.6	70.5	82.1	73.0	61.6	57.7	42.8	32.2	54.0
1975	29.9	27.8	36.5	52.8	67.1	74.2	80.9	79.8	63.5	58.9	45.9	34.1	54.3
1976	27.5	42.7	45.0	56.3	60.0	71.3	79.0	77.7	69.0	50.8	35.9	29.0	53.7
1977	15.6	35.1	48.0	59.7	69.0	74.5	77.4	74.4	69.0	55.7	42.3	29.5	54.4
1978	16.8	19.5	37.6	55.5	62.1	74.6	79.3	76.5	73.6	56.4	44.4	31.0	52.3
1979	12.5	20.9	42.2	51.6	63.5	72.2	75.9	75.7	68.6	57.7	40.7	33.9	51.4
1980	28.7	25.2	38.7	54.6	63.9	75.3	85.2	80.3	69.6	54.1	44.5	32.1	54.4
1981	30.3	33.4	45.2	61.1	60.0	74.1	78.3	72.9	68.4	55.3	45.6	29.0	54.5
1982	18.6	27.8	42.5	51.1	65.5	68.8	79.4	75.0	67.5	55.6	41.9	35.5	52.4
1983	30.1	35.9	43.1	46.3	59.6	70.8	81.5	83.5	71.2	57.2	44.3	13.2	53.1
1984	25.0	38.9	36.0	50.3	60.4	74.3	76.1	79.0	66.1	56.8	43.9	35.5	53.5
1985	18.7	25.3	47.4	57.9	66.0	68.8	77.0	72.1	66.7	56.8	36.8	22.9	51.3
1986	34.5	30.5	48.5	57.1	65.2	76.5	79.7	72.0	71.8	56.9	37.9	34.5	55.4
1987	29.7	39.4	47.1	56.8	70.6	76.0	79.9	76.4	67.8	52.0	46.7	35.1	56.5
1988	26.7	27.9	42.3	54.5	69.1	78.1	79.6	81.3	70.5	52.2	44.8	35.2	55.3
1989	37.7	22.8	43.8	56.9	63.2	71.1	77.8	75.5	63.2	57.9	42.3	21.1	52.8
1990	37.9	36.2	45.7	52.7	60.4	75.5	77.3	77.1	72.1	57.1	50.1	29.3	56.0
1991	22.9	39.4	47.2	56.7	67.6	76.1	80.6	77.7	68.8	57.6	36.7	36.1	55.6
1992	35.9	39.8	46.9	53.1	62.6	69.7	74.4	70.7	66.1	56.6	39.0	32.9	54.0
1993	25.9	28.6	39.8	50.4	63.5	72.9	77.6	77.8	62.8	53.6	39.4	34.5	52.2
1994	25.2	30.0	46.0	53.8	64.6	75.8	76.0	75.3	67.7	57.5	46.3	36.5	54.6
1995	28.2	35.9	45.0	52.5	59.4	72.7	78.2	78.9	65.2	57.1	40.1	31.1	53.7
1996	23.1	34.2	37.7	52.8	64.5	73.8	75.2	74.0	64.8	56.3	37.0	29.7	52.0
1997	24.4	33.8	45.3	48.8	59.9	73.1	77.7	74.5	69.2	59.7	40.7	33.5	53.2
1998	33.8	41.3	39.1	54.3	71.1	73.5	77.8	77.0	72.4	58.6	48.0	34.8	56.8
1999	27.8	41.2	42.6	54.6	63.4	71.5	81.0	76.3	65.7	51.9	35.8	35.5	53.9
2000	31.8	40.8	47.1	55.0	67.3	71.2	76.8	81.8	71.0	59.9	36.6	19.1	54.9
2001	29.2	30.0	40.0	60.3	66.1	72.1	80.7	77.4	65.7	56.0	51.2	37.3	55.5
POR= 112 YRS	29.0	33.1	43.3	55.3	65.1	74.4	79.3	77.7	69.6	58.5	44.2	33.0	55.2

REFERENCE NOTES:

PAGE 1:
THE TEMPERATURE GRAPH SHOWS NORMAL MAXIMUM AND NORMAL
MINIMUM DAILY TEMPERATURES (SOLID CURVES) AND THE
ACTUAL DAILY HIGH AND LOW TEMPERATURES (VERTICAL BARS).

PAGE 2 AND 3:
H/C INDICATES HEATING AND COOLING DEGREE DAYS.
RH INDICATES RELATIVE HUMIDITY
W/O INDICATES WEATHER AND OBSTRUCTIONS
S INDICATES SUNSHINE.
PR INDICATES PRESSURE.
CLOUDINESS ON PAGE 3 IS THE SUM OF THE CEILOMETER AND
SATELLITE DATA NOT TO EXCEED EIGHT EIGHTHS(OKTAS).

GENERAL:
T INDICATES TRACE PRECIPITATION, AN AMOUNT GREATER
THAN ZERO BUT LESS THAN THE LOWEST REPORTABLE VALUE.
+ INDICATES THE VALUE ALSO OCCURS ON EARLIER DATES.
BLANK ENTRIES DENOTE MISSING OR UNREPORTED DATA.
NORMALS ARE 30-YEAR AVERAGES (1961 - 1990).
ASOS INDICATES AUTOMATED SURFACE OBSERVING SYSTEM.
PM INDICATES THE LAST DAY OF THE PREVIOUS MONTH.
POR (PERIOD OF RECORD) BEGINS WITH THE JANUARY DATA
MONTH AND IS THE NUMBER OP YEARS USED TO COMPUTE
THE MEAN. INDIVIDUAL MONTHS WITHIN THE POR MAY
BE MISSING.
WHEN THE POR FOR A NORMAL IS LESS THAN 30 YEARS,
THE NORMAL IS PROVISIONAL AND IS BASED ON THE NUMBER
OF YEARS INDICATED.
0.* OR * INDICATES THE VALUE OR MEAN-DAYS-WITH
IS BETWEEN 0.00 AND 0.05.
CLOUDINESS FOR ASOS STATIONS DIFFERS FROM THE NON-ASOS
OBSERVATION TAKEN BY A HUMAN OBSERVER. ASOS STATION
CLOUDINESS IS BASED ON TIME-AVERAGED CEILOMETER DATA
FOR CLOUDS AT OR BELOW 12,000 FEET AND ON SATELLITE
DATA FOR CLOUDS ABOVE 12,000 FEET.
THE NUMBER OF DAYS WITH CLEAR, PARTLY CLOUDY, AND
CLOUDY CONDITIONS FOR ASOS STATIONS IS THE SUM
OF THE CEILOMETER AND SATELLITE DATA FOR THE
SUNRISE TO SUNSET PERIOD.

GENERAL CONTINUED:
CLEAR INDICATES 0 - 2 OKTAS, PARTLY CLOUDY INDICATES
3 - 6 OKTAS, AND CLOUDY INDICATES 7 OR 8 OKTAS.
WHEN AT LEAST ONE OF THE ELEMENTS (CEILOMETER OR
SATELLITE) IS MISSING, THE DAILY CLOUDINESS IS
NOT COMPUTED.
WIND DIRECTION IS RECORDED IN TENS OF DEGREES (2 DIGITS)
CLOCKWISE FROM TRUE NORTH. "00" INDICATES CALM. "36"
INDICATES TRUE NORTH.
RESULTANT WIND IS THE VECTOR AVERAGE OF THE SPEED AND
DIRECTION.
AVERAGE TEMPERATURE IS THE SUM OF THE MEAN DAILY MAXIMUM
AND MINIMUM TEMPERATURE DIVIDED BY 2.
SNOWFALL DATA COMPRISE ALL FORMS OF FROZEN
PRECIPITATION, INCLUDING HAIL.
A HEATING (COOLING) DEGREE DAY IS THE DIFFERENCE BETWEEN
THE AVERAGE DAILY TEMPERATURE AND 65 F.
DRY BULB IS THE TEMPERATURE OF THE AMBIENT AIR.
DEW POINT IS THE TEMPERATURE TO WHICH THE AIR MUST BE
COOLED TO ACHIEVE 100 PERCENT RELATIVE HUMIDITY.
WET BULB IS THE TEMPERATURE THE AIR WOULD HAVE IF THE
MOISTURE CONTENT WAS INCREASED TO 100 PERCENT RELATIVE
HUMIDITY.

ON JULY 1, 1996, THE NATIONAL WEATHER SERVICE BEGAN USING
THE "METAR" OBSERVATION CODE THAT WAS ALREADY EMPLOYED
BY MOST OTHER NATIONS OF THE WORLD. THE MOST NOTICEABLE
DIFFERENCE IN THIS ANNUAL PUBLICATION WILL BE THE CHANGE
IN UNITS FROM TENTHS TO EIGHTS(OKTAS) FOR REPORTING THE
AMOUNT OF SKY COVER.

HEATING DEGREE DAYS (base 65 F) 2001 INTERNATIONAL AIRPORT, MO (MCI)

YEAR	JUL	AUG	SEP	OCT	NOV	DEC	JAN	FEB	MAR	APR	MAY	JUN	TOTAL
1972-73	1	0	37	355	751	1155	1160	868	529	394	127	0	5377
1973-74	0	0	53	173	578	1077	1272	823	559	270	84	6	4895
1974-75	0	6	141	227	660	1009	1084	1036	878	381	42	5	5469
1975-76	0	0	142	226	567	954	1155	640	616	277	178	6	4761
1976-77	0	0	45	469	865	1108	1527	832	521	192	22	0	5581
1977-78	0	0	21	293	673	1094	1487	1266	848	291	172	6	6151
1978-79	0	0	28	272	618	1050	1624	1230	698	400	115	5	6040
1979-80	5	5	35	247	720	918	1118	1148	809	327	98	3	5433
1980-81	0	0	63	347	609	1011	1069	880	607	169	179	2	4936
1981-82	0	2	40	309	573	1112	1432	1037	690	416	51	32	5694
1982-83	0	2	78	307	688	911	1074	810	675	557	180	29	5311
1983-84	0	0	57	271	617	1602	1234	750	891	443	175	1	6041
1984-85	0	0	143	269	624	907	1431	1102	538	256	41	19	5330
1985-86	0	3	131	260	841	1297	940	960	528	267	60	0	5287
1986-87	0	12	23	251	805	938	1088	712	549	298	15	0	4691
1987-88	0	3	30	398	552	922	1180	1069	668	311	19	0	5152
1988-89	2	1	18	394	599	915	836	1176	658	319	135	7	5060
1989-90	0	1	138	267	675	1360	836	800	601	398	167	10	5253
1990-91	1	0	44	278	452	1104	1296	709	553	258	62	0	4757
1991-92	0	0	96	277	841	888	898	724	554	367	137	6	4788
1992-93	0	7	80	271	773	987	1203	1011	775	431	88	26	5652
1993-94	0	2	118	365	760	938	1227	974	581	353	91	1	5410
1994-95	0	4	69	252	552	878	1135	807	613	370	196	1	4877
1995-96	0	0	111	255	743	1047	1295	891	840	381	110	10	5683
1996-97	0	0	95	291	835	1089	1252	865	603	476	184	1	5691
1997-98	3	0	25	290	720	971	960	658	806	323	18	27	4801
1998-99	0	0	9	213	505	930	1145	656	690	313	84	12	4557
1999-00	0	0	94	261	392	898	1024	694	554	303	58	9	4287
2000-01	0	0	52	195	845	1416	1102	972	768	190	67	16	5623
2001-	0	0	71	280	407	854							

COOLING DEGREE DAYS (base 65 F) 2001 INTERNATIONAL AIRPORT, MO (MCI)

YEAR	JAN	FEB	MAR	APR	MAY	JUN	JUL	AUG	SEP	OCT	NOV	DEC	ANNUAL
1972	0	3	7	41	146	368	386	389	219	16	0	0	1575
1973	0	0	0	4	30	291	368	382	110	46	0	0	1231
1974	0	0	4	13	109	176	538	264	47	8	0	0	1159
1975	0	0	0	22	117	284	498	464	105	43	1	0	1534
1976	0	0	0	22	29	199	444	399	169	34	0	0	1296
1977	0	0	0	39	154	291	463	298	147	11	0	0	1403
1978	0	0	5	14	86	300	452	364	295	11	8	0	1535
1979	0	0	1	3	77	229	348	340	147	29	0	0	1174
1980	0	0	0	21	69	316	632	483	210	15	0	0	1746
1981	0	0	0	58	44	279	418	253	149	14	0	0	1215
1982	0	0	0	8	75	154	452	320	158	26	0	0	1193
1983	0	0	3	5	19	210	517	582	251	31	1	0	1619
1984	0	0	0	9	41	287	353	445	184	23	0	0	1342
1985	0	0	0	49	77	137	379	230	191	5	0	0	1068
1986	0	0	24	35	73	352	466	237	232	7	0	0	1426
1987	0	0	0	58	196	336	474	364	118	1	9	0	1556
1988	0	0	2	5	151	400	459	514	188	5	0	0	1724
1989	0	0	9	85	88	196	402	334	87	56	0	0	1257
1990	0	0	8	33	33	331	394	384	263	40	11	0	1497
1991	0	0	9	16	154	339	490	399	219	54	0	0	1680
1992	0	0	1	20	71	154	298	191	121	19	0	0	875
1993	0	0	0	0	49	271	398	406	55	19	0	0	1198
1994	0	0	0	23	86	331	345	333	154	28	0	0	1300
1995	0	0	0	1	27	239	417	438	123	18	0	0	1263
1996	0	0	0	19	105	282	325	313	94	29	0	0	1167
1997	0	0	0	0	33	248	403	300	159	75	0	0	1218
1998	0	0	10	10	213	291	407	378	237	20	0	0	1566
1999	0	0	0	7	42	215	501	358	123	24	4	0	1274
2000	0	0	4	10	134	202	369	528	239	41	0	0	1527
2001	0	0	0	58	109	239	495	392	100	8	1	0	1402

SNOWFALL (inches) 2001 INTERNATIONAL AIRPORT, MO (MCI)

YEAR	JUL	AUG	SEP	OCT	NOV	DEC	JAN	FEB	MAR	APR	MAY	JUN	TOTAL
1972-73	0.0	0.0	0.0	0.0	3.6	3.1	10.9	0.3	0.0	1.3	0.0	0.0	19.2
1973-74	0.0	0.0	0.0	0.0	T	8.0	4.8	0.3	0.5	0.3	0.0	0.0	13.9
1974-75	0.0	0.0	0.0	0.0	1.4	1.2	5.4	4.6	5.9	2.3	0.0	0.0	20.8
1975-76	0.0	0.0	0.0	0.0	7.1	2.8	5.2	5.2	1.5	0.0	0.0	0.0	21.8
1976-77	0.0	0.0	0.0	T	0.8	0.2	14.2	0.3	T	0.6	0.0	0.0	16.1
1977-78	0.0	0.0	0.0	0.0	T	1.2	3.9	12.7	11.4	0.0	0.0	0.0	29.2
1978-79	0.0	0.0	0.0	0.0	0.2	11.7	13.3	1.5	4.7	2.0	0.0	0.0	33.4
1979-80	0.0	0.0	0.0	0.0	T	T	5.4	12.7	5.4	0.0	0.0	0.0	23.5
1980-81	0.0	0.0	0.0	T	T	3.2	4.0	2.9	0.1	0.0	0.0	0.0	10.2
1981-82	0.0	0.0	0.0	0.0	0.1	5.3	6.0	12.7	4.0	1.3	0.0	0.0	29.4
1982-83	0.0	0.0	0.0	0.0	0.5	0.7	6.3	7.4	1.3	7.2	0.0	0.0	23.4
1983-84	0.0	0.0	0.0	0.0	0.7	13.2	1.3	0.5	8.7	0.0	0.0	0.0	24.4
1984-85	0.0	0.0	0.0	0.0	0.4	7.0	11.8	6.9	0.3	0.0	0.0	0.0	26.4
1985-86	0.0	0.0	0.0	0.0	3.5	5.4	T	4.5	T	T	0.0	0.0	13.4
1986-87	0.0	0.0	0.0	T	0.6	1.2	10.5	5.0	T	0.0	0.0	0.0	17.3
1987-88	0.0	0.0	0.0	T	2.0	11.9	0.9	9.3	2.2	0.0	0.0	0.0	26.3
1988-89	0.0	0.0	0.0	0.0	0.1	0.1	0.2	6.5	T	0.0	T	T	6.9
1989-90	0.0	0.0	0.0	0.0	T	6.8	1.0	2.1	9.6	0.0	T	0.0	19.5
1990-91	0.0	0.0	0.0	0.0	1.7	1.6	12.1	T	1.2	T	0.0	0.0	16.6
1991-92	0.0	0.0	0.0	0.0	4.6	0.2	T	2.0	0.5	2.8	0.0	0.0	10.1
1992-93	T	0.0	T	0.0	4.1	0.8	12.0	15.7	1.1	0.6	0.0	T	34.3
1993-94	0.0	0.0	0.0	T	0.5	2.7	1.4	10.3	0.5	2.6	0.0	T	18.0
1994-95	0.0	0.0	0.0	0.0	0.0	2.5	1.3	0.7	2.4	T	0.0	0.0	6.9
1995-96	0.0	0.0	0.0	0.0	0.7	5.3	11.4	T	0.7	1.0	0.0		
1996-97				6.5	4.8	0.5	9.8	6.3	0.0	1.3	T	0.0	
1997-98	0.0	0.0	0.0	1.0	0.5	10.9	0.6	1.0	5.6	0.0	T	T	19.6
1998-99	0.0	0.0	0.0	0.0	0.0	1.4	4.3	4.7	2.5	T	0.0	0.0	12.9
1999-00	0.0	0.0	0.0	0.0	0.0	3.9	4.9	2.1	2.0	0.0	0.0	0.0	12.9
2000-01	0.0	0.0	0.0	0.0	0.0	11.8	2.0	7.7	1.3	0.0	T	T	22.8
2001-	0.0	T	T	0.0	0.0	T							
POR= 66 YRS	T	0.0	T	T	1.1	4.5	5.5	4.4	3.3	0.9	T	T	19.7

2001
ST. LOUIS,
MISSOURI (STL)

St Louis is located at the confluence of the Missouri and Mississippi Rivers and near the geographical center of the United States. Its position in the middle latitudes allows the area to be affected by warm moist air that originates in the Gulf of Mexico, as well as cold air masses that originate in Canada. The alternate invasion of these airmasses produces a wide variety of weather conditions, and allows the region to enjoy a true four-season climate.

During the summer months, air originating from the Gulf of Mexico tends to dominate the area, producing warm and humid conditions. Since 1870, records indicate that temperatures of 90 degrees or higher occur on about 35-40 days per year. Extrmely hot days (100 degrees or more) are expected on no more than five days per year.

Winters are brisk and stimulating, but prolonged periods of extremely cold weather are rare. Records show that temperatures drop to zero or below an average of 2 or 3 days per year, and temperatures as cold as 32 degrees or lower occur less than 25 days in most years. Snowfall has averaged a little over 18 inches per winter season, and snowfall of an inch or less is received on 5 to 10 days in most years.

Normal annual precipitation for St. Louis is a little less than 34 inches. The three winter months are the driest, with an average total of about 6 inches of precipitation. The spring months of March through May are normally the wettest with normal total rainfall of just under 10.5 inches. It is not unusual to have extended dry periods of one to two weeks during growing season.

Thunderstorms normally occur on between 40 and 50 days per year. During any year, there are usually a few of these thunderstorms that are severe, and produce large hail and damaging winds. Tornadoes have produced extensive damage and loss of life in the St. Louis area.

NORMALS, MEANS, AND EXTREMES
ST. LOUIS, MO (STL)

LATITUDE: 38 45' 09" N　　LONGITUDE: 90 22' 25" W　　ELEVATION (FT): GRND: 707　BARO: 710　　TIME ZONE: CENTRAL (UTC + 6)　　WBAN: 13994

	ELEMENT	POR	JAN	FEB	MAR	APR	MAY	JUN	JUL	AUG	SEP	OCT	NOV	DEC	YEAR
TEMPERATURE °F	NORMAL DAILY MAXIMUM	30	37.7	42.6	54.6	66.9	76.1	85.2	89.3	87.3	79.9	68.5	54.7	41.7	65.4
	MEAN DAILY MAXIMUM	56	38.6	44.1	54.0	66.8	76.1	85.1	88.9	87.4	80.1	69.1	54.3	42.7	65.6
	HIGHEST DAILY MAXIMUM	44	76	85	89	93	94	102	107	107	104	94	85	76	107
	YEAR OF OCCURRENCE		1970	1972	1985	1989	1996	1988	1980	1984	1984	1963	1989	1970	AUG 1984
	MEAN OF EXTREME MAXS.	56	63.5	68.6	79.1	86.6	89.4	95.4	98.6	98.3	93.4	86.1	75.0	66.4	83.4
	NORMAL DAILY MINIMUM	30	20.8	25.1	35.5	46.4	56.0	65.7	70.4	67.9	60.5	48.3	37.7	26.0	46.7
	MEAN DAILY MINIMUM	56	21.6	25.9	34.4	45.9	55.7	65.2	69.4	67.6	59.4	47.9	36.3	26.3	46.3
	LOWEST DAILY MINIMUM	44	-18	-12	-5	22	31	43	51	47	36	23	1	-16	-18
	YEAR OF OCCURRENCE		1985	1996	1960	1975	1976	1969	1972	1986	1974	1976	1964	1989	JAN 1985
	MEAN OF EXTREME MINS.	56	0.2	6.3	16.1	29.9	40.5	52.0	58.3	56.3	43.8	32.5	19.1	6.5	30.1
	NORMAL DRY BULB	30	29.3	33.9	45.1	56.7	66.1	75.4	79.8	77.6	70.2	58.4	46.2	33.9	56.1
	MEAN DRY BULB	56	30.1	35.0	44.2	56.3	65.9	75.2	79.2	77.6	69.8	58.5	45.1	34.5	56.0
	MEAN WET BULB	17	28.5	32.9	40.2	49.9	59.6	67.6	71.4	70.2	62.3	52.2	41.5	31.9	50.7
	MEAN DEW POINT	17	22.8	26.4	32.8	42.6	53.5	62.8	66.9	65.9	57.2	45.8	35.1	26.1	44.8
	NORMAL NO. DAYS WITH:														
	MAXIMUM 90	30	0.0	0.0	0.0	0.4	1.3	8.0	15.1	11.6	4.4	0.2	0.0	0.0	41.0
	MAXIMUM 32	30	11.4	7.0	1.3	0.0	0.0	0.0	0.0	0.0	0.0	0.0	0.6	7.0	27.3
	MINIMUM 32	30	26.7	21.3	13.5	2.8	0.1	0.0	0.0	0.0	0.0	1.7	11.0	23.3	100.4
	MINIMUM 0	30	2.3	0.5	*	0.0	0.0	0.0	0.0	0.0	0.0	0.0	0.0	1.1	3.9
H/C	NORMAL HEATING DEG. DAYS	30	1107	871	617	266	111	0	0	0	21	237	564	964	4758
	NORMAL COOLING DEG. DAYS	30	0	0	0	17	145	312	459	391	177	33	0	0	1534
RH	NORMAL (PERCENT)	30	73	72	68	64	66	67	68	70	72	69	72	76	70
	HOUR 00 LST	30	77	77	74	70	75	76	77	80	81	76	77	79	77
	HOUR 06 LST	30	82	82	81	78	82	83	85	88	89	84	83	83	83
	HOUR 12 LST	30	65	64	59	54	55	56	56	57	58	56	63	68	59
	HOUR 18 LST	30	68	66	59	53	55	55	56	58	60	60	67	73	61
S	PERCENT POSSIBLE SUNSHINE	37	50	52	54	56	59	66	69	64	63	60	46	43	57
W/O	MEAN NO. DAYS WITH:														
	HEAVY FOG(VISBY 1/4 MI)	44	2.2	1.6	1.5	0.6	0.6	0.3	0.2	0.3	0.5	0.7	1.1	2.0	11.6
	THUNDERSTORMS	44	0.8	0.8	3.0	5.8	6.7	7.4	7.2	6.5	3.6	2.5	1.7	0.6	46.6
CLOUDINESS	MEAN:														
	SUNRISE-SUNSET (OKTAS)	48	5.4	5.3	5.4	5.2	5.1	4.8	4.4	4.2	4.1	4.0	4.9	5.3	4.8
	MIDNIGHT-MIDNIGHT (OKTAS)	32	5.0	5.1	5.1	4.9	4.7	4.3	3.8	3.8	3.9	3.9	4.8	5.1	4.5
	MEAN NO. DAYS WITH:														
	CLEAR	48	7.3	6.6	6.5	6.9	7.1	7.1	9.3	10.1	11.4	12.3	8.6	7.1	100.3
	PARTLY CLOUDY	48	6.6	6.5	8.0	8.2	9.5	11.0	11.3	11.1	8.2	7.5	6.6	6.6	101.1
	CLOUDY	48	17.1	15.1	16.5	14.9	14.4	11.9	10.4	9.8	10.4	11.2	14.9	17.3	163.9
PR	MEAN STATION PRESSURE(IN)	29	29.53	29.50	29.40	29.37	29.35	29.37	29.39	29.42	29.45	29.46	29.47	29.51	29.43
	MEAN SEA-LEVEL PRES. (IN)	17	30.14	30.12	30.04	29.95	29.95	29.94	29.98	30.01	30.03	30.07	30.08	30.16	30.04
WINDS	MEAN SPEED (MPH)	51	10.5	10.7	11.6	11.3	9.4	8.7	7.9	7.4	8.1	8.7	10.2	10.3	9.6
	PREVAIL.DIR(TENS OF DEGS)	32	30	30	30	30	30	18	18	18	18	18	30	30	30
	MAXIMUM 2-MINUTE:														
	SPEED (MPH)	5	36	39	37	44	41	43	37	39	33	43	41	38	44
	DIR. (TENS OF DEGS)		31	23	26	25	29	27	30	27	32	31	24	31	25
	YEAR OF OCCURRENCE		1997	1999	1998	1997	2000	2001	2001	2000	2000	2001	1998	1998	APR 1997
	MAXIMUM 5-SECOND:														
	SPEED (MPH)	5	43	54	52	59	51	51	48	48	40	51	53	47	59
	DIR. (TENS OF DEGS)		35	23	30	27	27	26	34	27	34	30	24	30	27
	YEAR OF OCCURRENCE		2000	1999	2001	1998	2000	2001	1999	2000	2000	2001	1998	2000	APR 1998
PRECIPITATION	NORMAL (IN)	30	1.81	2.12	3.58	3.50	3.97	3.72	3.85	2.85	3.12	2.68	3.28	3.03	37.51
	MAXIMUM MONTHLY (IN)	44	5.38	4.68	6.67	10.32	12.92	9.43	10.71	6.44	9.16	7.12	9.95	7.82	12.92
	YEAR OF OCCURRENCE		1975	1986	1978	1994	1995	1985	1981	1970	1993	1984	1985	1982	MAY 1995
	MINIMUM MONTHLY (IN)	44	0.10	0.25	1.09	0.99	1.02	0.44	0.60	0.08	T	0.21	0.44	0.32	T
	YEAR OF OCCURRENCE		1986	1963	1966	1977	1972	1991	1970	1971	1979	1975	1969	1958	SEP 1979
	MAXIMUM IN 24 HOURS (IN)	44	2.43	2.60	2.95	4.91	6.55	3.59	3.47	3.17	3.50	2.70	3.71	4.03	6.55
	YEAR OF OCCURRENCE		1975	1999	1977	1979	1995	2000	1982	1993	1986	1986	1985	1982	MAY 1995
	NORMAL NO. DAYS WITH:														
	PRECIPITATION 0.01	30	8.4	8.4	11.4	11.2	10.5	9.5	8.5	7.9	7.9	8.4	9.5	9.7	111.3
	PRECIPITATION 1.00	30	0.3	0.4	0.6	0.6	1.1	1.1	1.2	0.8	0.8	0.6	0.9	0.8	9.2
SNOWFALL	NORMAL (IN)	30	6.5	5.3	3.7	0.6	0.0	0.0	0.0	0.0	0.0	T	1.5	4.6	22.2
	MAXIMUM MONTHLY (IN)	65	23.9	20.8	22.3	6.5	0.6	T	T	0.0	0.0	T	11.3	26.3	26.3
	YEAR OF OCCURRENCE		1977	1993	1960	1971	1993	1993	1997			1993	1951	1973	DEC 1973
	MAXIMUM IN 24 HOURS (IN)	65	13.9	11.7	10.7	6.1	0.6	T	T	0.0	0.0	T	10.3	12.0	13.9
	YEAR OF OCCURRENCE		1982	1993	1989	1971	1993	1993	1997			1993	1951	1973	JAN 1982
	MAXIMUM SNOW DEPTH (IN)	53	13	20	17	6	0	0	0	0	0	0	10	12	20
	YEAR OF OCCURRENCE		1978	1982	1978	1971							1951	1973	FEB 1982
	NORMAL NO. DAYS WITH:														
	SNOWFALL 1.0	30	1.9	1.5	0.8	0.2	0.0	0.0	0.0	0.0	0.0	0.0	0.4	1.4	6.2

PRECIPITATION (inches) 2001 ST. LOUIS, MO (STL)

YEAR	JAN	FEB	MAR	APR	MAY	JUN	JUL	AUG	SEP	OCT	NOV	DEC	ANNUAL
1972	0.77	0.74	2.93	4.49	1.02	1.19	3.10	2.69	6.21	1.47	5.59	3.54	33.74
1973	1.40	1.04	5.81	4.25	3.92	4.23	2.85	2.46	3.52	2.33	3.65	4.36	39.82
1974	3.51	4.17	2.58	2.40	5.90	3.45	0.90	5.05	2.50	1.51	3.15	1.71	36.83
1975	5.38	3.59	4.08	4.56	3.23	3.78	2.56	5.44	2.48	0.21	1.51	3.62	40.21
1976	0.83	1.08	4.28	1.37	3.90	2.32	2.32	1.27	0.90	3.37	0.73	1.13	23.46
1977	2.38	2.47	6.28	0.99	2.13	5.47	4.28	5.34	3.64	3.76	4.33	2.34	43.41
1978	1.70	1.60	6.67	3.21	3.69	2.39	6.03	0.76	3.10	2.28	4.47	1.81	37.71
1979	1.95	1.48	3.63	1.62	1.62	1.67	3.67	2.26	T	1.81	2.07	1.85	29.48
1980	0.63	1.54	3.98	1.54	3.40	2.19	3.56	2.72	3.12	2.89	1.25	0.66	27.48
1981	0.64	2.18	2.97	3.40	6.79	5.82	10.71	3.31	1.17	3.81	2.71	2.01	45.52
1982	4.90	1.37	2.88	2.55	4.85	5.96	7.91	5.27	5.27	2.30	3.89	7.82	54.97
1983	0.72	0.95	3.54	7.30	6.32	4.32	1.23	2.24	1.24	5.40	7.79	3.75	44.80
1984	0.84	3.43	5.37	6.29	5.19	2.74	0.76	0.64	8.88	7.12	5.50	4.89	51.65
1985	0.53	3.77	5.18	3.60	3.30	9.43	5.23	3.66	0.43	1.96	9.95	3.69	50.73
1986	0.10	4.68	1.22	1.23	2.42	4.43	2.61	2.22	7.99	5.34	1.58	1.06	34.88
1987	1.98	1.40	2.16	1.74	2.00	3.59	5.04	5.56	1.62	1.74	4.09	7.46	38.38
1988	3.30	2.27	4.73	1.15	1.44	1.97	3.02	2.31	1.99	1.86	6.65	3.24	33.93
1989	2.58	1.43	4.53	2.10	4.11	2.34	4.59	3.00	1.69	0.95	0.59	0.69	28.60
1990	1.42	3.53	2.66	3.07	9.59	3.02	3.34	2.84	0.78	4.96	3.36	6.52	45.09
1991	1.52	0.98	3.20	3.27	3.87	0.44	5.18	0.98	2.98	5.70		2.10	33.48
1992	1.12	1.89	3.45	2.46	1.45	1.19	4.31	3.45	2.98	1.21	6.32	3.66	33.49
1993	3.54	2.75	3.31	6.16	3.94	7.12	5.06	4.78	9.16	2.61	4.85	1.48	54.76
1994	2.09	1.51	1.27	10.32	1.72	2.16	1.42	3.76	1.18	2.85	4.90	1.52	34.70
1995	4.39	1.33	3.19	3.33	12.92	2.96	2.16	4.52	0.74	2.01	1.28	2.85	41.68
1996	3.27	0.52	3.06	7.97	4.34	3.72	6.33	1.57	2.86	2.67	6.50	0.86	43.67
1997	2.74	4.14	2.85	2.66	3.05	2.00	1.44	3.36	2.73	2.05	2.36	1.85	31.23
1998	2.88	2.93	6.00	4.63	3.62	6.90	6.39	2.35	1.86	2.51	2.72	0.83	43.62
1999	5.10	3.52	2.40	3.72	2.20	5.26	4.22	1.95	1.09	2.04	0.72	1.84	34.06
2000	1.23	3.11	1.88	1.84	5.84	8.22	2.25	3.64	2.62	2.60	2.79	1.35	37.37
2001	1.12	2.48	1.45	3.01	2.81	3.60	4.00	1.99	2.81	5.50	3.06	3.46	35.29
POR= 131 YRS	2.20	2.31	3.32	3.75	4.13	4.06	3.44	2.89	3.07	2.71	2.82	2.29	36.99

AVERAGE TEMPERATURE (F) 2001 ST. LOUIS, MO (STL)

YEAR	JAN	FEB	MAR	APR	MAY	JUN	JUL	AUG	SEP	OCT	NOV	DEC	ANNUAL
1972	29.9	34.3	45.0	56.1	66.4	73.4	77.5	76.3	71.1	54.8	39.8	30.4	54.6
1973	32.6	34.5	50.9	53.7	61.7	74.6	76.9	76.0	60.7	47.0	30.0	35.3	55.9
1974	29.8	36.1	48.1	57.5	65.0	69.5	79.8	74.5	62.3	57.8	44.0	34.0	54.9
1975	33.2	32.0	39.0	53.4	67.6	74.9	77.8	77.9	64.7	59.0	48.3	35.5	55.3
1976	28.1	43.5	48.8	56.5	60.8	72.7	79.5	74.3	68.2	50.9	37.0	28.5	54.1
1977	15.1	34.8	49.6	60.8	71.1	74.7	81.2	76.4	70.7	55.5	44.7	30.6	55.4
1978	19.6	21.1	37.9	56.2	63.6	74.4	78.5	76.4	73.1	55.6	47.5	35.0	53.2
1979	16.6	23.1	44.1	52.9	65.5	76.5	79.2	78.4	70.8	59.2	44.5	38.7	54.1
1980	31.4	27.9	41.0	54.5	66.9	75.5	85.0	83.5	72.5	55.9	46.4	36.6	56.4
1981	31.2	36.8	46.7	63.1	60.7	75.7	78.7	75.1	69.2	55.7	48.7	31.1	56.1
1982	22.5	28.6	45.3	51.5	70.7	70.6	79.3	75.2	68.0	58.3	46.3	41.6	54.8
1983	32.3	38.1	44.5	50.3	62.3	75.3	83.5	84.2	72.0	59.7	48.2	20.5	55.9
1984	28.3	40.4	37.1	54.1	63.2	79.5	78.3	80.7	68.2	61.8	44.3	40.7	56.4
1985	22.6	30.5	49.5	60.4	67.6	71.6	79.2	74.7	70.9	61.4	46.5	27.3	55.2
1986	34.9	34.5	49.2	60.8	68.2	78.3	82.8	74.0	71.3	58.3	41.5	35.4	57.6
1987	30.6	40.1	48.6	56.9	72.6	77.9	81.0	78.9	70.5	53.8	49.1	38.0	58.2
1988	29.2	30.5	45.2	57.1	69.0	77.7	81.6	82.7	72.5	58.7	47.2	37.2	57.0
1989	41.2	28.2	45.0	57.7	64.3	74.8	79.3	77.8	67.4	61.3	47.1	24.1	55.7
1990	42.9	41.3	49.8	55.7	63.6	77.2	80.2	77.9	74.1	58.1	52.7	34.7	59.0
1991	29.3	41.7	50.1	61.5	73.0	79.9	80.9	79.7	72.4	60.5	42.4	39.2	59.2
1992	37.0	42.7	48.5	57.8	65.1	73.8	79.0	73.4	69.2	59.4	44.3	35.7	57.2
1993	32.1	31.7	41.5	54.2	66.4	75.1	81.9	80.4	66.4	56.1	43.9	34.2	55.6
1994	26.8	35.0	47.7	57.7	65.0	79.0	79.5	76.6	70.2	61.5	52.1	41.5	57.7
1995	31.1	36.4	49.0	56.8	64.7	75.6	81.3	83.9	68.0	60.9	42.7	33.8	57.0
1996	29.4	37.4	41.0	54.2	68.6	74.6	75.8	77.6	67.2	58.1	39.0	35.5	54.9
1997	26.7	38.1	47.6	50.9	61.7	73.6	80.2	76.0	70.2	59.1	42.0	35.1	55.1
1998	36.4	42.6	43.2	55.3	71.2	75.5	78.9	79.1	75.0	60.2	49.5	37.2	58.7
1999	31.3	42.5	42.8	58.6	66.8	74.9	82.9	76.5	69.7	58.8	52.6	38.3	58.0
2000	33.0	42.9	49.1	55.2	69.0	72.9	77.6	80.4	69.6	61.3	41.3	21.6	56.2
2001	30.5	35.1	41.5	63.1	68.7	74.1	80.7	79.6	68.8	57.7	53.1	39.9	57.7
POR= 128 YRS	31.2	35.0	44.5	56.2	66.0	75.2	79.5	77.6	70.1	58.8	45.5	35.0	56.2

REFERENCE NOTES:

PAGE 1:
THE TEMPERATURE GRAPH SHOWS NORMAL MAXIMUM AND NORMAL
MINIMUM DAILY TEMPERATURES (SOLID CURVES) AND THE
ACTUAL DAILY HIGH AND LOW TEMPERATURES (VERTICAL BARS).

PAGE 2 AND 3:
H/C INDICATES HEATING AND COOLING DEGREE DAYS.
RH INDICATES RELATIVE HUMIDITY
W/O INDICATES WEATHER AND OBSTRUCTIONS
S INDICATES SUNSHINE.
PR INDICATES PRESSURE.
CLOUDINESS ON PAGE 3 IS THE SUM OF THE CEILOMETER AND
SATELLITE DATA NOT TO EXCEED EIGHT EIGHTHS(OKTAS).

GENERAL:
T INDICATES TRACE PRECIPITATION, AN AMOUNT GREATER
THAN ZERO BUT LESS THAN THE LOWEST REPORTABLE VALUE.
+ INDICATES THE VALUE ALSO OCCURS ON EARLIER DATES.
BLANK ENTRIES DENOTE MISSING OR UNREPORTED DATA.
NORMALS ARE 30-YEAR AVERAGES (1961 - 1990).
ASOS INDICATES AUTOMATED SURFACE OBSERVING SYSTEM.
PM INDICATES THE LAST DAY OF THE PREVIOUS MONTH.
POR (PERIOD OF RECORD) BEGINS WITH THE JANUARY DATA
MONTH AND IS THE NUMBER OF YEARS USED TO COMPUTE
THE MEAN. INDIVIDUAL MONTHS WITHIN THE POR MAY
BE MISSING.
WHEN THE POR FOR A NORMAL IS LESS THAN 30 YEARS,
THE NORMAL IS PROVISIONAL AND IS BASED ON THE NUMBER
OF YEARS INDICATED.
0.* OR * INDICATES THE VALUE OR MEAN-DAYS-WITH
IS BETWEEN 0.00 AND 0.05.
CLOUDINESS FOR ASOS STATIONS DIFFERS FROM THE NON-ASOS
OBSERVATION TAKEN BY A HUMAN OBSERVER. ASOS STATION
CLOUDINESS IS BASED ON TIME-AVERAGED CEILOMETER DATA
FOR CLOUDS AT OR BELOW 12,000 FEET AND ON SATELLITE
DATA FOR CLOUDS ABOVE 12,000 FEET.
THE NUMBER OF DAYS WITH CLEAR, PARTLY CLOUDY, AND
CLOUDY CONDITIONS FOR ASOS STATIONS IS THE SUM
OF THE CEILOMETER AND SATELLITE DATA FOR THE
SUNRISE TO SUNSET PERIOD.

GENERAL CONTINUED:
CLEAR INDICATES 0 - 2 OKTAS, PARTLY CLOUDY INDICATES
3 - 6 OKTAS, AND CLOUDY INDICATES 7 OR 8 OKTAS.
WHEN AT LEAST ONE OF THE ELEMENTS (CEILOMETER OR
SATELLITE) IS MISSING, THE DAILY CLOUDINESS IS
NOT COMPUTED.
WIND DIRECTION IS RECORDED IN TENS OF DEGREES (2 DIGITS)
CLOCKWISE FROM TRUE NORTH. "00" INDICATES CALM. "36"
INDICATES TRUE NORTH.
RESULTANT WIND IS THE VECTOR AVERAGE OF THE SPEED AND
DIRECTION.
AVERAGE TEMPERATURE IS THE SUM OF THE MEAN DAILY MAXIMUM
AND MINIMUM TEMPERATURE DIVIDED BY 2.
SNOWFALL DATA COMPRISE ALL FORMS OF FROZEN
PRECIPITATION, INCLUDING HAIL.
A HEATING (COOLING) DEGREE DAY IS THE DIFFERENCE BETWEEN
THE AVERAGE DAILY TEMPERATURE AND 65 F.
DRY BULB IS THE TEMPERATURE OF THE AMBIENT AIR.
DEW POINT IS THE TEMPERATURE TO WHICH THE AIR MUST BE
COOLED TO ACHIEVE 100 PERCENT RELATIVE HUMIDITY.
WET BULB IS THE TEMPERATURE THE AIR WOULD HAVE IF THE
MOISTURE CONTENT WAS INCREASED TO 100 PERCENT RELATIVE
HUMIDITY.

ON JULY 1, 1996, THE NATIONAL WEATHER SERVICE BEGAN USING
THE "METAR" OBSERVATION CODE THAT WAS ALREADY EMPLOYED
BY MOST OTHER NATIONS OF THE WORLD. THE MOST NOTICEABLE
DIFFERENCE IN THIS ANNUAL PUBLICATION WILL BE THE CHANGE
IN UNITS FROM TENTHS TO EIGHTS(OKTAS) FOR REPORTING THE
AMOUNT OF SKY COVER.

HEATING DEGREE DAYS (base 65 F) 2001 ST. LOUIS, MO (STL)

YEAR	JUL	AUG	SEP	OCT	NOV	DEC	JAN	FEB	MAR	APR	MAY	JUN	TOTAL
1972-73	2	0	29	317	751	1069	997	849	430	348	121	0	4913
1973-74	0	0	31	182	538	1077	1083	804	539	253	101	21	4629
1974-75	0	0	127	242	625	954	979	919	803	353	48	1	5051
1975-76	2	0	110	228	498	910	1137	619	505	288	158	0	4455
1976-77	0	0	24	456	832	1125	1541	839	471	190	36	3	5517
1977-78	0	0	11	291	601	1059	1401	1223	840	275	158	5	5864
1978-79	0	0	24	292	528	923	1496	1167	644	364	80	0	5518
1979-80	0	0	16	223	610	810	1035	1071	740	331	54	0	4890
1980-81	0	0	30	305	553	877	1039	784	569	127	168	0	4452
1981-82	0	0	35	298	483	1048	1308	1015	603	407	9	11	5217
1982-83	0	0	49	261	569	721	1008	745	632	437	117	7	4546
1983-84	0	0	58	192	498	1376	1133	705	860	342	115	0	5279
1984-85	0	0	103	151	616	746	1308	960	487	200	42	10	4623
1985-86	0	0	64	145	550	1159	929	850	505	194	44	0	4441
1986-87	0	11	12	221	699	910	1062	691	501	267	10	0	4384
1987-88	0	0	12	346	490	830	1102	995	610	241	17	3	4646
1988-89	0	0	5	354	528	854	730	1029	625	293	128	4	4550
1989-90	0	0	73	183	536	1261	679	657	496	327	85	9	4306
1990-91	3	0	24	250	375	934	1101	648	474	154	21	0	3984
1991-92	0	0	67	191	674	796	859	642	509	264	95	5	4102
1992-93	0	3	46	204	615	902	1014	927	726	327	52	13	4829
1993-94	0	0	58	292	628	852	1179	833	531	261	84	1	4719
1994-95	0	0	33	155	388	722	1045	794	488	260	88	0	3973
1995-96	0	0	65	167	664	961	1096	797	735	322	67	14	4888
1996-97	0	0	57	235	776	908	1180	746	530	418	142	5	4997
1997-98	0	0	13	271	684	922	880	621	689	293	19	24	4416
1998-99	0	0	1	175	460	859	1038	626	679	214	41	7	4100
1999-00	0	0	41	226	370	826	985	638	490	297	36	8	3917
2000-01	0	0	57	176	709	1338	1063	831	721	154	47	13	5109
2001-	0	0	50	246	352	772							

COOLING DEGREE DAYS (base 65 F) 2001 ST. LOUIS, MO (STL)

YEAR	JAN	FEB	MAR	APR	MAY	JUN	JUL	AUG	SEP	OCT	NOV	DEC	ANNUAL
1972	0	0	5	35	129	268	394	358	219	9	0	0	1417
1973	0	0	1	17	25	294	435	375	187	57	4	0	1395
1974	0	0	24	36	109	164	463	300	52	25	2	0	1175
1975	0	0	2	12	133	308	405	406	110	51	4	0	1431
1976	0	0	8	38	34	239	458	298	129	25	0	0	1229
1977	0	0	2	69	231	302	509	360	190	4	0	0	1667
1978	0	0	7	17	120	295	426	360	276	10	8	0	1519
1979	0	0	2	9	102	354	446	420	195	50	0	0	1578
1980	0	0	0	23	120	320	626	580	262	31	2	0	1964
1981	0	0	7	77	42	327	431	353	166	13	1	0	1417
1982	0	0	0	7	191	186	453	322	146	63	15	4	1387
1983	0	0	3	3	41	322	578	603	274	36	2	0	1862
1984	0	0	0	24	67	442	423	493	202	57	2	1	1711
1985	0	0	14	70	128	214	451	310	245	43	2	0	1477
1986	0	0	25	75	150	407	561	298	267	21	0	0	1804
1987	0	0	0	32	251	393	501	439	183	5	20	0	1824
1988	0	0	4	10	144	389	521	556	238	16	0	0	1878
1989	0	0	11	80	111	305	450	403	151	75	6	0	1592
1990	0	0	30	55	47	382	480	408	304	41	12	0	1759
1991	0	0	19	56	277	452	498	463	295	58	3	0	2121
1992	0	0	7	54	104	277	440	268	182	36	0	0	1368
1993	0	0	1	9	102	320	529	483	107	23	0	0	1574
1994	0	0	4	48	90	428	453	365	194	53	7	0	1642
1995	0	0	0	22	85	322	510	595	164	47	0	0	1745
1996	0	5	0	16	188	309	342	402	132	25	0	0	1419
1997	0	0	1	1	44	268	477	349	173	99	0	0	1412
1998	0	0	21	7	217	347	439	446	305	33	2	3	1820
1999	0	0	0	28	104	309	563	364	190	37	5	0	1600
2000	0	5	6	8	167	251	398	483	202	69	5	0	1594
2001	0	0	0	106	166	293	492	457	167	29	2	1	1713

SNOWFALL (inches) 2001 ST. LOUIS, MO (STL)

YEAR	JUL	AUG	SEP	OCT	NOV	DEC	JAN	FEB	MAR	APR	MAY	JUN	TOTAL
1972-73	0.0	0.0	0.0	0.0	5.2	1.0	2.2	3.0	0.2	0.2	0.0	0.0	11.8
1973-74	0.0	0.0	0.0	0.0	0.0	26.3	4.2	4.5	7.4	0.0	0.0	0.0	42.4
1974-75	0.0	0.0	0.0	0.0	1.2	1.5	4.1	12.1	6.3	T	0.0	0.0	25.2
1975-76	0.0	0.0	0.0	0.0	7.6	3.9	4.5	4.5	4.8	0.0	0.0	0.0	25.3
1976-77	0.0	0.0	0.0	0.0	0.3	5.2	23.9	6.7	0.1	0.1	0.0	0.0	36.3
1977-78	0.0	0.0	0.0	0.0	6.7	11.7	22.9	9.3	15.4	0.0	0.0	0.0	66.0
1978-79	0.0	0.0	0.0	0.0	0.0	1.4	18.4	4.8	2.0	0.0	0.0	0.0	26.6
1979-80	0.0	0.0	0.0	0.0	0.3	T	4.2	7.4	8.7	5.0	0.0	0.0	25.6
1980-81	0.0	0.0	0.0	0.0	8.0	1.1	0.6	8.2	0.2	0.0	0.0	0.0	18.1
1981-82	0.0	0.0	0.0	0.0	T	7.9	16.6	9.0	0.4	2.7	0.0	0.0	36.6
1982-83	0.0	0.0	0.0	0.0	T	T	3.3	0.3	1.4	2.4	0.0	0.0	7.4
1983-84	0.0	0.0	0.0	0.0	T	6.5	2.3	9.9	5.2	0.0	0.0	0.0	23.9
1984-85	0.0	0.0	0.0	0.0	1.7	1.8	5.1	1.3	T	0.0	0.0	0.0	9.9
1985-86	0.0	0.0	0.0	0.0	T	5.7	1.0	6.1	0.2	T	0.0	0.0	13.0
1986-87	0.0	0.0	0.0	0.0	T	T	23.6	0.6	T	0.0	0.0	0.0	24.2
1987-88	0.0	0.0	0.0	0.0	T	7.3	1.4	6.7	2.8	0.0	0.0	0.0	18.2
1988-89	0.0	0.0	0.0	0.0	2.9	5.9	0.1	3.9	11.0	0.0	0.0	0.0	23.8
1989-90	0.0	0.0	0.0	T	T	9.1	0.2	6.9	8.4	T	T	T	24.6
1990-91	0.0	0.0	0.0	0.0	0.0	T	13.2	1.9	T	1.7	0.0	0.0	16.8
1991-92	0.0	0.0	0.0	T	3.9	0.4	3.8	2.3	3.1	0.0	0.0	0.0	13.5
1992-93	0.0	0.0	0.0	0.0	T	0.7	5.5	20.8	3.0	0.3	0.6	T	30.9
1993-94	0.0	0.0	0.0	T	1.6	1.1	5.0	1.0	0.2	1.4	0.0	0.0	10.3
1994-95	0.0	0.0	0.0	0.0	0.0	0.0	0.1	3.5	1.6	0.5	T	0.0	5.7
1995-96	0.0	0.0	0.0	0.0	0.9	7.9	13.4	2.8	1.0	T	0.0	0.0	26.0
1996-97	T	0.0	0.0	T		1.5	13.1	0.8	T	4.2	0.0	0.0	
1997-98	T	0.0	0.0	0.0	2.6	4.1	5.7	0.5	5.9	T	0.2	0.0	19.0
1998-99	T	0.0	0.0	0.0	0.0	1.8	12.0	1.5	0.5	T	0.0	0.0	15.8
1999-00	0.0	0.0	0.0	0.0	0.0	0.3	5.0	T	4.8	T	0.0	0.0	10.1
2000-01	0.0	0.0	0.0	0.0	0.0	15.9	0.7	1.6	T	0.4	0.0	0.0	18.6
2001-	0.0	0.0	0.0	T	0.2	T							
POR= 64 YRS	0.0	0.0	0.0	T	1.4	3.8	5.4	4.1	3.7	0.5	T	T	18.9

2001
HELENA,
MONTANA (HLN)

Helena is located on the south side of an intermountain valley bounded on the west and south by the main chain of the Continental Divide. The valley is approximately 25 miles in width from north to south and 35 miles long from east to west. The average height of the mountains above the valley floor is about 3,000 feet.

The climate of Helena may be described as modified continental. Several factors enter into modifying the continental climate characteristics. Some of these are invasion by Pacific Ocean air masses, drainage of cool air into the valley from the surrounding mountains, and the protecting mountain shield in all directions.

The mountains to the north and east sometimes deflect shallow masses of invading cold Arctic air to the east. Following periods of extreme cold, when the return circulation of maritime air has brought warming to most of the eastern part of the state, cold air may remain trapped in the valley for several days before being replaced by warmer air. During these periods of transition from cold-to-warm temperatures, inversions are often quite pronounced.

As may be expected in a northern latitude, cold waves may occur from November through February, with temperatures occasionally dropping to zero or lower.

Summertime temperatures are moderate, with maximum readings generally under 90 degrees and very seldom reaching 100 degrees. Like all mountain stations, there is usually a marked change in temperature from day to night. During the summer this tends to produce an agreeable combination of fairly warm days and cool nights.

Most of the precipitation falls from April through July from frequent showers or thunderstorms, but usually with some steady rains in June, the wettest month of the year. Like summer, fall and winter months are relatively dry. During the April to September growing season, precipitation varies considerably.

Thunderstorms are rather frequent from May through August. Snow can be expected from September through May, but amounts during the spring and fall are usually light, and snow on the ground ordinarily lasts only a day or two. During the winter months snow may remain on the ground for several weeks at a time. There is little drifting of snow in the valley, and blizzard conditions are very infrequent.

Severe ice, sleet, and hailstorms are very seldom observed. Since 1880, only a few hailstorms have caused extensive damage in the city of Helena.

In winter, hours of sunshine are more than would be expected at a mountain location.

Due to the sheltering influence of the mountains, Foehn (Chinook) winds are not as pronounced as might be expected for a location on the eastern slopes of the Rocky Mountains. Strong winds can occur at any time throughout the year, but generally do not last more than a few hours at a time.

Based on the 1951-1980 period, the average first occurrence of 32 degrees Fahrenheit in the fall is September 18 and the average last occurrence in the spring is May 18.

NORMALS, MEANS, AND EXTREMES
HELENA, MT (HLN)

LATITUDE:	LONGITUDE:	ELEVATION (FT):		TIME ZONE:	WBAN: 24144
46 36' 20" N	111 57' 49" W	GRND: 3864	BARO: 3867	MOUNTAIN (UTC + 7)	

	ELEMENT	POR	JAN	FEB	MAR	APR	MAY	JUN	JUL	AUG	SEP	OCT	NOV	DEC	YEAR
TEMPERATURE °F	NORMAL DAILY MAXIMUM	30	29.6	36.9	44.8	56.1	65.4	75.8	85.0	83.2	69.8	58.5	42.4	31.3	56.6
	MEAN DAILY MAXIMUM	50	29.9	36.7	44.8	55.8	65.8	74.1	83.5	82.6	71.0	58.5	42.1	32.4	56.4
	HIGHEST DAILY MAXIMUM	61	72	69	77	86	93	100	102	105	99	87	75	64	105
	YEAR OF OCCURRENCE		2000	1995	1978	1992	2001	1988	1981	1969	1967	2001	1999	1980	AUG 1969
	MEAN OF EXTREME MAXS.	54	52.3	55.4	64.2	75.3	83.6	90.7	95.8	95.5	88.8	77.7	61.9	52.9	74.5
	NORMAL DAILY MINIMUM	30	9.6	15.9	22.3	30.6	39.6	48.3	53.4	51.7	41.0	31.6	20.7	11.2	31.3
	MEAN DAILY MINIMUM	50	9.5	15.6	22.2	30.0	39.9	47.4	52.6	50.7	41.3	31.5	20.5	12.5	31.1
	LOWEST DAILY MINIMUM	61	-42	-42	-30	1	17	30	36	28	18	-8	-39	-38	-42
	YEAR OF OCCURRENCE		1957	1996	1955	1954	1954	1999	1971	1992	1970	1991	1959	1964	FEB 1996
	MEAN OF EXTREME MINS.	54	-18.6	-9.3	-.2	16.8	27.8	36.4	42.8	40.3	27.7	15.7	-1.0	-13.8	13.7
	NORMAL DRY BULB	30	19.6	26.4	33.6	43.4	52.5	62.1	69.2	67.4	55.4	45.1	31.6	21.2	44.0
	MEAN DRY BULB	54	19.0	25.8	33.0	43.2	52.6	60.6	68.0	66.6	56.0	45.1	31.4	22.0	43.6
	MEAN WET BULB	49	16.7	22.5	27.3	35.6	43.6	50.2	54.5	53.4	46.0	37.5	27.5	19.5	36.2
	MEAN DEW POINT	49	9.6	15.3	19.2	25.9	34.3	41.6	44.5	42.8	36.9	29.2	20.6	13.2	27.8
	NORMAL NO. DAYS WITH:														
	MAXIMUM 90	30	0.0	0.0	0.0	0.0	0.2	2.4	7.6	6.6	1.1	0.0	0.0	0.0	17.9
	MAXIMUM 32	30	14.4	7.9	4.1	0.5	0.0	0.0	0.0	0.0	0.0	0.5	5.3	14.8	47.5
	MINIMUM 32	30	29.4	26.0	27.3	18.0	3.7	0.1	0.0	0.0	3.6	16.6	26.7	29.5	180.9
	MINIMUM 0	30	8.9	4.5	1.7	0.0	0.0	0.0	0.0	0.0	0.0	0.1	1.6	6.2	23.0
H/C	NORMAL HEATING DEG. DAYS	30	1407	1081	973	648	388	137	34	65	321	617	1002	1358	8031
	NORMAL COOLING DEG. DAYS	30	0	0	0	0	0	50	164	139	33	0	0	0	386
RH	NORMAL (PERCENT)	30	66	64	60	54	54	52	46	48	54	58	65	68	57
	HOUR 05 LST	30	70	72	72	70	72	71	66	66	73	73	73	72	71
	HOUR 11 LST	30	66	62	55	46	45	44	39	41	48	52	62	67	52
	HOUR 17 LST	30	62	54	46	38	38	36	29	30	36	42	57	66	44
	HOUR 23 LST	30	69	68	65	60	60	58	50	52	60	64	69	71	62
S	PERCENT POSSIBLE SUNSHINE	55	47	56	61	59	61	64	78	75	68	60	44	42	60
W/O	MEAN NO. DAYS WITH:														
	HEAVY FOG(VISBY 1/4 MI)	61	1.9	1.2	1.8	0.3	0.2	0.1	0.1	0.1	0.1	0.5	1.4	2.3	10.0
	THUNDERSTORMS	61	0.1	0.1	0.1	1.0	4.1	7.4	8.6	7.5	1.8	0.3	0.1	0.0	31.1
CLOUDINESS	MEAN:														
	SUNRISE-SUNSET (OKTAS)	1			9.6		8.8	4.8							
	MIDNIGHT-MIDNIGHT (OKTAS)	1		3.2			8.8	4.8							
	MEAN NO. DAYS WITH:														
	CLEAR	1	1.5	2.0	2.0			8.0	2.0	11.0	7.0	1.0		6.0	
	PARTLY CLOUDY	1	1.0	5.0	1.0		6.0	5.0	1.0	3.0	5.0	7.0		2.0	
	CLOUDY	1	3.0	3.0	12.0		23.0	6.0			4.0	7.0		3.0	
PR	MEAN STATION PRESSURE(IN)	29	26.02	26.00	25.94	25.97	25.96	25.99	26.04	26.05	26.07	26.06	26.02	26.02	26.01
	MEAN SEA-LEVEL PRES. (IN)	50	30.17	30.12	30.03	29.98	29.95	29.93	29.96	29.96	30.03	30.09	30.12	30.15	30.04
WINDS	MEAN SPEED (MPH)	38	9.0	9.4	10.7	11.0	11.2	12.0	12.0	11.7	10.5	9.8	9.9	9.6	10.6
	PREVAIL.DIR(TENS OF DEGS)	21	27	27	27	27	27	27	27	27	27	27	27	27	27
	MAXIMUM 2-MINUTE:														
	SPEED (MPH)	7	46	51	47	41	44	52	47	46	52	51	39	44	52
	DIR. (TENS OF DEGS)		29	26	28	28	29	18	26	26	26	28	27	28	26
	YEAR OF OCCURRENCE		1997	1999	2001	1997	1996	1999	2001	1999	1999	1999	1995	1995	SEP 1999
	MAXIMUM 5-SECOND:														
	SPEED (MPH)	7	55	64	56	49	55	60	58	59	64	61	47	57	64
	DIR. (TENS OF DEGS)		28	26	28	27	25	19	26	26	26	26	27	29	26
	YEAR OF OCCURRENCE		2000	1999	2000	1997	2001	1999	2001	1999	1999	1999	1995	1995	SEP 1999
PRECIPITATION	NORMAL (IN)	30	0.63	0.41	0.73	0.97	1.78	1.87	1.10	1.29	1.15	0.60	0.48	0.59	11.60
	MAXIMUM MONTHLY (IN)	61	2.78	1.20	1.62	3.00	6.09	4.74	4.70	4.23	3.37	2.68	1.50	1.48	6.09
	YEAR OF OCCURRENCE		1969	1986	1982	1975	1981	1944	1993	1974	1965	1975	1950	1977	MAY 1981
	MINIMUM MONTHLY (IN)	61	T	0.02	0.02	0.10	0.29	0.08	0.08	0.02	0.08	0.02	0.04	0.01	T
	YEAR OF OCCURRENCE		1987	1991	1959	1977	1979	1985	1973	1988	1972	1978	1969	1997	JAN 1987
	MAXIMUM IN 24 HOURS (IN)	61	0.77	0.58	1.01	1.25	2.31	1.78	2.26	1.86	1.61	1.01	0.82	0.51	2.31
	YEAR OF OCCURRENCE		1969	1953	1957	1951	1981	1979	1983	1974	1980	2000	1959	1982	MAY 1981
	NORMAL NO. DAYS WITH:														
	PRECIPITATION 0.01	30	8.2	6.5	8.6	8.1	11.1	10.1	7.1	7.6	7.1	5.2	6.9	8.2	94.7
	PRECIPITATION 1.00	30	0.0	0.0	0.0	0.0	0.2	0.2	0.1	0.1	0.1	0.0	0.0	0.0	0.7
SNOWFALL	NORMAL (IN)	30	9.1	5.6	7.4	5.5	1.7	0.1	T	0.0	1.6	2.2	5.0	8.5	46.7
	MAXIMUM MONTHLY (IN)	57	35.6	19.7	21.6	20.6	12.7	2.7	T	6.2	13.7	11.0	32.9	22.8	35.6
	YEAR OF OCCURRENCE		1969	1959	1955	1967	1967	1969	1993	1992	1965	1969	1959	1967	JAN 1969
	MAXIMUM IN 24 HOURS (IN)	57	11.5	11.7	8.7	12.9	12.5	2.7	T	6.2	13.3	7.4	21.5	10.7	21.5
	YEAR OF OCCURRENCE		1969	1993	1955	1960	1967	1969		1993	1992	1957	1969	1991	NOV 1959
	MAXIMUM SNOW DEPTH (IN)	46	23	15	15	8	8	T	0	0	10	6	19	14	23
	YEAR OF OCCURRENCE		1969	1975	1969	1986	1967	1969			1965	1969	1978	1985	JAN 1969
	NORMAL NO. DAYS WITH:														
	SNOWFALL 1.0	30	3.0	1.8	2.3	1.5	0.4	0.*	0.0	0.0	0.5	0.7	1.5	2.3	14.0

PRECIPITATION (inches) 2001 HELENA, MT (HLN)

YEAR	JAN	FEB	MAR	APR	MAY	JUN	JUL	AUG	SEP	OCT	NOV	DEC	ANNUAL
1972	1.12	0.54	0.63	0.41	0.77	1.12	0.56	1.63	0.08	0.57	0.33	0.46	8.22
1973	0.22	0.13	0.05	0.66	1.08	0.73	0.08	0.56	0.43	0.66	1.03	0.63	6.26
1974	0.66	0.23	0.38	0.76	2.07	0.34	0.49	4.23	0.22	0.51	0.30	0.26	10.45
1975	1.26	0.72	0.88	3.00	1.95	2.83	3.89	2.47	0.47	2.68	0.48	0.31	20.94
1976	0.26	0.38	0.41	1.34	0.87	2.74	0.29	1.58	1.82	0.04	0.30	0.04	10.07
1977	0.65	0.13	1.11	0.10	1.82	1.37	1.37	0.72	1.93	0.17	0.48	1.48	11.33
1978	0.96	0.61	0.31	0.94	1.20	0.44	2.83	0.59	1.11	0.02	1.19	0.76	10.96
1979	0.77	0.72	1.34	2.26	0.29	2.75	0.32	0.79	0.12	0.38	0.06	0.59	10.39
1980	0.62	0.74	0.88	0.63	4.32	3.16	1.92	0.28	2.57	1.21	0.32	0.40	17.05
1981	0.15	0.10	1.10	0.75	6.09	1.15	1.78	0.10	0.90	0.82	0.54	0.33	13.81
1982	0.80	0.58	1.62	0.54	1.77	2.99	0.49	0.74	2.74	0.35	0.31	1.05	13.98
1983	0.24	0.07	0.36	0.29	1.79	2.20	3.48	2.67	1.56	0.35	0.26	0.76	14.03
1984	0.17	0.15	0.49	1.45	1.03	2.14	0.11	1.11	0.73	0.74	0.47	0.41	9.00
1985	0.16	0.38	0.32	0.46	0.75	0.08	0.10	2.64	2.11	0.76	0.84	0.35	8.95
1986	0.32	1.20	0.49	1.08	0.83	1.56	1.37	1.84	2.45	0.03	0.54	0.38	12.09
1987	T	0.03	1.19	0.76	1.90	1.50	2.88	0.38	0.80	0.05	0.12	0.42	10.03
1988	0.27	0.50	0.45	1.32	1.82	1.50	0.36	0.02	2.09	0.69	0.69	0.32	10.03
1989	1.42	0.82	1.35	0.72	1.00	1.43	1.55	1.61	1.31	0.54	0.26	0.48	12.49
1990	0.47	0.14	0.91	0.43	1.54	0.92	0.40	2.57	0.11	0.11	0.36	0.47	8.43
1991	0.27	0.02	0.90	0.75	1.71	3.27	0.72	0.70	1.26	0.65	0.88	0.79	11.92
1992	0.29	0.10	0.60	0.55	0.64	2.36	1.06	1.01	0.09	1.87	0.19	0.57	9.33
1993	0.80	1.03	0.56	1.63	1.71	3.14	4.70	2.79	1.25	0.71	0.36	0.13	18.81
1994	0.20	0.40	0.32	1.45	1.23	0.84	0.71	0.47	0.09	1.14	0.55	0.07	7.47
1995	0.20	0.08	0.49	1.15	3.09	2.93	1.51	0.33	1.59	0.10	0.62	0.28	12.37
1996	0.55	0.11	0.58	0.70	2.42	1.20	1.27	0.89	0.51	0.04	0.84	0.61	9.72
1997	0.28	0.10	0.10	0.20	2.35	2.43	1.79	0.31	1.62	0.13	0.01	0.57	10.57
1998	0.49	0.12	0.39	0.64	2.27	3.03	2.96	0.50	0.82	0.14	1.07	0.14	12.57
1999	0.38	0.26	0.02	1.05	2.19	2.15	0.41	1.92	0.54	0.39	0.13	0.10	9.54
2000	0.26	0.32	0.26	0.73	0.98	1.42	0.73	0.43	0.54	2.12	0.36	0.23	8.38
2001	0.27	0.17	0.44	1.39	1.23	2.11	1.94	0.43	1.38	0.54	0.13	0.28	10.31
POR= 120 YRS	0.69	0.51	0.71	0.98	1.87	2.17	1.15	0.97	1.13	0.76	0.61	0.61	12.16

AVERAGE TEMPERATURE (F) 2001 HELENA, MT (HLN)

YEAR	JAN	FEB	MAR	APR	MAY	JUN	JUL	AUG	SEP	OCT	NOV	DEC	ANNUAL
1972	12.4	26.8	39.1	39.8	50.9	61.8	61.7	66.0	50.9	39.0	31.1	13.6	41.1
1973	17.7	22.8	36.0	40.3	52.9	61.2	68.9	54.4	45.8	24.5	28.2	43.3	43.3
1974	18.1	32.8	33.9	46.0	48.6	64.9	70.4	61.5	53.5	45.8	34.7	27.7	44.8
1975	21.1	13.7	28.8	32.9	48.6	56.9	69.2	61.2	54.1	43.3	29.9	26.0	40.5
1976	24.9	29.5	30.4	43.2	54.1	56.6	67.6	64.8	57.9	44.9	33.8	28.7	44.7
1977	18.0	34.1	32.4	46.9	50.4	63.6	66.3	63.6	56.3	45.6	31.2	20.7	44.1
1978	16.9	22.8	38.8	47.7	53.2	63.2	67.2	65.5	57.0	46.5	22.7	15.2	43.1
1979	1.1	20.1	34.7	42.7	53.2	62.0	69.2	67.7	61.3	47.8	29.0	28.2	43.1
1980	14.3	25.3	31.3	49.0	55.4	59.8	67.3	62.9	56.8	45.4	26.9	44.1	43.1
1981	28.4	29.7	38.0	46.4	52.8	58.9	67.7	69.5	58.8	43.8	36.4	25.3	46.3
1982	16.5	23.7	33.9	40.5	50.9	61.2	68.6	66.1	54.9	44.9	27.4	22.7	42.9
1983	30.6	35.3	38.3	43.1	51.5	60.2	66.0	70.8	53.5	46.0	34.7	5.5	44.6
1984	27.3	32.4	37.2	43.5	52.8	60.0	70.0	69.1	52.8	41.7	33.1	11.6	44.3
1985	12.3	18.8	33.4	46.9	56.2	63.3	75.0	63.1	49.6	42.3	12.6	15.0	40.7
1986	25.5	21.8	42.9	43.3	53.6	66.4	64.2	68.0	51.2	45.3	29.0	18.4	44.2
1987	23.3	31.9	37.0	50.2	55.9	64.4	66.2	62.8	59.9	46.8	34.7	24.7	46.5
1988	18.8	29.1	36.1	47.1	55.5	68.4	71.3	68.6	56.4	50.3	33.8	23.1	46.5
1989	24.5	6.1	27.6	44.5	51.7	62.2	72.0	64.3	55.6	45.1	36.8	23.8	42.9
1990	28.8	26.9	34.9	45.8	51.0	61.5	68.4	68.4	63.6	45.3	37.4	14.0	45.6
1991	19.0	37.5	35.3	43.8	52.8	59.5	70.8	72.8	58.4	43.3	29.8	22.9	45.3
1992	23.7	34.2	42.7	47.7	57.5	64.9	64.3	64.7	56.7	46.7	32.8	14.7	45.9
1993	12.3	15.0	36.4	44.3	57.3	58.5	59.5	61.5	54.8	43.8	26.8	30.0	41.6
1994	29.8	20.7	39.4	45.6	56.2	62.7	69.0	69.9	61.6	44.3	28.3	23.6	45.9
1995	22.7	30.5	30.7	42.0	49.7	57.4	65.8	65.2	57.7	42.8	34.3	22.7	43.3
1996	9.9	21.1	25.7		48.2	61.3	68.3	66.6	55.5	43.8	23.1	16.6	
1997	13.3	28.2	36.9	38.4	53.2	60.1	65.8	65.4	58.8	44.1	31.2	24.4	43.3
1998	21.2	31.0	32.8	44.8	53.9	55.1	69.8	68.6	61.9	43.8	34.5	24.5	45.2
1999	26.8	32.4	37.1	40.8	50.2	58.8	65.3	68.4	52.9	45.8	39.0	28.1	45.5
2000	25.0	28.8	38.6	47.0	54.9	62.4	72.1	69.6	56.4	44.6	22.0	16.2	44.8
2001	20.1	17.6	35.4	44.0	59.2	64.1	72.0	74.7	64.6	48.4	39.2	24.5	47.0
POR= 121 YRS	19.8	24.7	32.7	43.6	52.4	60.0	67.8	66.4	55.9	45.4	32.3	23.6	43.7

REFERENCE NOTES:

PAGE 1:
THE TEMPERATURE GRAPH SHOWS NORMAL MAXIMUM AND NORMAL
MINIMUM DAILY TEMPERATURES (SOLID CURVES) AND THE
ACTUAL DAILY HIGH AND LOW TEMPERATURES (VERTICAL BARS).

PAGE 2 AND 3:
H/C INDICATES HEATING AND COOLING DEGREE DAYS.
RH INDICATES RELATIVE HUMIDITY
W/O INDICATES WEATHER AND OBSTRUCTIONS
S INDICATES SUNSHINE.
PR INDICATES PRESSURE.
CLOUDINESS ON PAGE 3 IS THE SUM OF THE CEILOMETER AND
SATELLITE DATA NOT TO EXCEED EIGHT EIGHTHS(OKTAS).

GENERAL:
T INDICATES TRACE PRECIPITATION, AN AMOUNT GREATER
THAN ZERO BUT LESS THAN THE LOWEST REPORTABLE VALUE.
+ INDICATES THE VALUE ALSO OCCURS ON EARLIER DATES.
BLANK ENTRIES DENOTE MISSING OR UNREPORTED DATA.
NORMALS ARE 30-YEAR AVERAGES (1961 - 1990).
ASOS INDICATES AUTOMATED SURFACE OBSERVING SYSTEM.
PM INDICATES THE LAST DAY OF THE PREVIOUS MONTH.
POR (PERIOD OF RECORD) BEGINS WITH THE JANUARY DATA
MONTH AND IS THE NUMBER OF YEARS USED TO COMPUTE
THE MEAN. INDIVIDUAL MONTHS WITHIN THE POR MAY
BE MISSING.
WHEN THE POR FOR A NORMAL IS LESS THAN 30 YEARS,
THE NORMAL IS PROVISIONAL AND IS BASED ON THE NUMBER
OF YEARS INDICATED.
0.* OR * INDICATES THE VALUE OR MEAN-DAYS-WITH
IS BETWEEN 0.00 AND 0.05.
CLOUDINESS FOR ASOS STATIONS DIFFERS FROM THE NON-ASOS
OBSERVATION TAKEN BY A HUMAN OBSERVER. ASOS STATION
CLOUDINESS IS BASED ON TIME-AVERAGED CEILOMETER DATA
FOR CLOUDS AT OR BELOW 12,000 FEET AND ON SATELLITE
DATA FOR CLOUDS ABOVE 12,000 FEET.
THE NUMBER OF DAYS WITH CLEAR, PARTLY CLOUDY, AND
CLOUDY CONDITIONS FOR ASOS STATIONS IS THE SUM
OF THE CEILOMETER AND SATELLITE DATA FOR THE
SUNRISE TO SUNSET PERIOD.

GENERAL CONTINUED:
CLEAR INDICATES 0 - 2 OKTAS, PARTLY CLOUDY INDICATES
3 - 6 OKTAS, AND CLOUDY INDICATES 7 OR 8 OKTAS.
WHEN AT LEAST ONE OF THE ELEMENTS (CEILOMETER OR
SATELLITE) IS MISSING, THE DAILY CLOUDINESS IS
NOT COMPUTED.
WIND DIRECTION IS RECORDED IN TENS OF DEGREES (2 DIGITS)
CLOCKWISE FROM TRUE NORTH. "00" INDICATES CALM. "36"
INDICATES TRUE NORTH.
RESULTANT WIND IS THE VECTOR AVERAGE OF THE SPEED AND
DIRECTION.
AVERAGE TEMPERATURE IS THE SUM OF THE MEAN DAILY MAXIMUM
AND MINIMUM TEMPERATURE DIVIDED BY 2.
SNOWFALL DATA COMPRISE ALL FORMS OF FROZEN
PRECIPITATION, INCLUDING HAIL.
A HEATING (COOLING) DEGREE DAY IS THE DIFFERENCE BETWEEN
THE AVERAGE DAILY TEMPERATURE AND 65 F.
DRY BULB IS THE TEMPERATURE OF THE AMBIENT AIR.
DEW POINT IS THE TEMPERATURE TO WHICH THE AIR MUST BE
COOLED TO ACHIEVE 100 PERCENT RELATIVE HUMIDITY.
WET BULB IS THE TEMPERATURE THE AIR WOULD HAVE IF THE
MOISTURE CONTENT WAS INCREASED TO 100 PERCENT RELATIVE
HUMIDITY.

ON JULY 1, 1996, THE NATIONAL WEATHER SERVICE BEGAN USING
THE "METAR" OBSERVATION CODE THAT WAS ALREADY EMPLOYED
BY MOST OTHER NATIONS OF THE WORLD. THE MOST NOTICEABLE
DIFFERENCE IN THIS ANNUAL PUBLICATION WILL BE THE CHANGE
IN UNITS FROM TENTHS TO EIGHTS(OKTAS) FOR REPORTING THE
AMOUNT OF SKY COVER.

HEATING DEGREE DAYS (base 65 F) 2001 HELENA, MT (HLN)

YEAR	JUL	AUG	SEP	OCT	NOV	DEC	JAN	FEB	MAR	APR	MAY	JUN	TOTAL
1972-73	136	49	418	798	1010	1588	1465	1174	892	732	374	155	8791
1973-74	23	47	317	588	1208	1136	1452	897	956	564	500	99	7787
1974-75	16	130	338	588	905	1149	1355	1429	1114	954	501	235	8714
1975-76	14	119	322	666	1045	1199	1236	1023	1064	649	331	257	7925
1976-77	15	45	219	615	928	1120	1449	862	1005	535	443	90	7326
1977-78	52	92	270	593	1008	1367	1485	1175	806	512	361	87	7808
1978-79	32	60	244	564	1263	1540	1979	1250	930	665	359	128	9014
1979-80	11	15	127	528	1072	1138	1566	1148	1039	473	304	164	7585
1980-81	25	81	242	602	899	1175	1127	986	832	552	371	191	7083
1981-82	21	16	195	650	853	1227	1497	1153	959	726	428	136	7861
1982-83	30	16	304	618	1120	1306	1059	828	823	649	417	152	7322
1983-84	76	0	351	584	901	1842	1164	941	856	538	266	97	7909
1984-85	2	7	377	716	954	1654	1625	1291	973	538	266	97	8500
1985-86	3	105	455	696	1571	1545	1218	1202	677	645	380	42	8539
1986-87	66	23	409	602	1077	1432	1288	923	862	437	276	77	7472
1987-88	75	104	163	556	901	1241	1426	1034	889	529	297	63	7278
1988-89	10	13	282	449	934	1292	1251	1650	1156	610	407	107	8161
1989-90	0	92	274	611	839	1268	1116	1058	925	573	426	177	7359
1990-91	15	31	78	604	823	1579	1579	884	914	630	373	159	7393
1991-92	2	0	220	666	1053	1297	1273	884	687	513	237	94	6926
1992-93	68	141	246	564	960	1552	1631	1397	879	617	228	214	8497
1993-94	179	122	323	652	1139	1080	1080	1236	783	576	268	128	7566
1994-95	18	17	110	633	1095	1275	1307	963	1059	681	465	222	7845
1995-96	37	61	293	683	912	1303	1703	1268	1210		510	127	8543
1996-97	15	53	286	651	1251	1494	1596	1026	865	793	361	152	
1997-98	49	53	189	642	1006	1252	1352	942	990	600	336	293	7704
1998-99	2	8	158	650	909	1248	1176	905	859	718	451	188	7272
1999-00	79	20	354	590	767	1136	1232	1043	814	535	308	119	6997
2000-01	16	13	279	625	1283	1504	1385	1319	912	625	212	113	8286
2001-	11	0	66	511	766	1249							

COOLING DEGREE DAYS (base 65 F) 2001 HELENA, MT (HLN)

YEAR	JAN	FEB	MAR	APR	MAY	JUN	JUL	AUG	SEP	OCT	NOV	DEC	ANNUAL
1972	0	0	0	0	2	30	42	89	0	0	0	0	163
1973	0	0	0	0	4	45	151	108	3	0	0	0	311
1974	0	0	0	0	0	102	190	31	0	0	0	0	323
1975	0	0	0	0	0	1	154	12	0	0	0	0	167
1976	0	0	0	0	0	14	102	45	10	0	0	0	171
1977	0	0	0	0	0	57	101	54	13	0	0	0	225
1978	0	0	0	0	3	37	109	85	37	0	0	0	271
1979	0	0	0	0	1	45	152	103	21	0	0	0	322
1980	0	0	0	0	14	14	104	25	4	0	0	0	161
1981	0	0	0	0	0	15	109	165	16	0	0	0	305
1982	0	0	0	0	0	30	147	151	25	0	0	0	353
1983	0	0	0	0	4	16	115	186	12	0	0	0	333
1984	0	0	0	0	10	31	165	163	4	0	0	0	373
1985	0	0	0	0	2	55	318	54	0	0	0	0	429
1986	0	0	0	0	35	91	45	123	1	0	0	0	295
1987	0	0	0	0	3	66	122	41	15	0	0	0	247
1988	0	0	0	0	8	170	211	132	30	0	0	0	551
1989	0	0	0	0	0	30	222	82	0	0	0	0	334
1990	0	0	0	0	0	77	159	142	42	0	0	0	420
1991	0	0	0	0	0	2	188	250	29	0	0	0	469
1992	0	0	0	0	10	96	55	140	3	0	0	0	304
1993	0	0	0	0	2	24	15	23	0	0	0	0	64
1994	0	0	0	0	1	67	146	176	16	0	0	0	406
1995	0	0	0	0	0	5	71	73	24	0	0	0	173
1996	0	0	0	0	0	25	125	109	9	0	0	0	
1997	0	0	0	0	1	13	81	71	12	0	0	0	178
1998	0	0	0	0	0	0	161	124	72	0	0	0	357
1999	0	0	0	0	0	10	93	132	0	0	0	0	235
2000	0	0	0	0	2	50	245	165	27	0	0	0	489
2001	0	0	0	0	41	94	237	308	61	0	0	0	741

SNOWFALL (inches) 2001 HELENA, MT (HLN)

YEAR	JUL	AUG	SEP	OCT	NOV	DEC	JAN	FEB	MAR	APR	MAY	JUN	TOTAL
1972-73	T	0.0	0.3	4.7	0.7	7.8	3.2	1.8	0.1	6.5	T	0.0	25.1
1973-74	0.0	0.0	1.3	7.2	12.5	7.2	9.9	5.9	2.2	1.5	0.2	0.0	47.9
1974-75	0.0	0.0	T	1.5	0.8	2.7	15.2	10.7	12.3	15.4	0.2	0.0	58.8
1975-76	0.0	0.0	0.0	6.3	4.9	3.9	3.3	4.4	6.7	10.9	0.0	0.0	40.4
1976-77	0.0	0.0	0.0	T	2.9	0.9	13.8	0.8	14.1	0.5	1.0	0.0	34.0
1977-78	0.0	0.0	0.0	0.4	6.8	19.5	15.7	13.3	2.6	0.9	T	0.0	59.2
1978-79	0.0	0.0	T	T	22.1	13.8	11.7	7.2	12.4	9.3	0.0	T	76.5
1979-80	0.0	0.0	0.0	0.2	0.6	6.5	9.3	10.5	10.1	3.1	T	0.0	40.3
1980-81	0.0	0.0	0.0	3.7	1.2	3.8	2.7	2.1	3.3	0.1	0.8	0.0	16.9
1981-82	0.0	0.0	0.0	5.2	3.4	6.1	18.4	4.8	13.9	4.1	0.8	0.0	56.7
1982-83	0.0	0.0	6.5	0.5	0.0	4.1	11.3	3.2	0.2	2.2	9.9	0.0	39.0
1983-84	0.0	0.0	6.4	0.0	1.3	13.0	1.3	1.5	5.7	2.3	0.8	0.0	32.3
1984-85	0.0	0.0	6.3	9.0	5.7	7.5	3.9	6.2	4.0	0.8	0.2	0.0	43.4
1985-86	0.0	0.0	2.9	8.8	10.4	8.5	4.2	15.6	1.2	11.6	0.2	0.0	63.4
1986-87	0.0	0.0	0.0	T	7.6	5.0	0.2	0.2	9.1	4.3	3.8	0.0	30.2
1987-88	0.0	0.0	0.0	0.3	0.9	1.2	4.4	8.0	5.2	2.9	0.0	0.0	22.9
1988-89	0.0	0.0	5.9	1.5	6.0	5.0	23.0	13.0	20.7	7.9	3.5	T	86.5
1989-90	0.0	T	T	2.6	1.8	9.4	4.3	1.8	14.0	0.8	0.1	0.0	34.8
1990-91	0.0	0.5	0.0	0.5	8.6	11.6	5.9	0.8	12.2	5.6	0.4	0.5	46.6
1991-92	0.0	T	0.0	7.3	8.8	15.7	4.7	1.3	T	2.2	T	0.0	40.0
1992-93	T	6.2	T	8.6	1.5	11.8	10.2	17.9	2.3	2.5	T	0.2	61.2
1993-94	T	0.0	0.0	4.6	7.3	1.1	4.4	5.0	4.4	8.0	T	0.0	34.8
1994-95	0.0	T	0.0	T	8.6	2.0	0.3	1.6	8.1	0.0	0.0		
1995-96	T	0.0	T		7.4	2.7	14.2						
1996-97													
1997-98													
1998-99													
1999-00				T			3.9	5.7	2.3				
2000-01													
2001-													
POR= 56 YRS	T	0.1	1.5	2.3	6.3	8.2	8.6	6.0	7.3	4.8	1.4	0.1	46.6

2001
NORTH PLATTE,
NEBRASKA (LBF)

The climate of North Platte is characterized throughout the year by frequent rapid changes in the weather. During the winter, most North Pacific lows cross the country north of North Platte. The passage usually brings little or no snowfall, and only a moderate drop in temperature. Only when there is a major outbreak of cold air from Canada does the temperature fall to zero or below. The duration of below-zero temperature is hardly more than two mornings, and by the third or fourth day the temperature is ordinarily rising to the 40s or higher. Snowfall at the onset of a cold outbreak is usually less than 2 inches.

Only when a low moves from the middle Rockies through Nebraska, allowing easterly winds to draw moist air into the low circulation, does snowfall of appreciable amounts occur. Few of these storms move slowly enough, or are intense enough, to deposit much precipitation in the North Platte area. However, during some winters the cold outbreak and intense low from the mid-Rockies combine to produce severe cold and snow several inches in depth, with blizzard conditions following. During and after these snowfalls and blizzards, rail and highway traffic may be stalled until the snow is cleared. Widespread loss of unsheltered livestock and wild life results from such conditions.

The sudden and frequent weather changes of the winter continue through spring with decreasing intensity of temperature changes but increasing precipitation. The summer and fall months bring frequent changes from hot to cool weather. Most summer and fall precipitation is associated with thunderstorms, so the amounts are extremely variable. The surrounding area is occasionally damaged by locally severe winds and hailstorms.

Temperatures may reach into the upper 90s and lower 100s frequently during the summer months, but the elevation and clear skies bring rapid cooling after sunset to lows in the 60s or below by daybreak. Since the humidity is generally low, the extremely hot days of summer are not uncomfortable.

Based on the 1951-1980 period, the average first occurrence of 32 degrees Fahrenheit in the fall is September 24 and the average last occurrence in the spring is May 11.

NORMALS, MEANS, AND EXTREMES
NORTH PLATTE, NE (LBF)

LATITUDE: 41 07' 19" N LONGITUDE: 100 40' 06" W ELEVATION (FT): GRND: 2778 BARO: 2781 TIME ZONE: CENTRAL (UTC + 6) WBAN: 24023

	ELEMENT	POR	JAN	FEB	MAR	APR	MAY	JUN	JUL	AUG	SEP	OCT	NOV	DEC	YEAR
TEMPERATURE °F	NORMAL DAILY MAXIMUM	30	34.6	40.9	49.9	62.5	71.8	81.7	87.8	86.0	76.6	65.7	49.5	37.2	62.0
	MEAN DAILY MAXIMUM	54	35.7	41.5	49.4	61.7	71.6	81.5	87.6	86.1	77.4	66.0	49.6	39.3	62.3
	HIGHEST DAILY MAXIMUM	50	73	79	86	98	97	107	112	105	102	94	82	75	112
	YEAR OF OCCURRENCE		1990	1962	1986	1992	1953	1952	1954	1954	1990	1990	1980	1980	JUL 1954
	MEAN OF EXTREME MAXS.	54	59.3	65.0	75.6	84.7	89.2	96.3	99.9	98.4	94.2	86.3	72.6	62.4	82.0
	NORMAL DAILY MINIMUM	30	8.6	14.3	23.0	33.9	44.6	54.1	60.1	57.6	46.2	33.5	21.3	11.2	34.0
	MEAN DAILY MINIMUM	54	9.6	15.1	22.6	33.5	45.1	54.8	60.6	58.6	47.0	34.0	21.1	12.7	34.6
	LOWEST DAILY MINIMUM	50	-23	-22	-22	7	19	29	39	35	17	10	-13	-34	-34
	YEAR OF OCCURRENCE		1979	1981	1962	1975	1989	1969	1997	1976	1984	1993	1976	1989	DEC 1989
	MEAN OF EXTREME MINS.	54	-11.0	-5.7	2.3	17.4	29.2	41.1	48.5	45.9	29.7	17.5	3.9	-7.9	17.6
	NORMAL DRY BULB	30	21.6	27.6	36.5	48.2	58.2	67.9	74.0	71.8	61.4	49.7	35.4	24.2	48.0
	MEAN DRY BULB	54	22.6	28.1	35.9	47.7	58.2	68.2	74.0	72.5	62.0	50.0	35.3	25.8	48.4
	MEAN WET BULB	18	21.7	25.3	32.7	40.9	52.1	60.4	65.1	64.0	54.3	42.4	28.6	21.2	42.4
	MEAN DEW POINT	18	15.9	19.1	25.4	32.9	46.2	55.1	60.0	59.5	48.4	35.4	22.8	15.7	36.4
	NORMAL NO. DAYS WITH:														
	MAXIMUM 90	30	0.0	0.0	0.0	0.3	1.0	6.0	13.6	11.3	3.8	0.3	0.0	0.0	36.3
	MAXIMUM 32	30	12.1	8.3	4.1	0.2	*	0.0	0.0	0.0	0.0	0.1	3.6	10.8	39.2
	MINIMUM 32	30	31.0	27.7	26.9	13.1	2.3	*	0.0	0.0	2.2	14.0	27.7	30.8	175.7
	MINIMUM 0	30	8.5	3.5	0.6	0.0	0.0	0.0	0.0	0.0	0.0	0.0	0.8	5.1	18.5
H/C	NORMAL HEATING DEG. DAYS	30	1345	1047	884	504	234	49	0	11	158	474	888	1265	6859
	NORMAL COOLING DEG. DAYS	30	0	0	0	0	23	136	282	222	50	0	0	0	713
RH	NORMAL (PERCENT)	30	69	68	64	60	63	64	63	64	64	62	67	70	65
	HOUR 00 LST	30	75	76	73	70	73	74	73	75	74	72	75	76	74
	HOUR 06 LST	30	78	79	79	79	82	83	83	84	83	80	80	78	81
	HOUR 12 LST	30	61	59	53	46	50	51	50	50	49	45	54	60	52
	HOUR 18 LST	30	62	56	48	42	47	47	46	46	46	46	56	62	50
S	PERCENT POSSIBLE SUNSHINE	49	63	62	62	64	65	71	77	75	72	70	60	61	67
W/O	MEAN NO. DAYS WITH:														
	HEAVY FOG(VISBY 1/4 MI)	49	1.4	2.4	2.2	1.1	1.1	0.8	1.1	2.0	2.0	1.9	2.3	1.7	20.0
	THUNDERSTORMS	49	0.0	0.1	0.8	2.6	6.5	9.9	9.9	8.1	3.9	1.5	0.2	0.0	43.5
CLOUDINESS	MEAN:														
	SUNRISE-SUNSET (OKTAS)	44	5.0	5.0	5.2	5.1	5.1	4.2	3.7	3.8	3.7	3.9	4.7	4.5	4.5
	MIDNIGHT-MIDNIGHT (OKTAS)	32	4.4	4.3	4.7	4.6	4.7	4.2	3.8	3.7	3.5	3.6	4.3	4.1	4.2
	MEAN NO. DAYS WITH:														
	CLEAR	44	8.4	7.2	7.5	7.0	6.8	10.1	12.1	11.7	12.8	12.5	8.8	9.8	114.7
	PARTLY CLOUDY	44	8.7	7.5	7.7	9.0	10.0	10.7	12.0	11.4	8.4	8.4	8.0	7.8	109.6
	CLOUDY	44	13.9	13.6	15.8	14.0	14.3	9.2	6.9	8.0	8.8	10.1	13.2	13.4	141.2
PR	MEAN STATION PRESSURE(IN)	29	27.10	27.09	27.00	27.00	27.00	27.10	27.10	27.10	27.10	27.10	27.11	27.12	27.08
	MEAN SEA-LEVEL PRES. (IN)	18	30.11	30.10	30.00	29.93	29.91	29.99	29.93	29.95	29.99	30.03	30.06	30.14	30.01
WINDS	MEAN SPEED (MPH)	46	9.3	9.8	11.6	12.6	11.4	10.4	9.6	9.1	9.7	9.7	9.9	9.3	10.2
	PREVAIL.DIR(TENS OF DEGS)	30	31	31	31	11	16	11	11	11	18	30	31	31	11
	MAXIMUM 2-MINUTE:														
	SPEED (MPH)	5	44	52	47	46	48	49	55	56	46	45	46	45	56
	DIR. (TENS OF DEGS)		31	36	31	36	31	34	24	31	30	31	29	36	31
	YEAR OF OCCURRENCE		1997	1999	2000	1999	1998	1999	1999	1999	1999	1997	1998	2000	AUG 1999
	MAXIMUM 5-SECOND:														
	SPEED (MPH)	5	55	62	56	54	59	62	70	75	53	61	53	53	75
	DIR. (TENS OF DEGS)		32	36	31	01	31	33	24	30	28	31	32	36	30
	YEAR OF OCCURRENCE		1997	1999	2000	1999	1998	1999	1999	1999	1999	1997	1997	2000	AUG 1999
PRECIPITATION	NORMAL (IN)	30	0.36	0.43	1.20	1.99	3.43	3.37	3.06	1.74	1.61	0.98	0.66	0.47	19.30
	MAXIMUM MONTHLY (IN)	50	1.12	1.98	2.98	5.94	8.01	6.81	7.05	6.30	6.03	3.24	2.89	1.22	8.01
	YEAR OF OCCURRENCE		1960	1978	1992	2001	1962	1965	1979	1992	1963	2000	1979	1977	MAY 1962
	MINIMUM MONTHLY (IN)	49	T	T	T	0.05	0.10	0.77	0.33	0.42	0.06	T	0.05	0.02	T
	YEAR OF OCCURRENCE		1964	1996	1994	1989	1966	1952	1955	1967	1953	1988	1989	1988	FEB 1996
	MAXIMUM IN 24 HOURS (IN)	50	0.75	1.15	2.26	2.42	2.95	3.80	3.15	2.93	2.53	2.24	1.48	0.79	3.80
	YEAR OF OCCURRENCE		1992	1971	1959	1971	1962	1965	1964	1957	1963	2000	1979	1978	JUN 1965
	NORMAL NO. DAYS WITH:														
	PRECIPITATION 0.01	30	5.0	5.0	6.8	7.9	11.3	9.2	10.0	7.6	7.4	4.8	4.7	4.5	84.2
	PRECIPITATION 1.00	30	0.0	*	0.2	0.4	0.9	0.7	0.9	0.3	0.3	0.1	*	0.0	3.8
SNOWFALL	NORMAL (IN)	30	5.3	4.2	6.6	2.6	0.1	0.0	0.0	0.0	0.1	1.4	4.1	5.2	29.6
	MAXIMUM MONTHLY (IN)	50	17.1	20.6	21.9	14.5	3.6	T	T	T	3.1	15.7	17.5	14.1	21.9
	YEAR OF OCCURRENCE		1976	1978	1980	1984	1967	1995	1996	1992	1985	1969	1979	1973	MAR 1980
	MAXIMUM IN 24 HOURS (IN)	50	11.9	9.7	15.1	8.5	2.3	T	T	T	3.1	8.8	9.7	8.6	15.1
	YEAR OF OCCURRENCE		1976	1955	1980	1984	1967	1995	1995	1992	1985	1969	2000	1968	MAR 1980
	MAXIMUM SNOW DEPTH (IN)	53	18	13	18	9	1	0	0	0	3	5	13	10	18
	YEAR OF OCCURRENCE		1949	1993	1980	1949	1967				1985	1969	1979	1968	MAR 1980
	NORMAL NO. DAYS WITH:														
	SNOWFALL 1.0	30	1.7	1.4	1.7	0.8	0.1	0.0	0.0	0.0	0.*	0.5	1.1	1.4	8.7

PRECIPITATION (inches) 2001 NORTH PLATTE, NE (LBF)

YEAR	JAN	FEB	MAR	APR	MAY	JUN	JUL	AUG	SEP	OCT	NOV	DEC	ANNUAL
1972	0.16	0.08	0.65	1.19	3.18	2.96	3.58	0.95	1.46	0.62	1.12	0.42	16.37
1973	0.40	0.10	2.45	1.45	3.85	0.88	2.87	2.82	3.98	0.54	1.13	21.73
1974	0.27	0.08	0.42	1.17	1.64	3.86	2.27	1.15	0.24	0.61	0.04	0.42	12.17
1975	0.26	0.17	0.92	1.77	2.12	6.12	2.51	0.25	0.56	0.14	1.15	0.22	16.19
1976	0.99	0.14	2.03	2.80	2.88	2.43	0.92	2.87	2.03	1.18	0.08	0.01	18.36
1977	0.18	0.27	2.89	4.85	5.90	1.31	4.41	1.79	1.48	0.18	0.39	1.22	24.87
1978	0.52	1.98	0.40	1.96	4.84	1.75	3.70	1.92	0.33	0.43	1.03	0.99	19.85
1979	0.86	0.09	2.78	1.46	2.96	3.37	7.05	1.49	0.45	1.32	2.89	0.28	25.00
1980	0.52	0.82	2.56	0.77	2.59	1.89	0.64	3.04	0.34	1.00	0.13	0.02	14.32
1981	0.07	0.05	2.72	2.47	5.37	2.32	5.09	0.25	0.60	1.94	0.43	23.79
1982	0.20	0.15	0.99	1.42	6.32	2.35	1.78	1.18	1.26	2.44	0.73	1.08	19.90
1983	0.33	0.25	1.54	2.12	3.20	3.32	3.74	1.98	0.14	0.56	1.56	0.46	19.20
1984	0.36	0.87	1.20	5.01	2.82	4.37	0.94	1.38	0.39	2.41	0.69	0.72	21.16
1985	0.55	0.44	0.44	1.84	4.01	0.87	3.98	1.16	3.23	1.24	1.09	0.79	19.34
1986	0.02	1.10	0.70	3.77	2.80	1.70	2.57	1.22	1.02	1.58	0.19	0.27	16.94
1987	0.16	1.55	1.65	1.01	3.19	3.95	2.81	1.19	1.16	1.67	1.26	0.81	20.41
1988	0.72	0.03	0.37	2.02	3.59	3.12	3.03	3.93	1.59	0.05	0.40	T	18.85
1989	0.55	0.73	0.38	0.10	3.02	3.51	1.86	2.37	1.11	0.08	0.02	0.28	14.01
1990	0.27	0.18	1.75	1.52	3.65	1.90	1.99	1.79	0.31	1.48	0.87	0.09	15.80
1991	0.35	0.21	1.00	3.00	5.39	2.78	1.81	0.53	1.75	2.14	0.92	0.67	20.55
1992	0.89	1.42	2.98	0.18	3.18	2.61	3.75	6.30	0.25	0.92	0.20	0.33	23.01
1993	0.74	1.37	0.61	1.80	2.47	6.12	5.47	3.78	0.71	1.59	1.32	0.22	26.20
1994	0.53	0.29	0.05	1.47	1.16	4.92	4.52	1.24	0.60	2.63	0.78	0.66	18.85
1995	0.12	0.09	1.21	3.09	4.51	2.59	2.01	0.73	1.98	0.87	0.08	0.02	17.30
1996	0.50	T	0.33	0.84	4.12	3.87	5.57	3.25	5.55	0.33	0.40	0.03	24.79
1997	T	0.73	0.07	1.00	1.87	2.58	3.19	3.40	1.76	2.77	0.04	0.20	17.61
1998	0.27	0.40	1.30	0.67	2.93	4.84	5.81	1.85	1.13	1.92	1.31	0.01	22.44
1999	0.34	0.26	0.64	2.83	1.92	5.32	0.93	5.49	1.20	0.22	0.14	0.05	19.34
2000	0.31	0.46	1.22	1.65	1.21	1.53	2.86	2.05	1.22	3.24	0.53	0.04	16.32
2001	0.48	0.40	0.60	5.94	2.19	1.71	2.52	5.26	2.76	0.80	0.96	0.07	23.69
POR= 127 YRS	0.42	0.50	0.97	2.07	3.00	3.25	2.74	2.18	1.54	1.02	0.53	0.40	18.62

AVERAGE TEMPERATURE (F) 2001 NORTH PLATTE, NE (LBF)

YEAR	JAN	FEB	MAR	APR	MAY	JUN	JUL	AUG	SEP	OCT	NOV	DEC	ANNUAL
1972	20.7	29.9	40.7	46.6	58.1	68.4	70.4	70.8	61.2	46.4	29.8	17.2	46.7
1973	21.5	30.3	40.4	45.8	56.0	67.7	72.8	73.7	58.4	51.7	35.0	23.2	48.0
1974	16.0	33.0	40.1	50.0	59.5	66.5	77.5	67.1	57.5	52.9	34.8	24.2	48.3
1975	25.7	23.4	31.5	46.6	57.1	65.7	74.6	73.2	58.5	50.9	30.2	26.4	47.0
1976	17.8	32.7	34.0	47.4	53.1	64.6	72.8	70.6	59.8	45.2	30.5	28.2	46.4
1977	16.6	32.0	37.0	51.7	63.1	70.5	74.5	68.5	63.8	50.0	35.8	24.9	49.0
1978	11.2	14.7	35.1	48.4	56.6	68.2	74.5	71.0	65.4	49.2	32.1	14.6	45.1
1979	6.0	17.2	37.0	48.4	57.4	68.1	74.1	72.9	67.1	53.8	32.2	34.0	47.4
1980	25.1	26.8	35.5	50.5	60.2	72.5	78.0	74.0	63.7	48.9	38.5	32.0	50.5
1981	29.7	29.1	40.2	55.8	54.8	68.6	73.8	70.1	63.0	49.0	39.9	26.7	50.1
1982	17.1	28.1	36.3	44.8	57.3	63.4	75.0	73.1	61.7	49.2	33.3	28.1	47.3
1983	27.1	35.1	37.2	42.3	54.0	65.2	75.3	78.0	64.8	51.3	36.7	7.5	47.9
1984	20.4	33.2	34.1	43.3	58.1	68.0	73.7	75.5	58.4	48.0	37.2	22.5	47.7
1985	18.3	23.1	40.5	52.2	60.8	65.4	75.3	70.1	59.8	48.6	24.6	19.2	46.5
1986	31.6	27.0	44.3	48.9	58.0	71.4	75.6	71.3	63.0	50.4	35.5	30.2	50.6
1987	29.5	35.8	36.7	51.3	63.1	70.2	75.7	70.3	61.3	47.0	38.2	27.7	50.6
1988	16.4	26.5	37.8	48.5	60.7	74.6	74.5	73.4	62.0	48.7	38.0	30.3	49.3
1989	30.2	17.8	35.7	51.1	59.0	65.8	74.3	71.7	61.0	50.7	37.6	20.6	48.0
1990	30.6	30.9	40.0	48.5	56.7	71.1	73.7	74.0	67.6	50.2	38.1	21.9	50.3
1991	23.1	37.5	40.5	49.3	61.8	71.0	74.9	73.7	63.6	48.4	32.2	33.4	50.8
1992	32.1	37.3	41.8	50.6	58.7	65.7	68.5	66.5	63.0	49.9	32.4	22.5	49.1
1993	18.5	18.0	37.0	45.5	58.7	65.3	71.5	70.3	58.3	48.5	32.7	30.4	46.2
1994	24.0	23.7	41.5	47.8	62.6	71.7	70.5	72.6	65.4	52.0	37.4	29.1	49.9
1995	26.8	34.0	37.8	43.8	52.6	66.7	73.8	79.3	62.6	48.2	38.6	28.8	49.4
1996	20.9	30.1	31.3	47.3	56.6	68.5	71.2	70.0	59.1	48.6	30.7	25.0	46.6
1997	23.3	29.1	39.5	42.3	55.1	69.5	73.9	70.9	64.2	51.1	33.9	28.2	48.4
1998	26.0	34.7	32.6	46.3	59.6	64.1	75.3	72.4	68.5	49.7	39.3	27.7	49.7
1999	27.6	36.7	36.7	45.3	57.1	66.9	75.6	70.7	58.2	50.0	41.8	30.7	49.9
2000	27.3	33.9	40.9	48.0	60.7	67.9	76.5	77.5	63.6	51.1	24.6	21.6	49.5
2001	27.7	23.4	36.9	49.1	57.0	67.5	77.5	71.1	61.7	48.4	38.9	27.4	48.9
POR= 127 YRS	23.3	28.0	36.5	48.6	58.7	68.5	74.6	72.9	63.3	51.0	36.4	27.0	49.1

REFERENCE NOTES:

PAGE 1:
THE TEMPERATURE GRAPH SHOWS NORMAL MAXIMUM AND NORMAL MINIMUM DAILY TEMPERATURES (SOLID CURVES) AND THE ACTUAL DAILY HIGH AND LOW TEMPERATURES (VERTICAL BARS).

PAGE 2 AND 3:
H/C INDICATES HEATING AND COOLING DEGREE DAYS.
RH INDICATES RELATIVE HUMIDITY
W/O INDICATES WEATHER AND OBSTRUCTIONS
S INDICATES SUNSHINE.
PR INDICATES PRESSURE.
CLOUDINESS ON PAGE 3 IS THE SUM OF THE CEILOMETER AND SATELLITE DATA NOT TO EXCEED EIGHT EIGHTHS(OKTAS).

GENERAL:
T INDICATES TRACE PRECIPITATION, AN AMOUNT GREATER THAN ZERO BUT LESS THAN THE LOWEST REPORTABLE VALUE.
+ INDICATES THE VALUE ALSO OCCURS ON EARLIER DATES.
BLANK ENTRIES DENOTE MISSING OR UNREPORTED DATA.
NORMALS ARE 30-YEAR AVERAGES (1961 - 1990).
ASOS INDICATES AUTOMATED SURFACE OBSERVING SYSTEM.
PM INDICATES THE LAST DAY OF THE PREVIOUS MONTH.
POR (PERIOD OF RECORD) BEGINS WITH THE JANUARY DATA MONTH AND IS THE NUMBER OF YEARS USED TO COMPUTE THE MEAN. INDIVIDUAL MONTHS WITHIN THE POR MAY BE MISSING.
WHEN THE POR FOR A NORMAL IS LESS THAN 30 YEARS, THE NORMAL IS PROVISIONAL AND IS BASED ON THE NUMBER OF YEARS INDICATED.
0.* OR * INDICATES THE VALUE OR MEAN-DAYS-WITH IS BETWEEN 0.00 AND 0.05.
CLOUDINESS FOR ASOS STATIONS DIFFERS FROM THE NON-ASOS OBSERVATION TAKEN BY A HUMAN OBSERVER. ASOS STATION CLOUDINESS IS BASED ON TIME-AVERAGED CEILOMETER DATA FOR CLOUDS AT OR BELOW 12,000 FEET AND ON SATELLITE DATA FOR CLOUDS ABOVE 12,000 FEET.
THE NUMBER OF DAYS WITH CLEAR, PARTLY CLOUDY, AND CLOUDY CONDITIONS FOR ASOS STATIONS IS THE SUM OF THE CEILOMETER AND SATELLITE DATA FOR THE SUNRISE TO SUNSET PERIOD.

GENERAL CONTINUED:
CLEAR INDICATES 0 - 2 OKTAS, PARTLY CLOUDY INDICATES 3 - 6 OKTAS, AND CLOUDY INDICATES 7 OR 8 OKTAS. WHEN AT LEAST ONE OF THE ELEMENTS (CEILOMETER OR SATELLITE) IS MISSING, THE DAILY CLOUDINESS IS NOT COMPUTED.
WIND DIRECTION IS RECORDED IN TENS OF DEGREES (2 DIGITS) CLOCKWISE FROM TRUE NORTH. "00" INDICATES CALM. "36" INDICATES TRUE NORTH.
RESULTANT WIND IS THE VECTOR AVERAGE OF THE SPEED AND DIRECTION.
AVERAGE TEMPERATURE IS THE SUM OF THE MEAN DAILY MAXIMUM AND MINIMUM TEMPERATURE DIVIDED BY 2.
SNOWFALL DATA COMPRISE ALL FORMS OF FROZEN PRECIPITATION, INCLUDING HAIL.
A HEATING (COOLING) DEGREE DAY IS THE DIFFERENCE BETWEEN THE AVERAGE DAILY TEMPERATURE AND 65 F.
DRY BULB IS THE TEMPERATURE OF THE AMBIENT AIR.
DEW POINT IS THE TEMPERATURE TO WHICH THE AIR MUST BE COOLED TO ACHIEVE 100 PERCENT RELATIVE HUMIDITY.
WET BULB IS THE TEMPERATURE THE AIR WOULD HAVE IF THE MOISTURE CONTENT WAS INCREASED TO 100 PERCENT RELATIVE HUMIDITY.

ON JULY 1, 1996, THE NATIONAL WEATHER SERVICE BEGAN USING THE "METAR" OBSERVATION CODE THAT WAS ALREADY EMPLOYED BY MOST OTHER NATIONS OF THE WORLD. THE MOST NOTICEABLE DIFFERENCE IN THIS ANNUAL PUBLICATION WILL BE THE CHANGE IN UNITS FROM TENTHS TO EIGHTS(OKTAS) FOR REPORTING THE AMOUNT OF SKY COVER.

HEATING DEGREE DAYS (base 65 F) 2001 NORTH PLATTE, NE (LBF)

YEAR	JUL	AUG	SEP	OCT	NOV	DEC	JAN	FEB	MAR	APR	MAY	JUN	TOTAL
1972-73	24	15	169	567	1047	1479	1343	967	756	567	283	34	7251
1973-74	10	0	219	407	892	1290	1518	889	765	445	190	62	6687
1974-75	0	40	256	373	900	1262	1216	1160	1032	558	247	62	7106
1975-76	6	0	228	437	1035	1191	1460	929	956	521	363	70	7196
1976-77	0	13	178	608	1028	1133	1493	920	858	395	81	2	6709
1977-78	2	34	96	458	869	1236	1662	1400	924	491	275	71	7518
1978-79	5	24	99	488	982	1560	1828	1335	862	491	259	56	7989
1979-80	4	11	52	341	975	957	1233	1102	909	439	166	10	6199
1980-81	0	6	107	491	790	1019	1089	1000	762	283	318	26	5891
1981-82	9	4	101	492	749	1179	1479	1030	885	601	239	111	6879
1982-83	0	18	160	484	946	1138	1167	833	854	672	343	90	6705
1983-84	2	0	128	419	840	1780	1379	915	953	647	236	33	7332
1984-85	0	0	247	519	829	1312	1440	1168	752	393	156	83	6899
1985-86	0	23	252	502	1205	1416	1029	1060	634	479	219	2	6821
1986-87	0	14	98	446	878	1074	1093	810	868	420	102	15	5818
1987-88	13	36	139	551	796	1152	1501	1109	839	490	170	3	6799
1988-89	0	13	128	498	803	1067	1072	1316	902	430	211	67	6507
1989-90	2	7	180	437	815	1374	1061	948	771	502	259	15	6371
1990-91	15	1	84	457	797	1331	1290	762	754	466	149	5	6111
1991-92	3	1	148	508	977	971	1010	797	714	436	219	45	5829
1992-93	18	60	113	466	970	1310	1436	1311	862	578	204	73	7401
1993-94	2	24	218	513	965	1066	1263	1151	724	517	131	5	6579
1994-95	3	8	90	395	820	1106	1180	863	840	631	377	67	6380
1995-96	17	2	167	517	786	1116	1359	1004	1037	527	283	40	6855
1996-97	3	1	214	503	1021	1233	1284	1001	782	673	306	14	7035
1997-98	10	17	113	452	926	1133	1202	841	998	551	195	114	6552
1998-99	2	0	39	467	764	1149	1151	785	815	586	248	54	6060
1999-00	8	9	227	458	690	1057	1164	894	738	503	169	49	5966
2000-01	0		142	424	1204	1337	1152	1156	868	473	255	87	
2001-	0	5	146	511	773	1160							

COOLING DEGREE DAYS (base 65 F) 2001 NORTH PLATTE, NE (LBF)

YEAR	JAN	FEB	MAR	APR	MAY	JUN	JUL	AUG	SEP	OCT	NOV	DEC	ANNUAL
1972	0	0	0	0	43	137	199	202	64	0	0	0	645
1973	0	0	0	0	11	121	260	278	29	3	0	0	702
1974	0	0	0	5	24	117	394	115	40	3	0	0	698
1975	0	0	0	10	11	89	311	260	39	7	0	0	727
1976	0	0	0	0	5	62	247	192	27	0	0	0	533
1977	0	0	0	2	26	174	305	149	64	0	0	0	720
1978	0	0	0	1	21	174	307	218	117	5	0	0	843
1979	0	0	0	0	27	156	294	262	123	0	0	0	862
1980	0	0	0	10	27	243	411	289	74	0	0	0	1054
1981	0	0	0	10	9	141	288	168	47	1	0	0	664
1982	0	0	0	0	8	70	314	276	68	0	0	0	736
1983	0	0	0	0	8	103	331	412	128	1	0	0	983
1984	0	0	0	0	27	129	281	331	55	2	0	0	825
1985	0	0	0	14	32	100	326	189	100	0	0	0	761
1986	0	0	0	3	11	201	334	217	43	0	0	0	809
1987	0	0	0	16	48	176	352	208	35	0	0	0	835
1988	0	0	0	1	41	293	301	282	46	0	0	0	964
1989	0	0	0	21	35	99	295	220	67	2	0	0	739
1990	0	0	0	15	10	205	291	289	165	5	0	0	980
1991	0	0	0	3	59	194	317	276	112	2	0	0	963
1992	0	0	0	13	31	68	135	116	60	1	0	0	424
1993	0	0	0	0	15	88	212	194	24	6	0	0	539
1994	0	0	0	10	63	213	183	251	111	0	0	0	831
1995	0	0	0	0	0	125	298	449	103	4	0	0	979
1996	0	0	0	0	33	149	204	165	44	3	0	0	598
1997	0	0	0	0	5	156	295	206	96	29	0	0	787
1998	0	0	0	0	37	94	326	237	150	0	0	0	844
1999	0	0	0	0	9	119	343	192	28	0	0	0	691
2000	0	0	0	0	42	140	362	393	107	0	0	0	1044
2001	0	0	0	4	13	168	394	201	52	3	0	0	835

SNOWFALL (inches) 2001 NORTH PLATTE, NE (LBF)

YEAR	JUL	AUG	SEP	OCT	NOV	DEC	JAN	FEB	MAR	APR	MAY	JUN	TOTAL
1972-73	0.0	0.0	0.0	0.8	8.0	8.7	3.8	0.4	1.6	1.7	0.0	0.0	25.0
1973-74	0.0	0.0	0.0	T	4.3	14.1	4.2	1.0	2.3	0.7	0.0	0.0	26.6
1974-75	0.0	0.0	0.0	0.0	T	4.5	2.2	2.6	3.9	3.0	0.0	0.0	16.2
1975-76	0.0	0.0	0.0	0.8	10.9	1.2	17.1	1.6	5.2	T	0.0	0.0	36.8
1976-77	0.0	0.0	0.0	1.0	0.6	T	2.5	2.2	9.9	8.1	0.0	0.0	24.3
1977-78	0.0	0.0	0.0	T	0.3	6.4	6.2	20.6	2.0	T	T	0.0	35.5
1978-79	0.0	0.0	0.0	T	7.6	10.3	6.3	0.3	5.9	2.4	0.0	0.0	32.7
1979-80	0.0	0.0	0.0	2.9	17.5	2.3	5.2	9.1	21.9	7.4	0.0	0.0	66.3
1980-81	0.0	0.0	0.0	0.6	1.2	T	0.7	0.5	0.5	0.4	0.0	0.0	3.9
1981-82	0.0	0.0	0.0	T	5.3	6.4	4.4	1.7	6.4	0.9	0.0	0.0	25.1
1982-83	0.0	0.0	0.0	1.0	2.0	9.7	1.6	0.1	5.9	5.4	0.0	0.0	25.7
1983-84	0.0	0.0	T	0.0	12.1	7.3	5.3	9.6	8.8	14.5	0.2	0.0	57.8
1984-85	0.0	0.0	T	0.3	0.9	8.7	8.5	0.8	2.1	0.0	0.0	0.0	21.3
1985-86	0.0	0.0	3.1	T	13.0	8.1	0.2	8.7	3.8	T	0.0	0.0	36.9
1986-87	0.0	0.0	0.0	2.0	1.0	2.6	1.6	9.2	7.2	0.1	0.0	0.0	23.7
1987-88	0.0	0.0	0.0	1.3	8.2	7.2	12.6	0.6	3.8	2.1	0.0	0.0	35.8
1988-89	0.0	0.0	0.0	T	2.5	T	6.1	10.6	4.3	0.3	T	0.0	23.8
1989-90	T	0.0	T	T	T	2.2	5.9	1.9	5.3	0.3	T	T	15.6
1990-91	0.0	0.0	0.0	2.0	9.7	0.8	4.1	1.1	5.6	0.3	T	0.0	22.7
1991-92	0.0	0.0	0.0	7.3	1.6	1.8	5.4	1.2	5.1	T	0.0	0.0	22.7
1992-93	T	T	0.0	T	2.2	3.9	10.1	17.2	2.7	2.0	T	T	38.1
1993-94	T	0.0	T	T	3.9	3.3	10.6	6.0	0.4	11.5	T	T	35.7
1994-95	T	0.0	0.0	0.0	4.6	9.0	0.1	2.7	7.4	9.6	0.0	T	33.4
1995-96	T	0.0	0.7	4.1	0.8	0.5	4.9	T	3.4	4.0	0.0	0.0	18.4
1996-97	T	0.0	0.0	T	5.2	1.0	0.9	12.1	0.5	4.4	0.0		
1997-98	0.0	T	0.0	5.7	1.6	3.2	4.4	4.2	8.8	2.7	T	T	27.9
1998-99	0.0	T	0.0	T	5.7	0.7	4.5	5.2	2.7	1.2	0.0	T	20.0
1999-00	T	T	0.0	0.0	1.4	1.1	8.1	3.6	0.8	T	T	T	15.0
2000-01	T	T	0.0	2.8	0.0	10.6	1.0	7.4	5.3	4.2	6.3	T	37.6
2001-	T	T	0.0	0.0	1.6	3.4							
POR= 49 YRS	T	T	0.2	1.5	3.8	4.4	5.4	5.1	6.3	3.0	0.2	T	29.9

2001
OMAHA (EPPLEY AIRFIELD),
NEBRASKA (OMA)

Omaha, Nebraska, is situated on the west bank of the Missouri River. The river level at Omaha is normally about 965 feet above sea level and the rolling hills in and around Omaha rise to about 1,300 feet above sea level. The climate is typically continental with relatively warm summers and cold, dry winters. It is situated midway between two distinctive climatic zones, the humid east and the dry west. Fluctuations between these two zones produce weather conditions for periods that are characteristic of either zone, or combinations of both. Omaha is also affected by most low pressure systems that cross the country. This causes periodic and rapid changes in weather, especially during the winter months.

Most of the precipitation in Omaha falls during sharp showers or thunderstorms, and these occur mostly during the growing season from April to September. Of the total precipitation, about 75 percent falls during this six-month period. The rain occurs mostly as evening or nighttime showers and thunderstorms. Although winters are relatively cold, precipitation is light, with only 10 percent of the total annual precipitation falling during the winter months.

Sunshine is fairly abundant, ranging around 50 percent of the possible in the winter to 75 percent of the possible in the summer.

NORMALS, MEANS, AND EXTREMES
OMAHA, NE (OMA)

LATITUDE:	LONGITUDE:	ELEVATION (FT):		TIME ZONE:	WBAN: 14942
41 18' 37" N	95 53' 57" W	GRND: 1025	BARO: 1028	CENTRAL (UTC + 6)	

	ELEMENT	POR	JAN	FEB	MAR	APR	MAY	JUN	JUL	AUG	SEP	OCT	NOV	DEC	YEAR
TEMPERATURE °F	NORMAL DAILY MAXIMUM	30	31.3	37.1	49.4	63.8	74.0	83.7	87.9	85.2	76.5	65.6	49.3	34.6	61.5
	MEAN DAILY MAXIMUM	54	30.8	36.7	48.5	63.2	74.2	83.5	87.7	85.3	77.0	66.1	49.1	36.0	61.5
	HIGHEST DAILY MAXIMUM	64	69	78	89	97	99	105	114	110	104	96	83	72	114
	YEAR OF OCCURRENCE		1944	1972	1986	1989	1939	1953	1936	1936	1939	1938	1999	1939	JUL 1936
	MEAN OF EXTREME MAXS.	54	52.2	59.4	74.9	86.0	89.8	96.6	99.1	97.2	92.7	84.8	70.3	58.5	80.1
	NORMAL DAILY MINIMUM	30	10.9	16.7	27.7	39.9	50.9	60.4	65.9	62.9	53.6	41.2	28.7	15.6	39.5
	MEAN DAILY MINIMUM	54	11.5	17.3	27.2	39.8	51.4	61.2	66.4	64.0	53.9	42.1	28.8	17.2	40.1
	LOWEST DAILY MINIMUM	64	-23	-21	-16	5	27	38	44	43	25	13	-9	-23	-23
	YEAR OF OCCURRENCE		1982	1981	1948	1975	1980	1983	1972	1967	1984	1972	1964	1989	DEC 1989
	MEAN OF EXTREME MINS.	54	-10.1	-3.8	7.0	23.9	36.4	46.9	54.8	51.8	37.1	25.1	10.9	-3.0	23.1
	NORMAL DRY BULB	30	21.1	26.9	38.6	51.9	62.4	72.1	76.9	74.1	65.1	53.4	39.0	25.1	50.5
	MEAN DRY BULB	54	21.1	27.1	37.7	51.6	62.8	72.4	77.0	74.7	65.5	54.1	39.0	26.7	50.8
	MEAN WET BULB	16	20.7	24.6	33.2	45.1	56.0	64.8	65.6	64.0	58.4	47.0	34.3	22.7	44.7
	MEAN DEW POINT	16	15.5	19.3	27.9	37.7	50.2	60.3	62.5	60.9	54.0	40.8	28.7	18.2	39.7
	NORMAL NO. DAYS WITH:														
	MAXIMUM 90	30	0.0	0.0	0.0	0.5	1.5	7.3	13.6	9.2	2.9	0.3	0.0	0.0	35.3
	MAXIMUM 32	30	15.2	10.6	3.5	0.1	0.0	0.0	0.0	0.0	0.0	0.0	2.2	12.4	44.0
	MINIMUM 32	30	30.2	26.0	21.0	6.7	0.4	0.0	0.0	0.0	0.4	5.8	20.2	29.3	140.0
	MINIMUM 0	30	7.7	3.8	0.4	0.0	0.0	0.0	0.0	0.0	0.0	*	0.2	4.0	16.1
H/C	NORMAL HEATING DEG. DAYS	30	1361	1067	818	398	164	14	0	9	80	372	780	1237	6300
	NORMAL COOLING DEG. DAYS	30	0	0	0	5	84	227	369	292	83	12	0	0	1072
RH	NORMAL (PERCENT)	30	71	71	66	61	64	66	68	71	72	67	71	74	68
	HOUR 00 LST	30	75	77	73	68	72	75	78	80	81	76	77	78	76
	HOUR 06 LST	30	78	79	78	76	79	82	84	86	87	82	81	80	81
	HOUR 12 LST	30	65	63	58	51	54	55	57	59	59	54	62	67	59
	HOUR 18 LST	30	66	64	55	47	50	51	55	58	60	56	65	70	58
S	PERCENT POSSIBLE SUNSHINE														
W/O	MEAN NO. DAYS WITH:														
	HEAVY FOG(VISBY 1/4 MI)	65	2.0	2.0	1.4	0.7	0.9	0.4	0.5	1.6	1.4	1.5	1.5	2.2	16.1
	THUNDERSTORMS	65	0.1	0.4	1.6	4.0	7.4	9.4	8.4	7.6	5.1	2.3	0.8	0.2	47.3
CLOUDINESS	MEAN:														
	SUNRISE-SUNSET (OKTAS)	49	4.9	5.0	5.3	5.1	5.0	4.6	3.8	3.8	3.8	3.8	4.8	5.0	4.6
	MIDNIGHT-MIDNIGHT (OKTAS)	22	4.5	4.5	4.8	4.7	4.6	4.4	3.8	3.7	3.8	3.8	4.6	4.7	4.3
	MEAN NO. DAYS WITH:														
	CLEAR	49	8.8	7.3	7.1	7.3	7.4	8.0	11.2	12.2	12.3	12.9	8.6	8.0	111.1
	PARTLY CLOUDY	49	7.9	7.4	8.0	8.5	9.6	11.4	12.1	10.1	7.6	8.1	7.4	7.5	105.6
	CLOUDY	49	14.3	13.5	15.9	14.2	14.1	10.6	7.7	8.7	10.2	10.1	13.9	15.4	148.6
PR	MEAN STATION PRESSURE(IN)	16	29.03	29.03	28.94	28.89	28.86	28.86	28.92	28.94	28.96	28.99	28.99	29.06	28.96
	MEAN SEA-LEVEL PRES. (IN)	12	30.12	30.14	30.04	29.94	29.91	29.90	29.93	29.98	30.00	30.05	30.04	30.15	30.02
WINDS	MEAN SPEED (MPH)	46	10.8	10.9	12.1	12.4	10.8	9.9	8.8	8.7	9.3	9.8	10.9	10.8	10.4
	PREVAIL.DIR(TENS OF DEGS)	31	33	35	35	16	16	16	16	16	16	16	16	16	16
	MAXIMUM 2-MINUTE:														
	SPEED (MPH)	5	45	39	40	51	57	37	56	39	36	48	45	39	57
	DIR. (TENS OF DEGS)		33	33	19	20	16	36	03	01	19	11	28	33	16
	YEAR OF OCCURRENCE		1997	2000	1999	2001	1998	2000	2001	1999	2000	1998	1998	2000	MAY 1998
	MAXIMUM 5-SECOND:														
	SPEED (MPH)	5	53	47	47	64	86	46	70	47	41	57	55	48	86
	DIR. (TENS OF DEGS)		33	30	30	19	17	16	02	01	19	11	28	33	17
	YEAR OF OCCURRENCE		1997	1999	2000	2001	1998	1998	2001	1999	2000	1998	1998	1997	MAY 1998
PRECIPITATION	NORMAL (IN)	30	0.74	0.77	2.04	2.66	4.52	3.87	3.51	3.24	3.72	2.28	1.49	1.02	29.86
	MAXIMUM MONTHLY (IN)	64	3.70	2.97	5.96	8.48	10.33	10.81	10.34	12.26	13.75	5.25	4.70	5.42	13.75
	YEAR OF OCCURRENCE		1949	1965	1973	1999	1959	1947	1993	1999	1965	1997	1983	1984	SEP 1965
	MINIMUM MONTHLY (IN)	64	T	0.09	0.12	0.23	0.56	1.03	0.39	0.61	0.41	T	0.03	T	T
	YEAR OF OCCURRENCE		1986	1981	1956	1936	1948	1972	1983	1984	1953	1952	1976	1943	JAN 1986
	MAXIMUM IN 24 HOURS (IN)	58	1.52	2.24	1.45	2.82	4.16	4.27	3.37	10.48	6.47	3.13	2.53	3.03	10.48
	YEAR OF OCCURRENCE		1967	1954	1990	1999	1987	1994	1958	1999	1965	1968	1948	1984	AUG 1999
	NORMAL NO. DAYS WITH:														
	PRECIPITATION 0.01	30	6.1	5.9	8.9	9.7	11.7	10.2	9.3	8.6	9.1	6.9	6.2	7.1	99.7
	PRECIPITATION 1.00	30	0.1	0.1	0.5	0.6	1.1	1.2	1.0	1.0	1.1	0.5	0.4	0.1	7.7
SNOWFALL	NORMAL (IN)	30	6.3	6.3	5.9	0.7	0.*	0.0	0.0	0.0	T	0.3	2.7	6.2	28.4
	MAXIMUM MONTHLY (IN)	63	25.7	25.4	27.2	10.0	2.0	T	T	0.0	T	7.2	12.0	19.9	27.2
	YEAR OF OCCURRENCE		1936	1965	1948	1992	1945	1994	1995		1985	1941	1957	1969	MAR 1948
	MAXIMUM IN 24 HOURS (IN)	56	13.1	18.3	13.0	9.9	2.0	T	T	0.0	T	7.2	8.7	10.2	18.3
	YEAR OF OCCURRENCE		1949	1965	1948	1992	1945	1994	1995		1985	1941	1957	1969	FEB 1965
	MAXIMUM SNOW DEPTH (IN)	52	17	18	27	8	0	0	0	0	0	5	9	17	27
	YEAR OF OCCURRENCE		1984	1965	1960	1992						1997	1957	1983	MAR 1960
	NORMAL NO. DAYS WITH:														
	SNOWFALL 1.0	30	2.0	1.9	1.9	0.3	0.*	0.0	0.0	0.0	0.0	0.2	1.0	2.1	9.4

```
PRECIPITATION (inches)  2001   OMAHA (EPPLEY AIRFIELD), NE (OMA)
```

YEAR	JAN	FEB	MAR	APR	MAY	JUN	JUL	AUG	SEP	OCT	NOV	DEC	ANNUAL
1972	0.38	0.36	1.14	4.63	5.13	1.03	7.28	2.60	4.93	3.21	3.25	1.62	35.56
1973	1.44	0.87	5.96	3.60	4.94	1.56	4.98	1.27	8.04	2.60	1.43	1.65	38.34
1974	0.63	0.16	0.80	1.72	2.65	1.79	0.79	4.17	2.54	2.74	1.41	0.81	20.21
1975	2.01	1.06	1.78	3.37	3.72	4.30	0.46	1.80	2.16	0.01	2.85	0.46	23.98
1976	0.11	1.39	2.13	2.98	3.55	2.84	1.52	0.62	1.64	1.35	0.03	0.21	18.37
1977	0.93	0.38	3.72	3.28	6.05	2.40	4.69	8.63	5.05	4.38	2.90	0.35	42.76
1978	0.15	0.76	1.04	4.61	5.05	2.19	5.89	3.87	6.01	0.96	1.09	0.50	32.12
1979	1.11	0.30	4.59	2.58	2.84			2.07	3.36	3.12	1.23	0.14	
1980	0.93	0.52	1.40	1.72	2.50	8.99	3.63	6.98	0.82	2.70	0.11	0.04	30.34
1981	0.20	0.09	0.88	1.33	4.13	2.14	1.87	4.80	1.51	1.92	2.60	0.86	22.33
1982	1.83	0.26	1.90	1.22	9.92	4.16	2.46	3.21	2.27	1.10	1.81	1.17	31.31
1983	0.86	0.68	3.65	1.00	2.81	6.52	0.39	1.24	2.45	2.16	4.70	0.63	27.09
1984	0.38	0.62	2.32	4.77	4.92	5.56	1.58	0.61	2.55	3.87	0.52	5.42	33.12
1985	0.56	1.88	1.36	3.16	2.46	1.73	3.27	1.50	2.71	1.36	0.85	0.37	21.21
1986	T	1.00	2.51	4.96	4.88	2.37	2.77	3.86	8.11	4.86	0.99	0.89	37.20
1987	0.08	0.55	4.14	2.24	8.64	3.29	6.72	10.16	1.56	1.33	1.60	1.01	41.32
1988	0.42	0.18	0.14	1.57	4.68	1.60	2.68	1.78	2.63	0.14	2.55	0.95	19.32
1989	1.10	0.86	0.40	1.80	0.83	5.05	3.06	1.80	6.46	1.55	0.15	0.74	23.80
1990	0.59	0.34	4.01	0.36	5.08	3.88	6.36	0.81	0.81	1.71	1.15	1.18	26.28
1991	1.08	0.26	2.85	4.46	4.07	7.79	2.96	3.67	1.37	3.76	3.51	1.75	37.53
1992	1.41	1.18	3.08	3.19	2.27	1.44	7.31	1.57	6.86	2.22	3.01	1.15	34.69
1993	1.42	0.93	2.67	2.26	4.90	8.03	10.34	7.53	2.29	1.18	0.66	0.51	42.72
1994	0.50	1.01	0.15	1.46	1.73	8.54	3.60	1.97	3.32	1.37	1.64	1.21	26.50
1995	0.80	0.47	2.50	4.26	7.07	1.28	3.14	2.52	2.75				
1996			0.83	2.36	7.57	2.96	2.39	2.19	4.90	1.64	2.46	0.32	
1997	0.29	0.69	1.08	3.66	1.54	4.51	4.69	1.33	4.30	5.25	2.30	0.57	30.21
1998	1.13	1.27	4.13	3.53	4.71	8.23	7.77	3.85	0.85	2.65	1.48	0.13	39.73
1999	0.59	1.40	1.31	8.48	4.50	3.75	3.07	12.26	1.64	0.03	1.11	0.57	38.71
2000	0.17	1.95	0.81	2.85	2.69	5.52	5.03	1.30	0.64	1.93	3.27	0.95	27.11
2001	1.61	1.53	1.38	2.37	8.78	2.29	2.06	1.95	2.39	2.10	1.55	0.67	28.68
POR= 129 YRS	0.75	0.89	1.56	2.72	3.96	4.50	3.73	3.42	3.08	2.02	1.25	0.90	28.78

```
AVERAGE TEMPERATURE ( F) 2001      OMAHA (EPPLEY AIRFIELD), NE (OMA)
```

YEAR	JAN	FEB	MAR	APR	MAY	JUN	JUL	AUG	SEP	OCT	NOV	DEC	ANNUAL
1972	20.0	25.4	40.8	51.2	62.4	72.6	74.6	73.6	65.4	49.1	37.1	21.1	49.4
1973	22.6	28.6	44.3	50.4	59.4	73.3	75.5	77.6	64.2	57.3	39.8	22.7	51.3
1974	18.8	30.6	42.1	52.6	61.9	69.8	82.2	70.5	59.7	55.3	40.7	28.9	51.1
1975	22.5	22.4	31.7	49.4	66.2	72.7	78.9	79.7	62.6	58.2	41.6	30.9	51.4
1976	25.5	37.5	40.4	56.8	60.2	72.7	79.0	76.3	67.4	48.5	33.3	24.1	51.8
1977	13.3	33.2	41.6	59.0	70.2	75.1	80.6	72.5	67.4	53.1	39.7	26.2	52.7
1978	12.0	15.7	35.7	52.6	61.7	73.7	77.3	75.5	71.1	53.7	39.6	24.4	49.4
1979	10.7	17.3	39.8	50.0	61.8	72.7	76.0	75.7	66.7		35.8	30.3	
1980	22.4	20.3	32.9	50.7	61.6	73.1	79.6	76.6	64.7	49.0	40.3	26.1	49.8
1981	24.1	28.4	40.8	57.4	58.6	72.6	76.6	71.0	64.8	50.2	40.6	22.9	50.7
1982	9.4	22.6	34.8	47.7	62.9	65.5	77.1	72.6	64.7	55.0	37.1	28.3	48.1
1983	24.9	30.1	37.3	43.5	56.5	69.6	79.4	81.5	67.0	52.2	38.5	7.3	49.0
1984	19.6	33.2	30.3	46.6	57.7	71.8	76.5	76.1	61.9	52.2	39.4	27.1	49.3
1985	19.1	23.6	43.4	54.9	63.4	67.2	74.1	69.4	62.1	52.7	28.5	16.4	47.9
1986	29.4	22.8	42.8	51.9	61.3	73.9	77.9	70.1	67.9	53.7	34.4	29.4	51.3
1987	28.6	36.8	42.8	55.3	67.3	74.2	77.8	70.9	64.8	48.3	43.4	30.9	53.4
1988	21.1	23.9	40.7	50.5	67.3	76.3	76.4	77.3	66.2	49.1	40.0	29.5	51.5
1989	32.4	16.0	37.9	54.6	62.4	69.4	77.4	74.4	63.0	53.9	36.2	17.7	49.6
1990	33.5	31.3	42.7	50.8	58.5	73.5	74.8	75.3	69.0	53.3	42.9	21.2	52.2
1991	16.3	34.2	42.5	54.2	67.3	74.6	75.8	75.8	74.6	52.1	30.6	31.8	51.6
1992	32.6	36.7	43.6	50.0	61.8	69.4	71.1	68.4	64.0	53.5	35.4	27.9	51.2
1993	19.5	22.2	35.6	47.7	61.4	70.0	75.1	75.2	59.9	50.9	35.9	48.5	49.4
1994	18.1	22.9	42.1	51.3	64.5	74.1	73.2	72.6	67.1	55.3	41.4	28.7	50.9
1995	23.0	31.5	39.3	47.9	58.1	71.8	78.7	79.6					
1996			33.0	49.2	59.4	72.8	73.4	72.8	63.1	54.0	32.9	22.2	
1997	19.0	29.5	41.6	45.7	58.0	73.4	77.2	73.4	66.5	54.0	35.3	29.9	50.3
1998	25.8	36.0	32.5	52.0	66.5	69.6	76.8	75.6	71.6	55.8	43.6	30.9	53.1
1999	22.6	35.5	39.5	51.6	62.3	70.8	80.4	72.8	63.8	53.3	47.0	31.0	52.6
2000	27.1	36.2	44.6	51.9	66.0	70.5	75.1	77.1	67.8	57.5	33.4	15.6	51.9
2001	26.6	21.4	35.0	55.5	63.5	71.6	79.0	75.7	65.0	53.5	49.4	32.0	52.4
POR= 128 YRS	21.9	26.8	37.8	51.6	62.6	72.2	77.3	75.0	66.2	54.5	39.2	27.0	51.0

REFERENCE NOTES:

PAGE 1:
THE TEMPERATURE GRAPH SHOWS NORMAL MAXIMUM AND NORMAL
MINIMUM DAILY TEMPERATURES (SOLID CURVES) AND THE
ACTUAL DAILY HIGH AND LOW TEMPERATURES (VERTICAL BARS).

PAGE 2 AND 3:
H/C INDICATES HEATING AND COOLING DEGREE DAYS.
RH INDICATES RELATIVE HUMIDITY
W/O INDICATES WEATHER AND OBSTRUCTIONS
S INDICATES SUNSHINE.
PR INDICATES PRESSURE.
CLOUDINESS ON PAGE 3 IS THE SUM OF THE CEILOMETER AND
SATELLITE DATA NOT TO EXCEED EIGHT EIGHTHS(OKTAS).

GENERAL:
T INDICATES TRACE PRECIPITATION, AN AMOUNT GREATER
THAN ZERO BUT LESS THAN THE LOWEST REPORTABLE VALUE.
+ INDICATES THE VALUE ALSO OCCURS ON EARLIER DATES.
BLANK ENTRIES DENOTE MISSING OR UNREPORTED DATA.
NORMALS ARE 30-YEAR AVERAGES (1961 - 1990).
ASOS INDICATES AUTOMATED SURFACE OBSERVING SYSTEM.
PM INDICATES THE LAST DAY OF THE PREVIOUS MONTH.
POR (PERIOD OF RECORD) BEGINS WITH THE JANUARY DATA
MONTH AND IS THE NUMBER OF YEARS USED TO COMPUTE
THE MEAN.   INDIVIDUAL MONTHS WITHIN THE POR MAY
BE MISSING.
WHEN THE POR FOR A NORMAL IS LESS THAN 30 YEARS,
THE NORMAL IS PROVISIONAL AND IS BASED ON THE NUMBER
OF YEARS INDICATED.
0.* OR * INDICATES THE VALUE OR MEAN-DAYS-WITH
IS BETWEEN 0.00 AND 0.05.
CLOUDINESS FOR ASOS STATIONS DIFFERS FROM THE NON-ASOS
OBSERVATION TAKEN BY A HUMAN OBSERVER. ASOS STATION
CLOUDINESS IS BASED ON TIME-AVERAGED CEILOMETER DATA
FOR CLOUDS AT OR BELOW 12,000 FEET AND ON SATELLITE
DATA FOR CLOUDS ABOVE 12,000 FEET.
THE NUMBER OF DAYS WITH CLEAR, PARTLY CLOUDY, AND
CLOUDY CONDITIONS FOR ASOS STATIONS IS THE SUM
OF THE CEILOMETER AND SATELLITE DATA FOR THE
SUNRISE TO SUNSET PERIOD.

GENERAL CONTINUED:
CLEAR INDICATES 0 - 2 OKTAS, PARTLY CLOUDY INDICATES
3 - 6 OKTAS, AND CLOUDY INDICATES 7 OR 8 OKTAS.
WHEN AT LEAST ONE OF THE ELEMENTS (CEILOMETER OR
SATELLITE) IS MISSING, THE DAILY CLOUDINESS IS
NOT COMPUTED.
WIND DIRECTION IS RECORDED IN TENS OF DEGREES (2 DIGITS)
CLOCKWISE FROM TRUE NORTH. "00" INDICATES CALM. "36"
INDICATES TRUE NORTH.
RESULTANT WIND IS THE VECTOR AVERAGE OF THE SPEED AND
DIRECTION.
AVERAGE TEMPERATURE IS THE SUM OF THE MEAN DAILY MAXIMUM
AND MINIMUM TEMPERATURE DIVIDED BY 2.
SNOWFALL DATA COMPRISE ALL FORMS OF FROZEN
PRECIPITATION, INCLUDING HAIL.
A HEATING (COOLING) DEGREE DAY IS THE DIFFERENCE BETWEEN
THE AVERAGE DAILY TEMPERATURE AND 65 F.
DRY BULB IS THE TEMPERATURE OF THE AMBIENT AIR.
DEW POINT IS THE TEMPERATURE TO WHICH THE AIR MUST BE
COOLED TO ACHIEVE 100 PERCENT RELATIVE HUMIDITY.
WET BULB IS THE TEMPERATURE THE AIR WOULD HAVE IF THE
MOISTURE CONTENT WAS INCREASED TO 100 PERCENT RELATIVE
HUMIDITY.

ON JULY 1, 1996, THE NATIONAL WEATHER SERVICE BEGAN USING
THE "METAR" OBSERVATION CODE THAT WAS ALREADY EMPLOYED
BY MOST OTHER NATIONS OF THE WORLD.   THE MOST NOTICEABLE
DIFFERENCE IN THIS ANNUAL PUBLICATION WILL BE THE CHANGE
IN UNITS FROM TENTHS TO EIGHTS(OKTAS) FOR REPORTING THE
AMOUNT OF SKY COVER.

HEATING DEGREE DAYS (base 65 F) 2001 OMAHA (EPPLEY AIRFIELD), NE (OMA)

YEAR	JUL	AUG	SEP	OCT	NOV	DEC	JAN	FEB	MAR	APR	MAY	JUN	TOTAL
1972-73	6	7	106	488	831	1357	1307	1014	636	437	191	0	6380
1973-74	0	0	90	254	750	1302	1427	955	702	379	140	31	6030
1974-75	0	15	191	300	726	1115	1311	1189	1024	469	72	11	6423
1975-76	0	0	141	251	695	1051	1219	791	757	261	177	4	5347
1976-77	0	0	61	522	947	1265	1598	883	579	219	10	1	6085
1977-78	0	1	28	361	754	1196	1637	1375	910	372	160	17	6811
1978-79	0	0	39	350	754	1255	1676	1333	775	451	156	12	6801
1979-80	1	6	65		867	1070	1318	1290	987	440	158	4	
1980-81	0	3	108	491	735	1198	1259	1018	743	241	221	0	6017
1981-82	7	3	85	452	723	1299	1721	1183	930	518	102	56	7079
1982-83	0	13	115	315	829	1131	1240	971	854	638	278	37	6421
1983-84	0	0	102	405	789	1786	1401	916	1071	552	243	7	7272
1984-85	0	3	184	391	766	1166	1416	1153	666	325	88	45	6203
1985-86	0	13	217	378	1089	1501	1095	1176	689	389	134	1	6682
1986-87	0	15	40	338	913	1096	1122	784	685	322	67	7	5389
1987-88	1	32	67	512	639	1048	1353	1185	748	433	29	6	6053
1988-89	1	7	56	488	744	1095	1002	1368	844	380	143	23	6151
1989-90	0	7	140	356	855	1460	973	935	684	460	206	15	6091
1990-91	4	1	75	371	662	1350	1506	859	695	338	108	0	5969
1991-92	0	0	123	402	1027	1022	999	816	656	449	154	11	5659
1992-93	2	26	114	359	881	1141	1403	1192	904	514	140	34	6710
1993-94	0	1	171	448	895	1097	1448	1174	702	432	120	8	6496
1994-95	0	5	83	302	698	1116	1296	931	791	507	213	21	5963
1995-96	1	0							981	471	211	22	
1996-97	0	0	130	347	956	1321	1417	986	719	572	226	0	6674
1997-98	1	7	57	399	884	1082	1207	806	1000	390	52	58	5943
1998-99	0	0	22	287	634	1050	1307	818	781	396	122	27	5444
1999-00	0	2	112	366	535	1047	1168	828	625	391	72	16	5162
2000-01	0	1	87	242	940	1526	1185	1214	925	301	120	20	6561
2001-	0	0	87	360	462	1014							

COOLING DEGREE DAYS (base 65 F) 2001 OMAHA (EPPLEY AIRFIELD), NE (OMA)

YEAR	JAN	FEB	MAR	APR	MAY	JUN	JUL	AUG	SEP	OCT	NOV	DEC	ANNUAL
1972	0	0	0	3	74	249	314	280	124	1	0	0	1045
1973	0	0	0	3	23	257	332	395	74	22	0	0	1106
1974	0	0	0	11	52	182	540	193	39	4	0	0	1021
1975	0	0	0	7	115	242	441	464	76	44	0	0	1389
1976	0	0	0	21	34	240	440	358	139	17	0	0	1249
1977	0	0	0	45	179	310	489	236	105	0	0	0	1364
1978	0	0	7	5	67	287	386	333	231	5	0	0	1321
1979	0	0	0	8	64	249	344	345	122		0	0	
1980	0	0	0	15	61	254	459	368	107	0	0	0	1264
1981	0	0	0	24	29	235	372	196	85	0	0	0	941
1982	0	0	0	5	43	78	383	252	113	12	0	0	886
1983	0	0	0	0	20	183	453	519	167	17	0	0	1359
1984	0	0	0	6	22	220	320	366	96	4	0	0	1034
1985	0	0	0	30	44	116	290	156	137	1	0	0	774
1986	0	0	10	5	26	276	408	181	133	0	0	0	1039
1987	0	0	0	39	145	292	407	221	69	2	1	0	1176
1988	0	0	0	5	109	351	364	394	99	3	0	0	1325
1989	0	0	10	77	68	159	395	306	89	19	0	0	1123
1990	0	0	0	41	9	277	316	327	199	12	4	0	1185
1991	0	0	5	21	184	295	345	295	161	9	0	0	1315
1992	0	0	0	7	63	150	198	136	90	11	0	0	655
1993	0	0	0	0	35	188	322	324	21	19	0	0	909
1994	0	0	0	28	108	287	262	248	151	9	0	0	1093
1995	0	0	0	0	8	230	431	458					
1996			0	8	46	264	266	250	79	14	0		
1997	0	0	0	0	18	260	388	273	110	62	0	0	1111
1998	0	0	0	1	6	108	205	373	336	227	9	0	1265
1999	0	0	0	0	43	208	484	248	84	12	0	0	1079
2000	0	0	0	2	110	188	317	385	179	18	0	0	1199
2001	0	0	0	27	80	225	439	342	93	10	1	0	1217

SNOWFALL (inches) 2001 OMAHA (EPPLEY AIRFIELD), NE (OMA)

YEAR	JUL	AUG	SEP	OCT	NOV	DEC	JAN	FEB	MAR	APR	MAY	JUN	TOTAL
1972-73	0.0	0.0	0.0	T	9.7	5.9	15.8	4.6	T	2.3	0.0	0.0	38.3
1973-74	0.0	0.0	0.0	0.0	4.1	10.7	11.5	2.4	4.2	0.8	0.0	0.0	33.7
1974-75	0.0	0.0	0.0	0.0	5.4	8.1	22.7	11.9	5.1	4.6	0.0	0.0	57.8
1975-76	0.0	0.0	0.0	0.0	6.3	0.6	1.8	7.1	4.2	1.6	0.0	0.0	20.0
1976-77	0.0	0.0	0.0	1.4	0.5	2.8	14.4	0.1	3.7		0.0	0.0	24.5
1977-78	0.0	0.0	0.0	0.0	T	4.0	1.4	17.0	6.5	T	0.0	0.0	28.9
1978-79	0.0	0.0	0.0	0.0	2.0	5.6	10.0	1.4	6.0	1.1	0.0	0.0	26.1
1979-80	0.0	0.0	0.0	2.0	T	1.3	5.0	3.0	9.2	T	0.0	0.0	20.5
1980-81	0.0	0.0	0.0	2.0	T	0.5	3.6	3.0	T	0.0	0.0	0.0	9.1
1981-82	0.0	0.0	0.0	T	1.0	7.0	2.4	3.0	8.0	2.9	0.0	0.0	24.3
1982-83	0.0	0.0	0.0	T	T	4.0	6.5	9.0	9.0	3.0	0.0	0.0	31.5
1983-84	0.0	0.0	0.0	0.0	6.9	13.5	3.3	2.0	14.2	T	0.0	0.0	39.9
1984-85	0.0	0.0	0.0	T	T	5.0	4.0	4.0	6.0	T	0.0	0.0	19.0
1985-86	0.0	0.0	T	0.0	5.0	4.5	T	7.7	T	0.3	0.0	0.0	17.5
1986-87	0.0	0.0	0.0	0.0	1.8	5.5	1.2	1.6	10.5	T	0.0	0.0	20.6
1987-88	0.0	0.0	0.0	T	9.0	2.0	2.0	2.2	0.8	0.0	0.0	0.0	16.0
1988-89	0.0	0.0	0.0	0.0	4.3	2.7	1.4	11.8	3.3	T	T	0.0	23.5
1989-90	0.0	0.0	0.0	T	1.2	5.3	5.2	4.0	6.7	0.1	0.0	T	22.5
1990-91	0.0	0.0	0.0	T	1.1	10.1	14.9	0.3	4.1	1.1	0.0	T	31.6
1991-92	0.0	0.0	0.0	2.5	8.8	T	0.3	1.1	0.4	10.0	0.0	0.0	23.1
1992-93	0.0	0.0	0.0	0.0	5.8	3.9	13.0	8.5	4.1	1.5	0.0	T	36.8
1993-94	0.0	0.0	0.0	T	2.8	3.1	3.5	8.8	1.4	1.2	0.0	T	20.8
1994-95	0.0	0.0	0.0	0.0	2.0	12.1	5.5	3.0	4.8	0.4	T	0.0	27.8
1995-96	T	0.0	0.0										
1996-97						6.2			T				
1997-98													
1998-99						4.0	6.0	12.2	5.8	0.4	0.0	0.0	
1999-00					T	4.1	1.6	7.7	T	T	0.0	T	
2000-01	0.0	0.0	0.0	0.0	2.7	18.1	6.9	10.4	0.7	T	T	T	38.8
2001-	0.0	0.0	0.0	T	T	1.8							
POR= 62 YRS	T	0.0	T	0.3	2.6	5.7	7.1	6.7	5.9	1.0	0.1	T	29.4

2001
LAS VEGAS,
NEVADA (LAS)

Las Vegas is situated near the center of a broad desert valley, which is almost surrounded by mountains ranging from 2,000 to 10,000 feet higher than the floor of the valley. This Vegas Valley, comprising about 600 square miles, runs from northwest to southeast, and slopes gradually upward on each side toward the surrounding mountains. Weather observations are taken at McCarran Airport, 7 miles south of downtown Las Vegas, and about 5 miles southwest and 300 feet higher than the lower portions of the valley. Since mountains encircle the valley, drainage winds are usually downslope toward the center, or lowest portion of the valley. This condition also affects minimum temperatures, which in lower portions of the valley can be from 15 to 25 degrees colder than recorded at the airport on clear, calm nights.

The four seasons are well defined. Summers display desert conditions, with maximum temperatures usually in the 100 degree range. The proximity of the mountains contributes to the relatively cool summer nights, with the majority of minimum temperatures in the mid 70s. During about 2 weeks almost every summer warm, moist air predominates in this area, and causes scattered thunderstorms, occasionally quite severe, together with higher than average humidity. Soil erosion, especially near the mountains and foothills surrounding the valley, is evidence of the intensity of some of the thunderstorm activity. Winters, on the whole, are mild and pleasant. Daytime temperatures average near 60 degrees with mostly clear skies. The spring and fall seasons are generally considered most ideal, although rather sharp temperature changes can occur during these months. There are very few days during the spring and fall months when outdoor activities are affected in any degree by the weather.

The Sierra Nevada Mountains of California and the Spring Mountains immediately west of the Vegas Valley, the latter rising to elevations over 10,000 feet above the valley floor, act as effective barriers to moisture moving eastward from the Pacific Ocean. It is mainly these barriers that result in a minimum of dark overcast and rainy days. Rainy days average less than one in June to three per month in the winter months. Snow rarely falls in this valley and it usually melts as it falls, or shortly thereafter. Notable exceptions have occurred.

Strong winds, associated with major storms, usually reach this valley from the southwest or through the pass from the northwest. Winds over 50 mph are infrequent but, when they do occur, are probably the most provoking of the elements experienced in the Vegas Valley, because of the blowing dust and sand associated with them.

Based on the 1951-1980 period, the average first occurrence of 32 degrees Fahrenheit in the fall is November 21 and the average last occurrence in the spring is March 7.

NORMALS, MEANS, AND EXTREMES
LAS VEGAS, NV (LAS)

LATITUDE: 36 04' 44" N   LONGITUDE: 115 09' 19" W   ELEVATION (FT): GRND: 2127 BARO: 2091   TIME ZONE: PACIFIC (UTC + 8)   WBAN: 23169

	ELEMENT	POR	JAN	FEB	MAR	APR	MAY	JUN	JUL	AUG	SEP	OCT	NOV	DEC	YEAR
TEMPERATURE F	NORMAL DAILY MAXIMUM	30	57.3	63.3	68.8	77.5	87.8	100.3	105.9	103.2	94.7	82.1	67.4	57.5	80.5
	MEAN DAILY MAXIMUM	53	56.4	62.4	69.2	78.1	87.7	98.4	104.0	102.0	94.3	81.2	66.2	57.1	79.7
	HIGHEST DAILY MAXIMUM	53	77	87	91	99	109	115	116	116	113	103	87	77	116
	YEAR OF OCCURRENCE		1975	1986	1966	2000	1951	1994	1998	1979	1950	1978	1988	1980	JUL 1998
	MEAN OF EXTREME MAXS.	53	67.9	74.5	82.7	92.2	100.9	110.0	112.2	110.0	104.8	94.3	79.2	68.1	91.4
	NORMAL DAILY MINIMUM	30	33.6	38.8	43.8	50.7	60.2	69.4	76.2	74.2	66.2	54.3	42.6	33.9	53.7
	MEAN DAILY MINIMUM	53	34.3	38.9	44.1	51.5	60.7	69.9	76.4	74.9	66.6	54.3	42.0	34.5	54.0
	LOWEST DAILY MINIMUM	53	8	16	23	31	40	48	60	56	46	26	21	11	8
	YEAR OF OCCURRENCE		1963	1989	1971	1975	1964	1993	1987	1968	1965	1971	1952	1990	JAN 1963
	MEAN OF EXTREME MINS.	53	23.3	27.5	32.5	39.6	47.6	56.7	66.7	65.9	55.4	42.2	30.1	24.2	42.6
	NORMAL DRY BULB	30	45.5	51.1	56.3	64.1	74.0	84.9	91.1	88.7	80.5	68.3	55.0	45.7	67.1
	MEAN DRY BULB	53	45.4	50.7	56.6	64.6	74.2	84.1	90.3	88.5	80.4	67.9	54.1	45.7	66.9
	MEAN WET BULB	18	38.2	41.5	45.5	54.3	58.8	63.8	66.6	60.6	58.6	48.0	42.4	34.7	49.6
	MEAN DEW POINT	18	25.2	27.3	28.3	27.7	32.3	35.0	43.6	43.0	39.3	29.8	26.6	21.1	31.6
	NORMAL NO. DAYS WITH:														
	MAXIMUM ≥ 90	30	0.0	0.0	*	3.3	15.3	25.8	30.5	29.9	21.8	5.6	0.0	0.0	132.2
	MAXIMUM ≤ 32	30	0.1	0.0	0.0	0.0	0.0	0.0	0.0	0.0	0.0	0.0	0.0	*	0.1
	MINIMUM ≤ 32	30	13.0	4.7	1.3	0.1	0.0	0.0	0.0	0.0	0.0	0.1	2.2	11.4	32.8
	MINIMUM ≤ 0	30	0.0	0.0	0.0	0.0	0.0	0.0	0.0	0.0	0.0	0.0	0.0	0.0	0.0
H/C	NORMAL HEATING DEG. DAYS	30	605	389	292	143	14	0	0	0	0	62	304	598	2407
	NORMAL COOLING DEG. DAYS	30	0	0	22	116	293	597	809	735	465	164	0	0	3201
RH	NORMAL (PERCENT)	30	45	40	33	25	21	16	21	26	25	29	37	45	30
	HOUR 04 LST	30	55	50	44	35	31	24	29	35	34	38	46	54	40
	HOUR 10 LST	30	41	36	29	22	19	15	19	24	22	25	33	40	27
	HOUR 16 LST	30	30	26	21	16	13	10	15	17	17	19	26	32	20
	HOUR 22 LST	30	49	42	35	26	22	17	22	26	26	31	40	49	32
S	PERCENT POSSIBLE SUNSHINE	47	77	81	83	87	88	93	88	88	91	87	81	78	85
W/O	MEAN NO. DAYS WITH:														
	HEAVY FOG(VISBY ≤ 1/4 MI)	54	0.3	0.1	0.1	0.0	0.0	0.0	0.0	0.0	0.0	0.0	0.1	0.1	0.7
	THUNDERSTORMS	54	0.0	0.2	0.4	0.5	1.0	1.0	3.8	3.8	1.5	0.5	0.2	0.0	12.9
CLOUDINESS	MEAN:														
	SUNRISE-SUNSET (OKTAS)	1			1.6		0.8	0.0							
	MIDNIGHT-MIDNIGHT (OKTAS)	1					0.8								
	MEAN NO. DAYS WITH:														
	CLEAR	1	4.0	1.0	9.0		27.0	16.0							
	PARTLY CLOUDY	1	2.0	3.0	6.0		1.0								
	CLOUDY	1	1.0	3.0			1.0	1.0							
PR	MEAN STATION PRESSURE(IN)	29	27.80	27.79	27.70	27.60	27.60	27.60	27.60	27.60	27.60	27.70	27.80	27.81	27.68
	MEAN SEA-LEVEL PRES. (IN)	18	30.13	30.04	29.94	29.87	29.79	29.97	29.80	29.83	29.85	29.95	30.06	30.15	29.95
WINDS	MEAN SPEED (MPH)	37	7.4	8.6	10.3	11.0	11.4	11.4	10.4	9.8	9.0	8.1	7.7	7.0	9.3
	PREVAIL.DIR(TENS OF DEGS)	22	25	25	24	22	22	20	18	18	24	24	24	25	24
	MAXIMUM 2-MINUTE:														
	SPEED (MPH)	6	45	46	46	44	56	47	45	43	41	46	40	48	56
	DIR. (TENS OF DEGS)		23	22	33	21	22	33	03	32	16	33	24	34	22
	YEAR OF OCCURRENCE		1996	2001	2001	1999	2000	1998	1998	1998	1998	1996	2001	2000	MAY 2000
	MAXIMUM 5-SECOND:														
	SPEED (MPH)	6	52	54	56	51	64	54	54	61	49	52	49	56	64
	DIR. (TENS OF DEGS)		24	20	34	34	26	33	04	16	23	33	24	33	26
	YEAR OF OCCURRENCE		1999	2001	2000	1997	2000	1998	1998	1998	1998	1996	2001	1999	MAY 2000
PRECIPITATION	NORMAL (IN)	30	0.48	0.48	0.42	0.21	0.28	0.12	0.35	0.49	0.28	0.21	0.43	0.38	4.13
	MAXIMUM MONTHLY (IN)	53	3.00	2.89	4.80	2.44	0.96	0.97	2.48	2.59	2.06	1.22	2.22	1.71	4.80
	YEAR OF OCCURRENCE		1995	1998	1992	1965	1969	1990	1984	1957	1997	1992	1965	1992	MAR 1992
	MINIMUM MONTHLY (IN)	53	T	0.00	0.00	0.00	0.00	0.00	0.00	0.00	0.00	0.00	0.00	0.00	0.00
	YEAR OF OCCURRENCE		1984	1977	1972	1962	1970	1982	1981	1980	1971	1979	1980	1981	JUN 1982
	MAXIMUM IN 24 HOURS (IN)	53	1.09	1.30	1.27	0.97	0.83	0.97	1.36	2.59	1.07	1.09	1.78	0.95	2.59
	YEAR OF OCCURRENCE		1990	1993	1992	1965	1987	1990	1984	1957	1963	1992	1960	1977	AUG 1957
	NORMAL NO. DAYS WITH:														
	PRECIPITATION ≥ 0.01	30	2.9	2.8	3.3	1.9	1.5	0.8	2.4	3.1	1.9	1.8	2.1	2.6	27.1
	PRECIPITATION ≥ 1.00	30	0.0	*	0.0	0.0	0.0	0.0	0.1	0.1	*	0.0	0.1	0.0	0.3
SNOWFALL	NORMAL (IN)	30	0.8	0.1	0.*	0.0	0.0	0.0	0.0	0.0	0.0	0.0	0.1	0.1	1.1
	MAXIMUM MONTHLY (IN)	48	16.7	1.4	0.1	T	0.0	0.0	0.0	T	0.0	T	4.0	2.0	16.7
	YEAR OF OCCURRENCE		1949	1990	1976	1970				1989		1956	1964	1967	JAN 1949
	MAXIMUM IN 24 HOURS (IN)	48	9.0	6.9	0.1	T	0.0	0.0	0.0	T	0.0	T	4.0	2.0	9.0
	YEAR OF OCCURRENCE		1974	1979	1976	1970				1989		1956	1964	1967	JAN 1974
	MAXIMUM SNOW DEPTH (IN)	47	8	6	0	0	0	0	0	0	0	0	3	2	8
	YEAR OF OCCURRENCE		1974	1979									1964	1967	JAN 1974
	NORMAL NO. DAYS WITH:														
	SNOWFALL ≥ 1.0	30	0.2	0.*	0.0	0.0	0.0	0.0	0.0	0.0	0.0	0.0	0.1	0.*	0.3

PRECIPITATION (inches) 2001 LAS VEGAS, NV (LAS)

YEAR	JAN	FEB	MAR	APR	MAY	JUN	JUL	AUG	SEP	OCT	NOV	DEC	ANNUAL
1972	0.00	T	0.00	0.07	0.46	0.32	0.13	0.84	0.63	1.12	1.09	0.19	4.85
1973	0.49	1.64	1.83	0.35	0.09	0.03	T	0.08	T	0.02	0.14	0.01	4.68
1974	2.00	0.11	0.16	T	T	0.00	0.58	0.08	0.16	0.61	0.23	0.59	4.52
1975	0.01	0.05	1.07	0.42	0.35	T	0.26	0.06	1.17	0.03	T	0.05	3.47
1976	0.00	2.49	0.02	0.13	0.34	0.00	1.95	0.00	1.09	0.70	0.02	0.03	6.77
1977	0.21	0.00	0.28	0.01	0.72	0.05	T	1.38	0.19	0.06	0.01	1.06	3.97
1978	1.00	1.51	1.13	0.36	0.54	0.00	0.19	0.53	0.03	0.62	0.59	1.15	7.65
1979	2.18	0.07	0.96	0.06	0.35	0.00	0.78	2.12	T	0.00	0.03	0.24	6.79
1980	1.45	2.25	0.94	0.18	0.15	T	0.43	0.00	0.18	0.04	0.00	0.01	5.63
1981	0.09	0.20	1.44	0.02	0.50	T	0.00	0.20	0.25	0.15	0.29	0.00	3.14
1982	0.09	1.10	0.29	0.01	0.31	0.00	0.05	0.71	0.07	0.04	0.60	0.72	3.99
1983	0.43	0.32	0.90	0.45	0.16	T	0.06	1.25	0.50	0.26	0.10	0.43	4.86
1984	T	0.03	T	0.04	0.00	0.22	2.48	0.99	0.47	T	0.94	1.68	6.85
1985	0.19	0.02	0.06	0.31	T	0.02	0.13	0.00	0.08	0.07	0.37	0.02	1.27
1986	0.23	0.15	0.32	0.10	0.28	T	0.13	0.04	0.05	0.07	0.81	0.47	2.65
1987	1.13	0.45	0.49	0.17	0.90	0.13	0.13	0.01	T	0.49	1.80	0.89	6.59
1988	0.65	0.26	0.00	0.76	T	0.04	0.04	0.46	T	0.00	T	0.08	2.29
1989	0.51	0.06	0.05	T	0.64	T	0.05	0.80	T	T	0.00	T	2.11
1990	1.18	0.37	T	0.18	T	0.97	0.59	T	0.19	0.17	0.10	T	3.75
1991	0.21	0.54	1.01	T	0.05	0.19	0.54	0.78	0.06	0.06	0.38	0.24	4.06
1992	0.45	1.30	4.80	0.02	0.05	0.09	0.03	0.21	0.00	1.22	0.00	1.71	9.88
1993	1.63	2.52	0.14	0.01	0.01	0.08	0.00	0.26	0.00	0.02	0.17	0.21	5.05
1994	0.04	0.48	0.13	T	0.01	0.00	0.11	0.08	0.35	T	0.28	1.08	2.56
1995	3.00	0.03	0.39	0.03	0.16	0.02	T	0.05	T	T	0.00	0.01	3.69
1996	0.13	0.14	0.10	0.00	0.13	T	1.18	T	T	0.11	0.79	0.18	2.76
1997	0.30	T	0.00	0.04	T	T	0.60	0.33	2.06	T	0.23	0.07	3.63
1998	0.17	2.89	1.03	0.14	0.13	0.03	0.46	0.23	1.29	0.22	0.33	0.43	7.35
1999	T	0.08	T	0.73	T	0.14	2.18	0.25	0.35	T	0.00	T	3.73
2000	T	1.59	0.21	T	T	T	T	0.71	0.00	0.92	T	0.04	3.47
2001	0.87	2.21	0.16	0.04	0.02	T	T	0.39	0.05	T	0.09	0.11	3.94
POR= 65 YRS	0.54	0.54	0.49	0.22	0.18	0.08	0.44	0.40	0.35	0.22	0.31	0.38	4.15

AVERAGE TEMPERATURE (F) 2001   LAS VEGAS, NV (LAS)

YEAR	JAN	FEB	MAR	APR	MAY	JUN	JUL	AUG	SEP	OCT	NOV	DEC	ANNUAL
1972	42.3	52.0	63.7	65.1	74.5	84.7	93.1	86.5	78.0	63.5	49.7	41.3	66.2
1973	40.9	49.6	50.7	62.2	76.7	85.2	91.7	87.6	78.9	67.7	53.4	46.2	65.9
1974	41.0	48.9	59.5	63.4	77.0	89.1	88.8	87.7	83.4	69.3	54.8	44.4	67.3
1975	45.3	48.8	53.9	56.6	72.5	83.8	90.3	87.5	81.7	66.1	53.0	48.2	65.6
1976	46.9	53.2	53.4	62.6	77.8	81.5	86.9	85.5	78.7	66.5	58.0	46.4	66.5
1977	45.7	54.2	52.6	68.6	67.7	88.0	92.4	90.1	80.6	71.4	57.2	51.9	68.4
1978	47.9	52.1	59.9	63.1	73.1	87.1	91.9	89.0	79.0	73.5	54.2	42.9	67.8
1979	41.1	48.4	56.0	66.1	75.4	85.5	91.1	85.9	85.3	70.7	51.6	47.2	67.0
1980	49.5	53.2	54.2	63.5	69.0	83.9	92.0	90.2	81.4	68.9	56.8	52.7	67.9
1981	51.1	52.5	56.4	70.6	74.3	88.8	92.7	90.0	82.5	64.4	58.0	48.8	69.2
1982	45.6	50.5	55.1	63.8	73.6	81.5	88.1	87.3	77.9	63.0	50.5	44.5	65.1
1983	46.6	51.7	56.4	58.5	72.8	82.8	88.5	83.8	82.5	67.8	55.3	47.9	66.2
1984	47.1	50.1	57.9	63.1	80.7	83.5	88.2	85.4	81.7	63.0	52.7	44.0	66.5
1985	44.4	47.4	54.9	68.2	76.9	87.4	92.0	89.9	75.4	67.3	51.7	48.3	67.0
1986	51.7	55.8	63.0	66.2	76.6	87.8	87.6	91.2	75.4	65.0	55.8	46.0	68.5
1987	44.7	51.4	54.6	68.4	74.5	86.3	86.9	88.2	81.2	71.0	53.4	42.5	66.9
1988	45.1	52.4	58.1	64.2	73.4	85.3	92.6	86.9	79.1	74.9	56.0	46.0	67.8
1989	43.9	50.0	63.4	72.7	75.7	85.3	93.4	86.9	80.0	67.2	57.3	48.0	68.7
1990	45.2	48.8	60.5	68.8	74.5	85.9	90.8	87.8	82.0	69.2	55.1	40.2	67.4
1991	45.5	55.9	52.7	64.2	69.9	82.1	90.2	87.8	81.9	72.2	55.2	47.0	67.1
1992	45.9	54.1	56.8	70.5	77.7	83.2	88.7	90.5	83.7	70.9	52.7	43.6	68.2
1993	45.7	50.1	60.9	67.5	77.0	82.5	89.4	88.5	81.3	69.1	51.5	46.3	67.5
1994	49.3	48.5	62.7	67.6	76.6	89.3	92.9	89.3	81.2	67.2	49.4	47.5	68.5
1995	47.5	58.7	57.9	64.8	71.0	80.9	92.4	93.1	83.7	69.4	59.8	48.9	69.0
1996	48.5	54.8	59.8	68.3	77.3	87.0	93.2	91.9	80.4	66.8	56.5	47.9	69.4
1997	48.3	51.7	62.7	65.4	81.6	84.3	88.2	90.7	81.3	67.3	56.2	45.9	68.6
1998	48.7	49.4	56.6	61.1	70.0	80.0	91.7	92.0	80.0	66.6	54.8	47.8	66.6
1999	50.5	52.7	60.6	60.9	75.3	85.2	88.2	88.0	81.6	71.6	58.8	48.7	68.5
2000	51.4	53.6	58.6	71.2	80.8	88.7	92.3	90.5	81.7	67.3	50.2	49.5	69.7
2001	46.4	49.8	60.6	65.0	82.2	87.9	90.3	91.9	85.1	72.1	58.6	45.4	69.6
POR= 65 YRS	45.2	50.1	56.3	64.5	74.1	83.6	89.9	88.2	80.1	67.6	53.7	45.9	66.6

REFERENCE NOTES:

PAGE 1:
THE TEMPERATURE GRAPH SHOWS NORMAL MAXIMUM AND NORMAL
   MINIMUM DAILY TEMPERATURES (SOLID CURVES) AND THE
   ACTUAL DAILY HIGH AND LOW TEMPERATURES (VERTICAL BARS).

PAGE 2 AND 3:
H/C INDICATES HEATING AND COOLING DEGREE DAYS.
RH INDICATES RELATIVE HUMIDITY
W/O INDICATES WEATHER AND OBSTRUCTIONS
S INDICATES SUNSHINE.
PR INDICATES PRESSURE.
CLOUDINESS ON PAGE 3 IS THE SUM OF THE CEILOMETER AND
   SATELLITE DATA NOT TO EXCEED EIGHT EIGHTHS (OKTAS).

GENERAL:
T INDICATES TRACE PRECIPITATION, AN AMOUNT GREATER
   THAN ZERO BUT LESS THAN THE LOWEST REPORTABLE VALUE.
+ INDICATES THE VALUE ALSO OCCURS ON EARLIER DATES.
BLANK ENTRIES DENOTE MISSING OR UNREPORTED DATA.
NORMALS ARE 30-YEAR AVERAGES (1961 - 1990).
ASOS INDICATES AUTOMATED SURFACE OBSERVING SYSTEM.
PM INDICATES THE LAST DAY OF THE PREVIOUS MONTH.
POR (PERIOD OF RECORD) BEGINS WITH THE JANUARY DATA
   MONTH AND IS THE NUMBER OF YEARS USED TO COMPUTE
   THE MEAN. INDIVIDUAL MONTHS WITHIN THE POR MAY
   BE MISSING.
WHEN THE POR FOR A NORMAL IS LESS THAN 30 YEARS,
   THE NORMAL IS PROVISIONAL AND IS BASED ON THE NUMBER
   OF YEARS INDICATED.
0.* OR * INDICATES THE VALUE OR MEAN-DAYS-WITH
   IS BETWEEN 0.00 AND 0.05.
CLOUDINESS FOR ASOS STATIONS DIFFERS FROM THE NON-ASOS
   OBSERVATION TAKEN BY A HUMAN OBSERVER. ASOS STATION
   CLOUDINESS IS BASED ON TIME-AVERAGED CEILOMETER DATA
   FOR CLOUDS AT OR BELOW 12,000 FEET AND ON SATELLITE
   DATA FOR CLOUDS ABOVE 12,000 FEET.
THE NUMBER OF DAYS WITH CLEAR, PARTLY CLOUDY, AND
   CLOUDY CONDITIONS FOR ASOS STATIONS IS THE SUM
   OF THE CEILOMETER AND SATELLITE DATA FOR THE
   SUNRISE TO SUNSET PERIOD.

GENERAL CONTINUED:
CLEAR INDICATES 0 - 2 OKTAS, PARTLY CLOUDY INDICATES
   3 - 6 OKTAS, AND CLOUDY INDICATES 7 OR 8 OKTAS.
   WHEN AT LEAST ONE OF THE ELEMENTS (CEILOMETER OR
   SATELLITE) IS MISSING, THE DAILY CLOUDINESS IS
   NOT COMPUTED.
WIND DIRECTION IS RECORDED IN TENS OF DEGREES (2 DIGITS)
   CLOCKWISE FROM TRUE NORTH. "00" INDICATES CALM. "36"
   INDICATES TRUE NORTH.
RESULTANT WIND IS THE VECTOR AVERAGE OF THE SPEED AND
   DIRECTION.
AVERAGE TEMPERATURE IS THE SUM OF THE MEAN DAILY MAXIMUM
   AND MINIMUM TEMPERATURE DIVIDED BY 2.
SNOWFALL DATA COMPRISE ALL FORMS OF FROZEN
   PRECIPITATION, INCLUDING HAIL.
A HEATING (COOLING) DEGREE DAY IS THE DIFFERENCE BETWEEN
   THE AVERAGE DAILY TEMPERATURE AND 65 F.
DRY BULB IS THE TEMPERATURE OF THE AMBIENT AIR.
DEW POINT IS THE TEMPERATURE TO WHICH THE AIR MUST BE
   COOLED TO ACHIEVE 100 PERCENT RELATIVE HUMIDITY.
WET BULB IS THE TEMPERATURE THE AIR WOULD HAVE IF THE
   MOISTURE CONTENT WAS INCREASED TO 100 PERCENT RELATIVE
   HUMIDITY.

ON JULY 1, 1996, THE NATIONAL WEATHER SERVICE BEGAN USING
   THE "METAR" OBSERVATION CODE THAT WAS ALREADY EMPLOYED
   BY MOST OTHER NATIONS OF THE WORLD. THE MOST NOTICEABLE
   DIFFERENCE IN THIS ANNUAL PUBLICATION WILL BE THE CHANGE
   IN UNITS FROM TENTHS TO EIGHTS (OKTAS) FOR REPORTING THE
   AMOUNT OF SKY COVER.

HEATING DEGREE DAYS (base 65 F) 2001 LAS VEGAS, NV (LAS)

YEAR	JUL	AUG	SEP	OCT	NOV	DEC	JAN	FEB	MAR	APR	MAY	JUN	TOTAL
1972-73	0	0	0	108	453	727	744	428	437	132	12	0	3041
1973-74	0	0	0	42	349	576	738	443	188	82	13	0	2431
1974-75	0	0	0	55	300	634	607	446	340	249	37	0	2668
1975-76	0	0	0	73	354	516	553	339	357	124	1	0	2317
1976-77	0	0	0	39	212	569	593	297	374	45	56	0	2185
1977-78	0	0	0	3	226	399	522	356	168	91	16	0	1781
1978-79	0	0	1	2	324	676	737	458	270	66	18	0	2552
1979-80	0	0	0	44	395	546	474	335	328	108	32	0	2262
1980-81	0	0	0	82	255	374	426	344	263	29	2	0	1775
1981-82	0	0	0	74	214	497	594	398	301	98	9	0	2185
1982-83	0	0	10	84	429	631	564	364	263	198	21	0	2564
1983-84	0	0	0	3	297	524	548	424	216	111	0	0	2123
1984-85	0	0	0	127	363	641	629	487	308	41	0	0	2596
1985-86	0	0	1	31	393	512	404	270	125	57	11	0	1804
1986-87	0	0	14	53	268	586	622	375	316	40	1	0	2275
1987-88	0	0	0	18	342	689	612	357	225	83	33	0	2359
1988-89	0	0	0	0	291	581	647	425	118	23	16	0	2101
1989-90	0	0	0	70	224	519	606	449	172	12	0	0	2052
1990-91	0	0	0	23	290	761	597	247	376	57	25	2	2378
1991-92	0	0	0	77	297	552	584	308	248	7	0	0	2073
1992-93	0	0	0	16	364	655	591	410	143	32	3	8	2222
1993-94	0	0	0	33	398	573	480	455	93	60	1	0	2093
1994-95	0	0	0	35	465	537	537	170	230	90	18	6	2088
1995-96	0	0	0	22	151	490	504	287	169	22	11	0	1656
1996-97	0	0	0	138	249	524	512	368	115	117	0	0	2023
1997-98	0	0	0	57	259	583	500	426	270	166	23	0	2284
1998-99	0	0	0	39	296	525	444	337	143	188	5	6	1983
1999-00	0	0	0	11	184	499	412	325	195	16	3	0	1645
2000-01	0	0	0	80	436	474	570	421	175	114	2	0	2272
2001-	0	0	0	0	204	602							

COOLING DEGREE DAYS (base 65 F) 2001 LAS VEGAS, NV (LAS)

YEAR	JAN	FEB	MAR	APR	MAY	JUN	JUL	AUG	SEP	OCT	NOV	DEC	ANNUAL
1972	0	2	66	80	308	597	876	675	398	69	0	0	3071
1973	0	0	0	54	382	612	833	708	424	134	8	0	3155
1974	0	0	24	43	394	731	744	713	559	195	0	0	3403
1975	0	0	2	2	276	570	792	704	508	117	2	0	2973
1976	0	0	2	57	404	500	687	641	419	93	6	0	2809
1977	0	0	0	161	149	694	858	781	476	210	3	0	3332
1978	0	0	17	40	277	672	841	752	425	268	8	0	3300
1979	0	0	0	104	346	625	813	656	614	229	0	0	3387
1980	0	0	0	68	160	575	842	788	498	211	15	0	3157
1981	0	0	5	205	296	721	866	781	531	64	12	0	3481
1982	0	0	2	70	281	501	721	699	404	30	0	0	2708
1983	0	0	2	9	269	541	735	589	534	94	10	0	2783
1984	0	0	3	61	496	563	724	641	508	74	1	0	3071
1985	0	0	0	143	377	678	844	778	319	110	2	0	3251
1986	0	20	69	98	379	693	707	821	332	59	0	0	3178
1987	0	0	0	148	302	645	685	729	495	211	0	0	3215
1988	0	0	16	64	300	615	864	685	434	312	31	0	3321
1989	0	11	74	259	351	614	887	687	456	143	0	0	3482
1990	0	0	42	134	302	634	810	713	516	163	0	0	3314
1991	0	0	0	42	187	524	788	714	515	307	12	0	3089
1992	0	0	0	180	402	552	742	798	571	206	3	0	3454
1993	0	0	21	114	381	537	765	737	494	166	0	0	3215
1994	0	0	31	145	369	768	883	870	551	108	0	0	3725
1995	0	0	13	91	211	490	883	880	570	168	2	0	3281
1996	0	3	16	129	402	665	883	841	471	201	0	0	3611
1997	0	0	48	136	522	584	727	805	494	137	4	0	3457
1998	0	0	18	55	186	457	834	842	456	95	0	0	2943
1999	0	0	13	74	333	620	726	719	508	221	3	0	3217
2000	0	0	3	208	498	719	851	799	509	158	0	0	3745
2001	0	0	47	122	541	692	792	843	609	231	17	0	3894

REFERENCE NOTES:

PAGE 1:
THE TEMPERATURE GRAPH SHOWS NORMAL MAXIMUM AND NORMAL
MINIMUM DAILY TEMPERATURES (SOLID CURVES) AND THE
ACTUAL DAILY HIGH AND LOW TEMPERATURES (VERTICAL BARS).

PAGE 2 AND 3:
H/C INDICATES HEATING AND COOLING DEGREE DAYS.
RH INDICATES RELATIVE HUMIDITY
W/O INDICATES WEATHER AND OBSTRUCTIONS
S INDICATES SUNSHINE.
PR INDICATES PRESSURE.
CLOUDINESS ON PAGE 3 IS THE SUM OF THE CEILOMETER AND
SATELLITE DATA NOT TO EXCEED EIGHT EIGHTHS(OKTAS).

GENERAL:
T INDICATES TRACE PRECIPITATION, AN AMOUNT GREATER
THAN ZERO BUT LESS THAN THE LOWEST REPORTABLE VALUE.
+ INDICATES THE VALUE ALSO OCCURS ON EARLIER DATES.
BLANK ENTRIES DENOTE MISSING OR UNREPORTED DATA.
NORMALS ARE 30-YEAR AVERAGES (1961 - 1990).
ASOS INDICATES AUTOMATED SURFACE OBSERVING SYSTEM.
PM INDICATES THE LAST DAY OF THE PREVIOUS MONTH.
POR (PERIOD OF RECORD) BEGINS WITH THE JANUARY DATA
MONTH AND IS THE NUMBER OF YEARS USED TO COMPUTE
THE MEAN. INDIVIDUAL MONTHS WITHIN THE POR MAY
BE MISSING.
WHEN THE POR FOR A NORMAL IS LESS THAN 30 YEARS,
THE NORMAL IS PROVISIONAL AND IS BASED ON THE NUMBER
OF YEARS INDICATED.
0.+ OR * INDICATES THE VALUE OR MEAN-DAYS-WITH
IS BETWEEN 0.00 AND 0.05.
CLOUDINESS FOR ASOS STATIONS DIFFERS FROM THE NON-ASOS
OBSERVATION TAKEN BY A HUMAN OBSERVER. ASOS STATION
CLOUDINESS IS BASED ON TIME-AVERAGED CEILOMETER DATA
FOR CLOUDS AT OR BELOW 12,000 FEET AND ON SATELLITE
DATA FOR CLOUDS ABOVE 12,000 FEET.
THE NUMBER OF DAYS WITH CLEAR, PARTLY CLOUDY, AND
CLOUDY CONDITIONS FOR ASOS STATIONS IS THE SUM
OF THE CEILOMETER AND SATELLITE DATA FOR THE
SUNRISE TO SUNSET PERIOD.

GENERAL CONTINUED:
CLEAR INDICATES 0 - 2 OKTAS, PARTLY CLOUDY INDICATES
3 - 6 OKTAS, AND CLOUDY INDICATES 7 OR 8 OKTAS.
WHEN AT LEAST ONE OF THE ELEMENTS (CEILOMETER OR
SATELLITE) IS MISSING, THE DAILY CLOUDINESS IS
NOT COMPUTED.
WIND DIRECTION IS RECORDED IN TENS OF DEGREES (2 DIGITS)
CLOCKWISE FROM TRUE NORTH. "00" INDICATES CALM. "36"
INDICATES TRUE NORTH.
RESULTANT WIND IS THE VECTOR AVERAGE OF THE SPEED AND
DIRECTION.
AVERAGE TEMPERATURE IS THE SUM OF THE MEAN DAILY MAXIMUM
AND MINIMUM TEMPERATURE DIVIDED BY 2.
SNOWFALL DATA COMPRISE ALL FORMS OF FROZEN
PRECIPITATION, INCLUDING HAIL.
A HEATING (COOLING) DEGREE DAY IS THE DIFFERENCE BETWEEN
THE AVERAGE DAILY TEMPERATURE AND 65 F.
DRY BULB IS THE TEMPERATURE OF THE AMBIENT AIR.
DEW POINT IS THE TEMPERATURE TO WHICH THE AIR MUST BE
COOLED TO ACHIEVE 100 PERCENT RELATIVE HUMIDITY.
WET BULB IS THE TEMPERATURE THE AIR WOULD HAVE IF THE
MOISTURE CONTENT WAS INCREASED TO 100 PERCENT RELATIVE
HUMIDITY.

ON JULY 1, 1996, THE NATIONAL WEATHER SERVICE BEGAN USING
THE "METAR" OBSERVATION CODE THAT WAS ALREADY EMPLOYED
BY MOST OTHER NATIONS OF THE WORLD. THE MOST NOTICEABLE
DIFFERENCE IN THIS ANNUAL PUBLICATION WILL BE THE CHANGE
IN UNITS FROM TENTHS TO EIGHTS(OKTAS) FOR REPORTING THE
AMOUNT OF SKY COVER.

2001
CONCORD,
NEW HAMPSHIRE (CON)

Concord, the Capital of New Hampshire, is situated near the geographical center of New England at an altitude of approximately 300 feet above sea level on the Merrimack River.   Its surroundings are hilly with many lakes and ponds. The countryside is generously wooded, mostly on land reclaimed from fields which were formerly cleared for farming. From the coast about 50 miles to the southeast, the terrain slopes gently upward to the city. West of the city, the land rises some 2,000 feet higher in only half that distance.   Mount Washington, at an elevation of 6,288 feet is in the White Mountains 75 miles north of town.

Northwesterly winds are prevalent.   They bring cold, dry air during the winter and pleasantly cool, dry air in the summer. Stronger southerly winds occur during July and August, and easterly winds usually accompany summer and winter storms. Winter breezes are somewhat lighter, and winds are frequently calm during the night and early morning hours. Low temperatures, as a rule, do not interrupt normal out-of-doors activity because winds are calm or light, producing a low wind chill factor.

Very hot summer weather is infrequent. During any month, temperatures considerably above the average maxima and much below the normal minima are observed.

The average amount of precipitation for the warmer half of the year differs little from that for the colder half.   Precipitation occurrences average approximately one day of three for the year, with a somewhat higher frequency for the April-May period, offsetting the lower frequency of August-October. The more significant rains and heavier snowfalls are associated with easterly winds, especially northeasterly winds. The first snowfall of an inch or more is likely to come between the middle of November and the middle of December. The snow cover normally lasts from mid-December until the last week of March, but bare ground is not rare in the winter, nor is a snowscape rare earlier or later in the season. Rain, sleet, or freezing rain may also occur.

Agriculture is neither intensive nor large-scale in the vicinity of the station.   Potatoes and other frost-resistant vegetables, hardy fruits such as apples, forage for the dairy industry, and maple sugar are the principal crops.

Based on the 1951-1980 period, the average first occurrence of 32 degrees Fahrenheit in the fall is September 22 and the average last occurrence in the spring is May 23.   Freezing temperatures have occurred as late as June and as early as August.

NORMALS, MEANS, AND EXTREMES
CONCORD, NH (CON)

LATITUDE:	LONGITUDE:	ELEVATION (FT):		TIME ZONE:	WBAN: 14745
43 11' 43" N	71 30' 04" W	GRND: 340	BARO: 343	EASTERN (UTC + 5)	

	ELEMENT	POR	JAN	FEB	MAR	APR	MAY	JUN	JUL	AUG	SEP	OCT	NOV	DEC	YEAR
TEMPERATURE °F	NORMAL DAILY MAXIMUM	30	29.8	33.0	42.8	56.3	68.9	77.3	82.4	79.8	71.6	60.7	47.1	34.2	57.0
	MEAN DAILY MAXIMUM	54	31.0	34.0	42.7	56.7	69.0	77.8	82.4	80.4	72.0	61.1	47.6	35.3	57.5
	HIGHEST DAILY MAXIMUM	60	68	67	89	95	97	98	102	101	98	90	80	73	102
	YEAR OF OCCURRENCE		1950	1997	1998	1976	1962	1995	1966	1975	1953	1963	1950	1998	JUL 1966
	MEAN OF EXTREME MAXS.	54	50.6	52.3	64.8	78.8	88.3	92.4	94.0	92.1	88.2	79.5	67.9	55.4	75.4
	NORMAL DAILY MINIMUM	30	7.4	10.4	22.1	31.5	41.4	51.2	56.5	54.7	46.0	34.9	27.0	14.4	33.1
	MEAN DAILY MINIMUM	54	9.7	12.0	22.2	32.1	41.9	51.7	56.7	54.9	46.4	35.6	27.7	15.8	33.9
	LOWEST DAILY MINIMUM	60	-33	-37	-16	8	21	30	35	29	21	10	-5	-22	-37
	YEAR OF OCCURRENCE		1984	1943	1967	1969	1966	1972	1965	1965	1947	1972	1989	1951	FEB 1943
	MEAN OF EXTREME MINS.	54	-13.7	-11.6	1.0	18.0	27.5	36.6	43.3	40.3	29.6	19.4	10.2	-6.5	16.2
	NORMAL DRY BULB	30	18.6	21.8	32.4	43.9	55.2	64.2	69.5	67.3	58.8	47.8	37.1	24.3	45.1
	MEAN DRY BULB	54	20.2	22.9	32.3	44.3	55.4	64.7	69.7	67.6	59.0	48.3	37.7	25.5	45.6
	MEAN WET BULB	17	19.8	21.8	29.2	39.3	50.1	58.9	63.7	62.4	54.7	41.7	34.6	24.6	41.7
	MEAN DEW POINT	17	13.0	13.8	20.7	31.0	43.3	53.9	59.4	58.5	50.8	36.8	28.8	18.5	35.7
	NORMAL NO. DAYS WITH:														
	MAXIMUM 90	30	0.0	0.0	0.0	0.2	0.9	2.2	4.6	2.5	0.7	*	0.0	0.0	11.1
	MAXIMUM 32	30	17.8	13.1	4.1	0.2	0.0	0.0	0.0	0.0	0.0	*	1.5	13.6	50.3
	MINIMUM 32	30	30.5	27.1	26.1	16.9	5.5	0.2	0.0	0.1	2.4	13.6	21.7	29.2	173.3
	MINIMUM 0	30	9.6	6.9	1.1	0.0	0.0	0.0	0.0	0.0	0.0	0.0	0.2	5.2	23.0
H/C	NORMAL HEATING DEG. DAYS	30	1438	1210	1011	633	312	70	13	39	196	533	837	1262	7554
	NORMAL COOLING DEG. DAYS	30	0	0	0	0	8	46	153	111	10	0	0	0	328
RH	NORMAL (PERCENT)	30	68	66	65	62	65	71	72	74	76	73	73	72	70
	HOUR 01 LST	30	74	73	74	76	82	88	89	90	90	85	81	78	82
	HOUR 07 LST	30	74	75	76	75	76	82	84	88	90	87	83	79	81
	HOUR 13 LST	30	58	55	52	46	46	52	51	53	54	52	59	62	53
	HOUR 19 LST	30	66	62	59	54	56	63	64	70	75	72	72	70	65
S	PERCENT POSSIBLE SUNSHINE	58	52	55	53	52	54	57	63	60	56	53	43	47	54
W/O	MEAN NO. DAYS WITH:														
	HEAVY FOG(VISBY 1/4 MI)	61	2.5	2.1	3.0	1.9	3.1	3.7	4.9	6.4	8.5	6.1	3.6	2.8	48.6
	THUNDERSTORMS	61	0.0	0.1	0.2	0.9	2.4	4.6	5.4	3.7	1.8	0.6	0.1	0.0	19.8
CLOUDINESS	MEAN:														
	SUNRISE-SUNSET (OKTAS)	55	5.0	5.0	5.1	5.3	5.4	5.0	4.9	4.6	4.7	4.6	5.3	5.0	5.0
	MIDNIGHT-MIDNIGHT (OKTAS)	32	4.7	4.6	4.7	4.9	5.0	4.7	4.6	4.4	4.6	4.4	5.0	4.8	4.7
	MEAN NO. DAYS WITH:														
	CLEAR	55	9.0	7.5	7.7	7.0	6.1	6.1	6.5	8.1	8.4	9.0	6.2	7.7	89.3
	PARTLY CLOUDY	55	7.1	7.6	8.0	8.1	9.9	11.6	12.2	11.1	8.9	8.7	7.5	7.7	108.4
	CLOUDY	55	14.9	13.2	15.3	14.9	15.0	12.2	11.7	11.3	12.2	12.7	15.8	15.1	164.3
PR	MEAN STATION PRESSURE(IN)	29	29.61	29.60	29.60	29.60	29.60	29.59	29.60	29.70	29.69	29.70	29.69	29.59	29.63
	MEAN SEA-LEVEL PRES. (IN)	17	30.04	30.03	29.99	29.96	29.97	29.94	29.96	30.02	30.07	30.09	30.05	30.03	30.01
WINDS	MEAN SPEED (MPH)	37	7.0	7.5	8.0	7.7	6.8	6.4	5.7	5.2	5.4	5.9	6.5	6.6	6.6
	PREVAIL.DIR(TENS OF DEGS)	21	31	31	31	31	32	32	31	32	32	31	31	31	31
	MAXIMUM 2-MINUTE:														
	SPEED (MPH)	5	38	40	39	38	33	31	43	26	36	37	36	39	43
	DIR. (TENS OF DEGS)		33	31	30	20	29	35	32	26	06	31	34	32	32
	YEAR OF OCCURRENCE		2000	2001	1997	2000	2001	1997	1999	1997	1999	1998	1997	1999	JUL 1999
	MAXIMUM 5-SECOND:														
	SPEED (MPH)	5	47	49	51	49	43	46	52	33	44	49	44	52	52
	DIR. (TENS OF DEGS)		29	28	34	20	30	35	32	24	32	29	34	32	32
	YEAR OF OCCURRENCE		1997	2001	1999	2000	1997	1997	1999	1997	1999	1998	1997	1999	DEC 1999
PRECIPITATION	NORMAL (IN)	30	2.51	2.53	2.72	2.91	3.14	3.15	3.23	3.32	2.81	3.23	3.66	3.16	36.37
	MAXIMUM MONTHLY (IN)	60	8.09	7.77	7.81	6.11	9.52	10.10	6.53	7.26	9.44	8.78	7.36	7.52	10.10
	YEAR OF OCCURRENCE		1979	1981	1953	1996	1984	1944	1988	1991	1999	1962	1983	1973	JUN 1944
	MINIMUM MONTHLY (IN)	60	0.40	0.03	0.86	0.83	0.60	0.64	0.96	0.06	0.41	0.59	0.75	0.58	0.03
	YEAR OF OCCURRENCE		1970	1987	1981	1999	1965	1979	1955	1996	1948	1947	1976	1943	FEB 1987
	MAXIMUM IN 24 HOURS (IN)	60	2.12	2.26	2.48	2.40	2.59	4.47	2.54	3.97	4.12	4.87	2.89	3.31	4.87
	YEAR OF OCCURRENCE		1979	1981	2001	1996	1984	1944	1971	1991	1960	1996	1947	1969	OCT 1996
	NORMAL NO. DAYS WITH:														
	PRECIPITATION 0.01	30	10.0	9.0	10.6	11.2	11.7	11.0	10.1	10.5	8.6	8.7	11.6	11.2	124.2
	PRECIPITATION 1.00	30	0.4	0.5	0.5	0.5	0.5	0.5	1.0	0.8	0.8	1.0	0.8	0.8	8.1
SNOWFALL	NORMAL (IN)	30	17.7	15.6	10.0	2.6	0.*	0.0	0.0	0.0	0.0	0.1	4.5	14.6	65.1
	MAXIMUM MONTHLY (IN)	60	45.4	49.8	38.3	15.3	5.0		0.0	0.0	T	2.1	18.4	38.1	49.8
	YEAR OF OCCURRENCE		1987	1969	1956	1982	1945	1993			1992	1969	1971	1956	FEB 1969
	MAXIMUM IN 24 HOURS (IN)	60	19.0	14.2	18.1	13.9	5.0	T	0.0	0.0	T	2.1	9.5	14.6	19.0
	YEAR OF OCCURRENCE		1944	1972	2001	1982	1945	1993			1992	1969	1961	1946	JAN 1944
	MAXIMUM SNOW DEPTH (IN)	53	25	37	34	17	1	0			0	2	10	26	37
	YEAR OF OCCURRENCE		1952	1969	1969	1956	1966					1969	1971	1970	FEB 1969
	NORMAL NO. DAYS WITH:														
	SNOWFALL 1.0	30	4.5	3.7	2.9	0.6	0.*	0.0	0.0	0.0	0.0	0.1	1.4	4.5	17.7

PRECIPITATION (inches) 2001   CONCORD, NH (CON)

YEAR	JAN	FEB	MAR	APR	MAY	JUN	JUL	AUG	SEP	OCT	NOV	DEC	ANNUAL
1972	1.44	2.60	4.16	2.71	4.20	3.54	5.40	2.12	2.55	2.23	6.57	4.55	42.07
1973	2.44	1.91	2.58	4.55	4.20	4.86	1.05	6.88	1.77	2.46	1.82	7.52	42.04
1974	2.80	2.32	3.98	2.58	3.74	1.82	1.41	2.20	4.74	1.64	3.20	4.02	34.45
1975	4.12	2.36	3.12	2.47	1.22	3.87	3.71	3.94	5.15	4.29	4.91	3.12	42.28
1976	3.40	2.36	2.02	2.43	3.90	2.74	3.20	2.66	2.73	4.05	0.75	2.27	32.51
1977	2.16	2.02	4.51	4.04	2.44	3.47	1.26	3.51	5.64	5.52	3.07	4.00	41.64
1978	6.32	0.67	2.16	2.06	2.67	3.18	1.08	2.87	2.72	1.77	2.91	2.89	28.87
1979	8.09	2.29	2.85	3.10	4.86	0.64	3.45	4.20	3.15	3.79	2.92	1.93	41.27
1980	0.43	0.78	3.37	3.72	0.86	2.35	2.83	3.99	2.19	2.63	3.12	0.79	27.06
1981	0.48	7.77	0.86	3.12	3.21	2.81	5.54	3.25	4.61	6.51	3.51	4.17	45.84
1982	3.98	2.88	2.47	3.08	1.91	7.84	2.83	2.54	1.85	1.52	2.93	0.91	34.74
1983	3.92	2.17	7.07	5.88	5.19	2.52	2.07	2.07	1.21	3.28	7.36	5.35	48.09
1984	1.89	5.06	2.92	3.74	9.52	2.83	4.44	0.97	1.08	4.42	2.67	2.70	42.24
1985	0.95	1.99	2.86	1.02	2.05	3.05	2.83	2.51	3.78	3.62	4.58	1.65	30.89
1986	4.78	2.23	3.58	1.85	1.44	4.95	4.77	3.72	2.27	1.71	4.48	4.50	40.28
1987	3.00	0.03	3.47	4.71	1.08	5.77	3.77	2.84	3.94	4.14	2.50	1.55	36.80
1988	1.97	2.24	1.32	2.75	3.35	0.80	6.53	5.44	1.56	1.23	5.06	1.05	33.30
1989	0.74	2.05	2.18	3.40	5.11	4.25	3.62	3.55	4.22	4.86	3.34	0.91	38.23
1990	2.82	2.63	1.64	3.00	5.09	2.51	1.79	7.19	2.31	4.93	3.25	4.12	41.28
1991	1.85	1.42	3.01	2.61	2.51	1.72	1.80	7.26	5.52	3.32	4.84	3.62	39.48
1992	1.97	1.30	2.77	1.78	2.39	3.28	3.50	2.13	2.48	2.41	3.54	2.23	29.78
1993	1.51	1.63	3.04	3.77	0.97	1.95	2.10	1.97	4.01	3.58	3.72	3.36	31.61
1994	3.48	0.80	5.03	2.38	4.33	1.94	4.02	2.85	3.55	0.91	2.67	4.19	36.15
1995	3.39	2.03	2.46	1.64	3.56	1.36	4.01	2.42	2.70	7.31	5.58	1.95	38.41
1996	4.87	2.58	1.55	6.11	4.81	3.09	5.17	0.06	3.05	8.11	2.06	5.84	47.30
1997	3.29	2.20	4.04	3.66	2.29	0.70	3.80	3.67	1.68	1.44	5.63	1.95	34.35
1998	3.62	2.85	3.70	2.21	3.97	7.95	1.90	1.41	1.63	3.58	2.20	0.92	35.94
1999	5.38	2.84	2.82	0.83	2.79	1.94	4.35	3.45	9.44	2.47	2.66	1.35	40.32
2000	2.26	2.83	3.34	4.82	2.89	4.33	1.89	3.54	2.29	3.42	2.99	2.24	37.74
2001	1.44	2.57	6.44	0.84	2.33	5.74	3.29	0.67	3.51	0.99	1.10		31.16
POR= 147 YRS	2.91	2.58	3.10	2.99	3.17	3.29	3.52	3.35	3.33	3.19	3.38	2.99	37.80

AVERAGE TEMPERATURE (F) 2001   CONCORD, NH (CON)

YEAR	JAN	FEB	MAR	APR	MAY	JUN	JUL	AUG	SEP	OCT	NOV	DEC	ANNUAL
1972	22.0	21.2	27.9	40.2	56.8	62.8	69.3	64.1	57.5	42.3	31.2	23.4	43.2
1973	21.0	20.2	35.5	45.0	52.9	66.9	70.3	72.3	58.8	48.2	36.1	28.9	46.3
1974	21.4	21.1	31.8	45.8	51.1	62.8	67.5	67.0	58.9	42.5	35.9	26.7	44.4
1975	21.6	21.5	30.1	40.4	61.3	65.0	72.9	66.5	56.1	47.6	39.9	21.9	45.4
1976	10.9	24.7	31.9	46.7	53.6	68.9	66.8	65.3	57.2	45.0	31.7	16.1	43.2
1977	10.6	20.5	36.7	45.1	58.3	62.9	69.0	68.0	58.5	47.0	39.4	21.0	44.8
1978	17.5	13.5	28.0	40.6	57.5	65.2	69.7	69.6	55.9	46.7	35.3	22.7	43.5
1979	23.3	15.1	37.6	44.2	56.3	63.8	71.2	67.5	59.8	47.4	42.2	29.4	46.5
1980	22.4	19.1	31.8	44.4	55.6	62.8	70.6	68.4	58.2	45.1	34.8	19.2	44.4
1981	12.5	30.8	34.2	47.1	57.5	66.1	69.9	66.6	58.7	45.2	37.8	25.4	46.0
1982	10.9	20.8	30.2	41.6	57.3	60.9	69.5	65.4	60.2	47.6	41.7	32.4	44.9
1983	23.1	26.1	35.9	45.2	53.1	65.1	70.1	69.3	62.3	48.2	39.4	23.4	46.8
1984	16.0	30.5	28.2	44.6	52.9	65.9	68.7	69.3	57.8	50.4	38.3	30.3	46.1
1985	15.9	25.8	35.7	45.2	56.1	62.2	70.3	66.7	60.5	49.2	38.5	22.0	45.7
1986	23.0	21.3	35.5	48.2	57.5	61.3	67.3	66.1	57.3	47.4	34.1	29.0	45.7
1987	20.0	22.0	34.5	46.8	56.7	64.4	70.6	65.1	58.7	45.8	36.9	28.1	45.8
1988	18.5	23.5	33.4	43.9	57.3	62.9	72.6	70.5	57.8	44.9	39.1	23.2	45.6
1989	25.7	22.6	31.8	41.4	58.3	65.0	69.5	67.7	60.8	49.1	35.8	11.9	45.0
1990	28.6	24.6	34.7	45.6	52.8	65.3	70.8	69.8	59.7	51.8	40.1	30.9	47.9
1991	20.2	28.3	36.0	47.6	60.2	65.6	69.2	69.9	57.7	50.8	39.7	25.5	47.6
1992	23.3	26.0	30.2	42.5	55.0	64.0	66.0	67.1	59.7	44.9	35.6	26.5	45.1
1993	23.5	15.7	30.5	45.7	56.9	64.9	71.4	70.0	58.8	45.3	36.4	26.8	45.5
1994	11.3	16.7	31.9	45.8	54.4	67.9	73.4	66.4	58.2	49.3	41.5	30.6	45.6
1995	28.1	20.7	36.0	41.5	55.1	66.1	73.0	69.4	56.9	52.0	34.2	22.9	46.3
1996	21.2	23.4	29.6	43.9	53.7	65.2	68.4	69.3	60.6	46.3	33.4	31.3	45.5
1997	21.8	28.3	29.8	42.3	51.6	65.9	69.1	68.1	59.8	46.8	36.2	28.0	45.6
1998	27.3	30.9	36.8	46.3	59.2	64.3	69.8	69.6	61.2	49.4	39.0	31.1	48.7
1999	20.7	27.8	34.5	44.9	57.0	68.4	71.7	67.2	63.4	46.3	41.9	29.7	47.8
2000	20.1	25.4	38.5	45.4	55.2	65.1	67.2	67.4	58.6	49.2	38.4	22.5	46.1
2001	19.5	23.5	29.4	44.0	56.4	66.7	66.6	72.1	60.3	49.2	40.8	32.1	46.7
POR= 131 YRS	21.1	22.9	32.2	44.4	56.1	64.8	70.0	67.3	59.6	48.8	37.5	25.5	45.8

REFERENCE NOTES:

PAGE 1:
THE TEMPERATURE GRAPH SHOWS NORMAL MAXIMUM AND NORMAL
  MINIMUM DAILY TEMPERATURES (SOLID CURVES) AND THE
  ACTUAL DAILY HIGH AND LOW TEMPERATURES (VERTICAL BARS).

PAGE 2 AND 3:
H/C INDICATES HEATING AND COOLING DEGREE DAYS.
RH INDICATES RELATIVE HUMIDITY
W/O INDICATES WEATHER AND OBSTRUCTIONS
S INDICATES SUNSHINE.
PR INDICATES PRESSURE.
CLOUDINESS ON PAGE 3 IS THE SUM OF THE CEILOMETER AND
  SATELLITE DATA NOT TO EXCEED EIGHT EIGHTHS(OKTAS).

GENERAL:
T INDICATES TRACE PRECIPITATION, AN AMOUNT GREATER
  THAN ZERO BUT LESS THAN THE LOWEST REPORTABLE VALUE.
+ INDICATES THE VALUE ALSO OCCURS ON EARLIER DATES.
BLANK ENTRIES DENOTE MISSING OR UNREPORTED DATA.
NORMALS ARE 30-YEAR AVERAGES (1961 - 1990).
ASOS INDICATES AUTOMATED SURFACE OBSERVING SYSTEM.
PM INDICATES THE LAST DAY OF THE PREVIOUS MONTH.
POR (PERIOD OF RECORD) BEGINS WITH THE JANUARY DATA
  MONTH AND IS THE NUMBER OF YEARS USED TO COMPUTE
  THE MEAN. INDIVIDUAL MONTHS WITHIN THE POR MAY
  BE MISSING.
WHEN THE POR FOR A NORMAL IS LESS THAN 30 YEARS,
  THE NORMAL IS PROVISIONAL AND IS BASED ON THE NUMBER
  OF YEARS INDICATED.
0.* OR * INDICATES THE VALUE OR MEAN-DAYS-WITH
  IS BETWEEN 0.00 AND 0.05.
CLOUDINESS FOR ASOS STATIONS DIFFERS FROM THE NON-ASOS
  OBSERVATION TAKEN BY A HUMAN OBSERVER. ASOS STATION
  CLOUDINESS IS BASED ON TIME-AVERAGED CEILOMETER DATA
  FOR CLOUDS AT OR BELOW 12,000 FEET AND ON SATELLITE
  DATA FOR CLOUDS ABOVE 12,000 FEET.
THE NUMBER OF DAYS WITH CLEAR, PARTLY CLOUDY, AND
  CLOUDY CONDITIONS FOR ASOS STATIONS IS THE SUM
  OF THE CEILOMETER AND SATELLITE DATA FOR THE
  SUNRISE TO SUNSET PERIOD.

GENERAL CONTINUED:
CLEAR INDICATES 0 - 2 OKTAS, PARTLY CLOUDY INDICATES
  3 - 6 OKTAS, AND CLOUDY INDICATES 7 OR 8 OKTAS.
  WHEN AT LEAST ONE OF THE ELEMENTS (CEILOMETER OR
  SATELLITE) IS MISSING, THE DAILY CLOUDINESS IS
  NOT COMPUTED.
WIND DIRECTION IS RECORDED IN TENS OF DEGREES (2 DIGITS)
  CLOCKWISE FROM TRUE NORTH. "00" INDICATES CALM. "36"
  INDICATES TRUE NORTH.
RESULTANT WIND IS THE VECTOR AVERAGE OF THE SPEED AND
  DIRECTION.
AVERAGE TEMPERATURE IS THE SUM OF THE MEAN DAILY MAXIMUM
  AND MINIMUM TEMPERATURE DIVIDED BY 2.
SNOWFALL DATA COMPRISE ALL FORMS OF FROZEN
  PRECIPITATION, INCLUDING HAIL.
A HEATING (COOLING) DEGREE DAY IS THE DIFFERENCE BETWEEN
  THE AVERAGE DAILY TEMPERATURE AND 65 F.
DRY BULB IS THE TEMPERATURE OF THE AMBIENT AIR.
DEW POINT IS THE TEMPERATURE TO WHICH THE AIR MUST BE
  COOLED TO ACHIEVE 100 PERCENT RELATIVE HUMIDITY.
WET BULB IS THE TEMPERATURE THE AIR WOULD HAVE IF THE
  MOISTURE CONTENT WAS INCREASED TO 100 PERCENT RELATIVE
  HUMIDITY.

ON JULY 1, 1996, THE NATIONAL WEATHER SERVICE BEGAN USING
  THE "METAR" OBSERVATION CODE THAT MOST OTHER NATIONS OF THE WORLD ALREADY EMPLOYED
  BY MOST OTHER NATIONS OF THE WORLD. THE MOST NOTICEABLE
  DIFFERENCE IN THIS ANNUAL PUBLICATION WILL BE THE CHANGE
  IN UNITS FROM TENTHS TO EIGHTS(OKTAS) FOR REPORTING THE
  AMOUNT OF SKY COVER.

HEATING DEGREE DAYS (base 65 F) 2001 CONCORD, NH (CON)

YEAR	JUL	AUG	SEP	OCT	NOV	DEC	JAN	FEB	MAR	APR	MAY	JUN	TOTAL
1972-73	27	82	223	695	1007	1284	1357	1250	905	596	370	78	7874
1973-74	15	9	244	518	860	1112	1345	1223	1025	573	432	99	7455
1974-75	34	26	213	694	865	1182	1339	1218	1075	730	152	98	7626
1975-76	10	78	260	532	747	1330	1672	1162	1019	565	356	60	7791
1976-77	37	84	234	615	992	1506	1683	1242	870	594	259	119	8235
1977-78	37	58	222	551	760	1360	1466	1435	1138	725	270	72	8094
1978-79	45	34	275	563	882	1304	1284	1392	841	617	280	99	7616
1979-80	33	64	199	546	675	1098	1317	1324	1022	610	290	123	7301
1980-81	13	33	245	611	899	1417	1626	953	951	530	267	40	7585
1981-82	12	43	192	608	810	1222	1674	1233	1072	695	246	136	7943
1982-83	25	66	169	535	692	1007	1291	1086	895	588	364	82	6800
1983-84	14	33	167	521	760	1283	1516	993	1135	607	382	85	7496
1984-85	19	27	238	446	792	1072	1514	1092	901	588	286	106	7081
1985-86	9	38	166	485	785	1326	1295	1216	907	499	267	140	7133
1986-87	41	67	251	538	919	1109	1390	1199	939	542	296	77	7368
1987-88	18	89	201	589	837	1138	1436	971	626	254	137	7490
1988-89	19	60	219	622	769	1289	1211	1182	1022	703	219	80	7395
1989-90	6	53	169	484	865	1639	1121	1121	933	585	369	67	7412
1990-91	22	15	183	409	737	1049	1381	1024	893	517	186	58	6474
1991-92	19	7	238	438	754	1216	1286	1123	1070	669	319	90	7229
1992-93	52	33	203	617	875	1190	1277	1372	1062	574	248	85	7588
1993-94	6	15	229	600	851	1177	1661	1345	1021	568	338	50	7861
1994-95	1	54	210	479	699	1059	1137	1235	893	697	309	46	6819
1995-96	6	17	257	397	918	1298	1352	1198	1093	625	358	54	7573
1996-97	10	6	171	572	943	1038	1334	1023	1086	675	409	83	7350
1997-98	16	18	178	560	858	1138	1161	950	870	554	186	91	6580
1998-99	5	15	144	475	773	1044	1365	1037	940	598	249	51	6696
1999-00	10	38	116	571	687	1087	1387	1142	815	583	316	98	6850
2000-01	20	26	217	486	790	1310	1403	1157	1095	627	276	68	7475
2001-	43	1	167	484	719	1011							

COOLING DEGREE DAYS (base 65 F) 2001 CONCORD, NH (CON)

YEAR	JAN	FEB	MAR	APR	MAY	JUN	JUL	AUG	SEP	OCT	NOV	DEC	ANNUAL
1972	0	0	0	0	17	52	167	60	5	0	0	0	301
1973	0	0	0	3	2	145	184	242	67	2	0	0	645
1974	0	0	0	3	9	40	118	92	40	0	0	0	302
1975	0	0	0	0	48	108	263	134	0	0	0	0	553
1976	0	0	0	18	9	184	100	99	8	1	0	0	419
1977	0	0	0	5	57	64	168	156	38	0	0	0	488
1978	0	0	0	0	46	83	198	186	8	0	0	0	521
1979	0	0	0	1	15	69	232	150	46	6	0	0	519
1980	0	0	0	0	5	64	193	145	51	0	0	0	458
1981	0	0	0	0	38	80	172	101	11	0	0	0	402
1982	0	0	0	0	11	17	171	87	31	0	0	0	317
1983	0	0	0	0	2	93	179	172	93	5	0	0	544
1984	0	0	0	0	0	15	117	139	165	28	0	0	464
1985	0	0	0	0	16	27	184	93	36	1	0	0	357
1986	0	0	0	0	44	37	118	109	23	0	0	0	331
1987	0	0	0	1	45	64	198	96	21	0	0	0	425
1988	0	0	0	0	19	81	262	238	7	4	0	0	611
1989	0	0	0	0	0	17	87	153	141	50	0	0	448
1990	0	0	0	10	0	80	210	170	30	8	0	0	508
1991	0	0	0	2	44	84	157	166	29	4	0	0	486
1992	0	0	0	2	15	66	90	105	52	0	0	0	330
1993	0	0	0	0	0	7	90	212	176	49	0	0	534
1994	0	0	0	0	0	18	143	269	104	11	0	0	545
1995	0	0	0	0	9	84	264	159	22	0	0	0	538
1996	0	0	0	2	15	65	122	149	48	0	0	0	401
1997	0	0	0	0	2	113	149	119	28	3	0	0	414
1998	0	0	5	0	13	80	157	165	36	0	0	0	456
1999	0	0	0	0	9	157	228	112	76	0	0	0	582
2000	0	0	0	0	20	106	93	107	32	0	0	0	358
2001	0	0	0	5	18	123	101	226	32	0	0	0	505

SNOWFALL (inches) 2001 CONCORD, NH (CON)

YEAR	JUL	AUG	SEP	OCT	NOV	DEC	JAN	FEB	MAR	APR	MAY	JUN	TOTAL
1972-73	0.0	0.0	0.0	0.0	12.9	23.2	13.1	5.6	0.1	3.4	0.0	0.0	58.3
1973-74	0.0	0.0	0.0	0.0	T	6.6	15.4	7.0	6.5	3.9	0.0	0.0	39.4
1974-75	0.0	0.0	0.0	0.0	2.9	18.3	19.5	17.1	4.7	5.5	0.0	0.0	68.0
1975-76	0.0	0.0	0.0	0.0	2.2	25.1	14.7	8.4	24.3	T	0.0	0.0	74.7
1976-77	0.0	0.0	0.0	T	3.1	11.6	37.1	11.7	22.3	0.5	T	0.0	86.3
1977-78	0.0	0.0	0.0	0.0	1.5	20.2	37.1	13.5	11.5	0.4	T	0.0	84.2
1978-79	0.0	0.0	0.0	0.0	10.5	16.2	42.3	4.5	0.2	5.1	0.0	0.0	78.8
1979-80	0.0	0.0	0.0	1.3	T	2.1	3.1	11.9	8.6	T	0.0	0.0	27.0
1980-81	0.0	0.0	0.0	T	9.4	9.8	9.2	20.9	5.4	T	0.0	0.0	54.7
1981-82	0.0	0.0	0.0	0.0	T	33.0	26.2	9.0	6.5	15.3	0.0	0.0	90.0
1982-83	0.0	0.0	0.0	0.0	1.3	3.6	9.0	20.8	4.0	T	T	0.0	38.7
1983-84	0.0	0.0	0.0	0.0	T	17.5	20.4	12.7	25.0	T	T	0.0	75.6
1984-85	0.0	0.0	0.0	0.0	T	16.5	11.6	11.0	12.4	1.0	0.0	0.0	52.5
1985-86	0.0	0.0	0.0	0.0	8.3	11.2	15.1	11.5	4.4	T	T	0.0	50.5
1986-87	0.0	0.0	0.0	0.0	14.4	7.7	45.4	0.6	7.0	9.4	0.0	0.0	84.5
1987-88	0.0	0.0	0.0	0.0	5.8	12.0	19.3	23.7	4.6	0.1	0.0	0.0	65.5
1988-89	0.0	0.0	0.0	T	0.5	5.0	5.6	7.0	10.2	0.8	T	0.0	29.1
1989-90	0.0	0.0	0.0	0.0	4.7	12.0	23.1	22.0	1.3	T	0.0	0.0	63.1
1990-91	0.0	0.0	0.0	0.0	0.6	8.8	11.2	6.8	6.0	0.2	0.0	0.0	33.6
1991-92	0.0	0.0	0.0	0.0	1.3	13.5	3.8	5.3	6.8	4.5	0.0	0.0	35.2
1992-93	0.0	0.0	T	T	1.1	10.9	19.7	24.0	35.8	6.6	0.0	T	98.1
1993-94	0.0	0.0	0.0	T	0.2	10.0	36.8	14.4	23.6	T	0.0	0.0	85.0
1994-95	0.0	0.0	0.0	0.0	4.3	3.2	8.1	18.2	0.7	T	0.0	0.0	34.5
1995-96	0.0	0.0	0.0	0.0	7.0	28.0	27.3	15.6		14.3	T	0.0	
1996-97	0.0	0.0	0.0	0.0	2.2	20.0	14.3	14.1	22.7	3.4	0.0	0.0	76.7
1997-98	0.0	0.0	0.0	0.0	12.4	17.3	16.3	3.3	9.6	0.0	0.0	0.0	58.9
1998-99	0.0	0.0	0.0	0.0	0.6	5.5	18.0	1.9	12.5	T	0.0	0.0	38.5
1999-00	0.0	0.0	0.0	0.0	T	T	16.6	15.1					
2000-01							13.5	20.9	37.5	T	0.0	0.0	
2001-	0.0	0.0	0.0	0.0	0.0	6.0							
POR= 58 YRS	0.0	0.0	T	0.1	3.8	13.3	17.6	14.1	11.5	2.5	0.1	T	63.0

2001
ATLANTIC CITY, NEW JERSEY
STATE MARINA (0325)

The Atlantic City State Marina is located on Abescon Island on the southeast coast of New Jersey. Surrounding terrain, composed of tidal marshes and beach sand, is flat and lies slightly above sea level. The climate is principally continental in character. However, the moderating influence of the Atlantic Ocean is apparent throughout the year, being more marked in the city than at the airport. As a result, summers are relatively cooler and winters milder than elsewhere at the same latitude.

Land and sea breezes, local circulations resulting from the differential heating and cooling of the land and sea, often prevail. These winds occur when moderate or intense storms are not present in the area, thus enabling the local circulation to overcome the general wind pattern. During the warm season sea breezes in the late morning and afternoon hours prevent excessive heating. Frequently, the temperature at Atlantic City during the afternoon hours in the summer averages several degrees lower than at the airport and the airport averages several degrees lower than localities farther inland. On occasions, sea breezes have lowered the temperature as much as 15 to 20 degrees within a half hour. However, the major effect of the sea breeze at the airport is preventing the temperature from rising above the 80s. Because the change in ocean temperature lags behind the air temperature from season to season, the weather tends to remain comparatively mild late into the fall, but on the other hand, warming is retarded in the spring. Normal ocean temperatures range from an average near 37 degrees in January to near 72 degrees in August.

Precipitation is moderate and well distributed throughout the year, with June the driest month and August the wettest. Tropical storms or hurricanes occasionally bring excessive rainfall to the area. The bulk of winter precipitation results from storms which move northeastward along or near the east coast of the United States. Snowfall is considerably less than elsewhere at the same latitude and does not remain long on the ground. Precipitation, often beginning as snow, will frequently become mixed with or change to rain while continuing as snow over more interior sections. In addition, ice storms and resultant glaze are relatively infrequent.

NORMALS, MEANS, AND EXTREMES
ATLANTIC CITY C.O., NJ (0325)

LATITUDE:	LONGITUDE:	ELEVATION (FT):		TIME ZONE:	WBAN: 13724
39 23' 0 " N	74 26' 0 " W	GRND: 11	BARO: 11	EASTERN (UTC + 5)	

	ELEMENT	POR	JAN	FEB	MAR	APR	MAY	JUN	JUL	AUG	SEP	OCT	NOV	DEC	YEAR
TEMPERATURE F	NORMAL DAILY MAXIMUM	30	39.5	41.5	48.8	57.3	65.7	74.4	80.1	79.8	73.8	64.3	54.8	45.5	60.5
	MEAN DAILY MAXIMUM	48	41.1	42.8	48.7	57.5	66.2	74.8	80.6	79.7	72.4	64.7	55.1	46.2	60.8
	HIGHEST DAILY MAXIMUM	41	70	72	82	91	94	97	101	102	92	89	78	74	102
	YEAR OF OCCURRENCE		1967	1976	1998	1960	1969	1991	1999	1983	1985	1997	1993	1998	AUG 1983
	MEAN OF EXTREME MAXS.	48	57.8	58.8	65.4	75.8	83.3	89.4	92.4	90.7	84.1	78.3	69.6	60.9	75.5
	NORMAL DAILY MINIMUM	30	26.8	28.8	36.0	44.1	53.6	62.3	68.3	68.0	61.8	51.0	41.7	32.5	47.9
	MEAN DAILY MINIMUM	48	28.4	29.8	36.0	44.4	53.8	63.0	68.9	68.4	60.7	51.5	41.8	33.1	48.3
	LOWEST DAILY MINIMUM	41	-3	1	2	22	36	45	53	50	42	27	8	4	-3
	YEAR OF OCCURRENCE		1994	1961	1991	1982	1966	1972	1988	1986	1974	1974	2000	1983	JAN 1994
	MEAN OF EXTREME MINS.	48	12.2	14.6	21.0	32.3	43.7	52.5	60.2	58.5	47.6	37.3	26.7	17.5	35.3
	NORMAL DRY BULB	30	33.1	35.2	42.5	50.7	59.7	68.4	74.2	73.9	67.8	57.6	48.2	39.1	54.2
	MEAN DRY BULB	48	34.7	36.3	42.5	50.9	60.0	68.8	74.8	74.1	66.5	58.1	48.4	39.7	54.6
	MEAN WET BULB														
	MEAN DEW POINT														
	NORMAL NO. DAYS WITH:														
	MAXIMUM 90	30	0.0	0.0	0.0	*	0.4	0.6	2.2	1.4	0.4	0.0	0.0	0.0	5.0
	MAXIMUM 32	30	7.2	4.3	0.4	0.0	0.0	0.0	0.0	0.0	0.0	0.0	*	2.4	14.3
	MINIMUM 32	30	22.5	18.5	9.5	1.3	0.0	0.0	0.0	0.0	0.0	0.4	4.4	15.7	72.3
	MINIMUM 0	30	*	0.0	0.0	0.0	0.0	0.0	0.0	0.0	0.0	0.0	*	0.0	0.0
H/C	NORMAL HEATING DEG. DAYS	30	989	834	698	429	179	19	0	0	30	243	504	803	4728
	NORMAL COOLING DEG. DAYS	30	0	0	0	0	15	121	285	276	114	13	0	0	824
RH	NORMAL (PERCENT)														
	HOUR 01 LST														
	HOUR 07 LST														
	HOUR 13 LST														
	HOUR 19 LST														
S	PERCENT POSSIBLE SUNSHINE														
W/O	MEAN NO. DAYS WITH:														
	HEAVY FOG(VISBY 1/4 MI)														
	THUNDERSTORMS														
CLOUDINESS	MEAN:														
	SUNRISE-SUNSET (OKTAS)														
	MIDNIGHT-MIDNIGHT (OKTAS)														
	MEAN NO. DAYS WITH:														
	CLEAR														
	PARTLY CLOUDY														
	CLOUDY														
PR	MEAN STATION PRESSURE(IN)														
	MEAN SEA-LEVEL PRES. (IN)														
WINDS	MEAN SPEED (MPH)														
	PREVAIL.DIR(TENS OF DEGS)														
	MAXIMUM 2-MINUTE:														
	SPEED (MPH)	11	43	43	63	37	37	30	44	30	32	44	36	45	63
	DIR. (TENS OF DEGS)		04	04	07	01	04	35	19	03	07	04	04	01	07
	YEAR OF OCCURRENCE		1996	1998	1984	1996	1998	2001	1996	1991	1992	1996	1997	1992	MAR 1984
	PEAK GUST :														
	SPEED (MPH)	14	58	60	87	63	52	64	52	67	78	77	64	67	87
	DIR. (TENS OF DEGS)		SSW	WNW	ENE	NE	NE	SW	SE	ENE	WSW	SE	NW	E	ENE
	YEAR OF OCCURRENCE		1992	1985	1984	1987	1985	1986	1984	1986	1985	1991	1995	1992	MAR 1984
PRECIPITATION	NORMAL (IN)	30	3.27	3.02	3.36	3.25	2.96	2.56	3.31	3.84	2.65	2.40	3.19	3.29	37.10
	MAXIMUM MONTHLY (IN)	41	8.40	6.29	8.48	6.68	6.01	7.28	15.69	14.77	6.87	5.47	7.57	6.87	15.69
	YEAR OF OCCURRENCE		1987	1979	1994	1987	1984	1973	1959	1967	1996	1959	1972	1996	JUL 1959
	MINIMUM MONTHLY (IN)	41	0.35	0.79	0.68	0.82	0.28	0.61	0.57	0.93	0.39	0.01	0.82	0.71	0.01
	YEAR OF OCCURRENCE		1981	1980	1966	1976	1986	1971	1981	1964	1959	2000	1965	1985	OCT 2000
	MAXIMUM IN 24 HOURS (IN)	41	3.40	2.91	3.30	2.73	3.13	4.10	6.62	8.60	3.77	2.67	3.90	3.32	8.60
	YEAR OF OCCURRENCE		1962	1998	1993	1991	1978	1968	1969	1967	2000	1972	1977	1993	AUG 1967
	NORMAL NO. DAYS WITH:														
	PRECIPITATION 0.01	30	10.5	9.4	10.2	10.3	10.2	8.9	8.0	8.6	7.5	7.2	9.4	10.2	110.4
	PRECIPITATION 1.00	30	0.6	0.6	0.7	0.7	0.5	0.5	1.0	1.2	0.8	0.6	0.7	0.6	8.5
SNOWFALL	NORMAL (IN)														
	MAXIMUM MONTHLY (IN)														
	YEAR OF OCCURRENCE														
	MAXIMUM IN 24 HOURS (IN)														
	YEAR OF OCCURRENCE														
	MAXIMUM SNOW DEPTH (IN)	11	11	7	7	0	0	0	0	0	0	0	0	2	11
	YEAR OF OCCURRENCE		1958	1958	1952									1957	JAN 1958
	NORMAL NO. DAYS WITH:														
	SNOWFALL 1.0														

PRECIPITATION (inches) 2001 ATLANTIC CITY, NEW JERSEY NJ (0325)

YEAR	JAN	FEB	MAR	APR	MAY	JUN	JUL	AUG	SEP	OCT	NOV	DEC	ANNUAL
1971	2.89	4.31	1.84	1.35	2.14	0.61	4.20	10.37	5.47	3.60	4.29	2.85	43.92
1972	3.25	3.76	3.87	3.97	5.10	5.21	3.73	1.46	4.35	4.43	7.57	4.98	51.68
1973	2.87	3.51	2.64	4.03	2.83	7.28	3.21	1.06	2.96	1.80	1.33	5.23	38.75
1974	2.78	2.35	4.33	1.75	2.44	2.01	2.57	6.31	2.56	1.59	1.00	3.31	33.00
1975	4.29	3.08	3.39	3.94	2.73	3.72	5.63	2.05	3.82	1.74	2.93	2.57	39.89
1976	3.66	2.98	0.91	0.82	2.19	1.59	1.74	5.01	3.08	4.94	0.95	2.42	30.29
1977	2.48	2.10	2.30	1.89	0.90	1.78	2.17	4.10	1.81	3.05	6.77	5.12	34.47
1978	5.59	1.11	3.82	1.33	6.00	2.07	4.20	4.01	0.87	1.35	2.48	3.95	36.78
1979	6.96	6.29	3.28	4.00	3.55	3.12	2.21	5.93	3.68	2.07	3.80	1.56	46.45
1980	2.97	0.79	6.39	4.38	1.47	2.73	2.25	1.41	2.08	3.63	3.11	1.13	32.34
1981	0.35	4.08	1.98	4.96	2.16	2.49	0.57	3.75	2.09	2.98	1.28	3.83	30.52
1982	2.81	2.03	2.48	3.98	2.24	2.88	4.14	2.08	1.16	1.11	4.05	3.04	32.00
1983	2.22	2.82	5.63	5.17	3.75	2.50	0.71	2.43	1.88	3.41	6.64	3.47	40.63
1984	2.38	4.32	6.50	4.46	6.01	1.44	4.24	1.03	2.08	1.53	2.28	1.53	37.80
1985	2.04	1.73	2.58	0.97	5.01	2.39	3.37	3.43	2.07	1.08	2.16	0.71	27.54
1986	2.46	2.83	1.46	2.79	0.28			6.11	1.88	3.55	4.19	5.01	
1987	8.40	1.55	4.10	6.68	3.84	0.71	3.11	2.91	3.27	2.81	2.41	1.99	41.78
1988	3.07	3.74	2.00	2.37	3.28	1.55	2.95	3.73	2.22				
1990						1.65	4.06	3.70	1.44	1.98	1.95	3.67	
1991	5.81	1.36	4.90	5.41	0.34	0.81	2.59	4.86	2.86	2.70	1.38	4.29	37.31
1992	1.20	2.40	2.97	1.43	2.81	1.75	5.89	5.11	5.01	1.42	3.25	4.25	37.49
1993	2.75	2.39	8.20	1.96	2.24	1.17	1.89	4.53	1.24	3.81	1.29	5.21	36.68
1994	4.00	3.25	8.48	3.37	1.86	3.68	4.03	2.80	1.07	2.59	2.32	39.54	
1995	2.79	2.30	0.96	2.54	4.49	2.86	2.07	2.40	3.87	5.09	4.55	2.14	36.06
1996	2.68	2.17	3.09	3.18	3.90	2.73	3.49	4.75	6.87	5.03	1.09	6.87	45.93
1997	2.40	3.36	3.88	3.13	3.21	1.73	8.38	6.01	1.06	3.45	3.74	4.77	45.12
1998	5.17	5.99	6.29	3.77	5.79	3.61	1.58	1.84	1.88	2.50	0.89	2.45	41.76
1999	6.40	2.91	3.93	2.90	0.71	2.09	1.20	5.24	5.81	3.90	0.92	2.34	38.35
2000	3.38	2.14	4.67	3.05	2.21	4.30	6.04	11.77	6.42	0.01	2.13	1.42	47.54
2001	2.68	2.31	4.04	1.42	0.78	4.59	2.52	2.88	1.27	0.99	0.85	1.26	25.59
POR= 42 YRS	3.32	3.07	3.80	3.15	2.93	2.55	3.81	4.20	2.88	2.62	2.81	3.29	38.43

AVERAGE TEMPERATURE (F) 2001 ATLANTIC CITY, NEW JERSEY NJ (0325)

YEAR	JAN	FEB	MAR	APR	MAY	JUN	JUL	AUG	SEP	OCT	NOV	DEC	ANNUAL
1971	31.5	38.1	40.9	48.7	56.9	67.9	71.9	71.0	67.9	60.0	46.2	44.9	53.8
1972	38.6	36.6	40.4	48.3	58.1	64.5	73.1	72.0	67.5	54.3	46.4	43.6	53.6
1973	39.0	38.4	48.4	54.0	57.2	68.8	73.7	75.1	69.8	60.4	48.9	40.8	56.2
1974	39.1	34.1	43.6	52.5	59.8	67.5	75.6	73.4	65.8	53.3	47.4	39.5	54.3
1975	37.4	36.0	39.4	45.9	61.3	68.8	73.5	74.9	64.7	59.3	50.8	38.5	54.2
1976	32.7	42.0	45.0	53.4	58.0	69.1	71.8	71.4	65.9	53.1	41.6	34.5	53.2
1977	24.4	35.6	45.7	52.2	60.9	67.1	75.3	76.1	71.1	56.6	48.9	36.9	54.2
1978	31.2	27.9	39.8	49.6	55.3	65.8	70.6	74.4	65.0	55.2	48.2	39.3	51.9
1979	33.1	27.4	43.3	50.5	59.3	66.6	72.8	73.1	67.5	56.2	51.4	41.9	53.6
1980	35.5	32.7	40.7	52.2	62.3	67.0	72.9	75.1	69.4	55.6		36.7	
1981	29.9	39.1	41.4	53.2	60.7	71.4	77.5	74.7	67.3	54.6	46.9	37.9	54.6
1982	29.1	35.8	42.2	48.7	60.2	66.1	72.8	69.6	66.0	56.3	48.8	40.7	53.0
1983	34.7	39.0	46.0	51.3	59.0	67.5	76.0	74.6	68.7	57.6	48.3	35.4	54.8
1984	30.4	41.3	39.6	51.1	60.7	72.8	73.4	77.4	67.4	62.8	47.6	47.2	56.0
1985	30.6	36.9	46.4	54.1	62.0	68.4	75.0	74.4	70.4	61.7	54.4	36.4	55.9
1986	35.5	34.0	43.6	51.5	62.4	69.4	77.0	73.3	68.3	59.8	48.4	41.1	55.4
1987	35.1	34.8	44.1	49.8		70.4	76.3	74.5	69.0	54.4	49.4	40.6	
1988	30.5	35.2	43.0	49.4	59.4	68.4	73.6	73.8	66.2				
1990						69.9	75.0	75.4	67.9	62.2	50.1	44.0	
1991	37.5	40.9	44.9	53.0	66.0	71.8	76.4	76.7	68.3	59.1	48.6	41.5	57.1
1992	36.9	37.9	41.0	48.7	56.8	66.7	74.9	72.6	67.8	55.7	48.7	40.3	54.0
1993	39.8	32.9	39.2	49.2	61.4	69.6	78.1	75.1	68.8	57.9	49.3	38.4	55.0
1994	29.5	34.3	41.8	54.5	58.5	70.6	75.5	71.0	67.4	57.6	53.2	43.7	54.8
1995	39.2	33.5	44.1	50.8	60.5	69.2		75.6	68.9	62.8	45.4	34.8	
1996	33.6	35.5	38.3	50.3	58.9	70.9	73.3	73.6		58.5	43.6	43.4	
1997	35.1	41.5	44.5	51.1	58.6	68.4	75.2	74.4	69.2	59.2	47.6	41.6	55.5
1998	43.1	43.0	44.7	53.3	62.5	70.5	76.8	76.6	72.2	59.6	50.3	44.1	58.1
1999	38.7	39.1	42.7	52.5	61.6	70.1	78.6	76.0	70.2	59.7	52.7	43.8	57.1
2000	34.7	39.8	47.2	51.4	62.6	70.9	73.1	74.4	68.3	59.5	46.9	33.7	55.2
2001	35.5	38.7	42.0	52.6	62.5	73.5	72.9	76.7	68.3	59.6	53.9	46.3	56.9
POR= 42 YRS	33.9	35.9	42.2	50.8	60.1	68.9	74.6	74.3	66.5	57.9	48.6	39.2	54.4

REFERENCE NOTES:

PAGE 1:
THE TEMPERATURE GRAPH SHOWS NORMAL MAXIMUM AND NORMAL MINIMUM DAILY TEMPERATURES (SOLID CURVES) AND THE ACTUAL DAILY HIGH AND LOW TEMPERATURES (VERTICAL BARS).

PAGE 2 AND 3:
H/C INDICATES HEATING AND COOLING DEGREE DAYS.
RH INDICATES RELATIVE HUMIDITY
W/O INDICATES WEATHER AND OBSTRUCTIONS
S INDICATES SUNSHINE.
PR INDICATES PRESSURE.
CLOUDINESS ON PAGE 3 IS THE SUM OF THE CEILOMETER AND SATELLITE DATA NOT TO EXCEED EIGHT EIGHTHS(OKTAS).

GENERAL:
T INDICATES TRACE PRECIPITATION, AN AMOUNT GREATER THAN ZERO BUT LESS THAN THE LOWEST REPORTABLE VALUE.
+ INDICATES THE VALUE ALSO OCCURS ON EARLIER DATES.
BLANK ENTRIES DENOTE MISSING OR UNREPORTED DATA.
NORMALS ARE 30-YEAR AVERAGES (1961 - 1990).
ASOS INDICATES AUTOMATED SURFACE OBSERVING SYSTEM.
PM INDICATES THE LAST DAY OF THE PREVIOUS MONTH.
POR (PERIOD OF RECORD) BEGINS WITH THE JANUARY DATA MONTH AND IS THE NUMBER OF YEARS USED TO COMPUTE THE MEAN. INDIVIDUAL MONTHS WITHIN THE POR MAY BE MISSING.
WHEN THE POR FOR A NORMAL IS LESS THAN 30 YEARS, THE NORMAL IS PROVISIONAL AND IS BASED ON THE NUMBER OF YEARS INDICATED.
0.* OR * INDICATES THE VALUE OR MEAN-DAYS-WITH IS BETWEEN 0.00 AND 0.05.
CLOUDINESS FOR ASOS STATIONS DIFFERS FROM THE NON-ASOS OBSERVATION TAKEN BY A HUMAN OBSERVER. ASOS STATION CLOUDINESS IS BASED ON TIME-AVERAGED CEILOMETER DATA FOR CLOUDS AT OR BELOW 12,000 FEET AND ON SATELLITE DATA FOR CLOUDS ABOVE 12,000 FEET.
THE NUMBER OF DAYS WITH CLEAR, PARTLY CLOUDY, AND CLOUDY CONDITIONS FOR ASOS STATIONS IS THE SUM OF THE CEILOMETER AND SATELLITE DATA FOR THE SUNRISE TO SUNSET PERIOD.

GENERAL CONTINUED:
CLEAR INDICATES 0 - 2 OKTAS, PARTLY CLOUDY INDICATES 3 - 6 OKTAS, AND CLOUDY INDICATES 7 OR 8 OKTAS. WHEN AT LEAST ONE OF THE ELEMENTS (CEILOMETER OR SATELLITE) IS MISSING, THE DAILY CLOUDINESS IS NOT COMPUTED.
WIND DIRECTION IS RECORDED IN TENS OF DEGREES (2 DIGITS) CLOCKWISE FROM TRUE NORTH. "00" INDICATES CALM. "36" INDICATES TRUE NORTH.
RESULTANT WIND IS THE VECTOR AVERAGE OF THE SPEED AND DIRECTION.
AVERAGE TEMPERATURE IS THE SUM OF THE MEAN DAILY MAXIMUM AND MINIMUM TEMPERATURE DIVIDED BY 2.
SNOWFALL DATA COMPRISE ALL FORMS OF FROZEN PRECIPITATION, INCLUDING HAIL.
A HEATING (COOLING) DEGREE DAY IS THE DIFFERENCE BETWEEN THE AVERAGE DAILY TEMPERATURE AND 65 F.
DRY BULB IS THE TEMPERATURE OF THE AMBIENT AIR.
DEW POINT IS THE TEMPERATURE TO WHICH THE AIR MUST BE COOLED TO ACHIEVE 100 PERCENT RELATIVE HUMIDITY.
WET BULB IS THE TEMPERATURE THE AIR WOULD HAVE IF THE MOISTURE CONTENT WAS INCREASED TO 100 PERCENT RELATIVE HUMIDITY.

ON JULY 1, 1996, THE NATIONAL WEATHER SERVICE BEGAN USING THE "METAR" OBSERVATION CODE THAT WAS ALREADY EMPLOYED BY MOST OTHER NATIONS OF THE WORLD. THE MOST NOTICEABLE DIFFERENCE IN THIS ANNUAL PUBLICATION WILL BE THE CHANGE IN UNITS FROM TENTHS TO EIGHTS(OKTAS) FOR REPORTING THE AMOUNT OF SKY COVER.

HEATING DEGREE DAYS (base 65 F) 2001 ATLANTIC CITY, NEW JERSEY NJ (0325)

YEAR	JUL	AUG	SEP	OCT	NOV	DEC	JAN	FEB	MAR	APR	MAY	JUN	TOTAL
1972-73	2	0	23	328	553	658	796	738	508	330	245	9	4190
1973-74	0	0	16	163	476	742	798	860	655	379	173	24	4286
1974-75	0	0	57	358	524	784	848	805	786	567	151	11	4891
1975-76	0	0	49	188	422	814	994	664	612	356	212	29	4340
1976-77	0	3	38	366	696	936	1253	816	590	384	154	36	5272
1977-78	0	0	11	267	477	862	1043	1031	776	455	302	41	5265
1978-79	6	0	62	297	497	791	979	1045	666	427	173	22	4965
1979-80	5	3	27	275	404	708	906	931	746	376	115	43	4539
1980-81	0	0	19	288		871	1080	720	723	350	160	3	
1981-82	0	0	51	315	538	834	1106	813	698	482	155	39	5031
1982-83	1	10	36	280	486	749	931	722	584	405	179	17	4400
1983-84	1	9	63	236	496	911	1065	681	781	406	139	5	4793
1984-85	0	0	56	96	514	543	1059	783	571	336	119	16	4093
1985-86	0	0	19	132	313	876	910	860	654	399	144	15	4322
1986-87	0	9	20	201	491	735	922	842	641	452		3	
1987-88	0	0	14	324	460	747	1062	859	675	465	200	51	4857
1988-89	1	2	40									13	
1989-90													
1990-91	0	0	42	154	437	646	847	668	616	357	79	8	3854
1991-92	0	0	36	205	484	722	865	780	738	484	267	23	4604
1992-93	0	0	37	292	482	759	776	890	792	467	136	19	4650
1993-94	0	0	33	214	469	819	1093	851	712	313	206	3	4713
1994-95	0	3	16	221	346	652	791	875	641	420	159	5	4129
1995-96		0	21	115	583	928	967	850	820	432	226	10	
1996-97	0	0		201	638	663	923	655	624	411	199	74	
1997-98	0	0	25	221	518	719	672	609	626	344	138	10	3882
1998-99	0	0	9	171	434	641	808	720	684	368	140	4	3979
1999-00	0	0	3	181	363	651	933	722	546	401	132	19	3951
2000-01	0	0	41	192	538	964	908	730	705	371	105	4	4558
2001-	0	0	39	195	326	574							

COOLING DEGREE DAYS (base 65 F) 2001 ATLANTIC CITY, NEW JERSEY NJ (0325)

YEAR	JAN	FEB	MAR	APR	MAY	JUN	JUL	AUG	SEP	OCT	NOV	DEC	ANNUAL
1971	0	0	0	0	2	122	219	194	336	7	2	0	882
1972	0	0	0	0	0	39	258	224	103	2	0	0	626
1973	0	0	0	3	9	233	278	319	168	29	0	0	1039
1974	0	0	0	10	18	125	334	266	91	0	1	0	845
1975	0	0	0	0	42	131	269	313	50	16	0	0	821
1976	0	0	0	0	14	2	160	219	210	69	2	0	0	676
1977	0	0	0	6	37	105	324	355	200	10	0	0	1037
1978	0	0	0	0	9	72	186	297	69	0	0	0	633
1979	0	0	0	0	5	74	252	263	111	8	0	0	713
1980	0	0	0	0	39	109	250	319	159	2		0	
1981	0	0	0	3	33	200	392	305	125	1	0	0	1059
1982	0	0	0	0	13	77	247	160	70	20	3	0	590
1983	0	0	0	2	0	96	349	313	181	14	0	0	955
1984	0	0	0	0	12	246	269	393	135	36	0	0	1091
1985	0	0	2	12	34	124	315	298	188	35	0	0	1008
1986	0	0	0	0	70	154	378	274	126	47	0	0	1049
1987	0	0	0	0	0		173	357	299	142	0	0	
1988	0	0	0	0	0	34	160	275	280	83			
1990							166	318	330	136	74	1	
1991	0	0	0	3	116	219	359	370	143	30	0	0	1240
1992	0	0	0	0	21	83	315	243	126	8	0	0	796
1993	0	0	0	0	30	164	415	319	155	2	4	0	1089
1994	0	0	0	3	12	176	331	197	94	0	0	0	813
1995	0	0	0	0	24	139		338	145	56	6	0	
1996	0	0	0	0	43	192	263	275		6	0		
1997	0	0	0	0	7	179	326	299	159	50	0	0	1020
1998	0	0	2	0	66	184	376	368	236	13	0	0	1245
1999	0	0	0	0	44	164	427	348	168	25	1	0	1177
2000	0	0	0	1	65	201	261	297	146	29	0	0	1000
2001	0	0	0	0	5	33	264	253	370	147	32	0	0	1104

SNOWFALL (inches) 2001 ATLANTIC CITY, NEW JERSEY NJ (0325)

YEAR	JUL	AUG	SEP	OCT	NOV	DEC	JAN	FEB	MAR	APR	MAY	JUN	TOTAL
POR=													

2001
NEWARK,
NEW JERSEY (EWR)

Terrain in vicinity of the station is flat and rather marshy. To the northwest are ridges oriented roughly in a south-southwest to north-northeast direction. They rise to an elevation of about 200 feet at 4.5 to 5 miles and to 500 to 600 feet at 7 to 8 miles. All winds between west-northwest and north-northwest are downslope and therefore are subject to some adiabatic temperature increase. This effect is evident in the rapid improvement which normally occurs with shift of wind to westerly, following a coastal storm or frontal passage. The drying effect of the downslope winds accounts for the relatively few local thunderstorms occurring at the station, compared to areas to the west. Easterly winds, particularly southeasterly, moderate the temperature because of the influence of the Atlantic Ocean.

Temperature falls of 5 to 15 degrees, depending on the season, are not uncommon when the wind backs from southwesterly to southeasterly. Periods of very hot weather, lasting as long as a week, are associated with a west-southwest air flow which has a long trajectory over land. Extremes of cold are related to rapidly moving outbreaks of cold air traveling southeastward from the

Hudson Bay region. Temperatures of zero or below occur in one winter out of four, but are much more common several miles to the west of the station. Average dates of the last occurrence in spring and the first occurrence in autumn of temperatures as low as 32 degrees are in mid-April and the end of October or early November. Areas to the west of the station experience a growing season at least a month shorter than that at the airport.

A considerable amount of precipitation is realized from the Northeasters of the Atlantic coast. These storms, more typical of the fall and winter, generally last for a period of two days and commonly produce between 1 and 2 inches of precipitation. Storms producing 4 inches or more of snow occur from two to five times a winter. Snowstorms producing 8 inches or more have occurred in about one-half the winters. As many as three such storms have been experienced in one winter. The frequency and intensity of snow storms and the duration of snow cover increase dramatically within a few miles to the west of the station.

NORMALS, MEANS, AND EXTREMES

NEWARK, NJ (EWR)

LATITUDE:	LONGITUDE:	ELEVATION (FT):		TIME ZONE:	WBAN: 14734
40 42' 57" N	74 10' 10" W	GRND: 25	BARO: 28	EASTERN (UTC + 5)	

	ELEMENT	POR	JAN	FEB	MAR	APR	MAY	JUN	JUL	AUG	SEP	OCT	NOV	DEC	YEAR
TEMPERATURE F	NORMAL DAILY MAXIMUM	30	37.7	40.5	50.8	61.9	72.4	82.3	87.0	85.4	77.6	66.7	55.4	42.9	63.4
	MEAN DAILY MAXIMUM	66	38.6	41.1	49.9	61.3	72.0	81.0	84.7	84.1	76.7	66.2	54.3	42.8	62.7
	HIGHEST DAILY MAXIMUM	60	74	76	89	94	99	102	105	105	105	92	85	76	105
	YEAR OF OCCURRENCE		1950	1949	1945	1990	1996	1994	1993	2001	1953	1949	1950	1998	AUG 2001
	MEAN OF EXTREME MAXS.	66	58.4	60.0	71.7	82.8	89.9	95.2	96.2	95.1	91.1	82.2	72.1	61.9	79.7
	NORMAL DAILY MINIMUM	30	23.4	25.4	33.4	42.7	53.2	62.8	68.6	67.4	59.9	48.2	39.2	29.1	46.1
	MEAN DAILY MINIMUM	66	24.3	25.8	33.2	42.7	52.8	62.3	67.0	66.6	58.8	47.9	38.7	28.7	45.7
	LOWEST DAILY MINIMUM	60	-8	-7	6	16	33	43	52	45	35	28	15	-1	-8
	YEAR OF OCCURRENCE		1985	1943	1943	1982	1947	1945	1952	1982	1947	1969	1955	1980	JAN 1985
	MEAN OF EXTREME MINS.	66	7.7	9.5	18.2	30.8	41.0	51.4	58.0	56.3	49.9	34.5	24.9	13.0	32.5
	NORMAL DRY BULB	30	30.6	33.0	42.1	52.3	62.8	72.6	77.8	76.4	68.8	57.5	47.3	36.0	54.8
	MEAN DRY BULB	66	31.5	33.5	41.6	52.0	62.4	71.7	75.9	75.4	67.8	57.1	46.5	35.7	54.3
	MEAN WET BULB	18	29.4	30.6	37.0	45.8	55.9	64.5	65.0	68.1	61.7	51.5	42.2	33.4	48.8
	MEAN DEW POINT	18	21.8	22.1	28.0	37.3	49.0	58.6	60.5	63.5	56.8	45.4	35.2	25.6	42.0
	NORMAL NO. DAYS WITH:														
	MAXIMUM 90	30	0.0	0.0	0.0	0.0	0.2	1.5	4.8	8.7	6.3	1.6	0.0	0.0	23.1
	MAXIMUM 32	30	9.2	5.7	0.9	0.0	0.0	0.0	0.0	0.0	0.0	0.0	0.1	4.5	20.4
	MINIMUM 32	30	23.8	21.1	12.6	1.5	0.0	0.0	0.0	0.0	0.0	0.6	5.8	19.9	85.3
	MINIMUM 0	30	0.5	0.2	0.0	0.0	0.0	0.0	0.0	0.0	0.0	0.0	0.0	0.1	0.8
H/C	NORMAL HEATING DEG. DAYS	30	1066	896	710	381	127	0	0	0	26	252	531	899	4888
	NORMAL COOLING DEG. DAYS	30	0	0	0	0	59	232	397	353	140	20	0	0	1201
RH	NORMAL (PERCENT)	30	65	63	60	58	62	63	63	66	68	66	66	67	64
	HOUR 01 LST	30	70	68	66	65	71	72	73	75	77	75	72	71	71
	HOUR 07 LST	30	73	72	69	66	70	70	72	75	78	78	76	74	73
	HOUR 13 LST	30	58	55	50	47	50	51	51	53	54	52	56	58	53
	HOUR 19 LST	30	63	60	56	54	58	58	58	62	64	63	64	64	60
S	PERCENT POSSIBLE SUNSHINE														
W/O	MEAN NO. DAYS WITH:														
	HEAVY FOG(VISBY 1/4 MI)	61	2.1	1.7	1.4	1.1	1.6	1.1	0.4	0.5	0.8	1.9	1.7	1.7	16.0
	THUNDERSTORMS	61	0.3	0.3	1.1	1.6	3.6	4.9	5.8	4.5	2.4	1.1	0.5	0.2	26.3
CLOUDINESS	MEAN:														
	SUNRISE-SUNSET (OKTAS)	50	5.2	5.0	5.0	5.1	5.2	5.0	4.9	4.7	4.5	4.3	5.0	5.1	4.9
	MIDNIGHT-MIDNIGHT (OKTAS)	51	4.8	4.7	4.9	4.8	5.0	4.9	4.7	4.5	4.4	4.2	4.9	4.9	4.7
	MEAN NO. DAYS WITH:														
	CLEAR	54	7.7	7.3	8.0	7.2	6.3	6.7	6.5	7.7	9.5	10.8	7.5	7.9	93.1
	PARTLY CLOUDY	54	7.7	7.6	8.4	8.9	10.6	10.9	12.2	11.7	8.9	8.4	8.2	8.0	111.5
	CLOUDY	54	15.7	13.4	14.6	14.0	14.1	14.1	12.4	12.3	11.6	11.6	14.3	15.1	160.9
PR	MEAN STATION PRESSURE(IN)	28	30.01	30.02	30.00	29.90	29.99	29.91	30.00	30.00	30.02	30.06	30.10	30.09	30.01
	MEAN SEA-LEVEL PRES. (IN)	18	30.08	30.08	30.02	29.97	29.99	29.96	30.00	30.00	30.02	30.06	30.10	30.09	30.04
WINDS	MEAN SPEED (MPH)	48	11.0	11.3	11.9	11.3	10.0	9.6	9.0	8.8	9.1	9.5	10.2	10.6	10.2
	PREVAIL.DIR(TENS OF DEGS)	31	30	31	31	32	23	23	23	23	23	23	23	25	23
	MAXIMUM 2-MINUTE:														
	SPEED (MPH)	5	44	44	45	37	43	34	47	40	44	34	40	48	48
	DIR. (TENS OF DEGS)		28	27	30	01	30	30	31	33	03	32	33	27	27
	YEAR OF OCCURRENCE		1999	1997	1997	1997	1997	1998	1997	1997	1999	1999	1997	2000	DEC 2000
	MAXIMUM 5-SECOND:														
	SPEED (MPH)	5	54	56	59	48	51	58	59	53	55	47	53	62	62
	DIR. (TENS OF DEGS)		28	28	30	33	32	32	32	30	35	32	16	27	27
	YEAR OF OCCURRENCE		2000	1997	1997	1999	1997	2000	1997	1997	1998	1999	1999	2000	DEC 2000
PRECIPITATION	NORMAL (IN)	30	3.39	3.04	3.87	3.84	4.13	3.22	4.50	3.91	3.66	3.05	3.91	3.45	43.97
	MAXIMUM MONTHLY (IN)	60	10.10	4.94	11.14	11.14	10.22	6.40	9.98	11.84	10.28	8.20	11.53	9.47	11.84
	YEAR OF OCCURRENCE		1979	1979	1983	1983	1984	1975	1988	1955	1944	1943	1977	1983	AUG 1955
	MINIMUM MONTHLY (IN)	60	0.45	1.22	1.10	0.90	0.52	0.07	0.89	0.36	0.95	0.21	0.51	0.27	0.07
	YEAR OF OCCURRENCE		1981	1968	1963	1963	1964	1949	1966	1995	1951	1963	1976	1955	JUN 1949
	MAXIMUM IN 24 HOURS (IN)	48	3.59	2.45	2.83	3.73	4.22	2.97	4.64	7.84	6.41	4.04	7.22	2.77	7.84
	YEAR OF OCCURRENCE		1979	1961	1991	1984	1979	1992	1997	1971	1999	1996	1977	1983	AUG 1971
	NORMAL NO. DAYS WITH:														
	PRECIPITATION 0.01	30	10.8	9.7	11.3	11.1	11.8	10.3	9.8	9.2	8.1	7.8	10.5	10.9	121.3
	PRECIPITATION 1.00	30	0.7	0.7	1.0	1.0	0.9	0.9	1.2	1.0	1.0	0.9	0.8	0.8	10.9
SNOWFALL	NORMAL (IN)	30	8.9	9.2	3.7	0.7	T	0.0	0.0	0.0	0.0	T	0.6	3.9	27.0
	MAXIMUM MONTHLY (IN)	60	31.6	33.4	26.0	13.8	T	T	0.0	0.0	T	0.3	5.7	29.1	33.4
	YEAR OF OCCURRENCE		1996	1994	1956	1982	1995	2001			1998	1952	1989	1947	FEB 1994
	MAXIMUM IN 24 HOURS (IN)	60	27.4	20.0	17.6	12.8	T	T	0.0	0.0	T	0.3	5.7	26.0	27.4
	YEAR OF OCCURRENCE		1996	1961	1956	1982	1995	2001			1998	1952	1989	1947	JAN 1996
	MAXIMUM SNOW DEPTH (IN)	15	17	25	18	11	0	0	0	0	0	0	9	22	25
	YEAR OF OCCURRENCE		1978	1961	1956	1982							1938	1947	FEB 1961
	NORMAL NO. DAYS WITH:														
	SNOWFALL 1.0	30	2.2	1.9	1.3	0.2	0.0	0.0	0.0	0.0	0.0	0.0	0.2	1.2	7.0

PRECIPITATION (inches) 2001 NEWARK, NJ (EWR)

YEAR	JAN	FEB	MAR	APR	MAY	JUN	JUL	AUG	SEP	OCT	NOV	DEC	ANNUAL
1972	2.26	4.01	3.09	3.08	6.02	6.02	4.70	2.30	1.03	4.83	8.42	4.10	49.86
1973	3.65	3.39	3.63	5.77	3.56	4.03	3.63	3.36	3.39	3.35	1.29	7.24	46.29
1974	2.84	1.44	4.11	2.37	3.49	3.60	1.31	7.17	5.76	1.85	0.80	4.02	38.76
1975	3.99	2.56	2.94	2.29	3.27	6.40	8.02	4.36	9.00	3.24	3.67	2.91	52.65
1976	5.04	2.52	2.33	2.50	4.12	1.54	3.91	2.98	2.50	5.07	0.51	2.17	35.19
1977	1.55	2.77	5.67	3.16	1.31	3.89	1.51	4.29	3.99	3.53	11.53	4.77	47.97
1978	7.76	2.26	4.58	2.60	7.97	2.05	4.99	7.30	4.23	1.64	2.66	5.37	53.41
1979	10.10	4.94	3.65	3.66	7.78	2.73	3.39	4.38	5.72	4.58	3.09	2.08	56.10
1980	1.66	1.28	9.13	7.28	2.61	3.27	2.78	0.92	1.87	3.37	3.71	0.63	38.51
1981	0.45	4.81	1.10	3.15	3.88	2.61	4.51	0.57	3.42	3.47	1.75	5.32	35.04
1982	6.77	2.36	2.82	6.20	2.96	5.28	2.86	2.78	2.39	1.68	3.16	1.32	40.58
1983	4.37	3.03	11.14	11.14	4.22	2.81	1.59	3.46	2.93	5.80	5.54	9.47	65.50
1984	2.78	4.57	6.96	6.36	10.22	4.77	8.65	1.74	2.46	3.93	2.88	3.69	59.01
1985	1.22	2.58	1.59	1.17	4.23	4.29	4.52	2.58	4.19	1.29	8.32	1.31	37.29
1986	4.44	3.88	1.95	5.88	1.41	1.71	6.62	4.16	1.96	1.93	6.78	5.23	45.95
1987	6.21	1.30	3.81	5.06	2.55	4.13	4.66	5.26	3.87	3.37	2.94	2.37	45.53
1988	3.74	4.15	2.13	1.97	5.86	1.06	9.98	1.82	1.66	2.45	7.71	0.98	43.51
1989	1.98	2.70	4.42	3.25	8.80	5.41	5.23	7.03	6.45	5.40	2.57	0.75	53.99
1990	4.72	1.71	2.81	3.98	6.87	3.68	4.98	4.72	5.11	2.82	2.82	5.19	52.30
1991	3.72	1.81	5.49	3.91	4.80	2.95	5.21	5.63	3.24	1.29	2.04	3.67	43.76
1992	1.27	1.37	3.48	1.35	3.46	4.67	4.79	3.37	2.60	0.73	5.02	4.63	36.74
1993	2.75	2.87	7.22	4.59	1.77	1.21	2.15	2.84	6.29	3.98	1.95	4.89	42.51
1994	6.09	4.77	6.90	2.98	3.64	3.58	3.57	5.01	2.26	1.04	4.36	3.12	47.32
1995	3.29	3.36	1.30	2.24	3.27	1.64	5.98	0.36	3.64	4.77	5.79	2.03	37.67
1996	5.24	2.34	4.40	5.63	2.59	5.06	8.27	2.39	6.05	6.92	2.31	6.87	58.07
1997	3.50	2.18	5.19	3.08	3.12	2.42	7.05	2.89	2.20	2.02	4.54	4.16	42.35
1998	4.93	4.77	4.14	6.17	6.52	5.98	1.34	3.20	2.72	1.81	0.86	1.03	43.47
1999	6.87	3.10	3.63	1.90	4.19	0.41	1.01	5.51	9.38	2.90	2.90	2.95	44.75
2000	3.39	1.60	3.43	3.57	5.66	3.42	6.30	4.73	4.58	0.54	2.71	3.42	43.35
2001	2.57	1.79	6.69	1.71	2.88	3.97	2.29	1.97	4.29	0.46	0.81	2.01	31.44
POR= 70 YRS	3.47	2.89	4.09	3.64	3.88	3.32	4.02	4.05	3.81	2.95	3.50	3.31	42.93

AVERAGE TEMPERATURE (F) 2001   NEWARK, NJ (EWR)

YEAR	JAN	FEB	MAR	APR	MAY	JUN	JUL	AUG	SEP	OCT	NOV	DEC	ANNUAL
1972	35.4	31.3	40.5	50.0	63.0	68.8	77.9	75.9	69.8	53.3	44.8	39.7	54.2
1973	35.5	33.3	48.6	54.2	60.4	74.6	78.7	79.6	71.0	60.3	48.8	39.4	57.0
1974	35.4	31.9	43.4	56.5	62.7	70.1	77.1	76.5	66.6	53.9	47.5	38.9	55.0
1975	36.9	35.1	39.7	47.3	65.8	71.6	76.9	75.1	64.3	59.1	51.7	35.4	54.9
1976	26.8	39.3	44.0	55.2	61.1	73.6	74.9	74.4	66.5	52.6	39.9	29.1	53.1
1977	20.9	32.8	46.8	53.7	65.4	70.3	78.2	75.1	68.0	54.5	47.1	33.3	53.8
1978	27.2	25.5	38.5	51.0	60.5	71.6	75.1	76.7	66.1	57.5	48.8	38.1	53.1
1979	32.5	23.5	46.2	52.0	64.5	69.4	77.0	76.6	69.1	56.5	51.8	40.2	54.9
1980	34.0	30.8	38.9	52.6	65.9	70.2	78.9	78.6	70.8	55.0	42.9	30.4	54.1
1981	24.1	37.6	40.2	55.3	64.0	74.6	79.3	75.1	67.2	53.1	46.0	34.6	54.3
1982	24.2	36.2	41.8	50.6	63.2	67.9	78.4	72.5	66.7	56.9	48.8	42.8	54.2
1983	35.0	35.9	44.7	52.2	60.8	73.5	79.6	77.6	70.6	57.8	47.8	34.2	55.8
1984	27.8	40.8	36.5	52.7	62.2	75.0	76.6	77.3	65.4	62.3	45.3	40.8	55.2
1985	24.9	33.5	44.5	57.0	67.1	69.4	76.3	75.6	70.2	58.5	49.5	33.3	55.0
1986	33.0	31.1	44.2	53.4	66.7	72.7	76.9	74.2	68.6	58.0	45.0	38.1	55.2
1987	31.5	33.0	45.0	53.9	63.9	74.5	79.4	75.3	68.7	53.7	47.6	38.4	55.4
1988	28.7	34.4	43.9	51.1	63.4	73.0	80.5	79.8	68.0	52.6	48.9	35.5	55.0
1989	37.0	34.2	42.4	52.5	63.2	74.3	77.2	76.3	69.9	59.1	45.0	25.6	54.7
1990	40.4	39.8	44.9	53.3	61.1	73.4	77.8	76.6	68.6	62.4	50.0	42.3	57.6
1991	33.6	38.6	44.4	54.8	68.9	74.2	77.9	77.7	68.0	58.3	47.6	38.8	56.9
1992	35.2	36.0	39.3	50.2	61.7	72.6	76.9	75.1	69.6	55.7	47.8	38.8	54.9
1993	37.6	31.0	40.2	54.3	67.0	75.8	82.6	79.2	69.2	56.4	47.8	37.2	56.5
1994	25.4	30.4	41.6	57.4	63.7	77.8	81.9	75.7	69.7	58.7	52.0	41.4	56.3
1995	37.5	30.8	45.5	52.6	62.7	73.1	79.6	78.5	68.6	61.0	42.9	31.7	55.4
1996	29.7	33.6	38.8	53.1	61.6	72.9	73.9	74.0	68.0	55.5	41.9	40.2	53.6
1997	31.1	39.4	41.8	50.9	59.2	70.9	76.8	73.6	66.9	56.5	43.7	37.6	54.0
1998	40.1	40.8	45.2	53.9	64.9	70.1	77.6	77.0	70.4	57.6	47.6	41.9	57.3
1999	33.5	37.8	42.9	53.4	63.3	74.2	80.9	76.2	69.3	55.5	50.1	39.4	56.4
2000	31.5	37.4	47.8	51.4	64.2	72.4	73.7	73.3	66.4	57.0	45.2	30.7	54.3
2001	32.2	35.7	40.0	53.4	64.0	73.9	74.1	79.1	67.4	57.8	51.9	43.6	56.1
POR= 71 YRS	31.7	33.3	41.3	51.7	62.4	71.6	75.7	75.2	67.8	56.9	46.4	35.7	54.1

REFERENCE NOTES:

PAGE 1:
THE TEMPERATURE GRAPH SHOWS NORMAL MAXIMUM AND NORMAL MINIMUM DAILY TEMPERATURES (SOLID CURVES) AND THE ACTUAL DAILY HIGH AND LOW TEMPERATURES (VERTICAL BARS).

PAGE 2 AND 3:
H/C INDICATES HEATING AND COOLING DEGREE DAYS.
RH INDICATES RELATIVE HUMIDITY
W/O INDICATES WEATHER AND OBSTRUCTIONS
S INDICATES SUNSHINE.
PR INDICATES PRESSURE.
CLOUDINESS ON PAGE 3 IS THE SUM OF THE CEILOMETER AND SATELLITE DATA NOT TO EXCEED EIGHT EIGHTHS(OKTAS).

GENERAL:
T INDICATES TRACE PRECIPITATION, AN AMOUNT GREATER THAN ZERO BUT LESS THAN THE LOWEST REPORTABLE VALUE.
+ INDICATES THE VALUE ALSO OCCURS ON EARLIER DATES.
BLANK ENTRIES DENOTE MISSING OR UNREPORTED DATA.
NORMALS ARE 30-YEAR AVERAGES (1961 - 1990).
ASOS INDICATES AUTOMATED SURFACE OBSERVING SYSTEM.
PM INDICATES THE LAST DAY OF THE PREVIOUS MONTH.
POR (PERIOD OF RECORD) BEGINS WITH THE JANUARY DATA MONTH AND IS THE NUMBER OF YEARS USED TO COMPUTE THE MEAN. INDIVIDUAL MONTHS WITHIN THE POR MAY BE MISSING.
WHEN THE POR FOR A NORMAL IS LESS THAN 30 YEARS, THE NORMAL IS PROVISIONAL AND IS BASED ON THE NUMBER OF YEARS INDICATED.
0.* OR * INDICATES THE VALUE OR MEAN-DAYS-WITH IS BETWEEN 0.00 AND 0.05.
CLOUDINESS FOR ASOS STATIONS DIFFERS FROM THE NON-ASOS OBSERVATION TAKEN BY A HUMAN OBSERVER. ASOS STATION CLOUDINESS IS BASED ON TIME-AVERAGED CEILOMETER DATA FOR CLOUDS AT OR BELOW 12,000 FEET AND ON SATELLITE DATA FOR CLOUDS ABOVE 12,000 FEET.
THE NUMBER OF DAYS WITH CLEAR, PARTLY CLOUDY, AND CLOUDY CONDITIONS FOR ASOS STATIONS IS THE SUM OF THE CEILOMETER AND SATELLITE DATA FOR THE SUNRISE TO SUNSET PERIOD.

GENERAL CONTINUED:
CLEAR INDICATES 0 - 2 OKTAS, PARTLY CLOUDY INDICATES 3 - 6 OKTAS, AND CLOUDY INDICATES 7 OR 8 OKTAS.
WHEN AT LEAST ONE OF THE ELEMENTS (CEILOMETER OR SATELLITE) IS MISSING, THE DAILY CLOUDINESS IS NOT COMPUTED.
WIND DIRECTION IS RECORDED IN TENS OF DEGREES (2 DIGITS) CLOCKWISE FROM TRUE NORTH. "00" INDICATES CALM. "36" INDICATES TRUE NORTH.
RESULTANT WIND IS THE VECTOR AVERAGE OF THE SPEED AND DIRECTION.
AVERAGE TEMPERATURE IS THE SUM OF THE MEAN DAILY MAXIMUM AND MINIMUM TEMPERATURE DIVIDED BY 2.
SNOWFALL DATA COMPRISE ALL FORMS OF FROZEN PRECIPITATION, INCLUDING HAIL.
A HEATING (COOLING) DEGREE DAY IS THE DIFFERENCE BETWEEN THE AVERAGE DAILY TEMPERATURE AND 65 F.
DRY BULB IS THE TEMPERATURE OF THE AMBIENT AIR.
DEW POINT IS THE TEMPERATURE TO WHICH THE AIR MUST BE COOLED TO ACHIEVE 100 PERCENT RELATIVE HUMIDITY.
WET BULB IS THE TEMPERATURE THE AIR WOULD HAVE IF THE MOISTURE CONTENT WAS INCREASED TO 100 PERCENT RELATIVE HUMIDITY.

ON JULY 1, 1996, THE NATIONAL WEATHER SERVICE BEGAN USING THE "METAR" OBSERVATION CODE THAT WAS ALREADY EMPLOYED BY MOST OTHER NATIONS OF THE WORLD. THE MOST NOTICEABLE DIFFERENCE IN THIS ANNUAL PUBLICATION WILL BE THE CHANGE IN UNITS FROM TENTHS TO EIGHTS(OKTAS) FOR REPORTING THE AMOUNT OF SKY COVER.

HEATING DEGREE DAYS (base 65 F) 2001 NEWARK, NJ (EWR)

YEAR	JUL	AUG	SEP	OCT	NOV	DEC	JAN	FEB	MAR	APR	MAY	JUN	TOTAL
1972-73	0	0	22	356	599	776	906	882	504	339	163	1	4548
1973-74	0	0	18	166	479	787	909	921	661	273	127	12	4353
1974-75	0	0	62	341	521	802	864	832	775	524	84	6	4811
1975-76	0	1	59	195	400	913	1177	738	645	338	141	17	4624
1976-77	0	4	56	381	745	1107	1361	895	563	352	89	24	5577
1977-78	0	0	50	319	527	975	1168	1099	814	411	190	13	5566
1978-79	6	0	66	239	481	830	1001	1155	577	386	68	11	4820
1979-80	2	4	28	289	393	763	953	987	802	366	62	24	4673
1980-81	0	0	28	314	654	1066	1261	762	764	290	96	0	5235
1981-82	0	0	52	360	563	934	1258	802	712	433	85	42	5241
1982-83	0	13	36	267	493	679	923	810	622	395	162	5	4405
1983-84	0	0	52	249	510	949	1144	696	874	366	128	9	4977
1984-85	0	0	83	114	584	745	1235	877	641	268	62	15	4624
1985-86	0	0	21	212	462	971	985	942	642	341	89	7	4672
1986-87	0	11	22	240	594	826	1030	893	616	331	140	3	4706
1987-88	0	1	25	342	518	818	1117	880	647	410	120	28	4906
1988-89	1	0	18	386	476	906	859	853	698	366	132	6	4701
1989-90	0	0	37	190	594	1215	756	699	622	369	122	2	4606
1990-91	1	1	50	163	446	697	734	630	330	63	4	4086
1991-92	0	0	55	227	513	804	917	834	790	441	148	4	4733
1992-93	0	0	38	295	510	807	842	946	765	318	42	4	4567
1993-94	0	0	48	263	513	853	1219	964	718	242	104	0	4924
1994-95	0	0	7	195	387	724	848	952	596	371	112	0	4192
1995-96	0	0	32	163	657	1026	1091	906	809	369	176	7	5236
1996-97	0	0	46	291	685	763	1043	714	713	418	179	43	4895
1997-98	1	0	51	294	635	842	765	672	633	327	94	22	4336
1998-99	0	0	19	227	516	711	970	754	674	345	92	2	4310
1999-00	0	2	22	290	439	786	1031	793	521	399	106	22	4411
2000-01	0	0	80	251	586	1056	1010	813	768	353	106	7	5030
2001-	0	0	53	241	389	658							

COOLING DEGREE DAYS (base 65 F) 2001 NEWARK, NJ (EWR)

YEAR	JAN	FEB	MAR	APR	MAY	JUN	JUL	AUG	SEP	OCT	NOV	DEC	ANNUAL
1972	0	0	3	4	41	142	406	347	175	3	0	0	1121
1973	0	0	0	20	26	296	432	459	205	28	0	0	1466
1974	0	0	0	28	64	172	381	361	115	1	3	0	1125
1975	0	0	0	0	117	211	375	321	46	20	10	0	1100
1976	0	0	0	50	30	281	317	305	110	6	0	0	1099
1977	0	0	6	18	111	191	414	321	146	1	0	0	1208
1978	0	0	0	0	59	217	325	367	105	15	0	0	1088
1979	0	0	0	2	59	147	381	372	158	34	3	0	1156
1980	0	0	0	0	97	187	435	427	209	10	0	0	1365
1981	0	0	0	6	75	293	446	319	124	0	0	0	1263
1982	0	0	0	6	39	136	421	249	95	24	12	0	982
1983	0	0	0	19	39	268	458	396	226	36	0	0	1442
1984	0	0	0	2	47	316	365	388	102	36	0	0	1256
1985	0	0	11	36	134	152	357	335	183	19	3	0	1230
1986	0	0	2	2	149	243	380	303	136	30	0	0	1245
1987	0	0	0	6	116	293	453	327	143	0	1	0	1339
1988	0	0	0	0	75	274	488	465	115	10	0	0	1427
1989	0	0	3	1	81	294	385	360	194	16	0	0	1334
1990	0	0	7	23	11	262	403	365	165	89	2	0	1327
1991	0	0	0	28	190	288	406	399	151	28	0	0	1490
1992	0	0	0	4	52	242	373	323	185	15	0	0	1194
1993	0	0	0	5	113	340	553	450	182	9	4	0	1656
1994	0	0	0	23	74	389	530	338	155	5	4	0	1518
1995	0	0	0	3	49	247	460	425	147	44	0	0	1375
1996	0	0	0	21	77	252	282	288	144	2	2	0	1068
1997	0	0	0	0	6	229	375	278	115	37	0	0	1040
1998	0	0	29	0	95	182	398	377	188	4	0	0	1273
1999	0	0	0	2	47	283	499	360	158	3	0	0	1352
2000	0	0	0	0	89	251	278	265	129	11	0	0	1023
2001	0	0	0	11	84	282	291	442	131	27	1	0	1269

SNOWFALL (inches) 2001 NEWARK, NJ (EWR)

YEAR	JUL	AUG	SEP	OCT	NOV	DEC	JAN	FEB	MAR	APR	MAY	JUN	TOTAL
1972-73	0.0	0.0	0.0	T	T	T	0.7	0.6	0.6	T	0.0	0.0	1.9
1973-74	0.0	0.0	0.0	0.0	0.0	2.1	6.8	8.1	3.1	0.3	0.0	0.0	20.4
1974-75	0.0	0.0	0.0	0.0	T	1.2	1.4	12.7	1.1	T	0.0	0.0	16.4
1975-76	0.0	0.0	0.0	0.0	T	2.4	7.2	6.1	4.2	T	0.0	0.0	19.9
1976-77	0.0	0.0	0.0	0.0	T	6.7	10.8	5.8	1.7	T	T	0.0	25.0
1977-78	0.0	0.0	0.0	0.0	1.5	0.2	27.4	25.3	10.5	T	0.0	0.0	64.9
1978-79	0.0	0.0	0.0	0.0	2.6	T	9.4	26.1	T	T	0.0	0.0	38.1
1979-80	0.0	0.0	0.0	T	0.0	3.7	2.5	1.8	6.3	T	0.0	0.0	14.3
1980-81	0.0	0.0	0.0	0.0	0.4	3.1	6.9	T	9.1	0.0	0.0	0.0	19.5
1981-82	0.0	0.0	0.0	0.0	T	3.4	12.3	0.5	0.8	13.8	0.0	0.0	30.8
1982-83	0.0	0.0	0.0	0.0	T	2.9	2.3	21.5	0.2	4.1	0.0	0.0	31.0
1983-84	0.0	0.0	0.0	0.0	1.2	2.4	13.7	0.3	11.3	T	0.0	0.0	28.9
1984-85	0.0	0.0	0.0	0.0	T	6.8	8.9	7.4	0.1	T	0.0	0.0	23.2
1985-86	0.0	0.0	0.0	0.0	0.6	4.6	2.8	13.9	T	0.1	0.0	0.0	22.0
1986-87	0.0	0.0	0.0	0.0	T	2.3	21.4	6.5	2.4	0.0	0.0	0.0	32.6
1987-88	0.0	0.0	0.0	0.0	1.5	2.3	15.4	2.7	0.9	T	0.0		7.5
1988-89	0.0	0.0	0.0	0.0	0.0	0.1	4.1	0.6	2.7	0.0	0.0	0.0	7.5
1989-90	0.0	0.0	0.0	0.0	5.7	0.5	2.4	2.8	2.5	0.6	0.0	0.0	14.5
1990-91	0.0	0.0	0.0	0.0	T	7.6	8.5	5.2	0.2	0.0	0.0	0.0	21.5
1991-92	0.0	0.0	0.0	0.0	T	0.5	1.0	1.0	11.4	T	0.0	0.0	13.9
1992-93	0.0	0.0	0.0	0.0	T	0.5	0.8	10.7	16.8	0.0	0.0	0.0	28.8
1993-94	0.0	0.0	0.0	0.0	T	3.9	18.5	33.4	8.7	0.0	T	0.0	64.5
1994-95	0.0	0.0	0.0	0.0	T	T	0.1	10.2	T	0.0	0.0	0.0	10.3
1995-96	0.0	0.0	0.0	0.0	3.0	12.8	31.6	18.4	11.9	0.7	0.0	0.0	78.4
1996-97					T	T	3.4	4.4	7.1	1.4	0.0	0.0	
1997-98	0.0	0.0	0.0	0.0	0.2	1.4	2.2	T	3.1	T	0.0	0.0	6.9
1998-99	0.0	0.0	T	0.0	0.0	1.2	4.1	2.0	5.5	0.0	0.0	0.0	12.8
1999-00	0.0	0.0	0.0	0.0	T	T	12.2	5.3	T	0.9	0.0	0.0	18.4
2000-01	0.0	0.0	0.0	0.0	0.0	14.9	6.1	11.1	7.2	0.0	T	T	39.3
2001-	0.0	0.0	0.0	0.0	0.0	0.0							
POR= 59 YRS	0.0	0.0	0.0	0.0	0.6	5.3	7.7	8.2	4.8	0.7	T	0.0	27.3

2001
ALBUQUERQUE,
NEW MEXICO (ABQ)

The Albuquerque metropolitan area is largely situated in the Rio Grande Valley and on the mesas and piedmont slopes which rise either side of the valley floor. The Rio Grande flows from north to south through the area. The Sandia and Manzano Mountains rise abruptly at the eastern edge of the city with Tijeras Canyon separating the two ranges. West of the city the land gradually rises to the Continental Divide, some 90 miles away.

The climate of Albuquerque is best described as arid continental with abundant sunshine, low humidity, scant precipitation, and a wide yet tolerable seasonal range of temperatures. Sunny days and low humidity are renowned features of the climate. More than three-fourths of the daylight hours have sunshine, even in the winter months. The air is normally dry and muggy days are rare. The combination of dry air and plentiful solar radiation allows widespread use of energy-efficient devices such as evaporative coolers and solar collectors.

Precipitation within the valley area is adequate only for native desert vegetation and deep-rooted imports. However, irrigation supports successful farming and fruit growing in the Rio Grande Valley. On the east slopes of the Sandias and Manzanos, precipitation is sufficient for thick stands of timber and good grass cover.

Meager amounts of precipitation fall in the winter, much of it as snow. Snowfalls of an inch or more occur about four times a year in the Rio Grande Valley, while the mountains receive substantial snowfall on occasion. Snow seldom remains on the ground more than 24 hours in the city proper. However, snow cover on the east slopes of the Sandias is sufficient for skiing during most winters.

Nearly half of the annual precipitation in Albuquerque results from afternoon and evening thunderstorms during the summer. Thunderstorm frequency increases rapidly around July 1st, peaks during August, then tapers off by the end of September. Thunderstorms are usually brief, sometimes produce heavy rainfall, and often lower afternoon temperatures noticeably. Hailstorms are infrequent and tornadoes rare.

Temperatures in Albuquerque are those characteristic of a dry, high altitude, continental climate. The average daily range of temperature is relatively high, but extreme temperatures are rare. High temperatures during the winter are near 50 degrees with only a few days on which the temperature fails to rise above the freezing mark. In the summer, daytime maxima are about 90 degrees, but with the large daily range, the nights usually are comfortably cool.

The average number of days between the last freezing temperature in spring and the first freeze in fall varies widely across the Albuquerque metropolitan area. The growing season in Albuquerque and adjacent suburbs ranges from around 170 days in the Rio Grande Valley to about 200 days in parts of the northeast section of the city.

Sustained winds of 12 mph or less occur approximately 80 percent of the time at the Albuquerque International Airport, while sustained winds greater than 25 mph have a frequency less than 3 percent. Late winter and spring storms along with occasional east winds out of Tijeras Canyon are the main sources of strong wind conditions. Blowing dust, the least attractive feature of the climate, often accompanies the occasional strong winds of winter and spring.

NORMALS, MEANS, AND EXTREMES
ALBUQUERQUE, NM (ABQ)

LATITUDE:	LONGITUDE:	ELEVATION (FT):	TIME ZONE:	WBAN: 23050
35 02' 32" N	106 36' 23" W	GRND: 5305　BARO: 5308	MOUNTAIN (UTC + 7)	

	ELEMENT	POR	JAN	FEB	MAR	APR	MAY	JUN	JUL	AUG	SEP	OCT	NOV	DEC	YEAR
TEMPERATURE F	NORMAL DAILY MAXIMUM	30	46.8	53.5	61.4	70.8	79.7	90.0	92.5	89.0	81.9	71.0	57.3	47.5	70.1
	MEAN DAILY MAXIMUM	54	47.3	53.6	61.1	70.5	80.0	90.0	92.2	89.4	82.9	71.3	57.0	47.8	70.3
	HIGHEST DAILY MAXIMUM	62	69	76	85	89	98	107	105	101	100	91	77	72	107
	YEAR OF OCCURRENCE		1994	1986	1971	1989	1951	1994	1980	1979	1979	1979	1975	1958	JUN 1994
	MEAN OF EXTREME MAXS.	54	60.9	67.5	75.3	83.5	91.1	99.5	99.9	96.3	92.4	83.5	70.4	60.7	81.8
	NORMAL DAILY MINIMUM	30	21.7	26.4	32.2	39.6	48.6	58.3	64.4	62.6	55.2	43.0	31.2	23.1	42.2
	MEAN DAILY MINIMUM	54	23.6	27.7	33.1	40.9	50.2	59.8	65.0	63.3	56.4	44.2	31.9	24.4	43.4
	LOWEST DAILY MINIMUM	62	-17	-5	8	19	28	40	52	50	37	21	-7	-7	-17
	YEAR OF OCCURRENCE		1971	1951	1948	1980	1975	1980	1985	1992	1971	1991	1976	1990	JAN 1971
	MEAN OF EXTREME MINS.	54	9.3	14.1	20.1	28.1	37.3	48.4	58.6	56.9	45.2	31.7	19.0	11.3	31.7
	NORMAL DRY BULB	30	34.2	40.0	46.9	55.2	64.2	74.2	78.5	75.9	68.6	57.0	44.3	35.3	56.2
	MEAN DRY BULB	54	35.6	40.7	47.1	55.7	65.0	74.9	78.7	76.3	69.6	57.9	44.5	36.1	56.8
	MEAN WET BULB	18	29.8	33.6	37.1	41.9	48.9	55.4	60.4	61.2	54.7	45.6	36.0	29.8	44.5
	MEAN DEW POINT	18	19.5	21.1	21.4	23.6	30.7	38.5	48.8	51.9	43.4	33.0	24.3	20.0	31.3
	NORMAL NO. DAYS WITH:														
	MAXIMUM　90	30	0.0	0.0	0.0	0.0	2.6	17.2	23.2	15.9	3.9	0.1	0.0	0.0	62.9
	MAXIMUM　32	30	2.3	0.7	0.1	0.0	0.0	0.0	0.0	0.0	0.0	0.0	0.2	1.8	5.1
	MINIMUM　32	30	29.0	22.6	15.8	4.5	0.2	0.0	0.0	0.0	0.0	2.0	16.1	28.5	118.7
	MINIMUM　0	30	0.4	0.0	0.0	0.0	0.0	0.0	0.0	0.0	0.0	0.0	0.1	0.1	0.6
H/C	NORMAL HEATING DEG. DAYS	30	955	700	561	301	89	0	0	0	18	259	621	921	4425
	NORMAL COOLING DEG. DAYS	30	0	0	0	7	64	279	419	338	126	11	0	0	1244
RH	NORMAL (PERCENT)	30	56	50	40	32	31	30	42	47	47	45	50	57	44
	HOUR 05 LST	30	70	65	56	49	48	46	60	66	66	62	65	70	60
	HOUR 11 LST	30	50	44	34	26	25	24	34	40	40	38	42	50	37
	HOUR 17 LST	30	40	32	24	19	18	18	27	30	31	30	36	43	29
	HOUR 23 LST	30	61	53	43	36	34	32	48	53	52	50	54	61	48
S	PERCENT POSSIBLE SUNSHINE	62	72	72	73	77	79	83	76	76	79	79	76	71	76
W/O	MEAN NO. DAYS WITH:														
	HEAVY FOG(VISBY 1/4 MI)	63	1.2	1.0	0.6	0.2	0.0	0.0	0.1	0.0	0.1	0.3	0.6	1.7	5.8
	THUNDERSTORMS	63	0.1	0.3	0.9	1.5	4.0	5.1	10.9	10.8	4.6	2.2	0.6	0.2	41.2
CLOUDINESS	MEAN:														
	SUNRISE-SUNSET (OKTAS)	57	3.9	4.1	4.1	3.7	3.4	2.7	3.5	3.5	2.8	2.8	3.1	3.6	3.4
	MIDNIGHT-MIDNIGHT (OKTAS)	32	3.5	3.8	3.6	3.2	3.1	2.7	3.7	3.7	2.9	2.6	2.9	3.3	3.2
	MEAN NO. DAYS WITH:														
	CLEAR	58	12.8	10.8	11.3	12.5	13.9	17.4	11.9	13.1	16.3	17.0	14.8	13.7	165.5
	PARTLY CLOUDY	58	7.8	7.7	9.6	9.5	10.4	8.7	13.9	12.3	7.7	7.7	7.6	7.2	110.1
	CLOUDY	58	10.4	9.8	10.1	8.0	6.6	3.9	4.7	5.1	5.5	5.9	7.1	9.6	86.7
PR	MEAN STATION PRESSURE(IN)	29	24.80	24.70	24.70	24.70	24.70	24.70	24.80	24.80	24.80	24.80	24.80	24.80	24.76
	MEAN SEA-LEVEL PRES. (IN)	17	30.11	30.02	29.92	29.85	29.81	29.82	29.88	29.92	29.93	29.98	30.04	30.12	29.95
WINDS	MEAN SPEED (MPH)	45	7.9	8.6	9.8	10.6	10.3	9.5	8.8	7.9	8.1	7.9	7.8	7.6	8.7
	PREVAIL.DIR(TENS OF DEGS)	30	36	36	36	36	18	09	10	10	10	36	36	36	36
	MAXIMUM 2-MINUTE:														
	SPEED (MPH)	5	41	47	49	47	48	48	46	51	43	41	44	51	51
	DIR. (TENS OF DEGS)		10	27	09	11	25	27	09	08	34	16	24	06	08
	YEAR OF OCCURRENCE		2001	2000	2000	1999	1999	2001	2000	2000	2000	1997	2001	1997	AUG 2000
	MAXIMUM 5-SECOND:														
	SPEED (MPH)	5	53	59	58	58	64	61	56	61	53	49	49	57	64
	DIR. (TENS OF DEGS)		09	29	09	18	28	28	06	09	33	16	24	07	28
	YEAR OF OCCURRENCE		2001	2000	2000	2001	2001	2001	2000	2000	2000	1997	2001	1997	MAY 2001
PRECIPITATION	NORMAL (IN)	30	0.44	0.46	0.54	0.52	0.50	0.59	1.37	1.64	1.00	0.89	0.43	0.50	8.88
	MAXIMUM MONTHLY (IN)	62	1.32	1.82	2.34	1.97	3.07	2.86	3.33	3.30	2.63	3.08	1.93	1.85	3.33
	YEAR OF OCCURRENCE		1978	1993	1998	1942	1941	1996	1968	1967	1988	1972	1991	1959	JUL 1968
	MINIMUM MONTHLY (IN)	62	T	T	T	T	T	T	0.08	T	T	0.00	0.00	0.00	0.00
	YEAR OF OCCURRENCE		1970	1984	1966	1996	1945	1975	1980	1962	1957	1952	1949	1981	DEC 1981
	MAXIMUM IN 24 HOURS (IN)	62	0.87	0.80	1.45	1.66	1.14	1.64	1.77	2.13	1.92	1.80	1.67	1.35	2.13
	YEAR OF OCCURRENCE		1962	1993	1998	1969	1969	1952	1961	1994	1955	1969	1991	1958	AUG 1994
	NORMAL NO. DAYS WITH:														
	PRECIPITATION　0.01	30	4.0	4.2	4.7	3.1	4.4	4.2	9.0	9.4	6.4	4.7	4.0	4.2	62.3
	PRECIPITATION　1.00	30	0.0	0.0	0.0	0.1	0.0	0.1	0.1	0.2	0.0	0.1	0.0	0.0	0.6
SNOWFALL	NORMAL (IN)	30	2.9	2.7	2.0	0.7	0.*	0.0	0.0	0.0	T	0.2	0.7	2.2	11.4
	MAXIMUM MONTHLY (IN)	62	9.5	10.3	13.9	8.1	1.0	T	T	T	T	3.2	9.3	14.7	14.7
	YEAR OF OCCURRENCE		1973	1986	1973	1973	1979	1996	1996	1996	1971	1986	1940	1959	DEC 1959
	MAXIMUM IN 24 HOURS (IN)	62	5.1	6.0	10.7	10.9	1.0	T	T	T	T	3.2	5.5	14.2	14.2
	YEAR OF OCCURRENCE		1973	1986	1973	1988	1979	1996	1990	1993	1971	1986	1946	1958	DEC 1958
	MAXIMUM SNOW DEPTH (IN)	53	47	16	8	11	0	0	0	0	0	3	12	25	47
	YEAR OF OCCURRENCE		1977	1986	1973	1988						1986	1992	1958	JAN 1977
	NORMAL NO. DAYS WITH:														
	SNOWFALL　1.0	30	1.0	1.2	0.6	0.2	0.0	0.0	0.0	0.0	0.0	0.*	0.3	0.8	4.1

PRECIPITATION (inches) 2001 ALBUQUERQUE, NM (ABQ)

YEAR	JAN	FEB	MAR	APR	MAY	JUN	JUL	AUG	SEP	OCT	NOV	DEC	ANNUAL
1972	0.12	0.12	0.08	T	0.18	0.55	1.00	2.93	1.00	3.08	0.69	0.36	10.11
1973	0.85	0.33	2.18	0.91	0.66	1.37	1.80	1.19	1.13	0.35	0.08	0.03	10.88
1974	0.88	0.11	0.85	0.14	0.01	0.22	2.40	0.79	1.58	1.96	0.38	0.51	9.83
1975	0.26	0.99	0.95	0.10	0.66	T	1.43	1.40	1.66	T	0.28	0.28	8.01
1976	0.00	0.40	0.09	0.31	0.82	0.60	1.32	0.73	0.45	0.03	0.24	0.20	5.19
1977	0.88	0.13	0.63	1.07	0.10	0.04	0.69	2.28	0.78	0.76	0.42	0.13	7.91
1978	1.32	1.02	0.54	0.05	0.69	1.05	0.24	2.49	0.59	1.22	1.00	0.76	10.97
1979	1.07	0.62	0.14	0.24	2.48	1.02	0.80	1.53	0.40	0.27	0.91	0.87	10.35
1980	0.87	0.58	0.60	0.60	0.56	0.01	0.08	2.61	1.83	0.09	0.30	0.74	8.87
1981	0.05	0.67	0.80	0.30	0.53	0.35	1.07	1.68	0.41	1.43	0.37	0.00	7.66
1982	0.32	0.20	0.84	0.05	0.52	0.09	1.32	1.09	1.34	0.26	0.60	0.78	7.41
1983	1.10	0.71	0.61	0.02	0.32	1.21	0.55	0.27	0.91	1.20	0.44	0.42	7.76
1984	0.33	T	0.62	0.50	0.16	0.48	1.13	2.70	1.13	3.04	0.63	1.36	12.08
1985	0.49	0.54	0.70	1.69	1.12	0.53	1.16	0.49	1.53	2.15	0.19	0.16	10.75
1986	0.22	1.01	0.17	0.33	1.11	2.57	1.51	2.26	0.53	1.54	1.29	0.44	12.98
1987	0.66	0.61	0.07	1.00	0.58	0.13	0.91	2.98	0.20	0.44	0.42	0.34	8.34
1988	0.15	0.07	0.85	1.42	0.62	1.25	2.26	3.29	2.63	0.32	0.22	0.03	13.11
1989	0.57	0.35	0.48	T	0.02	0.02	1.51	0.48	0.31	0.97	T	0.28	4.99
1990	0.21	0.49	0.41	1.71	0.45	0.27	2.36	1.79	0.96	0.15	0.86	0.59	10.25
1991	0.60	0.06	0.14	T	1.14	0.65	2.63	1.26	1.43	0.26	1.93	1.49	11.59
1992	0.60	0.20	0.63	0.22	1.81	0.67	2.01	2.17	0.79	0.70	1.12	1.16	12.08
1993	0.94	1.82	0.22	T	0.20	0.44	0.23	3.05	0.49	0.64	0.97	0.03	9.03
1994	0.02	0.26	0.59	0.07	1.87	0.28	0.61	2.70	1.21	1.38	0.62	1.15	11.15
1995	0.55	0.39	0.16	0.69	0.08	0.20	0.35	0.74	2.32	T	0.03	0.17	5.68
1996	0.17	0.19	0.02	T	0.02	2.86	1.03	1.54	1.45	1.52	0.95	T	9.75
1997	0.55	0.12	0.11	1.65	0.42	1.03	2.04	1.96	2.43	0.32	0.73	1.00	12.36
1998	0.14	0.66	2.34	0.64	T	0.17	2.37	0.88	0.15	1.80	0.46	0.22	9.83
1999	0.12	T	1.10	0.59	0.54	0.60	1.47	3.04	0.54	0.26	T	0.03	8.29
2000	0.30	0.30	1.27	T	0.07	0.72	0.83	0.57	0.37	2.66	0.91	0.24	8.24
2001	0.28	0.27	0.27	0.51	0.38	0.26	1.37	1.59	0.51	0.14	0.68	0.24	6.50
POR= 109 YRS	0.40	0.38	0.48	0.55	0.63	0.62	1.33	1.44	0.91	0.83	0.43	0.41	8.41

AVERAGE TEMPERATURE (F) 2001 ALBUQUERQUE, NM (ABQ)

YEAR	JAN	FEB	MAR	APR	MAY	JUN	JUL	AUG	SEP	OCT	NOV	DEC	ANNUAL
1972	36.1	42.5	53.6	56.9	64.0	73.7	78.6	74.1	68.1	57.6	40.1	35.0	56.7
1973	31.8	35.9	45.1	50.2	62.7	73.5	78.4	78.0	67.5	56.4	44.6	34.0	54.8
1974	33.6	37.9	52.8	56.4	68.5	80.1	77.0	72.7	66.1	58.1	45.0	32.0	56.7
1975	30.8	38.0	45.0	49.9	61.0	73.0	76.8	76.1	66.3	56.5	42.6	35.6	54.3
1976	33.2	43.3	44.3	54.6	62.8	73.4	77.0	75.0	68.0	53.1	40.6	33.0	54.9
1977	29.8	40.7	43.2	56.5	64.2	75.5	78.6	77.4	69.4	58.9	46.4	40.4	56.8
1978	36.8	39.3	50.2	57.7	60.5	75.5	81.6	75.5	69.1	60.3	47.5	34.3	57.4
1979	32.9	41.1	48.4	56.9	63.7	73.3	80.6	77.1	72.3	61.5	41.0	37.7	57.2
1980	40.2	44.2	46.1	52.1	61.1	77.2	82.7	77.4	69.9	54.5	43.5	40.5	57.5
1981	38.0	42.9	46.2	59.0	64.5	77.0	79.8	76.4	69.7	55.7	47.0	40.5	58.1
1982	35.9	39.4	47.4	56.1	63.0	74.8	79.1	77.4	69.5	54.8	42.9	34.4	56.2
1983	35.0	39.7	46.9	50.2	63.0	73.4	80.4	79.4	73.4	58.3	45.1	36.7	56.8
1984	34.1	40.1	46.8	52.8	69.9	73.6	78.9	75.7	68.8	51.6	43.7	35.6	56.0
1985	33.8	38.3	47.5	57.4	64.0	74.1	77.1	76.6	65.9	57.5	45.4	37.6	56.3
1986	41.3	43.0	50.9	56.5	63.7	72.7	74.7	76.0	66.5	54.6	42.0	36.3	56.5
1987	32.3	39.2	43.7	54.8	62.8	73.0	77.8	74.7	68.8	61.3	45.2	35.3	55.7
1988	34.6	43.9	47.0	55.1	64.3	74.4	78.1	75.0	66.3	61.1	45.4	33.9	56.6
1989	35.5	41.9	52.8	61.4	68.8	75.6	78.6	74.3	69.4	56.7	46.4	35.1	58.0
1990	34.6	38.5	48.6	57.3	63.6	79.0	76.8	73.8	70.9	58.3	45.0	32.1	56.5
1991	35.7	44.6	46.1	56.0	65.5	73.4	76.9	75.5	68.1	59.6	43.4	37.3	56.8
1992	32.7	42.3	48.9	60.0	64.6	72.4	76.2	75.0	70.3	60.8	39.7	32.8	56.3
1993	39.7	42.5	48.8	57.1	65.7	75.1	79.9	75.6	69.1	56.2	43.3	37.3	57.5
1994	38.1	40.7	50.2	58.5	66.9	80.4	81.3	79.4	71.0	57.0	44.5	40.9	59.1
1995	39.2	49.3	50.7	54.2	64.5	74.8	80.0	79.8	69.5	59.5	50.8	40.9	59.4
1996	39.3	46.1	46.8	57.5	71.5	76.5	79.5	76.2	66.2	55.9	45.6	38.7	58.3
1997	33.4	40.7	51.8	52.6	65.9	73.0	77.6	76.3	71.5	56.5	43.6	32.7	56.3
1998	37.9	38.8	46.7	52.2	65.6	74.4	77.1	77.1	74.4	57.9	46.5	38.5	57.3
1999	40.7	44.0	50.7	53.7	63.7	72.8	76.7	74.7	68.2	58.6	49.8	35.6	57.4
2000	40.9	45.0	47.7	59.1	70.5	76.2	79.5	78.1	72.4	55.7	39.4	37.1	58.5
2001	33.8	42.7	48.4	57.8	68.9	76.9	79.4	75.9	72.4	60.8	47.9	36.3	58.4
POR= 109 YRS	34.9	40.0	46.7	55.1	64.0	73.6	77.4	75.3	68.6	56.9	44.0	35.3	56.0

REFERENCE NOTES:

PAGE 1:
THE TEMPERATURE GRAPH SHOWS NORMAL MAXIMUM AND NORMAL
MINIMUM DAILY TEMPERATURES (SOLID CURVES) AND THE
ACTUAL DAILY HIGH AND LOW TEMPERATURES (VERTICAL BARS).

PAGE 2 AND 3:
H/C INDICATES HEATING AND COOLING DEGREE DAYS.
RH INDICATES RELATIVE HUMIDITY
W/O INDICATES WEATHER AND OBSTRUCTIONS
S INDICATES SUNSHINE.
PR INDICATES PRESSURE.
CLOUDINESS ON PAGE 3 IS THE SUM OF THE CEILOMETER AND
SATELLITE DATA NOT TO EXCEED EIGHT EIGHTHS(OKTAS).

GENERAL:
T INDICATES TRACE PRECIPITATION, AN AMOUNT GREATER
THAN ZERO BUT LESS THAN THE LOWEST REPORTABLE VALUE.
+ INDICATES THE VALUE ALSO OCCURS ON EARLIER DATES.
BLANK ENTRIES DENOTE MISSING OR UNREPORTED DATA.
NORMALS ARE 30-YEAR AVERAGES (1961 - 1990).
ASOS INDICATES AUTOMATED SURFACE OBSERVING SYSTEM.
PM INDICATES THE LAST DAY OF THE PREVIOUS MONTH.
POR (PERIOD OF RECORD) BEGINS WITH THE JANUARY DATA
MONTH AND IS THE NUMBER OF YEARS USED TO COMPUTE
THE MEAN. INDIVIDUAL MONTHS WITHIN THE POR MAY
BE MISSING.
WHEN THE POR FOR A NORMAL IS LESS THAN 30 YEARS,
THE NORMAL IS PROVISIONAL AND IS BASED ON THE NUMBER
OF YEARS INDICATED.
0.* OR * INDICATES THE VALUE OR MEAN-DAYS-WITH
IS BETWEEN 0.00 AND 0.05.
CLOUDINESS FOR ASOS STATIONS DIFFERS FROM THE NON-ASOS
OBSERVATION TAKEN BY A HUMAN OBSERVER. ASOS STATION
CLOUDINESS IS BASED ON TIME-AVERAGED CEILOMETER DATA
FOR CLOUDS AT OR BELOW 12,000 FEET AND ON SATELLITE
DATA FOR CLOUDS ABOVE 12,000 FEET.
THE NUMBER OF DAYS WITH CLEAR, PARTLY CLOUDY, AND
CLOUDY CONDITIONS FOR ASOS STATIONS IS THE SUM
OF THE CEILOMETER AND SATELLITE DATA FOR THE
SUNRISE TO SUNSET PERIOD.

GENERAL CONTINUED:
CLEAR INDICATES 0 - 2 OKTAS, PARTLY CLOUDY INDICATES
3 - 6 OKTAS, AND CLOUDY INDICATES 7 OR 8 OKTAS.
WHEN AT LEAST ONE OF THE ELEMENTS (CEILOMETER OR
SATELLITE) IS MISSING, THE DAILY CLOUDINESS IS
NOT COMPUTED.
WIND DIRECTION IS RECORDED IN TENS OF DEGREES (2 DIGITS)
CLOCKWISE FROM TRUE NORTH. "00" INDICATES CALM. "36"
INDICATES TRUE NORTH.
RESULTANT WIND IS THE VECTOR AVERAGE OF THE SPEED AND
DIRECTION.
AVERAGE TEMPERATURE IS THE SUM OF THE MEAN DAILY MAXIMUM
AND MINIMUM TEMPERATURE DIVIDED BY 2.
SNOWFALL DATA COMPRISE ALL FORMS OF FROZEN
PRECIPITATION, INCLUDING HAIL.
A HEATING (COOLING) DEGREE DAY IS THE DIFFERENCE BETWEEN
THE AVERAGE DAILY TEMPERATURE AND 65 F.
DRY BULB IS THE TEMPERATURE OF THE AMBIENT AIR.
DEW POINT IS THE TEMPERATURE TO WHICH THE AIR MUST BE
COOLED TO ACHIEVE 100 PERCENT RELATIVE HUMIDITY.
WET BULB IS THE TEMPERATURE THE AIR WOULD HAVE IF THE
MOISTURE CONTENT WAS INCREASED TO 100 PERCENT RELATIVE
HUMIDITY.

ON JULY 1, 1996, THE NATIONAL WEATHER SERVICE BEGAN USING
THE "METAR" OBSERVATION CODE THAT WAS ALREADY EMPLOYED
BY MOST OTHER NATIONS OF THE WORLD. THE MOST NOTICEABLE
DIFFERENCE IN THIS ANNUAL PUBLICATION WILL BE THE CHANGE
IN UNITS FROM TENTHS TO EIGHTS(OKTAS) FOR REPORTING THE
AMOUNT OF SKY COVER.

HEATING DEGREE DAYS (base 65 F) 2001 ALBUQUERQUE, NM (ABQ)

YEAR	JUL	AUG	SEP	OCT	NOV	DEC	JAN	FEB	MAR	APR	MAY	JUN	TOTAL
1972-73	0	3	14	244	740	925	1020	811	607	440	113	3	4920
1973-74	0	0	43	257	606	955	963	754	373	255	29	4	4239
1974-75	0	2	68	212	593	1020	1051	748	614	449	143	6	4906
1975-76	0	0	47	256	664	905	979	622	634	304	99	1	4511
1976-77	0	0	35	367	726	985	1084	675	669	250	61	0	4852
1977-78	0	0	1	192	551	757	870	713	454	215	175	2	3930
1978-79	0	0	20	167	521	945	988	665	509	241	100	12	4168
1979-80	0	0	23	148	715	840	763	595	577	379	139	2	4181
1980-81	0	0	6	335	640	752	827	611	575	197	62	2	4007
1981-82	0	0	3	280	534	754	895	709	538	268	94	0	4075
1982-83	0	0	23	314	658	941	922	703	556	439	127	0	4683
1983-84	0	0	11	198	592	875	948	714	559	362	22	3	4284
1984-85	0	0	51	411	631	903	960	744	536	220	74	7	4537
1985-86	0	0	61	228	581	842	727	610	431	249	80	8	3817
1986-87	0	0	51	313	680	882	1004	717	653	300	81	2	4683
1987-88	0	0	2	133	589	914	937	605	551	290	103	2	4126
1988-89	0	5	39	118	579	959	909	640	373	133	31	0	3786
1989-90	0	0	10	260	551	918	934	735	501	233	103	0	4245
1990-91	0	0	14	202	595	1013	903	563	581	263	60	12	4206
1991-92	0	0	21	188	645	851	994	651	493	170	53	5	4071
1992-93	0	0	8	128	752	991	778	624	496	238	69	3	4087
1993-94	0	0	15	284	642	853	827	676	453	218	52	0	4020
1994-95	0	0	1	251	610	741	793	435	433	320	67	0	3651
1995-96	0	0	37	165	419	741	787	540	558	239	12	0	3498
1996-97	0	0	66	297	575	809	973	676	403	366	48	12	4225
1997-98	0	0	9	278	635	994	834	727	558	376	51	4	4466
1998-99	0	0	0	225	549	816	744	584	435	332	98	5	3788
1999-00	0	0	23	203	447	902	740	571	528	184	32	0	3630
2000-01	0	0	16	302	760	858	961	619	509	218	35	3	4281
2001-	0	0	1	134	508	881							

COOLING DEGREE DAYS (base 65 F) 2001 ALBUQUERQUE, NM (ABQ)

YEAR	JAN	FEB	MAR	APR	MAY	JUN	JUL	AUG	SEP	OCT	NOV	DEC	ANNUAL
1972	0	0	0	5	52	267	428	294	113	23	0	0	1182
1973	0	0	0	0	48	267	422	409	124	0	0	0	1270
1974	0	0	0	5	144	464	380	247	107	6	0	0	1353
1975	0	0	0	0	25	256	372	351	96	0	0	0	1100
1976	0	0	0	0	38	260	382	319	137	5	0	0	1141
1977	0	0	0	0	44	324	427	392	141	7	0	0	1335
1978	0	0	0	4	41	324	521	330	151	27	0	0	1398
1979	0	0	0	5	67	269	491	382	249	45	0	0	1508
1980	0	0	0	0	27	375	557	392	160	15	0	0	1526
1981	0	0	0	28	51	368	470	360	152	1	0	0	1430
1982	0	0	0	6	38	301	441	394	163	4	0	0	1347
1983	0	0	0	1	72	260	484	450	267	1	0	0	1535
1984	0	0	0	4	179	266	441	340	169	1	0	0	1400
1985	0	0	0	0	51	289	383	368	97	0	0	0	1188
1986	0	0	0	1	50	245	310	349	103	0	0	0	1058
1987	0	0	0	0	17	251	404	308	120	25	0	0	1125
1988	0	0	0	1	85	288	411	322	86	3	0	0	1196
1989	0	0	0	31	154	323	426	295	150	10	0	0	1389
1990	0	0	0	10	66	426	374	281	200	2	0	0	1359
1991	0	0	0	0	87	269	375	331	120	25	0	0	1207
1992	0	0	0	27	49	235	354	318	171	5	0	0	1159
1993	0	0	0	8	101	312	470	337	145	16	0	0	1389
1994	0	0	0	29	115	469	512	455	188	10	0	0	1778
1995	0	0	0	4	55	302	472	467	182	1	0	0	1483
1996	0	0	0	19	218	352	457	354	109	21	0	0	1530
1997	0	0	0	0	85	261	398	354	212	21	0	0	1331
1998	0	0	0	1	78	292	383	383	288	11	0	0	1436
1999	0	0	0	0	65	246	367	306	122	12	0	0	1118
2000	0	0	0	13	208	346	457	411	244	21	0	0	1700
2001	0	0	0	8	162	369	453	346	228	12	0	0	1578

SNOWFALL (inches) 2001 ALBUQUERQUE, NM (ABQ)

YEAR	JUL	AUG	SEP	OCT	NOV	DEC	JAN	FEB	MAR	APR	MAY	JUN	TOTAL
1972-73	0.0	0.0	0.0	T	2.9	1.2	9.5	1.8	13.9	8.1	0.0	0.0	37.4
1973-74	0.0	0.0	0.0	0.3	0.6	0.1	9.3	0.6	2.0	0.0	0.0	0.0	12.9
1974-75	0.0	0.0	0.0	0.0	T	4.9	0.9	6.7	3.8	0.2	0.0	0.0	16.5
1975-76	0.0	0.0	0.0	0.0	0.2	2.9	0.0	T	0.5	0.2	0.0	0.0	3.8
1976-77	0.0	0.0	0.0	T	2.4	1.2	8.4	1.4	2.3	2.6	0.0	0.0	18.3
1977-78	0.0	0.0	0.0	0.0	0.0	T	6.0	3.4	2.0	0.0	0.1	0.0	11.5
1978-79	0.0	0.0	0.0	0.0	T	1.0	2.6	6.0	T	0.5	1.0	0.0	11.1
1979-80	0.0	0.0	0.0	0.9	0.8	2.7	T	0.9	3.1	T	T	0.0	8.4
1980-81	0.0	0.0	0.0	T	2.8	7.4	0.5	2.6	0.9	T	0.0	0.0	14.2
1981-82	0.0	0.0	0.0	0.0	0.0	0.0	3.6	1.2	0.7	T	0.0	0.0	5.5
1982-83	0.0	0.0	0.0	0.0	0.9	3.3	7.3	4.2	1.0	T	T	0.0	16.7
1983-84	0.0	0.0	0.0	0.0	0.8	0.8	4.1	T	0.1	3.0	0.0	0.0	8.8
1984-85	0.0	0.0	0.0	T	T	3.4	2.0	2.9	0.6	0.0	0.0	0.0	8.9
1985-86	0.0	0.0	0.0	0.0	0.7	0.9	2.9	10.3	0.3	0.0	T	0.0	15.1
1986-87	0.0	0.0	0.0	3.2	0.6	0.2	4.9	4.9	0.2	2.2	0.0	0.0	16.2
1987-88	0.0	0.0	0.0	0.0	1.1	1.7	1.2	T	7.9	4.2	0.0	0.0	16.1
1988-89	0.0	0.0	0.0	0.0	1.7	0.3	3.4	3.2	3.1	0.0	0.0	0.0	11.7
1989-90	0.0	0.0	0.0	0.0	T	2.5	1.8	4.8	T	0.3	T	0.0	9.4
1990-91	T	0.0	0.0	0.0	2.2	6.3	0.9	T	0.8	T	0.0	0.0	10.2
1991-92	0.0	0.0	0.0	2.5	1.5	2.1	5.6	1.0	0.0	T	T	0.0	12.7
1992-93	0.0	T	0.0	0.0	5.9	7.6	0.8	2.0	0.2	0.0	T	0.0	16.5
1993-94	0.0	T	0.0	T	4.1	0.2	T	T	1.2	0.0	T	0.0	5.5
1994-95	0.0	0.0	0.0	0.0	T	0.5	5.3	0.0	0.4	3.2	0.0	0.0	9.4
1995-96	0.0	0.0	0.0	0.0	0.0	0.9	0.6	1.7	0.2	0.0	0.0	T	3.4
1996-97	T	T	0.0	1.1	2.5	0.0	4.7	0.9	0.3	2.9	0.0	T	12.4
1997-98	0.0	0.0	0.0	T	1.1	8.8	0.2	1.0	0.9	0.4	0.0	0.0	12.4
1998-99	T	0.0	0.0	0.0	1.3	2.7	T	0.0	3.3	T	T	0.0	7.3
1999-00	0.0	0.0	0.0	T	0.0	0.1	0.7	0.8	2.9	T	0.0	0.0	4.5
2000-01	T	0.0	0.0	0.0	0.1	6.3	2.7	0.8	0.1	0.0	0.0	0.0	10.0
2001-	0.0	0.0	0.0	0.0	T	0.7							
POR= 61 YRS	T	T	T	0.1	1.2	2.7	2.5	2.1	1.8	0.6	T	T	11.0

2001
ALBANY,
NEW YORK (ALB)

Albany is located on the west bank of the Hudson River some 150 miles north of New York City, and 8 miles south of the confluence of the Mohawk and Hudson Rivers. The river-front portion of the city is only a few feet above sea level, and there is a tidal effect upstream to Troy. Eleven miles west of Albany the Helderberg escarpment rises to 1,800 feet. Between it and the Hudson River the valley floor is gently rolling, ranging some 200 to 500 feet above sea level. East of the city there is more rugged terrain 5 or 6 miles wide with elevations of 300 to 600 feet. Farther to the east the terrain rises more sharply. It reaches a north-south range of hills 12 miles east of Albany with elevations ranging to 2,000 feet.

The climate at Albany is primarily continental in character, but is subjected to some modification by the Atlantic Ocean. The moderating effect on temperatures is more pronounced during the warmer months than in winter when outbursts of cold air sweep down from Canada. In the warmer seasons, temperatures rise rapidly in the daytime. However, temperatures also fall rapidly after sunset so that the nights are relatively cool. Occasionally there are extended periods of oppressive heat up to a week or more in duration.

Winters are usually cold and sometimes fairly severe. Maximum temperatures during the colder winters are often below freezing and nighttime lows are frequently below 10 degrees. Sub-zero readings occur about twelve times a year. Snowfall throughout the area is quite variable and snow flurries are quite frequent during the winter. Precipitation is sufficient to serve the economy of the region in most years, and only occasionally do periods of drought exist. Most of the rainfall in the summer is from thunderstorms. Tornadoes are quite rare and hail is not usually of any consequence.

Wind velocities are moderate. The north-south Hudson River Valley has a marked effect on the lighter winds and in the warm months, average wind direction is usually southerly. Destructive winds rarely occur.

The area enjoys one of the highest percentages of sunshine in the entire state. Seldom does the area experience long periods of cloudy days and long periods of smog are rare.

Based on the 1951-1980 period, the average first occurrence of 32 degrees Fahrenheit in the fall is September 29 and the average last occurrence in the spring is May 7.

NORMALS, MEANS, AND EXTREMES
ALBANY, NY (ALB)

LATITUDE:	LONGITUDE:	ELEVATION (FT):		TIME ZONE:	WBAN: 14735
42 44' 53" N	73 48' 12" W	GRND: 278	BARO: 281	EASTERN (UTC + 5)	

	ELEMENT	POR	JAN	FEB	MAR	APR	MAY	JUN	JUL	AUG	SEP	OCT	NOV	DEC	YEAR
TEMPERATURE F	NORMAL DAILY MAXIMUM	30	30.2	33.2	44.0	57.5	69.7	79.0	84.0	81.4	73.2	61.8	48.7	34.9	58.1
	MEAN DAILY MAXIMUM	56	30.9	33.5	43.5	57.4	69.4	78.2	82.9	80.7	72.4	61.6	48.0	35.3	57.8
	HIGHEST DAILY MAXIMUM	55	65	68	89	92	94	99	100	99	100	89	82	71	100
	YEAR OF OCCURRENCE		1995	1997	1998	1990	1981	1952	1953	1955	1963	1963	1950	1984	SEP 1953
	MEAN OF EXTREME MAXS.	56	51.1	51.9	66.7	79.0	86.7	91.7	93.3	91.6	87.2	78.7	67.6	54.7	75.0
	NORMAL DAILY MINIMUM	30	11.0	13.8	24.5	35.1	45.4	54.6	59.6	57.8	49.4	38.6	30.7	18.2	36.6
	MEAN DAILY MINIMUM	56	13.2	15.1	24.7	35.7	46.0	55.2	60.0	58.1	49.9	39.4	31.0	19.3	37.3
	LOWEST DAILY MINIMUM	55	-28	-21	-21	10	26	36	40	34	24	16	5	-22	-28
	YEAR OF OCCURRENCE		1971	1973	1948	1965	1968	1986	1978	1982	1947	1969	1972	1969	JAN 1971
	MEAN OF EXTREME MINS.	56	-9.1	-7.8	5.4	21.8	31.9	40.4	47.5	44.5	33.6	24.3	14.8	-3.0	20.4
	NORMAL DRY BULB	30	20.6	23.5	34.3	46.4	57.6	66.9	71.8	69.6	61.3	50.2	39.7	26.5	47.4
	MEAN DRY BULB	56	21.9	24.2	34.0	46.7	57.7	66.7	71.4	69.4	61.2	50.5	39.5	27.3	47.5
	MEAN WET BULB	18	21.8	23.7	30.9	41.7	52.3	60.7	65.2	64.3	56.9	46.1	36.7	25.0	43.8
	MEAN DEW POINT	18	16.0	16.7	23.4	33.8	46.0	56.0	61.2	60.6	53.4	41.6	31.5	19.8	38.3
	NORMAL NO. DAYS WITH:														
	MAXIMUM 90	30	0.0	0.0	0.0	0.2	0.5	1.9	4.5	2.0	0.6	0.0	0.0	0.0	9.7
	MAXIMUM 32	30	17.3	13.4	3.9	0.2	0.0	0.0	0.0	0.0	0.0	0.0	1.2	12.2	48.2
	MINIMUM 32	30	29.6	25.9	24.4	12.5	1.7	0.0	0.0	0.0	0.7	8.4	18.1	27.7	149.0
	MINIMUM 0	30	6.9	4.4	0.4	0.0	0.0	0.0	0.0	0.0	0.0	0.0	0.0	2.4	14.1
H/C	NORMAL HEATING DEG. DAYS	30	1376	1162	952	558	247	34	0	12	141	459	759	1194	6894
	NORMAL COOLING DEG. DAYS	30	0	0	0	0	18	91	213	155	30	0	0	0	507
RH	NORMAL (PERCENT)	30	71	68	65	61	66	70	70	74	76	72	73	74	70
	HOUR 01 LST	30	76	74	72	71	77	82	84	87	87	82	79	78	79
	HOUR 07 LST	30	77	76	75	72	75	78	80	86	88	85	81	80	79
	HOUR 13 LST	30	63	59	54	49	51	55	54	57	58	56	62	66	57
	HOUR 19 LST	30	70	66	61	56	58	62	63	69	74	71	72	73	66
S	PERCENT POSSIBLE SUNSHINE	58	46	52	54	54	56	60	64	60	57	52	37	39	53
W/O	MEAN NO. DAYS WITH:														
	HEAVY FOG(VISBY 1/4 MI)	64	1.3	1.0	1.4	0.8	1.3	1.2	1.4	2.5	3.4	3.9	1.6	1.8	21.6
	THUNDERSTORMS	64	0.1	0.1	0.5	1.3	3.2	5.2	5.8	4.4	2.2	0.9	0.3	0.1	24.1
CLOUDINESS	MEAN:														
	SUNRISE-SUNSET (OKTAS)	1			5.6		6.4	6.4							
	MIDNIGHT-MIDNIGHT (OKTAS)	1			6.4										
	MEAN NO. DAYS WITH:														
	CLEAR														
	PARTLY CLOUDY														
	CLOUDY														
PR	MEAN STATION PRESSURE(IN)	29	29.71	29.71	29.70	29.70	29.70	29.69	29.70	29.70	29.78	29.80	29.79	29.80	29.73
	MEAN SEA-LEVEL PRES. (IN)	17	30.09	30.08	30.05	29.98	29.98	29.96	29.99	30.04	30.07	30.10	30.09	30.10	30.04
WINDS	MEAN SPEED (MPH)	63	9.8	10.2	10.5	10.4	9.0	8.2	7.6	7.1	7.5	8.1	9.2	9.4	8.9
	PREVAIL.DIR(TENS OF DEGS)	30	30	29	29	29	18	18	18	18	18	18	18	30	18
	MAXIMUM 2-MINUTE:														
	SPEED (MPH)	6	40	44	46	35	55	38	41	47	34	39	37	43	55
	DIR. (TENS OF DEGS)		30	29	30	28	32	29	35	27	28	29	29	29	32
	YEAR OF OCCURRENCE		2000	1997	1997	2001	1998	2000	1999	1999	1997	1998	1997	2000	MAY 1998
	MAXIMUM 5-SECOND:														
	SPEED (MPH)	6	52	56	56	47	82	47	52	59	49	53	51	53	82
	DIR. (TENS OF DEGS)		28	29	27	28	32	29	28	31	01	30	28	28	32
	YEAR OF OCCURRENCE		2000	2001	1997	1999	1998	2000	1999	1999	1999	1998	1997	2000	MAY 1998
PRECIPITATION	NORMAL (IN)	30	2.36	2.27	2.93	2.99	3.41	3.62	3.18	3.47	2.95	2.83	3.23	2.93	36.17
	MAXIMUM MONTHLY (IN)	55	6.44	5.02	5.90	7.95	8.96	7.36	6.96	7.33	11.06	8.83	8.07	6.73	11.06
	YEAR OF OCCURRENCE		1978	1981	1977	1983	1953	1973	1975	1950	1999	1955	1972	1973	SEP 1999
	MINIMUM MONTHLY (IN)	55	0.42	0.24	0.26	0.60	1.05	0.65	0.49	0.73	0.40	0.20	0.91	0.64	0.20
	YEAR OF OCCURRENCE		1980	1987	1981	1999	1980	1964	1968	1947	1964	1963	1978	1958	OCT 1963
	MAXIMUM IN 24 HOURS (IN)	55	1.91	1.74	2.38	2.20	2.17	3.48	3.49	4.52	6.00	3.31	2.26	4.02	6.00
	YEAR OF OCCURRENCE		1978	1990	1986	1968	1968	1952	1996	1971	1999	1987	1991	1948	SEP 1999
	NORMAL NO. DAYS WITH:														
	PRECIPITATION 0.01	30	11.7	10.3	11.8	11.6	13.2	11.8	10.0	10.6	9.6	8.9	12.4	12.7	134.6
	PRECIPITATION 1.00	30	0.3	0.3	0.4	0.6	0.4	0.7	0.7	0.6	0.8	0.7	0.5	0.3	6.3
SNOWFALL	NORMAL (IN)	30	16.8	14.1	10.2	2.8	0.1	0.0	0.0	0.0	0.0	0.2	5.0	16.7	65.9
	MAXIMUM MONTHLY (IN)	55	47.8	34.5	34.7	17.7	1.6	T	T	0.0	T	6.5	24.6	57.5	57.5
	YEAR OF OCCURRENCE		1987	1962	1956	1982	1977	1991	1995		1989	1987	1972	1969	DEC 1969
	MAXIMUM IN 24 HOURS (IN)	55	21.2	17.9	26.6	17.5	1.6	T	T	0.0	T	6.5	21.9	18.3	26.6
	YEAR OF OCCURRENCE		1983	1958	1993	1982	1977	1991	1995		1989	1987	1971	1966	MAR 1993
	MAXIMUM SNOW DEPTH (IN)	53	36	22	28	13	0	0	0	0	0	2	18	36	36
	YEAR OF OCCURRENCE		1970	1971	1993	1982						1987	1971	1969	DEC 1969
	NORMAL NO. DAYS WITH:														
	SNOWFALL 1.0	30	4.0	3.2	2.3	0.6	0.1	0.0	0.0	0.0	0.0	0.*	1.1	4.5	15.8

PRECIPITATION (inches) 2001 ALBANY, NY (ALB)

YEAR	JAN	FEB	MAR	APR	MAY	JUN	JUL	AUG	SEP	OCT	NOV	DEC	ANNUAL
1972	1.21	3.04	4.05	3.63	5.98	6.84	3.10	1.48	1.99	3.60	8.07	4.19	47.18
1973	2.16	1.34	1.99	4.47	5.45	7.36	1.68	2.89	1.33	2.07	1.27	6.73	38.74
1974	2.04	2.12	3.10	2.80	3.47	3.31	4.84	3.53	5.37	1.49	3.83	2.57	38.47
1975	2.75	3.58	2.72	2.18	2.96	3.80	6.96	5.98	4.57	5.88	2.89	2.78	47.05
1976	3.78	2.60	3.57	3.63	4.89	5.37	2.60	5.04	2.61	5.65	1.41	1.39	42.54
1977	1.51	2.63	5.90	3.41	2.29	2.87	2.31	3.66	6.66	4.00	4.85	4.21	44.30
1978	6.44	0.88	1.99	1.68	1.96	4.60	4.04	3.06	1.87	2.95	0.91	3.08	33.46
1979	6.37	1.71	1.83	3.89	4.13	1.94	2.78	2.67	4.05	3.42	0.94	0.94	37.14
1980	0.42	0.89	4.44	3.02	1.05	4.90	2.69	6.45	2.24	2.27	2.99	1.23	32.59
1981	0.59	5.02	0.26	1.99	2.44	2.78	3.50	1.76	3.55	1.56	3.54	3.54	30.44
1982	3.18	2.14	3.23	2.46	2.60	6.48	2.43	2.01	1.42	0.99	3.80	1.33	32.07
1983	3.73	2.03	5.33	7.95	6.26	1.95	1.34	3.41	2.28	2.18	4.73	5.10	46.29
1984	1.28	2.98	3.04	4.29	7.92	1.74	1.86	3.97	3.25	1.53	2.50	2.15	2.48	37.13
1985	0.81	1.18	3.67	1.44	2.71	4.12	1.86	2.23	3.07	1.81	5.00	2.05	29.95
1986	3.17	3.00	3.72	1.49	3.11	5.43	6.68	4.09	2.61	2.12	4.62	3.92	43.96
1987	4.23	0.24	1.99	4.25	1.57	3.54	2.50	3.67	6.98	6.90	1.78	1.64	39.29
1988	1.95	3.00	1.62	2.22	2.95	1.42	3.12	4.77	1.50	1.40	4.58	1.02	29.55
1989	0.46	1.60	2.69	2.68	5.92	6.52	5.91	2.90	2.81	5.53	1.90	0.75	39.67
1990	3.84	3.94	3.66	3.87	6.12	2.66	1.68	6.66	1.81	4.60	3.67	3.50	46.01
1991	2.15	1.67	2.53	4.14	2.74	1.69	1.65	4.32	3.33	3.82	4.76	2.92	35.72
1992	1.86	1.30	1.66	2.77	3.61	1.96	4.26	2.05	2.43	2.80	3.66	3.02	31.38
1993	2.14	2.86	5.12	5.39	1.37	2.87	6.55	1.54	3.22	3.31	3.80	3.08	41.25
1994	3.20	1.80	4.27	3.45	3.27	3.26	4.25	4.13	2.15	0.83	1.53	2.58	34.72
1995	2.11	1.95	2.20	1.94	1.35	2.27	2.23	3.66	2.28	8.03	3.76	2.30	34.08
1996	5.08	1.49	2.10	5.76	4.24	3.60	6.46	3.15	5.07	2.03	2.91	4.50	46.39
1997	1.67	2.00	4.41	2.30	2.60	0.74	2.34	4.64	4.10	1.91	5.91	2.10	34.72
1998	3.80	2.58	2.86	3.49	5.87	6.58	2.74	2.21	1.98	4.14	1.65	1.04	38.94
1999	4.78	1.59	4.15	0.60	2.77	2.20	2.24	3.45	11.06	2.42	2.07	1.42	38.63
2000	3.43	2.83	3.80	4.23	4.95	6.69	4.48	4.69	3.06	2.48	1.90	4.38	46.92
2001	1.00	1.85	5.50	1.33	3.21	3.78	3.59	2.10	1.64	1.26	1.38	1.95	28.59
POR= 176 YRS	2.52	2.34	2.82	2.87	3.35	3.68	3.60	3.51	3.26	2.99	2.90	2.60	36.44

AVERAGE TEMPERATURE (F) 2001 ALBANY, NY (ALB)

YEAR	JAN	FEB	MAR	APR	MAY	JUN	JUL	AUG	SEP	OCT	NOV	DEC	ANNUAL
1972	22.9	21.1	30.5	41.2	59.5	63.6	70.9	67.2	60.7	45.7	35.1	28.9	45.6
1973	27.0	22.0	41.9	48.8	55.3	68.7	72.8	72.9	60.5	51.0	39.9	28.2	49.1
1974	23.3	21.3	32.4	48.1	54.1	65.0	69.3	67.9	58.3	44.4	38.6	28.9	46.0
1975	25.7	24.9	30.8	40.7	61.9	65.1	72.8	70.0	59.4	53.3	45.5	26.1	48.0
1976	16.0	31.5	36.7	49.7	55.0	69.4	68.5	67.4	59.0	46.5	34.9	21.4	46.3
1977	15.5	24.5	40.0	46.8	60.2	64.6	71.7	67.8	64.6	49.7	42.6	26.7	47.6
1978	21.5	18.2	30.8	43.4	58.4	64.4	68.9	69.2	56.8	48.6	38.6	28.7	45.6
1979	22.1	14.4	38.9	45.4	60.0	66.0	72.5	69.3	61.6	50.2	44.1	31.4	47.9
1980	24.1	19.8	33.3	48.0	59.5	63.3	72.2	70.7	62.6	47.4	34.8	19.9	46.3
1981	14.0	33.1	34.7	48.1	58.9	66.7	69.3	68.5	58.4	44.8	37.7	25.7	46.7
1982	14.3	23.4	32.8	44.3	59.5	62.9	70.1	65.5	60.5	50.6	43.0	33.7	46.7
1983	24.3	26.8	37.6	46.7	54.9	67.2	72.2	69.8	62.6	49.6	39.2	24.0	47.9
1984	18.1	32.4	29.0	47.6	53.2	66.4	68.9	71.8	60.2	53.8	40.3	34.7	48.0
1985	19.9	26.8	37.3	49.7	60.0	62.2	70.7	68.7	63.3	50.2	40.1	24.5	47.8
1986	23.0	22.0	37.2	50.5	61.3	64.6	71.3	67.8	61.3	48.9	35.7	30.8	47.8
1987	21.7	21.7	37.7	50.4	60.0	68.3	73.5	67.2	60.6	46.6	40.1	30.7	48.2
1988	20.6	24.1	34.2	46.6	59.5	65.1	75.0	72.3	59.0	41.0	26.6	27.6	47.6
1989	27.8	24.2	33.5	44.6	59.5	68.0	71.6	69.8	62.5	51.5	39.3	13.7	47.2
1990	32.8	28.2	37.8	48.9	55.3	67.3	73.0	70.1	61.7	53.1	41.8	33.6	50.4
1991	23.2	30.0	37.4	51.2	63.2	69.0	71.6	71.2	59.9	53.2	40.1	28.9	49.9
1992	24.5	26.9	31.5	44.7	58.5	65.2	67.6	67.4	61.4	46.5	38.9	29.8	46.9
1993	26.6	18.3	31.4	48.4	59.5	66.3	73.1	71.7	60.5	49.3	38.4	27.4	47.5
1994	12.7	19.2	33.1	48.2	56.4	68.9	74.0	67.1	60.9	50.1	43.3	31.6	47.1
1995	31.3	22.8	40.0	43.9	57.0	66.9	74.0	70.9	59.1	53.4	35.7	23.9	48.2
1996	20.6	25.3	31.1	46.2	55.2	68.6	69.7	70.1	62.3	49.1	34.6	33.8	47.2
1997	22.7	30.4	33.2	44.2	53.6	67.9	70.6	68.6	60.7	48.0	35.8	29.8	47.1
1998	29.0	31.8	38.4	48.8	62.9	66.3	70.9	71.1	63.2	50.7	39.9	33.7	50.6
1999	21.8	28.2	34.4	46.6	59.4	69.7	74.2	69.2	66.8	48.8	44.1	31.0	49.4
2000	20.7	27.6	40.2	45.3	59.4	65.9	67.6	68.5	59.4	50.0	38.0	22.2	47.1
2001	24.6	26.9	30.9	47.4	58.9	68.4	68.9	73.7	62.4	51.9	44.8	34.2	49.4
POR= 128 YRS	22.8	24.0	33.8	46.6	58.4	67.4	72.3	70.0	62.4	51.1	39.6	27.9	48.0

REFERENCE NOTES:

PAGE 1:
THE TEMPERATURE GRAPH SHOWS NORMAL MAXIMUM AND NORMAL
MINIMUM DAILY TEMPERATURES (SOLID CURVES) AND THE
ACTUAL DAILY HIGH AND LOW TEMPERATURES (VERTICAL BARS).

PAGE 2 AND 3:
H/C INDICATES HEATING AND COOLING DEGREE DAYS.
RH INDICATES RELATIVE HUMIDITY
W/O INDICATES WEATHER AND OBSTRUCTIONS
S INDICATES SUNSHINE.
PR INDICATES PRESSURE.
CLOUDINESS ON PAGE 3 IS THE SUM OF THE CEILOMETER AND
SATELLITE DATA NOT TO EXCEED EIGHT EIGHTHS(OKTAS).

GENERAL:
T INDICATES TRACE PRECIPITATION, AN AMOUNT GREATER
THAN ZERO BUT LESS THAN THE LOWEST REPORTABLE VALUE.
+ INDICATES THE VALUE ALSO OCCURS ON EARLIER DATES.
BLANK ENTRIES DENOTE MISSING OR UNREPORTED DATA.
NORMALS ARE 30-YEAR AVERAGES (1961 - 1990).
ASOS INDICATES AUTOMATED SURFACE OBSERVING SYSTEM.
PM INDICATES THE LAST DAY OF THE PREVIOUS MONTH.
POR (PERIOD OF RECORD) BEGINS WITH THE JANUARY DATA
MONTH AND IS THE NUMBER OF YEARS USED TO COMPUTE
THE MEAN. INDIVIDUAL MONTHS WITHIN THE POR MAY
BE MISSING.
WHEN THE POR FOR A NORMAL IS LESS THAN 30 YEARS,
THE NORMAL IS PROVISIONAL AND IS BASED ON THE NUMBER
OF YEARS INDICATED.
0.* OR * INDICATES THE VALUE OR MEAN-DAYS-WITH
IS BETWEEN 0.00 AND 0.05.
CLOUDINESS FOR ASOS STATIONS DIFFERS FROM THE NON-ASOS
OBSERVATION TAKEN BY A HUMAN OBSERVER. ASOS STATION
CLOUDINESS IS BASED ON TIME-AVERAGED CEILOMETER DATA
FOR CLOUDS AT OR BELOW 12,000 FEET AND ON SATELLITE
DATA FOR CLOUDS ABOVE 12,000 FEET.
THE NUMBER OF DAYS WITH CLEAR, PARTLY CLOUDY, AND
CLOUDY CONDITIONS FOR ASOS STATIONS IS THE SUM
OF THE CEILOMETER AND SATELLITE DATA FOR THE
SUNRISE TO SUNSET PERIOD.

GENERAL CONTINUED:
CLEAR INDICATES 0 - 2 OKTAS, PARTLY CLOUDY INDICATES
3 - 6 OKTAS, AND CLOUDY INDICATES 7 OR 8 OKTAS.
WHEN AT LEAST ONE OF THE ELEMENTS (CEILOMETER OR
SATELLITE) IS MISSING, THE DAILY CLOUDINESS IS
NOT COMPUTED.
WIND DIRECTION IS RECORDED IN TENS OF DEGREES (2 DIGITS)
CLOCKWISE FROM TRUE NORTH. "00" INDICATES CALM. "36"
INDICATES TRUE NORTH.
RESULTANT WIND IS THE VECTOR AVERAGE OF THE SPEED AND
DIRECTION.
AVERAGE TEMPERATURE IS THE SUM OF THE MEAN DAILY MAXIMUM
AND MINIMUM TEMPERATURE DIVIDED BY 2.
SNOWFALL DATA COMPRISE ALL FORMS OF FROZEN
PRECIPITATION, INCLUDING HAIL.
A HEATING (COOLING) DEGREE DAY IS THE DIFFERENCE BETWEEN
THE AVERAGE DAILY TEMPERATURE AND 65 F.
DRY BULB IS THE TEMPERATURE OF THE AMBIENT AIR.
DEW POINT IS THE TEMPERATURE TO WHICH THE AIR MUST BE
COOLED TO ACHIEVE 100 PERCENT RELATIVE HUMIDITY.
WET BULB IS THE TEMPERATURE THE AIR WOULD HAVE IF THE
MOISTURE CONTENT WAS INCREASED TO 100 PERCENT RELATIVE
HUMIDITY.

ON JULY 1, 1996, THE NATIONAL WEATHER SERVICE BEGAN USING
THE "METAR" OBSERVATION CODE THAT WAS ALREADY EMPLOYED
BY MOST OTHER NATIONS OF THE WORLD. THE MOST NOTICEABLE
DIFFERENCE IN THIS ANNUAL PUBLICATION WILL BE THE CHANGE
IN UNITS FROM TENTHS TO EIGHTS(OKTAS) FOR REPORTING THE
AMOUNT OF SKY COVER.

HEATING DEGREE DAYS (base 65 F) 2001 ALBANY, NY (ALB)

YEAR	JUL	AUG	SEP	OCT	NOV	DEC	JAN	FEB	MAR	APR	MAY	JUN	TOTAL
1972-73	16	38	154	590	890	1113	1168	1198	709	486	299	47	6708
1973-74	2	3	200	431	750	1136	1285	1216	1005	511	343	54	6936
1974-75	17	14	227	631	786	1113	1212	1115	1053	722	145	88	7123
1975-76	0	19	173	357	580	1199	1511	964	871	472	315	43	6504
1976-77	7	40	196	564	895	1345	1526	1127	764	545	205	85	7299
1977-78	7	51	156	471	666	1179	1340	1306	1051	642	245	84	7198
1978-79	43	19	256	503	784	1119	1324	1414	803	579	188	63	7095
1979-80	19	37	163	468	619	1036	1259	1303	974	503	190	106	6677
1980-81	0	7	140	539	900	1393	1575	885	930	502	235	30	7136
1981-82	8	22	204	622	816	1209	1564	1160	992	617	182	87	7483
1982-83	20	65	156	436	657	969	1255	1062	843	539	312	58	6372
1983-84	5	24	150	479	766	1265	1448	939	1109	517	363	60	7125
1984-85	12	8	170	344	737	959	1389	1062	852	458	184	106	6281
1985-86	7	16	123	452	740	1246	1295	1177	859	432	154	75	6576
1986-87	17	46	173	495	872	1053	1332	1207	842	433	210	29	6709
1987-88	2	56	154	567	741	1056	1370	1181	946	546	198	99	6916
1988-89	8	30	160	584	714	1185	1146	1133	968	607	194	35	6764
1989-90	0	22	134	413	766	1584	990	1026	839	500	298	44	6616
1990-91	5	6	148	388	689	964	1248	973	850	417	141	22	5893
1991-92	6	0	197	372	740	1111	1248	1098	1034	605	210	56	6677
1992-93	17	27	167	565	773	1082	1183	1300	1034	492	185	67	6892
1993-94	0	11	185	500	791	1161	1619	1272	983	502	283	33	7340
1994-95	0	47	138	457	644	1027	1037	1177	766	627	252	41	6213
1995-96	2	12	196	355	872	1266	1369	1146	1046	559	316	18	7157
1996-97	1	2	133	488	903	961	1306	961	977	616	350	35	6733
1997-98	3	11	152	521	872	1083	1112	924	828	478	98	84	6166
1998-99	2	10	99	435	746	964	1334	1021	940	542	180	22	6295
1999-00	2	14	93	493	621	1048	1367	1077	762	585	209	74	6345
2000-01	14	25	207	456	801	1323	1246	1060	1048	522	202	46	6950
2001-	21	0	116	405	604	949							

COOLING DEGREE DAYS (base 65 F) 2001 ALBANY, NY (ALB)

YEAR	JAN	FEB	MAR	APR	MAY	JUN	JUL	AUG	SEP	OCT	NOV	DEC	ANNUAL
1972	0	0	0	0	12	58	208	112	31	0	0	0	421
1973	0	0	0	7	6	164	248	255	71	2	0	0	753
1974	0	0	0	11	12	59	157	111	35	0	1	0	386
1975	0	0	0	0	58	97	248	180	12	0	2	0	597
1976	0	0	0	19	11	184	120	120	22	0	0	0	476
1977	0	0	0	8	66	79	222	146	53	0	0	0	574
1978	0	0	0	0	47	70	169	154	16	0	0	0	456
1979	0	0	0	0	39	99	258	168	55	17	0	0	636
1980	0	0	0	0	28	63	230	189	73	0	0	0	583
1981	0	0	0	2	53	87	149	137	25	0	0	0	453
1982	0	0	0	0	19	31	184	88	29	0	4	0	355
1983	0	0	0	0	8	134	236	179	86	6	0	0	649
1984	0	0	0	0	3	107	140	226	35	3	0	0	514
1985	0	0	0	5	37	27	191	140	80	2	0	0	482
1986	0	0	6	4	46	69	220	140	33	1	0	0	519
1987	0	0	0	4	62	136	271	133	29	0	0	0	635
1988	0	0	0	0	36	110	326	263	16	4	0	0	755
1989	0	0	1	0	31	132	213	178	63	0	0	0	618
1990	0	0	2	22	1	119	261	197	55	24	0	0	681
1991	0	0	0	9	92	147	221	198	50	14	0	0	731
1992	0	0	0	2	15	70	106	112	65	0	0	0	370
1993	0	0	0	0	17	116	259	224	55	2	0	0	673
1994	0	0	0	6	24	160	290	119	21	0	1	0	621
1995	0	0	0	0	11	102	289	200	27	0	0	0	629
1996	0	0	0	1	19	132	153	168	57	0	0	0	530
1997	0	0	0	0	3	128	186	129	28	1	0	0	475
1998	0	0	10	0	38	128	192	206	54	0	0	0	628
1999	0	0	0	0	15	170	291	151	94	0	0	0	721
2000	0	0	0	0	42	107	100	141	45	1	0	0	436
2001	0	0	0	3	16	152	150	279	45	8	0	0	653

SNOWFALL (inches) 2001 ALBANY, NY (ALB)

YEAR	JUL	AUG	SEP	OCT	NOV	DEC	JAN	FEB	MAR	APR	MAY	JUN	TOTAL
1972-73	0.0	0.0	0.0	T	24.6	22.5	11.2	12.5	T	0.1	0.0	0.0	70.9
1973-74	0.0	0.0	0.0	0.0	0.1	18.9	10.0	12.4	5.6	11.3	0.0	0.0	58.3
1974-75	0.0	0.0	0.0	T	2.2	12.5	14.0	21.2	2.9	1.8	0.0	0.0	54.6
1975-76	0.0	0.0	0.0	0.0	3.6	16.4	15.0	4.4	14.8	T	1.6	0.0	54.2
1976-77	0.0	0.0	0.0	T	5.7	7.8	22.1	17.9	15.2	0.3	1.6	0.0	70.6
1977-78	0.0	0.0	0.0	0.0	8.4	19.8	40.8	15.8	7.4	0.2	T	0.0	92.4
1978-79	0.0	0.0	0.0	0.0	3.4	19.9	26.5	4.6	0.9	8.2	0.0	0.0	63.5
1979-80	0.0	0.0	0.0	T	0.0	5.8	0.6	10.2	10.8	0.0	0.0	0.0	27.4
1980-81	0.0	0.0	0.0	0.0	11.8	12.8	11.9	6.9	1.5	T	0.0	0.0	44.9
1981-82	0.0	0.0	0.0	0.0	1.1	31.4	18.2	9.6	19.1	17.7	0.0	0.0	97.1
1982-83	0.0	0.0	0.0	0.0	0.6	5.5	27.5	17.4	9.2	14.7	0.1	0.0	75.0
1983-84	0.0	0.0	0.0	0.0	1.7	11.6	16.5	7.2	28.2	T	0.0	0.0	65.2
1984-85	0.0	0.0	0.0	0.0	2.2	11.7	8.4	10.1	8.7	0.2	0.0	0.0	41.3
1985-86	0.0	0.0	0.0	0.0	11.8	11.5	18.0	16.1	3.4	1.7	T	0.0	62.5
1986-87	0.0	0.0	0.0	0.0	8.3	20.3	47.8	2.8	0.8	0.6	0.0	0.0	80.6
1987-88	0.0	0.0	0.0	6.5	6.2	11.4	21.7	26.0	4.8	0.1	0.0	0.0	76.7
1988-89	0.0	0.0	0.0	T	T	7.8	1.3	5.1	4.7	0.1	0.0	0.0	19.0
1989-90	T	0.0	T	0.0	1.9	8.0	20.3	22.8	4.9	T	0.0	0.0	57.9
1990-91	0.0	0.0	0.0	T	0.4	8.5	11.2	5.3	3.3	0.0	0.0	T	28.7
1991-92	0.0	0.0	0.0	0.0	1.5	12.7	3.4	6.3	4.9	1.9	0.0	0.0	30.7
1992-93	0.0	0.0	0.0	T	2.8	12.6	14.3	28.6	34.3	1.6	0.0	0.0	94.2
1993-94	0.0	0.0	0.0	T	0.7	6.1	42.0	20.2	19.1	T	T	0.0	88.1
1994-95	0.0	0.0	0.0	0.0	4.1	2.9	3.9	15.4	4.6	T	0.0	0.0	30.9
1995-96	T	0.0	0.0	0.0	5.8	25.1	28.4		20.3	1.1	0.0	0.0	
1996-97				0.0	4.0	11.1	16.7	8.2	23.6	3.0	0.0	0.0	
1997-98	0.0	0.0	0.0	T	11.8	14.7	13.5	6.0	6.1	0.0	0.0	0.0	52.1
1998-99	0.0	0.0	0.0	0.0	T	3.2	20.4	5.8	14.7	0.0	0.0	0.0	44.1
1999-00	0.0	0.0	0.0	0.0	0.4	1.1	31.0	12.4	3.9	13.3	0.0	0.0	62.1
2000-01	0.0	0.0	0.0	0.4	2.5	20.0	7.6	16.0	30.6	T	T	0.0	77.1
2001-	T	0.0	0.0	T	T	7.8							
POR= 54 YRS	T	0.0	T	0.2	4.1	14.4	16.3	13.6	11.7	2.7	0.2	T	63.2

2001
BUFFALO,
NEW YORK (BUF)

The Niagara Frontier experiences a fairly humid, continental type climate, but with a definite "maritime" flavor due to a strong modification from the Great Lakes (especially Lake Erie). Buffalo's weather repution is highly exaggerated, and due mainly to its propensity for localized heavy Lake-effect snowstorms in late fall and early winter. Summers, on the other hand, are among the most pleasant in the Northeast.

Winters in general are cloudy, cold and snowy...but are changeable and include frequent thaws and rain as well. Snow covers the ground more often than not from Christmas into early March...but periods of bare ground are not uncommon. Over half of the annual snowfall comes from "Lake-effect" process and is very localized. This feature develops when cold air crosses the warmer lake waters and becomes saturated.. creating clouds and precipitation downwind. The exact location of these snowbands are determined by the direction of the wind. Areas south of Buffalo derive much more snow from this process than the more densely populated northern suburbs. This snow machine can start as early as mid-November, peaks in December, then virtually shuts down after Lake Erie freezes in mid to late January. The Buffalo area is not subject to heavy general or "synoptic" snowstorms. Most of them pass by to the east. Total season snowfall ranges from about 60 inches in the far northern suburbs to 80-90 inches in the city and eastern suburbs to as much as 120 inches south of the city. The lakes do modify any extreme cold as the mercury falls below zero on only about four nights in an average winter...with anything below -10 extremely rare.

Spring comes slowly to the Niagara Frontier. The ice pack in lake Erie does not usually disappear until mid-April and the Lake remains chilly through most of May. As the prevailing flow is southwesterly, areas near the lake are often as much as 20 degrees colder than inland locations. Conversely, the cool Lake acts as a strong stabilizing influence so areas near the city and lakeshore experience fewer thunderstorms and more sunshine then inland areas in spring. The slow start to the growing season also diminishes the threat of damaging late season frosts. The average date of the last frost is around April 30 in the metro area...but mid-May well inland.

Summer is beautiful in the Buffalo area. Sunshine is plentiful, temperature are warm but seldom hot, and humidity levels moderate. Rainfall is adequate, but does show an overnight maximum and seldom is a problem for outdoor activities. The stabilizing effect of Lake Erie continues to inhibit thunderstorms and ehance sunshine in the immediate Buffalo area..at least through most of July. It also moderates most extreme heat approaching from the Ohio Valley. There usually are several periods of uncomfortably warm and muggy weather in an average summer...but 90-degree readings are relatively rare (only 3 per year). August usually turns a bit more humid and showery as the Lake is warmer and loses its stabilizing influence. In fact, a good nighttime thunderstorm or two is often a feature of late summmer in Buffalo. Overall though...Buffalo has the sunniest and driest summers of any major city in the Northeast.

Autumn is pleasant, but rather brief. September is usually very tame, and much of October as well. The first frosts can be expected in late September over interior sections, but not until mid-October in the metro area. The warm lake can extend the growing season into early November during some years close to the Lakeshore. The growing season is relatively long for the latitude...about 180 days...and is conducive to the many Fruit orchards and wineries, especially near Lake Ontario and along the Lake Erie shore. Cold air surges from Canada become more common starting in late October...with their passage over the warmer Great Lakes resulting in a drastic increase in cloud cover in late October and early November as the Lake-effect season begins. The first measurable snows can be expected in mid to late November, but ground cover is only sporadic until mid December. Many of Buffalo's greatest snowstorms however, have occurred in late November and early December, all due to the Lake effect phenomenon.

NORMALS, MEANS, AND EXTREMES
BUFFALO, NY (BUF)

LATITUDE:	LONGITUDE:	ELEVATION (FT):	TIME ZONE:	WBAN: 14733
42 56′ 27″ N	78 44′ 09″ W	GRND: 714 BARO: 717	EASTERN (UTC + 5)	

	ELEMENT	POR	JAN	FEB	MAR	APR	MAY	JUN	JUL	AUG	SEP	OCT	NOV	DEC	YEAR
TEMPERATURE °F	NORMAL DAILY MAXIMUM	30	30.2	31.6	41.7	54.2	66.1	75.3	80.2	77.9	70.8	59.4	47.1	35.3	55.8
	MEAN DAILY MAXIMUM	56	31.3	33.0	41.4	54.7	66.3	76.0	80.4	78.7	71.1	60.5	47.4	36.0	56.4
	HIGHEST DAILY MAXIMUM	58	72	71	81	94	90	96	97	99	98	87	80	74	99
	YEAR OF OCCURRENCE		1950	2000	1945	1990	1991	1988	1995	1948	1953	1951	1961	1982	AUG 1948
	MEAN OF EXTREME MAXS.	56	53.1	54.7	67.1	77.1	83.5	88.7	89.9	88.6	85.9	78.1	67.9	56.9	74.3
	NORMAL DAILY MINIMUM	30	17.0	17.4	25.9	36.2	47.0	56.5	61.9	60.1	53.0	42.7	33.9	22.9	39.5
	MEAN DAILY MINIMUM	56	18.1	18.6	25.9	36.2	46.6	56.5	61.7	60.1	52.7	43.1	33.9	23.5	39.7
	LOWEST DAILY MINIMUM	58	-16	-20	-7	12	26	35	43	38	32	20	9	-10	-20
	YEAR OF OCCURRENCE		1982	1961	1984	1982	1947	1945	1945	1982	1991	1965	1971	1980	FEB 1961
	MEAN OF EXTREME MINS.	56	-.6	0.2	8.0	23.2	33.9	43.4	50.7	47.6	37.6	29.4	18.9	4.2	24.7
	NORMAL DRY BULB	30	23.6	24.5	33.8	45.2	56.6	65.9	71.1	69.0	61.9	51.1	40.5	29.1	47.7
	MEAN DRY BULB	56	24.7	25.8	33.8	45.4	56.6	66.2	71.1	69.5	61.9	51.8	40.6	29.7	48.1
	MEAN WET BULB	18	23.8	24.5	30.6	40.9	51.3	60.0	64.1	63.4	56.7	46.6	37.2	28.1	43.9
	MEAN DEW POINT	18	19.0	19.4	24.6	34.5	45.4	55.4	59.6	59.2	52.5	41.6	32.1	23.6	38.9
	NORMAL NO. DAYS WITH:														
	MAXIMUM 90	30	0.0	0.0	0.0	*	0.1	0.6	1.6	0.7	*	0.0	0.0	0.0	3.0
	MAXIMUM 32	30	17.5	15.4	7.2	0.6	0.0	0.0	0.0	0.0	0.0	0.0	2.0	12.3	55.0
	MINIMUM 32	30	28.6	25.7	23.9	10.5	0.7	0.0	0.0	0.0	*	3.1	13.9	25.9	132.3
	MINIMUM 0	30	2.2	1.4	0.2	0.0	0.0	0.0	0.0	0.0	0.0	0.0	0.0	0.6	4.4
H/C	NORMAL HEATING DEG. DAYS	30	1283	1134	967	594	279	59	5	17	130	431	735	1113	6747
	NORMAL COOLING DEG. DAYS	30	0	0	0	0	19	86	194	141	37	0	0	0	477
RH	NORMAL (PERCENT)	30	76	76	73	68	67	69	68	72	74	73	76	78	72
	HOUR 01 LST	30	77	79	78	75	76	79	79	83	83	80	79	79	79
	HOUR 07 LST	30	79	80	80	76	75	77	78	83	84	82	80	81	80
	HOUR 13 LST	30	72	70	65	58	56	56	55	58	60	61	69	73	63
	HOUR 19 LST	30	76	75	72	64	62	62	60	66	72	73	76	77	70
S	PERCENT POSSIBLE SUNSHINE	57	31	38	46	50	58	64	67	64	57	50	29	27	48
W/O	MEAN NO. DAYS WITH:														
	HEAVY FOG(VISBY 1/4 MI)	59	1.6	1.8	2.6	2.1	2.3	1.3	0.8	0.9	1.0	1.3	1.4	1.4	18.5
	THUNDERSTORMS	59	0.2	0.2	1.2	2.3	3.4	5.3	5.8	5.9	3.7	1.6	1.1	0.4	31.1
CLOUDINESS	MEAN:														
	SUNRISE-SUNSET (OKTAS)	1												8.8	
	MIDNIGHT-MIDNIGHT (OKTAS)														
	MEAN NO. DAYS WITH:														
	CLEAR	1			5.0		4.0	4.0							
	PARTLY CLOUDY	1			4.0		8.0	5.0							
	CLOUDY	1	3.0	4.0	9.0		7.0	8.0							
PR	MEAN STATION PRESSURE(IN)	29	29.29	29.30	29.20	29.20	29.20	29.20	29.20	29.30	29.30	29.30	29.30	29.29	29.26
	MEAN SEA-LEVEL PRES. (IN)	18	30.05	30.07	30.04	29.98	29.98	29.95	29.99	30.04	30.05	30.08	30.06	30.06	30.03
WINDS	MEAN SPEED (MPH)	53	13.8	13.0	12.8	12.1	11.1	10.6	10.1	9.5	10.0	10.7	12.2	12.9	11.6
	PREVAIL.DIR(TENS OF DEGS)	34	24	24	24	23	23	24	24	24	24	24	27	27	24
	MAXIMUM 2-MINUTE:														
	SPEED (MPH)	6	44	54	44	38	43	38	41	32	39	46	46	51	54
	DIR. (TENS OF DEGS)		23	24	26	24	24	24	31	24	26	23	23	23	24
	YEAR OF OCCURRENCE		1996	1997	1998	1997	1997	2000	1999	2001	1997	2001	1998	2000	FEB 1997
	MAXIMUM 5-SECOND:														
	SPEED (MPH)	6	57	70	57	48	55	48	54	44	47	60	61	66	70
	DIR. (TENS OF DEGS)		24	25	27	26	23	26	31	27	22	24	22	22	25
	YEAR OF OCCURRENCE		1997	1997	1998	1997	1997	2000	1999	1997	1999	2001	1998	2000	FEB 1997
PRECIPITATION	NORMAL (IN)	30	2.70	2.31	2.68	2.87	3.14	3.55	3.08	4.17	3.49	3.09	3.83	3.67	38.58
	MAXIMUM MONTHLY (IN)	58	6.88	5.90	5.97	5.90	7.22	8.36	8.93	10.67	8.99	9.13	9.75	8.71	10.67
	YEAR OF OCCURRENCE		1982	1990	1991	1961	1989	1987	1992	1977	1977	1954	1985	1990	AUG 1977
	MINIMUM MONTHLY (IN)	58	1.03	0.81	1.20	1.27	1.21	0.11	0.73	1.10	0.77	0.30	1.44	0.69	0.11
	YEAR OF OCCURRENCE		1946	1968	1967	1946	1965	1955	2001	1948	1964	1963	1944	1943	JUN 1955
	MAXIMUM IN 24 HOURS (IN)	58	2.57	2.31	2.14	2.09	3.52	5.01	3.38	3.88	4.94	3.49	2.51	2.33	5.01
	YEAR OF OCCURRENCE		1982	1954	1954	1991	1986	1987	1963	1963	1979	1945	1949	1990	JUN 1987
	NORMAL NO. DAYS WITH:														
	PRECIPITATION 0.01	30	19.8	17.7	16.1	14.0	12.3	11.2	9.8	11.3	11.3	12.8	16.0	20.1	172.4
	PRECIPITATION 1.00	30	0.2	0.2	0.2	0.2	0.5	0.8	0.7	1.0	0.7	0.5	0.5	0.4	5.9
SNOWFALL	NORMAL (IN)	30	25.7	18.2	10.3	3.8	0.3	0.0	0.0	0.0	0.0	0.2	9.9	23.8	92.2
	MAXIMUM MONTHLY (IN)	58	68.3	54.2	32.8	15.0	7.9	T	T	T	T	3.1	45.6	82.7	82.7
	YEAR OF OCCURRENCE		1977	1958	2001	1975	1989	1980	1993	1991	1994	1972	2000	2001	DEC 2001
	MAXIMUM IN 24 HOURS (IN)	58	25.3	19.4	17.2	6.8	7.9	T	T	T	T	2.8	24.9	37.9	37.9
	YEAR OF OCCURRENCE		1982	1984	1993	1975	1989	1980	1993	1991	1994	1993	2000	1995	DEC 1995
	MAXIMUM SNOW DEPTH (IN)	53	38	42	20	12	4	0	0	0	0	2	25	44	44
	YEAR OF OCCURRENCE		1977	1977	1984	1975	1989					1972	2000	2001	DEC 2001
	NORMAL NO. DAYS WITH:														
	SNOWFALL 1.0	30	7.6	5.8	3.3	1.1	0.1	0.0	0.0	0.0	0.0	0.1	2.8	6.7	27.5

PRECIPITATION (inches) 2001 BUFFALO, NY (BUF)

YEAR	JAN	FEB	MAR	APR	MAY	JUN	JUL	AUG	SEP	OCT	NOV	DEC	ANNUAL
1972	2.17	3.44	3.99	2.99	3.64	6.06	0.99	4.19	3.06	2.96	4.28	3.86	41.63
1973	2.03	1.98	3.27	3.56	2.99	1.68	3.68	2.98	1.44	4.27	4.07	4.89	36.84
1974	2.44	2.19	3.19	3.15	3.36	3.86	1.80	2.42	1.75	5.38	3.13	3.63	36.31
1975	2.11	2.93	2.92	1.86	3.31	3.65	2.34	8.49	2.44	1.13	2.77	4.58	38.53
1976	3.19	3.43	5.59	4.01	4.70	3.36	5.65	1.65	5.39	3.61	2.11	3.83	46.52
1977	3.38	1.59	2.42	3.60	1.39	2.79	3.64	10.67	8.99	2.61	4.45	8.02	53.55
1978	6.29	1.36	1.72	1.84	3.95	2.42	1.48	3.51	4.40	3.72	1.55	3.50	35.74
1979	5.43	2.03	2.48	3.16	1.63	2.18	3.51	6.26	5.61	3.88	3.43	43.74	43.74
1980	1.97	1.08	4.05	2.43	1.60	5.82	3.55	3.58	4.53	4.69	2.36	2.65	38.31
1981	1.11	3.50	1.70	3.09	2.56	3.68	5.05	4.24	3.31	2.22	2.87	36.46	36.46
1982	6.88	1.28	2.64	2.33	3.66	3.14	1.50	4.62	3.37	2.06	6.31	3.32	41.11
1983	1.44	1.30	3.20	2.55	3.28	2.99	2.01	3.51	2.11	4.62	5.19	7.30	39.50
1984	1.54	3.59	1.77	2.53	4.67	6.86	1.37	4.16	3.73	0.87	2.66	3.67	37.42
1985	4.27	3.34	4.42	1.33	3.46	3.21	1.81	4.63	1.20	3.73	9.75	4.85	46.00
1986	2.31	2.60	1.95	3.33	4.42	4.15	2.82	2.73	3.88	4.34	3.11	4.02	39.66
1987	2.90	0.85	3.66	3.40	1.35	8.36	3.09	3.38	5.32	2.62	4.44	2.78	42.15
1988	1.58	4.07	2.99	2.96	2.74	1.56	6.35	2.69	2.07	6.08	3.37	2.15	38.61
1989	1.77	2.54	3.15	1.88	7.22	7.83	0.93	1.84	3.85	2.98	4.83	2.34	41.16
1990	2.69	5.90	1.50	5.22	6.08	3.55	3.14	3.25	3.65	4.59	2.61	8.71	50.89
1991	2.07	2.06	5.97	5.83	3.10	0.86	3.34	2.84	3.19	3.11	4.02	3.81	40.20
1992	2.01	2.45	2.93	4.68	3.48	2.21	8.93	3.79	5.56	2.80	4.92	3.80	47.56
1993	4.35	1.92	3.02	2.55	1.79	4.99	1.78	3.86	5.53	3.69	3.58	3.60	40.66
1994	2.90	1.40	2.61	4.02	3.54	4.27	2.08	4.09	3.19	1.87	4.08	2.67	36.72
1995	4.89	2.62	1.33	1.41	2.40	1.33	3.53	2.07	1.32	6.07	4.14	2.88	33.99
1996	3.42	2.09	2.37	5.63	4.08	5.20	5.15	2.14	7.51	4.22	2.99	3.42	48.22
1997	4.25	2.97	4.47	1.65	3.61	3.06	1.85	4.67	5.06	2.29	4.32	2.88	41.08
1998	5.61	2.28	3.86	2.54	3.73	2.87	4.39	1.74	2.43	2.10	1.61	1.54	34.70
1999	5.78	1.10	2.43	2.21	2.82	1.93	1.00	4.38	3.95	2.95	3.33	2.20	34.08
2000	2.65	1.75	2.12	4.07	4.38	6.51	2.90	3.21	3.92	1.11	5.82	3.76	42.20
2001	2.15	2.33	3.31	1.27	4.28	1.36	0.73	2.13	3.45	4.34	3.35	6.48	35.18
POR= 131 YRS	3.13	2.66	2.79	2.76	3.00	2.95	2.90	3.20	3.17	3.08	3.33	3.31	36.28

AVERAGE TEMPERATURE (F) 2001 BUFFALO, NY (BUF)

YEAR	JAN	FEB	MAR	APR	MAY	JUN	JUL	AUG	SEP	OCT	NOV	DEC	ANNUAL
1972	25.5	22.0	30.1	41.1	59.1	62.6	71.0	67.7	62.8	46.2	36.0	30.8	46.2
1973	27.6	22.9	42.4	46.9	54.5	68.2	72.3	71.8	61.7	54.3	40.8	29.0	49.4
1974	27.1	22.3	33.0	46.2	53.1	65.6	69.9	69.9	59.6	49.2	40.2	31.7	47.3
1975	30.1	29.1	30.8	39.3	62.1	68.0	72.3	69.7	58.3	53.1	46.9	28.3	49.0
1976	19.7	31.8	37.2	46.5	53.4	68.4	67.8	67.5	60.1	46.3	34.1	22.0	46.2
1977	13.8	24.6	39.8	47.0	60.3	64.4	72.0	68.1	62.6	49.6	43.3	27.9	47.8
1978	20.4	15.5	28.2	42.5	57.4	65.1	70.4	70.3	60.8	49.5	40.4	30.4	45.9
1979	20.5	14.9	38.2	44.3	56.9	66.5	71.3	67.5	61.9	50.7	43.5	33.4	47.5
1980	25.8	21.2	31.8	46.1	58.1	61.9	71.7	72.6	62.4	48.7	39.4	25.3	47.1
1981	19.3	32.9	33.9	47.2	56.4	66.2	71.8	70.0	61.6	48.2	40.4	29.0	47.0
1982	17.2	23.2	32.5	41.6	61.0	62.2	71.8	65.0	61.6	52.6	43.0	37.5	47.4
1983	27.0	29.6	36.7	43.6	53.9	67.6	74.2	71.2	63.7	51.7	40.8	22.7	48.6
1984	20.4	33.8	27.1	47.7	52.9	67.8	70.3	70.3	58.5	53.2	39.0	35.6	48.1
1985	21.1	24.8	35.6	49.5	59.5	62.7	69.7	69.2	64.2	52.5	42.0	25.6	48.0
1986	25.5	24.5	36.2	47.8	59.7	64.1	71.1	67.9	61.8	50.9	37.7	32.4	48.3
1987	26.1	25.0	37.7	50.0	60.5	68.9	74.2	68.9	63.4	47.9	42.5	34.3	50.0
1988	26.6	24.3	35.2	46.1	59.7	64.0	74.8	72.4	62.1	46.9	43.0	30.0	48.8
1989	31.3	22.7	33.0	41.9	55.1	65.9	71.5	68.5	60.8	51.5	37.9	17.4	46.5
1990	33.4	29.3	36.9	48.5	54.9	66.7	71.4	70.4	61.7	52.5	43.4	34.4	50.3
1991	26.0	30.6	37.8	50.5	64.3	69.1	71.9	71.0	62.0	53.1	39.3	31.3	50.6
1992	27.1	27.7	31.6	43.8	57.3	63.4	66.8	66.3	61.6	47.9	40.2	31.9	47.1
1993	29.5	20.7	30.7	47.3	57.0	66.0	73.4	72.0	59.4	49.2	39.6	29.6	47.9
1994	17.2	22.8	33.4	48.2	54.7	69.0	73.3	68.0	61.9	52.2	45.1	34.0	48.3
1995	29.8	21.9	37.8	42.3	56.8	69.9	72.7	73.0	60.0	54.2	36.4	24.5	48.3
1996	22.5	24.2	29.0	42.2	54.5	67.8	68.5	70.5	62.7	51.7	35.4	33.5	46.9
1997	24.7	30.1	33.1	42.3	50.6	66.7	68.6	66.8	60.5	50.5	37.6	31.8	46.9
1998	31.1	34.1	36.5	46.8	62.8	65.3	69.6	71.2	63.7	52.6	42.0	35.3	50.9
1999	23.5	31.0	31.0	46.0	59.7	68.4	74.3	67.9	64.3	43.9	42.0	29.3	49.3
2000	23.6	29.9	40.0	44.2	57.5	64.9	67.6	68.0	61.2	52.3	38.8	22.1	47.5
2001	27.0	28.2	31.1	47.3	58.8	67.0	69.8	73.0	62.7	53.0	46.9	35.9	50.1
POR≈ 128 YRS	25.0	24.8	32.7	43.7	55.1	64.8	70.5	69.0	62.4	51.5	40.2	29.6	47.4

REFERENCE NOTES:

PAGE 1:
THE TEMPERATURE GRAPH SHOWS NORMAL MAXIMUM AND NORMAL
MINIMUM DAILY TEMPERATURES (SOLID CURVES) AND THE
ACTUAL DAILY HIGH AND LOW TEMPERATURES (VERTICAL BARS).

PAGE 2 AND 3:
H/C INDICATES HEATING AND COOLING DEGREE DAYS.
RH INDICATES RELATIVE HUMIDITY
W/O INDICATES WEATHER AND OBSTRUCTIONS
S INDICATES SUNSHINE.
PR INDICATES PRESSURE.
CLOUDINESS ON PAGE 3 IS THE SUM OF THE CEILOMETER AND
SATELLITE DATA NOT TO EXCEED EIGHT EIGHTHS(OKTAS).

GENERAL:
T INDICATES TRACE PRECIPITATION, AN AMOUNT GREATER
THAN ZERO BUT LESS THAN THE LOWEST REPORTABLE VALUE.
+ INDICATES THE VALUE ALSO OCCURS ON EARLIER DATES.
BLANK ENTRIES DENOTE MISSING OR UNREPORTED DATA.
NORMALS ARE 30-YEAR AVERAGES (1961 - 1990).
ASOS INDICATES AUTOMATED SURFACE OBSERVING SYSTEM.
PM INDICATES THE LAST DAY OF THE PREVIOUS MONTH.
POR (PERIOD OF RECORD) BEGINS WITH THE JANUARY DATA
MONTH AND IS THE NUMBER OF YEARS USED TO COMPUTE
THE MEAN. INDIVIDUAL MONTHS WITHIN THE POR MAY
BE MISSING.
WHEN THE POR FOR A NORMAL IS LESS THAN 30 YEARS,
THE NORMAL IS PROVISIONAL AND IS BASED ON THE NUMBER
OF YEARS INDICATED.
0.* OR * INDICATES THE VALUE OR MEAN-DAYS-WITH
IS BETWEEN 0.00 AND 0.05.
CLOUDINESS FOR ASOS STATIONS DIFFERS FROM THE NON-ASOS
OBSERVATION TAKEN BY A HUMAN OBSERVER. ASOS STATION
CLOUDINESS IS BASED ON TIME-AVERAGED CEILOMETER DATA
FOR CLOUDS AT OR BELOW 12,000 FEET AND ON SATELLITE
DATA FOR CLOUDS ABOVE 12,000 FEET.
THE NUMBER OF DAYS WITH CLEAR, PARTLY CLOUDY, AND
CLOUDY CONDITIONS FOR ASOS STATIONS IS THE SUM
OF THE CEILOMETER AND SATELLITE DATA FOR THE
SUNRISE TO SUNSET PERIOD.

GENERAL CONTINUED:
CLEAR INDICATES 0 - 2 OKTAS, PARTLY CLOUDY INDICATES
3 - 6 OKTAS, AND CLOUDY INDICATES 7 OR 8 OKTAS.
WHEN AT LEAST ONE OF THE ELEMENTS (CEILOMETER OR
SATELLITE) IS MISSING, THE DAILY CLOUDINESS IS
NOT COMPUTED.
WIND DIRECTION IS RECORDED IN TENS OF DEGREES (2 DIGITS)
CLOCKWISE FROM TRUE NORTH. "00" INDICATES CALM. "36"
INDICATES TRUE NORTH.
RESULTANT WIND IS THE VECTOR AVERAGE OF THE SPEED AND
DIRECTION.
AVERAGE TEMPERATURE IS THE SUM OF THE MEAN DAILY MAXIMUM
AND MINIMUM TEMPERATURE DIVIDED BY 2.
SNOWFALL DATA COMPRISE ALL FORMS OF FROZEN
PRECIPITATION, INCLUDING HAIL.
A HEATING (COOLING) DEGREE DAY IS THE DIFFERENCE BETWEEN
THE AVERAGE DAILY TEMPERATURE AND 65 F.
DRY BULB IS THE TEMPERATURE OF THE AMBIENT AIR.
DEW POINT IS THE TEMPERATURE TO WHICH THE AIR MUST BE
COOLED TO ACHIEVE 100 PERCENT RELATIVE HUMIDITY.
WET BULB IS THE TEMPERATURE THE AIR WOULD HAVE IF THE
MOISTURE CONTENT WAS INCREASED TO 100 PERCENT RELATIVE
HUMIDITY.

ON JULY 1, 1996, THE NATIONAL WEATHER SERVICE BEGAN USING
THE "METAR" OBSERVATION CODE THAT WAS ALREADY EMPLOYED
BY MOST OTHER NATIONS OF THE WORLD. THE MOST NOTICEABLE
DIFFERENCE IN THIS ANNUAL PUBLICATION WILL BE THE CHANGE
IN UNITS FROM TENTHS TO EIGHTS(OKTAS) FOR REPORTING THE
AMOUNT OF SKY COVER.

HEATING DEGREE DAYS (base 65 F) 2001 BUFFALO, NY (BUF)

YEAR	JUL	AUG	SEP	OCT	NOV	DEC	JAN	FEB	MAR	APR	MAY	JUN	TOTAL
1972-73	16	33	113	574	860	1054	1152	1173	696	542	318	24	6555
1973-74	2	14	171	326	720	1107	1167	1187	989	553	365	51	6652
1974-75	2	0	187	483	738	1024	1077	1001	1053	764	175	32	6536
1975-76	3	15	197	368	535	1134	1400	958	853	557	358	40	6418
1976-77	15	35	180	573	921	1328	1580	1123	775	544	207	90	7371
1977-78	5	40	110	473	646	1146	1376	1378	1130	670	282	81	7337
1978-79	14	3	154	472	732	1067	1371	1400	823	619	285	65	7005
1979-80	16	35	134	455	636	973	1208	1265	1022	559	240	142	6685
1980-81	2	0	128	498	759	1224	1411	895	956	527	269	33	6702
1981-82	6	11	170	514	732	1108	1476	1163	1002	698	147	95	7122
1982-83	4	65	140	382	656	848	1172	987	868	636	342	71	6171
1983-84	5	10	125	418	722	1304	1378	899	1167	519	385	35	6967
1984-85	11	22	210	360	774	905	1354	1120	902	476	196	95	6425
1985-86	8	12	114	378	685	1215	1215	1128	885	519	197	80	6436
1986-87	4	42	137	430	811	1003	1199	1115	837	447	213	28	6266
1987-88	3	25	91	527	665	948	1184	1174	916	560	186	113	6392
1988-89	5	17	122	560	654	1078	1038	1177	985	687	321	60	6704
1989-90	1	28	170	411	806	1466	970	995	866	518	311	46	6588
1990-91	5	2	141	395	640	941	1203	956	836	431	141	22	5713
1991-92	1	1	166	376	762	1037	1169	1076	1027	633	254	93	6595
1992-93	28	41	148	525	738	1021	1095	1235	1053	526	257	60	6727
1993-94	0	8	212	486	752	1089	1476	1174	972	502	327	48	7046
1994-95	0	26	123	390	591	955	1085	1201	835	674	247	22	6149
1995-96	14	3	164	329	851	1250	1310	1178	1107	677	333	22	7238
1996-97	15	1	130	406	881	969	1241	970	983	673	438	40	6747
1997-98	17	25	150	457	814	1023	1045	862	878	538	96	104	6009
1998-99	0	9	88	378	682	912	1280	949	1048	566	193	58	6163
1999-00	0	17	97	454	628	1014	1276	1012	770	617	246	73	6204
2000-01	20	26	176	385	780	1323	1171	1023	1042	528	190	61	6725
2001-	18	0	127	371	535	893							

COOLING DEGREE DAYS (base 65 F) 2001 BUFFALO, NY (BUF)

YEAR	JAN	FEB	MAR	APR	MAY	JUN	JUL	AUG	SEP	OCT	NOV	DEC	ANNUAL
1972	0	0	0	0	12	48	210	123	57	0	0	0	450
1973	0	0	0	6	2	126	233	230	78	3	0	0	678
1974	0	0	0	0	7	71	163	158	29	0	0	0	428
1975	0	0	0	0	90	129	238	171	3	3	0	0	634
1976	0	0	0	8	7	149	109	119	40	0	0	0	432
1977	0	0	0	12	68	78	228	142	45	0	1	0	574
1978	0	0	0	0	52	91	189	173	35	0	0	0	540
1979	0	0	0	6	40	118	217	120	49	20	0	0	570
1980	0	0	0	0	32	56	217	242	58	2	0	0	607
1981	0	0	0	2	13	78	225	173	55	0	0	0	546
1982	0	0	0	3	31	18	221	74	45	2	0	2	396
1983	0	0	0	0	5	157	300	214	90	15	0	0	781
1984	0	0	0	5	16	123	183	193	23	1	0	0	544
1985	0	0	0	18	32	32	161	151	96	0	1	0	491
1986	0	0	0	7	38	60	200	137	46	0	0	0	488
1987	0	0	0	4	79	151	298	152	49	0	0	0	733
1988	0	0	0	0	29	88	315	255	41	8	0	0	736
1989	0	0	0	0	0	21	97	207	143	50	0	0	518
1990	0	0	3	29	4	104	208	176	47	14	0	0	585
1991	0	0	0	3	125	153	221	193	83	13	0	0	791
1992	0	0	0	1	24	53	90	90	55	0	0	0	313
1993	0	0	0	0	14	97	267	231	51	3	0	0	663
1994	0	0	0	0	5	14	175	267	125	36	2	0	0	624
1995	0	0	0	0	2	176	262	256	21	1	0	0	718
1996	0	0	0	0	12	108	131	177	65	2	0	0	495
1997	0	0	0	0	0	99	135	84	22	12	0	0	352
1998	0	0	1	0	34	118	148	207	57	1	0	0	566
1999	0	0	0	0	33	165	297	112	81	0	0	0	688
2000	0	0	0	0	17	76	108	126	69	0	0	0	396
2001	0	0	0	3	8	129	174	255	64	5	0	0	638

SNOWFALL (inches) 2001 BUFFALO, NY (BUF)

YEAR	JUL	AUG	SEP	OCT	NOV	DEC	JAN	FEB	MAR	APR	MAY	JUN	TOTAL
1972-73	0.0	0.0	0.0	3.1	18.9	19.8	9.9	16.1	8.5	2.4	0.1	0.0	78.8
1973-74	0.0	0.0	0.0	0.0	3.0	23.1	19.7	22.8	12.9	7.1	0.1	0.0	88.7
1974-75	0.0	0.0	0.0	T	22.1	23.6	11.0	16.3	7.6	15.0	0.0	0.0	95.6
1975-76	0.0	0.0	0.0	T	5.5	27.3	21.6	8.3	17.3	2.5	T	0.0	82.5
1976-77	0.0	0.0	0.0	0.2	31.3	60.7	68.3	22.7	13.5	2.2	0.5	0.0	199.4
1977-78	0.0	0.0	0.0	T	15.0	53.4	56.5	21.7	5.8	1.8	0.1	0.0	154.3
1978-79	0.0	0.0	0.0	T	3.0	10.1	42.6	28.3	4.6	8.7	0.0	0.0	97.3
1979-80	0.0	0.0	0.0	T	12.6	19.7	10.2	11.7	13.9	0.3	T	T	68.4
1980-81	0.0	0.0	0.0	T	6.7	21.6	14.4	5.0	13.2	T	0.0	0.0	60.9
1981-82	0.0	0.0	0.0	T	1.8	24.8	53.2	12.7	9.0	10.9	0.0	0.0	112.4
1982-83	0.0	0.0	0.0	0.0	15.8	12.9	9.0	5.5	6.9	2.3	T	0.0	52.4
1983-84	0.0	0.0	0.0	T	17.7	52.0	13.4	32.5	16.0	0.9	T	0.0	132.5
1984-85	0.0	0.0	0.0	0.0	1.4	11.2	65.9	20.9	6.3	1.5	0.0	0.0	107.2
1985-86	0.0	0.0	0.0	0.0	5.2	68.4	17.3	17.3	4.8	1.7	0.0	0.0	114.7
1986-87	0.0	0.0	0.0	0.0	13.7	4.8	28.5	7.7	10.8	2.0	0.0	0.0	67.5
1987-88	0.0	0.0	0.0	T	0.9	9.8	6.9	31.9	6.1	0.8	0.0	0.0	56.4
1988-89	0.0	0.0	0.0	0.5	0.6	10.8	5.4	29.6	10.1	2.5	7.9	0.0	67.4
1989-90	0.0	0.0	0.0	T	7.8	34.8	11.8	28.0	1.4	9.9	T	0.0	93.7
1990-91	0.0	0.0	0.0	T	0.7	15.4	16.6	16.1	8.5	0.2	T	0.0	57.5
1991-92	0.0	T	0.0	0.2	18.0	21.4	18.4	7.0	22.8	5.0	0.0	0.0	92.8
1992-93	0.0	0.0	0.0	0.6	13.7	16.5	13.1	19.5	29.3	0.5	T	0.0	93.2
1993-94	T	0.0	T	2.9	4.8	27.9	35.4	21.6	13.2	6.9	0.0	0.0	112.7
1994-95	0.0	0.0	T	0.0	0.9	7.8	23.1	34.6	4.3	3.9	T	0.0	74.6
1995-96	0.0	0.0	0.0	0.0	15.7	61.2	25.3	11.9	24.1	3.2	T		
1996-97	0.0	0.0	0.0	0.0	11.5	18.9	42.4	9.3	13.4	2.1	0.0	0.0	97.6
1997-98	0.0	0.0	0.0	0.2	16.5	16.8	13.6	1.8	25.3	T	T	T	74.2
1998-99	0.0	0.0	0.0	0.0	0.2	11.6	65.1	6.9	15.8	1.0	0.0	0.0	100.6
1999-00	0.0	0.0	0.0	T	0.9	12.7	19.4	16.2	10.7	3.7	0.0	0.0	63.6
2000-01	0.0	0.0	0.0	T	45.6	50.3	19.6	9.8	32.8	0.6	0.0	0.0	158.7
2001-	0.0	0.0	0.0	0.4	0.0	82.7							
POR= 57 YRS	T	T	T	0.3	11.4	24.2	24.0	17.4	12.4	3.0	0.4	T	93.1

2001
NEW YORK, CENTRAL PARK,
NEW YORK (NYC)

New York City, in area exceeding 300 square miles, is located on the Atlantic coastal plain at the mouth of the Hudson River. The terrain is laced with numerous waterways, all but one of the five boroughs in the city are situated on islands. Elevations range from less than 50 feet over most of Manhattan, Brooklyn, and Queens to almost 300 feet in northern Manhattan and the Bronx, and over 400 feet in Staten Island. Extensive suburban areas on Long Island, and in Connecticut, New York State and New Jersey border the city on the east, north, and west. About 30 miles to the west and northwest, hills rise to about 1,500 feet and to the north in upper Westchester County to 800 feet. To the southwest and to the east are the low-lying land areas of the New Jersey coastal plain and of Long Island, bordering on the Atlantic.

The New York Metropolitan area is close to the path of most storm and frontal systems which move across the North American continent. Therefore, weather conditions affecting the city most often approach from a westerly direction. New York City can thus experience higher temperatures in summer and lower ones in winter than would otherwise be expected in a coastal area. However, the frequent passage of weather systems often helps reduce the length of both warm and cold spells, and is also a major factor in keeping periods of prolonged air stagnation to a minimum.

Although continental influence predominates, oceanic influence is by no means absent. During the summer local sea breezes, winds blowing onshore from the cool water surface, often moderate the afternoon heat. The effect of the sea breeze diminishes inland. On winter mornings, ocean temperatures which are warm relative to the land reinforce the effect of the city heat island and low temperatures are often 10-20 degrees lower in the inland suburbs than in the central city. The relatively warm water temperatures also delay the advent of winter snows. Conversely, the lag in warming of water temperatures keeps spring temperatures relatively cool. One year-round measure of the ocean influence is the small average daily variation in temperature.

Precipitation is moderate and distributed fairly evenly throughout the year. Most of the rainfall from May through October comes from thunderstorms, usually of brief duration and sometimes intense. Heavy rains of long duration associated with tropical storms occur infrequently in late summer or fall. For the other months of the year precipitation is more likely to be associated with widespread storm areas, so that day-long rain, snow or a mixture of both is more common. Coastal storms, occurring most often in the fall and winter months, produce on occasion considerable amounts of precipitation and have been responsible for record rains, snows, and high winds.

The average annual precipitation is reasonably uniform within the city but is higher in the northern and western suburbs and less on eastern Long Island. Annual snowfall totals also show a consistent increase to the north and west of the city with lesser amounts along the south shores and the eastern end of Long Island, reflecting the influence of the ocean waters.

Local Climatological Data is published for three locations in New York City, Central Park, La Guardia Airport, and John F. Kennedy International Airport. Other nearby locations for which it is published are Newark, New Jersey, and Bridgeport, Connecticut.

Based on the 1951-1980 period, the average first occurrence of 32 degrees Fahrenheit in the fall is November 11 and the average last occurrence in the spring is April 1.

NORMALS, MEANS, AND EXTREMES

NEW YORK C.PARK, NY (NYC)

LATITUDE:	LONGITUDE:	ELEVATION (FT):		TIME ZONE:	WBAN: 94728
40 47' 0 " N	73 58' 0 " W	GRND: 158　BARO: 161		EASTERN (UTC + 5)	

	ELEMENT	POR	JAN	FEB	MAR	APR	MAY	JUN	JUL	AUG	SEP	OCT	NOV	DEC	YEAR
TEMPERATURE F	NORMAL DAILY MAXIMUM	30	37.6	40.3	50.0	61.2	71.7	80.1	85.2	83.7	76.2	65.3	54.0	42.5	62.3
	MEAN DAILY MAXIMUM	39	38.4	41.2	49.9	61.4	71.8	80.2	85.1	83.7	76.0	65.1	54.0	43.1	62.5
	HIGHEST DAILY MAXIMUM	133	72	75	86	96	99	101	106	104	102	94	84	75	106
	YEAR OF OCCURRENCE		1950	1985	1998	1976	1962	1966	1936	1918	1953	1941	1950	1998	JUL 1936
	MEAN OF EXTREME MAXS.	39	58.1	59.6	70.8	81.7	88.9	92.9	96.0	93.5	89.5	79.4	71.3	62.0	78.6
	NORMAL DAILY MINIMUM	30	25.3	26.9	34.8	43.8	53.7	63.0	68.4	67.3	60.1	49.7	41.1	30.7	47.1
	MEAN DAILY MINIMUM	39	26.0	27.4	34.8	43.9	53.8	63.2	68.4	67.5	60.2	49.6	41.1	31.3	47.3
	LOWEST DAILY MINIMUM	133	-6	-15	3	12	32	44	52	50	39	28	5	-13	-15
	YEAR OF OCCURRENCE		1882	1934	1872	1923	1891	1945	1943	1986	1912	1936	1875	1917	FEB 1934
	MEAN OF EXTREME MINS.	39	8.1	10.9	18.9	31.1	42.7	52.2	59.5	57.6	47.3	36.6	27.1	15.0	33.9
	NORMAL DRY BULB	30	31.5	33.6	42.4	52.5	62.7	71.6	76.8	75.5	68.2	57.5	47.6	36.6	54.7
	MEAN DRY BULB	39	32.1	34.2	42.3	52.7	62.7	71.6	76.8	75.6	68.0	57.4	47.5	37.2	54.8
	MEAN WET BULB	4	27.5	33.0	36.2	44.9	54.8	63.7	67.2	68.0	60.5	51.2	41.4	31.8	48.4
	MEAN DEW POINT	4	19.6	24.2	27.2	36.5	48.3	58.3	62.2	63.9	55.7	45.1	34.7	24.2	41.7
	NORMAL NO. DAYS WITH:														
	MAXIMUM 90	30	0.0	0.0	0.0	0.2	1.0	3.3	7.3	4.9	1.4	0.0	0.0	0.0	18.1
	MAXIMUM 32	30	10.1	5.7	0.9	*	0.0	0.0	0.0	0.0	0.0	0.0	0.1	4.8	21.6
	MINIMUM 32	30	22.6	19.9	10.8	1.4	0.0	0.0	0.0	0.0	0.0	0.4	3.8	16.8	75.7
	MINIMUM 0	30	0.2	0.1	0.0	0.0	0.0	0.0	0.0	0.0	0.0	0.0	0.0	*	0.3
H/C	NORMAL HEATING DEG. DAYS	30	1039	879	701	375	125	0	0	0	34	250	522	880	4805
	NORMAL COOLING DEG. DAYS	30	0	0	0	0	54	203	366	326	130	17	0	0	1096
RH	NORMAL (PERCENT)	30	62	60	58	55	63	65	64	66	68	66	64	64	63
	HOUR 01 LST	30	64	63	62	61	69	72	71	73	75	72	68	67	68
	HOUR 07 LST	30	67	67	66	64	72	74	74	76	78	75	72	69	71
	HOUR 13 LST	30	57	55	52	46	52	55	53	54	56	54	57	58	54
	HOUR 19 LST	30	59	57	56	52	59	61	60	63	65	63	62	62	60
S	PERCENT POSSIBLE SUNSHINE	107	51	55	57	58	61	64	65	64	62	61	52	49	58
W/O	MEAN NO. DAYS WITH:														
	HEAVY FOG(VISBY 1/4 MI)	14	1.4	0.7	0.7	0.2	0.5	0.6	0.5	0.4	0.3	0.2	0.4	0.4	6.3
	THUNDERSTORMS	38	0.2	0.2	0.8	1.1	2.0	3.3	3.7	3.1	1.2	0.7	0.3	0.1	16.7
CLOUDINESS	MEAN:														
	SUNRISE-SUNSET (OKTAS)														
	MIDNIGHT-MIDNIGHT (OKTAS)														
	MEAN NO. DAYS WITH:														
	CLEAR														
	PARTLY CLOUDY														
	CLOUDY														
PR	MEAN STATION PRESSURE(IN)	16	29.94	29.94	29.88	29.85	29.86	29.90	29.89	29.92	29.94	29.98	29.94	29.94	29.92
	MEAN SEA-LEVEL PRES. (IN)	6	30.07	30.27	29.98	29.93	29.95	29.98	29.96	30.00	30.01	30.08	30.07	30.01	30.03
WINDS	MEAN SPEED (MPH)	7	8.6	8.4	9.0	8.1	6.9	6.2	5.9	5.7	6.3	6.8	7.0	7.4	7.2
	PREVAIL.DIR(TENS OF DEGS)	9	32	32	32	32	05	23	23	23	27	27	32	32	32
	MAXIMUM 2-MINUTE:														
	SPEED (MPH)	6	40	34	37	35	29	28	24	33	29	28	30	34	40
	DIR. (TENS OF DEGS)		07	08	08	08	08	08	17	30	09	33	32	07	07
	YEAR OF OCCURRENCE		1996	1998	1996	2000	1999	1998	1996	1997	1999	1999	1997	1997	JAN 1996
	MAXIMUM 5-SECOND:														
	SPEED (MPH)	6	53	52	52	51	40	38	41	57	41	46	41	51	57
	DIR. (TENS OF DEGS)		06	07	07	07	30	06	16	31	09	02	31	06	31
	YEAR OF OCCURRENCE		1996	1998	1996	1998	1997	2000	1996	1997	1999	1996	1997	1997	AUG 1997
PRECIPITATION	NORMAL (IN)	30	3.42	3.27	4.08	4.20	4.42	3.67	4.35	4.01	3.89	3.56	4.47	3.91	47.25
	MAXIMUM MONTHLY (IN)	132	10.52	6.87	10.41	8.77	10.24	9.78	11.89	12.36	16.85	13.31	12.41	9.98	16.85
	YEAR OF OCCURRENCE		1979	1869	1980	1874	1989	1903	1889	1990	1882	1903	1972	1973	SEP 1882
	MINIMUM MONTHLY (IN)	132	0.58	0.46	0.90	0.95	0.30	0.02	0.44	0.18	0.21	0.14	0.34	0.25	0.02
	YEAR OF OCCURRENCE		1981	1895	1885	1881	1903	1949	1999	1995	1884	1963	1976	1955	JUN 1949
	MAXIMUM IN 24 HOURS (IN)	89	3.91	3.04	4.25	4.22	4.88	4.74	4.39	5.78	8.30	11.17	8.09	3.21	11.17
	YEAR OF OCCURRENCE		1979	1973	1876	1984	1968	1884	1997	1971	1882	1903	1977	1909	OCT 1903
	NORMAL NO. DAYS WITH:														
	PRECIPITATION 0.01	30	10.0	8.7	10.4	10.7	11.2	11.1	9.9	9.6	8.4	7.6	9.9	10.7	118.2
	PRECIPITATION 1.00	30	0.8	0.9	1.0	1.1	1.1	0.9	1.4	0.9	1.1	1.0	1.2	1.1	12.5
SNOWFALL	NORMAL (IN)	30	7.5	8.0	3.5	0.5	T	0.0	0.0	0.0	0.0	T	0.4	3.2	23.1
	MAXIMUM MONTHLY (IN)	133	27.4	27.9	30.5	13.5	T	0.0	T	0.0	0.0	0.8	19.0	29.6	30.5
	YEAR OF OCCURRENCE		1925	1934	1896	1875	1995		1990			1925	1898	1947	MAR 1896
	MAXIMUM IN 24 HOURS (IN)	133	19.2	17.6	18.1	10.2	T	0.0	T	0.0	0.0	0.8	10.0	26.4	26.4
	YEAR OF OCCURRENCE		1996	1983	1941	1915	1995		1990			1925	1898	1947	DEC 1947
	MAXIMUM SNOW DEPTH (IN)	128	15	22	9	9	0	0	0	0	0	0	5	7	22
	YEAR OF OCCURRENCE		1978	1994	1967	1982							1989	1995	FEB 1994
	NORMAL NO. DAYS WITH:														
	SNOWFALL 1.0	30	2.4	2.0	1.2	0.1	0.0	0.0	0.0	0.0	0.0	0.0	0.1	1.1	6.9

PRECIPITATION (inches) 2001 NEW YORK, CENTRAL PARK, NY (NYC)

YEAR	JAN	FEB	MAR	APR	MAY	JUN	JUL	AUG	SEP	OCT	NOV	DEC	ANNUAL
1972	2.41	5.90	4.55	3.92	8.39	9.30	4.54	1.92	1.33	6.27	12.41	6.09	67.03
1973	4.53	4.55	3.60	8.05	4.51	4.55	5.89	2.75	3.92	1.82	9.98	57.23
1974	3.80	1.49	5.76	3.83	4.29	3.29	1.33	5.99	8.05	2.59	0.94	6.33	47.69
1975	4.76	3.33	3.32	3.04	3.38	7.58	11.77	9.32	3.70	4.33	3.63	61.21
1976	5.78	3.13	2.99	2.80	4.77	2.78	1.42	6.52	3.15	5.31	0.34	2.29	41.28
1977	2.25	2.51	7.41	3.75	1.71	3.83	1.60	4.57	4.75	5.03	12.26	5.06	54.73
1978	8.27	1.59	2.73	2.38	9.15	1.69	4.48	5.50	4.06	1.50	2.85	5.61	49.81
1979	10.52	4.58	4.40	4.04	6.23	1.56	1.76	4.27	4.83	3.87	3.38	2.69	52.13
1980	1.72	1.04	10.41	8.26	2.33	3.84	5.26	1.16	1.98	3.86	4.11	0.58	44.55
1981	0.58	6.04	1.19	3.42	3.56	2.71	6.21	0.59	3.45	3.49	1.69	5.18	38.11
1982	6.46	2.37	2.56	5.67	2.43	5.12	3.14	4.66	1.77	2.31	3.44	1.47	41.40
1983													
1984	1.87	4.86	6.30	6.62	9.74	5.76	7.03	1.38	2.51	3.63	4.07	3.26	57.03
1985	1.00	2.41	1.91	1.41	5.72	4.41	4.41	2.58	4.75	1.30	8.09	0.83	38.82
1986	4.23	2.86	1.46	3.93	1.68	1.86	5.56	4.24	2.20	1.92	6.85	6.16	42.95
1987	5.81	1.01	4.93	5.90	1.45	3.94	4.12	4.89	5.25	3.89	3.08	2.17	46.44
1988	3.64	3.91	2.10	2.20	5.27	1.29	8.14	2.19	2.34	3.56	8.90	1.13	44.67
1989	2.29	3.03	4.93	4.26	10.24	8.79	5.13	8.44	6.90	7.48	2.79	0.83	65.11
1990	5.34	2.33	3.64	5.12	9.10	2.50	3.51	12.36	2.24	6.38	2.82	5.58	60.92
1991	3.38	1.93	5.16	3.68	3.11	4.16	4.57	7.13	3.71	2.13	1.96	4.26	45.18
1992	1.68	1.87	4.08	1.76	4.02	4.77	4.49	3.49	4.89	1.16	5.64	5.50	43.35
1993	3.44	2.81	6.64	4.28	1.56	1.49	1.70	5.41	5.25	4.55	2.20	4.95	44.28
1994	5.62	3.44	6.33	2.42	4.26	3.21	3.86	6.33	3.33	1.35	4.34	2.90	47.39
1995	3.75	3.13	1.26	2.29	2.84	2.09	6.13	0.18	3.03	7.82	5.78	2.12	40.42
1996	5.64	2.59	3.81	6.33	2.64	5.71	5.76	1.87	4.97	7.52	2.87	6.48	56.19
1997	3.65	2.54	5.18	2.86	3.05	1.93	3.21	2.10	2.10	4.68	4.27	43.93
1998	5.20	5.81	5.08	7.05	6.94	5.94	1.09	2.78	3.44	2.76	1.48	1.12	48.69
1999	7.02	3.49	4.01	1.93	4.04	0.19	0.44	2.89	8.81	2.73	2.33	3.23	41.11
2000	3.23	1.66	3.34	3.53	4.50	4.87	7.28	3.82	5.82	0.67	3.54	3.19	45.45
2001	3.16	1.95	7.48	1.58	2.03	5.29	2.04	2.56	5.30	0.66	1.36	2.24	35.65
POR= 132 YRS	3.50	3.33	3.93	3.54	3.66	3.49	4.21	4.24	3.78	3.46	3.48	3.41	44.03

AVERAGE TEMPERATURE (F) 2001 NEW YORK, CENTRAL PARK, NY (NYC)

YEAR	JAN	FEB	MAR	APR	MAY	JUN	JUL	AUG	SEP	OCT	NOV	DEC	ANNUAL
1972	36.1	31.4	39.8	50.1	63.3	67.9	77.2	75.6	69.5	53.5	44.4	38.5	53.9
1973	35.5	32.5	46.4	53.4	59.5	73.4	77.6	69.5	60.2	48.3	39.0	56.1
1974	35.3	31.7	42.1	55.2	61.0	69.0	77.2	76.4	66.7	54.1	48.2	39.4	54.7
1975	37.3	35.8	40.2	47.9	65.8	70.5	75.8	74.4	64.2	59.2	52.3	35.9	54.9
1976	27.4	39.9	44.4	55.0	60.2	73.2	74.8	74.3	66.6	52.9	41.7	29.9	53.4
1977	22.1	33.5	46.8	53.7	65.0	70.2	79.0	75.7	68.2	54.9	47.3	35.7	54.3
1978	28.0	27.2	39.0	51.6	61.5	71.3	74.4	76.0	65.0	54.9	47.8	38.9	53.6
1979	33.6	25.5	46.9	52.6	65.3	69.2	76.9	76.8	70.5	57.3	52.5	41.1	55.7
1980	33.7	31.4	41.2	54.5	65.6	70.3	79.3	80.3	70.8	55.2	44.6	32.5	55.0
1981	26.3	39.3	42.3	56.2	64.8	73.0	78.5	76.0	67.6	54.4	47.7	36.5	55.2
1982	26.1	35.3	42.0	51.2	64.1	68.6	77.9	73.2	68.3	58.5	50.4	42.8	54.9
1983	34.5	36.4	44.0	52.3	60.2	73.4	79.5	77.7	71.8	57.9	48.9	35.2	56.0
1984	29.9	40.6	36.7	51.9	61.6	74.5	74.7	76.7	65.9	61.8	47.3	43.8	55.5
1985	28.8	36.6	45.8	55.5	65.3	68.6	76.2	75.4	70.5	59.5	50.0	34.2	55.5
1986	34.1	32.0	45.1	54.5	66.0	71.6	76.0	73.1	67.6	58.0	45.7	39.0	55.3
1987	32.3	33.2	45.2	53.4	63.6	72.8	78.0	74.2	67.7	53.8	47.7	39.5	55.1
1988	29.5	35.0	43.6	51.2	62.7	71.8	79.3	78.8	67.4	52.8	49.4	35.9	54.8
1989	37.4	34.5	42.4	52.2	62.1	72.0	75.0	74.0	68.1	58.2	45.7	25.9	54.0
1990	41.4	39.8	45.1	53.5	60.2	72.1	76.8	75.3	67.5	57.1	50.4	42.6	57.2
1991	34.9	40.0	44.6	55.7	68.7	74.1	77.7	77.1	67.5	58.4	48.3	39.6	57.2
1992	35.7	36.4	40.0	50.5	61.0	70.3	74.2	73.0	67.2	54.5	46.5	37.9	53.9
1993	36.3	30.8	39.7	53.3	66.7	73.3	80.2	77.2	67.3	56.0	48.8	37.3	55.5
1994	25.6	30.6	40.7	55.6	61.8	75.2	79.4	74.0	67.6	58.0	52.0	42.2	55.2
1995	37.5	31.6	45.0	51.9	61.9	71.8	79.2	78.6	68.3	61.6	43.6	32.4	55.2
1996	30.5	33.8	38.9	52.2	61.1	71.4	73.3	74.5	68.0	56.4	43.0	41.3	53.7
1997	32.2	40.0	41.9	51.7	59.4	70.9	75.8	73.3	67.0	56.7	44.5	34.3	54.3
1998	40.0	40.6	45.4	54.0	64.3	69.2	76.5	76.7	70.2	57.6	48.1	43.2	57.2
1999	33.9	38.9	42.5	53.5	63.1	73.2	81.4	75.5	69.1	56.0	50.8	40.0	56.5
2000	31.3	37.3	47.2	51.0	63.5	71.3	72.3	72.5	66.0	57.0	45.3	31.1	53.8
2001	33.7	35.9	39.6	54.0	63.6	73.0	73.2	78.7	67.7	58.5	52.7	44.1	56.2
POR= 90 YRS	32.2	33.4	41.4	51.6	62.2	71.1	76.5	74.8	68.0	57.7	47.0	36.2	54.3

REFERENCE NOTES:

PAGE 1:
THE TEMPERATURE GRAPH SHOWS NORMAL MAXIMUM AND NORMAL MINIMUM DAILY TEMPERATURES (SOLID CURVES) AND THE ACTUAL DAILY HIGH AND LOW TEMPERATURES (VERTICAL BARS).

PAGE 2 AND 3:
H/C INDICATES HEATING AND COOLING DEGREE DAYS.
RH INDICATES RELATIVE HUMIDITY
W/O INDICATES WEATHER AND OBSTRUCTIONS
S INDICATES SUNSHINE.
PR INDICATES PRESSURE.
CLOUDINESS ON PAGE 3 IS THE SUM OF THE CEILOMETER AND SATELLITE DATA NOT TO EXCEED EIGHT EIGHTHS (OKTAS).

GENERAL:
T INDICATES TRACE PRECIPITATION, AN AMOUNT GREATER THAN ZERO BUT LESS THAN THE LOWEST REPORTABLE VALUE.
+ INDICATES THE VALUE ALSO OCCURS ON EARLIER DATES.
BLANK ENTRIES DENOTE MISSING OR UNREPORTED DATA.
NORMALS ARE 30-YEAR AVERAGES (1961 - 1990).
ASOS INDICATES AUTOMATED SURFACE OBSERVING SYSTEM.
PM INDICATES THE LAST DAY OF THE PREVIOUS MONTH.
POR (PERIOD OF RECORD) BEGINS WITH THE JANUARY DATA MONTH AND IS THE NUMBER OF YEARS USED TO COMPUTE THE MEAN. INDIVIDUAL MONTHS WITHIN THE POR MAY BE MISSING.
WHEN THE POR FOR A NORMAL IS LESS THAN 30 YEARS, THE NORMAL IS PROVISIONAL AND IS BASED ON THE NUMBER OF YEARS INDICATED.
0.* OR * INDICATES THE VALUE OR MEAN-DAYS-WITH IS BETWEEN 0.00 AND 0.05.
CLOUDINESS FOR ASOS STATIONS DIFFERS FROM THE NON-ASOS OBSERVATION TAKEN BY A HUMAN OBSERVER. ASOS STATION CLOUDINESS IS BASED ON TIME-AVERAGED CEILOMETER DATA FOR CLOUDS AT OR BELOW 12,000 FEET AND ON SATELLITE DATA FOR CLOUDS ABOVE 12,000 FEET.
THE NUMBER OF DAYS WITH CLEAR, PARTLY CLOUDY, AND CLOUDY CONDITIONS FOR ASOS STATIONS IS THE SUM OF THE CEILOMETER AND SATELLITE DATA FOR THE SUNRISE TO SUNSET PERIOD.

GENERAL CONTINUED:
CLEAR INDICATES 0 - 2 OKTAS, PARTLY CLOUDY INDICATES 3 - 6 OKTAS, AND CLOUDY INDICATES 7 OR 8 OKTAS. WHEN AT LEAST ONE OF THE ELEMENTS (CEILOMETER OR SATELLITE) IS MISSING, THE DAILY CLOUDINESS IS NOT COMPUTED.
WIND DIRECTION IS RECORDED IN TENS OF DEGREES (2 DIGITS) CLOCKWISE FROM TRUE NORTH. "00" INDICATES CALM. "36" INDICATES TRUE NORTH.
RESULTANT WIND IS THE VECTOR AVERAGE OF THE SPEED AND DIRECTION.
AVERAGE TEMPERATURE IS THE SUM OF THE MEAN DAILY MAXIMUM AND MINIMUM TEMPERATURE DIVIDED BY 2.
SNOWFALL DATA COMPRISE ALL FORMS OF FROZEN PRECIPITATION, INCLUDING HAIL.
A HEATING (COOLING) DEGREE DAY IS THE DIFFERENCE BETWEEN THE AVERAGE DAILY TEMPERATURE AND 65 F.
DRY BULB IS THE TEMPERATURE OF THE AMBIENT AIR.
DEW POINT IS THE TEMPERATURE TO WHICH THE AIR MUST BE COOLED TO ACHIEVE 100 PERCENT RELATIVE HUMIDITY.
WET BULB IS THE TEMPERATURE THE AIR WOULD HAVE IF THE MOISTURE CONTENT WAS INCREASED TO 100 PERCENT RELATIVE HUMIDITY.

ON JULY 1, 1996, THE NATIONAL WEATHER SERVICE BEGAN USING THE "METAR" OBSERVATION CODE THAT WAS ALREADY EMPLOYED BY MOST OTHER NATIONS OF THE WORLD. THE MOST NOTICEABLE DIFFERENCE IN THIS ANNUAL PUBLICATION WILL BE THE CHANGE IN UNITS FROM TENTHS TO EIGHTS (OKTAS) FOR REPORTING THE AMOUNT OF SKY COVER.

HEATING DEGREE DAYS (base 65 F) 2001 NEW YORK, CENTRAL PARK, NY (NYC)

YEAR	JUL	AUG	SEP	OCT	NOV	DEC	JAN	FEB	MAR	APR	MAY	JUN	TOTAL
1972-73	2	0	25	355	611	812	907	903	572	362	188	2	4739
1973-74	0	0	29	162	493	800	913	925	704	309	165	27	4527
1974-75	1	0	59	333	502	789	852	812	764	507	86	11	4716
1975-76	0	3	62	193	387	898	1163	723	630	360	167	18	4604
1976-77	0	4	44	373	692	1082	1322	877	560	354	100	27	5435
1977-78	0	0	56	307	524	903	1140	1051	797	394	179	13	5364
1978-79	5	0	75	311	510	802	969	1100	554	369	55	14	4764
1979-80	4	4	20	271	373	734	963	969	731	310	67	22	4468
1980-81	0	0	31	305	602	1000	1194	715	698	264	78	3	4890
1981-82	0	0	48	320	513	876	1198	825	707	413	74	36	5010
1982-83	0	5	24	229	446	679	936	793	644	393	161	3	4313
1983-84	0	0	34	249	480	914	1082	698	870	389	137	9	4862
1984-85	0	0	69	114	525	654	1113	789	596	305	79	24	4268
1985-86	0	0	17	188	448	947	950	917	615	312	89	11	4494
1986-87	0	10	27	236	572	797	1008	883	608	348	146	8	4643
1987-88	0	2	29	343	512	780	1093	867	656	409	133	31	4855
1988-89	3	0	23	385	459	896	844	849	696	376	143	14	4688
1989-90	0	1	54	217	572	1205	724	702	612	366	150	4	4607
1990-91	3	2	57	166	436	686	927	696	625	311	61	3	3973
1991-92	0	0	60	222	496	782	902	827	767	434	160	12	4662
1992-93	0	3	54	324	547	834	882	953	779	347	57	14	4794
1993-94	0	0	65	275	483	852	1215	958	749	282	142	0	5021
1994-95	0	0	18	212	388	700	846	931	614	386	130	2	4227
1995-96	0	0	31	146	637	1001	1065	894	801	389	183	8	5155
1996-97	0	0	46	263	656	726	1010	691	712	393	174	40	4711
1997-98	2	0	48	284	611	822	768	676	635	322	99	29	4296
1998-99	0	0	20	222	499	670	955	725	687	340	98	4	4220
1999-00	0	3	23	271	418	769	1038	795	544	411	118	31	4421
2000-01	0	0	81	256	586	1041	965	809	780	340	124	6	4988
2001-	0	0	47	228	364	639							

COOLING DEGREE DAYS (base 65 F) 2001 NEW YORK, CENTRAL PARK, NY (NYC)

YEAR	JAN	FEB	MAR	APR	MAY	JUN	JUL	AUG	SEP	OCT	NOV	DEC	ANNUAL
1972	0	0	0	5	47	118	384	338	169	3	0	0	1064
1973	0	0	0	20	23	260	390	401	171	22	2	0	1289
1974	0	0	0	19	47	155	385	360	115	1	6	0	1088
1975	0	0	0	0	120	185	341	299	43	22	15	0	1025
1976	0	0	0	65	24	270	310	299	103	5	0	0	1076
1977	0	0	3	22	110	189	442	338	159	0	0	0	1263
1978	0	0	0	0	77	209	301	348	81	4	0	0	1020
1979	0	0	0	4	71	149	378	376	192	43	5	0	1218
1980	0	0	0	1	94	188	448	480	213	11	0	0	1435
1981	0	0	0	4	78	252	425	347	129	0	0	0	1235
1982	0	0	0	7	55	152	405	266	129	36	16	0	1066
1983	0	0	0	19	16	259	460	404	244	35	0	0	1437
1984	0	0	0	3	39	301	306	367	106	26	0	0	1148
1985	0	0	8	28	95	139	353	329	189	21	5	0	1167
1986	0	0	0	5	4	127	214	348	269	120	27	0	1114
1987	0	0	0	5	110	251	406	295	118	0	2	0	1187
1988	0	0	0	0	66	243	455	435	104	12	0	0	1315
1989	0	0	4	0	61	231	313	287	151	10	0	0	1057
1990	0	0	4	25	8	225	375	328	140	77	4	0	1186
1991	0	0	0	38	182	280	403	382	142	24	1	0	1452
1992	0	0	0	5	46	174	292	256	127	8	0	0	908
1993	0	0	0	0	82	269	474	386	140	3	4	0	1358
1994	0	0	0	7	51	316	454	286	102	2	3	0	1221
1995	0	0	0	0	40	212	445	428	137	48	0	0	1310
1996	0	0	0	13	67	209	267	300	142	4	0	0	1002
1997	0	0	0	0	7	222	343	265	113	32	0	0	982
1998	0	0	0	36	89	162	366	368	184	1	0	2	1208
1999	0	0	0	3	46	258	517	336	152	3	0	0	1315
2000	0	0	0	0	81	227	234	240	117	12	0	0	911
2001	0	0	0	15	89	250	262	430	137	32	1	0	1216

SNOWFALL (inches) 2001 NEW YORK, CENTRAL PARK, NY (NYC)

YEAR	JUL	AUG	SEP	OCT	NOV	DEC	JAN	FEB	MAR	APR	MAY	JUN	TOTAL
1972-73	0.0	0.0	0.0	T	T	T	1.8	0.8	0.2	T	0.0	0.0	2.8
1973-74	0.0	0.0	0.0	0.0	0.0	2.8	7.8	9.4	3.2	0.3	0.0	0.0	23.5
1974-75	0.0	0.0	0.0	0.0	0.1	0.1	2.0	10.6	0.3	T	0.0	0.0	13.1
1975-76	0.0	0.0	0.0	0.0	T	2.3	5.6	5.0	4.4	T	0.0	0.0	17.3
1976-77	0.0	0.0	0.0	0.0	T	5.1	13.0	5.8	0.6	T	T	0.0	24.5
1977-78	0.0	0.0	0.0	0.0	0.2	0.4	20.3	23.0	6.8	T	0.0	0.0	50.7
1978-79	0.0	0.0	0.0	0.0	2.2	0.5	6.6	20.1	T	T	0.0	0.0	29.4
1979-80	0.0	0.0	0.0	T	0.0	3.5	2.0	2.7	4.6	T	0.0	0.0	12.8
1980-81	0.0	0.0	0.0	0.0	T	2.8	8.0	T	8.6	T	0.0	0.0	19.4
1981-82	0.0	0.0	0.0	0.0	0.0	2.1	11.8	0.4	0.7	9.6	0.0	0.0	24.6
1982-83	0.0	0.0	0.0	0.0	0.0	3.0	1.9	23.5	T	0.8	0.0	0.0	29.2
1983-84	0.0	0.0	0.0	0.0	T	1.6	11.7	0.2	11.9	0.0	0.0	0.0	25.4
1984-85	0.0	0.0	0.0	0.0	T	5.5	8.4	10.0	0.2	T	0.0	0.0	24.1
1985-86	0.0	0.0	0.0	0.0	T	0.9	2.2	9.9	T	T	0.0	0.0	13.0
1986-87	0.0	0.0	0.0	0.0	T	0.6	13.6	7.0	1.9	0.0	0.0	0.0	23.1
1987-88	0.0	0.0	0.0	0.0	1.1	2.6	13.9	1.5	T	0.0	0.0	0.0	19.1
1988-89	0.0	0.0	0.0	0.0	0.0	0.3	5.0	0.3	2.5	0.0	0.0	0.0	8.1
1989-90	0.0	0.0	0.0	0.0	4.7	1.4	1.8	1.8	3.1	0.6	0.0	0.0	13.4
1990-91	T	0.0	0.0	0.0	0.0	7.2	8.4	9.1	0.2	0.0	0.0	0.0	24.9
1991-92	0.0	0.0	0.0	0.0	T	0.7	1.5	1.0	9.4	T	0.0	0.0	12.6
1992-93	0.0	0.0	0.0	0.0	0.0	0.4	1.5	10.7	11.9	0.0	0.0	0.0	24.5
1993-94	0.0	0.0	0.0	0.0	T	6.9	12.0	26.4	8.1	0.0	0.0	0.0	53.4
1994-95	0.0	0.0	0.0	0.0	T	T	0.2	11.6	T	T	T	0.0	11.8
1995-96	0.0	0.0	0.0	0.0	2.9	11.5	26.1	21.2	13.2	0.7	0.0	0.0	75.6
1996-97	0.0	0.0	0.0	0.0	0.1	T	4.4	3.8	1.7	T	0.0	0.0	10.0
1997-98	0.0	0.0	0.0	0.0	T	T	0.5	0.0	5.0	0.0	0.0	0.0	5.5
1998-99	0.0	0.0	0.0	0.0	0.0	2.0	4.5	1.7	4.5	0.0	0.0	0.0	12.7
1999-00	0.0	0.0	0.0	0.0	T	9.5	5.2	0.4	1.2	0.0	0.0	0.0	16.3
2000-01	0.0	0.0	0.0	T	0.0	13.4	8.3	9.5	3.8	0.0	0.0	0.0	35.0
2001-	0.0	0.0	0.0	0.0	0.0	T							
POR= 132 YRS	T	0.0	0.0	0.0	0.9	5.5	7.5	8.3	5.0	0.9	T	0.0	28.1

2001
ROCHESTER,
NEW YORK (ROC)

Rochester and the Genesee Valley experience a fairly humid, continental type climate, which is strongly modified by the proximity of the Great Lakes. Precipitation is rather evenly distributed throughout the year in quanity, but frequency is much higher during the cloudy winter months than in the sunny ones. Snowfall is heavy, but is highly variable over short distances.

Winters in general are cloudy, cold and snowy..but are changeable and include frequent thaws and rain as well. Snow covers the ground more often than not from christmas into early March..but periods of bare ground are not uncommon. About half of the annual snowfall comes from "lake-effect" process and is very localized. This feature develops when cold air crosses the warmer lake waters and becomes saturated.. creating clouds and precipitation downwind. The exact location of these snowbands are determined by the direction of the wind. Areas east of Rochester receive the most snow from this process..as northwest winds have a longer "fetch" off Lake Ontario..while areas south of the city get somewhat less. Lake Erie can even contribute some snow from this process if a west or southwest wind is storng enough. Since Lake Ontario does not freeze in most winters..this Lake effect machine can remain active throughout the winter. The Rochester area is also subject to occasional general or "synoptic" snowfalls..but the worst effects from these usually pass by to the east. Total season snowfall ranges from 70 inches south of the city to about 90 inches in Rochester to over 120 inches along the lakeshore east of the city. About 50 inches of this total results from general snows..the rest is due to the Lake effect machine. The lake does modify any extreme cold as the mercury falls below zero on only anout six nights in an average winter..with anything below -10 extremely rare.

Spring comes slowly to the region. The last frosts usually occur by April 30 near Lake Ontario..but as late as mid-May south of the Thruway. The spring months are actually the driest months statistically, due in part to the stabilizing effects of the Great Lakes, although soils are wet. Sunshine increases markedly in May.

Summers are warm and sunny across the region. The average temperature is in the 70 to 72 degree range. Rain can be expected on every third or fourth day.. almost always in the form of showers and thunderstorms. This activity is more common inland than near the Lake. Completely overcast days in summer are rare. Severe weather is not common..but a few cases of damaging winds and small tornadoes occur each year. The greatest risk of this type of activity is south of the Thruway. There usually are several periods of uncomfortably warm and muggy weather in an average summer..but only about nine days reach 90-degree mark in an average year. Still, the area usually experiences some of the most delightful summer weather in the East.

Autumn is pleasant, but rather brief. Mild and dry conditions predominate through September and much of October, but colder airmasses cross the Great Lakes with increasing frequency starting in late October, and result in a drastic increase in cloud cover across the region in late October and early November. Although the first frosts may not occur until late October near Lake Ontario, the firstlake effect snows of the season follow soon after...usually by mid November. These early snows melt off quickly, with a general snow cover seldom established before mid December. The growing season is relatively long for the latitude...average about 180 days. The long growing season...combined with ample spring moisture and abundant summer sunshine...is beneficial for the many fruit orchards and wineries...especially near the Lake Ontario shore and Finger Lakes.

NORMALS, MEANS, AND EXTREMES
ROCHESTER, NY (ROC)

LATITUDE: LONGITUDE: ELEVATION (FT):   TIME ZONE: WBAN: 14768
43 07' 00" N   77 40' 36" W   GRND: 585   BARO: 588   EASTERN (UTC + 5)

	ELEMENT	POR	JAN	FEB	MAR	APR	MAY	JUN	JUL	AUG	SEP	OCT	NOV	DEC	YEAR
	NORMAL DAILY MAXIMUM	30	30.9	32.5	42.7	55.9	67.8	75.8	80.7	78.1	71.8	60.5	47.8	35.8	56.7
	MEAN DAILY MAXIMUM	54	31.4	33.3	41.8	55.8	67.8	77.6	81.9	79.8	72.0	61.2	47.8	36.2	57.2
	HIGHEST DAILY MAXIMUM	61	74	73	84	93	94	100	98	99	99	91	81	72	100
	YEAR OF OCCURRENCE		1950	1997	1945	1990	1987	1953	1993	1948	1953	1951	1950	1982	JUN 1953
F	MEAN OF EXTREME MAXS.	54	53.2	54.0	68.3	78.9	85.9	91.1	92.6	90.8	87.9	80.1	69.0	57.1	75.7
	NORMAL DAILY MINIMUM	30	16.3	16.6	25.7	35.9	46.3	54.3	59.6	57.8	51.7	41.6	33.3	22.4	38.5
TEMPERATURE	MEAN DAILY MINIMUM	54	17.1	17.7	25.3	36.1	46.2	55.6	60.4	58.9	51.5	41.6	33.1	23.0	38.9
	LOWEST DAILY MINIMUM	61	-17	-19	-7	13	26	35	42	36	28	20	5	-16	-19
	YEAR OF OCCURRENCE		1994	1979	1999	1982	1979	1949	1963	1965	1947	1972	1971	1942	FEB 1979
	MEAN OF EXTREME MINS.	54	-2.2	-1.4	6.6	22.4	32.8	41.8	48.5	46.4	37.1	28.1	18.1	3.3	23.5
	NORMAL DRY BULB	30	23.6	24.6	34.3	45.9	57.1	65.1	70.2	68.0	61.7	51.1	40.5	29.1	47.6
	MEAN DRY BULB	54	24.2	25.5	33.4	46.0	57.1	66.6	71.2	69.4	61.8	51.4	40.5	29.5	48.0
	MEAN WET BULB	17	23.6	24.3	30.6	41.0	51.6	60.1	64.3	63.3	56.6	46.4	36.5	27.9	43.9
	MEAN DEW POINT	17	18.8	19.1	24.5	34.5	45.8	55.5	60.3	59.8	53.1	41.8	31.8	23.2	39.0
	NORMAL NO. DAYS WITH:														
	MAXIMUM 90	30	0.0	0.0	0.0	0.1	0.4	1.6	4.3	2.0	0.7	0.0	0.0	0.0	9.1
	MAXIMUM 32	30	17.0	14.5	6.1	0.4	0.0	0.0	0.0	0.0	0.0	0.0	1.5	11.5	51.0
	MINIMUM 32	30	28.9	25.4	23.0	11.1	1.2	0.0	0.0	0.0	0.1	4.1	15.1	26.0	134.9
	MINIMUM 0	30	3.0	2.3	0.2	0.0	0.0	0.0	0.0	0.0	0.0	0.0	0.0	0.9	6.4
H/C	NORMAL HEATING DEG. DAYS	30	1283	1131	952	573	270	62	10	33	137	435	735	1113	6734
	NORMAL COOLING DEG. DAYS	30	0	0	0	0	25	65	171	126	38	0	0	0	425
RH	NORMAL (PERCENT)	30	74	74	71	67	67	69	70	74	77	74	76	78	73
	HOUR 01 LST	30	76	78	76	76	78	82	83	87	87	82	80	79	80
	HOUR 07 LST	30	77	79	78	77	77	80	82	87	88	84	82	81	81
	HOUR 13 LST	30	69	68	62	55	54	55	54	58	60	60	67	72	61
	HOUR 19 LST	30	74	73	69	62	60	60	60	67	74	74	76	78	69
S	PERCENT POSSIBLE SUNSHINE	57	35	41	49	53	59	66	69	66	59	49	31	31	51
W/O	MEAN NO. DAYS WITH:														
	HEAVY FOG(VISBY 1/4 MI)	61	0.9	0.6	1.5	1.0	1.1	1.1	0.6	0.8	1.3	1.7	0.8	1.0	12.4
	THUNDERSTORMS	61	0.1	0.1	0.9	2.0	3.7	5.2	6.1	5.7	3.1	0.9	0.4	0.2	28.4
CLOUDINESS	MEAN:														
	SUNRISE-SUNSET (OKTAS)	56	6.6	6.3	5.8	5.4	5.3	4.9	4.6	4.7	4.9	5.3	6.4	6.6	5.6
	MIDNIGHT-MIDNIGHT (OKTAS)	32	6.5	6.2	5.7	5.3	5.0	4.7	4.3	4.5	4.7	5.2	6.3	6.5	5.4
	MEAN NO. DAYS WITH:														
	CLEAR	56	2.0	2.4	4.6	6.1	6.0	7.2	7.8	7.7	7.0	6.4	2.1	2.0	61.3
	PARTLY CLOUDY	56	6.6	6.7	8.1	7.7	9.7	10.8	12.5	11.7	10.5	8.4	6.1	5.6	104.4
	CLOUDY	56	22.4	19.2	18.3	16.3	15.4	12.1	10.7	11.7	12.6	16.3	21.8	23.4	200.2
PR	MEAN STATION PRESSURE(IN)	28	29.43	29.45	29.42	29.37	29.38	29.37	29.40	29.44	29.47	29.47	29.45	29.44	29.42
	MEAN SEA-LEVEL PRES. (IN)	17	30.06	30.08	30.05	29.99	29.97	29.97	29.98	30.04	30.05	30.08	30.06	30.08	30.03
WINDS	MEAN SPEED (MPH)	43	12.0	11.5	11.7	11.3	9.9	9.2	8.5	8.3	8.7	9.2	10.9	11.3	10.2
	PREVAIL.DIR(TENS OF DEGS)	27	25	25	25	25	25	24	23	22	21	21	25	25	25
	MAXIMUM 2-MINUTE:														
	SPEED (MPH)	5	45	59	45	41	44	39	52	36	68	38	41	45	68
	DIR. (TENS OF DEGS)		24	25	26	25	27	36	20	30	27	26	17	24	27
	YEAR OF OCCURRENCE		2000	1997	1998	1997	1998	1998	1999	1999	1998	2001	1999	2000	SEP 1998
	MAXIMUM 5-SECOND:														
	SPEED (MPH)	5	55	75	61	62	63	48	63	53	89	48	52	54	89
	DIR. (TENS OF DEGS)		25	24	26	25	28	36	20	31	29	25	17	25	29
	YEAR OF OCCURRENCE		2000	1997	1998	1997	1998	1998	1999	1999	1998	2001	1999	2000	SEP 1998
PRECIPITATION	NORMAL (IN)	30	2.08	2.10	2.28	2.61	2.72	3.00	2.71	3.40	2.97	2.44	2.92	2.73	31.96
	MAXIMUM MONTHLY (IN)	61	5.79	5.07	5.42	4.90	6.62	7.11	9.70	6.00	6.07	7.85	6.99	5.05	9.70
	YEAR OF OCCURRENCE		1978	1950	1942	1944	1974	1998	1947	1984	1977	1955	1985	1944	JUL 1947
	MINIMUM MONTHLY (IN)	61	0.72	0.66	0.47	1.18	0.36	0.22	0.61	0.76	0.28	0.23	0.44	0.62	0.22
	YEAR OF OCCURRENCE		1988	1987	1958	1995	1977	1963	1994	1951	1960	1963	1976	1958	JUN 1963
	MAXIMUM IN 24 HOURS (IN)	61	2.24	2.43	2.21	2.22	3.85	2.86	3.25	2.52	3.54	3.13	3.13	1.60	3.85
	YEAR OF OCCURRENCE		1998	1950	1942	1991	1974	1950	1987	1998	1979	1995	1945	1978	MAY 1974
	NORMAL NO. DAYS WITH:														
	PRECIPITATION 0.01	30	16.6	15.5	14.0	13.0	11.4	10.8	9.8	10.9	11.1	12.5	15.4	18.1	159.1
	PRECIPITATION 1.00	30	0.1	0.2	0.1	0.2	0.3	0.5	0.5	0.8	0.7	0.2	0.3	0.2	4.1
SNOWFALL	NORMAL (IN)	30	25.1	22.6	12.3	4.6	0.4	0.0	0.0	0.0	0.0	0.1	6.3	21.6	93.0
	MAXIMUM MONTHLY (IN)	61	60.4	64.8	45.0	20.2	10.9	T	T	T	T	2.6	24.9	46.1	64.8
	YEAR OF OCCURRENCE		1978	1958	1999	1979	1989	1998	1996	1965	1994	1993	1996	1981	FEB 1958
	MAXIMUM IN 24 HOURS (IN)	61	23.0	22.8	23.3	10.4	10.8	T	T	T	T	2.6	14.1	19.1	23.3
	YEAR OF OCCURRENCE		1996	1978	1999	1990	1989	1998	1990	1965	1994	1993	1995	1978	MAR 1999
	MAXIMUM SNOW DEPTH (IN)	53	32	34	34	10	4	0	0	0	0	1	8	40	40
	YEAR OF OCCURRENCE		1978	1966	1999	1975	1989					1957	1972	1959	DEC 1959
	NORMAL NO. DAYS WITH:														
	SNOWFALL 1.0	30	7.7	7.2	3.6	1.2	0.1	0.0	0.0	0.0	0.0	0.0	2.0	6.9	28.7

PRECIPITATION (inches) 2001 ROCHESTER, NY (ROC)

YEAR	JAN	FEB	MAR	APR	MAY	JUN	JUL	AUG	SEP	OCT	NOV	DEC	ANNUAL
1972	1.50	3.96	2.19	2.68	3.32	6.56	1.43	3.14	3.84	2.25	4.83	2.58	38.28
1973	1.28	1.70	2.92	3.21	2.68	2.84	1.14	1.94	1.41	2.67	3.82	3.62	29.23
1974	1.75	2.06	3.61	2.60	6.62	2.59	2.82	3.64	3.48	1.34	3.23	2.86	36.60
1975	1.83	2.82	2.74	1.43	2.85	5.35	1.18	2.31	3.15	1.83	1.35	3.76	30.60
1976	2.33	1.67	3.54	3.81	2.63	3.37	5.15	3.04	2.13	4.73	0.44	1.48	34.32
1977	1.49	0.97	2.18	2.49	0.36	1.33	3.26	5.65	6.30	2.64	3.78	4.65	35.10
1978	5.79	2.40	1.48	2.25	2.03	1.30	2.17	2.66	3.63	2.56	1.14	4.35	31.76
1979	4.18	2.40	1.76	3.78	3.14	1.85	3.16	2.05	5.32	2.60	1.80	2.86	34.90
1980	1.11	1.16	3.83	2.35	1.49	6.77	1.90	3.44	3.57	3.73	2.52	2.45	34.32
1981	1.24	3.13	1.04	1.95	2.27	2.70	4.60	4.44	5.37	2.18	2.78	3.49	34.99
1982	4.16	1.01	1.73	1.63	1.77	3.92	3.13	3.00	3.57	1.79	3.95	2.17	31.83
1983	1.43	1.23	2.45	3.50	3.44	2.40	1.13	5.43	1.56	3.26	4.91	4.47	35.21
1984	1.62	2.97	2.08	3.05	5.47	1.67	1.90	6.00	3.34	0.76	1.47	3.31	33.64
1985	2.49	1.78	3.47	1.30	2.08	2.63	1.86	1.11	2.49	2.34	6.99	1.46	30.00
1986	1.63	2.46	1.90	3.80	1.64	4.27	3.13	3.29	5.11	3.56	1.93	3.56	36.28
1987	1.89	0.66	1.98	3.68	1.19	3.94	5.85	3.92	4.60	1.65	2.74	1.98	34.08
1988	0.72	2.18	1.62	2.32	1.73	1.10	4.30	3.81	1.69	2.34	1.68	1.11	24.60
1989	1.18	1.55	3.69	1.62	5.99	5.65	0.98	2.46	2.82	3.13	2.01	1.58	32.66
1990	1.61	3.93	1.56	3.58	5.76	2.88	3.05	3.59	3.36	4.37	2.27	4.18	40.14
1991	1.69	1.16	4.70	4.07	2.43	1.19	2.37	1.80	2.86	1.65	2.39	2.92	29.23
1992	1.46	1.87	3.53	3.43	2.83	1.98	6.03	4.45	3.02	1.78	2.90	2.98	36.26
1993	2.32	1.52	2.44	3.07	1.24	2.76	1.67	1.67	4.37	3.21	3.27	1.60	29.14
1994	2.68	1.63	1.70	4.08	2.56	2.43	0.61	4.27	2.68	1.34	3.24	2.32	29.54
1995	2.46	1.58	1.15	1.18	1.75	2.07	3.85	3.05	1.50	5.70	4.21	1.50	30.00
1996	3.18	1.72	2.07	4.84	3.51	6.65	2.18	3.33	5.09	5.40	4.12	2.97	45.06
1997	2.03	2.40	3.88	1.33	2.12	3.01	1.94	4.22	5.36	1.94	3.57	2.88	34.68
1998	5.63	2.34	3.50	1.81	2.63	7.11	6.09	5.39	3.00	1.45	1.41	1.60	41.96
1999	3.92	0.69	3.28	2.07	2.72	2.52	1.78	5.71	3.41	2.12	2.86	2.06	33.14
2000	2.98	1.97	2.04	4.35	4.70	4.47	3.66	4.11	3.53	1.36	2.19	2.47	37.83
2001	1.95	2.26	4.13	1.19	2.66	1.84	1.80	4.30	3.15	2.28	1.90	1.72	29.18
POR= 173 YRS	2.49	2.33	2.63	2.63	2.91	3.03	3.01	2.96	2.75	2.71	2.71	2.49	32.65

AVERAGE TEMPERATURE (F) 2001 ROCHESTER, NY (ROC)

YEAR	JAN	FEB	MAR	APR	MAY	JUN	JUL	AUG	SEP	OCT	NOV	DEC	ANNUAL
1972	26.0	23.4	30.3	42.2	60.4	65.2	73.0	69.8	64.3	47.6	37.0	33.1	47.7
1973	28.7	22.2	42.5	48.0	56.1	70.7	73.4	73.0	62.5	55.0	43.6	30.0	50.5
1974	27.1	22.5	33.0	49.4	53.9	65.7	71.3	70.6	59.2	47.6	39.5	31.4	47.6
1975	29.5	28.4	31.6	39.3	63.2	67.1	73.0	70.0	58.4	53.1	47.2	27.8	49.1
1976	19.8	33.3	37.2	48.6	55.4	69.8	69.2	68.4	60.9	47.5	35.4	23.6	47.4
1977	15.5	25.4	39.8	47.9	60.7	64.8	72.9	68.7	62.7	49.4	43.6	28.4	48.3
1978	22.9	16.2	29.7	43.6	60.2	67.4	72.6	71.6	62.4	51.0	41.0	30.0	47.4
1979	21.5	13.7	38.6	44.0	56.4	66.2	72.3	67.1	61.3	50.3	43.0	32.3	47.2
1980	24.0	19.7	32.4	47.8	60.0	63.1	72.9	74.3	63.7	48.8	38.8	24.7	47.5
1981	15.7	32.3	34.5	48.0	57.2	67.3	71.9	69.4	59.7	43.9	28.7	47.7	47.7
1982	16.1	23.0	33.5	43.2	60.9	63.6	72.0	66.1	62.8	52.7	43.4	37.4	47.9
1983	27.4	29.1	37.2	43.9	53.8	66.7	73.8	70.7	63.9	52.9	40.7	25.1	48.8
1984	20.4	33.2	26.5	47.5	52.6	66.8	69.2	72.0	60.6	54.9	40.7	35.9	48.4
1985	21.9	25.6	36.7	49.6	58.6	61.7	68.8	68.7	63.8	51.0	41.4	25.0	47.7
1986	25.0	24.5	37.0	47.9	59.8	63.3	69.8	65.7	59.8	49.9	36.9	31.7	47.6
1987	25.3	23.6	37.1	49.7	59.9	67.9	72.7	67.3	61.6	47.1	40.6	32.6	48.8
1988	25.0	23.7	34.7	45.0	58.7	64.2	73.7	71.1	60.1	45.8	42.6	29.4	47.8
1989	30.3	22.5	32.3	42.1	56.3	67.4	72.8	68.5	61.7	52.6	38.1	17.1	46.8
1990	33.6	29.3	37.3	48.8	54.4	67.2	70.7	69.9	60.7	52.1	42.4	33.8	50.0
1991	25.1	30.5	37.0	50.0	62.8	68.3	72.3	70.3	60.5	52.1	39.0	30.7	49.9
1992	26.2	27.2	30.2	44.1	57.1	63.5	66.6	66.3	60.9	46.4	38.9	30.2	46.5
1993	27.5	18.7	30.0	46.9	56.6	65.5	72.4	71.4	59.0	48.0	39.0	28.4	47.0
1994	14.9	21.1	32.0	48.0	54.2	67.8	73.5	69.2	62.4	45.8	34.8	44.0	48.0
1995	32.1	23.8	38.5	41.6	57.2	69.5	73.0	73.3	60.4	55.2	35.2	25.3	48.8
1996	23.7	24.6	29.7	43.3	54.6	68.0	68.1	69.3	62.0	51.2	34.3	33.6	46.9
1997	24.3	30.5	32.9	43.8	50.4	67.1	67.8	66.3	59.3	49.1	37.4	31.2	46.7
1998	31.6	32.6	38.1	47.7	62.9	65.7	69.6	69.8	62.7	51.2	41.8	34.6	50.7
1999	22.9	30.6	30.8	45.3	59.6	68.3	74.3	67.1	64.0	54.9	44.9	32.2	49.2
2000	23.2	30.2	41.2	45.1	59.5	65.8	67.1	67.5	60.9	51.7	38.4	22.6	47.8
2001	26.6	28.6	30.3	47.7	59.5	66.7	69.0	72.2	61.2	52.4	47.1	35.9	49.8
POR= 130 YRS	24.7	24.8	33.0	45.1	56.9	66.4	71.3	69.3	62.5	51.3	39.9	29.1	47.9

REFERENCE NOTES:

PAGE 1:
THE TEMPERATURE GRAPH SHOWS NORMAL MAXIMUM AND NORMAL
MINIMUM DAILY TEMPERATURES (SOLID CURVES) AND THE
ACTUAL DAILY HIGH AND LOW TEMPERATURES (VERTICAL BARS).

PAGE 2 AND 3:
H/C INDICATES HEATING AND COOLING DEGREE DAYS.
RH INDICATES RELATIVE HUMIDITY
W/O INDICATES WEATHER AND OBSTRUCTIONS
S INDICATES SUNSHINE.
PR INDICATES PRESSURE.
CLOUDINESS ON PAGE 3 IS THE SUM OF THE CEILOMETER AND
SATELLITE DATA NOT TO EXCEED EIGHT EIGHTHS(OKTAS).

GENERAL:
T INDICATES TRACE PRECIPITATION, AN AMOUNT GREATER
THAN ZERO BUT LESS THAN THE LOWEST REPORTABLE VALUE.
+ INDICATES THE VALUE ALSO OCCURS ON EARLIER DATES.
BLANK ENTRIES DENOTE MISSING OR UNREPORTED DATA.
NORMALS ARE 30-YEAR AVERAGES (1961 - 1990).
ASOS INDICATES AUTOMATED SURFACE OBSERVING SYSTEM.
PM INDICATES THE LAST DAY OF THE PREVIOUS MONTH.
POR (PERIOD OF RECORD) BEGINS WITH THE JANUARY DATA
MONTH AND IS THE NUMBER OF YEARS USED TO COMPUTE
THE MEAN. INDIVIDUAL MONTHS WITHIN THE POR MAY
BE MISSING.
WHEN THE POR FOR A NORMAL IS LESS THAN 30 YEARS,
THE NORMAL IS PROVISIONAL AND IS BASED ON THE NUMBER
OF YEARS INDICATED.
0.* OR * INDICATES THE VALUE OR MEAN-DAYS-WITH
IS BETWEEN 0.00 AND 0.05.
CLOUDINESS FOR ASOS STATIONS DIFFERS FROM THE NON-ASOS
OBSERVATION TAKEN BY A HUMAN OBSERVER. ASOS STATION
CLOUDINESS IS BASED ON TIME-AVERAGED CEILOMETER DATA
FOR CLOUDS AT OR BELOW 12,000 FEET AND ON SATELLITE
DATA FOR CLOUDS ABOVE 12,000 FEET.
THE NUMBER OF DAYS WITH CLEAR, PARTLY CLOUDY, AND
CLOUDY CONDITIONS FOR ASOS STATIONS IS THE SUM
OF THE CEILOMETER AND SATELLITE DATA FOR THE
SUNRISE TO SUNSET PERIOD.

GENERAL CONTINUED:
CLEAR INDICATES 0 - 2 OKTAS, PARTLY CLOUDY INDICATES
3 - 6 OKTAS, AND CLOUDY INDICATES 7 OR 8 OKTAS.
WHEN AT LEAST ONE OF THE ELEMENTS (CEILOMETER OR
SATELLITE) IS MISSING, THE DAILY CLOUDINESS IS
NOT COMPUTED.
WIND DIRECTION IS RECORDED IN TENS OF DEGREES (2 DIGITS)
CLOCKWISE FROM TRUE NORTH. "00" INDICATES CALM. "36"
INDICATES TRUE NORTH.
RESULTANT WIND IS THE VECTOR AVERAGE OF THE SPEED AND
DIRECTION.
AVERAGE TEMPERATURE IS THE SUM OF THE MEAN DAILY MAXIMUM
AND MINIMUM TEMPERATURE DIVIDED BY 2.
SNOWFALL DATA COMPRISE ALL FORMS OF FROZEN
PRECIPITATION, INCLUDING HAIL.
A HEATING (COOLING) DEGREE DAY IS THE DIFFERENCE BETWEEN
THE AVERAGE DAILY TEMPERATURE AND 65 F.
DRY BULB IS THE TEMPERATURE OF THE AMBIENT AIR.
DEW POINT IS THE TEMPERATURE TO WHICH THE AIR MUST BE
COOLED TO ACHIEVE 100 PERCENT RELATIVE HUMIDITY.
WET BULB IS THE TEMPERATURE THE AIR WOULD HAVE IF THE
MOISTURE CONTENT WAS INCREASED TO 100 PERCENT RELATIVE
HUMIDITY.

ON JULY 1, 1996, THE NATIONAL WEATHER SERVICE BEGAN USING
THE "METAR" OBSERVATION CODE THAT WAS ALREADY EMPLOYED
BY MOST OTHER NATIONS OF THE WORLD. THE MOST NOTICEABLE
DIFFERENCE IN THIS ANNUAL PUBLICATION WILL BE THE CHANGE
IN UNITS FROM TENTHS TO EIGHTS(OKTAS) FOR REPORTING THE
AMOUNT OF SKY COVER.

HEATING DEGREE DAYS (base 65 F) 2001 ROCHESTER, NY (ROC)

YEAR	JUL	AUG	SEP	OCT	NOV	DEC	JAN	FEB	MAR	APR	MAY	JUN	TOTAL
1972-73	7	24	92	534	833	982	1118	1189	690	519	279	17	6284
1973-74	2	14	162	305	653	1040	1167	1187	983	475	352	59	6399
1974-75	1	1	209	535	755	1034	1096	1017	1031	764	139	52	6634
1975-76	4	14	194	365	525	1146	1395	914	858	507	300	35	6257
1976-77	11	27	173	538	879	1279	1524	1103	777	523	204	89	7127
1977-78	9	44	113	477	634	1127	1298	1360	1087	634	220	63	7066
1978-79	5	1	136	428	711	1077	1342	1432	813	626	310	79	6960
1979-80	13	37	155	468	655	1006	1264	1306	1003	510	195	125	6737
1980-81	1	0	108	498	782	1243	1522	908	938	507	260	26	6793
1981-82	6	12	201	546	748	1119	1510	1171	972	648	162	67	7162
1982-83	10	54	113	377	643	847	1161	998	854	627	347	78	6109
1983-84	9	8	121	387	723	1228	1376	917	1187	520	395	50	6921
1984-85	14	7	162	307	724	897	1330	1097	869	471	217	119	6214
1985-86	15	23	121	429	700	1231	1235	1129	864	506	206	100	6559
1986-87	16	62	175	462	840	1026	1223	1153	858	454	234	39	6542
1987-88	7	50	139	547	722	997	1232	1192	933	594	220	126	6759
1988-89	6	40	164	596	664	1095	1070	1184	1009	682	288	33	6831
1989-90	0	33	149	383	801	1478	967	993	853	520	327	46	6550
1990-91	7	6	171	406	669	959	1230	957	862	458	170	29	5924
1991-92	2	1	196	408	776	1057	1195	1088	1069	621	259	89	6761
1992-93	26	46	172	571	774	1071	1158	1289	1077	538	263	66	7051
1993-94	0	10	214	525	775	1127	1550	1221	1015	505	345	56	7343
1994-95	1	16	106	386	568	928	1015	1147	811	694	239	33	5944
1995-96	13	7	162	300	885	1224	1271	1163	1086	647	347	16	7121
1996-97	19	6	140	421	915	968	1255	960	989	628	448	45	6794
1997-98	32	29	181	492	822	1041	1029	901	837	513	109	99	6085
1998-99	4	14	115	424	690	935	1295	955	1054	583	194	57	6320
1999-00	2	25	100	431	595	1008	1289	1003	732	591	211	75	6062
2000-01	23	35	178	404	793	1308	1186	1014	1067	522	192	63	6785
2001-	24	3	152	391	529	896							

COOLING DEGREE DAYS (base 65 F) 2001 ROCHESTER, NY (ROC)

YEAR	JAN	FEB	MAR	APR	MAY	JUN	JUL	AUG	SEP	OCT	NOV	DEC	ANNUAL
1972	0	0	0	0	24	94	261	179	79	0	0	0	637
1973	0	0	0	15	6	194	267	269	96	4	0	0	851
1974	0	0	0	13	14	88	204	181	40	0	0	0	540
1975	0	0	0	0	89	121	257	178	5	6	0	0	656
1976	0	0	0	24	9	189	150	138	55	1	0	0	566
1977	0	0	3	16	80	88	260	164	50	0	1	0	662
1978	0	0	0	0	77	141	245	212	66	3	0	0	744
1979	0	0	0	1	52	121	244	112	49	18	0	0	597
1980	0	0	0	0	46	73	253	294	76	2	0	0	744
1981	0	0	0	5	23	102	228	156	50	0	0	0	564
1982	0	0	0	3	40	30	232	95	52	3	1	0	456
1983	0	0	0	0	7	136	289	192	96	20	0	0	740
1984	0	0	0	1	14	113	152	233	35	1	0	0	549
1985	0	0	0	15	23	27	139	145	90	0	0	0	439
1986	0	0	1	0	50	53	168	94	28	0	0	0	394
1987	0	0	0	1	82	131	254	127	42	0	0	0	637
1988	0	0	0	0	34	107	284	232	29	7	0	0	693
1989	0	0	0	0	26	111	248	153	60	3	0	0	601
1990	0	0	3	41	5	122	192	164	45	14	0	0	586
1991	0	0	0	14	108	135	234	175	68	14	0	0	748
1992	0	0	0	1	19	51	84	96	57	0	0	0	308
1993	0	0	0	0	9	86	239	214	42	5	0	0	595
1994	0	0	0	0	16	145	271	152	36	2	1	0	623
1995	0	0	0	0	3	175	266	269	33	5	0	0	751
1996	0	0	0	1	31	115	120	147	58	0	0	0	472
1997	0	0	0	1	0	114	126	75	15	8	0	0	339
1998	0	0	9	0	48	125	152	168	53	4	0	0	559
1999	0	0	0	0	38	161	296	93	77	0	0	0	665
2000	0	0	0	2	49	105	105	95	120	63	0	0	434
2001	0	0	0	9	28	119	154	233	44	6	0	0	593

SNOWFALL (inches) 2001 ROCHESTER, NY (ROC)

YEAR	JUL	AUG	SEP	OCT	NOV	DEC	JAN	FEB	MAR	APR	MAY	JUN	TOTAL
1972-73	0.0	0.0	0.0	0.2	16.9	22.7	8.9	18.4	4.4	1.5	T	0.0	73.0
1973-74	0.0	0.0	0.0	0.0	4.2	23.4	14.4	26.6	22.3	8.2	T	0.0	99.1
1974-75	0.0	0.0	0.0	0.3	4.6	26.5	10.8	23.2	10.9	14.9	0.0	0.0	91.2
1975-76	0.0	0.0	0.0	T	1.8	28.3	29.9	8.8	15.2	1.8	0.4	0.0	86.2
1976-77	0.0	0.0	0.0	0.5	6.5	24.5	30.2	15.0	13.0	1.8	0.6	0.0	92.1
1977-78	0.0	0.0	0.0	T	12.7	35.2	60.4	40.7	7.5	4.2	0.2	0.0	160.9
1978-79	0.0	0.0	0.0	T	3.3	30.9	36.8	39.1	8.2	20.2	0.0	0.0	138.5
1979-80	0.0	0.0	0.0	0.2	1.2	12.2	13.1	24.0	21.2	0.3	0.0	0.0	72.2
1980-81	0.0	0.0	0.0	T	8.4	31.8	31.5	9.3	12.0	1.4	0.0	0.0	94.4
1981-82	0.0	0.0	0.0	0.1	2.4	46.1	43.6	14.9	8.9	12.4	0.0	0.0	128.4
1982-83	0.0	0.0	0.0	T	3.0	11.6	10.2	13.6	9.3	12.2	T	0.0	59.9
1983-84	0.0	0.0	0.0	0.0	17.6	19.6	23.4	27.8	29.1	0.5	T	0.0	118.0
1984-85	0.0	0.0	0.0	0.0	1.6	11.6	36.8	26.1	8.4	2.6	0.0	0.0	87.1
1985-86	0.0	0.0	0.0	0.0	7.6	18.3	15.5	17.9	9.3	2.1	T	0.0	70.7
1986-87	0.0	0.0	0.0	0.0	7.4	9.3	9.8	29.6	13.0	5.3	2.5	0.0	67.1
1987-88	0.0	0.0	0.0	T	4.6	19.3	9.8	29.4	5.6	1.1	0.0	0.0	69.8
1988-89	0.0	0.0	0.0	0.1	0.2	10.3	15.0	30.6	15.6	3.9	10.9	0.0	86.6
1989-90	0.0	0.0	0.0	T	6.5	32.8	14.0	31.3	5.4	15.8	T	0.0	105.8
1990-91	T	0.0	0.0	T	4.4	18.2	26.5	16.1	2.0	1.1	0.0	0.0	68.3
1991-92	0.0	0.0	0.0	0.0	13.7	23.9	18.3	12.8	38.1	3.8	0.0	0.0	110.6
1992-93	0.0	0.0	0.0	T	9.5	29.3	22.4	31.2	37.1	2.0	T	0.0	131.5
1993-94	0.0	0.0	T	2.6	9.8	14.0	43.0	35.1	12.1	9.6	0.0	0.0	126.2
1994-95	0.0	0.0	T	0.0	2.8	7.6	12.8	23.6	5.3	4.1	0.0	0.0	56.2
1995-96	0.0	0.0	0.0	T	23.4	20.5	36.9	14.1	28.6	5.3	1.5	0.0	130.3
1996-97	T	0.0	0.0	0.0	24.9	14.0	24.8	13.3	26.4	1.3	T	0.0	104.7
1997-98	0.0	0.0	0.0	0.1	21.6	26.4	14.6	9.1	27.9	0.0	0.0	T	99.7
1998-99	0.0	0.0	0.0	0.0	0.1	10.1	48.8	4.7	45.0	2.9	0.0	0.0	111.6
1999-00	0.0	0.0	T	0.0	4.7	19.1	42.0	25.7	13.6	5.6	0.0	0.0	110.7
2000-01	0.0	0.0	0.0	T	7.3	39.3	21.6	23.3	41.4	0.1	0.0	0.0	133.0
2001-	0.0	0.0	0.0	T	0.1	7.1							
POR= 60 YRS	T	T	T	0.2	7.0	19.4	23.3	21.5	15.4	3.4	0.7	0.0	90.9

2001
SYRACUSE,
NEW YORK (SYR)

Syracuse is located approximately at the geographical center of the state. Gently rolling terrain stretches northward for about 30 miles to the eastern end of Lake Ontario. Oneida Lake is about 8 miles northeast of Syracuse. Approximately 5 miles south of the city, hills rise to 1,500 feet. Immediately to the west, the terrain is gently rolling with elevations 500 to 800 feet above sea level.

The climate of Syracuse is primarily continental in character and comparatively humid. Nearly all cyclonic systems moving from the interior of the country through the St. Lawrence Valley will affect the Syracuse area. Seasonal and diurnal changes are marked and produce an invigorating climate.

In the summer and in portions of the transitional seasons, temperatures usually rise rapidly during the daytime to moderate levels and as a rule fall rapidly after sunset. The nights are relatively cool and comfortable. There are only a few days in a year when atmospheric humidity causes great personal discomfort.

Winters are usually cold and are sometimes severe in part. Daytime temperatures average in the low 30s with nighttime lows in the teens. Low winter temperatures below -25 degrees have been recorded. The autumn, winter, and spring seasons display marked variability.

Based on the 1951-1980 period, the average first occurrence of 32 degrees Fahrenheit in the fall is October 16 and the average last occurrence in the spring is April 28.

Precipitation in the Syracuse area is derived principally from cyclonic storms which pass from the interior of the country through the St. Lawrence Valley. Lake Ontario provides the source of significant winter precipitation. The lake is quite deep and never freezes so cold air flowing over the lake is quickly saturated and produces the cloudiness and snow squalls which are a well-known feature of winter weather in the Syracuse area.

The area enjoys sufficient precipitation in most years to meet the needs of agriculture and water supplies. The precipitation is uncommonly well distributed, averaging about 3 inches per month throughout the year. Snowfall is moderately heavy with an average just over 100 inches. There are about 30 days per year with thunderstorms, mostly during the warmer months.

Wind velocities are moderate, but during the winter months there are numerous days with sufficient winds to cause blowing and drifting snow.

During December, January, and February there is much cloudiness. Syracuse receives only about one-third of possible sunshine during winter months. Approximately two-thirds of possible sunshine is received during the warm months.

NORMALS, MEANS, AND EXTREMES
SYRACUSE, NY (SYR)

LATITUDE: 43 06'33" N LONGITUDE: 76 06'12" W ELEVATION (FT): GRND: 414 BARO: 417 TIME ZONE: EASTERN (UTC + 5) WBAN: 14771

ELEMENT	POR	JAN	FEB	MAR	APR	MAY	JUN	JUL	AUG	SEP	OCT	NOV	DEC	YEAR
TEMPERATURE (F)														
NORMAL DAILY MAXIMUM	30	30.6	32.5	42.7	56.0	68.3	76.7	81.7	79.0	71.6	60.3	48.0	35.4	56.9
MEAN DAILY MAXIMUM	52	31.0	33.5	42.1	56.5	68.3	77.5	81.8	79.8	72.0	60.9	48.1	36.0	57.3
HIGHEST DAILY MAXIMUM	51	70	69	87	92	96	98	97	100	97	87	81	72	100
YEAR OF OCCURRENCE		1967	1981	1986	1990	1977	1953	1990	2001	1953	1963	1950	2001	AUG 2001
MEAN OF EXTREME MAXS.	56	53.5	53.5	67.7	78.8	85.8	91.0	92.3	91.0	87.4	79.2	67.5	58.1	75.5
NORMAL DAILY MINIMUM	30	14.2	15.4	25.1	35.5	46.0	55.3	60.4	58.9	51.6	41.2	33.0	21.1	37.8
MEAN DAILY MINIMUM	52	14.9	16.8	25.0	36.2	46.2	55.3	60.4	58.9	51.6	41.2	33.2	21.9	38.5
LOWEST DAILY MINIMUM	51	-26	-26	-16	9	25	35	45	40	28	19	5	-22	-26
YEAR OF OCCURRENCE		1966	1979	1950	1972	1966	1966	1966	1965	1991	1976	1976	1980	FEB 1979
MEAN OF EXTREME MINS.	56	-7.9	-5.7	4.7	22.4	32.9	41.8	49.4	46.5	36.3	26.9	16.9	-.6	22.0
NORMAL DRY BULB	30	22.4	24.0	33.9	45.7	57.1	65.3	70.4	68.4	61.5	50.7	40.5	28.3	47.4
MEAN DRY BULB	56	23.2	24.9	33.5	46.2	57.3	66.3	71.1	69.5	61.9	51.2	39.9	28.6	47.8
MEAN WET BULB	52	21.5	23.2	30.6	40.9	51.0	59.8	64.3	63.2	56.7	46.7	37.5	26.8	43.5
MEAN DEW POINT	52	15.9	17.4	24.5	34.0	45.1	55.2	60.0	59.4	53.1	42.2	32.8	22.1	38.5
NORMAL NO. DAYS WITH:														
MAXIMUM 90	30	0.0	0.0	0.0	*	0.5	1.4	3.7	1.6	0.4	0.0	0.0	0.0	7.6
MAXIMUM 32	30	17.0	14.4	5.7	0.3	0.0	0.0	0.0	0.0	0.0	0.0	1.5	12.1	51.0
MINIMUM 32	30	28.8	25.2	23.7	11.8	1.0	0.0	0.0	0.0	0.2	4.8	14.8	26.6	136.9
MINIMUM 0	30	4.4	3.0	0.6	0.0	0.0	0.0	0.0	0.0	0.0	0.0	0.0	1.7	9.7
H/C														
NORMAL HEATING DEG. DAYS	30	1321	1148	964	579	268	61	10	28	139	443	735	1138	6834
NORMAL COOLING DEG. DAYS	30	0	0	0	0	23	70	178	133	34	0	0	0	438
RH														
NORMAL (PERCENT)	30	73	72	70	65	67	70	70	75	76	74	75	77	72
HOUR 01 LST	30	76	76	76	75	78	82	84	86	86	82	79	79	80
HOUR 07 LST	30	76	78	78	76	76	79	80	86	87	84	81	80	80
HOUR 13 LST	30	68	65	60	53	55	56	55	59	61	60	67	71	61
HOUR 19 LST	30	74	72	67	59	60	62	63	69	75	74	75	77	69
S PERCENT POSSIBLE SUNSHINE	50	35	39	45	49	54	59	63	58	52	44	26	26	46
W/O MEAN NO. DAYS WITH:														
HEAVY FOG(VISBY 1/4 MI)	51	0.9	0.8	0.9	0.6	0.8	0.5	0.5	0.7	0.9	1.2	0.6	0.8	9.2
THUNDERSTORMS	51	0.2	0.2	0.8	1.8	3.1	5.2	6.1	5.3	2.5	1.0	0.6	0.1	26.9
CLOUDINESS MEAN:														
SUNRISE-SUNSET (OKTAS)	1			8.0			5.6					9.6		
MIDNIGHT-MIDNIGHT (OKTAS)	1			8.0										
MEAN NO. DAYS WITH:														
CLEAR	1			3.0		5.0	3.0							
PARTLY CLOUDY	1			7.0		6.0	10.0							
CLOUDY	1	3.0	5.0	11.0		9.0	6.0							
PR														
MEAN STATION PRESSURE(IN)	27	29.59	29.60	29.57	29.52	29.53	29.52	29.54	29.59	29.61	29.63	29.60	29.61	29.58
MEAN SEA-LEVEL PRES. (IN)	52	30.06	30.05	30.01	29.98	29.97	29.96	29.99	30.01	30.06	30.08	30.03	30.06	30.02
WINDS														
MEAN SPEED (MPH)	45	10.5	10.6	10.5	10.4	9.1	8.2	7.9	7.5	8.1	8.5	10.1	10.2	9.3
PREVAIL.DIR(TENS OF DEGS)	25	25	25	25	27	29	30	26	26	25	26	25	25	25
MAXIMUM 2-MINUTE:														
SPEED (MPH)	8	46	49	43	38	43	33	54	37	59	38	41	48	59
DIR. (TENS OF DEGS)		16	26	25	26	29	28	28	25	32	29	26	25	32
YEAR OF OCCURRENCE		1996	2001	1996	2001	1998	1998	1999	1996	1998	2001	1994	2000	SEP 1998
MAXIMUM 5-SECOND:														
SPEED (MPH)	8	59	64	49	46	56	48	66	56	77	46	51	62	77
DIR. (TENS OF DEGS)		27	27	24	25	28	28	28	23	32	31	12	25	32
YEAR OF OCCURRENCE		1999	2001	1996	2001	1998	1998	1999	2001	1998	1998	1999	2000	SEP 1998
PRECIPITATION														
NORMAL (IN)	30	2.34	2.15	2.77	3.33	3.28	3.79	3.81	3.51	3.79	3.24	3.72	3.20	38.93
MAXIMUM MONTHLY (IN)	51	5.77	5.38	6.84	8.12	7.41	12.30	9.52	8.41	8.81	8.29	6.79	5.50	12.30
YEAR OF OCCURRENCE		1978	1951	1955	1976	1976	1972	1974	1956	1975	1955	1972	1983	JUN 1972
MINIMUM MONTHLY (IN)	51	1.02	0.63	1.01	1.22	0.75	1.10	0.90	1.02	0.75	0.21	1.25	1.40	0.21
YEAR OF OCCURRENCE		1970	1987	1981	1985	1977	1962	1969	1999	1964	1963	1978	1999	OCT 1963
MAXIMUM IN 24 HOURS (IN)	51	1.88	1.99	1.77	2.85	3.13	3.88	4.07	4.27	4.14	3.60	2.09	2.18	4.27
YEAR OF OCCURRENCE		1999	1961	1993	1976	1969	1972	1974	1954	1975	1955	1996	1952	AUG 1954
NORMAL NO. DAYS WITH:														
PRECIPITATION 0.01	30	18.5	15.6	16.3	13.4	12.9	11.7	10.6	11.3	11.3	12.9	16.7	19.1	170.3
PRECIPITATION 1.00	30	0.1	0.3	0.2	0.5	0.3	0.8	0.9	0.6	0.9	0.6	0.5	0.4	6.1
SNOWFALL														
NORMAL (IN)	30	29.8	23.3	14.8	4.4	0.1	0.0	0.0	0.0	0.0	0.6	9.1	28.0	110.1
MAXIMUM MONTHLY (IN)	51	72.2	72.6	54.4	16.4	2.1	T	T	0.0	T	5.7	25.9	70.3	72.6
YEAR OF OCCURRENCE		1978	1958	1993	1983	1996	1992	1992		1992	1988	1976	2000	FEB 1958
MAXIMUM IN 24 HOURS (IN)	51	24.5	21.4	35.6	7.1	2.1	T	T	0.0	T	2.9	12.1	18.8	35.6
YEAR OF OCCURRENCE		1966	1961	1993	1975	1996	1992	1992		1992	1988	1973	1997	MAR 1993
MAXIMUM SNOW DEPTH (IN)	51	39	48	25	8	1	0	0	0	0	2	14	23	48
YEAR OF OCCURRENCE		1966	1966	1971	1975	1996					1965	1973	1969	FEB 1966
NORMAL NO. DAYS WITH:														
SNOWFALL 1.0	30	8.5	7.1	4.6	1.4	0.1	0.0	0.0	0.0	0.0	0.2	2.9	8.1	32.9

PRECIPITATION (inches) 2001 SYRACUSE, NY (SYR)

YEAR	JAN	FEB	MAR	APR	MAY	JUN	JUL	AUG	SEP	OCT	NOV	DEC	ANNUAL
1972	1.10	2.87	2.49	4.03	6.19	12.30	3.45	3.76	4.12	4.36	6.79	3.95	55.41
1973	1.85	1.71	3.45	6.91	5.58	7.07	3.62	2.97	4.57	3.81	6.73	4.38	52.65
1974	2.08	1.70	4.34	3.09	5.78	4.67	9.52	4.60	4.45	1.58	4.95	3.47	50.23
1975	2.54	3.05	2.67	2.01	2.74	4.08	9.32	5.35	8.81	3.69	3.54	4.10	51.90
1976	2.79	2.71	4.62	8.12	7.41	7.42	5.24	3.27	6.53	1.53	1.80	58.17
1977	1.84	1.62	3.47	3.04	0.75	3.30	4.76	4.93	6.54	4.75	5.31	4.33	44.64
1978	5.77	0.80	3.08	1.87	1.90	3.58	2.78	3.31	3.93	2.68	1.25	4.12	35.07
1979	4.70	2.54	2.73	3.89	3.07	2.33	2.33	3.69	5.25	2.91	3.25	1.91	38.53
1980	1.47	1.38	4.34	3.33	1.34	4.45	2.57	1.33	3.40	2.56	2.64	3.27	32.08
1981	1.34	2.72	1.01	2.04	2.61	1.89	2.68	2.63	5.58	6.66	3.09	2.96	35.21
1982	3.59	1.26	2.63	1.71	2.87	4.64	3.83	2.60	4.22	0.72	4.52	2.55	35.14
1983	1.92	1.07	2.30	6.34	3.33	1.50	2.31	2.80	2.98	1.98	4.30	5.50	36.33
1984	1.30	2.88	2.39	3.16	4.97	2.02	3.66	5.17	2.61	1.95	3.48	4.38	37.97
1985	2.49	1.55	2.61	1.22	3.39	2.80	2.75	1.44	3.88	3.39	5.18	1.80	32.50
1986	2.41	2.27	2.82	3.42	2.67	4.89	5.23	3.36	5.47	3.32	3.74	3.33	42.93
1987	3.03	0.63	1.86	3.31	1.41	5.04	2.16	2.12	5.99	3.13	3.02	1.99	33.69
1988	1.50	2.13	1.79	2.70	3.05	2.46	5.72	3.77	1.88	3.57	3.95	1.92	34.44
1989	1.06	1.71	3.13	1.52	4.27	5.41	2.20	2.68	5.96	4.08	2.78	2.13	36.93
1990	2.13	3.95	3.70	4.09	5.62	2.92	3.72	5.33	3.45	6.09	3.23	5.24	49.47
1991	2.44	1.54	4.07	3.90	3.90	1.67	2.86	4.03	4.20	2.62	2.72	3.10	37.05
1992	2.62	2.46	3.80	3.54	5.21	1.78	8.00	2.64	4.55	2.69	3.75	2.57	43.61
1993	3.08	2.45	3.75	6.55	2.25	2.93	4.76	4.71	3.83	2.91	3.19	3.20	43.61
1994	3.37	1.92	5.14	3.62	3.02	2.38	2.64	5.19	2.43	1.61	3.50	2.52	37.34
1995	1.80	2.19	1.31	1.88	1.70	1.00	1.98	3.50	2.53	6.57	4.83	2.05	31.34
1996	3.35	1.25	1.74	4.28	3.02	3.05	4.24	1.71	4.38	2.14	5.78	4.45	39.39
1997	1.46	2.25	3.57	1.77	2.43	1.64	2.78	4.06	2.75	1.50	4.28	4.13	32.62
1998	4.76	3.14	2.94	2.09	2.37	4.62	3.63	4.77	2.41	2.53	2.06	1.74	37.06
1999	5.33	1.43	3.53	1.75	0.81	1.78	2.55	1.02	5.35	2.77	3.16	1.40	30.88
2000	2.80	2.46	2.37	4.24	4.75	4.46	2.73	2.48	3.13	2.25	2.98	2.36	37.01
2001	1.57	1.77	5.38	1.53	2.24	3.58	2.08	4.84	4.05	2.15	2.92	2.19	34.30
POR= 99 YRS	2.71	2.47	3.11	3.12	3.03	3.46	3.48	3.36	3.20	3.06	3.08	2.95	37.03

AVERAGE TEMPERATURE (F) 2001   SYRACUSE, NY (SYR)

YEAR	JAN	FEB	MAR	APR	MAY	JUN	JUL	AUG	SEP	OCT	NOV	DEC	ANNUAL
1972	26.4	22.9	29.4	40.5	58.5	64.5	72.9	69.2	63.5	46.5	37.0	30.8	46.8
1973	28.4	21.4	42.6	46.8	54.3	69.6	72.7	73.5	62.0	53.7	40.7	29.5	49.6
1974	26.0	21.6	32.3	48.8	54.1	65.6	69.1	68.0	59.1	46.5	40.6	30.4	46.9
1975	29.4	28.1	31.7	39.9	62.9	67.1	71.7	68.2	57.1	53.2	46.6	27.6	48.6
1976	18.1	32.5	36.6	48.4	54.2	67.9	66.7	66.0	59.8	46.9	35.8	22.6	46.3
1977	15.7	26.0	40.1	48.2	60.3	62.7	70.8	67.3	62.5	50.6	44.0	27.3	48.0
1978	21.3	17.6	29.4	42.2	58.3	64.8	71.9	71.7	59.9	49.6	40.3	30.6	46.5
1979	22.4	12.9	39.1	45.1	58.6	66.0	71.7	67.9	61.4	50.9	44.5	33.4	47.8
1980	25.6	19.8	32.4	47.8	59.8	63.0	72.5	73.8	63.4	48.8	37.6	22.6	47.3
1981	15.0	33.7	36.4	50.0	59.2	68.0	73.3	70.4	61.6	47.9	39.0	29.0	48.6
1982	14.8	25.1	33.2	43.9	59.4	63.1	70.4	65.3	60.6	50.4	43.9	34.1	47.0
1983	23.4	26.4	35.7	44.3	53.7	66.7	72.0	69.0	62.5	50.3	39.0	22.5	47.1
1984	18.7	32.0	24.5	46.0	52.4	65.4	68.0	68.8	57.7	52.2	38.3	33.5	46.5
1985	22.0	27.3	36.3	47.8	59.5	62.0	69.8	68.9	63.5	51.4	41.2	26.0	48.0
1986	23.9	23.4	37.4	49.2	61.0	64.3	71.0	66.8	60.5	49.7	36.8	31.6	48.0
1987	23.8	21.4	38.0	51.9	60.3	68.3	73.6	68.5	61.1	47.7	40.9	32.3	49.0
1988	23.1	24.6	34.4	45.7	59.7	64.1	74.0	71.8	60.8	46.6	43.0	27.8	48.0
1989	28.6	22.7	32.9	43.5	58.2	67.3	71.1	68.2	61.8	51.7	38.8	14.7	46.6
1990	33.2	29.0	37.5	49.3	54.5	67.3	71.8	70.3	61.2	52.8	42.2	33.5	50.2
1991	24.3	29.8	37.7	51.0	62.8	68.4	72.4	71.8	60.5	53.1	40.0	30.7	50.2
1992	24.7	26.5	29.3	44.4	57.5	64.0	67.3	67.5	61.3	46.6	39.7	31.0	46.7
1993	27.5	17.0	30.1	46.9	58.0	65.2	72.5	70.7	60.0	48.2	38.6	26.9	46.8
1994	12.7	19.2	30.8	47.9	54.1	68.0	72.9	67.5	60.5	44.0	39.1	31.5	46.7
1995	30.4	20.7	37.4	42.4	56.9	69.4	73.4	71.8	59.1	54.7	35.3	24.2	48.0
1996	21.5	24.6	29.8	43.2	55.0	68.2	69.4	70.3	63.1	50.7	34.7	34.9	47.1
1997	23.9	30.4	33.6	44.2	52.2	67.9	69.8	68.4	60.1	49.1	37.2	30.4	47.3
1998	29.6	31.2	37.9	48.1	62.9	66.3	70.1	71.1	64.0	51.9	41.7	35.4	50.9
1999	22.5	29.6	31.5	46.5	60.7	69.8	75.0	68.9	64.8	49.5	44.3	30.9	49.5
2000	21.3	28.8	40.1	44.3	59.1	65.6	67.0	68.2	60.7	50.9	38.5	21.7	47.2
2001	25.6	27.6	29.9	47.8	59.3	67.2	69.4	73.7	62.3	53.3	47.3	36.8	50.0
POR= 99 YRS	24.0	24.4	33.7	45.6	57.2	66.1	71.1	69.3	62.0	51.2	40.1	28.4	47.8

REFERENCE NOTES:

PAGE 1:
THE TEMPERATURE GRAPH SHOWS NORMAL MAXIMUM AND NORMAL
MINIMUM DAILY TEMPERATURES (SOLID CURVES) AND THE
ACTUAL DAILY HIGH AND LOW TEMPERATURES (VERTICAL BARS).

PAGE 2 AND 3:
H/C INDICATES HEATING AND COOLING DEGREE DAYS.
RH INDICATES RELATIVE HUMIDITY
W/O INDICATES WEATHER AND OBSTRUCTIONS
S INDICATES SUNSHINE.
PR INDICATES PRESSURE.
CLOUDINESS ON PAGE 3 IS THE SUM OF THE CEILOMETER AND
SATELLITE DATA NOT TO EXCEED EIGHT EIGHTHS(OKTAS).

GENERAL:
T INDICATES TRACE PRECIPITATION, AN AMOUNT GREATER
THAN ZERO BUT LESS THAN THE LOWEST REPORTABLE VALUE.
+ INDICATES THE VALUE ALSO OCCURS ON EARLIER DATES.
BLANK ENTRIES DENOTE MISSING OR UNREPORTED DATA.
NORMALS ARE 30-YEAR AVERAGES (1961 - 1990).
ASOS INDICATES AUTOMATED SURFACE OBSERVING SYSTEM.
PM INDICATES THE LAST DAY OF THE PREVIOUS MONTH.
POR (PERIOD OF RECORD) BEGINS WITH THE JANUARY DATA
MONTH AND IS THE NUMBER OF YEARS USED TO COMPUTE
THE MEAN. INDIVIDUAL MONTHS WITHIN THE POR MAY
BE MISSING.
WHEN THE POR FOR A NORMAL IS LESS THAN 30 YEARS,
THE NORMAL IS PROVISIONAL AND IS BASED ON THE NUMBER
OF YEARS INDICATED.
0.* OR * INDICATES THE VALUE OR MEAN-DAYS-WITH
IS BETWEEN 0.00 AND 0.05.
CLOUDINESS FOR ASOS STATIONS DIFFERS FROM THE NON-ASOS
OBSERVATION TAKEN BY A HUMAN OBSERVER. ASOS STATION
CLOUDINESS IS BASED ON TIME-AVERAGED CEILOMETER DATA
FOR CLOUDS AT OR BELOW 12,000 FEET AND ON SATELLITE
DATA FOR CLOUDS ABOVE 12,000 FEET.
THE NUMBER OF DAYS WITH CLEAR, PARTLY CLOUDY, AND
CLOUDY CONDITIONS FOR ASOS STATIONS IS THE SUM
OF THE CEILOMETER AND SATELLITE DATA FOR THE
SUNRISE TO SUNSET PERIOD.

GENERAL CONTINUED:
CLEAR INDICATES 0 - 2 OKTAS, PARTLY CLOUDY INDICATES
3 - 6 OKTAS, AND CLOUDY INDICATES 7 OR 8 OKTAS.
WHEN AT LEAST ONE OF THE ELEMENTS (CEILOMETER OR
SATELLITE) IS MISSING, THE DAILY CLOUDINESS IS
NOT COMPUTED.
WIND DIRECTION IS RECORDED IN TENS OF DEGREES (2 DIGITS)
CLOCKWISE FROM TRUE NORTH. "00" INDICATES CALM. "36"
INDICATES TRUE NORTH.
RESULTANT WIND IS THE VECTOR AVERAGE OF THE SPEED AND
DIRECTION.
AVERAGE TEMPERATURE IS THE SUM OF THE MEAN DAILY MAXIMUM
AND MINIMUM TEMPERATURE DIVIDED BY 2.
SNOWFALL DATA COMPRISE ALL FORMS OF FROZEN
PRECIPITATION, INCLUDING HAIL.
A HEATING (COOLING) DEGREE DAY IS THE DIFFERENCE BETWEEN
THE AVERAGE DAILY TEMPERATURE AND 65 F.
DRY BULB IS THE TEMPERATURE OF THE AMBIENT AIR.
DEW POINT IS THE TEMPERATURE TO WHICH THE AIR MUST BE
COOLED TO ACHIEVE 100 PERCENT RELATIVE HUMIDITY.
WET BULB IS THE TEMPERATURE THE AIR WOULD HAVE IF THE
MOISTURE CONTENT WAS INCREASED TO 100 PERCENT RELATIVE
HUMIDITY.

ON JULY 1, 1996, THE NATIONAL WEATHER SERVICE BEGAN USING
THE "METAR" OBSERVATION CODE THAT WAS ALREADY EMPLOYED
BY MOST OTHER NATIONS OF THE WORLD. THE MOST NOTICEABLE
DIFFERENCE IN THIS ANNUAL PUBLICATION WILL BE THE CHANGE
IN UNITS FROM TENTHS TO EIGHTS(OKTAS) FOR REPORTING THE
AMOUNT OF SKY COVER.

HEATING DEGREE DAYS (base 65 F) 2001 SYRACUSE, NY (SYR)

YEAR	JUL	AUG	SEP	OCT	NOV	DEC	JAN	FEB	MAR	APR	MAY	JUN	TOTAL
1972-73	9	23	98	567	833	1053	1128	1217	687	547	325	31	6518
1973-74	2	12	164	344	723	1094	1200	1206	1004	493	138	52	6633
1974-75	16	3	202	565	726	1069	1100	1026	1026	749	138	46	6666
1975-76	3	32	230	357	545	1154	1449	936	872	509	329	47	6463
1976-77	24	45	179	556	869	1303	1520	1086	767	511	209	111	7180
1977-78	14	60	121	444	624	1162	1348	1322	1097	677	252	92	7213
1978-79	10	1	184	470	735	1062	1315	1457	796	591	242	74	6937
1979-80	19	39	146	454	607	971	1215	1302	1007	511	194	115	6580
1980-81	3	0	120	496	814	1307	1544	869	882	446	221	27	6729
1981-82	2	4	145	523	775	1110	1552	1114	978	626	183	79	7091
1982-83	13	57	152	449	628	951	1280	1073	902	615	351	67	6538
1983-84	11	25	140	457	769	1312	949	1246	1054	563	386	68	7358
1984-85	16	33	227	390	797	971	1329	1048	882	514	193	109	6509
1985-86	10	18	121	415	702	1200	1266	1156	856	471	172	76	6463
1986-87	12	50	155	468	838	1027	1270	1208	831	395	211	35	6500
1987-88	7	27	138	529	717	1007	1290	1167	942	571	187	131	6713
1988-89	9	33	150	574	653	1148	1120	1175	989	639	242	38	6770
1989-90	3	36	151	406	779	1554	976	1001	849	496	319	43	6613
1990-91	4	4	160	386	675	967	1253	980	839	428	153	24	5873
1991-92	1	0	189	378	743	1056	1240	1112	1099	617	245	79	6759
1992-93	15	33	164	562	753	1047	1156	1337	1074	537	230	68	6976
1993-94	2	10	190	515	785	1172	1618	1274	1054	507	345	53	7525
1994-95	0	32	146	439	621	1019	1065	1235	850	671	248	36	6362
1995-96	5	8	194	313	884	1256	1344	1160	1085	648	328	21	7246
1996-97	5	2	123	438	903	929	1267	966	964	613	389	31	6630
1997-98	12	7	156	491	828	1065	1089	941	844	500	102	96	6131
1998-99	3	13	89	403	692	911	1310	986	1032	545	161	41	6186
1999-00	0	14	96	473	613	1049	1349	1042	765	614	224	69	6308
2000-01	22	29	184	427	791	1337	1215	1038	1080	508	190	62	6883
2001-	16	0	126	366	527	867							

COOLING DEGREE DAYS (base 65 F) 2001 SYRACUSE, NY (SYR)

YEAR	JAN	FEB	MAR	APR	MAY	JUN	JUL	AUG	SEP	OCT	NOV	DEC	ANNUAL
1973	0	0	0	7	0	177	249	281	79	2	0	0	795
1974	0	0	0	14	6	77	148	128	31	1	0	0	405
1975	0	0	0	0	80	114	221	138	1	1	0	0	555
1976	0	0	0	16	2	141	84	83	31	0	0	0	357
1977	0	0	1	12	71	47	202	138	49	0	0	0	520
1978	0	0	0	0	49	92	231	215	36	0	0	0	623
1979	0	0	0	2	50	109	232	134	46	22	0	0	595
1980	0	0	0	0	41	62	243	279	80	1	0	0	706
1981	0	0	3	4	47	125	264	180	49	0	0	0	672
1982	0	0	0	0	18	25	186	72	25	0	3	0	329
1983	0	0	0	0	2	125	236	155	70	7	0	0	595
1984	0	0	0	0	4	88	119	154	14	1	0	0	380
1985	0	0	0	7	30	26	165	144	87	0	0	0	459
1986	0	0	5	1	52	62	201	112	28	0	0	0	461
1987	0	0	0	7	73	142	280	143	29	0	0	0	674
1988	0	0	0	0	33	112	296	251	32	9	0	0	733
1989	0	0	0	0	37	112	198	144	59	0	0	0	550
1990	0	0	5	33	2	118	222	177	51	16	0	0	624
1991	0	0	0	16	89	136	237	218	61	16	0	0	773
1992	0	0	0	0	5	21	54	94	118	60	0	0	352
1993	0	0	0	0	17	81	241	195	48	0	0	0	582
1993	0	0	0	0	17	81	241	195	48	0	0	0	582
1994	0	0	0	1	15	150	251	120	28	0	0	0	565
1995	0	0	0	0	8	178	275	226	26	2	0	0	715
1996	0	0	0	0	22	126	150	175	73	0	0	0	546
1997	0	0	0	0	0	123	167	122	15	3	0	0	430
1998	0	0	9	0	41	140	167	209	65	1	0	0	632
1999	0	0	0	0	34	190	315	141	99	0	0	0	779
2000	0	0	0	1	46	94	91	132	58	0	0	0	422
2001	0	0	0	2	22	137	158	276	51	10	1	0	657

SNOWFALL (inches) 2001 SYRACUSE, NY (SYR)

YEAR	JUL	AUG	SEP	OCT	NOV	DEC	JAN	FEB	MAR	APR	MAY	JUN	TOTAL
1972-73	0.0	0.0	0.0	0.3	15.8	29.8	11.9	13.3	3.6	5.3	1.2	0.0	81.2
1973-74	0.0	0.0	0.0	T	20.6	24.4	15.5	23.7	31.2	7.8	T	0.0	123.2
1974-75	0.0	0.0	0.0	2.8	4.8	26.2	11.8	27.3	20.6	12.0	0.0	0.0	105.5
1975-76	0.0	0.0	0.0	T	2.8	27.0	35.8	12.7	16.6	0.9	T	0.0	95.8
1976-77	0.0	0.0	0.0	0.3	25.9	25.7	52.3	24.4	13.5	1.9	1.0	0.0	145.0
1977-78	0.0	0.0	0.0	0.3	11.3	40.1	72.2	26.1	11.1	0.4	T	0.0	161.2
1978-79	0.0	0.0	0.0	T	3.9	40.9	27.9	20.7	14.9	10.2	0.0	0.0	118.5
1979-80	0.0	0.0	0.0	0.1	1.5	13.8	24.5	32.8	20.5	0.2	0.0	0.0	93.4
1980-81	0.0	0.0	0.0	T	7.3	28.8	23.4	8.5	10.6	0.4	0.0	0.0	79.0
1981-82	0.0	0.0	0.0	0.5	12.1	37.3	48.2	11.6	14.4	13.0	0.0	0.0	137.1
1982-83	0.0	0.0	0.0	T	1.9	10.9	20.3	8.2	8.3	16.4	T	0.0	66.0
1983-84	0.0	0.0	0.0	0.0	7.6	24.2	21.8	19.7	40.3	T	0.0	0.0	113.6
1984-85	0.0	0.0	0.0	0.0	5.0	23.4	57.3	21.6	7.1	2.0	0.0	0.0	116.4
1985-86	0.0	0.0	0.0	0.0	8.0	28.2	29.9	26.1	11.0	1.7	T	0.0	104.9
1986-87	0.0	0.0	0.0	0.0	16.1	8.8	49.2	15.1	3.0	1.3	0.0	0.0	93.5
1987-88	0.0	0.0	0.0	T	10.8	20.7	18.0	46.1	10.2	5.6	0.0	0.0	111.4
1988-89	0.0	0.0	0.0	5.7	0.2	34.4	19.4	21.7	9.9	6.5	0.0	0.0	97.8
1989-90	0.0	0.0	T	T	12.9	64.6	27.4	33.3	15.2	8.6	0.0	0.0	162.0
1990-91	0.0	0.0	0.0	0.2	7.8	24.5	30.9	27.7	2.8	3.0	0.0	0.0	96.9
1991-92	0.0	0.0	0.0	0.0	5.5	37.9	50.5	27.6	41.3	4.1	0.0	T	166.9
1992-93	T	0.0	T	1.4	10.1	19.8	42.9	51.3	54.4	12.2	0.0	0.0	192.1
1993-94	0.0	0.0	0.0	1.0	17.1	34.0	57.0	30.8	25.3	3.7	0.0	0.0	168.9
1994-95	0.0	0.0	0.0	0.0	3.5	5.9	13.4	32.3	7.1	0.0	0.0	0.0	62.2
1995-96	0.0	0.0	0.0	0.0	34.2	45.1	36.0	16.5	32.2	4.8	2.1	0.0	170.9
1996-97	0.0	0.0	0.0	T	25.9	21.2	38.7	19.1	23.4	2.8	0.0	0.0	131.1
1997-98	0.0	0.0	0.0	1.2	19.3	47.8	31.8	14.7	19.9	0.0	0.0	0.0	134.7
1998-99	0.0	0.0	0.0	0.0	T	13.5	50.7	5.7	28.4	0.0	0.0	0.0	98.3
1999-00	0.0	0.0	0.0	0.0	3.8	15.7	29.9	27.4	7.1	1.9	T	0.0	85.8
2000-01	0.0	0.0	0.0	T	20.2	70.3	28.4	27.8	45.0	0.2	T	0.0	191.9
2001-	0.0	0.0	0.0	0.2	0.5	7.3							
POR= 52 YRS	T	0.0	T	T	9.9	27.1	30.5	25.5	18.3	3.7	0.1	T	115.1

2001
ASHEVILLE,
NORTH CAROLINA (AVL)

The city of Asheville is located on both banks of the French Broad River, near the center of the French Broad Basin. Upstream from Asheville, the valley runs south for 18 miles and then curves toward the south-southwest. Downstream from the city, the valley is oriented toward the north-northwest. Two miles upstream from the principal section of Asheville, the Swannanoa River joins the French Broad from the east. The entire valley is known as the Asheville Plateau, having an average elevation near 2,200 feet above sea level, and is flanked by mountain ridges to the east and west, whose peaks range from 2,000 to 4,400 feet above the valley floor. At the Carolina-Tennessee border, about 25 miles north-northwest of Asheville, a relatively high ridge of mountains blocks the northern end of the valley. Thirty miles south, the Blue Ridge Mountains form an escarpment, having a general elevation of about 2,700 feet above sea level. The tallest peaks near Asheville are Mt. Mitchell, 6,684 feet above sea level, 20 miles northeast of the city, and Big Pisgah Mountain, 5,721 feet above sea level, 16 miles to the southwest.

Asheville has a temperate, but invigorating, climate. Considerable variation in temperature often occurs from day to day in summer, as well as during the other seasons.

While the office was located in the city, the combination of roof exposure conditions and a smoke blanket, caused by inversions in temperature in the valley on quiet nights, resulted in higher early morning temperatures at City Office sites than were experienced nearer ground level in nearby rural areas. The growing season in this area is of sufficient length for commercial crops, the average length of freeze-free period being about 195 days. The average last occurrence in spring of a temperature 32 degrees or lower is mid-April and the average first occurrence in fall of 32 degrees is late October.

The orientation of the French Broad Valley appears to have a pronounced influence on the wind direction. Prevailing winds are from the northwest during all months of the year. Also, the shielding effect of the nearby mountain barriers apparently has a direct bearing on the annual amount of precipitation received in this vicinity. In an area northwest of Asheville, the average annual precipitation is the lowest in North Carolina. Precipitation increases sharply in all other directions, especially to the south and southwest.

Destructive events caused directly by meteorological conditions are infrequent. The most frequent, occurring at approximately 12-year intervals, are floods on the French Broad River. These floods are usually associated with heavy rains caused by storms moving out of the Gulf of Mexico. Snowstorms which have seriously disrupted normal life in this community are infrequent. Hailstorms that cause property damage are extremely rare.

NORMALS, MEANS, AND EXTREMES
ASHEVILLE, NC (AVL)

LATITUDE:	LONGITUDE:	ELEVATION (FT):		TIME ZONE:	WBAN: 03812
35 26' 09" N	82 32' 21" W	GRND: 2171	BARO: 2174	EASTERN (UTC + 5)	

	ELEMENT	POR	JAN	FEB	MAR	APR	MAY	JUN	JUL	AUG	SEP	OCT	NOV	DEC	YEAR
TEMPERATURE F	NORMAL DAILY MAXIMUM	30	46.5	50.0	59.2	67.8	75.0	80.4	83.0	82.1	76.9	68.3	59.3	50.3	66.6
	MEAN DAILY MAXIMUM	44	47.7	51.3	58.8	68.2	75.3	81.6	84.4	83.2	77.2	68.7	58.4	50.5	67.1
	HIGHEST DAILY MAXIMUM	37	80	78	83	89	93	96	96	100	92	86	81	78	100
	YEAR OF OCCURRENCE		1999	1996	1985	1972	1996	1969	1988	1983	1998	1986	1974	1971	AUG 1983
	MEAN OF EXTREME MAXS.	44	65.5	69.5	76.5	83.1	85.9	90.0	92.0	90.6	87.0	80.9	73.7	67.6	80.2
	NORMAL DAILY MINIMUM	30	24.8	27.4	35.4	42.6	50.9	58.3	62.7	61.9	55.5	43.5	35.7	28.6	43.9
	MEAN DAILY MINIMUM	44	26.6	28.6	34.7	42.3	50.5	58.4	62.8	61.8	55.0	43.0	34.4	28.8	43.9
	LOWEST DAILY MINIMUM	37	-16	-2	2	22	28	35	44	42	30	21	8	-7	-16
	YEAR OF OCCURRENCE		1985	1967	1993	1987	1989	1966	1988	1986	1967	1976	1970	1983	JAN 1985
	MEAN OF EXTREME MINS.	44	8.1	11.5	19.1	27.5	35.1	45.5	54.2	52.0	40.6	27.7	19.1	11.9	29.4
	NORMAL DRY BULB	30	35.7	38.7	47.4	55.2	63.0	69.4	72.8	72.0	66.2	55.9	47.5	39.5	55.3
	MEAN DRY BULB	44	37.3	40.0	46.8	55.3	62.9	70.0	73.7	72.4	66.3	55.9	46.5	39.6	55.6
	MEAN WET BULB	15	32.9	36.1	41.0	48.1	57.0	64.4	67.7	66.8	60.8	50.6	42.4	32.6	50.0
	MEAN DEW POINT	15	27.0	29.5	33.9	41.0	52.8	61.4	65.0	64.5	58.0	46.0	36.9	27.1	45.3
	NORMAL NO. DAYS WITH:														
	MAXIMUM 90	30	0.0	0.0	0.0	0.0	0.1	2.1	4.4	2.5	0.3	0.0	0.0	0.0	9.4
	MAXIMUM 32	30	3.4	1.7	0.2	0.0	0.0	0.0	0.0	0.0	0.0	0.0	0.1	1.0	6.4
	MINIMUM 32	30	23.9	20.8	13.3	4.3	0.3	0.0	0.0	0.0	*	4.4	13.0	20.7	100.7
	MINIMUM 0	30	0.5	*	0.0	0.0	0.0	0.0	0.0	0.0	0.0	0.0	0.0	0.1	0.6
H/C	NORMAL HEATING DEG. DAYS	30	908	736	546	298	117	18	5	5	64	295	525	791	4308
	NORMAL COOLING DEG. DAYS	30	0	0	0	0	55	150	247	222	100	13	0	0	787
RH	NORMAL (PERCENT)	30	73	70	68	66	75	79	82	84	84	78	75	74	76
	HOUR 01 LST	30	80	78	79	78	89	93	95	96	96	91	86	82	87
	HOUR 07 LST	30	85	84	85	85	92	94	96	98	97	94	88	86	90
	HOUR 13 LST	30	59	56	53	50	56	59	63	64	64	57	57	59	58
	HOUR 19 LST	30	67	62	60	56	66	70	74	78	81	74	70	70	69
S	PERCENT POSSIBLE SUNSHINE	32	55	59	61	66	61	62	60	54	56	62	58	55	59
W/O	MEAN NO. DAYS WITH:														
	HEAVY FOG(VISBY 1/4 MI)	38	4.1	2.9	2.1	2.5	5.2	8.3	8.8	12.3	11.2	7.9	4.2	4.5	74.0
	THUNDERSTORMS	38	0.5	1.0	2.2	3.2	6.4	7.9	8.8	7.6	3.1	0.8	0.8	0.3	42.6
CLOUDINESS	MEAN:														
	SUNRISE-SUNSET (OKTAS)	32	5.0	4.8	4.9	4.6	5.0	5.0	5.0	5.0	4.8	4.0	4.3	4.7	4.8
	MIDNIGHT-MIDNIGHT (OKTAS)	32	4.9	4.7	4.7	4.4	4.8	4.7	4.8	5.0	4.7	4.0	4.1	4.5	4.6
	MEAN NO. DAYS WITH:														
	CLEAR	33	9.1	8.7	8.8	9.6	7.2	6.2	5.0	4.8	6.7	11.6	10.2	9.3	97.2
	PARTLY CLOUDY	33	7.3	6.4	8.1	8.5	10.4	12.0	13.4	13.1	10.2	7.8	6.9	7.0	111.1
	CLOUDY	33	14.6	13.2	14.1	11.8	13.3	11.8	11.7	12.2	12.2	11.7	11.9	13.7	151.2
PR	MEAN STATION PRESSURE(IN)	28	27.80	27.79	27.80	27.80	27.80	27.80	27.90	27.90	27.90	27.90	27.89	27.80	27.84
	MEAN SEA-LEVEL PRES. (IN)	16	30.12	30.10	30.05	30.02	30.03	30.03	30.07	30.07	30.10	30.14	30.15	30.15	30.09
WINDS	MEAN SPEED (MPH)	37	9.3	9.2	9.2	8.7	6.9	5.9	5.9	5.5	5.6	6.8	8.1	8.8	7.5
	PREVAIL.DIR(TENS OF DEGS)	30	34	34	34	34	34	34	34	34	34	34	34	34	34
	MAXIMUM 2-MINUTE:														
	SPEED (MPH)	5	45	43	47	38	33	32	41	41	45	33	36	33	47
	DIR. (TENS OF DEGS)		34	33	32	32	32	32	33	30	36	32	32	34	32
	YEAR OF OCCURRENCE		2000	2001	2001	1997	2000	1998	2001	2000	1999	1997	1999	2001	MAR 2001
	MAXIMUM 5-SECOND:														
	SPEED (MPH)	5	53	49	56	47	45	46	49	53	54	39	43	40	56
	DIR. (TENS OF DEGS)		31	32	31	01	27	10	31	30	36	33	36	33	31
	YEAR OF OCCURRENCE		2000	2001	2001	2001	2000	2000	2001	2000	1999	2001	1999	2001	MAR 2001
PRECIPITATION	NORMAL (IN)	30	3.25	3.91	4.63	3.36	4.43	4.23	4.52	4.69	3.87	3.59	3.59	3.52	47.59
	MAXIMUM MONTHLY (IN)	37	9.96	8.07	9.86	8.70	8.83	10.73	9.92	11.28	9.12	8.82	7.76	8.48	11.28
	YEAR OF OCCURRENCE		1998	1990	1975	1998	1973	1989	1982	1967	1977	1990	1979	1973	AUG 1967
	MINIMUM MONTHLY (IN)	37	0.45	0.44	0.77	0.25	1.06	0.90	0.46	0.52	0.16	0.00	1.19	0.16	0.00
	YEAR OF OCCURRENCE		1981	1978	1985	1976	1988	1990	1986	1981	1984	2000	1981	1965	OCT 2000
	MAXIMUM IN 24 HOURS (IN)	37	4.67	3.47	5.13	3.06	4.95	4.36	4.02	5.10	3.41	4.22	4.03	2.66	5.13
	YEAR OF OCCURRENCE		1998	1982	1968	1973	1973	1997	1969	1990	1975	1995	1977	1973	MAR 1968
	NORMAL NO. DAYS WITH:														
	PRECIPITATION 0.01	30	10.1	9.5	11.3	9.5	11.8	10.8	12.1	12.3	9.3	8.2	9.2	9.8	123.9
	PRECIPITATION 1.00	30	0.6	0.8	1.2	1.0	1.2	0.9	1.1	1.0	1.0	0.9	1.1	1.0	11.8
SNOWFALL	NORMAL (IN)	30	5.0	5.3	2.5	0.7	0.0	0.0	0.0	0.0	0.0	T	0.8	1.9	16.2
	MAXIMUM MONTHLY (IN)	37	17.6	25.5	18.2	11.5	T	T	T	T	0.0	T	9.6	16.3	25.5
	YEAR OF OCCURRENCE		1966	1969	1993	1987	1993	1995	1994	1990		1993	1968	1971	FEB 1969
	MAXIMUM IN 24 HOURS (IN)	37	14.0	11.7	16.5	11.5	T	T	T	T	0.0	T	5.7	16.3	16.5
	YEAR OF OCCURRENCE		1988	1969	1993	1987	1993	1995	1994	1990		1993	1968	1971	MAR 1993
	MAXIMUM SNOW DEPTH (IN)	42	14	13	18	12	0	0	0	0	0	0	5	14	18
	YEAR OF OCCURRENCE		1988	1969	1993	1987							1968	1971	MAR 1993
	NORMAL NO. DAYS WITH:														
	SNOWFALL 1.0	30	1.4	1.5	0.7	0.2	0.0	0.0	0.0	0.0	0.0	0.0	0.2	0.5	4.5

PRECIPITATION (inches) 2001 ASHEVILLE, NC (AVL)

YEAR	JAN	FEB	MAR	APR	MAY	JUN	JUL	AUG	SEP	OCT	NOV	DEC	ANNUAL
1972	3.57	2.02	3.19	1.49	6.63	6.54	4.66	1.88	5.29	4.44	4.42	3.89	48.02
1973	4.26	4.23	8.91	5.71	8.83	3.87	6.95	4.57	3.12	2.41	3.57	8.48	64.91
1974	3.44	4.24	3.18	4.99	5.58	3.73	3.93	7.34	4.13	1.28	4.22	2.38	48.44
1975	3.86	4.56	9.86	0.61	8.17	2.12	3.31	3.63	7.53	3.94	4.89	4.44	56.92
1976	3.51	2.20	4.96	0.25	8.67	5.51	3.18	4.23	3.50	5.59	1.58	4.05	47.23
1977	2.09	1.02	7.29	4.05	3.96	5.11	1.03	3.68	9.12	3.79	6.88	2.43	50.45
1978	7.47	0.44	5.22	2.97	4.65	2.29	0.63	6.91	2.57	0.30	2.49	4.32	40.26
1979	6.81	5.14	5.72	7.26	5.35	2.20	5.52	3.63	5.60	1.40	7.76	1.05	57.44
1980	2.85	0.53	8.26	4.77	4.54	4.68	2.21	2.38	4.36	2.62	3.04	0.59	40.83
1981	0.45	4.80	3.24	2.07	7.50	4.41	2.06	0.52	1.36	2.19	1.19	4.79	34.58
1982	5.41	7.02	1.92	3.62	3.78	3.98	9.92	1.73	1.33	3.48	4.59	4.04	50.82
1983	3.39	5.63	6.27	5.27	3.48	3.71	1.06	0.95	5.66	4.43	4.77	8.30	52.92
1984	2.36	6.43	4.82	4.05	6.62	3.69	5.88	5.02	0.16	2.73	2.61	1.34	45.71
1985	2.95	4.74	0.77	2.74	1.59	1.47	4.37	7.04	1.25	3.41	4.91	0.70	35.94
1986	1.11	1.85	2.75	0.57	3.55	1.28	0.46	6.10	3.15	4.19	5.28	4.28	34.57
1987	3.49	6.17	2.85	3.67	1.87	8.94	1.86	1.79	6.79	0.36	3.09	2.33	43.21
1988	3.71	0.88	1.31	3.46	1.06	0.94	2.65	1.78	2.79	3.12	3.47	1.41	26.58
1989	1.65	4.61	2.91	3.17	5.54	10.73	8.33	4.98	8.17	2.98	4.27	3.29	60.63
1990	3.27	8.07	5.95	1.96	5.09	0.90	6.55	7.78	1.43	8.82	1.55	4.50	55.87
1991	3.25	1.66	6.13	5.38	2.41	5.27	6.07	3.83	1.27	0.19	3.34	4.86	43.66
1992	3.08	3.66	3.52	3.99	6.18	6.62	1.10	7.64	3.15	4.15	7.24	3.71	54.04
1993	3.82	2.03	6.16	3.21	4.59	1.12	2.07	5.29	1.56	1.21	3.32	3.59	37.97
1994	5.35	5.11	7.52	3.30	1.74	5.89	6.76	6.01	5.33	4.27	3.15	3.03	57.46
1995	7.03	2.93	2.42	0.98	6.04	8.89	3.61	9.22	1.95	7.23	3.66	1.43	55.39
1996	7.22	2.71	3.36	2.00	2.55	3.54	4.83	6.68	5.22	0.68	4.45	3.92	47.16
1997	4.44	5.29	5.48	5.26	2.91	8.29	2.97	1.37	4.89	3.90	1.60	2.98	49.38
1998	9.96	6.38	3.71	8.70	2.22	3.64	1.97	2.23	1.62	1.79	2.76	3.04	48.02
1999	6.38	3.29	2.82	2.44	2.53	4.39	3.85	3.37	2.20	3.29	3.31	1.98	39.85
2000	3.10	2.33	3.82	5.11	1.27	2.78	2.84	4.45	3.27	0.00	4.25	2.37	35.59
2001	2.63	2.73	5.00	1.32	2.47	2.91	5.50	3.20	4.37	0.60	1.42	2.34	34.49
POR= 36 YRS	3.70	3.78	4.56	3.40	4.31	4.25	4.21	4.56	3.65	3.20	3.52	3.35	46.49

AVERAGE TEMPERATURE (F) 2001 ASHEVILLE, NC (AVL)

YEAR	JAN	FEB	MAR	APR	MAY	JUN	JUL	AUG	SEP	OCT	NOV	DEC	ANNUAL
1972	42.1	37.6	46.5	55.8	61.2	66.5	72.3	72.5	69.0	55.0	45.5	45.2	55.8
1973	37.5	38.5	52.7	52.8	60.3	71.0	74.1	74.2	70.0	58.8	49.1	39.8	56.6
1974	48.2	40.5	51.1	54.9	64.2	66.7	72.9	72.3	65.7	54.6	47.4	40.3	56.6
1975	41.7	42.5	44.8	54.1	66.0	68.8	72.4	72.9	65.2	57.3	48.2	38.6	56.0
1976	33.6	45.3	50.6	54.9	59.5	68.1	71.2	70.2	63.1	51.5	41.2	36.2	53.8
1977	24.8	37.4	50.7	58.2	64.6	69.7	75.7	73.8	69.1	54.3	49.3	36.8	55.4
1978	29.3	33.4	45.9	56.8	62.0	71.1	73.4	74.1	70.0	55.7	51.8	41.0	55.4
1979	34.2	35.8	50.0	55.9	64.3	68.8	72.2	73.2	66.5	55.3	49.2	42.0	55.6
1980	40.5	35.1	46.2	56.5	64.8	71.7	77.5	74.8	70.2	54.7	47.1	39.6	56.6
1981	33.3	39.9	44.9	60.1	60.7	74.3	75.0	71.7	66.1	54.5	48.2	35.8	55.4
1982	32.3	41.2	50.0	53.6	67.3	71.5	74.6	71.7	64.5	56.3	47.1	44.9	56.3
1983	36.7	38.8	46.7	51.1	61.6	69.0	75.7	76.5	66.6	57.5	47.3	36.4	55.3
1984	34.0	40.5	44.8	51.7	59.9	70.0	70.6	71.6	62.8	62.7	43.1	46.3	54.8
1985	30.5	38.3	48.1	56.6	62.6	69.8	72.2	70.9	64.2	60.5	56.0	34.7	55.4
1986	35.0	42.2	46.0	56.0	63.3	71.7	76.1	70.9	68.0	57.4	50.7	39.8	56.4
1987	35.3	38.9	46.5	52.6	66.7	71.2	74.7	74.7	66.5	50.1	47.0	42.2	55.5
1988	32.1	37.1	47.1	54.6	61.1	69.3	73.8	74.6	66.3	50.2	46.7	38.7	54.3
1989	42.1	39.8	50.3	54.5	59.4	70.0	73.3	71.9	65.8	56.1	46.2	31.6	55.1
1990	42.8	45.6	50.4	54.2	63.3	70.9	73.9	73.9	67.6	57.8	49.9	45.5	58.0
1991	39.2	42.5	49.6	58.4	67.3	70.3	75.2	72.1	67.3	57.0	45.5	43.4	57.3
1992	40.7	44.2	46.3	55.9	59.9	68.1	74.9	70.4	67.8	54.4	46.2	39.2	55.7
1993	42.0	38.0	43.0	52.6	63.2	71.4	78.0	73.9	67.8	55.0	46.0	36.8	55.6
1994	32.4	40.6	47.3	58.5	60.2	72.6	73.1	71.4	64.6	55.9	50.1	43.6	55.9
1995	38.9	38.4	49.4	56.4	64.1	68.5	75.0	75.3	65.4	56.8	43.3	37.1	55.7
1996	35.3	39.5	41.4	53.6	66.1	69.6	72.0	71.5	64.4	55.8	42.3	40.5	54.3
1997	36.5	43.2	51.5	51.5	58.4	66.1	73.4	70.6	66.1	55.4	42.3	36.9	54.4
1998	41.3	42.2	43.9	54.1	66.5	71.7	74.9	73.4	70.3	58.3	49.2	43.9	57.5
1999	41.7	42.4	42.9	58.5	61.9	69.3	74.3	74.1	65.6	56.0	51.3	41.9	56.7
2000	36.0	43.3	49.9	53.0	66.2	70.6	72.8	71.7	65.2	57.2	44.6	32.5	55.3
2001	36.7	44.5	44.6	58.2	64.0	70.5	72.9	73.8	64.5	53.9	51.7	44.8	56.7
POR= 37 YRS	36.6	39.6	46.9	55.3	62.9	69.9	73.6	72.5	66.5	55.9	47.0	39.9	55.6

REFERENCE NOTES:

PAGE 1:
THE TEMPERATURE GRAPH SHOWS NORMAL MAXIMUM AND NORMAL MINIMUM DAILY TEMPERATURES (SOLID CURVES) AND THE ACTUAL DAILY HIGH AND LOW TEMPERATURES (VERTICAL BARS).

PAGE 2 AND 3:
H/C INDICATES HEATING AND COOLING DEGREE DAYS.
RH INDICATES RELATIVE HUMIDITY
W/O INDICATES WEATHER AND OBSTRUCTIONS
S INDICATES SUNSHINE.
PR INDICATES PRESSURE.
CLOUDINESS ON PAGE 3 IS THE SUM OF THE CEILOMETER AND SATELLITE DATA NOT TO EXCEED EIGHT EIGHTHS(OKTAS).

GENERAL:
T INDICATES TRACE PRECIPITATION, AN AMOUNT GREATER THAN ZERO BUT LESS THAN THE LOWEST REPORTABLE VALUE.
+ INDICATES THE VALUE ALSO OCCURS ON EARLIER DATES.
BLANK ENTRIES DENOTE MISSING OR UNREPORTED DATA.
NORMALS ARE 30-YEAR AVERAGES (1961 - 1990).
ASOS INDICATES AUTOMATED SURFACE OBSERVING SYSTEM.
PM INDICATES THE LAST DAY OF THE PREVIOUS MONTH.
POR (PERIOD OF RECORD) BEGINS WITH THE JANUARY DATA MONTH AND IS THE NUMBER OF YEARS USED TO COMPUTE THE MEAN. INDIVIDUAL MONTHS WITHIN THE POR MAY BE MISSING.
WHEN THE POR FOR A NORMAL IS LESS THAN 30 YEARS, THE NORMAL IS PROVISIONAL AND IS BASED ON THE NUMBER OF YEARS INDICATED.
0.* OR * INDICATES THE VALUE OR MEAN-DAYS-WITH IS BETWEEN 0.00 AND 0.05.
CLOUDINESS FOR ASOS STATIONS DIFFERS FROM THE NON-ASOS OBSERVATION TAKEN BY A HUMAN OBSERVER. ASOS STATION CLOUDINESS IS BASED ON TIME-AVERAGED CEILOMETER DATA FOR CLOUDS AT OR BELOW 12,000 FEET AND ON SATELLITE DATA FOR CLOUDS ABOVE 12,000 FEET.
THE NUMBER OF DAYS WITH CLEAR, PARTLY CLOUDY, AND CLOUDY CONDITIONS FOR ASOS STATIONS IS THE SUM OF THE CEILOMETER AND SATELLITE DATA FOR THE SUNRISE TO SUNSET PERIOD.

GENERAL CONTINUED:
CLEAR INDICATES 0 - 2 OKTAS, PARTLY CLOUDY INDICATES 3 - 6 OKTAS, AND CLOUDY INDICATES 7 OR 8 OKTAS. WHEN AT LEAST ONE OF THE ELEMENTS (CEILOMETER OR SATELLITE) IS MISSING, THE DAILY CLOUDINESS IS NOT COMPUTED.
WIND DIRECTION IS RECORDED IN TENS OF DEGREES (2 DIGITS) CLOCKWISE FROM TRUE NORTH. "00" INDICATES CALM. "36" INDICATES TRUE NORTH.
RESULTANT WIND IS THE VECTOR AVERAGE OF THE SPEED AND DIRECTION.
AVERAGE TEMPERATURE IS THE SUM OF THE MEAN DAILY MAXIMUM AND MINIMUM TEMPERATURE DIVIDED BY 2.
SNOWFALL DATA COMPRISE ALL FORMS OF FROZEN PRECIPITATION, INCLUDING HAIL.
A HEATING (COOLING) DEGREE DAY IS THE DIFFERENCE BETWEEN THE AVERAGE DAILY TEMPERATURE AND 65 F.
DRY BULB IS THE TEMPERATURE OF THE AMBIENT AIR.
DEW POINT IS THE TEMPERATURE TO WHICH THE AIR MUST BE COOLED TO ACHIEVE 100 PERCENT RELATIVE HUMIDITY.
WET BULB IS THE TEMPERATURE THE AIR WOULD HAVE IF THE MOISTURE CONTENT WAS INCREASED TO 100 PERCENT RELATIVE HUMIDITY.

ON JULY 1, 1996, THE NATIONAL WEATHER SERVICE BEGAN USING THE "METAR" OBSERVATION CODE THAT WAS ALREADY EMPLOYED BY MOST OTHER NATIONS OF THE WORLD. THE MOST NOTICEABLE DIFFERENCE IN THIS ANNUAL PUBLICATION WILL BE THE CHANGE IN UNITS FROM TENTHS TO EIGHTS(OKTAS) FOR REPORTING THE AMOUNT OF SKY COVER.

HEATING DEGREE DAYS (base 65 F) 2001 ASHEVILLE, NC (AVL)

YEAR	JUL	AUG	SEP	OCT	NOV	DEC	JAN	FEB	MAR	APR	MAY	JUN	TOTAL
1972-73	3	0	8	304	578	605	846	737	374	362	158	0	3975
1973-74	0	0	7	205	473	772	516	680	423	299	83	24	3482
1974-75	0	0	65	316	519	760	715	624	619	331	46	7	4002
1975-76	0	0	77	232	498	812	966	566	439	296	168	33	4087
1976-77	2	3	83	411	706	884	1239	768	437	198	66	25	4822
1977-78	0	0	14	331	466	868	1101	878	586	241	139	0	4624
1978-79	0	0	12	283	390	741	951	810	457	268	71	18	4001
1979-80	5	0	44	299	468	707	753	861	573	258	65	2	4035
1980-81	0	0	37	315	533	778	778	696	615	152	152	0	4256
1981-82	0	1	57	326	499	897	1006	659	458	333	38	0	4274
1982-83	0	0	74	274	531	616	872	725	562	410	127	13	4204
1983-84	0	0	84	229	527	882	955	706	618	391	176	9	4577
1984-85	1	0	107	91	648	576	1064	737	520	249	109	19	4121
1985-86	0	6	111	156	266	932	923	633	581	273	91	2	3974
1986-87	0	32	16	268	419	774	913	725	567	369	40	1	4124
1987-88	0	0	47	452	532	702	1013	802	545	308	132	31	4564
1988-89	5	0	33	453	544	808	702	698	454	331	200	4	4232
1989-90	1	8	74	279	558	1028	679	535	446	321	91	3	4023
1990-91	0	0	55	229	445	601	793	627	472	204	53	16	3495
1991-92	0	1	64	242	578	663	748	596	573	274	169	25	3933
1992-93	0	0	34	324	558	794	708	751	677	365	82	9	4302
1993-94	0	0	44	310	563	866	1005	676	544	206	165	0	4379
1994-95	0	1	52	276	441	656	803	736	474	264	87	17	3807
1995-96	0	0	50	255	645	856	912	735	725	341	78	20	4617
1996-97	0	0	76	284	673	753	876	601	412	400	212	64	4351
1997-98	0	5	35	292	675	863	727	631	647	322	71	23	4291
1998-99	0	0	11	214	465	649	718	627	680	210	100	7	3681
1999-00	6	0	56	274	405	705	893	621	460	353	41	11	3825
2000-01	1	1	88	238	607	998	872	566	628	222	63	1	4285
2001-	0	0	97	340	391	620							

COOLING DEGREE DAYS (base 65 F) 2001 ASHEVILLE, NC (AVL)

YEAR	JAN	FEB	MAR	APR	MAY	JUN	JUL	AUG	SEP	OCT	NOV	DEC	ANNUAL
1972	0	0	0	24	6	84	236	237	134	0	1	0	722
1973	0	0	0	1	16	190	288	292	163	19	0	0	969
1974	0	0	0	3	65	82	254	234	92	1	0	0	731
1975	0	0	0	11	82	124	237	252	89	0	0	0	795
1976	0	0	0	0	5	135	198	170	35	2	0	0	545
1977	0	0	0	2	59	173	340	279	146	7	1	0	1007
1978	0	0	0	2	53	188	266	292	168	4	0	0	973
1979	0	0	0	1	55	141	234	261	96	4	0	0	792
1980	0	0	0	8	64	210	396	311	198	4	0	0	1191
1981	0	0	0	10	25	286	316	213	98	7	0	0	955
1982	0	0	0	0	117	206	305	215	64	16	0	0	923
1983	0	0	0	0	25	139	335	362	141	5	0	0	1007
1984	0	0	0	0	25	165	180	211	49	27	0	0	657
1985	0	0	5	2	43	170	229	194	90	25	4	0	762
1986	0	0	0	8	43	209	353	222	112	38	0	0	985
1987	0	0	0	7	97	192	310	309	97	0	0	0	1012
1988	0	0	0	0	18	168	282	304	79	3	0	0	854
1989	0	0	5	23	34	159	264	229	107	11	0	0	832
1990	0	0	0	3	48	187	279	283	141	11	0	0	952
1991	0	0	0	3	13	132	181	324	227	139	3	0	1022
1992	0	0	0	7	20	125	313	174	125	0	0	0	764
1993	0	0	0	0	37	210	411	285	133	5	0	0	1081
1994	0	0	0	15	21	233	259	205	43	2	0	0	778
1995	0	0	1	12	65	127	315	325	68	8	0	0	921
1996	0	0	0	6	120	164	224	209	64	5	0	0	792
1997	0	0	0	0	13	116	265	184	76	1	0	0	655
1998	0	0	0	1	124	230	315	269	178	14	0	0	1131
1999	0	0	0	23	9	143	304	291	78	3	0	0	851
2000	0	0	0	0	83	187	251	215	103	4	1	0	844
2001	0	0	0	24	42	172	249	282	87	1	0	0	857

SNOWFALL (inches) 2001 ASHEVILLE, NC (AVL)

YEAR	JUL	AUG	SEP	OCT	NOV	DEC	JAN	FEB	MAR	APR	MAY	JUN	TOTAL
1972-73	0.0	0.0	0.0	0.0	0.6	T	7.1	0.5	1.0	T	0.0	0.0	9.2
1973-74	0.0	0.0	0.0	0.0	0.0	3.0	T	0.3	1.1	T	0.0	0.0	4.4
1974-75	0.0	0.0	0.0	0.0	3.1	3.0	0.4	4.3	3.7	T	0.0	0.0	14.5
1975-76	0.0	0.0	0.0	0.0	5.0	0.4	1.6	3.5	T	0.0	0.0	0.0	10.5
1976-77	0.0	0.0	0.0	0.0	0.1	0.3	11.9	0.7	0.0	0.0	0.0	0.0	13.0
1977-78	0.0	0.0	0.0	T	T	1.5	9.7	5.3	5.3	T	0.0	0.0	21.8
1978-79	0.0	0.0	0.0	0.0	0.0	0.0	5.2	17.8	T	0.0	0.0	0.0	23.0
1979-80	0.0	0.0	0.0	0.0	T	T	2.1	6.3	5.4	T	0.0	0.0	13.8
1980-81	0.0	0.0	0.0	0.0	T	T	4.7	T	9.9	0.0	0.0	0.0	14.6
1981-82	0.0	0.0	0.0	0.0	T	2.0	8.6	8.1	0.1	3.0	0.0	0.0	21.8
1982-83	0.0	0.0	0.0	0.0	0.0	0.4	10.5	9.3	4.5	2.0	0.0	0.0	26.7
1983-84	0.0	0.0	0.0	0.0	T	T	0.2	2.9	T	0.0	0.0	0.0	3.1
1984-85	0.0	0.0	0.0	0.0	0.0	0.0	4.5	3.1	0.1	0.4	0.0	0.0	8.1
1985-86	0.0	0.0	0.0	0.0	0.0	0.4	0.8	3.7	0.1	T	0.0	0.0	5.0
1986-87	0.0	0.0	0.0	0.0	T	T	15.0	2.7	0.3	11.5	0.0	0.0	29.5
1987-88	0.0	0.0	0.0	0.0	0.3	0.5	14.2	T	T	1.2	0.0	0.0	16.2
1988-89	0.0	0.0	0.0	0.0	0.0	T	1.2	6.0	T	1.0	0.0	0.0	8.2
1989-90	0.0	0.0	0.0	T	T	3.0	T	T	T	0.0	0.0	0.0	3.0
1990-91	0.0	T	0.0	0.0	0.0	T	T	0.4	3.1	0.0	0.0	T	3.5
1991-92	0.0	0.0	0.0	0.0	1.0	0.0	T	0.5	T	0.0	T	0.0	1.5
1992-93	0.0	0.0	0.0	0.0	T	T	T	3.0	18.2	T	T	0.0	21.2
1993-94	0.0	0.0	0.0	T	T	8.0	2.6	0.3	0.1	T	0.0	0.0	11.0
1994-95	T	0.0	0.0	0.0	0.0	0.0	T	0.2	3.2	T	0.0	T	3.4
1995-96	0.0	0.0	0.0	0.0	T	0.1	15.7	4.4	1.5	T	0.0	T	21.7
1996-97	0.0	0.0	0.0	0.0	0.1	2.5	3.8	1.9	T	T	0.0	0.0	8.3
1997-98	0.0	0.0	0.0	0.0	T	4.5	12.7	T	0.4	0.0	0.0	0.0	17.6
1998-99	0.0	0.0	0.0	0.0	0.0	0.0	0.4	1.6	T	4.2	0.0	0.0	6.2
1999-00	0.0	0.0	0.0	0.0	0.1	T	6.2	T	T	0.1	0.0	0.0	6.4
2000-01	0.0	0.0	0.0	0.0	2.0	6.1	T	0.8	6.4	0.2	0.0	0.0	15.5
2001-	0.0	0.0	0.0	0.0	0.0	0.0							
POR= 36 YRS	T	T	0.0	T	0.7	1.9	4.7	3.8	2.5	0.6	T	T	14.2

2001
RALEIGH,
NORTH CAROLINA (RDU)

The Raleigh-Durham Airport is located in the zone of transition between the Coastal Plain and the Piedmont Plateau. The surrounding terrain is rolling, with an average elevation of around 400 feet, the range over a 10-mile radius is roughly between 200 and 550 feet. Being centrally located between the mountains on the west and the coast on the south and east, the Raleigh-Durham area enjoys a favorable climate. The mountains form a partial barrier to cold air masses moving eastward from the interior of the nation. As a result, there are few days in the heart of the winter season when the temperature falls below 20 degrees. Tropical air is present over the eastern and central sections of North Carolina during much of the summer season, bringing warm temperatures and rather high humidities to the Raleigh-Durham area. Afternoon temperatures reach 90 degrees or higher on about one-fourth of the days in the middle of summer, but reach 100 degrees less than once per year. Even in the hottest weather, early morning temperatures almost always drop into the lower 70s.

Rainfall is well distributed throughout the year as a whole. July and August have the greatest amount of rainfall, and October and November the least. There are times in spring and summer when soil moisture is scanty. This usually results from too many days between rains rather than from a shortage of total rainfall, but occasionally the accumulated total during the growing season falls short of plant needs. Most summer rain is produced by thunderstorms, which may occasionally be accompanied by strong winds, intense rains, and hail. The Raleigh-Durham area is far enough from the coast so that the bad weather effects of coastal storms are reduced. While snow and sleet usually occur each year, excessive accumulations of snow are rare.

From September 1887 to December 1950, the office was located in the downtown areas of Raleigh. The various buildings occupied were within an area of three blocks. All thermometers were exposed on the roof, and this, plus the smoke over the city, had an effect on the temperature record of that period. Lowest temperatures at the city office were frequently from 2 to 5 degrees higher than those recorded in surrounding rural areas. Maximum temperatures in the city were generally a degree or two lower. These observations are supported by a period of simultaneous record from the Municipal Airport and the city office location between 1937 and 1940.

From September 1946 to May 1954, simultaneous records were kept at a surface location on the North Carolina State College campus in Raleigh, and at the Raleigh-Durham Airport 10 1/2 air miles to the northwest.

Based on the 1951-1980 period, the average first occurrence of 32 degrees Fahrenheit in the fall is October 27 and the average last occurrence in the spring is April 11.

NORMALS, MEANS, AND EXTREMES
RALEIGH, NC (RDU)

LATITUDE:	LONGITUDE:	ELEVATION (FT):	TIME ZONE:	WBAN: 13722
35 52' 14" N	78 47' 11" W	GRND: 427 BARO: 430	EASTERN (UTC + 5)	

	ELEMENT	POR	JAN	FEB	MAR	APR	MAY	JUN	JUL	AUG	SEP	OCT	NOV	DEC	YEAR
TEMPERATURE F	NORMAL DAILY MAXIMUM	30	48.9	52.6	62.1	71.7	78.6	85.0	88.0	86.8	81.1	71.6	62.6	52.7	70.1
	MEAN DAILY MAXIMUM	54	50.5	53.9	61.7	71.9	79.0	85.6	88.7	87.4	81.3	71.9	62.4	53.1	70.6
	HIGHEST DAILY MAXIMUM	57	79	84	92	95	97	104	105	105	104	98	88	80	105
	YEAR OF OCCURRENCE		1952	1977	1945	1980	1953	1954	1952	1988	1954	1954	1950	1998	AUG 1988
	MEAN OF EXTREME MAXS.	54	70.7	73.6	81.0	87.5	90.4	95.6	96.5	95.5	91.8	85.3	78.5	72.1	84.9
	NORMAL DAILY MINIMUM	30	28.8	31.3	38.7	46.2	55.3	63.6	68.1	67.5	61.1	48.4	39.7	32.4	48.4
	MEAN DAILY MINIMUM	54	29.8	31.8	38.1	46.6	55.4	63.6	68.0	66.8	60.4	48.1	38.8	32.0	48.3
	LOWEST DAILY MINIMUM	57	-9	0	11	23	31	38	48	46	37	19	11	4	-9
	YEAR OF OCCURRENCE		1985	1996	1980	1985	1977	1977	1975	1965	1983	1962	1970	1983	JAN 1985
	MEAN OF EXTREME MINS.	54	11.7	15.1	22.3	30.4	40.5	50.8	57.1	55.5	45.5	31.5	22.5	14.7	33.1
	NORMAL DRY BULB	30	38.9	42.0	50.4	59.0	67.0	74.3	78.1	77.1	71.1	60.1	51.2	42.6	59.3
	MEAN DRY BULB	54	40.1	43.0	50.0	59.2	67.2	74.5	78.5	77.1	71.0	60.0	50.6	42.5	59.5
	MEAN WET BULB	18	36.3	39.4	44.4	52.1	60.6	68.2	68.1	66.7	61.2	54.9	46.4	36.3	52.9
	MEAN DEW POINT	18	28.9	31.5	36.3	45.0	52.5	64.6	65.2	64.0	58.1	50.3	40.3	29.6	47.2
	NORMAL NO. DAYS WITH:														
	MAXIMUM 90	30	0.0	0.0	*	0.6	1.5	6.9	11.9	10.1	3.0	0.2	0.0	0.0	34.2
	MAXIMUM 32	30	2.4	0.6	0.1	0.0	0.0	0.0	0.0	0.0	0.0	0.0	*	0.8	3.9
	MINIMUM 32	30	20.8	17.4	9.5	2.3	0.1	0.0	0.0	0.0	0.0	1.6	9.1	17.8	78.6
	MINIMUM 0	30	0.2	0.0	0.0	0.0	0.0	0.0	0.0	0.0	0.0	0.0	0.0	0.0	0.2
H/C	NORMAL HEATING DEG. DAYS	30	809	644	458	193	47	0	0	0	9	189	414	694	3457
	NORMAL COOLING DEG. DAYS	30	0	0	6	13	109	279	406	375	192	37	0	0	1417
RH	NORMAL (PERCENT)	30	66	64	63	62	71	74	76	78	77	73	69	68	70
	HOUR 01 LST	30	73	71	71	73	84	87	88	90	88	85	78	75	80
	HOUR 07 LST	30	79	78	80	80	85	87	89	92	92	89	84	80	85
	HOUR 13 LST	30	54	52	49	45	53	56	58	59	58	52	52	55	54
	HOUR 19 LST	30	63	59	56	53	65	67	70	74	76	74	67	66	66
S	PERCENT POSSIBLE SUNSHINE	42	52	56	60	63	59	60	60	58	58	60	57	53	58
W/O	MEAN NO. DAYS WITH:														
	HEAVY FOG(VISBY 1/4 MI)	52	3.5	2.8	2.1	1.4	2.3	1.9	2.6	3.2	3.3	3.3	3.2	3.4	33.0
	THUNDERSTORMS	57	0.4	0.9	2.1	3.5	6.2	7.2	10.4	7.7	3.3	1.4	0.8	0.3	44.2
CLOUDINESS	MEAN:														
	SUNRISE-SUNSET (OKTAS)	47	5.0	4.7	4.7	4.4	4.7	4.6	4.8	4.7	4.7	3.9	4.2	4.6	4.6
	MIDNIGHT-MIDNIGHT (OKTAS)	32	4.7	4.4	4.3	4.1	4.6	4.4	4.7	4.6	4.2	3.7	4.0	4.3	4.3
	MEAN NO. DAYS WITH:														
	CLEAR	48	9.0	8.7	9.4	9.6	8.2	7.6	7.2	7.3	9.5	12.8	11.3	10.0	110.6
	PARTLY CLOUDY	48	6.7	6.1	7.3	9.1	10.0	11.8	12.2	12.3	9.1	7.1	7.3	7.0	106.0
	CLOUDY	48	15.2	13.5	14.3	11.3	12.8	10.7	11.6	11.3	11.4	11.1	11.4	14.0	148.6
PR	MEAN STATION PRESSURE(IN)	29	29.64	29.61	29.57	29.54	29.54	29.55	29.57	29.60	29.62	29.65	29.66	29.67	29.60
	MEAN SEA-LEVEL PRES. (IN)	18	30.13	30.12	30.05	30.00	30.01	30.00	30.04	30.04	30.07	30.12	30.15	30.17	30.08
WINDS	MEAN SPEED (MPH)	47	8.1	8.4	9.1	8.7	7.5	6.9	6.6	6.2	6.6	6.5	7.0	7.4	7.4
	PREVAIL.DIR(TENS OF DEGS)	32	23	23	22	22	23	22	23	22	04	04	23	23	23
	MAXIMUM 2-MINUTE:														
	SPEED (MPH)	5	38	45	46	33	33	30	37	36	40	23	32	40	46
	DIR. (TENS OF DEGS)		28	24	24	23	27	16	06	24	03	22	22	22	24
	YEAR OF OCCURRENCE		1999	1997	1999	2000	2001	2000	1997	2001	1999	2001	1999	2000	MAR 1999
	MAXIMUM 5-SECOND:														
	SPEED (MPH)	5	70	64	61	45	44	43	52	43	61	32	38	49	70
	DIR. (TENS OF DEGS)		24	23	23	30	28	22	05	03	25	25	23	20	24
	YEAR OF OCCURRENCE		1999	1997	1999	2000	2001	1998	1997	1998	1998	2001	1999	2000	JAN 1999
PRECIPITATION	NORMAL (IN)	30	3.48	3.69	3.77	2.59	3.92	3.68	4.01	4.02	3.19	2.86	2.98	3.24	41.43
	MAXIMUM MONTHLY (IN)	57	7.52	6.42	7.78	6.10	7.67	9.38	10.27	12.18	21.79	9.10	8.22	6.65	21.79
	YEAR OF OCCURRENCE		1954	1989	1983	1978	1974	1973	1991	1986	1999	1995	1948	1983	SEP 1999
	MINIMUM MONTHLY (IN)	57	0.87	0.69	1.03	0.23	0.58	0.33	0.80	0.81	0.23	0.44	0.50	0.25	0.23
	YEAR OF OCCURRENCE		1981	1991	1985	1976	1999	1993	1993	1950	1985	2000	2001	1965	SEP 1985
	MAXIMUM IN 24 HOURS (IN)	57	3.11	3.22	3.70	4.04	4.40	3.44	4.27	5.20	5.41	4.24	4.70	3.18	5.41
	YEAR OF OCCURRENCE		1984	1973	1983	1978	1957	1967	1997	5020	1999	1995	1963	1958	SEP 1999
	NORMAL NO. DAYS WITH:														
	PRECIPITATION 0.01	30	9.8	9.4	10.1	8.7	10.5	9.1	10.7	9.8	7.0	6.6	7.8	9.3	108.8
	PRECIPITATION 1.00	30	1.0	1.0	0.7	0.4	0.9	1.1	0.9	1.3	0.9	0.9	0.6	0.6	10.3
SNOWFALL	NORMAL (IN)	30	2.4	3.2	1.5	0.1	0.0	0.0	0.0	0.0	0.0	0.0	0.2	0.5	7.9
	MAXIMUM MONTHLY (IN)	57	25.8	17.2	14.0	1.8	T	T	T	0.0	0.0	0.0	2.6	10.6	25.8
	YEAR OF OCCURRENCE		2000	1979	1960	1983	1996	1996	1993				1975	1958	JAN 2000
	MAXIMUM IN 24 HOURS (IN)	57	17.9	10.4	9.3	1.8	T	T	T	0.0	0.0	0.0	2.6	9.1	17.9
	YEAR OF OCCURRENCE		2000	1979	1969	1983	1995	1996	1993				1975	1958	JAN 2000
	MAXIMUM SNOW DEPTH (IN)	53	20	10	11	0	0	0	0	0	0	0	2	9	20
	YEAR OF OCCURRENCE		2000	1979	1980								2000	1958	JAN 2000
	NORMAL NO. DAYS WITH:														
	SNOWFALL 1.0	30	0.9	0.8	0.4	0.*	0.0	0.0	0.0	0.0	0.0	0.0	0.1	0.2	2.4

PRECIPITATION (inches) 2001 RALEIGH, NC (RDU)

YEAR	JAN	FEB	MAR	APR	MAY	JUN	JUL	AUG	SEP	OCT	NOV	DEC	ANNUAL
1972	1.97	4.13	2.50	1.92	5.34	4.16	6.80	4.17	5.80	3.96	5.98	5.01	51.74
1973	2.67	5.50	4.06	4.40	3.99	9.38	3.12	1.13	0.60	0.61	6.38	4.44	46.44
1974	4.39	2.87	3.34	1.32	7.67	4.02	1.56	4.82	3.71	1.23	1.79	4.02	40.74
1975	6.09	2.85	6.26	1.64	3.84	1.66	6.74	2.11	5.77	1.23	4.60	4.04	46.83
1976	3.07	1.54	3.17	0.23	4.74	2.55	1.00	1.52	5.99	3.97	1.89	4.04	33.71
1977	2.82	2.13	5.63	1.89	3.94	0.84	0.89	4.12	3.86	5.06	2.22	3.70	37.10
1978	7.03	1.43	4.40	6.10	4.20	4.06	3.63	1.86	1.37	1.46	4.17	3.26	42.97
1979	5.71	5.55	2.69	2.63	4.71	3.27	4.84	1.66	6.76	1.88	4.73	0.94	45.37
1980	4.39	1.91	5.87	1.97	2.33	4.89	2.11	1.87	3.76	2.25	2.87	1.42	35.64
1981	0.87	3.02	2.35	1.03	4.28	0.55	5.69	5.34	2.70	4.64	0.95	4.96	36.38
1982	3.43	4.97	3.02	3.33	4.20	8.39	3.34	1.83	1.55	3.93	2.34	4.02	44.35
1983	1.79	6.00	7.78	3.54	5.89	3.09	1.10	1.81	2.13	3.59	3.86	6.65	47.23
1984	4.93	5.65	5.40	4.45	5.43	3.08	9.20	1.13	2.31	0.73	1.64	2.32	46.27
1985	4.83	4.44	1.03	0.64	3.95	2.87	6.28	3.73	0.23	1.75	7.61	0.81	38.17
1986	1.88	1.65	3.06	1.01	2.98	1.92	4.32	12.18	0.95	1.28	2.77	2.95	36.95
1987	6.53	5.52	2.88	4.68	1.19	2.11	1.78	5.80	5.48	1.71	1.39	3.02	42.09
1988	3.15	2.42	1.76	3.56	2.85	2.88	2.69	3.40	4.90	5.67	3.34	1.04	37.66
1989	1.35	6.42	5.40	4.91	3.88	7.30	5.46	5.08	3.96	3.44	3.94	3.01	54.15
1990	3.07	3.82	5.02	2.19	6.97	1.03	2.22	2.65	0.30	5.69	1.51	3.08	37.55
1991	4.12	0.69	4.59	1.04	2.89	2.05	10.27	1.87	3.16	1.40	0.73	2.65	35.46
1992	3.80	2.23	2.95	1.93	2.60	5.12	3.45	7.63	2.22	3.79	5.02	2.44	43.18
1993	4.50	2.22	6.13	4.84	3.32	0.33	2.11	1.77	3.50	2.95	2.66	3.72	38.05
1994	3.55	2.97	5.91	0.86	2.85	2.20	4.67	4.20	1.99	4.62	1.32	1.27	36.41
1995	4.50	4.52	2.49	1.32	3.91	7.75	3.29	2.70	2.46	9.10	4.67	1.88	48.59
1996	4.24	2.94	3.39	3.98	3.26	3.28	6.98	3.72	16.65	3.48	4.33	2.89	59.14
1997	3.13	2.83	3.41	4.75	2.21	3.83	6.51	1.01	3.07	3.26	4.05	2.75	40.81
1998	7.49	5.79	7.36	3.12	3.79	3.45	4.84	4.66	3.55	2.79	2.40	3.44	52.68
1999	5.77	1.96	3.69	3.53	0.58	1.16	3.00	3.20	21.79	2.45	1.20	2.31	50.64
2000	6.27	2.20	1.76	4.66	1.23	2.50	6.19	6.64	3.82	T	2.56	1.51	39.34
2001	1.30	2.34	7.11	1.72	3.53	4.54	4.13	4.88	0.86	1.86	0.50	2.01	34.78
POR= 115 YRS	3.53	3.63	3.81	3.20	3.81	4.03	5.06	4.85	3.77	2.89	2.61	3.17	44.36

AVERAGE TEMPERATURE (F) 2001 RALEIGH, NC (RDU)

YEAR	JAN	FEB	MAR	APR	MAY	JUN	JUL	AUG	SEP	OCT	NOV	DEC	ANNUAL
1972	44.7	40.0	49.5	58.0	64.3	69.9	77.1	75.6	70.4	57.4	48.1	46.2	58.4
1973	39.3	40.0	54.8	57.9	64.5	75.1	76.5	76.6	73.2	62.2	54.6	42.5	59.8
1974	49.3	42.9	54.4	60.2	67.4	71.8	76.5	76.0	69.0	56.5	48.7	43.2	59.7
1975	43.8	43.7	47.1	55.8	68.2	73.7	75.6	78.3	71.1	62.2	53.5	42.0	59.6
1976	37.6	50.1	56.2	60.7	66.8	74.7	78.6	75.4	69.7	55.7	42.5	37.1	58.8
1977	26.6	39.3	53.3	62.6	68.4	73.3	80.6	78.1	72.8	56.4	51.8	39.9	58.6
1978	35.3	33.1	48.2	59.2	66.0	75.8	78.0	79.9	73.7	59.5	55.1	44.9	59.1
1979	39.1	36.5	52.2	59.7	66.8	69.8	75.4	77.2	71.2	59.9	52.0	43.4	58.6
1980	40.6	36.5	46.5	62.0	69.8	75.0	78.9	79.6	74.9	58.7	49.1	40.7	59.4
1981	33.4	43.9	46.2	61.7	64.0	78.9	80.8	74.6	68.3	57.2	50.7	39.7	58.3
1982	35.5	45.5	51.7	57.4	71.0	74.7	79.1	76.5	70.5	60.6	51.9	47.5	60.2
1983	38.1	40.7	50.7	55.1	65.4	72.5	79.1	79.1	70.7	60.4	50.9	39.5	58.5
1984	36.3	45.7	47.2	55.9	65.5	75.5	74.9	76.6	67.5	66.3	47.1	49.7	59.0
1985	34.0	41.9	52.7	62.0	67.3	73.9	76.8	75.2	69.7	63.7	58.4	39.4	59.6
1986	38.5	44.5	51.7	61.2	67.4	78.4	81.7	75.6	72.3	63.0	52.9	42.6	60.8
1987	38.3	40.4	49.2	56.9	69.3	76.3	81.2	79.2	73.0	54.6	52.8	44.5	59.6
1988	34.7	41.9	50.6	57.8	66.0	72.4	79.0	80.3	70.3	54.4	52.0	42.4	58.5
1989	44.8	42.9	50.7	58.0	65.0	77.0	78.1	76.2	71.7	61.3	51.4	34.6	59.3
1990	48.0	51.0	54.9	60.3	67.4	75.2	80.0	78.0	72.2	64.1	54.2	48.1	62.8
1991	41.9	46.8	54.3	62.3	72.3	75.8	80.6	77.8	71.6	61.1	50.0	47.0	61.8
1992	43.4	46.2	50.5	59.2	63.2	72.0	80.4	74.1	71.5	58.1	51.9	42.5	59.4
1993	43.3	40.5	47.9	56.9	69.0	76.1	82.5	78.2	73.7	59.6	51.8	40.3	60.0
1994	36.9	44.2	52.2	63.6	64.2	77.0	79.6	76.0	69.1	59.3	54.0	47.9	60.3
1995	42.9	41.0	52.7	61.0	68.5	73.8	80.4	80.6	70.5	63.4	47.1	39.5	60.1
1996	38.3	42.5	45.3	58.4	67.7	76.4	78.6	75.2	70.3	61.0	45.7	45.8	58.8
1997	41.3	46.7	54.6	55.2	63.9	71.5	79.5	76.6	70.9	59.6	47.3	40.9	59.0
1998	44.7	46.1	50.2	58.8	69.2	77.3	79.5	77.8	74.6	61.1	52.0	46.6	61.5
1999	45.7	45.7	47.5	61.5	66.8	74.5	81.3	80.0	69.1	58.6	54.4	43.9	60.8
2000	38.4	45.8	53.4	57.6	70.0	77.2	77.0	76.1	69.4	60.2	48.0	35.5	59.1
2001	41.6	47.2	49.1	60.8	68.0	76.9	76.2	79.6	69.0	59.4	55.8	47.6	60.9
POR= 115 YRS	41.1	42.9	50.4	59.3	67.9	75.2	77.7	76.4	71.6	60.7	50.8	42.7	59.7

REFERENCE NOTES:

PAGE 1:
THE TEMPERATURE GRAPH SHOWS NORMAL MAXIMUM AND NORMAL MINIMUM DAILY TEMPERATURES (SOLID CURVES) AND THE ACTUAL DAILY HIGH AND LOW TEMPERATURES (VERTICAL BARS).

PAGE 2 AND 3:
H/C INDICATES HEATING AND COOLING DEGREE DAYS.
RH INDICATES RELATIVE HUMIDITY
W/O INDICATES WEATHER AND OBSTRUCTIONS
S INDICATES SUNSHINE.
PR INDICATES PRESSURE.
CLOUDINESS ON PAGE 3 IS THE SUM OF THE CEILOMETER AND SATELLITE DATA NOT TO EXCEED EIGHT EIGHTHS(OKTAS).

GENERAL:
T INDICATES TRACE PRECIPITATION, AN AMOUNT GREATER THAN ZERO BUT LESS THAN THE LOWEST REPORTABLE VALUE.
+ INDICATES THE VALUE ALSO OCCURS ON EARLIER DATES.
BLANK ENTRIES DENOTE MISSING OR UNREPORTED DATA.
NORMALS ARE 30-YEAR AVERAGES (1961 - 1990).
ASOS INDICATES AUTOMATED SURFACE OBSERVING SYSTEM.
PM INDICATES THE LAST DAY OF THE PREVIOUS MONTH.
POR (PERIOD OF RECORD) BEGINS WITH THE JANUARY DATA MONTH AND IS THE NUMBER OF YEARS USED TO COMPUTE THE MEAN. INDIVIDUAL MONTHS WITHIN THE POR MAY BE MISSING.
WHEN THE POR FOR A NORMAL IS LESS THAN 30 YEARS, THE NORMAL IS PROVISIONAL AND IS BASED ON THE NUMBER OF YEARS INDICATED.
0.* OR * INDICATES THE VALUE OR MEAN-DAYS-WITH IS BETWEEN 0.00 AND 0.05.
CLOUDINESS FOR ASOS STATIONS DIFFERS FROM THE NON-ASOS OBSERVATION TAKEN BY A HUMAN OBSERVER. ASOS STATION CLOUDINESS IS BASED ON TIME-AVERAGED CEILOMETER DATA FOR CLOUDS AT OR BELOW 12,000 FEET AND ON SATELLITE DATA FOR CLOUDS ABOVE 12,000 FEET.
THE NUMBER OF DAYS WITH CLEAR, PARTLY CLOUDY, AND CLOUDY CONDITIONS FOR ASOS STATIONS IS THE SUM OF THE CEILOMETER AND SATELLITE DATA FOR THE SUNRISE TO SUNSET PERIOD.

GENERAL CONTINUED:
CLEAR INDICATES 0 - 2 OKTAS, PARTLY CLOUDY INDICATES 3 - 6 OKTAS, AND CLOUDY INDICATES 7 OR 8 OKTAS. WHEN AT LEAST ONE OF THE ELEMENTS (CEILOMETER OR SATELLITE) IS MISSING, THE DAILY CLOUDINESS IS NOT COMPUTED.
WIND DIRECTION IS RECORDED IN TENS OF DEGREES (2 DIGITS) CLOCKWISE FROM TRUE NORTH. "00" INDICATES CALM. "36" INDICATES TRUE NORTH.
RESULTANT WIND IS THE VECTOR AVERAGE OF THE SPEED AND DIRECTION.
AVERAGE TEMPERATURE IS THE SUM OF THE MEAN DAILY MAXIMUM AND MINIMUM TEMPERATURE DIVIDED BY 2.
SNOWFALL DATA COMPRISE ALL FORMS OF FROZEN PRECIPITATION, INCLUDING HAIL.
A HEATING (COOLING) DEGREE DAY IS THE DIFFERENCE BETWEEN THE AVERAGE DAILY TEMPERATURE AND 65 F.
DRY BULB IS THE TEMPERATURE OF THE AMBIENT AIR.
DEW POINT IS THE TEMPERATURE TO WHICH THE AIR MUST BE COOLED TO ACHIEVE 100 PERCENT RELATIVE HUMIDITY.
WET BULB IS THE TEMPERATURE THE AIR WOULD HAVE IF THE MOISTURE CONTENT WAS INCREASED TO 100 PERCENT RELATIVE HUMIDITY.

ON JULY 1, 1996, THE NATIONAL WEATHER SERVICE BEGAN USING THE "METAR" OBSERVATION CODE THAT WAS ALREADY EMPLOYED BY MOST OTHER NATIONS OF THE WORLD. THE MOST NOTICEABLE DIFFERENCE IN THIS ANNUAL PUBLICATION WILL BE THE CHANGE IN UNITS FROM TENTHS TO EIGHTS(OKTAS) FOR REPORTING THE AMOUNT OF SKY COVER.

HEATING DEGREE DAYS (base 65 F) 2001 RALEIGH, NC (RDU)

YEAR	JUL	AUG	SEP	OCT	NOV	DEC	JAN	FEB	MAR	APR	MAY	JUN	TOTAL
1972-73	0	0	9	238	504	576	790	692	334	231	88	0	3462
1973-74	0	0	2	126	312	690	481	614	346	187	48	0	2806
1974-75	0	0	44	268	501	668	651	589	553	293	34	0	3601
1975-76	0	0	17	117	351	705	843	426	300	194	52	6	3011
1976-77	0	0	7	302	668	857	1183	715	358	132	49	14	4285
1977-78	0	0	4	283	411	768	914	883	514	196	83	0	4056
1978-79	0	0	7	184	292	627	793	792	398	183	43	8	3327
1979-80	0	0	13	196	394	661	753	820	564	130	33	0	3564
1980-81	0	0	16	225	477	747	973	583	579	149	99	0	3848
1981-82	0	4	31	253	425	776	907	538	411	244	15	0	3604
1982-83	0	0	14	182	392	542	828	675	438	305	79	7	3462
1983-84	0	0	59	180	417	784	882	553	545	283	83	5	3791
1984-85	0	0	63	42	530	468	954	644	395	146	42	4	3288
1985-86	0	0	36	96	207	789	812	569	415	157	59	0	3140
1986-87	0	11	12	149	370	687	820	681	484	248	29	0	3491
1987-88	0	0	1	319	362	631	932	665	444	228	62	22	3666
1988-89	0	0	8	336	386	695	619	623	459	257	102	0	3485
1989-90	0	3	30	167	404	934	518	390	357	186	37	0	3026
1990-91	0	0	18	124	323	520	709	501	354	153	18	0	2720
1991-92	0	0	24	156	451	562	659	537	446	226	114	3	3178
1992-93	0	0	29	224	390	691	666	670	524	244	15	0	3453
1993-94	0	0	19	198	405	758	863	576	394	113	101	0	3427
1994-95	0	0	8	190	326	527	679	665	380	165	36	0	2976
1995-96	0	0	18	123	532	787	819	646	601	227	72	2	3827
1996-97	0	0	9	142	569	590	729	510	326	300	97	45	3317
1997-98	0	0	14	223	524	742	624	525	475	207	21	7	3362
1998-99	0	0	7	146	383	573	593	533	536	160	36	2	2969
1999-00	2	0	29	216	314	646	818	548	365	235	24	0	3197
2000-01	0	0	40	181	511	910	718	492	490	200	36	0	3578
2001-	0	0	50	211	287	532							

COOLING DEGREE DAYS (base 65 F) 2001 RALEIGH, NC (RDU)

YEAR	JAN	FEB	MAR	APR	MAY	JUN	JUL	AUG	SEP	OCT	NOV	DEC	ANNUAL
1972	0	0	5	37	36	170	382	336	177	6	6	2	1157
1973	0	0	24	26	81	310	363	254	48	7	0	1478
1974	0	0	25	51	130	210	363	347	169	9	21	0	1325
1975	0	0	3	22	141	269	337	421	209	38	12	0	1452
1976	0	3	31	71	116	304	428	330	157	19	0	0	1459
1977	0	2	4	68	162	272	490	414	245	25	19	0	1701
1978	0	0	2	30	120	330	412	468	275	21	3	10	1671
1979	0	0	6	28	105	159	332	384	205	46	10	0	1275
1980	0	0	0	45	190	306	441	460	321	38	6	0	1807
1981	0	0	2	56	75	425	497	309	139	19	0	0	1522
1982	0	0	3	24	208	299	443	363	183	53	5	7	1588
1983	0	0	0	16	97	238	441	447	239	42	0	0	1520
1984	0	0	0	16	108	324	311	366	143	90	0	0	1358
1985	0	3	20	65	121	277	373	323	181	64	14	0	1441
1986	0	0	7	51	142	408	526	349	237	96	15	0	1831
1987	0	0	0	11	170	347	508	447	250	0	2	0	1735
1988	0	3	5	17	98	249	438	482	172	14	3	0	1481
1989	0	11	23	54	110	367	412	359	237	59	4	0	1636
1990	0	3	49	51	117	312	472	410	239	102	4	5	1764
1991	0	0	28	78	253	333	493	403	230	40	9	13	1880
1992	0	0	3	58	65	218	487	285	228	15	3	0	1362
1993	0	0	0	12	147	338	550	417	286	34	17	0	1801
1994	0	0	6	78	86	367	462	350	139	21	3	1	1513
1995	3	0	3	53	154	270	486	490	191	81	3	0	1734
1996	0	0	0	35	164	352	427	324	176	26	0	0	1504
1997	1	3	14	11	68	249	451	368	199	64	0	0	1428
1998	3	0	23	29	158	382	457	401	302	32	0	12	1799
1999	3	0	0	64	100	292	513	474	161	24	0	0	1631
2000	0	0	9	19	185	374	380	351	176	38	8	0	1540
2001	0	0	3	83	137	364	356	461	176	46	17	2	1645

SNOWFALL (inches) 2001 RALEIGH, NC (RDU)

YEAR	JUL	AUG	SEP	OCT	NOV	DEC	JAN	FEB	MAR	APR	MAY	JUN	TOTAL
1972-73	0.0	0.0	0.0	0.0	T	0.0	6.4	4.5	0.4	0.0	0.0	0.0	11.3
1973-74	0.0	0.0	0.0	0.0	0.0	2.8	0.0	T	2.9	0.0	0.0	0.0	5.7
1974-75	0.0	0.0	0.0	0.0	0.0	0.0	T	T	0.6	0.0	0.0	0.0	0.6
1975-76	0.0	0.0	0.0	0.0	2.6	T	0.4	T	0.0	0.0	0.0	0.0	3.0
1976-77	0.0	0.0	0.0	0.0	T	T	2.1	1.5	0.0	0.0	0.0	0.0	3.6
1977-78	0.0	0.0	0.0	0.0	T	T	T	9.0	1.6	0.0	0.0	0.0	10.6
1978-79	0.0	0.0	0.0	0.0	0.0	0.0	0.4	17.2	T	0.0	0.0	0.0	17.6
1979-80	0.0	0.0	0.0	0.0	0.0	0.0	2.2	5.0	11.1	0.0	0.0	0.0	18.3
1980-81	0.0	0.0	0.0	0.0	0.0	3.1	2.6	0.0	T	0.0	0.0	0.0	5.7
1981-82	0.0	0.0	0.0	0.0	0.0	T	6.0	0.6	0.0	0.0	0.0	0.0	6.6
1982-83	0.0	0.0	0.0	0.0	0.0	T	2.7	7.3	1.8	0.0	0.0	0.0	11.8
1983-84	0.0	0.0	0.0	0.0	0.0	0.0	T	6.9	T	0.0	0.0	0.0	6.9
1984-85	0.0	0.0	0.0	0.0	0.0	0.0	4.1	T	0.0	0.0	0.0	0.0	4.1
1985-86	0.0	0.0	0.0	0.0	0.0	T	T	0.9	T	0.0	0.0	0.0	0.9
1986-87	0.0	0.0	0.0	0.0	T	0.6	0.0	10.2	T	T	0.0	0.0	10.8
1987-88	0.0	0.0	0.0	0.0	0.6	0.0	7.3	T	0.0	0.0	0.0	0.0	7.9
1988-89	0.0	0.0	0.0	0.0	0.0	0.1	0.0	11.1	0.5	0.3	0.0	0.0	12.0
1989-90	0.0	0.0	0.0	0.0	0.0	2.7	0.0	T	T	0.0	0.0	0.0	2.7
1990-91	0.0	0.0	0.0	0.0	0.0	T	T	T	T	0.0	0.0	0.0	T
1991-92	0.0	0.0	0.0	0.0	T	0.0	T	T	T	0.0	0.0	0.0	T
1992-93	0.0	0.0	0.0	0.0	0.0	T	T	1.6	0.9	0.0	0.0	0.0	2.5
1993-94	T	0.0	0.0	0.0	0.0	0.0	3.1	1.1	0.2	0.0	0.0	0.0	4.4
1994-95	0.0	0.0	0.0	0.0	0.0	0.0	0.0	0.8	1.4	T	0.0	0.0	2.2
1995-96	0.0	0.0	0.0	0.0	T	T	5.6	8.1	0.9	0.0	T	T	14.6
1996-97							0.4	T	0.0	0.0	T	T	
1997-98	0.0	0.0	0.0	0.0	0.0	0.4	2.0	0.0	T	0.0	0.0	T	2.4
1998-99	0.0	0.0	0.0	0.0	0.0	0.0	T	T	T	0.0	0.0	0.0	T
1999-00	0.0	0.0	0.0	0.0	T	0.0	25.8	0.0	0.0	0.0	0.0	T	25.8
2000-01	0.0	0.0	0.0	0.0	2.2	0.1	T	0.3	T	T	T	0.0	2.6
2001-	0.0	0.0	0.0	0.0	0.0	0.0							
POR= 56 YRS	T	0.0	0.0	0.0	0.1	0.8	2.6	2.5	1.3	0.0	T	T	7.3

2001
FARGO,
NORTH DAKOTA (FAR)

Moorhead, Minnesota, and Fargo are twin cities in the Red River Valley of the north. The Red River of the north flows northward between the two cities and is a part of the Hudson Bay drainage area. The Red River is approximately 2 miles east of the airport at its nearest point and has no significant effect on the weather. In recent years, spring floods due to melting snow have been common. Summer floods caused by heavy rains are infrequent.

The surrounding terrain is flat and open. Northerly winds blowing up the valley occasionally causing low cloudiness and fog. However, this upslope cloudiness is very infrequent. Aside from this, there are no pronounced climatic differences due to geographical features in the immediate area.

The summers are generally comfortable with very few days of hot and humid weather. Nights, with few exceptions, are comfortably cool. The winter months are cold and dry with temperatures rising above freezing only on an average of six days each month, and nighttime lows dropping below zero approximately half of the time.

Precipitation is the most important climatic factor in the area. The Red River Valley lies in an area where lighter amounts fall to the west and heavier amounts to the east. Seventy-five percent of the precipitation occurs during the growing season (April to September) and is often accompanied by electrical storms and heavy falls in a short time. Winter precipitation is light, indicating that heavy snowfall is the exception rather than the rule. The first light snow in the fall occasionally falls in September, but usually very little, if any, occurs until October or November. The latest fall is generally in April.

With the flat terrain, surface friction has little effect on the wind in the area and this fact has led to the legendary Dakota blizzards. Strong winds with even light snowfall cause much drifting and blowing snow, reducing visibility to near zero. Fortunately, these conditions occur only several times during the winter months.

NORMALS, MEANS, AND EXTREMES
FARGO, ND (FAR)

LATITUDE:	LONGITUDE:	ELEVATION (FT):	TIME ZONE:	WBAN: 14914
46 55' 31" N	96 48' 40" W	GRND: 908　　BARO: 911	CENTRAL (UTC + 6)	

	ELEMENT	POR	JAN	FEB	MAR	APR	MAY	JUN	JUL	AUG	SEP	OCT	NOV	DEC	YEAR
TEMPERATURE F	NORMAL DAILY MAXIMUM	30	15.4	21.1	34.6	53.8	68.5	77.4	83.4	81.3	69.4	56.7	36.8	20.1	51.5
	MEAN DAILY MAXIMUM	54	15.1	21.6	34.1	53.5	68.6	77.0	82.4	81.4	70.1	57.0	36.4	21.2	51.5
	HIGHEST DAILY MAXIMUM	49	52	66	78	100	98	100	106	106	102	93	74	57	106
	YEAR OF OCCURRENCE		1981	1958	1967	1980	1964	1995	1988	1976	1959	1963	1990	1962	JUL 1988
	MEAN OF EXTREME MAXS.	54	37.3	41.4	54.1	77.9	88.1	92.0	95.0	95.5	89.6	79.3	58.8	41.7	70.9
	NORMAL DAILY MINIMUM	30	-3.6	2.7	17.3	32.1	43.8	53.6	58.8	56.4	45.9	34.6	19.4	3.1	30.3
	MEAN DAILY MINIMUM	54	-3.4	3.3	16.8	32.0	44.1	54.1	58.8	57.0	46.4	35.1	19.6	4.7	30.7
	LOWEST DAILY MINIMUM	49	-35	-39	-23	-7	20	30	36	33	19	7	-24	-32	-39
	YEAR OF OCCURRENCE		1977	1996	1980	1975	1966	1969	1967	1982	1965	1976	1985	1967	FEB 1996
	MEAN OF EXTREME MINS.	54	-25.4	-21.1	-8.6	14.8	27.2	39.5	46.0	43.2	30.2	18.9	-1.8	-18.7	12.0
	NORMAL DRY BULB	30	5.9	12.0	25.9	43.0	56.2	65.5	71.1	68.8	57.7	45.7	28.1	11.6	41.0
	MEAN DRY BULB	54	5.8	12.5	25.4	42.6	56.2	65.5	70.6	69.2	58.1	46.0	28.0	12.9	41.1
	MEAN WET BULB	16	8.8	14.1	25.8	38.0	50.4	59.3	63.4	61.9	52.3	40.2	24.1	12.7	37.6
	MEAN DEW POINT	16	4.3	9.7	21.1	30.1	42.1	53.8	58.8	56.9	46.8	33.7	19.7	8.7	32.1
	NORMAL NO. DAYS WITH:														
	MAXIMUM 90	30	0.0	0.0	0.0	0.1	0.7	2.4	5.5	5.3	1.0	*	0.0	0.0	15.0
	MAXIMUM 32	30	27.1	21.2	12.5	1.2	*	0.0	0.0	0.0	0.0	0.5	10.5	24.9	97.9
	MINIMUM 32	30	31.0	27.9	27.6	16.4	4.0	*	0.0	0.0	1.8	12.8	26.7	30.9	179.1
	MINIMUM 0	30	18.5	12.8	4.3	0.1	0.0	0.0	0.0	0.0	0.0	0.0	2.2	13.4	51.3
H/C	NORMAL HEATING DEG. DAYS	30	1832	1484	1212	660	307	93	19	48	239	598	1107	1655	9254
	NORMAL COOLING DEG. DAYS	30	0	0	0	0	35	108	209	165	20	0	0	0	537
RH	NORMAL (PERCENT)	30	73	75	76	66	60	66	67	66	69	69	76	76	70
	HOUR 00 LST	30	74	77	80	73	68	75	77	76	78	75	79	78	76
	HOUR 06 LST	30	74	76	82	79	76	82	85	86	85	81	82	78	80
	HOUR 12 LST	30	71	72	70	57	50	56	54	54	58	59	70	74	62
	HOUR 18 LST	30	73	74	70	53	46	52	51	50	55	60	73	76	61
S	PERCENT POSSIBLE SUNSHINE	54	50	56	58	60	61	62	71	69	60	54	40	43	57
W/O	MEAN NO. DAYS WITH:														
	HEAVY FOG(VISBY 1/4 MI)	59	1.1	1.6	2.0	0.6	0.4	0.6	0.8	1.1	0.9	1.0	1.5	1.9	13.5
	THUNDERSTORMS	59	0.0	0.0	0.2	1.2	3.5	6.9	7.9	6.9	2.9	0.9	0.1	0.0	30.5
CLOUDINESS	MEAN:														
	SUNRISE-SUNSET (OKTAS)	1			6.4			4.8							
	MIDNIGHT-MIDNIGHT (OKTAS)	1			5.6										
	MEAN NO. DAYS WITH:														
	CLEAR	1	2.0	2.0	7.0		3.0	9.0							
	PARTLY CLOUDY	1		1.0	3.0		3.0	5.0							
	CLOUDY	1	1.0	4.0	12.0		7.0	5.0							
PR	MEAN STATION PRESSURE(IN)	29	29.10	29.09	29.00	29.00	28.99	28.90	29.00	29.00	29.00	29.00	29.01	29.11	29.02
	MEAN SEA-LEVEL PRES. (IN)	17	30.12	30.14	30.07	29.97	29.91	29.99	29.93	29.97	29.99	29.99	30.05	30.11	30.02
WINDS	MEAN SPEED (MPH)	46	12.6	12.5	13.0	13.7	12.8	11.8	10.3	10.8	11.7	12.8	12.8	12.4	12.3
	PREVAIL.DIR(TENS OF DEGS)	30	35	16	35	35	16	16	16	16	16	16	16	16	16
	MAXIMUM 2-MINUTE:														
	SPEED (MPH)	6	49	51	49	49	45	43	74	51	39	49	47	43	74
	DIR. (TENS OF DEGS)		34	33	34	32	24	24	33	34	16	33	31	16	33
	YEAR OF OCCURRENCE		1996	1996	1999	2000	1996	2000	1999	2001	2001	1996	1999	1996	JUL 1999
	MAXIMUM 5-SECOND:														
	SPEED (MPH)	6	56	56	55	59	61	54	91	58	48	61	56	52	91
	DIR. (TENS OF DEGS)		34	32	34	32	25	13	34	34	16	33	30	16	34
	YEAR OF OCCURRENCE		1997	1996	1999	2000	1996	2000	1999	2001	2001	1996	1999	1996	JUL 1999
PRECIPITATION	NORMAL (IN)	30	0.67	0.45	1.06	1.82	2.45	2.82	2.70	2.43	1.99	1.68	0.73	0.65	19.45
	MAXIMUM MONTHLY (IN)	60	1.85	1.74	2.62	5.28	7.34	11.72	8.42	8.52	6.50	7.03	4.58	2.19	11.72
	YEAR OF OCCURRENCE		1989	1979	1995	1986	1998	2000	1952	1944	1999	1982	1977	1951	JUN 2000
	MINIMUM MONTHLY (IN)	60	0.09	0.03	0.03	0.01	0.46	0.58	0.42	0.18	0.13	0.05	0.02	0.04	0.01
	YEAR OF OCCURRENCE		1961	1954	1958	1988	1976	1972	1960	1984	1974	1986	1999	1958	APR 1988
	MAXIMUM IN 24 HOURS (IN)	60	1.19	1.22	1.16	1.91	4.10	4.64	5.10	4.72	3.97	3.22	1.99	0.87	5.10
	YEAR OF OCCURRENCE		1996	1946	1950	1963	1977	2000	1993	1943	1957	1982	1977	1960	JUL 1993
	NORMAL NO. DAYS WITH:														
	PRECIPITATION 0.01	30	8.3	6.8	8.1	8.4	9.7	10.3	9.5	8.9	7.8	6.8	5.6	8.1	98.3
	PRECIPITATION 1.00	30	0.0	0.0	0.0	0.2	0.5	0.4	0.7	0.5	0.4	0.3	0.1	0.0	3.1
SNOWFALL	NORMAL (IN)	30	10.0	5.8	7.4	3.1	0.*	0.0	0.0	0.0	T	0.7	5.3	7.8	40.1
	MAXIMUM MONTHLY (IN)	60	31.5	19.5	26.2	12.8	1.0	T	T	T	0.6	8.1	26.4	20.4	31.5
	YEAR OF OCCURRENCE		1989	1979	1997	1970	1950	1994	1999	1994	1942	1951	1996	1996	JAN 1989
	MAXIMUM IN 24 HOURS (IN)	60	19.4	11.2	12.0	8.6	1.0	T	T	T	0.6	7.8	12.6	9.3	19.4
	YEAR OF OCCURRENCE		1989	1951	1997	1970	1950	1994	1999	1994	1942	1951	1977	1988	JAN 1989
	MAXIMUM SNOW DEPTH (IN)	53	30	24	32	8	1	0	0	0	0	5	17	19	32
	YEAR OF OCCURRENCE		1989	1994	1997	1975	1979					1951	1985	1985	MAR 1997
	NORMAL NO. DAYS WITH:														
	SNOWFALL 1.0	30	2.9	1.8	2.5	1.1	0.0	0.0	0.0	0.0	0.0	0.3	1.6	2.4	12.6

PRECIPITATION (inches) 2001 FARGO, ND (FAR)

YEAR	JAN	FEB	MAR	APR	MAY	JUN	JUL	AUG	SEP	OCT	NOV	DEC	ANNUAL
1972	0.94	0.61	0.74	0.96	3.52	0.58	2.78	3.45	1.22	1.25	0.22	1.51	17.78
1973	0.12	0.13	1.25	0.70	1.65	1.78	3.60	3.85	4.98	1.54	0.90	1.02	21.52
1974	0.35	0.36	0.71	3.40	4.03	0.90	4.75	6.46	0.13	3.10	0.48	0.32	24.99
1975	1.32	0.27	1.48	3.24	1.45	9.40	2.42	2.90	1.24	1.76	0.64	0.18	26.30
1976	1.25	0.35	1.00	1.19	0.46	2.34	0.63	0.41	0.55	0.16	0.26	0.24	8.84
1977	0.65	1.24	1.72	0.84	7.30	1.64	5.36	2.53	3.21	2.46	4.58	0.75	32.28
1978	0.16	0.18	0.43	1.15	1.78	4.40	2.92	3.79	0.92	0.13	1.11	0.47	17.44
1979	0.44	1.74	2.00	3.04	2.02	2.92	3.38	0.90	0.31	2.60	0.48	0.14	19.97
1980	1.23	0.57	0.62	0.02	0.64	2.68	0.76	4.24	2.52	1.06	0.47	0.30	15.11
1981	0.11	0.49	0.67	0.61	3.46	2.56	3.21	1.76	1.11	2.36	0.40	0.85	17.59
1982	1.32	0.54	1.25	0.45	1.82	1.61	2.64	1.12	1.12	7.03	1.13	0.17	20.20
1983	0.46	0.21	2.27	0.42	2.00	2.34	4.16	2.56	1.63	1.62	1.04	0.96	19.67
1984	0.79	0.90	1.12	1.68	0.61	5.38	0.64	0.18	1.23	6.76	0.18	0.90	20.37
1985	0.20	0.18	1.35	0.60	5.03	1.44	3.91	2.30	1.39	1.12	1.06	0.59	19.17
1986	0.85	0.27	0.19	5.28	1.00	3.98	4.78	1.72	3.67	0.05	1.43	0.29	23.51
1987	0.27	0.86	0.49	0.12	3.46	0.66	2.86	3.23	1.70	0.18	0.48	0.69	15.00
1988	1.62	0.22	1.02	0.01	1.82	1.24	0.46	2.14	3.22	0.49	1.18	1.11	14.53
1989	1.85	0.21	1.49	1.03	2.60	1.51	0.62	6.07	2.10	0.31	1.18	0.24	19.21
1990	0.13	0.58	1.54	1.78	1.52	6.05	0.78	0.99	1.75	1.22	0.02	0.77	17.13
1991	0.29	1.27	0.97	3.15	2.38	6.26	1.86	1.87	1.28	0.71	0.46	0.37	20.87
1992	0.89	0.51	1.05	0.89	2.32	6.47	0.83	2.35	2.55	0.26	1.73	0.56	20.41
1993	0.79	0.19	0.83	0.74	2.67	4.28	7.71	1.13	0.49	0.19	1.88	1.00	21.90
1994	0.67	0.64	0.97	2.56	0.82	2.53	5.76	2.85	2.06	3.15	0.89	0.20	23.10
1995	0.76	0.62	2.62	0.69	2.07	1.41	5.27	1.75	2.58	2.04	0.99	0.73	21.53
1996	1.82	0.94	0.41	0.21	3.00	1.33	1.36	2.11	3.18	2.41	0.07	0.69	17.53
1997	1.79	0.59	1.89	3.12	2.54	4.86	2.73	2.60	2.31	2.89	0.45	0.07	25.84
1998	0.81	1.51	0.97	0.60	7.34	6.62	2.74	1.93	2.44	4.73	1.75	0.31	31.75
1999	1.15	0.20	1.83	1.04	3.50	2.83	2.34	4.43	6.50	1.04	T	0.45	25.31
2000	0.33	0.99	1.77	1.33	2.69	11.72	2.44	3.07	3.64	1.96	4.13	0.69	34.76
2001	0.20	0.74	0.26	2.70	2.88	2.73	3.14	2.19	1.45	2.74	1.00	0.22	20.25
POR= 121 YRS	0.66	0.63	0.57	1.41	3.56	2.57	1.72	0.90	2.37	1.19	0.41	0.08	16.07

AVERAGE TEMPERATURE (F) 2001 FARGO, ND (FAR)

YEAR	JAN	FEB	MAR	APR	MAY	JUN	JUL	AUG	SEP	OCT	NOV	DEC	ANNUAL
1972	2.7	4.1	23.9	41.0	59.8	66.9	68.4	70.5	56.9	42.4	28.6	3.8	39.1
1973	10.3	16.3	36.0	41.4	54.2	64.7	68.3	71.7	54.8	50.2	25.1	10.1	41.9
1974	1.7	9.6	22.9	42.2	51.1	64.4	73.7	64.3	53.4	47.5	29.2	20.9	40.1
1975	12.3	10.1	18.4	35.9	56.6	65.2	74.3	68.1	55.4	49.4	31.1	14.8	41.0
1976	7.7	21.4	23.0	47.0	55.8	68.5	71.8	73.6	60.0	39.5	23.2	6.9	41.5
1977	-3.3	17.5	32.0	49.5	66.5	66.6	72.2	62.5	57.0	47.1	25.6	6.5	41.7
1978	-1.4	3.4	23.5	42.5	59.1	64.7	69.5	69.1	63.6	46.4	22.8	7.3	39.2
1979	-4.2	-1.5	20.4	36.0	50.4	65.4	71.9	67.3	62.6	42.6	24.5	20.7	38.0
1980	6.6	8.3	20.7	49.0	61.4	65.7	71.9	67.5	56.9	42.4	33.1	12.7	41.4
1981	11.8	19.6	33.5	45.6	55.5	62.8	71.1	66.9	57.4	44.5	35.4	8.7	43.0
1982	-7.0	8.9	22.9	40.7	58.1	59.1	70.9	68.5	57.5	45.7	24.1	20.9	39.2
1983	16.1	21.8	29.9	40.2	52.1	66.1	73.5	72.9	54.7	44.4	31.3	-.3	42.1
1984	9.7	24.9	23.4	45.6	54.2	65.8	70.6	73.3	54.4	47.4	29.7	9.6	42.4
1985	5.1	10.9	32.9	46.6	60.2	60.0	69.4	64.5	53.3	45.4	15.4	3.9	38.9
1986	13.8	10.5	31.6	43.9	57.5	67.5	71.5	65.6	56.1	45.3	23.1	20.8	42.3
1987	18.2	27.5	31.4	51.5	61.7	69.1	74.0	66.8	59.6	42.6	33.4	20.6	46.4
1988	5.9	9.3	29.5	44.5	63.9	73.8	75.8	72.2	58.5	42.9	27.5	15.2	43.3
1989	11.4	1.7	20.1	42.2	58.2	64.1	75.9	70.8	58.5	45.8	24.0	4.3	39.8
1990	21.8	17.6	31.4	43.6	55.0	67.0	70.0	71.1	62.3	46.5	32.1	12.2	44.1
1991	6.4	20.5	30.4	48.0	61.5	70.1	70.2	72.7	58.8	42.0	22.0	18.8	43.5
1992	17.3	23.5	32.6	41.4	58.7	62.0	64.3	64.8	56.8	45.1	27.3	10.5	42.0
1993	7.5	9.8	25.6	43.3	56.7	63.1	67.0	69.2	54.7	43.9	26.6	16.0	40.3
1994	-3.9	6.3	30.5	43.7	59.9	68.2	67.6	66.5	61.9	50.4	34.0	20.7	42.2
1995	10.8	12.4	28.4	38.9	54.8	71.4	70.0	72.0	58.8	44.1	21.2	11.5	41.2
1996	-1.8	11.2	17.4	37.8	53.6	67.0	67.9	71.1	59.1	45.3	17.7	5.9	37.7
1997	1.8	12.4	20.1	37.8	53.0	69.0	69.3	67.9	61.9	47.0	23.2	23.5	40.6
1998	11.3	28.0	26.6	49.2	60.9	63.4	71.7	72.7	63.9	47.6	29.3	17.3	45.2
1999	6.3	22.5	31.2	45.1	58.0	66.4	71.5	68.0	55.4	44.3	37.1	22.9	44.1
2000	9.8	21.6	35.2	42.4	57.3	62.7	70.6	69.6	58.0	48.3	26.0	-.2	41.8
2001	14.3	3.9	23.0	44.4	58.5	65.9	72.6	70.5	59.4	44.4	39.7	20.1	43.1
POR= 121 YRS	5.6	10.5	24.8	42.6	55.3	64.8	70.1	68.1	58.1	45.5	27.5	12.8	40.5

REFERENCE NOTES:

PAGE 1:
THE TEMPERATURE GRAPH SHOWS NORMAL MAXIMUM AND NORMAL
MINIMUM DAILY TEMPERATURES (SOLID CURVES) AND THE
ACTUAL DAILY HIGH AND LOW TEMPERATURES (VERTICAL BARS).

PAGE 2 AND 3:
H/C INDICATES HEATING AND COOLING DEGREE DAYS.
RH INDICATES RELATIVE HUMIDITY
W/O INDICATES WEATHER AND OBSTRUCTIONS
S INDICATES SUNSHINE.
PR INDICATES PRESSURE.
CLOUDINESS ON PAGE 3 IS THE SUM OF THE CEILOMETER AND
SATELLITE DATA NOT TO EXCEED EIGHT EIGHTHS(OKTAS).

GENERAL:
T INDICATES TRACE PRECIPITATION, AN AMOUNT GREATER
THAN ZERO BUT LESS THAN THE LOWEST REPORTABLE VALUE.
+ INDICATES THE VALUE ALSO OCCURS ON EARLIER DATES.
BLANK ENTRIES DENOTE MISSING OR UNREPORTED DATA.
NORMALS ARE 30-YEAR AVERAGES (1961 - 1990).
ASOS INDICATES AUTOMATED SURFACE OBSERVING SYSTEM.
PM INDICATES THE LAST DAY OF THE PREVIOUS MONTH.
POR (PERIOD OF RECORD) BEGINS WITH THE JANUARY DATA
MONTH AND IS THE NUMBER OF YEARS USED TO COMPUTE
THE MEAN. INDIVIDUAL MONTHS WITHIN THE POR MAY
BE MISSING.
WHEN THE POR FOR A NORMAL IS LESS THAN 30 YEARS,
THE NORMAL IS PROVISIONAL AND IS BASED ON THE NUMBER
OF YEARS INDICATED.
0.* OR * INDICATES THE VALUE OR MEAN-DAYS-WITH
IS BETWEEN 0.00 AND 0.05.
CLOUDINESS FOR ASOS STATIONS DIFFERS FROM THE NON-ASOS
OBSERVATION TAKEN BY A HUMAN OBSERVER. ASOS STATION
CLOUDINESS IS BASED ON TIME-AVERAGED CEILOMETER DATA
FOR CLOUDS AT OR BELOW 12,000 FEET AND ON SATELLITE
DATA FOR CLOUDS ABOVE 12,000 FEET.
THE NUMBER OF DAYS WITH CLEAR, PARTLY CLOUDY, AND
CLOUDY CONDITIONS FOR ASOS STATIONS IS THE SUM
OF THE CEILOMETER AND SATELLITE DATA FOR THE
SUNRISE TO SUNSET PERIOD.

GENERAL CONTINUED:
CLEAR INDICATES 0 - 2 OKTAS, PARTLY CLOUDY INDICATES
3 - 6 OKTAS, AND CLOUDY INDICATES 7 OR 8 OKTAS.
WHEN AT LEAST ONE OF THE ELEMENTS (CEILOMETER OR
SATELLITE) IS MISSING, THE DAILY CLOUDINESS IS
NOT COMPUTED.
WIND DIRECTION IS RECORDED IN TENS OF DEGREES (2 DIGITS)
CLOCKWISE FROM TRUE NORTH. "00" INDICATES CALM. "36"
INDICATES TRUE NORTH.
RESULTANT WIND IS THE VECTOR AVERAGE OF THE SPEED AND
DIRECTION.
AVERAGE TEMPERATURE IS THE SUM OF THE MEAN DAILY MAXIMUM
AND MINIMUM TEMPERATURE DIVIDED BY 2.
SNOWFALL DATA COMPRISE ALL FORMS OF FROZEN
PRECIPITATION, INCLUDING HAIL.
A HEATING (COOLING) DEGREE DAY IS THE DIFFERENCE BETWEEN
THE AVERAGE DAILY TEMPERATURE AND 65 F.
DRY BULB IS THE TEMPERATURE OF THE AMBIENT AIR.
DEW POINT IS THE TEMPERATURE TO WHICH THE AIR MUST BE
COOLED TO ACHIEVE 100 PERCENT RELATIVE HUMIDITY.
WET BULB IS THE TEMPERATURE THE AIR WOULD HAVE IF THE
MOISTURE CONTENT WAS INCREASED TO 100 PERCENT RELATIVE
HUMIDITY.

ON JULY 1, 1996, THE NATIONAL WEATHER SERVICE BEGAN USING
THE "METAR" OBSERVATION CODE THAT WAS ALREADY EMPLOYED
BY MOST OTHER NATIONS OF THE WORLD. THE MOST NOTICEABLE
DIFFERENCE IN THIS ANNUAL PUBLICATION WILL BE THE CHANGE
IN UNITS FROM TENTHS TO EIGHTS(OKTAS) FOR REPORTING THE
AMOUNT OF SKY COVER.

HEATING DEGREE DAYS (base 65 F) 2001 FARGO, ND (FAR)

YEAR	JUL	AUG	SEP	OCT	NOV	DEC	JAN	FEB	MAR	APR	MAY	JUN	TOTAL
1972-73	25	41	261	695	1089	1897	1688	1361	893	701	328	79	9058
1973-74	32	3	309	451	1187	1698	1963	1550	1298	676	431	86	9684
1974-75	3	91	345	537	1066	1362	1630	1535	1438	867	265	79	9218
1975-76	14	22	284	492	1012	1550	1550	1257	1296	533	285	55	8574
1976-77	13	9	227	788	1247	1797	2119	1327	1015	466	74	30	9112
1977-78	7	95	211	549	1178	1817	2061	1721	1284	668	209	90	9890
1978-79	15	39	179	571	1262	1788	2147	1863	1377	861	457	64	10623
1979-80	3	45	139	689	1209	1367	1808	1644	1363	493	206	61	9027
1980-81	3	35	267	696	951	1616	1645	1266	971	574	298	84	8406
1981-82	14	10	250	627	881	1742	2236	1570	1298	725	222	187	9762
1982-83	0	66	257	589	1219	1359	1513	1206	1082	738	390	74	8493
1983-84	16	2	301	631	1004	2023	1714	1154	1280	576	344	52	9097
1984-85	15	13	339	541	1053	1715	1853	1514	988	550	172	179	8932
1985-86	13	72	329	625	1487	1895	1585	1527	1027	627	266	45	9498
1986-87	0	69	268	602	1251	1360	1447	1047	1036	415	163	39	7697
1987-88	15	59	177	688	940	1369	1832	1614	1092	609	131	8	8534
1988-89	3	25	207	677	1118	1537	1658	1771	1386	677	224	96	9379
1989-90	0	17	224	599	1224	1881	1332	1324	1034	666	314	58	8673
1990-91	8	18	173	594	982	1637	1813	1242	1066	505	211	6	8255
1991-92	3	3	234	708	1284	1425	1473	1198	998	709	247	137	8419
1992-93	66	84	252	613	1125	1686	1780	1542	1214	643	269	111	9385
1993-94	35	19	310	652	1144	1514	2137	1640	1062	637	211	18	9379
1994-95	19	58	146	446	925	1366	1674	1469	1131	777	316	42	8369
1995-96	10	0	241	641	1308	1654	2071	1557	1472	812	355	65	10186
1996-97	12	6	234	604	1412	1826	1952	1466	1382	810	368	4	10076
1997-98	48	43	126	560	1249	1281	1659	1031	1185	467	151	111	7911
1998-99	5	0	124	530	1064	1471	1814	1186	1042	588	234	66	8124
1999-00	4	27	285	635	831	1297	1704	1251	918	672	252	119	7995
2000-01	26	23	226	511	1166	2014	1567	1705	1296	617	215	81	9447
2001-	15	26	203	631	751	1385							

COOLING DEGREE DAYS (base 65 F) 2001 FARGO, ND (FAR)

YEAR	JAN	FEB	MAR	APR	MAY	JUN	JUL	AUG	SEP	OCT	NOV	DEC	ANNUAL
1972	0	0	0	0	74	125	135	217	23	0	0	0	574
1973	0	0	0	0	0	76	140	219	13	0	0	0	448
1974	0	0	0	0	9	75	281	75	3	1	0	0	444
1975	0	0	0	0	11	92	308	126	1	15	0	0	553
1976	0	0	0	0	4	164	228	283	83	4	0	0	766
1977	0	0	0	6	129	86	235	23	8	0	0	0	487
1978	0	0	0	0	31	86	176	146	0	0	0	0	604
1979	0	0	0	0	12	85	225	124	58	0	0	0	504
1980	0	0	0	18	102	89	222	119	31	1	0	0	582
1981	0	0	0	0	9	25	212	159	26	0	0	0	431
1982	0	0	0	2	11	20	189	179	39	0	0	0	440
1983	0	0	0	0	2	113	288	252	55	0	0	0	710
1984	0	0	0	0	18	81	196	279	24	4	0	0	602
1985	0	0	0	6	31	35	143	63	4	0	0	0	282
1986	0	0	0	0	41	126	208	92	10	0	0	0	477
1987	0	0	0	17	66	169	303	121	25	0	0	0	701
1988	0	0	0	0	102	280	346	252	22	0	0	0	1002
1989	0	0	0	0	19	76	345	201	34	11	0	0	686
1990	0	0	0	29	10	123	172	214	98	1	0	0	647
1991	0	0	0	2	107	166	171	250	54	0	0	0	750
1992	0	0	0	9	58	52	49	85	11	3	0	0	267
1993	0	0	0	0	19	61	107	155	9	4	0	0	355
1994	0	0	0	4	59	122	105	109	60	0	0	0	459
1995	0	0	0	0	5	243	175	227	62	3	0	0	715
1996	0	0	0	0	8	131	107	204	63	0	0	0	513
1997	0	0	0	0	3	131	186	136	38	9	0	0	503
1998	0	0	0	0	32	71	218	245	96	0	0	0	662
1999	0	0	0	0	23	117	214	127	4	0	0	0	485
2000	0	0	0	0	19	57	208	172	19	0	0	0	475
2001	0	0	0	9	20	112	257	202	44	0	0	0	644

SNOWFALL (inches) 2001 FARGO, ND (FAR)

YEAR	JUL	AUG	SEP	OCT	NOV	DEC	JAN	FEB	MAR	APR	MAY	JUN	TOTAL
1972-73	0.0	0.0	T	3.8	1.7	18.5	1.7	1.4	1.4	2.4	0.0	0.0	30.9
1973-74	0.0	0.0	0.0	T	3.9	12.3	6.1	7.1	10.5	2.7	0.0	0.0	42.6
1974-75	0.0	0.0	0.0	0.4	1.0	5.1	18.3	5.9	18.7	3.7	T	0.0	53.1
1975-76	0.0	0.0	0.0	0.4	3.8	1.9	14.0	6.2	14.0	T	0.1	0.0	40.4
1976-77	0.0	0.0	0.0	0.1	2.7	5.5	12.6	10.7	4.6	2.1	0.0	0.0	38.3
1977-78	0.0	0.0	0.0	T	24.2	7.2	4.6	3.9	7.1	2.8	0.0	0.0	49.8
1978-79	0.0	0.0	0.0	T	8.5	11.7	7.8	19.5	4.3	2.7	0.8	0.0	55.3
1979-80	0.0	0.0	0.0	1.4	6.0	1.5	17.3	7.2	6.5	T	0.0	0.0	39.9
1980-81	0.0	0.0	0.0	0.5	1.1	4.6	2.1	4.5	0.3	T	0.0	0.0	13.1
1981-82	0.0	0.0	T	2.3	2.2	9.9	30.0	10.9	14.0	0.2	0.0	0.0	69.5
1982-83	0.0	0.0	0.0	0.0	6.8	0.3	3.8	2.0	7.4	2.9	T	0.0	23.2
1983-84	0.0	0.0	T	T	5.3	11.8	11.5	3.1	7.7	0.5	0.0	0.0	39.9
1984-85	0.0	0.0	T	T	1.4	7.4	3.7	3.1	12.6	T	0.0	0.0	28.2
1985-86	0.0	0.0	0.0	T	24.3	10.4	11.2	6.7	0.7	3.7	T	0.0	57.0
1986-87	0.0	0.0	0.0	T	5.3	3.8	2.8	10.4	1.2	T	0.0	0.0	23.5
1987-88	0.0	0.0	0.0	T	3.0	6.6	24.3	4.4	6.2	T	0.0	0.0	44.5
1988-89	0.0	0.0	0.0	T	11.6	14.9	31.5	2.3	12.4	0.9	T	0.0	73.6
1989-90	0.0	T	T	T	16.3	2.6	0.8	7.9	11.5	7.2	T	0.0	46.3
1990-91	0.0	0.0	0.0	1.3	0.2	12.4	4.0	15.3	10.9	4.2	T	T	48.3
1991-92	0.0	0.0	T	0.3	5.2	5.9	10.5	2.1	0.2	3.3	T	0.0	27.5
1992-93	0.0	0.0	T	1.8	16.4	9.2	16.7	3.3	6.4	T	0.0	0.0	53.8
1993-94	0.0	0.0	0.0	T	21.5	13.8	18.0	12.8	12.1	10.9	0.0	T	89.1
1994-95	0.0	T	0.0	0.0	4.0	3.0	10.8	9.5	19.0	4.0	0.0	0.0	50.3
1995-96	0.0	0.0	T	1.0	9.6	12.7	27.2	8.9	15.0	0.2	0.0	0.0	74.6
1996-97	0.0	0.0	0.0	T	26.4	20.4	28.6	8.0	26.2	7.4	T	0.0	117.0
1997-98	0.0	0.0	0.0	0.3	11.1	7.4	12.6	3.6	5.4	0.7	0.0	T	41.1
1998-99	0.0	0.0	0.0	0.0	12.3	4.5	19.7	2.1	10.0	T	0.0	0.0	48.6
1999-00	T	0.0	0.0	T	0.0	4.3	6.5	11.4	5.6	6.2	0.0	0.0	34.0
2000-01	0.0	0.0	0.0	T	15.3	13.4	2.7	11.6	1.5	8.0	0.0	0.0	52.5
2001-	0.0	0.0	0.0	5.4	11.2	5.0							
POR= 59 YRS	0.0	T	0.0	0.7	6.3	7.2	9.4	6.1	7.3	3.3	0.1	T	40.4

2001
COVINGTON/CINCINNATI,
KENTUCKY (CVG)

Greater Cincinnati Airport is located on a gently rolling plateau about 12 miles southwest of downtown Cincinnati and 2 miles south of the Ohio River at its nearest point. The river valley is rather narrow and steep-sided varying from 1 to 3 miles in width and the river bed is 500 feet below the level of the airport.

The climate is continental with a rather wide range of temperatures from winter to summer. A precipitation maximum occurs during winter and spring with a late summer and fall minimum. On the average, the maximum snowfall occurs during January, although the heaviest 24-hour amounts have been recorded during late November and February.

The heaviest precipitation, as well as the precipitation of the longest duration, is normally associated with low pressure disturbances moving in a general southwest to northeast direction through the Ohio valley and south of the Cincinnati area.

Summers are warm and rather humid. The temperature will reach 100 degrees or more in 1 year out of 3. However, the temperature will reach 90 degrees or higher on about 19 days each year. Winters are moderately cold with frequent periods of extensive cloudiness.

The freeze free period lasts on the average 187 days from mid-April to the latter part of October.

NORMALS, MEANS, AND EXTREMES
COVINGTON/CINCINNATI, OH (CVG)

LATITUDE:	LONGITUDE:	ELEVATION (FT):		TIME ZONE:	WBAN: 93814
39 02' 35" N	84 40' 18" W	GRND: 882	BARO: 885	EASTERN (UTC + 5)	

	ELEMENT	POR	JAN	FEB	MAR	APR	MAY	JUN	JUL	AUG	SEP	OCT	NOV	DEC	YEAR
TEMPERATURE F	NORMAL DAILY MAXIMUM	30	36.6	40.8	53.0	64.2	74.0	82.0	85.5	84.1	77.9	66.0	53.3	41.5	63.2
	MEAN DAILY MAXIMUM	54	37.9	42.5	51.9	64.2	74.1	82.4	85.9	84.6	77.9	66.6	52.8	42.0	63.6
	HIGHEST DAILY MAXIMUM	40	69	75	84	89	93	102	103	102	98	88	81	75	103
	YEAR OF OCCURRENCE		1967	2000	1986	1976	1962	1988	1988	1962	1964	1963	1987	1982	JUL 1988
	MEAN OF EXTREME MAXS.	54	60.3	64.5	74.7	82.1	87.1	92.5	94.7	93.5	90.6	82.3	72.1	62.6	79.7
	NORMAL DAILY MINIMUM	30	19.5	22.7	33.1	42.2	51.8	60.0	64.8	62.9	56.6	44.2	35.3	25.3	43.2
	MEAN DAILY MINIMUM	54	21.6	24.7	32.6	42.7	52.3	61.0	65.4	63.5	56.3	44.7	35.0	26.1	43.8
	LOWEST DAILY MINIMUM	40	-25	-11	-11	15	27	39	47	43	31	16	1	-20	-25
	YEAR OF OCCURRENCE		1977	1996	1980	1997	1963	1972	1963	1986	1993	1962	1976	1989	JAN 1977
	MEAN OF EXTREME MINS.	54	-2.1	3.0	14.2	26.2	36.2	47.5	53.8	52.3	40.1	27.6	17.5	5.2	26.8
	NORMAL DRY BULB	30	28.1	31.8	43.0	53.2	62.9	71.0	75.1	73.5	67.3	55.1	44.3	33.5	53.2
	MEAN DRY BULB	54	29.8	33.6	42.3	53.4	63.2	71.6	75.6	74.1	67.3	55.6	44.0	34.0	53.7
	MEAN WET BULB	18	28.3	32.1	38.0	47.5	57.5	65.3	68.8	67.3	60.4	50.2	40.6	31.8	49.0
	MEAN DEW POINT	18	23.5	26.2	31.1	40.7	52.3	61.5	65.1	63.7	56.0	45.1	35.1	27.0	43.9
	NORMAL NO. DAYS WITH:														
	MAXIMUM 90	30	0.0	0.0	0.0	0.0	0.7	4.3	7.8	5.6	2.1	0.0	0.0	0.0	20.5
	MAXIMUM 32	30	12.0	7.6	1.3	0.0	0.0	0.0	0.0	0.0	0.0	0.0	0.6	6.6	28.1
	MINIMUM 32	30	26.5	22.3	15.6	4.6	0.3	0.0	0.0	0.0	0.0	3.5	12.7	22.4	107.9
	MINIMUM 0	30	3.2	1.6	0.1	0.0	0.0	0.0	0.0	0.0	0.0	0.0	0.0	1.3	6.2
H/C	NORMAL HEATING DEG. DAYS	30	1144	930	682	354	151	11	0	0	51	327	621	977	5248
	NORMAL COOLING DEG. DAYS	30	0	0	0	0	86	191	313	266	120	20	0	0	996
RH	NORMAL (PERCENT)	30	72	70	67	63	67	69	72	72	73	69	71	74	70
	HOUR 01 LST	30	75	74	72	70	77	80	83	83	82	77	75	76	77
	HOUR 07 LST	30	78	78	77	76	80	82	85	88	88	83	80	80	81
	HOUR 13 LST	30	67	64	59	53	54	56	57	57	57	55	62	68	59
	HOUR 19 LST	30	69	65	60	54	58	59	61	63	66	64	68	72	63
S	PERCENT POSSIBLE SUNSHINE	13	33	40	48	56	57	61	62	61	61	54	36	31	50
W/O	MEAN NO. DAYS WITH:														
	HEAVY FOG(VISBY 1/4 MI)	39	2.5	2.0	1.6	0.9	1.2	1.5	1.7	2.7	2.8	2.1	1.6	2.5	23.1
	THUNDERSTORMS	55	0.7	0.8	2.5	4.3	5.9	7.3	8.0	7.1	3.0	1.4	1.2	0.5	42.7
CLOUDINESS	MEAN:														
	SUNRISE-SUNSET (OKTAS)														
	MIDNIGHT-MIDNIGHT (OKTAS)														
	MEAN NO. DAYS WITH:														
	CLEAR	1			5.0										
	PARTLY CLOUDY	1			1.0		1.0								
	CLOUDY	1	1.0	4.0	7.0		2.0	1.0							
PR	MEAN STATION PRESSURE(IN)	29	29.20	29.10	29.10	29.10	29.09	29.10	29.10	29.10	29.10	29.20	29.19	29.19	29.13
	MEAN SEA-LEVEL PRES. (IN)	18	30.12	30.11	30.06	29.98	29.99	29.97	30.01	30.04	30.06	30.11	30.11	30.14	30.06
WINDS	MEAN SPEED (MPH)	47	10.3	10.4	11.0	10.6	8.6	7.9	7.2	6.8	7.4	8.1	9.9	9.9	9.0
	PREVAIL.DIR(TENS OF DEGS)	31	21	21	21	21	22	21	21	21	22	21	21	21	21
	MAXIMUM 2-MINUTE:														
	SPEED (MPH)	6	41	39	40	40	35	41	45	41	36	48	40	40	48
	DIR. (TENS OF DEGS)		17	30	27	29	30	27	32	29	31	29	17	27	29
	YEAR OF OCCURRENCE		1996	1996	2001	1997	2001	2001	2001	2000	1997	2001	2001	2000	OCT 2001
	MAXIMUM 5-SECOND:														
	SPEED (MPH)	6	51	47	55	59	54	56	61	53	43	60	55	52	61
	DIR. (TENS OF DEGS)		01	21	28	31	25	27	31	29	31	29	17	28	31
	YEAR OF OCCURRENCE		1996	2001	1997	1998	1999	2001	2001	2000	1997	2001	2001	2000	JUL 2001
PRECIPITATION	NORMAL (IN)	30	2.59	2.69	4.24	3.75	4.28	3.84	4.24	3.35	2.88	2.86	3.46	3.15	41.33
	MAXIMUM MONTHLY (IN)	54	9.43	6.72	12.18	9.77	9.48	9.61	8.70	7.71	8.61	8.60	7.51	7.90	12.18
	YEAR OF OCCURRENCE		1950	1955	1964	1998	1968	1998	2001	1982	1979	1983	1985	1990	MAR 1964
	MINIMUM MONTHLY (IN)	54	0.57	0.25	1.14	1.04	1.13	0.95	0.63	0.31	0.18	0.25	0.43	0.51	0.18
	YEAR OF OCCURRENCE		1981	1978	1960	1971	1964	1965	1999	1997	1963	1963	1949	1976	SEP 1963
	MAXIMUM IN 24 HOURS (IN)	54	4.33	2.84	5.21	3.31	3.71	3.45	4.28	3.52	4.54	4.47	3.36	2.96	5.21
	YEAR OF OCCURRENCE		1959	1990	1964	1996	1956	1974	1988	1995	1979	1985	1948	1948	MAR 1964
	NORMAL NO. DAYS WITH:														
	PRECIPITATION 0.01	30	11.3	11.0	13.3	12.1	11.1	10.4	10.1	9.2	8.3	8.4	10.8	12.5	128.5
	PRECIPITATION 1.00	30	0.5	0.6	0.7	0.6	1.0	1.0	1.1	0.8	0.6	0.7	0.7	0.5	8.8
SNOWFALL	NORMAL (IN)	30	7.1	6.1	4.1	0.5	0.*	0.0	0.0	0.0	0.0	0.3	1.8	3.7	23.6
	MAXIMUM MONTHLY (IN)	54	31.5	19.9	13.0	3.7	0.2	T	T	T	0.0	6.2	12.1	12.5	31.5
	YEAR OF OCCURRENCE		1978	1993	1968	1977	1989	1993	1994	2000		1993	1966	1989	JAN 1978
	MAXIMUM IN 24 HOURS (IN)	54	12.8	12.6	9.8	3.6	0.2	T	T	T	0.0	5.9	9.0	7.5	12.8
	YEAR OF OCCURRENCE		1996	1998	1968	1977	1989	1993	1994	2000		1993	1966	1990	JAN 1996
	MAXIMUM SNOW DEPTH (IN)	53	14	19	11	5	0	0	0	0	0	4	8	8	19
	YEAR OF OCCURRENCE		1978	1998	1980	1987						1993	1966	1990	FEB 1998
	NORMAL NO. DAYS WITH:														
	SNOWFALL 1.0	30	2.2	2.0	1.1	0.2	0.0	0.0	0.0	0.0	0.0	0.1	0.4	1.1	7.1

PRECIPITATION (inches) 2001 COVINGTON/CINCINNATI, OH (CVG)

YEAR	JAN	FEB	MAR	APR	MAY	JUN	JUL	AUG	SEP	OCT	NOV	DEC	ANNUAL
1972	1.96	2.20	3.68	5.89	6.02	2.41	1.50	2.64	5.96	2.55	6.26	4.23	45.30
1973	1.79	1.58	6.11	5.81	3.46	6.27	7.16	2.62	2.63	4.39	4.95	2.66	49.43
1974	3.65	1.63	4.39	5.08	5.53	4.38	3.82	5.75	4.44	1.07	4.19	2.83	46.76
1975	4.05	3.38	6.76	4.16	3.11	5.09	1.62	1.97	3.64	4.59	2.50	3.36	44.23
1976	3.00	2.37	2.14	1.21	1.80	5.94	2.33	1.95	3.85	0.43	0.83	0.51	30.29
1977	1.90	1.29	4.52	4.16	1.53	7.36	1.90	5.45	1.80	3.74	3.90	4.00	41.55
1978	4.52	0.25	1.99	2.28	5.30	6.63	6.86	4.41	0.43	5.03	2.67	6.46	46.83
1979	3.68	3.77	2.05	4.90	4.00	5.92	5.49	4.80	8.61	1.77	4.86	2.91	52.76
1980	2.26	1.04	4.50	1.96	4.59	4.13	5.51	4.19	1.83	3.28	2.58	1.26	37.13
1981	0.57	3.86	1.72	5.05	5.07	3.34	3.66	2.15	1.47	2.33	2.94	2.39	34.55
1982	7.17	1.17	4.67	2.18	4.60	3.61	2.44	7.71	1.27	0.99	5.08	4.25	45.14
1983	1.56	1.14	2.02	4.84	8.89	2.22	1.96	3.23	1.22	8.60	4.20	2.84	42.72
1984	0.75	2.40	3.61	4.88	4.82	2.11	2.57	3.30	3.50	3.85	6.00	4.21	42.00
1985	1.68	2.25	6.90	1.34	6.18	4.55	3.59	2.02	0.76	5.83	7.51	1.52	44.13
1986	1.01	2.85	3.07	1.57	3.59	1.46	3.33	3.78	3.53	3.08	3.79	2.58	33.64
1987	0.92	1.62	4.65	2.88	2.73	4.62	1.17	2.27	1.17	1.82	3.43	3.43	32.60
1988	2.75	4.94	3.42	3.92	1.99	1.19	6.85	2.44	3.05	1.86	4.78	2.78	39.97
1989	3.21	4.67	6.40	5.19	4.64	3.18	5.97	5.33	2.97	3.18	3.05	1.96	49.61
1990	2.59	5.82	2.75	3.22	9.41	5.01	3.68	5.67	4.13	5.09	2.31	7.90	57.58
1991	2.84	3.99	6.20	3.62	3.41	1.39	2.66	5.04	2.60	1.37	1.89	5.08	40.09
1992	2.99	0.93	4.19	2.71	2.84	3.65	7.00	3.17	3.23	1.11	4.31	1.36	37.49
1993	3.83	3.43	3.60	3.13	2.33	4.80	1.26	4.20	2.68	2.61	4.31	2.53	38.71
1994	3.22	1.68	2.22	6.46	2.06	4.08	5.64	5.14	1.49	2.87	2.88	3.35	38.29
1995	3.51	1.80	2.58	4.26	8.57	2.65	2.37	5.59	2.43	4.28	2.15	3.43	43.62
1996	4.36	1.98	5.58	8.20	9.20	5.83	2.62	0.76	5.41	1.74	3.40	4.33	53.41
1997	2.79	2.13	6.00	1.98	6.33	8.34	0.63	3.95	0.55	1.68	2.97	2.77	40.12
1998	3.27	3.04	3.52	9.77	5.12	9.61	4.75	2.67	0.67	2.82	2.33	3.82	51.39
1999	4.76	3.66	1.89	2.88	1.98	3.16	3.16	2.61	0.86	2.49	1.42	3.60	32.47
2000	4.45	5.71	3.34	4.27	5.21	4.74	3.53	2.90	4.79	1.37	2.33	3.18	45.82
2001	1.33	1.81	1.42	1.46	5.15	4.45	8.70	5.00	3.13	6.73	3.31	4.08	46.57
POR= 54 YRS	3.24	2.88	3.88	3.73	4.20	4.09	4.08	3.28	2.71	2.72	3.23	3.17	41.21

AVERAGE TEMPERATURE (F) 2001 COVINGTON/CINCINNATI, OH (CVG)

YEAR	JAN	FEB	MAR	APR	MAY	JUN	JUL	AUG	SEP	OCT	NOV	DEC	ANNUAL
1972	30.2	29.5	40.6	52.5	63.1	65.3	75.5	73.4	68.7	51.7	40.3	36.1	52.2
1973	31.7	33.0	50.6	51.1	58.4	72.3	74.8	73.7	69.6	59.0	45.4	33.6	54.4
1974	35.7	33.9	45.9	54.3	61.9	67.7	75.2	74.3	61.6	53.2	44.4	34.3	53.5
1975	34.0	37.1	39.8	50.6	67.6	72.3	74.6	76.9	62.4	57.5	49.1	37.4	54.9
1976	27.1	41.7	47.0	56.1	60.0	71.5	75.1	71.2	63.6	48.8	34.9	27.6	51.9
1977	12.0	29.8	46.5	56.3	68.2	68.7	77.9	73.9	69.8	52.3	45.4	28.7	52.5
1978	18.4	18.2	36.4	53.4	60.1	72.2	75.0	73.0	70.8	52.6	46.4	36.0	51.0
1979	21.3	21.4	46.6	50.9	60.2	69.4	73.4	72.3	66.1	54.3	44.2	36.2	51.4
1980	29.9	24.0	38.5	50.3	64.9	70.1	76.6	76.5	68.6	50.6	41.5	32.9	52.0
1981	24.1	34.2	40.1	58.1	59.8	72.3	75.9	73.6	65.2	53.9	43.7	29.0	52.5
1982	23.9	30.6	44.3	49.6	68.1	67.4	77.0	71.3	66.9	59.3	48.4	42.9	54.1
1983	31.6	35.3	44.7	49.4	59.0	71.6	79.2	78.3	67.2	55.7	44.7	24.6	53.4
1984	23.7	38.2	34.4	51.1	58.9	74.1	72.2	74.3	65.7	61.5	42.0	42.4	53.2
1985	22.7	29.4	47.5	57.9	65.4	69.9	75.1	72.5	67.5	59.3	49.9	26.2	53.6
1986	30.9	35.2	45.2	55.3	64.5	72.9	77.6	72.0	70.1	56.4	42.7	34.0	54.7
1987	30.7	37.3	45.0	53.0	69.3	73.6	76.1	75.2	68.3	49.3	48.0	36.8	55.2
1988	27.5	30.5	42.2	52.4	64.4	72.4	78.5	77.5	67.2	48.5	45.0	34.1	53.4
1989	38.6	30.8	45.4	52.9	60.1	71.3	76.7	73.5	66.4	55.7	43.9	21.6	53.1
1990	40.0	40.8	48.2	52.8	61.6	71.8	74.7	73.7	67.7	55.9	48.9	38.3	56.2
1991	31.0	37.3	45.4	56.9	70.4	75.4	77.9	74.8	68.6	58.6	41.4	37.3	56.3
1992	32.9	39.7	43.6	54.1	61.2	68.0	73.5	69.6	64.5	54.3	44.2	35.6	53.4
1993	34.5	29.3	40.2	52.0	63.3	70.9	79.1	76.2	63.9	53.0	43.5	32.8	53.2
1994	23.3	33.2	41.4	55.5	59.3	75.0	71.9	65.7	57.1	49.9	44.0	45.0	54.0
1995	31.3	31.0	45.7	53.3	62.7	72.9	76.6	79.5	65.7	55.3	37.5	29.3	53.4
1996	27.9	32.4	36.7	49.2	63.2	71.0	72.1	74.3	65.4	55.7	37.8	37.6	51.9
1997	28.2	37.6	44.2	48.1	57.2	69.1	75.1	71.6	65.6	55.4	41.3	35.7	52.4
1998	38.6	40.5	43.6	53.0	66.6	71.3	74.3	75.1	71.6	56.2	45.5	38.3	56.2
1999	32.4	37.9	38.3	55.0	63.6	73.0	79.0	72.7	66.8	55.3	48.5	34.8	54.8
2000	28.6	40.4	46.8	52.3	65.4	71.4	72.6	72.3	64.9	57.8	41.7	23.2	53.1
2001	29.7	37.0	39.7	58.0	64.2	69.7	74.2	74.9	64.6	55.2	50.0	39.3	54.7
POR= 54 YRS	29.8	33.6	42.3	53.4	63.2	71.7	75.6	74.1	67.3	55.6	44.0	34.0	53.7

REFERENCE NOTES:

PAGE 1:
THE TEMPERATURE GRAPH SHOWS NORMAL MAXIMUM AND NORMAL
  MINIMUM DAILY TEMPERATURES (SOLID CURVES) AND THE
  ACTUAL DAILY HIGH AND LOW TEMPERATURES (VERTICAL BARS).

PAGE 2 AND 3:
H/C INDICATES HEATING AND COOLING DEGREE DAYS.
RH INDICATES RELATIVE HUMIDITY.
W/O INDICATES WEATHER AND OBSTRUCTIONS
S INDICATES SUNSHINE.
PR INDICATES PRESSURE.
CLOUDINESS ON PAGE 3 IS THE SUM OF THE CEILOMETER AND
  SATELLITE DATA NOT TO EXCEED EIGHT EIGHTHS (OKTAS).

GENERAL:
T INDICATES TRACE PRECIPITATION, AN AMOUNT GREATER
  THAN ZERO BUT LESS THAN THE LOWEST REPORTABLE VALUE.
+ INDICATES THE VALUE ALSO OCCURS ON EARLIER DATES.
BLANK ENTRIES DENOTE MISSING OR UNREPORTED DATA.
NORMALS ARE 30-YEAR AVERAGES (1961 - 1990).
ASOS INDICATES AUTOMATED SURFACE OBSERVING SYSTEM.
PM INDICATES THE LAST DAY OF THE PREVIOUS MONTH.
POR (PERIOD OF RECORD) BEGINS WITH THE JANUARY DATA
  MONTH AND IS THE NUMBER OF YEARS USED TO COMPUTE
  THE MEAN. INDIVIDUAL MONTHS WITHIN THE POR MAY
  BE MISSING.
WHEN THE POR FOR A NORMAL IS LESS THAN 30 YEARS,
  THE NORMAL IS PROVISIONAL AND IS BASED ON THE NUMBER
  OF YEARS INDICATED.
0.* OR * INDICATES THE VALUE OR MEAN-DAYS-WITH
  IS BETWEEN 0.00 AND 0.05.
CLOUDINESS FOR ASOS STATIONS DIFFERS FROM THE NON-ASOS
  OBSERVATION TAKEN BY A HUMAN OBSERVER. ASOS STATION
  CLOUDINESS IS BASED ON TIME-AVERAGED CEILOMETER DATA
  FOR CLOUDS AT OR BELOW 12,000 FEET AND ON SATELLITE
  DATA FOR CLOUDS ABOVE 12,000 FEET.
THE NUMBER OF DAYS WITH CLEAR, PARTLY CLOUDY, AND
  CLOUDY CONDITIONS FOR ASOS STATIONS IS THE SUM
  OF THE CEILOMETER AND SATELLITE DATA FOR THE
  SUNRISE TO SUNSET PERIOD.

GENERAL CONTINUED:
CLEAR INDICATES 0 - 2 OKTAS, PARTLY CLOUDY INDICATES
  3 - 6 OKTAS, AND CLOUDY INDICATES 7 OR 8 OKTAS.
  WHEN AT LEAST ONE OF THE ELEMENTS (CEILOMETER OR
  SATELLITE) IS MISSING, THE DAILY CLOUDINESS IS
  NOT COMPUTED.
WIND DIRECTION IS RECORDED IN TENS OF DEGREES (2 DIGITS)
  CLOCKWISE FROM TRUE NORTH. "00" INDICATES CALM. "36"
  INDICATES TRUE NORTH.
RESULTANT WIND IS THE VECTOR AVERAGE OF THE SPEED AND
  DIRECTION.
AVERAGE TEMPERATURE IS THE SUM OF THE MEAN DAILY MAXIMUM
  AND MINIMUM TEMPERATURE DIVIDED BY 2.
SNOWFALL DATA COMPRISE ALL FORMS OF FROZEN
  PRECIPITATION, INCLUDING HAIL.
A HEATING (COOLING) DEGREE DAY IS THE DIFFERENCE BETWEEN
  THE AVERAGE DAILY TEMPERATURE AND 65 F.
DRY BULB IS THE TEMPERATURE OF THE AMBIENT AIR.
DEW POINT IS THE TEMPERATURE TO WHICH THE AIR MUST BE
  COOLED TO ACHIEVE 100 PERCENT RELATIVE HUMIDITY.
WET BULB IS THE TEMPERATURE THE AIR WOULD HAVE IF THE
  MOISTURE CONTENT WAS INCREASED TO 100 PERCENT RELATIVE
  HUMIDITY.

ON JULY 1, 1996, THE NATIONAL WEATHER SERVICE BEGAN USING
  THE "METAR" OBSERVATION CODE THAT WAS ALREADY EMPLOYED
  BY MOST OTHER NATIONS OF THE WORLD. THE MOST NOTICEABLE
  DIFFERENCE IN THIS ANNUAL PUBLICATION WILL BE THE CHANGE
  IN UNITS FROM TENTHS TO EIGHTS (OKTAS) FOR REPORTING THE
  AMOUNT OF SKY COVER.

HEATING DEGREE DAYS (base 65 F) 2001 COVINGTON/CINCINNATI, OH (CVG)

YEAR	JUL	AUG	SEP	OCT	NOV	DEC	JAN	FEB	MAR	APR	MAY	JUN	TOTAL
1972-73	2	3	24	404	733	891	1025	888	445	425	206	3	5049
1973-74	0	1	27	217	583	964	901	863	591	332	161	31	4671
1974-75	0	0	147	371	614	942	953	774	778	435	50	10	5074
1975-76	5	0	142	244	474	848	1168	670	558	321	176	5	4611
1976-77	0	4	72	498	893	1157	1640	980	571	276	67	36	6194
1977-78	0	2	32	391	586	1118	1440	1303	880	346	207	10	6315
1978-79	0	0	21	381	552	891	1348	1216	563	425	179	15	5591
1979-80	1	14	60	346	616	887	1080	1182	814	434	92	24	5550
1980-81	0	0	48	446	697	988	1261	858	768	230	191	6	5493
1981-82	0	0	87	344	634	1107	1268	956	635	460	28	19	5538
1982-83	0	1	56	244	505	682	1029	825	627	466	199	21	4655
1983-84	1	0	89	288	600	1247	1274	773	939	425	219	4	5859
1984-85	0	0	101	128	684	692	1306	992	543	256	72	22	4796
1985-86	0	0	78	212	450	1195	1056	828	613	305	105	3	4845
1986-87	0	21	25	292	664	955	1058	766	612	365	52	2	4812
1987-88	0	1	39	477	505	868	1156	991	699	374	84	22	5216
1988-89	1	0	38	509	595	949	811	949	608	380	211	14	5065
1989-90	0	4	77	297	630	1335	770	671	531	390	127	21	4853
1990-91	0	1	66	296	477	821	1046	773	602	250	44	0	4376
1991-92	0	0	81	232	700	853	988	727	658	339	172	27	4777
1992-93	0	8	97	358	617	907	937	997	762	384	103	40	5210
1993-94	0	0	101	370	640	992	1291	885	724	298	211	5	5517
1994-95	0	6	56	250	447	766	1037	948	589	353	109	2	4563
1995-96	0	0	68	298	818	1099	1142	939	873	478	132	10	5857
1996-97	0	0	93	307	811	847	1136	765	640	498	252	36	5385
1997-98	1	7	53	339	703	900	808	679	679	355	51	47	4622
1998-99	0	0	17	284	576	826	1007	754	820	297	90	5	4676
1999-00	0	0	63	294	489	931	1119	709	560	375	73	13	4626
2000-01	0	0	107	248	693	1290	1084	778	777	268	96	28	5369
2001-	1	0	96	316	442	786							

COOLING DEGREE DAYS (base 65 F) 2001 COVINGTON/CINCINNATI, OH (CVG)

YEAR	JAN	FEB	MAR	APR	MAY	JUN	JUL	AUG	SEP	OCT	NOV	DEC	ANNUAL
1972	0	0	0	7	65	96	334	268	142	0	0	0	912
1973	0	0	4	13	7	228	310	278	170	36	0	0	1046
1974	0	0	8	17	70	121	323	297	50	12	2	0	900
1975	0	0	0	8	138	236	309	376	72	19	3	0	1161
1976	0	0	6	59	28	207	276	202	39	5	0	0	822
1977	0	0	7	22	171	152	407	285	181	4	5	0	1234
1978	0	0	0	4	63	231	315	255	200	2	0	0	1070
1979	0	0	2	8	38	154	271	248	102	22	0	0	845
1980	0	0	0	0	98	187	364	363	166	5	0	0	1183
1981	0	0	1	31	34	234	343	275	99	9	0	0	1026
1982	0	0	0	5	129	99	381	203	120	73	13	8	1031
1983	0	0	4	4	18	225	448	417	161	8	0	0	1285
1984	0	0	0	13	38	289	233	295	130	29	0	0	1027
1985	0	0	6	47	93	174	318	241	162	41	5	0	1087
1986	0	0	4	22	97	247	399	243	183	30	0	0	1225
1987	0	0	0	12	193	266	353	325	147	0	4	0	1300
1988	0	0	2	3	70	251	425	392	111	6	0	0	1260
1989	0	0	7	26	67	210	369	275	125	17	0	0	1096
1990	0	0	17	32	27	230	309	276	155	21	3	0	1070
1991	0	0	0	14	218	317	408	310	195	42	0	0	1504
1992	0	0	0	19	59	126	273	158	88	2	0	0	725
1993	0	0	0	0	0	56	224	443	353	75	5	0	1156
1994	0	0	0	19	40	314	325	228	83	14	0	0	1023
1995	0	0	0	9	45	242	365	454	97	1	0	0	1213
1996	0	0	0	7	79	198	226	296	109	5	0	0	920
1997	0	0	0	0	16	163	320	218	77	48	0	0	842
1998	0	0	23	0	104	242	299	319	222	16	0	5	1230
1999	0	0	0	6	52	253	440	247	125	2	0	0	1125
2000	0	0	0	2	89	211	240	235	107	30	0	0	914
2001	0	0	0	64	78	174	296	315	90	16	0	0	1033

SNOWFALL (inches) 2001 COVINGTON/CINCINNATI, OH (CVG)

YEAR	JUL	AUG	SEP	OCT	NOV	DEC	JAN	FEB	MAR	APR	MAY	JUN	TOTAL
1972-73	0.0	0.0	0.0	T	6.5	1.9	0.5	3.1	3.9	1.8	0.0	0.0	17.7
1973-74	0.0	0.0	0.0	0.0	0.4	2.0	1.7	3.2	3.4	0.5	0.0	0.0	11.2
1974-75	0.0	0.0	0.0	T	5.0	5.6	2.0	2.2	6.8	0.2	0.0	0.0	21.8
1975-76	0.0	0.0	0.0	0.0	3.7	1.7	8.0	0.1	0.6	0.0	0.0	0.0	14.1
1976-77	0.0	0.0	0.0	0.0	2.7	1.0	30.3	4.2	5.4	3.7	0.0	0.0	47.3
1977-78	0.0	0.0	0.0	0.0	4.0	6.3	31.5	4.6	7.5	0.0	0.0	0.0	53.9
1978-79	0.0	0.0	0.0	0.0	T	0.7	17.5	11.7	0.6	0.1	0.0	0.0	30.6
1979-80	0.0	0.0	0.0	0.0	1.0	1.0	8.3	11.9	7.9	T	0.0	0.0	30.1
1980-81	0.0	0.0	0.0	T	1.2	3.7	4.0	2.6	2.5	0.0	0.0	0.0	14.0
1981-82	0.0	0.0	0.0	T	0.3	10.9	7.1	3.9	0.5	1.5	0.0	0.0	24.2
1982-83	0.0	0.0	0.0	0.0	T	T	0.8	5.5	0.3	T	0.0	0.0	6.6
1983-84	0.0	0.0	0.0	0.0	T	1.7	4.1	6.7	4.1	0.0	0.0	0.0	16.6
1984-85	0.0	0.0	0.0	0.0	1.4	7.3	12.2	9.5	0.4	1.7	0.0	0.0	32.5
1985-86	0.0	0.0	0.0	0.0	0.0	5.0	2.8	11.3	0.8	T	0.0	0.0	19.9
1986-87	0.0	0.0	0.0	0.0	T	0.8	1.6	2.4	8.8	2.3	0.0	0.0	15.9
1987-88	0.0	0.0	0.0	0.0	0.1	0.2	4.3	4.7	2.3	T	0.0	0.0	11.6
1988-89	0.0	0.0	0.0	0.0	0.7	2.9	T	3.0	1.2	0.3	0.2	0.0	8.3
1989-90	0.0	0.0	0.0	5.9	0.2	12.5	1.3	3.6	5.6	T	0.0	0.0	29.1
1990-91	0.0	0.0	0.0	0.0	0.0	8.6	4.3	2.6	T	0.0	0.0	0.0	15.5
1991-92	T	0.0	0.0	0.0	1.9	0.5	3.6	1.2	3.6	2.9	0.1	0.0	13.8
1992-93	0.0	0.0	0.0	T	1.6	3.8	0.3	19.9	3.9	T	0.0	T	29.5
1993-94	0.0	0.0	0.0	6.2	0.8	5.4	13.3	0.5	6.6	0.2	0.0	0.0	33.0
1994-95	T	0.0	0.0	0.0	0.0	0.2	16.4	7.6	3.3	0.0	0.0	0.0	27.5
1995-96	0.0	0.0	0.0	0.0	0.9	4.1	27.0	1.7	8.4	2.5	0.0	0.0	44.6
1996-97	0.0	0.0	0.0	0.0	2.1	1.3	4.3	4.5	T	T	0.0	0.0	12.2
1997-98	0.0	0.0	0.0	0.0	0.4	4.2	1.2	18.5	7.1	T	0.0	0.0	31.4
1998-99	0.0	0.0	0.0	0.0	0.0	3.4	9.6	3.9	9.5	T	0.0	0.0	26.4
1999-00	0.0	0.0	0.0	T	2.3	7.6	0.6	0.1	0.2	0.0	0.0	0.0	10.8
2000-01	0.0	T	0.0	0.0	0.4	8.5	5.8	0.7	1.2	0.7	0.0	T	17.3
2001-	0.0	0.0	0.0	T	0.0	0.7							
POR= 53 YRS	T	0.0	0.0	0.3	2.0	3.8	7.1	5.4	4.1	0.5	0.0	T	23.2

2001
CLEVELAND,
OHIO (CLE)

Cleveland is on the south shore of Lake Erie in northeast Ohio. The metropolitan area has a lake frontage of 31 miles. The surrounding terrain is generally level except for an abrupt ridge on the eastern edge of the city which rises some 500 feet above the shore terrain. The Cuyahoga River, which flows through a rather deep but narrow north-south valley, bisects the city.

Local climate is continental in character but with strong modifying influences by Lake Erie. West to northerly winds blowing off Lake Erie tend to lower daily high temperatures in summer and raise temperatures in winter. Temperatures at Hopkins Airport which is 5 miles south of the lakeshore average from 2-4 degrees higher than the lakeshore in summer, while overnight low temperatures average from 2-4 degrees lower than the lakefront during all seasons.

In this area, summers are moderately warm and humid with occasional days when temperatures exceed 90 degrees. Winters are relatively cold and cloudy with an average of 5 days with sub-zero temperatures. Weather changes occur every few days from the passing of cold fronts.

The daily range in temperature is usually greatest in late summer and least in winter. Annual extremes in temperature normally occur soon after late June and December. Maximum temperatures below freezing occur most often in December, January, and February. Temperatures of 100 degrees or higher are rare. On the average, freezing temperatures in fall are first recorded in October while the last freezing temperature in spring normally occurs in April.

As is characteristic of continental climates, precipitation varies widely from year to year. However, it is normally abundant and well distributed throughout the year with spring being the wettest season. Showers and thunderstorms account for most of the rainfall during the growing season. Thunderstorms are most frequent from April through August. Snowfall may fluctuate widely. Mean annual snowfall increases from west to east in Cuyahoga County ranging from about 45 inches in the west to more than 90 inches in the extreme east.

Damaging winds of 50 mph or greater are usually associated with thunderstorms. Tornadoes, one of the most destructive of all atmospheric storms, occasionally occur in Cuyahoga County.

NORMALS, MEANS, AND EXTREMES
CLEVELAND, OH (CLE)

LATITUDE: 41 24' 18" N LONGITUDE: 81 51' 10" W ELEVATION (FT): GRND: 802 BARO: 805 TIME ZONE: EASTERN (UTC + 5) WBAN: 14820

	ELEMENT	POR	JAN	FEB	MAR	APR	MAY	JUN	JUL	AUG	SEP	OCT	NOV	DEC	YEAR
TEMPERATURE °F	NORMAL DAILY MAXIMUM	30	31.9	35.0	46.3	57.9	68.6	78.3	82.4	80.5	73.6	62.1	50.0	37.4	58.7
	MEAN DAILY MAXIMUM	54	33.6	36.6	45.5	58.4	69.3	78.7	82.5	80.7	73.9	63.0	49.7	38.0	59.2
	HIGHEST DAILY MAXIMUM	60	73	74	83	88	92	104	103	102	101	90	82	77	104
	YEAR OF OCCURRENCE		1950	2000	1945	1986	1959	1988	1941	1948	1953	1946	1950	1982	JUN 1988
	MEAN OF EXTREME MAXS.	54	56.2	59.1	72.1	80.3	85.6	91.5	93.2	91.3	88.5	80.1	70.1	59.9	77.3
	NORMAL DAILY MINIMUM	30	17.6	19.3	28.2	37.3	47.3	56.8	61.4	60.3	54.2	43.5	35.0	24.5	40.5
	MEAN DAILY MINIMUM	54	19.4	21.1	28.6	38.4	48.4	57.8	62.3	61.0	54.4	44.0	34.8	24.8	41.2
	LOWEST DAILY MINIMUM	60	-20	-15	-5	10	25	31	41	38	32	19	3	-15	-20
	YEAR OF OCCURRENCE		1994	1963	1984	1964	1966	1972	1968	1982	1942	1988	1976	1989	JAN 1994
	MEAN OF EXTREME MINS.	54	-1.1	1.2	11.2	23.4	33.6	43.2	50.0	48.4	39.1	29.7	19.3	6.2	25.4
	NORMAL DRY BULB	30	24.8	27.2	37.3	47.6	58.0	67.6	71.9	70.4	63.9	52.8	42.6	30.9	49.6
	MEAN DRY BULB	54	26.5	28.9	37.0	48.4	58.9	68.2	72.4	70.9	64.2	53.5	42.4	31.4	50.2
	MEAN WET BULB	18	25.6	27.5	33.8	43.7	53.8	62.4	66.0	65.1	58.6	48.6	39.1	29.7	46.2
	MEAN DEW POINT	18	21.1	22.4	27.9	37.5	48.5	57.8	62.0	61.6	54.5	43.7	34.2	25.2	41.4
	NORMAL NO. DAYS WITH:														
	MAXIMUM 90	30	0.0	0.0	0.0	0.0	0.2	1.8	3.9	2.0	0.6	0.0	0.0	0.0	8.5
	MAXIMUM 32	30	15.6	12.7	4.9	0.2	0.0	0.0	0.0	0.0	0.0	0.0	1.1	10.3	44.8
	MINIMUM 32	30	27.9	24.3	21.0	9.3	0.9	*	0.0	0.0	0.0	0.0	2.8	12.5	24.8	123.5
	MINIMUM 0	30	3.2	2.1	0.1	0.0	0.0	0.0	0.0	0.0	0.0	0.0	0.0	0.9	6.3
H/C	NORMAL HEATING DEG. DAYS	30	1246	1058	859	522	250	40	0	11	99	387	672	1057	6201
	NORMAL COOLING DEG. DAYS	30	0	0	0	0	33	118	218	178	66	8	0	0	621
RH	NORMAL (PERCENT)	30	73	73	70	66	67	69	70	73	74	71	72	74	71
	HOUR 01 LST	30	75	76	74	73	76	79	81	83	82	77	75	76	77
	HOUR 07 LST	30	77	78	78	76	77	79	81	85	84	80	77	77	79
	HOUR 13 LST	30	69	68	63	57	57	57	57	60	60	59	65	70	62
	HOUR 19 LST	30	72	71	68	61	60	61	61	66	70	69	71	74	67
S	PERCENT POSSIBLE SUNSHINE	58	30	37	45	52	58	65	67	63	60	52	31	26	49
W/O	MEAN NO. DAYS WITH:														
	HEAVY FOG(VISBY 1/4 MI)	60	1.4	1.8	1.9	1.2	1.3	0.7	0.5	0.9	0.6	0.9	0.7	1.2	13.1
	THUNDERSTORMS	60	0.1	0.5	1.7	3.4	4.8	6.6	6.3	5.0	3.2	1.7	1.0	0.2	34.5
CLOUDINESS	MEAN:														
	SUNRISE-SUNSET (OKTAS)	0			6.4			5.6						8.8	
	MIDNIGHT-MIDNIGHT (OKTAS)	0			6.4										
	MEAN NO. DAYS WITH:														
	CLEAR	0			7.0		5.0	6.0							
	PARTLY CLOUDY	1	2.0	1.0	4.0		3.0	9.0							
	CLOUDY	1	9.0	6.0	11.0		11.0	9.0							
PR	MEAN STATION PRESSURE(IN)	29	29.20	29.21	29.20	29.10	29.10	29.10	29.20	29.20	29.20	29.21	29.21	29.20	29.18
	MEAN SEA-LEVEL PRES. (IN)	18	30.10	30.12	30.07	30.00	30.01	29.98	30.02	30.06	30.07	30.11	30.10	30.12	30.06
WINDS	MEAN SPEED (MPH)	50	12.3	11.7	12.1	11.5	10.0	9.3	8.7	8.3	9.0	9.9	11.8	12.0	10.6
	PREVAIL.DIR(TENS OF DEGS)	34	24	23	20	36	20	20	22	22	22	20	24	23	22
	MAXIMUM 2-MINUTE:														
	SPEED (MPH)	6	41	45	38	40	41	41	43	39	35	46	39	46	46
	DIR. (TENS OF DEGS)		25	24	04	23	29	28	23	24	29	24	24	25	25
	YEAR OF OCCURRENCE		1996	1997	1996	2001	1998	1998	1999	2000	2000	1996	1998	2000	DEC 2000
	MAXIMUM 5-SECOND:														
	SPEED (MPH)	6	52	55	46	53	51	58	68	47	45	57	47	60	68
	DIR. (TENS OF DEGS)		24	25	21	29	29	31	21	26	29	26	24	26	21
	YEAR OF OCCURRENCE		1996	1997	1998	1996	1998	1998	1999	2000	2000	1996	1998	1996	JUL 1999
PRECIPITATION	NORMAL (IN)	30	2.04	2.19	2.91	3.14	3.49	3.70	3.52	3.40	3.44	2.54	3.17	3.09	36.63
	MAXIMUM MONTHLY (IN)	60	7.01	4.70	6.07	6.61	9.14	9.06	9.12	8.96	11.05	9.50	8.80	8.59	11.05
	YEAR OF OCCURRENCE		1950	1990	1954	1961	1989	1972	1992	1975	1996	1954	1985	1990	SEP 1996
	MINIMUM MONTHLY (IN)	60	0.36	0.48	0.78	1.18	1.00	0.65	0.68	0.53	0.74	0.61	0.80	0.71	0.36
	YEAR OF OCCURRENCE		1961	1978	1958	1946	1963	1988	2001	1969	1964	1952	1976	1958	JAN 1961
	MAXIMUM IN 24 HOURS (IN)	60	2.53	2.33	2.76	2.24	3.73	4.00	2.87	3.65	5.24	3.44	2.73	2.81	5.24
	YEAR OF OCCURRENCE		1995	1959	1948	1961	1955	1972	1969	1994	1996	1954	1985	1992	SEP 1996
	NORMAL NO. DAYS WITH:														
	PRECIPITATION 0.01	30	16.3	14.2	15.3	13.8	12.8	11.0	10.2	10.1	10.1	11.7	14.0	16.9	156.4
	PRECIPITATION 1.00	30	0.1	0.2	0.2	0.3	0.6	0.8	1.2	0.8	0.8	0.2	0.4	0.4	6.0
SNOWFALL	NORMAL (IN)	30	13.8	12.9	9.7	2.2	0.1	0.0	0.0	0.0	0.0	0.7	4.2	12.6	56.2
	MAXIMUM MONTHLY (IN)	60	42.8	39.1	26.5	14.5	2.1	T	T	0.0	T	8.0	23.4	30.3	42.8
	YEAR OF OCCURRENCE		1978	1993	2001	1943	1974	1996	1993		1993	1962	1996	1962	JAN 1978
	MAXIMUM IN 24 HOURS (IN)	60	10.8	14.8	16.0	11.6	2.1	T	T	0.0	T	6.7	15.0	12.2	16.0
	YEAR OF OCCURRENCE		1996	1993	1987	1982	1974	1996	1993		1993	1962	1950	1974	MAR 1987
	MAXIMUM SNOW DEPTH (IN)	53	21	21	15	14	0	0	0	0	0	6	20	19	21
	YEAR OF OCCURRENCE		1978	1993	1984	1987						1962	1950	1962	FEB 1993
	NORMAL NO. DAYS WITH:														
	SNOWFALL 1.0	30	4.5	4.0	3.1	0.7	0.*	0.0	0.0	0.0	0.0	0.2	1.5	4.1	18.1

PRECIPITATION (inches) 2001 CLEVELAND, OH (CLE)

YEAR	JAN	FEB	MAR	APR	MAY	JUN	JUL	AUG	SEP	OCT	NOV	DEC	ANNUAL
1972	1.95	2.01	2.97	3.40	3.74	9.06	4.44	6.38	4.91	1.64	4.58	3.26	48.34
1973	1.62	2.40	3.48	3.40	4.79	6.72	2.94	3.11	2.69	3.95	2.62	3.53	41.25
1974	2.56	2.43	3.88	3.64	4.78	3.57	1.90	3.29	3.06	1.19	4.72	4.86	39.88
1975	3.06	3.20	3.47	1.31	3.23	4.10	2.54	8.96	3.35	1.73	2.09	3.77	40.81
1976	3.38	3.97	3.11	2.17	2.94	3.64	3.48	3.50	3.71	2.54	0.80	1.57	34.81
1977	1.29	1.38	4.49	3.56	1.02	4.91	3.94	3.92	2.52	1.93	3.62	3.51	36.09
1978	3.67	0.48	2.17	3.02	3.01	3.30	2.40	3.58	3.68	3.23	1.19	2.96	32.69
1979	2.61	2.74	2.33	3.09	4.77	3.47	3.76	4.46	3.66	1.79	3.16	4.00	39.84
1980	1.18	1.27	3.66	2.65	3.13	2.69	4.77	4.38	3.11	2.38	1.29	2.10	32.61
1981	0.76	2.72	1.61	4.62	2.19	4.68	5.31	2.61	2.33	1.99	5.34	3.44	39.01
1982	4.00	1.41	3.77	1.62	2.65	5.01	1.21	2.66	4.82	0.93	5.17	3.68	36.93
1983	1.08	0.77	3.54	4.48	4.17	3.45	4.16	3.15	2.87	4.14	5.89	2.92	40.62
1984	1.25	3.82	3.80	2.29	5.95	3.40	3.35	5.51	2.43	2.20	3.95	3.38	41.33
1985	1.78	2.60	4.97	1.38	3.45	2.93	3.23	4.01	2.05	3.45	8.80	2.63	41.28
1986	2.23	3.08	2.44	3.90	4.34	2.97	3.10	3.58	6.41	2.83	3.01	2.82	40.71
1987	1.98	0.49	3.84	2.97	2.40	7.94	3.36	5.51	2.07	3.41	1.02	2.96	37.95
1988	1.03	2.84	2.20	3.47	1.33	0.65	3.42	3.35	1.77	2.51	4.63	2.49	29.69
1989	2.07	1.73	3.46	3.73	9.14	5.22	3.02	1.09	4.61	4.50	3.61	1.72	43.90
1990	2.35	4.70	0.86	4.57	6.10	1.72	5.62	4.79	7.33	4.92	2.28	8.59	53.83
1991	2.18	2.31	3.64	4.22	3.24	1.37	1.69	2.79	3.40	2.65	2.92	2.26	32.67
1992	3.32	2.65	3.05	3.77	3.01	2.66	9.12	4.58	3.25	2.27	6.54	4.31	48.53
1993	4.44	2.61	3.85	3.16	1.56	5.18	2.58	1.52	5.94	3.52	4.06	2.21	40.63
1994	2.66	0.83	1.30	2.73	1.67	3.35	2.46	5.35	1.73	1.05	2.52	2.94	29.56
1995	5.81	1.73	1.72	4.33	3.96	3.67	5.39	2.00	1.03	4.08	3.88	1.45	39.05
1996	2.69	1.63	2.81	5.61	2.08	3.89	3.18	0.79	11.05	4.65	5.03	3.03	46.44
1997	1.77	2.93	3.26	2.20	4.21	3.34	1.51	5.26	4.25	1.63	2.58	2.42	35.36
1998	3.92	1.89	3.25	6.07	1.92	2.97	2.72	3.02	1.20	2.36	1.59	1.92	32.83
1999	3.64	2.36	1.65	3.89	1.54	1.43	4.66	1.80	1.93	3.06	3.31	2.70	31.97
2000	2.63	2.05	1.57	3.72	5.46	5.72	2.57	4.72	3.29	3.56	2.55	2.75	40.59
2001	1.59	1.63	2.43	2.33	3.84	3.96	0.68	3.31	3.90	5.56	2.62	2.53	34.38
POR= 131 YRS	2.53	2.32	2.86	2.93	3.22	3.38	3.36	3.01	3.15	2.65	2.71	2.51	34.63

AVERAGE TEMPERATURE (F) 2001 CLEVELAND, OH (CLE)

YEAR	JAN	FEB	MAR	APR	MAY	JUN	JUL	AUG	SEP	OCT	NOV	DEC	ANNUAL
1972	27.3	25.7	34.8	46.0	58.6	62.7	71.4	68.9	63.8	49.1	39.6	34.5	48.5
1973	30.4	27.9	46.5	50.2	56.7	70.4	72.6	73.2	66.3	57.7	44.6	34.3	52.6
1974	32.0	27.8	39.6	51.3	56.4	66.2	72.2	70.4	59.9	51.2	42.9	31.7	50.1
1975	31.9	30.4	34.6	41.8	62.3	69.8	71.3	72.3	58.6	53.8	47.0	32.0	50.5
1976	21.6	36.0	45.0	49.1	55.3	69.5	71.6	68.4	61.1	48.1	33.7	23.3	48.6
1977	11.0	25.0	42.7	51.4	61.8	63.3	73.1	69.6	65.6	52.6	45.4	29.2	49.2
1978	20.1	16.8	32.4	47.0	59.4	69.0	72.2	73.0	69.2	53.2	44.2	33.7	49.2
1979	22.0	19.1	42.9	46.6	56.9	66.9	71.1	71.5	65.0	52.4	42.3	33.7	49.0
1980	25.5	21.9	33.6	46.1	58.5	64.0	72.3	73.2	64.7	47.9	39.4	28.5	48.0
1981	20.1	31.5	36.0	50.6	55.7	68.2	71.3	70.0	62.0	42.6	30.6	49.1	49.1
1982	19.8	25.2	37.1	44.6	64.9	64.1	73.6	67.9	62.7	55.3	45.4	40.5	50.1
1983	30.7	33.9	40.8	47.1	55.7	69.0	75.2	73.7	65.1	53.4	43.9	23.2	51.0
1984	20.7	34.5	28.4	46.8	54.0	69.5	68.7	70.6	61.1	56.3	40.9	36.5	49.0
1985	20.8	25.2	40.3	53.6	60.4	62.7	71.1	68.9	64.9	54.0	46.0	24.3	49.4
1986	26.7	28.8	39.5	49.8	60.8	67.2	73.1	69.0	67.0	54.3	40.3	32.6	50.8
1987	27.4	30.5	39.0	49.1	63.0	70.2	75.2	70.8	63.5	47.5	46.1	34.8	51.4
1988	25.6	25.8	37.5	47.9	59.7	68.9	75.9	74.2	64.0	47.1	43.8	31.3	50.1
1989	35.0	26.1	38.1	45.3	57.6	68.3	73.4	71.0	64.0	54.0	41.0	19.2	49.4
1990	35.8	34.1	42.0	49.4	56.3	67.6	71.2	69.8	63.4	53.7	45.3	35.6	52.0
1991	27.3	32.8	40.7	52.6	66.9	71.1	74.7	72.7	64.5	55.7	40.2	34.8	52.8
1992	30.2	32.7	36.6	47.9	57.9	64.1	70.7	67.6	63.3	49.9	42.0	34.1	49.8
1993	32.3	25.6	33.6	47.6	59.2	67.9	75.0	73.2	62.2	51.3	41.8	30.5	50.0
1994	19.3	26.5	36.3	50.5	54.7	69.7	73.3	69.3	54.5	47.8	36.9	36.9	50.2
1995	30.1	27.0	40.0	46.6	59.3	71.5	75.5	77.8	63.2	56.4	38.5	26.1	51.0
1996	25.9	27.6	30.9	46.3	56.7	69.6	70.9	69.9	70.0	54.0	36.3	34.7	48.9
1997	25.7	34.4	38.8	45.2	52.9	68.2	70.6	67.5	62.6	53.1	39.0	33.4	49.3
1998	35.2	37.6	41.2	49.3	64.4	68.5	71.4	72.1	67.0	53.5	44.9	37.0	53.5
1999	27.1	34.7	34.3	50.4	61.0	70.1	76.2	69.1	63.5	52.5	45.9	33.8	51.7
2000	26.6	34.3	43.2	47.0	61.5	68.6	68.0	68.9	63.1	54.7	39.8	22.3	49.8
2001	27.7	32.2	34.0	51.4	60.0	68.0	71.9	72.6	61.4	54.4	48.8	37.1	51.6
POR= 131 YRS	27.1	28.2	36.3	47.2	58.4	67.8	72.2	70.6	64.5	53.5	41.8	31.3	49.9

REFERENCE NOTES:

PAGE 1:
THE TEMPERATURE GRAPH SHOWS NORMAL MAXIMUM AND NORMAL MINIMUM DAILY TEMPERATURES (SOLID CURVES) AND THE ACTUAL DAILY HIGH AND LOW TEMPERATURES (VERTICAL BARS).

PAGE 2 AND 3:
H/C INDICATES HEATING AND COOLING DEGREE DAYS.
RH INDICATES RELATIVE HUMIDITY
W/O INDICATES WEATHER AND OBSTRUCTIONS
S INDICATES SUNSHINE.
PR INDICATES PRESSURE.
CLOUDINESS ON PAGE 3 IS THE SUM OF THE CEILOMETER AND SATELLITE DATA NOT TO EXCEED EIGHT EIGHTHS(OKTAS).

GENERAL:
T INDICATES TRACE PRECIPITATION, AN AMOUNT GREATER THAN ZERO BUT LESS THAN THE LOWEST REPORTABLE VALUE.
+ INDICATES THE VALUE ALSO OCCURS ON EARLIER DATES.
BLANK ENTRIES DENOTE MISSING OR UNREPORTED DATA.
NORMALS ARE 30-YEAR AVERAGES (1961 - 1990).
ASOS INDICATES AUTOMATED SURFACE OBSERVING SYSTEM.
PM INDICATES THE LAST DAY OF THE PREVIOUS MONTH.
POR (PERIOD OF RECORD) BEGINS WITH THE JANUARY DATA MONTH AND IS THE NUMBER OF YEARS USED TO COMPUTE THE MEAN. INDIVIDUAL MONTHS WITHIN THE POR MAY BE MISSING.
WHEN THE POR FOR A NORMAL IS LESS THAN 30 YEARS, THE NORMAL IS PROVISIONAL AND IS BASED ON THE NUMBER OF YEARS INDICATED.
0.* OR * INDICATES THE VALUE OR MEAN-DAYS-WITH IS BETWEEN 0.00 AND 0.05.
CLOUDINESS FOR ASOS STATIONS DIFFERS FROM THE NON-ASOS OBSERVATION TAKEN BY A HUMAN OBSERVER. ASOS STATION CLOUDINESS IS BASED ON TIME-AVERAGED CEILOMETER DATA FOR CLOUDS AT OR BELOW 12,000 FEET AND ON SATELLITE DATA FOR CLOUDS ABOVE 12,000 FEET.
THE NUMBER OF DAYS WITH CLEAR, PARTLY CLOUDY, AND CLOUDY CONDITIONS FOR ASOS STATIONS IS THE SUM OF THE CEILOMETER AND SATELLITE DATA FOR THE SUNRISE TO SUNSET PERIOD.

GENERAL CONTINUED:
CLEAR INDICATES 0 - 2 OKTAS, PARTLY CLOUDY INDICATES 3 - 6 OKTAS, AND CLOUDY INDICATES 7 OR 8 OKTAS. WHEN AT LEAST ONE OF THE ELEMENTS (CEILOMETER OR SATELLITE) IS MISSING, THE DAILY CLOUDINESS IS NOT COMPUTED.
WIND DIRECTION IS RECORDED IN TENS OF DEGREES (2 DIGITS) CLOCKWISE FROM TRUE NORTH. "00" INDICATES CALM. "36" INDICATES TRUE NORTH.
RESULTANT WIND IS THE VECTOR AVERAGE OF THE SPEED AND DIRECTION.
AVERAGE TEMPERATURE IS THE SUM OF THE MEAN DAILY MAXIMUM AND MINIMUM TEMPERATURE DIVIDED BY 2.
SNOWFALL DATA COMPRISE ALL FORMS OF FROZEN PRECIPITATION, INCLUDING HAIL.
A HEATING (COOLING) DEGREE DAY IS THE DIFFERENCE BETWEEN THE AVERAGE DAILY TEMPERATURE AND 65 F.
DRY BULB IS THE TEMPERATURE OF THE AMBIENT AIR.
DEW POINT IS THE TEMPERATURE TO WHICH THE AIR MUST BE COOLED TO ACHIEVE 100 PERCENT RELATIVE HUMIDITY.
WET BULB IS THE TEMPERATURE THE AIR WOULD HAVE IF THE MOISTURE CONTENT WAS INCREASED TO 100 PERCENT RELATIVE HUMIDITY.

ON JULY 1, 1996, THE NATIONAL WEATHER SERVICE BEGAN USING THE "METAR" OBSERVATION CODE THAT WAS ALREADY EMPLOYED BY MOST OTHER NATIONS OF THE WORLD. THE MOST NOTICEABLE DIFFERENCE IN THIS ANNUAL PUBLICATION WILL BE THE CHANGE IN UNITS FROM TENTHS TO EIGHTS(OKTAS) FOR REPORTING THE AMOUNT OF SKY COVER.

HEATING DEGREE DAYS (base 65 F) 2001   CLEVELAND, OH (CLE)

YEAR	JUL	AUG	SEP	OCT	NOV	DEC	JAN	FEB	MAR	APR	MAY	JUN	TOTAL
1972-73	32	27	95	485	752	937	1067	1033	569	450	254	3	5704
1973-74	3	9	73	234	605	946	1015	1035	777	419	280	49	5445
1974-75	2	5	176	423	660	1026	1021	962	934	691	154	38	6092
1975-76	5	4	187	345	532	1015	1336	836	614	493	309	25	5701
1976-77	0	25	150	519	932	1286	1672	1113	689	423	166	115	7090
1977-78	4	26	60	378	592	1103	1387	1343	1005	534	218	43	6693
1978-79	7	2	43	362	620	965	1328	1281	680	552	290	60	6190
1979-80	20	11	87	403	670	967	1218	1244	967	561	223	103	6474
1980-81	3	2	97	521	763	1125	1385	935	894	430	298	30	6483
1981-82	11	11	145	458	664	1059	1393	1109	860	608	78	75	6471
1982-83	5	42	136	310	586	760	1056	864	742	533	294	56	5384
1983-84	7	0	116	362	628	1291	1366	878	1126	544	347	19	6684
1984-85	16	17	174	270	716	877	1364	1110	757	370	187	99	5957
1985-86	2	7	118	338	565	1255	1180	1009	785	459	172	52	5942
1986-87	3	40	63	332	736	999	1158	958	795	473	170	23	5750
1987-88	3	22	90	535	562	929	1213	1129	848	506	208	60	6105
1988-89	8	5	83	557	629	1040	922	1084	831	585	272	33	6049
1989-90	0	6	108	350	716	1416	898	858	718	492	270	56	5888
1990-91	7	3	121	350	585	906	1163	897	748	379	111	11	5281
1991-92	0	0	123	310	738	930	1074	929	872	513	243	90	5822
1992-93	8	26	118	462	682	952	1009	1097	967	519	192	56	6088
1993-94	0	3	134	420	691	1063	1414	1073	880	443	330	57	6508
1994-95	4	10	83	322	507	865	1077	1061	769	546	190	13	5447
1995-96	3	0	103	271	787	1200	1203	1077	1052	556	297	19	6568
1996-97	8	0	98	333	850	932	1213	849	805	584	368	46	6086
1997-98	11	30	103	404	773	972	916	763	758	465	94	73	5362
1998-99	1	4	52	350	597	863	1171	840	942	433	155	56	5464
1999-00	3	3	83	384	569	961	1180	884	674	533	169	54	5497
2000-01	10	21	136	315	750	1317	1152	914	955	427	177	58	6232
2001-	11	0	150	332	480	857							

COOLING DEGREE DAYS (base 65 F) 2001   CLEVELAND, OH (CLE)

YEAR	JAN	FEB	MAR	APR	MAY	JUN	JUL	AUG	SEP	OCT	NOV	DEC	ANNUAL
1972	0	0	0	1	5	63	239	157	64	0	0	0	529
1973	0	0	0	13	7	168	244	273	119	17	0	0	841
1974	0	0	0	14	18	91	231	180	30	3	2	0	569
1975	0	0	0	0	75	187	206	241	6	5	0	0	720
1976	0	0	3	23	14	167	214	138	39	2	0	0	600
1977	0	0	4	22	74	73	262	175	84	0	9	0	703
1978	0	0	0	0	53	170	237	256	177	3	0	0	896
1979	0	0	0	6	42	122	213	218	93	21	0	0	715
1980	0	0	0	0	27	83	235	263	97	0	0	0	705
1981	0	0	0	4	16	132	214	175	73	0	0	0	614
1982	0	0	0	3	84	54	278	140	73	17	6	6	661
1983	0	0	0	5	12	185	327	277	127	12	0	0	945
1984	0	0	0	3	13	159	139	197	60	5	0	0	576
1985	0	0	0	38	52	34	201	131	122	4	2	0	584
1986	0	0	1	9	48	128	259	168	131	8	0	0	752
1987	0	0	0	0	114	183	322	209	53	0	3	0	884
1988	0	0	0	0	47	185	348	297	58	9	0	0	944
1989	0	0	4	0	46	138	268	199	83	14	0	0	752
1990	0	0	10	31	8	141	208	158	80	8	1	0	645
1991	0	0	1	14	176	200	307	245	114	26	0	0	1083
1992	0	0	0	8	28	68	191	114	74	0	0	0	483
1993	0	0	0	0	17	147	316	262	62	2	0	0	806
1994	0	0	0	15	17	204	269	149	58	5	1	0	718
1995	0	0	0	0	21	216	336	404	58	11	0	0	1046
1996	0	0	0	2	48	155	160	193	76	1	0	0	635
1997	0	0	0	0	0	147	194	113	38	41	0	0	533
1998	0	0	25	0	82	183	207	231	118	4	0	0	850
1999	0	0	0	0	39	218	357	138	88	1	0	0	841
2000	0	0	4	1	69	169	114	145	86	7	0	0	595
2001	0	0	0	23	31	156	231	240	48	11	0	0	740

SNOWFALL (inches) 2001 CLEVELAND, OH (CLE)

YEAR	JUL	AUG	SEP	OCT	NOV	DEC	JAN	FEB	MAR	APR	MAY	JUN	TOTAL
1972-73	0.0	0.0	0.0	5.5	7.8	15.2	9.8	20.4	8.3	0.9	0.6	0.0	68.5
1973-74	0.0	0.0	0.0	T	3.3	13.8	8.9	16.9	7.1	6.4	2.1	0.0	58.5
1974-75	0.0	0.0	0.0	1.6	5.3	24.1	9.7	9.9	15.2	1.2	0.0	0.0	67.0
1975-76	0.0	0.0	0.0	0.0	5.6	13.1	21.5	6.8	5.8	1.6	T	0.0	54.4
1976-77	0.0	0.0	T	1.6	8.9	16.3	21.1	9.6	4.2	1.7	0.0	0.0	63.4
1977-78	0.0	0.0	0.0	T	9.7	23.1	42.8	10.8	3.5	0.2	0.0	0.0	90.1
1978-79	0.0	0.0	0.0	0.0	1.9	2.5	15.1	16.0	2.4	0.4	0.0	0.0	38.3
1979-80	0.0	0.0	0.0	0.2	0.5	4.0	11.3	19.2	3.5	T	T	0.0	38.7
1980-81	0.0	0.0	0.0	T	5.4	13.5	15.0	9.7	16.9	T	0.0	0.0	60.5
1981-82	0.0	0.0	0.0	4.0	2.9	27.1	28.1	7.6	17.6	13.2	0.0	0.0	100.5
1982-83	0.0	0.0	0.0	T	2.2	6.3	6.5	8.3	11.3	3.4	0.0	0.0	38.0
1983-84	0.0	0.0	0.0	0.0	7.1	13.0	12.9	27.1	19.3	T	0.0	0.0	79.4
1984-85	0.0	0.0	0.0	0.0	4.0	8.9	25.5	18.2	1.2	5.9	0.0	0.0	63.7
1985-86	0.0	0.0	0.0	0.0	T	23.4	17.2	10.8	6.7	0.2	0.0	0.0	58.3
1986-87	0.0	0.0	0.0	0.0	3.1	1.1	16.4	5.0	26.2	4.0	0.0	0.0	55.8
1987-88	0.0	0.0	0.0	T	1.0	16.4	8.7	22.9	20.4	1.9	0.0	0.0	71.3
1988-89	0.0	0.0	0.0	T	1.7	17.9	6.6	13.8	9.9	4.9	T	0.0	54.8
1989-90	0.0	0.0	0.0	T	9.1	24.0	10.5	9.9	4.4	4.7	0.0	0.0	62.6
1990-91	0.0	0.0	0.0	T	T	7.4	16.6	18.9	4.2	T	0.0	0.0	47.1
1991-92	0.0	0.0	0.0	0.0	3.5	9.4	23.8	6.2	18.4	4.4	0.0	0.0	65.7
1992-93	0.0	0.0	0.0	T	7.1	7.1	8.7	39.1	25.4	1.1	0.0	T	88.5
1993-94	T	0.0	T	0.2	3.0	19.0	27.4	12.3	7.0	3.6	0.0	0.0	72.5
1994-95	0.0	0.0	0.0	0.0	T	1.0	23.4	14.7	4.3	0.2	0.0	0.0	43.6
1995-96	0.0	0.0	0.0	0.0	9.9	29.6	21.9	10.1	19.4	10.2	0.0	T	101.1
1996-97	0.0	0.0	0.0	0.0	23.4	5.0	13.0	8.4	5.3	0.8	0.0	0.0	55.9
1997-98	0.0	0.0	0.0	T	8.6	10.7	5.0	0.2	9.5	T	0.0	0.0	34.0
1998-99	0.0	0.0	0.0	0.0	0.0	6.9	29.6	14.2	11.6	T	0.0	0.0	62.4
1999-00	0.0	0.0	0.0	T	1.6	10.3	24.7	13.9	8.0	1.0	0.0	0.0	59.5
2000-01	0.0	0.0	0.0	T	11.2	21.9	14.9	3.2	26.5	0.3	0.0	0.0	78.1
2001-	0.0	0.0	0.0	1.0	T	3.5							

POR=													
59 YRS	T	0.0	T	0.6	4.9	12.0	13.4	11.7	10.6	2.4	0.5	T	56.1

2001
COLUMBUS,
OHIO (CMH)

Columbus is located in the center of the state and in the drainage area of the Ohio River. The airport is located at the eastern boundary of the city approximately 7 miles from the center of the business district.

Four nearly parallel streams run through or adjacent to the city. The Scioto River is the principal stream and flows from the northwest into the center of the city and then flows straight south toward the Ohio River. The Olentangy River runs almost due south and empties into the Scioto just west of the business district. Two minor streams run through portions of Columbus or skirt the eastern and southern fringes of the area. They are Alum Creek and Big Walnut Creek. Alum Creek empties into the Big Walnut southeast of the city and the Big Walnut empties into the Scioto a few miles downstream. The Scioto and Olentangy are gorge-like in character with very little flood plain and the two creeks have only a little more flood plain or bottomland.

The narrow valleys associated with the streams flowing through the city supply the only variation in the micro-climate of the area. The city proper shows the typical metropolitan effect with shrubs and flowers blossoming earlier than in the immediate surroundings and in retarding light frost on clear quiet nights. Many small areas to the southeast and to the north and northeast show marked effects of air drainage as evidenced by the frequent formation of shallow ground fog at daybreak during the summer and fall months and the higher frequency of frost in the spring and fall.

The average occurrence of the last freezing temperature in the spring within the city proper is mid-April, and the first freeze in the fall is very late October, but in the immediate surroundings there is much variation. For example, at Valley Crossing located at the southeastern outskirts of the city, the average occurrence of the last 32 degree temperature in the spring is very early May, while the first 32 degree temperature in the fall is mid-October.

The records show a high frequency of calm or very low wind speeds during the late evening and early morning hours, from June through September. The rolling landscape is conducive to air drainage and from the Weather Service location at the airport the air drainage is toward the northwest with the wind direction indicated as southeast. Air drainage takes place at speeds generally 4 mph or less and frequently provides the only perceptible breeze during the night.

Columbus is located in the area of changeable weather. Air masses from central and northwest Canada frequently invade this region. Air from the Gulf of Mexico often reachs central Ohio during the summer and to a much lesser extent in the fall and winter. There are also occasional weather changes brought about by cool outbreaks from the Hudson Bay region of Canada, especially during the spring months. At infrequent intervals the general circulation will bring showers or snow to Columbus from the Atlantic. Although Columbus does not have a wet or dry season as such, the month of October usually has the least amount of precipitation.

NORMALS, MEANS, AND EXTREMES
COLUMBUS, OH　(CMH)

LATITUDE:	LONGITUDE:	ELEVATION (FT):	TIME ZONE:	WBAN: 14821
39 59' 29" N	82 52' 51" W	GRND: 846　BARO: 849	EASTERN　(UTC + 5)	

	ELEMENT	POR	JAN	FEB	MAR	APR	MAY	JUN	JUL	AUG	SEP	OCT	NOV	DEC	YEAR
TEMPERATURE F	NORMAL DAILY MAXIMUM	30	34.1	38.0	50.5	62.0	72.3	80.4	83.7	82.1	76.2	64.5	51.4	39.2	61.2
	MEAN DAILY MAXIMUM	54	35.9	39.9	49.9	62.5	72.9	81.7	85.0	83.4	76.9	65.3	51.5	39.9	62.1
	HIGHEST DAILY MAXIMUM	62	74	75	85	89	94	102	100	101	100	90	80	76	102
	YEAR OF OCCURRENCE		1950	2000	1945	1948	1941	1944	1999	1983	1951	1951	1987	1982	JUN 1944
	MEAN OF EXTREME MAXS.	54	58.1	62.3	73.9	81.9	87.5	92.9	94.3	92.9	89.9	81.8	71.4	61.8	79.1
	NORMAL DAILY MINIMUM	30	18.5	21.2	31.2	40.0	50.1	58.0	62.7	60.8	54.8	42.9	34.3	24.6	41.6
	MEAN DAILY MINIMUM	54	20.3	22.9	30.7	40.5	50.5	59.5	63.8	62.0	54.6	43.2	34.1	25.0	42.3
	LOWEST DAILY MINIMUM	62	-22	-13	-6	14	25	35	43	39	31	20	5	-17	-22
	YEAR OF OCCURRENCE		1994	1977	1984	1982	1966	1972	1972	1965	1963	1962	1976	1989	JAN 1994
	MEAN OF EXTREME MINS.	54	-1.5	3.0	12.8	24.8	35.5	45.5	51.6	49.6	38.1	27.9	17.8	4.8	25.8
	NORMAL DRY BULB	30	26.4	29.6	40.9	51.0	61.2	69.2	73.2	71.5	65.5	53.7	42.9	31.9	51.4
	MEAN DRY BULB	54	28.2	31.4	40.4	51.5	61.8	70.5	74.4	72.8	65.7	54.3	42.8	32.4	52.2
	MEAN WET BULB	17	27.3	30.1	36.4	46.1	55.8	64.0	67.8	66.3	59.6	49.1	39.8	30.8	47.8
	MEAN DEW POINT	17	22.2	24.4	29.4	39.2	50.3	59.2	63.4	62.4	55.2	43.8	34.4	26.2	42.5
	NORMAL NO. DAYS WITH:														
	MAXIMUM　90	30	0.0	0.0	0.0	0.0	0.5	3.5	5.9	3.4	1.2	0.0	0.0	0.0	14.5
	MAXIMUM　32	30	13.6	9.4	2.4	0.1	0.0	0.0	0.0	0.0	0.0	0.0	1.1	8.9	35.5
	MINIMUM　32	30	27.2	23.5	18.3	6.9	0.7	0.0	0.0	0.0	0.1	4.1	13.6	23.9	118.3
	MINIMUM　0	30	3.0	1.6	0.1	0.0	0.0	0.0	0.0	0.0	0.0	0.0	0.0	0.0	5.6
H/C	NORMAL HEATING DEG. DAYS	30	1197	991	747	420	187	23	0	12	81	361	663	1026	5708
	NORMAL COOLING DEG. DAYS	30	0	0	0	0	69	149	258	214	96	11	0	0	797
RH	NORMAL (PERCENT)	30	71	70	64	62	66	68	71	73	73	69	72	74	69
	HOUR 01 LST	30	74	73	70	70	77	80	82	84	83	78	76	77	77
	HOUR 07 LST	30	76	76	74	75	79	81	84	87	87	82	80	79	80
	HOUR 13 LST	30	67	64	56	52	54	55	56	58	58	55	63	69	59
	HOUR 19 LST	30	69	66	60	54	56	58	58	60	63	65	64	73	63
S	PERCENT POSSIBLE SUNSHINE	45	36	42	44	50	56	60	60	60	61	56	37	31	49
W/O	MEAN NO. DAYS WITH:														
	HEAVY FOG(VISBY 1/4 MI)	52	1.9	1.6	1.1	0.6	0.9	0.9	1.1	1.6	1.7	1.4	1.2	1.6	15.6
	THUNDERSTORMS	62	0.4	0.6	2.0	4.1	6.3	8.0	8.0	6.1	3.0	1.3	1.0	0.3	41.1
CLOUDINESS	MEAN:														
	SUNRISE-SUNSET (OKTAS)	47	6.2	6.1	5.9	5.6	5.3	5.0	4.8	4.6	4.5	4.5	5.8	6.2	5.4
	MIDNIGHT-MIDNIGHT (OKTAS)	32	6.0	5.5	5.5	5.1	4.8	4.7	4.4	4.3	4.3	4.3	5.6	6.0	5.0
	MEAN NO. DAYS WITH:														
	CLEAR	1	2.0												
	PARTLY CLOUDY	1	6.0												
	CLOUDY	1	23.0												
PR	MEAN STATION PRESSURE(IN)	28	29.20	29.19	29.10	29.10	29.10	29.10	29.10	29.20	29.19	29.20	29.20	29.19	29.16
	MEAN SEA-LEVEL PRES. (IN)	17	30.12	30.11	30.06	29.98	29.99	29.97	30.02	30.05	30.06	30.11	30.12	30.14	30.06
WINDS	MEAN SPEED (MPH)	42	9.7	9.4	10.0	9.5	8.0	6.9	6.3	5.8	6.3	7.2	8.9	9.1	8.1
	PREVAIL.DIR(TENS OF DEGS)	27	27	27	28	36	18	19	18	36	36	18	18	18	18
	MAXIMUM 2-MINUTE:														
	SPEED (MPH)	6	40	41	43	47	47	36	47	43	44	40	45	47	47
	DIR. (TENS OF DEGS)		24	24	24	26	26	22	33	29	23	24	27	22	22
	YEAR OF OCCURRENCE		2000	2001	1996	1996	2001	2000	1997	1996	1999	1996	1996	2001	DEC 2001
	MAXIMUM 5-SECOND:														
	SPEED (MPH)	6	52	53	51	58	55	52	60	56	61	52	56	59	61
	DIR. (TENS OF DEGS)		24	18	23	26	26	23	28	29	23	25	25	26	23
	YEAR OF OCCURRENCE		2000	1997	1996	1996	2001	1996	1997	1996	1999	2001	1998	2000	SEP 1999
PRECIPITATION	NORMAL (IN)	30	2.18	2.24	3.27	3.21	3.93	4.04	4.31	3.72	2.96	2.15	3.22	2.86	38.09
	MAXIMUM MONTHLY (IN)	62	8.29	5.15	9.59	6.51	9.11	9.75	12.36	8.63	6.76	5.24	10.67	6.98	12.36
	YEAR OF OCCURRENCE		1950	1990	1964	1998	1968	1958	1992	1979	1979	1954	1985	1990	JUL 1992
	MINIMUM MONTHLY (IN)	62	0.53	0.29	0.61	0.67	0.95	0.65	0.48	0.58	0.51	0.11	0.60	0.46	0.11
	YEAR OF OCCURRENCE		1944	1978	1941	1971	1977	1999	1940	1951	1963	1963	1976	1955	OCT 1963
	MAXIMUM IN 24 HOURS (IN)	54	4.81	2.15	3.40	2.37	2.72	2.93	5.16	3.79	4.86	2.21	2.47	2.56	5.16
	YEAR OF OCCURRENCE		1959	1975	1964	1957	1968	1958	1992	1972	1979	1986	1985	1998	JUL 1992
	NORMAL NO. DAYS WITH:														
	PRECIPITATION　0.01	30	12.8	11.1	13.8	12.4	12.3	10.5	10.9	10.0	8.4	9.3	11.7	13.5	136.7
	PRECIPITATION　1.00	30	0.2	0.3	0.3	0.6	0.8	1.1	1.1	1.0	0.7	0.3	0.7	0.3	7.4
SNOWFALL	NORMAL (IN)	30	9.2	7.0	4.4	1.1	0.*	0.0	0.0	0.0	T	0.1	1.9	5.3	29.0
	MAXIMUM MONTHLY (IN)	53	34.4	16.4	13.5	12.6	0.8	T	T	0.0	T	4.6	15.2	17.3	34.4
	YEAR OF OCCURRENCE		1978	1979	1962	1987	1989	1995	1995		1994	1993	1950	1960	JAN 1978
	MAXIMUM IN 24 HOURS (IN)	53	8.8	8.9	8.6	12.3	0.8	T	T	0.0	T	4.6	8.2	8.7	12.3
	YEAR OF OCCURRENCE		1996	1971	1962	1987	1989	1995	1995		1994	1993	1950	1960	APR 1987
	MAXIMUM SNOW DEPTH (IN)	52	17	13	9	10	0	0	0	0	0	0	13	10	17
	YEAR OF OCCURRENCE		1978	1979	1984	1987							1950	1960	JAN 1978
	NORMAL NO. DAYS WITH:														
	SNOWFALL　1.0	30	3.0	2.4	1.4	0.2	0.0	0.0	0.0	0.0	0.0	0.*	0.6	1.9	9.5

PRECIPITATION (inches) 2001 COLUMBUS, OH (CMH)

YEAR	JAN	FEB	MAR	APR	MAY	JUN	JUL	AUG	SEP	OCT	NOV	DEC	ANNUAL
1972	1.40	1.74	2.86	3.74	6.56	3.98	2.60	7.96	5.13	1.74	4.40	3.49	45.60
1973	2.46	1.29	3.43	3.72	3.36	8.77	4.07	4.97	2.82	3.29	5.37	2.70	46.25
1974	2.40	2.30	4.38	2.66	3.29	5.04	1.14	4.88	3.32	1.51	3.39	2.68	36.99
1975	3.21	3.47	4.10	2.71	3.17	3.53	2.04	4.51	5.46	2.29	1.54	3.01	39.04
1976	3.15	2.03	2.17	1.44	1.41	4.52	5.12	5.08	2.54	2.86	0.60	0.93	31.85
1977	1.57	1.02	3.88	4.04	0.95	4.02	2.52	4.76	2.57	3.77	3.54	36.12	
1978	5.89	0.29	2.98	3.02	4.15	3.65	1.81	5.23	1.16	2.39	1.56	5.01	37.14
1979	3.32	2.88	1.01	4.01	3.27	4.23	8.06	8.63	6.76	1.26	3.91	1.83	49.17
1980	1.69	1.38	3.77	1.59	4.56	5.17	4.58	6.26	1.86	2.53	2.07	1.96	37.42
1981	0.70	4.60	1.11	5.38	6.50	5.73	4.14	1.41	2.28	1.40	1.65	2.88	37.78
1982	4.77	1.49	3.99	1.90	4.68	3.37	3.90	1.02	4.25	0.92	5.19	3.84	39.32
1983	1.20	0.74	1.69	5.58	5.06	4.59	2.80	2.23	1.91	4.45	5.00	3.16	38.41
1984	1.07	1.97	3.89	3.10	4.93	0.71	3.15	2.96	1.48	2.91	4.41	2.84	33.42
1985	1.31	1.67	3.78	0.73	4.96	1.41	6.88	2.34	1.18	1.93	10.67	1.81	38.67
1986	1.54	2.96	2.61	1.31	2.47	5.53	3.60	1.61	3.44	4.16	3.00	2.81	35.04
1987	1.14	0.59	2.04	2.02	2.85	3.60	3.89	2.96	1.53	1.57	1.63	2.88	26.70
1988	2.14	4.26	2.54	2.24	2.27	1.34	7.80	2.68	3.52	1.70	3.59	2.49	36.57
1989	1.97	3.10	4.16	3.30	4.69	6.36	6.79	4.30	2.16	2.49	2.65	1.79	43.76
1990	2.43	5.15	1.32	2.82	7.01	5.25	8.00	1.86	5.26	5.05	2.03	6.98	53.16
1991	1.97	2.30	3.97	4.15	2.47	2.81	2.14	2.02	4.05	1.76	1.31	3.79	32.74
1992	1.79	0.85	3.40	2.83	3.40	2.33	12.36	3.75	2.14	1.40	4.03	1.32	39.60
1993	4.14	1.82	3.50	4.49	2.47	3.33	5.95	0.74	1.75	3.05	4.45	2.16	37.85
1994	3.79	1.56	1.94	3.64	1.69	1.93	6.02	3.29	1.68	0.92	2.94	2.22	31.62
1995	4.54	1.64	1.61	3.17	4.86	5.30	6.99	7.56	1.15	4.04	2.47	1.97	45.30
1996	3.73	2.14	3.40	6.39	5.81	3.82	5.09	1.58	5.50	1.44	3.20	3.46	45.56
1997	2.19	1.50	3.96	1.65	5.58	6.62	2.91	5.76	1.36	1.58	2.92	2.13	38.16
1998	2.32	2.48	1.88	6.51	3.09	6.99	2.75	1.99	1.27	3.05	1.99	3.25	37.57
1999	2.87	2.77	1.88	4.65	1.80	0.65	3.02	2.40	1.91	1.00	1.95	2.69	27.59
2000	3.53	2.79	2.70	4.15	5.42	3.50	4.10	4.10	4.18	2.70	2.13	3.59	42.89
2001	1.31	1.37	1.03	3.39	7.03	2.30	4.66	4.14	1.60	3.32	3.69	3.01	36.85
POR= 123 YRS	2.86	2.42	3.79	3.23	3.74	3.71	3.89	3.21	2.60	2.19	2.78	2.64	37.06

AVERAGE TEMPERATURE (F) 2001 COLUMBUS, OH (CMH)

YEAR	JAN	FEB	MAR	APR	MAY	JUN	JUL	AUG	SEP	OCT	NOV	DEC	ANNUAL
1972	28.2	27.7	37.0	48.8	60.8	63.6	71.9	70.1	64.6	49.6	40.5	36.1	49.9
1973	31.2	31.4	50.4	51.1	59.5	72.6	74.3	74.2	68.9	58.6	45.1	34.2	54.3
1974	33.2	31.2	44.6	54.3	60.8	67.7	74.4	74.0	62.2	52.8	44.5	34.0	52.8
1975	32.5	33.4	37.3	46.7	66.6	72.4	75.1	77.3	62.7	54.7	47.5	33.5	53.3
1976	24.0	37.4	46.5	50.9	58.1	70.5	72.0	68.3	61.7	47.5	33.9	24.8	49.6
1977	11.4	26.5	45.6	54.8	66.8	67.5	76.2	72.0	68.2	52.0	45.1	29.5	51.3
1978	19.0	16.6	34.5	50.6	59.6	70.4	73.5	73.2	69.8	51.5	44.4	34.4	49.8
1979	21.4	19.3	44.3	50.1	60.5	69.6	71.8	71.9	65.1	53.3	43.6	35.1	50.5
1980	29.3	25.2	37.2	49.5	62.4	67.4	75.9	75.9	68.3	50.8	40.8	32.5	51.3
1981	23.3	34.0	40.2	55.8	59.5	70.9	71.9	70.4	62.3	51.1	40.9	30.6	50.9
1982	21.2	29.2	40.4	46.4	66.8	65.8	74.4	69.2	63.5	56.2	45.4	40.4	51.6
1983	29.9	34.0	43.3	48.4	57.6	69.4	76.7	76.2	67.1	54.5	44.0	24.8	52.2
1984	23.3	37.4	32.3	50.0	57.6	73.1	71.2	72.9	63.1	59.4	40.6	39.5	51.7
1985	21.7	26.0	43.7	56.3	62.6	66.9	72.7	71.2	66.6	57.3	48.2	26.0	51.6
1986	30.1	32.7	42.5	54.5	64.3	70.6	75.7	71.0	69.2	56.3	41.3	33.5	53.5
1987	29.9	34.9	44.3	52.1	66.0	72.7	76.6	74.3	66.9	49.1	47.6	35.7	54.2
1988	26.5	29.3	40.2	50.3	62.6	69.6	77.5	75.3	65.2	47.4	43.9	31.6	51.6
1989	36.6	28.7	42.0	48.2	57.2	68.8	73.9	71.2	65.2	54.2	42.1	19.8	50.7
1990	37.7	37.5	45.3	50.7	59.1	70.3	73.6	72.5	66.4	55.1	46.2	37.2	54.3
1991	29.7	35.7	43.9	56.1	70.9	75.0	77.6	75.0	66.2	55.9	41.0	36.4	55.3
1992	32.2	36.8	40.7	51.8	59.9	67.3	73.5	69.4	64.7	51.9	44.8	34.7	52.3
1993	34.3	27.8	38.6	50.3	62.2	69.8	76.2	75.7	64.9	53.0	43.5	32.9	52.4
1994	21.3	30.0	39.5	53.9	58.4	73.9	75.2	71.7	65.4	55.5	48.2	38.8	52.7
1995	29.4	27.9	43.6	50.8	60.9	72.9	76.0	78.4	64.2	56.1	37.7	28.8	52.2
1996	27.8	30.5	35.6	50.2	60.9	72.3	72.8	74.0	65.8	55.0	37.5	37.1	51.6
1997	28.1	36.3	42.7	48.4	56.6	70.2	74.2	70.3	65.1	55.2	40.1	34.7	51.8
1998	37.6	40.5	43.6	53.1	67.3	71.7	74.8	76.3	71.6	55.7	45.9	38.1	56.4
1999	31.1	37.1	37.5	55.0	64.8	74.5	80.2	73.1	67.9	55.3	47.5	34.6	54.9
2000	27.0	37.5	45.9	51.4	64.9	71.6	72.5	71.3	64.8	57.2	41.0	23.3	52.4
2001	28.7	35.3	38.1	56.8	63.6	71.1	74.3	75.2	64.4	55.8	49.6	38.4	54.3
POR= 123 YRS	29.0	31.1	40.2	51.2	61.9	70.7	74.8	72.8	66.4	54.7	42.5	32.5	52.3

REFERENCE NOTES:

PAGE 1:
THE TEMPERATURE GRAPH SHOWS NORMAL MAXIMUM AND NORMAL
MINIMUM DAILY TEMPERATURES (SOLID CURVES) AND THE
ACTUAL DAILY HIGH AND LOW TEMPERATURES (VERTICAL BARS).

PAGE 2 AND 3:
H/C INDICATES HEATING AND COOLING DEGREE DAYS.
RH INDICATES RELATIVE HUMIDITY
W/O INDICATES WEATHER AND OBSTRUCTIONS
S INDICATES SUNSHINE.
PR INDICATES PRESSURE.
CLOUDINESS ON PAGE 3 IS THE SUM OF THE CEILOMETER AND
SATELLITE DATA NOT TO EXCEED EIGHT EIGHTHS(OKTAS).

GENERAL:
T INDICATES TRACE PRECIPITATION, AN AMOUNT GREATER
THAN ZERO BUT LESS THAN THE LOWEST REPORTABLE VALUE.
+ INDICATES THE VALUE ALSO OCCURS ON EARLIER DATES.
BLANK ENTRIES DENOTE MISSING OR UNREPORTED DATA.
NORMALS ARE 30-YEAR AVERAGES (1961 - 1990).
ASOS INDICATES AUTOMATED SURFACE OBSERVING SYSTEM.
PM INDICATES THE LAST DAY OF THE PREVIOUS MONTH.
POR (PERIOD OF RECORD) BEGINS WITH THE JANUARY DATA
MONTH AND IS THE NUMBER OF YEARS USED TO COMPUTE
THE MEAN. INDIVIDUAL MONTHS WITHIN THE POR MAY
BE MISSING.
WHEN THE POR FOR A NORMAL IS LESS THAN 30 YEARS,
THE NORMAL IS PROVISIONAL AND IS BASED ON THE NUMBER
OF YEARS INDICATED.
0.* OR * INDICATES THE VALUE OR MEAN-DAYS-WITH
IS BETWEEN 0.00 AND 0.05.
CLOUDINESS FOR ASOS STATIONS DIFFERS FROM THE NON-ASOS
OBSERVATION TAKEN BY A HUMAN OBSERVER. ASOS STATION
CLOUDINESS IS BASED ON TIME-AVERAGED CEILOMETER DATA
FOR CLOUDS AT OR BELOW 12,000 FEET AND ON SATELLITE
DATA FOR CLOUDS ABOVE 12,000 FEET.
THE NUMBER OF DAYS WITH CLEAR, PARTLY CLOUDY, AND
CLOUDY CONDITIONS FOR ASOS STATIONS IS THE SUM
OF THE CEILOMETER AND SATELLITE DATA FOR THE
SUNRISE TO SUNSET PERIOD.

GENERAL CONTINUED:
CLEAR INDICATES 0 - 2 OKTAS, PARTLY CLOUDY INDICATES
3 - 6 OKTAS, AND CLOUDY INDICATES 7 OR 8 OKTAS.
WHEN AT LEAST ONE OF THE ELEMENTS (CEILOMETER OR
SATELLITE) IS MISSING, THE DAILY CLOUDINESS IS
NOT COMPUTED.
WIND DIRECTION IS RECORDED IN TENS OF DEGREES (2 DIGITS)
CLOCKWISE FROM TRUE NORTH. "00" INDICATES CALM. "36"
INDICATES TRUE NORTH.
RESULTANT WIND IS THE VECTOR AVERAGE OF THE SPEED AND
DIRECTION.
AVERAGE TEMPERATURE IS THE SUM OF THE MEAN DAILY MAXIMUM
AND MINIMUM TEMPERATURE DIVIDED BY 2.
SNOWFALL DATA COMPRISE ALL FORMS OF FROZEN
PRECIPITATION, INCLUDING HAIL.
A HEATING (COOLING) DEGREE DAY IS THE DIFFERENCE BETWEEN
THE AVERAGE DAILY TEMPERATURE AND 65 F.
DRY BULB IS THE TEMPERATURE OF THE AMBIENT AIR.
DEW POINT IS THE TEMPERATURE TO WHICH THE AIR MUST BE
COOLED TO ACHIEVE 100 PERCENT RELATIVE HUMIDITY.
WET BULB IS THE TEMPERATURE THE AIR WOULD HAVE IF THE
MOISTURE CONTENT WAS INCREASED TO 100 PERCENT RELATIVE
HUMIDITY.

ON JULY 1, 1996, THE NATIONAL WEATHER SERVICE BEGAN USING
THE "METAR" OBSERVATION CODE THAT WAS ALREADY EMPLOYED
BY MOST OTHER NATIONS OF THE WORLD. THE MOST NOTICEABLE
DIFFERENCE IN THIS ANNUAL PUBLICATION WILL BE THE CHANGE
IN UNITS FROM TENTHS TO EIGHTS(OKTAS) FOR REPORTING THE
AMOUNT OF SKY COVER.

HEATING DEGREE DAYS (base 65 F) 2001 COLUMBUS, OH (CMH)

YEAR	JUL	AUG	SEP	OCT	NOV	DEC	JAN	FEB	MAR	APR	MAY	JUN	TOTAL
1972-73	22	18	77	473	727	889	1041	934	444	427	184	0	5236
1973-74	0	3	35	219	589	963	977	940	628	332	178	31	4895
1974-75	0	0	130	374	609	954	999	878	850	542	73	18	5427
1975-76	0	0	110	321	520	973	1263	791	570	440	229	4	5221
1976-77	1	25	118	537	925	1241	1659	1071	601	324	91	64	6657
1977-78	1	17	36	394	594	1091	1420	1346	938	424	223	23	6507
1978-79	0	0	38	411	610	943	1346	1270	637	449	185	18	5907
1979-80	11	16	83	376	632	920	1099	1148	855	458	133	53	5784
1980-81	0	0	46	435	717	1000	1286	864	761	287	195	14	5605
1981-82	8	5	141	429	713	1061	1351	997	758	556	45	33	6097
1982-83	3	19	107	304	585	759	1081	863	669	493	239	30	5152
1983-84	6	0	83	325	626	1236	1284	796	1007	447	254	3	6067
1984-85	6	3	143	182	727	782	1339	1086	654	286	134	35	5377
1985-86	0	2	96	249	500	1202	1076	901	694	328	113	19	5180
1986-87	0	26	41	287	702	974	1083	838	637	393	103	9	5093
1987-88	0	4	53	489	521	900	1187	1029	762	433	119	49	5546
1988-89	3	7	57	547	624	1032	873	1009	711	499	274	28	5664
1989-90	0	11	90	345	680	1394	840	766	613	444	190	26	5399
1990-91	0	3	83	310	558	857	1089	817	649	282	42	0	4690
1991-92	0	0	105	296	714	878	1011	814	747	402	190	35	5192
1992-93	0	8	101	403	600	932	942	1034	811	434	130	51	5446
1993-94	0	1	84	366	637	989	1351	973	787	340	233	12	5773
1994-95	0	8	51	295	497	805	1098	1031	657	427	150	6	5025
1995-96	0	0	78	274	810	1112	1147	992	905	448	191	11	5968
1996-97	2	0	74	307	821	859	1136	798	687	494	264	15	5457
1997-98	0	10	63	346	741	933	843	679	681	352	38	41	4727
1998-99	0	0	22	292	567	829	1045	776	848	299	68	9	4755
1999-00	0	0	46	295	517	936	1172	789	586	403	84	21	4849
2000-01	1	3	104	256	717	1285	1117	822	826	289	92	24	5536
2001-	2	0	97	297	455	815							

COOLING DEGREE DAYS (base 65 F) 2001 COLUMBUS, OH (CMH)

YEAR	JAN	FEB	MAR	APR	MAY	JUN	JUL	AUG	SEP	OCT	NOV	DEC	ANNUAL
1972	0	0	0	1	24	67	245	183	71	0	0	0	591
1973	0	0	3	14	17	236	295	292	160	25	0	0	1042
1974	0	0	4	20	58	117	296	286	52	3	0	0	836
1975	0	0	0	1	130	248	320	389	48	10	1	0	1147
1976	0	0	3	23	23	174	223	135	25	2	0	0	608
1977	0	0	8	24	151	148	354	242	139	0	7	0	1073
1978	0	0	0	0	59	190	270	261	188	0	0	0	968
1979	0	0	0	7	54	163	230	239	93	22	0	0	808
1980	0	0	0	0	61	132	343	344	151	3	0	0	1034
1981	0	0	0	16	32	198	231	181	64	4	0	0	726
1982	0	0	0	4	111	66	301	154	67	39	7	4	753
1983	0	0	1	2	17	167	377	355	152	9	0	0	1080
1984	0	0	0	8	30	253	205	254	94	14	0	0	858
1985	0	0	2	32	64	97	245	201	152	19	2	0	814
1986	0	0	2	19	95	194	339	221	171	25	0	0	1066
1987	0	0	0	11	142	246	366	299	116	0	5	0	1185
1988	0	0	0	0	54	194	396	333	70	5	0	0	1052
1989	0	0	5	2	40	149	282	211	106	12	0	0	807
1990	0	0	11	21	13	191	273	244	133	9	3	0	898
1991	0	0	0	21	232	307	402	317	147	23	0	0	1449
1992	0	0	0	13	37	115	272	152	99	2	0	0	690
1993	0	0	0	0	48	204	352	343	89	2	0	0	1038
1994	0	0	0	15	39	286	322	224	71	8	0	0	965
1995	0	0	0	6	32	251	347	424	61	4	0	0	1125
1996	0	0	0	11	72	238	251	285	102	2	0	0	961
1997	0	0	0	0	10	181	291	182	73	48	0	0	785
1998	0	0	27	0	118	248	313	357	227	10	0	3	1303
1999	0	0	0	4	69	301	476	258	139	0	0	0	1247
2000	0	0	0	2	87	225	239	209	105	18	0	0	885
2001	0	0	0	50	56	215	299	324	85	19	0	0	1048

SNOWFALL (inches) 2001 COLUMBUS, OH (CMH)

YEAR	JUL	AUG	SEP	OCT	NOV	DEC	JAN	FEB	MAR	APR	MAY	JUN	TOTAL
1972-73	0.0	0.0	0.0	T	6.3	2.8	4.4	1.8	2.1	7.1	0.0	0.0	24.5
1973-74	0.0	0.0	0.0	0.0	T	6.4	2.3	5.0	4.5	0.3	0.0	0.0	18.5
1974-75	0.0	0.0	0.0	T	0.3	7.4	8.1	3.7	2.6	T	0.0	0.0	22.1
1975-76	0.0	0.0	0.0	0.0	1.1	2.9	12.4	1.8	1.0	T	0.0	0.0	19.2
1976-77	0.0	0.0	0.0	T	3.1	4.6	18.1	6.7	0.3	0.1	0.0	0.0	32.9
1977-78	0.0	0.0	0.0	0.0	2.2	7.5	34.4	4.5	5.5	T	0.0	0.0	54.1
1978-79	0.0	0.0	0.0	T	1.3	1.8	17.3	16.4	0.8	0.3	0.0	0.0	37.9
1979-80	0.0	0.0	0.0	0.0	0.1	0.2	7.0	8.1	1.2	T	0.0	0.0	16.6
1980-81	0.0	0.0	0.0	T	8.0	7.3	7.8	3.7	3.3	0.0	0.0	0.0	30.1
1981-82	0.0	0.0	0.0	0.0	1.9	9.8	11.8	3.7	3.2	4.7	0.0	0.0	35.1
1982-83	0.0	0.0	0.0	T	1.5	2.6	4.5	2.8	0.1	0.0	0.0		11.5
1983-84	0.0	0.0	0.0	0.0	0.5	5.7	9.0	10.8	9.8	0.3	0.0	0.0	36.1
1984-85	0.0	0.0	0.0	0.0	0.9	7.3	21.9	12.5	T	0.8	0.0	0.0	43.4
1985-86	0.0	0.0	0.0	0.0	0.0	8.6	4.8	9.8	1.8	T	0.0	0.0	25.0
1986-87	0.0	0.0	0.0	0.0	0.4	0.4	2.7	1.2	1.2	5.9	12.6	0.0	23.2
1987-88	0.0	0.0	0.0	T	0.6	4.6	8.4	6.5	3.8	T	0.0	0.0	23.9
1988-89	0.0	0.0	0.0	T	0.8	5.9	0.6	3.9	6.6	0.1	0.8	0.0	18.7
1989-90	0.0	0.0	0.0	0.4	0.3	9.4	3.3	6.0	1.4	0.4	0.0	0.0	21.2
1990-91	0.0	0.0	0.0	0.0	0.0	3.7	3.4	4.5	4.0	T	0.0	0.0	15.6
1991-92	0.0	0.0	0.0	T	0.6	1.6	12.2	1.8	1.6	1.1	0.0	0.0	18.9
1992-93	0.0	0.0	0.0	T	3.0	2.4	1.5	14.6	8.9	0.2	0.0	0.0	30.6
1993-94	0.0	0.0	0.0	4.6	0.8	4.2	19.5	2.9	4.6	1.1	0.0	T	37.7
1994-95	T	0.0	T	0.0	0.0	0.3	12.6	5.3	2.5	T	0.0	T	20.7
1995-96	T	0.0	0.0	0.0	2.7	11.8	24.5	4.3	7.6	3.2	0.0		
1996-97					1.9								
1997-98					0.9	2.9	1.2	2.2	2.8	T	0.0	T	
1998-99	0.0	0.0	0.0	0.0	0.0	2.8	20.6	7.3	9.8	0.1	0.0	0.0	40.5
1999-00	0.0	0.0	0.0	0.0	1.5	4.6	13.8	7.9	1.9	0.1	0.0	0.0	29.8
2000-01	T	0.0	0.0	T	1.3	13.4	6.3	3.4	1.1	0.8	T	0.0	26.3
2001-	0.0	0.0	0.0	T	0.0	1.7							
POR= 52 YRS	T	0.0	T	0.1	2.2	5.4	8.7	5.9	4.3	1.0	0.1	T	27.7

2001
TOLEDO,
OHIO (TOL)

Toledo is located on the western end of Lake Erie at the mouth of the Maumee River. Except for a bank up from the river about 30 feet, the terrain is generally level with only a slight slope toward the river and Lake Erie. The city has quite a diversified industrial section and excellent harbor facilities, making it a large transportation center for rail, water, and motor freight. Generally rich agricultural land is found in the surrounding area, especially up the Maumee Valley toward the Indiana state line.

Rainfall is usually sufficient for general agriculture. The terrain is level and drainage rather poor, therefore, a little less than the normal precipitation during the growing season is better than excessive amounts. Snowfall is generally light in this area, distributed throughout the winter from November to March with frequent thaws.

The nearness of Lake Erie and the other Great Lakes has a moderating effect on the temperature, and extremes are seldom recorded. On average, only fifteen days a year experience temperatures of 90 degrees or higher, and only eight days when it drops to zero or lower. The growing season averages 160 days, but has ranged from over 220 to less than 125 days.

Humidity is rather high throughout the year in this area, and there is an excessive amount of cloudiness. In the winter months the sun shines during only about 30 percent of the daylight hours. December and January, the cloudiest months, sometimes have as little as 16 percent of the possible hours of sunshine.

Severe windstorms, causing more than minor damage, occur infrequently. There are on the average twenty-three days per year having a sustained wind velocity of 32 mph or more.

Flooding in the Toledo area is produced by several factors. Heavy rains of 1 inch or more will cause a sudden rise in creeks and drainage ditches to the point of overflow. The western shores of Lake Erie are subject to flooding when the lake level is high and prolonged periods of east to northeast winds prevail.

NORMALS, MEANS, AND EXTREMES

TOLEDO, OH (TOL)

LATITUDE:	LONGITUDE:	ELEVATION (FT):	TIME ZONE:	WBAN: 94830
41 35' 19" N	83 48' 05" W	GRND: 690 BARO: 693	EASTERN (UTC + 5)	

	ELEMENT	POR	JAN	FEB	MAR	APR	MAY	JUN	JUL	AUG	SEP	OCT	NOV	DEC	YEAR
TEMPERATURE °F	NORMAL DAILY MAXIMUM	30	30.2	33.4	45.5	58.8	70.5	79.8	83.4	81.3	74.4	62.4	48.5	35.2	58.6
	MEAN DAILY MAXIMUM	46	30.8	34.6	45.6	59.1	70.7	79.8	83.7	81.8	74.8	62.9	48.5	35.9	59.0
	HIGHEST DAILY MAXIMUM	46	65	71	81	88	95	104	104	99	98	91	78	70	104
	YEAR OF OCCURRENCE		1995	2000	1998	1990	1962	1988	1995	1993	1978	1963	1987	2001	JUL 1995
	MEAN OF EXTREME MAXS.	46	51.3	55.9	70.3	80.6	87.3	92.9	94.4	91.9	89.6	80.4	68.5	57.5	76.7
	NORMAL DAILY MINIMUM	30	14.9	17.0	26.8	36.4	46.7	56.0	60.6	58.4	51.5	40.0	31.5	20.5	38.4
	MEAN DAILY MINIMUM	46	15.9	18.7	26.8	37.3	47.5	56.8	61.1	59.3	51.7	40.7	31.8	21.4	39.1
	LOWEST DAILY MINIMUM	46	−20	−14	−6	8	25	32	40	34	26	15	2	−19	−20
	YEAR OF OCCURRENCE		1984	1982	1984	1982	1974	1972	1988	1982	1974	1976	1958	1989	JAN 1984
	MEAN OF EXTREME MINS.	46	−5.5	−1.4	9.0	21.1	32.4	42.5	48.5	46.4	35.1	24.8	15.9	0.6	22.5
	NORMAL DRY BULB	30	22.5	25.2	36.2	47.6	58.6	67.9	72.1	69.9	63.0	51.2	40.0	27.9	48.5
	MEAN DRY BULB	46	23.4	26.6	36.3	48.3	59.2	68.3	72.5	70.5	63.1	51.7	40.1	28.8	49.1
	MEAN WET BULB	15	23.5	25.9	33.4	43.4	53.8	62.5	66.6	65.3	57.8	47.6	37.6	28.3	45.5
	MEAN DEW POINT	15	19.1	20.8	26.9	36.5	47.9	57.4	62.5	61.9	54.1	43.1	33.0	24.5	40.6
	NORMAL NO. DAYS WITH:														
	MAXIMUM 90	30	0.0	0.0	0.0	0.0	0.9	3.5	5.3	2.8	1.1	*	0.0	0.0	13.6
	MAXIMUM 32	30	17.5	13.2	4.5	0.2	0.0	0.0	0.0	0.0	0.0	0.0	1.8	12.3	49.5
	MINIMUM 32	30	29.2	25.5	22.5	11.4	1.7	*	0.0	0.0	0.4	7.1	17.4	26.8	142.0
	MINIMUM 0	30	4.9	3.3	0.2	0.0	0.0	0.0	0.0	0.0	0.0	0.0	0.0	1.8	10.2
H/C	NORMAL HEATING DEG. DAYS	30	1318	1114	893	522	238	33	0	16	109	436	750	1150	6579
	NORMAL COOLING DEG. DAYS	30	0	0	0	0	40	120	225	168	49	8	0	0	610
RH	NORMAL (PERCENT)	30	74	73	70	66	66	69	72	76	76	72	76	79	72
	HOUR 01 LST	30	77	77	77	76	78	82	84	88	88	82	80	81	81
	HOUR 07 LST	30	79	79	80	79	80	82	86	91	91	85	83	82	83
	HOUR 13 LST	30	69	66	60	54	52	54	55	59	58	57	66	73	60
	HOUR 19 LST	30	73	70	65	58	56	58	61	67	72	70	74	78	67
S	PERCENT POSSIBLE SUNSHINE	40	41	46	50	52	60	64	65	63	61	54	37	33	52
W/O	MEAN NO. DAYS WITH:														
	HEAVY FOG(VISBY 1/4 MI)	46	1.8	1.8	1.8	0.8	0.7	1.0	0.8	1.8	1.8	2.0	1.6	2.3	18.2
	THUNDERSTORMS	46	0.2	0.5	2.0	3.7	4.8	6.5	17.6	14.8	7.3	2.0	4.6	5.9	69.9
CLOUDINESS	MEAN:														
	SUNRISE-SUNSET (OKTAS)														
	MIDNIGHT-MIDNIGHT (OKTAS)														
	MEAN NO. DAYS WITH:														
	CLEAR	0			2.0		2.0								
	PARTLY CLOUDY	0			1.0										
	CLOUDY	1	1.0	1.0	2.0										
PR	MEAN STATION PRESSURE(IN)	29	29.31	29.33	29.27	29.23	29.25	29.25	29.27	29.31	29.33	29.33	29.31	29.32	29.29
	MEAN SEA-LEVEL PRES. (IN)	16	30.09	30.10	30.06	29.97	29.99	30.01	30.00	30.04	30.06	30.10	30.08	30.11	30.05
WINDS	MEAN SPEED (MPH)	33	11.1	10.4	11.1	10.9	9.6	8.2	7.7	7.1	7.8	8.7	10.3	10.2	9.4
	PREVAIL.DIR(TENS OF DEGS)	27	24	24	07	07	23	24	24	24	24	24	24	24	24
	MAXIMUM 2-MINUTE:														
	SPEED (MPH)	6	43	46	46	48	46	41	38	43	38	45	45	48	48
	DIR. (TENS OF DEGS)		24	26	26	25	25	24	03	26	24	24	24	30	30
	YEAR OF OCCURRENCE		1996	2001	1996	1997	2000	1997	2001	1998	2001	1996	1998	1998	DEC 1998
	MAXIMUM 5-SECOND:														
	SPEED (MPH)	6	56	56	53	59	68	48	47	54	47	59	66	56	68
	DIR. (TENS OF DEGS)		25	26	26	25	27	28	29	26	23	25	24	31	27
	YEAR OF OCCURRENCE		1996	2001	1996	1997	1999	2001	1998	1998	2001	1996	1998	1998	MAY 1999
PRECIPITATION	NORMAL (IN)	30	1.75	1.73	2.66	2.96	2.91	3.75	3.27	3.25	2.85	2.10	2.81	2.93	32.97
	MAXIMUM MONTHLY (IN)	46	4.61	5.39	5.70	6.80	8.48	6.75	8.47	8.10	6.26	6.86	6.81	JUN 1981
	YEAR OF OCCURRENCE		1965	1990	1985	1977	2000	1981	1969	1965	1972	2001	1982	1967	JUN 1981
	MINIMUM MONTHLY (IN)	46	0.27	0.27	0.58	0.88	0.96	0.27	0.34	0.40	0.58	0.28	0.55	0.54	0.27
	YEAR OF OCCURRENCE		1961	1969	1958	1962	1964	1988	1995	1976	1963	1964	1976	1958	JUN 1988
	MAXIMUM IN 24 HOURS (IN)	46	1.78	2.59	2.60	3.43	2.34	3.21	4.39	2.42	3.97	3.21	3.17	3.53	4.39
	YEAR OF OCCURRENCE		1959	1990	1985	1977	1991	1978	1969	1972	1972	1988	1982	1967	JUL 1969
	NORMAL NO. DAYS WITH:														
	PRECIPITATION 0.01	30	13.5	10.8	13.2	12.7	11.9	10.4	9.5	9.1	9.9	9.5	11.8	14.7	137.0
	PRECIPITATION 1.00	30	0.1	0.1	0.2	0.3	0.4	0.7	0.8	0.8	0.6	0.2	0.4	0.4	5.0
SNOWFALL	NORMAL (IN)	30	9.5	8.6	6.0	1.5	0.*	0.0	0.0	0.0	T	0.1	3.0	9.3	38.0
	MAXIMUM MONTHLY (IN)	42	30.8	16.6	17.7	12.0	1.3	T	T	T	T	2.0	17.9	24.2	30.8
	YEAR OF OCCURRENCE		1978	1994	1993	1957	1989	1995	1992	1994	1993	1989	1966	1977	JAN 1978
	MAXIMUM IN 24 HOURS (IN)	42	10.4	7.7	9.7	9.8	1.3	T	T	T	T	1.8	8.3	13.9	13.9
	YEAR OF OCCURRENCE		1978	1981	1993	1957	1989	1995	1992	1994	1993	1989	1966	1974	DEC 1974
	MAXIMUM SNOW DEPTH (IN)	40	17	19	8	10	1	0	0	0	0	1	8	16	19
	YEAR OF OCCURRENCE		1978	1978	1993	1957	1989					1989	1966	1977	FEB 1978
	NORMAL NO. DAYS WITH:														
	SNOWFALL 1.0	30	2.9	2.7	2.0	0.5	0.0	0.0	0.0	0.0	0.0	0.*	1.2	2.9	12.2

PRECIPITATION (inches) 2001 TOLEDO, OH (TOL)

YEAR	JAN	FEB	MAR	APR	MAY	JUN	JUL	AUG	SEP	OCT	NOV	DEC	ANNUAL
1972	1.42	0.77	2.33	3.74	2.63	4.09	2.77	4.47	8.10	1.46	3.55	3.08	38.41
1973	1.63	1.05	4.20	1.79	2.85	6.51	1.18	1.18	1.09	2.76	3.27	3.17	32.67
1974	2.27	2.00	2.93	2.55	4.18	3.31	0.68	1.61	1.41	0.70	3.57	3.41	28.62
1975	2.57	2.57	1.90	2.34	3.83	4.21	4.99	5.52	2.70	2.42	2.17	3.35	38.57
1976	2.80	4.43	3.56	2.79	1.72	3.70	2.08	0.40	3.68	2.14	0.55	0.93	28.78
1977	1.29	1.99	4.43	6.10	1.53	3.48	1.83	5.79	4.27	1.77	2.72	3.56	38.76
1978	3.14	0.54	2.34	3.74	2.48	5.34	1.86	1.67	3.19	1.65	2.48	3.31	31.74
1979	1.24	0.70	2.55	4.03	3.15	4.23	3.96	4.71	2.90	2.02	4.25	2.46	36.20
1980	0.74	0.96	3.65	3.13	2.93	3.26	4.49	5.89	1.63	1.79	0.97	2.48	31.92
1981	0.48	3.27	0.63	3.54	2.38	8.48	3.72	2.28	6.05	3.79	2.93	38.39
1982	3.61	1.15	3.74	1.53	2.61	2.01	1.97	1.38	2.03	1.14	6.86	3.48	31.51
1983	0.88	0.59	1.86	4.28	3.98	4.06	3.39	2.15	1.42	3.59	5.56	3.91	35.67
1984	0.99	1.18	2.95	5.15	3.48	1.49	2.30	3.87	2.02	1.75	2.74	3.22	31.14
1985	2.02	3.23	5.70	1.40	1.85	2.90	3.86	4.30	2.53	3.05	5.89	1.62	38.35
1986	0.99	2.46	2.16	2.81	2.72	5.32	3.37	5.93	4.75	4.78	1.66	1.87	38.82
1987	1.87	0.53	1.78	1.72	2.32	5.62	1.51	4.45	2.31	2.21	2.59	3.80	30.71
1988	1.17	1.33	1.69	1.45	1.37	0.27	3.76	5.11	1.80	4.37	4.27	1.96	28.55
1989	1.80	0.74	2.03	3.50	4.87	6.74	6.31	3.59	3.30	1.36	1.89	1.29	37.42
1990	2.18	5.39	3.46	2.09	4.63	3.14	1.89	3.32	1.72	2.63	2.27	5.69	38.41
1991	1.41	1.42	1.42	4.29	4.82	1.51	0.52	1.94	0.73	5.53	2.15	1.51	27.25
1992	1.70	1.68	3.05	3.41	3.18	1.28	6.51	2.40	4.01	1.77	4.45	3.60	37.04
1993	3.17	1.71	3.46	3.06	1.13	4.60	1.60	1.15	4.50	1.51	2.73	1.25	29.87
1994	2.83	1.88	2.06	4.86	1.11	3.63	2.14	3.05	0.93	1.00	2.69	3.01	29.19
1995	3.07	0.57	1.59	4.52	2.96	4.46	0.34	2.72	1.41	3.71	2.72	0.89	28.96
1996	2.22	0.95	2.67	3.85	2.62	4.91	1.81	0.74	2.74	1.75	2.79	2.92	29.97
1997	2.35	4.27	2.53	1.55	6.76	3.70	2.63	4.07	4.74	1.24	2.16	2.07	38.07
1998	2.96	3.77	3.32	4.54	2.07	1.73	2.70	5.44	0.96	2.13	1.63	0.61	31.86
1999	3.17	1.67	1.42	4.89	4.93	1.86	2.87	1.40	1.50	1.92	1.46	1.71	28.80
2000	1.19	1.08	1.84	3.55	6.80	5.52	2.29	4.15	4.98	2.83	1.36	2.53	38.12
2001	0.52	2.45	0.64	2.47	5.06	2.87	1.87	2.48	4.72	6.26	2.11	1.96	33.41
POR= 131 YRS	2.15	1.93	2.57	2.92	3.21	3.49	2.89	2.91	2.64	2.34	2.38	2.39	31.82

AVERAGE TEMPERATURE (F) 2001   TOLEDO, OH (TOL)

YEAR	JAN	FEB	MAR	APR	MAY	JUN	JUL	AUG	SEP	OCT	NOV	DEC	ANNUAL
1972	23.4	24.4	34.1	46.1	60.4	63.9	71.4	68.4	62.2	47.2	37.7	30.3	47.5
1973	28.2	25.2	44.1	48.3	55.7	70.1	72.3	71.3	64.4	55.7	41.9	27.5	50.4
1974	26.1	23.2	36.2	48.8	56.0	65.4	72.5	71.5	59.6	49.5	40.4	28.9	48.2
1975	29.2	28.3	33.3	42.7	62.5	69.3	70.8	72.0	57.4	51.9	45.3	29.7	49.3
1976	19.8	32.8	41.6	49.3	56.0	69.3	72.2	68.2	60.5	45.6	32.3	19.9	47.3
1977	9.6	24.3	41.6	53.3	63.6	65.0	74.6	69.3	65.0	41.0	24.7	24.8	44.8
1978	16.7	11.8	28.7	45.8	58.9	67.6	70.9	70.4	68.0	49.8	40.3	30.1	46.6
1979	17.6	15.1	38.7	45.5	57.9	67.7	70.1	68.3	63.0	51.3	40.6	32.1	47.4
1980	24.3	21.4	32.4	46.8	59.5	65.5	73.6	73.3	63.8	46.8	37.4	26.0	47.6
1981	17.6	28.5	36.5	49.9	55.4	68.4	71.7	69.8	61.3	47.7	39.6	27.4	47.8
1982	15.8	20.2	33.4	42.7	64.4	64.3	72.6	67.5	61.9	52.7	41.8	36.6	47.8
1983	27.6	30.5	37.9	44.2	54.8	67.9	74.7	73.8	64.2	51.9	41.3	20.0	49.1
1984	16.6	33.0	27.6	46.8	54.4	71.2	69.8	71.2	60.8	55.2	38.7	34.4	48.3
1985	19.5	22.6	39.3	53.5	61.6	64.8	73.2	69.1	64.0	53.3	43.9	22.3	48.9
1986	25.6	25.0	39.2	50.0	60.3	66.8	73.6	67.0	65.3	53.2	37.2	31.6	49.6
1987	25.8	30.0	39.7	50.3	62.5	70.8	74.9	71.0	63.8	45.4	44.4	33.0	51.0
1988	23.8	23.3	37.5	48.1	61.0	69.3	75.9	73.9	62.5	45.2	41.8	28.0	49.2
1989	33.1	24.5	36.7	45.5	57.2	68.2	73.2	69.8	61.8	52.2	38.5	16.8	48.1
1990	34.3	32.4	41.1	49.4	56.6	69.1	71.8	70.0	63.7	51.8	43.3	33.1	51.5
1991	25.2	31.6	40.3	52.6	67.0	72.6	74.6	73.0	62.9	55.0	37.9	33.0	52.1
1992	28.8	31.9	36.1	47.4	57.9	65.1	70.1	67.8	61.9	49.6	40.8	32.9	49.2
1993	30.2	24.7	34.3	48.3	60.6	68.1	76.1	74.3	61.1	49.8	39.7	29.5	49.7
1994	17.1	23.0	36.7	50.8	57.2	71.0	72.6	67.1	64.1	53.6	45.9	35.7	49.6
1995	28.2	25.7	40.5	46.8	60.1	72.3	76.5	78.5	62.8	55.8	36.6	25.4	50.8
1996	23.7	26.9	31.7	46.1	57.5	70.8	70.5	72.3	64.3	53.1	34.4	31.6	48.6
1997	21.9	31.4	38.6	46.2	52.5	68.8	71.7	67.3	62.4	52.1	36.6	31.3	48.4
1998	33.0	36.3	40.0	50.0	66.0	70.2	73.3	72.2	67.2	53.5	42.9	35.1	53.3
1999	24.4	33.6	34.4	50.7	62.6	70.8	77.3	69.7	65.1	51.5	44.8	31.0	51.7
2000	23.8	33.3	43.8	48.2	61.8	69.0	70.3	70.2	62.5	55.4	39.9	18.3	49.7
2001	26.0	30.4	35.2	51.5	61.4	68.7	72.7	73.4	63.1	53.6	48.5	37.0	51.8
POR= 128 YRS	25.6	27.1	36.0	47.6	59.1	68.7	73.2	71.0	64.2	52.7	40.5	29.5	49.6

REFERENCE NOTES:

PAGE 1:
THE TEMPERATURE GRAPH SHOWS NORMAL MAXIMUM AND NORMAL
  MINIMUM DAILY TEMPERATURES (SOLID CURVES) AND THE
  ACTUAL DAILY HIGH AND LOW TEMPERATURES (VERTICAL BARS).

PAGE 2 AND 3:
H/C INDICATES HEATING AND COOLING DEGREE DAYS.
RH INDICATES RELATIVE HUMIDITY
W/O INDICATES WEATHER AND OBSTRUCTIONS
S INDICATES SUNSHINE.
PR INDICATES PRESSURE.
CLOUDINESS ON PAGE 3 IS THE SUM OF THE CEILOMETER AND
  SATELLITE DATA NOT TO EXCEED EIGHT EIGHTHS(OKTAS).

GENERAL:
T INDICATES TRACE PRECIPITATION, AN AMOUNT GREATER
  THAN ZERO BUT LESS THAN THE LOWEST REPORTABLE VALUE.
+ INDICATES THE VALUE ALSO OCCURS ON EARLIER DATES.
BLANK ENTRIES DENOTE MISSING OR UNREPORTED DATA.
NORMALS ARE 30-YEAR AVERAGES (1961 - 1990).
ASOS INDICATES AUTOMATED SURFACE OBSERVING SYSTEM.
PM INDICATES THE LAST DAY OF THE PREVIOUS MONTH.
POR (PERIOD OF RECORD) BEGINS WITH THE JANUARY DATA
  MONTH AND IS THE NUMBER OF YEARS USED TO COMPUTE
  THE MEAN. INDIVIDUAL MONTHS WITHIN THE POR MAY
  BE MISSING.
WHEN THE POR FOR A NORMAL IS LESS THAN 30 YEARS,
  THE NORMAL IS PROVISIONAL AND IS BASED ON THE NUMBER
  OF YEARS INDICATED.
0.* OR * INDICATES THE VALUE OR MEAN-DAYS-WITH
  IS BETWEEN 0.00 AND 0.05.
CLOUDINESS FOR ASOS STATIONS DIFFERS FROM THE NON-ASOS
  OBSERVATION TAKEN BY A HUMAN OBSERVER. ASOS STATION
  CLOUDINESS IS BASED ON TIME-AVERAGED CEILOMETER DATA
  FOR CLOUDS AT OR BELOW 12,000 FEET AND ON SATELLITE
  DATA FOR CLOUDS ABOVE 12,000 FEET.
THE NUMBER OF DAYS WITH CLEAR, PARTLY CLOUDY, AND
  CLOUDY CONDITIONS FOR ASOS STATIONS IS THE SUM
  OF THE CEILOMETER AND SATELLITE DATA FOR THE
  SUNRISE TO SUNSET PERIOD.

GENERAL CONTINUED:
CLEAR INDICATES 0 - 2 OKTAS, PARTLY CLOUDY INDICATES
  3 - 6 OKTAS, AND CLOUDY INDICATES 7 OR 8 OKTAS.
  WHEN AT LEAST ONE OF THE ELEMENTS (CEILOMETER OR
  SATELLITE) IS MISSING, THE DAILY CLOUDINESS IS
  NOT COMPUTED.
WIND DIRECTION IS RECORDED IN TENS OF DEGREES (2 DIGITS)
  CLOCKWISE FROM TRUE NORTH. "00" INDICATES CALM. "36"
  INDICATES TRUE NORTH.
RESULTANT WIND IS THE VECTOR AVERAGE OF THE SPEED AND
  DIRECTION.
AVERAGE TEMPERATURE IS THE SUM OF THE MEAN DAILY MAXIMUM
  AND MINIMUM TEMPERATURE DIVIDED BY 2.
SNOWFALL DATA COMPRISE ALL FORMS OF FROZEN
  PRECIPITATION, INCLUDING HAIL.
A HEATING (COOLING) DEGREE DAY IS THE DIFFERENCE BETWEEN
  THE AVERAGE DAILY TEMPERATURE AND 65 F.
DRY BULB IS THE TEMPERATURE OF THE AMBIENT AIR.
DEW POINT IS THE TEMPERATURE TO WHICH THE AIR MUST BE
  COOLED TO ACHIEVE 100 PERCENT RELATIVE HUMIDITY.
WET BULB IS THE TEMPERATURE THE AIR WOULD HAVE IF THE
  MOISTURE CONTENT WAS INCREASED TO 100 PERCENT RELATIVE
  HUMIDITY.

ON JULY 1, 1996, THE NATIONAL WEATHER SERVICE BEGAN USING
  THE "METAR" OBSERVATION CODE THAT WAS ALREADY EMPLOYED
  BY MOST OTHER NATIONS OF THE WORLD.   THE MOST NOTICEABLE
  DIFFERENCE IN THIS ANNUAL PUBLICATION WILL BE THE CHANGE
  IN UNITS FROM TENTHS TO EIGHTS(OKTAS) FOR REPORTING THE
  AMOUNT OF SKY COVER.

HEATING DEGREE DAYS (base 65 F) 2001 TOLEDO, OH (TOL)

YEAR	JUL	AUG	SEP	OCT	NOV	DEC	JAN	FEB	MAR	APR	MAY	JUN	TOTAL
1972-73	28	36	134	543	810	1073	1135	1106	639	499	285	3	6291
1973-74	3	16	114	289	686	1157	1197	1166	885	483	295	71	6362
1974-75	2	0	190	478	730	1108	1104	1021	974	664	148	45	6464
1975-76	7	6	227	406	585	1110	1393	927	717	497	277	16	6168
1976-77	1	33	162	596	976	1393	1708	1135	718	381	135	91	7329
1977-78	3	29	71	481	713	1241	1490	1484	1121	573	243	43	7492
1978-79	11	11	74	466	732	1076	1461	1390	808	577	259	42	6907
1979-80	16	33	121	440	724	1009	1258	1256	1005	542	199	83	6686
1980-81	0	3	113	560	822	1206	1464	1015	879	450	309	24	6845
1981-82	7	15	169	529	754	1160	1522	1250	972	665	81	76	7200
1982-83	3	47	148	386	690	871	1154	958	833	624	311	55	6080
1983-84	8	0	127	407	705	1389	1494	920	1151	545	341	9	7096
1984-85	11	15	173	297	782	951	1404	1182	791	368	158	58	6190
1985-86	0	16	138	356	626	1316	1216	1113	793	449	185	54	6262
1986-87	2	54	87	365	828	1027	1209	972	778	439	173	20	5954
1987-88	5	34	89	601	611	986	1269	1202	845	498	159	53	6352
1988-89	4	5	104	613	691	1141	979	1127	869	578	270	29	6410
1989-90	0	14	159	396	789	1488	947	907	742	492	262	31	6227
1990-91	4	3	125	415	612	981	1228	928	758	377	115	7	5553
1991-92	0	0	167	315	806	986	1116	953	889	525	245	62	6064
1992-93	7	25	146	473	719	987	1072	1123	943	493	156	48	6192
1993-94	0	3	151	465	756	1095	1479	1170	868	442	272	34	6735
1994-95	0	34	87	344	566	897	1137	1091	753	537	160	6	5612
1995-96	3	0	124	287	846	1221	1272	1099	1027	559	279	11	6728
1996-97	6	0	100	365	911	1027	1329	937	813	557	382	44	6471
1997-98	4	22	112	430	848	1037	985	795	783	447	63	72	5598
1998-99	0	3	52	363	655	920	1255	871	942	425	113	39	5638
1999-00	0	7	85	411	599	1047	1272	911	652	495	163	43	5685
2000-01	1	12	149	296	746	1445	1201	963	919	409	144	54	6339
2001-	7	0	117	352	490	860							

COOLING DEGREE DAYS (base 65 F) 2001 TOLEDO, OH (TOL)

YEAR	JAN	FEB	MAR	APR	MAY	JUN	JUL	AUG	SEP	OCT	NOV	DEC	ANNUAL
1972	0	0	0	0	22	67	236	148	55	0	0	0	528
1973	0	0	0	5	3	163	237	222	103	9	0	0	742
1974	0	0	0	4	25	91	243	206	34	5	0	0	608
1975	0	0	0	0	79	172	197	230	7	7	0	0	692
1976	0	0	0	31	10	155	230	137	34	2	0	0	599
1977	0	0	0	37	95	99	309	167	77	0	0	0	784
1978	0	0	0	0	58	128	200	184	170	.1	0	0	741
1979	0	0	0	0	46	127	182	158	67	22	0	0	602
1980	0	0	0	3	35	106	275	265	84	4	0	0	772
1981	0	0	1	2	17	132	220	170	64	0	0	0	606
1982	0	0	0	0	68	61	245	132	62	11	0	0	579
1983	0	0	0	4	2	148	311	279	109	11	0	0	864
1984	0	0	0	5	17	203	168	214	51	1	0	0	659
1985	0	0	0	29	60	58	263	147	116	0	0	0	673
1986	0	0	1	4	48	113	282	125	103	4	0	0	680
1987	0	0	0	5	105	202	318	225	59	0	4	0	918
1988	0	0	0	0	43	190	350	286	39	5	0	0	913
1989	0	0	2	0	34	132	259	168	69	5	0	0	669
1990	0	0	7	32	11	164	222	164	91	14	0	0	705
1991	0	0	0	14	185	244	305	256	111	13	0	0	1128
1992	0	0	0	3	32	66	170	120	59	2	0	0	452
1993	0	0	0	0	26	148	351	297	41	1	0	0	864
1994	0	0	0	22	39	222	245	104	66	1	0	0	699
1995	0	0	0	0	16	230	367	426	64	8	0	0	1111
1996	0	0	0	2	53	191	184	234	85	2	0	0	751
1997	0	0	0	0	0	163	215	101	41	34	0	0	554
1998	0	0	13	0	100	233	263	236	126	11	0	0	982
1999	0	0	0	2	46	220	386	161	95	2	0	0	912
2000	0	0	2	0	69	168	170	182	82	6	0	0	679
2001	0	0	0	10	40	171	254	268	64	6	0	0	813

SNOWFALL (inches) 2001 TOLEDO, OH (TOL)

YEAR	JUL	AUG	SEP	OCT	NOV	DEC	JAN	FEB	MAR	APR	MAY	JUN	TOTAL
1972-73	0.0	0.0	0.0	0.2	5.0	7.7	3.0	11.6	4.0	T	0.0	0.0	31.5
1973-74	0.0	0.0	0.0	0.0	0.2	13.8	7.5	11.6	2.9	1.1	T	0.0	37.1
1974-75	0.0	0.0	0.0	T	2.8	23.9	5.4	5.5	5.3	1.8	0.0	0.0	44.7
1975-76	0.0	0.0	0.0	0.0	5.7	12.2	14.5	8.4	4.0	1.3	0.0	0.0	46.1
1976-77	0.0	0.0	0.0	T	1.3	11.1	17.2	8.7	15.0	0.6	0.0	0.0	53.9
1977-78	0.0	0.0	0.0	0.0	6.6	24.2	30.8	9.0	2.5	T	0.0	0.0	73.1
1978-79	0.0	0.0	0.0	0.0	2.8	2.3	7.6	5.1	1.2	4.0	0.0	0.0	23.0
1979-80	0.0	0.0	0.0	T	1.6	1.5	4.1	6.4	3.4	0.5	T	0.0	17.5
1980-81	0.0	0.0	0.0	0.9	3.5	11.6	6.9	11.2	3.6	0.0	0.0	0.0	37.7
1981-82	0.0	0.0	0.0	T	0.8	14.9	18.4	14.3	10.7	9.1	0.0	0.0	68.2
1982-83	0.0	0.0	0.0	T	2.2	1.2	0.7	4.1	3.6	0.7	0.0	0.0	12.5
1983-84	0.0	0.0	0.0	0.0	3.4	13.4	12.2	6.3	9.8	T	T	0.0	45.1
1984-85	0.0	0.0	0.0	0.0	2.4	5.1	14.0	12.4	2.6	2.0	0.0	0.0	38.5
1985-86	0.0	0.0	0.0	0.0	2.5	8.7	6.6	10.2	2.2	0.2	0.0	0.0	30.4
1986-87	0.0	0.0	0.0	T	4.5	1.3	20.5	0.5	10.0	2.4	0.0	0.0	39.2
1987-88	0.0	0.0	0.0	T	0.1	11.1	8.3	14.3	4.2	T	0.0	0.0	38.0
1988-89	0.0	0.0	0.0	T	2.3	6.6	2.4	4.8	2.6	0.7	1.3	0.0	20.7
1989-90	0.0	0.0	0.0	2.0	2.3	6.5	2.5	10.4	3.5	0.3	0.0	0.0	27.5
1990-91	0.0	0.0	0.0	0.0	T	8.2	5.0	10.1	T	T	0.0	0.0	23.3
1991-92	0.0	0.0	0.0	T	1.7	2.5	10.5	3.0	12.5	0.1	0.0	0.0	30.3
1992-93	T	0.0	0.0	1.0	0.2	5.2	6.3	10.2	17.7	0.8	0.0	0.0	41.4
1993-94	0.0	T	T	0.8	1.1	6.9	20.2	16.6	4.2	7.0	0.0	T	56.8
1994-95	0.0	T	0.0	0.0	T	4.8	13.6	1.2	2.6	0.1	0.0	T	22.3
1995-96	0.0	0.0	0.0	0.0	6.8	7.0	11.0	2.5	4.4	T	0.0	0.0	31.7
1996-97													
1997-98									2.1				
1998-99						0.5	22.5	7.3	12.7				
1999-00													
2000-01													
2001-													
POR= 40 YRS	T	T	T	0.1	2.9	8.2	9.8	7.9	6.1	1.6	0.0	T	36.6

2001
OKLAHOMA CITY,
OKLAHOMA (OKC)

Oklahoma City is located along the North Canadian River, a frequently nearly-dry stream, at the geographic center of the state. It is not quite 1,000 miles south of the Canadian Border and a little less than 500 miles north of the Gulf of Mexico. The surrounding country is gently rolling with the nearest hills or low mountains, the Arbuckles, 80 miles south. The elevation ranges around 1,250 feet above sea level.

Although some influence is exerted at times by warm, moist air currents from the Gulf of Mexico, the climate of Oklahoma City falls mainly under continental controls characteristic of the Great Plains Region. The continental effect produces pronounced daily and seasonal temperature changes and considerable variation in seasonal and annual precipitation. Summers are long and usually hot. Winters are comparatively mild and short.

During the year, temperatures of 100 degrees or more occur on an average of 10 days, but have occurred on as many as 50 days or more. While summers are usually hot, the discomforting effect of extreme heat is considerably mitigated by low humidity and the prevalence of a moderate southerly breeze. Approximately one winter in three has temperatures of zero or lower.

The length of the growing season varies from 180 to 251 days. Average date of last freeze is early April and average date of first freeze is early November. Freezes have occurred in early October.

During an average year, skies are clear approximately 40 percent of the time, partly cloudy 25 percent, and cloudy 35 percent of the time. The city is almost smoke-free as a result of favorable atmospheric conditions and the almost exclusive use of natural gas for heating. Flying conditions are generally very good with flight by visual flight rules possible about 96 percent of the time.

Summer rainfall comes mainly from showers and thunderstorms. Winter precipitation is generally associated with frontal passages. Measurable precipitation has occurred on as many as 122 days and as few as 55 days during the year. The seasonal distribution of precipitation is normally 12 percent in winter, 34 percent in spring, 30 percent in summer, and 24 percent in fall. The The period with the least number of days with precipitation is November through January, and the month with the most rainy days is May. Thunderstorms occur most often in late spring and early summer. Large hail and/or destructive winds on occasion accompany these thunderstorms.

Snowfall averages less than 10 inches per year and seldom remains on the ground very long. Occasional brief periods of freezing rain and sleet storms occur.

Heavy fogs are infrequent. Prevailing winds are southerly except in January and February when northerly breezes predominate.

NORMALS, MEANS, AND EXTREMES
OKLAHOMA CITY, OK (OKC)

LATITUDE:	LONGITUDE:	ELEVATION (FT):	TIME ZONE:	WBAN: 13967
35 23' 19" N	97 36' 01" W	GRND: 1281 BARO: 1284	CENTRAL (UTC + 6)	

	ELEMENT	POR	JAN	FEB	MAR	APR	MAY	JUN	JUL	AUG	SEP	OCT	NOV	DEC	YEAR
TEMPERATURE °F	NORMAL DAILY MAXIMUM	30	46.7	52.1	62.0	71.9	79.1	87.3	93.4	92.5	83.8	73.6	60.4	49.9	71.1
	MEAN DAILY MAXIMUM	62	46.9	52.7	61.3	71.5	79.1	87.3	93.1	92.7	84.4	74.1	59.8	50.3	71.1
	HIGHEST DAILY MAXIMUM	47	80	92	93	100	104	105	110	110	108	96	87	86	110
	YEAR OF OCCURRENCE		1986	1996	1967	1972	1985	1998	1996	1980	2000	1972	1980	1955	JUL 1996
	MEAN OF EXTREME MAXS.	53	69.2	74.9	82.2	87.5	91.6	96.8	101.4	101.4	96.8	89.1	77.6	70.5	86.6
	NORMAL DAILY MINIMUM	30	25.2	29.6	38.5	48.8	57.7	66.1	70.6	69.6	62.2	50.4	38.6	28.6	48.8
	MEAN DAILY MINIMUM	62	26.4	30.8	38.0	48.8	58.0	66.7	70.8	69.8	62.0	51.1	38.2	29.6	49.2
	LOWEST DAILY MINIMUM	47	-4	-3	3	20	37	47	53	51	36	16	11	-8	-8
	YEAR OF OCCURRENCE		1988	1996	1960	1957	1981	1954	1971	1956	1989	1993	1991	1989	DEC 1989
	MEAN OF EXTREME MINS.	53	7.7	12.9	20.2	32.3	44.6	55.6	62.4	60.3	46.4	33.9	21.8	12.3	34.2
	NORMAL DRY BULB	30	35.9	40.9	50.3	60.4	68.4	76.7	82.0	81.1	73.0	62.0	49.6	39.3	60.0
	MEAN DRY BULB	62	36.8	41.8	49.7	60.2	68.6	77.1	82.0	81.4	73.2	62.6	49.1	39.9	60.2
	MEAN WET BULB	59	32.2	36.5	43.0	52.4	61.3	68.5	71.1	69.8	64.1	54.6	43.0	35.2	52.6
	MEAN DEW POINT	59	25.9	29.7	35.5	45.8	57.0	64.8	66.3	65.0	59.2	48.6	36.8	28.9	47.0
	NORMAL NO. DAYS WITH:														
	MAXIMUM 90	30	0.0	0.0	0.1	0.5	2.2	12.0	22.4	22.4	8.8	1.2	0.0	0.0	69.6
	MAXIMUM 32	30	5.2	2.4	0.2	0.0	0.0	0.0	0.0	0.0	0.0	0.0	0.1	2.9	10.8
	MINIMUM 32	30	23.1	17.0	8.4	0.9	0.0	0.0	0.0	0.0	0.0	0.4	7.9	20.0	77.7
	MINIMUM 0	30	0.5	0.1	0.0	0.0	0.0	0.0	0.0	0.0	0.0	0.0	0.0	0.2	0.8
H/C	NORMAL HEATING DEG. DAYS	30	902	675	464	176	31	0	0	0	15	137	462	797	3659
	NORMAL COOLING DEG. DAYS	30	0	0	9	38	136	351	527	499	255	44	0	0	1859
RH	NORMAL (PERCENT)	30	67	66	61	61	68	67	61	62	67	64	67	68	65
	HOUR 00 LST	30	72	72	68	69	76	77	70	70	76	72	74	73	72
	HOUR 06 LST	30	77	77	75	76	83	84	80	80	83	79	79	78	79
	HOUR 12 LST	30	58	58	52	51	57	56	49	50	55	52	56	58	54
	HOUR 18 LST	30	58	54	48	48	54	53	46	46	53	54	60	62	53
S	PERCENT POSSIBLE SUNSHINE	41	60	61	65	66	66	76	80	80	74	70	62	59	68
W/O	MEAN NO. DAYS WITH:														
	HEAVY FOG(VISBY 1/4 MI)	52	3.8	3.3	1.8	1.0	0.8	0.5	0.3	0.5	0.8	1.6	2.1	3.3	19.8
	THUNDERSTORMS	61	0.5	1.4	3.3	5.5	8.8	8.7	6.0	6.3	5.1	3.4	1.3	0.7	51.0
CLOUDINESS	MEAN:														
	SUNRISE-SUNSET (OKTAS)	2	4.8	5.1	5.3	4.8	5.2	2.4	2.0	2.7	4.8	3.6	3.2	4.8	4.1
	MIDNIGHT-MIDNIGHT (OKTAS)	1	4.8	5.2	5.1	4.8	5.2	2.4	2.0	2.4	2.4	3.2	3.6	4.0	3.8
	MEAN NO. DAYS WITH:														
	CLEAR	2	4.3	10.0	8.3	10.5	8.7	13.0	16.0	15.0	7.0	12.5	5.5	8.5	119.3
	PARTLY CLOUDY	2	4.7	2.7	2.7	4.0	6.0	5.3	4.5	4.5	3.0	4.0	4.5	5.0	48.9
	CLOUDY	2	7.3	10.0	7.0	8.0	6.3	3.3	2.5	3.0	2.0	5.0	6.5	7.5	68.4
PR	MEAN STATION PRESSURE(IN)	28	28.73	28.69	28.59	28.58	28.55	28.58	28.63	28.64	28.66	28.69	28.68	28.73	28.65
	MEAN SEA-LEVEL PRES. (IN)	59	30.15	30.09	29.98	29.94	29.92	29.90	29.96	29.97	30.00	30.04	30.07	30.12	30.01
WINDS	MEAN SPEED (MPH)	40	13.1	13.8	14.9	14.6	13.1	12.4	11.4	10.6	11.5	12.3	12.9	12.8	12.8
	PREVAIL.DIR(TENS OF DEGS)	21	36	36	16	16	16	16	16	16	16	16	16	36	16
	MAXIMUM 2-MINUTE:														
	SPEED (MPH)	8	45	45	52	46	47	41	48	46	38	43	46	44	52
	DIR. (TENS OF DEGS)		34	32	24	32	01	25	34	05	33	03	19	33	24
	YEAR OF OCCURRENCE		1996	1997	1996	1999	2001	1993	2000	1996	1998	1994	1994	2000	MAR 1996
	MAXIMUM 5-SECOND:														
	SPEED (MPH)	8	53	54	62	55	63	63	64	52	48	51	56	52	64
	DIR. (TENS OF DEGS)		33	33	23	32	01	22	33	05	29	03	20	33	33
	YEAR OF OCCURRENCE		1996	1997	1996	1999	2001	1998	2000	1996	1998	1994	1994	2000	JUL 2000
PRECIPITATION	NORMAL (IN)	30	1.13	1.56	2.71	2.77	5.22	4.31	2.61	2.60	3.84	3.23	1.98	1.40	33.36
	MAXIMUM MONTHLY (IN)	61	5.68	4.63	7.85	10.78	12.07	14.66	11.90	6.77	11.85	13.18	5.72	8.14	14.66
	YEAR OF OCCURRENCE		1949	1990	1988	1947	1982	1989	1996	1966	1991	1983	1994	1984	JUN 1989
	MINIMUM MONTHLY (IN)	61	0.00	T	T	0.17	0.33	0.55	T	0.00	T	T	T	0.03	0.00
	YEAR OF OCCURRENCE		1985	1947	1940	1989	1942	2001	1983	2000	1948	1958	1949	1996	AUG 2000
	MAXIMUM IN 24 HOURS (IN)	61	3.10	2.21	3.44	4.48	7.56	4.56	5.75	3.56	7.68	8.95	2.89	2.89	8.95
	YEAR OF OCCURRENCE		1982	1978	1944	1999	1993	1989	1981	1989	1970	1983	1994	1991	OCT 1983
	NORMAL NO. DAYS WITH:														
	PRECIPITATION 0.01	30	5.1	6.1	7.2	7.4	9.6	8.1	6.1	6.7	7.7	6.5	5.7	5.6	81.8
	PRECIPITATION 1.00	30	0.3	0.3	0.8	0.8	1.5	1.4	0.7	0.9	1.1	0.9	0.6	0.4	9.7
SNOWFALL	NORMAL (IN)	30	2.8	2.7	1.2	0.*	0.0	0.0	0.0	0.0	0.0	0.0	0.6	1.8	9.1
	MAXIMUM MONTHLY (IN)	61	17.3	12.0	13.9	0.7	T	T	T	T	T	0.1	7.5	8.3	17.3
	YEAR OF OCCURRENCE		1949	1978	1968	1957	1992	1992	1997	1997	1992	1993	1972	1987	JAN 1949
	MAXIMUM IN 24 HOURS (IN)	61	8.9	6.5	8.4	0.7	T	T	T	T	T	0.1	5.5	8.3	8.9
	YEAR OF OCCURRENCE		1988	1986	1948	1957	1992	1992	1997	1997	1992	1993	1972	1987	JAN 1988
	MAXIMUM SNOW DEPTH (IN)	53	12	8	8	T	0	0	0	0	0	T	3	7	12
	YEAR OF OCCURRENCE		1988	1951	1948	1973						1993	1980	1987	JAN 1988
	NORMAL NO. DAYS WITH:														
	SNOWFALL 1.0	30	0.9	1.1	0.3	0.0	0.0	0.0	0.0	0.0	0.0	0.0	0.2	0.6	3.1

PRECIPITATION (inches) 2001 OKLAHOMA CITY, OK (OKC)

YEAR	JAN	FEB	MAR	APR	MAY	JUN	JUL	AUG	SEP	OCT	NOV	DEC	ANNUAL
1972	0.21	0.43	1.13	3.10	4.03	1.36	3.22	1.82	2.04	7.17	2.28	0.84	27.63
1973	3.39	0.31	6.76	2.32	3.61	6.31	3.38	1.36	8.00	3.05	2.81	0.47	41.77
1974	0.10	2.68	3.12	4.66	5.01	3.36	0.48	4.42	6.24	5.57	2.34	1.47	39.45
1975	1.99	1.90	1.72	1.92	8.76	4.82	7.71	0.60	1.92	0.84	1.77	1.30	35.25
1976	T	0.33	3.09	2.94	4.36	0.88	1.38	1.46	1.53	1.78	0.12	0.19	18.06
1977	0.32	1.40	1.30	2.88	7.97	2.00	4.10	3.08	1.20	2.41	1.59	0.34	28.59
1978	1.26	3.23	1.32	1.65	10.12	4.04	3.75	0.25	0.96	1.02	2.88	0.70	31.18
1979	1.55	0.63	2.73	2.78	7.29	9.94	5.62	3.78	0.72	1.58	1.93	2.57	41.12
1980	1.69	1.29	1.38	2.16	9.00	2.52	0.42	0.60	2.21	0.99	0.51	1.58	24.35
1981	0.19	1.15	2.87	2.97	2.73	7.49	6.45	3.61	1.48	7.70	2.11	0.20	38.95
1982	3.68	0.98	1.63	1.92	12.07	4.06	2.11	1.13	2.86	1.03	2.78	1.94	36.19
1983	2.62	1.71	2.51	2.34	6.88	3.18	T	3.18	0.90	13.18	1.90	0.70	39.10
1984	0.35	1.16	4.70	1.79	1.62	3.48	0.30	2.35	1.01	6.64	2.05	8.14	33.59
1985	0.92	3.71	6.60	5.35	1.49	8.34	1.33	2.63	4.59	5.23	3.73	0.26	44.18
1986	0.00	0.68	1.75	4.42	8.21	3.11	0.38	3.29	9.54	8.00	4.63	1.16	45.17
1987	2.45	4.05	2.33	0.41	11.86	6.50	2.99	1.83	4.58	1.82	1.92	3.75	44.49
1988	1.24	0.41	7.85	3.19	1.07	3.59	1.92	1.60	5.19	2.04	2.45	1.39	31.94
1989	1.17	2.20	2.72	0.17	4.33	14.66	1.91	5.55	4.51	3.26	0.09	0.32	40.89
1990	1.85	4.63	4.43	5.11	5.79	1.25	2.65	3.16	7.35	1.27	1.59	1.46	40.54
1991	0.89	0.03	1.59	2.10	6.39	3.85	1.98	3.24	11.85	3.98	1.94	5.90	43.74
1992	1.15	1.28	1.08	3.64	4.88	6.35	4.01	5.82	2.92	1.13	4.51	3.08	39.85
1993	1.90	3.21	2.82	2.50	10.90	2.65	1.24	1.86	7.05	0.47	1.34	1.27	37.21
1994	0.21	2.56	3.18	3.38	2.69	1.70	2.17	1.81	2.17	1.88	5.72	1.63	29.10
1995	1.28	0.04	2.21	3.76	7.39	6.06	1.94	3.15	6.66	1.54	0.39	2.35	36.77
1996	0.08	0.02	2.17	2.00	1.90	1.16	11.90	5.85	5.88	2.53	3.36	T	36.85
1997	0.52	2.59	0.60	4.39	3.68	3.01	4.60	4.04	1.66	3.93	1.11	2.96	33.09
1998	4.09	0.32	6.45	3.34	2.12	2.67	0.02	0.48	4.39	6.76	3.09	1.62	35.35
1999	1.81	1.20	3.45	6.92	3.10	8.61	1.94	1.35	4.88	2.22	0.06	3.71	39.25
2000	0.75	1.47	3.12	5.17	1.36	6.71	5.25	0.00	1.73	8.39	2.79	2.30	39.04
2001	2.23	2.25	1.01	1.04	7.70	0.55	1.27	1.95	5.55	3.56	1.08	0.91	29.10
POR= 63 YRS	1.24	1.49	2.40	3.13	5.27	4.31	2.80	2.54	3.70	3.03	1.79	1.53	33.23

AVERAGE TEMPERATURE (F) 2001 OKLAHOMA CITY, OK (OKC)

YEAR	JAN	FEB	MAR	APR	MAY	JUN	JUL	AUG	SEP	OCT	NOV	DEC	ANNUAL
1972	34.9	42.1	53.4	63.2	67.6	79.0	79.8	80.4	75.8	61.1	43.4	34.4	59.6
1973	33.3	39.8	52.5	56.0	66.8	75.2	79.8	79.7	70.6	64.3	53.1	34.4	59.2
1974	35.0	44.4	54.8	60.0	71.5	74.1	82.7	78.5	65.5	63.5	49.3	39.6	59.9
1975	40.3	36.5	46.1	58.7	67.4	75.1	78.0	80.1	68.3	63.4	50.7	41.8	58.9
1976	39.0	52.2	52.4	61.6	63.6	74.8	79.8	81.3	72.6	56.5	43.9	38.8	59.7
1977	29.2	45.9	54.1	62.5	70.0	79.6	83.0	80.7	78.0	62.7	50.9	40.0	61.4
1978	26.3	29.4	49.1	64.5	68.1	77.3	87.0	82.6	79.7	64.7	50.4	36.9	59.7
1979	25.4	31.5	51.2	58.1	65.8	75.2	81.0	80.0	73.1	65.7	46.5	43.3	58.1
1980	38.2	38.2	46.3	56.7	69.0	81.4	88.3	88.0	76.3	61.1	50.3	41.9	61.3
1981	37.7	43.9	51.9	65.6	65.7	78.4	84.2	78.8	74.1	60.1	50.3	39.1	59.8
1982	35.3	37.7	52.7	57.5	68.2	72.2	81.0	84.1	74.5	62.7	48.6	43.2	59.8
1983	38.6	42.6	48.8	54.0	64.6	73.4	81.6	84.0	74.9	62.7	50.4	25.8	58.5
1984	34.0	45.4	46.4	56.5	68.4	78.6	81.6	82.6	71.5	61.6	49.7	43.0	59.9
1985	30.6	37.2	53.0	62.7	70.0	76.0	81.0	81.3	73.1	61.2	35.1	40.8	58.9
1986	43.6	44.8	55.5	62.8	69.0	79.0	85.9	80.0	74.8	61.6	44.8	40.8	61.9
1987	35.1	45.9	50.3	61.8	72.6	77.1	80.1	82.2	72.4	60.0	50.5	40.6	60.7
1988	34.2	40.3	49.5	58.9	70.3	78.4	81.6	82.8	73.5	59.3	51.2	43.9	60.3
1989	42.8	33.1	51.1	63.4	69.4	74.3	79.6	78.3	67.8	63.1	52.2	32.7	59.0
1990	45.9	46.0	52.6	59.2	68.6	82.0	80.7	81.6	77.0	60.9	54.9	37.1	62.2
1991	34.9	49.0	54.3	62.5	72.3	78.0	82.2	81.2	70.9	62.6	45.0	44.1	61.4
1992	42.0	49.9	54.8	61.3	66.5	74.1	81.1	74.8	72.5	62.3	45.9	39.8	60.4
1993	36.5	38.8	48.0	56.2	66.0	76.8	83.6	82.3	69.8	57.1	44.2	42.0	58.4
1994	36.0	37.4	52.7	59.4	66.8	79.6	79.7	79.7	70.7	62.6	50.1	42.5	59.8
1995	38.6	44.9	49.8	56.8	64.3	73.0	81.0	81.5	70.9	61.6	49.3	39.9	59.3
1996	35.8	44.3	46.0	58.7	73.8	77.9	81.3	81.3	70.1	56.0	46.3	46.3	59.6
1997	37.8	44.0	52.5	54.7	66.7	75.4	81.6	78.6	75.3	62.3	46.5	39.1	59.5
1998	40.6	45.5	47.4	57.4	72.5	81.1	88.0	85.0	81.2	64.4	53.2	41.6	63.2
1999	40.5	50.7	49.8	61.3	68.1	75.7	82.2	84.8	71.1	62.7	56.8	43.3	62.3
2000	40.7	49.1	53.4	59.0	71.0	74.6	80.8	85.4	76.1	64.1	43.3	30.5	60.7
2001	36.3	40.8	46.6	63.6	69.5	76.4	85.7	82.9	70.7	60.2	53.9	42.2	60.7
POR= 63 YRS	36.8	41.9	49.9	60.1	68.5	76.9	82.0	81.4	73.3	62.6	49.1	39.9	60.2

REFERENCE NOTES:

PAGE 1:
THE TEMPERATURE GRAPH SHOWS NORMAL MAXIMUM AND NORMAL MINIMUM DAILY TEMPERATURES (SOLID CURVES) AND THE ACTUAL DAILY HIGH AND LOW TEMPERATURES (VERTICAL BARS).

PAGE 2 AND 3:
H/C INDICATES HEATING AND COOLING DEGREE DAYS.
RH INDICATES RELATIVE HUMIDITY
W/O INDICATES WEATHER AND OBSTRUCTIONS
S INDICATES SUNSHINE.
PR INDICATES PRESSURE.
CLOUDINESS ON PAGE 3 IS THE SUM OF THE CEILOMETER AND SATELLITE DATA NOT TO EXCEED EIGHT EIGHTHS(OKTAS).

GENERAL:
T INDICATES TRACE PRECIPITATION, AN AMOUNT GREATER THAN ZERO BUT LESS THAN THE LOWEST REPORTABLE VALUE.
+ INDICATES THE VALUE ALSO OCCURS ON EARLIER DATES.
BLANK ENTRIES DENOTE MISSING OR UNREPORTED DATA.
NORMALS ARE 30-YEAR AVERAGES (1961 - 1990).
ASOS INDICATES AUTOMATED SURFACE OBSERVING SYSTEM.
PM INDICATES THE LAST DAY OF THE PREVIOUS MONTH.
POR (PERIOD OF RECORD) BEGINS WITH THE JANUARY DATA MONTH AND IS THE NUMBER OF YEARS USED TO COMPUTE THE MEAN. INDIVIDUAL MONTHS WITHIN THE POR MAY BE MISSING.
WHEN THE POR FOR A NORMAL IS LESS THAN 30 YEARS, THE NORMAL IS PROVISIONAL AND IS BASED ON THE NUMBER OF YEARS INDICATED.
0.* OR * INDICATES THE VALUE OR MEAN-DAYS-WITH IS BETWEEN 0.00 AND 0.05.
CLOUDINESS FOR ASOS STATIONS DIFFERS FROM THE NON-ASOS OBSERVATION TAKEN BY A HUMAN OBSERVER. ASOS STATION CLOUDINESS IS BASED ON TIME-AVERAGED CEILOMETER DATA FOR CLOUDS AT OR BELOW 12,000 FEET AND ON SATELLITE DATA FOR CLOUDS ABOVE 12,000 FEET.
THE NUMBER OF DAYS WITH CLEAR, PARTLY CLOUDY, AND CLOUDY CONDITIONS FOR ASOS STATIONS IS THE SUM OF THE CEILOMETER AND SATELLITE DATA FOR THE SUNRISE TO SUNSET PERIOD.

GENERAL CONTINUED:
CLEAR INDICATES 0 - 2 OKTAS, PARTLY CLOUDY INDICATES 3 - 6 OKTAS, AND CLOUDY INDICATES 7 OR 8 OKTAS. WHEN AT LEAST ONE OF THE ELEMENTS (CEILOMETER OR SATELLITE) IS MISSING, THE DAILY CLOUDINESS IS NOT COMPUTED.
WIND DIRECTION IS RECORDED IN TENS OF DEGREES (2 DIGITS) CLOCKWISE FROM TRUE NORTH. "00" INDICATES CALM. "36" INDICATES TRUE NORTH.
RESULTANT WIND IS THE VECTOR AVERAGE OF THE SPEED AND DIRECTION.
AVERAGE TEMPERATURE IS THE SUM OF THE MEAN DAILY MAXIMUM AND MINIMUM TEMPERATURE DIVIDED BY 2.
SNOWFALL DATA COMPRISE ALL FORMS OF FROZEN PRECIPITATION, INCLUDING HAIL.
A HEATING (COOLING) DEGREE DAY IS THE DIFFERENCE BETWEEN THE AVERAGE DAILY TEMPERATURE AND 65 F.
DRY BULB IS THE TEMPERATURE OF THE AMBIENT AIR.
DEW POINT IS THE TEMPERATURE TO WHICH THE AIR MUST BE COOLED TO ACHIEVE 100 PERCENT RELATIVE HUMIDITY.
WET BULB IS THE TEMPERATURE THE AIR WOULD HAVE IF THE MOISTURE CONTENT WAS INCREASED TO 100 PERCENT RELATIVE HUMIDITY.

ON JULY 1, 1996, THE NATIONAL WEATHER SERVICE BEGAN USING THE "METAR" OBSERVATION CODE THAT WAS ALREADY EMPLOYED BY MOST OTHER NATIONS OF THE WORLD. THE MOST NOTICEABLE DIFFERENCE IN THIS ANNUAL PUBLICATION WILL BE THE CHANGE IN UNITS FROM TENTHS TO EIGHTS(OKTAS) FOR REPORTING THE AMOUNT OF SKY COVER.

HEATING DEGREE DAYS (base 65 F) 2001 OKLAHOMA CITY, OK (OKC)

YEAR	JUL	AUG	SEP	OCT	NOV	DEC	JAN	FEB	MAR	APR	MAY	JUN	TOTAL
1972-73	0	0	23	225	640	940	975	701	380	283	55	0	4222
1973-74	0	0	37	99	362	787	922	573	330	168	8	0	3286
1974-75	0	0	56	88	463	784	763	792	583	235	29	0	3793
1975-76	0	0	64	126	430	713	801	367	406	128	100	0	3135
1976-77	0	0	19	306	629	805	1103	529	338	107	7	0	3843
1977-78	0	0	0	115	420	766	1192	990	493	90	64	0	4130
1978-79	0	0	2	89	437	866	1221	932	434	217	81	0	4279
1979-80	0	0	2	92	551	669	823	771	572	249	24	0	3753
1980-81	0	0	23	180	444	710	839	587	400	69	69	0	3321
1981-82	0	0	22	189	434	797	913	759	382	248	25	13	3782
1982-83	0	0	14	156	490	671	809	622	496	345	96	9	3708
1983-84	0	0	25	117	439	1207	955	561	572	263	45	0	4184
1984-85	0	0	75	162	462	676	1059	773	377	108	10	0	3702
1985-86	0	0	63	146	562	921	656	562	308	122	17	0	3357
1986-87	0	0	2	137	599	742	918	528	450	177	3	0	3556
1987-88	0	0	1	165	442	748	948	712	473	204	14	0	3707
1988-89	0	0	8	196	408	644	679	887	441	140	38	0	3441
1989-90	0	0	78	135	386	993	583	525	387	202	52	0	3341
1990-91	0	0	9	169	307	860	925	444	339	110	25	0	3188
1991-92	0	0	37	150	594	642	704	430	332	154	59	2	3104
1992-93	0	1	5	115	563	774	878	725	525	265	53	0	3904
1993-94	0	1	27	269	619	706	896	767	394	204	53	0	3936
1994-95	0	0	31	138	451	690	810	554	477	253	84	0	3488
1995-96	0	0	75	129	465	767	898	602	584	209	10	0	3739
1996-97	0	0	29	151	556	697	839	583	385	310	43	0	3593
1997-98	0	1	2	188	549	798	750	542	554	238	9	3	3634
1998-99	0	0	0	75	347	719	752	398	463	144	28	0	2926
1999-00	0	0	34	115	249	669	746	457	354	192	40	1	2857
2000-01	0	0	35	113	648	1063	882	672	561	105	22	0	4101
2001-	0	0	18	169	338	698							

COOLING DEGREE DAYS (base 65 F) 2001 OKLAHOMA CITY, OK (OKC)

YEAR	JAN	FEB	MAR	APR	MAY	JUN	JUL	AUG	SEP	OCT	NOV	DEC	ANNUAL
1972	0	2	11	97	133	429	470	483	351	109	0	0	2085
1973	0	0	0	19	119	312	465	462	216	83	11	0	1687
1974	0	0	22	26	217	280	553	426	80	47	0	0	1651
1975	0	0	1	53	108	310	410	476	170	83	4	0	1615
1976	0	1	23	33	62	300	468	512	253	50	0	0	1702
1977	0	1	8	37	170	445	565	491	395	49	2	0	2163
1978	0	0	8	80	165	378	690	553	450	87	7	0	2418
1979	0	0	10	18	112	314	505	471	252	121	2	0	1805
1980	0	0	0	7	155	498	729	721	366	65	11	2	2554
1981	0	4	0	94	98	409	603	435	304	47	0	0	1994
1982	0	0	9	28	130	234	503	598	305	90	3	1	1901
1983	0	0	0	20	91	266	523	599	329	54	8	0	1890
1984	0	0	0	16	159	414	521	551	279	64	5	0	2009
1985	0	0	12	43	172	336	501	512	313	38	0	0	1927
1986	0	2	21	63	147	425	653	473	301	40	0	0	2125
1987	0	0	0	88	242	371	475	543	230	18	12	0	1979
1988	0	0	1	29	186	410	525	558	270	25	1	0	2005
1989	0	0	16	100	179	285	459	419	170	83	8	0	1719
1990	0	0	12	33	169	517	495	522	378	48	13	0	2187
1991	0	0	15	45	257	398	542	507	219	85	1	0	2069
1992	0	0	3	51	114	283	508	312	239	36	0	1	1547
1993	0	0	4	9	89	362	584	545	177	32	0	0	1802
1994	0	0	20	44	116	446	470	464	208	70	9	0	1847
1995	0	0	11	14	72	250	506	521	262	33	0	0	1669
1996	0	7	4	28	288	382	514	408	170	42	0	0	1843
1997	2	0	5	7	104	316	520	429	317	112	0	0	1812
1998	0	0	13	16	252	496	719	627	497	62	2	0	2684
1999	0	4	0	40	131	327	540	619	225	51	10	0	1947
2000	0	0	3	20	232	295	498	639	372	94	0	0	2153
2001	0	0	0	70	168	348	650	563	196	27	8	1	2031

SNOWFALL (inches) 2001 OKLAHOMA CITY, OK (OKC)

YEAR	JUL	AUG	SEP	OCT	NOV	DEC	JAN	FEB	MAR	APR	MAY	JUN	TOTAL
1972-73	0.0	0.0	0.0	0.0	7.5	1.4	8.3	0.2	T	T	0.0	0.0	17.4
1973-74	0.0	0.0	0.0	0.0	0.0	0.6	0.7	1.0	0.5	0.0	0.0	0.0	2.8
1974-75	0.0	0.0	0.0	0.0	1.0	2.0	0.6	0.9	0.1	0.0	0.0	0.0	4.6
1975-76	0.0	0.0	0.0	0.0	0.7	3.9	T	0.3	0.0	0.0	0.0	0.0	4.9
1976-77	0.0	0.0	0.0	0.0	0.3	T	2.8	0.4	0.0	0.0	0.0	0.0	3.5
1977-78	0.0	0.0	0.0	0.0	T	0.0	8.4	12.0	T	0.0	0.0	0.0	20.4
1978-79	0.0	0.0	0.0	0.0	0.0	3.3	4.0	6.1	0.0	0.0	0.0	0.0	13.4
1979-80	0.0	0.0	0.0	0.0	T	T	T	1.8	T	0.0	0.0	0.0	1.8
1980-81	0.0	0.0	0.0	0.0	4.0	0.0	T	T	0.0	0.0	0.0	0.0	4.0
1981-82	0.0	0.0	0.0	0.0	0.0	T	1.0	3.9	2.5	0.0	0.0	0.0	7.4
1982-83	0.0	0.0	0.0	0.0	T	T	5.1	4.3	T	0.0	0.0	0.0	9.4
1983-84	0.0	0.0	0.0	0.0	T	1.9	5.6	2.0	T	0.0	0.0	0.0	9.5
1984-85	0.0	0.0	0.0	0.0	T	6.1	1.5	2.3	0.0	0.0	0.0	0.0	9.9
1985-86	0.0	0.0	0.0	0.0	T	2.9	0.0	10.9	0.0	0.0	0.0	0.0	13.8
1986-87	0.0	0.0	0.0	0.0	0.0	T	10.0	1.0	T	0.0	0.0	0.0	11.0
1987-88	0.0	0.0	0.0	0.0	2.0	8.3	12.1	0.2	0.9	0.0	0.0	0.0	23.5
1988-89	0.0	0.0	0.0	0.0	0.0	0.6	2.0	4.8	T	4.0	0.6	T	12.0
1989-90	0.0	0.0	0.0	0.0	T	1.7	0.0	1.7	0.1	0.0	0.0	0.0	3.5
1990-91	0.0	0.0	0.0	0.0	0.0	4.2	T	0.0	T	0.0	0.0	0.0	4.2
1991-92	0.0	0.0	0.0	T	2.1	1.0	5.0	0.0	T	0.0	T	T	8.1
1992-93	0.0	0.0	T		T	3.3	0.4	1.8	T	T		0.0	
1993-94	0.0	0.0	0.0	0.0	T	0.0	0.5	T	6.0	0.0	0.0		
1994-95	0.0	0.0	0.0	0.0	0.0	0.0	4.9	T	4.5	T	T	T	9.4
1995-96	0.0	0.0		0.0	0.5	4.1	0.3	T		0.0	T		
1996-97						T	6.5		T	0.0	T		
1997-98	T	T	0.0	0.0	0.1	2.0	T	T	0.0	T	T	T	2.1
1998-99	0.0	0.0	0.0	T	0.0	1.0	T	0.0	1.3	T	T	0.0	2.3
1999-00	0.0	0.0	T	0.0	0.0	0.0	9.1	0.0	T	T	0.0	0.0	9.1
2000-01	0.0	0.0	0.0	T	T	8.2	3.4	T	0.0	T	0.0		11.6
2001-	0.0	0.0	T	T	3.2	1.5							
POR= 60 YRS	0.0	0.0	T	T	0.1	1.9	3.0	2.3	T	T	T	T	7.3

2001
TULSA,
OKLAHOMA (TUL)

The city of Tulsa lies along the Arkansas River at an elevation of 700 feet above sea level. The surrounding terrain is gently rolling.

At latitude 36 degrees, Tulsa is far enough north to escape the long periods of heat in summer, yet far enough south to miss the extreme cold of winter. The influence of warm moist air from the Gulf of Mexico is often noted, due to the high humidity, but the climate is essentially continental characterized by rapid changes in temperature. Generally the winter months are mild. Temperatures occasionally fall below zero but only last a very short time. Temperatures of 100 degrees or higher are often experienced from late July to early September, but are usually accompanied by low relative humidity and a good southerly breeze. The fall season is long with a great number of pleasant, sunny days and cool, bracing nights.

Rainfall is ample for most agricultural pursuits and is distributed favorably throughout the year. Spring is the wettest season, having an abundance of rain in the form of showers and thunderstorms.

The steady rains of fall are a contrast to the spring and summer showers and provide a good supply of moisture and more ideal conditions for the growth of winter grains and pastures. The greatest amounts of snow are received in January and early March. The snow is usually light and only remains on the ground for brief periods.

The average date of the last 32 degree temperature occurrence is late March and the average date of the first 32 degree occurrence is early November. The average growing season is 216 days.

The Tulsa area is occasionally subjected to large hail and violent windstorms which occur mostly during spring and early summer, although occurrences have been noted throughout the year.

Prevailing surface winds are southerly during most of the year. Heavy fogs are infrequent. Sunshine is abundant. The prevalence of good flying weather throughout the year has contributed to the development of Tulsa as an aviation center.

NORMALS, MEANS, AND EXTREMES
TULSA, OK (TUL)

LATITUDE:	LONGITUDE:	ELEVATION (FT):		TIME ZONE:	WBAN: 13968
36 11' 51" N	95 53' 11" W	GRND: 739	BARO: 742	CENTRAL (UTC + 6)	

	ELEMENT	POR	JAN	FEB	MAR	APR	MAY	JUN	JUL	AUG	SEP	OCT	NOV	DEC	YEAR
TEMPERATURE °F	NORMAL DAILY MAXIMUM	30	45.4	51.0	62.1	73.0	79.7	87.7	93.7	92.5	83.6	73.8	60.3	48.8	71.0
	MEAN DAILY MAXIMUM	65	46.7	52.4	61.0	72.1	79.3	87.4	93.3	92.8	84.6	74.7	60.2	50.1	71.2
	HIGHEST DAILY MAXIMUM	62	79	90	96	102	96	103	112	110	109	98	87	80	112
	YEAR OF OCCURRENCE		1950	1996	1974	1972	1985	1953	1954	1970	1939	1979	1945	1966	JUL 1954
	MEAN OF EXTREME MAXS.	53	68.7	74.5	83.0	88.0	90.7	96.1	102.0	102.0	96.7	89.4	78.5	71.0	86.7
	NORMAL DAILY MINIMUM	30	24.9	29.5	39.1	49.9	58.8	67.7	72.8	70.6	63.0	50.7	39.5	28.9	49.6
	MEAN DAILY MINIMUM	65	26.4	30.7	38.5	49.6	58.7	67.5	72.2	70.5	62.3	50.9	38.6	30.0	49.7
	LOWEST DAILY MINIMUM	62	-8	-11	-3	22	35	49	51	52	35	18	10	-8	-11
	YEAR OF OCCURRENCE		1947	1996	1948	1957	1961	1954	1971	1988	1984	1993	1976	1989	FEB 1996
	MEAN OF EXTREME MINS.	53	6.8	12.4	20.4	33.3	44.4	55.5	62.3	59.5	46.3	34.2	22.3	11.4	34.1
	NORMAL DRY BULB	30	35.2	40.3	50.6	61.5	69.3	77.7	83.3	81.5	73.3	62.2	49.9	38.9	60.3
	MEAN DRY BULB	65	36.6	41.5	49.9	60.9	69.1	77.5	82.8	81.7	73.4	62.8	49.3	40.0	60.5
	MEAN WET BULB	55	32.2	36.5	43.4	52.9	62.3	69.8	72.7	71.2	65.0	54.9	43.6	35.4	53.3
	MEAN DEW POINT	55	25.7	29.6	35.7	46.1	58.0	66.1	68.4	66.7	60.3	49.2	37.4	29.1	47.7
	NORMAL NO. DAYS WITH:														
	MAXIMUM 90	30	0.0	0.0	0.3	0.8	2.1	13.3	24.0	22.0	9.4	1.6	0.0	0.0	73.5
	MAXIMUM 32	30	5.8	2.6	0.3	0.0	0.0	0.0	0.0	0.0	0.0	0.0	0.2	3.1	12.0
	MINIMUM 32	30	23.9	17.2	8.5	0.5	0.0	0.0	0.0	0.0	0.0	0.3	7.6	20.1	78.1
	MINIMUM 0	30	0.6	0.1	0.0	0.0	0.0	0.0	0.0	0.0	0.0	0.0	0.0	0.4	1.1
H/C	NORMAL HEATING DEG. DAYS	30	924	692	457	151	41	0	0	0	20	144	453	809	3691
	NORMAL COOLING DEG. DAYS	30	0	0	11	46	174	381	567	512	269	57	0	0	2017
RH	NORMAL (PERCENT)	30	67	65	62	61	69	69	64	64	70	66	67	68	66
	HOUR 00 LST	30	72	70	68	68	78	78	72	73	79	75	74	73	73
	HOUR 06 LST	30	78	77	76	78	85	86	82	84	86	82	80	79	81
	HOUR 12 LST	30	59	57	53	51	58	58	53	53	58	53	57	60	56
	HOUR 18 LST	30	58	55	49	48	56	56	49	50	57	55	59	61	54
S	PERCENT POSSIBLE SUNSHINE	53	54	55	58	59	59	69	76	74	68	65	56	53	62
W/O	MEAN NO. DAYS WITH:														
	HEAVY FOG(VISBY 1/4 MI)	62	2.1	1.6	0.9	0.2	0.4	0.3	0.2	0.1	0.6	1.1	1.2	1.7	10.4
	THUNDERSTORMS	62	0.7	1.3	3.4	6.1	8.7	8.7	8.2	5.6	5.9	5.1	3.1	1.6	0.9	50.6
CLOUDINESS	MEAN:														
	SUNRISE-SUNSET (OKTAS)	1	6.4	5.6	5.2	6.0	5.2	2.8	3.2	2.7	5.6	3.5	4.8	4.0	4.6
	MIDNIGHT-MIDNIGHT (OKTAS)	1	6.4	5.6	6.4	6.0	4.8	2.4	2.8	2.8	4.0	3.5	4.8	4.8	4.5
	MEAN NO. DAYS WITH:														
	CLEAR	3	5.0	8.3	8.0	9.5	9.7	10.7	13.5	15.0	7.5	10.0	7.0	9.0	113.2
	PARTLY CLOUDY	3	3.7	3.3	1.0	1.0	3.7	8.0	6.5	5.0	3.5	3.5	1.0	5.0	45.2
	CLOUDY	3	8.3	7.3	8.7	10.5	8.7	3.7	4.0	4.5	4.5	6.5	9.5	7.0	83.2
PR	MEAN STATION PRESSURE(IN)	27	29.42	29.37	29.25	29.23	29.20	29.22	29.26	29.28	29.31	29.34	29.34	29.41	29.30
	MEAN SEA-LEVEL PRES. (IN)	55	30.15	30.10	29.99	29.95	29.93	29.92	29.96	29.98	30.00	30.05	30.08	30.13	30.02
WINDS	MEAN SPEED (MPH)	42	10.3	10.6	11.8	11.8	10.7	9.7	9.3	8.6	8.9	9.6	10.4	10.2	10.2
	PREVAIL.DIR(TENS OF DEGS)	28	18	18	18	18	18	18	18	18	18	18	18	18	18
	MAXIMUM 2-MINUTE:														
	SPEED (MPH)	9	37	41	46	55	41	49	51	38	39	38	44	36	55
	DIR. (TENS OF DEGS)		18	20	18	34	30	28	19	29	27	35	29	33	34
	YEAR OF OCCURRENCE		1996	2000	1997	1993	2001	1998	1993	2001	1999	2001	1998	2000	APR 1993
	MAXIMUM 5-SECOND:														
	SPEED (MPH)	9	47	53	55	63	48	61	55	51	49	46	52	44	63
	DIR. (TENS OF DEGS)		19	21	16	34	29	28	19	29	27	30	29	34	34
	YEAR OF OCCURRENCE		1996	2000	1997	1993	2001	1998	1993	2001	1999	1994	1998	2000	APR 1993
PRECIPITATION	NORMAL (IN)	30	1.54	1.97	3.46	3.72	5.60	4.44	3.09	3.12	4.70	3.66	3.13	2.16	40.59
	MAXIMUM MONTHLY (IN)	62	6.65	5.73	11.94	9.23	18.00	11.17	11.39	7.89	18.81	16.51	7.57	8.70	18.81
	YEAR OF OCCURRENCE		1949	1985	1973	1947	1943	1948	1994	1997	1971	1941	1946	1984	SEP 1971
	MINIMUM MONTHLY (IN)	62	0.00	0.16	0.08	0.34	1.17	0.53	0.03	0.01	T	T	0.01	0.10	0.00
	YEAR OF OCCURRENCE		1993	1996	1971	1989	1988	1963	1954	2000	1948	1952	1949	1996	JAN 1993
	MAXIMUM IN 24 HOURS (IN)	62	2.25	4.34	2.72	4.58	9.27	5.01	7.54	5.37	6.39	5.80	5.14	3.27	9.27
	YEAR OF OCCURRENCE		1946	1985	1998	1964	1984	1941	1963	1989	1940	1983	1974	1984	MAY 1984
	NORMAL NO. DAYS WITH:														
	PRECIPITATION 0.01	30	6.1	6.7	8.4	8.8	10.1	8.5	5.9	6.8	8.0	6.5	6.5	6.7	89.0
	PRECIPITATION 1.00	30	0.2	0.4	1.1	1.0	1.6	1.3	0.9	1.1	1.5	1.3	1.0	0.6	12.0
SNOWFALL	NORMAL (IN)	30	2.8	2.5	1.5	0.*	0.0	0.0	0.0	0.0	0.0	0.0	0.4	1.9	9.1
	MAXIMUM MONTHLY (IN)	60	12.7	10.1	14.1	1.7	T	T	T	0.0	T	0.3	5.6	11.4	14.1
	YEAR OF OCCURRENCE		1979	1960	1994	1957	1991	1994	1994		1990	1993	1972	2000	MAR 1994
	MAXIMUM IN 24 HOURS (IN)	60	9.0	6.5	12.9	1.7	T	T	T	0.0	T	0.3	4.0	8.8	12.9
	YEAR OF OCCURRENCE		1944	1993	1994	1957	1991	1994	1994		1990	1993	1972	1954	MAR 1994
	MAXIMUM SNOW DEPTH (IN)	53	11	6	10	T	0	0	0	0	0	0	3	8	11
	YEAR OF OCCURRENCE		1988	1949	1968	1993							2001	1954	JAN 1988
	NORMAL NO. DAYS WITH:														
	SNOWFALL 1.0	30	1.0	0.9	0.3	0.0	0.0	0.0	0.0	0.0	0.0	0.0	0.2	0.8	3.2

PRECIPITATION (inches)   2001   TULSA, OK (TUL)

YEAR	JAN	FEB	MAR	APR	MAY	JUN	JUL	AUG	SEP	OCT	NOV	DEC	ANNUAL
1972	0.17	0.49	0.91	4.45	2.43	2.69	2.68	5.16	2.95	7.58	5.00	1.03	35.54
1973	3.39	0.74	11.94	7.22	5.30	7.69	6.47	6.56	6.16	6.32	3.39	69.88
1974	0.79	3.17	2.62	3.65	6.94	7.88	0.55	5.30	11.78	6.40	7.30	2.88	59.26
1975	2.61	3.44	5.45	2.20	7.22	6.75	2.14	3.52	3.34	1.47	3.53	3.04	44.71
1976	0.21	0.84	3.95	8.27	6.75	1.87	4.37	1.17	2.60	2.65	0.68	0.55	33.91
1977	1.43	1.57	5.58	2.05	5.72	6.69	2.00	4.86	5.57	2.75	2.31	0.93	41.46
1978	0.81	2.84	2.99	7.14	9.28	6.06	0.36	1.37	0.13	0.95	5.48	0.78	38.19
1979	2.07	0.81	3.97	4.47	6.15	8.90	2.68	4.77	0.28	2.20	5.60	0.45	42.35
1980	2.07	1.32	3.59	3.44	7.23	5.57	0.09	2.34	3.47	2.05	0.79	1.37	33.33
1981	0.69	1.63	1.67	1.90	6.70	3.31	1.65	2.47	3.11	6.73	2.25	0.20	36.88
1982	3.58	0.67	1.04	1.28	9.30	4.13	1.65	1.42	2.95	1.22	4.61	3.39	35.24
1983	2.95	1.98	2.19	3.88	6.85	1.47	0.58	0.65	2.11	9.33	2.14	0.61	34.74
1984	1.00	1.95	6.72	2.44	11.25	1.72	0.48	1.96	2.77	6.98	2.80	8.70	48.77
1985	1.24	5.74	5.39	5.62	4.19	7.63	2.38	1.91	3.29	6.26	6.27	1.39	51.31
1986	0.00	1.22	2.28	5.10	6.97	4.23	1.15	3.96	8.36	5.53	2.99	0.97	42.76
1987	2.21	4.72	2.20	0.70	10.02	2.31	4.20	3.72	3.52	1.27	5.17	5.87	45.91
1988	1.11	1.03	6.52	3.18	1.17	0.58	4.20	2.43	5.37	1.43	4.38	1.82	33.22
1989	2.94	2.26	3.14	0.34	3.95	5.16	4.09	6.69	3.32	2.80	0.15	0.26	35.10
1990	2.93	4.14	6.51	5.31	5.21	1.08	0.24	1.83	4.19	2.15	2.41	2.94	38.94
1991	1.47	0.38	1.02	2.58	5.11	3.64	0.35	1.17	6.15	5.12	1.98	4.57	33.54
1992	0.48	1.32	1.37	4.75	5.65	8.41	2.12	3.09	2.66	3.53	4.83	5.21	43.42
1993		2.86	2.76	4.59	6.86	3.79	2.42	2.29	6.90	1.13	1.69	1.76	
1994	0.68	2.21	3.35	6.57	2.81	2.73	11.39	4.12	3.60	3.68	7.10	1.21	49.45
1995	0.93	0.57	1.83	5.92	10.73	9.84	2.55	1.44	4.96	1.05	0.25	1.77	41.84
1996	0.47	0.16	2.07	1.40	2.14	3.64	3.22	1.34	5.04	5.60	7.16	0.10	32.34
1997	0.27	3.41	1.39	4.09	1.66	5.77	5.64	7.89	3.06	2.07	1.63	4.32	41.20
1998	3.49	0.30	7.30	4.54	2.52	3.36	4.31	1.67	5.13	9.14	3.26	1.58	46.60
1999	3.03	1.25	3.55	7.20	9.55	5.21	0.40	0.42	9.70	1.75	1.32	5.10	48.48
2000	0.89	1.33	3.77	2.71	7.01	6.25	6.58	0.01	1.10	6.32	3.51	1.62	41.10
2001	2.09	2.62	0.77	1.19	6.32	3.04	0.51	2.26	1.95	2.81	3.33	2.25	29.14
POR= 65 YRS	1.57	1.82	2.86	3.99	5.52	4.60	3.22	2.97	4.21	3.55	2.68	2.05	39.04

AVERAGE TEMPERATURE (F) 2001   TULSA, OK (TUL)

YEAR	JAN	FEB	MAR	APR	MAY	JUN	JUL	AUG	SEP	OCT	NOV	DEC	ANNUAL
1972	34.8	41.8	53.0	62.8	68.0	79.5	80.4	81.7	75.5	60.9	43.6	33.7	59.6
1973	34.0	39.8	54.3	58.2	67.5	76.7	81.2	79.3	72.3	65.0	53.4	38.1	60.0
1974	34.1	43.7	55.2	61.8	72.1	73.8	85.4	78.3	64.7	63.0	49.1	39.7	60.1
1975	39.9	36.9	45.3	60.4	69.1	76.0	81.2	82.2	69.2	63.2	50.8	40.1	59.5
1976	37.2	51.1	51.9	61.5	63.0	75.0	81.4	79.7	72.8	56.1	43.1	37.1	59.2
1977	26.9	46.8	55.0	64.4	72.6	81.0	84.8	81.7	75.6	62.2	51.1	39.0	61.7
1978	24.9	29.4	47.5	63.5	68.3	77.6	87.8	84.3	80.6	63.5	51.6	38.0	59.8
1979	23.1	30.2	52.4	61.0	68.7	77.8	83.4	81.8	74.7	66.2	47.5	44.4	59.3
1980	38.6	37.1	48.3	61.1	70.6	82.5	91.7	89.7	78.3	61.5	50.5	42.3	62.7
1981	37.6	43.6	53.3	68.0	65.9	80.0	85.9	79.4	73.9	60.9	51.4	38.5	61.5
1982	33.6	38.2	55.3	59.3	72.9	74.7	84.2	85.3	74.6	63.4	50.6	44.4	61.4
1983	39.1	42.9	49.0	55.4	67.0	76.6	84.7	88.1	77.4	64.5	52.9	26.7	60.4
1984	34.4	46.4	48.3	58.0	67.5	80.1	82.0	82.7	71.5	63.8	50.4	44.7	60.8
1985	30.2	35.9	54.7	63.3	70.6	75.8	82.9	81.7	74.6	63.1	47.8	34.5	59.6
1986	42.8	43.2	55.0	62.6	69.4	79.7	86.6	78.2	74.7	61.0	43.6	40.0	61.4
1987	36.0	45.4	51.5	63.2	74.1	78.9	81.9	83.1	72.4	59.3	51.6	41.4	61.6
1988	34.8	39.3	49.3	59.5	71.0	79.9	82.6	83.0	73.2	58.5	51.7	43.4	60.5
1989	43.4	31.9	49.3	63.3	69.2	74.8	80.2	80.4	68.7	64.0	52.7	31.6	59.1
1990	46.1	46.1	53.2	59.6	67.4	82.1	83.2	83.5	78.3	61.2	56.4	38.5	63.0
1991	34.7	48.3	55.1	63.8	73.7	80.0	84.9	82.9	72.7	64.3	45.8	44.6	62.6
1992	42.8	50.1	54.9	61.6	67.6	74.7	81.8	76.6	72.8	60.8	45.9	38.6	60.7
1993	35.7	37.8	46.8	55.8	66.0	76.8	84.4	83.5	68.6	56.2	44.5	42.3	58.2
1994	35.2	39.0	52.9	60.4	67.4	80.5	79.3	78.4	70.9	63.4	52.0	42.7	60.2
1995	39.4	44.2	51.5	58.3	65.5	74.1	82.3	84.6	70.5	62.8	48.8	39.4	60.1
1996	35.4	43.0	45.4	59.2	72.9	78.5	81.6	79.8	70.0	61.4	44.9	42.1	59.5
1997	35.9	44.1	52.4	55.9	66.9	75.5	81.7	78.3	73.9	62.0	46.0	39.3	59.3
1998	40.1	44.8	46.7	57.8	72.9	79.9	85.4	84.1	80.8	63.0	53.4	40.8	62.5
1999	39.0	50.0	48.8	61.5	68.1	75.4	84.4	84.6	69.8	62.3	57.7	43.4	62.1
2000	40.0	48.0	53.2	59.5	70.8	74.4	81.4	86.8	75.7	66.2	43.8	28.6	60.7
2001	35.3	41.3	47.1	66.4	70.6	78.2	87.4	85.2	71.9	62.0	55.2	43.6	62.0
POR= 66 YRS	36.7	41.6	49.9	60.9	69.0	77.5	82.7	81.8	73.3	62.8	49.3	40.0	60.5

REFERENCE NOTES:

PAGE 1:
THE TEMPERATURE GRAPH SHOWS NORMAL MAXIMUM AND NORMAL MINIMUM DAILY TEMPERATURES (SOLID CURVES) AND THE ACTUAL DAILY HIGH AND LOW TEMPERATURES (VERTICAL BARS).

PAGE 2 AND 3:
H/C INDICATES HEATING AND COOLING DEGREE DAYS.
RH INDICATES RELATIVE HUMIDITY
W/O INDICATES WEATHER AND OBSTRUCTIONS
S INDICATES SUNSHINE.
PR INDICATES PRESSURE.
CLOUDINESS ON PAGE 3 IS THE SUM OF THE CEILOMETER AND SATELLITE DATA NOT TO EXCEED EIGHT EIGHTHS(OKTAS).

GENERAL:
T INDICATES TRACE PRECIPITATION, AN AMOUNT GREATER THAN ZERO BUT LESS THAN THE LOWEST REPORTABLE VALUE.
+ INDICATES THE VALUE ALSO OCCURS ON EARLIER DATES.
BLANK ENTRIES DENOTE MISSING OR UNREPORTED DATA.
NORMALS ARE 30-YEAR AVERAGES (1961 - 1990).
ASOS INDICATES AUTOMATED SURFACE OBSERVING SYSTEM.
PM INDICATES THE LAST DAY OF THE PREVIOUS MONTH.
POR (PERIOD OF RECORD) BEGINS WITH THE JANUARY DATA MONTH AND IS THE NUMBER OF YEARS USED TO COMPUTE THE MEAN. INDIVIDUAL MONTHS WITHIN THE POR MAY BE MISSING.
WHEN THE POR FOR A NORMAL IS LESS THAN 30 YEARS, THE NORMAL IS PROVISIONAL AND IS BASED ON THE NUMBER OF YEARS INDICATED.
0.* OR * INDICATES THE VALUE OR MEAN-DAYS-WITH IS BETWEEN 0.00 AND 0.05.
CLOUDINESS FOR ASOS STATIONS DIFFERS FROM THE NON-ASOS OBSERVATION TAKEN BY A HUMAN OBSERVER. ASOS STATION CLOUDINESS IS BASED ON TIME-AVERAGED CEILOMETER DATA FOR CLOUDS AT OR BELOW 12,000 FEET AND ON SATELLITE DATA FOR CLOUDS ABOVE 12,000 FEET.
THE NUMBER OF DAYS WITH CLEAR, PARTLY CLOUDY, AND CLOUDY CONDITIONS FOR ASOS STATIONS IS THE SUM OF THE CEILOMETER AND SATELLITE DATA FOR THE SUNRISE TO SUNSET PERIOD.

GENERAL CONTINUED:
CLEAR INDICATES 0 - 2 OKTAS, PARTLY CLOUDY INDICATES 3 - 6 OKTAS, AND CLOUDY INDICATES 7 OR 8 OKTAS. WHEN AT LEAST ONE OF THE ELEMENTS (CEILOMETER OR SATELLITE) IS MISSING, THE DAILY CLOUDINESS IS NOT COMPUTED.
WIND DIRECTION IS RECORDED IN TENS OF DEGREES (2 DIGITS) CLOCKWISE FROM TRUE NORTH. "00" INDICATES CALM. "36" INDICATES TRUE NORTH.
RESULTANT WIND IS THE VECTOR AVERAGE OF THE SPEED AND DIRECTION.
AVERAGE TEMPERATURE IS THE SUM OF THE MEAN DAILY MAXIMUM AND MINIMUM TEMPERATURE DIVIDED BY 2.
SNOWFALL DATA COMPRISE ALL FORMS OF FROZEN PRECIPITATION, INCLUDING HAIL.
A HEATING (COOLING) DEGREE DAY IS THE DIFFERENCE BETWEEN THE AVERAGE DAILY TEMPERATURE AND 65 F.
DRY BULB IS THE TEMPERATURE OF THE AMBIENT AIR.
DEW POINT IS THE TEMPERATURE TO WHICH THE AIR MUST BE COOLED TO ACHIEVE 100 PERCENT RELATIVE HUMIDITY.
WET BULB IS THE TEMPERATURE THE AIR WOULD HAVE IF THE MOISTURE CONTENT WAS INCREASED TO 100 PERCENT RELATIVE HUMIDITY.

ON JULY 1, 1996, THE NATIONAL WEATHER SERVICE BEGAN USING THE "METAR" OBSERVATION CODE THAT WAS ALREADY EMPLOYED BY MOST OTHER NATIONS OF THE WORLD. THE MOST NOTICEABLE DIFFERENCE IN THIS ANNUAL PUBLICATION WILL BE THE CHANGE IN UNITS FROM TENTHS TO EIGHTS(OKTAS) FOR REPORTING THE AMOUNT OF SKY COVER.

HEATING DEGREE DAYS (base 65 F) 2001 TULSA, OK (TUL)

YEAR	JUL	AUG	SEP	OCT	NOV	DEC	JAN	FEB	MAR	APR	MAY	JUN	TOTAL
1972-73	0	0	19	183	634	964	954	700	321	233	42	0	4050
1973-74	0	0	24	95	343	824	951	591	341	137	5	0	3311
1974-75	0	0	74	94	473	777	773	780	610	205	19	0	3805
1975-76	0	0	57	146	429	762	855	402	407	126	109	0	3293
1976-77	0	0	16	317	648	858	1173	511	309	99	1	0	3932
1977-78	0	0	1	118	412	801	1236	989	541	110	67	0	4275
1978-79	0	0	0	121	406	834	1293	972	391	164	47	0	4228
1979-80	0	0	0	90	525	632	812	801	513	154	22	0	3549
1980-81	0	0	13	172	438	703	843	598	360	48	58	0	3233
1981-82	0	0	23	178	402	817	967	747	322	208	11	5	3680
1982-83	0	0	23	146	437	635	794	611	492	321	50	0	3509
1983-84	0	0	19	89	378	1179	941	533	509	229	47	0	3924
1984-85	0	0	73	130	438	628	1073	809	330	103	7	0	3591
1985-86	0	0	46	111	510	936	680	602	322	127	13	0	3347
1986-87	0	0	5	148	632	771	893	544	413	149	0	0	3555
1987-88	0	0	1	189	416	727	928	739	483	187	9	0	3679
1988-89	0	0	8	218	393	662	663	921	487	155	53	0	3560
1989-90	0	0	67	126	375	1029	580	527	376	194	54	0	3328
1990-91	0	0	8	172	271	813	933	459	327	83	17	0	3083
1991-92	0	0	35	121	570	628	682	423	311	156	53	0	2979
1992-93	0	0	9	151	565	812	903	755	556	280	57	0	4088
1993-94	0	0	40	294	611	695	917	721	387	186	56	0	3907
1994-95	0	0	25	126	390	683	783	574	436	219	70	0	3306
1995-96	0	0	79	112	480	786	911	640	604	204	19	0	3835
1996-97	0	0	26	152	594	701	896	579	393	275	40	0	3656
1997-98	0	0	3	195	563	791	764	558	581	225	12	6	3698
1998-99	0	0	0	109	344	747	798	415	496	134	18	0	3061
1999-00	0	0	38	130	238	664	770	487	363	183	23	2	2898
2000-01	0	0	26	100	628	1121	914	657	546	93	18	0	4103
2001-	0	0	20	142	298	660							

COOLING DEGREE DAYS (base 65 F) 2001 TULSA, OK (TUL)

YEAR	JAN	FEB	MAR	APR	MAY	JUN	JUL	AUG	SEP	OCT	NOV	DEC	ANNUAL
1972	0	6	11	99	144	446	487	524	339	64	0	0	2120
1973	0	0	0	35	124	357	508	452	249	101	5	0	1831
1974	0	0	47	48	232	270	641	419	71	40	2	0	1770
1975	0	0	9	77	156	335	509	542	192	97	12	0	1929
1976	0	6	7	28	52	307	520	461	256	48	0	0	1685
1977	0	1	6	84	248	486	619	525	327	38	0	0	2334
1978	0	0	7	73	180	388	713	605	476	79	14	0	2535
1979	0	0	9	48	167	388	577	527	298	137	6	0	2157
1980	0	0	0	43	200	533	833	774	419	69	6	4	2881
1981	0	5	4	145	96	456	658	452	296	57	1	0	2170
1982	0	0	28	44	266	300	601	637	319	106	10	5	2316
1983	0	0	3	40	120	353	615	725	396	80	20	0	2352
1984	0	0	0	25	132	464	534	556	272	100	9	2	2094
1985	0	0	19	59	185	333	564	523	340	57	0	0	2080
1986	0	0	20	60	157	448	676	415	303	31	0	0	2110
1987	0	0	2	102	290	421	532	567	230	18	19	0	2181
1988	0	0	2	30	200	454	555	564	262	23	1	0	2091
1989	0	0	6	107	191	300	475	483	183	105	14	0	1864
1990	0	0	17	38	137	521	571	581	416	63	21	0	2365
1991	0	0	29	53	293	458	622	562	274	108	3	0	2402
1992	0	0	5	63	140	298	526	369	251	29	0	0	1681
1993	0	0	1	7	95	360	609	579	153	27	0	0	1831
1994	0	0	21	55	135	470	452	424	212	82	8	0	1859
1995	0	0	24	26	97	282	545	618	252	50	0	0	1894
1996	0	6	0	37	273	410	522	463	183	49	0	0	1943
1997	3	0	9	10	106	321	524	420	278	108	0	0	1779
1998	0	0	18	16	264	459	640	598	480	53	5	2	2535
1999	0	2	0	35	118	318	607	616	190	57	25	0	1968
2000	0	0	2	26	211	290	517	684	353	143	1	0	2227
2001	0	2	0	139	197	401	700	630	236	56	10	3	2374

SNOWFALL (inches) 2001 TULSA, OK (TUL)

YEAR	JUL	AUG	SEP	OCT	NOV	DEC	JAN	FEB	MAR	APR	MAY	JUN	TOTAL
1972-73	0.0	0.0	0.0	0.0	5.6	1.7	4.3	2.2	0.0	0.3	0.0	0.0	14.1
1973-74	0.0	0.0	0.0	0.0	T	1.8	T	T	T	T	0.0	0.0	1.8
1974-75	0.0	0.0	0.0	0.0	1.7	T	T	3.0	1.8	T	0.0	0.0	6.5
1975-76	0.0	0.0	0.0	0.0	0.8	1.3	T	T	T	0.0	0.0	0.0	2.1
1976-77	0.0	0.0	0.0	0.0	0.5	T	10.5	0.3	0.0	0.0	0.0	0.0	11.3
1977-78	0.0	0.0	0.0	0.0	T	0.0	5.4	6.3	T	0.0	0.0	0.0	11.7
1978-79	0.0	0.0	0.0	0.0	0.0	2.8	12.7	3.4	0.0	T	0.0	0.0	18.9
1979-80	0.0	0.0	0.0	0.0	T	0.0	0.4	3.8	T	0.0	0.0	0.0	4.2
1980-81	0.0	0.0	0.0	0.0	T	0.0	T	0.9	T	0.0	0.0	0.0	0.9
1981-82	0.0	0.0	0.0	0.0	0.0	T	0.3	5.6	T	0.0	0.0	0.0	5.9
1982-83	0.0	0.0	0.0	0.0	T	T	3.8	1.4	T	0.0	0.0	0.0	5.2
1983-84	0.0	0.0	0.0	0.0	T	3.0	4.6	0.2	T	0.0	0.0	0.0	7.8
1984-85	0.0	0.0	0.0	0.0	0.0	6.6	3.3	4.3	0.0	0.0	0.0	0.0	14.2
1985-86	0.0	0.0	0.0	0.0	T	2.5	0.0	4.9	0.0	0.0	0.0	0.0	7.4
1986-87	0.0	0.0	0.0	0.0	0.0	8.7	4.6	0.0	0.0	0.0	0.0	0.0	13.3
1987-88	0.0	0.0	0.0	0.0	T	6.7	11.0	T	0.5	0.0	0.0	0.0	18.2
1988-89	0.0	0.0	0.0	0.0	0.4	2.7	3.4	0.3	9.7	0.0	0.0	0.0	16.5
1989-90	0.0	0.0	0.0	0.0	0.0	2.0	0.0	0.0	0.2	0.0	0.0	0.0	2.2
1990-91	0.0	0.0	0.0	0.0	0.0	4.6	T	1.4	T	0.0	T	0.0	6.0
1991-92	0.0	0.0	0.0	0.0	0.2	0.1	0.8	0.0	T	0.0	0.0	0.0	1.1
1992-93	0.0	0.0	0.0	0.0	3.5	1.1	0.8	6.7	T	T	0.0	0.0	12.1
1993-94	0.0	0.0	0.0	0.3	0.0	0.0	T	T	14.1	T	0.0	T	14.4
1994-95	T	0.0	0.0	0.0	T	0.0	1.8	T	6.3	0.0	0.0	0.0	8.1
1995-96	0.0	0.0	0.0	T	1.8	T	1.0	5.0	0.0	0.0	0.0	0.0	7.8
1996-97	0.0	0.0	0.0	T	T	T	5.9	0.3	0.0	T	0.0	0.0	6.2
1997-98	T	0.0	0.0	0.0	T	0.6	4.0	0.0	T	0.0	0.0	0.0	4.6
1998-99	0.0	0.0	0.0	0.0	0.0	T	3.3	0.0	5.9	0.0	0.0	0.0	9.2
1999-00	0.0	0.0	0.0	0.0	0.0	T	7.1	0.0	2.2	0.0	0.0	0.0	9.3
2000-01	0.0	0.0	0.0	T	2.1	11.4	1.4	T	0.0	0.0	0.0	0.0	14.9
2001-	0.0	0.0	0.0	0.0	3.0	T							
POR= 66 YRS	T	0.0	0.0	T	0.4	1.7	3.1	2.3	1.8	T	T	T	9.3

2001
PENDLETON,
OREGON (PDT)

Pendleton is located in the southeastern part of the Columbia Basin, that low country of northern Oregon and central and eastern Washington which is almost entirely surrounded by mountains. This Basin is bounded on the south by the high country of central Oregon, on the north by the mountains of western Canada, on the west by the Cascade Range and on the east by the Blue Mountains and the north Idaho plateau. The gorge in the Cascades through which the Columbia River reaches the Pacific is the most important break in the barriers surrounding this basin. These physical features have important influences on the general climate of Pendleton and the surrounding territory.

The Weather Service Office at Pendleton Airport is located in rolling country which slopes generally upward toward the Blue Mountains about 15 miles to the east and southeast. The Columbia River approaches the area from the northwest to its junction with the Walla Walla River at an elevation of 351 feet and some 25 miles north of Pendleton, then turns southwestward to be joined a few miles below by the Umatilla River. Both the Walla Walla and Umatilla Rivers have their sources in the Blue Mountains and flow westward to the Columbia. The observation station is at an elevation of nearly 1,500 feet, about 3 miles northwest of downtown Pendleton. The city of Pendleton lies in the shallow east-west valley of the Umatilla River, approximately 400 feet lower than the airport.

Precipitation in the Pendleton area is definitely seasonal in occurrence with an average of only 10 percent of the annual total occurring in the three-month period, July-September. Most precipitation reaching this area accompanies cyclonic storms moving in from the Pacific Ocean. These storms reach their greatest intensity and frequency from October through April. The Cascade Range west of the Columbia Basin reduces the amount of precipitation received from the Pacific cyclonic storms. This influence is felt, particularly, in the desert area of the central part of the Basin. A gradual rise in elevation from the Columbia River to the foothills of the Blue Mountains again results in increased precipitation. This increase supplies sufficient moisture for productive wheat, pea, and stock raising activity in the area surrounding Pendleton.

The lighter summertime precipitation usually accompanies thunderstorms which often move into the area from the south or southwest. On occasion, these storms are quite intense, causing flash flooding with resultant heavy property damage and even loss of life.

Seasonal temperature extremes are usually quite moderate for the latitude. The last occurrence in spring of temperatures as low as 32 degrees is mid-April, and the average last occurrence in the fall of 32 degrees is late October. At the city station, where cool air settles in the valley on still nights, temperatures of 32 degrees have been recorded later in the spring and earlier in the fall. Under usual atmospheric conditions, air from the Pacific, with moderate temperature characteristics, moves across the Cascades or through the Columbia Gorge resulting in mild temperatures in the Pendleton area. When this flow of air from the west is impeded by slow-moving high pressure systems over the interior of the continent, temperature conditions sometimes become rather severe, hot in summer and cold in winter. During the summer or early fall, if a stagnant high predominates to the north or east of Pendleton, the hot, dry conditions may prove detrimental to crops during late May and June, and cause fire danger in the forest and grassland areas during late summer and early fall. During winter, coldest temperatures occur when air from a cold high pressure system in central Canada moves southwestward across the Rockies and flows down into the Columbia Basin. Under this condition the heavy cold air sometimes remains at low levels in the Basin for several days while warmer air from the Pacific flows above it, causing comparatively mild temperatures at higher elevations. Extreme winter temperatures are not particularly common in the Pendleton area. Below zero readings are recorded in approximately 60 percent of winters. Maximum temperatures usually reach 100 degrees or slightly higher on a few days during the summer.

NORMALS, MEANS, AND EXTREMES
PENDLETON, OR   (PDT)

LATITUDE:	LONGITUDE:	ELEVATION (FT):		TIME ZONE:	WBAN: 24155
45 41' 54" N	118 50' 03" W	GRND: 1504	BARO: 1504	PACIFIC (UTC + 8)	

	ELEMENT	POR	JAN	FEB	MAR	APR	MAY	JUN	JUL	AUG	SEP	OCT	NOV	DEC	YEAR
TEMPERATURE °F	NORMAL DAILY MAXIMUM	30	39.7	46.9	54.2	61.3	70.0	79.5	87.8	86.2	76.3	63.7	48.9	40.5	62.9
	MEAN DAILY MAXIMUM	54	39.5	46.3	53.7	61.5	70.1	78.5	87.7	85.9	76.9	63.4	48.9	40.9	62.8
	HIGHEST DAILY MAXIMUM	66	70	75	79	91	100	108	110	113	102	92	80	67	113
	YEAR OF OCCURRENCE		1995	1996	1964	1977	1986	1961	1939	1961	1955	1980	1999	1980	AUG 1961
	MEAN OF EXTREME MAXS.	54	58.4	62.2	68.1	77.7	88.0	94.6	101.4	99.7	92.4	79.9	65.9	59.3	79.0
	NORMAL DAILY MINIMUM	30	27.2	31.6	35.4	39.4	45.8	52.9	58.0	57.7	49.9	41.0	34.1	27.9	41.7
	MEAN DAILY MINIMUM	54	26.7	31.0	34.8	39.5	46.0	52.4	57.9	57.3	50.2	41.0	33.7	28.5	41.6
	LOWEST DAILY MINIMUM	66	-22	-18	1	18	25	35	42	40	30	11	-12	-19	-22
	YEAR OF OCCURRENCE		1957	1950	1993	1936	1954	1991	1971	1980	2000	1935	1985	1983	JAN 1957
	MEAN OF EXTREME MINS.	54	7.3	15.0	23.8	30.1	35.1	42.4	47.9	47.4	38.7	28.9	20.8	11.8	29.1
	NORMAL DRY BULB	30	33.5	39.2	44.8	50.3	57.9	66.2	72.9	72.0	63.1	52.4	41.5	34.3	52.3
	MEAN DRY BULB	54	33.0	38.6	44.2	50.5	58.0	65.5	72.8	71.6	63.6	52.3	41.3	34.6	52.2
	MEAN WET BULB	18	31.9	34.2	39.7	44.2	49.3	53.6	56.9	56.3	51.7	44.3	37.6	28.8	44.0
	MEAN DEW POINT	18	28.1	29.5	33.3	36.4	40.9	43.3	44.0	43.3	40.7	36.3	33.1	25.6	36.2
	NORMAL NO. DAYS WITH:														
	MAXIMUM 90	30	0.0	0.0	0.0	*	0.8	5.3	14.4	11.7	2.6	0.1	0.0	0.0	34.9
	MAXIMUM 32	30	8.6	2.7	0.2	0.0	0.0	0.0	0.0	0.0	0.0	0.0	2.1	8.1	21.7
	MINIMUM 32	30	20.1	14.5	8.8	3.2	0.1	0.0	0.0	0.0	0.2	3.1	10.8	19.7	80.5
	MINIMUM 0	30	1.4	0.4	0.0	0.0	0.0	0.0	0.0	0.0	0.0	0.0	0.1	1.2	3.1
H/C	NORMAL HEATING DEG. DAYS	30	977	722	626	441	226	71	15	23	145	391	705	952	5294
	NORMAL COOLING DEG. DAYS	30	0	0	0	0	6	107	260	240	88	0	0	0	701
RH	NORMAL (PERCENT)	30	77	73	63	58	52	46	36	38	47	59	74	78	58
	HOUR 04 LST	30	79	78	73	71	68	63	53	53	61	70	78	80	69
	HOUR 10 LST	30	77	71	60	52	47	41	33	36	42	54	72	78	55
	HOUR 16 LST	30	73	63	49	42	37	31	23	26	32	44	68	76	47
	HOUR 22 LST	30	79	76	68	62	56	49	37	40	51	64	77	80	62
S	PERCENT POSSIBLE SUNSHINE														
W/O	MEAN NO. DAYS WITH:														
	HEAVY FOG(VISBY 1/4 MI)	64	7.4	4.9	1.9	0.3	0.2	0.1	0.0	0.0	0.2	1.0	6.1	8.4	30.5
	THUNDERSTORMS	64	0.0	0.0	0.2	0.9	1.8	2.0	2.0	2.1	1.2	0.3	0.1	0.0	10.6
CLOUDINESS	MEAN:														
	SUNRISE-SUNSET (OKTAS)	1					6.4				3.2				
	MIDNIGHT-MIDNIGHT (OKTAS)	1										3.2			
	MEAN NO. DAYS WITH:														
	CLEAR	1	1.0	3.0	4.0		5.0	7.0							
	PARTLY CLOUDY	1		3.0	2.0		5.0	3.0							
	CLOUDY	1	3.0	2.0	9.0		10.0	3.0							
PR	MEAN STATION PRESSURE(IN)	28	28.52	28.47	28.40	28.42	28.40	28.39	28.40	28.39	28.44	28.49	28.47	28.54	28.44
	MEAN SEA-LEVEL PRES. (IN)	18	30.13	30.10	30.02	30.01	29.96	29.96	29.96	29.95	29.98	30.06	30.09	30.19	30.03
WINDS	MEAN SPEED (MPH)	37	7.2	7.9	8.8	9.5	9.2	9.1	8.8	8.4	8.0	7.5	7.7	7.5	8.3
	PREVAIL.DIR(TENS OF DEGS)	23	14	14	26	26	26	26	26	27	14	14	15	16	26
	MAXIMUM 2-MINUTE:														
	SPEED (MPH)	6	47	49	55	48	43	41	40	43	45	44	41	54	55
	DIR. (TENS OF DEGS)		24	15	25	25	24	25	27	23	27	24	26	22	25
	YEAR OF OCCURRENCE		2000	1999	1997	1997	1996	1998	2000	1997	1999	2001	2000	1998	MAR 1997
	MAXIMUM 5-SECOND:														
	SPEED (MPH)	6	56	58	63	55	51	51	47	59	53	52	48	66	66
	DIR. (TENS OF DEGS)		23	16	25	21	24	25	27	23	27	24	26	22	22
	YEAR OF OCCURRENCE		2000	1999	1997	1998	1996	1998	2000	1997	1999	2001	2000	1998	DEC 1998
PRECIPITATION	NORMAL (IN)	30	1.51	1.14	1.16	1.04	0.99	0.64	0.35	0.53	0.59	0.86	1.58	1.63	12.02
	MAXIMUM MONTHLY (IN)	66	3.92	3.03	2.82	2.78	3.18	2.70	1.45	2.58	2.34	2.79	3.76	4.68	4.68
	YEAR OF OCCURRENCE		1970	1940	1983	1978	1991	1947	1993	1977	1941	1947	1973	1973	DEC 1973
	MINIMUM MONTHLY (IN)	66	0.21	0.07	0.24	0.01	0.03	0.03	T	0.00	T	T	0.04	0.21	0.00
	YEAR OF OCCURRENCE		1949	1964	1941	1956	1964	1986	1967	1969	1993	1987	1939	1989	AUG 1969
	MAXIMUM IN 24 HOURS (IN)	66	1.29	1.41	1.33	1.24	1.52	1.49	1.19	2.19	1.23	1.88	1.35	1.25	2.19
	YEAR OF OCCURRENCE		1956	1994	1983	1990	1972	1947	1948	1993	1981	1982	1971	1978	AUG 1993
	NORMAL NO. DAYS WITH:														
	PRECIPITATION 0.01	30	12.0	10.9	10.6	8.7	7.2	5.8	2.9	3.6	4.7	6.1	11.7	12.2	96.4
	PRECIPITATION 1.00	30	*	0.0	*	0.1	*	0.0	0.0	*	*	*	*	*	0.1
SNOWFALL	NORMAL (IN)	30	6.1	2.1	1.0	0.1	T	0.0	0.0	0.0	0.0	0.2	2.2	5.2	16.9
	MAXIMUM MONTHLY (IN)	64	41.6	16.8	4.9	2.2	T	T	T	0.0	0.0	3.2	14.9	26.6	41.6
	YEAR OF OCCURRENCE		1950	1994	1971	1975	1993	1994	1993			1973	1985	1983	JAN 1950
	MAXIMUM IN 24 HOURS (IN)	64	13.3	16.1	4.0	2.2	T	T	T	0.0	0.0	3.2	8.0	9.9	16.1
	YEAR OF OCCURRENCE		1950	1994	1970	1975	1993	1994	1993			1973	1977	1948	FEB 1994
	MAXIMUM SNOW DEPTH (IN)	52	16	12	6	0	0	0	0	0	0	2	8	11	16
	YEAR OF OCCURRENCE		1957	1994	1993							1971	1978	1985	JAN 1957
	NORMAL NO. DAYS WITH:														
	SNOWFALL 1.0	30	2.2	0.6	0.4	0.1	0.0	0.0	0.0	0.0	0.0	0.1	0.6	1.9	5.9

PRECIPITATION (inches) 2001 PENDLETON, OR (PDT)

YEAR	JAN	FEB	MAR	APR	MAY	JUN	JUL	AUG	SEP	OCT	NOV	DEC	ANNUAL
1972	0.96	1.08	1.47	0.68	1.97	0.80	0.58	0.36	0.16	0.58	0.70	2.31	11.65
1973	0.50	1.09	0.43	0.27	0.67	0.15	0.01	0.08	1.34	1.71	3.76	4.68	14.69
1974	0.79	1.57	0.81	2.13	0.26	0.19	0.90	T	T	0.29	1.00	1.59	9.53
1975	3.53	1.30	0.65	0.97	0.30	0.28	0.73	0.67	0.00	1.80	0.84	1.98	13.05
1976	1.77	1.00	1.65	1.09	0.92	0.33	0.16	1.77	0.18	0.54	0.19	0.44	10.04
1977	0.48	0.64	1.51	0.18	1.87	0.37	0.06	2.58	1.17	0.51	2.00	2.42	13.79
1978	2.82	1.60	1.03	2.78	0.63	0.76	0.77	2.21	0.92	T	2.37	1.86	17.75
1979	1.43	1.72	1.18	1.17	0.39	0.21	0.09	1.40	0.30	1.68	1.83	0.62	12.02
1980	2.48	1.39	1.60	0.59	2.14	1.12	0.77	0.03	0.53	1.22	0.84	1.20	13.97
1981	0.89	1.35	1.43	1.20	1.59	1.53	0.94	0.03	1.31	0.86	1.91	2.31	15.35
1982	1.54	0.77	1.22	0.84	0.31	0.63	0.51	0.24	1.47	2.67	0.34	2.20	12.74
1983	0.86	1.57	2.82	0.70	0.73	1.44	0.52	0.56	0.46	0.84	1.67	3.42	15.59
1984	0.53	1.74	1.83	1.70	1.02	1.13	0.06	0.44	0.39	1.02	2.14	0.92	12.92
1985	0.44	1.33	1.13	0.37	0.44	0.69	0.34	0.26	2.10	0.89	2.11	1.27	11.37
1986	1.66	2.58	1.13	0.43	1.18	0.03	0.48	0.02	1.28	0.80	2.12	0.82	12.53
1987	1.48	0.64	1.39	0.47	0.85	0.38	0.34	0.05	0.03	T	0.76	1.23	7.62
1988	1.86	0.12	0.95	2.47	1.56	0.31	0.01	T	0.31	0.10	2.16	0.37	10.22
1989	1.86	1.36	1.72	1.57	1.47	0.57	0.09	1.25	0.12	0.84	1.27	0.21	12.33
1990	0.77	0.28	1.14	1.54	1.83	0.58	0.18	0.62	T	0.78	0.87	0.84	9.43
1991	0.98	0.57	1.00	0.71	3.18	2.14	0.24	0.42	T	0.92	2.68	0.67	13.51
1992	0.41	1.04	0.26	1.21	0.07	0.94	0.70	0.43	0.42	1.32	1.15	0.73	8.68
1993	1.79	0.80	1.49	1.85	1.51	0.71	1.45	2.19	T	0.22	0.93	0.92	13.86
1994	1.57	1.71	0.56	0.45	2.55	0.77	0.38	T	0.36	1.28	1.98	0.85	12.46
1995	2.53	1.07	1.93	2.28	0.97	2.30	0.24	0.29	0.55	1.21	2.18	1.73	17.28
1996	1.88	1.80	1.00	1.08	2.00	0.47	0.06	0.05	0.61	1.22	1.96	2.32	14.45
1997	1.84	0.39	1.16	1.56	0.33	0.76	0.66	0.07	0.77	1.43	1.64	1.05	11.66
1998	2.61	1.19	1.01	1.28	1.53	0.76	0.68	T	1.11	0.60	2.31	1.37	14.45
1999	0.81	1.22	0.74	0.50	1.27	0.51	T	0.54	0.01	1.51	1.23	1.01	9.35
2000	1.99	2.98	2.42	0.69	1.60	0.72	0.07	T	2.01	2.06	1.22	0.57	16.33
2001	0.95	0.62	1.31	1.89	0.45	1.12	0.52	0.08	0.09	1.54	1.15	0.70	10.42
POR= 102 YRS	1.57	1.31	1.24	1.08	1.15	0.96	0.31	0.40	0.62	1.04	1.51	1.50	12.69

AVERAGE TEMPERATURE (F) 2001 PENDLETON, OR (PDT)

YEAR	JAN	FEB	MAR	APR	MAY	JUN	JUL	AUG	SEP	OCT	NOV	DEC	ANNUAL
1972	34.0	37.4	47.8	47.6	60.9	68.3	74.7	76.2	61.1	51.1	42.6	27.1	52.4
1973	31.3	38.4	45.8	50.3	61.3	66.9	75.3	71.7	64.0	52.8	42.6	41.5	53.5
1974	30.4	43.8	46.4	51.7	57.3	71.1	73.3	75.5	67.5	54.8	44.8	40.6	54.8
1975	37.1	39.0	45.2	47.5	59.3	65.8	78.4	70.1	67.0	54.3	42.3	40.5	53.9
1976	39.2	37.9	42.8	50.2	58.8	63.7	73.4	67.7	67.3	53.1	42.7	35.9	52.7
1977	26.3	41.5	44.2	55.3	55.1	69.1	70.3	74.9	58.5	50.0	38.3	34.8	51.5
1978	32.2	39.3	45.7	48.0	54.4	66.3	72.2	69.4	60.5	51.7	33.5	29.5	50.2
1979	15.3	37.7	46.0	50.4	59.5	66.6	72.8	70.6	65.5	54.3	34.7	38.2	51.0
1980	25.6	36.1	41.3	51.9	56.4	60.4	72.1	66.9	63.3	51.3	42.0	39.2	50.5
1981	36.2	38.9	45.7	50.4	56.0	61.6	69.2	74.3	63.8	50.6	44.2	37.2	52.3
1982	35.0	38.1	43.5	47.6	56.8	67.6	71.1	71.5	60.7	50.7	37.3	35.7	51.3
1983	40.8	43.8	47.8	49.0	58.9	62.7	68.4	72.7	58.9	52.5	45.9	23.2	52.1
1984	34.6	39.7	46.8	48.2	54.7	62.1	72.9	72.2	60.4	49.1	41.8	30.4	51.1
1985	26.3	33.5	43.2	53.1	58.5	65.6	77.4	68.1	57.0	50.3	26.5	19.5	48.3
1986	35.9	39.0	48.8	50.0	58.6	70.0	67.6	75.8	58.9	54.0	42.2	31.5	52.7
1987	30.4	39.1	46.4	53.9	59.7	67.2	68.9	70.6	66.2	54.1	42.6	32.7	52.7
1988	32.4	41.1	44.1	51.9	56.8	63.9	72.0	70.0	63.4	58.4	44.3	33.9	52.7
1989	38.3	25.1	42.5	52.9	55.9	65.9	70.3	68.8	63.6	51.8	44.6	33.2	51.1
1990	39.6	37.9	45.7	54.8	56.4	64.7	75.2	72.2	68.2	51.3	45.4	25.8	53.1
1991	31.2	44.7	42.6	49.6	53.9	59.6	71.5	73.3	64.9	51.6	41.2	36.9	51.8
1992	38.7	42.4	47.7	52.8	62.4	70.4	71.9	72.6	61.9	54.2	39.8	32.0	53.9
1993	25.1	28.2	41.9	50.3	61.8	62.9	65.4	68.0	64.3	54.6	34.6	35.6	49.4
1994	41.0	34.9	46.7	54.4	60.1	64.4	75.3	71.7	67.2	51.6	40.4	35.3	53.6
1995	34.9	41.7	45.3	49.0	57.7	62.1	72.0	67.4	66.0	50.2	46.1	34.5	52.2
1996	34.4	35.7	43.3	52.0	54.5	64.0	74.7	72.1	61.0	51.1	40.3	35.4	51.5
1997	32.6	38.6	45.4	48.6	59.8	63.1	70.4	73.1	64.9	51.4	42.7	34.9	52.1
1998	37.9	42.9	45.9	50.3	56.2	65.4	77.4	74.6	67.8	50.9	45.5	36.2	54.3
1999	41.0	42.3	43.4	47.3	54.1	63.7	70.0	72.5	62.5	50.8	46.0	38.1	52.6
2000	35.5	39.2	44.1	54.0	58.3	65.9	71.6	70.7	61.6	49.7	35.1	31.1	51.4
2001	32.7	35.5	44.4	47.3	58.7	62.1	70.5	73.2	66.2	51.2	42.2	36.6	51.7
POR= 96 YRS	32.8	38.0	44.9	51.3	58.4	65.5	73.1	71.4	63.2	52.5	41.4	35.1	52.3

REFERENCE NOTES:

PAGE 1:
THE TEMPERATURE GRAPH SHOWS NORMAL MAXIMUM AND NORMAL MINIMUM DAILY TEMPERATURES (SOLID CURVES) AND THE ACTUAL DAILY HIGH AND LOW TEMPERATURES (VERTICAL BARS).

PAGE 2 AND 3:
H/C INDICATES HEATING AND COOLING DEGREE DAYS.
RH INDICATES RELATIVE HUMIDITY
W/O INDICATES WEATHER AND OBSTRUCTIONS
S INDICATES SUNSHINE.
PR INDICATES PRESSURE.
CLOUDINESS ON PAGE 3 IS THE SUM OF THE CEILOMETER AND SATELLITE DATA NOT TO EXCEED EIGHT EIGHTHS(OKTAS).

GENERAL:
T INDICATES TRACE PRECIPITATION, AN AMOUNT GREATER THAN ZERO BUT LESS THAN THE LOWEST REPORTABLE VALUE.
+ INDICATES THE VALUE ALSO OCCURS ON EARLIER DATES.
BLANK ENTRIES DENOTE MISSING OR UNREPORTED DATA.
NORMALS ARE 30-YEAR AVERAGES (1961 - 1990).
ASOS INDICATES AUTOMATED SURFACE OBSERVING SYSTEM.
PM INDICATES THE LAST DAY OF THE PREVIOUS MONTH.
POR (PERIOD OF RECORD) BEGINS WITH THE JANUARY DATA MONTH AND IS THE NUMBER OF YEARS USED TO COMPUTE THE MEAN. INDIVIDUAL MONTHS WITHIN THE POR MAY BE MISSING.
WHEN THE POR FOR A NORMAL IS LESS THAN 30 YEARS, THE NORMAL IS PROVISIONAL AND IS BASED ON THE NUMBER OF YEARS INDICATED.
0.* OR * INDICATES THE VALUE OR MEAN-DAYS-WITH IS BETWEEN 0.00 AND 0.05.
CLOUDINESS FOR ASOS STATIONS DIFFERS FROM THE NON-ASOS OBSERVATION TAKEN BY A HUMAN OBSERVER. ASOS STATION CLOUDINESS IS BASED ON TIME-AVERAGED CEILOMETER DATA FOR CLOUDS AT OR BELOW 12,000 FEET AND ON SATELLITE DATA FOR CLOUDS ABOVE 12,000 FEET.
THE NUMBER OF DAYS WITH CLEAR, PARTLY CLOUDY, AND CLOUDY CONDITIONS FOR ASOS STATIONS IS THE SUM OF THE CEILOMETER AND SATELLITE DATA FOR THE SUNRISE TO SUNSET PERIOD.

GENERAL CONTINUED:
CLEAR INDICATES 0 - 2 OKTAS, PARTLY CLOUDY INDICATES 3 - 6 OKTAS, AND CLOUDY INDICATES 7 OR 8 OKTAS. WHEN AT LEAST ONE OF THE ELEMENTS (CEILOMETER OR SATELLITE) IS MISSING, THE DAILY CLOUDINESS IS NOT COMPUTED.
WIND DIRECTION IS RECORDED IN TENS OF DEGREES (2 DIGITS) CLOCKWISE FROM TRUE NORTH. "00" INDICATES CALM. "36" INDICATES TRUE NORTH.
RESULTANT WIND IS THE VECTOR AVERAGE OF THE SPEED AND DIRECTION.
AVERAGE TEMPERATURE IS THE SUM OF THE MEAN DAILY MAXIMUM AND MINIMUM TEMPERATURE DIVIDED BY 2.
SNOWFALL DATA COMPRISE ALL FORMS OF FROZEN PRECIPITATION, INCLUDING HAIL.
A HEATING (COOLING) DEGREE DAY IS THE DIFFERENCE BETWEEN THE AVERAGE DAILY TEMPERATURE AND 65 F.
DRY BULB IS THE TEMPERATURE OF THE AMBIENT AIR.
DEW POINT IS THE TEMPERATURE TO WHICH THE AIR MUST BE COOLED TO ACHIEVE 100 PERCENT RELATIVE HUMIDITY.
WET BULB IS THE TEMPERATURE THE AIR WOULD HAVE IF THE MOISTURE CONTENT WAS INCREASED TO 100 PERCENT RELATIVE HUMIDITY.

ON JULY 1, 1996, THE NATIONAL WEATHER SERVICE BEGAN USING THE "METAR" OBSERVATION CODE THAT WAS ALREADY EMPLOYED BY MOST OTHER NATIONS OF THE WORLD. THE MOST NOTICEABLE DIFFERENCE IN THIS ANNUAL PUBLICATION WILL BE THE CHANGE IN UNITS FROM TENTHS TO EIGHTS(OKTAS) FOR REPORTING THE AMOUNT OF SKY COVER.

HEATING DEGREE DAYS (base 65 F) 2001 PENDLETON, OR (PDT)

YEAR	JUL	AUG	SEP	OCT	NOV	DEC	JAN	FEB	MAR	APR	MAY	JUN	TOTAL
1972-73	5	4	165	422	663	1170	1036	738	588	434	169	73	5467
1973-74	1	16	97	372	666	721	1064	589	573	391	241	29	4760
1974-75	8	0	39	313	600	750	857	721	609	517	194	57	4665
1975-76	0	12	43	332	673	751	791	782	679	436	206	89	4794
1976-77	4	42	31	363	660	896	1192	653	639	299	301	26	5106
1977-78	20	35	200	461	792	927	1011	714	593	504	322	46	5625
1978-79	7	41	146	403	936	1094	1533	757	582	432	184	62	6177
1979-80	12	0	43	326	902	823	1210	829	728	388	267	141	5669
1980-81	4	33	88	438	681	794	886	724	593	435	275	126	5077
1981-82	20	1	128	440	617	855	919	747	662	515	256	72	5232
1982-83	22	7	171	435	825	901	741	588	528	470	242	95	5025
1983-84	42	1	180	381	569	1292	935	729	558	496	316	134	5633
1984-85	4	0	182	490	692	1065	1196	876	665	351	224	65	5810
1985-86	4	22	242	452	1149	1402	898	722	497	446	277	25	6136
1986-87	33	0	213	335	675	1031	1065	717	571	332	201	71	5244
1987-88	25	12	65	334	668	995	1004	689	637	387	264	126	5206
1988-89	22	4	120	208	616	957	821	1113	691	354	279	42	5227
1989-90	11	17	76	403	607	978	781	752	591	299	262	89	4866
1990-91	9	13	11	419	583	1211	1039	564	689	454	338	162	5492
1991-92	4	2	52	418	707	865	810	649	527	362	127	36	4559
1992-93	11	28	129	333	752	1015	1231	1025	709	432	153	98	5916
1993-94	27	35	114	318	908	903	736	838	559	321	174	83	5016
1994-95	15	0	30	406	731	915	928	644	602	473	237	126	5107
1995-96	1	26	51	452	560	941	938	840	664	387	321	64	5245
1996-97	8	4	153	423	733	909	998	733	599	488	185	84	5317
1997-98	13	0	73	415	662	927	832	611	585	442	273	37	4870
1998-99	0	3	56	429	578	887	738	628	662	522	336	108	4947
1999-00	17	20	117	433	565	830	905	740	640	324	219	57	4867
2000-01	7	12	137	468	892	1046	995	819	631	523	238	121	5889
2001-	14	5	48	419	676	874							

COOLING DEGREE DAYS (base 65 F) 2001 PENDLETON, OR (PDT)

YEAR	JAN	FEB	MAR	APR	MAY	JUN	JUL	AUG	SEP	OCT	NOV	DEC	ANNUAL
1972	0	0	0	0	50	134	314	358	55	0	0	0	911
1973	0	0	0	0	63	137	327	232	72	0	0	0	831
1974	0	0	0	0	9	219	272	332	122	4	0	0	958
1975	0	0	0	0	27	88	423	179	109	8	0	0	834
1976	0	0	0	0	20	53	270	129	103	3	0	0	578
1977	0	0	0	16	3	152	190	348	16	0	0	0	725
1978	0	0	0	0	1	93	236	182	16	0	0	0	528
1979	0	0	0	0	21	114	261	186	65	3	0	0	650
1980	0	0	0	2	5	13	232	101	44	20	0	0	417
1981	0	0	0	4	2	28	155	297	101	0	0	0	587
1982	0	0	0	0	7	158	219	215	47	0	0	0	646
1983	0	0	0	0	60	32	155	246	6	0	0	0	499
1984	0	0	0	0	7	55	256	231	51	3	0	0	603
1985	0	0	0	28	91	394	127	7	0	0	0	0	647
1986	0	0	0	2	88	184	121	341	35	1	0	0	772
1987	0	0	0	8	41	145	152	194	108	4	0	0	652
1988	0	0	0	0	16	98	246	164	78	9	0	0	611
1989	0	0	0	0	5	76	182	143	41	0	0	0	447
1990	0	0	0	0	4	92	330	245	114	3	0	0	788
1991	0	0	0	0	0	8	214	267	56	9	0	0	554
1992	0	0	0	1	52	204	229	275	45	4	0	0	810
1993	0	0	0	0	59	42	47	136	99	4	0	0	387
1994	0	0	0	8	29	72	341	214	103	0	0	0	767
1995	0	0	0	0	17	44	230	108	89	0	0	0	488
1996	0	0	0	0	0	42	312	230	39	0	0	0	623
1997	0	0	0	0	31	34	191	258	79	5	0	0	598
1998	0	0	0	6	5	56	390	309	144	0	0	0	910
1999	0	0	0	0	7	78	180	260	50	0	5	0	580
2000	0	0	0	0	17	89	215	196	40	0	0	0	557
2001	0	0	0	0	48	43	193	265	93	0	0	0	642

SNOWFALL (inches) 2001 PENDLETON, OR (PDT)

YEAR	JUL	AUG	SEP	OCT	NOV	DEC	JAN	FEB	MAR	APR	MAY	JUN	TOTAL
1972-73	0.0	0.0	0.0	T	T	12.6	2.2	5.9	T	0.1	0.0	0.0	20.8
1973-74	0.0	0.0	0.0	3.2	9.1	5.3	2.6	0.5	T	T	0.0	0.0	20.7
1974-75	0.0	0.0	0.0	0.0	T	16.6	3.3	0.3	2.2	T	0.0	0.0	22.1
1975-76	0.0	0.0	0.0	0.0	5.2	3.0	0.3	0.3	0.1	0.0	0.0	0.0	8.9
1976-77	0.0	0.0	0.0	0.0	1.0	3.1	0.5	0.4	0.0	0.0	0.0	0.0	5.0
1977-78	0.0	0.0	0.0	0.0	8.5	11.5	6.1	T	3.9	0.0	T	0.0	30.0
1978-79	0.0	0.0	0.0	0.0	9.0	7.4	14.7	2.2	T	0.0	0.0	0.0	33.3
1979-80	0.0	0.0	0.0	0.0	4.3	T	16.6	0.9	3.9	0.0	0.0	0.0	25.7
1980-81	0.0	0.0	0.0	0.0	2.0	2.7	3.6	1.2	0.0	0.0	0.0	0.0	9.5
1981-82	0.0	0.0	0.0	0.0	0.6	5.1	5.7	1.5	1.9	T	0.0	0.0	14.8
1982-83	0.0	0.0	0.0	0.0	T	1.6	0.2	0.9	0.0	0.0	0.0	0.0	2.7
1983-84	0.0	0.0	0.0	0.0	T	26.6	1.0	1.2	T	T	0.0	0.0	28.8
1984-85	0.0	0.0	0.0	0.0	T	6.2	0.8	12.7	0.6	T	0.0	0.0	20.3
1985-86	0.0	0.0	0.0	0.0	14.9	9.1	T	7.6	0.0	0.0	T	0.0	31.6
1986-87	0.0	0.0	0.0	0.0	1.2	6.8	5.8	0.0	T	0.0	0.0	0.0	13.8
1987-88	0.0	0.0	0.0	0.0	0.3	2.3	10.6	0.0	1.5	0.0	0.0	0.0	14.7
1988-89	0.0	0.0	0.0	0.0	T	T	4.3	4.9	4.0	0.0	T	0.0	13.2
1989-90	0.0	0.0	0.0	0.0	0.0	1.0	T	2.0	1.3	0.0	0.0	0.0	4.3
1990-91	0.0	0.0	0.0	0.0	T	6.4	1.6	0.0	0.6	T	0.0	0.0	8.6
1991-92	0.0	0.0	0.0	2.3	1.0	T	0.8	T	T	0.0	0.0	0.0	4.1
1992-93	0.0	0.0	0.0	0.0	0.2	7.6	25.1	14.8	1.8	T	T	0.0	49.5
1993-94	T	0.0	0.0	0.0	0.7	0.4	T	16.8	0.2	0.0	0.0	T	18.1
1994-95	0.0	0.0	0.0	0.0	0.6	3.8	2.0	7.2	T	0.0	0.0	0.0	13.6
1995-96	0.0	0.0	0.0		0.0								
1996-97						10.1	3.2			T			
1997-98						4.2		0.0					
1998-99			0.0	0.0		0.8	0.0	2.4	T	T	T	0.0	
1999-00	0.0	0.0	0.0	0.0	0.0	0.5	5.7	4.5	1.0	0.0	T	0.0	11.7
2000-01	0.0	0.0	0.0	0.0	0.7	2.5	3.1	0.5	T	T	T	T	6.8
2001-	0.0	0.0	0.0	0.0	T	3.0							
POR= 63 YRS	T	0.0	0.0	0.1	1.7	3.8	6.7	3.6	0.9	0.1	T	T	16.9

2001
PORTLAND,
OREGON (PDX)

The Portland Weather Service Office is located 6 miles north-northeast of downtown Portland. Portland is situated about 65 miles inland from the Pacific Coast and midway between the northerly oriented low coast range on the west and the higher Cascade range on the east, each about 30 miles distant. The airport lies on the south bank of the Columbia River. The coast range provides limited shielding from the Pacific Ocean. The Cascade range provides a steep slope for orographic lift of moisture-laden westerly winds and consequent moderate rainfall, and also forms a barrier from continental air masses originating over the interior Columbia Basin. Airflow is usually northwesterly in Portland in spring and summer and southeasterly in fall and winter. The Portland Airport location is drier than most surrounding localities.

Portland has a very definite winter rainfall climate. Approximately 88 percent of the annual total occurs in the months of October through May, 9 percent in June and September, while only 3 percent comes in July and August. Precipitation is mostly rain, as on the average there are only five days each year with measurable snow. Snowfalls are seldom more than a couple of inches, and generally last only a few days.

The winter season is marked by relatively mild temperatures, cloudy skies and rain with southeasterly surface winds predominating. Summer produces pleasantly mild temperatures, northwesterly winds and very little precipitation. Fall and spring are transitional in nature. Fall and early winter are times with most frequent fog.

At all times, incursions of marine air are a frequent moderating influence. Outbreaks of continental high pressure from east of the Cascade Mountains produce strong easterly flow through the Columbia Gorge into the Portland area. In winter this brings the coldest weather with the extremes of low temperature registered in the cold air mass. Freezing rain and ice glaze are sometimes transitional effects. Temperatures below zero are very infrequent. In summer, hot, dry continental air brings the highest temperatures. Temperatures above 100 degrees are infrequent, but 90 degrees or higher are reached every year, but seldom persist for more than two or three days.

Destructive storms are infrequent in the Portland area. Surface winds seldom exceed gale force and rarely in the period of record have winds reached higher than 75 mph. Thunderstorms occur about once a month through the spring and summer months. Heavy downpours are infrequent but gentle rains occur almost daily during winter months.

Most rural areas around Portland are farmed for berries, green beans, and vegetables for fresh market and processing. The long growing season with mild temperatures and ample moisture favors local nursery and seed industries.

Based on the 1951-1980 period, the average first occurrence of 32 degrees Fahrenheit in the fall is November 7 and the average last occurrence in the spring is April 3.

NORMALS, MEANS, AND EXTREMES
PORTLAND, OR (PDX)

LATITUDE:	LONGITUDE:	ELEVATION (FT):		TIME ZONE:	WBAN: 24229
45 35' 27" N	122 36' 01" W	GRND: 220	BARO: 223	PACIFIC (UTC + 8)	

	ELEMENT	POR	JAN	FEB	MAR	APR	MAY	JUN	JUL	AUG	SEP	OCT	NOV	DEC	YEAR
TEMPERATURE °F	NORMAL DAILY MAXIMUM	30	45.4	51.0	56.0	60.6	67.1	74.0	79.9	80.3	74.6	64.0	52.6	45.6	62.6
	MEAN DAILY MAXIMUM	71	44.7	50.1	55.7	61.3	67.6	73.1	79.6	79.5	74.5	64.0	52.6	46.3	62.4
	HIGHEST DAILY MAXIMUM	61	63	71	80	90	100	100	107	107	105	92	73	65	107
	YEAR OF OCCURRENCE		1986	1988	1947	1998	1983	1992	1965	1981	1988	1987	1975	1993	AUG 1981
	MEAN OF EXTREME MAXS.	71	56.8	60.6	68.7	77.7	85.7	90.5	95.5	94.6	90.2	78.6	63.6	57.7	76.7
	NORMAL DAILY MINIMUM	30	33.7	36.1	38.6	41.3	47.0	52.9	56.5	56.9	52.0	44.9	39.5	34.8	44.5
	MEAN DAILY MINIMUM	71	34.1	36.4	39.0	42.4	47.8	53.1	56.7	56.9	52.4	45.9	39.9	35.9	45.0
	LOWEST DAILY MINIMUM	61	-2	-3	19	29	29	39	43	44	34	26	13	6	-3
	YEAR OF OCCURRENCE		1950	1950	1989	1955	1954	1966	1955	1980	1965	1971	1985	1964	FEB 1950
	MEAN OF EXTREME MINS.	71	21.4	24.5	29.4	33.8	38.4	45.7	49.5	49.2	42.9	34.7	28.5	24.1	35.2
	NORMAL DRY BULB	30	39.6	43.6	47.3	51.0	57.1	63.5	68.2	68.6	63.3	54.5	46.1	40.2	53.6
	MEAN DRY BULB	71	39.4	43.2	47.4	51.9	57.8	63.2	68.2	68.2	63.4	54.9	46.2	41.1	53.7
	MEAN WET BULB	17	38.8	40.1	44.0	47.8	52.2	56.2	56.3	57.1	54.0	50.6	44.2	35.9	48.1
	MEAN DEW POINT	17	35.3	35.7	39.5	43.1	47.4	50.9	50.8	51.7	48.9	46.6	41.2	32.6	43.6
	NORMAL NO. DAYS WITH:														
	MAXIMUM 90	30	0.0	0.0	0.0	0.0	0.4	1.4	4.0	4.5	1.6	0.1	0.0	0.0	12.0
	MAXIMUM 32	30	1.7	0.1	0.0	0.0	0.0	0.0	0.0	0.0	0.0	0.0	0.2	1.4	3.4
	MINIMUM 32	30	12.4	7.7	4.2	1.1	0.1	0.0	0.0	0.0	0.0	0.6	4.6	10.4	41.1
	MINIMUM 0	30	0.0	0.0	0.0	0.0	0.0	0.0	0.0	0.0	0.0	0.0	0.0	0.0	0.0
H/C	NORMAL HEATING DEG. DAYS	30	787	599	549	420	249	91	28	35	102	326	567	769	4522
	NORMAL COOLING DEG. DAYS	30	0	0	0	0	0	46	127	147	51	0	0	0	371
RH	NORMAL (PERCENT)	30	81	78	75	72	69	66	63	65	69	78	82	83	73
	HOUR 04 LST	30	85	85	85	85	84	82	80	82	86	90	87	86	85
	HOUR 10 LST	30	82	80	75	70	66	64	61	64	67	78	82	83	73
	HOUR 16 LST	30	74	66	59	55	52	49	44	44	48	60	73	78	58
	HOUR 22 LST	30	82	80	77	74	72	69	66	68	74	82	83	84	76
S	PERCENT POSSIBLE SUNSHINE	46	28	38	48	52	57	56	69	66	62	44	28	23	48
W/O	MEAN NO. DAYS WITH:														
	HEAVY FOG(VISBY 1/4 MI)	59	4.3	3.7	2.3	1.1	0.1	0.1	0.1	0.2	2.6	6.7	5.7	4.7	31.6
	THUNDERSTORMS	61	0.1	0.1	0.5	0.9	1.4	0.9	0.8	1.0	0.7	0.4	0.3	0.1	7.2
CLOUDINESS	MEAN:														
	SUNRISE-SUNSET (OKTAS)	47	6.7	6.6	6.4	6.2	5.8	5.4	3.8	4.0	4.3	5.6	6.4	6.8	5.7
	MIDNIGHT-MIDNIGHT (OKTAS)	31	6.3	6.0	5.7	5.7	5.3	4.9	3.6	3.6	3.9	5.0	6.3	6.4	5.2
	MEAN NO. DAYS WITH:														
	CLEAR	47	2.9	2.9	3.3	3.5	5.0	6.2	12.6	11.4	10.3	5.4	2.8	2.1	68.4
	PARTLY CLOUDY	47	3.6	3.8	4.9	5.8	7.2	7.7	8.5	9.6	8.1	7.6	4.3	3.3	74.4
	CLOUDY	47	24.4	21.5	22.8	20.7	18.9	16.1	9.9	10.1	11.6	18.0	22.9	25.5	222.4
PR	MEAN STATION PRESSURE(IN)	29	30.09	29.99	30.00	30.00	30.00	30.00	30.00	30.00	30.00	30.02	30.00	30.10	30.02
	MEAN SEA-LEVEL PRES. (IN)	17	30.09	30.04	30.06	30.06	30.03	30.04	30.05	30.02	30.02	30.08	30.07	30.14	30.06
WINDS	MEAN SPEED (MPH)	47	9.9	9.4	8.3	7.5	7.2	7.3	7.6	7.1	6.5	6.6	8.6	9.6	8.0
	PREVAIL.DIR(TENS OF DEGS)	31	12	12	12	32	32	32	32	33	32	12	12	12	32
	MAXIMUM 2-MINUTE:														
	SPEED (MPH)	6	43	40	43	37	33	34	24	26	31	37	43	38	43
	DIR. (TENS OF DEGS)		20	10	21	18	24	20	18	32	07	11	20	19	20
	YEAR OF OCCURRENCE		2000	2000	1999	1996	2000	1997	1996	1999	2000	1999	1998	2001	JAN 2000
	MAXIMUM 5-SECOND:														
	SPEED (MPH)	6	59	51	51	46	40	40	29	33	40	43	51	51	59
	DIR. (TENS OF DEGS)		19	17	20	09	24	20	18	17	08	12	21	22	19
	YEAR OF OCCURRENCE		2000	1996	1999	1999	2000	1997	1996	1997	2000	1999	1998	1996	JAN 2000
PRECIPITATION	NORMAL (IN)	30	5.35	3.85	3.56	2.39	2.06	1.48	0.63	1.09	1.75	2.67	5.34	6.13	36.30
	MAXIMUM MONTHLY (IN)	61	12.83	10.03	7.52	5.26	5.55	4.06	2.68	4.53	4.30	8.41	11.57	13.35	13.35
	YEAR OF OCCURRENCE		1953	1996	1957	1993	1998	1984	1983	1968	1986	1994	1942	1996	DEC 1996
	MINIMUM MONTHLY (IN)	61	0.06	0.72	1.10	0.53	0.10	0.03	0.00	T	T	0.19	0.77	1.38	0.00
	YEAR OF OCCURRENCE		1985	1993	1965	1956	1992	1951	1967	1970	1993	1988	1976	1976	JUL 1967
	MAXIMUM IN 24 HOURS (IN)	61	2.61	2.46	1.83	1.47	1.47	1.82	1.09	1.54	2.38	4.44	4.10	2.59	4.44
	YEAR OF OCCURRENCE		1974	1994	1943	1962	1968	1958	1978	1977	1982	1994	1996	1977	OCT 1994
	NORMAL NO. DAYS WITH:														
	PRECIPITATION 0.01	30	17.5	15.5	17.1	14.3	11.8	8.5	4.2	5.3	8.0	12.0	18.5	18.7	151.4
	PRECIPITATION 1.00	30	1.1	0.4	0.2	0.2	0.1	0.1	0.0	0.1	0.2	0.2	0.8	1.1	4.5
SNOWFALL	NORMAL (IN)	30	1.8	0.9	0.1	T	T	0.0	0.0	0.0	0.0	0.0	0.5	2.0	5.3
	MAXIMUM MONTHLY (IN)	55	41.4	13.2	12.9	T	0.6	T	0.0	T	T	0.2	8.2	15.7	41.4
	YEAR OF OCCURRENCE		1950	1949	1951	1995	1953	1995		1989	1949	1950	1955	1968	JAN 1950
	MAXIMUM IN 24 HOURS (IN)	55	10.6	6.4	7.7	T	0.5	T	0.0	T	T	0.2	7.4	8.0	10.6
	YEAR OF OCCURRENCE		1950	1993	1951	1995	1953	1995		1989	1949	1950	1977	1964	JAN 1950
	MAXIMUM SNOW DEPTH (IN)	48	15	14	5	0	0	0	0	0	0	0	5	11	15
	YEAR OF OCCURRENCE		1950	1950	1951								1977	1964	JAN 1950
	NORMAL NO. DAYS WITH:														
	SNOWFALL 1.0	30	0.6	0.3	0.*	0.0	0.0	0.0	0.0	0.0	0.0	0.0	0.1	0.7	1.7

PRECIPITATION (inches) 2001 PORTLAND, OR (PDX)

YEAR	JAN	FEB	MAR	APR	MAY	JUN	JUL	AUG	SEP	OCT	NOV	DEC	ANNUAL
1972	5.71	4.08	5.41	2.98	2.23	0.68	0.56	0.67	3.06	0.87	3.78	8.79	38.82
1973	3.69	1.94	2.45	1.33	1.43	1.45	0.06	1.41	3.29	3.14	11.55	9.93	41.67
1974	8.51	4.61	5.65	1.76	1.74	0.80	2.01	0.07	0.21	2.14	6.73	6.05	40.28
1975	8.43	4.75	3.45	1.88	1.35	1.13	0.43	2.10	T	4.76	4.10	6.68	39.06
1976	5.14	4.92	2.93	2.34	2.29	0.78	0.66	3.29	0.73	1.48	0.77	1.38	26.71
1977	1.07	2.49	3.50	1.04	4.30	0.83	0.39	3.26	3.33	2.28	5.56	8.98	37.03
1978	4.85	3.28	1.49	3.96	3.17	1.69	1.36	2.05	2.07	0.36	3.83	2.51	30.62
1979	2.55	6.53	2.51	2.47	2.41	0.64	0.25	1.18	1.75	4.85	3.38	7.23	35.75
1980	8.51	4.01	3.11	2.58	2.19	2.50	0.19	0.39	1.56	1.18	6.47	9.72	42.41
1981	1.47	3.86	2.33	1.79	2.25	3.23	0.24	0.15	1.86	4.12	4.62	8.37	34.29
1982	6.31	5.98	2.38	3.56	0.46	1.66	0.94	1.66	3.98	4.44	3.51	8.16	43.04
1983	6.23	7.78	6.80	1.87	1.30	1.95	2.68	2.29	0.39	1.95	5.30	8.65	47.19
1984	2.01	3.93	3.19	3.20	3.41	4.06	T	0.09	1.46	3.85	9.74	2.56	37.50
1985	0.06	1.79	3.08	1.07	1.52	2.34	0.55	0.48	2.76	2.75	3.89	2.19	22.48
1986	4.65	5.31	2.60	1.91	2.19	0.23	1.20	0.10	4.30	1.99	6.26	4.30	35.04
1987	6.93	2.45	4.91	1.94	1.63	0.14	1.03	0.35	0.30	0.27	1.96	8.00	29.91
1988	4.95	1.17	3.13	4.57	2.53	2.34	0.69	0.10	1.76	0.19	7.92	2.37	31.72
1989	3.30	2.84	6.73	2.08	2.87	0.78	0.91	1.07	1.48	1.73	3.18	3.08	30.05
1990	7.95	3.43	2.52	2.31	2.37	1.94	0.32	0.95	0.34	4.65	3.68	2.40	32.86
1991	2.56	3.65	4.64	4.05	3.34	2.31	0.07	0.70	0.02	1.51	6.36	4.34	33.55
1992	4.31	4.12	1.87	3.82	0.10	0.60	0.67	0.49	1.12	2.87	4.55	4.98	29.50
1993	3.06	0.72	4.39	5.26	4.36	1.69	2.41	0.37	T	1.59	1.50	5.01	30.36
1994	3.56	4.92	1.84	1.91	0.56	1.67	0.07	0.13	1.13	8.41	5.91	4.85	34.96
1995	5.56	3.19	3.82	3.49	1.65	2.62	1.23	0.81	1.31	3.15	11.15	5.35	43.33
1996	7.15	10.03	3.24	5.12	4.88	0.44	0.73	0.25	3.05	5.39	9.58	13.35	63.21
1997	7.32	1.63	7.14	3.73	3.63	2.83	0.52	1.58	1.98	6.40	4.02	3.03	43.81
1998	6.77	5.27	4.06	1.04	5.55	1.73	0.59	T	1.09	2.16	11.02	6.74	46.02
1999	6.63	8.73	4.03	1.56	1.97	1.73	0.51	0.75	0.10	2.44	6.81	3.62	38.88
2000	5.66	4.50	3.21	1.82	2.70	1.19	0.15	0.12	1.67	3.25	2.46	3.47	30.20
2001	1.47	1.29	3.11	2.85	0.91	1.79	0.95	0.74	0.70	3.12	6.89	6.62	30.44
POR= 61 YRS	5.42	4.08	3.69	2.44	2.23	1.60	0.61	0.87	1.60	3.18	5.63	5.84	37.19

AVERAGE TEMPERATURE (F) 2001 PORTLAND, OR (PDX)

YEAR	JAN	FEB	MAR	APR	MAY	JUN	JUL	AUG	SEP	OCT	NOV	DEC	ANNUAL
1972	39.2	43.8	49.8	48.0	60.2	64.0	70.9	71.7	61.2	53.0	48.2	37.4	54.0
1973	39.0	44.9	47.9	52.3	59.4	63.9	70.3	65.9	64.3	54.3	44.2	44.7	54.3
1974	38.0	43.0	47.2	51.3	55.7	64.4	67.1	68.9	67.3	55.2	48.1	44.1	54.2
1975	41.5	41.2	45.0	47.3	57.5	61.9	69.0	65.3	65.7	53.5	46.0	42.7	53.1
1976	42.2	42.1	44.4	50.3	56.6	60.4	67.2	65.5	64.2	54.7	47.0	39.5	52.8
1977	35.7	44.6	45.5	52.9	53.8	63.9	66.3	71.7	60.9	53.8	43.3	42.0	52.9
1978	40.1	44.7	49.1	50.5	54.7	65.1	68.4	67.6	60.9	54.7	39.1	35.3	52.5
1979	30.7	42.9	50.8	53.1	60.1	65.1	70.5	68.6	66.3	58.1	45.0	44.4	54.6
1980	35.1	42.5	46.3	53.8	57.3	60.7	68.9	66.4	63.8	56.0	48.5	44.0	53.6
1981	43.9	44.0	48.8	52.5	57.5	61.8	67.5	72.2	64.9	53.3	48.8	42.7	54.8
1982	39.7	43.6	48.5	49.0	57.6	66.0	67.5	68.6	63.2	54.9	44.4	41.7	53.7
1983	44.4	47.3	50.7	52.7	60.4	62.8	66.5	69.1	61.5	54.2	49.3	36.4	54.6
1984	42.2	45.9	51.1	50.4	56.4	62.2	69.1	69.4	63.7	52.9	46.7	38.3	54.0
1985	36.1	41.1	45.8	53.9	58.3	64.4	74.1	69.3	60.8	52.7	37.3	33.0	52.2
1986	42.5	43.7	51.3	50.2	57.6	66.3	65.3	72.3	61.5	57.0	47.7	40.6	54.7
1987	39.6	45.2	48.7	54.2	60.4	66.5	67.2	70.5	65.5	58.2	48.8	39.1	55.3
1988	39.0	44.7	47.2	52.2	56.4	62.4	68.4	68.0	64.0	58.3	47.5	42.0	54.2
1989	42.2	36.0	45.6	56.0	58.0	64.3	65.5	66.1	65.3	54.9	48.6	40.3	53.6
1990	43.4	41.9	49.4	54.5	56.7	63.6	71.2	70.0	67.4	53.9	49.4	34.7	54.6
1991	38.9	48.8	46.2	50.8	54.7	59.9	69.9	70.1	67.4	55.4	47.5	42.7	54.4
1992	44.5	48.1	52.3	55.4	63.1	67.4	70.2	69.8	62.3	55.9	46.3	39.3	56.2
1993	36.5	40.1	47.9	52.6	61.1	62.5	64.3	68.5	65.7	57.4	41.5	41.4	53.3
1994	44.5	40.9	50.2	54.2	60.7	63.8	71.1	70.2	67.6	54.1	42.7	42.3	55.1
1995	43.4	46.5	48.3	51.3	60.7	63.8	70.6	67.3	67.1	54.5	51.7	42.4	55.6
1996	40.7	41.6	47.7	53.5	55.3	63.2	72.0	70.4	61.8	54.0	45.5	41.6	53.9
1997	41.0	42.9	46.7	51.0	62.2	62.7	68.9	71.6	65.9	53.7	49.8	41.1	54.8
1998	43.0	46.1	48.8	52.8	56.3	63.4	71.0	71.0	66.9	54.3	48.9	40.5	55.3
1999	42.8	43.7	45.8	50.7	54.9	61.2	67.2	69.0	66.4	55.1	50.1	43.9	54.3
2000	39.8	43.7	45.6	54.0	57.6	65.3	67.8	67.8	63.9	55.2	42.8	40.4	53.7
2001	41.2	42.0	48.0	49.6	59.4	61.0	66.5	69.2	65.1	53.6	48.0	42.0	53.8
POR= 61 YRS	39.3	43.3	46.8	51.3	57.4	61.5	66.8	66.9	63.4	54.4	46.0	40.8	53.2

REFERENCE NOTES:

PAGE 1:
THE TEMPERATURE GRAPH SHOWS NORMAL MAXIMUM AND NORMAL MINIMUM DAILY TEMPERATURES (SOLID CURVES) AND THE ACTUAL DAILY HIGH AND LOW TEMPERATURES (VERTICAL BARS).

PAGE 2 AND 3:
H/C INDICATES HEATING AND COOLING DEGREE DAYS.
RH INDICATES RELATIVE HUMIDITY
W/O INDICATES WEATHER AND OBSTRUCTIONS
S INDICATES SUNSHINE.
PR INDICATES PRESSURE.
CLOUDINESS ON PAGE 3 IS THE SUM OF THE CEILOMETER AND SATELLITE DATA NOT TO EXCEED EIGHT EIGHTHS(OKTAS).

GENERAL:
T INDICATES TRACE PRECIPITATION, AN AMOUNT GREATER THAN ZERO BUT LESS THAN THE LOWEST REPORTABLE VALUE.
+ INDICATES THE VALUE ALSO OCCURS ON EARLIER DATES.
BLANK ENTRIES DENOTE MISSING OR UNREPORTED DATA.
NORMALS ARE 30-YEAR AVERAGES (1961 - 1990).
ASOS INDICATES AUTOMATED SURFACE OBSERVING SYSTEM.
PM INDICATES THE LAST DAY OF THE PREVIOUS MONTH.
POR (PERIOD OF RECORD) BEGINS WITH THE JANUARY DATA MONTH AND IS THE NUMBER OF YEARS USED TO COMPUTE THE MEAN. INDIVIDUAL MONTHS WITHIN THE POR MAY BE MISSING.
WHEN THE POR FOR A NORMAL IS LESS THAN 30 YEARS, THE NORMAL IS PROVISIONAL AND IS BASED ON THE NUMBER OF YEARS INDICATED.
0.* OR * INDICATES THE VALUE OR MEAN-DAYS-WITH IS BETWEEN 0.00 AND 0.05.
CLOUDINESS FOR ASOS STATIONS DIFFERS FROM THE NON-ASOS OBSERVATION TAKEN BY A HUMAN OBSERVER. ASOS STATION CLOUDINESS IS BASED ON TIME-AVERAGED CEILOMETER DATA FOR CLOUDS AT OR BELOW 12,000 FEET AND ON SATELLITE DATA FOR CLOUDS ABOVE 12,000 FEET.
THE NUMBER OF DAYS WITH CLEAR, PARTLY CLOUDY, AND CLOUDY CONDITIONS FOR ASOS STATIONS IS THE SUM OF THE CEILOMETER AND SATELLITE DATA FOR THE SUNRISE TO SUNSET PERIOD.

GENERAL CONTINUED:
CLEAR INDICATES 0 - 2 OKTAS, PARTLY CLOUDY INDICATES 3 - 6 OKTAS, AND CLOUDY INDICATES 7 OR 8 OKTAS. WHEN AT LEAST ONE OF THE ELEMENTS (CEILOMETER OR SATELLITE) IS MISSING, THE DAILY CLOUDINESS IS NOT COMPUTED.
WIND DIRECTION IS RECORDED IN TENS OF DEGREES (2 DIGITS) CLOCKWISE FROM TRUE NORTH. "00" INDICATES CALM. "36" INDICATES TRUE NORTH.
RESULTANT WIND IS THE VECTOR AVERAGE OF THE SPEED AND DIRECTION.
AVERAGE TEMPERATURE IS THE SUM OF THE MEAN DAILY MAXIMUM AND MINIMUM TEMPERATURE DIVIDED BY 2.
SNOWFALL DATA COMPRISE ALL FORMS OF FROZEN PRECIPITATION, INCLUDING HAIL.
A HEATING (COOLING) DEGREE DAY IS THE DIFFERENCE BETWEEN THE AVERAGE DAILY TEMPERATURE AND 65 F.
DRY BULB IS THE TEMPERATURE OF THE AMBIENT AIR.
DEW POINT IS THE TEMPERATURE TO WHICH THE AIR MUST BE COOLED TO ACHIEVE 100 PERCENT RELATIVE HUMIDITY.
WET BULB IS THE TEMPERATURE THE AIR WOULD HAVE IF THE MOISTURE CONTENT WAS INCREASED TO 100 PERCENT RELATIVE HUMIDITY.

ON JULY 1, 1996, THE NATIONAL WEATHER SERVICE BEGAN USING THE "METAR" OBSERVATION CODE THAT WAS ALREADY EMPLOYED BY MOST OTHER NATIONS OF THE WORLD. THE MOST NOTICEABLE DIFFERENCE IN THIS ANNUAL PUBLICATION WILL BE THE CHANGE IN UNITS FROM TENTHS TO EIGHTS(OKTAS) FOR REPORTING THE AMOUNT OF SKY COVER.

HEATING DEGREE DAYS (base 65 F) 2001 PORTLAND, OR (PDX)

YEAR	JUL	AUG	SEP	OCT	NOV	DEC	JAN	FEB	MAR	APR	MAY	JUN	TOTAL
1972-73	10	6	153	363	497	848	799	560	525	378	202	89	4430
1973-74	6	47	59	326	618	624	832	610	545	403	282	72	4424
1974-75	32	16	29	301	500	640	722	660	615	523	240	127	4405
1975-76	24	41	48	354	565	686	698	658	632	437	258	155	4556
1976-77	15	41	47	319	536	783	901	564	596	358	340	68	4568
1977-78	40	19	131	339	644	707	764	561	485	430	317	58	4495
1978-79	29	26	134	312	772	915	1058	615	434	351	162	57	4865
1979-80	8	2	19	214	592	631	920	647	575	329	232	125	4294
1980-81	15	25	64	284	485	644	650	583	494	372	229	108	3953
1981-82	23	5	76	355	478	687	780	596	502	472	229	71	4274
1982-83	22	10	99	307	614	715	635	492	435	363	184	81	3957
1983-84	27	2	109	325	463	880	701	546	425	430	269	115	4292
1984-85	9	2	80	377	539	820	893	664	588	327	213	62	4574
1985-86	0	7	124	373	826	982	691	591	417	437	265	43	4756
1986-87	37	0	148	242	510	750	780	550	495	321	173	51	4057
1987-88	22	2	54	214	479	798	801	581	544	380	272	109	4256
1988-89	33	15	91	208	518	705	699	805	594	263	219	77	4227
1989-90	32	27	44	306	486	759	664	641	476	308	251	78	4072
1990-91	10	5	14	336	492	933	802	446	575	420	310	156	4499
1991-92	4	13	25	295	517	685	627	483	387	282	108	35	3461
1992-93	5	9	107	274	556	789	877	692	522	366	135	94	4426
1993-94	37	17	57	231	698	726	630	668	451	316	143	85	4059
1994-95	16	0	23	330	662	696	666	514	510	405	149	95	4066
1995-96	3	15	20	318	392	693	747	671	530	340	294	70	4093
1996-97	16	8	113	340	581	717	737	612	563	413	116	79	4295
1997-98	7	0	42	341	452	734	676	523	498	376	268	72	3989
1998-99	12	4	32	323	475	753	677	592	587	424	312	150	4341
1999-00	38	16	73	298	441	648	774	612	593	323	229	62	4107
2000-01	19	13	75	296	661	759	733	640	517	455	201	140	4509
2001-	22	5	55	347	504	706							

COOLING DEGREE DAYS (base 65 F) 2001 PORTLAND, OR (PDX)

YEAR	JAN	FEB	MAR	APR	MAY	JUN	JUL	AUG	SEP	OCT	NOV	DEC	ANNUAL
1972	0	0	0	0	27	39	200	221	44	0	0	0	531
1973	0	0	0	0	34	65	178	81	45	0	0	0	403
1974	0	0	0	0	1	60	102	144	102	0	0	0	409
1975	0	0	0	0	12	39	157	57	75	2	0	0	342
1976	0	0	0	0	4	23	89	66	30	4	0	0	216
1977	0	0	0	0	0	42	90	233	10	0	0	0	375
1978	0	0	0	0	3	69	141	112	18	0	0	0	343
1979	0	0	0	0	18	65	183	124	65	7	0	0	462
1980	0	0	0	1	0	2	141	75	35	12	0	0	266
1981	0	0	0	3	4	16	109	232	82	0	0	0	446
1982	0	0	0	0	4	107	103	127	50	0	0	0	391
1983	0	0	0	0	48	23	80	137	12	0	0	0	300
1984	0	0	0	0	10	34	140	144	47	6	0	0	381
1985	0	0	0	0	11	53	291	145	5	0	0	0	505
1986	0	0	0	0	40	87	52	235	50	0	0	0	464
1987	0	0	0	4	37	102	95	177	77	12	0	0	504
1988	0	0	0	0	10	39	147	115	67	8	0	0	386
1989	0	0	0	0	9	62	53	66	60	0	0	0	250
1990	0	0	0	2	3	45	206	193	83	0	0	0	532
1991	0	0	0	0	0	7	164	176	102	2	0	0	451
1992	0	0	0	1	57	114	174	164	31	0	0	0	541
1993	0	0	0	0	21	24	22	132	80	3	0	0	282
1994	0	0	0	0	15	31	210	168	109	0	0	0	533
1995	0	0	0	0	23	63	184	92	91	0	0	0	453
1996	0	0	0	0	1	36	240	181	25	7	0	0	490
1997	0	0	0	0	37	16	134	210	75	0	0	0	472
1998	0	0	0	14	4	30	206	197	96	0	0	0	547
1999	0	0	0	0	9	41	111	149	68	0	0	0	378
2000	0	0	0	0	3	78	113	107	50	0	0	0	351
2001	0	0	0	1	35	28	75	144	64	0	0	0	347

SNOWFALL (inches) 2001 PORTLAND, OR (PDX)

YEAR	JUL	AUG	SEP	OCT	NOV	DEC	JAN	FEB	MAR	APR	MAY	JUN	TOTAL
1972-73	0.0	0.0	0.0	0.0	0.0	6.1	0.4	T	T	T	0.0	0.0	6.5
1973-74	0.0	0.0	0.0	0.0	T	0.0	T	T	T	0.0	T	0.0	T
1974-75	0.0	0.0	0.0	0.0	0.0	T	T	0.1	T	T	T	0.0	0.1
1975-76	0.0	0.0	0.0	0.0	T	T	T	T	T	T	T	0.0	T
1976-77	0.0	0.0	0.0	0.0	0.0	0.0	T	0.0	T	0.0	T	0.0	T
1977-78	0.0	0.0	0.0	0.0	7.6	T	0.0	0.0	0.1	T	T	0.0	7.7
1978-79	0.0	0.0	0.0	0.0	3.0	2.4	1.9	1.1	T	T	0.0	0.0	8.4
1979-80	0.0	0.0	0.0	0.0	0.0	0.0	T	12.4	T	T	T	0.0	12.4
1980-81	0.0	0.0	0.0	0.0	T	T	0.0	T	T	T	T	0.0	T
1981-82	0.0	0.0	0.0	0.0	0.0	2.0	2.1	T	T	T	T	0.0	4.1
1982-83	0.0	0.0	0.0	T	T	0.0	0.0	0.0	T	T	T	0.0	T
1983-84	0.0	0.0	0.0	0.0	0.0	2.3	0.1	T	T	0.0	T	0.0	2.4
1984-85	0.0	0.0	0.0	T	0.0	2.8	T	4.8	T	T	0.0	0.0	7.6
1985-86	0.0	0.0	0.0	0.0	3.4	1.6	T	5.8	T	T	T	0.0	10.8
1986-87	0.0	0.0	0.0	0.0	T	0.1	0.0	T	T	T	T	0.0	0.1
1987-88	0.0	0.0	0.0	0.0	0.0	2.9	0.6	0.0	0.0	T	T	0.0	3.5
1988-89	0.0	0.0	0.0	0.0	T	T	0.9	0.3	2.0	T	0.0	0.0	3.2
1989-90	0.0	0.0	T	0.0	T	0.0	0.0	0.0	8.3	T	0.0	0.0	8.3
1990-91	0.0	0.0	0.0	0.0	0.0	0.0	1.3	0.6	0.0	T	0.0	0.0	1.9
1991-92	0.0	0.0	0.0	0.0	T	0.0	0.0	0.0	0.0	T	0.0	0.0	T
1992-93	0.0	0.0	0.0	0.0	T	4.6	2.9	6.6	0.0	T	T	0.0	14.1
1993-94	0.0	0.0	0.0	0.0	T	T	0.0	2.6	T	T	0.0	T	2.6
1994-95	0.0	0.0	0.0	0.0	0.3	1.1	T	3.6	0.4	T	T	0.0	5.4
1995-96	0.0	0.0	0.0	0.0	0.0								
1996-97													
1997-98													
1998-99													
1999-00													
2000-01													
2001-													
POR= 55 YRS	0.0	T	T	0.0	0.4	1.4	3.1	1.1	0.4	T	0.0	T	6.4

2001
AVOCA, WILKES-BARRE - SCRANTON
PENNSYLVANIA (AVP)

The Wilkes-Barre Scranton National Weather Service Office is located about midway between the two cities, at the southwest end of the crescent-shaped Lackawanna River Valley. The river flows through this valley and empties into the Susquehanna River and the Wyoming Valley a few miles west of the airport. The surrounding mountains protect both cities and the airport from high winds. They influence the temperature and precipitation during both summer and winter, causing wide departures in both within a few miles of the station. Because of the proximity of the mountains, the climate is relatively cool in summer with frequent shower and thunderstorm activity, usually of brief duration. The winter temperatures in the valley are not severe. The occurrence of sub-zero temperatures and severe snowstorms is infrequent. A high percentage of the winter precipitation occurs as rain.

Although severe snowstorms are infrequent, when they do occur they approach blizzard conditions. High winds cause huge drifts and normal routines are disrupted for several days.

While the incidence of tornadoes is very low, Wilkes-Barre has occasionally been hit with these storms which caused loss of life and great property damage.

The area has felt the effects of tropical storms. Considerable wind damage has occasionally occurred, but the most devastating damage has come from flooding caused by the large amounts of precipitation deposited by the storms. The worst natural disaster to hit the region was the result of the flooding caused by a hurricane.

NORMALS, MEANS, AND EXTREMES
AVOCA, PA (AVP)

LATITUDE: 41 20' 20" N   LONGITUDE: 75 43' 36" W   ELEVATION (FT): GRND: 955 BARO: 958   TIME ZONE: EASTERN (UTC + 5)   WBAN: 14777

	ELEMENT	POR	JAN	FEB	MAR	APR	MAY	JUN	JUL	AUG	SEP	OCT	NOV	DEC	YEAR
TEMPERATURE °F	NORMAL DAILY MAXIMUM	30	31.8	34.5	45.5	57.8	69.3	77.5	81.8	79.7	72.4	61.0	48.8	36.6	58.1
	MEAN DAILY MAXIMUM	52	33.3	35.8	44.9	58.2	69.6	78.0	82.2	80.2	72.2	61.6	48.6	37.1	58.5
	HIGHEST DAILY MAXIMUM	46	67	71	85	92	93	97	101	98	95	84	80	69	101
	YEAR OF OCCURRENCE		1967	1985	1998	1976	1962	1964	1988	2001	1983	1959	1982	1998	JUL 1988
	MEAN OF EXTREME MAXS.	52	54.3	55.2	67.9	80.2	86.1	90.3	92.2	89.9	86.4	78.2	68.1	57.2	75.5
	NORMAL DAILY MINIMUM	30	17.5	19.0	28.3	38.1	48.3	56.8	61.6	60.0	52.8	42.1	33.9	23.4	40.1
	MEAN DAILY MINIMUM	52	18.7	20.1	27.9	38.2	48.0	56.8	61.4	59.9	52.3	41.9	33.4	23.4	40.2
	LOWEST DAILY MINIMUM	46	-21	-16	-4	14	27	34	43	38	29	19	9	-9	-21
	YEAR OF OCCURRENCE		1994	1979	1967	1982	1974	1972	1979	1982	2000	1972	2000	1989	JAN 1994
	MEAN OF EXTREME MINS.	52	-.4	1.1	9.9	23.6	34.2	43.1	49.2	46.7	36.7	27.7	18.1	4.7	24.6
	NORMAL DRY BULB	30	24.7	26.8	36.9	48.0	58.8	67.2	71.7	69.9	62.6	51.5	41.4	30.0	49.1
	MEAN DRY BULB	52	26.0	28.0	36.2	48.1	58.7	67.4	71.8	70.1	62.3	51.7	41.0	30.3	49.3
	MEAN WET BULB	15	24.3	26.2	32.8	42.5	52.2	61.0	65.2	63.6	57.0	46.9	37.3	26.7	44.6
	MEAN DEW POINT	15	18.2	19.1	24.9	34.5	42.5	56.3	61.1	59.7	53.4	41.7	31.5	21.0	38.7
	NORMAL NO. DAYS WITH:														
	MAXIMUM 90	30	0.0	0.0	0.0	0.1	0.2	1.4	3.3	1.7	0.5	0.0	0.0	0.0	7.2
	MAXIMUM 32	30	15.7	11.7	3.4	0.2	0.0	0.0	0.0	0.0	0.0	0.0	1.4	10.3	42.7
	MINIMUM 32	30	28.0	24.3	20.8	8.2	0.6	0.0	0.0	0.0	0.0	*	4.3	13.6	25.1	124.9
	MINIMUM 0	30	2.3	1.1	0.1	0.0	0.0	0.0	0.0	0.0	0.0	0.0	0.0	0.6	4.1
H/C	NORMAL HEATING DEG. DAYS	30	1249	1070	871	510	217	38	0	11	109	423	708	1085	6291
	NORMAL COOLING DEG. DAYS	30	0	0	0	0	25	104	210	163	37	0	0	0	539
RH	NORMAL (PERCENT)	30	70	68	63	60	65	70	71	74	75	72	72	72	69
	HOUR 01 LST	30	72	70	67	66	72	80	81	83	83	78	75	75	75
	HOUR 07 LST	30	75	74	73	72	77	83	84	87	88	84	79	77	79
	HOUR 13 LST	30	66	63	57	52	53	57	57	59	62	60	65	68	60
	HOUR 19 LST	30	67	63	58	53	56	61	62	66	70	66	68	70	63
S	PERCENT POSSIBLE SUNSHINE	41	41	47	50	53	57	61	62	61	55	52	36	34	51
W/O	MEAN NO. DAYS WITH:														
	HEAVY FOG(VISBY 1/4 MI)	46	1.9	2.1	1.9	1.4	1.1	1.1	1.6	1.8	2.7	2.0	1.7	2.2	21.5
	THUNDERSTORMS	46	0.1	0.2	0.8	2.0	3.4	5.6	6.3	4.7	2.4	1.0	0.4	0.2	27.1
CLOUDINESS	MEAN:														
	SUNRISE-SUNSET (OKTAS)	41	6.0	5.8	5.7	5.4	5.4	5.0	4.8	4.8	4.8	4.8	5.9	6.0	5.4
	MIDNIGHT-MIDNIGHT (OKTAS)	32	5.7	5.4	5.4	5.0	5.0	4.6	4.5	4.4	4.6	4.6	5.7	5.9	5.1
	MEAN NO. DAYS WITH:														
	CLEAR	42	4.3	4.5	5.6	6.2	6.0	6.8	6.1	7.1	7.0	8.2	3.8	3.7	69.3
	PARTLY CLOUDY	42	7.1	7.0	7.7	7.5	9.5	10.9	12.5	11.3	9.4	8.2	6.7	6.5	104.3
	CLOUDY	42	19.6	16.7	17.7	16.3	15.5	12.3	11.7	11.8	13.0	13.9	18.7	19.9	187.1
PR	MEAN STATION PRESSURE(IN)	28	29.00	29.00	29.00	29.00	29.00	28.99	29.00	29.10	29.09	29.10	29.09	29.10	29.04
	MEAN SEA-LEVEL PRES. (IN)	16	30.07	30.06	30.04	29.96	29.98	29.97	30.00	30.04	30.08	30.10	30.09	30.10	30.04
WINDS	MEAN SPEED (MPH)	40	8.8	8.9	9.5	9.2	8.2	7.7	7.1	6.9	7.2	7.5	8.3	8.4	8.1
	PREVAIL.DIR(TENS OF DEGS)	26	23	23	32	22	22	22	24	23	22	23	22	23	23
	MAXIMUM 2-MINUTE:														
	SPEED (MPH)	5	36	38	39	34	45	36	39	46	45	34	37	43	46
	DIR. (TENS OF DEGS)		23	26	28	26	31	29	36	25	32	26	31	26	25
	YEAR OF OCCURRENCE		1999	1997	2000	2000	1998	2000	2001	1997	1998	2001	1997	2000	AUG 1997
	MAXIMUM 5-SECOND:														
	SPEED (MPH)	5	49	46	53	45	55	45	47	55	51	43	51	54	55
	DIR. (TENS OF DEGS)		13	24	25	25	31	31	01	23	35	27	15	26	31
	YEAR OF OCCURRENCE		1999	1997	1997	2000	1998	2000	2001	1997	2001	2001	1999	2000	MAY 1998
PRECIPITATION	NORMAL (IN)	30	2.10	2.15	2.55	2.97	3.65	3.98	3.79	3.32	3.31	2.79	3.06	2.51	36.18
	MAXIMUM MONTHLY (IN)	46	6.48	8.06	4.83	9.56	8.02	7.22	7.25	6.78	9.76	8.12	7.69	6.58	9.76
	YEAR OF OCCURRENCE		1979	1981	1977	1983	1989	1982	1986	1994	1999	1976	1972	1983	SEP 1999
	MINIMUM MONTHLY (IN)	46	0.39	0.30	0.49	0.96	0.77	0.27	1.04	0.95	0.82	0.03	0.80	0.35	0.03
	YEAR OF OCCURRENCE		1980	1968	1981	1997	1959	1966	1993	1995	1964	1963	1976	1958	OCT 1963
	MAXIMUM IN 24 HOURS (IN)	46	2.07	3.11	3.02	3.80	2.58	3.61	2.83	3.18	6.52	4.28	3.60	2.86	6.52
	YEAR OF OCCURRENCE		1996	1981	1986	1983	1972	1973	2000	1966	1985	1995	1996	1983	SEP 1985
	NORMAL NO. DAYS WITH:														
	PRECIPITATION 0.01	30	12.0	10.7	12.1	12.2	13.1	12.4	11.3	10.9	9.9	9.9	12.4	13.0	139.9
	PRECIPITATION 1.00	30	0.2	0.3	0.3	0.5	0.8	0.8	0.8	0.7	0.8	0.7	0.6	0.3	6.8
SNOWFALL	NORMAL (IN)	30	12.4	10.6	8.0	2.8	0.1	0.0	0.0	0.0	0.0	0.1	4.1	8.6	46.7
	MAXIMUM MONTHLY (IN)	41	42.3	22.0	32.0	26.7	2.4	0.0	T	T	T	4.4	22.5	33.9	42.3
	YEAR OF OCCURRENCE		1994	1964	1993	1983	1977		1995	1993	1993	1962	1971	1969	JAN 1994
	MAXIMUM IN 24 HOURS (IN)	41	20.6	13.3	20.4	12.2	2.4	0.0	T	T	T	4.4	20.5	12.4	20.6
	YEAR OF OCCURRENCE		1996	1961	1993	1983	1977		1995	1993	1993	1962	1971	1969	JAN 1996
	MAXIMUM SNOW DEPTH (IN)	46	27	25	21	14	0	0	0	0	0	2	17	22	27
	YEAR OF OCCURRENCE		1994	1961	1993	1983						1962	1971	1969	JAN 1994
	NORMAL NO. DAYS WITH:														
	SNOWFALL 1.0	30	3.3	3.0	2.4	0.6	0.*	0.0	0.0	0.0	0.0	0.0	1.0	2.2	12.5

PRECIPITATION (inches)   2001   AVOCA, WILKES-BARRE – SCRANTON PA (AVP)

YEAR	JAN	FEB	MAR	APR	MAY	JUN	JUL	AUG	SEP	OCT	NOV	DEC	ANNUAL
1972	2.05	2.42	4.00	3.31	7.33	7.04	1.23	1.64	1.57	3.30	7.69	3.61	45.19
1973	2.13	1.28	1.79	4.38	3.80	5.99	3.87	2.61	3.62	1.97	1.50	6.07	39.01
1974	2.66	1.48	4.75	2.71	1.89	3.85	2.80	3.50	6.85	1.07	2.26	3.40	37.22
1975	2.78	3.26	2.52	1.17	4.01	5.64	3.85	2.78	6.10	3.29	3.00	1.84	40.24
1976	3.25	2.14	2.18	2.27	3.24	5.43	3.20	2.57	3.81	8.12	0.80	1.50	38.51
1977	0.88	1.82	4.83	3.98	1.72	3.16	3.44	4.23	5.97	5.27	3.98	3.44	42.72
1978	5.33	0.93	2.30	1.67	4.30	2.48	2.16	3.28	3.06	3.35	1.02	3.09	32.97
1979	6.48	2.44	1.52	3.69	5.16	2.54	2.97	2.05	5.84	3.68	3.17	1.70	41.24
1980	0.39	0.69	3.72	2.35	2.37	4.36	3.76	1.23	1.43	2.17	2.83	1.24	26.54
1981	0.63	8.06	0.49	3.54	3.00	3.45	4.27	1.75	2.74	3.50	1.84	2.13	35.40
1982	2.71	2.28	2.55	3.48	3.52	7.22	3.32	3.42	1.10	0.84	3.44	1.52	35.40
1983	1.17	1.46	3.28	9.56	3.28	4.81	2.76	1.77	2.12	2.73	3.71	6.58	43.23
1984	1.11	2.92	2.42	4.09	6.70	4.75	5.12	2.81	1.36	2.30	2.63	2.36	38.57
1985	0.61	1.58	2.24	2.00	6.10	3.00	2.62	2.62	7.83	1.92	4.47	1.96	40.42
1986	2.59	2.58	4.25	2.98	2.24	6.77	7.25	3.94	3.07	2.61	3.94	2.04	44.26
1987	2.60	0.68	1.18	4.38	2.22	4.35	5.80	4.16	8.15	2.77	2.24	0.99	39.52
1988	1.41	2.32	1.97	2.65	4.24	0.82	6.26	5.03	1.89	1.93	3.33	1.08	32.93
1989	1.02	1.73	2.23	0.97	8.02	6.10	2.76	2.92	3.92	4.73	3.57	0.96	38.93
1990	3.81	2.70	1.88	2.48	5.27	4.78	4.36	5.69	3.16	4.33	3.33	4.30	46.09
1991	1.54	1.35	2.91	2.69	2.84	1.72	2.45	3.28	2.55	2.41	3.81	2.73	30.28
1992	1.23	1.41	2.65	1.67	4.21	1.45	3.83	2.13	2.65	1.91	3.47	2.91	29.52
1993	1.97	1.24	4.02	7.47	1.38	1.85	1.04	3.82	6.55	3.99	3.32	3.09	39.74
1994	3.29	1.00	3.80	4.90	2.68	3.54	3.22	6.78	5.12	0.76	4.54	2.29	41.92
1995	2.73	1.45	1.34	2.15	1.40	1.45	3.16	0.95	2.31	7.15	3.65	1.22	28.96
1996	6.40	1.46	2.55	5.26	4.03	3.90	6.01	1.24	3.92	4.40	4.45	5.37	48.99
1997	1.57	1.13	3.20	0.96	2.46	2.96	1.33	5.30	1.91	1.17	3.48	2.25	27.72
1998	2.96	2.91	2.54	5.41	4.38	4.16	2.43	2.69	2.46	2.99	0.95	0.86	34.74
1999	4.85	1.41	2.73	2.16	2.62	2.89	1.63	2.12	9.76	1.52	2.31	1.24	35.24
2000	2.08	2.40	2.84	2.90	2.86	6.09	6.21	1.93	3.07	1.46	1.44	2.77	36.05
2001	1.13	1.14	2.54	1.97	2.92	3.04	3.67	3.02	3.95	1.01	1.89	1.11	27.39
POR= 101 YRS	2.38	2.20	2.81	3.13	3.30	3.73	3.97	3.46	3.27	2.87	2.79	2.49	36.40

AVERAGE TEMPERATURE (F) 2001   AVOCA, WILKES-BARRE – SCRANTON PA (AVP)

YEAR	JAN	FEB	MAR	APR	MAY	JUN	JUL	AUG	SEP	OCT	NOV	DEC	ANNUAL
1972	27.6	24.3	32.8	43.5	60.0	62.9	71.5	69.6	61.9	45.3	36.0	32.9	47.4
1973	28.5	24.5	41.9	47.5	53.4	68.3	71.5	71.6	62.4	53.2	41.5	31.3	49.6
1974	27.8	24.2	34.7	49.1	56.2	64.4	70.8	68.5	61.2	48.5	43.0	34.4	48.6
1975	31.8	32.1	35.8	43.9	64.9	68.8	73.7	71.2	59.9	55.8	47.8	31.1	51.4
1976	22.1	35.0	41.0	51.3	56.7	70.4	69.2	69.1	60.9	48.0	37.0	22.9	48.6
1977	15.0	26.9	40.5	48.9	60.0	63.8	71.1	68.7	62.7	48.5	43.1	29.6	48.2
1978	24.4	19.2	33.1	46.0	58.7	65.2	69.4	71.0	60.5	50.7	40.5	29.1	47.3
1979	24.2	16.0	40.7	46.4	58.6	65.1	70.7	70.7	63.2	51.8	45.9	35.6	49.1
1980	27.8	24.2	35.9	51.0	61.8	65.3	73.2	75.2	66.1	49.8	37.6	25.7	49.5
1981	19.5	34.9	36.2	51.0	59.8	68.5	72.2	70.1	62.1	49.1	41.1	29.1	49.5
1982	18.7	27.8	36.0	46.3	61.2	64.3	71.0	66.2	62.4	52.3	44.3	36.7	48.9
1983	27.3	29.4	38.8	45.9	55.7	67.6	72.4	71.6	64.8	52.5	42.8	27.1	49.7
1984	23.0	35.9	30.9	48.3	57.5	69.3	71.6	72.6	61.7	58.0	40.8	37.3	50.6
1985	21.5	29.8	39.1	51.4	60.6	63.8	70.1	69.0	64.0	52.7	44.5	26.7	49.4
1986	27.2	26.1	39.8	49.3	62.7	66.2	71.0	67.4	61.6	51.2	37.3	32.8	49.4
1987	24.7	25.1	39.7	50.7	60.0	68.7	73.5	68.3	61.8	47.4	41.3	32.7	49.5
1988	22.0	27.4	38.1	46.6	60.1	65.5	75.8	72.9	60.4	46.3	43.3	29.8	49.0
1989	31.3	28.0	37.2	45.6	57.6	67.0	70.4	68.5	62.2	52.8	39.7	18.6	48.2
1990	35.3	33.3	40.6	50.2	56.1	67.6	71.7	69.5	61.6	55.3	44.2	36.0	51.8
1991	27.5	33.4	40.4	51.7	65.7	69.5	73.4	72.2	61.8	53.3	41.4	32.9	51.9
1992	28.6	30.3	34.4	47.4	58.7	66.3	71.1	68.4	62.8	49.0	42.2	32.8	49.3
1993	32.0	23.8	33.3	49.9	61.7	68.8	75.5	72.7	61.2	50.0	41.4	30.0	50.0
1994	18.8	24.4	35.4	52.2	56.6	69.9	73.8	67.3	60.9	50.9	47.4	36.6	49.5
1995	32.9	26.5	40.5	46.4	59.2	69.9	75.2	74.7	62.4	56.5	35.7	25.7	50.5
1996	24.1	27.6	33.2	47.4	55.9	68.2	68.9	69.6	62.5	50.8	36.4	34.5	48.3
1997	25.3	32.9	36.8	45.5	54.3	65.9	70.4	67.5	60.3	50.2	37.1	31.3	48.1
1998	34.1	35.3	41.1	49.5	63.3	65.7	69.8	70.4	64.1	51.6	41.8	35.9	51.9
1999	27.9	31.9	35.5	48.1	60.2	68.5	74.7	68.9	64.5	49.0	45.3	33.1	50.6
2000	24.7	30.5	42.8	47.8	59.9	66.8	67.2	67.7	60.0	51.6	38.3	23.0	48.4
2001	26.6	30.1	33.1	48.6	59.6	68.5	68.3	73.1	60.4	52.9	46.9	36.9	50.4
POR= 101 YRS	26.8	27.6	36.9	48.0	59.1	67.4	72.1	70.0	63.0	52.1	41.2	30.4	49.6

REFERENCE NOTES:

PAGE 1:
THE TEMPERATURE GRAPH SHOWS NORMAL MAXIMUM AND NORMAL
MINIMUM DAILY TEMPERATURES (SOLID CURVES) AND THE
ACTUAL DAILY HIGH AND LOW TEMPERATURES (VERTICAL BARS).

PAGE 2 AND 3:
H/C INDICATES HEATING AND COOLING DEGREE DAYS.
RH INDICATES RELATIVE HUMIDITY
W/O INDICATES WEATHER AND OBSTRUCTIONS
S INDICATES SUNSHINE.
PR INDICATES PRESSURE.
CLOUDINESS ON PAGE 3 IS THE SUM OF THE CEILOMETER AND
SATELLITE DATA NOT TO EXCEED EIGHT EIGHTHS(OKTAS).

GENERAL:
T INDICATES TRACE PRECIPITATION, AN AMOUNT GREATER
THAN ZERO BUT LESS THAN THE LOWEST REPORTABLE VALUE.
+ INDICATES THE VALUE ALSO OCCURS ON EARLIER DATES.
BLANK ENTRIES DENOTE MISSING OR UNREPORTED DATA.
NORMALS ARE 30-YEAR AVERAGES (1961 - 1990).
ASOS INDICATES AUTOMATED SURFACE OBSERVING SYSTEM.
PM INDICATES THE LAST DAY OF THE PREVIOUS MONTH.
POR (PERIOD OF RECORD) BEGINS WITH THE JANUARY DATA
MONTH AND IS THE NUMBER OF YEARS USED TO COMPUTE
THE MEAN. INDIVIDUAL MONTHS WITHIN THE POR MAY
BE MISSING.
WHEN THE POR FOR A NORMAL IS LESS THAN 30 YEARS,
THE NORMAL IS PROVISIONAL AND IS BASED ON THE NUMBER
OF YEARS INDICATED.
0.* OR * INDICATES THE VALUE OR MEAN-DAYS-WITH
IS BETWEEN 0.00 AND 0.05.
CLOUDINESS FOR ASOS STATIONS DIFFERS FROM THE NON-ASOS
OBSERVATION TAKEN BY A HUMAN OBSERVER. ASOS STATION
CLOUDINESS IS BASED ON TIME-AVERAGED CEILOMETER DATA
FOR CLOUDS AT OR BELOW 12,000 FEET AND ON SATELLITE
DATA FOR CLOUDS ABOVE 12,000 FEET.
THE NUMBER OF DAYS WITH CLEAR, PARTLY CLOUDY, AND
CLOUDY CONDITIONS FOR ASOS STATIONS IS THE SUM
OF THE CEILOMETER AND SATELLITE DATA FOR THE
SUNRISE TO SUNSET PERIOD.

GENERAL CONTINUED:
CLEAR INDICATES 0 - 2 OKTAS, PARTLY CLOUDY INDICATES
3 - 6 OKTAS, AND CLOUDY INDICATES 7 OR 8 OKTAS.
WHEN AT LEAST ONE OF THE ELEMENTS (CEILOMETER OR
SATELLITE) IS MISSING, THE DAILY CLOUDINESS IS
NOT COMPUTED.
WIND DIRECTION IS RECORDED IN TENS OF DEGREES (2 DIGITS)
CLOCKWISE FROM TRUE NORTH. "00" INDICATES CALM. "36"
INDICATES TRUE NORTH.
RESULTANT WIND IS THE VECTOR AVERAGE OF THE SPEED AND
DIRECTION.
AVERAGE TEMPERATURE IS THE SUM OF THE MEAN DAILY MAXIMUM
AND MINIMUM TEMPERATURE DIVIDED BY 2.
SNOWFALL DATA COMPRISE ALL FORMS OF FROZEN
PRECIPITATION, INCLUDING HAIL.
A HEATING (COOLING) DEGREE DAY IS THE DIFFERENCE BETWEEN
THE AVERAGE DAILY TEMPERATURE AND 65 F.
DRY BULB IS THE TEMPERATURE OF THE AMBIENT AIR.
DEW POINT IS THE TEMPERATURE TO WHICH THE AIR MUST BE
COOLED TO ACHIEVE 100 PERCENT RELATIVE HUMIDITY.
WET BULB IS THE TEMPERATURE THE AIR WOULD HAVE IF THE
MOISTURE CONTENT WAS INCREASED TO 100 PERCENT RELATIVE
HUMIDITY.

ON JULY 1, 1996, THE NATIONAL WEATHER SERVICE BEGAN USING
THE "METAR" OBSERVATION CODE THAT WAS ALREADY EMPLOYED
BY MOST OTHER NATIONS OF THE WORLD. THE MOST NOTICEABLE
DIFFERENCE IN THIS ANNUAL PUBLICATION WILL BE THE CHANGE
IN UNITS FROM TENTHS TO EIGHTS(OKTAS) FOR REPORTING THE
AMOUNT OF SKY COVER.

HEATING DEGREE DAYS (base 65 F) 2001 AVOCA, WILKES-BARRE - SCRANTON PA (AVP)

YEAR	JUL	AUG	SEP	OCT	NOV	DEC	JAN	FEB	MAR	APR	MAY	JUN	TOTAL
1972-73	21	20	125	603	860	988	1124	1131	704	521	354	25	6476
1973-74	2	11	140	368	699	1036	1145	1135	934	480	291	65	6306
1974-75	5	2	155	503	655	941	1024	913	902	627	88	25	5840
1975-76	0	8	158	291	509	1043	1322	864	737	451	265	31	5679
1976-77	8	25	155	519	834	1297	1546	1058	756	487	206	90	6981
1977-78	14	37	119	505	653	1090	1252	1279	984	562	240	73	6808
1978-79	38	2	153	436	728	1103	1257	1370	747	552	221	66	6673
1979-80	34	31	120	420	568	900	1144	1175	895	414	137	94	5932
1980-81	1	0	82	466	813	1211	1407	835	886	416	195	19	6331
1981-82	2	5	132	485	706	1105	1426	1034	896	554	147	68	6560
1982-83	17	55	112	390	619	870	1158	992	805	569	292	41	5920
1983-84	7	11	119	392	659	1169	1297	837	1052	493	247	34	6317
1984-85	7	6	148	219	719	852	1342	981	799	421	162	78	5734
1985-86	4	11	127	376	610	1181	1163	1083	777	467	140	61	6000
1986-87	16	50	139	428	823	990	1243	1111	779	425	208	20	6232
1987-88	2	34	119	539	706	995	1326	1082	823	546	176	91	6439
1988-89	13	12	156	574	643	1083	1037	1031	853	575	251	39	6267
1989-90	6	31	133	377	750	1433	915	881	757	465	269	44	6061
1990-91	10	13	152	320	619	894	1153	877	757	402	113	23	5333
1991-92	0	0	160	370	699	989	1121	1001	942	521	209	40	6052
1992-93	4	9	132	488	676	993	1017	1148	975	447	136	39	6064
1993-94	0	4	174	460	703	1077	1428	1131	910	390	270	33	6580
1994-95	0	31	143	427	522	875	986	1072	755	552	189	15	5567
1995-96	0	0	123	264	872	1210	1260	1075	979	525	312	16	6636
1996-97	9	5	140	432	852	938	1222	893	866	577	327	60	6321
1997-98	7	23	150	466	828	949	1036	828	754	460	104	97	5702
1998-99	2	11	90	408	689	896	1143	919	910	500	159	32	5759
1999-00	4	18	104	491	586	982	1243	995	683	510	196	59	5871
2000-01	26	32	202	410	794	1289	1181	973	982	491	187	45	6612
2001-	28	0	156	370	537	863							

COOLING DEGREE DAYS (base 65 F) 2001 AVOCA, WILKES-BARRE - SCRANTON PA (AV

YEAR	JAN	FEB	MAR	APR	MAY	JUN	JUL	AUG	SEP	OCT	NOV	DEC	ANNUAL
1972	0	0	0	0	20	53	232	167	42	0	0	0	514
1973	0	0	0	4	0	132	212	223	72	7	0	0	650
1974	0	0	0	10	28	52	194	117	46	0	0	0	447
1975	0	0	0	0	91	146	278	207	13	14	0	0	749
1976	0	0	0	46	16	198	145	159	37	0	0	0	601
1977	0	0	2	13	57	62	208	162	59	0	2	0	565
1978	0	0	0	0	52	84	181	194	26	0	0	0	537
1979	0	0	0	2	32	78	218	214	75	15	0	0	634
1980	0	0	0	0	42	107	263	322	122	3	0	0	859
1981	0	0	0	4	42	131	231	172	55	0	0	0	635
1982	0	0	0	1	34	55	208	98	41	3	5	0	445
1983	0	0	0	4	12	125	243	224	118	9	0	0	735
1984	0	0	0	0	20	165	218	248	58	12	0	0	721
1985	0	0	0	20	32	47	169	142	104	0	0	0	514
1986	0	0	2	0	76	104	205	131	44	8	0	0	570
1987	0	0	0	3	62	138	273	141	30	0	0	0	647
1988	0	0	0	0	34	111	356	266	23	3	0	0	793
1989	0	0	0	0	31	109	179	148	57	4	0	0	528
1990	0	0	0	8	29	2	127	225	160	56	24	0	631
1991	0	0	0	12	142	164	264	230	69	14	0	0	895
1992	0	0	0	0	21	85	201	123	71	0	0	0	501
1993	0	0	0	5	41	158	334	249	66	0	1	0	853
1994	0	0	0	13	18	188	280	109	27	0	0	0	636
1995	0	0	0	0	16	167	323	308	54	6	0	0	874
1996	0	0	0	0	38	117	132	155	71	0	0	0	513
1997	0	0	0	0	4	97	178	105	16	11	0	0	411
1998	0	0	17	0	59	128	157	184	71	0	0	0	616
1999	0	0	0	0	19	145	312	146	95	0	0	0	717
2000	0	0	0	0	45	117	101	126	60	0	0	0	449
2001	0	0	0	6	25	156	137	258	24	2	0	0	608

SNOWFALL (inches) 2001 AVOCA, WILKES-BARRE - SCRANTON PA (AVP)

YEAR	JUL	AUG	SEP	OCT	NOV	DEC	JAN	FEB	MAR	APR	MAY	JUN	TOTAL
1972-73	0.0	0.0	0.0	0.8	7.9	3.9	5.0	3.1	1.9	0.2	0.4	0.0	23.2
1973-74	0.0	0.0	0.0	T	0.4	16.0	12.8	4.5	15.7	2.8	0.0	0.0	52.2
1974-75	0.0	0.0	0.0	0.2	2.2	5.2	13.7	15.2	5.5	1.2	0.0	0.0	43.2
1975-76	0.0	0.0	0.0	0.0	1.3	3.7	13.0	7.5	10.2	0.5	T	0.0	36.2
1976-77	0.0	0.0	0.0	0.0	6.0	6.7	15.7	13.0	11.2	1.4	2.4	0.0	56.4
1977-78	0.0	0.0	0.0	0.6	8.7	9.8	28.8	18.2	6.5	0.9	0.0	0.0	73.5
1978-79	0.0	0.0	0.0	T	4.1	7.9	12.7	14.3	1.1	4.4	0.0	0.0	44.5
1979-80	0.0	0.0	0.0	T	T	5.5	1.4	8.1	10.5	T	0.0	0.0	25.5
1980-81	0.0	0.0	0.0	T	8.6	8.0	11.1	7.0	5.8	T	0.0	0.0	40.5
1981-82	0.0	0.0	0.0	T	1.0	14.2	14.1	13.5	8.7	8.1	0.0	0.0	59.6
1982-83	0.0	0.0	0.0	0.0	0.5	7.4	8.4	12.3	3.8	26.7	0.0	0.0	59.1
1983-84	0.0	0.0	0.0	0.0	3.1	2.7	11.2	4.0	18.4	0.0	0.0	0.0	39.4
1984-85	0.0	0.0	0.0	0.0	3.0	9.2	10.8	9.1	1.4	1.8	0.0	0.0	35.3
1985-86	0.0	0.0	0.0	0.0	1.7	13.4	12.9	11.6	1.1	8.6	0.0	0.0	49.3
1986-87	0.0	0.0	0.0	0.0	8.6	1.4	29.6	6.4	0.9	0.6	0.0	0.0	47.5
1987-88	0.0	0.0	0.0	T	6.4	6.4	13.0	14.9	4.3	0.7	0.0	0.0	45.7
1988-89	0.0	0.0	0.0	T	T	1.1	2.1	3.0	1.1	T	0.0	0.0	7.3
1989-90	0.0	0.0	0.0	T	2.6	8.3	10.8	7.3	6.2	2.1	0.0	0.0	37.3
1990-91	0.0	0.0	T	T	0.4	8.4	6.6	7.6	7.2	1.1	0.0	0.0	31.3
1991-92	0.0	0.0	0.0	0.0	0.1	3.5	5.4	6.3	8.5	0.7	0.0	0.0	24.5
1992-93	0.0	0.0	T	T	0.5	10.0	3.8	12.1	32.0	1.9	0.0	0.0	60.3
1993-94	0.0	T	T	1.9	0.7	8.0	42.3	16.9	20.6	T	T	0.0	90.4
1994-95	0.0	0.0	0.0	0.0	1.3	0.4	7.6	9.7	5.2	0.8	0.0	0.0	25.0
1995-96	T	0.0	0.0	T	18.6	16.0	37.5	6.5	14.2				
1996-97													
1997-98													
1998-99													
1999-00													
2000-01													
2001-													
POR= 41 YRS	T	T	T	0.2	3.7	8.8	12.2	10.5	9.5	2.8	0.1	0.0	47.8

2001
MIDDLETOWN/HARRISBURG INTL APT
PENNSYLVANIA (MDT)

Harrisburg, the capital of Pennsylvania, is situated on the east bank of the Susquehanna River. It is in the Great Valley formed by the eastern foothills of the Appalachian Chain, and about 60 miles southeast of the Commonwealths geographic center. It is nested in a saucer-like bowl, 10 miles south of Blue Mountain, which serves as a barrier to provide a modifying influence upon the severe winter climate experienced 50 to 100 miles to the north and west. Although the severity of the winter climate is lessened, the city lies a little too far inland to derive the full benefits of the coastal Climate.

Air masses change with some regularity, and any one condition does not persist for many days in succession. The mountain barrier occasionally prevents cold waves from reaching the Great Valley. The city is favorably located to receive precipitation produced when warm, maritime air from the Atlantic Ocean is forced upslope to cross the Blue Ridge Mountains.

The Growing Season is 192 days.

During June 1992, Hurricane Agnes produced 15.11 inches of rain from the 20th to the 23rd. Prolonged dry spells occur occasionally. During September and October 1947 there were 35 consecutive days with less then .01 inches of precipitation.

Flood stage on the Susquehanna River occurs on the average of about every three years in Harrisburg, but serious flooding is much less frequent. About one-third of all floods have occurred during March. Tropical hurricanes rarely reach Harrisburg with destructive winds.

NORMALS, MEANS, AND EXTREMES
MIDDLETOWN/HARRISBURG, PA　(MDT)

LATITUDE:	LONGITUDE:	ELEVATION (FT):		TIME ZONE:	WBAN: 14711
40 11' 37" N	76 45' 48" W	GRND: 311　BARO: 314		EASTERN (UTC + 5)	

	ELEMENT	POR	JAN	FEB	MAR	APR	MAY	JUN	JUL	AUG	SEP	OCT	NOV	DEC	YEAR
TEMPERATURE °F	NORMAL DAILY MAXIMUM	30	35.9	39.2	50.3	62.0	72.5	81.2	85.8	83.8	76.3	64.7	52.6	40.6	62.1
	MEAN DAILY MAXIMUM	28	38.0	41.2	49.0	62.5	72.9	81.9	86.0	83.8	76.1	66.0	52.4	41.2	62.6
	HIGHEST DAILY MAXIMUM	63	73	78	87	93	97	100	107	101	102	97	84	75	107
	YEAR OF OCCURRENCE		1950	1997	1998	1985	1942	1966	1966	1944	1953	1941	1950	1998	JUL 1966
	MEAN OF EXTREME MAXS.	28	57.6	60.8	71.1	83.3	89.3	94.2	96.0	94.1	90.1	81.6	70.4	60.7	79.1
	NORMAL DAILY MINIMUM	30	21.2	23.3	32.0	41.2	51.1	60.6	65.6	64.3	56.5	44.6	36.1	26.6	43.6
	MEAN DAILY MINIMUM	28	23.7	25.4	31.8	42.2	52.1	61.6	66.4	64.1	56.6	45.1	35.4	26.8	44.3
	LOWEST DAILY MINIMUM	63	-22	-5	5	19	31	40	49	45	30	23	13	-8	-22
	YEAR OF OCCURRENCE		1994	1979	1984	1982	1966	1980	1945	1976	1963	2000	1955	1960	JAN 1994
	MEAN OF EXTREME MINS.	28	7.2	9.0	18.3	29.0	38.7	49.4	56.3	53.0	40.8	31.0	21.6	10.4	30.4
	NORMAL DRY BULB	30	28.6	31.3	41.2	51.6	61.8	70.9	75.7	74.1	66.4	54.7	44.4	33.6	52.9
	MEAN DRY BULB	28	31.0	33.3	40.4	52.4	62.5	71.8	76.2	74.0	66.4	55.4	43.8	33.7	53.4
	MEAN WET BULB	6	27.1	27.5	34.9	46.2	55.0	65.5	68.2	54.2	50.1	42.6	34.8	27.7	43.7
	MEAN DEW POINT	6	21.6	20.2	27.8	38.8	49.0	61.2	55.5	51.4	46.9	38.4	29.8	22.6	38.6
	NORMAL NO. DAYS WITH:														
	MAXIMUM　90	30	0.0	0.0	0.0	0.0	0.8	3.2	5.6	2.6	3.4	0.0	0.0	0.0	15.6
	MAXIMUM　32	30	10.8	7.8	0.6	0.0	0.0	0.0	0.0	0.0	0.0	0.0	0.0	7.6	26.8
	MINIMUM　32	30	27.7	23.8	14.0	3.4	0.0	0.0	0.0	0.0	0.0	1.8	8.8	24.2	103.7
	MINIMUM　0	30	1.2	0.5	0.0	0.0	0.0	0.0	0.0	0.0	0.0	0.0	0.0	0.0	1.7
H/C	NORMAL HEATING DEG. DAYS	30	1128	944	738	402	147	9	0	0	59	329	618	973	5347
	NORMAL COOLING DEG. DAYS	30	0	0	0	0	48	186	332	285	101	10	0	0	962
RH	NORMAL (PERCENT)														
	HOUR 01 LST														
	HOUR 07 LST														
	HOUR 13 LST														
	HOUR 19 LST														
S	PERCENT POSSIBLE SUNSHINE	62	48	54	57	58	59	64	67	66	61	58	47	44	57
W/O	MEAN NO. DAYS WITH:														
	HEAVY FOG(VISBY 1/4 MI)	58	2.3	2.3	1.8	1.0	0.9	0.6	0.6	0.8	1.5	2.7	1.8	2.5	18.8
	THUNDERSTORMS	58	0.2	0.2	1.0	2.4	4.8	6.0	6.5	4.9	2.9	0.8	0.5	0.2	30.4
CLOUDINESS	MEAN:														
	SUNRISE-SUNSET (OKTAS)	49	5.4	5.3	5.3	5.2	5.2	4.9	4.9	4.7	4.7	4.6	5.2	5.6	5.1
	MIDNIGHT-MIDNIGHT (OKTAS)	8	5.4	5.1	5.2	5.0	4.8	5.3	4.9	4.6	4.6	3.5	4.6	5.3	4.9
	MEAN NO. DAYS WITH:														
	CLEAR	61	6.5	6.6	7.1	6.5	5.9	6.4	7.0	7.7	8.5	9.8	6.3	6.1	84.4
	PARTLY CLOUDY	61	7.8	7.4	8.3	8.7	10.9	11.7	11.7	11.3	9.5	8.4	8.2	7.5	111.4
	CLOUDY	61	16.7	14.3	15.4	14.9	14.3	11.9	11.6	11.6	11.9	11.6	12.4	15.2	167.0
PR	MEAN STATION PRESSURE(IN)	25	29.72	29.73	29.67	29.62	29.63	29.63	29.65	29.69	29.73	29.75	29.73	29.74	29.69
	MEAN SEA-LEVEL PRES. (IN)	10	30.10	30.09	30.02	29.98	29.99	29.99	29.98	30.04	30.05	30.12	30.12	30.12	30.05
WINDS	MEAN SPEED (MPH)	28	7.8	8.3	9.3	8.4	7.1	6.3	5.5	5.0	5.2	5.5	7.0	7.3	6.9
	PREVAIL.DIR(TENS OF DEGS)	14	31	31	31	31	30	30	30	30	31	31	31	30	31
	MAXIMUM 2-MINUTE:														
	SPEED (MPH)	1	35	36	37	38	28	31	35	28	23	32	31	38	38
	DIR. (TENS OF DEGS)		30	31	30	34	12	31	31	30	30	29	31	30	30
	YEAR OF OCCURRENCE		2001	2001	2001	2001	2001	2001	2001	2001	2001	2001	2001	2001	DEC 2001
	MAXIMUM 5-SECOND:														
	SPEED (MPH)	1	39	51	45	48	32	38	41	32	30	40	43	44	51
	DIR. (TENS OF DEGS)		31	28	30	34	12	31	31	30	30	26	31	31	28
	YEAR OF OCCURRENCE		2001	2001	2001	2001	2001	2001	2001	2001	2001	2001	2001	2001	FEB 2001
PRECIPITATION	NORMAL (IN)	30	2.84	2.93	3.28	3.24	4.26	3.85	3.59	3.31	3.51	2.93	3.52	3.24	40.50
	MAXIMUM MONTHLY (IN)	23	8.01	5.93	6.32	7.96	9.71	8.12	8.09	6.26	9.11	5.59	6.23	7.57	9.71
	YEAR OF OCCURRENCE		1979	1981	1993	1983	1989	1982	1994	1986	1999	1989	1985	1983	MAY 1989
	MINIMUM MONTHLY (IN)	23	0.43	0.82	0.95	0.45	1.39	1.00	0.97	0.53	0.65	0.47	0.92	0.31	0.31
	YEAR OF OCCURRENCE		1981	1980	1995	1985	1999	1988	1983	1995	1986	2000	1998	1998	DEC 1998
	MAXIMUM IN 24 HOURS (IN)	21	2.09	1.84	2.63	2.06	2.91	2.32	3.84	2.78	3.46	2.78	3.27	2.58	3.84
	YEAR OF OCCURRENCE		1979	1985	2000	1992	1984	1987	1994	1991	1999	1995	1993	1993	JUL 1994
	NORMAL NO. DAYS WITH:														
	PRECIPITATION　0.01	30	9.6	11.0	12.5	13.0	9.7	10.8	9.2	10.2	6.4	5.6	9.6	9.0	116.6
	PRECIPITATION　1.00	30	0.6	0.6	0.7	0.5	0.2	0.5	0.6	0.6	0.8	1.0	0.7	0.5	7.3
SNOWFALL	NORMAL (IN)	30	15.7	15.2	9.6	0.3	T	0.0	0.0	0.0	0.0	T	2.0	11.0	53.8
	MAXIMUM MONTHLY (IN)	23	38.9	28.8	22.8	10.2	0.0	T	0.0	T	0.0	T	9.7	17.4	38.9
	YEAR OF OCCURRENCE		1996	1983	1993	1982		1993		1993		1982	1987	1995	JAN 1996
	MAXIMUM IN 24 HOURS (IN)	21	21.7	14.2	20.4	4.0	0.0	T	0.0	T	0.0	T	7.9	9.1	21.7
	YEAR OF OCCURRENCE		1996	1979	1993	1996		1993		1993		1979	1987	1990	JAN 1996
	MAXIMUM SNOW DEPTH (IN)	28	20	23	20	1	0	0	0	0	0	0	12	13	23
	YEAR OF OCCURRENCE		1961	1961	1993	1959							1953	1951	FEB 1961
	NORMAL NO. DAYS WITH:														
	SNOWFALL　1.0	30	3.0	4.2	2.3	0.2	0.0	0.0	0.0	0.0	0.0	0.0	0.4	2.7	12.8

PRECIPITATION (inches) 2001 MIDDLETOWN/HARRISBURG INTL APT PA (MDT)

YEAR	JAN	FEB	MAR	APR	MAY	JUN	JUL	AUG	SEP	OCT	NOV	DEC	ANNUAL
1972	2.65	5.00	2.68	4.10	5.56	18.55	2.26	2.52	1.41	2.03	7.20	5.31	59.27
1973	3.24	2.50	2.00	6.23	6.37	3.34	2.18	2.19	5.73	2.47	1.04	6.52	43.81
1974	3.82	1.36	4.64	3.21	4.38	3.69	2.79	4.13	6.79	1.25	2.30	4.59	42.95
1975	4.12	3.10	3.78	2.80	5.25	6.51	3.13	1.83	14.97	2.62	2.92	3.19	54.22
1976	4.34	1.88	3.43	1.63	5.42	2.42	2.42	5.50	4.79	9.87	0.79	1.96	45.31
1977	1.44	1.75	6.10	4.48	1.00	3.17	3.01	0.93	3.73	3.66	5.61	4.82	39.70
1978	7.44	1.35	3.94	1.97	5.67	5.16	4.35	3.60	1.64	2.51	2.13	3.95	43.71
1979	8.01	4.74	1.93	3.60	4.66	2.62	3.14	3.24	6.62	3.91	2.67	1.46	46.60
1980	0.90	0.82	5.47	4.27	4.58	2.50	1.59	1.51	1.06	2.94	3.65	0.77	30.06
1981	0.43	5.93	1.02	2.77	1.86	4.66	4.67	4.11	2.20	3.76	0.96	2.41	34.78
1982	3.63	1.92	2.20	4.17	4.89	8.12	2.90	2.47	2.87	1.82	3.37	1.56	39.92
1983	2.26	3.38	4.86	7.96	5.36	2.81	0.97	2.50	1.40	4.21	5.29	7.57	48.57
1984	1.12	4.51	5.36	4.46	6.20	6.36	3.76	2.75	1.49	1.98	3.78	2.28	44.05
1985	1.06	2.91	2.78	0.45	6.29	3.07	2.50	2.14	3.76	1.34	6.23	1.28	33.81
1986	2.24	4.50	3.16	4.10	2.29	1.48	5.17	6.26	0.65	2.59	4.58	4.90	41.92
1987	3.69	1.59	1.43	2.93	3.73	3.46	1.96	2.89	8.41	2.63	4.96	1.84	39.52
1988	2.18	3.28	1.98	2.65	5.79	1.00	4.40	2.67	2.42	1.81	3.67	0.90	32.75
1989	2.29	1.90	3.60	1.10	9.71	6.02	7.20	3.03	2.63	5.59	2.17	1.27	46.51
1990	3.77	2.73	1.76	2.60	7.20	1.10	3.62	6.14	1.65	4.92	2.58	6.05	44.12
1991	2.61	1.39	3.54	2.00	3.15	1.08	1.99	5.29	1.35	3.15	2.08	3.49	31.12
1992	1.62	1.56	5.13	2.62	3.17	1.90	3.54	1.45	5.65	1.64	4.82	2.42	35.52
1993	2.39	2.32	6.32	6.49	1.96	3.20	3.65	3.45	7.84	2.66	4.17	3.95	48.40
1994	5.00	3.24	6.22	2.96	2.73	1.81	8.09	4.94	2.33	0.74	5.18	2.92	46.16
1995	3.52	1.52	0.95	2.22	3.52	4.16	5.81	0.53	1.95	5.43	3.67	2.53	35.81
1996	5.87	1.60	3.26	4.69	3.93	6.33	7.06	2.22	3.35	4.26	3.88	5.98	52.43
1997	2.00	1.47	3.44	0.92	3.66	2.41	4.82	3.60	2.52	1.50	3.97	2.01	32.32
1998	4.49	5.05	4.36	5.17	6.37	5.82	4.86	3.88	1.81	2.92	0.92	0.31	45.96
1999	4.94	1.90	2.55	2.80	1.39	1.87	2.96	3.92	9.11	2.65	1.66	2.57	38.32
2000	2.01	2.33	6.06	2.63	4.03	4.09	2.32	4.14	8.61	0.47	1.55	3.99	42.23
2001	2.44	1.48	4.19	1.72	1.66	2.01	1.90	3.79	2.18	1.01	1.51	1.87	25.76
POR= 112 YRS	2.87	2.65	3.31	3.07	3.76	3.58	3.62	3.61	3.18	2.87	2.77	2.91	38.20

AVERAGE TEMPERATURE (F) 2001   MIDDLETOWN/HARRISBURG INTL APT PA (MDT)

YEAR	JAN	FEB	MAR	APR	MAY	JUN	JUL	AUG	SEP	OCT	NOV	DEC	ANNUAL
1972	34.5	30.2	40.0	49.6	62.3	67.7	76.2	74.0	68.4	52.1	41.8	36.8	52.8
1973	33.7	31.7	45.2	50.9	57.8	73.4	75.5	76.1	68.6	57.4	46.9	36.2	54.5
1974	34.8	32.3	42.8	56.2	63.0	70.5	77.6	77.1	64.5	51.6	44.9	35.4	54.2
1975	33.3	32.4	38.3	47.4	64.8	70.7	75.3	76.2	63.3	57.7	50.0	34.2	53.6
1976	27.6	39.6	44.2	54.5	59.4	73.4	72.3	72.4	64.8	51.5	39.6	30.0	52.4
1977	20.1	30.4	46.0	54.1	64.5	68.6	75.9	74.3	68.3	52.7	46.1	31.6	52.7
1978	26.2	22.8	38.6	51.0	61.5	69.5	73.1	76.9	67.6	54.4	46.7	36.5	52.1
1979	29.2	22.6	45.1	50.4	62.2	68.5	73.4	72.9	65.5	52.3	46.8	37.6	52.2
1980	30.3	29.1	38.9	52.8	63.3	67.8	76.3	76.1	67.7	51.5	39.4	29.6	51.9
1981	23.7	34.6	38.7	53.7	61.9	71.7	75.7	72.2	63.9	54.7	44.7	31.9	52.0
1982	22.8	30.9	38.6	47.6	62.2	65.1	74.4	70.5	65.3	55.1	47.6	41.4	51.8
1983	33.0	33.4	42.7	49.3	58.4	69.1	75.9	74.9	66.2	53.6	43.9	28.7	52.4
1984	24.8	36.6	33.7	48.0	58.2	72.9	74.3	75.8	64.4	61.5	43.9	41.5	53.0
1985	27.9	34.4	44.5	56.9	65.2	69.4	75.9	74.1	69.2	57.2	47.9	31.0	54.5
1986	31.4	30.0	43.5	53.5	65.6	71.4	76.3	72.0	66.1	56.0	41.2	36.1	53.6
1987	30.0	32.3	44.1	52.3	63.2	72.6	78.2	73.1	65.7	49.4	43.9	36.6	53.5
1988	24.4	31.8	42.4	50.0	62.3	70.8	78.8	76.2	63.9	49.5	43.7	33.3	52.2
1989	34.8	32.4	40.4	50.4	60.0	70.6	73.7	72.6	66.0	55.6	42.7	22.6	51.8
1990	38.2	38.2	44.9	53.1	59.4	71.2	75.2	72.7	65.0	58.2	46.9	38.4	55.1
1991	31.7	37.4	43.7	53.7	69.1	74.2	78.7	76.2	65.4	56.9	43.7	36.5	55.6
1992	33.0	35.7	40.3	51.8	60.2	69.3	75.0	71.6	65.5	51.3	44.3	35.3	52.8
1993	34.8	28.8	36.6	51.3	64.6	72.1	78.0	75.9	65.6	52.8	43.9	34.0	53.2
1994	20.3	26.7	38.1	56.8	59.8	76.2	78.7	71.6	65.9	53.8	49.0	38.5	53.0
1995	34.8	29.1	44.7	50.9	61.9	71.6	77.9	77.4	66.9	59.0	39.6	29.6	53.6
1996	26.2	31.8	37.7	52.5	60.0	73.2	74.1	74.0	67.2	54.9	39.6	37.3	52.4
1997	29.8	38.3	42.5	51.1	59.6	71.4	77.4	74.0	67.1	56.2	41.6	36.0	53.8
1998	38.2	40.4	44.5	54.6	66.8	70.8	75.3	75.1	70.6	55.9	45.9	40.7	56.6
1999	30.7	36.5	41.2	53.9	65.0	73.2	81.9	75.2	68.3	54.0	49.3	37.4	55.6
2000	30.2	34.7	47.5	52.4	65.1	72.2	73.2	73.7	65.3	55.6	43.5	26.5	53.3
2001	28.4	34.8	38.1	51.5	62.4	72.6	73.2	76.8	65.1	55.7	48.8	39.8	53.9
POR= 113 YRS	30.2	31.5	40.4	50.7	62.1	70.8	75.4	73.4	66.4	55.0	43.8	33.3	52.8

REFERENCE NOTES:

PAGE 1:
THE TEMPERATURE GRAPH SHOWS NORMAL MAXIMUM AND NORMAL
MINIMUM DAILY TEMPERATURES (SOLID CURVES) AND THE
ACTUAL DAILY HIGH AND LOW TEMPERATURES (VERTICAL BARS).

PAGE 2 AND 3:
H/C INDICATES HEATING AND COOLING DEGREE DAYS.
RH INDICATES RELATIVE HUMIDITY
W/O INDICATES WEATHER AND OBSTRUCTIONS
S INDICATES SUNSHINE.
PR INDICATES PRESSURE.
CLOUDINESS ON PAGE 3 IS THE SUM OF THE CEILOMETER AND
SATELLITE DATA NOT TO EXCEED EIGHT EIGHTHS(OKTAS).

GENERAL:
T INDICATES TRACE PRECIPITATION, AN AMOUNT GREATER
THAN ZERO BUT LESS THAN THE LOWEST REPORTABLE VALUE.
+ INDICATES THE VALUE ALSO OCCURS ON EARLIER DATES.
BLANK ENTRIES DENOTE MISSING OR UNREPORTED DATA.
NORMALS ARE 30-YEAR AVERAGES (1961 - 1990).
ASOS INDICATES AUTOMATED SURFACE OBSERVING SYSTEM.
PM INDICATES THE LAST DAY OF THE PREVIOUS MONTH.
POR (PERIOD OF RECORD) BEGINS WITH THE JANUARY DATA
MONTH AND IS THE NUMBER OF YEARS USED TO COMPUTE
THE MEAN. INDIVIDUAL MONTHS WITHIN THE POR MAY
BE MISSING.
WHEN THE POR FOR A NORMAL IS LESS THAN 30 YEARS,
THE NORMAL IS PROVISIONAL AND IS BASED ON THE NUMBER
OF YEARS INDICATED.
0.* OR * INDICATES THE VALUE OR MEAN-DAYS-WITH
IS BETWEEN 0.00 AND 0.05.
CLOUDINESS FOR ASOS STATIONS DIFFERS FROM THE NON-ASOS
OBSERVATION TAKEN BY A HUMAN OBSERVER. ASOS STATION
CLOUDINESS IS BASED ON TIME-AVERAGED CEILOMETER DATA
FOR CLOUDS AT OR BELOW 12,000 FEET AND ON SATELLITE
DATA FOR CLOUDS ABOVE 12,000 FEET.
THE NUMBER OF DAYS WITH CLEAR, PARTLY CLOUDY, AND
CLOUDY CONDITIONS FOR ASOS STATIONS IS THE SUM
OF THE CEILOMETER AND SATELLITE DATA FOR THE
SUNRISE TO SUNSET PERIOD.

GENERAL CONTINUED:
CLEAR INDICATES 0 - 2 OKTAS, PARTLY CLOUDY INDICATES
3 - 6 OKTAS, AND CLOUDY INDICATES 7 OR 8 OKTAS.
WHEN AT LEAST ONE OF THE ELEMENTS (CEILOMETER OR
SATELLITE) IS MISSING, THE DAILY CLOUDINESS IS
NOT COMPUTED.
WIND DIRECTION IS RECORDED IN TENS OF DEGREES (2 DIGITS)
CLOCKWISE FROM TRUE NORTH. "00" INDICATES CALM. "36"
INDICATES TRUE NORTH.
RESULTANT WIND IS THE VECTOR AVERAGE OF THE SPEED AND
DIRECTION.
AVERAGE TEMPERATURE IS THE SUM OF THE MEAN DAILY MAXIMUM
AND MINIMUM TEMPERATURE DIVIDED BY 2.
SNOWFALL DATA COMPRISE ALL FORMS OF FROZEN
PRECIPITATION, INCLUDING HAIL.
A HEATING (COOLING) DEGREE DAY IS THE DIFFERENCE BETWEEN
THE AVERAGE DAILY TEMPERATURE AND 65 F.
DRY BULB IS THE TEMPERATURE OF THE AMBIENT AIR.
DEW POINT IS THE TEMPERATURE TO WHICH THE AIR MUST BE
COOLED TO ACHIEVE 100 PERCENT RELATIVE HUMIDITY.
WET BULB IS THE TEMPERATURE THE AIR WOULD HAVE IF THE
MOISTURE CONTENT WAS INCREASED TO 100 PERCENT RELATIVE
HUMIDITY.

ON JULY 1, 1996, THE NATIONAL WEATHER SERVICE BEGAN USING
THE "METAR" OBSERVATION CODE THAT WAS ALREADY EMPLOYED
BY MOST OTHER NATIONS OF THE WORLD. THE MOST NOTICEABLE
DIFFERENCE IN THIS ANNUAL PUBLICATION WILL BE THE CHANGE
IN UNITS FROM TENTHS TO EIGHTS(OKTAS) FOR REPORTING THE
AMOUNT OF SKY COVER.

HEATING DEGREE DAYS (base 65 F) 2001 MIDDLETOWN/HARRISBURG INTL APT PA (MDT)

YEAR	JUL	AUG	SEP	OCT	NOV	DEC	JAN	FEB	MAR	APR	MAY	JUN	TOTAL
1972-73	0	1	25	395	686	865	964	926	607	422	227	7	5125
1973-74	0	1	34	238	534	887	931	910	683	289	133	2	4642
1974-75	0	0	94	414	600	911	977	903	818	520	97	13	5347
1975-76	0	0	87	232	445	951	1150	730	639	354	193	23	4804
1976-77	0	5	75	418	756	1075	1387	966	588	340	106	32	5748
1977-78	0	5	35	377	562	1029	1196	1175	810	414	173	17	5793
1978-79	14	0	48	321	544	876	1104	1182	611	435	123	26	5284
1979-80	12	14	71	393	536	844	1070	1033	799	361	103	48	5284
1980-81	0	0	57	411	761	1091	1277	844	809	339	147	6	5742
1981-82	0	1	94	437	599	1021	1304	948	812	518	128	61	5923
1982-83	7	12	67	318	520	725	985	876	686	468	221	25	4910
1983-84	0	2	103	362	628	1117	1238	817	962	502	240	11	5982
1984-85	0	0	105	131	627	724	1143	849	627	292	87	16	4601
1985-86	0	0	41	237	508	1049	1038	974	664	349	89	9	4958
1986-87	2	17	46	300	705	890	1080	907	643	380	142	2	5114
1987-88	0	8	51	477	627	873	1252	961	693	445	131	41	5559
1988-89	4	5	88	475	633	975	931	912	760	433	196	9	5421
1989-90	1	6	81	292	663	1306	824	744	629	385	175	13	5119
1990-91	5	8	96	248	535	816	1026	769	651	345	69	2	4570
1991-92	0	0	103	279	634	877	986	842	761	392	170	16	5060
1992-93	0	3	86	420	615	914	931	1005	875	405	67	15	5336
1993-94	0	2	99	371	628	954	1382	1067	828	260	195	2	5788
1994-95	0	3	41	340	475	815	928	999	623	420	126	3	4773
1995-96	0	0	64	218	755	1090	1196	960	840	376	208	3	5710
1996-97	0	0	54	307	756	852	1083	741	688	410	175	32	5098
1997-98	1	1	46	318	696	892	824	680	653	305	67	29	4512
1998-99	0	1	24	278	565	748	1059	792	733	329	61	6	4596
1999-00	0	0	36	334	464	850	1075	872	534	375	101	16	4657
2000-01	0	4	99	293	638	1187	1127	836	826	402	112	16	5540
2001-	1	0	82	288	477	774							

COOLING DEGREE DAYS (base 65 F) 2001 MIDDLETOWN/HARRISBURG INTL APT PA (MD

YEAR	JAN	FEB	MAR	APR	MAY	JUN	JUL	AUG	SEP	OCT	NOV	DEC	ANNUAL
1972	0	0	0	0	33	133	357	287	137	3	0	0	950
1973	0	0	0	8	13	266	334	352	148	9	0	0	1130
1974	0	0	0	34	79	176	401	381	88	1	3	0	1163
1975	0	0	0	0	97	192	325	354	44	12	2	0	1026
1976	0	0	0	47	22	283	233	240	73	3	0	0	901
1977	0	0	4	19	95	147	347	297	140	1	3	0	1053
1978	0	0	0	0	69	162	273	377	133	1	0	0	1015
1979	0	0	0	5	43	138	279	264	92	7	0	0	828
1980	0	0	0	0	57	138	355	350	145	0	0	0	1045
1981	0	0	0	6	60	213	339	232	66	0	0	0	916
1982	0	0	0	2	48	70	307	191	83	19	4	0	724
1983	0	0	0	6	22	154	343	315	146	13	0	0	999
1984	0	0	0	0	34	256	292	342	95	35	0	1	1055
1985	0	0	0	55	99	157	345	290	173	4	0	0	1123
1986	0	0	5	12	116	205	360	238	88	27	0	0	1051
1987	0	0	0	9	94	237	418	266	76	0	0	0	1100
1988	0	0	0	0	53	219	439	355	52	4	0	0	1122
1989	0	0	5	0	49	182	279	249	114	9	1	0	888
1990	0	0	14	34	12	205	330	254	102	43	0	0	994
1991	0	0	0	16	202	284	429	355	118	31	0	0	1435
1992	0	0	0	1	29	151	314	214	108	0	0	0	817
1993	0	0	0	0	63	236	410	348	127	0	0	0	1184
1994	0	0	0	20	40	343	428	213	76	0	1	0	1121
1995	0	0	0	6	41	208	409	392	124	36	0	0	1216
1996	0	0	0	9	61	253	286	288	126	1	0	0	1024
1997	0	0	0	0	14	230	392	283	117	50	0	0	1086
1998	0	0	24	1	130	209	326	323	199	0	0	0	1212
1999	0	0	0	0	72	260	530	323	142	1	0	0	1328
2000	0	0	0	5	110	235	262	280	115	8	0	0	1015
2001	0	0	0	5	39	250	260	371	94	8	0	0	1027

SNOWFALL (inches) 2001 MIDDLETOWN/HARRISBURG INTL APT PA (MDT)

YEAR	JUL	AUG	SEP	OCT	NOV	DEC	JAN	FEB	MAR	APR	MAY	JUN	TOTAL
1972-73	0.0	0.0	0.0	1.2	5.9	0.5	T	5.7	T	T	0.0	0.0	13.3
1973-74	0.0	0.0	0.0	0.0	T	15.3	7.0	5.5	T	T	0.0	0.0	27.8
1974-75	0.0	0.0	0.0	0.0	T	0.5	11.3	13.1	6.1	T	0.0	0.0	31.0
1975-76	0.0	0.0	0.0	0.0	T	2.2	3.6	2.5	10.0	0.0	0.0	0.0	18.3
1976-77	0.0	0.0	0.0	0.0	1.4	5.1	12.2	4.5	0.2	T	T	0.0	23.4
1977-78	0.0	0.0	0.0	T	1.5	5.5	33.5	21.1	9.0	0.0	0.0	0.0	70.6
1978-79	0.0	0.0	0.0	0.0	4.0	0.3	9.2	26.0	T	0.0	0.0	0.0	39.5
1979-80	0.0	0.0	0.0	T	T	0.2	3.8	2.7	7.9	0.0	0.0	0.0	14.6
1980-81	0.0	0.0	0.0	0.0	4.0	4.3	5.5	4.4	6.7	0.0	0.0	0.0	24.9
1981-82	0.0	0.0	0.0	0.0	0.8	12.5	18.8	8.4	7.8	10.2	0.0	0.0	58.5
1982-83	0.0	0.0	0.0	T	T	1.1	4.4	28.8	0.4	1.3	0.0	0.0	36.0
1983-84	0.0	0.0	0.0	0.0	T	4.6	9.7	2.3	14.9	T	0.0	0.0	31.5
1984-85	0.0	0.0	0.0	0.0	1.9	2.6	10.6	11.4	T	3.6	0.0	0.0	30.1
1985-86	0.0	0.0	0.0	0.0	T	5.6	7.8	23.1	T	T	0.0	0.0	36.5
1986-87	0.0	0.0	0.0	0.0	T	1.9	31.5	10.1	1.4	1.0	0.0	0.0	45.9
1987-88	0.0	0.0	0.0	0.0	9.7	3.6	9.6	2.8	1.0	T	0.0	0.0	26.7
1988-89	0.0	0.0	0.0	0.0	T	T	6.4	2.2	11.3	0.0	0.0	0.0	19.9
1989-90	0.0	0.0	0.0	0.0	1.8	6.7	4.9	1.3	3.5	1.1	0.0	0.0	19.3
1990-91	0.0	0.0	0.0	0.0	0.0	9.3	5.2	0.7	5.9	T	0.0	0.0	21.1
1991-92	0.0	0.0	0.0	0.0	0.0	T	1.7	4.6	6.6	0.0	0.0	T	12.9
1992-93	0.0	0.0	0.0	0.0	T	3.9	1.4	18.5	22.8	0.6	0.0	T	47.2
1993-94	0.0	T	0.0	0.0	T	0.9	34.2	22.0	18.8	0.0	0.0	0.0	75.9
1994-95	0.0	0.0	0.0	0.0	0.5	0.5	0.6	5.6	1.8	0.0	0.0	0.0	9.0
1995-96	0.0	0.0	0.0	0.0	8.1	17.4	38.9	6.3	2.3	4.6			
1996-97	0.0	0.0	0.0	0.0	0.2	6.1	3.7	6.9	5.9	0.0	0.0		22.8
1997-98	0.0	0.0	0.0	0.0	0.3	4.8	5.3	0.4	0.6	0.0	0.0	0.0	11.4
1998-99	0.0	0.0	0.0	0.0	0.0	0.8	7.4	1.3	10.6	0.6	0.0	0.0	20.1
1999-00	0.0	0.0	0.0	0.0	T	0.1	15.0	7.2	0.0	0.6	0.0	0.0	22.9
2000-01	0.0	0.0	0.0	0.0	0.0	5.6	9.4	9.5	3.5	T	0.0	0.0	28.0
2001-	0.0	0.0	0.0	0.0	0.0	T							
POR= 61 YRS	0.0	T	0.0	0.0	2.0	6.4	9.9	8.9	6.3	0.6	T	T	34.1

2001
PHILADELPHIA,
PENNSYLVANIA (PHL)

The Appalachian Mountains to the west and the Atlantic Ocean to the east have a moderating effect on climate. Periods of very high or very low temperatures seldom last for more than three or four days. Temperatures below zero or above 100 degrees are a rarity. On occasion, the area becomes engulfed with maritime air during the summer months, and high humidity adds to the discomfort of seasonably warm temperatures.

Precipitation is fairly evenly distributed throughout the year with maximum amounts during the late summer months. Much of the summer rainfall is from local thunderstorms and amounts vary in different areas of the city. This is due, in part, to the higher elevations to the west and north. Snowfall amounts are often considerably larger in the northern suburbs than in the central and southern parts of the city. In many cases, the precipitation will change from snow to rain within the city. Single storms of 10 inches or more occur about every five years.

The prevailing wind direction for the summer months is from the southwest, while northwesterly winds prevail during the winter. The annual prevailing direction is from the west-southwest. Destructive velocities are comparatively rare and occur mostly in gustiness during summer thunderstorms. High winds occurring in the winter months, as a rule, come with the advance of cold air after the passage of a deep low pressure system. Only rarely have hurricanes in the vicinity caused widespread damage, primarily because of flooding.

Flood stages in the Schuylkill River normally occur about twice a year. Flood stages seldom last over 12 hours and usually occur after excessive thunderstorms. Flooding rarely occurs on the Delaware River.

NORMALS, MEANS, AND EXTREMES
PHILADELPHIA, PA (PHL)

LATITUDE: 39 52' 06" N LONGITUDE: 75 13' 52" W ELEVATION (FT): GRND: 59 BARO: 62 TIME ZONE: EASTERN (UTC + 5) WBAN: 13739

ELEMENT	POR	JAN	FEB	MAR	APR	MAY	JUN	JUL	AUG	SEP	OCT	NOV	DEC	YEAR
TEMPERATURE F														
NORMAL DAILY MAXIMUM	30	37.9	41.0	51.6	62.6	73.1	81.7	86.1	84.6	77.6	66.3	55.1	43.4	63.4
MEAN DAILY MAXIMUM	54	39.6	42.5	51.2	63.3	73.3	82.2	86.5	84.8	77.5	66.7	55.2	43.9	63.9
HIGHEST DAILY MAXIMUM	60	74	74	87	94	97	100	104	101	100	96	81	73	104
YEAR OF OCCURRENCE		1950	1997	1945	1976	1991	1994	1966	2001	1953	1941	1993	1998	JUL 1966
MEAN OF EXTREME MAXS.	54	59.8	61.9	73.0	82.6	88.6	93.5	95.9	93.9	90.3	81.8	72.5	63.3	79.8
NORMAL DAILY MINIMUM	30	22.8	24.8	33.2	42.1	52.7	61.8	67.2	66.3	58.7	46.4	37.6	28.1	45.1
MEAN DAILY MINIMUM	54	24.5	26.5	33.4	42.9	52.9	62.2	67.8	66.5	58.9	47.3	37.8	28.9	45.8
LOWEST DAILY MINIMUM	60	-7	-4	7	19	28	44	51	44	35	25	15	1	-7
YEAR OF OCCURRENCE		1984	1961	1984	1982	1966	1984	1966	1986	1963	1969	1976	1983	JAN 1984
MEAN OF EXTREME MINS.	54	8.7	10.5	18.7	30.0	40.7	50.2	57.5	55.5	44.2	33.3	23.8	14.0	32.3
NORMAL DRY BULB	30	30.4	33.0	42.4	52.4	62.9	71.8	76.7	75.5	68.2	56.4	46.4	35.8	54.3
MEAN DRY BULB	54	32.1	34.5	42.3	53.0	63.0	72.2	77.1	75.7	68.2	57.0	46.6	36.4	54.8
MEAN WET BULB	18	30.3	32.2	38.2	47.0	56.8	65.4	69.8	68.6	62.3	52.3	42.9	34.1	50.0
MEAN DEW POINT	18	23.3	24.5	30.0	39.7	50.6	60.3	65.4	64.4	57.7	47.0	36.6	27.0	43.9
NORMAL NO. DAYS WITH:														
MAXIMUM 90	30	0.0	0.0	0.0	0.4	0.8	4.0	8.8	6.4	1.9	0.0	0.0	0.0	22.3
MAXIMUM 32	30	9.4	5.7	0.7	0.0	0.0	0.0	0.0	0.0	0.0	0.0	0.1	4.2	20.1
MINIMUM 32	30	26.2	22.3	14.0	2.6	*	0.0	0.0	0.0	0.0	1.5	8.5	21.2	96.3
MINIMUM 0	30	0.5	0.1	0.0	0.0	0.0	0.0	0.0	0.0	0.0	0.0	0.0	0.0	0.6
H/C														
NORMAL HEATING DEG. DAYS	30	1073	896	701	378	123	5	0	0	32	283	558	905	4954
NORMAL COOLING DEG. DAYS	30	0	0	0	0	58	209	363	326	128	17	0	0	1101
RH														
NORMAL (PERCENT)	30	66	64	62	60	65	68	70	70	72	71	68	68	67
HOUR 01 LST	30	70	68	68	69	76	80	81	82	82	81	75	72	75
HOUR 07 LST	30	73	71	72	70	74	77	79	81	83	83	78	74	76
HOUR 13 LST	30	59	55	52	49	53	54	54	55	56	54	56	59	55
HOUR 19 LST	30	65	61	57	54	59	61	63	65	69	69	66	67	63
S PERCENT POSSIBLE SUNSHINE	58	49	53	55	55	56	62	61	62	59	59	52	49	56
W/O MEAN NO. DAYS WITH:														
HEAVY FOG(VISBY 1/4 MI)	62	2.6	2.3	1.7	1.2	1.3	1.0	0.8	1.0	1.5	3.0	2.3	2.5	21.2
THUNDERSTORMS	62	0.3	0.3	1.0	2.0	4.1	5.1	5.5	4.8	2.4	0.8	0.6	0.2	27.1
CLOUDINESS MEAN:														
SUNRISE-SUNSET (OKTAS)	1			6.4			5.6							
MIDNIGHT-MIDNIGHT (OKTAS)	1			5.6										
MEAN NO. DAYS WITH:														
CLEAR	1	3.0	2.0	8.0		8.0	11.0							
PARTLY CLOUDY	1		1.0	4.0		5.0	3.0							
CLOUDY	1	3.0	6.0	8.0		7.0	9.0							
PR														
MEAN STATION PRESSURE(IN)	29	30.05	30.05	30.00	29.95	29.95	29.95	29.97	30.01	30.05	30.07	30.06	30.06	30.01
MEAN SEA-LEVEL PRES. (IN)	18	30.10	30.09	30.04	29.98	29.98	29.97	29.99	30.03	30.06	30.10	30.11	30.10	30.05
WINDS														
MEAN SPEED (MPH)	61	10.4	10.9	11.4	10.9	9.6	8.8	8.2	7.9	8.4	8.9	9.7	10.1	9.6
PREVAIL.DIR(TENS OF DEGS)	38	29	30	29	23	23	23	23	23	23	23	23	29	23
MAXIMUM 2-MINUTE:														
SPEED (MPH)	6	44	43	40	37	40	51	41	41	39	36	40	45	51
DIR. (TENS OF DEGS)		28	29	27	31	31	30	33	25	32	35	28	27	30
YEAR OF OCCURRENCE		1999	1996	1996	2001	1997	1998	1999	1997	1998	1996	1996	2000	JUN 1998
MAXIMUM 5-SECOND:														
SPEED (MPH)	6	57	52	52	51	51	71	46	52	49	46	52	53	71
DIR. (TENS OF DEGS)		18	31	26	28	29	30	32	24	31	35	27	28	30
YEAR OF OCCURRENCE		1996	1996	1996	2000	1997	1998	1999	1997	1996	1996	1996	2000	JUN 1998
PRECIPITATION														
NORMAL (IN)	30	3.21	2.79	3.46	3.62	3.75	3.74	4.28	3.80	3.42	2.62	3.34	3.38	41.41
MAXIMUM MONTHLY (IN)	59	8.86	6.44	7.01	8.12	7.41	7.88	10.42	9.70	13.07	5.99	9.06	8.47	13.07
YEAR OF OCCURRENCE		1978	1979	1980	1983	1948	1973	1994	1955	1999	1995	1972	1996	SEP 1999
MINIMUM MONTHLY (IN)	59	0.45	0.75	0.68	0.52	0.47	0.11	0.64	0.49	0.44	0.09	0.32	0.25	0.09
YEAR OF OCCURRENCE		1955	1991	1966	1985	1964	1949	1957	1964	1968	1963	1976	1955	OCT 1963
MAXIMUM IN 24 HOURS (IN)	55	2.70	1.96	3.08	2.76	3.18	4.62	4.49	5.68	6.77	3.85	3.99	3.03	6.77
YEAR OF OCCURRENCE		1979	1966	2000	1970	1984	1973	1989	1971	1999	1980	1977	1992	SEP 1999
NORMAL NO. DAYS WITH:														
PRECIPITATION 0.01	30	10.4	9.8	10.0	10.6	11.2	10.0	9.1	8.7	7.9	7.3	9.9	10.5	115.4
PRECIPITATION 1.00	30	0.8	0.6	1.0	0.9	0.9	1.0	1.4	1.2	1.0	0.7	0.9	0.8	11.2
SNOWFALL														
NORMAL (IN)	30	7.8	7.5	3.3	0.5	T	0.0	0.0	0.0	0.0	0.1	0.6	3.4	23.2
MAXIMUM MONTHLY (IN)	58	23.4	27.6	13.4	4.3	T	T	0.0	0.0	0.0	2.1	8.8	18.8	27.6
YEAR OF OCCURRENCE		1978	1979	1958	1971	1963	1993				1979	1953	1966	FEB 1979
MAXIMUM IN 24 HOURS (IN)	58	13.2	21.3	12.0	4.3	T	T	0.0	0.0	0.0	2.1	8.7	14.6	21.3
YEAR OF OCCURRENCE		1961	1983	1993	1971	1963	1993				1979	1953	1960	FEB 1983
MAXIMUM SNOW DEPTH (IN)	53	12	22	12	3	0	0	0	0	0	0	8	12	22
YEAR OF OCCURRENCE		1961	1983	1993	1997							1953	1966	FEB 1983
NORMAL NO. DAYS WITH:														
SNOWFALL 1.0	30	2.3	1.8	1.0	0.2	0.0	0.0	0.0	0.0	0.0	0.*	0.2	0.9	6.4

PRECIPITATION (inches) 2001 PHILADELPHIA, PA (PHL)

YEAR	JAN	FEB	MAR	APR	MAY	JUN	JUL	AUG	SEP	OCT	NOV	DEC	ANNUAL
1972	2.34	5.09	2.69	4.08	4.11	5.79	2.62	3.76	1.12	3.77	9.06	5.20	49.63
1973	3.93	2.96	3.52	6.68	4.14	7.88	2.39	2.03	3.39	2.16	0.64	6.34	46.06
1974	2.95	2.14	4.91	2.77	3.21	4.43	2.08	3.83	4.68	1.93	0.81	4.04	37.78
1975	4.00	2.91	4.68	2.97	4.99	7.57	6.32	2.21	7.21	3.24	3.14	2.89	52.13
1976	4.50	1.66	2.38	2.06	4.35	3.42	4.04	2.17	2.44	4.30	0.32	1.63	33.27
1977	2.61	1.33	4.19	5.59	0.70	5.33	1.47	8.70	3.44	3.11	7.76	5.19	49.42
1978	8.86	1.35	4.31	1.76	6.01	1.75	5.27	6.04	1.59	1.20	2.20	5.61	45.95
1979	8.74	6.44	2.43	4.08	3.98	4.34	3.95	5.95	4.89	3.84	2.48	1.67	52.79
1980	2.27	0.96	7.01	4.79	3.22	1.73	6.58	0.80	2.79	5.03	2.85	0.77	38.80
1981	0.50	2.94	1.61	3.60	4.53	4.40	4.54	5.11	2.83	2.68	0.95	4.14	37.83
1982	4.45	3.16	2.66	6.06	4.47	5.76	1.94	2.20	2.32	1.94	3.67	1.80	40.43
1983	2.81	3.53	6.70	8.12	7.03	2.75	0.68	2.57	3.45	3.69	5.71	7.37	54.41
1984	2.22	2.81	6.14	4.25	6.87	2.85	6.99	3.28	1.96	2.56	1.56	2.17	43.66
1985	1.55	2.44	1.95	0.52	4.99	1.88	4.66	2.82	5.78	1.54	6.09	0.98	35.20
1986	4.13	3.38	1.25	4.46	0.70	1.99	4.10	3.70	2.33	2.22	6.27	5.89	40.42
1987	4.58	1.17	1.16	3.63	3.15	2.01	4.82	3.72	2.78	2.62	2.08	1.68	33.40
1988	2.72	4.11	2.24	2.92	3.67	0.57	8.07	3.16	2.62	2.16	5.17	1.00	38.41
1989	2.41	3.25	4.41	2.27	6.76	4.73	9.44	3.92	5.03	3.44	1.79	1.21	48.66
1990	4.09	1.44	2.59	3.16	6.08	3.39	2.62	4.07	1.71	1.68	1.17	3.79	35.79
1991	4.10	0.75	4.13	2.81	1.82	3.36	4.79	3.86	3.58	1.61	1.55	3.86	36.22
1992	0.88	1.31	3.19	1.26	2.74	1.84	5.05	2.00	3.04	1.23	3.26	4.61	30.41
1993	1.97	3.03	6.61	4.20	2.42	1.52	1.98	5.18	6.66	2.69	2.23	3.69	42.18
1994	4.27	3.27	6.44	2.86	3.66	1.74	10.42	4.54	1.64	0.94	3.03	2.11	44.92
1995	3.10	2.41	1.67	1.96	2.67	0.62	2.92	1.15	3.55	5.99	3.34	2.15	31.53
1996	4.39	2.12	4.27	4.48	3.25	4.73	8.17	4.29	4.95	4.30	3.03	8.47	56.45
1997	2.80	2.48	3.91	2.58	2.32	1.49	2.38	4.56	1.59	1.83	3.49	3.09	32.52
1998	4.24	3.25	3.93	2.70	3.87	4.91	1.79	1.26	1.86	1.84	1.18	0.82	31.65
1999	4.89	2.95	4.02	3.31	3.70	1.16	1.22	5.32	13.07	3.55	2.31	2.99	48.49
2000	3.22	2.02	6.32	3.05	3.03	3.82	5.54	2.90	8.28	1.51	2.21	2.82	44.72
2001	2.77	3.04	5.44	1.49	3.99	5.93	1.30	0.97	2.58	0.83	0.56	2.11	31.01
POR= 130 YRS	3.24	3.02	3.60	3.32	3.44	3.54	4.20	4.25	3.33	2.68	3.07	3.12	40.81

AVERAGE TEMPERATURE (F) 2001 PHILADELPHIA, PA (PHL)

YEAR	JAN	FEB	MAR	APR	MAY	JUN	JUL	AUG	SEP	OCT	NOV	DEC	ANNUAL
1972	35.1	32.4	40.7	49.7	63.6	68.7	77.1	76.0	69.2	52.7	43.6	39.9	54.1
1973	34.4	33.6	47.2	53.4	60.3	74.6	77.9	78.8	70.7	59.2	48.0	38.6	56.4
1974	35.9	31.7	43.3	55.8	62.4	70.3	76.9	76.8	68.1	54.8	48.5	39.4	55.3
1975	37.3	35.8	41.2	48.7	66.6	72.2	76.6	77.1	66.6	61.2	52.7	36.9	56.1
1976	28.7	40.9	46.3	56.6	62.7	75.2	75.3	74.8	67.3	52.5	39.9	30.3	54.2
1977	20.0	33.6	48.8	57.2	65.8	68.6	77.8	76.2	69.9	54.3	46.4	32.6	54.3
1978	28.0	24.7	39.0	50.6	61.4	72.6	75.6	79.2	68.5	55.5	47.9	38.6	53.5
1979	32.5	23.0	47.0	52.3	66.4	69.1	76.2	75.5	68.5	54.9	50.1	38.2	54.5
1980	31.8	29.7	40.2	54.7	65.4	70.6	78.5	80.0	72.2	54.9	43.2	34.5	54.5
1981	25.3	37.9	40.0	54.7	62.6	72.0	76.9	74.9	66.8	53.1	45.6	34.6	53.7
1982	24.7	34.4	41.7	50.2	65.9	68.7	76.9	73.5	67.6	56.9	48.4	41.3	54.2
1983	34.1	34.0	43.7	51.0	62.1	72.0	77.9	77.1	69.0	56.6	46.7	33.2	54.8
1984	26.2	38.7	35.5	50.2	63.0	73.0	73.9	75.2	64.7	61.2	44.4	41.9	53.8
1985	27.3	35.3	44.6	55.5	64.5	68.8	75.4	74.1	69.1	59.3	51.3	33.3	54.9
1986	32.8	32.1	44.5	53.3	66.8	73.8	78.1	74.0	68.3	57.8	44.5	37.9	55.3
1987	31.9	32.5	45.7	53.1	63.9	74.6	79.5	75.4	68.8	52.5	48.0	39.2	55.4
1988	27.3	34.6	44.7	51.3	63.6	72.3	80.7	78.3	66.7	51.8	47.7	35.4	54.5
1989	36.5	34.8	42.3	52.4	62.4	74.7	76.3	75.6	69.7	58.3	44.9	25.5	54.5
1990	40.3	41.2	46.1	53.3	61.3	72.2	78.0	75.8	68.0	61.9	49.7	42.1	57.5
1991	35.2	40.0	46.1	55.5	70.8	75.7	79.0	79.0	69.5	58.9	47.3	39.6	58.1
1992	35.7	37.5	41.6	53.2	62.5	71.3	77.1	73.5	68.4	55.2	48.0	38.8	55.2
1993	38.2	31.9	39.8	54.2	66.4	74.4	81.4	78.9	69.7	58.0	49.0	38.4	56.7
1994	27.4	33.2	42.8	59.5	62.5	78.1	82.1	74.9	68.3	57.5	51.7	41.9	56.7
1995	38.2	31.5	47.3	54.5	64.3	74.3	81.5	79.9	70.4	61.4	43.0	31.6	56.5
1996	30.2	33.9	38.7	52.7	60.6	73.0	74.4	74.5	68.7	56.3	41.3	40.2	53.7
1997	32.5	40.0	44.1	51.4	59.5	71.1	77.5	73.9	67.1	57.3	44.4	38.5	54.8
1998	41.0	41.8	45.5	55.3	66.2	71.5	77.5	78.0	71.8	58.3	48.2	42.0	58.1
1999	35.0	38.0	42.4	53.5	64.0	72.9	81.2	77.4	69.9	56.1	50.9	39.9	56.8
2000	32.1	37.5	48.0	52.6	64.2	72.6	74.1	74.1	66.4	57.7	45.4	31.3	54.7
2001	32.5	37.4	41.1	54.9	64.7	75.2	75.4	79.9	68.4	59.1	52.9	43.7	57.1
POR= 128 YRS	32.8	34.0	41.9	52.4	63.1	71.9	76.8	74.9	68.4	57.3	46.4	36.4	54.7

REFERENCE NOTES:

PAGE 1:
THE TEMPERATURE GRAPH SHOWS NORMAL MAXIMUM AND NORMAL
MINIMUM DAILY TEMPERATURES (SOLID CURVES) AND THE
ACTUAL DAILY HIGH AND LOW TEMPERATURES (VERTICAL BARS).

PAGE 2 AND 3:
H/C INDICATES HEATING AND COOLING DEGREE DAYS.
RH INDICATES RELATIVE HUMIDITY
W/O INDICATES WEATHER AND OBSTRUCTIONS
S INDICATES SUNSHINE.
PR INDICATES PRESSURE.
CLOUDINESS ON PAGE 3 IS THE SUM OF THE CEILOMETER AND
SATELLITE DATA NOT TO EXCEED EIGHT EIGHTHS(OKTAS).

GENERAL:
T INDICATES TRACE PRECIPITATION, AN AMOUNT GREATER
THAN ZERO BUT LESS THAN THE LOWEST REPORTABLE VALUE.
+ INDICATES THE VALUE ALSO OCCURS ON EARLIER DATES.
BLANK ENTRIES DENOTE MISSING OR UNREPORTED DATA.
NORMALS ARE 30-YEAR AVERAGES (1961 - 1990).
ASOS INDICATES AUTOMATED SURFACE OBSERVING SYSTEM.
PM INDICATES THE LAST DAY OF THE PREVIOUS MONTH.
POR (PERIOD OF RECORD) BEGINS WITH THE JANUARY DATA
MONTH AND IS THE NUMBER OF YEARS USED TO COMPUTE
THE MEAN. INDIVIDUAL MONTHS WITHIN THE POR MAY
BE MISSING.
WHEN THE POR FOR A NORMAL IS LESS THAN 30 YEARS,
THE NORMAL IS PROVISIONAL AND IS BASED ON THE NUMBER
OF YEARS INDICATED.
0.* OR * INDICATES THE VALUE OR MEAN-DAYS-WITH
IS BETWEEN 0.00 AND 0.05.
CLOUDINESS FOR ASOS STATIONS DIFFERS FROM THE NON-ASOS
OBSERVATION TAKEN BY A HUMAN OBSERVER. ASOS STATION
CLOUDINESS IS BASED ON TIME-AVERAGED CEILOMETER DATA
FOR CLOUDS AT OR BELOW 12,000 FEET AND ON SATELLITE
DATA FOR CLOUDS ABOVE 12,000 FEET.
THE NUMBER OF DAYS WITH CLEAR, PARTLY CLOUDY, AND
CLOUDY CONDITIONS FOR ASOS STATIONS IS THE SUM
OF THE CEILOMETER AND SATELLITE DATA FOR THE
SUNRISE TO SUNSET PERIOD.

GENERAL CONTINUED:
CLEAR INDICATES 0 - 2 OKTAS, PARTLY CLOUDY INDICATES
3 - 6 OKTAS, AND CLOUDY INDICATES 7 OR 8 OKTAS.
WHEN AT LEAST ONE OF THE ELEMENTS (CEILOMETER OR
SATELLITE) IS MISSING, THE DAILY CLOUDINESS IS
NOT COMPUTED.
WIND DIRECTION IS RECORDED IN TENS OF DEGREES (2 DIGITS)
CLOCKWISE FROM TRUE NORTH. "00" INDICATES CALM. "36"
INDICATES TRUE NORTH.
RESULTANT WIND IS THE VECTOR AVERAGE OF THE SPEED AND
DIRECTION.
AVERAGE TEMPERATURE IS THE SUM OF THE MEAN DAILY MAXIMUM
AND MINIMUM TEMPERATURE DIVIDED BY 2.
SNOWFALL DATA COMPRISE ALL FORMS OF FROZEN
PRECIPITATION, INCLUDING HAIL.
A HEATING (COOLING) DEGREE DAY IS THE DIFFERENCE BETWEEN
THE AVERAGE DAILY TEMPERATURE AND 65 F.
DRY BULB IS THE TEMPERATURE OF THE AMBIENT AIR.
DEW POINT IS THE TEMPERATURE TO WHICH THE AIR MUST BE
COOLED TO ACHIEVE 100 PERCENT RELATIVE HUMIDITY.
WET BULB IS THE TEMPERATURE THE AIR WOULD HAVE IF THE
MOISTURE CONTENT WAS INCREASED TO 100 PERCENT RELATIVE
HUMIDITY.

ON JULY 1, 1996, THE NATIONAL WEATHER SERVICE BEGAN USING
THE "METAR" OBSERVATION CODE THAT WAS ALREADY EMPLOYED
BY MOST OTHER NATIONS OF THE WORLD. THE MOST NOTICEABLE
DIFFERENCE IN THIS ANNUAL PUBLICATION WILL BE THE CHANGE
IN UNITS FROM TENTHS TO EIGHTS(OKTAS) FOR REPORTING THE
AMOUNT OF SKY COVER.

HEATING DEGREE DAYS (base 65 F) 2001 PHILADELPHIA, PA (PHL)

YEAR	JUL	AUG	SEP	OCT	NOV	DEC	JAN	FEB	MAR	APR	MAY	JUN	TOTAL
1972-73	0	0	22	378	635	775	940	874	547	359	176	1	4707
1973-74	0	0	18	196	507	810	897	926	667	292	128	11	4452
1974-75	0	0	46	313	500	786	852	812	732	483	66	4	4594
1975-76	0	0	45	152	372	866	1120	692	572	307	119	13	4258
1976-77	0	2	42	387	743	1069	1390	873	505	258	73	36	5378
1977-78	0	0	24	328	558	998	1139	1121	797	423	161	10	5559
1978-79	5	0	41	296	507	811	999	1170	556	378	38	17	4818
1979-80	4	7	28	324	439	823	1021	1016	763	301	72	17	4815
1980-81	0	0	22	320	646	999	1222	752	768	309	129	4	5171
1981-82	0	0	58	364	576	936	1243	850	714	440	50	25	5256
1982-83	0	8	31	277	497	730	951	861	653	423	128	2	4561
1983-84	0	0	70	283	540	981	1196	756	911	438	181	13	5369
1984-85	0	0	92	138	613	709	1161	824	627	306	89	9	4568
1985-86	0	0	38	187	407	975	990	914	628	345	77	6	4567
1986-87	0	21	23	255	609	838	1017	904	591	359	129	1	4747
1987-88	0	0	20	379	504	796	1162	876	624	404	105	32	4902
1988-89	0	0	35	408	513	908	876	840	700	371	138	0	4789
1989-90	0	0	43	220	594	1219	757	662	588	375	127	6	4591
1990-91	2	1	55	171	453	701	920	694	576	296	44	0	3913
1991-92	0	0	42	215	527	778	903	789	720	356	124	9	4463
1992-93	0	1	49	310	504	804	824	920	773	314	47	6	4552
1993-94	0	0	34	219	479	820	1156	881	681	193	132	0	4595
1994-95	0	0	20	231	392	710	824	932	542	318	82	0	4051
1995-96	0	0	21	159	656	1029	1072	894	809	370	191	7	5208
1996-97	0	0	34	265	704	761	999	695	643	401	172	42	4716
1997-98	2	0	45	278	612	811	738	644	619	287	67	13	4116
1998-99	0	0	14	207	498	706	920	751	691	336	77	3	4203
1999-00	0	0	22	275	415	772	1012	792	519	371	107	15	4300
2000-01	0	0	73	230	584	1039	999	768	734	320	74	4	4825
2001-	0	0	48	211	358	652							

COOLING DEGREE DAYS (base 65 F) 2001 PHILADELPHIA, PA (PHL)

YEAR	JAN	FEB	MAR	APR	MAY	JUN	JUL	AUG	SEP	OCT	NOV	DEC	ANNUAL
1972	0	0	3	0	47	143	381	344	153	3	0	0	1074
1973	0	0	0	16	35	294	404	435	193	23	0	0	1400
1974	0	0	0	24	55	179	373	372	145	5	12	0	1165
1975	0	0	0	0	121	224	366	380	98	42	12	0	1243
1976	0	0	0	64	58	326	326	315	115	7	0	0	1211
1977	0	0	10	32	104	150	402	355	175	3	6	0	1237
1978	0	0	0	0	57	244	338	447	153	8	0	0	1247
1979	0	0	6	5	90	146	357	339	137	16	1	0	1097
1980	0	0	0	0	89	194	428	470	244	10	0	0	1435
1981	0	0	0	9	62	224	373	315	119	1	0	0	1103
1982	0	0	0	3	85	142	376	280	115	31	5	0	1037
1983	0	0	0	11	43	217	409	380	199	27	0	0	1286
1984	0	0	0	0	39	260	283	324	90	30	0	0	1026
1985	0	0	0	27	81	133	330	291	166	19	0	0	1047
1986	0	0	0	0	139	278	413	307	129	40	0	0	1306
1987	0	0	0	7	101	295	456	332	142	0	0	0	1333
1988	0	0	0	0	1	70	259	495	418	93	7	0	1343
1989	0	0	1	1	62	298	357	332	192	18	1	0	1262
1990	0	0	0	9	29	226	413	341	152	83	1	0	1274
1991	0	0	1	21	230	327	443	437	183	30	0	0	1672
1992	0	0	0	10	54	205	380	268	158	13	0	0	1088
1993	0	0	0	0	95	298	517	438	181	11	6	0	1546
1994	0	0	0	34	63	401	540	313	125	4	0	0	1480
1995	0	0	0	8	66	284	519	468	189	53	1	0	1588
1996	0	0	0	9	61	257	301	305	151	2	1	0	1087
1997	0	0	0	0	9	230	397	280	113	49	0	0	1078
1998	0	0	20	3	114	216	397	410	223	3	0	0	1386
1999	0	0	0	0	51	246	508	394	177	4	0	0	1380
2000	0	0	0	3	89	251	286	286	123	12	0	0	1050
2001	0	0	0	22	72	316	330	468	158	37	1	0	1404

SNOWFALL (inches) 2001 PHILADELPHIA, PA (PHL)

YEAR	JUL	AUG	SEP	OCT	NOV	DEC	JAN	FEB	MAR	APR	MAY	JUN	TOTAL
1972-73	0.0	0.0	0.0	T	T	T	T	T	T	T	0.0	0.0	T
1973-74	0.0	0.0	0.0	0.0	0.0	4.6	4.1	12.1	T	T	0.0	0.0	20.8
1974-75	0.0	0.0	0.0	0.0	T	0.8	3.9	6.6	2.3	T	0.0	0.0	13.6
1975-76	0.0	0.0	0.0	0.0	0.0	1.1	6.4	3.1	6.9	0.0	0.0	0.0	17.5
1976-77	0.0	0.0	0.0	0.0	T	2.8	15.7	0.2	T	0.0	0.0	0.0	18.7
1977-78	0.0	0.0	0.0	0.0	0.2	0.2	23.4	19.0	12.1	T	0.0	0.0	54.9
1978-79	0.0	0.0	0.0	0.0	2.5	T	10.1	27.6	T	T	0.0	0.0	40.2
1979-80	0.0	0.0	0.0	2.1	T	4.9	6.1	0.4	7.4	T	0.0	0.0	20.9
1980-81	0.0	0.0	0.0	0.0	0.2	1.4	5.0	T	8.8	0.0	0.0	0.0	15.4
1981-82	0.0	0.0	0.0	0.0	T	2.8	14.0	3.5	1.1	4.0	0.0	0.0	25.4
1982-83	0.0	0.0	0.0	0.0	0.0	6.8	0.2	26.1	0.9	1.9	0.0	0.0	35.9
1983-84	0.0	0.0	0.0	0.0	0.8	T	10.5	T	10.3	T	0.0	0.0	21.6
1984-85	0.0	0.0	0.0	0.0	T	0.2	11.9	4.4	T	T	0.0	0.0	16.5
1985-86	0.0	0.0	0.0	0.0	0.0	1.5	3.4	11.5	T	T	0.0	0.0	16.4
1986-87	0.0	0.0	0.0	0.0	T	0.4	15.2	10.1	T	T	0.0	0.0	25.7
1987-88	0.0	0.0	0.0	0.0	1.4	1.5	10.6	1.5	T	0.0	0.0	0.0	15.0
1988-89	0.0	0.0	0.0	0.0	0.0	0.4	6.0	2.4	2.4	0.0	0.0	0.0	11.2
1989-90	0.0	0.0	0.0	0.0	4.6	5.3	1.4	0.9	2.4	2.4	0.0	0.0	17.0
1990-91	0.0	0.0	0.0	0.0	0.0	6.4	6.5	1.0	0.7	0.0	0.0	0.0	14.6
1991-92	0.0	0.0	0.0	0.0	T	T	1.2	1.0	2.5	T	0.0	0.0	4.7
1992-93	0.0	0.0	0.0	0.0	T	T	1.0	10.9	12.4	T	0.0	T	24.3
1993-94	0.0	0.0	0.0	0.0	T	0.9	4.1	13.2	4.9	0.0	0.0	0.0	23.1
1994-95	0.0	0.0	0.0	0.0	T	0.0	T	9.8	T	0.0	0.0	0.0	9.8
1995-96	0.0	0.0	0.0	0.0	1.9	7.3	33.8	12.9	7.2	2.4	0.0	0.0	65.5
1996-97	0.0	0.0	0.0	0.0	T	T	1.7	4.1	5.5	1.6	0.0	0.0	12.9
1997-98	0.0	0.0	0.0	0.0	0.0	0.2	0.5	T	0.1	0.0	0.0	T	0.8
1998-99	0.0	0.0	0.0	0.0	0.0	2.0	4.9	0.7	4.9	0.0	0.0	0.0	12.5
1999-00	0.0	0.0	0.0	0.0	0.0	T	13.7	5.7	0.0	1.6	T	0.0	21.0
2000-01	0.0	0.0	0.0	0.0	0.0	10.5	3.8	10.0	1.8	T	0.0	0.0	26.1
2001-	0.0	0.0	0.0	0.0	0.0	0.0							
POR= 58 YRS	0.0	0.0	0.0	0.0	0.7	3.1	6.0	6.4	3.4	0.3	T	T	19.9

2001
PITTSBURGH, GRTR. PITT. AIRPORT
PENNSYLVANIA (PIT)

Pittsburgh lies at the foothills of the Allegheny Mountains at the confluence of the Allegheny and Monongahela Rivers which form the Ohio. The city is a little over 100 miles southeast of Lake Erie. It has a humid continental type of climate modified only slightly by its nearness to the Atlantic Seaboard and the Great Lakes.

The predominant winter air masses influencing the climate of Pittsburgh have a polar continental source in Canada and move in from the Hudson Bay region or the Canadian Rockies. During the summer, frequent invasions of air from the Gulf of Mexico bring warm humid weather. Occasionally, Gulf air reaches as far north as Pittsburgh during the winter and produces intermittent periods of thawing. The last spring temperature of 32 degrees usually occurs in late April and the first in late October. The average growing season is about 180 days. There is a wide variation in the time of the first and last frosts over a radius of 25 miles from the center of Pittsburgh due to terrain differences.

Precipitation is distributed well throughout the year. During the winter months about a fourth of the precipitation occurs as snow and there is about a 50 percent chance of measurable precipitation on any day. Thunderstorms occur normally during all months, except midwinter, and have a maximum frequency in midsummer. The first appreciable snowfall generally occurs in late November and usually the last occurs early in April. Snow lies on the ground in the suburbs on an average of about 33 days during the year.

Seven months of the year, April through October, have sunshine more than 50 percent of the possible time. During the remaining five months cloudiness is heavier because the track of migratory storms from west to east is closer to the area and because of the frequent periods of cloudy, showery weather associated with northwest winds from across the Great Lakes. Cold air drainage induced by the many hills leads to the frequent formation of early morning fog which may be quite persistent in the river valleys during the colder months.

The Allegheny River flowing south and the Monongahela River flowing north meet to form the Ohio River at Pittsburgh. Heavier rainfall and steeper topography cause the Monongahela River to flood more frequently than the Allegheny River.

Both rivers combine to cause the Ohio River at Pittsburgh to reach the 25 foot flood stage approximately once every four years. The serious flood level of 30 feet is reached much less frequently.

NORMALS, MEANS, AND EXTREMES
PITTSBURGH, PA (PIT)

LATITUDE: 40 30' 05" N　　LONGITUDE: 80 13' 52" W　　ELEVATION (FT): GRND: 1172　BARO: 1175　　TIME ZONE: EASTERN (UTC + 5)　　WBAN: 94823

	ELEMENT	POR	JAN	FEB	MAR	APR	MAY	JUN	JUL	AUG	SEP	OCT	NOV	DEC	YEAR
TEMPERATURE °F	NORMAL DAILY MAXIMUM	30	33.7	36.9	49.0	60.3	70.6	78.9	82.6	80.8	74.3	62.5	50.4	38.6	59.9
	MEAN DAILY MAXIMUM	54	35.5	38.7	48.2	60.8	70.9	79.3	82.8	81.2	74.4	63.1	50.5	39.2	60.4
	HIGHEST DAILY MAXIMUM	49	69	76	82	89	91	98	103	100	97	87	82	74	103
	YEAR OF OCCURRENCE		1999	2000	1998	1990	1987	1988	1988	1988	1954	1959	1961	1982	JUL 1988
	MEAN OF EXTREME MAXS.	54	58.3	61.2	73.4	80.8	85.6	90.3	91.9	90.3	87.7	79.2	71.2	60.9	77.6
	NORMAL DAILY MINIMUM	30	18.5	20.3	29.8	38.8	48.4	56.9	61.6	60.2	53.5	42.3	34.1	24.4	40.7
	MEAN DAILY MINIMUM	54	20.3	22.3	29.6	39.6	49.1	57.8	62.2	60.9	53.7	42.9	34.2	25.0	41.5
	LOWEST DAILY MINIMUM	49	-22	-12	-1	14	26	34	42	39	31	16	-1	-12	-22
	YEAR OF OCCURRENCE		1994	1979	1980	1982	1970	1972	1963	1982	1959	1965	1958	1989	JAN 1994
	MEAN OF EXTREME MINS.	54	-.6	1.7	11.2	23.0	34.3	43.7	50.2	48.6	38.0	28.1	17.3	6.1	25.1
	NORMAL DRY BULB	30	26.1	28.7	39.4	49.6	59.5	67.9	72.1	70.5	63.9	52.4	42.3	31.5	50.3
	MEAN DRY BULB	54	27.8	30.6	38.9	50.2	60.0	68.6	72.5	71.0	64.2	53.0	42.2	32.0	50.9
	MEAN WET BULB	18	26.5	29.1	34.9	44.4	53.7	62.1	65.8	64.6	58.2	47.6	38.6	30.2	46.3
	MEAN DEW POINT	18	21.0	22.7	27.6	36.6	47.7	57.3	61.5	60.4	54.0	42.2	32.9	24.8	40.7
	NORMAL NO. DAYS WITH:														
	MAXIMUM　90	30	0.0	0.0	0.0	0.0	0.3	1.6	3.4	1.7	0.6	0.0	0.0	0.0	7.6
	MAXIMUM　32	30	14.5	10.3	3.1	0.1	0.0	0.0	0.0	0.0	0.0	0.0	1.3	9.9	39.2
	MINIMUM　32	30	27.3	23.9	19.5	8.4	0.9	0.0	0.0	0.0	0.0	4.3	14.1	24.5	122.9
	MINIMUM　0	30	2.6	1.5	0.1	0.0	0.0	0.0	0.0	0.0	0.0	0.0	*	0.8	5.0
H/C	NORMAL HEATING DEG. DAYS	30	1206	1016	794	462	214	36	6	14	100	400	681	1039	5968
	NORMAL COOLING DEG. DAYS	30	0	0	0	0	44	123	227	184	67	9	0	0	654
RH	NORMAL (PERCENT)	30	70	67	64	60	63	66	69	71	72	68	70	72	68
	HOUR 01 LST	30	72	71	69	66	72	77	80	82	82	77	74	75	75
	HOUR 07 LST	30	75	74	74	72	76	79	82	86	86	81	78	77	78
	HOUR 13 LST	30	65	62	57	50	52	52	54	56	57	55	62	67	57
	HOUR 19 LST	30	67	63	58	52	55	57	60	63	66	63	67	70	62
S	PERCENT POSSIBLE SUNSHINE	49	32	36	43	46	50	55	57	55	55	51	36	29	45
W/O	MEAN NO. DAYS WITH:														
	HEAVY FOG(VISBY 1/4 MI)	50	1.4	1.2	1.0	0.8	1.3	1.1	1.6	2.1	2.3	1.8	1.4	1.8	17.8
	THUNDERSTORMS	50	0.2	0.5	1.7	3.4	5.1	6.9	6.6	5.4	3.0	1.2	0.6	0.3	34.9
CLOUDINESS	MEAN:														
	SUNRISE-SUNSET (OKTAS)	44	6.5	6.2	6.0	5.8	5.5	5.2	4.9	4.9	4.8	4.9	6.0	6.4	5.6
	MIDNIGHT-MIDNIGHT (OKTAS)	32	6.3	5.9	5.7	5.4	5.1	4.9	4.6	4.4	4.5	4.6	5.8	6.2	5.3
	MEAN NO. DAYS WITH:														
	CLEAR	44	2.9	3.3	4.5	4.4	5.3	4.8	5.4	6.6	7.3	7.8	3.7	2.7	58.7
	PARTLY CLOUDY	44	6.0	5.9	6.8	8.3	9.0	11.8	12.8	11.6	10.1	8.8	6.3	5.7	103.1
	CLOUDY	44	22.2	19.1	19.8	17.3	16.7	13.4	12.7	12.9	12.5	14.4	20.0	22.6	203.6
PR	MEAN STATION PRESSURE(IN)	28	28.74	28.75	28.70	28.68	28.69	28.71	28.74	28.75	28.80	28.79	28.77	28.76	28.74
	MEAN SEA-LEVEL PRES. (IN)	18	30.09	30.09	30.05	29.97	29.99	29.99	30.02	30.06	30.08	30.12	30.11	30.12	30.06
WINDS	MEAN SPEED (MPH)	44	10.5	10.3	10.8	10.5	8.8	8.0	7.4	7.0	7.6	8.3	9.6	10.0	9.1
	PREVAIL.DIR(TENS OF DEGS)	33	26	27	27	27	27	23	23	24	23	26	26	26	26
	MAXIMUM 2-MINUTE:														
	SPEED (MPH)	5	41	38	38	51	40	53	45	46	38	39	36	37	53
	DIR. (TENS OF DEGS)		27	20	30	30	31	34	29	33	20	31	32	26	34
	YEAR OF OCCURRENCE		1999	2001	1999	1998	2000	1998	2001	2001	1999	1999	2001	2000	JUN 1998
	MAXIMUM 5-SECOND:														
	SPEED (MPH)	5	48	55	45	61	59	62	61	60	47	47	47	52	62
	DIR. (TENS OF DEGS)		27	26	28	31	22	34	33	34	21	31	31	23	34
	YEAR OF OCCURRENCE		1999	2001	1997	1998	1997	1998	1997	2001	1999	1999	2001	2000	JUN 1998
PRECIPITATION	NORMAL (IN)	30	2.54	2.39	3.41	3.15	3.59	3.71	3.75	3.21	2.97	2.36	2.85	2.92	36.85
	MAXIMUM MONTHLY (IN)	49	6.25	5.98	6.10	7.61	6.56	10.29	8.71	7.86	6.00	8.20	11.05	8.51	11.05
	YEAR OF OCCURRENCE		1978	1956	1967	1964	1989	1989	1992	1987	1990	1954	1985	1990	NOV 1985
	MINIMUM MONTHLY (IN)	49	0.77	0.51	1.14	0.48	1.21	0.64	1.62	0.78	0.28	0.16	0.90	0.40	0.16
	YEAR OF OCCURRENCE		1981	1969	1969	1971	1965	1992	1997	1985	1963	1963	1976	1955	OCT 1963
	MAXIMUM IN 24 HOURS (IN)	49	1.69	2.30	2.00	2.15	2.90	3.11	4.41	3.06	2.59	3.56	1.97	2.76	4.41
	YEAR OF OCCURRENCE		1986	1975	1964	1964	1997	1996	1999	1956	1990	1954	1985	1990	JUL 1999
	NORMAL NO. DAYS WITH:														
	PRECIPITATION　0.01	30	16.4	14.0	15.4	13.6	12.9	11.8	10.4	9.5	10.0	10.5	13.3	16.7	154.5
	PRECIPITATION　1.00	30	0.3	0.3	0.3	0.4	0.7	0.8	0.9	0.9	0.5	0.2	0.3	0.3	5.9
SNOWFALL	NORMAL (IN)	30	12.6	9.9	7.7	1.7	0.2	0.0	0.0	0.0	0.0	0.2	3.2	8.1	43.6
	MAXIMUM MONTHLY (IN)	48	40.2	24.2	34.1	8.1	3.1	T	T	T	T	8.5	13.9	21.2	40.2
	YEAR OF OCCURRENCE		1978	1972	1993	1987	1966	1990	1991	1994	1989	1993	1995	1974	JAN 1978
	MAXIMUM IN 24 HOURS (IN)	48	14.0	12.3	23.8	7.7	3.1	T	T	T	T	6.6	10.5	12.5	23.8
	YEAR OF OCCURRENCE		1966	1960	1993	1987	1966	1990	1991	1994	1989	1993	1958	1974	MAR 1993
	MAXIMUM SNOW DEPTH (IN)	53	26	16	25	7	1	0	0	0	0	1	22	13	26
	YEAR OF OCCURRENCE		1978	1961	1993	1987	1963					1992	1950	1974	JAN 1978
	NORMAL NO. DAYS WITH:														
	SNOWFALL　1.0	30	3.9	3.1	2.3	0.5	0.1	0.0	0.0	0.0	0.0	0.1	0.9	2.5	13.4

PRECIPITATION (inches) 2001 PITTSBURGH, GRTR. PITT. AIRPORT PA (PIT)

YEAR	JAN	FEB	MAR	APR	MAY	JUN	JUL	AUG	SEP	OCT	NOV	DEC	ANNUAL
1972	1.84	3.64	3.68	4.37	1.38	5.08	2.98	1.79	5.42	2.15	4.70	3.04	40.07
1973	2.03	1.80	3.86	4.69	5.87	3.12	2.16	3.40	3.56	4.45	2.65	2.15	39.74
1974	3.47	2.10	3.72	3.26	5.35	5.08	3.30	2.93	4.42	1.12	3.06	4.02	41.83
1975	3.34	4.64	4.62	2.27	1.84	4.58	4.38	7.56	5.06	3.46	1.77	2.90	46.42
1976	3.25	1.74	4.45	1.24	1.99	3.37	4.72	1.25	3.30	3.76	0.90	1.81	31.78
1977	2.06	0.87	4.12	3.26	2.57	2.85	3.38	2.66	3.13	2.44	2.59	3.27	33.20
1978	6.25	0.54	1.65	2.25	4.26	4.11	2.15	3.65	2.64	3.42	1.62	5.24	37.78
1979	4.80	3.12	1.32	3.17	4.49	1.73	4.31	6.84	3.60	2.46	2.43	2.29	40.56
1980	1.56	1.32	5.65	2.94	4.32	4.34	6.76	5.10	1.29	2.42	2.38	1.38	39.46
1981	0.77	4.20	2.12	4.92	2.04	8.20	3.82	0.98	4.13	1.82	1.50	3.00	37.50
1982	4.44	1.93	3.52	1.44	3.98	3.05	2.36	1.97	2.80	0.40	3.33	2.79	32.01
1983	1.19	1.58	3.50	4.33	5.24	4.82	3.32	3.13	2.42	3.67	3.94	4.27	41.41
1984	1.40	2.05	2.32	3.72	5.22	1.98	3.01	5.15	0.84	3.45	3.14	3.04	35.32
1985	1.43	1.45	3.37	1.64	5.80	2.26	4.06	2.64	0.28	2.27	11.05	2.26	38.51
1986	2.49	3.43	1.38	1.94	1.67	5.24	5.66	3.04	2.33	2.83	3.92	3.47	37.40
1987	2.23	0.71	2.65	5.30	2.41	6.30	2.42	7.86	3.97	0.92	2.02	2.41	39.20
1988	1.49	3.46	2.56	1.97	2.78	1.26	2.82	2.04	2.34	1.40	2.80	2.17	27.09
1989	1.99	3.42	5.52	1.43	6.56	10.29	1.62	1.12	4.57	2.04	1.56	2.39	42.51
1990	3.30	3.31	1.47	3.48	6.19	4.24	6.59	3.59	6.00	3.51	2.05	8.51	52.24
1991	2.55	1.88	2.92	2.56	3.29	3.82	3.74	1.63	3.45	0.55	1.97	3.66	32.02
1992	2.13	1.73	3.54	2.30	2.31	0.64	8.71	4.77	2.91	1.47	3.31	2.83	36.65
1993	2.99	2.92	4.14	3.66	2.85	3.35	2.85	2.44	3.87	2.77	4.30	2.12	38.26
1994	3.90	2.13	5.00	3.72	2.54	2.91	3.27	7.75	3.59	0.88	3.64	2.01	41.34
1995	2.23	1.73	1.56	1.70	3.72	3.74	3.06	1.75	1.80	3.24	2.74	1.62	28.89
1996	3.68	2.54	4.54	4.42	2.95	7.95	4.01	1.99	5.63	2.96	2.83	1.98	45.48
1997	1.58	1.15	3.22	1.57	6.33	3.95	1.82	3.80	2.90	0.95	5.98	1.29	34.54
1998	3.63	2.57	1.91	5.00	2.39	6.71	2.02	3.32	1.09	2.27	1.50	1.81	34.22
1999	4.88	2.43	1.24	4.19	4.12	1.67	6.25	2.21	1.97	1.55	3.46	2.24	36.21
2000	1.70	2.53	2.26	3.15	5.69	5.64	6.28	3.66	3.07	2.09	1.38	2.64	40.09
2001	1.35	1.09	3.28	3.75	2.11	3.43	3.15	7.12	2.23	2.33	3.47	2.43	35.74
POR= 130 YRS	2.86	2.45	3.26	3.12	3.41	3.79	3.91	3.23	2.62	2.38	2.44	2.67	36.14

AVERAGE TEMPERATURE (F) 2001   PITTSBURGH, GRTR. PITT. AIRPORT PA (PIT)

YEAR	JAN	FEB	MAR	APR	MAY	JUN	JUL	AUG	SEP	OCT	NOV	DEC	ANNUAL
1972	29.6	26.5	36.4	48.5	61.8	63.8	71.2	70.6	65.3	48.4	39.3	37.2	49.9
1973	29.7	28.8	48.3	49.3	56.4	70.9	73.2	73.2	66.5	56.1	44.1	33.3	52.5
1974	34.0	29.9	41.2	51.8	58.3	65.2	73.1	72.8	62.2	52.4	43.9	32.5	51.4
1975	32.6	32.1	36.3	44.3	63.0	67.8	72.8	73.0	58.8	53.3	46.3	32.9	51.1
1976	23.5	37.2	45.2	50.6	55.6	68.4	67.4	65.3	59.9	45.9	33.1	23.9	48.0
1977	11.4	26.9	43.7	50.8	63.0	63.8	71.8	68.1	50.5	45.6	31.1	26.9	49.3
1978	22.6	20.9	36.9	51.0	60.2	69.4	73.0	71.4	66.2	49.1	43.0	32.7	49.7
1979	21.4	18.0	43.1	49.7	59.1	67.7	70.3	69.6	63.4	50.9	44.7	34.6	49.4
1980	26.9	24.2	35.6	48.1	60.3	66.2	75.0	74.5	67.1	49.5	38.6	28.6	49.6
1981	20.5	31.4	35.6	51.9	58.4	68.8	72.1	69.7	61.9	49.4	40.3	29.4	49.1
1982	20.9	28.4	38.4	45.3	64.7	63.7	72.4	68.2	63.4	54.4	44.7	39.9	50.4
1983	30.0	32.6	40.7	47.1	55.8	67.8	73.0	72.8	64.4	53.0	43.5	25.4	50.5
1984	23.2	36.4	32.2	49.2	55.3	69.7	68.5	70.8	61.4	58.3	40.2	39.3	50.4
1985	22.1	27.7	42.1	55.0	60.6	64.2	70.5	69.6	65.3	55.2	47.1	27.4	50.6
1986	28.3	31.3	41.1	53.1	62.0	68.3	73.3	68.6	66.6	54.2	40.4	33.1	51.7
1987	28.0	32.6	41.9	50.0	63.0	70.9	75.7	71.8	65.1	47.8	46.2	35.1	52.3
1988	26.6	29.0	39.3	49.4	61.4	68.5	76.9	75.1	63.5	46.6	44.2	31.9	51.0
1989	35.5	27.8	41.1	47.0	58.0	69.0	74.1	71.6	64.8	53.3	40.6	19.2	50.2
1990	36.8	36.9	44.0	51.3	57.7	68.3	71.7	70.5	63.7	55.0	45.5	38.0	53.3
1991	29.7	35.4	43.0	54.5	68.7	72.6	75.4	74.4	64.6	55.7	41.6	35.3	54.2
1992	30.5	34.3	38.8	51.3	59.1	66.0	72.4	67.9	63.7	50.1	42.9	33.9	50.9
1993	35.1	27.8	37.6	50.0	61.9	68.9	75.7	75.4	63.3	51.7	43.0	31.7	51.8
1994	21.1	29.5	37.8	53.9	56.7	72.9	74.5	70.0	63.8	53.4	47.7	38.2	51.8
1995	31.0	26.5	42.8	48.6	60.4	72.0	75.8	77.8	64.3	56.3	38.1	27.7	51.8
1996	27.5	30.1	34.7	50.7	60.0	72.3	69.6	70.6	63.7	52.0	36.3	36.5	50.3
1997	28.1	35.1	40.5	46.4	54.3	68.1	71.7	68.2	62.1	52.6	39.0	33.5	50.0
1998	37.1	38.5	42.4	51.0	64.4	67.1	71.2	73.1	67.2	53.1	44.4	37.8	53.9
1999	30.1	34.0	35.1	51.8	61.0	69.4	76.0	69.0	64.2	52.3	46.7	34.6	52.0
2000	27.7	36.0	44.9	50.2	62.5	69.7	69.8	69.1	62.3	54.7	39.5	23.1	50.7
2001	28.4	35.1	35.3	54.3	60.0	68.4	70.2	73.1	62.1	54.2	48.2	37.5	52.2
POR= 126 YRS	30.0	31.2	39.9	50.9	61.6	70.2	74.2	72.5	66.2	54.7	43.0	33.2	52.3

REFERENCE NOTES:

PAGE 1:
THE TEMPERATURE GRAPH SHOWS NORMAL MAXIMUM AND NORMAL
MINIMUM DAILY TEMPERATURES (SOLID CURVES) AND THE
ACTUAL DAILY HIGH AND LOW TEMPERATURES (VERTICAL BARS).

PAGE 2 AND 3:
H/C INDICATES HEATING AND COOLING DEGREE DAYS.
RH INDICATES RELATIVE HUMIDITY
W/O INDICATES WEATHER AND OBSTRUCTIONS
S INDICATES SUNSHINE.
PR INDICATES PRESSURE.
CLOUDINESS ON PAGE 3 IS THE SUM OF THE CEILOMETER AND
SATELLITE DATA NOT TO EXCEED EIGHT EIGHTHS(OKTAS).

GENERAL:
T INDICATES TRACE PRECIPITATION, AN AMOUNT GREATER
THAN ZERO BUT LESS THAN THE LOWEST REPORTABLE VALUE.
+ INDICATES THE VALUE ALSO OCCURS ON EARLIER DATES.
BLANK ENTRIES DENOTE MISSING OR UNREPORTED DATA.
NORMALS ARE 30-YEAR AVERAGES (1961 - 1990).
ASOS INDICATES AUTOMATED SURFACE OBSERVING SYSTEM.
PM INDICATES THE LAST DAY OF THE PREVIOUS MONTH.
POR (PERIOD OF RECORD) BEGINS WITH THE JANUARY DATA
MONTH AND IS THE NUMBER OF YEARS USED TO COMPUTE
THE MEAN. INDIVIDUAL MONTHS WITHIN THE POR MAY
BE MISSING.
WHEN THE POR FOR A NORMAL IS LESS THAN 30 YEARS,
THE NORMAL IS PROVISIONAL AND IS BASED ON THE NUMBER
OF YEARS INDICATED.
0.* OR * INDICATES THE VALUE OR MEAN-DAYS-WITH
IS BETWEEN 0.00 AND 0.05.
CLOUDINESS FOR ASOS STATIONS DIFFERS FROM THE NON-ASOS
OBSERVATION TAKEN BY A HUMAN OBSERVER. ASOS STATION
CLOUDINESS IS BASED ON TIME-AVERAGED CEILOMETER DATA
FOR CLOUDS AT OR BELOW 12,000 FEET AND ON SATELLITE
DATA FOR CLOUDS ABOVE 12,000 FEET.
THE NUMBER OF DAYS WITH CLEAR, PARTLY CLOUDY, AND
CLOUDY CONDITIONS FOR ASOS STATIONS IS THE SUM
OF THE CEILOMETER AND SATELLITE DATA FOR THE
SUNRISE TO SUNSET PERIOD.

GENERAL CONTINUED:
CLEAR INDICATES 0 - 2 OKTAS, PARTLY CLOUDY INDICATES
3 - 6 OKTAS, AND CLOUDY INDICATES 7 OR 8 OKTAS.
WHEN AT LEAST ONE OF THE ELEMENTS (CEILOMETER OR
SATELLITE) IS MISSING, THE DAILY CLOUDINESS IS
NOT COMPUTED.
WIND DIRECTION IS RECORDED IN TENS OF DEGREES (2 DIGITS)
CLOCKWISE FROM TRUE NORTH. "00" INDICATES CALM. "36"
INDICATES TRUE NORTH.
RESULTANT WIND IS THE VECTOR AVERAGE OF THE SPEED AND
DIRECTION.
AVERAGE TEMPERATURE IS THE SUM OF THE MEAN DAILY MAXIMUM
AND MINIMUM TEMPERATURE DIVIDED BY 2.
SNOWFALL DATA COMPRISE ALL FORMS OF FROZEN
PRECIPITATION, INCLUDING HAIL.
A HEATING (COOLING) DEGREE DAY IS THE DIFFERENCE BETWEEN
THE AVERAGE DAILY TEMPERATURE AND 65 F.
DRY BULB IS THE TEMPERATURE OF THE AMBIENT AIR.
DEW POINT IS THE TEMPERATURE TO WHICH THE AIR MUST BE
COOLED TO ACHIEVE 100 PERCENT RELATIVE HUMIDITY.
WET BULB IS THE TEMPERATURE THE AIR WOULD HAVE IF THE
MOISTURE CONTENT WAS INCREASED TO 100 PERCENT RELATIVE
HUMIDITY.

ON JULY 1, 1996, THE NATIONAL WEATHER SERVICE BEGAN USING
THE "METAR" OBSERVATION CODE THAT WAS ALREADY EMPLOYED
BY MOST OTHER NATIONS OF THE WORLD. THE MOST NOTICEABLE
DIFFERENCE IN THIS ANNUAL PUBLICATION WILL BE THE CHANGE
IN UNITS FROM TENTHS TO EIGHTS(OKTAS) FOR REPORTING THE
AMOUNT OF SKY COVER.

HEATING DEGREE DAYS (base 65 F) 2001 PITTSBURGH, GRTR. PITT. AIRPORT PA (PIT)

YEAR	JUL	AUG	SEP	OCT	NOV	DEC	JAN	FEB	MAR	APR	MAY	JUN	TOTAL
1972-73	20	11	63	508	767	853	1087	1006	508	474	264	2	5563
1973-74	2	8	55	274	621	978	978	957	729	403	223	54	5282
1974-75	0	0	124	384	630	1001	997	916	881	617	116	48	5714
1975-76	0	0	192	362	554	989	1278	801	605	453	301	24	5559
1976-77	15	59	159	587	953	1268	1655	1060	658	436	138	102	7090
1977-78	11	41	78	442	583	1043	1307	1229	860	412	209	38	6253
1978-79	4	3	80	485	656	993	1346	1311	671	458	219	38	6264
1979-80	23	26	111	438	601	935	1175	1177	906	500	172	71	6135
1980-81	0	5	48	476	787	1117	1372	936	904	391	223	18	6277
1981-82	3	10	159	475	736	1098	1361	1017	819	586	82	67	6413
1982-83	9	23	119	336	605	770	1080	904	746	535	280	44	5451
1983-84	10	2	126	365	639	1223	1293	823	1008	471	305	16	6281
1984-85	12	7	165	214	734	790	1322	1038	701	334	163	65	5545
1985-86	3	9	116	300	531	1160	1131	936	737	368	148	37	5476
1986-87	1	40	65	346	733	983	1139	904	710	451	145	22	5539
1987-88	4	20	61	529	560	920	1181	1040	792	461	149	64	5781
1988-89	5	3	83	570	619	1018	905	1033	739	532	260	25	5792
1989-90	1	14	102	364	723	1414	869	781	657	439	229	49	5642
1990-91	4	1	116	314	577	829	1085	820	674	337	63	5	4825
1991-92	0	0	127	308	698	913	1063	880	805	417	210	50	5471
1992-93	1	17	116	457	657	956	920	1037	841	445	135	43	5625
1993-94	0	0	118	407	654	1028	1357	988	836	345	278	21	6032
1994-95	0	10	80	351	516	824	1047	1071	678	487	163	5	5232
1995-96	0	0	79	269	799	1150	1155	1005	930	436	215	5	6043
1996-97	14	1	104	397	853	873	1136	832	755	554	329	35	5883
1997-98	2	21	109	401	772	973	858	736	716	415	78	75	5156
1998-99	0	0	52	369	616	837	1076	861	917	387	140	40	5295
1999-00	1	6	101	386	541	935	1148	833	620	442	135	33	5181
2000-01	8	14	158	317	761	1291	1123	833	912	350	171	43	5981
2001-	17	2	125	341	497	846							

COOLING DEGREE DAYS (base 65 F) 2001 PITTSBURGH, GRTR. PITT. AIRPORT PA (P

YEAR	JAN	FEB	MAR	APR	MAY	JUN	JUL	AUG	SEP	OCT	NOV	DEC	ANNUAL
1972	0	0	0	0	34	68	219	192	76	0	0	0	589
1973	0	0	0	10	5	185	264	269	108	5	0	0	846
1974	0	0	0	13	19	66	258	247	45	5	4	0	657
1975	0	0	0	0	60	137	248	257	12	7	0	0	721
1976	0	0	1	25	14	134	99	73	12	0	0	0	358
1977	0	0	3	14	83	75	231	141	72	0	4	0	623
1978	0	0	0	0	69	178	260	207	122	0	0	0	836
1979	0	0	0	9	41	125	193	175	70	7	0	0	620
1980	0	0	0	0	34	115	317	306	118	0	0	0	890
1981	0	0	0	5	25	139	230	160	72	0	0	0	631
1982	0	0	0	0	79	33	246	127	77	15	3	0	580
1983	0	0	0	3	3	135	263	251	115	0	0	0	770
1984	0	0	0	3	12	165	127	194	63	13	0	0	577
1985	0	0	0	41	33	49	181	160	130	4	0	0	598
1986	0	0	3	20	65	144	265	157	121	20	0	0	795
1987	0	0	0	6	93	204	342	240	72	0	1	0	958
1988	0	0	0	0	44	174	381	322	47	7	0	0	975
1989	0	0	5	0	49	154	291	225	100	9	0	0	833
1990	0	0	14	37	9	153	218	179	83	13	0	0	706
1991	0	0	0	27	184	239	333	298	124	27	0	0	1232
1992	0	0	0	9	33	87	233	112	86	3	0	0	563
1993	0	0	0	2	45	166	337	327	73	1	1	0	952
1994	0	0	0	0	21	28	264	302	175	51	0	2	843
1995	0	0	0	0	27	220	342	402	65	7	0	0	1063
1996	0	0	0	11	67	230	162	186	71	0	0	0	727
1997	0	0	0	2	4	134	217	128	29	24	0	0	538
1998	0	0	22	0	67	147	199	254	125	6	0	0	820
1999	0	0	0	0	24	178	350	136	83	0	0	0	771
2000	0	0	5	2	67	181	131	147	83	6	0	0	622
2001	0	0	0	34	20	152	184	258	44	12	0	0	704

SNOWFALL (inches) 2001 PITTSBURGH, GRTR. PITT. AIRPORT PA (PIT)

YEAR	JUL	AUG	SEP	OCT	NOV	DEC	JAN	FEB	MAR	APR	MAY	JUN	TOTAL
1972-73	0.0	0.0	0.0	1.8	6.1	2.9	3.4	6.1	4.6	1.4	T	0.0	26.3
1973-74	0.0	0.0	0.0	0.0	0.8	4.8	4.9	2.2	2.3	1.6	T	0.0	16.6
1974-75	0.0	0.0	0.0	T	2.6	21.2	10.1	13.9	9.8	1.1	0.0	0.0	58.7
1975-76	0.0	0.0	0.0	0.0	1.9	3.8	21.8	3.3	4.3	0.5	0.0	0.0	35.6
1976-77	0.0	0.0	0.0	T	6.6	7.9	26.5	6.4	0.9	1.3	T	0.0	49.6
1977-78	0.0	0.0	0.0	T	3.3	9.1	40.2	5.4	4.0	0.2	0.0	0.0	62.2
1978-79	0.0	0.0	0.0	0.0	2.3	3.2	18.2	13.7	2.0	1.4	0.0	0.0	40.8
1979-80	0.0	0.0	0.0	T	1.1	1.1	7.8	6.2	7.9	T	0.0	0.0	24.1
1980-81	0.0	0.0	0.0	T	9.7	6.3	12.5	11.9	7.6	T	0.0	0.0	48.0
1981-82	0.0	0.0	0.0	T	0.6	11.5	13.4	3.6	12.2	3.8	0.0	0.0	45.1
1982-83	0.0	0.0	0.0	T	0.1	8.8	3.9	12.0	4.3	1.0	0.0	0.0	30.1
1983-84	0.0	0.0	0.0	0.0	6.1	10.5	10.8	11.4	10.4	T	0.0	0.0	49.2
1984-85	0.0	0.0	0.0	0.0	1.5	4.8	14.6	8.1	0.2	7.2	0.0	0.0	36.4
1985-86	0.0	0.0	0.0	0.0	T	15.3	11.1	12.4	4.8	2.7	0.0	0.0	46.3
1986-87	0.0	0.0	0.0	0.0	1.0	0.9	11.6	1.1	7.3	8.1	0.0	0.0	30.0
1987-88	0.0	0.0	0.0	T	4.1	7.9	5.5	6.9	9.8	0.9	0.0	0.0	35.1
1988-89	0.0	0.0	0.0	0.2	1.1	4.0	4.2	4.1	7.5	0.6	T	0.0	21.7
1989-90	0.0	0.0	T	0.2	1.6	12.5	7.7	2.5	0.6	3.3	0.0	T	28.4
1990-91	0.0	0.0	0.0	0.0	T	4.6	4.8	3.6	4.2	T	0.0	0.0	17.2
1991-92	T	0.0	0.0	0.0	2.1	3.1	12.9	2.4	10.6	2.8	T	0.0	33.9
1992-93	0.0	0.0	0.0	1.3	1.5	14.1	2.1	18.5	34.1	0.5	0.0	0.0	72.1
1993-94	0.0	0.0	0.0	8.5	2.6	10.4	30.1	11.0	13.6	0.6	0.0	0.0	76.8
1994-95	0.0	T	0.0	0.0	0.2	T	7.6	9.0	6.4	0.2	0.0	0.0	23.4
1995-96	0.0	0.0	0.0	0.0	13.9	12.6	23.4	9.6	13.4	1.6	0.0	0.0	74.5
1996-97	0.0	0.0	0.0	0.0	4.4	6.1					0.0	0.0	
1997-98	0.0	0.0	0.0	0.0		12.6	1.7	2.5	4.9	0.0	0.0	0.0	
1998-99	0.0	0.0	0.0	T	T	1.8	17.1	4.7	15.6	T	0.0	0.0	39.2
1999-00	0.0	0.0	0.0	T	2.1	3.2	12.0	8.4	1.1	0.3	T	0.0	27.1
2000-01	0.0	0.0	0.0	T	2.1	8.3	13.7	2.7	8.0	1.1	0.0	0.0	35.9
2001-	T	0.0	0.0	0.2	T	5.0							
POR= 48 YRS	T	T	T	0.4	3.3	8.0	11.5	8.8	8.5	1.6	0.2	T	42.3

2001
SAN JUAN,
PUERTO RICO (SJU)

San Juan, located on the north coast of the island of Puerto Rico, is surrounded by the waters of the Atlantic Ocean and San Juan Bay. Local custom assigns the name San Juan to the old city which lies right on the coast, but the modern metropolitan area extends inland about 12 miles. These inland sections have a temperature and rainfall regime significantly different from the coastal area. Isla Verde Airport, where weather observations are made, lies on the coast about 7 miles east of old San Juan. The surrounding terrain is level with a gradual upslope inland. Mountain ranges, with peak elevations of 4,000 feet, extend east and west through the central portion of Puerto Rico, and are located 15 to 20 miles east and south of San Juan. These mountain ranges have a decided influence on the rainfall of the San Juan metropolitan area, and on the entire island in general.

The climate is tropical maritime, characteristic of all tropical islands. The predominant easterly trade winds, modified by local effects such as the land and sea breeze and the particular island topography, are a primary feature of the climate of San Juan and have a significant influence on the temperature and rainfall. During daylight hours the wind blows almost constantly off the ocean. Usually, after sunset the wind shifts to the south or southeast, off land. This daily wind variation is a contributing factor to the delightful climate of the city. The annual temperature range is small with about a 5-6 degree difference between the temperatures of the warmest and coldest months. The inland sectors have warmer afternoons and cooler nights. In the interior mountain and valley regions even greater daily and annual ranges of temperature occur. The highest temperatures recorded in Puerto Rico have exceeded 105 degrees and the lowest have been near 40. Sea water temperatures range from 78 degrees in March to about 83 degrees in September.

Although rainfall in San Juan is nearly 60 inches, the geographical distribution of rainfall over the island shows the heaviest rainfall, of about 180 inches per year, in the Luquillo Range, only 23 miles distant from San Juan. The driest area, with annual rainfall of 30 to 35 inches, is located in the southwest corner of the island. Rain showers occur mostly in the afternoon and at night. The nocturnal showers, usually light, are a characteristic feature of the San Juan rainfall pattern. Rainfall is generally of the brief showery type except for the continuous rains occuring with the passage of tropical disturbances, or when the trailing edge of a cold front out of the United States reaches Puerto Rico. This normally occurs from about November to April.

Puerto Rico is in the tropical hurricane region of the eastern Caribbean. The hurricane season begins June 1 and ends November 30. Only a few hurricanes have passed close enough to San Juan to produce hurricane force winds or damage.

Mild temperatures, refreshing sea breezes in the daytime, plenty of sunshine, and adequate rainfall make the climate of San Juan most enjoyable for tourists and residents alike.

NORMALS, MEANS, AND EXTREMES
SAN JUAN, PR (SJU)

LATITUDE: 18 26' 29" N LONGITUDE: 66 00' 08" W ELEVATION (FT): GRND: 7 BARO: 10 TIME ZONE: ATLANTIC (UTC + 4) WBAN: 11641

	ELEMENT	POR	JAN	FEB	MAR	APR	MAY	JUN	JUL	AUG	SEP	OCT	NOV	DEC	YEAR
TEMPERATURE °F	NORMAL DAILY MAXIMUM	30	83.2	83.6	84.4	85.8	87.2	88.6	88.5	88.7	88.8	88.3	85.9	83.8	86.4
	MEAN DAILY MAXIMUM	46	83.2	83.6	84.6	85.8	87.1	88.5	88.4	88.8	88.8	88.4	85.9	83.9	86.4
	HIGHEST DAILY MAXIMUM	47	92	96	96	97	96	97	95	97	97	98	96	94	98
	YEAR OF OCCURRENCE		1983	1983	1983	1983	1980	1988	1981	1980	1981	1981	1981	1989	OCT 1981
	MEAN OF EXTREME MAXS.	46	87.2	87.8	89.7	91.2	92.1	92.7	91.8	92.2	92.6	92.6	90.0	87.6	90.6
	NORMAL DAILY MINIMUM	30	70.8	70.6	71.6	72.9	74.5	76.1	76.8	76.7	76.2	75.5	74.0	72.4	74.0
	MEAN DAILY MINIMUM	46	70.5	70.4	71.1	72.7	74.3	75.8	76.5	76.5	76.0	75.3	73.7	72.0	73.7
	LOWEST DAILY MINIMUM	47	61	62	60	64	66	69	69	70	69	46	66	63	46
	YEAR OF OCCURRENCE		1962	1968	1957	1968	1962	1957	1959	1956	1960	2001	1969	1964	OCT 2001
	MEAN OF EXTREME MINS.	46	66.0	66.1	66.9	68.9	71.1	72.9	73.1	73.4	73.2	71.7	70.1	68.2	70.1
	NORMAL DRY BULB	30	77.0	77.1	78.0	79.4	80.9	82.3	82.6	82.7	82.5	81.9	80.0	78.1	80.2
	MEAN DRY BULB	46	76.8	77.0	77.9	79.3	80.7	82.2	82.4	82.6	82.4	81.8	79.9	78.1	80.1
	MEAN WET BULB	16	71.3	70.7	71.1	72.9	70.8	71.5	72.5	77.1	76.6	76.0	74.5	72.6	73.1
	MEAN DEW POINT	16	68.6	67.7	67.9	69.8	68.3	69.1	70.1	74.8	74.3	73.9	72.3	70.1	70.6
	NORMAL NO. DAYS WITH:														
	MAXIMUM 90	30	0.4	0.9	1.9	3.6	6.2	9.4	8.8	10.3	11.1	9.2	1.9	0.6	64.3
	MAXIMUM 32	30	0.0	0.0	0.0	0.0	0.0	0.0	0.0	0.0	0.0	0.0	0.0	0.0	0.0
	MINIMUM 32	30	0.0	0.0	0.0	0.0	0.0	0.0	0.0	0.0	0.0	0.0	0.0	0.0	0.0
	MINIMUM 0	30	0.0	0.0	0.0	0.0	0.0	0.0	0.0	0.0	0.0	0.0	0.0	0.0	0.0
H/C	NORMAL HEATING DEG. DAYS	30	0	0	0	0	0	0	0	0	0	0	0	0	0
	NORMAL COOLING DEG. DAYS	30	372	339	403	432	493	519	546	549	525	524	450	406	5558
RH	NORMAL (PERCENT)	30	74	72	71	71	75	76	76	76	76	77	76	75	75
	HOUR 02 LST	30	80	79	78	79	83	84	83	84	84	84	83	81	82
	HOUR 08 LST	30	80	79	76	74	76	77	78	79	79	80	80	80	78
	HOUR 14 LST	30	63	62	60	61	65	66	66	67	67	66	67	65	65
	HOUR 20 LST	30	75	73	72	73	76	77	78	78	78	79	77	76	76
S	PERCENT POSSIBLE SUNSHINE	46	69	70	76	72	64	65	69	68	62	64	60	62	67
W/O	MEAN NO. DAYS WITH:														
	HEAVY FOG(VISBY 1/4 MI)	46	0.0	0.0	0.0	0.0	0.0	0.0	0.0	0.0	0.0	0.0	0.0	0.0	0.0
	THUNDERSTORMS	46	0.3	0.3	0.3	1.0	4.4	4.9	5.1	6.1	8.3	7.7	3.2	0.9	42.5
CLOUDINESS	MEAN:														
	SUNRISE-SUNSET (OKTAS)	41	3.9	4.0	3.9	4.3	5.1	4.8	4.6	4.5	4.8	4.7	4.4	4.2	4.4
	MIDNIGHT-MIDNIGHT (OKTAS)	32	3.5	3.6	3.6	3.7	4.5	4.1	4.2	4.0	4.2	4.0	4.0	3.8	3.9
	MEAN NO. DAYS WITH:														
	CLEAR	41	9.0	7.5	9.4	7.5	3.7	4.4	5.2	5.5	4.0	4.7	5.4	6.7	73.0
	PARTLY CLOUDY	41	18.0	16.3	17.2	16.4	15.5	15.8	17.2	17.4	16.6	17.0	17.6	18.2	203.2
	CLOUDY	41	4.0	4.4	4.4	6.1	11.8	9.8	8.6	8.1	9.4	9.4	6.9	6.1	89.0
PR	MEAN STATION PRESSURE(IN)	27	29.97	29.97	29.95	29.92	29.92	29.96	29.98	29.94	29.90	29.87	29.89	29.94	29.93
	MEAN SEA-LEVEL PRES. (IN)	16	30.04	30.03	30.01	29.97	29.97	30.02	30.04	30.00	29.95	29.92	29.93	30.01	29.99
WINDS	MEAN SPEED (MPH)	33	7.9	8.3	8.6	8.5	8.0	8.7	9.4	8.5	7.4	6.5	7.3	7.7	8.1
	PREVAIL.DIR(TENS OF DEGS)	25	09	09	07	07	07	09	09	07	07	09	07	07	07
	MAXIMUM 2-MINUTE:														
	SPEED (MPH)	5	28	29	32	26	26	32	28	34	79	37	33	30	79
	DIR. (TENS OF DEGS)		05	06	07	07	07	07	07	04	05	04	36	07	05
	YEAR OF OCCURRENCE		2000	1997	1997	2000	1997	1998	1997	1998	1998	2001	1999	2000	SEP 1998
	MAXIMUM 5-SECOND:														
	SPEED (MPH)	5	37	40	36	32	30	45	37	47	93	43	39	36	93
	DIR. (TENS OF DEGS)		02	06	06	06	14	06	07	07	07	04	36	07	07
	YEAR OF OCCURRENCE		2000	2001	1997	1999	2001	1998	1997	1998	1998	2001	1999	2000	SEP 1998
PRECIPITATION	NORMAL (IN)	30	2.81	2.15	2.35	3.76	5.93	4.00	4.37	5.32	5.28	5.71	5.94	4.72	52.34
	MAXIMUM MONTHLY (IN)	47	7.60	6.69	5.41	10.37	14.99	10.96	9.35	11.76	15.15	15.06	15.96	16.81	16.81
	YEAR OF OCCURRENCE		1977	1982	1958	1988	1965	1965	1961	1955	1996	1970	1979	1981	DEC 1981
	MINIMUM MONTHLY (IN)	47	0.61	0.20	0.72	0.08	0.44	0.29	1.12	1.83	1.73	1.17	1.91	0.68	0.08
	YEAR OF OCCURRENCE		1978	1983	1970	1997	1972	1985	1974	1994	1987	1979	1980	1963	APR 1997
	MAXIMUM IN 24 HOURS (IN)	47	5.08	2.75	3.91	7.20	4.74	3.55	2.91	5.30	8.84	5.04	7.07	8.40	8.84
	YEAR OF OCCURRENCE		1969	1989	1969	1988	1986	1965	1993	2000	1989	1985	1979	1981	SEP 1989
	NORMAL NO. DAYS WITH:														
	PRECIPITATION 0.01	30	17.3	13.0	12.6	12.1	16.1	14.9	18.8	18.6	16.7	17.5	17.6	18.7	193.9
	PRECIPITATION 1.00	30	0.2	0.2	0.4	0.8	1.7	0.8	1.0	1.0	1.2	1.2	1.3	1.2	11.0
SNOWFALL	NORMAL (IN)	30	0.0	0.0	0.0	0.0	0.0	0.0	0.0	0.0	0.0	0.0	0.0	0.0	0.0
	MAXIMUM MONTHLY (IN)	47	0.0	0.0	0.0	0.0	0.0	0.0	0.0	0.0	T	0.0	0.0	0.0	T
	YEAR OF OCCURRENCE										1989				SEP 1989
	MAXIMUM IN 24 HOURS (IN)	47	0.0	0.0	0.0	0.0	0.0	0.0	0.0	0.0	T	0.0	0.0	0.0	T
	YEAR OF OCCURRENCE										1989				SEP 1989
	MAXIMUM SNOW DEPTH (IN)	45	0	0	0	0	0	0	0	0	0	0	0	0	0
	YEAR OF OCCURRENCE														
	NORMAL NO. DAYS WITH:														
	SNOWFALL 1.0	30	0.0	0.0	0.0	0.0	0.0	0.0	0.0	0.0	0.0	0.0	0.0	0.0	0.0

PRECIPITATION (inches) 2001 SAN JUAN, PR (SJU)

YEAR	JAN	FEB	MAR	APR	MAY	JUN	JUL	AUG	SEP	OCT	NOV	DEC	ANNUAL
1972	2.76	2.00	3.40	2.79	0.44	1.58	2.24	3.06	3.68	5.46	2.78	7.53	37.72
1973	2.27	0.92	4.66	8.48	0.48	4.71	2.44	7.00	3.13	3.29	3.01	4.16	44.55
1974	2.92	0.82	1.92	1.20	2.42	2.34	1.12	6.57	8.23	6.55	3.92	41.68
1975	2.69	0.71	1.13	1.01	1.04	2.64	3.35	4.08	9.29	6.60	10.90	7.82	51.26
1976	1.50	2.18	2.05	3.94	2.96	2.48	5.12	11.44	7.69	2.77	2.11	47.20
1977	7.60	1.02	1.73	0.96	4.04	1.49	4.64	4.42	4.71	5.94	12.44	3.82	52.81
1978	0.61	1.56	3.52	8.27	7.14	2.86	3.46	3.21	6.34	4.88	5.40	2.61	49.86
1979	1.29	1.80	2.25	4.28	12.13	5.76	6.61	9.38	10.11	1.17	15.96	3.81	74.55
1980	1.75	1.67	1.47	2.55	5.19	1.31	2.19	3.17	4.85	6.71	1.91	3.18	35.95
1981	2.55	2.72	4.39	2.89	11.02	5.48	7.04	3.32	2.98	9.32	4.94	16.81	73.46
1982	2.53	6.69	0.98	1.01	10.26	5.24	2.33	1.93	2.87	2.06	4.34	4.76	45.00
1983	0.69	0.20	1.47	8.54	3.85	1.91	6.53	5.15	2.75	4.06	3.25	3.50	41.90
1984	1.96	3.13	0.82	0.28	3.75	6.85	2.66	6.04	3.16	5.10	5.65	4.69	44.09
1985	2.80	2.40	1.84	1.02	5.95	0.29	2.85	4.33	5.44	11.10	4.54	2.80	45.36
1986	2.18	1.13	1.61	8.93	12.80	1.52	1.94	5.19	1.98	8.54	5.87	3.59	55.28
1987	2.16	1.20	5.17	8.88	12.17	7.07	3.26	2.48	1.73	2.72	7.49	7.69	62.00
1988	3.83	2.27	1.76	10.37	6.06	1.45	4.02	11.31	5.49	4.12	5.68	4.07	60.43
1989	2.96	6.05	3.39	2.63	4.88	2.97	5.54	7.88	14.83	2.09	4.95	2.50	60.67
1990	4.56	3.02	3.14	1.05	2.44	4.32	5.76	3.42	2.23	8.65	5.33	5.03	48.95
1991	2.57	2.26	1.99	1.76	3.23	2.77	3.30	1.94	5.00	1.84	6.16	2.71	35.53
1992	4.03	1.19	1.47	2.12	8.76	5.55	4.38	4.00	5.35	1.74	11.98	4.72	55.29
1993	2.35	0.51	0.78	6.55	4.48	5.46	7.34	3.01	4.36	2.78	4.34	3.00	44.96
1994	3.39	1.69	1.38	2.75	1.69	4.49	3.58	1.83	5.30	3.52	8.32	3.04	40.98
1995	3.69	3.57	1.29	2.98	9.47	4.66	4.93	5.15	7.17	6.11	3.09	3.84	55.95
1996	5.74	2.44	2.06	5.04	2.39	6.02	7.64	6.94	15.15	2.55	7.98	3.29	67.24
1997	4.01	4.58	1.83	0.08	3.38	0.98	4.75	6.86	3.56		6.82	1.02	
1998	7.29	3.72	3.48	4.14	3.67	3.25	7.18	7.80	10.47	6.82	6.74	7.99	72.55
1999	3.24	2.74	0.75	2.53	3.99	5.53	5.81	7.34	5.82	7.38	9.84	6.15	61.12
2000	2.37	1.01	0.78	1.20	4.72	2.97	3.67	9.52	2.96	3.73	3.76	3.08	39.77
2001	2.91	3.27	1.59	2.27	5.90	2.01	3.57	6.44	5.05	5.00	6.25	11.79	56.05
POR= 51 YRS	3.15	2.38	0.65	1.46	3.63	2.49	3.78	3.33	6.50	1.90	3.27	1.63	34.17

AVERAGE TEMPERATURE (F) 2001 SAN JUAN, PR (SJU)

YEAR	JAN	FEB	MAR	APR	MAY	JUN	JUL	AUG	SEP	OCT	NOV	DEC	ANNUAL
1972	77.7	77.8	78.7	80.1	81.7	83.8	83.9	83.4	82.9	82.6	81.1	79.2	81.1
1973	78.9	78.3	78.6	80.0	82.9	83.2	83.4	83.2	82.7	83.5	80.7	77.9	81.1
1974	77.3	78.0	78.5	79.6	81.4	83.4	83.7	83.2	82.9	82.5	80.0	78.3	80.7
1975	76.9	77.2	78.4	79.6	81.0	82.9	83.0	82.7	82.1	81.6	79.8	77.5	80.2
1976	76.3	76.6	76.8	78.5	80.5	81.7	83.1	83.5	83.3	82.5	80.9	78.3	80.2
1977	77.0	77.6	78.3	80.2	80.9	81.9	80.6	81.0	81.1	81.9	80.3	79.3	80.0
1978	78.3	79.2	79.8	80.1	80.8	82.3	83.1	83.2	83.9	82.7	81.8	79.6	81.2
1979	78.5	78.7	77.8	79.4	81.3	83.8	83.5	83.0	81.8	83.0	80.6	78.7	80.8
1980	78.2	78.6	79.2	80.6	83.8	85.1	85.2	85.1	84.5	84.4	82.2	80.7	82.3
1981	79.8	79.4	80.9	80.3	83.4	84.2	85.0	83.8	84.3	83.3	81.6	78.7	82.1
1982	78.5	77.8	78.0	80.4	81.0	82.9	83.3	84.6	84.6	83.8	80.2	78.0	81.1
1983	78.5	79.9	82.2	81.8	83.2	85.4	84.2	84.0	84.3	83.6	81.4	79.9	82.4
1984	78.1	77.8	80.2	81.8	81.1	82.4	82.6	82.6	82.2	81.3	78.7	77.3	80.5
1985	76.1	77.3	76.7	78.3	80.4	83.0	83.5	83.4	81.7	79.6	79.0	76.8	79.7
1986	75.5	76.0	77.4	79.0	79.1	81.6	82.0	82.2	82.5	81.4	79.1	77.7	79.5
1987	76.7	77.4	78.0	81.2	81.6	81.7	82.7	83.7	83.8	83.3	80.6	79.7	80.9
1988	76.9	76.6	77.8	80.4	82.4	84.3	83.6	82.8	82.2	81.6	79.8	76.9	80.4
1989	76.2	76.0	76.1	78.8	80.3	81.2	81.8	82.3	82.3	81.7	80.0	79.2	79.7
1990	77.1	76.4	76.5	79.1	81.8	82.4	82.6	82.9	83.4	82.1	80.7	77.5	80.2
1991	77.0	77.8	78.6	79.2	81.0	83.2	83.1	83.9	83.3	82.4	79.9	77.0	80.5
1992	77.1	78.2	78.9	80.8	81.3	83.4	82.4	82.7	82.3	84.0	80.8	79.3	80.9
1993	77.3	78.5	79.3	80.3	82.1	83.3	82.8	84.2	83.0	82.7	80.9	78.9	81.1
1994	77.9	78.0	79.3	79.7	82.8	82.8	82.9	83.8	83.3	82.8	81.1	79.7	81.2
1995	78.0	78.7	77.2	80.6	81.5	83.3	84.2	84.5	84.9	82.3	80.9	80.0	81.3
1996	78.2	77.9	79.1	79.1	80.3	81.3	81.6	81.8	81.8	81.6	78.8	76.7	79.9
1997	76.2	76.0	76.6	80.3	81.5	83.4	83.1	83.1	83.2	82.4	80.9	80.1	80.6
1998	78.7	78.7	78.2	79.6	81.7	83.1	82.6	83.7	82.8	82.2	80.2	78.6	80.8
1999	76.9	75.4	77.7	79.5	82.4	82.1	81.5	82.7	83.1	81.5	79.8	76.9	80.0
2000	75.7	76.7	76.6	79.4	80.6	82.0	82.9	82.8	82.0	81.8	79.5	78.4	79.9
2001	77.3	76.9	79.2	80.2	82.7	82.8	82.9	83.2	82.9	82.1	79.4	78.3	80.7
POR= 51 YRS	76.6	76.8	77.6	79.0	80.6	82.0	82.2	82.5	82.2	81.6	79.7	77.9	79.9

REFERENCE NOTES:

PAGE 1:
THE TEMPERATURE GRAPH SHOWS NORMAL MAXIMUM AND NORMAL
MINIMUM DAILY TEMPERATURES (SOLID CURVES) AND THE
ACTUAL DAILY HIGH AND LOW TEMPERATURES (VERTICAL BARS).

PAGE 2 AND 3:
H/C INDICATES HEATING AND COOLING DEGREE DAYS.
RH INDICATES RELATIVE HUMIDITY
W/O INDICATES WEATHER AND OBSTRUCTIONS
S INDICATES SUNSHINE.
PR INDICATES PRESSURE.
CLOUDINESS ON PAGE 3 IS THE SUM OF THE CEILOMETER AND
SATELLITE DATA NOT TO EXCEED EIGHT EIGHTHS(OKTAS).

GENERAL:
T INDICATES TRACE PRECIPITATION, AN AMOUNT GREATER
THAN ZERO BUT LESS THAN THE LOWEST REPORTABLE VALUE.
+ INDICATES THE VALUE ALSO OCCURS ON EARLIER DATES.
BLANK ENTRIES DENOTE MISSING OR UNREPORTED DATA.
NORMALS ARE 30-YEAR AVERAGES (1961 - 1990).
ASOS INDICATES AUTOMATED SURFACE OBSERVING SYSTEM.
PM INDICATES THE LAST DAY OF THE PREVIOUS MONTH.
POR (PERIOD OF RECORD) BEGINS WITH THE JANUARY DATA
MONTH AND IS THE NUMBER OF YEARS USED TO COMPUTE
THE MEAN. INDIVIDUAL MONTHS WITHIN THE POR MAY
BE MISSING.
WHEN THE POR FOR A NORMAL IS LESS THAN 30 YEARS,
THE NORMAL IS PROVISIONAL AND IS BASED ON THE NUMBER
OF YEARS INDICATED.
0.* OR * INDICATES THE VALUE OR MEAN-DAYS-WITH
IS BETWEEN 0.00 AND 0.05.
CLOUDINESS FOR ASOS STATIONS DIFFERS FROM THE NON-ASOS
OBSERVATION TAKEN BY A HUMAN OBSERVER. ASOS STATION
CLOUDINESS IS BASED ON TIME-AVERAGED CEILOMETER DATA
FOR CLOUDS AT OR BELOW 12,000 FEET AND ON SATELLITE
DATA FOR CLOUDS ABOVE 12,000 FEET.
THE NUMBER OF DAYS WITH CLEAR, PARTLY CLOUDY, AND
CLOUDY CONDITIONS FOR ASOS STATIONS IS THE SUM
OF THE CEILOMETER AND SATELLITE DATA FOR THE
SUNRISE TO SUNSET PERIOD.

GENERAL CONTINUED:
CLEAR INDICATES 0 - 2 OKTAS, PARTLY CLOUDY INDICATES
3 - 6 OKTAS, AND CLOUDY INDICATES 7 OR 8 OKTAS.
WHEN AT LEAST ONE OF THE ELEMENTS (CEILOMETER OR
SATELLITE) IS MISSING, THE DAILY CLOUDINESS IS
NOT COMPUTED.
WIND DIRECTION IS RECORDED IN TENS OF DEGREES (2 DIGITS)
CLOCKWISE FROM TRUE NORTH. "00" INDICATES CALM. "36"
INDICATES TRUE NORTH.
RESULTANT WIND IS THE VECTOR AVERAGE OF THE SPEED AND
DIRECTION.
AVERAGE TEMPERATURE IS THE SUM OF THE MEAN DAILY MAXIMUM
AND MINIMUM TEMPERATURE DIVIDED BY 2.
SNOWFALL DATA COMPRISE ALL FORMS OF FROZEN
PRECIPITATION, INCLUDING HAIL.
A HEATING (COOLING) DEGREE DAY IS THE DIFFERENCE BETWEEN
THE AVERAGE DAILY TEMPERATURE AND 65 F.
DRY BULB IS THE TEMPERATURE OF THE AMBIENT AIR.
DEW POINT IS THE TEMPERATURE TO WHICH THE AIR MUST BE
COOLED TO ACHIEVE 100 PERCENT RELATIVE HUMIDITY.
WET BULB IS THE TEMPERATURE THE AIR WOULD HAVE IF THE
MOISTURE CONTENT WAS INCREASED TO 100 PERCENT RELATIVE
HUMIDITY.

ON JULY 1, 1996, THE NATIONAL WEATHER SERVICE BEGAN USING
THE "METAR" OBSERVATION CODE THAT WAS ALREADY EMPLOYED
BY MOST OTHER NATIONS OF THE WORLD. THE MOST NOTICEABLE
DIFFERENCE IN THIS ANNUAL PUBLICATION WILL BE THE CHANGE
IN UNITS FROM TENTHS TO EIGHTS(OKTAS) FOR REPORTING THE
AMOUNT OF SKY COVER.

HEATING DEGREE DAYS (base 65 F) 2001 SAN JUAN, PR (SJU)

YEAR	JUL	AUG	SEP	OCT	NOV	DEC	JAN	FEB	MAR	APR	MAY	JUN	TOTAL
1983-84	0	0	0	0	0	0	0	0	0	0	0	0	0
1984-85	0	0	0	0	0	0	0	0	0	0	0	0	0
1985-86	0	0	0	0	0	0	0	0	0	0	0	0	0
1986-87	0	0	0	0	0	0	0	0	0	0	0	0	0
1987-88	0	0	0	0	0	0	0	0	0	0	0	0	0
1988-89	0	0	0	0	0	0	0	0	0	0	0	0	0
1989-90	0	0	0	0	0	0	0	0	0	0	0	0	0
1990-91	0	0	0	0	0	0	0	0	0	0	0	0	0
1991-92	0	0	0	0	0	0	0	0	0	0	0	0	0
1992-93	0	0	0	0	0	0	0	0	0	0	0	0	0
1993-94	0	0	0	0	0	0	0	0	0	0	0	0	0
1994-95	0	0	0	0	0	0	0	0	0	0	0	0	0
1995-96	0	0	0	0	0	0	0	0	0	0	0	0	0
1996-97	0	0	0	0	0	0	0	0	0	0	0	0	0
1997-98	0	0	0	0	0	0	0	0	0	0	0	0	0
1998-99	0	0	0	0	0	0	0	0	0	0	0	0	0
1999-00	0	0	0	0	0	0	0	0	0	0	0	0	0
2000-01	0	0	0	0	0	0	0	0	0	0	0	0	0
2001-	0	0	0	0	0	0							

COOLING DEGREE DAYS (base 65 F) 2001 SAN JUAN, PR (SJU)

YEAR	JAN	FEB	MAR	APR	MAY	JUN	JUL	AUG	SEP	OCT	NOV	DEC	ANNUAL
1972	400	376	431	458	525	569	592	578	543	554	490	446	5962
1973	438	379	428	458	562	553	578	570	539	579	475	406	5965
1974	390	368	424	442	511	558	586	575	544	549	455	418	5820
1975	375	346	420	446	501	543	565	556	521	519	450	395	5637
1976	357	342	375	413	485	506	566	579	557	548	485	419	5632
1977	382	360	419	463	499	514	491	502	487	532	469	452	5570
1978	420	404	464	462	499	526	566	570	575	555	513	459	6013
1979	426	393	404	436	512	571	582	564	511	569	477	432	5877
1980	414	402	446	479	590	610	633	628	591	608	522	494	6417
1981	467	407	499	468	578	584	626	587	588	574	505	429	6312
1982	423	364	412	470	504	543	573	615	593	590	466	405	5958
1983	426	424	541	509	573	621	604	593	583	582	502	468	6426
1984	415	377	479	508	503	528	553	553	522	517	417	389	5761
1985	352	349	368	403	483	547	579	577	508	459	429	371	5425
1986	329	315	390	428	444	501	535	538	534	517	429	401	5361
1987	366	354	407	490	521	508	556	588	571	575	477	464	5877
1988	377	341	403	470	547	587	581	558	523	522	453	375	5737
1989	354	313	353	421	482	495	529	543	525	526	456	443	5440
1990	383	325	362	430	526	526	551	561	559	539	478	392	5632
1991	379	363	428	430	503	551	571	593	558	547	455	380	5758
1992	383	393	435	479	512	556	547	555	524	596	482	454	5916
1993	391	383	450	469	535	557	561	603	550	557	483	440	5979
1994	404	370	449	448	559	541	563	592	556	562	493	461	5998
1995	412	389	387	475	518	557	602	611	603	545	486	472	6057
1996	414	381	444	430	480	495	519	529	510	524	421	372	5519
1997	353	316	369	467	518	561	569	567	555	545	485	473	5778
1998	430	391	415	446	523	550	555	589	538	541	462	428	5868
1999	374	297	402	443	547	515	519	558	549	516	449	379	5548
2000	339	345	366	443	492	517	560	557	516	530	443	424	5532
2001	388	342	448	464	555	540	563	569	544	535	437	421	5806

SNOWFALL (inches) 2001 SAN JUAN, PR (SJU)

YEAR	JUL	AUG	SEP	OCT	NOV	DEC	JAN	FEB	MAR	APR	MAY	JUN	TOTAL
1972-73	0.0	0.0	0.0	0.0	0.0	0.0	0.0	0.0	0.0	0.0	0.0	0.0	0.0
1973-74	0.0	0.0	0.0	0.0	0.0	0.0	0.0	0.0	0.0	0.0	0.0	0.0	0.0
1974-75	0.0	0.0	0.0	0.0	0.0	0.0	0.0	0.0	0.0	0.0	0.0	0.0	0.0
1975-76	0.0	0.0	0.0	0.0	0.0	0.0	0.0	0.0	0.0	0.0	0.0	0.0	0.0
1976-77	0.0	0.0	0.0	0.0	0.0	0.0	0.0	0.0	0.0	0.0	0.0	0.0	0.0
1977-78	0.0	0.0	0.0	0.0	0.0	0.0	0.0	0.0	0.0	0.0	0.0	0.0	0.0
1978-79	0.0	0.0	0.0	0.0	0.0	0.0	0.0	0.0	0.0	0.0	0.0	0.0	0.0
1979-80	0.0	0.0	0.0	0.0	0.0	0.0	0.0	0.0	0.0	0.0	0.0	0.0	0.0
1980-81	0.0	0.0	0.0	0.0	0.0	0.0	0.0	0.0	0.0	0.0	0.0	0.0	0.0
1981-82	0.0	0.0	0.0	0.0	0.0	0.0	0.0	0.0	0.0	0.0	0.0	0.0	0.0
1982-83	0.0	0.0	0.0	0.0	0.0	0.0	0.0	0.0	0.0	0.0	0.0	0.0	0.0
1983-84	0.0	0.0	0.0	0.0	0.0	0.0	0.0	0.0	0.0	0.0	0.0	0.0	0.0
1984-85	0.0	0.0	0.0	0.0	0.0	0.0	0.0	0.0	0.0	0.0	0.0	0.0	0.0
1985-86	0.0	0.0	0.0	0.0	0.0	0.0	0.0	0.0	0.0	0.0	0.0	0.0	0.0
1986-87	0.0	0.0	0.0	0.0	0.0	0.0	0.0	0.0	0.0	0.0	0.0	0.0	0.0
1987-88	0.0	0.0	0.0	0.0	0.0	0.0	0.0	0.0	0.0	0.0	0.0	0.0	0.0
1988-89	0.0	0.0	0.0	0.0	0.0	0.0	0.0	0.0	0.0	0.0	0.0	0.0	0.0
1989-90	0.0	0.0	T	0.0	0.0	0.0	0.0	0.0	0.0	0.0	0.0	0.0	T
1990-91	0.0	0.0	0.0	0.0	0.0	0.0	0.0	0.0	0.0	0.0	0.0	0.0	0.0
1991-92	0.0	0.0	0.0	0.0	0.0	0.0	0.0	0.0	0.0	0.0	0.0	0.0	0.0
1992-93	0.0	0.0	0.0	0.0	0.0	0.0	0.0	0.0	0.0	0.0	0.0	0.0	0.0
1993-94	0.0	0.0	0.0	0.0	0.0	0.0	0.0	0.0	0.0	0.0	0.0	0.0	0.0
1994-95	0.0	0.0	0.0	0.0	0.0	0.0	0.0	0.0	0.0	0.0	0.0	0.0	0.0
1995-96	0.0	0.0	0.0	0.0	0.0	0.0	0.0	0.0	0.0	0.0	0.0	0.0	0.0
1996-97	0.0	0.0	0.0	0.0	0.0	0.0	0.0	0.0	0.0	0.0	0.0	0.0	0.0
1997-98	0.0	0.0	0.0	0.0	0.0	0.0	0.0	0.0	0.0	0.0	0.0	0.0	0.0
1998-99	0.0	0.0	0.0	0.0	0.0	0.0	0.0	0.0	0.0	0.0	0.0	0.0	0.0
1999-00	0.0	0.0	0.0	0.0	0.0	0.0	0.0	0.0	0.0	0.0	0.0	0.0	0.0
2000-01	0.0	0.0	0.0	0.0	0.0	0.0	0.0	0.0	0.0	0.0	0.0	0.0	0.0
2001-	0.0	0.0	0.0	0.0	0.0	0.0							
POR= 45 YRS	0.0	0.0	T	0.0	0.0	0.0	0.0	0.0	0.0	0.0	0.0	0.0	T

2001
PROVIDENCE,
RHODE ISLAND (PVD)

The proximity to Narragansett Bay and the Atlantic Ocean plays an important part in determining the climate for Providence and vicinity. In winter, the temperatures are modified considerably, and many major snowstorms change to rain before reaching the area. In summer, many days that could be uncomfortably warm are cooled by refreshing sea breezes. At other times of the year, sea fog may be advected in over land by onshore winds. In fact, most cases of dense fog are produced this way, but the number of such days is few, averaging two or three days per month. In early fall, severe coastal storms of tropical origin sometimes bring destructive winds to this area. Even at other times of the year, it is usually coastal storms which produce the severest weather.

The temperature for the entire year averages around 50 degrees with 70 degree temperatures common from near the end of May to the latter part of September. During this period, there may be several days reaching 90 degrees or more. Temperatures of 100 degrees and more are rare.

Freezing temperatures occur on the average about 125 days per year. They become a common daily occurrence in the latter part of November, and become less frequent near the end of March. The average date for the last freeze in spring is mid-April, while the average date for the first freeze in fall is late October, making the growing season about 195 days in length. Sub-zero weather in winter seldom occurs, averaging less than one day for December and one or two days each for January and February.

Measurable precipitation occurs on about one day out of every three, and is fairly evenly distributed throughout the year. There is usually no definite dry season, but occasionally droughts do occur.

Thunderstorms are responsible for much of the rainfall from May through August. They usually produce heavy, and sometimes even excessive amounts of rainfall. However, since their duration is relatively short, damage is ordinarily light. The thunderstorms of summer are frequently accompanied by extremely gusty winds, which may result in some damage to property.

The first measurable snowfall of winter usually comes toward the end of November, and the last in spring is about the middle of March. Winters with over 50 inches of snow are not common. The area normally receives less than 25 inches. The month of greatest snowfall is usually February, but January and March are close seconds. It is unusual for the ground to remain well covered with snow for any long period of time.

NORMALS, MEANS, AND EXTREMES
PROVIDENCE, RI (PVD)

LATITUDE:	LONGITUDE:	ELEVATION (FT):		TIME ZONE:	WBAN: 14765
41 43' 19" N	71 25' 57" W	GRND: 50	BARO: 53	EASTERN (UTC + 5)	

	ELEMENT	POR	JAN	FEB	MAR	APR	MAY	JUN	JUL	AUG	SEP	OCT	NOV	DEC	YEAR
TEMPERATURE F	NORMAL DAILY MAXIMUM	30	36.6	38.3	46.1	57.0	67.3	76.9	82.1	80.7	74.3	64.1	53.0	41.2	59.8
	MEAN DAILY MAXIMUM	54	37.0	38.8	46.3	57.7	67.9	77.0	80.7	80.6	73.3	63.5	52.5	41.3	59.7
	HIGHEST DAILY MAXIMUM	48	69	72	85	98	95	97	102	104	100	86	78	77	104
	YEAR OF OCCURRENCE		1995	1985	1998	1976	1996	1988	1991	1975	1983	1979	1993	1998	AUG 1975
	MEAN OF EXTREME MAXS.	54	56.1	56.2	66.7	78.1	85.8	91.1	92.1	91.5	87.1	79.4	69.5	60.4	76.2
	NORMAL DAILY MINIMUM	30	19.1	20.9	28.8	37.7	47.3	56.8	63.2	61.9	53.8	43.0	34.9	24.4	41.0
	MEAN DAILY MINIMUM	54	20.5	22.1	29.4	38.7	48.0	57.4	62.4	62.2	54.1	43.4	35.2	25.1	41.5
	LOWEST DAILY MINIMUM	48	-13	-7	1	14	29	41	0	40	33	20	6	-10	-13
	YEAR OF OCCURRENCE		1976	1979	1967	1954	1956	1980	1996	1965	1980	1976	1989	1980	JAN 1976
	MEAN OF EXTREME MINS.	54	2.3	4.3	14.0	27.2	36.6	46.2	52.5	50.4	39.2	29.4	20.9	7.4	27.5
	NORMAL DRY BULB	30	27.9	29.7	37.4	47.4	57.3	66.9	72.7	71.3	64.1	53.6	44.0	32.8	50.4
	MEAN DRY BULB	54	28.6	30.4	37.8	48.2	57.9	67.3	71.5	71.4	63.7	53.4	43.8	33.2	50.6
	MEAN WET BULB	17	26.5	28.5	33.9	42.7	52.5	61.6	62.9	62.2	58.8	49.1	40.0	31.2	45.8
	MEAN DEW POINT	17	18.7	20.2	25.9	34.9	46.5	57.0	59.1	62.2	54.9	43.6	33.7	24.1	40.1
	NORMAL NO. DAYS WITH:														
	MAXIMUM 90	30	0.0	0.0	0.0	0.1	0.5	1.8	3.6	2.1	0.9	0.0	0.0	0.0	9.0
	MAXIMUM 32	30	11.6	7.7	1.4	*	0.0	0.0	0.0	0.0	0.0	0.0	0.2	6.8	27.7
	MINIMUM 32	30	28.0	24.2	19.6	5.7	0.2	0.0	0.0	0.0	0.0	3.6	12.7	25.1	119.1
	MINIMUM 0	30	1.4	0.7	0.0	0.0	0.0	0.0	0.0	0.0	0.0	0.0	0.0	0.4	2.5
H/C	NORMAL HEATING DEG. DAYS	30	1150	988	856	528	246	31	0	8	90	359	630	998	5884
	NORMAL COOLING DEG. DAYS	30	0	0	0	0	7	88	239	203	63	6	0	0	606
RH	NORMAL (PERCENT)	30	64	63	63	61	67	70	71	72	73	70	69	67	68
	HOUR 01 LST	30	69	68	70	71	78	82	83	84	84	80	75	72	76
	HOUR 07 LST	30	70	70	71	69	73	76	78	80	82	80	77	74	75
	HOUR 13 LST	30	56	54	52	48	53	56	56	56	56	53	57	58	55
	HOUR 19 LST	30	63	62	61	59	64	67	68	71	73	71	69	66	66
S	PERCENT POSSIBLE SUNSHINE	42	56	58	58	57	58	61	63	62	62	61	50	52	58
W/O	MEAN NO. DAYS WITH:														
	HEAVY FOG(VISBY 1/4 MI)	49	2.1	2.2	2.1	2.1	2.1	2.5	1.9	1.5	1.7	2.7	2.1	2.0	25.0
	THUNDERSTORMS	49	0.2	0.3	0.8	1.3	2.7	3.6	4.2	3.5	1.8	1.0	0.9	0.2	20.5
CLOUDINESS	MEAN:														
	SUNRISE-SUNSET (OKTAS)	42	5.0	5.0	5.3	5.3	5.4	5.1	5.0	4.8	4.6	4.3	5.0	4.9	5.0
	MIDNIGHT-MIDNIGHT (OKTAS)	31	4.6	4.5	4.7	4.8	5.0	4.9	4.9	4.6	4.5	4.2	4.8	4.7	4.7
	MEAN NO. DAYS WITH:														
	CLEAR	42	9.5	8.0	8.4	7.3	6.5	6.6	6.7	8.4	9.4	11.0	8.3	8.3	98.4
	PARTLY CLOUDY	42	6.8	7.1	7.7	8.2	9.9	10.3	11.9	10.3	8.1	7.9	6.9	7.8	102.9
	CLOUDY	42	14.7	13.2	14.9	14.5	14.6	13.1	12.4	12.3	12.4	12.1	14.9	14.9	164.0
PR	MEAN STATION PRESSURE(IN)	29	29.97	29.96	29.93	29.89	29.91	29.89	29.91	29.97	30.00	30.01	29.99	29.98	29.95
	MEAN SEA-LEVEL PRES. (IN)	17	30.06	30.04	30.01	29.97	29.97	30.01	29.99	30.02	30.07	30.09	30.07	30.05	30.03
WINDS	MEAN SPEED (MPH)	45	10.7	11.3	11.9	11.6	10.4	9.8	9.3	9.0	9.2	9.3	10.1	10.3	10.2
	PREVAIL.DIR(TENS OF DEGS)	29	30	30	30	30	16	16	18	22	23	22	36	30	30
	MAXIMUM 2-MINUTE:														
	SPEED (MPH)	6	41	41	46	40	38	38	39	28	34	40	38	41	46
	DIR. (TENS OF DEGS)		20	30	29	03	22	36	01	01	02	06	20	27	29
	YEAR OF OCCURRENCE		1996	1996	1997	1997	2000	1997	1996	1999	1996	1996	1998	2000	MAR 1997
	MAXIMUM 5-SECOND:														
	SPEED (MPH)	6	56	54	59	51	48	51	51	37	44	47	53	53	59
	DIR. (TENS OF DEGS)		21	20	28	03	24	36	01	01	18	05	20	26	28
	YEAR OF OCCURRENCE		1997	2000	1997	1997	2000	1997	1996	1999	1999	1996	1998	2000	MAR 1997
PRECIPITATION	NORMAL (IN)	30	3.88	3.61	4.05	4.11	3.76	3.33	3.18	3.63	3.48	3.69	4.43	4.38	45.53
	MAXIMUM MONTHLY (IN)	48	11.66	7.20	8.84	12.74	8.38	11.08	8.08	11.12	7.92	11.89	11.01	10.75	12.74
	YEAR OF OCCURRENCE		1979	1984	1983	1983	1984	1982	1976	1955	1961	1962	1983	1969	APR 1983
	MINIMUM MONTHLY (IN)	48	0.50	0.39	0.56	1.48	0.71	0.17	0.82	0.71	0.77	0.40	0.41	0.58	0.17
	YEAR OF OCCURRENCE		1970	1987	1981	1966	1964	1999	1999	1984	1994	2001	1955	JUN 1999	
	MAXIMUM IN 24 HOURS (IN)	48	3.34	3.14	4.53	4.45	5.17	5.03	4.83	6.71	4.89	6.63	4.18	3.85	6.71
	YEAR OF OCCURRENCE		1962	1978	1968	1983	1984	1984	1976	1979	1961	1962	1983	1969	AUG 1979
	NORMAL NO. DAYS WITH:														
	PRECIPITATION 0.01	30	10.8	9.8	11.4	10.7	11.4	10.7	8.6	9.4	8.1	8.5	10.7	12.2	122.3
	PRECIPITATION 1.00	30	1.1	0.9	1.2	1.2	1.0	0.8	0.9	0.8	0.9	1.0	1.3	1.2	12.3
SNOWFALL	NORMAL (IN)	30	10.0	10.7	6.0	0.7	0.*	0.0	0.0	0.0	0.0	0.2	1.2	7.2	36.0
	MAXIMUM MONTHLY (IN)	47	37.2	30.9	31.6	7.6	7.0	0.0	0.0	0.0	0.0	2.5	8.0	19.8	37.2
	YEAR OF OCCURRENCE		1996	1962	1956	1982	1977					1979	1989	1963	JAN 1996
	MAXIMUM IN 24 HOURS (IN)	47	20.8	27.6	16.9	7.6	7.0	0.0	0.0	0.0	0.0	2.5	8.0	11.9	27.6
	YEAR OF OCCURRENCE		1996	1978	1960	1982	1977					1979	1989	1981	FEB 1978
	MAXIMUM SNOW DEPTH (IN)	52	19	30	20	10	2	0	0	0	0	1	7	12	30
	YEAR OF OCCURRENCE		1948	1961	1956	1970	1977					1979	1989	1981	FEB 1961
	NORMAL NO. DAYS WITH:														
	SNOWFALL 1.0	30	2.8	2.6	1.7	0.2	0.0	0.0	0.0	0.0	0.0	0.1	0.4	2.2	10.0

PRECIPITATION (inches) 2001 PROVIDENCE, RI (PVD)

YEAR	JAN	FEB	MAR	APR	MAY	JUN	JUL	AUG	SEP	OCT	NOV	DEC	ANNUAL
1972	1.85	5.19	6.70	3.71	5.73	6.83	4.25	2.98	7.31	4.36	8.45	7.70	65.06
1973	3.06	3.55	2.78	7.16	3.99	3.48	2.92	5.17	3.04	3.17	2.29	7.63	48.24
1974	4.45	3.04	4.51	2.86	2.74	3.28	1.64	3.10	6.15	2.79	1.56	4.54	40.66
1975	6.78	3.29	3.07	2.99	2.06	4.73	2.19	6.15	5.11	4.66	6.29	5.11	50.83
1976	6.38	2.91	3.44	2.00	2.53	1.60	8.08	7.01	1.57	6.52	0.81	3.47	46.32
1977	3.90	2.87	5.62	3.35	3.43	3.92	2.04	2.12	5.60	6.90	3.24	5.85	48.84
1978	9.01	3.20	3.10	2.53	5.27	1.97	2.63	6.46	1.82	3.22	2.61	5.19	47.01
1979	11.66	4.08	2.21	5.12	7.62	1.44	1.65	10.09	4.08	3.94	4.49	1.81	58.19
1980	1.40	1.16	8.11	6.18	1.78	3.85	2.03	1.99	0.90	3.41	3.73	1.57	36.11
1981	0.77	4.79	0.56	4.10	1.92	2.31	3.75	2.65	2.58	3.38	3.20	6.36	36.37
1982	6.09	3.08	3.76	3.64	1.61	11.08	3.51	3.67	3.61	3.08	4.32	1.81	49.26
1983	4.32	4.81	8.84	12.74	4.67	1.91	2.14	2.71	2.16	4.50	11.01	7.71	67.52
1984	2.00	7.20	5.77	4.30	8.38	4.09	5.16	1.77	1.77	4.25	1.95	3.16	48.74
1985	1.18	1.57	3.08	1.65	4.76	4.70	2.88	8.57	1.69	1.78	7.14	1.42	40.42
1986	5.88	3.18	2.86	2.10	2.29	3.27	5.95	3.29	0.97	2.48	5.77	8.09	46.13
1987	4.73	0.39	5.62	6.91	1.80	2.00	1.20	2.58	7.47	2.28	3.40	2.29	40.67
1988	2.69	5.29	4.09	3.11	2.83	0.91	5.73	2.38	1.77	7.60	1.03	3.38	38.37
1989	1.17	2.69	4.13	5.30	6.07	5.84	5.59	6.14	4.75	8.37	4.35	1.66	56.06
1990	5.01	2.93	2.01	5.57	5.70	1.13	3.52	3.74	2.28	4.96	2.45	5.48	44.78
1991	3.44	2.31	6.61	4.80	3.30	0.93	2.76	5.98	5.09	2.65	4.65	3.17	45.69
1992	4.82	2.10	4.04	2.34	1.42	4.61	3.59	6.06	5.09	1.53	5.05	6.83	47.48
1993	2.42	5.06	6.99	5.02	1.12	1.40	2.18	1.23	4.08	3.55	3.35	5.76	42.16
1994	5.53	2.10	7.19	2.07	2.98	2.70	1.34	6.34	4.12	0.40	5.34	4.58	44.69
1995	3.67	3.14	2.03	3.34	2.83	2.89	1.17	1.80	4.06	6.37	4.76	2.18	38.24
1996	5.02	2.19	2.71	4.88	2.44	2.17	5.57	2.19	5.72	6.20	2.38	6.59	48.06
1997	4.27	1.89	4.68	3.25	2.68	2.23	0.96	6.32	0.99	1.80	6.06	2.84	37.97
1998	6.55	5.85	5.86	4.91	6.05	9.61	1.37	2.39	2.69	3.78	2.76	1.27	52.70
1999	6.70	5.45	3.33	1.54	4.25	0.17	0.92	3.25	7.00	2.85	2.39	4.26	42.26
2000	4.19	2.74	5.37	5.06	3.72	4.78	3.64	2.41	3.79	1.31	4.73	4.26	46.00
2001	2.40	1.96	8.78	2.04	3.96	6.72	1.92	4.50	4.40	0.64	0.41	2.46	40.19
POR= 97 YRS	3.83	3.30	3.97	3.73	3.29	3.10	3.07	3.70	3.36	3.20	3.76	3.78	42.09

AVERAGE TEMPERATURE (F) 2001 PROVIDENCE, RI (PVD)

YEAR	JAN	FEB	MAR	APR	MAY	JUN	JUL	AUG	SEP	OCT	NOV	DEC	ANNUAL
1972	30.8	28.0	36.4	44.3	57.7	64.9	72.6	70.6	65.1	49.6	40.9	34.3	49.6
1973	31.1	29.6	43.7	50.0	56.7	70.3	73.6	75.0	63.4	54.2	43.8	38.3	52.5
1974	31.6	29.0	38.7	50.6	55.6	65.3	72.6	72.6	63.2	48.2	43.7	35.7	50.6
1975	34.1	30.4	35.5	44.5	61.4	66.0	74.3	71.4	61.0	55.3	48.0	32.1	51.2
1976	23.5	35.5	39.0	52.6	58.0	70.0	70.0	70.0	61.6	48.7	37.9	25.4	49.4
1977	20.9	29.8	43.6	50.6	61.2	66.7	74.3	70.0	64.1	52.9	45.9	31.6	51.2
1978	25.1	22.1	33.8	46.7	57.8	68.0	71.7	71.3	59.5	51.5	42.3	33.4	48.6
1979	30.0	19.7	40.4	46.8	60.3	65.1	73.5	70.2	64.0	53.0	48.3	37.3	50.7
1980	29.7	26.8	37.1	49.5	60.3	64.5	74.8	73.4	60.9	49.8	41.0	28.5	50.0
1981	20.3	37.4	38.7	51.4	58.5	69.4	75.6	70.0	62.4	49.1	43.0	31.1	50.6
1982	21.5	31.5	38.8	47.8	58.9	63.9	73.6	69.2	63.6	53.2	47.5	38.6	50.7
1983	31.4	32.9	40.4	49.9	56.9	70.2	76.6	74.3	69.6	55.3	46.0	32.5	53.0
1984	26.4	37.1	33.8	47.6	57.4	69.1	71.5	73.5	62.1	56.3	43.6	37.9	51.4
1985	22.5	32.1	40.8	51.0	60.2	64.8	73.0	71.1	65.2	54.6	45.9	30.4	51.0
1986	31.1	29.0	39.9	49.4	59.4	66.4	71.0	69.3	62.3	53.0	41.6	35.4	50.7
1987	29.0	28.6	39.8	48.4	59.3	68.4	72.2	69.6	64.3	51.4	43.0	35.1	50.8
1988	26.8	31.8	39.4	47.0	58.0	66.9	74.3	75.3	63.0	48.9	45.2	32.4	50.8
1989	33.8	29.9	37.5	46.2	59.3	68.7	72.3	72.1	65.3	54.1	42.5	21.8	50.3
1990	36.3	34.3	40.1	48.1	56.0	67.7	73.0	73.5	63.7	58.6	46.5	39.5	53.1
1991	29.6	35.1	41.3	51.8	63.9	69.3	74.2	73.6	63.1	56.1	45.4	36.3	53.3
1992	31.4	33.0	36.6	46.5	57.6	67.3	70.3	70.1	64.0	51.7	43.0	34.2	50.5
1993	31.4	26.3	35.8	49.6	61.8	69.4	74.5	73.8	65.1	51.5	44.0	33.7	51.4
1994	22.7	25.8	38.8	51.4	56.5	69.4	76.2	69.9	63.0	54.2	48.5	38.4	51.2
1995	36.0	29.4	41.2	48.4	57.3	68.3	75.8	73.5	62.8	57.0	40.7	29.8	51.7
1996	28.7	29.6	35.0	48.3	57.4	68.1		71.0	64.1	52.1	40.0	38.6	
1997	29.3	36.5	37.8	47.1	55.4	68.1	73.7	70.6	66.0	51.8	41.0	34.2	50.8
1998	35.1	37.0	41.2	49.8	61.1	66.0	73.5	73.3	66.3	54.1	43.5	38.5	53.3
1999	30.2	34.8	39.9	50.0	59.8	70.5	76.6	72.0	66.5	52.4	47.4	36.8	53.1
2000	27.7	34.0	43.5	47.3	58.5	67.8	70.5	70.2	63.4	53.4	43.4	28.9	50.7
2001	29.0	31.9	36.3	49.1	60.1	70.5	69.9	74.3	64.9	54.2	47.6	39.3	52.3
POR= 97 YRS	29.1	29.8	37.9	47.8	58.1	67.1	72.1	71.1	63.8	53.7	43.5	33.0	50.6

REFERENCE NOTES:

PAGE 1:
THE TEMPERATURE GRAPH SHOWS NORMAL MAXIMUM AND NORMAL
   MINIMUM DAILY TEMPERATURES (SOLID CURVES) AND THE
   ACTUAL DAILY HIGH AND LOW TEMPERATURES (VERTICAL BARS).

PAGE 2 AND 3:
H/C INDICATES HEATING AND COOLING DEGREE DAYS.
RH INDICATES RELATIVE HUMIDITY
W/O INDICATES WEATHER AND OBSTRUCTIONS
S INDICATES SUNSHINE.
PR INDICATES PRESSURE.
CLOUDINESS ON PAGE 3 IS THE SUM OF THE CEILOMETER AND
   SATELLITE DATA NOT TO EXCEED EIGHT EIGHTHS(OKTAS).

GENERAL:
T INDICATES TRACE PRECIPITATION, AN AMOUNT GREATER
   THAN ZERO BUT LESS THAN THE LOWEST REPORTABLE VALUE.
+ INDICATES THE VALUE ALSO OCCURS ON EARLIER DATES.
BLANK ENTRIES DENOTE MISSING OR UNREPORTED DATA.
NORMALS ARE 30-YEAR AVERAGES (1961 - 1990).
ASOS INDICATES AUTOMATED SURFACE OBSERVING SYSTEM.
PM INDICATES THE LAST DAY OF THE PREVIOUS MONTH.
POR (PERIOD OF RECORD) BEGINS WITH THE JANUARY DATA
   MONTH AND IS THE NUMBER OF YEARS USED TO COMPUTE
   THE MEAN. INDIVIDUAL MONTHS WITHIN THE POR MAY
   BE MISSING.
WHEN THE POR FOR A NORMAL IS LESS THAN 30 YEARS,
   THE NORMAL IS PROVISIONAL AND IS BASED ON THE NUMBER
   OF YEARS INDICATED.
0.* OR * INDICATES THE VALUE OR MEAN-DAYS-WITH
   IS BETWEEN 0.00 AND 0.05.
CLOUDINESS FOR ASOS STATIONS DIFFERS FROM THE NON-ASOS
   OBSERVATION TAKEN BY A HUMAN OBSERVER. ASOS STATION
   CLOUDINESS IS BASED ON TIME-AVERAGED CEILOMETER DATA
   FOR CLOUDS AT OR BELOW 12,000 FEET AND ON SATELLITE
   DATA FOR CLOUDS ABOVE 12,000 FEET.
THE NUMBER OF DAYS WITH CLEAR, PARTLY CLOUDY, AND
   CLOUDY CONDITIONS FOR ASOS STATIONS IS THE SUM
   OF THE CEILOMETER AND SATELLITE DATA FOR THE
   SUNRISE TO SUNSET PERIOD.

GENERAL CONTINUED:
CLEAR INDICATES 0 - 2 OKTAS, PARTLY CLOUDY INDICATES
   3 - 6 OKTAS, AND CLOUDY INDICATES 7 OR 8 OKTAS.
WHEN AT LEAST ONE OF THE ELEMENTS (CEILOMETER OR
   SATELLITE) IS MISSING, THE DAILY CLOUDINESS IS
   NOT COMPUTED.
WIND DIRECTION IS RECORDED IN TENS OF DEGREES (2 DIGITS)
   CLOCKWISE FROM TRUE NORTH. "00" INDICATES CALM. "36"
   INDICATES TRUE NORTH.
RESULTANT WIND IS THE VECTOR AVERAGE OF THE SPEED AND
   DIRECTION.
AVERAGE TEMPERATURE IS THE SUM OF THE MEAN DAILY MAXIMUM
   AND MINIMUM TEMPERATURE DIVIDED BY 2.
SNOWFALL DATA COMPRISE ALL FORMS OF FROZEN
   PRECIPITATION, INCLUDING HAIL.
A HEATING (COOLING) DEGREE DAY IS THE DIFFERENCE BETWEEN
   THE AVERAGE DAILY TEMPERATURE AND 65 F.
DRY BULB IS THE TEMPERATURE OF THE AMBIENT AIR.
DEW POINT IS THE TEMPERATURE TO WHICH THE AIR MUST BE
   COOLED TO ACHIEVE 100 PERCENT RELATIVE HUMIDITY.
WET BULB IS THE TEMPERATURE THE AIR WOULD HAVE IF THE
   MOISTURE CONTENT WAS INCREASED TO 100 PERCENT RELATIVE
   HUMIDITY.

ON JULY 1, 1996, THE NATIONAL WEATHER SERVICE BEGAN USING
   THE "METAR" OBSERVATION CODE THAT WAS ALREADY EMPLOYED
   BY MOST OTHER NATIONS OF THE WORLD. THE MOST NOTICEABLE
   DIFFERENCE IN THIS ANNUAL PUBLICATION WILL BE THE CHANGE
   IN UNITS FROM TENTHS TO EIGHTS(OKTAS) FOR REPORTING THE
   AMOUNT OF SKY COVER.

HEATING DEGREE DAYS (base 65 F) 2001 PROVIDENCE, RI (PVD)

YEAR	JUL	AUG	SEP	OCT	NOV	DEC	JAN	FEB	MAR	APR	MAY	JUN	TOTAL
1972-73	8	10	64	473	717	945	1044	984	653	451	268	16	5633
1973-74	2	3	125	331	632	819	1028	1003	808	433	313	62	5559
1974-75	0	0	114	512	634	899	951	962	907	606	160	64	5809
1975-76	0	13	132	298	506	1013	1283	850	798	403	223	39	5558
1976-77	2	23	124	501	806	1219	1361	983	653	434	176	51	6333
1977-78	0	6	103	368	568	1030	1231	1192	964	542	238	26	6268
1978-79	8	8	180	412	673	970	1075	1261	755	540	162	52	6096
1979-80	11	25	94	380	496	849	1088	1104	857	459	158	93	5614
1980-81	0	1	120	465	715	1125	1379	769	808	405	228	13	6028
1981-82	0	20	119	486	651	1044	1343	932	802	510	190	91	6188
1982-83	1	26	78	363	518	809	1038	892	755	449	254	13	5196
1983-84	0	4	62	323	563	1001	1190	802	961	513	236	36	5691
1984-85	1	0	125	270	637	832	1309	914	743	417	177	63	5488
1985-86	0	6	78	321	567	1065	1045	999	772	460	216	57	5586
1986-87	14	25	113	380	697	911	1111	1014	772	494	228	23	5782
1987-88	2	25	70	414	653	921	1177	954	787	532	238	67	5840
1988-89	8	10	89	491	587	1003	960	975	847	557	181	22	5730
1989-90	2	9	89	332	668	1329	882	854	761	511	275	24	5736
1990-91	6	0	107	242	549	781	1090	829	726	400	121	29	4880
1991-92	1	0	125	275	581	884	1034	919	876	549	246	27	5517
1992-93	11	4	100	404	652	951	1036	1077	901	455	118	32	5741
1993-94	1	0	102	413	623	966	1307	1092	805	401	263	17	5990
1994-95	0	5	85	326	487	815	892	990	731	493	244	22	5090
1995-96	0	0	111	245	721	1082	1116	1018	921	494	265	18	5991
1996-97	0	4	83	389	743	814	1098	794	836	529	289	77	5656
1997-98	3	2	83	410	713	948	919	776	738	451	149	61	5253
1998-99	0	0	45	329	636	816	1072	837	773	441	172	14	5135
1999-00	1	5	43	384	519	866	1148	892	659	524	219	59	5319
2000-01	2	7	120	355	642	1115	1111	922	883	473	193	15	5838
2001-	3	0	72	334	515	790							

COOLING DEGREE DAYS (base 65 F) 2001 PROVIDENCE, RI (PVD)

YEAR	JAN	FEB	MAR	APR	MAY	JUN	JUL	AUG	SEP	OCT	NOV	DEC	ANNUAL
1972	0	0	0	1	7	60	248	190	76	1	0	0	583
1973	0	0	0	8	17	181	272	318	84	3	0	0	883
1974	0	0	0	7	27	79	242	244	66	0	1	0	666
1975	0	0	0	0	55	100	300	218	16	4	1	0	694
1976	0	0	0	40	13	196	163	183	33	3	0	0	631
1977	0	0	0	5	68	108	295	260	85	0	0	0	821
1978	0	0	0	0	25	126	224	211	24	0	0	0	610
1979	0	0	0	0	26	59	279	190	74	12	0	0	640
1980	0	0	0	0	21	84	312	272	122	0	0	0	811
1981	0	0	0	2	33	152	335	183	47	0	0	0	752
1982	0	0	0	0	11	64	276	165	59	3	2	0	580
1983	0	0	0	1	8	177	367	298	206	30	0	0	1087
1984	0	0	0	0	6	164	206	272	47	7	0	0	702
1985	0	0	0	5	34	65	256	203	90	5	0	0	658
1986	0	0	0	0	51	105	207	164	38	14	0	0	579
1987	0	0	0	0	57	130	231	177	53	0	0	0	648
1988	0	0	0	0	26	131	302	336	37	2	0	0	834
1989	0	0	0	0	10	141	237	237	103	0	0	0	728
1990	0	0	0	8	1	114	262	272	74	49	2	0	782
1991	0	0	0	12	96	166	295	276	73	8	0	0	926
1992	0	0	0	0	27	103	183	169	75	1	0	0	558
1993	0	0	0	0	26	167	303	281	110	1	1	0	889
1994	0	0	0	0	7	155	352	163	30	0	0	0	707
1995	0	0	0	0	12	128	344	272	53	4	0	0	813
1996	0	0	0	3	36	117		200	65	0	0	0	
1997	0	0	0	0	0	177	281	184	63	10	0	0	715
1998	0	0	8	0	36	96	270	266	93	0	0	0	769
1999	0	0	0	0	18	185	366	229	96	0	0	0	894
2000	0	0	0	0	27	147	179	173	77	1	0	0	604
2001	0	0	0	1	48	190	161	293	74	10	0	0	777

SNOWFALL (inches) 2001 PROVIDENCE, RI (PVD)

YEAR	JUL	AUG	SEP	OCT	NOV	DEC	JAN	FEB	MAR	APR	MAY	JUN	TOTAL
1972-73	0.0	0.0	0.0	0.7	0.3	4.1	2.0	3.2	0.6	0.4	0.0	0.0	11.3
1973-74	0.0	0.0	0.0	0.0	T	T	15.1	11.1	0.4	1.3	0.0	0.0	27.9
1974-75	0.0	0.0	0.0	0.0	0.3	2.1	2.0	18.2	2.2	0.2	0.0	0.0	25.0
1975-76	0.0	0.0	0.0	T	1.2	7.5	15.6	3.7	9.5	0.0	0.0	0.0	37.5
1976-77	0.0	0.0	0.0	0.0	T	9.8	14.0	11.4	4.4	T	7.0	0.0	46.6
1977-78	0.0	0.0	0.0	0.0	1.3	3.9	20.5	28.6	15.9	T	0.0	0.0	70.2
1978-79	0.0	0.0	0.0	0.0	2.3	2.4	6.0	5.5	1.1	0.0	0.0	0.0	17.3
1979-80	0.0	0.0	0.0	2.5	0.0	T	0.6	3.8	5.3	0.0	0.0	0.0	12.2
1980-81	0.0	0.0	0.0	0.0	4.1	3.6	12.9	0.6	0.3	0.0	0.0	0.0	21.5
1981-82	0.0	0.0	0.0	T	T	16.4	13.4	4.3	5.7	7.6	0.0	0.0	47.4
1982-83	0.0	0.0	0.0	0.0	0.0	7.3	3.8	21.3	T	T	0.0	0.0	32.4
1983-84	0.0	0.0	0.0	0.0	T	4.5	17.9	T	13.7	T	0.0	0.0	36.1
1984-85	0.0	0.0	0.0	0.0	T	2.0	9.8	10.0	0.6	T	0.0	0.0	22.4
1985-86	0.0	0.0	0.0	0.0	1.8	2.6	0.7	13.0	0.5	T	T	0.0	18.6
1986-87	0.0	0.0	0.0	0.0	4.4	8.0	21.5	4.7	1.6	1.1	0.0	0.0	41.3
1987-88	0.0	0.0	0.0	0.0	8.0	7.8	13.5	6.7	2.7	T	0.0	0.0	38.7
1988-89	0.0	0.0	0.0	0.0	T	1.2	0.2	7.3	1.9	0.3	0.0	0.0	10.9
1989-90	0.0	0.0	0.0	0.0	8.0	15.8	10.8	10.5	9.3	1.8	0.0	0.0	56.2
1990-91	0.0	0.0	0.0	0.0	T	6.9	6.4	6.0	5.3	0.0	0.0	0.0	24.6
1991-92	0.0	0.0	0.0	0.0	T	4.8	2.4	4.9	8.2	2.0	0.0	0.0	22.3
1992-93	0.0	0.0	0.0	T	T	3.6	5.4	12.7	17.8	0.2	0.0	0.0	39.7
1993-94	0.0	0.0	0.0	0.0	T	10.1	18.0	25.8	9.6	0.0	T	0.0	63.5
1994-95	0.0	0.0	0.0	0.0	0.7	0.3	3.0	8.4	T	0.1	0.0	0.0	12.5
1995-96	0.0	0.0	0.0	0.0	4.0		37.2						
1996-97													
1997-98					0.2		1.6	0.1	0.3	T			
1998-99							6.5	12.8	12.2	T			
1999-00							6.9	6.8	2.6				
2000-01							9.7	10.8	10.3	T	0.0	0.0	
2001-	0.0	0.0	0.0	0.0	0.0	1.5							
POR= 43 YRS	0.0	0.0	0.0	0.1	1.1	6.7	9.6	9.7	7.3	0.7	0.2	0.0	35.4

2001
CHARLESTON,
SOUTH CAROLINA (CHS)

Charleston is a peninsula city bounded on the west and south by the Ashley River, on the east by the Cooper River, and on the southeast by a spacious harbor. Weather records for the airport are from a site some 10 miles inland. The terrain is generally level, ranging in elevation from sea level to 20 feet on the peninsula, with gradual increases in elevation toward inland areas. The soil is sandy to sandy loam with lesser amounts of loam. The drainage varies from good to poor. Because of the very low elevation, a considerable portion of this community and the nearby coastal islands are vulnerable to tidal flooding.

The climate is temperate, modified considerably by the nearness to the ocean. The marine influence is noticeable during winter when the low temperatures are sometimes 10-15 degrees higher on the peninsula than at the airport. By the same token, high temperatures are generally a few degrees lower on the peninsula. The prevailing winds are northerly in the fall and winter, southerly in the spring and summer.

Summer is warm and humid. Temperatures of 100 degrees or more are infrequent. High temperatures are generally several degrees lower along the coast than inland due to the cooling effect of the sea breeze. Summer is the rainiest season with 41 percent of the annual total. The rain, except during occasional tropical storms, generally occurs as showers or thunderstorms.

The fall season passes through the warm Indian Summer period to the pre-winter cold spells which begin late in November. From late September to early November the weather is mostly sunny and temperature extremes are rare. Late summer and early fall is the period of maximum threat to the South Carolina coast from hurricanes.

The winter months, December through February, are mild with periods of rain. However, the winter rainfall is generally of a more uniform type. There is some chance of a snow flurry, with the best probability of its occurrence in January, but a significant amount is rarely measured. An average winter would experience less than one cold wave and severe freeze. Temperatures of 20 degrees or less on the peninsula and along the coast are very unusual.

The most spectacular time of the year, weatherwise, is spring with its rapid changes from windy and cold in March to warm and pleasant in May. Severe local storms are more likely to occur in spring than in summer.

The average occurrence of the first freeze in the fall is early December, and the average last freeze is late February, giving an average growing season of about 294 days.

NORMALS, MEANS, AND EXTREMES
CHARLESTON, SC (CHS)

LATITUDE:	LONGITUDE:	ELEVATION (FT):	TIME ZONE:	WBAN: 13880
32 53' 55" N	80 02' 27" W	GRND: 45 BARO: 48	EASTERN (UTC + 5)	

	ELEMENT	POR	JAN	FEB	MAR	APR	MAY	JUN	JUL	AUG	SEP	OCT	NOV	DEC	YEAR
TEMPERATURE F	NORMAL DAILY MAXIMUM	30	57.8	61.0	68.6	75.8	82.7	87.6	90.2	89.0	84.9	77.2	69.5	61.6	75.5
	MEAN DAILY MAXIMUM	56	59.3	62.3	68.5	76.2	82.9	87.6	90.1	89.1	84.6	76.9	69.0	61.3	75.6
	HIGHEST DAILY MAXIMUM	59	83	87	90	94	98	103	104	105	99	94	88	83	105
	YEAR OF OCCURRENCE		1950	1989	1974	1989	1989	1944	1986	1999	1944	1986	1961	1972	AUG 1999
	MEAN OF EXTREME MAXS.	56	75.5	78.1	82.8	88.3	92.6	96.4	96.6	95.9	92.4	86.9	81.6	76.9	87.0
	NORMAL DAILY MINIMUM	30	37.7	40.0	47.5	53.9	62.9	69.1	72.7	72.2	67.9	56.3	47.2	40.7	55.7
	MEAN DAILY MINIMUM	56	37.9	40.0	46.1	53.1	61.8	68.7	72.2	71.5	67.0	55.6	46.2	39.4	55.0
	LOWEST DAILY MINIMUM	59	6	12	15	29	36	50	58	56	42	27	15	8	6
	YEAR OF OCCURRENCE		1985	1973	1980	1944	1963	1972	1952	1979	1967	1976	1950	1962	JAN 1985
	MEAN OF EXTREME MINS.	56	21.1	23.8	29.4	37.8	48.2	58.6	65.7	64.3	55.1	38.9	29.4	22.7	41.2
	NORMAL DRY BULB	30	47.8	50.5	58.1	64.9	72.8	78.3	81.5	80.6	76.4	66.8	58.4	51.2	65.6
	MEAN DRY BULB	56	48.6	51.1	57.3	64.7	72.5	78.1	81.1	80.4	75.8	66.3	57.7	50.4	65.3
	MEAN WET BULB	18	44.5	47.5	51.8	57.9	66.0	72.3	75.8	74.8	70.3	61.5	54.3	46.6	60.3
	MEAN DEW POINT	18	38.6	41.5	45.5	52.1	61.7	69.4	73.2	72.4	67.4	57.5	49.9	41.2	55.9
	NORMAL NO. DAYS WITH:														
	MAXIMUM 90	30	0.0	0.0	0.1	1.2	4.1	10.7	18.1	15.3	5.7	0.4	0.0	0.0	55.6
	MAXIMUM 32	30	0.3	0.1	*	0.0	0.0	0.0	0.0	0.0	0.0	0.0	0.0	0.1	0.5
	MINIMUM 32	30	12.2	8.6	2.5	0.1	0.0	0.0	0.0	0.0	0.0	0.1	2.9	9.2	35.6
	MINIMUM 0	30	0.0	0.0	0.0	0.0	0.0	0.0	0.0	0.0	0.0	0.0	0.0	0.0	0.0
H/C	NORMAL HEATING DEG. DAYS	30	548	414	239	66	0	0	0	0	0	74	233	439	2013
	NORMAL COOLING DEG. DAYS	30	15	8	25	63	242	399	512	484	342	130	35	11	2266
RH	NORMAL (PERCENT)	30	70	67	68	68	72	75	77	79	78	74	73	72	73
	HOUR 01 LST	30	78	77	80	81	86	87	88	89	89	86	84	80	84
	HOUR 07 LST	30	80	79	82	83	85	85	86	89	89	87	85	82	84
	HOUR 13 LST	30	55	52	50	47	53	58	60	63	61	54	52	55	55
	HOUR 19 LST	30	68	64	64	62	68	71	74	77	77	75	74	71	70
S	PERCENT POSSIBLE SUNSHINE	39	56	60	66	71	70	66	67	64	61	63	59	56	63
W/O	MEAN NO. DAYS WITH:														
	HEAVY FOG(VISBY 1/4 MI)	52	4.1	2.2	2.4	2.1	2.2	1.5	0.7	1.2	1.9	2.5	3.6	3.7	28.1
	THUNDERSTORMS	59	0.9	1.1	2.2	2.9	6.4	10.1	13.3	11.2	5.4	1.5	0.8	0.6	56.4
CLOUDINESS	MEAN:														
	SUNRISE-SUNSET (OKTAS)	1			5.6		2.4	4.0							
	MIDNIGHT-MIDNIGHT (OKTAS)	1			5.6			4.0							
	MEAN NO. DAYS WITH:														
	CLEAR	1	3.0	4.0	11.0		13.0	9.0							
	PARTLY CLOUDY	1	1.0	2.0	3.0		4.0	11.0							
	CLOUDY	1	4.0	3.0	9.0		3.0	2.0							
PR	MEAN STATION PRESSURE(IN)	29	30.10	30.10	30.00	30.00	30.00	30.00	30.00	30.00	30.00	30.00	30.09	30.10	30.03
	MEAN SEA-LEVEL PRES. (IN)	18	30.14	30.13	30.07	30.03	30.03	30.01	30.05	30.04	30.03	30.09	30.13	30.17	30.08
WINDS	MEAN SPEED (MPH)	49	8.9	9.7	10.0	9.7	8.4	8.1	7.7	7.3	7.7	7.9	8.0	8.4	8.5
	PREVAIL.DIR(TENS OF DEGS)	30	27	20	20	20	20	20	20	20	20	04	02	02	20
	MAXIMUM 2-MINUTE:														
	SPEED (MPH)	6	36	33	39	38	33	44	36	38	51	39	32	38	51
	DIR. (TENS OF DEGS)		31	19	26	27	24	17	36	27	36	21	20	26	36
	YEAR OF OCCURRENCE		1996	2000	1997	2001	2000	1997	2000	1997	1999	1996	1996	2000	SEP 1999
	MAXIMUM 5-SECOND:														
	SPEED (MPH)	6	45	47	52	49	45	51	55	48	67	56	39	48	67
	DIR. (TENS OF DEGS)		18	25	24	26	02	18	11	26	36	21	20	25	36
	YEAR OF OCCURRENCE		2001	1997	1997	2001	1999	1997	1999	1997	1999	1996	1996	2000	SEP 1999
PRECIPITATION	NORMAL (IN)	30	3.45	3.30	4.34	2.67	4.01	6.43	6.84	7.22	4.73	2.90	2.49	3.15	51.53
	MAXIMUM MONTHLY (IN)	59	8.92	10.17	11.11	9.50	9.28	27.24	18.46	16.99	17.31	12.11	7.35	7.09	27.24
	YEAR OF OCCURRENCE		1993	1998	1983	1958	1957	1973	1964	1974	1945	1994	1972	1953	JUN 1973
	MINIMUM MONTHLY (IN)	59	0.63	0.33	0.70	0.01	0.57	0.96	1.76	0.73	0.18	0.08	0.16	0.66	0.01
	YEAR OF OCCURRENCE		1950	1947	1995	1972	2000	1970	1972	1980	1990	2000	1998	1984	APR 1972
	MAXIMUM IN 24 HOURS (IN)	59	3.90	5.93	6.63	4.10	6.23	10.10	5.81	5.77	10.52	5.77	5.24	3.40	10.52
	YEAR OF OCCURRENCE		1993	1998	1959	1958	1967	1973	1960	1964	1998	1944	1969	1978	SEP 1998
	NORMAL NO. DAYS WITH:														
	PRECIPITATION 0.01	30	10.0	8.9	9.4	7.1	8.7	11.1	12.4	13.2	9.3	5.7	6.7	8.4	110.9
	PRECIPITATION 1.00	30	0.8	0.6	1.3	0.7	1.1	2.0	2.1	2.3	1.3	1.0	0.6	0.9	14.7
SNOWFALL	NORMAL (IN)	30	0.1	0.4	0.1	0.0	0.0	0.0	0.0	0.0	0.0	0.0	0.0	0.4	1.0
	MAXIMUM MONTHLY (IN)	54	1.0	7.1	2.0	T	0.0	T	T	0.0	0.0	0.0	T	8.0	8.0
	YEAR OF OCCURRENCE		1977	1973	1969	1985		1995	1993				1995	1989	DEC 1989
	MAXIMUM IN 24 HOURS (IN)	54	0.8	5.9	2.0	T	0.0	T	T	0.0	0.0	0.0	T	6.6	6.6
	YEAR OF OCCURRENCE		1966	1973	1969	1985		1995	1993				1995	1989	DEC 1989
	MAXIMUM SNOW DEPTH (IN)	48	1	7	1	0	0	0	0	0	0	0	0	8	8
	YEAR OF OCCURRENCE		1966	1973	1980									1989	DEC 1989
	NORMAL NO. DAYS WITH:														
	SNOWFALL 1.0	30	0.0	0.1	0.1	0.0	0.0	0.0	0.0	0.0	0.0	0.0	0.0	0.1	0.3

PRECIPITATION (inches) 2001 CHARLESTON, SC (CHS)

YEAR	JAN	FEB	MAR	APR	MAY	JUN	JUL	AUG	SEP	OCT	NOV	DEC	ANNUAL
1972	4.13	5.18	2.52	0.01	5.67	5.29	1.76	4.52	1.82	0.25	7.35	4.36	42.86
1973	4.59	5.57	6.15	2.55	1.83	27.24	3.60	7.93	0.63	0.84	4.58	72.17
1974	1.42	2.96	3.04	0.86	4.82	9.45	3.09	16.99	4.80	0.40	3.78	3.00	54.61
1975	4.92	3.54	4.54	3.74	5.06	5.96	9.34	7.18	5.16	1.97	1.43	3.35	56.19
1976	1.62	0.95	2.33	0.62	8.87	5.59	4.48	5.22	6.03	4.10	3.57	5.12	48.50
1977	2.72	1.38	5.31	0.45	4.66	2.12	3.86	8.13	2.49	1.76	5.88	41.24
1978	4.31	1.82	3.25	1.97	4.68	3.42	6.19	4.01	5.06	0.18	1.87	4.13	40.89
1979	3.43	3.04	3.01	3.81	8.09	2.23	8.35	0.88	15.36	3.87	3.29	2.62	57.98
1980	3.99	1.25	7.99	3.43	5.85	3.15	6.97	0.73	2.60	1.52	2.19	1.25	40.92
1981	0.93	2.23	2.38	1.87	4.02	6.04	12.66	9.30	1.27	1.95	1.06	5.73	49.44
1982	2.18	3.64	1.26	6.51	3.04	9.16	5.40	4.10	3.92	2.42	1.19	4.20	47.02
1983	4.86	6.35	11.11	3.57	0.75	2.37	8.89	2.90	3.50	2.36	3.08	4.35	54.09
1984	5.12	3.51	5.63	6.30	6.89	2.96	4.87	1.96	5.27	1.67	1.39	0.66	46.23
1985	0.87	2.70	1.50	1.12	2.79	7.02	12.06	8.48	2.53	4.58	5.49	1.21	50.35
1986	2.05	4.17	2.67	0.83	0.93	2.51	5.07	13.41	4.60	2.95	4.03	5.21	48.43
1987	7.17	4.58	5.55	1.31	2.29	5.64	2.92	6.97	14.49	0.56	3.65	1.57	56.70
1988	2.76	2.38	1.78	3.21	1.86	2.32	4.13	11.88	9.72	0.73	1.08	0.72	42.57
1989	2.31	1.17	2.87	4.84	2.14	7.26	1.93	9.18	13.35	4.08	1.85	4.74	55.72
1990	3.96	1.68	6.63	1.65	1.91	3.12	5.95	6.32	0.18	7.29	3.75	2.69	45.13
1991	7.78	0.94	4.66	4.59	5.37	4.54	7.38	8.09	2.29	0.77	1.64	1.62	49.67
1992	4.93	2.23	3.59	2.75	5.07	6.22	4.36	9.55	3.04	4.87	5.76	1.50	53.87
1993	8.92	3.08	5.80	2.72	2.67	3.70	4.21	7.69	5.01	3.00	3.59	2.30	52.69
1994	7.50	1.23	4.44	0.39	2.35	11.71	8.07	5.39	8.08	12.11	2.92	6.35	70.54
1995	3.94	3.73	0.70	1.77	1.31	6.72	5.81	11.07	7.98	3.52	2.02	1.02	49.59
1996	1.05	1.36	4.04	2.70	1.72	4.04	7.34	5.73	8.77	5.07	1.74	2.14	45.70
1997	2.68	2.86	1.81	6.61	2.04	13.76	8.51	2.15	9.58	4.12	3.26	5.19	62.57
1998	7.58	10.17	5.51	4.01	4.63	3.41	6.74	4.44	14.74	1.99	0.16	4.34	67.72
1999	4.96	2.01	2.15	2.90	3.95	2.32	3.19	3.68	10.81	6.20	1.70	2.54	46.41
2000	4.04	2.01	3.66	1.78	0.57	4.35	10.81	4.47	8.88	T	2.72	2.65	45.94
2001	1.07	2.31	6.30	0.94	1.36	8.83	9.73	1.65	4.90	0.65	0.53	1.71	39.98
POR= 131 YRS	3.20	3.23	3.73	2.77	3.47	5.29	7.04	6.66	5.31	3.11	2.28	2.82	48.91

AVERAGE TEMPERATURE (F) 2001 CHARLESTON, SC (CHS)

YEAR	JAN	FEB	MAR	APR	MAY	JUN	JUL	AUG	SEP	OCT	NOV	DEC	ANNUAL
1972	54.8	48.9	56.9	64.0	69.8	74.0	80.1	80.4	76.4	67.3	56.8	55.7	65.4
1973	48.0	46.1	61.0	61.5	71.7	78.4	82.6	81.0	79.6	69.6	61.4	51.3	66.0
1974	61.8	51.5	62.0	63.6	74.0	75.4	78.2	79.3	75.0	61.8	55.5	51.0	65.8
1975	53.8	54.7	56.9	62.3	75.1	78.5	79.2	81.5	76.8	69.0	59.3	49.7	66.4
1976	44.8	55.9	62.3	64.0	70.3	75.8	81.0	77.2	73.9	61.4	50.9	48.8	63.9
1977	38.7	46.3	60.6	66.4	72.8	81.2	83.8	81.4	78.7	63.5	61.3	5.0	61.6
1978	43.5	42.7	55.2	66.5	72.0	78.6	81.1	81.3	77.4	65.7	63.3	52.7	65.0
1979	45.4	46.8	57.4	64.9	72.4	75.9	82.0	81.4	76.5	66.0	59.4	48.7	64.7
1980	48.7	45.9	54.6	64.3	71.4	78.4	82.4	82.1	79.8	65.0	55.4	47.5	64.6
1981	41.6	50.8	54.3	67.5	70.8	82.7	83.5	80.3	74.8	64.1	55.4	46.2	64.3
1982	45.1	51.5	59.2	61.8	72.2	78.8	81.2	80.0	74.5	60.9	57.0	65.6
1983	45.6	49.0	56.4	61.0	71.7	76.9	82.8	82.9	75.5	68.9	57.4	48.8	64.7
1984	46.1	52.8	57.6	64.0	71.7	78.8	79.9	81.1	73.0	71.3	54.2	57.2	65.6
1985	42.6	50.5	60.7	67.8	73.6	79.6	80.9	79.9	75.8	72.2	67.3	47.9	66.6
1986	45.8	55.5	58.0	66.1	74.3	81.4	86.1	79.9	78.6	68.8	63.1	52.8	67.5
1987	47.2	48.8	56.8	62.6	73.3	80.1	83.0	83.5	77.8	61.0	60.1	53.5	65.6
1988	43.2	49.2	57.4	64.4	71.9	76.7	81.8	82.0	76.3	62.7	61.0	50.6	64.8
1989	55.6	55.0	59.7	65.3	72.3	80.4	82.8	80.7	76.6	68.7	60.6	43.2	66.7
1990	55.4	59.2	62.5	66.0	74.4	81.0	83.6	82.5	79.2	70.5	60.4	56.4	69.3
1991	50.8	54.9	60.5	67.6	76.3	78.9	83.6	82.0	77.4	67.6	56.0	54.3	67.5
1992	49.5	55.0	57.9	63.6	70.3	77.5	83.9	80.6	76.2	65.1	60.4	51.0	65.9
1993	53.7	49.7	55.9	61.6	72.5	79.5	85.5	82.0	78.3	67.3	59.3	48.5	66.2
1994	46.5	53.4	61.3	68.3	71.1	79.8	82.0	80.0	75.7	66.4	62.5	54.4	66.8
1995	49.7	50.2	60.4	68.5	76.0	78.5	82.5	75.6	69.8	53.8	47.5	66.3
1996	48.0	51.8	53.7	64.2	74.4	78.3	81.9	78.7	75.4	65.8	54.6	52.3	64.9
1997	49.7	55.1	64.3	63.1	69.2	74.8	80.8	79.0	76.0	66.0	55.2	49.9	65.3
1998	52.5	53.7	56.2	65.2	75.1	82.8	83.8	81.2	77.3	68.6	62.4	55.7	67.9
1999	53.6	53.3	54.9	67.8	70.4	76.4	82.5	83.2	74.7	66.4	59.8	50.5	66.2
2000	47.4	53.2	60.9	62.8	75.6	79.7	81.5	80.3	75.2	64.9	55.3	42.6	65.0
2001	46.4	54.6	56.6	65.2	72.6	79.0	79.8	80.9	73.6	64.8	63.6	56.4	66.1
POR= 128 YRS	49.6	51.5	57.6	64.6	72.5	78.7	81.3	80.6	76.5	67.2	57.9	51.0	65.8

REFERENCE NOTES:

PAGE 1:
THE TEMPERATURE GRAPH SHOWS NORMAL MAXIMUM AND NORMAL
MINIMUM DAILY TEMPERATURES (SOLID CURVES) AND THE
ACTUAL DAILY HIGH AND LOW TEMPERATURES (VERTICAL BARS).

PAGE 2 AND 3:
H/C INDICATES HEATING AND COOLING DEGREE DAYS.
RH INDICATES RELATIVE HUMIDITY
W/O INDICATES WEATHER AND OBSTRUCTIONS
S INDICATES SUNSHINE.
PR INDICATES PRESSURE.
CLOUDINESS ON PAGE 3 IS THE SUM OF THE CEILOMETER AND
SATELLITE DATA NOT TO EXCEED EIGHT EIGHTHS(OKTAS).

GENERAL:
T INDICATES TRACE PRECIPITATION, AN AMOUNT GREATER
THAN ZERO BUT LESS THAN THE LOWEST REPORTABLE VALUE.
+ INDICATES THE VALUE ALSO OCCURS ON EARLIER DATES.
BLANK ENTRIES DENOTE MISSING OR UNREPORTED DATA.
NORMALS ARE 30-YEAR AVERAGES (1961 - 1990).
ASOS INDICATES AUTOMATED SURFACE OBSERVING SYSTEM.
PM INDICATES THE LAST DAY OF THE PREVIOUS MONTH.
POR (PERIOD OF RECORD) BEGINS WITH THE JANUARY DATA
MONTH AND IS THE NUMBER OF YEARS USED TO COMPUTE
THE MEAN. INDIVIDUAL MONTHS WITHIN THE POR MAY
BE MISSING.
WHEN THE POR FOR A NORMAL IS LESS THAN 30 YEARS,
THE NORMAL IS PROVISIONAL AND IS BASED ON THE NUMBER
OF YEARS INDICATED.
0.* OR * INDICATES THE VALUE OR MEAN-DAYS-WITH
IS BETWEEN 0.00 AND 0.05.
CLOUDINESS FOR ASOS STATIONS DIFFERS FROM THE NON-ASOS
OBSERVATION TAKEN BY A HUMAN OBSERVER. ASOS STATION
CLOUDINESS IS BASED ON TIME-AVERAGED CEILOMETER DATA
FOR CLOUDS AT OR BELOW 12,000 FEET AND ON SATELLITE
DATA FOR CLOUDS ABOVE 12,000 FEET.
THE NUMBER OF DAYS WITH CLEAR, PARTLY CLOUDY, AND
CLOUDY CONDITIONS FOR ASOS STATIONS IS THE SUM
OF THE CEILOMETER AND SATELLITE DATA FOR THE
SUNRISE TO SUNSET PERIOD.

GENERAL CONTINUED:
CLEAR INDICATES 0 - 2 OKTAS, PARTLY CLOUDY INDICATES
3 - 6 OKTAS, AND CLOUDY INDICATES 7 OR 8 OKTAS.
WHEN AT LEAST ONE OF THE ELEMENTS (CEILOMETER OR
SATELLITE) IS MISSING, THE DAILY CLOUDINESS IS
NOT COMPUTED.
WIND DIRECTION IS RECORDED IN TENS OF DEGREES (2 DIGITS)
CLOCKWISE FROM TRUE NORTH. "00" INDICATES CALM. "36"
INDICATES TRUE NORTH.
RESULTANT WIND IS THE VECTOR AVERAGE OF THE SPEED AND
DIRECTION.
AVERAGE TEMPERATURE IS THE SUM OF THE MEAN DAILY MAXIMUM
AND MINIMUM TEMPERATURE DIVIDED BY 2.
SNOWFALL DATA COMPRISE ALL FORMS OF FROZEN
PRECIPITATION, INCLUDING HAIL.
A HEATING (COOLING) DEGREE DAY IS THE DIFFERENCE BETWEEN
THE AVERAGE DAILY TEMPERATURE AND 65 F.
DRY BULB IS THE TEMPERATURE OF THE AMBIENT AIR.
DEW POINT IS THE TEMPERATURE TO WHICH THE AIR MUST BE
COOLED TO ACHIEVE 100 PERCENT RELATIVE HUMIDITY.
WET BULB IS THE TEMPERATURE THE AIR WOULD HAVE IF THE
MOISTURE CONTENT WAS INCREASED TO 100 PERCENT RELATIVE
HUMIDITY.

ON JULY 1, 1996, THE NATIONAL WEATHER SERVICE BEGAN USING
THE "METAR" OBSERVATION CODE THAT WAS ALREADY EMPLOYED
BY MOST OTHER NATIONS OF THE WORLD. THE MOST NOTICEABLE
DIFFERENCE IN THIS ANNUAL PUBLICATION WILL BE THE CHANGE
IN UNITS FROM TENTHS TO EIGHTS(OKTAS) FOR REPORTING THE
AMOUNT OF SKY COVER.

HEATING DEGREE DAYS (base 65 F) 2001   CHARLESTON, SC (CHS)

YEAR	JUL	AUG	SEP	OCT	NOV	DEC	JAN	FEB	MAR	APR	MAY	JUN	TOTAL
1972-73	0	0	0	33	268	302	520	524	167	141	18	0	1973
1973-74	0	0	0	34	158	428	131	378	150	114	2	0	1395
1974-75	0	0	5	136	299	432	350	294	273	152	0	0	1941
1975-76	0	0	0	40	221	466	624	265	146	94	15	3	1874
1976-77	0	0	0	159	418	501	808	516	186	58	17	0	2663
1977-78	0	0	0	112	175	459	663	616	309	52	18	0	2404
1978-79	0	0	0	57	83	399	602	505	241	70	2	0	1959
1979-80	0	0	0	68	203	500	495	555	321	82	17	0	2241
1980-81	0	0	0	80	287	537	719	393	333	55	16	0	2420
1981-82	0	0	3	88	291	577	611	372	214	132	3	0	2291
1982-83	0	0	0	102	154	276	596	440	264	146	2	0	1980
1983-84	0	0	4	24	230	500	578	347	240	92	16	0	2031
1984-85	0	0	9	13	337	249	692	418	183	47	4	0	1952
1985-86	0	0	2	16	54	526	586	261	244	74	4	0	1767
1986-87	0	6	0	56	128	376	545	446	272	131	7	0	1967
1987-88	0	0	0	135	188	358	669	458	239	85	7	2	2141
1988-89	0	0	0	107	145	442	286	312	220	121	14	0	1647
1989-90	0	0	1	50	169	669	294	189	137	67	0	0	1576
1990-91	0	0	0	65	152	280	432	293	185	34	0	0	1441
1991-92	0	0	0	41	281	349	472	293	234	119	28	0	1817
1992-93	0	0	3	70	197	430	353	421	279	132	1	0	1886
1993-94	0	0	3	49	206	504	568	327	152	30	11	0	1850
1994-95	0	0	0	46	119	338	468	414	167	35	1	0	1588
1995-96	0	0	8	44	357	535	518	383	350	107	15	0	2317
1996-97	0	0	0	54	323	391	467	288	94	106	20	9	1752
1997-98	0	0	0	87	293	463	391	318	288	66	0	0	1906
1998-99	0	0	0	38	113	306	358	315	314	61	20	0	1525
1999-00	0	0	4	71	175	446	543	339	148	96	0	0	1822
2000-01	0	0	1	80	306	687	571	300	272	94	1	0	2312
2001-	0	0	8	110	94	277							

COOLING DEGREE DAYS (base 65 F) 2001   CHARLESTON, SC (CHS)

YEAR	JAN	FEB	MAR	APR	MAY	JUN	JUL	AUG	SEP	OCT	NOV	DEC	ANNUAL
1972	7	3	5	93	163	275	475	488	351	110	30	22	2022
1973	1	0	50	42	233	412	554	501	445	184	56	10	2488
1974	41	7	63	80	288	319	417	450	312	46	18	3	2044
1975	8	13	26	74	318	414	449	516	361	171	58	0	2408
1976	2	9	73	70	187	329	502	384	274	52	2	1	1885
1977	0	1	54	107	263	493	588	518	417	71	71	1	2584
1978	0	0	13	106	242	414	505	514	378	86	40	21	2319
1979	0	2	9	71	241	335	533	514	354	105	40	0	2204
1980	0	9	7	69	221	407	549	539	451	87	5	1	2345
1981	0	0	9	138	199	539	582	481	307	66	9	0	2330
1982	0	2	42	42	232	420	510	475	293	111	36	34	2197
1983	0	0	6	32	217	362	559	567	322	149	10	4	2228
1984	0	1	19	67	228	420	471	509	254	212	21	10	2212
1985	7	17	57	136	276	445	501	470	332	245	129	5	2620
1986	0	2	36	114	300	499	662	474	414	182	78	5	2766
1987	0	0	26	62	269	459	567	580	389	18	48	9	2427
1988	0	2	12	74	229	359	529	534	349	43	30	4	2165
1989	5	37	64	136	246	470	561	493	358	173	44	0	2587
1990	4	34	65	105	302	487	583	548	430	238	24	21	2841
1991	0	20	52	118	356	426	584	531	379	128	18	25	2637
1992	0	7	20	83	199	381	592	495	345	79	68	5	2274
1993	7	0	4	37	241	440	644	536	410	127	42	0	2488
1994	0	7	42	133	210	449	537	473	327	93	53	17	2341
1995	1	6	29	146	349	410	577	552	334	198	26	0	2628
1996	0	8	6	88	313	405	531	432	320	87	18	6	2214
1997	0	17	77	56	158	309	496	444	336	124	7	0	2024
1998	10	6	23	77	320	542	592	509	378	158	43	26	2684
1999	10	1	6	153	193	351	548	571	303	122	25	1	2284
2000	1	2	29	36	336	448	522	481	315	84	25	0	2279
2001	0	13	16	107	242	423	467	499	269	114	58	17	2225

SNOWFALL (inches) 2001   CHARLESTON, SC (CHS)

YEAR	JUL	AUG	SEP	OCT	NOV	DEC	JAN	FEB	MAR	APR	MAY	JUN	TOTAL
1972-73	0.0	0.0	0.0	0.0	0.0	0.0	T	7.1	0.0	0.0	0.0	0.0	7.1
1973-74	0.0	0.0	0.0	0.0	0.0	0.0	T	0.0	0.0	0.0	0.0	0.0	T
1974-75	0.0	0.0	0.0	0.0	0.0	0.0	0.0	0.4	0.0	0.0	0.0	0.0	0.4
1975-76	0.0	0.0	0.0	0.0	0.0	0.0	1.0	0.3	0.0	0.0	0.0	0.0	1.3
1976-77	0.0	0.0	0.0	0.0	0.0	0.0	0.0	0.0	0.0	0.0	0.0	0.0	0.0
1977-78	0.0	0.0	0.0	0.0	0.0	0.0	T	0.4	T	0.0	0.0	0.0	0.4
1978-79	0.0	0.0	0.0	0.0	0.0	0.0	0.0	1.8	0.0	0.0	0.0	0.0	1.8
1979-80	0.0	0.0	0.0	0.0	0.0	0.0	0.0	T	1.3	0.0	0.0	0.0	1.3
1980-81	0.0	0.0	0.0	0.0	0.0	3.8	0.0	0.0	0.0	0.0	0.0	0.0	3.8
1981-82	0.0	0.0	0.0	0.0	0.0	0.0	T	0.0	0.0	0.0	0.0	0.0	T
1982-83	0.0	0.0	0.0	0.0	0.0	0.0	0.0	0.0	T	0.0	0.0	0.0	T
1983-84	0.0	0.0	0.0	0.0	0.0	0.0	T	0.0	0.0	0.0	0.0	0.0	T
1984-85	0.0	0.0	0.0	0.0	0.0	0.0	T	0.0	0.0	T	0.0	0.0	T
1985-86	0.0	0.0	0.0	0.0	0.0	0.0	0.5	0.0	0.0	0.0	0.0	0.0	0.5
1986-87	0.0	0.0	0.0	0.0	0.0	T	0.0	0.0	T	0.0	0.0	0.0	T
1987-88	0.0	0.0	0.0	0.0	0.0	0.0	0.4	0.0	0.0	0.0	0.0	0.0	0.4
1988-89	0.0	0.0	0.0	0.0	0.0	T	0.0	0.9	T	0.0	0.0	T	0.9
1989-90	0.0	0.0	0.0	0.0	0.0	8.0	0.0	0.0	0.0	0.0	0.0	0.0	8.0
1990-91	0.0	0.0	0.0	0.0	0.0	0.0	T	0.0	0.0	0.0	0.0	0.0	T
1991-92	0.0	0.0	0.0	0.0	0.0	0.0	0.0	0.0	0.0	0.0	0.0	0.0	0.0
1992-93	0.0	0.0	0.0	0.0	0.0	0.0	0.0	0.0	T	0.0	0.0	0.0	T
1993-94	T	0.0	0.0	0.0	0.0	T	0.0	0.0	0.0	0.0	0.0	0.0	T
1994-95	0.0	0.0	0.0	0.0	0.0	0.0	0.0	0.0	0.0	0.0	T		T
1995-96	0.0	0.0	0.0	0.0	T	0.0	0.0						
1996-97													
1997-98													
1998-99													
1999-00													
2000-01													
2001-				0.0	0.0	0.0							
POR= 53 YRS	T	0.0	0.0	0.0	T	0.3	0.1	0.2	0.1	T	0.0	T	0.7

2001
RAPID CITY,
SOUTH DAKOTA (RAP)

Rapid City, which is not far from the geographical center of North America, experiences the large temperature ranges, both daily and seasonal, that are typical of semi-arid continental climates.

The city is surrounded by contrasting landforms, with the forested Black Hills rising immediately west of the city, and rolling prairie extending out in the other directions. From 40 to 70 miles southeast lie the eroded Badlands. The Black Hills, many of which are more than 5,000 feet above sea level, with a number of peaks above 7,000 feet, exert a pronounced influence on the climate of this area. The rolling land to the east of the city is cut by the valleys of the Box Elder and Rapid Creeks, which flow generally east-southeastward. The station is located on the north slope of the irrigated Rapid Valley. An east-west ridge 200 to 300 feet higher than the airport separates the station from the Box Elder Creek Valley.

The principal agricultural products in the area are cattle and wheat, and ranchers and farmers are dependent on the current weather forecasts, which are at times of vital interest in the protection of livestock.

Although the annual precipitation is light at lower elevations, the distribution is beneficial to agriculture with the greatest amounts occurring during the growing season. The heaviest snows are expected in the spring, which helps to furnish moisture for the early maturing crops such as wheat, while heavy winter snows at the higher elevations provide irrigation water for the fertile valleys.

Summer days are normally warm with cool, comfortable nights. Nearly all of the summer precipitation occurs as thunderstorms. Hail is often associated with the more severe thunderstorms, with resultant damage to vegetation as well as other fragile material in the path of the storms. Autumn, which begins soon after the first of September, is characterized by mild, balmy days, and cool, invigorating mornings and evenings. Autumn weather usually extends into November and often into December.

Temperatures for the winter months of December, January, and February are among the warmest in South Dakota due to the protection of the Black Hills, the frequent occurrence of Chinook winds, and the fact that the winter tracks of arctic air masses usually pass east of Rapid City. Rapid City has become the retirement home for many farmers and ranchers from the western half of the state because of the cool summer nights and the relatively mild winters.

Snowfall is normally light with the greatest monthly average of about 8 inches occurring in March. Cold waves can be expected occasionally, and one or more blizzards may occur each winter.

Spring is characterized by unsettled conditions. Wide variations usually occur in temperatures, and snows may fall as late as May.

Based on the 1951-1980 period, the average first occurrence of 32 degrees Fahrenheit in the fall is September 29 and the average last occurrence in the spring is May 7.

NORMALS, MEANS, AND EXTREMES
RAPID CITY, SD (RAP)

LATITUDE: LONGITUDE: ELEVATION (FT):   TIME ZONE: WBAN: 24090
44 02' 44" N   103 03' 14" W   GRND: 3150   BARO: 3153   MOUNTAIN (UTC + 7)

	ELEMENT	POR	JAN	FEB	MAR	APR	MAY	JUN	JUL	AUG	SEP	OCT	NOV	DEC	YEAR
TEMPERATURE F	NORMAL DAILY MAXIMUM	30	33.8	38.2	45.9	57.9	68.1	77.8	86.2	85.1	74.4	62.5	46.7	35.6	59.4
	MEAN DAILY MAXIMUM	53	33.8	38.4	45.4	57.4	67.8	77.6	86.2	86.0	75.1	61.4	46.6	37.1	59.4
	HIGHEST DAILY MAXIMUM	59	76	75	82	93	98	106	110	106	104	94	83	75	110
	YEAR OF OCCURRENCE		1987	1995	1993	1989	1969	1988	1989	1988	1978	1993	1999	1965	JUL 1989
	MEAN OF EXTREME MAXS.	53	60.2	63.0	71.7	81.7	87.5	94.3	100.6	100.0	94.9	83.8	71.1	62.7	81.0
	NORMAL DAILY MINIMUM	30	10.7	15.2	22.2	32.2	42.3	51.7	58.2	56.1	45.5	34.9	22.8	12.7	33.7
	MEAN DAILY MINIMUM	53	10.5	15.2	21.8	32.0	43.0	52.0	58.3	57.0	46.2	34.9	22.8	14.0	34.0
	LOWEST DAILY MINIMUM	59	-27	-31	-21	1	18	31	39	38	18	-2	-19	-30	-31
	YEAR OF OCCURRENCE		1950	1996	1996	1975	1950	1951	1987	1992	1985	1991	1959	1990	FEB 1996
	MEAN OF EXTREME MINS.	53	-12.3	-7.2	1.2	17.1	29.5	40.0	48.1	46.2	30.4	18.8	3.3	-9.4	17.1
	NORMAL DRY BULB	30	22.3	26.7	34.1	45.1	55.2	64.8	72.2	70.6	60.0	48.7	34.8	24.2	46.6
	MEAN DRY BULB	53	22.1	26.7	33.6	44.8	55.4	65.0	72.3	71.5	60.8	48.2	34.7	25.6	46.7
	MEAN WET BULB	16	20.8	23.5	30.5	38.5	48.9	57.1	61.3	59.9	50.4	38.0	28.7	21.5	39.9
	MEAN DEW POINT	16	13.8	16.1	23.1	29.8	42.1	51.2	54.4	52.0	41.5	29.6	21.4	14.1	32.4
	NORMAL NO. DAYS WITH:														
	MAXIMUM 90	30	0.0	0.0	0.0	0.1	0.5	3.7	11.4	11.2	3.3	0.2	0.0	0.0	30.4
	MAXIMUM 32	30	13.3	9.3	5.7	0.7	0.0	0.0	0.0	0.0	0.0	0.3	5.0	11.8	46.1
	MINIMUM 32	30	30.0	26.6	26.8	14.8	2.5	*	0.0	0.0	2.0	10.9	24.9	30.0	168.5
	MINIMUM 0	30	8.1	4.2	1.3	0.0	0.0	0.0	0.0	0.0	0.0	0.0	1.0	5.0	19.6
H/C	NORMAL HEATING DEG. DAYS	30	1324	1072	958	597	311	103	18	31	211	505	906	1265	7301
	NORMAL COOLING DEG. DAYS	30	0	0	0	0	7	97	241	205	61	0	0	0	611
RH	NORMAL (PERCENT)	30	64	65	64	59	61	62	56	53	54	54	62	65	60
	HOUR 05 LST	30	66	71	74	73	76	77	74	71	70	66	70	69	71
	HOUR 11 LST	30	58	58	55	48	50	51	45	41	42	42	52	59	50
	HOUR 17 LST	30	62	59	52	45	47	47	40	36	39	45	58	64	50
	HOUR 23 LST	30	67	70	71	68	71	72	66	62	63	63	68	68	67
S	PERCENT POSSIBLE SUNSHINE	54	57	60	63	62	60	65	73	74	70	66	55	55	63
W/O	MEAN NO. DAYS WITH:														
	HEAVY FOG(VISBY 1/4 MI)	60	1.7	2.2	2.8	1.9	1.2	1.2	0.6	0.6	0.6	0.8	2.1	2.0	17.7
	THUNDERSTORMS	60	0.0	0.0	0.1	1.2	5.6	10.4	11.4	8.3	3.2	0.4	0.0	0.0	40.6
CLOUDINESS	MEAN:														
	SUNRISE-SUNSET (OKTAS)	1					9.6	3.2			4.8	5.6		5.6	
	MIDNIGHT-MIDNIGHT (OKTAS)	1					9.6	3.2				6.4			
	MEAN NO. DAYS WITH:														
	CLEAR	1		7.0	3.0		1.0	10.0							
	PARTLY CLOUDY	1	4.0	3.0	6.0		7.0	9.0							
	CLOUDY	1	3.0	6.0	9.0		21.0	4.0							
PR	MEAN STATION PRESSURE(IN)	29	26.71	26.71	26.67	26.66	26.67	26.68	26.73	26.74	26.75	26.74	26.70	26.70	26.71
	MEAN SEA-LEVEL PRES. (IN)	17	30.07	30.08	30.01	29.95	29.92	29.90	29.93	29.95	29.98	30.02	30.05	30.11	30.00
WINDS	MEAN SPEED (MPH)	34	10.5	11.1	12.3	12.9	11.9	10.8	10.1	10.0	10.7	11.0	11.0	10.5	11.1
	PREVAIL.DIR(TENS OF DEGS)	22	33	33	33	33	33	33	33	33	33	33	33	33	33
	MAXIMUM 2-MINUTE:														
	SPEED (MPH)	6	51	59	54	61	57	54	53	49	47	55	57	52	61
	DIR. (TENS OF DEGS)		33	33	33	32	32	32	24	32	33	33	33	33	33
	YEAR OF OCCURRENCE		2001	1998	2000	1997	1999	2001	2001	2000	1999	1999	1998	1996	APR 1997
	MAXIMUM 5-SECOND:														
	SPEED (MPH)	6	60	70	62	69	68	69	66	58	56	67	67	61	70
	DIR. (TENS OF DEGS)		32	31	33	31	32	33	23	32	34	34	33	33	31
	YEAR OF OCCURRENCE		1997	1996	2000	1997	1999	2001	2001	2000	1999	1999	1998	1996	FEB 1996
PRECIPITATION	NORMAL (IN)	30	0.39	0.52	1.03	1.89	2.68	3.06	2.04	1.67	1.23	1.10	0.56	0.47	16.64
	MAXIMUM MONTHLY (IN)	59	1.77	2.46	3.02	5.16	8.18	7.00	6.13	4.83	3.94	5.60	2.22	1.65	8.18
	YEAR OF OCCURRENCE		1944	1953	1945	1967	1996	1968	1969	1982	1946	1998	1985	1975	MAY 1996
	MINIMUM MONTHLY (IN)	59	0.01	0.02	0.12	0.02	0.33	0.64	0.38	0.10	0.03	T	0.03	0.01	T
	YEAR OF OCCURRENCE		1952	1999	1981	1987	1966	1973	1988	1943	1975	1960	1945	2001	OCT 1960
	MAXIMUM IN 24 HOURS (IN)	59	1.26	1.00	2.19	3.19	3.40	4.01	2.51	2.60	2.13	2.49	1.09	1.04	4.01
	YEAR OF OCCURRENCE		1944	1953	1945	1997	1965	1963	1944	1982	1966	1982	1944	1975	JUN 1963
	NORMAL NO. DAYS WITH:														
	PRECIPITATION 0.01	30	6.7	7.2	8.4	9.6	11.7	11.7	9.3	7.8	7.2	5.2	5.6	6.8	97.2
	PRECIPITATION 1.00	30	0.0	0.0	0.1	0.2	0.4	0.7	0.4	0.2	0.2	0.2	0.0	*	2.4
SNOWFALL	NORMAL (IN)	30	4.3	6.5	9.3	6.3	0.7	0.*	0.0	0.0	0.2	1.6	5.2	5.8	39.9
	MAXIMUM MONTHLY (IN)	57	24.0	23.7	30.7	30.6	11.6	T	T	T	2.0	10.2	33.6	17.9	33.6
	YEAR OF OCCURRENCE		1949	1953	1950	1970	1950	1951	1993	1994	1970	1995	1985	1975	NOV 1985
	MAXIMUM IN 24 HOURS (IN)	57	16.3	10.0	14.9	16.0	13.4	3.6	T	T	2.0	7.8	9.4	9.8	16.3
	YEAR OF OCCURRENCE		1944	1953	1973	1970	1967	1951	1993	1994	1970	1995	1977	1975	JAN 1944
	MAXIMUM SNOW DEPTH (IN)	49	16	14	16	17	13	2	0	0	1	6	15	11	17
	YEAR OF OCCURRENCE		1993	1987	1977	1970	1967	1951			1965	1954	1985	1985	APR 1970
	NORMAL NO. DAYS WITH:														
	SNOWFALL 1.0	30	1.3	2.3	2.7	1.8	0.2	0.0	0.0	0.0	0.1	0.5	1.7	1.9	12.5

PRECIPITATION (inches) 2001 RAPID CITY, SD (RAP)

YEAR	JAN	FEB	MAR	APR	MAY	JUN	JUL	AUG	SEP	OCT	NOV	DEC	ANNUAL
1972	0.22	0.44	0.47	2.78	3.28	4.11	1.67	2.49	0.24	0.77	0.38	0.34	17.19
1973	0.11	0.31	2.71	2.69	2.37	0.64	1.46	0.74	1.44	1.38	0.73	0.54	15.12
1974	0.16	0.30	0.34	1.55	1.32	1.10	0.68	1.37	0.88	1.18	0.12	0.12	9.12
1975	1.05	0.35	2.45	1.37	1.23	5.63	1.57	0.87	0.03	0.69	0.57	1.65	17.46
1976	0.28	0.47	0.33	2.70	2.74	4.81	1.05	1.31	0.28	0.21	0.61	0.41	15.20
1977	0.83	0.25	2.63	1.57	2.49	1.76	2.98	1.79	2.86	1.06	0.82	0.36	19.40
1978	0.19	0.84	0.40	2.19	3.12	2.01	4.08	1.42	0.18	0.26	0.63	0.25	15.57
1979	0.49	0.33	0.47	0.31	1.17	3.60	4.11	2.32	0.07	0.90	0.15	0.07	13.99
1980	0.20	0.51	0.86	1.13	1.58	4.75	1.78	2.38	0.48	2.28	0.57	0.66	17.18
1981	0.14	0.09	0.12	0.32	2.81	1.89	4.47	1.74	0.16	1.81	0.23	0.35	14.13
1982	0.39	0.37	1.35	0.69	6.50	2.89	1.81	4.83	2.69	3.82	0.27	0.36	25.97
1983	0.34	0.18	0.84	1.00	2.18	3.01	1.94	2.39	0.33	1.74	1.07	0.47	15.49
1984	0.10	0.18	0.69	3.10	1.57	4.72	1.57	1.00	0.74	0.67	0.51	0.38	15.23
1985	0.46	0.06	1.55	0.32	1.24	1.58	1.03	1.86	1.57	0.98	2.22	0.77	13.64
1986	0.49	0.92	0.88	4.74	1.43	4.56	0.91	1.32	3.14	1.64	1.40	0.01	21.44
1987	0.04	1.71	1.14	0.02	3.39	1.37	0.83	2.37	0.68	0.26	0.30	0.31	12.42
1988	0.17	0.34	0.52	0.60	3.25	1.09	0.38	1.98	0.56	0.76	0.81	0.46	10.92
1989	0.02	0.34	0.96	1.46	1.40	1.04	0.82	1.70	3.09	1.49	0.43	0.82	13.57
1990	0.22	0.37	1.17	0.77	4.87	1.42	1.94	1.87	2.44	0.61	0.44	0.33	16.45
1991	0.32	0.77	0.63	2.99	4.40	3.27	1.97	0.58	0.59	1.00	0.73	0.04	17.29
1992	0.29	0.16	1.92	0.71	2.47	2.17	3.25	0.47	0.42	0.68	0.39	0.57	13.50
1993	0.68	0.61	0.82	3.05	2.16	3.39	4.31	1.18	1.46	0.90	0.70	0.53	19.79
1994	0.45	0.66	0.37	1.20	1.47	0.67	0.64	0.92	0.27	2.84	0.66	0.35	10.50
1995	0.09	0.55	0.79	2.57	4.03	4.50	2.87	0.46	0.82	0.42	0.13		19.65
1996	0.85	0.10	1.06	1.63	8.18	1.24	0.52	1.85	1.55		0.07		
1997	0.65	0.28	0.20	4.80	5.35	3.43	3.67	3.93	0.78	0.47	0.19	0.08	23.83
1998	0.15	1.24	1.32	0.28	2.34	5.59	1.26	1.42	1.50	5.60	1.13	0.06	21.89
1999	0.21	0.02	1.32	2.45	4.49	5.24	3.68	0.47	0.85	0.11	0.43	0.21	19.48
2000	0.23	0.15	1.37	3.95	2.40	1.60	2.06	0.70	0.45	1.54	0.47	0.11	15.03
2001	0.24	0.17	0.42	2.16	1.73	3.57	2.46	1.59	0.91	0.97	0.07	T	14.29
POR= 102 YRS	0.42	0.48	0.99	1.96	3.11	3.22	2.21	1.52	1.20	1.03	0.60	0.49	17.23

AVERAGE TEMPERATURE (F) 2001 RAPID CITY, SD (RAP)

YEAR	JAN	FEB	MAR	APR	MAY	JUN	JUL	AUG	SEP	OCT	NOV	DEC	ANNUAL
1972	17.3	24.0	36.5	43.6	55.2	64.7	65.6	69.1	59.4	43.8	31.0	18.6	44.1
1973	26.8	28.9	37.5	42.9	53.7	64.7	70.7	74.3	57.1	51.3	32.3	25.4	47.1
1974	21.9	32.5	37.6	47.2	53.8	66.6	77.3	67.7	57.6	51.7	36.4	28.9	48.3
1975	23.7	17.6	27.6	40.5	54.0	62.2	74.7	70.3	58.5	49.2	33.8	29.4	45.1
1976	23.9	34.6	34.4	46.9	55.4	64.3	72.8	72.6	63.3	45.4	31.7	26.8	47.7
1977	12.7	34.6	36.4	49.5	60.3	69.0	73.4	66.0	61.0	48.8	33.3	21.9	47.2
1978	11.0	15.3	35.4	45.2	55.1	65.1	71.1	69.7	66.2	50.5	29.2	17.1	44.2
1979	7.4	16.8	35.4	44.4	53.6	65.4	70.4	68.5	66.3	50.9	33.0	33.2	45.4
1980	21.0	27.1	31.5	48.9	57.4	67.1	74.9	68.6	61.6	48.9	39.3	30.3	48.1
1981	32.6	29.4	40.0	51.5	54.9	64.9	72.0	70.0	63.5	47.7	40.4	25.8	49.4
1982	11.9	23.9	33.0	42.1	53.3	59.7	70.7	70.2	58.7	47.2	32.7	28.6	44.3
1983	32.1	37.3	36.4	40.7	52.0	63.1	73.6	78.0	60.8	49.4	34.9	8.1	47.2
1984	28.0	36.1	34.3	43.8	53.6	62.8	72.2	74.8	57.2	47.2	37.5	21.4	47.4
1985	21.6	23.8	35.9	52.0	61.8	62.1	74.6	69.0	55.6	47.3	16.0	21.0	45.1
1986	29.8	21.5	43.0	44.2	54.9	70.9	69.6	55.0	48.7	30.6	30.5	47.2
1987	31.1	32.5	32.6	51.6	59.5	67.1	75.4	68.1	61.4	47.1	40.3	28.9	49.6
1988	21.7	26.9	35.6	47.1	60.0	75.6	76.1	72.5	60.4	49.7	36.4	28.4	49.2
1989	28.7	14.4	31.1	45.8	55.4	64.0	77.0	73.1	61.0	48.8	36.4	36.4	46.3
1990	32.4	28.9	36.4	45.0	53.2	66.6	71.7	73.8	65.9	48.2	40.5	17.8	48.4
1991	18.4	36.0	38.5	46.3	56.6	67.2	72.9	74.0	61.1	46.4	30.0	32.7	48.3
1992	33.6	36.3	40.9	47.1	57.4	62.6	64.3	65.9	62.6	49.3	32.4	19.0	47.6
1993	16.0	15.2	37.6	43.5	56.1	60.3	65.1	68.6	56.3	48.2	32.4	31.5	44.3
1994	22.4	19.8	40.5	46.0	60.2	68.5	70.7	73.3	65.5	49.7	35.5	30.3	48.5
1995	29.2	31.9	33.2	40.3	51.8	62.9	70.6	73.6	60.3	46.5	33.7	25.4	46.6
1996	14.9	27.3	26.2	43.3	50.6	65.5	70.8	73.2	59.8		24.3	17.9	
1997	18.1	28.8	36.7	38.7	52.8	66.3	70.9	68.7	62.5	48.6	33.2	30.3	46.3
1998	24.3	34.9	28.0	46.2	56.2	58.7	72.7	72.3	67.1	47.8	36.9	26.8	47.7
1999	25.6	36.6	36.3	42.6	53.0	62.2	71.3	71.4	55.7	49.9	44.4	33.3	48.6
2000	26.2	33.2	38.7	43.1	56.0	62.5	73.6	74.4	63.4	49.8	27.0	16.7	47.1
2001	29.7	16.9	34.4	46.0	56.2	64.1	74.8	74.4	62.5	47.9	39.9	28.9	48.0
POR= 102 YRS	23.0	25.7	33.5	44.9	55.1	64.5	72.2	70.8	60.8	48.7	35.6	26.2	46.8

REFERENCE NOTES:

PAGE 1:
THE TEMPERATURE GRAPH SHOWS NORMAL MAXIMUM AND NORMAL
MINIMUM DAILY TEMPERATURES (SOLID CURVES) AND THE
ACTUAL DAILY HIGH AND LOW TEMPERATURES (VERTICAL BARS).

PAGE 2 AND 3:
H/C INDICATES HEATING AND COOLING DEGREE DAYS.
RH INDICATES RELATIVE HUMIDITY
W/O INDICATES WEATHER AND OBSTRUCTIONS
S INDICATES SUNSHINE.
PR INDICATES PRESSURE.
CLOUDINESS ON PAGE 3 IS THE SUM OF THE CEILOMETER AND
SATELLITE DATA NOT TO EXCEED EIGHT EIGHTHS(OKTAS).

GENERAL:
T INDICATES TRACE PRECIPITATION, AN AMOUNT GREATER
THAN ZERO BUT LESS THAN THE LOWEST REPORTABLE VALUE.
+ INDICATES THE VALUE ALSO OCCURS ON EARLIER DATES.
BLANK ENTRIES DENOTE MISSING OR UNREPORTED DATA.
NORMALS ARE 30-YEAR AVERAGES (1961 - 1990).
ASOS INDICATES AUTOMATED SURFACE OBSERVING SYSTEM.
PM INDICATES THE LAST DAY OF THE PREVIOUS MONTH.
POR (PERIOD OF RECORD) BEGINS WITH THE JANUARY DATA
MONTH AND IS THE NUMBER OF YEARS USED TO COMPUTE
THE MEAN. INDIVIDUAL MONTHS WITHIN THE POR MAY
BE MISSING.
WHEN THE POR FOR A NORMAL IS LESS THAN 30 YEARS,
THE NORMAL IS PROVISIONAL AND IS BASED ON THE NUMBER
OF YEARS INDICATED.
0.* OR * INDICATES THE VALUE OR MEAN-DAYS-WITH
IS BETWEEN 0.00 AND 0.05.
CLOUDINESS FOR ASOS STATIONS DIFFERS FROM THE NON-ASOS
OBSERVATION TAKEN BY A HUMAN OBSERVER. ASOS STATION
CLOUDINESS IS BASED ON TIME-AVERAGED CEILOMETER DATA
FOR CLOUDS AT OR BELOW 12,000 FEET AND ON SATELLITE
DATA FOR CLOUDS ABOVE 12,000 FEET.
THE NUMBER OF DAYS WITH CLEAR, PARTLY CLOUDY, AND
CLOUDY CONDITIONS FOR ASOS STATIONS IS THE SUM
OF THE CEILOMETER AND SATELLITE DATA FOR THE
SUNRISE TO SUNSET PERIOD.

GENERAL CONTINUED:
CLEAR INDICATES 0 - 2 OKTAS, PARTLY CLOUDY INDICATES
3 - 6 OKTAS, AND CLOUDY INDICATES 7 OR 8 OKTAS.
WHEN AT LEAST ONE OF THE ELEMENTS (CEILOMETER OR
SATELLITE) IS MISSING, THE DAILY CLOUDINESS IS
NOT COMPUTED.
WIND DIRECTION IS RECORDED IN TENS OF DEGREES (2 DIGITS)
CLOCKWISE FROM TRUE NORTH. "00" INDICATES CALM. "36"
INDICATES TRUE NORTH.
RESULTANT WIND IS THE VECTOR AVERAGE OF THE SPEED AND
DIRECTION.
AVERAGE TEMPERATURE IS THE SUM OF THE MEAN DAILY MAXIMUM
AND MINIMUM TEMPERATURE DIVIDED BY 2.
SNOWFALL DATA COMPRISE ALL FORMS OF FROZEN
PRECIPITATION, INCLUDING HAIL.
A HEATING (COOLING) DEGREE DAY IS THE DIFFERENCE BETWEEN
THE AVERAGE DAILY TEMPERATURE AND 65 F.
DRY BULB IS THE TEMPERATURE OF THE AMBIENT AIR.
DEW POINT IS THE TEMPERATURE TO WHICH THE AIR MUST BE
COOLED TO ACHIEVE 100 PERCENT RELATIVE HUMIDITY.
WET BULB IS THE TEMPERATURE THE AIR WOULD HAVE IF THE
MOISTURE CONTENT WAS INCREASED TO 100 PERCENT RELATIVE
HUMIDITY.

ON JULY 1, 1996, THE NATIONAL WEATHER SERVICE BEGAN USING
THE "METAR" OBSERVATION CODE THAT WAS ALREADY EMPLOYED
BY MOST OTHER NATIONS OF THE WORLD. THE MOST NOTICEABLE
DIFFERENCE IN THIS ANNUAL PUBLICATION WILL BE THE CHANGE
IN UNITS FROM TENTHS TO EIGHTS(OKTAS) FOR REPORTING THE
AMOUNT OF SKY COVER.

HEATING DEGREE DAYS (base 65 F) 2001 RAPID CITY, SD (RAP)

YEAR	JUL	AUG	SEP	OCT	NOV	DEC	JAN	FEB	MAR	APR	MAY	JUN	TOTAL
1972-73	74	43	193	649	1010	1436	1181	1008	847	659	350	83	7533
1973-74	16	0	246	416	972	1223	1329	905	840	530	343	87	6907
1974-75	1	42	242	407	849	1112	1274	1318	1151	728	343	119	7586
1975-76	3	16	206	493	929	1096	1269	878	940	535	295	98	6758
1976-77	3	6	132	606	991	1177	1616	846	877	459	165	17	6895
1977-78	1	48	163	494	944	1330	1669	1383	912	588	312	91	7935
1978-79	17	40	111	443	1068	1480	1781	1348	910	614	362	82	8256
1979-80	3	25	64	433	952	982	1359	1094	1032	483	251	54	6732
1980-81	1	18	144	510	763	1070	998	993	765	402	311	65	6040
1981-82	21	7	108	531	730	1209	1646	1146	985	682	358	170	7593
1982-83	7	21	226	545	962	1119	1012	772	880	723	407	113	6787
1983-84	8	0	208	474	896	1762	1139	832	948	626	366	101	7360
1984-85	0	0	268	546	820	1344	1341	1148	895	393	146	144	7045
1985-86	8	27	327	544	1466	1358	1083	1211	672	617	317	35	7665
1986-87	5	12	296	497	1025	1059	1045	907	997	408	199	46	6496
1987-88	10	49	147	545	736	1111	1340	1103	905	533	195	17	6691
1988-89	3	18	163	470	850	1127	1120	1414	1047	586	303	116	7217
1989-90	3	6	182	495	847	1410	1004	1004	880	597	363	68	6859
1990-91	10	5	112	514	730	1462	1440	807	815	556	269	16	6736
1991-92	2	6	183	581	1045	997	964	828	742	539	262	107	6256
1992-93	67	83	138	493	973	1418	1515	1390	841	636	269	158	7981
1993-94	56	26	256	522	972	1035	1312	1260	749	565	171	27	6951
1994-95	9	17	79	468	876	1068	1104	919	983	734	400	130	6787
1995-96	11	5	215	568	933	1221	1551	1087	1200	643	447	66	7947
1996-97	6	0	203		1215	1453	1446	1008	872	783	373	30	6962
1997-98	23	26	127	510	947	1067	1254	837	1143	558	272	198	6962
1998-99	3	0	91	522	840	1178	1214	790	885	666	366	104	6659
1999-00	22	6	284	460	609	976	1196	914	808	652	281	139	6347
2000-01	10	1	152	462	1134	1490	1086	1339	944	565	278	127	7588
2001-	9	0	138	523	746	1111							

COOLING DEGREE DAYS (base 65 F) 2001 RAPID CITY, SD (RAP)

YEAR	JAN	FEB	MAR	APR	MAY	JUN	JUL	AUG	SEP	OCT	NOV	DEC	ANNUAL
1972	0	0	0	0	21	70	98	178	34	0	0	0	401
1973	0	0	0	0	7	80	202	295	15	0	0	0	599
1974	0	0	0	1	3	143	390	132	28	0	0	0	697
1975	0	0	0	0	9	41	314	188	19	12	0	0	583
1976	0	0	0	0	4	83	251	248	87	3	0	0	676
1977	0	0	0	0	25	143	270	85	51	0	0	0	574
1978	0	0	0	0	9	99	214	193	154	0	0	0	669
1979	0	0	0	4	15	99	179	141	110	2	0	0	550
1980	0	0	0	6	25	123	315	136	48	14	0	0	667
1981	0	0	0	3	5	67	243	170	74	0	0	0	562
1982	0	0	0	0	3	18	189	190	41	0	0	0	441
1983	0	0	0	0	9	62	282	407	88	0	0	0	848
1984	0	0	0	0	18	41	234	309	42	2	0	0	646
1985	0	0	0	10	53	64	312	158	51	0	0	0	648
1986	0	0	0	0	11	124	192	164	0	0	0	0	491
1987	0	0	0	13	33	118	341	152	45	0	0	0	702
1988	0	0	0	2	46	341	355	255	33	2	0	0	1034
1989	0	0	0	15	9	95	380	265	70	0	0	0	834
1990	0	0	0	2	4	120	226	282	147	3	0	0	784
1991	0	0	0	3	16	89	256	294	76	15	0	0	749
1992	0	0	0	9	36	41	51	118	72	11	0	0	338
1993	0	0	0	0	0	25	66	146	9	7	0	0	253
1994	0	0	0	0	28	141	193	280	97	0	0	0	739
1995	0	0	0	0	0	72	192	280	72	1	0	0	617
1996	0	0	0	0	5	86	194	262	52	0	0		
1997	0	0	0	0	4	76	211	148	58	7	0	0	504
1998	0	0	0	0	8	15	250	232	160	0	0	0	665
1999	0	0	0	0	2	40	225	212	9	0	0	0	488
2000	0	0	0	0	10	71	282	299	108	0	0	0	770
2001	0	0	0	4	11	107	320	300	69	1	0	0	812

SNOWFALL (inches) 2001 RAPID CITY, SD (RAP)

YEAR	JUL	AUG	SEP	OCT	NOV	DEC	JAN	FEB	MAR	APR	MAY	JUN	TOTAL
1972-73	0.0	0.0	0.0	0.7	3.6	6.0	2.0	3.1	16.9	1.9	0.0	0.0	34.2
1973-74	0.0	0.0	0.0	1.4	9.7	5.6	2.0	4.0	1.7	4.7	0.0	0.0	29.1
1974-75	0.0	0.0	0.0	T	1.5	1.4	13.7	6.1	27.4	1.3	0.0	0.0	51.4
1975-76	0.0	0.0	0.0	4.6	8.4	17.9	2.0	6.5	4.3	3.5	0.0	0.0	47.2
1976-77	0.0	0.0	0.0	0.5	7.0	6.1	11.1	2.6	26.0	3.4	0.0	0.0	56.7
1977-78	0.0	0.0	0.0	0.8	10.6	7.4	3.1	15.0	3.0	2.6	T	0.0	42.5
1978-79	0.0	0.0	0.0	T	9.8	4.0	6.1	4.4	2.6	1.8	2.8	0.0	31.5
1979-80	0.0	0.0	0.0	T	1.8	0.3	3.0	10.1	8.6	5.4	0.0	0.0	29.2
1980-81	0.0	0.0	0.0	1.4	6.9	6.1	1.2	1.3	T	T	0.0	0.0	16.9
1981-82	0.0	0.0	0.0	1.6	1.2	3.8	6.2	5.0	11.5	5.5	0.0	0.0	34.8
1982-83	0.0	0.0	0.0	1.4	1.2	4.0	2.9	0.3	6.5	4.3	4.3	0.0	24.9
1983-84	0.0	0.0	0.3	0.9	6.9	7.1	1.9	2.5	6.1	22.1	0.2	0.0	48.0
1984-85	0.0	0.0	1.3	0.7	2.0	4.9	3.8	0.7	16.2	0.4	0.0	T	30.0
1985-86	0.0	0.0	1.4	0.6	33.6	10.2	5.7	10.7	6.0	12.7	0.0	0.0	80.9
1986-87	0.0	0.0	0.0	T	12.6	0.1	0.5	21.5	10.9	0.3	0.0	0.0	45.9
1987-88	0.0	0.0	0.0	1.7	T	4.7	2.7	3.6	10.6	6.1	0.0	0.0	29.4
1988-89	0.0	0.0	0.0	0.0	2.2	9.0	0.4	7.3	10.7	6.4	0.0	0.0	36.0
1989-90	0.0	T	0.0	3.9	4.6	10.9	3.1	5.0	9.2	0.5	T	0.0	40.2
1990-91	0.0	T	0.0	1.5	1.0	6.2	6.3	8.2	3.6	11.1	4.9	T	42.8
1991-92	T	0.0	T	4.0	6.7	0.8	5.2	1.2	8.2	3.5	0.0	0.0	29.6
1992-93	T	0.0	0.0	0.9	5.0	12.4	13.6	9.0	1.0	12.2	0.0	0.0	54.1
1993-94	T	0.0	T	2.4	11.4	5.5	7.5	11.1	5.6	12.8	0.0	T	56.3
1994-95	0.0	T	0.0	0.0	6.8	4.6	0.9	6.3	9.2	10.1	T	T	37.9
1995-96	0.0	0.0	T	10.2	3.1	1.4	19.4	4.4	16.2	10.3	0.9	T	65.9
1996-97	0.0	0.0	T								0.0	T	
1997-98	T	T	0.0	3.1									
1998-99	0.0	T			15.9	0.5	2.0	0.2	18.5	6.3	0.0	T	
1999-00	0.0	0.0	0.0	T	0.3	3.7	2.9	1.5	5.5	19.0	0.0	0.0	32.9
2000-01	0.0	0.0	1.4	0.0	2.9	3.0	2.7						
2001-													
POR= 55 YRS	T	T	0.1	1.7	5.2	4.9	5.0	6.2	8.9	6.4	0.8	0.1	39.3

2001
SIOUX FALLS,
SOUTH DAKOTA (FSD)

Sioux Falls is located in the Big Sioux River Valley in southeast South Dakota. The surrounding terrain is gently rolling. The land slopes upward for about 100 miles north and northwest to an elevation about 400 feet higher than the city. To the southeast, the land slopes downward 200 to 300 feet over the same distance. Little change in elevation occurs in the remaining directions.

The climate is of the continental type. There are frequent weather changes from day to day or week to week as the locality is visited by differing air masses. Cold air masses arrive from the interior of Canada, cool, dry air from the northern Pacific, warm, moist air from the Gulf of Mexico, or hot, dry air from the southwest.

Temperatures fluctuate frequently as cold air masses move in very rapidly. During the late fall and winter, cold fronts accompanied by strong, gusty winds drop temperatures by 20 to 30 degrees in a 24-hour period. Severe cold spells usually last only a few days. The winter months of December through February have experienced cold spells with average temperatures under 8 degrees and more than 60 consecutive days below 32 degrees.

Temperatures of 100 degrees and above occur about one in every three years, and will most likely happen in July. Summer nights are usually comfortable with temperatures below 70 degrees.

Rainfall is heavier during the spring and summer with lighter amounts in winter. Nearly 64 percent of the normal yearly precipitation falls during the growing season of April through August.

One or two very heavy snows usually fall each winter. Eight to 12 inches of snow may fall in 24 hours. There have been a few snows in excess of 15 inches and almost 30 inches have fallen during a severe winter storm. Strong winds often cause drifting snow, and blizzard conditions may block highways for a day or so.

Southerly winds prevail from late spring to early fall with northwest winds the remainder of the year. Strong winds of 70 mph with gusts to 90 mph have occurred.

Thunderstorms are frequent during the late spring and summer with June and July the most active months. The thunderstorms usually occur during the late afternoon and evening with a secondary peak of activity between 2 and 5 in the morning. Some of the most severe thunderstorms with damaging winds, hail and an occasional tornado, occur most frequently June.

There is occasional flooding in the lower areas of Sioux Falls along the Big Sioux River and Skunk Creek. Runoff from the melting snow in the spring often causes substantial rises in the rivers. A diversion canal around Sioux Falls has reduced the threat of damaging floods.

Based on the 1951-1980 period, the average first occurrence of 32 degrees Fahrenheit in the fall is October 1 and the average last occurrence in the spring is May 10.

NORMALS, MEANS, AND EXTREMES
SIOUX FALLS, SD   (FSD)

LATITUDE: LONGITUDE: ELEVATION (FT):   TIME ZONE: WBAN: 14944
43 34' 37" N   96 45' 13" W   GRND: 1425   BARO: 1428   CENTRAL   (UTC +   6)

	ELEMENT	POR	JAN	FEB	MAR	APR	MAY	JUN	JUL	AUG	SEP	OCT	NOV	DEC	YEAR
TEMPERATURE °F	NORMAL DAILY MAXIMUM	30	24.3	29.6	42.3	59.0	70.7	80.5	86.3	83.3	73.1	61.2	43.4	28.0	56.8
	MEAN DAILY MAXIMUM	70	24.7	30.2	41.8	58.3	70.9	80.0	85.7	83.5	73.7	61.8	42.9	29.5	56.9
	HIGHEST DAILY MAXIMUM	56	66	70	87	94	100	110	108	108	104	94	81	63	110
	YEAR OF OCCURRENCE		1981	1982	1968	1962	1967	1988	1989	1973	1976	1963	1999	1998	JUN 1988
	MEAN OF EXTREME MAXS.	70	45.3	51.1	67.2	81.9	88.5	94.8	98.4	96.7	90.6	81.9	65.2	49.8	76.0
	NORMAL DAILY MINIMUM	30	3.3	9.7	22.6	34.8	45.9	56.1	62.3	59.4	48.7	36.0	22.6	8.6	34.2
	MEAN DAILY MINIMUM	70	4.7	10.3	22.2	34.8	46.7	56.6	62.3	60.0	49.4	37.4	22.9	10.7	34.8
	LOWEST DAILY MINIMUM	56	-36	-31	-23	5	17	33	38	34	22	9	-17	-28	-36
	YEAR OF OCCURRENCE		1970	1962	1948	1982	1967	1969	1971	1950	1974	1972	1964	1990	JAN 1970
	MEAN OF EXTREME MINS.	70	-18.8	-15.0	-.5	18.7	30.4	41.8	49.3	45.9	31.9	19.8	2.5	-12.6	16.1
	NORMAL DRY BULB	30	13.8	19.7	32.5	46.9	58.4	68.3	74.3	71.4	60.9	48.6	33.0	18.3	45.5
	MEAN DRY BULB	70	14.8	20.1	31.8	46.5	58.7	68.4	74.1	71.8	61.7	49.6	32.9	20.2	45.9
	MEAN WET BULB	18	15.9	20.6	29.8	40.9	52.8	61.4	66.0	64.9	55.1	43.0	29.1	18.8	41.5
	MEAN DEW POINT	18	11.4	15.9	24.6	33.9	46.7	56.2	62.0	61.0	50.2	36.7	24.1	14.6	36.4
	NORMAL NO. DAYS WITH:														
	MAXIMUM 90	30	0.0	0.0	0.0	0.2	0.7	4.3	10.4	7.6	1.8	0.1	0.0	0.0	25.1
	MAXIMUM 32	30	20.9	15.5	7.2	0.3	0.0	0.0	0.0	0.0	0.0	*	5.8	19.0	68.7
	MINIMUM 32	30	30.9	27.3	25.7	12.6	2.1	0.0	0.0	0.0	1.3	11.2	25.4	30.5	167.0
	MINIMUM 0	30	13.4	7.9	1.6	0.0	0.0	0.0	0.0	0.0	0.0	0.0	1.2	8.3	32.4
H/C	NORMAL HEATING DEG. DAYS	30	1587	1268	1008	543	240	50	10	22	165	508	960	1448	7809
	NORMAL COOLING DEG. DAYS	30	0	0	0	0	0	35	149	298	220	42	0	0	744
RH	NORMAL (PERCENT)	30	72	73	72	64	64	65	65	68	70	67	73	76	69
	HOUR 00 LST	30	75	77	78	73	72	75	76	78	79	75	79	78	76
	HOUR 06 LST	30	75	78	82	80	80	82	83	85	86	81	82	80	81
	HOUR 12 LST	30	67	67	64	54	53	55	53	55	57	55	64	70	60
	HOUR 18 LST	30	70	69	63	51	50	51	50	53	56	57	69	75	60
S	PERCENT POSSIBLE SUNSHINE	5	56	46	55	50	48	67	68	70	70	68	57	52	59
W/O	MEAN NO. DAYS WITH:														
	HEAVY FOG(VISBY 1/4 MI)	56	2.5	3.0	2.6	1.0	0.7	0.5	0.7	1.2	1.4	1.5	2.6	3.7	21.4
	THUNDERSTORMS	56	0.0	0.1	0.9	2.9	6.0	8.7	9.4	7.7	5.2	2.0	0.4	0.1	43.4
CLOUDINESS	MEAN:														
	SUNRISE-SUNSET (OKTAS)														
	MIDNIGHT-MIDNIGHT (OKTAS)														
	MEAN NO. DAYS WITH:														
	CLEAR	1					2.0	11.0							
	PARTLY CLOUDY	1					3.0	3.0							
	CLOUDY	1					11.0	7.0							
PR	MEAN STATION PRESSURE(IN)	29	28.54	28.54	28.45	28.43	28.42	28.42	28.46	28.49	28.50	28.50	28.49	28.54	28.48
	MEAN SEA-LEVEL PRES. (IN)	18	30.12	30.12	30.04	29.95	29.92	29.91	29.95	29.99	29.99	30.02	30.05	30.12	30.01
WINDS	MEAN SPEED (MPH)	42	10.9	11.0	12.1	12.7	11.6	10.4	9.6	9.5	10.1	10.6	11.5	10.7	10.9
	PREVAIL.DIR(TENS OF DEGS)	27	31	31	34	36	18	18	18	18	18	18	31	30	18
	MAXIMUM 2-MINUTE:														
	SPEED (MPH)	5	44	41	44	51	41	52	51	44	43	41	43	43	52
	DIR. (TENS OF DEGS)		31	34	30	32	18	08	32	03	34	19	30	32	08
	YEAR OF OCCURRENCE		2000	1999	1999	2000	1997	2001	1998	1998	1997	1997	1998	2000	JUN 2001
	MAXIMUM 5-SECOND:														
	SPEED (MPH)	5	52	48	55	62	53	56	64	54	49	52	54	52	64
	DIR. (TENS OF DEGS)		30	33	30	32	28	08	32	03	34	29	31	32	32
	YEAR OF OCCURRENCE		2000	1999	1999	2000	1998	2001	1998	1998	1997	1997	1998	2000	JUL 1998
PRECIPITATION	NORMAL (IN)	30	0.51	0.64	1.64	2.52	3.03	3.40	2.68	2.85	3.02	1.78	1.09	0.70	23.86
	MAXIMUM MONTHLY (IN)	56	1.71	4.05	4.08	6.97	8.26	8.43	8.41	9.09	9.26	6.28	4.76	2.62	9.26
	YEAR OF OCCURRENCE		1969	1962	1998	2001	1993	1984	1992	1975	1986	1998	2001	1968	SEP 1986
	MINIMUM MONTHLY (IN)	56	0.05	0.05	0.14	0.17	0.61	0.91	0.25	0.53	0.29	T	0.02	T	T
	YEAR OF OCCURRENCE		1958	1986	1967	1969	1981	1988	1947	1970	1956	1952	1980	1986	DEC 1986
	MAXIMUM IN 24 HOURS (IN)	56	1.61	2.00	2.53	3.72	3.92	4.32	3.39	4.59	4.02	4.54	2.68	1.44	4.59
	YEAR OF OCCURRENCE		1960	1962	1995	2001	1972	1957	1992	1975	1966	1973	2001	1955	AUG 1975
	NORMAL NO. DAYS WITH:														
	PRECIPITATION 0.01	30	6.8	6.8	8.7	10.0	10.8	10.5	9.6	8.6	8.5	6.2	6.6	6.3	99.4
	PRECIPITATION 1.00	30	*	0.1	0.2	0.4	0.6	0.8	0.6	0.6	1.0	0.4	0.3	0.0	5.0
SNOWFALL	NORMAL (IN)	30	6.7	7.7	8.0	2.2	0.*	0.0	0.0	0.0	0.*	0.6	5.1	7.8	38.1
	MAXIMUM MONTHLY (IN)	56	19.6	48.4	31.5	18.4	0.2	T	T	T	0.9	10.0	21.9	41.1	48.4
	YEAR OF OCCURRENCE		1969	1962	1951	1983	1954	1994	1995	2001	1985	1991	1985	1968	FEB 1962
	MAXIMUM IN 24 HOURS (IN)	56	11.8	26.0	18.9	10.5	0.2	T	T	T	0.9	8.8	12.6	16.6	26.0
	YEAR OF OCCURRENCE		1960	1962	1956	1994	1954	1994	1995	2001	1985	1991	1998	1968	FEB 1962
	MAXIMUM SNOW DEPTH (IN)	58	33	34	33	10	1	0	0	0	0	3	13	34	34
	YEAR OF OCCURRENCE		1969	1969	1969	1969	1947					1982	1983	1968	DEC 1968
	NORMAL NO. DAYS WITH:														
	SNOWFALL 1.0	30	2.2	2.1	2.2	0.6	0.0	0.0	0.0	0.0	0.0	0.2	1.6	2.2	11.1

PRECIPITATION (inches) 2001 SIOUX FALLS, SD (FSD)

YEAR	JAN	FEB	MAR	APR	MAY	JUN	JUL	AUG	SEP	OCT	NOV	DEC	ANNUAL
1972	0.18	0.40	0.97	2.73	7.25	2.09	3.49	2.65	1.75	1.78	1.89	1.25	26.43
1973	0.43	0.43	3.52	2.12	1.93	2.38	3.50	1.05	5.61	5.73	1.01	0.48	28.19
1974	0.13	0.30	1.65	1.33	3.11	2.79	1.27	5.16	0.58	0.34	0.27	0.10	17.03
1975	1.35	0.22	1.95	2.45	1.66	4.48	0.62	9.09	1.35	0.49	2.25	0.19	26.10
1976	0.41	0.48	1.60	2.15	1.02	1.02	1.53	1.31	0.76	0.71	0.07	0.36	11.42
1977	0.19	0.83	3.60	2.17	3.17	1.73	3.64	5.63	5.63	2.36	1.80	0.87	31.62
1978	0.47	0.33	0.56	3.98	3.47	2.91	4.79	3.08	2.45	0.14	0.48	0.51	23.17
1979	1.14	0.41	3.47	2.75	4.90	3.01	3.13	4.35	4.03	3.30	1.72	0.04	32.25
1980	0.18	0.47	0.70	0.77	2.52	2.17	1.63	2.92	0.79	1.36	0.02	0.29	13.82
1981	0.12	0.33	1.86	0.58	0.58	3.90	3.89	2.28	0.50	2.45	1.21	0.38	18.11
1982	0.76	0.13	1.17	1.87	4.72	1.18	4.60	5.23	3.49	5.18	2.94	1.99	33.26
1983	0.52	0.22	3.35	2.88	2.92	6.75	1.82	2.00	1.92	0.71	2.95	0.73	26.77
1984	0.37	1.10	1.83	5.79	2.95	8.43	1.63	0.76	1.62	4.11	0.03	1.02	29.64
1985	0.45	0.05	2.37	5.18	3.29	2.52	2.70	4.07	3.34	0.75	1.97	0.47	27.16
1986	0.72	0.05	1.50	5.15	2.42	3.93	2.59	2.77	9.26	1.22	0.89	T	30.50
1987	0.19	0.26	3.27	0.28	2.94	1.78	3.16	1.36	2.05	0.31	1.66	1.40	18.66
1988	1.54	0.25	0.63	3.00	1.54	0.91	0.49	4.02	4.39	0.02	1.98	0.37	19.14
1989	0.23	0.51	1.07	1.59	1.42	2.50	1.37	2.46	3.38	0.10	0.91	0.25	15.79
1990	0.08	0.31	1.57	1.86	4.07	4.86	1.77	1.17	0.47	1.82	0.61	0.61	19.20
1991	0.22	0.34	0.86	2.21	6.20	6.36	2.26	1.41	3.95	1.65	1.78	0.20	27.44
1992	0.75	1.76	2.36	2.01	1.80	2.44	8.41	5.29	3.06	2.72	1.04	0.83	32.47
1993	0.70	0.81	2.04	2.61	8.26	6.43	7.86	3.10	1.88	0.62	1.50	0.30	36.11
1994	0.97	0.63	0.20	3.34	1.26	6.03	1.70	2.66	2.36	2.36	1.03	0.33	22.87
1995	0.18	0.13	4.06	5.83	4.76	2.70	2.55	5.11	1.86	2.76	0.38	0.10	30.42
1996	0.99	0.16	0.82	0.55	5.27	1.14	0.98	1.79	2.82	1.63	2.91	0.78	19.84
1997	0.41	1.39	0.23	2.43	3.58	3.77	2.94	1.58	1.59	1.75	0.35	0.24	20.26
1998	0.50	0.67	4.08	3.57	1.92	4.52	2.66	3.29	1.19	6.28	2.20	0.24	31.12
1999	0.35	0.28	1.15	4.32	6.20	2.57	4.81	0.80	0.84	0.37	0.05	0.17	21.91
2000	0.68	1.04	0.91	2.27	5.56	3.26	3.22	3.17	1.34	1.79	2.52	0.35	26.11
2001	1.50	0.65	0.78	6.97	1.92	3.13	5.88	1.37	2.25	0.86	4.76	0.11	30.18
POR= 111 YRS	0.62	0.77	1.47	2.58	3.55	4.10	3.03	3.06	2.65	1.54	1.07	0.70	25.14

AVERAGE TEMPERATURE (F) 2001 SIOUX FALLS, SD (FSD)

YEAR	JAN	FEB	MAR	APR	MAY	JUN	JUL	AUG	SEP	OCT	NOV	DEC	ANNUAL
1972	8.9	11.5	31.5	43.8	59.1	66.7	70.2	70.5	59.5	44.4	31.9	14.8	42.7
1973	18.5	23.1	39.7	46.4	56.9	68.9	74.4	76.7	60.1	52.9	34.7	17.1	47.5
1974	12.8	24.3	34.5	48.9	55.7	65.9	79.2	67.6	57.7	51.4	33.9	23.6	46.3
1975	17.1	16.5	25.5	41.4	61.1	67.6	78.5	72.7	57.0	52.1	33.5	20.4	45.3
1976	15.6	29.5	34.8	51.0	57.1	71.5	76.7	75.2	62.9	44.3	27.6	16.6	46.9
1977	4.7	25.6	36.0	53.8	66.8	71.1	77.0	68.1	61.9	47.7	31.3	16.3	46.7
1978	1.9	8.6	30.5	44.3	57.7	66.6	70.9	71.1	66.7	47.8	31.3	14.6	42.7
1979	1.8	7.5	28.2	44.1	55.2	67.8	74.0	69.9	64.3	48.2	30.2	27.1	43.2
1980	18.1	18.1	31.7	50.1	58.4	68.8	74.2	71.1	62.0	45.5	36.7	21.2	46.3
1981	22.3	26.0	38.4	53.5	58.3	70.4	75.5	71.1	63.0	48.7	39.5	18.0	48.7
1982	3.8	20.2	32.8	43.9	59.6	62.6	74.4	71.3	60.0	48.8	30.3	24.5	44.4
1983	20.1	26.2	33.1	41.5	55.1	66.7	77.1	78.3	63.5	49.6	34.9	2.1	45.7
1984	17.4	27.8	24.6	45.8	55.9	68.1	73.6	74.2	57.2	50.4	36.2	20.4	46.0
1985	13.4	19.9	37.8	52.8	62.7	64.2	71.5	66.5	59.6	48.6	20.7	9.5	43.6
1986	20.7	16.9	36.9	48.2	58.8	69.6	75.0	66.5	59.6	48.6	28.7	25.2	46.2
1987	24.0	32.8	37.8	52.6	64.9	71.4	77.0	68.8	62.8	43.9	38.1	24.4	49.9
1988	9.6	15.0	36.4	46.4	65.1	76.3	77.4	74.8	62.6	44.6	34.2	22.1	47.0
1989	25.5	9.2	29.8	47.8	58.0	66.9	77.3	72.0	60.2	49.5	29.5	11.5	44.8
1990	28.2	24.9	36.7	46.4	56.4	70.1	71.2	72.9	66.5	48.0	35.6	15.3	47.7
1991	13.7	29.8	37.0	50.0	62.5	73.5	73.1	72.8	62.0	46.1	25.7	26.0	47.7
1992	26.5	30.0	36.7	43.7	60.5	67.0	65.5	66.1	60.3	48.7	30.4	19.2	46.2
1993	14.0	15.0	29.0	44.1	57.1	65.1	71.4	71.4	56.8	47.1	29.9	21.9	43.6
1994	6.2	13.3	36.6	46.5	63.9	71.6	70.8	69.5	65.2	52.4	36.5	21.8	46.2
1995	17.7	23.5	34.8	41.8	55.7	70.0	75.1	76.9	60.4	48.3	29.2	23.7	46.4
1996	10.9	22.6	27.4	42.4	54.9	68.8	69.1	70.0	59.4	47.7	23.5	10.3	42.3
1997	8.6	17.9	30.6	40.6	51.5	68.8	72.4	68.9	62.4	49.4	27.9	26.1	43.8
1998	19.6	31.8	28.0	46.5	61.7	63.1	71.9	71.0	66.3	50.6	35.5	24.0	47.5
1999	14.0	29.9	33.8	45.7	58.5	66.9	75.1	70.6	59.7	48.5	42.1	25.1	47.5
2000	17.3	30.4	39.4	46.0	59.2	65.7	71.7	71.6	61.4	51.1	25.0	7.7	45.5
2001	18.8	11.1	26.7	48.1	59.1	68.7	75.7	72.2	61.3	48.2	43.6	25.0	46.5
POR= 81 YRS	15.2	20.7	32.2	46.9	58.9	68.4	73.2	71.7	61.8	49.8	33.0	20.4	46.0

REFERENCE NOTES:

PAGE 1:
THE TEMPERATURE GRAPH SHOWS NORMAL MAXIMUM AND NORMAL MINIMUM DAILY TEMPERATURES (SOLID CURVES) AND THE ACTUAL DAILY HIGH AND LOW TEMPERATURES (VERTICAL BARS).

PAGE 2 AND 3:
H/C INDICATES HEATING AND COOLING DEGREE DAYS.
RH INDICATES RELATIVE HUMIDITY
W/O INDICATES WEATHER AND OBSTRUCTIONS
S INDICATES SUNSHINE.
PR INDICATES PRESSURE.
CLOUDINESS ON PAGE 3 IS THE SUM OF THE CEILOMETER AND SATELLITE DATA NOT TO EXCEED EIGHT EIGHTHS(OKTAS).

GENERAL:
T INDICATES TRACE PRECIPITATION, AN AMOUNT GREATER THAN ZERO BUT LESS THAN THE LOWEST REPORTABLE VALUE.
+ INDICATES THE VALUE ALSO OCCURS ON EARLIER DATES.
BLANK ENTRIES DENOTE MISSING OR UNREPORTED DATA.
NORMALS ARE 30-YEAR AVERAGES (1961 - 1990).
ASOS INDICATES AUTOMATED SURFACE OBSERVING SYSTEM.
PM INDICATES THE LAST DAY OF THE PREVIOUS MONTH.
POR (PERIOD OF RECORD) BEGINS WITH THE JANUARY DATA MONTH AND IS THE NUMBER OF YEARS USED TO COMPUTE THE MEAN. INDIVIDUAL MONTHS WITHIN THE POR MAY BE MISSING.
WHEN THE POR FOR A NORMAL IS LESS THAN 30 YEARS, THE NORMAL IS PROVISIONAL AND IS BASED ON THE NUMBER OF YEARS INDICATED.
0.* OR * INDICATES THE VALUE OR MEAN-DAYS-WITH IS BETWEEN 0.00 AND 0.05.
CLOUDINESS FOR ASOS STATIONS DIFFERS FROM THE NON-ASOS OBSERVATION TAKEN BY A HUMAN OBSERVER. ASOS STATION CLOUDINESS IS BASED ON TIME-AVERAGED CEILOMETER DATA FOR CLOUDS AT OR BELOW 12,000 FEET AND ON SATELLITE DATA FOR CLOUDS ABOVE 12,000 FEET.
THE NUMBER OF DAYS WITH CLEAR, PARTLY CLOUDY, AND CLOUDY CONDITIONS FOR ASOS STATIONS IS THE SUM OF THE CEILOMETER AND SATELLITE DATA FOR THE SUNRISE TO SUNSET PERIOD.

GENERAL CONTINUED:
CLEAR INDICATES 0 - 2 OKTAS, PARTLY CLOUDY INDICATES 3 - 6 OKTAS, AND CLOUDY INDICATES 7 OR 8 OKTAS. WHEN AT LEAST ONE OF THE ELEMENTS (CEILOMETER OR SATELLITE) IS MISSING, THE DAILY CLOUDINESS IS NOT COMPUTED.
WIND DIRECTION IS RECORDED IN TENS OF DEGREES (2 DIGITS) CLOCKWISE FROM TRUE NORTH. "00" INDICATES CALM. "36" INDICATES TRUE NORTH.
RESULTANT WIND IS THE VECTOR AVERAGE OF THE SPEED AND DIRECTION.
AVERAGE TEMPERATURE IS THE SUM OF THE MEAN DAILY MAXIMUM AND MINIMUM TEMPERATURE DIVIDED BY 2.
SNOWFALL DATA COMPRISE ALL FORMS OF FROZEN PRECIPITATION, INCLUDING HAIL.
A HEATING (COOLING) DEGREE DAY IS THE DIFFERENCE BETWEEN THE AVERAGE DAILY TEMPERATURE AND 65 F.
DRY BULB IS THE TEMPERATURE OF THE AMBIENT AIR.
DEW POINT IS THE TEMPERATURE TO WHICH THE AIR MUST BE COOLED TO ACHIEVE 100 PERCENT RELATIVE HUMIDITY.
WET BULB IS THE TEMPERATURE THE AIR WOULD HAVE IF THE MOISTURE CONTENT WAS INCREASED TO 100 PERCENT RELATIVE HUMIDITY.

ON JULY 1, 1996, THE NATIONAL WEATHER SERVICE BEGAN USING THE "METAR" OBSERVATION CODE THAT WAS ALREADY EMPLOYED BY MOST OTHER NATIONS OF THE WORLD. THE MOST NOTICEABLE DIFFERENCE IN THIS ANNUAL PUBLICATION WILL BE THE CHANGE IN UNITS FROM TENTHS TO EIGHTS(OKTAS) FOR REPORTING THE AMOUNT OF SKY COVER.

HEATING DEGREE DAYS (base 65 F) 2001 SIOUX FALLS, SD (FSD)

YEAR	JUL	AUG	SEP	OCT	NOV	DEC	JAN	FEB	MAR	APR	MAY	JUN	TOTAL
1972-73	29	43	194	631	984	1555	1440	1166	775	549	256	19	7641
1973-74	1	0	178	373	902	1481	1616	939	477	303	85		7488
1974-75	2	39	240	419	927	1279	1476	1351	1217	701	157	51	7859
1975-76	3	5	271	407	937	1377	1528	1023	929	418	261	23	7182
1976-77	1	0	142	643	1114	1495	1867	1097	891	341	38	4	7633
1977-78	0	22	125	530	1000	1503	1957	1576	1063	613	252	78	8719
1978-79	11	18	104	525	1004	1554	1960	1607	1134	622	321	40	8900
1979-80	0	28	103	512	1038	1168	1448	1349	1027	464	243	32	7412
1980-81	1	14	157	602	841	1354	1318	1087	816	353	230	5	6778
1981-82	5	5	114	497	758	1452	1899	1247	991	632	174	105	7879
1982-83	0	28	188	494	1035	1249	1385	1082	982	699	317	67	7526
1983-84	1	0	160	476	894	1947	1468	1072	1247	569	285	23	8142
1984-85	2	12	265	451	857	1376	1594	1260	836	378	124	101	7256
1985-86	5	44	269	562	1327	1721	1363	1341	865	498	204	17	8216
1986-87	0	54	180	504	1082	1227	1265	896	835	387	96	23	6549
1987-88	5	54	110	649	801	1252	1715	1448	878	554	83	1	7550
1988-89	0	22	126	628	916	1321	1219	1559	1083	514	235	61	7684
1989-90	0	5	192	481	1056	1655	1131	1117	871	586	268	39	7401
1990-91	11	7	109	527	875	1538	1587	977	859	455	192	7	7144
1991-92	6	3	180	583	1171	1200	1186	1009	871	636	187	35	7067
1992-93	35	63	159	500	1030	1414	1576	1397	1109	621	244	74	8222
1993-94	2	19	257	557	1046	1331	1821	1443	872	563	140	8	8059
1994-95	5	15	100	386	849	1332	1461	1157	931	689	281	55	7261
1995-96	3	0	198	523	1068	1275	1674	1224	1159	672	334	44	8174
1996-97	8	4	228	535	1237	1691	1738	1313	1060	727	413	7	8961
1997-98	20	26	130	497	1105	1202	1402	922	1138	549	147	117	7255
1998-99	4	2	79	440	881	1262	1574	976	961	572	208	72	7031
1999-00	0	4	199	505	686	1229	1471	997	789	565	205	80	6730
2000-01	23	8	167	425	1195	1771	1424	1501	1185	516	203	43	8461
2001-	7	7	161	522	637	1234							

COOLING DEGREE DAYS (base 65 F) 2001 SIOUX FALLS, SD (FSD)

YEAR	JAN	FEB	MAR	APR	MAY	JUN	JUL	AUG	SEP	OCT	NOV	DEC	ANNUAL
1972	0	0	0	0	60	108	197	219	35	0	0	0	619
1973	0	0	0	0	13	143	300	370	38	8	0	0	872
1974	0	0	0	2	23	118	450	126	26	6	0	0	751
1975	0	0	0	0	44	134	428	252	36	16	0	0	910
1976	0	0	0	2	24	226	370	324	85	9	0	0	1040
1977	0	0	0	11	101	197	382	123	40	0	0	0	854
1978	0	0	0	0	32	133	202	213	163	0	0	0	743
1979	0	0	0	3	25	131	289	189	87	0	0	0	724
1980	0	0	0	23	46	153	290	207	74	6	0	0	799
1981	0	0	0	14	27	175	342	203	61	0	0	0	822
1982	0	0	0	4	12	42	298	230	45	0	0	0	631
1983	0	0	0	0	15	122	381	420	120	2	0	0	1060
1984	0	0	0	0	10	120	276	305	36	4	0	0	751
1985	0	0	0	19	59	85	212	93	72	0	0	0	540
1986	0	0	0	0	20	163	318	108	25	0	0	0	634
1987	0	0	0	20	100	219	381	178	53	0	0	0	951
1988	0	0	0	1	94	349	393	330	61	0	0	0	1228
1989	0	0	0	6	25	125	387	228	56	7	0	0	834
1990	0	0	0	31	11	198	209	258	160	7	0	0	874
1991	0	0	0	11	119	265	264	249	100	5	0	0	1013
1992	0	0	0	4	53	58	103	103	25	5	0	0	351
1993	0	0	0	0	5	85	208	223	16	9	0	0	546
1994	0	0	0	13	113	212	191	160	113	2	0	0	804
1995	0	0	0	0	0	214	324	375	68	11	0	0	992
1996	0	0	0	0	27	163	140	167	63	4	0	0	564
1997	0	0	0	0	1	128	255	152	62	20	0	0	618
1998	0	0	0	0	54	69	225	196	123	0	0	0	667
1999	0	0	0	0	14	136	319	185	48	3	0	0	705
2000	0	0	0	0	31	106	240	219	64	4	0	0	664
2001	0	0	0	14	29	163	345	237	58	8	0	0	854

SNOWFALL (inches) 2001 SIOUX FALLS, SD (FSD)

YEAR	JUL	AUG	SEP	OCT	NOV	DEC	JAN	FEB	MAR	APR	MAY	JUN	TOTAL
1972-73	0.0	0.0	0.0	0.2	2.5	6.2	5.5	7.4	0.3	T	0.0	0.0	22.1
1973-74	0.0	0.0	0.0	T	0.1	11.6	1.9	2.9	5.1	3.3	0.0	0.0	24.9
1974-75	0.0	0.0	0.0	0.0	2.5	1.1	18.3	3.1	17.9	T	0.0	0.0	42.9
1975-76	0.0	0.0	0.0	T	13.2	2.5	6.5	6.6	7.9	T	0.1	0.0	36.8
1976-77	0.0	0.0	0.0	3.4	0.7	6.7	2.7	1.9	13.5	T	0.0	0.0	28.9
1977-78	0.0	0.0	0.0	0.9	8.6	8.0	7.6	6.5	5.5	0.6	0.0	0.0	37.7
1978-79	0.0	0.0	0.0	0.0	2.9	9.5	19.0	7.3	13.2	1.5	0.0	0.0	53.4
1979-80	0.0	0.0	0.0	T	15.2	0.3	2.9	5.5	5.8	T	0.0	0.0	29.7
1980-81	0.0	0.0	0.0	0.0	0.8	0.4	4.6	1.4	3.1	T	0.5	0.0	10.8
1981-82	0.0	0.0	0.0	0.8	8.9	5.5	16.9	1.0	1.8	7.5	0.0	0.0	42.4
1982-83	0.0	0.0	0.0	3.3	4.1	17.6	4.7	3.6	18.8	18.4	0.0	0.0	70.5
1983-84	0.0	0.0	0.0	T	19.0	13.7	5.0	11.9	19.4	6.0	0.0	0.0	75.0
1984-85	0.0	0.0	T	T	T	4.7	7.4	0.7	16.1	2.5	0.0	0.0	31.4
1985-86	0.0	0.0	0.9	0.2	21.9	9.1	9.1	0.9	8.2	0.3	0.0	0.0	50.6
1986-87	0.0	0.0	0.0	T	5.4	T	2.4	0.3	T	T	0.0	0.0	8.1
1987-88	0.0	0.0	0.0	0.1	7.5	13.3	17.9	7.3	2.5	11.3	0.0	0.0	59.9
1988-89	0.0	0.0	0.0	T	10.9	2.2	2.0	10.0	16.0	0.7	T	0.0	41.8
1989-90	0.0	0.0	0.0	0.0	2.4	4.0	0.2	5.8	0.7	T	0.0	T	13.1
1990-91	0.0	0.0	0.0	1.2	8.4	8.8	5.4	5.4	3.2	T	T	T	32.4
1991-92	0.0	0.0	0.0	10.0	10.8	0.2	3.8	11.2	8.4	3.5	T	T	47.9
1992-93	0.0	0.0	0.0	T	4.7	8.4	8.6	13.1	14.9	2.2	T	T	51.9
1993-94	T	0.0	T	T	8.8	4.6	16.8	13.0	1.1	14.9	0.0	0.0	59.2
1994-95	0.0	0.0	0.0	0.0	9.7	7.0	0.5	2.4	8.8	12.5	0.0	0.0	40.9
1995-96	T	0.0	0.0	7.4	5.1	1.9	15.4	1.7	6.9				
1996-97			0.0	T	11.3	19.8	8.8	16.5	1.3	6.0	T	0.0	
1997-98	0.0	0.0	0.0	0.4	5.0	3.3	11.7	6.9	21.4	0.3	0.0	T	49.0
1998-99	0.0	0.0	0.0	0.0	14.8	9.0	9.9	7.9	11.2	2.4	T	0.0	55.2
1999-00	0.0	T	0.0	2.7	T	3.2	6.0	1.2	4.5	5.6	T	0.0	23.2
2000-01	T	0.0	0.0	0.0	19.6	11.4	13.9	8.4	3.2	1.4	0.0	T	57.9
2001-	0.0	T	0.0	T	13.8	2.9							
POR= 55 YRS	T	0.0	0.0	0.8	6.0	6.9	6.9	7.8	9.1	2.9	0.2	T	40.6

2001
KNOXVILLE,
TENNESSEE (TYS)

Knoxville is located in a broad valley between the Cumberland Mountains, which lie northwest of the city, and the Great Smoky Mountains, which lie southeast of the city. These two mountain ranges exercise a marked influence upon the climate of the valley. The Cumberland Mountains, to the northwest, serve to retard and weaken the force of the cold winter air which frequently penetrates far south of the latitude of Knoxville over the plains areas to the west of the mountains.

The mountains also serve to modify the hot summer winds which are common to the plains to the west. In addition, they serve as a fixed incline plane which lifts the warm, moist air flowing northward from the Gulf of Mexico and thereby increases the frequency of afternoon thunderstorms. Relief from extremely high temperatures which such thunderstorms produce serves to reduce the number of extremely warm days in the valley.

July is usually the warmest month of the year. The coldest weather usually occurs during the month of January. Sudden great temperature changes occur infrequently. This again is due mainly to the retarding effect of the mountains. Summer nights are nearly always comfortable.

Rainfall is ample for agricultural purposes and is favorably distributed during the year for most crops. Precipitation is greatest in the wintertime. Another peak period occurs during the late spring and summer months. The period of lowest rainfall occurs during the fall. A cumulative total of approximately 12 inches of snow falls annually. However, this usually comes in amounts of less than 4 inches at one time. It is unusual for snow to remain on the ground in measurable amounts longer than one week.

The topography also has a pronounced effect upon the prevailing wind direction. Daytime winds usually have a southwesterly component, while nighttime winds usually move from the northeast. The winds are relatively light and tornadoes are extremely rare.

NORMALS, MEANS, AND EXTREMES
KNOXVILLE, TN (TYS)

LATITUDE:	LONGITUDE:	ELEVATION (FT):		TIME ZONE:	WBAN: 13891
35 49' 05" N	83 59' 09" W	GRND: 979	BARO: 982	EASTERN (UTC + 5)	

	ELEMENT	POR	JAN	FEB	MAR	APR	MAY	JUN	JUL	AUG	SEP	OCT	NOV	DEC	YEAR
TEMPERATURE F	NORMAL DAILY MAXIMUM	30	45.9	50.9	61.3	70.4	77.6	84.5	87.1	86.7	81.2	70.6	59.9	50.1	68.8
	MEAN DAILY MAXIMUM	54	47.3	52.3	60.5	70.3	78.1	85.0	88.0	87.2	81.5	71.3	59.4	50.0	69.2
	HIGHEST DAILY MAXIMUM	60	77	83	86	92	94	102	103	102	103	91	84	80	103
	YEAR OF OCCURRENCE		1950	1996	1963	1942	1962	1988	1952	1944	1954	1953	1948	1982	SEP 1954
	MEAN OF EXTREME MAXS.	54	67.2	70.9	78.5	85.0	88.1	92.8	94.9	94.1	91.6	83.4	75.9	68.5	82.6
	NORMAL DAILY MINIMUM	30	26.0	29.1	36.6	44.6	53.1	61.8	66.0	65.3	59.0	46.0	37.5	30.0	46.2
	MEAN DAILY MINIMUM	54	29.4	32.1	38.7	47.4	56.2	64.0	68.1	67.1	60.6	48.0	38.2	31.8	48.5
	LOWEST DAILY MINIMUM	60	-24	-8	1	22	32	43	49	49	36	25	5	-6	-24
	YEAR OF OCCURRENCE		1985	1996	1980	1987	1986	1956	1988	1946	1967	1987	1950	1983	JAN 1985
	MEAN OF EXTREME MINS.	54	9.7	12.9	22.2	31.4	41.4	52.6	59.7	58.8	47.1	32.7	22.7	14.7	33.8
	NORMAL DRY BULB	30	36.0	41.0	49.0	57.5	65.4	73.2	76.6	76.0	70.1	58.4	48.7	40.1	57.6
	MEAN DRY BULB	54	38.3	42.3	49.6	58.9	67.2	74.4	78.1	77.2	71.1	59.5	48.8	41.0	58.9
	MEAN WET BULB	18	35.0	38.7	44.1	51.8	60.9	67.7	67.1	66.2	60.4	53.5	45.0	37.6	52.3
	MEAN DEW POINT	18	29.7	32.6	37.3	45.6	56.4	64.4	64.2	63.2	56.8	49.2	39.8	32.9	47.7
	NORMAL NO. DAYS WITH:														
	MAXIMUM 90	30	0.0	0.0	0.0	*	0.8	5.1	10.4	8.5	2.8	0.0	0.0	0.0	27.6
	MAXIMUM 32	30	3.5	1.3	0.1	0.0	0.0	0.0	0.0	0.0	0.0	0.0	0.1	1.4	6.4
	MINIMUM 32	30	20.8	16.5	8.6	1.8	*	0.0	0.0	0.0	0.0	1.0	7.6	17.7	74.0
	MINIMUM 0	30	0.5	*	0.0	0.0	0.0	0.0	0.0	0.0	0.0	0.0	0.0	0.1	0.6
H/C	NORMAL HEATING DEG. DAYS	30	899	697	496	234	94	0	0	0	18	238	489	772	3937
	NORMAL COOLING DEG. DAYS	30	0	0	0	9	106	246	360	341	171	33	0	0	1266
RH	NORMAL (PERCENT)	30	72	68	65	63	71	74	76	76	76	73	72	73	72
	HOUR 01 LST	30	77	74	71	72	82	86	87	88	87	85	80	78	81
	HOUR 07 LST	30	81	80	80	81	86	88	90	92	92	89	84	82	85
	HOUR 13 LST	30	63	59	54	51	57	58	61	61	60	56	59	64	59
	HOUR 19 LST	30	66	60	56	52	60	62	65	66	67	65	66	68	63
S	PERCENT POSSIBLE SUNSHINE	57	40	47	53	63	64	65	64	63	61	61	49	40	56
W/O	MEAN NO. DAYS WITH:														
	HEAVY FOG(VISBY 1/4 MI)	59	2.9	1.8	1.6	1.2	2.3	2.1	2.1	3.3	3.6	4.4	3.1	2.6	31.0
	THUNDERSTORMS	59	0.8	1.5	3.1	4.4	6.9	8.4	9.5	6.8	3.1	1.3	1.0	0.5	47.3
CLOUDINESS	MEAN:														
	SUNRISE-SUNSET (OKTAS)	1					4.0	3.2							
	MIDNIGHT-MIDNIGHT (OKTAS)	1						3.2							
	MEAN NO. DAYS WITH:														
	CLEAR	1	4.0	1.0	4.0		13.0	11.0							
	PARTLY CLOUDY	1			1.0		3.0	5.0							
	CLOUDY	1	2.0	3.0	10.0		4.0	4.0							
PR	MEAN STATION PRESSURE(IN)	29	29.10	29.00	29.00	29.00	28.99	29.00	29.00	29.00	29.00	29.10	29.10	29.09	29.03
	MEAN SEA-LEVEL PRES. (IN)	18	30.13	30.11	30.05	29.99	30.00	30.01	30.03	30.02	30.06	30.10	30.13	30.15	30.07
WINDS	MEAN SPEED (MPH)	44	7.6	8.0	8.5	8.5	6.9	6.4	6.1	5.5	5.6	5.5	6.7	7.2	6.9
	PREVAIL.DIR(TENS OF DEGS)	28	24	24	24	23	23	24	23	23	05	04	24	23	24
	MAXIMUM 2-MINUTE:														
	SPEED (MPH)	6	43	39	43	64	40	34	43	38	28	43	49	39	64
	DIR. (TENS OF DEGS)		27	22	24	28	20	15	24	03	26	26	25	25	28
	YEAR OF OCCURRENCE		1996	2000	1996	1996	1999	1997	1996	2000	2000	2001	2000	1996	APR 1996
	MAXIMUM 5-SECOND:														
	SPEED (MPH)	6	53	48	48	76	52	51	68	48	36	56	61	57	76
	DIR. (TENS OF DEGS)		22	23	25	27	20	14	25	03	23	26	24	28	27
	YEAR OF OCCURRENCE		1996	1998	1996	1996	1999	1997	1997	2000	2000	2001	2000	1996	APR 1996
PRECIPITATION	NORMAL (IN)	30	4.17	4.06	5.09	3.72	4.13	3.97	4.67	3.13	3.07	2.84	3.75	4.54	47.14
	MAXIMUM MONTHLY (IN)	60	11.74	9.38	11.81	11.07	10.98	8.21	12.66	8.88	9.19	6.67	10.36	11.63	12.66
	YEAR OF OCCURRENCE		1954	1944	1994	1998	1974	1989	1999	1942	1989	1949	1948	1961	JUL 1999
	MINIMUM MONTHLY (IN)	60	0.95	0.74	1.69	0.39	0.74	0.20	0.33	0.77	0.42	T	0.97	0.45	T
	YEAR OF OCCURRENCE		1986	1968	1986	1976	1970	1944	1995	1954	1985	1963	1942	1965	OCT 1963
	MAXIMUM IN 24 HOURS (IN)	60	3.89	3.42	5.77	3.69	3.40	3.57	4.69	3.25	5.08	2.44	4.06	4.89	5.77
	YEAR OF OCCURRENCE		1946	1991	1994	1998	1984	1972	1942	1959	1944	1961	1948	1969	MAR 1994
	NORMAL NO. DAYS WITH:														
	PRECIPITATION 0.01	30	11.9	11.0	12.5	10.9	11.3	10.1	11.5	9.3	8.2	8.0	10.5	11.0	126.2
	PRECIPITATION 1.00	30	0.7	0.9	1.4	0.7	1.0	1.0	1.4	0.7	0.8	0.8	0.7	1.2	11.3
SNOWFALL	NORMAL (IN)	30	5.4	3.9	1.3	0.7	0.0	0.0	0.0	0.0	0.0	T	0.2	1.9	13.4
	MAXIMUM MONTHLY (IN)	59	15.1	23.3	20.2	10.7	T	T	0.0	T	0.0	T	18.2	12.2	23.3
	YEAR OF OCCURRENCE		1962	1960	1960	1987	1945	1998		1995		1993	1952	1963	FEB 1960
	MAXIMUM IN 24 HOURS (IN)	59	12.0	17.5	14.1	10.7	T	T	0.0	T	0.0	T	18.2	8.9	18.2
	YEAR OF OCCURRENCE		1962	1960	1993	1987	1945	1998		1995		1993	1952	1969	NOV 1952
	MAXIMUM SNOW DEPTH (IN)	52	10	15	15	7	0	0	0	0	0	0	10	6	15
	YEAR OF OCCURRENCE		1966	1960	1993	1987							1952	1963	MAR 1993
	NORMAL NO. DAYS WITH:														
	SNOWFALL 1.0	30	1.7	1.2	0.6	0.1	0.0	0.0	0.0	0.0	0.0	0.0	0.1	0.6	4.3

PRECIPITATION (inches) 2001 KNOXVILLE, TN (TYS)

YEAR	JAN	FEB	MAR	APR	MAY	JUN	JUL	AUG	SEP	OCT	NOV	DEC	ANNUAL
1972	7.35	4.19	4.98	2.54	4.49	5.02	6.76	1.61	4.70	5.99	3.36	7.02	58.01
1973	3.24	2.59	10.24	5.15	5.71	5.26	4.38	2.31	3.28	3.48	5.01	7.38	58.03
1974	7.05	5.24	6.15	5.77	10.98	2.70	2.92	3.14	3.33	2.35	5.18	4.52	59.33
1975	4.66	4.68	10.42	2.43	2.98	2.43	2.25	1.61	3.28	4.02	2.92	3.59	45.27
1976	3.86	2.18	5.22	0.39	5.53	3.46	3.75	1.98	2.87	5.33	3.45	4.42	42.44
1977	2.55	1.52	6.08	6.96	1.16	6.49	1.08	5.78	6.91	4.04	5.06	3.30	50.93
1978	5.22	1.01	4.42	4.10	3.44	5.27	5.06	2.44	1.26	0.82	3.62	5.91	42.57
1979	6.18	4.17	4.21	4.30	7.21	3.80	9.47	2.29	2.64	1.97	5.73	1.92	53.89
1980	5.54	1.78	8.72	3.30	3.80	1.94	3.57	2.24	2.38	1.53	3.78	1.78	40.46
1981	1.05	3.62	2.83	4.84	3.02	5.53	2.03	3.48	6.09	4.15	3.01	4.14	43.79
1982	6.03	4.88	6.36	3.26	5.52	3.93	6.60	2.68	2.66	5.21	4.89	54.70
1983	1.58	2.90	1.99	5.88	5.42	3.26	3.18	3.89	0.95	3.34	4.40	5.69	42.48
1984	2.26	4.42	3.79	3.37	10.14	4.34	9.03	1.72	0.85	3.26	2.87	2.49	48.54
1985	3.17	4.11	1.98	2.86	1.60	4.77	2.63	4.07	0.42	3.04	5.39	2.36	36.40
1986	0.95	3.90	1.69	2.25	2.40	0.69	1.89	3.37	3.59	3.84	3.83	4.08	32.48
1987	4.68	4.63	2.91	2.18	4.62	2.66	4.67	1.08	1.93	0.60	1.21	3.49	34.66
1988	4.29	2.94	2.42	2.34	2.35	0.51	3.60	3.20	2.68	1.52	4.82	3.99	34.66
1989	4.96	6.26	3.82	3.50	5.31	8.21	2.68	3.16	9.19	1.47	4.92	2.74	56.22
1990	5.88	6.90	5.73	2.56	4.71	1.72	7.56	3.02	2.68	3.64	1.77	8.99	55.16
1991	2.53	6.99	6.36	3.97	3.10	6.02	3.45	6.13	3.14	1.10	5.24	10.23	58.26
1992	3.87	3.36	3.99	2.11	3.20	3.07	4.62	4.22	1.77	3.05	4.29	6.68	44.23
1993	4.09	2.20	6.16	2.50	3.78	0.90	2.03	6.04	4.77	2.25	3.33	7.04	45.09
1994	7.08	8.81	11.01	7.90	3.04	5.97	5.94	4.19	1.20	2.38	2.93	2.02	63.27
1995	4.90	4.54	3.54	1.74	7.05	3.90	0.33	2.08	2.46	3.58	6.21	2.50	42.83
1996	7.53	2.74	4.72	4.50	4.57	2.96	4.77	1.33	3.45	1.17	7.71	5.44	50.89
1997	5.20	5.17	6.34	3.82	6.76	5.70	3.75	2.02	3.32	2.90	2.78	2.37	50.13
1998	4.62	2.71	4.75	11.07	3.84	7.96	5.69	2.07	1.81	1.42	2.51	5.95	53.90
1999	6.62	3.50	4.85	3.40	4.92	5.58	12.66	0.85	0.82	2.84	2.40	1.70	50.14
2000	5.14	3.42	4.37	6.69	5.90	3.36	6.12	1.40	3.82	T	4.35	2.45	47.02
2001	4.74	6.46	2.63	2.82	3.46	3.88	4.14	3.74	3.56	0.86	1.55	4.66	42.50
POR= 131 YRS	4.63	4.55	5.12	4.12	3.90	4.06	4.48	3.54	2.80	2.56	3.40	4.38	47.54

AVERAGE TEMPERATURE (F) 2001 KNOXVILLE, TN (TYS)

YEAR	JAN	FEB	MAR	APR	MAY	JUN	JUL	AUG	SEP	OCT	NOV	DEC	ANNUAL
1972	42.5	39.4	48.2	59.5	64.9	70.8	75.5	75.9	71.9	56.8	47.5	44.8	58.1
1973	38.0	39.9	55.8	56.4	63.5	74.8	76.7	76.2	73.8	62.4	52.4	41.1	59.3
1974	49.3	43.1	55.2	59.8	68.1	70.5	77.7	76.5	69.8	58.6	50.8	41.9	60.1
1975	43.3	45.5	46.6	57.5	70.1	74.6	77.6	78.8	68.9	60.1	50.2	41.1	59.5
1976	34.3	48.2	52.6	58.9	64.3	73.9	76.2	74.9	67.7	54.5	43.2	37.5	57.1
1977	27.2	41.4	55.0	62.9	69.8	75.3	80.3	78.9	74.0	57.4	52.6	40.2	59.6
1978	29.4	34.6	49.0	61.1	66.6	75.4	78.7	77.1	74.8	58.2	54.4	42.8	58.5
1979	33.0	36.7	53.1	59.9	66.9	72.8	75.0	77.6	72.2	58.3	51.4	42.2	58.3
1980	41.6	36.6	47.6	59.1	68.1	75.2	82.1	81.7	73.7	56.0	47.3	40.1	59.1
1981	33.4	41.9	46.8	63.5	63.9	77.1	81.4	78.3	70.2	58.9	49.4	38.7	58.6
1982	33.8	43.1	53.0	55.2	69.3	72.6	77.6	76.7	70.0	59.5	50.9	46.4	59.0
1983	38.0	41.1	49.7	53.2	65.4	72.5	78.5	79.6	70.3	60.2	47.4	36.4	57.7
1984	35.3	43.46	47.7	57.3	62.8	74.8	73.3	74.1	67.1	67.6	46.3	47.5	58.1
1985	29.4	37.1	50.2	58.6	65.7	72.5	76.5	75.1	68.5	63.7	56.8	34.8	57.4
1986	35.0	43.9	48.7	58.6	68.1	77.2	81.4	76.7	72.8	60.2	52.4	39.1	59.5
1987	36.8	42.2	49.2	55.0	71.3	75.5	78.5	79.6	70.1	52.8	50.2	43.5	58.7
1988	33.9	39.4	50.1	57.2	65.0	74.4	78.7	79.6	71.3	52.7	49.7	40.4	57.7
1989	42.8	41.3	53.5	57.8	62.4	73.3	77.5	76.8	70.9	59.4	48.1	32.2	58.0
1990	44.7	48.9	53.6	57.9	65.9	75.4	78.0	77.9	72.1	59.9	52.1	45.1	61.0
1991	41.2	43.8	52.0	62.8	72.0	74.9	79.3	76.5	71.4	61.0	47.1	43.5	60.5
1992	40.5	45.8	48.3	59.0	64.0	72.2	77.9	73.7	71.3	57.4	48.3	40.0	58.2
1993	43.5	39.5	45.7	56.2	67.0	76.0	83.0	78.7	70.6	57.9	48.4	40.2	58.9
1994	32.0	43.4	49.5	61.9	63.9	76.5	76.8	75.8	69.0	60.1	53.6	44.8	58.9
1995	39.2	40.7	53.1	61.0	67.3	73.8	80.2	81.5	70.5	58.6	43.6	37.1	58.9
1996	36.4	39.6	44.8	55.1	69.4	74.2	76.8	76.2	68.9	59.2	44.1	43.3	57.3
1997	39.4	46.6	54.2	54.0	62.2	71.9	78.4	75.7	71.7	58.9	43.9	39.2	58.0
1998	42.8	46.0	49.4	57.3	70.7	75.9	79.4	77.7	76.0	61.9	50.1	44.4	61.0
1999	43.6	44.1	45.3	61.2	66.2	75.0	78.6	77.7	69.9	58.5	52.7	42.4	59.6
2000	37.5	46.1	53.2	55.9	69.9	74.8	77.3	76.9	70.7	61.0	47.4	34.0	58.7
2001	35.5	45.8	46.3	61.7	68.4	72.9	78.0	77.5	68.2	56.9	54.2	45.6	59.3
POR= 131 YRS	38.9	41.8	49.4	58.7	67.2	74.7	77.8	76.8	71.3	59.7	48.4	40.6	58.8

REFERENCE NOTES:

PAGE 1:
THE TEMPERATURE GRAPH SHOWS NORMAL MAXIMUM AND NORMAL
MINIMUM DAILY TEMPERATURES (SOLID CURVES) AND THE
ACTUAL DAILY HIGH AND LOW TEMPERATURES (VERTICAL BARS).

PAGE 2 AND 3:
H/C INDICATES HEATING AND COOLING DEGREE DAYS.
RH INDICATES RELATIVE HUMIDITY
W/O INDICATES WEATHER AND OBSTRUCTIONS
S INDICATES SUNSHINE.
PR INDICATES PRESSURE.
CLOUDINESS ON PAGE 3 IS THE SUM OF THE CEILOMETER AND
SATELLITE DATA NOT TO EXCEED EIGHT EIGHTHS(OKTAS) .

GENERAL:
T INDICATES TRACE PRECIPITATION, AN AMOUNT GREATER
THAN ZERO BUT LESS THAN THE LOWEST REPORTABLE VALUE.
+ INDICATES THE VALUE ALSO OCCURS ON EARLIER DATES.
BLANK ENTRIES DENOTE MISSING OR UNREPORTED DATA.
NORMALS ARE 30-YEAR AVERAGES (1961 - 1990).
ASOS INDICATES AUTOMATED SURFACE OBSERVING SYSTEM.
PM INDICATES THE LAST DAY OF THE PREVIOUS MONTH.
POR (PERIOD OF RECORD) BEGINS WITH THE JANUARY DATA
MONTH AND IS THE NUMBER OF YEARS USED TO COMPUTE
THE MEAN. INDIVIDUAL MONTHS WITHIN THE POR MAY
BE MISSING.
WHEN THE POR FOR A NORMAL IS LESS THAN 30 YEARS,
THE NORMAL IS PROVISIONAL AND IS BASED ON THE NUMBER
OF YEARS INDICATED.
0.* OR * INDICATES THE VALUE OR MEAN-DAYS-WITH
IS BETWEEN 0.00 AND 0.05.
CLOUDINESS FOR ASOS STATIONS DIFFERS FROM THE NON-ASOS
OBSERVATION TAKEN BY A HUMAN OBSERVER. ASOS STATION
CLOUDINESS IS BASED ON TIME-AVERAGED CEILOMETER DATA
FOR CLOUDS AT OR BELOW 12,000 FEET AND ON SATELLITE
DATA FOR CLOUDS ABOVE 12,000 FEET.
THE NUMBER OF DAYS WITH CLEAR, PARTLY CLOUDY, AND
CLOUDY CONDITIONS FOR ASOS STATIONS IS THE SUM
OF THE CEILOMETER AND SATELLITE DATA FOR THE
SUNRISE TO SUNSET PERIOD.

GENERAL CONTINUED:
CLEAR INDICATES 0 - 2 OKTAS, PARTLY CLOUDY INDICATES
3 - 6 OKTAS, AND CLOUDY INDICATES 7 OR 8 OKTAS.
WHEN AT LEAST ONE OF THE ELEMENTS (CEILOMETER OR
SATELLITE) IS MISSING, THE DAILY CLOUDINESS IS
NOT COMPUTED.
WIND DIRECTION IS RECORDED IN TENS OF DEGREES (2 DIGITS)
CLOCKWISE FROM TRUE NORTH. "00" INDICATES CALM. "36"
INDICATES TRUE NORTH.
RESULTANT WIND IS THE VECTOR AVERAGE OF THE SPEED AND
DIRECTION.
AVERAGE TEMPERATURE IS THE SUM OF THE MEAN DAILY MAXIMUM
AND MINIMUM TEMPERATURE DIVIDED BY 2.
SNOWFALL DATA COMPRISE ALL FORMS OF FROZEN
PRECIPITATION, INCLUDING HAIL.
A HEATING (COOLING) DEGREE DAY IS THE DIFFERENCE BETWEEN
THE AVERAGE DAILY TEMPERATURE AND 65 F.
DRY BULB IS THE TEMPERATURE OF THE AMBIENT AIR.
DEW POINT IS THE TEMPERATURE TO WHICH THE AIR MUST BE
COOLED TO ACHIEVE 100 PERCENT RELATIVE HUMIDITY.
WET BULB IS THE TEMPERATURE THE AIR WOULD HAVE IF THE
MOISTURE CONTENT WAS INCREASED TO 100 PERCENT RELATIVE
HUMIDITY.

ON JULY 1, 1996, THE NATIONAL WEATHER SERVICE BEGAN USING
THE "METAR" OBSERVATION CODE THAT WAS ALREADY EMPLOYED
BY MOST OTHER NATIONS OF THE WORLD. THE MOST NOTICEABLE
DIFFERENCE IN THIS ANNUAL PUBLICATION WILL BE THE CHANGE
IN UNITS FROM TENTHS TO EIGHTS(OKTAS) FOR REPORTING THE
AMOUNT OF SKY COVER.

HEATING DEGREE DAYS (base 65 F) 2001 KNOXVILLE, TN (TYS)

YEAR	JUL	AUG	SEP	OCT	NOV	DEC	JAN	FEB	MAR	APR	MAY	JUN	TOTAL
1972-73	0	0	9	251	522	622	828	697	283	272	97	0	3581
1973-74	0	0	10	135	373	734	481	606	304	190	41	10	2884
1974-75	0	0	23	215	439	713	666	542	563	257	9	0	3427
1975-76	0	0	42	172	437	733	944	485	380	203	91	0	3487
1976-77	0	0	19	320	651	845	1166	658	319	110	33	2	4123
1977-78	0	0	3	242	374	764	1097	846	487	148	74	0	4035
1978-79	0	0	0	210	310	681	985	786	376	167	46	0	3561
1979-80	0	0	0	220	407	700	715	815	529	185	32	1	3604
1980-81	0	0	23	284	523	761	974	641	556	104	94	0	3960
1981-82	0	0	32	196	461	809	959	606	384	295	20	0	3762
1982-83	0	0	30	228	416	577	829	662	472	356	50	3	3623
1983-84	0	0	51	163	523	878	912	619	530	240	139	5	4060
1984-85	0	0	44	30	556	536	1095	776	459	208	51	13	3768
1985-86	0	0	44	86	250	927	922	582	499	206	51	0	3567
1986-87	0	2	0	197	377	797	863	631	483	313	13	0	3676
1987-88	0	0	15	370	437	660	956	734	458	247	66	9	3952
1988-89	0	0	3	384	454	755	681	660	360	258	148	1	3704
1989-90	0	1	36	204	499	1011	622	443	358	239	68	0	3481
1990-91	0	0	26	182	382	612	730	589	405	93	16	0	3035
1991-92	0	0	46	159	538	660	753	549	512	225	113	6	3561
1992-93	0	0	13	229	494	769	659	708	592	270	38	2	3774
1993-94	0	0	37	232	495	761	1016	597	473	139	93	0	3843
1994-95	0	0	11	158	336	620	794	674	364	162	54	0	3173
1995-96	0	0	22	216	638	857	877	732	622	302	45	0	4311
1996-97	0	0	30	196	623	663	786	512	332	333	124	8	3607
1997-98	0	0	1	226	625	794	681	528	500	234	30	12	3631
1998-99	0	0	1	133	438	632	656	578	604	149	24	0	3215
1999-00	0	0	28	206	359	694	846	542	354	265	9	6	3309
2000-01	0	0	34	149	522	955	908	530	571	155	25	1	3850
2001-	0	0	57	263	320	594							

COOLING DEGREE DAYS (base 65 F) 2001 KNOXVILLE, TN (TYS)

YEAR	JAN	FEB	MAR	APR	MAY	JUN	JUL	AUG	SEP	OCT	NOV	DEC	ANNUAL
1972	0	0	0	53	51	193	332	347	223	4	2	0	1205
1973	0	0	4	19	56	298	369	355	282	60	2	0	1445
1974	0	0	9	39	142	179	399	365	171	19	17	0	1340
1975	0	0	0	39	173	293	397	434	168	26	0	0	1530
1976	0	3	1	26	48	273	356	314	107	5	0	0	1133
1977	0	5	14	57	188	317	483	440	283	14	10	0	1811
1978	0	0	0	37	130	319	432	384	302	8	0	0	1612
1979	0	0	15	20	111	242	317	399	224	23	4	0	1355
1980	0	0	0	16	136	315	538	525	290	12	0	0	1832
1981	0	0	0	62	65	373	512	421	193	14	1	0	1641
1982	0	0	17	6	157	233	396	372	187	65	2	8	1443
1983	0	0	4	6	71	235	425	462	217	23	0	0	1443
1984	0	0	0	12	77	306	263	290	116	120	2	0	1186
1985	0	0	11	20	81	247	363	322	155	53	11	0	1263
1986	0	0	0	21	156	374	517	373	241	55	6	0	1743
1987	0	0	0	18	215	321	427	460	172	0	0	0	1613
1988	0	0	3	19	74	297	431	458	200	9	0	0	1491
1989	0	2	10	47	76	257	395	374	219	35	0	0	1415
1990	0	1	13	32	101	316	410	406	245	32	1	0	1557
1991	0	0	9	32	239	306	453	364	244	42	6	2	1697
1992	0	0	0	49	91	230	410	279	206	0	0	0	1265
1993	0	0	0	10	106	339	566	432	212	21	3	0	1689
1994	0	0	0	54	67	349	372	342	137	15	0	0	1336
1995	0	0	3	50	134	268	480	521	192	26	1	0	1675
1996	0	4	0	10	187	285	371	353	152	22	3	0	1387
1997	0	1	5	9	44	222	423	338	209	46	0	0	1297
1998	0	0	24	8	212	346	454	402	338	43	0	1	1828
1999	0	0	0	41	67	304	429	403	181	13	0	0	1438
2000	0	0	0	2	167	309	390	376	211	33	4	0	1492
2001	0	0	0	62	137	247	411	393	160	19	1	0	1430

SNOWFALL (inches) 2001 KNOXVILLE, TN (TYS)

YEAR	JUL	AUG	SEP	OCT	NOV	DEC	JAN	FEB	MAR	APR	MAY	JUN	TOTAL
1972-73	0.0	0.0	0.0	0.0	T	T	9.0	T	1.9	T	0.0	0.0	10.9
1973-74	0.0	0.0	0.0	0.0	0.0	2.0	T	T	1.8	0.0	0.0	0.0	3.8
1974-75	0.0	0.0	0.0	0.0	T	1.6	1.3	T	2.5	0.0	0.0	0.0	5.4
1975-76	0.0	0.0	0.0	0.0	T	T	0.3	2.8	0.0	0.0	0.0	0.0	3.1
1976-77	0.0	0.0	0.0	0.0	1.0	2.1	7.9	0.2	0.0	0.0	0.0	0.0	11.2
1977-78	0.0	0.0	0.0	0.0	0.1	0.6	11.3	5.5	1.8	0.0	0.0	0.0	19.3
1978-79	0.0	0.0	0.0	0.0	0.0	T	4.7	18.4	0.0	0.0	0.0	0.0	23.1
1979-80	0.0	0.0	0.0	0.0	T	T	0.5	11.0	3.5	T	0.0	0.0	15.0
1980-81	0.0	0.0	0.0	0.0	T	T	5.0	2.5	T	0.0	0.0	0.0	7.5
1981-82	0.0	0.0	0.0	0.0	T	0.1	4.4	0.1	1.1	T	0.0	0.0	5.7
1982-83	0.0	0.0	0.0	0.0	0.0	3.0	1.1	3.4	0.7	2.0	0.0	0.0	10.2
1983-84	0.0	0.0	0.0	0.0	0.0	0.7	2.8	3.5	T	0.0	0.0	0.0	7.0
1984-85	0.0	0.0	0.0	0.0	0.0	T	14.2	8.3	0.0	0.0	0.0	0.0	22.5
1985-86	0.0	0.0	0.0	0.0	0.0	0.4	3.6	5.0	T	0.0	0.0	0.0	9.0
1986-87	0.0	0.0	0.0	0.0	T	T	7.5	T	1.6	10.7	0.0	0.0	19.8
1987-88	0.0	0.0	0.0	0.0	T	T	9.2	0.9	T	0.0	0.0	0.0	10.1
1988-89	0.0	0.0	0.0	0.0	T	2.6	1.8	2.8	0.0	0.0	0.0	0.0	7.2
1989-90	0.0	0.0	0.0	T	T	1.1	T	T	T	0.0	0.0	0.0	1.1
1990-91	0.0	0.0	0.0	0.0	0.0	T	T	1.2	0.7	T	0.0	0.0	1.9
1991-92	0.0	0.0	0.0	0.0	0.2	T	0.2	T	T	0.0	0.0	0.0	0.2
1992-93	0.0	0.0	0.0	0.0	0.0	0.3	T	1.0	15.1	T	0.0	0.0	16.4
1993-94	0.0	0.0	0.0	T	0.0	0.3	3.3	0.6	0.6	0.0	0.0	0.0	4.2
1994-95	0.0	0.0	0.0	0.0	0.0	0.0	2.0	1.1	0.6	0.0	0.0	0.0	3.7
1995-96	0.0	T	0.0	0.0	T	10.0							
1996-97							1.5						
1997-98					T	3.3	0.4	0.1	T	0.0	T	T	4.1
1998-99	0.0	0.0	0.0	0.0	0.0	T	T	2.1	2.0	0.0	T	0.0	2.0
1999-00	0.0	0.0	0.0	0.0	0.0	T	2.0	T	T	0.0	T	0.0	2.0
2000-01													
2001-													
POR= 55 YRS	0.0	T	0.0	T	0.6	1.5	3.7	3.3	1.7	0.4	T	0.0	11.2

2001
MEMPHIS,
TENNESSEE (MEM)

Topography varies from the level alluvial area in east-central Arkansas to the slightly rolling area in northwestern Mississippi and southwestern Tennessee.

Agricultural interests are varied, with major crops being cotton, corn, hay, soybeans, peaches, apples, and a considerable number of vegetables. The climate is quite favorable for dairy interests, and for the raising of cattle and hogs.

The growing season is about 230 days in length. The average date for the last occurrence of temperatures as low as 32 degrees is late March. The average date of the first temperature of 32 degrees or below is early November.

Precipitation of nearly 50 inches per year is fairly well distributed. Crops and pastures receive, on the average, an adequate supply of moisture during the growing season, with lesser amounts during the fall harvesting period.

Sunshine averages slightly over 70 percent of the possible amount during the growing season. Relative humidity averages about 70 percent for the year.

Memphis, although not in the normal paths of storms coming from the Gulf or from western Canada, is affected by both, and thereby has comparatively frequent changes in weather. Extremely high or low temperatures, however, are relatively rare.

NORMALS, MEANS, AND EXTREMES
MEMPHIS, TN (MEM)

LATITUDE: 35 03' 40" N LONGITUDE: 89 59' 06" W ELEVATION (FT): GRND: 283 BARO: 286 TIME ZONE: CENTRAL (UTC + 6) WBAN: 13893

ELEMENT	POR	JAN	FEB	MAR	APR	MAY	JUN	JUL	AUG	SEP	OCT	NOV	DEC	YEAR
TEMPERATURE F														
NORMAL DAILY MAXIMUM	30	48.5	53.5	63.2	73.3	81.0	89.3	92.3	90.8	83.9	74.3	62.3	52.5	72.1
MEAN DAILY MAXIMUM	54	49.0	53.9	62.1	72.8	80.9	88.6	91.6	90.5	84.5	74.6	61.7	52.2	71.9
HIGHEST DAILY MAXIMUM	60	78	81	85	94	99	104	108	107	103	95	86	81	108
YEAR OF OCCURRENCE		1972	1962	1986	1987	1977	1954	1980	1954	1954	2000	1982		JUL 1980
MEAN OF EXTREME MAXS.	54	70.3	73.5	79.6	85.7	90.8	96.2	98.1	97.5	94.2	87.5	79.0	71.1	85.3
NORMAL DAILY MINIMUM	30	30.9	34.8	43.0	52.4	61.2	68.9	72.9	71.1	64.5	51.9	42.7	34.8	52.4
MEAN DAILY MINIMUM	54	31.6	35.3	42.6	52.3	61.4	69.3	73.0	71.3	64.1	52.2	42.0	34.7	52.5
LOWEST DAILY MINIMUM	60	-4	-11	12	29	38	48	52	48	36	25	9	-13	-13
YEAR OF OCCURRENCE		1985	1951	1943	1987	1944	1966	1947	1946	1949	1952	1950	1963	DEC 1963
MEAN OF EXTREME MINS.	54	13.4	17.9	26.1	35.7	47.6	57.6	64.2	62.0	49.2	36.5	25.1	17.1	37.7
NORMAL DRY BULB	30	39.7	44.2	53.1	62.9	71.2	79.1	82.6	81.0	74.2	63.1	52.5	43.7	62.3
MEAN DRY BULB	54	40.3	44.7	52.4	62.6	71.3	78.9	82.3	80.8	74.2	63.4	51.8	43.4	62.2
MEAN WET BULB	17	36.9	41.4	46.9	54.9	63.9	70.9	70.1	68.9	66.4	53.4	44.9	39.8	54.9
MEAN DEW POINT	17	31.5	35.4	40.2	48.5	59.1	66.9	70.6	69.1	62.1	48.3	39.6	34.3	50.5
NORMAL NO. DAYS WITH:														
MAXIMUM 90	30	0.0	0.0	0.0	0.2	3.2	13.6	20.9	17.4	7.3	0.6	0.0	0.0	63.2
MAXIMUM 32	30	3.8	1.4	0.1	0.0	0.0	0.0	0.0	0.0	0.0	0.0	0.1	1.7	7.1
MINIMUM 32	30	19.1	13.4	4.8	0.4	0.0	0.0	0.0	0.0	0.0	0.0	4.2	14.2	56.2
MINIMUM 0	30	0.2	0.0	0.0	0.0	0.0	0.0	0.0	0.0	0.0	0.0	0.0	0.2	0.4
H/C														
NORMAL HEATING DEG. DAYS	30	784	582	383	127	25	0	0	0	10	131	380	660	3082
NORMAL COOLING DEG. DAYS	30	0	0	14	64	217	423	546	496	286	72	0	0	2118
RH														
NORMAL (PERCENT)	30	68	66	63	62	66	67	69	70	71	66	68	69	67
HOUR 00 LST	30	73	71	68	69	74	75	77	78	80	75	73	72	74
HOUR 06 LST	30	76	76	75	77	81	82	84	85	86	82	79	77	80
HOUR 12 LST	30	61	58	55	53	56	56	58	58	58	52	57	61	57
HOUR 18 LST	30	63	60	55	53	56	56	58	59	62	58	62	65	59
S PERCENT POSSIBLE SUNSHINE	35	50	54	56	64	69	74	74	75	69	70	58	50	64
W/O MEAN NO. DAYS WITH:														
HEAVY FOG(VISBY 1/4 MI)	51	1.9	1.4	0.7	0.3	0.2	0.2	0.3	0.3	0.6	1.0	1.2	1.8	9.9
THUNDERSTORMS	51	2.0	2.4	4.5	6.3	7.0	7.4	8.3	6.0	3.4	2.2	2.5	1.6	53.6
CLOUDINESS MEAN:														
SUNRISE-SUNSET (OKTAS)	45	5.4	5.1	5.2	4.8	4.7	4.3	4.2	3.8	3.9	3.6	4.5	5.0	4.5
MIDNIGHT-MIDNIGHT (OKTAS)	29	4.9	4.6	4.9	4.3	4.3	3.9	3.6	3.4	3.6	3.3	4.2	4.7	4.1
MEAN NO. DAYS WITH:														
CLEAR	43	7.7	7.9	7.9	8.6	8.4	9.6	10.2	11.8	12.5	14.4	10.2	8.9	118.1
PARTLY CLOUDY	43	5.8	5.6	6.5	7.3	9.6	11.1	11.9	11.7	7.5	7.0	6.2	5.6	95.8
CLOUDY	43	17.5	14.8	16.6	14.1	13.0	9.3	8.9	7.4	10.0	9.7	13.6	16.5	151.4
PR MEAN STATION PRESSURE(IN)	26	29.86	29.82	29.72	29.70	29.70	29.68	29.71	29.72	29.75	29.79	29.81	29.89	29.76
MEAN SEA-LEVEL PRES. (IN)	17	30.16	30.11	30.05	29.99	29.97	30.01	30.01	30.02	30.04	30.10	30.12	30.16	30.06
WINDS														
MEAN SPEED (MPH)	52	10.0	10.1	10.7	10.2	8.8	7.9	7.5	6.9	7.5	7.7	9.0	9.6	8.8
PREVAIL.DIR(TENS OF DEGS)	27	18	19	18	20	18	21	21	21	03	15	18	18	18
MAXIMUM 2-MINUTE:														
SPEED (MPH)	2	31	37	32	35	37	39	36	31	31	35	33	31	39
DIR. (TENS OF DEGS)		29	18	28	30	24	33	06	06	36	31	17	31	33
YEAR OF OCCURRENCE		2001	2001	2001	2001	2001	2000	2001	2000	2000	2001	2001	2000	JUN 2000
MAXIMUM 5-SECOND:														
SPEED (MPH)	2	37	44	45	48	48	46	46	39	38	40	45	39	48
DIR. (TENS OF DEGS)		21	18	27	30	25	33	03	03	34	31	19	31	25
YEAR OF OCCURRENCE		2001	2001	2001	2001	2001	2000	2001	2000	2000	2001	2001	2000	MAY 2001
PRECIPITATION														
NORMAL (IN)	30	3.73	4.35	5.41	5.46	4.98	3.57	3.79	3.43	3.53	3.01	5.10	5.74	52.10
MAXIMUM MONTHLY (IN)	51	12.21	10.51	12.08	17.13	11.58	10.17	9.96	9.65	7.61	7.75	11.60	13.81	17.13
YEAR OF OCCURRENCE		1951	1989	1975	1991	1953	1996	1998	1978	1958	1984	2001	1982	APR 1991
MINIMUM MONTHLY (IN)	51	0.57	1.12	1.50	1.39	0.83	0.04	0.43	0.43	0.19	T	0.75	1.05	T
YEAR OF OCCURRENCE		1986	1980	1966	1992	1977	1953	1954	1953	1963	1963	1965	1955	OCT 1963
MAXIMUM IN 24 HOURS (IN)	51	3.89	4.24	5.95	4.35	4.94	4.76	4.71	4.04	4.63	3.71	7.30	5.42	7.30
YEAR OF OCCURRENCE		1974	1989	1975	1985	1958	1980	1980	1978	1957	2001	2001	1978	NOV 2001
NORMAL NO. DAYS WITH:														
PRECIPITATION 0.01	30	9.8	9.5	10.6	10.1	9.6	8.5	8.5	7.7	7.7	6.5	9.0	9.7	107.2
PRECIPITATION 1.00	30	0.9	1.3	1.8	1.9	1.5	1.0	1.3	1.0	1.2	1.0	1.5	1.9	16.3
SNOWFALL														
NORMAL (IN)	30	2.6	1.4	0.9	T	0.0	0.0	0.0	0.0	0.0	0.0	0.1	0.7	5.7
MAXIMUM MONTHLY (IN)	49	12.4	8.3	17.3	T	T	T	0.0	0.0	0.0	T	1.5	14.3	17.3
YEAR OF OCCURRENCE		1985	1985	1968	1993	1995	1995					1976	1963	MAR 1968
MAXIMUM IN 24 HOURS (IN)	49	8.1	5.8	16.1	T	T	T	0.0	0.0	0.0	T	1.2	14.3	16.1
YEAR OF OCCURRENCE		1985	1960	1968	1993	1995	1995					1976	1963	MAR 1968
MAXIMUM SNOW DEPTH (IN)	51	12	6	12	0	0	0	0	0	0	0	1	13	13
YEAR OF OCCURRENCE		1948	1985	1968								1991	1963	DEC 1963
NORMAL NO. DAYS WITH:														
SNOWFALL 1.0	30	0.9	0.5	0.3	0.0	0.0	0.0	0.0	0.0	0.0	0.0	0.*	0.1	1.8

PRECIPITATION (inches) 2001 MEMPHIS, TN (MEM)

YEAR	JAN	FEB	MAR	APR	MAY	JUN	JUL	AUG	SEP	OCT	NOV	DEC	ANNUAL
1972	4.73	2.23	4.80	3.51	4.55	5.50	4.89	1.94	5.46	3.92	8.05	9.37	58.95
1973	4.62	3.62	7.63	9.44	6.23	1.00	4.49	4.88	5.06	3.37	8.49	5.35	64.18
1974	8.90	4.65	3.40	6.34	7.76	6.30	6.33	4.78	3.45	2.67	4.96	5.03	64.57
1975	4.65	5.53	12.08	4.98	8.72	2.42	2.26	2.03	2.62	2.69	7.77	2.93	58.68
1976	2.85	4.41	7.68	2.41	4.73	4.06	3.82	0.86	5.40	5.66	1.83	1.79	45.50
1977	2.57	1.99	4.13	5.42	0.83	3.38	3.41	1.62	6.43	2.02	6.01	3.39	41.20
1978	8.13	1.31	4.05	2.14	8.14	4.45	3.89	9.65	1.52	1.82	5.56	13.12	63.78
1979	5.98	5.66	6.60	11.47	7.78	4.93	3.12	5.92	4.49	2.60	7.42	4.92	70.89
1980	3.23	1.12	10.86	7.53	4.43	5.75	4.73	1.23	5.32	3.14	5.23	1.86	54.43
1981	1.38	3.66	4.98	3.67	7.06	2.93	1.71	0.61	5.83	2.12	1.84	40.00
1982	6.61	4.16	4.47	6.76	5.50	6.68	4.13	3.11	1.92	5.23	6.43	13.81	68.81
1983	2.32	2.61	3.66	8.84	9.58	3.50	3.83	0.61	1.52	2.94	9.56	8.68	57.65
1984	1.88	4.37	6.07	5.24	9.06	1.12	4.59	5.00	1.96	7.75	5.85	4.35	57.24
1985	3.78	4.10	4.96	6.51	2.23	4.55	3.50	3.50	4.03	3.36	3.87	3.27	47.66
1986	0.57	2.50	1.90	3.72	4.63	3.80	1.21	2.74	1.21	3.75	8.67	3.92	38.62
1987	1.76	5.81	3.38	3.78	2.96	3.66	2.06	4.12	2.01	1.96	10.45	11.39	53.34
1988	4.25	3.49	4.20	2.85	2.38	2.15	5.21	0.85	4.73	3.62	10.52	5.99	50.24
1989	7.91	10.51	5.50	2.13	2.36	7.20	7.55	1.43	6.08	2.37	3.65	2.20	58.89
1990	3.97	8.99	5.65	6.93	4.55	2.68	2.21	1.18	5.21	4.37	3.44	10.61	59.79
1991	2.90	6.46	3.68	17.13	5.10	1.42	1.92	2.06	1.47	4.39	5.54	7.04	59.11
1992	1.78	2.18	7.07	1.39	3.68	7.50	5.38	2.44	3.62	4.01	4.82	3.28	47.15
1993	3.59	2.46	3.14	6.20	4.56	4.20	0.86	3.69	3.73	1.91	4.07	5.59	44.00
1994	5.53	4.67	6.65	4.04	3.00	5.03	4.03	2.00	1.44	2.72	4.34	6.04	49.49
1995	7.07	2.12	3.35	4.38	5.90	6.66	8.42	3.63	0.39	1.57	7.62	5.79	56.90
1996	5.53	2.77	5.31	3.37	6.28	10.17	9.89	2.07	5.99	7.12	11.51	6.18	76.19
1997	4.37	7.63	11.28	8.91	8.00	6.75	4.35	4.01	6.31	4.06	1.69	4.52	71.88
1998	6.73	5.55	5.14	8.24	1.71	1.09	9.96	2.54	1.70	2.34	2.56	4.25	51.81
1999	6.15	2.10	6.59	8.92	4.79	2.42	3.63	1.18	1.11	1.53	2.37	4.73	45.52
2000	1.37	5.37	4.42	4.41	4.02	3.79	2.23	0.77	1.08	0.45	6.89	2.47	37.27
2001	3.32	6.89	3.54	3.94	6.35	2.12	5.53	2.25	3.35	6.93	11.60	10.19	66.01
POR= 130 YRS	4.84	4.39	5.25	5.21	4.42	3.77	3.53	3.16	2.93	2.86	4.49	4.85	49.70

AVERAGE TEMPERATURE (F) 2001 MEMPHIS, TN (MEM)

YEAR	JAN	FEB	MAR	APR	MAY	JUN	JUL	AUG	SEP	OCT	NOV	DEC	ANNUAL
1972	42.3	44.7	52.2	63.1	69.7	77.5	79.4	79.7	75.9	61.4	45.5	40.0	61.0
1973	38.6	40.5	57.3	59.8	68.3	81.0	83.2	79.6	76.1	67.6	57.3	43.7	62.7
1974	45.7	45.6	58.7	61.8	72.1	74.7	82.5	79.2	68.5	62.4	53.3	45.2	62.5
1975	45.9	46.2	49.9	61.9	73.5	78.8	81.1	81.2	70.9	65.8	53.8	44.1	62.8
1976	39.5	53.8	58.5	63.6	65.6	76.4	81.5	78.9	73.0	58.9	45.5	41.9	61.4
1977	30.7	45.1	50.8	66.9	76.4	81.9	82.6	79.0	62.2	55.1	44.1	63.9
1978	32.7	35.0	50.3	66.3	70.9	79.8	83.8	80.9	77.7	62.5	57.7	44.0	61.8
1979	30.9	38.5	54.3	63.0	70.0	77.9	82.6	80.9	73.4	65.8	50.7	45.4	61.1
1980	43.2	39.5	49.4	60.9	72.5	80.9	88.8	87.2	80.5	62.7	53.3	45.9	63.7
1981	40.9	47.3	54.3	70.2	70.0	82.5	84.6	81.8	74.0	62.5	53.8	40.9	63.6
1982	36.6	40.5	55.5	58.5	74.5	78.0	85.0	82.9	73.3	63.7	53.4	49.5	62.6
1983	40.4	45.3	51.9	56.8	68.2	77.2	83.6	84.9	75.2	66.2	53.1	34.7	61.5
1984	35.9	47.6	51.1	61.0	69.5	80.9	79.8	71.9	68.4	50.9	53.8	62.0
1985	32.4	40.6	57.7	65.0	71.8	78.9	82.2	80.2	73.6	67.2	57.6	36.9	62.0
1986	41.9	48.0	55.2	64.4	72.5	81.0	86.5	79.2	79.1	64.2	51.1	42.6	63.8
1987	39.6	47.1	54.7	62.2	76.5	80.0	82.5	83.6	75.3	59.2	54.4	46.7	63.5
1988	36.8	42.2	52.3	62.8	71.8	80.3	81.6	83.7	76.3	59.1	54.6	44.9	62.2
1989	47.3	40.0	53.8	62.6	69.5	77.6	80.7	81.2	72.4	64.2	54.2	33.6	61.4
1990	48.5	52.0	55.3	61.4	68.3	80.7	82.5	82.0	77.9	61.3	57.1	45.8	64.4
1991	41.2	48.0	55.4	65.0	74.6	80.3	83.3	80.9	75.3	65.5	49.2	46.8	63.8
1992	42.2	49.9	53.7	63.0	70.4	76.9	82.5	76.8	73.2	63.8	50.7	43.8	62.2
1993	42.7	42.6	51.1	59.3	69.9	79.2	86.4	83.6	73.5	62.2	50.2	44.8	62.1
1994	37.8	46.3	54.2	67.0	69.4	82.3	81.1	79.4	72.5	63.6	56.6	46.7	63.1
1995	42.2	44.2	56.2	62.9	71.1	77.0	81.1	84.5	73.0	65.2	48.3	43.2	62.4
1996	39.2	45.0	48.0	59.7	74.2	78.4	80.5	79.6	72.3	63.1	49.1	46.8	61.3
1997	39.7	47.6	56.7	57.7	67.9	75.7	82.5	79.3	74.7	62.6	48.5	42.3	61.3
1998	45.8	48.5	51.9	61.4	75.3	83.2	84.1	81.8	80.5	66.6	55.9	45.1	65.0
1999	46.0	51.4	50.7	67.0	71.2	79.7	83.7	82.7	75.2	64.3	57.0	45.8	64.6
2000	42.8	50.2	56.4	60.6	73.9	78.6	83.4	86.3	76.7	68.1	50.2	32.7	63.3
2001	38.0	47.7	49.3	67.7	72.1	77.7	83.2	81.8	73.6	61.7	57.3	48.2	63.2
POR= 127 YRS	40.9	44.3	52.5	62.2	70.6	78.4	81.5	80.3	74.3	63.5	51.8	43.5	62.0

REFERENCE NOTES:

PAGE 1:
THE TEMPERATURE GRAPH SHOWS NORMAL MAXIMUM AND NORMAL
   MINIMUM DAILY TEMPERATURES (SOLID CURVES) AND THE
   ACTUAL DAILY HIGH AND LOW TEMPERATURES (VERTICAL BARS).

PAGE 2 AND 3:
H/C INDICATES HEATING AND COOLING DEGREE DAYS.
RH INDICATES RELATIVE HUMIDITY
W/O INDICATES WEATHER AND OBSTRUCTIONS
S INDICATES SUNSHINE.
PR INDICATES PRESSURE.
CLOUDINESS ON PAGE 3 IS THE SUM OF THE CEILOMETER AND
   SATELLITE DATA NOT TO EXCEED EIGHT EIGHTHS(OKTAS).

GENERAL:
T INDICATES TRACE PRECIPITATION, AN AMOUNT GREATER
   THAN ZERO BUT LESS THAN THE LOWEST REPORTABLE VALUE.
+ INDICATES THE VALUE ALSO OCCURS ON EARLIER DATES.
BLANK ENTRIES DENOTE MISSING OR UNREPORTED DATA.
NORMALS ARE 30-YEAR AVERAGES (1961 - 1990).
ASOS INDICATES AUTOMATED SURFACE OBSERVING SYSTEM.
PM INDICATES THE LAST DAY OF THE PREVIOUS MONTH.
POR (PERIOD OF RECORD) BEGINS WITH THE JANUARY DATA
   MONTH AND IS THE NUMBER OF YEARS USED TO COMPUTE
   THE MEAN. INDIVIDUAL MONTHS WITHIN THE POR MAY
   BE MISSING.
WHEN THE POR FOR A NORMAL IS LESS THAN 30 YEARS,
   THE NORMAL IS PROVISIONAL AND IS BASED ON THE NUMBER
   OF YEARS INDICATED.
0.* OR * INDICATES THE VALUE OR MEAN-DAYS-WITH
   IS BETWEEN 0.00 AND 0.05.
CLOUDINESS FOR ASOS STATIONS DIFFERS FROM THE NON-ASOS
   OBSERVATION TAKEN BY A HUMAN OBSERVER. ASOS STATION
   CLOUDINESS IS BASED ON TIME-AVERAGED CEILOMETER DATA
   FOR CLOUDS AT OR BELOW 12,000 FEET AND ON SATELLITE
   DATA FOR CLOUDS ABOVE 12,000 FEET.
THE NUMBER OF DAYS WITH CLEAR, PARTLY CLOUDY, AND
   CLOUDY CONDITIONS FOR ASOS STATIONS IS THE SUM
   OF THE CEILOMETER AND SATELLITE DATA FOR THE
   SUNRISE TO SUNSET PERIOD.

GENERAL CONTINUED:
CLEAR INDICATES 0 - 2 OKTAS, PARTLY CLOUDY INDICATES
   3 - 6 OKTAS, AND CLOUDY INDICATES 7 OR 8 OKTAS.
   WHEN AT LEAST ONE OF THE ELEMENTS (CEILOMETER OR
   SATELLITE) IS MISSING, THE DAILY CLOUDINESS IS
   NOT COMPUTED.
WIND DIRECTION IS RECORDED IN TENS OF DEGREES (2 DIGITS)
   CLOCKWISE FROM TRUE NORTH. "00" INDICATES CALM. "36"
   INDICATES TRUE NORTH.
RESULTANT WIND IS THE VECTOR AVERAGE OF THE SPEED AND
   DIRECTION.
AVERAGE TEMPERATURE IS THE SUM OF THE MEAN DAILY MAXIMUM
   AND MINIMUM TEMPERATURE DIVIDED BY 2.
SNOWFALL DATA COMPRISE ALL FORMS OF FROZEN
   PRECIPITATION, INCLUDING HAIL.
A HEATING (COOLING) DEGREE DAY IS THE DIFFERENCE BETWEEN
   THE AVERAGE DAILY TEMPERATURE AND 65 F.
DRY BULB IS THE TEMPERATURE OF THE AMBIENT AIR.
DEW POINT IS THE TEMPERATURE TO WHICH THE AIR MUST BE
   COOLED TO ACHIEVE 100 PERCENT RELATIVE HUMIDITY.
WET BULB IS THE TEMPERATURE THE AIR WOULD HAVE IF THE
   MOISTURE CONTENT WAS INCREASED TO 100 PERCENT RELATIVE
   HUMIDITY.

ON JULY 1, 1996, THE NATIONAL WEATHER SERVICE BEGAN USING
   THE "METAR" OBSERVATION CODE THAT WAS ALREADY EMPLOYED
   BY MOST OTHER NATIONS OF THE WORLD.   THE MOST NOTICEABLE
   DIFFERENCE IN THIS ANNUAL PUBLICATION WILL BE THE CHANGE
   IN UNITS FROM TENTHS TO EIGHTS(OKTAS) FOR REPORTING THE
   AMOUNT OF SKY COVER.

HEATING DEGREE DAYS (base 65 F) 2001 MEMPHIS, TN (MEM)

YEAR	JUL	AUG	SEP	OCT	NOV	DEC	JAN	FEB	MAR	APR	MAY	JUN	TOTAL
1972-73	0	0	12	172	583	766	809	679	237	200	32	0	3490
1973-74	0	0	8	67	244	665	599	535	235	150	1	0	2504
1974-75	0	0	28	121	367	607	591	521	463	180	2	0	2880
1975-76	0	0	40	90	352	643	783	326	238	100	58	0	2630
1976-77	0	0	0	231	581	708	1056	547	212	61	4	0	3400
1977-78	0	0	0	123	313	640	995	835	454	74	47	0	3481
1978-79	0	0	0	116	230	643	1049	734	345	121	23	0	3261
1979-80	0	0	0	76	426	598	669	733	478	156	7	0	3143
1980-81	0	0	5	146	362	586	739	492	342	18	23	0	2713
1981-82	0	0	9	153	331	739	873	680	324	215	2	0	3326
1982-83	0	0	20	134	352	500	759	543	406	273	25	0	3012
1983-84	0	0	27	73	368	935	894	499	426	162	24	0	3408
1984-85	0	0	37	48	423	367	1004	683	254	100	6	0	2922
1985-86	0	0	17	54	257	864	708	475	307	102	8	0	2792
1986-87	0	0	0	102	413	687	782	492	322	154	0	0	2952
1987-88	0	0	0	186	324	559	867	657	393	108	0	0	3094
1988-89	0	0	1	202	314	619	544	691	369	174	41	0	2958
1989-90	0	0	24	102	337	966	503	363	321	181	38	0	2835
1990 91	0	0	11	182	249	593	728	471	329	57	7	0	2627
1991-92	0	0	23	90	481	556	699	433	347	144	30	0	2803
1992-93	0	0	15	87	428	650	685	617	433	200	15	2	3132
1993-94	0	0	14	164	447	618	839	524	337	78	38	0	3059
1994-95	0	0	16	117	261	559	703	574	292	123	32	0	2677
1995-96	0	0	18	79	495	666	792	577	525	190	9	0	3351
1996-97	0	0	12	115	475	561	776	479	267	214	36	0	2935
1997-98	0	0	0	181	490	696	587	457	439	133	4	0	2987
1998-99	0	0	0	58	275	622	586	381	432	68	0	0	2422
1999-00	0	0	6	112	241	591	679	430	269	150	0	0	2478
2000-01	0	0	11	81	455	996	829	479	479	72	5	0	3407
2001-	0	0	20	150	235	513							

COOLING DEGREE DAYS (base 65 F) 2001 MEMPHIS, TN (MEM)

YEAR	JAN	FEB	MAR	APR	MAY	JUN	JUL	AUG	SEP	OCT	NOV	DEC	ANNUAL
1972	0	0	3	95	179	383	449	464	346	66	6	0	1991
1973	0	0	4	48	143	486	571	458	350	156	19	0	2235
1974	6	0	46	59	228	299	550	445	138	46	23	0	1840
1975	8	0	3	93	272	421	507	510	224	121	23	2	2184
1976	0	7	44	64	84	349	519	438	247	48	0	0	1800
1977	0	0	23	123	362	516	619	551	426	41	20	0	2681
1978	0	0	6	122	235	452	590	501	387	46	18	0	2357
1979	0	0	19	68	184	394	553	499	259	108	4	0	2088
1980	0	0	0	40	249	480	744	695	476	80	18	2	2784
1981	0	5	20	181	184	532	614	527	285	80	2	0	2430
1982	0	1	36	26	305	399	623	563	275	100	14	25	2367
1983	0	0	7	32	131	373	584	622	338	116	17	0	2220
1984	0	1	4	51	169	482	502	462	249	162	5	23	2110
1985	0	6	30	107	224	425	540	478	285	129	42	0	2266
1986	0	3	12	91	247	487	673	448	427	81	4	0	2473
1987	0	0	9	78	366	458	549	584	315	12	13	0	2384
1988	0	0	7	52	221	469	518	586	347	24	8	0	2232
1989	0	2	27	109	186	386	496	510	254	84	20	0	2074
1990	0	7	27	83	146	477	550	534	404	71	18	2	2319
1991	0	0	34	62	309	464	576	500	339	113	13	0	2410
1992	0	0	5	92	204	363	551	375	268	55	4	0	1917
1993	0	0	8	38	173	436	669	584	275	84	13	0	2280
1994	0	7	8	145	179	525	502	451	247	80	15	0	2159
1995	4	0	26	66	229	365	507	612	264	91	2	1	2167
1996	0	6	5	40	302	408	486	458	237	62	4	3	2011
1997	1	1	14	7	132	327	550	446	297	114	0	0	1889
1998	0	0	39	33	330	550	598	528	471	114	9	13	2685
1999	5	6	0	136	199	447	587	555	319	99	10	1	2364
2000	0	7	12	24	284	416	579	667	371	184	18	0	2562
2001	0	2	0	159	232	387	569	526	285	56	8	2	2226

SNOWFALL (inches) 2001 MEMPHIS, TN (MEM)

YEAR	JUL	AUG	SEP	OCT	NOV	DEC	JAN	FEB	MAR	APR	MAY	JUN	TOTAL
1972-73	0.0	0.0	0.0	0.0	T	T	1.4	T	0.0	0.0	0.0	0.0	1.4
1973-74	0.0	0.0	0.0	0.0	0.0	0.2	0.9	0.5	T	0.0	0.0	0.0	1.6
1974-75	0.0	0.0	0.0	0.0	T	0.2	3.9	0.5	1.4	0.0	0.0	0.0	6.0
1975-76	0.0	0.0	0.0	0.0	0.0	0.1	0.3	T	0.0	0.0	0.0	0.0	0.4
1976-77	0.0	0.0	0.0	0.0	1.5	0.3	3.5	T	0.0	0.0	0.0	0.0	5.3
1977-78	0.0	0.0	0.0	0.0	0.0	T	4.3	3.2	T	0.0	0.0	0.0	7.5
1978-79	0.0	0.0	0.0	0.0	0.0	T	3.0	7.4	0.0	0.0	0.0	0.0	10.4
1979-80	0.0	0.0	0.0	0.0	T	0.0	1.3	1.5	0.8	0.0	0.0	0.0	3.6
1980-81	0.0	0.0	0.0	0.0	T	T	T	T	0.0	0.0	0.0	0.0	T
1981-82	0.0	0.0	0.0	0.0	0.0	T	4.5	0.7	1.2	0.0	0.0	0.0	6.4
1982-83	0.0	0.0	0.0	0.0	0.0	T	7.3	T	0.2	0.0	0.0	0.0	7.5
1983-84	0.0	0.0	0.0	0.0	0.0	0.8	2.0	0.5	T	0.0	0.0	0.0	3.3
1984-85	0.0	0.0	0.0	0.0	0.0	T	12.4	8.3	0.0	0.0	0.0	0.0	20.7
1985-86	0.0	0.0	0.0	0.0	0.0	T	T	2.0	0.0	0.0	0.0	0.0	2.0
1986-87	0.0	0.0	0.0	0.0	0.0	0.0	T	T	0.4	0.0	0.0	0.0	
1987-88	0.0	0.0	0.0	0.0	0.0	T	8.2	3.0	T	0.0	0.0	0.0	11.2
1988-89	0.0	0.0	0.0	0.0	0.0	T	T	0.3	T	0.0	T	0.0	0.3
1989-90	0.0	0.0	0.0	T	T	0.4	0.0	0.0	0.0	T	0.0	0.0	0.4
1990-91	0.0	0.0	0.0	0.0	0.0	0.4	T	T	1.0	0.0	0.0	0.0	1.4
1991-92	0.0	0.0	0.0	0.0	0.6	0.0	T	0.0	T	0.0	0.0	0.0	0.6
1992-93	0.0	0.0	0.0	0.0	T	T	T	T	T	T	T	0.0	T
1993-94	0.0	0.0	0.0	T	0.0	0.0	0.7	2.4	T	0.0	0.0	0.0	3.1
1994-95	0.0	0.0	0.0	0.0	0.0	0.0	0.5	2.5	0.7	0.0	T	T	3.7
1995-96	0.0	0.0	0.0	0.0	T	T	0.4	0.2	T	0.0	0.0	0.0	0.6
1996-97	0.0	0.0	0.0	0.0	0.0	T	3.8	0.7	0.0	0.0	T	0.0	4.5
1997-98	0.0	0.0	0.0	0.0	0.0	T	3.0	0.0	T	0.0	0.0	0.0	3.0
1998-99	0.0	0.0	0.0	0.0	0.0	0.8	T	T	T	0.0			
1999-00													
2000-01													
2001-													
POR= 47 YRS	0.0	0.0	0.0	T	0.1	0.6	2.2	1.4	0.8	T	T	T	5.1

2001
NASHVILLE,
TENNESSEE (BNA)

The city of Nashville is located on the Cumberland River, in the northwestern corner of the Central Basin of middle Tennessee near the escarpment of the Highland Rim. The Rim, as it is called, rises to the height of 300 to 400 feet above the mean elevation of the basin, forming an amphitheater about the city from the southwest to the southeast, with the south being more or less open but undulating.

Temperatures are moderate, with great extremes of either heat or cold rarely occurring, yet there are changes of sufficient amplitude and frequency to give variety.

Based on the 1951-1980 period, the average first occurrence of 32 degrees Fahrenheit in the fall is October 29 and the average last occurrence in the spring is April 5.

Humidity is an important phase of climate in relation to bodily health and comfort. The Nashville records show that the average relative humidity is moderate as compared with the general conditions east of the Mississippi River and south of the Ohio.

Nashville is not in the most frequented path of general storms that cross the country, however, it is in the zone of moderate frequency of thunderstorms. The thunderstorm season usually begins in the latter part of March and continues through September.

NORMALS, MEANS, AND EXTREMES
NASHVILLE, TN (BNA)

LATITUDE: 36 07' 08" N
LONGITUDE: 86 41' 21" W
ELEVATION (FT): GRND: 571 BARO: 574
TIME ZONE: CENTRAL (UTC + 6)
WBAN: 13897

	ELEMENT	POR	JAN	FEB	MAR	APR	MAY	JUN	JUL	AUG	SEP	OCT	NOV	DEC	YEAR
TEMPERATURE °F	NORMAL DAILY MAXIMUM	30	45.9	50.8	61.2	70.8	78.8	86.5	89.5	88.4	82.5	72.5	60.4	50.2	69.8
	MEAN DAILY MAXIMUM	54	46.9	51.8	60.3	70.8	79.1	86.5	89.8	88.8	82.8	72.5	59.7	50.3	69.9
	HIGHEST DAILY MAXIMUM	62	78	84	86	91	97	106	107	104	105	94	84	79	107
	YEAR OF OCCURRENCE		1972	1962	1982	1989	1941	1952	1952	1954	1954	1953	2000	1982	JUL 1952
	MEAN OF EXTREME MAXS.	54	68.0	72.2	78.8	85.3	89.4	94.9	97.2	96.3	93.6	85.8	77.0	69.1	84.0
	NORMAL DAILY MINIMUM	30	26.5	29.9	39.1	47.5	56.6	64.7	68.9	67.7	61.1	48.3	39.6	30.9	48.4
	MEAN DAILY MINIMUM	54	28.4	31.4	38.7	47.8	57.1	65.2	69.5	67.8	60.9	48.5	38.7	31.5	48.8
	LOWEST DAILY MINIMUM	62	-17	-13	2	23	34	42	51	47	36	26	-1	-10	-17
	YEAR OF OCCURRENCE		1985	1951	1980	1982	1976	1966	1947	1946	1983	1987	1950	1989	JAN 1985
	MEAN OF EXTREME MINS.	54	7.6	11.5	21.9	31.2	42.1	52.9	60.0	58.0	45.7	32.4	20.9	12.5	33.1
	NORMAL DRY BULB	30	36.2	40.4	50.2	59.2	67.7	75.6	79.3	78.1	71.8	60.4	50.0	40.5	59.1
	MEAN DRY BULB	54	37.6	41.7	49.5	59.4	68.1	75.9	79.7	78.4	71.9	60.6	49.2	41.0	59.4
	MEAN WET BULB	18	34.7	38.8	44.2	52.3	61.5	68.7	72.0	70.7	64.5	54.3	43.0	37.7	53.5
	MEAN DEW POINT	18	28.9	32.1	36.5	45.2	56.7	64.8	68.3	67.0	60.0	48.8	37.6	32.3	48.2
	NORMAL NO. DAYS WITH:														
	MAXIMUM 90	30	0.0	0.0	0.0	0.1	1.5	9.2	16.6	13.1	5.8	0.1	0.0	0.0	46.4
	MAXIMUM 32	30	5.1	2.3	0.1	0.0	0.0	0.0	0.0	0.0	0.0	0.0	0.2	2.4	10.1
	MINIMUM 32	30	22.2	17.1	9.2	1.8	0.0	0.0	0.0	0.0	0.0	1.2	7.7	17.8	77.0
	MINIMUM 0	30	0.9	0.2	0.0	0.0	0.0	0.0	0.0	0.0	0.0	0.0	0.0	0.3	1.4
H/C	NORMAL HEATING DEG. DAYS	30	893	689	469	193	59	0	0	0	21	195	450	760	3729
	NORMAL COOLING DEG. DAYS	30	0	0	10	19	143	318	443	406	225	52	0	0	1616
RH	NORMAL (PERCENT)	30	70	68	65	63	70	70	73	73	74	69	70	71	70
	HOUR 00 LST	30	75	74	71	72	81	83	85	85	85	81	77	76	79
	HOUR 06 LST	30	80	80	79	80	86	87	89	90	90	86	82	80	84
	HOUR 12 LST	30	63	60	54	51	55	55	57	57	58	53	59	63	57
	HOUR 18 LST	30	65	61	55	52	58	59	62	62	64	60	64	67	61
S	PERCENT POSSIBLE SUNSHINE	54	41	47	52	59	60	65	63	63	62	62	50	42	56
W/O	MEAN NO. DAYS WITH:														
	HEAVY FOG (VISBY 1/4 MI)	61	2.4	1.3	1.0	0.6	1.0	0.9	1.1	1.5	1.8	2.1	1.6	1.8	17.1
	THUNDERSTORMS	61	1.2	1.7	4.0	5.1	7.2	8.2	9.2	7.4	3.5	1.6	1.6	1.0	51.7
CLOUDINESS	MEAN:														
	SUNRISE-SUNSET (OKTAS)	56	5.7	5.4	5.3	4.9	4.8	4.5	4.4	4.2	4.2	3.9	4.8	5.3	4.8
	MIDNIGHT-MIDNIGHT (OKTAS)	32	5.3	5.1	5.0	4.6	4.6	4.1	4.0	3.8	4.0	3.6	4.7	5.0	4.5
	MEAN NO. DAYS WITH:														
	CLEAR	55	6.3	6.9	7.5	8.3	8.0	8.0	8.1	9.9	10.5	12.8	8.8	7.1	102.2
	PARTLY CLOUDY	55	6.1	5.9	7.2	8.5	10.1	12.5	13.2	12.2	9.1	8.0	6.8	7.0	106.6
	CLOUDY	55	18.5	15.5	16.3	13.2	13.0	9.5	9.7	8.9	10.4	10.2	14.4	17.0	156.6
PR	MEAN STATION PRESSURE (IN)	29	29.50	29.46	29.39	29.37	29.35	29.36	29.40	29.41	29.43	29.47	29.48	29.50	29.43
	MEAN SEA-LEVEL PRES. (IN)	18	30.16	30.13	30.06	30.00	30.00	30.01	30.02	30.03	30.06	30.12	30.12	30.18	30.07
WINDS	MEAN SPEED (MPH)	49	8.9	9.1	9.7	9.0	7.4	6.9	6.5	6.0	6.3	6.7	8.3	8.8	7.8
	PREVAIL.DIR (TENS OF DEGS)	33	18	18	18	18	18	18	18	18	18	18	18	18	18
	MAXIMUM 2-MINUTE:														
	SPEED (MPH)	5	38	36	39	40	33	38	38	36	28	31	36	30	40
	DIR. (TENS OF DEGS)		17	02	17	25	33	01	36	33	08	27	26	18	25
	YEAR OF OCCURRENCE		2000	1998	1999	1998	1998	1999	2000	1999	2000	2001	1997	2001	APR 1998
	MAXIMUM 5-SECOND:														
	SPEED (MPH)	5	61	47	48	59	47	47	47	49	33	45	45	41	61
	DIR. (TENS OF DEGS)		26	19	16	23	32	01	35	36	17	30	25	23	26
	YEAR OF OCCURRENCE		1999	1997	1999	1998	2000	1999	2000	1999	2001	2001	1997	2001	JAN 1999
PRECIPITATION	NORMAL (IN)	30	3.58	3.81	4.85	4.37	4.88	3.57	3.97	3.46	3.46	2.62	4.12	4.61	47.30
	MAXIMUM MONTHLY (IN)	62	13.92	10.31	12.35	8.41	11.04	11.95	7.75	8.31	11.44	6.13	9.04	13.63	13.92
	YEAR OF OCCURRENCE		1950	1956	1975	1984	1983	1998	1950	1942	1979	1959	1945	1978	JAN 1950
	MINIMUM MONTHLY (IN)	62	0.19	0.64	1.18	0.52	0.69	0.45	0.71	0.69	0.28	T	0.54	0.98	T
	YEAR OF OCCURRENCE		1986	1968	1987	1986	1941	1988	1954	1968	1956	1963	1949	1985	OCT 1963
	MAXIMUM IN 24 HOURS (IN)	62	4.40	4.73	4.66	3.29	4.39	5.24	4.32	5.34	6.68	3.75	4.20	5.12	6.68
	YEAR OF OCCURRENCE		1946	1989	1975	1979	2000	1998	1992	1963	1979	1975	1997	1978	SEP 1979
	NORMAL NO. DAYS WITH:														
	PRECIPITATION 0.01	30	10.3	10.3	11.6	10.8	10.9	9.3	10.4	8.3	8.2	7.2	10.1	11.0	118.4
	PRECIPITATION 1.00	30	0.7	0.7	1.1	1.2	1.5	0.9	1.2	1.0	0.7	0.7	1.0	1.5	12.2
SNOWFALL	NORMAL (IN)	30	4.4	3.5	1.1	0.*	0.0	0.0	0.0	0.0	0.0	T	0.3	1.7	11.0
	MAXIMUM MONTHLY (IN)	57	18.8	18.9	16.1	1.1	0.0	T	0.0	T	0.0	0.4	9.2	13.2	18.9
	YEAR OF OCCURRENCE		1948	1979	1960	1971		1994		1989		1993	1950	1963	FEB 1979
	MAXIMUM IN 24 HOURS (IN)	57	8.1	8.3	8.8	1.1	0.0	T	0.0	T	0.0	0.4	9.2	10.2	10.2
	YEAR OF OCCURRENCE		1988	1979	1951	1971		1994		1989		1993	1950	1963	DEC 1963
	MAXIMUM SNOW DEPTH (IN)	48	70	8	7	0	0	0	0	0	0	0	5	7	70
	YEAR OF OCCURRENCE		1948	1979	1968								1966	1963	JAN 1948
	NORMAL NO. DAYS WITH:														
	SNOWFALL 1.0	30	1.2	1.4	0.3	0.*	0.0	0.0	0.0	0.0	0.0	0.0	0.*	0.5	3.4

PRECIPITATION (inches) 2001 NASHVILLE, TN (BNA)

YEAR	JAN	FEB	MAR	APR	MAY	JUN	JUL	AUG	SEP	OCT	NOV	DEC	ANNUAL
1972	5.15	3.45	4.34	3.58	3.52	2.54	6.40	4.30	3.71	4.06	5.22	8.14	54.41
1973	3.40	3.63	9.88	7.00	5.72	4.80	7.67	1.79	1.56	3.32	7.78	3.23	59.78
1974	9.45	3.01	5.25	3.97	5.04	6.80	2.10	4.13	10.44	1.47	6.23	2.81	60.70
1975	4.67	5.22	12.35	3.55	6.52	2.22	2.96	4.69	5.42	5.86	3.00	4.12	60.58
1976	4.11	2.28	5.32	1.53	6.19	4.72	4.01	8.05	5.08	5.17	1.30	1.81	49.57
1977	2.53	3.27	5.83	7.87	1.65	4.29	1.15	4.65	5.04	4.22	5.96	4.25	50.71
1978	5.95	1.57	4.88	2.42	8.03	1.46	4.03	3.81	1.37	2.28	4.01	13.63	53.44
1979	7.13	4.01	4.92	7.80	8.18	2.79	4.27	4.59	11.44	3.97	5.98	5.04	70.12
1980	2.59	1.38	7.27	3.67	6.14	2.89	3.53	1.24	1.09	1.17	2.55	1.40	34.92
1981	1.60	3.83	3.38	4.78	3.05	3.49	3.10	8.05	1.37	2.82	3.83	2.38	41.68
1982	6.50	4.80	3.00	4.36	4.19	2.28	5.47	3.46	3.23	1.91	3.87	6.36	49.43
1983	2.56	2.93	3.44	6.80	11.04	3.93	1.71	1.36	0.45	2.77	6.98	7.75	51.72
1984	1.79	2.38	5.14	8.41	9.68	4.49	6.63	2.42	0.97	6.00	6.20	2.38	56.49
1985	3.02	3.30	2.70	2.91	2.65	1.53	2.00	3.91	2.52	1.59	3.81	0.98	30.92
1986	0.19	3.59	2.29	0.52	3.36	2.38	0.77	3.38	2.19	2.19	7.43	3.31	31.60
1987	1.61	4.87	1.18	1.03	4.41	2.82	0.73	1.95	0.21	3.40	5.46		30.23
1988	3.73	2.02	2.18	2.09	1.86	0.45	3.26	2.39	2.45	1.54	5.49	3.95	31.41
1989	4.52	9.36	5.31	2.68	4.61	7.87	3.18	3.67	6.30	3.62	3.94	1.97	57.03
1990	2.76	4.73	3.26	1.60	2.80	2.37	4.86	3.12	2.13	4.41	4.29	10.76	47.09
1991	2.92	5.44	4.25	3.35	5.63	1.25	2.82	1.79	5.47	3.88	2.87	7.27	46.94
1992	2.97	2.60	4.50	0.77	3.12	4.31	5.89	3.25	3.45	1.62	4.48	2.88	39.84
1993	2.76	3.33	5.50	3.33	4.50	5.31	3.64	1.76	2.90	2.20	2.53	6.62	44.38
1994	4.36	6.18	7.56	5.72	3.76	8.08	4.82	5.05	4.20	3.31	4.04	2.69	59.77
1995	5.61	1.81	3.87	3.95	7.66	3.69	1.95	3.40	5.00	5.60	3.98	2.32	48.84
1996	3.82	2.46	5.15	3.68	4.48	3.68	5.45	1.09	4.89	3.16	6.00	4.77	48.63
1997	4.19	3.10	9.64	2.42	4.92	6.66	3.26	3.52	5.75	2.71	6.59	2.19	54.95
1998	3.68	4.11	3.13	6.31	4.46	11.95	4.63	2.93	1.39	1.59	1.30	6.53	52.01
1999	9.28	2.33	4.27	2.29	4.35	3.56	3.19	3.05	1.97	2.04	2.99	2.50	41.82
2000	3.52	3.75	3.34	6.23	7.66	1.74	2.25	1.95	1.90	0.26	6.39	3.44	42.43
2001	3.21	8.54	2.73	2.42	5.54	4.47	2.77	4.07	1.79	4.61	5.09	3.32	48.56
POR= 131 YRS	4.57	4.13	5.06	4.15	4.16	3.89	3.86	3.28	3.28	2.51	3.67	4.01	46.57

AVERAGE TEMPERATURE (F) 2001   NASHVILLE, TN (BNA)

YEAR	JAN	FEB	MAR	APR	MAY	JUN	JUL	AUG	SEP	OCT	NOV	DEC	ANNUAL
1972	41.8	41.8	50.2	60.5	67.4	73.0	77.3	77.2	75.7	60.2	47.3	42.8	59.6
1973	38.0	39.7	56.8	56.4	64.2	76.1	78.7	78.1	76.2	66.2	54.7	40.5	60.5
1974	45.4	41.8	54.9	58.6	70.0	71.4	78.0	77.6	67.5	59.4	50.0	42.6	59.8
1975	43.4	44.6	47.3	58.5	70.4	76.0	78.5	79.1	67.9	62.4	52.1	42.6	60.3
1976	36.7	50.5	55.9	59.9	64.0	73.3	76.4	74.4	66.8	53.9	40.9	36.6	57.4
1977	24.5	40.6	53.9	63.0	71.9	77.2	82.2	79.5	74.0	57.0	50.8	38.6	59.4
1978	27.6	29.2	46.9	61.0	66.7	76.2	80.5	78.7	75.5	57.4	53.6	42.4	58.0
1979	29.7	33.4	50.7	57.8	66.3	73.7	77.6	77.0	70.5	60.3	48.6	41.5	57.3
1980	39.7	35.7	46.3	57.3	67.8	75.5	82.8	81.7	76.0	57.8	48.5	41.0	59.2
1981	35.5	42.6	47.5	64.0	64.2	77.5	79.8	76.6	68.1	60.4	49.9	38.3	58.7
1982	34.0	39.5	52.5	54.6	71.0	73.3	79.8	76.1	69.6	61.1	51.4	48.2	59.3
1983	38.8	42.7	50.3	54.5	64.8	75.5	80.5	83.2	73.7	62.4	49.9	34.0	59.2
1984	32.2	43.4	46.1	58.2	64.2	77.4	76.1	76.5	68.6	66.7	46.0	49.6	58.8
1985	27.8	36.5	53.2	61.9	68.4	75.7	80.2	77.2	70.8	64.4	56.9	34.2	58.9
1986	37.2	44.7	50.8	60.8	68.6	76.5	82.4	76.7	74.9	61.0	49.9	39.9	60.3
1987	36.1	43.1	51.8	57.7	73.4	77.5	80.2	81.1	72.2	54.6	52.4	44.1	60.4
1988	34.4	38.7	49.3	57.1	67.3	77.3	81.4	81.9	72.8	54.2	51.1	42.4	59.0
1989	44.9	39.0	52.6	59.3	65.7	74.7	79.1	78.0	70.5	61.0	51.4	29.5	58.8
1990	45.8	49.9	53.6	58.4	66.4	78.2	80.4	79.6	74.7	60.1	54.3	43.7	62.1
1991	39.2	43.9	52.5	63.8	74.2	78.2	81.1	78.3	72.3	61.2	47.2	44.5	61.4
1992	40.0	45.9	50.1	59.6	65.8	72.4	79.9	74.9	70.9	59.4	49.5	41.2	59.1
1993	41.6	39.3	47.1	56.7	67.6	75.9	83.3	81.0	71.0	58.6	47.4	40.3	59.2
1994	33.4	44.0	50.7	62.5	64.1	78.1	78.5	77.1	69.1	61.0	54.5	45.3	59.9
1995	38.6	40.4		60.9	68.5	74.7	80.8	83.3	70.7	60.0	44.0	39.5	
1996	36.3	40.6	44.6	55.8	71.5	75.6	77.6	77.5	69.5	60.8	45.7	44.4	58.3
1997	37.3	45.6	53.6	54.4	63.2	72.0	79.8	76.9	71.9	59.8	45.5	39.5	58.3
1998	44.7	46.0	49.4	57.9	71.5	77.6	79.6	79.2	77.1	63.5	52.3	43.1	61.8
1999	42.5	45.8	45.6	62.6	67.6	76.5	81.9	79.4	71.9	59.9	54.0	43.7	61.0
2000	39.2	46.7	53.1	56.7	69.9	76.4	80.3	79.8	71.3	63.8	48.0	30.8	59.7
2001	35.4	44.7	45.3	63.7	68.8	73.4	79.8	78.8	70.2	58.5	54.3	44.9	59.8
POR= 131 YRS	38.6	41.0	49.6	59.4	68.1	76.2	78.9	78.3	72.1	60.8	49.1	40.9	59.4

REFERENCE NOTES:

PAGE 1:
THE TEMPERATURE GRAPH SHOWS NORMAL MAXIMUM AND NORMAL MINIMUM DAILY TEMPERATURES (SOLID CURVES) AND THE ACTUAL DAILY HIGH AND LOW TEMPERATURES (VERTICAL BARS).

PAGE 2 AND 3:
H/C INDICATES HEATING AND COOLING DEGREE DAYS.
RH INDICATES RELATIVE HUMIDITY
W/O INDICATES WEATHER AND OBSTRUCTIONS
S INDICATES SUNSHINE.
PR INDICATES PRESSURE.
CLOUDINESS ON PAGE 3 IS THE SUM OF THE CEILOMETER AND SATELLITE DATA NOT TO EXCEED EIGHT EIGHTHS(OKTAS).

GENERAL:
T INDICATES TRACE PRECIPITATION, AN AMOUNT GREATER THAN ZERO BUT LESS THAN THE LOWEST REPORTABLE VALUE.
+ INDICATES THE VALUE ALSO OCCURS ON EARLIER DATES.
BLANK ENTRIES DENOTE MISSING OR UNREPORTED DATA.
NORMALS ARE 30-YEAR AVERAGES (1961 - 1990).
ASOS INDICATES AUTOMATED SURFACE OBSERVING SYSTEM.
PM INDICATES THE LAST DAY OF THE PREVIOUS MONTH.
POR (PERIOD OF RECORD) BEGINS WITH THE JANUARY DATA MONTH AND IS THE NUMBER OF YEARS USED TO COMPUTE THE MEAN. INDIVIDUAL MONTHS WITHIN THE POR MAY BE MISSING.
WHEN THE POR FOR A NORMAL IS LESS THAN 30 YEARS, THE NORMAL IS PROVISIONAL AND IS BASED ON THE NUMBER OF YEARS INDICATED.
0.* OR * INDICATES THE VALUE OR MEAN-DAYS-WITH IS BETWEEN 0.00 AND 0.05.
CLOUDINESS FOR ASOS STATIONS DIFFERS FROM THE NON-ASOS OBSERVATION TAKEN BY A HUMAN OBSERVER. ASOS STATION CLOUDINESS IS BASED ON TIME-AVERAGED CEILOMETER DATA FOR CLOUDS AT OR BELOW 12,000 FEET AND ON SATELLITE DATA FOR CLOUDS ABOVE 12,000 FEET.
THE NUMBER OF DAYS WITH CLEAR, PARTLY CLOUDY, AND CLOUDY CONDITIONS FOR ASOS STATIONS IS THE SUM OF THE CEILOMETER AND SATELLITE DATA FOR THE SUNRISE TO SUNSET PERIOD.

GENERAL CONTINUED:
CLEAR INDICATES 0 - 2 OKTAS, PARTLY CLOUDY INDICATES 3 - 6 OKTAS, AND CLOUDY INDICATES 7 OR 8 OKTAS. WHEN AT LEAST ONE OF THE ELEMENTS (CEILOMETER OR SATELLITE) IS MISSING, THE DAILY CLOUDINESS IS NOT COMPUTED.
WIND DIRECTION IS RECORDED IN TENS OF DEGREES (2 DIGITS) CLOCKWISE FROM TRUE NORTH. "00" INDICATES CALM. "36" INDICATES TRUE NORTH.
RESULTANT WIND IS THE VECTOR AVERAGE OF THE SPEED AND DIRECTION.
AVERAGE TEMPERATURE IS THE SUM OF THE MEAN DAILY MAXIMUM AND MINIMUM TEMPERATURE DIVIDED BY 2.
SNOWFALL DATA COMPRISE ALL FORMS OF FROZEN PRECIPITATION, INCLUDING HAIL.
A HEATING (COOLING) DEGREE DAY IS THE DIFFERENCE BETWEEN THE AVERAGE DAILY TEMPERATURE AND 65 F.
DRY BULB IS THE TEMPERATURE OF THE AMBIENT AIR.
DEW POINT IS THE TEMPERATURE TO WHICH THE AIR MUST BE COOLED TO ACHIEVE 100 PERCENT RELATIVE HUMIDITY.
WET BULB IS THE TEMPERATURE THE AIR WOULD HAVE IF THE MOISTURE CONTENT WAS INCREASED TO 100 PERCENT RELATIVE HUMIDITY.

ON JULY 1, 1996, THE NATIONAL WEATHER SERVICE BEGAN USING THE "METAR" OBSERVATION CODE THAT WAS ALREADY EMPLOYED BY MOST OTHER NATIONS OF THE WORLD. THE MOST NOTICEABLE DIFFERENCE IN THIS ANNUAL PUBLICATION WILL BE THE CHANGE IN UNITS FROM TENTHS TO EIGHTS(OKTAS) FOR REPORTING THE AMOUNT OF SKY COVER.

HEATING DEGREE DAYS (base 65 F) 2001 NASHVILLE, TN (BNA)

YEAR	JUL	AUG	SEP	OCT	NOV	DEC	JAN	FEB	MAR	APR	MAY	JUN	TOTAL
1972-73	0	0	10	168	533	682	830	702	261	275	83	0	3544
1973-74	0	0	8	84	316	753	601	641	320	227	28	3	2981
1974-75	0	0	48	196	464	685	665	567	547	241	6	0	3419
1975-76	0	0	68	138	398	683	870	417	303	183	94	0	3154
1976-77	0	0	31	349	718	872	1250	679	350	129	28	1	4407
1977-78	0	0	3	255	425	813	1152	996	556	164	92	0	4456
1978-79	0	0	1	240	338	695	1088	877	449	213	57	0	3958
1979-80	0	0	5	180	487	723	777	848	571	240	38	0	3869
1980-81	0	0	9	259	487	739	909	621	537	97	96	0	3754
1981-82	0	0	42	175	445	820	956	707	416	309	8	0	3878
1982-83	0	0	30	194	413	537	806	620	458	322	71	0	3451
1983-84	0	0	45	121	447	956	1009	621	578	220	106	0	4103
1984-85	0	0	59	63	564	473	1146	794	383	145	25	6	3658
1985-86	0	0	30	91	264	948	854	561	432	171	55	0	3406
1986-87	0	3	0	175	447	773	889	608	401	242	6	0	3544
1987-88	0	0	7	317	376	640	941	756	485	242	43	2	3809
1988-89	0	0	5	343	408	693	618	721	397	258	90	0	3533
1989-90	0	0	36	158	408	1095	590	422	373	245	65	1	3393
1990-91	0	0	21	195	323	654	791	586	402	80	9	0	3061
1991-92	0	0	42	166	535	628	768	544	456	217	85	2	3443
1992-93	0	0	26	181	461	731	717	713	552	252	32	4	3669
1993-94	0	0	27	227	528	759	974	585	437	134	90	0	3761
1994-95	0	0	21	144	316	605	814	683		175	42	0	
1995-96	0	0	31	184	624	783	886	702	626	292	32	0	4160
1996-97	0	0	25	160	572	634	851	538	357	319	107	4	3567
1997-98	0	0	0	227	576	785	622	527	505	209	19	5	3475
1998-99	0	0	0	111	375	671	686	531	595	125	10	0	3104
1999-00	0	0	18	185	322	654	796	523	364	248	14	0	3124
2000-01	0	0	32	123	518	1051	912	565	602	124	17	7	3951
2001-	0	0	47	218	315	617							

COOLING DEGREE DAYS (base 65 F) 2001 NASHVILLE, TN (BNA)

YEAR	JAN	FEB	MAR	APR	MAY	JUN	JUL	AUG	SEP	OCT	NOV	DEC	ANNUAL
1972	0	0	1	62	117	250	385	387	341	24	6	0	1573
1973	0	0	14	25	61	339	432	412	351	128	8	0	1770
1974	0	0	16	39	191	203	410	399	130	30	22	0	1440
1975	3	0	3	55	183	341	424	444	164	62	19	0	1698
1976	0	1	28	36	68	257	363	299	92	10	0	0	1154
1977	0	0	11	74	253	371	543	458	281	13	4	0	2008
1978	0	0	1	50	152	344	489	432	324	13	2	0	1807
1979	0	0	11	5	103	264	393	381	175	44	0	0	1376
1980	0	0	0	17	131	322	562	527	344	44	1	0	1948
1981	0	0	1	71	81	383	464	366	145	42	0	0	1553
1982	0	0	37	4	199	256	470	352	177	84	12	21	1612
1983	0	0	9	12	69	320	488	568	315	49	2	0	1832
1984	0	0	0	21	87	382	352	364	173	121	0	1	1501
1985	0	2	24	59	137	335	479	386	206	79	29	0	1736
1986	0	0	1	52	174	352	551	371	304	59	0	0	1864
1987	0	0	0	31	272	381	479	507	227	3	7	0	1907
1988	0	0	5	17	120	380	515	531	246	17	0	0	1831
1989	0	0	21	93	120	298	446	408	208	39	8	0	1641
1990	0	4	26	51	115	401	485	458	315	52	10	0	1917
1991	0	0	22	50	300	403	507	419	268	57	4	0	2030
1992	0	0	0	60	115	233	471	311	208	15	2	0	1415
1993	0	0	1	9	121	336	573	506	215	33	7	0	1801
1994	0	3	0	66	67	400	426	381	152	26	8	0	1529
1995	0	0	0	60	158	297	496	573	210	34	1	0	
1996	0	4	0	22	240	324	397	395	166	35	0	2	1585
1997	0	0	13	6	61	222	465	374	211	72	0	0	1424
1998	0	0	27	5	227	391	459	447	370	69	1	1	1997
1999	0	0	0	60	100	354	533	453	229	35	1	0	1765
2000	2	0	6	6	172	346	483	467	227	95	18	0	1822
2001	0	0	0	93	146	264	468	437	209	27	2	0	1646

SNOWFALL (inches) 2001 NASHVILLE, TN (BNA)

YEAR	JUL	AUG	SEP	OCT	NOV	DEC	JAN	FEB	MAR	APR	MAY	JUN	TOTAL
1972-73	0.0	0.0	0.0	0.0	0.1	0.6	4.8	T	0.2	0.1	0.0	0.0	5.8
1973-74	0.0	0.0	0.0	0.0	0.0	2.4	T	0.3	T	0.0	0.0	0.0	2.7
1974-75	0.0	0.0	0.0	0.0	T	2.1	4.2	T	T	T	0.0	0.0	6.3
1975-76	0.0	0.0	0.0	0.0	T	1.1	T	2.3	T	0.0	0.0	0.0	3.4
1976-77	0.0	0.0	0.0	0.0	1.2	1.8	18.5	T	0.0	T	0.0	0.0	21.5
1977-78	0.0	0.0	0.0	0.0	T	0.1	12.9	9.8	2.4	0.0	0.0	0.0	25.2
1978-79	0.0	0.0	0.0	0.0	0.0	T	8.0	18.9	0.6	0.0	0.0	0.0	27.5
1979-80	0.0	0.0	0.0	0.0	T	T	0.3	6.6	3.1	0.0	0.0	0.0	10.0
1980-81	0.0	0.0	0.0	0.0	T	T	1.2	1.7	T	0.0	0.0	0.0	2.9
1981-82	0.0	0.0	0.0	0.0	0.0	0.2	4.8	3.7	1.0	0.0	0.0	0.0	9.7
1982-83	0.0	0.0	0.0	0.0	0.0	0.4	0.3	0.8	T	0.0	0.0	0.0	1.5
1983-84	0.0	0.0	0.0	0.0	0.0	0.7	5.3	3.7	T	0.0	0.0	0.0	9.7
1984-85	0.0	0.0	0.0	0.0	0.0	0.8	9.8	8.0	0.0	0.0	0.0	0.0	18.6
1985-86	0.0	0.0	0.0	0.0	0.0	0.5	0.4	2.1	T	0.0	0.0	0.0	3.0
1986-87	0.0	0.0	0.0	0.0	0.0	T	1.4	1.3	1.6	T	0.0	0.0	4.3
1987-88	0.0	0.0	0.0	0.0	T	T	8.6	1.4	T	0.0	0.0	0.0	10.0
1988-89	0.0	0.0	0.0	0.0	0.0	1.6	T	5.2	T	0.0	0.0	0.0	6.8
1989-90	0.0	0.0	T	0.0	T	T	0.4	T	T	0.4	0.0	0.0	0.8
1990-91	0.0	0.0	0.0	0.0	0.0	0.3	T	0.6	1.1	0.0	0.0	0.0	2.0
1991-92	0.0	0.0	0.0	0.0	T	0.0	T	T	0.0	1.0	0.0	0.0	1.0
1992-93	0.0	0.0	0.0	0.0	T	0.3	T	5.9	2.8	T	0.0	T	9.0
1993-94	0.0	0.0	0.0	0.4	T	0.3	2.3	1.0	T	T	0.0	T	4.0
1994-95	0.0	0.0	0.0	0.0	0.0	0.0	1.5	1.3	0.1	0.0	0.0	0.0	2.9
1995-96	0.0	0.0	0.0	0.0	T	0.4	6.2	7.8	9.3	0.0	0.0		
1996-97						0.2							
1997-98													
1998-99									1.0				
1999-00													
2000-01													
2001-													
POR= 54 YRS	0.0	T	0.0	0.0	0.4	1.4	3.7	2.9	1.5	0.0	0.0	T	9.9

2001
AMARILLO,
TEXAS (AMA)

The station is located 7 statute miles east northeast of the downtown post office in a region of rather flat topography. The Canadian River flows eastward 18 miles north of the station, with its bed about 800 feet below the plains. The Prairie Dog Town Fork of the Red River flows southeastward about 15 miles south of the station where it enters the Palo Duro Canyon, which is about 1,000 feet deep. There are numerous shallow Playa lakes, often dry, over the area, and the nearly treeless grasslands slope downward to the east. The terrain gradually rises to the west and northwest.

Three-fourths of the total annual precipitation falls from April through September, occurring from thunderstorm activity. Snow usually melts within a few days after it falls. Heavier snowfalls of 10 inches or more, usually with near blizzard conditions, average once every 5 years and last 2 to 3 days.

The Amarillo area is subject to rapid and large temperature changes, especially during the winter months when cold fronts from the northern Rocky Mountain and Plains states sweep across the area. Temperature drops of 50 to 60 degrees within a 12-hour period are not uncommon. Temperature drops of 40 degrees have occurred within a few minutes.

Humidity averages are low, occasionally dropping below 20 percent in the spring. Low humidity moderates the effect of high summer afternoon temperatures, permits evaporative cooling systems to be very effective, and provides many pleasant evenings and nights.

Severe local storms are infrequent, although a few thunderstorms with damaging hail, lightning, and wind in a very localized area occur most years, usually in spring and summer. These storms are often accompanied by very heavy rain, which produces local flooding, particularly of roads and streets. Tornadoes are rare.

Based on the 1951-1980 period, the average first occurrence of 32 degrees Fahrenheit in the fall is October 29 and the average last occurrence in the spring is April 14.

NORMALS, MEANS, AND EXTREMES
AMARILLO, TX (AMA)

	LATITUDE:	LONGITUDE:	ELEVATION (FT):	TIME ZONE:	WBAN: 23047
	35 13' 10" N	101 42' 20" W	GRND: 3586　BARO: 3589	CENTRAL (UTC + 6)	

	ELEMENT	POR	JAN	FEB	MAR	APR	MAY	JUN	JUL	AUG	SEP	OCT	NOV	DEC	YEAR
TEMPERATURE °F	NORMAL DAILY MAXIMUM	30	49.0	52.8	61.6	71.5	79.1	87.6	91.7	89.1	81.8	72.5	59.7	50.1	70.5
	MEAN DAILY MAXIMUM	61	49.0	53.5	61.0	70.8	78.9	87.7	91.3	91.0	83.9	73.8	60.0	51.6	71.0
	HIGHEST DAILY MAXIMUM	60	81	88	94	98	103	108	105	106	102	99	87	81	108
	YEAR OF OCCURRENCE		1950	1963	1971	1989	1996	1998	1994	1944	2000	2000	1980	1955	JUN 1998
	MEAN OF EXTREME MAXS.	53	71.2	75.7	82.2	89.0	94.2	100.0	99.9	98.5	94.7	89.3	78.6	72.0	87.1
	NORMAL DAILY MINIMUM	30	21.2	25.5	32.7	42.1	51.6	60.7	65.5	63.8	56.4	44.5	32.3	23.7	43.3
	MEAN DAILY MINIMUM	61	22.0	26.3	32.0	41.8	51.7	61.1	65.6	64.1	56.7	45.3	32.1	24.5	43.6
	LOWEST DAILY MINIMUM	60	-11	-14	-3	14	28	41	51	49	30	12	0	-8	-14
	YEAR OF OCCURRENCE		1984	1951	1948	1945	1954	1998	1990	1956	1984	1993	1976	1989	FEB 1951
	MEAN OF EXTREME MINS.	53	3.7	8.7	14.9	27.1	37.7	50.1	58.1	56.5	42.4	30.1	15.9	7.7	29.4
	NORMAL DRY BULB	30	35.1	39.2	47.1	56.8	65.4	74.1	78.6	76.5	69.1	58.5	46.0	36.9	56.9
	MEAN DRY BULB	61	35.6	39.8	46.6	56.4	65.3	74.5	78.6	76.9	69.5	58.9	45.7	37.6	57.1
	MEAN WET BULB	55	29.0	32.7	37.6	45.3	54.4	62.1	65.1	64.4	58.6	48.6	37.3	30.6	47.1
	MEAN DEW POINT	55	19.5	23.0	26.2	33.8	45.7	55.2	58.5	58.4	51.9	39.8	28.6	21.8	38.5
	NORMAL NO. DAYS WITH:														
	MAXIMUM 90	30	0.0	0.0	0.1	1.0	4.7	12.8	21.1	16.5	6.4	1.0	0.0	0.0	63.6
	MAXIMUM 32	30	4.5	2.8	0.9	*	0.0	0.0	0.0	0.0	0.0	*	0.8	3.6	12.6
	MINIMUM 32	30	27.5	21.8	14.7	3.5	0.1	0.0	0.0	0.0	0.2	1.8	14.5	26.7	110.8
	MINIMUM 0	30	1.1	0.4	0.0	0.0	0.0	0.0	0.0	0.0	0.0	0.0	*	0.7	2.2
H/C	NORMAL HEATING DEG. DAYS	30	927	722	555	266	89	6	0	0	26	226	570	871	4258
	NORMAL COOLING DEG. DAYS	30	0	0	0	20	102	279	422	357	149	25	0	0	1354
RH	NORMAL (PERCENT)	30	58	59	52	48	54	56	53	58	61	56	59	59	56
	HOUR 00 LST	30	64	65	58	56	63	65	61	67	70	64	66	66	64
	HOUR 06 LST	30	70	72	68	68	74	77	74	78	80	73	72	70	73
	HOUR 12 LST	30	49	50	42	38	43	45	42	46	49	43	47	49	45
	HOUR 18 LST	30	46	43	35	31	37	39	38	43	45	42	48	50	41
S	PERCENT POSSIBLE SUNSHINE	59	69	69	74	76	72	78	79	77	74	76	72	68	74
W/O	MEAN NO. DAYS WITH:														
	HEAVY FOG(VISBY 1/4 MI)	59	3.2	4.0	3.4	1.9	2.1	0.7	0.5	0.7	2.0	2.5	2.8	2.8	26.6
	THUNDERSTORMS	59	0.2	0.5	1.7	3.3	8.0	9.2	9.1	8.6	4.0	2.6	0.6	0.2	48.0
CLOUDINESS	MEAN:														
	SUNRISE-SUNSET (OKTAS)	1	3.2	4.4	4.0	3.7	4.0	2.0	2.0	2.9	4.0	3.6	3.2	2.4	3.3
	MIDNIGHT-MIDNIGHT (OKTAS)	1	3.2	4.4	4.0	3.6	4.0	1.6	2.0	3.2	1.6	3.2	2.8	2.4	3.0
	MEAN NO. DAYS WITH:														
	CLEAR	3	7.7	9.7	11.0	12.0	11.0	17.8	17.0	15.5	9.5	12.0	10.0	13.0	146.2
	PARTLY CLOUDY	3	5.0	3.7	3.3	4.3	5.3	4.7	3.0	2.5	1.0	2.5	3.0	5.0	43.3
	CLOUDY	3	4.7	6.0	3.3	4.7	5.7	1.5	2.0	3.5	1.5	4.0	3.5	3.5	43.9
PR	MEAN STATION PRESSURE(IN)	28	26.35	26.31	26.23	26.25	26.25	26.30	26.37	26.38	26.37	26.37	26.33	26.35	26.32
	MEAN SEA-LEVEL PRES. (IN)	54	30.07	30.02	29.91	29.87	29.85	29.84	29.91	29.92	29.95	29.99	30.03	30.06	29.95
WINDS	MEAN SPEED (MPH)	37	12.6	13.4	14.6	14.7	14.1	13.7	12.4	11.3	12.4	12.5	12.8	12.6	13.1
	PREVAIL.DIR(TENS OF DEGS)	23	23	36	21	18	18	18	18	18	18	20	22	36	19
	MAXIMUM 2-MINUTE:														
	SPEED (MPH)	9	45	48	47	53	46	60	48	44	40	41	45	51	60
	DIR. (TENS OF DEGS)		25	25	03	25	11	35	30	34	24	17	30	32	35
	YEAR OF OCCURRENCE		2001	1993	1997	2001	1999	1994	1994	1996	1994	2000	1994	1993	JUN 1994
	MAXIMUM 5-SECOND:														
	SPEED (MPH)	9	54	55	60	74	52	68	58	59	52	61	52	57	74
	DIR. (TENS OF DEGS)		26	24	23	24	04	33	09	21	26	23	30	31	24
	YEAR OF OCCURRENCE		1996	1993	2000	2001	2001	1994	1999	2000	1994	1996	1994	1993	APR 2001
PRECIPITATION	NORMAL (IN)	30	0.50	0.61	0.96	0.99	2.48	3.70	2.62	3.22	1.99	1.37	0.69	0.43	19.56
	MAXIMUM MONTHLY (IN)	60	2.67	2.08	4.14	6.45	9.81	10.73	7.59	7.55	5.02	7.64	2.26	4.52	10.73
	YEAR OF OCCURRENCE		1999	1998	2000	1997	1951	1965	1960	1974	1960	1941	1961	1959	JUN 1965
	MINIMUM MONTHLY (IN)	60	0.00	T	T	T	0.04	0.01	0.04	0.28	0.03	0.00	0.00	T	0.00
	YEAR OF OCCURRENCE		1986	1991	1950	1964	1984	1953	2001	1983	1977	1952	1989	1976	NOV 1989
	MAXIMUM IN 24 HOURS (IN)	60	1.74	1.28	2.27	2.65	6.75	6.15	4.74	4.26	3.42	3.45	1.53	3.11	6.75
	YEAR OF OCCURRENCE		1968	1971	1973	1999	1951	1960	1982	1945	1941	1948	1980	1943	MAY 1951
	NORMAL NO. DAYS WITH:														
	PRECIPITATION 0.01	30	3.9	4.4	5.2	4.9	7.9	8.9	7.7	8.7	6.5	4.7	3.6	4.1	70.5
	PRECIPITATION 1.00	30	*	0.1	0.1	0.1	0.5	0.9	0.7	0.9	0.5	0.2	*	0.0	4.0
SNOWFALL	NORMAL (IN)	30	3.8	4.6	2.9	0.6	0.*	0.0	0.0	0.0	0.*	0.3	1.9	2.8	16.9
	MAXIMUM MONTHLY (IN)	60	14.5	17.3	14.7	6.5	0.5	T	T	T	0.3	3.9	13.6	21.2	21.2
	YEAR OF OCCURRENCE		1983	1971	1961	1997	1978	1992	1994	1999	1984	1976	1952	2000	DEC 2000
	MAXIMUM IN 24 HOURS (IN)	60	10.2	13.5	9.8	6.5	0.5	T	T	T	0.3	3.2	12.2	16.8	16.8
	YEAR OF OCCURRENCE		1994	1971	1957	1997	1978	1992	1994	1999	1984	1976	1952	2000	DEC 2000
	MAXIMUM SNOW DEPTH (IN)	56	10	14	10	6	T	T	0	0	T	1	9	15	15
	YEAR OF OCCURRENCE		1987	1983	1957	1973	1988	1949			1984	1970	1952	2000	DEC 2000
	NORMAL NO. DAYS WITH:														
	SNOWFALL 1.0	30	1.4	1.3	1.0	0.2	0.0	0.0	0.0	0.0	0.0	0.1	0.6	0.8	5.4

PRECIPITATION (inches) 2001 AMARILLO, TX (AMA)

YEAR	JAN	FEB	MAR	APR	MAY	JUN	JUL	AUG	SEP	OCT	NOV	DEC	ANNUAL
1972	0.21	0.11	0.11	0.03	2.81	3.87	2.59	1.73	0.71	1.66	1.19	0.32	15.34
1973	0.56	0.42	3.99	1.88	1.43	0.84	2.31	1.22	1.05	0.10	0.17	18.05
1974	0.33	0.24	0.60	0.04	4.06	3.33	1.31	7.55	1.65	3.44	0.12	0.42	23.09
1975	0.28	1.33	0.51	1.02	2.47	4.15	5.19	0.76	0.33	0.92	0.15	21.08
1976	T	0.10	0.79	1.65	1.36	2.94	1.77	1.78	4.28	1.14	0.43	T	16.24
1977	0.64	0.53	0.24	2.74	4.01	2.06	3.14	4.94	0.03	0.26	0.32	0.27	19.18
1978	0.63	0.80	0.21	0.55	5.76	6.50	1.82	1.61	2.42	0.97	0.47	0.27	22.01
1979	0.92	0.28	1.46	1.29	3.94	1.34	2.03	5.08	0.52	1.28	0.40	0.07	20.46
1980	0.85	0.55	1.38	0.82	2.88	1.30	0.65	1.80	1.55	0.42	0.84	0.35	13.39
1981	0.11	0.23	1.87	0.90	2.11	1.04	2.73	5.22	3.47	1.79	1.50	0.03	21.00
1982	0.15	0.39	0.52	0.43	1.96	4.75	6.23	0.55	1.37	0.71	0.75	0.79	18.60
1983	1.78	1.19	0.98	0.83	2.85	1.76	0.74	0.28	0.37	3.23	0.33	0.64	14.98
1984	0.56	0.37	0.98	1.18	0.04	6.76	0.83	2.28	0.95	3.19	1.09	1.00	19.23
1985	0.99	0.77	1.49	2.79	0.86	3.08	2.07	1.67	4.96	3.07	0.39	0.26	22.40
1986	0.00	1.02	0.60	0.30	3.28	3.70	3.52	7.04	1.45	1.94	1.82	0.66	25.33
1987	1.26	0.84	0.92	0.57	4.28	3.29	0.83	3.28	3.40	1.17	0.43	1.75	22.02
1988	0.33	0.04	1.19	2.22	6.02	3.68	3.30	3.59	3.15	0.71	0.29	0.17	24.69
1989	0.16	0.55	0.52	0.75	2.51	6.07	2.74	3.22	1.80	0.74	0.00	0.49	19.55
1990	1.22	1.61	2.56	1.10	0.90	0.14	3.28	2.79	2.72	0.46	0.50	0.23	17.51
1991	0.86	T	0.41	0.04	3.08	2.47	2.20	1.28	2.04	0.64	0.66	2.24	15.92
1992	0.50	0.30	1.11	1.60	3.10	7.57	2.36	2.27	0.16	0.31	0.80	0.55	20.63
1993	0.76	0.36	1.29	0.35	1.92	2.76	3.36	4.64	1.00	0.53	0.51	0.95	18.43
1994	1.01	0.07	1.22	1.15	1.41	1.26	5.01	2.86	2.02	0.46	0.58	0.30	17.35
1995	0.32	T	0.80	0.83	4.94	2.71	2.85	2.18	2.62	0.45	0.06	0.59	18.35
1996	0.07	0.26	0.24	T	1.67	2.92	4.95	6.43	1.86	0.84	1.30	0.05	20.59
1997	0.64	0.47	0.01	6.45	2.16	2.93	5.51	1.40	1.35	0.74	1.17	2.12	24.95
1998	0.68	2.08	2.46	0.97	0.53	0.12	1.09	0.76	1.23	6.48	0.34	0.41	17.15
1999	2.67	T	1.35	6.30	4.29	3.61	2.87	2.04	2.54	0.00	0.52	0.26	26.57
2000	0.24	0.04	4.14	0.43	1.14	5.54	0.16	0.29	0.03	3.95	0.96	1.47	18.39
2001	1.67	0.93	3.96	0.49	3.05	1.99	0.04	1.39	3.03	0.05	1.86	0.23	18.69
POR= 62 YRS	0.61	0.57	1.01	1.28	2.77	3.32	2.75	2.92	1.88	1.61	0.66	0.68	20.06

AVERAGE TEMPERATURE (F) 2001   AMARILLO, TX (AMA)

YEAR	JAN	FEB	MAR	APR	MAY	JUN	JUL	AUG	SEP	OCT	NOV	DEC	ANNUAL
1972	35.7	40.9	51.7	59.5	63.2	73.6	74.5	73.9	69.2	57.9	36.9	32.9	55.8
1973	33.4	39.5	47.1	50.3	62.9	75.1	78.1	78.8	68.7	62.0	50.8	38.6	57.1
1974	35.0	42.2	53.1	60.2	71.5	75.0	79.3	73.6	62.9	59.0	45.7	36.1	57.8
1975	37.1	35.3	45.0	54.9	64.5	73.3	75.5	76.7	65.8	61.1	45.8	40.4	56.3
1976	36.9	47.8	46.5	56.8	60.2	72.2	74.8	75.1	66.5	49.9	38.5	37.5	55.2
1977	30.1	43.8	48.7	57.8	67.3	78.1	80.7	77.5	74.3	60.7	47.4	39.9	58.9
1978	29.0	30.1	47.2	61.4	63.7	75.4	80.7	76.1	70.7	59.4	45.6	32.7	56.0
1979	24.9	40.0	46.6	54.5	62.2	70.9	77.1	73.6	69.8	60.1	40.5	38.8	54.9
1980	34.9	37.7	43.9	52.4	61.9	78.3	82.9	78.5	70.7	56.6	42.7	41.4	56.8
1981	37.9	42.2	48.8	63.1	65.6	78.5	81.3	74.4	69.0	56.6	49.1	40.1	58.9
1982	37.1	35.9	47.3	53.8	63.3	72.2	78.8	78.7	71.1	57.3	45.6	36.1	56.4
1983	33.4	36.2	45.3	50.9	60.3	70.4	80.0	81.0	73.6	60.5	47.9	24.7	55.4
1984	31.6	40.3	44.0	51.8	66.9	74.6	75.6	75.3	65.7	56.9	47.0	40.5	55.9
1985	31.5	37.2	49.5	60.0	68.1	75.1	80.1	79.6	68.9	56.9	43.5	34.0	57.0
1986	42.5	40.4	52.7	59.4	65.1	73.3	80.6	74.4	68.8	55.7	42.7	37.1	57.7
1987	34.0	42.0	44.6	54.8	64.5	72.1	77.2	75.3	67.6	57.9	45.4	34.6	55.8
1988	32.2	38.3	44.3	54.1	63.6	73.3	75.8	76.4	67.7	59.0	47.6	38.5	55.9
1989	40.7	32.3	51.2	59.9	67.3	69.4	76.2	76.0	66.3	60.4	47.9	31.4	56.6
1990	39.2	40.8	47.0	55.9	63.6	81.3	76.4	76.3	72.0	57.7	49.7	33.2	57.8
1991	32.6	45.9	49.7	58.2	68.5	74.7	76.5	76.2	67.1	58.5	41.2	39.0	57.3
1992	38.0	45.0	50.6	58.7	63.3	70.8	76.6	73.5	70.2	60.1	39.3	34.2	56.7
1993	33.1	36.0	46.1	54.4	64.0	74.0	79.1	74.9	67.9	54.6	40.9	38.7	55.4
1994	36.5	37.6	48.9	55.0	65.2	79.4	77.4	75.9	69.1	58.2	46.5	41.7	57.6
1995	39.2	43.9	47.4	53.9	61.6	70.8	77.2	78.2	67.9	58.2	48.2	38.4	57.1
1996	34.9	43.1	44.2	56.9	72.5	75.6	77.1	74.4	65.9	57.6	45.7	39.7	57.3
1997	35.3	38.8	50.3	48.9	63.2	71.6	77.9	75.0	71.9	58.2	42.8	33.5	55.6
1998	39.8	41.0	43.5	53.5	69.2	77.4	82.4	78.0	75.8	60.4	49.2	38.1	59.0
1999	40.1	47.3	46.3	54.8	62.3	71.8	77.5	78.0	66.2	58.1	52.3	37.6	57.7
2000	39.5	47.1	49.2	57.8	69.7	72.1	80.1	81.8	73.7	59.6	39.6	33.1	58.6
2001	34.8	41.2	45.0	61.0	65.2	76.5	83.9	78.7	70.4	60.2	50.7	40.4	59.0
POR= 62 YRS	35.7	40.0	46.7	56.5	65.4	74.5	78.6	77.0	69.5	58.9	45.7	37.7	57.2

REFERENCE NOTES:

PAGE 1:
THE TEMPERATURE GRAPH SHOWS NORMAL MAXIMUM AND NORMAL MINIMUM DAILY TEMPERATURES (SOLID CURVES) AND THE ACTUAL DAILY HIGH AND LOW TEMPERATURES (VERTICAL BARS).

PAGE 2 AND 3:
H/C INDICATES HEATING AND COOLING DEGREE DAYS.
RH INDICATES RELATIVE HUMIDITY
W/O INDICATES WEATHER AND OBSTRUCTIONS
S INDICATES SUNSHINE.
PR INDICATES PRESSURE.
CLOUDINESS ON PAGE 3 IS THE SUM OF THE CEILOMETER AND SATELLITE DATA NOT TO EXCEED EIGHT EIGHTHS (OKTAS).

GENERAL:
T INDICATES TRACE PRECIPITATION, AN AMOUNT GREATER THAN ZERO BUT LESS THAN THE LOWEST REPORTABLE VALUE.
+ INDICATES THE VALUE ALSO OCCURS ON EARLIER DATES.
BLANK ENTRIES DENOTE MISSING OR UNREPORTED DATA.
NORMALS ARE 30-YEAR AVERAGES (1961 - 1990).
ASOS INDICATES AUTOMATED SURFACE OBSERVING SYSTEM.
PM INDICATES THE LAST DAY OF THE PREVIOUS MONTH.
POR (PERIOD OF RECORD) BEGINS WITH THE JANUARY DATA MONTH AND IS THE NUMBER OF YEARS USED TO COMPUTE THE MEAN. INDIVIDUAL MONTHS WITHIN THE POR MAY BE MISSING.
WHEN THE POR FOR A NORMAL IS LESS THAN 30 YEARS, THE NORMAL IS PROVISIONAL AND IS BASED ON THE NUMBER OF YEARS INDICATED.
0.* OR * INDICATES THE VALUE OR MEAN-DAYS-WITH IS BETWEEN 0.00 AND 0.05.
CLOUDINESS FOR ASOS STATIONS DIFFERS FROM THE NON-ASOS OBSERVATION TAKEN BY A HUMAN OBSERVER. ASOS STATION CLOUDINESS IS BASED ON TIME-AVERAGED CEILOMETER DATA FOR CLOUDS AT OR BELOW 12,000 FEET AND ON SATELLITE DATA FOR CLOUDS ABOVE 12,000 FEET.
THE NUMBER OF DAYS WITH CLEAR, PARTLY CLOUDY, AND CLOUDY CONDITIONS FOR ASOS STATIONS IS THE SUM OF THE CEILOMETER AND SATELLITE DATA FOR THE SUNRISE TO SUNSET PERIOD.

GENERAL CONTINUED:
CLEAR INDICATES 0 - 2 OKTAS, PARTLY CLOUDY INDICATES 3 - 6 OKTAS, AND CLOUDY INDICATES 7 OR 8 OKTAS.
WHEN AT LEAST ONE OF THE ELEMENTS (CEILOMETER OR SATELLITE) IS MISSING, THE DAILY CLOUDINESS IS NOT COMPUTED.
WIND DIRECTION IS RECORDED IN TENS OF DEGREES (2 DIGITS) CLOCKWISE FROM TRUE NORTH. "00" INDICATES CALM. "36" INDICATES TRUE NORTH.
RESULTANT WIND IS THE VECTOR AVERAGE OF THE SPEED AND DIRECTION.
AVERAGE TEMPERATURE IS THE SUM OF THE MEAN DAILY MAXIMUM AND MINIMUM TEMPERATURE DIVIDED BY 2.
SNOWFALL DATA COMPRISE ALL FORMS OF FROZEN PRECIPITATION, INCLUDING HAIL.
A HEATING (COOLING) DEGREE DAY IS THE DIFFERENCE BETWEEN THE AVERAGE DAILY TEMPERATURE AND 65 F.
DRY BULB IS THE TEMPERATURE OF THE AMBIENT AIR.
DEW POINT IS THE TEMPERATURE TO WHICH THE AIR MUST BE COOLED TO ACHIEVE 100 PERCENT RELATIVE HUMIDITY.
WET BULB IS THE TEMPERATURE THE AIR WOULD HAVE IF THE MOISTURE CONTENT WAS INCREASED TO 100 PERCENT RELATIVE HUMIDITY.

ON JULY 1, 1996, THE NATIONAL WEATHER SERVICE BEGAN USING THE "METAR" OBSERVATION CODE THAT WAS ALREADY EMPLOYED BY MOST OTHER NATIONS OF THE WORLD.   THE MOST NOTICEABLE DIFFERENCE IN THIS ANNUAL PUBLICATION WILL BE THE CHANGE IN UNITS FROM TENTHS TO EIGHTS (OKTAS) FOR REPORTING THE AMOUNT OF SKY COVER.

HEATING DEGREE DAYS (base 65 F) 2001 AMARILLO, TX (AMA)

YEAR	JUL	AUG	SEP	OCT	NOV	DEC	JAN	FEB	MAR	APR	MAY	JUN	TOTAL
1972-73	13	2	48	274	833	987	972	706	548	438	113	0	4934
1973-74	0	0	56	154	420	813	922	633	368	190	17	0	3573
1974-75	0	0	119	199	571	890	853	826	612	323	67	16	4476
1975-76	0	0	104	158	569	757	861	492	567	253	171	0	3932
1976-77	0	0	59	464	790	846	1075	592	499	215	22	0	4562
1977-78	0	0	1	150	522	768	1107	972	551	136	139	6	4352
1978-79	0	0	33	197	574	992	1236	697	561	319	144	30	4783
1979-80	0	2	28	186	727	806	926	788	649	373	146	0	4631
1980-81	0	0	35	280	662	723	832	630	496	111	65	0	3834
1981-82	0	0	26	271	469	765	849	803	534	340	96	2	4155
1982-83	0	0	23	252	575	888	972	800	603	421	171	32	4737
1983-84	0	0	40	175	506	1241	1028	709	642	390	56	0	4787
1984-85	0	0	125	262	531	752	1034	769	474	169	37	5	4158
1985-86	0	0	111	249	640	957	691	681	379	203	72	0	3983
1986-87	0	2	26	290	665	858	954	634	624	315	70	9	4447
1987-88	0	8	18	226	584	936	1010	765	633	323	102	4	4609
1988-89	0	7	32	197	517	815	747	909	429	219	59	27	3958
1989-90	0	0	91	185	507	1037	795	672	551	276	140	0	4254
1990-91	0	0	11	234	454	981	994	529	467	208	53	1	3932
1991-92	0	0	71	258	709	798	833	576	437	203	98	10	3993
1992-93	0	0	21	171	765	949	984	808	577	317	82	5	4679
1993-94	0	2	39	332	717	810	875	760	499	309	78	0	4421
1994-95	0	0	31	228	553	716	792	583	547	333	146	13	3942
1995-96	0	1	89	217	497	813	926	634	636	250	21	0	4084
1996-97	3	0	70	251	571	775	917	728	447	476	99	2	4339
1997-98	0	4	35	259	661	967	774	667	664	343	39	21	4434
1998-99	0	0	13	180	469	828	764	490	573	305	132	9	3763
1999-00	0	0	93	226	373	844	785	510	482	220	60	9	3602
2000-01	0	0	50	195	755	981	929	660	611	139	75	0	4395
2001-	0	0	12	180	418	755							

COOLING DEGREE DAYS (base 65 F) 2001 AMARILLO, TX (AMA)

YEAR	JAN	FEB	MAR	APR	MAY	JUN	JUL	AUG	SEP	OCT	NOV	DEC	ANNUAL
1972	0	0	4	44	59	265	313	285	182	60	0	0	1212
1973	0	0	0	3	53	310	414	435	173	67	1	0	1456
1974	0	0	6	52	228	306	449	274	61	20	0	0	1396
1975	0	0	0	24	61	273	330	367	134	46	0	0	1235
1976	0	0	0	13	32	223	312	318	110	5	0	0	1013
1977	0	0	0	10	100	399	493	389	287	22	0	0	1700
1978	0	0	6	33	108	324	497	351	209	28	0	0	1556
1979	0	0	0	12	64	218	380	276	177	41	0	0	1168
1980	0	0	0	4	58	408	562	429	211	26	0	0	1698
1981	0	0	1	59	87	410	512	299	154	16	0	0	1538
1982	0	0	0	9	52	225	437	432	213	22	0	0	1390
1983	0	0	0	2	31	201	473	502	306	41	3	0	1559
1984	0	0	0	1	121	298	331	325	151	17	0	0	1244
1985	0	0	2	27	143	315	473	458	235	6	0	0	1659
1986	0	0	3	45	87	256	489	299	144	10	0	0	1333
1987	0	0	0	19	64	227	386	334	105	11	0	0	1146
1988	0	0	0	3	67	263	340	366	120	19	0	0	1178
1989	0	0	10	75	136	164	354	347	138	51	0	0	1275
1990	0	0	0	9	101	494	359	358	229	17	1	0	1568
1991	0	0	0	12	166	298	364	358	144	66	0	0	1408
1992	0	0	0	19	52	188	366	271	186	25	0	0	1107
1993	0	0	0	6	77	283	443	315	134	15	0	0	1273
1994	0	0	6	18	91	439	392	345	159	24	4	0	1478
1995	0	0	9	7	47	194	388	419	182	13	0	0	1259
1996	0	5	0	14	260	325	386	297	103	27	0	0	1417
1997	0	0	0	0	49	205	407	320	247	55	0	0	1283
1998	0	0	1	6	177	399	547	407	342	43	0	0	1922
1999	0	0	0	5	56	221	396	410	135	16	0	0	1239
2000	0	0	0	14	214	225	475	529	320	39	0	0	1816
2001	0	0	0	27	88	353	594	433	180	38	0	0	1713

SNOWFALL (inches) 2001 AMARILLO, TX (AMA)

YEAR	JUL	AUG	SEP	OCT	NOV	DEC	JAN	FEB	MAR	APR	MAY	JUN	TOTAL
1972-73	0.0	0.0	0.0	0.4	9.9	2.8	7.0	4.2	0.5	5.7	0.0	0.0	30.5
1973-74	0.0	0.0	0.0	0.0	0.4	1.5	3.6	1.2	1.2	T	0.0	0.0	6.7
1974-75	0.0	0.0	0.0	0.0	T	1.5	3.8	11.7	1.2	T	0.0	0.0	18.2
1975-76	0.0	0.0	0.0	0.0	0.4	1.8	T	4.2	0.0	0.0	0.0	0.0	6.4
1976-77	0.0	0.0	0.0	3.9	4.3	T	6.4	3.7	T	T	0.0	0.0	18.3
1977-78	0.0	0.0	0.0	0.0	1.3	0.2	8.2	9.4	1.5	T	0.5	0.0	21.1
1978-79	0.0	0.0	0.0	0.0	2.0	3.3	5.9	1.9	T	T	0.0	0.0	13.1
1979-80	0.0	0.0	0.0	1.6	2.2	0.3	1.4	5.2	1.0	T	0.0	0.0	11.7
1980-81	0.0	0.0	0.0	0.0	8.6	T	1.1	0.1	0.1	0.0	0.0	0.0	9.9
1981-82	0.0	0.0	0.0	0.0	T	0.3	1.1	4.7	1.7	T	0.0	0.0	7.8
1982-83	0.0	0.0	0.0	0.0	4.5	1.9	14.5	13.0	8.1	5.9	0.0	0.0	47.9
1983-84	0.0	0.0	0.0	0.0	T	5.2	7.0	3.5	2.5	0.0	0.0	0.0	18.2
1984-85	0.0	0.0	0.3	0.0	0.1	0.3	7.4	0.3	0.2	0.0	0.0	0.0	8.6
1985-86	0.0	0.0	T	0.0	T	2.8	0.0	10.9	1.0	0.0	0.0	0.0	14.7
1986-87	0.0	0.0	0.0	T	0.2	3.3	12.1	3.1	6.4	0.1	0.0	0.0	25.2
1987-88	0.0	0.0	0.0	0.0	0.7	15.3	4.3	0.5	8.5	4.2	0.0	0.0	33.5
1988-89	0.0	0.0	0.0	0.0	2.4	2.2	T	0.1	4.2	0.1	T	T	9.0
1989-90	0.0	0.0	T	0.0	0.0	5.4	8.5	3.0	0.0	0.0	0.0	0.0	16.9
1990-91	0.0	0.0	0.0	0.0	2.2	3.8	5.0	T	1.0	0.0	0.0	0.0	12.0
1991-92	0.0	0.0	T	3.2	T	2.0	1.0	0.2	T	0.3	T	T	6.7
1992-93	0.0	0.0	0.0	T	9.0	10.0	2.5	2.2	2.1	T	T	0.0	25.8
1993-94	0.0	0.0	0.0	T	T	0.5	10.2	T	3.0	T	0.0	0.0	13.7
1994-95	T	0.0	0.0	0.0	T	0.0	3.7	T	2.3	0.0	0.0	0.0	6.0
1995-96	T	0.0	T	0.0	0.5	2.7	1.1	4.6	0.1	0.0	T	0.0	9.0
1996-97	T	0.0	0.0	1.5	8.0	1.1	10.8	2.0	0.0	6.5	0.0	T	29.9
1997-98	T	0.0	0.0	T	5.4	10.9	T	10.4	4.4	T	0.0	0.0	31.1
1998-99	0.0	0.0	0.0	0.0	0.0	0.0	1.0	13.1	1.3	T	0.0	T	15.4
1999-00	0.0	T	0.0	T	0.0	8.6	2.8	T	0.5	0.5	T	0.0	12.4
2000-01	0.0	0.0	0.0	T	8.9	21.2	13.0	1.4	1.8	T	T	0.0	46.3
2001-	0.0	0.0	T	0.0	2.3	2.0							
POR= 62 YRS	0.3	0.1	0.0	0.2	1.9	3.2	4.2	3.3	2.5	0.7	0.1	0.0	16.5

2001
DALLAS - FORT WORTH, TEXAS (DFW)

The Dallas-Fort Worth Metroplex is located in North Central Texas, approximately 250 miles north of the Gulf of Mexico. It is near the headwaters of the Trinity River, which lie in the upper margins of the Coastal Plain. The rolling hills in the area range from 500 to 800 feet in elevation.

The Dallas-Fort Worth climate is humid subtropical with hot summers. It is also continental, characterized by a wide annual temperature range. Precipitation also varies considerably, ranging from less than 20 to more than 50 inches.

Winters are mild, but northers occur about three times each month, and often are accompanied by sudden drops in temperature. Periods of extreme cold that occasionally occur are short-lived, so that even in January mild weather occurs frequently.

The highest temperatures of summer are associated with fair skies, westerly winds and low humidities. Characteristically, hot spells in summer are broken into three-to-five day periods by thunderstorm activity. There are only a few nights each summer when the low temperature exceeds 80 degrees. Summer daytime temperatures frequently exceed 100 degrees. Air conditioners are recommended for maximum comfort indoors and while traveling via automobile.

Throughout the year, rainfall occurs more frequently during the night. Usually, periods of rainy weather last for only a day or two, and are followed by several days with fair skies. A large part of the annual precipitation results from thunderstorm activity, with occasional heavy rainfall over brief periods of time. Thunderstorms occur throughout the year, but are most frequent in the spring. Hail falls on about two or three days a year, ordinarily with only slight and scattered damage. Windstorms occurring during thunderstorm activity are sometimes destructive. Snowfall is rare.

The average length of the warm season (freeze-free period) in the Dallas-Fort Worth Metroplex is about 249 days. The average last occurrence of 32 degrees or below is mid March and the average first occurrence of 32 degrees or below is in late November.

NORMALS, MEANS, AND EXTREMES
DALLAS-FORT WORTH, TX (DFW)

LATITUDE: 32 53' 47" N   LONGITUDE: 97 02' 28" W   ELEVATION (FT): GRND: 559   BARO: 562   TIME ZONE: CENTRAL (UTC + 6)   WBAN: 03927

ELEMENT	POR	JAN	FEB	MAR	APR	MAY	JUN	JUL	AUG	SEP	OCT	NOV	DEC	YEAR
TEMPERATURE F														
NORMAL DAILY MAXIMUM	30	54.1	58.9	67.8	76.3	82.9	91.9	96.5	96.2	87.8	78.5	66.8	57.5	76.3
MEAN DAILY MAXIMUM	54	54.8	60.0	67.5	76.2	83.3	91.6	96.0	95.8	88.5	78.7	66.3	57.8	76.4
HIGHEST DAILY MAXIMUM	48	88	95	96	95	103	113	110	108	111	102	89	88	113
YEAR OF OCCURRENCE		1969	1996	1991	1990	1985	1980	1998	2000	2000	1979	1989	1955	JUN 1980
MEAN OF EXTREME MAXS.	54	76.1	80.0	85.1	89.1	94.1	98.8	102.7	103.3	98.7	92.4	82.9	77.2	90.0
NORMAL DAILY MINIMUM	30	32.7	36.9	45.6	54.7	62.6	70.0	74.1	73.6	66.9	55.8	45.4	36.3	54.6
MEAN DAILY MINIMUM	54	33.9	38.4	45.3	54.5	63.1	70.9	74.7	74.2	67.0	56.2	44.8	37.0	55.0
LOWEST DAILY MINIMUM	48	4	7	15	29	41	51	59	56	43	29	20	-1	-1
YEAR OF OCCURRENCE		1964	1985	1980	1989	1978	1964	1972	1967	1984	1993	1959	1989	DEC 1989
MEAN OF EXTREME MINS.	54	16.2	21.6	27.8	37.7	49.7	60.4	67.7	65.9	52.5	40.4	28.7	20.6	40.8
NORMAL DRY BULB	30	43.4	47.9	56.7	65.5	72.8	81.0	85.3	84.9	77.4	67.2	56.2	46.9	65.4
MEAN DRY BULB	54	44.4	49.2	56.3	65.3	73.1	81.2	85.3	85.1	77.7	67.3	55.4	47.3	65.6
MEAN WET BULB	18	40.5	45.0	50.5	57.8	66.6	72.3	73.9	73.1	68.1	59.6	50.1	42.5	58.3
MEAN DEW POINT	18	34.4	38.2	44.0	51.5	62.3	68.2	68.6	67.3	62.6	53.9	44.4	36.6	52.7
NORMAL NO. DAYS WITH:														
MAXIMUM 90	30	0.0	0.0	0.2	1.0	4.5	19.5	27.5	26.8	14.4	3.1	0.0	0.0	97.0
MAXIMUM 32	30	1.9	0.7	0.1	0.0	0.0	0.0	0.0	0.0	0.0	0.0	0.0	0.9	3.6
MINIMUM 32	30	15.7	9.3	2.8	0.2	0.0	0.0	0.0	0.0	0.0	*	2.3	10.7	41.0
MINIMUM 0	30	0.0	0.0	0.0	0.0	0.0	0.0	0.0	0.0	0.0	0.0	0.0	*	0.0
H/C														
NORMAL HEATING DEG. DAYS	30	670	484	286	75	0	0	0	0	0	51	275	566	2407
NORMAL COOLING DEG. DAYS	30	0	5	29	90	246	480	629	617	372	119	11	5	2603
RH														
NORMAL (PERCENT)	30	68	66	64	65	70	66	60	60	66	66	67	68	66
HOUR 00 LST	30	72	72	69	72	78	74	67	66	74	73	74	73	72
HOUR 06 LST	30	79	80	79	82	87	85	80	80	84	82	81	79	82
HOUR 12 LST	30	60	58	56	56	59	55	49	49	55	54	56	59	56
HOUR 18 LST	30	57	54	50	52	56	50	44	44	52	54	58	59	52
S PERCENT POSSIBLE SUNSHINE	17	52	54	58	61	57	67	75	73	67	63	57	52	61
W/O MEAN NO. DAYS WITH:														
HEAVY FOG(VISBY 1/4 MI)	48	2.5	1.5	1.0	0.6	0.3	0.1	0.0	0.0	0.1	0.8	1.5	2.4	10.8
THUNDERSTORMS	48	1.4	1.9	4.5	6.0	7.6	6.2	4.5	4.3	3.4	3.0	2.1	1.3	46.2
CLOUDINESS MEAN:														
SUNRISE-SUNSET (OKTAS)	1				4.0		4.0	3.2					4.8	
MIDNIGHT-MIDNIGHT (OKTAS)	1				4.0									
MEAN NO. DAYS WITH:														
CLEAR	1	2.0	6.0	15.0		10.0	11.0							
PARTLY CLOUDY	1		2.0			4.0	8.0							
CLOUDY	1	2.0		7.0		6.0	2.0							
PR MEAN STATION PRESSURE(IN)	29	29.49	29.49	29.40	29.30	29.30	29.30	29.40	29.40	29.39	29.40	29.40	29.50	29.40
MEAN SEA-LEVEL PRES. (IN)	18	30.13	30.08	30.01	29.93	29.90	29.91	29.96	29.96	29.98	30.04	30.08	30.13	30.01
WINDS														
MEAN SPEED (MPH)	48	11.0	11.7	12.6	12.4	11.1	10.6	10.0	9.1	9.4	9.9	11.0	11.1	10.8
PREVAIL.DIR(TENS OF DEGS)	5	18	18	18	18	18	18	18	18	16	16	18	18	18
MAXIMUM 2-MINUTE:														
SPEED (MPH)	6	41	51	48	47	43	47	36	47	33	46	40	39	51
DIR. (TENS OF DEGS)		29	23	30	30	34	32	11	33	24	23	28	33	23
YEAR OF OCCURRENCE		1996	2000	2000	2000	1998	1996	1999	1996	1996	2001	2001	2000	FEB 2000
MAXIMUM 5-SECOND:														
SPEED (MPH)	6	51	78	74	64	55	57	45	47	39	54	47	47	78
DIR. (TENS OF DEGS)		19	23	27	26	28	34	12	34	19	23	30	26	23
YEAR OF OCCURRENCE		1996	2000	2000	2000	2000	1996	1999	1996	2001	2001	1998	1997	FEB 2000
PRECIPITATION														
NORMAL (IN)	30	1.83	2.18	2.77	3.50	4.88	2.98	2.31	2.21	3.39	3.52	2.29	1.84	33.70
MAXIMUM MONTHLY (IN)	48	5.07	7.40	6.69	12.19	13.66	8.75	11.13	6.85	9.52	14.18	6.95	8.75	14.18
YEAR OF OCCURRENCE		1998	1997	1995	1957	1982	1989	1973	1970	1964	1981	2000	1991	OCT 1981
MINIMUM MONTHLY (IN)	48	T	0.15	0.10	0.11	0.95	0.40	0.00	0.00	0.09	T	0.20	0.17	0.00
YEAR OF OCCURRENCE		1986	1963	1972	1987	1996	1964	1993	2000	1984	1975	1970	1981	AUG 2000
MAXIMUM IN 24 HOURS (IN)	48	3.15	4.06	4.39	4.55	5.34	3.15	3.83	4.05	4.76	5.91	2.83	4.22	5.91
YEAR OF OCCURRENCE		1998	1965	1977	1957	1989	1989	2001	1976	1965	1959	1964	1991	OCT 1959
NORMAL NO. DAYS WITH:														
PRECIPITATION 0.01	30	6.7	6.3	7.3	7.6	8.7	6.4	4.7	4.6	7.1	6.2	6.0	6.5	78.1
PRECIPITATION 1.00	30	0.3	0.5	0.7	1.2	1.4	0.9	0.7	0.8	1.1	1.4	0.6	0.4	10.0
SNOWFALL														
NORMAL (IN)	30	1.4	1.0	0.2	0.0	0.0	0.0	0.0	0.0	0.0	0.0	0.2	0.3	3.1
MAXIMUM MONTHLY (IN)	43	12.1	13.5	2.5	T	T	0.0	0.0	0.0	0.0	T	5.0	2.6	13.5
YEAR OF OCCURRENCE		1964	1978	1962	1995	1995						1993	1976	1963	FEB 1978
MAXIMUM IN 24 HOURS (IN)	43	12.1	7.5	2.5	T	T	0.0	0.0	0.0	0.0	T	4.8	2.5	12.1
YEAR OF OCCURRENCE		1964	1978	1962	1995	1995						1993	1976	1963	JAN 1964
MAXIMUM SNOW DEPTH (IN)	48	6	8	2	0	0	0	0	0	0		3	2	8
YEAR OF OCCURRENCE		1964	1978	1971								1976	1983	FEB 1978
NORMAL NO. DAYS WITH:														
SNOWFALL 1.0	30	0.5	0.4	0.1	0.0	0.0	0.0	0.0	0.0	0.0	0.0	0.*	0.1	1.1

PRECIPITATION (inches) 2001 DALLAS - FORT WORTH, TX (DFW)

YEAR	JAN	FEB	MAR	APR	MAY	JUN	JUL	AUG	SEP	OCT	NOV	DEC	ANNUAL
1972	1.09	0.26	0.10	3.25	2.35	1.50	0.59	0.81	2.42	6.89	2.36	0.61	22.23
1973	3.26	1.92	2.28	6.06	3.18	5.88	11.13	0.01	7.16	6.85	2.06	0.83	50.62
1974	1.79	1.01	0.80	2.51	6.00	5.44	0.67	4.19	6.04	5.93	3.32	1.93	39.63
1975	3.34	3.72	1.67	3.40	6.88	1.95	0.30	0.87	T	3.46	0.42	1.49	29.10
1976	0.13	0.52	2.29	5.71	6.03	1.40	3.83	4.75	5.02	3.46	0.50	1.99	35.63
1977	2.39	1.68	5.88	4.31	0.99	0.69	2.20	2.33	1.72	2.96	1.79	0.25	27.19
1978	1.41	3.33	2.66	1.34	8.01	0.77	0.33	1.53	0.93	0.55	2.73	0.78	24.37
1979	3.35	1.52	6.33	2.03	5.90	1.36	1.94	2.47	0.99	3.38	0.43	2.72	32.42
1980	2.52	0.84	1.24	2.23	3.01	1.25	0.71	T	6.54	1.08	1.23	1.43	22.08
1981	0.58	1.44	3.39	2.69	6.24	7.85	1.81	2.32	2.40	14.18	1.53	0.17	44.60
1982	2.33	1.89	1.71	2.71	13.66	4.28	2.73	0.52	0.58	3.36	4.22	2.76	40.75
1983	2.55	1.25	4.36	0.59	5.83	2.07	1.56	5.55	0.22	4.04	2.22	0.83	31.07
1984	1.07	3.11	4.92	1.41	3.04	2.79	0.43	1.47	0.09	6.50	2.97	6.09	33.89
1985	0.81	2.62	3.70	3.75	2.13	3.78	2.40	0.53	3.35	3.91	3.11	0.61	30.70
1986	T	2.49	1.08	5.30	5.52	3.92	0.41	1.63	4.60	1.81	3.25	2.44	32.45
1987	1.22	3.67	1.70	0.11	5.95	3.45	1.77	0.81	1.38	0.12	4.17	2.90	27.25
1988	0.88	1.23	2.03	2.21	2.11	3.23	2.47	0.44	4.04	1.64	2.28	2.48	25.04
1989	2.56	3.70	3.72	1.86	9.62	8.75	2.61	1.89	2.40	2.02	0.49	0.33	39.95
1990	4.54	4.72	5.89	6.90	7.16	1.89	2.60	2.37	1.12	2.81	3.81	1.46	45.27
1991	2.72	2.60	1.35	3.63	6.97	4.26	3.99	4.30	4.61	9.32	1.04	8.75	53.54
1992	3.25	2.40	3.24	2.46	6.93	5.23	2.48	2.08	3.25	3.05	3.56	4.26	42.19
1993	1.74	5.78	3.03	3.49	1.75	3.75	0.00	0.75	3.28	5.10	1.62	2.54	32.83
1994	1.43	2.01	1.69	3.62	5.80	2.05	4.58	4.89	1.39	8.19	6.03	2.42	44.10
1995	2.11	0.44	6.69	6.83	7.50	2.41	3.45	0.86	1.54	0.75	0.74	2.08	35.40
1996	0.97	0.35	2.36	2.14	0.95	3.42	3.85	5.02	1.51	6.56	5.54	0.47	33.14
1997	0.33	7.40	2.21	6.73	3.92	3.99	1.68	3.13	2.01	5.66	1.01	6.93	45.00
1998	5.07	3.22	4.45	1.25	2.38	1.75	0.11	0.35	0.68	5.64	4.91	4.43	34.24
1999	1.44	0.48	2.84	2.74	6.91	0.99	0.77	T	2.30	2.26	0.31	2.55	23.59
2000	1.59	3.30	2.92	4.28	3.17	5.93	T	0.00	0.16	4.38	6.95	3.57	36.25
2001	2.44	6.17	5.27	0.89	5.58	1.28	3.85	2.72	3.72	1.86	1.11	3.24	38.13
POR= 103 YRS	1.03	0.54	2.42	2.21	1.13	3.40	3.74	4.84	1.54	6.44	5.41	0.66	33.36

AVERAGE TEMPERATURE (F) 2001 DALLAS - FORT WORTH, TX (DFW)

YEAR	JAN	FEB	MAR	APR	MAY	JUN	JUL	AUG	SEP	OCT	NOV	DEC	ANNUAL
1972	45.0	51.5	62.1	70.1	72.7	81.4	83.1	84.7	80.8	67.5	50.1	44.0	66.1
1973	42.5	47.9	60.0	60.7	71.7	79.3	83.9	82.9	76.1	68.3	59.8	48.4	65.1
1974	43.6	52.3	62.9	65.8	75.7	78.7	86.1	82.9	70.9	69.2	55.5	47.2	65.9
1975	49.0	46.6	53.8	64.7	72.4	80.9	83.6	84.8	75.6	69.3	57.3	45.0	65.6
1976	45.0	58.4	59.4	64.9	68.6	78.8	82.1	84.2	76.1	60.2	49.5	45.0	64.4
1977	34.7	49.4	57.2	66.8	77.4	84.1	87.1	84.9	81.6	66.7	56.4	47.6	66.2
1978	33.8	36.7	54.1	67.1	73.1	82.3	88.4	84.6	80.2	68.9	57.7	46.1	64.4
1979	35.4	42.2	56.7	64.4	69.7	81.0	84.5	82.5	77.0	70.8	52.9	49.4	63.9
1980	45.5	46.6	54.2	63.1	75.0	87.0	92.0	88.5	80.3	65.4	54.9	49.4	66.8
1981	44.6	48.9	55.7	69.2	70.5	80.3	85.9	83.4	76.2	66.1	57.5	47.3	65.5
1982	44.6	44.5	59.8	62.5	72.5	79.2	84.6	86.7	78.1	67.0	55.6	49.2	65.4
1983	43.4	48.5	54.5	60.6	69.5	77.3	83.6	84.9	77.1	67.8	57.3	34.8	63.3
1984	39.3	50.9	56.3	63.7	73.7	82.5	85.5	85.8	76.1	67.0	54.6	52.7	65.7
1985	37.8	45.0	60.8	67.2	74.0	80.2	84.4	87.6	77.7	67.6	56.3	42.3	65.1
1986	48.8	51.2	60.2	67.2	71.5	80.8	86.4	83.4	80.2	65.7	52.4	46.1	66.2
1987	44.5	50.8	53.9	65.0	75.1	79.6	83.4	86.5	77.1	66.5	55.7	47.3	65.5
1988	42.2	47.1	56.0	64.5	72.8	80.4	85.3	87.9	79.2	65.7	58.1	49.1	65.7
1989	50.0	42.2	56.7	66.4	74.3	77.9	82.8	82.3	74.7	69.0	58.2	39.0	64.5
1990	51.8	53.9	57.7	64.0	73.4	84.0	82.5	84.6	80.0	66.4	59.8	44.0	66.8
1991	42.8	53.7	59.8	67.4	75.4	81.0	85.0	82.5	75.2	68.1	51.7	50.3	66.1
1992	46.9	54.4	59.1	66.0	71.1	79.4	84.3	80.2	77.7	69.5	52.7	49.9	65.9
1993	45.1	49.0	56.1	63.3	71.9	81.7	87.3	87.5	78.2	63.8	51.6	49.5	65.4
1994	45.7	48.6	59.0	65.9	71.5	84.1	83.9	84.0	76.3	67.3	57.9	49.0	66.1
1995	48.2	52.5	56.9	64.2	73.2	79.9	85.5	85.5	76.6	67.8	54.8	47.5	66.1
1996	43.0	52.1	53.3	64.2	79.7	82.5	86.2	82.5	74.5	66.9	54.9	49.2	65.8
1997	44.0	49.5	58.3	60.4	70.1	78.8	84.9	83.1	80.2	67.2	52.1	45.6	64.5
1998	48.4	51.1	54.9	63.8	78.5	85.5	91.6	87.8	83.6	69.5	57.6	47.0	68.3
1999	48.6	55.5	56.4	67.9	73.8	82.1	86.1	90.2	79.0	62.9	51.5	48.6	68.6
2000	50.6	57.3	61.0	65.3	76.6	80.7	87.3	90.2	80.4	69.7	49.8	39.4	67.4
2001	42.7	50.2	51.8	67.8	74.2	80.3	86.7	84.9	74.6	65.0	59.7	49.5	65.6
POR= 103 YRS	45.2	49.0	56.9	65.1	72.8	81.0	84.7	84.8	77.9	67.6	55.9	47.4	65.7

REFERENCE NOTES:

PAGE 1:
THE TEMPERATURE GRAPH SHOWS NORMAL MAXIMUM AND NORMAL MINIMUM DAILY TEMPERATURES (SOLID CURVES) AND THE ACTUAL DAILY HIGH AND LOW TEMPERATURES (VERTICAL BARS).

PAGE 2 AND 3:
H/C INDICATES HEATING AND COOLING DEGREE DAYS.
RH INDICATES RELATIVE HUMIDITY
W/O INDICATES WEATHER AND OBSTRUCTIONS
S INDICATES SUNSHINE.
PR INDICATES PRESSURE.
CLOUDINESS ON PAGE 3 IS THE SUM OF THE CEILOMETER AND SATELLITE DATA NOT TO EXCEED EIGHT EIGHTHS(OKTAS).

GENERAL:
T INDICATES TRACE PRECIPITATION, AN AMOUNT GREATER THAN ZERO BUT LESS THAN THE LOWEST REPORTABLE VALUE.
+ INDICATES THE VALUE ALSO OCCURS ON EARLIER DATES.
BLANK ENTRIES DENOTE MISSING OR UNREPORTED DATA.
NORMALS ARE 30-YEAR AVERAGES (1961 - 1990).
ASOS INDICATES AUTOMATED SURFACE OBSERVING SYSTEM.
PM INDICATES THE LAST DAY OF THE PREVIOUS MONTH.
POR (PERIOD OF RECORD) BEGINS WITH THE JANUARY DATA MONTH AND IS THE NUMBER OF YEARS USED TO COMPUTE THE MEAN. INDIVIDUAL MONTHS WITHIN THE POR MAY BE MISSING.
WHEN THE POR FOR A NORMAL IS LESS THAN 30 YEARS, THE NORMAL IS PROVISIONAL AND IS BASED ON THE NUMBER OF YEARS INDICATED.
0.* OR * INDICATES THE VALUE OR MEAN-DAYS-WITH IS BETWEEN 0.00 AND 0.05.
CLOUDINESS FOR ASOS STATIONS DIFFERS FROM THE NON-ASOS OBSERVATION TAKEN BY A HUMAN OBSERVER. ASOS STATION CLOUDINESS IS BASED ON TIME-AVERAGED CEILOMETER DATA FOR CLOUDS AT OR BELOW 12,000 FEET AND ON SATELLITE DATA FOR CLOUDS ABOVE 12,000 FEET.
THE NUMBER OF DAYS WITH CLEAR, PARTLY CLOUDY, AND CLOUDY CONDITIONS FOR ASOS STATIONS IS THE SUM OF THE CEILOMETER AND SATELLITE DATA FOR THE SUNRISE TO SUNSET PERIOD.

GENERAL CONTINUED:
CLEAR INDICATES 0 - 2 OKTAS, PARTLY CLOUDY INDICATES 3 - 6 OKTAS, AND CLOUDY INDICATES 7 OR 8 OKTAS. WHEN AT LEAST ONE OF THE ELEMENTS (CEILOMETER OR SATELLITE) IS MISSING, THE DAILY CLOUDINESS IS NOT COMPUTED.
WIND DIRECTION IS RECORDED IN TENS OF DEGREES (2 DIGITS) CLOCKWISE FROM TRUE NORTH. "00" INDICATES CALM. "36" INDICATES TRUE NORTH.
RESULTANT WIND IS THE VECTOR AVERAGE OF THE SPEED AND DIRECTION.
AVERAGE TEMPERATURE IS THE SUM OF THE MEAN DAILY MAXIMUM AND MINIMUM TEMPERATURE DIVIDED BY 2.
SNOWFALL DATA COMPRISE ALL FORMS OF FROZEN PRECIPITATION, INCLUDING HAIL.
A HEATING (COOLING) DEGREE DAY IS THE DIFFERENCE BETWEEN THE AVERAGE DAILY TEMPERATURE AND 65 F.
DRY BULB IS THE TEMPERATURE OF THE AMBIENT AIR.
DEW POINT IS THE TEMPERATURE TO WHICH THE AIR MUST BE COOLED TO ACHIEVE 100 PERCENT RELATIVE HUMIDITY.
WET BULB IS THE TEMPERATURE THE AIR WOULD HAVE IF THE MOISTURE CONTENT WAS INCREASED TO 100 PERCENT RELATIVE HUMIDITY.

ON JULY 1, 1996, THE NATIONAL WEATHER SERVICE BEGAN USING THE "METAR" OBSERVATION CODE THAT WAS ALREADY EMPLOYED BY MOST OTHER NATIONS OF THE WORLD. THE MOST NOTICEABLE DIFFERENCE IN THIS ANNUAL PUBLICATION WILL BE THE CHANGE IN UNITS FROM TENTHS TO EIGHTS(OKTAS) FOR REPORTING THE AMOUNT OF SKY COVER.

HEATING DEGREE DAYS (base 65 F) 2001 DALLAS - FORT WORTH, TX (DFW)

YEAR	JUL	AUG	SEP	OCT	NOV	DEC	JAN	FEB	MAR	APR	MAY	JUN	TOTAL
1972-73	0	0	3	96	446	644	690	475	155	182	12	0	2703
1973-74	0	0	1	36	182	509	656	352	173	70	1	0	1980
1974-75	0	0	20	16	296	546	489	355	112	0	0	0	2342
1975-76	0	0	4	33	266	500	616	217	222	48	20	0	1926
1976-77	0	0	0	214	459	614	931	431	241	37	0	0	2927
1977-78	0	0	0	55	257	536	962	786	346	54	41	0	3037
1978-79	0	0	0	27	247	578	911	635	261	78	29	0	2766
1979-80	0	0	0	34	370	478	597	530	339	102	6	0	2456
1980-81	0	0	18	99	330	486	625	448	284	26	23	0	2339
1981-82	0	0	10	116	228	541	625	569	232	140	9	0	2470
1982-83	0	0	1	94	316	495	663	454	324	186	21	2	2556
1983-84	0	0	12	52	269	933	789	401	281	89	11	0	2837
1984-85	0	0	38	66	322	389	837	558	171	37	0	0	2418
1985-86	0	0	19	53	285	696	495	400	164	41	5	0	2158
1986-87	0	0	0	61	376	580	632	387	342	109	0	0	2487
1987-88	0	0	0	55	297	540	703	512	301	70	0	0	2478
1988-89	0	0	0	51	240	487	460	630	294	102	4	0	2268
1989-90	0	0	14	80	251	799	401	306	251	102	19	0	2223
1990-91	0	0	0	100	190	646	681	314	198	37	7	0	2173
1991-92	0	0	12	69	405	456	555	302	182	73	14	0	2068
1992-93	0	0	0	14	366	474	612	445	290	120	4	0	2325
1993-94	0	0	2	145	414	476	594	451	229	83	27	0	2421
1994-95	0	0	6	71	226	487	521	348	289	86	13	0	2047
1995-96	0	0	22	41	304	549	671	403	384	104	0	0	2478
1996-97	0	0	14	51	308	483	650	429	213	155	16	0	2319
1997-98	0	0	0	89	383	593	512	383	347	86	0	0	2393
1998-99	0	0	0	31	226	559	505	266	263	39	3	0	1892
1999-00	0	0	0	44	118	421	447	239	169	77	0	0	1515
2000-01	0	0	12	64	458	785	685	417	402	41	0	0	2864
2001-	0	0	3	77	191	489							

COOLING DEGREE DAYS (base 65 F) 2001 DALLAS - FORT WORTH, TX (DFW)

YEAR	JAN	FEB	MAR	APR	MAY	JUN	JUL	AUG	SEP	OCT	NOV	DEC	ANNUAL
1972	2	14	57	185	249	498	618	480	183	4	0	2859	
1973	1	0	6	60	230	435	593	559	342	146	33	0	2405
1974	0	2	115	101	341	419	660	563	202	153	20	2	2578
1975	0	0	15	107	236	483	580	620	331	189	39	9	2609
1976	0	32	59	52	138	421	537	602	338	72	0	0	2251
1977	0	0	7	94	391	581	693	626	505	112	6	2	3017
1978	0	0	16	125	301	524	733	614	462	153	37	0	2965
1979	0	1	9	67	179	489	613	551	366	220	14	0	2509
1980	0	0	11	52	320	668	844	737	485	117	35	10	3279
1981	0	5	5	158	200	467	654	577	352	155	8	0	2581
1982	1	2	77	71	252	433	614	679	403	160	40	10	2742
1983	0	0	7	61	171	382	582	626	381	145	46	0	2401
1984	0	0	20	60	288	531	644	652	376	135	16	12	2734
1985	0	5	51	108	287	460	608	706	408	139	29	0	2801
1986	0	19	24	112	212	480	673	578	464	91	3	0	2656
1987	0	0	6	114	318	442	576	674	370	111	23	0	2634
1988	4	0	28	61	247	467	639	714	433	78	39	1	2711
1989	1	0	45	154	297	393	561	542	314	208	52	0	2567
1990	1	2	30	79	286	575	551	617	457	152	41	2	2793
1991	0	3	42	115	335	484	624	550	324	174	14	5	2670
1992	0	2	4	109	213	437	606	480	386	161	4	13	2415
1993	0	0	19	75	223	507	697	701	406	115	20	4	2767
1994	3	0	50	119	238	578	591	596	350	149	19	1	2694
1995	5	5	44	67	271	456	641	643	374	132	6	12	2656
1996	0	37	27	87	464	531	663	549	302	120	10	2	2792
1997	6	0	12	24	181	423	622	566	461	163	3	0	2461
1998	0	0	45	56	424	623	831	710	563	177	9	9	3447
1999	4	5	4	134	282	520	662	785	425	181	63	9	3074
2000	7	23	49	93	364	475	698	787	482	215	7	0	3200
2001	0	6	0	134	293	465	682	624	300	86	39	13	2642

SNOWFALL (inches) 2001 DALLAS - FORT WORTH, TX (DFW)

YEAR	JUL	AUG	SEP	OCT	NOV	DEC	JAN	FEB	MAR	APR	MAY	JUN	TOTAL
1972-73	0.0	0.0	0.0	0.0	T	1.4	2.3	T	0.0	0.0	0.0	0.0	3.7
1973-74	0.0	0.0	0.0	0.0	0.0	0.0	T	0.0	0.0	0.0	0.0	0.0	T
1974-75	0.0	0.0	0.0	0.0	T	T	T	3.7	T	0.0	0.0	0.0	3.7
1975-76	0.0	0.0	0.0	0.0	0.0	0.4	0.0	0.0	T	0.0	0.0	0.0	0.4
1976-77	0.0	0.0	0.0	0.0	5.0	5.4	0.0	0.0	0.0	0.0	0.0	0.0	10.4
1977-78	0.0	0.0	0.0	0.0	0.0	0.0	4.1	13.5	T	0.0	0.0	0.0	17.6
1978-79	0.0	0.0	0.0	0.0	0.0	0.8	1.8	0.7	0.0	0.0	0.0	0.0	3.3
1979-80	0.0	0.0	0.0	0.0	0.0	T	0.0	1.6	0.0	0.0	0.0	0.0	1.6
1980-81	0.0	0.0	0.0	0.0	0.0	0.0	T	T	0.0	0.0	0.0	0.0	T
1981-82	0.0	0.0	0.0	0.0	0.0	0.0	0.8	T	0.0	0.0	0.0	0.0	0.8
1982-83	0.0	0.0	0.0	0.0	0.0	0.0	T	T	T	0.0	0.0	0.0	T
1983-84	0.0	0.0	0.0	0.0	0.0	2.0	0.0	0.0	0.0	0.0	0.0	0.0	2.0
1984-85	0.0	0.0	0.0	0.0	0.0	0.0	3.4	1.7	0.0	0.0	0.0	0.0	5.1
1985-86	0.0	0.0	0.0	0.0	0.0	T	0.0	0.8	0.0	0.0	0.0	0.0	0.8
1986-87	0.0	0.0	0.0	0.0	0.0	1.7	T	T	0.5	0.0	0.0	0.0	2.2
1987-88	0.0	0.0	0.0	0.0	0.0	T	0.8	2.7	0.0	0.0	0.0	0.0	3.5
1988-89	0.0	0.0	0.0	0.0	0.0	T	T	0.7	1.1	0.0	T	0.0	1.8
1989-90	0.0	0.0	0.0	0.0	0.0	T	0.0	0.0	0.0	T	T	0.0	T
1990-91	0.0	0.0	0.0	0.0	0.0	0.3	0.3	0.0	0.0	0.0	0.0	0.0	0.6
1991-92	0.0	0.0	0.0	0.0	0.0	0.0	T	0.0	0.0	0.0	0.0	0.0	T
1992-93	0.0	0.0	0.0	0.0	0.0	T	0.0	0.0	0.0	0.0	0.0	0.0	T
1993-94	0.0	0.0	0.0	T	0.3	T	T	0.1	T	T	T	0.0	0.4
1994-95	0.0	0.0	0.0	0.0	0.0	0.0	T	0.0	T	T	T	0.0	T
1995-96	0.0				T	0.0							
1996-97					T	T							
1997-98													
1998-99													
1999-00													
2000-01													
2001-													
POR= 42 YRS	0.0	0.0	0.0	T	0.1	0.2	1.1	0.9	0.2	T	T	0.0	2.5

2001
EL PASO,
TEXAS (ELP)

The city of El Paso is located in the extreme west point of Texas at an elevation of about 3,700 feet . The National Weather Service station is located on a mesa about 200 feet higher than the city. The climate of the region is characterized by an abundance of sunshine throughout the year, high daytime summer temperatures, very low humidity, scanty rainfall, and a relatively mild winter season. The Franklin Mountains begin within the city limits and extend northward for about 16 miles. Peaks of these mountains range from 4,687 to 7,152 feet above sea level.

Rainfall throughout the year is light, insufficient for any growth except desert vegetation. Irrigation is necessary for crops, gardens, and lawns. Dry periods lasting several months are not unusual. Almost half of the precipitation occurs in the three-month period, July through September, from brief but often heavy thunderstorms. Small amounts of snow fall nearly every winter, but snow cover rarely amounts to more than an inch and seldom remains on the ground for more than a few hours.

Daytime summer temperatures are high, frequently above 90 degrees and occasionally above 100 degrees. Summer nights are usually comfortable, with temperatures in the 60s. It should be noted that when temperatures are high the relative humidity is generally quite low. A 20-year tabulation of observations with temperatures above 90 degrees shows that in April, May, and June the humidity averaged from 10 to 14 percent, while in July, August, and September it averaged 22 to 24 percent. This low humidity aids the efficiency of evaporative air coolers, which are widely used in homes and public buildings and are quite effective in cooling the air to comfortable temperatures.

Winter daytime temperatures are mild. At night they drop below freezing about half the time in December and January. The flat, irrigated land of the Rio Grande Valley in the vicinity of El Paso is noticeably cooler, particularly at night, than the airport or the city proper, both in summer and winter. This results in more comfortable temperatures in summer but increases the severity of freezes in winter. The cooler air in the Valley also causes marked short-period fluctuations of temperature and dewpoint at the airport with changes in wind direction, especially during the early morning hours.

Dust and sandstorms are the most unpleasant features of the weather in El Paso. While wind velocities are not excessively high, the soil surface is dry and loose and natural vegetation is sparse, so moderately strong winds raise considerable dust and sand. A tabulation of duststorms for a period of 20 years shows that they are most frequent in March and April, and comparatively rare in the period July through December. prevailing winds are from the north in winter and the south in summer.

NORMALS, MEANS, AND EXTREMES
EL PASO, TX (ELP)

LATITUDE: 31° 48′ 40″ N LONGITUDE: 106° 22′ 33″ W ELEVATION (FT): GRND: 3942 BARO: 3945 TIME ZONE: MOUNTAIN (UTC + 7) WBAN: 23044

	ELEMENT	POR	JAN	FEB	MAR	APR	MAY	JUN	JUL	AUG	SEP	OCT	NOV	DEC	YEAR
TEMPERATURE °F	NORMAL DAILY MAXIMUM	30	56.1	62.2	69.9	78.7	87.1	96.5	96.1	93.5	87.1	78.4	66.4	57.5	77.5
	MEAN DAILY MAXIMUM	54	57.7	63.3	69.9	78.7	87.3	95.7	95.2	93.0	88.0	78.7	64.9	58.2	77.6
	HIGHEST DAILY MAXIMUM	62	80	83	89	98	104	114	112	108	104	96	87	80	114
	YEAR OF OCCURRENCE		1970	1986	1989	1989	1951	1994	1979	1980	1982	1994	1983	1973	JUN 1994
	MEAN OF EXTREME MAXS.	54	70.8	76.4	82.6	90.5	97.6	104.4	103.4	100.3	96.7	89.8	77.1	70.9	88.4
	NORMAL DAILY MINIMUM	30	29.4	33.9	40.2	48.0	56.5	64.3	68.4	66.6	61.6	49.6	38.4	30.7	49.0
	MEAN DAILY MINIMUM	54	31.1	35.3	41.3	49.3	58.0	66.6	69.7	68.2	62.1	50.3	37.3	31.7	50.1
	LOWEST DAILY MINIMUM	62	-8	8	14	23	31	46	57	56	41	25	0	5	-8
	YEAR OF OCCURRENCE		1962	1985	1971	1983	1967	1988	1988	1945	1970	1996	1953		JAN 1962
	MEAN OF EXTREME MINS.	54	16.7	21.1	26.2	34.7	44.6	55.4	63.3	61.8	51.4	37.1	23.4	17.9	37.8
	NORMAL DRY BULB	30	42.8	48.1	55.1	63.4	71.8	80.4	82.3	80.1	74.4	64.0	52.4	44.1	63.2
	MEAN DRY BULB	54	44.4	49.4	55.6	64.0	72.7	81.2	82.5	80.6	75.0	64.5	51.1	44.9	63.8
	MEAN WET BULB	18	36.5	40.0	43.0	47.8	54.4	61.1	65.5	66.0	61.1	51.9	40.3	36.6	50.4
	MEAN DEW POINT	18	24.6	25.5	25.8	27.6	35.2	45.6	55.5	57.9	51.4	40.2	29.0	26.0	37.0
	NORMAL NO. DAYS WITH:														
	MAXIMUM 90	30	0.0	0.0	0.0	2.0	12.8	26.0	26.9	23.4	11.8	1.6	0.0	0.0	104.5
	MAXIMUM 32	30	0.4	0.1	0.0	0.0	0.0	0.0	0.0	0.0	0.0	0.0	0.1	0.2	0.8
	MINIMUM 32	30	19.6	12.3	5.1	0.9	*	0.0	0.0	0.0	0.0	0.4	7.8	18.8	64.9
	MINIMUM 0	30	0.1	0.0	0.0	0.0	0.0	0.0	0.0	0.0	0.0	0.0	0.0	0.0	0.1
H/C	NORMAL HEATING DEG. DAYS	30	688	473	316	110	7	0	0	0	0	88	378	648	2708
	NORMAL COOLING DEG. DAYS	30	0	0	10	62	217	462	536	468	282	57	0	0	2094
RH	NORMAL (PERCENT)	30	50	42	32	27	27	30	44	48	50	47	46	52	41
	HOUR 05 LST	30	65	56	46	40	41	46	63	67	69	65	61	65	57
	HOUR 11 LST	30	44	37	28	23	23	25	37	41	43	38	38	44	35
	HOUR 17 LST	30	34	27	20	17	16	18	29	33	34	30	32	37	27
	HOUR 23 LST	30	55	44	34	29	29	32	48	53	56	53	52	56	45
S	PERCENT POSSIBLE SUNSHINE	54	78	82	86	89	90	90	82	81	83	84	83	77	84
W/O	MEAN NO. DAYS WITH:														
	HEAVY FOG(VISBY 1/4 MI)	63	0.6	0.2	0.1	0.0	0.0	0.0	0.0	0.0	0.1	0.1	0.3	0.6	2.0
	THUNDERSTORMS	63	0.2	0.4	0.5	1.1	3.0	4.5	10.3	9.9	4.0	1.8	0.3	0.2	36.2
CLOUDINESS	MEAN:														
	SUNRISE-SUNSET (OKTAS)	53	3.7	3.4	3.4	2.8	2.6	2.2	3.5	3.4	2.7	2.5	2.7	3.4	3.0
	MIDNIGHT-MIDNIGHT (OKTAS)	31	3.3	3.1	2.8	2.4	2.4	2.2	3.6	3.6	2.9	2.2	2.5	3.0	2.8
	MEAN NO. DAYS WITH:														
	CLEAR	54	13.9	13.6	15.1	16.7	18.5	19.8	12.0	13.6	17.4	18.5	17.2	14.8	191.1
	PARTLY CLOUDY	54	7.5	7.3	8.4	8.0	8.2	7.4	13.1	11.9	7.1	6.7	6.1	7.2	98.9
	CLOUDY	54	9.7	7.3	7.5	5.2	4.2	2.7	5.4	4.9	5.0	5.2	6.2	8.5	71.8
PR	MEAN STATION PRESSURE(IN)	29	26.10	26.09	26.00	26.00	26.00	26.00	26.10	26.10	26.10	26.10	26.11	26.10	26.07
	MEAN SEA-LEVEL PRES. (IN)	17	30.06	29.98	29.91	29.83	29.77	29.97	29.84	29.87	29.90	29.94	30.02	30.07	29.93
WINDS	MEAN SPEED (MPH)	59	8.3	9.1	10.9	11.0	10.3	9.3	8.4	7.8	7.7	7.7	8.0	7.9	8.9
	PREVAIL.DIR(TENS OF DEGS)	30	02	27	27	27	26	27	14	14	36	36	01	36	36
	MAXIMUM 2-MINUTE:														
	SPEED (MPH)	6	64	51	52	56	45	51	45	47	41	41	52	43	64
	DIR. (TENS OF DEGS)		26	24	28	26	25	32	30	36	35	28	23	25	26
	YEAR OF OCCURRENCE		1996	1998	2001	2001	1999	2000	1996	1998	1996	2001	1998	2000	JAN 1996
	MAXIMUM 5-SECOND:														
	SPEED (MPH)	6	75	60	63	70	58	59	62	54	48	58	64	49	75
	DIR. (TENS OF DEGS)		26	26	26	24	26	32	29	36	08	26	24	26	26
	YEAR OF OCCURRENCE		1996	1998	1999	2001	1999	2000	1996	1998	1999	1998	1998	2000	JAN 1996
PRECIPITATION	NORMAL (IN)	30	0.40	0.41	0.29	0.20	0.25	0.67	1.54	1.58	1.70	0.76	0.44	0.57	8.81
	MAXIMUM MONTHLY (IN)	62	1.84	1.69	2.26	1.42	4.22	3.18	5.53	5.57	6.68	4.31	1.63	3.29	6.68
	YEAR OF OCCURRENCE		1949	1973	1958	1983	1992	1984	1968	1984	1974	1945	1961	1991	SEP 1974
	MINIMUM MONTHLY (IN)	62	0.00	0.00	T	0.00	0.00	T	0.04	T	T	0.00	0.00	0.00	0.00
	YEAR OF OCCURRENCE		1967	1943	1996	1978	1962	1990	1978	1962	1959	1952	1964	1955	APR 1978
	MAXIMUM IN 24 HOURS (IN)	62	0.61	0.87	1.72	1.08	2.40	1.56	2.63	2.30	2.52	1.77	1.19	1.76	2.63
	YEAR OF OCCURRENCE		1960	1956	1941	1966	1992	1986	1968	1984	1958	1945	1943	1987	JUL 1968
	NORMAL NO. DAYS WITH:														
	PRECIPITATION 0.01	30	4.3	3.0	2.4	1.5	2.1	3.1	7.8	8.5	6.5	4.4	3.1	3.9	50.6
	PRECIPITATION 1.00	30	0.0	0.0	0.0	0.0	0.0	0.1	0.3	0.3	0.4	0.0	0.0	*	1.1
SNOWFALL	NORMAL (IN)	30	1.1	1.0	0.3	0.6	0.0	0.0	0.0	0.0	0.0	0.*	1.4	2.2	6.6
	MAXIMUM MONTHLY (IN)	57	8.3	8.9	7.3	16.5	T	T	T	0.0	T	1.0	12.7	25.9	25.9
	YEAR OF OCCURRENCE		1949	1956	1958	1983	1992	1992	1990		1993	1980	1976	1987	DEC 1987
	MAXIMUM IN 24 HOURS (IN)	57	5.2	7.2	7.3	8.8	T	T	T	0.0	T	1.0	7.8	16.8	16.8
	YEAR OF OCCURRENCE		1992	1956	1958	1983	1992	1992	1990		1993	1980	1961	1987	DEC 1987
	MAXIMUM SNOW DEPTH (IN)	48	6	7	7	9	0	0	0	0	0	1	7	14	14
	YEAR OF OCCURRENCE		1983	1956	1958	1983						1993	1968	1987	DEC 1987
	NORMAL NO. DAYS WITH:														
	SNOWFALL 1.0	30	0.4	0.4	0.1	0.1	0.0	0.0	0.0	0.0	0.0	0.*	0.5	0.6	2.1

PRECIPITATION (inches) 2001 EL PASO, TX (ELP)

YEAR	JAN	FEB	MAR	APR	MAY	JUN	JUL	AUG	SEP	OCT	NOV	DEC	ANNUAL
1972	0.44	T	T	0.00	0.04	1.62	0.71	2.59	1.60	1.25	0.33	0.42	9.00
1973	1.23	1.69	0.60	0.00	0.29	0.71	2.12	0.73	0.01	0.07	0.08	T	7.53
1974	0.27	T	0.36	0.12	0.05	0.36	2.21	0.63	6.68	1.90	0.50	0.87	13.95
1975	0.70	0.59	0.19	T	0.03	T	1.11	0.45	2.18	0.25	T	0.71	6.21
1976	0.26	0.52	T	0.30	0.74	0.50	3.17	0.23	1.70	1.20	1.20	0.32	10.14
1977	0.57	T	0.17	0.09	0.06	0.04	1.09	1.36	0.16	1.65	0.05	0.26	5.50
1978	0.44	0.47	0.07	0.00	0.57	1.46	0.04	2.18	4.14	2.28	0.45	0.47	12.57
1979	0.77	0.68	T	0.28	0.24	0.03	0.98	2.16	0.41	T	0.04	0.25	5.84
1980	0.54	0.73	0.25	0.31	0.08	T	0.21	1.76	1.90	0.95	0.54	0.04	7.31
1981	1.10	0.36	0.39	0.65	0.72	0.64	2.08	5.26	0.52	0.53	0.30	0.08	12.63
1982	0.34	0.55	T	0.05	0.19	0.18	1.00	0.48	5.28	T	0.29	2.61	10.97
1983	0.35	0.60	0.45	1.42	0.05	0.23	0.43	0.97	1.51	1.48	0.34	0.16	7.99
1984	0.31	0.00	0.44	0.01	0.59	3.18	0.69	5.57	0.58	3.12	0.51	1.17	16.17
1985	0.95	0.19	0.59	0.07	0.01	0.10	1.32	1.46	1.47	1.82	0.13	0.05	8.16
1986	0.01	0.39	0.39	T	0.83	3.05	2.66	0.70	0.85	0.45	1.42	1.42	12.17
1987	0.29	0.30	0.49	0.32	0.24	2.24	0.64	2.22	0.89	0.15	0.29	2.87	10.94
1988	0.25	0.70	0.10	0.23	0.15	0.03	3.35	3.46	1.52	0.59	0.24	0.44	11.06
1989	0.11	0.72	0.62	T	0.65	T	1.23	3.06	0.48	0.23	T	0.16	7.26
1990	0.29	0.14	0.41	0.25	0.10	T	3.96	1.98	3.46	0.58	1.34	0.34	12.85
1991	0.82	0.66	0.10	T	0.23	0.01	2.69	2.06	1.82	0.20	0.50	3.29	12.38
1992	1.14	0.16	0.50	0.30	4.22	0.27	0.65	2.11	0.15	0.27	0.28	1.35	11.40
1993	1.34	0.32	0.01	0.12	T	1.47	0.95	2.73	1.32	0.17	0.49	0.71	9.63
1994	0.03	0.23	0.37	0.65	0.80	0.67	0.18	0.02	0.03	0.35	0.54	1.61	5.48
1995	0.26	0.88	0.42	0.04	0.01	1.74	0.28	0.76	3.18	T	0.26	0.03	7.86
1996	0.11	0.19	T	0.49	0.00	2.36	1.87	1.24	T	T	0.16	0.00	8.39
1997	0.38	0.29	0.64	0.43	0.52	1.11	0.91	1.41	1.55	0.19	0.79	1.41	9.63
1998	0.05	0.15	0.18	0.04	T	0.27	2.07	0.53	0.66	2.14	0.34	0.34	6.77
1999	0.10	0.00	0.04	T	0.02	1.44	2.00	1.43	1.94	0.56	0.00	0.63	8.16
2000	0.00	0.03	0.06	0.28	T	2.45	1.59	0.70	T	0.82	1.06	0.42	7.41
2001	0.06	0.24	0.40	T	0.18	0.30	0.36	1.72	0.30	T	0.60	0.13	4.29
POR= 123 YRS	0.44	0.41	0.34	0.26	0.37	0.68	1.59	1.50	1.29	0.71	0.42	0.51	8.52

AVERAGE TEMPERATURE (F) 2001 EL PASO, TX (ELP)

YEAR	JAN	FEB	MAR	APR	MAY	JUN	JUL	AUG	SEP	OCT	NOV	DEC	ANNUAL
1972	45.2	52.3	61.2	65.2	69.8	78.3	82.2	77.7	72.9	65.7	48.8	46.6	63.8
1973	42.9	47.0	52.4	57.7	68.7	76.6	79.7	79.2	75.9	63.3	53.8	45.7	61.9
1974	44.3	44.8	59.6	65.0	75.6	82.8	79.6	77.0	69.4	63.0	41.3	62.7	62.7
1975	43.1	48.9	55.1	59.7	69.8	80.9	79.8	80.9	72.2	64.0	51.5	44.2	62.5
1976	42.3	52.6	55.9	64.1	69.5	79.4	78.2	78.7	70.5	58.5	44.8	41.9	61.4
1977	44.7	47.3	49.7	61.5	70.7	81.5	82.2	82.5	77.4	64.3	53.7	49.6	63.8
1978	45.4	48.8	58.6	66.1	73.6	83.4	84.5	80.0	72.3	63.6	55.8	44.6	64.7
1979	41.0	47.1	53.1	63.6	70.1	78.4	85.1	78.8	74.2	66.2	47.9	43.1	62.4
1980	46.8	50.6	54.1	60.6	70.5	86.3	87.2	82.4	75.6	60.3	49.2	48.5	64.3
1981	45.2	50.3	57.2	64.8	73.6	82.6	83.6	79.5	75.9	64.6	54.3	48.7	65.0
1982	42.3	48.7	57.7	64.4	69.6	80.9	84.2	83.5	77.1	64.6	53.2	43.3	64.1
1983	41.6	49.5	54.6	56.3	68.9	77.3	82.9	81.8	78.7	66.6	54.3	45.5	63.2
1984	44.4	47.0	55.7	62.0	75.0	79.5	81.1	80.4	72.9	61.4	51.6	45.7	63.1
1985	40.0	45.6	55.2	64.2	72.1	79.0	79.4	80.6	72.8	61.4	52.9	43.1	62.2
1986	44.7	52.1	55.7	67.3	71.5	77.7	80.0	80.4	74.1	62.4	49.4	42.6	63.2
1987	41.3	46.2	51.2	59.7	68.6	78.1	81.6	79.1	72.1	67.0	51.4	40.5	61.4
1988	42.6	48.4	53.4	61.1	70.3	79.0	80.3	77.6	72.4	66.7	54.0	42.6	62.4
1989	43.6	51.4	58.7	67.4	74.2	81.3	81.8	79.1	73.4	63.2	53.3	41.9	64.1
1990	44.1	49.1	56.1	66.4	73.1	87.1	80.2	76.8	73.9	63.4	52.9	44.8	64.0
1991	44.3	51.1	53.9	64.0	72.6	79.0	78.6	79.5	70.7	65.4	50.3	46.3	63.0
1992	44.0	50.2	58.1	67.9	70.6	81.5	85.1	81.3	77.8	66.5	48.0	44.7	64.6
1993	48.5	51.4	57.9	66.7	74.6	82.7	85.2	82.5	75.8	64.4	52.3	46.9	65.7
1994	45.6	50.4	58.3	66.7	75.7	89.0	88.1	86.2	78.3	66.6	54.4	49.4	67.4
1995	47.1	56.3	59.7	65.2	74.3	80.2	83.1	82.8	74.1	65.6	55.0	47.6	65.9
1996	46.5	54.5	54.6	65.5	80.1	83.5	82.9	79.6	73.5	65.5		46.3	
1997	44.2	48.2	58.6	61.5	74.3	81.0	82.7	81.6	78.4	63.6	52.2	40.6	63.9
1998	47.1	47.6	54.0	59.5	74.4	82.2	83.5	80.6	79.2	66.0	55.2	45.8	64.6
1999	48.1	53.6	58.0	63.2	74.2	80.0	81.2	81.2	75.4	63.9	56.5	43.4	64.9
2000	49.7	54.0	57.2	68.3	79.0	80.3	84.1	81.5	78.6	62.5	47.6	44.0	65.6
2001	41.8	50.1	56.3	65.5	76.0	83.3	84.3	81.0	76.9	67.6	54.7	42.7	65.0
POR= 115 YRS	44.6	49.3	55.6	63.8	72.3	80.8	82.0	80.2	74.7	64.5	51.9	45.1	63.7

REFERENCE NOTES:

PAGE 1:
THE TEMPERATURE GRAPH SHOWS NORMAL MAXIMUM AND NORMAL
MINIMUM DAILY TEMPERATURES (SOLID CURVES) AND THE
ACTUAL DAILY HIGH AND LOW TEMPERATURES (VERTICAL BARS).

PAGE 2 AND 3:
H/C INDICATES HEATING AND COOLING DEGREE DAYS.
RH INDICATES RELATIVE HUMIDITY
W/O INDICATES WEATHER AND OBSTRUCTIONS
S INDICATES SUNSHINE.
PR INDICATES PRESSURE.
CLOUDINESS ON PAGE 3 IS THE SUM OF THE CEILOMETER AND
SATELLITE DATA NOT TO EXCEED EIGHT EIGHTHS(OKTAS).

GENERAL:
T INDICATES TRACE PRECIPITATION, AN AMOUNT GREATER
THAN ZERO BUT LESS THAN THE LOWEST REPORTABLE VALUE.
+ INDICATES THE VALUE ALSO OCCURS ON EARLIER DATES.
BLANK ENTRIES DENOTE MISSING OR UNREPORTED DATA.
NORMALS ARE 30-YEAR AVERAGES (1961 - 1990).
ASOS INDICATES AUTOMATED SURFACE OBSERVING SYSTEM.
PM INDICATES THE LAST DAY OF THE PREVIOUS MONTH.
POR (PERIOD OF RECORD) BEGINS WITH THE JANUARY DATA
MONTH AND IS THE NUMBER OF YEARS USED TO COMPUTE
THE MEAN. INDIVIDUAL MONTHS WITHIN THE POR MAY
BE MISSING.
WHEN THE POR FOR A NORMAL IS LESS THAN 30 YEARS,
THE NORMAL IS PROVISIONAL AND IS BASED ON THE NUMBER
OF YEARS INDICATED.
0.* OR * INDICATES THE VALUE OR MEAN-DAYS-WITH
IS BETWEEN 0.00 AND 0.05.
CLOUDINESS FOR ASOS STATIONS DIFFERS FROM THE NON-ASOS
OBSERVATION TAKEN BY A HUMAN OBSERVER. ASOS STATION
CLOUDINESS IS BASED ON TIME-AVERAGED CEILOMETER DATA
FOR CLOUDS AT OR BELOW 12,000 FEET AND ON SATELLITE
DATA FOR CLOUDS ABOVE 12,000 FEET.
THE NUMBER OF DAYS WITH CLEAR, PARTLY CLOUDY, AND
CLOUDY CONDITIONS FOR ASOS STATIONS IS THE SUM
OF THE CEILOMETER AND SATELLITE DATA FOR THE
SUNRISE TO SUNSET PERIOD.

GENERAL CONTINUED:
CLEAR INDICATES 0 - 2 OKTAS, PARTLY CLOUDY INDICATES
3 - 6 OKTAS, AND CLOUDY INDICATES 7 OR 8 OKTAS.
WHEN AT LEAST ONE OF THE ELEMENTS (CEILOMETER OR
SATELLITE) IS MISSING, THE DAILY CLOUDINESS IS
NOT COMPUTED.
WIND DIRECTION IS RECORDED IN TENS OF DEGREES (2 DIGITS)
CLOCKWISE FROM TRUE NORTH. "00" INDICATES CALM. "36"
INDICATES TRUE NORTH.
RESULTANT WIND IS THE VECTOR AVERAGE OF THE SPEED AND
DIRECTION.
AVERAGE TEMPERATURE IS THE SUM OF THE MEAN DAILY MAXIMUM
AND MINIMUM TEMPERATURE DIVIDED BY 2.
SNOWFALL DATA COMPRISE ALL FORMS OF FROZEN
PRECIPITATION, INCLUDING HAIL.
A HEATING (COOLING) DEGREE DAY IS THE DIFFERENCE BETWEEN
THE AVERAGE DAILY TEMPERATURE AND 65 F.
DRY BULB IS THE TEMPERATURE OF THE AMBIENT AIR.
DEW POINT IS THE TEMPERATURE TO WHICH THE AIR MUST BE
COOLED TO ACHIEVE 100 PERCENT RELATIVE HUMIDITY.
WET BULB IS THE TEMPERATURE THE AIR WOULD HAVE IF THE
MOISTURE CONTENT WAS INCREASED TO 100 PERCENT RELATIVE
HUMIDITY.

ON JULY 1, 1996, THE NATIONAL WEATHER SERVICE BEGAN USING
THE "METAR" OBSERVATION CODE THAT WAS ALREADY EMPLOYED
BY MOST OTHER NATIONS OF THE WORLD. THE MOST NOTICEABLE
DIFFERENCE IN THIS ANNUAL PUBLICATION WILL BE THE CHANGE
IN UNITS FROM TENTHS TO EIGHTS(OKTAS) FOR REPORTING THE
AMOUNT OF SKY COVER.

HEATING DEGREE DAYS (base 65 F) 2001　EL PASO, TX (ELP)

YEAR	JUL	AUG	SEP	OCT	NOV	DEC	JAN	FEB	MAR	APR	MAY	JUN	TOTAL
1972-73	0	0	3	87	480	563	679	499	384	218	31	0	2944
1973-74	0	0	7	90	336	592	636	558	178	79	5	0	2481
1974-75	0	0	41	107	445	728	672	445	309	188	13	0	2948
1975-76	0	0	20	66	399	640	696	351	278	91	26	0	2567
1976-77	0	0	7	214	601	709	623	492	469	138	3	0	3256
1977-78	0	0	0	56	328	472	603	449	200	57	22	0	2187
1978-79	0	0	16	106	272	625	735	494	362	118	26	1	2755
1979-80	0	0	24	56	505	670	555	410	331	157	19	0	2727
1980-81	0	0	2	203	467	503	607	405	233	82	2	0	2504
1981-82	0	0	0	93	313	499	697	449	237	82	17	0	2387
1982-83	0	0	0	88	344	668	720	430	316	284	23	0	2873
1983-84	0	0	0	52	317	599	633	514	285	126	8	0	2534
1984-85	0	0	18	144	404	592	768	537	302	71	5	0	2841
1985-86	0	0	10	125	358	670	621	356	283	47	22	0	2492
1986-87	0	0	1	116	460	687	725	521	420	173	5	0	3108
1987-88	0	0	0	13	405	750	686	474	360	121	16	0	2825
1988-89	0	2	2	17	337	684	661	377	208	52	9	0	2349
1989-90	0	0	2	108	344	708	640	439	271	45	20	0	2577
1990-91	0	2	8	88	354	617	636	383	336	78	3	0	2505
1991-92	0	0	26	66	434	576	644	423	209	49	3	0	2430
1992-93	0	0	0	21	502	624	504	373	222	53	4	0	2303
1993-94	0	0	0	145	375	551	596	401	218	38	0	0	2324
1994-95	0	0	0	41	313	475	548	240	185	84	6	0	1892
1995-96	0	0	15	31	292	533	562	305	322	89	0	0	2149
1996-97	0	0	0	103		571	635	467	202	155	0	0	
1997-98	0	0	0	150	378	746	550	481	342	187	0	0	2834
1998-99	0	0	0	80	289	590	518	315	218	114	1	0	2125
1999-00	0	0	4	108	253	662	466	314	242	49	2	0	2100
2000-01	0	0	3	145	513	644	711	412	265	71	0	0	2764
2001-	0	0	0	18	315	684							

COOLING DEGREE DAYS (base 65 F) 2001　EL PASO, TX (ELP)

YEAR	JAN	FEB	MAR	APR	MAY	JUN	JUL	AUG	SEP	OCT	NOV	DEC	ANNUAL
1972	0	4	15	70	159	404	543	401	247	120	0	0	1963
1973	0	0	0	7	152	355	459	448	340	44	7	0	1812
1974	0	0	19	84	338	540	459	378	181	54	0	0	2053
1975	0	0	9	35	170	482	469	502	241	41	1	0	1950
1976	0	0	4	71	170	441	418	434	179	18	0	0	1735
1977	0	0	0	39	186	500	540	552	380	43	0	0	2240
1978	0	0	8	98	295	559	612	474	238	70	2	0	2356
1979	0	0	0	84	190	414	630	432	308	99	0	0	2157
1980	0	0	0	34	198	646	693	546	329	63	0	0	2509
1981	0	2	2	84	275	534	586	455	333	89	0	2	2362
1982	0	0	14	70	167	484	602	583	371	82	0	0	2373
1983	0	0	0	32	151	374	564	527	417	108	6	0	2179
1984	0	0	2	44	324	441	507	482	260	38	8	0	2106
1985	0	0	9	55	233	428	457	488	252	18	0	0	1940
1986	0	3	0	122	228	391	474	485	281	41	0	0	2025
1987	0	0	0	22	121	399	521	446	222	83	2	0	1816
1988	0	0	9	15	186	429	478	399	226	76	16	0	1834
1989	0	2	20	130	300	494	530	442	261	60	0	0	2239
1990	0	1	0	96	278	667	480	374	282	43	0	0	2221
1991	0	0	0	55	245	424	428	455	204	88	0	0	1899
1992	0	0	3	144	182	500	628	511	389	75	0	0	2432
1993	0	0	11	110	306	538	632	550	331	132	2	0	2612
1994	0	0	17	92	338	725	722	666	404	100	3	0	3067
1995	0	0	27	97	304	462	567	559	292	58	1	0	2367
1996	0	6	7	112	475	560	561	462	263	125		0	
1997	0	0	10	53	298	489	556	518	408	116	0	0	2448
1998	0	0	8	30	299	527	581	491	434	116	0	0	2486
1999	0	0	8	67	291	457	506	511	324	81	5	0	2250
2000	0	0	4	159	443	463	600	517	417	77	0	0	2680
2001	0	0	4	95	351	554	606	506	365	108	13	0	2602

SNOWFALL (inches) 2001　EL PASO, TX (ELP)

YEAR	JUL	AUG	SEP	OCT	NOV	DEC	JAN	FEB	MAR	APR	MAY	JUN	TOTAL
1972-73	0.0	0.0	0.0	0.0	0.0	T	5.3	4.6	T	0.0	0.0	0.0	9.9
1973-74	0.0	0.0	0.0	0.0	T	0.0	T	T	0.0	0.0	0.0	0.0	T
1974-75	0.0	0.0	0.0	0.0	0.0	5.3	2.2	0.2	2.0	0.0	0.0	0.0	9.7
1975-76	0.0	0.0	0.0	0.0	0.0	T	1.0	0.0	T	T	0.0	0.0	1.0
1976-77	0.0	0.0	0.0	T	12.7	2.0	T	T	0.0	T	0.0	0.0	14.7
1977-78	0.0	0.0	0.0	0.0	0.0	0.0	T	0.0	0.0	0.0	0.0	0.0	T
1978-79	0.0	0.0	0.0	0.0	T	T	T	1.2	0.0	0.0	0.0	0.0	1.2
1979-80	0.0	0.0	0.0	0.0	0.0	0.0	T	3.6	0.0	2.0	0.0	0.0	5.6
1980-81	0.0	0.0	0.0	1.0	4.0	0.0	4.8	0.0	0.0	0.0	0.0	0.0	9.8
1981-82	0.0	0.0	0.0	0.0	0.0	0.0	T	T	0.0	0.0	0.0	0.0	T
1982-83	0.0	0.0	0.0	0.0	0.3	18.2	T	0.0	T	16.5	0.0	0.0	35.0
1983-84	0.0	0.0	0.0	0.0	T	T	T	6.1	0.0	0.0	0.0	0.0	6.1
1984-85	0.0	0.0	0.0	0.0	T	2.9	5.4	1.1	0.0	0.0	0.0	0.0	9.4
1985-86	0.0	0.0	0.0	0.0	0.0	0.9	T	0.9	T	0.0	0.0	0.0	1.8
1986-87	0.0	0.0	0.0	0.0	0.0	0.6	3.4	2.8	0.6	0.0	0.0	0.0	7.4
1987-88	0.0	0.0	0.0	0.0	0.0	25.9	T	6.6	0.0	0.0	0.0	0.0	32.5
1988-89	0.0	0.0	0.0	0.0	0.0	0.3	0.0	0.0	T	0.0	T	0.0	0.3
1989-90	0.0	0.0	0.0	0.0	0.0	T	T	T	T	0.0	T	0.0	T
1990-91	T	0.0	0.0	0.0	4.3	T	1.5	T	T	0.0	T	0.0	5.8
1991-92	0.0	0.0	0.0	0.0	0.0	0.5	6.9	0.0	T	0.0	T	T	7.4
1992-93	0.0	0.0	0.0	0.0	0.0	1.6	0.0	0.0	0.0	0.0	0.0	0.0	1.6
1993-94	0.0	0.0	T	0.8	0.0	1.6	0.3	0.0	0.0	0.0	0.0	0.0	2.7
1994-95	0.0	0.0	0.0	0.0	0.0	0.0	0.0	0.0	0.0	0.0	0.0	0.0	0.0
1995-96	0.0	0.0	0.0	0.0	0.0	T	1.6	0.0	T	0.0			
1996-97													
1997-98													
1998-99													
1999-00													
2000-01													
2001-													
POR= 56 YRS	T	0.0	T	0.0	0.9	1.6	1.3	0.8	0.4	0.3	T	T	5.3

2001
HOUSTON,
TEXAS (IAH)

Houston, the largest city in Texas, is located in the flat Coastal Plains, about 50 miles from the Gulf of Mexico and about 25 miles from Galveston Bay. The climate is predominantly marine. The terrain includes numerous small streams and bayous which, together with the nearness to Galveston Bay, favor the development of both ground and advective fogs. Prevailing winds are from the southeast and south, except in January, when frequent passages of high pressure areas bring invasions of polar air and prevailing northerly winds.

Temperatures are moderated by the influence of winds from the Gulf, which result in mild winters. Another effect of the nearness of the Gulf is abundant rainfall, except for rare extended dry periods. Polar air penetrates the area frequently enough to provide variability in the weather.

Records of sky cover for daylight hours indicate about one-fourth of the days per year as clear, with a high number of clear days in October and November. Cloudy days are relatively frequent from December to May and partly cloudy days are the more frequent for June through September. Sunshine averages nearly 60 percent of the possible amount for the year ranging from 42 percent in January to 67 percent in June.

Heavy fog occurs on an average of 16 days a year and light fog occurs about 62 days a year in the city. The frequency of heavy fog is considerably higher at William P. Hobby Airport and at Intercontinental Airport.

Destructive windstorms are fairly infrequent, but both thundersqualls and tropical storms occasionally pass through the area.

NORMALS, MEANS, AND EXTREMES

HOUSTON, TX (IAH)

LATITUDE: 29 59' 33" N   LONGITUDE: 95 21' 50" W   ELEVATION (FT): GRND: 118   BARO: 121   TIME ZONE: CENTRAL (UTC + 6)   WBAN: 12960

	ELEMENT	POR	JAN	FEB	MAR	APR	MAY	JUN	JUL	AUG	SEP	OCT	NOV	DEC	YEAR
TEMPERATURE °F	NORMAL DAILY MAXIMUM	30	61.0	65.3	71.1	78.4	84.6	90.1	92.7	92.5	88.4	81.6	72.4	64.7	78.6
	MEAN DAILY MAXIMUM	32	61.8	66.3	72.7	79.1	85.4	90.8	93.9	93.6	88.9	81.4	71.6	64.7	79.2
	HIGHEST DAILY MAXIMUM	32	84	91	91	95	99	103	104	107	109	96	89	85	109
	YEAR OF OCCURRENCE		1975	1986	1989	1987	1996	1980	1980	2000	2000	1991	1989	1995	SEP 2000
	MEAN OF EXTREME MAXS.	32	78.3	81.5	85.4	88.5	92.7	96.7	98.6	99.3	96.6	91.0	85.0	80.2	89.5
	NORMAL DAILY MINIMUM	30	39.7	42.6	50.0	58.1	64.4	70.6	72.4	72.0	67.9	57.6	49.6	42.2	57.3
	MEAN DAILY MINIMUM	32	41.2	44.7	50.9	57.9	65.6	71.3	73.2	72.8	68.4	58.6	49.7	43.4	58.1
	LOWEST DAILY MINIMUM	32	12	20	22	31	44	52	62	60	48	29	19	7	7
	YEAR OF OCCURRENCE		1982	1997	1980	1987	1978	1970	1990	1992	1975	1993	1976	1989	DEC 1989
	MEAN OF EXTREME MINS.	32	24.8	27.4	32.8	39.8	52.1	62.4	68.2	67.6	54.1	41.9	32.0	26.5	44.1
	NORMAL DRY BULB	30	50.4	53.9	60.6	68.3	74.5	80.4	82.6	82.3	78.2	69.6	61.0	53.5	67.9
	MEAN DRY BULB	32	51.6	55.4	61.7	68.4	75.4	81.0	83.6	83.2	78.6	70.0	60.6	54.1	68.6
	MEAN WET BULB	18	48.1	51.7	56.4	62.6	70.0	74.5	75.7	75.6	71.4	64.2	56.3	47.1	62.8
	MEAN DEW POINT	18	43.0	46.8	51.5	58.2	66.7	71.5	72.7	72.3	67.9	60.4	52.3	43.0	58.9
	NORMAL NO. DAYS WITH:														
	MAXIMUM 90	30	0.0	*	0.1	1.0	5.1	19.4	26.5	26.0	14.7	2.7	0.0	0.0	95.5
	MAXIMUM 32	30	0.1	0.2	0.0	0.0	0.0	0.0	0.0	0.0	0.0	0.0	0.0	0.3	0.6
	MINIMUM 32	30	7.8	4.8	1.4	0.1	0.0	0.0	0.0	0.0	0.0	*	1.6	5.6	21.3
	MINIMUM 0	30	0.0	0.0	0.0	0.0	0.0	0.0	0.0	0.0	0.0	0.0	0.0	0.0	0.0
H/C	NORMAL HEATING DEG. DAYS	30	468	322	187	36	0	0	0	0	0	31	181	374	1599
	NORMAL COOLING DEG. DAYS	30	16	11	50	135	295	462	546	536	396	174	61	18	2700
RH	NORMAL (PERCENT)	30	75	73	73	73	75	75	74	75	77	75	76	76	75
	HOUR 00 LST	30	82	83	83	85	87	87	87	88	89	88	86	84	86
	HOUR 06 LST	30	85	86	87	89	91	92	93	93	93	91	89	86	90
	HOUR 12 LST	30	63	61	59	58	59	59	58	58	60	56	59	62	59
	HOUR 18 LST	30	67	61	60	60	63	62	62	63	67	68	72	70	65
S	PERCENT POSSIBLE SUNSHINE	27	45	50	54	58	62	68	70	68	66	64	52	51	59
W/O	MEAN NO. DAYS WITH:														
	HEAVY FOG(VISBY 1/4 MI)	32	4.5	3.5	2.9	2.4	1.7	0.7	0.2	0.3	1.2	2.7	3.5	4.0	27.6
	THUNDERSTORMS	32	2.3	2.3	3.6	3.9	7.0	8.8	9.9	10.2	7.2	3.7	3.0	1.9	63.8
CLOUDINESS	MEAN:														
	SUNRISE-SUNSET (OKTAS)	27	5.5	5.3	5.4	5.3	5.2	4.6	4.4	4.4	4.2	4.0	4.6	5.3	4.8
	MIDNIGHT-MIDNIGHT (OKTAS)	26	5.3	5.0	5.1	5.0	4.9	3.9	3.8	3.7	3.6	3.5	4.3	5.0	4.4
	MEAN NO. DAYS WITH:														
	CLEAR	27	7.2	7.0	7.0	7.2	5.6	7.3	6.9	6.1	8.6	11.2	9.0	7.2	90.3
	PARTLY CLOUDY	27	5.4	5.6	6.4	7.1	11.0	13.3	15.9	16.9	11.5	8.9	7.0	5.5	114.5
	CLOUDY	27	18.4	15.6	17.6	15.7	14.3	9.3	8.2	8.0	10.0	10.9	14.0	18.3	160.3
PR	MEAN STATION PRESSURE(IN)	29	30.00	29.99	29.90	29.90	29.80	29.80	29.90	29.90	29.89	29.90	30.00	30.00	29.91
	MEAN SEA-LEVEL PRES. (IN)	17	30.13	30.08	30.02	29.97	29.93	29.94	30.00	29.98	29.97	30.04	30.09	30.14	30.02
WINDS	MEAN SPEED (MPH)	29	8.1	8.5	9.1	9.0	8.1	7.6	6.8	6.1	6.6	7.1	7.7	7.7	7.7
	PREVAIL.DIR(TENS OF DEGS)	29	36	36	13	13	16	16	18	16	13	13	36	36	13
	MAXIMUM 2-MINUTE:														
	SPEED (MPH)	5	28	40	33	37	36	38	36	31	36	28	30	32	40
	DIR. (TENS OF DEGS)		09	10	11	11	02	08	34	09	12	31	15	32	10
	YEAR OF OCCURRENCE		2000	2000	2001	1997	1999	2001	1998	1998	2000	1998	1997	2000	FEB 2000
	MAXIMUM 5-SECOND:														
	SPEED (MPH)	5	37	44	39	43	47	66	49	44	48	47	43	41	66
	DIR. (TENS OF DEGS)		33	10	32	09	35	15	33	23	10	13	31	33	15
	YEAR OF OCCURRENCE		1998	2000	2001	1997	1997	2000	1998	1997	2000	1998	2000	2000	JUN 2000
PRECIPITATION	NORMAL (IN)	30	3.29	2.96	2.92	3.21	5.24	4.96	3.60	3.49	4.89	4.27	3.79	3.45	46.07
	MAXIMUM MONTHLY (IN)	32	9.78	5.99	8.52	10.92	14.39	19.21	8.10	10.58	11.35	16.05	10.21	9.34	19.21
	YEAR OF OCCURRENCE		1991	1992	1972	1976	1970	2001	1979	1996	1976	1984	1998	1991	JUN 2001
	MINIMUM MONTHLY (IN)	32	0.36	0.38	0.12	0.43	0.04	0.26	0.47	0.31	0.80	0.05	0.41	0.64	0.04
	YEAR OF OCCURRENCE		1971	1976	1996	1983	1998	1990	1993	2000	1975	1978	1988	1973	MAY 1998
	MAXIMUM IN 24 HOURS (IN)	32	2.73	2.22	7.47	8.16	10.36	11.02	3.99	6.83	7.98	9.31	6.33	4.14	11.02
	YEAR OF OCCURRENCE		1995	1985	1972	1976	1989	2001	1973	1981	1976	1984	1998	1995	JUN 2001
	NORMAL NO. DAYS WITH:														
	PRECIPITATION 0.01	30	10.2	8.4	9.3	6.9	8.0	8.5	9.2	8.9	9.4	7.4	8.3	8.9	103.4
	PRECIPITATION 1.00	30	0.9	0.8	0.8	1.0	1.9	1.6	1.1	1.0	1.3	1.5	1.0	0.9	13.8
SNOWFALL	NORMAL (IN)	30	0.2	0.2	0.0	0.0	0.0	0.0	0.0	0.0	0.0	0.0	T	0.1	0.5
	MAXIMUM MONTHLY (IN)	32	2.0	2.8	T	T	T	T	0.0	0.0	0.0	0.0	T	1.7	2.8
	YEAR OF OCCURRENCE		1973	1973	1992	1993	1993	1996					1979	1989	FEB 1973
	MAXIMUM IN 24 HOURS (IN)	32	2.0	1.4	T	T	T	T	0.0	0.0	0.0	0.0	T	1.7	2.0
	YEAR OF OCCURRENCE		1973	1980	1992	1993	1993	1996					1979	1989	JAN 1973
	MAXIMUM SNOW DEPTH (IN)	31	1	1	0	0	0	0	0	0	0	0	0	0	1
	YEAR OF OCCURRENCE		1973	1973											FEB 1973
	NORMAL NO. DAYS WITH:														
	SNOWFALL 1.0	30	0.1	0.1	0.0	0.0	0.0	0.0	0.0	0.0	0.0	0.0	0.0	0.*	0.2

PRECIPITATION (inches) 2001 HOUSTON, TX (IAH)

YEAR	JAN	FEB	MAR	APR	MAY	JUN	JUL	AUG	SEP	OCT	NOV	DEC	ANNUAL
1972	3.30	1.20	8.52	2.85	6.99	3.02	2.76	3.90	6.23	3.34	6.49	2.20	50.80
1973	5.00	3.40	3.68	7.15	4.22	13.46	6.77	3.73	9.38	9.31	1.59	2.47	70.16
1974	7.68	0.55	4.20	1.68	5.61	0.59	1.75	6.94	4.51	4.53	7.90	3.35	49.29
1975	1.97	2.63	3.19	4.80	7.57	7.50	5.48	5.72	0.80	5.62	2.08	3.61	50.97
1976	1.39	0.38	1.53	10.92	5.80	2.63	3.93	1.59	11.35	5.83	3.05	6.22	54.62
1977	2.67	1.70	1.95	4.34	0.79	3.55	2.69	4.45	3.92	0.82	5.17	2.89	34.94
1978	7.15	3.07	1.70	0.57	4.15	9.37	2.35	3.66	4.27	0.05	5.99	2.60	44.93
1979	6.30	5.23	2.88	7.79	3.78	1.88	8.10	4.57	9.83	2.80	1.78	4.03	58.97
1980	6.09	2.54	5.39	2.05	5.63	0.92	1.57	1.40	6.00	4.03	2.12	1.25	38.99
1981	2.32	2.21	1.74	2.69	8.75	9.65	4.43	7.01	2.91	6.96	5.26	2.05	55.98
1982	1.82	1.59	1.55	2.28	6.87	1.10	4.32	1.90	0.98	6.64	8.91	4.91	42.87
1983	2.00	3.97	3.85	0.43	7.29	5.37	5.23	9.42	7.23	1.56	3.17	3.69	53.21
1984	3.99	4.37	2.41	0.56	3.13	1.99	3.43	3.52	3.87	16.05	2.28	2.59	48.19
1985	2.10	5.38	4.52	4.31	1.57	5.29	4.93	1.14	4.67	6.54	4.84	3.85	49.14
1986	0.71	2.74	1.44	2.63	4.29	6.34	0.61	3.27	3.70	6.83	6.66	5.71	44.93
1987	2.42	4.26	0.88	0.47	5.39	9.31	4.79	1.48	3.46	0.17	3.41	4.56	40.60
1988	1.27	1.29	4.88	1.26	1.32	2.00	3.23	3.52	1.20	1.29	0.41	1.26	22.93
1989	4.80	0.90	3.96	1.48	13.56	16.28	1.92	2.74	2.69	1.76	1.84	0.80	52.73
1990	3.96	4.54	5.11	6.21	2.23	2.98	4.85	0.31	1.57	3.79	3.01	1.81	40.37
1991	9.78	5.79	1.77	8.06	4.02	7.69	1.31	2.97	2.76	2.57	5.03	9.34	61.09
1992	7.70	5.99	6.28	3.74	7.05	3.38	3.85	2.78	1.08	1.03	5.99	3.46	52.33
1993	5.79	2.67	6.41	7.88	8.50	12.08	0.47	1.82	1.10	5.32	3.27	2.68	57.99
1994	2.08	2.79	2.39	2.11	5.02	3.40	1.60	5.45	1.12	10.62	1.67	4.90	43.15
1995	5.95	2.55	4.11	2.59	3.83	4.11	2.68	4.90	2.52	2.77	3.63	4.99	44.63
1996	0.88	1.29	0.12	2.05	0.56	8.37	1.11	10.58	6.96	2.60	4.55	3.74	42.81
1997	3.26	5.35	7.96	7.17	6.69	4.46	2.30	2.26	4.86	7.11	3.38	5.42	60.22
1998	4.35	5.85	2.32	1.21	0.04	2.87	1.65	4.38	10.16	7.79	10.21	4.01	54.84
1999	2.12	0.80	3.44	1.06	4.10	5.26	5.11	0.50	1.36	0.56	1.53	2.20	28.04
2000	1.25	2.32	1.35	5.52	12.35	3.29	0.64	2.11	4.34	3.27	8.50	2.69	47.63
2001	4.25	0.82	7.97	2.00	3.53	19.21	2.05	4.83	8.82	8.95	2.58	6.18	71.19
POR= 67 YRS	3.70	3.11	2.97	3.42	4.88	5.00	3.77	3.94	4.46	4.11	4.01	4.02	47.39

AVERAGE TEMPERATURE (F) 2001 HOUSTON, TX (IAH)

YEAR	JAN	FEB	MAR	APR	MAY	JUN	JUL	AUG	SEP	OCT	NOV	DEC	ANNUAL
1972	56.5	55.2	64.3	71.2	73.7	80.8	80.3	79.6	69.8	54.7	52.0	68.2	
1973	47.4	51.4	63.7	64.6	72.8	79.2	83.0	79.5	78.1	71.8	67.3	53.5	67.7
1974	55.0	56.2	66.8	67.2	76.9	79.9	82.8	81.6	74.6	70.6	60.2	54.6	68.9
1975	56.9	55.4	61.1	68.3	75.9	80.0	81.5	81.1	74.9	69.5	59.7	52.7	68.1
1976	50.6	60.1	62.2	67.7	70.5	78.4	80.5	81.4	76.2	60.6	51.8	49.2	65.8
1977	42.7	53.8	60.9	66.9	74.5	81.0	82.4	83.1	80.0	69.2	61.8	53.7	67.5
1978	40.8	45.1	57.3	67.6	76.0	80.4	83.7	83.1	79.3	68.9	64.7	52.9	66.7
1979	44.1	51.7	62.4	68.7	73.1	79.8	82.6	81.5	75.6	70.7	55.7	52.4	66.5
1980	55.0	53.7	60.9	66.2	77.3	85.1	87.5	86.6	83.2	67.8	58.0	55.2	69.7
1981	51.4	55.4	60.9	74.3	75.3	82.7	84.4	84.4	78.6	72.3	64.4	54.5	69.9
1982	52.9	52.1	64.9	67.8	75.3	83.0	85.4	84.1	79.3	69.5	60.9	55.4	69.2
1983	50.1	52.5	58.3	64.0	73.4	79.0	82.2	82.6	76.6	70.1	63.1	45.7	66.5
1984	47.0	54.0	61.9	67.8	74.9	78.6	81.8	82.9	77.4	74.2	60.0	63.4	68.7
1985	45.7	49.6	64.7	70.0	75.6	81.0	81.6	84.2	79.8	72.5	67.0	51.0	68.0
1986	54.4	59.9	63.3	71.7	75.8	82.0	85.9	82.6	81.8	68.9	62.0	51.7	70.0
1987	51.4	56.1	58.9	67.2	77.1	81.3	83.5	86.2	78.9	68.7	60.5	55.6	68.8
1988	48.1	54.1	61.3	67.6	73.6	80.5	84.4	85.3	80.8	72.0	65.7	55.4	68.9
1989	57.5	52.7	61.3	69.4	77.8	79.9	82.4	81.7	77.0	70.2	62.9	44.4	68.1
1990	57.0	59.1	62.9	69.4	78.1	84.8	82.1	85.1	80.1	68.7	63.4	53.6	70.4
1991	50.4	57.4	63.5	72.2	78.0	82.0	84.0	83.0	77.4	72.3	56.7	56.2	69.4
1992	51.0	58.5	64.0	68.7	73.7	81.7	83.6	80.1	79.3	71.4	56.8	56.7	68.8
1993	53.6	56.7	61.1	65.9	73.4	81.6	85.8	86.5	80.2	69.5	56.9	54.6	68.8
1994	52.6	55.2	62.7	69.6	76.0	83.5	85.5	83.1	78.3	71.9	65.7	57.2	70.1
1995	54.3	58.7	62.9	68.6	77.9	80.6	84.8	84.9	81.6	70.4	61.4	57.1	70.3
1996	52.0	58.7	58.1	69.4	81.4	80.7	83.8	82.0	77.5	70.6	62.0	57.4	69.5
1997	50.8	55.3	65.3	64.2	73.6	79.2	83.1	83.2	79.1	68.9	55.7	50.1	67.4
1998	57.1	55.1	60.1	65.9	78.7	85.5	86.6	84.7	82.2	72.6	64.3	55.1	70.7
1999	57.1	61.5	63.7	73.0	76.6	81.9	83.1	86.8	78.0	69.0	62.2	53.7	70.6
2000	56.5	61.7	66.4	67.9	78.1	81.4	85.2	84.8	79.4	70.9	57.6	47.6	69.8
2001	49.3	59.3	56.4	71.7	75.9	80.5	83.6	83.5	77.0	66.9	63.4	56.0	68.6
POR= 32 YRS	51.5	55.3	61.9	68.5	75.4	81.0	83.5	83.3	78.6	70.0	60.6	54.1	68.6

REFERENCE NOTES:

PAGE 1:
THE TEMPERATURE GRAPH SHOWS NORMAL MAXIMUM AND NORMAL
MINIMUM DAILY TEMPERATURES (SOLID CURVES) AND THE
ACTUAL DAILY HIGH AND LOW TEMPERATURES (VERTICAL BARS).

PAGE 2 AND 3:
H/C INDICATES HEATING AND COOLING DEGREE DAYS.
RH INDICATES RELATIVE HUMIDITY
W/O INDICATES WEATHER AND OBSTRUCTIONS
S INDICATES SUNSHINE.
PR INDICATES PRESSURE.
CLOUDINESS ON PAGE 3 IS THE SUM OF THE CEILOMETER AND
SATELLITE DATA NOT TO EXCEED EIGHT EIGHTHS (OKTAS).

GENERAL:
T INDICATES TRACE PRECIPITATION, AN AMOUNT GREATER
THAN ZERO BUT LESS THAN THE LOWEST REPORTABLE VALUE.
+ INDICATES THE VALUE ALSO OCCURS ON EARLIER DATES.
BLANK ENTRIES DENOTE MISSING OR UNREPORTED DATA.
NORMALS ARE 30-YEAR AVERAGES (1961 - 1990).
ASOS INDICATES AUTOMATED SURFACE OBSERVING SYSTEM.
PM INDICATES THE LAST DAY OF THE PREVIOUS MONTH.
POR (PERIOD OF RECORD) BEGINS WITH THE JANUARY DATA
MONTH AND IS THE NUMBER OF YEARS USED TO COMPUTE
THE MEAN. INDIVIDUAL MONTHS WITHIN THE POR MAY
BE MISSING.
WHEN THE POR FOR A NORMAL IS LESS THAN 30 YEARS,
THE NORMAL IS PROVISIONAL AND IS BASED ON THE NUMBER
OF YEARS INDICATED.
0.* OR * INDICATES THE VALUE OR MEAN-DAYS-WITH
IS BETWEEN 0.00 AND 0.05.
CLOUDINESS FOR ASOS STATIONS DIFFERS FROM THE NON-ASOS
OBSERVATION TAKEN BY A HUMAN OBSERVER. ASOS STATION
CLOUDINESS IS BASED ON TIME-AVERAGED CEILOMETER DATA
FOR CLOUDS AT OR BELOW 12,000 FEET AND ON SATELLITE
DATA FOR CLOUDS ABOVE 12,000 FEET.
THE NUMBER OF DAYS WITH CLEAR, PARTLY CLOUDY, AND
CLOUDY CONDITIONS FOR ASOS STATIONS IS THE SUM
OF THE CEILOMETER AND SATELLITE DATA FOR THE
SUNRISE TO SUNSET PERIOD.

GENERAL CONTINUED:
CLEAR INDICATES 0 - 2 OKTAS, PARTLY CLOUDY INDICATES
3 - 6 OKTAS, AND CLOUDY INDICATES 7 OR 8 OKTAS.
WHEN AT LEAST ONE OF THE ELEMENTS (CEILOMETER OR
SATELLITE) IS MISSING, THE DAILY CLOUDINESS IS
NOT COMPUTED.
WIND DIRECTION IS RECORDED IN TENS OF DEGREES (2 DIGITS)
CLOCKWISE FROM TRUE NORTH. "00" INDICATES CALM. "36"
INDICATES TRUE NORTH.
RESULTANT WIND IS THE VECTOR AVERAGE OF THE SPEED AND
DIRECTION.
AVERAGE TEMPERATURE IS THE SUM OF THE MEAN DAILY MAXIMUM
AND MINIMUM TEMPERATURE DIVIDED BY 2.
SNOWFALL DATA COMPRISE ALL FORMS OF FROZEN
PRECIPITATION, INCLUDING HAIL.
A HEATING (COOLING) DEGREE DAY IS THE DIFFERENCE BETWEEN
THE AVERAGE DAILY TEMPERATURE AND 65 F.
DRY BULB IS THE TEMPERATURE OF THE AMBIENT AIR.
DEW POINT IS THE TEMPERATURE TO WHICH THE AIR MUST BE
COOLED TO ACHIEVE 100 PERCENT RELATIVE HUMIDITY.
WET BULB IS THE TEMPERATURE THE AIR WOULD HAVE IF THE
MOISTURE CONTENT WAS INCREASED TO 100 PERCENT RELATIVE
HUMIDITY.

ON JULY 1, 1996, THE NATIONAL WEATHER SERVICE BEGAN USING
THE "METAR" OBSERVATION CODE THAT WAS ALREADY EMPLOYED
BY MOST OTHER NATIONS OF THE WORLD. THE MOST NOTICEABLE
DIFFERENCE IN THIS ANNUAL PUBLICATION WILL BE THE CHANGE
IN UNITS FROM TENTHS TO EIGHTS(OKTAS) FOR REPORTING THE
AMOUNT OF SKY COVER.

HEATING DEGREE DAYS (base 65 F) 2001 HOUSTON, TX (IAH)

YEAR	JUL	AUG	SEP	OCT	NOV	DEC	JAN	FEB	MAR	APR	MAY	JUN	TOTAL
1972-73	0	0	2	50	320	410	540	379	75	117	5	0	1898
1973-74	0	0	0	8	74	364	330	273	95	60	0	0	1204
1974-75	0	0	0	15	196	336	290	270	179	48	0	0	1334
1975-76	0	0	0	26	217	399	441	178	155	26	7	0	1449
1976-77	0	0	0	173	398	484	687	312	166	25	0	0	2245
1977-78	0	0	0	34	150	365	752	553	250	33	17	0	2154
1978-79	0	0	0	22	111	393	646	376	135	23	2	0	1708
1979-80	0	0	0	27	297	389	308	350	169	45	0	0	1585
1980-81	0	0	0	67	255	323	416	291	144	6	1	0	1503
1981-82	0	0	0	50	82	326	409	363	143	79	1	0	1453
1982-83	0	0	0	53	175	328	457	346	219	96	0	0	1674
1983-84	0	0	6	27	138	606	549	325	150	45	2	0	1848
1984-85	0	0	6	12	204	144	591	432	91	22	0	0	1502
1985-86	0	0	5	17	76	434	326	209	99	11	0	0	1177
1986-87	0	0	0	28	175	411	421	245	196	82	0	0	1558
1987-88	0	0	0	16	185	301	525	331	171	35	0	0	1564
1988-89	0	0	0	5	120	309	260	379	210	56	0	0	1339
1989-90	0	0	0	47	160	637	264	177	122	34	0	0	1441
1990-91	0	0	0	61	129	395	448	222	115	8	0	0	1378
1991-92	0	0	0	15	289	303	428	197	95	37	4	0	1368
1992-93	0	0	0	1	270	268	351	235	157	62	0	0	1344
1993-94	0	0	0	76	269	343	391	291	136	40	1	0	1547
1994-95	0	0	0	21	75	268	347	192	155	28	0	0	1086
1995-96	0	0	4	8	145	303	408	267	259	54	0	0	1448
1996-97	0	0	1	29	159	280	458	287	77	70	0	0	1361
1997-98	0	0	0	58	282	454	254	276	212	57	0	0	1593
1998-99	0	0	0	8	92	349	276	153	106	25	0	0	1009
1999-00	0	0	0	51	114	355	298	152	72	50	0	0	1092
2000-01	0	0	5	51	267	532	480	207	262	19	0	0	1823
2001-	0	0	0	57	111	311							

COOLING DEGREE DAYS (base 65 F) 2001 HOUSTON, TX (IAH)

YEAR	JAN	FEB	MAR	APR	MAY	JUN	JUL	AUG	SEP	OCT	NOV	DEC	ANNUAL
1972	58	20	71	208	275	480	480	482	447	206	17	12	2756
1973	1	4	41	111	253	434	564	458	401	225	151	12	2655
1974	24	33	158	132	374	454	558	519	295	196	60	18	2821
1975	47	8	61	155	342	455	514	505	303	174	68	24	2656
1976	5	43	75	110	182	408	490	520	341	42	9	0	2225
1977	0	5	44	91	302	487	547	565	456	173	58	23	2751
1978	10	5	19	120	369	471	584	568	437	150	108	25	2866
1979	7	13	62	142	261	454	552	519	324	211	26	6	2577
1980	4	31	49	86	388	610	705	677	553	162	52	26	3343
1981	1	28	23	295	330	538	606	609	413	285	71	7	3206
1982	39	11	147	170	329	547	641	599	437	199	60	40	3219
1983	0	0	18	76	268	427	541	554	362	196	87	18	2547
1984	0	13	64	135	315	415	527	562	384	302	62	100	2879
1985	0	6	87	180	335	487	521	602	456	257	143	9	3083
1986	4	71	52	220	341	518	654	553	510	157	92	4	3176
1987	4	4	14	154	383	497	580	661	423	137	54	15	2926
1988	7	20	65	121	274	472	609	637	478	229	144	20	3076
1989	33	44	105	194	405	454	547	526	363	218	105	5	2999
1990	20	19	65	174	413	603	536	630	456	181	87	47	3231
1991	0	14	76	231	408	514	593	565	376	248	47	37	3109
1992	0	17	74	155	281	508	584	476	437	210	32	20	2794
1993	4	10	42	92	267	506	652	674	463	221	31	27	2989
1994	15	25	74	186	347	561	644	569	406	243	102	33	3205
1995	24	22	98	142	406	476	622	622	508	183	42	65	3210
1996	15	89	53	192	513	475	587	532	384	205	78	51	3174
1997	23	21	94	52	274	432	567	570	426	187	9	0	2655
1998	16	6	64	90	429	621	678	616	524	250	75	50	3419
1999	37	60	73	270	365	515	568	685	395	182	35	11	3196
2000	41	64	126	143	413	495	632	622	443	242	51	0	3272
2001	2	51	3	230	344	474	585	581	370	123	70	40	2873

SNOWFALL (inches) 2001 HOUSTON, TX (IAH)

YEAR	JUL	AUG	SEP	OCT	NOV	DEC	JAN	FEB	MAR	APR	MAY	JUN	TOTAL
1972-73	0.0	0.0	0.0	0.0	0.0	0.0	2.0	2.8	0.0	0.0	0.0	0.0	4.8
1973-74	0.0	0.0	0.0	0.0	0.0	0.0	0.0	0.0	0.0	0.0	0.0	0.0	0.0
1974-75	0.0	0.0	0.0	0.0	0.0	0.0	T	0.0	0.0	0.0	0.0	0.0	T
1975-76	0.0	0.0	0.0	0.0	0.0	0.0	0.0	0.0	0.0	0.0	0.0	0.0	0.0
1976-77	0.0	0.0	0.0	0.0	T	0.0	0.0	0.0	0.0	0.0	0.0	0.0	T
1977-78	0.0	0.0	0.0	0.0	0.0	0.0	0.4	0.0	0.0	0.0	0.0	0.0	0.4
1978-79	0.0	0.0	0.0	0.0	0.0	0.0	T	0.0	0.0	0.0	0.0	0.0	T
1979-80	0.0	0.0	0.0	0.0	T	0.0	0.0	1.4	0.0	0.0	0.0	0.0	1.4
1980-81	0.0	0.0	0.0	0.0	0.0	0.0	T	T	0.0	0.0	0.0	0.0	T
1981-82	0.0	0.0	0.0	0.0	0.0	0.0	T	0.0	0.0	0.0	0.0	0.0	T
1982-83	0.0	0.0	0.0	0.0	0.0	0.0	0.0	0.0	0.0	0.0	0.0	0.0	0.0
1983-84	0.0	0.0	0.0	0.0	0.0	0.0	0.0	0.0	0.0	0.0	0.0	0.0	0.0
1984-85	0.0	0.0	0.0	0.0	0.0	0.0	1.4	0.3	0.0	0.0	0.0	0.0	1.7
1985-86	0.0	0.0	0.0	0.0	0.0	0.0	0.0	0.0	0.0	0.0	0.0	0.0	0.0
1986-87	0.0	0.0	0.0	0.0	0.0	0.0	0.0	0.0	0.0	0.0	0.0	0.0	0.0
1987-88	0.0	0.0	0.0	0.0	0.0	0.0	0.0	T	0.0	0.0	0.0	0.0	T
1988-89	0.0	0.0	0.0	0.0	0.0	0.0	0.0	T	0.0	0.0	0.0	0.0	T
1989-90	0.0	0.0	0.0	0.0	0.0	1.7	0.0	0.0	0.0	0.0	T	0.0	1.7
1990-91	0.0	0.0	0.0	0.0	0.0	0.0	0.0	0.0	0.0	0.0	0.0	0.0	0.0
1991-92	0.0	0.0	0.0	0.0	0.0	0.0	0.0	0.0	T	0.0	T	0.0	T
1992-93	0.0	0.0	0.0	0.0	0.0	0.0	0.0	0.0	0.0	T	T	0.0	T
1993-94	0.0	0.0	0.0	0.0	0.0	0.0	0.0	0.1	0.0	0.0	0.0	0.0	0.1
1994-95	0.0	0.0	0.0	0.0	0.0	0.0	T	0.0	0.0	0.0	0.0	0.0	T
1995-96	0.0	0.0	0.0	0.0	0.0	0.0	0.0	T	0.0	0.0	T	0.0	T
1996-97	0.0	0.0	0.0	0.0	0.0	T	0.0	0.0	0.0	0.0	0.0	0.0	T
1997-98	0.0	0.0	0.0	0.0	0.0	0.0	0.0	0.0	0.0	0.0	0.0	0.0	0.0
1998-99	0.0	0.0	0.0	0.0	0.0	0.0	0.0	0.0	T	0.0	0.0	0.0	T
1999-00	0.0	0.0	0.0	0.0	0.0	0.0	0.0	0.0	0.0	0.0	T	0.0	T
2000-01	0.0	0.0	0.0	0.0	0.0	0.0	0.0	0.0	0.0	T	0.0	0.0	T
2001-	0.0	0.0	0.0	0.0	0.0								
POR= 66 YRS	0.0	0.0	0.0	0.0	T	0.0	0.2	0.2	0.0	T	T	T	0.4

2001
SAN ANTONIO,
TEXAS (SAT)

The city of San Antonio is located in the south-central portion of Texas on the Balcones escarpment. Northwest of the city, the terrain slopes upward to the Edwards Plateau and to the southeast it slopes downward to the Gulf Coastal Plains. Soils are blackland clay and silty loam on the Plains and thin limestone soils on the Edwards Plateau.

The location of San Antonio on the edge of the Gulf Coastal Plains is influenced by a modified subtropical climate, predominantly continental during the winter months and marine during the summer months. Temperatures range from 50 degrees in January to the middle 80s in July and August. While the summer is hot, with daily temperatures above 90 degrees over 80 percent of the time, extremely high temperatures are rare. Mild weather prevails during much of the winter months, with below-freezing temperatures occurring on an average of about 20 days each year.

San Antonio is situated between a semi-arid area to the west and the coastal area of heavy precipitation to the east. The normal annual rainfall of nearly 28 inches is sufficient for the production of most crops. Precipitation is fairly well distributed throughout the year with the heaviest amounts occurring during May and September. The precipitation from April through September usually occurs from thunderstorms. Large amounts of precipitation may fall during short periods of time. Most of the winter precipitation occurs as light rain or drizzle. Thunderstorms and heavy rains have occurred in all months of the year. Hail of damaging intensity seldom occurs but light hail is frequent with the springtime thunderstorms. Measurable snow occurs only once in three or four years. Snowfall of 2 to 4 inches occurs about every ten years.

Northerly winds prevail during most of the winter, and strong northerly winds occasionally occur during storms called northers. Southeasterly winds from the Gulf of Mexico also occur frequently during winter and are predominant in summer.

Since San Antonio is located only 140 miles from the Gulf of Mexico, tropical storms occasionally affect the city with strong winds and heavy rains. One of the fastest winds recorded, 74 mph, occurred as a tropical storm moved inland east of the city in August 1942.

Relative humidity is above 80 percent during the early morning hours most of the year, dropping to near 50 percent in the late afternoon.

San Antonio has about 50 percent of the possible amount of sunshine during the winter months and more than 70 percent during the summer months. Skies are clear to partly cloudy more than 60 percent of the time and cloudy less than 40 percent. Air carried over San Antonio by southeasterly winds is lifted orographically, causing low stratus clouds to develop frequently during the later part of the night. These clouds usually dissipate around noon, and clear skies prevail a high percentage of the time during the afternoon.

The first occurrence of 32 degrees Fahrenheit is in late November and the average last occurrence is in early March.

NORMALS, MEANS, AND EXTREMES
SAN ANTONIO, TX (SAT)

LATITUDE: LONGITUDE: ELEVATION (FT):   TIME ZONE: WBAN: 12921
29 31' 58" N   98 27' 49" W   GRND: 818   BARO: 821   CENTRAL (UTC + 6)

ELEMENT	POR	JAN	FEB	MAR	APR	MAY	JUN	JUL	AUG	SEP	OCT	NOV	DEC	YEAR
TEMPERATURE °F														
NORMAL DAILY MAXIMUM	30	60.8	65.7	73.5	80.3	85.3	91.8	95.0	95.3	89.3	81.7	71.9	63.5	79.5
MEAN DAILY MAXIMUM	54	61.9	66.5	73.4	80.4	86.0	91.9	95.1	95.3	89.9	81.9	71.2	64.3	79.8
HIGHEST DAILY MAXIMUM	60	89	100	100	101	103	107	106	108	111	99	94	90	111
YEAR OF OCCURRENCE		1971	1996	1991	1996	1989	1998	1989	1986	2000	1991	1988	1955	SEP 2000
MEAN OF EXTREME MAXS.	54	79.6	83.4	88.3	91.6	94.4	97.9	100.1	100.4	97.5	92.0	85.2	80.2	90.9
NORMAL DAILY MINIMUM	30	37.9	41.3	49.7	58.4	65.7	72.6	75.0	74.5	69.2	58.8	48.8	40.8	57.7
MEAN DAILY MINIMUM	54	39.4	43.4	50.0	58.3	66.1	72.1	74.4	73.8	69.2	59.2	48.7	41.6	58.0
LOWEST DAILY MINIMUM	60	0	6	19	31	43	53	62	61	41	27	21	6	0
YEAR OF OCCURRENCE		1949	1951	1980	1987	1984	1964	1967	1992	1942	1993	1976	1989	JAN 1949
MEAN OF EXTREME MINS.	54	22.3	26.5	31.3	40.7	52.1	62.5	68.8	67.5	55.5	42.2	30.8	24.6	43.7
NORMAL DRY BULB	30	49.3	53.5	61.7	69.3	75.5	82.2	85.0	84.9	79.3	70.2	60.4	52.2	68.6
MEAN DRY BULB	54	50.8	54.9	61.7	69.2	76.1	82.0	84.8	84.5	79.5	70.5	60.0	52.9	68.9
MEAN WET BULB	18	45.9	49.8	54.6	61.2	68.8	73.1	73.8	73.7	70.0	63.0	51.5	47.7	61.1
MEAN DEW POINT	18	39.2	43.3	48.0	55.0	64.5	69.1	69.0	68.9	64.9	57.8	46.3	41.4	55.6
NORMAL NO. DAYS WITH:														
MAXIMUM 90	30	0.0	0.2	0.8	2.1	7.7	20.7	27.4	28.0	16.8	4.1	0.1	0.0	107.9
MAXIMUM 32	30	0.2	0.2	0.0	0.0	0.0	0.0	0.0	0.0	0.0	0.0	0.0	0.2	0.6
MINIMUM 32	30	9.5	5.7	1.7	*	0.0	0.0	0.0	0.0	0.0	0.0	1.8	6.6	25.3
MINIMUM 0	30	0.0	0.0	0.0	0.0	0.0	0.0	0.0	0.0	0.0	0.0	0.0	0.0	0.0
H/C														
NORMAL HEATING DEG. DAYS	30	494	332	167	32	0	0	0	0	0	30	180	409	1644
NORMAL COOLING DEG. DAYS	30	8	10	64	161	326	516	620	617	429	191	42	12	2996
RH														
NORMAL (PERCENT)	30	67	65	63	66	70	69	65	65	68	67	68	68	67
HOUR 00 LST	30	74	74	72	75	80	79	74	74	78	77	78	76	76
HOUR 06 LST	30	79	80	79	82	87	87	87	86	86	84	82	80	83
HOUR 12 LST	30	58	56	54	57	60	58	53	52	56	55	56	57	56
HOUR 18 LST	30	55	50	47	50	55	52	46	46	52	52	56	56	51
S PERCENT POSSIBLE SUNSHINE	53	47	50	57	56	56	67	74	74	67	64	54	48	60
W/O MEAN NO. DAYS WITH:														
HEAVY FOG(VISBY 1/4 MI)	60	5.0	2.8	2.4	1.3	0.8	0.1	0.1	0.1	0.2	1.4	2.9	4.3	21.4
THUNDERSTORMS	60	1.0	1.6	2.7	4.0	6.6	4.8	3.6	4.3	4.1	2.7	1.7	0.9	38.0
CLOUDINESS MEAN:														
SUNRISE-SUNSET (OKTAS)	1					5.0	3.0							
MIDNIGHT-MIDNIGHT (OKTAS)														
MEAN NO. DAYS WITH:														
CLEAR	1	4.0	5.0	8.0		6.0	12.0							
PARTLY CLOUDY	1	1.0	4.0	2.0		11.0	5.0							
CLOUDY	1	1.0		6.0		3.0	2.0							
PR MEAN STATION PRESSURE(IN)	29	29.28	29.24	29.14	29.11	29.07	29.11	29.16	29.15	29.15	29.20	29.23	29.27	29.18
MEAN SEA-LEVEL PRES. (IN)	18	30.12	30.06	30.00	29.92	29.89	29.91	29.97	29.96	29.96	30.02	30.07	30.12	30.00
WINDS MEAN SPEED (MPH)	39	8.5	9.0	9.6	9.7	9.6	9.6	9.1	8.2	8.0	8.2	8.3	8.1	8.8
PREVAIL.DIR(TENS OF DEGS)	24	36	02	14	14	14	14	16	16	16	16	02	36	16
MAXIMUM 2-MINUTE:														
SPEED (MPH)	59	39	32	36	38	46	38	39	39	43	35	32	32	46
DIR. (TENS OF DEGS)		29	15	01	31	28	09	30	11	25	31	31	02	28
YEAR OF OCCURRENCE		1999	2000	1996	1997	1997	1999	1996	2000	1996	2001	2000	2000	MAY 1997
MAXIMUM 5-SECOND:														
SPEED (MPH)	59	46	40	47	51	55	44	54	45	71	40	39	43	71
DIR. (TENS OF DEGS)		28	26	32	29	29	08	29	11	26	29	29	33	26
YEAR OF OCCURRENCE		1999	1998	1998	1997	1997	1999	1996	2000	1996	2001	2000	2000	SEP 1996
PRECIPITATION NORMAL (IN)	30	1.71	1.81	1.52	2.50	4.22	3.81	2.16	2.54	3.41	3.17	2.62	1.51	30.98
MAXIMUM MONTHLY (IN)	59	8.52	6.43	6.12	9.32	12.85	11.95	8.29	11.14	15.78	18.07	8.58	13.96	18.07
YEAR OF OCCURRENCE		1968	1965	1992	1957	1987	1986	1990	1974	1946	1998	2000	1991	OCT 1998
MINIMUM MONTHLY (IN)	59	T	0.01	0.03	0.05	0.17	0.01	T	0.00	0.05	T	T	0.03	0.00
YEAR OF OCCURRENCE		1996	1999	1961	1998	1961	1967	1993	1952	1999	1952	1966	1950	AUG 1952
MAXIMUM IN 24 HOURS (IN)	59	3.18	2.44	3.59	4.88	6.53	6.30	6.97	5.57	7.28	13.35	4.87	6.90	13.35
YEAR OF OCCURRENCE		1968	1986	1992	1977	1972	1986	1958	1950	1973	1998	1977	1991	OCT 1998
NORMAL NO. DAYS WITH:														
PRECIPITATION 0.01	30	7.7	7.0	7.1	7.3	8.9	6.4	4.8	5.1	7.2	6.8	6.6	7.1	82.0
PRECIPITATION 1.00	30	0.4	0.4	0.4	0.8	1.2	1.2	0.7	0.8	0.8	1.1	0.7	0.3	8.8
SNOWFALL NORMAL (IN)	30	0.6	0.3	T	0.0	0.0	0.0	0.0	0.0	0.0	0.0	T	0.*	0.9
MAXIMUM MONTHLY (IN)	56	15.9	3.5	T	T	T	T	0.0	0.0	0.0	T	0.3	0.2	15.9
YEAR OF OCCURRENCE		1985	1966	1994	1993	1993	1989				1993	1957	1964	JAN 1985
MAXIMUM IN 24 HOURS (IN)	56	13.2	3.5	T	T	T	T	0.0	0.0	0.0	T	0.3	0.2	13.2
YEAR OF OCCURRENCE		1985	1966	1994	1993	1993	1989				1993	1957	1964	JAN 1985
MAXIMUM SNOW DEPTH (IN)	50	9	3	0	0	0	0	0	0	0	0	0	0	9
YEAR OF OCCURRENCE		1985	1966											JAN 1985
NORMAL NO. DAYS WITH:														
SNOWFALL 1.0	30	0.1	0.1	0.0	0.0	0.0	0.0	0.0	0.0	0.0	0.0	0.0	0.0	0.2

PRECIPITATION (inches) 2001 SAN ANTONIO, TX (SAT)

YEAR	JAN	FEB	MAR	APR	MAY	JUN	JUL	AUG	SEP	OCT	NOV	DEC	ANNUAL
1972	1.35	0.40	0.13	1.94	11.24	2.86	3.13	4.24	1.40	1.99	2.37	0.44	31.49
1973	2.77	2.76	1.58	5.41	2.73	10.44	6.91	1.29	13.09	4.85	0.29	0.16	52.28
1974	1.36	0.04	0.94	2.18	4.28	1.02	1.28	11.14	3.85	4.09	5.39	1.43	37.00
1975	1.04	3.30	0.52	2.69	6.91	4.60	1.06	0.51	2.25	0.03	1.48	25.67
1976	0.56	0.13	1.20	5.67	5.80	1.61	5.39	2.09	3.79	8.48	2.46	1.95	39.13
1977	3.10	0.91	0.88	8.80	1.62	2.26	0.10	0.06	2.11	3.47	6.01	0.32	29.64
1978	0.68	1.76	1.71	3.62	2.45	3.96	1.43	4.97	8.86	0.55	4.91	1.09	35.99
1979	4.07	1.38	3.55	5.34	1.98	5.59	7.38	2.09	0.86	0.11	1.43	2.86	36.64
1980	0.72	0.74	0.98	1.67	6.42	0.52	0.26	2.64	5.05	1.09	3.53	0.61	24.23
1981	2.06	0.96	1.96	2.21	6.43	8.71	0.25	2.41	1.36	8.61	0.72	0.69	36.37
1982	0.72	1.28	0.69	1.23	6.42	1.37	0.14	0.55	0.87	2.84	4.54	2.31	22.96
1983	1.48	1.54	3.89	0.18	4.37	1.27	2.43	2.00	3.86	1.64	3.06	0.39	26.11
1984	1.87	0.54	1.91	0.11	3.76	1.40	T	3.04	1.06	5.94	2.91	3.41	25.95
1985	2.68	1.91	2.85	3.27	2.47	8.20	5.80	0.45	4.80	3.91	5.00	0.09	41.43
1986	0.76	2.52	0.35	0.60	6.29	11.95	0.05	1.86	2.83	6.58	1.83	7.11	42.73
1987	1.13	4.78	1.10	1.48	12.85	7.69	1.21	0.33	2.24	0.44	2.53	2.18	37.96
1988	0.39	0.92	0.86	1.23	0.41	5.50	5.58	1.98	0.83	0.62	0.02	0.67	19.01
1989	2.96	0.29	1.24	2.55	0.33	3.96	0.69	0.48	1.54	5.81	1.93	0.36	22.14
1990	1.17	2.68	5.17	4.52	3.28	1.18	8.29	1.30	3.70	3.71	3.11	0.20	38.31
1991	5.08	2.34	1.06	4.91	5.30	2.28	2.23	2.84	1.42	0.87	0.47	13.96	42.76
1992	5.64	6.37	6.12	3.03	8.15	5.67	1.28	2.56	1.12	0.92	3.47	2.16	46.49
1993	1.31	3.72	1.56	1.81	12.47	6.43	T	0.01	0.52	3.07	0.66	0.44	32.00
1994	1.55	0.64	5.06	2.21	7.01	1.66	0.50	2.54	5.52	9.75	0.71	3.28	40.43
1995	0.28	1.19	1.58	1.07	5.36	4.81	0.71	2.03	4.49	0.23	0.82	0.09	22.66
1996	T	0.70	0.30	0.89	1.26	2.12	1.31	2.86	3.66	0.36	2.79	1.56	17.81
1997	0.44	2.44	2.24	5.72	3.91	7.30	T	0.62	1.86	4.08	1.76	3.55	33.92
1998	3.21	3.37	2.85	0.05	0.34	0.81	0.21	7.78	1.57	18.07	3.40	0.39	42.05
1999	0.04	0.01	3.48	0.91	2.78	3.37	1.80	2.11	0.05	1.29	0.05	0.52	16.41
2000	1.40	2.20	0.91	1.22	3.59	7.61	0.34	0.16	2.65	5.62	8.58	1.57	35.85
2001	2.85	0.70	2.77	2.29	2.48	3.39	0.50	7.83	4.05	2.06	4.37	3.43	36.72
POR= 117 YRS	1.66	1.73	1.77	2.83	3.65	3.16	1.96	2.39	3.10	2.75	2.03	1.78	28.81

AVERAGE TEMPERATURE (F) 2001 SAN ANTONIO, TX (SAT)

YEAR	JAN	FEB	MAR	APR	MAY	JUN	JUL	AUG	SEP	OCT	NOV	DEC	ANNUAL
1972	52.8	56.7	66.3	73.7	72.8	80.3	82.2	82.1	82.0	71.9	54.0	50.3	68.8
1973	47.2	51.9	66.1	66.0	74.7	79.2	83.2	82.1	79.3	65.8	52.2	68.4
1974	51.0	56.5	67.9	69.7	77.3	79.4	83.0	81.2	72.3	68.2	57.3	50.9	67.9
1975	53.2	53.5	61.4	68.4	73.5	80.0	80.9	81.7	76.0	71.1	60.3	53.1	67.8
1976	49.6	61.2	63.8	68.9	71.3	79.8	79.8	81.6	77.5	61.1	52.1	49.9	66.4
1977	44.1	52.8	61.8	66.9	74.8	81.5	84.9	84.7	82.3	71.2	61.4	53.4	68.3
1978	43.4	46.4	59.6	68.9	77.1	82.7	86.1	83.1	78.5	69.3	62.4	51.7	67.4
1979	43.7	52.4	63.3	69.7	73.9	80.9	84.7	83.1	78.7	74.7	58.2	55.4	68.2
1980	52.6	53.7	61.5	67.6	76.1	85.1	88.1	85.3	83.7	70.7	58.3	55.0	69.8
1981	50.8	53.7	60.7	72.9	75.3	81.5	84.2	84.7	78.9	71.9	62.4	53.0	69.2
1982	50.8	49.7	63.1	66.9	74.5	81.6	85.5	86.0	80.1	69.3	59.4	52.4	68.3
1983	48.9	52.1	58.7	65.2	73.6	79.2	82.9	84.5	78.5	70.9	62.5	43.0	66.7
1984	46.7	54.1	64.2	69.7	77.1	82.8	85.0	84.7	77.6	71.2	58.8	56.9	69.3
1985	44.2	50.5	64.1	69.4	76.7	80.2	82.2	85.5	79.4	71.7	64.4	49.9	68.2
1986	53.4	58.0	62.9	72.6	74.6	81.5	85.8	85.7	83.7	69.7	59.4	56.9	69.9
1987	50.7	55.9	57.8	66.1	75.8	80.5	83.8	86.0	79.2	71.2	60.6	54.2	68.5
1988	47.6	54.3	61.3	69.1	76.1	81.2	84.6	86.4	80.7	73.2	65.1	56.0	69.6
1989	56.2	51.6	61.9	70.4	81.7	83.3	86.6	86.0	79.1	71.3	61.8	43.4	69.4
1990	56.4	58.9	61.5	69.7	79.3	87.5	83.4	85.3	80.0	69.3	63.0	51.9	70.5
1991	48.9	56.6	64.0	72.4	77.7	82.8	84.5	85.8	77.8	73.3	57.4	55.5	69.7
1992	50.8	59.1	63.3	69.0	73.7	82.5	84.7	82.2	81.7	73.4	57.3	56.3	69.5
1993	51.2	55.5	61.5	67.3	73.9	81.6	86.1	87.3	81.5	70.7	56.3	55.1	69.6
1994	52.3	56.2	63.9	69.8	76.0	84.5	87.9	86.1	78.4	72.7	64.8	57.0	70.8
1995	53.5	57.4	61.9	69.8	78.6	79.3	84.3	85.5	80.1	69.8	59.5	55.6	69.6
1996	51.0	57.9	57.6	69.5	81.9	84.1	87.3	84.4	78.4	71.1	61.3	54.5	69.9
1997	49.2	53.1	63.3	63.9	74.0	79.8	85.1	86.1	82.2	70.2	57.4	50.2	67.9
1998	56.4	55.3	59.8	66.7	79.8	86.3	88.1	83.6	80.5	71.4	62.4	52.7	70.3
1999	54.6	61.8	62.7	71.2	76.2	81.9	82.9	86.1	80.3	63.1	54.0	70.4
2000	55.2	62.6	67.0	70.7	78.6	81.0	85.9	86.3	81.0	71.1	56.9	46.4	70.2
2001	49.2	57.5	56.6	70.8	76.3	82.6	85.4	85.6	76.9	67.9	62.9	53.8	68.8
POR= 117 YRS	51.6	55.4	62.1	69.3	75.5	81.6	84.0	84.2	79.4	70.9	60.4	53.5	69.0

REFERENCE NOTES:

PAGE 1:
THE TEMPERATURE GRAPH SHOWS NORMAL MAXIMUM AND NORMAL
MINIMUM DAILY TEMPERATURES (SOLID CURVES) AND THE
ACTUAL DAILY HIGH AND LOW TEMPERATURES (VERTICAL BARS).

PAGE 2 AND 3:
H/C INDICATES HEATING AND COOLING DEGREE DAYS.
RH INDICATES RELATIVE HUMIDITY
W/O INDICATES WEATHER AND OBSTRUCTIONS
S INDICATES SUNSHINE.
PR INDICATES PRESSURE.
CLOUDINESS ON PAGE 3 IS THE SUM OF THE CEILOMETER AND
SATELLITE DATA NOT TO EXCEED EIGHT EIGHTHS(OKTAS).

GENERAL:
T INDICATES TRACE PRECIPITATION, AN AMOUNT GREATER
THAN ZERO BUT LESS THAN THE LOWEST REPORTABLE VALUE.
+ INDICATES THE VALUE ALSO OCCURS ON EARLIER DATES.
BLANK ENTRIES DENOTE MISSING OR UNREPORTED DATA.
NORMALS ARE 30-YEAR AVERAGES (1961 - 1990).
ASOS INDICATES AUTOMATED SURFACE OBSERVING SYSTEM.
PM INDICATES THE LAST DAY OF THE PREVIOUS MONTH.
POR (PERIOD OF RECORD) BEGINS WITH THE JANUARY DATA
MONTH AND IS THE NUMBER OF YEARS USED TO COMPUTE
THE MEAN. INDIVIDUAL MONTHS WITHIN THE POR MAY
BE MISSING.
WHEN THE POR FOR A NORMAL IS LESS THAN 30 YEARS,
THE NORMAL IS PROVISIONAL AND IS BASED ON THE NUMBER
OF YEARS INDICATED.
0.* OR * INDICATES THE VALUE OR MEAN-DAYS-WITH
IS BETWEEN 0.00 AND 0.05.
CLOUDINESS FOR ASOS STATIONS DIFFERS FROM THE NON-ASOS
OBSERVATION TAKEN BY A HUMAN OBSERVER. ASOS STATION
CLOUDINESS IS BASED ON TIME-AVERAGED CEILOMETER DATA
FOR CLOUDS AT OR BELOW 12,000 FEET AND ON SATELLITE
DATA FOR CLOUDS ABOVE 12,000 FEET.
THE NUMBER OF DAYS WITH CLEAR, PARTLY CLOUDY, AND
CLOUDY CONDITIONS FOR ASOS STATIONS IS THE SUM
OF THE CEILOMETER AND SATELLITE DATA FOR THE
SUNRISE TO SUNSET PERIOD.

GENERAL CONTINUED:
CLEAR INDICATES 0 - 2 OKTAS, PARTLY CLOUDY INDICATES
3 - 6 OKTAS, AND CLOUDY INDICATES 7 OR 8 OKTAS.
WHEN AT LEAST ONE OF THE ELEMENTS (CEILOMETER OR
SATELLITE) IS MISSING, THE DAILY CLOUDINESS IS
NOT COMPUTED.
WIND DIRECTION IS RECORDED IN TENS OF DEGREES (2 DIGITS)
CLOCKWISE FROM TRUE NORTH. "00" INDICATES CALM. "36"
INDICATES TRUE NORTH.
RESULTANT WIND IS THE VECTOR AVERAGE OF THE SPEED AND
DIRECTION.
AVERAGE TEMPERATURE IS THE SUM OF THE MEAN DAILY MAXIMUM
AND MINIMUM TEMPERATURE DIVIDED BY 2.
SNOWFALL DATA COMPRISE ALL FORMS OF FROZEN
PRECIPITATION, INCLUDING HAIL.
A HEATING (COOLING) DEGREE DAY IS THE DIFFERENCE BETWEEN
THE AVERAGE DAILY TEMPERATURE AND 65 F.
DRY BULB IS THE TEMPERATURE OF THE AMBIENT AIR.
DEW POINT IS THE TEMPERATURE TO WHICH THE AIR MUST BE
COOLED TO ACHIEVE 100 PERCENT RELATIVE HUMIDITY.
WET BULB IS THE TEMPERATURE THE AIR WOULD HAVE IF THE
MOISTURE CONTENT WAS INCREASED TO 100 PERCENT RELATIVE
HUMIDITY.

ON JULY 1, 1996, THE NATIONAL WEATHER SERVICE BEGAN USING
THE "METAR" OBSERVATION CODE THAT WAS ALREADY EMPLOYED
BY MOST OTHER NATIONS OF THE WORLD. THE MOST NOTICEABLE
DIFFERENCE IN THIS ANNUAL PUBLICATION WILL BE THE CHANGE
IN UNITS FROM TENTHS TO EIGHTS(OKTAS) FOR REPORTING THE
AMOUNT OF SKY COVER.

HEATING DEGREE DAYS (base 65 F) 2001 SAN ANTONIO, TX (SAT)

YEAR	JUL	AUG	SEP	OCT	NOV	DEC	JAN	FEB	MAR	APR	MAY	JUN	TOTAL
1972-73	0	0	0	29	334	457	551	362	29	94	1	0	1857
1973-74	0	0	0	4	85	391	437	257	74	39	0	0	1287
1974-75	0	0	2	19	260	433	389	316	152	41	0	0	1612
1975-76	0	0	1	21	214	394	472	166	143	11	2	0	1424
1976-77	0	0	0	160	382	461	643	336	144	32	0	0	2158
1977-78	0	0	0	19	138	360	667	521	192	27	4	0	1928
1978-79	0	0	0	12	152	413	657	356	109	20	4	0	1723
1979-80	0	0	0	15	243	306	386	333	163	42	0	0	1488
1980-81	0	0	0	62	245	331	437	332	157	10	0	0	1574
1981-82	0	0	2	52	112	368	445	430	171	77	2	0	1659
1982-83	0	0	0	49	237	404	490	356	208	99	1	0	1844
1983-84	0	0	5	20	154	681	563	315	120	21	2	0	1881
1984-85	0	0	9	28	228	203	635	406	109	26	0	0	1644
1985-86	0	0	10	9	112	467	354	232	106	8	1	0	1299
1986-87	0	0	0	14	204	413	443	254	233	98	0	0	1659
1987-88	0	0	0	1	194	339	538	323	179	38	0	0	1612
1988-89	0	0	0	0	122	291	292	392	187	55	0	0	1339
1989-90	0	0	0	42	165	663	283	190	154	32	0	0	1529
1990-91	0	0	0	50	142	422	494	240	96	7	0	0	1451
1991-92	0	0	5	30	271	306	435	188	91	31	5	0	1362
1992-93	0	0	0	0	260	287	421	269	147	42	0	0	1426
1993-94	0	0	0	85	287	323	391	273	130	28	10	0	1527
1994-95	0	0	0	19	99	267	359	215	180	29	0	0	1168
1995-96	0	0	9	12	187	324	435	280	277	63	0	0	1587
1996-97	0	0	0	27	165	325	498	338	108	84	0	0	1545
1997-98	0	0	0	44	251	449	269	270	214	30	0	0	1527
1998-99	0	0	0	22	120	404	328	137	109	32	0	0	1152
1999-00	0	0	0	56	98	346	300	135	79	29	0	0	1043
2000-01	0	0	2	76	269	570	484	234	255	15	0	0	1905
2001-	0	0	0	38	133	367							

COOLING DEGREE DAYS (base 65 F) 2001 SAN ANTONIO, TX (SAT)

YEAR	JAN	FEB	MAR	APR	MAY	JUN	JUL	AUG	SEP	OCT	NOV	DEC	ANNUAL
1972	11	31	105	276	252	465	542	539	515	249	12	6	3003
1973	8	0	69	129	310	431	570	536	437	242	114	5	2846
1974	11	22	171	188	387	439	568	506	229	124	34	5	2684
1975	29	1	51	151	273	457	502	524	337	217	80	30	2652
1976	3	62	113	136	202	451	467	521	383	45	0	0	2383
1977	0	3	52	98	311	502	620	618	525	218	38	5	2990
1978	3	7	30	152	384	537	660	567	410	154	79	11	2994
1979	3	13	65	166	285	482	619	570	418	322	42	13	2998
1980	11	14	61	127	355	614	725	635	567	245	51	26	3431
1981	3	24	30	255	324	502	603	619	424	273	41	3	3101
1982	11	5	117	142	304	504	645	659	459	191	72	19	3128
1983	0	0	21	111	276	435	560	611	417	207	84	8	2730
1984	0	8	101	169	383	541	625	618	394	230	46	44	3159
1985	0	8	85	165	368	462	539	641	450	223	101	5	3047
1986	2	45	49	244	304	500	652	646	568	166	40	4	3220
1987	4	5	17	135	340	471	589	658	434	199	67	12	2931
1988	6	19	71	166	352	492	617	671	480	264	131	20	3289
1989	24	23	99	222	524	557	678	656	429	244	75	0	3531
1990	22	26	53	177	450	681	578	635	459	192	91	23	3387
1991	0	10	70	234	402	541	612	654	396	295	49	20	3283
1992	0	23	47	158	281	531	618	542	508	267	34	24	3033
1993	1	7	45	117	283	503	660	698	503	267	31	22	3137
1994	8	34	102	183	357	593	715	659	410	261	97	27	3446
1995	7	9	90	177	429	436	603	644	467	167	28	40	3097
1996	9	79	57	203	530	578	700	610	411	220	60	8	3465
1997	12	11	61	60	286	449	628	660	522	212	27	0	2928
1998	11	3	56	84	467	645	724	582	470	227	49	32	3350
1999	12	56	45	225	352	512	561	664	467	206	46	11	3157
2000	7	73	147	207	428	487	654	668	487	270	33	0	3461
2001	1	30	1	195	357	530	640	642	362	135	76	26	2995

SNOWFALL (inches) 2001 SAN ANTONIO, TX (SAT)

YEAR	JUL	AUG	SEP	OCT	NOV	DEC	JAN	FEB	MAR	APR	MAY	JUN	TOTAL
1972-73	0.0	0.0	0.0	0.0	0.0	T	0.8	2.1	0.0	0.0	0.0	0.0	2.9
1973-74	0.0	0.0	0.0	0.0	0.0	0.0	0.0	0.0	0.0	0.0	0.0	0.0	0.0
1974-75	0.0	0.0	0.0	0.0	0.0	T	T	0.0	0.0	0.0	0.0	0.0	T
1975-76	0.0	0.0	0.0	0.0	0.0	0.0	T	0.0	0.0	0.0	0.0	0.0	T
1976-77	0.0	0.0	0.0	0.0	T	T	0.0	0.0	0.0	0.0	0.0	0.0	T
1977-78	0.0	0.0	0.0	0.0	0.0	0.0	0.0	0.0	T	0.0	0.0	0.0	T
1978-79	0.0	0.0	0.0	0.0	0.0	0.0	T	0.0	0.0	0.0	0.0	0.0	T
1979-80	0.0	0.0	0.0	0.0	T	0.0	0.0	0.0	T	0.0	0.0	0.0	T
1980-81	0.0	0.0	0.0	0.0	T	0.0	T	0.0	0.0	0.0	0.0	0.0	T
1981-82	0.0	0.0	0.0	0.0	0.0	0.0	0.5	0.0	0.0	0.0	0.0	0.0	0.5
1982-83	0.0	0.0	0.0	0.0	0.0	0.0	0.0	0.0	0.0	0.0	0.0	0.0	0.0
1983-84	0.0	0.0	0.0	0.0	0.0	0.0	0.0	0.0	0.0	0.0	0.0	0.0	0.0
1984-85	0.0	0.0	0.0	0.0	0.0	0.0	15.9	T	0.0	0.0	0.0	0.0	15.9
1985-86	0.0	0.0	0.0	0.0	0.0	0.0	T	0.0	0.0	0.0	0.0	0.0	T
1986-87	0.0	0.0	0.0	0.0	0.0	0.0	1.3	0.0	0.0	0.0	0.0	0.0	1.3
1987-88	0.0	0.0	0.0	0.0	0.0	0.0	0.0	0.1	0.0	0.0	0.0	0.0	0.1
1988-89	0.0	0.0	0.0	0.0	0.0	0.0	0.0	0.0	T	0.0	0.0	T	T
1989-90	0.0	0.0	0.0	0.0	0.0	T	0.0	T	T	0.0	0.0		T
1990-91	0.0	0.0	0.0	0.0	0.0	T	0.0	T	0.0	T	T	0.0	T
1991-92	0.0	0.0	0.0	0.0	0.0	0.0	0.0	T	T	0.0	0.0	0.0	T
1992-93	0.0	0.0	0.0	0.0	T	0.0	T	0.0	0.0	T	T	0.0	T
1993-94	0.0	0.0	0.0	T	0.0	0.0	T	T	0.0	0.0	0.0	0.0	T
1994-95	0.0	0.0	0.0	0.0	0.0	0.0	0.0	0.0	0.0	0.0	0.0		
1995-96	0.0	0.0	0.0	0.0	0.0	0.0	0.0						
1996-97													
1997-98													
1998-99							0.0	T	0.0	T	0.0	0.0	
1999-00							0.0	T	0.0	T	0.0	0.0	0.0
2000-01	0.0	0.0	0.0	0.0	0.0	0.0	0.0	0.0	0.0	0.0	0.0	0.0	0.0
2001-	0.0	0.0	0.0	T	0.0	0.0							
POR= 55 YRS	0.0	0.0	0.0	T	0.0	0.0	0.5	0.2	T	T	T	T	0.7

2001
SALT LAKE CITY,
UTAH (SLC)

Salt Lake City is located in a northern Utah valley surrounded by mountains on three sides and the Great Salt Lake to the northwest. The city varies in altitude from near 4,200 to 5,000 feet above sea level.

The Wasatch Mountains to the east have peaks to nearly 12,000 feet above sea level. Their orographic effects cause more precipitation in the eastern part of the city than over the western part.

The Oquirrh Mountains to the southwest of the city have several peaks to above 10,000 feet above sea level. The Traverse Mountain Range at the south end of the Salt Lake Valley rises to above 6,000 feet above sea level. These mountain ranges help to shelter the valleys from storms from the southwest in the winter, but are instrumental in developing thunderstorms which can drift over the valley in the summer.

Besides the mountain ranges, the most influential natural condition affecting the climate of Salt Lake City is the Great Salt Lake. This large inland body of water, which never freezes over due to its high salt content, can moderate the temperatures of cold winter winds blowing from the northwest and helps drive a lake/valley wind system. The warmer lake water during the winter and spring also contributes to increased precipitation in the valley downwind from the lake. The combination of the Great Salt Lake and the Wasatch Mountains often enhances storm precipitation in the valley.

Salt Lake City normally has a semi-arid continental climate with four well-defined seasons. Summers are characterized by hot, dry weather, but the high temperatures are usually not oppressive since the relative humidity is generally low and the nights usually cool. July is the hottest month with temperature readings in the 90s.

The mean diurnal temperature range is about 30 degrees in the summer and 18 degrees during the winter. Temperatures above 102 degrees in the summer or colder than -10 degrees in the winter are likely to occur one season out of four.

Winters are cold, but usually not severe. Mountains to the north and east act as a barrier to frequent invasions of cold continental air. The average annual snowfall is under 60 inches at the airport but much higher amounts fall in higher bench locations. Heavy fog can develop under temperature inversions in the winter and persist for several days.

Precipitation, generally light during the summer and early fall, is heavy in the spring when storms from the Pacific Ocean are moving through the area more frequently than at any other season of the year.

Winds are usually light, although occasional high winds have occurred in every month of the year, particularly in March.

The growing season is over five months in length. Yard and garden foilage generally are making good growth by mid-April. The last freezing temperature in the spring averages late April and the first freeze of the fall is mid-October.

NORMALS, MEANS, AND EXTREMES
SALT LAKE CITY, UT   (SLC)

LATITUDE: LONGITUDE: ELEVATION (FT):   TIME ZONE: WBAN: 24127
40 47' 13" N   111 58' 05" W   GRND: 4221   BARO: 4224   MOUNTAIN   (UTC + 7)

	ELEMENT	POR	JAN	FEB	MAR	APR	MAY	JUN	JUL	AUG	SEP	OCT	NOV	DEC	YEAR
TEMPERATURE F	NORMAL DAILY MAXIMUM	30	36.4	43.6	52.2	61.3	71.9	82.8	92.2	89.4	79.2	66.1	50.8	37.8	63.6
	MEAN DAILY MAXIMUM	54	37.4	43.7	52.3	61.5	72.1	83.0	92.5	90.3	79.5	65.8	50.1	38.5	63.9
	HIGHEST DAILY MAXIMUM	73	62	69	78	86	95	104	107	106	100	89	75	69	107
	YEAR OF OCCURRENCE		1982	1972	1960	1992	1997	1979	1960	1994	1979	1963	1999	1995	JUL 1960
	MEAN OF EXTREME MAXS.	54	52.0	58.7	69.7	79.0	88.2	97.6	101.2	99.4	93.7	82.4	66.5	54.7	78.6
	NORMAL DAILY MINIMUM	30	19.3	24.6	31.4	37.9	45.6	55.4	63.7	61.8	51.0	40.2	30.9	21.6	40.3
	MEAN DAILY MINIMUM	54	20.2	24.8	31.4	38.1	46.1	54.4	62.6	61.4	51.3	40.0	29.9	21.9	40.2
	LOWEST DAILY MINIMUM	73	-22	-30	2	14	25	35	40	37	27	16	-14	-21	-30
	YEAR OF OCCURRENCE		1949	1933	1966	1936	1965	1962	1968	1965	1965	1971	1955	1932	FEB 1933
	MEAN OF EXTREME MINS.	54	2.2	7.1	18.1	27.5	33.5	41.8	51.3	49.9	37.8	27.9	15.4	6.1	26.5
	NORMAL DRY BULB	30	27.9	34.1	41.8	49.7	58.8	69.1	77.9	75.6	65.2	53.2	40.8	29.7	52.0
	MEAN DRY BULB	54	28.8	34.2	42.0	49.8	59.1	68.7	77.5	75.9	65.6	52.9	40.0	30.2	52.1
	MEAN WET BULB	18	26.6	30.5	37.3	42.7	49.6	55.2	60.0	59.6	53.1	44.2	34.8	27.1	43.4
	MEAN DEW POINT	18	22.2	25.2	29.1	33.1	39.1	43.2	47.4	47.3	42.3	34.9	28.4	22.5	34.6
	NORMAL NO. DAYS WITH:														
	MAXIMUM 90	30	0.0	0.0	0.0	0.0	0.6	9.0	23.6	19.1	3.8	0.0	0.0	0.0	56.1
	MAXIMUM 32	30	10.8	3.8	0.6	0.0	0.0	0.0	0.0	0.0	0.0	*	0.5	9.0	24.7
	MINIMUM 32	30	27.4	22.6	16.3	6.5	0.8	0.0	0.0	0.0	0.4	4.7	17.8	27.5	124.0
	MINIMUM 0	30	1.7	0.4	0.0	0.0	0.0	0.0	0.0	0.0	0.0	0.0	0.0	0.9	3.0
H/C	NORMAL HEATING DEG. DAYS	30	1150	865	719	464	215	51	0	0	108	373	726	1094	5765
	NORMAL COOLING DEG. DAYS	30	0	0	0	0	23	174	400	329	114	7	0	0	1047
RH	NORMAL (PERCENT)	30	74	70	60	53	49	41	36	38	46	56	66	74	55
	HOUR 05 LST	30	78	77	70	66	65	59	52	54	61	68	74	79	67
	HOUR 11 LST	30	70	64	52	44	38	31	27	30	35	43	58	70	47
	HOUR 17 LST	30	69	59	47	39	33	26	22	23	29	41	58	71	43
	HOUR 23 LST	30	78	76	67	61	57	50	43	45	54	66	73	78	62
S	PERCENT POSSIBLE SUNSHINE	63	45	54	64	68	72	80	84	83	81	72	53	40	66
W/O	MEAN NO. DAYS WITH:														
	HEAVY FOG(VISBY 1/4 MI)	73	4.3	2.3	0.3	0.1	0.0	0.0	0.0	0.0	0.0	0.0	0.9	3.6	11.5
	THUNDERSTORMS	73	0.3	0.7	1.4	2.2	5.3	5.4	6.8	7.6	4.2	1.9	0.5	0.3	36.6
CLOUDINESS	MEAN:														
	SUNRISE-SUNSET (OKTAS)	62	5.9	5.7	5.3	5.1	4.7	3.4	2.9	2.9	2.9	3.8	5.0	5.8	4.5
	MIDNIGHT-MIDNIGHT (OKTAS)	32	6.0	5.4	5.3	5.1	4.6	3.5	3.1	3.1	3.0	3.7	5.2	5.8	4.5
	MEAN NO. DAYS WITH:														
	CLEAR	69	5.4	5.2	7.0	6.7	9.0	13.8	15.0	14.3	14.8	12.5	7.6	5.6	116.9
	PARTLY CLOUDY	69	6.5	6.9	8.3	9.4	10.2	9.9	8.8	9.6	7.5	7.0	6.4	5.9	96.4
	CLOUDY	69	19.0	16.1	15.7	14.0	11.8	6.3	4.0	4.0	4.0	4.7	8.5	13.1	133.4
PR	MEAN STATION PRESSURE(IN)	29	25.82	25.78	25.69	25.68	25.66	25.69	25.73	25.74	25.76	25.79	25.79	25.83	25.75
	MEAN SEA-LEVEL PRES. (IN)	18	30.21	30.12	29.98	29.93	29.86	29.84	29.87	29.88	29.94	30.05	30.10	30.22	30.00
WINDS	MEAN SPEED (MPH)	54	7.6	8.2	9.5	9.7	9.4	9.4	9.5	9.7	9.1	8.4	8.1	7.7	8.9
	PREVAIL.DIR(TENS OF DEGS)	38	14	14	16	15	15	16	16	16	15	14	14	14	14
	MAXIMUM 2-MINUTE:														
	SPEED (MPH)	3	35	47	38	37	43	48	45	43	40	37	38	38	48
	DIR. (TENS OF DEGS)		17	18	33	16	28	32	25	10	27	35	29	17	32
	YEAR OF OCCURRENCE		1999	1999	2001	2000	2001	2001	2001	2001	2001	2001	2000	2001	JUN 2001
	MAXIMUM 5-SECOND:														
	SPEED (MPH)	3	41	62	45	46	54	55	54	52	53	44	55	46	62
	DIR. (TENS OF DEGS)		28	18	33	15	33	18	26	16	14	32	28	18	18
	YEAR OF OCCURRENCE		1999	1999	2001	2001	2001	2001	2001	1999	2001	2001	2000	2001	FEB 1999
PRECIPITATION	NORMAL (IN)	30	1.11	1.23	1.91	2.12	1.80	0.93	0.81	0.86	1.28	1.44	1.29	1.40	16.18
	MAXIMUM MONTHLY (IN)	73	3.23	4.89	3.97	4.90	4.76	3.84	2.57	3.66	7.04	3.91	3.34	4.37	7.04
	YEAR OF OCCURRENCE		1993	1998	1983	1944	1977	1998	1982	1968	1982	1981	2001	1983	SEP 1982
	MINIMUM MONTHLY (IN)	73	0.09	0.12	0.10	0.45	T	T	T	T	T	0.00	0.01	0.08	0.00
	YEAR OF OCCURRENCE		1961	1946	1956	1987	1934	1994	1963	1944	1951	1952	1939	1976	OCT 1952
	MAXIMUM IN 24 HOURS (IN)	73	1.36	1.76	1.83	2.41	2.03	1.88	2.35	1.96	2.30	1.76	1.53	1.82	2.41
	YEAR OF OCCURRENCE		1953	1998	1944	1957	1942	1948	1962	1932	1982	1984	2001	1972	APR 1957
	NORMAL NO. DAYS WITH:														
	PRECIPITATION 0.01	30	9.7	8.6	10.3	10.0	8.7	5.9	4.8	5.7	6.1	6.3	8.2	9.8	94.1
	PRECIPITATION 1.00	30	0.0	0.0	0.0	0.2	0.2	0.1	0.1	*	0.2	*	0.0	0.1	0.9
SNOWFALL	NORMAL (IN)	30	12.3	9.8	11.7	6.8	1.1	T	0.0	0.0	0.2	2.1	6.5	13.7	64.2
	MAXIMUM MONTHLY (IN)	73	50.3	32.1	41.9	26.4	7.5	T	T	T	4.0	20.4	33.3	35.2	50.3
	YEAR OF OCCURRENCE		1993	1998	1977	1974	1975	1995	1996	1993	1971	1984	1994	1972	JAN 1993
	MAXIMUM IN 24 HOURS (IN)	73	16.5	18.0	15.4	16.2	6.4	T	T	T	4.0	18.4	11.0	18.1	18.4
	YEAR OF OCCURRENCE		1996	1998	1944	1974	1975	1995	1991	1993	1971	1984	1930	1972	OCT 1984
	MAXIMUM SNOW DEPTH (IN)	53	25	17	14	8	4	0	0	0	0	9	11	14	25
	YEAR OF OCCURRENCE		1993	1949	1998	1974	1978					1984	1985	1972	JAN 1993
	NORMAL NO. DAYS WITH:														
	SNOWFALL 1.0	30	3.7	3.3	3.3	1.9	0.3	0.0	0.0	0.0	0.1	0.5	2.1	4.2	19.4

PRECIPITATION (inches) 2001 SALT LAKE CITY, UT (SLC)

YEAR	JAN	FEB	MAR	APR	MAY	JUN	JUL	AUG	SEP	OCT	NOV	DEC	ANNUAL
1972	1.22	0.48	1.18	3.62	0.14	0.15	0.06	0.21	1.36	2.74	1.36	3.22	15.74
1973	1.49	0.91	2.67	1.64	1.74	0.19	1.07	1.16	4.07	0.67	2.52	2.26	20.39
1974	1.80	1.65	0.97	4.57	0.39	0.28	0.18	0.32	0.03	2.03	0.90	1.34	14.46
1975	1.28	1.24	3.44	2.46	2.58	1.81	0.28	0.10	0.08	1.91	1.71	1.03	17.92
1976	0.63	1.90	1.90	2.47	0.99	1.24	1.55	0.82	0.16	0.57	0.03	0.08	12.34
1977	0.76	0.64	3.10	0.59	4.76	0.06	0.61	1.85	1.85	0.83	1.20	1.42	17.67
1978	2.33	1.96	3.47	2.90	1.57	0.06	0.06	0.92	2.51	T	1.73	0.58	18.09
1979	0.72	1.05	0.80	1.04	0.84	0.35	0.40	0.63	0.05	1.29	0.98	0.55	8.70
1980	2.87	2.25	2.46	0.89	2.70	0.42	1.34	0.26	0.72	1.74	1.17	0.37	17.19
1981	0.64	0.81	2.11	0.45	3.68	1.03	0.33	0.33	0.48	3.91	1.03	1.89	16.59
1982	1.08	0.53	2.39	1.63	1.86	0.66	2.57	0.56	7.04	1.87	0.75	1.92	22.86
1983	1.19	1.36	3.97	1.63	2.58	0.62	1.02	2.64	1.03	1.62	2.23	4.37	24.26
1984	0.50	0.95	1.76	4.43	1.17	1.86	1.72	1.49	1.72	3.70	1.45	0.80	21.55
1985	0.91	0.85	1.80	0.64	2.95	1.30	0.85	0.03	1.98	1.61	2.63	1.42	16.97
1986	0.86	1.28	2.32	4.55	3.39	0.42	0.85	1.32	2.75	0.39	1.17	0.10	19.40
1987	1.53	1.41	1.52	0.79	2.41	0.19	0.79	0.36	0.05	1.18	1.17	1.10	12.50
1988	1.06	0.13	0.94	1.84	2.16	0.03	0.04	0.22	0.07	0.01	2.17	0.62	9.29
1989	0.56	1.57	1.77	0.46	1.83	0.22	0.39	0.90	0.49	1.82	0.73	0.13	10.87
1990	0.57	0.35	2.17	1.14	1.65	0.66	0.64	0.46	0.56	0.69	1.24	0.56	10.69
1991	1.11	0.61	1.11	2.71	2.76	1.09	0.32	0.86	2.55	2.10	2.17	0.40	17.79
1992	0.78	1.24	1.11	0.96	1.86	0.45	0.29	0.35	0.47	1.03	2.46	1.07	12.07
1993	3.23	1.35	1.37		3.99	1.14	1.38	0.46	0.22	2.77	0.54	0.88	
1994	0.62	1.53	1.28	2.94	1.29	T	0.06	0.61	0.32	2.24	2.96	1.43	15.28
1995	1.81	1.08	2.35	2.07	3.68	1.49	0.32	0.21	1.33	0.53	0.85	1.20	16.92
1996	3.09	1.54	2.71	2.20	1.32	0.09	0.41	0.02	1.03	1.45	1.72	1.73	17.31
1997	2.27	1.62	0.97	2.22	1.77	1.73	0.84	0.63	1.50	1.87	0.87	0.64	16.93
1998	1.63	4.89	2.97	2.09	1.04	3.84	1.57	0.46	1.53	1.25	1.27	1.27	23.81
1999	1.29	0.96	0.80	3.09	2.59	0.82	0.25	0.70	0.45	0.02	0.70	1.84	13.51
2000	2.17	1.80	0.80	0.76	1.62	1.62	0.36	2.00	1.86	2.00	1.31	1.24	16.34
2001	0.78	1.50	1.55	2.46	0.22	1.12	1.13	0.53	0.05	0.92	3.34	1.44	15.04
POR= 127 YRS	1.31	1.37	1.84	2.01	1.82	0.90	0.62	0.80	0.92	1.40	1.33	1.30	15.62

AVERAGE TEMPERATURE (F) 2001 SALT LAKE CITY, UT (SLC)

YEAR	JAN	FEB	MAR	APR	MAY	JUN	JUL	AUG	SEP	OCT	NOV	DEC	ANNUAL
1972	29.8	37.8	46.9	48.1	60.5	71.9	77.2	75.8	63.9	53.6	39.4	22.7	52.3
1973	19.6	32.3	41.8	47.6	61.6	70.2	76.6	76.6	61.5	54.1	40.5	33.4	51.3
1974	26.7	31.4	45.2	48.1	58.8	73.4	79.2	74.2	66.5	54.7	43.4	31.7	52.8
1975	27.4	35.5	41.1	44.3	54.3	64.8	78.8	73.4	65.4	53.4	37.3	32.9	50.7
1976	27.9	34.1	38.1	49.3	62.2	67.6	78.7	72.3	66.4	51.0	41.8	29.4	51.6
1977	26.8	35.8	37.7	54.1	55.0	73.2	77.3	75.0	66.4	55.6	42.5	37.9	53.1
1978	36.3	39.8	48.0	50.2	56.0	69.2	78.0	74.0	64.0	55.5	41.0	26.8	53.2
1979	22.1	32.5	43.2	51.1	60.2	70.1	78.9	75.6	71.4	56.7	36.5	32.9	52.6
1980	33.7	36.0	41.5	52.7	57.0	67.5	77.6	74.1	66.3	52.6	41.3	33.6	52.8
1981	32.1	38.3	44.1	53.4	57.6	69.6	78.2	78.0	68.5	50.5	44.3	36.4	54.3
1982	29.8	32.3	43.3	46.5	56.7	68.0	75.5	78.4	64.0	48.8	38.1	29.9	50.9
1983	35.2	39.4	44.6	45.9	55.8	67.7	76.6	77.8	67.8	56.0	43.0	31.9	53.5
1984	23.8	25.8	40.1	48.5	61.6	67.3	78.5	77.2	66.5	49.5	42.7	29.9	51.0
1985	24.2	25.6	40.8	55.7	63.9	72.5	80.7	76.5	62.7	53.1	37.4	27.7	51.7
1986	29.0	41.4	47.7	48.8	57.2	73.5	74.2	77.9	60.2	51.3	40.9	29.8	52.7
1987	26.5	36.1	42.8	55.9	62.7	71.6	75.7	74.7	66.5	56.4	40.8	30.5	53.4
1988	25.0	34.8	41.4	52.0	59.6	75.7	80.9	76.5	63.8	60.0	41.1	28.1	53.2
1989	22.3	25.3	45.8	54.8	59.9	69.2	81.1	75.1	66.4	53.4	40.5	31.4	52.1
1990	33.4	32.8	45.0	54.9	57.8	72.0	78.9	76.2	72.0	54.0	41.4	21.0	53.3
1991	24.4	36.5	42.8	47.9	55.6	68.3	79.1	78.4	64.4	53.6	40.0	30.0	51.8
1992	25.6	39.3	49.3	57.1	65.6	70.4	75.4	77.3	66.5	56.0	34.1	27.1	53.6
1993	24.9	29.5	45.5	48.5	63.4	63.7	69.9	72.5	65.5	52.6	34.8	31.5	50.2
1994	36.8	35.3	45.6	52.1	63.0	74.3	80.7	80.8	70.5	51.3	32.6	31.6	54.6
1995	34.9	42.3	44.1	48.6	55.5	64.3	76.0	78.0	67.5	51.5	46.1	37.1	53.8
1996	33.0	30.8	45.3	50.6	59.9	73.4	80.5	77.8	64.7	55.9	43.4	37.1	54.2
1997	32.3	35.3	46.0	48.3	63.4	70.1	75.1	78.7	67.8	52.8	42.2	27.9	53.3
1998	37.9	37.5	40.4	48.1	58.6	62.6	79.7	77.6	68.1	51.0	42.9	29.6	52.8
1999	35.8	37.4	44.8	45.5	56.0	68.4	78.3	77.3	65.3	54.5	45.7	31.0	53.2
2000	35.1	39.8	42.0	54.5	61.7	72.1	80.8	78.9	64.6	52.3	31.4	30.7	53.7
2001	27.3	34.4	45.4	50.1	63.6	70.9	79.4	79.1	70.2	55.0	42.6	26.3	53.7
POR= 127 YRS	28.4	33.4	41.1	49.4	58.4	68.3	77.3	75.5	65.2	53.2	40.3	31.2	51.8

REFERENCE NOTES:

PAGE 1:
THE TEMPERATURE GRAPH SHOWS NORMAL MAXIMUM AND NORMAL
MINIMUM DAILY TEMPERATURES (SOLID CURVES) AND THE
ACTUAL DAILY HIGH AND LOW TEMPERATURES (VERTICAL BARS).

PAGE 2 AND 3:
H/C INDICATES HEATING AND COOLING DEGREE DAYS.
RH INDICATES RELATIVE HUMIDITY
W/O INDICATES WEATHER AND OBSTRUCTIONS
S INDICATES SUNSHINE.
PR INDICATES PRESSURE.
CLOUDINESS ON PAGE 3 IS THE SUM OF THE CEILOMETER AND
SATELLLITE DATA NOT TO EXCEED EIGHT EIGHTHS(OKTAS).

GENERAL:
T INDICATES TRACE PRECIPITATION, AN AMOUNT GREATER
THAN ZERO BUT LESS THAN THE LOWEST REPORTABLE VALUE.
+ INDICATES THE VALUE ALSO OCCURS ON EARLIER DATES.
BLANK ENTRIES DENOTE MISSING OR UNREPORTED DATA.
NORMALS ARE 30-YEAR AVERAGES (1961 - 1990).
ASOS INDICATES AUTOMATED SURFACE OBSERVING SYSTEM.
PM INDICATES THE LAST DAY OF THE PREVIOUS MONTH.
POR (PERIOD OF RECORD) BEGINS WITH THE JANUARY DATA
MONTH AND IS THE NUMBER OF YEARS USED TO COMPUTE
THE MEAN. INDIVIDUAL MONTHS WITHIN THE POR MAY
BE MISSING.
WHEN THE POR FOR A NORMAL IS LESS THAN 30 YEARS,
THE NORMAL IS PROVISIONAL AND IS BASED ON THE NUMBER
OF YEARS INDICATED.
0.* OR * INDICATES THE VALUE OR MEAN-DAYS-WITH
IS BETWEEN 0.00 AND 0.05.
CLOUDINESS FOR ASOS STATIONS DIFFERS FROM THE NON-ASOS
OBSERVATION TAKEN BY A HUMAN OBSERVER. ASOS STATION
CLOUDINESS IS BASED ON TIME-AVERAGED CEILOMETER DATA
FOR CLOUDS AT OR BELOW 12,000 FEET AND ON SATELLITE
DATA FOR CLOUDS ABOVE 12,000 FEET.
THE NUMBER OF DAYS WITH CLEAR, PARTLY CLOUDY, AND
CLOUDY CONDITIONS FOR ASOS STATIONS IS THE SUM
OF THE CEILOMETER AND SATELLITE DATA FOR THE
SUNRISE TO SUNSET PERIOD.

GENERAL CONTINUED:
CLEAR INDICATES 0 - 2 OKTAS, PARTLY CLOUDY INDICATES
3 - 6 OKTAS, AND CLOUDY INDICATES 7 OR 8 OKTAS.
WHEN AT LEAST ONE OF THE ELEMENTS (CEILOMETER OR
SATELLITE) IS MISSING, THE DAILY CLOUDINESS IS
NOT COMPUTED.
WIND DIRECTION IS RECORDED IN TENS OF DEGREES (2 DIGITS)
CLOCKWISE FROM TRUE NORTH. "00" INDICATES CALM. "36"
INDICATES TRUE NORTH.
RESULTANT WIND IS THE VECTOR AVERAGE OF THE SPEED AND
DIRECTION.
AVERAGE TEMPERATURE IS THE SUM OF THE MEAN DAILY MAXIMUM
AND MINIMUM TEMPERATURE DIVIDED BY 2.
SNOWFALL DATA COMPRISE ALL FORMS OF FROZEN
PRECIPITATION, INCLUDING HAIL.
A HEATING (COOLING) DEGREE DAY IS THE DIFFERENCE BETWEEN
THE AVERAGE DAILY TEMPERATURE AND 65 F.
DRY BULB IS THE TEMPERATURE OF THE AMBIENT AIR.
DEW POINT IS THE TEMPERATURE TO WHICH THE AIR MUST BE
COOLED TO ACHIEVE 100 PERCENT RELATIVE HUMIDITY.
WET BULB IS THE TEMPERATURE THE AIR WOULD HAVE IF THE
MOISTURE CONTENT WAS INCREASED TO 100 PERCENT RELATIVE
HUMIDITY.

ON JULY 1, 1996, THE NATIONAL WEATHER SERVICE BEGAN USING
THE "METAR" OBSERVATION CODE THAT WAS ALREADY EMPLOYED
BY MOST OTHER NATIONS OF THE WORLD. THE MOST NOTICEABLE
DIFFERENCE IN THIS ANNUAL PUBLICATION WILL BE THE CHANGE
IN UNITS FROM TENTHS TO EIGHTS(OKTAS) FOR REPORTING THE
AMOUNT OF SKY COVER.

HEATING DEGREE DAYS (base 65 F) 2001 SALT LAKE CITY, UT (SLC)

YEAR	JUL	AUG	SEP	OCT	NOV	DEC	JAN	FEB	MAR	APR	MAY	JUN	TOTAL
1972-73	0	0	110	347	761	1307	1400	909	711	515	135	67	6262
1973-74	1	0	140	333	732	975	1181	935	603	502	214	41	5657
1974-75	0	5	54	316	638	1025	1157	819	734	613	334	92	5787
1975-76	0	1	62	365	825	989	1144	890	826	464	112	67	5745
1976-77	0	7	37	432	689	1096	1175	813	838	333	304	0	5724
1977-78	0	11	73	282	670	835	880	697	522	433	293	36	4732
1978-79	0	12	144	284	714	1178	1327	902	666	414	196	57	5894
1979-80	0	0	7	270	846	987	964	835	723	371	250	77	5330
1980-81	0	10	57	379	704	965	1013	742	641	346	233	46	5136
1981-82	0	0	34	444	614	879	1087	909	668	548	259	62	5504
1982-83	7	0	134	495	800	1080	916	710	624	569	314	36	5685
1983-84	6	0	49	276	650	1018	1269	1130	763	493	157	76	5887
1984-85	0	0	98	480	662	1084	1260	1097	740	285	109	17	5832
1985-86	0	0	140	360	821	1151	1110	655	527	477	283	14	5538
1986-87	6	0	203	416	720	1085	1186	803	679	291	123	17	5529
1987-88	0	0	51	260	719	1060	1235	870	723	381	222	3	5524
1988-89	0	0	142	158	711	1138	1318	1105	587	313	193	35	5700
1989-90	0	15	44	355	729	1036	971	895	612	297	232	30	5216
1990-91	0	0	17	347	704	1359	1249	792	682	508	289	32	5979
1991-92	0	0	78	347	742	1078	1220	741	481	247	57	33	5024
1992-93	11	14	42	285	921	1169	1237	986	596	489	151	108	6009
1993-94	19	3	82	391	902	1031	863	823	590	397	100	12	5213
1994-95	3	0	19	416	963	1028	927	628	642	486	288	104	5504
1995-96	0	0	75	415	559	856	986	986	607	428	175	5	5092
1996-97	0	0	123	384	642	858	1009	827	583	495	115	17	5053
1997-98	6	0	61	382	677	1145	835	763	754	504	202	118	5447
1998-99	0	0	57	428	657	1089	898	766	619	578	295	63	5450
1999-00	0	0	95	322	573	1047	920	723	704	311	154	15	4864
2000-01	0	0	99	388	1002	1056	1162	850	598	450	123	62	5790
2001-	0	0	27	310	662	1194							

COOLING DEGREE DAYS (base 65 F) 2001 SALT LAKE CITY, UT (SLC)

YEAR	JAN	FEB	MAR	APR	MAY	JUN	JUL	AUG	SEP	OCT	NOV	DEC	ANNUAL
1972	0	0	0	0	34	213	386	340	85	0	0	0	1058
1973	0	0	0	0	38	226	370	367	44	3	0	0	1048
1974	0	0	0	2	31	303	446	298	108	3	0	0	1191
1975	0	0	0	0	9	89	439	269	80	14	0	0	900
1976	0	0	0	0	34	151	431	237	87	3	0	0	943
1977	0	0	0	12	2	254	389	328	123	0	0	0	1108
1978	0	0	0	0	21	167	411	299	120	0	0	0	1018
1979	0	0	0	2	54	214	439	336	208	21	0	0	1274
1980	0	0	0	9	10	159	399	301	99	1	0	0	978
1981	0	0	0	3	12	190	412	409	145	2	0	0	1173
1982	0	0	0	0	11	158	338	423	109	0	0	0	1039
1983	0	0	0	0	37	123	370	405	138	4	0	0	1077
1984	0	0	0	3	58	153	426	383	147	4	0	0	1174
1985	0	0	0	11	78	249	493	364	79	0	0	0	1274
1986	0	0	0	0	47	277	296	407	66	0	0	0	1093
1987	0	0	0	25	60	222	338	309	103	0	0	0	1057
1988	0	0	0	0	61	334	501	363	112	9	0	0	1380
1989	0	0	0	0	13	43	171	506	337	92	0	0	1162
1990	0	0	0	2	19	247	438	351	235	11	0	0	1303
1991	0	0	0	1	3	138	442	425	63	4	0	0	1076
1992	0	0	0	15	81	204	340	401	93	14	0	0	1148
1993	0	0	0	0	108	75	178	246	102	13	0	0	722
1994	0	0	0	16	46	300	498	500	189	0	0	0	1549
1995	0	0	0	0	2	89	347	409	158	5	0	0	1010
1996	0	0	0	2	24	264	490	405	125	48	0	0	1358
1997	0	0	0	0	71	178	324	431	151	15	0	0	1170
1998	0	0	0	3	13	53	462	399	156	0	0	0	1086
1999	0	0	0	0	22	172	416	388	58	6	0	0	1062
2000	0	0	0	4	61	233	497	438	91	2	0	0	1326
2001	0	0	0	9	87	249	452	445	191	8	0	0	1441

SNOWFALL (inches) 2001 SALT LAKE CITY, UT (SLC)

YEAR	JUL	AUG	SEP	OCT	NOV	DEC	JAN	FEB	MAR	APR	MAY	JUN	TOTAL
1972-73	0.0	0.0	0.0	6.0	1.1	35.2	20.9	3.6	17.8	2.6	0.0	0.0	87.2
1973-74	0.0	0.0	0.0	1.3	19.5	19.6	20.1	17.2	6.7	26.4	T	T	110.8
1974-75	0.0	0.0	0.0	T	T	8.8	12.5	7.9	22.8	13.1	7.5	0.0	72.6
1975-76	0.0	0.0	0.0	0.1	18.0	11.8	8.6	15.8	18.7	3.5	0.0	T	76.5
1976-77	0.0	0.0	0.0	0.0	T	1.2	8.6	3.2	41.9	4.8	0.6	0.0	60.3
1977-78	0.0	0.0	0.0	0.2	8.5	8.2	15.6	15.5	6.2	2.5	4.6	0.0	61.3
1978-79	0.0	0.0	1.0	0.0	17.4	8.7	13.8	12.4	3.6	7.7	T	0.0	64.6
1979-80	0.0	0.0	0.0	0.0	4.6	8.5	24.5	2.9	19.9	1.2	T	0.0	61.6
1980-81	0.0	0.0	0.0	T	3.9	3.3	8.9	2.7	11.1	0.3	T	T	30.2
1981-82	0.0	0.0	0.0	4.4	2.4	11.5	15.3	4.5	10.2	9.5	T	T	57.8
1982-83	0.0	0.0	0.0	0.2	1.0	20.1	6.2	1.0	13.3	9.0	5.0	0.0	55.8
1983-84	0.0	0.0	0.0	0.0	5.9	34.2	7.6	18.5	6.7	25.1	T	T	98.0
1984-85	0.0	0.0	T	20.4	6.6	12.9	12.7	11.4	8.0	0.7	0.0	0.0	72.7
1985-86	0.0	0.0	0.0	T	27.2	14.7	3.9	1.7	1.0	5.5	T	0.0	54.0
1986-87	0.0	0.0	T	0.0	4.4	1.7	16.4	9.9	3.0	2.1	0.0	0.0	37.5
1987-88	0.0	0.0	0.0	0.0	0.6	11.0	16.3	0.4	6.1	T	0.9	0.0	35.3
1988-89	0.0	0.0	T	0.0	8.5	12.5	9.4	27.5	2.1	T	0.0	0.0	60.0
1989-90	0.0	0.0	0.0	2.7	2.4	1.7	8.2	8.5	11.8	0.7	T	T	36.0
1990-91	0.0	0.0	T	0.0	4.8	14.3	10.9	1.2	1.9	13.7	T	0.0	46.8
1991-92	T	T	T	2.8	11.4	6.8	12.2	4.7	0.2	0.4	0.0	0.0	38.5
1992-93	0.0	T	0.0	0.0	14.2	16.9	50.3	13.2	T	2.4	1.7	T	98.7
1993-94	0.0	T	0.0	0.0	0.0	8.3	6.2	15.0	3.1	0.1	0.0	0.0	38.8
1994-95	0.0	0.0	0.0	T	33.3	13.6	11.1	13.3	7.3	6.6	T	T	85.2
1995-96	0.0	0.0	0.0	0.5	0.1	1.4	45.0	22.6	11.2	4.9	T	T	85.7
1996-97	T	0.0	0.0	5.1	3.5	16.0	16.2	11.0	7.1	4.4	T	T	63.3
1997-98	0.0	0.0	0.0	1.7	T	6.6	6.3	32.1	14.8	3.7	0.0	T	65.2
1998-99	0.0	0.0	T	0.0	2.1	6.6	8.4	8.6	3.0	3.5	T	T	32.2
1999-00	0.0	0.0	T	0.0	5.5	18.0	15.0	5.1	6.2	T	0.5	0.0	50.3
2000-01	0.0	0.0	0.0	0.0	17.0	12.9	6.5	13.1	5.3	9.9	0.0	0.0	64.7
2001-	T	0.0	0.0	T	18.8	19.9							
POR= 72 YRS	T	T	0.1	1.2	7.0	11.8	13.5	9.8	9.1	4.9	0.6	T	58.0

2001
BURLINGTON,
VERMONT (BTV)

Burlington is located on the eastern shore of Lake Champlain at the widest part of the lake. About 35 miles to the west lie the highest peaks of the Adirondacks, while the foothills of the Green Mountains begin 10 miles to the east and southeast.

Its northerly latitude assures the variety and vigor of a true New England climate, while thanks to the modifying influence of the lake, the many rapid and marked weather changes are tempered in severity. Due to its location in the path of the St. Lawrence Valley storm track and the lake effects, the city is one of the cloudiest in the United States.

Lake Champlain exercises a tempering influence on the local temperature. During the winter months and prior to the lake freezing, temperatures along the lake shore are often 5-10 degrees warmer than at the airport 3 1/2 miles inland. At the airport the average occurrence of the last freeze in spring is around May 10th and that of the first in fall is early October, giving a growing season of 145 days. This location is justly proud of its delightful summer weather. On average, there are few days a year with maxima of 90 degrees or higher. This moderate summer heat gives way to a cooler, but none the less pleasant fall period, usually extending well into October. High pressure systems moving down rapidly from central Canada or Hudson Bay produce the coldest temperatures during the winter months, but extended periods of very cold weather are rare.

Precipitation, although generally plentiful and well distributed throughout the year, is less in the Champlain Valley than in other areas of Vermont due to the shielding effect of the mountain barriers to the east and west. The heaviest rainfall usually occurs during summer thunderstorms, but excessively heavy rainfall is quite uncommon. Droughts are infrequent.

Because of the trend of the Champlain Valley between the Adirondack and Green Mountain ranges, most winds have a northerly or southerly component. The prevailing direction most of the year is from the south. Winds of damaging force are very uncommon.

Smoke pollution is nearly non-existent since there is no concentration of heavy industry here, however, haze has been on the increase over the years due to the large increase in industry to the north and south. During the spring and fall months, fog occasionally forms along the Winooski River to the north and east and may drift over the airport with favorable winds. In spite of the high percentage of cloudiness, periods of low aircraft ceilings and visibilities are usually of short duration, allowing this area to have one of the highest percentages of flying weather in New England.

NORMALS, MEANS, AND EXTREMES
BURLINGTON, VT (BTV)

LATITUDE:	LONGITUDE:	ELEVATION (FT):		TIME ZONE:	WBAN: 14742
44 28' 05" N	73 09' 01" W	GRND: 345	BARO: 348	EASTERN (UTC + 5)	

	ELEMENT	POR	JAN	FEB	MAR	APR	MAY	JUN	JUL	AUG	SEP	OCT	NOV	DEC	YEAR
TEMPERATURE °F	NORMAL DAILY MAXIMUM	30	25.1	27.5	39.3	53.6	67.2	75.8	81.2	77.9	69.0	57.0	44.0	30.4	54.0
	MEAN DAILY MAXIMUM	54	26.3	28.8	38.4	53.4	67.0	76.2	80.8	78.2	69.0	57.5	44.4	31.7	54.3
	HIGHEST DAILY MAXIMUM	58	66	62	84	91	93	100	100	101	94	85	75	67	101
	YEAR OF OCCURRENCE		1995	1981	1998	1976	1977	1995	1995	1944	1945	1949	1948	1998	AUG 1944
	MEAN OF EXTREME MAXS.	54	48.0	47.0	61.0	74.9	85.0	90.4	92.4	90.0	84.8	75.0	64.5	52.1	72.1
	NORMAL DAILY MINIMUM	30	7.5	8.9	22.0	34.2	45.4	54.6	59.7	57.9	48.8	38.6	29.6	15.5	35.2
	MEAN DAILY MINIMUM	54	8.7	10.1	20.9	33.3	44.7	54.5	59.3	57.3	49.3	38.9	30.1	16.6	35.3
	LOWEST DAILY MINIMUM	58	-30	-30	-20	2	24	33	39	35	25	15	-2	-26	-30
	YEAR OF OCCURRENCE		1957	1979	1948	1972	1966	1986	1962	1976	1963	1972	1958	1980	FEB 1979
	MEAN OF EXTREME MINS.	54	-14.9	-13.5	-1.3	18.8	30.3	39.7	46.9	44.1	33.3	24.2	11.6	-7.9	17.6
	NORMAL DRY BULB	30	16.3	18.2	30.7	43.9	56.3	65.2	70.5	67.9	58.9	47.8	36.8	23.0	44.6
	MEAN DRY BULB	54	17.5	19.4	29.7	43.4	55.8	65.3	70.1	67.7	59.2	48.3	37.3	24.2	44.8
	MEAN WET BULB	18	18.1	19.2	27.2	39.1	50.4	59.1	63.4	62.3	54.9	44.3	34.5	24.0	41.4
	MEAN DEW POINT	18	12.1	12.5	20.0	31.3	43.4	53.6	59.0	58.1	50.9	39.0	28.8	18.4	35.6
	NORMAL NO. DAYS WITH:														
	MAXIMUM 90	30	0.0	0.0	0.0	0.1	0.5	1.2	2.9	1.0	0.1	0.0	0.0	0.0	5.8
	MAXIMUM 32	30	21.3	18.3	8.6	0.6	0.0	0.0	0.0	0.0	0.0	*	4.3	17.4	70.5
	MINIMUM 32	30	30.1	26.4	26.0	15.9	3.2	0.0	0.0	0.0	0.9	8.9	19.1	28.1	158.6
	MINIMUM 0	30	10.3	8.7	2.0	0.0	0.0	0.0	0.0	0.0	0.0	0.0	0.0	5.0	26.0
H/C	NORMAL HEATING DEG. DAYS	30	1510	1310	1063	633	282	58	6	29	199	533	846	1302	7771
	NORMAL COOLING DEG. DAYS	30	0	0	0	0	13	64	176	119	16	0	0	0	388
RH	NORMAL (PERCENT)														
	HOUR 01 LST	30	71	72	73	74	77	81	83	85	85	79	76	75	78
	HOUR 07 LST	30	72	74	75	73	74	77	79	83	86	81	78	76	77
	HOUR 13 LST	30	64	61	58	53	51	54	53	57	61	61	66	67	59
	HOUR 19 LST	30	68	65	63	58	57	60	61	67	73	71	72	72	66
S	PERCENT POSSIBLE SUNSHINE	58	41	48	51	49	55	59	64	60	54	47	31	33	49
W/O	MEAN NO. DAYS WITH:														
	HEAVY FOG(VISBY 1/4 MI)	58	1.0	1.2	1.4	1.3	0.9	1.1	0.8	1.4	2.4	1.9	1.2	1.1	15.7
	THUNDERSTORMS	58	0.0	0.0	0.4	1.0	2.5	5.0	6.0	5.0	2.1	0.6	0.3	0.0	22.9
CLOUDINESS	MEAN:														
	SUNRISE-SUNSET (OKTAS)	53	6.0	5.8	5.7	5.8	5.6	5.4	5.0	5.0	5.1	5.4	6.4	6.3	5.6
	MIDNIGHT-MIDNIGHT (OKTAS)	32	5.9	5.4	5.3	5.3	5.3	5.0	4.8	4.8	5.0	5.3	6.2	6.2	5.4
	MEAN NO. DAYS WITH:														
	CLEAR	54	4.6	4.5	4.5	4.6	4.0	4.9	4.8	5.1	5.9	5.9	2.5	2.9	57.8
	PARTLY CLOUDY	54	6.4	6.9	6.9	6.9	7.5	9.3	10.8	12.7	11.6	9.8	7.6	5.2	5.7	100.4
	CLOUDY	54	20.1	16.9	18.3	17.5	16.7	14.4	12.7	13.1	13.8	17.1	21.7	21.7	204.0
PR	MEAN STATION PRESSURE(IN)	29	29.61	29.70	29.60	29.60	29.59	29.59	29.60	29.70	29.68	29.70	29.69	29.69	29.65
	MEAN SEA-LEVEL PRES. (IN)	18	30.06	30.06	30.04	29.97	29.95	29.94	29.96	30.02	30.05	30.08	30.06	30.06	30.02
WINDS	MEAN SPEED (MPH)	47	9.8	9.4	9.5	9.5	9.0	8.5	8.1	7.8	8.5	8.9	9.8	10.0	9.1
	PREVAIL.DIR(TENS OF DEGS)	31	18	18	18	18	18	18	18	18	18	18	18	18	18
	MAXIMUM 2-MINUTE:														
	SPEED (MPH)	5	38	37	32	36	33	32	34	35	36	31	33	35	38
	DIR. (TENS OF DEGS)		18	25	18	33	19	19	32	27	32	29	19	27	18
	YEAR OF OCCURRENCE		1997	2001	1997	2000	1997	2000	1999	2000	1999	1998	1999	2000	JAN 1997
	MAXIMUM 5-SECOND:														
	SPEED (MPH)	5	51	47	44	43	40	41	43	46	48	41	43	45	51
	DIR. (TENS OF DEGS)		19	19	19	32	19	28	31	26	32	27	18	18	19
	YEAR OF OCCURRENCE		1997	1997	1997	2000	1997	1998	1999	2000	1999	1998	1998	1998	JAN 1997
PRECIPITATION	NORMAL (IN)	30	1.82	1.63	2.23	2.76	3.12	3.47	3.65	4.06	3.30	2.88	3.13	2.42	34.47
	MAXIMUM MONTHLY (IN)	58	5.15	5.38	4.14	6.55	6.31	8.66	9.31	11.54	10.26	6.22	6.85	5.95	11.54
	YEAR OF OCCURRENCE		1998	1981	2001	1983	1983	1998	1998	1955	1999	1959	1983	1973	AUG 1955
	MINIMUM MONTHLY (IN)	58	0.42	0.21	0.38	0.73	0.29	0.82	0.77	0.72	0.87	0.50	0.63	0.37	0.21
	YEAR OF OCCURRENCE		1989	1978	1965	1999	1977	1995	2001	1957	1948	1963	1952	1998	FEB 1978
	MAXIMUM IN 24 HOURS (IN)	58	2.11	1.93	1.62	2.16	2.26	2.83	2.69	3.62	3.96	2.17	2.48	2.60	3.96
	YEAR OF OCCURRENCE		1998	1981	1971	1968	1955	1972	1985	1998	1999	1983	1990	1950	SEP 1999
	NORMAL NO. DAYS WITH:														
	PRECIPITATION 0.01	30	14.0	10.8	13.0	12.3	14.1	13.2	11.9	13.4	11.5	12.4	14.5	15.3	156.4
	PRECIPITATION 1.00	30	0.1	0.1	0.2	0.1	0.3	0.5	0.8	0.8	0.7	0.5	0.2	0.2	4.5
SNOWFALL	NORMAL (IN)	30	19.0	15.1	13.0	4.8	0.3	0.0	0.0	0.0	0.0	0.3	7.4	21.3	81.2
	MAXIMUM MONTHLY (IN)	58	42.4	34.3	47.6	21.3	3.9	0.0	T	T	0.1	5.1	19.2	56.7	56.7
	YEAR OF OCCURRENCE		1978	1958	2001	1983	1966		1989	2001	1992	1969	1971	1970	DEC 1970
	MAXIMUM IN 24 HOURS (IN)	58	14.5	17.7	22.4	15.6	3.5	0.0	T	T	0.1	5.1	10.1	17.0	22.4
	YEAR OF OCCURRENCE		1961	1995	1993	1983	1966		1989	2001	1992	1969	1958	1978	MAR 1993
	MAXIMUM SNOW DEPTH (IN)	53	40	33	31	16	1	0	0	0	0	5	10	33	40
	YEAR OF OCCURRENCE		1964	1958	1993	2001	1967					1969	1958	1969	JAN 1964
	NORMAL NO. DAYS WITH:														
	SNOWFALL 1.0	30	5.2	4.3	3.9	1.5	0.1	0.0	0.0	0.0	0.0	0.1	2.2	5.8	23.1

PRECIPITATION (inches) 2001 BURLINGTON, VT (BTV)

YEAR	JAN	FEB	MAR	APR	MAY	JUN	JUL	AUG	SEP	OCT	NOV	DEC	ANNUAL
1972	0.93	1.69	3.58	2.26	2.83	6.52	6.12	2.35	1.69	2.60	4.10	3.43	38.10
1973	2.13	1.55	2.09	3.80	5.38	7.69	3.02	5.41	5.02	1.93	2.31	5.95	46.28
1974	1.90	1.54	2.73	3.47	4.61	4.45	3.70	2.60	3.23	0.78	3.60	2.08	34.69
1975	2.20	2.01	2.86	1.71	1.17	2.47	3.77	2.85	4.12	3.85	3.14	2.36	32.51
1976	2.99	2.85	2.35	2.54	5.86	4.04	3.05	4.69	3.77	4.34	1.63	1.97	40.08
1977	1.61	1.78	2.97	3.13	0.29	2.06	3.34	6.27	6.33	5.02	4.22	3.42	40.44
1978	4.69	0.21	2.98	2.51	2.16	4.36	3.50	1.82	2.07	3.72	0.95	2.11	31.08
1979	4.50	0.60	2.15	3.61	3.12	1.39	1.23	3.42	3.84	2.31	3.89	1.50	31.56
1980	0.61	0.67	2.44	2.39	1.61	1.92	6.11	3.83	4.41	2.48	2.92	1.50	30.89
1981	0.49	5.38	1.32	3.05	3.76	3.07	3.22	5.58	6.24	5.26	2.73	2.03	42.13
1982	2.74	1.43	2.31	2.63	1.95	4.95	3.07	3.55	2.12	2.31	3.59	1.69	32.34
1983	3.09	1.66	2.60	6.55	6.31	1.49	3.92	4.31	3.77	4.38	6.85	5.23	50.16
1984	0.81	2.73	1.72	4.25	5.27	1.70	5.11	3.30	2.81	1.89	3.08	3.14	35.81
1985	1.46	1.26	2.46	1.90	3.53	3.76	4.42	2.67	3.30	3.31	3.68	1.59	33.34
1986	3.69	1.68	3.17	0.95	4.11	4.40	4.53	5.82	4.86	2.50	2.99	1.32	40.02
1987	1.91	0.49	1.33	1.42	2.69	4.42	2.79	2.09	3.58	3.28	2.24	1.17	27.41
1988	0.69	1.69	1.55	1.91	1.80	3.26	2.55	4.27	1.50	2.05	4.51	0.90	26.68
1989	0.42	0.67	2.60	1.89	3.19	3.68	3.65	7.30	5.98	2.98	2.41	1.26	36.03
1990	2.36	2.82	1.81	2.97	3.66	3.08	5.12	4.85	2.03	5.99	3.91	3.58	42.18
1991	1.65	0.51	2.55	3.41	3.15	1.28	2.83	4.00	5.14	5.07	1.58	1.35	32.52
1992	1.65	1.56	2.13	2.58	2.38	1.72	4.58	1.89	4.73	3.00	3.67	0.96	30.85
1993	2.17	1.90	1.54	3.76	2.19	3.35	3.34	4.46	3.38	2.93	2.27	1.57	32.86
1994	2.19	1.21	2.93	3.37	4.58	3.65	5.30	4.50	1.74	1.25	2.48	1.66	34.86
1995	1.88	1.26	1.60	2.35	1.41	0.82	3.49	4.64	2.97	5.81	3.33	2.63	32.19
1996	3.91	0.83	0.80	6.12	5.33	4.54	4.74	1.47	2.75	3.64	3.30	0.64	38.07
1997	1.71	1.38	2.59	1.54	2.24	2.62	3.89	4.63	2.98	1.23	4.16	1.65	30.62
1998	5.15	1.84	3.81	1.79	3.61	8.66	9.31	6.80	5.64	2.42	1.02	0.37	50.42
1999	3.51	1.13	2.22	0.73	2.40	1.79	1.97	2.41	10.26	3.18	1.86	1.12	32.58
2000	2.30	2.67	1.63	5.01	6.13	3.55	1.97	2.96	3.02	1.80	2.96	3.36	39.26
2001	0.98	1.54	4.14	0.85	2.28	2.32	0.77	4.32	1.40	1.37	1.81	1.49	23.27
POR= 118 YRS	1.88	1.65	2.18	2.54	3.06	3.47	3.63	3.52	3.37	2.88	2.70	2.01	32.89

AVERAGE TEMPERATURE (F) 2001 BURLINGTON, VT (BTV)

YEAR	JAN	FEB	MAR	APR	MAY	JUN	JUL	AUG	SEP	OCT	NOV	DEC	ANNUAL
1972	21.1	17.0	24.8	35.6	56.2	63.1	69.5	65.3	58.7	42.4	32.1	22.5	42.4
1973	21.5	14.6	37.1	44.6	53.6	66.9	70.6	72.1	58.5	49.3	37.2	27.4	46.1
1974	18.7	15.6	29.2	44.4	51.3	66.5	70.2	69.1	58.2	43.4	36.2	28.5	44.3
1975	23.6	20.7	28.0	37.1	62.3	66.4	74.6	69.1	58.0	50.4	42.1	20.1	46.0
1976	11.1	24.6	33.4	47.4	54.7	69.2	68.5	67.7	58.0	43.7	33.0	16.3	43.7
1977	11.1	20.5	37.6	45.3	60.0	64.7	69.6	67.5	58.7	46.6	40.0	22.4	45.3
1978	15.1	9.5	26.0	38.7	60.1	63.9	69.4	68.3	55.2	46.4	34.8	25.2	42.7
1979	18.0	7.5	36.9	43.5	58.5	65.3	72.2	65.9	58.6	48.1	41.3	29.0	45.4
1980	21.2	17.6	31.1	46.5	58.9	64.4	70.6	70.7	57.9	45.0	32.3	15.0	44.3
1981	8.9	32.9	33.5	46.7	58.2	66.1	71.1	67.1	59.3	44.9	36.9	25.3	45.9
1982	9.6	19.1	30.3	43.4	57.3	60.7	69.5	65.9	62.3	50.1	42.3	31.9	45.2
1983	21.0	22.3	33.0	42.3	52.9	66.3	71.3	68.9	62.9	48.2	38.1	22.4	45.8
1984	16.5	28.7	21.9	44.7	52.3	66.0	70.3	71.1	57.2	50.0	38.4	30.3	45.6
1985	13.4	22.5	31.6	44.3	55.8	61.7	69.6	67.5	60.3	49.1	36.9	24.3	44.5
1986	18.5	16.2	33.7	48.5	58.3	62.3	68.5	66.1	58.1	46.9	34.5	27.8	45.0
1987	18.1	15.0	33.3	48.6	55.5	66.3	71.5	66.5	55.9		37.0	28.5	45.5
1988	19.9	21.4	29.7	44.3	57.9	63.4	73.2	70.7	58.2	44.4	39.6	22.9	45.5
1989	23.7	19.7	28.4	41.6	59.6	67.2	71.7	67.7	58.6	50.3	36.4	7.6	44.0
1990	29.8	23.5	33.8	46.2	52.9	65.9	70.2	69.8	59.4	49.4	39.5	30.1	47.5
1991	18.9	26.5	34.0	49.1	59.3	65.9	70.4	70.5	57.8	50.4	37.6	24.0	47.0
1992	18.6	19.1	26.5	42.3	56.5	64.4	66.1	67.6	60.2	45.5	36.7	28.1	44.3
1993	21.7	10.6	27.4	45.3	56.6	64.7	72.2	70.8	59.1	46.1	36.8	24.7	44.7
1994	7.1	15.4	30.2	44.5	54.8	68.6	74.2	66.5	59.6	49.8	41.1	28.7	45.0
1995	27.9	19.0	35.0	40.4	56.4	69.6	74.7	70.0	57.5	54.1	35.2	22.0	46.8
1996	17.5	21.3	28.6	42.8	54.4	66.3	68.6	68.9	62.0	47.4	32.8	32.7	45.3
1997	19.1	25.0	26.9	41.4	51.4	67.3	68.7	66.8	58.6	46.6	35.1	25.8	44.4
1998	22.7	27.7	34.3	46.4	62.0	66.1	68.9	68.5	61.5	49.6	39.7	32.0	48.3
1999	19.1	24.8	30.9	44.7	59.8	70.5	74.2	68.1	64.5	46.2	42.6	28.7	47.8
2000	18.1	22.1	36.5	42.7	56.7	63.6	67.5	67.7	58.8	48.8	37.2	19.6	44.9
2001	20.1	22.1	27.2	43.2	58.7	67.2	68.0	72.6	61.3	51.1	42.4	32.7	47.2
POR= 109 YRS	17.9	18.9	29.7	43.0	55.6	65.0	69.9	67.6	59.6	48.5	36.9	23.8	44.7

REFERENCE NOTES:

PAGE 1:
THE TEMPERATURE GRAPH SHOWS NORMAL MAXIMUM AND NORMAL MINIMUM DAILY TEMPERATURES (SOLID CURVES) AND THE ACTUAL DAILY HIGH AND LOW TEMPERATURES (VERTICAL BARS).

PAGE 2 AND 3:
H/C INDICATES HEATING AND COOLING DEGREE DAYS.
RH INDICATES RELATIVE HUMIDITY
W/O INDICATES WEATHER AND OBSTRUCTIONS
S INDICATES SUNSHINE.
PR INDICATES PRESSURE.
CLOUDINESS ON PAGE 3 IS THE SUM OF THE CEILOMETER AND SATELLITE DATA NOT TO EXCEED EIGHT EIGHTHS(OKTAS).

GENERAL:
T INDICATES TRACE PRECIPITATION, AN AMOUNT GREATER THAN ZERO BUT LESS THAN THE LOWEST REPORTABLE VALUE.
+ INDICATES THE VALUE ALSO OCCURS ON EARLIER DATES.
BLANK ENTRIES DENOTE MISSING OR UNREPORTED DATA.
NORMALS ARE 30-YEAR AVERAGES (1961 - 1990).
ASOS INDICATES AUTOMATED SURFACE OBSERVING SYSTEM.
PM INDICATES THE LAST DAY OF THE PREVIOUS MONTH.
POR (PERIOD OF RECORD) BEGINS WITH THE JANUARY DATA MONTH AND IS THE NUMBER OF YEARS USED TO COMPUTE THE MEAN. INDIVIDUAL MONTHS WITHIN THE POR MAY BE MISSING.
WHEN THE POR FOR A NORMAL IS LESS THAN 30 YEARS, THE NORMAL IS PROVISIONAL AND IS BASED ON THE NUMBER OF YEARS INDICATED.
0.* OR * INDICATES THE VALUE OR MEAN-DAYS-WITH IS BETWEEN 0.00 AND 0.05.
CLOUDINESS FOR ASOS STATIONS DIFFERS FROM THE NON-ASOS OBSERVATION TAKEN BY A HUMAN OBSERVER. ASOS STATION CLOUDINESS IS BASED ON TIME-AVERAGED CEILOMETER DATA FOR CLOUDS AT OR BELOW 12,000 FEET AND ON SATELLITE DATA FOR CLOUDS ABOVE 12,000 FEET.
THE NUMBER OF DAYS WITH CLEAR, PARTLY CLOUDY, AND CLOUDY CONDITIONS FOR ASOS STATIONS IS THE SUM OF THE CEILOMETER AND SATELLITE DATA FOR THE SUNRISE TO SUNSET PERIOD.

GENERAL CONTINUED:
CLEAR INDICATES 0 - 2 OKTAS, PARTLY CLOUDY INDICATES 3 - 6 OKTAS, AND CLOUDY INDICATES 7 OR 8 OKTAS. WHEN AT LEAST ONE OF THE ELEMENTS (CEILOMETER OR SATELLITE) IS MISSING, THE DAILY CLOUDINESS IS NOT COMPUTED.
WIND DIRECTION IS RECORDED IN TENS OF DEGREES (2 DIGITS) CLOCKWISE FROM TRUE NORTH. "00" INDICATES CALM. "36" INDICATES TRUE NORTH.
RESULTANT WIND IS THE VECTOR AVERAGE OF THE SPEED AND DIRECTION.
AVERAGE TEMPERATURE IS THE SUM OF THE MEAN DAILY MAXIMUM AND MINIMUM TEMPERATURE DIVIDED BY 2.
SNOWFALL DATA COMPRISE ALL FORMS OF FROZEN PRECIPITATION, INCLUDING HAIL.
A HEATING (COOLING) DEGREE DAY IS THE DIFFERENCE BETWEEN THE AVERAGE DAILY TEMPERATURE AND 65 F.
DRY BULB IS THE TEMPERATURE OF THE AMBIENT AIR.
DEW POINT IS THE TEMPERATURE TO WHICH THE AIR MUST BE COOLED TO ACHIEVE 100 PERCENT RELATIVE HUMIDITY.
WET BULB IS THE TEMPERATURE THE AIR WOULD HAVE IF THE MOISTURE CONTENT WAS INCREASED TO 100 PERCENT RELATIVE HUMIDITY.

ON JULY 1, 1996, THE NATIONAL WEATHER SERVICE BEGAN USING THE "METAR" OBSERVATION CODE THAT WAS ALREADY EMPLOYED BY MOST OTHER NATIONS OF THE WORLD. THE MOST NOTICEABLE DIFFERENCE IN THIS ANNUAL PUBLICATION WILL BE THE CHANGE IN UNITS FROM TENTHS TO EIGHTS(OKTAS) FOR REPORTING THE AMOUNT OF SKY COVER.

HEATING DEGREE DAYS (base 65 F) 2001 BURLINGTON, VT (BTV)

YEAR	JUL	AUG	SEP	OCT	NOV	DEC	JAN	FEB	MAR	APR	MAY	JUN	TOTAL
1972-73	26	69	212	694	982	1310	1344	1410	855	608	345	86	7941
1973-74	10	17	256	480	825	1160	1431	1378	1101	618	430	37	7743
1974-75	2	6	224	665	858	1128	1276	1236	1141	831	152	82	7601
1975-76	0	45	208	448	681	1385	1669	1168	973	545	331	50	7503
1976-77	20	68	254	654	954	1505	1667	1240	842	590	223	89	8106
1977-78	24	53	207	564	740	1314	1539	1547	1202	781	225	90	8286
1978-79	49	38	295	571	897	1227	1452	1610	866	641	224	90	7960
1979-80	23	65	213	528	703	1107	1350	1371	1043	550	204	91	7248
1980-81	10	3	240	611	976	1545	1738	894	969	544	239	43	7812
1981-82	13	36	204	617	837	1224	1716	1277	1069	643	255	133	8024
1982-83	30	54	124	455	676	1021	1356	1188	983	675	367	77	7006
1983-84	19	36	148	518	803	1317	1500	1044	1331	602	395	68	7781
1984-85	6	24	241	460	792	1068	1592	1185	1029	615	296	118	7426
1985-86	11	42	169	489	835	1344	1436	1361	966	492	219	113	7477
1986-87	40	60	215	553	906	1144	1446	1397	975	488	328	48	7600
1987-88	19	66	185	584	833	1125	1389	1260	1088	614	236	136	7535
1988-89	15	52	212	635	755	1298	1273	1265	1128	691	188	45	7557
1989-90	2	43	164	451	849	1776	1084	1156	961	577	370	63	7496
1990-91	19	10	180	480	758	1074	1424	1072	954	475	206	59	6711
1991-92	7	11	240	451	813	1266	1434	1327	1187	674	277	83	7770
1992-93	49	33	197	597	843	1137	1335	1517	1159	584	256	80	7787
1993-94	3	12	211	579	839	1243	1793	1385	1073	609	328	48	8123
1994-95	1	57	168	467	711	1118	1145	1283	925	733	268	36	6912
1995-96	0	21	232	330	885	1326	1466	1262	1123	659	336	27	7667
1996-97	11	10	138	542	960	995	1416	1113	1173	701	412	31	7502
1997-98	17	20	205	564	889	1208	1306	1039	947	552	115	81	6943
1998-99	8	19	129	473	754	1017	1418	1120	1053	606	184	31	6812
1999-00	1	20	113	578	666	1119	1448	1235	879	663	263	108	7093
2000-01	28	28	209	498	828	1399	1385	1193	1169	649	206	49	7641
2001-	32	10	152	431	670	998							

COOLING DEGREE DAYS (base 65 F) 2001 BURLINGTON, VT (BTV)

YEAR	JAN	FEB	MAR	APR	MAY	JUN	JUL	AUG	SEP	OCT	NOV	DEC	ANNUAL
1972	0	0	0	0	14	64	169	81	30	0	0	0	358
1973	0	0	0	3	0	149	187	243	68	0	0	0	650
1974	0	0	0	5	9	89	171	140	27	1	0	0	442
1975	0	0	0	0	75	131	306	181	5	1	0	0	699
1976	0	0	0	24	19	185	135	97	23	0	0	0	483
1977	0	0	0	7	75	86	174	138	27	0	0	0	507
1978	0	0	0	0	79	64	194	146	6	0	0	0	489
1979	0	0	0	2	29	106	253	101	27	13	0	0	531
1980	0	0	0	0	24	78	189	184	34	0	0	0	509
1981	0	0	0	2	35	85	211	110	39	0	0	0	482
1982	0	0	0	1	24	11	179	90	51	0	0	0	356
1983	0	0	0	0	0	121	223	155	92	6	0	0	597
1984	0	0	0	0	7	106	175	217	15	3	0	0	523
1985	0	0	0	0	15	25	160	123	34	0	0	0	357
1986	0	0	0	4	19	38	156	104	14	0	0	0	335
1987	0	0	0	3	42	92	228	126	30	0	0	0	521
1988	0	0	0	0	19	96	274	238	15	3	0	0	645
1989	0	0	0	0	28	117	216	134	63	0	0	0	558
1990	0	0	0	16	1	95	189	165	18	6	0	0	490
1991	0	0	0	5	35	92	182	186	32	6	0	0	538
1992	0	0	0	3	21	71	91	121	61	0	0	0	368
1993	0	0	0	0	2	79	235	198	39	0	0	0	553
1994	0	0	0	2	17	165	293	110	14	0	0	0	601
1995	0	0	0	0	8	179	306	182	12	0	0	0	687
1996	0	0	0	0	17	72	124	139	56	0	0	0	408
1997	0	0	0	0	0	108	142	82	17	0	0	0	349
1998	0	0	0	2	27	121	141	134	33	1	0	0	459
1999	0	0	0	0	29	207	295	125	106	0	0	0	762
2000	0	0	0	0	12	74	112	118	31	1	0	0	348
2001	0	0	0	0	17	122	132	251	49	6	0	0	577

SNOWFALL (inches) 2001 BURLINGTON, VT (BTV)

YEAR	JUL	AUG	SEP	OCT	NOV	DEC	JAN	FEB	MAR	APR	MAY	JUN	TOTAL
1972-73	0.0	0.0	0.0	T	12.2	39.0	11.4	18.5	2.3	6.3	0.0	0.0	89.7
1973-74	0.0	0.0	0.0	0.1	2.6	24.1	21.5	9.9	20.9	16.8	0.0	0.0	95.9
1974-75	0.0	0.0	0.0	0.1	11.5	16.8	14.8	22.0	12.4	13.3	0.0	0.0	90.9
1975-76	0.0	0.0	0.0	T	5.3	16.0	28.3	20.4	18.8	0.9	T	0.0	89.7
1976-77	0.0	0.0	0.0	0.9	13.3	11.5	24.2	16.4	9.6	1.8	T	0.0	77.7
1977-78	0.0	0.0	0.0	0.0	16.0	22.6	42.4	4.0	12.5	1.9	T	0.0	99.4
1978-79	0.0	0.0	0.0	T	5.7	24.1	37.9	6.6	1.6	8.4	0.0	0.0	84.3
1979-80	0.0	0.0	0.0	1.5	0.4	6.0	3.0	11.6	16.8	0.3	0.0	0.0	39.6
1980-81	0.0	0.0	0.0	T	12.2	17.5	8.7	11.9	13.3	1.1	0.0	0.0	64.7
1981-82	0.0	0.0	0.0	T	3.9	32.8	19.4	8.3	13.0	4.1	0.0	0.0	81.5
1982-83	0.0	0.0	0.0	T	0.8	5.0	22.5	18.3	11.9	21.3	0.7	0.0	80.5
1983-84	0.0	0.0	0.0	T	4.7	14.4	15.2	13.7	16.1	0.4	T	0.0	64.5
1984-85	0.0	0.0	0.0	0.0	6.0	29.3	25.9	10.9	16.6	2.7	0.0	0.0	91.4
1985-86	0.0	0.0	0.0	T	4.6	21.3	33.6	18.3	8.4	T	T	0.0	86.2
1986-87	0.0	0.0	0.0	T	10.5	7.7	34.4	7.0	6.0	2.1	0.0	0.0	67.7
1987-88	0.0	0.0	0.0	0.6	6.5	12.4	9.2	26.9	6.4	2.4	0.0	0.0	64.4
1988-89	0.0	0.0	0.0	0.3	0.6	12.4	6.6	8.5	9.7	2.3	0.0	0.0	40.4
1989-90	T	0.0	0.0	T	5.6	20.7	17.6	20.5	10.2	2.1	0.0	0.0	76.7
1990-91	0.0	0.0	T	T	7.3	10.3	17.8	3.9	3.2	T	0.0	0.0	42.5
1991-92	0.0	0.0	T	T	2.3	14.9	12.2	27.2	14.0	8.6	0.0	0.0	79.2
1992-93	0.0	0.0	0.1	T	2.9	2.6	24.8	33.8	39.9	12.8	0.0	0.0	116.9
1993-94	0.0	0.0	0.0	1.3	9.1	38.6	15.9	26.6	7.8	T	0.0	0.0	107.2
1994-95	0.0	0.0	0.0	0.0	5.1	4.3	8.7	26.8	10.7	4.9	0.0	0.0	60.5
1995-96	0.0	0.0	0.0	0.1	7.3	44.0	19.0	4.5	11.5	12.4	0.3	0.0	99.1
1996-97	0.0	0.0	0.0	T	14.3	13.7	22.0	8.8	27.0	9.1	T	0.0	94.9
1997-98	0.0	0.0	0.0	0.6	15.5	20.6	25.1	10.3	21.5	0.3	0.0	0.0	93.9
1998-99	0.0	0.0	0.0	T	2.4	4.4	30.4	10.8	22.7	T	0.0	0.0	70.7
1999-00	0.0	0.0	0.0	T	0.7	2.3	21.9	23.1	9.3	19.1	0.0	0.0	76.4
2000-01	T	0.0	0.0	3.0	8.7	32.8	15.7	14.4	47.6	0.2	T	0.0	122.4
2001-	0.0	T	0.0	T	1.8	13.9							
POR= 57 YRS	T	0.0	0.0	0.2	6.5	18.0	19.0	16.5	13.6	4.2	0.5	0.0	78.5

2001
NORFOLK,
VIRGINIA (ORF)

The city of Norfolk, Virginia, is located near the coast and the southern border of the state. It is almost surrounded by water, with the Chesapeake Bay immediately to the north, Hampton Roads to the west, and the Atlantic Ocean only 18 miles to the east. It is traversed by numerous rivers and waterways and its average elevation above sea level is 13 feet. There are no nearby hilly areas and the land is low and level throughout the city. The climate is generally marine. The geographic location of the city with respect to the principal storm tracks, is especially favorable, being south of the average path of storms originating in the higher latitudes and north of the usual tracks of hurricanes and other tropical storms.

The winters are usually mild, while the autumn and spring seasons usually are delightful. Summers, though warm and long, frequently are tempered by cool periods, often associated with northeasterly winds off the Atlantic. Temperatures of 100 degrees or higher occur infrequently. Extreme cold waves seldom penetrate the area and temperatures of zero or below are almost nonexistent. Winters pass, on occasion, without a measurable amount of snowfall. Most of the snowfall in Norfolk is light and generally melts within 24 hours.

Based on the 1951-1980 period, the average first occurrence of 32 degrees Fahrenheit in the fall is November 17 and the average last occurrence in the spring is March 23.

NORMALS, MEANS, AND EXTREMES
NORFOLK, VA (ORF)

LATITUDE:	LONGITUDE:	ELEVATION (FT):	TIME ZONE:	WBAN: 13737
36 54' 13" N	76 11' 31" W	GRND: 66　　BARO: 69	EASTERN (UTC + 5)	

	ELEMENT	POR	JAN	FEB	MAR	APR	MAY	JUN	JUL	AUG	SEP	OCT	NOV	DEC	YEAR
TEMPERATURE F	NORMAL DAILY MAXIMUM	30	47.3	49.7	57.9	66.9	75.3	82.9	86.4	85.1	79.6	69.5	61.2	52.2	67.8
	MEAN DAILY MAXIMUM	54	48.7	50.9	58.1	68.0	75.9	83.5	87.2	85.5	80.1	70.1	61.3	52.3	68.5
	HIGHEST DAILY MAXIMUM	53	78	82	88	97	100	101	103	104	99	95	86	80	104
	YEAR OF OCCURRENCE		1970	1997	1990	1960	1991	1964	1993	1980	1983	1954	1974	1991	AUG 1980
	MEAN OF EXTREME MAXS.	54	70.4	72.4	79.8	86.6	90.9	95.3	96.8	95.4	91.8	84.6	78.4	72.2	84.6
	NORMAL DAILY MINIMUM	30	30.9	32.3	39.3	47.1	56.8	65.2	70.0	69.4	64.2	52.9	43.8	35.4	50.6
	MEAN DAILY MINIMUM	54	32.3	33.5	39.8	48.0	57.3	65.8	70.9	69.8	64.3	53.1	43.6	35.5	51.2
	LOWEST DAILY MINIMUM	53	-3	8	18	28	36	45	54	49	45	27	20	7	-3
	YEAR OF OCCURRENCE		1985	1965	1980	1982	1966	1967	1979	1982	1967	1976	1950	1983	JAN 1985
	MEAN OF EXTREME MINS.	54	17.0	19.9	26.9	35.1	44.4	54.4	61.6	60.3	52.1	38.5	29.4	21.0	38.4
	NORMAL DRY BULB	30	39.1	41.0	48.6	57.0	66.1	74.1	78.2	77.2	71.9	61.2	52.5	43.8	59.2
	MEAN DRY BULB	54	40.5	42.2	49.0	58.0	66.6	74.6	78.9	77.6	72.2	61.6	52.5	43.9	59.8
	MEAN WET BULB	18	37.5	39.3	44.1	51.8	60.3	68.4	68.8	71.7	62.8	57.0	48.7	40.7	54.3
	MEAN DEW POINT	18	31.0	32.6	37.2	45.4	55.1	64.5	65.7	68.7	59.5	52.3	43.1	34.6	49.1
	NORMAL NO. DAYS WITH:														
	MAXIMUM　90	30	0.0	0.0	0.0	0.4	1.5	5.9	10.9	8.6	2.8	0.1	0.0	0.0	30.2
	MAXIMUM　32	30	3.3	1.3	0.1	0.0	0.0	0.0	0.0	0.0	0.0	0.0	0.0	1.0	5.7
	MINIMUM　32	30	18.0	15.5	6.0	0.4	0.0	0.0	0.0	0.0	0.0	0.2	3.0	13.1	56.2
	MINIMUM　0	30	*	0.0	0.0	0.0	0.0	0.0	0.0	0.0	0.0	0.0	0.0	0.0	0.0
H/C	NORMAL HEATING DEG. DAYS	30	803	672	508	249	51	0	0	0	11	164	380	657	3495
	NORMAL COOLING DEG. DAYS	30	0	0	0	9	85	277	409	378	218	46	0	0	1422
RH	NORMAL (PERCENT)	30	66	66	65	63	69	71	73	75	74	72	68	67	69
	HOUR 01 LST	30	71	71	71	72	80	82	84	85	84	81	75	72	77
	HOUR 07 LST	30	73	73	74	72	77	79	82	84	83	82	78	74	78
	HOUR 13 LST	30	58	56	54	50	55	56	59	61	61	58	56	58	57
	HOUR 19 LST	30	65	64	62	59	65	66	70	73	74	73	68	66	67
S	PERCENT POSSIBLE SUNSHINE	32	53	56	60	63	62	67	62	62	61	58	56	54	60
W/O	MEAN NO. DAYS WITH:														
	HEAVY FOG(VISBY 1/4 MI)	53	2.1	2.5	2.0	1.5	1.8	1.0	0.5	1.0	1.2	2.1	1.9	2.1	19.7
	THUNDERSTORMS	53	0.4	0.6	1.9	2.7	5.0	5.6	8.1	6.5	2.7	1.3	0.5	0.4	35.7
CLOUDINESS	MEAN:														
	SUNRISE-SUNSET (OKTAS)	48	5.0	5.0	4.9	4.6	4.9	4.7	4.7	4.6	4.5	4.2	4.4	4.8	4.7
	MIDNIGHT-MIDNIGHT (OKTAS)	32	4.7	4.7	4.4	4.2	4.6	4.4	4.5	4.4	4.2	4.0	4.1	4.4	4.4
	MEAN NO. DAYS WITH:														
	CLEAR	48	8.8	8.1	8.9	8.8	7.7	7.5	7.4	7.9	9.1	11.6	10.3	9.4	105.5
	PARTLY CLOUDY	48	6.5	6.2	7.6	9.3	10.0	11.7	12.0	11.9	9.5	7.2	7.9	6.8	106.6
	CLOUDY	48	15.7	13.9	14.5	11.9	13.3	10.8	11.6	11.3	11.5	12.2	11.7	14.8	153.2
PR	MEAN STATION PRESSURE(IN)	29	30.10	30.09	30.00	30.00	29.99	29.99	30.01	30.00	30.00	30.10	30.09	30.09	30.04
	MEAN SEA-LEVEL PRES. (IN)	18	30.11	30.11	30.05	30.00	30.01	30.01	30.02	30.03	30.07	30.11	30.12	30.12	30.06
WINDS	MEAN SPEED (MPH)	49	11.3	11.8	12.2	11.8	10.3	9.6	9.0	8.7	9.7	10.1	10.4	10.8	10.5
	PREVAIL.DIR(TENS OF DEGS)	33	36	36	04	22	22	23	23	22	05	05	23	23	23
	MAXIMUM 2-MINUTE:														
	SPEED (MPH)	5	43	41	41	36	40	36	36	46	36	29	31	37	46
	DIR. (TENS OF DEGS)		02	07	22	06	07	28	03	11	06	01	20	26	11
	YEAR OF OCCURRENCE		2000	1998	1999	1998	2000	1998	1997	1998	1999	1999	1999	2000	AUG 1998
	MAXIMUM 5-SECOND:														
	SPEED (MPH)	5	54	53	49	44	52	58	51	67	46	40	43	47	67
	DIR. (TENS OF DEGS)		02	34	22	03	06	29	04	22	29	33	17	25	22
	YEAR OF OCCURRENCE		1998	1999	1999	2000	2000	1998	1997	2000	1999	1999	2000	2000	AUG 2000
PRECIPITATION	NORMAL (IN)	30	3.78	3.47	3.70	3.06	3.81	3.82	5.06	4.81	3.90	3.15	2.85	3.23	44.64
	MAXIMUM MONTHLY (IN)	53	9.93	8.21	10.36	7.25	10.12	9.72	14.37	14.32	13.80	10.12	7.01	6.10	14.37
	YEAR OF OCCURRENCE		1987	1998	1994	1984	1979	1963	1994	1992	1979	1971	1951	1983	JUL 1994
	MINIMUM MONTHLY (IN)	53	1.05	0.84	0.75	0.43	0.64	0.37	0.36	0.74	0.26	0.01	0.08	0.67	0.01
	YEAR OF OCCURRENCE		1981	1991	1986	1985	1991	1954	1993	1975	1986	2000	2001	1988	OCT 2000
	MAXIMUM IN 24 HOURS (IN)	53	3.80	4.78	4.02	5.90	3.41	6.85	5.64	11.40	6.79	7.29	3.35	2.76	11.40
	YEAR OF OCCURRENCE		1967	1998	1994	1991	1980	1963	1969	1964	1959	1999	1952	1983	AUG 1964
	NORMAL NO. DAYS WITH:														
	PRECIPITATION　0.01	30	10.7	10.3	10.4	9.8	9.9	9.7	11.1	10.1	7.7	7.4	7.7	9.5	114.3
	PRECIPITATION　1.00	30	0.8	0.7	0.7	0.8	0.9	1.1	1.5	1.3	1.2	0.8	0.5	0.6	10.9
SNOWFALL	NORMAL (IN)	30	3.1	3.9	1.2	0.1	0.0	0.0	0.0	0.0	0.0	0.0	0.*	0.5	8.8
	MAXIMUM MONTHLY (IN)	49	14.2	24.4	13.7	1.2	T	T	0.0	0.0	0.0	0.0	0.6	14.7	24.4
	YEAR OF OCCURRENCE		1966	1989	1980	1964	1994	1990		1991			1950	1958	FEB 1989
	MAXIMUM IN 24 HOURS (IN)	49	9.1	14.2	9.9	1.2	T	T	0.0	T	0.0	0.0	0.6	11.4	14.2
	YEAR OF OCCURRENCE		1973	1989	1980	1964	1994	1990		1991			1950	1958	FEB 1989
	MAXIMUM SNOW DEPTH (IN)	49	9	14	14	0	T	0	0	T	0	0	0	11	14
	YEAR OF OCCURRENCE		1966	1980	1980									1958	MAR 1980
	NORMAL NO. DAYS WITH:														
	SNOWFALL　1.0	30	0.9	0.8	0.2	0.*	0.0	0.0	0.0	0.0	0.0	0.0	0.0	0.2	2.1

PRECIPITATION (inches) 2001 NORFOLK, VA (ORF)

YEAR	JAN	FEB	MAR	APR	MAY	JUN	JUL	AUG	SEP	OCT	NOV	DEC	ANNUAL
1972	2.94	3.50	2.55	2.15	3.35	4.93	4.65	1.60	6.91	4.09	5.44	4.12	46.23
1973	2.54	3.21	4.69	3.44	3.62	5.93	4.19	7.92	0.86	1.37	1.90	5.83	45.50
1974	3.52	2.98	5.16	3.34	3.74	4.76	5.47	8.33	4.40	1.23	1.22	3.81	47.96
1975	4.18	4.18	5.72	4.19	3.37	1.16	13.73	0.74	3.19	1.63	3.62		50.53
1976	2.51	1.50	2.21	0.99	3.74	1.59	5.19	2.62	3.51	2.90	2.38	3.22	32.36
1977	3.33	2.23	4.05	2.20	3.86	2.41	2.70	4.57	3.00	6.09	5.41	3.92	43.77
1978	6.32	1.91	7.80	2.90	5.64	7.84	4.19	1.66	1.17	1.50	4.40	2.31	47.64
1979	6.47	5.01	5.13	7.00	10.12	2.97	4.69	1.79	13.80	1.74	5.26	0.98	64.96
1980	4.54	2.91	4.40	3.25	5.17	1.39	1.85	4.54	1.47	4.21	2.01	2.64	38.38
1981	1.05	2.26	1.88	2.26	2.75	5.00	5.10	6.87	3.18	3.28	1.78	5.77	41.18
1982	3.35	5.81	3.04	1.71	3.07	4.22	5.83	6.51	3.63	4.25	3.43	4.30	49.15
1983	2.21	6.23	4.55	6.13	3.52	3.84	0.77	3.07	4.52	5.29	3.24	6.10	49.47
1984	2.77	4.66	5.09	7.25	6.23	1.50	7.66	2.25	1.94	0.57	2.68	2.22	44.82
1985	3.98	3.53	2.02	0.43	3.23	6.81	6.14	1.89	6.36	3.92	5.71	0.79	44.81
1986	2.52	2.71	0.75	3.31	1.41	1.51	2.59	4.80	1.26	3.74			26.48
1987	9.93	3.11	2.30	3.83	2.65	2.98	3.20	2.04	7.00	1.81	3.51	2.33	44.69
1988	3.12	2.70	2.11	3.53	5.49	3.83	2.93	5.69	1.74	2.85	4.02	0.67	38.68
1989	2.70	5.80	8.50	3.62	2.97	5.10	4.86	7.49	5.10	2.94	3.69	3.86	56.63
1990	3.26	2.93	3.49	3.55	3.79	3.51	4.06	11.85	1.00	3.73	1.68	2.67	45.52
1991	4.74	0.84	4.70	6.39	0.64	4.54	6.46	3.77	2.04	4.65	1.72	2.43	42.92
1992	4.48	2.07	2.63	1.26	3.46	2.22	4.52	14.32	2.06	2.85	4.26	3.15	47.28
1993	4.89	2.36	5.91	3.59	2.88	2.79	0.36	1.45	4.14	3.40	0.97	3.29	36.03
1994	4.06	3.64	10.36	0.56	3.52	1.27	14.37	3.65	2.02	2.50	5.25	1.20	52.40
1995	2.40	2.80	2.95	2.76	2.42	4.54	1.78	1.50	5.05	4.82	2.94	1.86	35.82
1996	5.49	3.04	3.46	4.94	3.59	6.60	7.46	5.19	2.72	4.71	2.87	3.86	53.93
1997	2.09	2.94	3.21	3.01	1.66	1.10	7.85	1.76	1.97	4.31	4.94	2.64	37.48
1998	6.02	8.21	4.15	4.31	3.99	4.56	3.91	8.47	2.25	1.73	1.83	5.33	54.76
1999	3.51	2.33	3.29	3.66	3.85	3.60	6.17	4.47	13.16	8.24	1.40	1.71	55.39
2000	5.07	1.13	2.40	3.70	4.05	8.31	7.52	8.35	6.25	0.01	1.67	0.97	49.43
2001	1.46	2.16	4.73	1.48	2.89	6.91	4.40	4.21	2.46	0.75	0.08	1.83	33.36
POR= 131 YRS	3.40	3.36	3.80	3.24	3.66	4.03	5.55	5.31	3.85	3.12	2.59	3.11	45.02

AVERAGE TEMPERATURE (F) 2001   NORFOLK, VA (ORF)

YEAR	JAN	FEB	MAR	APR	MAY	JUN	JUL	AUG	SEP	OCT	NOV	DEC	ANNUAL
1972	46.4	43.2	49.0	56.4	63.5	70.5	77.6	75.8	71.9	59.2	51.6	49.2	59.5
1973	40.5	39.7	52.3	58.5	66.9	76.9	78.3	78.5	75.0	64.2	53.5	46.2	60.9
1974	48.6	43.4	53.1	60.8	66.8	72.8	78.3	77.4	71.4	58.7	53.5	46.0	60.9
1975	46.0	45.4	47.4	52.7	68.3	77.0	78.6	79.6	72.3	63.4	55.7	43.2	60.8
1976	38.9	49.9	53.4	61.9	66.3	75.9	78.2	75.9	71.1	57.7	45.9	41.4	59.7
1977	29.2	41.5	54.7	61.9	68.2	74.3	81.4	81.0	76.3	60.5	54.8	43.5	60.6
1978	37.0	32.6	46.1	57.2	65.6	74.1	76.1	80.5	73.2	60.8	56.0	45.3	58.7
1979	39.4	33.3	49.1	58.1	66.7	70.4	77.1	78.5	72.8	60.4	54.4	44.9	58.9
1980	40.3	34.7	46.5	58.6	67.8	73.9	80.9	80.9	76.1	60.4	49.9	42.3	59.4
1981	32.7	43.1	45.4	61.2	65.1	78.3	79.8	75.1	70.7	59.6	50.7	41.0	58.6
1982	35.4	42.0	48.8	55.0	69.4	73.4	78.6	75.3	70.0	60.2	54.4	48.8	59.3
1983	40.2	40.8	51.0	55.7	65.8	73.0	80.3	79.0	72.8	62.7	52.6	41.6	59.6
1984	35.5	46.7	45.4	55.6	67.8	76.2	76.7	78.4	70.5	66.9	49.9	50.9	60.0
1985	34.9	40.4	51.8	62.0	68.8	74.2	78.2	77.2	73.4	65.9	60.3	41.2	60.7
1986	39.3	42.1	49.9	57.3	67.6	76.1	77.1	76.6	72.4	65.4	54.9	44.8	60.7
1987	39.6	38.7	47.5	54.6	68.3	77.0	82.4	79.6	74.3	56.6	54.3	46.0	59.9
1988	37.3	42.5	49.5	56.5	65.8	73.6	80.1	80.8	70.5	56.9	54.2	42.4	59.2
1989	45.3	43.6	50.1	56.5	65.6	78.5	79.2	77.7	73.9	62.7	53.3	34.8	60.1
1990	47.3	50.2	53.2	58.7	66.6	75.5	80.6	78.0	71.6	65.9	55.0	50.5	62.8
1991	43.5	46.0	52.7	61.6	72.8	76.2	82.0	79.5	72.3	61.9	52.7	47.6	62.4
1992	42.7	45.0	48.8	58.2	62.4	72.0	81.8	75.6	73.5	59.7	53.8	44.6	59.8
1993	44.2	39.1	46.8	57.3	68.5	76.1	83.4	78.9	74.6	61.3	53.5	41.5	60.4
1994	36.5	43.4	51.7	64.7	64.9	79.6	83.0	77.6	72.4	62.0	57.5	50.1	62.0
1995	45.1	41.5	51.7	60.4	67.9	75.8	83.3	80.1	72.2	65.5	48.4	40.0	61.0
1996	38.7	41.3	45.0	58.8	65.2	75.2	77.7	74.9	72.4	61.8	46.5	46.7	58.7
1997	41.3	46.0	51.5	54.6	64.0	72.0	79.1	76.4	72.0	62.1	50.3	43.5	59.4
1998	46.5	47.2	50.8	58.9	67.5	75.9	79.0	78.9	75.3	61.9	52.6	48.7	61.9
1999	46.1	44.5	47.6	58.0	66.5	73.9	81.1	79.5	72.1	61.5	55.5	45.8	61.0
2000	40.2	45.1	53.0	57.5	70.3	76.1	76.3	76.6	71.5	62.1	49.3	38.2	59.7
2001	39.3	44.0	46.5	58.2	66.5	75.8	76.3	78.8	70.4	61.3	56.0	49.7	60.2
POR= 127 YRS	41.1	42.2	48.9	57.5	66.7	74.7	78.8	77.6	72.4	62.1	52.3	43.8	59.8

REFERENCE NOTES:

PAGE 1:
THE TEMPERATURE GRAPH SHOWS NORMAL MAXIMUM AND NORMAL
  MINIMUM DAILY TEMPERATURES (SOLID CURVES) AND THE
  ACTUAL DAILY HIGH AND LOW TEMPERATURES (VERTICAL BARS).

PAGE 2 AND 3:
  H/C INDICATES HEATING AND COOLING DEGREE DAYS.
  RH INDICATES RELATIVE HUMIDITY
  W/O INDICATES WEATHER AND OBSTRUCTIONS
  S INDICATES SUNSHINE.
  PR INDICATES PRESSURE.
  CLOUDINESS ON PAGE 3 IS THE SUM OF THE CEILOMETER AND
    SATELLITE DATA NOT TO EXCEED EIGHT EIGHTHS(OKTAS).

GENERAL:
  T INDICATES TRACE PRECIPITATION, AN AMOUNT GREATER
    THAN ZERO BUT LESS THAN THE LOWEST REPORTABLE VALUE.
  + INDICATES THE VALUE ALSO OCCURS ON EARLIER DATES.
  BLANK ENTRIES DENOTE MISSING OR UNREPORTED DATA.
  NORMALS ARE 30-YEAR AVERAGES (1961 - 1990).
  ASOS INDICATES AUTOMATED SURFACE OBSERVING SYSTEM.
  PM INDICATES THE LAST DAY OF THE PREVIOUS MONTH.
  POR (PERIOD OF RECORD) BEGINS WITH THE JANUARY DATA
    MONTH AND IS THE NUMBER OF YEARS USED TO COMPUTE
    THE MEAN.   INDIVIDUAL MONTHS WITHIN THE POR MAY
    BE MISSING.
  WHEN THE POR FOR A NORMAL IS LESS THAN 30 YEARS,
    THE NORMAL IS PROVISIONAL AND IS BASED ON THE NUMBER
    OF YEARS INDICATED.
  0.* OR * INDICATES THE VALUE OR MEAN-DAYS-WITH
    IS BETWEEN 0.00 AND 0.05.
  CLOUDINESS FOR ASOS STATIONS DIFFERS FROM THE NON-ASOS
    OBSERVATION TAKEN BY A HUMAN OBSERVER. ASOS STATION
    CLOUDINESS IS BASED ON TIME-AVERAGED CEILOMETER DATA
    FOR CLOUDS AT OR BELOW 12,000 FEET AND ON SATELLITE
    DATA FOR CLOUDS ABOVE 12,000 FEET.
  THE NUMBER OF DAYS WITH CLEAR, PARTLY CLOUDY, AND
    CLOUDY CONDITIONS FOR ASOS STATIONS IS THE SUM
    OF THE CEILOMETER AND SATELLITE DATA FOR THE
    SUNRISE TO SUNSET PERIOD.

GENERAL CONTINUED:
  CLEAR INDICATES 0 - 2 OKTAS, PARTLY CLOUDY INDICATES
    3 - 6 OKTAS, AND CLOUDY INDICATES 7 OR 8 OKTAS.
    WHEN AT LEAST ONE OF THE ELEMENTS (CEILOMETER OR
    SATELLITE) IS MISSING, THE DAILY CLOUDINESS IS
    NOT COMPUTED.
  WIND DIRECTION IS RECORDED IN TENS OF DEGREES (2 DIGITS)
    CLOCKWISE FROM TRUE NORTH. "00" INDICATES CALM. "36"
    INDICATES TRUE NORTH.
  RESULTANT WIND IS THE VECTOR AVERAGE OF THE SPEED AND
    DIRECTION.
  AVERAGE TEMPERATURE IS THE SUM OF THE MEAN DAILY MAXIMUM
    AND MINIMUM TEMPERATURE DIVIDED BY 2.
  SNOWFALL DATA COMPRISE ALL FORMS OF FROZEN
    PRECIPITATION, INCLUDING HAIL.
  A HEATING (COOLING) DEGREE DAY IS THE DIFFERENCE BETWEEN
    THE AVERAGE DAILY TEMPERATURE AND 65 F.
  DRY BULB IS THE TEMPERATURE OF THE AMBIENT AIR.
  DEW POINT IS THE TEMPERATURE TO WHICH THE AIR MUST BE
    COOLED TO ACHIEVE 100 PERCENT RELATIVE HUMIDITY.
  WET BULB IS THE TEMPERATURE THE AIR WOULD HAVE IF THE
    MOISTURE CONTENT WAS INCREASED TO 100 PERCENT RELATIVE
    HUMIDITY.

  ON JULY 1, 1996, THE NATIONAL WEATHER SERVICE BEGAN USING
    THE "METAR" OBSERVATION CODE THAT WAS ALREADY EMPLOYED
    BY MOST OTHER NATIONS OF THE WORLD.   THE MOST NOTICEABLE
    DIFFERENCE IN THIS ANNUAL PUBLICATION WILL BE THE CHANGE
    IN UNITS FROM TENTHS TO EIGHTS(OKTAS) FOR REPORTING THE
    AMOUNT OF SKY COVER.

HEATING DEGREE DAYS (base 65 F) 2001 NORFOLK, VA (ORF)

YEAR	JUL	AUG	SEP	OCT	NOV	DEC	JAN	FEB	MAR	APR	MAY	JUN	TOTAL
1972-73	0	0	4	197	406	486	752	703	403	217	47	0	3215
1973-74	0	0	0	83	353	575	504	599	377	183	63	0	2737
1974-75	0	0	16	213	371	584	584	547	541	382	47	0	3285
1975-76	0	0	6	98	290	671	804	443	362	186	62	6	2928
1976-77	0	0	0	245	566	726	1104	657	330	150	40	1	3819
1977-78	0	0	0	158	321	661	860	902	580	235	72	3	3792
1978-79	0	0	3	162	268	614	787	879	499	213	52	5	3482
1979-80	0	0	0	190	272	616	759	872	564	196	58	2	3529
1980-81	0	0	11	181	449	699	994	610	605	159	96	0	3804
1981-82	0	0	12	189	423	739	907	636	495	303	21	0	3725
1982-83	0	4	6	177	334	498	762	674	426	295	85	3	3264
1983-84	0	0	27	126	370	718	908	522	601	281	54	3	3610
1984-85	0	0	16	37	450	432	928	686	421	172	21	0	3163
1985-86	0	0	6	61	162	731	790	637	465	228	69	1	3150
1986-87	0	1	8	88	311	620	779	730	538	306	58	0	3439
1987-88	0	0	0	252	320	582	851	646	474	266	86	15	3492
1988-89	0	0	2	265	324	692	602	601	486	282	80	0	3334
1989-90	0	0	12	134	356	928	541	417	410	234	39	3	3074
1990-91	0	0	13	102	301	444	657	527	386	166	19	2	2617
1991-92	0	0	22	132	377	542	686	575	496	262	125	1	3218
1992-93	0	0	9	179	337	623	638	722	555	246	22	5	3336
1993-94	0	0	10	140	353	726	877	599	414	111	89	0	3319
1994-95	0	0	0	113	239	463	614	651	411	190	45	0	2726
1995-96	0	0	4	80	497	769	810	681	614	237	99	7	3798
1996-97	0	0	1	123	559	561	728	532	418	314	82	44	3362
1997-98	0	0	1	148	434	663	573	492	466	211	60	0	3048
1998-99	0	0	2	127	367	513	583	567	531	226	60	0	2976
1999-00	0	0	4	146	282	588	762	570	374	227	28	1	2982
2000-01	0	0	13	133	466	825	793	583	571	239	52	0	3675
2001-	0	0	19	163	275	468							

COOLING DEGREE DAYS (base 65 F) 2001 NORFOLK, VA (ORF)

YEAR	JAN	FEB	MAR	APR	MAY	JUN	JUL	AUG	SEP	OCT	NOV	DEC	ANNUAL
1972	0	0	8	20	40	183	398	343	217	22	10	2	1243
1973	0	0	16	27	112	363	420	424	307	64	17	1	1751
1974	3	0	16	64	124	244	419	390	213	26	32	0	1531
1975	2	3	0	22	157	366	429	460	233	55	17	0	1744
1976	1	13	11	102	110	337	417	347	193	27	0	0	1558
1977	0	4	16	66	145	289	515	502	347	24	22	0	1930
1978	0	0	0	9	96	286	352	487	257	36	3	9	1535
1979	0	0	11	13	112	171	385	426	239	54	22	0	1433
1980	0	0	0	11	153	274	499	497	351	45	1	1	1832
1981	0	0	0	51	103	407	468	320	189	29	0	0	1567
1982	0	0	1	8	166	257	428	331	164	39	21	4	1419
1983	0	0	0	21	115	250	481	440	265	62	4	0	1638
1984	0	0	0	5	146	345	368	426	188	102	5	2	1587
1985	0	5	20	91	146	284	419	382	267	97	28	0	1739
1986	0	0	2	2	153	343	537	367	237	109	15	0	1765
1987	0	0	0	2	168	364	544	461	285	0	7	0	1831
1988	0	0	1	18	118	280	477	498	173	17	10	0	1592
1989	0	9	30	31	106	412	447	399	286	69	10	0	1799
1990	0	8	52	51	98	324	489	407	218	137	8	5	1797
1991	0	0	13	71	269	347	534	456	248	43	18	11	2010
1992	0	0	0	63	51	220	527	336	270	23	9	0	1499
1993	0	0	0	23	136	345	579	438	305	31	18	0	1875
1994	0	0	11	111	94	447	562	400	229	30	20	8	1912
1995	2	0	4	59	143	330	574	475	226	102	6	0	1921
1996	0	0	1	55	116	320	401	312	227	31	10	0	1473
1997	0	7	7	7	59	262	443	360	218	63	0	0	1426
1998	7	0	33	37	143	338	440	440	317	39	0	12	1806
1999	4	0	0	25	112	275	509	458	222	42	4	0	1651
2000	0	0	7	10	196	342	358	368	216	51	0	0	1548
2001	0	0	3	40	108	331	357	435	187	57	15	3	1536

SNOWFALL (inches) 2001 NORFOLK, VA (ORF)

YEAR	JUL	AUG	SEP	OCT	NOV	DEC	JAN	FEB	MAR	APR	MAY	JUN	TOTAL
1972-73	0.0	0.0	0.0	0.0	T	0.0	9.1	4.7	T	0.0	0.0	0.0	13.8
1973-74	0.0	0.0	0.0	0.0	0.0	1.4	T	0.9	7.5	0.0	0.0	0.0	9.8
1974-75	0.0	0.0	0.0	0.0	0.0	T	0.3	T	0.8	T	0.0	0.0	1.1
1975-76	0.0	0.0	0.0	0.0	0.0	T	T	T	0.0	0.0	0.0	0.0	T
1976-77	0.0	0.0	0.0	0.0	T	1.0	4.7	1.4	0.0	0.0	0.0	0.0	7.1
1977-78	0.0	0.0	0.0	0.0	0.0	T	1.3	9.2	2.3	0.0	0.0	0.0	12.8
1978-79	0.0	0.0	0.0	0.0	0.0	0.0	0.0	1.0	12.7	T	0.0	0.0	13.7
1979-80	0.0	0.0	0.0	0.0	0.0	0.0	0.0	9.3	18.9	13.7	0.0	0.0	41.9
1980-81	0.0	0.0	0.0	0.0	0.0	0.0	T	T	0.0	0.3	0.0	0.0	0.3
1981-82	0.0	0.0	0.0	0.0	0.0	1.8	4.2	0.1	T	T	0.0	0.0	6.1
1982-83	0.0	0.0	0.0	0.0	0.0	0.0	0.4	T	3.0	T	T	0.0	3.4
1983-84	0.0	0.0	0.0	0.0	0.0	T	T	5.2	T	0.0	0.0	0.0	5.2
1984-85	0.0	0.0	0.0	0.0	0.0	T	4.3	0.0	T	0.0	0.0	0.0	4.3
1985-86	0.0	0.0	0.0	0.0	0.0	T	3.6	1.1	T	0.0	0.0	0.0	4.7
1986-87	0.0	0.0	0.0	0.0	0.0	T	1.6	1.0	1.2	T	0.0	0.0	3.8
1987-88	0.0	0.0	0.0	0.0	0.3	T	4.4	T	T	0.0	0.0	0.0	4.7
1988-89	0.0	0.0	0.0	0.0	0.0	T	T	24.4	T	0.5	0.0	0.0	24.9
1989-90	0.0	0.0	0.0	0.0	0.0	0.5	0.0	0.0	T	0.0	0.0	T	0.5
1990-91	0.0	0.0	0.0	0.0	0.0	T	0.0	T	T	T	0.0	0.0	T
1991-92	0.0	T	0.0	0.0	0.0	T	T	0.0	T	T	0.0	0.0	T
1992-93	0.0	0.0	0.0	0.0	0.0	0.0	0.3	1.9	1.7	0.0	0.0	0.0	3.9
1993-94	0.0	0.0	0.0	0.0	0.0	3.9	7.4	1.2	0.3	0.0	T	0.0	12.8
1994-95	0.0	0.0	0.0	0.0	0.0	0.0	T	0.3	T	0.0	0.0	0.0	0.3
1995-96	0.0	0.0	0.0	0.0	0.0	T	6.2	11.0					
1996-97													
1997-98													
1998-99								T					
1999-00							9.0						
2000-01									0.3	0.0	0.0	0.0	
2001-	0.0	0.0	0.0	0.0	0.0	0.0							
POR= 47 YRS	0.0	T	0.0	0.0	0.0	0.9	2.7	2.8	1.0	0.0	T	T	7.4

2001
RICHMOND,
VIRGINIA (RIC)

Richmond is located in east-central Virginia at the head of navigation on the James River and along a line separating the Coastal Plains (Tidewater Virginia) from the Piedmont. The Blue Ridge Mountains lie about 90 miles to the west and the Chesapeake Bay 60 miles to the east. Elevations range from a few feet above sea level along the river to a little over 300 feet in parts of the western section of the city.

The climate might be classified as modified continental. Summers are warm and humid and winters generally mild. The mountains to the west act as a partial barrier to outbreaks of cold, continental air in winter. The cold winter air is delayed long enough to be modified, then further warmed as it subsides in its approach to Richmond. The open waters of the Chesapeake Bay and Atlantic Ocean contribute to the humid summers and mild winters. The coldest weather normally occurs in late December and January, when low temperatures usually average in the upper 20s, and the high temperatures in the upper 40s. Temperatures seldom lower to zero, but there have been several occurrences of below zero temperatures. Summertime high temperatures above 100 degrees are not uncommon, but do not occur every year.

Precipitation is rather uniformly distributed throughout the year. However, dry periods lasting several weeks do occur, especially in autumn when long periods of pleasant, mild weather are most common. There is considerable variability in total monthly amounts from year to year. Snow usually remains on the ground only one or two days at a time. Ice storms (freezing rain or glaze) are not uncommon, but they are seldom severe enough to do any considerable damage. A notable exception was the spectacular glaze storm of January 27-28, 1943, when nearly 1 inch of ice accumulation caused heavy damage to trees and overhead transmission lines.

The James River reaches tidewater at Richmond where flooding may occur in every month of the year, most frequently in March and least in July. Hurricanes and tropical storms have been responsible for most of the flooding during the summer and early fall months. Hurricanes passing near Richmond have produced record rainfalls. In 1955, three hurricanes brought record rainfall to Richmond within a six-week period. The most noteworthy of these were Hurricanes Connie and Diane that brought heavy rains five days apart.

Damaging storms occur mainly from snow and freezing rain in winter and from hurricanes, tornadoes, and severe thunderstorms in other seasons. Damage may be from wind, flooding, or rain, or from any combination of these. Tornadoes are infrequent but some notable occurrences have been observed within the Richmond area.

Based on the 1951-1980 period, the average first occurrence of 32 degrees Fahrenheit in the fall is October 26 and the average last occurrence in the spring is April 10.

NORMALS, MEANS, AND EXTREMES
RICHMOND, VA (RIC)

LATITUDE:	LONGITUDE:	ELEVATION (FT):	TIME ZONE:	WBAN: 13740
37 30' 40" N	77 19' 24" W	GRND: 164 BARO: 167	EASTERN (UTC + 5)	

	ELEMENT	POR	JAN	FEB	MAR	APR	MAY	JUN	JUL	AUG	SEP	OCT	NOV	DEC	YEAR
TEMPERATURE °F	NORMAL DAILY MAXIMUM	30	45.7	49.2	59.5	70.0	77.8	85.1	88.4	87.1	80.9	70.7	61.3	50.2	68.8
	MEAN DAILY MAXIMUM	81	47.4	50.3	58.9	69.3	77.6	85.1	88.3	86.5	80.8	70.6	60.4	50.0	68.8
	HIGHEST DAILY MAXIMUM	72	80	83	93	96	100	104	105	102	103	99	86	81	105
	YEAR OF OCCURRENCE		1950	1932	1938	1990	1941	1952	1977	1983	1954	1941	1993	1998	JUL 1977
	MEAN OF EXTREME MAXS.	81	69.4	71.2	80.2	87.8	91.1	96.0	97.5	95.9	93.1	85.5	77.8	70.1	84.6
	NORMAL DAILY MINIMUM	30	25.7	28.1	36.3	44.6	54.2	62.7	67.5	66.4	59.0	46.5	37.9	29.9	46.6
	MEAN DAILY MINIMUM	81	28.4	30.0	36.6	45.4	54.7	63.5	68.0	66.6	60.1	47.8	38.4	30.9	47.5
	LOWEST DAILY MINIMUM	72	-12	-10	11	23	31	40	51	46	35	21	10	-1	-12
	YEAR OF OCCURRENCE		1940	1936	1960	1985	1956	1967	1965	1934	1974	1962	1933	1942	JAN 1940
	MEAN OF EXTREME MINS.	81	10.2	14.2	21.6	30.9	41.0	51.1	57.8	55.6	45.4	32.4	22.7	14.6	33.1
	NORMAL DRY BULB	30	35.7	38.7	48.0	57.3	66.0	73.9	78.0	76.8	70.0	58.6	49.6	40.1	57.7
	MEAN DRY BULB	81	38.0	40.1	47.8	57.4	66.2	74.3	78.2	76.6	70.5	59.2	49.5	40.4	58.2
	MEAN WET BULB	17	34.0	36.7	41.9	50.5	59.4	67.3	71.5	66.0	63.6	53.7	44.9	36.8	52.2
	MEAN DEW POINT	17	27.0	29.1	33.9	43.0	54.3	63.2	68.1	63.1	59.9	48.7	38.6	30.0	46.6
	NORMAL NO. DAYS WITH:														
	MAXIMUM 90	30	0.0	0.0	0.1	0.8	2.3	8.7	13.8	11.0	4.1	0.3	0.0	0.0	41.1
	MAXIMUM 32	30	4.3	1.7	0.1	0.0	0.0	0.0	0.0	0.0	0.0	0.0	0.0	1.5	7.6
	MINIMUM 32	30	23.0	19.5	10.8	2.3	0.1	0.0	0.0	0.0	0.0	2.1	9.4	19.2	86.4
	MINIMUM 0	30	0.3	0.1	0.0	0.0	0.0	0.0	0.0	0.0	0.0	0.0	0.0	0.0	0.4
H/C	NORMAL HEATING DEG. DAYS	30	908	736	527	241	61	0	0	0	23	233	462	772	3963
	NORMAL COOLING DEG. DAYS	30	0	0	0	10	92	270	403	366	173	34	0	0	1348
RH	NORMAL (PERCENT)	30	68	66	63	61	70	72	75	77	77	74	69	69	70
	HOUR 01 LST	30	75	73	72	73	83	86	88	90	90	86	79	76	81
	HOUR 07 LST	30	79	78	78	76	81	83	86	90	90	89	84	80	83
	HOUR 13 LST	30	56	53	49	45	52	54	57	58	57	52	51	55	53
	HOUR 19 LST	30	65	62	57	53	63	66	69	74	77	75	68	68	66
S	PERCENT POSSIBLE SUNSHINE	46	54	58	62	66	65	69	68	66	65	63	59	54	62
W/O	MEAN NO. DAYS WITH:														
	HEAVY FOG(VISBY 1/4 MI)	73	2.7	2.1	1.7	1.6	1.8	1.5	2.0	2.4	2.9	3.3	2.3	2.8	27.1
	THUNDERSTORMS	65	0.2	0.4	1.6	2.4	5.3	6.6	8.2	6.2	2.9	1.0	0.6	0.2	35.6
CLOUDINESS	MEAN:														
	SUNRISE-SUNSET (OKTAS)	1			5.6		5.6	3.2							
	MIDNIGHT-MIDNIGHT (OKTAS)	1			5.6										
	MEAN NO. DAYS WITH:														
	CLEAR	1	2.0	1.0	9.0		9.0	10.0							
	PARTLY CLOUDY	1		1.0	5.0		4.0	8.0							
	CLOUDY	1	7.0	4.0	10.0		8.0	1.0							
PR	MEAN STATION PRESSURE(IN)	28	29.91	29.89	29.85	29.81	29.81	29.81	29.83	29.86	29.89	29.92	29.92	29.93	29.87
	MEAN SEA-LEVEL PRES. (IN)	17	30.12	30.11	30.05	29.99	30.00	29.99	30.01	30.04	30.08	30.12	30.14	30.14	30.07
WINDS	MEAN SPEED (MPH)	43	8.3	8.7	9.3	9.2	7.9	7.5	7.1	6.6	7.0	7.2	7.7	7.9	7.9
	PREVAIL.DIR(TENS OF DEGS)	26	01	01	36	19	19	20	19	19	36	36	19	36	19
	MAXIMUM 2-MINUTE:														
	SPEED (MPH)	6	38	39	37	46	41	45	33	44	40	37	36	40	46
	DIR. (TENS OF DEGS)		31	23	22	33	30	26	29	36	01	10	16	15	33
	YEAR OF OCCURRENCE		2000	1997	1996	1999	1997	2000	1999	1996	1999	1996	1999	1996	APR 1999
	MAXIMUM 5-SECOND:														
	SPEED (MPH)	6	48	49	49	56	60	55	44	59	53	46	46	48	60
	DIR. (TENS OF DEGS)		31	26	25	28	28	27	29	32	08	10	16	16	28
	YEAR OF OCCURRENCE		2000	1997	1996	1998	1996	2000	1999	2000	1996	1996	1999	1996	MAY 1996
PRECIPITATION	NORMAL (IN)	30	3.24	3.16	3.61	2.96	3.84	3.62	5.03	4.40	3.34	3.53	3.17	3.26	43.16
	MAXIMUM MONTHLY (IN)	64	7.97	5.97	8.65	7.31	8.87	9.24	18.87	14.10	16.60	9.39	7.64	7.07	18.87
	YEAR OF OCCURRENCE		1978	1979	1984	1987	1972	1938	1945	1955	1999	1971	1959	1973	JUL 1945
	MINIMUM MONTHLY (IN)	64	0.64	0.48	0.94	0.64	0.87	0.38	0.51	0.52	0.26	0.01	0.17	0.40	0.01
	YEAR OF OCCURRENCE		1981	1978	1966	1963	1965	1980	1983	1943	1978	2000	2001	1980	OCT 2000
	MAXIMUM IN 24 HOURS (IN)	64	3.31	2.67	3.43	2.97	3.08	4.61	5.73	8.79	6.52	6.50	4.07	3.16	8.79
	YEAR OF OCCURRENCE		1962	1979	1992	1987	1981	1963	1969	1955	1999	1961	1956	1958	AUG 1955
	NORMAL NO. DAYS WITH:														
	PRECIPITATION 0.01	30	10.4	9.4	10.2	9.0	10.7	9.6	10.4	9.5	7.6	7.0	8.0	9.1	110.9
	PRECIPITATION 1.00	30	0.8	0.7	0.8	0.6	1.0	0.9	1.4	1.3	1.0	1.2	0.8	0.7	11.2
SNOWFALL	NORMAL (IN)	30	5.8	5.5	2.4	0.1	0.0	0.0	0.0	0.0	0.0	T	0.3	2.2	16.3
	MAXIMUM MONTHLY (IN)	62	28.5	21.4	19.7	2.0	T	0.0	0.0	0.0	0.0	T	7.3	12.5	28.5
	YEAR OF OCCURRENCE		1940	1983	1960	1940	1994					1979	1953	1958	JAN 1940
	MAXIMUM IN 24 HOURS (IN)	62	21.6	16.8	12.1	2.0	T	0.0	0.0	0.0	0.0	T	7.3	7.5	21.6
	YEAR OF OCCURRENCE		1940	1983	1962	1940	1994					1979	1953	1966	JAN 1940
	MAXIMUM SNOW DEPTH (IN)	77	18	20	13	1	0	0	0	0	0	0	6	9	20
	YEAR OF OCCURRENCE		1922	1922	1980	1964							1938	1958	FEB 1922
	NORMAL NO. DAYS WITH:														
	SNOWFALL 1.0	30	1.5	1.4	0.6	0.*	0.0	0.0	0.0	0.0	0.0	0.0	0.1	0.7	4.3

```
PRECIPITATION (inches)   2001   RICHMOND, VA (RIC)
```

YEAR	JAN	FEB	MAR	APR	MAY	JUN	JUL	AUG	SEP	OCT	NOV	DEC	ANNUAL
1972	1.43	5.15	2.11	3.35	8.87	8.82	5.80	3.84	3.35	7.89	5.82	2.91	59.34
1973	2.66	3.11	3.44	4.58	3.56	2.45	3.64	4.34	1.82	2.56	1.27	7.07	40.50
1974	3.21	2.54	3.79	1.58	3.02	1.80	6.84	4.83	0.39	1.23	4.22		35.70
1975	5.71	2.96	8.04	2.78	2.59	4.00	12.29	2.31	10.98	3.10	2.04	4.51	61.31
1976	3.39	1.35	2.14	1.08	3.76	2.85	2.63	1.35	4.98	6.99	1.88	2.56	34.76
1977	2.22	1.34	2.67	2.33	3.99	1.25	4.20	6.15	2.16	7.88	4.32	5.57	44.08
1978	7.97	0.48	5.67	4.31	3.92	5.26	4.24	5.93	0.26	1.21	4.57	3.80	47.62
1979	6.16	5.97	2.59	3.97	3.80	2.42	4.36	7.08	9.76	3.87	5.50	1.64	57.12
1980	6.05	1.01	5.49	4.28	4.68	0.38	5.18	2.15	2.37	6.96	2.18	0.40	41.13
1981	0.64	2.76	1.52	2.96	6.62	3.69	4.01	2.89	2.70	2.36	0.68	5.04	35.87
1982	2.76	4.44	3.74	2.97	3.48	3.97	9.21	4.39	2.55	2.90	2.70	3.37	46.48
1983	1.59	3.95	6.04	5.21	2.50	5.46	0.51	0.97	3.05	4.02	5.63	4.50	43.43
1984	3.98	3.97	8.65	5.92	4.52	2.01	3.55	4.58	1.86	2.14	3.34	1.52	46.04
1985	3.54	3.20	1.80	0.65	2.36	4.01	5.31	10.58	4.97	5.09	6.99	0.58	49.08
1986	2.69	2.67	1.16	1.16	3.15	1.30	7.01	6.75	0.63	2.43	2.46	5.15	36.56
1987	5.53	2.57	1.65	7.31	2.94	6.29	1.20	1.11	4.43	1.25	2.86		40.27
1988	2.53	3.08	1.98	2.55	4.81	2.25	7.50	2.95	1.74	2.74	4.34	0.79	37.26
1989	1.88	4.34	5.00	4.27	5.02	5.85	4.00	4.89	5.33	3.54	3.00	2.62	49.74
1990	2.84	2.38	2.54	2.81	6.85	0.97	6.74	5.76	1.92	3.90	1.70	3.52	41.93
1991	3.62	1.09	5.87	0.87	0.91	6.24	3.47	3.32	2.69	2.50	0.67	4.53	35.78
1992	1.57	2.89	5.87	2.21	4.95	2.28	5.68	6.40	2.35	1.94	2.62	2.79	41.55
1993	4.48	2.88	7.24	3.23	4.66	1.75	1.91	3.89	2.97	2.23	3.24	3.77	42.25
1994	3.10	4.38	7.92	2.70	2.49	1.73	7.46	2.54	3.99	2.56	3.99	0.95	43.81
1995	3.15	1.14	3.00	1.98	4.33	1.85	2.89	2.59	3.82	5.11	2.87	1.71	34.44
1996	4.65	2.97	2.71	2.88	3.18	4.35	6.51	4.40	6.87	7.18	3.52	4.91	54.13
1997	1.93	3.71	2.96	3.94	1.36	2.21	4.85	1.41	0.82	3.25	5.33	2.36	34.13
1998	6.85	5.76	6.72	4.32	3.72	4.41	2.37	1.89	3.94	0.47	1.30	5.00	46.75
1999	4.70	1.47	4.04	2.60	2.75	6.29	2.77	2.00	16.60	2.25	1.01	1.72	48.20
2000	3.96	1.60	3.67	4.78	3.03	6.07	4.05	8.28	3.60	0.01	1.72	2.38	43.15
2001	2.06	2.55	3.77	2.14	2.03	6.53	2.73	5.08	2.14	0.65	0.17	1.67	31.52
POR= 64 YRS	3.22	2.99	3.75	2.99	3.66	3.71	5.18	4.62	3.71	3.21	3.03	3.10	43.17

```
AVERAGE TEMPERATURE ( F) 2001    RICHMOND, VA (RIC)
```

YEAR	JAN	FEB	MAR	APR	MAY	JUN	JUL	AUG	SEP	OCT	NOV	DEC	ANNUAL
1972	40.7	37.6	47.2	56.2	64.6	70.1	77.1	75.2	70.1	55.8	45.9		57.4
1973	37.6	38.5	52.6	57.9	65.1	76.0	77.4	77.5	72.3	60.6	51.3	40.8	59.0
1974	45.8	40.1	50.4	59.9	65.8	70.6	76.9	75.7	67.4	55.4	48.5	41.7	58.2
1975	40.7	41.4	45.3	52.9	67.7	73.6	76.0	78.8	69.3	62.5	53.6	40.0	58.5
1976	35.1	48.5	52.6	60.5	65.2	74.6	77.3	75.7	68.7	54.4	42.7	36.7	57.7
1977	25.3	40.5	53.7	61.1	68.2	73.0	81.4	79.8	74.2	57.3	52.3	39.5	58.9
1978	33.4	30.3	44.5	57.3	65.5	74.7	77.5	80.1	72.9	58.3	52.5	42.5	57.5
1979	36.4	28.6	51.1	58.4	67.1	70.8	76.9	77.8	71.0	58.3	53.3	42.3	57.7
1980	38.8	36.0	47.4	61.1	68.3	72.8	80.0	80.7	74.7	56.9	46.2	38.6	58.5
1981	31.2	42.2	44.6	60.6	64.1	77.9	79.6	75.1	69.4	56.4	49.1	38.0	57.4
1982	31.6	41.7	49.1	55.9	70.4	73.4	78.6	75.0	69.8	59.2	51.9	46.1	58.6
1983	37.8	39.1	50.4	56.1	66.1	75.6	79.4	77.7	68.8	58.1	49.0	36.2	57.9
1984	32.6	44.5	43.6	55.8	65.4	77.7	76.0	77.0	67.5	66.1	46.6	47.7	58.4
1985	32.6	40.2	49.7	62.0	68.0	74.3	79.0	77.5	70.8	62.6	56.6	37.8	59.3
1986	36.2	39.3	50.0	59.2	66.9	76.1	80.9	74.2	70.8	61.8	49.1	40.9	58.8
1987	34.7	37.0	47.1	54.3	67.3	75.8	81.3	78.5	72.3	52.9	51.3	43.0	58.0
1988	32.3	39.1	47.9	56.0	65.8	72.8	79.9	79.8	68.4	53.6	50.6	39.2	57.1
1989	42.3	39.6	47.9	55.8	64.1	76.1	77.7	75.3	70.8	60.2	49.3	31.3	57.5
1990	46.3	48.0	52.1	57.9	65.8	75.0	79.9	76.3	69.5	63.1	52.6	46.3	61.1
1991	40.2	44.0	50.8	60.2	72.1	75.3	80.7	78.6	71.6	59.9	49.6	44.3	60.6
1992	40.7	42.7	47.2	57.8	61.7	70.8	79.5	73.7	70.3	56.3	50.6	40.9	57.7
1993	41.4	37.8	45.9	55.9	67.7	75.0	82.5	78.5	72.5	58.6	50.9	39.0	58.8
1994	33.4	40.2	48.9	63.2	63.2	78.0	81.1	75.4	69.2	57.6	52.6	45.6	59.0
1995	40.7	38.7	50.2	58.1	65.8	74.1	80.8	78.9	69.1	61.1	44.2	35.6	58.1
1996	34.1	37.8	43.2	57.9	64.8	75.1	76.4	74.3	70.1	58.9	43.2	43.8	56.6
1997	37.7	43.7	49.6	53.6	62.9	71.4	77.8	75.8	70.3	58.9	46.5	39.9	57.3
1998	43.2	44.2	48.5	58.0	67.4	74.0	78.3	78.3	74.2	59.7	49.5	44.2	60.0
1999	42.0	42.2	45.2	57.7	65.5	72.3	79.6	77.9	68.5	57.2	52.9	42.4	58.6
2000	36.3	43.4	52.0	56.4	68.0	75.4	74.7	74.9	67.5	59.7	46.1	33.0	57.3
2001	37.1	42.3	45.2	58.2	65.2	75.0	74.9	78.2	67.8	59.1	54.6	46.0	58.6
POR= 72 YRS	37.6	39.8	47.4	57.2	66.2	74.2	78.1	76.4	70.2	58.8	49.2	40.1	57.9

REFERENCE NOTES:

PAGE 1:
THE TEMPERATURE GRAPH SHOWS NORMAL MAXIMUM AND NORMAL MINIMUM DAILY TEMPERATURES (SOLID CURVES) AND THE ACTUAL DAILY HIGH AND LOW TEMPERATURES (VERTICAL BARS).

PAGE 2 AND 3:
H/C INDICATES HEATING AND COOLING DEGREE DAYS.
RH INDICATES RELATIVE HUMIDITY
W/O INDICATES WEATHER AND OBSTRUCTIONS
S INDICATES SUNSHINE.
PR INDICATES PRESSURE.
CLOUDINESS ON PAGE 3 IS THE SUM OF THE CEILOMETER AND SATELLITE DATA NOT TO EXCEED EIGHT EIGHTHS(OKTAS).

GENERAL:
T INDICATES TRACE PRECIPITATION, AN AMOUNT GREATER THAN ZERO BUT LESS THAN THE LOWEST REPORTABLE VALUE.
+ INDICATES THE VALUE ALSO OCCURS ON EARLIER DATES.
BLANK ENTRIES DENOTE MISSING OR UNREPORTED DATA.
NORMALS ARE 30-YEAR AVERAGES (1961 - 1990).
ASOS INDICATES AUTOMATED SURFACE OBSERVING SYSTEM.
PM INDICATES THE LAST DAY OF THE PREVIOUS MONTH.
POR (PERIOD OF RECORD) BEGINS WITH THE JANUARY DATA MONTH AND IS THE NUMBER OF YEARS USED TO COMPUTE THE MEAN. INDIVIDUAL MONTHS WITHIN THE POR MAY BE MISSING.
WHEN THE POR FOR A NORMAL IS LESS THAN 30 YEARS, THE NORMAL IS PROVISIONAL AND IS BASED ON THE NUMBER OF YEARS INDICATED.
0.* OR * INDICATES THE VALUE OR MEAN-DAYS-WITH IS BETWEEN 0.00 AND 0.05.
CLOUDINESS FOR ASOS STATIONS DIFFERS FROM THE NON-ASOS OBSERVATION TAKEN BY A HUMAN OBSERVER. ASOS STATION CLOUDINESS IS BASED ON TIME-AVERAGED CEILOMETER DATA FOR CLOUDS AT OR BELOW 12,000 FEET AND ON SATELLITE DATA FOR CLOUDS ABOVE 12,000 FEET.
THE NUMBER OF DAYS WITH CLEAR, PARTLY CLOUDY, AND CLOUDY CONDITIONS FOR ASOS STATIONS IS THE SUM OF THE CEILOMETER AND SATELLITE DATA FOR THE SUNRISE TO SUNSET PERIOD.

GENERAL CONTINUED:
CLEAR INDICATES 0 - 2 OKTAS, PARTLY CLOUDY INDICATES 3 - 6 OKTAS, AND CLOUDY INDICATES 7 OR 8 OKTAS. WHEN AT LEAST ONE OF THE ELEMENTS (CEILOMETER OR SATELLITE) IS MISSING, THE DAILY CLOUDINESS IS NOT COMPUTED.
WIND DIRECTION IS RECORDED IN TENS OF DEGREES (2 DIGITS) CLOCKWISE FROM TRUE NORTH. "00" INDICATES CALM. "36" INDICATES TRUE NORTH.
RESULTANT WIND IS THE VECTOR AVERAGE OF THE SPEED AND DIRECTION.
AVERAGE TEMPERATURE IS THE SUM OF THE MEAN DAILY MAXIMUM AND MINIMUM TEMPERATURE DIVIDED BY 2.
SNOWFALL DATA COMPRISE ALL FORMS OF FROZEN PRECIPITATION, INCLUDING HAIL.
A HEATING (COOLING) DEGREE DAY IS THE DIFFERENCE BETWEEN THE AVERAGE DAILY TEMPERATURE AND 65 F.
DRY BULB IS THE TEMPERATURE OF THE AMBIENT AIR.
DEW POINT IS THE TEMPERATURE TO WHICH THE AIR MUST BE COOLED TO ACHIEVE 100 PERCENT RELATIVE HUMIDITY.
WET BULB IS THE TEMPERATURE THE AIR WOULD HAVE IF THE MOISTURE CONTENT WAS INCREASED TO 100 PERCENT RELATIVE HUMIDITY.

ON JULY 1, 1996, THE NATIONAL WEATHER SERVICE BEGAN USING THE "METAR" OBSERVATION CODE THAT WAS ALREADY EMPLOYED BY MOST OTHER NATIONS OF THE WORLD. THE MOST NOTICEABLE DIFFERENCE IN THIS ANNUAL PUBLICATION WILL BE THE CHANGE IN UNITS FROM TENTHS TO EIGHTS(OKTAS) FOR REPORTING THE AMOUNT OF SKY COVER.

HEATING DEGREE DAYS (base 65 F) 2001 RICHMOND, VA (RIC)

YEAR	JUL	AUG	SEP	OCT	NOV	DEC	JAN	FEB	MAR	APR	MAY	JUN	TOTAL
1972-73	0	0	17	285	513	588	843	735	394	247	79	0	3701
1973-74	0	0	5	163	414	744	589	691	455	204	75	5	3345
1974-75	0	0	62	310	513	715	746	654	604	368	44	1	4017
1975-76	0	0	27	121	356	770	917	480	386	227	78	11	3373
1976-77	0	1	15	332	660	869	1227	680	366	176	42	7	4375
1977-78	0	0	4	259	401	784	974	964	627	235	88	5	4341
1978-79	0	0	16	214	366	694	876	1011	439	218	44	4	3882
1979-80	0	0	8	242	353	698	806	835	541	135	47	2	3667
1980-81	0	0	14	267	557	813	1042	633	626	171	107	0	4230
1981-82	0	1	29	273	473	834	1029	645	486	280	6	1	4057
1982-83	0	6	10	213	399	585	836	718	445	282	69	2	3565
1983-84	0	1	86	236	475	887	994	589	657	282	93	3	4303
1984-85	0	0	73	57	546	531	997	692	484	177	35	5	3597
1985-86	0	0	31	114	257	838	886	713	465	187	78	3	3572
1986-87	0	16	24	172	476	741	931	777	550	317	57	0	4061
1987-88	0	0	5	370	409	677	1008	746	527	279	79	32	4132
1988-89	0	0	27	361	425	794	696	709	546	293	108	0	3959
1989-90	0	3	38	181	468	1036	574	472	436	258	50	3	3519
1990-91	0	0	33	146	365	574	762	582	443	193	27	1	3126
1991-92	0	0	25	190	458	637	749	640	545	261	142	10	3657
1992-93	0	0	37	275	427	739	724	757	585	280	24	0	3848
1993-94	0	0	26	213	436	799	973	687	497	130	120	0	3881
1994-95	0	0	11	232	365	595	750	731	453	236	72	1	3446
1995-96	0	0	28	166	620	905	952	782	669	247	122	8	4499
1996-97	0	0	10	193	652	654	840	589	472	337	117	45	3909
1997-98	0	0	19	247	547	771	667	575	536	232	50	7	3651
1998-99	0	0	10	173	457	652	710	631	606	237	60	7	3543
1999-00	0	0	29	249	362	693	882	620	404	257	45	2	3543
2000-01	0	0	61	190	561	983	858	631	605	244	67	0	4200
2001-	0	0	57	213	312	582							

COOLING DEGREE DAYS (base 65 F) 2001 RICHMOND, VA (RIC)

YEAR	JAN	FEB	MAR	APR	MAY	JUN	JUL	AUG	SEP	OCT	NOV	DEC	ANNUAL
1972	0	0	7	30	52	180	381	326	178	9	8	0	1171
1973	0	0	13	42	91	338	391	395	231	32	9	2	1544
1974	0	0	10	58	106	180	377	340	141	21	26	0	1259
1975	0	0	0	16	135	267	348	433	165	51	18	0	1433
1976	0	8	9	99	91	307	389	337	133	12	0	0	1385
1977	0	0	22	66	148	258	513	467	289	24	27	0	1814
1978	0	0	0	12	112	302	393	475	263	15	0	1	1573
1979	0	0	16	30	117	188	374	404	195	42	9	0	1375
1980	0	0	1	25	157	243	472	494	313	23	1	0	1729
1981	0	0	1	45	89	395	458	319	169	16	0	0	1492
1982	0	0	0	13	181	259	428	323	157	43	13	7	1424
1983	0	0	0	23	108	325	452	405	207	27	0	0	1547
1984	0	0	0	10	114	392	346	381	154	100	0	2	1499
1985	0	4	20	94	139	290	441	392	213	51	10	0	1654
1986	0	0	8	19	142	344	498	308	205	79	6	0	1609
1987	0	0	0	2	136	329	513	427	227	0	3	0	1637
1988	0	0	3	16	108	269	466	465	137	12	0	0	1476
1989	0	3	24	27	88	341	403	331	218	42	4	0	1481
1990	0	1	43	51	81	312	470	356	177	96	0	2	1589
1991	0	0	10	53	254	316	494	428	227	40	3	0	1825
1992	0	0	0	51	48	192	457	275	202	12	3	0	1240
1993	0	0	0	14	112	307	547	426	261	20	18	0	1705
1994	0	0	5	80	70	396	507	332	142	9	2	2	1545
1995	1	0	0	35	105	281	495	437	157	52	3	0	1566
1996	0	0	0	40	123	315	360	293	173	12	0	0	1316
1997	0	0	4	2	59	244	406	341	182	65	0	0	1303
1998	2	0	32	26	131	282	420	418	292	15	0	12	1630
1999	1	0	0	24	82	231	460	409	140	13	2	0	1362
2000	0	0	5	6	143	320	305	313	141	33	0	0	1266
2001	0	0	0	45	79	307	314	414	149	38	7	0	1353

SNOWFALL (inches) 2001 RICHMOND, VA (RIC)

YEAR	JUL	AUG	SEP	OCT	NOV	DEC	JAN	FEB	MAR	APR	MAY	JUN	TOTAL
1972-73	0.0	0.0	0.0	T	0.6	0.0	4.3	0.4	1.4	0.0	0.0	0.0	6.7
1973-74	0.0	0.0	0.0	0.0	0.0	9.9	T	5.0	T	0.0	0.0	0.0	14.9
1974-75	0.0	0.0	0.0	0.0	0.0	T	2.7	2.9	0.4	0.0	0.0	0.0	6.0
1975-76	0.0	0.0	0.0	0.0	0.0	0.2	T	1.0	0.0	0.0	0.0	0.0	1.2
1976-77	0.0	0.0	0.0	0.0	1.0	1.7	11.1	T	0.0	0.0	0.0	0.0	13.8
1977-78	0.0	0.0	0.0	0.0	T	T	1.3	5.1	5.0	0.0	0.0	0.0	11.4
1978-79	0.0	0.0	0.0	0.0	0.0	0.0	0.7	19.5	T	0.0	0.0	0.0	20.2
1979-80	0.0	0.0	0.0	T	0.0	0.0	16.6	7.0	15.0	0.2	0.0	0.0	38.6
1980-81	0.0	0.0	0.0	0.0	0.0	0.2	0.6	0.0	0.2	0.0	0.0	0.0	1.0
1981-82	0.0	0.0	0.0	0.0	0.0	T	1.9	8.3	10.8	T	0.2	0.0	21.2
1982-83	0.0	0.0	0.0	0.0	0.0	7.9	0.1	21.4	0.0	T	0.0	0.0	29.4
1983-84	0.0	0.0	0.0	0.0	0.1	T	1.1	2.8	0.3	0.0	0.0	0.0	4.3
1984-85	0.0	0.0	0.0	0.0	0.0	0.0	8.3	T	T	0.0	0.0	0.0	8.3
1985-86	0.0	0.0	0.0	0.0	0.0	1.3	3.3	4.4	T	T	0.0	0.0	9.0
1986-87	0.0	0.0	0.0	0.0	0.0	0.0	15.8	5.3	0.7	T	0.0	0.0	21.8
1987-88	0.0	0.0	0.0	0.0	4.5	T	8.1	T	T	T	0.0	0.0	12.6
1988-89	0.0	0.0	0.0	0.0	0.0	1.8	T	13.6	T	T	T	0.0	15.4
1989-90	0.0	0.0	0.0	0.0	1.1	9.9	0.0	T	T	0.2	0.0	0.0	11.2
1990-91	0.0	0.0	0.0	0.0	0.0	T	T	0.0	1.9	T	0.0	0.0	1.9
1991-92	0.0	0.0	0.0	0.0	0.0	0.1	T	0.0	0.8	T	T	0.0	0.9
1992-93	0.0	0.0	0.0	0.0	0.0	0.0	0.5	5.3	3.5	T	0.0	0.0	9.3
1993-94	0.0	0.0	0.0	0.0	0.0	5.1	2.2	1.7	0.7	0.0	T	0.0	9.7
1994-95	0.0	0.0	0.0	0.0	0.0	0.0	T	3.1	T	0.8	0.0	0.0	3.9
1995-96	0.0	0.0	0.0	0.0	T	12.3							
1996-97					T	1.0	0.7	1.0	T	T	0.0	0.0	
1997-98	0.0	0.0	0.0	0.0									
1998-99													
1999-00							15.2	0.2					
2000-01							0.3	2.5	0.5	0.0	0.0	0.0	
2001-	0.0	0.0	0.0	0.0	0.0	0.0							
POR= 60 YRS	0.0	0.0	0.0	T	0.4	2.0	4.6	3.9	2.4	0.1	T	0.0	13.4

2001
ROANOKE,
VIRGINIA (ROA)

The climate of Roanoke is relatively mild. Roanoke is nestled among mountains which interrupt the Great Valley, extending from northernmost Virginia southwestward into east Tennessee. This location, at a point where the valley is pinched between the Blue Ridges and the Alleghenies, offers a natural barrier to the winter cold as it moves southward. It is also far enough inland that hurricanes lose much of their destructive force before reaching Roanoke. Finally, the rough terrain is an inhospitable breeding ground for tornadic activity. The elevation in the vicinity usually produces cool summer nights that make a light cover comfortable for sleeping. Although past records show extremes over 100 degrees and below zero, many years pass without either extreme being threatened.

Roanoke is located near the headwaters of the Roanoke River, which flows in a general southeasterly direction. Numerous creeks and small streams from nearby mountainous areas empty into the Roanoke River. The usual low water stage is 1 to 1.5 feet, and flood stage is 10 feet. Some low-lying streets in Roanoke and nearby Salem have to be blocked off during 7 to 8 foot stages, but damage is minor until the river overflows its banks. The highest stage on record exceeds 19 feet. Damage has been widespread on occasion and has amounted to several million dollars in the city of Roanoke alone.

The growing season averages 190 days. The average date of the last freezing temperature in spring is mid-April and the average date of the first freezing date in the fall is late October.

Rainfall is well apportioned throughout the year. Droughts are so infrequent that quoting actual records would be difficult. Snow usually falls each winter, ranging from only a trace to more than 60 inches.

NORMALS, MEANS, AND EXTREMES
ROANOKE, VA (ROA)

LATITUDE:	LONGITUDE:	ELEVATION (FT):		TIME ZONE:	WBAN: 13741
37 19' 01" N	79 58' 27" W	GRND: 1189	BARO: 1192	EASTERN (UTC + 5)	

	ELEMENT	POR	JAN	FEB	MAR	APR	MAY	JUN	JUL	AUG	SEP	OCT	NOV	DEC	YEAR
TEMPERATURE °F	NORMAL DAILY MAXIMUM	30	43.8	47.3	57.8	67.3	75.7	82.9	86.4	85.3	78.5	68.1	58.0	47.6	66.6
	MEAN DAILY MAXIMUM	54	45.1	48.6	56.8	67.7	76.1	83.4	87.0	85.4	78.7	68.8	57.5	47.7	66.9
	HIGHEST DAILY MAXIMUM	54	78	80	87	95	96	100	104	105	101	93	83	80	105
	YEAR OF OCCURRENCE		1952	1985	1986	1957	1962	1959	1954	1983	1954	1951	1950	1998	AUG 1983
	MEAN OF EXTREME MAXS.	54	65.8	69.6	77.8	85.5	89.3	93.4	95.8	94.5	91.3	83.3	75.0	67.2	82.4
	NORMAL DAILY MINIMUM	30	25.0	27.2	35.7	43.8	52.5	60.2	64.8	63.8	56.8	44.8	37.0	28.9	45.0
	MEAN DAILY MINIMUM	54	26.9	28.8	35.6	44.6	53.0	60.9	65.5	64.1	57.0	45.6	36.8	29.4	45.7
	LOWEST DAILY MINIMUM	54	-11	-1	9	20	31	39	47	42	34	22	9	-4	-11
	YEAR OF OCCURRENCE		1985	1996	1996	1985	1966	1977	1988	1986	1983	1976	1950	1983	JAN 1985
	MEAN OF EXTREME MINS.	54	8.8	12.4	19.7	28.7	38.5	47.8	54.9	53.5	41.6	30.1	20.7	12.2	30.7
	NORMAL DRY BULB	30	34.5	37.3	46.8	55.6	64.1	71.5	75.6	74.6	67.7	56.5	47.5	38.3	55.8
	MEAN DRY BULB	54	36.0	38.8	46.2	56.2	64.7	72.0	76.3	74.8	67.9	57.1	47.0	38.6	56.3
	MEAN WET BULB	17	31.6	34.6	40.1	48.4	57.4	65.4	68.9	67.9	58.2	51.2	42.2	31.9	49.8
	MEAN DEW POINT	17	23.1	25.4	30.5	39.9	51.9	61.3	65.0	64.4	54.4	45.1	35.0	24.4	43.4
	NORMAL NO. DAYS WITH:														
	MAXIMUM 90	30	0.0	0.0	0.0	0.4	0.8	5.6	10.2	7.8	2.5	0.0	0.0	0.0	27.3
	MAXIMUM 32	30	5.1	2.7	0.1	0.0	0.0	0.0	0.0	0.0	0.0	0.0	0.1	2.9	10.9
	MINIMUM 32	30	23.3	20.0	12.2	2.6	0.1	0.0	0.0	0.0	0.0	2.8	9.8	20.0	90.8
	MINIMUM 0	30	0.4	0.0	0.0	0.0	0.0	0.0	0.0	0.0	0.0	0.0	0.0	0.2	0.6
H/C	NORMAL HEATING DEG. DAYS	30	946	776	564	289	105	5	0	0	39	283	525	828	4360
	NORMAL COOLING DEG. DAYS	30	0	0	0	7	78	200	329	298	120	20	0	0	1052
RH	NORMAL (PERCENT)	30	61	60	58	57	66	69	71	73	74	68	64	64	65
	HOUR 01 LST	30	66	65	64	64	76	83	84	86	86	80	72	69	75
	HOUR 07 LST	30	69	70	70	70	79	81	83	87	88	83	76	72	77
	HOUR 13 LST	30	52	51	48	46	52	53	54	56	56	52	52	54	52
	HOUR 19 LST	30	57	55	51	48	58	62	64	66	68	63	60	61	59
S	PERCENT POSSIBLE SUNSHINE														
W/O	MEAN NO. DAYS WITH:														
	HEAVY FOG(VISBY 1/4 MI)	54	2.6	2.7	2.0	1.0	1.7	1.0	1.3	1.5	2.4	1.9	2.1	2.3	22.5
	THUNDERSTORMS	54	0.1	0.2	1.1	2.9	6.1	6.6	8.3	6.3	2.6	0.9	0.4	0.1	35.6
CLOUDINESS	MEAN:														
	SUNRISE-SUNSET (OKTAS)	48	5.1	5.0	5.0	4.8	5.0	4.8	4.7	4.6	4.5	3.9	4.6	4.8	4.7
	MIDNIGHT-MIDNIGHT (OKTAS)	32	4.7	4.6	4.7	4.3	4.5	4.3	4.3	4.2	4.2	3.7	4.3	4.6	4.4
	MEAN NO. DAYS WITH:														
	CLEAR	49	8.1	7.6	8.1	8.5	7.4	7.1	6.9	7.8	9.6	12.8	8.8	8.7	101.4
	PARTLY CLOUDY	49	7.7	7.1	8.4	8.8	10.5	11.9	13.0	12.4	8.8	7.5	8.4	7.9	112.4
	CLOUDY	49	15.2	13.6	14.4	12.7	13.0	11.1	11.1	10.8	11.6	10.7	12.8	14.4	151.4
PR	MEAN STATION PRESSURE(IN)	28	28.82	28.81	28.77	28.75	28.77	28.78	28.81	28.84	28.86	28.87	28.86	28.84	28.82
	MEAN SEA-LEVEL PRES. (IN)	17	30.11	30.10	30.03	29.99	29.99	30.00	30.02	30.05	30.07	30.12	30.13	30.13	30.06
WINDS	MEAN SPEED (MPH)	42	9.3	9.2	9.7	9.5	7.5	6.7	6.4	6.0	5.9	6.7	7.7	8.4	7.8
	PREVAIL.DIR(TENS OF DEGS)	26	30	30	30	30	30	27	27	27	31	30	31	30	30
	MAXIMUM 2-MINUTE:														
	SPEED (MPH)	5	44	36	38	39	39	34	40	34	34	33	37	37	44
	DIR. (TENS OF DEGS)		28	30	28	30	26	35	04	32	30	30	25	27	28
	YEAR OF OCCURRENCE		2000	1999	1997	1999	2001	1998	1997	1997	1999	2000	1999	2000	JAN 2000
	MAXIMUM 5-SECOND:														
	SPEED (MPH)	5	53	47	53	51	51	41	47	49	45	39	47	47	53
	DIR. (TENS OF DEGS)		28	20	29	29	26	36	04	28	29	29	24	27	28
	YEAR OF OCCURRENCE		2000	1998	1997	1999	2001	1998	1997	1997	1999	2000	1999	2000	JAN 2000
PRECIPITATION	NORMAL (IN)	30	2.62	3.04	3.48	3.25	3.98	3.19	3.91	4.15	3.50	3.85	3.19	2.97	41.13
	MAXIMUM MONTHLY (IN)	54	7.97	8.00	7.91	11.35	8.42	10.32	10.09	9.54	11.09	9.89	12.36	7.10	12.36
	YEAR OF OCCURRENCE		1998	1998	1993	1987	1950	1995	1989	1984	1987	1990	1985	1948	NOV 1985
	MINIMUM MONTHLY (IN)	54	0.29	0.56	0.43	0.48	1.04	0.62	0.45	0.74	0.15	0.02	0.44	0.18	0.02
	YEAR OF OCCURRENCE		1981	1968	1966	1976	1997	1986	1977	1995	1991	2000	1960	1965	OCT 2000
	MAXIMUM IN 24 HOURS (IN)	54	3.64	3.32	3.09	5.57	3.99	3.98	2.74	5.22	6.60	6.41	6.63	3.40	6.63
	YEAR OF OCCURRENCE		1995	1994	1998	1978	1973	1972	1989	1985	1987	1968	1985	1948	NOV 1985
	NORMAL NO. DAYS WITH:														
	PRECIPITATION 0.01	30	9.6	9.3	10.5	10.0	12.0	9.7	11.6	10.6	8.3	7.8	8.6	9.2	117.2
	PRECIPITATION 1.00	30	0.5	0.6	0.7	0.7	0.9	0.7	0.9	1.2	0.9	1.3	0.7	0.5	9.6
SNOWFALL	NORMAL (IN)	30	7.3	8.1	3.2	0.6	T	0.0	0.0	0.0	0.0	0.*	1.7	4.8	25.7
	MAXIMUM MONTHLY (IN)	49	41.2	27.6	30.3	7.3	T	T	0.0	0.0	T	1.0	13.8	22.6	41.2
	YEAR OF OCCURRENCE		1966	1960	1960	1971	1990	1989			1953	1957	1968	1966	JAN 1966
	MAXIMUM IN 24 HOURS (IN)	49	22.2	18.4	17.4	7.3	T	T	0.0	0.0	T	1.0	10.0	16.4	22.2
	YEAR OF OCCURRENCE		1996	1983	1960	1971	1990	1989			1953	1957	1968	1969	JAN 1996
	MAXIMUM SNOW DEPTH (IN)	48	23	18	17	3	0	0	0	0	0	0	8	15	23
	YEAR OF OCCURRENCE		1987	1983	1960	1987							1950	1969	JAN 1987
	NORMAL NO. DAYS WITH:														
	SNOWFALL 1.0	30	1.7	1.9	0.7	0.2	0.0	0.0	0.0	0.0	0.0	0.0	0.5	1.1	6.1

PRECIPITATION (inches) 2001 ROANOKE, VA (ROA)

YEAR	JAN	FEB	MAR	APR	MAY	JUN	JUL	AUG	SEP	OCT	NOV	DEC	ANNUAL
1972	2.49	4.80	1.76	3.31	6.00	7.55	4.89	2.62	4.79	3.18	5.63	4.62	51.64
1973	2.60	2.95	5.92	5.39	5.58	3.65	5.10	1.84	5.02	4.28	1.79	5.60	48.04
1974	3.33	2.13	3.12	1.86	3.76	2.93	3.71	4.93	3.04	0.77	1.28	3.16	34.02
1975	3.59	3.05	7.80	2.04	6.65	1.54	5.15	5.68	6.46	3.01	1.77	3.67	50.41
1976	2.16	1.27	4.54	0.48	6.13	5.20	1.24	2.20	3.17	9.72	1.31	2.59	40.01
1977	1.46	0.73	2.61	3.50	1.52	2.41	0.45	2.29	2.71	4.70	6.46	2.49	31.33
1978	6.12	0.65	5.92	7.54	4.85	2.05	4.83	6.33	0.52	0.78	2.55	3.15	45.29
1979	5.27	5.37	3.38	3.99	2.65	5.78	3.97	3.37	9.18	3.56	3.77	1.13	51.42
1980	4.10	0.67	5.41	5.51	2.66	1.81	5.18	2.87	1.66	2.30	1.78	0.60	34.55
1981	0.29	2.43	2.30	1.75	4.56	2.49	2.86	1.32	4.52	3.90	0.68	3.79	30.89
1982	3.76	4.75	2.33	2.01	4.83	4.99	3.98	5.20	2.67	4.13	3.65	2.53	44.83
1983	1.28	4.12	6.41	7.95	3.17	2.38	1.67	2.23	1.52	7.73	4.26	5.61	48.33
1984	1.35	4.85	4.30	3.97	4.49	2.34	4.17	9.54	2.69	1.42	2.67	1.84	43.63
1985	2.45	3.64	1.80	1.75	6.89	2.08	4.18	8.67	1.26	3.77	12.36	0.85	49.70
1986	0.93	2.87	1.36	1.67	4.15	0.62	2.83	4.31	3.04	2.76	3.73	5.48	33.75
1987	4.53	4.55	4.11	11.35	2.68	0.71	3.21	1.08	11.09	1.10	5.00	2.16	51.57
1988	1.87	1.07	0.88	3.40	2.76	3.66	3.75	4.30	3.01	1.26	2.42	1.28	29.66
1989	1.31	2.04	2.96	2.54	6.46	7.76	10.09	1.65	8.94	4.13	3.86	2.60	54.34
1990	2.33	2.76	3.42	2.07	7.45	0.83	3.80	4.42	1.86	9.89	1.08	3.79	43.70
1991	3.55	2.10	7.58	2.49	2.88	2.42	7.22	2.31	0.15	0.04	2.51	3.81	37.06
1992	2.51	3.75	2.54	4.89	6.06	6.87	1.73	2.38	2.07	1.90	5.60	2.62	42.92
1993	3.77	3.44	7.91	2.77	2.37	2.49	1.24	1.04	4.58	2.15	2.60	5.27	39.63
1994	4.52	5.34	5.67	5.59	1.69	1.99	7.07	4.22	0.76	1.84	1.53	2.41	42.63
1995	7.28	1.88	1.31	0.79	4.55	10.32	2.34	0.74	2.04	3.49	3.39	2.32	40.45
1996	6.87	2.12	3.75	1.76	4.58	7.65	2.63	6.40	10.14	1.57	4.75	2.72	54.94
1997	2.18	2.21	3.12	2.51	1.04	5.43	3.38	2.22	3.43	1.50	2.65	2.37	32.04
1998	7.97	8.00	5.20	4.58	4.47	2.03	0.86	6.17	1.10	1.47	0.85	2.32	45.02
1999	3.70	2.14	2.77	2.70	2.27	0.85	5.96	2.63	7.38	1.24	2.09	2.46	36.19
2000	2.08	1.69	2.86	5.71	2.67	4.57	7.16	3.16	5.88	0.02	2.08	1.67	39.55
2001	1.79	0.90	4.48	0.81	4.67	1.51	3.13	2.02	2.12	0.43	0.60	2.48	24.94
POR= 54 YRS	3.00	3.15	3.67	3.63	3.85	3.52	3.72	3.89	3.41	3.06	2.82	2.94	40.66

AVERAGE TEMPERATURE (F) 2001   ROANOKE, VA (ROA)

YEAR	JAN	FEB	MAR	APR	MAY	JUN	JUL	AUG	SEP	OCT	NOV	DEC	ANNUAL
1972	39.6	36.0	45.6	55.5	62.0	67.7	74.3	73.2	67.6	52.2	45.4	44.5	55.3
1973	37.3	35.8	51.1	57.8	61.1	73.8	75.5	76.0	70.8	59.2	49.3	38.2	56.8
1974	45.3	39.0	50.8	57.2	64.3	68.1	74.5	73.5	64.8	55.0	47.3	38.9	56.6
1975	39.6	40.4	42.2	52.9	66.5	71.5	74.5	77.0	66.6	60.0	51.1	38.8	56.8
1976	33.4	46.7	50.8	57.4	62.3	70.0	74.0	72.1	65.2	50.3	40.2	34.3	54.7
1977	23.6	36.6	52.5	58.9	68.1	71.1	79.7	77.8	70.4	53.6	48.5	35.9	56.4
1978	27.8	29.5	44.2	56.3	63.8	72.7	76.0	77.5	71.4	54.0	49.5	39.5	55.2
1979	31.4	29.4	48.6	55.7	63.8	69.3	73.5	74.0	66.3	54.6	49.6	40.9	54.8
1980	37.5	34.0	43.5	56.7	64.8	70.0	78.1	76.8	71.5	55.5	45.4	38.7	56.0
1981	32.0	38.2	43.1	58.6	60.9	74.0	76.1	73.4	66.1	53.7	45.2	32.8	54.5
1982	28.7	38.1	44.8	50.8	67.5	70.1	75.5	72.3	65.8	57.4	47.8	42.4	55.1
1983	35.6	36.5	47.0	52.4	61.0	70.1	77.0	77.9	68.1	57.1	47.1	34.9	55.4
1984	33.6	43.6	43.3	54.2	63.1	74.2	73.2	74.4	64.1	64.3	46.0	47.1	56.8
1985	31.3	38.6	50.3	61.9	67.1	72.7	76.5	73.8	68.4	60.3	55.4	35.1	57.6
1986	35.2	38.3	47.2	58.8	63.9	74.2	78.8	72.4	68.9	57.4	38.5	55.9
1987	34.2	36.9	46.6	53.2	67.1	75.0	79.3	77.8	68.9	51.5	49.6	41.4	56.8
1988	30.8	37.9	47.5	55.4	63.4	70.9	77.1	77.5	66.0	51.2	46.7	38.9	55.3
1989	41.3	38.2	47.4	54.3	61.2	73.7	75.9	74.1	68.1	58.5	46.0	29.6	55.7
1990	43.7	45.9	51.5	56.0	64.3	72.6	76.8	74.6	68.4	57.2	52.2	43.8	59.1
1991	39.3	42.7	49.2	57.5	69.6	73.5	77.9	74.8	69.3	59.4	46.8	42.4	58.5
1992	39.6	41.5	46.5	55.5	60.2	68.4	77.0	72.1	68.0	55.5	46.3	38.5	55.8
1993	39.7	36.0	42.1	54.5	65.6	71.9	80.3	76.8	68.6	55.3	47.3	36.2	56.2
1994	30.6	39.4	46.7	60.9	60.6	75.1	77.6	73.9	66.9	56.7	51.4	43.0	56.9
1995	38.5	36.9	50.1	56.8	64.4	72.0	76.9	76.9	67.5	58.5	43.2	35.5	56.6
1996	33.7	38.2	41.7	56.3	65.2	72.8	72.8	74.4	73.3	66.1	58.0	41.0	40.2	55.1
1997	35.8	43.0	49.7	52.5	60.3	69.2	76.1	73.4	67.7	56.3	43.6	38.3	55.5
1998	41.2	42.2	45.8	56.5	67.1	73.5	77.5	75.9	73.6	59.0	49.0	43.4	58.7
1999	39.8	42.0	44.0	58.2	64.3	73.2	79.2	75.5	66.4	55.7	52.8	42.5	57.7
2000	35.3	43.5	51.7	54.8	67.6	73.8	73.3	72.9	65.9	59.4	44.2	31.1	56.1
2001	37.0	42.9	43.4	59.5	64.8	73.3	74.3	76.9	66.5	56.9	53.2	44.3	57.8
POR= 54 YRS	36.0	38.9	46.3	56.6	64.7	72.0	76.3	74.8	67.9	57.1	47.1	38.6	56.4

REFERENCE NOTES:

PAGE 1:
THE TEMPERATURE GRAPH SHOWS NORMAL MAXIMUM AND NORMAL
  MINIMUM DAILY TEMPERATURES (SOLID CURVES) AND THE
  ACTUAL DAILY HIGH AND LOW TEMPERATURES (VERTICAL BARS).

PAGE 2 AND 3:
H/C INDICATES HEATING AND COOLING DEGREE DAYS.
RH INDICATES RELATIVE HUMIDITY
W/O INDICATES WEATHER AND OBSTRUCTIONS
S INDICATES SUNSHINE.
PR INDICATES PRESSURE.
CLOUDINESS ON PAGE 3 IS THE SUM OF THE CEILOMETER AND
  SATELLITE DATA NOT TO EXCEED EIGHT EIGHTHS(OKTAS).

GENERAL:
T INDICATES TRACE PRECIPITATION, AN AMOUNT GREATER
  THAN ZERO BUT LESS THAN THE LOWEST REPORTABLE VALUE.
+ INDICATES THE VALUE ALSO OCCURS ON EARLIER DATES.
BLANK ENTRIES DENOTE MISSING OR UNREPORTED DATA.
NORMALS ARE 30-YEAR AVERAGES (1961 - 1990).
ASOS INDICATES AUTOMATED SURFACE OBSERVING SYSTEM.
PM INDICATES THE LAST DAY OF THE PREVIOUS MONTH.
POR (PERIOD OF RECORD) BEGINS WITH THE JANUARY DATA
  MONTH AND IS THE NUMBER OF YEARS USED TO COMPUTE
  THE MEAN. INDIVIDUAL MONTHS WITHIN THE POR MAY
  BE MISSING.
WHEN THE POR FOR A NORMAL IS LESS THAN 30 YEARS,
  THE NORMAL IS PROVISIONAL AND IS BASED ON THE NUMBER
  OF YEARS INDICATED.
0.* OR * INDICATES THE VALUE OR MEAN-DAYS-WITH
  IS BETWEEN 0.00 AND 0.05.
CLOUDINESS FOR ASOS STATIONS DIFFERS FROM THE NON-ASOS
  OBSERVATION TAKEN BY A HUMAN OBSERVER. ASOS STATION
  CLOUDINESS IS BASED ON TIME-AVERAGED CEILOMETER DATA
  FOR CLOUDS AT OR BELOW 12,000 FEET AND ON SATELLITE
  DATA FOR CLOUDS ABOVE 12,000 FEET.
THE NUMBER OF DAYS WITH CLEAR, PARTLY CLOUDY, AND
  CLOUDY CONDITIONS FOR ASOS STATIONS IS THE SUM
  OF THE CEILOMETER AND SATELLITE DATA FOR THE
  SUNRISE TO SUNSET PERIOD.

GENERAL CONTINUED:
CLEAR INDICATES 0 - 2 OKTAS, PARTLY CLOUDY INDICATES
  3 - 6 OKTAS, AND CLOUDY INDICATES 7 OR 8 OKTAS.
  WHEN AT LEAST ONE OF THE ELEMENTS (CEILOMETER OR
  SATELLITE) IS MISSING, THE DAILY CLOUDINESS IS
  NOT COMPUTED.
WIND DIRECTION IS RECORDED IN TENS OF DEGREES (2 DIGITS)
  CLOCKWISE FROM TRUE NORTH. "00" INDICATES CALM. "36"
  INDICATES TRUE NORTH.
RESULTANT WIND IS THE VECTOR AVERAGE OF THE SPEED AND
  DIRECTION.
AVERAGE TEMPERATURE IS THE SUM OF THE MEAN DAILY MAXIMUM
  AND MINIMUM TEMPERATURE DIVIDED BY 2.
SNOWFALL DATA COMPRISE ALL FORMS OF FROZEN
  PRECIPITATION, INCLUDING HAIL.
A HEATING (COOLING) DEGREE DAY IS THE DIFFERENCE BETWEEN
  THE AVERAGE DAILY TEMPERATURE AND 65 F.
DRY BULB IS THE TEMPERATURE OF THE AMBIENT AIR.
DEW POINT IS THE TEMPERATURE TO WHICH THE AIR MUST BE
  COOLED TO ACHIEVE 100 PERCENT RELATIVE HUMIDITY.
WET BULB IS THE TEMPERATURE THE AIR WOULD HAVE IF THE
  MOISTURE CONTENT WAS INCREASED TO 100 PERCENT RELATIVE
  HUMIDITY.

ON JULY 1, 1996, THE NATIONAL WEATHER SERVICE BEGAN USING
  THE "METAR" OBSERVATION CODE THAT WAS ALREADY EMPLOYED
  BY MOST OTHER NATIONS OF THE WORLD. THE MOST NOTICEABLE
  DIFFERENCE IN THIS ANNUAL PUBLICATION WILL BE THE CHANGE
  IN UNITS FROM TENTHS TO EIGHTS(OKTAS) FOR REPORTING THE
  AMOUNT OF SKY COVER.

HEATING DEGREE DAYS (base 65 F) 2001 ROANOKE, VA (ROA)

YEAR	JUL	AUG	SEP	OCT	NOV	DEC	JAN	FEB	MAR	APR	MAY	JUN	TOTAL
1972-73	10	1	33	391	582	628	852	812	428	344	144	1	4226
1973-74	0	0	12	196	461	826	607	722	440	255	96	14	3629
1974-75	0	0	84	308	539	801	783	683	699	361	60	12	4330
1975-76	0	0	59	173	415	809	973	523	438	271	126	22	3809
1976-77	0	3	47	452	735	945	1275	786	385	203	56	23	4910
1977-78	0	0	18	350	496	896	1147	989	637	261	112	4	4910
1978-79	0	0	29	335	461	784	1037	992	512	279	88	15	4532
1979-80	3	9	49	329	458	738	848	893	656	266	78	14	4341
1980-81	0	0	30	301	582	807	1016	744	672	212	158	3	4525
1981-82	0	0	58	357	589	991	1121	746	618	421	51	3	4955
1982-83	0	6	61	264	509	695	904	792	551	381	157	14	4334
1983-84	1	0	87	246	531	924	966	614	664	336	123	6	4498
1984-85	0	0	120	74	565	549	1041	734	471	168	39	11	3772
1985-86	0	0	55	171	282	918	917	739	551	206	115	1	3955
1986-87	0	22	22	231	523	813	950	782	562	352	58	0	4315
1987-88	0	0	18	412	455	723	1055	778	535	289	101	46	4412
1988-89	1	0	53	423	543	802	726	743	556	340	172	0	4359
1989-90	0	3	63	234	560	1091	654	528	441	297	77	3	3951
1990-91	0	0	38	200	381	652	787	618	486	245	36	4	3447
1991-92	0	0	49	194	539	695	777	677	567	290	172	18	3978
1992-93	0	2	55	290	552	813	777	807	698	310	52	12	4368
1993-94	0	0	52	300	529	886	1061	709	559	172	176	4	4448
1994-95	0	0	22	261	401	675	815	779	454	258	93	0	3758
1995-96	0	0	54	226	647	908	965	771	717	283	118	6	4695
1996-97	0	0	52	221	712	762	898	610	467	369	175	49	4315
1997-98	1	0	32	294	634	819	730	629	615	255	46	22	4077
1998-99	0	0	10	184	473	662	775	638	646	223	61	8	3680
1999-00	5	3	38	289	357	689	913	619	414	301	44	5	3677
2000-01	2	0	86	190	617	1046	863	612	666	221	66	4	4373
2001-	4	0	70	263	345	638							

COOLING DEGREE DAYS (base 65 F) 2001 ROANOKE, VA (ROA)

YEAR	JAN	FEB	MAR	APR	MAY	JUN	JUL	AUG	SEP	OCT	NOV	DEC	ANNUAL
1972	0	0	0	26	13	118	305	262	118	0	2	0	844
1973	0	0	3	15	29	269	331	346	192	23	0	0	1208
1974	0	0	7	27	82	117	303	267	83	3	12	0	901
1975	0	0	0	5	108	213	301	379	114	27	5	0	1152
1976	0	0	8	53	49	179	285	230	58	1	0	0	863
1977	0	0	7	28	158	214	461	400	185	4	7	0	1464
1978	0	0	0	7	79	243	347	391	228	2	0	0	1297
1979	0	0	10	7	56	150	273	296	97	12	1	0	902
1980	0	0	0	21	78	171	412	374	231	13	0	0	1300
1981	0	0	2	26	39	278	350	267	97	13	0	0	1072
1982	0	0	0	0	137	165	332	242	90	35	0	0	1001
1983	0	0	0	11	41	172	382	407	188	10	0	0	1211
1984	0	0	0	20	71	290	260	301	101	56	0	0	1099
1985	0	1	18	78	112	247	365	280	163	34	3	0	1301
1986	0	0	4	24	84	282	438	257	145	48	0	0	1282
1987	0	0	0	4	130	306	450	403	140	0	0	0	1433
1988	0	0	0	7	58	232	386	395	90	1	0	0	1169
1989	0	0	17	27	61	265	344	293	163	38	0	0	1208
1990	0	0	29	34	61	238	377	305	145	27	3	0	1219
1991	0	0	2	29	188	267	407	311	185	28	1	0	1418
1992	0	0	2	12	29	125	377	230	152	4	0	0	931
1993	0	0	0	2	78	226	481	374	166	8	6	0	1341
1994	0	0	0	57	45	317	397	286	84	9	3	0	1198
1995	0	0	0	20	80	220	375	427	134	30	0	0	1286
1996	0	1	0	30	130	247	300	263	93	10	0	0	1074
1997	0	0	0	1	36	181	353	264	121	31	0	0	987
1998	0	0	24	5	120	282	394	344	276	7	0	0	1452
1999	0	0	0	26	47	233	451	336	86	9	0	0	1188
2000	0	0	10	4	131	277	269	253	121	25	0	0	1090
2001	0	0	0	64	67	259	301	377	121	20	0	0	1209

SNOWFALL (inches) 2001 ROANOKE, VA (ROA)

YEAR	JUL	AUG	SEP	OCT	NOV	DEC	JAN	FEB	MAR	APR	MAY	JUN	TOTAL
1972-73	0.0	0.0	0.0	0.0	T	T	2.4	1.4	5.5	0.2	0.0	0.0	9.5
1973-74	0.0	0.0	0.0	T	T	7.9	T	9.7	T	0.0	0.0	0.0	17.6
1974-75	0.0	0.0	0.0	T	3.3	6.6	4.6	6.3	6.2	T	0.0	0.0	27.0
1975-76	0.0	0.0	0.0	0.0	T	0.1	T	T	2.2	0.0	0.0	0.0	2.3
1976-77	0.0	0.0	0.0	0.0	2.4	6.4	8.9	1.5	T	T	0.0	0.0	19.2
1977-78	0.0	0.0	0.0	T	2.4	0.4	14.5	9.6	10.4	0.0	0.0	0.0	37.3
1978-79	0.0	0.0	0.0	0.0	1.6	T	3.0	19.3	T	0.0	0.0	0.0	23.9
1979-80	0.0	0.0	0.0	0.3	T	0.2	15.1	4.2	12.0	T	0.0	0.0	31.8
1980-81	0.0	0.0	0.0	T	T	1.4	T	T	10.4	0.0	0.0	0.0	11.8
1981-82	0.0	0.0	0.0	0.0	4.0	3.9	8.4	12.2	0.5	1.9	0.0	0.0	30.9
1982-83	0.0	0.0	0.0	T	0.0	6.2	3.8	24.3	0.3	0.4	0.0	0.0	35.0
1983-84	0.0	0.0	0.0	0.0	T	0.6	7.2	1.5	0.3	0.2	0.0	0.0	9.8
1984-85	0.0	0.0	0.0	0.0	T	0.8	2.7	1.2	1.3	T	0.0	0.0	6.0
1985-86	0.0	0.0	0.0	0.0	0.0	1.2	1.7	7.1	T	T	0.0	0.0	10.0
1986-87	0.0	0.0	0.0	0.0	T	T	27.9	19.1	2.7	6.3	0.0	0.0	56.0
1987-88	0.0	0.0	0.0	0.0	1.7	T	6.7	T	T	T	0.0	0.0	8.4
1988-89	0.0	0.0	0.0	T	0.0	3.8	0.7	9.1	0.2	0.2	T	T	14.0
1989-90	0.0	0.0	0.0	0.0	3.3	11.1	T	0.2	1.5	0.0	T	0.0	16.1
1990-91	0.0	0.0	0.0	0.0	0.0	0.8	T	T	0.4	0.0	0.0	0.0	1.2
1991-92	0.0	0.0	0.0	0.0	T	T	0.4	2.2	0.5	1.9	0.0	0.0	5.0
1992-93	0.0	0.0	0.0	0.0	T	0.9	0.1	10.7	16.0	0.3	0.0	0.0	28.0
1993-94	0.0	0.0	0.0	T	T	6.2	5.2	1.8	3.1	T	0.0	0.0	16.3
1994-95	0.0	0.0	0.0	0.0	T	0.0	9.6	1.2	1.0	0.0	0.0		11.8
1995-96	0.0	0.0	0.0	0.0	0.4	10.2	28.2	17.1	0.1	T			
1996-97													
1997-98													
1998-99													
1999-00													
2000-01													
2001-													
POR= 48 YRS	0.0	0.0	T	0.0	1.5	3.9	6.6	6.8	3.6	0.4	T	T	22.8

2001
SEATTLE, WASHINGTON
SEATTLE - TACOMA AIRPORT (SEA)

The Seattle-Tacoma International Airport is located 6 miles south of the Seattle city limits and 14 miles north of Tacoma. It is situated on a low ridge lying between Puget Sound on the west and the Green River valley on the east with terrain sloping moderately to the shores of Puget Sound some 2 miles to the west. The Olympic Mountains, rising sharply from Puget Sound, are about 50 miles to the northwest. Rather steep bluffs border the Green River Valley about 2.5 miles to the east and the foothills of the Cascade Range begin 10 to 15 miles to the east of the airport.

The mild climate of the Pacific Coast is modified by the Cascade Mountains and, to a lesser extent, by the Olympic Mountains. The climate is characterized by mild temperatures, a pronounced though not sharply defined rainy season, and considerable cloudiness, particularly during the winter months. The Cascades are very effective in shielding the Seattle-Tacoma area from the cold, dry continental air during the winter and the hot, dry continental air during the summer months. The extremes of temperature that occur in western Washington are the result of the occasional pressure distributions that force the continental air into the Puget Sound area. But the prevailing southwesterly circulation keeps the average winter daytime temperatures in the 40s and the nighttime readings in the 30s. During the summer, daytime temperatures are usually in the 70s with nighttime lows in the 50s. Extremes of temperatures, both in the winter and summer, are usually of short duration. The dry season is centered around July and early August with July being the driest month of the year. The rainy season extends from October to March with December normally the wettest month, however, precipitation is rather evenly distributed through the winter and early spring months with more than 75 percent of the yearly precipitation falling during the winter wet season. Most of the rainfall in the Seattle area comes from storms common to the middle latitudes. These disturbances are most vigorous during the winter as they move through western Washington. The storm track shifts to the north during the summer and those that reach the State are not the wind and rain producers of the winter months. Local summer afternoon showers and a few thunderstorms occur in the Seattle-Tacoma area but they do not contribute materially to the precipitation.

The occurrence of snow in the Seattle-Tacoma area is extremely variable and usually melts before accumulating measurable depths. There are winters on record with only a trace of snow, but at the other extreme, over 21 inches has fallen in a 24-hour period. Usually, winter storms do not produce snow unless the storm moves in such a way to bring cold air out of Canada directly or with only a short over water trajectory.

The highest winds recorded in the Seattle-Tacoma area were associated with strong storms crossing the state from the southwest. Prevailing winds are from the southwest but occasional severe winter storms will produce strong northerly winds. Winds during the summer months are relatively light with occasional land-sea breeze effects creating afternoon northerly winds of 8 to 15 miles an hour. Fog or low clouds that form over the southern Puget Sound area in the late summer, fall, and early winter months, often dominate the weather conditions during the late night and early morning hours with visibilities occasionally lower for a few hours near sunrise. Most of the summer clouds form along the coast and move into the Seattle area from the southwest.

Based on the 1951-1980 period, the average first occurrence of 32 degrees Fahrenheit in the fall is November 11 and the average last occurrence in the spring is March 24.

NORMALS, MEANS, AND EXTREMES

SEATTLE, WA (SEA)

LATITUDE: 47 27' 41" N　　LONGITUDE: 122 18' 49" W　　ELEVATION (FT): GRND: 447　BARO: 450　　TIME ZONE: PACIFIC (UTC + 8)　　WBAN: 24233

	ELEMENT	POR	JAN	FEB	MAR	APR	MAY	JUN	JUL	AUG	SEP	OCT	NOV	DEC	YEAR
TEMPERATURE F	NORMAL DAILY MAXIMUM	30	45.0	49.5	52.7	57.2	63.9	69.9	75.2	75.2	69.3	59.7	50.5	45.1	59.4
	MEAN DAILY MAXIMUM	54	44.6	48.8	52.2	57.3	64.0	69.3	75.0	74.8	69.5	59.4	50.6	45.4	59.2
	HIGHEST DAILY MAXIMUM	57	64	70	75	85	93	96	100	99	98	89	74	64	100
	YEAR OF OCCURRENCE		1981	1968	1987	1976	1963	1995	1994	1981	1988	1987	1949	1993	JUL 1994
	MEAN OF EXTREME MAXS.	54	55.2	59.7	64.4	73.1	81.1	84.8	89.1	88.1	84.3	73.1	61.0	55.5	72.4
	NORMAL DAILY MINIMUM	30	35.2	37.4	38.5	41.2	46.3	51.9	55.2	55.7	51.9	45.8	40.1	35.8	44.6
	MEAN DAILY MINIMUM	54	34.7	36.6	37.9	41.1	46.4	51.3	54.6	54.8	51.2	45.2	39.6	35.8	44.1
	LOWEST DAILY MINIMUM	57	0	1	11	29	28	38	43	44	35	28	6	6	0
	YEAR OF OCCURRENCE		1950	1950	1955	1975	1954	1952	1954	1955	1972	1949	1955	1968	JAN 1950
	MEAN OF EXTREME MINS.	54	21.9	25.4	29.2	33.8	38.4	45.2	49.0	49.0	43.0	34.7	28.2	24.1	35.2
	NORMAL DRY BULB	30	40.1	43.5	45.6	49.2	55.1	60.9	65.2	65.5	60.6	52.8	45.3	40.5	52.0
	MEAN DRY BULB	54	39.6	42.8	45.0	49.2	55.3	60.3	64.9	64.7	60.4	52.3	45.1	40.6	51.7
	MEAN WET BULB	18	39.1	39.8	42.7	45.9	50.4	54.1	57.8	58.2	55.2	49.1	42.9	38.3	47.8
	MEAN DEW POINT	18	35.6	35.5	38.4	41.2	45.6	49.3	52.5	53.3	50.7	45.9	40.0	35.1	43.6
	NORMAL NO. DAYS WITH:														
	MAXIMUM　90	30	0.0	0.0	0.0	0.0	0.2	0.3	1.1	1.2	0.3	0.0	0.0	0.0	3.1
	MAXIMUM　32	30	1.2	0.2	0.0	0.0	0.0	0.0	0.0	0.0	0.0	0.0	0.2	1.3	2.9
	MINIMUM　32	30	9.6	5.3	3.0	0.2	0.0	0.0	0.0	0.0	0.0	0.2	3.6	8.6	30.5
	MINIMUM　0	30	0.0	0.0	0.0	0.0	0.0	0.0	0.0	0.0	0.0	0.0	0.0	0.0	0.0
H/C	NORMAL HEATING DEG. DAYS	30	772	602	601	474	307	144	58	65	156	378	591	760	4908
	NORMAL COOLING DEG. DAYS	30	0	0	0	0	0	21	64	81	24	0	0	0	190
RH	NORMAL (PERCENT)	30	78	75	74	71	69	67	65	68	73	79	80	80	73
	HOUR 04 LST	30	81	80	82	83	82	81	81	83	86	86	84	82	83
	HOUR 10 LST	30	79	77	74	71	68	66	65	68	73	79	80	81	73
	HOUR 16 LST	30	74	67	61	57	54	53	49	50	57	67	75	77	62
	HOUR 22 LST	30	78	76	75	74	72	70	68	70	75	81	81	80	75
S	PERCENT POSSIBLE SUNSHINE	30	28	40	50	52	56	56	64	65	62	43	28	23	47
W/O	MEAN NO. DAYS WITH:														
	HEAVY FOG(VISBY 1/4 MI)	57	4.8	3.2	2.3	1.1	0.7	0.7	1.5	2.5	4.9	6.9	5.0	5.9	39.5
	THUNDERSTORMS	57	0.2	0.3	0.6	1.0	0.9	0.7	0.7	0.8	0.7	0.4	0.8	0.3	7.4
CLOUDINESS	MEAN:														
	SUNRISE-SUNSET (OKTAS)	52	6.7	6.5	6.3	6.2	5.7	5.6	4.2	4.5	4.8	5.9	6.6	6.7	5.8
	MIDNIGHT-MIDNIGHT (OKTAS)	32	6.6	6.1	5.8	5.9	5.5	5.3	4.1	4.2	4.4	5.3	6.3	6.4	5.5
	MEAN NO. DAYS WITH:														
	CLEAR	52	2.8	3.0	3.4	2.7	4.4	5.1	10.2	9.0	8.1	4.0	2.5	2.3	57.5
	PARTLY CLOUDY	52	3.9	4.2	5.8	7.3	8.8	7.9	9.5	9.7	8.5	7.4	4.2	3.9	81.1
	CLOUDY	52	24.3	21.1	21.9	20.0	17.7	16.9	10.5	11.9	13.0	19.5	23.3	24.9	225.0
PR	MEAN STATION PRESSURE(IN)	29	29.55	29.52	29.50	29.55	29.55	29.56	29.57	29.54	29.55	29.57	29.53	29.58	29.55
	MEAN SEA-LEVEL PRES. (IN)	18	30.04	30.02	30.01	30.04	30.02	30.03	30.05	30.03	30.03	30.06	30.02	30.08	30.04
WINDS	MEAN SPEED (MPH)	51	9.6	9.6	9.5	9.4	8.9	8.6	8.2	7.9	8.1	8.4	9.2	9.6	8.9
	PREVAIL.DIR(TENS OF DEGS)	35	19	19	19	19	21	21	21	22	21	01	19	18	19
	MAXIMUM 2-MINUTE:														
	SPEED (MPH)	5	40	37	44	29	25	26	22	22	36	32	37	37	44
	DIR. (TENS OF DEGS)		20	23	19	22	21	22	22	01	20	18	20	22	19
	YEAR OF OCCURRENCE		2000	1999	1999	2001	2001	1999	2001	1997	1999	1997	1998	2000	MAR 1999
	MAXIMUM 5-SECOND:														
	SPEED (MPH)	5	52	49	60	40	36	32	31	29	43	44	48	48	60
	DIR. (TENS OF DEGS)		20	21	18	21	27	23	23	34	21	18	21	22	18
	YEAR OF OCCURRENCE		2000	1999	1999	2001	1999	1999	1999	1998	1999	1997	1998	2000	MAR 1999
PRECIPITATION	NORMAL (IN)	30	5.38	3.99	3.54	2.33	1.70	1.50	0.76	1.14	1.88	3.23	5.83	5.91	37.19
	MAXIMUM MONTHLY (IN)	57	12.92	9.11	8.40	6.53	4.76	3.90	2.39	4.59	5.95	8.95	11.62	11.85	12.92
	YEAR OF OCCURRENCE		1953	1961	1950	1991	1948	1946	1983	1975	1978	1947	1998	1979	JAN 1953
	MINIMUM MONTHLY (IN)	57	0.58	0.35	0.57	0.33	0.12	0.13	T	0.01	T	0.31	0.74	1.37	T
	YEAR OF OCCURRENCE		1985	1993	1965	1956	1992	1951	1960	1974	1991	1987	1976	1978	SEP 1991
	MAXIMUM IN 24 HOURS (IN)	57	3.22	3.41	2.86	3.32	1.83	2.08	0.85	1.75	2.23	3.74	3.58	2.61	3.74
	YEAR OF OCCURRENCE		1986	1951	1972	1991	1969	1985	1981	1968	1978	1981	1990	1979	OCT 1981
	NORMAL NO. DAYS WITH:														
	PRECIPITATION　0.01	30	18.0	15.4	16.8	13.6	10.6	8.4	5.0	6.1	9.1	12.2	17.6	18.7	151.5
	PRECIPITATION　1.00	30	0.8	0.4	0.2	0.1	0.1	0.2	0.0	0.1	0.2	0.4	1.2	1.1	4.8
SNOWFALL	NORMAL (IN)	30	4.1	1.2	0.7	0.1	T	0.0	0.0	0.0	0.0	0.1	1.1	3.1	10.4
	MAXIMUM MONTHLY (IN)	52	57.2	13.1	18.2	2.3	T	0.0	T	0.0	T	2.0	17.5	22.1	57.2
	YEAR OF OCCURRENCE		1950	1949	1951	1972	1993		1980		1972	1971	1985	1968	JAN 1950
	MAXIMUM IN 24 HOURS (IN)	52	21.4	9.8	7.4	2.3	T	0.0	T	0.0	T	2.0	9.4	13.0	21.4
	YEAR OF OCCURRENCE		1950	1990	1989	1972	1993		1980		1972	1971	1946	1968	JAN 1950
	MAXIMUM SNOW DEPTH (IN)	47	21	11	7	2	0	0	0	0	0	2	8	10	21
	YEAR OF OCCURRENCE		1969	1969	1989	1972						1971	1985	1974	JAN 1969
	NORMAL NO. DAYS WITH:														
	SNOWFALL　1.0	30	1.4	0.3	0.3	0.1	0.0	0.0	0.0	0.0	0.0	0.*	0.3	1.0	3.4

PRECIPITATION (inches)   2001   SEATTLE, WASHINGTON WA (SEA)

YEAR	JAN	FEB	MAR	APR	MAY	JUN	JUL	AUG	SEP	OCT	NOV	DEC	ANNUAL
1972	7.24	8.11	6.74	4.12	0.69	1.81	1.34	1.13	4.10	0.72	3.38	8.98	48.36
1973	4.29	1.89	1.62	1.35	1.60	2.50	0.08	0.27	1.81	3.31	7.99	8.33	35.04
1974	7.78	4.01	5.84	2.39	1.37	1.25	1.51	0.01	0.21	1.99	5.06	6.45	37.87
1975	6.01	5.80	2.87	2.49	1.13	0.84	0.27	4.59	T	7.75	5.07	7.66	44.48
1976	5.55	4.74	2.71	1.67	1.61	0.63	1.17	2.71	1.25	2.06	0.74	1.86	26.70
1977	1.77	1.58	3.80	0.55	3.70	0.54	0.42	3.59	2.55	2.60	5.27	6.47	32.84
1978	4.30	3.59	2.43	4.19	1.79	0.75	1.40	1.19	5.95	0.98	6.05	1.37	33.99
1979	2.25	5.32	1.55	0.81	0.88	0.46	0.73	1.02	2.07	3.38	1.94	11.85	32.26
1980	4.09	5.04	2.10	3.23	0.97	1.77	0.46	0.64	1.43	1.32	7.16	7.39	35.60
1981	2.42	4.45	2.23	1.58	1.33	2.31	1.38	0.25	3.42	6.40	4.07	5.56	35.40
1982	5.35	7.57	3.73	2.07	0.63	1.03	0.59	0.62	1.49	4.07	5.31	6.86	39.32
1983	7.07	4.57	3.81	1.06	2.10	1.85	2.39	1.90	1.85	1.34	7.97	5.02	40.93
1984	3.62	3.91	3.91	2.87	3.38	2.81	0.17	0.13	1.01	2.14	8.09	4.95	36.99
1985	0.58	2.63	2.56	1.30	0.85	2.80	0.10	0.55	1.98	5.74	4.26	1.78	25.13
1986	8.54	4.41	2.67	1.38	1.71	0.68	1.10	0.10	1.89	4.21	7.98	3.67	38.34
1987	5.98	2.05	5.53	2.61	2.38	0.16	0.39	0.29	0.91	0.31	3.21	6.11	29.93
1988	4.07	0.71	3.75	3.20	3.01	1.56	0.50	0.28	1.75	2.24	8.43	3.48	32.98
1989	2.78	3.43	5.79	2.80	2.78	1.14	0.64	0.89	0.54	2.98	6.13	4.79	34.69
1990	9.41	3.72	2.58	2.54	1.98	3.05	0.58	0.71	0.05	5.79	10.71	3.63	44.75
1991	4.46	4.69	4.66	6.53	1.39	1.29	0.28	2.17	T	1.31	5.33	3.31	35.42
1992	7.82	3.09	1.68	4.12	0.12	1.14	0.89	0.66	1.15	2.45	5.57	4.09	32.78
1993	4.09	0.35	4.80	4.54	2.86	2.48	1.27	0.16	0.03	1.54	2.20	4.48	28.80
1994	2.51	4.47	3.17	2.27	1.43	1.25	0.28	0.30	1.69	3.51	5.79	8.15	34.82
1995	4.48	4.97	4.07	2.05	0.81	1.46	1.34	1.81	0.91	3.93	10.40	6.37	42.60
1996	7.34	8.35	2.06	5.37	2.07	0.59	0.77	1.32	1.85	5.54	5.23	10.18	50.67
1997	7.02	1.99	8.15	4.32	1.87	1.64	1.20	1.27	3.41	5.83	3.93	2.63	43.26
1998	7.15	3.31	3.96	0.99	1.98	1.11	0.41	0.35	0.72	3.48	11.62	8.98	44.06
1999	6.84	6.95	3.66	1.49	2.12	1.86	1.18	0.92	0.17	2.26	9.60	5.06	42.11
2000	3.77	5.25	2.82	1.48	3.27	1.61	0.23	0.33	1.12	3.00	3.27	2.51	28.66
2001	2.70	2.07	2.73	3.16	1.39	3.05	1.03	2.32	0.83	3.13	9.26	5.89	37.56
POR= 58 YRS	5.58	4.27	3.73	2.57	1.72	1.53	0.72	1.02	1.76	3.56	5.96	5.83	38.25

AVERAGE TEMPERATURE (F) 2001   SEATTLE, WASHINGTON WA (SEA)

YEAR	JAN	FEB	MAR	APR	MAY	JUN	JUL	AUG	SEP	OCT	NOV	DEC	ANNUAL
1972	37.0	41.4	46.9	47.0	58.3	60.1	66.0	66.7	55.4	50.1	46.7	38.1	51.1
1973	38.7	43.9	44.1	48.6	56.5	59.3	64.7	61.6	61.9	52.2	43.7	44.4	51.6
1974	38.7	43.2	46.3	50.3	54.9	62.6	64.0	64.6	64.4	52.5	45.1	42.4	52.4
1975	38.8	40.8	42.9	45.8	54.6	60.7	67.5	63.2	63.0	51.4	44.9	41.5	51.3
1976	41.8	40.9	41.3	49.5	56.4	60.0	65.9	64.1	62.6	54.9	47.8	44.7	52.5
1977	39.4	48.7	45.7	53.6	54.5	63.0	65.1	68.5	58.9	52.2	43.9	42.2	53.0
1978	44.4	46.0	48.6	49.9	54.5	64.3	65.8	65.5	58.8	54.3	41.2	37.5	52.6
1979	37.8	42.3	49.3	50.8	57.2	62.5	67.4	64.0	62.1	54.2	43.9	44.1	53.0
1980	34.8	43.8	44.3	51.6	54.2	57.5	63.8	61.9	59.6	53.9	46.7	44.1	51.4
1981	44.4	44.2	48.8	49.6	54.7	57.5	63.3	68.1	61.1	50.9	47.2	41.7	52.6
1982	39.3	42.1	44.1	47.4	54.7	63.1	62.8	65.1	60.6	52.7	43.2	40.8	51.3
1983	45.0	46.9	49.4	50.7	57.7	59.9	63.3	65.6	58.3	51.7	47.8	36.1	52.7
1984	43.2	44.8	48.5	48.7	52.9	58.8	65.0	64.9	59.9	49.7	44.6	36.8	51.5
1985	37.1	39.0	43.3	49.2	54.8	60.8	68.6	65.2	58.1	51.4	35.8	36.2	49.9
1986	44.9	42.8	49.2	48.1	55.7	62.7	61.7	68.4	59.1	54.3	45.3	42.0	52.9
1987	40.5	46.3	48.9	52.0	56.9	62.6	64.2	66.1	62.6	55.8	48.5	39.2	53.6
1988	40.1	44.4	45.6	50.3	54.9	59.6	65.3	65.4	60.5	55.4	41.9	42.4	52.4
1989	40.5	35.9	43.7	53.4	56.0	63.2	64.5	65.3	64.1	53.1	47.0	42.9	52.5
1990	42.5	40.0	47.1	52.1	54.7	59.8	68.0	67.3	63.4	51.2	46.6	35.3	52.3
1991	40.0	47.7	44.1	49.1	54.3	58.9	66.8	66.6	62.9	52.9	47.3	43.7	52.9
1992	43.9	47.3	50.3	53.1	59.8	65.0	66.7	66.8	60.9	54.4	45.5	38.8	54.3
1993	37.9	42.3	48.1	50.6	59.6	60.6	61.2	65.5	61.9	55.4	42.0	41.4	52.2
1994	44.9	40.4	48.5	52.3	58.2	60.6	67.9	67.3	64.4	52.8	42.4	41.7	53.5
1995	46.4	45.9	47.8	51.8	60.1	62.7	67.0	63.0	64.0	52.5	49.1	42.4	54.4
1996	39.7	43.2	47.2	51.5	51.5	60.5	67.9	66.5	57.9	51.1	43.1	39.3	51.9
1997	41.1	41.9	44.8	49.2	58.2	59.5	64.5	67.7	62.2	51.3	48.4	41.5	52.5
1998	42.2	45.8	46.6	50.4	54.9	59.9	67.6	66.8	63.1	52.6	46.7	39.9	53.0
1999	42.0	42.5	44.3	48.5	51.8	58.2	62.4	64.8	60.8	52.2	47.9	42.0	51.5
2000	40.3	43.7	44.5	50.9	53.8	60.6	64.3	63.5	60.2	52.8	42.5	40.8	51.5
2001	42.0	40.7	45.4	48.0	55.4	57.6	62.5	64.8	59.8	50.9	46.7	41.5	51.3
POR= 57 YRS	39.6	42.8	44.9	49.1	55.3	60.3	64.9	64.6	60.2	52.3	45.0	40.5	51.6

REFERENCE NOTES:

PAGE 1:
THE TEMPERATURE GRAPH SHOWS NORMAL MAXIMUM AND NORMAL
MINIMUM DAILY TEMPERATURES (SOLID CURVES) AND THE
ACTUAL DAILY HIGH AND LOW TEMPERATURES (VERTICAL BARS).

PAGE 2 AND 3:
H/C INDICATES HEATING AND COOLING DEGREE DAYS.
RH INDICATES RELATIVE HUMIDITY.
W/O INDICATES WEATHER AND OBSTRUCTIONS
S INDICATES SUNSHINE.
PR INDICATES PRESSURE.
CLOUDINESS ON PAGE 3 IS THE SUM OF THE CEILOMETER AND
SATELLITE DATA NOT TO EXCEED EIGHT EIGHTHS(OKTAS).

GENERAL:
T INDICATES TRACE PRECIPITATION, AN AMOUNT GREATER
THAN ZERO BUT LESS THAN THE LOWEST REPORTABLE VALUE.
+ INDICATES THE VALUE ALSO OCCURS ON EARLIER DATES.
BLANK ENTRIES DENOTE MISSING OR UNREPORTED DATA.
NORMALS ARE 30-YEAR AVERAGES (1961 - 1990).
ASOS INDICATES AUTOMATED SURFACE OBSERVING SYSTEM.
PM INDICATES THE LAST DAY OF THE PREVIOUS MONTH.
POR (PERIOD OF RECORD) BEGINS WITH THE JANUARY DATA
MONTH AND IS THE NUMBER OF YEARS USED TO COMPUTE
THE MEAN. INDIVIDUAL MONTHS WITHIN THE POR MAY
BE MISSING.
WHEN THE POR FOR A NORMAL IS LESS THAN 30 YEARS,
THE NORMAL IS PROVISIONAL AND IS BASED ON THE NUMBER
OF YEARS INDICATED.
0.* OR * INDICATES THE VALUE OR MEAN-DAYS-WITH
IS BETWEEN 0.00 AND 0.05.
CLOUDINESS FOR ASOS STATIONS DIFFERS FROM THE NON-ASOS
OBSERVATION TAKEN BY A HUMAN OBSERVER. ASOS STATION
CLOUDINESS IS BASED ON TIME-AVERAGED CEILOMETER DATA
FOR CLOUDS AT OR BELOW 12,000 FEET AND ON SATELLITE
DATA FOR CLOUDS ABOVE 12,000 FEET.
THE NUMBER OF DAYS WITH CLEAR, PARTLY CLOUDY, AND
CLOUDY CONDITIONS FOR ASOS STATIONS IS THE SUM
OF THE CEILOMETER AND SATELLITE DATA FOR THE
SUNRISE TO SUNSET PERIOD.

GENERAL CONTINUED:
CLEAR INDICATES 0 - 2 OKTAS, PARTLY CLOUDY INDICATES
3 - 6 OKTAS, AND CLOUDY INDICATES 7 OR 8 OKTAS.
WHEN AT LEAST ONE OF THE ELEMENTS (CEILOMETER OR
SATELLITE) IS MISSING, THE DAILY CLOUDINESS IS
NOT COMPUTED.
WIND DIRECTION IS RECORDED IN TENS OF DEGREES (2 DIGITS)
CLOCKWISE FROM TRUE NORTH. "00" INDICATES CALM. "36"
INDICATES TRUE NORTH.
RESULTANT WIND IS THE VECTOR AVERAGE OF THE SPEED AND
DIRECTION.
AVERAGE TEMPERATURE IS THE SUM OF THE MEAN DAILY MAXIMUM
AND MINIMUM TEMPERATURE DIVIDED BY 2.
SNOWFALL DATA COMPRISE ALL FORMS OF FROZEN
PRECIPITATION, INCLUDING HAIL.
A HEATING (COOLING) DEGREE DAY IS THE DIFFERENCE BETWEEN
THE AVERAGE DAILY TEMPERATURE AND 65 F.
DRY BULB IS THE TEMPERATURE OF THE AMBIENT AIR.
DEW POINT IS THE TEMPERATURE TO WHICH THE AIR MUST BE
COOLED TO ACHIEVE 100 PERCENT RELATIVE HUMIDITY.
WET BULB IS THE TEMPERATURE THE AIR WOULD HAVE IF THE
MOISTURE CONTENT WAS INCREASED TO 100 PERCENT RELATIVE
HUMIDITY.

ON JULY 1, 1996, THE NATIONAL WEATHER SERVICE BEGAN USING
THE "METAR" OBSERVATION CODE THAT WAS ALREADY EMPLOYED
BY MOST OTHER NATIONS OF THE WORLD. THE MOST NOTICEABLE
DIFFERENCE IN THIS ANNUAL PUBLICATION WILL BE THE CHANGE
IN UNITS FROM TENTHS TO EIGHTS(OKTAS) FOR REPORTING THE
AMOUNT OF SKY COVER.

HEATING DEGREE DAYS (base 65 F) 2001 SEATTLE, WASHINGTON WA (SEA)

YEAR	JUL	AUG	SEP	OCT	NOV	DEC	JAN	FEB	MAR	APR	MAY	JUN	TOTAL
1972-73	48	32	295	455	544	825	807	586	639	484	272	183	5170
1973-74	70	114	111	388	633	632	809	606	573	433	306	99	4774
1974-75	60	66	74	380	591	690	804	671	678	570	317	144	5045
1975-76	23	73	93	413	594	723	712	693	731	465	265	157	4942
1976-77	24	52	81	307	510	625	786	451	591	335	320	79	4161
1977-78	34	43	178	390	625	701	631	525	498	447	323	78	4473
1978-79	44	42	180	324	706	846	837	630	479	420	235	96	4839
1979-80	27	40	86	327	628	642	929	610	634	395	329	218	4865
1980-81	66	104	158	343	543	639	633	577	494	455	316	220	4548
1981-82	80	28	138	430	530	715	790	636	640	521	312	103	4923
1982-83	93	42	141	373	647	745	613	502	479	422	244	149	4450
1983-84	72	19	196	406	511	890	672	577	507	482	372	183	4887
1984-85	54	42	159	467	603	867	857	719	666	469	310	160	5373
1985-86	8	48	199	413	870	888	618	616	479	502	305	90	5036
1986-87	105	12	196	323	586	707	754	522	491	384	253	105	4438
1987-88	58	37	102	284	485	792	767	590	593	435	316	165	4624
1988-89	60	38	162	291	583	708	749	807	654	340	273	93	4758
1989-90	41	29	68	362	534	677	689	696	547	379	312	158	4492
1990-91	29	23	61	420	546	913	767	475	639	472	322	179	4846
1991-92	24	40	90	368	526	654	645	507	447	351	171	63	3886
1992-93	26	22	151	322	580	805	834	629	517	427	170	135	4618
1993-94	113	48	120	290	685	723	618	682	503	373	209	132	4496
1994-95	23	5	43	369	671	713	570	527	527	390	160	118	4116
1995-96	19	77	60	382	471	693	779	630	543	399	362	135	4550
1996-97	42	32	167	422	649	793	732	640	622	468	211	159	4937
1997-98	38	11	95	416	489	718	700	533	562	433	307	152	4454
1998-99	21	21	79	376	543	771	706	623	634	487	400	218	4879
1999-00	101	42	141	393	504	706	760	611	628	415	340	154	4795
2000-01	60	65	146	373	668	745	707	675	599	504	305	218	5065
2001-	86	44	150	430	547	723							

COOLING DEGREE DAYS (base 65 F) 2001 SEATTLE, WASHINGTON WA (SEA)

YEAR	JAN	FEB	MAR	APR	MAY	JUN	JUL	AUG	SEP	OCT	NOV	DEC	ANNUAL
1972	0	0	0	0	22	3	85	91	11	0	0	0	212
1973	0	0	0	0	16	19	67	17	21	0	0	0	140
1974	0	0	0	0	0	36	38	62	60	0	0	0	196
1975	0	0	0	0	0	21	108	29	39	0	0	0	197
1976	0	0	0	8	4	14	59	29	15	0	0	0	129
1977	0	0	0	0	0	26	44	158	4	0	0	0	232
1978	0	0	0	0	4	66	76	64	0	0	0	0	210
1979	0	0	0	0	2	27	106	15	21	0	0	0	171
1980	0	0	0	0	0	0	34	15	3	2	0	0	54
1981	0	0	0	0	1	3	35	131	24	0	0	0	194
1982	0	0	0	0	0	53	31	55	15	0	0	0	154
1983	0	0	0	0	24	2	24	44	0	0	0	0	94
1984	0	0	0	0	1	5	62	45	11	0	0	0	124
1985	0	0	0	0	3	17	125	59	0	0	0	0	204
1986	0	0	0	0	22	27	10	124	26	0	0	0	209
1987	0	0	0	0	11	42	39	80	35	5	0	0	212
1988	0	0	0	0	7	10	79	56	36	1	0	0	189
1989	0	0	0	0	2	47	32	45	46	0	0	0	172
1990	0	0	0	0	0	10	129	100	21	0	0	0	260
1991	0	0	0	0	0	0	0	85	96	34	0	0	215
1992	0	0	0	0	18	68	83	85	8	0	0	0	262
1993	0	0	0	0	10	8	1	72	32	0	0	0	123
1994	0	0	0	0	1	5	120	86	29	0	0	0	241
1995	0	0	0	0	15	55	88	20	34	0	0	0	212
1996	0	0	0	1	0	2	137	81	1	0	0	0	222
1997	0	0	0	0	4	0	29	100	17	0	0	0	150
1998	0	0	0	2	1	6	108	84	29	0	0	0	230
1999	0	0	0	0	0	18	29	43	22	0	0	0	112
2000	0	0	0	0	0	28	44	27	8	0	0	0	107
2001	0	0	0	0	11	3	14	44	0	0	0	0	72

SNOWFALL (inches) 2001 SEATTLE, WASHINGTON WA (SEA)

YEAR	JUL	AUG	SEP	OCT	NOV	DEC	JAN	FEB	MAR	APR	MAY	JUN	TOTAL
1972-73	0.0	0.0	T	0.0	T	5.6	2.7	T	0.8	T	0.0	0.0	9.1
1973-74	0.0	0.0	0.0	0.0	0.2	0.3	3.7	T	T	0.0	T	0.0	4.2
1974-75	0.0	0.0	0.0	0.0	0.0	9.8	1.3	T	T	0.2	0.0	0.0	11.3
1975-76	0.0	0.0	0.0	0.0	1.6	2.6	T	0.5	0.2	T	0.0	0.0	4.9
1976-77	0.0	0.0	0.0	0.0	0.0	T	1.0	T	0.9	0.0	0.0	0.0	1.9
1977-78	0.0	0.0	0.0	0.0	3.5	T	T	0.0	T	T	0.0	0.0	3.5
1978-79	0.0	0.0	0.0	0.0	4.9	0.2	0.5	0.4	0.0	0.0	0.0	0.0	6.0
1979-80	0.0	0.0	0.0	0.0	0.0	1.2	8.8	2.5	0.1	T	0.0	0.0	12.6
1980-81	T	0.0	0.0	0.0	T	0.3	0.0	1.1	0.0	0.0	0.0	0.0	1.4
1981-82	0.0	0.0	0.0	0.0	0.0	T	7.0	T	2.0	T	0.0	0.0	9.0
1982-83	0.0	0.0	0.0	T	0.0	T	0.0	0.0	0.0	T	0.0	0.0	T
1983-84	0.0	0.0	0.0	0.0	T	0.3	T	0.0	0.0	0.0	0.0	0.0	0.3
1984-85	0.0	0.0	0.0	T	T	2.4	T	5.7	T	T	0.0	0.0	8.1
1985-86	0.0	0.0	0.0	T	17.5	1.7	0.0	1.1	0.0	T	0.0	0.0	20.3
1986-87	0.0	0.0	0.0	0.0	T	1.4	0.0	0.0	0.0	0.0	0.0	0.0	1.4
1987-88	0.0	0.0	0.0	0.0	0.0	T	T	1.0	T	T	0.0	0.0	T
1988-89	0.0	0.0	0.0	0.0	T	T	1.0	5.8	7.4	T	T	0.0	14.2
1989-90	0.0	0.0	0.0	0.0	0.0	0.0	T	9.8	T	0.0	T	0.0	9.8
1990-91	0.0	0.0	0.0	0.0	0.0	3.8	0.4	0.0	2.5	T	0.0	0.0	6.7
1991-92	0.0	0.0	0.0	0.0	0.0	0.0	0.0	0.0	0.0	0.0	0.0	0.0	0.0
1992-93	0.0	0.0	0.0	0.0	T	6.7	1.3	1.4	0.0	0.0	T	0.0	9.4
1993-94	0.0	0.0	0.0	0.0	T	0.0	0.0	2.1	0.2	0.0	0.0	0.0	2.3
1994-95	0.0	0.0	0.0	0.0	T	1.9	0.0	0.2	T	0.0	0.0	0.0	2.1
1995-96	0.0	0.0	0.0	0.0	0.0	0.0	10.5	0.5	0.0	0.0	0.0	0.0	11.0
1996-97	0.0	0.0	0.0										
1997-98													
1998-99													
1999-00													
2000-01													
2001-													
POR= 51 YRS	T	0.0	T	0.0	1.1	2.4	4.8	1.6	1.3	0.1	T	0.0	11.3

2001
SPOKANE
WASHINGTON (GEG)

Spokane lies on the eastern edge of the broad Columbia Basin area of Washington which is bounded by the Cascade Range on the west and the Rocky Mountains on the east. The elevations in eastern Washington vary from less than 400 feet above sea level near Pasco where the Columbia River flows out of Washington to over 5,000 feet in the mountain areas of the extreme eastern edge of the State. Spokane is located on the upper plateau area where the long gradual slope from the Columbia River meets the sharp rise of the Rocky Mountain Ranges.

Much of the urban area of Spokane lies along both sides of the Spokane River at an elevation of approximately 2,000 feet, but the residential areas have spread to the crests of the plateaus on either side of the river with elevations up to 2,500 feet above sea level. Spokane International Airport is situated on the plateau area 6 miles west-southwest and some 400 feet higher than the downtown business district.

The climate of Spokane combines some of the characteristics of damp coastal type weather and arid interior conditions. Most of the air masses which reach Spokane are brought in by the prevailing westerly and southwesterly circulations. Frequently, much of the moisture in the storms that move eastward and southeastward from the Gulf of Alaska and the eastern Pacific Ocean is precipitated out as the storms are lifted across the Coast and Cascade Ranges. Annual precipitation totals in the Spokane area are generally less than 20 inches and less than 50 percent of the amounts received west of the Cascades. However, the precipitation and total cloudiness in the Spokane vicinity is greater than that of the desert areas of south-central Washington. The lifting action of the air masses as they move up the east slope of the Columbia Basin frequently produces the cooling and condensation necessary for formation of clouds and precipitation.

Infrequently, the Spokane area comes under the influence of dry continental air masses from the north or east. On occasions when these air masses penetrate into eastern Washington the result is high temperatures and very low humidity in the summer and sub-zero temperatures in the winter. In the winter most of the severe arctic outbursts of cold air move southward on the east side of the Continental Divide and do not affect Spokane.

In general, Spokane weather has the characteristics of a mild, arid climate during the summer months and a cold, coastal type in the winter. Approximately 70 percent of the total annual precipitation falls between the first of October and the end of March and about half of that falls as snow. The growing season usually extends over nearly six months from mid-April to mid-October. Irrigation is required for all crops except dry-land type grains. The summer weather is ideal for full enjoyment of the many mountain and lake recreational areas in the immediate vicinity. Winter weather includes many cloudy or foggy days and below freezing temperatures with occasional snowfall of several inches in depth. Sub-zero temperatures and traffic-stopping snowfalls are infrequent.

Based on the 1951-1980 period, the average first occurrence of 32 degrees Fahrenheit in the fall is October 6 and the average last occurrence in the spring is May 4.

NORMALS, MEANS, AND EXTREMES
SPOKANE, WA (GEG)

LATITUDE: LONGITUDE: ELEVATION (FT):   TIME ZONE: WBAN: 24157
47 37' 17" N   117 31' 40" W   GRND: 2381   BARO: 2384   PACIFIC (UTC + 8)

	ELEMENT	POR	JAN	FEB	MAR	APR	MAY	JUN	JUL	AUG	SEP	OCT	NOV	DEC	YEAR
TEMPERATURE F	NORMAL DAILY MAXIMUM	30	33.2	40.6	47.7	57.0	65.8	74.7	83.1	82.5	72.0	58.6	41.4	33.8	57.5
	MEAN DAILY MAXIMUM	54	32.0	39.0	47.4	57.1	66.3	74.0	83.2	82.3	72.6	58.2	41.7	33.2	57.2
	HIGHEST DAILY MAXIMUM	54	59	63	71	90	96	101	103	108	98	86	67	56	108
	YEAR OF OCCURRENCE		1971	1995	1960	1977	1986	1992	1998	1961	1988	1997	1999	1980	AUG 1961
	MEAN OF EXTREME MAXS.	54	45.9	51.2	61.8	73.6	84.3	90.2	96.6	96.0	88.7	75.1	55.7	47.2	72.2
	NORMAL DAILY MINIMUM	30	20.8	25.9	29.6	34.7	41.9	49.2	54.4	54.3	45.8	36.0	28.8	21.7	36.9
	MEAN DAILY MINIMUM	54	20.6	25.3	29.6	35.2	42.7	49.3	54.8	54.3	46.4	36.5	28.7	22.4	37.1
	LOWEST DAILY MINIMUM	54	-22	-24	-7	17	24	33	37	35	22	10	-21	-25	-25
	YEAR OF OCCURRENCE		1979	1996	1989	1966	1954	1984	1981	1965	2000	1991	1985	1968	DEC 1968
	MEAN OF EXTREME MINS.	54	0.0	7.5	16.3	25.7	31.1	38.9	44.3	43.5	33.6	23.9	13.8	2.5	23.4
	NORMAL DRY BULB	30	27.1	33.3	38.7	45.9	53.9	62.0	68.8	68.4	58.9	47.3	35.1	27.8	47.3
	MEAN DRY BULB	54	26.3	32.1	38.5	46.1	54.4	61.7	69.0	68.4	59.6	47.4	35.2	27.8	47.2
	MEAN WET BULB	17	27.6	29.7	35.7	40.9	46.9	52.0	55.9	51.8	49.4	41.2	33.4	26.3	40.9
	MEAN DEW POINT	17	25.2	26.0	30.0	33.6	38.8	43.3	45.0	41.2	39.8	34.6	30.7	24.0	34.4
	NORMAL NO. DAYS WITH:														
	MAXIMUM 90	30	0.0	0.0	0.0	*	0.3	2.1	8.4	7.2	1.0	0.0	0.0	0.0	19.0
	MAXIMUM 32	30	14.2	4.6	0.9	0.0	0.0	0.0	0.0	0.0	0.0	0.1	4.1	13.8	37.7
	MINIMUM 32	30	26.5	22.4	20.8	10.7	1.7	0.0	0.0	0.0	0.0	0.8	9.5	19.9	26.6	138.9
	MINIMUM 0	30	2.3	0.5	*	0.0	0.0	0.0	0.0	0.0	0.0	0.0	0.3	2.1	5.2
H/C	NORMAL HEATING DEG. DAYS	30	1175	888	815	573	344	139	30	56	223	549	897	1153	6842
	NORMAL COOLING DEG. DAYS	30	0	0	0	0	0	49	148	161	40	0	0	0	398
RH	NORMAL (PERCENT)	30	82	79	70	61	58	54	44	45	54	67	83	86	65
	HOUR 04 LST	30	85	84	81	77	76	74	64	63	71	79	87	87	77
	HOUR 10 LST	30	83	80	69	57	53	49	41	43	51	66	83	86	63
	HOUR 16 LST	30	78	69	55	44	41	36	27	28	35	49	76	82	52
	HOUR 22 LST	30	84	81	74	65	63	58	45	46	56	70	85	87	68
S	PERCENT POSSIBLE SUNSHINE	48	28	41	55	61	65	67	80	78	72	55	29	23	54
W/O	MEAN NO. DAYS WITH:														
	HEAVY FOG(VISBY 1/4 MI)	54	9.5	6.9	3.0	1.2	0.8	0.5	0.2	0.3	0.8	4.0	8.5	8.4	44.1
	THUNDERSTORMS	54	0.0	0.0	0.3	0.7	1.6	2.8	2.4	2.1	0.9	0.3	0.1	0.0	11.2
CLOUDINESS	MEAN:														
	SUNRISE-SUNSET (OKTAS)	1			7.2										
	MIDNIGHT-MIDNIGHT (OKTAS)														
	MEAN NO. DAYS WITH:														
	CLEAR	1		2.0	3.0		3.0	6.0							
	PARTLY CLOUDY	1		3.0	2.0		3.0	1.0							
	CLOUDY	1	4.0	3.0	10.0		10.0	4.0							
PR.	MEAN STATION PRESSURE(IN)	27	27.57	27.54	27.47	27.49	27.48	27.49	27.52	27.51	27.55	27.58	27.53	27.58	27.53
	MEAN SEA-LEVEL PRES. (IN)	17	30.13	30.09	30.01	29.98	29.93	29.94	29.96	29.95	29.99	30.06	30.08	30.15	30.02
WINDS	MEAN SPEED (MPH)	39	8.4	9.1	9.7	10.0	9.3	9.3	8.7	8.2	8.1	8.1	8.6	8.3	8.8
	PREVAIL.DIR(TENS OF DEGS)	23	05	05	22	22	22	22	22	22	22	05	05	05	22
	MAXIMUM 2-MINUTE:														
	SPEED (MPH)	6	43	44	46	46	44	37	41	31	38	40	37	41	46
	DIR. (TENS OF DEGS)		26	21	23	25	26	25	25	20	26	25	22	22	23
	YEAR OF OCCURRENCE		2000	1999	2000	1997	1997	2000	1998	2001	1999	1999	1998	1998	MAR 2000
	MAXIMUM 5-SECOND:														
	SPEED (MPH)	6	52	53	52	54	48	45	51	39	47	47	47	49	54
	DIR. (TENS OF DEGS)		24	22	24	23	26	27	25	27	24	25	23	22	23
	YEAR OF OCCURRENCE		2000	1999	2000	2000	1997	1996	1996	1996	1999	1999	1998	1998	APR 2000
PRECIPITATION	NORMAL (IN)	30	1.98	1.49	1.49	1.18	1.41	1.26	0.67	0.72	0.73	0.99	2.15	2.42	16.49
	MAXIMUM MONTHLY (IN)	54	4.96	3.94	3.81	3.08	5.71	3.06	2.33	1.83	2.05	4.05	5.10	5.13	5.71
	YEAR OF OCCURRENCE		1959	1961	1995	1948	1948	1964	1990	1976	1959	1950	1973	1964	MAY 1948
	MINIMUM MONTHLY (IN)	54	0.38	0.35	0.31	0.08	0.20	0.16	T	T	T	0.03	0.22	0.60	T
	YEAR OF OCCURRENCE		1985	1988	1965	1956	1982	1960	1994	1988	1990	1987	1976	1976	JUL 1994
	MAXIMUM IN 24 HOURS (IN)	54	1.48	1.11	1.08	1.51	1.67	2.07	1.80	1.09	1.12	1.23	1.41	1.60	2.07
	YEAR OF OCCURRENCE		1954	1963	1995	2000	1948	1964	1990	1959	1973	1994	1960	1951	JUN 1964
	NORMAL NO. DAYS WITH:														
	PRECIPITATION 0.01	30	13.1	10.8	11.1	8.9	9.2	7.7	4.5	5.1	5.7	7.1	12.8	14.7	110.7
	PRECIPITATION 1.00	30	*	0.0	0.0	0.0	0.0	0.1	*	0.0	*	0.0	*	0.1	0.2
SNOWFALL	NORMAL (IN)	30	14.2	6.7	3.6	0.9	0.2	0.0	0.0	0.0	0.0	0.3	6.4	15.1	47.4
	MAXIMUM MONTHLY (IN)	53	56.9	28.5	15.3	6.6	3.5	T	0.0	0.0	T	6.1	24.7	42.0	56.9
	YEAR OF OCCURRENCE		1950	1975	1962	1964	1967	1994			1991	1957	1955	1964	JAN 1950
	MAXIMUM IN 24 HOURS (IN)	53	13.0	11.0	6.1	4.9	3.5	T	0.0	0.0	T	6.1	9.0	12.1	13.0
	YEAR OF OCCURRENCE		1950	1993	1989	1964	1967	1994			1991	1957	1973	1951	JAN 1950
	MAXIMUM SNOW DEPTH (IN)	52	39	42	16	2	0		0		0	4	12	23	42
	YEAR OF OCCURRENCE		1969	1969	1969	1990						1957	1985	1951	FEB 1969
	NORMAL NO. DAYS WITH:														
	SNOWFALL 1.0	30	4.7	2.5	1.4	0.3	0.*	0.0	0.0	0.0	0.0	0.1	2.1	5.0	16.1

PRECIPITATION (inches) 2001 SPOKANE WA (GEG)

YEAR	JAN	FEB	MAR	APR	MAY	JUN	JUL	AUG	SEP	OCT	NOV	DEC	ANNUAL
1972	1.74	1.13	1.05	1.09	1.99	1.56	0.25	0.87	0.86	0.19	0.88	1.92	13.53
1973	2.05	0.48	0.77	0.42	1.34	0.57	T	0.19	1.44	0.97	5.10	3.78	17.11
1974	3.79	1.79	2.22	0.80	1.03	0.23	0.71	0.04	0.18	0.12	2.59	2.54	16.04
1975	2.53	3.12	1.83	1.78	1.41	1.45	1.60	0.93	0.03	2.23	1.94	2.42	21.27
1976	1.28	2.04	0.83	0.97	1.24	0.78	0.79	0.05	0.05	0.59	0.22	0.60	11.22
1977	0.75	0.52	1.15	0.13	1.71	1.45	0.11	1.25	1.42	0.44	2.12	4.52	15.57
1978	2.53	1.64	0.77	2.62	2.81	1.22	1.76	1.71	0.93	0.13	2.02	1.05	19.19
1979	1.11	2.19	1.03	0.69	1.60	0.78	0.85	1.01	0.78	1.22	1.15	1.94	14.35
1980	1.96	1.90	0.91	1.06	2.34	0.99	0.21	0.79	0.84	0.64	1.67	3.72	17.03
1981	1.00	1.41	1.57	0.85	2.02	1.92	0.51	0.04	0.59	1.53	0.96	2.51	14.91
1982	1.61	1.67	1.49	2.23	0.20	0.85	1.05	0.25	1.77	1.48	1.86	2.79	17.25
1983	1.89	2.07	2.20	0.61	0.92	2.84	1.85	0.96	0.79	1.33	4.80	2.38	22.64
1984	0.99	1.37	1.80	1.75	2.01	1.89	0.07	0.27	0.56	0.76	4.26	2.28	18.01
1985	0.38	0.93	1.39	0.28	1.13	0.67	0.26	0.19	1.64	1.40	2.23	0.71	11.21
1986	3.08	2.02	1.58	1.33	1.08	0.48	0.44	0.15	1.65	0.46	2.25	1.03	15.55
1987	1.59	0.88	2.18	1.12	0.90	0.59	2.27	1.81	0.01	0.03	1.37	4.93	17.68
1988	1.76	0.35	1.57	2.15	1.50	1.12	0.23	T	1.63	0.11	4.35	1.75	16.52
1989	0.82	1.34	2.87	0.72	2.17	0.41	0.40	1.61	0.18	1.58	1.66	0.95	14.71
1990	2.45	1.01	0.85	1.34	3.11	1.91	2.33	1.03	T	3.05	0.84	1.69	19.61
1991	1.72	0.81	2.31	1.35	1.72	1.13	0.58	0.17	0.01	0.34	3.08	1.23	14.45
1992	2.12	1.76	0.43	0.65	0.28	1.51	1.09	0.33	0.36	0.81	3.02	2.16	14.52
1993	1.40	0.86	1.13	1.90	1.36	0.48	2.08	1.24	0.28	0.42	0.68	1.80	13.63
1994	1.43	0.83	0.49	1.64	1.37	1.37	T	0.10	0.45	2.79	2.24	1.57	13.81
1995	2.74	1.60	3.81	0.93	1.33	2.17	1.08	0.63	0.66	1.50	0.77	2.63	19.85
1996	2.44	2.95	1.61	2.15	1.78	1.19	0.34	0.80	0.79	3.27	4.04	4.10	25.46
1997	1.67	1.40	2.40	2.56	2.27	0.63	0.80	0.14	0.92	1.67	1.99	1.00	17.45
1998	2.08	1.59	1.21	0.89	3.09	0.84	0.26	0.27	0.21	0.27	3.78	3.28	17.77
1999	1.89	3.27	0.69	0.44	0.73	1.36	0.13	1.07	T	0.89	2.06	2.26	14.79
2000	1.96	1.61	1.64	2.16	2.22	0.91	0.35	T	1.12	0.64	1.13	0.93	14.67
2001	0.63	0.66	1.37	1.71	0.79	1.10	0.28	0.26	0.17	2.10	2.61	2.03	13.71
POR= 120 YRS	2.04	1.58	1.38	1.13	1.39	1.26	0.50	0.59	0.78	1.22	2.04	2.21	16.12

AVERAGE TEMPERATURE (F) 2001   SPOKANE WA (GEG)

YEAR	JAN	FEB	MAR	APR	MAY	JUN	JUL	AUG	SEP	OCT	NOV	DEC	ANNUAL
1972	22.6	30.7	41.4	42.0	56.9	62.0	68.1	71.1	55.4	47.2	38.3	25.4	46.8
1973	27.0	34.9	41.1	46.2	56.5	62.0	71.2	69.1	59.7	47.2	33.7	33.3	48.5
1974	24.1	35.4	38.5	46.4	50.2	66.0	67.8	68.2	60.5	48.0	36.4	30.5	47.7
1975	23.6	24.7	34.0	41.7	52.7	59.2	72.4	64.1	61.0	46.9	33.8	30.9	45.4
1976	29.6	32.1	35.1	45.2	54.5	58.5	68.8	65.4	63.4	46.8	35.8	29.6	47.1
1977	22.0	35.1	38.2	50.9	51.6	65.0	67.0	71.2	55.1	46.5	34.0	26.2	46.9
1978	27.6	34.0	42.2	45.7	51.4	62.7	68.3	65.9	56.6	46.6	28.7	19.0	45.7
1979	10.5	28.8	40.4	45.5	54.7	62.7	70.4	70.0	63.1	51.1	30.5	35.2	46.9
1980	20.7	34.5	38.6	51.7	55.8	57.8	69.2	64.1	58.4	47.4	36.3	33.2	47.3
1981	32.8	33.9	40.9	45.7	52.0	57.0	65.1	71.5	59.7	45.9	39.9	29.7	47.8
1982	26.0	32.1	40.3	43.5	54.2	66.5	67.6	69.8	59.5	46.1	31.7	27.3	47.1
1983	35.8	38.1	43.0	46.3	57.1	61.9	65.5	72.3	57.1	49.7	39.3	16.2	48.5
1984	30.5	34.5	41.7	44.0	50.1	59.2	69.1	70.1	56.7	43.4	35.8	20.4	46.3
1985	21.4	24.9	35.9	48.0	56.2	61.8	75.0	64.9	53.3	44.7	19.5	19.3	43.7
1986	30.1	31.6	42.8	44.9	55.3	66.2	64.0	72.6	54.8	49.0	34.8	26.3	47.7
1987	26.5	35.1	41.8	51.1	57.2	65.1	66.6	66.2	62.8	49.5	38.1	25.9	48.8
1988	24.7	35.4	39.7	48.9	54.6	61.1	68.7	68.4	58.9	53.3	36.3	27.0	48.1
1989	28.8	21.8	36.6	48.9	53.1	64.3	68.7	64.8	60.1	47.0	38.0	31.0	46.9
1990	33.4	30.2	40.9	49.7	52.8	60.7	70.4	68.5	65.3	45.1	39.0	21.1	48.1
1991	25.7	39.2	36.8	45.8	51.6	56.6	68.7	70.2	61.8	46.2	34.2	32.8	47.5
1992	31.8	38.9	45.5	48.8	58.9	68.0	67.8	69.6	57.4	49.5	34.3	22.9	49.5
1993	21.9	25.4	37.8	45.5	59.8	60.2	60.2	64.2	58.7	50.0	29.4	30.9	45.3
1994	35.6	29.1	41.8	49.1	56.7	60.8	73.0	69.4	63.4	46.8	32.4	30.3	49.0
1995	31.0	37.3	39.9	45.5	56.8	60.1	67.9	63.9	61.2	43.9	40.2	28.6	48.0
1996	25.4	28.7	36.4	46.3	49.6	60.5	70.0	66.1	56.0	45.3	33.2	24.8	46.3
1997	28.4	31.7	39.2	43.3	56.7	59.9	67.6	71.0	61.9	47.3	38.6	29.3	47.9
1998	30.7	38.1	41.5	44.0	56.1	62.5	75.3	71.7	65.1	46.5	39.9	28.6	50.3
1999	32.2	34.9	39.9	44.9	50.6	59.9	66.2	70.3	59.1	47.4	41.4	31.6	48.2
2000	27.9	33.5	39.0	48.2	53.0	61.0	67.9	67.6	55.8	46.3	26.9	24.7	46.0
2001	27.2	26.8	39.2	43.7	55.4	58.7	68.4	71.1	63.3	45.9	39.9	28.1	47.3
POR= 120 YRS	26.9	31.9	39.5	47.6	55.6	62.4	69.3	68.6	59.5	48.5	36.4	29.7	48.0

REFERENCE NOTES:

PAGE 1:
THE TEMPERATURE GRAPH SHOWS NORMAL MAXIMUM AND NORMAL
  MINIMUM DAILY TEMPERATURES (SOLID CURVES) AND THE
  ACTUAL DAILY HIGH AND LOW TEMPERATURES (VERTICAL BARS).

PAGE 2 AND 3:
H/C INDICATES HEATING AND COOLING DEGREE DAYS.
RH INDICATES RELATIVE HUMIDITY
W/O INDICATES WEATHER AND OBSTRUCTIONS
S INDICATES SUNSHINE.
PR INDICATES PRESSURE.
CLOUDINESS ON PAGE 3 IS THE SUM OF THE CEILOMETER AND
  SATELLITE DATA NOT TO EXCEED EIGHT EIGHTHS(OKTAS).

GENERAL:
T INDICATES TRACE PRECIPITATION, AN AMOUNT GREATER
  THAN ZERO BUT LESS THAN THE LOWEST REPORTABLE VALUE.
+ INDICATES THE VALUE ALSO OCCURS ON EARLIER DATES.
BLANK ENTRIES DENOTE MISSING OR UNREPORTED DATA.
NORMALS ARE 30-YEAR AVERAGES (1961 - 1990).
ASOS INDICATES AUTOMATED SURFACE OBSERVING SYSTEM.
PM INDICATES THE LAST DAY OF THE PREVIOUS MONTH.
POR (PERIOD OF RECORD) BEGINS WITH THE JANUARY DATA
  MONTH AND IS THE NUMBER OF YEARS USED TO COMPUTE
  THE MEAN. INDIVIDUAL MONTHS WITHIN THE POR MAY
  BE MISSING.
WHEN THE POR FOR A NORMAL IS LESS THAN 30 YEARS,
  THE NORMAL IS PROVISIONAL AND IS BASED ON THE NUMBER
  OF YEARS INDICATED.
0.* OR * INDICATES THE VALUE OR MEAN-DAYS-WITH
  IS BETWEEN 0.00 AND 0.05.
CLOUDINESS FOR ASOS STATIONS DIFFERS FROM THE NON-ASOS
  OBSERVATION TAKEN BY A HUMAN OBSERVER. ASOS STATION
  CLOUDINESS IS BASED ON TIME-AVERAGED CEILOMETER DATA
  FOR CLOUDS AT OR BELOW 12,000 FEET AND ON SATELLITE
  DATA FOR CLOUDS ABOVE 12,000 FEET.
THE NUMBER OF DAYS WITH CLEAR, PARTLY CLOUDY, AND
  CLOUDY CONDITIONS FOR ASOS STATIONS IS THE SUM
  OF THE CEILOMETER AND SATELLITE DATA FOR THE
  SUNRISE TO SUNSET PERIOD.

GENERAL CONTINUED:
CLEAR INDICATES 0 - 2 OKTAS, PARTLY CLOUDY INDICATES
  3 - 6 OKTAS, AND CLOUDY INDICATES 7 OR 8 OKTAS.
  WHEN AT LEAST ONE OF THE ELEMENTS (CEILOMETER OR
  SATELLITE) IS MISSING, THE DAILY CLOUDINESS IS
  NOT COMPUTED.
WIND DIRECTION IS RECORDED IN TENS OF DEGREES (2 DIGITS)
  CLOCKWISE FROM TRUE NORTH. "00" INDICATES CALM. "36"
  INDICATES TRUE NORTH.
RESULTANT WIND IS THE VECTOR AVERAGE OF THE SPEED AND
  DIRECTION.
AVERAGE TEMPERATURE IS THE SUM OF THE MEAN DAILY MAXIMUM
  AND MINIMUM TEMPERATURE DIVIDED BY 2.
SNOWFALL DATA COMPRISE ALL FORMS OF FROZEN
  PRECIPITATION, INCLUDING HAIL.
A HEATING (COOLING) DEGREE DAY IS THE DIFFERENCE BETWEEN
  THE AVERAGE DAILY TEMPERATURE AND 65 F.
DRY BULB IS THE TEMPERATURE OF THE AMBIENT AIR.
DEW POINT IS THE TEMPERATURE TO WHICH THE AIR MUST BE
  COOLED TO ACHIEVE 100 PERCENT RELATIVE HUMIDITY.
WET BULB IS THE TEMPERATURE THE AIR WOULD HAVE IF THE
  MOISTURE CONTENT WAS INCREASED TO 100 PERCENT RELATIVE
  HUMIDITY.

ON JULY 1, 1996, THE NATIONAL WEATHER SERVICE BEGAN USING
  THE "METAR" OBSERVATION CODE THAT WAS ALREADY EMPLOYED
  BY MOST OTHER NATIONS OF THE WORLD. THE MOST NOTICEABLE
  DIFFERENCE IN THIS ANNUAL PUBLICATION WILL BE THE CHANGE
  IN UNITS FROM TENTHS TO EIGHTS(OKTAS) FOR REPORTING THE
  AMOUNT OF SKY COVER.

HEATING DEGREE DAYS (base 65 F) 2001 SPOKANE WA (GEG)

YEAR	JUL	AUG	SEP	OCT	NOV	DEC	JAN	FEB	MAR	APR	MAY	JUN	TOTAL
1972-73	36	18	292	545	795	1219	1171	838	734	558	286	152	6644
1973-74	17	47	193	546	933	978	1265	824	814	554	455	97	6723
1974-75	41	22	134	519	852	1062	1276	1122	953	694	375	173	7223
1975-76	22	75	136	554	933	1048	1091	946	922	588	317	213	6845
1976-77	20	71	74	556	871	1089	1324	832	824	436	409	66	6572
1977-78	57	56	289	563	921	1197	1154	862	701	576	412	101	6889
1978-79	37	97	252	562	1083	1424	1684	1011	756	577	313	134	7930
1979-80	41	4	91	423	1029	918	1365	880	809	392	283	211	6446
1980-81	19	77	195	543	854	977	992	867	741	570	395	243	6473
1981-82	73	7	209	584	747	1088	1202	912	761	639	328	76	6626
1982-83	62	17	193	582	996	1163	897	747	672	558	285	113	6285
1983-84	55	2	230	468	765	1508	1065	880	715	621	460	194	6963
1984-85	21	18	264	662	870	1381	1345	1117	895	501	280	128	7482
1985-86	0	64	343	622	1363	1409	1076	927	680	595	357	67	7503
1986-87	81	4	311	488	902	1193	1186	831	710	417	253	86	6462
1987-88	51	50	116	474	799	1206	1240	850	775	477	330	173	6541
1988-89	47	16	240	361	856	1171	1113	1205	873	473	364	65	6784
1989-90	22	76	149	554	805	1048	976	968	739	454	373	166	6330
1990-91	37	42	54	610	774	1356	1212	716	866	568	406	248	6889
1991-92	15	16	108	574	918	992	1024	750	598	477	206	61	5739
1992-93	32	60	232	481	916	1297	1331	1102	834	578	192	165	7220
1993-94	151	83	217	457	1063	1051	904	998	713	469	262	160	6528
1994-95	26	13	81	558	970	1071	1045	771	771	578	262	170	6316
1995-96	21	88	146	648	742	1120	1217	1045	880	556	471	143	7077
1996-97	35	49	281	603	949	1241	1130	928	794	642	264	154	7070
1997-98	35	15	116	549	785	1098	1058	747	721	505	276	90	5995
1998-99	0	20	101	565	748	1119	1010	836	769	594	448	186	6396
1999-00	75	36	181	540	703	1030	1143	908	799	496	363	142	6416
2000-01	51	43	285	572	1134	1245	1168	1060	795	634	320	201	7508
2001-	33	20	100	588	744	1136							

COOLING DEGREE DAYS (base 65 F) 2001 SPOKANE WA (GEG)

YEAR	JAN	FEB	MAR	APR	MAY	JUN	JUL	AUG	SEP	OCT	NOV	DEC	ANNUAL
1972	0	0	0	0	28	41	138	213	10	0	0	0	430
1973	0	0	0	0	31	67	216	177	39	0	0	0	530
1974	0	0	0	0	0	137	134	127	7	0	0	0	405
1975	0	0	0	0	0	7	256	57	20	0	0	0	340
1976	0	0	0	0	0	24	143	93	33	0	0	0	293
1977	0	0	0	18	2	72	126	254	0	0	0	0	472
1978	0	0	0	0	0	42	144	131	9	0	0	0	326
1979	0	0	0	0	1	73	217	166	39	0	0	0	496
1980	0	0	0	1	3	2	156	56	6	3	0	0	227
1981	0	0	0	0	0	9	82	213	60	0	0	0	364
1982	0	0	0	0	2	128	148	171	32	0	0	0	481
1983	0	0	0	0	46	26	77	235	1	0	0	0	385
1984	0	0	0	0	3	28	155	181	23	1	0	0	391
1985	0	0	0	0	15	36	317	68	0	0	0	0	436
1986	0	0	0	0	65	109	57	247	8	0	0	0	486
1987	0	0	0	8	20	94	110	97	53	1	0	0	383
1988	0	0	0	0	12	63	169	128	67	0	0	0	439
1989	0	0	0	0	0	49	145	78	9	0	0	0	281
1990	0	0	0	0	0	42	213	157	68	0	0	0	480
1991	0	0	0	0	0	0	139	187	20	0	0	0	346
1992	0	0	0	0	25	159	124	209	11	8	0	0	536
1993	0	0	0	0	36	27	11	64	34	0	0	0	172
1994	0	0	0	0	9	37	280	159	43	0	0	0	528
1995	0	0	0	0	14	29	119	59	38	0	0	0	259
1996	0	0	0	0	0	16	198	150	17	0	0	0	381
1997	0	0	0	0	14	9	122	209	30	6	0	0	390
1998	0	0	0	0	6	22	325	234	110	0	0	0	697
1999	0	0	0	0	7	41	118	210	14	0	0	0	390
2000	0	0	0	0	0	29	146	129	16	0	0	0	320
2001	0	0	0	0	29	19	146	213	54	0	0	0	461

SNOWFALL (inches) 2001 SPOKANE WA (GEG)

YEAR	JUL	AUG	SEP	OCT	NOV	DEC	JAN	FEB	MAR	APR	MAY	JUN	TOTAL
1972-73	0.0	0.0	0.0	0.8	T	4.7	6.5	3.5	0.5	T	0.0	0.0	16.0
1973-74	0.0	0.0	0.0	0.8	23.6	9.1	15.0	4.4	2.5	0.4	0.4	0.0	56.2
1974-75	0.0	0.0	0.0	0.0	0.3	16.6	30.9	28.5	7.6	5.1	T	0.0	89.0
1975-76	0.0	0.0	0.0	3.9	11.4	6.9	15.3	6.3	4.6	0.4	0.0	0.0	48.8
1976-77	0.0	0.0	0.0	0.0	0.1	4.2	6.8	2.5	2.7	T	T	0.0	16.3
1977-78	0.0	0.0	0.0	0.0	11.2	30.3	19.1	6.6	2.2	T	T	0.0	69.4
1978-79	0.0	0.0	0.0	0.0	15.4	14.8	16.5	10.6	3.4	T	T	0.0	60.7
1979-80	0.0	0.0	0.0	0.0	3.9	10.4	16.6	5.9	1.1	0.4	0.0	0.0	38.3
1980-81	0.0	0.0	0.0	0.0	1.2	6.8	2.6	3.3	T	T	0.3	0.0	14.2
1981-82	0.0	0.0	0.0	T	0.8	13.0	23.3	2.2	2.1	6.0	T	0.0	47.4
1982-83	0.0	0.0	0.0	T	5.4	17.4	8.1	5.5	T	0.2	T	0.0	36.6
1983-84	0.0	0.0	0.0	0.0	5.7	24.8	5.3	8.0	1.9	1.3	0.8	0.0	47.8
1984-85	0.0	0.0	0.0	1.1	12.0	24.7	4.6	14.8	9.6	T	T	0.0	66.8
1985-86	0.0	0.0	0.0	0.4	23.7	8.3	14.7	13.8	T	0.2	T	0.0	61.1
1986-87	0.0	0.0	0.0	0.0	5.0	7.9	11.7	1.1	1.1	T	T	0.0	25.7
1987-88	0.0	0.0	0.0	0.0	1.5	20.3	9.1	1.2	1.6	T	T	0.0	33.7
1988-89	0.0	0.0	0.0	0.0	10.9	16.3	10.5	19.0	9.4	T	T	0.0	66.1
1989-90	0.0	0.0	0.0	T	5.2	1.1	10.3	18.0	2.6	3.5	T	0.0	40.7
1990-91	0.0	0.0	0.0	0.0	1.2	14.3	15.9	1.1	9.5	0.2	0.0	0.0	42.2
1991-92	0.0	0.0	T	0.8	4.9	2.4	9.0	1.4	0.0	T	0.0	0.0	18.5
1992-93	0.0	0.0	0.0	0.0	11.1	40.2	18.8	15.1	2.1	T	T	T	87.3
1993-94	0.0	0.0	0.0	T	3.7	6.4	6.3	0.9	8.2	0.5	T	0.0	19.7
1994-95	0.0	0.0	0.0	0.8	13.7	6.3	3.9	4.4	0.7	T	0.0	0.0	29.8
1995-96	0.0	0.0	0.0				22.7						
1996-97													
1997-98						6.4	8.5	1.6	1.8	T	0.0	0.0	
1998-99	0.0	0.0	0.0	T	0.8	11.2	8.7	14.7	3.8	2.7	0.6	0.0	42.5
1999-00	0.0	0.0	0.0	0.0	2.1	9.7	21.3	6.7	1.1	T	0.3		
2000-01	0.0	0.0	0.0	T	10.9	15.1	9.6	8.3	2.2	2.5	0.0	0.0	48.6
2001-	0.0	0.0	0.0	0.7	12.5	21.9							
POR= 52 YRS	0.0	0.0	T	0.4	6.4	14.6	15.3	7.4	3.7	0.6	0.1	T	48.5

2001
CHARLESTON,
WEST VIRGINIA (CRW)

Charleston lies at the junction of the Kanawha and Elk Rivers in the western foothills of the Appalachian Mountains. The main urban and business areas have developed along the two river valleys, while some residential areas are in nearby valleys and on the surrounding hills. The hilltops are around 1,100 feet above sea level, about 500 feet higher than the valleys. The Kanawha Airport is just over 2 miles northeast of the center-city area, on an artificial plateau constructed from several hilltops.

Weather records are maintained at the Kanawha Airport by National Weather Service personnel. This site tends to be slightly cooler than the river valleys during the afternoons. Conversely, the valleys can become cooler than the hilltops during clear, calm nights. The weather at Charleston is highly changeable, especially from mid-autumn through the spring.

Winters can vary greatly from one season to the next. Snow does not favor any given winter month, heavy snowstorms are infrequent, and most snowfalls are in the 4-inch or less category. Snow and ice usually do not persist on valley roads, but can linger longer on nearby hills and outlying rural roads.

Afternoon temperatures in the 40s and morning readings in the 20s are common during the winter. Yet, every winter typically has two or three extended cold spells when temperatures stay below freezing for a few consecutive days. Northwesterly winds are associated with the cold weather. Air reaching Charleston from the northwest can cause cloudiness and flurries, even when there is no nearby organized storm system. Winter conditions are much more severe over the higher mountains less than 50 miles to the northeast through the southeast. Temperatures warm rapidly in the spring and are accompanied by low daytime humidities.

Summer and early autumn have more day-to-day consistency in the weather. Sunshine is more abundant than in winter. Summer precipitation falls mostly in brief, but sometimes heavy, showers. Flash flooding can occur along small streams, but flooding is rare on the dam-controlled Kanawha and Elk Rivers.

Afternoon summer temperatures are mostly in the 80s. Readings above 95 degrees are rare. However, during a hot spell, haze and humidity can add to the unpleasantness and indoor air conditioning is recommended. Cooler and less humid air often penetrates the area from the north to end a hot spell.

Early morning fog is common from late June into October. Industrial and vehicular pollutants can contribute to limited visibility any time of the year, especially when cooler air becomes trapped in the valleys. Autumn foliage is generally at its peak during the second and third weeks of October. By the end of October, the first 32 degree temperature has usually arrived.

Ample precipitation is well distributed throughout the year. July is quite often the wettest month of the year, while October averages the least rain. Droughts severe enough to limit water use are scarce. Any dry spells during the spring or autumn can cause conditions favorable for brush fires in outlying areas.

NORMALS, MEANS, AND EXTREMES
CHARLESTON, WV (CRW)

LATITUDE:	LONGITUDE:	ELEVATION (FT):	TIME ZONE:	WBAN: 13866
38 22' 46" N	81 35' 29" W	GRND: 1023　BARO: 1026	EASTERN (UTC + 5)	

	ELEMENT	POR	JAN	FEB	MAR	APR	MAY	JUN	JUL	AUG	SEP	OCT	NOV	DEC	YEAR
TEMPERATURE F	NORMAL DAILY MAXIMUM	30	41.2	45.3	56.7	66.8	75.5	83.1	85.7	84.4	78.8	68.2	57.3	46.0	65.8
	MEAN DAILY MAXIMUM	48	42.5	46.7	56.0	67.5	75.9	82.6	85.5	84.3	78.4	68.0	56.2	46.1	65.8
	HIGHEST DAILY MAXIMUM	54	79	79	89	94	93	98	104	101	102	92	85	80	104
	YEAR OF OCCURRENCE		1950	2000	1990	1990	1991	1999	1988	1988	1953	1951	1993	1982	JUL 1988
	MEAN OF EXTREME MAXS.	53	66.8	69.3	79.3	86.3	89.1	92.3	94.3	92.7	90.4	82.9	76.8	68.4	82.4
	NORMAL DAILY MINIMUM	30	23.0	25.7	35.0	42.8	51.5	59.8	64.4	63.4	56.5	44.2	36.3	28.0	44.2
	MEAN DAILY MINIMUM	48	24.5	27.1	34.3	43.5	51.9	59.8	64.2	63.3	56.1	44.3	35.5	28.2	44.4
	LOWEST DAILY MINIMUM	54	-16	-12	0	19	26	33	46	41	34	17	6	-12	-16
	YEAR OF OCCURRENCE		1994	1996	1980	1982	1966	1972	1963	1965	1983	1962	1950	1989	JAN 1994
	MEAN OF EXTREME MINS.	53	2.9	7.4	16.9	26.4	35.5	46.1	52.7	51.4	40.8	28.5	18.6	8.8	28.0
	NORMAL DRY BULB	30	32.1	35.5	45.9	54.8	63.5	71.4	75.1	73.9	67.7	56.2	46.8	37.0	55.0
	MEAN DRY BULB	53	34.0	37.1	44.9	55.4	63.8	71.4	75.0	73.8	67.3	56.4	46.0	37.3	55.2
	MEAN WET BULB	49	30.5	32.7	38.9	47.6	56.8	64.3	68.3	67.5	61.2	50.5	40.9	33.4	49.4
	MEAN DEW POINT	49	24.1	25.8	31.0	39.4	51.5	60.6	65.1	64.6	57.7	45.7	34.7	27.6	44.0
	NORMAL NO. DAYS WITH:														
	MAXIMUM 90	30	0.0	0.0	0.0	0.4	1.1	4.6	7.5	4.8	1.9	0.0	0.0	0.0	20.3
	MAXIMUM 32	30	8.4	5.2	0.7	*	0.0	0.0	0.0	0.0	0.0	0.0	0.3	4.9	19.5
	MINIMUM 32	30	24.0	20.7	13.7	4.8	0.5	0.0	0.0	0.0	0.0	3.8	12.2	21.0	100.7
	MINIMUM 0	30	1.2	0.4	*	0.0	0.0	0.0	0.0	0.0	0.0	0.0	0.0	0.4	2.0
H/C	NORMAL HEATING DEG. DAYS	30	1020	826	592	312	129	10	0	0	44	299	546	868	4646
	NORMAL COOLING DEG. DAYS	30	0	0	0	6	83	202	313	276	125	26	0	0	1031
RH	NORMAL (PERCENT)	30	69	67	61	59	67	72	76	77	78	72	69	70	70
	HOUR 01 LST	30	74	72	67	67	79	86	90	91	90	83	76	75	79
	HOUR 07 LST	30	77	77	74	76	83	86	90	92	92	87	81	78	83
	HOUR 13 LST	30	62	59	52	47	51	54	58	59	58	54	57	62	56
	HOUR 19 LST	30	63	60	52	47	55	61	66	69	73	65	62	66	62
S	PERCENT POSSIBLE SUNSHINE														
W/O	MEAN NO. DAYS WITH:														
	HEAVY FOG(VISBY 1/4 MI)	54	3.7	3.1	2.6	2.8	7.4	10.8	13.8	17.9	15.6	10.3	4.2	3.6	95.8
	THUNDERSTORMS	54	0.5	0.9	2.3	4.1	6.7	8.2	9.6	7.4	3.3	1.1	0.7	0.3	45.1
CLOUDINESS	MEAN:														
	SUNRISE-SUNSET (OKTAS)	1					7.2	4.0						8.0	
	MIDNIGHT-MIDNIGHT (OKTAS)														
	MEAN NO. DAYS WITH:														
	CLEAR	1	1.0		5.0		7.0	12.0	1.0	7.0	4.0	6.0		3.0	
	PARTLY CLOUDY	1		1.0			2.0	3.0	1.0	3.0	1.0	1.0		2.0	
	CLOUDY	1	8.0	10.0	8.0		12.0	5.0	2.0	1.0	7.0	4.0		14.0	
PR	MEAN STATION PRESSURE(IN)	29	29.07	29.04	28.99	28.97	28.97	28.99	29.02	29.05	29.06	29.08	29.07	29.07	29.03
	MEAN SEA-LEVEL PRES. (IN)	49	30.12	30.09	30.03	30.00	30.00	30.00	30.03	30.04	30.08	30.11	30.10	30.13	30.06
WINDS	MEAN SPEED (MPH)	35	6.8	6.6	7.4	6.8	5.6	4.9	4.5	4.0	4.2	4.6	5.9	6.1	5.6
	PREVAIL.DIR(TENS OF DEGS)	22	23	23	23	22	22	22	22	23	23	06	23	23	23
	MAXIMUM 2-MINUTE:														
	SPEED (MPH)	7	34	31	32	30	36	35	36	29	23	28	33	31	36
	DIR. (TENS OF DEGS)		15	14	30	15	28	35	30	27	29	21	27	21	28
	YEAR OF OCCURRENCE		1996	1997	2000	1996	2001	1998	2000	2001	1999	2001	2000	2000	MAY 2001
	MAXIMUM 5-SECOND:														
	SPEED (MPH)	7	46	46	43	44	54	52	66	61	30	39	46	47	66
	DIR. (TENS OF DEGS)		13	31	31	29	28	35	20	31	21	22	27	26	20
	YEAR OF OCCURRENCE		1996	1999	2001	1998	2001	1998	1997	2000	2000	1996	2000	2001	JUL 1997
PRECIPITATION	NORMAL (IN)	30	2.91	3.04	3.63	3.31	3.94	3.59	4.99	4.01	3.24	2.89	3.59	3.39	42.53
	MAXIMUM MONTHLY (IN)	54	9.11	6.89	8.35	6.46	8.76	10.56	13.54	10.45	7.61	6.49	8.45	8.02	13.54
	YEAR OF OCCURRENCE		1950	1956	1997	1965	2001	1998	1961	1958	1971	1983	1985	1978	JUL 1961
	MINIMUM MONTHLY (IN)	54	1.09	0.64	1.30	0.50	0.84	0.70	1.98	0.66	0.65	0.09	0.64	0.45	0.09
	YEAR OF OCCURRENCE		1981	1968	1987	1976	1977	1966	1993	1957	1959	1963	1965	1965	OCT 1963
	MAXIMUM IN 24 HOURS (IN)	54	2.45	2.70	2.86	2.72	3.31	2.24	5.60	4.17	2.40	2.48	2.88	2.47	5.60
	YEAR OF OCCURRENCE		1994	2000	1967	1948	1982	1962	1961	1958	1956	1961	1991	1978	JUL 1961
	NORMAL NO. DAYS WITH:														
	PRECIPITATION 0.01	30	14.9	13.6	14.4	13.8	13.1	11.4	12.9	11.2	10.0	9.6	12.1	14.2	151.2
	PRECIPITATION 1.00	30	0.4	0.3	0.5	0.4	0.6	0.7	1.3	0.8	0.8	0.6	0.9	0.5	7.8
SNOWFALL	NORMAL (IN)	30	12.2	10.3	4.6	1.0	0.*	0.0	0.0	0.0	0.0	0.2	1.6	6.1	36.0
	MAXIMUM MONTHLY (IN)	49	39.5	21.8	20.4	20.7	0.6	T	T	T	T	2.8	25.8	21.9	39.5
	YEAR OF OCCURRENCE		1978	1964	1993	1987	1989	1994	1990	1989	1994	1961	1950	1995	JAN 1978
	MAXIMUM IN 24 HOURS (IN)	49	15.8	11.2	17.1	11.3	0.6	T	T	T	T	2.8	15.1	11.2	17.1
	YEAR OF OCCURRENCE		1978	1983	1993	1987	1989	1994	1990	1989	1994	1961	1950	1967	MAR 1993
	MAXIMUM SNOW DEPTH (IN)	44	23	13	9	17	T	0	0	0	0	2	19	10	23
	YEAR OF OCCURRENCE		1978	1985	1980	1987	1989					1961	1950	1967	JAN 1978
	NORMAL NO. DAYS WITH:														
	SNOWFALL 1.0	30	3.9	3.1	1.5	0.2	0.0	0.0	0.0	0.0	0.0	0.*	0.7	2.1	11.5

PRECIPITATION (inches) 2001 CHARLESTON, WV (CRW)

YEAR	JAN	FEB	MAR	APR	MAY	JUN	JUL	AUG	SEP	OCT	NOV	DEC	ANNUAL
1972	5.47	5.51	2.17	5.16	2.55	4.33	4.13	4.13	3.61	2.48	5.26	6.35	51.15
1973	1.52	2.41	3.40	5.44	5.36	4.48	6.88	2.07	3.91	4.75	5.42	3.68	49.32
1974	4.67	2.50	4.54	3.05	6.06	5.07	2.64	2.64	1.64	3.72	3.19	43.46
1975	4.84	3.10	6.01	4.03	6.44	4.25	2.71	5.14	4.99	3.08	2.66	3.74	50.99
1976	2.89	2.11	4.21	0.50	3.66	4.24	6.93	2.23	5.37	5.44	1.02	2.18	40.78
1977	1.90	1.08	3.16	4.06	0.84	5.93	4.92	6.58	1.14	4.16	3.78	2.07	39.62
1978	5.59	1.31	2.67	3.31	3.99	2.96	9.83	8.21	1.45	2.68	2.26	8.02	52.28
1979	6.48	3.76	3.00	3.82	3.87	3.54	5.17	4.78	3.95	4.02	2.81	4.87	48.87
1980	2.85	2.25	5.32	4.49	2.67	2.17	8.47	10.32	2.37	2.03	3.02	1.85	47.81
1981	1.09	4.59	1.80	4.04	3.78	6.46	3.02	2.24	2.36	2.43	1.29	2.71	35.81
1982	3.74	3.23	4.96	1.14	6.19	7.00	2.68	2.65	2.58	1.65	4.65	2.71	43.18
1983	1.24	2.72	3.15	3.96	5.98	2.77	4.19	2.54	1.33	6.49	4.80	3.19	42.36
1984	1.67	2.56	2.72	4.00	3.71	2.56	4.37	4.57	2.95	3.28	4.73	3.78	40.90
1985	3.07	2.32	4.23	1.84	5.88	3.07	3.22	2.02	0.71	3.65	8.45	2.71	41.17
1986	2.12	4.35	1.87	1.39	4.86	2.36	7.61	4.71	3.51	2.20	6.88	3.89	45.75
1987	3.23	3.34	1.30	4.05	2.49	3.38	4.23	3.56	3.89	1.10	2.71	4.13	37.41
1988	1.62	2.50	2.71	2.17	2.59	0.94	3.00	2.86	3.46	1.87	5.02	2.66	31.40
1989	2.92	6.05	5.81	4.13	6.79	7.54	3.04	5.62	7.28	4.09	2.87	1.83	57.97
1990	2.86	3.74	1.94	2.89	4.87	3.01	5.35	2.54	4.26	3.51	2.07	7.01	44.05
1991	2.68	2.98	6.07	3.49	1.47	2.49	2.84	2.95	5.51	1.10	5.00	5.89	42.47
1992	1.94	2.72	4.79	2.93	4.66	3.21	6.41	4.41	1.38	0.94	3.15	3.50	40.04
1993	1.87	2.98	6.68	1.78	1.98	5.01	1.98	2.71	5.99	3.50	3.95	3.23	41.66
1994	6.42	5.56	7.73	3.78	3.98	4.43	3.71	6.20	1.95	1.13	1.95	2.52	49.36
1995	6.02	2.98	2.73	2.59	6.15	4.93	2.91	5.81	2.70	2.61	3.31	2.79	45.53
1996	5.18	2.82	4.32	3.77	7.40	3.59	8.50	2.82	7.37	2.49	4.36	2.04	54.66
1997	1.76	1.76	8.35	2.77	3.60	5.24	5.83	4.14	1.94	0.84	2.96	1.57	40.76
1998	3.43	4.23	3.41	4.77	5.27	10.56	3.65	3.70	2.50	1.67	1.89	3.18	48.26
1999	4.81	2.67	3.70	2.20	1.90	1.30	5.37	2.97	1.81	3.43	4.53	2.55	37.24
2000	1.41	4.25	2.26	4.67	4.75	3.38	6.06	4.35	2.87	0.87	1.27	2.10	38.24
2001	2.43	1.90	3.28	1.30	8.76	4.19	10.06	2.74	1.85	1.36	1.43	2.47	41.77
POR= 100 YRS	3.61	3.27	4.07	3.54	3.90	3.89	4.97	4.12	3.06	2.69	3.18	3.20	43.50

AVERAGE TEMPERATURE (F) 2001   CHARLESTON, WV (CRW)

YEAR	JAN	FEB	MAR	APR	MAY	JUN	JUL	AUG	SEP	OCT	NOV	DEC	ANNUAL
1972	38.7	35.2	43.8	54.5	63.7	64.9	73.4	72.7	68.7	52.5	43.4	42.1	54.5
1973	34.3	35.1	52.2	53.8	60.8	73.3	74.5	74.6	70.1	58.9	46.8	38.0	56.0
1974	43.5	37.1	49.1	56.9	64.2	67.8	74.2	73.6	63.2	52.4	45.2	36.8	55.3
1975	35.7	38.0	39.8	50.2	66.1	71.7	74.2	76.7	63.9	57.6	50.2	38.6	55.2
1976	31.7	45.8	51.6	55.2	61.9	71.9	72.4	70.2	63.3	49.5	37.6	31.0	53.5
1977	18.6	33.2	49.8	58.4	67.0	68.6	76.9	73.6	70.2	53.4	49.1	35.1	54.5
1978	24.4	24.2	42.6	56.7	63.1	70.9	73.8	74.9	71.0	53.8	49.4	39.0	53.7
1979	28.1	27.9	50.5	55.1	63.0	68.8	72.9	73.4	67.0	54.6	47.5	38.3	53.9
1980	34.1	29.7	42.0	53.3	63.6	68.8	76.6	76.3	69.8	53.6	43.2	36.3	53.9
1981	28.0	37.2	41.4	59.1	60.4	73.2	75.7	72.8	66.5	54.2	45.3	34.6	54.0
1982	29.8	36.1	47.2	51.5	68.6	68.8	76.2	71.1	65.9	57.7	49.0	44.8	55.6
1983	34.0	37.7	47.0	52.1	61.1	71.6	77.0	78.0	68.4	58.1	47.4	32.0	55.4
1984	30.6	41.5	41.1	54.2	61.4	75.3	73.2	74.9	65.4	64.4	44.6	46.9	56.1
1985	27.2	34.0	49.4	60.8	66.3	71.0	75.8	74.0	69.6	62.3	55.5	33.8	56.6
1986	34.1	40.5	47.1	57.9	65.3	72.2	77.2	71.9	69.5	57.9	46.4	36.4	56.4
1987	33.0	37.2	47.0	52.7	68.4	73.5	77.1	77.0	67.1	50.3	49.0	39.8	56.0
1988	31.1	35.2	46.1	54.1	63.2	71.0	78.6	77.4	66.6	49.3	47.0	37.4	54.8
1989	41.1	34.9	47.8	52.8	59.4	71.7	75.7	73.2	67.0	56.7	46.4	26.0	54.4
1990	42.3	45.2	51.7	55.1	62.9	72.3	75.8	74.1	68.7	58.2	50.7	43.6	58.4
1991	36.5	40.3	47.4	60.3	71.7	73.9	77.9	75.2	68.4	59.1	45.8	41.3	58.2
1992	35.6	41.8	45.4	55.7	61.2	68.8	76.0	70.9	66.4	54.1	47.4	37.9	55.2
1993	40.2	34.0	41.8	55.0	64.9	71.4	79.0	76.4	66.6	55.1	47.1	36.6	55.7
1994	28.1	38.7	45.3	60.5	59.7	74.2	76.1	72.4	65.4	54.8	50.6	41.2	55.6
1995	34.7	34.0	46.9	55.2	61.9	71.4	76.4	77.6	64.6	56.9	40.8	32.5	54.4
1996	32.1	35.5	39.6	54.0	64.7	71.9	71.6	72.8	65.9	55.8	40.2	41.4	53.8
1997	35.6	43.8	47.1	50.8	58.7	70.1	74.5	70.8	65.0	55.1	42.2	36.5	54.2
1998	41.0	42.0	45.9	55.4	65.8	70.3	73.7	73.9	70.2	56.1	46.7	40.4	56.8
1999	38.3	39.4	40.1	58.5	64.0	72.9	79.1	72.3	65.8	54.8	48.9	38.2	56.0
2000	32.1	43.2	49.2	54.1	66.0	72.3	72.3	71.3	65.7	57.7	43.3	28.6	54.6
2001	33.2	40.6	40.3	59.1	64.0	70.6	72.4	74.8	64.6	55.1	50.4	42.3	55.6
POR= 94 YRS	35.7	38.0	45.7	55.8	64.5	72.1	76.0	74.7	68.9	57.4	46.6	37.9	56.1

REFERENCE NOTES:

PAGE 1:
THE TEMPERATURE GRAPH SHOWS NORMAL MAXIMUM AND NORMAL
MINIMUM DAILY TEMPERATURES (SOLID CURVES) AND THE
ACTUAL DAILY HIGH AND LOW TEMPERATURES (VERTICAL BARS).

PAGE 2 AND 3:
H/C INDICATES HEATING AND COOLING DEGREE DAYS.
RH INDICATES RELATIVE HUMIDITY
W/O INDICATES WEATHER AND OBSTRUCTIONS
S INDICATES SUNSHINE.
PR INDICATES PRESSURE.
CLOUDINESS ON PAGE 3 IS THE SUM OF THE CEILOMETER AND
SATELLITE DATA NOT TO EXCEED EIGHT EIGHTHS (OKTAS).

GENERAL:
T INDICATES TRACE PRECIPITATION, AN AMOUNT GREATER
THAN ZERO BUT LESS THAN THE LOWEST REPORTABLE VALUE.
+ INDICATES THE VALUE ALSO OCCURS ON EARLIER DATES.
BLANK ENTRIES DENOTE MISSING OR UNREPORTED DATA.
NORMALS ARE 30-YEAR AVERAGES (1961 - 1990).
ASOS INDICATES AUTOMATED SURFACE OBSERVING SYSTEM.
PM INDICATES THE LAST DAY OF THE PREVIOUS MONTH.
POR (PERIOD OF RECORD) BEGINS WITH THE JANUARY DATA
MONTH AND IS THE NUMBER OF YEARS USED TO COMPUTE
THE MEAN. INDIVIDUAL MONTHS WITHIN THE POR MAY
BE MISSING.
WHEN THE POR FOR A NORMAL IS LESS THAN 30 YEARS,
THE NORMAL IS PROVISIONAL AND IS BASED ON THE NUMBER
OF YEARS INDICATED.
0.* OR * INDICATES THE VALUE OR MEAN-DAYS-WITH
IS BETWEEN 0.00 AND 0.05.
CLOUDINESS FOR ASOS STATIONS DIFFERS FROM THE NON-ASOS
OBSERVATION TAKEN BY A HUMAN OBSERVER. ASOS STATION
CLOUDINESS IS BASED ON TIME-AVERAGED CEILOMETER DATA
FOR CLOUDS AT OR BELOW 12,000 FEET AND ON SATELLITE
DATA FOR CLOUDS ABOVE 12,000 FEET.
THE NUMBER OF DAYS WITH CLEAR, PARTLY CLOUDY, AND
CLOUDY CONDITIONS FOR ASOS STATIONS IS THE SUM
OF THE CEILOMETER AND SATELLITE DATA FOR THE
SUNRISE TO SUNSET PERIOD.

GENERAL CONTINUED:
CLEAR INDICATES 0 - 2 OKTAS, PARTLY CLOUDY INDICATES
3 - 6 OKTAS, AND CLOUDY INDICATES 7 OR 8 OKTAS.
WHEN AT LEAST ONE OF THE ELEMENTS (CEILOMETER OR
SATELLITE) IS MISSING, THE DAILY CLOUDINESS IS
NOT COMPUTED.
WIND DIRECTION IS RECORDED IN TENS OF DEGREES (2 DIGITS)
CLOCKWISE FROM TRUE NORTH. "00" INDICATES CALM. "36"
INDICATES TRUE NORTH.
RESULTANT WIND IS THE VECTOR AVERAGE OF THE SPEED AND
DIRECTION.
AVERAGE TEMPERATURE IS THE SUM OF THE MEAN DAILY MAXIMUM
AND MINIMUM TEMPERATURE DIVIDED BY 2.
SNOWFALL DATA COMPRISE ALL FORMS OF FROZEN
PRECIPITATION, INCLUDING HAIL.
A HEATING (COOLING) DEGREE DAY IS THE DIFFERENCE BETWEEN
THE AVERAGE DAILY TEMPERATURE AND 65 F.
DRY BULB IS THE TEMPERATURE OF THE AMBIENT AIR.
DEW POINT IS THE TEMPERATURE TO WHICH THE AIR MUST BE
COOLED TO ACHIEVE 100 PERCENT RELATIVE HUMIDITY.
WET BULB IS THE TEMPERATURE THE AIR WOULD HAVE IF THE
MOISTURE CONTENT WAS INCREASED TO 100 PERCENT RELATIVE
HUMIDITY.

ON JULY 1, 1996, THE NATIONAL WEATHER SERVICE BEGAN USING
THE "METAR" OBSERVATION CODE THAT WAS ALREADY EMPLOYED
BY MOST OTHER NATIONS OF THE WORLD.   THE MOST NOTICEABLE
DIFFERENCE IN THIS ANNUAL PUBLICATION WILL BE THE CHANGE
IN UNITS FROM TENTHS TO EIGHTS (OKTAS) FOR REPORTING THE
AMOUNT OF SKY COVER.

HEATING DEGREE DAYS (base 65 F) 2001 CHARLESTON, WV (CRW)

YEAR	JUL	AUG	SEP	OCT	NOV	DEC	JAN	FEB	MAR	APR	MAY	JUN	TOTAL
1972-73	16	2	14	378	642	701	945	832	394	351	157	4	4436
1973-74	0	1	19	202	541	833	659	775	277	115	25	4	3947
1974-75	0	0	110	388	590	869	899	749	772	445	59	4	4885
1975-76	0	0	106	227	441	813	1025	549	427	342	142	4	4076
1976-77	0	9	84	475	814	1047	1432	888	482	242	81	52	5606
1977-78	0	2	19	357	482	919	1249	1138	691	258	137	23	5275
1978-79	0	0	18	344	462	797	1137	1031	456	308	125	19	4697
1979-80	5	10	39	331	519	820	951	1017	707	349	106	27	4881
1980-81	0	0	33	356	650	882	1138	774	727	207	175	2	4944
1981-82	0	1	76	335	585	936	1086	801	545	405	36	2	4808
1982-83	1	2	69	268	480	626	955	757	554	388	153	16	4269
1983-84	4	0	66	227	521	1019	1059	674	734	346	171	5	4826
1984-85	1	0	98	74	613	563	1164	860	488	192	54	18	4125
1985-86	0	0	51	127	294	960	954	679	554	249	83	7	3958
1986-87	0	23	23	255	550	880	989	770	549	374	63	4	4480
1987-88	0	0	37	447	473	774	1043	859	577	326	112	38	4686
1988-89	2	0	37	484	534	849	735	837	536	367	221	2	4604
1989-90	0	7	72	270	553	1203	697	549	446	323	111	8	4239
1990-91	0	0	59	230	428	655	876	685	558	192	21	1	3705
1991-92	0	0	80	229	576	729	904	670	602	319	170	24	4303
1992-93	0	1	67	335	522	834	760	862	713	306	73	24	4497
1993-94	0	0	64	307	540	873	1136	732	607	191	197	3	4650
1994-95	0	4	34	308	424	729	932	861	555	306	138	7	4298
1995-96	0	0	72	262	721	999	1008	848	779	357	105	4	5155
1996-97	2	0	59	278	743	727	908	592	547	429	213	23	4521
1997-98	0	6	53	324	676	877	733	639	626	283	60	38	4315
1998-99	0	0	29	290	541	758	821	711	763	213	74	13	4213
1999-00	0	0	56	312	475	821	1014	627	490	320	51	14	4181
2000-01	0	0	94	250	644	1122	978	681	759	249	82	13	4872
2001-	0	0	95	309	433	699							

COOLING DEGREE DAYS (base 65 F) 2001 CHARLESTON, WV (CRW)

YEAR	JAN	FEB	MAR	APR	MAY	JUN	JUL	AUG	SEP	OCT	NOV	DEC	ANNUAL
1972	0	0	0	24	56	85	283	247	132	0	0	0	827
1973	0	0	4	22	34	256	304	305	181	22	2	0	1130
1974	0	0	14	43	99	118	292	275	62	6	1	0	910
1975	0	0	0	7	99	212	291	372	82	7	4	0	1074
1976	0	1	17	58	54	218	238	175	39	1	0	0	801
1977	0	0	18	50	148	165	373	277	180	4	12	0	1227
1978	0	0	0	16	88	207	279	314	205	4	1	0	1114
1979	0	0	13	18	69	138	257	277	105	17	0	0	894
1980	0	0	0	6	71	147	370	358	182	9	0	0	1143
1981	0	0	2	38	41	256	340	251	126	5	0	0	1059
1982	0	0	0	6	154	122	355	196	101	47	5	6	992
1983	0	0	2	6	39	222	385	407	177	18	0	0	1256
1984	0	0	0	27	64	318	261	312	116	65	7	8	1178
1985	0	0	9	72	105	204	339	285	194	52	14	0	1274
1986	0	0	4	41	100	227	384	244	167	43	0	0	1210
1987	0	0	0	13	177	268	381	379	108	0	2	0	1328
1988	0	0	3	9	64	225	430	392	91	4	3	0	1221
1989	0	0	11	6	55	211	339	273	140	23	2	0	1060
1990	0	0	41	33	54	232	342	286	174	28	7	0	1197
1991	0	0	17	60	236	273	408	324	190	53	5	0	1566
1992	0	0	2	47	57	143	347	192	145	4	0	0	937
1993	0	0	0	14	76	222	444	364	119	7	8	0	1254
1994	0	0	4	62	37	288	353	240	51	0	1	0	1036
1995	1	0	1	20	51	209	363	401	69	23	2	0	1140
1996	0	0	0	34	105	214	213	247	93	2	5	0	913
1997	1	5	0	11	23	182	302	195	61	23	0	0	803
1998	0	0	39	3	93	205	275	283	194	20	0	4	1116
1999	0	0	0	24	48	254	443	236	89	1	0	0	1095
2000	0	0	5	3	93	241	201	218	122	27	0	0	910
2001	0	0	0	80	56	185	235	310	93	12	0	0	971

SNOWFALL (inches) 2001 CHARLESTON, WV (CRW)

YEAR	JUL	AUG	SEP	OCT	NOV	DEC	JAN	FEB	MAR	APR	MAY	JUN	TOTAL
1972-73	0.0	0.0	0.0	0.9	6.9	1.4	3.4	2.9	4.8	0.9	0.0	0.0	21.2
1973-74	0.0	0.0	0.0	0.0	T	6.5	0.3	13.0	0.9	1.2	0.0	0.0	21.9
1974-75	0.0	0.0	0.0	0.6	2.7	8.3	18.1	2.4	8.1	0.1	0.0	0.0	40.3
1975-76	0.0	0.0	0.0	0.0	T	4.0	14.3	2.6	5.4	0.0	0.0	0.0	26.3
1976-77	0.0	0.0	0.0	T	4.7	7.4	22.2	11.1	1.0	1.5	0.0	0.0	47.9
1977-78	0.0	0.0	0.0	0.0	4.4	5.4	39.5	15.6	11.7	0.0	0.0	0.0	76.6
1978-79	0.0	0.0	0.0	0.0	T	1.5	27.5	20.1	5.5	T	0.0	0.0	54.6
1979-80	0.0	0.0	0.0	T	0.9	0.4	11.7	12.7	10.5	T	0.0	0.0	36.2
1980-81	0.0	0.0	0.0	T	0.5	2.6	9.1	6.8	7.5	T	0.0	0.0	26.5
1981-82	0.0	0.0	0.0	T	0.5	8.2	12.1	6.4	7.4	1.0	0.0	0.0	35.6
1982-83	0.0	0.0	0.0	0.0	T	2.9	5.8	15.0	5.2	0.1	0.0	0.0	29.0
1983-84	0.0	0.0	0.0	0.0	0.3	3.8	12.8	9.7	2.4	0.0	0.0	0.0	29.0
1984-85	0.0	0.0	0.0	0.0	T	3.7	17.6	20.1	0.9	1.7	0.0	0.0	44.0
1985-86	0.0	0.0	0.0	0.0	0.0	8.9	13.1	17.7	3.7	T	0.0	0.0	43.4
1986-87	0.0	0.0	0.0	0.0	0.1	16.3	9.7	3.9	20.7	0.0	0.0	0.0	50.9
1987-88	0.0	0.0	0.0	T	2.4	5.7	8.3	7.8	4.6	T	0.0	0.0	28.8
1988-89	0.0	0.0	0.0	T	T	6.9	1.7	4.6	1.4	0.6	0.0	0.0	15.2
1989-90	0.0	T	T	T	2.0	14.1	11.0	3.8	6.6	1.1	0.0	0.0	38.6
1990-91	T	0.0	0.0	0.0	0.0	1.2	3.5	6.5	5.3	T	0.0	0.0	16.5
1991-92	0.0	0.0	0.0	0.0	4.1	0.7	5.6	1.1	8.6	3.6	T	T	23.7
1992-93	0.0	0.0	0.0	T	2.5	3.7	0.4	12.0	20.4	T	0.0	T	39.0
1993-94	0.0	0.0	0.0	1.5	0.4	12.4	34.2	7.0	3.1	0.0	0.0	T	58.6
1994-95	0.0	0.0	T	0.0	T	T	9.1	7.9	8.7	0.0	0.0	0.0	25.7
1995-96	0.0	0.0	0.0	0.0	13.6	21.9	35.1	14.2	20.4	0.8	0.0	T	106.0
1996-97	0.0	0.0	0.0										
1997-98													
1998-99													
1999-00													
2000-01													
2001-													
POR= 49 YRS	T	T	T	0.1	2.4	5.2	11.1	8.7	5.4	0.9	T	T	33.8

2001
MADISON,
WISCONSIN (MSN)

Madison is set on a narrow isthmus of land between Lakes Mendota and Monona. Lake Mendota (15 square miles) lies northwest of Lake Monona (5 square miles) and the lakes are only two-thirds of a mile apart at one point. Drainage at Madison is southeast through two other lakes into the Rock River, which flows south into Illinois, and then west to the Mississippi. The westward flowing Wisconsin River is only 20 miles northwest of Madison. Madison lakes are normally frozen from mid-December to early April.

Madison has the typical continental climate of interior North America with a large annual temperature range and with frequent short period temperature changes. The range of extreme temperatures is from about 110 to -40 degrees. Winter temperatures (December-February) average near 20 degrees and the summer average (June-August) is in the upper 60s. Daily temperatures average below 32 degrees about 120 days and above 40 degrees for about 210 days of the year.

Madison lies in the path of the frequent cyclones and anticyclones which move eastward over this area during fall, winter and spring. In summer, the cyclones have diminished intensity and tend to pass farther north. The most frequent air masses are of polar origin. Occasional outbreaks of arctic air affect this area during the winter months. Although northward moving tropical air masses contribute considerable cloudiness and precipitation, the true Gulf air mass does not reach this area in winter, and only occasionally at other seasons. Summers are pleasant, with only occasional periods of extreme heat or high humidity.

There are no dry and wet seasons, but about 60 percent of the annual precipitation falls in the five months of May through September. Cold season precipitation is lighter, but lasts longer. Soil moisture is usually adequate in the first part of the growing season. During July, August, and September, the crops depend on current rainfall, which is mostly from thunderstorms and tends to be erratic and variable. Average occurrence of thunderstorms is just under 7 days per month during this period.

March and November are the windiest months. Tornadoes are infrequent. Dane County has about one tornado in every three to five years.

The ground is covered with 1 inch or more of snow about 60 percent of the time from about December 10 to near February 25 in an average winter. The soil is usually frozen from the first of December through most of March with an average frost penetration of 25 to 30 inches. The growing season averages 175 days.

Farming is diversified with the main emphasis on dairying. Field crops are mainly corn, oats, clover, and alfalfa, but barley, wheat, rye, and tobacco are also raised. Canning factories pack peas, sweet corn, and lima beans. Fruits are mainly apples, strawberries, and raspberries.

NORMALS, MEANS, AND EXTREMES

MADISON, WI (MSN)

LATITUDE:	LONGITUDE:	ELEVATION (FT):		TIME ZONE:	WBAN: 14837
43 08' 26" N	89 20' 43" W	GRND: 857	BARO: 860	CENTRAL (UTC + 6)	

	ELEMENT	POR	JAN	FEB	MAR	APR	MAY	JUN	JUL	AUG	SEP	OCT	NOV	DEC	YEAR
TEMPERATURE °F	NORMAL DAILY MAXIMUM	30	24.8	30.1	41.5	56.7	68.9	78.2	82.4	79.6	71.5	59.9	44.0	29.8	55.6
	MEAN DAILY MAXIMUM	54	25.9	30.9	42.1	57.5	69.9	79.0	83.0	80.8	72.3	60.9	44.1	31.0	56.4
	HIGHEST DAILY MAXIMUM	62	56	64	82	94	93	101	104	102	99	90	76	64	104
	YEAR OF OCCURRENCE		1989	2000	1986	1980	1975	1988	1976	1988	1953	1976	1964	2001	JUL 1976
	MEAN OF EXTREME MAXS.	54	44.2	48.4	66.5	78.9	86.1	91.6	93.8	92.2	87.5	79.2	64.7	50.5	73.6
	NORMAL DAILY MINIMUM	30	7.2	11.1	23.0	34.1	44.2	54.2	59.5	56.9	48.2	37.7	26.7	13.5	34.7
	MEAN DAILY MINIMUM	54	8.3	12.8	22.9	34.9	45.2	54.7	59.6	57.6	48.7	38.4	26.8	14.9	35.4
	LOWEST DAILY MINIMUM	62	-37	-29	-29	0	19	31	36	35	25	13	-11	-25	-37
	YEAR OF OCCURRENCE		1951	1996	1962	1982	1978	1972	1965	1968	1974	1988	1947	1983	JAN 1951
	MEAN OF EXTREME MINS.	54	-15.0	-10.1	2.6	19.5	29.9	39.7	46.3	43.8	32.4	23.0	9.0	-7.6	17.8
	NORMAL DRY BULB	30	16.0	20.6	32.3	45.4	56.5	66.2	71.0	68.3	59.8	48.9	35.4	21.7	45.2
	MEAN DRY BULB	54	17.1	21.9	32.4	46.0	57.4	67.0	71.4	69.1	60.5	49.6	35.6	22.9	45.9
	MEAN WET BULB	17	18.1	22.4	30.9	41.4	52.7	61.3	62.1	64.3	56.2	44.9	32.9	22.4	42.4
	MEAN DEW POINT	17	13.3	16.9	24.5	34.1	46.1	56.4	58.5	60.9	52.3	39.6	27.9	17.7	37.3
	NORMAL NO. DAYS WITH:														
	MAXIMUM ≥ 90	30	0.0	0.0	0.0	*	0.3	3.1	5.4	2.8	0.5	0.1	0.0	0.0	12.2
	MAXIMUM ≤ 32	30	21.2	15.1	5.7	0.4	0.0	0.0	0.0	0.0	0.0	*	3.8	17.2	63.4
	MINIMUM ≤ 32	30	30.2	26.9	25.8	14.1	3.3	*	0.0	0.0	1.2	10.2	21.8	29.2	162.7
	MINIMUM ≤ 0	30	11.3	6.9	0.9	*	0.0	0.0	0.0	0.0	0.0	0.0	0.2	5.7	25.0
H/C	NORMAL HEATING DEG. DAYS	30	1519	1243	1014	588	294	68	12	38	168	499	888	1342	7673
	NORMAL COOLING DEG. DAYS	30	0	0	0	0	30	104	198	141	12	0	0	0	485
RH	NORMAL (PERCENT)	30	74	73	71	66	66	68	71	74	77	73	77	78	72
	HOUR 00 LST	30	78	78	78	75	76	80	84	87	88	82	82	82	81
	HOUR 06 LST	30	79	80	81	81	80	82	86	91	92	86	85	83	84
	HOUR 12 LST	30	69	66	62	54	53	55	56	59	61	59	67	73	61
	HOUR 18 LST	30	73	70	65	56	54	56	58	62	68	68	75	77	65
S	PERCENT POSSIBLE SUNSHINE	50	47	51	52	52	58	64	67	64	60	54	39	40	54
W/O	MEAN NO. DAYS WITH:														
	HEAVY FOG(VISBY 1/4 MI)	55	2.3	2.0	2.6	1.4	1.4	1.1	1.4	2.2	1.9	1.7	1.9	2.8	22.7
	THUNDERSTORMS	53	0.2	0.2	1.9	3.7	5.1	7.0	7.5	6.3	4.5	2.0	0.8	0.3	39.5
CLOUDINESS	MEAN:														
	SUNRISE-SUNSET (OKTAS)	1						6.4							
	MIDNIGHT-MIDNIGHT (OKTAS)														
	MEAN NO. DAYS WITH:														
	CLEAR	1	6.0	7.0	8.0										
	PARTLY CLOUDY	1	4.0	6.0	4.0										
	CLOUDY	1	21.0	16.0	19.0										
PR	MEAN STATION PRESSURE(IN)	29	29.11	29.13	29.06	29.03	29.03	29.03	29.06	29.10	29.11	29.10	29.08	29.12	29.08
	MEAN SEA-LEVEL PRES. (IN)	17	30.07	30.10	30.04	29.94	29.94	29.94	29.97	30.02	30.03	30.04	30.04	30.09	30.02
WINDS	MEAN SPEED (MPH)	46	10.2	10.1	11.0	11.3	9.8	8.8	8.0	7.7	8.3	9.3	10.3	9.8	9.6
	PREVAIL.DIR(TENS OF DEGS)	30	30	30	30	31	18	18	18	18	18	18	18	30	18
	MAXIMUM 2-MINUTE:														
	SPEED (MPH)	5	34	30	34	36	38	41	40	28	43	31	38	28	43
	DIR. (TENS OF DEGS)		06	16	01	02	16	36	33	31	15	17	18	34	15
	YEAR OF OCCURRENCE		1999	2000	1998	2000	2001	1998	1997	2000	2000	1997	1998	1998	SEP 2000
	MAXIMUM 5-SECOND:														
	SPEED (MPH)	5	45	41	41	53	44	53	53	41	74	40	52	37	74
	DIR. (TENS OF DEGS)		07	25	18	26	16	31	22	31	15	17	21	31	15
	YEAR OF OCCURRENCE		1999	2001	1998	1997	2001	2000	1997	2000	2000	1997	1998	1999	SEP 2000
PRECIPITATION	NORMAL (IN)	30	1.07	1.08	2.17	2.86	3.14	3.66	3.39	4.04	3.37	2.17	2.09	1.84	30.88
	MAXIMUM MONTHLY (IN)	62	2.53	2.77	5.46	7.11	9.63	9.95	10.93	9.49	9.51	5.63	5.13	4.09	10.93
	YEAR OF OCCURRENCE		1996	1953	1998	1973	2000	1978	1950	1980	1941	1984	1985	1987	JUL 1950
	MINIMUM MONTHLY (IN)	62	0.14	0.06	0.28	0.96	0.64	0.81	1.38	0.70	0.11	0.06	0.11	0.25	0.06
	YEAR OF OCCURRENCE		1981	1995	1978	1946	1981	1973	1946	1948	1979	1952	1976	1960	FEB 1995
	MAXIMUM IN 24 HOURS (IN)	53	1.27	1.59	3.01	2.83	4.11	4.51	5.25	3.95	3.57	2.78	2.36	2.19	5.25
	YEAR OF OCCURRENCE		1960	2001	1998	1975	2000	1996	1950	2001	1961	1984	1985	1990	JUL 1950
	NORMAL NO. DAYS WITH:														
	PRECIPITATION ≥ 0.01	30	10.5	7.8	10.6	11.6	11.5	10.1	9.6	9.5	9.5	9.3	10.0	10.2	120.2
	PRECIPITATION ≥ 1.00	30	0.1	0.1	0.1	0.5	0.7	0.9	0.8	1.0	0.9	0.3	0.4	0.3	6.1
SNOWFALL	NORMAL (IN)	30	9.9	7.1	7.9	2.6	0.1	0.0	0.0	0.0	T	0.2	3.4	12.2	43.4
	MAXIMUM MONTHLY (IN)	53	27.5	37.0	25.4	17.4	3.0	T	T	T	T	3.9	18.3	35.0	37.0
	YEAR OF OCCURRENCE		1995	1994	1959	1973	1990	1992	1994	1994	1994	1997	1985	2000	FEB 1994
	MAXIMUM IN 24 HOURS (IN)	53	13.0	14.2	13.6	12.9	3.0	T	T	T	T	3.8	9.0	17.3	17.3
	YEAR OF OCCURRENCE		1996	1994	1971	1973	1990	1992	1994	1994	1994	1997	1985	1990	DEC 1990
	MAXIMUM SNOW DEPTH (IN)	53	32	28	16	14	4	0	0	0	0	4	9	17	32
	YEAR OF OCCURRENCE		1979	1979	1986	1973	1994					1997	1985	1990	JAN 1979
	NORMAL NO. DAYS WITH:														
	SNOWFALL ≥ 1.0	30	2.8	2.5	2.2	0.7	0.*	0.0	0.0	0.0	0.0	0.*	1.2	3.5	12.9

PRECIPITATION (inches) 2001 MADISON, WI (MSN)

YEAR	JAN	FEB	MAR	APR	MAY	JUN	JUL	AUG	SEP	OCT	NOV	DEC	ANNUAL
1972	0.40	0.42	2.23	2.02	2.83	1.65	3.49	7.47	5.26	2.42	0.86	1.91	30.96
1973	1.54	1.20	5.04	7.11	5.27	0.81	2.68	2.53	3.59	2.30	1.48	1.98	35.53
1974	2.45	1.17	3.43	4.24	5.77	3.86	2.69	4.60	1.08	3.18	1.79	1.80	36.06
1975	0.98	1.54	3.09	4.19	4.57	4.30	6.05	5.25	0.84	0.64	2.79	0.29	34.53
1976	0.56	1.72	4.75	4.80	1.95	1.38	1.46	1.99	0.50	1.49	0.11	0.37	21.08
1977	0.53	1.44	3.03	2.59	2.52	2.63	6.63	5.19	2.84	1.41	2.12	1.60	32.53
1978	1.03	0.24	0.28	3.50	3.96	9.95	4.54	1.63	5.44	1.11	3.05	1.71	36.44
1979	1.69	0.90	2.67	2.46	2.70	2.53	2.80	4.96	0.11	3.10	2.27	1.93	28.12
1980	1.11	0.64	0.68	2.36	2.08	3.43	2.67	9.49	7.84	1.13	1.33	1.62	34.38
1981	0.14	2.47	0.33	3.42	0.64	4.99	4.81	7.06	3.10	2.68	1.71	0.75	32.10
1982	1.42	0.17	2.11	3.26	4.34	3.40	3.47	2.67	1.42	1.46	4.21	3.65	31.58
1983	0.53	2.26	2.70	2.23	4.21	1.85	1.92	5.05	2.85	2.59	3.18	2.30	31.67
1984	0.36	1.26	1.15	3.86	3.32	7.01	1.96	1.89	2.79	5.63	1.83	2.66	33.72
1985	1.43	1.89	3.13	1.52	3.35	3.06	4.48	2.98	5.00	4.58	5.13	2.39	38.94
1986	1.02	2.72	1.55	2.27	1.97	3.24	4.31	4.38	6.82	1.85	1.03	0.69	31.85
1987	0.68	0.62	1.99	2.46	3.90	1.17	3.26	7.16	3.61	1.24	3.24	4.09	33.42
1988	1.82	0.46	1.20	2.65	0.92	2.06	2.44	2.95	3.33	1.60	3.58	1.56	24.57
1989	0.61	0.57	1.69	1.69	1.72	1.67	4.97	6.46	0.89	1.88	0.98	0.26	23.39
1990	1.60	0.99	4.18	1.90	5.35	4.88	2.61	6.03	1.64	2.25	1.65	3.46	36.54
1991	1.17	0.44	4.24	4.89	2.20	3.75	5.18	2.34	3.96	5.35	3.86	1.71	39.09
1992	0.78	1.34	1.90	3.17	1.12	1.53	5.54	2.48	5.99	1.06	4.83	2.39	32.13
1993	1.60	1.18	3.29	5.33	3.81	6.67	9.34	5.57	3.74	0.91	1.55	0.35	43.34
1994	1.46	2.76	0.46	2.57	1.33	5.66	4.10	4.56	6.14	0.65	2.77	1.08	33.54
1995	2.12	0.06	2.17	4.14	3.92	1.22	4.36	5.58	1.78	4.29	3.17	0.77	33.58
1996	2.53	0.53	0.80	2.76	2.95	9.69	4.08	1.84	1.07	3.14	1.01	1.27	31.67
1997	1.36	2.52	1.54	2.50	1.94	5.23	6.23	2.33	1.38	1.23	1.25	1.25	28.76
1998	2.24	1.44	5.46	4.10	4.58	7.46	2.50	4.24	2.48	3.20	1.95	0.29	39.94
1999	2.10	0.91	0.47	6.91	3.72	5.57	4.49	3.26	1.55	0.88	1.21	0.86	31.93
2000	0.91	1.95	1.17	3.18	9.63	8.63	3.27	3.94	3.59	0.68	2.00	1.39	40.34
2001	0.99	2.64	0.59	3.07	4.16	5.40	3.09	7.64	5.53	2.62	1.59	1.13	38.45
POR= 62 YRS	1.25	1.16	2.07	3.01	3.34	4.23	3.82	3.73	3.14	2.19	2.06	1.55	31.55

AVERAGE TEMPERATURE (F) 2001   MADISON, WI (MSN)

YEAR	JAN	FEB	MAR	APR	MAY	JUN	JUL	AUG	SEP	OCT	NOV	DEC	ANNUAL
1972	12.7	16.5	28.7	41.3	59.2	62.8	68.3	69.2	59.4	45.8	34.6	17.3	43.0
1973	23.4	24.0	41.6	44.9	54.4	67.9	71.6	70.6	60.7	54.1	36.6	21.5	47.6
1974	19.2	18.4	33.1	48.7	54.1	64.0	72.1	66.8	57.4	50.5	37.1	26.8	45.7
1975	21.9	21.3	26.1	41.0	62.5	69.2	72.4	70.6	57.5	52.2	41.9	25.5	46.8
1976	15.7	28.4	36.5	49.3	54.3	68.3	73.4	68.8	58.0	43.7	28.1	13.2	44.8
1977	3.7	22.4	39.8	51.9	65.2	64.9	73.6	64.6	59.7	47.6	34.0	19.6	45.6
1978	10.5	12.4	29.4	44.5	57.9	65.9	69.7	69.5	63.8	47.3	33.5	21.3	43.8
1979	6.9	11.7	32.1	42.4	56.7	66.0	69.8	66.6	61.1	47.5	35.1	28.8	43.7
1980	17.3	15.7	28.0	45.5	57.8	65.3	73.4	70.3	59.9	43.7	35.4	22.6	44.6
1981	20.5	25.3	36.9	48.7	55.3	67.4	70.6	68.7	59.1	46.6	36.7	22.0	46.5
1982	8.0	19.1	30.6	41.7	60.8	59.6	70.9	66.3	59.0	50.6	34.2	28.8	44.1
1983	21.4	26.3	33.1	41.6	51.9	67.5	75.0	72.2	60.1	48.2	37.3	10.8	45.5
1984	14.8	30.2	26.7	45.6	53.4	67.5	70.2	71.3	59.3	52.0	33.9	26.4	45.9
1985	12.2	19.0	37.7	52.2	60.7	63.8	70.0	66.4	61.6	49.4	31.0	11.3	44.6
1986	18.2	19.4	36.2	49.8	58.4	65.9	73.2	64.8	61.6	49.7	31.2	25.5	46.2
1987	22.6	30.5	37.2	49.9	60.8	70.4	74.5	68.7	60.6	43.4	40.0	28.4	48.9
1988	13.8	17.4	34.6	46.0	60.5	69.5	74.1	74.5	63.0	43.5	38.8	24.6	46.6
1989	27.6	14.6	30.1	44.7	56.1	65.7	72.3	68.6	58.7	50.8	33.1	14.2	44.7
1990	28.6	25.8	37.7	48.5	53.6	67.6	70.6	69.9	63.7	48.3	41.0	21.4	48.1
1991	15.1	26.5	36.8	49.4	63.6	71.1	72.3	70.2	60.0	49.4	31.4	26.3	47.7
1992	25.5	29.3	34.8	43.6	58.0	64.9	67.2	65.3	60.4	48.4	34.7	24.9	46.4
1993	21.8	21.2	31.5	43.7	59.6	65.9	72.0	72.1	56.7	47.9	35.2	26.6	46.2
1994	8.8	15.8	35.8	48.3	58.0	70.3	70.7	66.5	64.9	52.4	40.2	28.8	46.7
1995	20.4	22.9	36.9	43.8	57.6	72.0	74.8	76.9	58.7	50.2	28.7	20.6	47.0
1996	15.2	21.6	28.6	42.2	53.2	66.7	67.5	68.6	60.3	48.5	28.7	22.6	43.6
1997	15.7	24.1	34.1	43.3	51.3	67.3	69.3	65.4	60.2	50.2	33.3	27.9	45.2
1998	23.7	33.5	34.2	48.1	62.5	66.3	71.2	70.6	64.7	51.6	40.6	31.3	49.9
1999	16.5	30.8	33.8	48.1	59.9	68.4	74.9	66.9	60.3	48.9	42.0	26.4	48.1
2000	20.6	29.6	40.7	45.3	59.1	65.8	69.4	69.8	61.1	52.5	34.9	11.2	46.7
2001	21.7	19.7	31.7	51.1	58.4	66.0	72.3	70.9	58.7	48.9	46.1	30.4	48.0
POR= 62 YRS	17.4	21.9	32.4	46.3	57.3	66.9	71.5	69.4	60.7	49.9	35.6	22.9	46.0

REFERENCE NOTES:

PAGE 1:
THE TEMPERATURE GRAPH SHOWS NORMAL MAXIMUM AND NORMAL MINIMUM DAILY TEMPERATURES (SOLID CURVES) AND THE ACTUAL DAILY HIGH AND LOW TEMPERATURES (VERTICAL BARS).

PAGE 2 AND 3:
H/C INDICATES HEATING AND COOLING DEGREE DAYS.
RH INDICATES RELATIVE HUMIDITY
W/O INDICATES WEATHER AND OBSTRUCTIONS
S INDICATES SUNSHINE.
PR INDICATES PRESSURE.
CLOUDINESS ON PAGE 3 IS THE SUM OF THE CEILOMETER AND SATELLITE DATA NOT TO EXCEED EIGHT EIGHTHS(OKTAS).

GENERAL:
T INDICATES TRACE PRECIPITATION, AN AMOUNT GREATER THAN ZERO BUT LESS THAN THE LOWEST REPORTABLE VALUE.
+ INDICATES THE VALUE ALSO OCCURS ON EARLIER DATES.
BLANK ENTRIES DENOTE MISSING OR UNREPORTED DATA.
NORMALS ARE 30-YEAR AVERAGES (1961 - 1990).
ASOS INDICATES AUTOMATED SURFACE OBSERVING SYSTEM.
PM INDICATES THE LAST DAY OF THE PREVIOUS MONTH.
POR (PERIOD OF RECORD) BEGINS WITH THE JANUARY DATA MONTH AND IS THE NUMBER OF YEARS USED TO COMPUTE THE MEAN. INDIVIDUAL MONTHS WITHIN THE POR MAY BE MISSING.
WHEN THE POR FOR A NORMAL IS LESS THAN 30 YEARS, THE NORMAL IS PROVISIONAL AND IS BASED ON THE NUMBER OF YEARS INDICATED.
0.* OR * INDICATES THE VALUE OR MEAN-DAYS-WITH IS BETWEEN 0.00 AND 0.05.
CLOUDINESS FOR ASOS STATIONS DIFFERS FROM THE NON-ASOS OBSERVATION TAKEN BY A HUMAN OBSERVER. ASOS STATION CLOUDINESS IS BASED ON TIME-AVERAGED CEILOMETER DATA FOR CLOUDS AT OR BELOW 12,000 FEET AND ON SATELLITE DATA FOR CLOUDS ABOVE 12,000 FEET.
THE NUMBER OF DAYS WITH CLEAR, PARTLY CLOUDY, AND CLOUDY CONDITIONS FOR ASOS STATIONS IS THE SUM OF THE CEILOMETER AND SATELLITE DATA FOR THE SUNRISE TO SUNSET PERIOD.

GENERAL CONTINUED:
CLEAR INDICATES 0 - 2 OKTAS, PARTLY CLOUDY INDICATES 3 - 6 OKTAS, AND CLOUDY INDICATES 7 OR 8 OKTAS.
WHEN AT LEAST ONE OF THE ELEMENTS (CEILOMETER OR SATELLITE) IS MISSING, THE DAILY CLOUDINESS IS NOT COMPUTED.
WIND DIRECTION IS RECORDED IN TENS OF DEGREES (2 DIGITS) CLOCKWISE FROM TRUE NORTH. "00" INDICATES CALM. "36" INDICATES TRUE NORTH.
RESULTANT WIND IS THE VECTOR AVERAGE OF THE SPEED AND DIRECTION.
AVERAGE TEMPERATURE IS THE SUM OF THE MEAN DAILY MAXIMUM AND MINIMUM TEMPERATURE DIVIDED BY 2.
SNOWFALL DATA COMPRISE ALL FORMS OF FROZEN PRECIPITATION, INCLUDING HAIL.
A HEATING (COOLING) DEGREE DAY IS THE DIFFERENCE BETWEEN THE AVERAGE DAILY TEMPERATURE AND 65 F.
DRY BULB IS THE TEMPERATURE OF THE AMBIENT AIR.
DEW POINT IS THE TEMPERATURE TO WHICH THE AIR MUST BE COOLED TO ACHIEVE 100 PERCENT RELATIVE HUMIDITY.
WET BULB IS THE TEMPERATURE THE AIR WOULD HAVE IF THE MOISTURE CONTENT WAS INCREASED TO 100 PERCENT RELATIVE HUMIDITY.

ON JULY 1, 1996, THE NATIONAL WEATHER SERVICE BEGAN USING THE "METAR" OBSERVATION CODE THAT WAS ALREADY EMPLOYED BY MOST OTHER NATIONS OF THE WORLD. THE MOST NOTICEABLE DIFFERENCE IN THIS ANNUAL PUBLICATION WILL BE THE CHANGE IN UNITS FROM TENTHS TO EIGHTS(OKTAS) FOR REPORTING THE AMOUNT OF SKY COVER.

HEATING DEGREE DAYS (base 65 F) 2001 MADISON, WI (MSN)

YEAR	JUL	AUG	SEP	OCT	NOV	DEC	JAN	FEB	MAR	APR	MAY	JUN	TOTAL
1972-73	44	42	188	587	905	1475	1279	1143	720	596	325	15	7319
1973-74	4	25	180	349	847	1342	1416	1298	979	494	347	90	7371
1974-75	1	37	253	443	829	1179	1329	1220	1198	714	150	43	7396
1975-76	18	11	236	412	687	1217	1520	1056	877	477	333	32	6876
1976-77	4	40	236	656	1102	1602	1898	1188	772	409	110	95	8112
1977-78	6	95	161	533	925	1404	1688	1466	1096	608	269	59	8310
1978-79	19	22	130	543	940	1348	1800	1489	1013	671	283	52	8310
1979-80	14	62	144	546	890	1112	1471	1424	1138	586	255	84	7726
1980-81	2	11	178	651	881	1303	1373	1107	864	482	307	30	7189
1981-82	16	27	193	566	842	1327	1765	1281	1059	688	155	172	8091
1982-83	5	66	230	444	918	1117	1346	1078	978	693	400	57	7332
1983-84	11	6	193	519	823	1678	1550	1006	1181	575	358	20	7920
1984-85	9	21	215	397	927	1191	1632	1287	839	418	155	96	7187
1985-86	12	36	198	475	1012	1661	1444	1272	888	462	220	73	7753
1986-87	7	59	145	471	1007	1218	1309	963	857	452	192	27	6707
1987-88	3	45	150	661	743	1127	1586	1377	938	565	176	53	7424
1988-89	4	18	107	661	777	1242	1153	1404	1076	602	290	68	7402
1989-90	5	22	207	437	952	1568	1122	1092	835	519	349	46	7154
1990-91	7	12	133	511	713	1349	1539	1072	868	467	173	22	6866
1991-92	8	11	222	476	1002	1195	1216	1031	929	634	244	73	7041
1992-93	26	68	176	514	903	1236	1333	1220	1032	633	196	74	7411
1993-94	0	9	260	525	887	1182	1739	1375	896	501	241	39	7654
1994-95	6	52	80	389	736	1116	1376	1173	867	629	226	20	6670
1995-96	9	0	228	452	1081	1370	1537	1256	1123	675	385	52	8168
1996-97	18	7	174	505	1083	1306	1521	1138	951	646	416	44	7809
1997-98	28	52	157	481	945	1141	1273	875	949	496	129	95	6621
1998-99	0	1	78	409	724	1040	1497	952	962	496	178	53	6390
1999-00	0	26	177	491	681	1190	1372	1021	747	587	206	64	6562
2000-01	17	10	187	386	895	1663	1334	1260	1027	417	231	89	7516
2001-	17	8	210	490	560	1063							

COOLING DEGREE DAYS (base 65 F) 2001 MADISON, WI (MSN)

YEAR	JAN	FEB	MAR	APR	MAY	JUN	JUL	AUG	SEP	OCT	NOV	DEC	ANNUAL
1972	0	0	0	0	41	61	156	180	27	0	0	0	465
1973	0	0	0	0	2	112	215	207	58	19	0	0	613
1974	0	0	0	9	17	68	228	102	31	2	0	0	457
1975	0	0	0	0	81	176	256	190	18	21	0	0	742
1976	0	0	0	14	6	136	270	165	34	2	0	0	627
1977	0	0	0	24	123	99	278	88	10	0	0	0	622
1978	0	0	0	0	56	92	171	168	102	0	0	0	589
1979	0	0	0	0	33	88	168	115	33	13	0	0	450
1980	0	0	0	8	39	100	268	183	31	0	0	0	629
1981	0	0	0	0	13	107	198	148	19	0	0	0	485
1982	0	0	0	0	29	16	194	114	53	3	0	0	409
1983	0	0	0	0	0	138	327	237	52	6	0	0	760
1984	0	0	0	1	5	102	177	224	50	0	0	0	559
1985	0	0	0	40	29	66	175	84	102	0	0	0	496
1986	0	0	0	13	24	105	269	59	49	0	0	0	519
1987	0	0	0	8	69	194	304	165	26	0	0	0	766
1988	0	0	0	0	43	194	296	315	54	0	0	0	902
1989	0	0	0	0	21	97	237	141	25	3	0	0	524
1990	0	0	0	32	2	132	191	171	100	1	0	0	629
1991	0	0	0	8	136	210	241	180	80	3	0	0	858
1992	0	0	0	0	33	76	100	86	42	5	0	0	342
1993	0	0	0	0	33	111	223	240	18	5	0	0	630
1994	0	0	0	10	32	207	192	108	85	3	0	0	637
1995	0	0	0	0	5	237	320	374	45	1	0	0	982
1996	0	0	0	0	28	110	103	125	41	0	0	0	407
1997	0	0	0	0	0	123	168	69	21	29	0	0	410
1998	0	0	0	0	57	140	199	182	76	0	0	0	654
1999	0	0	0	0	24	161	315	95	42	0	0	0	637
2000	0	0	0	0	32	96	159	165	78	5	0	0	535
2001	0	0	0	4	37	125	250	199	27	0	0	0	642

SNOWFALL (inches) 2001 MADISON, WI (MSN)

YEAR	JUL	AUG	SEP	OCT	NOV	DEC	JAN	FEB	MAR	APR	MAY	JUN	TOTAL
1972-73	0.0	0.0	0.0	T	1.3	16.3	1.9	6.0	1.1	17.4	T	0.0	44.0
1973-74	0.0	0.0	0.0	0.0	0.4	10.9	10.5	14.1	6.6	0.4	T	0.0	42.9
1974-75	0.0	0.0	0.0	0.0	3.0	15.4	5.2	20.9	10.0	5.9	0.0	0.0	60.4
1975-76	0.0	0.0	0.0	0.0	5.5	2.8	10.1	10.4	2.0	T	0.0	0.0	30.8
1976-77	0.0	0.0	0.0	T	1.1	5.8	8.7	2.6	5.8	2.3	0.0	0.0	26.3
1977-78	0.0	0.0	0.0	0.0	10.4	24.6	13.5	4.7	3.0	0.5	0.0	0.0	56.7
1978-79	0.0	0.0	0.0	0.0	6.2	23.0	26.9	8.7	4.0	7.3	0.0	0.0	76.1
1979-80	0.0	0.0	0.0	0.2	4.4	1.3	4.9	7.5	5.6	7.1	0.0	0.0	31.0
1980-81	0.0	0.0	0.0	T	3.5	9.2	2.9	9.2	1.7	0.0	0.0	0.0	26.5
1981-82	0.0	0.0	0.0	0.1	2.0	7.2	19.4	2.4	8.6	10.3	0.0	0.0	50.0
1982-83	0.0	0.0	0.0	0.0	0.3	3.3	6.5	13.0	14.1	4.2	0.0	0.0	41.4
1983-84	0.0	0.0	0.0	0.0	2.1	22.6	6.0	0.8	6.8	3.9	0.0	0.0	42.2
1984-85	0.0	0.0	0.0	0.0	0.5	15.8	19.9	7.4	8.2	1.9	0.0	0.0	53.7
1985-86	0.0	0.0	0.0	0.0	18.3	24.0	13.9	13.3	2.7	0.2	0.0	0.0	72.4
1986-87	0.0	0.0	0.0	T	8.6	8.0	8.7	0.3	8.9	T	0.0	0.0	34.5
1987-88	0.0	0.0	0.0	0.4	3.9	32.8	16.3	6.4	1.1	1.3	0.0	0.0	62.2
1988-89	0.0	0.0	0.0	0.2	5.5	8.2	2.6	9.7	9.3	0.2	0.5	T	36.2
1989-90	0.0	0.0	0.0	0.7	4.4	4.3	10.1	11.7	0.1	0.5	3.0	T	34.8
1990-91	0.0	0.0	0.0	3.1	4.5	23.0	14.5	5.0	3.6	1.3	0.0	0.0	55.0
1991-92	T	0.0	0.0	0.5	8.0	10.2	4.5	12.3	6.9	0.1	0.0	T	42.5
1992-93	0.0	T	0.0	2.1	3.9	10.5	12.5	12.1	21.6	8.5	0.0	0.0	71.2
1993-94	0.0	0.0	T	0.2	1.2	2.5	22.5	37.0	0.6	9.7	0.0	0.0	73.7
1994-95	T	T	T	0.0	0.3	12.5	27.5	0.7	11.2	0.6	0.0	0.0	52.8
1995-96	0.0	0.0	0.0	0.1	12.8	10.3	26.4	1.3	4.7	4.9	0.0	0.0	60.5
1996-97	0.0	0.0	0.0	0.0	5.9	6.7	13.1	14.4	2.7	7.1	0.1	0.0	50.0
1997-98	0.0	0.0	0.0	3.9	3.0	14.3	18.9	1.8	12.0	T	0.0	0.0	53.9
1998-99	0.0	0.0	0.0	0.0	0.4	2.2	23.9	3.8	7.8	T	0.0	0.0	38.1
1999-00	0.0	T	0.0	0.0	T	3.2	11.7	11.2	3.5	4.5	T	0.0	34.1
2000-01	0.0	0.0	0.0	T	5.3	35.0	1.6	7.8	1.9	0.6	0.0	0.0	52.2
2001-	0.0	0.0	0.0	T	T	2.3							
POR= 52 YRS	T	T	T	0.3	3.5	11.0	10.9	7.9	7.4	2.4	0.2	T	43.6

2001
MILWAUKEE,
WISCONSIN (MKE)

Milwaukee possesses a continental climate characterized by a wide range of temperatures between summer and winter. Precipitation is moderate and occurs mostly in the spring, less in the autumn, and very little in the wintertime. Rainfall is well distributed for agricultural purposes, although spring planting is sometimes delayed by wet ground and cold weather.

Milwaukee is in a region of frequently changeable weather and its climate is influenced by general easterly-moving storms which traverse the nations midsection. The most severe winter storms, which produce in excess of 10 inches of snow, develop in the southern Great Plains and move northeast across Illinois and Indiana.

Occasionally during the cold season, frigid air masses from Canada push southeast across the Great Lakes region. These arctic air masses account for the coldest winter temperatures. Very low temperatures, zero degrees or lower, most often occur in air that flows southward to the west of Lake Superior before reaching the Milwaukee area. If northwesterly wind circulation persists, repeated incursions of arctic air will result in a period of bitterly cold weather lasting several days.

Summer temperatures, which reach into the 90s but rarely exceed 100 degrees, occur with brisk southwest winds that carry hot air from the plains and lower Mississippi River Valley across the city. A combination of high temperatures and humidity occasionally develops, usually building up over a period of several days when persistent southerly winds transport moisture from the Gulf of Mexico into the area.

The Gulf is a major source of moisture for Milwaukee in all seasons, but the type of precipitation which results is dependent upon the time of year. Cold-season precipitation (rain, snow, or a mixture) is usually of relatively long duration and low intensity, and occasionally persists for two days or more, whereas in the warm season, relatively short-duration and high-intensity showery rainfall, usually lasting a few hours or less, predominates.

The Great Lakes significantly influence the local climate. Temperature extremes are modified by Lake Michigan and, to a lesser extent, the other Great Lakes. In late autumn and winter, air masses that are initially very cold often reach the city only after being tempered by passage over one or more of the lakes. Similarly, air masses that approach from the northeast in the spring and summer are cooler because of movement over the Great Lakes.

The influence of Lake Michigan is variable and occasionally dramatic, especially when the temperature of the lake water differs strongly from the air temperature. During the spring and early summer, a wind shift from a westerly to an easterly direction frequently causes a sudden 10 to 20 degree temperature drop. When the breeze off the lake is light, this effect reaches inland only a mile or two. With stronger on-shore winds, the entire city is cooled. In the winter the relatively warm water of the lake moderates the temperature during easterly wind situations. Lake-induced snows usually occur a few times each winter, but snow accumulation is rarely heavy.

Topography does not significantly affect air flow, except that lesser frictional drag over Lake Michigan causes winds to be frequently stronger along the lake shore, and often permits air masses approaching from the north to reach shore areas one hour or more before affecting inland portions of the city.

NORMALS, MEANS, AND EXTREMES
MILWAUKEE, WI (MKE)

LATITUDE:	LONGITUDE:	ELEVATION (FT):		TIME ZONE:	WBAN: 14839
42 56' 48" N	87 53' 49" W	GRND: 677	BARO: 680	CENTRAL (UTC + 6)	

	ELEMENT	POR	JAN	FEB	MAR	APR	MAY	JUN	JUL	AUG	SEP	OCT	NOV	DEC	YEAR
TEMPERATURE °F	NORMAL DAILY MAXIMUM	30	26.1	30.1	40.4	52.9	64.3	74.9	79.9	77.8	70.6	58.7	44.7	31.2	54.3
	MEAN DAILY MAXIMUM	54	27.6	31.7	40.7	53.7	65.1	75.7	80.3	78.7	71.2	60.0	45.3	32.5	55.2
	HIGHEST DAILY MAXIMUM	61	62	68	82	91	93	101	103	103	98	89	77	68	103
	YEAR OF OCCURRENCE		1944	1999	1986	1980	1991	1988	1995	1988	1953	1963	1950	2001	JUL 1995
	MEAN OF EXTREME MAXS.	54	46.0	48.3	65.4	77.7	85.4	91.6	93.3	91.9	87.5	78.7	64.5	51.9	73.5
	NORMAL DAILY MINIMUM	30	11.6	15.9	26.2	35.8	44.8	55.0	62.0	60.8	52.8	41.8	30.7	17.5	37.9
	MEAN DAILY MINIMUM	54	13.0	17.4	26.0	36.4	45.5	55.7	62.2	61.5	53.3	42.4	30.6	19.0	38.6
	LOWEST DAILY MINIMUM	61	-26	-26	-10	12	21	33	40	44	28	18	-5	-20	-26
	YEAR OF OCCURRENCE		1982	1996	1962	1982	1966	1945	1965	1982	1974	1981	1950	1983	FEB 1996
	MEAN OF EXTREME MINS.	54	-9.5	-3.0	8.4	23.3	33.6	43.5	51.2	50.6	38.9	27.9	14.0	-1.7	23.1
	NORMAL DRY BULB	30	18.9	23.0	33.3	44.4	54.6	65.0	70.9	69.3	61.7	50.3	37.7	24.4	46.1
	MEAN DRY BULB	54	20.3	24.5	33.5	44.9	55.4	65.6	71.2	70.1	62.3	51.3	37.8	25.7	46.9
	MEAN WET BULB	16	21.4	24.8	31.8	40.7	50.7	60.3	65.6	65.1	54.5	46.8	35.5	23.4	43.4
	MEAN DEW POINT	16	16.3	18.2	25.6	34.4	44.9	55.4	61.7	61.6	50.6	41.3	30.2	18.6	38.2
	NORMAL NO. DAYS WITH:														
	MAXIMUM 90	30	0.0	0.0	0.0	*	0.1	2.1	4.1	2.3	0.6	0.0	0.0	0.0	9.2
	MAXIMUM 32	30	20.1	15.6	6.4	0.5	0.0	0.0	0.0	0.0	0.0	0.0	2.6	15.4	60.6
	MINIMUM 32	30	29.7	26.1	23.6	9.9	1.1	0.0	0.0	0.0	0.1	4.3	18.0	28.0	140.8
	MINIMUM 0	30	7.6	3.5	0.2	0.0	0.0	0.0	0.0	0.0	0.0	0.0	0.1	3.3	14.7
H/C	NORMAL HEATING DEG. DAYS	30	1429	1176	983	618	338	82	14	27	123	456	819	1259	7324
	NORMAL COOLING DEG. DAYS	30	0	0	0	0	16	82	197	160	24	0	0	0	479
RH	NORMAL (PERCENT)	30	72	72	71	68	68	70	72	75	75	72	74	76	72
	HOUR 00 LST	30	75	74	75	74	75	77	80	84	83	78	78	78	78
	HOUR 06 LST	30	76	77	78	78	78	79	82	86	87	82	80	80	80
	HOUR 12 LST	30	68	67	64	60	60	60	61	63	63	62	67	72	64
	HOUR 18 LST	30	71	70	68	63	62	62	63	68	71	70	74	75	68
S	PERCENT POSSIBLE SUNSHINE	56	44	47	50	53	60	65	69	66	59	54	39	38	54
W/O	MEAN NO. DAYS WITH:														
	HEAVY FOG(VISBY 1/4 MI)	61	2.1	2.2	3.1	2.9	3.1	2.5	1.3	1.7	1.3	2.0	2.1	2.1	26.4
	THUNDERSTORMS	61	0.3	0.3	1.4	3.4	4.3	6.3	6.6	5.7	3.9	1.6	1.1	0.3	35.2
CLOUDINESS	MEAN:														
	SUNRISE-SUNSET (OKTAS)														
	MIDNIGHT-MIDNIGHT (OKTAS)														
	MEAN NO. DAYS WITH:														
	CLEAR	1		1.0	2.0		6.0								
	PARTLY CLOUDY	1		1.0	1.0		1.0								
	CLOUDY	1		1.0	3.0		10.0								
PR	MEAN STATION PRESSURE(IN)	28	29.29	29.32	29.25	29.22	29.22	29.21	29.25	29.29	29.30	29.30	29.27	29.31	29.27
	MEAN SEA-LEVEL PRES. (IN)	16	30.06	30.09	30.05	29.96	29.97	29.94	29.97	30.02	30.03	30.05	30.03	30.09	30.02
WINDS	MEAN SPEED (MPH)	49	12.4	12.2	12.5	12.7	11.5	10.2	9.7	9.5	10.3	11.3	12.2	12.1	11.4
	PREVAIL.DIR(TENS OF DEGS)	33	29	30	30	03	02	03	24	24	21	21	30	30	30
	MAXIMUM 2-MINUTE:														
	SPEED (MPH)	6	44	52	41	48	46	47	36	36	41	43	52	38	52
	DIR. (TENS OF DEGS)		08	27	26	24	30	31	31	23	07	24	23	23	27
	YEAR OF OCCURRENCE		1999	1999	1999	1997	1998	2001	2001	2000	2001	1996	1998	2001	FEB 1999
	MAXIMUM 5-SECOND:														
	SPEED (MPH)	6	51	70	53	62	61	69	44	48	47	48	68	47	70
	DIR. (TENS OF DEGS)		08	25	01	24	30	28	29	22	07	25	23	23	25
	YEAR OF OCCURRENCE		1999	1999	1996	1997	1998	2001	1999	2000	2001	2001	1998	2001	FEB 1999
PRECIPITATION	NORMAL (IN)	30	1.60	1.45	2.67	3.50	2.84	3.24	3.47	3.53	3.38	2.41	2.51	2.33	32.93
	MAXIMUM MONTHLY (IN)	61	4.38	3.94	6.93	7.31	8.42	9.98	7.66	9.05	9.87	7.03	7.11	5.42	9.98
	YEAR OF OCCURRENCE		1999	1986	1976	1973	2000	1997	1964	1987	1941	1991	1985	1987	JUN 1997
	MINIMUM MONTHLY (IN)	61	0.31	0.05	0.31	0.81	0.50	0.70	0.95	0.46	0.02	0.15	0.62	0.29	0.02
	YEAR OF OCCURRENCE		1981	1969	1968	1942	1988	1988	1946	1948	1979	1956	1949	1976	SEP 1979
	MAXIMUM IN 24 HOURS (IN)	61	1.73	1.92	2.57	3.11	3.11	4.23	4.42	6.84	5.28	2.60	2.69	2.24	6.84
	YEAR OF OCCURRENCE		1985	2001	1960	1976	1978	1997	2000	1986	1941	1959	1998	1982	AUG 1986
	NORMAL NO. DAYS WITH:														
	PRECIPITATION 0.01	30	11.6	9.8	12.3	12.4	11.3	10.5	9.7	9.3	9.5	9.7	10.6	12.0	128.7
	PRECIPITATION 1.00	30	0.2	0.2	0.4	0.8	0.6	0.8	1.0	0.8	0.9	0.4	0.3	0.4	6.8
SNOWFALL	NORMAL (IN)	30	12.9	11.1	8.8	2.5	0.1	0.0	0.0	0.0	0.0	0.4	2.7	11.4	49.9
	MAXIMUM MONTHLY (IN)	61	39.0	42.0	26.7	15.8	3.2	T	T	T	T	6.3	16.1	49.5	49.5
	YEAR OF OCCURRENCE		1999	1974	1965	1973	1990	1992	1990	1989	1993	1989	1977	2000	DEC 2000
	MAXIMUM IN 24 HOURS (IN)	61	13.8	16.7	11.2	11.6	3.2	T	T	T	T	6.3	10.6	13.6	16.7
	YEAR OF OCCURRENCE		1990	1960	1961	1973	1990	1992	1990	1989	1993	1989	1977	2000	FEB 1960
	MAXIMUM SNOW DEPTH (IN)	53	33	29	24	13	2	0	0	0	0	6	11	32	33
	YEAR OF OCCURRENCE		1979	1979	1960	1973	1990					1989	1977	2000	JAN 1979
	NORMAL NO. DAYS WITH:														
	SNOWFALL 1.0	30	3.8	2.9	2.4	0.5	0.*	0.0	0.0	0.0	0.0	0.1	0.7	3.3	13.7

PRECIPITATION (inches) 2001 MILWAUKEE, WI (MKE)

YEAR	JAN	FEB	MAR	APR	MAY	JUN	JUL	AUG	SEP	OCT	NOV	DEC	ANNUAL
1972	0.75	0.86	2.57	2.76	2.33	3.33	4.60	4.82	7.57	3.28	1.34	2.47	36.68
1973	1.12	1.51	2.86	7.31	3.39	1.96	1.55	0.95	4.50	2.97	1.83	3.80	33.75
1974	3.61	3.10	4.29	3.83	4.10	3.48	3.51	2.54	0.50	1.96	1.86	2.10	34.88
1975	2.25	2.53	3.01	4.08	2.01	3.99	1.14	2.05	1.00	0.72	2.83	1.70	29.15
1976	1.16	2.65	6.93	5.01	3.77	2.27	2.12	2.05	1.70	2.82	0.65	0.29	31.42
1977	0.90	0.59	4.56	2.09	0.90	5.78	5.99	3.82	4.11	2.02	2.56	3.27	36.59
1978	2.03	0.55	1.08	4.41	4.66	4.52	5.98	3.43	6.81	2.22	2.13	2.92	40.74
1979	3.00	0.97	4.17	5.43	1.82	2.84	1.06	4.85	0.02	1.77	2.67	2.27	30.87
1980	1.65	1.75	0.77	4.02	1.81	4.67	3.39	5.06	3.57	1.63	1.57	3.52	33.41
1981	0.31	2.88	0.51	4.87	3.05	2.39	4.35	4.26	5.47	2.71	2.05	1.03	33.88
1982	2.92	0.29	3.20	4.47	2.76	3.06	3.88	3.33	0.64	3.17	4.74	4.10	36.56
1983	0.75	2.23	4.12	4.66	5.83	1.41	1.34	4.70	2.79	2.65	4.10	2.89	37.47
1984	0.79	1.20	2.17	5.04	4.21	4.07	3.39	2.93	2.51	5.30	3.74	4.25	39.60
1985	1.94	2.34	4.11	1.93	2.73	1.27	2.18	2.23	3.44	5.39	7.11	2.62	37.29
1986	0.91	3.94	1.85	1.83	2.74	6.15	8.82	7.26	2.24	0.89	1.03	2.51	42.17
1987	1.22	1.22	1.74	4.26	3.76	2.23	4.20	9.05	2.22	1.09	2.73	5.42	39.14
1988	3.25	1.29	1.30	4.12	0.50	0.70	1.53	3.25	4.94	2.97	5.15	1.43	30.43
1989	0.86	0.69	3.03	1.33	2.86	1.89	6.16	5.19	3.25	2.67	1.90	0.47	30.30
1990	2.57	1.90	2.75	2.67	7.56	4.97	3.02	4.68	1.89	2.65	3.54	2.66	40.86
1991	1.55	0.38	4.06	3.70	4.25	2.13	4.34	2.27	4.34	7.03	3.36	1.94	39.35
1992	1.09	1.54	2.61	2.41	0.60	3.13	5.64	3.50	4.13	1.45	5.40	2.45	33.95
1993	2.63	0.98	3.19	6.64	1.56	6.39	4.22	4.20	3.91	0.44	1.98	0.70	36.84
1994	2.20	3.52	1.21	2.35	0.67	3.08	2.51	4.91	1.68	0.78	3.31	1.14	27.36
1995	2.14	0.25	1.76	3.86	3.41	1.46	2.80	5.83	1.24	4.64	3.42	0.53	31.34
1996	1.66	0.52	0.76	2.99	2.89	5.47	1.61	1.24	1.82	3.00	0.63	1.53	24.12
1997	1.59	2.47	0.63	2.16	1.95	9.98	3.59	3.95	2.91	1.11	1.11	1.30	32.75
1998	3.60	2.19	3.18	4.18	2.48	2.82	1.78	5.98	2.17	2.47	2.91	0.88	34.64
1999	4.38	0.98	1.35	6.14	3.74	6.96	5.58	1.69	4.16	0.94	0.70	1.26	37.88
2000	1.20	1.66	1.12	3.64	8.42	3.42	7.12	5.17	7.04	0.84	2.33	2.41	44.37
2001	1.11	3.48	0.67	3.45	4.68	4.13	2.70	5.41	4.76	4.29	1.19	0.86	36.73
POR= 55 YRS	1.87	1.64	2.39	3.02	3.26	3.65	3.12	3.16	3.24	2.34	2.12	1.80	31.61

AVERAGE TEMPERATURE (F) 2001 MILWAUKEE, WI (MKE)

YEAR	JAN	FEB	MAR	APR	MAY	JUN	JUL	AUG	SEP	OCT	NOV	DEC	ANNUAL
1972	15.9	19.8	27.7	39.4	55.0	61.6	69.3	69.1	61.3	47.5	36.2	20.2	43.6
1973	24.7	25.4	39.9	42.7	51.0	69.2	71.4	72.5	63.9	54.6	38.5	25.5	48.3
1974	21.6	22.9	33.8	46.1	50.6	62.4	71.5	67.3	58.0	49.9	38.6	29.1	46.0
1975	24.0	23.5	28.6	37.7	57.1	65.7	71.7	71.2	58.5	54.1	44.5	27.9	47.0
1976	18.5	31.1	38.6	48.7	52.6	68.2	72.6	69.9	62.6	45.9	29.5	16.3	46.2
1977	8.3	23.7	39.2	48.9	61.3	62.5	73.1	67.2	61.9	49.0	37.1	22.8	46.3
1978	15.4	16.4	29.8	42.5	55.2	65.2	68.6	65.8	49.5	38.2	23.8	45.0	45.0
1979	11.6	15.1	33.2	42.1	54.9	64.7	70.9	68.8	64.7	51.1	38.2	31.4	45.6
1980	20.7	20.3	30.5	45.3	57.2	61.3	71.2	67.8	61.1	45.7	37.7	24.4	45.4
1981	18.9	25.3	35.6	46.5	51.5	65.2	67.3	67.8	59.1	45.8	37.4	24.2	45.4
1982	9.7	19.4	31.5	41.2	58.5	59.8	71.1	67.3	60.8	52.7	38.0	33.2	45.3
1983	26.4	29.4	35.0	41.7	50.2	66.3	76.2	74.4	62.9	52.0	39.9	14.4	47.4
1984	18.9	33.4	29.2	45.5	54.9	68.7	71.7	73.3	60.9	52.9	37.7	29.1	48.0
1985	15.2	21.3	37.9	50.8	58.8	63.8	72.4	68.4	64.3	50.7	36.7	15.7	46.3
1986	21.9	23.3	38.1	48.5	56.1	63.3	72.5	63.8	51.6	51.6	35.1	29.2	47.5
1987	24.8	32.0	37.8	48.0	60.1	72.2	74.8	69.9	63.3	46.2	41.9	31.4	50.2
1988	18.3	20.1	34.8	45.7	58.7	70.2	75.4	75.7	63.5	45.8	40.7	26.6	48.0
1989	30.4	18.0	32.4	43.2	54.9	64.4	71.6	68.8	60.2	52.7	35.1	16.7	45.7
1990	31.1	28.9	38.8	49.3	52.8	67.6	70.5	71.2	66.0	51.5	44.5	26.9	49.9
1991	20.1	29.9	38.2	49.7	63.2	70.7	74.4	73.5	63.2	52.3	35.1	29.6	50.0
1992	27.9	31.9	35.4	42.9	56.7	63.0	67.7	67.4	61.6	49.8	37.3	28.2	47.5
1993	25.7	24.4	32.5	43.0	56.9	63.8	73.0	73.6	60.6	51.0	39.5	29.9	47.8
1994	14.8	21.6	35.8	48.1	57.0	70.2	73.7	70.1	67.5	55.9	43.7	34.5	49.4
1995	25.3	25.5	38.7	44.0	57.8	71.4	74.4	75.7	61.1	52.8	31.1	24.1	48.5
1996	20.7	24.0	29.8	42.4	52.0	65.2	68.7	72.4	63.8	52.3	33.1	27.1	46.0
1997	20.4	28.2	35.9	43.6	49.6	64.8	69.2	66.5	61.4	51.6	34.8	30.4	46.4
1998	26.9	34.6	35.9	45.8	59.9	66.7	72.4	72.5	67.0	54.0	43.0	32.0	50.9
1999	20.2	32.0	34.6	46.2	58.1	67.2	76.5	68.8	63.4	44.8	29.3	49.5	49.5
2000	23.6	32.2	41.9	44.3	58.0	65.9	68.2	70.7	62.2	54.7	36.9	16.6	47.9
2001	24.8	23.5	32.5	49.0	56.8	66.2	72.4	72.9	61.2	51.2	47.5	32.9	49.2
POR= 55 YRS	21.0	24.1	33.4	44.6	54.8	64.8	71.1	70.0	62.6	51.4	38.0	26.2	46.8

REFERENCE NOTES:

PAGE 1:
THE TEMPERATURE GRAPH SHOWS NORMAL MAXIMUM AND NORMAL
MINIMUM DAILY TEMPERATURES (SOLID CURVES) AND THE
ACTUAL DAILY HIGH AND LOW TEMPERATURES (VERTICAL BARS).

PAGE 2 AND 3:
H/C INDICATES HEATING AND COOLING DEGREE DAYS.
RH INDICATES RELATIVE HUMIDITY
W/O INDICATES WEATHER AND OBSTRUCTIONS
S INDICATES SUNSHINE.
PR INDICATES PRESSURE.
CLOUDINESS ON PAGE 3 IS THE SUM OF THE CEILOMETER AND
SATELLITE DATA NOT TO EXCEED EIGHT EIGHTHS(OKTAS).

GENERAL:
T INDICATES TRACE PRECIPITATION, AN AMOUNT GREATER
THAN ZERO BUT LESS THAN THE LOWEST REPORTABLE VALUE.
+ INDICATES THE VALUE ALSO OCCURS ON EARLIER DATES.
BLANK ENTRIES DENOTE MISSING OR UNREPORTED DATA.
NORMALS ARE 30-YEAR AVERAGES (1961 - 1990).
ASOS INDICATES AUTOMATED SURFACE OBSERVING SYSTEM.
PM INDICATES THE LAST DAY OF THE PREVIOUS MONTH.
POR (PERIOD OF RECORD) BEGINS WITH THE JANUARY DATA
MONTH AND IS THE NUMBER OF YEARS USED TO COMPUTE
THE MEAN. INDIVIDUAL MONTHS WITHIN THE POR MAY
BE MISSING.
WHEN THE POR FOR A NORMAL IS LESS THAN 30 YEARS,
THE NORMAL IS PROVISIONAL AND IS BASED ON THE NUMBER
OF YEARS INDICATED.
0.* OR * INDICATES THE VALUE OR MEAN-DAYS-WITH
IS BETWEEN 0.00 AND 0.05.
CLOUDINESS FOR ASOS STATIONS DIFFERS FROM THE NON-ASOS
OBSERVATION TAKEN BY A HUMAN OBSERVER. ASOS STATION
CLOUDINESS IS BASED ON TIME-AVERAGED CEILOMETER DATA
FOR CLOUDS AT OR BELOW 12,000 FEET AND ON SATELLITE
DATA FOR CLOUDS ABOVE 12,000 FEET.
THE NUMBER OF DAYS WITH CLEAR, PARTLY CLOUDY, AND
CLOUDY CONDITIONS FOR ASOS STATIONS IS THE SUM
OF THE CEILOMETER AND SATELLITE DATA FOR THE
SUNRISE TO SUNSET PERIOD.

GENERAL CONTINUED:
CLEAR INDICATES 0 - 2 OKTAS, PARTLY CLOUDY INDICATES
3 - 6 OKTAS, AND CLOUDY INDICATES 7 OR 8 OKTAS.
WHEN AT LEAST ONE OF THE ELEMENTS (CEILOMETER OR
SATELLITE) IS MISSING, THE DAILY CLOUDINESS IS
NOT COMPUTED.
WIND DIRECTION IS RECORDED IN TENS OF DEGREES (2 DIGITS)
CLOCKWISE FROM TRUE NORTH. "00" INDICATES CALM. "36"
INDICATES TRUE NORTH.
RESULTANT WIND IS THE VECTOR AVERAGE OF THE SPEED AND
DIRECTION.
AVERAGE TEMPERATURE IS THE SUM OF THE MEAN DAILY MAXIMUM
AND MINIMUM TEMPERATURE DIVIDED BY 2.
SNOWFALL DATA COMPRISE ALL FORMS OF FROZEN
PRECIPITATION, INCLUDING HAIL.
A HEATING (COOLING) DEGREE DAY IS THE DIFFERENCE BETWEEN
THE AVERAGE DAILY TEMPERATURE AND 65 F.
DRY BULB IS THE TEMPERATURE OF THE AMBIENT AIR.
DEW POINT IS THE TEMPERATURE TO WHICH THE AIR MUST BE
COOLED TO ACHIEVE 100 PERCENT RELATIVE HUMIDITY.
WET BULB IS THE TEMPERATURE THE AIR WOULD HAVE IF THE
MOISTURE CONTENT WAS INCREASED TO 100 PERCENT RELATIVE
HUMIDITY.

ON JULY 1, 1996, THE NATIONAL WEATHER SERVICE BEGAN USING
THE "METAR" OBSERVATION CODE THAT WAS ALREADY EMPLOYED
BY MOST OTHER NATIONS OF THE WORLD. THE MOST NOTICEABLE
DIFFERENCE IN THIS ANNUAL PUBLICATION WILL BE THE CHANGE
IN UNITS FROM TENTHS TO EIGHTS(OKTAS) FOR REPORTING THE
AMOUNT OF SKY COVER.

HEATING DEGREE DAYS (base 65 F) 2001 MILWAUKEE, WI (MKE)

YEAR	JUL	AUG	SEP	OCT	NOV	DEC	JAN	FEB	MAR	APR	MAY	JUN	TOTAL
1972-73	40	32	133	534	859	1381	1242	1101	769	659	426	6	7182
1973-74	10	5	111	324	788	1218	1340	1173	959	560	448	106	7042
1974-75	0	20	237	461	786	1103	1260	1157	1122	814	267	69	7296
1975-76	17	4	203	353	610	1144	1438	978	813	507	382	43	6492
1976-77	2	21	124	589	1056	1504	1754	1152	790	490	173	151	7806
1977-78	8	47	106	485	827	1302	1531	1356	1086	667	335	83	7833
1978-79	21	5	76	473	796	1273	1654	1391	980	681	322	91	7763
1979-80	20	25	70	436	797	1036	1368	1290	1063	594	259	154	7112
1980-81	8	9	140	590	812	1250	1423	1106	905	548	417	69	7277
1981-82	44	21	187	590	820	1257	1712	1272	1032	707	215	172	8029
1982-83	3	44	170	381	802	983	1186	990	925	692	453	81	6710
1983-84	10	0	148	405	748	1565	1424	910	1103	579	318	35	7245
1984-85	3	7	179	373	812	1103	1542	1215	831	461	222	96	6844
1985-86	2	13	139	436	843	1523	1328	1161	827	494	302	128	7196
1986-87	13	34	98	407	891	1106	1242	917	839	502	236	19	6304
1987-88	12	28	91	576	686	1037	1442	1294	930	571	245	55	6967
1988-89	3	7	87	587	720	1183	1065	1307	1006	649	324	85	7023
1989-90	0	16	166	381	890	1493	1040	1004	805	502	375	51	6723
1990-91	21	9	93	418	612	1173	1385	980	822	467	201	26	6207
1991-92	0	1	160	394	889	1091	1141	955	914	655	278	104	6582
1992-93	27	35	145	472	825	1132	1211	1132	1004	654	245	104	6986
1993-94	1	2	158	428	757	1080	1553	1211	899	502	286	55	6932
1994-95	0	22	49	284	633	937	1222	1096	807	623	230	31	5934
1995-96	1	0	171	379	1013	1260	1365	1180	1084	672	431	102	7658
1996-97	17	0	103	398	950	1167	1376	1026	896	635	470	100	7138
1997-98	23	37	129	441	899	1069	1176	847	897	571	203	100	6392
1998-99	0	0	47	337	653	1017	1380	919	936	558	239	78	6164
1999-00	0	9	113	388	598	1100	1275	945	711	616	252	84	6091
2000-01	25	8	151	323	836	1491	1239	1155	1001	481	267	99	7076
2001-	16	0	146	420	522	992							

COOLING DEGREE DAYS (base 65 F) 2001 MILWAUKEE, WI (MKE)

YEAR	JAN	FEB	MAR	APR	MAY	JUN	JUL	AUG	SEP	OCT	NOV	DEC	ANNUAL
1972	0	0	0	0	3	42	180	166	26	0	0	0	417
1973	0	0	0	0	0	140	216	247	84	6	0	0	693
1974	0	0	0	3	6	36	210	98	32	1	0	0	386
1975	0	0	0	0	30	98	230	203	16	21	0	0	598
1976	0	0	0	24	5	144	247	181	62	4	0	0	667
1977	0	0	0	12	65	81	264	122	20	0	0	0	564
1978	0	0	0	0	40	97	138	164	109	0	0	0	548
1979	0	0	0	0	16	87	209	147	68	11	0	0	538
1980	0	0	0	9	25	50	207	164	29	0	0	0	484
1981	0	0	0	2	3	84	121	112	16	0	0	0	338
1982	0	0	0	0	21	24	199	121	51	5	0	0	421
1983	0	0	0	0	0	127	364	299	92	9	0	0	891
1984	0	0	0	1	11	152	216	270	63	3	0	0	716
1985	0	0	0	42	35	68	240	127	127	0	0	0	639
1986	0	0	3	7	31	84	251	105	70	0	0	0	551
1987	0	0	0	2	87	244	323	189	47	0	1	0	893
1988	0	0	0	0	57	215	333	344	48	0	0	0	997
1989	0	0	2	0	16	76	214	144	29	4	0	0	485
1990	0	0	2	38	5	135	198	210	132	7	1	0	728
1991	0	0	0	13	150	204	300	268	114	7	0	0	1056
1992	0	0	0	0	25	49	117	119	50	4	0	0	364
1993	0	0	0	0	5	73	258	277	35	3	0	0	651
1994	0	0	0	5	46	218	278	187	133	10	0	0	877
1995	0	0	0	0	16	232	297	338	61	8	0	0	952
1996	0	0	0	0	37	112	140	237	75	7	0	0	608
1997	0	0	0	0	0	103	158	89	26	31	0	0	407
1998	0	0	3	0	51	157	235	238	114	2	0	0	800
1999	0	0	0	0	31	151	363	136	71	0	1	0	753
2000	0	0	1	0	44	116	131	188	76	10	0	0	566
2001	0	0	0	5	20	143	252	250	39	2	0	0	711

SNOWFALL (inches) 2001 MILWAUKEE, WI (MKE)

YEAR	JUL	AUG	SEP	OCT	NOV	DEC	JAN	FEB	MAR	APR	MAY	JUN	TOTAL
1972-73	0.0	0.0	0.0	T	3.4	13.7	0.2	9.9	1.8	15.8	T	0.0	44.8
1973-74	0.0	0.0	0.0	0.0	T	19.6	14.2	42.0	7.4	T	0.0	0.0	83.2
1974-75	0.0	0.0	0.0	T	2.0	9.1	3.5	12.2	15.1	10.4	0.0	0.0	52.3
1975-76	0.0	0.0	0.0	0.0	8.4	12.2	14.8	7.6	2.1	0.1	T	0.0	45.2
1976-77	0.0	0.0	0.0	4.0	3.6	5.3	15.6	5.6	12.4	2.1	0.0	0.0	48.6
1977-78	0.0	0.0	0.0	T	16.1	20.8	25.7	13.3	4.8	T	0.0	0.0	80.7
1978-79	0.0	0.0	0.0	0.0	5.3	27.9	33.6	9.1	6.2	0.8	0.0	0.0	82.9
1979-80	0.0	0.0	0.0	T	2.1	0.6	11.6	22.8	6.3	3.6	T	0.0	47.0
1980-81	0.0	0.0	0.0	T	2.3	17.5	4.9	15.7	1.5	T	0.0	0.0	41.9
1981-82	0.0	0.0	0.0	T	2.0	8.3	29.2	3.0	13.0	11.7	0.0	0.0	67.2
1982-83	0.0	0.0	0.0	T	4.1	3.1	6.3	13.5	13.8	1.0	0.0	0.0	38.1
1983-84	0.0	0.0	0.0	0.0	0.3	13.3	9.6	1.2	8.2	0.5	T	0.0	33.1
1984-85	0.0	0.0	0.0	0.0	T	19.0	20.8	15.3	9.0	2.5	0.0	0.0	66.6
1985-86	0.0	0.0	0.0	0.0	3.5	13.5	10.4	14.0	0.7	0.3	0.0	0.0	42.4
1986-87	0.0	0.0	0.0	T	2.4	2.5	11.4	T	5.2	0.4	0.0	0.0	21.9
1987-88	0.0	0.0	0.0	0.6	0.4	19.9	10.2	20.7	2.9	T	0.0	0.0	54.7
1988-89	0.0	0.0	0.0	T	2.7	7.1	2.7	13.1	13.3	0.4	0.6	0.0	39.9
1989-90	T	T	0.0	6.3	11.6	7.4	19.9	17.9	0.2	1.2	3.2	0.0	67.7
1990-91	T	0.0	0.0	0.0	0.4	10.5	15.2	2.4	1.1	0.4	0.0	0.0	30.0
1991-92	0.0	0.0	T	T	4.8	14.7	4.3	5.3	11.1	0.8	0.0	T	41.0
1992-93	0.0	0.0	0.0	1.2	0.4	8.4	12.2	11.6	12.7	3.4	T	0.0	49.9
1993-94	0.0	0.0	T	0.0	8.0	1.2	27.0	38.7	3.9	3.1	T	0.0	81.9
1994-95	0.0	0.0	0.0	T	0.2	10.4	15.2	2.1	7.6	0.2	0.0	0.0	35.7
1995-96	0.0	0.0	0.0	T	14.9	6.2	22.7	0.6	2.9	4.2	0.0	0.0	51.5
1996-97	0.0	0.0	0.0	0.0	1.8	9.2	23.6	10.7	0.5	5.4	0.0	T	51.2
1997-98	0.0	T	0.0	T	1.1	10.6	23.7	0.5	3.7	T	0.0	0.0	39.6
1998-99	0.0	0.0	0.0	0.0	0.3	3.4	39.0	4.4	13.6	0.0	T	0.0	60.7
1999-00	0.0	0.0	0.0	0.0	0.0	2.3	15.2	12.1	1.0	7.0	T	0.0	37.6
2000-01	0.0	0.0	0.0	0.2	2.8	49.5	1.3	4.2	0.8	0.5	T	0.0	59.3
2001-	0.0	0.0	0.0	T	0.0	2.6							
POR= 55 YRS	T	T	T	0.2	2.9	10.7	13.5	9.4	8.0	1.9	0.1	T	46.7

2001
CHEYENNE,
WYOMING (CYS)

The city of Cheyenne is located on a broad plateau between the North and South Platte Rivers in the extreme southeastern corner of Wyoming at an elevation of approximately 6,100 feet. The surrounding country is mostly rolling prairie which is used primarily for grazing. The ground level rises rapidly to a ridge approximately 9,000 feet in elevation about 30 miles west of the city. This ridge is known as the Laramie Mountains, one of the ranges of the Rockies, and extends in a north-south direction. Because of this ridge, winds from the northwest through west to southwest are downslope and produce a marked chinook effect in Cheyenne which is especially noticeable during the winter months. Also, winds from the north through east to south are upslope and may cause fog or low stratus clouds in the Cheyenne area throughout the year. Because of this terrain variation, the wind direction plays an important role in controlling the local temperature and weather.

Cheyenne experiences large diurnal and annual temperature ranges. This is due to the advent of both warm and cold air masses and the relatively high elevation of the city which permits rapid incoming and outgoing radiation. The daily temperature range averages about 30 degrees in the summer and 23 degrees in the winter. Many cold air masses from the north during the winter months miss Cheyenne. Because of the downslope of land to the east and the prevailing westerlies, some of the cold air masses do move over the city, but only about 13 percent of the days in an average January, the coldest month of the year, show temperatures dropping to zero or below. Temperatures during the winter months average a few degrees higher than over the Mississippi and Missouri Valleys at the same latitude.

Windy days are quire frequent during the winter and spring months. Since the wind is usually strongest during the daytime it is a very noticeable weather element. Usually the strong winds are from a westerly direction and this tends to raise the temperature because the air is moving downslope.

Most of the air masses reaching this area move in from the Pacific and since the mountains to the west are quite effective moisture barriers the climate is semi-arid. Fortunately, about 70 percent of normal annual precipitation occurs during the growing season. In the summer months, precipitation is mostly of the shower type and occurs mainly with thunderstorms. Hail is frequent and occasionally destructive in some thunderstorms. Most of the snow falls during the late winter and early spring months. It is not uncommon to have heavy snow in May.

The growing season in Cheyenne averages about 132 days a year and extends from around May 18th to September 27th. Freezing temperatures have occurred as late in the spring as mid-June, and as early in the fall as late August.

Relative humidity averages near 50 percent on an annual basis with large daily variations. Very seldom is the relative humidity above 30 percent when the temperature is above 80 degrees.

NORMALS, MEANS, AND EXTREMES
CHEYENNE, WY (CYS)

LATITUDE:	LONGITUDE:	ELEVATION (FT):		TIME ZONE:	WBAN: 24018
41 09' 28" N	104 48' 25" W	GRND: 6125	BARO: 6128	MOUNTAIN (UTC + 7)	

	ELEMENT	POR	JAN	FEB	MAR	APR	MAY	JUN	JUL	AUG	SEP	OCT	NOV	DEC	YEAR
TEMPERATURE °F	NORMAL DAILY MAXIMUM	30	37.7	40.5	44.9	54.7	64.6	74.4	82.2	80.0	71.1	60.0	46.8	38.8	58.0
	MEAN DAILY MAXIMUM	54	37.9	40.6	44.9	54.3	64.5	75.1	81.1	79.4	70.5	59.2	45.5	39.0	57.7
	HIGHEST DAILY MAXIMUM	66	66	71	74	83	91	100	100	96	95	83	75	69	100
	YEAR OF OCCURRENCE		1982	1962	1986	1992	2000	1954	1939	1979	1995	1992	1999	1939	JUN 1954
	MEAN OF EXTREME MAXS.	54	56.9	59.1	65.1	73.9	81.3	89.2	91.1	89.3	84.7	76.4	64.7	57.6	74.1
	NORMAL DAILY MINIMUM	30	15.2	18.1	22.1	30.1	39.4	48.3	54.6	52.8	43.7	33.9	23.7	16.7	33.2
	MEAN DAILY MINIMUM	54	15.4	18.1	21.8	29.9	39.6	48.4	53.5	52.2	43.1	33.1	22.9	17.1	32.9
	LOWEST DAILY MINIMUM	66	-29	-34	-21	-8	16	25	38	36	8	-1	-16	-28	-34
	YEAR OF OCCURRENCE		1984	1936	1943	1975	1947	1951	1952	1975	1985	1991	1993	1990	FEB 1936
	MEAN OF EXTREME MINS.	54	-8.8	-3.9	2.0	14.0	27.7	37.7	45.0	43.6	29.6	17.3	2.7	-5.4	16.8
	NORMAL DRY BULB	30	26.5	29.3	33.6	42.5	52.0	61.3	68.4	66.4	57.4	47.0	35.2	27.8	45.6
	MEAN DRY BULB	54	26.7	29.3	33.4	42.1	52.1	61.6	67.3	65.8	56.8	46.0	34.2	28.0	45.3
	MEAN WET BULB	18	22.8	23.7	29.4	35.2	44.6	51.8	52.9	52.6	44.6	35.5	26.2	21.4	36.7
	MEAN DEW POINT	18	12.9	14.8	20.1	26.2	36.3	43.4	44.3	44.7	35.8	26.0	17.3	12.1	27.8
	NORMAL NO. DAYS WITH:														
	MAXIMUM 90	30	0.0	0.0	0.0	0.0	*	1.3	5.2	2.1	0.1	0.0	0.0	0.0	8.7
	MAXIMUM 32	30	9.3	7.0	5.4	1.4	*	0.0	0.0	0.0	0.2	0.7	4.1	8.9	37.0
	MINIMUM 32	30	28.8	26.1	27.7	18.0	3.6	0.0	0.0	0.0	2.1	12.2	24.4	28.5	171.4
	MINIMUM 0	30	4.4	2.5	1.0	0.1	0.0	0.0	0.0	0.0	0.0	0.0	0.5	3.4	11.9
H/C	NORMAL HEATING DEG. DAYS	30	1194	1000	973	675	403	150	31	44	251	558	894	1153	7326
	NORMAL COOLING DEG. DAYS	30	0	0	0	0	0	39	136	87	23	0	0	0	285
RH	NORMAL (PERCENT)	30	52	55	56	54	56	54	51	51	52	50	54	54	53
	HOUR 05 LST	30	57	61	65	68	71	71	70	69	67	61	60	58	65
	HOUR 11 LST	30	45	46	46	43	42	40	36	36	38	38	43	46	42
	HOUR 17 LST	30	49	48	47	42	44	41	38	38	38	41	50	52	44
	HOUR 23 LST	30	58	61	64	65	67	64	63	63	62	59	60	59	62
S	PERCENT POSSIBLE SUNSHINE	62	64	67	67	63	61	67	69	68	70	69	61	60	66
W/O	MEAN NO. DAYS WITH:														
	HEAVY FOG(VISBY 1/4 MI)	67	1.0	1.9	3.2	3.3	3.0	1.9	1.2	1.4	1.9	2.0	1.9	1.4	24.1
	THUNDERSTORMS	67	0.0	0.0	0.2	2.1	7.8	11.3	13.2	10.5	4.5	0.9	0.0	0.0	50.5
CLOUDINESS	MEAN:														
	SUNRISE-SUNSET (OKTAS)	1		4.8	6.4		7.2	4.0						4.0	
	MIDNIGHT-MIDNIGHT (OKTAS)	1					7.2	4.0							
	MEAN NO. DAYS WITH:														
	CLEAR	1	4.0	7.0	5.0		4.0	13.0							
	PARTLY CLOUDY	1	3.0	6.0	5.0		12.0	8.0							
	CLOUDY	1	3.0	4.0	11.0		13.0	7.0							
PR	MEAN STATION PRESSURE(IN)	29	23.90	23.89	23.80	23.90	23.90	24.00	24.11	24.11	24.10	24.01	23.91	23.91	23.96
	MEAN SEA-LEVEL PRES. (IN)	18	30.06	30.04	29.97	29.92	29.88	29.98	29.96	29.99	30.00	30.03	30.03	30.08	29.99
WINDS	MEAN SPEED (MPH)	44	15.2	14.7	14.4	14.1	12.6	11.4	10.4	10.4	11.2	12.3	13.5	14.6	12.9
	PREVAIL.DIR(TENS OF DEGS)	22	28	29	28	28	28	28	28	28	28	28	28	28	28
	MAXIMUM 2-MINUTE:														
	SPEED (MPH)	6	55	59	48	55	44	51	46	56	48	53	59	63	63
	DIR. (TENS OF DEGS)		25	27	28	30	28	27	20	25	28	26	27	27	27
	YEAR OF OCCURRENCE		2000	1999	1999	1999	1997	1998	2001	1996	2000	2001	1999	1998	DEC 1998
	MAXIMUM 5-SECOND:														
	SPEED (MPH)	6	64	71	60	69	55	61	59	60	60	63	68	75	75
	DIR. (TENS OF DEGS)		26	27	29	32	25	28	32	25	22	26	27	26	26
	YEAR OF OCCURRENCE		1999	1999	2000	1999	1999	2001	1996	1996	2000	2001	1999	1998	DEC 1998
PRECIPITATION	NORMAL (IN)	30	0.40	0.39	1.03	1.37	2.39	2.08	2.09	1.69	1.27	0.74	0.53	0.42	14.40
	MAXIMUM MONTHLY (IN)	66	2.78	2.16	3.65	5.04	6.00	5.32	5.01	6.64	4.52	3.57	2.48	1.68	6.64
	YEAR OF OCCURRENCE		1949	1953	1990	1942	1995	1955	1973	1985	1973	1942	1979	1937	AUG 1985
	MINIMUM MONTHLY (IN)	66	T	T	0.12	0.34	0.11	0.07	0.58	0.03	0.01	0.03	T	0.03	T
	YEAR OF OCCURRENCE		1952	1983	1966	1992	1974	1980	1969	1944	1992	1964	1965	1959	FEB 1983
	MAXIMUM IN 24 HOURS (IN)	66	1.41	1.60	1.88	1.94	2.07	2.68	3.42	6.06	2.75	1.70	1.66	1.19	6.06
	YEAR OF OCCURRENCE		1949	1953	1946	1984	1991	1955	1973	1985	1973	1947	1979	1979	AUG 1985
	NORMAL NO. DAYS WITH:														
	PRECIPITATION 0.01	30	5.1	6.0	9.2	9.3	11.3	10.2	10.8	9.6	7.5	5.6	6.0	6.0	96.6
	PRECIPITATION 1.00	30	*	0.0	0.0	0.1	0.4	0.2	0.2	0.1	0.2	0.0	*	*	1.2
SNOWFALL	NORMAL (IN)	30	6.0	5.4	12.4	7.9	3.0	T	0.0	0.0	0.8	4.1	6.4	6.5	52.5
	MAXIMUM MONTHLY (IN)	66	35.5	23.3	39.2	31.8	30.4	8.7	1.0	0.5	11.8	21.3	31.1	21.3	39.2
	YEAR OF OCCURRENCE		1980	1995	1990	1984	1943	1947	1994	1993	2000	1969	1979	1958	MAR 1990
	MAXIMUM IN 24 HOURS (IN)	66	12.7	14.0	15.6	17.4	15.0	8.7	1.0	0.5	10.1	8.6	19.8	11.7	19.8
	YEAR OF OCCURRENCE		1992	1953	1973	1984	1942	1947	1994	1993	2000	1990	1979	1979	NOV 1979
	MAXIMUM SNOW DEPTH (IN)	53	23	12	19	15	12		0	0	8	10	26	17	26
	YEAR OF OCCURRENCE		1980	1989	1990	1984	1978				2000	1990	1979	1979	NOV 1979
	NORMAL NO. DAYS WITH:														
	SNOWFALL 1.0	30	1.9	1.7	3.6	2.1	0.7	0.0	0.0	0.0	0.3	1.3	2.0	2.0	15.6

PRECIPITATION (inches) 2001 CHEYENNE, WY (CYS)

YEAR	JAN	FEB	MAR	APR	MAY	JUN	JUL	AUG	SEP	OCT	NOV	DEC	ANNUAL
1972	0.36	0.02	0.79	0.80	2.76	1.71	1.35	1.83	1.01	0.42	0.59	0.40	12.04
1973	0.23	0.07	1.85	1.75	0.31	1.20	5.01	0.27	4.52	0.06	1.25	1.06	17.58
1974	0.48	0.03	1.24	0.50	0.11	2.81	1.41	1.29	0.50	0.91	0.49	0.10	9.87
1975	0.40	0.17	1.17	0.47	2.27	1.49	2.62	0.39	0.52	0.49	0.20	0.52	10.71
1976	0.32	0.71	0.32	1.79	2.07	0.68	2.39	1.40	0.77	0.15	0.28	0.10	10.98
1977	0.14	0.08	1.21	1.86	2.50	2.44	3.49	1.07	0.19	0.08	0.35	0.24	13.65
1978	0.58	0.78	0.35	0.52	3.98	0.63	0.98	1.38	0.12	0.50	0.45	0.54	10.81
1979	0.27	0.14	1.34	0.77	2.90	3.32	1.83	1.86	0.32	0.46	2.48	1.50	17.19
1980	2.71	0.73	1.36	0.93	2.39	0.07	2.00	1.55	0.97	0.51	0.46	0.08	13.76
1981	0.30	0.20	0.70	0.73	5.67	1.66	2.85	2.90	0.31	0.85	0.09	0.45	16.71
1982	0.41	0.19	0.17	0.53	3.56	4.52	2.71	1.81	2.87	1.20	0.43	0.83	19.23
1983	0.02	T	2.96	4.45	2.31	2.81	2.12	1.95	0.78	0.49	2.34	0.46	20.69
1984	0.54	0.84	1.28	3.71	0.78	2.43	2.84	0.65	1.55	0.11	0.34	1.77	17.64
1985	0.66	0.19	0.36	1.10	1.05	1.59	3.99	6.64	1.78	0.94	0.84	0.80	19.94
1986	0.13	0.50	0.54	2.26	1.03	2.42	1.04	1.55	2.47	1.78	0.66	0.18	14.56
1987	0.09	0.90	1.25	0.68	4.43	1.80	2.04	1.23	0.93	0.33	0.76	0.85	15.29
1988	0.52	0.65	1.34	1.84	3.09	2.03	1.79	1.79	1.66	0.09	0.42	0.53	15.75
1989	0.27	1.26	0.49	0.48	1.37	2.51	1.70	1.79	1.62	0.41	0.14	0.69	12.73
1990	0.35	0.69	3.65	1.66	3.37	1.03	3.64	1.98	0.80	1.35	0.72	0.39	19.63
1991	0.36	0.12	0.43	1.15	3.84	4.56	3.39	1.49	1.87	0.44	0.87	0.13	18.65
1992	1.18	0.11	1.75	0.34	1.86	2.11	1.87	1.90	0.01	0.46	1.69	0.49	13.77
1993	0.35	0.81	0.63	2.49	1.92	3.32	0.64	2.21	3.16	1.85	1.23	0.30	18.91
1994	0.66	0.88	0.55	1.45	2.05	1.44	2.13	1.49	0.38	1.55	0.34	0.57	13.49
1995	0.12	1.02	0.19	1.49	6.00	4.58	1.04	0.89	3.39	0.78	0.37	0.18	20.09
1996	0.50	0.07	1.50	1.51	2.24	1.79	3.10	1.79	2.34	0.50	0.45	0.01	15.80
1997	0.39	0.31	0.44	1.68	2.47	3.48	2.80	4.33	2.06	1.09	0.20	0.59	19.84
1998	0.07	0.29	0.69	1.09	2.35	1.63	1.21	0.96	0.49	1.27	0.31	0.46	10.82
1999	0.35	0.13	0.46	5.02	2.04	2.13	2.38	0.78	2.11	0.27	0.25	0.19	16.11
2000	0.35	0.59	1.48	0.58	1.38	0.46	2.64	1.87	2.64	0.69	0.30	0.75	13.73
2001	0.41	0.33	0.56	2.49	2.24	1.45	3.31	0.74	1.18	0.40	0.23	0.13	13.47
POR= 130 YRS	0.43	0.55	1.05	1.76	2.46	1.88	2.03	1.52	1.24	0.80	0.50	0.41	14.63

AVERAGE TEMPERATURE (F) 2001   CHEYENNE, WY (CYS)

YEAR	JAN	FEB	MAR	APR	MAY	JUN	JUL	AUG	SEP	OCT	NOV	DEC	ANNUAL
1972	25.3	33.2	39.3	42.5	51.2	63.2	64.3	64.6	56.6	45.5	29.5	21.0	44.7
1973	23.7	29.7	31.9	37.3	51.4	62.9	65.8	68.4	54.1	49.3	35.3	29.4	44.9
1974	24.1	31.9	37.1	43.3	55.0	64.1	70.1	65.2	55.3	48.3	35.6	26.6	46.4
1975	25.5	25.0	30.2	37.5	48.4	57.9	67.3	66.3	56.0	48.1	34.2	32.6	44.1
1976	26.9	32.9	30.9	42.7	51.1	59.7	69.2	65.0	57.7	43.4	33.5	30.6	45.3
1977	22.4	32.9	32.1	44.7	53.4	65.2	68.2	63.5	60.7	48.3	34.4	29.0	46.2
1978	22.1	25.0	37.7	43.9	49.0	61.0	68.7	64.3	59.6	47.9	33.5	21.1	44.5
1979	17.3	31.6	36.0	44.9	49.9	61.5	69.9	65.9	62.9	49.8	29.5	32.9	46.0
1980	22.2	29.4	33.0	42.3	51.2	65.5	71.4	66.5	60.4	46.7	36.8	38.5	47.0
1981	33.4	32.2	36.8	50.1	50.5	63.7	69.0	65.2	61.2	46.1	40.7	30.7	48.3
1982	25.5	28.2	36.1	41.8	50.5	57.7	67.7	69.1	56.7	45.1	31.4	28.3	44.8
1983	32.8	33.5	32.0	34.8	47.4	57.3	67.8	70.2	59.0	48.6	32.3	15.0	44.2
1984	24.3	28.4	32.4	35.7	53.6	59.4	68.4	66.6	53.0	40.0	35.0	27.0	43.7
1985	19.5	23.3	35.2	45.6	54.3	61.1	69.3	66.8	52.8	45.1	26.2	26.1	43.8
1986	37.0	30.1	42.2	43.9	50.6	64.5	68.5	67.0	55.3	44.6	34.3	28.7	47.2
1987	29.1	31.6	32.4	46.7	54.5	63.3	69.2	64.9	58.0	46.9	36.7	25.7	46.6
1988	21.5	28.4	32.3	44.4	53.3	67.4	69.7	68.7	57.5	50.3	35.3	28.7	46.5
1989	30.2	17.3	37.4	44.3	54.3	60.3	70.6	67.0	57.5	46.5	38.7	24.3	45.7
1990	31.5	28.7	31.9	42.8	49.3	64.1	65.0	66.4	62.2	46.7	38.7	20.8	45.7
1991	25.0	36.2	36.4	41.1	52.4	62.5	67.7	67.7	58.0	46.1	31.7	31.6	46.4
1992	30.5	36.7	38.6	48.6	55.0	60.8	64.6	63.8	60.0	49.2	30.9	25.5	47.0
1993	26.1	24.9	37.2	42.1	53.4	58.7	66.1	64.8	54.3	44.1	29.2	30.7	44.3
1994	29.4	27.2	38.6	43.3	56.9	66.3	68.0	69.4	61.7	46.9	34.6	33.3	48.0
1995	30.7	32.6	36.8	39.5	45.7	59.0	67.8	71.6	57.1	45.8	38.4	30.0	46.3
1996	23.0	30.2	31.6	42.7	51.7	63.6	68.3	66.3	56.6	45.8	34.7	30.1	45.4
1997	24.7	26.7	38.3	35.9	51.7	62.6	68.3	65.0	59.2	46.1	31.4	28.3	44.9
1998	30.6	30.6	32.7	41.2	53.3	56.8	70.3	68.7	64.0	45.8	39.5	26.6	46.7
1999	30.8	34.6	37.6	38.4	50.5	60.3	70.0	67.7	54.0	47.6	44.0	32.0	47.3
2000	29.8	34.7	37.1	45.4	55.5	62.5	72.2	70.5	58.2	46.3	25.9	25.3	47.0
2001	27.6	26.9	35.1	44.3	52.6	64.2	72.2	69.5	61.5	47.2	38.9	29.2	47.4
POR= 128 YRS	26.2	28.1	32.8	41.6	51.1	61.1	67.8	66.4	57.5	46.4	35.0	28.4	45.2

REFERENCE NOTES:

PAGE 1:
   THE TEMPERATURE GRAPH SHOWS NORMAL MAXIMUM AND NORMAL
   MINIMUM DAILY TEMPERATURES (SOLID CURVES) AND THE
   ACTUAL DAILY HIGH AND LOW TEMPERATURES (VERTICAL BARS).

PAGE 2 AND 3:
   H/C INDICATES HEATING AND COOLING DEGREE DAYS.
   RH INDICATES RELATIVE HUMIDITY
   W/O INDICATES WEATHER AND OBSTRUCTIONS
   S INDICATES SUNSHINE.
   PR INDICATES PRESSURE.
   CLOUDINESS ON PAGE 3 IS THE SUM OF THE CEILOMETER AND
   SATELLITE DATA NOT TO EXCEED EIGHT EIGHTHS(OKTAS).

GENERAL:
   T INDICATES TRACE PRECIPITATION, AN AMOUNT GREATER
   THAN ZERO BUT LESS THAN THE LOWEST REPORTABLE VALUE.
   + INDICATES THE VALUE ALSO OCCURS ON EARLIER DATES.
   BLANK ENTRIES DENOTE MISSING OR UNREPORTED DATA.
   NORMALS ARE 30-YEAR AVERAGES (1961 - 1990).
   ASOS INDICATES AUTOMATED SURFACE OBSERVING SYSTEM.
   PM INDICATES THE LAST DAY OF THE PREVIOUS MONTH.
   POR (PERIOD OF RECORD) BEGINS WITH THE JANUARY DATA
   MONTH AND IS THE NUMBER OF YEARS USED TO COMPUTE
   THE MEAN. INDIVIDUAL MONTHS WITHIN THE POR MAY
   BE MISSING.
   WHEN THE POR FOR A NORMAL IS LESS THAN 30 YEARS,
   THE NORMAL IS PROVISIONAL AND IS BASED ON THE NUMBER
   OF YEARS INDICATED.
   0.* OR * INDICATES THE VALUE OR MEAN-DAYS-WITH
   IS BETWEEN 0.00 AND 0.05.
   CLOUDINESS FOR ASOS STATIONS DIFFERS FROM THE NON-ASOS
   OBSERVATION TAKEN BY A HUMAN OBSERVER. ASOS STATION
   CLOUDINESS IS BASED ON TIME-AVERAGED CEILOMETER DATA
   FOR CLOUDS AT OR BELOW 12,000 FEET AND ON SATELLITE
   DATA FOR CLOUDS ABOVE 12,000 FEET.
   THE NUMBER OF DAYS WITH CLEAR, PARTLY CLOUDY, AND
   CLOUDY CONDITIONS FOR ASOS STATIONS IS THE SUM
   OF THE CEILOMETER AND SATELLITE DATA FOR THE
   SUNRISE TO SUNSET PERIOD.

GENERAL CONTINUED:
   CLEAR INDICATES 0 - 2 OKTAS, PARTLY CLOUDY INDICATES
   3 - 6 OKTAS, AND CLOUDY INDICATES 7 OR 8 OKTAS.
   WHEN AT LEAST ONE OF THE ELEMENTS (CEILOMETER OR
   SATELLITE) IS MISSING, THE DAILY CLOUDINESS IS
   NOT COMPUTED.
   WIND DIRECTION IS RECORDED IN TENS OF DEGREES (2 DIGITS)
   CLOCKWISE FROM TRUE NORTH. "00" INDICATES CALM. "36"
   INDICATES TRUE NORTH.
   RESULTANT WIND IS THE VECTOR AVERAGE OF THE SPEED AND
   DIRECTION.
   AVERAGE TEMPERATURE IS THE SUM OF THE MEAN DAILY MAXIMUM
   AND MINIMUM TEMPERATURE DIVIDED BY 2.
   SNOWFALL DATA COMPRISE ALL FORMS OF FROZEN
   PRECIPITATION, INCLUDING HAIL.
   A HEATING (COOLING) DEGREE DAY IS THE DIFFERENCE BETWEEN
   THE AVERAGE DAILY TEMPERATURE AND 65 F.
   DRY BULB IS THE TEMPERATURE OF THE AMBIENT AIR.
   DEW POINT IS THE TEMPERATURE TO WHICH THE AIR MUST BE
   COOLED TO ACHIEVE 100 PERCENT RELATIVE HUMIDITY.
   WET BULB IS THE TEMPERATURE THE AIR WOULD HAVE IF THE
   MOISTURE CONTENT WAS INCREASED TO 100 PERCENT RELATIVE
   HUMIDITY.

   ON JULY 1, 1996, THE NATIONAL WEATHER SERVICE BEGAN USING
   THE "METAR" OBSERVATION CODE THAT WAS ALREADY EMPLOYED
   BY MOST OTHER NATIONS OF THE WORLD. THE MOST NOTICEABLE
   DIFFERENCE IN THIS ANNUAL PUBLICATION WILL BE THE CHANGE
   IN UNITS FROM TENTHS TO EIGHTS(OKTAS) FOR REPORTING THE
   AMOUNT OF SKY COVER.

HEATING DEGREE DAYS (base 65 F) 2001 CHEYENNE, WY (CYS)

YEAR	JUL	AUG	SEP	OCT	NOV	DEC	JAN	FEB	MAR	APR	MAY	JUN	TOTAL
1972-73	85	75	248	599	1056	1358	1275	980	1017	824	414	122	8053
1973-74	80	4	323	482	883	1098	1264	922	862	643	304	110	6975
1974-75	4	55	302	509	873	1180	1215	1115	1070	819	506	212	7860
1975-76	11	39	274	515	920	998	1175	924	1048	661	425	158	7148
1976-77	11	44	224	664	937	1059	1314	894	1014	602	352	44	7159
1977-78	21	74	150	511	910	1108	1324	1115	840	627	491	157	7328
1978-79	28	73	200	523	937	1358	1471	933	893	597	459	139	7611
1979-80	2	62	105	468	1058	990	1321	1027	984	673	424	65	7179
1980-81	0	41	151	558	840	812	974	913	868	440	445	95	6137
1981-82	21	50	120	580	722	1058	1216	1025	892	687	446	227	7044
1982-83	29	7	264	608	1002	1131	992	875	1016	898	538	233	7593
1983-84	23	0	202	502	974	1547	1259	1058	1002	870	348	177	7962
1984-85	6	15	365	769	892	1171	1403	1163	917	576	323	162	7762
1985-86	11	37	364	612	1158	1199	862	970	698	629	440	71	7051
1986-87	4	20	286	630	914	1121	1102	928	1002	542	321	78	6948
1987-88	27	79	214	554	841	1213	1343	1056	1008	611	361	42	7349
1988-89	12	18	236	449	884	1116	1075	1332	851	615	334	180	7102
1989-90	4	18	236	566	779	1257	1037	1012	1021	662	480	106	7178
1990-91	75	28	127	558	784	1364	1231	799	879	712	385	95	7037
1991-92	26	17	217	586	993	1031	1062	813	811	488	307	131	6482
1992-93	63	92	161	484	1014	1216	1198	1118	856	680	351	202	7435
1993-94	45	58	322	642	1068	1059	1097	1053	809	644	249	43	7089
1994-95	24	16	131	552	904	978	1057	901	866	758	588	199	6974
1995-96	37	6	283	589	793	1076	1293	1002	1031	663	411	77	7261
1996-97	16	29	269	590	905	1076	1242	1065	820	867	400	102	7381
1997-98	34	58	181	577	1001	1130	1059	957	994	706	352	256	7305
1998-99	7	10	104	587	754	1183	1055	845	840	791	444	150	6770
1999-00	8	27	322	534	625	1015	1083	870	858	581	305	130	6358
2000-01	2	8	242	573	1167	1223	1155	1064	922	614	377	106	7453
2001-	0	11	135	545	776	1102							

COOLING DEGREE DAYS (base 65 F) 2001 CHEYENNE, WY (CYS)

YEAR	JAN	FEB	MAR	APR	MAY	JUN	JUL	AUG	SEP	OCT	NOV	DEC	ANNUAL
1972	0	0	0	0	0	11	69	69	2	0	0	0	151
1973	0	0	0	0	0	65	112	116	0	0	0	0	293
1974	0	0	0	0	4	88	173	67	17	0	0	0	349
1975	0	0	0	0	0	7	90	86	10	0	0	0	193
1976	0	0	0	0	0	7	145	50	15	0	0	0	217
1977	0	0	0	0	0	59	126	38	29	0	0	0	252
1978	0	0	0	0	1	43	150	59	44	0	0	0	297
1979	0	0	0	0	1	42	160	100	49	0	0	0	352
1980	0	0	0	0	0	88	205	94	21	0	0	0	408
1981	0	0	0	0	0	59	156	64	15	0	0	0	294
1982	0	0	0	0	0	8	120	140	21	0	0	0	289
1983	0	0	0	0	0	10	115	169	28	0	0	0	322
1984	0	0	0	0	1	14	118	72	8	0	0	0	213
1985	0	0	0	0	0	52	150	98	8	0	0	0	308
1986	0	0	0	0	0	62	118	89	0	0	0	0	269
1987	0	0	0	0	0	35	164	83	9	0	0	0	291
1988	0	0	0	0	4	122	166	140	18	0	0	0	450
1989	0	0	0	0	7	46	188	86	19	0	0	0	346
1990	0	0	0	0	0	86	84	79	49	0	0	0	298
1991	0	0	0	0	2	28	119	106	13	4	0	0	272
1992	0	0	0	1	2	15	58	63	16	0	0	0	155
1993	0	0	0	0	0	19	89	60	6	0	0	0	174
1994	0	0	0	0	3	91	125	156	38	0	0	0	413
1995	0	0	0	0	0	23	132	218	50	0	0	0	423
1996	0	0	0	0	2	40	124	73	24	0	0	0	263
1997	0	0	0	0	0	37	144	65	13	0	0	0	259
1998	0	0	0	0	0	17	182	131	78	0	0	0	408
1999	0	0	0	0	0	18	170	114	0	0	0	0	302
2000	0	0	0	0	19	59	231	187	44	0	0	0	540
2001	0	0	0	0	0	91	229	158	38	0	0	0	516

SNOWFALL (inches) 2001 CHEYENNE, WY (CYS)

YEAR	JUL	AUG	SEP	OCT	NOV	DEC	JAN	FEB	MAR	APR	MAY	JUN	TOTAL
1972-73	0.0	0.0	T	5.0	8.8	7.6	4.6	0.6	27.0		1.2	0.0	67.6
1973-74	0.0	0.0	T	0.5	17.4	13.9	5.0	0.8	16.3	8.0		0.0	61.9
1974-75	0.0	0.0	1.9	1.4	2.4	1.5	5.8	4.3	14.0	5.4	0.8	0.0	37.5
1975-76	0.0	0.0	0.6	5.5	4.7	8.4	6.0	8.5	9.0	6.7	T	T	49.4
1976-77	0.0	0.0	0.0	1.5	3.5	1.2	2.5	0.8	12.9	7.0	T		29.4
1977-78	0.0	0.0	0.0	0.4	4.3	5.6	7.0	12.0	2.5	1.0	18.3	0.0	51.1
1978-79	0.0	0.0	0.1	2.0	9.3	17.6	6.7	2.2	21.1	4.0	14.1	T	77.1
1979-80	0.0	0.0	0.0	3.6	31.1	15.6	35.5	10.7	17.8	3.4	3.8	0.0	121.5
1980-81	0.0	0.0	0.0	1.1	6.3	2.1	3.4	2.9	9.0	2.0	0.8	0.0	27.6
1981-82	0.0	0.0	0.0	4.6	0.8	5.8	5.6	2.4	1.7	2.0	4.0	0.0	26.9
1982-83	0.0	0.0	0.0	12.2	7.9	13.1	0.1	T	31.9	25.7	10.1	0.0	101.0
1983-84	0.0	0.0	T	0.2	27.2	7.0	7.9	12.1	13.0	31.8	T	0.0	99.2
1984-85	0.0	0.0	1.6	3.8	1.5	5.6	9.8	1.9	3.6	2.6	0.4	0.0	30.8
1985-86	0.0	0.0	7.4	6.0	12.4	13.0	1.3	4.9	5.6	11.3	3.8	0.0	65.7
1986-87	0.0	0.0	0.0	9.2	6.9	2.2	1.1	9.9	13.7	4.4	T	0.0	47.4
1987-88	0.0	0.0	0.0	1.3	4.5	16.1	9.0	7.5	16.0	7.5	2.2	0.0	64.1
1988-89	0.0	0.0	0.2	0.0	4.5	7.2	4.4	17.6	5.0	4.3	T	T	43.2
1989-90	T	T	2.6	3.4	1.9	9.7	5.6	11.6	39.2	5.6	1.3	T	80.9
1990-91	T	T	T	12.3	8.8	6.6	7.7	1.6	3.7	10.5	T	T	51.2
1991-92	T	T	T	7.5	9.1	2.0	18.2	0.8	11.3	0.7	4.4	0.5	54.5
1992-93	T	T	0.0	2.1	25.5	10.9	7.1	14.6	6.6	11.3	T	0.5	78.6
1993-94	0.0	0.5	5.5	3.1	18.0	5.7	11.5	16.1	4.2	9.1	T	T	73.7
1994-95	1.0	0.0	0.9	2.3	3.0	10.0	2.7	23.3	4.4	13.2	2.6	T	63.4
1995-96	0.0	0.0	3.0	6.4	6.4	3.6	11.9	2.1	21.9	8.7	0.1	T	64.1
1996-97	T	T	1.2	2.0	7.0	0.4	10.5	7.9	8.4	23.3	2.8	T	63.5
1997-98	T	T	T	8.8	2.5	10.9	1.3	3.4	8.4	10.1	1.1	0.7	47.2
1998-99	T	0.0	0.0	0.3	4.0	13.5	6.4	2.4	4.9	17.0	0.9	T	49.4
1999-00	0.0	0.0	3.7	3.3	4.3	2.9	2.9	10.7	16.2	1.7	0.2	0.0	45.9
2000-01	T	T	11.8	0.7	4.3	13.5	8.6	6.8	7.0	23.2	13.1	T	89.0
2001-	T	0.0	T	2.6	4.3	2.1							
POR= 65 YRS	0.0	0.0	1.1	3.7	7.1	6.3	6.4	6.3	11.5	9.3	3.5	0.2	55.4

Glossary of Weather Terms

A

absolute instability

A state of and layer within the atmosphere in which the vertical distribution of temperature is such that the air parcel, if given an upward or downward push, will move away from its initial level without further outside force being applied.

absolute temperature scale

See **kelvin temperature scale**

absolute vorticity

See **vorticity**

adiabatic process

The process by which fixed relationships are maintained during changes in temperature, volume, and pressure in a body of air without heat being added to or removed from the body.

advection

The horizontal transport of air or atmospheric probes. In meteorology, sometimes referred to as the horizontal component of convection.

advection fog

Fog resulting from the transport of warm, humid air over a cold surface.

air density

The mass density of the air in terms of weight per unit volume.

air mass

In meteorology, an extensive body of air within which the conditions of temperature and moisture in a horizontal plane are essentially uniform.

air mass classification

A system used to identify and characterize the different air masses according to a basic scheme. The system most commonly used classifies air masses primarily according to the thermal properties of their source regions: "tropical" (T); "polar" (P); and "arctic" or "antarctic" (A). They are further classified according to moisture characteristics as "continental" (c) or "maritime" (m).

air parcel

See **parcel**

albedo

The ratio of the amount of electromagnetic radiation reflected by a body to the amount incident upon it, commonly expressed in percentage; in meteorology, usually used in reference to insolation (solar radiation); i.e., the albedo of wet sand is 9, meaning that about 9% of the incident insolation is reflected; albedos of other surfaces range upward to 80–85 for fresh snow cover; average albedo for Earth and its atmosphere has been calculated to a range of 35–43.

All-hallow summer

Also called Allhallow summer, All Saints' summer. In English folklore, an old name for a period, like Indian summer, of unseasonable warmth, supposed to occur on the eve of All Hallows day (All Saints Day), November 1.

almwind

Local name for a foehn that blows from the south (Hungary) across the Tatra Mountains south of Krakow, Poland, and descends the northern valleys; similar to the Alpine south foehn.

alpenglow

The occasional reappearance of sunset colors on a snow-covered mountaintop soon after sunset and a similar phenomenon before sunrise.

altimeter

An instrument that determines the altitude of an object with respect to a fixed level.

altimeter setting

The value to which the scale of a pressure altimeter is set so as to read true altitude at field elevation.

altimeter setting indicator

A precision aneroid barometer calibrated to indicate directly the altimeter setting.

altitude

Height expressed in units of distance above a reference plane, usually above mean sea level or above ground.

altocumulus

White or gray layers or patches of cloud, often with a waved appearance; cloud elements appear as rounded masses or rolls; composed mostly of liquid water droplets which may be supercooled; may contain ice crystals at subfreezing temperatures.

altocumulus castellanus

A species of middle cloud of which at least a fraction of its upper part presents some vertically developed, cumuliform protuberances (some of which are taller than they are wide, as castles) that give the cloud a crenelated or turreted appearance; especially evident when seen from the side; elements usually have a common base arranged in lines. This cloud indicates instability and turbulence at the altitudes of occurrence.

altostratus

A principal cloud type in the form of a gray or bluish sheet of striated, fibrous, or uniform appearance.

anabatic wind

In mountain meteorology, an upslope wind driven by heating at the slope surface under fair-weather conditions.

anchor ice

Ice attached to the beds of streams, lakes, and shallow seas, irrespective of its nature of formation.

anemometer

An instrument for measuring wind speed.

aneroid barometer

A barometer that operates on the principle of having changing atmospheric pressure bend a metallic surface which, in turn, moves a pointer across a scale graduated in units of pressure.

anticyclone

An area of high atmospheric pressure that has a closed circulation that is anticyclonic. As viewed from above, the circulation is clockwise in the Northern Hemisphere, counterclockwise in the Southern Hemisphere, and undefined at the equator.

anvil cloud

Popular name given to the top portion of a cumulonimbus having an anvil-like form.

aphelion

The point on the Earth's orbit that is farthest from the Sun

arctic air

An air mass with characteristics developed mostly in winter over arctic surfaces of ice and snow. Arctic air extends to great heights, and the surface temperatures are basically, but not always, lower than those of polar air.

arctic front

The surface of discontinuity between very cold (arctic) air flowing directly from the arctic region and another less cold, and consequently less dense, air mass.

astronomical twilight

See **twilight**

atmosphere

The mass of air surrounding Earth.

atmospheric pressure (barometric pressure)

The pressure exerted by the atmosphere as a consequence of gravitational attraction exerted upon the "column" of air lying directly above the point in question.

atmospherics

Disturbing effects produced in radio receiving apparatus by atmospheric electrical phenomena such as an electrical storm. Static.

aurora

A luminous, radiant emission over middle and high latitudes confined to the thin air of high altitudes and centered over the Earth's magnetic poles. Called "aurora borealis" (northern lights) or "aurora australis" according to their occurrence in the Northern or Southern Hemisphere, respectively.

AWIPS

Advanced Weather Interactive Processing System. A high-speed communication network and of the National Weather Service that provides fast-response interactive analysis and display of data to assist meteorologists.

B

back-door cold front

The leading edge of a cold air mass moving toward the south and southwest along the United States' Atlantic seaboard.

backing

Shifting of the wind in a counterclockwise direction with respect to either space or time; opposite of veering. Commonly used by meteorologists to refer to a cyclonic shift (counterclockwise in the Northern Hemisphere and clockwise in the Southern Hemisphere).

backscatter

Pertaining to radar, the energy reflected or scattered by a target; an echo.

ball lightning (Also known as globe lightning)

A rare and randomly occurring bright ball of light observed floating or moving through the atmosphere close to the ground.

banner cloud (cloud banner)

A banner-like cloud streaming off from a mountain peak.

barb (also called feather)

A means of representing wind speed in the plotting of a synoptic map.

baroclinic

A state where areas of constant atmospheric pressure cross areas of constant atmospheric density.

barograph

A continuous-recording barometer.

barometer

An instrument for measuring the pressure of the atmosphere; the two principal types are mercurial and aneroid.

barometric altimeter

See **pressure altimeter**

barometric pressure

See **atmospheric pressure**

barometric tendency

The change of barometric pressure within a specified period of time. In aviation weather observations, routinely determined periodically, usually for a three-hour period.

barrier jet

A jet on the windward side of a mountain barrier, blowing parallel to the barrier.

Beaufort scale

A scale of wind speeds.

bird burst

Radar echoes caused by flocks of birds, or by roosting birds that take to the air as a group and fly away in different directions.

black blizzard

See **duststorm**

blizzard

A severe weather condition characterized by low temperatures and strong winds bearing a great amount of snow, either falling or picked up from the ground.

blowing dust

A type of lithometeor composed of dust particles picked up locally from the surface and blown about in clouds or sheets.

blowing sand

A type of lithometeor composed of sand picked up locally from the surface and blown about in clouds or sheets.

blowing snow

A type of hydrometeor composed of snow picked up from the surface by the wind and carried to a height of 6 ft or more.

blowing spray

A type of hydrometeor composed of water particles picked up by the wind from the surface of a large body of water.

Buys Ballot's law

If an observer in the Northern Hemisphere stands with his or her back to the wind, lower pressure is to the observer's left.

C

calm

The absence of wind or of apparent motion of the air.

cap cloud

A standing or stationary cap-like cloud crowning a mountain summit.

ceiling

In meteorology in the United States, the height above the surface of the base of the lowest layer of clouds or obscuring phenomenon aloft that hides more than half of the sky, or the vertical visibility into an obscuration. *See also* **summation principle**

Celsius temperature scale (°C)

A temperature scale with zero degrees as the melting point of pure ice and 100°C (212°F) as the boiling point of pure water at standard sea-level atmospheric pressure.

centigrade temperature scale

See **Celsius temperature scale**

change of state

In meteorology, the transformation of water from one form, i.e., solid (ice), liquid, or gaseous (water vapor), to any other form. There are six possible transformations designated by the five terms following: (1) condensation: the change of water vapor to liquid water; (2) evaporation: the change of liquid water to water vapor; (3) freezing: the change of liquid water to ice; (4) melting: the change of ice to liquid water; (5) sublimation: the change of ice to water vapor or water vapor to ice. *See also* **latent heat**

chinook

A warm, dry/foehn wind blowing down the eastern slopes of the Rocky Mountains over the adjacent plains in the United States and Canada.

chubasco

A severe thunderstorm with vivid lightning and violent squalls coming from the land on the west coast of Nicaragua and Costa Rica in Central America.

cirriform

All species and varieties of cirrus, cirrocumulus, and cirrostratus clouds; descriptive of clouds composed mostly or entirely of small ice crystals, usually transparent and white; often producing halo phenomena not observed with other cloud forms. Average height ranges upward from 20,000 ft in middle latitudes.

cirrocumulus

A cirriform cloud appearing as a thin sheet of small white puffs resembling flakes or patches of cotton without shadows; sometimes confused with altocumulus.

cirrostratus

A cirriform cloud appearing as a whitish veil, usually fibrous, sometimes smooth; often produces halo phenomena; may totally cover the sky.

cirrus

A cirriform cloud in the form of thin, white feather-like clouds in patches or narrow bands; has fibrous and/or silky sheen; large ice crystals often trail downward a considerable vertical distance in fibrous, slanted, or irregularly curved wisps called mare's tails.

civil twilight

See **twilight**

climate

The statistical collective of the weather conditions of a point or area during a specified interval of time (usually several decades); may be expressed in a variety of ways.

climatology

The study of climate.

clinometer

An instrument used in weather observing for measuring angles of inclination. It is used in conjunction with a ceiling light to determine cloud height at night.

cloud cap

See **cap cloud**

cloudburst

Any sudden and heavy fall of rain or hail, almost always of the shower type.

cold front

Any non-occluded front that moves in such a way that colder air replaces warmer air.

condensation

See **change of state**

condensation level

The height at which a rising parcel or layer of air would become saturated if lifted adiabatically.

condensation nuclei

Small particles in the air on which water vapor condenses or sublimates.

condensation trail (contrail or vapor trail)

A cloudlike streamer frequently observed to form behind aircraft flying in clear, cold, humid air.

conditionally unstable air

Unsaturated air that will become unstable on the condition it becomes saturated. *See also* **instability**

conduction

The transfer of heat by molecular action through a substance or from one substance in contact with another; transfer is always from warmer to colder temperature.

constant pressure chart

A chart of a constant pressure surface; may contain analyses of height, wind, temperature, humidity, and/or other elements.

continental polar air

See **polar air**

continental tropical air

See **tropical air**

contour

In meteorology, a line of equal height on a constant pressure chart; analogous to contours on a relief map; in radar meteorology, a line on a radar scope of equal echo intensity.

contouring circuit

On weather radar, a circuit that displays multiple contours of echo intensity simultaneously on the plan position indicator or range-height indicator scope. *See also* **contour**

convection

In general, mass motions within a fluid resulting in transport and mixing of the properties of that fluid. In meteorology, atmospheric motions that are predominantly vertical, resulting in vertical transport and mixing of atmospheric properties; distinguished from advection.

convective cloud

See **cumuliform**

convective condensation level (CCL)

The lowest level at which condensation will occur as a result of convection due to surface heating. When condensation occurs at this level, the layer between the surface and the CCL will be thoroughly mixed, temperature lapse rate will be dry adiabatic, and mixing ratio will be constant.

convective instability

The state of an unsaturated layer of air whose lapse rates of temperature and moisture are such that when lifted adiabatically until the layer becomes saturated, convection is spontaneous.

convergence

The condition that exists when the distribution of winds within a given area is such that there is a net horizontal inflow of air into the area. In convergence at lower levels, the removal of the resulting excess is accomplished by an upward movement of air; consequently, area of low-level convergent winds are regions favorable to the occurrence of clouds and precipitation. *See also* **divergence**

Coriolis force

A deflective force resulting from the Earth's rotation; it acts to the right of wind direction in the Northern Hemisphere and to the left in the Southern Hemisphere.

corona

A prismatically colored circle or arcs of a circle with the Sun or Moon at its center; coloration is from blue inside to red outside (opposite that of a halo); varies in size (much smaller) as opposed to the fixed diameter of the halo; characteristic of clouds composed of water droplets and valuable in differentiating between middle and cirriform clouds.

corposant

See **St. Elmo's Fire**

corrected altitude

See **altitude**

cumuliform

A term descriptive of all convective clouds exhibiting vertical development in contrast to the horizontally extended stratiform types.

cumulonimbus

A cumuliform cloud type; it is heavy and dense, with considerable vertical extent in the form of massive towers; often with tops in the shape of an anvil or massive plume; under the base of cumulonimbus, which often is very dark, there frequently exists virga, precipitation, and low ragged clouds (scud), either merged with it or not; frequently accompanied by lightning, thunder, and sometimes hail; occasionally produces a tornado or a waterspout; the ultimate manifestation of the growth of a cumulus cloud, occasionally extending well into the stratosphere.

cumulonimbus mamma

A cumulonimbus cloud having hanging protuberances, like pouches, festoons, or udders, on the underside of the cloud; usually indicative of severe turbulence.

cumulus

A cloud in the form of individual detached domes or towers that are usually dense and well defined; develops vertically in the form of rising mounds of which the bulging upper part often resembles a cauliflower; the sunlit parts of these clouds are mostly brilliant white; their bases are relatively dark and nearly horizontal.

cumulus fractus

See **fractus**

cyclogenesis

Any development or strengthening of cyclonic circulation in the atmosphere.

cyclone

An area of low atmospheric pressure that has a closed circulation that is cyclonic, i.e., as viewed from above, the circulation is counterclockwise in the Northern Hemisphere, clockwise in the Southern Hemisphere, and undefined at the equator. Because cyclonic circulation and relatively low atmospheric pressure usually co-exist, in common practice the terms cyclone and low are used interchangeably. Also, because cyclones often are accompanied by inclement (sometimes destructive) weather, they are frequently referred to simply as storms. The term is frequently misused to denote a tornado. In the Indian Ocean, a tropical cyclone of hurricane or typhoon force.

D

deepening

A decrease in the central pressure of a pressure system; usually applied to a low rather than to a high, although technically, it is acceptable in either sense.

density

The ratio of the mass of any substance to the volume it occupies—weight per unit volume. Also, the ratio of any quantity to the volume or area it occupies, i.e., population per unit area, power density.

density altitude

See **altitude**

depression

In meteorology, an area of low pressure; a low or trough. This is usually applied to a certain stage in the development of a tropical cyclone, to migratory lows and troughs, and to upper-level lows and troughs that are only weakly developed.

derecho

A widespread convectively induced straight-line windstorm.

dew

Water condensed onto grass and other objects near the ground, the temperatures of which have fallen below the initial dew-point temperature of the surface air, but are still above freezing. *See also* **frost**

dew point (dew-point temperature)

The temperature to which a sample of air must be cooled, while the mixing ratio and barometric pressure remain constant, in order to attain saturation with respect to water.

discontinuity

A zone with comparatively rapid transition of one or more meteorological elements.

disturbance

In meteorology, applied rather loosely: first, any low pressure or cyclone, but usually one that is relatively small in size; second, an area where weather, wind, pressure, etc., show signs of cyclonic development; third, any deviation in flow or pressure that is associated with a disturbed state of the weather, i.e., cloudiness and precipitation; and finally, any individual circulatory system within the primary circulation of the atmosphere.

diurnal

Daily, especially pertaining to a cycle completed within a 24-hour period, and that recurs every 24 hours.

divergence

The condition that exists when the distribution of winds within a given area is such that there is a net horizontal flow of air outward from the region. In divergence at lower levels, the resulting deficit is compensated for by subsidence of air from aloft; consequently the air is heated and the relative humidity lowered, making divergence a warming and drying process. Low-level divergent regions are areas unfavorable to the occurrence of clouds and precipitation. The opposite of convergence.

doldrums

The equatorial belt of calm or light and variable winds between the two trade wind belts. *See also* **intertropical convergence zone**

downburst

A strong and potentially destructive thunderstorm downdraft; depending on size, either a microburst or a macroburst. *See also* **microburst**

downdraft

A relatively small-scale downward current of air; often observed on the lee side of large objects restricting the smooth flow of the air or in precipitation areas in or near cumuliform clouds.

drifting snow

A type of hydrometeor composed of snow particles picked up from the surface, but carried to a height of less than 6 ft.

drizzle

A form of precipitation. Very small water drops that appear to float with the air currents while falling in an irregular path (unlike rain, which falls in a comparatively straight path, and unlike fog droplets, which remain suspended in the air).

dropsonde

A radiosonde dropped by parachute from an aircraft to obtain soundings (measurements) of the atmosphere below.

dry adiabatic lapse rate

The rate of decrease of temperature with height when unsaturated air is lifted adiabatically (due to expansion as it is lifted to lower pressure). *See also* **adiabatic process**

dry bulb

A name given to an ordinary thermometer used to determine temperature of the air.

dry-bulb temperature

The temperature of the air.

dry line

A boundary that separates warm, dry air from warm, moist air. It frequently represents a zone of instability along which possible severe thunderstorms form, particularly over the United States western plains, from eastern New Mexico and far west Texas northward, most notably in the spring.

dust

A type of lithometeor composed of small earthen particles suspended in the atmosphere.

dust devil

A small, vigorous whirlwind, usually of short duration, rendered visible by dust, sand, and debris picked up from the ground; extends from the ground upward. Caused by intense solar heating of dry surface areas.

duster

See **duststorm**

duststorm (duster, black blizzard)

An unusual, frequently severe weather condition characterized by strong winds and dust-filled air over an extensive area.

D-value

Departure of true altitude from pressure altitude; obtained by algebraically subtracting true altitude from pressure altitude; thus it may be plus or minus. On a constant pressure chart, the difference between actual height and standard atmospheric height of a constant pressure surface.

E

eddy

A local irregularity of wind in a larger scale wind flow. Small scale eddies produce turbulent conditions.

el Niño

Name given to the periodic warming of the ocean that occurs in the central and eastern tropical Pacific and can produce extreme weather in many locations of the world.

El Niño/Southern Oscillation (ENSO)

An episode of anomalously high sea-surface temperatures in the equatorial tropical eastern Pacific Ocean.

estimated ceiling

A ceiling classification applied when the ceiling height has been estimated by the observer or has been determined by some other method; but, because of the specified limits of time, distance, or precipitation conditions, a more descriptive classification cannot be applied.

evaporation

See **change of state**

extratropical low

Any cyclone that is not a tropical cyclone, usually referring to the migratory frontal cyclones of middle and high latitudes.

eye

The roughly circular area of calm or relatively light winds and comparatively fair weather at the center of a well-developed tropical cyclone. A wall cloud marks the outer boundary of the eye.

F

Fahrenheit temperature scale (°F)

A temperature scale with 32°F as the melting point of pure ice and 212°F as the boiling point of pure water at standard sea-level atmospheric pressure (29.92 in or 1013.2 mb).

fall wind

A cold wind blowing downslope. Fall wind differs from foehn in that the air is initially cold enough to remain relatively cold despite compressional heating during descent.

filling

An increase in the central pressure of a pressure system; opposite of deepening; more commonly applied to a low rather than a high.

first gust

The leading edge of the spreading downdraft, plow wind, from an approaching thunderstorm.

flow line

A streamline.

foehn

A warm, dry downslope wind; the warmness and dryness being due to adiabatic compression upon descent; characteristic of mountainous regions. *See also* **adiabatic process; chinook; Santa Ana**

fog

A hydrometeor consisting of numerous minute water droplets and based at Earth's surface; droplets are small enough to be suspended in the atmosphere indefinitely. Unlike drizzle, it does not fall to the surface; differs from cloud only in that a cloud is not based at the surface; distinguished from haze by its wetness and gray color.

fractus

Clouds in the form of irregular shreds, appearing as if torn; have a clearly ragged appearance; applies only to stratus and cumulus, i.e., cumulus fractus and stratus fractus.

freezing

See **change of state**

freezing level

A level in the atmosphere at which the temperature is 32°F (0°C).

freshet

The annual spring rise of streams in cold climates as a result of melting snow.

front

A surface, interface, or transition zone of discontinuity between two adjacent air masses of different densities; more simply the boundary between two different air masses. *See also* **frontal zone**

frontal zone

A front or zone with a marked increase of density gradient; used to denote that fronts are not truly a "surface" of discontinuity but rather a "zone" of rapid transition of meteorological elements.

frontogenesis

The initial formation of a front or frontal zone.

frontolysis

The dissipation of a front.

frost (hoarfrost)

Ice crystal deposits formed by sublimation when temperature and dew point are below freezing.

funnel cloud

A tornado cloud or vortex cloud extending downward from the parent cloud but not reaching the ground.

G

glaze

A coating of ice, generally clear and smooth, formed by freezing of supercooled water on a surface.

GOES

Geostationary Operational Environmental Satellites. Satellites that orbit the Earth at the same rate that the planet rotates so that the satellites remain over a fixed point above the equator at an altitude of about 22,238 mi.

gradient

In meteorology, a horizontal decrease in value per unit distance of a parameter in the direction of maximum decrease; most commonly used with pressure, temperature, and moisture.

graupel

Same as snow pellets.

ground fog

In the United States, a fog that conceals less than 60% of the sky and is not contiguous with the base of clouds.

gust

A sudden brief increase in wind; according to United States weather-observing practice, gusts are reported when the variation in wind speed between peaks and lulls is at least 10 knots.

gust front

A boundary or leading edge of a cool downdraft of a thunderstorm that spreads along the surface, usually in advance of the storm.

H

haboob

A dust or sand storm, caused either by the downdraft of a desert thunderstorm and lasting minutes, or by a larger (generally late) winter or early spring cyclone in the southwestern United States, and lasting for hours. Also found in the Sudan most months of the year.

hail

A form of precipitation composed of balls or irregular lumps of ice, always produced by convective clouds that are nearly always cumulonimbus.

halo

A prismatically colored or whitish circle or arcs of a circle with the Sun or Moon at its center; coloration, if not white, is from red inside to blue outside (opposite that of a corona); fixed in size with an angular diameter of 22° (common) or 46° (rare); characteristic of clouds composed of ice crystals; valuable in differentiating between cirriform and forms of lower clouds.

haze

A type of lithometeor composed of fine dust or salt particles dispersed through a portion of the atmosphere, particles are so small they cannot be felt or individually seen with the naked eye (as compared with the larger particles of dust), but diminish the visibility; distinguished from fog by its bluish or yellowish tinge.

high

An area of barometric pressure, with its attendant system of winds; an anticyclone. Also, high-pressure system.

hoarfrost

See **frost**

humidity

Water vapor content of the air; may be expressed as specific humidity, relative humidity, or mixing ratio.

hurricane

A tropical cyclone in the Western Hemisphere with winds in excess of 65 knots.

hydrograph

The record produced by a continuous-recording hygrometer.

hydrometeor

A general term for particles of liquid water or ice such as rain, fog, frost, etc., formed by modification of water vapor in the atmosphere; also water or ice particles lifted from the Earth by the wind such as sea spray or blowing snow.

hygrometer

An instrument for measuring the water vapor content of the air.

I

ice crystals

A type of precipitation composed of unbranched crystals in the form of needles, columns, or plates; usually having a very slight downward motion, may fall from a cloudless sky.

ice fog

A type of fog composed of minute suspended particles of ice; occurs at very low temperatures and may cause halo phenomena.

ice needles

A form of ice crystals.

ice pellets

Small, transparent, or translucent, round or irregularly shaped pellets of ice. They may be hard grains that rebound on striking a hard surface or pellets of snow encased in ice.

indefinite ceiling

A ceiling classification denoting vertical visibility into a surface-based obscuration.

indicated altitude

See **altitude**

insolation

Incoming solar radiation falling upon Earth and its atmosphere.

instability

A general term to indicate various states of the atmosphere in which spontaneous convection will occur when prescribed criteria are met; indicative of turbulence. *See also* **absolute instability; conditionally unstable air; convective instability**

intertropical convergence zone (ITCZ)

The boundary zone between the trade wind system of the Northern and Southern Hemispheres; it is characterized in maritime climates by showery precipitation with cumulonimbus clouds sometimes extending to great heights.

inversion

An increase in temperature with height—a reversal of the normal decrease with height in the troposphere; may also be applied to other meteorological properties.

isobar

A line of equal or constant barometric pressure.

isoheight

On a weather chart, a line of equal height.

isoline

A line of equal value of a variable quantity, i.e., an isoline of temperature is an isotherm, etc.

isoshear

A line of equal wind shear.

isotach

A line of equal or constant wind speed.

isotherm

A line of equal or constant temperature.

isothermal

Of equal or constant temperature, with respect to either space or time; more commonly, temperature with height; a zero lapse rate.

J

jet stream

A quasi-horizontal stream of winds 50 knots or more concentrated within a narrow band embedded in the westerlies in the high troposphere.

K

katabatic wind

Any wind blowing downslope. *See also* **fall wind; foehn**

kelvin temperature scale (K)

A temperature scale with zero equal to the temperature at which all molecular motion ceases, i.e., absolute zero (0K = −273°C); the kelvin degree is identical to the Celsius degree; hence at standard sea-level pressure, the melting point of water is 273K and its boiling point is 373K.

knot

A unit of speed equal to one nautical mile per hour.

L

la Niña

The condition where the central and eastern tropical Pacific Ocean sea-surface temperatures become cooler than normal.

land breeze

A coastal breeze blowing from land to sea, caused by temperature difference when the sea surface is warmer than the adjacent land. Therefore, it usually blows at night and alternates with a sea breeze, which blows in the opposite direction by day.

landspout

An informal but very common term among Great Plains storm observers and meteorologists for a non-supercell tornado over land.

lapse rate

The rate of decrease of an atmospheric variable with height; commonly refers to decrease of temperature with height.

latent heat

The amount of heat absorbed (converted to kinetic energy) during the processes of change of liquid water to water vapor, ice to water vapor, or ice to liquid water; or the amount released during the reverse processes. Four basic classifications are: (1) latent heat of condensation: heat released during change of water vapor to water; (2) latent heat of fusion: heat released during change of water to ice or the amount absorbed in change of ice to water; (3) latent heat of sublimation: heat released during change of water vapor to ice or the amount absorbed in the change of ice to water vapor; (4) latent heat of vaporization: heat absorbed in the change of water to water vapor; the negative of latent heat of condensation.

layer

In reference to sky cover, clouds, or other obscuring phenomena whose bases are approximately at the same level. The layer may be continuous or composed of detached elements. The term "layer" does not imply that a clear space exists between the layers or that the clouds or obscuring phenomena composing them are of the same type.

lee wave

Any stationary wave disturbance caused by a barrier in a fluid flow. In the atmosphere when sufficient moisture is present, this wave will be evidenced by lenticular clouds to the lee of mountain barriers; also called mountain wave or standing wave.

lenticular cloud (lenticularis)

A species of cloud whose elements have the form of more or less isolated, generally smooth lenses or almonds. These clouds appear most often in formations of orographic origin, the result of lee waves, in which case they remain nearly stationary with respect to the terrain (standing cloud), but they also occur in regions without marked orography.

level of free convection (LFC)

The level at which a parcel of air lifted dry-adiabatically until saturated, and moist-adiabatically thereafter, first becomes warmer than its surroundings in a conditionally unstable atmosphere. *See also* **conditional instability; adiabatic process**

lifting condensation level (LCL)

The level at which a parcel of unsaturated air lifted dry-adiabatically becomes saturated. *See also* **level of free convection; convective condensation level**

lightning

Generally, any and all forms of visible electrical discharge produced by a thunderstorm.

lithometeor

The general term for dry particles suspended in the atmosphere such as dust, haze, smoke, and sand.

low

An area of low barometric pressure, with its attendant system of winds. Also called a barometric depression or cyclone.

M

macroburst

A downburst that affects a path longer than 2.5 mi.

mammato cumulus

See **cumulonimbus mamma**

mare's tail

See **cirrus**

maritime polar air (mP)

See **polar air**

maritime tropical air (mT)

See **tropical air**

maximum wind axis

On a constant pressure chart, a line denoting the axis of maximum wind speeds at that constant pressure surface.

mean sea level

The average height of the surface of the sea for all stages of tide; used as reference for elevations throughout the United States.

melting

See **change of state**

mercurial barometer

A barometer in which pressure is determined by balancing air pressure against the weight of a column of mercury in an evacuated glass tube.

mesoscale convective complex (MCC)

A large, nearly circular cluster of individual thunderstorms. May cover an entire state, in excess of thousands of square miles.

meteorology

The science of the atmosphere.

microbarograph

An aneroid barograph designed to record atmospheric pressure changes of very small magnitudes.

microburst

A downburst that affects a path that is 2.5 mi or less.

millibar (mb)

An internationally used unit of pressure equal to 1,000 dynes per square centimeter. It is used for reporting atmospheric pressure.

mist

A popular expression for drizzle or heavy fog.

mixing ratio

The ratio by weight of the amount of water vapor in a volume of air to the amount of dry air; usually expressed as grams per kilogram (g/kg).

moist-adiabatic lapse rate

See **saturated adiabatic lapse rate**

moisture

An all-inclusive term denoting water in any or all of its three states.

monsoon

A wind that in summer blows from sea to a continental interior, bringing copious rain, and in winter blows from the interior to the sea, resulting in sustained dry weather.

mountain wave

A standing wave or lee wave to the lee of a mountain barrier.

N

National Climatic Data Center (NCDC)

Located in Asheville, North Carolina, houses archives of climatic data from the United States and around the world.

National Oceanic and Atmospheric Administration (NOAA)

The administrative unit within the United States Department of Commerce that oversees the National Weather Service.

nautical twilight

See **twilight**

negative vorticity

See **vorticity**

Neoglacial

A period of general expansion of glaciers variously defined as spanning from approximately 3,000 to

2,000 years ago or covering the last 4,000–5,000 years.

NEXRAD

Next Generation Weather Radar. Utilizing Doppler radar technology, NEXRAD (WSR-88D) systems observe the presence and calculate the speed and direction of motion of severe weather elements, such as tornadoes and violent thunderstorms.

nimbostratus

A principal cloud type, gray colored, often dark, the appearance of which is rendered diffuse by more or less continuously falling rain or snow, which in most cases reaches the ground. It is thick enough throughout to blot out the Sun.

noctilucent clouds

Clouds of unknown composition that occur at great heights, probably around 46.5–56 mi. They resemble thin cirrus, but usually with a bluish or silverish color, although sometimes orange to red, standing out against a dark night sky. Rarely observed.

normal

In meteorology, the value of an element averaged for a given location over a period of years and recognized as a standard.

northern lights

See **aurora**

numerical forecasting

See **numerical weather prediction**

numerical weather prediction

Forecasting by digital computers solving mathematical equations; used extensively in weather services worldwide.

O

obscuration

Denotes sky hidden by surface-based obscuring phenomena and vertical visibility restricted overhead.

obscuring phenomena

Any hydrometeor or lithometeor other than clouds; may be surface based or aloft.

occluded front (occlusion, frontal occlusion)

A composite of two fronts as a cold front overtakes a warm front or quasi-stationary front.

orographic

Of, pertaining to, or caused by, mountains as in orographic clouds, orographic lift, or orographic precipitation.

ozone (O_3)

An unstable form of oxygen; heaviest concentrations are in the stratosphere; corrosive to some metals; absorbs most ultraviolet solar radiation.

P

parcel

A small volume of air, small enough to contain uniform distribution of its meteorological properties, and large enough to remain relatively self-contained and respond to all meteorological processes. No specific dimensions have been defined, however, the order of magnitude of 1 cubic foot has been suggested.

parhelion

See **sun dog**

partial obscuration

A designation of sky cover when part of the sky is hidden by surface-based obscuring phenomena.

perihelion

The point on the Earth's orbit that is nearest the Sun.

pilot balloon

A small free–lift balloon used to determine the speed and direction of winds in the upper air.

pilot balloon observation (PIBAL)

A method of winds–aloft observation by visually tracking a pilot balloon.

plan position indicator (PPI) scope

A radar indicator scope displaying range and azimuth of targets in polar coordinates.

plow wind

The spreading downdraft of a thunderstorm; a strong, straight-line wind in advance of the storm. *See also* **first gust**

polar air

An air mass with characteristics developed over high latitudes, especially within the subpolar highs. Continental polar (cP) air has cold surface temperatures, low moisture content, and especially in its source regions, has great stability in the lower layers. It is shallow in comparison with arctic air. Maritime polar (mP) air initially possesses similar properties to those of continental polar air, but in passing over warmer water it becomes unstable with a higher moisture content. *See also* **tropical air**

polar front

The semipermanent, semicontinuous front separating air masses of tropical and polar origins.

positive vorticity

See **vorticity**

precipitation

Any or all forms of water particles, whether liquid or solid, that fall from the atmosphere and reach the surface. It is a major class of hydrometeor, distinguished from cloud and virga in that it must reach the surface.

pressure

See **atmospheric pressure**

pressure altimeter

An aneroid barometer with a scale graduated in altitude instead of pressure using standard atmospheric pressure-height relationships; shows indicated altitude (not necessarily true altitude); may be set to measure altitude (indicated) from any arbitrarily chosen level.

pressure gradient

The rate of decrease of pressure per unit distance at a fixed time.

pressure jump

A sudden, significant increase in station pressure.

pressure tendency

See **barometric tendency**

prevailing easterlies

The broad current or pattern of persistent easterly winds in the tropics and in polar regions.

prevailing visibility

In the United States, the greatest horizontal visibility that is equaled or exceeded throughout half of the horizon circle; it need not be a continuous half.

prevailing westerlies

The dominant west-to-east motion of the atmosphere, centered over middle latitudes of both hemispheres.

prevailing wind

Direction from which the wind blows most frequently.

prognostic chart (PROG)

A chart of expected or forecast conditions.

pseudo-adiabatic lapse rate

See **saturated-adiabatic lapse rate**

psychrometer

An instrument consisting of a wet-bulb and a dry-bulb thermometer for measuring wet-bulb and dry-bulb temperature; used to determine water vapor content of the air.

pulse

Pertaining to radar, a brief burst of electromagnetic radiation emitted by the radar; of very short time duration. *See also* **pulse length**

pulse length

Pertaining to radar, the dimension of a radar pulse; may be expressed as the time duration or the length in linear units. Linear dimension is equal to time duration multiplied by the speed of propagation (approximately the speed of light).

Q

quasi-stationary front (stationary front)

A front that is stationary or nearly so; conventionally, a front that is moving at a speed of less than five knots is generally considered to be quasi-stationary.

R

RADAR (radio detection and ranging)

An electronic instrument used for the detection and ranging of distant objects of such composition that they scatter or reflect radio energy. Since hydrometeors can scatter radio energy, weather radars, operating on certain frequency bands, can detect the presence of precipitation, clouds, or both.

radarsonde observation

A rawinsonde observation in which winds are determined by radar tracking a balloon-borne target.

radiation

The emission of energy by a medium and transferred, either through free space or another medium, in the form of electromagnetic waves.

radiation fog

Fog characteristically resulting when radiational cooling of Earth's surface lowers the air temperature near the ground to or below its initial dew point on calm, clear nights.

radiosonde

A balloon-borne instrument for measuring pressure, temperature, and humidity aloft. Radiosonde observation is a sounding made by the instrument.

rain

A form of precipitation; drops are larger than drizzle and fall in relatively straight, although not necessarily vertical, paths as compared to drizzle, which falls in irregular paths.

rain shower

See **shower**

RAOB

A radiosonde observation.

rawin

A rawinsonde observation.

rawinsonde observation

A combined winds aloft and radiosonde observation. Winds are determined by tracking the radiosonde by radio direction finder or radar.

relative humidity

The ratio of the existing amount of water vapor in the air at a given temperature to the maximum amount that could exist at that temperature; usually expressed in percent.

relative vorticity

See **vorticity**

Richter scale

Developed by Charles F. Richter in 1938. An open-ended logarithmic scale for expressing the magnitude of a seismic disturbance (earthquake or tremor).

ridge (ridge line)

In meteorology, an elongated area of relatively high atmospheric pressure; usually associated with and most clearly identified as an area of maximum anti-cyclonic curvature of the wind flow (isobars, contours, or streamlines).

River Forecast Centers (RFCs)

National Weather Service offices that provide hydro-logic forecasts and guidance information.

rocketsonde

A type of radiosonde launched by a rocket that makes its measurements during a parachute descent; capable of obtaining soundings to a much greater height than possible by balloon or aircraft.

roll cloud

A dense and horizontal roll-shaped accessory cloud located on the lower leading edge of a cumulonim-bus or less often, a rapidly developing cumulus; in-dicative of turbulence.

rotor cloud

A turbulent cloud formation found in the lee of some large mountain barriers, the air in the cloud rotates around an axis parallel to the range; indicative of possible violent turbulence.

S

St. Elmo's fire (corposant)

A luminous brush discharge of electricity from pro-truding objects, such as masts and yardarms of ships, aircraft, lightning rods, steeples, etc., occurring in stormy weather.

Santa Ana

A hot, dry, foehn wind, generally from the northeast or east, occurring west of the Sierra Nevada especially in the pass and river valley near Santa Ana, California.

saturated adiabatic lapse rate

The rate of decrease of temperature with height as saturated air is lifted with no gain or loss of heat from outside sources; varies with temperature, being greatest at low temperatures. *See also* **adiabatic process, dry-adiabatic lapse rate**

saturation

The condition of the atmosphere when actual water vapor present is the maximum possible at the exist-ing temperature.

scud

Small detached masses of stratus fractus clouds be-low a layer of higher clouds, usually nimbostratus.

sea breeze

A coastal breeze blowing from sea to land, caused by the temperature difference when the land surface is warmer than the sea surface. *See also* **land breeze**

sea fog

A type of advection fog formed when air that has been lying over a warm surface is transported over a colder water surface.

sea-level pressure

The atmospheric pressure at mean sea level, either directly measured by stations at sea level or empiri-cally determined from the station pressure and tem-perature by stations not at sea level; used as a com-mon reference for analyses of surface-pressure patterns.

sea smoke

See **steam fog**

sector visibility

Meteorological visibility within a specified sector of the horizon circle.

shear

See **wind shear**

shower

Precipitation from a cumuliform cloud; characterized by the suddenness of beginning and ending, by the rapid change of intensity, and usually by rapid change in the appearance of the sky; showery precip-itation may be in the form of rain, ice pellets, or snow.

slant visibility

For an airborne observer, the distance at which the pilot can see and distinguish objects on the ground.

sleet

See **ice pellets**

smog

A mixture of smoke and fog.

smoke

A restriction to visibility resulting from combustion.

snow

Precipitation composed of white or translucent ice crystals, chiefly in complex branched hexagonal form.

snow flurry

Popular term for snow shower, particularly of a very light and brief nature.

snow grains

Precipitation of very small, white, opaque grains of ice, similar in structure to snow crystals. The grains are fairly flat or elongated, with diameters generally less than 0.04 in.

snow pellets

Precipitation consisting of white, opaque, approxi-mately round (sometimes conical) ice particles hav-ing a snowlike structure, and about 0.08–0.2 in in di-ameter; crisp and easily crushed, differing in this respect from snow grains; rebound from a hard sur-face and often break up. Same as graupel.

snow roller

A mass of windblown snow, shaped somewhat like a lady's muff, and rather common in mountainous or hilly regions.

snow shower

See **shower**

solar radiation

The total electromagnetic radiation emitted by the Sun. *See also* **insolation**

sounding

In meteorology, an upper-air observation; a radiosonde observation.

source region

An extensive area of the Earth's surface characterized by relatively uniform surface conditions where large masses of air remain long enough to take on characteristic temperature and moisture properties imparted by that surface.

specific humidity

The ratio by weight of water vapor in a sample of air to the combined weight of water vapor and dry air. *See also* **mixing ratio**

squall

A sudden increase in wind speed by at least 15 knots to a peak of 20 knots or more and lasting for at least one minute. Essential difference between a gust and a squall is the duration of the peak speed.

squall line

Any nonfrontal line or narrow band of active thunderstorms (with or without squalls).

stability

A state of the atmosphere in which the vertical distribution of temperature is such that a parcel will resist displacement from its initial level. *See also* **instability**

standard atmosphere

A hypothetical atmosphere based on climatological averages comprised of numerous physical constants of which the most important are: (1) a surface temperature of 59°F (15°C) and a surface pressure of 29.92 in of mercury (1,013.2 mb) at sea level; (2) a lapse rate in the troposphere of 18.8°F per mi (6.5°C per km); (3) a tropopause of approximately 36,000 ft (11 km) with a temperature of −69.7°F (−56.5°C); and (4) an isothermal lapse rate in the stratosphere to an altitude of approximately 80,000 ft (24 km).

standing cloud (standing lenticular altocumulus)

See **lenticular cloud**

standing wave

A wave that remains stationary in a moving fluid. In aviation operations it is used most commonly to refer to a lee wave or mountain wave.

station pressure

The actual atmospheric pressure at the observing station.

stationary front

See **quasi-stationary front**

steam fog

Fog formed when cold air moves over relatively warm water or wet ground.

stepped leader

The initial leader of a lightning discharge

storm detection radar

A weather radar designed to detect hydrometeors of precipitation size; used primarily to detect storms with large drops or hailstones as opposed to clouds and light precipitation of small drop size.

stratiform

Descriptive of clouds of extensive horizontal development, as contrasted to vertically developed cumuliform clouds; characteristic of stable air and, therefore, composed of small water droplets.

stratocumulus

A low cloud, predominantly stratiform, in gray and/or whitish patches or layers, may or may not merge; elements are tessellated, rounded, or roll-shaped with relatively flat tops.

stratosphere

The atmospheric layer above the tropopause, average altitude of base and top is 7 and 22 mi respectively; characterized by a slight average increase of temperature from base to top and is very stable; also characterized by low moisture content and absence of clouds.

stratus

A low, gray cloud layer or sheet with a fairly uniform base; sometimes appears in ragged patches; seldom produces precipitation but may produce drizzle or snow grains. A stratiform cloud.

stratus fractus

See **fractus**

streamline

In meteorology, a line whose tangent is the wind direction at any point along the line. A flowline.

sublimation

See **change of state**

subsidence

A descending motion of air in the atmosphere over a rather broad area; usually associated with divergence.

summation principle

The principle states that the cover assigned to a layer is equal to the summation of the sky cover of the lowest layer plus the additional coverage at all successively higher layers up to and including the layer in question. Thus, no layer can be assigned a sky cover less than a lower layer, and no sky cover can be greater than 1.0 (10/10).

sun dog (parhelion)

Either of two colored, luminous spots that appear at points 22° (or more) on both sides of the Sun at the same elevation as the Sun. Caused by refraction of sunlight by ice crystals. Also called "mock Sun."

supercell thunderstorm

A large and intense severe thunderstorm with strong updrafts and downdrafts that may last for several hours; can produce large, damaging hail and tornadoes.

supera diabatic lapse rate

A lapse rate greater than the dry-adiabatic lapse rate. *See also* **absolute instability**

supercooled water

Liquid water at temperatures colder than freezing.

surface inversion

An inversion with its base at the surface, often caused by cooling of the air near the surface as a result of terrestrial radiation, especially at night.

surface visibility

Visibility observed from eye level above the ground.

synoptic chart

A chart, such as the familiar weather map, that depicts the distribution of meteorological conditions over an area at a given time.

T

target

In radar, any of the many types of objects detected by radar.

temperature

In general, the degree of hotness or coldness as measured on some definite temperature scale by means of any of various types of thermometers.

temperature inversion

See **inversion**

terrestrial radiation

The total infrared radiation emitted by the Earth and its atmosphere.

thermograph

A continuous-recording thermometer.

thermometer

An instrument for measuring temperature.

thunderstorm

In general, a local storm invariably produced by a cumulonimbus cloud, and always accompanied by lightning and thunder.

tornado (cyclone or twister)

A violently rotating column of air, pendant from a cumulonimbus cloud, and nearly always observable as "funnel-shaped." It is the most destructive of all small-scale atmospheric phenomena.

tower visibility

Prevailing visibility determined from the control tower.

towering cumulus

A rapidly growing cumulus in which height exceeds width.

trade winds

Prevailing, almost continuous winds blowing with an easterly component from the subtropical high-pressure belts toward the intertropical convergence zone northeast in the Northern Hemisphere, southeast in the Southern Hemisphere.

training

Individual storms that move over the same area and may produce flooding in a short period of time.

tropical air

An air mass with characteristics developed over low latitudes. Maritime tropical (mT) air, the principal type, is produced over the tropical and subtropical sea and is very warm and humid. Continental tropical (cT) air is produced over subtropical arid regions and is hot and very dry. *See also* **polar air**

tropical cyclone

A general term for cyclone that originates over tropical oceans. By international agreement, tropical cyclones have been classified according to their intensity, as follows: (1) tropical depression: winds up to 34 knots; (2) tropical storm: winds of 35–64 knots; (3) hurricane or typhoon: winds of 65 knots or higher.

tropical depression

See **tropical cyclone**

tropical storm

See **tropical cyclone**

tropopause

The transition zone between the troposphere and stratosphere, usually characterized by an abrupt change of lapse rate.

troposphere

That position of the atmosphere from Earth's surface to the tropopause; that is, the lowest 6–12 mi of the atmosphere. The troposphere is characterized by decreasing temperature with height, and by appreciable water vapor.

trough (trough line)

In meteorology, an elongated area of relatively low atmospheric pressure, usually associated with and most clearly identified as an area of maximum cyclonic curvature of the wind flow (isobars, contours, or streamlines). *See also* **ridge**

true altitude

See **altitude**

true wind direction

The direction, with respect to true north, from which the wind is blowing.

turbulence

In meteorology, any irregular or disturbed flow in the atmosphere.

twilight

The intervals of incomplete darkness following sunset and preceding sunrise. The time at which evening twilight ends or morning twilight begins is determined by arbitrary convention, and several kinds of twilight have been defined and used, most commonly civil, nautical, and astronomical twilight: (1) civil twilight: the period of time before sunrise and after sunset when the Sun is not more than 6 degrees below the horizon; (2) nautical twilight: the period of time before sunrise and after sunset when the Sun is not more than 12 degrees below the horizon; (3) astronomical twilight: the period of time before sunrise and after sunset when the Sun is not more than 18 degrees below the horizon.

twister

In the United States, a colloquial term for tornado.

typhoon

A tropical cyclone (hurricane) in the western tropical Pacific Ocean with winds in excess of 65 knots.

U

undercast

A cloud layer of 10/10 (1.0) coverage (to the nearest tenth) as viewed from an observation point above the layer.

unlimited ceiling

A clear sky or a sky cover that does not meet the criteria for a ceiling.

unstable

See **instability**

updraft

A localized upward current of air.

upper front

A front aloft not extending to the Earth's surface.

upslope fog

Fog formed when air flows upward over rising terrain and is, consequently, adiabatically cooled to or below its initial dew point.

V

vapor pressure

In meteorology, the pressure of water vapor in the atmosphere. Vapor pressure is that part of the total atmospheric pressure due to water vapor and is independent of the other atmospheric gases or vapors.

vapor trail

See **condensation trail**

veering

Shifting of the wind in a clockwise direction with respect to either space or time; opposite of backing. Commonly used by meteorologists to refer to an anticyclonic shift (clockwise in the Northern Hemisphere and counterclockwise in the Southern Hemisphere).

vertical visibility

The distance one can see upward into a surface-based obscuration; or the maximum height from which a pilot in flight can recognize the ground through a surface-based obscuration.

virga

Water or ice particles falling from a cloud, usually in wisps or streaks, and evaporating before reaching the ground.

visibility

The greatest distance one can see and identify prominent objects.

vortex

In meteorology, any rotary flow in the atmosphere.

vorticity

Turning of the atmosphere. Vorticity may be embedded in the total flow and not readily identified by a flow pattern. (1) absolute vorticity: the rotation of the Earth imparts vorticity to the atmosphere; absolute vorticity is the combined vorticity due to this rotation and vorticity due to circulation relative to the Earth (relative vorticity); (2) negative vorticity: vorticity caused by anticyclonic turning; it is associated with downward motion of the air; (3) positive vorticity: vorticity caused by cyclonic turning; it is associated with upward motion of the air; (4) relative vorticity: vorticity of the air relative to the Earth, disregarding the component of vorticity resulting from the Earth's rotation.

W

wake turbulence

Turbulence found to the rear of a solid body in motion relative to a fluid. In aviation terminology, the turbulence caused by a moving aircraft.

wall cloud

The well-defined bank of vertically developed clouds with a wall-like appearance that form the outer boundary of the eye of a well-developed tropical cyclone.

warm front

Any non-occluded front that moves in such a way that warmer air replaces colder air.

warm sector

The area covered by warm air at the surface and bounded by the warm front and cold front of a wave cyclone.

water equivalent

The depth of water that would result from the melting of snow or ice.

water vapor

Water in the invisible gaseous form.

waterspout

Usually a tornado occurring over warm water. Most common over tropical and subtropical waters.

wave cyclone

A cyclone that forms and moves along a front. The circulation about the cyclone center tends to produce a wavelike deformation of the front.

weather

The state of the atmosphere, mainly with respect to its effects on life and human activities; refers to instantaneous conditions or short-term changes as opposed to climate.

Weather Forecast Offices (WFOs)

National Weather Service offices (118) charged with providing area weather reports, forecasts, and warnings.

weather radar

Radar specifically designed for observing weather. *See also* **cloud detection radar; storm detection radar**

weather vane

A wind vane.

weather warning

Warnings that are issued for areas where severe weather is imminent.

weather watch

When conditions are favorable for severe weather to develop, a watch is issued.

wedge

See **ridge**

wet-bulb temperature

The lowest temperature that can be obtained on a wet-bulb thermometer in any given sample of air, by evaporation of water (or ice) from the muslin wick; used in computing dew point and relative humidity.

wet-bulb thermometer

A thermometer with a muslin-covered bulb used to measure wet-bulb temperature.

whirlwind

A small, rotating column of air; may be visible as a dust devil.

willy-willy

A tropical cyclone of hurricane strength near Australia. Also a dust devil.

wind

Air in motion relative to the surface of the Earth; generally used to denote horizontal movement.

wind direction

The direction from which wind is blowing.

wind shear

The rate of change of wind velocity (direction and/or speed) per unit distance; conventionally expressed as vertical or horizontal wind shear.

wind speed

Rate of wind movement in distance per unit time.

wind vane

An instrument to indicate wind direction.

wind velocity

A vector term to include both wind direction and wind speed.

X, Y, Z

zonal wind

A west wind; the westerly component of a wind. Conventionally used to describe large-scale flow that is neither cyclonic nor anticyclonic.

Further Resources

BOOKS

Abbott, Patrick L. *Natural Disasters.* Dubuque, IA: Wm. C. Brown Publishers, 1996.

Ahrens, C. Donald. *Meteorology Today: An Introduction to Weather, Climate, and the Environment.* 5th ed. Minneapolis/St. Paul: West Publishing Co., 1994.

Bluestein, Howard B. *Tornado Alley.* New York: Oxford University Press, 1999.

Bradley, Raymond S., and Philip D. Jones. *Climate Since 500 A.D.* New York: Routledge, 1995.

Danielson, Eric, James Levin, and Elliot Abrams. *Meteorology.* Columbus, OH: McGraw-Hill Higher Education, 1998.

Dashew, Steve, and Linda Dashew. *Mariner's Weather Handbook.* Pineville, NC: Beowulf Publishing, 1998.

Dudley, Walter C., and Min Lee. *Tsunami.* Honolulu: University of Hawaii Press, 1998.

Ellis, Beth R. *The World Almanac.* Published annually by World Almanac books.

Elsner, James B., and A. Birol Kara. *Hurricanes of the North Atlantic, Climate and Society.* New York: Oxford University Press, 1999.

Garoogian, David, ed. *Weather America, 2001.* Lakeville, CT: Grey House Publishing, 2000.

Glickman, Todd S. *Glossary of Meteorology.* Boston: American Meteorological Society, 2000

Kahl, Jonathan D. W. *National Audubon Society First Field Guide Weather.* New York: Scholastic, Inc., 1998.

Kahl, Jonathan D. W. *Weather Watch: Forecasting the Weather.* Minneapolis: Lerner Publishing Company, 1996.

Kingsbury, Stewart A., Mildred E. Kingsbury, and Wolfgang Mieder. *Weather Wisdom: Proverbs, Superstitions, and Signs.* New York: Peter Lang Publishing, 1996.

Longshore, David. *Hurricanes, Typhoons, and Cyclones.* New York: Checkmark Books, 2000.

Lutgens, Frederick K., and Edward J. Tarbuck. *The Atmosphere.* 6th ed. Englewood Cliffs, NJ: Prentice-Hall, 1995.

Lyons, Walter A. *The Handy Weather Answer Book.* Detroit: Visible Ink Press, 1997.

Miller, E. Willard, and Ruby M. Miller. *Natural Disasters: Floods.* Santa Barbara, CA: ABC-CLIO, Inc., 2000.

Monmonier, Mark. *Air Apparent: How Meteorologists Learned to Map, Predict, and Dramatize Weather.* Chicago: University of Chicago Press, 1999.

Moran, Joseph M., Michael D. Morgan, and Patricia M. Pauley. *Meteorology: The Atmosphere and the Science of Weather.* 5th ed. Upper Saddle River, NJ: Prentice-Hall, 1997.

Nese, Jon M., and Lee M. Grenci. *A World of Weather, Fundamentals of Meteorology.* Dubuque, IA: Kendall/Hunt Publishing Co., 2001.

Rinehart, Ronald E. *Radar for Meteorologists.* 3rd ed. Grand Forks, ND: Rinehart Publications, 1997.

Schneider, Stephen H. *Encyclopedia of Climate and Weather.* 2 vols. New York: Oxford University Press, 1996.

Science and Technology Department of the Carnegie Library of Pittsburgh. *The Handy Science Answer Book.* Detroit: Visible Ink Press, 1997.

Stein, Paul. *The Macmillan Encyclopedia of Weather.* Published by Macmillan Reference USA, 2001.

Uman, Martin A. *The Lightning Discharge.* Published by Dover Publications, 2001.

Watts, Alan. *The Weather Handbook.* Dobbs Ferry, NY: Sheridan House, 1999.

WEB SITES

American Meteorological Society.
<http://www.ametsoc.org/AMS/>

BBC Online Weather Centre.
<http://www.bbc.co.uk/weather/>

Climate Prediction Center.
<http://www.cpc.ncep.noaa.gov/>

Global Volcanism Program. National Museum of Natural History. Smithsonian Institution.
<http://www.volcano.si.edu/>

National Centers for Environmental Prediction.
<http://www.ncep.noaa.gov/>

National Climatic Data Center (NCDC).
<http://www.ncdc.noaa.gov/>

National Severe Storms Laboratory.
<http://www.nssl.noaa.gov/>

National Oceanic & Atmospheric Administration (NOAA).
<http://www.noaa.gov/>

National Weather Service (NWS).
    <http://www.nws.noaa.gov/>
NOAA Photo Library.
    <http://www.photolib.noaa.gov/>
Office of Oceanic and Atmospheric Research (OAR).
    <http://www.oar.noaa.gov/>
Storm Prediction Center.
    <http://www.spc.noaa.gov/>
Surfing the Internet for Earthquake Data.
    <http://www.geophys.washington.edu/seismosurf-ing.html>
Tropical Prediction Center/National Hurricane Center.
    <http://www.nhc.noaa.gov/>
UM Weather. University of Michigan.
    <http://cirrus.sprl.umich.edu/wxnet/>

United States Geological Survey (USGS).
    <http://www.usgs.gov/>
USGS Earthquake Hazards Program.
    <http://earthquake.usgs.gov/>
USGS National Earthquake Information Center.
    <http://wwwneic.cr.usgs.gov/>
USGS Volcano Hazards Program.
    <http://volcanoes.usgs.gov/>
Volcano World.
    <http://volcano.und.nodak.edu/>
World Meteorological Organization.
    <http://www.wmo.ch/>
World-Wide Earthquake Locator. University of Edinburgh, Scotland.
    <http://www.geo.ed.ac.uk/quakes/quakes.html>

General Index

Boldface page numbers indicate the location of a main chapter about the topic. Photographs are highlighted by an *italicized* page number; tables are indicated with an italicized *t* and figures with an *f*.

807